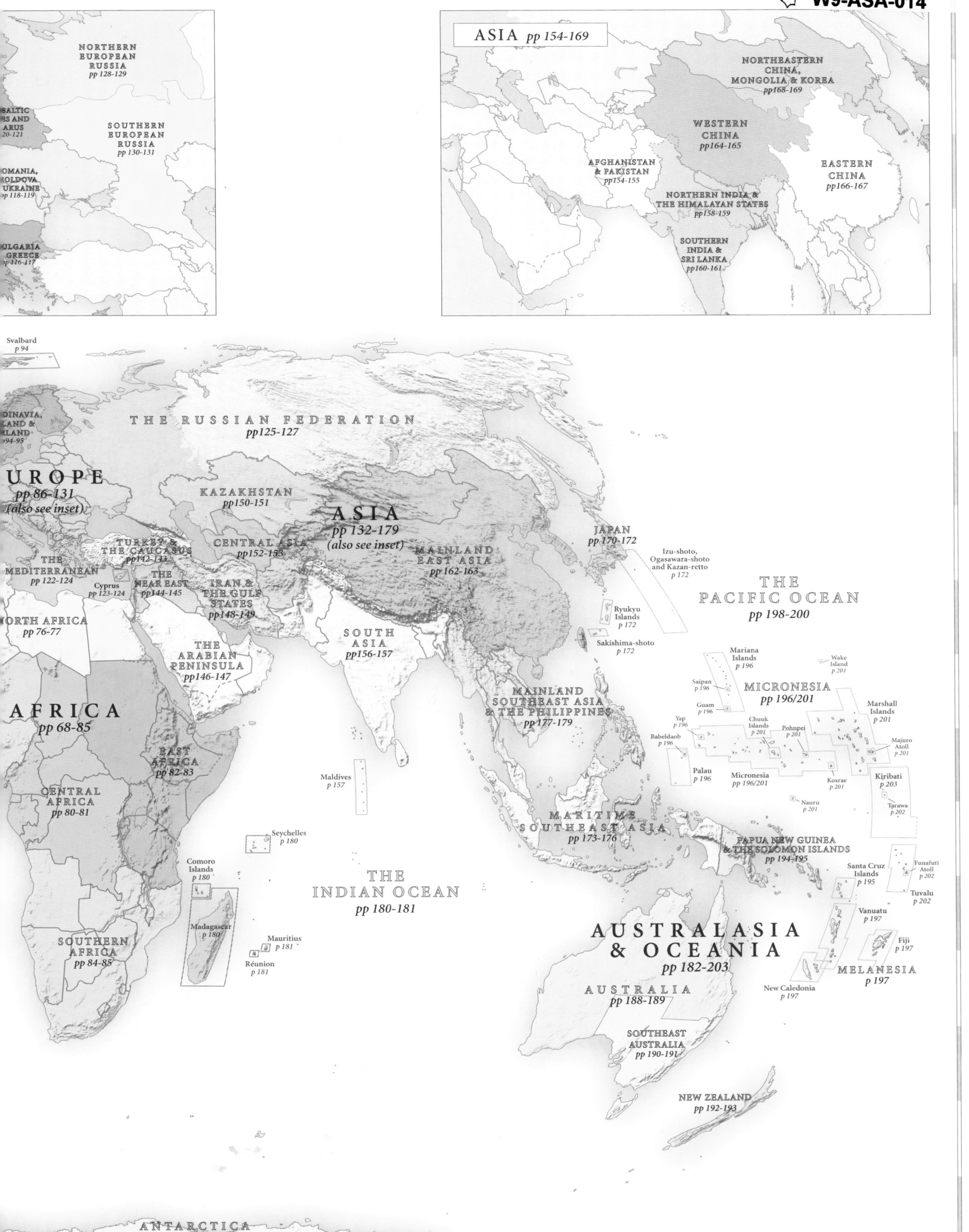

NORTHERN
EUROPEAN
RUSSIA
pp 128-129

BALTIC
ES AND
ARUS
20-121

SOUTHERN
EUROPEAN
RUSSIA
pp 130-131

OMANIA,
OLDOVA,
KRAINE
118-119

ULGARIA
GREECE
116-117

ASIA pp 154-169

NORTHEASTERN
CHINA,
MONGOLIA & KOREA
pp168-169

WESTERN
CHINA
pp164-165

AFGHANISTAN
& PAKISTAN
pp154-155

EASTERN
CHINA
pp166-167

NORTHERN INDIA &
THE HIMALAYAN STATES
pp158-159

SOUTHERN
INDIA &
SRI LANKA
pp160-161

Svalbard
p 94

DINAVIA,
LAND &
RLAND
94-95

THE RUSSIAN FEDERATION
pp125-127

KAZAKHSTAN
pp150-151

ASIA
pp 132-179
(also see inset)

JAPAN
pp 170-172

UROPE
pp 86-131
(also see inset)

TURKEY &
THE CAUCASUS
pp142-143

CENTRAL ASIA
pp152-153

MAINLAND
EAST ASIA
pp 162-163

Izu-shoto,
Ogasawara-shoto
and Kazan-retto
p 172

THE
PACIFIC OCEAN
pp 198-200

THE
MEDITERRANEAN
pp 122-124

Cyprus
pp 123-124

THE
NEAR EAST
pp144-145

IRAN &
THE GULF
STATES
pp148-149

Ryukyu
Islands
p 172

NORTH AFRICA
pp 76-77

Sakishima-shoto
p 172

Mariana
Islands
p 196

Wake
Island
p 201

THE
ARABIAN
PENINSULA
pp 146-147

SOUTH
ASIA
pp156-157

Saipan
p 196

MICRONESIA
pp 196/201

AFRICA
pp 68-85

Guam
p 196

Marshall
Islands
p 201

MAINLAND
SOUTHEAST ASIA
& THE PHILIPPINES
pp 177-179

Yap p 196

Chuuk
Islands
p 201

Pohnpei
p 201

EAST
AFRICA
pp 82-83

Babeldaob
p 196

Majuro
Atoll
p 201

Palau
p 196

Micronesia
pp 196/201

Kosrae
p 201

Kiribati
p 203

CENTRAL
AFRICA
pp 80-81

Maldives
p 157

Nauru
p 201

Tarawa
p 202

Seychelles
p 180

MARITIME
SOUTHEAST ASIA
pp 173-176

PAPUA NEW GUINEA
& THE SOLOMON ISLANDS
pp 194-195

Comoro
Islands
p 180

THE
INDIAN OCEAN
pp 180-181

Santa Cruz
Islands
p 195

Funafuti
Atoll
p 202

Madagascar
p 180

Tuvalu
p 202

Mauritius
p 181

Vanuatu
p 197

AUSTRALASIA
& OCEANIA
pp 182-203

SOUTHERN
AFRICA
pp 84-85

Réunion
p 181

Fiji
p 197

MELANESIA
p 197

New Caledonia
p 197

AUSTRALIA
pp 188-189

SOUTHEAST
AUSTRALIA
pp 190-191

NEW ZEALAND
pp 192-193

ANTARCTICA
pp 204-205

DORLING KINDERSLEY

WORLD ATLAS

DORLING KINDERSLEY

WORLD ATLAS

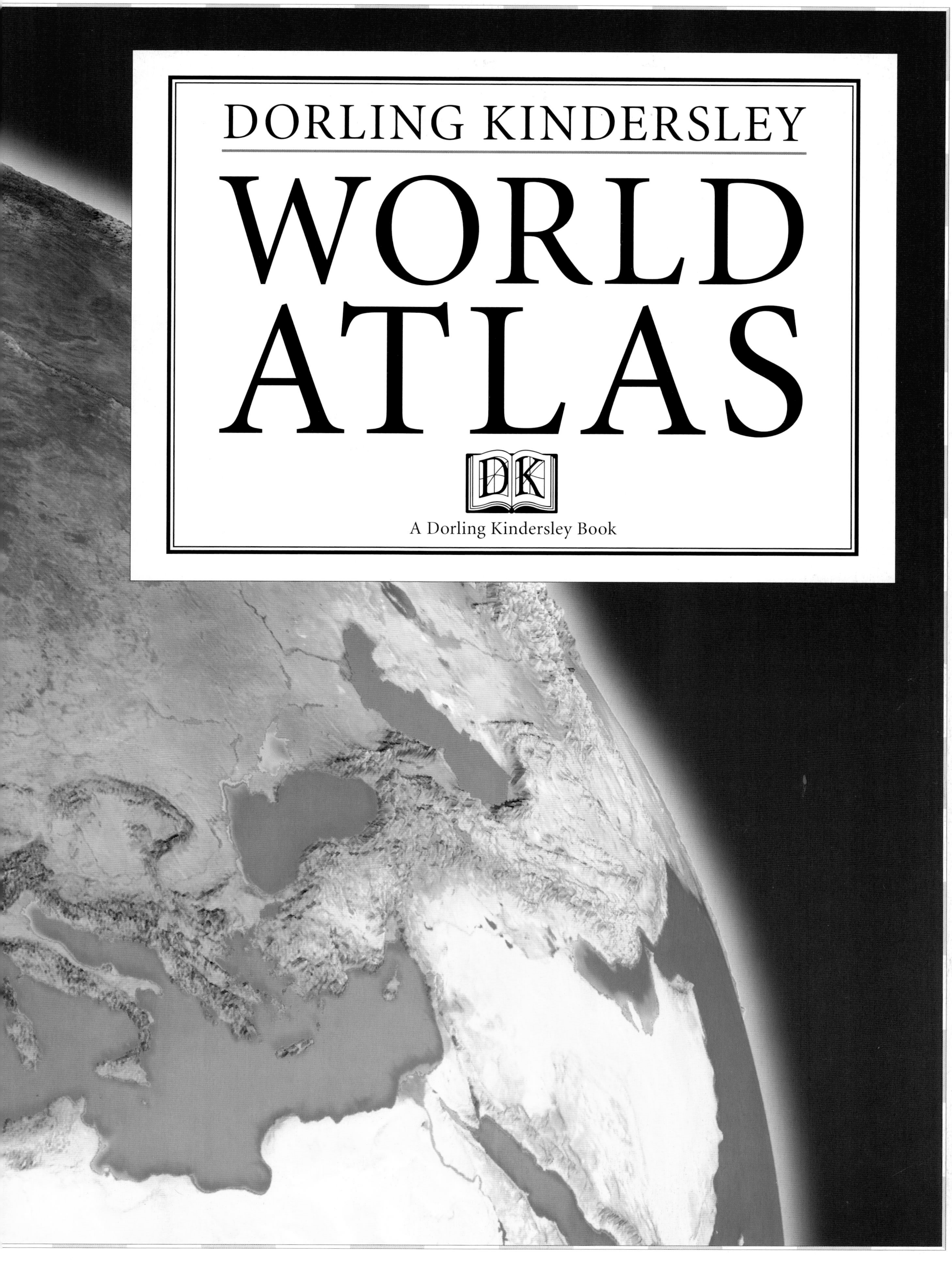

A Dorling Kindersley Book

Dorling DK Kindersley

LONDON, NEW YORK, SYDNEY, DELHI, PARIS, MUNICH, AND JOHANNESBURG

GENERAL GEOGRAPHICAL CONSULTANTS

PHYSICAL GEOGRAPHY • Denys Brunsden, Emeritus Professor, Department of Geography, King's College, London

HUMAN GEOGRAPHY • Professor J Malcolm Wagstaff, Department of Geography, University of Southampton

PLACE NAMES • Caroline Burgess, Permanent Committee on Geographical Names, London

BOUNDARIES • International Boundaries Research Unit, Mountjoy Research Centre, University of Durham

DIGITAL MAPPING CONSULTANTS

DK Cartopia developed by George Galfalvi and XMap Ltd, London

Professor Jan-Peter Muller, Department of Photogrammetry and Surveying, University College, London

Cover globes, planets and information on the Solar System provided by Philip Eales and Kevin Tildsley, Planetary Visions Ltd, London

REGIONAL CONSULTANTS

NORTH AMERICA • Dr David Green, Department of Geography, King's College, London
Jim Walsh, Head of Reference, Wessell Library, Tufts University, Medford, Massachussetts

SOUTH AMERICA • Dr David Preston, School of Geography, University of Leeds

EUROPE • Dr Edward M Yates, formerly of the Department of Geography, King's College, London

AFRICA • Dr Philip Amis, Development Administration Group, University of Birmingham
Dr Ieuan Ll Griffiths, Department of Geography, University of Sussex
Dr Tony Binns, Department of Geography, University of Sussex

CENTRAL ASIA • Dr David Turnock, Department of Geography, University of Leicester

SOUTH AND EAST ASIA • Dr Jonathan Rigg, Department of Geography, University of Durham

AUSTRALASIA AND OCEANIA • Dr Robert Allison, Department of Geography, University of Durham

ACKNOWLEDGMENTS

Digital terrain data created by Eros Data Center, Sioux Falls, South Dakota, USA. Processed by GVS Images Inc, California, USA and Planetary Visions Ltd, London, UK
• CIRCA Research and Reference Information, Cambridge, UK • Digitization by Robertson Research International, Swanley, UK • Peter Clark
British Isles maps generated from a dataset supplied by Map Marketing Ltd/European Map Graphics Ltd in combination with DK Cartopia copyright data

FOR THE SECOND EDITION

EDITOR-IN-CHIEF
Andrew Heritage

SENIOR CARTOGRAPHIC MANAGER
David Roberts

MANAGING CARTOGRAPHER   SENIOR CARTOGRAPHIC EDITOR
Roger Bullen   Simon Mumford

DIGITAL MAPPING SUPPLIERS
Encompass Graphics

CARTOGRAPHERS
Tony Chambers • Jan Clark • John Plumer • Rob Stokes • Julie Turner • Iorwerth Watkins • Peter Winfield

SENIOR EDITOR   SENIOR MANAGING ART EDITOR
Debra Clapson   Philip Lord

EDITORS   DESIGNERS
Wim Jenkins • Sam Atkinson   Karen Gregory, Carol Ann Davis, David Douglas

SYSTEMS COORDINATOR   INDEX GAZETTEER
Phil Rowles   Margaret Hynes

PRODUCTION
Michelle Thomas

DORLING KINDERSLEY CARTOGRAPHY

EDITOR-IN-CHIEF
Andrew Heritage

MANAGING CARTOGRAPHER   SENIOR CARTOGRAPHIC EDITOR
David Roberts   Roger Bullen

CARTOGRAPHERS
Pamela Alford • James Anderson • Sarah Baker-Ede • Caroline Bowie • Dale Buckton • Tony Chambers • Jan Clark • Bob Croser • Martin Darlison • Claire Ellam
Sally Gable • Jeremy Hepworth • Geraldine Horner • Chris Jackson • Christine Johnston • Julia Lunn • Michael Martin • James Mills-Hicks • Simon Mumford • John Plumer
John Scott • Ann Stephenson • Julie Turner • Jane Voss • Scott Wallace • Iorwerth Watkins • Bryony Webb • Alan Whitaker • Peter Winfield

DIGITAL MAPS CREATED IN DK CARTOPIA BY   PLACENAMES DATABASE TEAM
Tom Coulson • Thomas Robertshaw   Natalie Clarkson • Ruth Duxbury • Caroline Falce • John Featherstone • Dan Gardiner
Philip Rowles • Rob Stokes   Ciárán Hynes • Margaret Hynes • Helen Rudkin • Margaret Stevenson • Annie Wilson

DATABASE MANAGER
Simon Lewis

MANAGING EDITOR   SENIOR MANAGING ART EDITOR
Lisa Thomas   Philip Lord

EDITORS   DESIGNERS
Thomas Heath • Wim Jenkins • Jane Oliver • Siobhán Ryan • Elizabeth Wyse   Scott David • Carol Ann Davis • David Douglas • Rhonda Fisher • Karen Gregory • Nicola Liddiard

EDITORIAL RESEARCH   ILLUSTRATIONS
Helen Dangerfield • Andrew Rebeiro-Hargrave   Ciárán Hughes • Advanced Illustration, Congleton, UK

ADDITIONAL EDITORIAL ASSISTANCE   PICTURE RESEARCH
Debra Clapson • Robert Damon • Ailsa Heritage • Constance Novis • Jayne Parsons • Chris Whitwell   Melissa Albany • James Clarke • Anna Lord • Christine Rista • Sarah Moule • Louise Thomas

EDITORIAL DIRECTION • Louise Cavanagh   ART DIRECTION • Chez Picthall

First American Edition, 1997. Reprinted with revisions 1998, 1999. Third Edition (revised) 2000

Published in the United States by Dorling Kindersley Publishing, Inc., 95 Madison Avenue, New York, New York 10016

Copyright @ 1997, 1998, 1999, 2000 Dorling Kindersley Limited
see our complete catalog at
www.dk.com

DK Publishing, Inc.
DK world atlas.
p. cm.
Includes index.
ISBN 0-7894-5962-0 (alk. paper)
1. Atlases. I. Title: World atlas. II. Title
G1021 D625 2000
912--dc21

INTRODUCTION

FOR MANY, THE OUTSTANDING LEGACY OF THE TWENTIETH CENTURY was the way in which the
Earth shrank. As we enter the third millennium, it is increasingly important for us to have a clear vision
of the world in which we live. The human population has increased fourfold since 1900. The last scraps of
terra incognita – the polar regions and ocean depths – have been penetrated and mapped. New regions
have been colonized, and previously hostile realms claimed for habitation. The advent of aviation
technology and mass tourism allows many of us to travel farther, faster, and more frequently than ever
before. In doing so we are given a bird's-eye view of the Earth's surface denied to our forebears.

❧

AT THE SAME TIME, the amount of information about our world has grown enormously.
Telecommunications can span the greatest distances in fractions of a second: our multimedia
environment hurls uninterrupted streams of data at us, on the printed page, through the airwaves,
and across our television and computer screens; events from all corners of the globe reach us
instantaneously, and are witnessed as they unfold. Our sense of stability and certainty has
been eroded; instead, we are aware that the world is in a constant state of flux and change.
Natural disasters, man-made cataclysms, and conflicts between nations remind
us daily of the enormity and fragility of our domain.

❧

OUR CURRENT "GLOBAL" CULTURE has made the need greater than ever before for everyone to possess an
atlas. The *DK World Atlas* has been conceived to meet this need. At its core, like all atlases, it seeks to
define where places are, to describe their main characteristics, and to locate them in relation to other
places. Every attempt has been made to make the information on the maps as clear and accessible as
possible. In addition, each page of the atlas provides a wealth of further information, bringing the maps
to life. Using photographs, diagrams, "at-a-glance" maps, introductory texts, and captions, the atlas
builds up a detailed portait of those features – cultural, political, economic, and geomorphological –
which make each region unique, and which are also the main agents of change.

❧

THIS SECOND EDITION INCORPORATES thousands of revisions and updates affecting every map and every
page, and features a new typographic design for the maps; a further addition is the provision of
longitude and latitude coordinates for every site in the Index-Gazetteer. Since its first publication in
1997 the *DK World Atlas* has proved extremely popular – going into 22 editions around the world –
and has been translated into 13 languages, including Greek and Russian.

PETER KINDERSLEY

CONTENTS

INTRODUCTION v

CONTENTS............................ vi–vii

HOW TO USE THIS ATLAS......... viii–ix

THE WORLD TODAY

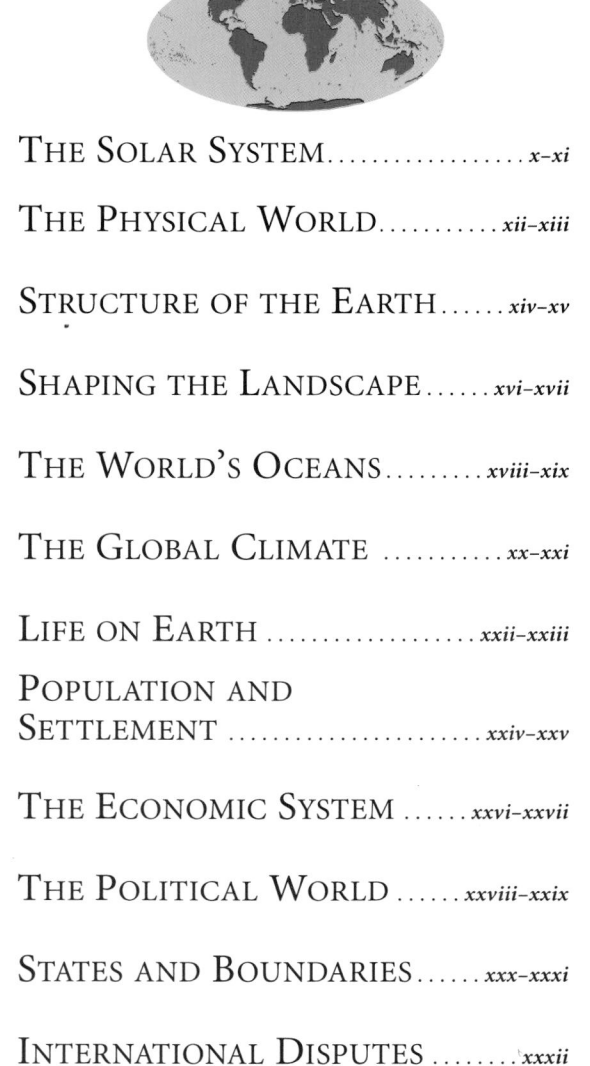

THE SOLAR SYSTEM.................. x–xi

THE PHYSICAL WORLD........... xii–xiii

STRUCTURE OF THE EARTH...... xiv–xv

SHAPING THE LANDSCAPE...... xvi–xvii

THE WORLD'S OCEANS......... xviii–xix

THE GLOBAL CLIMATE .... xx–xxi

LIFE ON EARTH .... xxii–xxiii

POPULATION AND
SETTLEMENT .... xxiv–xxv

THE ECONOMIC SYSTEM ...... xxvi–xxvii

THE POLITICAL WORLD ...... xxviii–xxix

STATES AND BOUNDARIES...... xxx–xxxi

INTERNATIONAL DISPUTES xxxii

ATLAS OF THE WORLD

NORTH AMERICA

NORTH AMERICA 1

PHYSICAL NORTH AMERICA 2–3

POLITICAL NORTH AMERICA........ 4–5

NORTH AMERICAN RESOURCES...... 6–7

CANADA: WESTERN PROVINCES........ 8–9
Alberta, British Columbia, Manitoba,
Saskatchewan, Yukon

CANADA: EASTERN PROVINCES........ 10–11
New Brunswick, Newfoundland & Labrador,
Nova Scotia, Ontario, Prince Edward Island, Québec,
St. Pierre & Miquelon

SOUTHEASTERN CANADA............... 12–13
Southern Ontario, Southern Québec

CANADA (gatefold)........................ 14–16

UNITED STATES OF AMERICA (gatefold) . 17–19

USA: NORTHEASTERN STATES......... 20–21
Connecticut, Maine, Massachussetts, New Hampshire,
New Jersey, New York, Pennsylvania, Rhode Island, Vermont

USA: MID-EASTERN STATES........... 22–23
Delaware, District of Columbia, Kentucky,
Maryland, North Carolina, South Carolina,
Tennessee, Virginia, West Virginia

USA: SOUTHERN STATES............... 24–25
Alabama, Florida, Georgia, Louisiana, Mississippi

USA: TEXAS............................ 26–27

USA: SOUTH MIDWESTERN STATES... 28–29
Arkansas, Kansas, Missouri, Oklahoma

USA: UPPER PLAINS STATES........... 30–31
Iowa, Minnesota, Nebraska, North Dakota, South Dakota

USA: GREAT LAKES STATES............ 32–33
Illinois, Indiana, Michigan, Ohio, Wisconsin

USA: NORTH MOUNTAIN STATES..... 34–35
Idaho, Montana, Oregon, Washington, Wyoming

USA: CALIFORNIA & NEVADA......... 36–37

USA: SOUTH MOUNTAIN STATES 38–39
Arizona, Colorado, New Mexico, Utah

USA: HAWAII & ALASKA 40–41

MEXICO 42–43

CENTRAL AMERICA.................... 44–45
Belize, Costa Rica, El Salvador, Guatemala, Honduras,
Nicaragua, Panama

THE CARIBBEAN 46–47
Anguilla, Antigua & Barbuda, Aruba, Bahamas,
Barbados, British Virgin Islands, Cayman Islands, Cuba,
Dominica, Dominican Republic, Grenada, Guadeloupe,
Haiti, Jamaica, Martinique, Montserrat, Navassa Island,
Netherlands Antilles, Puerto Rico, St. Kitts & Nevis,
St. Lucia, St. Vincent & the Grenadines, Trinidad &
Tobago, Turks & Caicos Islands, Virgin Islands (US)

SOUTH AMERICA

SOUTH AMERICA...................... 48–49

PHYSICAL SOUTH AMERICA 50–51

POLITICAL SOUTH AMERICA 52–53

SOUTH AMERICAN RESOURCES.... 54–55

NORTHERN SOUTH AMERICA.......... 56–57
Colombia, French Guiana, Guyana, Suriname, Venezuela

WESTERN SOUTH AMERICA 58–59
Bolivia, Ecuador, Peru

BRAZIL 60–61

EASTERN SOUTH AMERICA............. 62–63
Southeast Brazil, Northeast Argentina, Uruguay

SOUTHERN SOUTH AMERICA 64–65
Argentina, Chile, Paraguay

THE ATLANTIC OCEAN................ 66–67

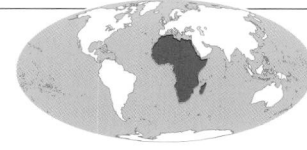

AFRICA

AFRICA 68–69

PHYSICAL AFRICA..................... 70–71

POLITICAL AFRICA 72–73

AFRICAN RESOURCES 74–75

NORTH AFRICA........................ 76–77
Algeria, Egypt, Libya, Morocco, Tunisia, Western Sahara

WEST AFRICA 78–79
Benin, Burkina, Cape Verde, Gambia, Ghana, Guinea,
Guinea-Bissau, Ivory Coast, Liberia, Mali, Mauritania,
Niger, Nigeria, Senegal, Sierra Leone, Togo

CENTRAL AFRICA 80–81
Cameroon, Central African Republic, Chad, Congo,
Dem. Rep. Congo (Zaire), Equatorial Guinea, Gabon,
Sao Tome & Principe

EAST AFRICA 82–83
Burundi, Djibouti, Eritrea, Ethiopia, Kenya,
Rwanda, Somalia, Sudan, Tanzania, Uganda

SOUTHERN AFRICA 84–85
Angola, Botswana, Lesotho, Malawi, Mozambique,
Namibia, South Africa, Swaziland, Zambia, Zimbabwe

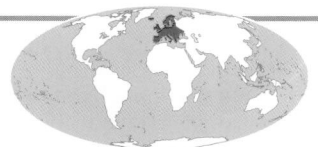

EUROPE

EUROPE *86–87*

PHYSICAL EUROPE *88–89*

POLITICAL EUROPE *90–91*

EUROPEAN RESOURCES *92–93*

SCANDINAVIA, FINLAND, & ICELAND . *94–95*
Denmark, Finland, Iceland, Norway, Svalbard, Sweden

SOUTHERN SCANDINAVIA *96–97*
Denmark, Faeroe Islands, Southern Norway, Southern Sweden

THE BRITISH ISLES *98–99*
Channel Islands, Isle of Man, Republic of Ireland, United Kingdom

THE LOW COUNTRIES *100–101*
Belgium, Luxembourg, Netherlands

GERMANY *102–103*

FRANCE *104–105*
France, Monaco

THE IBERIAN PENINSULA............. *106–107*
Andorra, Gibraltar, Portugal, Spain

THE ITALIAN PENINSULA............. *108–109*
Italy, San Marino, Vatican City

THE ALPINE STATES................... *110–111*
Austria, Liechtenstein, Slovenia, Switzerland

CENTRAL EUROPE..................... *112–113*
Czech Republic, Hungary, Poland, Slovakia

SOUTHEAST EUROPE *114–115*
Albania, Bosnia & Herzegovina, Croatia, Macedonia, Yugoslavia

BULGARIA & GREECE................. *116–117*
(including European Turkey)

ROMANIA, MOLDOVA, & UKRAINE......................... *118–119*

THE BALTIC STATES & BELARUS......................... *120–121*
Belarus, Estonia, Latvia, Lithuania, Kaliningrad

THE MEDITERRANEAN (gatefold)....... *122–124*

THE RUSSIAN FEDERATION (gatefold).. *125–127*

NORTHERN EUROPEAN RUSSIA *128–129*

SOUTHERN EUROPEAN RUSSIA....... *130–131*

ASIA

ASIA................................... *132–133*

ASIAN RESOURCES.................. *134–135*

POLITICAL ASIA..................... *136 / 141*

PHYSICAL ASIA (gatefold)............. *137–140*

TURKEY & THE CAUCASUS........... *142–143*
Armenia, Azerbaijan, Georgia, Turkey

THE NEAR EAST...................... *144–145*
Iraq, Israel, Jordan, Lebanon, Syria

THE ARABIAN PENINSULA *146–147*
Bahrain, Kuwait, Oman, Qatar, Saudi Arabia, United Arab Emirates, Yemen

IRAN & THE GULF STATES............ *148–149*
Bahrain, Iran, Kuwait, Qatar, United Arab Emirates

KAZAKHSTAN *150–151*

CENTRAL ASIA *152–153*
Kyrgyzstan, Tajikistan, Turkmenistan, Uzbekistan

AFGHANISTAN & PAKISTAN *154–155*

SOUTH ASIA *156–157*
Bangladesh, Bhutan, India, Maldives, Nepal, Pakistan, Sri Lanka

NORTHERN INDIA & THE HIMALAYAN STATES.............. *158–159*
Bangladesh, Bhutan, Nepal, Northern India

SOUTHERN INDIA & SRI LANKA..... *160–161*

MAINLAND EAST ASIA *162–163*
China, Mongolia, North Korea, South Korea, Taiwan

WESTERN CHINA.................... *164–165*

EASTERN CHINA..................... *166–167*
Eastern China, Taiwan

NORTHEASTERN CHINA, MONGOLIA & KOREA................ *168–169*
Mongolia, Northeastern China, North Korea, South Korea

JAPAN (gatefold) *170–172*

MARITIME SOUTHEAST ASIA (gatefold) . *173–176*
Brunei, East Timor, Indonesia, Malaysia, Singapore

MAINLAND SOUTHEAST ASIA & THE PHILIPPINES (gatefold) *177–179*
Myanmar, Cambodia, Laos, Paracel Islands, Philippines, Spratly Islands, Thailand, Vietnam

THE INDIAN OCEAN................. *180–181*

AUSTRALASIA AND OCEANIA

AUSTRALASIA & OCEANIA *182–183*

POLITICAL AUSTRALASIA & OCEANIA...................... *184–185*

AUSTRALASIAN & OCEANIAN RESOURCES *186–187*

AUSTRALIA *188–189*

SOUTHEAST AUSTRALIA *190–191*
New South Wales, South Australia, Tasmania, Victoria

NEW ZEALAND..................... *192–193*

PAPUA NEW GUINEA & THE SOLOMON ISLANDS........ *194–195*

MICRONESIA...................... *196 / 201*
Guam, Marshall Islands, Micronesia, Nauru, Northern Mariana Islands, Palau, Wake Island

MELANESIA *197*
Fiji, New Caledonia, Vanuatu

THE PACIFIC OCEAN (gatefold) *198–200*

POLYNESIA *202–203*
Cook Islands, Easter Island, French Polynesia, Kiribati, Niue, Pitcairn Islands, Tokelau, Tuvalu, Wallis & Futuna

ANTARCTICA *204–205*

THE ARCTIC....................... *206–207*

INDEX–GAZETTEER

GEOGRAPHICAL COMPARISONS *208–209*

COUNTRIES OF THE WORLD
including World Timezones *210–217*

GEOGRAPHICAL NAMES *218*

INDEX .... *219 - 353*

CREDITS/ACKNOWLEDGMENTS ... *353*

SELECTED GLOSSARY................. *354*

KEY TO REGIONAL MAPS

PHYSICAL FEATURES

elevation

6000m / 19,686ft
4000m / 13,124ft
3000m / 9843ft
2000m / 6562ft
1000m / 3281ft
500m / 1640ft
250m / 820ft
100m / 328ft
sea level
below sea level

▲ elevation above sea level (mountain height)
▲ volcano
✕ pass
▼ elevation below sea level (depression depth)

sand desert
lava flow
coastline
reef
atoll

sea depth

sea level
-250m / -820ft
-500m / -1640ft
-1000m / -3281ft
-2000m / -6562ft
-3000m / -9843ft

▲ seamount / guyot symbol
▼ undersea spot depth

DRAINAGE FEATURES

main river
secondary river
tertiary river
minor river
main seasonal river
secondary seasonal river
canal
waterfall
rapids
dam
perennial lake
seasonal lake
perennial salt lake
seasonal salt lake
reservoir
salt flat / salt pan
marsh / salt marsh
mangrove
wadi
⊙ spring / well / waterhole / oasis

ICE FEATURES

ice cap / sheet
ice shelf
glacier / snowfield
• • • • summer pack ice limit
winter pack ice limit

COMMUNICATIONS

━━━ highway
═ ═ ═ highway (under construction)
──── major road
──── minor road
⇥┈┈⇤ tunnel (road)
──── main line
──── minor line
⇥┈┈⇤ tunnel (railroad)
✈ international airport

BORDERS

━━━ full international border
■ ■ ■ ■ undefined international border
━ ━ ━ disputed de facto border
━ ∙ ━ ∙ ━ disputed territorial claim border
━ ━ ━ indication of country extent (Pacific only)
━ ━ ━ indication of dependent territory extent (Pacific only)
• • • • • • demarcation/ cease-fire line
──── autonomous / federal region border
──── 2nd order internal administrative border
──── 3rd order internal administrative border

SETTLEMENTS

built-up area

settlement population symbols

■ more than 5 million
■ 1 million to 5 million
◉ 500,000 to 1 million
◎ 100,000 to 500,000
⊕ 50,000 to 100,000
○ 10,000 to 50,000
○ fewer than 10,000

■ ◉ ● country/dependent territory capital city
■ ◉ ● autonomous / federal region / 2nd order internal administrative center
■ ◉ ● 3rd order internal administrative center

MISCELLANEOUS FEATURES

═══════ ancient wall
◇ site of interest
○ scientific station

GRATICULE FEATURES

lines of latitude and longitude / Equator
Tropics / Polar circles
45° degrees of longitude / latitude

TYPOGRAPHIC KEY

PHYSICAL FEATURES

landscape features .. *Namib Desert*
Massif Central
ANDES
headland .... *Nordkapp*
elevation / volcano / pass .... Mount Meru 4556 m
drainage features .... *Lake Rudolf*
rivers / canals spring / well / waterhole / oasis / waterfall / rapids / dam .... *Mekong*
ice features .... *Vatnajökull*
sea features........... *Golfe de Lion*
Andaman Sea
INDIAN OCEAN
undersea features ... *Barracuda Fracture Zone*

REGIONS

country................. **ARMENIA**
dependent territory with parent state..... NIUE (to NZ)
region outside feature area........... ANGOLA
autonomous / federal region........ MINAS GERAIS
2nd order internal administrative region................. MINSKAYA VOBLASTS'
3rd order internal administrative region................. Vaucluse
cultural region....... New England

SETTLEMENTS

capital city............ **BEIJING**
dependent territory capital city............ FORT-DE-FRANCE
other settlements.... **Chicago**
Adana
Tizi Ozou
Yonezawa
Farnham

MISCELLANEOUS

sites of interest / miscellaneous........ Valley of the Kings
Tropics / Polar circles........... *Antarctic Circle*

HOW TO USE THIS ATLAS

THE ATLAS IS ORGANIZED BY CONTINENT, moving eastward from the International Dateline. The opening section describes the world's structure, systems, and its main features. The Atlas of the World that follows, is a continent-by-continent guide to today's world, starting with a comprehensive insight into the physical, political, and economic structure of each continent, followed by integrated mapping and descriptions of each region or country.

THE WORLD

THE INTRODUCTORY SECTION of the Atlas deals with every aspect of the planet, from physical structure to human geography, providing an overall picture of the world we live in. Complex topics such as the landscape of the Earth, climate, oceans, population, and economic patterns are clearly explained with the aid of maps and diagrams drawn from the latest information.

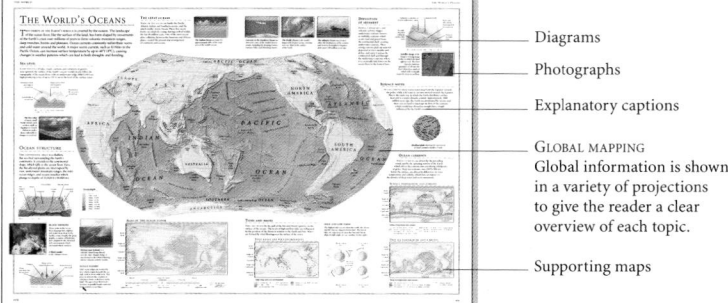

Diagrams
Photographs
Explanatory captions
GLOBAL MAPPING
Global information is shown in a variety of projections to give the reader a clear overview of each topic.
Supporting maps

THE POLITICAL CONTINENT

THE POLITICAL PORTRAIT of the continent is a vital reference point for every continental section, showing the position of countries relative to one another, and the relationship between human settlement and geographic location. The complex mosaic of languages spoken in each continent is mapped, as is the effect of communications networks on the pattern of settlement.

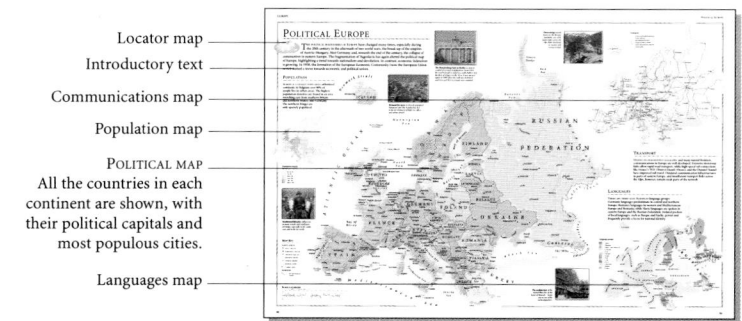

Locator map
Introductory text
Communications map
Population map
POLITICAL MAP
All the countries in each continent are shown, with their political capitals and most populous cities.
Languages map

CONTINENTAL RESOURCES

THE EARTH'S RICH NATURAL RESOURCES, including oil, gas, minerals, and fertile land, have played a key role in the development of society. These pages show the location of minerals and agricultural resources on each continent, and how they have been instrumental in dictating industrial growth and the varieties of economic activity across the continent.

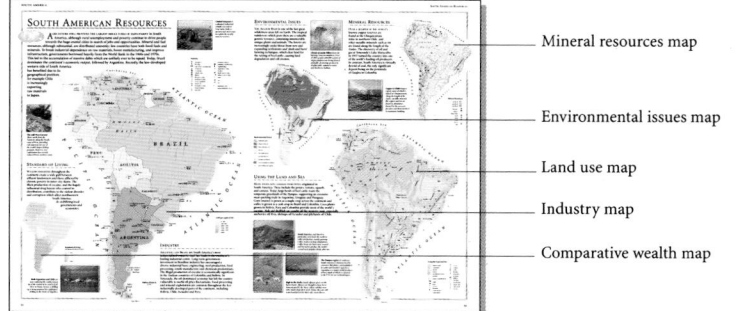

Mineral resources map
Environmental issues map
Land use map
Industry map
Comparative wealth map

THE PHYSICAL CONTINENT

THE ASTONISHING VARIETY of landforms, and the dramatic forces that created and continue to shape the landscape, are explained in the continental physical spread. Cross-sections, illustrations, and terrain maps highlight the different parts of the continent, showing how nature's forces have produced the landscapes we see today.

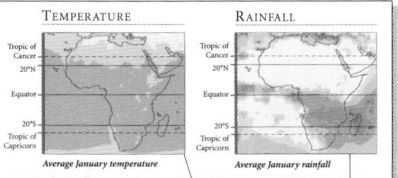

CLIMATE CHARTS
Rainfall and temperature charts clearly show the continental patterns of rainfall and temperature.

CLIMATE MAP
Climatic regions vary across each continent. The map displays the differing climatic regions, as well as daily hours of sunshine at selected weather stations.

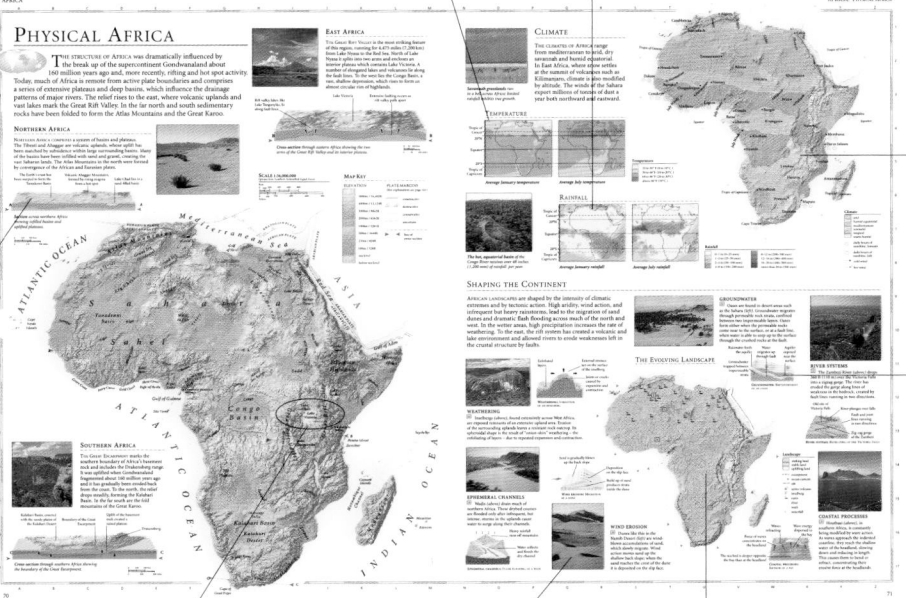

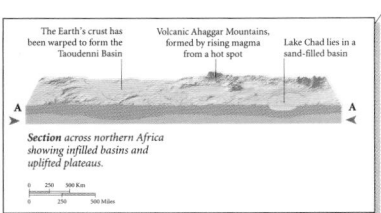

CROSS-SECTIONS
Detailed cross-sections through selected parts of the continent show the underlying geomorphic structure.

MAIN PHYSICAL MAP
Detailed satellite data has been used to create an accurate and visually striking picture of the surface of the continent.

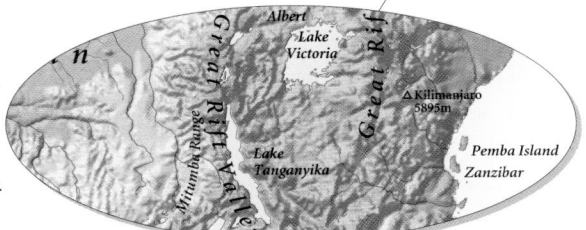

PHOTOGRAPHS
A wide range of beautiful photographs bring the world's regions to life.

Rainwater feeds the aquifer — Water migrates up through fault — Aquifer exposed near the surface. Groundwater trapped between impermeable strata.

GROUNDWATER: REPLENISHMENT OF AN OASIS

LANDFORM DIAGRAMS
The complex formation of many typical landforms is summarized in these easy-to-understand illustrations.

LANDSCAPE EVOLUTION MAP
The physical shape of each continent is affected by a variety of forces which continually sculpt and modify the landscape. This map shows the major processes which affect different parts of the continent.

REGIONAL MAPPING

THE MAIN BODY of the Atlas is a unique regional map set, with detailed information on the terrain, the human geography of the region and its infrastructure. Around the edge of the map, additional 'at-a-glance' maps, give an instant picture of regional industry, land use and agriculture. The detailed terrain map (shown in perspective), focuses on the main physical features of the region, and is enhanced by annotated illustrations, and photographs of the physical structure.

TRANSPORTATION NETWORK

340,090 miles (544,144 km)		4813 miles 7700 km	
12,872 miles (20,592 km)		2108 miles (3389 km)	

New York's commercial success is tied historically to its transportation connections. The Erie Canal, completed in 1825, opened up the Great Lakes and the interior to New York's markets and carried a stream of immigrants into the Midwest.

TRANSPORTATION NETWORK
The differing extent of the transportation network for each region is shown here, along with key facts about the transportation system.

REGIONAL LOCATOR
This small map shows the location of each country in relation to its continent.

KEY TO MAIN MAP
A key to the population symbols and land heights accompanies the main map.

WORLD LOCATOR
This locates the continent in which the region is found on a small world map.

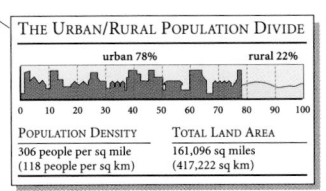

LAND USE MAP
This shows the different types of land use which characterize the region, as well as indicating the principal agricultural activities.

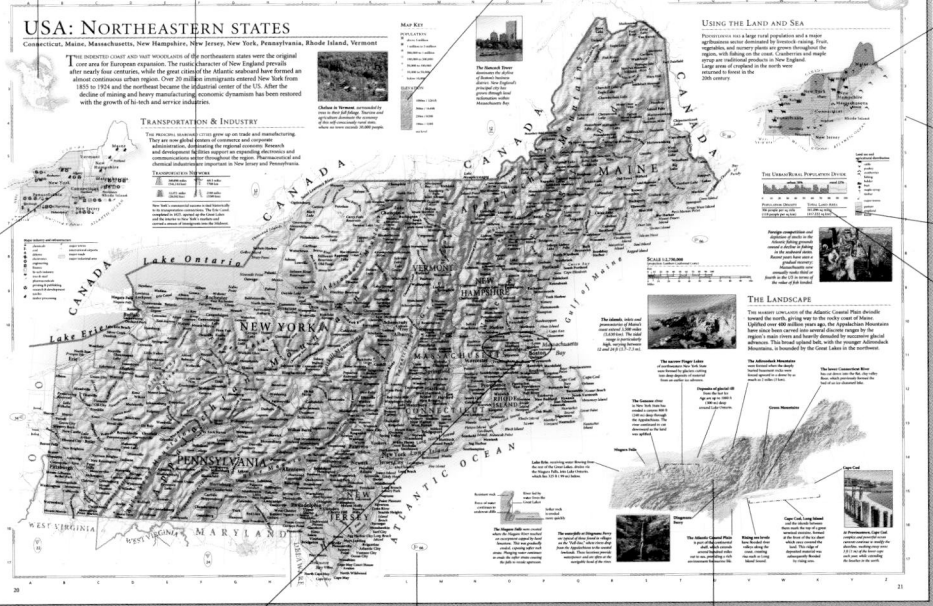

USA: NORTHEASTERN STATES
Connecticut, Maine, Massachusetts, New Hampshire, New Jersey, New York, Pennsylvania, Rhode Island, Vermont

GRID REFERENCE
The framing grid provides a location reference for each place listed in the Index.

MAP KEYS
Each supporting map has its own key.

THE URBAN/RURAL POPULATION DIVIDE

urban 78%   rural 22%

0 10 20 30 40 50 60 70 80 90 100

POPULATION DENSITY	TOTAL LAND AREA
306 people per sq mile (118 people per sq km)	161,096 sq miles (417,222 sq km)

URBAN/RURAL POPULATION DIVIDE
The proportion of people in the region who live in urban and rural areas, as well as the overall population density and land area are clearly shown in these simple graphics.

TRANSPORTATION AND INDUSTRY MAP
The main industrial areas are mapped, and the most important industrial and economic activities of the region are shown.

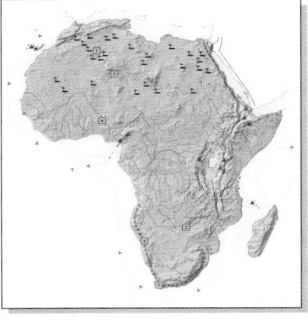

CONTINUATION SYMBOLS
These symbols indicate where adjacent maps can be found.

MAIN REGIONAL MAP
A wealth of information is displayed on the main map, building up a rich portrait of the interaction between the physical landscape and the human and political geography of each region. The key to the regional maps can be found on page viii.

LANDSCAPE MAP
The computer-generated terrain model accurately portrays an oblique view of the landscape. Annotations highlight the most important geographic features of the region.

JUPITER

- ⊖ **Diameter:** 88,846 miles (142,984 km)
- ⊙ **Mass:** 1,900,000 million million million tons
- ○ **Temperature:** -153°C (extremes not available)
- ▷◁ **Distance from Sun:** 483 million miles (778 million km)
- ◐ **Length of day:** 9.84 hours
- ◑ **Length of year:** 11.86 earth years
- ⊖ **Surface gravity:** 1 kg = 2.53 kg

MARS

- ⊖ **Diameter:** 4,217 miles (6,786 km)
- ○ **Mass:** 642 million million million tons
- ○ **Temperature:** -137 to 37°C
- ▷◁ **Distance from Sun:** 142 million miles (228 million km)
- ◐ **Length of day:** 24.623 hours
- ◑ **Length of year:** 1.88 earth years
- ⊖ **Surface gravity:** 1 kg = 0.38 kg

EARTH

- ⊖ **Diameter:** 7,926 miles (12,756 km)
- ○ **Mass:** 5,976 million million million tons
- ○ **Temperature:** -70 to 55°C
- ▷◁ **Distance from Sun:** 93 million miles (150 million km)
- ◐ **Length of day:** 23.92 hours
- ◑ **Length of year:** 365.25 earth days
- ⊖ **Surface gravity:** 1 kg = 1 kg

VENUS

- ⊖ **Diameter:** 7,520 miles (12,102 km)
- ○ **Mass:** 4,870 million million million tons
- ○ **Temperature:** 457°C (extremes not available)
- ▷◁ **Distance from Sun:** 67 million miles (108 million km)
- ◐ **Length of day:** 243.01 earth days
- ◑ **Length of year:** 224.7 earth days
- ⊖ **Surface gravity:** 1 kg = 0.88 kg

MERCURY

- ⊖ **Diameter:** 3,031 miles (4,878 km)
- ○ **Mass:** 330 million million million tons
- ○ **Temperature:** -173 to 427°C
- ▷◁ **Distance from Sun:** 36 million miles (58 million km)
- ◐ **Length of day:** 58.65 earth days
- ◑ **Length of year:** 87.97 earth days
- ⊖ **Surface gravity:** 1 kg = 0.38 kg

THE SOLAR SYSTEM

NINE MAJOR PLANETS, their satellites, and countless minor planets (asteroids) orbit the Sun to form the Solar System. The Sun, our nearest star, creates energy from nuclear reactions deep within its interior, providing all the light and heat which make life on Earth possible. The Earth is unique in the Solar System in that it supports life: its size, gravitational pull and distance from the Sun have all created the optimum conditions for the evolution of life. The planetary images seen here are composites derived from actual spacecraft images (not shown to scale).

THE SUN

- ⊖ **Diameter:** 864,948 miles (1,392,000 km)
- ⊙ **Mass:** 1990 million million million million tons

THE SUN was formed when a swirling cloud of dust and gas contracted, pulling matter into its center. When the temperature at the center rose to 1,000,000°C, nuclear fusion – the fusing of hydrogen into helium, creating energy – occurred, releasing a constant stream of heat and light.

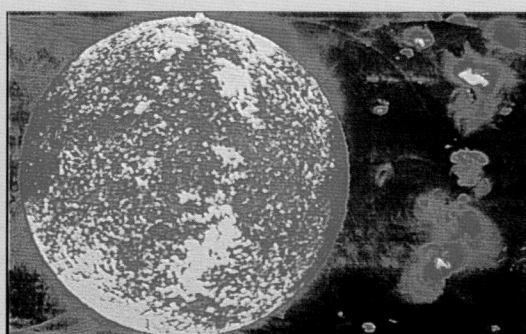

Solar flares are sudden bursts of energy from the Sun's surface. They can be 125,000 miles (200,000 km) long.

THE FORMATION OF THE SOLAR SYSTEM

The cloud of dust and gas thrown out by the Sun during its formation cooled to form the Solar System. The smaller planets nearest the Sun are formed of minerals and metals. The outer planets were formed at lower temperatures, and consist of swirling clouds of gases.

THE MILANKOVITCH CYCLE

The amount of radiation from the Sun which reaches the Earth is affected by variations in the Earth's orbit and the tilt of the Earth's axis, as well as by "wobbles" in the axis. These variations cause three separate cycles, corresponding with the durations of recent ice ages.

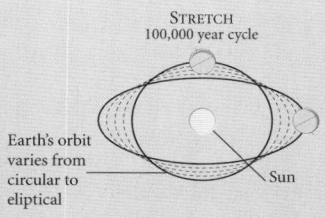

STRETCH
100,000 year cycle

Earth's orbit varies from circular to eliptical

Sun

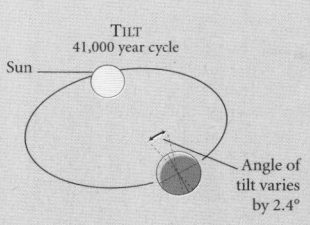

TILT
41,000 year cycle

Sun

Angle of tilt varies by 2.4°

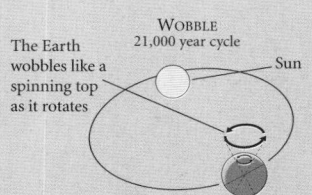

WOBBLE
21,000 year cycle

The Earth wobbles like a spinning top as it rotates

Sun

SATURN

- ⊖ **Diameter:** 74,974 miles (120,660 km)
- ○ **Mass:** 570,000 million million million tons
- ○ **Temperature:** -185°C (extremes not available)
- ◗ **Distance from Sun:** 887 million miles (1,427 million km)
- ◑ **Length of day:** 10.23 hours
- ◑ **Length of year:** 29.46 earth years
- ⊖ **Surface gravity:** 1 kg = 1.07 kg

URANUS

- ⊖ **Diameter:** 31,763 miles (51,118 km)
- ○ **Mass:** 86,800 million million million tons
- ○ **Temperature:** -214°C (extremes not available)
- ◗ **Distance from Sun:** 1,783 million miles (2,870 million km)
- ◑ **Length of day:** 17.9 hours
- ◑ **Length of year:** 84.01 earth years
- ⊖ **Surface gravity:** 1 kg = 0.92 kg

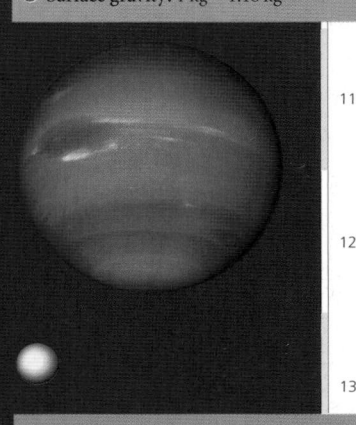

NEPTUNE

- ⊖ **Diameter:** 30,775 miles (49,528 km)
- ○ **Mass:** 102,000 million million million tons
- ○ **Temperature:** -225°C (extremes not available)
- ◗ **Distance from Sun:** 2794 million miles (4497 million km)
- ◑ **Length of day:** 19.2 hours
- ◑ **Length of year:** 164.79 earth years
- ⊖ **Surface gravity:** 1 kg = 1.18 kg

SPACE DEBRIS

MILLIONS OF OBJECTS, remnants of planetary formation, circle the Sun in a zone lying between Mars and Jupiter: the asteroid belt. Fragments of asteroids break off to form meteoroids, which can reach the Earth's surface. Comets, composed of ice and dust, originated outside our Solar System. Their elliptical orbit brings them close to the Sun and into the inner Solar System.

Meteor Crater in Arizona is 4200 ft (1300 m) wide and 660 ft (200 m) deep. It was formed over 10,000 years ago.

METEOROIDS

Meteoroids are fragments of asteroids which hurtle through space at great velocity. Although millions of meteoroids enter the Earth's atmosphere, the vast majority burn up on entry, and fall to the Earth as a meteor or shooting star. Large meteoroids traveling at speeds of 155,000 mph (250,000 kmph) can sometimes withstand the atmosphere and hit the Earth's surface with tremendous force, creating large craters on impact.

POSSIBLE AND ACTUAL METEORITE CRATERS

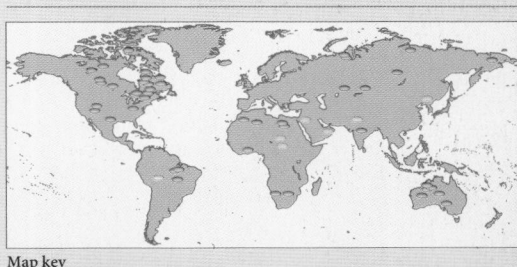

Map key

○ Possible impact craters   ○ Meteorite impact craters

The orbit of Halley's Comet brings it close to the Earth every 76 years. It last visited in 1986.

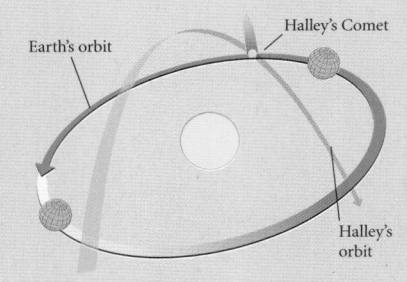

Earth's orbit

Halley's Comet

Halley's orbit

ORBIT OF HALLEY'S COMET AROUND THE SUN

THE EARTH'S ATMOSPHERE

DURING THE EARLY STAGES of the Earth's formation, ash, lava, carbon dioxide, and water vapor were discharged onto the surface of the planet by constant volcanic eruptions. The water formed the oceans, while carbon dioxide entered the atmosphere or was dissolved in the oceans. Clouds, formed of water droplets, reflected some of the Sun's radiation back into space. The Earth's temperature stabilized and early life forms began to emerge, converting carbon dioxide into life-giving oxygen.

It is thought that the gases that make up the Earth's atmosphere originated deep within the interior, and were released many millions of years ago during intense volcanic actvity, similar to this eruption at Mount St. Helens.

PLUTO

- ⊖ **Diameter:** 1,429 miles (2,300 km)
- ○ **Mass:** 13 million million million tons
- ○ **Temperature:** -236°C (extremes not available)
- ◗ **Distance from Sun:** 3,666 million miles (5,900 million km)
- ◑ **Length of day:** 6.39 hours
- ◑ **Length of year:** 248.54 earth years
- ⊖ **Surface gravity:** 1 kg = 0.30 kg

ORDER AND RELATIVE DISTANCE FROM THE SUN OF PLANETS

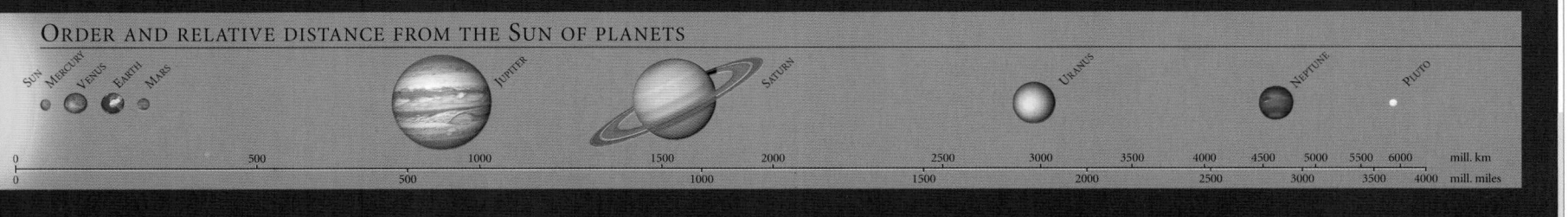

SUN MERCURY VENUS EARTH MARS JUPITER SATURN URANUS NEPTUNE PLUTO

| 0 | 500 | 1000 | 1500 | 2000 | 2500 | 3000 | 3500 | 4000 | 4500 | 5000 | 5500 | 6000 | mill. km |
| 0 | | 500 | | 1000 | | 1500 | | 2000 | | 2500 | 3000 | | 3500 | 4000 | mill. miles |

THE PHYSICAL WORLD

THE EARTH'S SURFACE is constantly being transformed: it is uplifted, folded and faulted by tectonic forces; weathered and eroded by wind, water, and ice. Sometimes change is dramatic, the spectacular results of earthquakes or floods. More often it is a slow process lasting millions of years. A physical map of the world represents a snapshot of the ever-evolving architecture of the Earth. This terrain map shows the whole surface of the Earth, both above and below the sea.

THE WORLD IN SECTION

These cross-sections around the Earth, one in the northern hemisphere; one straddling the Equator, reveal the limited areas of land above sea level in comparison with the extent of the sea floor. The greater erosive effects of weathering by wind and water limit the upward elevation of land above sea level, while the deep oceans retain their dramatic mountain and trench profiles.

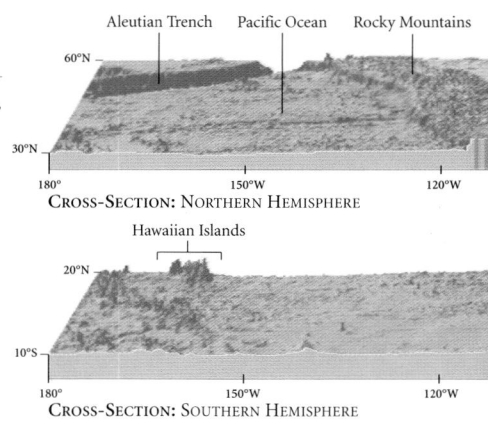

Aleutian Trench Pacific Ocean Rocky Mountains

CROSS-SECTION: NORTHERN HEMISPHERE

Hawaiian Islands

CROSS-SECTION: SOUTHERN HEMISPHERE

MAP KEY

GEOGRAPHICAL REGIONS

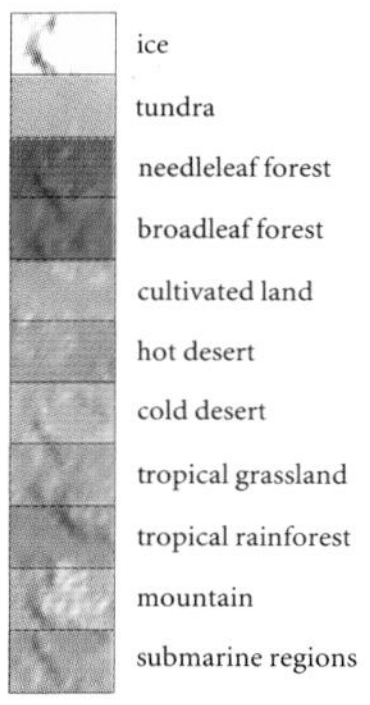

- ice
- tundra
- needleleaf forest
- broadleaf forest
- cultivated land
- hot desert
- cold desert
- tropical grassland
- tropical rainforest
- mountain
- submarine regions

SCALE 1:60,000,000
(projection: Wagner VII)

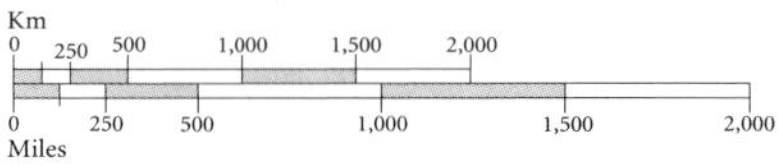

Km 0 250 500 1,000 1,500 2,000

Miles 0 250 500 1,000 1,500 2,000

NORTHERN HEMISPHERE

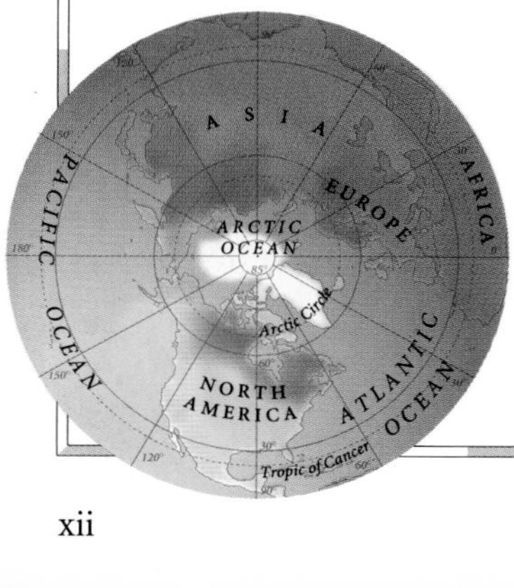

MOST OF the land on Earth is concentrated in the northern hemisphere, although Europe and North America are the only continents which lie wholly in the north.

ARCTIC OCEAN · ASIA · EUROPE · AMERICA · PACIFIC OCEAN · ATLANTIC OCEAN · NORTH AMERICA · Arctic Circle · Tropic of Cancer

NORTH AMERICA
Great Plains

SOUTH AMERICA

PACIFIC OCEAN

ATLANTIC OCEAN

SOUTHERN

ANT

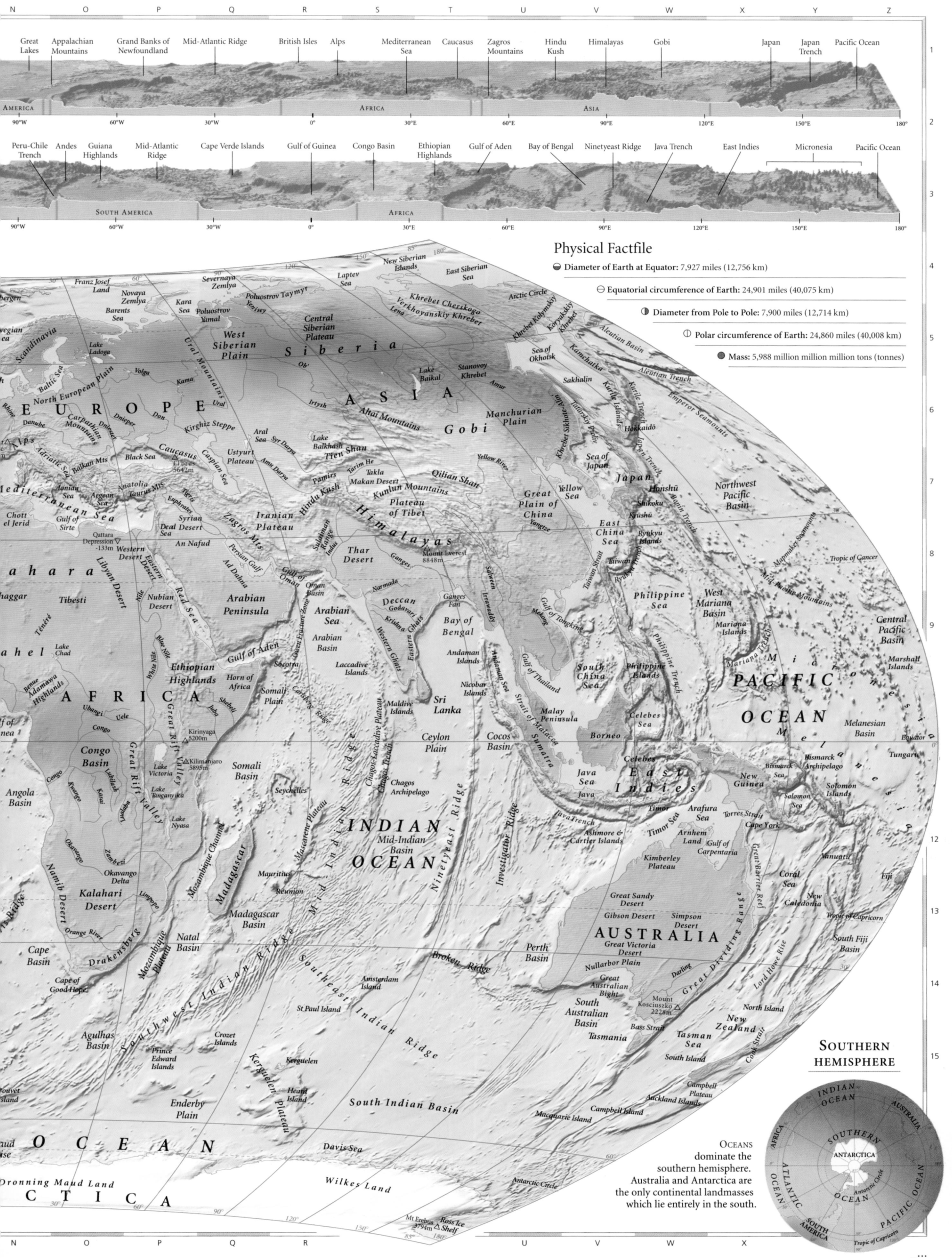

STRUCTURE OF THE EARTH

THE EARTH AS IT IS TODAY is just the latest phase in a constant process of evolution which has occurred over the past 4.5 billion years. The Earth's continents are neither fixed nor stable; over the course of the Earth's history, propelled by currents rising from the intense heat at its center, the great plates on which they lie have moved, collided, joined together, and separated. These processes continue to mold and transform the surface of the Earth, causing earthquakes and volcanic eruptions and creating oceans, mountain ranges, deep ocean trenches, and island chains.

INSIDE THE EARTH

THE EARTH'S HOT INNER CORE is made up of solid iron, while the outer core is composed of liquid iron and nickel. The mantle nearest the core is viscous, whereas the rocky upper mantle is fairly rigid. The crust is the rocky outer shell of the Earth. Together, the upper mantle and the crust form the lithosphere.

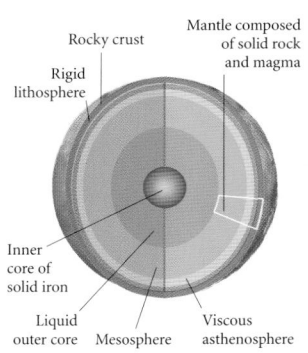

Rocky crust
Mantle composed of solid rock and magma
Rigid lithosphere
Inner core of solid iron
Liquid outer core
Mesosphere
Viscous asthenosphere

THE DYNAMIC EARTH

THE EARTH'S CRUST is made up of eight major (and several minor) rigid continental and oceanic tectonic plates, which fit closely together. The positions of the plates are not static. They are constantly moving relative to one another. The type of movement between plates affects the way in which they alter the structure of the Earth. The oldest parts of the plates, known as shields, are the most stable parts of the Earth and little tectonic activity occurs here.

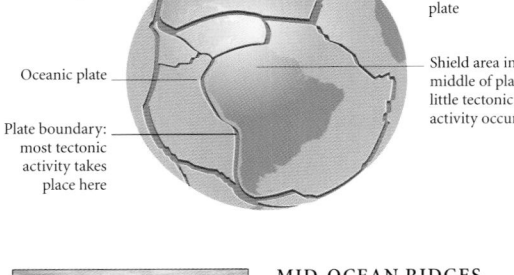

Continental plate
Oceanic plate
Plate boundary: most tectonic activity takes place here
Rigid tectonic plate
Shield area in middle of plate: little tectonic activity occurs here

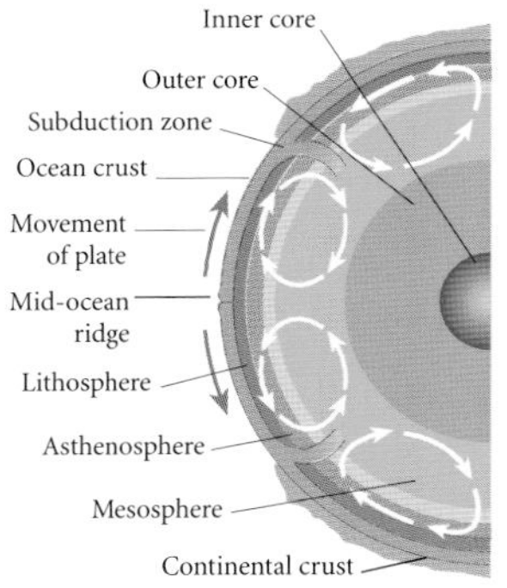

CONVECTION CURRENTS

DEEP WITHIN THE EARTH, at its inner core, temperatures may exceed 8,100°F (4,500°C). This heat warms rocks in the mesosphere which rise through the partially molten mantle, displacing cooler rocks just below the solid crust, which sink, and are warmed again by the heat of the mantle. This process is continuous, creating convection currents which form the moving force beneath the Earth's crust.

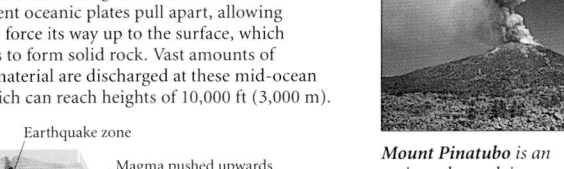

Inner core
Outer core
Subduction zone
Ocean crust
Movement of plate
Mid-ocean ridge
Lithosphere
Asthenosphere
Mesosphere
Continental crust

PLATE BOUNDARIES

THE BOUNDARIES BETWEEN THE PLATES are the areas where most tectonic activity takes place. Three types of movement occur at plate boundaries: the plates can either move toward each other, move apart, or slide past each other. The effect this has on the Earth's structure depends on whether the margin is between two continental plates, two oceanic plates, or an oceanic and continental plate.

MID-OCEAN RIDGES

Mid-ocean ridges are formed when two adjacent oceanic plates pull apart, allowing magma to force its way up to the surface, which then cools to form solid rock. Vast amounts of volcanic material are discharged at these mid-ocean ridges which can reach heights of 10,000 ft (3,000 m).

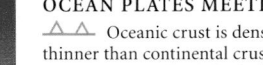

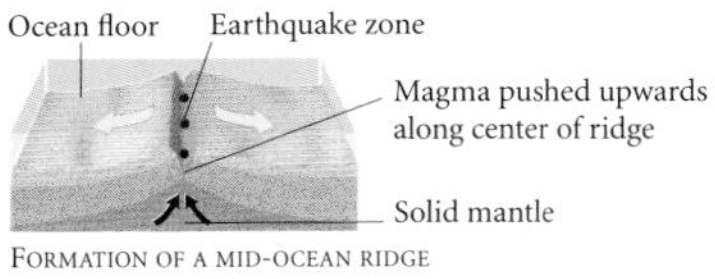

Ocean floor
Earthquake zone
Magma pushed upwards along center of ridge
Solid mantle

FORMATION OF A MID-OCEAN RIDGE

The Mid-Atlantic Ridge rises above sea level in Iceland, producing geysers and volcanoes.

OCEAN PLATES MEETING

Oceanic crust is denser and thinner than continental crust; on average it is 3 miles (5 km) thick, while continental crust averages 18–24 miles (30–40 km). When oceanic plates of similar density meet, the crust is contorted as one plate overrides the other, forming deep sea trenches and volcanic island arcs above sea level.

Mount Pinatubo is an active volcano, lying on the Pacific "Ring of Fire."

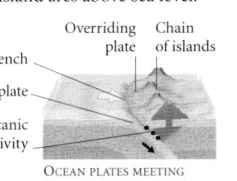

Overriding plate
Chain of islands
Ocean trench
Diving plate
Volcanic activity

OCEAN PLATES MEETING TO FORM AN ISLAND ARC

Tectonic Activity

- - - - -   uncertain plate boundary
▲   volcanic zone
●   earthquake zone
●   hot spot
ⱽⱽⱽⱽⱽ   rift valley

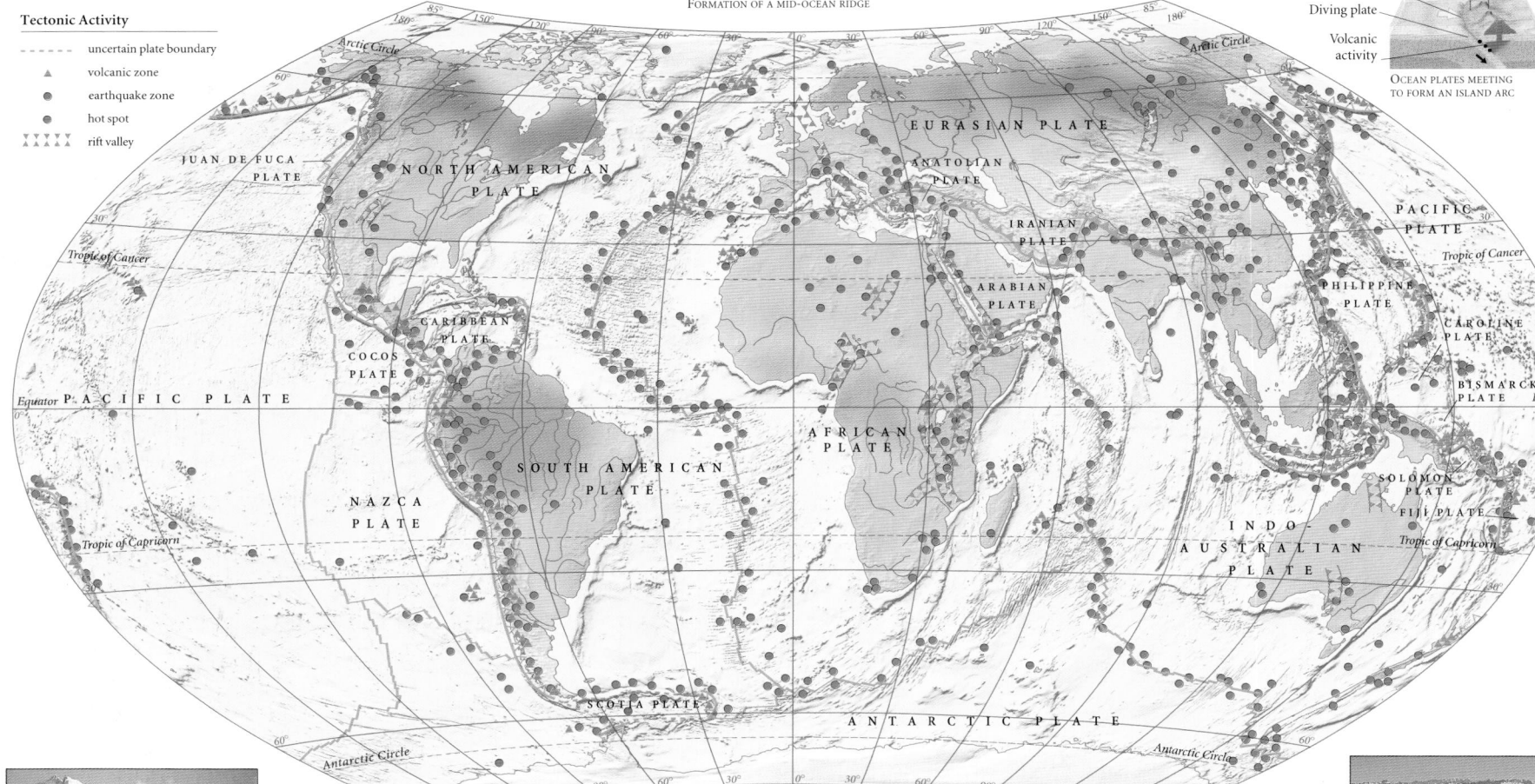

Arctic Circle

JUAN DE FUCA PLATE

NORTH AMERICAN PLATE

EURASIAN PLATE

ANATOLIAN PLATE

IRANIAN PLATE

ARABIAN PLATE

PACIFIC PLATE

PHILIPPINE PLATE

Tropic of Cancer

CARIBBEAN PLATE

COCOS PLATE

CAROLINE PLATE

BISMARCK PLATE

Equator   PACIFIC PLATE

Equator

AFRICAN PLATE

SOUTH AMERICAN PLATE

NAZCA PLATE

INDO-AUSTRALIAN PLATE

SOLOMON PLATE

FIJI PLATE

Tropic of Capricorn

Tropic of Capricorn

SCOTIA PLATE

ANTARCTIC PLATE

Antarctic Circle

Antarctic Circle

DIVING PLATES

When an oceanic and a continental plate meet, the denser oceanic plate is driven underneath the continental plate, which is crumpled by the collision to form mountain ranges. As the ocean plate plunges downward, it heats up, and molten rock (magma) is forced up to the surface.

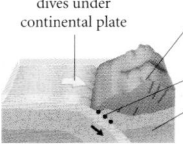

The Andean mountain chain is the typical result of the impact of a diving plate.

Oceanic plate dives under continental plate
Mountains thrust up by collision
Earthquake zone
Continental plate

DIVING PLATE

The deep fracture caused by the sliding plates of the San Andreas Fault can be clearly seen in parts of California.

SLIDING PLATES

When two plates slide past each other, friction is caused along the fault line which divides them. The plates do not move smoothly, and the uneven movement causes earthquakes.

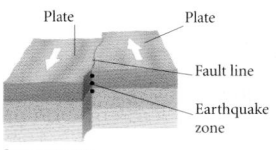

Plate
Plate
Fault line
Earthquake zone

SLIDING PLATES

The Alps were formed when the African plate collided with the Eurasian Plate, about 65 million years ago.

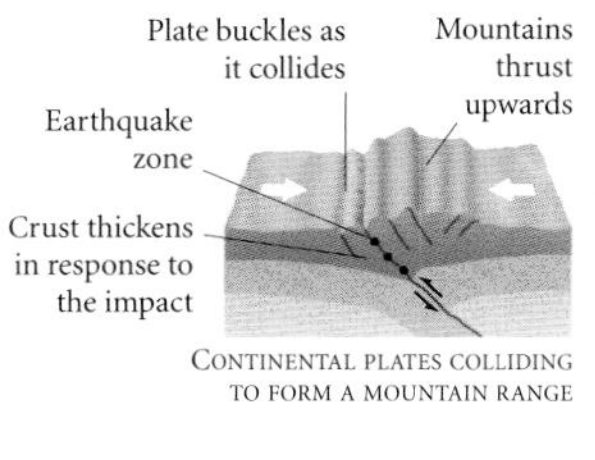

Plate buckles as it collides
Mountains thrust upwards
Earthquake zone
Crust thickens in response to the impact

CONTINENTAL PLATES COLLIDING TO FORM A MOUNTAIN RANGE

COLLIDING PLATES

When two continental plates collide, great mountain chains are thrust upward as the crust buckles and folds under the force of the impact.

CONTINENTAL DRIFT

ALTHOUGH THE PLATES which make up the Earth's crust move only a few inches in a year, over the millions of years of the Earth's history, its continents have moved many thousands of miles, to create new continents, oceans, and mountain chains.

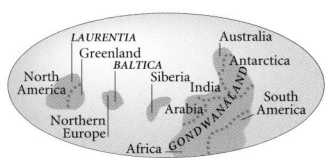

1: CAMBRIAN PERIOD

570–510 million years ago. Most continents are in tropical latitudes. The supercontinent of Gondwanaland reaches the South Pole.

2: DEVONIAN PERIOD

408–362 million years ago. The continents of Gondwanaland and Laurentia are drifting northward.

3: CARBONIFEROUS PERIOD

362–290 million years ago. The Earth is dominated by three continents; Laurentia, Angaraland, and Gondwanaland.

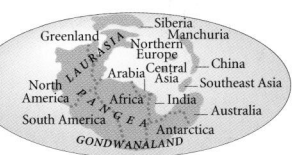

4: TRIASSIC PERIOD

245–208 million years ago. All three major continents have joined to form the supercontinent of Pangea.

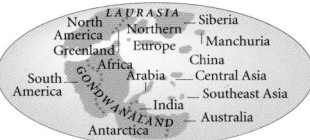

5: JURASSIC PERIOD

208–145 million years ago. The supercontinent of Pangea begins to break up, causing an overall rise in sea levels.

6: CRETACEOUS PERIOD

145–65 million years ago. Warm, shallow seas cover much of the land: sea levels are about 80 ft (25 m) above present levels.

7: TERTIARY PERIOD

65–2 million years ago. Although the world's geography is becoming more recognizable, major events such as the creation of the Himalayan mountain chain, are still to occur during this period.

CONTINENTAL SHIELDS

THE CENTERS OF THE EARTH'S CONTINENTS, known as shields, were established between 2500 and 500 million years ago; some contain rocks over three billion years old. They were formed by a series of turbulent events: plate movements, earthquakes, and volcanic eruptions. Since the Pre-Cambrian period, over 570 million years ago, they have experienced little tectonic activity, and today, these flat, low-lying slabs of solidified molten rock form the stable centers of the continents. They are bounded or covered by successive belts of younger sedimentary rock.

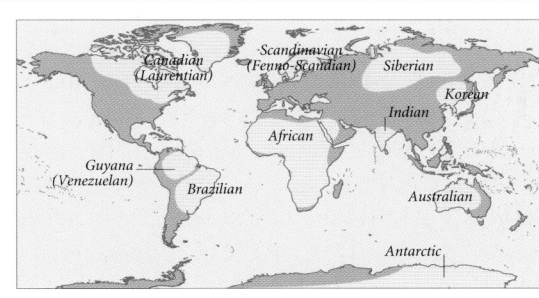

CREATION OF THE HIMALAYAS

BETWEEN 10 AND 20 MILLION YEARS AGO, the Indian subcontinent, part of the ancient continent of Gondwanaland, collided with the continent of Asia. The Indo-Australian Plate continued to move northward, displacing continental crust and uplifting the Himalayas, the world's highest mountain chain.

MOVEMENTS OF INDIA

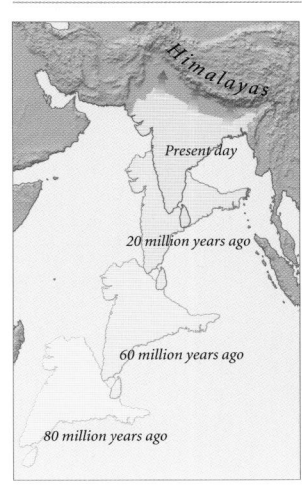

Present day

20 million years ago

60 million years ago

80 million years ago

Force of collision pushes up mountains

CROSS-SECTION THROUGH THE HIMALAYAS

The Himalayas were uplifted when the Indian subcontinent collided with Asia.

THE HAWAIIAN ISLAND CHAIN

A HOT SPOT lying deep beneath the Pacific Ocean pushes a plume of magma from the Earth's mantle up through the Pacific Plate to form volcanic islands. While the hot spot remains stationary, the plate on which the islands sit is moving slowly. A long chain of islands has been created as the plate passes over the hot spot.

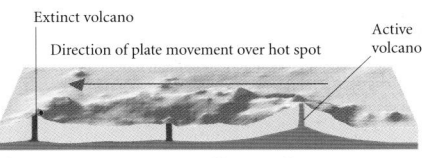

Extinct volcano

Direction of plate movement over hot spot

Active volcano

CROSS-SECTION THROUGH THE HAWAIIAN ISLANDS

EVOLUTION OF THE HAWAIIAN ISLANDS

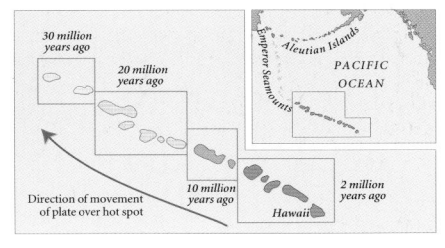

30 million years ago

20 million years ago

PACIFIC OCEAN

Direction of movement of plate over hot spot

10 million years ago

2 million years ago

Hawaii

THE EARTH'S GEOLOGY

THE EARTH'S ROCKS are created in a continual cycle. Exposed rocks are weathered and eroded by wind, water and chemicals and deposited as sediments. If they pass into the Earth's crust they will be transformed by high temperatures and pressures into metamorphic rocks or they will melt and solidify as igneous rocks.

GNEISS

1 Gneiss is a metamorphic rock made at great depth during the formation of mountain chains, when intense heat and pressure transform sedimentary or igneous rocks.

Gneiss formations in Norway's Jotunheimen Mountains.

Basalt columns at Giant's Causeway, Northern Ireland, UK.

BASALT

2 Basalt is an igneous rock, formed when small quantities of magma lying close to the Earth's surface cool rapidly.

LIMESTONE

3 Limestone is a sedimentary rock, which is formed mainly from the calcite skeletons of marine animals which have been compressed into rock.

Limestone hills, Guilin, China.

CORAL

4 Coral reefs are formed from the skeletons of millions of individual corals.

Great Barrier Reef, Australia.

SANDSTONE

8 Sandstones are sedimentary rocks formed mainly in deserts, beaches, and deltas. Desert sandstones are formed of grains of quartz which have been well rounded by wind erosion.

Rock stacks of desert sandstone, at Bryce Canyon National Park, Utah.

THE WORLD'S MAJOR GEOLOGICAL REGIONS

Extrusive igneous rocks are formed during volcanic eruptions, as here in Hawaii.

ANDESITE

7 Andesite is an extrusive igneous rock formed from magma which has solidified on the Earth's crust after a volcanic eruption.

Geological Regions

- continental shield
- sedimentary cover
- coral formation
- igneous rock types

Mountain Ranges

- Alpine (new)
- Hercynian (old)
- Caledonian (ancient)

SCHIST

6 Schist is a metamorphic rock formed during mountain building, when temperature and pressure are comparatively high. Both mudstones and shales reform into schist under these conditions.

Schist formations in the Atlas Mountains, northwestern Africa.

GRANITE

5 Granite is an intrusive igneous rock formed from magma which has solidified deep within the Earth's crust. The magma cools slowly, producing a coarse-grained rock.

Namibia's Namaqualand Plateau is formed of granite.

SHAPING THE LANDSCAPE

THE BASIC MATERIAL OF THE EARTH'S SURFACE is solid rock: valleys, deserts, soil, and sand are all evidence of the powerful agents of weathering, erosion, and deposition which constantly shape and transform the Earth's landscapes. Water, either flowing continually in rivers or seas, or frozen and compacted into solid sheets of ice, has the most clearly visible impact on the Earth's surface. But wind can transport fragments of rock over huge distances and strip away protective layers of vegetation, exposing rock surfaces to the impact of extreme heat and cold.

WATER

LESS THAN 2% of the world's water is on the land, but it is the most powerful agent of landscape change. Water, as rainfall, groundwater, and rivers, can transform landscapes through both erosion and deposition. Eroded material carried by rivers forms the world's most fertile soils.

Waterfalls such as the Iguaçu Falls on the border between Argentina and southern Brazil, erode the underlying rock, causing the falls to retreat.

COASTAL WATER

THE WORLD'S COASTLINES are constantly changing; every day, tides deposit, sift and sort sand and gravel on the shoreline. Over longer periods, powerful wave action erodes cliffs and headlands and carves out bays.

A low, wide sandy beach on South Africa's Cape Peninsula is continually re-shaped by the action of the Atlantic waves.

The sheer chalk cliffs at Seven Sisters in southern England are constantly under attack from waves.

GROUNDWATER

IN REGIONS where there are porous rocks such as chalk, water is stored underground in large quantities; these reservoirs of water are known as aquifers. Rain percolates through topsoil into the underlying bedrock, creating an underground store of water. The limit of the saturated zone is called the water table.

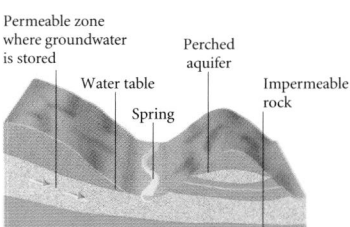

Permeable zone where groundwater is stored · Water table · Perched aquifer · Spring · Impermeable rock

STORAGE OF GROUNDWATER IN AN AQUIFER

World river systems:
Sediment deposited annually per drainage basin

tons per sq mile per year
9120   6080   1520   760

2400   1600   400   200 and less

tonnes per sq km per year

World river systems

drainage basin

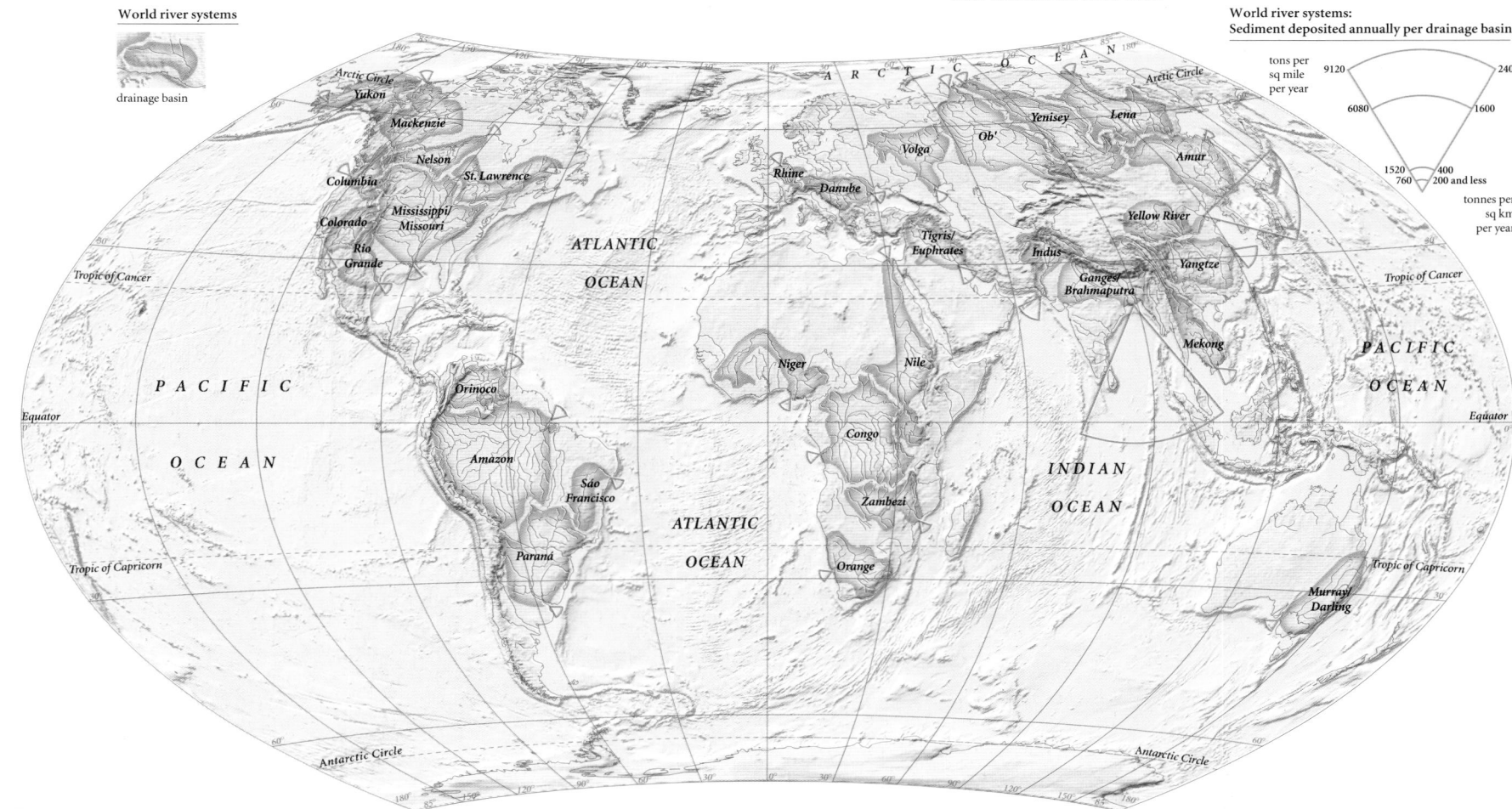

Yukon · Mackenzie · Nelson · St. Lawrence · Columbia · Colorado · Mississippi/Missouri · Rio Grande · Yenisey · Lena · Volga · Ob' · Amur · Rhine · Danube · Yellow River · Tigris/Euphrates · Indus · Yangtze · Ganges/Brahmaputra · Mekong · Niger · Nile · Orinoco · Amazon · Congo · São Francisco · Zambezi · Paraná · Orange · Murray/Darling

ARCTIC OCEAN · Arctic Circle · ATLANTIC OCEAN · PACIFIC OCEAN · INDIAN OCEAN · PACIFIC OCEAN · Tropic of Cancer · Equator · Tropic of Capricorn · Antarctic Circle

RIVERS

RIVERS ERODE THE LAND by grinding and dissolving rocks and stones. Most erosion occurs in the river's upper course as it flows through highland areas. Rock fragments are moved along the river bed by fast-flowing water and deposited in areas where the river slows down, such as flat plains, or where the river enters seas or lakes.

RIVER VALLEYS

Over long periods of time rivers erode uplands to form characteristic V-shaped valleys with smooth sides.

Resistant rock · River · Chemical erosion cuts valley in softer rock

RIVER VALLEY EROSION

DELTAS

When a river deposits its load of silt and sediment (alluvium) on entering the sea, it may form a delta. As this material accumulates, it chokes the mouth of the river, forcing it to create new channels to reach the sea.

The Nile forms a broad delta as it flows into the Mediterranean.

Watershed · Major trunk river · Alps · Apennines · Tributary river · Delta · River mouth · Po Valley · Dolomites

DRAINAGE BASINS

The drainage basin is the area of land drained by a major trunk river and its smaller branch rivers or tributaries. Drainage basins are separated from one another by natural boundaries known as watersheds.

The drainage basin of the Po River, northern Italy.

MEANDERS

In their lower courses, rivers flow slowly. As they flow across the lowlands, they form looping bends called meanders.

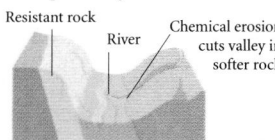

The Mississippi River forms meanders as it flows across the southern US.

The meanders of Utah's San Juan River have become deeply incised.

DEPOSITION

When rivers have deposited large quantities of fertile alluvium, they are forced to find new channels through the alluvium deposits, creating braided river systems.

Mud is deposited by China's Yellow River in its lower course.

LANDSLIDES

Heavy rain and associated flooding on slopes can loosen underlying rocks, which crumble, causing the top layers of rock and soil to slip.

A huge landslide in the Swiss Alps has left massive piles of rocks and pebbles called scree.

GULLIES

In areas where soil is thin, rainwater is not effectively absorbed, and may flow overland. The water courses downhill in channels, or gullies, and may lead to rapid erosion of soil.

A deep gully in the French Alps caused by the scouring of upper layers of turf.

ICE

DURING ITS LONG HISTORY, the Earth has experienced a number of glacial episodes when temperatures were considerably lower than today. During the last Ice Age, 18,000 years ago, ice covered an area three times larger than it does today. Over these periods, the ice has left a remarkable legacy of transformed landscapes.

GLACIERS

GLACIERS ARE FORMED by the compaction of snow into "rivers" of ice. As they move over the landscape, glaciers pick up and carry a load of rocks and boulders which erode the landscape they pass over, and are eventually deposited at the end of the glacier.

A massive glacier advancing down a valley in southern Argentina.

POST-GLACIAL FEATURES

WHEN A GLACIAL EPISODE ENDS, the retreating ice leaves many features. These include depositional ridges called moraines, which may be eroded into low hills known as drumlins; sinuous ridges called eskers; kames, which are rounded hummocks; depressions known as kettle holes; and windblown loess deposits.

GLACIAL VALLEYS

GLACIERS CAN ERODE much more powerfully than rivers. They form steep-sided, flat-bottomed valleys with a typical U-shaped profile. Valleys created by tributary glaciers, whose floors have not been eroded to the same depth as the main glacial valley floor, are called hanging valleys.

The U-shaped profile and piles of morainic debris are characteristic of a valley once filled by a glacier.

A series of hanging valleys high up in the Chilean Andes.

The profile of the Matterhorn has been formed by three cirques lying "back-to-back."

PAST AND PRESENT WORLD ICE-COVER AND GLACIAL FEATURES

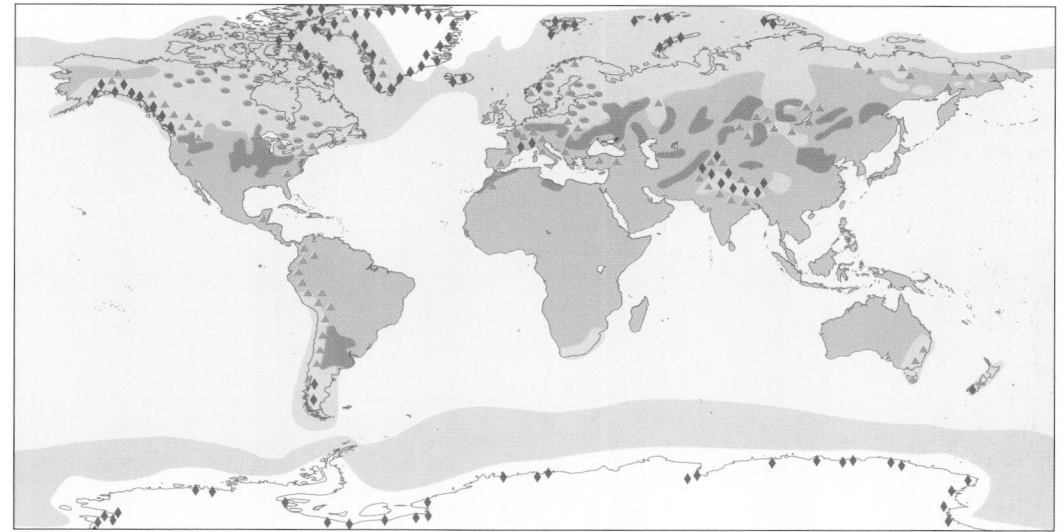

POST-GLACIAL LANDSCAPE FEATURES

Kame terrace — Retreating glacier
Kettle hole — Drumlin
Esker
Braided river — Terminal moraine
Windblown loess — Glacial till
— Bedrock

Past and present world ice cover and glacial features

☐ extent of last Ice Age / loess deposits
☐ post-glacial feature
▲ glacial feature
☐ present day ice cover
◆ glacial field

ICE SHATTERING

Water drips into fissures in rocks and freezes. The pressure weakens the rock, causing it to crack, and eventually to shatter into polygonal patterns.

Irregular polygons show through the sedge-grass tundra in the Yukon, Canada.

CIRQUES

Cirques are basin-shaped hollows which mark the head of a glaciated valley. Where neighboring cirques meet, they are divided by sharp rock ridges called arêtes. It is these arêtes which give the Matterhorn its characteristic profile.

FJORDS

Fjords are ancient glacial valleys flooded by the sea following the end of a period of glaciation. Beneath the water, the valley floor can be 4,000 ft (1,300 m) deep.

A fjord fills a former glacial valley in southern New Zealand.

PERIGLACIATION

Periglacial areas occur near to the edge of ice sheets. A layer of frozen ground lying just beneath the surface of the land is known as permafrost. When the surface melts in the summer, the water is unable to drain into the frozen ground, and so "creeps" downhill, a process known as solifluction.

WIND

STRONG WINDS can transport rock fragments great distances, especially where there is little vegetation to protect the rock. In desert areas, wind picks up loose, unprotected sand particles, carrying them over great distances. This powerfully abrasive debris is blasted at the surface by the wind, eroding the landscape into dramatic shapes.

PREVAILING WINDS AND DUST TRAJECTORIES

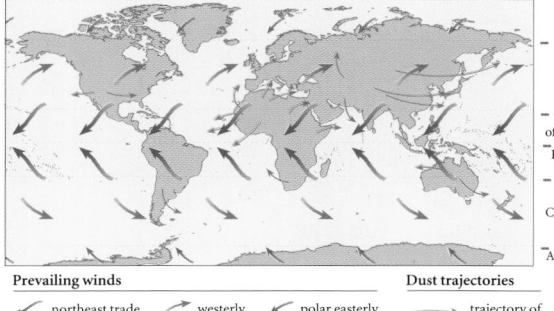

Prevailing winds
↙ northeast trade
↗ southeast trade
↘ westerly
↖ westerly
↖ polar easterly
↙ polar easterly

Dust trajectories
→ trajectory of aeolian dust

DEPOSITION

THE ROCKY, STONY FLOORS of the world's deserts are swept and scoured by strong winds. The smaller, finer particles of sand are shaped into surface ripples, dunes, or sand mountains, which rise to a height of 650 ft (200 m). Dunes usually form single lines, running perpendicular to the direction of the prevailing wind. These long, straight ridges can extend for over 100 miles (160 km).

Barchan dunes in the Arabian Desert.

Complex dune system in the Sahara.

DUNES

Dunes are shaped by wind direction and sand supply. Where sand supply is limited, crescent-shaped barchan dunes are formed.

— TYPES OF DUNE —

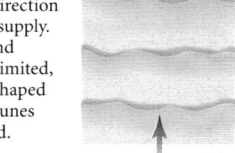
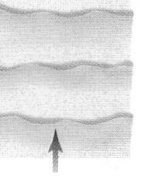
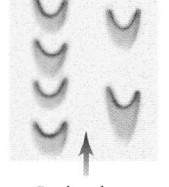

→ wind direction
Transverse dune   Barchan dune   Linear dune   Star dune

TEMPERATURE

HOT AND COLD DESERTS

— Arctic Circle
— Tropic of Cancer
— Equator
— Tropic of Capricorn
— Antarctic Circle

Main desert types
☐ hot arid   ☐ semiarid   ☐ cold polar

MOST OF THE WORLD'S deserts are in the tropics. The cold deserts which occur elsewhere are arid because they are a long way from the rain-giving sea. Rock in deserts is exposed because of lack of vegetation and is susceptible to changes in temperature; extremes of heat and cold can cause both cracks and fissures to appear in the rock.

HEAT

FIERCE SUN can heat the surface of rock, causing it to expand more rapidly than the cooler, underlying layers. This creates tensions which force the rock to crack or break up. In arid regions, the evaporation of water from rock surfaces dissolves certain minerals within the water, causing salt crystals to form in small openings in the rock. The hard crystals force the openings to widen into cracks and fissures.

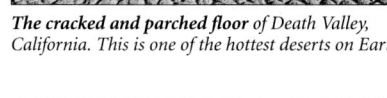
The cracked and parched floor of Death Valley, California. This is one of the hottest deserts on Earth.

DESERT ABRASION

Abrasion creates a wide range of desert landforms from faceted pebbles and wind ripples in the sand, to large-scale features such as yardangs (low, streamlined ridges), and scoured desert pavements.

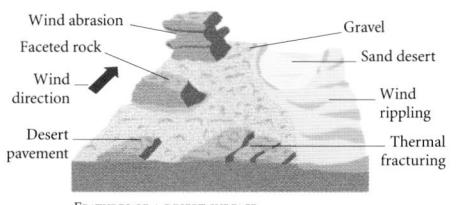
FEATURES OF A DESERT SURFACE

Wind abrasion — Gravel
Faceted rock — Sand desert
Wind direction — Wind rippling
Desert pavement — Thermal fracturing

This dry valley at Ellesmere Island in the Canadian Arctic is an example of a cold desert. The cracked floor and scoured slopes are features also found in hot deserts.

THE WORLD'S OCEANS

TWO-THIRDS OF THE EARTH'S SURFACE is covered by the oceans. The landscape of the ocean floor, like the surface of the land, has been shaped by movements of the Earth's crust over millions of years to form volcanic mountain ranges, deep trenches, basins, and plateaus. Ocean currents constantly redistribute warm and cold water around the world. A major warm current, such as El Niño in the Pacific Ocean, can increase surface temperature by up to 46°F (8°C), causing changes in weather patterns which can lead to both droughts and flooding.

THE GREAT OCEANS

THERE ARE FIVE OCEANS on Earth: the Pacific, Atlantic, Indian, and Southern oceans, and the much smaller Arctic Ocean. These five ocean basins are relatively young, having evolved within the last 80 million years. One of the most recent plate collisions, between the Eurasian and African plates, created the present-day arrangement of continents and oceans.

The Indian Ocean accounts for approximately 20% of the total area of the world's oceans.

SEA LEVEL

IF THE INFLUENCE of tides, winds, currents, and variations in gravity were ignored, the surface of the Earth's oceans would closely follow the topography of the ocean floor, with an underwater ridge 3,000 ft (915 m) high producing a rise of up to 3 ft (1 m) in the level of the surface water.

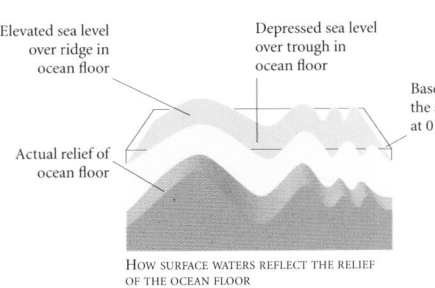

Elevated sea level over ridge in ocean floor

Depressed sea level over trough in ocean floor

Base level of the sea surface at 0 ft (0 m)

Actual relief of ocean floor

HOW SURFACE WATERS REFLECT THE RELIEF OF THE OCEAN FLOOR

The low relief of many small Pacific islands such as these atolls at Huahine in French Polynesia makes them vulnerable to changes in sea level.

OCEAN STRUCTURE

THE CONTINENTAL SHELF is a shallow, flat seabed surrounding the Earth's continents. It extends to the continental slope, which falls to the ocean floor. Here, the flat abyssal plains are interrupted by vast, underwater mountain ranges, the mid-ocean ridges, and ocean trenches which plunge to depths of 35,828 ft (10,920 m).

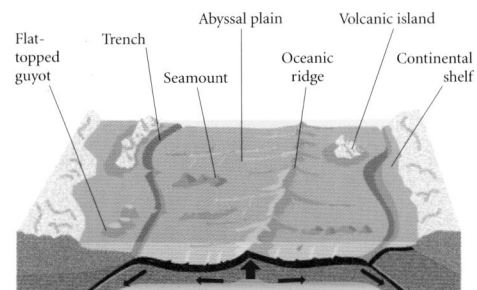

Flat-topped guyot

Trench

Abyssal plain

Seamount

Oceanic ridge

Volcanic island

Continental shelf

TYPICAL SEA-FLOOR FEATURES

Ocean depth

Sea level
200m / 656ft
1000m / 3281ft
2000m / 6562ft
3000m / 9843ft
4000m / 13,124ft
5000m / 16,400ft
6000m / 19,686ft

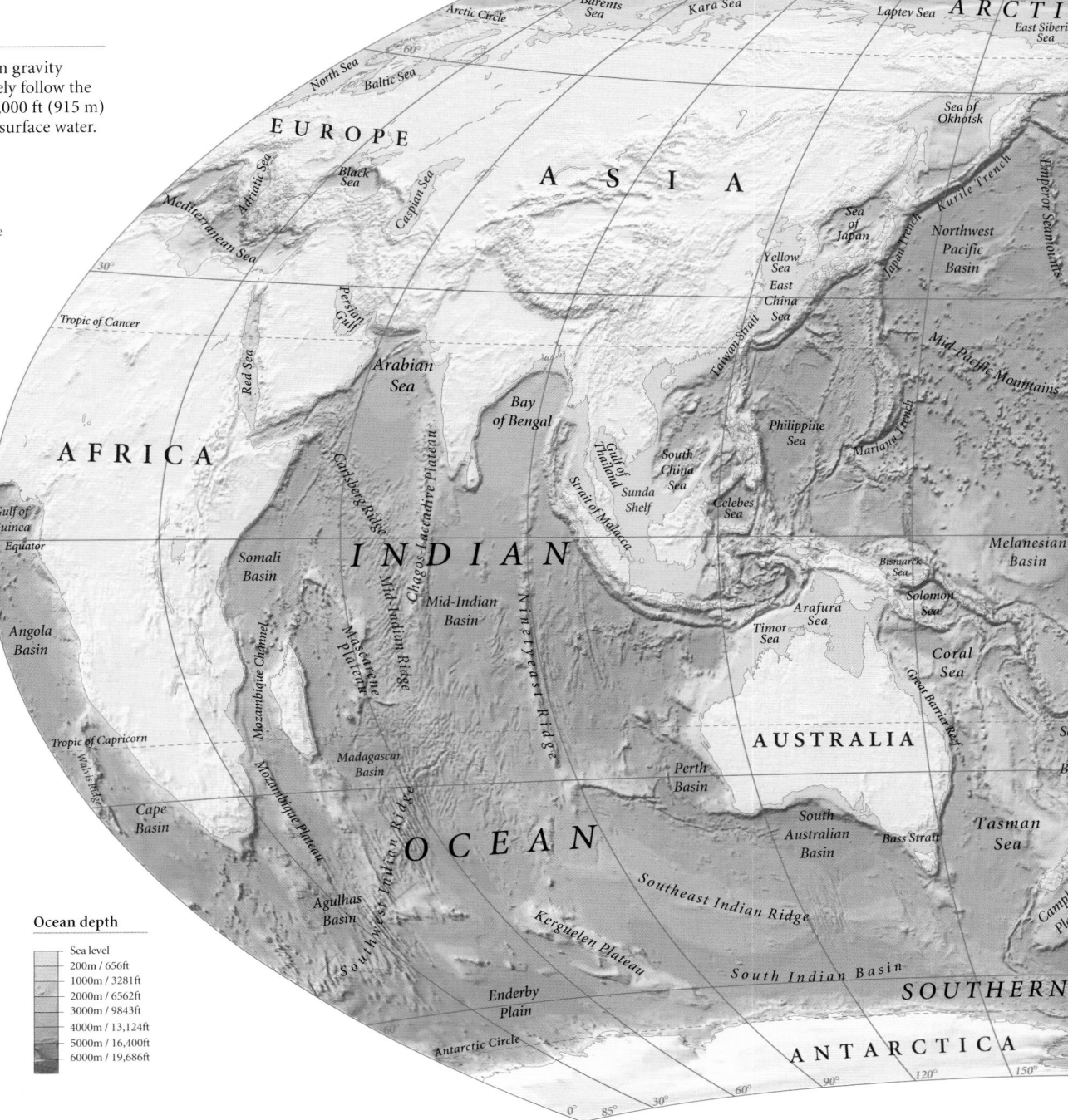

[Map of the Eastern Hemisphere oceans, showing: ARCTIC, Arctic Circle, Barents Sea, Kara Sea, Laptev Sea, East Siberia Sea, North Sea, Baltic Sea, Sea of Okhotsk, EUROPE, ASIA, Mediterranean Sea, Adriatic Sea, Black Sea, Caspian Sea, Sea of Japan, Kurile Trench, Emperor Seamounts, Tropic of Cancer, Red Sea, Persian Gulf, Arabian Sea, Yellow Sea, East China Sea, Japan Trench, Northwest Pacific Basin, Bay of Bengal, Philippine Sea, Mid-Pacific Mountains, AFRICA, Carlsberg Ridge, Gulf of Thailand, South China Sea, Mariana Trench, Gulf of Guinea, Equator, Chagos-Laccadive Plateau, Strait of Malacca, Sunda Shelf, Celebes Sea, Melanesian Basin, Somali Basin, Mid-Indian Ridge, Mascarene Plateau, INDIAN, Mid-Indian Basin, Bismarck Sea, Solomon Sea, Angola Basin, Mozambique Channel, Ninetyeast Ridge, Arafura Sea, Timor Sea, Coral Sea, Tropic of Capricorn, Madagascar Basin, AUSTRALIA, Great Barrier Reef, Cape Basin, Mozambique Plateau, Perth Basin, Tasman Sea, Bass Strait, Walvis Ridge, OCEAN, South Australian Basin, Agulhas Basin, Southwest Indian Ridge, Southeast Indian Ridge, Kerguelen Plateau, South Indian Basin, SOUTHERN, Enderby Plain, Antarctic Circle, ANTARCTICA]

BLACK SMOKERS

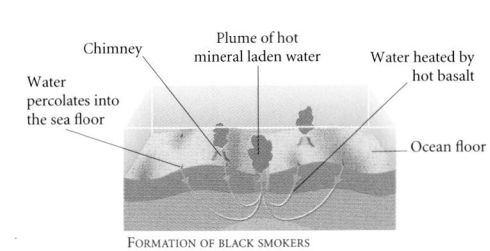

These vents in the ocean floor disgorge hot, sulfur-rich water from deep in the Earth's crust. Despite the great depths, a variety of lifeforms have adapted to the chemical-rich environment which surrounds black smokers.

A black smoker in the Atlantic Ocean.

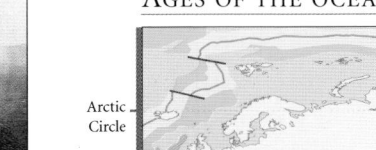

Surtsey, near Iceland, is a volcanic island lying directly over the Mid-Atlantic Ridge. It was formed in the 1960s following intense volcanic activity nearby.

OCEAN FLOORS

Mid-ocean ridges are formed by lava which erupts beneath the sea and cools to form solid rock. This process mirrors the creation of volcanoes from cooled lava on the land. The ages of sea floor rocks increase in parallel bands outward from central ocean ridges.

Chimney

Plume of hot mineral laden water

Water heated by hot basalt

Water percolates into the sea floor

Ocean floor

FORMATION OF BLACK SMOKERS

AGES OF THE OCEAN FLOOR

[World map showing ages of the ocean floor with labels: Arctic Circle, Tropic of Cancer, Equator, Tropic of Capricorn, Antarctic Circle]

Jurassic | Cretaceous | Tertiary (Paleogene) | Quaternary | Cretaceous | Jurassic

208 million years old | 145 | 65 | 23 | 0 | 23 | 65 | 145 | 208 million years old

Tertiary (Neogene)

Age uncertain
Continental shelf and island arcs

***Currents in the Southern Ocean** are driven by some of the world's fiercest winds, including the Roaring Forties, Furious Fifties, and Shrieking Sixties.*

***The Pacific Ocean** is the world's largest and deepest ocean, covering over one-third of the surface of the Earth.*

***The Atlantic Ocean** was formed when the landmasses of the eastern and western hemispheres began to drift apart 180 million years ago.*

DEPOSITION OF SEDIMENT

STORMS, EARTHQUAKES, and volcanic activity trigger underwater currents known as turbidity currents which scour sand and gravel from the continental shelf, creating underwater canyons. These strong currents pick up material deposited at river mouths and deltas, and carry it across the continental shelf and through the underwater canyons, where it is eventually laid down on the ocean floor in the form of fans.

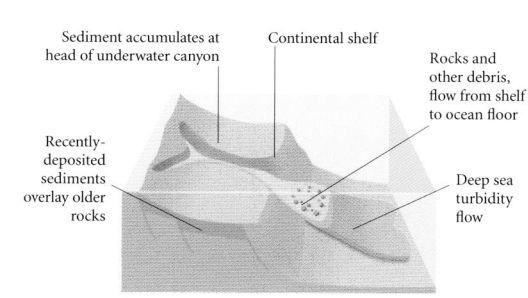

HOW SEDIMENT IS DEPOSITED ON THE OCEAN FLOOR

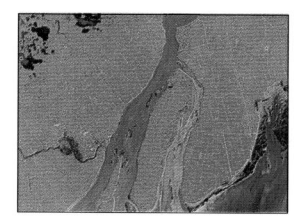

***Satellite image** of the Yangtze (Chang Jiang) Delta, in which the land appears red. The river deposits immense quantities of silt into the East China Sea, much of which will eventually reach the deep ocean floor.*

SURFACE WATER

OCEAN CURRENTS move warm water away from the Equator toward the poles, while cold water is, in turn, moved towards the Equator. This is the main way in which the Earth distributes surface heat and is a major climatic control. Approximately 4,000 million years ago, the Earth was dominated by oceans and there was no land to interrupt the flow of the currents, which would have flowed as straight lines, simply influenced by the Earth's rotation.

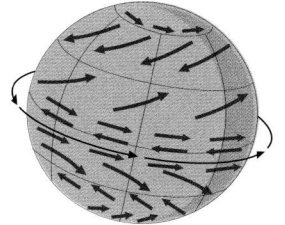

***Idealized globe** showing the movement of water around a landless Earth.*

OCEAN CURRENTS

SURFACE CURRENTS are driven by the prevailing winds and by the spinning motion of the Earth, which drives the currents into circulating whirlpools, or gyres. Deep sea currents, over 330 ft (100 m) below the surface, are driven by differences in water temperature and salinity, which have an impact on the density of deep water and on its movement.

SURFACE TEMPERATURE AND CURRENTS

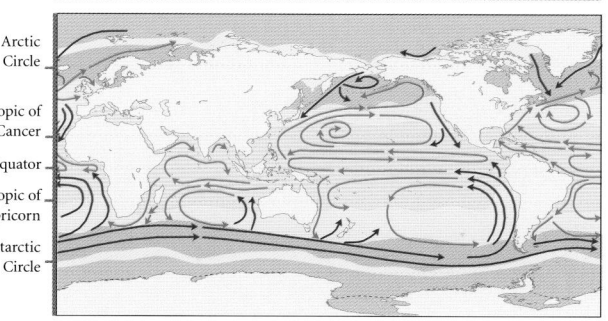

Surface temperature and currents

- ········ Ice-shelf (below 32°F / 0°C)
- Sea-ice* (average) below 28°F / -2°C
- Sea-water 28–32°F / -2–0°C
- *Sea-water freezes at 28.4°F / -1.9°C
- 32–50°F / 0–10°C
- 50–68°F / 10–20°C
- 68–86°F / 20–30°C
- → warm current
- → cold current

TIDES AND WAVES

TIDES ARE CREATED by the pull of the Sun and Moon's gravity on the surface of the oceans. The levels of high and low tides are influenced by the position of the Moon in relation to the Earth and Sun. Waves are formed by wind blowing over the surface of the water.

TIDAL RANGE AND WAVE ENVIRONMENTS

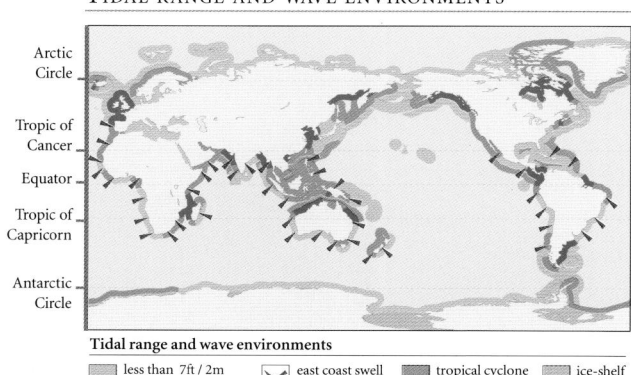

Tidal range and wave environments

- less than 7ft / 2m
- 7–13ft / 2–4m
- greater than 13ft / 4m
- east coast swell
- west coast swell
- tropical cyclone
- storm wave
- ice-shelf

HIGH AND LOW TIDES

The highest tides occur when the Earth, the Moon and the Sun are aligned *(below left)*. The lowest tides are experienced when the Sun and Moon align at right angles to one another *(below right)*.

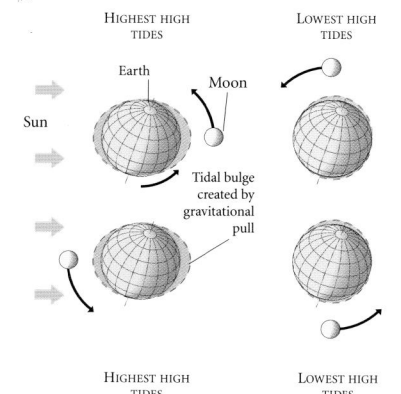

HIGHEST HIGH TIDES

LOWEST HIGH TIDES

Sun

Earth

Moon

Tidal bulge created by gravitational pull

HIGHEST HIGH TIDES

LOWEST HIGH TIDES

DEEP SEA TEMPERATURE AND CURRENTS

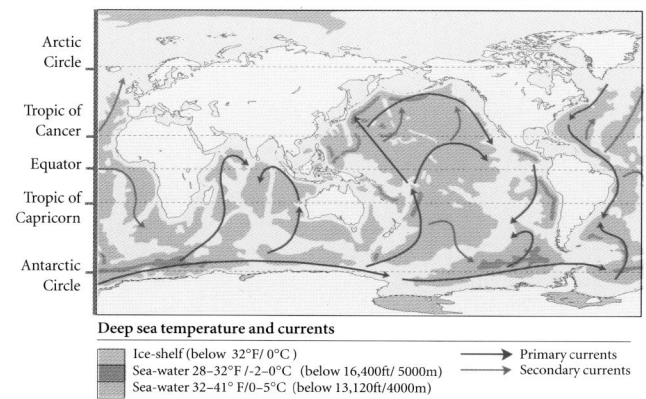

Deep sea temperature and currents

- Ice-shelf 32°F/ 0°C)
- Sea-water 28–32°F / -2–0°C (below 16,400ft/ 5000m)
- Sea-water 32–41° F/0–5°C (below 13,120ft/4000m)
- → Primary currents
- → Secondary currents

Map labels

OCEAN
Beaufort Sea
Gulf of Alaska
Aleutian Trench
Mendocino Fracture Zone
Murray Fracture Zone
Molokai Fracture Zone
Clarion Fracture Zone
Clipperton Fracture Zone
Central Pacific Basin
PACIFIC
OCEAN
Southwest Pacific Basin
Antarctic Ridge
OCEAN
Amundsen Sea
Bellingshausen Sea
Southeast Pacific Basin
NORTH AMERICA
Baffin Bay
Davis Strait
Greenland Sea
Arctic Circle
Hudson Strait
Hudson Bay
Labrador Sea
Gulf of Mexico
Yucatan Basin
Caribbean Sea
Guatemala Basin
Mid-America Trench
East Pacific Rise
Nazca Ridge
Sala y Gomez Ridge
Chile Basin
Peru Basin
Peru-Chile Trench
SOUTH AMERICA
Argentine Basin
Scotia Sea
South Sandwich Trench
Weddell Sea
Antarctic Circle
Newfoundland Basin
Mid-Atlantic Ridge
North American Basin
ATLANTIC
Sargasso Sea
Canary Basin
Tropic of Cancer
Barracuda Fracture Zone
Brazil Basin
OCEAN
Tropic of Capricorn
Rio Grande Rise
Mid-Atlantic Ridge
Equator

THE GLOBAL CLIMATE

THE EARTH'S CLIMATIC TYPES CONSIST of stable patterns of weather conditions averaged out over a long period of time. Different climates are categorized according to particular combinations of temperature and humidity. By contrast, weather consists of short-term fluctuations in wind, temperature, and humidity conditions. Different climates are determined by latitude, altitude, the prevailing wind, and circulation of ocean currents. Longer-term changes in climate, such as global warming or the onset of ice ages, are punctuated by shorter-term events which comprise the day-to-day weather of a region, such as frontal depressions, hurricanes, and blizzards.

THE ATMOSPHERE, WIND AND WEATHER

THE EARTH'S ATMOSPHERE has been compared to a giant ocean of air which surrounds the planet. Its circulation patterns are similar to the currents in the oceans and are influenced by three factors; the Earth's orbit around the Sun and rotation about its axis, and variations in the amount of heat radiation received from the Sun. If both heat and moisture were not redistributed between the Equator and the poles, large areas of the Earth would be uninhabitable.

Heavy fogs, as here in southern England, form as moisture-laden air passes over cold ground.

TEMPERATURE

THE WORLD CAN BE DIVIDED into three major climatic zones, stretching like large belts across the latitudes: the tropics which are warm; the cold polar regions and the temperate zones which lie between them. Temperatures across the Earth range from above 86°F (30°C) in the deserts to as low as -70°F (-55°C) at the poles. Temperature is also controlled by altitude; because air becomes cooler and less dense the higher it gets, mountainous regions are typically colder than those areas which are at, or close to, sea level.

AVERAGE JANUARY TEMPERATURES

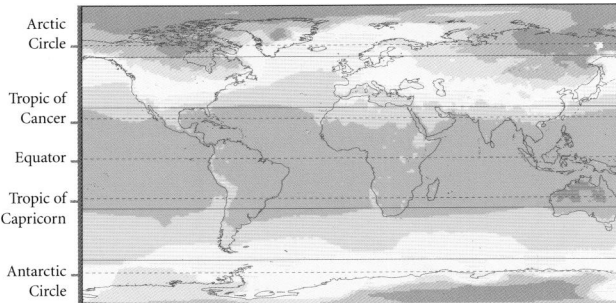

Arctic Circle
Tropic of Cancer
Equator
Tropic of Capricorn
Antarctic Circle

AVERAGE JULY TEMPERATURES

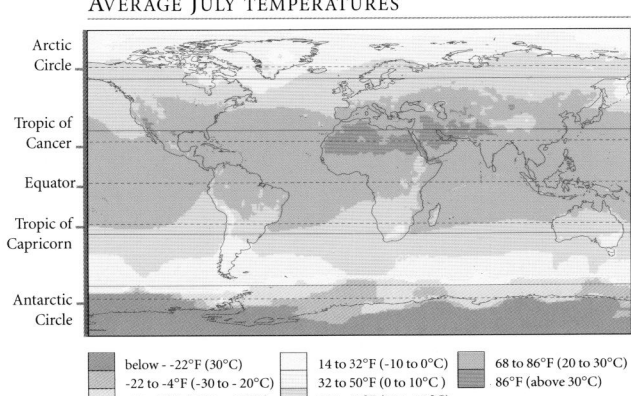

Arctic Circle
Tropic of Cancer
Equator
Tropic of Capricorn
Antarctic Circle

below - -22°F (30°C)	14 to 32°F (-10 to 0°C)	68 to 86°F (20 to 30°C)
-22 to -4°F (-30 to -20°C)	32 to 50°F (0 to 10°C)	86°F (above 30°C)
-4 to 14°F (-20 to -10°C)	50 to 68°F (10 to 20°C)	

GLOBAL AIR CIRCULATION

AIR DOES NOT SIMPLY FLOW FROM THE EQUATOR TO THE POLES, it circulates in giant cells known as Hadley and Ferrel cells. As air warms it expands, becoming less dense and rising; this creates areas of low pressure. As the air rises it cools and condenses, causing heavy rainfall over the tropics and slight snowfall over the poles. This cool air then sinks, forming high pressure belts. At surface level in the tropics these sinking currents are deflected poleward as the westerlies and toward the Equator as the trade winds. At the poles they become the polar easterlies.

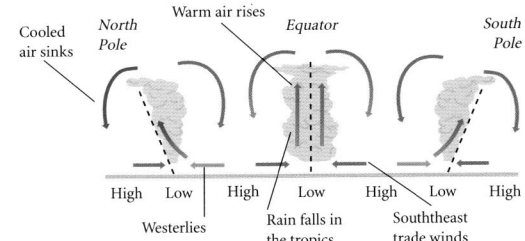

Cooled air sinks | North Pole | Warm air rises | Equator | South Pole
High Low | High | Low | High Low | High
Westerlies | Rain falls in the tropics | Souththheast trade winds

The Antarctic pack ice expands its area by almost seven times during the winter as temperatures drop and surrounding seas freeze.

CLIMATIC CHANGE

THE EARTH IS CURRENTLY IN A WARM PHASE between ice ages. Warmer temperatures result in higher sea levels as more of the polar ice caps melt. Most of the world's population lives near coasts, so any changes which might cause sea levels to rise, could have a potentially disastrous impact.

This ice fair, painted by Pieter Brueghel the Younger in the 17th century, shows the Little Ice Age which peaked around 300 years ago.

THE GREENHOUSE EFFECT

Gases such as carbon dioxide are known as "greenhouse gases" because they allow shortwave solar radiation to enter the Earth's atmosphere, but help to stop longwave radiation from escaping. This traps heat, raising the Earth's temperature. An excess of these gases, such as that which results from the burning of fossil fuels, helps trap more heat and can lead to global warming.

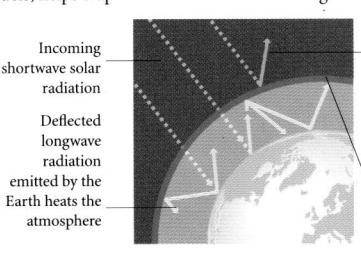

Incoming shortwave solar radiation
Deflected longwave radiation emitted by the Earth heats the atmosphere
Deflected shortwave solar radiation
Greenhouse gases prevent the escape of longwave radiation

The islands of the Caribbean, Mexico's Gulf coast and the southeastern US are often hit by hurricanes formed far out in the Atlantic.

OCEANIC WATER CIRCULATION

IN GENERAL, OCEAN CURRENTS parallel the movement of winds across the Earth's surface. Incoming solar energy is greatest at the Equator and least at the poles. So, water in the oceans heats up most at the Equator and flows poleward, cooling as it moves north or south toward the Arctic or Antarctic. The flow is eventually reversed and cold water currents move back toward the Equator. These ocean currents act as a vast system for moving heat from the Equator toward the poles and are a major influence on the distribution of the Earth's climates.

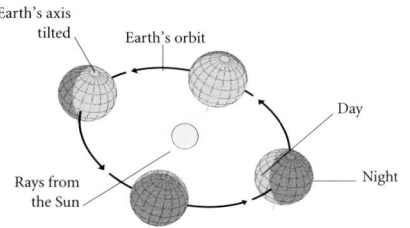

In marginal climatic zones years of drought can completely dry out the land and transform grassland to desert.

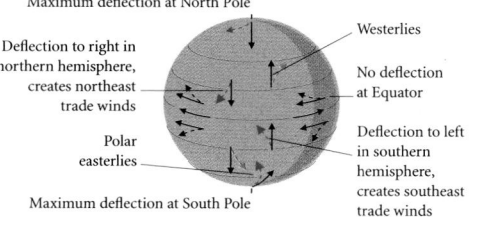

The wide range of environments found in the Andes is strongly related to their altitude, which modifies climatic influences. While the peaks are snow-capped, many protected interior valleys are semitropical.

TILT AND ROTATION

The tilt and rotation of the Earth during its annual orbit largely control the distribution of heat and moisture across its surface, which correspondingly controls its large-scale weather patterns. As the Earth annually rotates around the Sun, half its surface is receiving maximum radiation, creating summer and winter seasons. The angle of the Earth means that on average the tropics receive two and a half times as much heat from the Sun each day as the poles.

Earth's axis tilted
Earth's orbit
Day
Rays from the Sun
Night

THE CORIOLIS EFFECT

The rotation of the Earth influences atmospheric circulation by deflecting winds and ocean currents. Winds blowing in the northern hemisphere are deflected to the right and those in the southern hemisphere are deflected to the left, creating large-scale patterns of wind circulation, such as the northeast and southeast trade winds and the westerlies. This effect is greatest at the poles and least at the Equator.

Maximum deflection at North Pole
Deflection to right in northern hemisphere, creates northeast trade winds
Westerlies
No deflection at Equator
Polar easterlies
Deflection to left in southern hemisphere, creates southeast trade winds
Maximum deflection at South Pole

MAP KEY

Climate zones
ice cap
subarctic
tundra
continental
temperate
warm temperate
mediterranean
semiarid
arid
hot humid
humid equatorial
tropical

Ocean currents
warm
cold

Prevailing winds
warm
cold

Local winds
warm
cold
seasonal*
* (seasonal winds which can either be warm or cold)

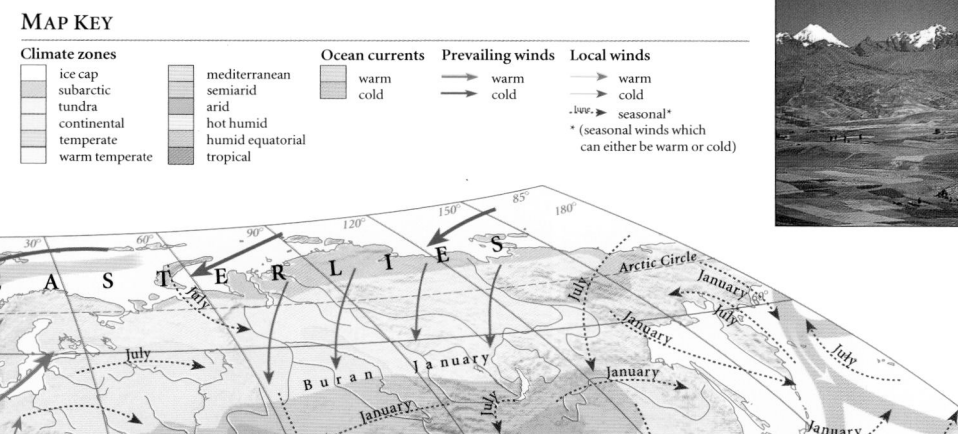

PRECIPITATION

WHEN WARM AIR EXPANDS, it rises and cools, and the water vapor it carries condenses to form clouds. Heavy, regular rainfall is characteristic of the equatorial region, while the poles are cold and receive only slight snowfall. Tropical regions have marked dry and rainy seasons, while in the temperate regions rainfall is relatively unpredictable.

Monsoon rains, which affect southern Asia from May to September, are caused by sea winds blowing across the warm land.

Heavy tropical rainstorms occur frequently in Papua New Guinea, often causing soil erosion and landslides in cultivated areas.

AVERAGE JANUARY RAINFALL

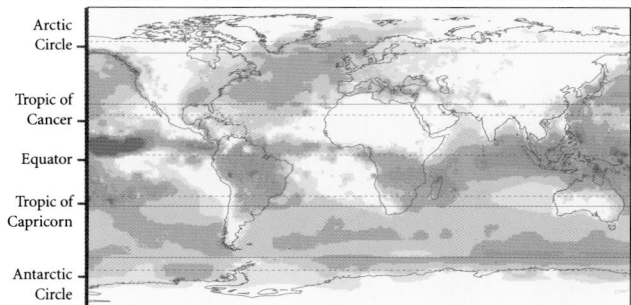

Arctic Circle
Tropic of Cancer
Equator
Tropic of Capricorn
Antarctic Circle

AVERAGE JULY RAINFALL

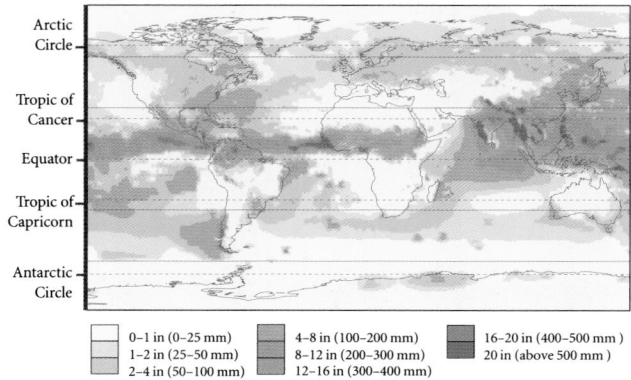

Arctic Circle
Tropic of Cancer
Equator
Tropic of Capricorn
Antarctic Circle

The intensity of some blizzards in Canada and the northern US can give rise to snowdrifts as high as 10 ft (3 m).

The Atacama Desert in Chile is one of the driest places on Earth, with an average rainfall of less than 2 inches (50 mm) per year.

Violent thunderstorms occur along advancing cold fronts, when cold, dry air masses meet warm, moist air, which rises rapidly, its moisture condensing into thunderclouds. Rain and hail become electrically charged, causing lightning.

THE RAINSHADOW EFFECT

When moist air is forced to rise by mountains, it cools and the water vapor falls as precipitation, either as rain or snow. Only the dry, cold air continues over the mountains, leaving inland areas with little or no rain. This is called the rainshadow effect and is one reason for the existence of the Mojave Desert in California, which lies east of the Coast Ranges.

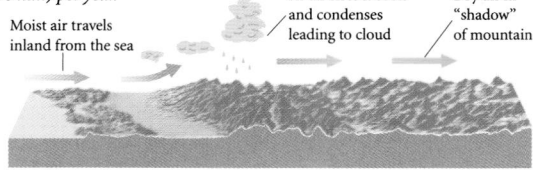

As air rises it cools and condenses leading to cloud
Dry air in "shadow" of mountain
Moist air travels inland from the sea
THE RAINSHADOW EFFECT

0–1 in (0–25 mm)
1–2 in (25–50 mm)
2–4 in (50–100 mm)
4–8 in (100–200 mm)
8–12 in (200–300 mm)
12–16 in (300–400 mm)
16–20 in (400–500 mm)
20 in (above 500 mm)

LIFE ON EARTH

A UNIQUE COMBINATION of an oxygen-rich atmosphere and plentiful water is the key to life on Earth. Apart from the polar ice caps, there are few areas which have not been colonized by animals or plants over the course of the Earth's history. Plants process sunlight to provide them with their energy, and ultimately all the Earth's animals rely on plants for survival. Because of this reliance, plants are known as primary producers, and the availability of nutrients and temperature of an area is defined as its primary productivity, which affects the quantity and type of animals which are able to live there. This index is affected by climatic factors – cold and aridity restrict the quantity of life, whereas warmth and regular rainfall allow a greater diversity of species.

BIOGEOGRAPHICAL REGIONS

THE EARTH CAN BE DIVIDED into a series of biogeographical regions, or biomes, ecological communities where certain species of plant and animal coexist within particular climatic conditions. Within these broad classifications, other factors including soil richness, altitude, and human activities such as urbanization, intensive agriculture, and deforestation, affect the local distribution of living species within each biome.

POLAR REGIONS
A layer of permanent ice at the Earth's poles covers both seas and land. Very little plant and animal life can exist in these harsh regions.

TUNDRA
A desolate region, with long, dark freezing winters and short, cold summers. With virtually no soil and large areas of permanently frozen ground known as permafrost, the tundra is largely treeless, though it is briefly clothed by small flowering plants in the summer months.

NEEDLELEAF FORESTS
With milder summers than the tundra and less wind, these areas are able to support large forests of coniferous trees.

BROADLEAF FORESTS
Much of the northern hemisphere was once covered by deciduous forests, which occurred in areas with marked seasonal variations. Most deciduous forests have been cleared for human settlement.

TEMPERATE RAIN FORESTS
In warmer wetter areas, such as southern China, temperate deciduous forests are replaced by evergreen forest.

DESERTS
Deserts are areas with negligible rainfall. Most hot deserts lie within the tropics; cold deserts are dry because of their distance from the moisture-providing sea.

MEDITERRANEAN
Hot, dry summers and short winters typify these areas, which were once covered by evergreen shrubs and woodland, but have now been cleared by humans for agriculture.

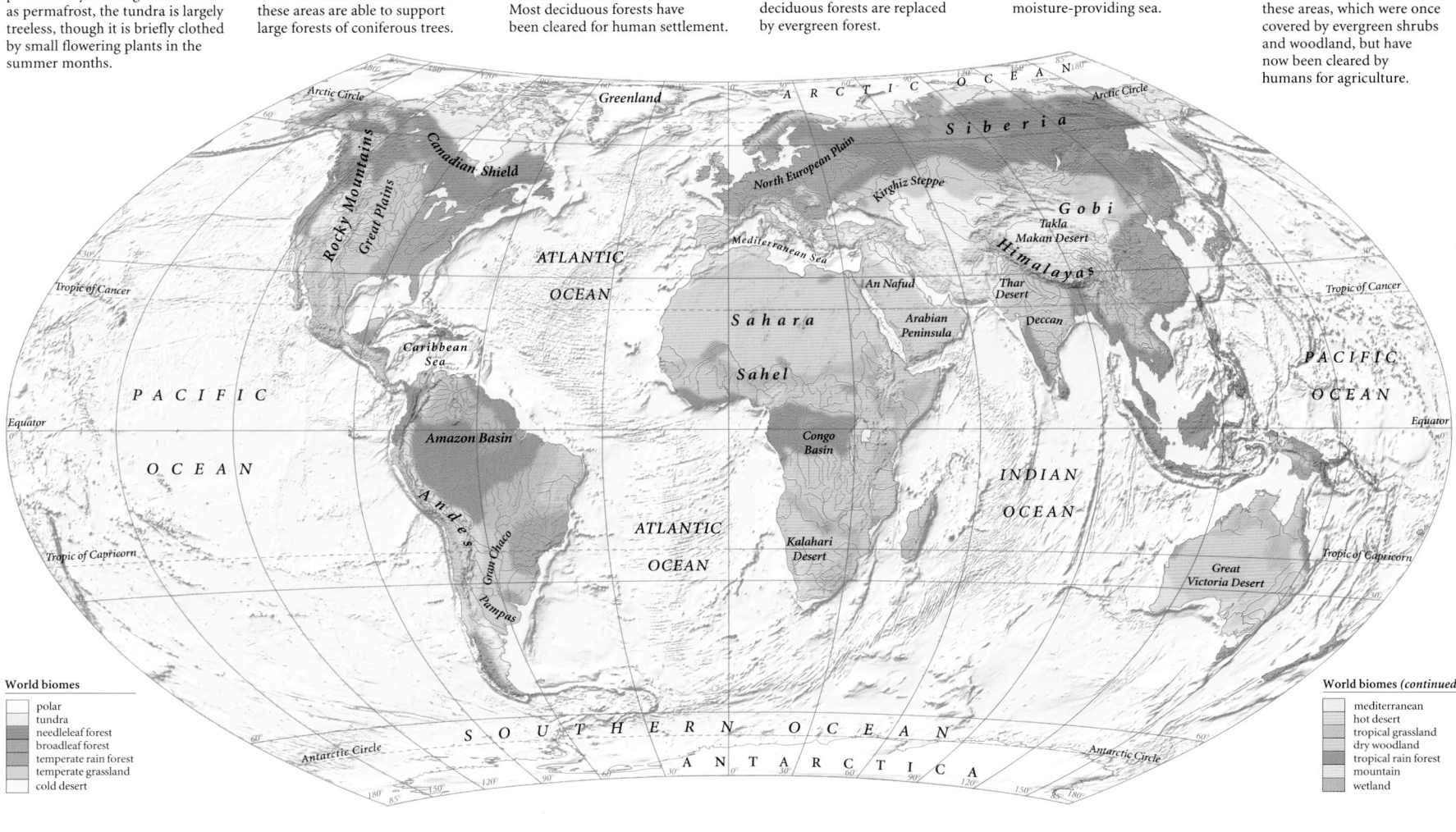

World biomes
- polar
- tundra
- needleleaf forest
- broadleaf forest
- temperate rain forest
- temperate grassland
- cold desert

World biomes (continued)
- mediterranean
- hot desert
- tropical grassland
- dry woodland
- tropical rain forest
- mountain
- wetland

TROPICAL AND TEMPERATE GRASSLANDS
The major grassland areas are found in the centers of the larger continental landmasses. In Africa's tropical savannah regions, seasonal rainfall alternates with drought. Temperate grasslands, also known as *steppes* and *prairies* are found in the northern hemisphere, and in South America, where they are known as the *pampas*.

DRY WOODLANDS
Trees and shrubs, adapted to dry conditions, grow widely spaced from one another, interspersed by savannah grasslands.

TROPICAL RAIN FORESTS
Characterized by year-round warmth and high rainfall, tropical rain forests contain the highest diversity of plant and animal species on Earth.

MOUNTAINS
Though the lower slopes of mountains may be thickly forested, only ground-hugging shrubs and other vegetation will grow above the tree line which varies according to both altitude and latitude.

WETLANDS
Rarely lying above sea level, wetlands are marshes, swamps and tidal flats. Some, with their moist, fertile soils, are rich feeding grounds for fish and breeding grounds for birds. Others have little soil structure and are too acidic to support much plant and animal life.

BIODIVERSITY

THE NUMBER OF PLANT AND ANIMAL SPECIES, and the range of genetic diversity within the populations of each species, make up the Earth's biodiversity. The plants and animals which are endemic to a region – that is, those which are found nowhere else in the world – are also important in determining levels of biodiversity. Human settlement and intervention have encroached on many areas of the world once rich in endemic plant and animal species. Increasing international efforts are being made to monitor and conserve the biodiversity of the Earth's remaining wild places.

ANIMAL ADAPTATION

THE DEGREE OF AN ANIMAL'S ADAPTABILITY to different climates and conditions is extremely important in ensuring its success as a species. Many animals, particularly the largest mammals, are becoming restricted to ever-smaller regions as human development and modern agricultural practices reduce their natural habitats. In contrast, humans have been responsible – both deliberately and accidentally – for the spread of some of the world's most successful species. Many of these introduced species are now more numerous than the indigenous animal populations.

POLAR ANIMALS

The frozen wastes of the polar regions are able to support only a small range of species which derive their nutritional requirements from the sea. Animals such as the walrus (left) have developed insulating fat, stocky limbs, and double-layered coats to enable them to survive in the freezing conditions.

DIVERSITY OF ANIMAL SPECIES

DESERT ANIMALS

Many animals which live in the extreme heat and aridity of the deserts are able to survive for days and even months with very little food or water. Their bodies are adapted to lose heat quickly and to store fat and water. The Gila monster (above) stores fat in its tail.

AMAZON RAINFOREST

The vast Amazon Basin is home to the world's greatest variety of animal species. Animals are adapted to live at many different levels from the treetops to the tangled undergrowth which lies beneath the canopy. The sloth (below) hangs upside down in the branches. Its fur grows from its stomach to its back to enable water to run off quickly.

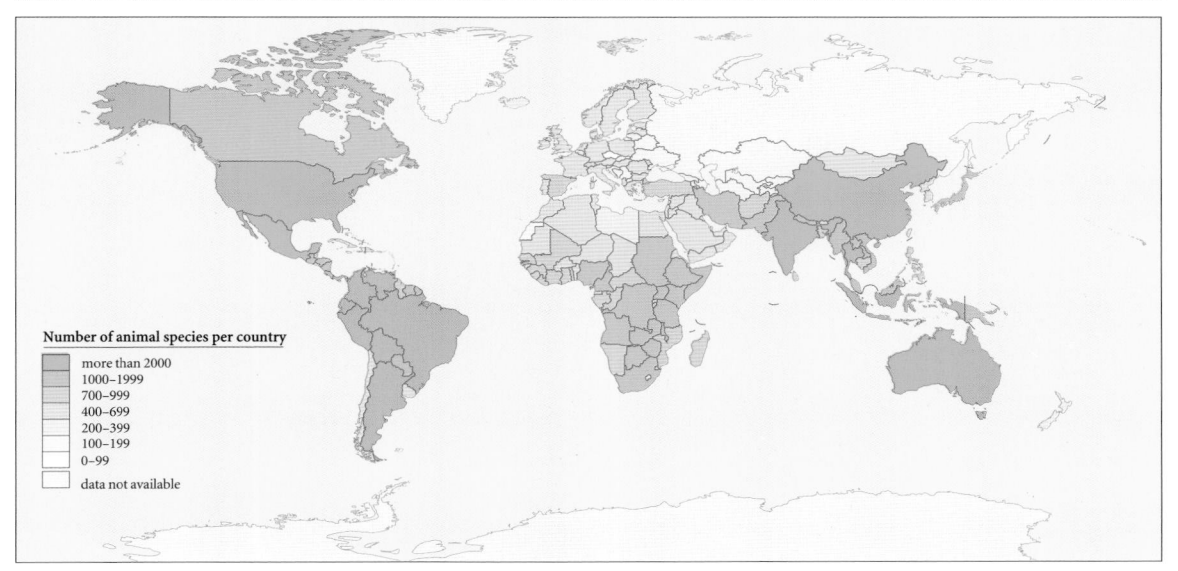

Number of animal species per country
- more than 2000
- 1000–1999
- 700–999
- 400–699
- 200–399
- 100–199
- 0–99
- data not available

MARINE BIODIVERSITY

The oceans support a huge variety of different species, from the world's largest mammals like whales and dolphins down to the tiniest plankton. The greatest diversities occur in the warmer seas of continental shelves, where plants are easily able to photosynthesize, and around coral reefs, where complex ecosystems are found. On the ocean floor, nematodes can exist at a depth of more than 10,000 ft (3,000 m) below sea level.

HIGH ALTITUDES

Few animals exist in the rarefied atmosphere of the highest mountains. However, birds of prey such as eagles and vultures (above), with their superb eyesight can soar as high as 23,000 ft (7,000 m) to scan for prey below.

URBAN ANIMALS

The growth of cities has reduced the amount of habitat available to many species. A number of animals are now moving closer into urban areas to scavenge from the detritus of the modern city (left). Rodents, particularly rats and mice, have existed in cities for thousands of years, and many insects, especially moths, quickly develop new coloring to provide them with camouflage.

ENDEMIC SPECIES

Isolated areas such as Australia and the island of Madagascar, have the greatest range of endemic species. In Australia, these include marsupials such as the kangaroo (below), which carry their young in pouches on their bodies. Destruction of habitat, pollution, hunting, and predators introduced by humans, are threatening this unique biodiversity.

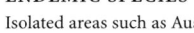

PLANT ADAPTATION

ENVIRONMENTAL CONDITIONS, particularly climate, soil type, and the extent of competition with other organisms, influence the development of plants into a number of distinctive forms. Similar conditions in quite different parts of the world create similar adaptations in the plants, which may then be modified by other, local, factors specific to the region.

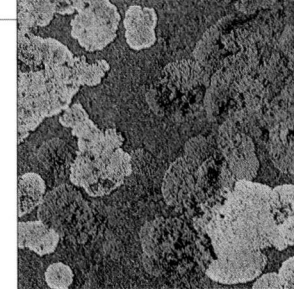

COLD CONDITIONS

In areas where temperatures rarely rise above freezing, plants such as lichens (left) and mosses grow densely, close to the ground.

RAIN FORESTS

Most of the world's largest and oldest plants are found in rain forests; warmth and heavy rainfall provide ideal conditions for vast plants like the world's largest flower, the rafflesia (left).

HOT, DRY CONDITIONS

Arid conditions lead to the development of plants whose surface area has been reduced to a minimum to reduce water loss. In cacti (above), which can survive without water for months, leaves are minimal or not present at all.

ANCIENT PLANTS

Some of the world's most primitive plants still exist today, including algae, cyclads, and many ferns (above), reflecting the success with which they have adapted to changing conditions.

DIVERSITY OF PLANT SPECIES

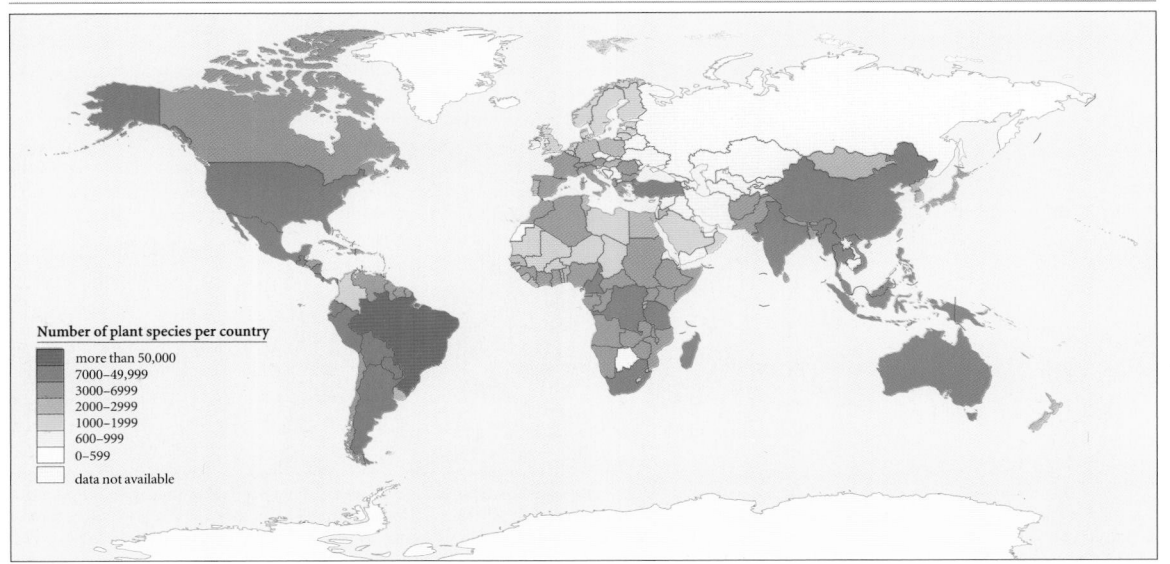

Number of plant species per country
- more than 50,000
- 7000–49,999
- 3000–6999
- 2000–2999
- 1000–1999
- 600–999
- 0–599
- data not available

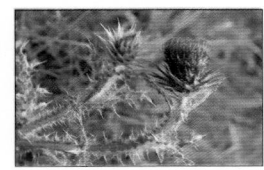

RESISTING PREDATORS

A great variety of plants have developed devices including spines (above), poisons, stinging hairs, and an unpleasant taste or smell to deter animal predators.

WEEDS

Weeds such as bindweed (above) are fast-growing, easily dispersed, and tolerant of a number of different environments, enabling them to quickly colonize suitable habitats. They are among the most adaptable of all plants.

POPULATION AND SETTLEMENT

THE EARTH'S POPULATION IS PROJECTED to rise from its current level of about 5.5 billion to reach some 10 billion by 2025. The global distribution of this rapidly growing population is very uneven, and is dictated by climate, terrain, and natural and economic resources. The great majority of the Earth's people live in coastal zones, and along river valleys. Deserts cover over 20% of the Earth's surface, but support less than 5% of the world's population. It is estimated that over half of the world's population live in cities – most of them in Asia – as a result of mass migration from rural areas in search of jobs. Many of these people live in the so-called "megacities," some with populations as great as 40 million.

PATTERNS OF SETTLEMENT

THE PAST 200 YEARS have seen the most radical shift in world population patterns in recorded history.

NOMADIC LIFE

ALL THE WORLD'S PEOPLES were hunter-gatherers 10,000 years ago. Today nomads, who live by following available food resources, account for less than 0.0001% of the world's population. They are mainly pastoral herders, moving their livestock from place to place in search of grazing land.

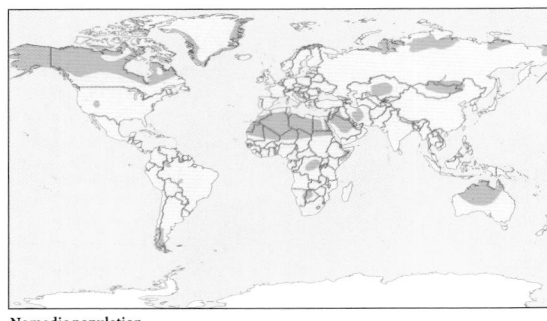

Nomadic population

Nomadic population area

THE GROWTH OF CITIES

IN 1900 there were only 14 cities in the world with populations of more than a million, mostly in the northern hemisphere. Today, as more and more people in the developing world migrate to towns and cities, there are 29 cities whose population exceeds 5 million, and around 200 "million-cities."

MILLION-CITIES IN 1900

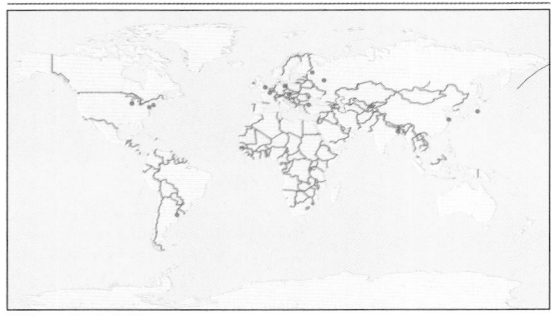

Million-cities in 1900

* Cities over 1 million population

MILLION-CITIES IN 1995

Million-cities in 1995

* Cities over 1 million population

NORTH AMERICA

THE EASTERN AND WESTERN SEABOARDS of the US, with huge expanses of interconnected cities, towns, and suburbs, are vast, densely-populated megalopolises. Central America and the Caribbean also have high population densities. Yet, away from the coasts and in the wildernesses of northern Canada the land is very sparsely settled.

Vancouver on Canada's west coast, grew up as a port city. In recent years it has attracted many Asian immigrants, particularly from the Pacific Rim.

North America's central plains, the continent's agricultural heartland, are thinly populated and highly productive.

EUROPE

WITH ITS TEMPERATE CLIMATE, and rich mineral and natural resources, Europe is generally very densely settled. The continent acts as a magnet for economic migrants from the developing world, and immigration is now widely restricted. Birthrates in Europe are generally low, and in some countries, such as Germany, the populations have stabilized at zero growth, with a fast-growing elderly population.

Many European cities, like Siena, once reflected the "ideal" size for human settlements. Modern technological advances have enabled them to grow far beyond the original walls.

Within the densely-populated Netherlands the reclamation of coastal wetlands is vital to provide much-needed land for agriculture and settlement.

Population density (inhabitants per sq mile)

More than 520
260–519
130–259
55–129
28–54
15–27
1–15
Less than 1

NORTH AMERICA

Population 9% World land area 17%

EUROPE

Population 14% World land area 7.1%

AFRICA

Population 12% World land area 20.2%

SOUTH AMERICA

Population 5.5% World land area 11.8%

SOUTH AMERICA

MOST SETTLEMENT IN SOUTH AMERICA is clustered in a narrow belt in coastal zones and in the northern Andes. During the 20th century, cities such as São Paulo and Buenos Aires grew enormously, acting as powerful economic magnets to the rural population. Shantytowns have grown up on the outskirts of many major cities to house these immigrants, often lacking basic amenities.

Many people in western South America live at high altitudes in the Andes, both in cities and in villages such as this one in Bolivia.

Venezuela is the most highly urbanized country in South America, with more than 90% of the population living in cities such as Caracas.

AFRICA

THE ARID CLIMATE of much of Africa means that settlement of the continent is sparse, focusing in coastal areas and fertile regions such as the Nile Valley. Africa still has a high proportion of nomadic agriculturalists, although many are now becoming settled, and the population is predominantly rural.

Cities such as Nairobi (above), Cairo and Johannesburg have grown rapidly in recent years, although only Cairo has a significant population on a global scale.

Traditional lifestyles and homes persist across much of Africa, which has a higher proportion of rural or village-based population than any other continent.

ASIA

MOST ASIAN SETTLEMENT originally centered around the great river valleys such as the Indus, the Ganges, and the Yangtze. Today, almost 60% of the world's population lives in Asia, many in burgeoning cities – particularly in the economically-buoyant Pacific Rim countries. Even rural population densities are high in many countries; practices such as terracing in Southeast Asia making the most of the available land.

Many of China's cities are now vast urban areas with populations of more than 5 million people.

This stilt village in Bangladesh is built to resist the regular flooding. Pressure on land, even in rural areas, forces many people to live in marginal areas.

POPULATION STRUCTURES

POPULATION PYRAMIDS are an effective means of showing the age structures of different countries, and highlighting changing trends in population growth and decline. The typical pyramid for a country with a growing, youthful population, is broad-based *(left)*, reflecting a high birthrate and a far larger number of young rather than elderly people. In contrast, countries with populations whose numbers are stabilizing have a more balanced distribution of people in each age band, and may even have lower numbers of people in the youngest age ranges, indicating both a high life expectancy, and that the population is now barely replacing itself *(right)*. The Russian Federation *(center)* still bears the scars of World War II, reflected in the dramatically lower numbers of men than women in the 60–80+ age range.

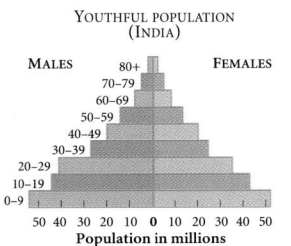
YOUTHFUL POPULATION
(INDIA)
MALES 80+ FEMALES
70–79
60–69
50–59
40–49
30–39
20–29
10–19
0–9
50 40 30 20 10 0 10 20 30 40 50
Population in millions

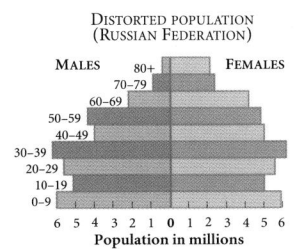
DISTORTED POPULATION
(RUSSIAN FEDERATION)
MALES 80+ FEMALES
70–79
60–69
50–59
40–49
30–39
20–29
10–19
0–9
6 5 4 3 2 1 0 1 2 3 4 5 6
Population in millions

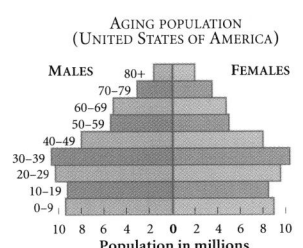
AGING POPULATION
(UNITED STATES OF AMERICA)
MALES 80+ FEMALES
70–79
60–69
50–59
40–49
30–39
20–29
10–19
0–9
10 8 6 4 2 0 2 4 6 8 10
Population in millions

POPULATION GROWTH

IMPROVEMENTS IN FOOD SUPPLY and advances in medicine have both played a major role in the remarkable growth in global population, which has increased five-fold over the last 150 years. Food supplies have risen with the mechanization of agriculture and improvements in crop yields. Better nutrition, together with higher standards of public health and sanitation, have led to increased longevity and higher birthrates.

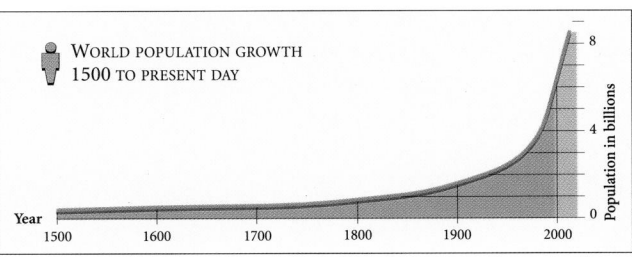
WORLD POPULATION GROWTH
1500 TO PRESENT DAY
Year 1500 1600 1700 1800 1900 2000
Population in billions 0 4 8

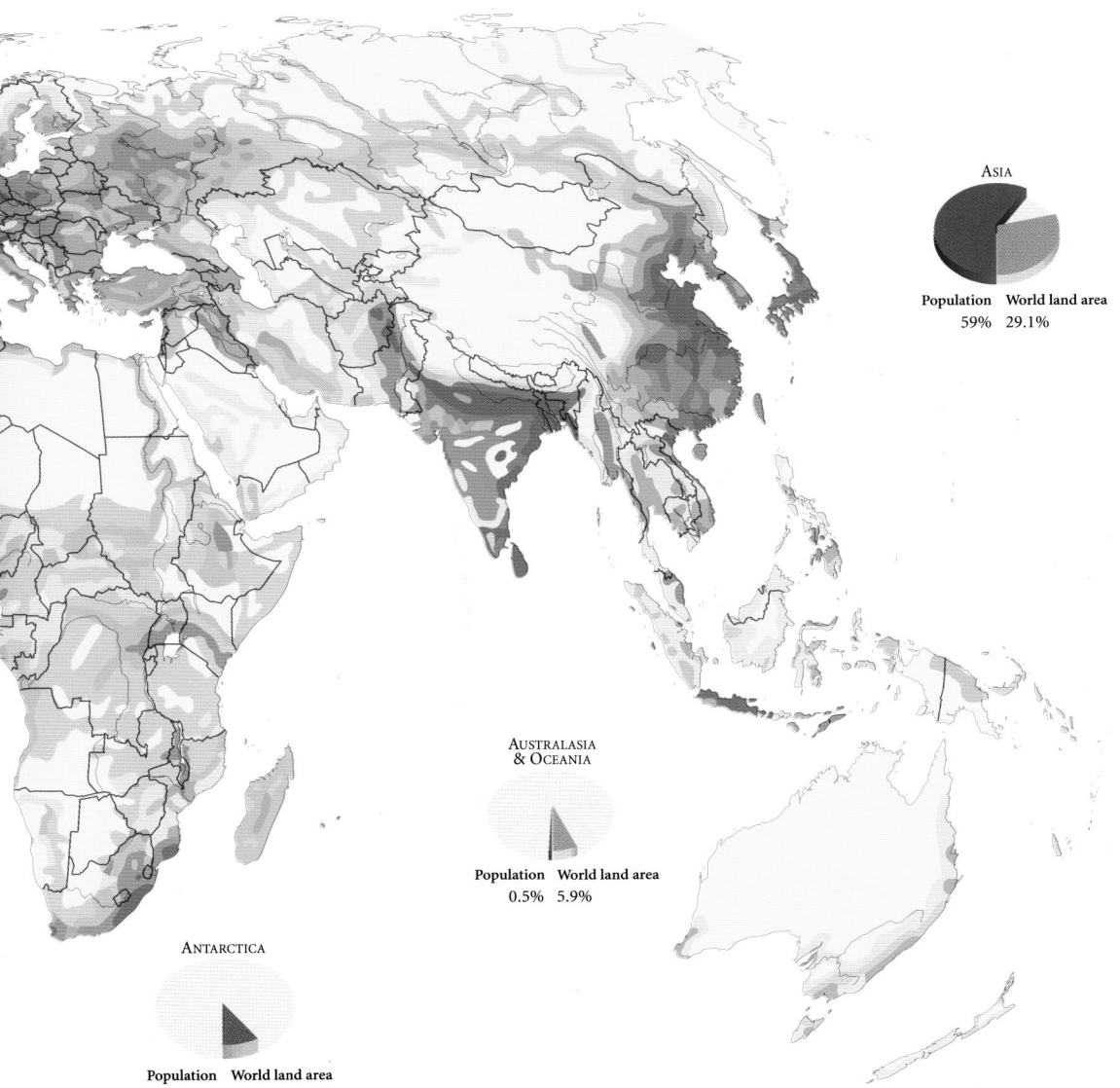
ASIA
Population World land area
59% 29.1%

AUSTRALASIA
& OCEANIA
Population World land area
0.5% 5.9%

ANTARCTICA
Population World land area
0% 8.9%

WORLD NUTRITION

TWO-THIRDS OF THE WORLD'S food supply is consumed by the industrialized nations, many of which have a daily calorific intake far higher than is necessary for their populations to maintain a healthy body weight. In contrast, in the developing world, about 800 million people do not have enough food to meet their basic nutritional needs.

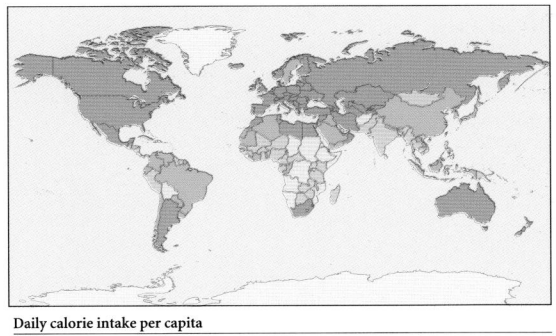
Daily calorie intake per capita
above 3,000   2,000–2,499   data not available
2,500–2,999   below 2,000

WORLD LIFE EXPECTANCY

IMPROVED PUBLIC HEALTH and living standards have greatly increased life expectancy in the developed world, where people can now expect to live twice as long as they did 100 years ago. In many of the world's poorest nations, inadequate nutrition and disease, means that the average life expectancy still does not exceed 45 years.

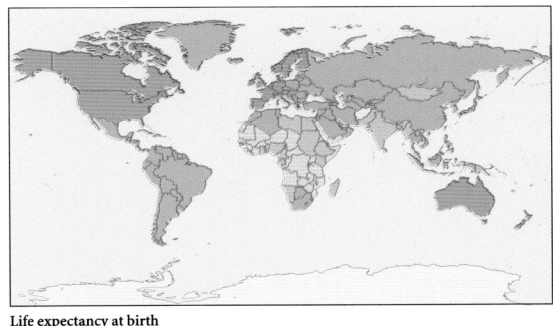
Life expectancy at birth
above 75 years   55–64 years   below 44 years
65–74 years   45–54 years   data not available

AUSTRALASIA & OCEANIA

THIS IS THE WORLD'S most sparsely settled region. The peoples of Australia and New Zealand live mainly in the coastal cities, with only scattered settlements in the arid interior. The Pacific islands can only support limited populations because of their remoteness and lack of resources.

Brisbane, on Australia's Gold Coast is the most rapidly expanding city in the country. The great majority of Australia's population lives in cities near the coasts.

The remote highlands of Papua New Guinea are home to a wide variety of peoples, many of whom still subsist by traditional hunting and gathering.

AVERAGE WORLD BIRTHRATES

BIRTHRATES ARE MUCH HIGHER in Africa, Asia, and South America than in Europe and North America. Increased affluence and easy access to contraception are both factors which can lead to a significant decline in a country's birthrate.

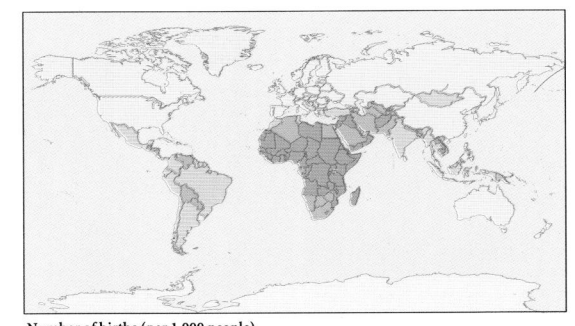
Number of births (per 1,000 people)
above 40   20–29   data not available
30–39   below 20

WORLD INFANT MORTALITY

IN PARTS OF THE DEVELOPING WORLD infant mortality rates are still high; access to medical services such as immunization, adequate nutrition, and the promotion of breast-feeding have been important in combating infant mortality.

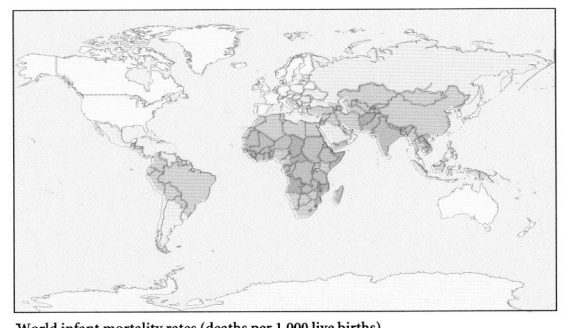
World infant mortality rates (deaths per 1,000 live births)
above 125   35–74   below 15
75–124   15–43   data not available

THE ECONOMIC SYSTEM

THE WEALTHY COUNTRIES OF THE DEVELOPED WORLD, with their aggressive, market-led economies and their access to productive new technologies and international markets, dominate the world economic system. At the other extreme, many of the countries of the developing world are locked in a cycle of national debt, rising populations, and unemployment. The state-managed economies of the former communist bloc began to be dismantled during the 1990s, and China is emerging as a major economic power following decades of isolation.

Trade blocs

▨ EU	▨ NAFTA
CACM	SADC
▨ ASEAN	▨ LAIA
ECOWAS	CEEAC

TRADE BLOCS

INTERNATIONAL TRADE BLOCS are formed when groups of countries, often already enjoying close military and political ties, join together to offer mutually preferential terms of trade for both imports and exports. Increasingly, global trade is dominated by three main blocs: the EU, NAFTA, and ASEAN. They are supplanting older trade blocs such as the Commonwealth, a legacy of colonialism.

INTERNATIONAL TRADE FLOWS

WORLD TRADE acts as a stimulus to national economies, encouraging growth. Over the last three decades, as heavy industries have declined, services – banking, insurance, tourism, airlines, and shipping – have taken an increasingly large share of world trade. Manufactured articles now account for nearly two-thirds of world trade; raw materials and food make up less than a quarter of the total.

SHIPPING
Ships carry 80% of international cargo, and extensive container ports, where cargo is stored, are vital links in the international transportation network.

MULTINATIONALS
Multinational companies are increasingly penetrating inaccessible markets. The reach of many American commodities is now global.

PRIMARY PRODUCTS
Many countries, particularly in the Caribbean and Africa, are still reliant on primary products such as rubber and coffee, which makes them vulnerable to fluctuating prices.

SERVICE INDUSTRIES
Service industries such as banking, tourism and insurance were the fastest-growing industrial sector in the last half of the 20th century. Lloyds of London is the center of the world insurance market.

Countries reliant on a single export
- 🍌 bananas
- ☕ coffee
- 🛢 oil/petroleum
- ⛏ copper

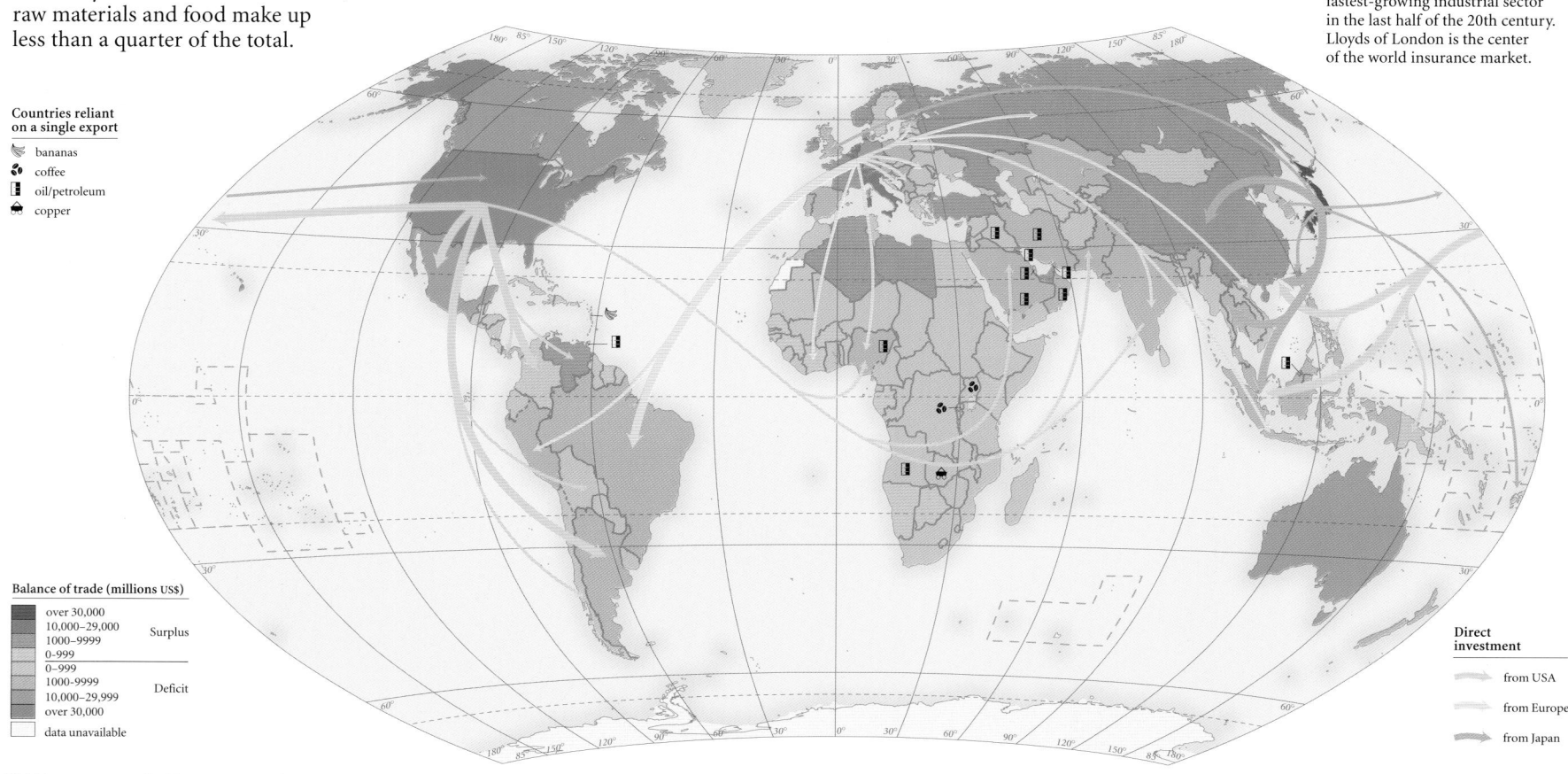

Balance of trade (millions US$)

over 30,000	
10,000–29,000	
1000–9999	Surplus
0–999	
0–999	
1000–9999	
10,000–29,999	Deficit
over 30,000	
data unavailable	

Direct investment
- from USA
- from Europe
- from Japan

WORLD MONEY MARKETS

THE FINANCIAL WORLD has traditionally been dominated by three major centers – Tokyo, New York and London, which house the headquarters of stock exchanges, multinational corporations and international banks. Their geographic location means that, at any one time in a 24-hour day, one major market is open for trading in shares, currencies, and commodities. Since the late 1980s, technological advances have enabled transactions between financial centers to occur at ever-greater speed, and new markets have sprung up throughout the world.

NEW STOCK MARKETS

NEW STOCK MARKETS are now opening in many parts of the world, where economies have recently emerged from state controls. In Moscow and Beijing, and several countries in eastern Europe, newly-opened stock exchanges reflect the transition to market-driven economies.

THE DEVELOPING WORLD

INTERNATIONAL TRADE in capital and currency is dominated by the rich nations of the northern hemisphere. In parts of Africa and Asia, where exports of any sort are extremely limited, home-produced commodities are simply sold in local markets.

MAJOR MONEY MARKETS

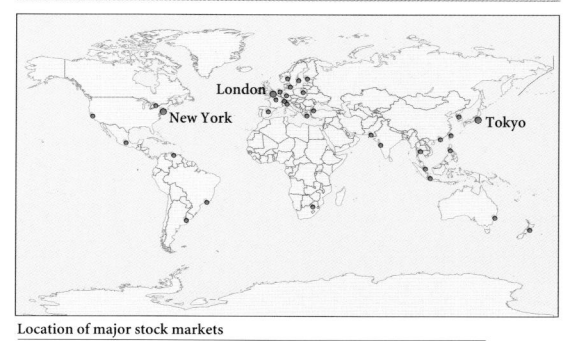

London
New York
Tokyo

Location of major stock markets
- ● Major stock markets

The Tokyo Stock Market *crashed in 1990, leading to a slow-down in the growth of the world's most powerful economy, and a refocusing on economic policy away from export-led growth and toward the domestic market.*

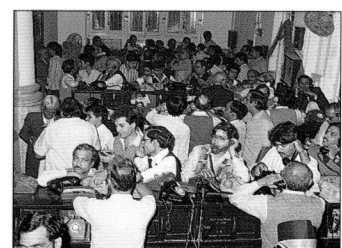

Dealers at the Calcutta Stock Market. *The Indian economy has been opened up to foreign investment and many multinationals now have bases there.*

Markets have thrived *in communist Vietnam since the introduction of a liberal economic policy.*

WORLD WEALTH DISPARITY

A GLOBAL ASSESSMENT of Gross Domestic Product (GDP) by nation reveals great disparities. The developed world, with only a quarter of the world's population, has 80% of the world's manufacturing income. Civil war, conflict, and political instability further undermine the economic self-sufficiency of many of the world's poorest nations.

Cities such as Detroit have been badly hit by the decline in heavy industry.

URBAN DECAY

ALTHOUGH THE US still dominates the global economy, it faces deficits in both the federal budget and the balance of trade. Vast discrepancies in personal wealth, high levels of unemployment, and the dismantling of welfare provisions throughout the 1980s have led to severe deprivation in several of the inner cities of North America's industrial heartland.

BOOMING CITIES

SINCE THE 1980s the Chinese government has set up special industrial zones, such as Shanghai, where foreign investment is encouraged through tax incentives. Migrants from rural China pour into these regions in search of work, creating "boomtown" economies.

Foreign investment has encouraged new infrastructure development in cities like Shanghai.

URBAN SPRAWL

CITIES ARE EXPANDING all over the developing world, attracting economic migrants in search of work and opportunities. In cities such as Rio de Janeiro, housing has not kept pace with the population explosion, and squalid shanty towns (*favelas*) rub shoulders with middle-class housing.

The favelas of Rio de Janeiro sprawl over the hills surrounding the city.

COMPARATIVE WORLD WEALTH

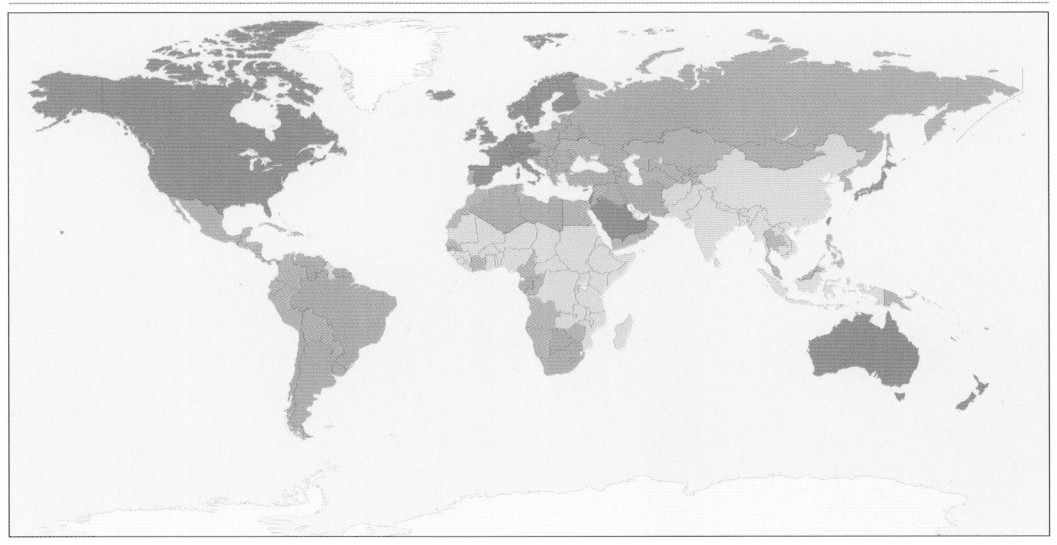

World economies

- high income
- upper-middle income
- lower-middle income
- low income
- data unavailable

ECONOMIC "TIGERS"

THE ECONOMIC "TIGERS" of the Pacific Rim – Taiwan, Singapore, and South Korea – have grown faster than Europe and the US over the last decade. Their export- and service-led economies have benefited from stable government, low labor costs, and foreign investment.

Hong Kong, with its fine natural harbor, is one of the most important ports in Asia.

AGRICULTURAL ECONOMIES

IN PARTS OF THE DEVELOPING WORLD, people survive by subsistence farming – only growing enough food for themselves and their families. With no surplus product, they are unable to exchange goods for currency, the only means of escaping the poverty trap. In other countries, farmers have been encouraged to concentrate on growing a single crop for the export market. This reliance on cash crops leaves farmers vulnerable to crop failure and to changes in the market price of the crop.

The Ugandan uplands are fertile, but poor infrastructure hampers the export of cash crops.

A shopping arcade in Paris displays a great profusion of luxury goods.

THE AFFLUENT WEST

THE CAPITAL CITIES of many countries in the developed world are showcases for consumer goods, reflecting the increasing importance of the service sector, and particularly the retail sector, in the world economy. The idea of shopping as a leisure activity is unique to the western world. Luxury goods and services attract visitors, who in turn generate tourist revenue.

TOURISM

IN 1995, THERE WERE 567 million tourists worldwide. Tourism is now the world's biggest single industry, employing 127 million people, though frequently in low-paid unskilled jobs. While tourists are increasingly exploring inaccessible and less-developed regions of the world, the benefits of the industry are not always felt at a local level. There are also worries about the environmental impact of tourism, as the world's last wildernesses increasingly become tourist attractions.

Botswana's Okavango Delta is an area rich in wildlife. Tourists go on safaris to the region, but the impact of tourism is controlled.

MONEY FLOWS

FOREIGN INVESTMENT in the developing world during the 1970s led to a global financial crisis in the 1980s, when many countries were unable to meet their debt repayments. The International Monetary Fund (IMF) was forced to reschedule the debts and, in some cases, write them off completely. Within the developing world, austerity programs have been initiated to cope with the debt, leading in turn to high unemployment and galloping inflation. In many parts of Africa, stricken economies are now dependent on international aid.

In rural Southeast Asia, babies are given medical checks by UNICEF as part of a global aid program sponsored by the un.

TOURIST ARRIVALS

Tourist arrivals

- over 20 million
- 10–20 million
- 5–10 million
- 2.5–5 million
- 1–2.5 million
- 700,000–999,000
- under 700,000
- data unavailable

INTERNATIONAL DEBT: DONORS AND RECEIVERS

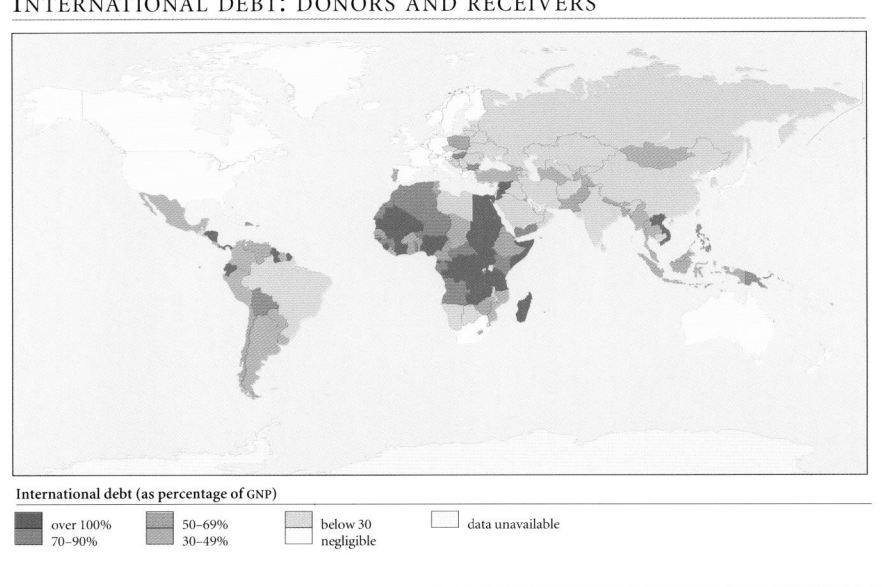

International debt (as percentage of GNP)

- over 100%
- 70–90%
- 50–69%
- 30–49%
- below 30
- negligible
- data unavailable

THE POLITICAL WORLD

THERE ARE 192 INDEPENDENT COUNTRIES in the world today. With the exception of Antarctica, where territorial claims have been deferred by international treaty, every land area of the Earth's surface either belongs to, or is claimed by, one country or another. The largest country in the world is the Russian Federation, the smallest is Vatican City. Some 60 overseas dependent territories remain, administered variously by France, Australia, Denmark, New Zealand, Norway, Portugal, the UK, the US, and the Netherlands.

INTERNATIONAL BORDERS

THE MAP SHOWS three main types of boundary between states. Full borders represent internationally agreed and recognized territorial boundaries. Undefined borders exist where no fixed boundary between states has been demarcated; the boundaries indicated in this way show approximate areas of sovereignty. A disputed border is indicated where a *de facto* territorial boundary exists, which is not agreed or is subject to arbitration.

MOST DENSELY POPULATED COUNTRY
Monaco: 15,897 people per sq mile (41,333 people per sq km)

SMALLEST COUNTRY
Vatican City: 0.17 sq miles (0.44 sq km)

LONGEST LAND BORDERS
Russian Federation: 12,427 miles (20,000 km)

LARGEST COUNTRY
Russian Federation: 6,592,863 sq miles (17,075,400 sq km)

LEAST DENSELY POPULATED COUNTRY
Mongolia: 5 people per sq mile (2 people per sq km)

LONGEST SINGLE LAND BORDER
Canada/US: 5,526 miles (8,893 km)

SMALLEST ISLAND COUNTRY
Nauru: 8.2 sq miles (21 sq km)

MOST POPULOUS CITY
Mexico City: 16,700,000 people

MOST POPULOUS COUNTRY
China: 1,255,100,000 people (estimated)

LARGEST ISLAND COUNTRY
Australia: 2,967,915 sq miles (7,686,850 sq km)

MAP KEY

BORDERS

- full borders
- undefined borders
- disputed borders
- indication of country extent (island territories only)
- indication of dependent territory extent (island territories only)

POLITICAL STATUS

MEXICO: independent state

Gibraltar (to UK): self-governing dependent territory

Laccadive Is (to India): non self-governing dependent territory, with parent state indicated

---

ARCTIC OCEAN

Arctic Circle

USA (Alaska)

Great Bear Lake

Great Slave Lake

CANADA

Baffin Bay

Greenland (to Denmark)

Jan M (to No

Bering Sea

Hudson Bay

ICELAND

Faeroe Islands (to Denmark)

Aleutian Is (to US)

Lake Winnipeg

REPUBLIC OF IRELAND

Isle of Man (to UK)

PACIFIC OCEAN

Lake Superior   Lake Huron   Montreal

Lake Michigan   Toronto   Lake Ontario

Chicago   Lake Erie   New York

UNITED STATES OF AMERICA

St Pierre & Miquelon (to France)

Channel Islands (to UK)

Azores (to Portugal)

Los Angeles

Bermuda (to UK)

ATLANTIC OCEAN

Gibraltar (to UK)

Ceuta (to Spain)

Melilla (to Spain)

Madeira (to Portugal)

Casablanca

Midway Islands (to US)

Tropic of Cancer

Guadalupe (to Mexico)

Gulf of Mexico

BAHAMAS

Turks & Caicos Is (to UK)

Canary Islands (to Spain)

Hawaii (to US)

Monterrey

MEXICO

Havana

CUBA

Puerto Rico (to US)

Virgin Is (to US)

British Virgin Is (to UK)

Anguilla (to UK)

ANTIGUA & BARBUDA

WESTERN SAHARA (occupied by Morocco)

Johnston Atoll (to US)

Guadalajara

Mexico City

Cayman Is (to UK)

JAMAICA

HAITI

DOM. REP.

Guadeloupe (to France)

DOMINICA

Martinique (to France)

ST LUCIA

MAURITANIA

Revillagigedo Islands (to Mexico)

BELIZE Navassa I. (to US)

GUATEMAL

ST KITTS & Montserrat (to UK)

CAPE VERDE

Caribbean Sea

Guatemala City   HONDURAS

EL SALVADOR

Netherlands Antilles (to Neth.)

ST VINCENT & THE GRENADINES

BARBADOS

GRENADA

SENEGAL

MAL

NICARAGUA

Aruba (to Neth.)

TRINIDAD & TOBAGO

GAMBIA

GUINEA-

GUINEA

Clipperton Island (to French Polynesia)

COSTA RICA

PANAMA

Caracas

VENEZUELA

SIERRA LEONE

IVOR COAS

Kingman Reef (to US)

Palmyra Atoll (to US)

Bogotá

COLOMBIA

GUYANA

SURINAME

LIBERIA

Abidja

Baker & Howland Is (to US)

French Guiana (to France)

Equator

ECUADOR

Jarvis I (to US)

Galapagos Is (to Ecuador)

Fernando de Noronha (to Brazil)

KIRIBATI

B R A Z I L

Ascension (to St Helena)

Tokelau (to NZ)

PERU

SAMOA

Lima

Salvador

ATLANTIC OCEAN

Wallis & Futuna (to France)

American Samoa (to US)

Cook Islands (to NZ)

French Polynesia (to France)

PACIFIC OCEAN

Lake Titicaca

BOLIVIA

St He (to

TONGA

Belo Horizonte

Trindade (to Brazil)

NIUE (to NZ)

PARAGUAY

São Paulo   Rio de Janeiro

Tropic of Capricorn

Pitcairn Islands (to UK)

Easter Island (to Chile)

Sala y Gomez (to Chile)

San Felix Island (to Chile)

San Ambrosio Island (to Chile)

Kermadec Islands (to NZ)

Gough Island (to Tristan da Cun

Juan Fernandez Islands (to Chile)

Santiago

URUGUAY

Buenos Aires

Tristan da Cunha (to St Helena)

Chatham Islands (to NZ)

CHILE   ARGENTINA

Falkland Islands (to UK)

South Georgia & South Sandwich Islands (to UK)

South Orkney Islands

South Shetland Islands

S O U T H E R

Peter I Island (to Norway)

Antarctic Circle

Ronne Ice Shelf

Ross Ice Shelf

THE WORLD IN 1914

THE EARLY YEARS OF the 20th century saw the mainly European colonial empires reaching their greatest extents by 1914. Two world wars inaugurated their disintegration, but even in 1950 there were only 82 independent countries. Since then, over 100 have gained their independence, culminating in the breakup of the Soviet Union and former Yugoslavia in the early 1990s.

PERCENTAGE OF EARTH'S LAND SURFACE CONTROLLED BY COLONIAL EMPIRES IN 1914

- Independent: 29.8%
- Chinese: 6%
- Ottoman: 1.5%
- Russian: 15%
- Portuguese: 1%
- Spanish: 1%
- British: 21.5%
- Dutch: 1.4%
- Danish: 1.5%
- United States: 7.6%
- Japanese: 0.4%
- German: 1.6%
- Italian: 1.8%
- Belgian: 1.6%
- French: 7.7%

COLONIAL EMPIRES IN 1914

Colonial Empires in 1914

Belgian	Japanese
British	Ottoman
Chinese	Portuguese
Danish	Russian
Dutch	Spanish
French	United States
German	Independent
Italian	Disputed

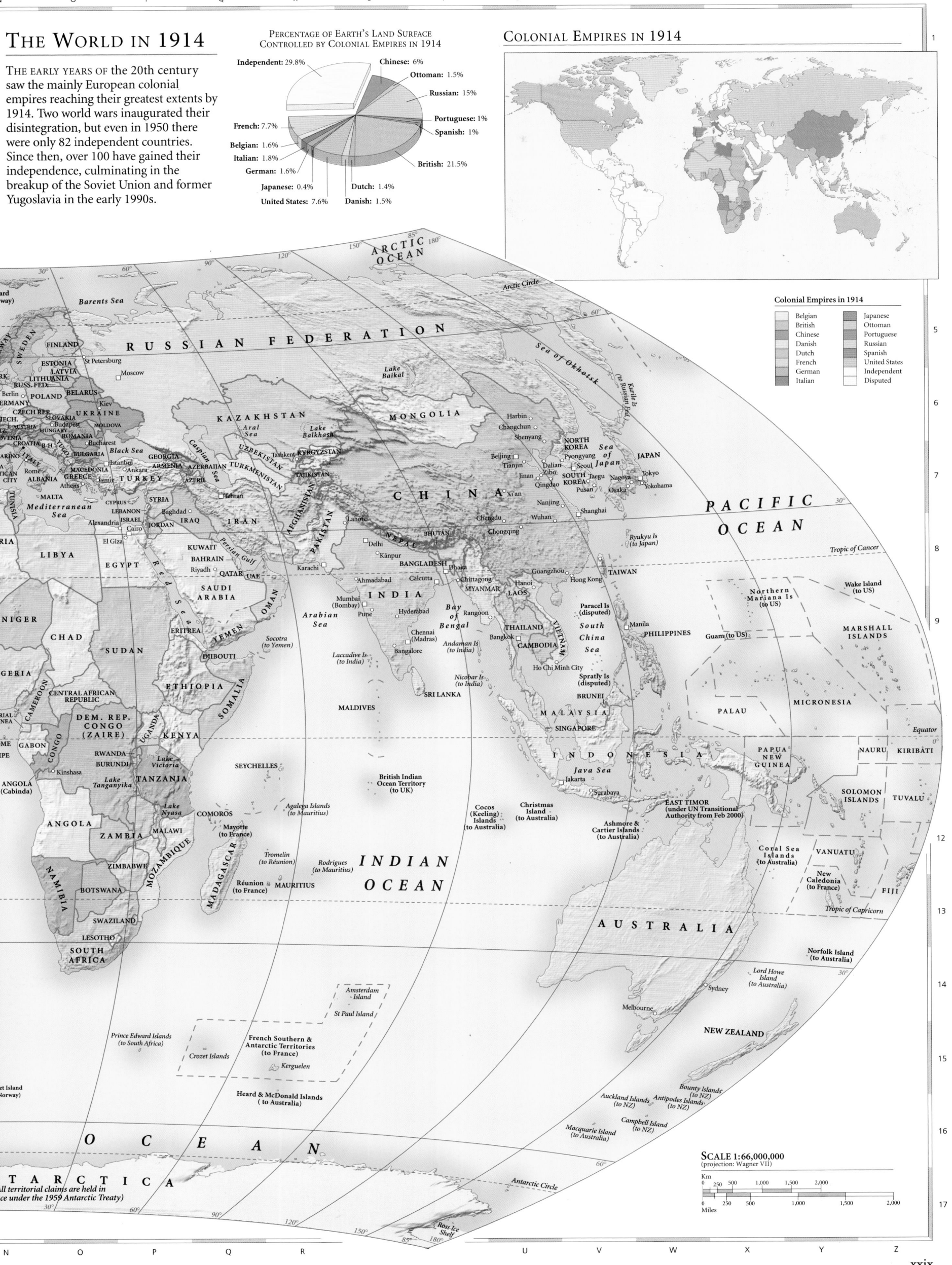

SCALE 1:66,000,000
(projection: Wagner VII)

STATES AND BOUNDARIES

THERE ARE OVER 190 SOVEREIGN STATES in the world today; in 1950 there were only 82. Over the last half-century national self-determination has been a driving force for many states with a history of colonialism and oppression. As more borders are added to the world map, the number of international border disputes increases.

In many cases, where the impetus toward independence has been religious or ethnic, disputes with minority groups have also caused violent internal conflict. While many newly-formed states have moved peacefully toward independence, successfully establishing government by multi-party democracy, dictatorship by military regime or individual despot is often the result of the internal power-struggles which characterize the early stages in the lives of new nations.

THE NATURE OF POLITICS

Democracy is a broad term: it can range from the ideal of multiparty elections and fair representation to, in countries such as Singapore and Indonesia, a thin disguise for single-party rule. In despotic regimes, on the other hand, a single, often personal authority has total power; institutions such as parliament and the military are mere instruments of the dictator.

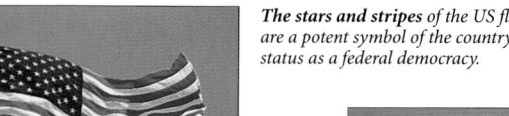

The stars and stripes of the US flag are a potent symbol of the country's status as a federal democracy.

Types of government

- Multiparty democracy for more than 10 yrs
- Multiparty/transitional democracy within last 10 yrs
- Single-party government
- Military regime
- Theocracy
- Absolute monarchy
- ♛ Current civil unrest

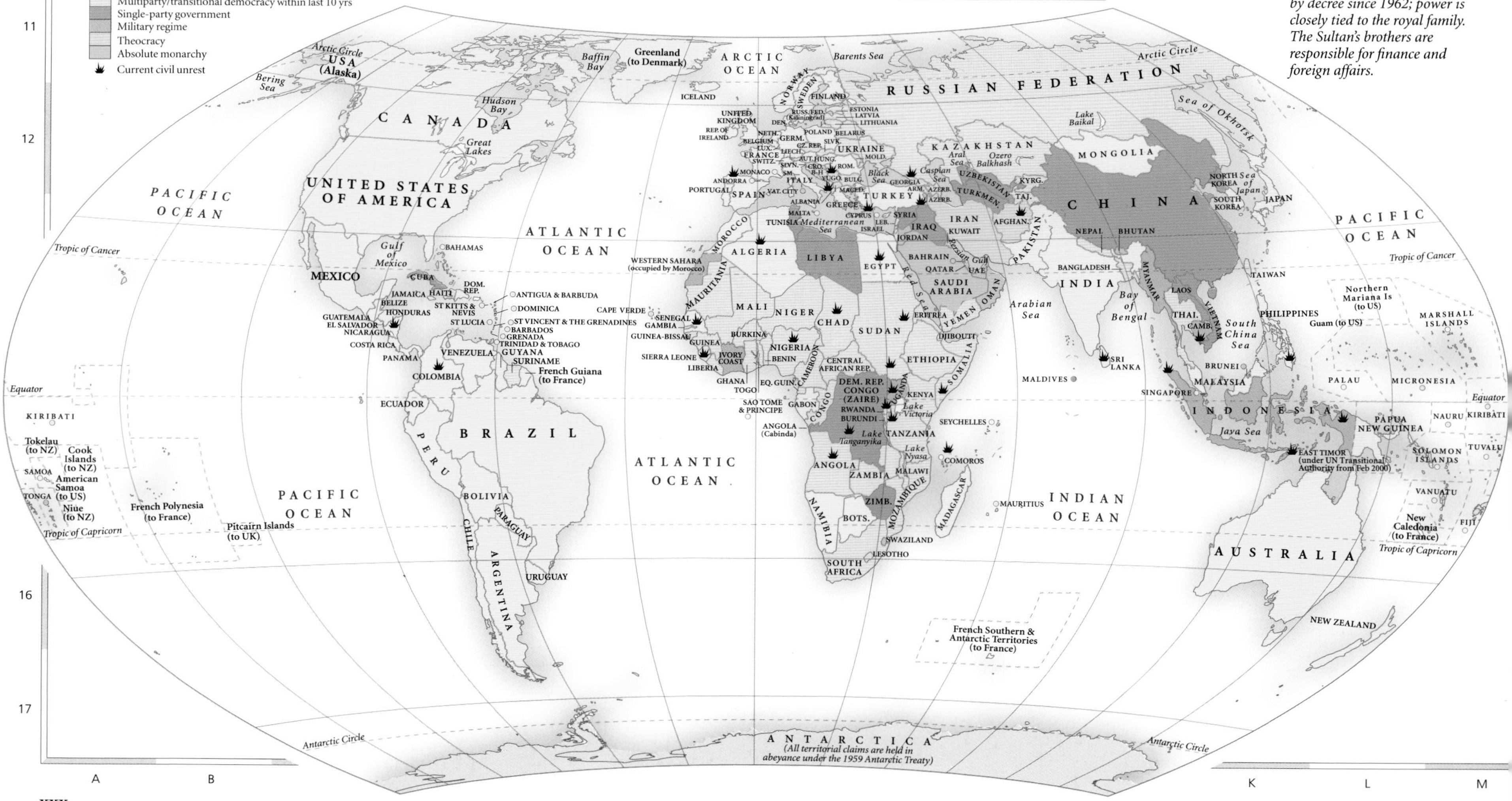

THE CHANGING WORLD MAP

DECOLONIZATION

In 1950, large areas of the world remained under the control of a handful of European countries (*page xxviii*). The process of decolonization had begun in Asia, where, following World War II, much of southern and southeastern Asia sought and achieved self-determination. In the 1960s, a host of African states achieved independence, so that by 1965, most of the larger tracts of the European overseas empires had been substantially eroded. The final major stage in decolonization came with the breakup of the Soviet Union and the Eastern bloc after 1990. The process continues today as the last toeholds of European colonialism, often tiny island nations, press increasingly for independence.

Icons of communism, including statues of former leaders such as Lenin and Stalin, were destroyed when the Soviet bloc was dismantled in 1989, creating several new nations.

Iran is one of the world's true theocracies; Islam has an impact on every aspect of political life.

North Korea is an independent communist republic. Power is concentrated in the hands of Kim Jong Il.

Saddam Hussein overthrew his predecessor in 1979. Since then he has promoted an extreme personality cult, with autocratic control over 21.8 million Iraqis.

NEW NATIONS 1945–1965

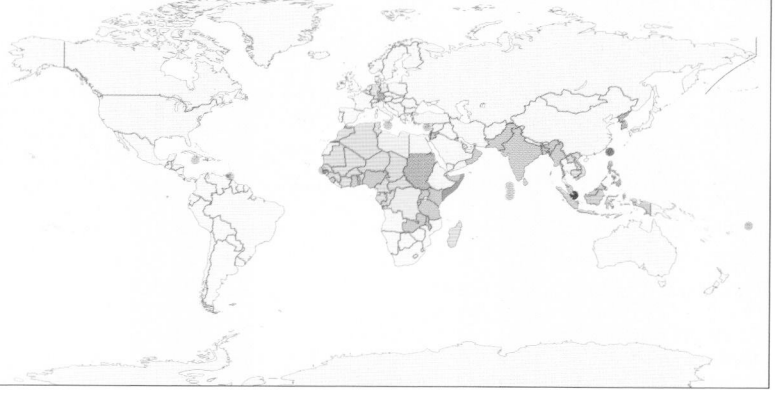

NEW NATIONS 1965–1996

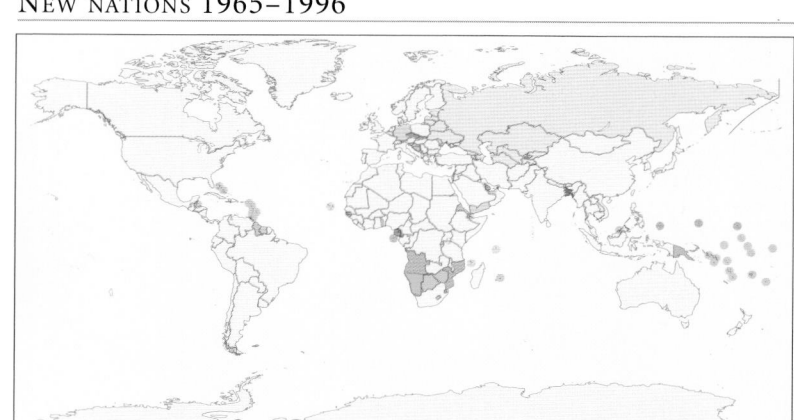

Administration at the time of independence

Australia	Netherlands
Aust/NZ/UK	New Zealand
Belgium	Pakistan
China	Portugal
Czechoslovakia	South Africa
Egypt/UK	Spain
Ethiopia	UK
France	Unified country
France/UK	USA
Italy	USSR
Japan	Yugoslavia
Malaysia	

South Africa became a democracy in 1994, when elections ended over a century of white minority rule.

In Brunei the Sultan has ruled by decree since 1962; power is closely tied to the royal family. The Sultan's brothers are responsible for finance and foreign affairs.

LINES ON THE MAP

THE DETERMINATION OF INTERNATIONAL BOUNDARIES can use a variety of criteria. Many of the borders between older states follow physical boundaries; some mirror religious and ethnic differences; others are the legacy of complex histories of conflict and colonialism, while others have been imposed by international agreements or arbitration.

POST-COLONIAL BORDERS

WHEN THE EUROPEAN COLONIAL EMPIRES IN AFRICA were dismantled during the second half of the 20th century, the outlines of the new African states mirrored colonial boundaries. These boundaries had been drawn up by colonial administrators, often based on inadequate geographical knowledge. Such arbitrary boundaries were imposed on people of different languages, racial groups, religions, and customs.
This confused legacy often led to civil and international war.

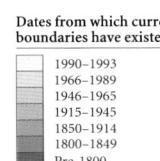

Dates from which current boundaries have existed
- 1990–1993
- 1966–1989
- 1946–1965
- 1915–1945
- 1850–1914
- 1800–1849
- Pre-1800

The conflict that has plagued many African countries since independence has caused millions of people to become refugees.

PHYSICAL BORDERS

MANY OF THE WORLD'S COUNTRIES are divided by physical borders: lakes, rivers, mountains. The demarcation of such boundaries can, however, lead to disputes. Control of waterways, water supplies, and fisheries are frequent causes of international friction.

ENCLAVES

THE SHIFTING POLITICAL MAP over the course of history has frequently led to anomalous situations. Parts of national territories may become isolated by territorial agreement, forming an enclave. The West German part of the city of Berlin, which until 1989 lay several hundred miles within East German territory, was a famous example.

Since the independence of Lithuania and Belarus, the peoples of the Russian enclave of Kaliningrad have become physically isolated.

ANTARCTICA

WHEN ANTARCTIC EXPLORATION began a century ago, seven nations, Australia, Argentina, Britain, Chile, France, New Zealand, and Norway, laid claim to the new territory. In 1961 the Antarctic Treaty, signed by 39 nations, agreed to hold all territorial claims in abeyance.

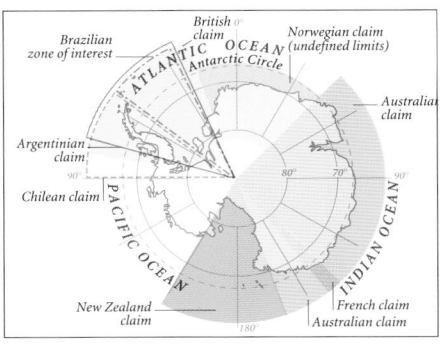

GEOMETRIC BORDERS

STRAIGHT LINES and lines of longitude and latitude have occasionally been used to determine international boundaries; and indeed the world's longest international boundary, between Canada and the USA follows the 49th Parallel for over one-third of its course. Many Canadian, American and Australian internal administrative boundaries are similarly determined using a geometric solution.

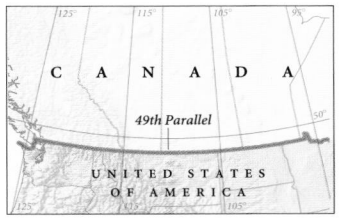

Different farming techniques in Canada and the US clearly mark the course of the international boundary in this satellite map.

WORLD BOUNDARIES

LAKE BORDERS
Countries which lie next to lakes usually fix their borders in the middle of the lake. Unusually the Lake Nyasa border between Malawi and Tanzania runs along Tanzania's shore.

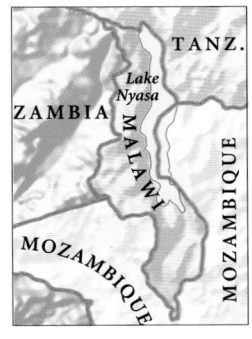

Complicated agreements between colonial powers led to the awkward division of Lake Nyasa.

RIVER BORDERS
Rivers alone account for one-sixth of the world's borders. Many great rivers form boundaries between a number of countries. Changes in a river's course and interruptions of its natural flow can lead to disputes, particularly in areas where water is scarce. The center of the river's course is the nominal boundary line.

The Danube forms all or part of the border between nine European nations.

MOUNTAIN BORDERS
Mountain ranges form natural barriers and are the basis for many major borders, particularly in Europe and Asia. The watershed is the conventional boundary demarcation line, but its accurate determination is often problematic.

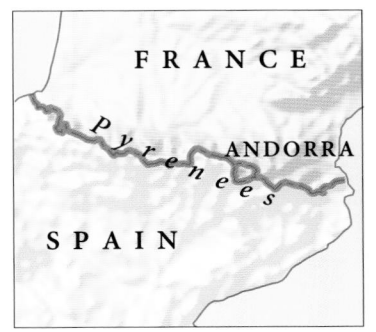

The Pyrenees form a natural mountain border between France and Spain.

SHIFTING BOUNDARIES – POLAND

BORDERS BETWEEN COUNTRIES can change dramatically over time. The nations of eastern Europe have been particularly affected by changing boundaries. Poland is an example of a country whose boundaries have changed so significantly that it has literally moved around Europe. At the start of the 16th century, Poland was the largest nation in Europe. Between 1772 and 1795, it was absorbed into Prussia, Austria, and Russia, and it effectively ceased to exist. After World War I, Poland became an independent country once more, but its borders changed again after World War II following invasions by both Soviet Russia and Nazi Germany.

In 1634, Poland was the largest nation in Europe, its eastern boundary reaching toward Moscow.

From 1772–1795, Poland was gradually partitioned between Austria, Russia, and Prussia. Its eastern boundary receded by over 100 miles (160 km).

Following World War I, Poland was reinstated as an independent state, but it was less than half the size it had been in 1634.

After World War II, the Baltic Sea border was extended westward, but much of the eastern territory was annexed by Russia.

INTERNATIONAL DISPUTES

THERE ARE MORE THAN 60 DISPUTED BORDERS or territories in the world today. Although many of these disputes can be settled by peaceful negotiation, some areas have become a focus for international conflict. Ethnic tensions have been a major source of territorial disagreement throughout history, as has the ownership of, and access to, valuable natural resources. The turmoil of the postcolonial era in many parts of Africa is partly a result of the 19th century "carve-up" of the continent, which created potential for conflict by drawing often arbitrary lines through linguistic and cultural areas.

JAMMU AND KASHMIR

DISPUTES OVER JAMMU AND KASHMIR have caused three serious wars between India and Pakistan since 1947. Pakistan wishes to annex the largely Muslim territory, while India refuses to cede any territory or to hold a referendum, and also lays claim to the entire territory. Most international maps show the "line of control" agreed in 1972 as the *de facto* border. In addition, both Pakistan and India have territorial disputes with neighboring China. The situation is further complicated by a Kashmiri independence movement, active since the late 1980s.

Indian army troops maintain their positions in the mountainous terrain of northern Kashmir.

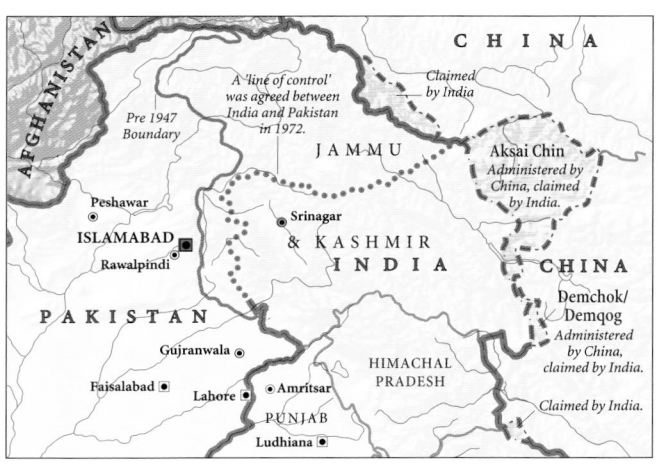

NORTH AND SOUTH KOREA

SINCE 1953, the *de facto* border between North and South Korea has been a ceasefire line which straddles the 38th Parallel and is designated as a demilitarized zone. Both countries have heavy fortifications and troop concentrations behind this zone.

CYPRUS

CYPRUS WAS PARTITIONED in 1974, following an invasion by Turkish troops. The south is now the Greek Cypriot Republic of Cyprus, while the self-proclaimed Turkish Republic of Northern Cyprus is recognized only by Turkey.

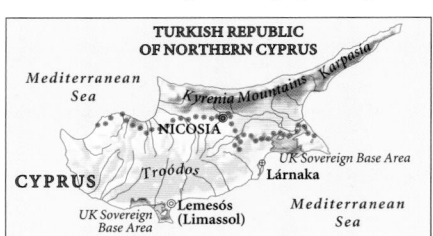
The so-called 'green line' divides Cyprus into Greek and Turkish sectors.

Heavy fortifications on the border between North and South Korea.

THE FALKLAND ISLANDS

THE BRITISH DEPENDENT TERRITORY of the Falkland Islands was invaded by Argentina in 1982, sparking a full-scale war with the UK. In 1995, the UK and Argentina reached an agreement on the exploitation of oil reserves around the islands.

British warships in Falkland Sound during the 1982 war with Argentina.

ISRAEL

ISRAEL WAS CREATED IN 1948 following the 1947 UN Resolution (147) on Palestine. Until 1979 Israel had no borders, only ceasefire lines from a series of wars in 1948, 1967 and 1973. Treaties with Egypt in 1979 and Jordan in 1994 led to these borders being defined and agreed. Negotiations over Israeli settlements in disputed territories such as the West Bank, and the issue of self-government for the Palestinians, continue.

Israeli settlement
Major settlement
Palestinian settlement
Area under Palestinian control

Barbed-wire fences surround a settlement in the Golan Heights.

YUGOSLAVIA

FOLLOWING THE DISINTEGRATION in 1991 of the communist state of Yugoslavia, the breakaway states of Croatia and Bosnia-Herzegovina came into conflict with the "parent" state (consisting of Serbia and Montenegro). Warfare focused on ethnic and territorial ambitions in Bosnia. The tenuous Dayton Accord of 1995 sought to recognize the post-1990 borders, whilst providing for ethnic partition and required international peace-keeping troops to maintain the terms of the peace.

Republika Srpska
Federacija Bosna i Hercegovina

Most claimant states have small military garrisons on the Spratly Islands.

THE SPRATLY ISLANDS

THE SITE OF POTENTIAL OIL and natural gas reserves, the Spratly Islands in the South China Sea have been claimed by China, Vietnam, Taiwan, Malaysia, and the Philippines since the Japanese gave up a wartime claim in 1951.

Occupied by Taiwan
Occupied by Philippines
Occupied by Malaysia
Occupied by China
Occupied by Vietnam

Disputed territories and borders
Countries involved in active territorial or border disputes
Disputed borders
Undefined borders
Disputed territories

ATLAS
OF THE
WORLD

THE MAPS IN THIS ATLAS ARE ARRANGED CONTINENT BY CONTINENT, STARTING FROM THE INTERNATIONAL DATE LINE, AND MOVING EASTWARD. THE MAPS PROVIDE A UNIQUE VIEW OF TODAY'S WORLD, COMBINING TRADITIONAL CARTOGRAPHIC TECHNIQUES WITH THE LATEST REMOTE-SENSED AND DIGITAL TECHNOLOGY.

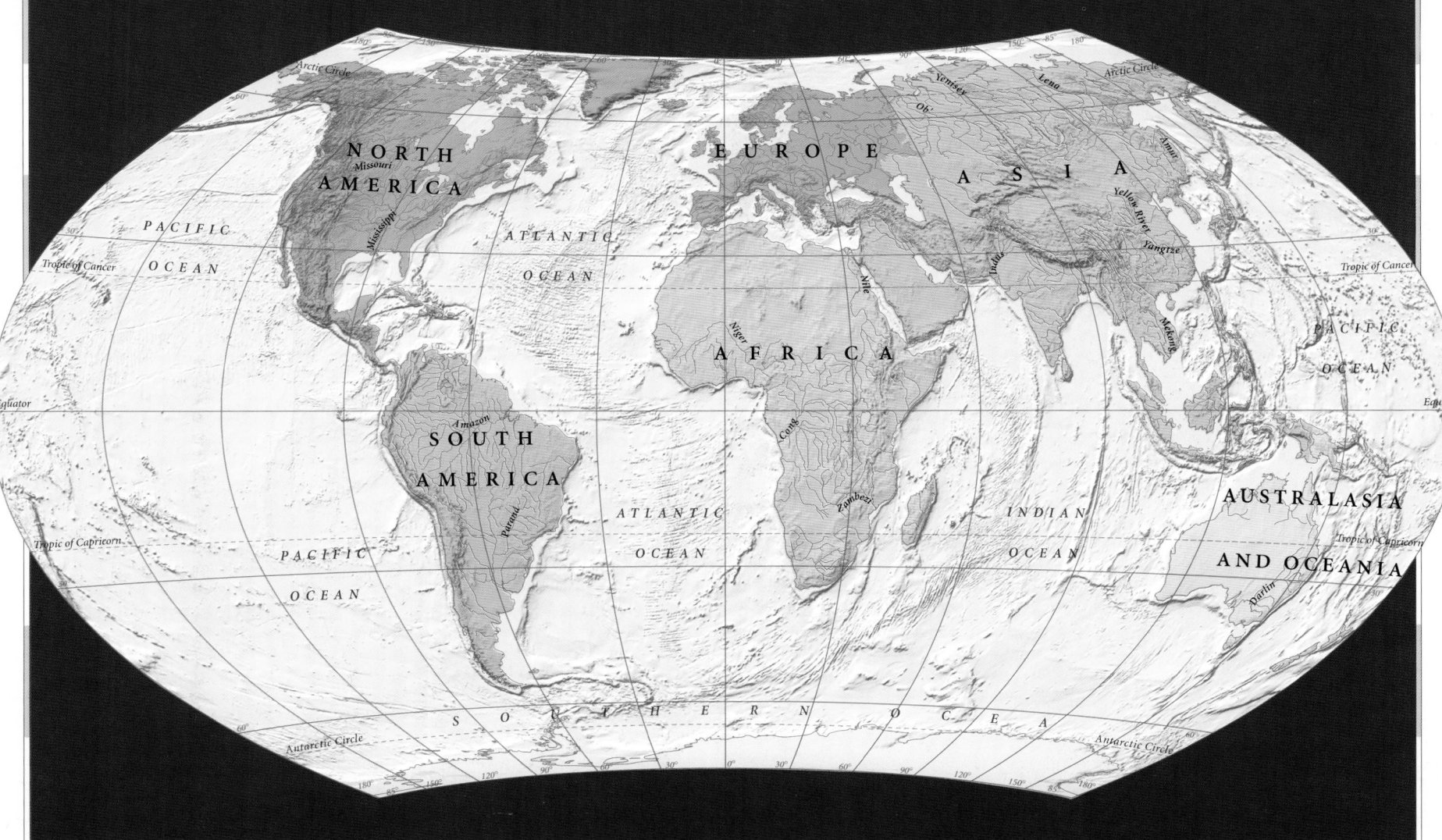

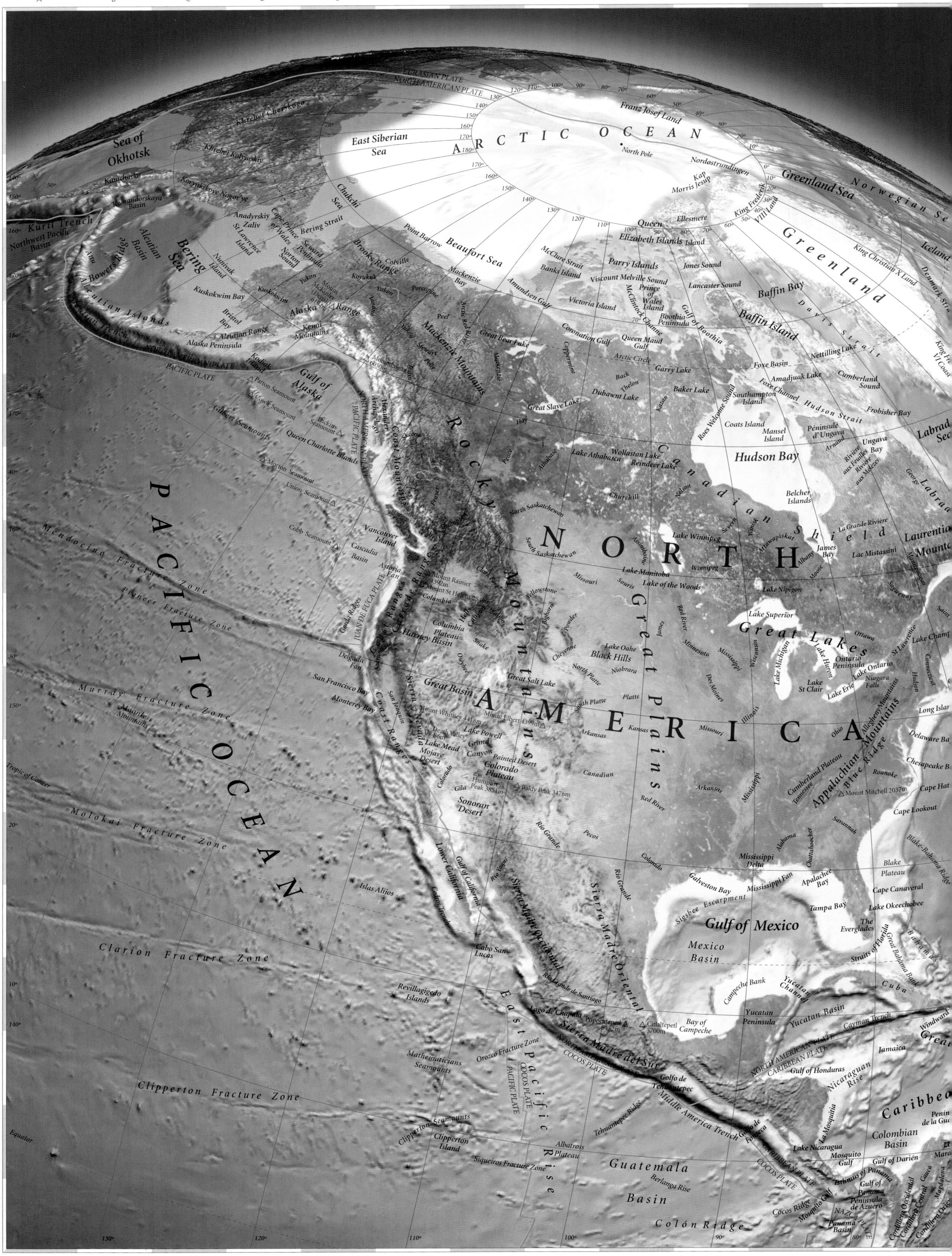

EURASIAN PLATE
NORTH AMERICAN PLATE

Khrebet Cherskogo
Sea of Okhotsk
Khrebet Kolymskiy
East Siberian Sea

ARCTIC OCEAN
North Pole
Nordostrundingen
Franz Josef Land

Greenland Sea
Norwegian Sea

Kamchatka
Koryakskoye Nagor'ye
Chukchi Sea
Morris Jesup
Kap
Queen
Ellesmere Island
King Frederik VIII Land
Greenland
King Christian X Land
Iceland
Denmark Strait

Kurit Trench
Northwest Pacific Basin
Komandorskaya Basin
Anadyrskiy
Cape Prince of Wales
St Lawrence Island
Bering Strait
Seward Peninsula
Point Barrow
Beaufort Sea
McClure Strait
Elizabeth Islands
Parry Islands
Banks Island
Jones Sound
Lancaster Sound
Baffin Bay

Aleutian Basin
Bering Sea
Bowers Ridge
Nunivak
Norton Sound
Brooks Range
Mackenzie Bay
Viscount Melville Sound
Prince of Wales Island
McClintock Channel
Gulf of Boothia
Baffin Island
Davis Strait

Aleutian Islands
Kuskokwim Bay
Bristol Bay
Yukon
Koyukuk
Yukon
Porcupine
Peel
Victoria Island
Coronation Gulf
Boothia Peninsula
Queen Maud Gulf
Nettilling Lake

Alaska Peninsula
Aleutian Range
Kodiak
Kenai Mountains
Alaska Range
Mackenzie Mountains
Great Bear Lake
Coppermine
Arctic Circle
Back
Garry Lake
Foxe Basin
Amadjuak Lake
Cumberland Sound

PACIFIC PLATE
Gulf of Alaska
PACIFIC PLATE
Patton Seamount
Cook Seamount
Gibbs Seamounts
Queen Charlotte Islands
Morton Seamount
Rocky
Great Slave Lake
Thelon
Dubawnt Lake
Baker Lake
Kazan
Foxe Channel
Southampton Island
Hudson Strait
Frobisher Bay

Union Seamount
Roes Welcome Sound
Coats Island
Mansel Island
Péninsule d'Ungava
Rivière Ungava
aux Feuilles Bay
Labrador Sea

Cobb Seamount
Vancouver Island
Churchill
NORTH
Nelson
Belcher Islands
La Grande Rivière
Rivière aux Mélèzes
Georg

Cascadia
North Saskatchewan
Athabasca
Lake Athabasca
Wollaston Lake
Reindeer Lake
Canadian Shield

Cascadia Basin
Astoria Fan
South Saskatchewan
Lake Winnipeg
James Bay
Lac Mistassini
Laurentian Mountains

Gorda Ridges
Mount Rainier 4392m
Mount St Helens
Columbia
Clark Fork
Lake Manitoba
Souris
Lake of the Woods
Lake Nipigon
Moose
Albany
Saguenay

JUAN DE FUCA PLATE
Yellowstone
Columbia Plateau
Snake
Powder
Missouri
Lake of the Woods
Winnipeg
Lake Superior
Great Lakes
Ottawa
Lake Champlain

Delgada Fan
Harney Basin
Cheyenne
Lake Oahe
Black Hills
Niobrara
James
Red River
Minnesota
Mississippi
Wisconsin
Lake Nipissing
Lake Huron
Lake Ontario
St Lawrence

San Francisco Bay
Great Basin
North Platte
Des Moines
Lake Michigan
Niagara Falls
Allegheny Mountains

Monterey Bay
Mount Whitney 4418m
Mount Elbert 4399m
South Platte
Platte
Illinois
Lake St Clair
Lake Erie
Connecticut
Hudson
Long Island

Death Valley
Lake Powell
AMERICA
Arkansas
Kansas
Missouri
Ohio
Cumberland Plateau
Appalachian Mountains
Blue Ridge
Delaware Bay

Coast Ranges
Sierra Nevada
Lake Mead
Grand Canyon
Colorado Plateau
Painted Desert
Canadian
Tennessee
Roanoke
Mount Mitchell 2037m
Cape Hatteras
Chesapeake Bay

San Joaquin
Mojave Desert
Humphreys Peak 3851m
Gila
Baldy Peak 3476m
Red River
Savannah
Cape Lookout

Sonoran Desert
Colorado
Pecos
Great Plains
Alabama
Chattahoochee
Blake Plateau

Rio Grande
Colorado
Mississippi Delta
Apalachee Bay
Cape Canaveral

Gulf of California
Rio Grande
Galveston Bay
Mississippi Fan
Tampa Bay
Lake Okeechobee

Islas Alijos
Lower California
Sigsbee Escarpment
Gulf of Mexico
Mexico Basin
The Everglades
Straits of Florida
Great Bahama Bank

Clarion Fracture Zone
Cabo San Lucas
Campeche Bank
Yucatan Channel
Cuba
Blake Bahama Ridge

Revillagigedo Islands
Sierra Madre Occidental
Rio Grande de Santiago
Sierra Madre Oriental
Lago de Chapala
Popocatépetl
Citlaltépetl 5700m
Yucatan Peninsula
Bay of Campeche
Yucatan Basin
Cayman Trench
Jamaica

Mathematicians Seamounts
Orozco Fracture Zone
East Pacific Rise
Sierra Madre del Sur
NORTH AMERICAN PLATE
CARIBBEAN PLATE
Gulf of Honduras
Nicaraguan Rise
Great

Clipperton Fracture Zone
COCOS PLATE
Golfo de Tehuantepec
Tehuantepec Ridge
Middle America Trench
Golfo de Fonseca
La Mosquitia
Caribbean
Colombian Basin

Clipperton Seamounts
Clipperton Island
PACIFIC PLATE
COCOS PLATE
Lake Nicaragua
Mosquito Gulf
Gulf of Darién
Mara

Equator
Albatross Plateau
Siqueiros Fracture Zone
PACIFIC PLATE
COCOS PLATE
Guatemala Basin
Berlanga Rise
Mosquito Gulf
Isthmus of Panama
Gulf of Panama
Peninsula de Azuero
Panama Basin
Cordillera Occidental
Cordillera Central
Cocos Ridge
Colón Ridge

Mendocino Fracture Zone
Pioneer Fracture Zone
Murray Fracture Zone
Molokai Fracture Zone
Tropic of Cancer
PACIFIC OCEAN

NORTH AMERICA

NORTH AMERICA IS THE WORLD'S THIRD LARGEST CONTINENT WITH A TOTAL AREA OF 9,358,340 SQ MILES (24,238,000 SQ KM) INCLUDING GREENLAND AND THE CARIBBEAN ISLANDS. IT LIES WHOLLY WITHIN THE NORTHERN HEMISPHERE.

○ GREATEST EXTENT, NORTH–SOUTH:
4,600 miles / 7,400 km

□ GREATEST EXTENT, EAST–WEST:
3,500 miles / 5,700 km

Most northerly point:
Kap Morris Jesup,
northern Greenland
83° 38' N

Most easterly point:
Nordøstrundingen,
northeast Greenland
12° 08' W

Most westerly point:
Attu,
Aleutian Islands,
USA 172° 30' E

CAPE PRINCE
OF WALES
(168° 4' W)

Lowest recorded temperature:
Northice, Greenland -
-87° F (-66° C)

Highest point:
Mount McKinley *(Denali)*,
Alaska, USA
20,322 ft (6,194 m)

BOOTHIA PENINSULA
(71° 59' N)

BATTLE HARBOUR
(55° 35' W)

Highest recorded temperature:
Death Valley,
California, USA
135°F (57°C)

Largest lake:
Lake Superior,
Canada/USA
32,142 sq miles
(83,270 sq km)

SAN FRANCISCO

WASHINGTON DC

Lowest point:
Death Valley,
California, USA
-282 ft (-86 m)
below sea level

Tropic of Cancer

Most southerly point:
Península de Azuero, southeast
Panama 7° 15' N

PENÍNSULA
DE AZUERO
(7° 15' N)

SAN FRANCISCO

Rocky Mountains

Great Plains

Great Lakes

Appalachian
Mountains

WASHINGTON DC

CROSS-SECTION FROM SAN FRANCISCO TO WASHINGTON DC

← line of cross-section

| 0 | 500 | 1000 Km |
| 0 | 500 | 1000 Miles |

PHYSICAL NORTH AMERICA

The North American continent can be divided into a number of major structural areas: the Western Cordillera, the Canadian Shield, the Great Plains, and Central Lowlands, and the Appalachians. Other smaller regions include the Gulf Atlantic Coastal Plain which borders the southern coast of North America from the southern Appalachians to the Great Plains. This area includes the expanding Mississippi Delta. A chain of volcanic islands, running in an arc around the margin of the Caribbean Plate, lie to the east of the Gulf of Mexico.

THE CANADIAN SHIELD

SPANNING NORTHERN CANADA and Greenland, this geologically stable plain forms the heart of the continent, containing rocks more than two billion years old. A long history of weathering and repeated glaciation has scoured the region, leaving flat plains, gentle hummocks, numerous small basins and lakes, and the bays and islands of the Arctic.

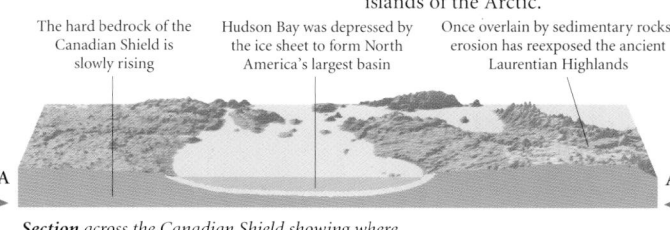

The hard bedrock of the Canadian Shield is slowly rising

Hudson Bay was depressed by the ice sheet to form North America's largest basin

Once overlain by sedimentary rocks, erosion has reexposed the ancient Laurentian Highlands

A — A

Section across the Canadian Shield showing where the ice sheet has depressed the underlying rock and formed bays and islands.

0 100 200 Km
0   100   200 Miles

THE WESTERN CORDILLERA

ABOUT 80 MILLION YEARS ago the Pacific and North American plates collided, uplifting the Western Cordillera. This consists of the Aleutian, Coast, Cascade and Sierra Nevada mountains, and the inland Rocky Mountains. These run parallel from the Arctic to Mexico.

The weight of the ice sheet, 1.8 miles (3 km) thick, has depressed the land to 0.6 miles (1 km) below sea level

Strata have been thrust eastward along fault lines

The Rocky Mountain Trench is the longest linear fault on the continent

This computer-generated view shows the ice-covered island of Greenland without its ice cap.

B — B

Volcanic rock

Cross-section through the Western Cordillera showing direction of mountain building. →

0 50 100 Km
0 50 100 Miles

MAP KEY

ELEVATION

3500m / 11,484ft
3000m / 9843ft
2500m / 8203ft
2000m / 6562ft
1500m / 4922ft
1000m / 3281ft
500m / 1640ft
250m / 820ft
100m / 328ft
sea level

PLATE MARGINS
(for explanation see page xiv)

constructive
destructive
conservative
uncertain
physiographic regions
line of cross-section

SCALE 1:38,000,000
(projection: Lambert Azimuthal Equal Area)

Km
0 100 200   400   600   800   1000
0 50 100 200 300 400 500 600 700 800 900 1000
Miles

THE GREAT PLAINS & CENTRAL LOWLANDS

DEPOSITS LEFT by retreating glaciers and rivers have made this vast flat area very fertile. In the north this is the result of glaciation, with deposits up to one mile (1.7 km) thick, covering the basement rock. To the south and west, the massive Missouri/Mississippi river system has for centuries deposited silt across the plains, creating broad, flat floodplains and deltas.

THE APPALACHIANS

THE APPALACHIAN MOUNTAINS, uplifted about 400 million years ago, are some of the oldest in the world. They have been lowered and rounded by erosion and now slope gently toward the Atlantic across a broad coastal plain.

Horizontal strata

Sedimentary strata folded and faulted into ridges and valleys

Softer strata has been crumpled against the harder basement rock

Hard basement rock

C — C

Cross-section through the Appalachians showing the numerous folds, which have subsequently been weathered to create a rounded relief.

0 50 100 Km
0 50 100 Miles

Sedimentary layers overlay domed basement rock

Upland rivers drain south toward the Mississippi Basin

Confluence of the Missouri and Mississippi Rivers

D — D

Section across the Great Plains and Central Lowlands showing river systems and structure.

0 200 400 Km
0 200 400 Miles

Map labels

ASIA
Bering Strait
Bering Sea
Aleutian Islands
Aleutian Range
Aleutian Range Alaska Range
Brooks Range
Mount McKinley 6194m
Beaufort Sea
Mackenzie Delta
Mackenzie Mountains
Mackenzie
Great Bear Lake
Gulf of Alaska
NORTH AMERICAN PLATE
PACIFIC PLATE
Coast Mountains
Great Slave Lake
Lake Athabasca
Reindeer Lake
Great Salt Lake
WESTERN ROCKY MOUNTAINS CORDILLERA
Cascade Range
Mount Rainier 4392m
Mount St Helens 2549m
Sierra Nevada
Coast Ranges
San Joaquin Valley
San Andreas Fault
Great Basin
Death Valley -86m
Mojave Desert
Grand Canyon
Colorado Plateau
Colorado
Sonoran Desert
Lower California
Gulf of California
Sierra Madre Occidental
Sierra Madre Oriental
Rio Grande
Volcán Pico de Orizaba 5760m
Yucatan Peninsula
Sierra Madre del Sur
CANADIAN SHIELD
CENTRAL LOWLANDS
GREAT PLAINS
Lake Winnipeg
Lake Manitoba
Missouri
Lake Superior
Lake Huron
Lake Michigan
Lake Ontario
Lake Erie
Great Lakes
Ohio
Arkansas
Mississippi
GULF ATLANTIC COASTAL PLAIN
Mississippi Delta
Gulf of Mexico
Laurentian Mountains
St Lawrence
Nova Scotia
Cape Cod
Newfoundland
Labrador
Labrador Sea
Hudson Bay
Hudson Strait
Foxe Basin
Baffin Island
Baffin Bay
Davis Strait
Greenland
ATLANTIC OCEAN
APPALACHIAN MOUNTAINS
APPALACHIANS
West Indies
Greater Antilles
Lesser Antilles
Caribbean Sea
NORTH AMERICAN PLATE
CARIBBEAN PLATE
COCOS PLATE
Isthmus of Panama
Lake Nicaragua
SOUTH AMERICA
SOUTH AMERICAN PLATE
PACIFIC OCEAN

CLIMATE

NORTH AMERICA'S climate includes extremes ranging from freezing Arctic conditions in Alaska and Greenland, to desert in the southwest, and tropical conditions in southeastern Florida, the Caribbean, and Central America. Central and southern regions are prone to severe storms including tornadoes and hurricanes.

"Tornado alley" in the Mississippi Valley suffers frequent tornadoes.

Much of the southwest is semi-desert; receiving less than 12 inches (300 mm) of rainfall a year.

Climate
- ice cap
- tundra
- subarctic
- cool continental
- warm humid
- semiarid
- arid
- humid equatorial
- tropical
- daily hours of sunshine, January
- daily hours of sunshine, July
- direction of hurricanes
- tornado zones

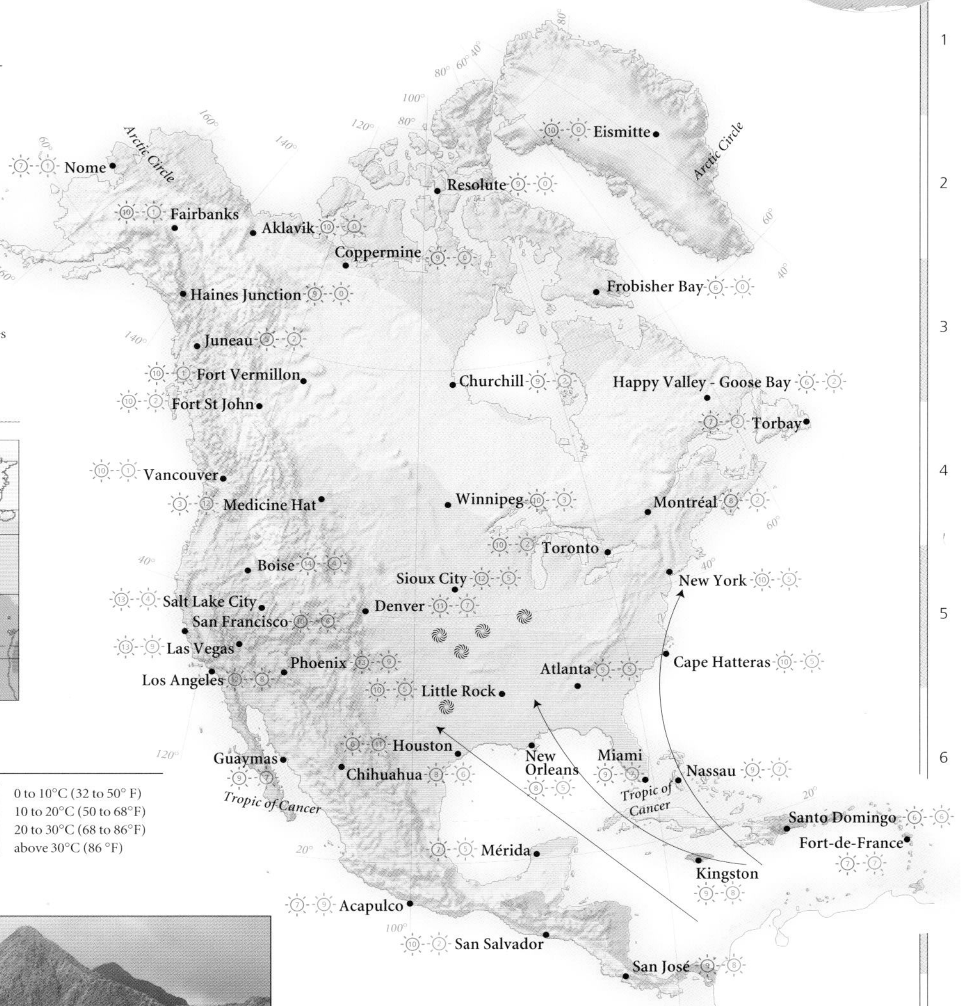

TEMPERATURE

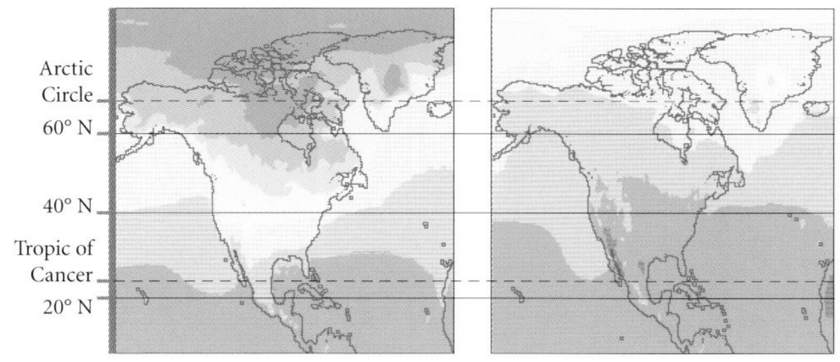

Average January temperature

Average July temperature

Temperature
- below -30°C (-22°F)
- -30 to -20°C (-22 to -4°F)
- -20 to -10°C (-4 to 14°F)
- -10 to 0°C (14 to 32°F)
- 0 to 10°C (32 to 50°F)
- 10 to 20°C (50 to 68°F)
- 20 to 30°C (68 to 86°F)
- above 30°C (86°F)

RAINFALL

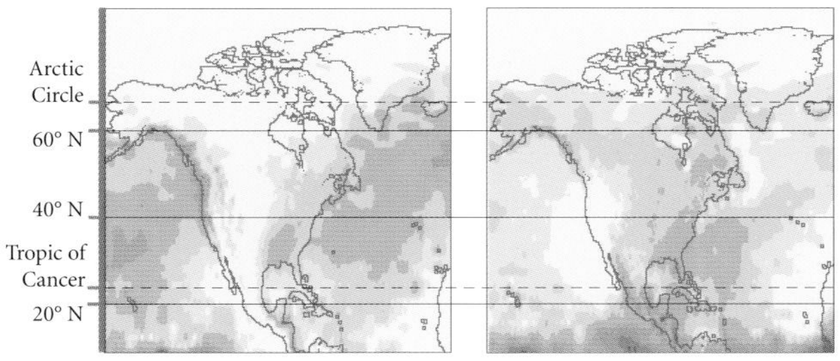

Average January rainfall

Average July rainfall

Rainfall
- 0–25 mm (0–1 in)
- 25–50 mm (1–2 in)
- 50–100 mm (2–4 in)
- 100–200 mm (4–8 in)
- 200–300 mm (8–12 in)
- 300–400 mm (12–16 in)
- 400–500 mm (16–20 in)
- more than 500 mm (20 in)

The lush, green mountains of the Lesser Antilles receive annual rainfalls of up to 360 inches (9,000 mm).

Map labels: Eismitte, Nome, Fairbanks, Aklavik, Coppermine, Resolute, Frobisher Bay, Haines Junction, Juneau, Churchill, Happy Valley - Goose Bay, Torbay, Fort Vermillon, Fort St John, Vancouver, Medicine Hat, Winnipeg, Montréal, Boise, Toronto, New York, Salt Lake City, Sioux City, Denver, San Francisco, Las Vegas, Phoenix, Atlanta, Cape Hatteras, Los Angeles, Little Rock, Guaymas, Houston, New Orleans, Miami, Nassau, Chihuahua, Santo Domingo, Fort-de-France, Mérida, Kingston, Acapulco, San José, San Salvador, Arctic Circle, Tropic of Cancer

SHAPING THE CONTINENT

GLACIAL PROCESSES affect much of northern Canada, Greenland and the Western Cordillera. Along the western coast of North America, Central America, and the Caribbean, underlying plates moving together lead to earthquakes and volcanic eruptions. The vast river systems, fed by mountain streams, constantly erode and deposit material along their paths.

VOLCANIC ACTIVITY

[1] Mount St. Helens volcano *(right)* in the Cascade Range erupted violently in May 1980, killing 57 people and leveling large areas of forest. The lateral blast filled a valley with debris for 15 miles (25 km).

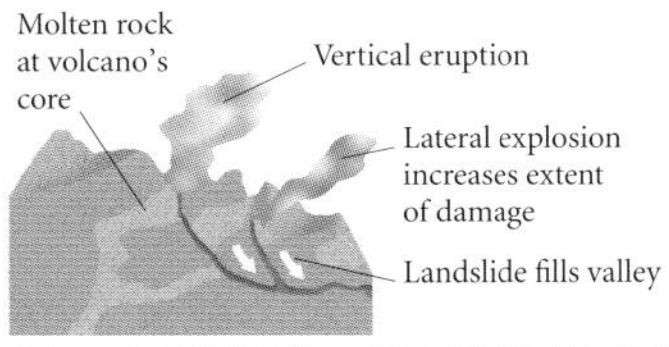

Diagram labels: Molten rock at volcano's core, Vertical eruption, Lateral explosion increases extent of damage, Landslide fills valley

VOLCANIC ACTIVITY: ERUPTION OF MOUNT ST.. HELENS

SEISMIC ACTIVITY

[5] The San Andreas Fault *(above)* places much of the North America's west coast under constant threat from earthquakes. It is caused by the Pacific Plate grinding past the North American Plate at a faster rate, though in the same direction.

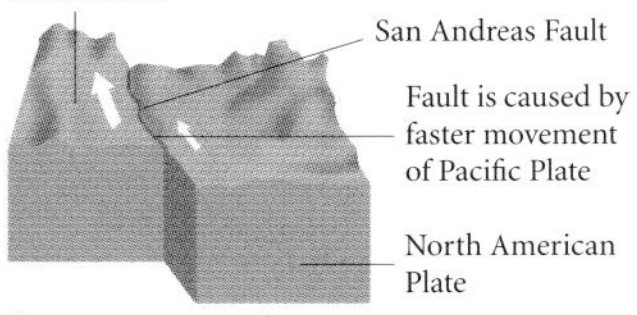

Diagram labels: Pacific Plate, San Andreas Fault, Fault is caused by faster movement of Pacific Plate, North American Plate

SEISMIC ACTIVITY: ACTION OF THE SAN ANDREAS FAULT

RIVER EROSION

[6] The Grand Canyon *(above)* in the Colorado Plateau was created by the downward erosion of the Colorado River, combined with the gradual uplift of the plateau, over the past 30 million years. The contours of the canyon formed as the softer rock layers eroded into gentle slopes, and the hard rock layers into cliffs. The depth varies from 3,855–6,560 ft (1,175–2,000 m).

PERIGLACIATION

[2] The ground in the far north is nearly always frozen: the surface thaws only in summer. This freeze-thaw process produces features such as pingos *(left)*; formed by the freezing of groundwater. With each successive winter ice accumulates producing a mound with a core of ice.

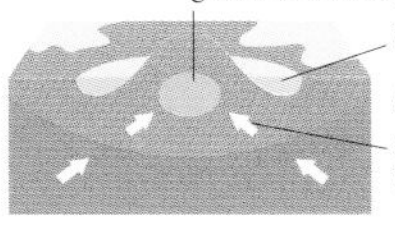

Diagram labels: Ice core pushes up ground to form pingo, Unfrozen lake, Groundwater attracted to ice core

PERIGLACIATION: FORMATION OF A PINGO IN THE MACKENZIE DELTA

THE EVOLVING LANDSCAPE

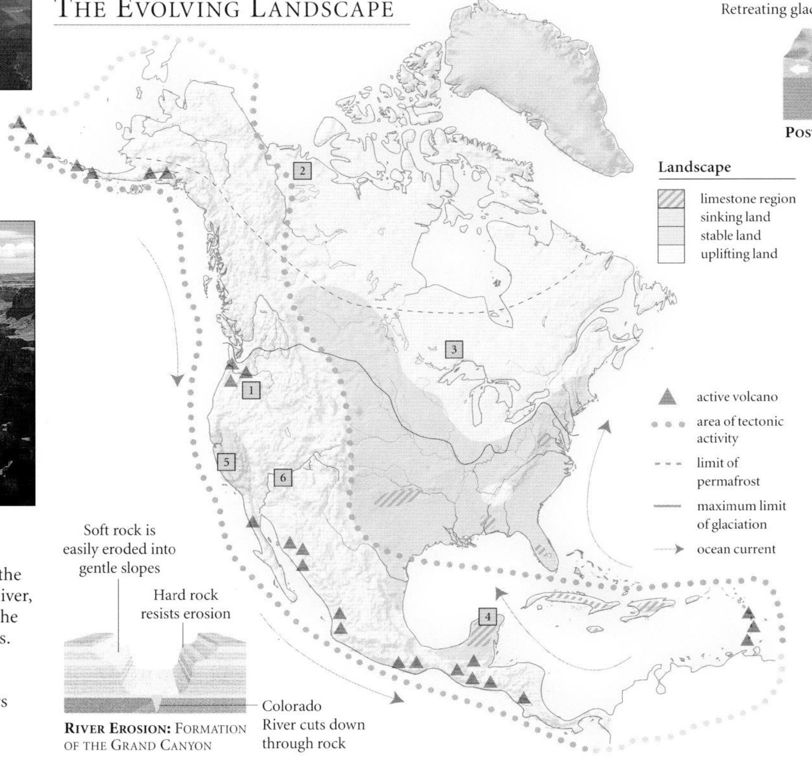

Landscape
- limestone region
- sinking land
- stable land
- uplifting land

- ▲ active volcano
- ⋯ area of tectonic activity
- - - limit of permafrost
- — maximum limit of glaciation
- → ocean current

Diagram labels (River Erosion): Soft rock is easily eroded into gentle slopes, Hard rock resists erosion, Colorado River cuts down through rock

RIVER EROSION: FORMATION OF THE GRAND CANYON

POST-GLACIAL LAKES

[3] A chain of lakes from Great Bear Lake to the Great Lakes *(above)* was created as the ice retreated northward. Glaciers scoured hollows in the softer lowland rock. Glacial deposits at the lip of the hollows, and ridges of harder rock, trapped water to form lakes.

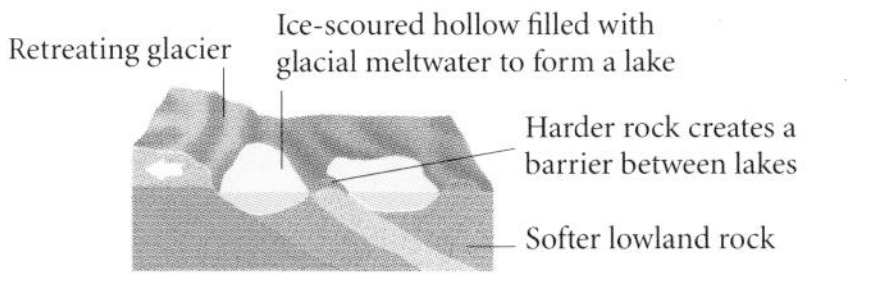

Diagram labels: Retreating glacier, Ice-scoured hollow filled with glacial meltwater, Harder rock creates a barrier between lakes, Softer lowland rock

POST-GLACIAL LAKES: FORMATION OF THE GREAT LAKES

WEATHERING

[4] The Yucatan Peninsula is a vast, flat limestone plateau in southern Mexico. Weathering action from both rainwater and underground streams has enlarged fractures in the rock to form caves and hollows, called sinkholes *(above)*.

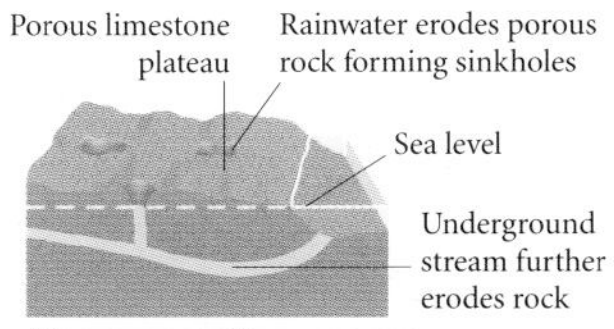

Diagram labels: Porous limestone plateau, Rainwater erodes porous rock forming sinkholes, Sea level, Underground stream further erodes rock

WEATHERING: WATER EROSION ON THE YUCATAN PENINSULA

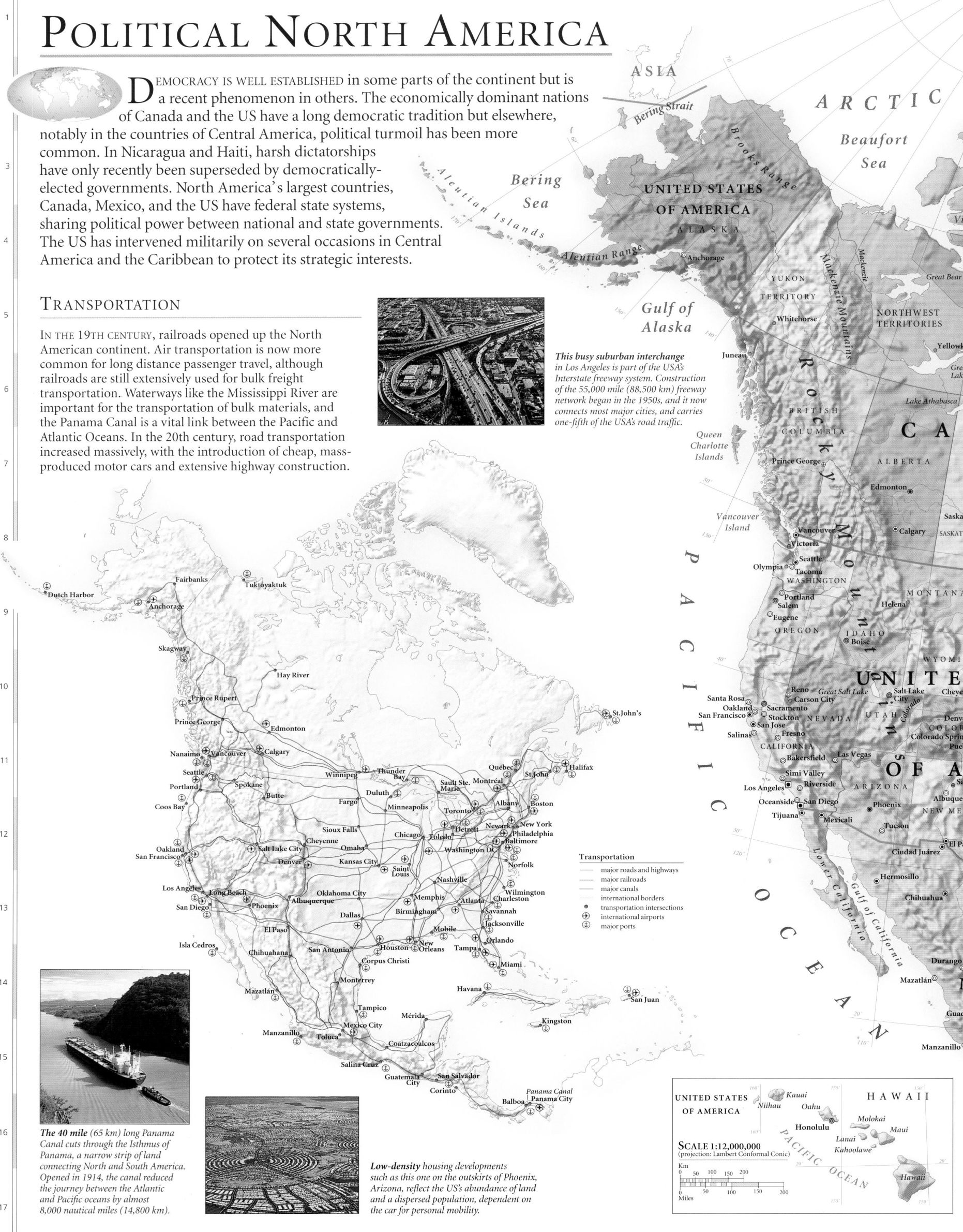

POLITICAL NORTH AMERICA

DEMOCRACY IS WELL ESTABLISHED in some parts of the continent but is a recent phenomenon in others. The economically dominant nations of Canada and the US have a long democratic tradition but elsewhere, notably in the countries of Central America, political turmoil has been more common. In Nicaragua and Haiti, harsh dictatorships have only recently been superseded by democratically-elected governments. North America's largest countries, Canada, Mexico, and the US have federal state systems, sharing political power between national and state governments. The US has intervened militarily on several occasions in Central America and the Caribbean to protect its strategic interests.

TRANSPORTATION

IN THE 19TH CENTURY, railroads opened up the North American continent. Air transportation is now more common for long distance passenger travel, although railroads are still extensively used for bulk freight transportation. Waterways like the Mississippi River are important for the transportation of bulk materials, and the Panama Canal is a vital link between the Pacific and Atlantic Oceans. In the 20th century, road transportation increased massively, with the introduction of cheap, mass-produced motor cars and extensive highway construction.

This busy suburban interchange in Los Angeles is part of the USA's Interstate freeway system. Construction of the 55,000 mile (88,500 km) freeway network began in the 1950s, and it now connects most major cities, and carries one-fifth of the USA's road traffic.

The 40 mile (65 km) long Panama Canal cuts through the Isthmus of Panama, a narrow strip of land connecting North and South America. Opened in 1914, the canal reduced the journey between the Atlantic and Pacific oceans by almost 8,000 nautical miles (14,800 km).

Low-density housing developments such as this one on the outskirts of Phoenix, Arizona, reflect the US's abundance of land and a dispersed population, dependent on the car for personal mobility.

Transportation

— major roads and highways
— major railroads
— major canals
— international borders
• transportation intersections
⊕ international airports
⊕ major ports

UNITED STATES OF AMERICA

SCALE 1:12,000,000
(projection: Lambert Conformal Conic)

HAWAII

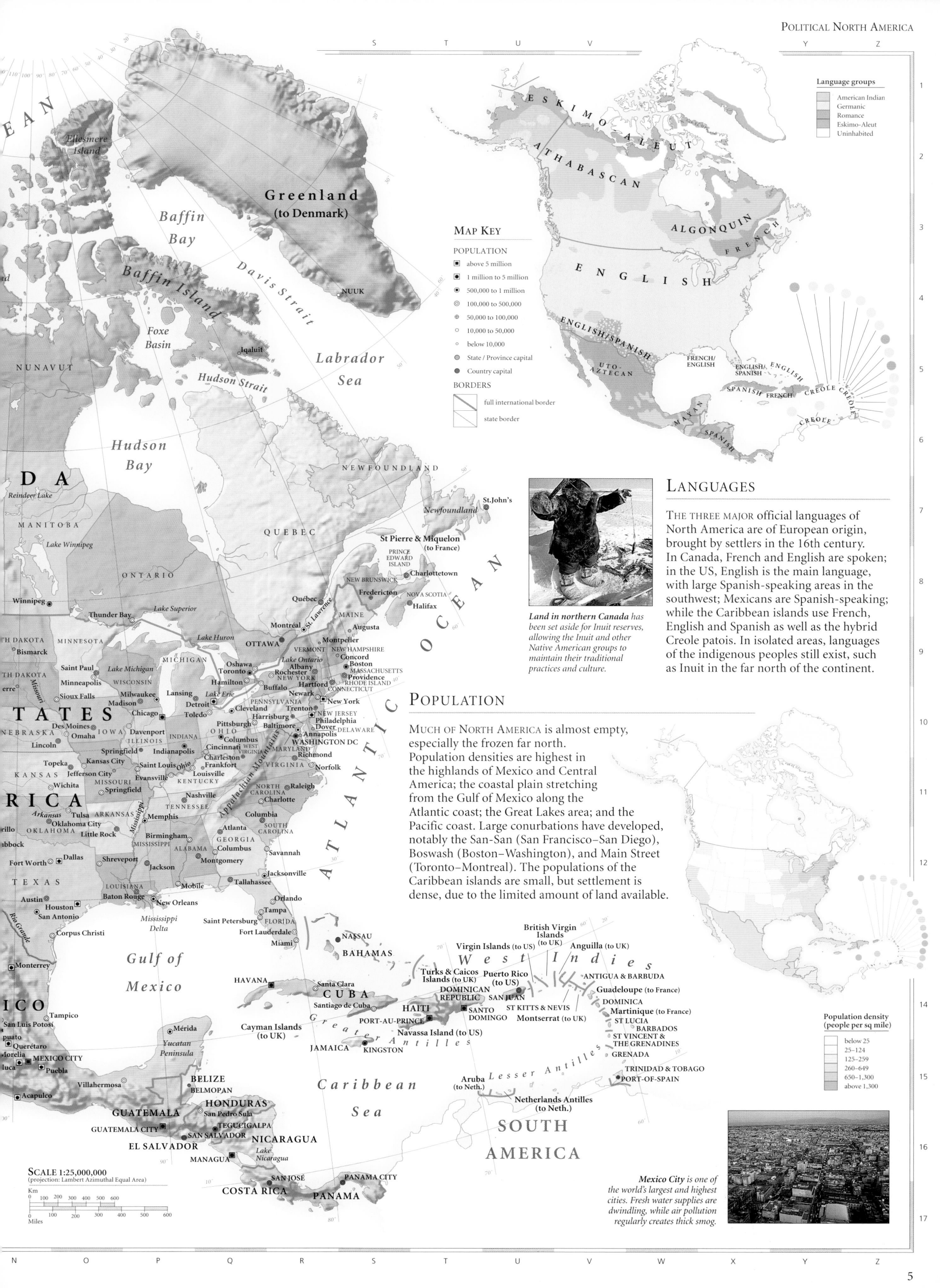

Language groups

- American Indian
- Germanic
- Romance
- Eskimo-Aleut
- Uninhabited

MAP KEY

POPULATION

- ▪ above 5 million
- ▪ 1 million to 5 million
- ● 500,000 to 1 million
- ⊙ 100,000 to 500,000
- ⊕ 50,000 to 100,000
- ○ 10,000 to 50,000
- ∘ below 10,000
- ● State / Province capital
- ● Country capital

BORDERS

- full international border
- state border

LANGUAGES

THE THREE MAJOR official languages of North America are of European origin, brought by settlers in the 16th century. In Canada, French and English are spoken; in the US, English is the main language, with large Spanish-speaking areas in the southwest; Mexicans are Spanish-speaking; while the Caribbean islands use French, English and Spanish as well as the hybrid Creole patois. In isolated areas, languages of the indigenous peoples still exist, such as Inuit in the far north of the continent.

Land in northern Canada has been set aside for Inuit reserves, allowing the Inuit and other Native American groups to maintain their traditional practices and culture.

POPULATION

MUCH OF NORTH AMERICA is almost empty, especially the frozen far north. Population densities are highest in the highlands of Mexico and Central America; the coastal plain stretching from the Gulf of Mexico along the Atlantic coast; the Great Lakes area; and the Pacific coast. Large conurbations have developed, notably the San-San (San Francisco–San Diego), Boswash (Boston–Washington), and Main Street (Toronto–Montreal). The populations of the Caribbean islands are small, but settlement is dense, due to the limited amount of land available.

Population density (people per sq mile)

- below 25
- 25–124
- 125–259
- 260–649
- 650–1,300
- above 1,300

Mexico City is one of the world's largest and highest cities. Fresh water supplies are dwindling, while air pollution regularly creates thick smog.

SCALE 1:25,000,000
(projection: Lambert Azimuthal Equal Area)

Km 0 100 200 300 400 500 600
Miles 0 100 200 300 400 500 600

NORTH AMERICAN RESOURCES

THE TWO NORTHERN COUNTRIES of Canada and the US are richly endowed with natural resources that have helped to fuel economic development. The US is the world's largest economy, although today it is facing stiff competition from the Far East. Mexico has relied on oil revenues but there are hopes that the North American Free Trade Agreement (NAFTA), will encourage trade growth with Canada and the US. The poorer countries of Central America and the Caribbean depend largely on cash crops and tourism.

INDUSTRY

THE MODERN, INDUSTRIALIZED economies of the US and Canada contrast sharply with those of Mexico, Central America, and the Caribbean. Manufacturing is especially important in the US; vehicle production is concentrated around the Great Lakes, while electronic and hi-tech industries are increasingly found in the western and southern states. Mexico depends on oil exports and assembly work, taking advantage of cheap labor. Many Central American and Caribbean countries rely heavily on agricultural exports.

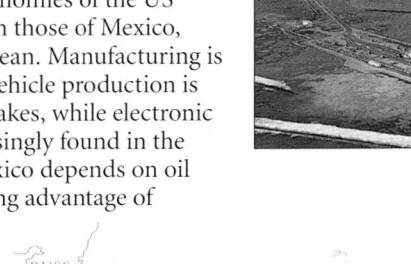

After its purchase from Russia in 1867, Alaska's frozen lands were largely ignored by the US. Oil reserves similar in magnitude to those in eastern Texas were discovered in Prudhoe Bay, Alaska in 1968. Freezing temperatures and a fragile environment hamper oil extraction.

STANDARD OF LIVING

THE US AND CANADA have one of the highest overall standards of living in the world. However, many people still live in poverty, especially in urban ghettos and some rural areas. Central America and the Caribbean are markedly poorer than their wealthier northern neighbors. Haiti is the poorest country in the western hemisphere.

Standard of Living
(UN Human Development Index)

high

low

Fish such as cod, flounder, and plaice are caught in the Grand Banks, off the Newfoundland coast, and processed in many North Atlantic coastal settlements.

South of San Francisco, "Silicon Valley" is both a national and international center for hi-tech industries, electronic industries, and research institutions.

Multinational companies rely on cheap labor and tax benefits to assemble vehicles in Mexican factories.

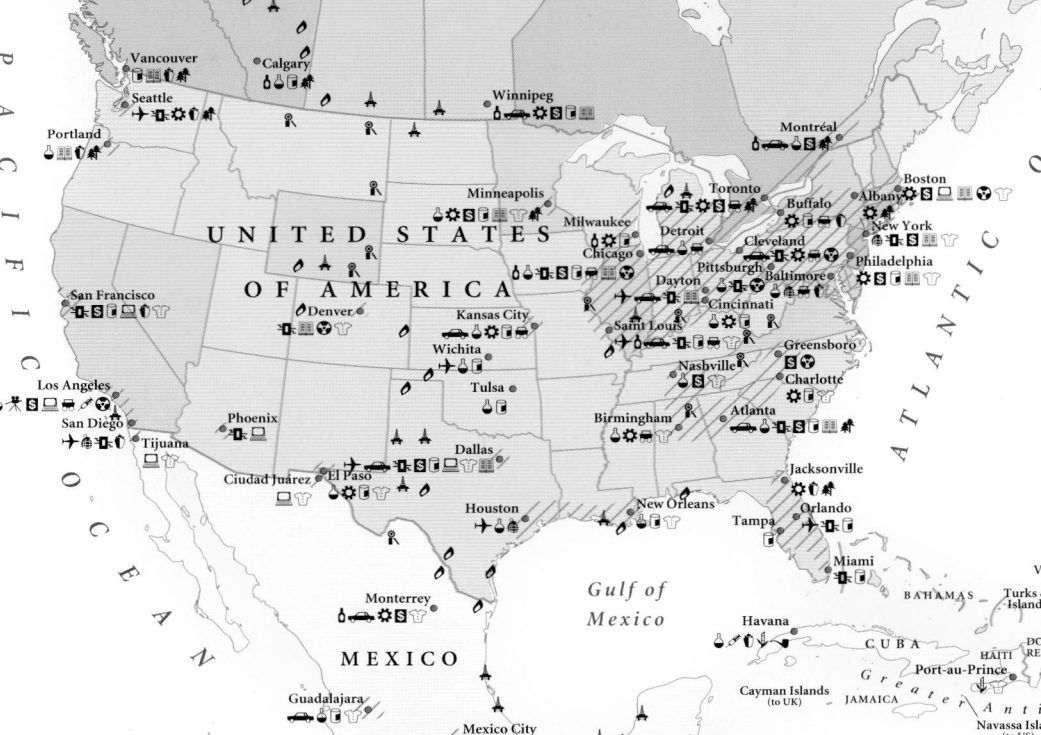

The twin towers of the World Trade Center dominate the Manhattan skyline. New York is one of the world's leading trade and finance centers.

Industry

✈ aerospace	printing & publishing
brewing	research & development
car/vehicle manufacture	shipbuilding
chemicals	sugar processing
defense	textiles
electronics	timber processing
engineering	tobacco processing
movie industry	
finance	coal
food processing	oil
hi-tech industry	gas
iron & steel	industrial cities
pharmaceuticals	major industrial areas

GNP per capita (US$)

	0–1999
	2000–4999
	5000–9999
	10,000–19,999
	20,000–24,999
	25,000+

ARCTIC OCEAN

Bering Strait

RUSS. FED.

Beaufort Sea

Greenland (to Denmark)

Baffin Bay

Bering Sea

USA

Prudhoe Bay

Labrador Sea

Hudson Strait

Gulf of Alaska

Hudson Bay

CANADA

PACIFIC OCEAN

Vancouver
Calgary
Seattle
Winnipeg
Portland
Montréal

Boston
Albany
Minneapolis
Toronto
Buffalo
New York
Milwaukee
Detroit
Cleveland
Philadelphia
Chicago
Pittsburgh
Baltimore
Dayton
UNITED STATES
Cincinnati
San Francisco
OF AMERICA
Denver
Kansas City
Saint Louis
Wichita
Greensboro
Nashville
Charlotte
Los Angeles
Tulsa
San Diego
Phoenix
Atlanta
Tijuana
Dallas
Birmingham
Ciudad Juárez El Paso
Jacksonville
Houston
Orlando
New Orleans
Monterrey
Tampa
Miami

ATLANTIC OCEAN

Gulf of Mexico

MEXICO

Guadalajara

Mexico City

West Indies

Virgin Islands (to US)
British Virgin Islands (to UK) Anguilla (to UK)
BAHAMAS
Turks & Caicos Islands (to UK)
ST KITTS & NEVIS
ANTIGUA & BARBUDA
Puerto Rico (to US)
Montserrat (to UK)
Guadeloupe (to Fra)
Havana
CUBA
DOMINICAN REPUBLIC
San Juan
DOMINICA
Cayman Islands (to UK)
HAITI
Santo Domingo
Martinique (to Fr)
Port-au-Prince
ST LUCIA
Greater Antilles
BARBADOS
JAMAICA
ST VINCENT &
THE GRENADIN
Navassa Island (to US)
GRENADA
Aruba (to Neth.)
Lesser Antilles
TRINIDAD &
TOBAGO
Port-of-Spa
Caribbean Sea

BELIZE
Netherlands Antilles (to Neth.)
GUATEMALA
HONDURAS
VENEZUELA
Guatemala City
Tegucigalpa
EL SALVADOR San Salvador NICARAGUA
Managua
San José
Panama City
COSTA RICA
PANAMA

ENVIRONMENTAL ISSUES

MANY FRAGILE ENVIRONMENTS ARE UNDER THREAT throughout the region. In Haiti, all the primary rain forest has been destroyed, while air pollution from factories and cars in Mexico City is among the worst in the world. Elsewhere, industry and mining pose threats, particularly in the delicate arctic environment of Alaska where oil spills have polluted coastlines and decimated fish stocks.

Environmental Issues
- national parks
- acid rain
- tropical forest
- forest destroyed
- desert
- desertification
- polluted rivers
- radioactive contamination
- marine pollution
- heavy marine pollution
- poor urban air quality

Wild bison graze in Yellowstone National Park, the world's first national park. Designated in 1872, geothermal springs and boiling mud are among its natural spectacles, making it a major tourist attraction.

MINERAL RESOURCES

FOSSIL FUELS ARE EXPLOITED in considerable quantities throughout the continent. Coal mining in the Appalachians is declining but vast open pits exist further west in Wyoming. Oil and natural gas are found in Alaska, Texas, the Gulf of Mexico, and the Canadian West. Canada has large quantities of nickel, while Jamaica has considerable deposits of bauxite, and Mexico has large reserves of silver.

Mineral Resources
- oil field
- gas field
- coal field
- bauxite
- copper
- gold
- iron
- lead
- nickel
- phosphates
- silver
- uranium

In addition to fossil fuels, North America is also rich in exploitable metallic ores. This vast, mile-deep (1.6 km) pit is a copper mine in New Mexico.

In agriculturally marginal areas where the soil is either too poor, or the climate too dry for crops, cattle ranching proliferates – especially in Mexico and the western reaches of the Great Plains.

USING THE LAND AND SEA

ABUNDANT LAND AND FERTILE SOILS stretch from the Canadian prairies to Texas creating North America's agricultural heartland. Cereals and cattle ranching form the basis of the farming economy, with corn and soybeans also important. Fruit and vegetables are grown in California using irrigation, while Florida is a leading producer of citrus fruits. Caribbean and Central American countries depend on cash crops such as bananas, coffee, and sugar cane, often grown on large plantations. This reliance on a single crop can leave these countries vulnerable to fluctuating world crop prices.

Sugar cane is Cuba's main agricultural crop, and is grown and processed throughout the Caribbean. Fermented sugar is used to make rum.

The Great Plains support large-scale arable farming throughout central North America. Corn is grown in a belt south and west of the Great Lakes, while farther west where the climate is drier, wheat is grown.

Using the Land and Sea
- cropland
- forest
- ice cap
- mountain region
- pasture
- tundra
- wetland
- desert
- major conurbations
- cattle
- goats
- pigs
- poultry
- reindeer
- sheep
- bananas
- citrus fruits
- coffee
- corn (maize)
- cotton
- fishing
- fruit
- maple syrup
- peanuts
- rice
- shellfish
- soybeans
- sugar cane
- timber
- tobacco
- vineyards
- wheat

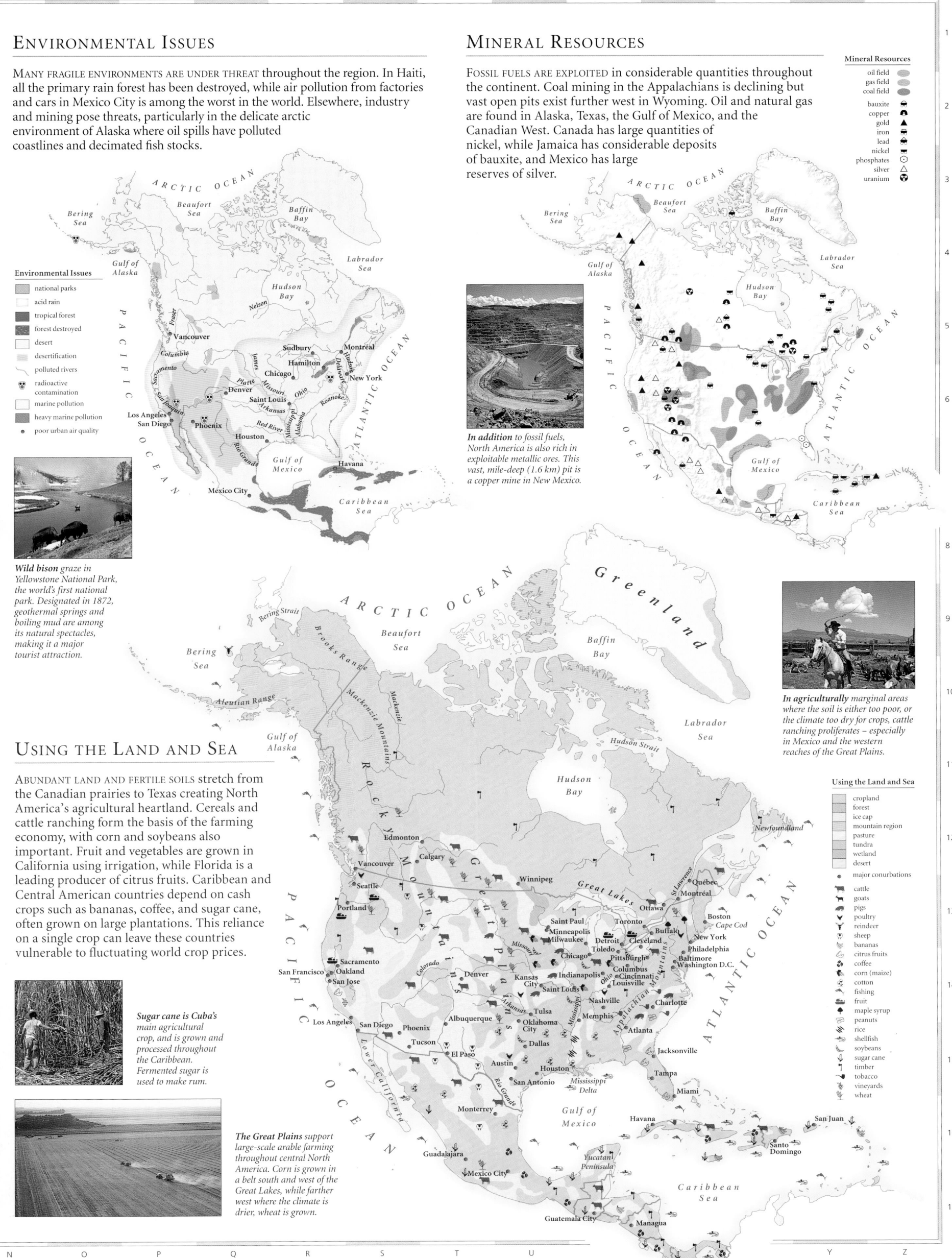

CANADA: WESTERN PROVINCES

Alberta, British Columbia, Manitoba, Saskatchewan, Yukon Territory

THE MOUNTAINS OF THE WEST COAST, incorporating British Columbia and the Yukon Territory, descend into the vast, flat prairies of Alberta, Saskatchewan, and Manitoba. The empty lands and fertile soils of the prairie provinces attracted migrants, and the descendants of early European immigrants still make up a large proportion of the population. The mechanization of agriculture has reduced the need for labor, and rural population densities remain low. The majority of the people live within 100 miles (160 km) of the southern Canada–US border, and in British Columbia, one of the leading Canadian provinces in terms of economic wealth. The Yukon Territory, in the far north, remains a relatively unspoiled wilderness, containing large, untapped mineral reserves. This province has a significant population of Native Americans people, many of whom maintain a traditional lifestyle.

USING THE LAND AND SEA

WHEAT FARMING IS THE ECONOMIC MAINSTAY of Alberta, Manitoba, and Saskatchewan, which contain 82% of farmland in Canada. Cattle are also raised on the prairies. Forestry and fishing are the most prominent resource-based industries in British Columbia. Despite the mountainous terrain, fruit and specialized grains can be grown in the Okanagan and Fraser valleys.

Land use and agricultural distribution

- cattle
- cereals
- fishing
- fruit
- timber
- ● major towns
- pasture
- cropland
- forest
- wetland
- barren
- tundra

THE URBAN/RURAL POPULATION DIVIDE

77% urban   23% rural

0 10 20 30 40 50 60 70 80 90 100

POPULATION DENSITY
7 people per sq mile
(3 people per sq km)

TOTAL LAND AREA
1,224,449 sq miles
(3,172,150 sq km)

Large, highly-mechanized and often very specialized farms, requiring huge investment but little labor, characterize modern farming in the prairies.

TRANSPORTATION & INDUSTRY

THE WESTERN PROVINCES contain a wealth of mineral resources. Alberta holds the bulk of Canada's fossil fuels; the other provinces contain reserves of metallic ores, such as zinc, lead, and silver. Isolation from markets has slowed the development of manufacturing, restricting it to the large cities like Vancouver, Winnipeg, and Calgary. Hydroelectric power is widely exploited, although there is increasing concern about potential ecological damage.

Major industry and infrastructure

- ✈ aerospace
- chemicals
- coal
- ⚙ engineering
- food processing
- hydroelectric power
- mining
- oil & gas
- timber processing
- ● major towns
- ✚ international airports
- — major roads
- ☐ major industrial areas

TRANSPORTATION NETWORK

82,438 miles (135,145 km)

6,459 miles (10,401 km)

10,811 miles (17,410 km)

None

The transportation network of the western provinces is dominated by east–west routes that weave through mountain passes and spread across the plains. Access to some northern areas is restricted to air travel.

The Fraser River valley is a major area of settlement in British Columbia. Railroads cross the Rocky Mountains via this valley.

Established in 1907, Jasper National Park lies in the heart of the Rocky Mountains. It is noted for its spectacular alpine scenery and contains part of the large Columbia Icefield.

Much of the Yukon Territory is uninhabited tundra. Industry is based on the extraction of mineral resources, and to a lesser extent, on the scattered forests of the south.

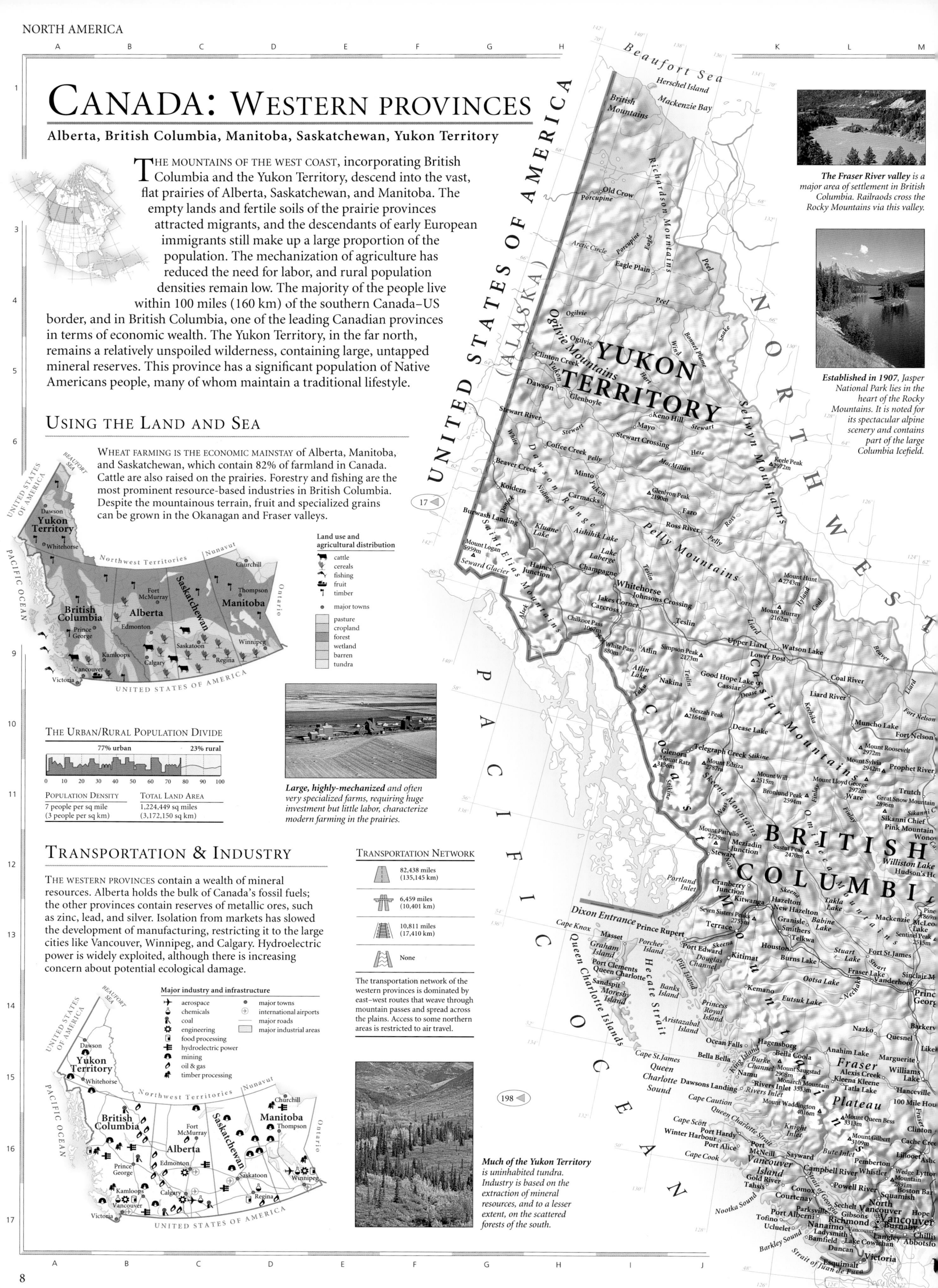

THE LANDSCAPE

THE MASSIVE ROCKY MOUNTAINS form a continental divide between rivers flowing eastward and westward. The interior plains lie east of the mountains, stretching from the Arctic Circle south into the US. Covered with glacial deposits from the last Ice Age, these are interspersed with hilly regions and long, steep escarpments.

MAP KEY

POPULATION

◉ 500,000 to 1 million
◎ 100,000 to 500,000
⊕ 50,000 to 100,000
○ 10,000 to 50,000
○ below 10,000

ELEVATION

6000m / 19,686ft
4000m / 13,124ft
3000m / 9843ft
2000m / 6562ft
1000m / 3281ft
500m / 1640ft
250m / 820ft
100m / 328ft
sea level

SCALE 1:7,500,000
(projection: Lambert Conformal Conic)

Km
0 25 50   100   150   200   250

Miles
0   25   50   100   150   200   250

Mount Logan rises 19,551 ft (5,959 m). It is the highest peak in Canada.

The Rocky Mountain Trench is the longest linear fault in the world. It has formed a straight, flat-bottomed valley between 2–9 miles (4–15 km) wide, and up to 3,280 ft (1,000 m) deep.

Hundreds of islands dot the fjord-indented coast of British Columbia; the largest is Vancouver Island.

Three major passes cut through the Rocky Mountains: Yellowhead, Kicking Horse, and Crowsnest. They are all used as transportation routes through the mountains.

The Columbia Icefield in the Rocky Mountains is the source of two major rivers, the Athabasca and the North Saskatchewan.

The badlands of Alberta were created when east-flowing rivers, swollen by meltwater at the end of the last Ice Age, cut deep, wide canyons producing eroded, barren landscapes.

South Saskatchewan River

Vegetated island — Bar
River flow is diverted by — Sand
deposited sediments — flat

Braided rivers are shallow and fast-flowing. The interlaced branches are formed when excess sediments, which can no longer be transported, are deposited. The sediments collect in the river channel forming bars and sand flats. Islands form when the bars are colonized by vegetation.

Across the tundra of northern Manitoba, widespread permafrost inhibits water from permeating the soil. This causes rivers like the Churchill to flow in many channels, which can be frozen for up to six months during the winter.

The Nelson and Churchill Rivers drain northward across the Canadian Shield to Hudson Bay. The shield covers three-fifths of Saskatchewan.

Setting Lake

Ancient granite outcrops, part of the Canadian Shield, rise above the surface of Setting Lake, which was initially formed by meltwater from the last Ice Age.

The Cypress Hills rise to 4,806 ft (1,465 m) above the surrounding plain. Having escaped the last glaciation they contain unique plant and animal life. The silvery lupine, bunchberry, and lodgepole pine all grow in the cool, moist climate of the hills.

The Alberta and Saskatchewan plains bear strong testament to past glaciations. The Assiniboine, Saskatchewan and Qu'Appelle Rivers occupy flat-bottomed, steep-sided valleys eroded during the last Ice Age by glacial meltwater.

The lowlands of Manitoba are a basin that once held the vast post-glacial Lake Agassiz, remnants of which include Lake Winnipeg, Lake Winnipegosis, and Lake Manitoba.

CANADA: EASTERN PROVINCES

New Brunswick, Newfoundland, Nova Scotia, Ontario,
Prince Edward Island, Quebec, *St. Pierre & Miquelon* (to France)

COLONIZED BY BOTH THE ENGLISH AND THE FRENCH during the 16th century, Canada's eastern provinces are still marked by their dual influences. They contain the last fragment of once-sizeable French territories, the islands of St. Pierre and Miquelon. French remains Canada's second official language and Quebec's first language. The population of the eastern provinces is highly concentrated in the south, especially along the border with the US. A recent decline in fishing in the Atlantic provinces has encouraged a steady flow of westerly migration to more properous regions. The north, around Hudson Bay, remains snow-covered for most of the year and the indigenous Inuit people make up the bulk of its sparse population.

Rocher Percé, is 290 ft (88 m) high. Lying off the southeastern coast of Quebec, it is a sanctuary for sea birds.

SCALE 1:7,000,000
(projection: Lambert Conformal Conic)

Km
0 25 50   100   150   200   250

Miles
0   25   50   100   150   200   250

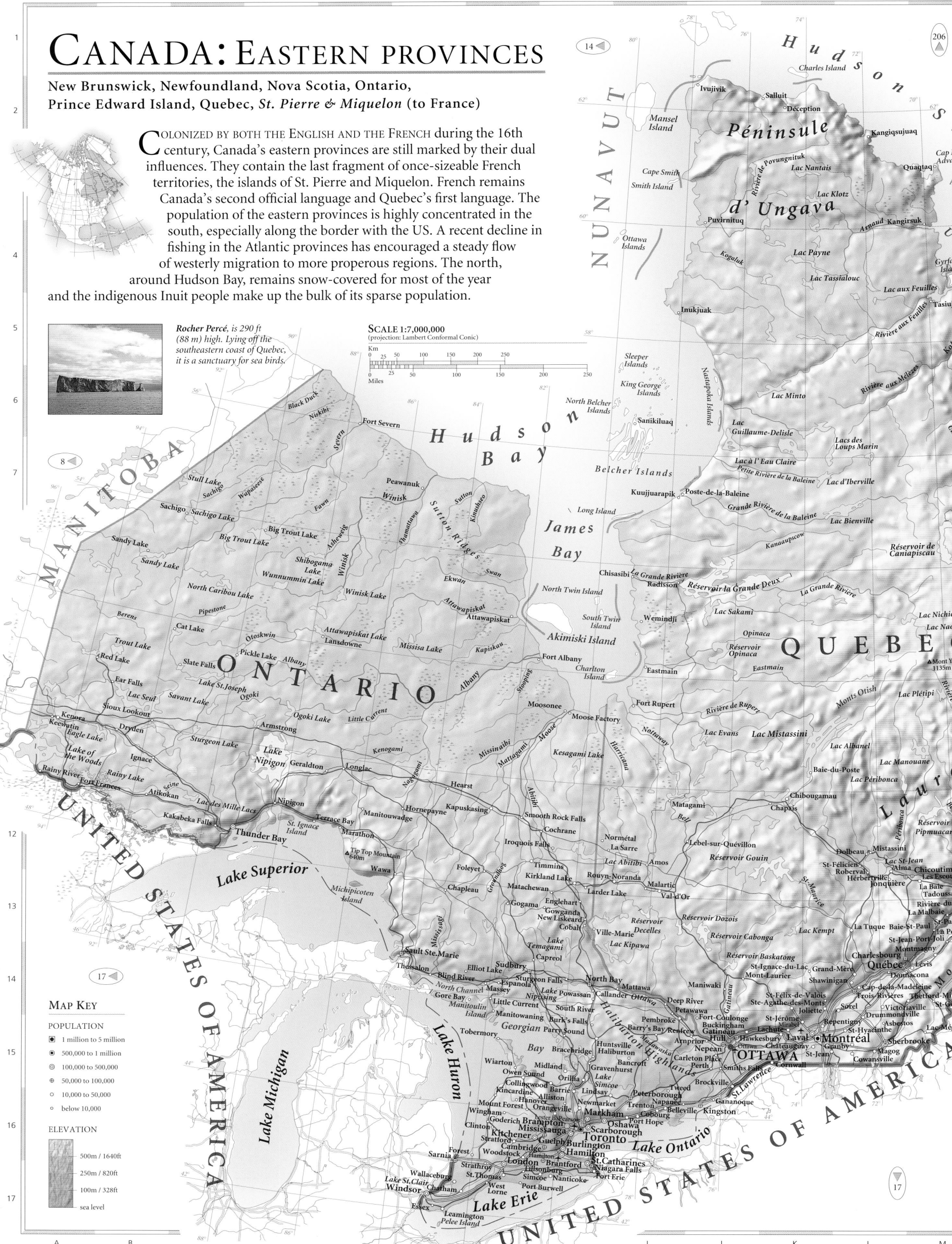

MAP KEY

POPULATION

- ◙ 1 million to 5 million
- ◉ 500,000 to 1 million
- ◎ 100,000 to 500,000
- ⊕ 50,000 to 100,000
- ○ 10,000 to 50,000
- ○ below 10,000

ELEVATION

- 500m / 1640ft
- 250m / 820ft
- 100m / 328ft
- sea level

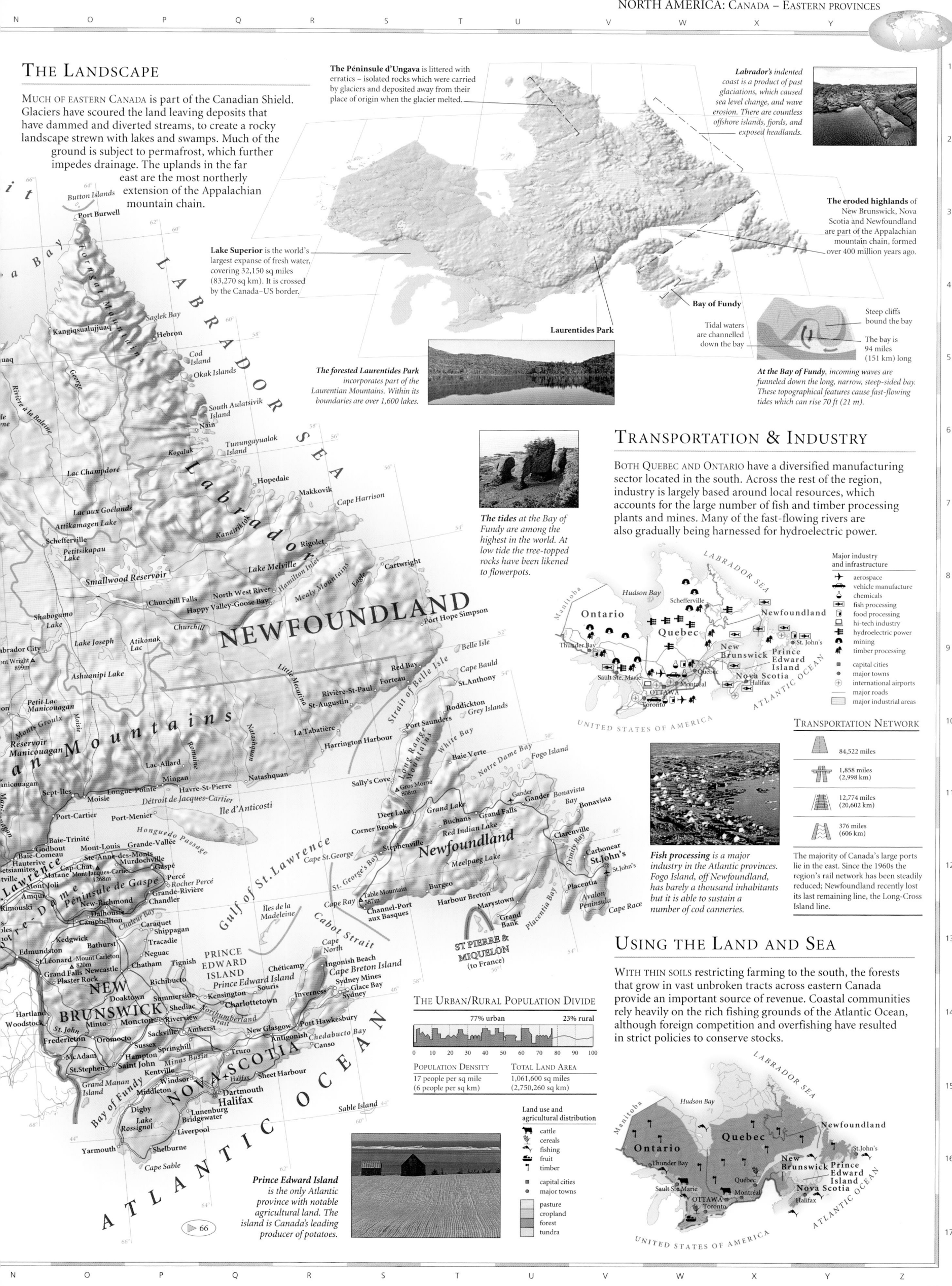

THE LANDSCAPE

MUCH OF EASTERN CANADA is part of the Canadian Shield. Glaciers have scoured the land leaving deposits that have dammed and diverted streams, to create a rocky landscape strewn with lakes and swamps. Much of the ground is subject to permafrost, which further impedes drainage. The uplands in the far east are the most northerly extension of the Appalachian mountain chain.

The Péninsule d'Ungava is littered with erratics – isolated rocks which were carried by glaciers and deposited away from their place of origin when the glacier melted.

Labrador's indented coast is a product of past glaciations, which caused sea level change, and wave erosion. There are countless offshore islands, fjords, and exposed headlands.

Lake Superior is the world's largest expanse of fresh water, covering 32,150 sq miles (83,270 sq km). It is crossed by the Canada–US border.

The eroded highlands of New Brunswick, Nova Scotia and Newfoundland are part of the Appalachian mountain chain, formed over 400 million years ago.

Bay of Fundy

Tidal waters are channelled down the bay

Steep cliffs bound the bay

The bay is 94 miles (151 km) long

Laurentides Park

The forested Laurentides Park incorporates part of the Laurentian Mountains. Within its boundaries are over 1,600 lakes.

At the Bay of Fundy, incoming waves are funneled down the long, narrow, steep-sided bay. These topographical features cause fast-flowing tides which can rise 70 ft (21 m).

The tides at the Bay of Fundy are among the highest in the world. At low tide the tree-topped rocks have been likened to flowerpots.

TRANSPORTATION & INDUSTRY

BOTH QUEBEC AND ONTARIO have a diversified manufacturing sector located in the south. Across the rest of the region, industry is largely based around local resources, which accounts for the large number of fish and timber processing plants and mines. Many of the fast-flowing rivers are also gradually being harnessed for hydroelectric power.

Major industry and infrastructure

- ✈ aerospace
- 🚗 vehicle manufacture
- chemicals
- fish processing
- food processing
- hi-tech industry
- hydroelectric power
- mining
- timber processing
- capital cities
- major towns
- + international airports
- major roads
- major industrial areas

TRANSPORTATION NETWORK

	84,522 miles
	1,858 miles (2,998 km)
	12,774 miles (20,602 km)
	376 miles (606 km)

The majority of Canada's large ports lie in the region's east. Since the 1960s the region's rail network has been steadily reduced; Newfoundland recently lost its last remaining line, the Long-Cross Island line.

Fish processing is a major industry in the Atlantic provinces. Fogo Island, off Newfoundland, has barely a thousand inhabitants but it is able to sustain a number of cod canneries.

USING THE LAND AND SEA

WITH THIN SOILS restricting farming to the south, the forests that grow in vast unbroken tracts across eastern Canada provide an important source of revenue. Coastal communities rely heavily on the rich fishing grounds of the Atlantic Ocean, although foreign competition and overfishing have resulted in strict policies to conserve stocks.

THE URBAN/RURAL POPULATION DIVIDE

77% urban   23% rural

0 10 20 30 40 50 60 70 80 90 100

POPULATION DENSITY
17 people per sq mile
(6 people per sq km)

TOTAL LAND AREA
1,061,600 sq miles
(2,750,260 sq km)

Land use and agricultural distribution

- cattle
- cereals
- fishing
- fruit
- timber
- capital cities
- major towns
- pasture
- cropland
- forest
- tundra

Prince Edward Island is the only Atlantic province with notable agricultural land. The island is Canada's leading producer of potatoes.

▶ 66

Map labels

Button Islands
Port Burwell
Torngat Mountains
Saglek Bay
Kangiqsualujjuaq
Hebron
George
Cod Island
Okak Islands
Rivière à la Baleine
South Aulatsivik Island
Nain
Tunungayualok Island
LABRADOR SEA
Labrador
Lac Champdoré
Kogaluk
Hopedale
Makkovik
Lac aux Goélands
Cape Harrison
Attikamagen Lake
Schefferville
Kanairiktok
Rigolet
Petitsikapau Lake
Smallwood Reservoir
Lake Melville
Cartwright
Shabogamo Lake
North West River
Hamilton Inlet
Churchill Falls
Happy Valley-Goose Bay
Mealy Mountains
Eagle
Lake Joseph
Atikonak Lac
Churchill
Port Hope Simpson
Labrador City
Mount Wright 899m
Ashuanipi Lake
NEWFOUNDLAND
Little Mecatina
Belle Isle
Red Bay
Cape Bauld
Petit Lac Manicouagan
Rivière-St-Paul
Forteau
St.Anthony
Monts Groulx
St-Augustin
Roddickton
Grey Islands
Réservoir Manicouagan
La Tabatière
Natashquan
Harrington Harbour
Port Saunders
White Bay
Romaine
Baie Verte
Sept-Iles
Lac-Allard
Mingan
Longue-Pointe
Havre-St-Pierre
Natashquan
Sally's Cove
Notre Dame Bay
Fogo Island
Moisie
Détroit de Jacques-Cartier
Gros Morne 808m
Gander
Bonavista Bay
Port-Cartier
Port-Menier
Île d'Anticosti
Deer Lake
Grand Lake
Gander
Bonavista
Baie-Trinité
Honguedo Passage
Corner Brook
Buchans
Grand Falls
Clarenville
Baie-Comeau
Mont-Louis
Grande-Vallée
Red Indian Lake
Trinity Bay
Ste-Anne-des-Monts
Murdochville
Newfoundland
Carbonear
Cap-Chat
Mont Jacques-Cartier 1268m
Gaspé
Meelpaeg Lake
St-John's
Mont-Joli
Percé
Rocher Percé
Grande-Rivière
Chandler
Burgeo
Amqui
Péninsule de Gaspé
New-Richmond
Table Mountain 587m
Harbour Breton
Marystown
Placentia Bay
Avalon Peninsula
Rimouski
Cape Ray
Channel-Port aux Basques
Grand Bank
Placentia
Gulf of St.Lawrence
Îles de la Madeleine
ST PIERRE & MIQUELON (to France)
Cape Race
Kedgwick
Dalhousie
Campbellton
Chaleur Bay
Caraquet
Bathurst
Shippagan
St-Léonard
Tracadie
Edmundston
Mount Carleton 820m
Neguac
PRINCE EDWARD ISLAND
Chéticamp
Ingonish Beach
Cape Breton Island
Grand Falls
Newcastle
Chatham
Tignish
Prince Edward Island
Souris
Sydney Mines
Plaster Rock
Doaktown
Richibucto
Summerside
Kensington
Inverness
Glace Bay
NEW BRUNSWICK
Minto
Shediac
Charlottetown
Sydney
Hartland
Chipman
Riverview
Northumberland Strait
Woodstock
St.John
Moncton
Port Hawkesbury
Fredericton
Oromocto
Sackville
Amherst
New Glasgow
McAdam
Hampton
Springhill
Antigonish
Chedabucto Bay
St.Stephen
Saint John
Sussex
Truro
Canso
Grand Manan Island
Kentville
Minas Basin
NOVA SCOTIA
Digby
Windsor
Dartmouth
Sheet Harbour
Bay of Fundy
Middleton
Halifax
Lake Rossignol
Bridgewater
Lunenburg
Sable Island
Yarmouth
Liverpool
Shelburne
Cape Sable
ATLANTIC OCEAN

Hudson Bay
Manitoba
Ontario
Quebec
Newfoundland
Thunder Bay
Schefferville
St.John's
New Brunswick
Prince Edward Island
Sault Ste. Marie
Nova Scotia
Halifax
OTTAWA
Montreal
Quebec
Toronto
UNITED STATES OF AMERICA

SOUTHEASTERN CANADA

Southern Ontario, Southern Quebec

THE SOUTHERN PARTS of Quebec and Ontario form the economic heart of Canada. The two provinces are divided by their language and culture; in Quebec, French is the main language, whereas English is spoken in Ontario. Separatist sentiment in Quebec has led to a provincial referendum on the question of a sovereignty association with Canada. The region contains Canada's capital, Ottawa and its two largest cities: Toronto, the center of commerce and Montréal, the cultural and administrative heart of French Canada.

The port at Montréal is situated on the St. Lawrence Seaway. A network of 16 locks allows sea-going vessels access to routes once plied by fur-trappers and early settlers.

Niagara Falls lies on the border between Canada and the US. It comprises a system of two falls: American Falls, in New York, is separated from Horseshoe Falls, in Ontario, by Goat Island. Horseshoe Falls, seen here, plunges 184 ft (56 m) and is 2,500 ft (762 m) wide.

TRANSPORTATION & INDUSTRY

THE CITIES OF SOUTHERN QUEBEC and ONTARIO, and their hinterlands, form the heart of Canadian manufacturing industry. Toronto is Canada's leading financial center, and Ontario's motor and aerospace industries have developed around the city. A major center for nickel mining lies to the north of Toronto. Most of Quebec's industry is located in Montréal, the oldest port in North America. Chemicals, paper manufacture, and the construction of transportation equipment are leading industrial activities.

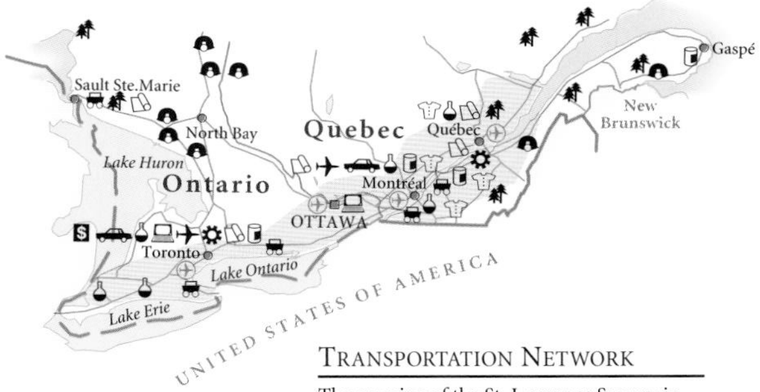

Major industry and infrastructure

- car manufacture
- chemicals
- engineering
- finance
- food processing
- hi-tech industry
- mining
- iron & steel
- textiles
- paper industry
- timber processing
- capital cities
- major towns
- international airports
- major roads
- major industrial areas

TRANSPORTATION NETWORK

The opening of the St. Lawrence Seaway in 1959 finally allowed ocean-going ships (up to 24,000 tons (tonnes)) access to the interior of Canada, creating a vital trading route.

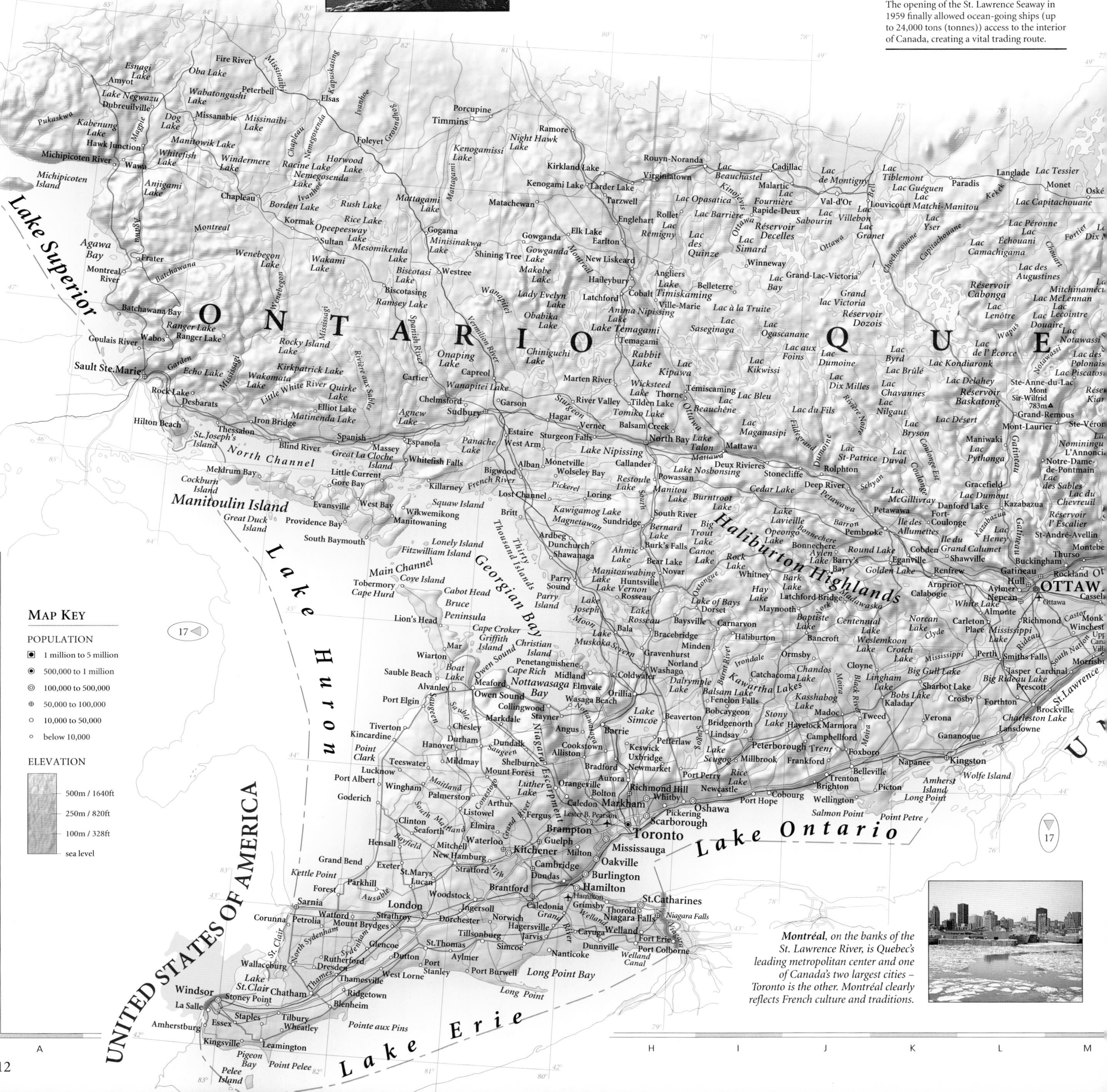

MAP KEY

POPULATION

- 1 million to 5 million
- 500,000 to 1 million
- 100,000 to 500,000
- 50,000 to 100,000
- 10,000 to 50,000
- below 10,000

ELEVATION

- 500m / 1640ft
- 250m / 820ft
- 100m / 328ft
- sea level

Montréal, on the banks of the St. Lawrence River, is Quebec's leading metropolitan center and one of Canada's two largest cities – Toronto is the other. Montréal clearly reflects French culture and traditions.

USING THE LAND AND SEA

THE PRODUCTIVE NIAGARA "FRUIT BELT" on the shores of Lake Erie and Lake Ontario is a major farming region, although available farmland is being challenged by urban expansion. Quebec is Canada's leading producer of maple syrup and dairy products. In the north, farmland gives way to extensive areas of forest, partly used for commercial logging. Fishing occurs in Atlantic waters and in the Great Lakes.

Land use and agricultural distribution

cattle	▪	capital cities
fish	●	major towns
cereals		pasture
fruit		cropland
maple syrup		forest
timber		
tobacco		

THE URBAN/RURAL POPULATION DIVIDE

urban 87%   rural 13%

0 10 20 30 40 50 60 70 80 90 100

POPULATION DENSITY
64 people per sq mile
(25 people per sq km)

TOTAL LAND AREA
214,230 sq miles
(555,000 sq km)

Pumpkins are just one of the crops grown in the Niagara "fruit belt." The mild climate, moderated by the lakes, allows the cultivation of a wide range of fruit and vegetables, including cherries, apples, peaches, grapes, and asparagus. Fruit and vegetable growing is confined to southern Canada, due to the colder climate and short growing season of the northern regions.

In contrast to the boreal forest which spans northern Canada, the Gaspé Peninsula (Peninsule de Gaspé) is covered with a band of mixed coniferous-deciduous woodland, including sugar and red maple, cedar, and eastern hemlock.

THE LANDSCAPE

THE HEART OF SOUTHEASTERN CANADA is the lowland area surrounding the St. Lawrence River, the principal outlet for the Great Lakes. The lowlands are bordered to the east by an extension of the Appalachian mountain chain and to the north by the Canadian Shield. The Champlain Sea, which flooded the area during the last glacial period, deposited clay over much of the area.

The wooded Gaspé Peninsula (Peninsule de Gaspé) includes the Notre Dame and Shickshock mountains (Monts Chic-Chocs). These are a northerly outcrop of the Appalachian mountain chain.

The flat plains of the St. Lawrence Valley were formed when the area was inundated by the Champlain Sea during the last glacial period.

The Laurentide Scarp, along the north shore of the St. Lawrence River, is a 2,000 ft (610 m) escarpment, marking the rim of the Canadian Shield.

In 1971, large quantities of marine clay liquefied and flowed into the Saguenay River, killing 30 people. Large landslides often occur on waterlogged slopes.

SCALE 1:3,000,000
(projection: Lambert Conformal Conic)

Km
0 5 10 20 30 40 50 60 70 80

Miles
0 5 10 20 30 40 50 60 70 80

Lake Superior

Lake Huron

Point Pelee is a world-famous site for bird migration. Over 250 species of bird have been sighted on the sandspit which forms the southern tip of the Canadian mainland.

Lake Erie

Lake Ontario

The Great Lakes moderate the climate of the area surrounding the St. Lawrence River. Their water, which cools more slowly than the land, acts as a reservoir for warmth, extending the growing season into the early autumn.

Mount Royal, around which the city of Montréal has developed, is the result of an igneous intrusion which occurred between 135 and 65 million years ago.

River bank or bluff

Earthflow

Sand

Clay

River

In the lowlands around the St. Lawrence, earthflows have developed along gentle river banks where sand overlies clay, making the surface layers very unstable. When the slope's natural equilibrium is disturbed, an earthflow can occur.

CANADA

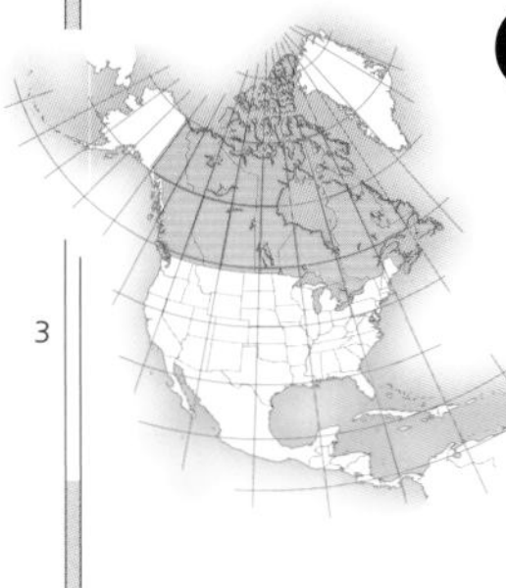

CANADA IS THE THIRD LARGEST COUNTRY in the world, and with only about one-tenth of its land area inhabited, it is one of the most sparsely populated. Canada became a confederation in 1867, though Newfoundland did not join until 1949. As a founding member of the UN and of the Commonwealth, Canada has played an important role in international affairs. A constitutional crisis, focusing on the French-speaking Québécois, and Inuit, and Native American land rights, dominated politics in the 1990s. In 1999, part of the Northwest Territories, Nunavut, became a self-governing homeland for the Inuit.

The Selwyn Mountains in northwestern Canada form part of the Rocky Mountains. The highest point, Keele Peak, rises to 9,750 ft (2,972 m).

TRANSPORTATION & INDUSTRY

ABUNDANT ENERGY in the form of coal, oil, natural gas, and hydroelectric power underpins Canadian industry. Over 75% of manufacturing is concentrated in the Great Lakes–St. Lawrence region, including prospering aerospace, transportation, and hi-tech industries. Across Canada as a whole, manufacturing has developed around a diversified, high-quality resource base and a wide range of metallic and nonmetallic minerals.

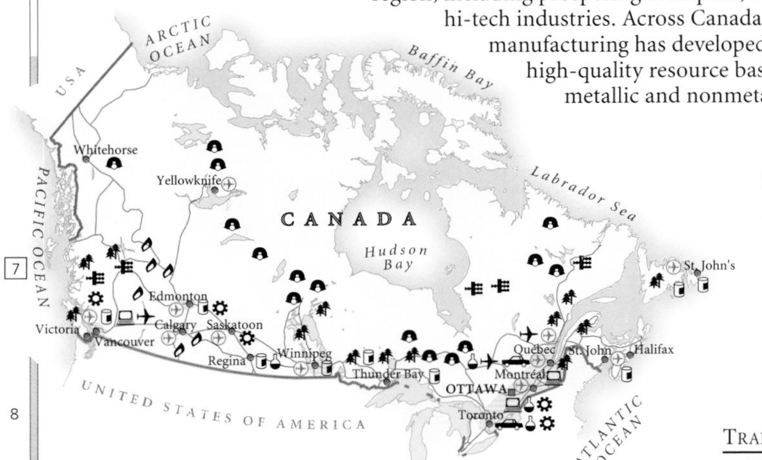

Major industry and infrastructure

aerospace	capital cities
car manufacture	major towns
chemicals	international airports
engineering	major roads
food processing	major industrial areas
hi-tech industry	
hydroelectric power	
oil & gas	
mining	
timber processing	

Canada has one of the world's highest rates of energy consumption per person. It is endowed with vast hydroelectric potential from which more than 60% of its electricity requirements are generated.

TRANSPORTATION NETWORK

566,352 miles (912,000 km)	15,189 miles (24,459 km)
8,755 miles (14,098 km)	2,341 miles (3,769 km)

In recent years the road network has been expanded, especially links to remote areas. Meanwhile, for long-distance travel, air transportation now supersedes the declining rail network, which focuses mainly on east–west routes.

THE LANDSCAPE

GLACIERS ON ISLANDS IN THE ARCTIC OCEAN are the last remnants of the ice sheet that once covered and shaped Canada. Hudson Bay is the center of the Canadian Shield, a huge, eroded plateau marked at its southern extremity by a string of lakes running southeastward from Great Bear Lake to the Great Lakes. In contrast to the rolling relief of the Shield and the central lowland region, the Rocky Mountains rise to peaks of over 13,000 ft (4,000 m), stretching 500 miles (800 km) along the west coast.

Along the northeastern coast of Baffin Island the mountains rise to 8,000 ft (2,440 m). Glaciers move down through the valleys to the sea, eroding wide U-shaped valleys.

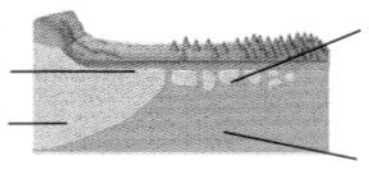

Top layer thaws in the summer

Permanently frozen ground

Marginal areas of permafrost thaw in summer

Unfrozen ground where temperature is more moderate

Permanently frozen ground known as permafrost is common in Canada's northern tundra. It thickens farther north, becoming hundreds of yards deep in parts of the Arctic.

The Mackenzie River, flowing north over the permafrost, forms a wide river channel with many tributaries. Together with the Peel River it has created a long, narrow delta at its mouth. The entire river freezes during the winter.

Great Bear Lake

Exposure to three phases of mountain-building and subsequent erosion over millions of years has molded the ancient Canadian Shield into a series of basins and ridges.

The Rocky Mountains were formed some 80 million years ago, when the Pacific Plate was driven under the North American Plate, forcing up the land.

Isolated pillars, known as hoodos *near Red Deer River in the badlands of Alberta are a product of wind and water erosion, especially flash floods. The badlands lie in the rain shadow of the Rocky Mountains, which creates a semiarid climate.*

Fertile prairies stretch from the southern rim of the Canadian Shield, south into the US.

The Great Lakes lie on the Canada–US border. The basins they now occupy were fashioned by repeated ice advance. Once, Lakes Superior, Huron, and Michigan formed one large lake, Lake Nipissing.

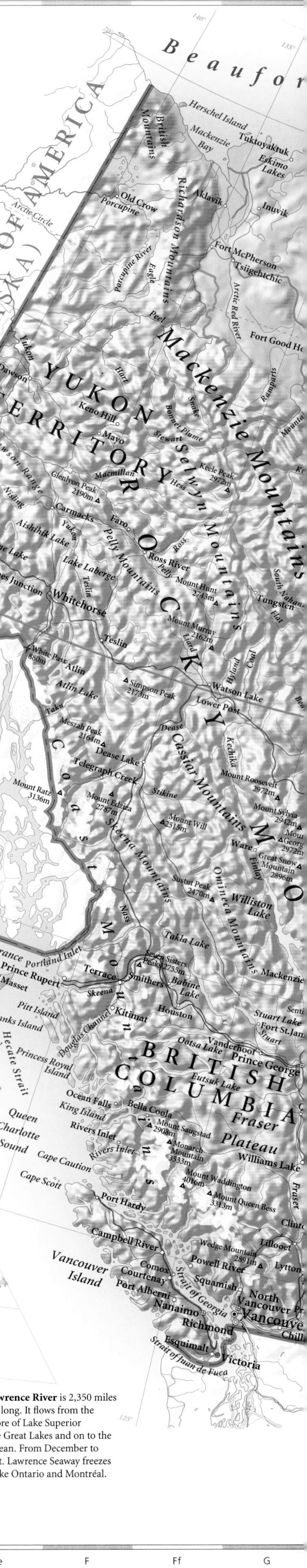

The St. Lawrence River is 2,350 miles (3,782 km) long. It flows from the western shore of Lake Superior through the Great Lakes and on to the Atlantic Ocean. From December to April, the St. Lawrence Seaway freezes between Lake Ontario and Montréal.

Cape Kellett
Banks
Island
Sachs Harbour
Cape Lambton
Cape Wollaston
Cape Parry
Cape Bathurst
Franklin Bay
Cape Lyon
Holman

Amundsen Gulf

Prince of Wales Strait
Prince Albert Peninsula
Prince Albert Sound
Wollaston Peninsula

Victoria Island

Dolphin & Union Strait

Viscount Melville Sound
Passage Point
Peel Point
Stefansson Island
Hadley Bay
206
McClintock Channel

Peel Sound
Somerset Island

Prince of Wales Island

Zeta Lake
Gateshead Island
Larsen Sound

Boothia Peninsula

Franklin Strait

Gulf of Boothia

Prince Regent Inlet

Brodeur Peninsula

Admiralty Inlet

Borden Peninsula

Baffin

Cape Henry Kater
Hantzsch

Cape Englefield
Cape Chapman
Melville Peninsula

Gifford
Igloolik
Hall Beach
Jens Munk Island
Baird Island
Rowley Island

Prince Charles Island
Air Force Island
Netilling Lake
Koukdjuak

Foxe
Basin

Taloyoak
Pelly Bay
Simpson Peninsula
Committee Bay
Wales Island

Vansittart Island
Cape Dorchester
Foxe Peninsula
Cape Dorset
Salisbury Island

Foxe Channel

Repulse Bay
Hayes
Roes Welcome Sound
Southampton Island
Coral Harbour

Evans Strait

Nottingham Island
Mansel Island

Ivujivik

Cape Low
Fisher Strait
Coats Island

Ottawa Islands

Cambridge Bay
Kent Peninsula
Cape Krusenstern
Coronation Gulf
Bathurst Inlet

King William Island
Jenny Lind Island
Gjoa Haven
Adelaide Peninsula
Queen Maud Gulf
Bowes Point

Rae Strait
Chantrey Inlet

Rae
Kugluktuk
Ellice
Hood
Burnside

Back

NUNAVUT

Wager Bay

Cape Kendall

Hudson
Bay

an Wells
Déline
Great Bear Lake
Echo Bay
Takijuq Lake
Coppermine

Garry Lake
Aberdeen Lake
Baker Lake
Baker Lake

Chesterfield Inlet
Chesterfield Inlet
Rankin Inlet

Tulita
Mackenzie
Wrigley
Lac La Martre

Hottah Lake

NORTH WEST TERRITORIES

Yellowknife
Snare
Aylmer Lake
Clinton-Colden Lake
Hanbury
Thelon

Quoich

Willowlake
Wha Ti
Edzo
Reliance

Dubawnt Lake
Yathkyed Lake

Whale Cove

Fort Simpson
Yellowknife
Łutselk'e
Snowdrift
Nonacho Lake

Kaza
Eskimo Point
Arviat

Trout
Fort Providence
Great Slave Lake
Fort Resolution

Tha-Anne

Hay River
Pine Point

Thoa
Dubawnt
Thlewiaza

t Liard
Petitot

Fort Smith

Wholdaia Lake
Kasba Lake
Nueltin Lake

Nejanilini Lake

t Nelson
Fontas

Bistcho Lake
Steen River

Caribou Mountains

Slave

Uranium City
Selwyn Lake

Phelps Lake

Seal
Tadoule Lake

Cape Churchill
Churchill

Cape Tatnam

High Level
Fort Vermilion
Lake Claire

Lake Athabasca
Black Lake
Lac Brochet

South Seal
Southern Indian Lake

Fort Severn

Chinchaga
Clear Hills
Manning
Peace

ALBERTA
Birch Mountains

Pasfield Lake
Cree Lake
MacFarlane
William

Wollaston Lake
Wollaston Lake
Reindeer Lake

Lynn Lake
Gillam
Waskaiowaka Lake

Winisk

Fort St.John
Dawson Creek
Grimshaw
Fairview
Peace River
Desmarais
Utikuma Lake

SASKATCHEWAN

Cree
Geikie
Foster Lakes
Macoun Lake

Leaf Rapids
Granville Lake
Split Lake

Nelson
Gods

Hayes
Big Trout Lake
Winisk Lake
Attawapiskat

twynd
Grande Prairie
Wapiti
High Prairie
Lesser Slave Lake

Clearwater
Turnor Lake
Frobisher Lake
La Loche
Churchill Lake

Reindeer
Churchill

Thompson
Burntwood
Sipiwesk Lake

Oxford Lake
Molson Lake
Gods Lake

Island Lake
Sandy Lake

ONTARIO

r Alexander
Grande Cache

Slave Lake
Wallace Mountain
1250m
Swan Hills

Primrose Lake
Athabasca
Peter Pond Lake
Buffalo Narrows
Pinehouse Lake
Missinipe

La Ronge
Beaver

Kississing Lake
Wabowden

MANITOBA

Sandy Lake
North Caribou Lake
Attawapiskat Lake

Mount Robson 3954m
Whitecourt
Edson
Hinton
Mount Sir Wilfrid Laurier 3505m

Cold Lake
Cold Lake

Meadow Lake
Tobin Lake

Deschambault Lake
Amisk Lake
Creighton
Flin Flon

Lac La Ronge
Montreal Lake

Trout Lake
Lake St.Joseph

Red Lake
Ogoki

St.Albert
Spruce Grove
Edmonton
Stony Plain
Drayton Valley
Devon
Leduc
Pembina

Morinville
Fort Saskatchewan
Vegreville
Vermilion

St.Paul
North Saskatchewan
St.Walburg

Prince Albert

The Pas
Cedar Lake
Grand Rapids

Porcupine Hills

Gypsumville

Pipestone

Lake Nipigon

Mount Columbia 3741m
North Saskatchewan
Rocky Mountain House
Camrose
Wetaskiwin
Ponoka
Lacombe

Lloydminster
Wainwright
Battleford
Unity
Martensville

Nipawin
Pasquia Hills
Hudson Bay

Lake Winnipegosis
Swan River
Duck Mountain

Eriksdale
Gimli

Ear Falls
Lac Seul
Sioux Lookout

Red Deer
Innisfail
Sylvan Lake
Stettler

North Battleford
Melfort
Tisdale

Saskatchewan

Lake Manitoba

Dryden

Kamloops
Okanagan
Kelowna
Penticton

Olds
Didsbury
Red Deer
Drumheller

Biggar
Saskatoon
Humboldt
Quill Lakes

Canora
Dauphin

Neepawa
Stonewall
Selkirk

Kenora
Eagle Lake
Lake of the Woods

Salmon Arm
Kinbasket Lake
Golden
Mount Assiniboine 3618m

Rosetown
Outlook
Lanigan
Watrous
Wynyard
Yorkton

Kamsack
Baldy Mountain 831m
Minnedosa
Portage la Prairie
Winnipeg
Steinbach

Rainy Lake
Fort Frances
Atikokan
Thunder Bay

Nelson
Castlegar

Banff
Canmore
Invermere
High River
Okotoks
Calgary

Oyen
Kindersley
Lake Diefenbaker
Last Mountain Lake
Fort Qu'Appelle
Melville
Esterhazy
Riding Mountain

Neepawa
Beausejour
Pinawa

Kicking Horse Pass
Crowsnest Pass

Claresholm
Travers Reservoir
Fort Macleod
Lethbridge
Coaldale
Taber

Medicine Hat
Redcliff
Maple Creek

Swift Current
Moose Jaw
Old Wives Lake

South Saskatchewan

Lumsden
Regina
Moosomin
Virden
Brandon
Assiniboine
Carman
Winkler
Morden
Altona

Rainy River
Lake Nipigon

Kimberley
Pincher Creek
Cardston
Raymond
Milk River

Cypress Hills
Val Marie
Wood Mountain

Assiniboia
Rockglen

Weyburn
Carlyle
Estevan
Melita
Killarney

Lac des Milles Lacs

UNITED STATES OF AMERICA

SCALE 1:9,250,000
(Projection: Lambert Azimuthal Equal Area)

17

H   Hh   I   Ii   J   Jj   K   Kk   L   Ll   M   Mm

A   Aa   A   B   Bb   C   Cc   D   Dd   E   Ee   F

THE UNITED STATES OF AMERICA

CONTERMINOUS US (FOR ALASKA AND HAWAII SEE PAGES 40–41)

THE US'S PROGRESSION FROM FRONTIER TERRITORY to economic and political superpower has taken less than 200 years. The 48 conterminous states, along with the outlying states of Alaska and Hawaii, are part of a federal union, held together by the guiding principles of the US Constitution, which embodies the ideals of democracy and liberty for all. Abundant fertile land and a rich resource-base fueled and sustained US economic development. With the spread of agriculture and the growth of trade and industry came the need for a larger workforce, which was supplied by millions of immigrants, many seeking an escape from poverty and political or religious persecution. Immigration continues today, particularly from Central America and Asia.

Mount Rainier is a dormant volcano in the Cascade Range, Washington. This 14,090 ft (4,392 m) peak is flanked by the most extensive glacier outside Alaska.

TRANSPORTATION & INDUSTRY

THE US HAS BEEN THE INDUSTRIAL POWERHOUSE of the world since the Second World War, pioneering mass-production and the consumer lifestyle. Initially, heavy engineering and manufacturing in the northeast led the economy. Today, heavy industry has declined and the US economy is driven by service and financial industries, with the most important being defense, hi-tech, and electronics.

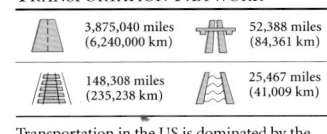

Washington D.C. was established as the nation's capital in 1790. It is home to the seat of national government, on Capitol Hill, as well as the President's official residence, the White House.

198 ◄

Major industry and infrastructure

🚗	aerospace	🌐	research & development
🚙	car manufacture		textiles
	chemicals		tourism
	coal		
	electronics	■	capital cities
⚙	engineering	◉	major towns
	food processing	✈	international airports
💻	hi-tech industry		major roads
	oil & gas		major industrial areas

TRANSPORTATION NETWORK

🛣 3,875,040 miles (6,240,000 km)		🛣 52,388 miles (84,361 km)	
🛤 148,308 miles (235,238 km)		🚢 25,467 miles (41,009 km)	

Transportation in the US is dominated by the car which, with the extensive Interstate Highway system, allows great personal mobility. Today, internal air flights between major cities provide the most rapid cross-country travel.

198 ◄

THE LANDSCAPE

THE HIGH, RUGGED MOUNTAIN RANGES of the west are about 80 million years old, geologically young compared to the old, eroded, Appalachian mountain chain, which dates from when North America and Europe were joined together as part of the supercontinent Pangaea, 400 million years ago. In contrast, the Great Plains and Mississippi Basin have a low relief and fertile soils.

Devils Tower, in Wyoming is a 1,280 ft (390 m) intrusion of basalt rock, which cooled to form octagonal pillars. In 1906 it became the first US National Monument.

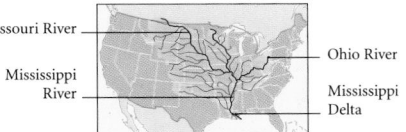

The massive drainage basin of the Mississippi covers 1,250,000 sq miles (3,200,000 sq km). It includes all areas drained by the Mississippi and its chief tributaries, the Missouri and Ohio Rivers, and drains the entire region from the Appalachians to the Rockies.

Mount Rainier

Hells Canyon running through part of Idaho and Oregon, is North America's deepest gorge. It was formed by the down-cutting of the Snake River through the thick basalt rocks of the Columbia–Snake Plateau.

The Rocky Mountains form the backbone of the US, running from Alaska to New Mexico. They contain the US's highest mountains and many active volcanoes.

The Great Lakes

The Hudson-Mohawk Gap, lying at the point where the two rivers join, allows passage from the Atlantic Ocean to the continental interior.

Death Valley, California, 282 ft (86 m) below sea level, is the lowest point in the western hemisphere, and one of the hottest places on Earth. Temperatures of 190° F (88° C) have been recorded here.

Niagara Falls

Barrier beaches, bars and spits are typical of the Atlantic coast. These sand formations around Cape Hatteras stretch along the coast for 200 miles (320 km).

The Great Smoky Mountains, part of the ancient Appalachian mountain chain formed a natural barrier to early settlers attempting to penetrate the country's interior.

Volcanically heated water erupts every 40–80 minutes from Old Faithful geyser in Yellowstone National Park, Wyoming. The 170 ft (50 m) column of water and steam persists for 4 minutes.

Monument Valley's striking sandstone spires and pillars (buttes) have been formed by the action of wind, water, heat, and cold.

The deep gullies of South Dakota's badlands are created by periodic, torrential rainfall, which erodes the soft soils and rocks. Their form has been greatly affected by changes in land use.

Great Plains

Most of the US is drained by the great Mississippi River system. At its mouth, where levées are breached, floodwaters are carried to the swamps through a series of channels. This region is known as the bayou.

The US Gulf Coast is seriously affected by hurricane erosion which reshapes its beaches and sandbanks.

The Everglades are a vast area of sawgrass swamp covering 4,000 sq miles (10,300 sq km) of southern Florida.

A   Aa   B   Bb   C   Cc   D   Dd   E   Ee   F   Ff

USING THE LAND AND SEA

THE MAJORITY OF CANADA'S agricultural land is found in the prairies, which cover 140 million acres (57 million ha) and support wheat and grain-fed cattle. More specialized crops, such as fruit and vegetables, are grown in pockets of agricultural land in the east and west. Of Canada's many islands, only Prince Edward Island has notable farmland. Further north, boreal forests, exploited for timber, run in an almost unbroken arc, giving way to uncultivable tundra and ice sheets in the far north.

THE URBAN/RURAL POPULATION DIVIDE

urban 77%   rural 23%

0 10 20 30 40 50 60 70 80 90 100

POPULATION DENSITY	TOTAL LAND AREA
8 people per sq mile (3 people per sq km)	3,559,294 sq miles (9,220,970 sq km)

Land use and agricultural distribution

- cattle
- cereals
- fishing
- fruit
- timber
- capital cities
- major towns

- pasture
- cropland
- forest
- wetland
- mountain region
- barren
- tundra

The climate and topography of the prairies makes them ideally suited to farming. Long summer days, moderate temperatures, limited rainfall, and flat plains provide excellent conditions for wheat farming.

Ottawa was selected by Queen Victoria as the Canadian capital in 1858. Prior to this date it was a notorious work camp centered around the lumber industry. Today, the city is known as "Silicon Valley North," due to its concentration of hi-tech industries.

MAP KEY

POPULATION

- ▣ 1 million to 5 million
- ◉ 500,000 to 1 million
- ◎ 100,000 to 500,000
- ⊕ 50,000 to 100,000
- ○ 10,000 to 50,000
- · below 10,000

ELEVATION

- 6000m / 19,686ft
- 4000m / 13,124ft
- 3000m / 9843ft
- 2000m / 6562ft
- 1000m / 3281ft
- 500m / 1640ft
- 250m / 820ft
- 100m / 328ft
- sea level

The Great Lakes are drained by the St. Lawrence River which flows down through a wide tectonic depression. It forms a broad estuary for much of its course, the width varying from 1.2 miles (1.9 km) in the upper reaches to 90 miles (145 km) at its mouth.

▶ 66

The clear waters of Niagara Falls cascade 190 ft (58 m) into the gorge below. It is one of America's most famous spectacles and a leading tourist attraction. The falls are slowly receding and the gorge may one day stretch from Lake Ontario to Lake Erie.

USING THE LAND AND SEA

OVER HALF OF THE US is used for agriculture, typified by the large cereal grain farms and cattle ranches of the Great Plains and Midwest prairie regions. Although wheat and corn are still primary crops, a diverse range of fruits and vegetables are grown in the fertile areas, particularly near the east and west coasts. Despite the abundance of cultivable land, inadequate soil management has resulted in a third of the topsoil being lost through wind and water erosion.

THE URBAN/RURAL POPULATION DIVIDE

urban 76%   rural 24%

POPULATION DENSITY
76 people per sq mile
(29 people per sq km)

TOTAL LAND AREA
3,538,307 sq miles
(9,166,600 sq km)

Land use and agricultural distribution

- cattle
- pigs
- poultry
- citrus fruits
- cotton
- fishing
- fruit
- corn (maize)
- peanuts
- shellfish
- soybeans
- timber
- tobacco
- wheat

- capital cities
- major towns
- pasture
- cropland
- forest
- wetland
- desert
- mountain region

Fakahatchee Strand is part of the extensive subtropical swamps in the Florida Everglades. The swamps support a wide variety of animal life, including many rare birds, fish, alligators, and crocodiles.

Farming on the Great Plains and in the Midwest is characterized by large-scale, mechanized wheat farms.

19

USA: NORTHEASTERN STATES

Connecticut, Maine, Massachusetts, New Hampshire, New Jersey, New York, Pennsylvania, Rhode Island, Vermont

THE INDENTED COAST AND VAST WOODLANDS of the northeastern states were the original core area for European expansion. The rustic character of New England prevails after nearly four centuries, while the great cities of the Atlantic seaboard have formed an almost continuous urban region. Over 20 million immigrants entered New York from 1855 to 1924 and the northeast became the industrial center of the US. After the decline of mining and heavy manufacturing, economic dynamism has been restored with the growth of hi-tech and service industries.

Chelsea in Vermont, surrounded by trees in their fall foliage. Tourism and agriculture dominate the economy of this self-consciously rural state, where no town exceeds 30,000 people.

MAP KEY

POPULATION

- above 5 million
- 1 million to 5 million
- 500,000 to 1 million
- 100,000 to 500,000
- 50,000 to 100,000
- 10,000 to 50,000
- below 10,000

ELEVATION

- 1000m / 3281ft
- 500m / 1640ft
- 250m / 820ft
- 100m / 328ft
- sea level

TRANSPORTATION & INDUSTRY

THE PRINCIPAL SEABOARD CITIES grew up on trade and manufacturing. They are now global centers of commerce and corporate administration, dominating the regional economy. Research and development facilities support an expanding electronics and communications sector throughout the region. Pharmaceutical and chemical industries are important in New Jersey and Pennsylvania.

TRANSPORTATION NETWORK

340,090 miles (544,144 km)	4813 miles 7700 km
12,872 miles (20,592 km)	2108 miles (3389 km)

New York's commercial success is tied historically to its transportation connections. The Erie Canal, completed in 1825, opened up the Great Lakes and the interior to New York's markets and carried a stream of immigrants into the Midwest.

Major industry and infrastructure

- chemicals
- coal
- defense
- electronics
- engineering
- finance
- hi-tech industry
- iron & steel
- pharmaceuticals
- printing & publishing
- research & development
- textiles
- timber processing
- major towns
- international airports
- major roads
- major industrial area

Map labels include: CANADA, Maine, Vermont, New Hampshire, New York, Massachusetts, Connecticut, Rhode Island, Pennsylvania, New Jersey, Ohio, West Virginia, Maryland, Delaware, Atlantic Ocean, Lake Ontario, Lake Erie, Buffalo, Rochester, Syracuse, Albany, Pittsburgh, Harrisburg, Philadelphia, Boston, Providence, Hartford, New York, Trenton, Scranton, Allentown, Atlantic City, Cape May.

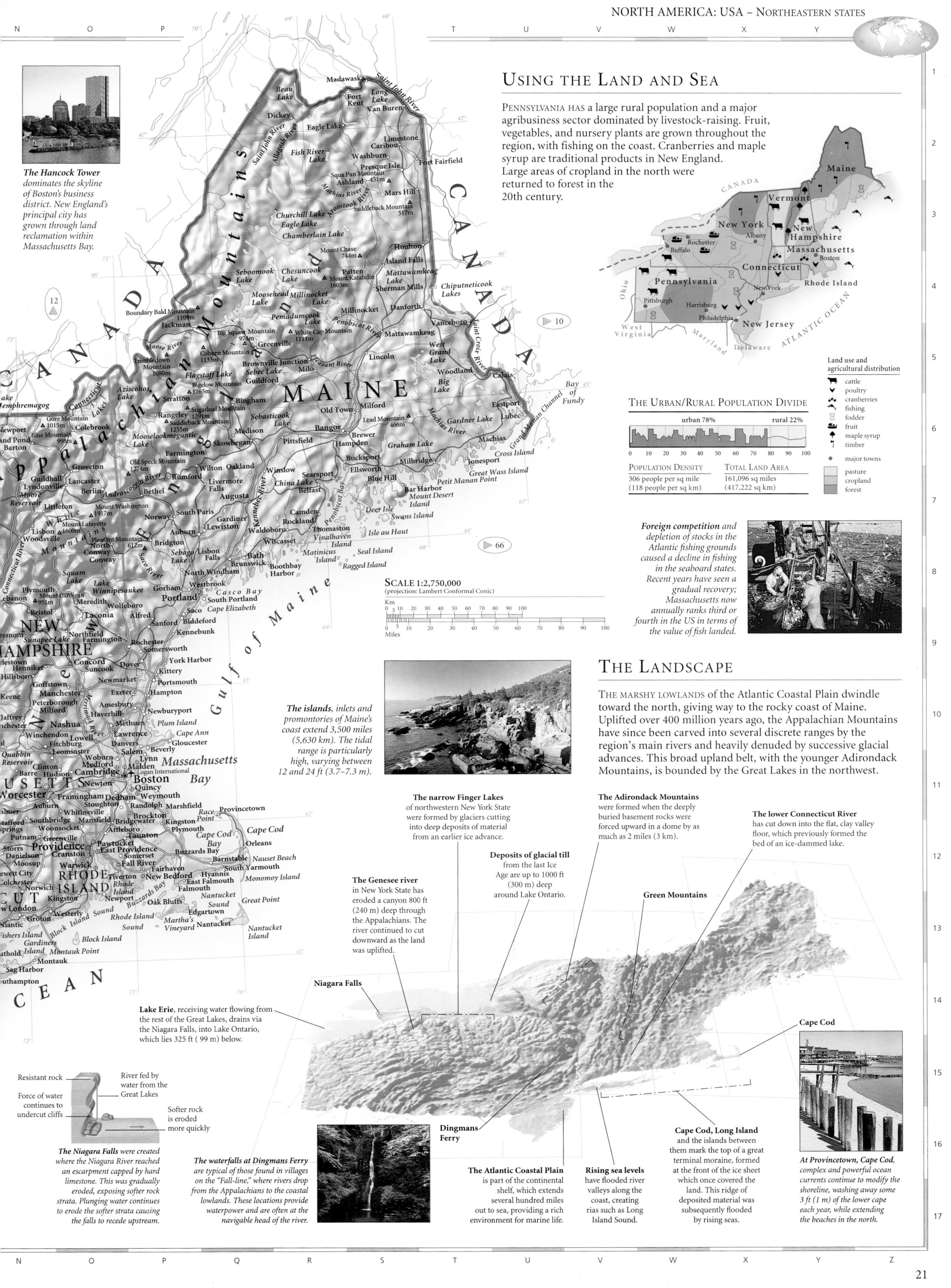

The Hancock Tower dominates the skyline of Boston's business district. New England's principal city has grown through land reclamation within Massachusetts Bay.

USING THE LAND AND SEA

PENNSYLVANIA HAS a large rural population and a major agribusiness sector dominated by livestock-raising. Fruit, vegetables, and nursery plants are grown throughout the region, with fishing on the coast. Cranberries and maple syrup are traditional products in New England. Large areas of cropland in the north were returned to forest in the 20th century.

Land use and agricultural distribution

- 🐄 cattle
- 🦃 poultry
- cranberries
- fishing
- fodder
- fruit
- 🍁 maple syrup
- timber
- • major towns
- pasture
- cropland
- forest

THE URBAN/RURAL POPULATION DIVIDE

urban 78%   rural 22%

0 10 20 30 40 50 60 70 80 90 100

POPULATION DENSITY
306 people per sq mile
(118 people per sq km)

TOTAL LAND AREA
161,096 sq miles
(417,222 sq km)

Foreign competition and depletion of stocks in the Atlantic fishing grounds caused a decline in fishing in the seaboard states. Recent years have seen a gradual recovery; Massachusetts now annually ranks third or fourth in the US in terms of the value of fish landed.

THE LANDSCAPE

THE MARSHY LOWLANDS of the Atlantic Coastal Plain dwindle toward the north, giving way to the rocky coast of Maine. Uplifted over 400 million years ago, the Appalachian Mountains have since been carved into several discrete ranges by the region's main rivers and heavily denuded by successive glacial advances. This broad upland belt, with the younger Adirondack Mountains, is bounded by the Great Lakes in the northwest.

The islands, inlets and promontories of Maine's coast extend 3,500 miles (5,630 km). The tidal range is particularly high, varying between 12 and 24 ft (3.7–7.3 m).

The narrow Finger Lakes of northwestern New York State were formed by glaciers cutting into deep deposits of material from an earlier ice advance.

The Adirondack Mountains were formed when the deeply buried basement rocks were forced upward in a dome by as much as 2 miles (3 km).

The lower Connecticut River has cut down into the flat, clay valley floor, which previously formed the bed of an ice-dammed lake.

Deposits of glacial till from the last Ice Age are up to 1000 ft (300 m) deep around Lake Ontario.

Green Mountains

The Genesee river in New York State has eroded a canyon 800 ft (240 m) deep through the Appalachians. The river continued to cut downward as the land was uplifted.

Niagara Falls

Lake Erie, receiving water flowing from the rest of the Great Lakes, drains via the Niagara Falls, into Lake Ontario, which lies 325 ft (99 m) below.

Cape Cod

Resistant rock

River fed by water from the Great Lakes

Force of water continues to undercut cliffs

Softer rock is eroded more quickly

The Niagara Falls were created where the Niagara River reached an escarpment capped by hard limestone. This was gradually eroded, exposing softer rock strata. Plunging water continues to erode the softer strata causing the falls to recede upstream.

The waterfalls at Dingmans Ferry are typical of those found in villages on the "Fall-line," where rivers drop from the Appalachians to the coastal lowlands. These locations provide waterpower and are often at the navigable head of the river.

Dingmans Ferry

The Atlantic Coastal Plain is part of the continental shelf, which extends several hundred miles out to sea, providing a rich environment for marine life.

Rising sea levels have flooded river valleys along the coast, creating rias such as Long Island Sound.

Cape Cod, Long Island and the islands between them mark the top of a great terminal moraine, formed at the front of the ice sheet which once covered the land. This ridge of deposited material was subsequently flooded by rising seas.

At Provincetown, Cape Cod, complex and powerful ocean currents continue to modify the shoreline, washing away some 3 ft (1 m) of the lower cape each year, while extending the beaches in the north.

SCALE 1:2,750,000
(projection: Lambert Conformal Conic)

USA: MID-EASTERN STATES

Delaware, District of Columbia, Kentucky, Maryland, North Carolina, South Carolina, Tennessee, Virginia, West Virginia

KEY EVENTS IN AMERICAN HISTORY took place in this diverse region, which became the front line between the North and the South during the Civil War of the 1860s. Strong regional contrasts exist between the fertile coastal plains, the isolated upcountry of the Appalachian Mountains, and the cotton-growing areas of the Mississippi lowlands to the west. While coal mining, a traditional industry in the Appalachians, has declined in recent years leaving much rural poverty, service industries elsewhere have increased, especially in Washington D.c, the nation's capital.

MAP KEY

POPULATION
- ◉ 500,000 to 1 million
- ◎ 100,000 to 500,000
- ⊕ 50,000 to 50,000
- ○ below 10,000

ELEVATION
- 6000m / 19,686ft
- 4000m / 13,124ft
- 3000m / 9843ft
- 2000m / 6562ft
- 1000m / 3281ft
- 500m / 1640ft
- 250m / 820ft
- 100m / 328ft
- sea level

SCALE 1:3,000,000
(projection: Lambert Conformal Conic)

Km 0 5 10 20 30 40 50 60 70 80
Miles 0 5 10 20 30 40 50 60 70 80

The Bluegrass region of Kentucky centers on the town of Lexington. This exceptionally fertile rolling plain is well known for its thoroughbred horse-breeding ranches.

TRANSPORTATION & INDUSTRY

IN THE URBANIZED NORTHEAST, manufacturing remains important, alongside a burgeoning service sector. North Carolina is a major center for industrial research and development. Traditional industries include Tennessee whiskey and textiles in South Carolina. The decline of open-cast coal mining in the Appalachians has been hastened by environmental controls, although adventure-tourism is a flourishing new industry.

Major industry and infrastructure
- ⚲ adventure-tourism
- 🚗 car manufacture
- coal
- ✿ electronics
- ⚙ engineering
- $ finance
- 🍴 food processing
- 💻 hi-tech industry
- mining
- ✿ research & development
- textiles
- ■ capital cities
- ⊞ major towns
- ✈ international airports
- — major roads
- ▭ major industrial areas

TRANSPORTATION NETWORK
- 452,218 miles (723,548 km)
- 5,737 miles (8,267 km)
- 18,336 miles (29,503 km)
- 4,404 miles (7,081 km)

Tennessee's rivers are part of an important inland bulk-transportation network. Memphis connects with New Orleans in the south, and with cities as distant as Minneapolis, Sioux City, Chicago, and Pittsburgh, via the Mississippi and its tributaries.

THE LANDSCAPE

THE EASTERN TRIBUTARIES OF THE MISSISSIPPI drain the interior lowlands. The Cumberland Plateau and the parallel ranges of the Appalachians have been successively uplifted and eroded over time, with the eastern side reduced to a series of foothills known as the Piedmont. The broad coastal plain gradually falls away into salt marshes, lagoons, and offshore bars, broken by flooded estuaries along the shores of the Atlantic.

The Mammoth Cave is part of an extensive cave system in the limestone region of southwestern Kentucky. It stretches for over 300 miles (485 km) on five different levels and contains three rivers and three lakes.

The Mississippi River and its tributary the Ohio River form the western border of the region.

Natural Bridge in eastern Kentucky is an arch 78 ft (26 m) long and 65 ft (20 m) high. It has been shaped from resistant sandstone by gradual weathering processes, which removed the softer rock lying underneath.

The Allegheny Mountains form the northwestern edge of the Appalachian mountain chain. Continuous folding has formed rich seams of bituminous coal.

Appalachian Mountains

The Cumberland Plateau is the most southwesterly part of the Appalachians. Big Black Mountain at 4,180 ft (1,274 m) is the highest point in the range.

The Great Smoky Mountains form the western escarpment of the Appalachians. The region is heavily forested, with over 130 species of tree.

The Blue Ridge Mountains are a steep ridge, culminating in Mount Mitchell, the highest point in the Appalachians, at 6,684 ft (2,037 m).

Farmland on the eastern shores of Chesapeake Bay is sustained by artificial drainage. The area also provides refuge for a variety of waterfowl.

The many inlets of Chesapeake Bay are the flooded tributaries of the main river valley, which have been inundated by rising sea levels.

Salt marshes such as Great Dismal Swamp, develop where the coast is sheltered. Vast areas of such marshland have been reclaimed for farmland and settlement.

Cape Hatteras is the easternmost point of an offshore barrier island; a wave-deposited sand-bar which has become permanent, establishing its own vegetation.

Barrier islands

These intertidal mudflats become submerged at high tide

Tidal inlet
Barrier island

Barrier islands are common along the coasts of North and South Carolina. As sea levels rise, wave action builds up ridges of sand and pebbles parallel to the coast, separated by lagoons or intertidal mudflats, which are flooded at high tide.

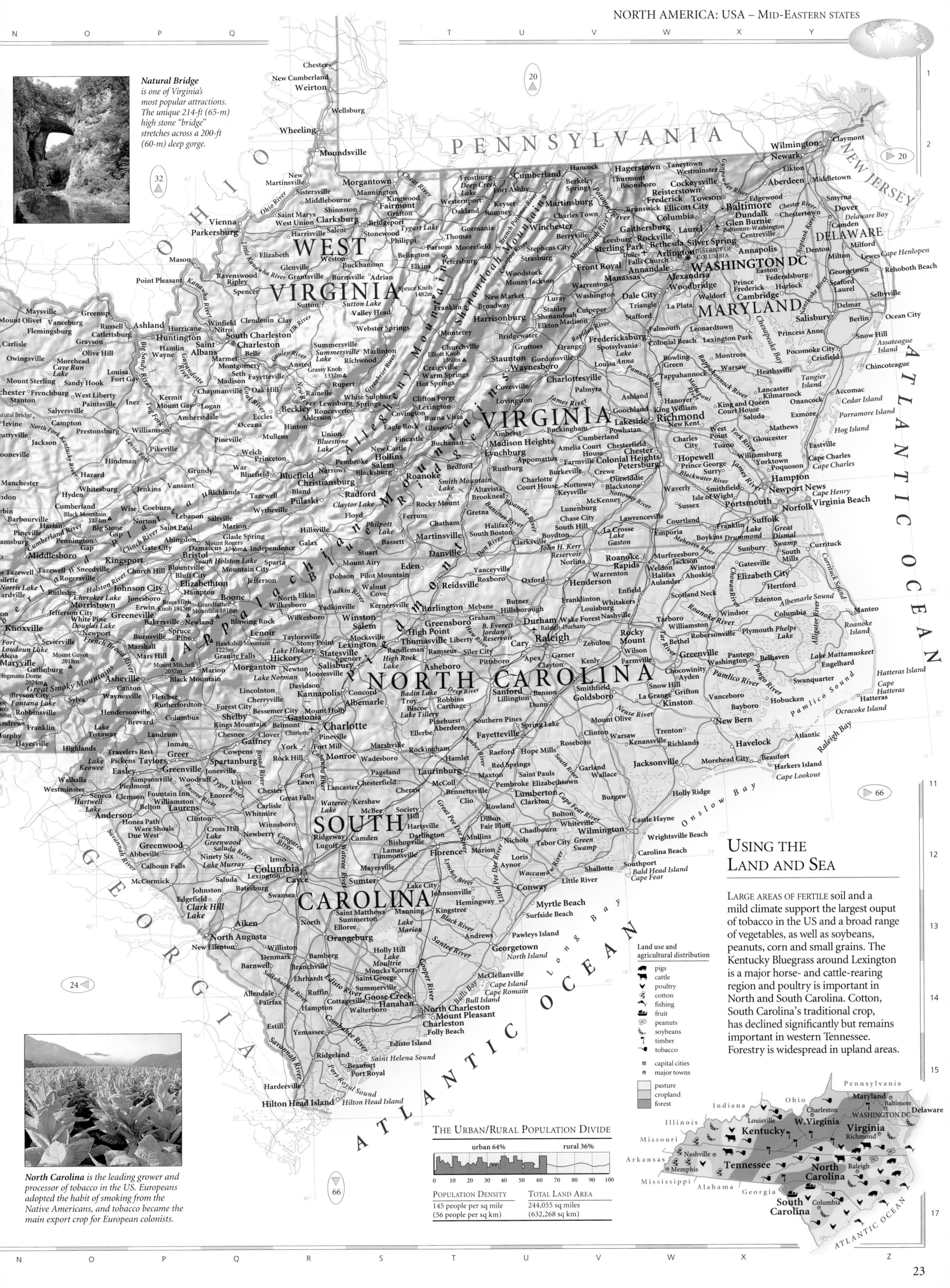

Natural Bridge is one of Virginia's most popular attractions. The unique 214-ft (65-m) high stone "bridge" stretches across a 200-ft (60-m) deep gorge.

North Carolina is the leading grower and processor of tobacco in the US. Europeans adopted the habit of smoking from the Native Americans, and tobacco became the main export crop for European colonists.

USING THE LAND AND SEA

LARGE AREAS OF FERTILE soil and a mild climate support the largest ouput of tobacco in the US and a broad range of vegetables, as well as soybeans, peanuts, corn and small grains. The Kentucky Bluegrass around Lexington is a major horse- and cattle-rearing region and poultry is important in North and South Carolina. Cotton, South Carolina's traditional crop, has declined significantly but remains important in western Tennessee. Forestry is widespread in upland areas.

Land use and agricultural distribution
- pigs
- cattle
- poultry
- cotton
- fishing
- fruit
- peanuts
- soybeans
- timber
- tobacco

■ capital cities
● major towns

pasture
cropland
forest

THE URBAN/RURAL POPULATION DIVIDE

urban 64%   rural 36%

0 10 20 30 40 50 60 70 80 90 100

POPULATION DENSITY
145 people per sq mile
(56 people per sq km)

TOTAL LAND AREA
244,055 sq miles
(632,268 sq km)

USA: SOUTHERN STATES

Alabama, Florida, Georgia, Louisiana, Mississippi

THE SOUTH HAS MAINTAINED a separate identity and outlook throughout the history of the US. Defeat in the Civil War (1861–65) brought chronic poverty to the former confederate states, while the subsequent liberation of four million slaves began a struggle not resolved until the 1960s, when the Civil Rights movement achieved an end to legal racial segregation. Many parts of the South have experienced rapid change. Tourism and retirement communities, together with agriculture, have fueled growth in Florida, while defense-related industries have boosted the growth of cities such as Miami and Atlanta. Many people retain a strong attachment to their history and culture, evidenced by Creole-speaking Cajuns in Louisiania and Hispanic communities in South Florida.

TRANSPORTATION & INDUSTRY

FLORIDA'S TOURIST TRADE is only part of a flourishing service sector, which has swelled the principal cities of he south. Petroleum and mineral extraction has made the Gulf Coast a major industrial region. Traditional textile production remains important in Georgia, while advanced new industries have grown from the NASA Space Program.

TRANSPORTATION NETWORK

441,625 miles (706,600 km)	
5,116 miles (8,186 km)	
16,597 miles (26,555 km)	
6,179 miles (9,942 km)	

Atlanta's Hartsfield International airport is one of the busiest in the world. A dramatic rise in the use of regional air transportation has helped to integrate the major cities of the southern states.

The French Quarter is the traditional cultural center of New Orleans, one of the historic Southern cities. The city once thrived on the cotton trade but now relies mainly on tourism and on oil from the Gulf of Mexico.

Major industry and infrastructure

- ✈ aerospace
- car manufacture
- chemicals
- coal
- defense
- electronics
- engineering
- food processing
- oil
- textiles
- tourism
- • major towns
- ⊕ international airports
- major roads
- major industrial areas

The cypress swamps of the Mississippi Delta form in the backswamps behind the levees of the river and in the multitude of subsiding delta basins.

THE LANDSCAPE

THE BLUE RIDGE MOUNTAINS in the north are skirted by the gentle hills of the Piedmont, whose rivers drain south on to the great flat expanse of the coastal plain. Sandy barrier beaches and islands dominate the sea shore, tracing round the swampy limestone arm of Florida. In the west, the Mississippi meanders toward its delta, crossing the thickly mantled alluvial plain of the interior lowlands.

The Yazoo River flows parallel to the Mississippi through a common floodplain. The confluence of the rivers is deferred downstream because flood deposition has built the Mississippi channel up above the level of the Yazoo.

The Mississippi is the world's third longest river and moves over a billion tons (tonnes) of sediment a year, creating deep alluvial plains. Flooding is a constant threat in lowland areas.

Cathedral Caverns near Huntsville in Alabama is a system of vast limestone caves, with a main opening 1000 ft (300 m) high and 150 ft (50 m) wide.

At De Soto Falls, Alabama, the Little River descends into the deepest canyon east of the Mississippi, with sheer cliff walls up to 700 ft (230 m) high.

Brasstown Bald in the Blue Ridge mountains of Georgia is the region's highest point, at 4,784 ft (1,458 m).

Piedmont

In Providence Canyon, Georgia, the Chattahoochee River has cut straight down through the sandy bedrock, to leave sheer rock faces and pinnacles, which have been smoothed by subsequent weathering.

Sandbars, deposited by waves breaking offshore, form barrier beaches along much of the coastline, creating sheltered lagoons and salt marshes behind them.

Atchafalaya Bay

Mississippi Delta

The delta of the Mississippi over 5,000 years ago

Delta lobe

Present-day delta

Over the last 5,000 years the lower course of the Mississippi has moved back and forth over great distances. These changes, caused by varying sediment loads and human modification, have resulted in a "bird's foot" delta with several lobes, each reflecting the river's different historic position.

Lake Okeechobee is actually a shallow, slow-moving river, 150 miles (240 km) long and 50 miles (80 km) wide.

The Everglades lie in a limestone hollow formed over two million years ago, which has gradually become in-filled with swamp deposits.

Across Florida the coastal plain is mostly less than 75 ft (25 m) above sea level. The land is underlain by limestone, pitted with hollows which have been filled by over 10,000 lakes.

Florida Keys

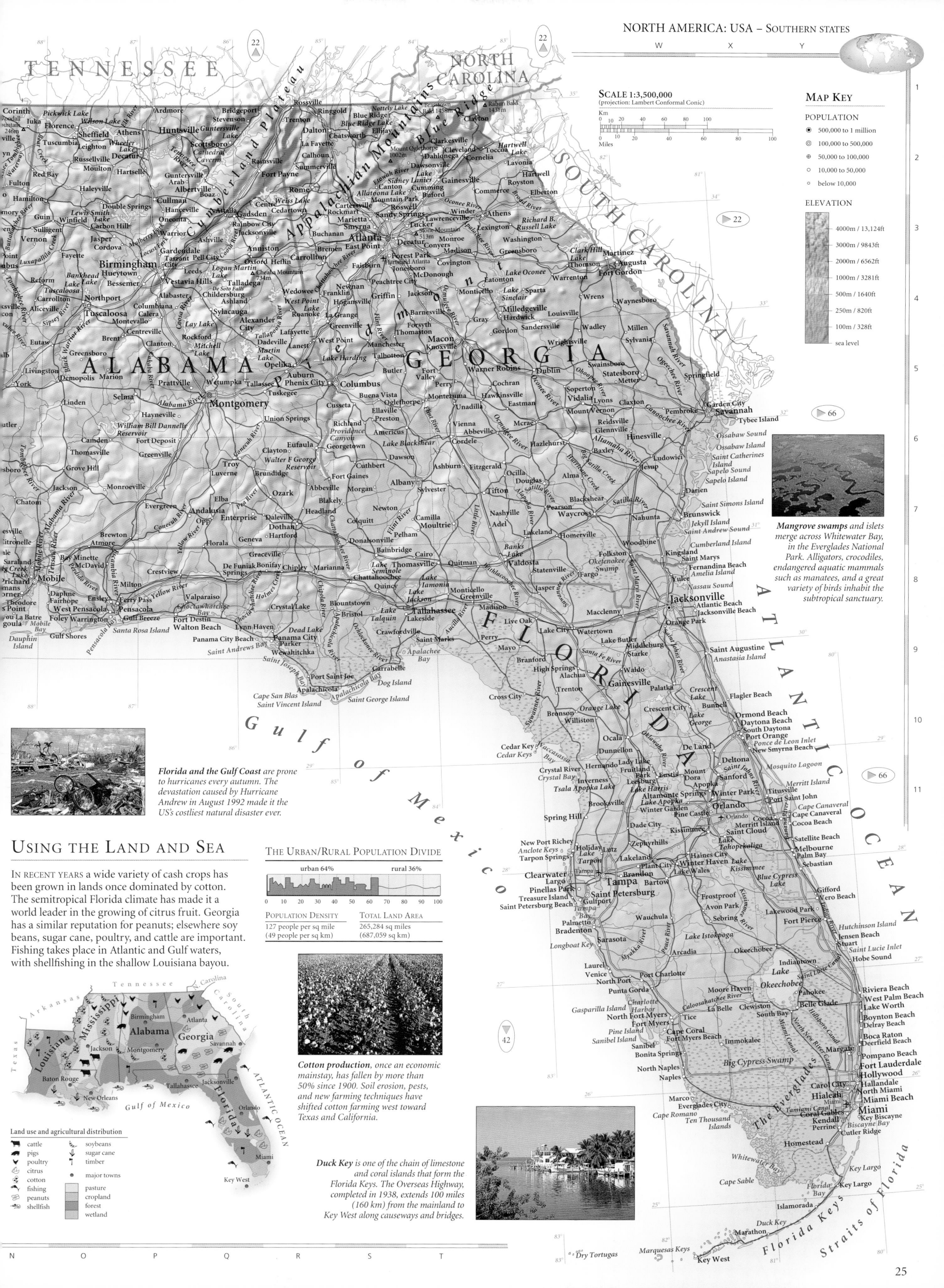

SCALE 1:3,500,000
(projection: Lambert Conformal Conic)

MAP KEY

POPULATION
- 500,000 to 1 million
- 100,000 to 500,000
- 50,000 to 100,000
- 10,000 to 50,000
- below 10,000

ELEVATION
- 4000m / 13,124ft
- 3000m / 9843ft
- 2000m / 6562ft
- 1000m / 3281ft
- 500m / 1640ft
- 250m / 820ft
- 100m / 328ft
- sea level

Mangrove swamps and islets merge across Whitewater Bay, in the Everglades National Park. Alligators, crocodiles, endangered aquatic mammals such as manatees, and a great variety of birds inhabit the subtropical sanctuary.

Florida and the Gulf Coast are prone to hurricanes every autumn. The devastation caused by Hurricane Andrew in August 1992 made it the US's costliest natural disaster ever.

USING THE LAND AND SEA

IN RECENT YEARS a wide variety of cash crops has been grown in lands once dominated by cotton. The semitropical Florida climate has made it a world leader in the growing of citrus fruit. Georgia has a similar reputation for peanuts; elsewhere soy beans, sugar cane, poultry, and cattle are important. Fishing takes place in Atlantic and Gulf waters, with shellfishing in the shallow Louisiana bayou.

THE URBAN/RURAL POPULATION DIVIDE

urban 64%   rural 36%

POPULATION DENSITY
127 people per sq mile
(49 people per sq km)

TOTAL LAND AREA
265,284 sq miles
(687,059 sq km)

Cotton production, once an economic mainstay, has fallen by more than 50% since 1900. Soil erosion, pests, and new farming techniques have shifted cotton farming west toward Texas and California.

Land use and agricultural distribution
- cattle
- pigs
- poultry
- citrus
- cotton
- fishing
- peanuts
- shellfish
- soybeans
- sugar cane
- timber
- major towns
- pasture
- cropland
- forest
- wetland

Duck Key is one of the chain of limestone and coral islands that form the Florida Keys. The Overseas Highway, completed in 1938, extends 100 miles (160 km) from the mainland to Key West along causeways and bridges.

USA: Texas

FIRST EXPLORED BY SPANIARDS moving north from Mexico in search of gold, Texas was controlled by Spain and then by Mexico, before becoming an independent republic in 1836, and joining the Union of States in 1845. During the 19th century, many migrants who came to Texas raised cattle on the abundant land; in the 20th century, they were joined by prospectors attracted by the promise of oil riches. Today, although natural resources, especially oil, still form the basis of its wealth, the diversified Texan economy includes thriving hi-tech and financial industries. The major urban centers, home to 80% of the population, lie in the south and east, and include Houston, the "oil-city," and Dallas Fort Worth. Hispanic influences remain strong, especially in southern and western Texas.

Dallas was founded in 1841 as a prairie trading post and its development was stimulated by the arrival of railroads. Cotton and then oil funded the town's early growth. Today, the modern, high-rise skyline of Dallas reflects the city's position as a leading center of banking, insurance, and the petroleum industry in the southwest.

USING THE LAND

COTTON PRODUCTION AND LIVESTOCK-RAISING, particularly cattle, dominate farming, although crop failures and the demands of local markets have led to some diversification. Following the introduction of modern farming techniques, cotton production spread out from the east to the plains of western Texas. Cattle ranches are widespread, while sheep and goats are raised on the dry Edwards Plateau.

The huge cattle ranches of Texas developed during the 19th century when land was plentiful and could be acquired cheaply. Today, more cattle and sheep are raised in Texas than in any other state.

Land use and agricultural distribution
- cattle
- goats
- sheep
- cereals
- cotton
- major towns
- pasture
- cropland
- forest
- barren

THE URBAN/RURAL POPULATION DIVIDE

urban 80%   rural 20%

0 10 20 30 40 50 60 70 80 90 100

POPULATION DENSITY: 73 people per sq mile (28 people per sq km)

TOTAL LAND AREA: 267,338 sq miles (692,402 sq km)

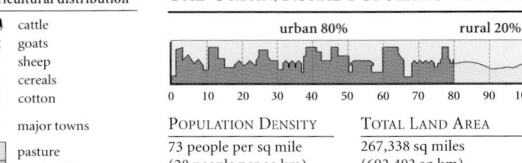

THE LANDSCAPE

TEXAS IS MADE UP OF A SERIES of massive steps descending from the mountains and high plains of the west and northwest to the coastal lowlands in the southeast. Many of the state's borders are delineated by water. The Rio Grande flows from the Rocky Mountains to the Gulf of Mexico, marking the border with Mexico.

Cap Rock Escarpment juts out from the plains, running 200 miles (320 km) from north to south. Its height varies from 300 ft (90 m) rising to sheer cliffs up to 1,000 ft (300 m).

The Llano Estacado or Staked Plain in northern Texas is known for its harsh environment. In the north, freezing winds carrying ice and snow sweep down from the Rocky Mountains. To the south, sandstorms frequently blow up, scouring anything in their paths. Flash floods, in the wide, flat riverbeds that remain dry for most of the year, are another hazard.

The Guadalupe Mountains lie in the southern Rocky Mountains. They incorporate Guadalupe Peak, the highest in Texas, rising 8,749 ft (2,667 m).

The Rio Grande flows from the Rocky Mountains through semi-arid land, supporting sparse vegetation. The river actually shrinks along its course, losing more water through evaporation and seepage than it gains from its tributaries and rainfall.

Big Bend National Park

Flowing through 1,500 ft (450 m) high gorges, the shallow, muddy Rio Grande makes a 90° bend. This marks the southern border of Big Bend National Park, and gives its name. The area is a mixture of forested mountains, deserts, and canyons.

Edwards Plateau is a limestone outcrop. It is part of the Great Plains, bounded to the southeast by the Balcones Escarpment, which marks the southerly limit of the plains.

The Red River flows for 1300 miles (2090 km), marking most of the northern border of Texas. A dam and reservoir along its course provide vital irrigation and hydro-electric power to the surrounding area.

Sabine River

Extensive forests of pine and cypress grow in the eastern corner of the coastal lowlands where the average rainfall is 45 inches (1145 mm) a year. This is higher than the rest of the state and over twice the average in the west.

In the coastal lowlands of southeastern Texas the Earth's crust is warping, causing the land to subside and allowing the sea to invade. Around Galveston, the rate of downward tilting is 6 inches (15 cm) per year. Erosion of the coast is also exacerbated by hurricanes.

Laguna Madre in southern Texas has been almost completely cut off from the sea by Padre Island. This sand bank was created by wave action, carrying and depositing material along the coast. The process is known as longshore drift.

Padre Island

Oil deposits

Oil trapped by fault

Oil deposits migrate through reservoir rocks such as shale

Oil accumulates beneath impermeable cap rock

Impermeable rock strata

Salt dome

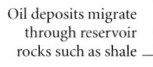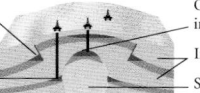

Oil deposits are found beneath much of Texas. They collect as oil migrates upward through porous layers of rock until it is trapped, either by a cap of rock above a salt dome, or by a fault line which exposes impermeable rock through which the oil cannot rise.

Map labels

Oklahoma • Arkansas • New Mexico • Louisiana • Amarillo • Dallas • El Paso • Texas • Austin • Houston • San Antonio • MEXICO

Texline • Dalhart • Hartley • Channing • Canadian River • High • Adrian • Wilde • Hereford • Friona • Bovina • Farwell • Running Water • Springlak • Earth • Muleshoe • Sudan • Enochs • Littl • Morton • Whiteface • Levell • Rope • Sundow • Mea • Brow • Plains • Tokio • Sulphur Spring • Denver City • Wellman • Seagraves • Cedar La • Seminole • Mustang L • Andre • Goldsmith • Kermit • Mentone • Wink • Penwell • Wickett • Monahans • Barstow • Pecos • Royalty • Grandfalls • Imperial • McCame • Girvin • Bakersfi • Big Ca • Fort Stockton • Stockton Plateau • Sand • Dry

Canutillo • El Paso • Dell City • Salt Basin • Guadalupe Mountains • Guadalupe Peak 2667m • Red Bluff Reservoir • Orla • Pecos River • San Elizario • Clint • Fabens • Tornillo • Salt Flat • Fort Hancock • McNary • Sierra Blanca 2100m • Sierra Blanca • Esperanza • Van Horn • Kent • Toyah • Saragosa • Balmorhea • Davis Mountains • Mount Livermore 2554m • Fort Davis • Valentine • Marfa • Alpine • Glass Mountains • Marathon • Candelaria • Cathedral Mountain 2093m • Ruidosa • Chinati Mountains • Shafter • Casa Piedra • Presidio • Redford • Terlingua • Emory Peak 2385m • Big Bend National Park • Rio Grande

See also page 38, page 42

TRANSPORTATION & INDUSTRY

INDUSTRY IN THE 20TH CENTURY was largely concentrated on the processing of local raw materials, especially oil – deposits were discovered under 65% of the state's area. The technological demands of the oil industry and defense-related institutions, particularly NASA, have stimulated the development of numerous electronics and hi-tech firms which, alongside many national corporate headquarters, are based in Dallas–Fort Worth and Houston.

Major industry and infrastructure

chemicals		mining	
defense		oil	
engineering		textiles	
finance		major towns	
food processing		international airports	
gas		major roads	
hi-tech industry		major industrial areas	

TRANSPORTATION NETWORK

293,509 miles (496,614 km)	3,229 miles (5,166 km)
10,681 miles (17,089 km)	845 miles (1,359 km)

The sheer size of Texas promoted the development of an extensive road and rail network. The highway system, although well-developed, is concentrated in the east.

The Texas hill country is the most southerly extension of the Great Plains. Although farming is the primary source of income, the beautiful hills, valleys, and lakes are a major tourist attraction.

Padre Island is a sand bank. It extends 113 miles (182 km) along the southern coast of Texas.

MAP KEY

POPULATION

- 1 million to 5 million
- 500,000 to 1 million
- 100,000 to 500,000
- 50,000 to 100,000
- 10,000 to 50,000
- below 10,000

ELEVATION

- 2000m / 6562ft
- 1000m / 3281ft
- 500m / 1640ft
- 250m / 820ft
- 100m / 328ft
- sea level

SCALE 1:3,250,000
(projection: Lambert Conformal Conic)

USA: South Midwestern states

Arkansas, Kansas, Missouri, Oklahoma

THE EXPANSION OF THE US focused on this region in the mid-19th century. Settlers spread from the confluence of the Missouri and Mississippi Rivers up onto the Great Plains. This treeless expanse, which early explorers had called the Great American Desert was turned into one of the world's richest agricultural regions. But periodic droughts, coupled with overintensive farming, led to the "dustbowl" soil erosion crisis of the 1930s, the abandonment of many farms, and a mass exodus to the west coast. The land has since recovered, although the mechanization of agriculture has led to a decline in the rural population. In recent years, suburban residential development has spread rapidly across the wooded Ozark Plateau in the east of the region.

TRANSPORTATION & INDUSTRY

THE PROCESSING OF AGRICULTURAL PRODUCTS, such as brewing and meatpacking, has been traditionally important in these states. In Kansas and Oklahoma, diversified manufacturing now supplements income from fossil fuels; Wichita has become a world center for aeronautical engineering, an industry which also employs many people in neighboring Missouri.

Major industry and infrastructure

- ✈ aerospace
- ✿ engineering
- $ finance
- 🗊 food processing
- 🛢 gas
- ⛏ mining
- ⚓ oil
- 🚚 vehicle manufacture
- ● major towns
- ✈ international airports
- — major roads
- ▨ major industrial areas

Agricultural produce from the plains is moved by barges along the Mississippi. The river now carries a far greater tonnage of freight than any other waterway system in the US.

TRANSPORTATION NETWORK

380,307 miles (608,491 km)	4068 miles (6508 km)
16,185 miles (25,896 km)	1994 miles (3208 km)

The Arkansas River and its tributaries allow access to over half of the US's navigable inland waterways. A system of locks and dams along the river provides Tulsa, in Oklahom, with a navigable water route to the Gulf of Mexico.

MAP KEY

POPULATION

- ◎ 100,000 to 500,000
- ⊕ 50,000 to 100,000
- ○ 10,000 to 50,000
- ○ below 10,000

ELEVATION

- 1000m / 3281ft
- 500m / 1640ft
- 250m / 820ft
- 100m / 328ft
- sea level

The Ozark Plateau is a wooded, hilly region of rivers and narrow, winding lakes. The Lake of the Ozarks was created by the damming of the Osage River in 1930.

THE LANDSCAPE

MOST OF THE REGION consists of high, treeless plains, which gradually descend east from the Rocky Mountains. Drainage follows this slope, with rivers flowing toward the alluvial lowlands of the Mississippi in the southeast. Between the plains and the lowlands lie various ranges of wooded hills, including the deeply incised Ozark Plateau.

The Mississippi, North America's longest river, is joined by the Missouri, its main tributary, on a flood plain which spreads south to the Gulf of Mexico.

Collapsed limestone caverns led to the formation of Big Basin in Kansas; a depression 100 ft (33 m) deep and 1 mile (1.6 km) wide.

Flint Hills is the region's easternmost major escarpment. Steep, grassy uplands are interspersed with rocky, wooded ravines and outcrops of limestone and chert.

Missouri River

The Great Salt Plains of northern Oklahoma cover 45 sq miles (116 sq km). The arid, white flats were left by the gradual evaporation of an ancient salt lake.

Lake Ouachita, in Arkansas is one of a number of irregularly-shaped lakes found among the ridges of the Ouachita Mountains.

Mississippi River

Underground water reserves

- Extent of the aquifer
- Kansas
- Oklahoma

The Ogallala Aquifer, beneath the Great Plains, is the largest known source of underground water in the world. There is concern about the rapid depletion of this finite water supply by irrigation schemes.

Red River

Devil's Den is a dry badland area. The rugged landscape, strewn with large boulders, is the eroded remnant of a spur extending from the Arbuckle mountains to the west.

Ouachita Mountains

Crowleys Ridge is a long, sandy ridge, rising from the Mississippi floodplain. It was formed over thousands of years by the deposition of sand blown eastward from the Great Plains.

SCALE 1:3,000,000
(projection: Lambert Conformal Conic)

Km
0 5 10 20 30 40 50 60 70

Miles
0 10 20 30 40 50 60 70

The landscape of northeast Kansas is interlaced by rivers which have cut broad wooded valleys through the gentle hills. All the rivers in Kansas form part of the massive Missouri/Mississippi drainage basin.

Gateway Arch, in Saint Louis, Missouri, is 634 ft (192 m) high. The huge steel arch symbolizes the city's historic role as the "Gateway to the West".

Map labels

States: IOWA · NEBRASKA · ILLINOIS · KANSAS · MISSOURI · OKLAHOMA · ARKANSAS · TENNESSEE · KENTUCKY · TEXAS · LOUISIANA · MISSISSIPPI

Physical features: Ozark Plateau · Boston Mountains · Ouachita Mountains · Kiamichi Mountains · Caddo Mountains · Flint Hills · Saint Francois Mountains · Crowley's Ridge · Missouri River · Mississippi River · Arkansas River · Red River · White River · Osage River

Towns (selection):
Tarkio, Burlington Junction, Grant City, Lancaster, Unionville, Memphis, Rock Port, Maryville, Princeton, Queen City, Kahoka, Mound City, Oregon, Bethany, Albany, Milan, Green City, Edina, Monticello, Canton, La Grange, Savannah, King City, Pattonsburg, Trenton, Kirksville, Linneus, Shelbyville, Palmyra, Hannibal, Saint Joseph, Stewartsville, Chillicothe, Hamilton, Brookfield, Macon, Monroe City, New London, Dearborn, Lawson, Brunswick, Keytesville, Huntsville, Paris, Louisiana, Bowling Green, Vandalia, Kansas City, Lexington, Waverly, Marshall, Fayette, Centralia, Mexico, Montgomery City, Troy, O'Fallon, Saint Charles, Florissant, Independence, Higginsville, Concordia, Boonville, Columbia, Ashland, Fulton, Jefferson City, Warrenton, Hermann, Washington, Union, Pacific, Webster Groves, Saint Louis, Clayton, Kirkwood, University City, New Haven, Gerald, Owensville, Belle, Sullivan, Hillsboro, Crystal City, Festus, Sainte Genevieve, Rolla, Steelville, Cuba, Salem, Potosi, Bonne Terre, Deslodge, Flat River, Farmington, Perryville, Bismarck, Fredericktown, Jackson, Cape Girardeau, Sikeston, Charleston, New Madrid, Kennett, Caruthersville, Springfield, Branson, Joplin, Neosho, Fayetteville, Fort Smith, Van Buren, Tulsa, Muskogee, McAlester, Little Rock, North Little Rock, Hot Springs, Pine Bluff, Jonesboro, West Memphis, Texarkana, El Dorado, Magnolia, Camden, Warren, Monticello, Topeka, Lawrence, Overland Park, Olathe, Lenexa, Leavenworth, Manhattan, Junction City, Salina, Emporia, Newton, El Dorado, Wichita, Derby, Winfield, Arkansas City, Ponca City, Bartlesville, Stillwater, Oklahoma City, Norman, Ada, Durant, Ardmore

Legend

Land use and agricultural distribution
- cattle
- poultry
- cereals
- corn (maize)
- cotton
- fodder
- rice
- soya beans
- ● major towns
- pasture
- cropland
- forest

USING THE LAND

THE PROBLEMS of a harsh continental climate, with severe winters and hot, dry summers, are partially offset by the rich soils of the plains. Kansas is a major cereal crop producer, ranking first in US production of wheat and sorghum. Rainfall increases toward the east, favoring the cultivation of soybeans, cotton, and rice, with corn concentrated in Missouri. Huge herds of cattle are raised in Oklahoma, Kansas, and Missouri.

A combine harvester works the land on the great plains. A hundred years ago this region, also known as the prairies – the French word for pasture – was covered with tall, wild grasses.

THE URBAN/RURAL POPULATION DIVIDE

urban 65% · rural 35%

0 10 20 30 40 50 60 70 80 90 100

POPULATION DENSITY
50 people per sq mile
(19 people per sq km)

TOTAL LAND AREA
274,900 sq miles
(712,177 sq km)

USA: UPPER PLAINS STATES

Iowa, Minnesota, Nebraska, North Dakota, South Dakota

LYING AT THE VERY HEART of the North American continent, much of this region was acquired from France as part of the Louisiana Purchase in 1803. The area was largely bypassed by the early waves of westward migrants. When Europeans did settle, during the 19th century, they displaced the Native Americans who lived on the plains. The settlers planted arable crops and raised cattle on the immensely fertile prairie land, founding an agrarian tradition which flourishes today. Most of this region remains rural; of the five states, only in Minnesota has there been significant diversification away from agriculture and resource-based industries into the hi-tech and service sectors.

USING THE LAND

THE POPULAR IMAGE of these states as agricultural is entirely justified; prairies stretch uninterrupted across most of the area. Croplands fall into two regions: the wheat belt of the plains, and the corn belt of the central US. Cash crops, such as soybeans, are grown to supplement incomes. Livestock, particularly pigs and cattle, are raised throughout this region.

Dark, fertile prairie soils in the southeast provide Minnesota's most productive farmland. Hot, humid summers create a long growing season for corn cultivation.

Land use and agricultural distribution
- cattle
- pigs
- corn (maize)
- soybeans
- wheat
- major towns
- pasture
- cropland
- forest
- wetland

THE URBAN/RURAL POPULATION DIVIDE

urban 64%   rural 36%

0 10 20 30 40 50 60 70 80 90 100

POPULATION DENSITY
29 people per sq mile
(11 people per sq km)

TOTAL LAND AREA
365,287 sq miles
(946,056 sq km)

TRANSPORTATION & INDUSTRY

FOOD PROCESSING and the production of farm machinery are supported by the large agricultural sector. Mineral exploitation is also an important activity: gold is mined in the ore-rich Black Hills of South Dakota, and both North Dakota and Nebraska are emerging as major petroleum producers.

Water erosion along the Little Missouri River has carried away sedimentary deposits, creating rugged landscapes known as badlands.

Major industry and infrastructure
- coal
- engineering
- electronics
- finance
- food processing
- oil & gas
- mining
- major towns
- international airports
- major roads
- major industrial areas

TRANSPORTATION NETWORK

504,522 miles (807,235 km)

3,422 miles (5,475 km)

16,940 miles (27,104 km)

683 miles (1,098 km)

Nebraska's central location has made it an important transportation artery for east–west traffic. Minnesota's road network radiates out from the hub of the twin cities, Minneapolis–Saint Paul.

THE LANDSCAPE

THESE STATES STRADDLE the Great Plains and the lowlands of the central US, with Minnesota lying in a transition zone between the eastern forests and the prairies. The region was shaped by repeated ice advances and retreats, leaving a flat relief, broken only by the numerous lakes and broad river networks that drain the prairies.

Escarpment   Ridge   In permeable strata hollows are formed by small mudslides

Water flowing into gullies erodes back the escarpment

Badlands are formed by stormwater run-off. This flows down the impermeable strata of the escarpment and saturates the permeable strata, leading to mudslides and the formation of gullies.
North Dakota Badlands

The Minnesota landscape contains many post-glacial features, including its numerous lakes, boulder-strewn hills, and mineral-rich deposits.

Although it escaped the last glaciation, the limestone bedrock of southeastern Minnesota has been eroded by surface and subterranean streams, leaving a network of underground caverns and steep-sided valleys.

In the badlands of North and South Dakota, horizontal layers of sandstone have been eroded by rivers, leaving a landscape of narrow gullies, sharp crests and pinnacles.
South Dakota Badlands

Chimney Rock is a remnant of an ancient land surface, eroded by the North Platte River. The tip of its spire stands 500 ft (150 m) above the plain.

Missouri River

Mississippi River

In northeastern Iowa, the Mississippi and its tributaries have deeply incised the underlying bedrock creating a hilly terrain, with bluffs standing 300 ft (90 m) above the valley.

Along the shores of Lake Superior in Minnesota, the average number of frost-free days can be as few as 90, and frosts may occur in any month of the year.

MAP KEY

POPULATION

◎ 100,000 to 500,000
⊕ 50,000 to 100,000
○ 10,000 to 50,000
∘ below 10,000

ELEVATION

2000m / 6562ft
1000m / 3281ft
500m / 1640ft
250m / 820ft
100m / 328ft
sea level

SCALE 1:3,250,000
(projection: Lambert Conformal Conic)

Km
0 10 20   40   60   80   100   120

Miles
0   20   40   60   80   100   120

USA: GREAT LAKES STATES

Illinois, Indiana, Michigan, Ohio, Wisconsin

THE STATES BORDERING THE GREAT LAKES developed rapidly in the second half of the 19th century as a result of improvements in communications: railroads to the west and waterways to the south and east. Fertile land and good links with growing eastern seaboard cities encouraged the development of agriculture and food processing. Migrants from Europe and other parts of the US flooded into the region and for much of the 20th century the region's economy boomed. However, in recent years heavy industry has declined, earning the region the unwanted label the "Rustbelt."

TRANSPORTATION & INDUSTRY

THE GREAT LAKES REGION IS THE CENTER of the US car industry. Since the early part of the 20th century, its prosperity has been closely linked to the fortunes of automobile manufacturing. Iron and steel production has expanded to meet demand from this industry. In the 1970s, nationwide recession, cheaper foreign competition in the automobile sector, pollution in and around the Great Lakes, and the collapse of the meatpacking industry, centered on Chicago, forced these states to diversify their industrial base. New industries have emerged, notably electronics, service, and finance industries.

TRANSPORTATION NETWORK

540,682 miles (865,091 km)		6,550 miles (10,480 km)	
24,928 miles (39,884 km)		2,330 miles (3,748 km)	

Few areas of the US have a comparable system. Chicago is a principal transportation terminus with a dense network of roads, railroads, and Interstate freeways that radiates out from the city.

Ever since Ransom Olds and Henry Ford started mass-producing automobiles in Detroit early in the 20th century, the city's name has become synonymous with the American automotive industry.

Major industry and infrastructure

- car manufacture
- coal
- electronics
- engineering
- finance
- food processing
- iron & steel
- oil
- research & development
- textiles
- major towns
- international airports
- major roads
- major industrial areas

THE LANDSCAPE

MUCH OF THIS REGION shows the impact of glaciation which lasted until about 10,000 years ago, and extended as far south as Illinois and Ohio. Although the relief of the region slopes toward the Great Lakes, because the ice sheets blocked northerly drainage, most of the rivers today flow southward, forming part of the massive Mississippi/Missouri drainage basin.

Lake Michigan

The dunes near Sleeping Bear Point rise 400 ft (120 m) from the banks of Lake Michigan. They are constantly being resculpted by wind action.

Lake Erie is the shallowest of the five Great Lakes. Its average depth is about 62 ft (19 m). Storms sweeping across from Canada erode its shores and cause the silting of its harbors.

The many lakes and marshes of Wisconsin and Michigan are the result of glacial erosion and deposition which occurred during the last Ice Age.

Southwestern Wisconsin is known as a "driftless" area. Unlike most of the region, low hills protected it from erosion by the advancing ice sheet.

Most of the water used in northern Illinois is pumped from underground reservoirs. Due to increased demand, many areas now face a water shortage. Around Joliet, the water table was lowered by more than 700 ft (210 m) over the last century.

Illinois plains

The plains of Illinois are characteristic of drift landscapes, scoured and flattened by glacial erosion and covered with fertile glacial deposits.

Mississippi River

Relict landforms from the last glaciation, such as shallow basins and ridges, cover all but the south of this region. Ridges, known as moraines, up to 300 ft (100 m) high, lie to the south of Lake Michigan.

Ohio River

Unlike the level prairie to the north, southern Indiana is relatively rugged. Limestone in the hills has been dissolved by water, producing features such as sinkholes and underground caves.

The Appalachian plateau stretches eastward from Ohio. It is dissected by streams flowing west into the Mississippi and Ohio Rivers.

Present-day river or stream

Channels caused by outwash from melting glacier

Glacial till

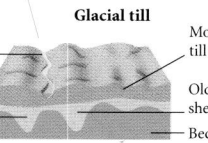

Most recent till deposits

Older till sheet

Bedrock

As a result of successive glacial depositions, the total depth of till along the former southern margin of the Laurentide ice sheet can exceed 1,300 ft (400 m).

MINNESOTA
WISCONSIN
ILLINOIS
IOWA
MISSOURI

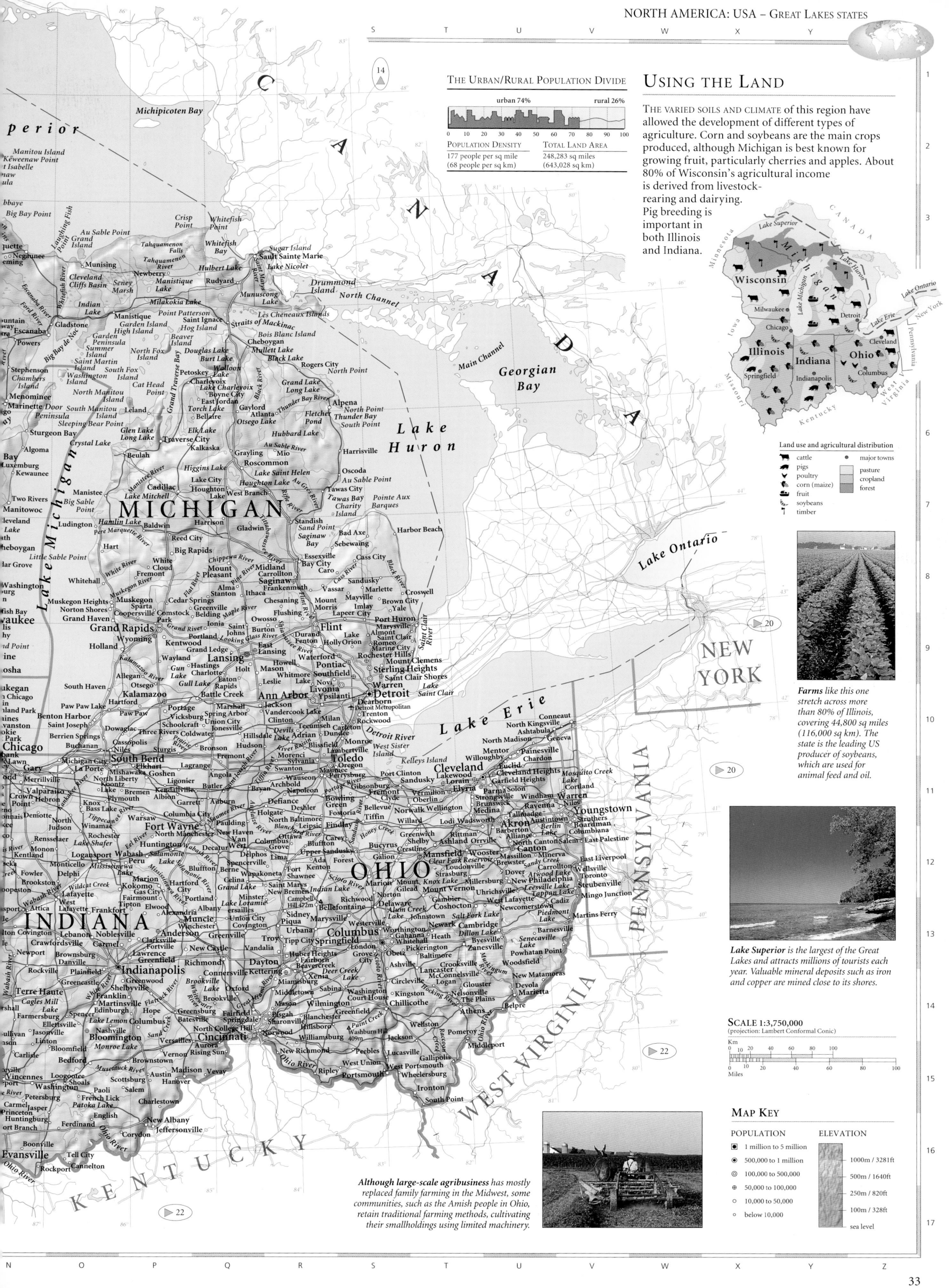

THE URBAN/RURAL POPULATION DIVIDE

urban 74% rural 26%

POPULATION DENSITY	TOTAL LAND AREA
177 people per sq mile (68 people per sq km)	248,283 sq miles (643,028 sq km)

USING THE LAND

THE VARIED SOILS AND CLIMATE of this region have allowed the development of different types of agriculture. Corn and soybeans are the main crops produced, although Michigan is best known for growing fruit, particularly cherries and apples. About 80% of Wisconsin's agricultural income is derived from livestock-rearing and dairying. Pig breeding is important in both Illinois and Indiana.

Land use and agricultural distribution

- cattle
- pigs
- poultry
- corn (maize)
- fruit
- soybeans
- timber
- major towns
- pasture
- cropland
- forest

Farms like this one stretch across more than 80% of Illinois, covering 44,800 sq miles (116,000 sq km). The state is the leading US producer of soybeans, which are used for animal feed and oil.

Lake Superior is the largest of the Great Lakes and attracts millions of tourists each year. Valuable mineral deposits such as iron and copper are mined close to its shores.

SCALE 1:3,750,000
(projection: Lambert Conformal Conic)

Km
Miles

MAP KEY

POPULATION

- 1 million to 5 million
- 500,000 to 1 million
- 100,000 to 500,000
- 50,000 to 100,000
- 10,000 to 50,000
- below 10,000

ELEVATION

- 1000m / 3281ft
- 500m / 1640ft
- 250m / 820ft
- 100m / 328ft
- sea level

Although large-scale agribusiness has mostly replaced family farming in the Midwest, some communities, such as the Amish people in Ohio, retain traditional farming methods, cultivating their smallholdings using limited machinery.

USA: NORTH MOUNTAIN STATES

Idaho, Montana, Oregon, Washington, Wyoming

THE REMOTENESS OF THE NORTHWESTERN STATES, coupled with the rugged landscape, ensured that this was one of the last areas settled by Europeans in the 19th century. Fur-trappers and gold-prospectors followed the Snake River westward as it wound its way through the Rocky Mountains. The states of the northwest have pioneered many conservationist policies, with the first US National Park opened at Yellowstone in 1872. More recently, the Cascades and Rocky Mountains have become havens for adventure tourism. The mountains still serve to isolate the western seaboard from the rest of the continent. This isolation has encouraged West Coast cities to expand their trade links with countries of the Pacific Rim.

The Snake River has cut down into the basalt of the Columbia Basin to form Hells Canyon, the deepest in the US, with cliffs up to 7,900 ft (2,408 m) high.

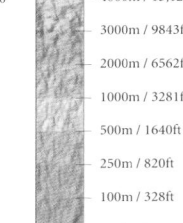
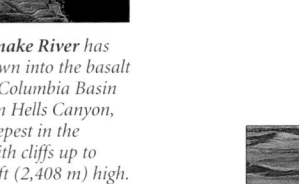

Fine-textured, volcanic soils in the hilly Palouse region of eastern Washington are susceptible to erosion.

USING THE LAND

WHEAT FARMING IN THE EAST gives way to cattle ranching as rainfall decreases. Irrigated farming in the Snake River valley produces large yields of potatoes and other vegetables. Dairying and fruit-growing take place in the wet western lowlands between the mountain ranges.

THE URBAN/RURAL POPULATION DIVIDE

urban 70%   rural 30%

0 10 20 30 40 50 60 70 80 90 100

POPULATION DENSITY
23 people per sq mile
(9 people per sq km)

TOTAL LAND AREA
493,782 sq miles
(1,278,846 sq km)

SCALE 1:3,750,000
(projection: Lambert Conformal Conic)

Km 0 10 20   40   60   80   100
Miles 0 10 20   40   60   80   100

Land use and agricultural distribution

- 🐄 cattle
- 🐓 poultry
- 🌾 cereals
- 🍎 fruit
- 🥔 potatoes
- 🌲 timber
- major towns
- pasture
- cropland
- forest

198 ◀

TRANSPORTATION & INDUSTRY

MINERALS AND TIMBER are extremely important in this region. Uranium, precious metals, copper, and coal are all mined, the latter in vast open-cast pits in Wyoming; oil and natural gas are extracted further north. Manufacturing, notably related to the aerospace and electronics industries, is important in western cities.

TRANSPORTATION NETWORK

- 347,857 miles (556,571 km)
- 4,200 miles (6,720 km)
- 12,354 miles (19,766 km)
- 1,108 miles (1,782 km)

Major industry and infrastructure

- ⛷ adventure tourism
- ✈ aerospace
- coal
- chemicals
- electronics
- food processing
- mining
- oil & gas
- timber processing
- · major towns
- ⊕ international airports
- — major roads
- major industrial areas

The Union Pacific Railroad has been in service across Wyoming since 1867. The route through the Rocky Mountains is now shared with the Interstate 80, a major east–west highway.

Seattle lies in one of Puget Sound's many inlets. The city receives oil and other resources from Alaska, and benefits from expanding trade across the Pacific.

Crater Lake, Oregon, is 6 miles (10 km) wide and 1,800 ft (600 m) deep. It marks the site of a volcanic cone, which collapsed after an eruption within the last 7,000 years.

THE LANDSCAPE

THE ROCKY MOUNTAINS are flanked by lower parallel ranges, which spread onto the Great Plains in the east and surmount the broad lava plateau which extends westward. The Cascade Range divides the Columbia Basin from the coastlands, where the low areas around Puget Sound are broken by the steep, volcanic Olympic Mountains and the wooded hills of the Coast Ranges.

Molten rock cools, forming parallel columns
Surrounding strata eroded away
Molten rock wells up from the Earth's core

Devil's Tower in Wyoming is an igneous intrusion, formed below the Earth's surface. Molten rock intruded through cracks in the overlying strata and cooled. Over time, the softer rock layers have been eroded away, leaving only the tower standing.

Glacial valleys on the seaward side of the Olympic Mountains receive about 142 inches (3,600 mm) of rain per year, supporting the only true rain forest of the northern hemisphere.

The Cascades are glacially scoured volcanic mountains, the highest of which is Mount Rainier, a dormant volcano at 14,409 ft (4,392 m).

Coast Ranges

Mount St. Helens erupted in 1980, killing 57 people and devastating a huge area.

Puget Sound

Columbia Basin

Grand Coulee and the lesser *coulées* (ravines) were cut by cataclysmic floods, from the release of an ice-dammed lake, at the end of the last ice age.

The Continental Divide, or watershed, crosses the Lewis Range. From here, rivers flow east to Hudson Bay, south to the Gulf of Mexico and west to the Pacific Ocean.

Piney Buttes are the remnants of an older, higher land surface gradually weathered and eroded into isolated outcrops with flat tops and steep sides.

Great Plains

Devil's Tower

The plateaus of the Columbia and Snake Rivers represent one of the world's largest accumulations of lava. Over 5 million years ago, successive flows of molten basalt buried the existing land surface by up to 450 ft (150 m).

The contorted rock shapes at "Craters of the Moon" National Monument in Idaho were left 2,000 years ago by the sporadic upwelling of viscous lava from fissures in the basalt plateau.

Rocky Mountains

Water from the hot springs in Yellowstone National Park deposits minerals as it cools in rock pools. Long periods of deposition have created these rock terraces.

USA: CALIFORNIA & NEVADA

THE GOLD RUSH OF 1849 attracted the first major wave of European settlers to the West Coast. The pleasant climate, beautiful scenery and dynamic economy continue to attract immigrants – despite the ever-present danger of earthquakes – and California has become the US's most populous state. The overwhelmingly urban population is concentrated in the vast conurbations of Los Angeles, San Francisco, and San Diego; new immigrants include people from South Korea, the Philippines, Vietnam, and Mexico. Nevada's arid lands were initially exploited for minerals; in recent years, revenue from mining has been superseded by income from the tourist and gambling centers of Las Vegas and Reno.

MAP KEY

POPULATION

- 1 million to 5 million
- 500,000 to 1 million
- 100,000 to 500,000
- 50,000 to 100,000
- 10,000 to 50,000
- below 10,000

ELEVATION

- 4000m / 13,124ft
- 3000m / 9843ft
- 2000m / 6562ft
- 1000m / 3281ft
- 500m / 1640ft
- 250m / 820ft
- 100m / 328ft
- sea level

SCALE 1:3,000,000
(projection: Lambert Conformal Conic)

Km 0 5 10 20 30 40 50 60 70 80

Miles 0 5 10 20 30 40 50 60 70 80

TRANSPORTATION & INDUSTRY

NEVADA'S RICH MINERAL RESERVES ushered in a period of mining wealth which has now been replaced by revenue generated from gambling. California supports a broad set of activities including defense-related industries and research and development facilities. "Silicon Valley," near San Francisco, is a world leading center for microelectronics, while tourism and the Los Angeles film industry also generate large incomes.

Gambling was legalized in Nevada in 1931. Las Vegas has since become the center of this multimillion dollar industry.

Major industry and infrastructure

- aerospace
- car manufacture
- defense
- movie industry
- finance
- food processing
- gambling
- hi-tech industry
- mining
- pharmaceuticals
- research & development
- textiles
- tourism
- major towns
- international airports
- major roads
- major industrial areas

TRANSPORTATION NETWORK

- 211,459 miles (338,334 km)
- 2,944 miles (4,710 km)
- 7,872 miles (12,595 km)
- 190 miles (306 km)

In California, the motor vehicle is a vital part of daily life, and an extensive freeway system runs throughout the state, which has a greater *per capita* car ownership than anywhere else in the world.

THE LANDSCAPE

THE BROAD CENTRAL VALLEY divides California's coastal mountains from the Sierra Nevada. The San Andreas Fault, running beneath much of the state, is the site of frequent earth tremors and sometimes more serious earthquakes. East of the Sierra Nevada, the landscape is characterized by the basin and range topography with stony deserts and many salt lakes.

Rising molten rock causes stretching of the Earth's crust

Extensive cracking (faulting) uplifted a series of ridges

As ridges are eroded they fill intervening valleys with sediments

Molten rock (magma) welling up to form a dome in the Earth's interior, causes the brittle surface rocks to stretch and crack. Some areas were uplifted to form mountains (ranges), while others sunk to form flat valleys (basins).

The General Sherman sequoia tree in Sequoia National Park is 3000 years old and at 275 ft (84 m) is one of the largest living things on earth.

Most of California's agriculture is confined to the fertile and extensively irrigated Central Valley, running between the Coast Ranges and the Sierra Nevada. It incorporates the San Joaquin and Sacramento valleys.

The dramatic granitic rock formations of Half Dome and El Capitan, and the verdant coniferous forests, attract millions of visitors annually to Yosemite National Park in the Sierra Nevada.

The Great Basin dominates most of Nevada's topography containing large open basins, punctuated by eroded features such as *buttes* and *mesas*. River flow tends to be seasonal, dependent upon spring showers and winter snow melt.

Sierra Nevada

USING THE LAND

CALIFORNIA is the leading agricultural producer in the US, although low rainfall makes irrigation essential. The long growing season and abundant sunshine allow many crops to be grown in the fertile Central Valley including grapes, citrus fruits, vegetables, and cotton. Almost 17 million acres (6.8 million hectares) of California's forests are used commercially. Nevada's arid climate and poor soil are largely unsuitable for agriculture; 85% of its land is state owned and large areas are used for underground testing of nuclear weapons.

Land use and agricultural distribution

- cattle
- citrus fruits
- fruit
- irrigation
- timber
- vineyards
- major towns
- pasture
- cropland
- forest
- desert

Wheeler Peak is home to some of the world's oldest trees, bristlecone pines, which live for up to 5,000 years.

When the Hoover Dam across the Colorado River was completed in 1936, it created Lake Mead, one of the largest artificial lakes in the world, extending for 115 miles (285 km) upstream.

The San Andreas Fault is a transverse fault which extends for 650 miles (1,050 km) through California. Major earthquakes occur when the land either side of the fault moves at different rates. San Francisco was devastated by an earthquake in 1906.

Named by migrating settlers in 1849, Death Valley is the driest, hottest place in North America, as well as being the lowest point on land in the western hemisphere, at 282 ft (86 m) below sea level.

Death Valley

The sparsely populated Mojave Desert receives less than 8 inches (200 mm) of rainfall a year. It is used extensively for testing weapons and other military purposes.

The Salton Sea was created accidentally between 1905 and 1907 when an irrigation channel from the Colorado River broke out of its banks and formed this salty 300 sq mile (777 sq km), landlocked lake.

Amargosa Desert

The Sierra Nevada create a "rainshadow," preventing rain from reaching most of Nevada. Pacific air masses, passing over the mountains, are stripped of their moisture.

Without considerable irrigation, this fertile valley at Palm Springs would still be part of the Sonoran Desert. California's farmers account for about 80% of the state's total water usage.

THE URBAN/RURAL POPULATION DIVIDE

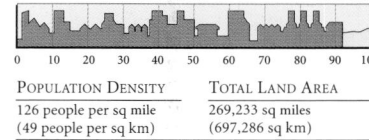

urban 92% | rural 8%

0 10 20 30 40 50 60 70 80 90 100

POPULATION DENSITY	TOTAL LAND AREA
126 people per sq mile (49 people per sq km)	269,233 sq miles (697,286 sq km)

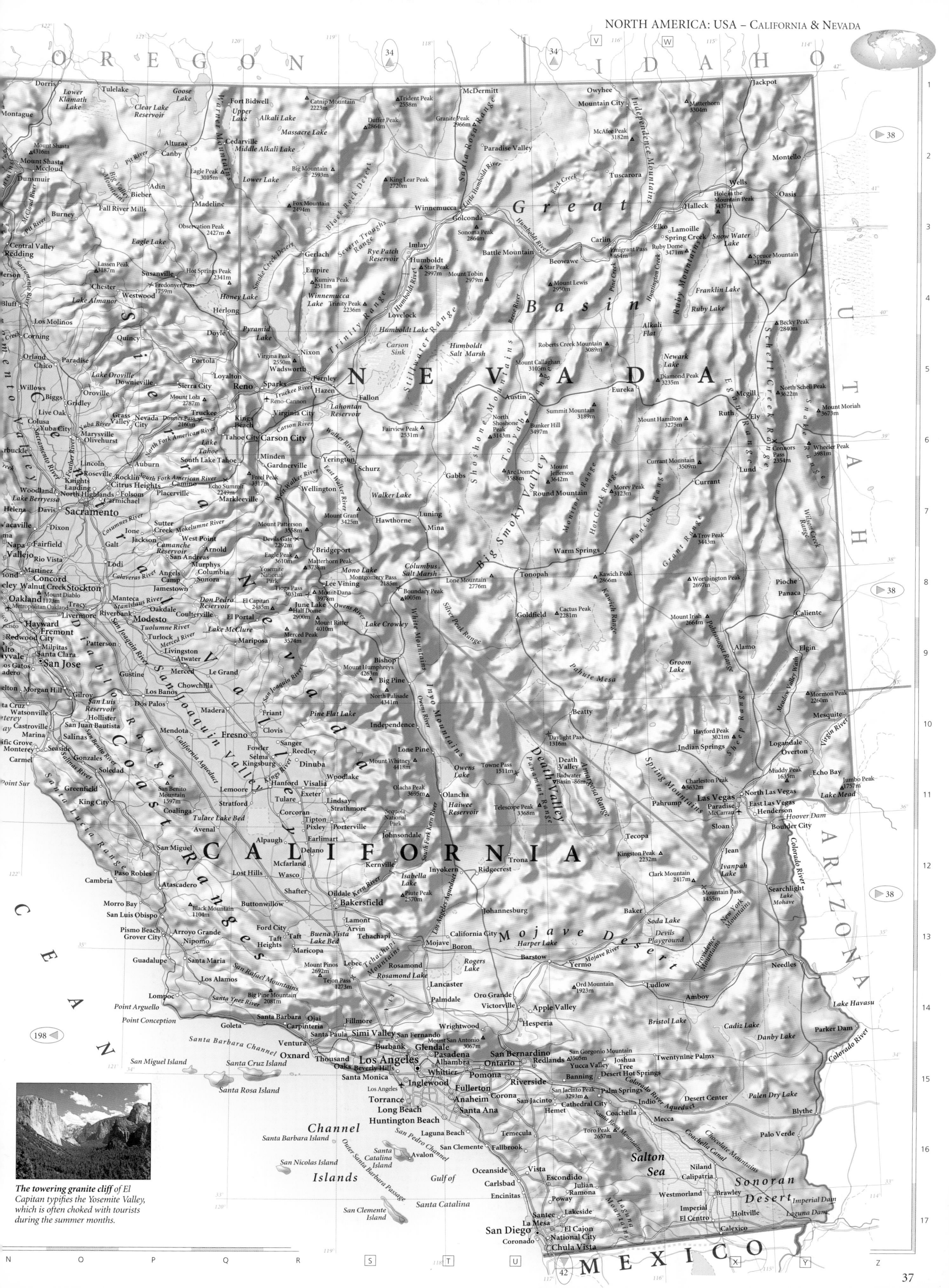

The towering granite cliff of El Capitan typifies the Yosemite Valley, which is often choked with tourists during the summer months.

USA: SOUTH MOUNTAIN STATES

Arizona, Colorado, New Mexico, Utah

THIS ARID REGION, CHARACTERIZED BY EXPANSIVE PLATEAUS and spectacular canyons is home to several distinct peoples. The ruins of cliff dwellings built a thousand years ago by the Anasazi people still exist today, and native Americans own one-third of the land in Arizona. Spanish and Mexican conquest and settlement left a Hispanic presence which is strongest in New Mexico. The Mormons, who came to the Great Salt Lake seeking religious freedom in 1847, were among the earliest Anglo-American settlers and now make up over 70% of Utah's population. The region's mineral wealth drove rapid development in the 20th century, yet the constraints of a fragile environment, including widespread water shortages, may limit prospects for growth.

When water evaporates it leaves a salt pan

Mudflats

Water level of lake varies according to quantity of run-off received from snow melt

Lake is fed by seasonal snow melt

The Great Salt Lake is an ephemeral lake; it can remain dry for extended periods, leaving a pan of evaporated mineral salts in its center.

THE LANDSCAPE

THE ARID, ROCKY EXPANSE of the Colorado Plateau is dissected by immense canyons of the Colorado River. Desert lies to the north and south and branches of the Rocky Mountains run east and west. The Great Salt Lake and Desert lie within the Great Basin, a barren region of parallel mountain ranges that extends into Arizona.

Over 13 million years of weathering has created thousands of spires and pinnacles from the alternating rock strata of Bryce Canyon.

Lake Powell

The Rio Grande has its source in several meltwater streams, which have cut deep valleys into the platform of the San Juan mountains.

Sand dunes, 600 ft (180 m) high, have been deposited in San Luis Valley, by winds funneled through the San Juan and Sangre de Cristo mountains in the Rockies.

The parallel basins and ridges, which run north–south along the Great Basin, reflect a major series of block-faults in the underlying bedrock.

Parts of the Grand Canyon, which cuts through the Colorado Plateau, are 16 miles (25 km) wide. The Colorado River has cut down 6262 ft (2000 m), exposing rock strata more than 2 billion years old.

Rainbow Bridge is the world's largest natural arch. The 309 ft (94 m) span probably began to grow when the sandstone spur of a meandering creek was breached during a flash flood.

The striking colour effects seen in the Painted Desert come from minerals such as gypsum and haematite, combined with ambient heat and dust.

Petrified Forest

In the arid landscape of Petrified Forest National Park in Arizona, the grain of prehistoric trees has been preserved as a fossil imprint in the rocks. The bog-preserved trees were gradually turned to stone by seeping mineral-rich water.

Shifting gypsum sands produce a constantly changing land surface, overwhelming plants and any other obstacles in Tularosa Valley.

Carlsbad Caverns

The intricate stalactites of Carlsbad Caverns have grown with the seepage of calcium-rich water over the last 100,000 years. The huge caves are home to around 100,000 Mexican freetail bats.

TRANSPORTATION & INDUSTRY

NEW INDUSTRIES HAVE HELPED reduce the region's dependence on the extraction of minerals and fossil fuels. Precision manufacture has grown rapidly, particularly in Arizona and Colorado. Salt Lake City and Denver are well-established financial centers and New Mexico, the main US producer of uranium, is a prominent region for nuclear research. Colorado is the most important US center for winter sports.

TRANSPORTATION NETWORK

232,434 miles (373,986 km)		4,059 miles (6,515 km)
8,627 miles (13,881 km)		none

The Colorado Rockies are crossed by 32 mountain passes, some as high as 12,183 ft (3,713 m). The Eisenhower Tunnel west of Denver carries Interstate Highway 70 straight through the Continental Divide.

Major industry and infrastructure

- chemicals
- coal
- defense
- finance
- food processing
- hi-tech industry
- oil & gas
- mining
- research & development
- winter sports
- major towns
- international airports
- major roads
- major industrial areas

Glen Canyon Dam on the Colorado river was completed in 1964. it provides hydroelectric power and irrigation water as part of a long-term federal project to harness the river.

The flat tablelands (mesas), and the isolated pinnacles (buttes) which rise from the floor of Monument Valley are the resistant remnants of an earlier land surface, gradually cut back by erosion under arid conditions.

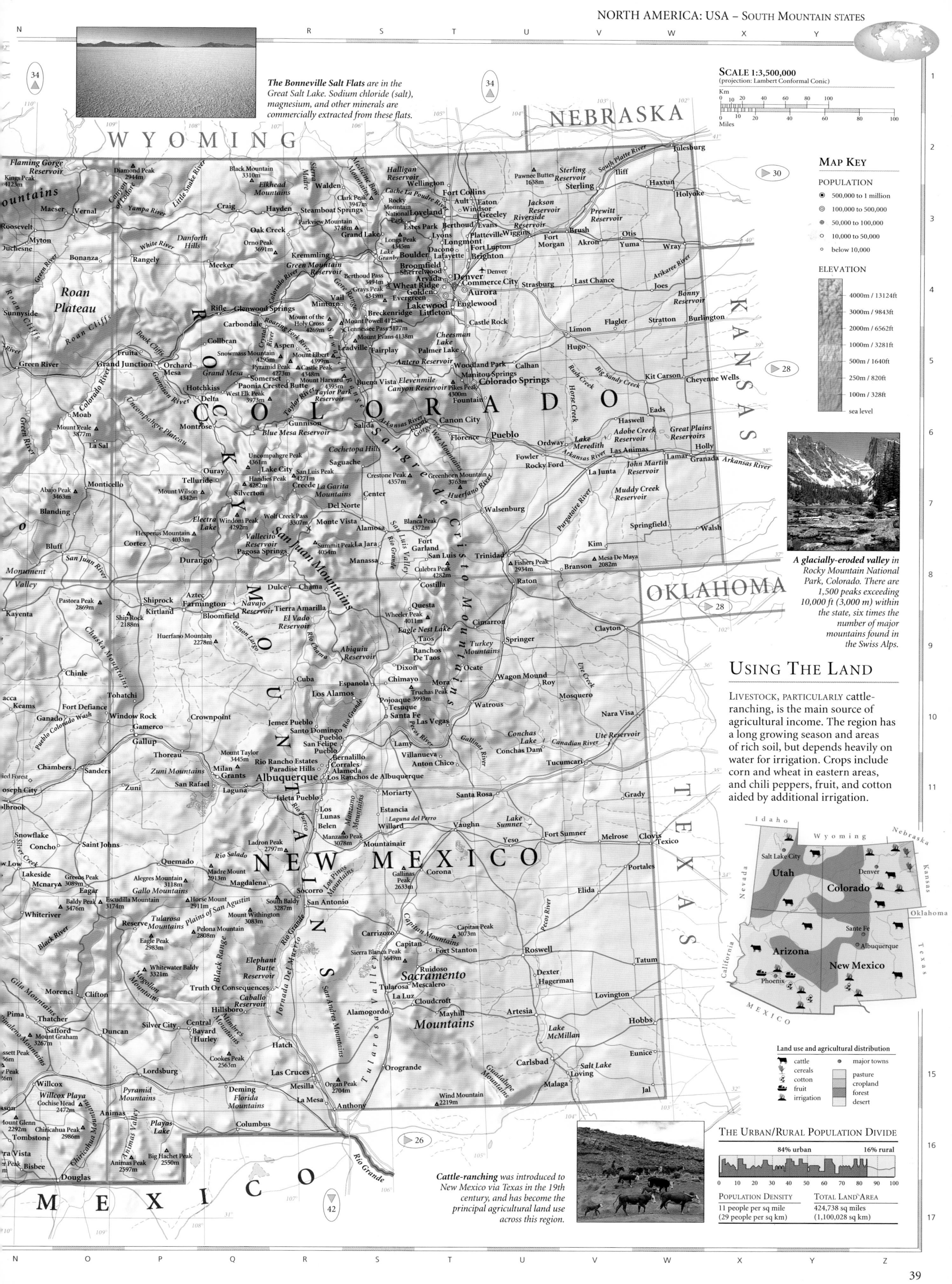

The Bonneville Salt Flats are in the Great Salt Lake. Sodium chloride (salt), magnesium, and other minerals are commercially extracted from these flats.

SCALE 1:3,500,000
(projection: Lambert Conformal Conic)

MAP KEY

POPULATION

- 500,000 to 1 million
- 100,000 to 500,000
- 50,000 to 100,000
- 10,000 to 50,000
- below 10,000

ELEVATION

- 4000m / 13124ft
- 3000m / 9843ft
- 2000m / 6562ft
- 1000m / 3281ft
- 500m / 1640ft
- 250m / 820ft
- 100m / 328ft
- sea level

A glacially-eroded valley in Rocky Mountain National Park, Colorado. There are 1,500 peaks exceeding 10,000 ft (3,000 m) within the state, six times the number of major mountains found in the Swiss Alps.

USING THE LAND

LIVESTOCK, PARTICULARLY cattle-ranching, is the main source of agricultural income. The region has a long growing season and areas of rich soil, but depends heavily on water for irrigation. Crops include corn and wheat in eastern areas, and chili peppers, fruit, and cotton aided by additional irrigation.

Land use and agricultural distribution

- cattle
- cereals
- cotton
- fruit
- irrigation
- major towns
- pasture
- cropland
- forest
- desert

Cattle-ranching was introduced to New Mexico via Texas in the 19th century, and has become the principal agricultural land use across this region.

THE URBAN/RURAL POPULATION DIVIDE

84% urban   16% rural

POPULATION DENSITY	TOTAL LAND AREA
11 people per sq mile (29 people per sq km)	424,738 sq miles (1,100,028 sq km)

USA: HAWAII

THE 122 ISLANDS of the Hawaiian archipelago – which are part of Polynesia – are the peaks of the world's largest volcanoes. They rise approximately 6 miles (9.7 km) from the floor of the Pacific Ocean. The largest, the island of Hawaii, remains highly active. Hawaii became the US's 50th state in 1959. A tradition of receiving immigrant workers is reflected in the islands' ethnic diversity, with peoples drawn from around the rim of the Pacific. Only 2% of the current population are native Polynesians.

The island of Molokai is formed from volcanic rock. Mature sand dunes cover the rocks in coastal areas.

TRANSPORTATION & INDUSTRY

TOURISM DOMINATES the economy, with over half of the population employed in services. The naval base at Pearl Harbor is also a major source of employment. Industry is concentrated on the island of Oahu and relies mostly on imported materials, while agricultural produce is processed locally.

Major industry and infrastructure
- food processing
- military base
- textiles
- tourism
- major towns
- international airports
- major roads
- major industrial areas

TRANSPORTATION NETWORK

4,102 miles (6,600 km)	43 miles (69 km)
none	none

Hawaii relies on ocean-surface transportation. Honolulu is the main focus of this network, bringing foreign trade and the markets of mainland US to Hawaii's outer islands.

Haleakala's extinct volcanic crater is the world's largest. The giant caldera, containing many secondary cones, is 2,000 ft (600 m) deep and 20 miles (32 km) in circumference.

SCALE 1:3,500,000
(projection: Lambert Conformal Conic)

MAP KEY

POPULATION
- ◎ 100,000 to 500,000
- ⊕ 50,000 to 100,000
- ⊙ 10,000 to 50,000
- ○ below 10,000

ELEVATION
- 4000m / 13,124ft
- 3000m / 9843ft
- 2000m / 6562ft
- 1000m / 3281ft
- 500m / 1640ft
- 250m / 820ft
- 100m / 328ft
- sea level

USING THE LAND AND SEA

THE VOLCANIC SOILS are extremely fertile and the climate hot and humid on the lower slopes, supporting large commercial plantations growing sugar cane, bananas, pineapples, and other tropical fruit, as well as nursery plants and flowers. Some land is given to pasture, particularly for beef and dairy cattle.

Land use and agricultural distribution
- cattle
- fishing
- fruit
- sugar cane
- major towns

- pasture
- cropland
- forest
- mountain region

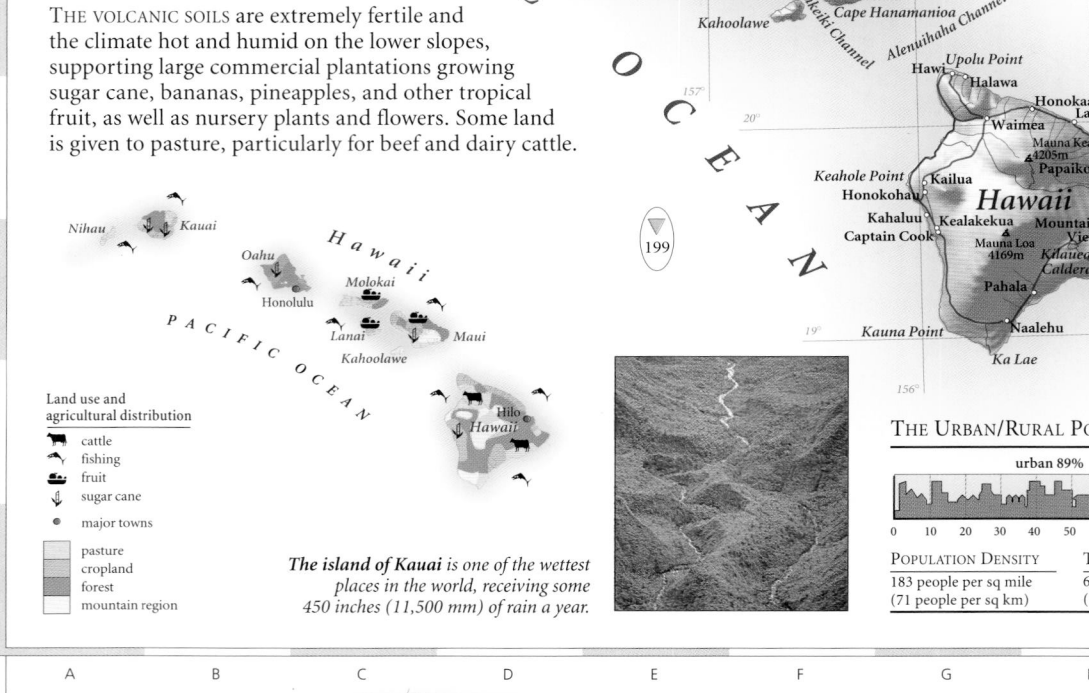

The island of Kauai is one of the wettest places in the world, receiving some 450 inches (11,500 mm) of rain a year.

THE URBAN/RURAL POPULATION DIVIDE

urban 89% rural 11%

0 10 20 30 40 50 60 70 80 90 100

POPULATION DENSITY	TOTAL LAND AREA
183 people per sq mile (71 people per sq km)	6,423 sq miles (16,636 sq km)

USING THE LAND AND SEA

THE ICE-FREE COASTLINE of Alaska provides access to salmon fisheries and more than 5.5 million acres (2.2 million ha) of forest. Most of Alaska is uncultivable, and around 90% of food is imported. Barley, hay, and hothouse products are grown around Anchorage, where dairy farming is also concentrated.

THE URBAN/RURAL POPULATION DIVIDE

urban 68% rural 32%

0 10 20 30 40 50 60 70 80 90 100

POPULATION DENSITY	TOTAL LAND AREA
1 person per sq mile (0.4 people per sq km)	586,412 sq miles (1,518,800 sq km)

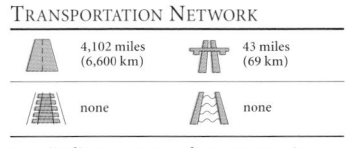

A raft of timber from the Tongass forest is hauled by a tug, bound for the pulp mills of the Alaskan coast between Juneau and Ketchikan.

MAP KEY

POPULATION
- ◎ 100,000 to 500,000
- ⊕ 50,000 to 100,000
- ⊙ 10,000 to 50,000
- ○ below 10,000

ELEVATION
- 4000m / 13,124ft
- 3000m / 9843ft
- 2000m / 6562ft
- 1000m / 3281ft
- 500m / 1640ft
- 250m / 820ft
- 100m / 328ft
- sea level

SCALE 1:8,000,000
(projection: Lambert Conformal Conic)

USA: ALASKA

JUST OVER HALF A MILLION people live in Alaska, a wilderness of ice, forest, mountains, and plains, purchased from Russia in 1867 and twice the size of Texas. The discovery of large oil reserves has brought prosperity to the US's "last frontier," while advancing the need to preserve natural habitats and the traditional livelihoods of indigenous peoples, such as the Aleuts and Inupiaq.

Land use and agricultural distribution
- fishing
- reindeer
- fruit
- major towns
- forest
- barren
- tundra

THE LANDSCAPE

THE MOUNTAINS OF THE PACIFIC COAST culminate in the heavily glaciated Alaska Range and extend west, to the Alaska Peninsula and the great volcanic arc of the Aleutian Islands. The interior plains are drained by the Yukon River and bounded by the bare, jagged peaks of the Brooks Range to the north.

The Yukon Delta is a fan of alluvial material eroded by the Yukon River and its tributaries. It is approximately twice the size of the Mississippi Delta.

Brooks Range

West Fork Glacier

The ten highest mountains in the US are all in the Alaska Range, Mount McKinley (Denali), at 20,321 ft (6,194 m) is the highest.

Yukon River

Alaska Range

The arc of the Aleutian Islands marks the boundary between the Eurasian and Pacific tectonic plates.

Fjords are found along the coast where valleys, deeply excavated by large glaciers, were inundated by rising seas.

By August, the Alaska Range is covered with autumnal tundra vegetation.

West Fork Glacier

The surging ice mass shears along the glacier margin

Deep crevasses divide the front of the surging glacier into large ice blocks

Surging glaciers make rapid and dramatic advances, normally after periods of snow accumulation. West Fork Glacier in the Susitna River Basin traveled 2.5 miles (4 km) in 1987.

TRANSPORTATION & INDUSTRY

LARGE AREAS OF ALASKA are undeveloped, and much of the existing infrastructure is a legacy of Cold War military investment. Mineral ores, including gold, have been mined for over a century, but the oil business now dominates the economy. Processing industries such as paper-pulp mills supply Japan and other markets on the Pacific Rim.

TRANSPORTATION NETWORK

- 13,524 miles (21,760 km)
- 49 miles (78 km)
- 482 miles (772 km)
- none

Nearly 80 million gallons of oil are pumped through the Trans-Alaska Pipeline every day. The oil takes six days to travel the 789 miles (1,262 km) from Prudhoe Bay to Valdez.

Major industry and infrastructure
- fish processing
- gold mining
- oil
- timber processing
- major towns
- international airports
- major roads

The Trans-Alaska Pipeline has carried crude oil from Prudhoe Bay since 1977. The oilfield is the US's largest and is estimated to be equal in size to the biggest oilfields of the Persian Gulf.

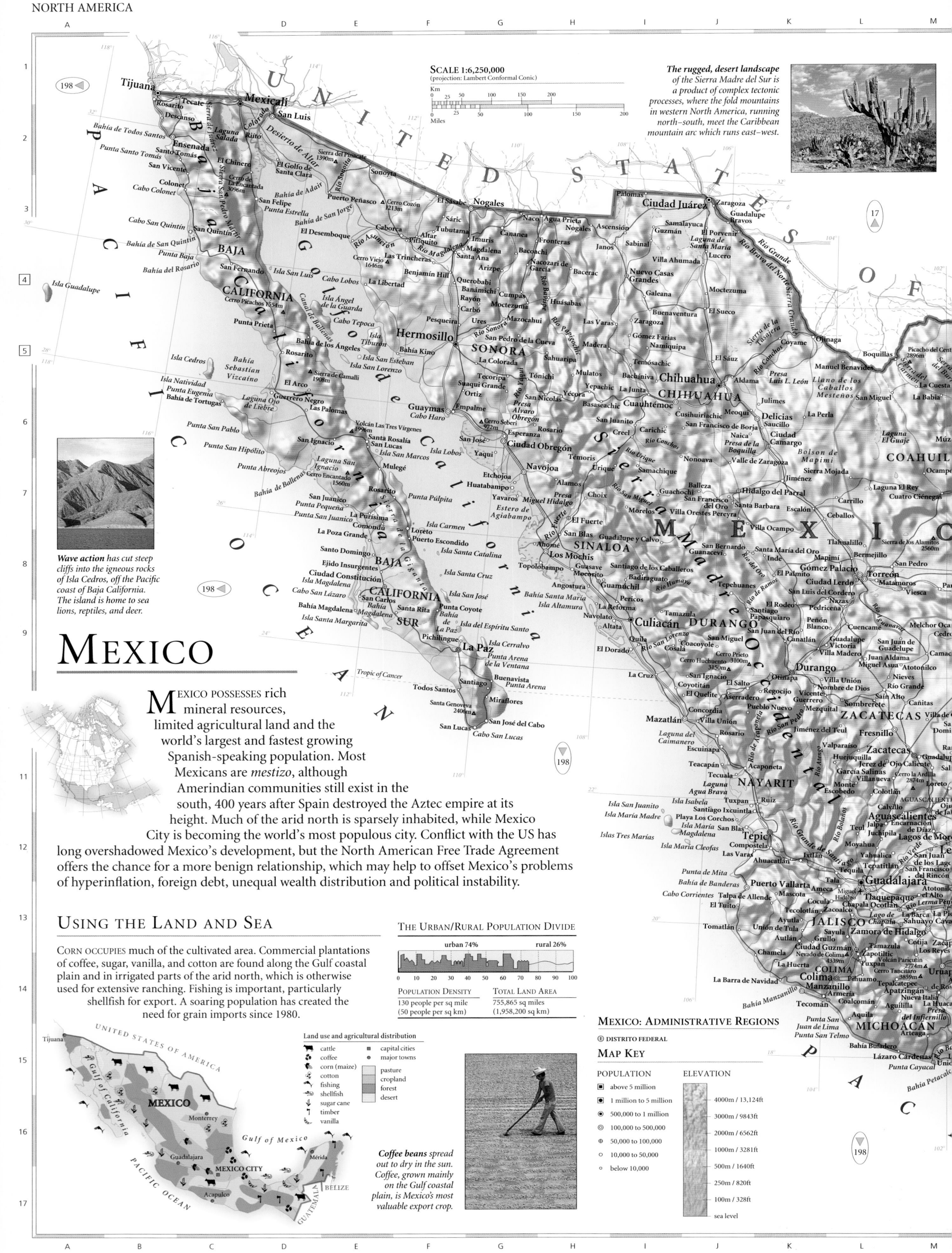

SCALE 1:6,250,000
(projection: Lambert Conformal Conic)

The rugged, desert landscape of the Sierra Madre del Sur is a product of complex tectonic processes, where the fold mountains in western North America, running north–south, meet the Caribbean mountain arc which runs east–west.

Wave action has cut steep cliffs into the igneous rocks of Isla Cedros, off the Pacific coast of Baja California. The island is home to sea lions, reptiles, and deer.

MEXICO

MEXICO POSSESSES rich mineral resources, limited agricultural land and the world's largest and fastest growing Spanish-speaking population. Most Mexicans are *mestizo*, although Amerindian communities still exist in the south, 400 years after Spain destroyed the Aztec empire at its height. Much of the arid north is sparsely inhabited, while Mexico City is becoming the world's most populous city. Conflict with the US has long overshadowed Mexico's development, but the North American Free Trade Agreement offers the chance for a more benign relationship, which may help to offset Mexico's problems of hyperinflation, foreign debt, unequal wealth distribution and political instability.

USING THE LAND AND SEA

CORN OCCUPIES much of the cultivated area. Commercial plantations of coffee, sugar, vanilla, and cotton are found along the Gulf coastal plain and in irrigated parts of the arid north, which is otherwise used for extensive ranching. Fishing is important, particularly shellfish for export. A soaring population has created the need for grain imports since 1980.

THE URBAN/RURAL POPULATION DIVIDE

urban 74%   rural 26%

0 10 20 30 40 50 60 70 80 90 100

POPULATION DENSITY	TOTAL LAND AREA
130 people per sq mile (50 people per sq km)	755,865 sq miles (1,958,200 sq km)

Land use and agricultural distribution

- cattle
- coffee
- corn (maize)
- cotton
- fishing
- shellfish
- sugar cane
- timber
- vanilla

- capital cities
- major towns

- pasture
- cropland
- forest
- desert

Coffee beans spread out to dry in the sun. Coffee, grown mainly on the Gulf coastal plain, is Mexico's most valuable export crop.

MEXICO: ADMINISTRATIVE REGIONS

ⓓ DISTRITO FEDERAL

MAP KEY

POPULATION	ELEVATION
▪ above 5 million	4000m / 13,124ft
▪ 1 million to 5 million	3000m / 9843ft
◉ 500,000 to 1 million	2000m / 6562ft
◎ 100,000 to 500,000	1000m / 3281ft
⊕ 50,000 to 100,000	500m / 1640ft
○ 10,000 to 50,000	250m / 820ft
∘ below 10,000	100m / 328ft
	sea level

N O P Q R S T U V W X Y

THE LANDSCAPE

THE GREAT CENTRAL PLATEAU rises gently southward from the Rio Grande, isolated from the coastal plains by the Sierra Madre Oriental and Occidental. The two ranges converge from east and west respectively, culminating in high volcanic peaks around Mexico City. Further ranges of the Sierra Madre rise to the south of the Balsas Basin, skirted by the low-lying Isthmus of Tehuantepec (*Istmo de Tehuantepec*) and Yucatan Peninsula.

The long, narrow, extremely arid peninsula of Baja (lower) California is an elongated granite block, separated from the mainland by the flooded rift valley of the Gulf of California (*Golfo de California*).

Wave action has constructed sand bars which shelter lagoons along the shore of the Gulf coastal plain.

Sierra Madre Oriental

Rio Grande

The dormant cone of Volcán Pico de Orizaba is, at 18,700 ft (5,700 m), the highest peak in Mexico. In North America, only Mount McKinley and Mount Logan are taller.

Tropical rain forest abounds in the Yucatan Peninsula, a broad, low limestone shelf. Rivers are rare due to the porous nature of limestone, so the forest is mostly fed by streams and underground water.

The heavily-forested Isthmus of Tehuantepec (*Istmo de Tehuantepec*) is a *graben*; a low-lying trough created by downward movement of the bedrock between two fault lines.

Formation of the Gulf of California

Direction of plate movement
Baja California
Gulf of California
Transform fault
Spreading oceanic ridge
Edge of continental crust

The Gulf of California (Golfo de California) began to open out about 4 million years ago as a result of rifting and plate displacement along transform faults.

Sierra Madre Occidental

Popocatépetl is a dormant volcano, part of the Pacific "Rim of Fire." The crater is over half a mile (1 km) wide.

Río Balsas

Popocatépetl

The unstable, earthquake-prone, upland basin around Mexico City was once a region of shallow lakes. Flood control measures and domestic consumption over the last four centuries have caused the virtual disappearance of this surface water.

The highlands of Chiapas are a series of *horsts*, blocks of land thrust upward between two fault lines. Volcanic cones have developed where lava has flowed out from the faults.

TRANSPORTATION & INDUSTRY

OIL AND GAS ON THE GULF COAST are Mexico's main sources of export income. Metal mining has declined but the country remains a leading global producer of silver. Manufacturing is heavily concentrated around the metropolitan area of Mexico City, while the duty-free movement of goods in the US border region, under the *Maquiladora* (twin plant) scheme, has created new hi-tech and service growth centers.

Major industry and infrastructure

- brewing
- car manufacture
- chemicals
- electronics
- fish processing
- maquiladoras
- mining
- oil & gas
- textiles
- capital cities
- major towns
- international airports
- major roads
- major industrial areas

TRANSPORTATION NETWORK

55,021 miles (88,601 km)

4,186 miles (6,740 km)

16,422 miles (26,445 km)

1,801 miles (2,900 km)

Fast, modern highways or *autopistas* now link Mexico City with Toluca, Puebla and other satellite cities, yet distant centers like Chihuahua are still served by narrow roads and an outdated railroad network.

A stone figure reclines by the Temple of Warriors, within the Mayan city of Chichén-Itzá. The Maya civilization flourished across the Yucatan Peninsula between 200 and 900 ad.

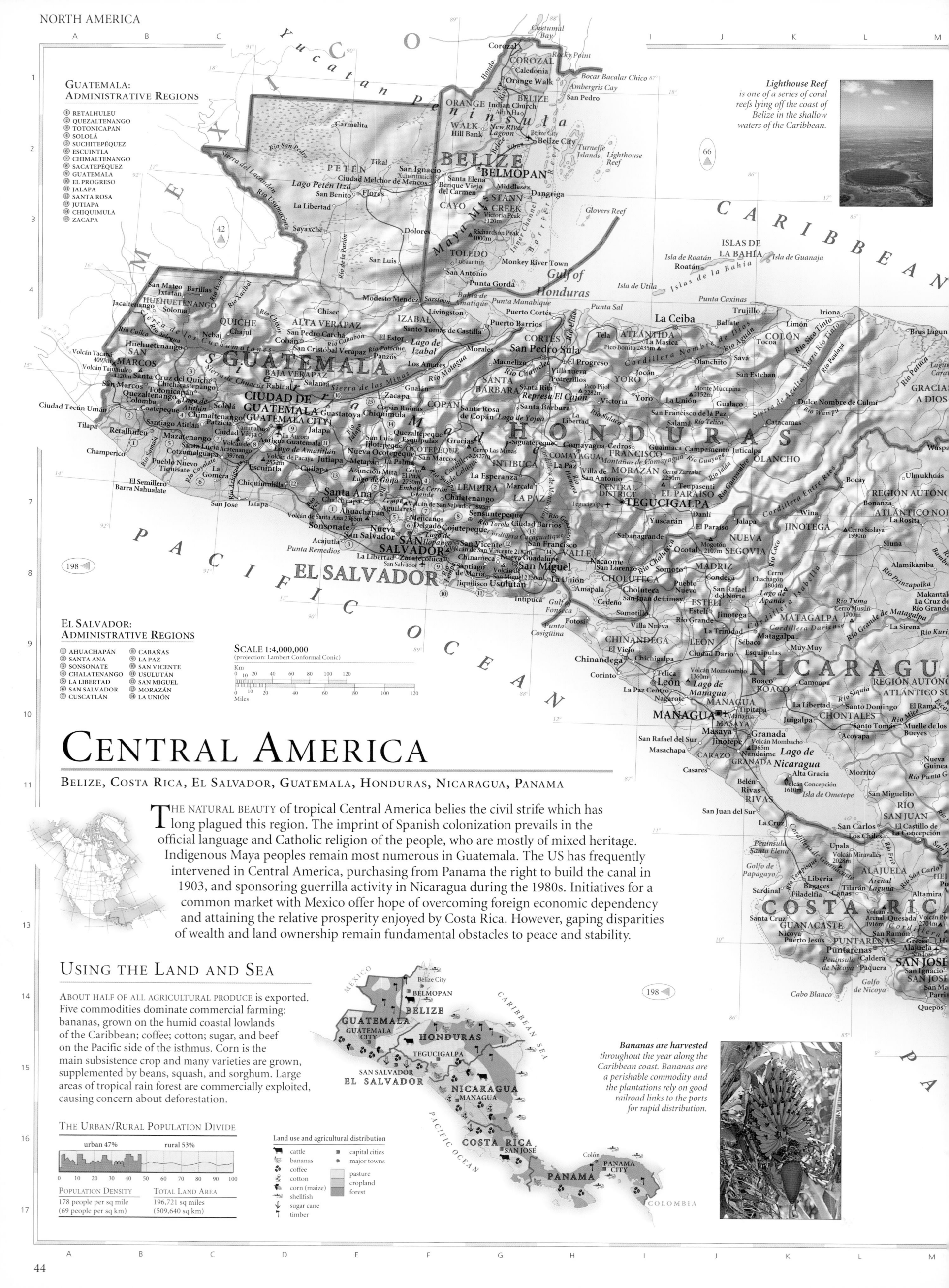

GUATEMALA: ADMINISTRATIVE REGIONS

① RETALHULEU
② QUEZALTENANGO
③ TOTONICAPÁN
④ SOLOLÁ
⑤ SUCHITEPÉQUEZ
⑥ ESCUINTLA
⑦ CHIMALTENANGO
⑧ SACATEPÉQUEZ
⑨ GUATEMALA
⑩ EL PROGRESO
⑪ JALAPA
⑫ SANTA ROSA
⑬ JUTIAPA
⑭ CHIQUIMULA
⑮ ZACAPA

Lighthouse Reef is one of a series of coral reefs lying off the coast of Belize in the shallow waters of the Caribbean.

EL SALVADOR: ADMINISTRATIVE REGIONS

① AHUACHAPÁN
② SANTA ANA
③ SONSONATE
④ CHALATENANGO
⑤ LA LIBERTAD
⑥ SAN SALVADOR
⑦ CUSCATLÁN
⑧ CABAÑAS
⑨ LA PAZ
⑩ SAN VICENTE
⑪ USULUTÁN
⑫ SAN MIGUEL
⑬ MORAZÁN
⑭ LA UNIÓN

SCALE 1:4,000,000
(projection: Lambert Conformal Conic)

Km
0 10 20   40   60   80   100 120

Miles
0 10   20   40   60   80   100   120

CENTRAL AMERICA

BELIZE, COSTA RICA, EL SALVADOR, GUATEMALA, HONDURAS, NICARAGUA, PANAMA

THE NATURAL BEAUTY of tropical Central America belies the civil strife which has long plagued this region. The imprint of Spanish colonization prevails in the official language and Catholic religion of the people, who are mostly of mixed heritage. Indigenous Maya peoples remain most numerous in Guatemala. The US has frequently intervened in Central America, purchasing from Panama the right to build the canal in 1903, and sponsoring guerrilla activity in Nicaragua during the 1980s. Initiatives for a common market with Mexico offer hope of overcoming foreign economic dependency and attaining the relative prosperity enjoyed by Costa Rica. However, gaping disparities of wealth and land ownership remain fundamental obstacles to peace and stability.

USING THE LAND AND SEA

ABOUT HALF OF ALL AGRICULTURAL PRODUCE is exported. Five commodities dominate commercial farming: bananas, grown on the humid coastal lowlands of the Caribbean; coffee; cotton; sugar, and beef on the Pacific side of the isthmus. Corn is the main subsistence crop and many varieties are grown, supplemented by beans, squash, and sorghum. Large areas of tropical rain forest are commercially exploited, causing concern about deforestation.

Bananas are harvested throughout the year along the Caribbean coast. Bananas are a perishable commodity and the plantations rely on good railroad links to the ports for rapid distribution.

THE URBAN/RURAL POPULATION DIVIDE

urban 47%   rural 53%

0 10 20 30 40 50 60 70 80 90 100

POPULATION DENSITY
178 people per sq mile
(69 people per sq km)

TOTAL LAND AREA
196,721 sq miles
(509,640 sq km)

Land use and agricultural distribution

cattle
bananas
coffee
cotton
corn (maize)
shellfish
sugar cane
timber

■ capital cities
• major towns

pasture
cropland
forest

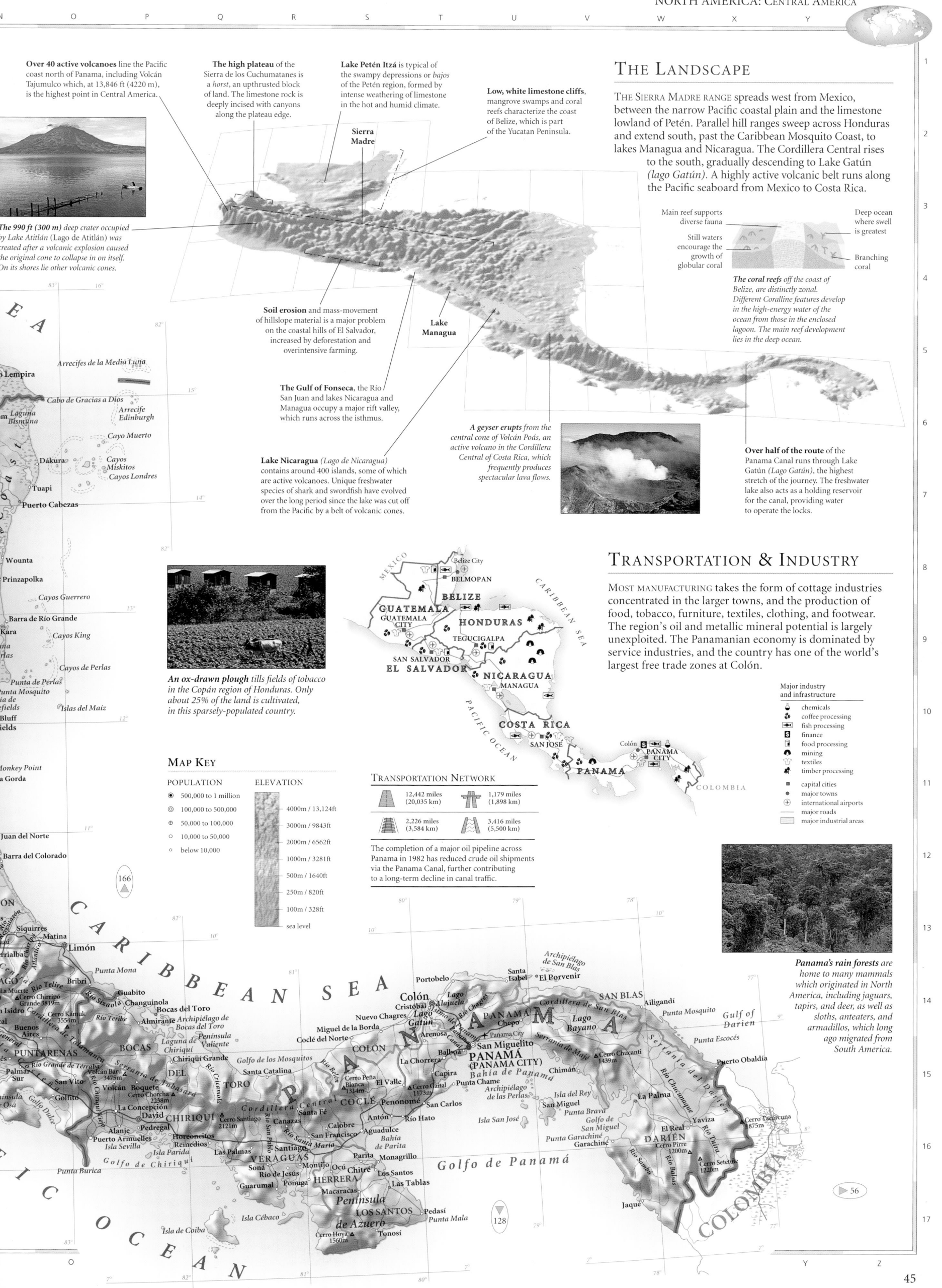

Over 40 active volcanoes line the Pacific coast north of Panama, including Volcán Tajumulco which, at 13,846 ft (4220 m), is the highest point in Central America.

The high plateau of the Sierra de los Cuchumatanes is a *horst*, an unthrusted block of land. The limestone rock is deeply incised with canyons along the plateau edge.

Lake Petén Itzá is typical of the swampy depressions or *bajos* of the Petén region, formed by intense weathering of limestone in the hot and humid climate.

Low, white limestone cliffs, mangrove swamps and coral reefs characterize the coast of Belize, which is part of the Yucatan Peninsula.

Sierra Madre

The 990 ft (300 m) deep crater occupied by Lake Atitlán (Lago de Atitlán) was created after a volcanic explosion caused the original cone to collapse in on itself. On its shores lie other volcanic cones.

Soil erosion and mass-movement of hillslope material is a major problem on the coastal hills of El Salvador, increased by deforestation and overintensive farming.

Lake Managua

The Gulf of Fonseca, the Río San Juan and lakes Nicaragua and Managua occupy a major rift valley, which runs across the isthmus.

Lake Nicaragua (*Lago de Nicaragua*) contains around 400 islands, some of which are active volcanoes. Unique freshwater species of shark and swordfish have evolved over the long period since the lake was cut off from the Pacific by a belt of volcanic cones.

A geyser erupts from the central cone of Volcán Poás, an active volcano in the Cordillera Central of Costa Rica, which frequently produces spectacular lava flows.

THE LANDSCAPE

THE SIERRA MADRE RANGE spreads west from Mexico, between the narrow Pacific coastal plain and the limestone lowland of Petén. Parallel hill ranges sweep across Honduras and extend south, past the Caribbean Mosquito Coast, to lakes Managua and Nicaragua. The Cordillera Central rises to the south, gradually descending to Lake Gatún (*lago Gatún*). A highly active volcanic belt runs along the Pacific seaboard from Mexico to Costa Rica.

Main reef supports diverse fauna

Deep ocean where swell is greatest

Still waters encourage the growth of globular coral

Branching coral

The coral reefs off the coast of Belize, are distinctly zonal. Different Coralline features develop in the high-energy water of the ocean from those in the enclosed lagoon. The main reef development lies in the deep ocean.

Over half of the route of the Panama Canal runs through Lake Gatún (*Lago Gatún*), the highest stretch of the journey. The freshwater lake also acts as a holding reservoir for the canal, providing water to operate the locks.

TRANSPORTATION & INDUSTRY

MOST MANUFACTURING takes the form of cottage industries concentrated in the larger towns, and the production of food, tobacco, furniture, textiles, clothing, and footwear. The region's oil and metallic mineral potential is largely unexploited. The Panamanian economy is dominated by service industries, and the country has one of the world's largest free trade zones at Colón.

An ox-drawn plough tills fields of tobacco in the Copán region of Honduras. Only about 25% of the land is cultivated, in this sparsely-populated country.

Major industry and infrastructure

- chemicals
- coffee processing
- fish processing
- finance
- food processing
- mining
- textiles
- timber processing

- ▪ capital cities
- ● major towns
- + international airports
- — major roads
- ▢ major industrial areas

MAP KEY

POPULATION
- ◉ 500,000 to 1 million
- ◎ 100,000 to 500,000
- ⊕ 50,000 to 100,000
- ⊙ 10,000 to 50,000
- ○ below 10,000

ELEVATION
- 4000m / 13,124ft
- 3000m / 9843ft
- 2000m / 6562ft
- 1000m / 3281ft
- 500m / 1640ft
- 250m / 820ft
- 100m / 328ft
- sea level

TRANSPORTATION NETWORK

12,442 miles (20,035 km)	1,179 miles (1,898 km)
2,226 miles (3,584 km)	3,416 miles (5,500 km)

The completion of a major oil pipeline across Panama in 1982 has reduced crude oil shipments via the Panama Canal, further contributing to a long-term decline in canal traffic.

Panama's rain forests are home to many mammals which originated in North America, including jaguars, tapirs, and deer, as well as sloths, anteaters, and armadillos, which long ago migrated from South America.

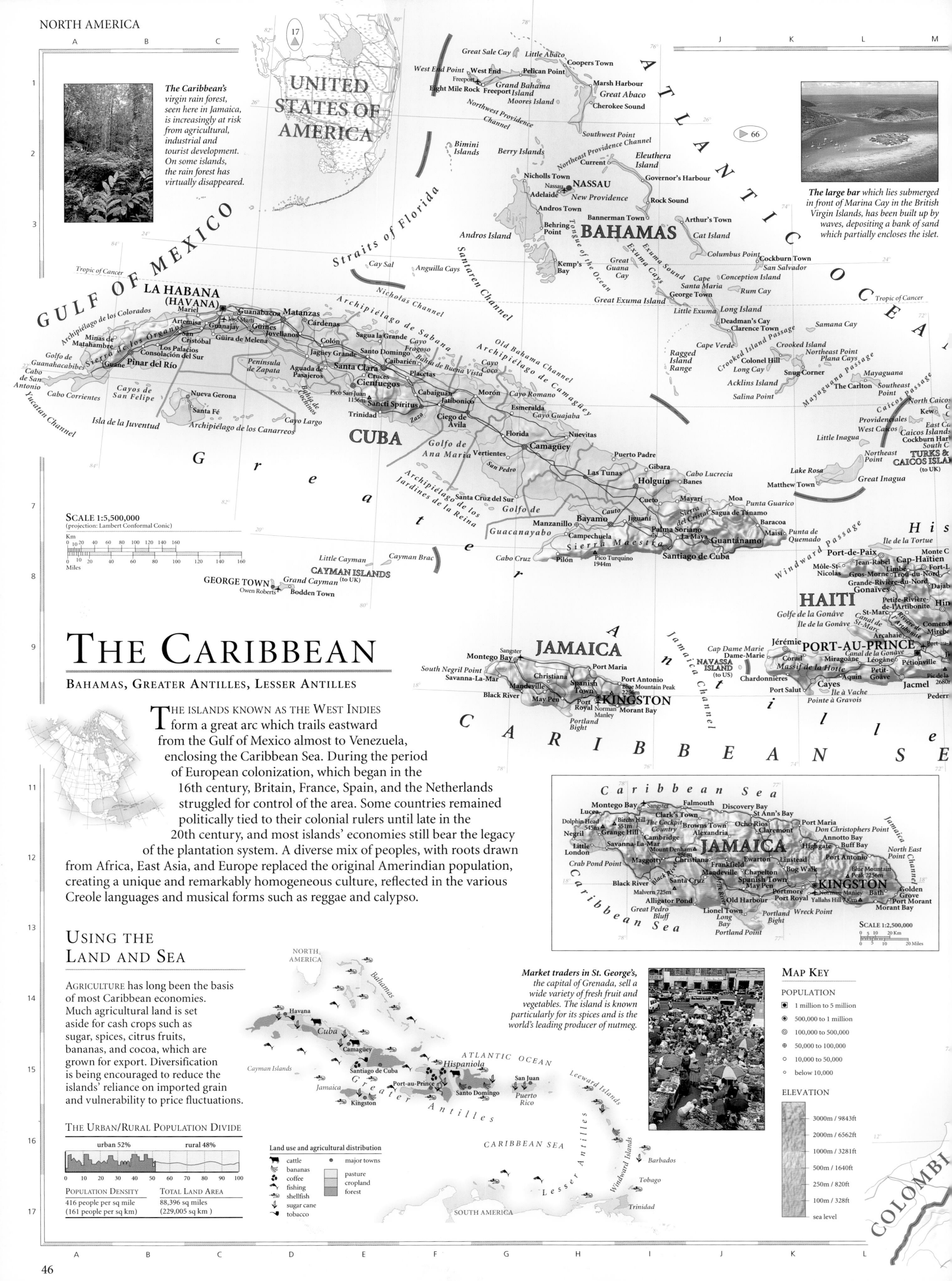

The Caribbean's virgin rain forest, seen here in Jamaica, is increasingly at risk from agricultural, industrial and tourist development. On some islands, the rain forest has virtually disappeared.

The large bar which lies submerged in front of Marina Cay in the British Virgin Islands, has been built up by waves, depositing a bank of sand which partially encloses the islet.

THE CARIBBEAN

BAHAMAS, GREATER ANTILLES, LESSER ANTILLES

THE ISLANDS KNOWN AS THE WEST INDIES form a great arc which trails eastward from the Gulf of Mexico almost to Venezuela, enclosing the Caribbean Sea. During the period of European colonization, which began in the 16th century, Britain, France, Spain, and the Netherlands struggled for control of the area. Some countries remained politically tied to their colonial rulers until late in the 20th century, and most islands' economies still bear the legacy of the plantation system. A diverse mix of peoples, with roots drawn from Africa, East Asia, and Europe replaced the original Amerindian population, creating a unique and remarkably homogeneous culture, reflected in the various Creole languages and musical forms such as reggae and calypso.

USING THE LAND AND SEA

AGRICULTURE has long been the basis of most Caribbean economies. Much agricultural land is set aside for cash crops such as sugar, spices, citrus fruits, bananas, and cocoa, which are grown for export. Diversification is being encouraged to reduce the islands' reliance on imported grain and vulnerability to price fluctuations.

THE URBAN/RURAL POPULATION DIVIDE

urban 52%	rural 48%

POPULATION DENSITY
416 people per sq mile
(161 people per sq km)

TOTAL LAND AREA
88,396 sq miles
(229,005 sq km)

SCALE 1:5,500,000
(projection: Lambert Conformal Conic)

Market traders in St. George's, the capital of Grenada, sell a wide variety of fresh fruit and vegetables. The island is known particularly for its spices and is the world's leading producer of nutmeg.

Land use and agricultural distribution

- cattle
- bananas
- coffee
- fishing
- shellfish
- sugar cane
- tobacco
- major towns
- pasture
- cropland
- forest

MAP KEY

POPULATION
- 1 million to 5 million
- 500,000 to 1 million
- 100,000 to 500,000
- 50,000 to 100,000
- 10,000 to 50,000
- below 10,000

ELEVATION
- 3000m / 9843ft
- 2000m / 6562ft
- 1000m / 3281ft
- 500m / 1640ft
- 250m / 820ft
- 100m / 328ft
- sea level

SCALE 1:2,500,000

TRANSPORTATION & INDUSTRY

CARIBBEAN INDUSTRY remains, with few exceptions, agricultural, and export-led, or service-based, supporting the flourishing tourist industry. However, several countries including Jamaica, Barbados, Trinidad and Tobago, and Puerto Rico have developed important mineral industries, and Cuba is attempting to diversify its economy by importing capital goods to start up new manufacturing businesses.

Cruise ships, such as this one moored at Castries in St. Lucia, have become a popular way for tourists to travel round the Caribbean islands, stopping off at several islands for sightseeing and shopping.

TRANSPORTATION NETWORK

21,197 miles (34,133 km)	369 miles (627 km)
9,100 miles (14,654 km)	211 miles (340 km)

Air links are well-developed between most of the Caribbean islands. The importance of the tourist trade has recently encouraged many countries to upgrade their paved roads.

Major industry and infrastructure

- fish processing
- finance
- mining
- oil refining
- sugar refining
- tourism
- major towns
- international airports
- major roads
- major industrial areas

This rock stack on the coast of St. Martin in the Leeward Islands has been created by wave action which undercut the cliffs, forming an arch. Continued wave action weakened the arch, which eventually collapsed leaving a single tower of rock.

The Pitons in St. Lucia are two volcanic domes; the tallest is 2,620 ft (798 m) high. Their steep slopes are covered in thick forest.

PUERTO RICO (to US)

Isabela · Aguadilla · Punta Higüero · Bahía de Mayagüez · Mayagüez · San Germán · Cabo Rojo · Punta Brea · Arecibo · Laguna Tortuguero · Manatí · Vega Baja · Bayamón · Cataño · SAN JUAN · Carolina · Río Grande · Cabezas de San Juan · Fajardo · Lago Dos Bocas · Río Grande de Arecibo · Utuado · Orocovis · Sierra de Luquillo · Caguas · Adjuntas · 1338m Embalse · Cayey · Humacao · Yauco · Juana Díaz · Loa Vaca · Salinas · Guayama · Yabucoa · Ponce · Punta Petrona · 1065m Monte Pirata 301m · Sonda de Vieques · Vieques · Pasaje de Vieques · Isla de Vieques · Punta Guayanés · Culebra · Isla Puerca Culebra · Passage de Vieques

ATLANTIC OCEAN · Caribbean Sea
SCALE 1:2,500,000

GUADELOUPE (to France)

Pointe de la Grande Vigie · Anse-Bertrand · Port-Louis · Morne-à-l'Eau · le Moule · Ste-Rose · Baie-Mahault · Lamentin · Pointe-à-Pitre · Grande Terre · les Abymes · Ste-Anne · Pointe des Colibris · Pointe Noire · Basse Terre · Petit-Bourg · Petit Cul-de-Sac Marin · Soufrière 1467m · Vieux Habitants · St-Claude · Capesterre-Belle-Eau · BASSE-TERRE · Ste-François · Canal de Marie-Galante · Canal des Saintes

Caribbean Sea · ATLANTIC OCEAN
SCALE 1:2,500,000

DOMINICAN REPUBLIC

Cabo Isabela · Puerto Plata · Cabo Francés Viejo · Cabrera · Bahía Escocesa · Cabo Frances · Puerto Plata · Moca · Nagua · Santiago · La Vega · Salcedo · San Francisco de Macorís · Samaná · Bahía de Samaná · Bonao · Monte Plata · El Seibo · Higüey · Cabo Engaño · Villa Altagracia · Hato Mayor · San Cristóbal · San Pedro de Macorís · La Romana · SANTO DOMINGO · Isla Saona · Baní · Barahona · Punta Palenque · Enriquillo · Cabo Beata · Beata

PUERTO RICO (to US) · Aguadilla · Arecibo · SAN JUAN · Carolina · Bayamón · Caguas · El Yunque · Fajardo · Vieques · Mayagüez · Utuado · Cordillera Central · Yauco · Ponce · Guayama · Vieques · Cabo Rojo · Isla Mona · Mona Passage

Leeward Islands · BRITISH VIRGIN ISLANDS · Anegada · ROAD TOWN · Virgin Gorda · Beef Island · Tortola · Virgin Passage · CHARLOTTE AMALIE · St Thomas · St John · VIRGIN ISLANDS (to US) · St Croix · Christiansted · Frederiksted · Sombrero (to Anguilla) · Anegada Passage

St Martin · ANGUILLA (to UK) · THE VALLEY · Anguilla · Wall Blake · Marigot · St-Martin (to France) · Philipsburg · Sint Maarten (to Netherlands) · St-Barthélémy (to France) · Barbuda · Codrington · NETHERLANDS ANTILLES (to Netherlands) · Saba · St Eustatius · Golden Rock · St Kitts · Nevis · ANTIGUA & BARBUDA · BASSETERRE · Newcastle · ST JOHN'S · V.C. Bird · Redonda · Blackburne · Falmouth · ST KITTS & NEVIS · Charlestown · PLYMOUTH · MONTSERRAT (to UK) · Guadeloupe Passage · Port-Louis · Grande Terre · la Désirade · GUADELOUPE (to France) · Ste-Rose · Pointe-à-Pitre · Marie-Galante · BASSE-TERRE · Basse Terre · Soufrière 1467m · Grand-Bourg · les Saintes · Dominica Passage · Portsmouth · Marigot · DOMINICA · Melville Hall · Canefield · La Plaine · ROSEAU · Martinique Passage · Montagne Pelée 1397m · St-Pierre · MARTINIQUE (to France) · Ste-Marie · le Lamentin · FORT-DE-FRANCE · Rivière-Pilote · St Lucia Channel · Vigie · CASTRIES · Mount Gimie 950m · ST LUCIA · Soufrière · Vieux Fort · Hewanorra · Saint Vincent Passage · La Soufrière 1234m · Georgetown · KINGSTOWN · St Vincent · Arnos Vale · ST VINCENT & THE GRENADINES · BARBADOS · Speightstown · Bathsheba · BRIDGETOWN · Port Elizabeth · Grantley Adams · Bequia · Mustique · Canouan · Union Island · GRENADA · Hillsborough · Carriacou · Victoria · Grenville · ST GEORGE'S · Point Salines

DOMINICA

Dominica Passage · Pointe Jaco · Vieille Case · Portsmouth · Melville Hall · Morne Diablotins 1447m · Marigot · DOMINICA · Salisbury · Castle Bruce · St.Joseph · Canefield · Rosalie · ROSEAU · La Plaine · Scotts Head Village · Berekua · Martinique Passage · ATLANTIC OCEAN · Caribbean Sea
SCALE 1:2,000,000

MARTINIQUE (to France)

Martinique Passage · Grand' Rivière · Basse-Pointe · le Prêcheur · Montagne Pelée 1397m · Ste-Marie · St-Pierre · la Trinité · Schœlcher · le Robert · MARTINIQUE · le François · FORT-DE-FRANCE · le Lamentin · Baie de Fort-de-France · Rivière-Pilote · les Anses-d'Arlets · le Diamant · Ste-Anne · ATLANTIC OCEAN · Caribbean Sea
SCALE 1:2,500,000

ST LUCIA

Pointe Du Cap · Gros Islet · Vigie · CASTRIES · Anse La Raye · Dennery · Soufrière · Petit Piton 743m · Mount Gimie 950m · Micoud · Gros Piton 798m · Laborie · Hewanorra · Vieux Fort · Ministre Point · Saint Vincent Passage · Caribbean Sea · ATLANTIC OCEAN
SCALE 1:2,000,000

BARBADOS

ATLANTIC OCEAN · Speightstown · Mount Hillaby 340m · Bathsheba · Holetown · Welchman Hall · BRIDGETOWN · The Crane · Oistins · Grantley Adams
SCALE 1:2,000,000

ST VINCENT

Saint Vincent Passage · Fancy · Porter Point · La Soufrière 1234m · Georgetown · Chateaubelair · St Vincent · Barrouallie · North Union · Layou · KINGSTOWN · Arnos Vale · Stubbs · Caribbean Sea · ATLANTIC OCEAN
SCALE 1:2,000,000

GRENADA

Sauteurs · Victoria · Mount St.Catherine 840m · Gouyave · Grenville · GRENADA · ST.GEORGE'S · St.David's · Grand Anse · Point Salines · Caribbean Sea · ATLANTIC OCEAN
SCALE 1:2,000,000

TRINIDAD & TOBAGO

Caribbean Sea · Galera Point · Matelot · Blanchisseuse · Redhead · VENEZUELA · The Dragon's Mouth · Tunapuna · PORT-OF-SPAIN · Arima · Sangre Grande · Chaguanas · Caroni River · Caroni · Couva · Dam · Guatuaro Point · Gulf of Paria · San Fernando · Rio Claro · Point Fortin · Princes Town · Wildbeer River · Rushville · La Brea · Siparia · Moruga · Galeota Point · The Serpent's Mouth · Bonasse · ATLANTIC OCEAN
SCALE 1:2,500,000

Lesser Antilles

ARUBA (to Netherlands) · ORANJESTAD · Reina Beatrix · Sint Nicholaas · NETHERLANDS ANTILLES (to Netherlands) · Noordpunt · Curaçao · Malmok · Bonaire · Hato Airport · Santa Catherina · Kralendijk · WILLEMSTAD · VENEZUELA

Windward Islands

DOMINICA · Portsmouth · Melville Hall · Marigot · Roseau · Martinique Passage · MARTINIQUE (to France) · Montagne Pelée 1397m · Ste-Marie · St-Pierre · le Lamentin · FORT-DE-FRANCE · Rivière-Pilote · St Lucia Channel · Vigie · CASTRIES · Mount Gimie 950m · ST LUCIA · Soufrière · Vieux Fort · Hewanorra · Saint Vincent Passage · La Soufrière 1234m · Georgetown · Chateaubelair · St Vincent · Arnos Vale · KINGSTOWN · ST VINCENT & THE GRENADINES · BARBADOS · Speightstown · Bathsheba · Bridgetown · Grantley Adams · Port Elizabeth · Bequia · Mustique · Canouan · Union Island · The Grenadines · GRENADA · Hillsborough · Carriacou · Victoria · Grenville · ST GEORGE'S · Point Salines

TRINIDAD & TOBAGO · Tobago · Charlotteville · Scarborough · Galera Point · PORT-OF-SPAIN · Arima · Sangre Grande · Piarco · Trinidad · Rio Claro · Gulf of Paria · San Fernando · Galeota Point · Point Fortin · Siparia · Bonasse

VENEZUELA

47

SOUTH AMERICA

REACHING FROM THE HUMID TROPICS DOWN INTO THE COLD SOUTH ATLANTIC, SOUTH AMERICA HAS AN AREA OF 6,886,000 SQ MILES (17,835,000 SQ KM). THERE ARE 12 SEPARATE COUNTRIES, WITH THE LARGEST, BRAZIL, COVERING ALMOST HALF THE CONTINENT.

○ Greatest extent, north–south:
(4,750 miles / 7,640 km)
□ Greatest extent, east-west:
(3,100 miles / 4,990 km)

Most northerly point:
Punta Gallinas,
Colombia 12° 28′ N

PUNTA GALLINAS,
COLOMBIA
12° 28′ N

Most westerly point:
Galapagos Islands,
Ecuador
92° W

Equator

PUNTA PARIÑAS,
PERU
81° 20′ W

CABO BRANCO /
PONTA DO SEIXAS,
BRAZIL 34° 50′ W

Largest lake:
Lake Titicaca, Bolivia/Peru
3,220 sq miles (8,340 sq km)

Highest recorded temperature:
Rivadavia, Argentina
120°F (49°C)

Tropic of Capricorn

ANTOFAGASTA

SÃO PAULO

Most easterly point:
Ilhas Martin Vaz,
Brazil (28° 51′ W)

Highest point:
Cerro Aconcagua,
Argentina
22,833 ft (6,959 m)

Lowest recorded temperature:
Sarmiento, Argentina
-27°F (-33°C)

Lowest point:
Peninsula Valdés,
Argentina
131 ft (40 m) below sea level

Most southerly point:
Cape Horn, Chile
55° 59′ S

FROWARD CAPE,
CHILE 53° 54′ S

ANTOFAGASTA,
CHILE

Atacama
Desert

Andes

Paraguay River

Planalto de Mato Grosso

SÃO PAULO, BRAZIL

CROSS-SECTION FROM ANTOFAGASTA, CHILE TO SÃO PAULO, BRAZIL

◄— line of cross-section

0 250 500 750 1000 Km

0   250   500   750   1000 Miles

NORT

Cedros Trench

East Pacific Rise

PACIFIC

NAZCA PLATE
PACIFIC PLATE

Union Reefs

Clipperton Seamounts

Clipperton
Island

COCOS PL.
NAZCA PL.

San Lucas
Cape

Gulf of California

Sierra Madre Oriental

Sierra Madre Occidental

Lower California

Río Grande

Río de Santiago

Lago de Chapala

Popoca
5452m

AMERICA

Cape Canaveral

Apalachee Bay

Lake Okeechobee

Sargasso
Sea

Cape Verde
Islands

Cape Verde
Basin

Mississippi Fan

Escarpment

Gulf of Mexico

Hatteras plain

Nares Plain

Straits of Florida

Bahamas

Great Bahama Bank

West Indies

Yucatan
Basin

Cuba

NORTH AMERICAN PLATE

SOUTH AMERICAN PLATE

Puerto Rico Trench

Cayman Trough

Greater Antilles

Leeward Islands

Barbuda

Antigua

Gambia
Plain

Gulf of
Honduras

Jamaica

Hispaniola

Puerto Rico

Nevis

Guadeloupe

Dominica

Caribbean Sea

Windward Passage

Martinique

AFRICAN PLATE

Sierra
Madre Sur

Nicaraguan Rise

Punta
Gallinas

Lesser Antilles

Saint Lucia

Barbados

Doldrums Fracture Zone

Gulf of
Fonseca

Peninsula
de la Guajira

Aruba

Bonaire

Isla de
Margarita

Grenada

Grenada Basin

Windward Islands

Tobago

Demerara
Plain

Trinidad

Four North Fracture Zone

Saint Paul Fracture Zone

Colombian
Basin

Curaçao

Gulf of Venezuela

Cordillera de la Costa

Orinoco

Lake
Maracaibo

Middle America Trench

Mosquito Coast

Colombian
Basin

Guiana
Basin

Equator

Lake
Nicaragua

Mosquito
Gulf

Gulf of
Darien

CARIBBEAN PLATE

Apure

Arauca

Caroní

Cordillera Occidental

SOUTH AMERICAN
PLATE

Meta

Orinoco

Guatemala
Basin

Isthmus of Panama

Cordillera Central

Cordillera Oriental

Guaviare

Vichada

Guiana Highlands

Orinoco

Ceara Plain

Peninsula
de Azuero

Gulf of
Panama

Serra
Parima

Tumuc-Humac Mountains

Amazon Fan

Colón Ridge

Panama
Basin

Caquetá

Uaupés

Branco

Negro

Baía de
Marajó

Baía de
São Marcos

Atol
das Rocas

Fernando
de Noronha

Chimborazo
6310m

Cordillera Real

Putumayo

Japurá

Río Negro

Ilha de
Marajó

Represa
de Tucuruí

Cabo de
São Roque

Galapagos
Islands

Napo

Içá

Amazon

SOUTH

Amazon Alto

Tocantins

Planalto da
Borborema

Cabo Branco

Pernambuco
Plain

Gulf of
Guayaquil

Marañón

Juruá

Amazon Basin

Madeira

Curuá

Xingu

Represa de
Itaparica

Punta
Parinas

Tapauá

AMERICA

Serra do Cachimbo

Serra Grande

Chapada das
Mangabeiras

Represa de
Sobradinho

Purus

Jurua

Roosevelt

Araguaia

São Manuel

São Francisco

Chapada Diamantina

Brazil
Basin

Peru
Basin

Mendaña Fracture Zone

Madre de Dios

Mamoré

Guaporé

Chapada dos Parecis

Planalto de
Mato Grosso

Serra Formosa

Serra Geral
de Goiás

Baía de
Todos os Santos

NAZCA PLATE

Cordillera Occidental

Cordillera Oriental

Rapulo

Río Grande

Paraguai

Paranaíba

Serra de Espinhaço

Abrolhos
Bank

Trindade Spur

Lake
Titicaca

Altiplano

Yungas

Doce

Nazca Ridge

Lago Poopó

Pilcomayo

Gran Chaco

Jacuí

Río Grande

Serra do Mar

Tropic of Capricorn

Chile
Basin

Pantanal

Aporé

Rio Grande

Serra da
Mantiqueira

Rio Grande
Rise

Islas de los
Desventurados

Atacama Desert

Peru-Chile Trench

ANDES

Salado

Paraná

Iguaçu

Ilha de
São Sebastião

Santos
Plateau

Sala y Gomez Fracture Zone

Represa
de Itaipú

Uruguay

Easter
Island

Roggeveen
Basin

Salinas
Grandes

Cerro
Aconcagua
6959m

Sierras de Córdoba

Mar Chiquita

Paraná

Mesopotamia

Embalse
de Río Negro

Lagoa
dos Patos

Mirim
Lagoon

Cuchilla Grande

Juan Fernandez
Islands

Neuquén

Colorado

Río Negro

Pampas

Río de la Plata

Argentine
Basin

East Pacific Rise

Limay

Chubut

Golfo San Matías

Bahía
Blanca

NAZCA PLATE

ANTARCTIC PLATE

Golfo Corcovado

Lago
Buenos
Aires

Patagonia

Chico

Deseado

Gulf of
San Jorge

Bahía
Grande

Argentine
Plain

Falkland Escarpment

Maurice Ewing
Bank

South Sandwich Trench

ANTARCTIC PLATE
PACIFIC PLATE

Archipiélago
de los Chonos

Falkland
Plateau

South Georgia

South Georgia Ridge

South
Sandwich
Islands

Strait of Magellan

Falkland Islands

SOUTH AMERICAN PLATE

Tierra
del Fuego

Scotia Ridge

SCOTIA PLATE

Scotia
Sea

Cape Horn

SCOTIA PLATE

South Shetland Trough

South Shetland
Islands

South Orkney
Islands

ANTARCTIC PLATE

Antarctic Circle

Weddell
Sea

ANTARCTICA

49

PHYSICAL SOUTH AMERICA

THREE MAJOR PHYSIOGRAPHIC REGIONS characterize South America. The oldest, the ancient Brazilian Shield and the smaller Guyana and Patagonian shields, form the stable core of the continent. Stretching along the entire west coast are the younger Andean fold mountains with many summits rising to 20,000 ft (6,100 m). These two diverse regions are separated by a number of sedimentary basins carrying South America's large river systems to the sea. These include the massive Amazon Basin and the basin of the Gran Chaco.

THE AMAZON BASIN AND GUYANA SHIELD

THE RIVER AMAZON occupies a large depression in the Earth's crust, formed by the uplift of the Andes. It is covered by thick volcanic deposits and layers of alluvium – these have been laid down by the Amazon's many tributaries. To the north is the smaller Guyana Shield.

Headwaters of the Amazon rise in the Andes
Thick alluvium deposits
Mouths of the Amazon

A — A

Section across northern South America showing Amazon Basin and its drainage pattern.

0   500   1000 Km
0   500   1000 Miles

SCALE 1:27,500,000
(projection: Lambert Azimuthal Equal Area)

Km
0   100   200   400   600   800
0   100   200   400   600   800
Miles

THE ANDEAN UPLANDS

THE ANDEAN UPLANDS run along the west coast of South America. They are being uplifted as the Nazca Plate is subducted beneath the South American Plate. They contain some of the world's largest volcanoes, such as Cotopaxi, and Lake Titicaca which occupies a dormant site. The far south has many large ice-sheets and a fragmented coastline.

Nazca Plate
South American Plate
Volcanic intrusions

B — B

Cross-section through the Andes showing the subduction of the Nazca Plate beneath the South American Plate.

0   200   400 Km
0   200   400 Miles

MAP KEY

ELEVATION

6000m / 19,686ft
4000m / 13,124ft
3000m / 9843ft
2000m / 6562ft
1500m / 4922ft
1000m / 3281ft
500m / 1640ft
250m / 820ft
100m / 328ft
sea level

PLATE MARGINS
(for explanation see page xiv)

— constructive
△ △ destructive
— conservative
······ uncertain

— physiographic regions
▶ ◀ line of cross-section

THE BRAZILIAN SHIELD AND GRAN CHACO

THE IMMENSE BRAZILIAN SHIELD underlies more than one-third of South America. It is pitted with numerous volcanic intrusions, and a large basaltic plateau exists between the Paraná River and the Atlantic Ocean. The flat Gran Chaco lies to the west of the shield, covered by sedimentary deposits eroded from the Andes, and transported by South America's mighty rivers.

Young, folded Andes Mountains
Volcanic intrusions
Major rivers drain to the south through the Gran Chaco
Ancient resistant shield

C — C

Section across central South America showing the flat basin of the Gran Chaco and the ancient Brazilian Shield.

0   200   400 Km
0   200   400 Miles

Map labels

Punta Gallinas
Gulf of Venezuela
Lake Maracaibo
Gulf of Darien
Gulf of Panama
COCOS PLATE
NAZCA PLATE
Cauca
Magdalena
Cordillera Occidental
Cordillera Central
Cordillera Oriental
Llanos
Orinoco
Rio Negro
Japurá
Pakaraima Mountains
GUYANA SHIELD
Guiana Highlands
Tumuc-Humac Mountains
ATLANTIC OCEAN
Putumayo
Cotopaxi 5897m
Chimborazo 6310m
Cordillera Real
Gulf of Guayaquil
Marañón
Amazon
Juruá
Amazon
Purus
Madeira
Represa Balbina
Ilha de Marajó
Amazon Basin
Tapajós
Xingu
Tocantins
Serra dos Carajás
Cabo de São Roque
Punta Negra
Ucayali
Nevado Huascarán 6768m
Madre de Dios
Guaporé
Chapada dos Parecis
Serra do Cachimbo
Serra Formosa
Araguaia
Planalto da Borborema
BRAZILIAN SHIELD
Serra do Roncador
Serra Dourada
Represa de Sobradinho
NAZCA PLATE
SOUTH AMERICAN PLATE
PACIFIC OCEAN
A
Lake Titicaca
Lago Poopó
Altiplano
Planalto de Mato Grosso
Pantanal
Serra do Caiapó
São Francisco
Brazilian Highlands
Serra do Espinhaço
B
Atacama Desert
Pilcomayo
Gran Chaco
Serra de Maracaju
Paraná
Serra Geral
Serra da Mantiqueira
Cerro Ojos del Salado 6880m
Andes System
Paraguay
Paraná
Uruguay
Serra do Mar
Lagoa dos Patos
Cerro Aconcagua 6959m
Salado
Mesopotamia
Mirim Lagoon
C
Pampas
Rio de la Plata
PATAGONIAN SHIELD
Colorado
Rio Negro
Isla de Chiloé
Península Valdés
Patagonia
Chico
Lago Colhué Huapí
Gulf of San Jorge
Golfo de Penas
Deseado
Bahía Grande
Strait of Magellan
Tierra del Fuego
Falkland Islands
SOUTH AMERICAN PLATE
SCOTIA PLATE
Cape Horn
ANTARCTIC PLATE

50

CLIMATE

THE CLIMATE OF SOUTH AMERICA is influenced by three principal factors: the seasonal shift of high pressure air masses over the tropics, cold ocean currents along the western coast, affecting temperature and precipitation, and the mountain barrier produced by by the Andes, which creates a rain shadow over much of the south.

Mild winters and cool summers typify the extensive Pampas grasslands of Argentina.

Chile's hyperarid Atacama Desert is renowned as one of the driest places on Earth.

Climate
- tundra
- cool continental
- warm humid
- semiarid
- arid
- humid equatorial
- tropical
- daily hours of sunshine, January
- daily hours of sunshine, July
- cold wind

TEMPERATURE

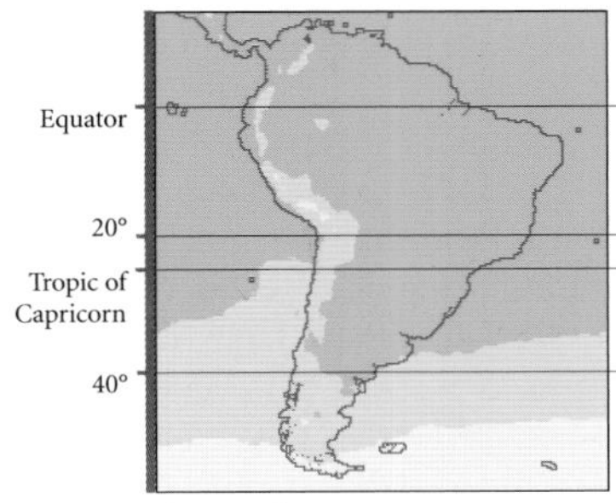

Equator
20°
Tropic of Capricorn
40°

Average January temperature

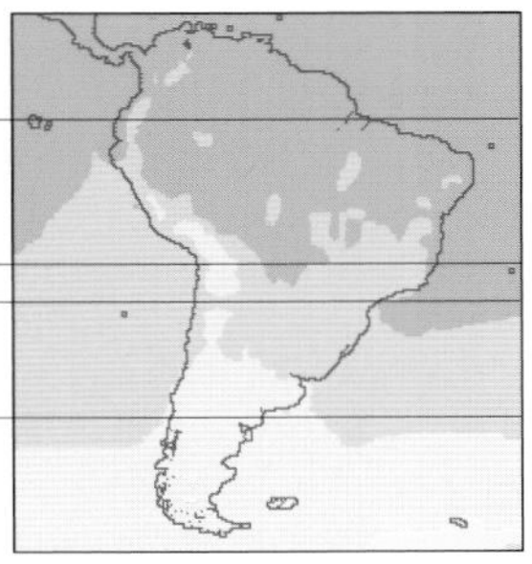

Average July temperature

Temperature
- below -22°F (-30°C)
- -22 to -4°F (-30 to -20°C)
- -4 to 14°F (-20 to -10°C)
- 14 to 32°F (-10 to 0°C)
- 32 to 50°F (0 to 10°C)
- 50°F (10 to 20°C)
- 68 to 86°F (20 to 30°C)
- above 86°F (30°C)

RAINFALL

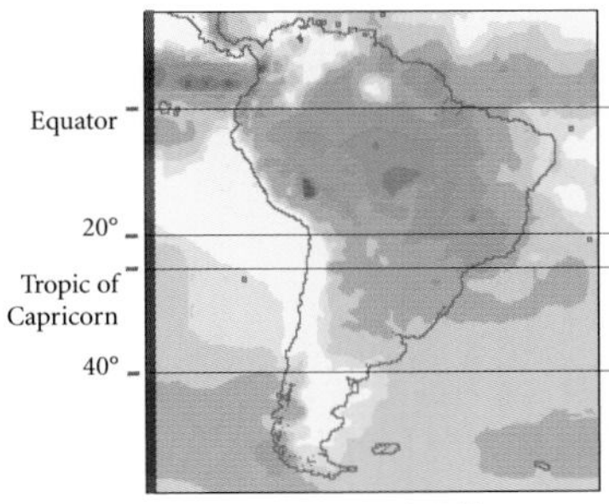

Equator
20°
Tropic of Capricorn
40°

Average January rainfall

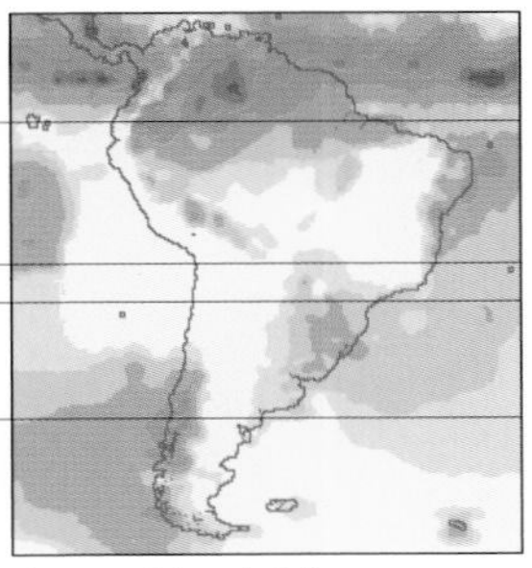

Average July rainfall

Rainfall
- 0–1 in (0–25 mm)
- 1–2 in (25–50 mm)
- 2–4 in (50–100 mm)
- 4–8 in (100–200 mm)
- 8–12 in (200–300 mm)
- 12–16 in (300–400 mm)
- 16–20 in (400–500 mm)
- more than 20 in (500 mm)

Map labels: Maracaibo, Caracas, Georgetown, Cayenne, Bogotá, Quito, Equator, Manaus, Belém, Altos, Recife, Lima, La Paz, Santa Cruz, Brasília, Belo Horizonte, La Quiaca, Rio de Janeiro, Tropic of Capricorn, Antofagasta, Asunción, Córdoba, Porto Alegre, Santiago, Buenos Aires, Montevideo, Concepción, Pamperos, Stanley

Tropical conditions are found across over half of South America. When both rainfall and temperatures are high, hot humid rain forests prevail.

SHAPING THE CONTINENT

SOUTH AMERICA'S ACTIVE TECTONIC BELT has been extensively folded over millions of years; landslides are still frequent in the mountains. The large river systems that erode the mountains flow across resistant shield areas, depositing sediment. Present-day glaciation affects the distinctive landscape of the far south.

MASS MOVEMENT

[6] Debris slides are common in the highlands of South America (left). They occur where soil on a slope is saturated by rainwater and therefore less stable. The actual slides are often triggered by earthquakes.

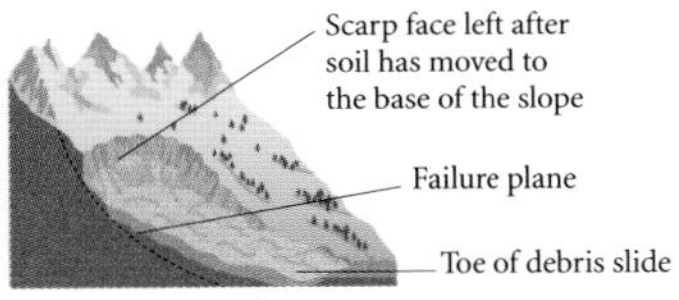

Scarp face left after soil has moved to the base of the slope
Failure plane
Toe of debris slide

MASS MOVEMENT: A SECTION OF A DEBRIS SLIDE

CHEMICAL WEATHERING

[1] Table mountains (left) are the eroded remnants of an ancient upland. As water percolates along cracks in these high, flat-topped mountains it forms intricate cave systems. Chemical weathering also isolates large blocks which then collapse, accumulating as rockfalls at the foot of scarp slopes.

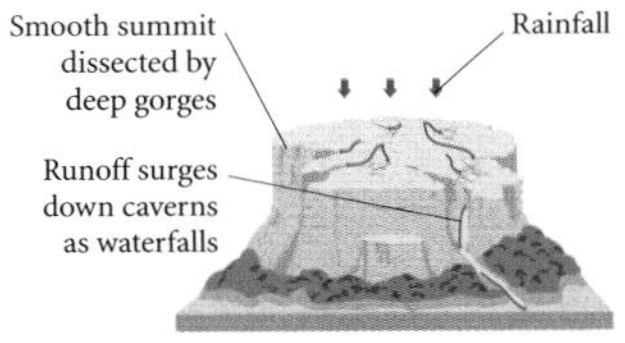

Smooth summit dissected by deep gorges
Rainfall
Runoff surges down caverns as waterfalls

CHEMICAL WEATHERING: EROSION OF THE GUYANA SHIELD

THE EVOLVING LANDSCAPE

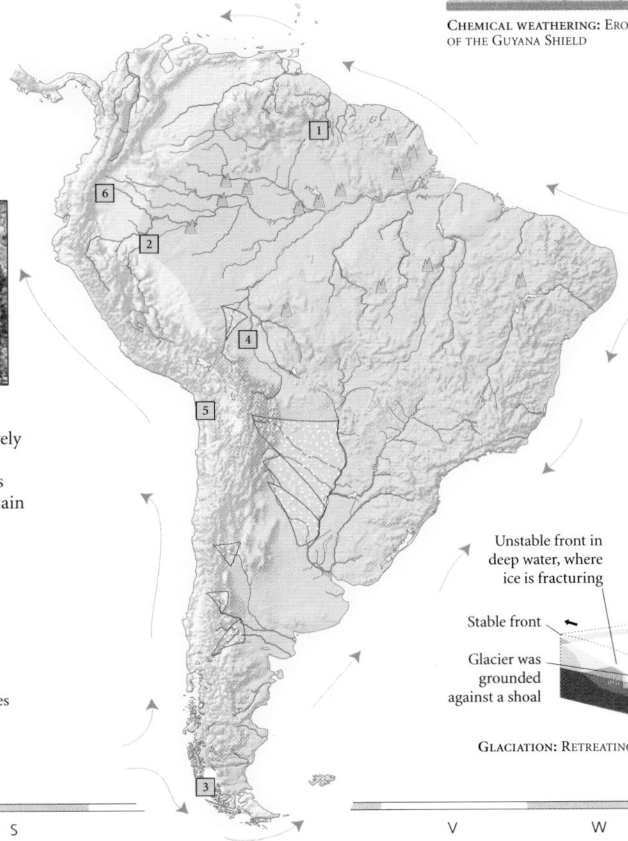

RIVER SYSTEMS

[2] Along the Amazon (above) there is a great variation in rates of erosion. As the headwaters of the Amazon flow down from the Andes, they erode and transport vast quantities of sediment, and are known as whitewaters. Across the shield areas erosion rates are very low. These rivers, carrying rotting vegetation, are called blackwaters.

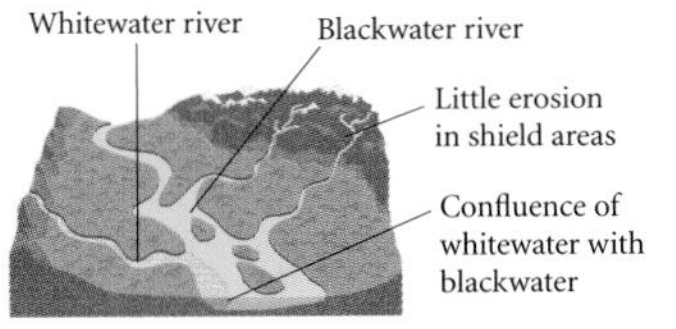

Whitewater river
Blackwater river
Little erosion in shield areas
Confluence of whitewater with blackwater

RIVER SYSTEMS: SUSPENDED SEDIMENTS IN THE AMAZON

Landscape
- uplifting land
- stable land
- sinking land
- glacier
- ocean current
- aluvial fan
- inselberg
- river

FOLDING

[5] Folding occurs beneath the surface under high temperatures and pressures. Rocks become sufficiently malleable to flow and not fracture as tectonic plates collide. In the Valley of the Moon in Chile (above), anticlines (or upfolds) and synclines (or troughs) have been exploited by erosion.

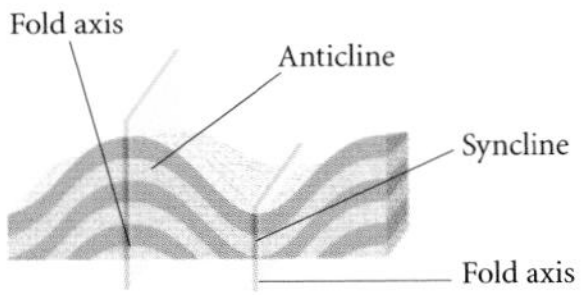

Fold axis
Anticline
Syncline
Fold axis

FOLDING: SYNCLINES AND ANTICLINES

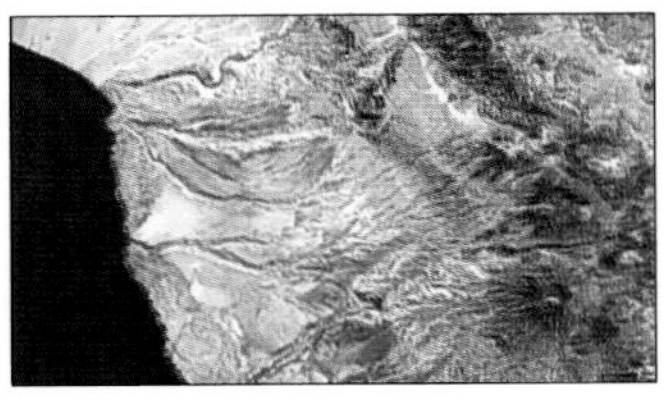

DEPOSITION

[4] Large alluvial fans are found extensively across South America (above). Confined mountain rivers, carrying large quantities of eroded material, emerge from a mountain gorge onto the plains, where they deposit their load in huge fans.

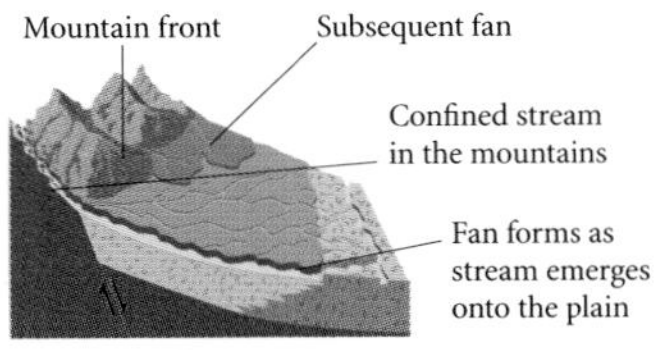

Mountain front
Subsequent fan
Confined stream in the mountains
Fan forms as stream emerges onto the plain

DEPOSITION: FORMATION OF AN ALLUVIAL FAN

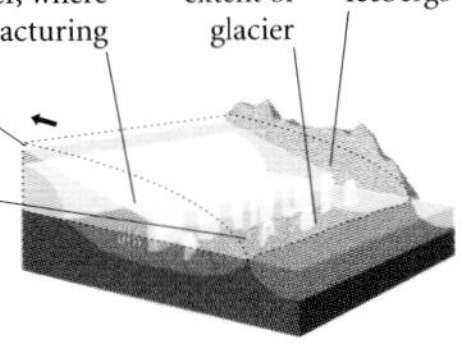

Unstable front in deep water, where ice is fracturing
Original extent of glacier
Icebergs
Stable front
Glacier was grounded against a shoal

GLACIATION: RETREATING GLACIER IN PATAGONIA

GLACIATION

[3] As fjord glaciers in Patagonia (above) retreat, they become grounded on shoals. In deeper water the base of the glacier becomes unstable, and icebergs break off (calve) until the glacier snout grounds once more.

POLITICAL SOUTH AMERICA

MODERN SOUTH AMERICA'S POLITICAL BOUNDARIES have their origins in the territorial endeavors of explorers during the 16th century, who claimed almost the entire continent for Portugal and Spain. The Portuguese land in the east later evolved into the federal state of Brazil, while the Spanish vice-royalties eventually emerged as separate independent nation-states in the early 19th century. South America's growing population has become increasingly urbanized, with the growth of coastal cities into large conurbations like Rio de Janeiro and Buenos Aires. In Brazil, Argentina, Chile and Uruguay, a succession of military dictatorships has given way to fragile, but strengthening, democracies.

Europe retains a small foothold in South America. Kourou in French Guiana was the site chosen by the European Space Agency to launch the Ariane rocket. As a result of its status as a French overseas department, French Guiana is actually part of the European Union.

SCALE 1:21,500,000
(projection: Lambert Azimuthal Equal Area)

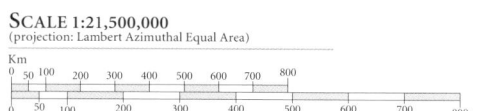

TRANSPORTATION

MOST MAJOR ROAD AND RAIL ROUTES are confined to the coastal regions by the forbidding natural barriers of the Andes Mountains and the Amazon Basin. Few major cross-continental routes exist, although Buenos Aires serves as a transportation center for the main rail links to La Paz and Valparaíso, while the construction of the Trans-Amazon and Pan-American Highways have made direct road travel possible from Recife to Lima and from Puerto Montt up the coast into central America. A new waterway project is proposed to transform the River Paraguay into a major shipping route, although it involves considerable wetland destruction.

South America's most extensive rail network is centered on the Argentinian capital, Buenos Aires. The construction of new rail lines ouward from this important port, allowed the colonization of the Pampas lands for agriculture.

LANGUAGES

PRIOR TO EUROPEAN EXPLORATION in the 16th century, a diverse range of indigenous languages were spoken across the continent. With the arrival of Iberian settlers, Spanish became the dominant language, with Portuguese spoken in Brazil, and Native American languages such as Quechua and Guaraní, becoming concentrated in the continental interior. Today this pattern persists, although successive European colonization has led to Dutch being spoken in Suriname, English in Guyana, and French in French Guiana, while in large urban areas, Japanese and Chinese are increasingly common.

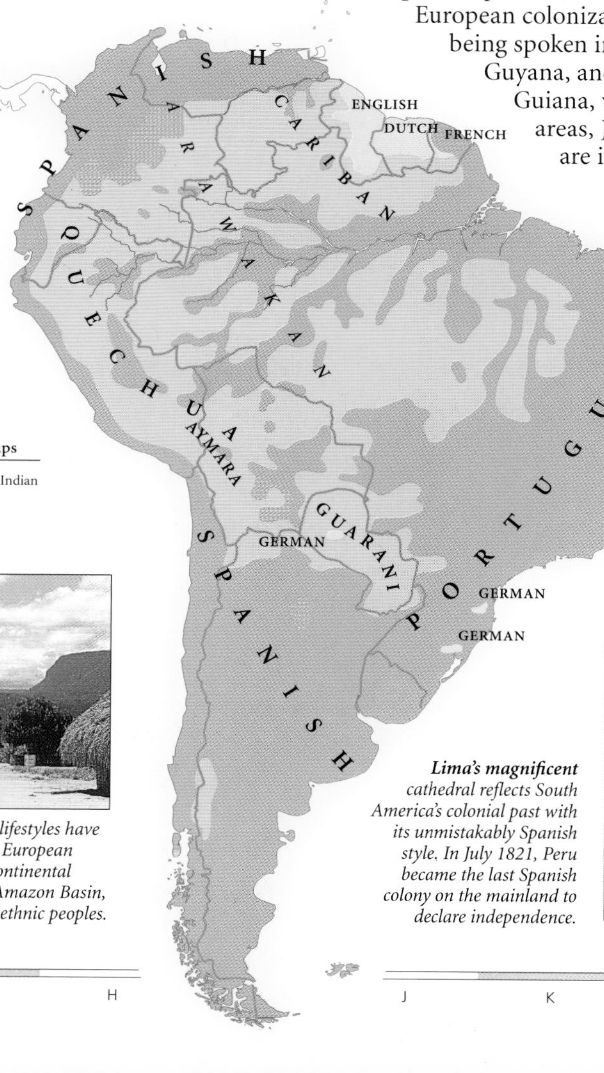

Transportation
- major roads and highways
- major railroads
- international borders
- • transportation intersections
- ⊕ international airports
- ⊕ major ports

Language groups
- American Indian
- Germanic
- Romance

Chile's main port, Valparaíso, is a vital national shipping center, in addition to playing a key role in the growing trade with Pacific nations. The country's awkward, elongated shape means that sea transportation is frequently used for internal travel and communications in Chile.

Indigenous South American lifestyles have not been totally submerged by European cultures and languages. The continental interior, and particularly the Amazon Basin, is still home to many different ethnic peoples.

Lima's magnificent cathedral reflects South America's colonial past with its unmistakably Spanish style. In July 1821, Peru became the last Spanish colony on the mainland to declare independence.

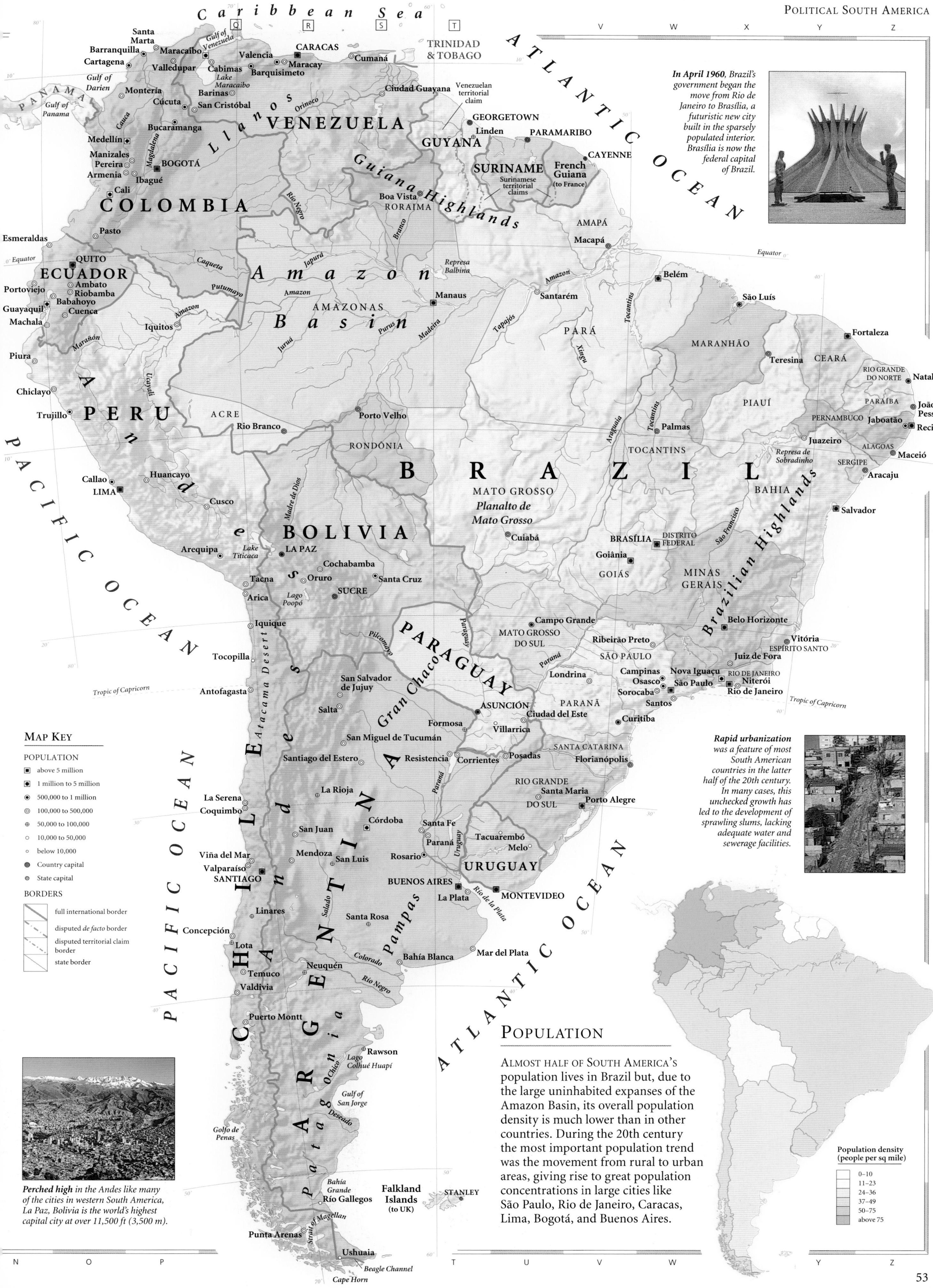

Map labels

Caribbean Sea

Water bodies and oceans:
ATLANTIC OCEAN
PACIFIC OCEAN
Gulf of Venezuela
Gulf of Darien
Gulf of Panama
Lake Maracaibo
Caribbean Sea

Countries:
VENEZUELA
COLOMBIA
ECUADOR
PERU
BOLIVIA
BRAZIL
PARAGUAY
URUGUAY
ARGENTINA
CHILE
GUYANA
SURINAME
French Guiana (to France)
TRINIDAD & TOBAGO
PANAMA

Cities and places:
Santa Marta, Barranquilla, Cartagena, Maracaibo, Valledupar, Cabimas, Valencia, Maracay, CARACAS, Cumaná, Montería, Cúcuta, San Cristóbal, Barinas, Barquisimeto, Ciudad Guayana, Venezuelan territorial claim, GEORGETOWN, Linden, PARAMARIBO, CAYENNE, Surinamese territorial claims

Medellín, Manizales, Pereira, Armenia, BOGOTÁ, Ibagué, Cali, Bucaramanga, Boa Vista, RORAIMA, AMAPÁ, Macapá

Esmeraldas, QUITO, Pasto, Represa Balbina

Portoviejo, Ambato, Riobamba, Babahoyo, Cuenca, Guayaquil, Machala, Manaus, Belém, São Luís

Piura, Iquitos, Santarém, Fortaleza, MARANHÃO

Chiclayo, ACRE, Porto Velho, Teresina, CEARÁ, Natal, RIO GRANDE DO NORTE

Trujillo, Rio Branco, RONDÔNIA, PIAUÍ, PARAÍBA, João Pessoa, Jaboatão, Recife, PERNAMBUCO, Juazeiro

Callao, LIMA, Huancayo, Cusco, MATO GROSSO, Planalto de Mato Grosso, Cuiabá, TOCANTINS, Palmas, Represa de Sobradinho, ALAGOAS, Maceió, SERGIPE, Aracaju

Arequipa, Lake Titicaca, LA PAZ, Cochabamba, Santa Cruz, BRASÍLIA, DISTRITO FEDERAL, Goiânia, GOIÁS, BAHIA, Salvador

Tacna, Oruro, SUCRE, MINAS GERAIS, Brazilian Highlands

Arica, Lago Poopó, Campo Grande, Belo Horizonte

Iquique, MATO GROSSO DO SUL, Ribeirão Preto, Vitória, ESPÍRITO SANTO

Tocopilla, Londrina, SÃO PAULO, Campinas, Nova Iguaçu, Juiz de Fora, RIO DE JANEIRO, Niterói

Antofagasta, San Salvador de Jujuy, Osasco, São Paulo, Santos, Rio de Janeiro, Sorocaba, PARANÁ

Salta, Formosa, ASUNCIÓN, Ciudad del Este, Curitiba, Villarrica

San Miguel de Tucumán, Santiago del Estero, Resistencia, Corrientes, Posadas, SANTA CATARINA, Florianópolis

La Rioja, RIO GRANDE DO SUL, Santa Maria, Porto Alegre

La Serena, Coquimbo, Córdoba, Santa Fe, Tacuarembó, Melo

San Juan, Paraná, Rosario, MONTEVIDEO

Viña del Mar, Valparaíso, SANTIAGO, Mendoza, San Luis, BUENOS AIRES, La Plata

Linares, Santa Rosa

Concepción, Lota, Temuco, Valdivia, Neuquén, Bahía Blanca, Mar del Plata

Puerto Montt, Rawson, Lago Colhué Huapí, Bahía Grande, Río Gallegos, Falkland Islands (to UK), STANLEY, Punta Arenas, Ushuaia

Regions and physical features:
Llanos, Orinoco, Guiana Highlands, Amazon, Amazon Basin, AMAZONAS, Rio Negro, Branco, Japurá, Caquetá, Putumayo, Marañón, Ucayali, Juruá, Purus, Madeira, Tapajós, Xingu, Tocantins, Araguaia, Madre de Dios, Pilcomayo, Gran Chaco, Andes, Atacama Desert, Paraguay, Paraná, Uruguay, Salado, Pampas, Patagonia, Colorado, Río Negro, Río de la Plata, Deseado, Gulf of San Jorge, Golfo de Penas, Strait of Magellan, Beagle Channel, Cape Horn, Tropic of Capricorn, Equator

Sidebar texts

In April 1960, Brazil's government began the move from Rio de Janeiro to Brasília, a futuristic new city built in the sparsely populated interior. Brasília is now the federal capital of Brazil.

Rapid urbanization was a feature of most South American countries in the latter half of the 20th century. In many cases, this unchecked growth has led to the development of sprawling slums, lacking adequate water and sewerage facilities.

Perched high in the Andes like many of the cities in western South America, La Paz, Bolivia is the world's highest capital city at over 11,500 ft (3,500 m).

POPULATION

ALMOST HALF OF SOUTH AMERICA'S population lives in Brazil but, due to the large uninhabited expanses of the Amazon Basin, its overall population density is much lower than in other countries. During the 20th century the most important population trend was the movement from rural to urban areas, giving rise to great population concentrations in large cities like São Paulo, Rio de Janeiro, Caracas, Lima, Bogotá, and Buenos Aires.

Population density (people per sq mile)
- 0–10
- 11–23
- 24–36
- 37–49
- 50–75
- above 75

MAP KEY

POPULATION
- ▪ above 5 million
- ▪ 1 million to 5 million
- ◉ 500,000 to 1 million
- ◎ 100,000 to 500,000
- ⊕ 50,000 to 100,000
- ⊙ 10,000 to 50,000
- ○ below 10,000
- ● Country capital
- ◉ State capital

BORDERS
- full international border
- disputed *de facto* border
- disputed territorial claim border
- state border

SOUTH AMERICAN RESOURCES

Ciudad Guayana is a planned industrial complex in eastern Venezuela, built as an iron and steel centre to exploit the nearby iron ore reserves.

Agriculture still provides the largest single form of employment in South America, although rural unemployment and poverty continue to drive people toward the huge coastal cities in search of jobs and opportunities. Mineral and fuel resources, although substantial, are distributed unevenly; few countries have both fossil fuels and minerals. To break industrial dependence on raw materials, boost manufacturing, and improve infrastructure, governments borrowed heavily from the World Bank in the 1960s and 1970s. This led to the accumulation of massive debts which are unlikely ever to be repaid. Today, Brazil dominates the continent's economic output, followed by Argentina. Recently, the less-developed western side of South America has benefited due to its geographical position; for example Chile is increasingly exporting raw materials to Japan.

Industry

aerospace	pharmaceuticals
brewing	printing & publishing
car/vehicle manufacture	shipbuilding
chemicals	sugar processing
electronics	textiles
engineering	timber processing
finance	tobacco processing
fish processing	wine
food processing	oil
hi-tech industry	gas
iron & steel	industrial cities
meat processing	major industrial areas
metal refining	
narcotics	

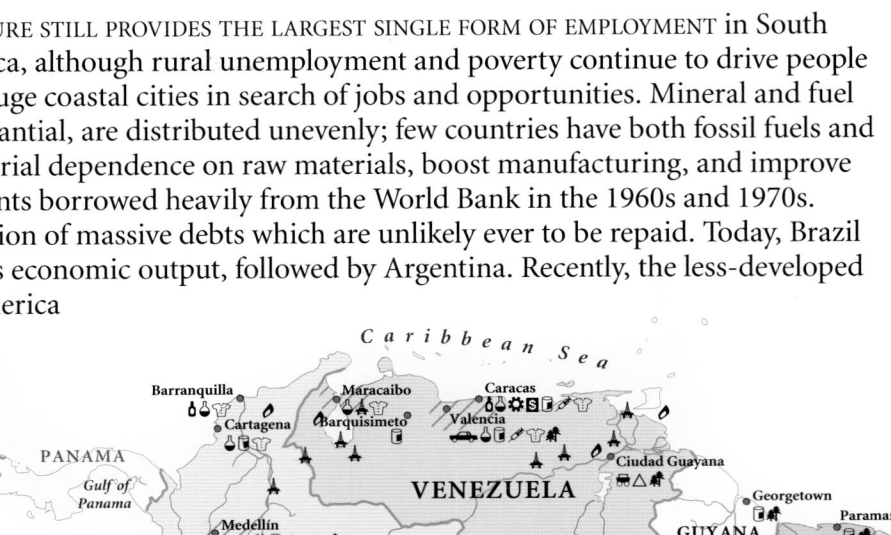

The cold Peru Current flows north from the Antarctic along the Pacific coast of Peru, providing rich nutrients for one of the world's largest fishing grounds. Overexploitation has severely reduced Peru's anchovy catch.

STANDARD OF LIVING

WEALTH DISPARITIES throughout the continent create a wide gulf between affluent landowners and the chronically poor in inner-city slums. The illicit production of cocaine, and the hugely influential drug barons who control its distribution, contribute to the violent disorder and corruption which affect northwestern South America, de-stabilizing local governments and economies.

Standard of Living
(UN Human Development Index)
low — high

Both Argentina and Chile are now exploring the southernmost tip of the continent in search of oil. Here in Punta Arenas, a drilling rig is being prepared for exploratory drilling in the Strait of Magellen.

GNP per capita (US$)
0–499
500–999
1000–1499
1500–2999
3000–5999
6000+

INDUSTRY

ARGENTINA AND BRAZIL are South America's most industrialized countries and São Paulo is the continent's leading industrial center. Long-term government investment in Brazilian industry has encouraged a diverse industrial base; engineering, steel production, food processing, textile manufacture, and chemicals predominate. The illegal production of cocaine is economically significant in the Andean countries of Colombia and Bolivia. In Venezuela, the oil-dominated economy has left the country vulnerable to world oil price fluctuations. Food processing and mineral exploitation are common throughout the less industrially developed parts of the continent, including Bolivia, Chile, Ecuador, and Peru.

ENVIRONMENTAL ISSUES

THE AMAZON BASIN is one of the last great wilderness areas left on Earth. The tropical rain forests which grow there are a valuable genetic resource, containing innumerable unique plants and animals. The forests are increasingly under threat from new and expanding settlements and "slash and burn" farming techniques, which clear land for the raising of beef cattle, causing land degradation and soil erosion.

Clouds of smoke billow from the burning Amazon rain forest. Over 25,000 sq miles (60,000 sq km) of virgin rain forest are being cleared annually, destroying an ancient, irreplaceable, natural resource and biodiverse habitat.

Environmental Issues

	national parks
	tropical forest
	forest destroyed
	desert
	desertification
	polluted rivers
	marine pollution
	heavy marine pollution
•	poor urban air quality

USING THE LAND AND SEA

MANY FOODS NOW COMMON WORLDWIDE originated in South America. These include the potato, tomato, squash, and cassava. Today, large herds of beef cattle roam the temperate grasslands of the Pampas, supporting an extensive meatpacking trade in Argentina, Uruguay and Paraguay. Corn (maize) is grown as a staple crop across the continent and coffee is grown as a cash crop in Brazil and Colombia. Coca plants grown in Bolivia, Peru, and Colombia provide most of the world's cocaine. Fish and shellfish are caught off the western coast, especially anchovies off Peru, shrimps off Ecuador and pilchards off Chile.

South America, and Brazil in particular, now leads the world in coffee production, mainly growing Coffea Arabica in large plantations. Coffee beans are harvested, roasted, and brewed to produce the world's second most popular drink, after tea.

The Pampas region of southeast South America is characterized by extensive, flat plains, and populated by cattle and ranchers (gauchos). Argentina is a major world producer of beef, much of which is exported to the US for use in hamburgers.

High in the Andes, hardy alpacas graze on the barren land. Alpacas are thought to have been domesticated by the Incas, whose nobility wore robes made from their wool. Today, they are still reared and prized for their soft, warm fleeces.

MINERAL RESOURCES

OVER A QUARTER OF THE WORLD'S known copper reserves are found at the Chuquicamata mine in northern Chile, and other metallic minerals such as tin are found along the length of the Andes. The discovery of oil and gas at Venezuela's Lake Maracaibo in 1917 turned the country into one of the world's leading oil producers. In contrast, South America is virtually devoid of coal, the only significant deposit being on the peninsula of Guajira in Colombia.

Copper is Chile's largest export, most of which is mined at Chuquicamata. Along the length of the Andes, metallic minerals like copper and tin are found in abundance, formed by the excessive pressures and heat involved in mountain-building.

Mineral Resources

	oil field
	gas field
	coal field
	bauxite
	copper
	diamonds
	gold
	iron
	lead
	silver
	tin

Using the Land and Sea

	barren land		cocoa
	cropland		cotton
	desert		coffee
	forest		fishing
	mountain region		oil palms
	pasture		peanuts
•	major conurbations		rubber
	cattle		shellfish
	pigs		soybeans
	sheep		sugar cane
	bananas		vineyards
	corn		wheat
	citrus fruits		

NORTHERN SOUTH AMERICA

COLOMBIA, GUYANA, SURINAME, VENEZUELA, *French Guiana* (to France)

Fringed by the Pacific and Atlantic oceans and the Caribbean Sea, South America's northern region has a rich range of natural resources, some exploited for centuries by colonial powers including the Spanish, French, Dutch, and British, others still to be fully explored. The prospects for further economic development in Colombia, Guyana and Suriname are blighted by drug-related violence and political instability. Venezuela, despite huge incomes from its oil reserves, remains less developed in other industrial sectors.

French Guiana is an overseas *département* of France, now seeking greater autonomy. Most of the major population centers, such as Bogotá, have grown up in the temperate conditions of the high Andes or, like Caracas, at strategic points along the Caribbean coast.

Flowers grown in Colombia are exported all over the world, and include fine carnations and roses. Here, workers are cutting roses which have been grown in plastic greenhouses.

MAP KEY

POPULATION

- 1 million to 5 million
- 500,000 to 1 million
- 100,000 to 500,000
- 50,000 to 100,000
- 10,000 to 50,000
- below 10,000

ELEVATION

- 4000m / 13,124ft
- 3000m / 9843ft
- 2000m / 6562ft
- 1000m / 3281ft
- 500m / 1640ft
- 250m / 820ft
- 100m / 328ft
- sea level

Large open squares like the Plaza Bolivia in Bogotá are characteristic of many cities founded by the Spanish.

SCALE 1:6,500,000
(projection: Lambert Azimuthal Equal Area)

Km
0 25 50   100   150   200

Miles
0 25 50   100   150   200

Scattered farms and villages have grown up on the gentle slopes of this Colombian river valley, utilizing the fertile soils for farming.

The River Orinoco flows from its source in the southern Guiana Highlands to form a broad delta on Venezuela's Atlantic coast. One of its distributary channels opens into a wide bay called the Serpent's Mouth.

TRANSPORTATION & INDUSTRY

MANY MINERAL RESOURCES are mined in Colombia, including fuels, gold, and precious and semiprecious stones. Revenues from coffee and exports of illegal narcotics are crucial to the economy. Venezuela's major economic activity is the oil industry around Lake Maracaibo (Lago de Maracaibo). Sugar and bauxite are exported from Guyana and Suriname.

TRANSPORTATION NETWORK

29,185 miles
(46,996 km)

1,795 miles
(2,890 km)

1,729 miles
(2,785 km)

17,947 miles
(28,900 km)

Rivers are an important means of transportation in Colombia; many are extensively navigable. The Pan-American Highway runs through Colombia. In Venezuela, much infrastructure investment is linked to the oil industry.

Major industry and infrastructure
- chemicals
- finance
- food processing
- iron & steel
- narcotics
- mining
- oil
- oil refining
- pharmaceuticals
- textiles
- timber processing
- capital cities
- major towns
- international airports
- major roads
- major industrial areas

Vast oil reserves around Lake Maracaibo (Lago de Maracaibo) form the focus of Venezuelan industry. Incomes from oil are used to invest in other industries and in the development of infrastructure.

USING THE LAND

THE ANDEAN BASINS support cereals and potatoes. Livestock graze at higher altitudes and on the drier tropical grasslands known as the *llanos*; hardy goats are reared in scrubland areas. Grown at higher elevations, coffee is an important cash crop, as is cotton, sugar cane, bananas, citrus fruits, cocoa, and rice, farmed on the Caribbean lowlands. Coca is the most widely-grown narcotic plant, with heroin poppies grown in Colombia and marijuana in lowland areas throughout the region.

THE URBAN/RURAL POPULATION DIVIDE

urban 80%   rural 20%

0 10 20 30 40 50 60 70 80 90 100

POPULATION DENSITY	TOTAL LAND AREA
56 people per sq mile (22 people per sq km)	1,111,317 sq miles (2,879,060 sq km)

Land use and agricultural distribution
- cattle
- goats
- bananas
- cereals
- coffee
- cotton
- sugar cane
- capital cities
- major towns
- pasture
- cropland
- forest
- wetlands
- mountain region

THE LANDSCAPE

AT ITS NORTHERNMOST REACHES, in western Colombia and Venezuela, the great Andean mountain chain splits into three distinct ranges: the Cordillera Oriental, Cordillera Central, and Cordillera Occidental, intercut by a complex series of lesser ranges and basins. The relief becomes lower toward the coast and the interior plains of the northern Amazon Basin, rising again into the tropical hills of the Guiana Highlands.

The Sierra Nevada de Santa Marta is a granite massif which rises sharply from the Caribbean lowlands to snow-covered peaks, the tallest of which is 18,947 ft (5,775 m) high.

Lake Maracaibo (Lago de Maracaibo) is not a true lake but a shallow inlet of the Caribbean Sea. It is the main source of Venezuela's oil.

The drainage basin of the Magdalena River and the Cauca, its main tributary, covers over 20% of Colombia's total surface area.

Cordillera Occidental

Cordillera Central

Cordillera Oriental

Colombia's eastern lowlands are known locally as *llanos*, meaning grasslands.

In the Guiana Highlands, Venezuela's most remote region, the ancient crystalline rocks contain deposits of iron ore, gold, and diamonds.

Angel Falls (Salto Ángel), at 3,212 ft (979 m), is the world's highest waterfall.

Igneous intrusions into the crystalline plateau which forms most of central Guyana have led to the formation of the many rapids that characterize Guyana's rivers.

Potaru river

The Potaru River descends 741 ft (226 m) over a sandstone ledge at the Kaieteur Falls in Guyana.

Guyana Shield

- Alluvial plains
- Inselbergs
- Table mountains

The Guyana Shield is one of the oldest land surfaces in the world – probably formed more than 4 billion years ago. Chemical weathering over millions of years has created flat-topped table mountains and large numbers of inselbergs.

Over 80% of Suriname is covered by tropical rain forest.

Most of the land in French Guiana is low-lying; here, the rocks of the Guiana Highlands have been eroded by rivers flowing toward the sea.

WESTERN SOUTH AMERICA

BOLIVIA, ECUADOR, PERU

THE THREE STATES OF WESTERN SOUTH AMERICA share a similar geography and recent history. Dominated by the Inca empire until Spanish conquest in the 16th century, they achieved independence from Spain in the early 19th century. The precipitous terrain of the Andes presents severe difficulties for overland transportation and continues to be a barrier to national unity and stability. Although Ecuador is now a relatively stable democracy, the military is highly influential in Peru and Bolivia, while the drug trade and associated corruption discourages external aid and economic progress. Wealth and power are still largely concentrated in the hands of a small elite of families, who attained their position during the Spanish colonial period. Land rights and political recognition for the indigenous peoples are becoming increasingly important issues, particularly in Ecuador.

THE LANDSCAPE

BOLIVIA, PERU, AND ECUADOR each possess a high Andean mountain region and an eastern region consisting of tropical lowlands and the Andean slope leading down to them. Toward the south of the region, the mountains widen to form the high plateau of the Altiplano. Peru and Ecuador also have fertile, lowland coastal plains. A wide variety of environments include *selva* (tropical rain forest), *montaña* (mountain forest), and grassland.

There are many large and active volcanoes in the Andes. Magma generated in the heart of the volcano erupts in a huge cloud of ash. Ash-fall deposits are common throughout the Andes and the rock produced is known as andesite. This is rapidly soaked by heavy rain, causing massive debris flows.

Eruption column
Subduction zone
Zone of magma generation
Magma chamber
Lava flows
Falling ash

Cotopaxi is the world's highest active volcano, with a peak 19,347 ft (5,897 m) high. A massive eruption in 1877 caused a mudflow which destroyed everything in its path for 150 miles (240 km).

Fast-flowing tributaries of the Amazon, which rise in the Andes, run eastward through the front ranges to reach the tropical lowlands. They cut valleys so deep that tropical environments can be found extending well into mountainous areas.

Much of eastern Ecuador is covered by the tropical rain forest of the Amazon Basin

Rolling hills and level plains typify the *montaña* and *selva* region, which makes up more than 65% of Peru.

The Bolivian oriente covers more than two-thirds of the country. It includes *llanos* – low alluvial plains, massive swamps, flooded bottomlands, savannah grassland, and tropical forests.

The coastal floodplains are the source of Ecuador's richest soils, enabling the cultivation of a wide range of crops.

The steepness of the Andean slopes means that avalanches and debris flows are an ever-present danger. A landslide starting from Nevado Huascarán in Peru in 1970 killed 20,000 people in 2.5 minutes when it engulfed an inhabited valley.

The Peruvian Andes are relatively young mountains which are continually being uplifted, making the area very unstable, with frequent earthquakes. The transportation difficulties that they present continue to form a barrier to national unity.

Ecuador's capital city, Quito, lies high in the Andes, nestling between snowcapped peaks. At 9,350 ft (2,850 m), Quito is the second highest capital in the world – La Paz in Bolivia is the highest.

The Altiplano is a flat, high plateau lying between the Cordillera Oriental and the Cordillera Occidental at a height of up to 12,500 ft (3,800 m). At its margins lie many spurs and alluvial fans.

Bolivian Andes

Nevado de Illampu and Nevado de Ancohuma, at 21,275 ft (6,485 m) and 21,490 ft (6,550 m) respectively, form Illampu, the highest mountain in the Bolivian Andes.

Lake Titicaca

Lake Titicaca, which forms part of the border between Peru and Bolivia, is the largest lake in South America and the highest significant body of water in the world at an altitude of 12,507 ft (3,812 m).

SCALE 1:7,750,000
(projection: Lambert Azimuthal Equal Area)

Km
0 25 50 100 150 200 250 300
Miles
0 25 50 100 150 200 250 300

MAP KEY

POPULATION

- above 5 million
- 1 million to 5 million
- 500,000 to 1 million
- 100,000 to 500,000
- 50,000 to 100,000
- 10,000 to 50,000
- below 10,000

ELEVATION

6000m / 19,686ft
4000m / 13,124ft
3000m / 9843ft
2000m / 6562ft
1000m / 3281ft
500m / 1640ft
250m / 820ft
100m / 328ft
sea level

ECUADOREAN ADMINISTRATIVE REGIONS

① CARCHI
② TUNGURAHUA
③ BOLÍVAR
④ ZAMORA CHINCHIPE
⑤ CHIMBORAZO

COLOMBIA

ECUADOR

PERU

BRAZIL

Equator

Valdéz
Esmeraldas
Punta Galera
Ensenada de Mompiche
Muisne
Rosa Zárate
Punta de Jome
Bahía de Manta
Manta
Jipijapa
Portoviejo
Punta de Santa Elena
Bahía de Santa Elena
Salinas
Golfo de Guayaquil
Guayaquil
Isla Puná
Machala
Tumbes
Zorritos
Punta Parinas
Punta Negro
Bahía de Sechura
Talara
Sullana
Piura
Paita

Tulcán
San Gabriel
Ibarra
QUITO
Santo Domingo de los Colorados
Machachi
Latacunga
Ambato
Riobamba
Guaranda
Babahoyo
Milagro
Naranjal
Cuenca
Azogues
Loja

Nueva Loja
Puerto Francisco de Orellana
Nuevo Rocafuerte
Puyo
Macas
Zamora

Río Putumayo
Río San Miguel
Río Aguarico
Río Napo
Río Coca
Río Nashiño
Río Curaray
Río Tigre
Río Pastaza
Río Morona
Río Santiago

Iquitos
Nauta
Requena
Contamana
Pucallpa

Río Amazon
Río Marañón
Río Huallaga
Río Ucayali
Río Tapiche

Loreto
San Martín
Amazonas
Cajamarca
Lambayeque
La Libertad
Áncash
Huánuco

Moyobamba
Lamas
Tarapoto
Chachapoyas
Bagua
Cajamarca
Chiclayo
Pimentel
Trujillo
Salaverry
Chimbote
Casma

Cordillera Azul
Cordillera Blanca

Amazon

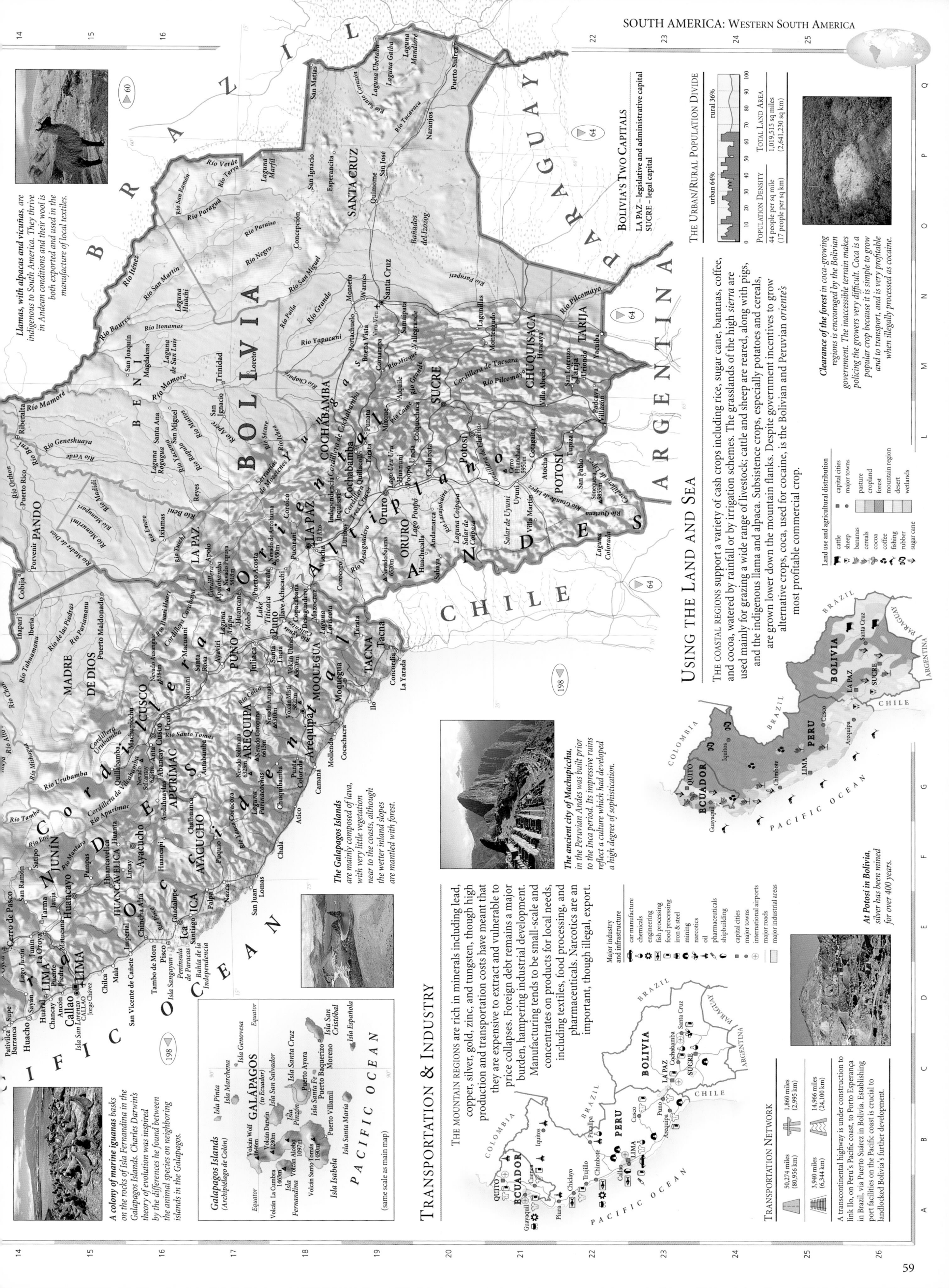

Llamas, with alpacas and vicuñas, are indigenous to South America. They thrive in Andean conditions and their wool is both exported and used in the manufacture of local textiles.

BOLIVIA'S TWO CAPITALS

LA PAZ – legislative and administrative capital
SUCRE – legal capital

THE URBAN/RURAL POPULATION DIVIDE

urban 64%   rural 36%

TOTAL LAND AREA
1,019,515 sq miles
(2,641,230 sq km)

POPULATION DENSITY
44 people per sq mile
(17 people per sq km)

Clearance of the forest in coca-growing regions is encouraged by the Bolivian government. The inaccessible terrain makes policing the growers very difficult. Coca is a popular crop because it is simple to grow and to transport, and is very profitable when illegally processed as cocaine.

USING THE LAND AND SEA

THE COASTAL REGIONS support a variety of cash crops including rice, sugar cane, bananas, coffee, and cocoa, watered by rainfall or by irrigation schemes. The grasslands of the high *sierra* are used mainly for grazing a wide range of livestock; cattle and sheep are reared, along with pigs, and the indigenous llama and alpaca. Subsistence crops, especially potatoes and cereals, are grown lower down the mountain flanks. Despite government incentives to grow alternative crops, coca, used for cocaine, is the Bolivian and Peruvian *oriente's* most profitable commercial crop.

Land use and agricultural distribution

- cattle
- sheep
- bananas
- cereals
- cocoa
- coffee
- fishing
- sugar cane

- capital cities
- major cities
- major towns

- pasture
- cropland
- forest
- mountain region
- desert
- wetlands

The ancient city of Machupicchu, in the Peruvian Andes was built prior to the Inca period. Its impressive ruins reflect a culture which had developed a high degree of sophistication.

At Potosí in Bolivia, silver has been mined for over 400 years.

The Galapagos Islands are mainly composed of lava, with very little vegetation near to the coasts, although the wetter inland slopes are mantled with forest.

A colony of marine iguanas basks on the rocks of Isla Fernandina in the Galapagos Islands. Charles Darwin's theory of evolution was inspired by the differences he found between the animal species on neighboring islands in the Galapagos.

Galápagos Islands
(Archipiélago de Colón)

(same scale as main map)

GALÁPAGOS
(to Ecuador)

TRANSPORTATION & INDUSTRY

THE MOUNTAIN REGIONS are rich in minerals including lead, copper, silver, gold, zinc, and tungsten, though high production and transportation costs have meant that they are expensive to extract and vulnerable to price collapses. Foreign debt remains a major burden, hampering industrial development. Manufacturing tends to be small-scale and concentrates on products for local needs, including textiles, food processing, and pharmaceuticals. Narcotics are an important, though illegal, export.

Major industry and infrastructure

- car manufacture
- chemicals
- engineering
- fish processing
- food processing
- iron & steel
- mining
- narcotics
- oil
- pharmaceuticals
- shipbuilding

- capital cities
- major towns
- international airports
- major roads
- major industrial areas

TRANSPORTATION NETWORK

50,274 miles
(80,956 km)

1,860 miles
(2,995 km)

3,940 miles
(6,344 km)

14,966 miles
(24,100 km)

A transcontinental highway is under construction to link Ilo, on Peru's Pacific coast, to Porto Esperança in Brazil, via Puerto Suárez in Bolivia. Establishing port facilities on the Pacific coast is crucial to landlocked Bolivia's further development.

BRAZIL

B RAZIL IS THE LARGEST COUNTRY in South America, with a population of over 165 million – greater than the combined total for the whole of the rest of the continent. The 26 states which make up the federal republic of Brazil are administered from the purpose-built capital, Brasília. Tropical rain forest, covering more than one-third of the country, contains rich natural resources, but great tracts are sacrificed to agriculture, industry and urban expansion on a daily basis. Most of Brazil's multiethnic population now live in cities, some of which are vast areas of urban sprawl; São Paulo is one of the world's biggest conurbations, with more than 17 million inhabitants. Although prosperity is a reality for some, many people still live in great poverty, and mounting foreign debts continue to damage Brazil's prospects of economic advancement.

USING THE LAND

BRAZIL HAS IMMENSE NATURAL RESOURCES, including minerals and hardwoods, many of which are found in the fragile rain forest. Brazil is the world's leading coffee grower and a major producer of livestock, sugar, and orange juice concentrate. Soybeans for animal feed, particularly for poultry feed, have become the country's most significant crop.

Land use and
agricultural distribution
- cattle
- pigs
- sheep
- citrus fruits
- coffee
- cotton
- soya beans
- sugar cane
- timber
- capital cities
- major towns
- pasture
- cropland
- forest

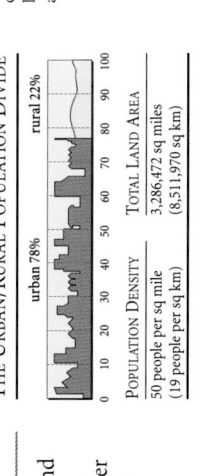

THE LANDSCAPE

THE AMAZON BASIN, containing the largest area of tropical rain forest on Earth, covers nearly half of Brazil. It is bordered by two shield areas: in the south by the Brazilian Highlands, and in the north by the Guiana Highlands. The east coast is dominated by a great escarpment which runs for 1,600 miles (2,565 km).

Brazil's highest mountain is the Pico da Neblina which was only discovered in 1962. It is 9,888 ft (3,014 m) high.

The floodplains which border the Amazon River are made up of a variety of different features including shallow lakes and swamps, mangrove forests in the tidal delta area, and fertile levees on river banks and point bars.

The Pantanal region in the south of Brazil is an extension of the Gran Chaco plain. The swamps and marshes of this area are renowned for their beauty, and abundant and unique wildlife, including wildfowl and these caimans, a type of crocodile.

Pantanal swamps

The fecundity of parts of Brazil's rain forest results from exceptionally high levels of rainfall and the quantities of silt deposited by the Amazon River system.

The ancient Brazilian Highlands have a varied topography. Their plateaus, hills, and deep valleys are bordered by highly-eroded mountains containing important mineral deposits. They are drained by three great river systems, the Amazon, the Paraguay–Paraná, and the São Francisco.

The São Francisco Basin has a climate unique in Brazil. Known as the "drought polygon," it has almost no rain during the dry season, leading to regular disastrous droughts.

The northeastern scrublands are known as the *caatinga*, a virtually impenetrable thorny woodland, sometimes intermixed with cacti where water is scarce.

The Amazon Basin is the largest river basin in the world. The Amazon River and over a thousand tributaries drain an area of 2,375,000 sq miles (6,150,000 sq km) and carry one-fifth of the world's fresh water out to sea.

Guiana Highlands

The famous Sugar Loaf Mountain (*Pão de Açúcar*) which overlooks Rio de Janeiro is a fine example of a volcanic plug – a domed core of solidified lava left after the slopes of the original volcano have eroded away.

Deep natural harbors such as Baía de Guanabara were created where the steep slopes of the Serra da Mantiqueira plunge directly into the ocean.

The Iguaçu River surges over the spectacular Iguaçu Falls (Saltos do Iguaçu) toward the Paraná River. Falls like these are increasingly under pressure from large-scale hydroelectric projects such as that at Itaipú.

Hillslope gullying

Direction of growth
Overland water flow
Gully

Rainfall
Water seeps through hillslope

Large-scale gullies are common in Brazil, particularly on hillslopes from which vegetation has been removed. Gullies grow headwards (up the slope), aided by a combination of erosion through water seepage and rainwater runoff.

THE URBAN/RURAL POPULATION DIVIDE

urban 78% rural 22%

POPULATION DENSITY
50 people per sq mile
(19 people per sq km)

TOTAL LAND AREA
3,286,472 sq miles
(8,511,970 sq km)

MAP KEY

POPULATION
- ■ above 5 million
- ● 1 million to 5 million
- ◉ 500,000 to 1 million
- ◎ 100,000 to 500,000
- ⊙ 50,000 to 100,000
- ⊙ 10,000 to 50,000
- ○ below 10,000

ELEVATION
| 3000m / 9843ft |
| 2000m / 6562ft |
| 1000m / 3281ft |
| 500m / 1640ft |
| 250m / 820ft |
| 100m / 328ft |
| sea level |

ATLANTIC OCEAN

Equator

FRENCH GUIANA (to France)

SURINAME

GUYANA

VENEZUELA

COLOMBIA

PERU

BOLIVIA

PARAGUAY

ARGENTINA

URUGUAY

BRAZIL

Guiana Highlands
Tumuc-Humac Mountains
Acaraí Mountains
Serra do Jatapu
Planalto
Amazon

Mouth of the Amazon
Ilha de Marajó

Manaus
Boa Vista
RORAIMA
AMAPÁ
Macapá
Belém
São Luís
Recife

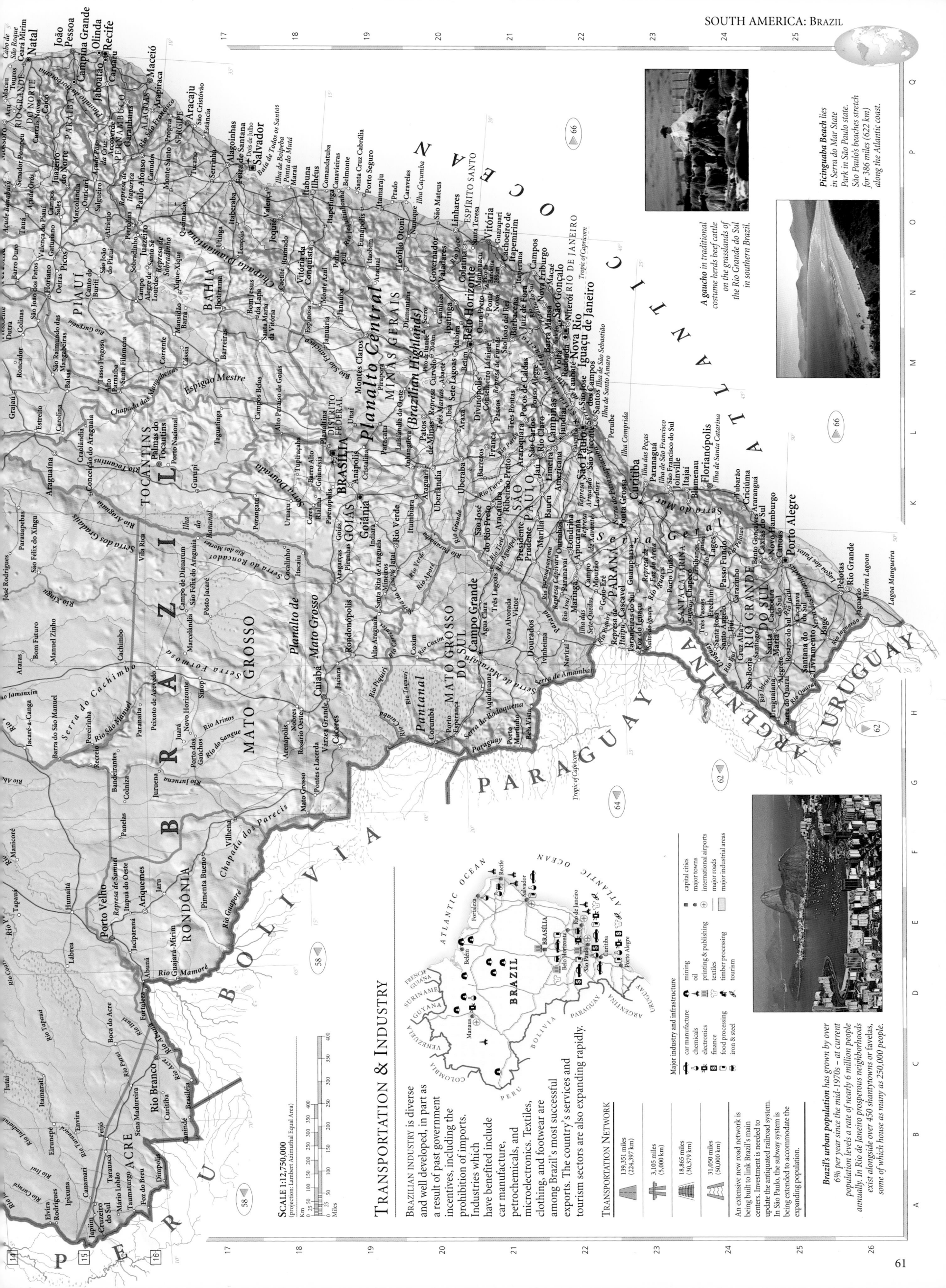

Picinguaba Beach lies in Serra do Mar State Park in São Paulo state. São Paulo's beaches stretch for 386 miles (622 km) along the Atlantic coast.

A gaucho in traditional costume herds beef cattle on the grasslands of the Rio Grande do Sul in southern Brazil.

TRANSPORTATION & INDUSTRY

BRAZILIAN INDUSTRY is diverse and well developed, in part as a result of past government incentives, including the prohibition of imports. Industries which have benefited include car manufacture, petrochemicals, and microelectronics. Textiles, clothing, and footwear are among Brazil's most successful exports. The country's services and tourism sectors are also expanding rapidly.

An extensive new road network is being built to link Brazil's main centers. Investment is needed to update the antiquated railroad system. In São Paulo, the subway system is being extended to accommodate the expanding population.

Brazil's urban population has grown by over 6% per year since the mid-1970s – at current population levels a rate of nearly 6 million people annually. In Rio de Janeiro prosperous neighborhoods exist alongside over 450 shantytowns or favelas, some of which house as many as 250,000 people.

TRANSPORTATION NETWORK

139,351 miles (224,397 km)

3,105 miles (5,000 km)

18,865 miles (30,379 km)

31,050 miles (50,000 km)

SCALE 1:12,750,000
(projection: Lambert Azimuthal Equal Area)

Major industry and infrastructure

- car manufacture
- chemicals
- electronics
- finance
- food processing
- iron & steel
- mining
- oil
- printing & publishing
- textiles
- timber processing
- tourism
- capital cities
- major towns
- international airports
- major roads
- major industrial areas

EASTERN SOUTH AMERICA

URUGUAY, NORTHEAST ARGENTINA, SOUTHEAST BRAZIL

The vast conurbations of Rio de Janeiro, São Paulo, and Buenos Aires form the core of South America's highly-urbanized eastern region. São Paulo state, with almost 35 million inhabitants, is among the world's 20 most powerful economies, and São Paulo is the fastest growing city on the continent. Rio de Janeiro and Buenos Aires, transformed in the last hundred years from port cities to great metropolitan areas each with more than 10 million inhabitants, typify the unstructured growth and wealth disparities of South America's great cities.

In Uruguay, over half of the population lives in the capital, Montevideo, which faces Buenos Aires across the Plate River (*Rio de la Plata*). Immigration from the countryside has created severe pressure on the urban infrastructure, particularly on available housing, leading to a profusion of crowded shanty settlements (*favelas or barrios*).

USING THE LAND

Most of Uruguay and the Pampas of northern Argentina are devoted to the rearing of livestock, especially cattle and sheep, which are central to both countries' economies. Soybeans, first produced in Brazil's Rio Grande do Sul, are now more widely grown for large-scale export, as are cereals, sugar cane, and grapes. Subsistence crops, including potatoes, corn and sugar beets, are grown on the remaining arable land.

Land use and agricultural distribution

- cattle
- sheep
- cereals
- coffee
- fruit
- soybeans
- sugar cane
- capital cities
- major towns
- pasture
- cropland
- forest
- wetlands
- barren land

TRANSPORTATION & INDUSTRY

Southeast Brazil is home to much of the important motor and capital goods industry, largely based around São Paulo; iron and steel production is also concentrated in this region. Uruguay's economy continues to be based mainly on the export of livestock products including meat and leather goods. Buenos Aires is Argentina's chief port, and the region has a varied and sophisticated economic base including service-based industries such as finance and publishing, as well as primary processing.

Major industry and infrastructure

- car manufacture
- chemicals
- engineering
- finance
- food processing
- iron & steel
- meat processing
- printing & publishing
- shipbuilding
- textiles
- timber processing
- capital cities
- major towns
- international airports
- major roads
- major industrial areas

TRANSPORTATION NETWORK

Throughout the region, road networks need to be expanded to cope with urban development. Plans are underway to build a bridge over the Plate River (*Rio de la Plata*) to link Colonia and Buenos Aires.

The Itaipú dam on the Paraná River is one of the largest hydroelectric projects in the world, jointly financed by Brazil and Paraguay.

Soybeans are harvested, pressed, and processed into soycake, which is used as animal feed. The cake is fed mainly to chickens on large-scale factory farms, and the growth in soy production has been an important factor in the expansion of the Brazilian poultry trade.

The rolling grasslands of Uruguay are ideally suited to the rearing of cattle, which are concentrated in great herds throughout the region.

Rio de Janeiro's annual carnival, Mardi Gras, which ushers in the start of Lent, is an extravagant, five-day parade through the city, characterized by fantastically decorated floats, exuberant dancing, and samba music.

MAP KEY

POPULATION
- ■ above 5 million
- ■ 1 million to 5 million
- ◉ 500,000 to 1 million
- ⊕ 100,000 to 500,000
- ○ 50,000 to 100,000
- ○ 10,000 to 50,000
- ○ below 10,000

ELEVATION
- 2000m / 6562ft
- 1000m / 3281ft
- 500m / 1640ft
- 250m / 820ft
- 100m / 328ft
- sea level

SCALE 1: 6,250,000
(projection: Lambert Azimuthal Equal Area)

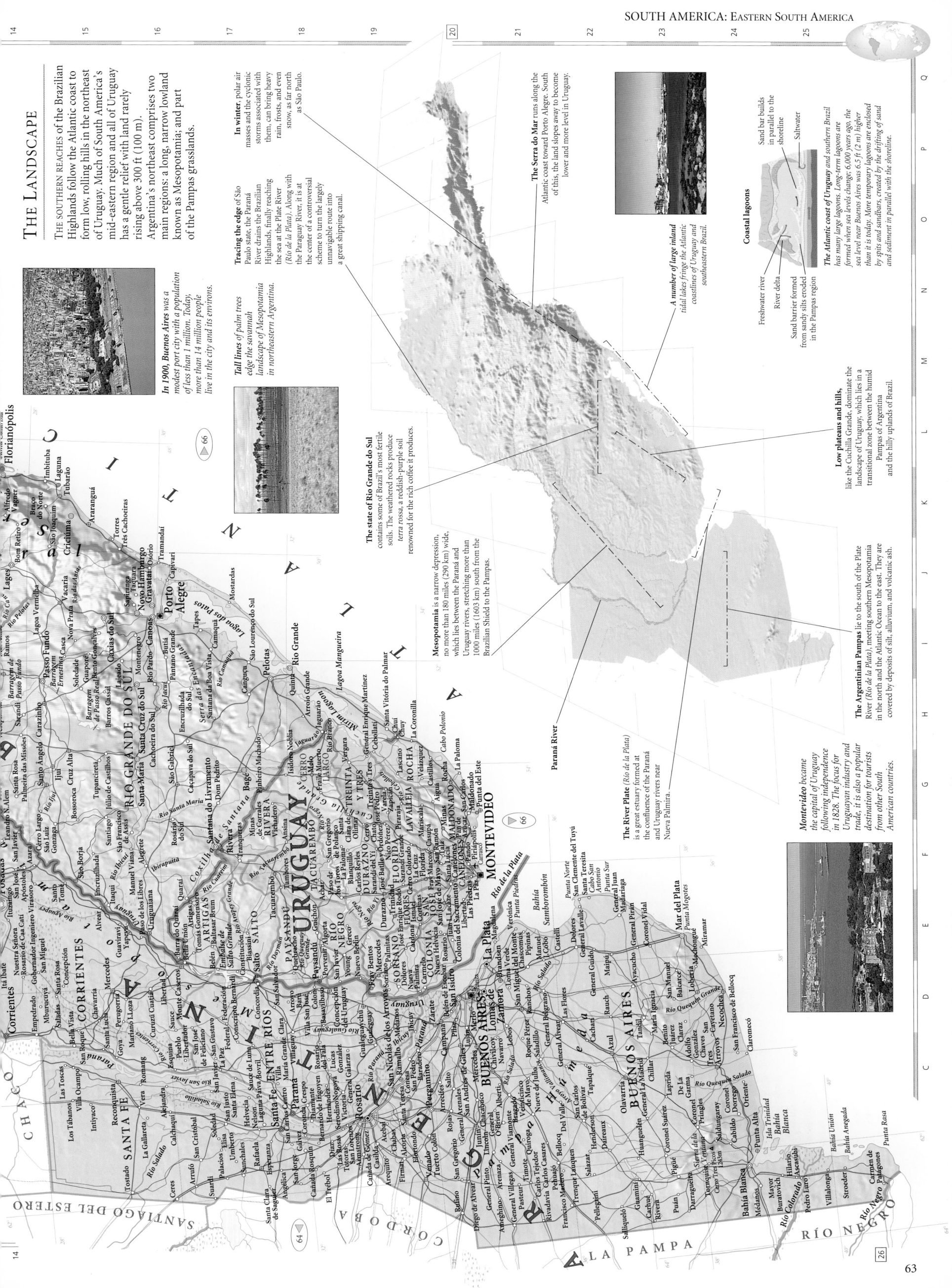

THE LANDSCAPE

THE SOUTHERN REACHES of the Brazilian Highlands follow the Atlantic coast to form low, rolling hills in the northeast of Uruguay. Much of South America's mid-eastern region and all of Uruguay has a gentle relief with land rarely rising above 300 ft (100 m). Argentina's northeast comprises two main regions: a long, narrow lowland known as Mesopotamia; and part of the Pampas grasslands.

In 1900, Buenos Aires was a modest port city with a population of less than 1 million. Today, more than 14 million people live in the city and its environs.

Tall lines of palm trees edge the savannah landscape of Mesopotamia in northeastern Argentina.

Tracing the edge of São Paulo state, the Paraná River drains the Brazilian Highlands, finally reaching the sea at the Plate River (*Río de la Plata*). Along with the Paraguay River, it is at the center of a controversial scheme to turn the largely unnavigable route into a great shipping canal.

In winter, polar air masses and the cyclonic storms associated with them, can bring heavy rain, frosts, and even snow, as far north as São Paulo.

The Serra do Mar runs along the Atlantic coast toward Porto Alegre. South of this, the land slopes away to become lower and more level in Uruguay.

A number of large inland tidal lakes fringe the Atlantic coastlines of Uruguay and southeastern Brazil.

The state of Rio Grande do Sul contains some of Brazil's most fertile soils. The weathered rocks produce *terra rossa*, a reddish-purple soil renowned for the rich coffee it produces.

Low plateaus and hills, like the Cuchilla Grande, dominate the landscape of Uruguay, which lies in a transitional zone between the humid Pampas of Argentina and the hilly uplands of Brazil.

Coastal lagoons

The Atlantic coast of Uruguay and southern Brazil has many large lagoons. Long-term lagoons are formed when sea levels change; 6,000 years ago, the sea level near Buenos Aires was 6.5 ft (2 m) higher than it is today. More temporary lagoons are enclosed by spits and sandbars, created by the drifting of sand and sediment in parallel with the shoreline.

Sand bar builds in parallel to the shoreline

Saltwater

Freshwater river

River delta

Sand barrier formed from sandy silts eroded in the Pampas region

Mesopotamia is a narrow depression, no more than 180 miles (290 km) wide, which lies between the Paraná and Uruguay rivers, stretching more than 1000 miles (1603 km) south from the Brazilian Shield to the Pampas.

The Argentinian Pampas lie to the south of the Plate River (*Río de la Plata*), meeting southern Mesopotamia in the north and the Atlantic Ocean to the east. They are covered by deposits of silt, alluvium, and volcanic ash.

Paraná River

Montevideo became the capital of Uruguay following independence in 1828. The focus for Uruguayan industry and trade, it is also a popular destination for tourists from other South American countries.

The River Plate (*Río de la Plata*) is a great estuary formed at the confluence of the Paraná and Uruguay rivers near Nueva Palmira.

SOUTHERN SOUTH AMERICA

ARGENTINA, CHILE, PARAGUAY

SOUTH AMERICA'S CONE-SHAPED SOUTHERN REGION is shared by Argentina and Chile, two overwhelmingly urbanized nations whose populations live mainly in or around the capital cities, Buenos Aires and Santiago. The people are largely *mestizo* or of European origin; in the early 20th century Argentina absorbed waves of new European immigrants, many from Italy and Germany. Paraguay is far less urbanized than its neighbors, with a homogeneous population of mixed Spanish and Guaraní origin, who retain their Indian roots through the Guaraní language. Though most Paraguayans live in the southeast, near Asunción, the indigenous Indians live in the sparsely populated Gran Chaco. The Gran Chaco is also home to some of Argentina's minority indigenous peoples, who otherwise live mainly in Andean regions. Chile's estimated 800,000 Mapauche Indians live almost exclusively in the south.

TRANSPORTATION & INDUSTRY

FOOD PROCESSING AND AGRICULTURAL EXPORTS remain a fundamental part of Argentina's economy. The growth of manufacturing is regularly hampered by hyperinflation and massive foreign debts. The world's most important copper-producer and one of the top ten gold producers, Chile also has a thriving wine and grape industry. Most Paraguayan exports involve primary processing, although domestic goods are produced for home markets.

Floodwaters cover the land in the Gran Chaco, partly submerging its vegetation of fan palms and hyacinths.

Boiling water and steam emerge from a volcanic vent, one of the Tatio geysers which lie at the foot of Cerro de Tocorpuri near Chile's border with Bolivia.

Chuquicamata copper mine, lies on a desert plateau near Calama in the Andes of northern Chile. It is the world's largest open-pit copper mine.

Transportation Network

89,104 miles (143,485 km)	2,809 miles (4,523 km)
23,107 miles (37,210 km)	9,206 miles (14,825 km)

Argentina's state transportation system is undergoing privatization, though the outmoded rail network requires updating. Paraguay requires foreign investment to upgrade its roads and railroads. Essential internal air routes, especially across the Andes, are well developed in all three countries.

Major industry and infrastructure

- chemicals
- engineering
- food processing
- meat processing
- mining
- oil
- textiles
- timber processing
- capital cities
- major towns
- international airports
- major roads
- major industrial areas

Map Key

POPULATION
- ◻ 1 million to 5 million
- ◉ 500,000 to 1 million
- ⊙ 100,000 to 500,000
- ⊚ 50,000 to 100,000
- ○ 10,000 to 50,000
- ∘ below 10,000

ELEVATION
- 6000m / 19,686ft
- 4000m / 13,124ft
- 3000m / 9843ft
- 2000m / 6562ft
- 1000m / 3281ft
- 500m / 1640ft
- 250m / 820ft
- 100m / 328ft
- sea level

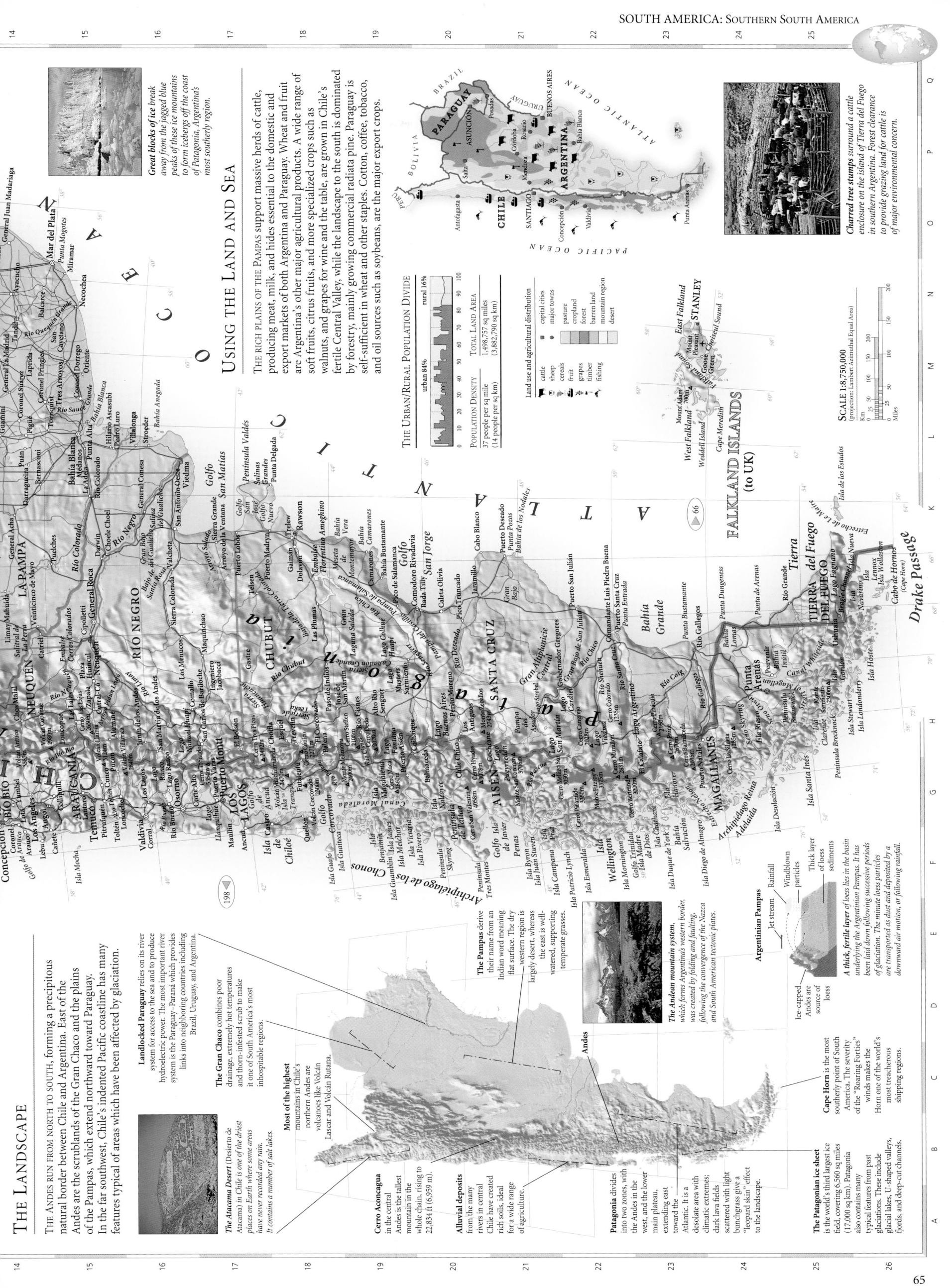

THE LANDSCAPE

THE ANDES RUN FROM NORTH TO SOUTH, forming a precipitous natural border between Chile and Argentina. East of the Andes are the scrublands of the Gran Chaco and the plains of the Pampas, which extend northward toward Paraguay. In the far southwest, Chile's indented Pacific coastline has many features typical of areas which have been affected by glaciation.

The Atacama Desert (Desierto de Atacama) in Chile is one of the driest places on Earth where some areas have never recorded any rain. It contains a number of salt lakes.

Cerro Aconcagua in the central Andes is the tallest mountain in the whole chain, rising to 22,834 ft (6,959 m).

Most of the highest mountains in Chile's northern Andes are volcanoes like Volcán Lascar and Volcán Rutana.

Alluvial deposits from the many rivers in central Chile have created rich soils, ideal for a wide range of agriculture.

Landlocked Paraguay relies on its river system for access to the sea and to produce hydroelectric power. The most important river system is the Paraguay–Paraná which provides links into neighboring countries including Brazil, Uruguay, and Argentina.

The Gran Chaco combines poor drainage, extremely hot temperatures and thorn-infested scrub to make it one of South America's most inhospitable regions.

The Pampas derive their name from an Indian word meaning flat surface. The dry western region is largely desert, whereas the east is well-watered, supporting temperate grasses.

The Andean mountain system, which forms Argentina's western border, was created by folding and faulting, following the convergence of the Nazca and South American tectonic plates.

Patagonia divides into two zones, with the Andes in the west, and the lower main plateau, extending east toward the Atlantic. It is a desolate area with dark lava fields scattered with light bunchgrass give a "leopard skin" effect to the landscape.

The Patagonian ice sheet is the world's third largest ice field, covering 6,560 sq miles (17,000 sq km). Patagonia also contains many typical features from past glaciations. These include glacial lakes, U-shaped valleys, fjords, and deep-cut channels.

Argentinian Pampas

Windblown particles

Jet stream

Rainfall

Thick layer of loess sediments

Ice-capped Andes are source of loess

A thick, fertile layer of loess lies in the basin underlying the Argentinian Pampas. It has been laid down following successive periods of glaciation. The minute loess particles are transported as dust and deposited by a downward air motion, or following rainfall.

Cape Horn is the most southerly point of South America. The severity of the "Roaring Forties" winds makes the Horn one of the world's most treacherous shipping regions.

Andes

USING THE LAND AND SEA

THE RICH PLAINS OF THE PAMPAS support massive herds of cattle, producing meat, milk, and hides essential to the domestic and export markets of both Argentina and Paraguay. Wheat and fruit are Argentina's other major agricultural products. A wide range of soft fruits, citrus fruits, and more specialized crops such as walnuts, and grapes for wine and the table, are grown in Chile's fertile Central Valley, while the landscape to the south is dominated by forestry, mainly growing commercial radiata pine. Paraguay is self-sufficient in wheat and other staples. Cotton, coffee, tobacco, and oil sources such as soybeans, are the major export crops.

Great blocks of ice break away from the jagged blue peaks of these ice mountains to form icebergs off the coast of Patagonia, Argentina's most southerly region.

Charred tree stumps surround a cattle enclosure on the island of Tierra del Fuego in southern Argentina. Forest clearance to provide grazing land for cattle is of major environmental concern.

THE URBAN/RURAL POPULATION DIVIDE

urban 84%   rural 16%

POPULATION DENSITY
37 people per sq mile
(14 people per sq km)

TOTAL LAND AREA
1,498,757 sq miles
(3,882,790 sq km)

Land use and agricultural distribution

- cattle
- sheep
- cereals
- fruit
- grapes
- timber
- fishing

- capital cities
- major towns
- pasture
- cropland
- forest
- barren land
- mountain region
- desert

SCALE 1:8,750,000
(projection: Lambert Azimuthal Equal Area)

Km
Miles

THE ATLANTIC OCEAN

THE ATLANTIC IS THE YOUNGEST OF THE WORLD'S OCEANS, formed about 180 million years ago when the landmasses of the eastern and western hemispheres separated. Its underwater topography is dominated by the Mid-Atlantic Ridge, a huge mountain system running north to south along the center of the ocean. Although most of the ridge's peaks lie below the sea, some emerge as volcanic islands, like Iceland and the Azores. The Atlantic contains a wealth of resources, including substantial oil and gas reserves and rich fishing grounds. Until the 1950s, the north Atlantic was the world's busiest shipping route; cheaper air transportation and alternative routes have shifted patterns of world trade.

RESOURCES

DEVELOPMENT OF THE OIL AND GAS RESERVES in the Atlantic began in the 1940s around the Gulf of Mexico. Since then other areas have been exploited, including the North Sea, the west coast of Africa and the area east of Newfoundland and Nova Scotia. There is also extensive mining of sand, gravel, and shell deposits by the US and UK. For centuries, the north Atlantic's fishing grounds have been utilized more heavily than other oceans, leading to a serious decline in many fish stocks.

Resources (including wildlife)
- fish
- whales
- aggregates
- oil & gas
- major towns
- major ports

Fishing in the seas around northwestern Europe dates back over 1,500 years. The high nutrient content of the seas makes them ideal breeding grounds for many species of fish.

Surtsey near Iceland, lies on the Mid-Atlantic Ridge. The island was formed in 1963 following a volcanic eruption caused by sea-floor spreading.

On January 5 1993, the oil tanker Braer ran aground in the Shetland Islands, spilling 83,660 tons (85,000 tonnes) of light crude oil into the ocean, devastating the local marine ecosystem.

Inset maps

AZORES (to Portugal)
SCALE 1:6,500,000

Corvo, Flores, Graciosa, Terceira, São Jorge, Faial, Pico, Ponta do Pico 2351m, Horta, Ponta Delgada, São Miguel, Ribeira Grande, Santa Maria, Vila do Porto

MADEIRA (to Portugal)
SCALE 1:2,500,000

Porto Santo, Camacha, Ilhéu de Baixo, Faial, Santana, Porto da Cruz, Machico, Santa Cruz, Funchal, Ilhas Desertas, Deserta Grande, Bugio, Ponta do Pargo, Porto do Moniz, Pico Ruivo de Santana 1862m, Ribeira Brava, Calheta, Câmara de Lobos, Deserta Grande

ISLAS CANARIAS (CANARY ISLANDS) (to Spain)
SCALE 1:6,500,000

Alegranza, Graciosa, Teguise, Arrecife, Tinajo, Lanzarote, Puerto del Rosario, Fuerteventura, Antigua, Punta de Jandía, Las Palmas, Las Palmas de Gran Canaria, Santa Cruz de Tenerife, La Orotava, Pico del Teide 3718m, Gomera, Garajonay 1487m, La Palma, Los Llanos, Santa Cruz de la Palma, Hierro, Valverde, Los Rodeos, Puerto de la Cruz, Santa Cruz de Tenerife

BERMUDA (to UK)
SCALE 1:500,000

Ireland Island North, Ireland Island South, Somerset, Gibbs Hill, Spanish Point, Great Sound, Harrington Sound, Hamilton, Tucker's Town, Harbour, St George, St George's Island, St Catherine Point, St David's Island, Commissioner's, Kindley Field

SCALE 1:43,000,000 (projection: Mollweide)

Map labels

ARCTIC CIRCLE, ATLANTIC OCEAN

EUROPE — ICELAND, Reykjavik, Faeroe Islands (to Denmark), Shetland Islands, North Sea, UNITED KINGDOM, Milford Haven, Southampton, English Channel, IRELAND, REPUBLIC OF IRELAND, Belfast, Cork, Celtic Sea, Rotterdam, FRANCE, Leire, Nantes, Bordeaux, Bay of Biscay, Bilbao, Gijón, SPAIN, PORTUGAL, Lisbon, Gibraltar, Strait of Gibraltar

AFRICA — MOROCCO, Casablanca, Safi, ALGERIA, Western Sahara (occupied by Morocco), MAURITANIA, Nouâdhibou, Nouakchott, SENEGAL, Dakar, GAMBIA, Banjul, GUINEA-BISSAU, Bissau, GUINEA, Conakry, SIERRA LEONE, Freetown, LIBERIA, Monrovia, IVORY COAST, Abidjan, GHANA, TOGO, BENIN, NIGERIA, Lagos, Porto-Novo, Niger, CAMEROON, Port Harcourt

NORTH AMERICA — Greenland (to Denmark), Nuuk, Baffin Bay, Baffin Island, Davis Strait, Labrador Sea, CANADA, Hudson Strait, Ungava Bay, Foxe Basin, Montréal, St Lawrence, Gulf of St Lawrence, Newfoundland, Halifax, Nova Scotia, Boston, New York, Baltimore, UNITED STATES OF AMERICA, Jacksonville, Savannah, Mobile, New Orleans, Gulf of Mexico, MEXICO, Tampico, Veracruz, BAHAMAS, CUBA, JAMAICA, HAITI, DOMINICAN REPUBLIC, PUERTO RICO (to USA), BELIZE, GUATEMALA, HONDURAS, NICARAGUA, COSTA RICA, PANAMA, Caribbean Sea

SOUTH AMERICA — COLOMBIA, VENEZUELA, Caracas, Maracaibo, TRINIDAD & TOBAGO, BARBADOS, GUYANA, Georgetown, Paramaribo, Cayenne, French Guiana, Orinoco

Ocean floor features
Mid-Atlantic Ridge, Reykjanes Ridge, Iceland Basin, Iceland-Faeroe Ridge, Hatton Bank, Rockall Bank, Rockall Trough, Porcupine Bank, Porcupine Plain, Biscay Plain, Iberian Plain, Tagus Plain, Madeira (to Portugal), Madeira Plain, Canary Islands (to Spain), Cape Verde Terrace, Cape Verde Plain, Cape Verde Basin, Sierra Leone Basin, Guinea Basin, Denmark Strait, Irminger Basin, Labrador Basin, Eirik Ridge, Flemish Cap, Orphan Knoll, Newfoundland Basin, Newfoundland Ridge, Newfoundland Seamounts, Grand Banks of Newfoundland, Georges Bank, Northwest Atlantic Mid-Ocean Canyon, New England Seamounts, Sohm Plain, Bermuda (to UK), Bermuda Rise, Nares Plain, Hatteras Plain, Blake Plateau, Blake-Bahama Ridge, Puerto Rico Trench, Venezuelan Basin, Colombian Basin, Aves Ridge, Barracuda Ridge, Demerara Plain, Demerara Plateau, Sargasso Sea, Charlie-Gibbs Fracture Zone, Oceanographer Fracture Zone, Atlantis Fracture Zone, Kane Fracture Zone, Vema Fracture Zone, Doldrums Fracture Zone, Azores-Biscay Rise, Azores Fracture Zone, East Azores Fracture Zone, Great Meteor Tablemount, Cruiser Tablemount, Madeira Seamounts, Saldanha Seamounts, West Thulean Rise, East Thulean Rise, Milne Seamounts, Hamilton Bank, Saglek Bank, Great Halibut Bank, Reykjanes Basin

Globe inset
NORTH AMERICA, EUROPE, AFRICA, SOUTH AMERICA, ANTARCTICA, Reykjavik, Rotterdam, Gibraltar, New York, New Orleans, Sargasso Sea, Caribbean Sea, Lagos, Rio de Janeiro, Buenos Aires, Cape Town, La Guaira, Cristóbal, Scotia Sea, Weddell Sea, ATLANTIC OCEAN, SOUTH ATLANTIC OCEAN

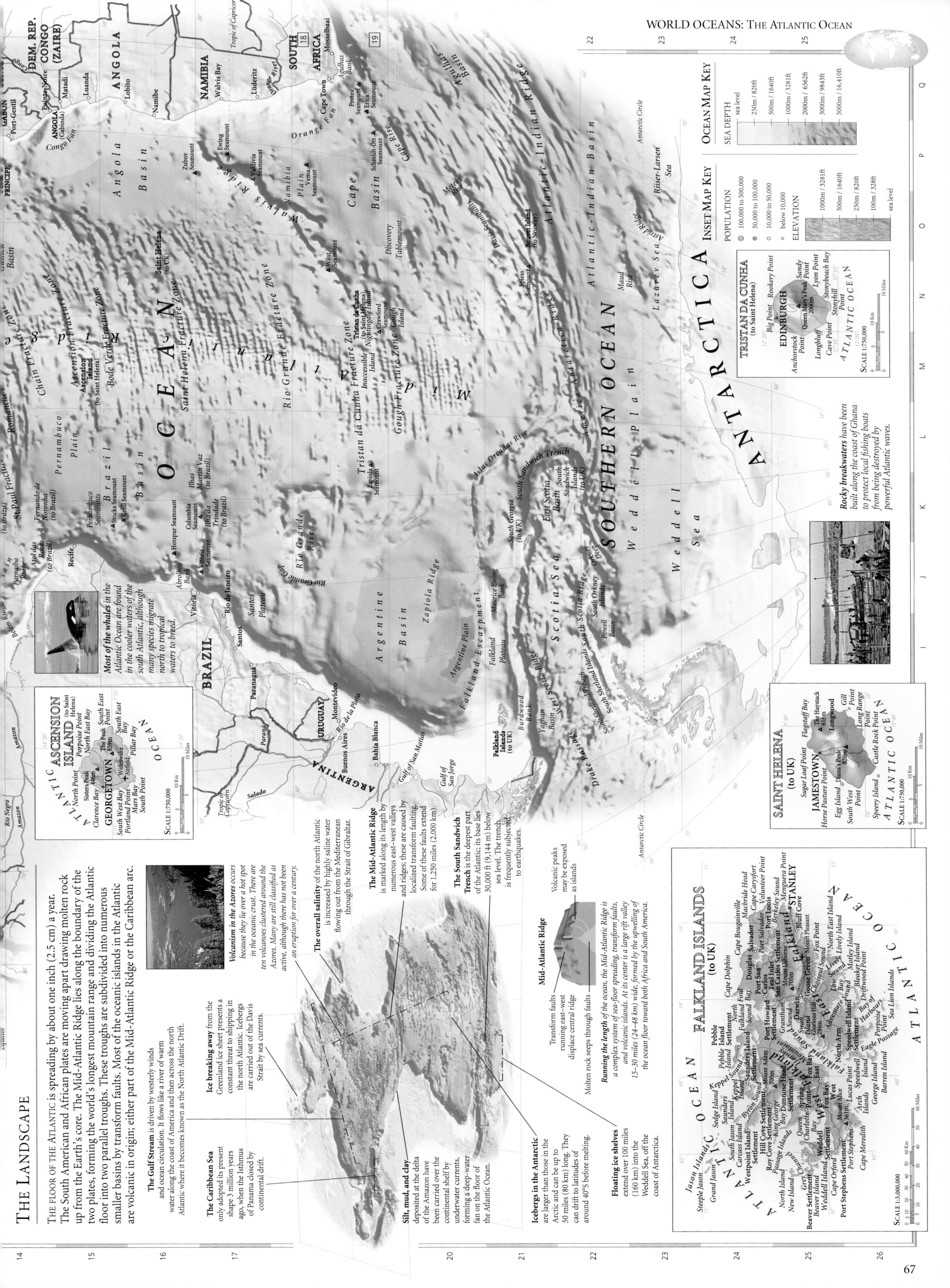

THE LANDSCAPE

THE FLOOR OF THE ATLANTIC is spreading by about one inch (2.5 cm) a year. The South American and African plates are moving apart drawing molten rock up from the Earth's core. The Mid-Atlantic Ridge lies along the boundary of the two plates, forming the world's longest mountain range and dividing the Atlantic floor into two parallel troughs. These troughs are subdivided into numerous smaller basins by transform faults. Most of the oceanic islands in the Atlantic are volcanic in origin; either part of the Mid-Atlantic Ridge or the Caribbean arc.

The Gulf Stream is driven by westerly winds and ocean circulation. It flows like a river of warm water along the coast of America and then across the north Atlantic where it becomes known as the North Atlantic Drift.

Ice breaking away from the Greenland ice sheet presents a constant threat to shipping in the north Atlantic. Icebergs are carried out of the Davis Strait by sea currents.

The Caribbean Sea only adopted its present shape 3 million years ago, when the Isthmus of Panama closed by continental drift.

Silt, mud, and clay, deposited at the delta of the Amazon have been carried over the continental shelf by underwater currents, forming a deep-water fan on the floor of the Atlantic Ocean.

Icebergs in the Antarctic are larger than those in the Arctic and can be up to 50 miles (80 km) long. They can drift to latitudes of around 40°S before melting.

Floating ice shelves extend over 100 miles (160 km) into the Weddell Sea, off the coast of Antarctica.

Volcanism in the Azores occurs because they lie over a hot spot in the oceanic crust. There are ten volcanoes clustered around the Azores. Many are still classified as active, although there has not been an eruption for over a century.

The overall salinity of the north Atlantic is increased by highly saline water flowing out from the Mediterranean through the Strait of Gibraltar.

The Mid-Atlantic Ridge is marked along its length by numerous east–west valleys and ridges; these are caused by localized transform faulting. Some of these faults extend for 1,250 miles (2,000 km).

The South Sandwich Trench is the deepest part of the Atlantic; its base lies 30,000 ft (9,144 m) below sea level. The trench is frequently subjected to earthquakes.

Volcanic peaks may be exposed as islands.

Mid-Atlantic Ridge

Transform faults running east–west displace central ridge

Molten rock seeps through faults

Running the length of the ocean, the Mid-Atlantic Ridge is a complex system of sea-floor spreading, transform faults, and volcanic islands. At its center is a large rift valley 15–30 miles (24–48 km) wide, formed by the upwelling of the ocean floor toward both Africa and South America.

Most of the whales in the Atlantic Ocean are found in the cooler waters of the south Atlantic, although many species migrate north to tropical waters to breed.

Rocky breakwaters have been built along the coast of Ghana to protect local fishing boats from being destroyed by powerful Atlantic waves.

OCEAN MAP KEY

SEA DEPTH

sea level
250m / 820ft
500m / 1640ft
1000m / 3281ft
2000m / 6562ft
3000m / 9843ft
5000m / 16,410ft

INSET MAP KEY

POPULATION
⊙ 100,000 to 500,000
◉ 50,000 to 100,000
○ 10,000 to 50,000
∘ below 10,000

ELEVATION
1000m / 3281ft
500m / 1640ft
250m / 820ft
100m / 328ft
sea level

TRISTAN DA CUNHA (to Saint Helena)
EDINBURGH
Big Point Rookery Point
Sandy Point
Queen Mary's Peak 2060m
Lyon Point
Anchorstock Point
Longbluff Point Stonyhill Point
Cave Point
ATLANTIC OCEAN
SCALE 1:750,000

SAINT HELENA (to UK)
JAMESTOWN
Sugar Loaf Point Flagstaff Bay
The Haystack
Horse Pasture Point Longwood
Egg Island Diana's Peak 820m
South West Point
Speery Island Castle Rock Point
ATLANTIC OCEAN
SCALE 1:750,000

ASCENSION ISLAND (to Saint Helena)
GEORGETOWN
North Point South East Bay
Sisters Peak Porpoise Point
The Peak 859m South East Bay
Clarence Bay Airfield Pillar Bay
Wideawake
South West Bay Mars Bay
Portland Point South Point
ATLANTIC OCEAN
SCALE 1:750,000

FALKLAND ISLANDS (to UK)
STANLEY
ATLANTIC OCEAN
SCALE 1:3,000,000

SCALE 1:750,000 (bottom 67)

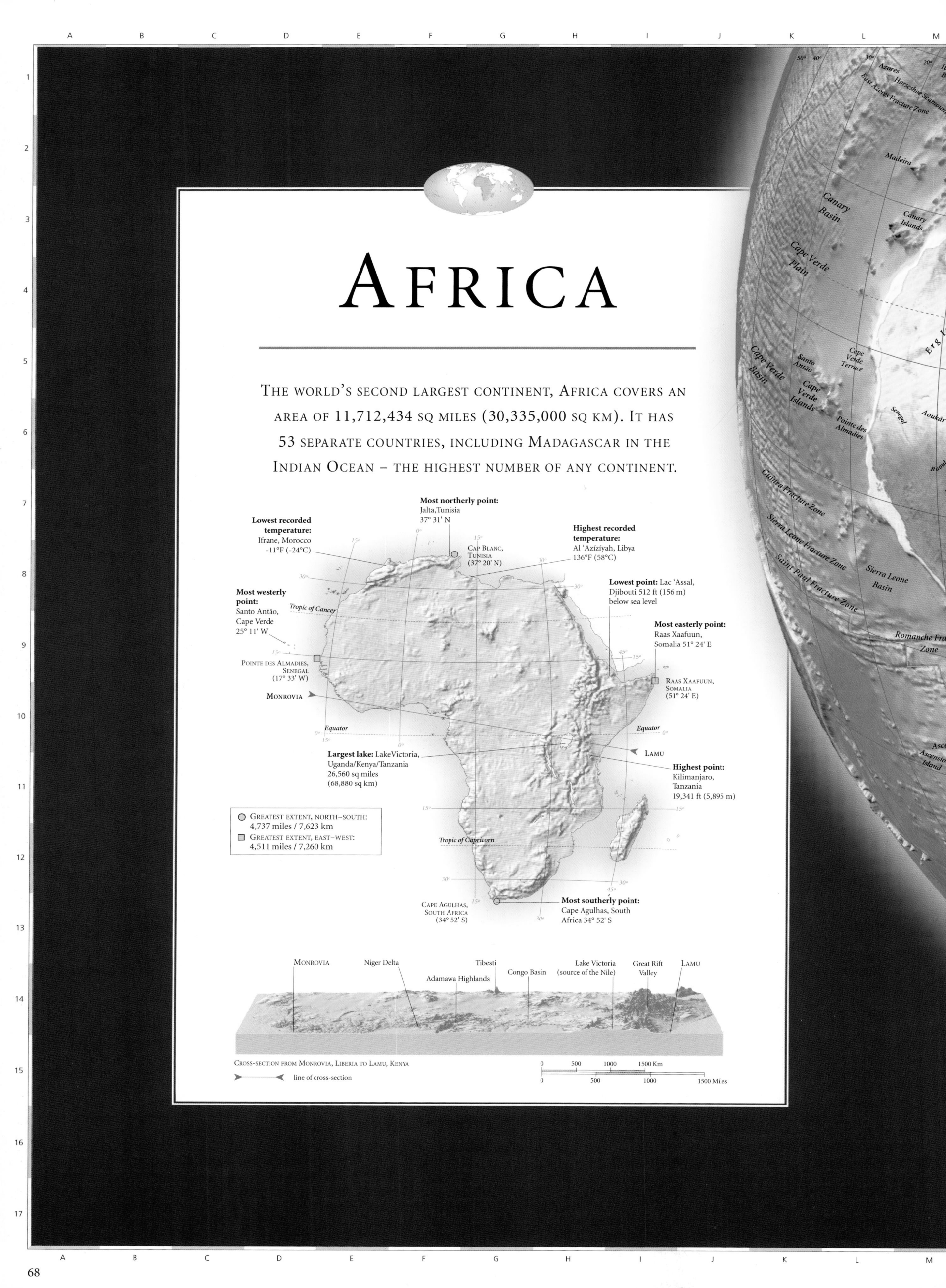

AFRICA

THE WORLD'S SECOND LARGEST CONTINENT, AFRICA COVERS AN AREA OF 11,712,434 SQ MILES (30,335,000 SQ KM). IT HAS 53 SEPARATE COUNTRIES, INCLUDING MADAGASCAR IN THE INDIAN OCEAN – THE HIGHEST NUMBER OF ANY CONTINENT.

Most northerly point:
Jalta, Tunisia
37° 31' N

Lowest recorded temperature:
Ifrane, Morocco
-11°F (-24°C)

Highest recorded temperature:
Al 'Azízíyah, Libya
136°F (58°C)

CAP BLANC, TUNISIA
(37° 20' N)

Lowest point: Lac 'Assal, Djibouti 512 ft (156 m) below sea level

Most westerly point:
Santo Antão, Cape Verde
25° 11' W

Tropic of Cancer

Most easterly point:
Raas Xaafuun,
Somalia 51° 24' E

POINTE DES ALMADIES, SENEGAL
(17° 33' W)

RAAS XAAFUUN, SOMALIA
(51° 24' E)

MONROVIA

Equator

Equator

LAMU

Largest lake: Lake Victoria, Uganda/Kenya/Tanzania 26,560 sq miles (68,880 sq km)

Highest point: Kilimanjaro, Tanzania 19,341 ft (5,895 m)

○ GREATEST EXTENT, NORTH–SOUTH: 4,737 miles / 7,623 km
□ GREATEST EXTENT, EAST–WEST: 4,511 miles / 7,260 km

Tropic of Capricorn

CAPE AGULHAS, SOUTH AFRICA
(34° 52' S)

Most southerly point:
Cape Agulhas, South Africa 34° 52' S

MONROVIA   Niger Delta   Tibesti   Congo Basin   Lake Victoria (source of the Nile)   Great Rift Valley   LAMU

Adamawa Highlands

CROSS-SECTION FROM MONROVIA, LIBERIA TO LAMU, KENYA

line of cross-section

| 0 | 500 | 1000 | 1500 Km |
| 0 | 500 | 1000 | 1500 Miles |

Azores
Horseshoe Seamounts
East Azores Fracture Zone
Madeira
Canary Basin
Canary Islands
Cape Verde Plain
Cape Verde Basin
Santo Antão
Cape Verde Islands
Cape Verde Terrace
Pointe des Almadies
Senegal
Guinea Fracture Zone
Sierra Leone Fracture Zone
Saint Paul Fracture Zone
Sierra Leone Basin
Romanche Fracture Zone
Ascension Island

PHYSICAL AFRICA

THE STRUCTURE OF AFRICA was dramatically influenced by the break up of the supercontinent Gondwanaland about 160 million years ago and, more recently, rifting and hot spot activity. Today, much of Africa is remote from active plate boundaries and comprises a series of extensive plateaus and deep basins, which influence the drainage patterns of major rivers. The relief rises to the east, where volcanic uplands and vast lakes mark the Great Rift Valley. In the far north and south sedimentary rocks have been folded to form the Atlas Mountains and the Great Karoo.

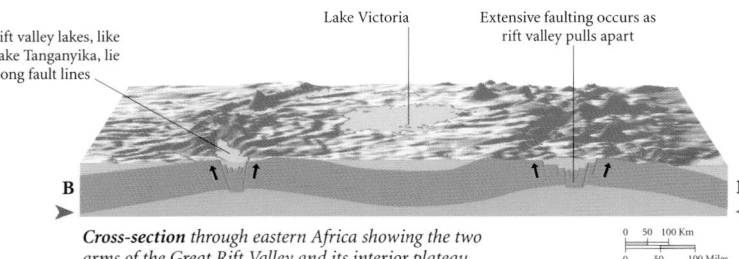

EAST AFRICA

THE GREAT RIFT VALLEY is the most striking feature of this region, running for 4,475 miles (7,200 km) from Lake Nyasa to the Red Sea. North of Lake Nyasa it splits into two arms and encloses an interior plateau which contains Lake Victoria. A number of elongated lakes and volcanoes lie along the fault lines. To the west lies the Congo Basin, a vast, shallow depression, which rises to form an almost circular rim of highlands.

Rift valley lakes, like Lake Tanganyika, lie along fault lines

Lake Victoria

Extensive faulting occurs as rift valley pulls apart

B

B

Cross-section through eastern Africa showing the two arms of the Great Rift Valley and its interior plateau.

0 50 100 Km
0 50 100 Miles

NORTHERN AFRICA

NORTHERN AFRICA COMPRISES a system of basins and plateaus. The Tibesti and Ahaggar are volcanic uplands, whose uplift has been matched by subsidence within large surrounding basins. Many of the basins have been infilled with sand and gravel, creating the vast Saharan lands. The Atlas Mountains in the north were formed by convergence of the African and Eurasian plates.

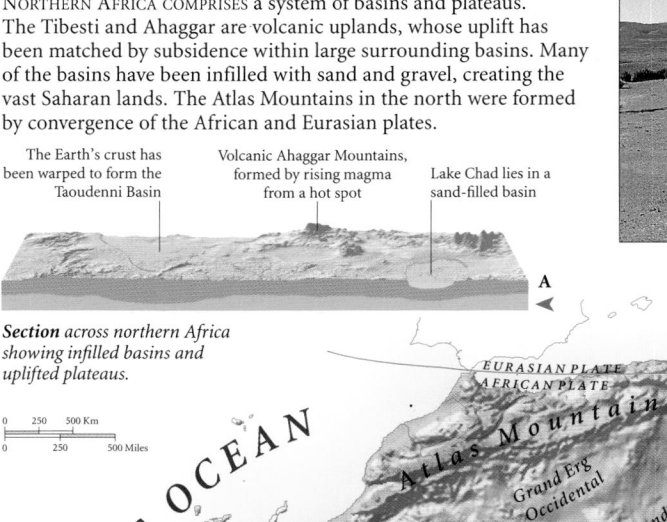

The Earth's crust has been warped to form the Taoudenni Basin

Volcanic Ahaggar Mountains, formed by rising magma from a hot spot

Lake Chad lies in a sand-filled basin

A

A

Section across northern Africa showing infilled basins and uplifted plateaus.

0 250 500 Km
0 250 500 Miles

SCALE 1:36,000,000
(projection: Lambert Azimuthal Equal Area)

Km
0 100 200 400 600 800
Miles
0 100 200 400 600 800

MAP KEY

ELEVATION

5000m / 16,405ft
4000m / 13,124ft
3000m / 9843ft
2000m / 6562ft
1000m / 3281ft
500m / 1640ft
250m / 820ft
100m / 328ft
sea level
below sea level

PLATE MARGINS
(for explanation see page xiv)

constructive
destructive
conservative
uncertain
line of cross-section

SOUTHERN AFRICA

THE GREAT ESCARPMENT marks the southern boundary of Africa's basement rock and includes the Drakensberg range. It was uplifted when Gondwanaland fragmented about 160 million years ago and it has gradually been eroded back from the coast. To the north, the relief drops steadily, forming the Kalahari Basin. In the far south are the fold mountains of the Great Karoo.

Kalahari Basin, covered with the sandy plains of the Kalahari Desert

Boundary of the Great Escarpment

Uplift of the basement rock created a raised plateau

Drakensberg

C

C

Cross-section through southern Africa showing the boundary of the Great Escarpment.

0 100 200 Km
0 100 200 Miles

Map labels

ATLANTIC OCEAN

Mediterranean Sea

EURASIAN PLATE
AFRICAN PLATE
ANATOLIAN PLATE
AFRICAN PLATE
ARABIAN PLATE

Atlas Mountains
Chott el Jerid
Gulf of Sirte
Grand Erg Occidental
Grand Erg Oriental
Erg Iguidi
Erg Chech
Ahaggar
Qattara Depression
Nile Delta
Western Desert
Great Sand Sea
Libyan Desert
Nile
Lake Nasser
Nubian Desert
Arabian Plate
Red Sea

ARABIAN PLATE
AFRICAN PLATE
ASIA

Sahara
Taoudenni Basin
Niger
Massif de l'Aïr
Tibesti
Ténéré

Cape Verde Islands
Senegal
Niger
Sahel
White Volta
Lake Volta
Niger
Benue
Adamawa Highlands
Grain Coast
Ivory Coast
Gold Coast
Slave Coast
Bight of Benin
Niger Delta
Cameroon Mountain 4070m
Gulf of Guinea
São Tomé
Ubangi
Massif des Bongo
Sudd
Congo
Congo Basin
Congo

Blue Nile
White Nile
Lake Tana
Gulf of Aden
Horn of Africa
Ethiopian Highlands
Shebeli
Lake Rudolf
Juba
Lake Albert
Lake Victoria
Great Rift Valley
Kilimanjaro 5895m
Lake Tanganyika
Pemba Island
Zanzibar
Seychelles

ATLANTIC OCEAN
Bié Plateau
Namib Desert
Okavango Delta
Kalahari Basin
Kalahari Desert
Zambezi
Lake Nyasa
Zambezi
Comoro Islands
Mozambique Channel
Madagascar
Mauritius
Réunion

Orange River
Great Karoo
Drakensberg
Cape of Good Hope

INDIAN OCEAN

CLIMATE

THE CLIMATES OF AFRICA range from mediterranean to arid, dry savannah and humid equatorial. In East Africa, where snow settles at the summit of volcanoes such as Kilimanjaro, climate is also modified by altitude. The winds of the Sahara export millions of tonnes of dust a year both northward and eastward.

Savannah grasslands run in a belt across Africa; limited rainfall inhibits tree growth.

TEMPERATURE

Average January temperature

Average July temperature

Temperature	
	32 to 50° F (0 to 10°C)
	50 to 68°F (10 to 20°C)
	68 to 86°F (20 to 30°C)
	above 86°F (30°C)

RAINFALL

Average January rainfall

Average July rainfall

The hot, equatorial basin of the Congo River receives over 48 inches (1,200 mm) of rainfall per year.

Rainfall	
0–1 in (0–25 mm)	8–12 in (200–300 mm)
1–2 in (25–50 mm)	12–16 in (300–400 mm)
2–4 in (50–100 mm)	16–20 in (400–500 mm)
4–8 in (100–200 mm)	more than 20 in (500 mm)

Climate	
	arid
	humid equatorial
	mediterranean
	semiarid
	tropical
	warm humid
☼	daily hours of sunshine, January
☼	daily hours of sunshine, July
→	cold wind
→	hot wind

SHAPING THE CONTINENT

AFRICAN LANDSCAPES are shaped by the intensity of climatic extremes and by tectonic action. High aridity, wind action, and infrequent but heavy rainstorms, lead to the migration of sand dunes and dramatic flash flooding across much of the north and west. In the wetter areas, high precipitation increases the rate of weathering. To the east, the rift system has created a volcanic and lake environment and allowed rivers to erode weaknesses left in the crustal structure by faults.

GROUNDWATER

1 Oases are found in desert areas such as the Sahara *(left)*. Groundwater migrates through permeable rock strata, confined between two impermeable layers. Oases form either when the permeable rocks come near to the surface, or at a fault line, when water is able to seep up to the surface through the crushed rocks at the fault.

Rainwater feeds the aquifer

Water migrates up through fault

Aquifer exposed near the surface

Groundwater trapped between impermeable strata

GROUNDWATER: REPLENISHMENT OF AN OASIS

RIVER SYSTEMS

2 The Zambezi River *(above)* drops 360 ft (110 m) over the Victoria Falls into a zigzag gorge. The river has eroded the gorge along lines of weakness in the bedrock, created by fault lines running in two directions.

Old site of Victoria Falls

River plunges over falls

Fault and joint lines running in two directions

Zig-zag gorge of the Zambezi

RIVER SYSTEMS: RETREATING OF THE VICTORIA FALLS

THE EVOLVING LANDSCAPE

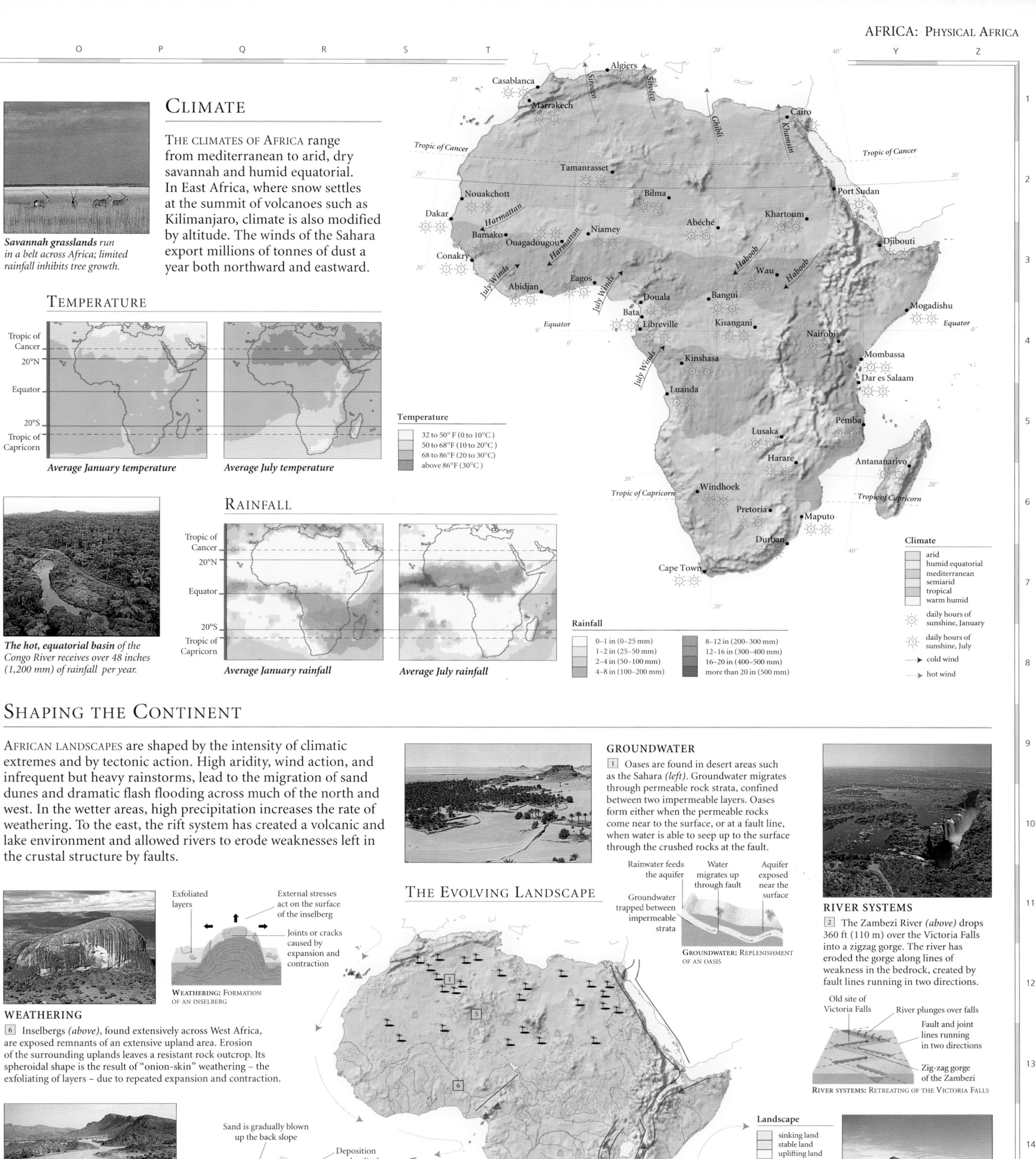

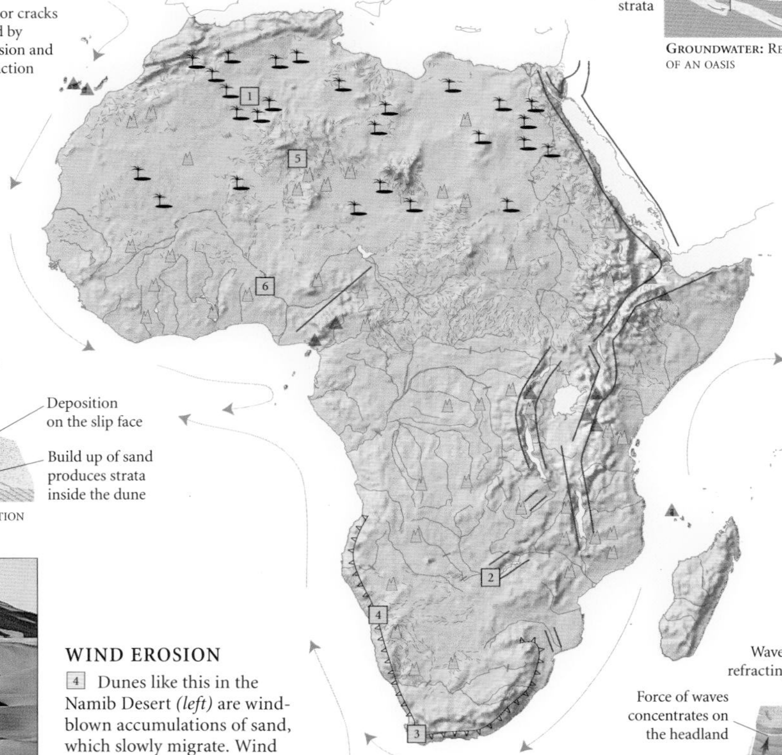

WEATHERING

Exfoliated layers

External stresses act on the surface of the inselberg

Joints or cracks caused by expansion and contraction

WEATHERING: FORMATION OF AN INSELBERG

6 Inselbergs *(above)*, found extensively across West Africa, are exposed remnants of an extensive upland area. Erosion of the surrounding uplands leaves a resistant rock outcrop. Its spheroidal shape is the result of "onion-skin" weathering – the exfoliating of layers – due to repeated expansion and contraction.

EPHEMERAL CHANNELS

5 Wadis *(above)* drain much of northern Africa. These drybed courses are flooded only after infrequent, but intense, storms in the uplands cause water to surge along their channels.

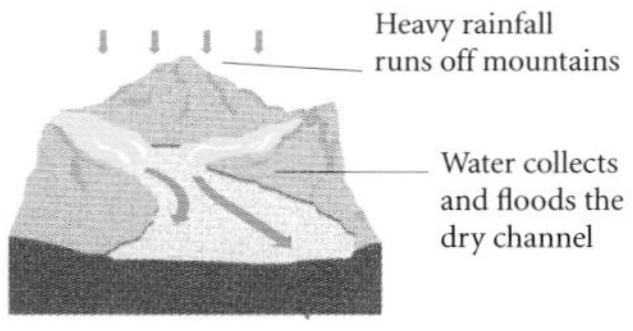

Heavy rainfall runs off mountains

Water collects and floods the dry channel

EPHEMERAL CHANNELS: FLASH FLOODING OF A WADI

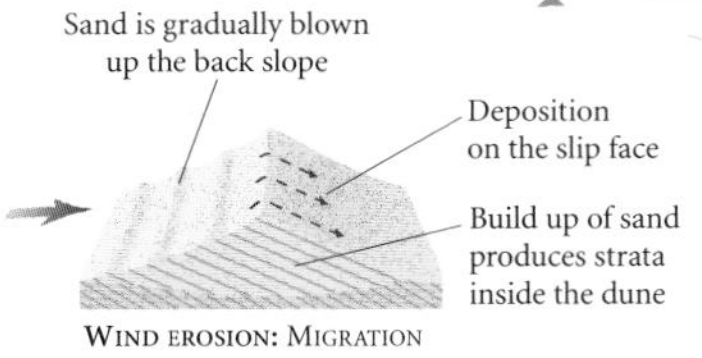

WIND EROSION

Sand is gradually blown up the back slope

Deposition on the slip face

Build up of sand produces strata inside the dune

WIND EROSION: MIGRATION OF A DUNE

4 Dunes like this in the Namib Desert *(left)* are wind-blown accumulations of sand, which slowly migrate. Wind action moves sand up the shallow back slope; when the sand reaches the crest of the dune it is deposited on the slip face.

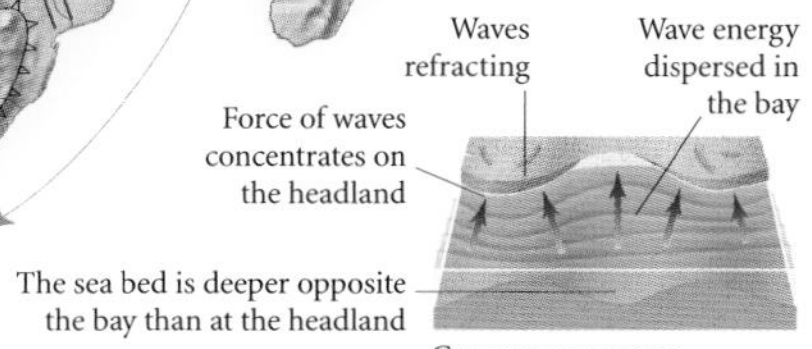

Landscape	
	sinking land
	stable land
	uplifting land
▽▽▽	escarpment
→	ocean current
—	rift
▲	active volcano
⛰	inselberg
○	oasis
	river
	wadi
	waterfall

COASTAL PROCESSES

Waves refracting

Wave energy dispersed in the bay

Force of waves concentrates on the headland

The sea bed is deeper opposite the bay than at the headland

COASTAL PROCESSES: EROSION OF A BAY

3 Houtbaai *(above)*, in southern Africa, is constantly being modified by wave action. As waves approach the indented coastline, they reach the shallow water of the headland, slowing down and reducing in length. This causes them to bend or refract, concentrating their erosive force at the headlands.

Map labels: Algiers, Casablanca, Marrakech, Sirocco, Sirocco, Cairo, Ghibli, Khamsin, Tropic of Cancer, Tamanrasset, Port Sudan, Nouakchott, Bilma, Khartoum, Dakar, Abéché, Harmattan, Djibouti, Bamako, Niamey, Ouagadougou, Harmattan, Wau, Haboob, Conakry, Lagos, Abidjan, Douala, Bangui, Haboob, Bata, Mogadishu, July Winds, Libreville, Kisangani, Nairobi, Equator, Kinshasa, Mombassa, July Winds, Dar es Salaam, Luanda, Pemba, Lusaka, Harare, Antananarivo, Windhoek, Tropic of Capricorn, Pretoria, Maputo, Durban, Cape Town

POLITICAL AFRICA

THE POLITICAL MAP OF MODERN AFRICA only emerged following the end of the Second World War. Over the next half-century, all of the countries formerly controlled by European powers gained independence from their colonial rulers – only Liberia and Ethiopia were never colonized. The postcolonial era has not been an easy period for many countries, but there have been moves toward multiparty democracy in much of West Africa, and in Zambia, Tanzania, and Kenya. In South Africa, democratic elections replaced the internationally-condemned apartheid system only in 1994. Other countries have still to find political stability; corruption in government, and ethnic tensions are serious problems. National infrastructures, based on the colonial transportation systems built to exploit Africa's resources, are often inappropriate for independent economic development.

LANGUAGES

THREE MAJOR WORLD LANGUAGES act as *lingua francas* across the African continent: Arabic in North Africa; English in southern and eastern Africa and Nigeria; and French in Central and West Africa, and in Madagascar. A huge number of African languages are spoken as well – over 2,000 have been recorded, with more than 400 in Nigeria alone – reflecting the continuing importance of traditional cultures and values. In the north of the continent, the extensive use of Arabic reflects Middle Eastern influences while Bantu is widely-spoken across much of southern Africa.

Language groups
- Afro-Asiatic (Hamito-Semitic)
- Niger-Congo
- Nilo-Saharan
- Khoisan
- Indo-European
- Austronesian

OFFICIAL AFRICAN LANGUAGES

Official languages
- French
- English
- Arabic
- Portuguese
- Swahili
- Amharic
- Spanish
- French/English
- French/Arabic
- French/Malagasy
- English/Swahili
- Arabic/Somali

Islamic influences are evident throughout North Africa. The Great Mosque at Kairouan, Tunisia, is Africa's holiest Islamic place.

In northeastern Nigeria, people speak Kanuri – a dialect of the Saharan language group.

TRANSPORTATION

AFRICAN RAILROADS WERE BUILT to aid the exploitation of natural resources, and most offer passage only from the interior to the coastal cities, leaving large parts of the continent untouched – five landlocked countries have no railroads at all. The Congo, Nile, and Niger River networks offer limited access to land within the continental interior, but have a number of waterfalls and cataracts which prevent navigation from the sea. Many roads were developed in the 1960s and 1970s, but economic difficulties are making the maintenance and expansion of the networks difficult.

South Africa has the largest concentration of railroads in Africa. Over 20,000 miles (32,000 km) of routes have been built since 1870.

Traditional means of transportation, such as the camel, are still widely used across the less accessible parts of Africa.

The Congo River, though not suitable for river transportation along its entire length, forms a vital link for people and goods in its navigable inland reaches.

Transportation
- major roads and highways
- major railroads
- major canal
- international borders
- transportation intersections
- international airports
- major ports

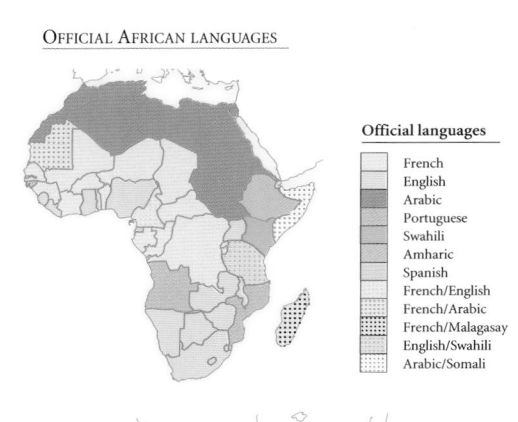

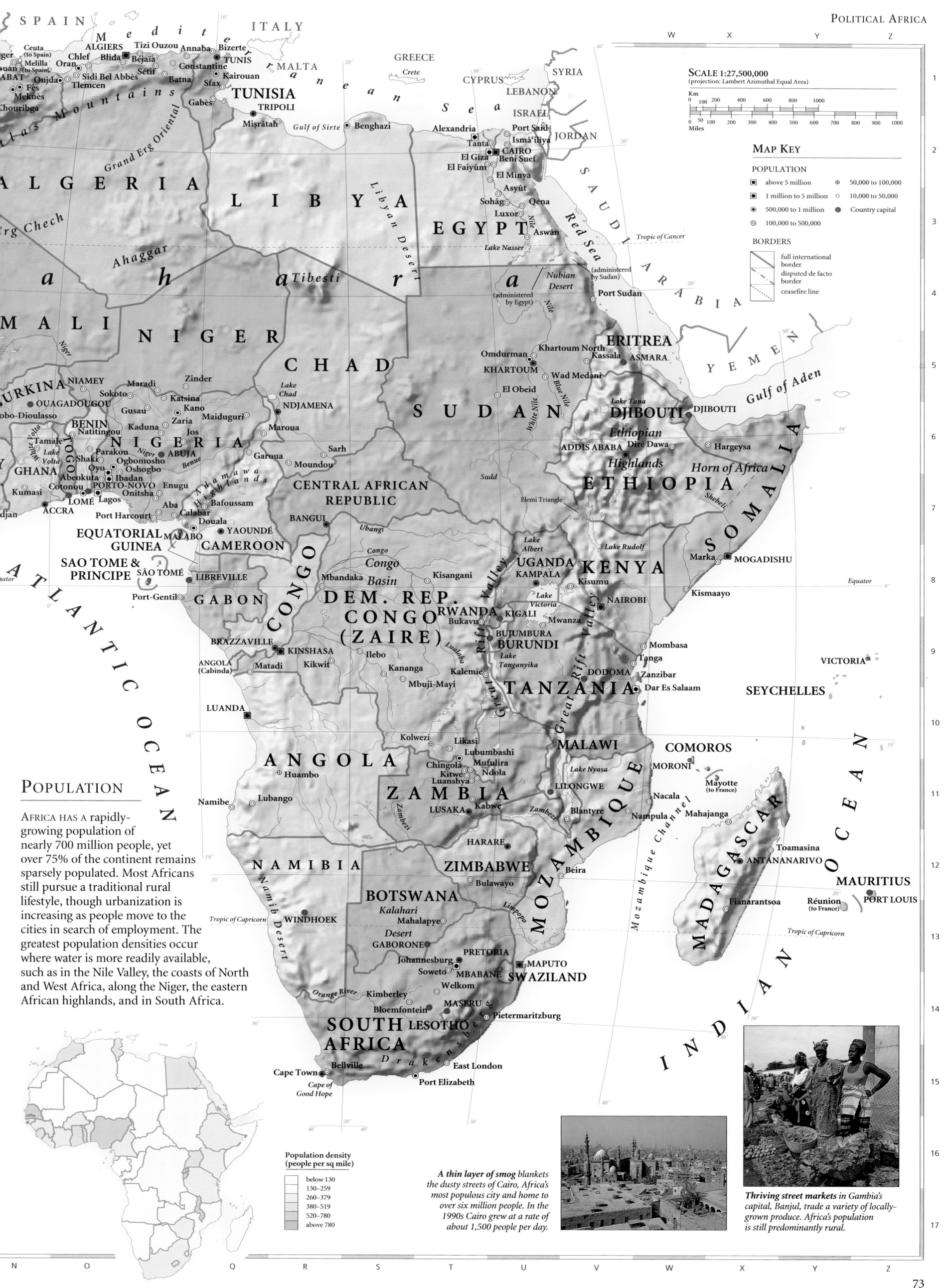

SPAIN
ITALY
Ceuta (to Spain)
Melilla (to Spain)
ALGIERS
Tizi Ouzou
Annaba
Bizerte
RABAT
Oujda
Blida
Béjaïa
Constantine
TUNIS
GREECE
Oran
Sidi Bel Abbès
Sétif
Batna
Kairouan
MALTA
CYPRUS
SYRIA
LEBANON
Fès
Meknès
Tlemcen
Sfax
Gabès
TUNISIA
TRIPOLI
ISRAEL
JORDAN
Mishrâtah
Gulf of Sirte
Benghazi
Alexandria
Port Said
Ismâ'îliya
Tanta
CAIRO
El Giza
Beni Suef
El Faiyûm
El Minya
Asyût
Sohâg
Qena
Luxor
Aswân

A L G E R I A
L I B Y A
E G Y P T
S A U D I   A R A B I A
Grand Erg Oriental
Ahaggar
Tibesti
Libyan Desert
Lake Nasser
Nubian Desert
(administered by Sudan)
(administered by Egypt)
Red Sea
Tropic of Cancer

MALI
NIGER
CHAD
SUDAN
ERITREA
YEMEN
Gulf of Aden

BURKINA
NIAMEY
Maradi
Zinder
Lake Chad
NDJAMENA
Omdurman
Khartoum North
Kassala
ASMARA
Sokoto
Katsina
Kano
Maiduguri
Maroua
KHARTOUM
Wad Medani
DJIBOUTI
DJIBOUTI
OUAGADOUGOU
Gusau
Kaduna
Zaria
El Obeid
Bobo-Dioulasso
BENIN
Natitingou
Jos
Sarh
Garoua
Moundou
ADDIS ABABA
Dire Dawa
Hargeysa
Shaki
Parakou
Oyo
Ogbomosho
ABUJA
Benue
Sudd
Ethiopian Highlands
Horn of Africa
GHANA
Ibadan
Oshogbo
CENTRAL AFRICAN
Shebeli
Abeokuta
PORTO-NOVO
Enugu
Onitsha
Aba
REPUBLIC
ETHIOPIA
SOMALIA
ACCRA
LOMÉ
Cotonou
Lagos
Calabar
BANGUI
Elemi Triangle
Port Harcourt
Douala
Ubangi
Lake Albert
Lake Rudolf
Marka
MOGADISHU
EQUATORIAL
MALABO
YAOUNDÉ
Congo
UGANDA
KENYA
GUINEA
CAMEROON
Mbandaka
Kisangani
KAMPALA
SAO TOME &
Congo Basin
Lake Victoria
Kisumu
PRINCIPE
SÃO TOMÉ
LIBREVILLE
KINSHASA
NAIROBI
Kismaayo
Port-Gentil
GABON
DEM. REP.
RWANDA
Mwanza
BRAZZAVILLE
CONGO
KIGALI
Bukavu
ANGOLA
(ZAIRE)
BUJUMBURA
Mombasa
(Cabinda)
Matadi
Ilebo
BURUNDI
Tanga
VICTORIA
Kikwit
Kananga
Kalemie
DODOMA
Zanzibar
LUANDA
Mbuji-Mayi
Lake Tanganyika
TANZANIA
Dar Es Salaam
SEYCHELLES
Kolwezi
Likasi
MALAWI
COMOROS
ANGOLA
Lubumbashi
MORONI
Chingola
Mufulira
Lake Nyasa
Mayotte (to France)
Huambo
Kitwe
Ndola
Nacala
Luanshya
Nampula
ZAMBIA
Blantyre
Mahajanga
Namibe
LUSAKA
Kabwe
LILONGWE
Lubango
Zambezi
MADAGASCAR
Toamasina
HARARE
ANTANANARIVO
MOZAMBIQUE
Beira
MAURITIUS
NAMIBIA
ZIMBABWE
Bulawayo
Fianarantsoa
Réunion (to France)
PORT LOUIS
BOTSWANA
Limpopo
Kalahari Desert
Mahalapye
Namib Desert
WINDHOEK
GABORONE
PRETORIA
Johannesburg
MAPUTO
Soweto
MBABANE
SWAZILAND
Welkom
MASERU
Kimberley
Bloemfontein
Pietermaritzburg
Orange River
SOUTH
LESOTHO
AFRICA
Drakensberg
Bellville
East London
Cape Town
Cape of Good Hope
Port Elizabeth

ATLANTIC OCEAN
INDIAN OCEAN
Mediterranean Sea
Niger
White Nile
Blue Nile
Lake Tana
Great Rift Valley
Lualaba
Mozambique Channel
Equator
Tropic of Capricorn

POPULATION

AFRICA HAS A rapidly-growing population of nearly 700 million people, yet over 75% of the continent remains sparsely populated. Most Africans still pursue a traditional rural lifestyle, though urbanization is increasing as people move to the cities in search of employment. The greatest population densities occur where water is more readily available, such as in the Nile Valley, the coasts of North and West Africa, along the Niger, the eastern African highlands, and in South Africa.

SCALE 1:27,500,000
(projection: Lambert Azimuthal Equal Area)

Km
0 50 100 200 300 400 500 600 700 800 900 1000
Miles

MAP KEY

POPULATION

▣ above 5 million	⊕ 50,000 to 100,000
▣ 1 million to 5 million	○ 10,000 to 50,000
◉ 500,000 to 1 million	● Country capital
○ 100,000 to 500,000	

BORDERS

full international border
disputed de facto border
ceasefire line

Population density
(people per sq mile)

	below 130
	130–259
	260–379
	380–519
	520–780
	above 780

A thin layer of smog blankets the dusty streets of Cairo, Africa's most populous city and home to over six million people. In the 1990s Cairo grew at a rate of about 1,500 people per day.

Thriving street markets in Gambia's capital, Banjul, trade a variety of locally-grown produce. Africa's population is still predominantly rural.

AFRICAN RESOURCES

THE ECONOMIES OF MOST AFRICAN COUNTRIES are dominated by subsistence and cash crop agriculture, with limited industrialization. Manufacturing is largely confined to South Africa. Many countries depend on a single resource, such as copper or gold, or a cash crop, such as coffee, for export income, which can leave them vulnerable to fluctuations in world commodity prices. In order to diversify their economies and develop a wider industrial base, investment from overseas is being actively sought by many African governments.

INDUSTRY

MANY AFRICAN INDUSTRIES concentrate on the extraction and processing of raw materials. These include the oil industry, food processing, mining, and textile production. South Africa accounts for over half of the continent's industrial output with much of the remainder coming from the countries along the northern coast. Over 60% of Africa's workforce is employed in agriculture.

The unspoiled natural splendor of wildlife reserves, like the Serengeti National Park in Tanzania, attract tourists to Africa from around the globe. The tourist industry in Kenya and Tanzania is particularly well developed, where it accounts for almost 10% of GNP.

STANDARD OF LIVING

SINCE THE 1960s most countries in Africa have seen significant improvements in life expectancy, healthcare and education. However, 18 of the 20 most deprived countries in the world are African, and the continent as a whole lies well behind the rest of the world in terms of meeting many basic human needs.

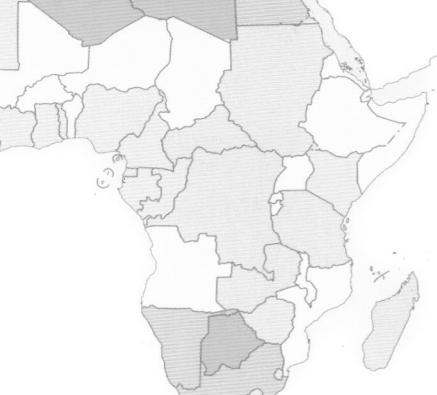

Standard of Living
(UN Human Development Index)

high

low

GNP per capita (US$)

0–199
200–399
400–599
600–899
900–1999
2000+

The discovery of *oil* in the swampy Niger Delta during the 1960s made Nigeria one of Africa's richer nations. As world oil prices fell in the 1980s, the Nigerian economy faltered.

Exotic rugs and brightly-colored textiles are sold in a street market along the banks of the Nile River in Luxor, Egypt.

The Rössing uranium mines in Namibia are the largest in the world. Africa and the US produce over half the world's uranium ore, used to fuel nuclear power plants. Elsewhere, South Africa and Niger also mine uranium on a large scale.

Industry

brewing	mining
car/vehicle manufacture	palm oil processing
cement	peanut processing
chemicals	pharmaceuticals
coffee processing	rice milling
electronics	shipbuilding
engineering	sugar processing
finance	tea processing
fish processing	textiles
food processing	timber processing
iron & steel	tobacco processing

coal
oil
gas

industrial cities
major industrial areas

PORTUGAL SPAIN *Mediterranean Sea* ITALY
Algiers Annaba Tunis CYPRUS SYRIA LEBANON
Oran TUNISIA Tripoli ISRAEL
Casablanca Rabat Benghazi Alexandria Port Said
Safi Cairo
MOROCCO
ALGERIA LIBYA EGYPT
Aswân
Western Sahara
(occupied by Morocco)
Port Sudan
MAURITANIA MALI NIGER CHAD ERITREA
CAPE VERDE Khartoum Asmara YEMEN
Dakar SENEGAL SUDAN DJIBOUTI *Gulf of Aden*
Banjul
GAMBIA Bamako BURKINA Katsina Kano
GUINEA-BISSAU BENIN Kaduna Addis Ababa
Conakry IVORY GHANA NIGERIA ETHIOPIA
Freetown COAST Kumasi Ibadan CENTRAL AFRICAN SOMALIA
SIERRA LEONE Lagos REPUBLIC
Monrovia LIBERIA Abidjan Accra CAMEROON Bangui Mogadishu
Sekondi-Takoradi Port Harcourt Douala UGANDA KENYA
EQUATORIAL Kisangani Kampala
GUINEA Libreville Nairobi
SAO TOME & GABON CONGO DEM. REP. RWANDA Mombasa
PRINCIPE Port-Gentil CONGO Bukavu BURUNDI
Brazzaville (ZAIRE)
Pointe-Noire Kinshasa Kananga Dodoma Zanzibar SEYCHELLES
Gulf of Guinea Dar es Salaam
Luanda TANZANIA
ATLANTIC OCEAN
Lobito MALAWI COMOROS
ANGOLA Lubumbashi Mayotte (to France)
Ndola
ZAMBIA Blantyre
Lusaka *Mozambique Channel*
NAMIBIA ZIMBABWE Harare Kwekwe Beira MADAGASCAR Antananarivo
Bulawayo MOZAMBIQUE MAURITIUS
Walvis Bay Windhoek BOTSWANA Réunion (to France)
Johannesburg Pretoria Maputo
Kimberley SWAZILAND *INDIAN OCEAN*
LESOTHO Durban
SOUTH AFRICA
Cape Town Port Elizabeth East London

Red Sea
SAUDI ARABIA

ENVIRONMENTAL ISSUES

ONE OF AFRICA'S most serious environmental problems occurs in marginal areas such as the Sahel where scrub and forest clearance, often for cooking fuel, combined with overgrazing, are causing desertification. Game reserves in southern and eastern Africa have helped to preserve many endangered animals, although the needs of growing populations have led to conflict over land use, and poaching is a serious problem.

Environmental Issues

- national parks
- tropical forest
- forest destroyed
- desert
- desertification
- polluted rivers
- radioactive contamination
- marine pollution
- heavy marine pollution
- • poor urban air quality

The Sahel's delicate natural equilibrium is easily destroyed by the clearing of vegetation, drought, and overgrazing. This causes the Sahara to advance south, engulfing the savannah grasslands.

MINERAL RESOURCES

AFRICA'S ANCIENT PLATEAUS contain some of the world's most substantial reserves of precious stones and metals. About 30% of the world's gold is mined in South Africa; Zambia has great copper deposits; and diamonds are mined in Botswana, Dem. Rep. Congo (Zaire), and South Africa. Oil has brought great economic benefits to Algeria, Libya, and Nigeria.

Mineral Resources

- oil field
- gas field
- coal field
- bauxite
- copper
- diamonds
- gold
- iron
- phosphates
- tin
- uranium

North and West Africa have large deposits of white phosphate minerals, which are used in making fertilizers. Morocco, Senegal, and Tunisia are the continent's leading producers.

Workers on a tea plantation gather one of Africa's most important cash crops, providing a valuable source of income. Coffee, rubber, bananas, cotton, and cocoa are also widely grown as cash crops.

Surrounded by desert, the fertile floodplains of the Nile Valley and Delta have been extensively irrigated, farmed, and settled since 3,000 BC.

USING THE LAND AND SEA

SOME OF AFRICA'S MOST PRODUCTIVE agricultural land is found in the eastern volcanic uplands, where fertile soils support a wide range of valuable export crops including vegetables, tea, and coffee. The most widely-grown grain is corn and peanuts are particularly important in West Africa. Without intensive irrigation, cultivation is not possible in desert regions and unreliable rainfall in other areas limits crop production. Pastoral herding is most commonly found in these marginal lands. Substantial local fishing industries are found along coasts and in vast lakes such as Lake Nyasa and Lake Victoria.

Using the Land and Sea

- cropland
- desert
- forest
- pasture
- wetland
- • major conurbations
- cattle
- goats
- cereals
- sheep
- bananas
- corn (maize)
- citrus fruits
- cocoa
- cotton
- coffee
- dates
- fishing
- fruit
- oil palms
- olives
- peanuts
- rice
- rubber
- shellfish
- sugar cane
- tea
- tobacco
- vineyards
- wheat

NORTH AFRICA

ALGERIA, EGYPT, LIBYA, MOROCCO, TUNISIA, WESTERN SAHARA

FRINGED BY THE MEDITERRANEAN along the northern coast and by the arid Sahara in the south, North Africa reflects the influence of many invaders, both European and, most importantly, Arab, giving the region an almost universal Islamic flavor and a common Arabic language. The countries lying to the west of Egypt are often referred to as the Maghreb, an Arabic term for "west." Today, Morocco and Tunisia exploit their culture and landscape for tourism, while rich oil and gas deposits aid development in Libya and Algeria, despite political turmoil. Egypt, with its fertile, Nile-watered agricultural land and varied industrial base, is the most populous nation.

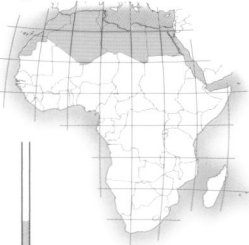

THE LANDSCAPE

These rock piles in Algeria's Ahaggar Mountains are the result of weathering caused by extremes of temperature. Great cracks or joints appear in the rocks, which are then worn and smoothed by the wind.

THE ATLAS MOUNTAINS, which extend across much of Morocco, northern Algeria, and Tunisia, are part of the fold mountain system which also runs through much of southern Europe. They recede to the south and east, becoming a steppe landscape before meeting the Sahara desert which covers more than 90% of the region. The sediments of the Sahara overlie an ancient plateau of crystalline rock, some of which is more than four billion years old.

MAP KEY

POPULATION

- ■ above 5 million
- ▪ 1 million to 5 million
- ◉ 500,000 to 1 million
- ◎ 100,000 to 500,000
- ⊕ 50,000 to 100,000
- ○ 10,000 to 50,000
- ∘ below 10,000

ELEVATION

- 4000m / 13,124ft
- 3000m / 9843ft
- 2000m / 6562ft
- 1000m / 3281ft
- 500m / 1640ft
- 250m / 820ft
- 100m / 328ft
- sea level

SCALE 1:11,000,000
(projection: Lambert Azimuthal Equal Area)

The town of Tiznit, Morocco, lies in an oasis in the desert. Crops and trees grow on the fertile land surrounding the town.

The Grand Erg Occidental is one of Algeria's great Saharan sand seas. Wind force and direction determines the nature of landforms such as the linear or seif dunes in the foreground.

USING THE LAND AND SEA

SHELTERED VALLEYS IN THE ATLAS MOUNTAINS, the Nile Valley and Delta, and the Mediterranean coast are the main sources of good farming land. A wide variety of valuable crops including cereals, rice, and cotton, and woods such as cedar and cork, are grown. Typical Mediterranean crops such as olives, figs, dates, and citrus fruits also thrive in these areas. The Nile Valley is particularly fertile, and most of Egypt's population lives close to the river. Elsewhere, irrigation is essential to improve crop yields on the desert margins.

Land use and agricultural distribution

- goats
- sheep
- cereals
- citrus fruits
- cork
- cotton
- dates
- fishing
- olives
- vineyards
- ■ capital cities
- ▲ major towns
- pasture
- cropland
- forest
- desert

THE URBAN/RURAL POPULATION DIVIDE

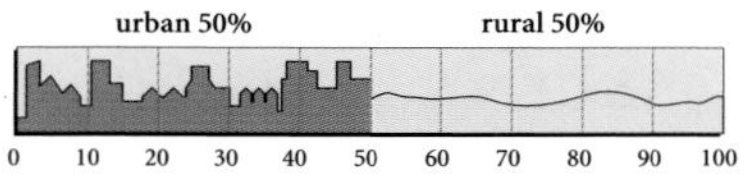

urban 50%   rural 50%

0 10 20 30 40 50 60 70 80 90 100

POPULATION DENSITY
62 people per sq mile
(24 people per sq km)

TOTAL LAND AREA
2,215,020 sq miles
(5,738,394 sq km)

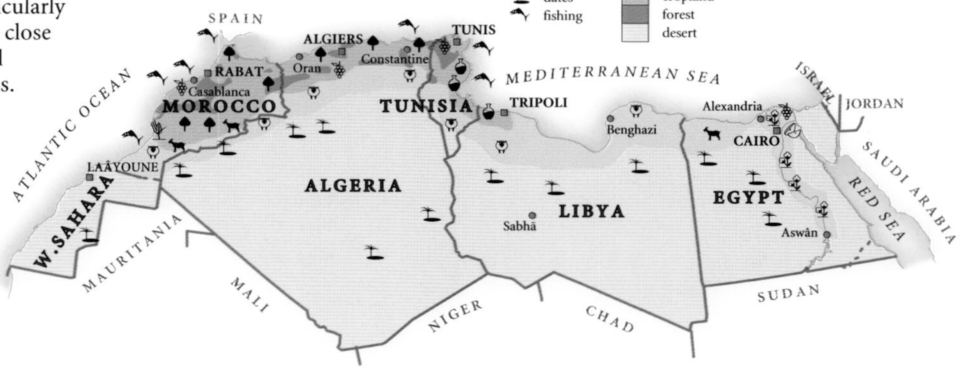

Many North African nomads, such as the Bedouin, maintain a traditional pastoral lifestyle on the desert fringes, moving their herds of sheep, goats, and camels from place to place – crossing country borders in order to find sufficient grazing land.

The Atlas Mountains run from Morocco to Tunisia, covering more than 1,200 miles (1,931 km). The northern Tell Atlas (Atlas Tellien) are well watered, with forested slopes; the drier southern High Atlas (Haut Atlas) (left) have the highest peaks, such as Jbel Toubkal, 13,665 ft (4,165 m) high.

The Tell Atlas (Atlas Tellien) are a range of recent, folded mountains. They are still being formed, and the region's frequent earth tremors reflect this.

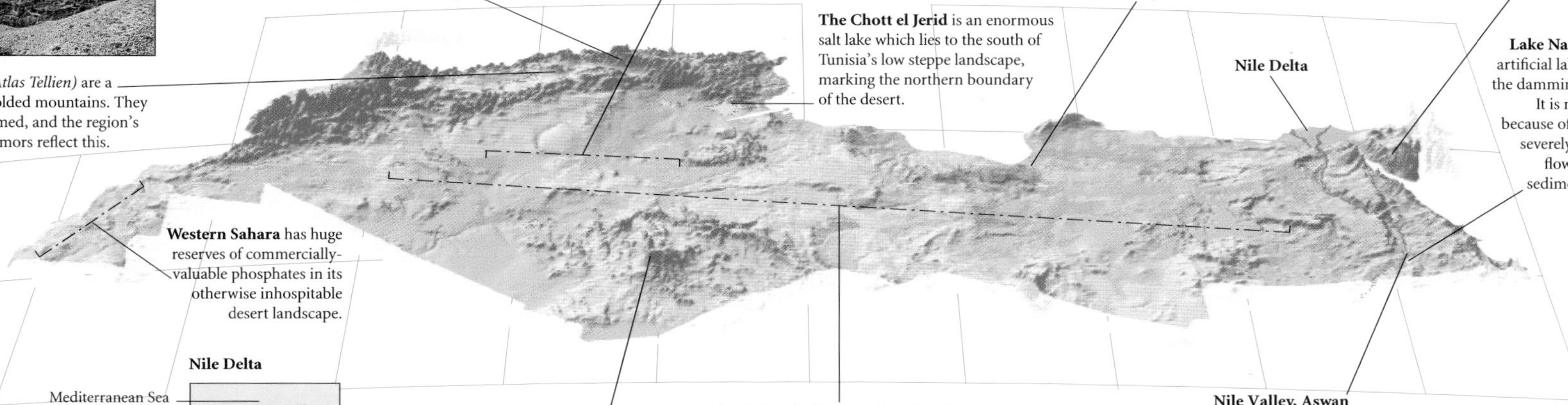

Western Sahara has huge reserves of commercially-valuable phosphates in its otherwise inhospitable desert landscape.

The spectacular sand seas of the Grand Ergs Occidental and Oriental in Algeria are only one of the varied landscapes of the Sahara. Hammadas, boulder-strewn rock plateaus, and reg, or desert pavements, plains strewn with gravel and small pebbles, are other important landforms.

The Chott el Jerid is an enormous salt lake which lies to the south of Tunisia's low steppe landscape, marking the northern boundary of the desert.

Despite its outward aridity, the Sahara has several underground aquifers. Libya has built an underground pipeline, the Great Man-made River Project, to enable fuller exploitation of this valuable resource.

Split from the rest of Egypt by the Suez Canal, the Sinai Peninsula is partially desert, dissected by countless wadis.

Lake Nasser is a huge artificial lake, created by the damming of the Nile. It is now silting up because of evaporation, severely affecting the flow of water and sediment to the sea.

Nile Delta

Mediterranean Sea

Fertile deposits of alluvium

Network of drainage channels

River Nile

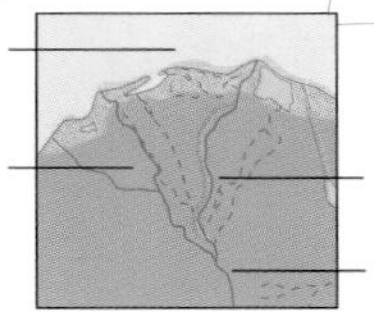

In its northernmost reaches, the Nile River has deposited huge quantities of silt and alluvium to form the fan-shaped Nile Delta. The Nile splits into two main channels at the base of the delta which are interlinked by a dense network of canals and drainage channels.

Ahaggar

The Sahara is the largest hot desert on Earth, covering nearly a third of Africa. The sandy parts of the desert contain a wide variety of sand dunes, created by differing wind directions and strengths.

Nile Valley, Aswan

Almost all of Egypt's people – more than 99% – live close to the Nile River, or on its massive delta. The river waters the only strip of fertile land in Egypt.

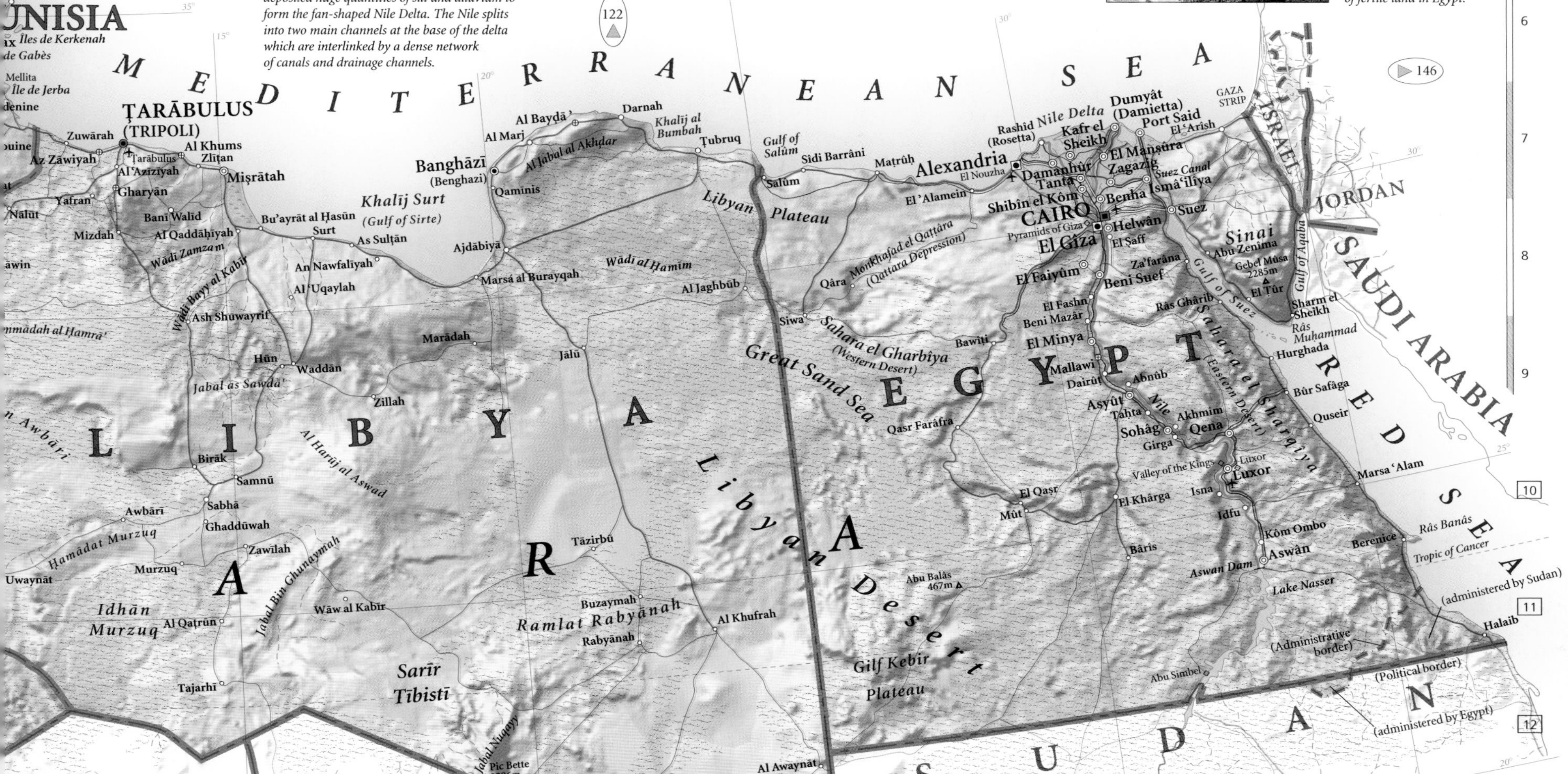

Built as great tombs for the pharaohs of ancient Egypt, the magnificent pyramids at Giza near Cairo have fascinated scholars, archaeologists, and tourists for centuries.

Oil rigs are scattered throughout the deserts of Libya and Algeria. Libyan oil is especially prized because of its low sulfur content, which means it produces much less pollution than other fuel oils.

TRANSPORTATION & INDUSTRY

THE ECONOMIES OF ALGERIA AND LIBYA were transformed by the discovery of oil and natural gas reserves in the deserts. Morocco's major exports are phosphates and agricultural produce, and as in Egypt and Tunisia, the tourist industry is essential to the economy. Egypt has the most varied industrial base, importing technology to develop electronics and engineering industries, and maintaining the reputation of its high-quality cotton textiles.

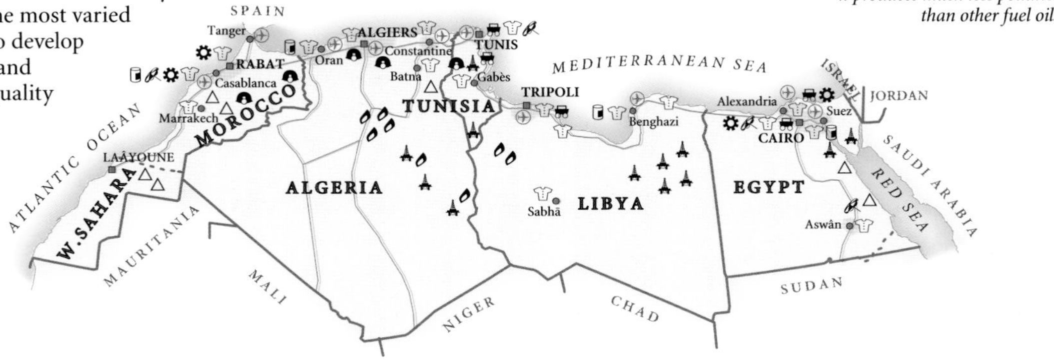

Major industry and infrastructure
- engineering
- food processing
- gas
- iron & steel
- iron ore
- oil
- phosphates
- textiles
- tourism
- capital cities
- major towns
- international airports
- major roads
- major industrial areas

TRANSPORTATION NETWORK

152,393 miles (245,400 km)	480 miles (773 km)	8025 miles (12,922 km)	121 miles (195 km)

Tourism and the oil industry have made improvements to the Maghreb's infrastructure both necessary and possible. The Suez Canal is a vital artery for shipping between Europe and Asia.

WEST AFRICA

Benin, Burkina, Cape Verde, Gambia, Ghana, Guinea, Guinea-Bissau, Ivory Coast, Liberia, Mali, Mauritania, Niger, Nigeria, Senegal, Sierra Leone, Togo

West Africa is an immensely diverse region, encompassing the desert landscapes and mainly Muslim populations of the southern Saharan countries, and the tropical rain forests of the more humid south, with a great variety of local languages and cultures. The rich natural resources and accessibility of the area were quickly exploited by Europeans; most of the Africans taken by slave traders came from this region, causing serious depopulation. The very different influences of West Africa's leading colonial powers, Britain and France, remain today, reflected in the languages and institutions of the countries they once governed.

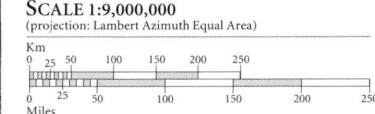

The dry scrub of the Sahel is only suitable for grazing herd animals like these cattle in Mali.

TRANSPORTATION & INDUSTRY

Abundant natural resources including oil and metallic minerals are found in much of West Africa, although investment is required for their further exploitation. Nigeria experienced an oil boom during the 1970s but subsequent growth has been sporadic. Most industry in other countries has a primary basis, including mining, logging, and food processing.

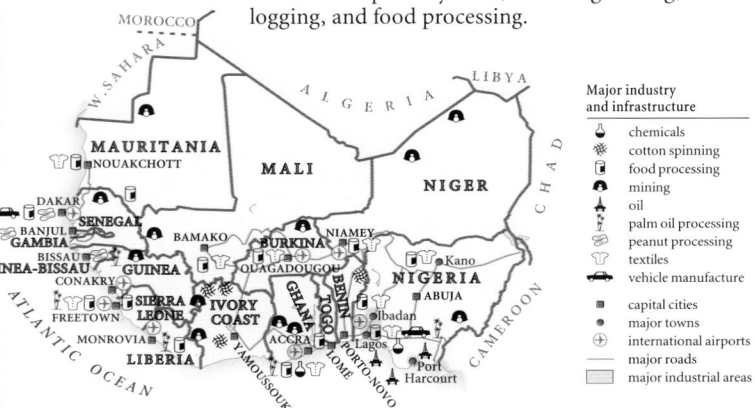

TRANSPORTATION NETWORK

163,769 miles (263,719 km)	1,554 miles (2,502 km)
6,819 miles (10,980 km)	9,470 miles (15,250 km)

The road and rail systems are most developed near the coasts. Some of the landlocked countries remain disadvantaged by the difficulty of access to ports, and their poor road networks.

Major industry and infrastructure
- chemicals
- cotton spinning
- food processing
- mining
- oil
- palm oil processing
- peanut processing
- textiles
- vehicle manufacture
- capital cities
- major towns
- international airports
- major roads
- major industrial areas

MAP KEY

POPULATION
- 1 million to 5 million
- 500,000 to 1 million
- 100,000 to 500,000
- 50,000 to 100,000
- 10,000 to 50,000
- below 10,000

ELEVATION
- 2000m / 6562ft
- 1000m / 3281ft
- 500m / 1640ft
- 250m / 820ft
- 100m / 328ft
- sea level

CAPE VERDE
Santo Antão, Pombas, Ilhas de Barlavento, Mindelo, Ribeira Brava, Pedra Lume, São Vicente, Amílcar Cabral, Sal, São Nicolau, Boa Vista, João Barrosa, Tarrafal, Maio, Maio, Fogo, Santiago, PRAIA, São Filipe, Ilhas de Sotavento

ATLANTIC OCEAN

(same scale as main map)

The southern regions of West Africa still contain great swaths of tropical rain forest, including some of the world's most prized hardwood trees, such as mahogany and iroko.

USING THE LAND AND SEA

The humid southern regions are most suitable for cultivation; in these areas, cash crops such as coffee, cotton, cocoa, and rubber are grown in large quantities. Peanuts are grown throughout West Africa. In the north, advancing desertification has made the Sahel increasingly uncultivable, and pastoral farming is more common. Great herds of sheep, cattle, and goats are grazed on the savannah grasses. Fishing is important in coastal and delta areas.

The Gambia, mainland Africa's smallest country, produces great quantities of peanuts. Winnowing is used to separate the nuts from their stalks.

Land use and agricultural distribution
- goats
- sheep
- cocoa
- coffee
- cotton
- oil palms
- peanuts
- rubber
- shellfish
- capital cities
- major towns
- pasture
- cropland
- forest
- desert

THE URBAN/RURAL POPULATION DIVIDE

urban 36%   rural 64%

POPULATION DENSITY	TOTAL LAND AREA
98 people per sq mile (38 people per sq km)	2,337,137 sq miles (6,054,760 sq km)

SCALE 1:9,000,000
(projection: Lambert Azimuth Equal Area)

Main map labels

WESTERN SAHARA (occupied by Morocco)

Yetti, 'Aïn Ben Tili, Bir Mogrein, 'Ayoûn 'Abd el Mâlek, TIRIS ZEMMOUR, Kâghet, El Ha, El Mreiti, Zouérat, El Hammâmi, Fdérik, Tourine, Touâjil, Châr, Maqteir, Erg, El Mrâyer, Ouarâne, Bou Lanouâr, Nouâdhibou, Ras Nouâdhibou, Nouâdhibou, Choûm, Dakhlet Nouâdhibou, DAKHLET NOUÂDHIBOU, Ouadâne, ADRAR, Et Tidra, Atâr, Chinguetti, INCHIRI, Akjoujt, Oujeft, MAURITANIA, El Mreyyé, Râs Timirist, Nouâmghâr, Bennichab, S, Sebkhet Te-n-Dghâmcha, Rachid, TAGANT, HODH, Beïla, Tidjikja, Tichit, ECH CHARGUI, NOUAKCHOTT, Nouakchott, TRARZA, Aoukâr, Moudjéria, Boûmdeïd, Tâmchekket, Ouâlata, Idini, Boutilimit, Mâgta' Lahjar, HODH EL, Néma, Mederdra, Rkiz, Aleg, Guérou, Kiffa, Tintâne, 'Ayoûn el 'Atroûs, GHARBI, Rosso, Senegal, Podor, Bogué, Bababé, Mônguel, Kaédi, ASSABA, Timbedgha, Amourj, Richard Toll, Dagana, Lac de Guier, Kankossa, Kobenni, Bassikounou, Saint Louis, Louga, Kébémèr, Matam, Maghama, GORGOL, Ould Yenjé, Sélibabi, Yélimané, Nioro, Ballé, Adel Bagrou, Mékhé, Tivaouane, Dara, Linguère, Ranérou, GUIDIMAKA, Ambidédi, Sandaré, Diéma, Nara, DAKAR, Thiès, Touba, Mbaké, Vélingara, Diamou, Bakel, Kidira, Maréna, Kayes, KAYES, Mourdiah, Sokolo, Rufisque, Diourbel, Bambey, Saloum, KOULIKORO, Nione, Mbour, Fatick, Kaolack, Goudiri, Tambacounda, Sadiola, Bafoulabé, Kolokani, Banamba, SÉGO, Joal-Fadiout, SENEGAL, Kaffrine, Koungheul, Maka, Georgetown, Dialakoto, Toukoto, Kita, Sébékoro, Koulikoro, Kati, Markala, Sokone, Nioro du Rip, Kédougou, Kéniéba, Satadougou, Kokofata, Bamako, BAMAKO, Baní, Dioïla, GAMBIA, BANJUL, Banjul, Gambie, Mansa Konko, Basse Santa Su, Saraya, Koundâra, Niagassola, Kangaba, Ouéléssébougou, Bagoé, Brikama, Mali, Malé, Kangare, Niéna, Diouloulou, Bignona, Kolda, Vélingara, Médina Gounas, Gambia, Labé, MOYENNE-GUINÉE, 1538m, Doko, Bougouni, SIKASSO, Ziguinchor, Farim, Sédhiou, Tamgué, Gaoual, Dinguiraye, Siguiri, Yanfolila, Garalo, Kolondiéba, Cacheu, Bissorã, Mansôa, Gabú, Rio Geba, Tougué, Fouta, Kadé, GUINÉE-BISSAU, Bafatá, Fulacunda, Rio Corubal, Djallon, Tikinso, Kouroussa, Mandiana, Manankoro, Kadi, Bissau, Quinhámel, Bolama, Buba, Boké, GUINÉE, HAUTE-GUINÉE, Dabola, Niger, Tengr, Arquipélago dos Bijagós, Catió, MARITIME, Télimélé, Kavendou, Dalaba, Kankan, Samatiguila, Odienné, Bound, Cap Verga, Boffa, Fria, 1421m, Mamou, Dabola, Kouto, Korh, Dubréka, Kindia, Mongo, Kabala, Faranah, Tokounou, Kérouane, Madinani, Bako, CONAKRY, Coyah, Forécariah, Falaba, Kissidougou, Koidu, Guéckedou, Macenta, Beyla, Pic de Tibé, 1504m, Borotou, Touba, Kani, Port Loko, Makeni, Sefa, Binimani, 1948m, Kamsar, Lungi, Pepel, Lunsar, Magburaka, GUINÉE, FORESTIÈRE, Monts Nimba, Nzérékoré, Lola, Séguéla, FREETOWN, Moyamba, Shenge, Boola, Yéképa, Zuénoula, Vavoua, SIERRA LEONE, Bo, Kenema, Mano, Zorzor, Yomou, Ganta, IVOR, Bonthe, Pujehun, Matru, Voinjama, Kolahun, Duékoué, Daloa, Sherbro Island, Sulima, Gbanga, Tubmanburg, Kakata, Tapeta, Toulépleu, Guiglo, Issia, Robertsport, MONROVIA, Monrovia, Marshall, Harbel, Saint John, Zwedru, Lac de Buyo, Buyo, Gagnoa, LIBERIA, Buchanan, Cestos, Taï, Soubré, River Cess, Greenville, Grabo, San-Pédro, Grand Cess, Plibo, Grand-Béréby, Harper, Tabou, Cap des Palmés

ATLANTIC OCEAN, Tropic of Cancer, ATLANTIC OCEAN

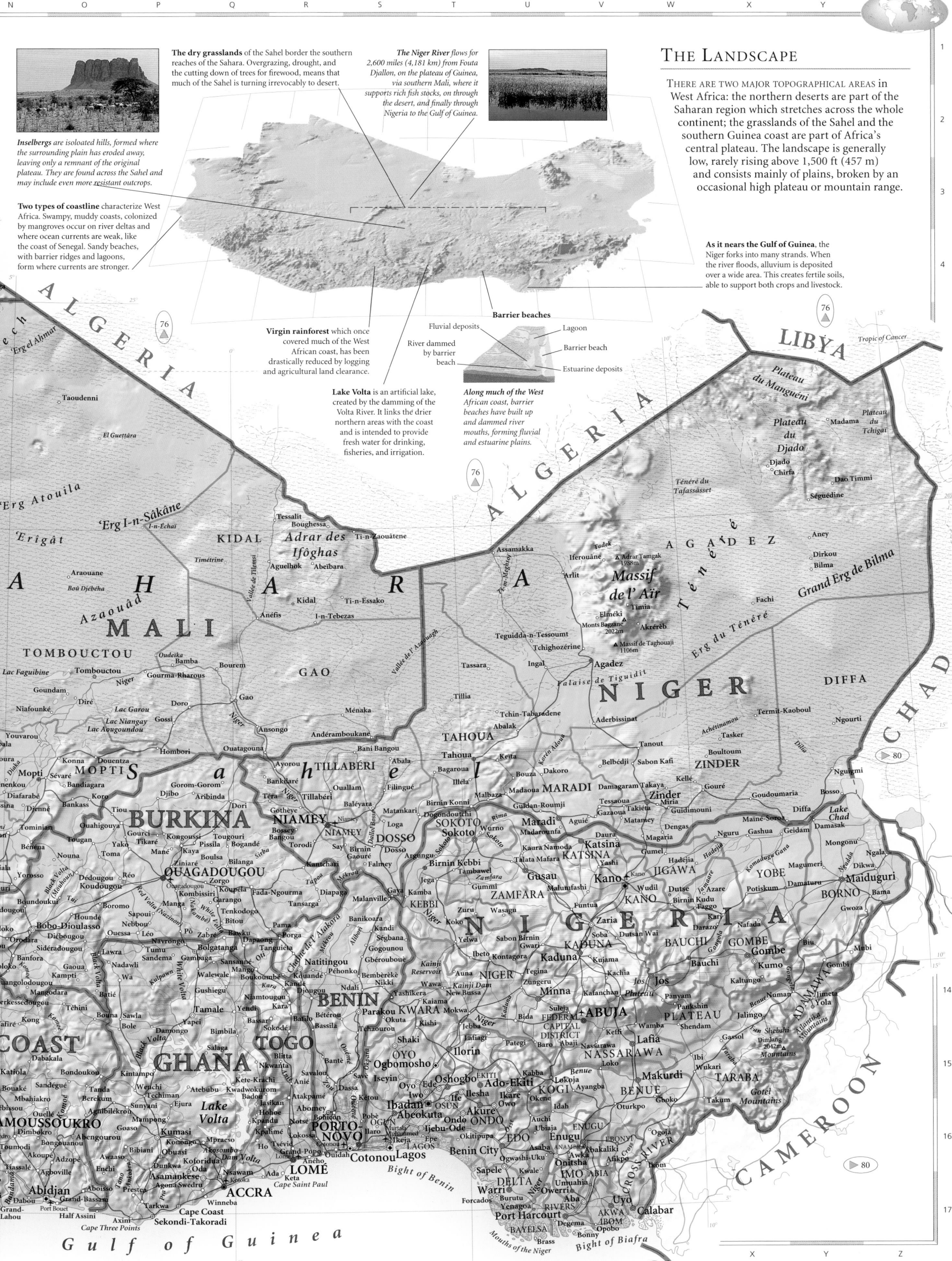

THE LANDSCAPE

THERE ARE TWO MAJOR TOPOGRAPHICAL AREAS in West Africa: the northern deserts are part of the Saharan region which stretches across the whole continent; the grasslands of the Sahel and the southern Guinea coast are part of Africa's central plateau. The landscape is generally low, rarely rising above 1,500 ft (457 m) and consists mainly of plains, broken by an occasional high plateau or mountain range.

The dry grasslands of the Sahel border the southern reaches of the Sahara. Overgrazing, drought, and the cutting down of trees for firewood, means that much of the Sahel is turning irrevocably to desert.

Inselbergs are isoloated hills, formed where the surrounding plain has eroded away, leaving only a remnant of the original plateau. They are found across the Sahel and may include even more resistant outcrops.

The Niger River flows for 2,600 miles (4,181 km) from Fouta Djallon, on the plateau of Guinea, via southern Mali, where it supports rich fish stocks, on through the desert, and finally through Nigeria to the Gulf of Guinea.

Two types of coastline characterize West Africa. Swampy, muddy coasts, colonized by mangroves occur on river deltas and where ocean currents are weak, like the coast of Senegal. Sandy beaches, with barrier ridges and lagoons, form where currents are stronger.

As it nears the Gulf of Guinea, the Niger forks into many strands. When the river floods, alluvium is deposited over a wide area. This creates fertile soils, able to support both crops and livestock.

Virgin rainforest which once covered much of the West African coast, has been drastically reduced by logging and agricultural land clearance.

Fluvial deposits
River dammed by barrier beach
Barrier beaches
Lagoon
Barrier beach
Estuarine deposits

Lake Volta is an artificial lake, created by the damming of the Volta River. It links the drier northern areas with the coast and is intended to provide fresh water for drinking, fisheries, and irrigation.

Along much of the West African coast, barrier beaches have built up and dammed river mouths, forming fluvial and estuarine plains.

CENTRAL AFRICA

CAMEROON, CENTRAL AFRICAN REPUBLIC, CHAD, CONGO,
DEM. REP. CONGO (ZAIRE), EQUATORIAL GUINEA, GABON,
SAO TOME & PRINCIPE

THE GREAT RAIN FOREST BASIN of the Congo River embraces most of remote Central Africa. The interior was largely unknown to Europeans until late in the 19th century, when its tribal kingdoms were split – principally between France and Belgium – with Sao Tome and Principe the lone Portuguese territory, and Equatorial Guinea controlled by Spain. Open democracy and regional economic integration are important goals for these nations – several of which have only recently emerged from restrictive regimes – and investment is needed to improve transportation infrastructures. Many of the small, but fast-growing and increasingly urban population, speak French, the regional *lingua franca*, along with several hundred Pygmy, Bantu, and Sudanic dialects.

TRANSPORTATION & INDUSTRY

LARGE RESERVES OF VALUABLE MINERALS are found in Central Africa: copper, cobalt, zinc, and tin are mined in Dem. Rep. Congo (Zaire) and Cameroon; diamonds in the Central African Republic, and manganese in Gabon. Congo, Cameroon, Gabon, and Dem. Rep. Congo (Zaire) have oil deposits and oil has also been recently discovered in Chad. Goods such as palm oil and rubber are processed for export.

The ancient rocks of Dem. Rep. Congo (Zaire) hold immense and varied mineral reserves. This open pit copper mine is at Kolwezi in the far south.

Major industry and infrastructure

- brewing
- chemicals
- cobalt
- copper
- diamonds
- food processing
- manganese
- oil
- palm oil processing
- textiles
- tin
- capital cities
- major towns
- international airports
- major roads
- major industrial areas

THE LANDSCAPE

LAKE CHAD LIES in a desert basin bounded by the volcanic Tibesti Mountains in the north, plateaus in the east and, in the south, the broad watershed of the Congo Basin. The vast circular depression of the Congo is isolated from the coastal plain by the granite Massif du Chaillu. To the northwest, the volcanoes and fold mountains of the Cameroon Ridge (*Dorsale Camerounaise*) extend as islands into the Gulf of Guinea. The high fold mountains fringing the east of the Congo Basin fall steeply to the lakes of the Great Rift Valley.

The Tibesti Mountains are the highest in the Sahara. They were pushed up by the movement of the African Plate over a hot spot, which first formed the northern Ahaggar Mountains and is now thought to lie under the Great Rift Valley.

The Congo River is second only to the Amazon in the volume of water it carries, and in the size of its drainage basin.

Lake Tanganyika, the world's second deepest lake, is the largest of a series of linear "ribbon" lakes occupying a trench within the Great Rift Valley.

Rich mineral deposits in the "Copper Belt" of Dem. Rep. Congo (Zaire) were formed under intense heat and pressure when the ancient African Shield was uplifted to form the region's mountains.

Virgin tropical rain forest covers the Ruwenzori range on the borders of Dem. Rep. Congo and Uganda.

The lake-like expansion of the Congo River at Stanley Pool is the lowest point of the interior basin, although the river still descends more than 1,000 ft (300 m) to reach the sea.

Waterfalls and cataracts

Submarine canyon

Broad, shallow basin

The Congo River flows sluggishly through the rain forest of the interior basin. Toward the coast, the river drops steeply in a series of waterfalls and cataracts. At this point, the erosional power of the river becomes so great that it has formed a deep submarine canyon offshore.

A plug of resistant lava, at the southwestern end of the Cameroon Ridge (Dorsale Camerounaise), is all that remains of an eroded volcano.

The volcanic massif of Cameroon Mountain occupies an area which remains volcanically active.

Gulf of Guinea

Massif du Chaillu

Lake Chad is the remnant of an inland sea, which once occupied much of the surrounding basin. A series of droughts since the 1970s has reduced the area of this shallow freshwater lake to about 1,000 sq miles (2,599 sq km).

The vast sandflats surrounding Lake Chad were once covered by water. Changing climatic patterns caused the lake to shrink, and desert now covers much of its previous area.

76

MAP KEY

POPULATION
- ◉ 1 million to 5 million
- ⊙ 500,000 to 1 million
- ⊕ 100,000 to 500,000
- ⊙ 50,000 to 100,000
- ○ 10,000 to 50,000
- ○ below 10,000

ELEVATION
- 4000m / 13,124ft
- 3000m / 9843ft
- 2000m / 6562ft
- 1000m / 3281ft
- 500m / 1640ft
- 250m / 820ft
- 100m / 328ft
- sea level

SCALE 1:9,500,000
(projection: Lambert Azimuthal Equal Area)

Km
Miles

Map labels

LIBYA

SUDAN

NIGER

NIGERIA

C H A D

BORKOU-ENNEDI-TIBESTI

Erdi

Erdi Ma

Rèdi Ma

Dépression du Mourdi

Enneri

Tibesti

Emi Koussi 3415m

Yebbi-Bou

Massif d'Abo

Bardai

Aozou

Zouar

Sherda

Bodélé

Erg du Djourab

Erg du Ténéré

Faya

Gouro

Koro Toro

Ounianga Kebir

Oum-Chalouba

Monou

Fada

Guéréda

Iriba

Arada

Biltine

BILTINE

Abéché

Adré

OUADDAÏ

Am Dam

Goz-Beïda

Mongororo

Am Timan

SALAMAT

Birao

Gordil

Ouanda Djallé

Garba

VAKAGA

BAMINGUI-RANGORAN

MOYEN-CHARI

Oum-Hadjer

Haraz-Djombo

BATHA

Djedaa

Ati

Oum-Hadjer

Mangalmé

Mongo

Bitkine

GUÉRA

Melfi

Zakouma

Singako

Kyabé

Sarh

Koumra

Gondey

Doba

Bahr Azoum

Bahr Salamat

Bahr Aouk

Bahr Aouk

Bahr Keita

KANEM

Salal

Moussoro

Tefsei

Ngoura

Bokoro

Bousso

TANDJILE

Nokou

Mao

Bol

Rig-Rig

Ngouri

Massakory

Massaguet

Moïto

Massalassé

Lai

Kélo

LOGONE OCCIDENTAL

Moundou

LAC

Lake Chad

Chari

NDJAMENA

CHARI-BAGUIRMI

Kousséri

Logone

Bongor

Guélengdeng

Bahr Erguig

Churi

Gounou-Gaya

Pala

Léré

KEBBI

MAYO-KEBBI

Garoua

EXTRÊME-NORD

Maroua

Mora

Mokolo

Guider

LOGONE ORIENTAL

Tropic of Cancer

SUDAN

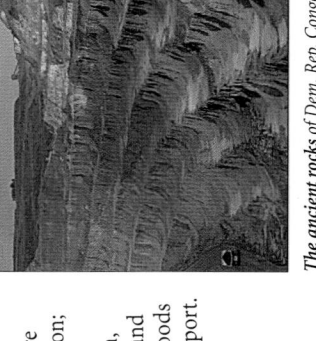

CHAD

CENTRAL AFRICAN REPUBLIC

DEM. REP. CONGO (ZAIRE)

CAMEROON

EQ. GUINEA

SAO TOME & PRINCIPE

GABON

CONGO

LIBYA

SUDAN

UGANDA

RWANDA

BURUNDI

TANZANIA

ZAMBIA

ANGOLA

ANGOLA (CABINDA)

NIGER

NIGERIA

ATLANTIC OCEAN

NDJAMENA

YAOUNDÉ

MALABO

Douala

BANGUI

Bangui

Kisangani

Bukavu

Kananga

Kolwezi

Lubumbashi

LIBREVILLE

Port-Gentil

BRAZZAVILLE

KINSHASA

TRANSPORTATION NETWORK

	miles	km
🚗	124,349 miles	(200,240 km)
✈	342 miles	(550 km)
🚂	3,830 miles	(6,167 km)
🚢	15,261 miles	(24,575 km)

The Trans-Gabon railroad, which began operating in 1987, has opened up new sources of timber and manganese. Elsewhere, much investment is needed to update and improve road, rail, and water transportation.

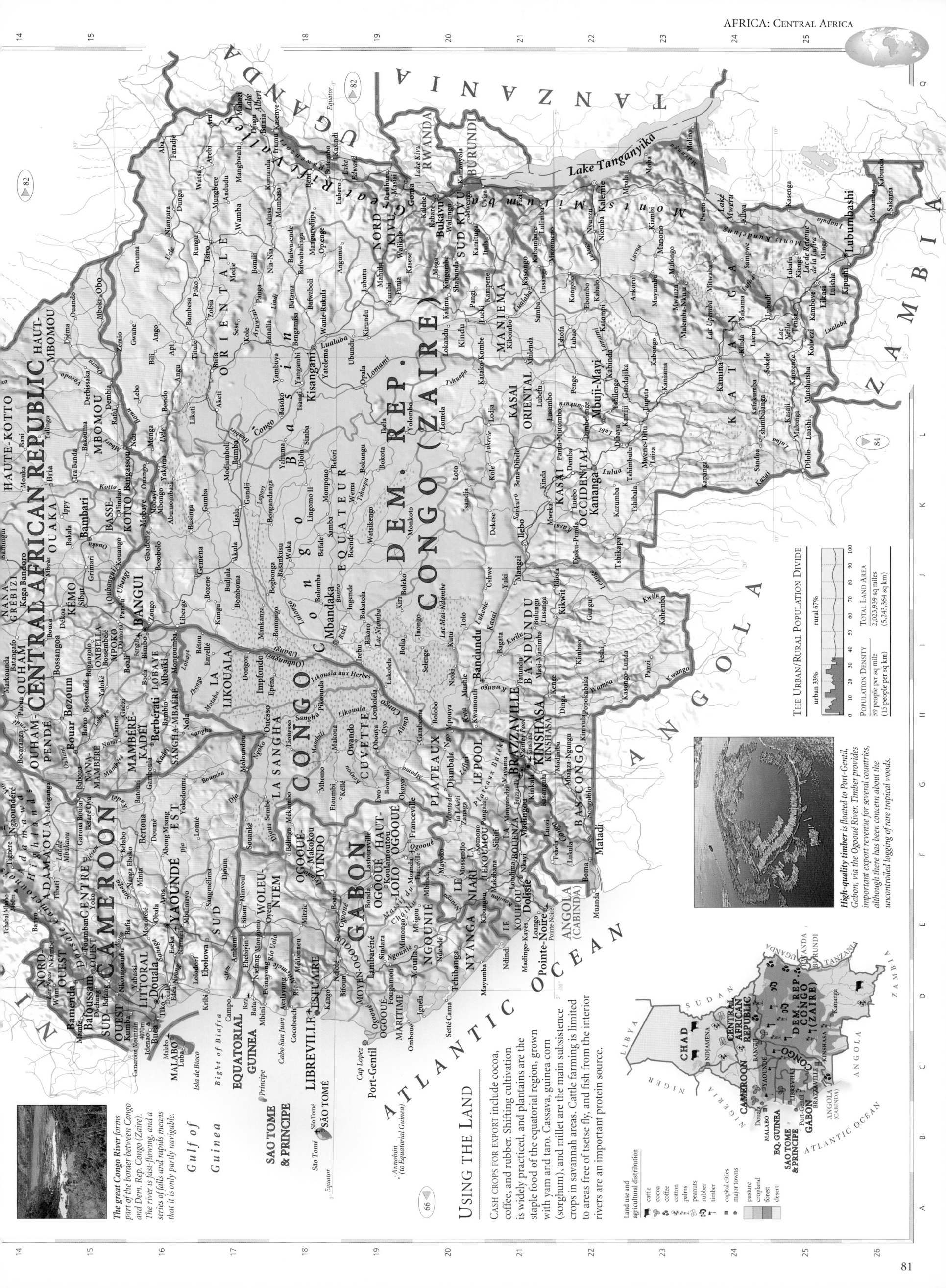

The great Congo River forms part of the border between Congo and Dem. Rep. Congo (Zaire). The river is fast-flowing, and a series of falls and rapids means that it is only partly navigable.

USING THE LAND

CASH CROPS FOR EXPORT include cocoa, coffee, and rubber. Shifting cultivation is widely practiced, and plantains are the staple food of the equatorial region, grown with yam and taro. Cassava, guinea corn (sorghum), and millet are the main subsistence crops in savannah areas. Cattle farming is limited to areas free of tsetse fly, and fish from the interior rivers are an important protein source.

High-quality timber is floated to Port-Gentil, Gabon, via the Ogooué River. Timber provides important export revenue for several countries, although there has been concern about the uncontrolled logging of rare tropical woods.

THE URBAN/RURAL POPULATION DIVIDE

urban 33%   rural 67%

TOTAL LAND AREA
2,023,939 sq miles
(5,243,364 sq km)

POPULATION DENSITY
39 people per sq mile
(15 people per sq km)

EAST AFRICA

BURUNDI, DJIBOUTI, ERITREA, ETHIOPIA, KENYA, RWANDA, SOMALIA, SUDAN, TANZANIA, UGANDA

THE COUNTRIES OF EAST AFRICA divide into two distinct cultural regions. Sudan and the "Horn" nations have been influenced by the Middle East; Ethiopia was the home of one of the earliest Christian civilizations, and Sudan reflects both Muslim and Christian influences. The southern countries share a closer cultural affinity with other sub-Saharan nations. Some of Africa's most densely populated countries lie in this region, and the needs of a growing number of people have put pressure on marginal lands and fragile environments. Although most East African economies remain strongly agricultural, Kenya has developed a varied industrial base.

THE LANDSCAPE

EAST AFRICA'S MOST SIGNIFICANT landscape feature is the Great Rift Valley, which formed during the most recent phase of continental movement when the rigid basement rocks cracked and buckled. Great blocks of land were raised and lowered, creating huge flat-bottomed valleys and steep escarpments, sometimes covered by volcanic extrusions in highland areas.

Central block slopes towards main fault

Ephemeral lake forms at far edge of slope

Boundary fault

The eastern arm of the Great Rift Valley is gradually being pulled apart; however the forces on one side are greater than the other causing the land to slope. This affects regional drainage which migrates down the slope.

This dome at Gonder, in Ethiopia, is a volcanic intrusion, formed when molten rock pushed up the surface of the Earth and then solidified, leaving an outcrop of igneous rock.

Lava flows on uplifted areas either side of the eastern branch of the Great Rift Valley gave the Ethiopian Highlands – a series of high, wide plateaus – their distinctive rounded appearance and fertile soils.

Kilimanjaro

An extinct volcano, Kilimanjaro is Africa's highest mountain, rising 19,340 ft (5,895 m). It is one of the few places in Africa where snow settles, allowing glacier ice to form.

A vast plateau lies between the eastern and western rift valleys in Kenya, Uganda, and western Tanzania. It has been leveled by long periods of erosion to form a peneplain, but is dotted with inselbergs – outcrops of more resistant rocks.

The Kassala region in eastern Sudan is watered by the Atbara River, an important tributary of the Nile. Most of the population is engaged in agriculture, growing cotton and cereals.

Lake Victoria occupies a vast basin between the two arms of the Great Rift Valley. It is the world's second largest lake in terms of surface area, extending 26,560 sq miles (68,880 sq km). The lake contains numerous islands and coral reefs.

Lake Tanganyika lies 8,202 ft (2,500 m) above sea level. It has a depth of nearly 4,700 ft (1,435 m). The lake traces the valley floor for some 400 miles (644 km) of the western arm of the Great Rift Valley.

The tiny countries of Rwanda and Burundi are mainly mountainous, with large areas of inaccessible tropical rain forest.

Much of northern Sudan is covered by desert. However, in the tropical wetlands of the southern Sudd region, annual rainfall can sometimes exceed 40 inches (1,000 mm).

(administered by Sudan)

(Political border)

(Administrative border)

(administered by Egypt)

82

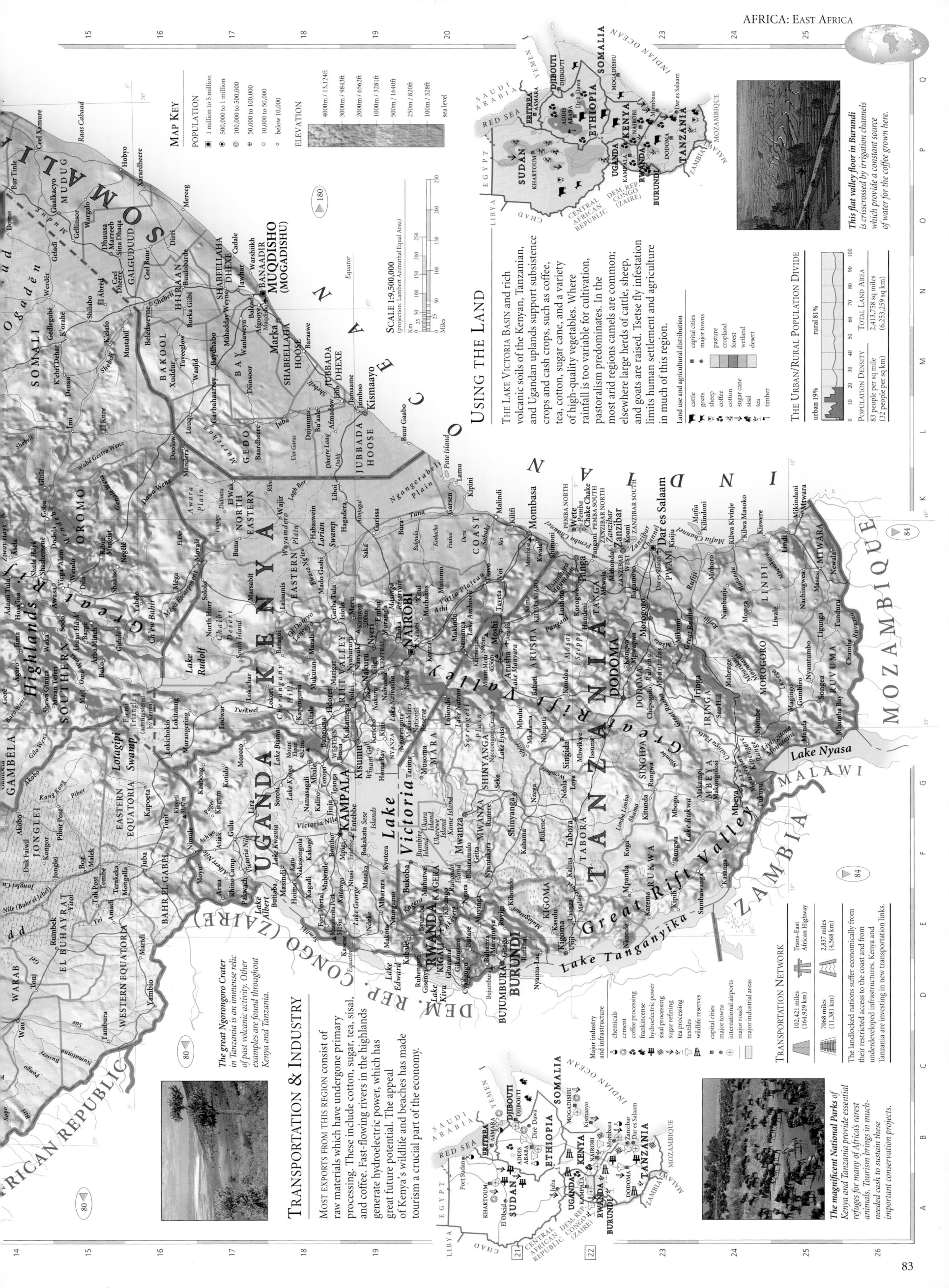

USING THE LAND

THE LAKE VICTORIA BASIN and rich volcanic soils of the Kenyan, Tanzanian, and Ugandan uplands support subsistence crops and cash crops, such as coffee, tea, cotton, sugar cane, and a variety of high-quality vegetables. Where rainfall is too variable for cultivation, pastoralism predominates. In the most arid regions camels are common; elsewhere large herds of cattle, sheep, and goats are raised. Tsetse fly infestation limits human settlement and agriculture in much of this region.

Land use and agricultural distribution

- ■ cattle
- ● goats
- ● sheep
- ● coffee
- ● cotton
- ● sugar cane
- ● sisal
- ● tea
- ● timber

- ■ capital cities
- ● major towns

- pasture
- cropland
- forest
- wetland
- desert

This flat valley floor in Burundi is crisscrossed by irrigation channels which provide a constant source of water for the coffee grown here.

THE URBAN/RURAL POPULATION DIVIDE

urban 19%   rural 81%

POPULATION DENSITY
83 people per sq mile
(32 people per sq km)

TOTAL LAND AREA
2,413,758 sq miles
(6,253,259 sq km)

TRANSPORTATION & INDUSTRY

MOST EXPORTS FROM THIS REGION consist of raw materials which have undergone primary processing. These include cotton, sugar, tea, sisal, and coffee. Fast-flowing rivers in the highlands generate hydroelectric power, which has great future potential. The appeal of Kenya's wildlife and beaches have made tourism a crucial part of the economy.

Major industry and infrastructure

- chemicals
- cement
- coffee processing
- frankincense
- hydroelectric power
- sisal processing
- sugar refining
- tea processing
- textiles
- wildlife reserves

- ■ capital cities
- ● major towns
- ✈ international airports
- major roads
- major industrial areas

TRANSPORTATION NETWORK

Trans-East African Highway		
102,421 miles (164,929 km)	7068 miles (11,381 km)	2,837 miles (4,568 km)

The landlocked nations suffer economically from their restricted access to the coast and from underdeveloped infrastructures. Kenya and Tanzania are investing in new transportation links.

The great Ngorongoro Crater in Tanzania is an immense relic of past volcanic activity. Other examples are found throughout Kenya and Tanzania.

The magnificent National Parks of Kenya and Tanzania provide essential refuges for many of Africa's rarest animals. Tourism brings in much-needed cash to sustain these important conservation projects.

MAP KEY

POPULATION

- ◻ 1 million to 5 million
- ◉ 500,000 to 1 million
- ◎ 100,000 to 500,000
- ⊕ 50,000 to 100,000
- ⊕ 10,000 to 50,000
- · below 10,000

ELEVATION

- 4000m / 13,124ft
- 3000m / 9843ft
- 2000m / 6562ft
- 1000m / 3281ft
- 500m / 1640ft
- 250m / 820ft
- 100m / 328ft
- sea level

▲ 180

SCALE 1:9,500,000

(projection: Lambert Azimuthal Equal Area)

SOUTHERN AFRICA

ANGOLA, BOTSWANA, LESOTHO, MALAWI, MOZAMBIQUE, NAMIBIA, SOUTH AFRICA, SWAZILAND, ZAMBIA, ZIMBABWE

A FRICA'S VAST SOUTHERN PLATEAU has been a contested homeland for disparate peoples for many centuries. The European incursion began with the slave trade and quickened in the 19th century, when the discovery of enormous mineral wealth secured South Africa's regional economic dominance. The struggle against white minority rule led to strife in Namibia, Zimbabwe, and the former Portuguese territories of Angola and Mozambique. South Africa's notorious apartheid laws, which denied basic human rights to more than 75% of the people, led to the state being internationally ostracized until 1994, when the first fully democratic elections inaugurated a new era of racial justice.

TRANSPORTATION & INDUSTRY

SOUTH AFRICA, the world's largest exporter of gold, has a varied economy which generates about 75% of the region's income and draws migrant labor from neighboring states. Angola exports petroleum; Botswana and Namibia rely on diamond mining; and Zambia is seeking to diversify its economy to compensate for declining copper reserves.

Almost all new mining ventures in Zimbabwe are now subject to government control. This mine at Bindura in northeastern Zimbabwe produces nickel, one of the country's top three minerals in terms of economic value.

THE LANDSCAPE

MOST OF SOUTHERN AFRICA rests on a concave plateau comprising the Kalahari basin and a mountainous fringe, skirted by a coastal plain which widens out in Mozambique. The plateau extends north, toward the Planalto de Bié in Angola, the Congo Basin and the lake-filled troughs of the Great Rift Valley. The eastern region is drained by the Zambezi and Limpopo Rivers, and the Orange is the major western river.

Lake Nyasa occupies one of the deep troughs of the Great Rift Valley, where the land has been displaced downward by as much as 3,000 ft (920 m).

The fast-flowing Zambezi River cuts a deep, wide channel as it flows along the Zimbabwe/Zambia border.

At Victoria Falls, the Zambezi River has cut a spectacular gorge taking advantage of large joints in the basalt, which were first formed as the lava cooled and contracted.

The Okavango/Cubango River flows from the Planalto de Bié to the swamplands of the Okavango Delta, one of the world's largest inland deltas, where it divides into countless distributary channels, feeding out into the desert.

Volcanic lava, over 250 million years old, caps the peaks of the Drakensberg range, which lie on the mountainous rim of southern Africa's interior plateau.

Broad, flat-topped mountains characterize the Great Karoo, which have been cut from level rock strata under extremely arid conditions.

The mountains of the Little Karoo are composed of sedimentary rocks which have been substantially folded and faulted.

The Orange River, one of the longest in Africa, rises in Lesotho and is the only major river in the south which flows westward, rather than to the east coast.

The Kalahari Desert is the largest continuous sand surface in the world. Iron oxide gives a distinctive red color to the windblown sand, which, in eastern areas covers the bedrock by over 200 ft (60 m).

Thousands of years of evaporating water have produced the Etosha Pan, one of the largest salt flats in the world. Lake and river sediments in the area indicate that the region was once less arid.

Finger Rock, near Khorixas, Namibia is a remnant of a former land surface, which has been denuded by erosion over the last 5 million years. These occasional stacks of partially weathered rocks interrupt the plains of the dry southern interior.

Khorixas, Namibia

Namib Desert

Planalto de Bié

Bushveld intrusion

Limpopo River

Great Rift Valley

Following a series of droughts, this baobab tree in Zimbabwe now stands alone in a field once filled by sugar cane. The thick trunk and small leaves of the baobab help it to conserve water, enabling it to survive even in drought conditions.

TRANSPORTATION NETWORK

84,213 miles (135,609 km)	746 miles (1,202 km)
23,208 miles (37,372 km)	3,815 miles (6,144 km)

Southern Africa's Cape-gauge rail network is by far the largest in the continent. About two-thirds of the 20,000 mile (32,000 km) system lies within South Africa. Lines such as the Harare–Bulawayo route have become corridors for industrial growth.

MAP KEY

POPULATION
- 1 million to 5 million
- 500,000 to 1 million
- 100,000 to 500,000
- 50,000 to 100,000
- 10,000 to 50,000
- below 10,000

ELEVATION
3000m / 9843ft	
2000m / 6562ft	
1000m / 3281ft	
500m / 1640ft	
250m / 820ft	
100m / 328ft	
sea level	

The Bushveld intrusion lies on South Africa's high "veld". Molten magma intruded into the Earth's crust creating a saucer-shaped feature, more than 180 miles (300 km) across, containing regular layers of precious minerals, overlain by a dome of granite.

Bushveld intrusion
- Granite
- Chromite
- Gabbro and peridotite
- Magnetite
- Platinum minerals

SOUTH AFRICA'S THREE CAPITALS
PRETORIA – administrative capital
CAPE TOWN – legislative capital
BLOEMFONTEIN – judicial capital

SCALE 1:9,500,000
(projection: Lambert Azimuthal Equal Area)

Major industry and infrastructure
- car manufacture
- coal
- copper
- diamonds
- gold
- oil
- textiles
- uranium
- food processing
- wildlife reserves
- capital cities
- major towns
- international airports
- major roads
- major industrial areas

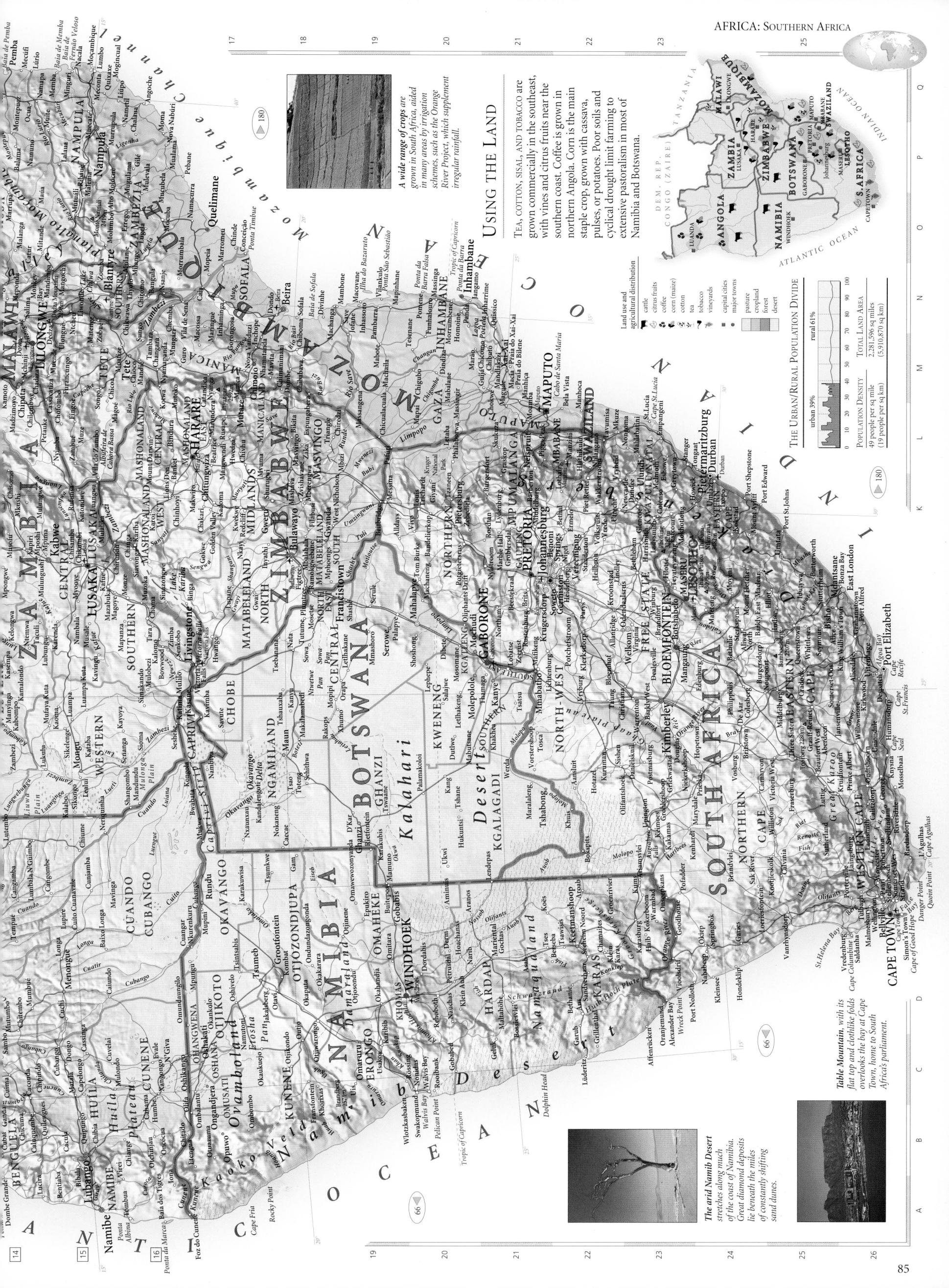

AFRICA: Southern Africa

Using the Land

Tea, cotton, sisal, and tobacco are grown commercially in the southeast, with vines and citrus fruits near the southern coast. Coffee is grown in northern Angola. Corn is the main staple crop, grown with cassava, pulses, or potatoes. Poor soils and cyclical drought limit farming to extensive pastoralism in most of Namibia and Botswana.

A wide range of crops are grown in South Africa, aided in many areas by irrigation schemes, such as the Orange River Project, which supplement irregular rainfall.

The Urban/Rural Population Divide

urban 39%   rural 61%

POPULATION DENSITY
49 people per sq mile
(19 people per sq km)

TOTAL LAND AREA
2,281,596 sq miles
(5,910,870 sq km)

Land use and agricultural distribution

cattle, citrus fruits, coffee, corn (maize), tea, tobacco, vineyards

pasture, cropland, forest, desert

capital cities, major towns

Table Mountain, with its flat top and clothlike folds overlooks the bay at Cape Town, home to South Africa's parliament.

The arid Namib Desert stretches along much of the coast of Namibia. Great diamond deposits lie beneath the miles of constantly shifting sand dunes.

85

ARCTIC OCEAN
North Pole
Laptev Sea
Ellesmere Island
Severnaya Zemlya
Poluostrov Taymyr
Kara Sea
Ostrov Rudol'fa
Franz Josef Land
Mys Zhelaniya
King Frederik VIII Land
Greenland
EURASIAN PLATE
NORTH AMERICAN PLATE
Spitsbergen
King Christian X Land
Novaya Zemlya
Poluostrov Yamal
Baydaratskaya Guba
Gulf of Ob
Yenisey
Greenland Sea
Barents Sea
Bjørnøya
Barents Trough
North Cape
Nordkinn
West Siberian Plain
ASIA

Arctic Circle
Denmark Strait
Bjargtangar
Reykjanes Ridge
Iceland
Vatnajökull
Iceland Plateau
Iceland Basin
Kolbeinsey Ridge
Jan Mayen Fracture Zone
Jan Mayen Ridge
Norwegian Sea
Norwegian Basin
Vøring Plateau
Tromsøflaket
Fugløya Bank
Vesterålen
Lofoten
Inarijärvi
Kebnekaise 2117m
Murmansk Rise
Kola Peninsula
Ozero Imandra
Ostrov Kolguyev
Poluostrov Kanin
Pechora
White Sea
Mezen'
Timanskiy Kryazh
URAL Mountains

Hatton Ridge
Rockall Rise
Feni Ridge
Rockall Trough
Faeroe-Iceland Ridge
Bill Baileys Bank
Faeroe-Shetland Trough
Faeroe Islands
Shetland Islands
Viking Bank
Orkney Islands
Outer Hebrides
Traena Bank
Scandinavia
Galdhøpiggen 2469m
Kjølen
Torneälven
Luleälven
Ljungan
Glåma
Ljusnan
Gulf of Bothnia
Oulujoki
Vänern
Åland
Onega Bay
Ozero Vygozero
Lake Ladoga
Lake Onega
Ozero Belaye
Northern Dvina
Mezen'
Gora Narodnaya
Yug
Vychegda
Sukhona
Vyatka
Kama

ATLANTIC OCEAN
Porcupine Plain
Ben Nevis 1343m
Grampian Mountains
British Isles
Ireland
Irish Sea
Snowdon 1085m
Britain
Pennines
North Channel
Celtic Sea
Celtic Shelf
St. George's Channel
Bristol Channel
Shannon
Trent
Severn
The Fens
Thames
Land's End
Channel Islands
Channel Islands
English Channel
Strait of Dover
North Sea
Great Fisher Bank
Jutland Bank
Dogger Bank
Frisian Islands
Skagerrak
Kattegat
Jylland
Sjælland
Gotland
Baltic Sea
Gulf of Riga
Lake Peipus
Lake Pskov
Gulf of Finland
Lake Ilmen
Vättern
Vättern
Jutland
Norwegian Trench

EUROPE
North European Plain
Harz
Elbe
Oder
Warta
Vistula
Bug
Western Dvina
Neman
Byelarus
Dnieper
Pripet Marshes
Central Russian Upland
Seym
Desna
Dnieper Lowlands
Dniester
Podil's'ka Vysochina
Pivdennyy Buh
Kiev Reservoir
Kremenchuk Reservoir
Don
Donets
Oka
Moskva
Volga Upland
Sura
Kuybyshev Reservoir
Samara
Ural
Kirghiz Steppe
Volga
Yergeni

Ardennes
Seine
Marne
Meuse
Moselle
Rhine
Vosges
Black Forest
Lake Constance
Danube
ALPS
Mont Blanc 4808m
Lake Geneva
Lake Garda
Loire
Vienne
Cher
Allier
Rhône
Garonne
Dordogne
Lot
Massif Central
Cévennes
Bay of Biscay
Biscay Plain
Charcot Seamounts
Azores-Biscay Rise
Theta Gap
Galicia Bank
Cordillera Cantábrica
Iberian Plain
Iberian Peninsula
Miño
Douro
Duero
Sistema Central
Sistema Ibérico
Aragón
Ebro
Aneto 3404m
Gulf of Lion
Pyrenees
Po
Apennines
Corno Grande 2912m
Ligurian Sea
Adriatic Sea
Dinaric Alps
Adriatic Basin
Corsica
Strait of Bonifacio
Sardinia
Tyrrhenian Sea
Tyrrhenian Basin
Gulf of Taranto
Strait of Otranto
Carpathian Mountains
Tisza
Prut
Bakony
Lake Balaton
Great Hungarian Plain
Drava
Sava
Tisza
Transylvanian Alps
Balkan Mountains
Danube
Scutari
Rhodope Mountains
Maritsa
Black Sea Lowland
Sea of Azov
Crimea
Kerch Strait
Black Sea
Bosporus
Sea of Marmara
EURASIAN PLATE
ANATOLIAN PLATE
Anatolia
Aegean Sea
Peloponnese
Mirtoan Sea
Sea of Crete
Crete
Mediterranean Ridge
Pontus Mountains
Karpathos
Rhodes Basin
Gulf of Antalya
Cyprus
Cyprus Basin
Levantine Basin
Nile Fan
Suez Canal
ARABIAN PLATE
AFRICAN PLATE

Madeira
Dacia Seamount
Ampère Seamount
Seine Plain
Seine Seamount
Horseshoe Seamounts
Gorringe Ridge
Cape Saint Vincent
Tagus Plain
Cabo da Roca
Tagus
Guadiana
Sierra Morena
Guadalquivir
Sistema Bético
Sistemas Béticos
Sierra Nevada
Punta de Tarifa
Strait of Gibraltar
Alboran Sea
Rif
Sebou
Oued Chelif
Tell Atlas
Middle Atlas
High Atlas
Atlas Mountains
Saharan Atlas
Oum er Rbia
Moulouya
Júcar
Segura
Gulf of Valencia
Balearic Islands
Algerian Basin
Mediterranean Sea
Sicily
Strait of Sicily
Mount Etna 3340m
Malta
Ionian Sea
Ionian Basin
Gávdos
Canary Islands
Agadir Canyon
Grand Erg Occidental
Grand Erg Oriental
Chott el Jerid
Gulf of Sirte
Western Desert
Qattara Depression -133m
Libyan Desert
'Erg Iguidi
Erg Chech
SAHARA
AFRICA

EUROPE

EUROPE IS THE WORLD'S SECOND SMALLEST CONTINENT, COVERING
4,053,309 SQ MILES (10,498,000 SQ KM). IT COMPRISES 44 SEPARATE
COUNTRIES, INCLUDING TURKEY AND THE RUSSIAN FEDERATION,
ALTHOUGH THE GREATER PARTS OF THESE NATIONS LIE IN ASIA.

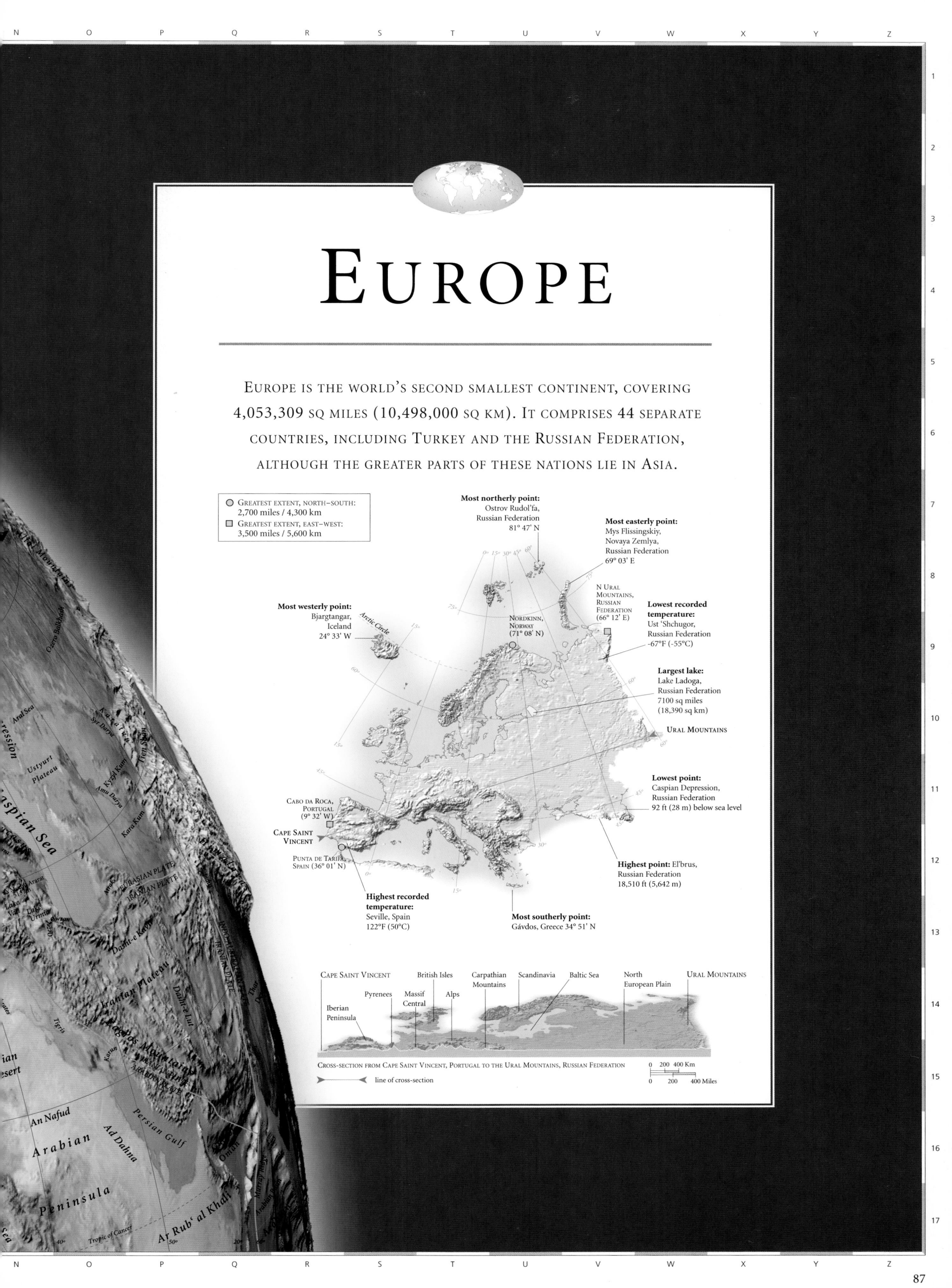

GREATEST EXTENT, NORTH–SOUTH:
2,700 miles / 4,300 km
GREATEST EXTENT, EAST–WEST:
3,500 miles / 5,600 km

Most northerly point:
Ostrov Rudol'fa,
Russian Federation
81° 47' N

Most easterly point:
Mys Flissingskiy,
Novaya Zemlya,
Russian Federation
69° 03' E

N URAL
MOUNTAINS,
RUSSIAN
FEDERATION
(66° 12' E)

**Lowest recorded
temperature:**
Ust 'Shchugor,
Russian Federation
-67°F (-55°C)

Most westerly point:
Bjargtangar, Iceland
24° 33' W

NORDKINN,
NORWAY
(71° 08' N)

Arctic Circle

Largest lake:
Lake Ladoga,
Russian Federation
7100 sq miles
(18,390 sq km)

URAL MOUNTAINS

Lowest point:
Caspian Depression,
Russian Federation
92 ft (28 m) below sea level

CABO DA ROCA,
PORTUGAL
(9° 32' W)

CAPE SAINT
VINCENT

PUNTA DE TARIFA,
SPAIN (36° 01' N)

Highest point: El'brus,
Russian Federation
18,510 ft (5,642 m)

**Highest recorded
temperature:**
Seville, Spain
122°F (50°C)

Most southerly point:
Gávdos, Greece 34° 51' N

CAPE SAINT VINCENT   British Isles   Carpathian   Scandinavia   Baltic Sea   North   URAL MOUNTAINS
                                          Mountains   European Plain
        Pyrenees   Massif   Alps
                     Central
Iberian
Peninsula

CROSS-SECTION FROM CAPE SAINT VINCENT, PORTUGAL TO THE URAL MOUNTAINS, RUSSIAN FEDERATION

line of cross-section

0   200   400 Km

0   200   400 Miles

PHYSICAL EUROPE

THE PHYSICAL DIVERSITY of Europe belies its relatively small size. To the northwest and south it is enclosed by mountains. The older, rounded Atlantic Highlands of Scandinavia and the British Isles lie to the north and the younger, rugged peaks of the Alpine Uplands to the south. In between lies the North European Plain, stretching 2,485 miles (4,000 km) from The Fens in England to the Ural Mountains in Russia. South of the plain lies a series of gently folded sedimentary rocks separated by ancient plateaus, known as massifs.

THE NORTH EUROPEAN PLAIN

RISING LESS THAN 1,000 ft (300 m) above sea level, the North European Plain strongly reflects past glaciation. Ridges of both coarse moraine and finer, wind-blown deposits have accumulated over much of the region. The ice sheet also diverted a number of river channels from their original courses.

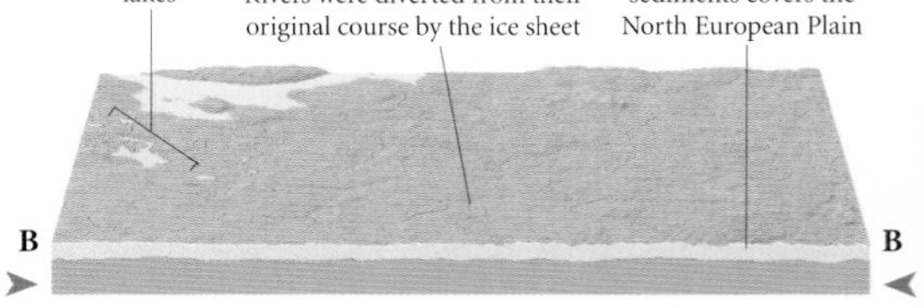

Glacial lakes

Rivers were diverted from their original course by the ice sheet

A layer of glacial sediments covers the North European Plain

Section across the North European Plain showing its low relief and drainage.

0   100   200 Km
0   100   200 Miles

THE ATLANTIC HIGHLANDS

THE ATLANTIC HIGHLANDS were formed by compression against the Scandinavian Shield during the Caledonian mountain-building period over 500 million years ago. The highlands were once part of a continuous mountain chain, now divided by the North Sea and a submerged rift valley.

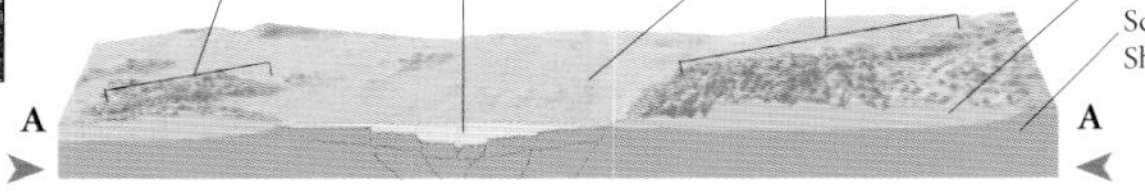

The Atlantic Highlands continue in the British Isles

Rift valley buried by sediments

North Sea

Atlantic Highlands in Norway

Rocks affected by ancient mountain-building

Scandinavian Shield

Cross-section through northeastern Europe showing the continuous mountain chain and rift valley system.

0   100   200 Km
0   100   200 Miles

SCALE 1:23,000,000
(projection: Lambert Azimuthal Equal Area)

Km
0   100   200   400   600
0 50 100   200   300   400   500   600
Miles

MAP KEY

ELEVATION

4000m / 13,124ft
3000m / 9843ft
2000m / 6562ft
1000m / 3281ft
500m / 1640ft
250m / 820ft
100m / 328ft
sea level

PLATE MARGINS
(for explanation see page xiv)

—————   constructive
△ △   destructive
—————   conservative
....   uncertain
—————   physiographic regions
▶—   line of cross-section

Map labels

NORTH AMERICAN PLATE
EURASIAN PLATE
Iceland
ATLANTIC OCEAN
Norwegian Sea
Faeroe Islands
Shetland Islands
Outer Hebrides
British Isles
Ireland
Shannon
Britain
The Fens
Thames
North Sea
English Channel
Seine
Loire
Garonne
Bay of Biscay
Ardennes
Rhine
Elbe
Weser
Oder
PLATEAUX AND LOWLANDS
Massif Central
Pyrenees
Iberian Peninsula
Douro
Tagus
Ebro
Guadalquivir
Balearic Islands
Corsica
Sardinia
ALPS
Mt Blanc 4807m
APENNINES
Po
Adriatic Sea
Tyrrhenian Sea
Sicily
Etna 3263m
Malta
Ionian Sea
Mediterranean Sea
ATLANTIC HIGHLANDS
Kölen
Novaya Zemlya
Kara Sea
Ostrov Kolguyev
Barents Sea
Kola Peninsula
White Sea
SCANDINAVIAN SHIELD
Gulf of Bothnia
Northern Dvina
Lake Onega
Lake Ladoga
Vänern
Vättern
Baltic Sea
Jylland
Gulf of Riga
Western Dvina
Ural Mountains
Central Russian Upland
Volga Uplands
Volga
Don
Dnieper
Dniester
Danube
Carpathian Mountains
Great Hungarian Plain
Danube
Balkan Mountains
Dinaric Alps
Vesuvius 1171m
Peloponnese
Crete
Aegean Sea
ANATOLIAN PLATE
EURASIAN PLATE
AFRICAN PLATE
Crimea
Sea of Azov
Black Sea
Caspian Sea
Caucasus
El'brus 5642m
ASIA
Vistula

THE PLATEAUS AND LOWLANDS

THE UPLIFTED PLATEAUS or massifs of southern central Europe are the result of long-term erosion, later followed by uplift. They are the source areas of many of the rivers which drain Europe's lowlands. In some of the higher reaches, fractures have enabled igneous rocks from deep in the Earth to reach the surface.

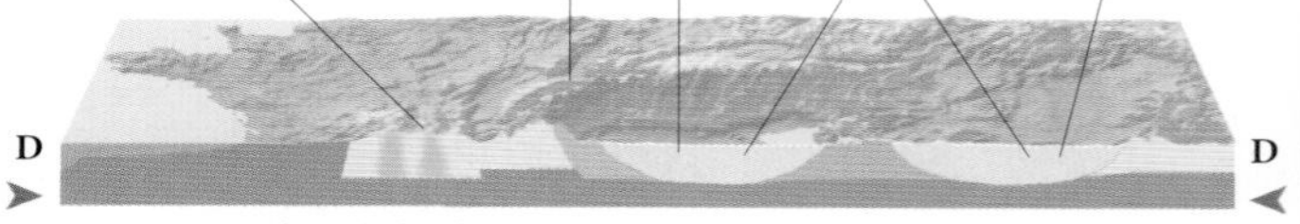

Igneous rocks have intruded into the Massif Central

Older, eroded massifs lie behind the arc of the Alps

Tectonically formed basins

Po Valley

Great Hungarian Plain

Cross-section through the plateaus and lowlands showing the lower elevation of the ancient massifs.

0   100   200 Km
0   100   200 Miles

THE ALPINE UPLANDS

THE COLLISION OF the African and European continents, which began about 65 million years ago, folded and then uplifted a series of mountain ranges running across southern Europe and into Asia. Two major lines of folding can be traced: one includes the Pyrenees, the Alps, and the Carpathian Mountains; the other incorporates the Apennines and the Dinaric Alps.

European basement rock

Alps

Weak sedimentary strata have been folded

African Plate moved northward

The Apennines

Cross-section through the Alps showing folding and faulting caused by plate tectonics.

0   50   100 Km
0   50   100 Miles

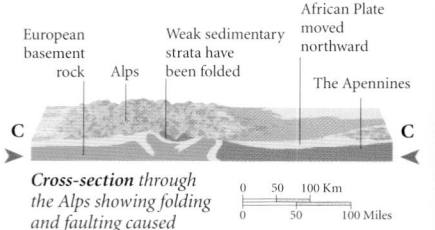

CLIMATE

EUROPE EXPERIENCES few extremes in either rainfall or temperature, with the exception of the far north and south. Along the west coast, the warm currents of the North Atlantic Drift moderate temperatures. Although east–west air movement is relatively unimpeded by relief, the Alpine Uplands halt the progress of north–south air masses, protecting most of the Mediterranean from cold, north winds.

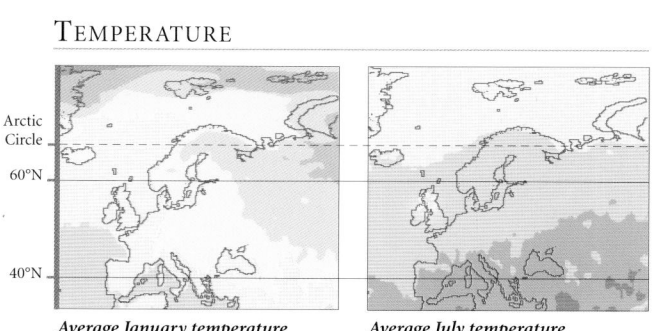

Frost grips northern and eastern Europe during the long cold winters. Lakes and rivers frequently freeze.

TEMPERATURE

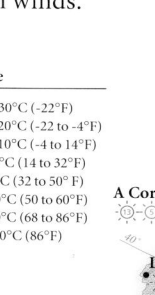

Arctic Circle
60°N
40°N

Average January temperature

Average July temperature

Temperature
	below -30°C (-22°F)
	-30 to -20°C (-22 to -4°F)
	-20 to -10°C (-4 to 14°F)
	-10 to 0°C (14 to 32°F)
	0 to 10°C (32 to 50°F)
	10 to 20°C (50 to 60°F)
	20 to 30°C (68 to 86°F)
	above 30°C (86°F)

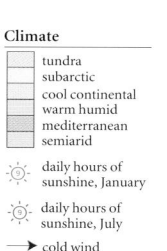

Mild temperatures and frequent rainfall contribute to the fertile farming land found over much of northwestern Europe.

RAINFALL

Arctic Circle
60°N
40°N

Average January rainfall

Average July rainfall

Rainfall
	0–25 mm (0–1 in)
	25–50 mm (1–2 in)
	50–100 mm (2–4 in)
	100–200 mm (4–8 in)
	200–300 mm (8–12 in)
	300–400 mm (12–16 in)
	400–500 mm (16–20 in)
	more than 500 mm (20 in)

Reykjavík · Karasjok · Murmansk · Pechora
Bodø · Pajala · Archangel
Hoyvík · Kajaani · Kirov
Bergen · Sveg · Härnösand · Ufa
Malin Head · Dundee · Oslo · Helsinki · St Petersburg · Kirov
Shannon · Morecambe · Vestervig · Stockholm · Tallinn · Moscow
Exeter · London · Malmö · Riga · Minsk
Brussels · Hamburg · Berlin · Warsaw · Kharkiv
Paris · Prague · Astrakhan'
A Coruña · Bordeaux · Zürich · Munich · Vienna · Bratislava · Rostov-na-Donu
Toulouse · Lyon · Milan · Zagreb · Belgrade · Bucharest · Simferopol'
Lisbon · Madrid · Monaco · Sarajevo · Sofia · Constanța
Barcelona · Naples · Tirana · Istanbul
Gibraltar · Palma · Cagliari · Messina · Salonica · Athens

Mistral · Bora · Sirocco

Climate
	tundra
	subarctic
	cool continental
	warm humid
	mediterranean
	semiarid
☼	daily hours of sunshine, January
☼	daily hours of sunshine, July
→	cold wind
→	hot wind

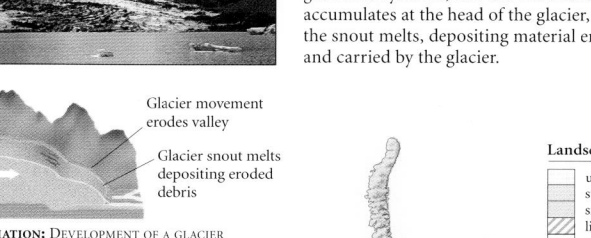

Dusty Sirocco winds from Africa help create the semiarid scrubland common across the Mediterranean coastlands of southern Europe.

SHAPING THE CONTINENT

SUCCESSIVE ICE AGES have left many relict landforms across Europe. Present glaciers continue to carve peaks and valleys in the northern Atlantic Highlands and Alpine Uplands. Tectonic activity, both past and present, has shaped southern Europe and Iceland. Active volcanoes and earthquakes still occur in Italy and Greece. Europe's extensive coastline, particularly in the northwest, is constantly modified by wave action and fluvial deposits.

GLACIATION

[1] Valley glaciers, such as this one *(left)* in Iceland, form in hollows at the top of valleys and flow downward, drawn by gravity. Their growth is dynamic; new snowfall constantly accumulates at the head of the glacier, while the snout melts, depositing material eroded and carried by the glacier.

COASTAL PROCESSES

[5] Spits are narrow bands of sand or shingle, formed by longshore drift; a process whereby waves carry material along the beach. They usually form where the coastline changes direction, and their growth is then halted by an opposing river current, as at Spurn Head, in the British Isles *(left)*. Coastal features such as these are constantly being created and destroyed.

Snow accumulates at the head of glacier
Glacier movement erodes valley
Glacier snout melts depositing eroded debris

GLACIATION: DEVELOPMENT OF A GLACIER

RIVER SYSTEMS

[2] Rivers are continuously transporting eroded material toward the sea. Slow-moving, low-gradient rivers, like this one in western Russia *(above)*, deposit their alluvium load, infilling valleys creating a floodplain. Subsequent climatic and tectonic fluctuations may erode the floodplain to form terraces.

Landscape
	uplifting land
	stable land
	sinking land
	limestone region
	glacier
▲	active volcano
→	ocean current
⋯	area of tectonic activity
—	maximum limit of glaciation

Terrace created by erosion
Floodplain
Deposited alluvium
River channel

RIVER SYSTEMS: FORMATION OF A FLOODPLAIN AND TERRACES

Sand and shingle spit
Original coastline
Opposing river current
Waves breaking at an angle

COASTAL PROCESSES: FORMATION OF A SPIT

THE EVOLVING LANDSCAPE

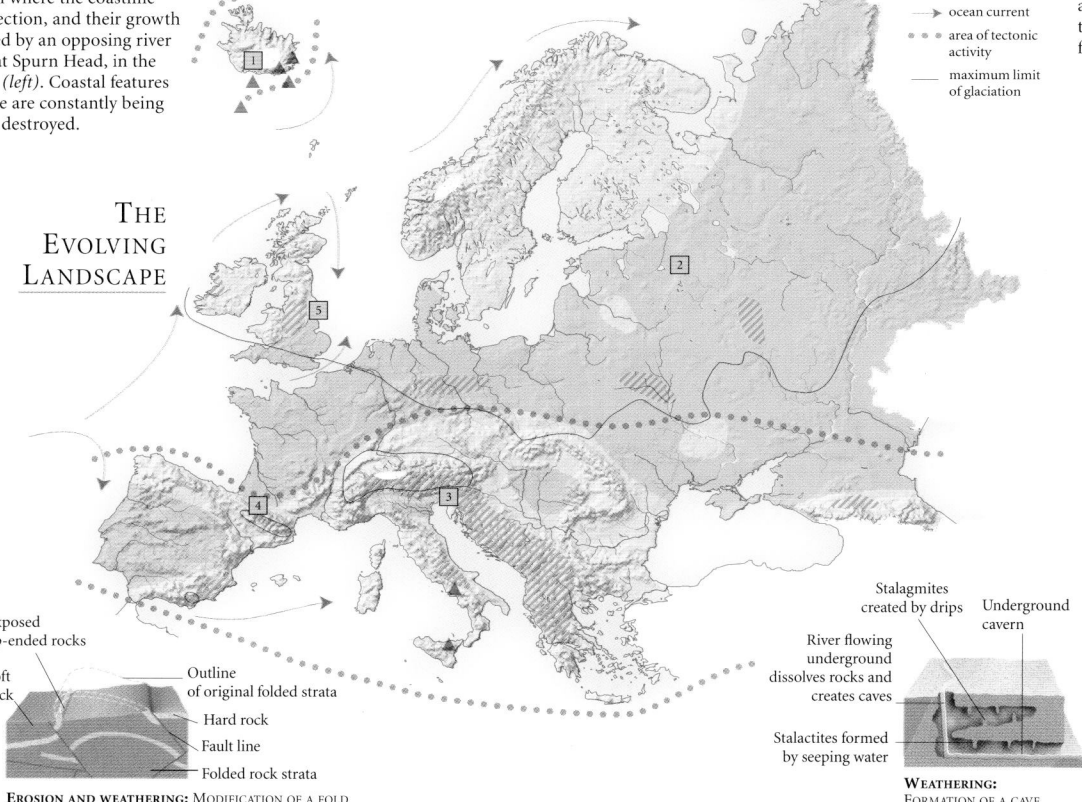

EROSION AND WEATHERING

[4] Much of Europe was once subjected to folding and faulting, exposing hard and soft rock layers. Subsequent erosion and weathering has worn away the softer strata, leaving up-ended layers of hard rock as in the French Pyrenees *(above)*.

Exposed up-ended rocks
Soft rock
Outline of original folded strata
Hard rock
Fault line
Folded rock strata

EROSION AND WEATHERING: MODIFICATION OF A FOLD

Stalagmites created by drips
Underground cavern
River flowing underground dissolves rocks and creates caves
Stalactites formed by seeping water

WEATHERING: FORMATION OF A CAVE

WEATHERING

[3] As surface water filters through permeable limestone, the rock dissolves to form underground caves, like Postojna in the Karst region of Slovenia *(above)*. Stalactites grow downward as lime-enriched water seeps from roof fractures; stalagmites grow upward where drips splash down.

POLITICAL EUROPE

THE POLITICAL BOUNDARIES OF EUROPE have changed many times, especially during the 20th century in the aftermath of two world wars, the breakup of the empires of Austria-Hungary, Nazi Germany and, toward the end of the century, the collapse of communism in eastern Europe. The fragmentation of Yugoslavia has again altered the political map of Europe, highlighting a trend toward nationalism and devolution. In contrast, economic federalism is growing. In 1958, the formation of the European Economic Community (now the European Union or EU) started a move toward economic and political union.

The Brandenburg Gate in Berlin is a potent symbol of German reunification. From 1961, the road beneath it ended in a wall, built to stop the flow of refugees to the West. It was opened again in 1989 when the wall was destroyed and East and West Germany were reunited.

POPULATION

EUROPE IS A DENSELY POPULATED, urbanized continent; in Belgium over 90% of people live in urban areas. The highest population densities are found in an area stretching east from southern Britain and northern France, into Germany. The northern fringes are only sparsely populated.

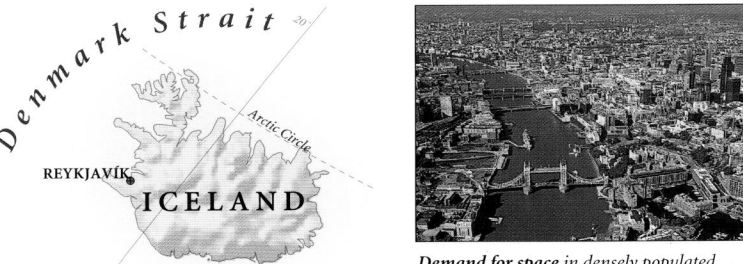

Demand for space in densely populated European cities like London has led to the development of high-rise offices and urban sprawl.

Population density (people per sq mile)
- below 130
- 130–259
- 260–379
- 380–519
- 520–780
- above 780

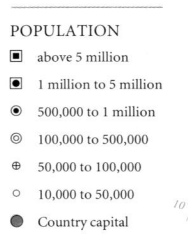

Traditional lifestyles still persist in many remote and rural parts of Europe, especially in the south, east, and in the far north.

MAP KEY

POPULATION
- ■ above 5 million
- ▪ 1 million to 5 million
- ◉ 500,000 to 1 million
- ◎ 100,000 to 500,000
- ⊕ 50,000 to 100,000
- ○ 10,000 to 50,000
- ● Country capital

BORDERS
- full international border

SCALE 1:15,500,000
(projection: Lambert Azimuthal Equal Area)

Km
0 50 100 200 300 400 500 600 700 800 900 1000
Miles
0 50 100 200 300 400 500 600 700

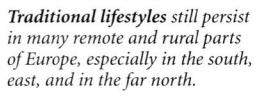

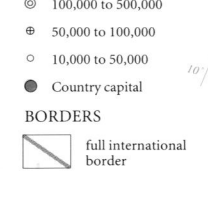
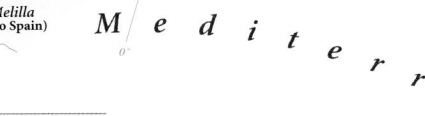

Overcoming natural barriers, the Brenner Autobahn, one of the main routes across the Alps, links Innsbruck in Austria with Verona in Italy.

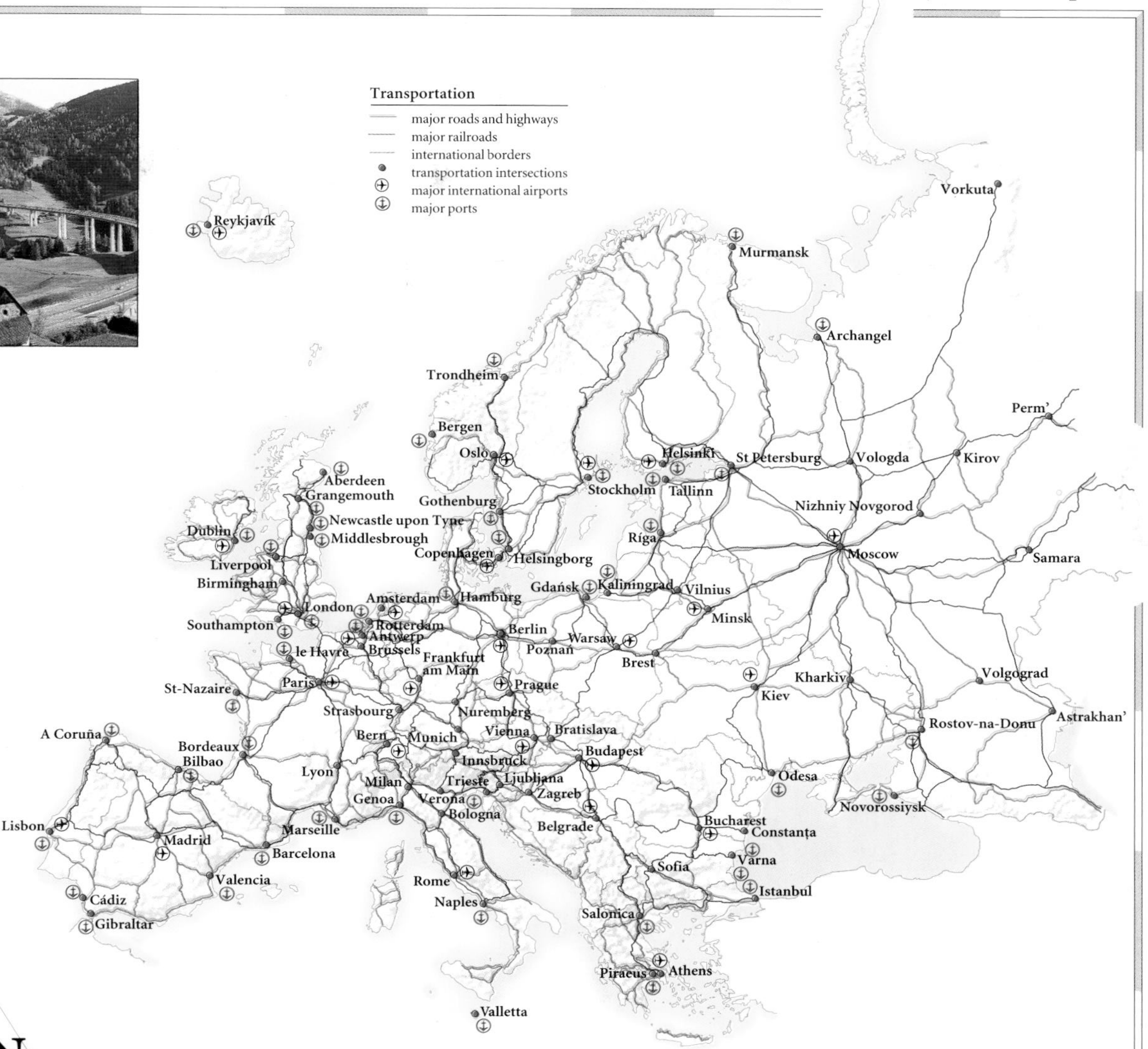

Transportation
— major roads and highways
— major railroads
— international borders
● transportation intersections
⊕ major international airports
⚓ major ports

Novaya Zemlya

Kara Sea

Barents Sea

White Sea

RUSSIAN

FEDERATION

Ural mountains

Vorkuta

Arkhangel'sk
Northern Dvina
Lake Onega
Vologda
Yaroslavl'
Nizhniy Novgorod
Kirov
Perm'
Ufa
Kazan'
MOSCOW
Ul'yanovsk
Tol'yatti
Samara
Orenburg
Tula
Kazakhstan
Saratov
Voronezh
Kharkiv
Volgograd
Volga
Astrakhan'
UKRAINE
Dnipropetrovs'k
Donets'k
Rostov-na-Donu
Caspian Sea
Sea of Azov
Stavropol'
Grozznyy
Novorossiysk
Simferopol'
Dnieper
Caucasus
Georgia
Azerbaijan
Black Sea
Turkey

Reykjavík
Murmansk
Archangel
Trondheim
Bergen
Oslo
Helsinki
St Petersburg
Vologda
Kirov
Perm'
Aberdeen
Grangemouth
Gothenburg
Stockholm
Tallinn
Nizhniy Novgorod
Samara
Dublin
Newcastle upon Tyne
Middlesbrough
Copenhagen
Helsingborg
Riga
Moscow
Liverpool
Birmingham
London
Amsterdam
Hamburg
Gdańsk
Kaliningrad
Vilnius
Minsk
Southampton
Rotterdam
Berlin
Warsaw
Brest
le Havre
Antwerp
Brussels
Frankfurt am Main
Poznań
Kharkiv
Volgograd
St-Nazaire
Paris
Prague
Kiev
A Coruña
Strasbourg
Nuremberg
Vienna
Bratislava
Odesa
Rostov-na-Donu
Astrakhan'
Bordeaux
Bilbao
Bern
Munich
Innsbruck
Budapest
Bucharest
Constanţa
Novorossiysk
Lyon
Milan
Trieste
Ljubljana
Zagreb
Lisbon
Genoa
Verona
Bologna
Belgrade
Sofia
Varna
Istanbul
Madrid
Marseille
Rome
Salonica
Barcelona
Valencia
Naples
Piraeus
Athens
Cádiz
Gibraltar
Valletta

TRANSPORTATION

DESPITE ITS FRAGMENTED GEOGRAPHY and many natural frontiers, communications in Europe are well developed. Extensive highway links allow rapid road transportation. High-speed rail connections like France's TGV *(Train à Grande Vitesse)*, and the Channel Tunnel have improved rail travel. Outdated communication infrastructures in parts of eastern Europe, and insufficient transportation links across the Alps, however, remain weak parts of the network.

LANGUAGES

THERE ARE THREE MAIN EUROPEAN language groups: Germanic languages predominate in central and northern Europe; Romance languages in western and Mediterranean Europe and Romania; while Slavic languages are spoken in eastern Europe and the Russian Federation. Isolated pockets of local languages, such as Basque and Gaelic, persist and frequently provide a focus for national identity.

Language groups
Turkic
Albanian
Finno-Ugric/Samoyed
Germanic
Slavic
Romance
Basque
Baltic
Celtic
Greek
Caucasian
Iranian
Mongol

ICELANDIC
FAEROESE
NORWEGIAN
SWEDISH
LAPPISH (SAMI)
NENETS
KOMI
FINNISH
KARELIAN
RUSSIAN
UDMURT
VEPSE
GAELIC
ENGLISH
IRISH
ENGLISH
ESTONIAN
SWEDISH
KARELIAN
MARI
CHUUASH
TARTAR
BASHKIR
LATVIAN
MORDUNIAN
WELSH
FRISIAN
DANISH
LITHUANIAN
RUSSIAN
BRETON
FRENCH
DUTCH
GERMAN
POLISH
BELARUSSIAN
FRENCH
GERMAN
CZECH
SLOVAK
UKRAINIAN
KALMYK
PORTUGUESE
GALICIAN
BASQUE
ITALIAN
SLOVENE
HUNGARIAN
KABARD
CIRCASSIAN
ADYGHE
KARACHAY
KUMYK
CHECHEN
AVAR
LEZGHIAN
OSSETIAN
BALKAR
SPANISH
CATALAN
SERBO-CROAT
ROMANIAN
CATALAN
ITALIAN
BULGARIAN
MACEDONIAN
TURKISH
SARDINIAN
ALBANIAN
GREEK
ITALIAN
MALTESE

The architecture of the Grand Place lies at the heart of Brussels – home city to one of the EU headquarters.

EUROPEAN RESOURCES

Europe's large tracts of fertile, accessible land, combined with its generally temperate climate, have allowed a greater percentage of land to be used for agricultural purposes than in any other continent. Extensive coal and iron ore deposits were used to create steel and manufacturing industries during the 19th and 20th centuries. Today, although natural resources have been widely exploited, and heavy industry is of declining importance, the growth of hi-tech and service industries has enabled Europe to maintain its wealth.

INDUSTRY

Europe's wealth was generated by the rise of industry and colonial exploitation during the 19th century. The mining of abundant natural resources made Europe the industrial center of the world. Adaptation has been essential in the changing world economy, and a move to service-based industries has been widespread except in eastern Europe, where heavy industry still dominates.

Countries like Hungary are still struggling to modernize inefficient factories left over from extensive, centrally-planned industrialization during the communist era.

Other power sources are becoming more attractive as fossil fuels run out; 16% of Europe's electricity is now provided by hydroelectric power.

Frankfurt am Main is an example of a modern service-based city. The skyline is dominated by headquarters from the worlds of banking and commerce.

STANDARD OF LIVING

Living standards in western Europe are among the highest in the world, although there is a growing sector of homeless, jobless people. Eastern Europeans have lower overall standards of living – a legacy of stagnated economies.

Standard of Living
(UN Human Development Index)
- low
- high

Skiing brings millions of tourists to the slopes each year, which means that even unproductive, marginal land is used to create wealth in the French, Swiss, Italian, and Austrian Alps.

GNP per capita (US$)
- below 1999
- 2000–4999
- 5000–9999
- 10,000–19,999
- 20,000–24,999
- above 25,000

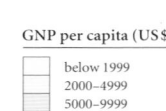

Industry
- ✈ aerospace
- brewing
- car/vehicle manufacture
- chemicals
- defense
- electronics
- engineering
- finance
- food processing
- hi-tech industry
- iron & steel
- pharmaceuticals
- printing & publishing
- shipbuilding
- textiles
- timber processing
- wine
- coal
- oil
- gas
- industrial cities
- major industrial areas

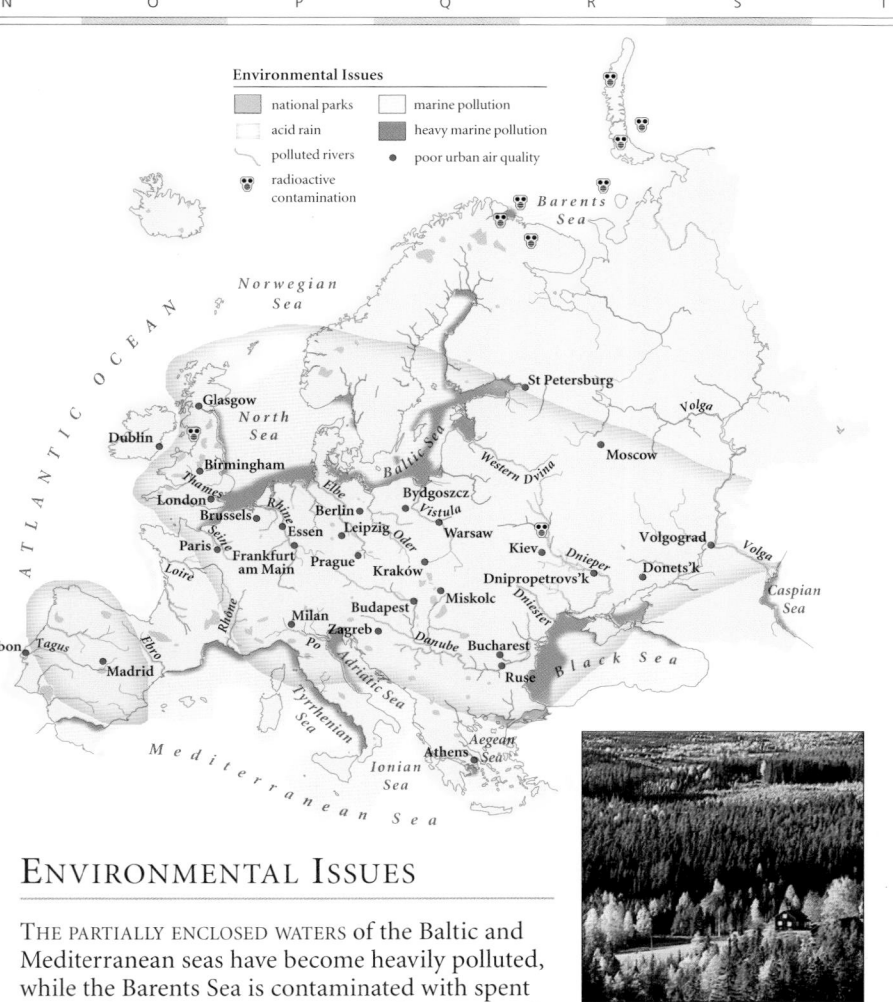

Environmental Issues

national parks   **marine pollution**
acid rain   **heavy marine pollution**
polluted rivers   • **poor urban air quality**
radioactive contamination

MINERAL RESOURCES

FOSSIL FUELS ARE EUROPE'S main mineral resource, although fuel demand far outstrips production. Sizeable coal reserves remain in the Donbass in Ukraine, Germany's Ruhr Valley, Poland, and in the British Isles. Oil and gas reserves are found mainly in the North Sea, and in the Volga Basin.

The valuable oil and gas reserves in the North Sea were first discovered in the early 1960s, and are exploited by the UK, Denmark, Germany, and Norway.

Mineral Resources

oil field
gas field
coal field

bauxite
iron
lead
mercury △
potassium ▲
uranium
zinc

ENVIRONMENTAL ISSUES

THE PARTIALLY ENCLOSED WATERS of the Baltic and Mediterranean seas have become heavily polluted, while the Barents Sea is contaminated with spent nuclear fuel from Russia's navy. Acid rain, caused by emissions from factories and power stations, is actively destroying northern forests. As a result, pressure is growing to safeguard Europe's natural environment and prevent further deterioration.

The Camargue in the Rhône Delta, southern France, is a protected wetland area, famous for its native population of white horses, and unique bird and plant life.

Coniferous forest covers vast swathes of northern Scandinavia and the Russian Federation. Pollutants from other parts of Europe mixing with rainfall are causing defoliation and serious damage to many forests.

USING THE LAND AND SEA

EUROPE'S SWELLING URBAN POPULATION and the outward expansion of many cities has created acute competition for land. Despite this, European resourcefulness has maximized land potential, and over half of Europe's land is still used for a wide variety of agricultural purposes. Land in northern Europe is used for cattle-rearing, pasture, and arable crops. Toward the Mediterranean, the mild climate allows the growing of grapes for wine; olives, sunflowers, tobacco, and citrus fruits. EU subsidies, however, have resulted in massive overproduction and a land "set-aside" policy has been introduced.

Using the Land and Sea

cropland
forest
ice cap
mountain region
pasture
tundra
wetland

• major conurbations

cattle
goats
pigs
poultry
reindeer
sheep
cereals
citrus fruits
cotton
fishing
fodder
fruit
olive oil
potatoes
rice
root crops
roses
shellfish
sunflowers
timber
tobacco
vineyards

Bulgarian roses are one of the many diverse crops grown in Europe. Rose oil, extracted from the petals, is used in perfume making.

Lowland pastures are used for dairy farming. Good transportation links and refrigeration allow fresh milk to be distributed throughout Europe.

SCANDINAVIA, FINLAND & ICELAND

DENMARK, NORWAY, SWEDEN, FINLAND, ICELAND

JUTTING INTO THE ARCTIC CIRCLE, this northern swath of Europe has some of the continent's harshest environments, but benefits from great reserves of oil, gas, and natural evergreen forests. While most early settlers came from the south, migrants to Finland came from the east, giving it a distinct language and culture. Since the late 19th century, the Scandinavian states have developed strong egalitarian traditions. Today, their welfare benefits systems are among the most extensive in the world, and standards of living are high. The Lapps, or Sami, maintain their traditional lifestyle in the northern regions of Norway, Sweden, and Finland.

THE LANDSCAPE

GLACIERS up to 10,000 ft (3,000 m) deep covered most of Scandinavia and Finland during the last Ice Age. The effects of glaciation mark the entire landscape, from the mountains to the lowlands, across the tundra landscape of Lapland, and the lake districts of Sweden and Finland.

Geysers are a by-product of Iceland's volcanic activity. Geysir, Iceland's largest spring, gives them their name.

Lapland, north of the Arctic Circle, is an area of undulating fells and plains known as tundra. The subsoil is permanently frozen and therefore impermeable. There are many peat bogs. Pools reappear in the summer when the surface thaws.

Halti Mountain is Finland's highest point, at 4,356 ft (1,328 m).

Finland's landscape was fashioned by ice action. Glaciers gouged out its distinctive shallow lake basins, such as Oulujärvi, and left debris called moraines in their wake.

The Lofoten Islands were one of the first areas exposed as the ice sheet melted.

Oulujärvi

Scandinavia is still recovering from the last Ice Age, when ice depressed the land by 2,000 ft (600 m). This gradual uplift is known as isostatic rebound.

Area of maximum yearly uplift 0.3 in/yr (9 mm/yr)

Slower rates of uplift 0.1 in/yr (3 mm/yr)

Sjelland coast

On the coast of Sjelland, these cliffs have been eroded by the sea, exposing layers of chalk and limestone.

Fjords

The fjords on the western coast of Norway were once gentle river valleys. Their deep floors and steep sides were carved out by glaciers during the last Ice Age, and they were later flooded by the sea.

USING THE LAND AND SEA

THE COLD CLIMATE, short growing season, poorly developed soil, steep slopes, and exposure to high winds across northern regions means that most agriculture is concentrated, with the population, in the south. Most of Finland and much of Norway and Sweden are covered by dense forests of pine, spruce and birch, which supply the timber industries.

Land use and agricultural distribution

- fishing
- pigs
- reindeer
- sheep
- timber
- capital cities
- major towns
- pasture
- cropland
- forest
- mountain region
- tundra

THE URBAN/RURAL POPULATION DIVIDE

urban 77%   rural 23%

TOTAL LAND AREA
473,970 sq miles
(1,227,610 sq km)

POPULATION DENSITY
20 people per sq mile
(51 people per sq km)

SCALE 1:8,000,000
(projection: Lambert Conformal Conic)

SCALE 1:5,000,000
(projection: Lambert Conformal Conic)

(same scale as main map)

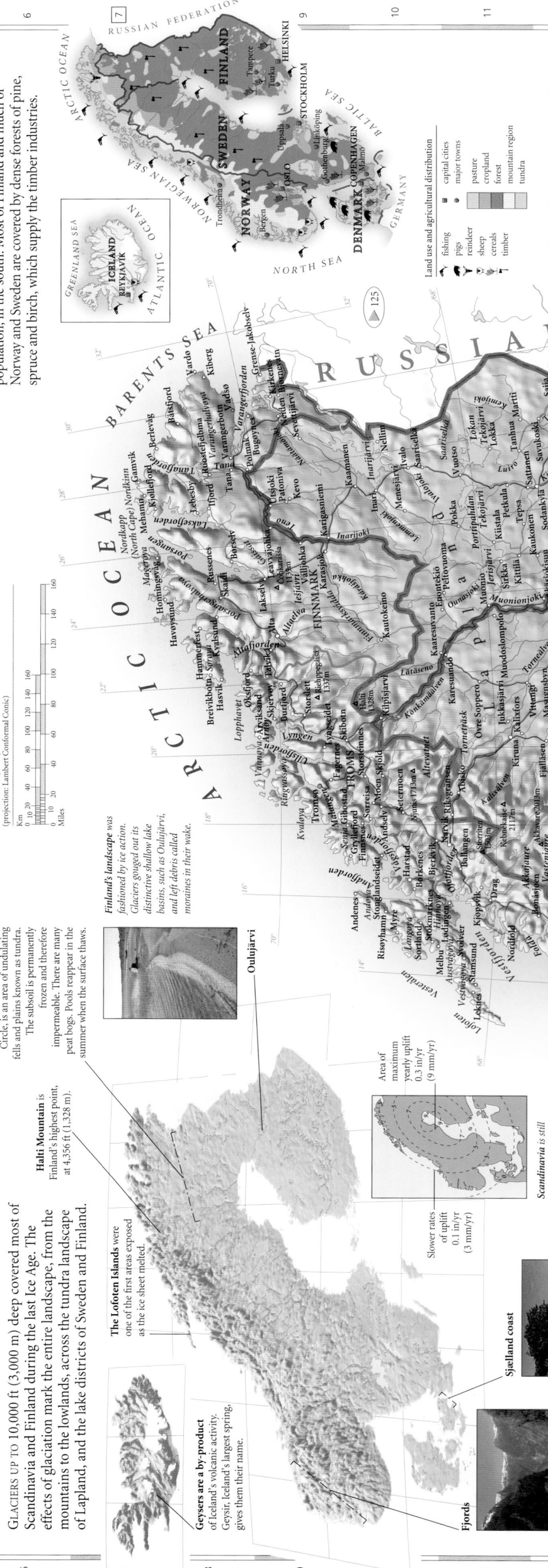

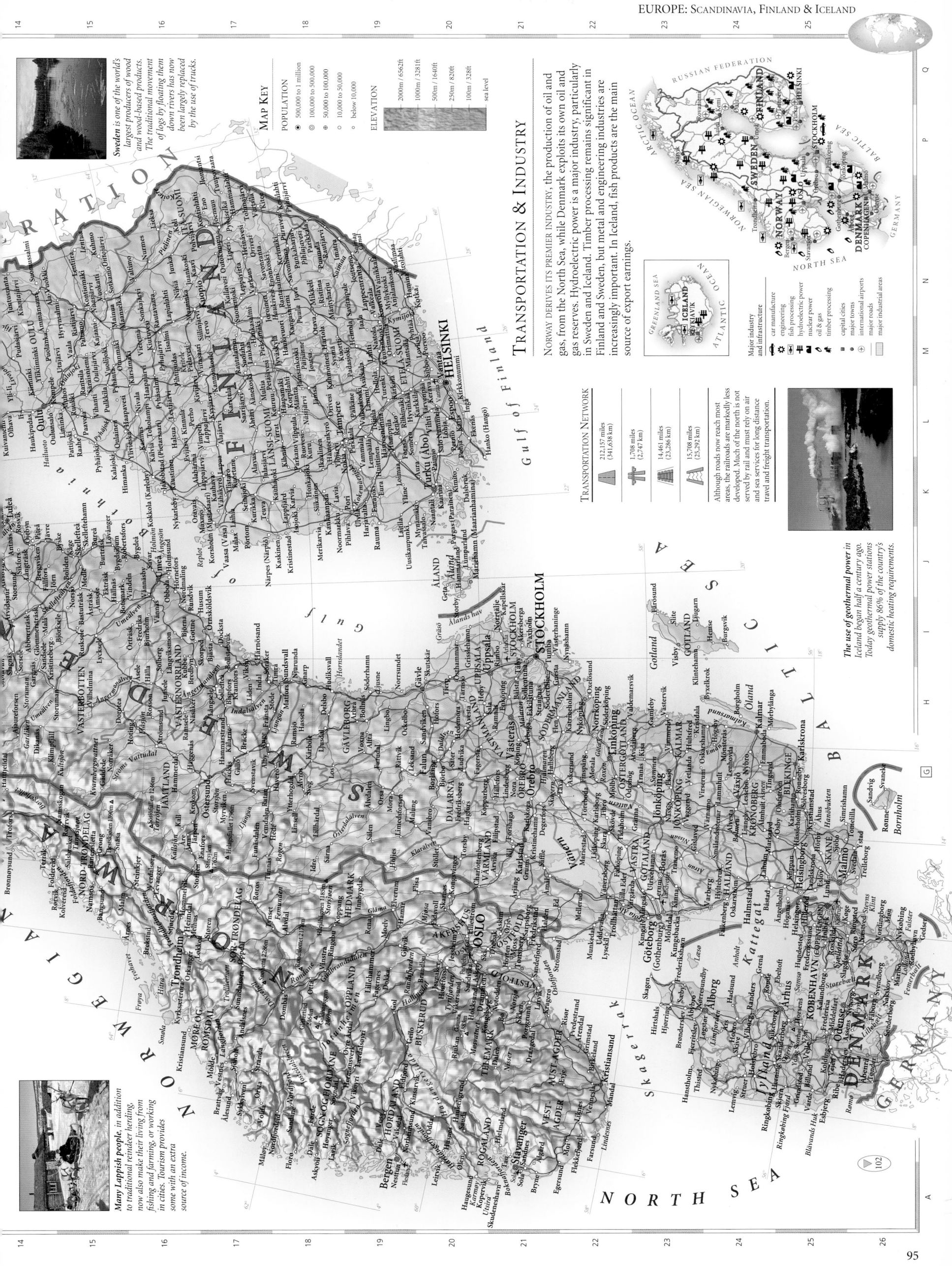

Sweden is one of the world's largest producers of wood and wood-based products. The traditional movement of logs by floating them down rivers has now been largely replaced by the use of trucks.

MAP KEY

POPULATION
- ◉ 500,000 to 1 million
- ◎ 100,000 to 500,000
- ⊕ 50,000 to 100,000
- ○ 10,000 to 50,000
- ○ below 10,000

ELEVATION
- 2000m / 6562ft
- 1000m / 3281ft
- 500m / 1640ft
- 250m / 820ft
- 100m / 328ft
- sea level

TRANSPORTATION & INDUSTRY

NORWAY DERIVES ITS PREMIER INDUSTRY, the production of oil and gas, from the North Sea, while Denmark exploits its own oil and gas reserves. Hydroelectric power is a major industry, particularly in Sweden and Iceland. Timber processing remains significant in Finland and Sweden, but metal and engineering industries are increasingly important. In Iceland, fish products are the main source of export earnings.

Major industry and infrastructure
- car manufacture
- engineering
- fish processing
- hydroelectric power
- nuclear power
- oil & gas
- timber processing
- capital cities
- major towns
- international airports
- major roads
- major industrial areas

TRANSPORTATION NETWORK

- 212,157 miles (341,638 km)
- 1,708 miles (2,747 km)
- 14,461 miles (23,286 km)
- 15,708 miles (25,292 km)

Although roads now reach most areas, the railroads are markedly less developed. Much of the north is not served by rail and must rely on air and sea services for long distance travel and freight transportation.

The use of geothermal power in Iceland began half a century ago. Today geothermal power stations supply 86% of the country's domestic heating requirements.

Many Lappish people, in addition to traditional reindeer herding, now also make their living from fishing and farming, or working in cities. Tourism provides some with an extra source of income.

SOUTHERN SCANDINAVIA

SOUTHERN NORWAY, SOUTHERN SWEDEN, DENMARK

SCANDINAVIA'S ECONOMIC AND POLITICAL HUB is the more habitable and accessible southern region. Many of the area's major cities are on the southern coasts, including Oslo and Stockholm, the capitals of Norway and Sweden. In Denmark, most of the population and the capital, Copenhagen, are located on its many islands. A cultural unity links the three Scandinavian countries. Their main languages, Danish, Swedish, and Norwegian, are mutually intelligible, and they all retain their monarchies, although the parliaments have legislative control.

USING THE LAND

AGRICULTURE IN SOUTHERN SCANDINAVIA is highly mechanized although farms are small. Denmark is the most intensively farmed country and its western pastureland is used mainly for pig farming. Cereal crops including wheat, barley, and oats, predominate in eastern Denmark and in the far south of Sweden. Southern Norway, and Sweden have large tracts of forest which are exploited for logging.

THE URBAN/RURAL POPULATION DIVIDE

urban 87% rural 13%

POPULATION DENSITY
152 people per sq mile
(61 people per sq km)

TOTAL LAND AREA
173,487 sq miles
(456,564 sq km)

Land use and agricultural distribution

- capital cities
- major towns
- pasture
- cropland
- forest
- mountain region

- cattle
- pigs
- sheep
- cereals
- fodder
- root crops
- timber

THE LANDSCAPE

SOUTHERN SCANDINAVIA, with the exception of Norway, has a flatter terrain than the rest of the region. Denmark and southern Sweden are both extensions of the North European Plain. In this area, because of glacial deposition rather than erosion, the soils are deeper and more fertile.

Acid rain, caused by industrial pollution carried north from elsewhere in Europe, harms plant and animal life in Scandinavian forests and lakes. The region's surface rocks lack lime to neutralize the acid, so making the problem more serious.

In the past, glaciers such as this one in Olden, Norway, were much larger. Today, many are retreating to yield the spectacular glacial scenery.

Olden

Distinctive low ridges, called eskers, are found across southern Sweden. They are formed from sand and gravel deposits left by retreating glaciers.

Limestone pillars eroded by the sea dot the coast of Gotland and surrounding islands.

The peak of Glittertind in the Jotunheimen Mountains is 8,044 ft (2,452 m) high.

The lakes of southern Sweden remain from a period when the land was completely flooded. As the ice which covered the area melted, the land rose, leaving lakes in shallow, ice-scoured depressions. Sweden has over 90,000 lakes.

Vänern in Sweden is the largest lake in Scandinavia. It covers an area of 2,080 sq miles (5,390 sq km).

Denmark's flat and fertile soils are formed on glacial deposits between 100–160 ft (30–50 m) deep.

When the ice retreated the valley was flooded by the sea

Old valley floor

Sea level

Erosion by glaciers deepened existing river valleys

Sognefjorden is the deepest of Norway's many fjords. It drops to 4,291 ft (1,308 m) below sea level.

Sognefjorden

MAP KEY

POPULATION
- ◉ 500,000 to 1 million
- ◉ 100,000 to 500,000
- ◎ 50,000 to 100,000
- ○ 10,000 to 50,000
- ○ below 10,000

ELEVATION
- 2000m / 6562ft
- 1000m / 3281ft
- 500m / 1640ft
- 250m / 820ft
- 100m / 328ft
- sea level

SCALE 1:2,900,000
(projection: Lambert Conformal Conic)

In Norway winters are longer and colder inland than in coastal areas, where the warm current of the North Atlantic Drift moderates the climate.

Gulf of Bothnia

NORTH SEA

BALTIC SEA

NORWEGIAN SEA

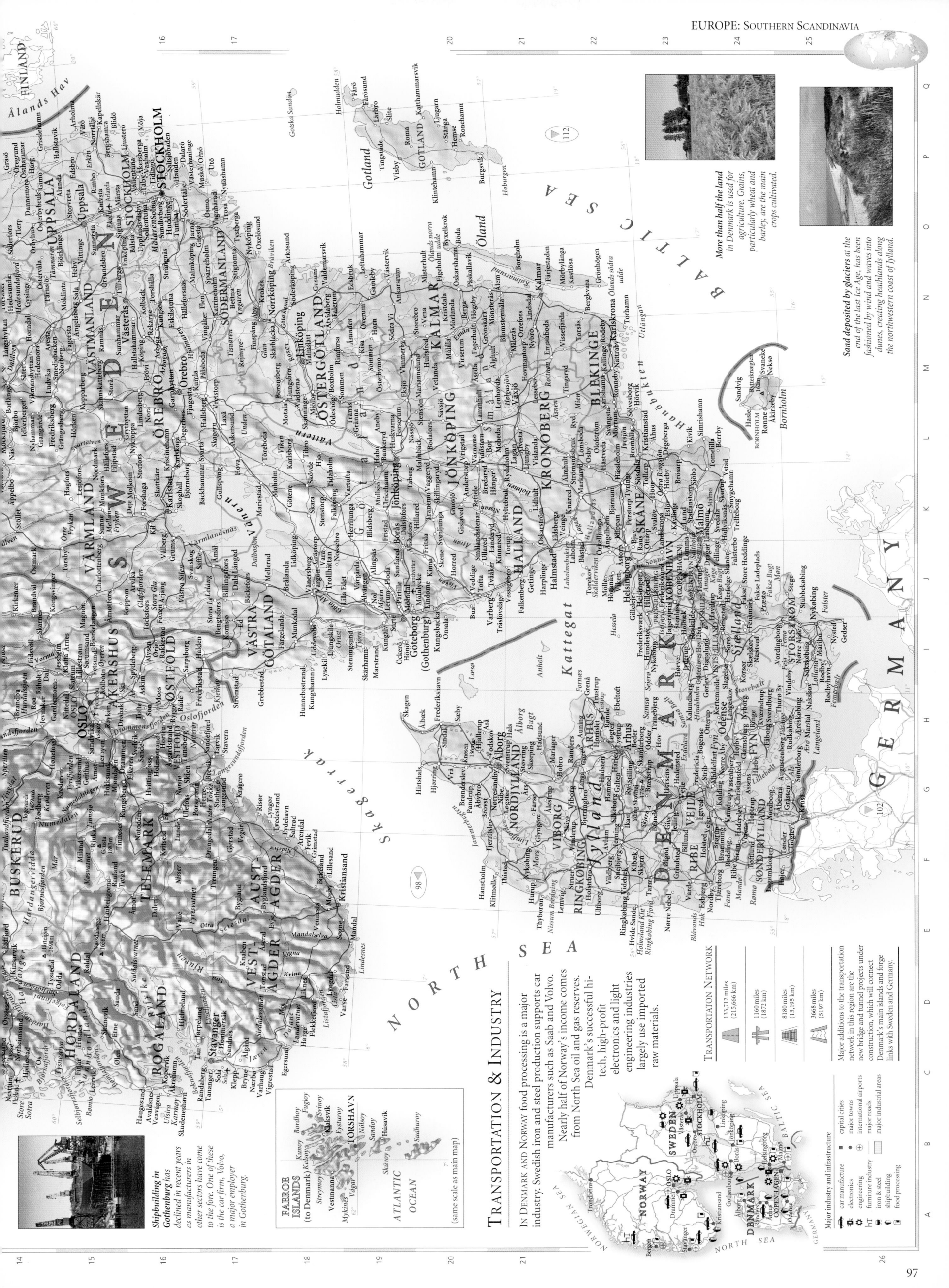

More than half the land in Denmark is used for agriculture. Grains, particularly wheat and barley, are the main crops cultivated.

Sand deposited by glaciers at the end of the last Ice Age, has been fashioned by wind and waves into dunes, creating heathlands along the northwestern coast of Jylland.

Shipbuilding in Gothenburg has declined in recent years as manufacturers in other sectors have come to the fore. One of these is the car firm, Volvo, a major employer in Gothenburg.

TRANSPORTATION & INDUSTRY

In Denmark and Norway food processing is a major industry. Swedish iron and steel production supports car manufacturers such as Saab and Volvo. Nearly half of Norway's income comes from North Sea oil and gas reserves. Denmark's successful hi-tech, high-profit electronics and light engineering industries largely use imported raw materials.

TRANSPORTATION NETWORK

133,712 miles (215,666 km)	
1160 miles (1872 km)	
8180 miles (13,195 km)	
3668 miles (5197 km)	

Major additions to the transportation network in this region are the new bridge and tunnel projects under construction, which will connect Denmark's main islands and forge links with Sweden and Germany.

Major industry and infrastructure

- car manufacture
- electronics
- engineering
- furniture industry
- iron & steel
- shipbuilding
- food processing
- capital cities
- major towns
- international airports
- major roads
- major industrial areas

FAERROE ISLANDS (to Denmark)

(same scale as main map)

97

THE BRITISH ISLES

UNITED KINGDOM, REPUBLIC OF IRELAND

THE BRITISH ISLES have for centuries played a central role in European and world history. England, Wales, Scotland, and Northern Ireland together form the United Kingdom (UK), while the southern portion of Ireland is an independent country, self-governing since 1921. Although England has tended to be the politically and economically dominant partner in the UK, the Scots, Welsh and Irish maintain independent cultures, distinct national identities and languages. Southeastern England is the most densely populated part of this crowded region, with over nine million people living in and around the London area.

TRANSPORTATION & INDUSTRY

THE BRITISH ISLES' INDUSTRIAL BASE was founded primarily on coal, iron and textiles, based largely in the north. Today, the most productive sectors include hi-tech industries clustered mainly in southeastern England, chemicals and the service sector, particularly tourism.

Major industry and infrastructure

- capital cities
- major towns
- international airports
- hi-tech industry
- major roads
- major industrial areas

car manufacture
chemicals
engineering
hi-tech industry
iron & steel
tourism

TRANSPORTATION NETWORK

288,330 miles (464,900 km)		2,046 miles (3,295 km)	
11,874 miles (19,121 km)		3,806 miles (6,129 km)	

The UK's congested roads have become a major focus of environmental concern in recent years. No longer an island, the UK was finally linked to continental Europe by the Channel Tunnel in 1994.

Clew Bay in western Ireland, is characteristic of the heavily indented west coast, where deep wide-mouthed bays separate the mountains of Mayo, Donegal, and Kerry as they thrust out into the Atlantic Ocean.

THE LANDSCAPE

RUGGED UPLANDS dominate the landscape of Scotland, Wales, and northern England. All the peaks in the British Isles over 4,000 ft (1,219 m) lie in highland Scotland. Lowland England rises into several ranges of rolling hills, including the older Mendips, and the Cotswolds and the Chilterns, which were formed at the same time as the Alps in southern Europe.

The valley of Glen Coe in the Scottish Highlands is a U-shaped valley, typical of the north and west of the British Isles, where glaciers shaped much of the landscape.

The Pennines, sometimes called "the backbone of England," are formed of limestones and grits.

Ullswater in the Lake District fills a deep valley formed by glacial erosion.

The Fens are a low-lying area reclaimed from the sea.

Chiltern Hills

The Cotswold Hills are characterized by a series of limestone ridges overlooking clay vales.

Durdle Door

Coastal erosion around the British Isles forms striking features such as this limestone arch, Durdle Door in Dorset.

The lowlands of Scotland, drained by the Tay, Forth, and Clyde Rivers, are centred on a rift valley. The region contains valuable coal reserves.

Lake District

Mendip Hills

Dartmoor, studded with tors, is an exposed part of a vast granite dome, formed when molten rock intruded into the Earth's crust.

Black Ven, Lyme Regis

Much of the south coast is subject to landslides. Following rain, porous sandstones feed water into the underlying, less permeable clays which then crumble and slide into the sea.

Ben Nevis at 4,409 ft (1,343 m) is the highest peak in the UK.

Over 600 islands, mostly uninhabited, lie west and north of the Scottish mainland.

Thousands of hexagonal basalt columns form Giant's Causeway on the north coast of Antrim. These were created by volcanic activity.

The British Isles have no large-scale river systems. The Shannon is the longest, at 230 miles (370 km).

Snowdon is the highest mountain in England and Wales reaching 3,556 ft (1,085 m).

Peat bogs dot the poorly-drained Irish lowlands.

MAP KEY

POPULATION
- above 5 million
- 1 million to 5 million
- 500,000 to 1 million
- 100,000 to 500,000
- 50,000 to 100,000
- 10,000 to 50,000
- below 10,000

ELEVATION
- 1000m / 3281ft
- 500m / 1640ft
- 250m / 820ft
- 100m / 328ft
- sea level

Cracks
Sandstone
Clay
Limestone

Water
Mudslide
Sea

Shetland Islands

Herma Ness
Unst
Fedlar
Yell
Yell Sound
Foula
Papa Stour
Hillswick
St Magnus Bay
Mainland
Scalloway
West Burra
Bressay
Whalsay
Sullom Voe
Out Skerries
Lerwick
Fitful Head
Sumburgh Head
Fair Isle

Orkney Islands

North Ronaldsay
Westray
Papa Westray
Rousay
Sanday
Eday
Stronsay
Shapinsay
Mainland
Kirkwall
Stromness
Scapa Flow
Burray
South Ronaldsay
St Margaret's Hope
The North Sound
Hoy
Sule Skerry
Stack Skerry

North Rona
Sula Sgeir

SCOTLAND

Cape Wrath
Durness
Loch Eriboll
Strathy Point
Tongue
Thurso
Dunnet Head
Duncansby Head
John o'Groats
Wick
Noss Head
Halkirk
Halladale
Helmsdale
Brora
Golspie
Lairg
Dornoch
Tarbat Ness
Tain
Dingwall
Beauly
Inverness
Nairn
Forres
Elgin
Lossiemouth
Buckie
Banff
Macduff
Buckie
Keith
Huntly
Turriff
Peterhead
Fraserburgh
Kinnaird Head
Buchan Ness
Aberdeen
Girdle Ness
Stonehaven
Inverurie
Ellon
Banchory
Braemar
Montrose
Brechin
Forfar
Kirriemuir
Blairgowrie
Arbroath
Carnoustie
Dundee
Perth
Firth of Tay
St Andrews
Cupar
Kinross
Dunfermline
Kirkcaldy
Firth of Forth
Edinburgh
Haddington
North Berwick
Dunbar

Cairn Gorm 1245m
Cairngorm Mountains
Ben Macdui 1309m
Grampian Mountains
Ben Lawers 1214m
Aviemore
Grantown-on-Spey
Kingussie
Pitlochry
Aberfeldy
Crieff
Callander
Stirling
Falkirk
Alloa
Glasgow

Lochinver
Ullapool
Stornoway
Isle of Lewis
Butt of Lewis
Port of Ness
Carloway
Tarbert
Isle of Harris
Sound of Harris
North Uist
Benbecula
South Uist
Barra
Barra Head
Monach Islands
Flannan Isles
St Kilda

Loch Ness
Fort Augustus
Fort William
Ben Nevis
Glen Coe
Oban
Isle of Mull
Ben More 966m
Tobermory
Iona
Colonsay
Islay
Jura
Port Askaig
Port Ellen
Mull of Oa

Gairloch
Loch Torridon
Loch Maree
Kyle of Lochalsh
Isle of Skye
Portree
Uig
Raasay
Broadford
Rhum
Eigg
Muck
Canna
Coll
Tiree
Ardnamurchan
Point of Ardnamurchan

North Rona

REPUBLIC OF IRELAND

ATLANTIC OCEAN

TRANSPORTATION & INDUSTRY map

NORTH SEA
Aberdeen
Dundee
Edinburgh
Glasgow
Belfast
UNITED KINGDOM
Newcastle upon Tyne
Middlesbrough
Leeds
Liverpool
Manchester
Sheffield
Nottingham
Norwich
Birmingham
Cardiff
Bristol
Oxford
LONDON
English Channel
REPUBLIC OF IRELAND
DUBLIN
Cork
ATLANTIC OCEAN

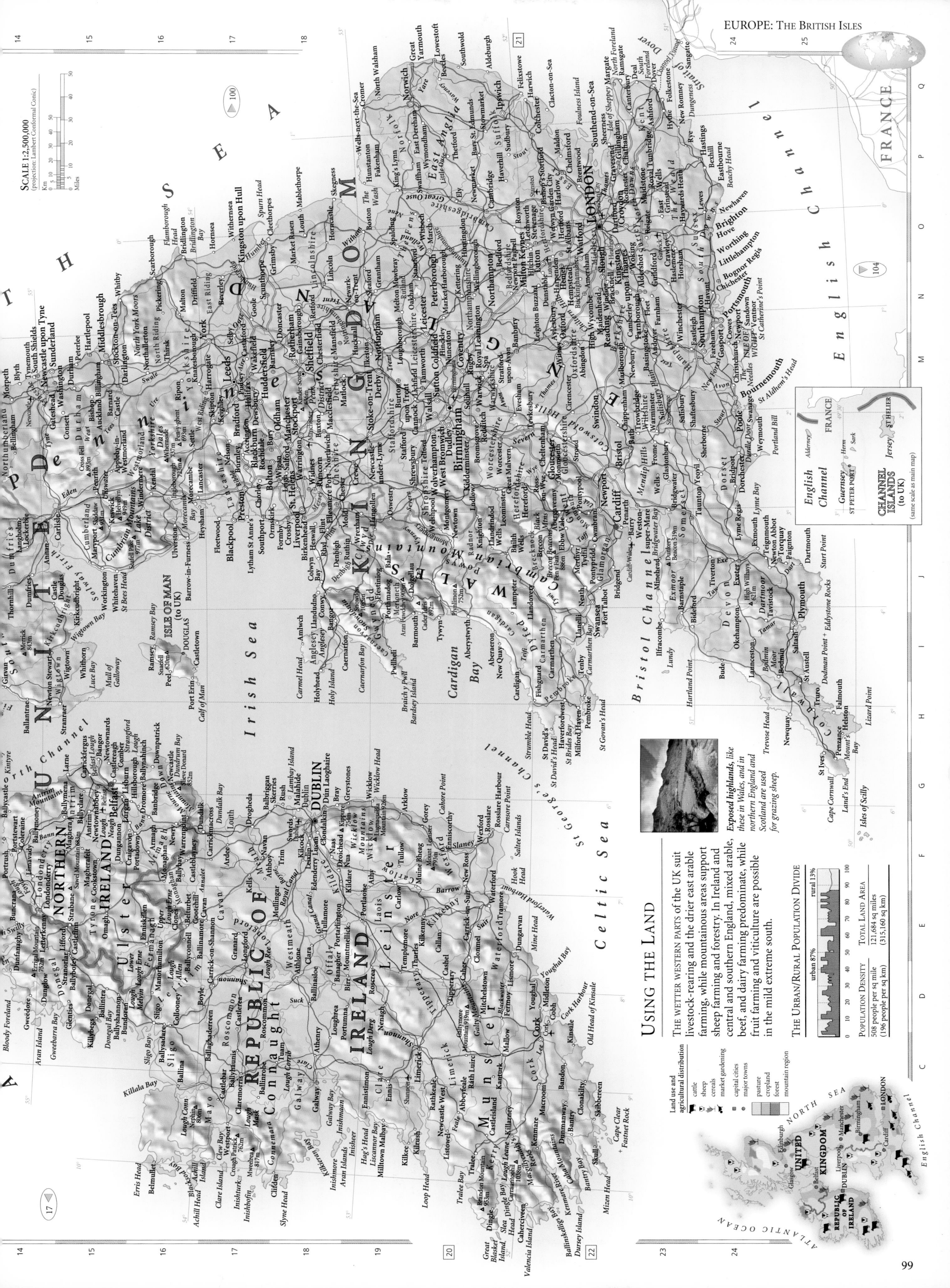

SCALE 1:2,500,000
(projection: Lambert Conformal Conic)

USING THE LAND

THE WETTER WESTERN PARTS of the UK suit livestock-rearing and the drier east arable farming, while mountainous areas support sheep farming and forestry. In Ireland and central and southern England, mixed arable, beef, and dairy farming predominate, while fruit farming and viticulture are possible in the mild extreme south.

Exposed highlands, like these in Wales, and in northern England and Scotland are used for grazing sheep.

THE URBAN/RURAL POPULATION DIVIDE

urban 87% rural 13%

POPULATION DENSITY	TOTAL LAND AREA
508 people per sq mile (196 people per sq km)	121,684 sq miles (315,160 sq km)

Land use and agricultural distribution
- cattle
- sheep
- cereals
- market gardening
- capital cities
- major towns
- pasture
- cropland
- forest
- mountain region

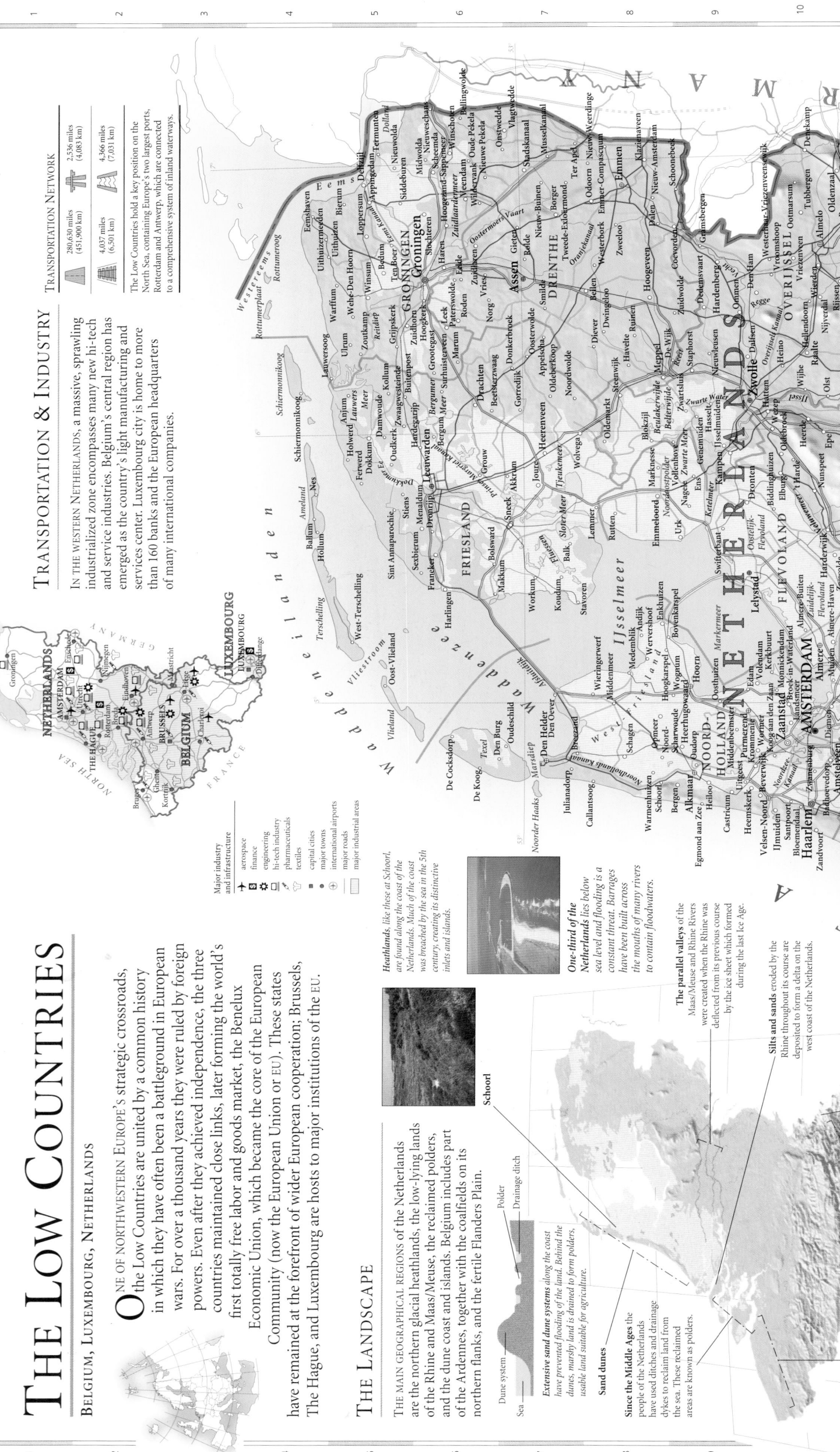

THE LOW COUNTRIES

BELGIUM, LUXEMBOURG, NETHERLANDS

ONE OF NORTHWESTERN EUROPE'S strategic crossroads, the Low Countries are united by a common history in which they have often been a battleground in European wars. For over a thousand years they were ruled by foreign powers. Even after they achieved independence, the three countries maintained close links, later forming the world's first totally free labor and goods market, the Benelux Economic Union, which became the core of the European Community (now the European Union or EU). These states have remained at the forefront of wider European cooperation; Brussels, The Hague, and Luxembourg are hosts to major institutions of the EU.

THE LANDSCAPE

THE MAIN GEOGRAPHICAL REGIONS of the Netherlands are the northern glacial heathlands, the low-lying lands of the Rhine and Maas/Meuse, the reclaimed polders, and the dune coast and islands. Belgium includes part of the Ardennes, together with the coalfields on its northern flanks, and the fertile Flanders Plain.

Extensive sand dune systems along the coast have prevented flooding of the land. Behind the dunes, marshy land is drained to form polders, usable land suitable for agriculture.

Sand dunes

Since the Middle Ages the people of the Netherlands have used ditches and drainage dykes to reclaim land from the sea. These reclaimed areas are known as polders.

Dune system
Polder
Drainage ditch
Sea

Schoorl

Heathlands, like these at Schoorl, are found along the coast of the Netherlands. Much of the coast was breached by the sea in the 5th century, creating its distinctive inlets and islands.

One-third of the Netherlands lies below sea level and flooding is a constant threat. Barrages have been built across the mouths of many rivers to contain floodwaters.

The parallel valleys of the Maas/Meuse and Rhine Rivers were created when the Rhine was deflected from its previous course by the ice sheet which formed during the last Ice Age.

Silts and sands eroded by the Rhine throughout its course are deposited to form a delta on the west coast of the Netherlands.

The loess soils of the Flanders Plain in western Belgium provide excellent conditions for arable farming.

Hautes Fagnes is the highest part of Belgium. The bogs and streams in this upland region result from high rainfall and low temperatures.

Ardennes

Uplifted and folded 220 million years ago, the Ardennes have since been reduced to relatively level plateaus, then sharply incised by rivers such as the Maas/Meuse.

TRANSPORTATION & INDUSTRY

IN THE WESTERN NETHERLANDS, a massive, sprawling industrialized zone encompasses many new hi-tech and service industries. Belgium's central region has emerged as the country's light manufacturing and services center. Luxembourg city is home to more than 160 banks and the European headquarters of many international companies.

TRANSPORTATION NETWORK

280,630 miles (451,900 km)	2,536 miles (4,083 km)
4,037 miles (6,501 km)	4,366 miles (7,031 km)

The Low Countries hold a key position on the North Sea, containing Europe's two largest ports, Rotterdam and Antwerp, which are connected to a comprehensive system of inland waterways.

Major industry and infrastructure
aerospace
finance
engineering
hi-tech industry
pharmaceuticals
textiles
capital cities
major towns
international airports
major roads
major industrial areas

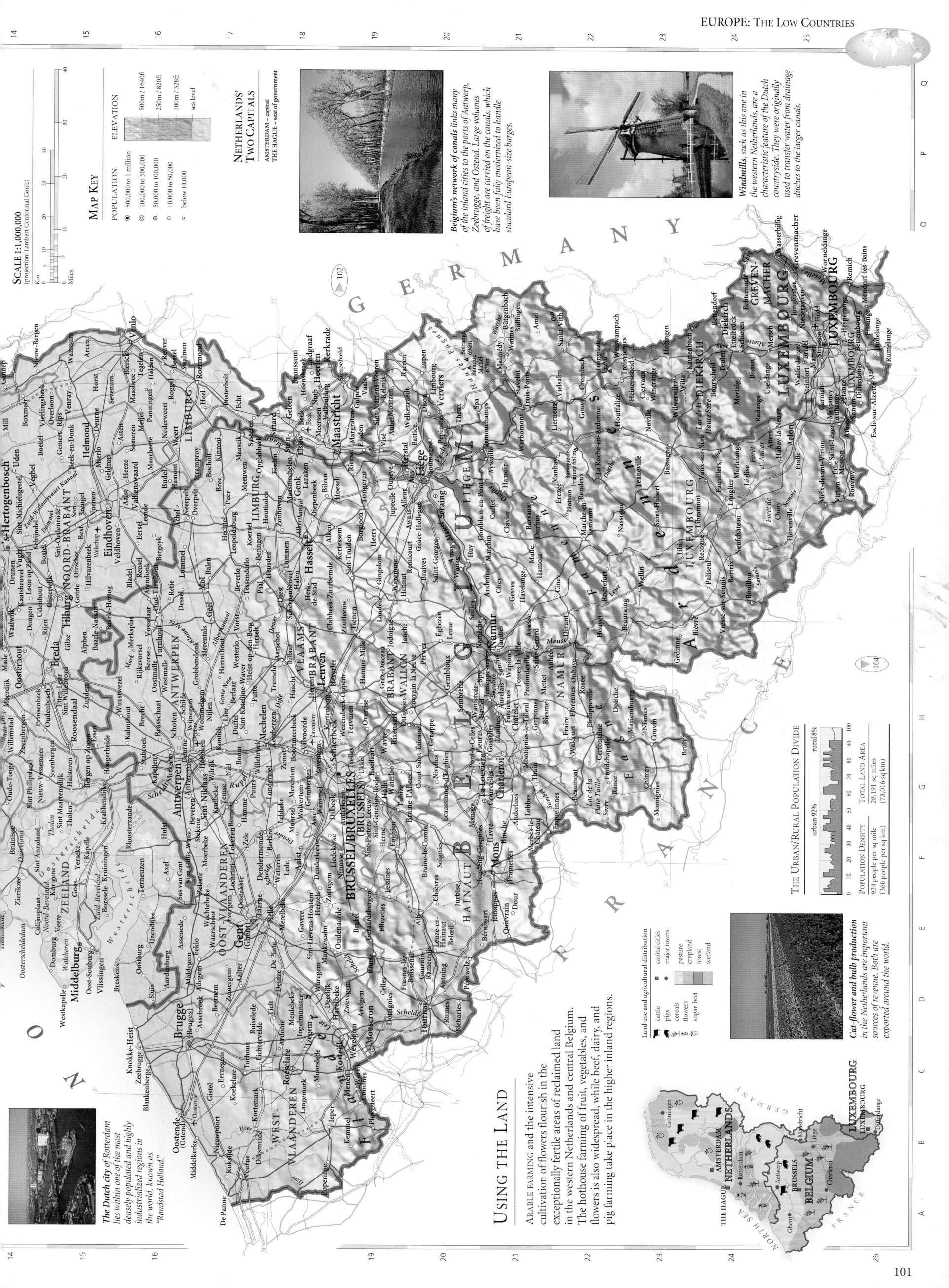

MAP KEY

SCALE 1:1,000,000
(projection: Lambert Conformal Conic)

ELEVATION
- 500m / 1640ft
- 250m / 820ft
- 100m / 328ft
- sea level

POPULATION
- ● 500,000 to 1 million
- ◉ 100,000 to 500,000
- ⊕ 50,000 to 100,000
- ⊙ 10,000 to 50,000
- ○ below 10,000

NETHERLANDS' TWO CAPITALS
AMSTERDAM – capital
THE HAGUE – seat of government

Belgium's network of canals links many of the inland cities to the ports of Antwerp, Zeebrugge, and Ostend. Large volumes of freight are carried on the canals, which have been fully modernized to handle standard European-size barges.

Windmills, such as this one in the western Netherlands, are a characteristic feature of the Dutch countryside. They were originally used to transfer water from drainage ditches to the larger canals.

USING THE LAND

ARABLE FARMING and the intensive cultivation of flowers flourish in the exceptionally fertile areas of reclaimed land in the western Netherlands and central Belgium. The hothouse farming of fruit, vegetables, and flowers is also widespread, while beef, dairy, and pig farming take place in the higher inland regions.

The Dutch city of Rotterdam lies within one of the most densely populated and highly industrialized regions in the world, known as "Randstad Holland."

Cut-flower and bulb production in the Netherlands are important sources of revenue. Both are exported around the world.

Land use and agricultural distribution
- capital cities
- major towns
- cattle
- pigs
- cereals
- flowers
- sugar beet
- pasture
- cropland
- forest
- wetland

THE URBAN/RURAL POPULATION DIVIDE

urban 92%   rural 8%

POPULATION DENSITY
934 people per sq mile
(360 people per sq km)

TOTAL LAND AREA
28,191 sq miles
(73,016 sq km)

101

GERMANY

D ESPITE THE DEVASTATION of its industry and infrastructure during the Second World War and its separation from eastern Germany during the Cold War, West Germany made a rapid recovery in the following generation to become Europe's most formidable economic power. When the Berlin Wall was dismantled in 1989, the two halves of Germany were politically united for the first time in 40 years. Complete social and economic unity remain a longer term goal, as East German industry and society adapt to a free market. Germany has been a key player in the creation of the European Union (EU) and in moves toward a single European currency.

USING THE LAND

GERMANY has a large, efficient agricultural sector, and produces more than three-quarters of its own food. The major crops grown are cereals and sugar beet on the more fertile soils, and root crops, rye, oats, and fodder on the poorer soils of the northern plains and central uplands. Southern Germany is also a principal producer of high quality wines. Vineyards cover the slopes surrounding the Rhine and its tributaries.

Land use and agricultural distribution

- cattle
- pigs
- cereals
- sugar beet
- vineyards

- capital cities
- major cities
- major towns

- pasture
- cropland
- forest

THE URBAN/RURAL POPULATION DIVIDE

urban 87%   rural 13%

POPULATION DENSITY
598 people per sq mile
(231 people per sq km)

TOTAL LAND AREA
13,804 sq miles
(356,910 sq km)

The Moselle River flows through the Rhine State Uplands (Rheinisches Schiefergebirge). During a period of uplift, preexisting river meanders were deeply incised, to form its present dramatic contours.

THE LANDSCAPE

THE PLAINS OF NORTHERN GERMANY, the volcanic plateaus and mountains of the central uplands, and the Bavarian Alps are the three principal geographic regions in Germany. North to south the land rises steadily from barely 300 ft (90 m) in the plains to 6,500 ft (2,000 m) in the Bavarian Alps, which are a small but distinct region in the far south.

The heathlands of northern Germany are covered by glacial deposits of sandy outwash soil which makes them largely infertile. They support only sheep and solitary trees.

Lüneburg Heath
(Lüneburger Heide)

Much of the landscape of northern Germany has been shaped by glaciation. During the last Ice Age, the ice sheet advanced as far the northern slopes of the central uplands.

Fault lines

Rhine

Downfaulted block

Part of the floor of the Rhine Rift Valley was let down between two parallel faults in the Earth's crust.

Rhine Rift Valley

Müritz lake covers 45 sq miles (117 sq km), but is only 1.08 ft (33 m) deep. It lies in a shallow valley formed by meltwater flowing out from a retreating ice sheet. These valleys are known as *Urstromtäler.*

The Harz Mountains were formed 300 million years ago. They are block-faulted mountains, formed when a section of the Earth's crust was thrust up between two faults.

Elbe River

The Elbe flows in wide meanders across the north German plain to the North Sea. At its mouth it is 10 miles (16 km) wide.

The Danube rises in the Black Forest (*Schwarzwald*) and flows east, across a wide valley, on its course to the Black Sea.

Zugspitze, the highest peak in Germany at 9,719 ft (2,962 m), was formed during the Alpine mountain-building period, 30 million years ago.

The Rhine is Germany's principal waterway and one of Europe's longest rivers, flowing 820 miles (1,320 km).

SCALE 1:2,250,000
(projection: Lambert Conformal Conic)

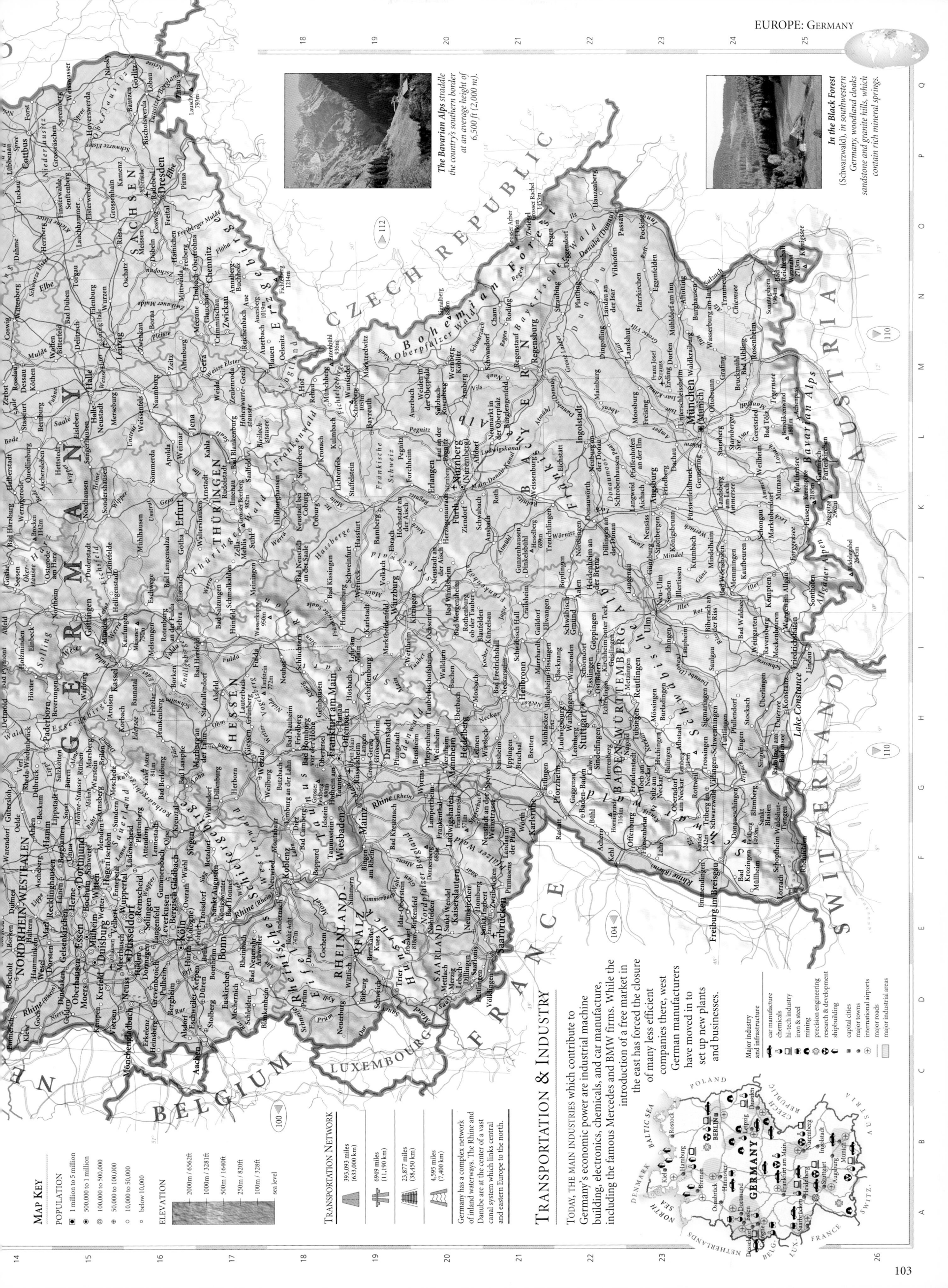

*The **Bavarian Alps** straddle the country's southern border at an average height of 6,500 ft (2,000 m).*

*In the **Black Forest** (Schwarzwald), in southwestern Germany, woodland cloaks sandstone and granite hills, which contain rich mineral springs.*

Transportation & Industry

Today, the main industries which contribute to Germany's economic power are industrial machine building, electronics, chemicals, and car manufacture, including the famous Mercedes and BMW firms. While the introduction of a free market in the east has forced the closure of many less efficient companies there, west German manufacturers have moved in to set up new plants and businesses.

Germany has a complex network of inland waterways. The Rhine and Danube are at the center of a vast canal system which links central and eastern Europe to the north.

Transportation Network

◄▬►	393,093 miles (633,000 km)
⟱	6949 miles (11,190 km)
◢◣	23,877 miles (38,450 km)
▲	4,595 miles (7,400 km)

Major industry and infrastructure

- ⚙ car manufacture
- 🔬 chemicals
- 💻 hi-tech industry
- ⚒ iron & steel
- ⚙ mining
- ⚙ precision engineering
- research & development
- ⚓ shipbuilding
- ■ capital cities
- □ major cities
- major towns
- ✈ international airports
- ▬ major roads
- ▭ major industrial areas

Map Key

POPULATION
- ◙ 1 million to 5 million
- ◉ 500,000 to 1 million
- ◎ 100,000 to 500,000
- ⊙ 50,000 to 100,000
- ○ 10,000 to 50,000
- ∘ below 10,000

ELEVATION
- 2000m / 6562ft
- 1000m / 3281ft
- 500m / 1640ft
- 250m / 820ft
- 100m / 328ft
- sea level

FRANCE

FRANCE, MONACO

EUROPE'S SECOND LARGEST nation and the founder of modern Republican government, France is a major center of culture and fashion, and a leading producer of both agricultural and industrial goods. It has played a leading role in European events for centuries, and remains a key player in the push toward European unity. The Paris Basin is the most highly populated area; Île de France is home to over nine million people. Large parts of France remain thinly populated, particularly the mountainous Massif Central, Pyrenees, and southern Alps.

The chalk cliffs of Normandy (Normandie) and southeastern England form part of a single geological region, now divided in two by the English Channel.

THE LANDSCAPE

FRANCE'S LANDSCAPE was fashioned by two phases of mountain-building. The northwestern peninsula, the Massif Central, and the Vosges date from 220 million years ago. The complex folds of the Alps and Pyrenees, the gently-folded Jura, and the low-lying sedimentary areas of the Paris, Garonne, and Rhône basins started to form 65 million years ago.

The coast of Brittany (Bretagne) is highly indented where deep valleys in the northwestern peninsula were drowned by the sea.

The Normandy (Normandie) coastline is characterized by high chalk cliffs.

The coastline of France is 2,141 miles (3,427 km) long.

The Paris Basin consists of a layered sequence of sedimentary rocks. Fertile soils over much of the area make good agricultural land.

The gently rounded summits of the Vosges are over 200 million years old.

The folded Jura form low ridges and long narrow valleys.

The Biscay coast, like the Mediterranean, is characterized by flat sandy beaches, interspersed with lagoons.

Garonne Basin

The Alps were forced up during several phases of mountain-building beginning 65 million years ago.

The Dordogne region contains spectacular examples of limestone scenery including caves and gorges.

The Pyrenees form a natural border between France and Spain.

The ancient Massif Central, disturbed by the formation of the Alps, was subject to volcanism that only ceased during the last 10,000 years.

Rhône Basin

Rhône Delta

Corsica's northeastern peninsula has dramatic cliffs of folded limestone.

The volcanic landscape of the Auvergne where the cones of its extinct volcanoes have worn away to leave "plugs" of lava.

Rhône / Delta plain / The marshes of the Camargue

Deposition in the Rhône Delta is wave-dominated. Sea currents carry river sediments extending the delta plain westwards.

TRANSPORTATION & INDUSTRY

TODAY THE MAIN FRENCH GROWTH INDUSTRIES are hi-tech, including microelectronics, telecommunications, and aerospace. Other important sectors are the nuclear industry, only rivalled in scale by that of the USA, car manufacture, dominated by the giants Renault and Peugeot and a highly diversified tourist industry.

Major industry and infrastructure

- aerospace industry
- car manufacture
- chemicals
- engineering
- hi-tech industry
- nuclear power
- tourism
- capital cities
- major towns
- international airports
- major roads
- major industrial areas

TRANSPORTATION NETWORK

599,017 miles (964,600 km)	5,900 miles (9,500 km)
19,761 miles (31,821 km)	5,279 miles (8,500 km)

The French TGV (*Train à Grande Vitesse*) leads the world in high-speed train technology, and provides a service which is faster, door-to-door, than air travel.

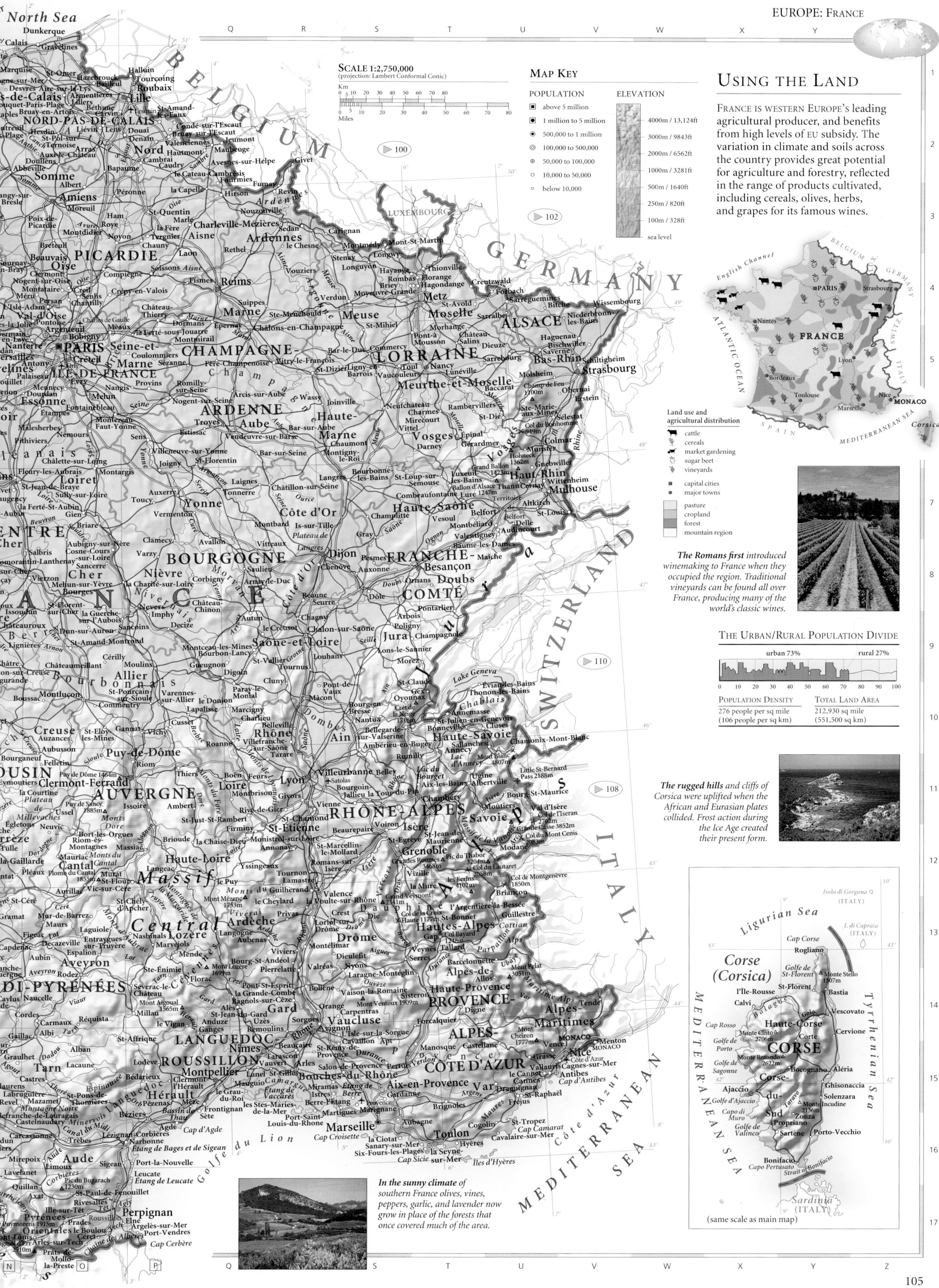

USING THE LAND

FRANCE IS WESTERN EUROPE's leading agricultural producer, and benefits from high levels of EU subsidy. The variation in climate and soils across the country provides great potential for agriculture and forestry, reflected in the range of products cultivated, including cereals, olives, herbs, and grapes for its famous wines.

Land use and agricultural distribution
- cattle
- cereals
- market gardening
- sugar beet
- vineyards
- ● capital cities
- • major towns
- pasture
- cropland
- forest
- mountain region

The Romans first introduced winemaking to France when they occupied the region. Traditional vineyards can be found all over France, producing many of the world's classic wines.

THE URBAN/RURAL POPULATION DIVIDE

urban 73%   rural 27%

0 10 20 30 40 50 60 70 80 90 100

POPULATION DENSITY
276 people per sq mile
(106 people per sq km)

TOTAL LAND AREA
212,930 sq mile
(551,500 sq km)

The rugged hills and cliffs of Corsica were uplifted when the African and Eurasian plates collided. Frost action during the Ice Age created their present form.

In the sunny climate of southern France olives, vines, peppers, garlic, and lavender now grow in place of the forests that once covered much of the area.

Corse (Corsica)

(same scale as main map)

SCALE 1:2,750,000
(projection: Lambert Conformal Conic)

MAP KEY

POPULATION
- ■ above 5 million
- ▣ 1 million to 5 million
- ◉ 500,000 to 1 million
- ◎ 100,000 to 500,000
- ⊕ 50,000 to 100,000
- ○ 10,000 to 50,000
- ○ below 10,000

ELEVATION
- 4000m / 13,124ft
- 3000m / 9843ft
- 2000m / 6562ft
- 1000m / 3281ft
- 500m / 1640ft
- 250m / 820ft
- 100m / 328ft
- sea level

THE IBERIAN PENINSULA

ANDORRA, GIBRALTAR, PORTUGAL, SPAIN *(Azores, Canary Islands, Madeira on p.66)*

THE IBERIAN PENINSULA is separated from the rest of Europe by the Pyrenees, and at its most southerly point is only 5 miles (8 km) from North Africa. The location of Iberia has been central to its diverse history. The Greeks, Carthaginians, Romans, Visigoths, and most recently the Moors, invaded Iberia at various times. For much of the 20th century, both Spain and Portugal were governed by right-wing dictators. Since the establishment of democratic governments in the mid-1970s, modernization has been rapid and both countries are now among the most popular of European holiday destinations.

USING THE LAND

THE PRINCIPAL CROPS grown in Iberia are cereals, especially wheat and barley. Both countries are major wine producers, most notably of Rioja, sherry, and port. Sheep are kept throughout the region, and citrus fruits thrive on the Mediterranean coast. The successful forest industry in Iberia produces two-thirds of the world's cork.

The steep, terraced slopes of the Douro Valley in northern Portugal, are used to cultivate vines. The grapes harvested produce Portugal's famous port wine.

Land use and agricultural distribution

- sheep
- cereals
- citrus fruit
- olives
- vineyards
- cork
- capital cities
- major towns

pasture
cropland
forest
mountain region

THE URBAN/RURAL POPULATION DIVIDE

urban 68%   rural 32%

POPULATION DENSITY	TOTAL LAND AREA
215 people per sq mile (83 people per sq km)	230,569 sq miles (597,170 sq km)

TRANSPORTATION & INDUSTRY

SINCE THE 1970s, the economies of Spain and Portugal have expanded and diversified. In both countries, tourism has outstripped agriculture in economic importance. Spain's resource base is varied, including coal, iron, and the world's largest reserves of mercury. Portugal is a leading producer of tungsten ore.

Major industry and infrastructure

- car manufacture
- chemicals
- engineering
- fish processing
- mining
- textiles
- tourism
- capital cities
- major towns
- international airports
- major roads
- major industrial areas

TRANSPORTATION NETWORK

241,720 miles (388,990 km)	1,552 miles (2,529 km)
11,793 miles (18,979 km)	1,159 miles (1,865 km)

Radiating from Madrid, the road network in Spain dates from the 18th century, but now includes many highways. Portugal's road system has been completely modernized in recent years.

The eroded cliffs of the Algarve in southern Portugal were carved by Atlantic waves. The numerous rocky bays and beaches, and the region's pleasant climate, have made it a popular tourist destination.

The climate in northwestern Spain is milder in both summer and winter than in the rest of the country, creating a verdant environment, more commonly associated with northwestern Europe.

MAP KEY

POPULATION

- ▣ 1 million to 5 million
- ◉ 500,000 to 1 million
- ◎ 100,000 to 500,000
- ⊕ 50,000 to 100,000
- ○ 10,000 to 50,000
- ○ below 10,000

ELEVATION

- 3000m / 9843ft
- 2000m / 6562ft
- 1000m / 3281ft
- 500m / 1640ft
- 250m / 820ft
- 100m / 328ft
- sea level

SCALE 1:2,750,000

(projection: Lambert Conformal Conic)

Km 0 5 10 20 30 40 50 60 70 80

Miles 0 10 20 30 40 50 60 70 80

THE LANDSCAPE

A VAST PLATEAU, the Meseta dominates the centre of the peninsula, enclosed by the Cordillera Cantábrica to the north and the Sierra Morena to the south. It is drained by three major rivers, the Douro/Duero, the Tagus, and the Guadalquivir. The peninsula experiences great variations in climate and rainfall, both regionally and locally.

The Pyrenees form Iberia's northeastern boundary, running for 270 miles (440 km), dividing the peninsula from the rest of Europe.

The Ebro River has formed the peninsula's largest delta. Recently, sediment flows have been seriously disturbed by nearby reservoirs.

On the northeastern coast sea level changes are evident from wave-cut beaches which rise up to 200 ft (60 m) above the present sea level.

Cordillera Cantábrica

Douro/Duero River

The Meseta plateau averages 1,970 ft (600 m) in height and is now largely dry and treeless.

Tagus River

The Balearic Islands (Islas Baleares) are characterized by jagged limestones and plains.

Mountain front — Pediment

Weathered material

Pediments are characteristic of semi-arid lands across Iberia. A pediment is a flat, low-lying, eroded platform, cut into the bedrock. Weathered material is transported by streams and deposited in broad fan shapes on the pediment.

The Guadalquivir River brings vital irrigation water to the plains, and like many of Iberia's rivers, is prone to flooding.

Sierra Morena

The Sierra Nevada in southern Spain contain Iberia's highest peak, Mulhacén, which rises 11,418 ft (3,481 m).

In the Sierra de los Filabres deforestation and overgrazing, which cause soil erosion, have created semidesert badlands.

THE ITALIAN PENINSULA

ITALY, SAN MARINO, VATICAN CITY

THE ITALIAN PENINSULA is a land of great contrasts. Until unification in 1861, Italy was a collection of independent states, whose competitiveness during the Renaissance resulted in the architectural and artistic magnificence of cities such as Rome, Florence, and Venice. The majority of Italy's population and economic activity is concentrated in the north, centered on the sophisticated industrial city of Milan. Southern Italy, the *Mezzogiorno*, has a harsh terrain, and remains far less developed than the north. Attempts to attract industry and investment in the south are frequently deterred by the entrenched network of organized crime and corruption.

THE LANDSCAPE

THE MAINLY MOUNTAINOUS and hilly Italian peninsula took its present form following a collision between the African and Eurasian tectonic plates. The Alps in the northwest rise to a high point of 15,772 ft (4,807 m) at Mont Blanc (*Monte Bianco*) on the French border, while the Apennines (*Appennino*) form a rugged backbone, running along the entire length of the country.

The island of Sardinia is an ancient land mass; an uplifted section of very old igneous rocks. Its rugged mountainous regions provide pasture for sheep and goats, while its valleys support some agriculture.

Mont Blanc (*Monte Bianco*)

Costa Smeralda

The Dolomites (*Alpi Dolomitiche*) *are formed of thick limestones, overlying weaker marine strata. They have distinctive serrated peaks and many massive landslides occur.*

The distinctive square shape of the Gulf of Taranto (*Golfo di Taranto*) was defined by numerous block faults. Earthquakes are common in this region.

Vesuvius (*Vesuvio*)

The Pontine Marshes (*Agro Pontino*) are bounded by low sand hills which prevent natural drainage.

The Apennines (*Appennino*) are the source of most of Italy's rivers. They run 823 miles (1324 km) down the length of the peninsula.

The Strait of Messina (*Stretto di Messina*) is between 2 and 12 miles (3–19 km) wide, and is a rich fishing ground.

Sicily is the largest island in the Mediterranean at 9,926 sq miles (25,708 sq km).

The southwestern tip of Sicily lies 95 miles (152 km) from the north African mainland and is part of the same geological region.

Sardinia is the second largest island in the Mediterranean Sea. The highest point is Punta La Marmora at 6,017 ft (1,834 m).

The Po Valley once formed part of the Adriatic Sea. Sediments of gravel, sand, and clay washed down from the Alps gradually filling the bay and forming a broad, cultivable plain.

Present-day crater has developed within the old crater of Monte Somma

Old crater

There have been four volcanoes on the site of Vesuvius since volcanic activity began here more than 10,000 years ago.

Vesuvius (*Vesuvio*)

Monte Somma

Old crater

USING THE LAND

ITALY PRODUCES 95% of its own food. The best farming land is in the Po Valley in northern Italy, where soft wheat and rice are grown. Irrigation is essential to agriculture in much of the south. Italy is a major producer and exporter of citrus fruits, olives, tomatoes, and wine.

THE URBAN/RURAL POPULATION DIVIDE

urban 67%   rural 33%

TOTAL LAND AREA
116,320 sq miles (301,270 sq km)

POPULATION DENSITY
492 people per sq mile
(190 people per sq km)

Land use and agricultural distribution

- cattle
- cereals
- citrus fruits
- olive oil
- rice
- vineyards

- capital cities
- major towns
- pasture
- cropland
- forest
- mountain region

SCALE 1:2,500,000
(projection: Lambert Conformal Conic)

Km 0 10 20 30 40 50 60 70
Miles 0 5 10 20 30 40 50

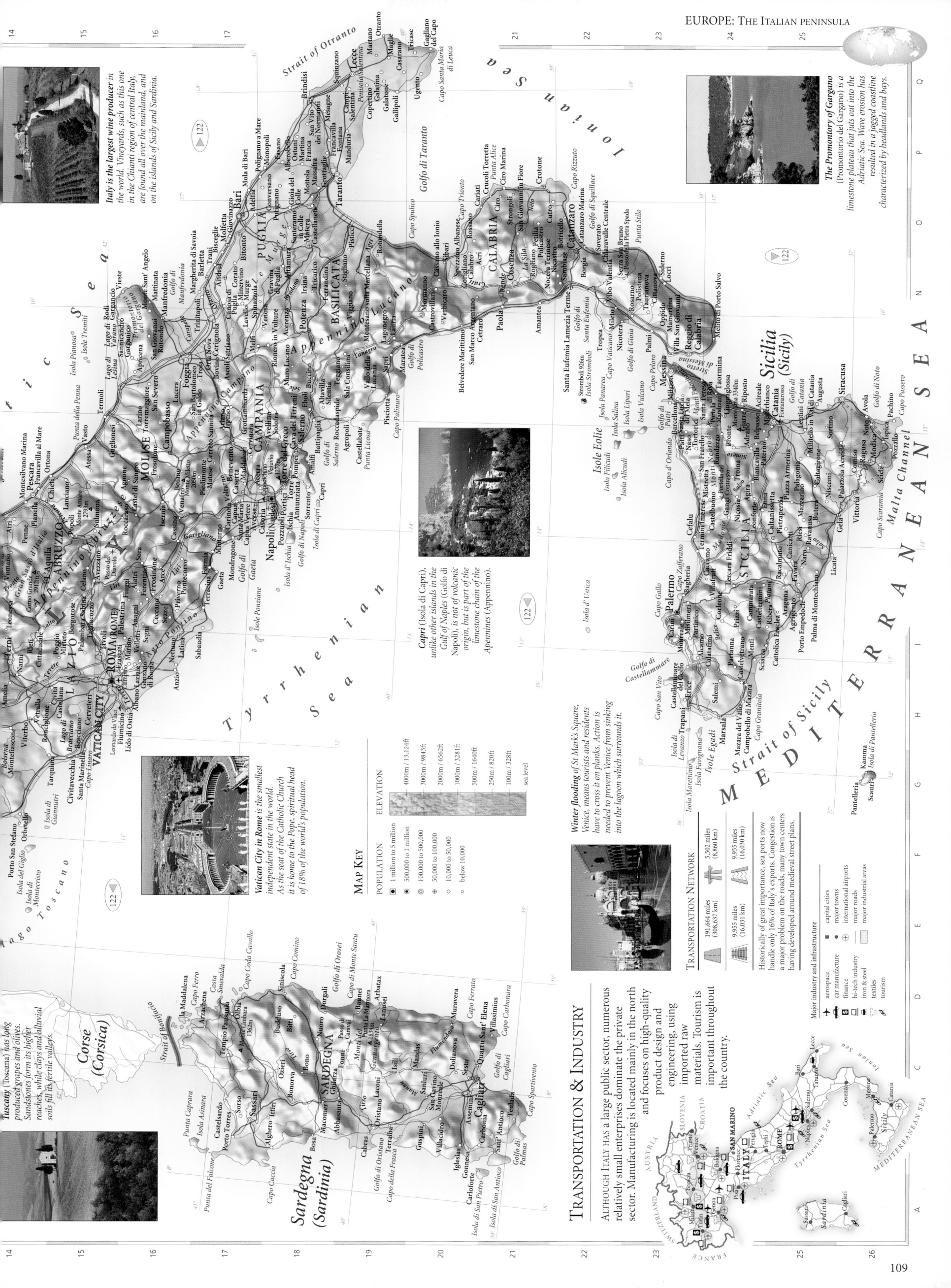

Italy is the largest wine producer in the world. Vineyards, such as this one in the Chianti region of central Italy, are found all over the mainland, and on the islands of Sicily and Sardinia.

The Promontory of Gargano (Promontorio del Gargano) is a limestone plateau that juts out into the Adriatic Sea. Wave erosion has resulted in a jagged coastline characterized by headlands and bays.

Capri (Isola di Capri), unlike other islands in the Gulf of Naples (Golfo di Napoli), is not of volcanic origin, but is part of the limestone chain of the Apennines (Appennino).

Vatican City in Rome is the smallest independent state in the world. As the seat of the Catholic Church it is home to the Pope, spiritual head of 18% of the world's population.

Winter flooding of St Mark's Square, Venice, means tourists and residents have to cross it on planks. Action is needed to prevent Venice from sinking into the lagoon which surrounds it.

Tuscany (Toscana) has long produced grapes and olives. Sandstones form its higher reaches, while clays and alluvial soils fill its fertile valleys.

MAP KEY

POPULATION
- ■ 1 million to 5 million
- ● 500,000 to 1 million
- ◉ 100,000 to 500,000
- ⊕ 50,000 to 100,000
- ○ 10,000 to 50,000
- ○ below 10,000

ELEVATION
- 4000m / 13,124ft
- 3000m / 9843ft
- 2000m / 6562ft
- 1000m / 3281ft
- 500m / 1640ft
- 250m / 820ft
- 100m / 328ft
- sea level

TRANSPORTATION & INDUSTRY

ALTHOUGH ITALY HAS a large public sector, numerous relatively small enterprises dominate the private sector. Manufacturing is located mainly in the north and focuses on high-quality product design and engineering, using imported raw materials. Tourism is important throughout the country.

TRANSPORTATION NETWORK

191,664 miles (308,637 km)	5,502 miles (8,860 km)		
9,955 miles (16,031 km)	9,955 miles (16,030 km)		

Historically of great importance, sea ports now handle only 16% of Italy's exports. Congestion is a major problem on the roads, many town centers having developed around medieval street plans.

Major industry and infrastructure
- aerospace
- car manufacture
- finance
- hi-tech industry
- iron & steel
- textiles
- tourism

- capital cities
- major towns
- international airports
- major roads
- major industrial areas

THE ALPINE STATES

AUSTRIA, LIECHTENSTEIN, SLOVENIA, SWITZERLAND

THE ALPINE COUNTRIES of Austria, Switzerland, Liechtenstein, and Slovenia form a narrow strip across western Europe's geographical core, lying on the main north–south trading routes across the Alps. Switzerland, politically neutral since 1815, is an important international meeting place and houses one of the headquarters of the United Nations, although not itself a member. Austria, once at the heart of the great Habsburg Empire has been a fully independent nation since 1955, and maintains a deserved reputation as an international center of culture. Slovenia declared independence from the former Yugoslavia in 1991 and despite initial economic hardship, is now starting to achieve the prosperity enjoyed by its Alpine neighbors.

USING THE LAND

THE ALPINE REGION's mountainous terrain discourages cultivation over much of the land area. The primary agricultural activity is the raising of dairy and beef cattle on the pasture land of the lower mountain slopes. Austria is self-supporting in grains, and crops such as wheat, barley, and grapes are grown on the east Austrian lowlands. Woodlands are more prevalent in the eastern Alps; both Austria and Slovenia have large tracts of forest.

Land use and agricultural distribution

- cattle
- pigs
- cereals
- vineyards
- capital cities
- major towns
- pasture
- cropland
- forest
- mountain region

The Matterhorn, on the Swiss-Italian border, is one of the highest mountains in the Alps, at 14,692 ft (4,478 m). The term "horn"refers to its distinctive peak, formed by three glaciers eroding hollows, known as cirques, in each of its sides.

THE LANDSCAPE

THE ALPS OCCUPY THREE-FIFTHS OF SWITZERLAND, most of southern Austria and the northwest of Slovenia. They were formed by the collision of the African and Eurasian tectonic plates, which began 65 million years ago. Their complex geology is reflected in the differing heights and rock types of the various ranges. The Rhine flows along Liechtenstein's border with Switzerland, creating a broad floodplain in the north and west of Liechtenstein. In the far northeast and east are a number of lowland regions, including the Vienna Basin, Burgenland, and the plain of the Danube. Slovenia's major rivers flow across the lower eastern regions; in the west, the rivers flow underground through the limestone Karst region.

Original height after uplift and folding

Folded strata are overturned creating a *nappe*

Eurasian Plate

Present-day height of Alps

African Plate

The convergence of the African and Eurasian plates compressed and folded huge masses of rock strata. As the plates continued to move together, the folded strata were overturned, creating complex nappes. Much of the rock strata has since been eroded, resulting in the current topography of the Alps.

Constricted as it cuts through ridges in the Alps, the Danube meanders across the lowlands, where uplift combined with river erosion has deepened meanders.

The Vienna Basin lies mainly below 390 ft (120 m). It gradually subsided and filled with sediment as the Alps were uplifted.

Neusiedler See straddles the border of Austria and Hungary; the area around it provides some of the best wine-growing land in Austria.

The mountains of the Jura form a natural border between Switzerland and France. Their marine limestones date from over 200 million years ago. When the Alps were formed the Jura were folded into a series of parallel ridges and troughs.

Tectonic activity has resulted in dramatic changes in land height over very short distances. Lake Geneva, lying at 1,221 ft (372 m) is only 43 miles (70 km) away from the 15,772 ft (4,807 m) peak of Mont Blanc, on the France–Italy border.

The Bernese Alps (*Berner Alpen*) contain the Aletsch, which at 15 miles (24 km) is the longest Alpine glacier.

The Rhine, like other major Alpine rivers, follows a broad, flat trough through the mountains. Along part of its course, the Rhine forms the boundary between Switzerland and Liechtenstein.

The first road through the Brenner Pass was built in 1772, although it has been used as a mountain route since Roman times. It is the lowest of the main Alpine passes at 4,298 ft (1374 m).

The deep, blue lakes of the Karst region are part of a drainage network which runs largely underground through this limestone area.

Karst region

The limestone cave system at Postojna extends for more than 10 miles (16 km) and includes caverns reaching 125 ft (40 m) in height and width.

The Austrian Alps comprise three distinct mountain ranges, separated by deep trenches. The northern and southern ranges are rugged limestones, while the Tauern range is formed of crystalline rocks.

The Tauern range in the central Austrian Alps contains the highest mountain in Austria, the towering Grossglockner, rising 12,461 ft (3,798 m).

THE URBAN/RURAL POPULATION DIVIDE

58% urban   42% rural

POPULATION DENSITY	TOTAL LAND AREA
318 people per sq mile (117 people per sq km)	20,687 sq miles (53,580 sq km)

In this mountainous region, the flatter, more accessible areas are often used for both cattle grazing and recreation.

CZECH REPUBLIC

SLOVAKIA

AUSTRIA

OBERÖSTERREICH

NIEDERÖSTERREICH

WIEN (VIENNA)

Linz · Wels · Salzburg · Graz

SALZBURG · Hohe Tauern · TIROL · KÄRNTEN · STEIERMARK · BURGENLAND

HUNGARY

These converging glaciers are marked by dark lines of moraine. This eroded material is carried by glaciers, and deposited as the ice melts.

SLOVENIA

LJUBLJANA

Maribor · Klagenfurt

Julian Alps · Gulf of Trieste

Koper · Piran · Izola

CROATIA

SCALE 1:1,750,000
(projection: Lambert Conformal Conic)

Km 0 10 20 30 40 50 60
Miles 0 5 10 20 30 40 50 60

The Austrian Tirol contains some of the most spectacular Alpine scenery. Snow cover is a permanent feature in the highest reaches.

MAP KEY

POPULATION

- 1 million to 5 million
- 500,000 to 1 million
- 100,000 to 500,000
- 50,000 to 100,000
- 10,000 to 50,000
- below 10,000

ELEVATION

- 4000m / 13,124ft
- 3000m / 9843ft
- 2000m / 6562ft
- 1000m / 3281ft
- 500m / 1640ft
- 250m / 820ft
- 100m / 328ft
- sea level

TRANSPORTATION & INDUSTRY

ALL FOUR NATIONS concentrate on high-quality manufacturing and services. Austrian iron and steel production is complemented by construction industries; and Slovenia, traditionally the industrial powerhouse of the western Balkans has increasingly diversified industries. Liechtenstein and Switzerland, lacking raw materials, produce pharmaceuticals and precision instruments, such as watches, and act as international banking centers. The spectacular scenery of the region encourages tourism all year round.

TRANSPORTATION NETWORK

119,805 miles (192,923 km)	2044 miles (3292 km)
6227 miles (10,028 km)	984 miles (1584 km)

Tunnels and passes through the Alps are an important feature of this region. The NEAT project, providing two new high-speed rail links between Basel and Milan, was given approval in 1992.

Major industry and infrastructure

- car manufacture
- chemicals
- engineering
- finance
- food processing
- iron & steel
- pharmaceuticals
- textiles
- tourism
- watch making
- winter sports

- capital cities
- major towns
- international airports
- major roads
- major industrial areas

SWITZERLAND · BERN · VADUZ · LIECHTENSTEIN · Geneva · Lausanne · Zürich · Basel · Biel · Luzern

GERMANY · FRANCE · ITALY

AUSTRIA · VIENNA · Linz · Sankt Pölten · Salzburg · Innsbruck · Kapfenberg · Donawitz · Leoben · Graz · Klagenfurt · Maribor

SLOVENIA · LJUBLJANA · Novo Mesto · Jesenice

The Schönbrunn Palace in Vienna was the summer residence of the Habsburg monarchy. Today, it is a major tourist attraction.

CENTRAL EUROPE

CZECH REPUBLIC, HUNGARY, POLAND, SLOVAKIA

WHEN SLOVAKIA AND THE CZECH REPUBLIC became separate countries in 1993, they joined Hungary and Poland in a new role as independent nation states, following centuries of shifting boundaries and imperial strife. This turbulent history bequeathed the region a rich cultural heritage, shared through the works of its many great writers and composers, and celebrated in the vibrant historic capitals of Prague, Budapest, and Warsaw. Having shaken off Soviet domination in 1989, these states are facing up to the challenge of winning commercial investment to modernize outmoded industry, while bearing the severe environmental impact from forty years of large-scale industrialization.

TRANSPORTATION & INDUSTRY

HEAVY INDUSTRY HAS DOMINATED POSTWAR LIFE in Central Europe. Poland has large coal reserves, having inherited the Silesian coalfield from Germany after the Second World War, allowing the export of large quantities of coal, along with other minerals. Hungary specializes in consumer goods and services, while Slovakia's industrial base is still relatively small. The Czech Republic's traditional glassworks and breweries bring some stability to its precarious Soviet-built manufacturing sector.

THE LANDSCAPE

THE FORESTED Carpathian Mountains, uplifted with the Alps, lie southeast of the older Bohemian massif, which contains the Sudeten and Krušné Hory (Erzgebirge) ranges. They divide the fertile plains of the Danube to the south and the Vistula (Wisła), which flows north across vast expanses of glacial deposits into the Baltic Sea.

The Biebrza River has left meanders and oxbow lakes as it flows across low-lying ground.

Gerlachovský štít, in the Tatra Mountains, is Slovakia's highest mountain, at 8,711ft (2,655 m).

Carpathian Mountains

Hot mineral springs occur where geothermally heated water wells up through faults and fractures in the rocks of the Sudeten Mountains.

Pomerania is a sandy coastal region of glacially-formed lakes stretching west from the Vistula (Wisła).

Longshore currents moving east along the Baltic coast have built a 40 mile (65 km) spit composed of material from the Vistula (Wisła) River.

Danube River

Slip-off slope

Bluff

Direction of flow

Meanders form as rivers flow across plains at a low gradient. A steep cliff or bluff, forms on the outside curve, and a gentler slip-off slope on the inside bend.

The Great Hungarian Plain formed by the floodplain of the Danube is a mixture of steppe and cultivated land, covering nearly half of Hungary's total area.

The Slovak Ore Mountains (Slovenské Rudohorie) are noted for their mineral resources, including high-grade iron ore.

Bohemian Massif

Krušné Hory (Erzgebirge)

The Berounka River cuts through the precipitous wooded landscape of the Bohemian massif, banked by a broad floodplain.

TRANSPORTATION NETWORK

✈ 817 miles (1,315 km)	🛫	
🚗 213,997 miles (344,600 km)		
🚗 22,479 miles (44,229 km)	🚆 3,784 miles (6,094 km)	

Major industry and infrastructure

- car manufacture
- chemicals
- engineering
- food processing
- mining
- shipbuilding
- tourism

○ ● capital cities
• ● major towns
✈ international airports
— major roads
major industrial areas

The huge growth of tourism and business has prompted major investment in the transportation infrastructure, with new roadbuilding schemes within and between the main cities of the region.

Budapest, the capital of Hungary, straddles the Danube. It comprises the historic towns of Buda, on the west bank, and Pest, which contains the Parliament Building, seen here on the far bank.

BELARUS

LITHUANIA

RUSSIAN FEDERATION (Kaliningrad)

POLAND

WARSZAWA (WARSAW)

GERMANY

BALTIC SEA

Gulf of Danzig

Pomeranian Bay

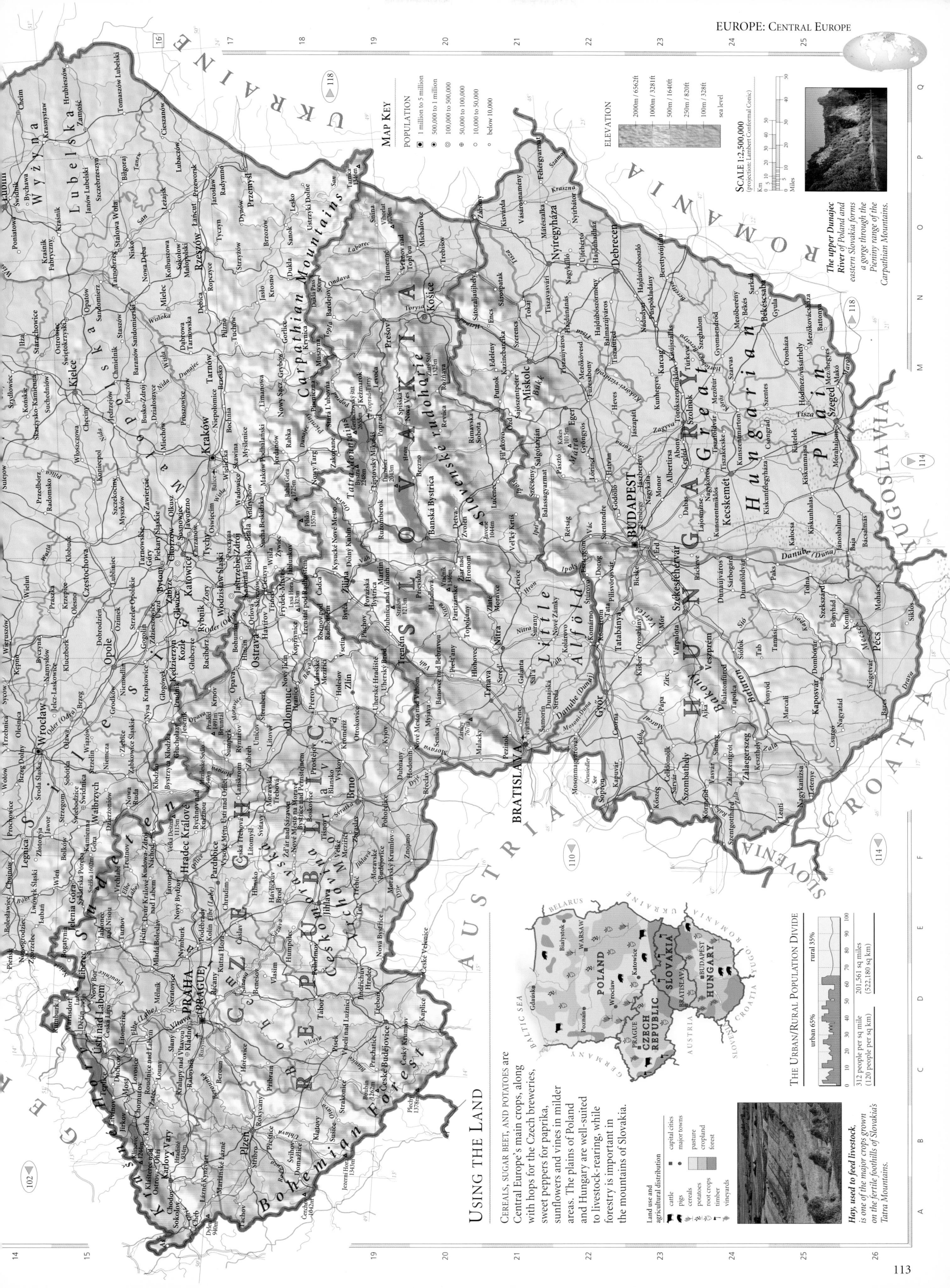

MAP KEY

POPULATION
- ■ 1 million to 5 million
- ● 500,000 to 1 million
- ◉ 100,000 to 500,000
- ⊕ 50,000 to 100,000
- ⊙ 10,000 to 50,000
- ○ below 10,000

ELEVATION
- 2000m / 6562ft
- 1000m / 3281ft
- 500m / 1640ft
- 250m / 820ft
- 100m / 328ft
- sea level

SCALE 1:2,500,000
(projection: Lambert Conformal Conic)

The upper Dunajec River of Poland and eastern Slovakia forms a gorge through the Pieniny range of the Carpathian Mountains.

USING THE LAND

CEREALS, SUGAR BEET, AND POTATOES are Central Europe's main crops, along with hops for the Czech breweries, sweet peppers for paprika, sunflowers and vines in milder areas. The plains of Poland and Hungary are well-suited to livestock-rearing, while forestry is important in the mountains of Slovakia.

Land use and agricultural distribution
- cattle
- pigs
- cereals
- potatoes
- root crops
- timber
- vineyards
- ● capital cities
- ■ major towns
- pasture
- cropland
- forest

Hay, used to feed livestock, is one of the major crops grown on the fertile foothills of Slovakia's Tatra Mountains.

THE URBAN/RURAL POPULATION DIVIDE

urban 65%   rural 35%

312 people per sq mile
(120 people per sq km)

201,561 sq miles
(522,180 sq km)

SOUTHEAST EUROPE

ALBANIA, BOSNIA & HERZEGOVINA, CROATIA, MACEDONIA, YUGOSLAVIA

FOR 46 YEARS THE FEDERATION of Yugoslavia held together the most diverse ethnic region in Europe, along the picturesque mountain hinterland of the Dalmatian coast. Economic collapse resulted in internal tensions. In the early 1990s, civil war broke out in both Croatia and Bosnia as the ethnic populations struggled to establish their own exclusive territories. Peace was only restored by the UN after NATO launched air strikes in 1995. In the province of Kosovo, attempts to gain autonomy from Yugoslavia in 1998 were crushed by the Serbian government. The slaughter of ethnic Albanians in Kosovo provoked the West to launch NATO air strikes yet again in the region, and Yugoslav forces withdrew. The flood of refugees from Kosovo has severely strained Albania.

THE LANDSCAPE

THE TISZA, SAVA, AND DRAVA RIVERS drain the broad northern lowland, meeting the Danube after it crosses the Hungarian border. In the west, the Dinaric Alps divide the Adriatic Sea from the interior. Mainland valleys and elongated islands run parallel to the steep Dalmatian (*Dalmacija*) coastline, following alternating bands of resistant limestone.

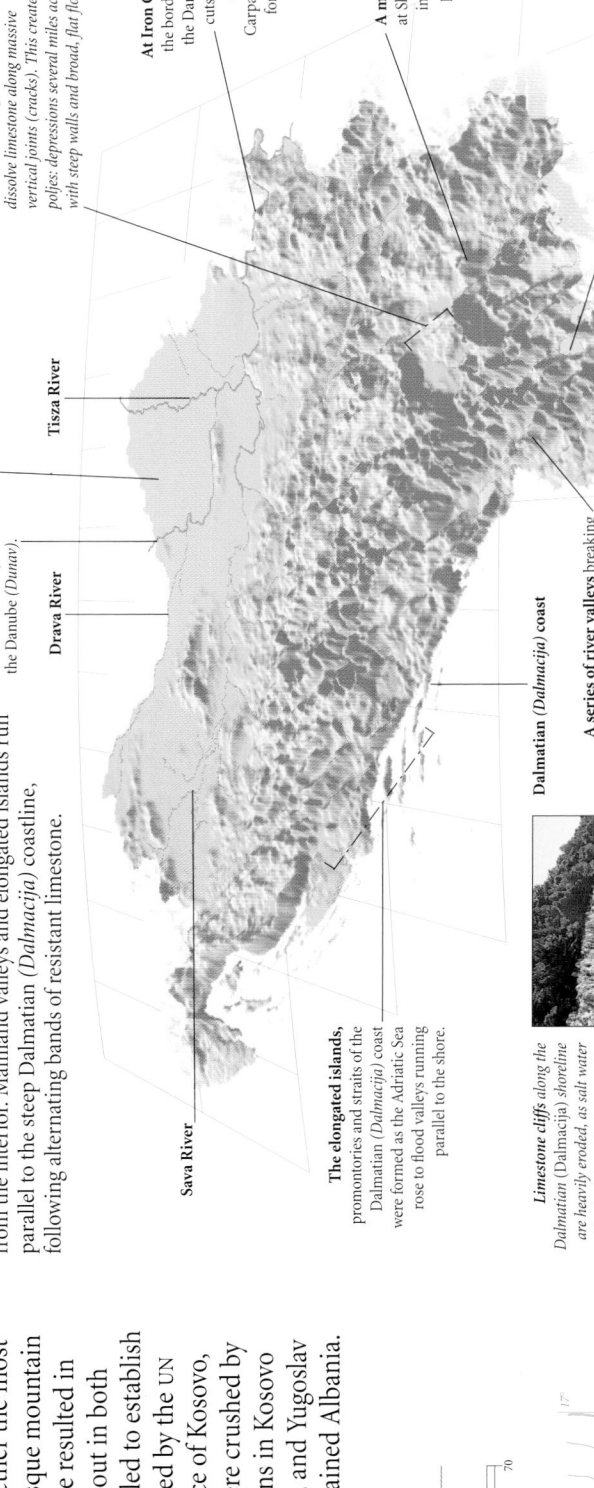

SCALE 1:2,500,000
(projection: Lambert Conformal Conic)

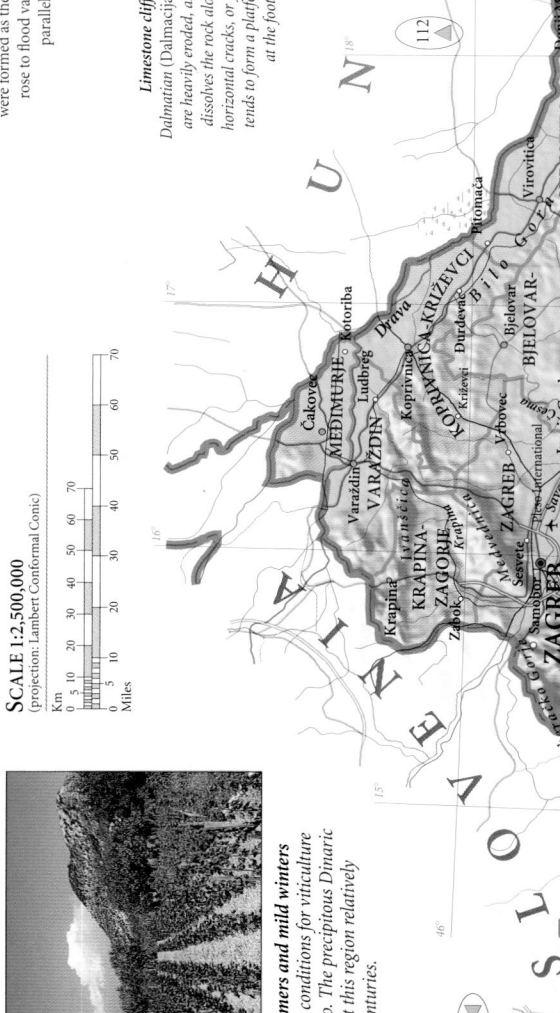

Hot, dry summers and mild winters offer excellent conditions for viticulture in Montenegro. The precipitous Dinaric Alps have kept this region relatively isolated for centuries.

Polvjes in the Kosovo region

Sheer limestone walls enclose all sides

Flat polje floor

Rain and underground water dissolve limestone along massive vertical joints (cracks). This creates poljes: depressions several miles across with steep walls and broad, flat floors.

Underground drainage along joints in the rock

Spring at foot of cliff

The river floodplains of the Pannonian Basin are flanked by terraces of gravel and wind-blown glacial deposits known as loess.

At least 70% of the fresh water in the Western Balkans drains eastward into the Black Sea, mostly via the Danube (*Dunav*).

At Iron Gate (*Ðerdap*), on the border with Romania, the Danube narrows and cuts through foothills of the Balkan and Carpathian mountains, forming the deepest gorge in Europe.

A major earthquake at Skopje, Macedonia, in 1963 killed 1,000 people. The whole region lies on an active crustal plate margin.

Tisza River

Drava River

Sava River

The elongated islands, promontories and straits of the Dalmatian (*Dalmacija*) coast were formed as the Adriatic Sea rose to flood valleys running parallel to the shore.

Dalmatian (*Dalmacija*) coast

Limestone cliffs along the Dalmatian (Dalmacija) shoreline are heavily eroded, as salt water dissolves the rock along existing horizontal cracks, or joints. This tends to form a platform of rock at the foot of the cliff.

A series of river valleys breaking through the Dinaric Alps from the lowlands of western Albania, give access to the interior.

Lake Ohrid

Lake Ohrid borders Albania and Macedonia. Ohrid is the deepest lake in the Western Balkans, reaching depths of 938 ft (286 m).

MAP KEY

POPULATION

- ⊚ 1 million to 5 million
- ◉ 500,000 to 1 million
- ⊕ 100,000 to 500,000
- ⊙ 50,000 to 100,000
- ○ 10,000 to 50,000
- ∘ below 10,000

ELEVATION

- 2000m / 6562ft
- 1000m / 3281ft
- 500m / 1640ft
- 250m / 820ft
- 100m / 328ft
- sea level

The Tara River is one of Montenegro's major rivers. It flows into the Danube via the Drina and Sava Rivers. Along its course the Tara has eroded spectacular gorges up to 3,280 ft (1,000 m) deep.

The ancient Croatian port of Dubrovnik was one of the former Yugoslavia's most popular tourist resorts and an important point of access to the sea along the Dalmatian (Dalmacija) coast. Shelling of the old city by Serb forces in 1991 provoked international condemnation.

TRANSPORTATION NETWORK

🛣	72,719 miles (117,100 km)
🚂	4,808 miles (7,743 km)
✈	415 miles (668 km)
⚓	1,911 miles (3,078 km)

The war has resulted in the destruction or disintegration of infrastructure for transportation, communications, and power supply, with essential provisions moved under armed UN convoy.

Industrial processing plants were established throughout Albania by the Hoxha regime, which collapsed in 1992. They remain incongruous among the villages of one of Europe's most conservative rural societies.

TRANSPORTATION & INDUSTRY

PROCESSING INDUSTRIES based on the region's wealth of mineral reserves predominate in Albania and Macedonia. In other regions, industrial plants have been commandeered, if not destroyed in the war and mineral extraction has severely declined. The fast-flowing rivers found throughout the Dinaric Alps are exploited to generate hydroelectric power.

Major industry and infrastructure

- △ aluminum refining
- ⚙ car manufacture
- ◆ chemicals
- ⚒ engineering
- 🍴 food processing
- ⚒ hydroelectric power
- ✕ mining
- ⚓ shipbuilding
- ⬥ textiles
- ■ timber processing

- ◼ capital cities
- ▪ major towns
- ⊕ international airports
- — major roads

The historic center of Mostar in southern Bosnia, with its famous 16th-century Turkish bridge, was destroyed by shelling during 1993. The town was formerly the capital of Herzegovina.

USING THE LAND

CROPS OF WHEAT, maize, sugar beet, vegetables, and fruit are widely grown. The hilly terrain is suited to forestry and livestock farming. The mild, Mediterranean climate of the coastal regions provides ideal conditions for growing vines and olives. Albania's largely agricultural economy has been adversely affected by the recent dismantling of state farms.

Land use and agricultural distribution

- pigs
- sheep
- cereals
- fruit
- olives
- sugar beet
- tobacco
- vineyards

- capital cities
- major towns
- pasture
- cropland
- forest
- mountain region

Sweet red peppers are dried in the sun, ready to make paprika. Macedonia's economy is mainly agricultural and its fertile soils support a broad range of crops.

THE URBAN/RURAL POPULATION DIVIDE

urban 44% rural 56%

POPULATION DENSITY
256 people per sq mile (99 people per sq km)

TOTAL LAND AREA
95,038 sq miles (246,278 sq km)

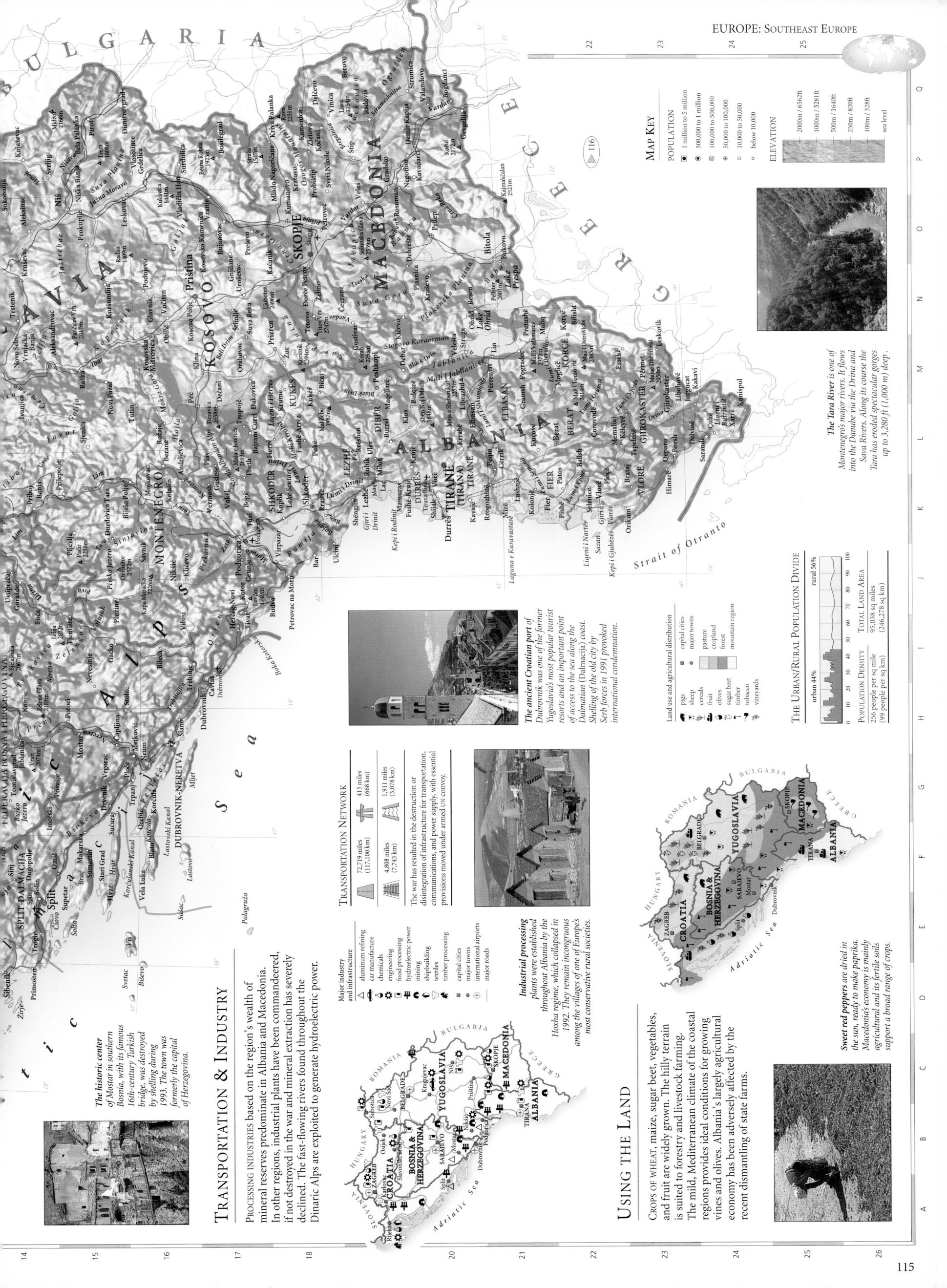

BULGARIA & GREECE

Including EUROPEAN TURKEY

G REECE IS RENOWNED as the original hearth of Western civilization. The rugged terrain and numerous islands have profoundly affected its development, creating a strong agricultural and maritime tradition. In the past 50 years, this formerly rural society has rapidly urbanized, with more than half the population now living in the capital, Athens, and in the northern city of Salonica. Bulgaria, dominated for centuries by the Ottoman Turks, became part of the eastern bloc after the Second World War, only slowly emerging from Soviet influence in 1989. Moves toward democracy have led to some political instability and Bulgaria has been slow to align its economy with the rest of Europe.

THE LANDSCAPE

BULGARIA'S BALKAN MOUNTAINS divide the Danubian Plain (*Dunavska Ravnina*) and Maritsa Basin, meeting the Black Sea in the east along sandy beaches. The steep Rhodope Mountains form a natural barrier with Greece, while the younger Pindus form a rugged central spine which descends into the Aegean Sea to give a vast archipelago of over 2000 islands, the largest of which is Crete.

Mount Olympus is the mythical home of the Greek Gods and, at 9,570 ft (2,917 m), is the highest mountain in Greece.

Ancient metamorphic rock, formed miles below the surface

Limestone rocks exposed by erosion of metamorphic rocks

Mount Olympus

Mount Olympus is a composite of rocks formed by two major tectonic events. First the older metamorphic rocks were thrust over the limestones, then two million years ago regional warping and subsequent erosion, reexposed the limestone.

Younger limestones created in shallow seas

The Peloponnese consist of several mountainous peninsulas, linked to the mainland by the Isthmus of Corinth. The Corinth Canal (*Dioryga Korinthou*), built in 1893, cuts through the isthmus, linking the Aegean and Ionian Seas.

TRANSPORTATION & INDUSTRY

SOVIET INVESTMENT introduced heavy industry into Bulgaria, and the processing of agricultural produce, such as tobacco, is important throughout the country. Both countries have substantial shipyards and Greece has one of the world's largest merchant fleets. Many small craft workshops, producing textiles and processed foods, are clustered around Greek cities. The service and construction sectors have profited from the successful tourist industry.

Major industry and infrastructure
- chemicals
- engineering
- food processing
- shipbuilding
- textiles
- tourism
- capital cities
- major towns
- international airports
- major roads
- major industrial areas

TRANSPORTATION NETWORK

103,930 miles (167,630 km)

345 miles (557 km)

4,346 miles (6,995 km)

294 miles (474 km)

Bulgaria's railroads require investment to revive an outdated infrastructure. In Greece, despite a developing road network, ferry-boats remain the most effective form of transportation in many areas.

Marginal captions

The Arda river cuts through the Rhodope mountains in rugged, rocky gorges.

The islands of Crete, Kythira, Karpathos, and Rhodes are part of an arc which bends southeastward from the southern Peloponnese, forming the southern boundary of the Aegean.

Layers of black volcanic ash still cover the island of Thíra. This volcano last erupted 3,500 years ago, but still shows signs of volcanic activity.

The Danube, Europe's second longest river, forms most of Bulgaria's northern border. The Danubian Plain (*Dunavska Ravnina*), extending from the southern bank, is extremely fertile.

Balkan Mountains

Maritsa Basin

Rhodes

Karpathos

Crete

Kythira

Corinth Canal (*Dioryga Korinthou*)

Rhodope Mountains

Pindus Mountains

SCALE 1:2,500,000
(projection: Lambert Conformal Conic)

A towering pinnacle at Metéora in central Greece is home to the monastery of Roussanou. The 24 rock towers which dominate the plain of Thessaly (Thessalía) are remnants of an old plateau. Long-term weathering along fissures in the rock has worn away the rest of the plateau.

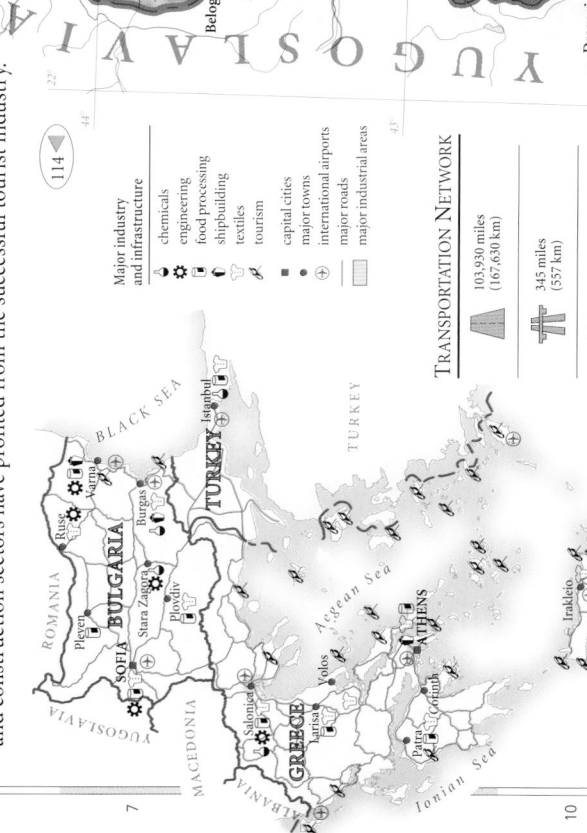

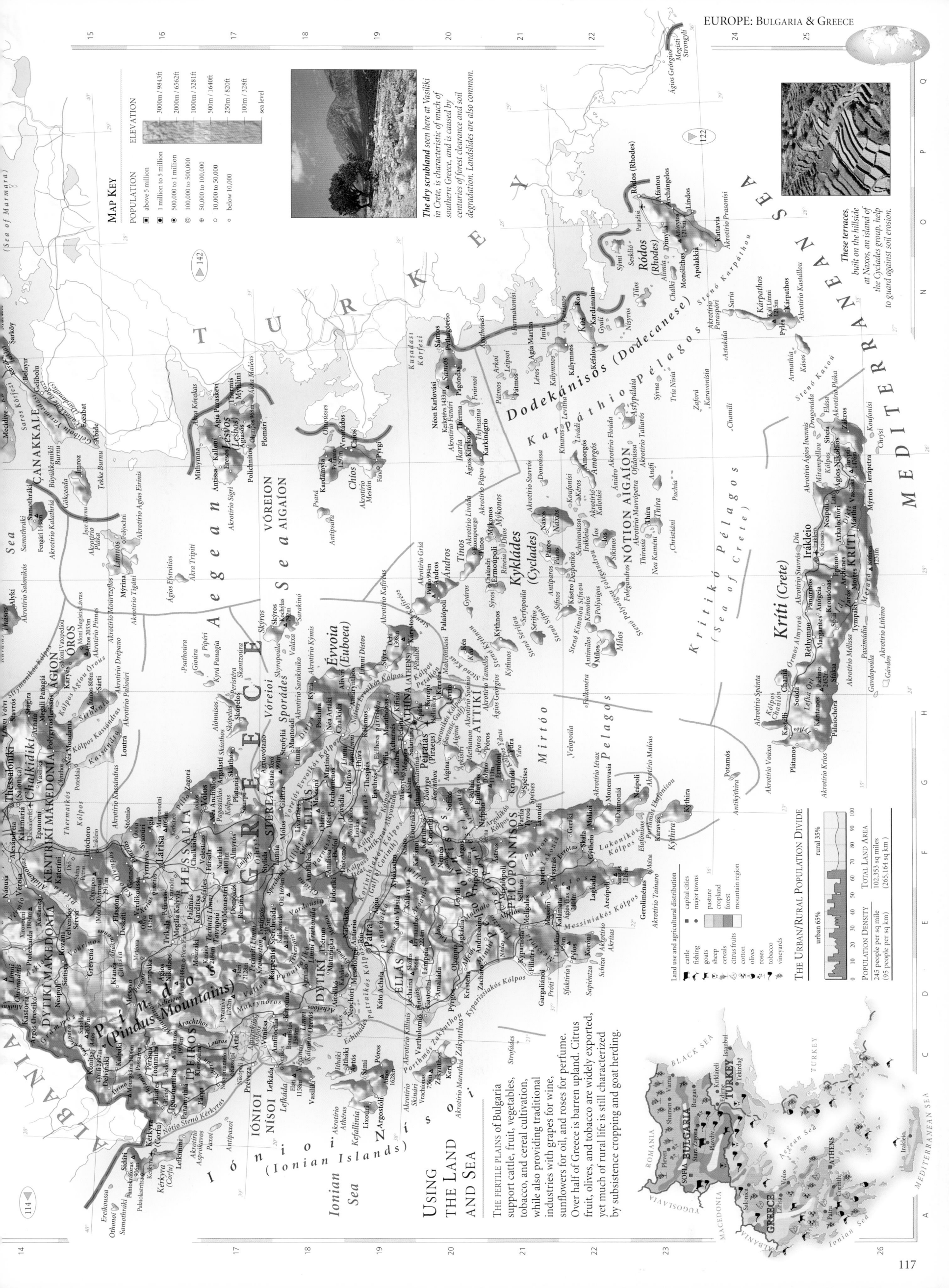

MAP KEY

POPULATION
- ■ above 5 million
- ■ 1 million to 5 million
- ● 500,000 to 1 million
- ◉ 100,000 to 500,000
- ⊕ 50,000 to 100,000
- ⊕ 10,000 to 50,000
- ○ below 10,000

ELEVATION
- 3000m / 9843ft
- 2000m / 6562ft
- 1000m / 3281ft
- 500m / 1640ft
- 250m / 820ft
- 100m / 328ft
- sea level

The dry scrubland seen here at Vasiliki in Crete, is characteristic of much of southern Greece, and is caused by centuries of forest clearance and soil degradation. Landslides are also common.

These terraces, built on the hillside at Naxos, an island of the Cyclades group, help to guard against soil erosion.

USING THE LAND AND SEA

THE FERTILE PLAINS of Bulgaria support cattle, fruit, vegetables, tobacco, and cereal cultivation, while also providing traditional industries with grapes for wine, sunflowers for oil, and roses for perfume. Over half of Greece is barren upland. Citrus fruit, olives, and tobacco are widely exported, yet much of rural life is still characterized by subsistence cropping and goat herding.

Land use and agricultural distribution
- ● capital cities
- ● major towns
- pasture
- cropland
- forest
- mountain region

cattle / fishing / goats / sheep / citrus fruits / cotton / olives / roses / tobacco / vineyards

THE URBAN/RURAL POPULATION DIVIDE

urban 65% rural 35%

TOTAL LAND AREA
102,353 sq miles
(265,164 sq km)

POPULATION DENSITY
245 people per sq mile
(95 people per sq km)

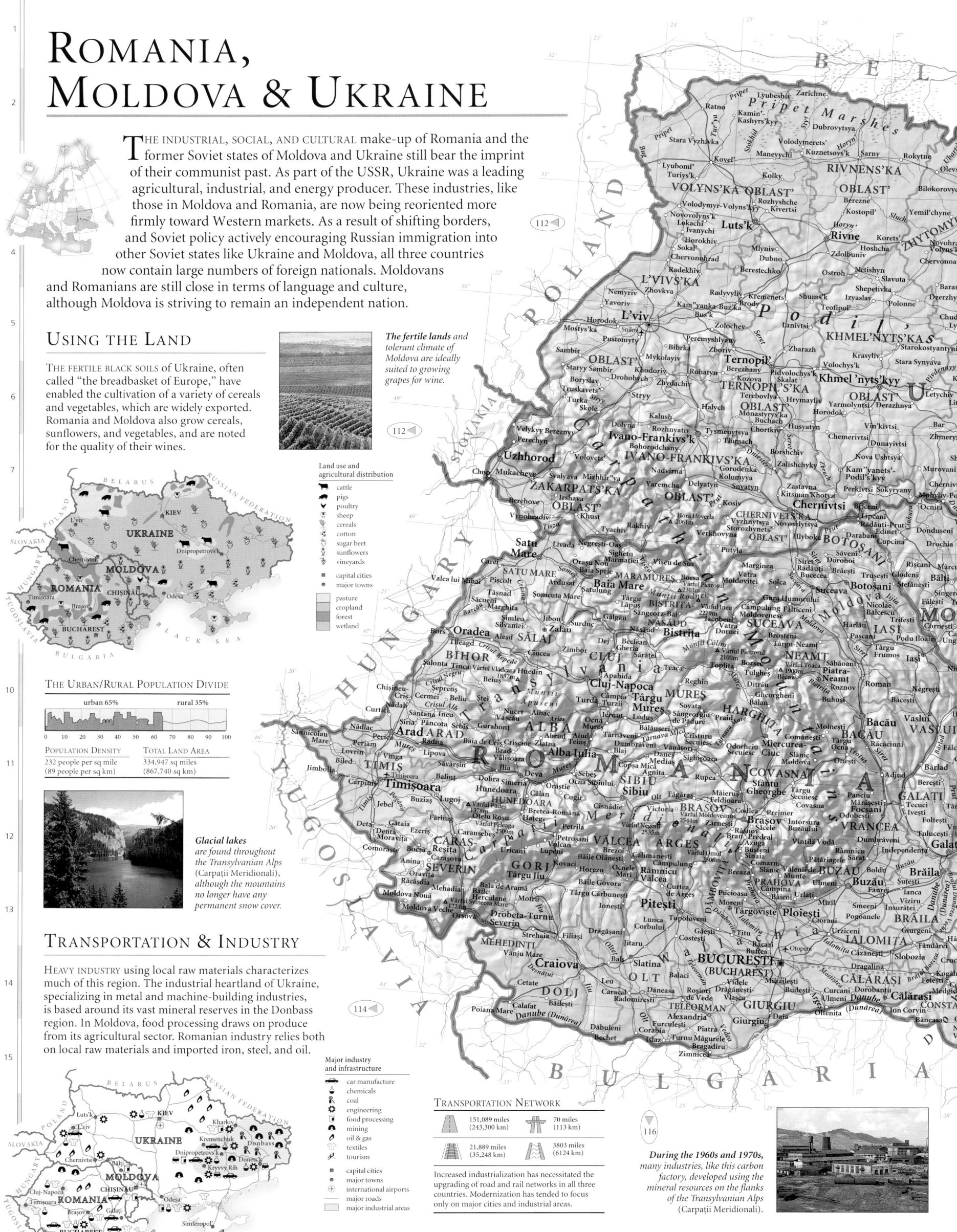

ROMANIA, MOLDOVA & UKRAINE

THE INDUSTRIAL, SOCIAL, AND CULTURAL make-up of Romania and the former Soviet states of Moldova and Ukraine still bear the imprint of their communist past. As part of the USSR, Ukraine was a leading agricultural, industrial, and energy producer. These industries, like those in Moldova and Romania, are now being reoriented more firmly toward Western markets. As a result of shifting borders, and Soviet policy actively encouraging Russian immigration into other Soviet states like Ukraine and Moldova, all three countries now contain large numbers of foreign nationals. Moldovans and Romanians are still close in terms of language and culture, although Moldova is striving to remain an independent nation.

USING THE LAND

THE FERTILE BLACK SOILS of Ukraine, often called "the breadbasket of Europe," have enabled the cultivation of a variety of cereals and vegetables, which are widely exported. Romania and Moldova also grow cereals, sunflowers, and vegetables, and are noted for the quality of their wines.

The fertile lands and tolerant climate of Moldova are ideally suited to growing grapes for wine.

Land use and agricultural distribution
- cattle
- pigs
- poultry
- sheep
- cereals
- cotton
- sugar beet
- sunflowers
- vineyards
- ▪ capital cities
- ▪ major towns
- pasture
- cropland
- forest
- wetland

THE URBAN/RURAL POPULATION DIVIDE

urban 65%　　rural 35%

0 10 20 30 40 50 60 70 80 90 100

POPULATION DENSITY
232 people per sq mile
(89 people per sq km)

TOTAL LAND AREA
334,947 sq miles
(867,740 sq km)

Glacial lakes are found throughout the Transylvanian Alps (Carpaţii Meridionali), although the mountains no longer have any permanent snow cover.

TRANSPORTATION & INDUSTRY

HEAVY INDUSTRY using local raw materials characterizes much of this region. The industrial heartland of Ukraine, specializing in metal and machine-building industries, is based around its vast mineral reserves in the Donbass region. In Moldova, food processing draws on produce from its agricultural sector. Romanian industry relies both on local raw materials and imported iron, steel, and oil.

Major industry and infrastructure
- car manufacture
- chemicals
- coal
- engineering
- food processing
- mining
- oil & gas
- textiles
- tourism
- ▪ capital cities
- ▪ major towns
- ⊕ international airports
- major roads
- major industrial areas

TRANSPORTATION NETWORK

151,089 miles (243,300 km)	70 miles (113 km)
21,889 miles (35,248 km)	3803 miles (6124 km)

Increased industrialization has necessitated the upgrading of road and rail networks in all three countries. Modernization has tended to focus only on major cities and industrial areas.

During the 1960s and 1970s, many industries, like this carbon factory, developed using the mineral resources on the flanks of the Transylvanian Alps (Carpaţii Meridionali).

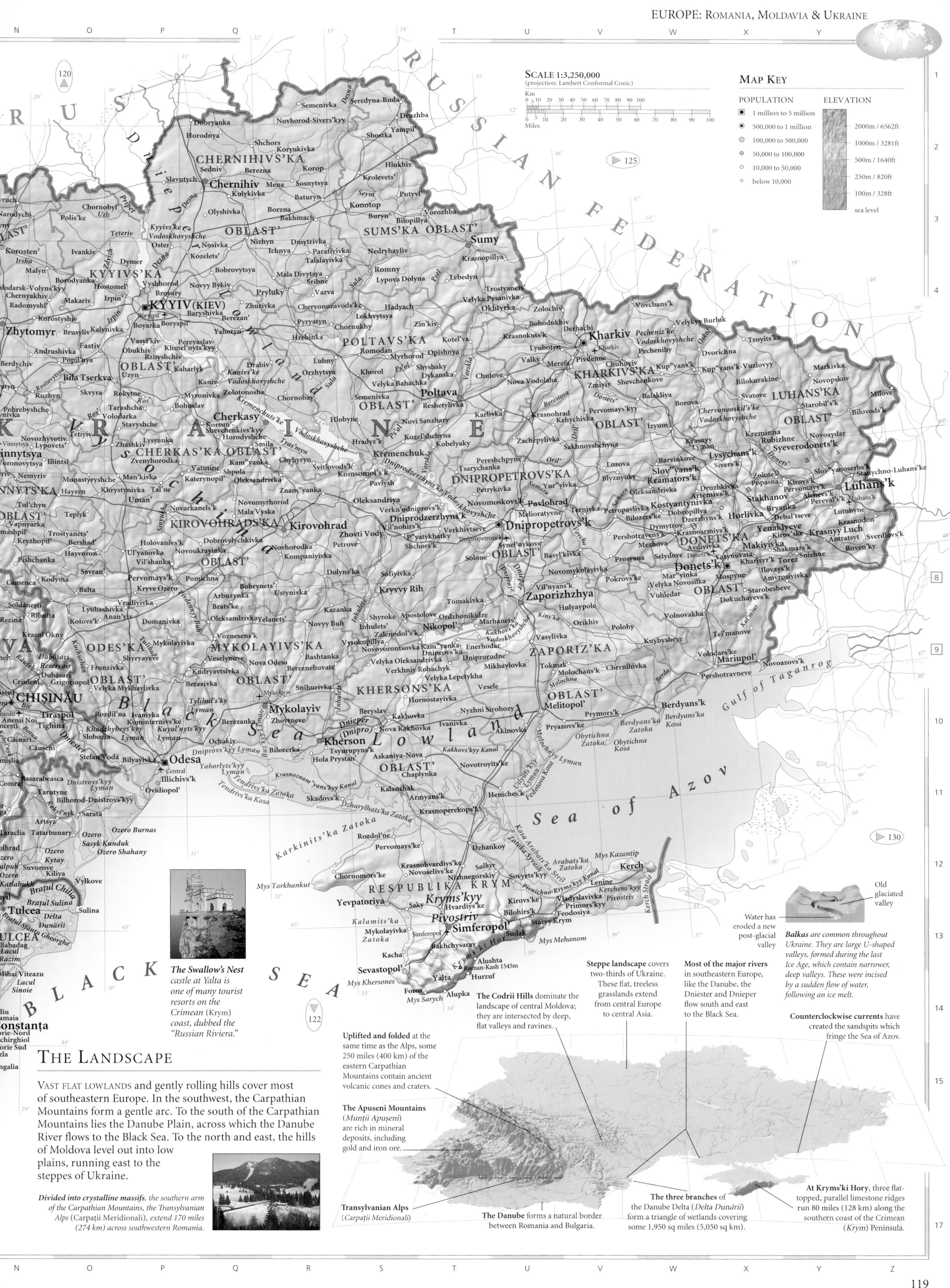

SCALE 1:3,250,000
(projection: Lambert Conformal Conic)

Km
0 10 20 30 40 50 60 70 80 90 100

Miles
0 10 20 30 40 50 60 70 80 90 100

MAP KEY

POPULATION
- 1 million to 5 million
- 500,000 to 1 million
- 100,000 to 500,000
- 50,000 to 100,000
- 10,000 to 50,000
- below 10,000

ELEVATION
2000m / 6562ft
1000m / 3281ft
500m / 1640ft
250m / 820ft
100m / 328ft
sea level

The Swallow's Nest castle at Yalta is one of many tourist resorts on the Crimean (Krym) coast, dubbed the "Russian Riviera."

THE LANDSCAPE

VAST FLAT LOWLANDS and gently rolling hills cover most of southeastern Europe. In the southwest, the Carpathian Mountains form a gentle arc. To the south of the Carpathian Mountains lies the Danube Plain, across which the Danube River flows to the Black Sea. To the north and east, the hills of Moldova level out into low plains, running east to the steppes of Ukraine.

Divided into crystalline massifs, the southern arm of the Carpathian Mountains, the Transylvanian Alps (Carpații Meridionali), extend 170 miles (274 km) across southwestern Romania.

Uplifted and folded at the same time as the Alps, some 250 miles (400 km) of the eastern Carpathian Mountains contain ancient volcanic cones and craters.

The Apuseni Mountains (*Munții Apuseni*) are rich in mineral deposits, including gold and iron ore.

The Codrii Hills dominate the landscape of central Moldova; they are intersected by deep, flat valleys and ravines.

Steppe landscape covers two-thirds of Ukraine. These flat, treeless grasslands extend from central Europe to central Asia.

Most of the major rivers in southeastern Europe, like the Danube, the Dniester and Dnieper flow south and east to the Black Sea.

Counterclockwise currents have created the sandspits which fringe the Sea of Azov.

Balkas are common throughout Ukraine. They are large U-shaped valleys, formed during the last Ice Age, which contain narrower, deep valleys. These were incised by a sudden flow of water, following an ice melt.

Water has eroded a new post-glacial valley

Old glaciated valley

Transylvanian Alps (*Carpații Meridionali*)

The Danube forms a natural border between Romania and Bulgaria.

The three branches of the Danube Delta (*Delta Dunării*) form a triangle of wetlands covering some 1,950 sq miles (5,050 sq km).

At Kryms'ki Hory, three flat-topped, parallel limestone ridges run 80 miles (128 km) along the southern coast of the Crimean (*Krym*) Peninsula.

THE BALTIC STATES & BELARUS

BELARUS, ESTONIA, LATVIA, LITHUANIA, Kaliningrad

OCCUPYING EUROPE's main corridor to Russia, the four distinct cultures of Estonia, Latvia, Lithuania, and Belarus share a history of struggle for nationhood against the interests of more powerful neighbors. As the first republics to declare their independence from the Soviet Union in 1990–91, the Baltic states of Estonia, Latvia, and Lithuania have sought an economic role in the EU, while reaffirming their European cultural roots through the church and a strong musical tradition. Meanwhile, Belarus has shown economic and political allegiance to Russia by joining the Commonwealth of Independent States.

The seaport of Riga is Latvia's capital and the center of economic and cultural life. With a 34% Russian minority in Latvia, language and the right to national citizenship are key issues.

USING THE LAND

ACROSS THE FOUR NATIONS cattle and pig farming are widespread, together with diverse arable crops, including flax for making linen, potatoes used to produce vodka, cereals, and other vegetables. Almost a third of the land is forested; demand for timber has increased the importance of forest management.

Land use and agricultural distribution
- cattle
- pigs
- flax
- cereals
- potatoes
- timber
- capital cities
- major towns
- pasture
- cropland
- forest
- wetland

THE URBAN/RURAL POPULATION DIVIDE

urban 69% / rural 31%

POPULATION DENSITY
122 people per sq mile
(47 people per sq km)

TOTAL LAND AREA
145,006 sq miles
(375,656 sq km)

A pine forest in northern Belarus. Conifers in the north give way to hardwood forest farther south. Timber mills are supplied with logs floated along the country's many navigable waterways.

The Western Dvina River provides hydro-electric power and, during the summer months, access to the Baltic Sea. The lower course of the river freezes from December to April.

MAP KEY

POPULATION
- ◻ 1 million to 5 million
- ◉ 500,000 to 1 million
- ⊚ 100,000 to 500,000
- ⊕ 50,000 to 100,000
- ○ 10,000 to 50,000
- ∘ below 10,000

ELEVATION
- 250m / 820ft
- 100m / 328ft
- sea level

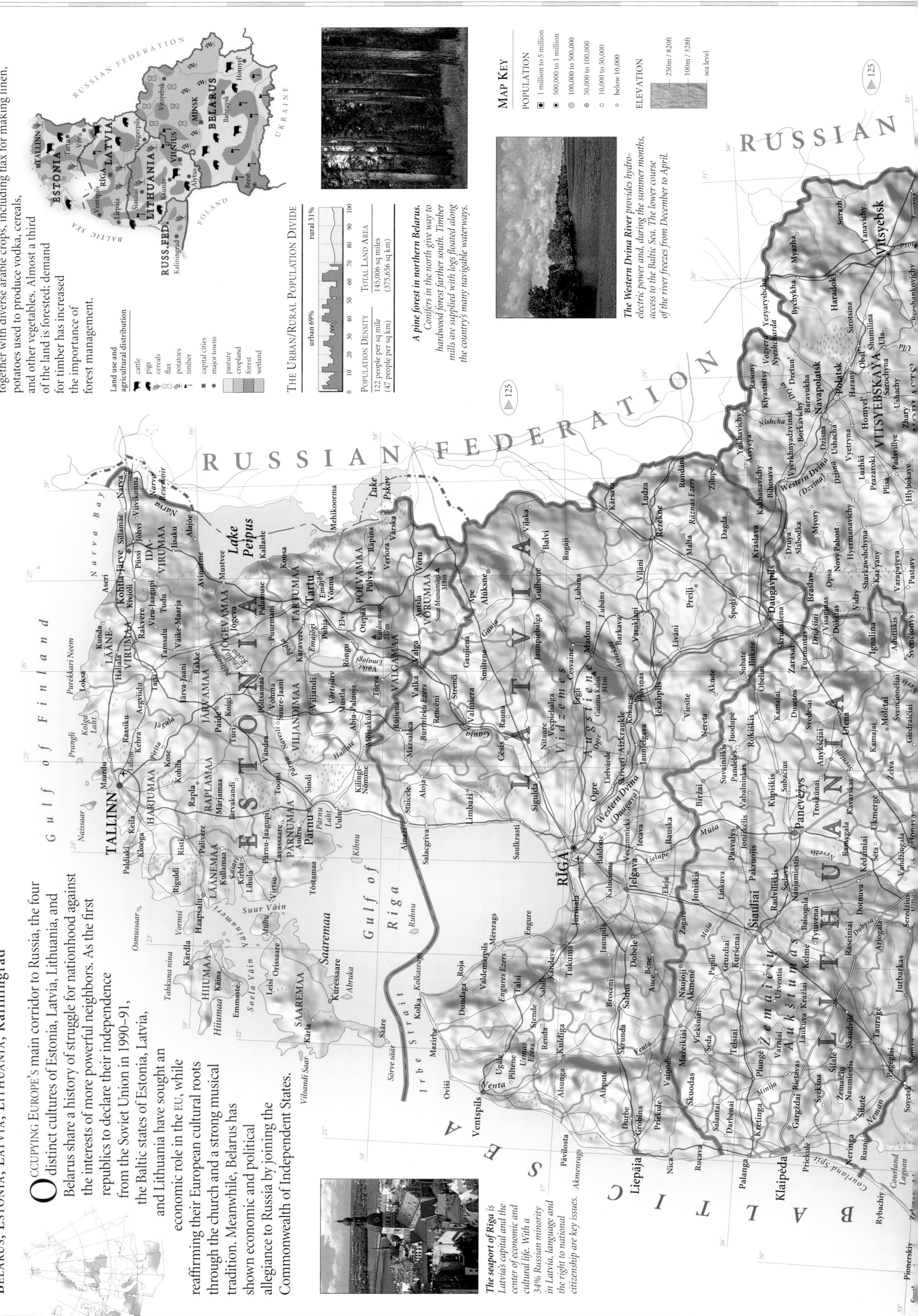

Major industry and infrastructure

- amber mining
- car manufacture
- chemicals
- electrical goods
- oil shale
- food processing
- light engineering
- paper industry

- ● capital cities
- ■ major towns
- ⊕ international airports
- — major roads
- ▢ major industrial areas

Rich oil shale deposits in northern Estonia are quarried, crushed, and heated to produce almost 32,000 barrels of oil a day.

TRANSPORTATION & INDUSTRY

RECENT ECONOMIC RESTRUCTURING has meant modernizing old Soviet industries such as vehicle production and the paper industry, and expanding the light engineering and electronics sectors. There has also been a revival of traditional crafts like carpentry and amber work. Although Estonia has oil shale reserves, the Baltic economies still rely heavily on Russian raw materials and energy.

TRANSPORTATION NETWORK

242,810 miles (391,630 km)		40 miles (64 km)	
6830 miles (11,016 km)		376 miles (606 km)	

Railroads are being superseded by roads linking the ports with eastern Europe and Russia. A highway connecting the three Baltic capitals with Warsaw has been proposed.

Nuclear fallout from the 1986 Chernobyl (Chornobyl') disaster in Ukraine has contaminated large areas of agricultural land in Belarus.

The Dnieper River is the third longest in Europe and forms the heart of Belarus's drainage system.

Pripet Marshes

A network of streams and creeks drains across the marshes

Peat deposits

Glacial deposits

Broad tectonic basin

This large area of marshland lies in a broad tectonic depression, mantled by glacial deposits. Peat deposits have developed below the marshes, which are prone to spring flooding.

The Pripet Marshes form the largest area of "unreclaimed" marshland in Europe. They also provide a network of navigable waterways across southern Belarus.

Suur Munamägi in southern Estonia is, at 1,088 ft (318 m), the highest point in the low-lying Baltic states.

The Vidzeme Uplands (Vidzemes Augstiene) is a region of mixed forest and pasture.

THE LANDSCAPE

ROCK-STREWN GLACIAL PLAINS meet the Baltic Sea along a coast of cliffs and sandy beaches. Hundreds of islands ranging from tiny, rocky outcrops to the large island of Saaremaa, lie scattered off the Estonian mainland, creating an archipelago. Lakes and marshes in low-lying areas give way to mixed woodland on fertile, undulating ground, with remnants of the primeval forest which once covered most of Europe preserved at Byelavyezhskaya Pushcha in western Belarus.

Saaremaa is the largest island in the Estonian archipelago. The southeastern parts are flat and fertile, giving way to numerous low hills and ridges toward the northwest.

There are many shallow depressions across Estonia. These formed as the ice sheet retreated and water from the melting ice was concentrated into lake basins, which eventually found outlets in the Baltic Sea.

A small delta has formed where the Neman River flows into the protected waters of Courland Lagoon, behind Courland Spit.

Saaremaa Island

Courland Spit

Courland Spit is one of the largest of its kind on the Baltic coast, created by longshore currents moving eastward.

Byelavyezhskaya Pushcha

SCALE 1:2,500,000
(projection: Lambert Conformal Conic)

Km 0 10 20 30 40 50 60 70 80 90 100
Miles 0 10 20 30 40 50 60 70

RUSSIAN FEDERATION

POLAND

UKRAINE

BELARUS

MINSKAYA VOBLASTS'

MAHILYOWSKAYA VOBLASTS'

HOMYEL'SKAYA VOBLASTS'

BRESTSKAYA VOBLASTS'

HRODZYENSKAYA VOBLASTS'

VILNIUS

Mahilyow

Minsk

Homyel'

Babruysk

Hrodna

Brest

Pripet Marshes

THE MEDITERRANEAN

THE MEDITERRANEAN SEA stretches over 2,500 miles (4,000 km) east to west, separating Europe from Africa. At its westernmost point it is connected to the Atlantic Ocean through the Strait of Gibraltar. In the east, the Suez Canal, opened in 1869, gives passage to the Indian Ocean. In the northeast, linked by the Sea of Marmara, lies the Black Sea. The Mediterranean is bordered by 28 states and territories, and more than 100 million people live on its shores and islands. Throughout history, the Mediterranean has been a focal area for many great empires and civilizations, reflected in the variety of cultures found on its shores. Since the 1960s, development along the southern coast of Europe has expanded rapidly to accommodate increasing numbers of tourists and to enable the exploitation of oil and gas reserves. This has resulted in rising levels of pollution, threatening the future of the sea.

USING THE LAND AND SEA

A QUARTER OF THE FISH SPECIES found in the Mediterranean are economically important. Sardines are the main catch in northern and western regions and aquaculture, including oyster farming, is becoming increasingly important in the eastern Mediterranean. Olives, citrus fruit, cork trees, and vines thrive in the Mediterranean climate, enjoying hot, dry summers and mild, wet winters. Italy and Spain are world leaders in commercial olive production.

The growing of citrus fruit such as lemons, limes, oranges, and grapefruit is common along the coasts surrounding the Mediterranean.

Land use and agricultural distribution
- goats
- sheep
- cereals
- citrus fruits
- cork
- fishing
- olives
- sunflowers
- tobacco
- vineyards
- major towns

pasture
cropland
forest
mountain region
wetland
desert

THE LANDSCAPE

THE MEDITERRANEAN SEA IS ALMOST TOTALLY LANDLOCKED, joined to the Atlantic Ocean through the Strait of Gibraltar, which is only 8 miles (13 km) wide. Lying on an active plate margin, sea floor movements have formed a variety of basins, troughs, and ridges. A submarine ridge running from Tunisia to the island of Sicily divides the Mediterranean into two distinct basins. The western basin is characterized by broad, smooth abyssal (or ocean) plains. In contrast, the eastern basin is dominated by a large ridge system, running east to west.

The narrow Strait of Gibraltar inhibits water exchange between the Mediterranean Sea and the Atlantic Ocean, producing a high degree of salinity and a low tidal range within the Mediterranean. The lack of tides has encouraged the build-up of pollutants in many semienclosed bays.

Main surface current

Denser, more saline currents flow back to Atlantic

Dense currents sink below surface

Because the Mediterranean is almost enclosed by land, its circulation is quite different to the oceans. There is one major current which flows in from the Atlantic and moves east. Currents flowing back to the Atlantic are denser and flow below the main current.

The Dalmatian (Dalmacija) coast has many long, elongated islands running parallel to the mainland. These resulted when rising sea levels drowned valleys running parallel with the coast.

TRANSPORTATION & INDUSTRY

THE OPENING OF THE SUEZ CANAL in 1869 made the Mediterranean a key shipping route to Asia. Oil and gas reserves, although comparatively small on a world scale, are being explored and exploited off the coasts of Libya, Greece, Italy, Spain, and Tunisia. The Mediterranean's greatest natural resources are its miles of beaches and warm sea. Over half the world's income from tourism is generated in the Mediterranean.

Benidorm is one of the most popular resorts on Spain's Costa Blanca. Many of the Mediterranean's coastal resorts have grown up since the 1950s, expanding from small fishing villages to large resorts catering almost exclusively for tourists.

The Ionian Basin is the deepest in the Mediterranean, reaching depths of 16,800 ft (5,121 m).

Industrial pollution flowing from the Dnieper and Danube Rivers has destroyed a large proportion of the fish population that used to inhabit the upper layers of the Black Sea.

The eastern basin of the Mediterranean contains many features which indicate the force of a colliding plate margin, including volcanoes, earthquake zones, ridges, and seamounts.

The Atlas Mountains are a range of fold mountains that lie in Morocco and Algeria. They run parallel to the Mediterranean, forming a topographical and climatic divide between the Mediterranean coast and the western Sahara.

The edge of the Eurasian Plate is edged by a continental shelf. In the Mediterranean Sea this is widest at the Ebro Fan where it extends 60 miles (96 km).

Beneath the Strait of Sicily lies a submarine ridge which rises to 1,200 ft (360 m) below sea level. It divides the eastern and western basins of the Mediterranean.

An arc of active submarine, island and mainland volcanoes, including Etna and Vesuvius, lie in and around southern Italy. The area is also susceptible to earthquakes and landslides.

The shallow basin of the Aegean contains numerous small islands, many of volcanic origin.

Nutrient flows into the eastern Mediterranean, and sediment flows to the Nile Delta have been severely lowered by the building of the Aswan Dam across the Nile in Egypt. This is causing the delta to shrink.

THE RUSSIAN FEDERATION

THE COLD WAR ERA OF GLOBAL RELATIONS was concluded in 1991 with the formal dissolution of the Soviet Union. The Russian Federation declared its separate sovereignty from the foundering communist empire following independence declarations from a number of former Soviet republics. As the leading member of the Commonwealth of Independent States, the Russian Federation has a central role in the development of post-Soviet Eurasia. Crossing 11 time zones, the Russian Federation is almost twice the size of the US, and with more than 150 ethnic minorities and 21 autonomous republics, regionalist dissent within its own territory remains a danger.

Summer beds of moss and lichen scatter a 90% surface cover of ice across the islands of Franz Josef Land (Zemlya Frantsa-Iosifa), the northernmost land in the eastern hemisphere.

MAP KEY

POPULATION

- above 5 million
- 1 million to 5 million
- 500,000 to 1 million
- 100,000 to 500,000
- 50,000 to 100,000
- 10,000 to 50,000
- below 10,000

ELEVATION

- 4000m / 13,124ft
- 3000m / 9843ft
- 2000m / 6562ft
- 1000m / 3281ft
- 500m / 1640ft
- 250m / 820ft
- 100m / 328ft
- sea level

USING THE LAND

THE MAIN AGRICULTURAL REGIONS follow the belt of rich, black *chernozem* soils between Ukraine and Novosibirsk, producing cereals, fodder, and a broad range of crops for industrial use. Small pockets of pastureland are also found in this region. Large areas of terrain are uncultivable, and the constraints of a severe climate force the Federation to be partly dependent on imported grain. The wilds of Siberia are given over to hunting and reindeer herding, and contain the world's largest timber reserves.

Land use and agricultural distribution

- cattle
- cereals
- root crops
- timber
- capital cities
- major towns
- pasture
- cropland
- forest
- desert
- mountain region
- barren

THE RUSSIAN FEDERATION: ADMINISTRATIVE REGIONS

① PSKOVSKAYA OBLAST'
② YAROSLAVSKAYA OBLAST'
③ IVANOVSKAYA OBLAST'
④ SMOLENSKAYA OBLAST'
⑤ MOSKOVSKAYA OBLAST'
⑥ VLADIMIRSKAYA OBLAST'
⑦ RESPUBLIKA MARIY EL
⑧ CHUVASHSKAYA RESPUBLIKA
⑨ KALUZHSKAYA OBLAST'
⑩ TUL'SKAYA OBLAST'
⑪ RYAZANSKAYA OBLAST'
⑫ RESPUBLIKA MORDOVIYA
⑬ UL'YANOVSKAYA OBLAST'
⑭ SAMARSKAYA OBLAST'
⑮ BRYANSKAYA OBLAST'
⑯ ORLOVSKAYA OBLAST'

⑰ LIPETSKAYA OBLAST'
⑱ TAMBOVSKAYA OBLAST'
⑲ KURSKAYA OBLAST'
⑳ BELGORODSKAYA OBLAST'
㉑ VORONEZHSKAYA OBLAST'
㉒ KRASNODARSKIY KRAY
㉓ RESPUBLIKA ADYGEYA
㉔ KARACHAYEVO-CHERKESSKAYA RESPUBLIKA
㉕ KABARDINO-BALKARSKAYA RESPUBLIKA
㉖ RESPUBLIKA SEVERNAYA OSETIYA - ALANIYA
㉗ INGUSHSKAYA RESPUBLIKA
㉘ CHECHENSKAYA RESPUBLIKA
㉙ YEVREYSKAYA AVTONOMNAYA OBLAST'

THE URBAN/RURAL POPULATION DIVIDE

urban 76%   rural 24%

0 10 20 30 40 50 60 70 80 90 100

POPULATION DENSITY
22 people per sq mile
(9 people per sq km)

TOTAL LAND AREA
65,592,800 sq miles
(17,075,400 sq km)

RUSSIAN FEDERATION

N   Nn   O   Oo   P   Pp   Q   Qq   R   Rr   S   Ss

SCALE 1:7,500,000
(projection: Lambert Conformal Conic)
Km 0 25 50   100   150   200   250   300
Miles 0 25 50   100   150   200   250   300

St. Peter's Castle at Bodrum in southwestern Turkey is a crusader's castle. It is one of many ancient ruins found along the shores of the Mediterranean, reflecting different civilizations and the strategic importance of many coastal towns.

TURKISH REPUBLIC OF NORTHERN CYPRUS
(recognised only by Turkey)

Zafer Burnu
(Akrotíri Apostólou Andréa)
Yenierenköy (Agialoúsa)
Dipkarpaz (Rizokárpason)
Lapta
Girne (Kerýnia)
Tatlısu (Akanthoú)
Besparmak Dağları (Kyrenia Mountains)
Değirmenlik (Kythréa)
İskele (Trikomon)
Geçitkale (Lefkónikon)
NICOSIA
Ercan
Gazimağusa Körfezi (Kólpos Ammóchostos)
Yenicboğazıçı (Agios Sergios)
Gazimağusa (Ammóchostos, Famagusta)
Vadili (Vatilí)
Akdoğan (Lýsi)
Paralímni
Athiénou
Agia Nápa
Aradippou
Akrotíri Gkréko
Dhekelia
Lárnaka
Lárnaka
Sovereign Base Area (to UK)
Kofinou
Lemesós (Limassol)
Kólpos Akrotíri Gátas

SCALE 1:2,000,000
(projection: Lambert Conformal Conic)
Km 0 5 10 20 30 40 50
Miles 0 5 10 20 30 40 50

TURKEY OCCUPIED the northern part of ... while Greek Cypriots remained in control ... ath. Cyprus was effectively partitioned ... buffer zone currently divides the two ... 1983 the north of the island proclaimed ... Turkish Republic of North Cyprus. ... recognized by Turkey.

Black Sea region

UKRAINE
Mykolayiv
Kherson
Odesa
MOLDOVA
Dniester
Danube
Kakhovs'ka Vodoskhovyshche
Nova Kakhovka
Melitopol'
Mariupol'
Berdyans'k
Yeysk
Gulf of Taganrog
Taganrog
RUSSIAN FEDERATION
Stavropol'
Nevinnomyssk
Krasnodar
Maykop
Cherkessk
Kislovodsk
Pyatigorsk
Nal'chik
Vladikavkaz
El'brus 5642m
Caucasus
K'ut'aisi
GEORGIA
Bat'umi
P'ot'i
Sokhumi
Tuapse
Sochi
Novorossiysk
Kerch Strait
Kerch
Mys Sarych
Sevastopol'
Simferopol'
Yevpatoriya
Kryms'kyy Pivostrov
Mys Tarkhankut
Karkinits'ka Zatoka
Dnieper
Sea of Azov

ROMANIA
Ploiești
BUCUREȘTI (BUCUREST)
Ialomița
Lacul Sinoie
Constanța
Ruse
Danube (Dunav)
Shumen
Pleven
Varna
Nos Kaliakra
BULGARIA
Stara Zagora
Sliven
Burgaski Zaliv
Burgas
Sofiya (SOFIA)
Plovdiv
Balkan Mountains
Rhodope Mountains
Maritsa
Edirne
İgneada Burnu

YUGOSLAVIA
Kruševac
Niš
Leskovac
Priština
Uroševac
SKOPJE
MACEDONIA
Bitola
Lake Prespa

BLACK SEA

Kerempe Burnu
Sinop Burnu
Sinop
Bafra Burnu
Civa Burnu
Yasun Burnu
Fener Burnu
Trabzon
Samsun
Ordu
Zonguldak
Kızıl Irmak
Yeşil Irmak
Kelkit Çayı
Erzurum
Sivas

İstanbul
Marmara Denizi (Sea of Marmara)
İzmit
Adapazarı
Köroğlu Dağları
ANKARA
Çorum
Bursa
İznik Gölü
Bandırma
Eskişehir
Kırıkkale
Gelibolu
Gökçeada
Kütahya
Polatlı
Çanakkale
Edremit
Balıkesir
Kayseri
Karakaya Barajı
Malatya

TURKEY

GREECE
Thássos
Samothráki
Thracian Sea
Strymonikós Kólpos
Thessaloníki (Salonica)
Thermaikós Kólpos
Lárisa
Stereá
Límnos
Lésvos (Lesbos)
Manisa
Gediz Nehri
Uşak
İzmir
Aydın
Denizli
Işparta
Konya
Eber Gölü
Akşehir Gölü
Tuz Gölü
Eğridir Gölü
Beyşehir Gölü
Seyhan Nehri
Ceyhan Nehri
Gaziantep
Osmaniye
Adana
İskenderun
Mersin
Halab (Aleppo)
Antakya
SYRIA
Ḥamāh
Ḥimṣ
Al Lādhiqīyah (Latakia)

Aegean Sea
Vóreioi Sporádes
Évvoia (Euboea)
Lamía
Pátra
Peiraiás (Piraeus)
ATHÍNA (ATHENS)
Kórinthos (Corinth)
Spárti
Chíos
Sámos
Kafiréas
Kykládes (Cyclades)
Náxos
Dodekánisa (Dodecanese)
Kos
Mírtoo Pélagos
Ródos (Rhodes)
Rhodes Basin
Prasonísi Basin
Akrotírio Prasonísi
Stenó Karpáthou
Kárpathos
Stenó Kásou
Kritikó Pélagos (Sea of Crete)
Kríti (Crete)
Akrotírio Lithíno
Irákleio
Akrotírio Spátha
Cretan Trough
Akrotírio Pláka
Akrotírio Kastállou
Akrotírio Krios
Mediterranean Ridge
Pliny Trench
Ptolemy Seamounts

Büyükmenderes Nehri
Acı Göl
Burdur Gölü
Bodrum
Antalya
Finike
Yardımcı Burnu
Antalya Körfezi
Antalya Basin
Cilicia Trough
İskenderun Körfezi
Göksu Nehri

TURKISH REPUBLIC OF NORTHERN CYPRUS (recognised only by Turkey)
Girne (Keryneia)
NICOSIA
Gazimağusa (Ammóchostos, Famagusta)
CYPRUS
Ólympos 1951m
Lárnaka
Lemesós (Limassol)
Cyprus Basin
Tripoli

MEDITERRANEAN SEA
Herodotus Trough
Al Bayḍā'
Ra's al Hilāl
Darnah
Jabal al Akhḍar
Khalīj al Bumbah
Ṭubruq
Ra's al Tīn
Ra's al Muraysah
Gulf of Salūm
Ra's 'Alam el Rūm
Ra's el-Kenāyis
Khalīg el 'Arab
El'Alamein
Libyan Plateau
Herodotus Basin
Levantine Basin
Eratosthenes Tablemount
Cyprus Basin

LEBANON
BEYROUTH (BEIRUT)
DIMASHQ (DAMASCUS)
Lake Tiberias
Hefa
Irbid
Az Zarqa'
AMMAN
WEST BANK
Petah Tiqwa
Tel Aviv-Yafo
JERUSALEM
Gaza
GAZA STRIP
Be'er Sheva
ISRAEL
Dead Sea
JORDAN
Jordan
Wadi el 'Araba

Nile Fan
Alexandria
Rashid (Rosetta)
Dumyât (Damietta)
Port Said
Nile Delta
Kafr el Sheikh
Damanhûr
El Mansûra
El Mahalla el Kubra
Ismâ'îliya
Tanta
Zagazig
Shibîn el Kôm
Benha
Suez
Shubrâ el Kheima
CAIRO
El Gîza
Helwân
El'Alamein
Monkhafad el Qattâra (Qattara Depression)
Siwa
EGYPT
El Faiyûm
Beni Suef
El Minya
Nile
Sinai
Gebel Mûsa 2285m
Gulf of Suez
Gulf of Aqaba
Elat
Al 'Aqabah
SAUDI ARABIA
El Tûr
Red Sea

MAP KEY

POPULATION
- ■ above 5 million
- ■ 1 million to 5 million
- ● 500,000 to 1 million
- ◉ 100,000 to 500,000
- ⊕ 50,000 to 100,000
- ○ 10,000 to 50,000
- ○ below 10,000

ELEVATION
- 4000m / 13,124ft
- 3000m / 9843ft
- 2000m / 6562ft
- 1000m / 3281ft
- 500m / 1640ft
- 250m / 820ft
- 100m / 328ft
- sea level

SEA DEPTH
- sea level
- 250m / 820ft
- 500m / 1640ft
- 1000m / 3281ft
- 2000m / 6562ft
- 3000m / 9843ft

The Suez Canal links the Mediterranean with the Red Sea providing an important shipping route between Europe and Asia.

Beirut is Lebanon's largest city. In the 1960s and 70s it was the chief financial, commercial, and transportation center for the Arab states. In 1975 civil war broke out. Rebuilding is under way, however many buildings bear the scars of the war, which only ended in 1990.

Major industry and infrastructure
- fishing port
- oil & gas
- tourism
- major towns
- international airports
- major roads
- major industrial areas

Monte Carlo is just one of the luxurious resorts scattered along the Riviera, which stretches along the coast from Cannes in France to La Spezia in Italy. The region's mild winters and hot summers have attracted wealthy tourists since the early 19th century.

CYPRUS

In 1974 Cyprus ...

Oxygen in the Black Sea is dissolved only in its upper layers; at depths below 230–300 ft (70–100 m) the sea is "dead" and can support no lifeforms other than specially-adapted bacteria.

The city of Venice is built on an archipelago of islands and mud-flats in the middle of a lagoon at the head of the Adriatic Sea. The city's numerous canals follow water routes between the original 118 islands.

Cyprus is the third largest Mediterranean island after Sardinia and Sicily. The island is mountainous; containing two main ranges, the Troodos and the Kyrenia mountains.

The Suez Canal, opened in 1869, extends 100 miles (160 km) from Port Said to the Gulf of Suez.

Both the Dead Sea in Jordan and the Gulf of Aqaba are extensions of the Great Rift Valley which runs through eastern Africa.

MALTA

SCALE 1:900,000
(projection: Lambert Conformal Conic)

0 10 20 Km
0 10 20 Miles

Commercial fisheries are found throughout the Mediterranean. Operations have traditionally been small-scale. As elsewhere, high demand has caused a decline in fish stocks.

A fishing trawler lies at anchor in the icy waters of Karaginskiy Zaliv, at the northern end of the Kamchatka Peninsula (Poluostrov Kamchatka) in eastern Siberia. The Russian Federation's fishing fleet is the largest in the world and operates worldwide.

The shores of Lake Baikal (Ozero Baykal) are a mixture of forest and the grassy steppe seen here. The lake freezes to a depth of 33 ft (10 m) in winter.

SCALE 1:13,800,000
(projection: Lambert Conformal Conic)

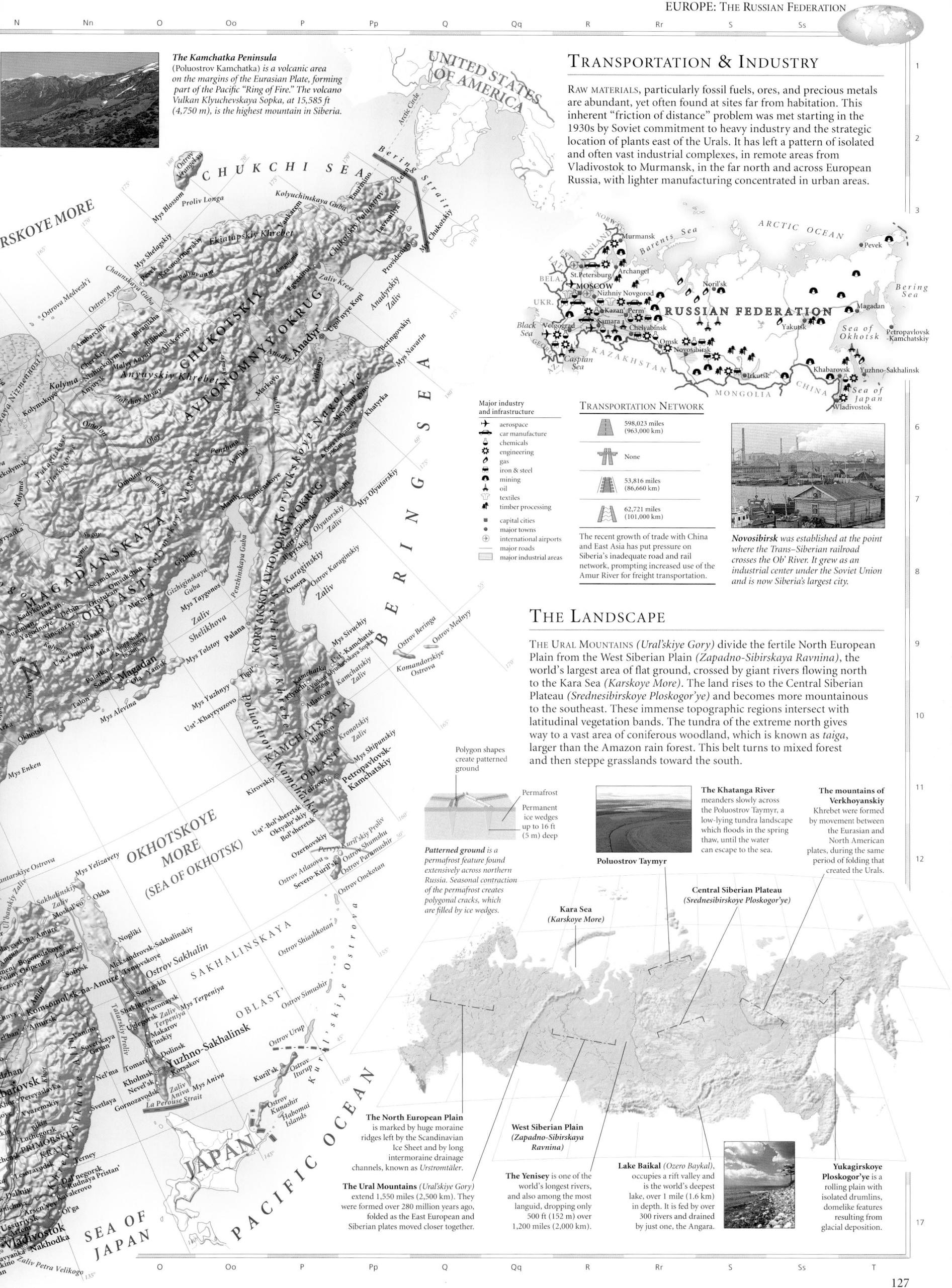

The Kamchatka Peninsula
(Poluostrov Kamchatka) *is a volcanic area on the margins of the Eurasian Plate, forming part of the Pacific "Ring of Fire." The volcano Vulkan Klyuchevskaya Sopka, at 15,585 ft (4,750 m), is the highest mountain in Siberia.*

TRANSPORTATION & INDUSTRY

RAW MATERIALS, particularly fossil fuels, ores, and precious metals are abundant, yet often found at sites far from habitation. This inherent "friction of distance" problem was met starting in the 1930s by Soviet commitment to heavy industry and the strategic location of plants east of the Urals. It has left a pattern of isolated and often vast industrial complexes, in remote areas from Vladivostok to Murmansk, in the far north and across European Russia, with lighter manufacturing concentrated in urban areas.

Major industry and infrastructure

- aerospace
- car manufacture
- chemicals
- engineering
- gas
- iron & steel
- mining
- oil
- textiles
- timber processing
- capital cities
- major towns
- international airports
- major roads
- major industrial areas

TRANSPORTATION NETWORK

598,023 miles (963,000 km)	
None	
53,816 miles (86,660 km)	
62,721 miles (101,000 km)	

The recent growth of trade with China and East Asia has put pressure on Siberia's inadequate road and rail network, prompting increased use of the Amur River for freight transportation.

Novosibirsk was established at the point where the Trans–Siberian railroad crosses the Ob' River. It grew as an industrial center under the Soviet Union and is now Siberia's largest city.

THE LANDSCAPE

THE URAL MOUNTAINS (Ural'skiye Gory) divide the fertile North European Plain from the West Siberian Plain (Zapadno-Sibirskaya Ravnina), the world's largest area of flat ground, crossed by giant rivers flowing north to the Kara Sea (Karskoye More). The land rises to the Central Siberian Plateau (Srednesibirskoye Ploskogor'ye) and becomes more mountainous to the southeast. These immense topographic regions intersect with latitudinal vegetation bands. The tundra of the extreme north gives way to a vast area of coniferous woodland, which is known as *taiga*, larger than the Amazon rain forest. This belt turns to mixed forest and then steppe grasslands toward the south.

Polygon shapes create patterned ground

Permafrost
Permanent ice wedges up to 16 ft (5 m) deep

Patterned ground is a permafrost feature found extensively across northern Russia. Seasonal contraction of the permafrost creates polygonal cracks, which are filled by ice wedges.

The Khatanga River meanders slowly across the Poluostrov Taymyr, a low-lying tundra landscape which floods in the spring thaw, until the water can escape to the sea.

Poluostrov Taymyr

The mountains of Verkhoyanskiy Khrebet were formed by movement between the Eurasian and North American plates, during the same period of folding that created the Urals.

Kara Sea (Karskoye More)

Central Siberian Plateau (Srednesibirskoye Ploskogor'ye)

The North European Plain is marked by huge moraine ridges left by the Scandinavian Ice Sheet and by long intermoraine drainage channels, known as *Urstromtäler.*

West Siberian Plain (Zapadno-Sibirskaya Ravnina)

The Ural Mountains (Ural'skiye Gory) extend 1,550 miles (2,500 km). They were formed over 280 million years ago, folded as the East European and Siberian plates moved closer together.

The Yenisey is one of the world's longest rivers, and also among the most languid, dropping only 500 ft (152 m) over 1,200 miles (2,000 km).

Lake Baikal (Ozero Baykal), occupies a rift valley and is the world's deepest lake, over 1 mile (1.6 km) in depth. It is fed by over 300 rivers and drained by just one, the Angara.

Yukagirskoye Ploskogor'ye is a rolling plain with isolated drumlins, domelike features resulting from glacial deposition.

NORTHERN EUROPEAN RUSSIA

Reaching into the Arctic Circle, this region of lakeland, forest, and tundra is historically bound to Europe by St. Petersburg, the old imperial capital of Tsarist Russia and home to a third of the region's population. Communist rule from Moscow left the north politically marginalized, contributing to the present problems of outmoded industry, poor infrastructure, and serious environmental neglect. However, with borders embracing Finland, Norway, the Baltic, and the northern sea route to the Atlantic, the region's success in foreign trade is now of prime importance to the Russian economy.

St. Peter and Paul Fortress is the oldest building in St. Petersburg, founded by Peter the Great in 1703 as a modern, European capital for Russia.

THE LANDSCAPE

The ancient bedrock of the Scandinavian Shield lies exposed across the glacially scoured Khibiny Mountains of the Kola Peninsula *(Kol'skiy Poluostrov)*, becoming mantled with till toward the North European Plain. The Valdai Hills *(Valdayskaya Vozvyshennost')* form an important watershed for the plain's rivers, while thick forest veils a complicated topography of moraines, lakes, and ground disturbed by frost action. The Ural Mountains *(Ural'skiye Gory)* form a border with Asia in the east.

The Kola Peninsula (Kol'skiy Poluostrov) *is part of the Scandinavian Shield, an area of ancient bedrock underlying Scandinavia. Rocks in excess of 2,500 million years old are exposed across the peninsula.*

The Khibiny Mountains were formed by volcanic intrusions into the Scandinavian Shield, over 570 million years ago.

Kola Peninsula
(Kol'skiy Poluostrov)

Karst features, including sinkholes, lakes, and caverns, are found in limestone outcrops across the plain of the Severnaya Dvina and Mezen' Rivers.

The low-lying plains of the Pechora, Mezen', and Severnaya Dvina Rivers were flooded by the sea while the land was still isostatically depressed following the last Ice Age, a process which has hidden the landforms created by glacial deposition.

Retreating glacier
Meltwater channels
Terminal moraine

Terminal moraines are crescent-shaped ridges of glacial deposits, widely found in central Russia. Detritus is carried by the glacier and deposited at its terminus (snout) as it melts, marking the limit of the ice advance.

Ural Mountains
(Ural'skiye Gory)

Lake Onega (Onezhskoye Ozero) *is the remnant of a body of water which, 12,000 years ago, connected the White Sea (Beloye More) with the Gulf of Finland and the Baltic Sea.*

Two of Europe's biggest rivers, the Volga and Western Dvina, rise in the swampy uplands of the Valdai Hills *(Valdayskaya Vozvyshennost')*.

USING THE LAND AND SEA

The cold climate confines agriculture mainly to southern and western provinces, where dairy farming predominates and arable land is given over to fodder crops as well as flax, potatoes, oats, and rye. Areas beyond the northern margins of cultivation are used for forestry, hunting, herding, and fishing, with some vegetables grown in hothouses around urban areas.

Land use and agricultural distribution
- cattle
- fishing
- reindeer
- timber
- fodder
- major towns
- pasture
- cropland
- forest
- mountain region
- wetland
- tundra
- barren
- ice

THE URBAN/RURAL POPULATION DIVIDE

urban 74% rural 26%

0 10 20 30 40 50 60 70 80 90 100

POPULATION DENSITY
26 people per sq mile
10 people per sq km

TOTAL LAND AREA
829,398 sq miles
(2,148,700 sq km)

Many rapids are found along the 175 mile (280 km) course of the Suna River.

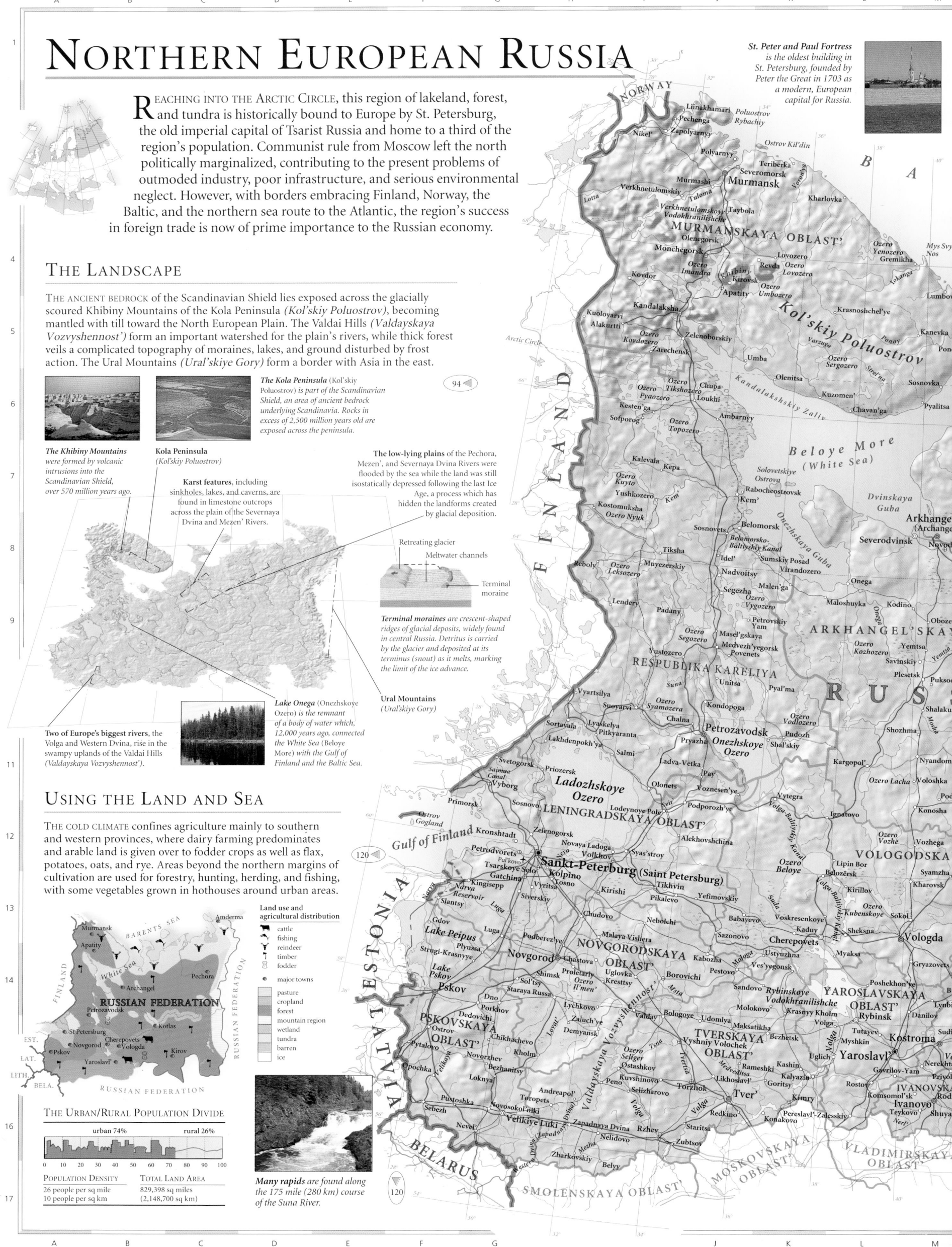

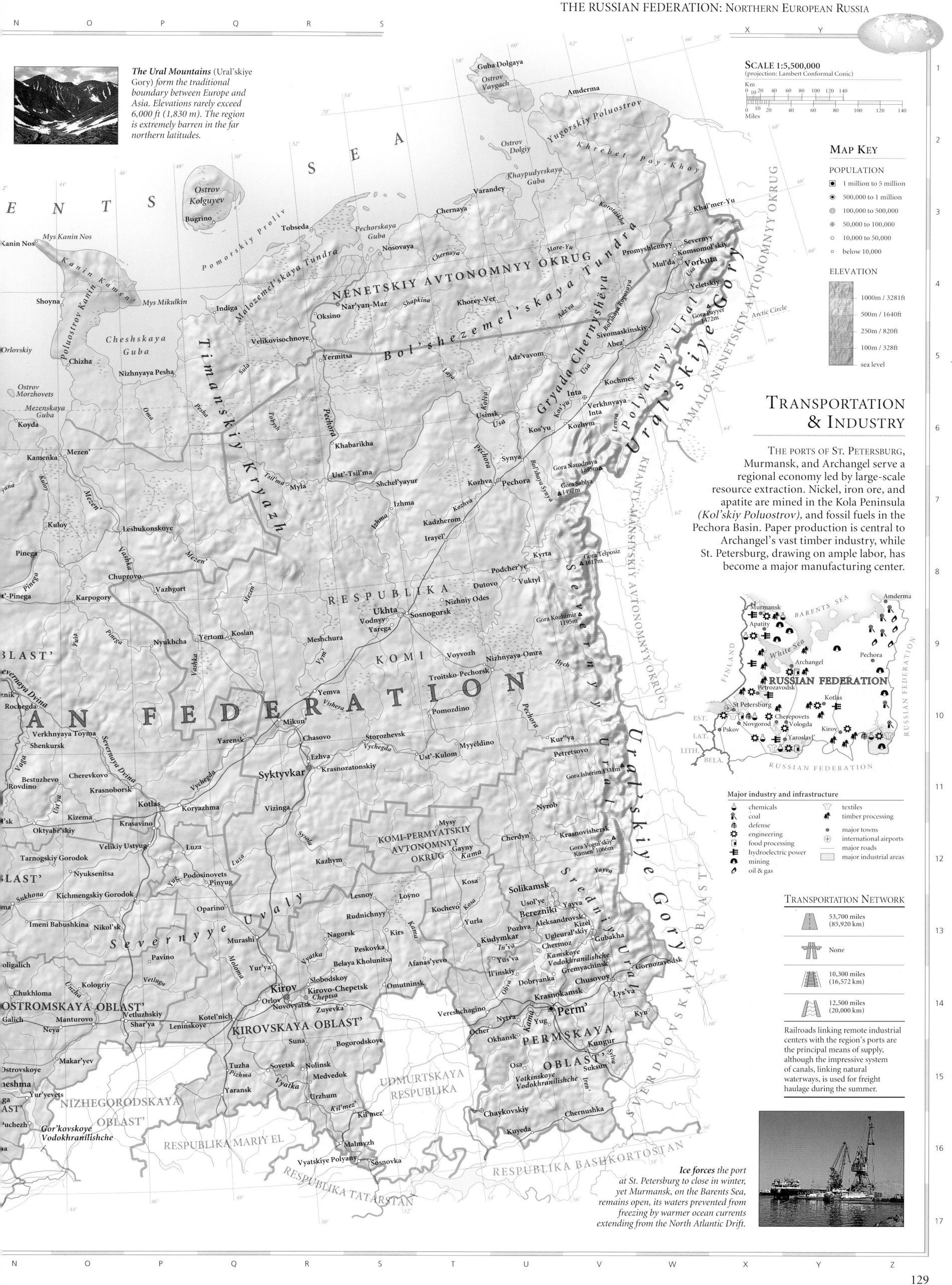

The Ural Mountains (Ural'skiye Gory) *form the traditional boundary between Europe and Asia. Elevations rarely exceed 6,000 ft (1,830 m). The region is extremely barren in the far northern latitudes.*

SCALE 1:5,500,000
(projection: Lambert Conformal Conic)

Km
Miles

MAP KEY

POPULATION

- 1 million to 5 million
- 500,000 to 1 million
- 100,000 to 500,000
- 50,000 to 100,000
- 10,000 to 50,000
- below 10,000

ELEVATION

- 1000m / 3281ft
- 500m / 1640ft
- 250m / 820ft
- 100m / 328ft
- sea level

TRANSPORTATION & INDUSTRY

THE PORTS OF ST. PETERSBURG, Murmansk, and Archangel serve a regional economy led by large-scale resource extraction. Nickel, iron ore, and apatite are mined in the Kola Peninsula (Kol'skiy Poluostrov), and fossil fuels in the Pechora Basin. Paper production is central to Archangel's vast timber industry, while St. Petersburg, drawing on ample labor, has become a major manufacturing center.

Major industry and infrastructure

- chemicals
- coal
- defense
- engineering
- food processing
- hydroelectric power
- mining
- oil & gas
- textiles
- timber processing
- major towns
- international airports
- major roads
- major industrial areas

TRANSPORTATION NETWORK

- 53,700 miles (85,920 km)
- None
- 10,300 miles (16,572 km)
- 12,500 miles (20,000 km)

Railroads linking remote industrial centers with the region's ports are the principal means of supply, although the impressive system of canals, linking natural waterways, is used for freight haulage during the summer.

Ice forces the port at St. Petersburg to close in winter, yet Murmansk, on the Barents Sea, remains open, its waters prevented from freezing by warmer ocean currents extending from the North Atlantic Drift.

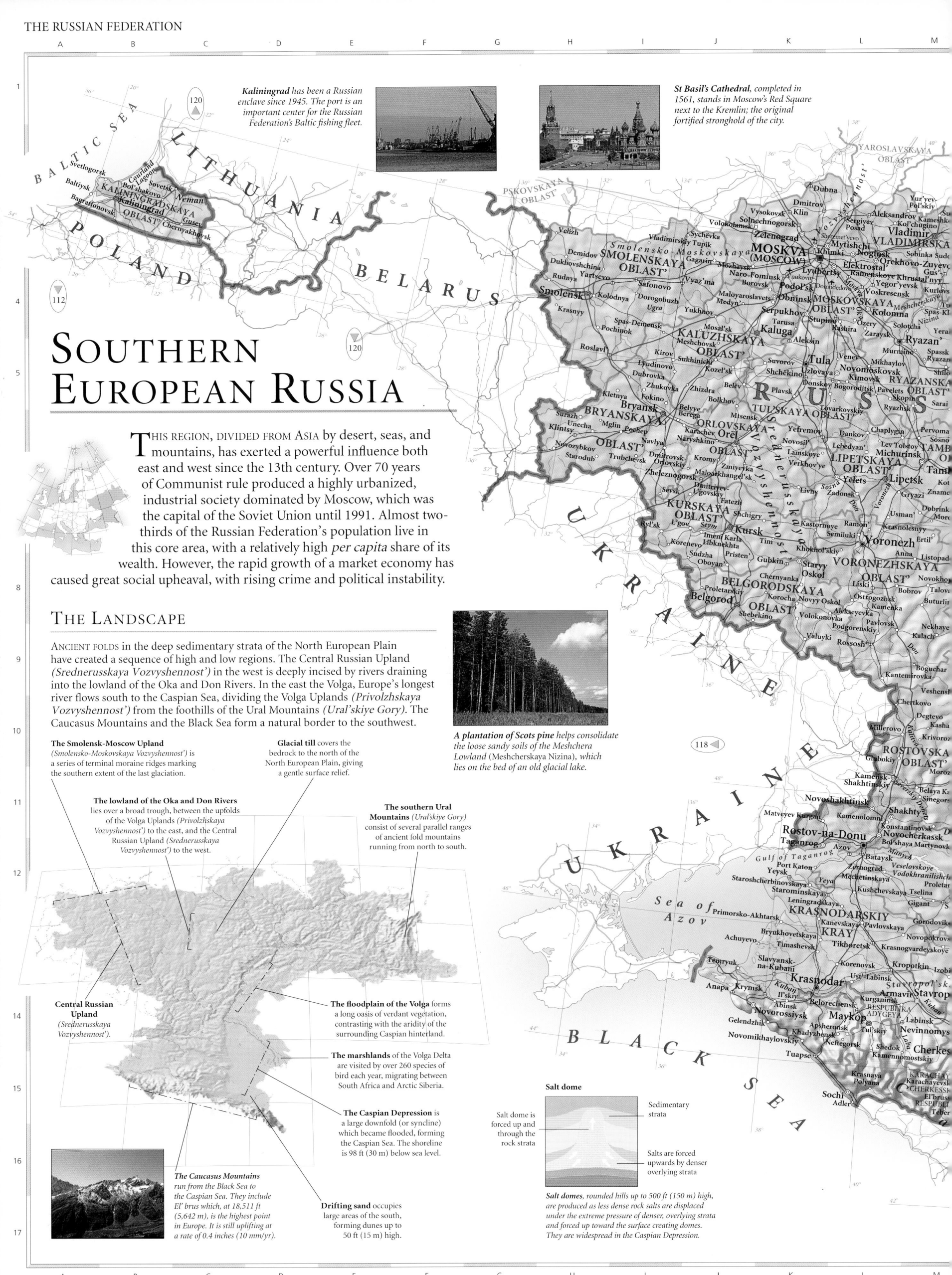

Kaliningrad has been a Russian enclave since 1945. The port is an important center for the Russian Federation's Baltic fishing fleet.

St Basil's Cathedral, completed in 1561, stands in Moscow's Red Square next to the Kremlin; the original fortified stronghold of the city.

SOUTHERN EUROPEAN RUSSIA

THIS REGION, DIVIDED FROM ASIA by desert, seas, and mountains, has exerted a powerful influence both east and west since the 13th century. Over 70 years of Communist rule produced a highly urbanized, industrial society dominated by Moscow, which was the capital of the Soviet Union until 1991. Almost two-thirds of the Russian Federation's population live in this core area, with a relatively high *per capita* share of its wealth. However, the rapid growth of a market economy has caused great social upheaval, with rising crime and political instability.

THE LANDSCAPE

ANCIENT FOLDS in the deep sedimentary strata of the North European Plain have created a sequence of high and low regions. The Central Russian Upland (*Srednerusskaya Vozvyshennost'*) in the west is deeply incised by rivers draining into the lowland of the Oka and Don Rivers. In the east the Volga, Europe's longest river flows south to the Caspian Sea, dividing the Volga Uplands (*Privolzhskaya Vozvyshennost'*) from the foothills of the Ural Mountains (*Ural'skiye Gory*). The Caucasus Mountains and the Black Sea form a natural border to the southwest.

A plantation of **Scots pine** helps consolidate the loose sandy soils of the Meshchera Lowland (Meshcherskaya Nizina), which lies on the bed of an old glacial lake.

The Smolensk-Moscow Upland (*Smolensko-Moskovskaya Vozvyshennost'*) is a series of terminal moraine ridges marking the southern extent of the last glaciation.

Glacial till covers the bedrock to the north of the North European Plain, giving a gentle surface relief.

The lowland of the Oka and Don Rivers lies over a broad trough, between the upfolds of the Volga Uplands (*Privolzhskaya Vozvyshennost'*) to the east, and the Central Russian Upland (*Srednerusskaya Vozvyshennost'*) to the west.

The southern Ural Mountains (*Ural'skiye Gory*) consist of several parallel ranges of ancient fold mountains running from north to south.

Central Russian Upland (*Srednerusskaya Vozvyshennost'*).

The floodplain of the Volga forms a long oasis of verdant vegetation, contrasting with the aridity of the surrounding Caspian hinterland.

The marshlands of the Volga Delta are visited by over 260 species of bird each year, migrating between South Africa and Arctic Siberia.

The Caspian Depression is a large downfold (or syncline) which became flooded, forming the Caspian Sea. The shoreline is 98 ft (30 m) below sea level.

The Caucasus Mountains run from the Black Sea to the Caspian Sea. They include El' brus which, at 18,511 ft (5,642 m), is the highest point in Europe. It is still uplifting at a rate of 0.4 inches (10 mm/yr).

Drifting sand occupies large areas of the south, forming dunes up to 50 ft (15 m) high.

Salt dome

Salt dome is forced up and through the rock strata

Sedimentary strata

Salts are forced upwards by denser overlying strata

Salt domes, rounded hills up to 500 ft (150 m) high, are produced as less dense rock salts are displaced under the extreme pressure of denser, overlying strata and forced up toward the surface creating domes. They are widespread in the Caspian Depression.

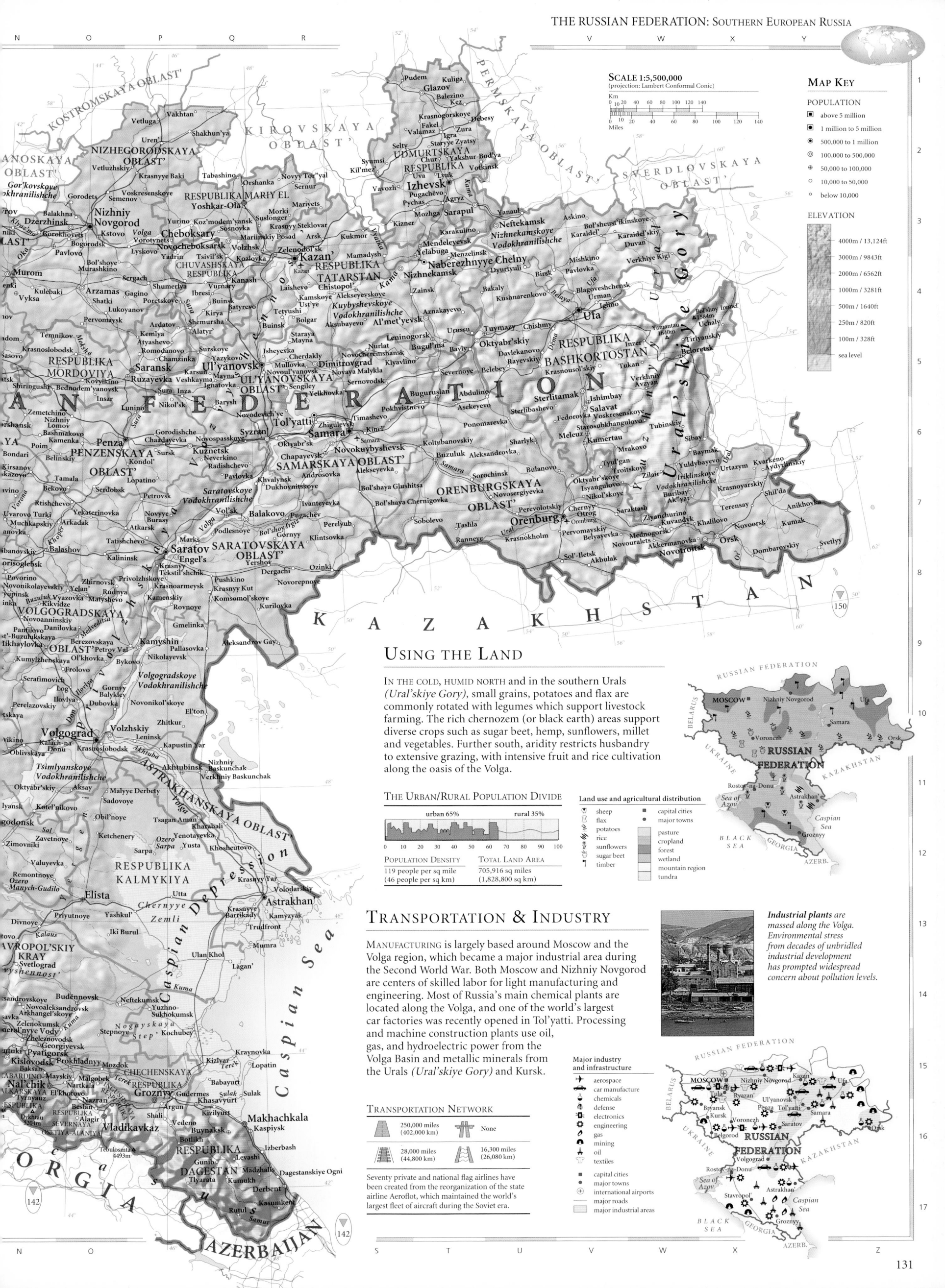

SCALE 1:5,500,000
(projection: Lambert Conformal Conic)

MAP KEY

POPULATION
- ■ above 5 million
- ■ 1 million to 5 million
- ◉ 500,000 to 1 million
- ◉ 100,000 to 500,000
- ⊕ 50,000 to 100,000
- ○ 10,000 to 50,000
- ○ below 10,000

ELEVATION
- 4000m / 13,124ft
- 3000m / 9843ft
- 2000m / 6562ft
- 1000m / 3281ft
- 500m / 1640ft
- 250m / 820ft
- 100m / 328ft
- sea level

USING THE LAND

IN THE COLD, HUMID NORTH and in the southern Urals (Ural'skiye Gory), small grains, potatoes and flax are commonly rotated with legumes which support livestock farming. The rich chernozem (or black earth) areas support diverse crops such as sugar beet, hemp, sunflowers, millet and vegetables. Further south, aridity restricts husbandry to extensive grazing, with intensive fruit and rice cultivation along the oasis of the Volga.

THE URBAN/RURAL POPULATION DIVIDE

urban 65%   rural 35%

0 10 20 30 40 50 60 70 80 90 100

POPULATION DENSITY
119 people per sq mile
(46 people per sq km)

TOTAL LAND AREA
705,916 sq miles
(1,828,800 sq km)

Land use and agricultural distribution
- sheep
- flax
- potatoes
- rice
- sunflowers
- sugar beet
- timber
- ● capital cities
- ● major towns
- pasture
- cropland
- forest
- wetland
- mountain region
- tundra

TRANSPORTATION & INDUSTRY

MANUFACTURING is largely based around Moscow and the Volga region, which became a major industrial area during the Second World War. Both Moscow and Nizhniy Novgorod are centers of skilled labor for light manufacturing and engineering. Most of Russia's main chemical plants are located along the Volga, and one of the world's largest car factories was recently opened in Tol'yatti. Processing and machine construction plants use oil, gas, and hydroelectric power from the Volga Basin and metallic minerals from the Urals (Ural'skiye Gory) and Kursk.

Industrial plants are massed along the Volga. Environmental stress from decades of unbridled industrial development has prompted widespread concern about pollution levels.

TRANSPORTATION NETWORK

- 250,000 miles (402,000 km)
- None
- 28,000 miles (44,800 km)
- 16,300 miles (26,080 km)

Seventy private and national flag airlines have been created from the reorganization of the state airline Aeroflot, which maintained the world's largest fleet of aircraft during the Soviet era.

Major industry and infrastructure
- ✈ aerospace
- 🚗 car manufacture
- chemicals
- defense
- electronics
- ⚙ engineering
- gas
- mining
- oil
- textiles
- ■ capital cities
- ● major towns
- ✈ international airports
- major roads
- major industrial areas

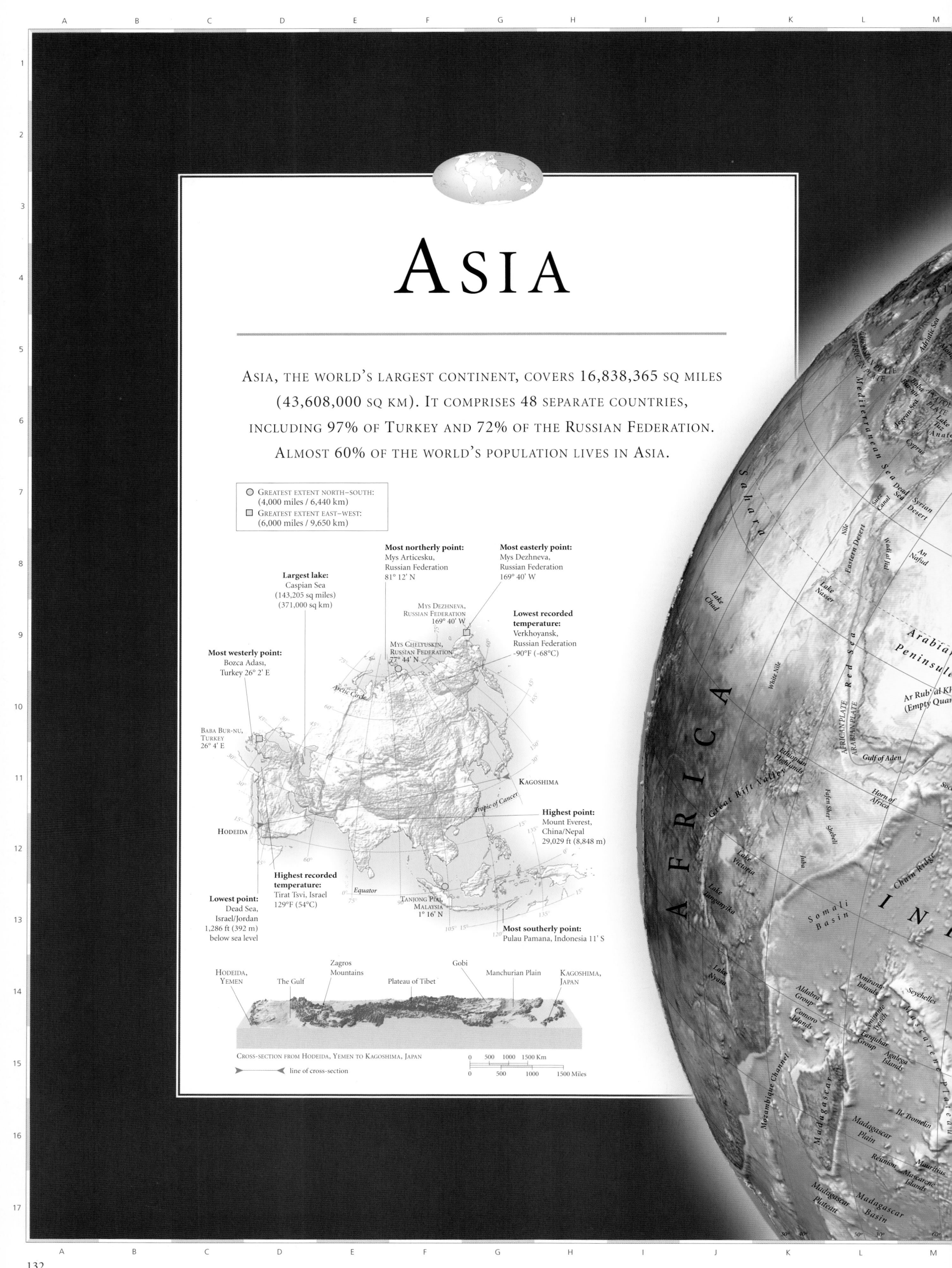

ASIA

ASIA, THE WORLD'S LARGEST CONTINENT, COVERS 16,838,365 SQ MILES (43,608,000 SQ KM). IT COMPRISES 48 SEPARATE COUNTRIES, INCLUDING 97% OF TURKEY AND 72% OF THE RUSSIAN FEDERATION. ALMOST 60% OF THE WORLD'S POPULATION LIVES IN ASIA.

○ GREATEST EXTENT NORTH–SOUTH:
(4,000 miles / 6,440 km)
□ GREATEST EXTENT EAST–WEST:
(6,000 miles / 9,650 km)

Most northerly point:
Mys Articesku,
Russian Federation
81° 12' N

Most easterly point:
Mys Dezhneva,
Russian Federation
169° 40' W

Largest lake:
Caspian Sea
(143,205 sq miles)
(371,000 sq km)

MYS DEZHNEVA,
RUSSIAN FEDERATION
169° 40' W

Lowest recorded temperature:
Verkhoyansk,
Russian Federation
-90°F (-68°C)

MYS CHELYUSKIN,
RUSSIAN FEDERATION
77° 44' N

Most westerly point:
Bozca Adası,
Turkey 26° 2' E

BABA BUR-NU,
TURKEY
26° 4' E

Arctic Circle

KAGOSHIMA

Tropic of Cancer

Highest point:
Mount Everest,
China/Nepal
29,029 ft (8,848 m)

HODEIDA

Highest recorded temperature:
Tirat Tsvi, Israel
129°F (54°C)

Equator

Lowest point:
Dead Sea,
Israel/Jordan
1,286 ft (392 m)
below sea level

TANJONG PIAI
MALAYSIA
1° 16' N

Most southerly point:
Pulau Pamana, Indonesia 11' S

HODEIDA,
YEMEN

The Gulf

Zagros
Mountains

Plateau of Tibet

Gobi

Manchurian Plain

KAGOSHIMA,
JAPAN

CROSS-SECTION FROM HODEIDA, YEMEN TO KAGOSHIMA, JAPAN

line of cross-section

0   500   1000 1500 Km

0   500   1000   1500 Miles

ASIA

EUROPE

ASIA

Norwegian Sea
Scandinavia
North Sea
Gulf of Bothnia
North Cape
Barents Sea
Kola Peninsula
Novaya Zemlya
Kara Sea
Severnaya Zemlya
Mys Chelyuskin
Laptev Sea
New Siberia Islands
East Siberian Sea
Wrangel Island
Long Strait
Bering Strait
Chukot Range
Bering Sea
Aleutian Basin
Baltic Sea
Gulf of Finland
Lake Ladoga
Lake Onega
White Sea
Poluostrov Kanin
Pechora
Poluostrov Yamal
Ob
Gydanskiy Poluostrov
Poluostrov Taymyr
North Siberian Lowland
Kheta
Kolyma
Khrebet Cherskogo
Koryak Range
Kolyma Range
North European Plain
Central Russian Upland
Ural Mountains
West Siberian Plain
Central Siberian Plateau
Verkhoyanskiy Khrebet
Indigirka
Kamchatka

Rhine
Dnieper
Dvina
Don
Khopёr
Volga
Desna
Oka
Kama
Belaya
Kazym
Nadym
Pur
Taz
Yenisey
Podkamennaya Tunguska
Lower Tunguska
Olenёk
Lena
Markha
Vilyuy
Aldan
Uchur
Maya
Sea of Okhotsk

Black Sea
Sea of Azov
Caspian Depression
Emba
Ozero Tengiz
Nura
Om
Lake Chany
Ob
Tomь
Angara
Buryatya
Lake Baikal
Vitim
Shilka
Amur
Stanovoy Khrebet
Zeya Reservoir
Karile Trench

Caucasus
Elbrus
Caspian Sea
Ustyurt Plateau
Aral Sea
Syr Darya
Kirghiz Steppe
Sarysu
Lake Balkhash
Ili
Sayanskiy Khrebet
Uvs Nuur
Har Us Nuur
Hyargas Nuur
Dzavhan Gol
Hövsgöl Nuur
Selenga
Orhon Gol
Chikoy
Sablonovyy Khrebet
Hulun Nur
Kerulen
Great Khingan Range
Manchurian Plain
Lake Khanka
Hokkaido

Lake Van
Lake Urmia
Turan Lowland
Amu Darya
Aydarkul
Chu
Naryn
Ozero Issyk-Kul
Ozero Alakol
Dzungaria
Altai Mountains
Gobi
Plateau of Mongolia
Xar Moron
Sea of Japan

Zagros Mountains
Elburz Mountains
Great Salt Desert
EURASIAN PLATE
IRANIAN PLATE
Pontus
Tien Shan
Tarim He
Tarim Basin
Konqi He
Shule He
Lop Nur
Ruo He
Yellow River
Yongding He
Ordos Desert
Wutai Shan
Bo Hai
Korea Bay

Iranian Plateau
Kuh-e-Baba 5143m
Hindu Kush
Karakoram Range 8611m
Takla Makan Desert
Keriya He
Qarqan He
Altun Shan
Kunlun Mountains
Nan Shan
Qilian Shan
Qinghai Hu
Xining Shan
Wei He
Taihang Shan
Yellow River
Huai He
Yellow Sea
Cheju-do

Rigestan
Indus
Jhelum
Plateau of Tibet
Dogai Coring
Bayan Har Shan
Min Shan
Huai He
East China Sea

Hamun
Jaz Murian
Sulaiman Range
Punjab Plains
Sutlej
Himalayas
Siling Co
Tangra Yumco
Nam Co
Xiqing Shan
Hong Hu
Dongting Hu
Tai Hu

Central Makran Range
Oman Basin
Indus
Thar Desert
Luni
Annapurna 8091m
Mount Everest 8848m
Kanchenjunga
Brahmaputra
Yangtze
Yuan Jiang
Lang Jiang
Xi Jiang
Taiwan

Strait of Hormuz
Gulf of Oman
IRANIAN PLATE
ARABIAN PLATE
INDO-AUSTRALIAN PLATE
Gulf of Kachchh
Sabarmati
Mahi
Chambal
Betwa
Yamuna
Ganges
Ghaghara
Damodar
Khasi Hills
Chindwin
Red River
Black River
Dongnan Qiuling

Arabian Sea
Arabian Basin
Gulf of Khambhat
Narmada
Vindhya Range
Satpura Range
Son
Ganges
Brahmani
Irrawaddy
Salween
Mekong
Gulf of Tongking
Hainan
Hainan Strait
Luzon Strait
Philippine Sea

Ajanta Range
Godavari
Mahanadi
Mouths of the Ganges
Arakan Yoma
Sittang
Nam Ma
Luzon
Philippine Basin

Deccan
Bhima
Krishna
Indravati
Godavari
Gulf of Martaban
Chi
Mun
Trung Phan
Mindoro

Laccadive Islands
Western Ghats
Tungabhadra
Eastern Ghats
Bay of Bengal
Chao Phraya
Tônlé Sap
Xe Bang Fai
South China Sea
Panay
Philippines

Malabar Coast
Kaveri
Penner
Coromandel Coast
Andaman Islands
Gulf of Thailand
South China Basin
Negros
Palawan

Cape Comorin
Gulf of Mannar
Sri Lanka
Andaman Sea
Isthmus of Kra
Mouths of the Mekong
Sulu Sea
Mindanao

Maldives
Ceylon Plain
Nicobar Islands
EURASIAN PLATE
INDO-AUSTRALIAN PLATE
Malay Peninsula
Anambas Islands
Sunda Natuna Islands Shelf
Gunung Kinabalu 4104m
Celebes Sea

Chagos-Laccadive Plateau
INDO-AUSTRALIAN PLATE
AFRICAN PLATE
Nikitin Seamount
Danau Toba
Strait of Malacca
Sumatra
Tanjong Piai
Borneo
Kapuas Sungai
Molucca Sea
Halmahera
Buru
Seram
New Guinea

INDIAN OCEAN

Mid-Indian Ridge
Chagos Bank
Chagos Trench
Ninetyeast Ridge
Investigator Ridge
Cocos Basin
Pulau Bangka
Greater Sunda Islands
Celebes
Banda Sea

Mid-Indian Basin
Gunung Kerinci 3806m
Mentawai Ridge
Selat Sunda
Java Sea
Flores Sea
Lesser Sunda Islands
Arafura Sea

Tropic of Capricorn
Cocos Islands
Christmas Island
Java Trench
Sunda Trough
Bali
Sumba Islands
Timor
Timor Trough

PACIFIC OCEAN
Shikotan Basin
Ryukyu Islands
Kyushu-Palau Ridge
PHILIPPINE PLATE
Philippine Trench
Palau
CAROLINE PLATE
New Guinea Trench
AUSTRALIA

Tropic of Cancer
Equator

A B C D E F G H I J K L M

ASIAN RESOURCES

Although agriculture remains the economic mainstay of most Asian countries, the number of people employed in agriculture has steadily declined, as new industries have been developed during the past 30 years. China, Indonesia, Malaysia, Thailand, and Turkey have all experienced far-reaching structural change in their economies, while the breakup of the Soviet Union has created a new economic challenge in the Central Asian republics. The countries of the Persian Gulf illustrate the rapid transformation from rural nomadism to modern, urban society which oil wealth has brought to parts of the continent. Asia's most economically dynamic countries, Japan, Singapore, South Korea, and Taiwan, fringe the Pacific Ocean and are known as the Pacific Rim. In contrast, other Southeast Asian countries like Laos and Cambodia remain both economically and industrially underdeveloped.

INDUSTRY

Japanese industry leads the continent in both productivity and efficiency; electronics, hi-tech industries, car manufacture and shipbuilding are important. In recent years, the so-called economic "tigers" of the Pacific Rim such as Taiwan and South Korea are now challenging Japan's economic dominance. Heavy industries such as engineering, chemicals, and steel typify the industrial complexes along the corridor created by the Trans-Siberian Railway, the Fergana Valley in Central Asia, and also much of the huge industrial plain of east China. The discovery of oil in the Persian Gulf has brought immense wealth to countries that previously relied on subsistence agriculture on marginal desert land.

STANDARD OF LIVING

Despite Japan's high standards of living, and Southwest Asia's oil-derived wealth, immense disparities exist across the continent. Afghanistan remains one of the world's most underdeveloped nations, as do the mountain states of Nepal and Bhutan. Further rapid population growth is exacerbating poverty and overcrowding in many parts of India and Bangladesh.

Standard of Living
(UN Human Development Index)

low

high

On a small island at the southern tip of the Malay Peninsula lies Singapore, one of the Pacific Rim's most vibrant economic centers. Multinational banking and finance form the core of the city's wealth.

GNP per capita (US$)

0–499
500–999
1000–4999
5000–9999
10000–19999
20000+

Industry

✈ aerospace	🖶 printing & publishing
🍺 brewing	⚓ shipbuilding
🚗 car/vehicle manufacture	sugar processing
cement	tea processing
⚗ chemicals	🧵 textiles
electronics	timber processing
⚙ engineering	tobacco processing
$ finance	
fish processing	coal
food processing	oil
🖥 hi-tech industry	gas
iron & steel	
pharmaceuticals	• industrial cities
	⫽ major industrial areas

ARCTIC OCEAN

PACIFIC OCEAN

Sea of Okhotsk

RUSSIAN FEDERATION

Yakutsk

Trans-Siberian Railway

Bratsk

Khabarovsk

Yekaterinburg

Chelyabinsk

Magnitogorsk

Omsk

Krasnoyarsk

Kemerovo

Vladivostok

Novosibirsk

Novokuznetsk

Irkutsk

Harbin

JAPAN

KAZAKHSTAN

Karaganda

Ulan Bator

Shenyang

NORTH KOREA

Tokyo

Istanbul

Izmir

Ankara

GEORGIA

Tbilisi

ARMENIA

Yerevan

AZERB.

Baku

Aral Sea

Urumqi

MONGOLIA

Beijing

Pyongyang

Seoul

Nagoya

Kobe

TURKEY

CYPRUS

LEBANON

Beirut

SYRIA

Damascus

Tel Aviv-Yafo

ISRAEL

JORDAN

Amman

Kirkuk

UZBEKISTAN

Caspian Sea

Tashkent

KYRGYZSTAN

Farghona

Dushanbe

TURKMENISTAN

Ashgabat

Alma-Ata

Tianjin

Taiyuan

Jinan

Qingdao

SOUTH KOREA

Pusan

Baghdad

Tehran

IRAQ

Basra

Isfahan

TAJIKISTAN

AFGHANISTAN

Lanzhou

Zhengzhou

Xi'an

Nanjing

Shanghai

SAUDI ARABIA

Kuwait

KUWAIT

IRAN

Rawalpindi

CHINA

Wuhan

Ad Damman

BAHRAIN

Jedda

Riyadh

QATAR

Abu Dhabi

Dubai

UAE

Persian Gulf

PAKISTAN

Lahore

Chengdu

Chongqing

Taipei

TAIWAN

Gulf of Oman

Delhi

NEPAL

BHUTAN

Kunming

Guangzhou

Hong Kong

Red Sea

OMAN

Karachi

INDIA

Kanpur

Indore

Jamshedpur

BANGLADESH

Dhaka

MYANMAR

Mandalay

Hanoi

Da Nang

Manila

PHILIPPINES

YEMEN

Ahmadabad

Nagpur

Chittagong

Calcutta

LAOS

Mumbai (Bombay)

Arabian Sea

Rangoon

THAILAND

Bangkok

VIETNAM

South China Sea

Gulf of Aden

Bangalore

Chennai (Madras)

CAMBODIA

Ho Chi Minh City

SRI LANKA

INDIAN OCEAN

MALAYSIA

BRUNEI

Kuala Lumpur

Singapore

SINGAPORE

INDONESIA

Jakarta

Surabaya

EAST TIMOR
(under UN Transitional Authority from Feb 2000)

Iron and steel, engineering, and shipbuilding typify the heavy industry found in eastern China's industrial cities, especially the nation's leading manufacturing center, Shanghai.

Traditional industries are still crucial to many rural economies across Asia. Here, on the Vietnamese coast, salt has been extracted from seawater by evaporation and is being loaded into a van to take to market.

ENVIRONMENTAL ISSUES

THE TRANSFORMATION OF UZBEKISTAN by the former Soviet Union into the world's second largest producer of cotton led to the diversion of several major rivers for irrigation. Starved of this water, the Aral Sea diminished in volume by over 50% in 30 years, irreversibly altering the ecology of the area. Heavy industries in eastern China have polluted coastal waters, rivers, and urban air, while in Myanmar, Malaysia, and Indonesia, ancient hardwood rain forests are felled faster than they can regenerate.

Although Siberia remains a quintessentially frozen, inhospitable wasteland, vast untapped mineral reserves – especially the oil and gas of the West Siberian Plain – have lured industrial development to the area since the 1950s and 1960s.

Environmental Issues
- tropical forest
- forest destroyed
- desert
- desertification
- acid rain
- polluted rivers
- marine pollution
- heavy marine pollution
- radioactive contamination
- poor urban air quality

The long-term environmental impact of the Gulf War (1991) is still uncertain. As Iraqi troops left Kuwait, equipment was abandoned to rust and thousands of oil wells were set alight, pouring crude oil into the Persian Gulf.

MINERAL RESOURCES

AT LEAST 60% OF THE WORLD's known oil and gas deposits are found in Asia; notably the vast oil fields of the Persian Gulf, and the less-exploited oil and gas fields of the Ob' Basin in west Siberia. Immense coal reserves in Siberia and China have been utilized to support large steel industries. Southeast Asia has some of the world's largest deposits of tin, found in a belt running down the Malay Peninsula to Indonesia.

Mineral Resources
- oil field
- gas field
- coal field
- chromite
- copper
- gold
- iron
- lead
- nickel
- platinum
- tin
- wolfram

USING THE LAND AND SEA

VAST AREAS OF ASIA REMAIN UNCULTIVATED as a result of unsuitable climatic and soil conditions. In favorable areas such as river deltas, farming is intensive. Rice is the staple crop of most Asian countries, grown in paddy fields on waterlogged alluvial plains and terraced hillsides, and often irrigated for higher yields. Across the black earth region of the Eurasian steppe in southern Siberia and Kazakhstan, wheat farming is the dominant activity. Cash crops, like tea in Sri Lanka and dates in the Arabian Peninsula, are grown for export, and provide valuable income. The sovereignty of the rich fishing grounds in the South China Sea is disputed by China, Malaysia, Taiwan, the Philippines, and Vietnam, because of potential oil reserves.

Using the Land and Sea
- cropland
- desert
- forest
- mountain region
- pasture
- tundra
- wetland
- major conurbations
- cattle
- pigs
- goats
- sheep
- coconuts
- corn
- cotton
- dates
- fishing
- fruit
- jute
- peanuts
- rice
- rubber
- shellfish
- soybeans
- sugar beet
- sugar cane
- tea
- timber
- wheat

Date palms have been cultivated in oases throughout the Arabian Peninsula since antiquity. In addition to the fruit, palms are used for timber, fuel, rope, and for making vinegar, syrup, and a liquor known as arrack.

Rice terraces blanket the landscape across the small Indonesian island of Bali. The large amounts of water needed to grow rice have resulted in Balinese farmers organizing water-control cooperatives.

SIBERIAN PLATEAU AND PLAIN

THE WEST SIBERIAN PLAIN is one of the largest in the world, and contains a vast system of marshes. The whole area is covered by glacial deposits, underlain by the Angara Shield, a remnant of the ancient continent of Laurasia. The flat relief of the region and thick surface deposits result in poor drainage; this, combined with the freezing and thawing of the extensive permafrost layer leads to the formation of the vast swamps which cover the area. Many of the north-flowing rivers are also frozen for up to half the year.

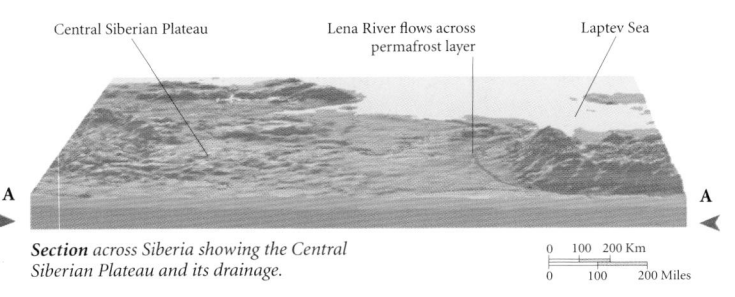

Section across Siberia showing the Central Siberian Plateau and its drainage.

Central Siberian Plateau • Lena River flows across permafrost layer • Laptev Sea

0 100 200 Km
0 100 200 Miles

THE ARABIAN SHIELD AND IRANIAN PLATEAU

APPROXIMATELY FIVE MILLION YEARS AGO, rifting of the continental crust split the Arabian Plate from the African Plate and flooded the Red Sea. As this rift spread, the Arabian Plate collided with the Eurasian Plate, transforming part of the Tethys seabed into the Zagros Mountains which run northwest-southeast across western Iran.

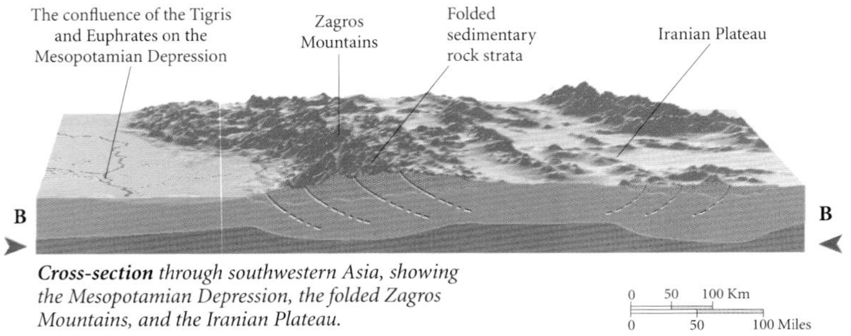

The confluence of the Tigris and Euphrates on the Mesopotamian Depression • Zagros Mountains • Folded sedimentary rock strata • Iranian Plateau

Cross-section through southwestern Asia, showing the Mesopotamian Depression, the folded Zagros Mountains, and the Iranian Plateau.

0 50 100 Km
0 50 100 Miles

THE TURAN BASIN AND KAZAKH UPLANDS

THE TURAN BASIN AND KAZAKH UPLANDS are a complex mixture of mountain foothills, an arid limestone plateau, and deserts including the Kyzl Kum and Kara Kum. In the center of the Turan Lowland – an area of inland drainage – is the desiccated Aral Sea, reduced to a fraction of its former size because of the diversion of its flow into irrigation channels. The only rivers with sufficient water to cross this arid region are the Syr Dayra and Amu Dayra.

THE INDIAN SHIELD AND HIMALAYAN SYSTEM

THE LARGE SHIELD AREA beneath the Indian subcontinent is between 2.5 and 3.5 billion years old. As the floor of the southern Indian Ocean spread, it pushed the Indian Shield north. This was eventually driven beneath the Plateau of Tibet. This process closed up the ancient Tethys Sea and uplifted the world's highest mountain chain, the Himalayas. Much of the uplifted rock strata was from the seabed of the Tethys Sea, partly accounting for the weakness of the rocks and the high levels of erosion found in the Himalayas.

Indo-Gangetic Depression • Crushed sediment from seabed of the Tethys Sea • Himalayas • Thrust zone • Plateau of Tibet

Cross-section through the Himalayas showing thrust faulting of the rock strata.

0 50 100 Km
0 50 100 Miles

CENTRAL ASIAN PLATEAUS AND BASINS

THE PLATEAU OF TIBET lies north of the Himalayas and covers 965,250 sq miles (2,500,000 sq km); its average elevation is 16,500 ft (5,000 m). The region is noted for its extreme aridity. In the south, the Himalayan mountain belt blocks moisture-bearing winds. The pressure from the Indo-Australian Plate against the plateau is causing both uplift and, when combined with the downward force caused by weight of the plateau, extension east and west of the of the more malleable underlying crust. The brittle upper rock layers are extensively faulted.

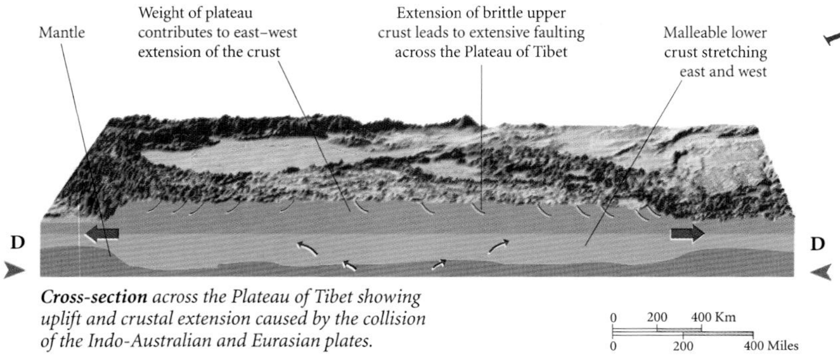

Mantle • Weight of plateau contributes to east–west extension of the crust • Extension of brittle upper crust leads to extensive faulting across the Plateau of Tibet • Malleable lower crust stretching east and west

Cross-section across the Plateau of Tibet showing uplift and crustal extension caused by the collision of the Indo-Australian and Eurasian plates.

0 200 400 Km
0 200 400 Miles

Map labels: Franz Josef Land, Kara Sea, Poluostrov Yamal, Gulf of Ob, Ob', Nadym, SIBERIA, West Siberian Plain, Tobol, Ishim, Irtysh, Ural Mountains, Kirghiz Steppe, Kulunda, Kazakh Uplands, Zaysan, Aral Sea, TURAN BASIN & KAZAKH UPLANDS, Lake Balkhash, Ili, EUROPE, Black Sea, Sea of Azov, Caucasus, Caspian Sea, Ustyurt Plateau, Turan Lowland, Kyzyl Kum, Syr Darya, Chatkal Range, Kirghiz Range, Ozero Issyk-Kul', Tien Shan, Tarim, EURASIAN PLATE, Kura Delta, ANATOLIAN PLATE, Anatolia, Taurus Mountains, Güney Dağı Toroslar, Kara Kum, Amu Darya, Turkestan Range, Pamir, Karakoram Range, Takla, AFRICAN PLATE, Gulf of Antalya, Mediterranean Sea, ARABIAN PLATE, Syrian Desert, Euphrates, Tigris, ARABIAN SHIELD & IRANIAN PLATEAU, Elburz Mountains, Great Salt Desert, Iranian Plateau, Zagros Mountains, IRANIAN PLATE, Hindu Kush, Kabul, K2, 8611m, Karakoram Pass 5568m, Khyber Pass 1080m, Himalaya, AFRICA, Red Sea, ARABIAN PLATE, AFRICAN PLATE, An Nafūd, Persian Gulf, As Summān, Ad Dahnā', Al Biyad, Strait of Hormuz, Gulf of Oman, INDO-AUSTRALIAN PLATE, Indus, Sutlej, Kashmir Range, Thar Desert, INDIAN SHIELD & HIMALAYAN SYSTEM, Arabian Peninsula, Ar Rub' al Khālī (Empty Quarter), Hadhramaut, Mouths of the Indus, Rann of Kachchh, Gulf of Kachchh, Vindhya Range, Narmada, Satpura Range, Gulf of Aden, ARABIAN PLATE, AFRICAN PLATE, Socotra, Arabian Sea, Gulf of Khambhat, Deccan, Western Ghats, Eastern Ghats, Krishna, Godavari, Sri Lanka, Gulf of Mannar, INDIAN OCEAN

PHYSICAL ASIA

THE STRUCTURE OF ASIA can be divided into two distinct regions. The landscape of northern Asia consists of old mountain chains, shields, plateaus, and basins, like the Ural Mountains in the west and the Central Siberian Plateau to the east. To the south of this region, are a series of plateaux and basins, including the vast Plateau of Tibet and the Tarim Basin. In contrast, the landscapes of southern Asia are much younger, formed by tectonic activity beginning about 65 million years ago, leading to an almost continuous mountain chain running from Europe, across much of Asia, and culminating in the mighty Himalayan mountain belt, formed when the Indo-Australian Plate collided with the Eurasian Plate. They are still being uplifted today. North of the mountains lies a belt of deserts, including the Gobi and the Takla Makan. In the far south, tectonic activity has formed narrow island arcs, extending over 4,000 miles (7,000 km). To the west lies the Arabian Shield, once part of the African Plate. As it was rifted apart from Africa, the Arabian Plate collided with the Eurasian Plate, uplifting the Zagros Mountains.

SHAPING THE LANDSCAPE

IN THE NORTH, melting of extensive permafrost leads to typical periglacial features such as thermokarst. In the arid areas wind action transports sand creating extensive dune systems. An active tectonic margin in the south causes continued uplift, and volcanic and seismic activity, but also high rates of weathering and erosion. Across the continent, huge rivers erode and transport vast quantities of sediment depositing it on the plains or forming large deltas.

PERIGLACIATION

[1] Permafrost is widespread across northern Siberia. When ground ice, which makes up a large proportion of the soil layer, melts, it contracts and extensive ground subsidence occurs. Over time this process leads to depressions in the landscape and the gradual movement of soil down slopes. Eventually the accumulation of water in the depressions leads to thermokarstic lakes (left).

PERIGLACIATION: FORMATION OF THERMOKARST

THE EVOLVING LANDSCAPE

Landscape
- limestone region
- sinking land
- stable land
- uplifting land
- active volcano
- area of tectonic activity
- limit of permafrost
- ocean current

TECTONIC ACTIVITY

[7] The Dead Sea (above) lies in a pull-apart basin. The sliding of the African Plate against the Arabian Plate, at unequal rates, led to the sinking of blocks of crust. This depression has been filled by the waters of the Dead Sea and Lake Tiberias (Sea of Galilee). The plates continue to move causing intermittent earthquakes.

Arabian Plate
Blocks of the Earth's crust sink, creating a basin
Dead Sea
African Plate

TECTONIC ACTIVITY: THE FORMATION OF A PULL-APART BASIN

RIVER SYSTEMS

[2] Vast river systems flow across Asia, many originating in the Himalayas and the Plateau of Tibet. Seasonal melting of snow and monsoon rains swell the river flow leading to flooding and erosion. The Yellow River (above) gets its color from the high level of eroded material from the loess plateau.

Snow melt
Monsoon rains
Yellow River dissects loess plateau
Carries large sediment load

RIVER SYSTEMS: EROSION OF THE LOESS PLATEAU BY THE YELLOW RIVER

CHEMICAL WEATHERING

[3] Tower karsts are widespread across south China (above) and Vietnam. It is thought the karstic towers were formed under a soil cover, where small depressions in the limestone bedrock began to be weathered by soil water acids, eventually creating larger hollows. This process continued over millions of years, deepening the hollows and leaving steep-sided limestone hills.

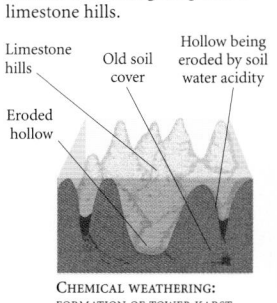

Limestone hills
Old soil cover
Hollow being eroded by soil water acidity
Eroded hollow

CHEMICAL WEATHERING: FORMATION OF TOWER KARST

SEDIMENTATION

[6] The Ganges/Brahmaputra is a tide-dominated delta (above). The two rivers transport huge quantities of mountain sediment, which is deposited on the delta plain. This debris is then redistributed by tidal currents, to form extensions to the bars, beach ridges, and deltaic deposits.

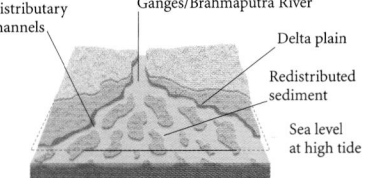

Distributary channels
Ganges/Brahmaputra River
Delta plain
Redistributed sediment
Sea level at high tide

SEDIMENTATION: THE DESTRUCTION OF A DELTA

COASTAL EROSION

[5] The erosion of cliffs along the coast of Indonesia (above) and Thailand occurs when waves and currents undermine the base leading to collapse of material. The surf then gradually erodes this material away, exposing the cliff to further undercutting. This process eventually creates shore platforms.

Undercutting by sea waves
Collapsed debris is eventually transported away by the surf
Shore platform showing how far cliffs have been eroded back

COASTAL EROSION: THE UNDERCUTTING OF A CLIFF

VOLCANIC ACTIVITY

[4] Volcanic eruptions occur frequently across Southeast Asia's island arcs (above). Low-level eruptions occur when groundwater, superheated by underlying magma, becomes pressurized, forcing hot fluid and rocks up through cracks in the volcanic cone. This is known as a phreatic eruption.

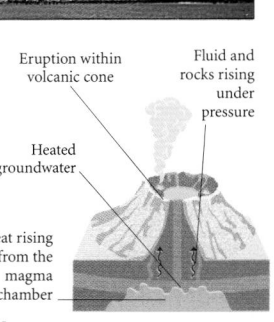

Eruption within volcanic cone
Fluid and rocks rising under pressure
Heated groundwater
Heat rising from the magma chamber

VOLCANIC ACTIVITY: A PHREATIC ERUPTION

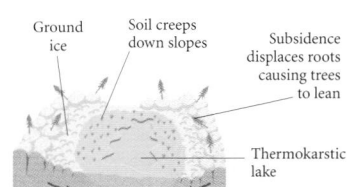
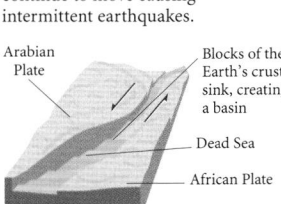
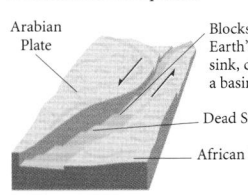
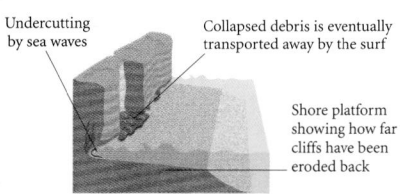

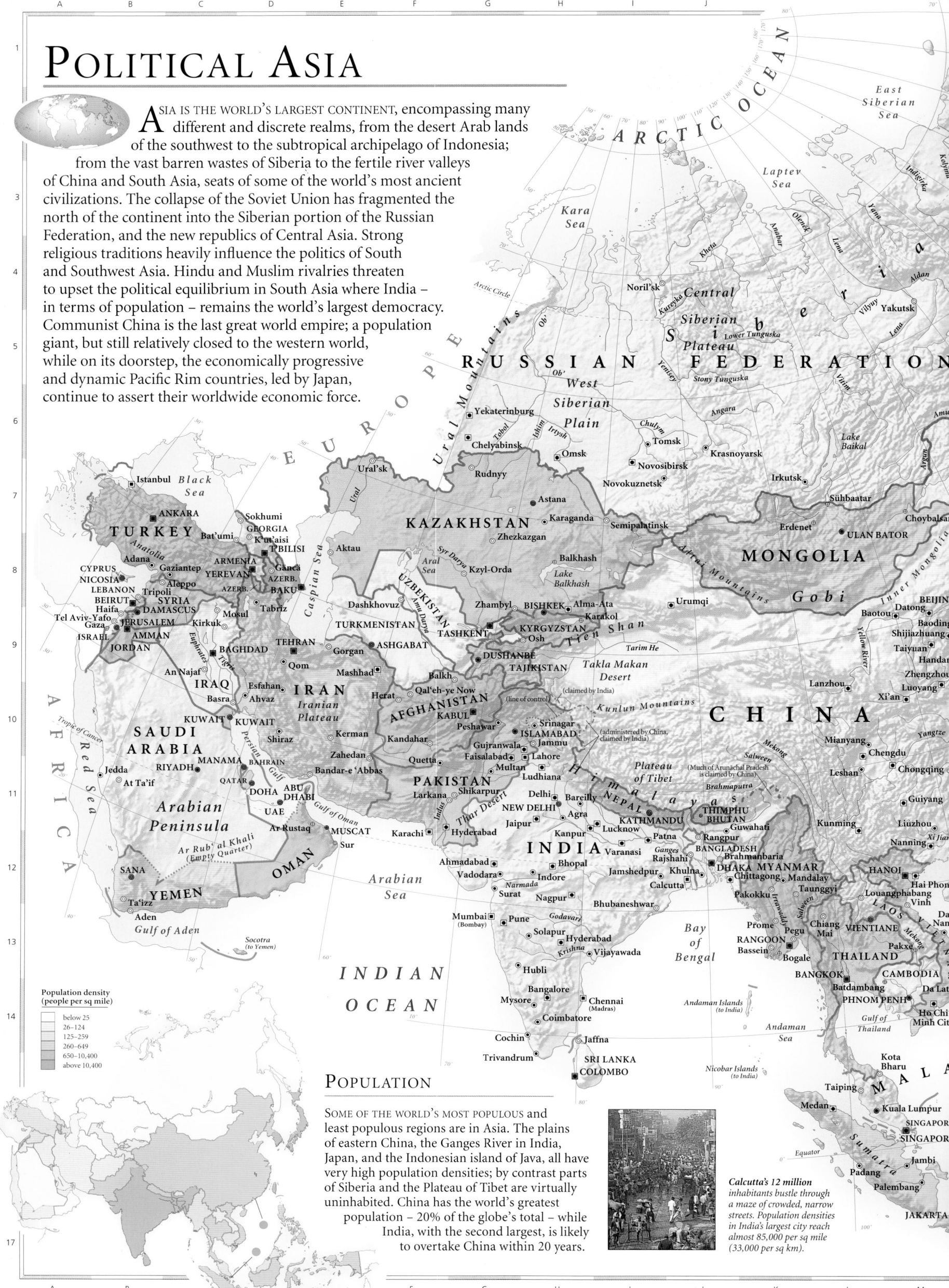

POLITICAL ASIA

ASIA IS THE WORLD'S LARGEST CONTINENT, encompassing many different and discrete realms, from the desert Arab lands of the southwest to the subtropical archipelago of Indonesia; from the vast barren wastes of Siberia to the fertile river valleys of China and South Asia, seats of some of the world's most ancient civilizations. The collapse of the Soviet Union has fragmented the north of the continent into the Siberian portion of the Russian Federation, and the new republics of Central Asia. Strong religious traditions heavily influence the politics of South and Southwest Asia. Hindu and Muslim rivalries threaten to upset the political equilibrium in South Asia where India – in terms of population – remains the world's largest democracy. Communist China is the last great world empire; a population giant, but still relatively closed to the western world, while on its doorstep, the economically progressive and dynamic Pacific Rim countries, led by Japan, continue to assert their worldwide economic force.

Population density
(people per sq mile)

- below 25
- 26–124
- 125–259
- 260–649
- 650–10,400
- above 10,400

POPULATION

SOME OF THE WORLD'S MOST POPULOUS and least populous regions are in Asia. The plains of eastern China, the Ganges River in India, Japan, and the Indonesian island of Java, all have very high population densities; by contrast parts of Siberia and the Plateau of Tibet are virtually uninhabited. China has the world's greatest population – 20% of the globe's total – while India, with the second largest, is likely to overtake China within 20 years.

Calcutta's 12 million inhabitants bustle through a maze of crowded, narrow streets. Population densities in India's largest city reach almost 85,000 per sq mile (33,000 per sq km).

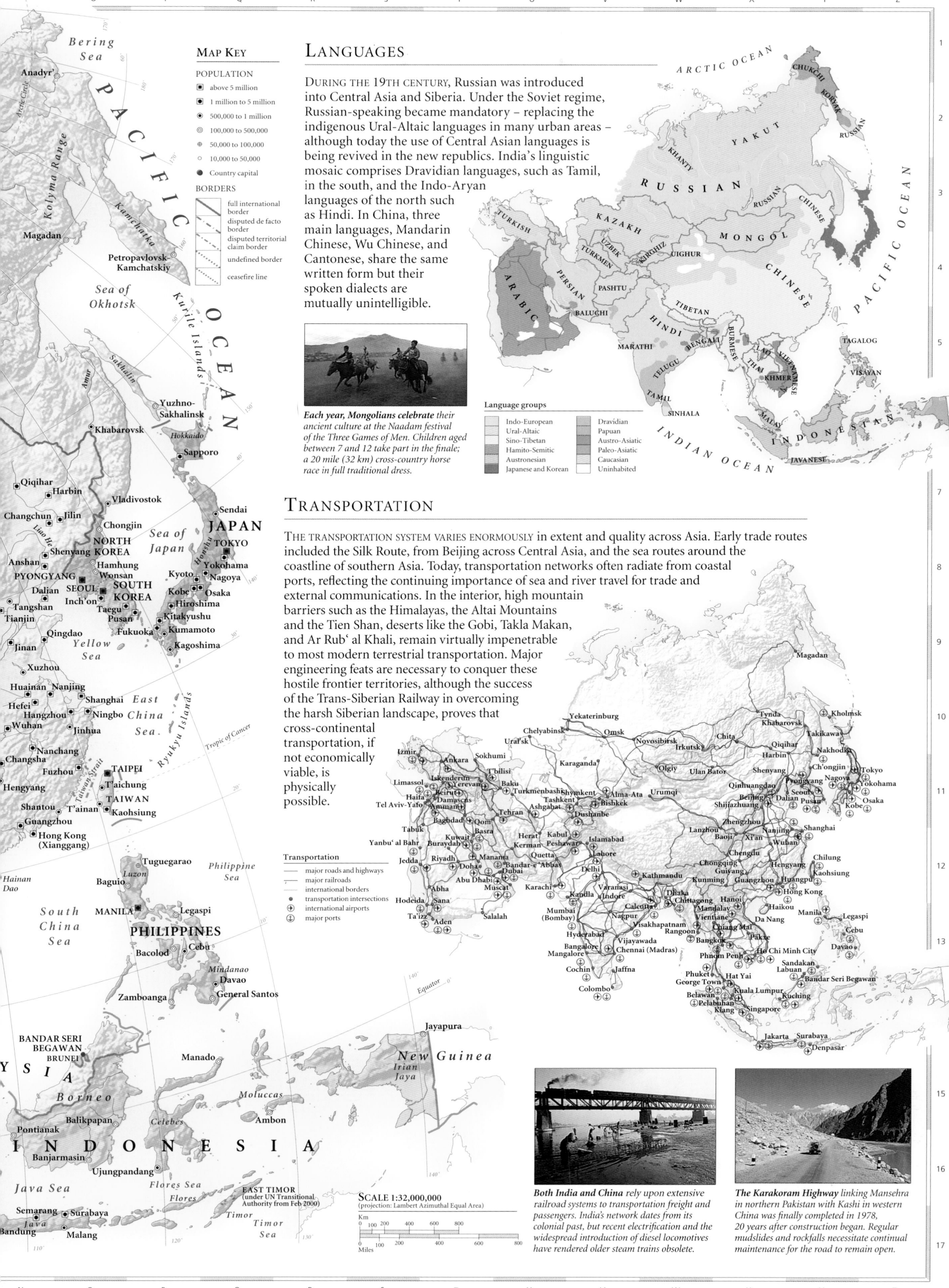

LANGUAGES

DURING THE 19TH CENTURY, Russian was introduced into Central Asia and Siberia. Under the Soviet regime, Russian-speaking became mandatory – replacing the indigenous Ural-Altaic languages in many urban areas – although today the use of Central Asian languages is being revived in the new republics. India's linguistic mosaic comprises Dravidian languages, such as Tamil, in the south, and the Indo-Aryan languages of the north such as Hindi. In China, three main languages, Mandarin Chinese, Wu Chinese, and Cantonese, share the same written form but their spoken dialects are mutually unintelligible.

Each year, Mongolians celebrate their ancient culture at the Naadam festival of the Three Games of Men. Children aged between 7 and 12 take part in the finale; a 20 mile (32 km) cross-country horse race in full traditional dress.

Language groups

- Indo-European
- Ural-Altaic
- Sino-Tibetan
- Hamito-Semitic
- Austronesian
- Japanese and Korean
- Dravidian
- Papuan
- Austro-Asiatic
- Paleo-Asiatic
- Caucasian
- Uninhabited

TRANSPORTATION

THE TRANSPORTATION SYSTEM VARIES ENORMOUSLY in extent and quality across Asia. Early trade routes included the Silk Route, from Beijing across Central Asia, and the sea routes around the coastline of southern Asia. Today, transportation networks often radiate from coastal ports, reflecting the continuing importance of sea and river travel for trade and external communications. In the interior, high mountain barriers such as the Himalayas, the Altai Mountains and the Tien Shan, deserts like the Gobi, Takla Makan, and Ar Rub' al Khali, remain virtually impenetrable to most modern terrestrial transportation. Major engineering feats are necessary to conquer these hostile frontier territories, although the success of the Trans-Siberian Railway in overcoming the harsh Siberian landscape, proves that cross-continental transportation, if not economically viable, is physically possible.

Transportation
- major roads and highways
- major railroads
- international borders
- transportation intersections
- international airports
- major ports

MAP KEY

POPULATION
- above 5 million
- 1 million to 5 million
- 500,000 to 1 million
- 100,000 to 500,000
- 50,000 to 100,000
- 10,000 to 50,000
- Country capital

BORDERS
- full international border
- disputed de facto border
- disputed territorial claim border
- undefined border
- ceasefire line

SCALE 1:32,000,000
(projection: Lambert Azimuthal Equal Area)

Both India and China rely upon extensive railroad systems to transportation freight and passengers. India's network dates from its colonial past, but recent electrification and the widespread introduction of diesel locomotives have rendered older steam trains obsolete.

The Karakoram Highway linking Mansehra in northern Pakistan with Kashi in western China was finally completed in 1978, 20 years after construction began. Regular mudslides and rockfalls necessitate continual maintenance for the road to remain open.

CLIMATE

THE CLIMATE OF ASIA exhibits marked differences from region to region, with freezing polar conditions in the north, hot and cold deserts in central regions and subtropical conditions throughout the south. Much of this variation can be attributed to enormous mountain barriers and internal depressions found across the continent. Monsoon winds, which reverse semiannually, cause alternate wet and dry seasons across southern Asia. These air masses moving north from the ocean are stripped of their moisture over the Himalayas causing arid conditions across the Plateau of Tibet. Both the south and east are susceptible to tropical cyclones or typhoons.

Treeless, frozen plains, with permanently frozen soil layers characterize much of Siberia. Even during the summer only the top 2–3 ft (1 m) of soil thaws.

Tundra-like marshes are found alongside vast sand dunes in the Takla Makan Desert in China. In the spring, windstorms of hurricane-force can send dust as high as 13,000 ft (4,000 m) in the air.

The Gobi Desert experiences major extremes in climate, with winter temperatures sometimes falling below -40°C (-40°F) and summer temperatures exceeding 45°C (113°F).

Climate

	tundra
	subarctic
	cool continental
	warm humid
	mediterranean
	semiarid
	arid
	humid equatorial
	tropical

☼ daily hours of sunshine, January

☼ daily hours of sunshine, July

→ cyclone
→ typhoon
→ cold/dry monsoon
→ warm/wet monsoon
→ cold wind

TEMPERATURE

Average January temperature

Average July temperature

Temperature

	below -22°F (-30°C)
	-22 to -4°F (-30 to -20°C)
	-4 to 14°F (-20 to -10°C)
	14 to 32°F (-10 to 0°C)
	32 to 50°F (0 to 10°C)
	50°F (10 to 20°C)
	68 to 86°F (20 to 30°C)
	above 86°F (30°C)

Tropical cyclones occur principally during late summer and early autumn. The intense winds and heavy rainfall can devastate entire villages.

Through India, the southwest monsoon, which brings heavy rainfall from May to September, accounts for 80% of annual precipitation.

RAINFALL

Average January rainfall

Average July rainfall

Rainfall

	0–1 in (0 –25 mm)
	1–2 in (25–50 mm)
	2–4 in (50–100 mm)
	4–8 in (100–200 mm)
	8–12 in (200–300 mm)
	12–16 in (300–400 mm)
	16–20 in (400–500 mm)
	more than 20 in (500 mm)

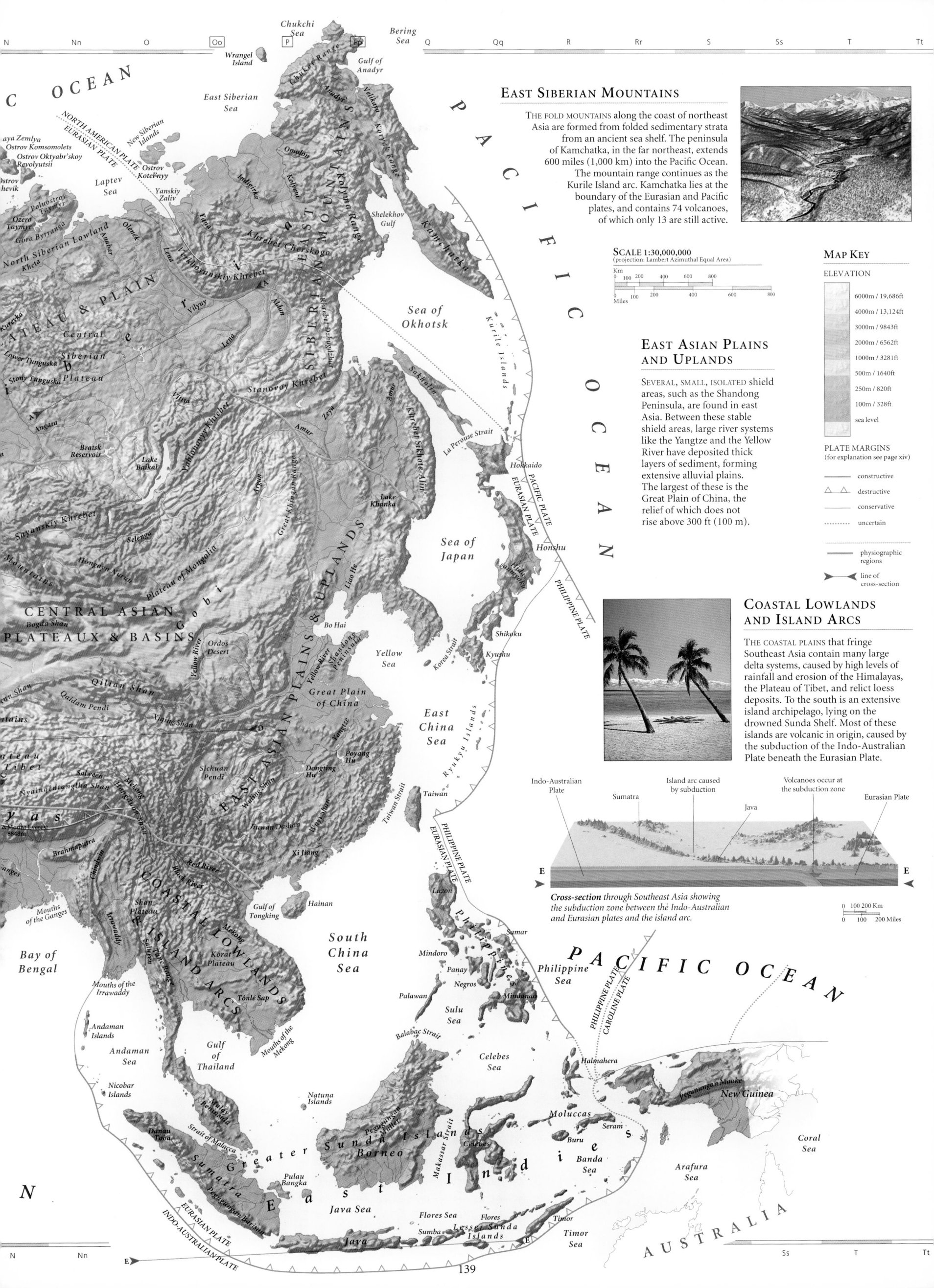

EAST SIBERIAN MOUNTAINS

THE FOLD MOUNTAINS along the coast of northeast Asia are formed from folded sedimentary strata from an ancient sea shelf. The peninsula of Kamchatka, in the far northeast, extends 600 miles (1,000 km) into the Pacific Ocean. The mountain range continues as the Kurile Island arc. Kamchatka lies at the boundary of the Eurasian and Pacific plates, and contains 74 volcanoes, of which only 13 are still active.

SCALE 1:30,000,000
(projection: Lambert Azimuthal Equal Area)

Km
0 100 200   400   600   800
0 100   200   400   600   800
Miles

EAST ASIAN PLAINS AND UPLANDS

SEVERAL, SMALL, ISOLATED shield areas, such as the Shandong Peninsula, are found in east Asia. Between these stable shield areas, large river systems like the Yangtze and the Yellow River have deposited thick layers of sediment, forming extensive alluvial plains. The largest of these is the Great Plain of China, the relief of which does not rise above 300 ft (100 m).

MAP KEY

ELEVATION

6000m / 19,686ft
4000m / 13,124ft
3000m / 9843ft
2000m / 6562ft
1000m / 3281ft
500m / 1640ft
250m / 820ft
100m / 328ft
sea level

PLATE MARGINS
(for explanation see page xiv)

——— constructive
△ △ destructive
——— conservative
········· uncertain

——— physiographic regions
►——◄ line of cross-section

COASTAL LOWLANDS AND ISLAND ARCS

THE COASTAL PLAINS that fringe Southeast Asia contain many large delta systems, caused by high levels of rainfall and erosion of the Himalayas, the Plateau of Tibet, and relict loess deposits. To the south is an extensive island archipelago, lying on the drowned Sunda Shelf. Most of these islands are volcanic in origin, caused by the subduction of the Indo-Australian Plate beneath the Eurasian Plate.

Indo-Australian Plate   Sumatra   Island arc caused by subduction   Java   Volcanoes occur at the subduction zone   Eurasian Plate

Cross-section through Southeast Asia showing the subduction zone between the Indo-Australian and Eurasian plates and the island arc.

0 100 200 Km
0 100   200 Miles

TURKEY & THE CAUCASUS

ARMENIA, AZERBAIJAN, GEORGIA, TURKEY

THIS REGION OCCUPIES THE FRAGMENTED JUNCTION between Europe, Asia, and the Russian Federation. Sunni Islam provides a common identity for the secular state of Turkey, which the revered leader Kemal Atatürk established from the remnants of the Ottoman Empire after the First World War. Turkey has a broad resource base and expanding trade links with Europe, but the east is relatively undeveloped and strife between the state and a large Kurdish minority has yet to be resolved. Georgia is similarly challenged by ethnic separatism, while the Christian state of Armenia and the mainly Muslim and oil-rich Azerbaijan are locked in conflict over the territory of Nagornyy Karabakh.

TRANSPORTATION & INDUSTRY

TURKEY LEADS THE REGION'S well-diversified economy. Petrochemicals, textiles, engineering, and food processing are the main industries. Azerbaijan is able to export oil, while the other states rely heavily on hydro-electric power and imported fuel. Georgia produces precision machinery. War and earthquake damage have devastated Armenia's infrastructure.

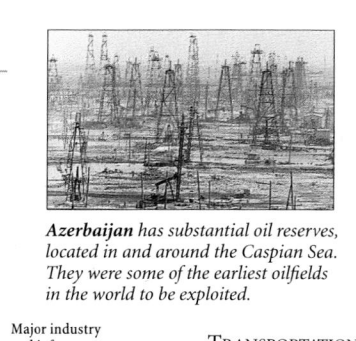

Azerbaijan has substantial oil reserves, located in and around the Caspian Sea. They were some of the earliest oilfields in the world to be exploited.

Major industry and infrastructure

- carpet weaving
- cement
- chemicals
- coal
- engineering
- food processing
- oil
- textiles
- tourism
- vehicle manufacture

- ■ capital cities
- ● major towns
- ⊕ international airports
- — major roads
- major industrial areas

TRANSPORTATION NETWORK

76,289 miles (122,849 km)	
7,74 miles (1,246 km)	
9,047 miles (14,569 km)	
745 miles (1,200 km)	

Physical and political barriers have severely limited communications between Armenia, Georgia and Azerbaijan. Turkey has a relatively well-developed transportation network.

USING THE LAND AND SEA

TURKEY IS LARGELY SELF-SUFFICIENT in food. The irrigated Black Sea coastlands have the world's highest yields of hazelnuts. Tobacco, cotton, sultanas, tea, and figs are the region's main cash crops and a great range of fruit and vegetables are grown. Wine grapes are among the labor-intensive crops which allow full use of limited agricultural land in the Caucasus. Sturgeon fishing is particularly important in Azerbaijan.

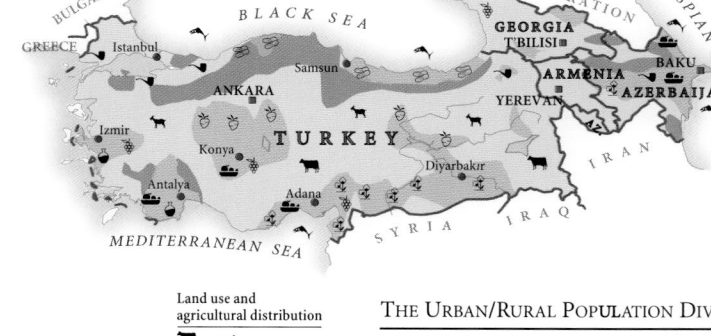

Land use and agricultural distribution

- cattle
- goats
- cotton
- fishing
- fruit
- hazelnuts
- olives
- sugar beet
- tobacco
- vineyards

- ■ capital cities
- major towns

- pasture
- cropland
- forest

THE URBAN/RURAL POPULATION DIVIDE

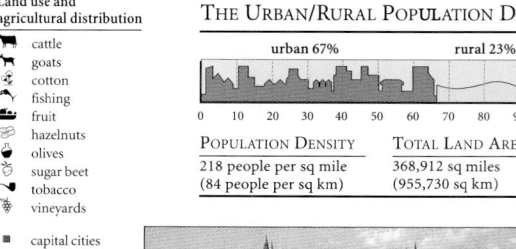

urban 67%   rural 23%

0 10 20 30 40 50 60 70 80 90 100

POPULATION DENSITY
218 people per sq mile
(84 people per sq km)

TOTAL LAND AREA
368,912 sq miles
(955,730 sq km)

For many centuries, Istanbul has held tremendous strategic importance as a crucial gateway between Europe and Asia. Founded by the Greeks as Byzantium, the city became the center of the East Roman Empire and was known as Constantinople to the Romans. From the 15th century onward the city became the center of the great Ottoman Empire.

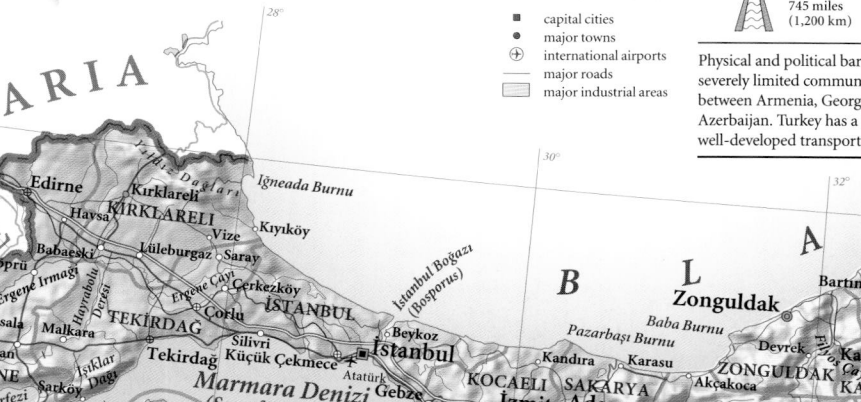

N O P Q R S T U V W X Y

THE LANDSCAPE

THE DEEPLY ERODED HILLS and salty basins of the Anatolian Plateau are bordered by several mountain ranges along the Black Sea coast, and the limestone Taurus Mountains (*Toros Dağlari*) in the south. A lowland trough divides the Caucasus and the Lesser Caucasus, which form a formidable barrier of peaks in the north.

Limestone weathering in the Anatolian Plateau
- Eroded gully
- High plateau
- Remnant landforms
- Layers of tephra

In central Turkey, rainwater has chemically weathered away numerous layers of limestone, leaving isolated outcrops and pinnacles and deep eroded gullies.

The Caucasus are fold mountains, which formed around the same time as the Taurus Mountains (*Toros Dağlari*) around 65 million years ago and have since been modified by volcanic erruptions.

The straits of the Bosporus and the Dardanelles, respectively linking the Black and Mediterranean seas with the Sea of Marmara, formed after the last Ice Age, when a rising sea level caused these former river valleys to be flooded.

Anatolian Plateau

Thick, temperate forest veils the seaward slopes of the Kaçkar Dağlari. The southern slopes, which lie in a rainshadow, are dry and barren.

Lava has flowed over large areas of the Lesser Caucasus within the last five million years, producing extensive basalt plateaus.

The white rock terraces at Pamukkale in western Turkey were formed when underground water, heated by volcanic activity, dissolved minerals in the rocks. When the water reached the surface and evaporated the minerals were left behind in these extraordinary formations.

The earthquake that struck Armenia in 1988 killed over 55,000 people and devastated the country's infrastructure.

Long, parallel mountain ranges run from east to west into the Aegean Sea, which has risen since the last Ice Age to form a drowned coastline of numerous islands and extended inlets.

Pamukkale

The volcanic cone of Mount Ararat is the highest peak in Turkey, with an altitude of 16,853 ft (5,137 m).

MAP KEY

POPULATION
- ■ above 5 million
- ■ 1 million to 5 million
- ◎ 500,000 to 1 million
- ◉ 100,000 to 500,000
- ⊕ 50,000 to 100,000
- ○ 10,000 to 50,000
- ○ below 10,000

The folded peaks of the Taurus Mountains (*Toros Dağlari*) were formed 60–65 million years ago, at the same time as the Alps. The rock is mainly limestone, with deep caves, gorges, and underground rivers.

The Cilician Gates (*Gůlek Boğazi*), a major pass through the Taurus Mountains (*Toros Dağlari*), is the point where streams flow from the interior plateau onto the lowland of Adana.

Many of the rivers crossing the Anatolian Plateau never reach the sea, but drain into salt marshes and shallow salt lakes such as Lake Tuz (*Tuz Gölü*), where much of the water is lost to evaporation.

The granite massif near Suram divides the lowlands of Georgia from the oil-rich basin of Azerbaijan's Kura River, which has built a large delta into the Caspian Sea.

The shallow, saline Lake Van (*Van Gölü*) is the largest lake in Turkey. Dry terraces mark a previous shoreline 181 ft (55 m) above the present water level.

ELEVATION
- 4000m / 13,124ft
- 3000m / 9843ft
- 2000m / 6562ft
- 1000m / 3281ft
- 500m / 1640ft
- 250m / 820ft
- 100m / 328ft
- sea level

Since the 6th century BC, the pinnacles and caves of east-central Anatolia have been utilized as dwellings. Many are still inhabited today.

SCALE 1:4,000,000
(projection: Lambert Conformal Conic)

Km
0 10 20 40 60 80 100 120

Miles
0 10 20 40 60 80 100 120

The fisheries of Azerbaijan are noted for their hauls of sturgeon, and the Caspian Sea accounts for 80% of the world's total catch. Sturgeon roe is used to make internationally-famed caviar.

Traditional steam baths are found throughout Turkey, and are used for socializing as well as for bathing.

THE NEAR EAST

IRAQ, ISRAEL, JORDAN, LEBANON, SYRIA

SOME OF THE WORLD'S OLDEST CIVILIZATIONS developed in this region – the Fertile Crescent – which is venerated by Jews, Muslims, and Christians, but torn by competing religious, ethnic, and national claims to the land. Turkish Ottoman rule ended with World War I and the region was divided into areas administered by Britain and France. The UN endorsed calls for a Jewish homeland in what was then Palestine and in 1948 the state of Israel was declared. Hostility towards the Jewish state led to a series of wars but since 1977, and especially since 1993, a peace process between Israel and her neighbors has been evolving. Since independence, Syria has played a leading role in Middle Eastern politics. The once-prosperous state of Lebanon is emerging from a ruinous factional war, while Iraq's great oil wealth has funded military campaigns against Iran and Kuwait, and the stifling of internal dissent, leading to international ostracization.

USING THE LAND AND SEA

WATER SCARCITY limits cropland to the north and to areas watered principally by the Tigris, Euphrates, and Jordan Rivers. In Israel, new irrigation techniques are allowing cultivation in the arid Negev. Wheat is the chief grain and large areas of scrub support livestock herding. Commercial produce includes dates, tobacco, citrus fruits, olives, grapes, and cotton, which is Syria's main export crop. Fishing is still important in the Mediterranean.

THE URBAN/RURAL POPULATION DIVIDE

urban 70%   rural 30%

0 10 20 30 40 50 60 70 80 90 100

POPULATION DENSITY
163 people per sq mile
(63 people per sq km)

TOTAL LAND AREA
325,460 sq miles
(843,160 sq km)

Land use and agricultural distribution
- sheep
- cereals
- citrus fruits
- cotton
- dates
- fishing
- rice
- tobacco
- capital cities
- major towns

pasture
cropland
wetland
desert

TRANSPORTATION & INDUSTRY

THE PETROCHEMICAL INDUSTRY is well established, and central to the economies of Syria and Iraq, which was the world's second largest oil exporter before the war with Iran which began in 1980. Lebanon has traditionally been a center for commerce, while Israel has a well-diversified economy with an expanding tourist industry, despite few natural resources.

TRANSPORTATION NETWORK

75,427 miles (121,461 km)

1,468 miles (2,364 km)

3,271 miles (5,267 km)

498 miles (802 km)

Jordan's seaport of Al 'Aqabah is connected to Damascus in Syria by road and rail. This route to the Red Sea provides for large exports of phosphate and trade with states in The Persian Gulf.

Major industry and infrastructure
- car manufacture
- cement
- chemicals
- electronics
- finance
- food processing
- iron & steel
- oil
- oil refining
- textiles
- capital cities
- major towns
- international airports
- major roads
- major industrial areas

The Dome of the Rock in Jerusalem is a magnificent mosque, revered by Muslims. Close by is the Wailing Wall, the city's most sacred Jewish landmark and the Church of the Holy Sepulchre, a famous Christian place of worship.

The city of Petra, carved from spectacular rose-colored limestone, lies deep within a canyon in southern Jordan. Revenues from the spice trade funded the construction of the city which was built by the Nabatean people in about 400 BC.

Water and wind erosion over thousands of years have created the Canyon of the Oasis at En 'Avedat in the Negev Desert (HaNegev). Extreme diurnal temperature fluctuations, coupled with wind erosion, have caused layers of rock to crack and peel away.

THE LANDSCAPE

THE AL JAZIRAH PLATEAU divides the Euphrates and Tigris Rivers, which cross the Mesopotamian plain to reach their confluence in the southeast. The rocky Syrian Desert extends west to the northern extremity of the Great Rift Valley, which runs from the mountains of Lebanon to the Gulf of Aqaba. The River Jordan flows south along this trough into the Dead Sea, divided from the Mediterranean coastal plain by a steep-sided plateau.

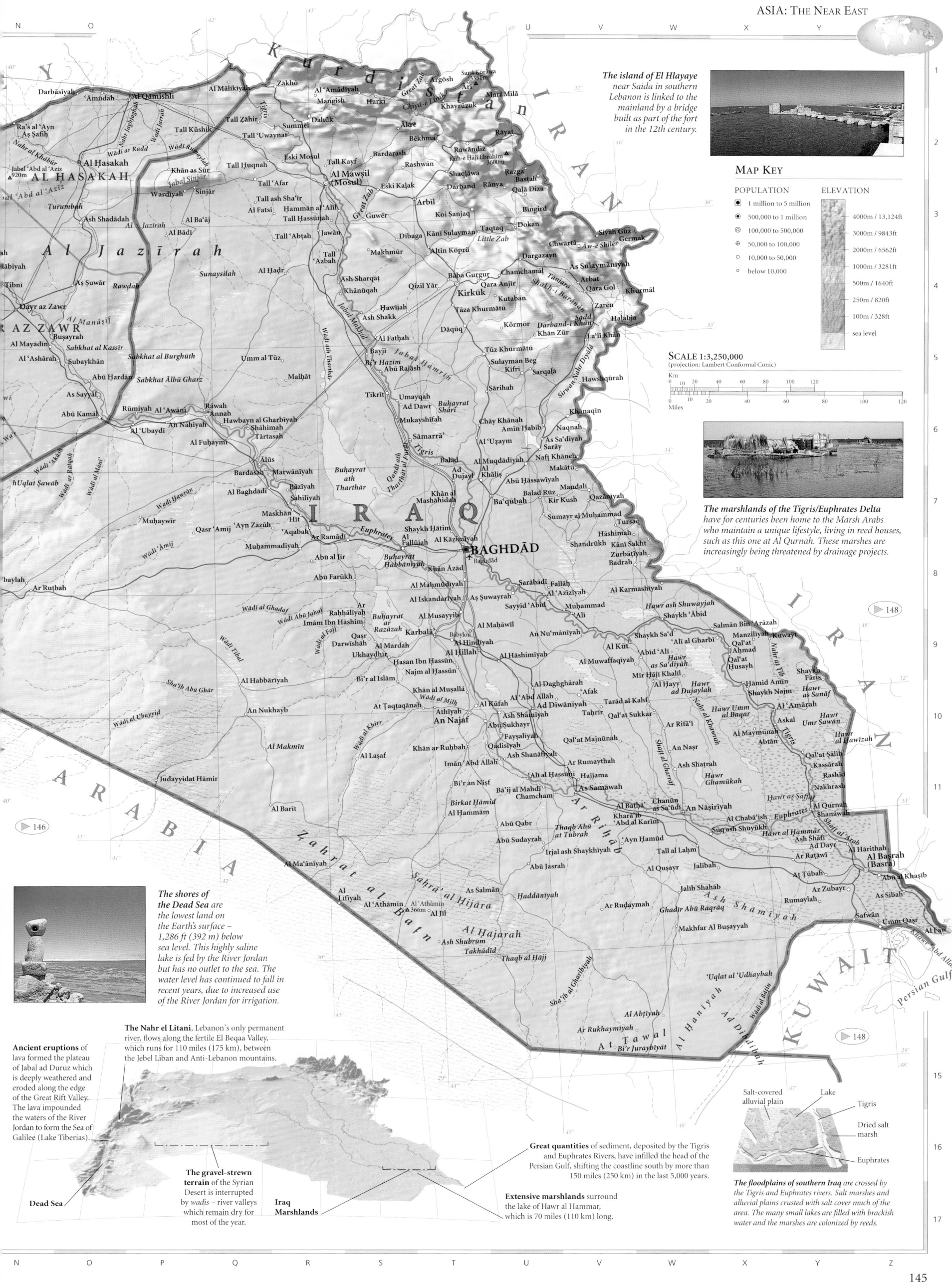

The island of El Hlayaye near Saida in southern Lebanon is linked to the mainland by a bridge built as part of the fort in the 12th century.

MAP KEY

POPULATION

- 1 million to 5 million
- 500,000 to 1 million
- 100,000 to 500,000
- 50,000 to 100,000
- 10,000 to 50,000
- below 10,000

ELEVATION

- 4000m / 13,124ft
- 3000m / 9843ft
- 2000m / 6562ft
- 1000m / 3281ft
- 500m / 1640ft
- 250m / 820ft
- 100m / 328ft
- sea level

SCALE 1:3,250,000
(projection: Lambert Conformal Conic)

Km
0 10 20   40   60   80   100   120

Miles
0   10   20   40   60   80   100   120

The marshlands of the Tigris/Euphrates Delta have for centuries been home to the Marsh Arabs who maintain a unique lifestyle, living in reed houses, such as this one at Al Qurnah. These marshes are increasingly being threatened by drainage projects.

The shores of the Dead Sea are the lowest land on the Earth's surface – 1,286 ft (392 m) below sea level. This highly saline lake is fed by the River Jordan but has no outlet to the sea. The water level has continued to fall in recent years, due to increased use of the River Jordan for irrigation.

Ancient eruptions of lava formed the plateau of Jabal ad Duruz which is deeply weathered and eroded along the edge of the Great Rift Valley. The lava impounded the waters of the River Jordan to form the Sea of Galilee (Lake Tiberias).

The Nahr el Litani, Lebanon's only permanent river, flows along the fertile El Beqaa Valley, which runs for 110 miles (175 km), between the Jebel Liban and Anti-Lebanon mountains.

Dead Sea

The gravel-strewn terrain of the Syrian Desert is interrupted by *wadis* – river valleys which remain dry for most of the year.

Iraq Marshlands

Great quantities of sediment, deposited by the Tigris and Euphrates Rivers, have infilled the head of the Persian Gulf, shifting the coastline south by more than 150 miles (250 km) in the last 5,000 years.

Extensive marshlands surround the lake of Hawr al Hammar, which is 70 miles (110 km) long.

Salt-covered alluvial plain
Lake
Tigris
Dried salt marsh
Euphrates

The floodplains of southern Iraq are crossed by the Tigris and Euphrates rivers. Salt marshes and alluvial plains crusted with salt cover much of the area. The many small lakes are filled with brackish water and the marshes are colonized by reeds.

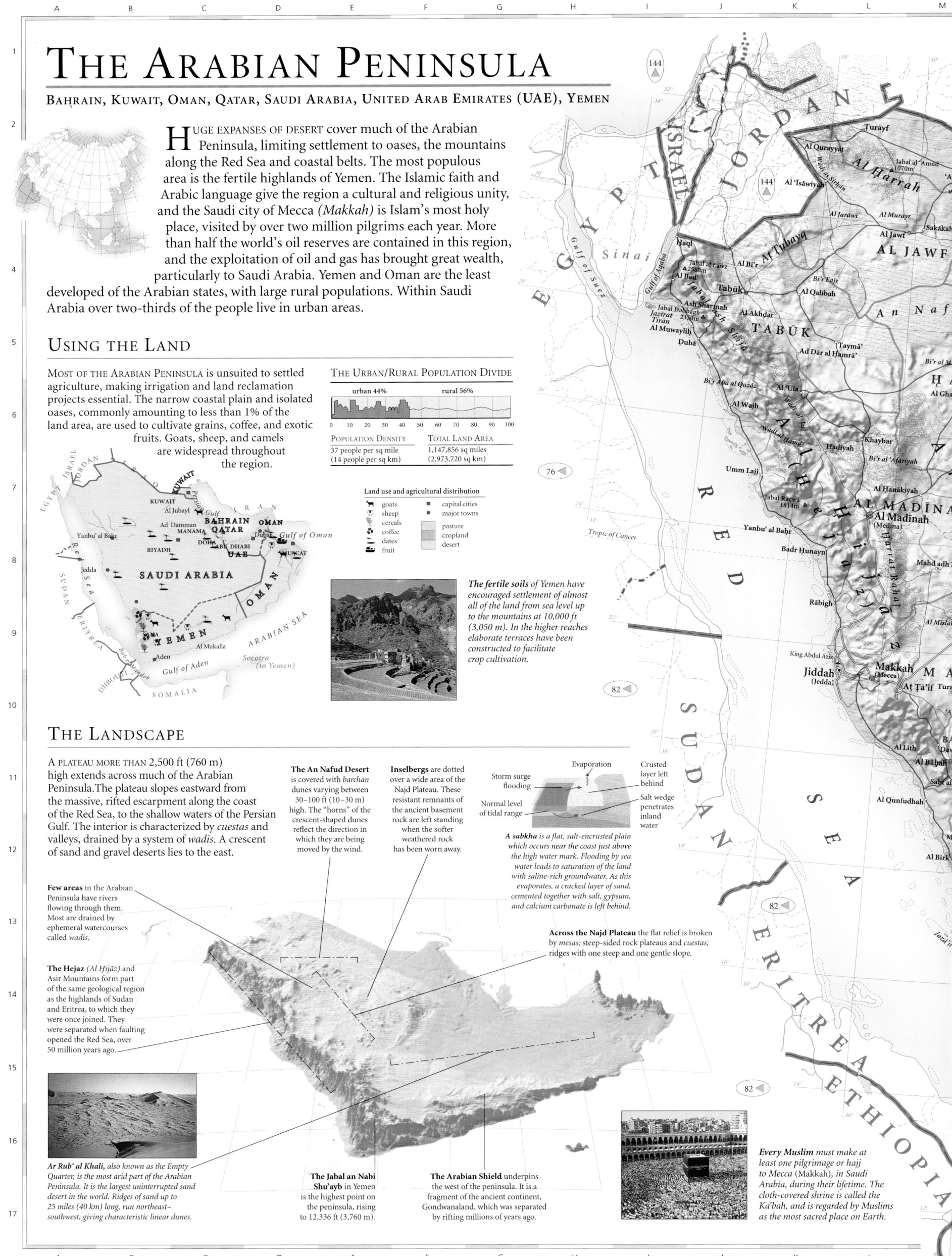

THE ARABIAN PENINSULA

BAHRAIN, KUWAIT, OMAN, QATAR, SAUDI ARABIA, UNITED ARAB EMIRATES (UAE), YEMEN

HUGE EXPANSES OF DESERT cover much of the Arabian Peninsula, limiting settlement to oases, the mountains along the Red Sea and coastal belts. The most populous area is the fertile highlands of Yemen. The Islamic faith and Arabic language give the region a cultural and religious unity, and the Saudi city of Mecca *(Makkah)* is Islam's most holy place, visited by over two million pilgrims each year. More than half the world's oil reserves are contained in this region, and the exploitation of oil and gas has brought great wealth, particularly to Saudi Arabia. Yemen and Oman are the least developed of the Arabian states, with large rural populations. Within Saudi Arabia over two-thirds of the people live in urban areas.

USING THE LAND

MOST OF THE ARABIAN PENINSULA is unsuited to settled agriculture, making irrigation and land reclamation projects essential. The narrow coastal plain and isolated oases, commonly amounting to less than 1% of the land area, are used to cultivate grains, coffee, and exotic fruits. Goats, sheep, and camels are widespread throughout the region.

THE URBAN/RURAL POPULATION DIVIDE

urban 44%   rural 56%

0 10 20 30 40 50 60 70 80 90 100

POPULATION DENSITY	TOTAL LAND AREA
37 people per sq mile (14 people per sq km)	1,147,856 sq miles (2,973,720 sq km)

Land use and agricultural distribution

- goats
- sheep
- cereals
- coffee
- dates
- fruit
- capital cities
- major towns
- pasture
- cropland
- desert

The fertile soils of Yemen have encouraged settlement of almost all of the land from sea level up to the mountains at 10,000 ft (3,050 m). In the higher reaches elaborate terraces have been constructed to facilitate crop cultivation.

THE LANDSCAPE

A PLATEAU MORE THAN 2,500 ft (760 m) high extends across much of the Arabian Peninsula. The plateau slopes eastward from the massive, rifted escarpment along the coast of the Red Sea, to the shallow waters of the Persian Gulf. The interior is characterized by *cuestas* and valleys, drained by a system of *wadis*. A crescent of sand and gravel deserts lies to the east.

The An Nafud Desert is covered with *barchan* dunes varying between 30–100 ft (10–30 m) high. The "horns" of the crescent-shaped dunes reflect the direction in which they are being moved by the wind.

Inselbergs are dotted over a wide area of the Najd Plateau. These resistant remnants of the ancient basement rock are left standing when the softer weathered rock has been worn away.

Evaporation
Storm surge flooding
Normal level of tidal range
Crusted layer left behind
Salt wedge penetrates inland water

A sabkha is a flat, salt-encrusted plain which occurs near the coast just above the high water mark. Flooding by sea water leads to saturation of the land with saline-rich groundwater. As this evaporates, a cracked layer of sand, cemented together with salt, gypsum, and calcium carbonate is left behind.

Few areas in the Arabian Peninsula have rivers flowing through them. Most are drained by ephemeral watercourses called *wadis*.

The Hejaz *(Al Ḥijāz)* and Asir Mountains form part of the same geological region as the highlands of Sudan and Eritrea, to which they were once joined. They were separated when faulting opened the Red Sea, over 50 million years ago.

Across the Najd Plateau the flat relief is broken by *mesas*; steep-sided rock plateaus and *cuestas*; ridges with one steep and one gentle slope.

Ar Rub' al Khali, also known as the Empty Quarter, is the most arid part of the Arabian Peninsula. It is the largest uninterrupted sand desert in the world. Ridges of sand up to 25 miles (40 km) long, run northeast–southwest, giving characteristic linear dunes.

The Jabal an Nabi Shu'ayb in Yemen is the highest point on the peninsula, rising to 12,336 ft (3,760 m).

The Arabian Shield underpins the west of the peninsula. It is a fragment of the ancient continent, Gondwanaland, which was separated by rifting millions of years ago.

Every Muslim must make at least one pilgrimage or hajj to Mecca (Makkah), in Saudi Arabia, during their lifetime. The cloth-covered shrine is called the Ka'bah, and is regarded by Muslims as the most sacred place on Earth.

TRANSPORTATION & INDUSTRY

THE EXTRACTION AND REFINING OF OIL AND GAS are the major industrial activities in the Arabian Peninsula. The region also has an active construction sector, with many Arab cities reflecting the wealth generated by the oil industry. The service sector is dominated by financial and technical institutions, which, like the construction sector, mainly serve the oil industry. Traditional handicrafts such as carpet-weaving are found in rural areas.

Saudi Arabia contains the world's largest oil reserves, lying mainly along the Persian Gulf coast. Each day the region produces 8.3 million barrels of oil. Here, in the desert, excess oil is being burnt off.

TRANSPORTATION NETWORK

65,239 miles (105,054 km)		2,071 miles (3,333 km)	
864 miles (1,392 km)		none	

Internal surface transportation is poorly developed across the peninsula. Along the coast, commercial routes have developed, but connections between bordering states rely on major airports.

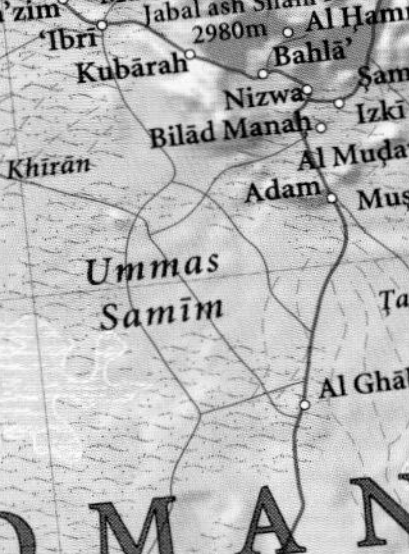

Major industry and infrastructure

- cement
- chemicals
- iron & steel
- oil
- oil refining
- food processing
- capital cities
- major towns
- international airports
- major roads
- major industrial areas

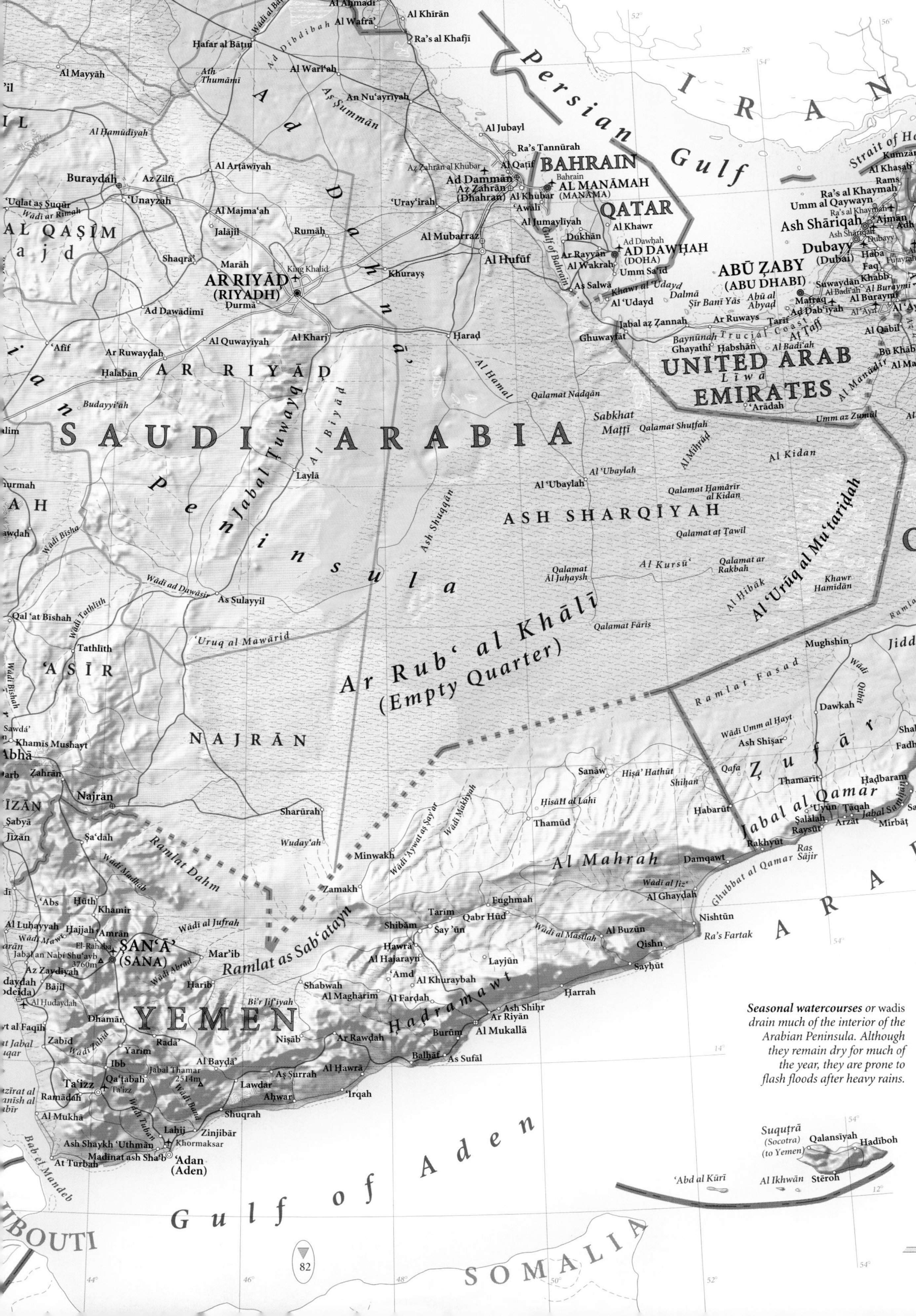

MAP KEY

POPULATION

- 1 million to 5 million
- 500,000 to 1 million
- 100,000 to 500,000
- 50,000 to 100,000
- 10,000 to 50,000
- below 10,000

ELEVATION

- 3000m / 9843ft
- 2000m / 6562ft
- 1000m / 3281ft
- 500m / 1640ft
- 250m / 820ft
- 100m / 328ft
- sea level

Seasonal watercourses or wadis drain much of the interior of the Arabian Peninsula. Although they remain dry for much of the year, they are prone to flash floods after heavy rains.

SCALE 1:7,500,000
(projection: Lambert Conformal Conic)

Km 0 25 50 75 100 150 200 250
Miles 0 25 50 75 100 150 200 250

IRAN & THE GULF STATES

BAHRAIN, IRAN, KUWAIT, QATAR, UNITED ARAB EMIRATES (UAE)

THE DISCOVERY OF OIL in the Persian Gulf in the 1930s brought great wealth to the surrounding states. The revenue was largely used to modernize industry and infrastructure, initiating great social change in these formerly agrarian countries. Today, over 80% of the people in the Gulf states live in urban areas, and foreign nationals make up a sizeable proportion of the population in Kuwait, Qatar ,and the United Arab Emirates. The importance of control of the oil reserves has led to a number of territorial disputes, including most recently the Iran–Iraq War and the Gulf War. Islam is practiced almost exclusively throughout the region and two distinct strands are found; Sunni Muslims in Qatar, Kuwait, and UAE, and Shi'a Muslims in Iran and Bahrain. In 1979 Iran became the world's largest theocracy.

THE LANDSCAPE

THE LAND RISES STEEPLY from the fragmented coastal lowlands bordering the Persian Gulf, to reach Iran's interior plateau, bounded by heavily-eroded mountain chains. An unstable plate boundary runs northwest to southeast across Iran causing frequent earthquakes. On the sandy west coast of the Persian Gulf, the relief is generally flat, with patches of salt marsh. Bahrain consists of two groups of islands, which are mostly small and rocky.

Pyroclastic layers

Lava flow

Lava flow layers

Qolleh-ye Damavand in the Elburz Mountains is a composite volcano. It comprises layers of lava and pyroclasts fragmentary rocks which accumulate on the slopes of the volcano after being ejected into the air.

Marine sediments from deep beneath the ancient Tethys Sea have been uplifted to form the Elburz Mountains, which stretch along the shores of the Caspian Sea, northern Iran.

Lava and ash from previous volcanic activity covers a 200-mile (320-km) stretch from the border with Azerbaijan to the Caspian Sea.

Iran's two mountain chains, the Zagros and Elburz, were uplifted at the same time as the Alps in Europe, when the African Plate collided with the Eurasian Plate.

Caspian Sea

Qolleh-ye Damavand

Dominated by a vast, semi-arid interior plateau, most of Iran lies above 1,640 ft (500 m). The region is poorly drained with many of its basins remaining dry for months at a time.

The Dasht-e Lut

The fierce Shamal wind affects much of this region. Every summer it blows dust south from the flood plains of the Tigris and Euphrates, reducing visibility to such an extent that Kuwait International Airport is frequently forced to close.

The oilfields of The Gulf are formed from marine shale deposits lying in sedimentary basins at the margins of the Zagros Mountains.

Autumn winds blowing across The Gulf can reach speeds of up to 95 mph (150 kmph) causing severe storms, squalls, and waterspouts.

Prolific springs tapping artesian water make cultivation possible across the north of Bahrain's main island. This provides a sharp contrast to the sandy plains in the south and west.

Numerous islands lie along the southern coast of the Persian Gulf. Some of these are salt domes, created when less dense salts were displaced and forced up to the surface by denser, overlying strata.

The Dasht-e Lut covers a large portion of eastern Iran with its dry, wind-eroded plain of scattered sandstone pillars and salty depressions. During the summer, temperatures soar, making it one of the world's hottest, driest places.

USING THE LAND AND SEA

ALONG THE COAST of the Caspian Sea, desalinated water allows fruits and vegetables to be produced, although water shortages and desert soils still limit farming. Sheep are the most important livestock raised in Iran and commercial forests cover the northwest of the country. Shrimp stocks were decimated by pollution during the Gulf War, but fishing remains important for domestic and export markets.

All of the Gulf states have commercial fishing fleets. Before the discovery of oil, fishing was the region's leading industry.

The Kuwait Towers in the centre of Kuwait are symbols of the vast wealth oil has brought to the country. Before 1960, the city had only one main street and was surrounded by a mud wall.

Land use and agricultural distribution

- goats
- sheep
- cereals
- citrus fruits
- cotton
- dates
- fishing
- timber
- capital cities
- major towns

- pasture
- cropland
- forest
- desert
- wetland

THE URBAN/RURAL POPULATION DIVIDE

urban 59%   rural 41%

0 10 20 30 40 50 60 70 80 90 100

POPULATION DENSITY
118 people per sq mile
(46 people per sq km)

TOTAL LAND AREA
642,883 sq miles
(1,665,500 sq km)

Many volcanoes lie in Iran's 1,200 mile (1930 km) volcanic belt, including the country's highest peak, the now-extinct Qolleh-ye Damavand at 18,600 ft (5,671 m).

Extensive oil and gas exploitation in the Gulf region has allowed the economic transformation of the Gulf states. Kuwait and the United Arab Emirates today have the highest per capita incomes in the world.

TRANSPORTATION & INDUSTRY

BOTH ONSHORE AND OFFSHORE oil reserves are exploited throughout the region. Kuwait not only extracts but also refines 80% of its oil. Bahrain has diversified its economy to become the main commercial and financial center in the Persian Gulf. Iran produces a wide range of products: textile mills are widespread and carpet weaving is an important export industry.

Major industry and infrastructure

- carpet manufacture
- chemicals
- finance
- food processing
- oil
- oil refining
- textiles
- capital city
- major towns
- international airports
- major roads
- major industrial areas

TRANSPORTATION NETWORK

50,340 miles (81,063 km)	466 miles (750 km)
3723 miles (5995 km)	81 miles (130 km)

Major towns and neighboring countries are linked by adequate road networks, although rural areas are less well served. Bahrain is linked to the mainland by a 15 mile (25 km) long causeway.

MAP KEY

POPULATION

- above 5 million
- 1 million to 5 million
- 500,000 to 1 million
- 100,000 to 500,000
- 50,000 to 100,000
- 10,000 to 50,000
- below 10,000

ELEVATION

- 4000m / 13,124ft
- 3000m / 9843ft
- 2000m / 6562ft
- 1000m / 3281ft
- 500m / 1640ft
- 250m / 820ft
- 100m / 328ft
- sea level

SCALE 1:5,500,000
(projection: Lambert Conformal Conic)

Km
0 10 20 40 60 80 100 120 140 160 180 200

Miles
0 10 20 40 60 80 100 120 140 160 180 200

Tropic of Cancer

Map labels

TURKMENISTAN

AFGHANISTAN

PAKISTAN

OMAN

UNITED ARAB EMIRATES

I R A N

Caspian Sea

Pian Sea

Gulf of Oman

Strait of Hormuz

Makran Coast

Dasht-e Kavir

Dasht-e Lut

Iranian Plateau

Zagros Mountains

KHORĀSĀN

GOLESTĀN

MĀZANDARĀN

SEMNĀN

ESFAHĀN

YAZD

KERMĀN

FĀRS

HORMOZGĀN

SISTĀN VA BALŪCHESTĀN

CHAHĀR MAHALL VA BAKHTĪĀRĪ

QOM

Mashhad
TEHRĀN
Esfahān
Shirāz
Kermān
Yazd
Zāhedān
Bandar-e 'Abbās
Qom

BAHRAIN
QATAR
AD DAWHAH (DOHA)
ABŪ ZABY (ABU DHABI)
Dubayy (Dubai)
Ash Shāriqah

Tabriz
Rasht
Gorgān
Bojnūrd
Gonbad-e Kāvūs
Bandar-e Torkaman
Sāri
Bābol
Āmol
Semnān
Sabzevār
Neyshābūr
Torbat-e Heydarīyeh
Torbat-e Jām
Birjand
Kāshmar
Ferdows
Qā'en
Tabas
Deyhūk
Zābol

KAZAKHSTAN

ABUNDANT NATURAL RESOURCES lie in the immense steppe grasslands, deserts, and central plateau of the former Soviet republic of Kazakhstan. An intensive program of industrial and agricultural development to exploit these resources during the Soviet era resulted in catastrophic industrial pollution, including fallout from nuclear testing and the shrinkage of the Aral Sea. Since independence, the government has encouraged foreign investment and liberalized the economy to promote growth. The adoption of Kazakh as the national language is intended to encourage a new sense of national identity in a state where living conditions for the majority remain harsh, both in cramped urban centers and impoverished rural areas.

TRANSPORTATION & INDUSTRY

THE SINGLE MOST IMPORTANT INDUSTRY in Kazakhstan is mining, based around extensive oil deposits near the Caspian Sea, the world's largest chromium mine, and vast reserves of iron ore. Recent foreign investment has helped to develop industries including food processing and steel manufacture, and to expand the exploitation of mineral resources. The Russian space program is still based at Baykonur, near Zhezkazgan in central Kazakhstan.

Major industry and infrastructure

- ⚗ chemicals
- ⚙ engineering
- 🐟 fish processing
- 🍴 food processing
- 🏭 iron & steel
- △ metallurgy
- ⛏ mining
- ⚓ oil
- ■ capital cities
- ● major towns
- ✈ international airports
- — major roads
- ▨ major industrial areas

TRANSPORTATION NETWORK

🛣	87,561 miles (141,000 km)
🛣	none
🚉	8,483 miles (13,660 km)
🚉	none

Industrial areas in the north and east are well-connected to Russia. Air and rail links with Germany and China have been established through foreign investment. Better access to Baltic ports is being sought.

An open-cast coal mine in Kazakhstan. Foreign investment is being actively sought by the Kazakh government in order to fully exploit the potential of the country's rich mineral reserves.

MAP KEY

POPULATION

- ▣ 1 million to 5 million
- ◉ 500,000 to 1 million
- ◎ 100,000 to 500,000
- ⊕ 50,000 to 100,000
- ○ 10,000 to 50,000
- · below 10,000

ELEVATION

- 4000m / 13,124ft
- 3000m / 9843ft
- 2000m / 6562ft
- 1000m / 3281ft
- 500m / 1640ft
- 250m / 820ft
- 100m / 328ft
- sea level

USING THE LAND AND SEA

THE REARING OF LARGE HERDS of sheep and goats on the steppe grasslands forms the core of Kazakh agriculture. Arable cultivation and cotton-growing in pasture and desert areas was encouraged during the Soviet era, but relative yields are low. The heavy use of fertilizers and the diversion of natural water sources for irrigation has degraded much of the land.

THE URBAN/RURAL POPULATION DIVIDE

urban 60%   rural 40%

0 10 20 30 40 50 60 70 80 90 100

POPULATION DENSITY	TOTAL LAND AREA
16 people per sq mile (6 people per sq km)	1,048,878 sq miles (2,717,300 sq km)

Land use and agricultural distribution

- 🐄 cattle
- 🐐 goats
- 🐑 sheep
- cotton
- 🐟 fishing
- wheat
- ■ capital cities
- ● major towns
- pasture
- cropland
- forest
- mountain region
- desert

The nomadic peoples who moved their herds around the steppe grasslands are now largely settled, although echoes of their traditional lifestyle, in particular their superb riding skills, remain.

SCALE 1:6,250,000
(projection: Lambert Conformal Conic)

THE LANDSCAPE

STRETCHING MORE THAN 1,250 MILES (2,000 km) from the Caspian Sea in the west to China in the east, more than 40% of Kazakhstan is covered by steppe grasslands which give way to barren desert in the south. The land rises eastward towards the mineral-rich central plateau, to form the Altai Mountains.

1960   *1996*   *2010*

Since 1960, the Aral Sea has shrunk by 40%, become extremely saline, and lost all but five of its once-abundant fish species. Factors in this ecological disaster include the excessive use of fertilizers, defoliants and the diversion of its main source rivers for the irrigation of desert lands.

The Caspian Sea is the largest body of inland water in the world.

The desert of Peski Bol'shiye Barsuki is mainly sandy, displaying a number of classic dune formations. Groundwater supports a small amount of vegetation.

A large number of salt lakes fill depressions in the rolling uplands of central Kazakhstan.

The Altai Mountains lie on Kazakhstan's eastern borders with China and the Russian Federation. Cold and largely barren, they are the source of many of the rivers which flow across the steppe.

Altai Mountains

Tien Shan

Khrebet Kanchingiz

Aral Sea

Its waters taken for industry and irrigation, the Syr Darya, one of Kazakhstan's major rivers, now barely reaches the Aral Sea which it used to fill. Like many Kazakh rivers it has been heavily polluted with chemicals and its flow has been restricted by up to 60%.

The waters of Lake Balkhash (*Ozero Balkhash*), unlike those of the Aral Sea, are still able to support a fishing industry.

The central Kazakh Uplands (*Kazakhskiy Melkosopochnik*) contain much of the country's mineral riches. The landscape is largely flat with occasional rocky outcrops and hillocks.

Immense stretches of steppe grasslands characterize much of the Kazakh landscape. These lowland areas have been used for arable cultivation in recent years, although problems with irrigation have meant that much of the land is being allowed to revert to its natural vegetation and pastoral usage.

Rows of pine trees edge this valley near Alma-Ata. The snow-covered slopes in the background are used for skiing.

CENTRAL ASIA

KYRGYZSTAN, TAJIKISTAN, TURKMENISTAN, UZBEKISTAN

THE FOUR REPUBLICS that declared independence in 1991 were created in the early years of the Soviet Union, promoting ethnic divisions in a region whose common focus, since the 8th century, has been Islam. Traditional rural, nomadic ways of life have survived the Soviet era, while the benefits of modern industry and grand irrigation schemes have resulted in severe pollution in the delicate, arid environment of the steppe, particularly in Uzbekistan. Many ethnic minority groups are scattered among the four republics, with isolated communities in the mountains of Kyrgyzstan. The current Islamic revival has brought hope of greater regional unity, in spite of religious factionalism which, in 1992, plunged Tajikistan into civil war.

The desert of the Kara Kum (Garagumy) occupies over 70% of Turkmenistan; its wind-scoured surface of dune ridges and depressions severely limits human settlement.

The southern shoreline of the Aral Sea has retreated over 30 miles (48 km) since 1960. A major cause is the diversion of water from the Amu Darya River for irrigation via the Kara Kum Canal (Garagumskiy Kanal).

MAP KEY

POPULATION

- 1 million to 5 million
- 500,000 to 1 million
- 100,000 to 500,000
- 50,000 to 100,000
- 10,000 to 50,000
- below 10,000

ELEVATION

- 6000m / 19,686ft
- 4000m / 13,124ft
- 3000m / 9843ft
- 2000m / 6562ft
- 1000m / 3281ft
- 500m / 1640ft
- 250m / 820ft
- 100m / 328ft
- sea level

TRANSPORTATION & INDUSTRY

FOSSIL FUELS ARE extracted and processed in all four states, with scope for further exploitation. Agriculture provides raw materials for many industries, including food and textiles processing, and the manufacture of leather goods, clothing, and carpets. Farm machinery is also produced.

TRANSPORTATION NETWORK

🛣 85,574 miles (137,800 km)		🛤 None	
🚆 4,184 miles (6,738 km)		🚄 1,180 miles (1,900 km)	

The Kara Kum Canal (*Garagumskiy Kanal*) runs for 870 miles (1,400 km) from the Amu Darya River to the Caspian Sea. The canal is principally used for irrigation but is navigable for 280 miles (450 km).

Major industry and infrastructure
- carpet weaving
- chemicals
- engineering
- food processing
- oil & gas
- textiles
- capital cities
- major towns
- international airports
- major roads
- major industrial areas

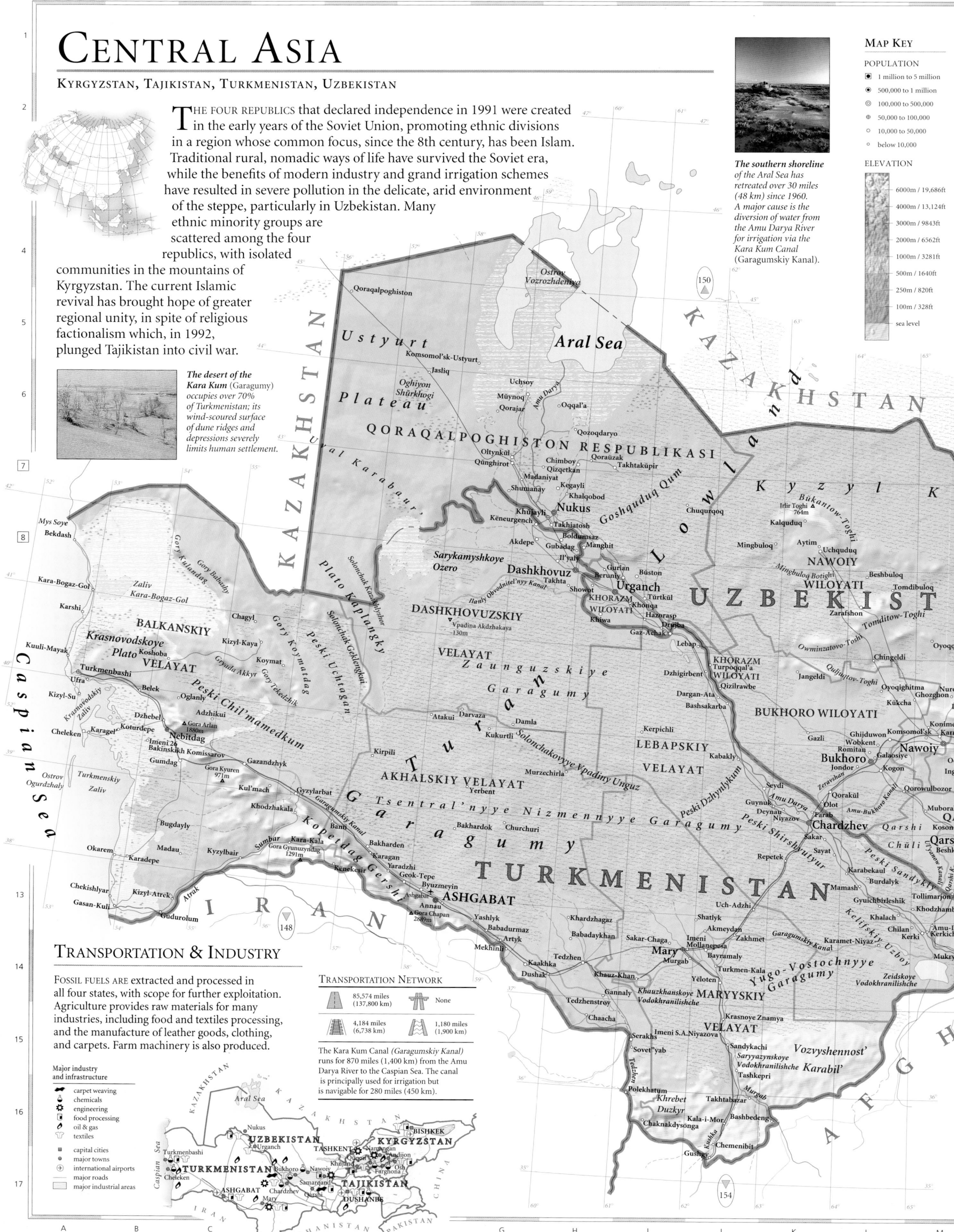

THE LANDSCAPE

THE GREAT TIEN SHAN and Pamir Ranges meet in a succession of high mountain chains. These mountains encircle the fertile Fergana Valley and reach west into the desert of the Kyzyl Kum, dividing the Syr Darya and Amu Darya Rivers. Sandy steppeland extends to the shores of the Caspian Sea, with the desert of the Kara Kum *(Garagumy)* in the south. The Amu Darya drains into the Aral Sea in the north.

Salt marshes fill many of the depressions in the Ustyurt Plateau, a barren, rocky tableland about 650 ft (200 m) above sea level.

Some of the world's largest deposits of marine salts are found in Zaliv Kara-Bogaz-Gol. This shallow, saline gulf has an average depth of only 33 ft (10 m), and a very high evaporation rate, producing the salty deposits.

The Kara Kum *(Garagumy)* is one of the world's largest expanses of sand. Wind action has created a terrain of shifting, crescent-shaped sand dunes known as *barchans*.

The Amu Darya is the only river in Central Asia with a sufficient volume of water to cross the desert of the Kara Kum *(Garagumy)* from the Pamirs to the Aral Sea, where it forms a delta largely vegetated by scrub grasses.

A series of major rock faults has created the Fergana Valley, a deep depression surrounded by high mountains. Water from the Syr Darya River and from underground sources supports intensive agriculture, despite minimal rainfall.

Shock waves travel through ground
Epicentre
Fault

In the heavily-fractured and faulted mountain region, earthquakes are common, caused by the sudden release of tension along active fault lines.

Earthquake zone

Kyzyl Kum

Mount Communism *(Qullai Kommunizm)*, in the northern Pamirs, was so named for being the highest point in the former Soviet Union, rising to 24,590 ft (7,495 m).

Syr Darya

Naryn River

Qarokŭl

Nestling high in the Pamir range, and fed by glacial meltwater, Qarokŭl is the largest of the lakes in this region.

Ozero Issyk-Kul' lies at an altitude of 5,193 ft (1,584 m). The lake remains ice-free throughout the year, due to the slight salinity of the water.

Tien Shan

The Tien Shan extend from China in the east, reaching heights over 24,400 ft (7,439 m) and branching into many parallel ranges in the west.

Bare mountains provide a stark background to the croplands along the Naryn River in Kyrgyzstan. Irrigation is essential for cultivation in this dry region.

SCALE 1:4,250,000
(projection: Lambert Conformal Conic)

Km
0 10 20 40 60 80 100 120

Miles
0 20 40 60 80 100 120

USING THE LAND

CROPLAND OUTSIDE Kyrgyzstan is restricted to irrigated areas such as the Fergana Valley. Central Asia is a leading global producer of cotton, and traditional silk-farming remains widespread. A wide range of fruits, vegetables, and grains are grown and livestock raised includes horses, goats, and karakul sheep.

Land use and agricultural distribution

- cattle
- goats
- sheep
- cereals
- cotton
- fruit
- capital cities
- major towns
- pasture
- cropland
- desert
- wetland

Plentiful sunshine, rich soils and massive irrigation schemes have made Uzbekistan the world's third largest cotton producer, although water shortages now prevent any further expansion of irrigated land.

THE URBAN/RURAL POPULATION DIVIDE

urban 40%   rural 60%

0 10 20 30 40 50 60 70 80 90 100

POPULATION DENSITY
79 people per sq mile
(31 people per sq km)

TOTAL LAND AREA
492,961 sq miles
(1,277,100 sq km)

AFGHANISTAN & PAKISTAN

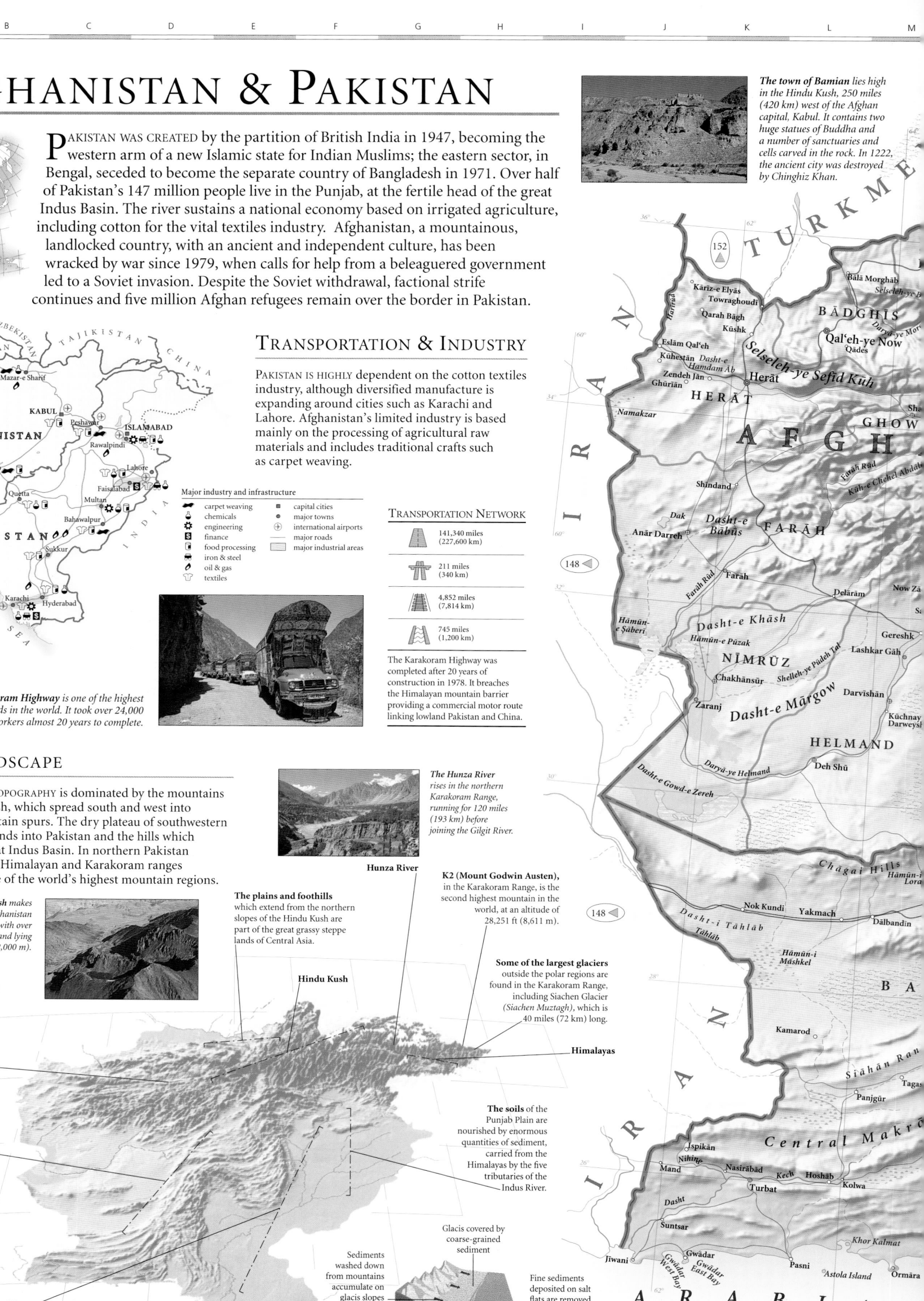

PAKISTAN WAS CREATED by the partition of British India in 1947, becoming the western arm of a new Islamic state for Indian Muslims; the eastern sector, in Bengal, seceded to become the separate country of Bangladesh in 1971. Over half of Pakistan's 147 million people live in the Punjab, at the fertile head of the great Indus Basin. The river sustains a national economy based on irrigated agriculture, including cotton for the vital textiles industry. Afghanistan, a mountainous, landlocked country, with an ancient and independent culture, has been wracked by war since 1979, when calls for help from a beleaguered government led to a Soviet invasion. Despite the Soviet withdrawal, factional strife continues and five million Afghan refugees remain over the border in Pakistan.

The town of Bamian lies high in the Hindu Kush, 250 miles (420 km) west of the Afghan capital, Kabul. It contains two huge statues of Buddha and a number of sanctuaries and cells carved in the rock. In 1222, the ancient city was destroyed by Chinghiz Khan.

TRANSPORTATION & INDUSTRY

PAKISTAN IS HIGHLY dependent on the cotton textiles industry, although diversified manufacture is expanding around cities such as Karachi and Lahore. Afghanistan's limited industry is based mainly on the processing of agricultural raw materials and includes traditional crafts such as carpet weaving.

Major industry and infrastructure

- carpet weaving
- chemicals
- engineering
- finance
- food processing
- iron & steel
- oil & gas
- textiles
- capital cities
- major towns
- international airports
- major roads
- major industrial areas

TRANSPORTATION NETWORK

141,340 miles (227,600 km)	
211 miles (340 km)	
4,852 miles (7,814 km)	
745 miles (1,200 km)	

The Karakoram Highway was completed after 20 years of construction in 1978. It breaches the Himalayan mountain barrier providing a commercial motor route linking lowland Pakistan and China.

The Karakoram Highway is one of the highest major roads in the world. It took over 24,000 workers almost 20 years to complete.

THE LANDSCAPE

AFGHANISTAN'S TOPOGRAPHY is dominated by the mountains of the Hindu Kush, which spread south and west into numerous mountain spurs. The dry plateau of southwestern Afghanistan extends into Pakistan and the hills which overlook the great Indus Basin. In northern Pakistan the Hindu Kush, Himalayan and Karakoram ranges meet to form one of the world's highest mountain regions.

The Hunza River rises in the northern Karakoram Range, running for 120 miles (193 km) before joining the Gilgit River.

Hunza River

K2 (Mount Godwin Austen), in the Karakoram Range, is the second highest mountain in the world, at an altitude of 28,251 ft (8,611 m).

The arid Hindu Kush makes much of Afghanistan uninhabitable, with over 50% of the land lying above 6,500 ft (2,000 m).

The plains and foothills which extend from the northern slopes of the Hindu Kush are part of the great grassy steppe lands of Central Asia.

Some of the largest glaciers outside the polar regions are found in the Karakoram Range, including Siachen Glacier (Siachen Muztagh), which is 40 miles (72 km) long.

Frequent earthquakes mean that mountain-building processes are continuing in this region, as the Indo-Australian Plate drifts northward, colliding with the Eurasian Plate.

Hindu Kush

Himalayas

Mountain chains running southwest from the Hindu Kush into Pakistan form a barrier to the humid winds which blow from the Indian Ocean, creating arid conditions across southern Afghanistan.

The soils of the Punjab Plain are nourished by enormous quantities of sediment, carried from the Himalayas by the five tributaries of the Indus River.

The Indus Basin is part of the Indus-Ganges lowland, a vast depression which has been filled with layers of sediment over the last 50 million years. These deposits are estimated to be over 16,400 ft (5,000 m) deep.

The Indus Delta is prone to heavy flooding and high levels of salinity. It remains a largely uncultivated wilderness area.

Sediments washed down from mountains accumulate on glacis slopes

Glacis covered by coarse-grained sediment

Fine sediments deposited on salt flats are removed by wind erosion

Bedrock

Glacis are gentle, debris-covered slopes which lead into saltflats or deserts. They typically occur at the base of mountains in arid regions such as Afghanistan.

SCALE 1:4,500,000
(projection: Lambert Conformal Conic)

Km
0 10 20 40 60 80 100 120 140 160 180 200

Miles
0 10 20 40 60 80 100 120 140 160 180 200

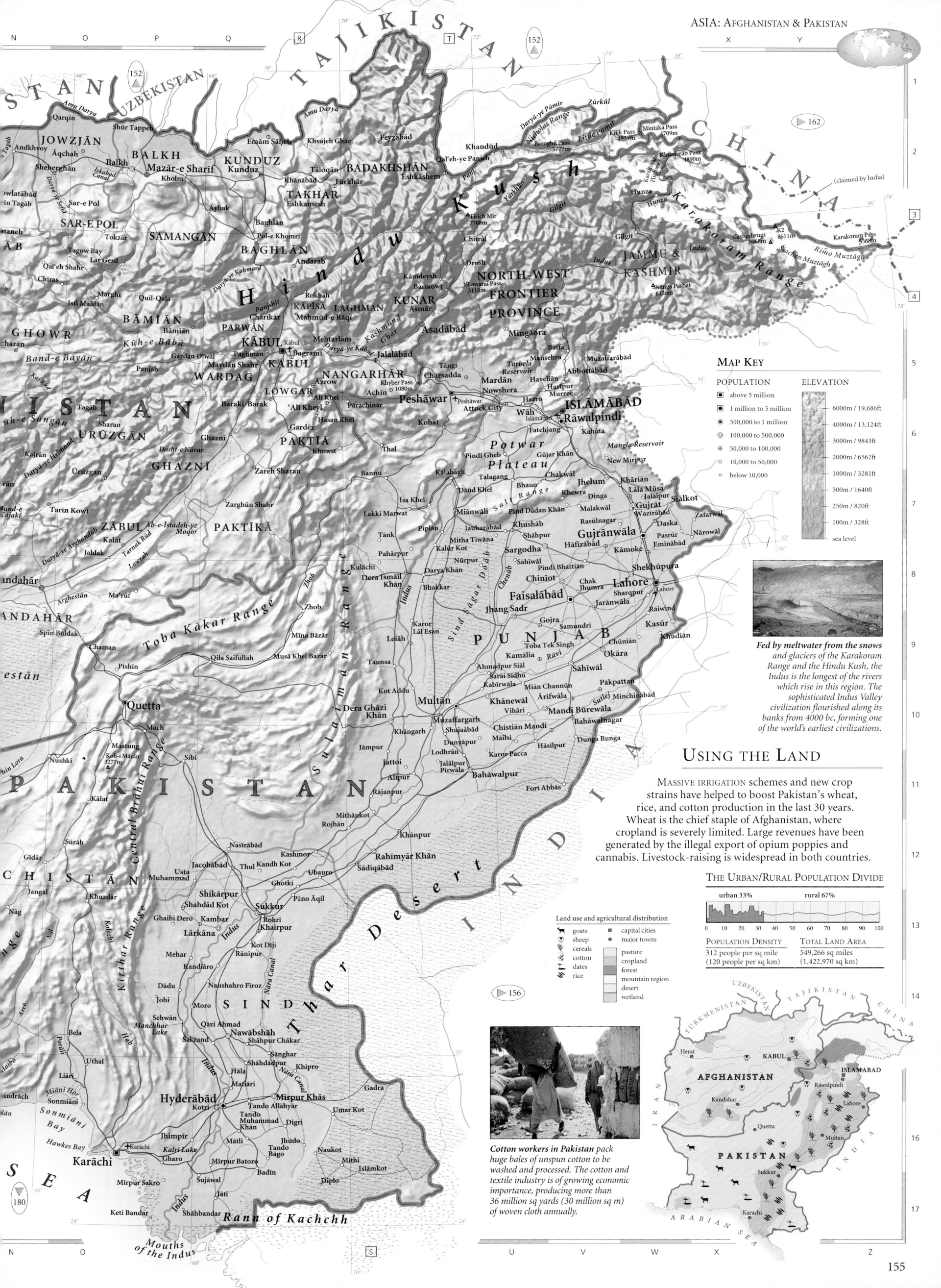

MAP KEY

POPULATION
- ■ above 5 million
- ■ 1 million to 5 million
- ▣ 500,000 to 1 million
- ▢ 100,000 to 500,000
- ▫ 50,000 to 100,000
- □ 10,000 to 50,000
- ○ below 10,000

ELEVATION
- 6000m / 19,686ft
- 4000m / 13,124ft
- 3000m / 9843ft
- 2000m / 6562ft
- 1000m / 3281ft
- 500m / 1640ft
- 250m / 820ft
- 100m / 328ft
- sea level

Fed by meltwater from the snows and glaciers of the Karakoram Range and the Hindu Kush, the Indus is the longest of the rivers which rise in this region. The sophisticated Indus Valley civilization flourished along its banks from 4000 bc, forming one of the world's earliest civilizations.

USING THE LAND

MASSIVE IRRIGATION schemes and new crop strains have helped to boost Pakistan's wheat, rice, and cotton production in the last 30 years. Wheat is the chief staple of Afghanistan, where cropland is severely limited. Large revenues have been generated by the illegal export of opium poppies and cannabis. Livestock-raising is widespread in both countries.

THE URBAN/RURAL POPULATION DIVIDE

urban 33%	rural 67%

0 10 20 30 40 50 60 70 80 90 100

POPULATION DENSITY	TOTAL LAND AREA
312 people per sq mile (120 people per sq km)	549,266 sq miles (1,422,970 sq km)

Land use and agricultural distribution
- goats
- sheep
- cereals
- cotton
- dates
- rice
- ■ capital cities
- ● major towns
- pasture
- cropland
- forest
- mountain region
- desert
- wetland

Cotton workers in Pakistan pack huge bales of unspun cotton to be washed and processed. The cotton and textile industry is of growing economic importance, producing more than 36 million sq yards (30 million sq m) of woven cloth annually.

SOUTH ASIA

BANGLADESH, BHUTAN, INDIA, MALDIVES, NEPAL, PAKISTAN, SRI LANKA

MORE THAN ONE-FIFTH of the world's population lives in the south Asian subcontinent. Great cultural diversity has come from a long succession of foreign invaders, including Hindu Aryans, Islamic Moguls, and the British, whose empire incorporated the princely states of the Maharajas and extended to the borders of Nepal and Bhutan in the Himalayas. Half a century after independence, India is the world's largest democracy, and at the current rate of growth, may overtake China as the world's most populous country within the next century. There are points of tension in the region over claims for independence by the Sikhs in the Indian Punjab and the Tamil separatists in Sri Lanka, and the long-standing dispute with Pakistan over Jammu and Kashmir in the north.

THE LANDSCAPE

SOUTH ASIA is effectively isolated from the rest of Asia by desert along the western flank of Pakistan, and a continuous wall of mountains, dominated by the Himalayas, to the north and east. The great basins of the Indus and Ganges separate this mountain fringe from the rolling plateau of the Indian peninsula, which is bordered by a line of coastal hills, the Eastern and Western Ghats.

The towering Karakoram and Hindu Kush ranges, formed at the same time as the Himalayas, dominate Pakistan's northern borders. K2 on the border of northern Pakistan is the second highest mountain on Earth, at 28,251 ft (8,611 m).

The Indus River flows more than 1,970 miles (3,180 km) from southwestern Tibet to its mouth on the Arabian Sea. It has an estimated catchment area of 450,000 sq miles (1,165,500 sq km).

The coast of western Pakistan is a staircase of folded rock strata caused by successive periods of rapid uplift.

The Indus Valley near Skardu in northern Pakistan has been partially infilled by great quantities of eroded sediment. Most of this is carried from the region's bare slopes by swollen rivers during the spring thaw and mass movement activity.

The Himalayas are the highest and most extensive mountain system in the world. They were formed when the Indo-Australian Plate collided with the Eurasian Plate about 40 million years ago, thrusting up huge masses of land and creating a "ripple" effect, which formed lesser mountain ranges in Tibet and Southeast Asia. Mount Everest is the world's tallest mountain at 29,028 ft (8,848 m).

Almost all of Bangladesh lies in the immense delta formed by the Ganges and the Brahmaputra which merge and flow out into the Bay of Bengal.

Ganges Delta

Deccan Plateau

Layers of volcanic basalt

Stepped valleys or 'traps'

The Deccan Plateau covers an area of more than 123,553 sq miles (320,000 sq km). It is formed of deep layers of volcanic basalt, reaching thicknesses of more than 9,800 ft (3,000 m) toward the coast. Distinctive stepped valleys cut in the basalt plateau by rivers are known as "traps."

Coastal deposition has formed many typical features along the western coast of Sri Lanka. These include spits and bars, sometimes enclosing lagoons.

Eastern Ghats

Trivandrum in southern India normally receives the first of the monsoon rains, which are essential to south Asian agriculture and moderate the extreme summer heat. The monsoon then moves northward over a period of about two months.

The Western Ghats are formed by a fault scarp which runs unbroken for more than 930 miles (1,500 km). They reach their highest point at the southern Cardamon Hills.

Rivers flowing from the Himalayas into a broad depression in northern India have formed marshes around Bharatpur. They are now a sanctuary for numerous bird species.

Bharatpur

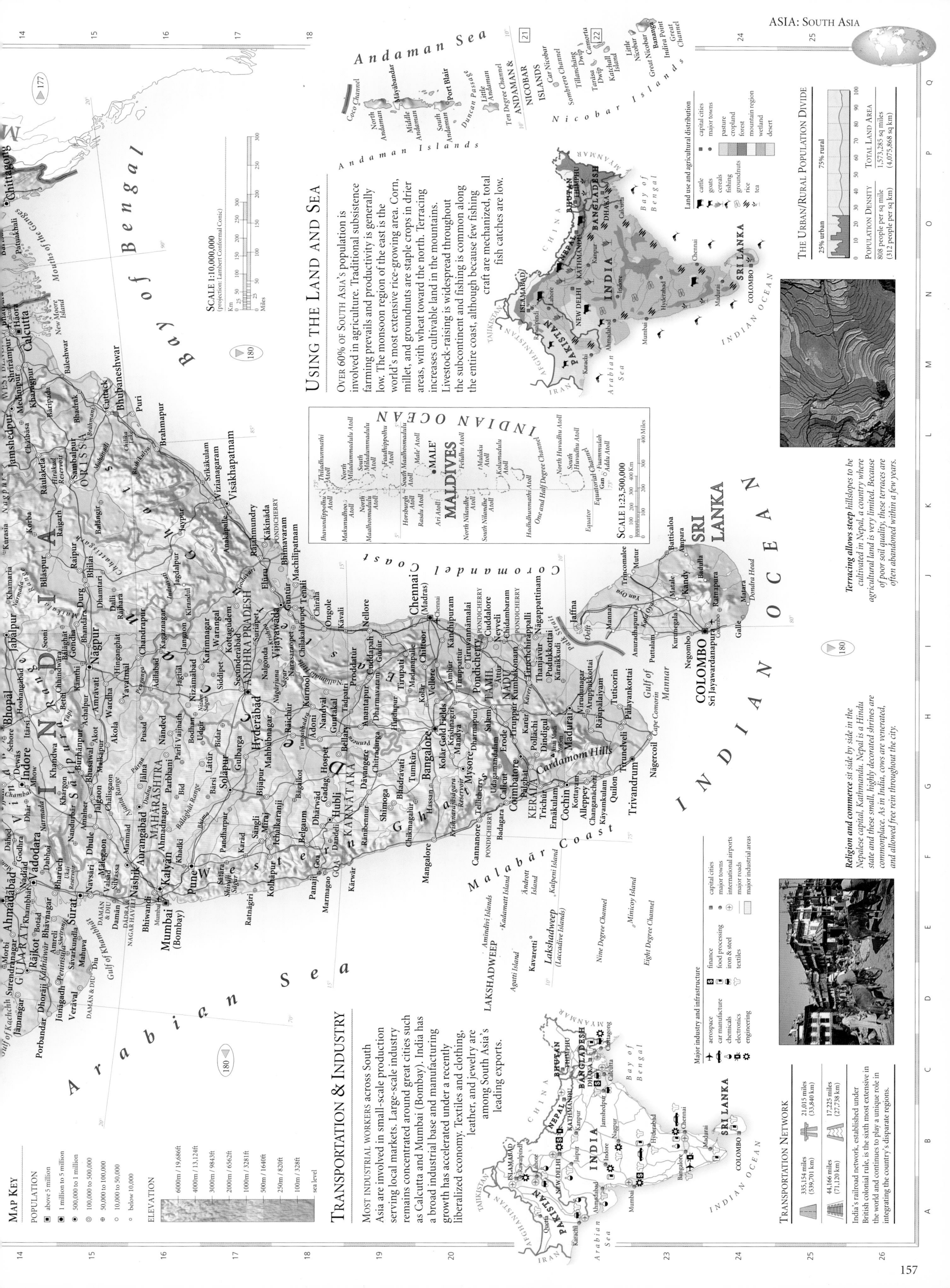

USING THE LAND AND SEA

OVER 60% OF SOUTH ASIA'S population is involved in agriculture. Traditional subsistence farming prevails and productivity is generally low. The monsoon region of the east is the world's most extensive rice-growing area. Corn, millet, and groundnuts are staple crops in drier areas, with wheat toward the north. Terracing increases cultivable land in the mountains. Livestock-raising is widespread throughout the subcontinent and fishing is common along the entire coast, although because few fishing craft are mechanized, total fish catches are low.

Terracing allows steep hillslopes to be cultivated in Nepal, a country where agricultural land is very limited. Because of poor soil quality, these terraces are often abandoned within a few years.

Religion and commerce sit side by side in the Nepalese capital, Kathmandu. Nepal is a Hindu state and these small, highly decorated shrines are commonplace. As in India, cows are venerated, and allowed free rein throughout the city.

TRANSPORTATION & INDUSTRY

MOST INDUSTRIAL WORKERS across South Asia are involved in small-scale production serving local markets. Large-scale industry remains concentrated around great cities such as Calcutta and Mumbai (Bombay). India has a broad industrial base and manufacturing growth has accelerated under a recently liberalized economy. Textiles and clothing, leather, and jewelry are among South Asia's leading exports.

Major industry and infrastructure

- ✈ aerospace
- 🚗 car manufacture
- chemicals
- ⚙ electronics
- ⚙ engineering
- S finance
- 🏭 food processing
- iron & steel
- 👔 textiles
- ● capital cities
- ■ major towns
- ✈ international airports
- major roads
- major industrial areas

THE URBAN/RURAL POPULATION DIVIDE

25% urban	75% rural

POPULATION DENSITY	TOTAL LAND AREA
808 people per sq mile (312 people per sq km)	1,573,285 sq miles (4,075,868 sq km)

TRANSPORTATION NETWORK

335,154 miles (539,701 km)	21,015 miles (33,840 km)	44,166 miles (71,120 km)
	17,225 miles (27,738 km)	

India's railroad network, established under British colonial rule, is the sixth most extensive in the world and continues to play a unique role in integrating the country's disparate regions.

MAP KEY

POPULATION

- ● above 5 million
- ■ 1 million to 5 million
- ● 500,000 to 1 million
- ⊕ 100,000 to 500,000
- ⊕ 50,000 to 100,000
- ○ 10,000 to 50,000
- ○ below 10,000

ELEVATION

- 6000m / 19,686ft
- 4000m / 13,124ft
- 3000m / 9843ft
- 2000m / 6562ft
- 1000m / 3281ft
- 500m / 1640ft
- 250m / 820ft
- 100m / 328ft
- sea level

SCALE 1:10,000,000
(projection: Lambert Conformal Conic)

SCALE 1:23,500,000

NORTHERN INDIA & THE HIMALAYAN STATES

Bangladesh, Bhutan, Nepal, Arunachal Pradesh, Assam, Bihar, Chandigarh, Delhi, Haryana, Himachal Pradesh, Jammu & Kashmir, Manipur, Meghalaya, Mizoram, Nagaland, Punjab, Rajasthan, Sikkim, Tripura, Uttar Pradesh, West Bengal

THE GANGES AND BRAHMAPUTRA river basins and the massive mountain barrier of the Himalayas define this region's landscape and have served to reinforce potent cultural and religious differences among its people. Hinduism pervades most aspects of national life and is a growing political force within India, a secular country which also encompasses the center of Sikhism at Amritsar and the world's largest Muslim minority. Nepal is a crowded mountain state, which faces severe ecological problems from deforestation, while the tiny Himalayan Buddhist kingdom of Bhutan is emerging from long-term isolation, to welcome selected visitors. The Muslim state of Bangladesh, formerly East Pakistan, is one of the world's most densely populated countries and one of the poorest, with more than 120 million people living largely on the massive Ganges/Brahmaputra Delta. Many Bangladeshis live under threat of repeated, catastrophic floods.

The Golden Temple in Amritsar, the most sacred shrine of the Sikh religion, was the scene of violent clashes between Sikh separatists and government forces in 1984.

MAP KEY

POPULATION

- ▣ 1 million to 5 million
- ◉ 500,000 to 1 million
- ◎ 100,000 to 500,000
- ⊕ 50,000 to 100,000
- ○ 10,000 to 50,000
- ∘ below 10,000

ELEVATION

- 6000m / 19,686ft
- 4000m / 13,124ft
- 3000m / 9843ft
- 2000m / 6562ft
- 1000m / 3281ft
- 500m / 1640ft
- 250m / 820ft
- 100m / 328ft
- sea level

TRANSPORTATION & INDUSTRY

TEXTILES, ENGINEERING, chemicals, and electronics are leading industries in north India. The plateau of Chota Nagpur provides ore for iron and steel production in the major industrial region northeast of Calcutta. Bangladesh processes jute and Nepal has a small manufacturing sector based on agricultural produce, while Bhutan's limited industry is concentrated in the southern lowland area.

Major industry and infrastructure

- ⚓ adventure tourism
- 🚗 car manufacture
- chemicals
- coal
- electronics
- engineering
- finance
- food processing
- iron & steel
- jute processing
- oil
- tea processing
- textiles
- ■ capital cities
- ● major towns
- ✈ international airports
- major roads
- major industrial areas

TRANSPORTATION NETWORK

Over 60% of Bangladesh's internal trade is carried by boat. The country has a very disjointed land transportation network, with no bridges over the Brahmaputra and few road crossings on the Ganges River.

THE LANDSCAPE

MOST OF THE REGION is drained by the Ganges River, which meets the Brahmaputra in Bangladesh to form an immense delta before flowing into the Bay of Bengal. The Himalayas extend eastward over 1,500 miles (2,400 km), from the parallel ranges running through Jammu and Kashmir. The Thar Desert occupies the southwest.

The Indian Punjab lies mainly to the west of the Ganges watershed and its rivers flow into the Indus. Control of this water resource has been a source of great friction with neighboring Pakistan.

The border between India and Pakistan runs through the Thar Desert, an area of sandy *seif* dunes 50–100 ft (15–30 m) in height. Fossils found in the desert indicate that the dunes, stabilized by vegetation, have been in their current position for about 3,000 years.

Sambhar Salt Lake in Rajasthan is India's largest lake. Unlike most of the Himalayan lakes which are glacial in origin – formed in ice-scoured basins or as the result of depositional damming – it is an ephemeral salt lake filled periodically by flash flooding.

The Pir Panjal Range in southwestern Kashmir rises to elevations of 12,500 ft (3,810 m). Despite the freezing conditions, settlements and extensive pastures are found above the tree line.

The Ganges River, sacred to the Hindu people, drains a vast lowland area at the base of the Himalayas. The northern plains are covered by sandy deposits, broken by mud-banks formed when the river floods.

The rapid deforestation of Himalayan valleys has led to acute soil erosion and increased rates of rainwater runoff, both cited as possible causes of the worsening floods downstream in the Ganges/Brahmaputra Delta, although natural rates are high and may be the real cause.

The northern ranges of the Himalayas contain the highest mountains in the world, with average heights of more than 23,000 ft (7,000 m) and many peaks higher than 26,000 ft (8,000 m).

In the last 40 million years, the course of the Brahmaputra has been diverted hundreds of miles to the east by the rising landmass of the Himalayas.

Over half of the great Ganges/Brahmaputra Delta floods each year during the monsoon as rivers, swollen by meltwater from the Himalayas and by excess rainwater, break their banks and fertilize the land with nutrient-rich sediment.

The Khasi Hills are an example of a *horst*, a fractured block of bedrock which has been thrust upward.

The summit of Machhapuchhre rises to 22,942 ft (6,993 m). It is also known as the "Fish's Tail" because of its distinctive peak.

Debris slides in the middle Himalayas

Soil blocks

Debris fans at base of slope

Slide plain

Soil loss in the middle Himalayas has largely been attributed to debris slides, where large blocks of soil are mobilized by saturation along a slide plane. Once mobile, the soil slides down the slope, gaining speed and thinning to form a fan at the base of the slope.

USING THE LAND

GRAIN PRODUCTION dominates land use. Rice is most widely grown in the east. Irrigation and new crop strains have dramatically increased yields in the Punjab, a major wheat-producing area. River floodplains are intensively farmed and livestock-herding is widespread, particularly in Bhutan. Regional crops include jute in Bangladesh, tea in Assam, cardamom in Sikkim, and saffron in Kashmir.

THE URBAN/RURAL POPULATION DIVIDE

urban 23%   rural 77%

0 10 20 30 40 50 60 70 80 90 100

POPULATION DENSITY
782 people per sq mile
(302 people per sq km)

TOTAL LAND AREA
665,104 sq miles
(1,723,068 sq km)

Land use and agricultural distribution

- cattle
- goats
- sheep
- cereals
- jute
- rice
- tea
- capital cities
- major towns
- pasture
- cropland
- forest
- mountain region
- wetland
- desert

An adverse climate, steep slopes, and poor soils limit crop cultivation in Bhutan, which is a largely agrarian economy. Rice, corn, and wheat are the main staples, although orchards are being established as the soil and climate suit this type of farming.

Flooded streets in Dhaka, Bangladesh are a testament to the region's vulnerability to flooding. In 1988 alone, 75% of the country was flooded, leaving thousands of people dead and over 25 million homeless.

Southern India & Sri Lanka

Sri Lanka, Andhra Pradesh, Dadra & Nagar Haveli, Daman & Diu, Goa, Gujarat, Karnataka, Kerala, Lakshadweep, Madhya Pradesh, Maharashtra, Orissa, Pondicherry, Tamil Nadu

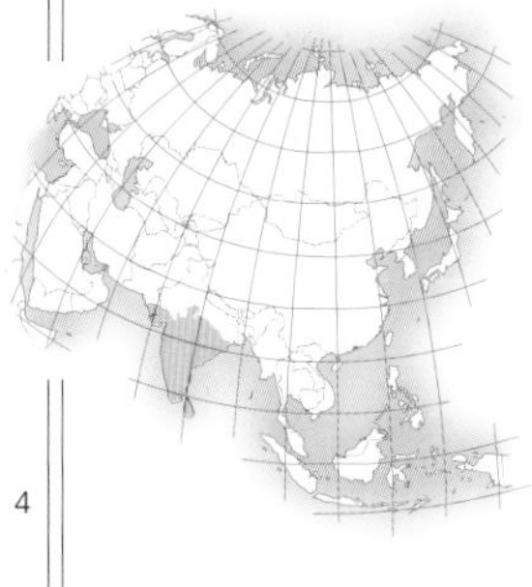

THE UNIQUE AND HIGHLY INDEPENDENT southern states reflect the diverse and decentralized nature of India, which has fourteen official languages. The southern half of the peninsula lay beyond the reach of early invaders from the north and retained the distinct and ancient culture of the Dravidian peoples such as the Tamils, whose language is spoken in preference to Hindi throughout southern India. The interior plateau of southern India is less densely populated than the coastal lowlands, where the European colonial imprint is strongest. Urban and industrial growth is accelerating, but southern India's vast population remains predominantly rural. The island of Sri Lanka has two distinct cultural groups; the mainly Buddhist Sinhalese majority, and the Tamil minority whose struggle for a homeland in the northeast has led to prolonged civil war.

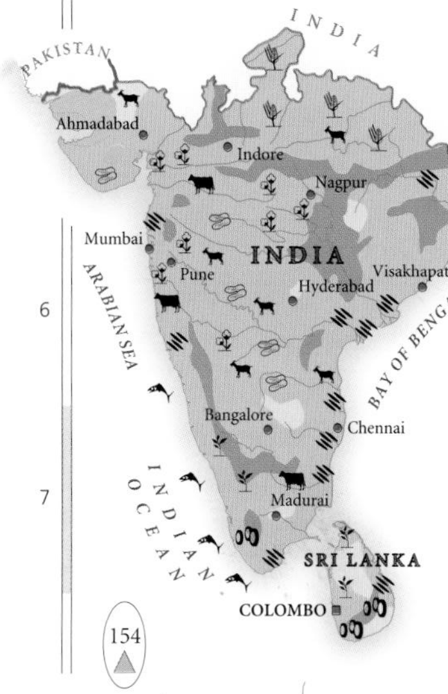

Using the Land and Sea

RICE IS THE MAIN staple in the east, in Sri Lanka and along the humid Malabar Coast. Peanuts are grown on the Deccan Plateau, with wheat, corn, and chickpeas, toward the north. Sri Lanka is a leading exporter of tea, coconuts and rubber. Cotton plantations supply local mills around Nagpur and Mumbai (Bombay). Fishing supports many communities in Kerala and the Laccadive Islands.

Commercial plantations, growing tea, (seen here), cardamom, coffee, coconuts, and rubber, occupy about half the agricultural land in Kerala, necessitating food imports for local consumption.

THE URBAN/RURAL POPULATION DIVIDE

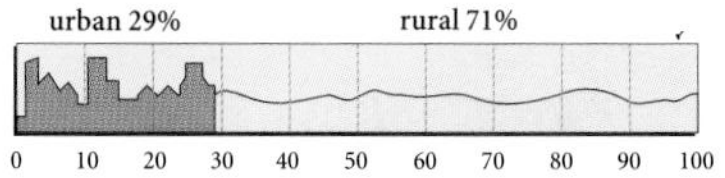

urban 29%   rural 71%

POPULATION DENSITY	TOTAL LAND AREA
715 people per sq mile (276 people per sq km)	698,295 sq miles (1,809,054 sq km)

Land use and agricultural distribution

- cattle
- goats
- cereals
- cotton
- fishing
- groundnuts
- rice
- rubber
- tea

- capital cities
- major towns
- pasture
- cropland
- forest
- wetland

The Landscape

THE UNDULATING DECCAN PLATEAU underlies most of southern India; it slopes gently down toward the east and is largely enclosed by the Ghats coastal hill ranges. The Western Ghats run continuously along the Arabian Sea coast, while the Eastern Ghats are interrupted by rivers which follow the slope of the plateau and flow across broad lowlands into the Bay of Bengal. The plateaus and basins of Sri Lanka's central highlands are surrounded by a broad plain.

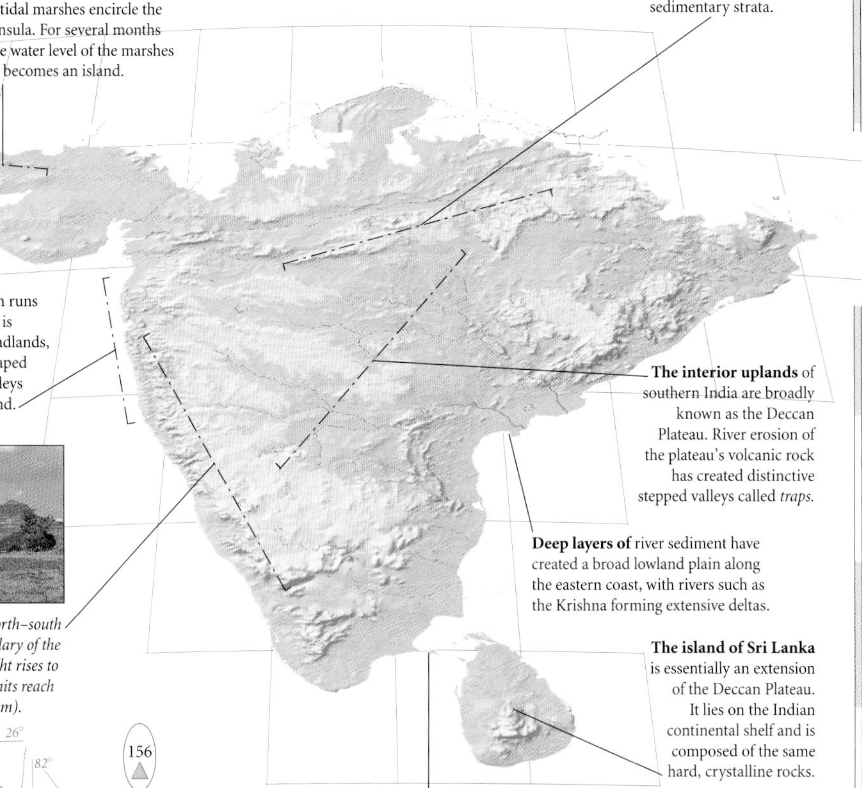

Along the northern boundary of the Deccan Plateau, old basement rocks are interspersed with younger sedimentary strata. This creates spectacular scarplands, cut by numerous waterfalls along the softer sedimentary strata.

The interior uplands of southern India are broadly known as the Deccan Plateau. River erosion of the plateau's volcanic rock has created distinctive stepped valleys called *traps*.

The island of Sri Lanka is essentially an extension of the Deccan Plateau. It lies on the Indian continental shelf and is composed of the same hard, crystalline rocks.

Deep layers of river sediment have created a broad lowland plain along the eastern coast, with rivers such as the Krishna forming extensive deltas.

The Rann of Kachchh tidal marshes encircle the low-lying Kachchh Peninsula. For several months during the rainy season the water level of the marshes rises and Kachchh becomes an island.

The Konkan coast, which runs between Daman and Goa, is characterized by rocky headlands, and bays with crescent-shaped beaches. Flooded river valleys known as *rías* extend inland.

The Western Ghats run north–south marking the western boundary of the Deccan Plateau. Their height rises to the south where their summits reach altitudes of 8,000 ft (2,500 m).

Adam's Bridge

Ocean currents cause sediment build up

Sri Lanka

Relict of ancient tombolo

Adam's Bridge

Adam's Bridge (Rama's Bridge) is a chain of sandy shoals lying about 4 ft (1.2 m) under the sea between India and Sri Lanka. They once formed the world's longest tombolo, or land bridge, before the sea level began to rise several thousand years ago.

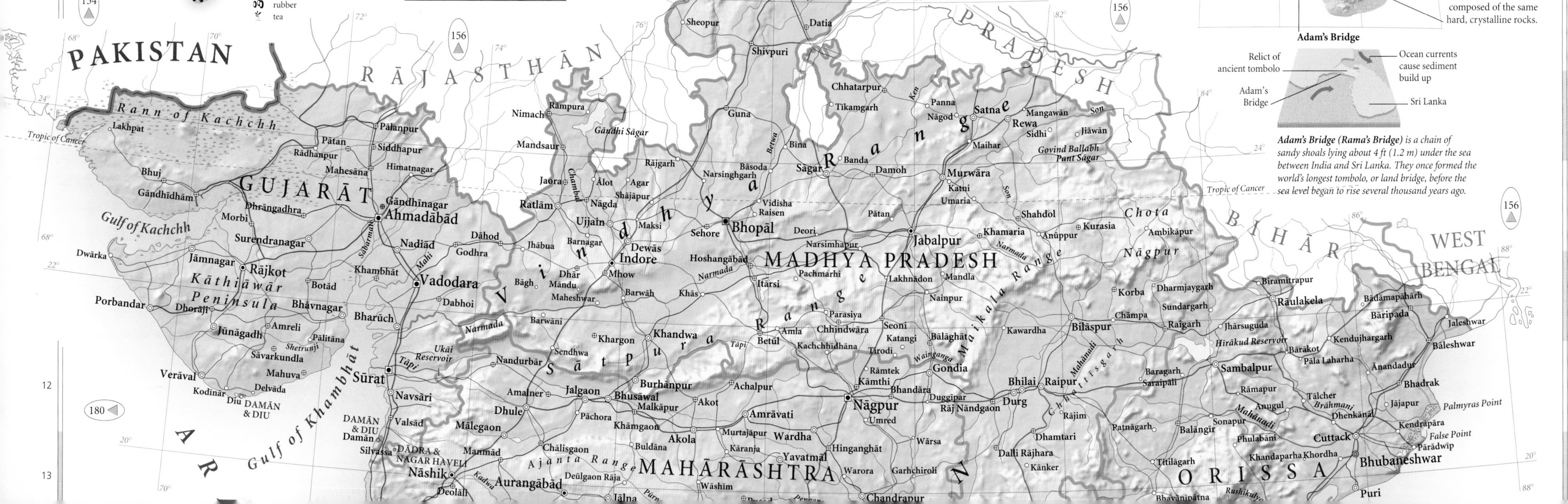

Tropic of Cancer

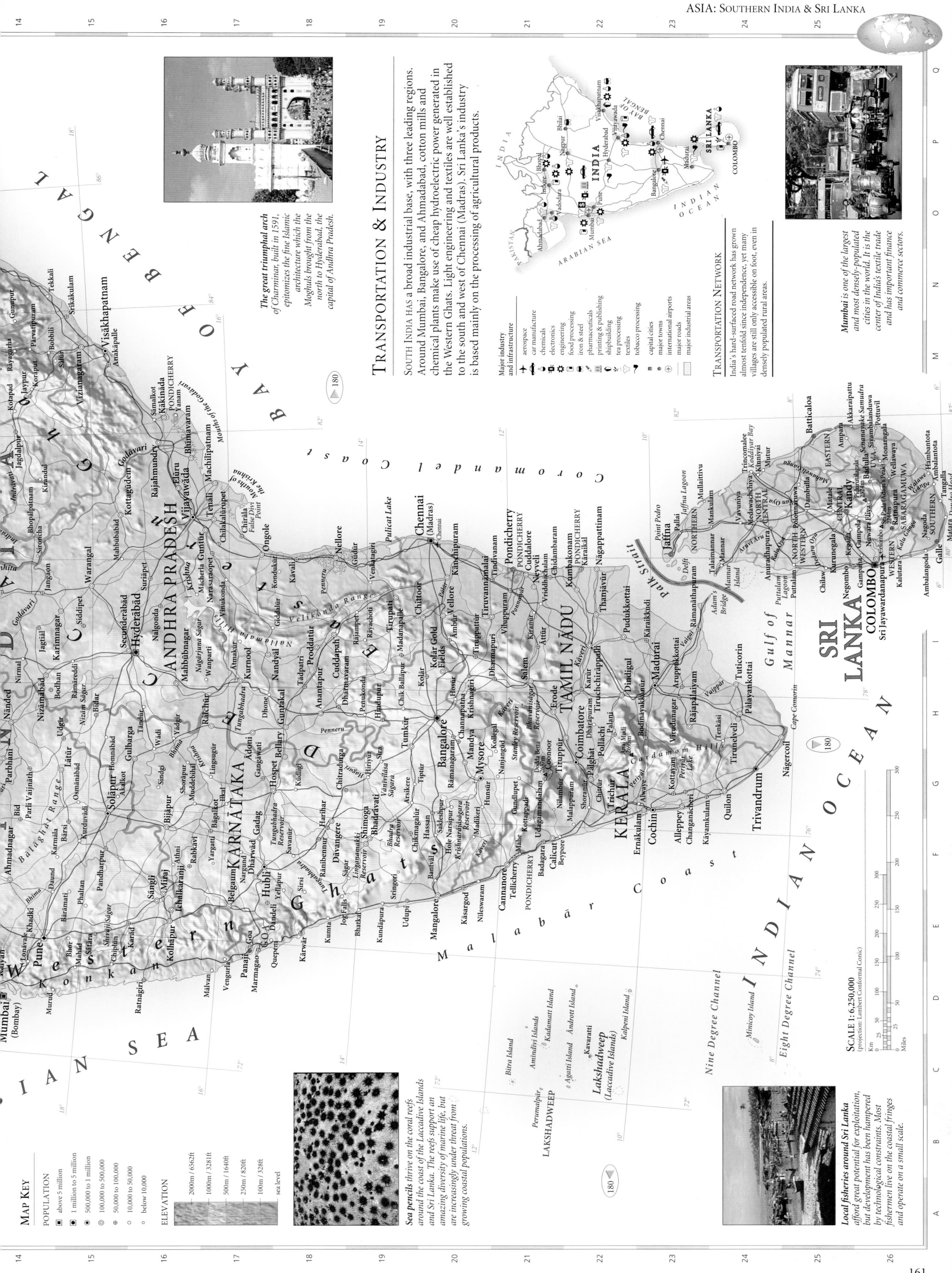

The great triumphal arch of Charminar, built in 1591, epitomizes the fine Islamic architecture which the Moghuls brought from the north to Hyderabad, the capital of Andhra Pradesh.

TRANSPORTATION & INDUSTRY

SOUTH INDIA HAS a broad industrial base, with three leading regions. Around Mumbai, Bangalore, and Ahmadabad, cotton mills and chemical plants make use of cheap hydroelectric power generated in the Western Ghats. Light engineering and textiles are well established to the south and west of Chennai (Madras). Sri Lanka's industry is based mainly on the processing of agricultural products.

Major industry and infrastructure

- aerospace
- car manufacture
- chemicals
- electronics
- engineering
- food processing
- iron & steel
- pharmaceuticals
- printing & publishing
- shipbuilding
- tea processing
- textiles
- tobacco processing
- capital cities
- major towns
- international airports
- major roads
- major industrial areas

TRANSPORTATION NETWORK

India's hard-surfaced road network has grown almost tenfold since independence, yet many villages are still only accessible on foot, even in densely populated rural areas.

Mumbai is one of the largest and most densely-populated cities in the world. It is the center of India's textile trade and has important finance and commerce sectors.

Sea pencils thrive on the coral reefs around the coast of the Laccadive Islands and Sri Lanka. The reefs support an amazing diversity of marine life, but are increasingly under threat from growing coastal populations.

Local fisheries around Sri Lanka afford great potential for exploitation, but development has been hampered by technological constraints. Most fishermen live on the coastal fringes and operate on a small scale.

MAP KEY

POPULATION
- above 5 million
- 1 million to 5 million
- 500,000 to 1 million
- 100,000 to 500,000
- 50,000 to 100,000
- 10,000 to 50,000
- below 10,000

ELEVATION
- 2000m / 6562ft
- 1000m / 3281ft
- 500m / 1640ft
- 250m / 820ft
- 100m / 328ft
- sea level

SCALE 1: 6,250,000
(projection: Lambert Conformal Conic)

Mainland East Asia

China, Mongolia, North Korea, South Korea, Taiwan

CHINA, THE WORLD'S MOST POPULOUS NATION, has an unbroken cultural history, longer than that of any other country, and is rapidly emerging as a leading world power. When Mao Zedong established Communist rule in 1949, China had become a backward feudal empire, stricken by civil war and over a century of European and Japanese incursions. The closed regime withstood the traumas of rapid industrialization, communal farming, and the brutal purges of the Cultural Revolution. Since the 1980s has introduced economic reforms, led by expanded foreign trade. China's population is heavily concentrated in the east and, despite accelerating urban growth, remains predominantly rural. One cultural group, the Han, make up over 90% of the people, while five "Autonomous Regions" have been established in the south and west for the main ethnic minorities.

Gansu province, through which the ancient Silk Route passes on its way to the west, is characterized by extensive loess deposits which are terraced and used for crop cultivation.

TRANSPORTATION & INDUSTRY

LARGE-SCALE INDUSTRIAL growth has always been a priority of the Communist government. Metals and machine production, chemicals, and engineering are among the leading industries, concentrated in the major cities of the east coast. Textiles and clothing manufacture, the main consumer goods sector, is relatively well dispersed, with a few significant centers such as Shanghai, Beijing, and Hong Kong.

Major industry and infrastructure

- car manufacture
- chemicals
- electronics
- engineering
- finance
- food processing
- iron & steel
- shipbuilding
- textiles
- capital cities
- major towns
- international airports
- major roads
- major industrial areas

TRANSPORTATION NETWORK

| 734,473 miles (1,182,727 km) | 1,182 miles (1,904 km) |
| 41,798 miles (67,308 km) | 70,495 miles (113,519 km) |

Steam trains use China's abundant coal and are still the main form of passenger and goods transportation. The railroad network is now struggling to meet an ever-growing demand.

Coal is China's most abundant mineral resource. This mine at Fuxin in Liaoning province is used to provide coal for a nearby power station.

THE LANDSCAPE

THE EAST ASIAN LANDMASS is arranged in three distinct levels, the highest of which is the Plateau of Tibet in the southwest. The arid uplands of northwestern China form a barren middle step. The main rivers flow eastward from these two platforms to the East China and South China sea coasts, across a broad region of alluvial lowlands and low hills.

Paektu-san, at 9,023 ft (2,750 m), is North Korea's highest peak; an extinct volcanic cone now filled by a crater lake.

The loess plateau of northern China is the world's greatest expanse of loess, a loose soil made up of wind-blown material. The plateau has been heavily eroded by tributaries of the Yellow River.

Shifting sand dunes are found in the arid west of the northeast China Plain, while the eastern part of this great expanse is wet and swampy.

River-eroded fine soils

Thick blanket of loess

Because of its very small grain-size, loess has been easily transported and deposited by winds which scour the plains, and in northern China, deposits of loess can be up to 3,000 ft (1,000 m) thick. Loess-based soils are very fertile, but clearing land for agriculture quickly destabilizes the soil and allows it to be eroded.

The Gobi Desert extends across the Nei Mongol Gaoyuan; a vast saucer-shaped upland surrounded by a rim of higher mountains.

Tarim Basin *(Tarim Pendi)*

Plateau of Tibet

The Plateau of Tibet occupies about a quarter of China's total area. The Yangtze, Mekong, Indus, and Brahmaputra Rivers all originate in the south and east of the plateau.

The Himalayas extend along the southwestern edge of the Plateau of Tibet, forming a continuous mountain barrier over 1,500 miles (2,500 km) long.

Warm, humid conditions have caused intensive erosion of south China's karst areas, producing spectacular jagged peaks and vast caves in the limestone.

North China Plain

Paektu-san

The Yangtze is China's longest river and the principal navigable waterway.

Sichuan Pendi

Although it is over 20 years since his death, the legacy of Chairman Mao Zedong, architect of the Great Proletariat Cultural Revolution, is still very much in evidence across China's landscape. In 1959 Mao launched a 20-year period of industrialization and socioeconomic realignment, rejecting western ideals and social codes.

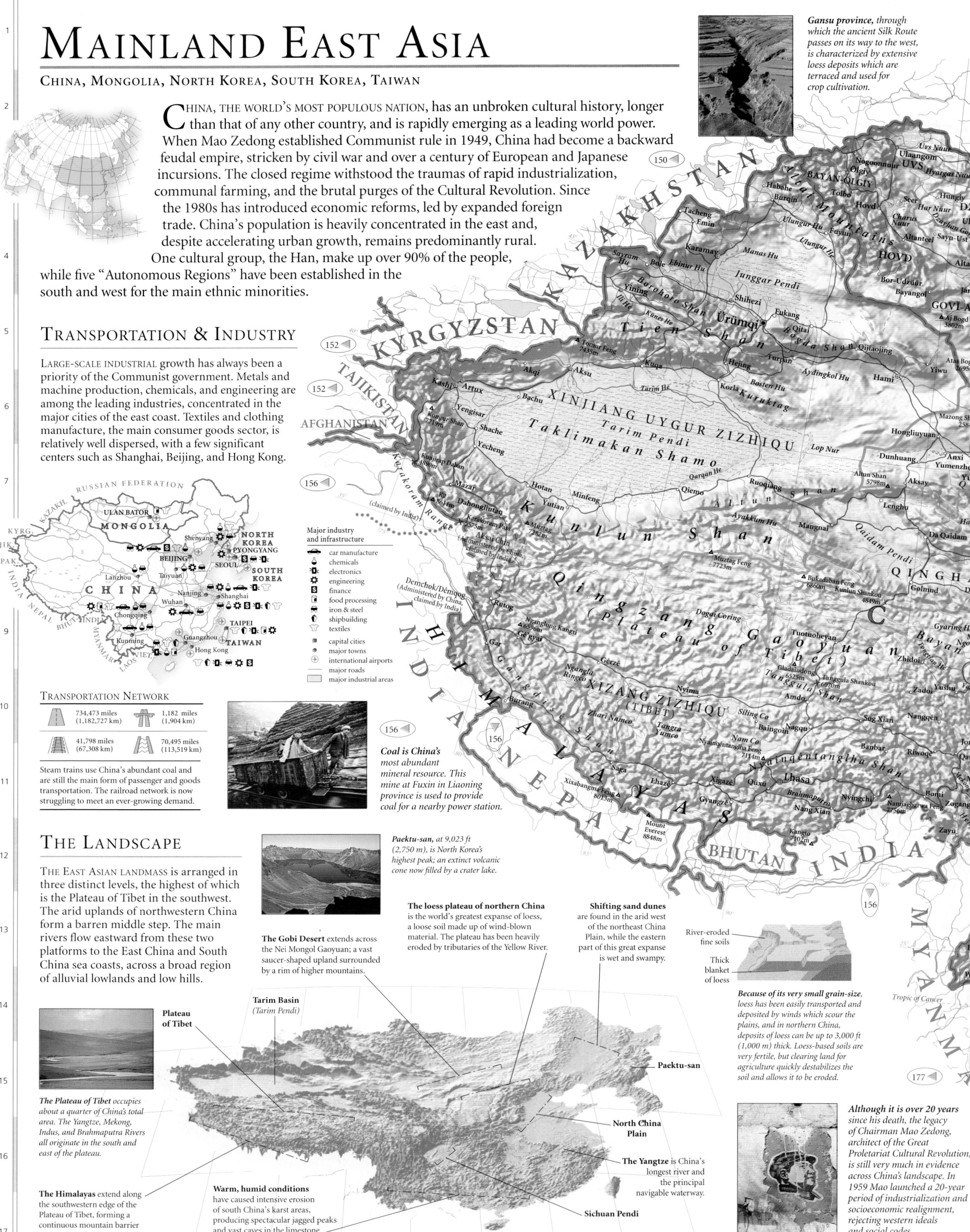

The Great Wall of China remains one of the world's largest-ever construction projects, and is so vast that it is visible from space. Finally completed in AD 214, it runs for over 4,000 miles (6,400 km) from the Yellow Sea, stretching into Central Asia.

USING THE LAND AND SEA

AROUND 90% OF China is unsuitable for cultivation, being either climactically or topographically adverse, or lacking sufficiently fertile soils. Most of the west is used for nomadic herding, while farmland is concentrated in the eastern monsoon region, with rice grown in the tropical and subtropical south. Cereals and soybeans predominate as rainfall and temperatures decline further north.

Beijing (formerly Peking), is China's capital city and, with Shanghai, one of its leading industrial and cultural centers. The morning and evening rush-hours are dominated by bicycles, which constitute the bulk of traffic.

THE URBAN/RURAL POPULATION DIVIDE
urban 32%   rural 68%

POPULATION DENSITY
297 people per sq mile
(115 people per sq km)

TOTAL LAND AREA
4,288,672 sq miles
(11,110,550 sq km)

A B C D E F G H I J

WESTERN CHINA

Gansu, Ningxia, Qinghai, Tibet, Xinjiang

THE PLATEAUS AND BASINS of China's dry, desolate western domain are sparsely populated and largely undeveloped, although they have rich mineral reserves; they also form a critical buffer zone for China, in a geographically important and culturally sensitive part of the Asian continent. Across most of the west, the Han Chinese are outnumbered by a range of cultural groups, including the Uygur, the largest group of the various seminomadic Muslim peoples from Central Asia. The remote, inhospitable Plateau of Tibet is the world's coldest and highest plateau. It has been occupied by the Chinese since 1950. Tibet is one of western China's five "Autonomous Regions," but its reclusive Buddhist culture has been systematically undermined by the Chinese government.

MAP KEY

POPULATION

- ▣ 1 million to 5 million
- ◉ 500,000 to 1 million
- ◎ 100,000 to 500,000
- ⊕ 50,000 to 100,000
- ⊙ 10,000 to 50,000
- ○ below 10,000

ELEVATION

- 6000m / 19,686ft
- 4000m / 13,124ft
- 3000m / 9843ft
- 2000m / 6562ft
- 1000m / 3281ft
- 500m / 1640ft
- 250m / 820ft
- 100m / 328ft
- sea level

SCALE 1:7,000,000
(projection: Lambert Conformal Conic)

Km 0 25 50 100 150 200 250 300

Miles 0 25 50 100 150 200 250 300

The Lhasa He is one of the many rivers that drain the vast Plateau of Tibet. From its source in the Nyainqêntanglha Shan range and fed by the spring meltwater, it eventually joins the upper Brahmaputra 40 miles (65 km) southwest of Lhasa.

USING THE LAND

AGRICULTURE IS CONSTRAINED by the cold, dry climate and lack of fertile soils in the region, although irrigation and glasshouse farming are increasing agricultural potential. Large quantities of fruit, like melons and grapes, are grown at the oases of Hami and Turpan in Xinjiang, and new irrigation schemes have greatly increased cotton and wheat production in the Tarim Basin *(Tarim Pendi)*. Most of the great area of Tibet and Qinghai is devoted to pastoralism. Sheep are the principal livestock.

Land use and agricultural distribution

- 🐐 goats
- 🐑 sheep
- 🌾 cereals
- ❀ cotton
- 🍇 grapes
- 🍈 melons
- oases
- • major towns
- pasture
- cropland
- forest
- mountain region
- desert

The Potala Palace, in Tibet's capital, Lhasa, was the former residence of the Dalai Lama, Tibetan Buddhism's spiritual leader. Tibet remains only sparsely populated; forming over 20% of China's landmass, it supports fewer than 1% of its population.

THE LANDSCAPE

THE HIMALAYAS MARK the southwestern edge of the Plateau of Tibet, an extreme mountain wilderness which occupies nearly a quarter of China's total area. A large structural depression, the Qaidam Pendi, lies at its northeastern edge. The Kunlun mountain chain isolates the plateau from the desert to the north, where the Tien Shan range forms a spur between the Tarim Basin (*Tarim Pendi*) and Dzungarian Basin (*Junggar Pendi*).

The Tien Shan reach elevations of over 24,400 ft (7435 m) and have permanent ice fields, from which large glaciers extend.

Dzungarian Basin (*Junggar Pendi*)

The Bogda Shan, an eastward arm of the Tien Shan range, rise high above the Turpan Depression (Turpan Pendi).

The Turpan Depression (*Turpan Pendi*) is the lowest and hottest place in China. Temperatures can exceed 117°F (47°C) around the lake of Aydingkol Hu, which lies 505 ft (154 m) below sea level.

Northwestern China is largely a region of internal drainage. The Tarim He flows only as far as Lop Nur, where its water is lost by evapotranspiration from the lake and land surface.

A vast glacial lake filled much of the Tarim Basin (*Tarim Pendi*) during the last Ice Age. This area is now occupied by the Takla Makan Desert (*Taklimakan Shamo*). A remnant of the lake, Lop Nur, forms the eastern margin, where it is fed by the Tarim He.

Sand dunes cover western parts of the the basin of Qaidam Pendi. Strong winds frequently carry the sands east, threatening the agricultural areas around the lake of Qinghai Hu.

The terrain of the Plateau of Tibet consists of mountain peaks and open plateaus, dotted with brackish lakes. These are probably remnants of the Tethys Sea, which covered the area before it was uplifted following the collision of the Indo-Australian and Eurasian plates.

Mount Everest is the world's highest peak, at 29,028 ft (8,848 m). The summit marks the border between China and Nepal.

Tarim Basin (*Tarim Pendi*)

Barchan sand dunes in Takla Makan Desert (*Taklimakan Shamo*)

Oases at edge of basin

Lop Nur

The Tarim Basin (Tarim Pendi) has no permanent rivers. Rainfall from the surrounding Plateau of Tibet and Tien Shan ranges drains into the basin's sand and gravel floor.

From its source, high in eastern Qinghai, the Yellow River starts on a 3,395 mile (5,464 km) journey to the Yellow Sea.

TRANSPORTATION & INDUSTRY

OIL EXTRACTION AT Yumen and in the Dzungarian and Qaidam basins has led to the growth of the petrochemical industry and a range of heavy manufacturing plants in the cities of Lanzhou and Urumqi. Tibet, and most of Xinjiang, have little industry beyond traditional handicrafts, especially textiles at Hotan and Kashi, located along the ancient Silk Route. Nuclear and space-research testing are carried out at Lop Nur in Xinjiang.

TRANSPORTATION NETWORK

The construction of roads connecting Lhasa in Tibet with Sichuan, Qinghai, and Xinjiang was achieved in the 1950s, in spite of the extreme physical conditions of the Plateau of Tibet.

Major industry and infrastructure

- agribusiness
- chemicals
- coal
- engineering
- food processing
- iron & steel
- nuclear testing
- oil
- textiles
- major towns
- major roads
- major industrial areas

EASTERN CHINA

TAIWAN, Anhui, Beijing, Fujian, Guangdong, Guangxi, Guizhou, Hainan, Hebei, Henan, Hubei, Hunan, Jiangsu, Jiangxi, Shaanxi, Shandong, Shanghai, Shanxi, Sichuan, Tianjin, Yunnan, Zhejiang

THE EAST IS CHINA'S HEARTLAND. Massive industrial development since 1949 has transformed much of the densely populated rural landscape, in a region still prone to flooding and drought. Over 20 cities have populations of over a million, including the giant metropolis of Shanghai and the capital Beijing, which has been China's cultural and political center since the 13th century. The ethnically diverse southwest and the oil-rich interior provinces of Sichuan and Shaanxi have largely missed out on the remarkable economic growth occurring in designated free-trade areas along the coasts of the South and East China seas. The republic of Taiwan was established in 1949 by Chinese nationalists ousted from the mainland by the victorious Communist forces. Taiwan now has one of the strongest economies in the world but its sovereignty is not recognized by China. Hong Kong provides a major international trade link for China; a 99-year "lease" period of British control was concluded in 1997.

USING THE LAND AND SEA

THIS IS A REGION of intensive cultivation. Wheat, millet, sorghum, and cotton are the main crops of the Yellow River basin. South from Sichuan, rice becomes the principal crop, grown with wheat, corn, and cotton along the Yangtze River. Tea is produced in the hills and sugar cane along the coast of the southeast, where flat land is limited. Pigs and poultry are raised in great numbers.

North of the Qin Ling range in Shaanxi province, is an agriculturally fertile region covered with fine, wind-blown deposits and known as the loess plateau. The loose sediments are vulnerable to water erosion.

On the hills above the North China Plain, slopes are terraced to utilize the rich loess soils of the Taihang Shan range.

The former Portuguese territory of Macao, with its colonial architecture, bars and casinos, reverted to Chinese rule in 1999.

Land use and agricultural distribution

cattle, pigs, cereals, corn (maize), cotton, fishing, peanuts, rice, sugar cane, tea, capital cities, major towns

pasture, cropland, forest, mountain region

MAP KEY

POPULATION
- above 5 million
- 1 million to 5 million
- 500,000 to 1 million
- 100,000 to 500,000
- 50,000 to 100,000
- 10,000 to 50,000
- below 10,000

ELEVATION
- 6000m / 19,686ft
- 4000m / 13,124ft
- 3000m / 9843ft
- 2000m / 6562ft
- 1000m / 3281ft
- 500m / 1640ft
- 250m / 820ft
- 100m / 328ft
- sea level

SCALE 1:7,750,000
(projection: Lambert Conformal Conic)

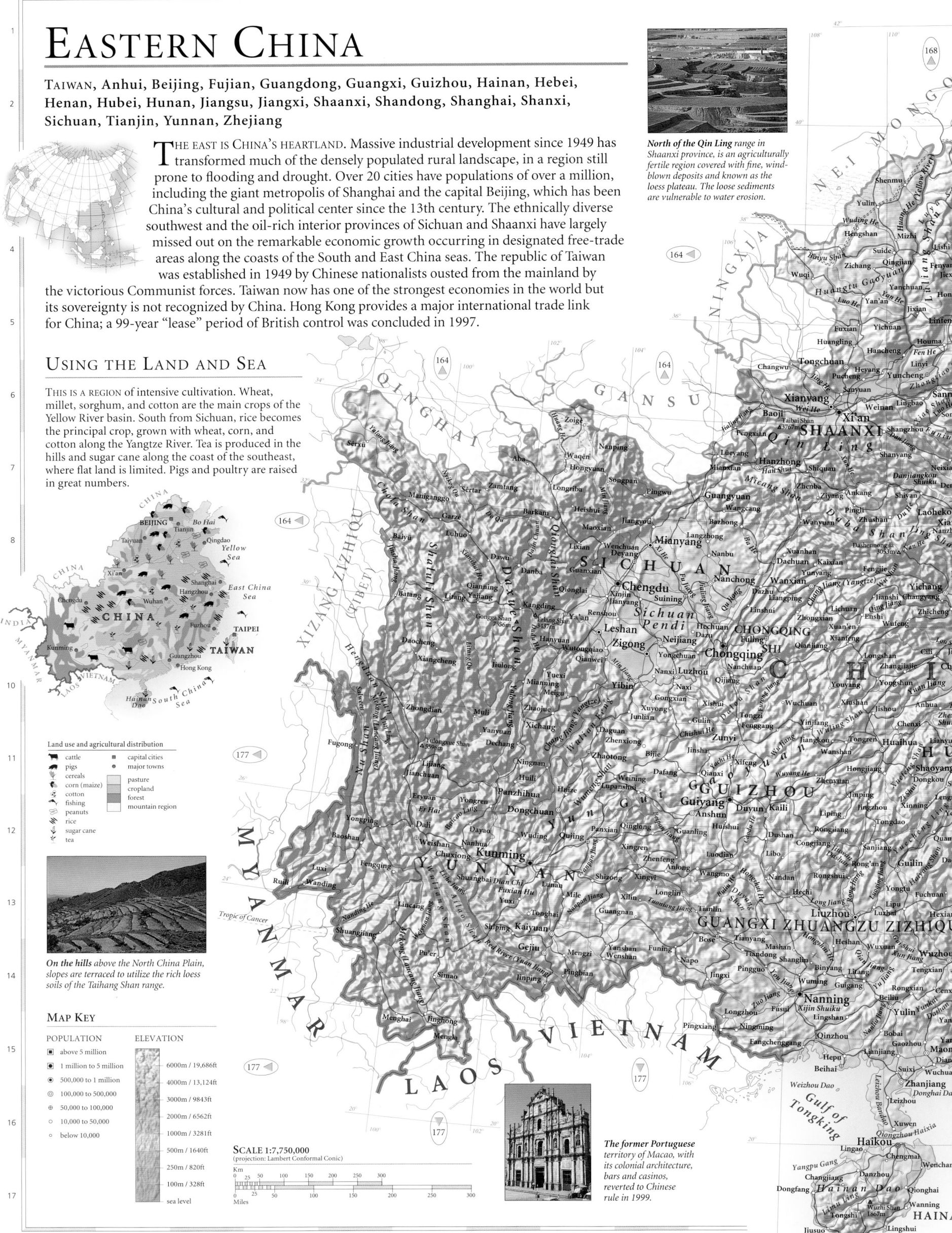

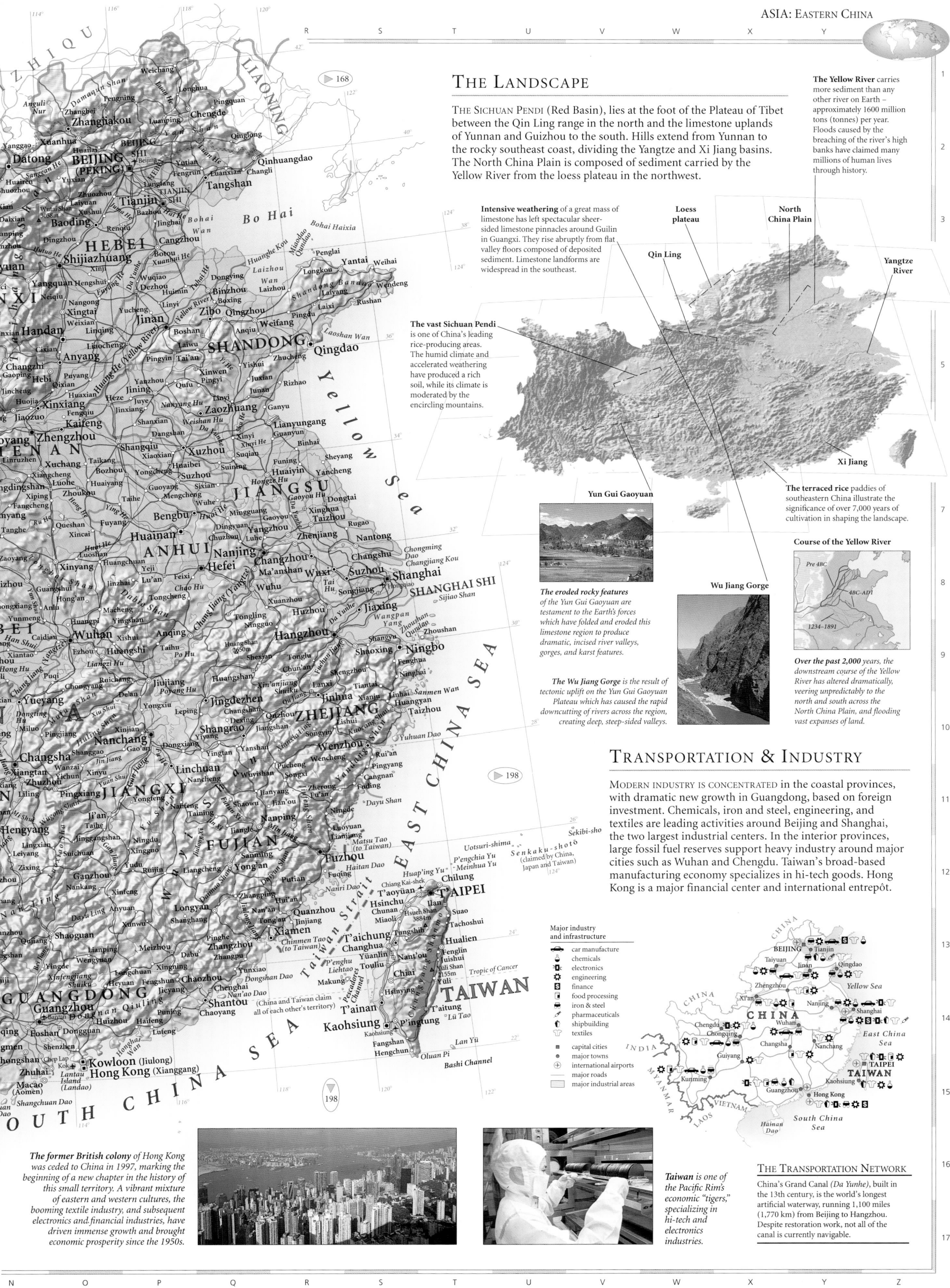

THE LANDSCAPE

THE SICHUAN PENDI (Red Basin), lies at the foot of the Plateau of Tibet between the Qin Ling range in the north and the limestone uplands of Yunnan and Guizhou to the south. Hills extend from Yunnan to the rocky southeast coast, dividing the Yangtze and Xi Jiang basins. The North China Plain is composed of sediment carried by the Yellow River from the loess plateau in the northwest.

The Yellow River carries more sediment than any other river on Earth – approximately 1600 million tons (tonnes) per year. Floods caused by the breaching of the river's high banks have claimed many millions of human lives through history.

Intensive weathering of a great mass of limestone has left spectacular sheer-sided limestone pinnacles around Guilin in Guangxi. They rise abruptly from flat valley floors composed of deposited sediment. Limestone landforms are widespread in the southeast.

Loess plateau

North China Plain

Qin Ling

Yangtze River

The vast Sichuan Pendi is one of China's leading rice-producing areas. The humid climate and accelerated weathering have produced a rich soil, while its climate is moderated by the encircling mountains.

Xi Jiang

Yun Gui Gaoyuan

The eroded rocky features of the Yun Gui Gaoyuan are testament to the Earth's forces which have folded and eroded this limestone region to produce dramatic, incised river valleys, gorges, and karst features.

Wu Jiang Gorge

The Wu Jiang Gorge is the result of tectonic uplift on the Yun Gui Gaoyuan Plateau which has caused the rapid downcutting of rivers across the region, creating deep, steep-sided valleys.

The terraced rice paddies of southeastern China illustrate the significance of over 7,000 years of cultivation in shaping the landscape.

Course of the Yellow River

Pre 4BC
4BC–AD1
1234–1891

Over the past 2,000 years, the downstream course of the Yellow River has altered dramatically, veering unpredictably to the north and south across the North China Plain, and flooding vast expanses of land.

TRANSPORTATION & INDUSTRY

MODERN INDUSTRY IS CONCENTRATED in the coastal provinces, with dramatic new growth in Guangdong, based on foreign investment. Chemicals, iron and steel, engineering, and textiles are leading activities around Beijing and Shanghai, the two largest industrial centers. In the interior provinces, large fossil fuel reserves support heavy industry around major cities such as Wuhan and Chengdu. Taiwan's broad-based manufacturing economy specializes in hi-tech goods. Hong Kong is a major financial center and international entrepôt.

Major industry and infrastructure
- car manufacture
- chemicals
- electronics
- engineering
- finance
- food processing
- iron & steel
- pharmaceuticals
- shipbuilding
- textiles
- capital cities
- major towns
- international airports
- major roads
- major industrial areas

The former British colony of Hong Kong was ceded to China in 1997, marking the beginning of a new chapter in the history of this small territory. A vibrant mixture of eastern and western cultures, the booming textile industry, and subsequent electronics and financial industries, have driven immense growth and brought economic prosperity since the 1950s.

Taiwan is one of the Pacific Rim's economic "tigers," specializing in hi-tech and electronics industries.

THE TRANSPORTATION NETWORK

China's Grand Canal (Da Yunhe), built in the 13th century, is the world's longest artificial waterway, running 1,100 miles (1,770 km) from Beijing to Hangzhou. Despite restoration work, not all of the canal is currently navigable.

Northeastern China, Mongolia & Korea

Mongolia, North Korea, South Korea, Heilongjiang, Inner Mongolia, Jilin, Liaoning

THIS NORTHERLY REGION has been a domain of shifting borders and competing colonial powers for centuries. Mongolia was the heartland of Chinghiz Khan's vast Mongol empire in the 13th century, while northeastern China was home to the Manchus, China's last ruling dynasty (1644–1911). The mineral and forest wealth of the northeast helped make this China's principal region of heavy industry, although the outdated state factories now face decline. South Korea's state-led market economy has grown dramatically and Seoul is now one of the world's largest cities. The austere communist regime of North Korea has isolated itself from the expanding markets of the Pacific Rim and faces continuing economic stagnation.

The Eurasian steppe stretches from the mouth of the Danube in Europe, to Mongolia. In Mongolia, nomadic people have lived in felt huts called yurts or gers, for thousands of years.

MAP KEY

POPULATION
- above 5 million
- 1 million to 5 million
- 500,000 to 1 million
- 100,000 to 500,000
- 50,000 to 100,000
- 10,000 to 50,000
- below 10,000

ELEVATION
- 4000m / 13,124ft
- 3000m / 9843ft
- 2000m / 6562ft
- 1000m / 3281ft
- 500m / 1640ft
- 250m / 820ft
- 100m / 328ft
- sea level

SCALE 1:7,000,000
(projection: Lambert Conformal Conic)

THE LANDSCAPE

THE GREAT NORTH CHINA PLAIN is largely enclosed by mountain ranges including the Great and Lesser Khingan Ranges (*Da Hinggan Ling* and *Xiao Hinggan Ling*) in the north, and the Changbai Shan, which extend south into the rugged peninsula of Korea. The broad steppeland plateau of Nei Mongol Gaoyuan borders the southeastern edge of the great cold desert of the Gobi which extends west across the southern reaches of Mongolia. In northwest Mongolia the Altai Mountains and various lesser ranges are interspersed with lakeland basins.

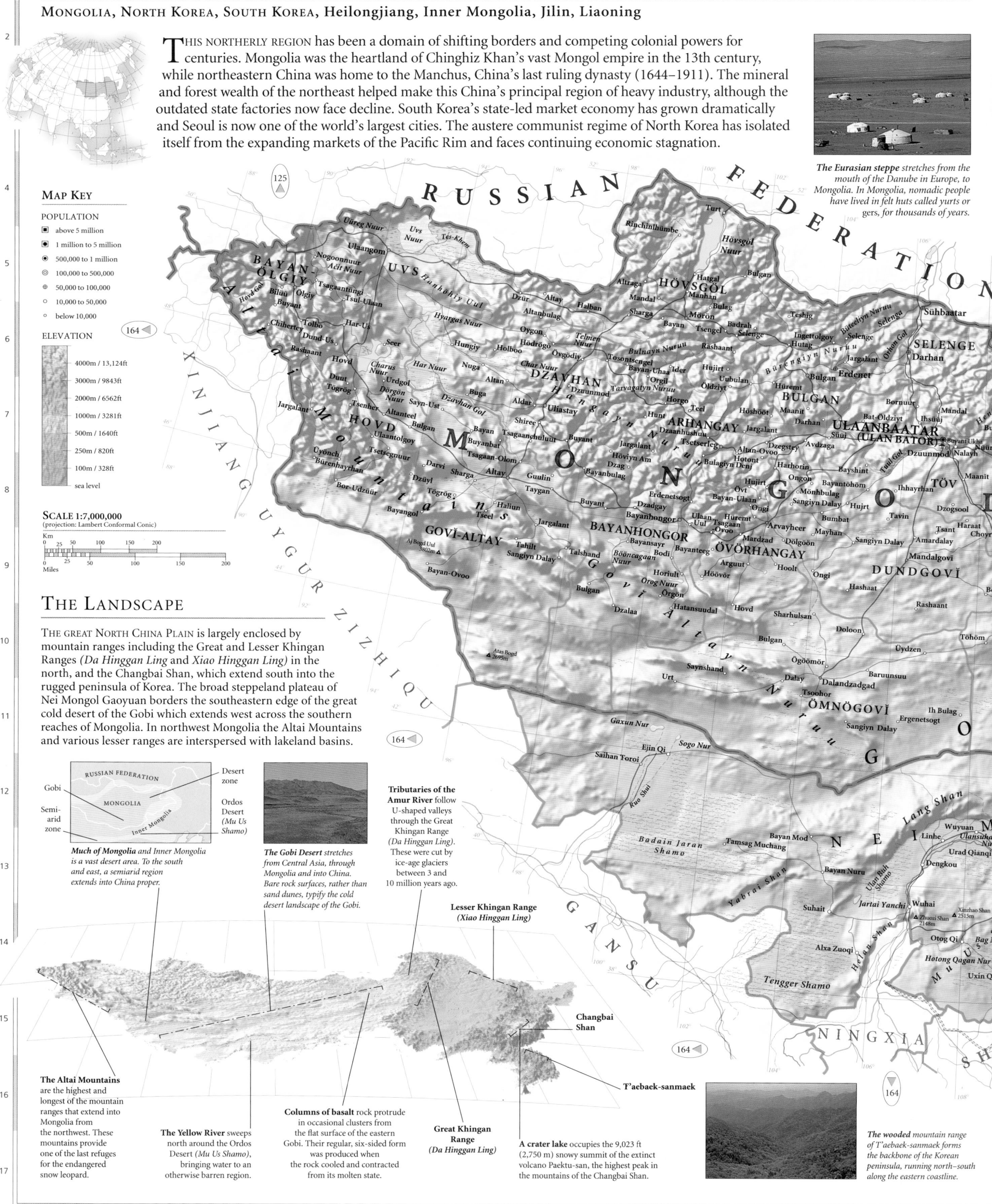

Desert zone
Ordos Desert (*Mu Us Shamo*)
Gobi
Semi-arid zone
RUSSIAN FEDERATION
MONGOLIA
Inner Mongolia

Much of Mongolia and Inner Mongolia is a vast desert area. To the south and east, a semiarid region extends into China proper.

The Gobi Desert stretches from Central Asia, through Mongolia and into China. Bare rock surfaces, rather than sand dunes, typify the cold desert landscape of the Gobi.

Tributaries of the Amur River follow U-shaped valleys through the Great Khingan Range (*Da Hinggan Ling*). These were cut by ice-age glaciers between 3 and 10 million years ago.

Lesser Khingan Range (*Xiao Hinggan Ling*)

Changbai Shan

T'aebaek-sanmaek

The Altai Mountains are the highest and longest of the mountain ranges that extend into Mongolia from the northwest. These mountains provide one of the last refuges for the endangered snow leopard.

The Yellow River sweeps north around the Ordos Desert (*Mu Us Shamo*), bringing water to an otherwise barren region.

Columns of basalt rock protrude in occasional clusters from the flat surface of the eastern Gobi. Their regular, six-sided form was produced when the rock cooled and contracted from its molten state.

Great Khingan Range (*Da Hinggan Ling*)

A crater lake occupies the 9,023 ft (2,750 m) snowy summit of the extinct volcano Paektu-san, the highest peak in the mountains of the Changbai Shan.

The wooded mountain range of T'aebaek-sanmaek forms the backbone of the Korean peninsula, running north–south along the eastern coastline.

TRANSPORTATION & INDUSTRY

NORTH KOREA'S CENTRALLY-PLANNED ECONOMY is strongly oriented toward heavy industry, while South Korea has a broad manufacturing base which includes textiles, steel, electronics, and one of the world's largest shipbuilding industries. Mongolia and Inner Mongolia's great mineral resource potential is largely undeveloped. The heavy industrial region around Shenyang produces iron, steel, chemicals, and cement on a massive scale.

Major industry and infrastructure

- car manufacture
- chemicals
- coal
- electronics
- engineering
- finance
- food processing
- iron & steel
- pharmaceuticals
- shipbuilding
- textiles
- capital cities
- major towns
- international airports
- major roads
- major industrial areas

TRANSPORTATION NETWORK

Liaoning has China's most comprehensive railroad network, the legacy of the Japanese occupation of Manchuria in the 20th century. The railroads are used primarily for freight transportation.

Ulan Bator, the Mongolian capital bears many of the hallmarks of Soviet-style central planning, the result of economic and industrial assistance from the Soviet Union following Mongolian independence in 1921.

While North Korea has remained politically and economically isolated from the rest of the world, South Korea has enjoyed immense economic growth. It has benefited considerably from US economic aid in the aftermath of the Korean war of 1950–1953.

USING THE LAND AND SEA

MONGOLIA AND INNER MONGOLIA rely heavily on livestock farming, with only about 1% of the land area cultivated. Northeastern China produces wheat, corn, soybeans, and sugar beet. The cool climate limits the range of crops and large upland areas of the northeast remain forested. Rice is the staple food of North and South Korea. The latter has become a leading ocean-fishing nation.

Land use and agricultural distribution

- goats
- pigs
- sheep
- corn
- fishing
- rice
- soybeans
- sugar beet
- wheat
- capital cities
- major towns
- pasture
- cropland
- forest
- mountain region
- desert

JAPAN

IN THE YEARS SINCE THE END of the Second World War, Japan has become the world's most dynamic industrial nation. The country comprises a string of over 4,000 islands which lie in a great northeast to southwest arc in the northwest Pacific. Four major islands: Hokkaido, Honshu, Shikoku, and Kyushu are home to the great majority of Japan's population of 125.9 million people, although the mountainous terrain of the central region means that most cities are situated on the coast. A densely populated industrial belt stretches along much of Honshu's southern coast, including Japan's crowded capital, Tokyo. Alongside its spectacular economic growth and the increasing westernization of its cities, Japan still maintains a highly individual culture, reflected in its traditional food, formal behavioral codes, unique Shinto religion, and the reverence for the emperor, who is officially regarded as a god.

TRANSPORTATION & INDUSTRY

JAPAN IS THE WORLD'S second largest market economy, outranked only by the US. Technological development, particularly of computers, electronic goods, cars, and motorcycles is second to none. Japanese industry invests in its workforce and in long-term research and development to maintain the high standard of its products and a reputation for innovation. Japanese businesses are now global both in their manufacturing bases and in the distribution of goods.

Major industry and infrastructure

- brewing
- car manufacture
- chemicals
- hi-tech industry
- engineering
- finance
- iron & steel
- research & development
- shipbuilding
- textiles
- winter sports
- capital cities
- major towns
- international airports
- major roads
- major industrial areas

TRANSPORTATION NETWORK

720,360 miles (1,160,000 km)	6,070 miles (12,529 km)
12,529 miles (20,175 km)	1,099 miles (1,770 km)

Japanese road construction traditionally lagged behind that of its extensive and technologically advanced railroad network. The road network's relative lack of development has led to severe urban congestion, although expressways have now been built in some cities.

Known in the west as the "bullet train," the Shinkansen is the second-fastest train in the world. It speeds past the snowcapped peak of Mount Fuji between the cities of Tokyo and Osaka.

USING THE LAND AND SEA

ALTHOUGH ONLY ABOUT 11% OF JAPAN is suitable for cultivation, substantial government support, a favorable climate and intensive farming methods enable the country to be virtually self-sufficient in rice production. Northern Hokkaido, the largest and most productive farming region, has an open terrain and climate similar to that of the American Midwest, and produces over half of Japan's cereal requirements. Farmers are being encouraged to diversify by growing fruit, vegetables, and wheat, as well as raising livestock.

Land use and agricultural distribution

- cattle
- pigs
- fishing
- cereals
- citrus fruits
- fruit
- herbs
- rice
- root crops
- tobacco
- capital cities
- major towns
- pasture
- cropland
- forest

Cutting terraces maximizes the limited agricultural land, enabling Japan to produce large quantities of rice.

THE URBAN/RURAL POPULATION DIVIDE

urban 78%   rural 22%

0 10 20 30 40 50 60 70 80 90 100

POPULATION DENSITY	TOTAL LAND AREA
863 people per sq mile (333 people per sq km)	145,869 sq miles (377,800 sq km)

The Kobe earthquake in January 1995 highlighted Japan's vulnerability to earthquakes, despite technological advances. It shattered much of the infrastructure of this important port. More than 5,000 people died as buildings and overhead highways collapsed and fires broke out.

A number of new volcanoes emerged in Japan during the 20th century. They exist alongside older cones like this one in Aso-Kuju National Park on Kyushu, now dormant and grass-covered.

THE LANDSCAPE

THE ISLANDS OF JAPAN LIE on the Pacific "Ring of Fire," and form a series of clearly defined arcs. The largely mountainous landscape was formed very recently in geological terms. Volcanic eruptions and earthquakes continue to reshape the terrain and shake the country's complex infrastructure. There is no single continuous mountain range; the mountains divide into many small land blocks separated by lowlands and dissected by numerous river valleys.

Active volcanic island / **Japan Trench (subduction zone)**

Japan is part of an arc of volcanic islands, formed by the Pacific Plate diving under the Eurasian Plate. This process generates intense stress which is periodically released as earthquakes.

A number of rivers which emerge from the volcanic parts of northeastern Honshu are so highly acidic that their water is unsuitable for irrigation and consumption.

Calderas are the wide, flat-bottomed craters of volcanoes. Many Japanese calderas are filled by lakes such as Towada-ko in northern Honshu.

Trees cling to the sheer slopes of the waterfalls on the northern island of Hokkaido. The island's climate is similar to that in northern Europe, with long, cold winters and short, warm summers.

The long, narrow, steep-sided islands which make up Japan give rise to numerous short, fast-flowing rivers. The river of Shinano-gawa is the longest, at 228 miles (367 km).

The Inland Sea (Seto-naikai) has resulted from the depression of faulted blocks which has allowed sea water to invade the region between northern Shikoku and western Honshu.

There are over 60 active volcanoes – like Asahi-dake, Hokkaido's highest peak – throughout Japan. This accounts for more than 10% of the world's total.

Rising land on the Pacific coast of Honshu leads to typical features such as raised beaches, some lying over 1,000 ft (300 m) above sea level.

In much of Kyushu the coast is subsiding, giving a highly indented coastline. In some places, former hilltops are barely visible above the current sea level.

Strong northwesterly winds blowing onshore during the winter create sand dunes which extend for miles along the western coasts.

Biwa-ko is the largest lake in Japan, covering 260 sq miles (673 sq km) in central Honshu. The depression in which it lies was created by recent faulting of the underlying rocks.

Mount Fuji

Japan experiences earthquakes on an almost daily basis. They can cause fast-moving landslides and immense sea waves called tsunami. One that hit Sagami-nada in 1923, reached heights of 40 ft (12 m).

Mount Fuji is Japan's highest mountain, rising 12,388 ft (3,776 m) above the Kanto Plain in the central region of Honshu. The flat land below is suitable for growing crops such as tea. Like many Japanese mountains, it is revered as a sacred site.

Autumnal trees near Gifu, on central Honshu, create a spectacular display. Native trees on this island include camphor, pasania, Japanese evergreen oak, camellia and holly.

Modern tower blocks overlook the docks in Tokyo, Japan's teeming capital. Nearly 8 million people live in the city, straining the infrastructure to its limits.

Malaysia exports a greater tonnage of tropical timber than anywhere else in the world. Much of it comes from Sarawak in Borneo. Although in principle logging is only allowed on a sustainable basis, environmentalists fear that the rainforest in Sarawak will have disappeared by the early 21st century.

This tiny island near Kota Kinabalu, in Sabah, eastern Malaysia, is a part of a designated national park. Thickly forested, it is surrounded by broad, sandy beaches and shallow inland seas.

Throughout Southeast Asia, where agricultural land is at a premium, terraces are cut into the slopes to maximize the area available for cultivation. These terraces on the Indonesian island of Bali are used to support rice paddies.

MAP KEY

POPULATION
- above 5 million
- 1 million to 5 million
- 500,000 to 1 million
- 100,000 to 500,000
- 50,000 to 100,000
- 10,000 to 50,000
- below 10,000

ELEVATION
- 4000m / 13,124ft
- 3000m / 9843ft
- 2000m / 6562ft
- 1000m / 3281ft
- 500m / 1640ft
- 250m / 820ft
- 100m / 328ft
- sea level

SCALE 1:6,250,000
(projection: Mercator)

MARITIME SOUTHEAST ASIA

BRUNEI, EAST TIMOR, INDONESIA, MALAYSIA, SINGAPORE

THE INTRICATE ARC OF ISLANDS which runs from peninsular Malaysia east to Irian Jaya in western New Guinea sustains a huge variety of peoples, languages, and cultures. Indonesia is by far the largest country in the region, and 87% of its huge, predominantly Muslim, population is crowded onto Java, the most habitable of Indonesia's 13,677 islands. Malaysia, split between the mainland and the east Malaysian states of Sabah and Sarawak on Borneo, has a diverse population, as well as a fast-growing economy, although the pace of its development is still far outstripped by that of Singapore. This small island nation is the financial and commercial capital of Southeast Asia, and an Asian "tiger" economy. The Sultanate of Brunei in northern Borneo, one of the world's last princely states, also has an extremely high standard of living, based on its oil revenues.

USING THE LAND AND SEA

RICE IS THE MOST IMPORTANT ARABLE CROP in Indonesia and Malaysia, and both countries manage to meet almost all of their domestic demand. Malaysian rubber accounts for 25% of world production and is the main cash crop, grown on plantations and small farms, along with oil palms and copra. Timber is exported from both Malaysia and Indonesia. Modern agricultural techniques enable Singapore to produce fruit and vegetables despite a shortage of suitable land.

Spiral cuts in the bark of this rubber palm show where it has been tapped. Sophisticated "cloning" techniques mean that trees which produce consistently high quantities of rubber can be easily reproduced.

THE URBAN/RURAL POPULATION DIVIDE

urban 38%   rural 62%

0 10 20 30 40 50 60 70 80 90 100

POPULATION DENSITY	TOTAL LAND AREA
262 people per sq mile (101 people per sq km)	828,356 sq miles (2,146,000 sq km)

Land use and agricultural distribution

- coconuts
- fishing
- oil palms
- rice
- rubber
- shellfish
- sugar cane
- timber
- capital cities
- major towns
- pasture
- cropland
- forest
- wetland

THE LANDSCAPE

FROM SUMATRA IN THE WEST, the volcanic islands of Indonesia run for nearly 3,100 miles (5,000 km). The Sunda Shelf, an extension of the Eurasian Plate, lies between Java, Bali, Sumatra, Lombok, and Borneo. Their volcanic mountains rise from a base below the sea and they were once joined together by dry land, which has since been submerged by rising sea levels.

The river of Sungai Mahakam cuts through the central highlands of Borneo, the third largest island in the world, with a total area of 290,000 sq miles (757,050 sq km). Although mountainous, Borneo is one of the most stable of the Indonesian islands, with little volcanic activity.

The Sunda Shelf underlies this whole region. It is one of the largest submarine shelves in the world, covering an area of 714,285 sq miles (1,850,000 sq km). During the early Quaternary period, when sea levels were lower, the shelf was exposed.

Malay Peninsula — Borneo — Broad, shallow valleys on sea floor — Present sea level — Quaternary sea level, 460 ft (140 m) below present sea level — Sumatra — Drowned rivers

Malay Peninsula has a rugged east coast, but the west coast, fronting the Strait of Malacca, has many sheltered beaches and bays. The two coasts are divided by the Banjaran Titiwangsa, which run the length of the peninsula.

Gunung Kinabalu is the highest peak in Malaysia, rising 13,455 ft (4,101 m).

The four-pronged island of Celebes is the product of complex tectonic activity which ruptured and then reattached small fragments of the Earth's crust to form the island's many peninsulas.

Irian Jaya contains some of the most dense and least explored tropical rain forests in the world, inhabited by many rare species of plants and animals.

The island of Krakatau (Pulau Rakata), lying between Sumatra and Java, was all but destroyed in 1883, when the volcano erupted. The release of gas and dust into the atmosphere disrupted cloud cover and global weather patterns for several years.

Gunung Semeru

The volcano of Gunung Semeru in eastern Java lies on the Pacific "Rim of Fire." It is part of the ancient Tennegger volcano and remains highly active.

Indonesia has more than 220 volcanoes, most of which are still active. They are strung out along the island arc from Sumatra through the Lesser Sunda Islands, into the Moluccas and Celebes.

Coral islands such as Timor in eastern Indonesia show evidence of very recent and dramatic movements of the Earth's plates. Reefs in Timor have risen by as much as 4,000 ft (1,300 m) in the last million years.

The Pegunungan Jayawijaya range in central Irian Jaya contains the world's highest range of limestone mountains, some with peaks more than 16,400 ft (5,000 m) high. Heavy rainfall and high temperatures, which promote rapid weathering, have led to the creation of large underground caves and river systems such as the river of Sungai Baliem.

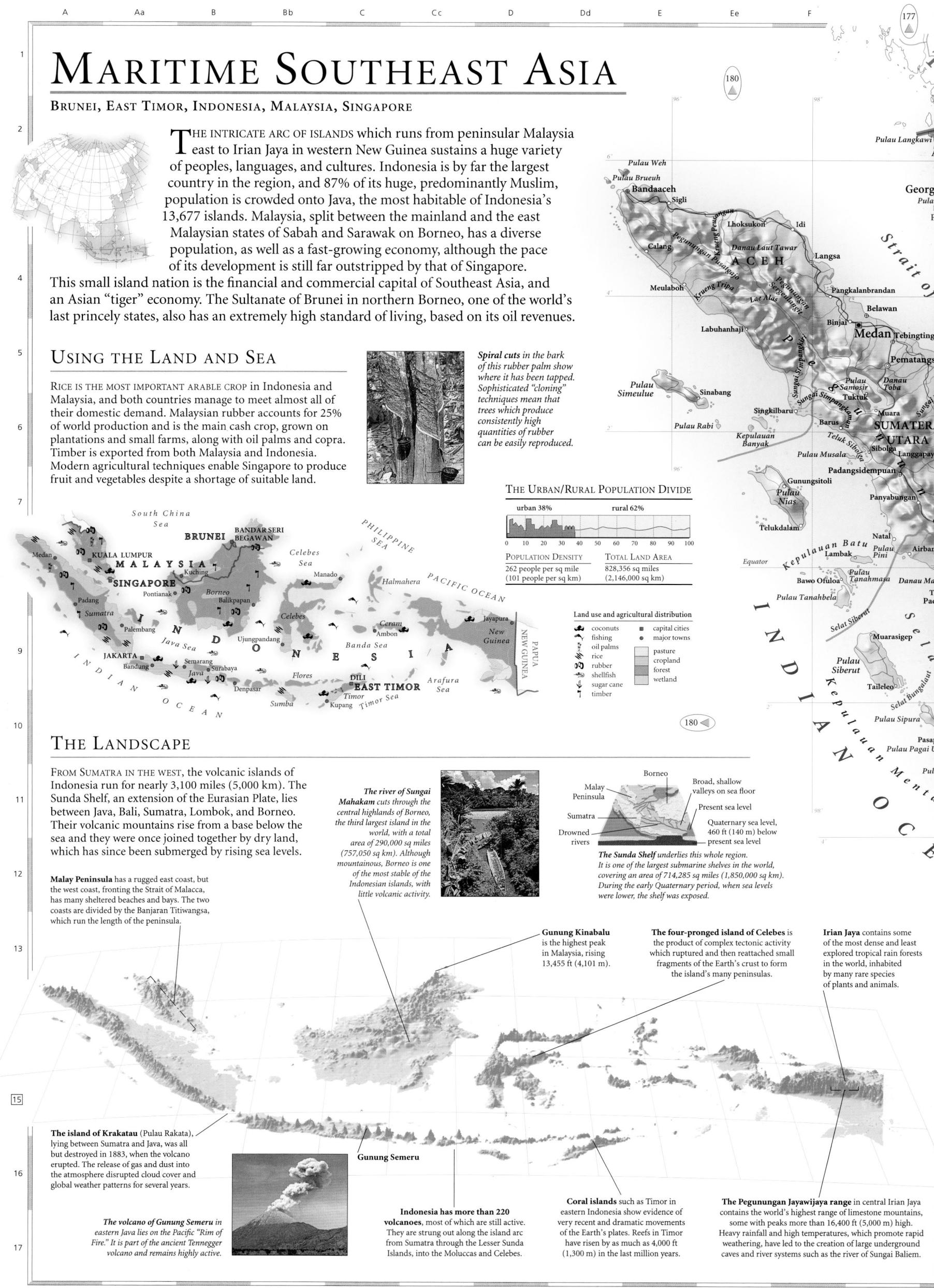

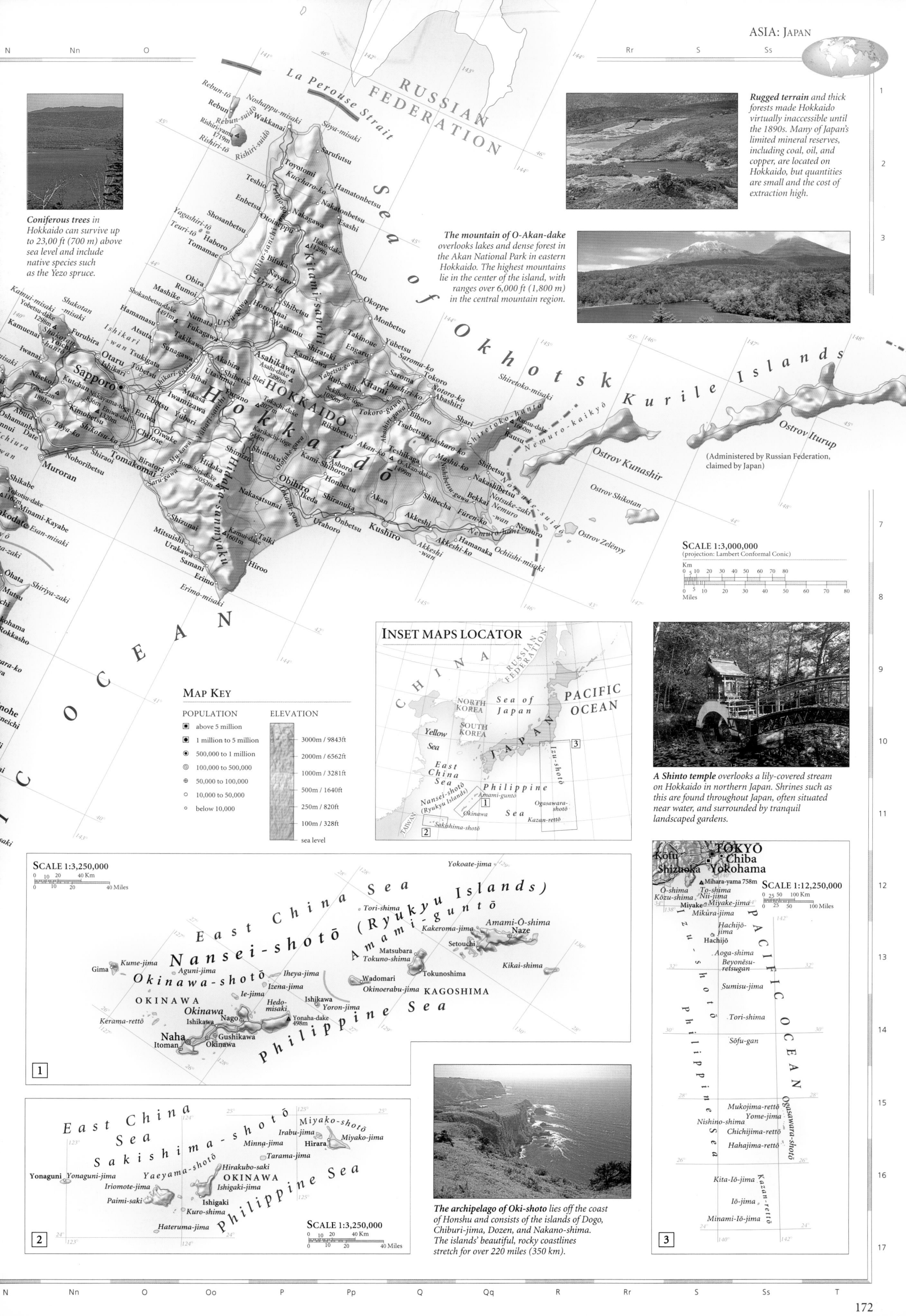

Rugged terrain and thick forests made Hokkaido virtually inaccessible until the 1890s. Many of Japan's limited mineral reserves, including coal, oil, and copper, are located on Hokkaido, but quantities are small and the cost of extraction high.

Coniferous trees in Hokkaido can survive up to 23,00 ft (700 m) above sea level and include native species such as the Yezo spruce.

The mountain of O-Akan-dake overlooks lakes and dense forest in the Akan National Park in eastern Hokkaido. The highest mountains lie in the center of the island, with ranges over 6,000 ft (1,800 m) in the central mountain region.

(Administered by Russian Federation, claimed by Japan)

SCALE 1:3,000,000
(projection: Lambert Conformal Conic)

Km
0 5 10 20 30 40 50 60 70 80
0 5 10 20 30 40 50 60 70 80
Miles

INSET MAPS LOCATOR

MAP KEY

POPULATION

▪ above 5 million
▪ 1 million to 5 million
◉ 500,000 to 1 million
◎ 100,000 to 500,000
⊕ 50,000 to 100,000
○ 10,000 to 50,000
○ below 10,000

ELEVATION

3000m / 9843ft
2000m / 6562ft
1000m / 3281ft
500m / 1640ft
250m / 820ft
100m / 328ft
sea level

A Shinto temple overlooks a lily-covered stream on Hokkaido in northern Japan. Shrines such as this are found throughout Japan, often situated near water, and surrounded by tranquil landscaped gardens.

SCALE 1:3,250,000
0 10 20 40 Km
0 10 20 40 Miles

1

SCALE 1:12,250,000
0 25 50 100 Km
0 25 50 100 Miles

2

SCALE 1:3,250,000
0 10 20 40 Km
0 10 20 40 Miles

The archipelago of Oki-shoto lies off the coast of Honshu and consists of the islands of Dogo, Chiburi-jima, Dozen, and Nakano-shima. The islands' beautiful, rocky coastlines stretch for over 220 miles (350 km).

3

A Aa B Bb C Cc D Dd E Ee F Ff

MAINLAND SOUTHEAST ASIA & THE PHILIPPINES

CAMBODIA, LAOS, MYANMAR, PHILIPPINES, THAILAND, VIETNAM

THICKLY FORESTED MOUNTAINS, intercut by the broad valleys of five great rivers characterize the landscape of Southeast Asia's mainland countries. Agriculture remains the main activity for much of the population, which is concentrated in the river floodplains and deltas. Linked ethnic and cultural roots give the region a distinct identity. Most people on the mainland are Theravada Buddhists, and the Philippines is the only predominantly Christian country in Southeast Asia. Foreign intervention began in the 16th century with the opening of the spice trade; Cambodia, Laos, and Vietnam were French colonies until the end of the Second World War, Myanmar was under British control; and the Philippines was controlled by Spain and the US in the 20th century. Only Thailand was never colonized. Today, Thailand and the Philippines are poised to play a leading role in the economic development of the Pacific Rim, and Laos and Vietnam have begun to mend the devastation of the Vietnam War, and to develop their economies. With continuing political instability and a shattered infrastructure, Cambodia faces an uncertain future, while Myanmar is seeking investment and the ending of its 38-year isolation from the world community.

The Irrawaddy River is Myanmar's vital central artery, watering the rice paddies and providing a rich source of fish, as well as an important transportation link, particularly for local traffic.

Commercial logging – still widespread in Myanmar – has now been stopped in Thailand because of overexploitation of the tropical rain forest.

THE LANDSCAPE

A SERIES OF MOUNTAIN RANGES runs north–south through the mainland, formed as the result of the collision between the Eurasian Plate and the Indian subcontinent, which created the Himalayas. They are interspersed by the valleys of a number of great rivers. On their passage to the sea these rivers have deposited sediment, forming huge, fertile floodplains and deltas. The Philippines' 7,000 islands are mountainous and volcanic, with narrow coastal plains.

Lake Taal on the Philippine island of Luzon lies within the crater of an immense volcano which erupted twice in the 20th century, first in 1911 and again in 1965, causing the deaths of more than 3,200 people.

The Irrawaddy River runs virtually north–south, draining the plains of northern Myanmar. The Irrawaddy Delta is the country's main rice-growing area.

Hkakabo Razi is the highest point in mainland Southeast Asia. It rises 19,300 ft (5,885 m) at the border between China and Myanmar.

Mountains dominate the Laotian landscape with more than 90% of the land lying more than 600 ft (180 m) above sea level. The mountains of the Chaine Annamitique form the country's eastern border.

The Red River Delta in northern Vietnam is fringed to the north by steep-sided, round-topped limestone hills, typical of karst scenery.

Mindanao has five mountain ranges, many of which have large numbers of active volcanoes. Lying just west of the Philippine Trench, which forms the boundary between the colliding Philippine and Eurasian plates, the entire island chain is subject to earthquakes and volcanic activity.

Salween River

The fast-flowing waters of the Mekong River cascade over this waterfall in Champasak province in Laos. The force of the water erodes rocks at the base of the fall.

The Mekong River flows through southern China and Myanmar, then for much of its length forms the border between Laos and Thailand, flowing through Cambodia before terminating in a vast delta on the southern Vietnamese coast.

Malay Peninsula

Tonle Sap, a freshwater lake, drains into the Mekong Delta via the Mekong River. It is the largest lake in Southeast Asia.

The coastline of the Isthmus of Kra

Longshore drift
Spit
Lagoon
Eroded coastline
Wave attack

The east and west coasts of the Isthmus of Kra differ greatly. The tectonically uplifting west coast is exposed to the harsh south-westerly monsoon and is heavily eroded. On the east coast, longshore currents produce depositional features such as spits and lagoons.

Bohol

Bohol in the southern Philippines is famous for its so-called "chocolate hills". There are more than 1,000 of these regular mounds on the island. The hills are limestone in origin, the smoothed remains of an earlier cycle of erosion. Their brown appearance in the dry season gives the hills their name.

Thailand

The coast of the Isthmus of Kra, in southeast Thailand has many small, precipitous islands like these, formed by chemical erosion on limestone, which is weathered along vertical cracks. The humidity of the climate in Southeast Asia increases the rate of weathering.

The Irrawaddy River map (Myanmar region — BANGLADESH, CHIN STATE, Sittwe, ARAKAN STATE, MAGWE, PEGU, IRRAWADDY, Bay of Bengal, Mouths of the Irrawaddy, Preparis Island, Great Coco Island, Little Coco Island, Andaman Sea)

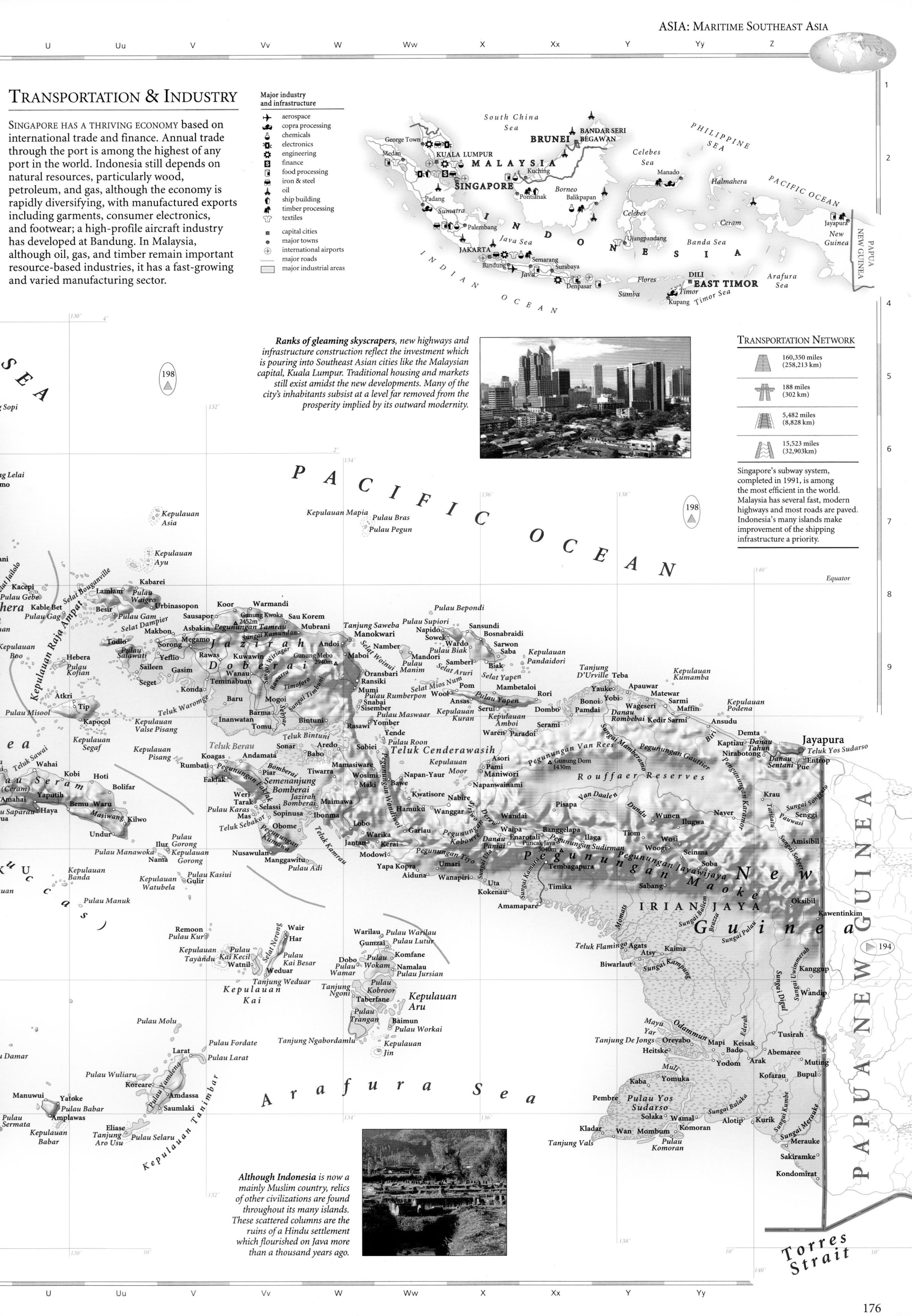

TRANSPORTATION & INDUSTRY

SINGAPORE HAS A THRIVING ECONOMY based on international trade and finance. Annual trade through the port is among the highest of any port in the world. Indonesia still depends on natural resources, particularly wood, petroleum, and gas, although the economy is rapidly diversifying, with manufactured exports including garments, consumer electronics, and footwear; a high-profile aircraft industry has developed at Bandung. In Malaysia, although oil, gas, and timber remain important resource-based industries, it has a fast-growing and varied manufacturing sector.

Major industry and infrastructure

- aerospace
- copra processing
- chemicals
- electronics
- engineering
- finance
- food processing
- iron & steel
- oil
- ship building
- timber processing
- textiles

- ■ capital cities
- ▪ major towns
- ⊕ international airports
- — major roads
- major industrial areas

Ranks of gleaming skyscrapers, new highways and infrastructure construction reflect the investment which is pouring into Southeast Asian cities like the Malaysian capital, Kuala Lumpur. Traditional housing and markets still exist amidst the new developments. Many of the city's inhabitants subsist at a level far removed from the prosperity implied by its outward modernity.

TRANSPORTATION NETWORK

- 160,350 miles (258,213 km)
- 188 miles (302 km)
- 5,482 miles (8,828 km)
- 15,523 miles (32,903km)

Singapore's subway system, completed in 1991, is among the most efficient in the world. Malaysia has several fast, modern highways and most roads are paved. Indonesia's many islands make improvement of the shipping infrastructure a priority.

Although Indonesia is now a mainly Muslim country, relics of other civilizations are found throughout its many islands. These scattered columns are the ruins of a Hindu settlement which flourished on Java more than a thousand years ago.

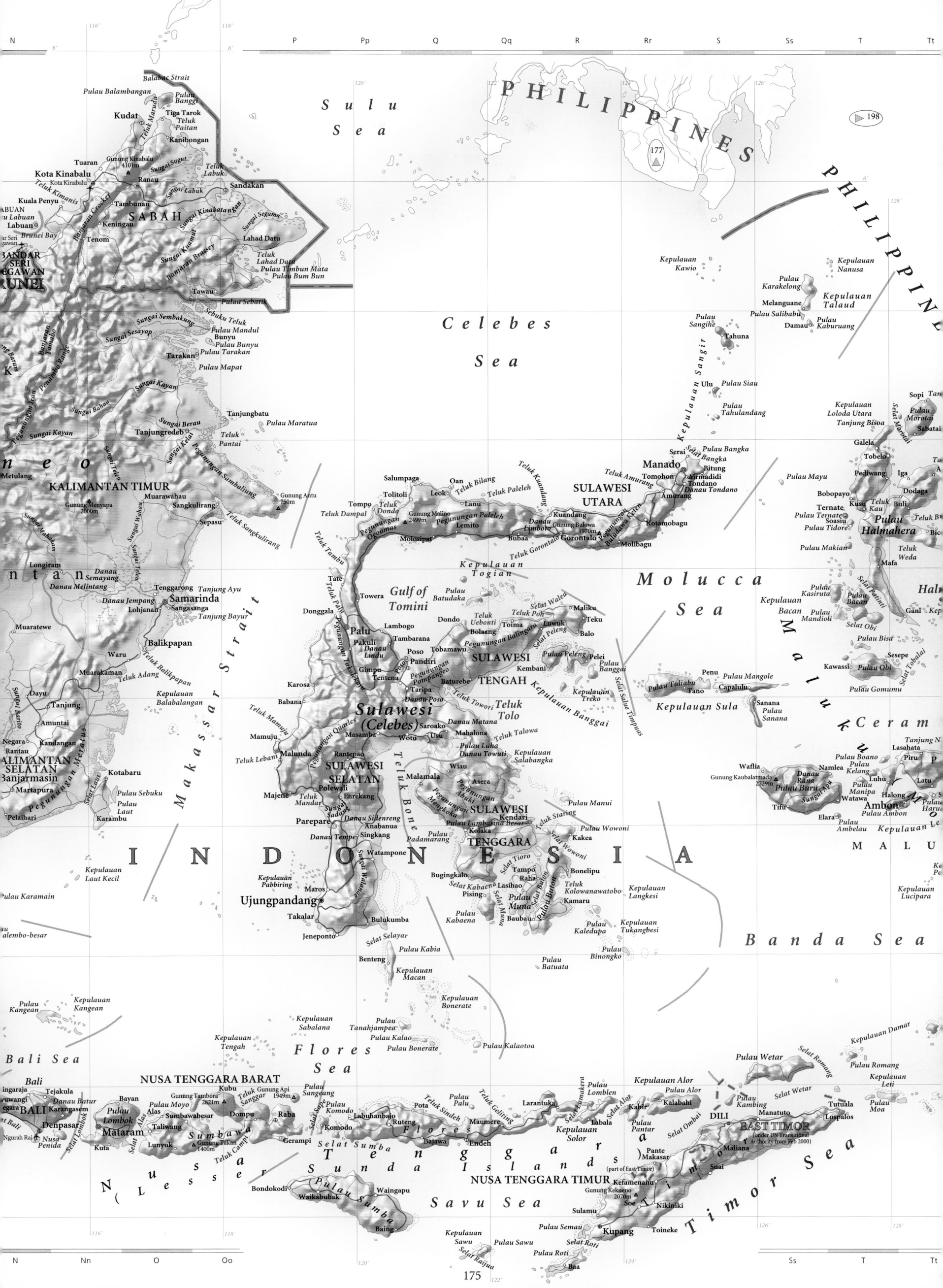

PHILIPPINES

PHILIPPINE

198

177

Sulu
Sea

Kudat
Pulau Balambangan
Pulau Banggi
Balabac Strait
Teluk Marudu
Tiga Tarok
Teluk
Paitan
Kanibongan

Tuaran
Gunung Kinabalu
4101m
Kota Kinabalu
Ranau
Sungai Sugut
Teluk
Labuk
Sandakan

Kota Kinabalu
Kuala Penyu
Tambunan
Sungai Kinabatangan
Sungai Labuk

Celebes
Sea

Kepulauan
Kawio
Pulau
Karakelong
Kepulauan
Nanusa

Melanguane
Kepulauan
Talaud

Kepulauan Sangir
Pulau
Sangihe
Pulau
Damau
Pulau
Kaburuang

Pulau Siau

Keningau
SABAH
Tenom
Sungai Kuamut
Lahad Datu
BANDAR
SERI
BEGAWAN
BRUNEI
Brunei Bay
ABUAN
Labuan
Banjaran Brassey
Teluk
Lahad Datu
Pulau Timbun Mata
Pulau Bum Bun
Tawau
Pulau Sebatik

Pulau Tahulandang

Kepulauan
Loloda Utara
Tanjung Bisoa
Sopi   Tan
Pulau
Morotai
Galela   Sabatai
Tobelo
Kepulauan Ulu
Pediwang   Iga

Sebuku Teluk
Sungai Sembakung
Pulau Mandul
Bunyu
Pulau Bunyu
Pulau Tarakan
Tarakan
Sungai Sesayap

Manado
Tomohon
Airmadidi
Tondano
Danau Tondano
Bitung
SULAWESI
UTARA
Teluk Amurang
Amurang
Kotamobagu
Serai
Pulau Bangka
Selat Bangka

Pulau Mayu
Bobopayo
Ternate
Pulau Ternate
Soasiu
Pulau Tidore
Pulau Makian
Mafa

Kusu
Teluk
Kau
Dodaga
Pulau Halmahera
Buli
Teluk Weda

Pulau Mapat

Sungai Kayan
Sungai Berau
Tanjungbatu
Tanjungredeb
Pulau Maratua
Teluk
Pantai

Salumpaga
Tolitoli
Leok
Oan
Teluk Bilang
Lanu
Kuandang
Bubaa
Gorontalo
Molibagu
Gunung Bulowa
1970m
Teluk Gorontalo

Molucca
Sea

Pulau
Kasiruta
Pulau Bacan
Selat Patinti
Gani   Kep

KALIMANTAN TIMUR
Gunung Menyapa
2000m
Muarawahau
Sangkulirang
Sepasu
Teluk Sangkulirang
Gunung Antu
750m
Tompo
Teluk Dampal
Pegunungan
Ogoamas
Tate
Towera
Teluk Tambu
Dondo
Molosipat
Lemito
Gunung Malino
2499m
Pegunungan Paleleh
Lembgo
Bolaang
Dandu
Limboto
Pegunungan
Buliohuto
Gorontalo
Kuandang
Pulau
Batudaka
Kepulauan
Togian
Maliku
Teluk Poh
Teku
Selat Walea
Luwuk
Kembani
Balo
Pulau Peleng
Pelei
Pegunungan Balingara

Kepulauan
Bacan
Pulau
Mandioli
Selat Obi
Pulau Bisa
Sesepe
Kawassi
Pulau Obi
Selat Tobalai

Longiram
Danau
Semayang
Danau Melintang
Tenggarong
Tanjung Ayu
Samarinda
Lohjanan
Sangasanga
Tanjung Bayur
Donggala
Palu
Pakuli
Danau
Lindu
Lambogo
Tambarana
Poso
Dondo
Tobamawu
Pandiri
Gimpu
Tentena
Pegunungan
Pompange
Baturebe
SULAWESI
TENGAH
Kepulauan Banggai
Pulau Banggai
Pulau Treko

Pulau Taliabu
Penu
Tano
Pulau Mangole
Capalulu
Kepulauan Sula
Sanana
Pulau
Sanana

Ceram

Danau Jempang
Muaratewe
Balikpapan
Waru
Teluk Balikpapan
Muarakaman
Karosa
Babana
Tenggarong
Teluk Adang
Kepulauan
Balabalangan
Teluk Mamuju
Mamuju
Pegunungan Quarles
Masamba
Sulawesi
(Celebes)
Saroako
Danau Matana
Wotu
Danau Towuti
Usu
Pulau Luha
Mahalona
Teluk Towori
Teluk
Tolo
Teluk Talowa
Kepulauan
Salabangka

Gunung Kaubalatmada
2729m
Waflia
Pulau Boano
Namlea
Pulau
Kelang
Piru
Lasahata
Danau
Rana
Pulau Buru
Luhu
Pulau Manipa
Watawa
Halong
Ambon
Tifu
Elara
Pulau
Ambelau
Pulau Ambon

Dayu
Tanjung
Amuntai
ALIMANTAN
SELATAN
Banjarmasin
Martapura
Negara
Rantau
Kandangan
Pegunungan Meratus
Selat Laut
Pulau Sebuku
Kotabaru
Pulau
Laut
Karambu
Pelaihari
Pulau Laut Kecil

Malunda
Rantepao
SULAWESI
SELATAN
Majene
Polewali
Enrekang
Malamala
Wiau
Asera
Pegunungan
Mekongka
Pegunungan
Mahu
SULAWESI
TENGGARA
Kendari
Kolaka
Pulau Manui
Pulau Lambasina Besar
Teluk Staring
Pulau Wowoni

MALU

Kepulauan
Lucipara

Parepare
Danau Sidenreng
Anabanua
Singkang
Sungai Walanae
Watampone
Danau Tempe
Teluk Mandar
Teluk
Bone
Kolaka
Padamarang
Selat Tioro
Tampo
Raha
Bonelipu
Selat Wowoni
Kakea

Makassar Strait

Banda Sea

Kepulauan
Laut Kecil
Pulau Karamain
Kepulauan
Pabbiring
Maros
Ujungpandang
Takalar
Bulukumba
Jeneponto
Bugingkalo
Lasihao
Pising
Selat Kabaena
Pulau
Kabaena
Pulau
Muna
Baubau
Pulau Buton
Kamaru
Teluk
Kolowanawatobo
Kepulauan
Langkesi
Kepulauan
Tukangbesi
Pulau
Kaledupa

alembo-besar

Benteng
Pulau Kabia
Kepulauan
Macan
Pulau
Batuata
Pulau
Binongko

Pulau
Kangean
Kepulauan
Kangean
Kepulauan
Sabalana
Pulau
Tanahjampea
Pulau Kalao
Pulau Bonerate
Pulau Kalaotoa
Kepulauan
Bonerate

Bali Sea

NUSA TENGGARA BARAT
Flores
Sea

Kepulauan Damar

Pulau Wetar
Selat Romang
Pulau Romang
Kepulauan
Leti

Bali
Tejakula
Danau Batur
Karangasem
Bayan
BALI
Denpasar
Pulau
Lombok
Mataram
Kuta
Ngurah Rai
Nusa
Penida
Gunung Tambora
2821m
Pulau Moyo
Alas
Taliwang
Lunyuk
Sumbawabesar
Dompu
Raba
Sumbawa
Gunung Takan
1400m
Teluk Saleh
Gerampi
Komodo
Pulau
Sangeang
Gunung Api
1949m
Teluk Komodo
Pota
Labuhanbajo
Ruteng
Teluk Palu
Bajawa
Endeh
Flores
Larantuka
Pulau
Lomblen
Kepulauan Alor
Pulau Alor
Kalabahi
Kabir
Pulau Pantar
Selat Ombai
DILI
Pante
Makasar
Maliana
Manatuto
EAST TIMOR
(under UN Transitional
Authority from Feb 2000)
Tutuala
Lospalos
Pulau
Moa

Selat
Alas
Selat Lombok
Selat Sumba
Selat Sumba
Lesser Sunda Islands
Nusa *Tenggara*
NUSA TENGGARA TIMUR
Savu Sea
Maumere
Kepulauan
Solor
Labala
Pante
Makasar
Snai
Gunung Kekneno
2070m
Soe
Nikiniki
Timor Sea

Bondokodi
Waikabubak
Waingapu
Pulau Sumba
Baing
Sulamu
Kefamenanu
Kupang
Toineke
Pulau Semau
Selat Roti
Pulau Roti
Baa
Kepulauan
Sawu
Pulau Sawu
Selat Raijua
Pulau Sawu

USING THE LAND AND SEA

THE FERTILE FLOODPLAINS of rivers such as the Mekong and Salween, and the humid climate, enable the production of rice throughout the region. Cambodia, Myanmar and Laos still have substantial forests, producing hardwoods such as teak and rosewood. Cash crops include tropical fruits such as coconuts, bananas, and pineapples, rubber, oil palm, sugar cane, and the jute substitute, kenaf. Pigs and cattle are the main livestock raised. Large quantities of marine and freshwater fish are caught throughout the region.

Land use and agricultural distribution
- cattle
- pigs
- bananas
- coconuts
- fishing
- oil palms
- rice
- rubber
- sugar cane
- timber
- ■ capital cities
- • major towns

pasture
cropland
forest
wetland

THE URBAN/RURAL POPULATION DIVIDE

urban 30% rural 70%

0 10 20 30 40 50 60 70 80 90 100

POPULATION DENSITY	TOTAL LAND AREA
322 people per sq mile (124 people per sq km)	733,828 sq miles (1,901,110 sq km)

The Paracel Islands and the Spratly Islands are two strategically sensitive island groups, disputed by several surrounding countries. The Paracels are claimed by China, Taiwan and Vietnam, though only China has actually occupied them. The Spratlys are claimed by China, Taiwan, Vietnam, Malaysia and the Philippines and are particularly important as they lie on oil and gas deposits.

The city of Hue in central Vietnam was the country's capital under the 13 emperors of the Nguyen dynasty from 1802 to 1945. It is the site of a number of religious monuments, including the Thien-Mu Pagoda.

Transportation & Industry

INDUSTRIAL MANUFACTURING has become increasingly important in Thailand, Vietnam, and the Philippines in recent years. The assembling of component-based electrical and electronic goods is becoming more common throughout this region, with foreign companies benefiting from low labor costs and the upgrading of technology. The economies of Myanmar and Cambodia are still based on agricultural produce and the processing of raw materials. Tin is the region's most important metal, and nickel, copper, and chromite are also mined, although the quantities produced are not significant on a global scale. Thailand's successful tourist industry is the country's highest earner of foreign exchange.

TRANSPORTATION NETWORK

131,566 miles (211,845 km)		267 miles (430 km)	
7,785 miles (12,536 km)		28,393 miles (45,722 km)	

Transportation development has concentrated on the building of road networks. Water and sea transportation remain important, although air links have improved, particularly in Thailand and the Philippines.

Major industry and infrastructure

chemicals		oil & gas	
electronics		mining	
engineering		shipbuilding	
finance		textiles	
food processing		timber processing	
iron & steel			

- capital cities
- major towns
- international airports
- major roads
- major industrial areas

Opium poppies are destroyed under army supervision in Thailand. This action is part of a government-sponsored initiative to reduce the trade in drugs such as heroin, which is derived from these plants. Drug trafficking is a major problem throughout the region; the area is known as the "Golden Triangle," and Laos is the third-largest producer of opium poppies in the world.

The terracing of land to restrict soil erosion and create flat surfaces for agriculture is a common practice throughout Southeast Asia, particularly where land is scarce. These terraces are on Luzon in the Philippines.

SCALE 1:7,750,000
(projection: Lambert Conformal Conic)

Km
0 25 50   100   150 200

Miles
0 25   50   100   150   200

MAP KEY

POPULATION

- ■ above 5 million
- ■ 1 million to 5 million
- ◉ 500,000 to 1 million
- ◎ 100,000 to 500,000
- ⊕ 50,000 to 100,000
- ○ 10,000 to 50,000
- ∘ below 10,000

ELEVATION

	4000m / 13,124ft
	3000m / 9843ft
	2000m / 6562ft
	1000m / 3281ft
	500m / 1640ft
	250m / 820ft
	100m / 328ft
	sea level

Straw and timber dwellings have been built close to the edge of the beach on this island near Palawan, one of the most westerly islands in the Philippines.

THE INDIAN OCEAN

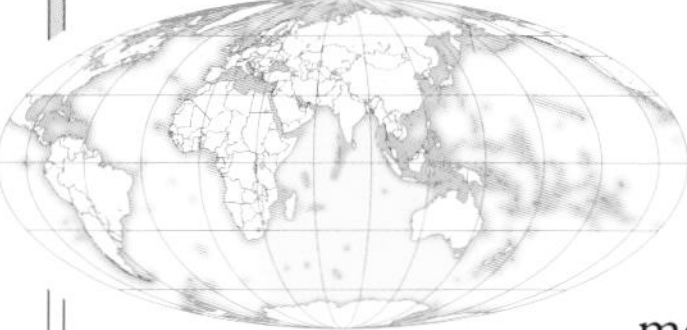

DESPITE BEING THE SMALLEST of the three major oceans, the evolution of the Indian Ocean was the most complex. The ocean basin was formed during the breakup of the supercontinent Gondwanaland, when the Indian subcontinent moved northeast, Africa moved west and Australia separated from Antarctica. Like the Pacific Ocean, the warm waters of the Indian Ocean are punctuated by coral atolls and islands. About one-fifth of the world's population – over a billion people – live on its shores. Those people living along the northern coasts are constantly threatened by flooding and typhoons caused by the monsoon winds.

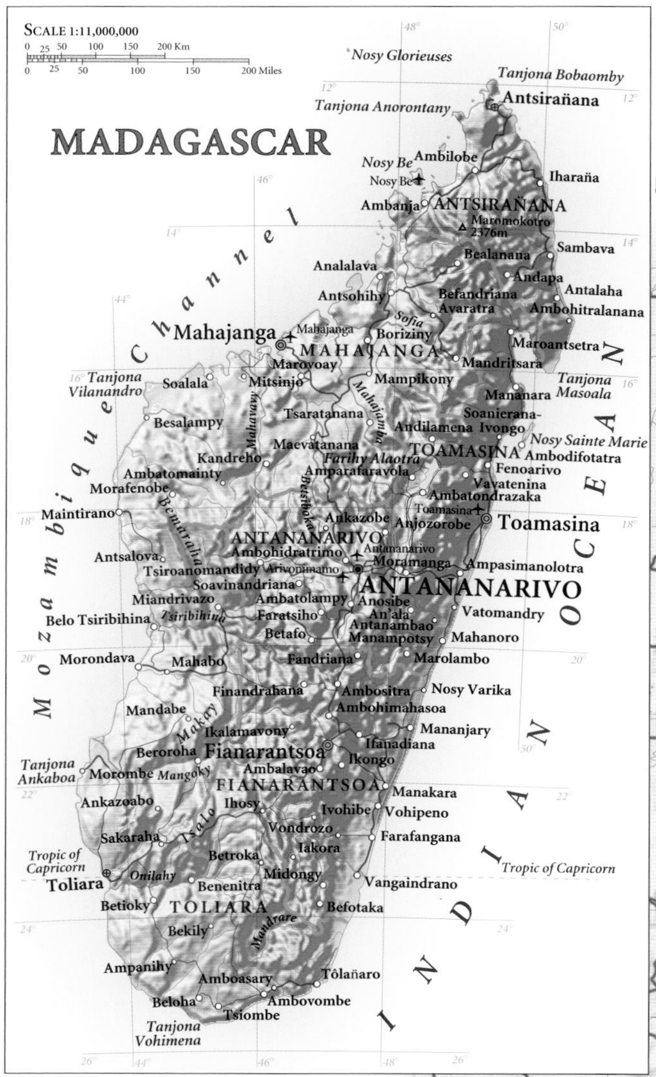

THE LANDSCAPE

THE INDIAN OCEAN BEGAN FORMING about 150 million years ago, but in its present form it is relatively young, only about 36 million years old. Along the three subterranean mountain chains of its mid-ocean ridge the seafloor is still spreading. The Indian Ocean has fewer trenches than other oceans and only a narrow continental shelf around most of its surrounding land.

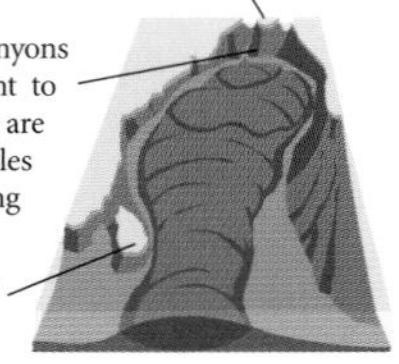

Sediments come from Ganges/Brahmaputra river system

Submarine canyons transport sediment to fan – some of these are more than 1,500 miles (2,500 km) long

Sri Lanka

The mid-oceanic ridge runs from the Arabian Sea. It diverges east of Madagascar. One arm runs southwest to join the Mid-Atlantic Ridge, the other branches southeast, joining the Pacific-Antarctic Ridge, southeast of Tasmania.

The Ninetyeast Ridge takes its name from the line of longitude it follows. It is the world's longest and straightest under-sea ridge.

Two of the world's largest rivers flow into the Indian Ocean; the Indus and the Ganges/Brahmaputra. Both have deposited enormous fans of sediment.

The Ganges Fan is one of the world's largest submarine accumulations of sediment, extending far beyond Sri Lanka. It is fed by the Ganges/Brahmaputra River system, whose sediment is carried through a network of underwater canyons at the edge of the continental shelf.

Indus River

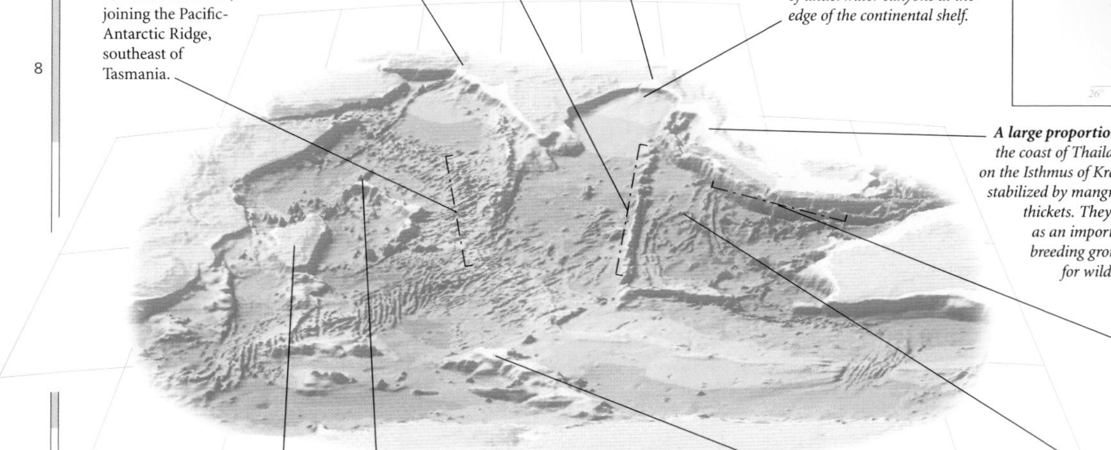

A large proportion of the coast of Thailand, on the Isthmus of Kra, is stabilized by mangrove thickets. They act as an important breeding ground for wildlife.

The Java Trench is the world's longest, it runs 1,600 miles (2,570 km) from the southwest of Java, but is only 50 miles (80 km) wide.

The relief of Madagascar rises from a low-lying coastal strip in the east, to the central plateau. The plateau is also a major watershed separating Madagascar's three main river basins.

The central group of the Seychelles are mountainous, granite islands. They have a narrow coastal belt and lush, tropical vegetation cloaks the highlands.

The Kerguelen Islands in the Southern Ocean were created by a hot spot in the Earth's crust. The islands were formed in succession as the Antarctic Plate moved slowly over the hot spot.

The circulation in the northern Indian Ocean is controlled by the monsoon winds. Biannually these winds reverse their pattern, causing a reversal in the surface currents and alternative high and low pressure conditions over Asia and Australia.

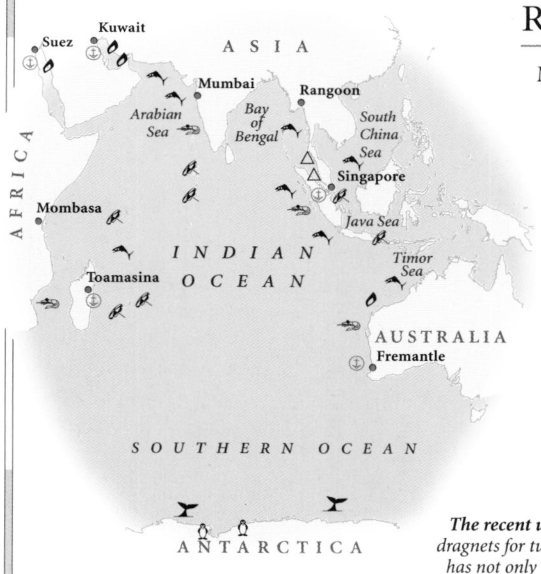

RESOURCES

MANY OF THE SMALL ISLANDS in the Indian Ocean rely exclusively on tuna-fishing and tourism to maintain their economies. Most fisheries are artisanal, although large-scale tuna-fishing does take place in the Seychelles, Mauritius and the western Indian Ocean. Nonliving resources include oil in the Persian Gulf, pearls in the Red Sea, and tin from deposits off the shores of Myanmar, Thailand, and Indonesia.

The recent use of large dragnets for tuna-fishing has not only threatened the livelihoods of many small-scale fisheries, but also caused widespread environmental concern about the potential impact on other marine species.

Resources (including wildlife)

⤚ fish	△ tin deposits	
🐧 penguins	🏖 tourism	
🐚 shellfish		
🐋 whales	⊕ major towns	
⬧ oil & gas	⊕ major ports	

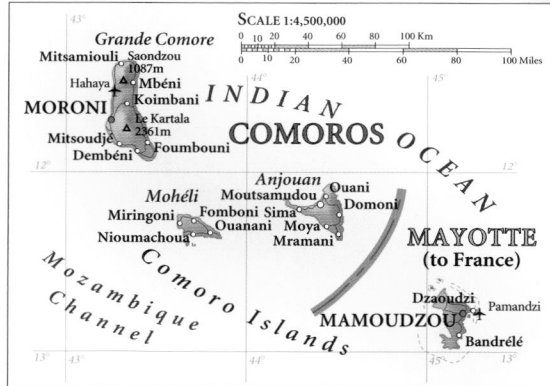

SCALE 1:4,500,000

MORONI
COMOROS
Grande Comore
Mitsamiouli Saondzou 1087m
Hahaya Mbéni
Mitsoudjé Koimbani
Dembéni Ile Kartala 2361m
Foumbouni

MAYOTTE (to France)
Mohéli Anjouan
Miringoni Moutsamoudou Ouani
Nioumachoua Fomboni Sima Domoni
Ouanani Moya
Mramani

MAMOUDZOU
Dzaoudzi Pamandzi
Bandrélé

Mozambique Channel
Comoro Islands
INDIAN OCEAN

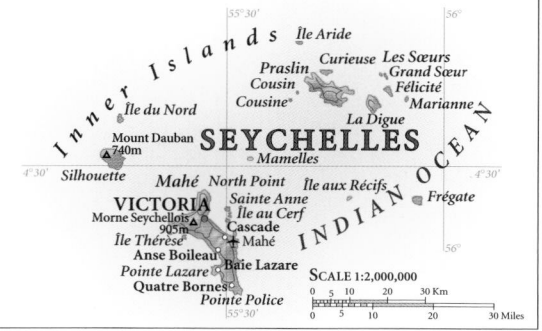

SEYCHELLES
Inner Islands
Ile Aride
Praslin Curieuse Les Sœurs
Cousin Grand Sœur
Cousine Félicité
La Digue
Marianne
Ile du Nord
Mount Dauban 740m
Mamelles
Silhouette
Frégate
Mahé North Point
Sainte Anne
VICTORIA Ile au Cerf
Morne Seychellois 905m Ile du Cerf
Ile Thérèse Cascade
Anse Boileau Mahé
Pointe Lazare Baie Lazare
Quatre Bornes
Pointe Police
SCALE 1:2,000,000

Coral reefs support an enormous diversity of animal and plant life. Many species of tiny tropical fish, like these squirrel fish, live and feed around the profusion of reefs and atolls in the Indian Ocean.

SCALE 1:11,000,000

MADAGASCAR

(map of Madagascar with numerous place names including Antsiranaña, Antananarivo, Toamasina, Mahajanga, Fianarantsoa, Toliara, Mozambique Channel, Indian Ocean, Tropic of Capricorn)

(map of eastern Africa and surrounding oceans including Red Sea, Saudi, Egypt, Sudan, Eritrea, Djibouti, Ethiopia, Kenya, Somalia, Tanzania, Mozambique, South Africa, Lesotho, Swaziland, Comoros, Mayotte, Agulhas Plateau, Agulhas Basin, Natal Basin, Atlantic-Indian Ridge, Antarctic Circle)

(small index map of the Indian Ocean region showing Asia, Africa, Australia, Antarctica, Southern Ocean, Arabian Sea, Bay of Bengal, South China Sea, Java Sea, Timor Sea, and cities: Suez, Kuwait, Mumbai, Rangoon, Singapore, Mombasa, Toamasina, Fremantle)

The steeper eastern side of Madagascar is drained by numerous short, fast-flowing rivers. In contrast, larger, more languid rivers flow across the west. Both erode huge quantities of Madagascar's reddish soil.

There are over 1,300 small coral islands in the Maldives, but only about 200 are inhabited. They are based around an ancient submerged volcanic mountain range and all the islands are low-lying, none rising more than 6 ft (1.8 m) above sea level.

SCALE 1:42,000,000
(projection: Mollweide)

The island of Mauritius is volcanic in origin. Its central plateau is bounded by mountains which may once have formed the rim of a volcanic crater.

RÉUNION (to France)
SCALE 1:2,000,000

INSET MAP KEY

POPULATION
- 500,000 to 1 million
- 100,000 to 500,000
- 50,000 to 100,000
- 10,000 to 50,000
- below 10,000

ELEVATION
- 3000m / 9843ft
- 2000m / 6562ft
- 1000m / 3281ft
- 500m / 1640ft
- 250m / 820ft
- 100m / 328ft
- sea level

OCEAN MAP KEY

SEA DEPTH
- sea level
- 250m / 820ft
- 500m / 1640ft
- 1000m / 3281ft
- 2000m / 6562ft
- 3000m / 9843ft

MAURITIUS
SCALE 1:2,000,000

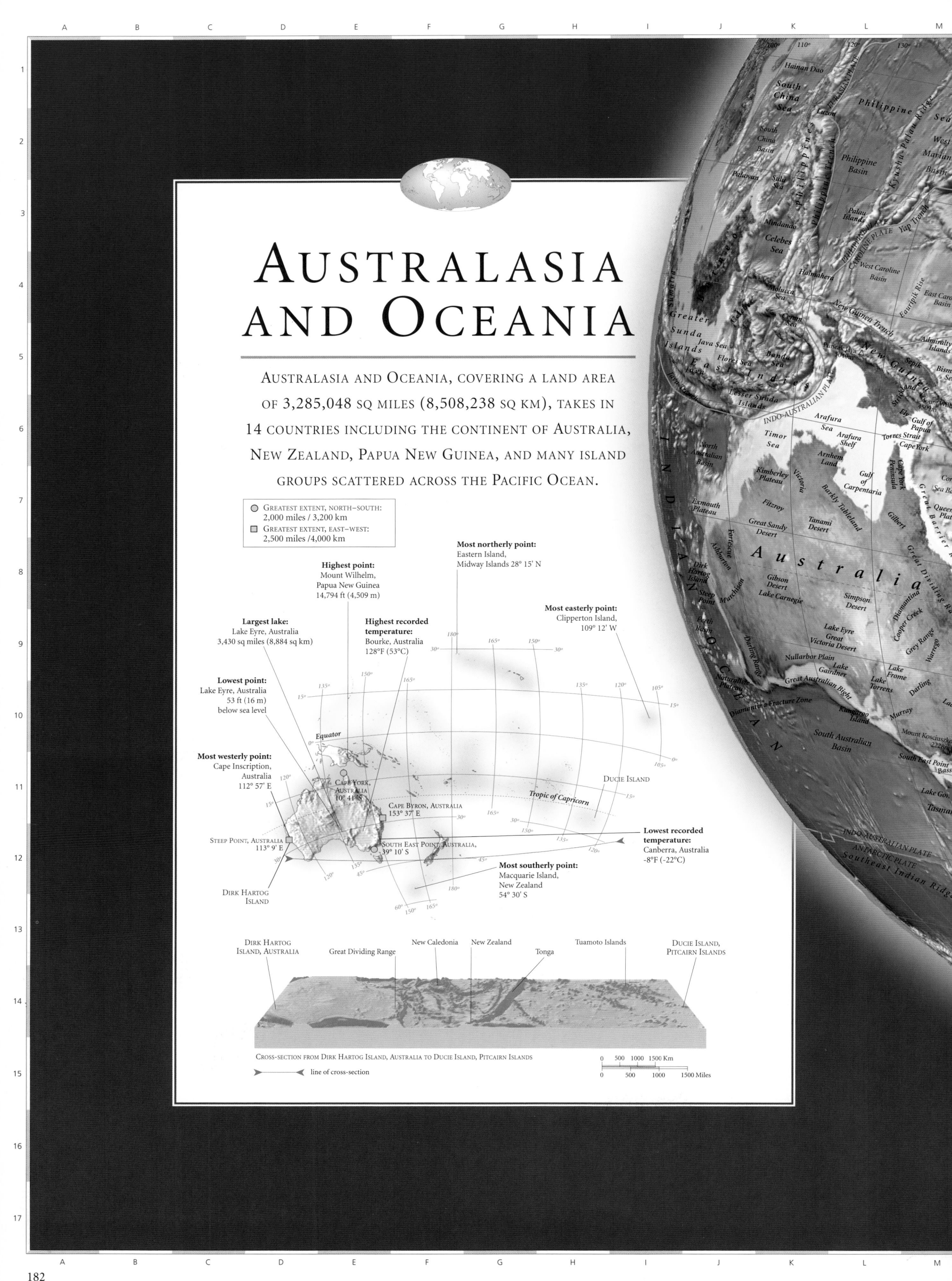

AUSTRALASIA AND OCEANIA

AUSTRALASIA AND OCEANIA, COVERING A LAND AREA OF 3,285,048 SQ MILES (8,508,238 SQ KM), TAKES IN 14 COUNTRIES INCLUDING THE CONTINENT OF AUSTRALIA, NEW ZEALAND, PAPUA NEW GUINEA, AND MANY ISLAND GROUPS SCATTERED ACROSS THE PACIFIC OCEAN.

- ⬤ GREATEST EXTENT, NORTH–SOUTH: 2,000 miles / 3,200 km
- ▪ GREATEST EXTENT, EAST–WEST: 2,500 miles /4,000 km

Most northerly point:
Eastern Island,
Midway Islands 28° 15' N

Highest point:
Mount Wilhelm,
Papua New Guinea
14,794 ft (4,509 m)

Most easterly point:
Clipperton Island,
109° 12' W

Largest lake:
Lake Eyre, Australia
3,430 sq miles (8,884 sq km)

Highest recorded temperature:
Bourke, Australia
128°F (53°C)

Lowest point:
Lake Eyre, Australia
53 ft (16 m)
below sea level

Most westerly point:
Cape Inscription,
Australia
112° 57' E

CAPE YORK, AUSTRALIA, 10° 41' N

CAPE BYRON, AUSTRALIA, 153° 37' E

DUCIE ISLAND

Tropic of Capricorn

STEEP POINT, AUSTRALIA, 113° 9' E

SOUTH EAST POINT, AUSTRALIA, 39° 10' S

Lowest recorded temperature:
Canberra, Australia
-8°F (-22°C)

DIRK HARTOG ISLAND

Most southerly point:
Macquarie Island,
New Zealand
54° 30' S

DIRK HARTOG ISLAND, AUSTRALIA

Great Dividing Range

New Caledonia

New Zealand

Tonga

Tuamoto Islands

DUCIE ISLAND, PITCAIRN ISLANDS

CROSS-SECTION FROM DIRK HARTOG ISLAND, AUSTRALIA TO DUCIE ISLAND, PITCAIRN ISLANDS

◄───◄ line of cross-section

| 0 | 500 | 1000 | 1500 Km |
| 0 | 500 | 1000 | 1500 Miles |

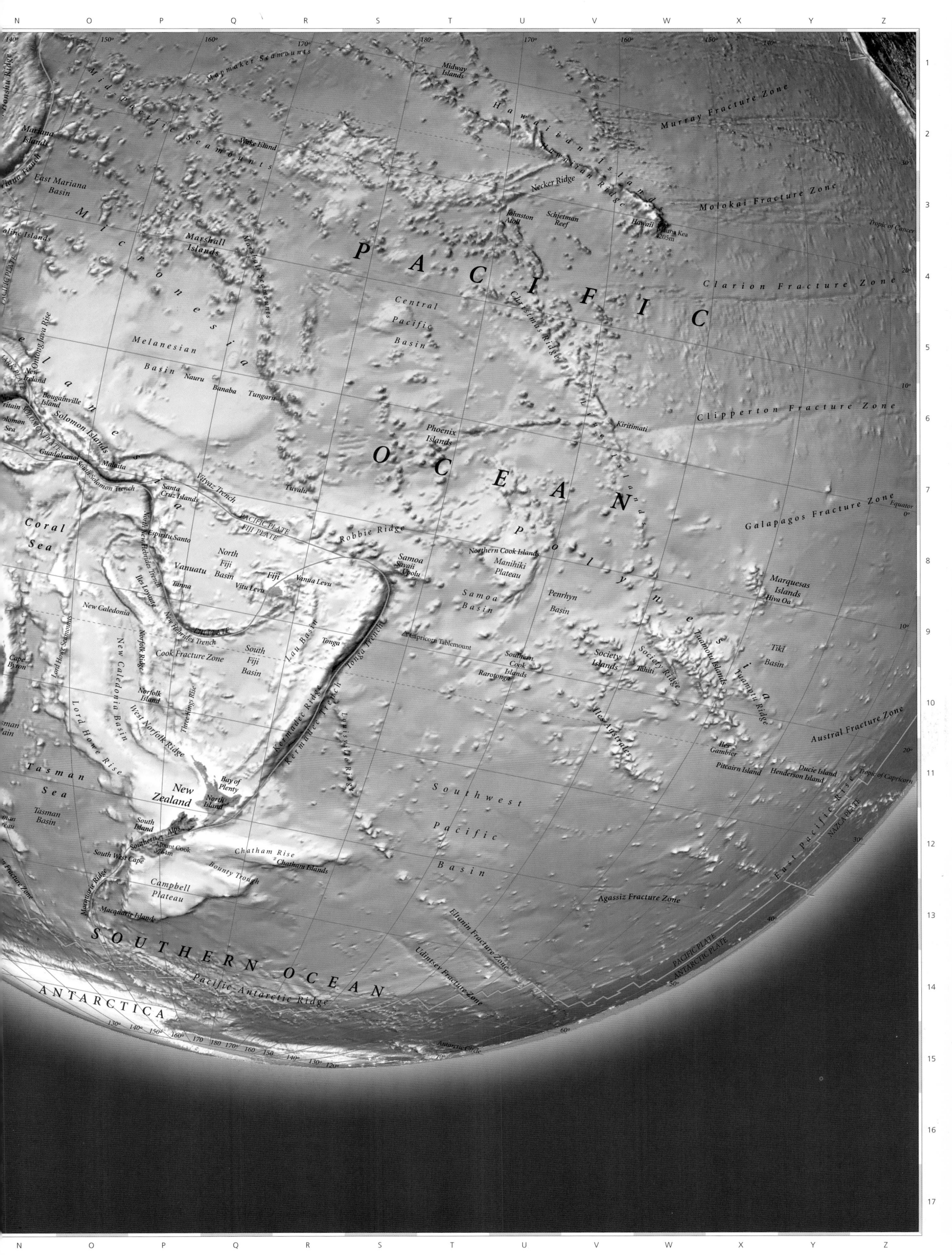

PACIFIC

OCEAN

Midway
Islands

Hawaiian Islands

Murray Fracture Zone

Manmaker Seamounts

Mariana
Islands

East Mariana
Basin

Micronesia

Marshall
Islands

Necker Ridge

Johnston
Atoll

Schjetman
Reef

Hawaii Mauna Kea
4205m

Molokai Fracture Zone

Tropic of Cancer

Clarion Fracture Zone

Central
Pacific
Basin

Christmas Ridge

Line Islands

Clipperton Fracture Zone

Caroline Islands

Melanesian
Basin

Nauru

Banaba

Tungaru

Phoenix
Islands

Kiritimati

New
Zealand

Kandavu Java Rise

Bougainville
Island

Solomon Islands

SOLOMON PLATE

Guadalcanal
Malaita

South Solomon Trench

Vityaz Trench

Santa
Cruz Islands

Tuvalu

PACIFIC PLATE
FIJI PLATE

Robbie Ridge

Samoa
Savaii
Upolu

Northern Cook Islands

Manihiki
Plateau

Galapagos Fracture Zone

Equator

Coral
Sea

North New Hebrides Trench

Espiritu Santo

North
Fiji
Basin

Fiji

Vanua Levu

Polynesia

Samoa
Basin

Penrhyn
Basin

Marquesas
Islands
Hiva Oa

Vanuatu

Tanna

Vitu Levu

Acapricorn Tablemount

Society
Islands

Society Ridge
Tahiti

Tuamotu Islands

Tiki
Basin

Iles Loyaute

New Caledonia

New Hebrides Trench

South
Fiji
Basin

Lau Basin

Tonga

Southern
Cook
Islands
Rarotonga

Tuamotu Ridge

Cape
Byron

Lord Howe Seamounts

New Caledonia Basin

Norfolk Ridge

Cook Fracture Zone

Tonga Trench

Austral Fracture Zone

Norfolk
Island

Kermadec Ridge

Louisville Ridge

Iles Australes

Iles
Gambier

Tasman
Sea

Lord Howe Rise

West Norfolk Ridge

Three Kings Rise

Kermadec Trench

Pitcairn Island

Ducie Island
Henderson Island

Tropic of Capricorn

Tasman
Basin

Bay of
Plenty

North
Island

Southwest

Pacific

East Pacific Rise

NAZCA PLATE

South
Island

Southern Alps
Mount Cook
3764m

Chatham Rise

Chatham Islands

Basin

Agassiz Fracture Zone

South West Cape

Macquarie Ridge

Bounty Trough

Campbell
Plateau

Macquarie Island

Eltanin Fracture Zone

PACIFIC PLATE
ANTARCTIC PLATE

SOUTHERN OCEAN

Udintsev Fracture Zone

Pacific-Antarctic Ridge

ANTARCTICA

Antarctic Circle

POLITICAL AUSTRALASIA AND OCEANIA

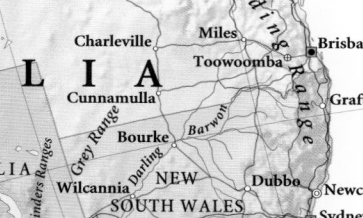

Western Australia's mineral wealth has transformed its state capital, Perth, into one of Australia's major cities. Perth is one of the world's most isolated cities – over 2,500 miles (4,000 km) from the population centers of the eastern seaboard.

VAST EXPANSES OF OCEAN separate this geographically fragmented realm, characterized more by each country's isolation than by any political unity. Australia's and New Zealand's traditional ties with the United Kingdom, as members of the Commonwealth, are now being called into question as Australasian and Oceanian nations are increasingly looking to forge new relationships with neighboring Asian countries like Japan. External influences have featured strongly in the politics of the Pacific Islands; the various territories of Micronesia were largely under US control until the late 1980s, and France, New Zealand, the US, and the UK still have territories under colonial rule in Polynesia. Nuclear weapons-testing by Western superpowers was widespread during the Cold War period, but has now been discontinued.

POPULATION

DENSITY OF SETTLEMENT in the region is generally low. Australia is one of the least densely populated countries on Earth with over 80% of its population living within 25 miles (40 km) of the coast – mostly in the southeast of the country. New Zealand, and the island groups of Melanesia, Micronesia, and Polynesia, are much more densely populated, although many of the smaller islands remain uninhabited.

Population density (people per sq mile)

	below 10
	10-62
	63-130
	131-259
	260-519
	520-780
	above 780

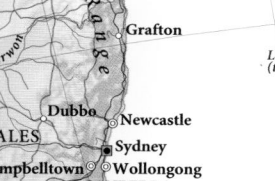

The myriad of small coral islands that are scattered across the Pacific Ocean are often uninhabited, as they offer little shelter from the weather, often no fresh water, and only limited food supplies.

The planes of the Australian Royal Flying Doctor Service are able to cover large expanses of barren land quickly, bringing medical treatment to the most inaccessible and far-flung places.

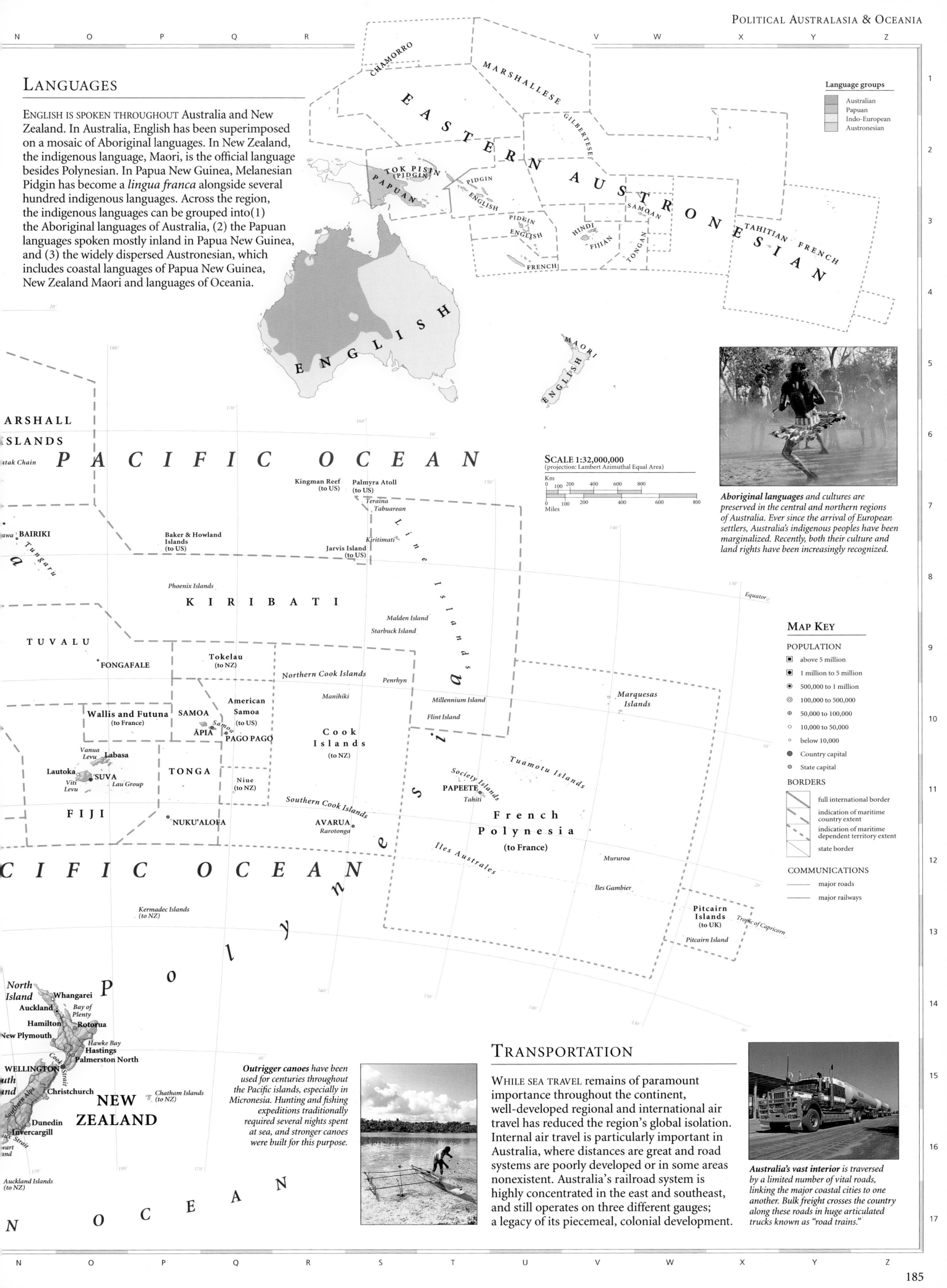

LANGUAGES

ENGLISH IS SPOKEN THROUGHOUT Australia and New Zealand. In Australia, English has been superimposed on a mosaic of Aboriginal languages. In New Zealand, the indigenous language, Maori, is the official language besides Polynesian. In Papua New Guinea, Melanesian Pidgin has become a *lingua franca* alongside several hundred indigenous languages. Across the region, the indigenous languages can be grouped into(1) the Aboriginal languages of Australia, (2) the Papuan languages spoken mostly inland in Papua New Guinea, and (3) the widely dispersed Austronesian, which includes coastal languages of Papua New Guinea, New Zealand Maori and languages of Oceania.

Language groups
- Australian
- Papuan
- Indo-European
- Austronesian

Aboriginal languages and cultures are preserved in the central and northern regions of Australia. Ever since the arrival of European settlers, Australia's indigenous peoples have been marginalized. Recently, both their culture and land rights have been increasingly recognized.

Scale 1:32,000,000
(projection: Lambert Azimuthal Equal Area)

MAP KEY

POPULATION
- ■ above 5 million
- ■ 1 million to 5 million
- ◉ 500,000 to 1 million
- ◎ 100,000 to 500,000
- ⊕ 50,000 to 100,000
- ⊕ 10,000 to 50,000
- ○ below 10,000
- ● Country capital
- ◉ State capital

BORDERS
- full international border
- indication of maritime country extent
- indication of maritime dependent territory extent
- state border

COMMUNICATIONS
- major roads
- major railways

TRANSPORTATION

WHILE SEA TRAVEL remains of paramount importance throughout the continent, well-developed regional and international air travel has reduced the region's global isolation. Internal air travel is particularly important in Australia, where distances are great and road systems are poorly developed or in some areas nonexistent. Australia's railroad system is highly concentrated in the east and southeast, and still operates on three different gauges; a legacy of its piecemeal, colonial development.

Outrigger canoes have been used for centuries throughout the Pacific islands, especially in Micronesia. Hunting and fishing expeditions traditionally required several nights spent at sea, and stronger canoes were built for this purpose.

Australia's vast interior is traversed by a limited number of vital roads, linking the major coastal cities to one another. Bulk freight crosses the country along these roads in huge articulated trucks known as "road trains."

AUSTRALASIAN AND OCEANIAN RESOURCES

NATURAL RESOURCES ARE OF MAJOR ECONOMIC IMPORTANCE throughout Australasia and Oceania. Australia in particular is a major world exporter of raw materials such as coal, iron ore, and bauxite, while New Zealand's agricultural economy is dominated by sheep-raising. Trade with western Europe has declined significantly in the last 20 years, and the Pacific Rim countries of Southeast Asia are now the main trading partners, as well as a source of new settlers to the region. Australasia and Oceania's greatest resources are its climate and environment; tourism increasingly provides a vital source of income for the whole continent.

The largely unpolluted waters of the Pacific Ocean support rich and varied marine life, much of which is farmed commercially. Here, oysters are gathered for market off the coast of New Zealand's South Island.

Huge flocks of sheep are a common sight in New Zealand, where they outnumber people by 20 to 1. New Zealand is one of the world's largest exporters of wool and frozen lamb.

STANDARD OF LIVING

IN MARKED CONTRAST TO ITS NEIGHBOR, Australia, with one of the world's highest life expectancies and standards of living, Papua New Guinea is one of the world's least developed countries. In addition, high population growth and urbanization rates throughout the Pacific islands contribute to overcrowding. In Australia and New Zealand, the Aboriginal and Maori people have been isolated, although recently their traditional land ownership rights have begun to be legally recognized in an effort to ease their social and economic isolation, and to improve living standards.

Standard of Living
(UN Human Development Index)

low

high

figures unavailable

ENVIRONMENTAL ISSUES

THE PROSPECT OF RISING SEA LEVELS poses a threat to many low-lying islands in the Pacific. The testing of nuclear weapons, once common throughout the region, was finally discontinued in 1996. Australia's ecological balance has been irreversibly altered by the introduction of alien species. Although it has the world's largest underground water reserve, the Great Artesian Basin, the availability of fresh water in Australia remains critical. Periodic droughts combined with overgrazing lead to desertification and increase the risk of devastating bush fires, and occasional flash floods.

Environmental Issues

national parks
tropical forest
forest destroyed
desert
desertification
polluted rivers
radioactive contamination
marine pollution
heavy marine pollution
poor urban air quality

In 1946 Bikini Atoll, in the Marshall Islands, was chosen as the site for Operation Crossroads – investigating the effects of atomic bombs upon naval vessels. Further nuclear tests continued until the early 1990s. The long-term environmental effects are unknown.

Northern Mariana Islands (to US)

Guam (to US)

Saipan

MICR

PALAU

Me l

PAPUA NEW GUINEA

New Guinea

Port Mores

Arafura Sea

Torres Strait

Timor Sea

Darwin

Gulf of Carpentaria

Great Barrier Re

Townsville

AUSTRALIA

INDIAN OCEAN

Adelaide

Geel

Perth

Bikini Atoll

Eniwetak Atoll

Malden Island

Fangataufa

Coral Sea

PACIFIC OCEAN

SOUTHERN

INDIAN OCEAN

Murchison

Mackenzie

Darling

Murray

Sydney

Tasman Sea

AGRICULTURE, INDUSTRY, AND MINERALS

MUCH OF THE REGION'S INDUSTRY IS RESOURCE-BASED: sheep farming for wool and meat in Australia and New Zealand; mining in Australia and Papua New Guinea and fishing throughout the Pacific islands. Manufacturing is mainly limited to the large coastal cities in Australia and New Zealand, like Sydney, Adelaide, Melbourne, Brisbane, Perth, and Auckland, although small-scale enterprises operate in the Pacific islands, concentrating on processing of fish and foods. Tourism continues to provide revenue to the area – in Fiji it accounts for 15% of GNP.

The massive Ok Tedi copper mine *was opened in 1988. It is situated in the midst of remote tropical jungle in Papua New Guinea.*

Plumes of steam *rise from the electricity turbines on New Zealand's North Island. New Zealand is one of the few countries in the world where geothermal energy makes a significant contribution to national energy production.*

Map labels

MARSHALL ISLANDS

ESIA
Pohnpei
Ralik Chain
Ratak Chain
Kosrae

PACIFIC OCEAN

Tungaru

Kiritimati

NAURU

KIRIBATI

Starbuck Island

TUVALU

Penrhyn

SOLOMON ISLANDS
Honiara

Tokelau (to NZ)

Marquesas Islands

Wallis and Futuna (to France)

SAMOA
Ápia

American Samoa (to US)
Pago Pago

Cook Islands (to NZ)

Tuamotu Islands

Society Islands
Tahiti

VANUATU
Port-Vila

Suva
FIJI

TONGA

Niue (to NZ)

French Polynesia (to France)

New Caledonia (to France)

Nuku'alofa

Avarua

Iles Australes

Iles Gambier

Pitcairn Islands (to UK)

Brisbane
Toowoomba

Newcastle
Sydney
Wollongong
Canberra

Auckland

NEW ZEALAND

Wellington

Tasman Sea

Launceston
Hobart

Christchurch

Dunedin

OCEAN

MAP KEY

Using the Land and Sea

	barren land
	cropland
	desert
	forest
	mountain region
	pasture

Industry

sheep	brewing	printing & publishing
coconuts	chemicals	shipbuilding
coffee	copra	sugar processing
fishing	engineering	textiles
fruit	finance	timber processing
shellfish	fish processing	
sugar cane	food processing	coal
vineyards	hi-tech industry	oil
whaling	iron & steel	gas
wheat	meat processing	industrial cities

Mineral Resources

bauxite
copper
gold
iron
lead
nickel

CLIMATE

SURROUNDED BY WATER, the climate of most areas is profoundly affected by the moderating effects of the oceans. Australia, however, is the exception. Its dry continental interior remains isolated from the ocean; temperatures soar during the day, and droughts are common. The coastal regions, where most people live, are cooler and wetter. The numerous islands scattered across the Pacific are generally hot and humid, subject to the different air circulation patterns and ocean currents that affect the area, including the El Niño ocean current anomaly, which produces extreme aridity.

The tourist trade *continues to bring valuable income to the region. Fiji, Guam, and the Cook Islands are favored destinations for Japanese, American, and Australian tourists. Surfers Paradise near Brisbane, Australia, is part of the fastest growing tourist area in the country; 40 years ago, the area was wild bushland.*

Climate map

Equator
Southeast Monsoon
Madang
Darwin
Suva
South East Trades
January Winds
Queensland
Townsville
Brisbane
Alice Springs
Tropic of Capricorn
Sydney
Auckland
Adelaide
Melbourne
Perth
Hobart
Dunedin

Climate

	arid
	cool continental
	humid subtropical
	mediterranean
	semiarid
	tropical
	warm humid

daily hours of sunshine, January
daily hours of sunshine, July
→ cold wind
→ hot wind

Coconuts are harvested *throughout the islands of the Pacific Ocean, and dried in the sun for their white meat which is known as copra. Dried copra is crushed in processing plants to produce valuable coconut oil, used in making soap, margarine, and cooking oil.*

AUSTRALIA

Australia is the world's smallest continent, a stable landmass lying between the Indian and Pacific oceans. Previously home to its aboriginal peoples only, since the end of the 18th century immigration has transformed the face of the country. Initially settlers came mainly from western Europe, particularly the UK, and for years Australia remained wedded to its British colonial past. More recent immigrants have come from eastern Europe, and from Asian countries such as Japan, South Korea, and Indonesia. Australia is now forging strong trading links with these "Pacific Rim" countries and its economic future seems to lie with Asia and the Americas, rather than Europe, its traditional partner.

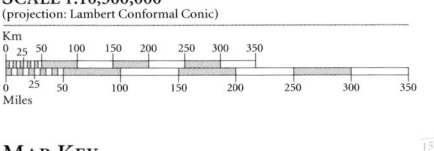

Uluru (Ayers Rock), the world's largest free-standing rock, is a massive outcrop of red sandstone in Australia's desert center. Wind and sandstorms have ground the rock into the smooth curves seen here. Uluru is revered as a sacred site by many aboriginal peoples.

SCALE 1:10,500,000
(projection: Lambert Conformal Conic)

MAP KEY

POPULATION

- ◉ 1 million to 5 million
- ◉ 500,000 to 1 million
- ⊙ 100,000 to 500,000
- ⊕ 50,000 to 100,000
- ○ 10,000 to 50,000
- ○ below 10,000

ELEVATION

- 2000m / 6562ft
- 1000m / 3281ft
- 500m / 1640ft
- 250m / 820ft
- 100m / 328ft
- sea level

USING THE LAND

Over 165 million sheep are dispersed in vast herds around the country, contributing to a major export industry. Cattle-ranching is important, particularly in the west. Wheat, and grapes for Australia's wine industry, are grown mainly in the south. Much of the country is desert, unsuitable for agriculture unless irrigation is used.

THE URBAN/RURAL POPULATION DIVIDE

urban 85%   rural 15%

0 10 20 30 40 50 60 70 80 90 100

POPULATION DENSITY	TOTAL LAND AREA
6 people per sq mile (2 people per sq km)	2,967,893 sq miles (7,686,850 sq km)

Land use and agricultural distribution

- cattle
- sheep
- cereals
- sugar cane
- timber
- vineyards
- ■ capital cities
- ● major towns
- pasture
- cropland
- forest
- desert
- mountain region

Lines of ripening vines stretch for miles in Barossa Valley, a major wine-growing region near Adelaide.

THE LANDSCAPE

Australia consists of many eroded plateaus, lying firmly in the middle of the Indo-Australian Plate. It is the world's flattest continent, and the driest, after Antarctica. The coasts tend to be more hilly and fertile, especially in the east. The mountains of the Great Dividing Range form a natural barrier between the eastern coastal areas and the flat, dry plains and desert regions of the Australian "outback."

The Great Barrier Reef is the world's largest area of coral islands and reefs. It runs for about 1,240 miles (2,000 km) along the Queensland coast.

The Pinnacles are a series of rugged sandstone pillars. Their strange shapes have been formed by water and wind erosion.

The ancient Kimberley Plateau is the source of some of Australia's richest mineral deposits, including diamonds.

Arnhem Land

Uluru (Ayers Rock)

The tropical rainforest of the Cape York Peninsula contains more than 600 different varieties of tree.

Great Artesian Basin

The Great Dividing Range forms a watershed between east- and west-flowing rivers. Erosion has created deep valleys, gorges, and waterfalls where rivers tumble over escarpments on their way to the sea.

Great Artesian Basin

Rainwater replenishes aquifer — Aquifers from which artesian water is obtained
Lake Eyre — Underground water movements

The Great Artesian Basin underlies nearly 20% of the total area of Australia, providing a valuable store of underground water, essential to Australian agriculture. The ephemeral rivers which drain the northern part of the basin have highly braided courses and, in consequence, the area is known as "channel country."

Australian Alps

More than half of Australia rests on a uniform shield over 600 million years old. It is one of the Earth's original geological plates.

The Nullarbor Plain is a low-lying limestone plateau which is so flat that the Trans-Australian Railway runs through it in a straight line for more than 300 miles (483 km).

The Simpson Desert has a number of large salt pans, created by the evaporation of past rivers and now sourced by seasonal rains. Some are crusted with gypsum, but most are covered by common salt crystals.

The Lake Eyre basin, lying 51 ft (16 m) below sea level, is one of the largest inland drainage systems in the world, covering an area of more than 500,000 sq miles (1,300,000 sq km).

Tasmania has the same geological structure as the Australian Alps. During the last period of glaciation, 18,000 years ago, sea levels were some 300 ft (100 m) lower and it was joined to the mainland.

Map labels

Cape Londonderry, Cape Bougainville, Kalumburu, Bigge Island, Bonaparte Archipelago, Heywood Islands, Adele Island, Mount Hann 779m, Kimberley, Collier Bay, King Sound, King Leopold Ran, Kupir, Plate, Lombadina, Derby, Fitzroy Crossi, Broome, Fitzroy River, Great Sandy Desert, De Grey River, Eighty Mile Beach, Percival Lakes, Port Hedland, Wickham, Whim Creek, Lake Dora, Lake Auld, Dampier Archipelago, Dampier, Karratha, Roebourne, Marble Bar, Barrow Island, Fortescue River, Hamersley Range, Wittenoom, Lake Disappointment, North West Cape, Onslow, Ashburton River, Tom Price, Mount Bruce, Little Sandy Desert, Gibson Des, Exmouth, Mount, Paraburdoo, McHarry 1251m, Newman, Learmonth, Kenneth Range, Coral Bay, Barlee Range, Mount Augustus 1105m, Kumarina Roadhouse, Carnarvon Range, Lake Gregory, Lake Carnegie, WESTERN, Minilya, Waldburg Range, Lake Macleod, Robinson Range, Lake Wells, Bernier Island, Gascoyne River, Gascoyne Junction, Wiluna, Lake Way, Dorre Island, Carnarvon, Murchison River, Lake Annean, Meekatharra, AUSTRALIA, Dirk Hartog Island, Denham, Lake Austin, Lake Throssell, Shark Bay, Lake Yeo, Kalbarri, Mount Magnet, Leonora, Lake Carey, Yalgoo, Lake Ballard, Geraldton, Mongers Lake, Lake Barlee, Menzies, Lake Rebecca, Lake Moore, Kalgoorlie, Rawl, Kitchener, Wubin, Coolgardie, Moora, Pithara, Kambalda, Lake Lefroy, The Pinnacles, Southern Cross, Lake Cowan, Balladonia, Gingin, Merredin, Lake Johnston, Norseman, Lake Dundas, Wanneroo, Northam, York, Perth, Fremantle, Brookton, Kondinin, Lake Hope, Rockingham, Narrogin, Lake King, Mandurah, Wagin, Ravensthorpe, Esperance, Bunbury, Collie, Katanning, Busselton, Stirling Range, Tower Peak 594m, Margaret River, Bridgetown, Manjimup, Mount Barker, Cape Leeuwin, Augusta, Pemberton, Albany, INDIAN OCEAN, PACIFIC OCEAN, Tropic of Capricorn

Using the Land map labels

Timor Sea, Darwin, INDIAN OCEAN, Townsville, Alice Springs, AUSTRALIA, Brisbane, Perth, Sydney, Adelaide, CANBERRA, Melbourne, Hobart, PACIFIC OCEAN, 160

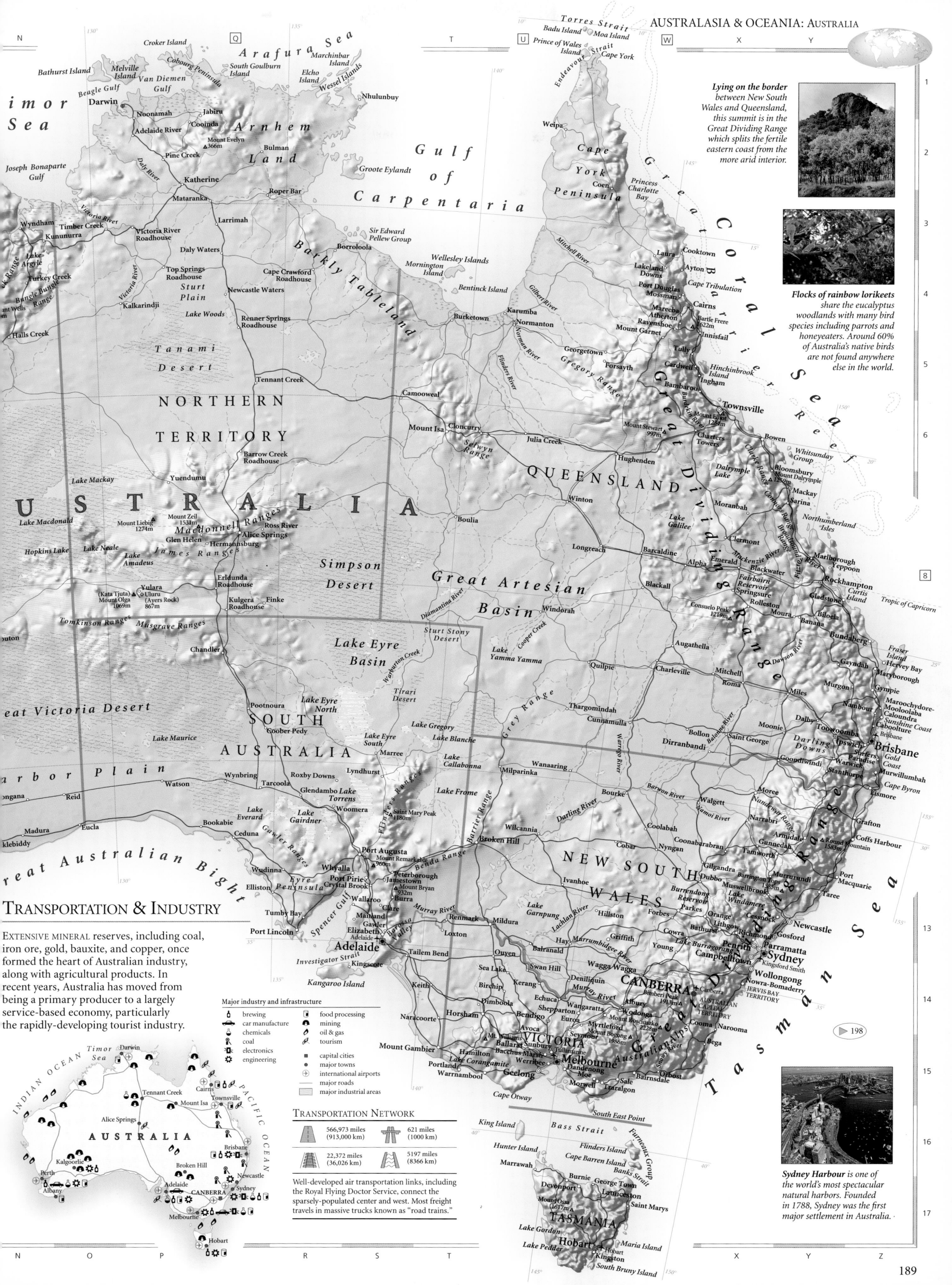

Lying on the border between New South Wales and Queensland, this summit is in the Great Dividing Range which splits the fertile eastern coast from the more arid interior.

Flocks of rainbow lorikeets share the eucalyptus woodlands with many bird species including parrots and honeyeaters. Around 60% of Australia's native birds are not found anywhere else in the world.

TRANSPORTATION & INDUSTRY

EXTENSIVE MINERAL reserves, including coal, iron ore, gold, bauxite, and copper, once formed the heart of Australian industry, along with agricultural products. In recent years, Australia has moved from being a primary producer to a largely service-based economy, particularly the rapidly-developing tourist industry.

Major industry and infrastructure

- brewing
- car manufacture
- chemicals
- coal
- electronics
- engineering
- food processing
- mining
- oil & gas
- tourism
- capital cities
- major towns
- international airports
- major roads
- major industrial areas

TRANSPORTATION NETWORK

566,973 miles (913,000 km)	621 miles (1000 km)
22,372 miles (36,026 km)	5197 miles (8366 km)

Well-developed air transportation links, including the Royal Flying Doctor Service, connect the sparsely-populated center and west. Most freight travels in massive trucks known as "road trains."

Sydney Harbour is one of the world's most spectacular natural harbors. Founded in 1788, Sydney was the first major settlement in Australia.

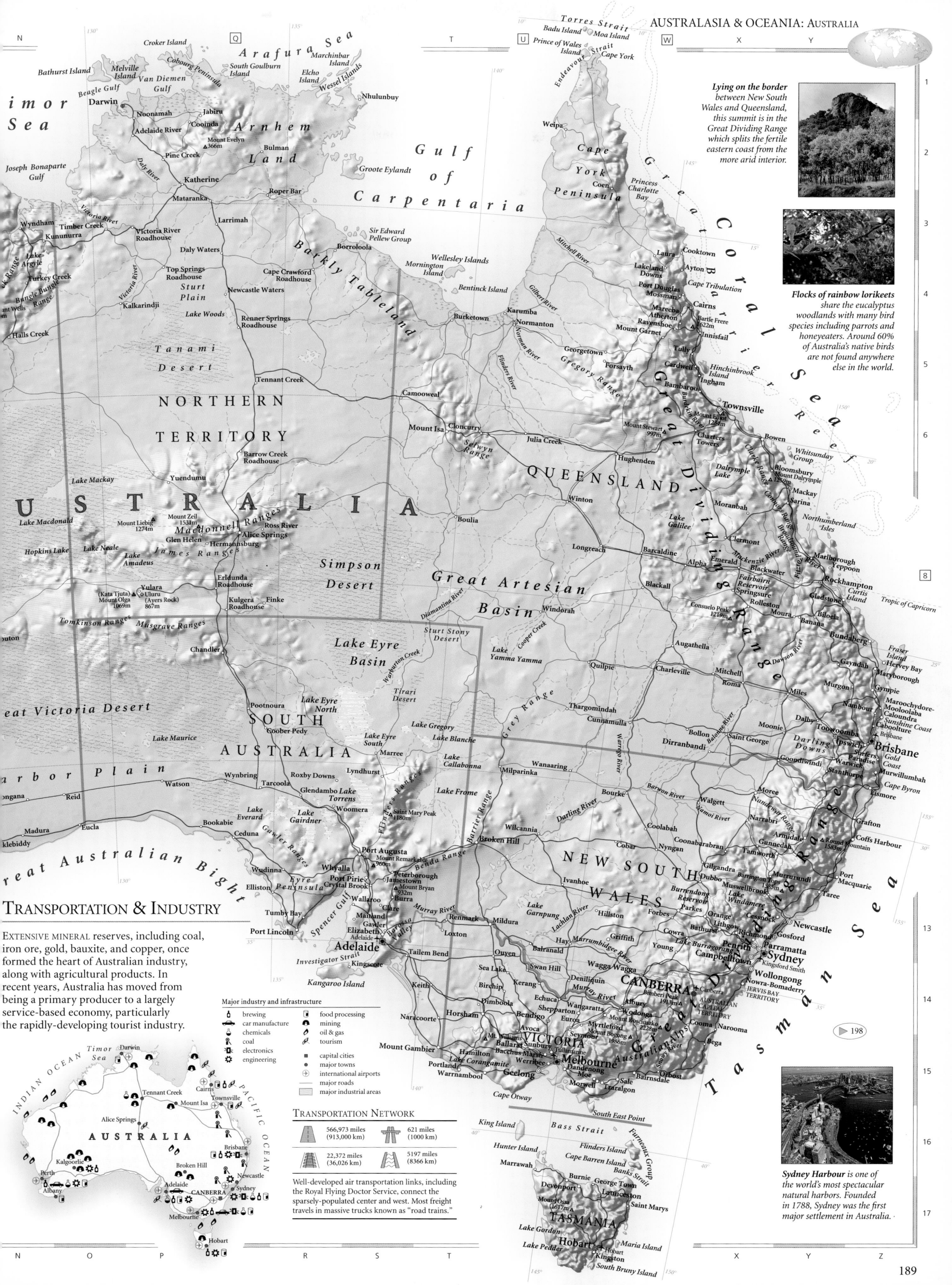

189

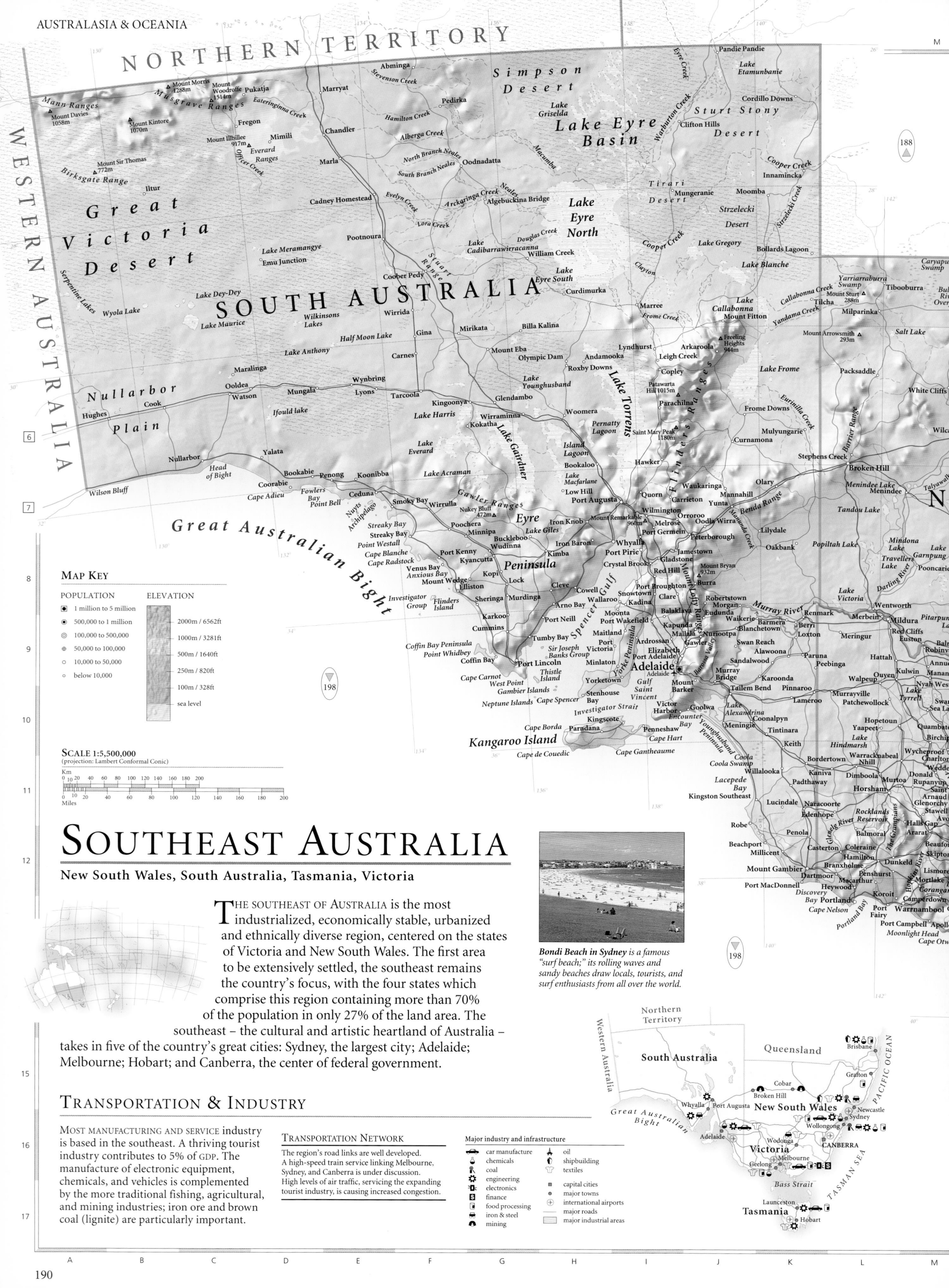

Map labels (as shown on map)

NORTHERN TERRITORY

WESTERN AUSTRALIA

Mann Ranges · Mount Davies 1058m · Mount Kintore 1070m · Mount Morris 1288m · Mount Woodroffe 1514m · Pukatja · Musgrave Ranges · Mount Illbillee 917m · Fregon · Mimili · Everard Ranges · Officer Creek · Mount Sir Thomas 772m · Iltur · Birksgate Range

Great Victoria Desert

Serpentine Lakes · Wyola Lake · Lake Dey-Dey · Lake Maurice · Lake Meramangye · Emu Junction · Maralinga · Ooldea · Watson · Mungala · Lyons · Tarcoola

Nullarbor Plain

Hughes · Cook · Ifould lake · Nullarbor · Head of Bight · Wilson Bluff · Cape Adieu · Coorabie · Bookabie · Penong · Koonibba · Ceduna · Fowlers Bay · Point Bell · Yalata

Simpson Desert

Abminga · Marryat · Stevenson Creek · Pedirka · Hamilton Creek · Alberga Creek · Oodnadatta · North Branch Neales · South Branch Neales · Cadney Homestead · Evelyn Creek · Arckaringa Creek · Algebuckina Bridge · Neales · Macumba · Coober Pedy · Stuart Range · Lora Creek · Cadibarrawirracanna · Douglas Creek · William Creek · Lake Cadibarrawirracanna

SOUTH AUSTRALIA

Pandie Pandie · Lake Etamunbanie · Cordillo Downs · Clifton Hills · Sturt Stony Desert · Lake Griselda · Warburton Creek · Eyre Creek · Mungeranie · Moomba · Innamincka · Cooper Creek · Tirari Desert · Strzelecki Desert · Strzelecki Creek

Lake Eyre Basin

Lake Eyre North · Marree · Frome Creek · Clayton · Lake Gregory · Lake Blanche · Bollards Lagoon · Lake Callabonna · Mount Fitton · Lake Frome · Frome Downs · Packsaddle · White Cliffs

Lake Eyre South · Curdimurka · Wirrida · Gina · Mirikata · Billa Kalina · Mount Eba · Olympic Dam · Andamooka · Roxby Downs · Lyndhurst · Leigh Creek · Copley · Arkaroola · Freeling Heights 944m · Patawarta Hill 1015m · Parachilna · Flinders Ranges · Mulyungarie · Curnamona · Mount Arrowsmith 293m · Salt Lake · Caryapundy Swamp · Yarriarraburra Swamp · Tibooburra · Milparinka · Bullo River Overfl.

Carnes · Wilkinsons Lakes · Half Moon Lake · Lake Anthony · Lake Harris · Wirraminna · Kokatha · Lake Gairdner · Pernatty Lagoon · Woomera · Saint Mary Peak 1180m · Saint Mary Peak · Hawker · Lake Torrens · Frome Creek · Eurinilla Creek · Yandama Creek · Tilcha · Mount Sturt 288m · Minona Lake · Travellers Lake · Lake Garnpung · Poancarie · Wilcan

Maralinga · Wynbring · Glendambo · Lake Younghusband · Lake Macfarlane · Low Hill · Port Augusta · Quorn · Waukaringa · Olary · Menindee Lake · Menindee · Tandou Lake · Stephens Creek · Broken Hill · Talyawalka Cr.

Great Australian Bight

Streaky Bay · Smoky Bay · Wirrulla · Nuyts Archipelago · Point Westall · Cape Blanche · Cape Radstock · Venus Bay · Anxious Bay · Mount Wedge · Elliston · Sheringa · Murdinga · Karkoo · Cummins · Coffin Bay Peninsula · Point Whidbey · Coffin Bay · Cape Carnot · West Point · Gambier Islands · Neptune Islands · Cape Spencer · Cape de Couedic

Eyre Peninsula

Nukey Bluff 472m · Poochera · Minnipa · Wudinna · Lake Giles · Buckleboo · Iron Knob · Iron Baron · Kimba · Kyancutta · Kopi · Lock · Cleve · Cowell · Arno Bay · Port Neill · Tumby Bay · Sir Joseph Banks Group · Port Lincoln · Thistle Island · Iron Monarch · Mount Remarkable 960m · Whyalla · Port Pirie · Crystal Brook · Port Germein · Melrose · Wilmington · Orroroo · Peterborough · Oodla Wirra · Yunta · Mannahill · Lilydale · Popiltah Lake · Pitarpung Lake

Port Wakefield · Wallaroo · Moonta · Kadina · Balaklava · Snowtown · Clare · Jamestown · Gladstone · Red Hill · Mount Bryan 932m · Burra · Robertstown · Eudunda · Kapunda · Nuriootpa · Waikerie · Barmera · Berri · Renmark · Wentworth · Mildura · Merbein · Meringur · Red Cliffs · Euston · Lake Victoria · Darling River

Yorke Peninsula

Maitland · Ardrossan · Port Victoria · Minlaton · Yorketown · Stenhouse Bay · Edithburgh · Gulf Saint Vincent · Stansbury

Spencer Gulf

Adelaide · Port Adelaide · Elizabeth · Gawler · Alawoona · Sandalwood · Paruna · Peebinga · Murray Bridge · Karoonda · Tailem Bend · Pinnaroo · Murrayville · Patchewollock · Ouyen · Kulwin · Walpeup · Nyah West · Swan Hill · Lake Tyrrell · Sea Lake

Mount Barker · Victor Harbor · Goolwa · Encounter Bay · Lake Alexandrina · Meningie · Coonalpyn · Tintinara · Keith · Hindmarsh · Bordertown · Kaniva · Nhill · Lake Hindmarsh · Dimboola · Jeparit · Warracknabeal · Wycheproof · Charlton · Donald · Rupanyup · Saint Arnaud · Avoca · Stawell · Glenorchy · Ararat · Beaufort

Kangaroo Island

Cape Borda · Parndana · Kingscote · Penneshaw · Cape Hart · Cape Gantheaume · Cape de Couedic · Youngusband Peninsula

Willalooka · Lucindale · Naracoorte · Rocklands Reservoir · Edenhope · Penola · Coleraine · Balmoral · Coola Swamp · Robe · Beachport · Millicent · Mount Gambier · Casterton · Hamilton · Penshurst · Dunkeld · Lismore · Camperdown · Port MacDonnell · Dartmoor · Macarthur · Heywood · Koroit · Corangamite · Discovery Bay · Portland · Port Fairy · Warrnambool · Moonlight Head · Cape Otway · Apollo · Port Campbell · Cape Nelson · Branxholme · The Grampians · Halls Gap · Mortlake · Glenelg River · Hotspur River

---

SOUTHEAST AUSTRALIA

New South Wales, South Australia, Tasmania, Victoria

THE SOUTHEAST OF AUSTRALIA is the most industrialized, economically stable, urbanized and ethnically diverse region, centered on the states of Victoria and New South Wales. The first area to be extensively settled, the southeast remains the country's focus, with the four states which comprise this region containing more than 70% of the population in only 27% of the land area. The southeast – the cultural and artistic heartland of Australia – takes in five of the country's great cities: Sydney, the largest city; Adelaide; Melbourne; Hobart; and Canberra, the center of federal government.

Bondi Beach in Sydney is a famous "surf beach;" its rolling waves and sandy beaches draw locals, tourists, and surf enthusiasts from all over the world.

TRANSPORTATION & INDUSTRY

MOST MANUFACTURING AND SERVICE industry is based in the southeast. A thriving tourist industry contributes to 5% of GDP. The manufacture of electronic equipment, chemicals, and vehicles is complemented by the more traditional fishing, agricultural, and mining industries; iron ore and brown coal (lignite) are particularly important.

TRANSPORTATION NETWORK

The region's road links are well developed. A high-speed train service linking Melbourne, Sydney, and Canberra is under discussion. High levels of air traffic, servicing the expanding tourist industry, is causing increased congestion.

Major industry and infrastructure

- car manufacture
- chemicals
- coal
- engineering
- electronics
- finance
- food processing
- iron & steel
- mining
- oil
- shipbuilding
- textiles
- ■ capital cities
- ● major towns
- ✈ international airports
- — major roads
- ▢ major industrial areas

MAP KEY

POPULATION
- ▪ 1 million to 5 million
- ◉ 500,000 to 1 million
- ◎ 100,000 to 500,000
- ⊕ 50,000 to 100,000
- ⊙ 10,000 to 50,000
- ○ below 10,000

ELEVATION
- 2000m / 6562ft
- 1000m / 3281ft
- 500m / 1640ft
- 250m / 820ft
- 100m / 328ft
- sea level

SCALE 1:5,500,000
(projection: Lambert Conformal Conic)

Km 0 20 40 60 80 100 120 140 160 180 200
Miles 0 20 40 60 80 100 120 140 160 180 200

Locator map
Northern Territory · Western Australia · South Australia · Queensland · New South Wales · Victoria · Tasmania · Brisbane · Grafton · Cobar · Broken Hill · Whyalla · Port Augusta · Adelaide · Newcastle · Sydney · Wollongong · CANBERRA · Wodonga · Melbourne · Geelong · Launceston · Hobart · Great Australian Bight · Bass Strait · Tasman Sea · PACIFIC OCEAN

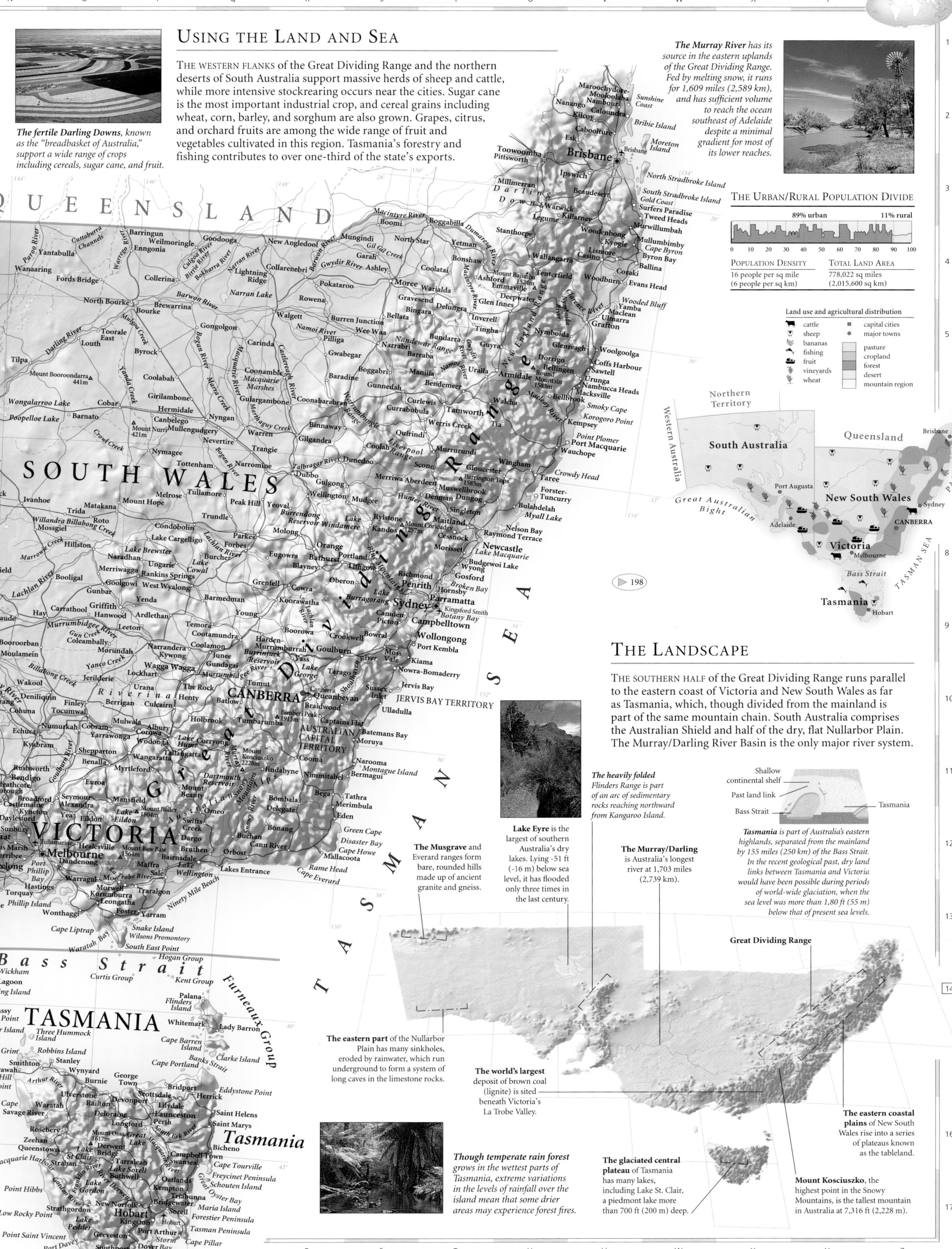

USING THE LAND AND SEA

THE WESTERN FLANKS of the Great Dividing Range and the northern deserts of South Australia support massive herds of sheep and cattle, while more intensive stockrearing occurs near the cities. Sugar cane is the most important industrial crop, and cereal grains including wheat, corn, barley, and sorghum are also grown. Grapes, citrus, and orchard fruits are among the wide range of fruit and vegetables cultivated in this region. Tasmania's forestry and fishing contributes to over one-third of the state's exports.

The fertile Darling Downs, known as the "breadbasket of Australia," support a wide range of crops including cereals, sugar cane, and fruit.

The Murray River has its source in the eastern uplands of the Great Dividing Range. Fed by melting snow, it runs for 1,609 miles (2,589 km), and has sufficient volume to reach the ocean southeast of Adelaide despite a minimal gradient for most of its lower reaches.

THE URBAN/RURAL POPULATION DIVIDE

89% urban   11% rural

0 10 20 30 40 50 60 70 80 90 100

POPULATION DENSITY
16 people per sq mile
(6 people per sq km)

TOTAL LAND AREA
778,022 sq miles
(2,015,600 sq km)

Land use and agricultural distribution

- cattle
- sheep
- bananas
- fishing
- fruit
- vineyards
- wheat

- capital cities
- major towns

- pasture
- cropland
- forest
- desert
- mountain region

THE LANDSCAPE

THE SOUTHERN HALF of the Great Dividing Range runs parallel to the eastern coast of Victoria and New South Wales as far as Tasmania, which, though divided from the mainland is part of the same mountain chain. South Australia comprises the Australian Shield and half of the dry, flat Nullarbor Plain. The Murray/Darling River Basin is the only major river system.

The heavily folded Flinders Range is part of an arc of sedimentary rocks reaching northward from Kangaroo Island.

The Musgrave and Everard ranges form bare, rounded hills made up of ancient granite and gneiss.

Lake Eyre is the largest of southern Australia's dry lakes. Lying -51 ft (-16 m) below sea level, it has flooded only three times in the last century.

The Murray/Darling is Australia's longest river at 1,703 miles (2,739 km).

Shallow continental shelf
Past land link
Bass Strait
Tasmania

Tasmania is part of Australia's eastern highlands, separated from the mainland by 155 miles (250 km) of the Bass Strait. In the recent geological past, dry land links between Tasmania and Victoria would have been possible during periods of world-wide glaciation, when the sea level was more than 1,80 ft (55 m) below that of present sea levels.

Great Dividing Range

The eastern part of the Nullarbor Plain has many sinkholes, eroded by rainwater, which run underground to form a system of long caves in the limestone rocks.

The world's largest deposit of brown coal (lignite) is sited beneath Victoria's La Trobe Valley.

Though temperate rain forest grows in the wettest parts of Tasmania, extreme variations in the levels of rainfall over the island mean that some drier areas may experience forest fires.

The glaciated central plateau of Tasmania has many lakes, including Lake St. Clair, a piedmont lake more than 700 ft (200 m) deep.

The eastern coastal plains of New South Wales rise into a series of plateaus known as the tableland.

Mount Kosciuszko, the highest point in the Snowy Mountains, is the tallest mountain in Australia at 7,316 ft (2,228 m).

▶ 198

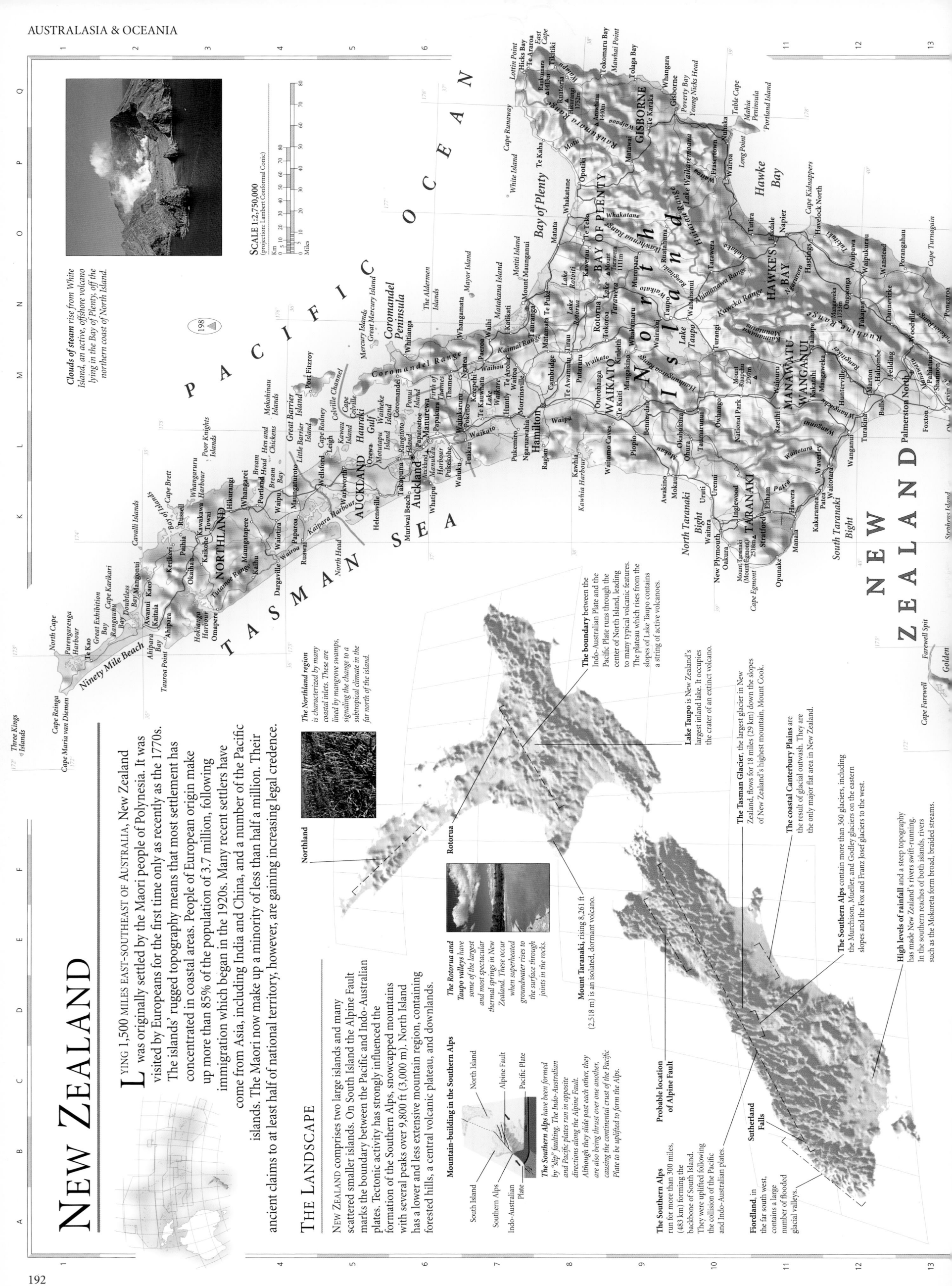

NEW ZEALAND

L YING 1,500 MILES EAST-SOUTHEAST OF AUSTRALIA, New Zealand was originally settled by the Maori people of Polynesia. It was visited by Europeans for the first time only as recently as the 1770s. The islands' rugged topography means that most settlement has concentrated in coastal areas. People of European origin make up more than 85% of the population of 3.7 million, following immigration which began in the 1920s. Many recent settlers have come from Asia, including India and China, and a number of the Pacific islands. The Maori now make up a minority of less than half a million. Their ancient claims to at least half of national territory, however, are gaining increasing legal credence.

THE LANDSCAPE

NEW ZEALAND comprises two large islands and many scattered smaller islands. On South Island the Alpine Fault marks the boundary between the Pacific and Indo-Australian plates. Tectonic activity has strongly influenced the formation of the Southern Alps, snowcapped mountains with several peaks over 9,800 ft (3,000 m). North Island has a lower and less extensive mountain region, containing forested hills, a central volcanic plateau, and downlands.

Mountain-building in the Southern Alps

North Island
Alpine Fault
Pacific Plate
South Island
Southern Alps
Indo-Australian Plate

The Southern Alps have been formed by "slip" faulting. The Indo-Australian and Pacific plates run in opposite directions along the Alpine Fault. Although they slide past each other, they are also being thrust over one another, causing the continental crust of the Pacific Plate to be uplifted to form the Alps.

The Southern Alps run for more than 300 miles, (483 km) forming the backbone of South Island. They were uplifted following the collision of the Pacific and Indo-Australian plates.

Probable location of Alpine Fault

Fiordland, in the far south west, contains a large number of flooded glacial valleys.

Sutherland Falls

The Northland region is characterized by many coastal inlets. These are lined by mangrove swamps, signaling the change to a subtropical climate in the far north of the island.

Northland

The Rotorua and Taupo valleys have some of the largest and most spectacular thermal springs in New Zealand. These occur when superheated groundwater rises to the surface through joints in the rocks.

Rotorua

Mount Taranaki, rising 8,261 ft (2,518 m) is an isolated, dormant volcano.

The boundary between the Indo-Australian Plate and the Pacific Plate runs through the center of North Island, leading to many typical volcanic features. The plateau which rises from the slopes of Lake Taupo contains a string of active volcanoes.

Lake Taupo is New Zealand's largest inland lake. It occupies the crater of an extinct volcano.

The Tasman Glacier, the largest glacier in New Zealand, flows for 18 miles (29 km) down the slopes of New Zealand's highest mountain, Mount Cook.

The coastal Canterbury Plains are the result of glacial outwash. They are the only major flat area in New Zealand.

The Southern Alps contain more than 360 glaciers, including the Murchison, Mueller, and Godley glaciers on the eastern slopes and the Fox and Franz Josef glaciers to the west.

High levels of rainfall and a steep topography has made New Zealand's rivers swift-running. In the southern reaches of both islands, rivers such as the Mokoreta form broad, braided streams.

Clouds of steam rise from White Island, an active, offshore volcano lying in the Bay of Plenty, off the northern coast of North Island.

198

SCALE 1:2,750,000
(projection: Lambert Conformal Conic)

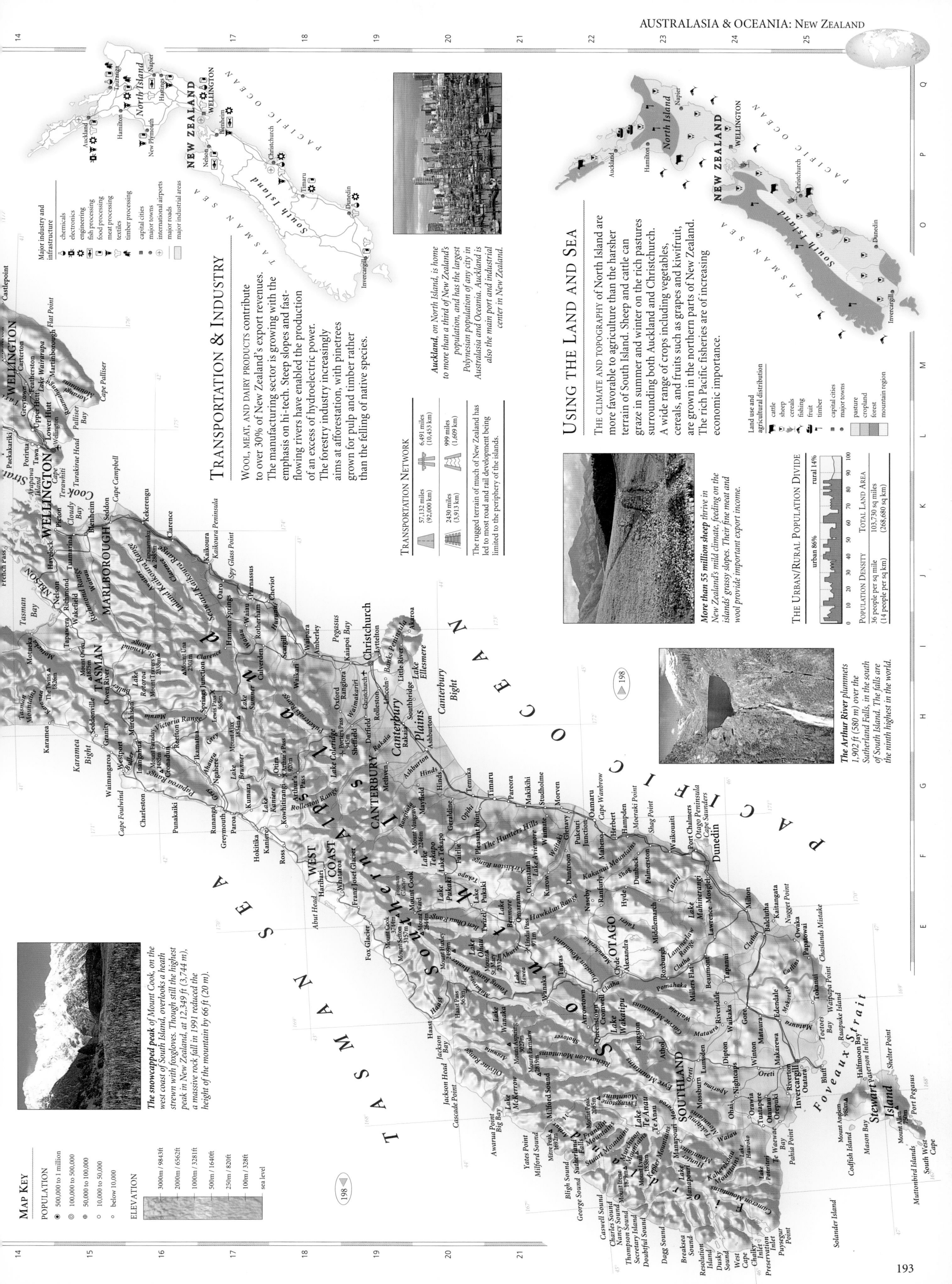

MAP KEY

POPULATION

- ⊙ 500,000 to 1 million
- ◎ 100,000 to 500,000
- ⊕ 50,000 to 100,000
- ⊙ 10,000 to 50,000
- ∘ below 10,000

ELEVATION

3000m / 9843ft	
2000m / 6562ft	
1000m / 3281ft	
500m / 1640ft	
250m / 820ft	
100m / 328ft	
sea level	

The snowcapped peak of Mount Cook, on the west coast of South Island, overlooks a heath strewn with foxgloves. Though still the highest peak in New Zealand, at 12,349 ft (3,744 m), a massive rock fall in 1991 reduced the height of the mountain by 66 ft (20 m).

TRANSPORTATION & INDUSTRY

WOOL, MEAT, AND DAIRY PRODUCTS contribute to over 30% of New Zealand's export revenues. The manufacturing sector is growing with the emphasis on hi-tech. Steep slopes and fast-flowing rivers have enabled the production of an excess of hydroelectric power. The forestry industry increasingly aims at afforestation, with pinetrees grown for pulp and timber rather than the felling of native species.

Major industry and infrastructure

- chemicals
- electronics
- engineering
- fish processing
- food processing
- meat processing
- textiles
- timber processing
- capital cities
- major towns
- international airports
- major roads
- major industrial areas

Auckland, on North Island, is home to more than a third of New Zealand's population, and has the largest Polynesian population of any city in Australasia and Oceania. Auckland is also the main port and industrial center in New Zealand.

TRANSPORTATION NETWORK

57,132 miles (92,000 km)	6,491 miles (10,453 km)
2430 miles (3,913 km)	999 miles (1,609 km)

The rugged terrain of much of New Zealand has led to most road and rail development being limited to the periphery of the islands.

USING THE LAND AND SEA

THE CLIMATE AND TOPOGRAPHY of North Island are more favorable to agriculture than the harsher terrain of South Island. Sheep and cattle can graze in summer and winter on the rich pastures surrounding both Auckland and Christchurch. A wide range of crops including vegetables, cereals, and fruits such as grapes and kiwifruit, are grown in the northern parts of New Zealand. The rich Pacific fisheries are of increasing economic importance.

Land use and agricultural distribution

- cattle
- sheep
- cereals
- fishing
- fruit
- timber
- capital cities
- major towns
- pasture
- cropland
- forest
- mountain region

More than 55 million sheep thrive in New Zealand's mild climate, feeding on the islands' grassy slopes. Their fine meat and wool provide important export income.

The Arthur River plummets 1,902 ft (580 m) over the Sutherland Falls, in the south of South Island. The falls are the ninth highest in the world.

THE URBAN/RURAL POPULATION DIVIDE

urban 86%	rural 14%

POPULATION DENSITY
36 people per sq mile
(14 people per sq km)

TOTAL LAND AREA
103,730 sq miles
(268,680 sq km)

A B C D E F G H I J K L M

PAPUA NEW GUINEA &
THE SOLOMON ISLANDS

CUT OFF BY INACCESSIBLE, largely mountainous terrain, the peoples of Papua New Guinea have maintained a remarkable diversity of language and culture. There are over 750 separate languages, and yet more distinct tribes. Much of the country remains isolated, with many of the indigenous inhabitants of the interior living as hunter-gatherers. To the east of Papua New Guinea, the Solomons form an archipelago of several hundred islands, scattered over an area of 252,897 sq miles (655,000 sq km). The Solomon Islanders, a mainly Melanesian people, live on the six largest islands.

USING THE LAND AND SEA

MOST AGRICULTURE IN Papua New Guinea is at a subsistence level, with more than two-thirds of the land used for rough grazing, particularly for pigs. The tropical rain forest is a rich timber resource. The Solomon Islanders rely heavily on coconuts for export revenue and fishing, mainly for tuna, is a staple industry.

TRANSPORTATION & INDUSTRY

PAPUA NEW GUINEA has substantial mineral resources including the world's largest copper reserves at Panguna on Bougainville Island; gold, and potential oil and natural gas. Political instability on Bougainville and an undeveloped infrastructure deters the investment necessary for exploition of these reserves. The Solomon Islanders rely mainly on copra and timber with some production of palm oil and cocoa. Traditional crafts are made for the tourist market and for export.

TRANSPORTATION NETWORK

🛣	460 miles (740 km)
🛤	None
🚆	None
✈	6,794 miles (10,940 km)

Much of Papua New Guinea and the Solomons is inaccessible by road. A network of airstrips serves even remote villages on the islands. The Solomons' airport has been extended to take jumbo jets to improve connections for tourism.

The slopes of this extinct volcano near Talasea on the island of New Britain have been almost entirely colonized by rain forest vegetation.

Major industry and infrastructure

- 🍹 beverages
- ☕ coffee processing
- 🥥 copra processing
- 🍴 food processing
- ⛏ mining
- 👕 textiles
- 🌲 timber processing

- ■ capital cities
- ● major towns
- ⊕ international airports
- — major roads

Land use and agricultural distribution

- 🍌 bananas
- 🍫 cocoa
- 🥥 coconuts
- 🎣 fishing
- 🌴 oil palms
- rubber
- timber

- ■ capital cities
- ● major towns
- cropland
- forest
- wetland

Over 70% of Papua New Guinea is covered by dense, tropical rain forest, sustained by high levels of rainfall. Uncontrolled logging in the formerly inaccessible rain forest has led to species loss and soil erosion on steep slopes.

THE URBAN/RURAL POPULATION DIVIDE

urban 16%   rural 84%

0 10 20 30 40 50 60 70 80 90 100

POPULATION DENSITY	TOTAL LAND AREA
17 people per sq mile (7 people per sq km)	290,210 sq miles (751,840 sq km)

MAP KEY

POPULATION
- ◉ 100,000 to 500,000
- ⊕ 50,000 to 100,000
- ○ 10,000 to 50,000
- ○ below 10,000

ELEVATION
- 4000m / 13,124ft
- 3000m / 9843ft
- 2000m / 6562ft
- 1000m / 3281ft
- 500m / 1640ft
- 250m / 820ft
- 100m / 328ft
- sea level

Huli tribesmen from Southern Highlands Province in Papua New Guinea parade in ceremonial dress, their powdered wigs decorated with exotic plumage and their faces and bodies painted with colored pigments.

SCALE 1:5,500,000
(projection: Mercator)

Km
0 10 20 40 60 80 100 120 140 160 180 200

Miles
0 10 20 40 60 80 100 120 140 160 180 200

173 ◀

NOPQRSTUVWXY

THE LANDSCAPE

THE PLATE MARGIN between the Pacific and Indo-Australian plates runs through the mainland of Papua New Guinea, which is dominated by steep and forested mountain ranges. The 600 or so outer islands are mainly high, volcanic islands, fringed by coral reefs. The Solomons comprise six large volcanic islands which form two parallel chains, and several hundred small islands and atolls.

The Sepik River drains the lowlands north of the Central Range, flowing eastward into the Bismarck Sea.

The Bismarck Range is precipitous, rugged and covered in dense vegetation, rising to 14,793 ft (4,509 m) at Mount Wilhelm in central Papua New Guinea.

Most of Papua New Guinea's outlying islands, including New Britain, Bougainville Island, and New Ireland, are precipitous and of volcanic origin.

The Star Mountains include some of the most remote terrain on Earth. The area is rich in gold and copper.

A series of coral reefs can be seen in the clear waters off Cape Esperance on the island of Guadalcanal in the Solomons.

Huon Peninsula

Cape Esperance

Kikori River

Kavachi is an active submarine volcano near New Georgia, which erupts every few years.

Southern Papua New Guinea is part of the Indo-Australian Plate. New Guinea only became separated physically from Australia about 8,000 years ago following the flooding of the Torres Strait.

The lowland plains in the south and north of the main island are swampy, and contain some fertile alluvial soils. This contrasts with the mountainous islands in the rest of Papua New Guinea where soils are generally thin and nutrients are retained in the existing vegetation.

Papua New Guinea's rivers, though fairly short, carry extremely high sediment loads, largely due to soil erosion. This is caused by a combination of very steep slopes and heavy rainfall, and is made worse by forest clearance, particularly "slash and burn" techniques and road or mine operations.

The Owen Stanley Range contains several of Papua New Guinea's highest peaks, the greatest of which is Mount Victoria at 13,200 ft (4,035 m).

The Louisiade Archipelago contains 10 volcanic islands and numerous coral islets. Tagula Island is the largest of the islands, containing the archipelago's highest peak at 2,645 ft (806 m).

Huon Peninsula

Caves and undercut cliffs mark former shoreline

Former level of beach

Stream cuts down through recently exposed land

Current beach

Uplift of the land in tectonically active regions can lead to former coastlines being lifted beyond the reach of the sea. New cliffs and caves are formed at a lower level, and rivers cut down through the lower land to reach sea level once more.

SOLOMON ISLANDS

PACIFIC OCEAN

TEMOTU

Duff Islands
Reef Islands
Tinakula
Nendö
Noka
Lata
Santa Cruz Islands
Utupua
Vanikolo

(same scale as main map)

Lying close to the banks of the Sepik River in northern Papua New Guinea, this building is known as the Spirit House. It is constructed from leaves and twigs, ornately woven and trimmed into geometric patterns. The house is decorated with a mask and topped by a carved statue.

Map labels

Matthias Group
Mirau Island
New Hanover
Taskul
North Cape
Kavieng
Tatau Island
Simberi Island
Tabar Islands
Tabar Island
Lihir Group
Lihir Island
Channel
Peteran
Dyaul Island
Konos
Konogogo
Schleinitz Range
Tanga Islands
Boang Island
Malendok Island

PACIFIC OCEAN

NEW IRELAND
New Ireland
Namatanai
Mount Konogaiang 1860m
Ambitle Island
Taron
Feni Islands
Babase Island

Nuguria Islands

Cape Lambert
Rabaul
Kokopo
Gazelle Peninsula
Toriu
Mount Sinewit 1360m
Verron Range
Cape St.George

St. George's Channel

Green Islands
Pinipel Island
Nissan Island

Tulun Islands

Takuu Islands

Nukumanu Islands

Lolobau Island
Open Bay
Mount Ulawun 1360m
Wide Bay
Sampun

Willaumez Peninsula
Talasea
Kimbe Bay
Hoskins
Kimbe
Ubai
Nakanai Mountains
Pomio
Jacquinot Bay
Bagabag
Lau
EAST NEW BRITAIN

Gasmata
New Britain

Lemankoa
Buka Island
Hutjena
NORTH SOLOMONS
Wakunai
Mount Balbi 2685m
Torokina
Empress Augusta Bay
Arawa
Kieta
Panguna
Buin
Bougainville Island

Solomon

Ontong Java Atoll

Roncador Reef

SOLOMON SEA

Shortland Island
Shortland Islands
Treasury Islands
Fauro
Bougainville Strait
Nukiki
Panggoe
Choiseul
Luti
Rob Roy
Vaghena
Kia
Baolo
ISABEL
Santa Isabel
Buala
Mount Sasari 1219m

WESTERN
New Georgia Sound
Manning Strait

Vella Lavella
Mongga
Kolombangara
New Georgia
Ranongga
Gizo
Ringgi
Munda
Rendova
New Georgia Islands
Blanche Channel
Vangunu
Nggatokae
Tetepare

Kaolo
San Jorge

Dai Island

MALAITA

Maluu
Kwailibesi
Auki
Malaita
Olomburi
Baunani

Sikaiana

Kiriwina Island
Losuia
Kitava Island
Vakuta Island
Madau Island
Woodlark Island
Gawa Island
Yanaba Island
Guasopa

D'Entrecasteaux Islands
Goodenough Island
Fergusson Island
Normanby Island
Esa'ala
Sehulea

Russell Islands
Yandina
CENTRAL
Cape Esperance
Tambea
Savo
Iron Bottom Sound
Florida Islands
Tulaghi
HONIARA
Tangarare
Guadalcanal
Nduindui
Henderson Field
Aola
Mount Popomanaseu 2330m
Avuavu

Tarapaina
Maramasike
Apio

Ulawa Island

Alotau
Milne Bay
Ahioma
Goschen Strait
Sideia Island
MILNE BAY
Samarai
Basilaki Island
Conflict Group
Bwagaoia
Misima Island
Louisiade Archipelago

GUADALCANAL

Heuru
Kirakira
Haurahu
San Cristobal
Star Harbour
Three Sisters Islands

Pocklington Reef

SOLOMON ISLANDS

CENTRAL

Bellona
Lavanggu
Rennell

MAKIRA

The Calvados Chain
Tagula
Tagula Island
Rossel Island

TUV

PACIFIC OCEAN

THE PACIFIC IS THE WORLD'S LARGEST AND DEEPEST OCEAN. It is nearly twice the area of the Atlantic and contains almost three times as much water. The ocean is dotted with islands and surrounded by some of the world's most populous states; over half the world's population lives on its shores. The Pacific is bordered by active plate margins known as the "Ring of Fire," causing earthquakes and tsunamis, and creating volcanic islands and subterranean mountain chains. The largest underwater mountains break the surface as island arcs. The fisheries of the Pacific are some of the most productive in the world and provide a vital resource for many of the Pacific islands. Since the Second World War there has been a shift in trading patterns, with a considerable growth in trade between the US and the countries of the Pacific Rim.

INSET MAP KEY

POPULATION
○ below 10,000

ELEVATION
1000m / 3281ft
500m / 1640ft
250m / 820ft
100m / 328ft
sea level

OCEAN MAP KEY

SEA DEPTH
sea level
250m / 820ft
500m / 1640ft
1000m / 3281ft
2000m / 6562ft
3000m / 9843ft
5000m / 16,410ft

SCALE 1:50,000,000
(projection: Mollweide)

AMERICAN SAMOA AND SAMOA

AMERICAN SAMOA AND SAMOA are part of the island archipelago of Polynesia. The two most populous islands are Tutuila in American Samoa and Upolu in Samoa. Although the economies of both these states remain predominantly resource-based, both are expanding their light manufacturing sectors, and the US administration is the primary employer in American Samoa. Tuna fishing is particularly important: 25% of all tuna consumed in the US is processed and canned in Pago Pago.

Japan is one of the major trading nations within the Pacific, importing iron and steel from Australia, and grain from the US. The major exports from the 'Pacific Rim' are electronics, precision equipment, and motor cars.

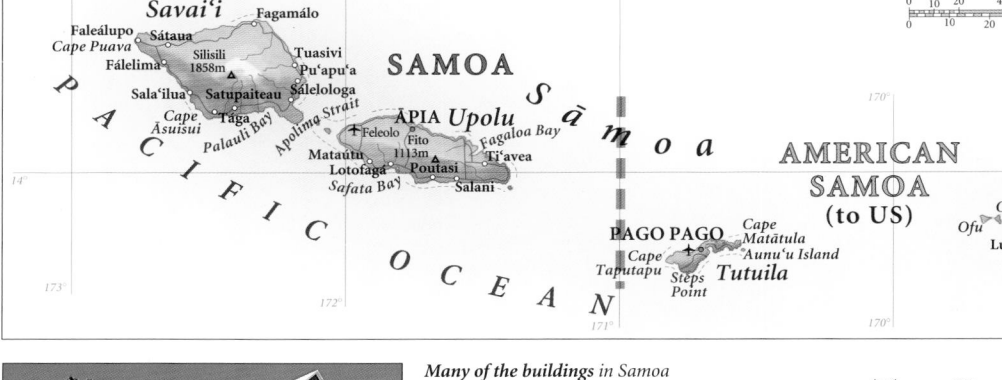

SCALE 1:3,000,000

Many of the buildings in Samoa reflect the country's colonial past. Once a colony of New Zealand, Samoa is now an independent state; American Samoa remains an unincorporated territory of the United States.

THE RING OF FIRE

THE ACTIVE PLATE MARGINS surrounding the Pacific have created numerous land and island volcanoes along its border. The actual basin of the Pacific is made up of a number of separate tectonic plates which move away from each other, colliding with other plates. When they collide, the oceanic plates, being thinner, are forced beneath the thicker continental plates, forming deep ocean trenches and high ridges. These collision zones are known as subduction zones and are characterized by intense seismic and volcanic activity.

RESOURCES

MANY OF THE SMALL ISLANDS in the Pacific rely heavily on marine resources to provide valuable export incomes. These fisheries tend to be small-scale and are forced to compete with the large commerical fleets from Japan and the Russian Federation. Although many metallic mineral deposits have been discovered in the Pacific, few are exploited. The major areas of oil and gas extraction are off the coast of Vietnam, along the Kamchatka Peninsula and off the coast of Alaska. The numerous reefs which fringe the islands of the Pacific are harvested for corals.

Farms such as this black pearl oyster farm in Tahiti are widespread throughout the Pacific. The culturing or farming of marine organisms, such as mollusks and crustaceans, has been practiced for hundreds of years.

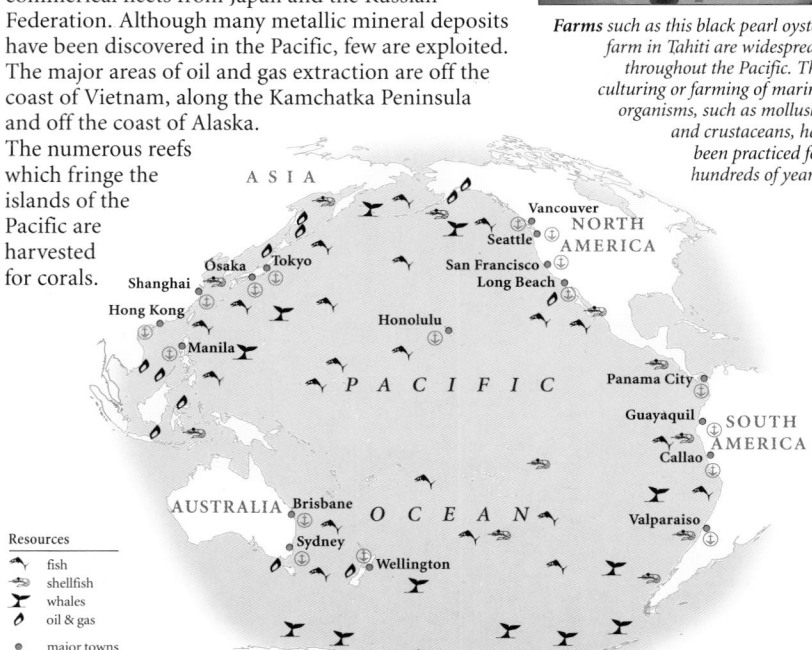

Resources
fish
shellfish
whales
oil & gas
● major towns
⊕ major ports

Ring of Fire
— plate boundaries
● major volcanoes

Mayon Volcano in the Philippines is one of many active volcanoes on the Pacific "Ring of Fire." It is noted for its perfect conical shape; the base of the cone is 80 miles (130 km) in circumference.

The Hawaiian volcanoes lie in the center of a plate, not on a plate margin, and are known as intraplate volcanoes. They are associated with hot spots, whereby a plume of hot molten rock rises to the surface as the plate moves over it.

A B C D E F G H I J K L

MELANESIA

FIJI, VANUATU, *New Caledonia* (to France)

THREE MAIN ISLAND groups make up the area of southern Melanesia in the southwestern Pacific: the independent countries of Fiji and Vanuatu and the French overseas territory of New Caledonia. The major Melanesian island group, the Solomon Islands, lies to the east of Papua New Guinea (pages 194–95). Most of the larger islands are volcanic in origin; the smaller ones are mainly coral atolls and are largely uninhabited. The economy in all three island groups is increasingly driven by tourism, not necessarily to the benefit of other economic activities.

VANUATU

A STRING OF MOUNTAINOUS VOLCANIC ISLANDS covering more than 4,706 sq miles (12,190 sq km) of the south Pacific, Vanuatu achieved independence from France and the UK in 1980. The majority of the population relies on subsistence fishing and agriculture. Once-important copra and cocoa exports are declining as a result of cost-effective substitutes from elsewhere, and alternatives are being explored. There is further resource potential in the forests and fishing grounds, and beef and arable farming are of growing importance. Tourism, accounting for 40% of GDP, is the fastest-growing sector of the economy, and further expansion is planned.

NEW CALEDONIA (to France)

NEW CALEDONIA, a French overseas territory known as Kanaky by its indigenous peoples, comprises a large main island, 260 miles (418 km) long, and many smaller islands and atolls. Socioeconomic inequality, unemployment, and the issue of independence have caused tension between the Kanaks and the French-speaking expatriate population. This resulted in a long history of political violence, although the Nouméa accord, signed in 1998, allowed for greater autonomy. New Caledonia produces 25% of the world's nickel, and improved incomes from tourism and agriculture have benefited the economy.

SCALE 1:6,000,000
(projection: Lambert Conformal Conic)

Km
0 10 20 40 60 80 100 120 140 160
Miles

On New Caledonia's main island, relatively high interior plateaus descend to coastal plains. Nickel is the most important mineral resource, but the hills also harbor metallic deposits including chrome, cobalt, iron, gold, silver, and copper.

MAP KEY

POPULATION
⊕ 50,000 to 100,000
○ 10,000 to 50,000
○ below 10,000

ELEVATION
1000m / 3281ft
500m / 1640ft
250m / 820ft
100m / 328ft
sea level

FIJI

FIJI IS A VOLCANIC ARCHIPELAGO in the southwestern Pacific consisting of two large islands and 880 smaller islets, and covering a total area of 7,054 sq miles (18,270 sq km). The majority of the population lives on the two largest islands. The people are split fairly evenly between Indo-Fijians, who arrived when Fiji was still a British colony, and the indigenous Fijians who have, since 1987, controlled the government. Sugar and copra are the most important crops in a diversified agricultural base and forestry is becoming increasingly important. A relatively varied economy has potential for mineral and hydroelectric exploitation, while Fiji's climate and location on the main Pacific air routes are an impetus to tourism.

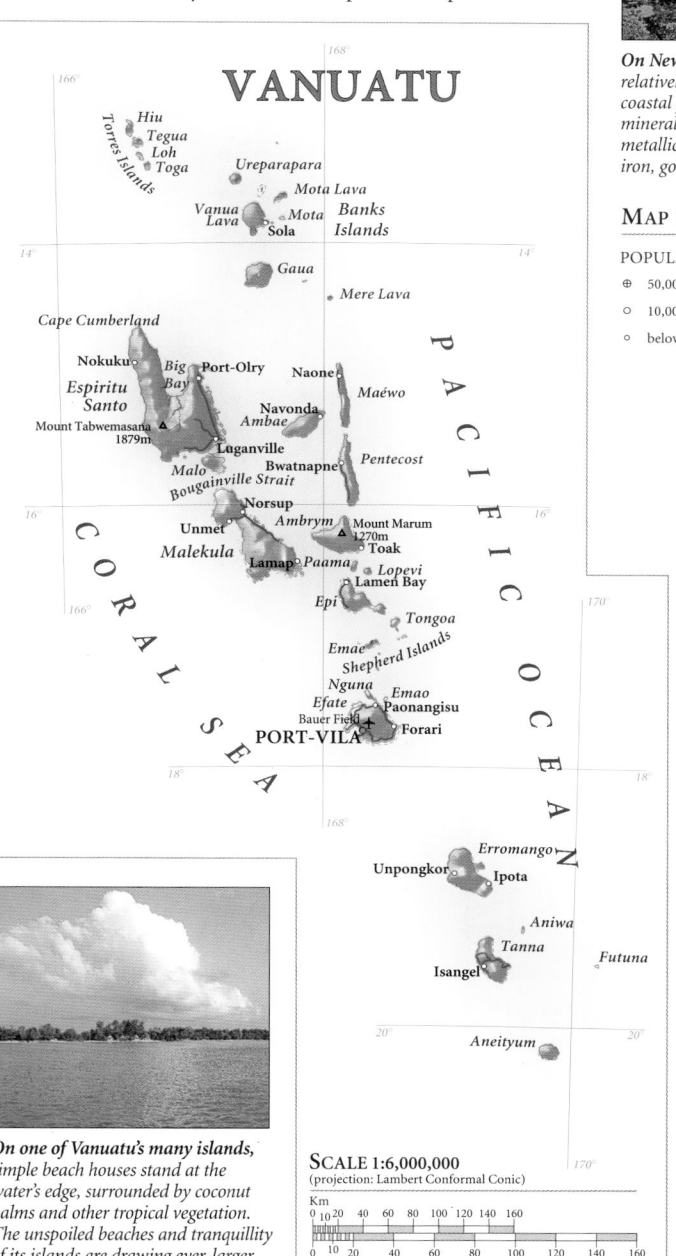

On one of Vanuatu's many islands, simple beach houses stand at the water's edge, surrounded by coconut palms and other tropical vegetation. The unspoiled beaches and tranquillity of its islands are drawing ever-larger numbers of tourists to Vanuatu.

SCALE 1:6,000,000
(projection: Lambert Conformal Conic)

Km
0 10 20 40 60 80 100 120 140 160
Miles

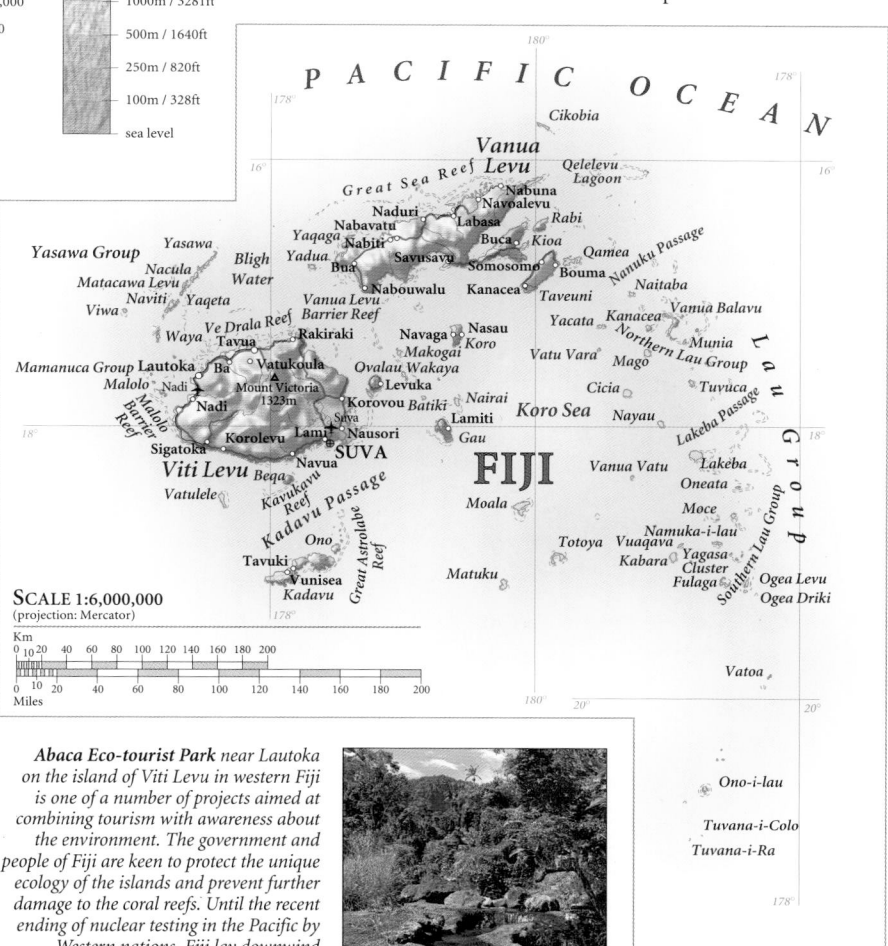

Abaca Eco-tourist Park near Lautoka on the island of Viti Levu in western Fiji is one of a number of projects aimed at combining tourism with awareness about the environment. The government and people of Fiji are keen to protect the unique ecology of the islands and prevent further damage to the coral reefs. Until the recent ending of nuclear testing in the Pacific by Western nations, Fiji lay downwind of some of the main testing sites.

SCALE 1:6,000,000
(projection: Mercator)

Km
0 10 20 40 60 80 100 120 140 160 180 200
Miles

A B C D E F G H I J K L M

MICRONESIA

MARSHALL ISLANDS, MICRONESIA, NAURU, PALAU, *Guam, Northern Mariana Islands, Wake Island*

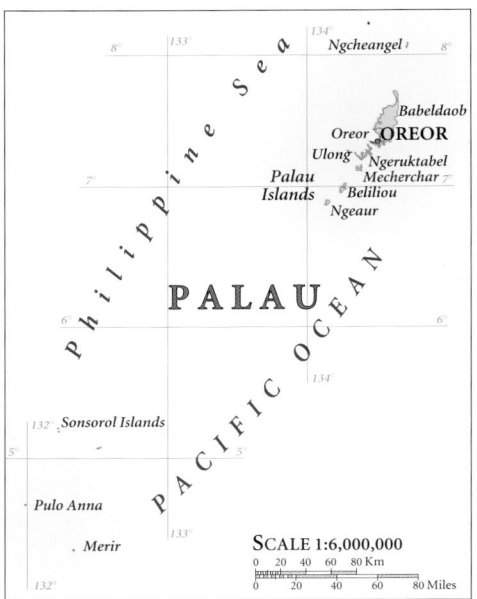

THE MICRONESIAN ISLANDS lie in the western reaches of the Pacific Ocean and are all part of the same volcanic zone. The Federated States of Micronesia is the largest group, with more than 600 atolls and forested volcanic islands in an area of more than 1,120 sq miles (2,900 sq km). Micronesia is a mixture of former colonies, overseas territories, and dependencies. Most of the region still relies on aid and subsidies to sustain economies limited by resources, isolation, and an emigrating population, drawn to New Zealand and Australia by the attractions of a western lifestyle.

PALAU

PALAU IS AN ARCHIPELAGO OF OVER 200 ISLANDS, only eight of which are inhabited. It was the last remaining UN trust territory in the Pacific, controlled by the US until 1994, when it became independent. The economy operates on a subsistence level, with coconuts and cassava the principal crops. Fishing licenses and tourism provide foreign currency.

SCALE 1:750,000

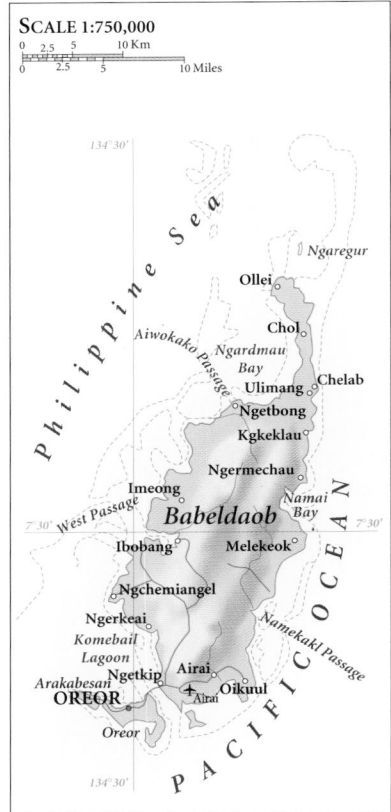

SCALE 1:6,000,000

GUAM (to US)

LYING AT THE SOUTHERN END of the Mariana Islands, Guam is an important US military base and tourist destination. Social and political life is dominated by the indigenous Chamorro, who make up just under half the population, although the increasing prevalence of western culture threatens Guam's traditional social stability.

The tranquillity of these coastal lagoons, at Inarajan in southern Guam, belies the fact that the island lies in a region where typhoons are common.

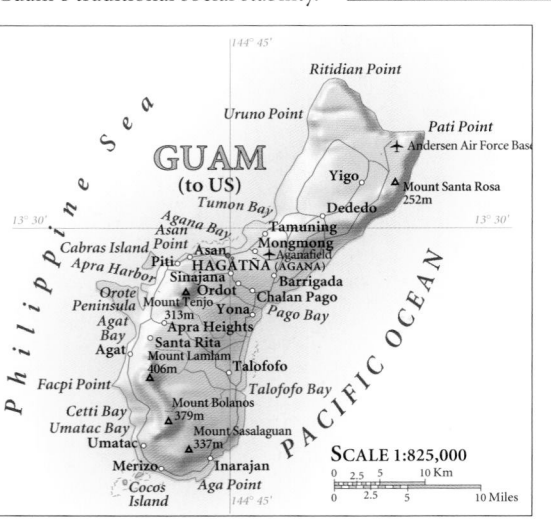

SCALE 1:825,000

NORTHERN MARIANA ISLANDS (to US)

A US COMMONWEALTH TERRITORY, the Northern Marianas comprise the whole of the Mariana archipelago except for Guam. The islands retain their close links with the US and continue to receive American aid. Tourism, though bringing in much-needed revenue, has speeded the decline of the traditional subsistence economy. Most of the population lives on Saipan.

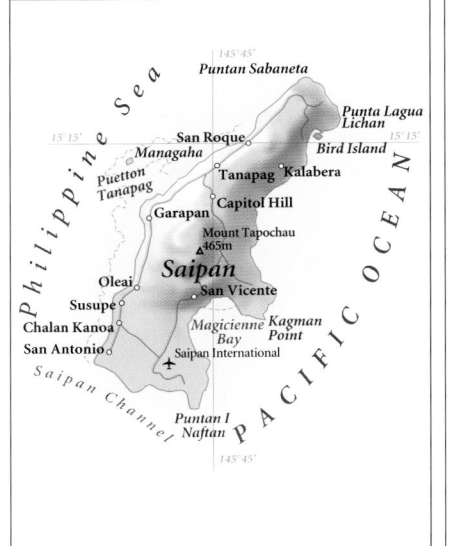

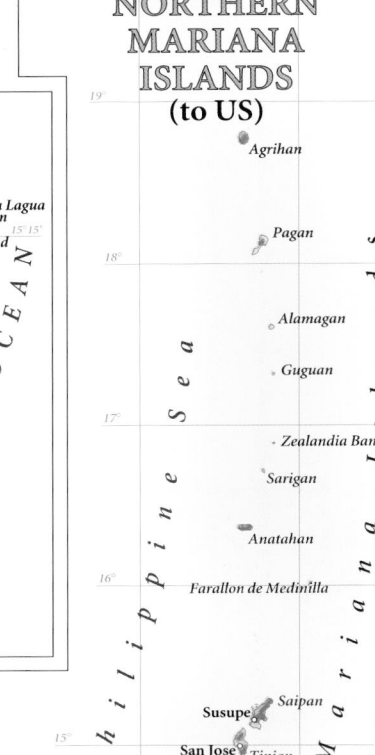

NORTHERN MARIANA ISLANDS (to US)

SCALE 1:5,000,000

The Palau Islands have numerous hidden lakes and lagoons. These sustain their own ecosystems which have developed in isolation. This has produced adaptations in the animals and plants that are often unique to each lake.

MICRONESIA

A MIXTURE OF HIGH VOLCANIC ISLANDS and low-lying coral atolls, the Federated St of Micronesia include all the Caroline Islands except Palau. Pohnpei, Kosrae, Ch and Yap are the four main island cluster states, each of which has its own langua with English remaining the official language. Nearly half the population is conce on Pohnpei, the largest island. Independent since 1986, the islands continue to receive considerable aid from the US which supplements an economy based primarily on fishing and copra processing.

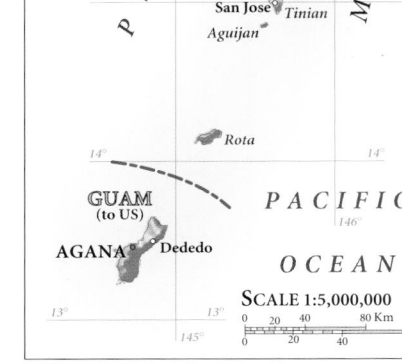

SCALE 1:825,000

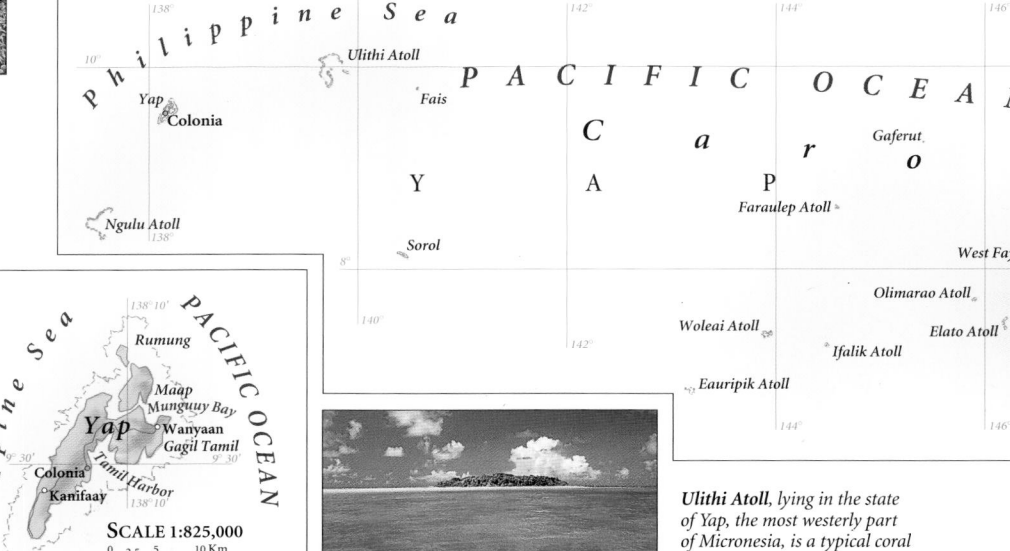

Ulithi Atoll, lying in the state of Yap, the most westerly part of Micronesia, is a typical coral island, with a series of reefs enclosing a large lagoon.

MARSHALL ISLANDS

A GROUP OF 34 WIDELY-SCATTERED ATOLLS in the central Pacific Ocean, the Marshall Islands include some of the largest atolls in the world, formed from low coral islands with sandy beaches and enclosing vast lagoons. Formerly under US protection as part of the UN Trust Territory of the Pacific Islands, and including the former US nuclear testing sites of Bikini Atoll and Enewetak Atoll, the Marshall Islands became self-governing in 1979. The economy is reliant on US aid and on the rent paid by the US for its missile base on Kwajalein Atoll.

SCALE 1:1,000,000

Majuro Atoll is the Marshall Islands' capital and commercial center. Almost half the population live on the narrow islands, often in overcrowded conditions.

NAURU

FORMER BRITISH COLONY, the tiny island of Nauru, with an area of only [ ] sq miles (21.2 sq km), has been exploited for its substantial phosphate deposits by the UK, Australia, and New Zealand. Since independence in 1968, the phosphate industry has made its citizens some of the wealthiest in the world, and scars from the vast mining operation pit the island's landscape. Phosphate reserves are now virtually exhausted and investment overseas will in future form the bulk of Nauru's income.

SCALE 1:200,000

A series of coral pinnacles stand exposed in the shallow water off the coast of Nauru. Much of the island has an extraordinary "unar" landscape, created by years of phosphate extraction.

WAKE ISLAND (to US)

AN UNINCORPORATED TERRITORY of the US with a tiny population, Wake Island remains strategically important to US forces, and has been used as a base in several conflicts. Formed by the rim of an extinct underwater volcano, it is now used as an emergency airstrip for trans-Pacific flights, and as a stopover for cargo planes.

SCALE 1:6,500,000

SCALE 1:650,000

Canoes, built following tradition, are still important in Micronesia, and are used for transportation and for fishing. This large canoe, on Satawal, in the state of Yap, needs nearly 20 people to return it to the boathouse.

SCALE 1:1,500,000

SCALE 1:250,000

SCALE 1:500,000

SCALE 1:8,000,000

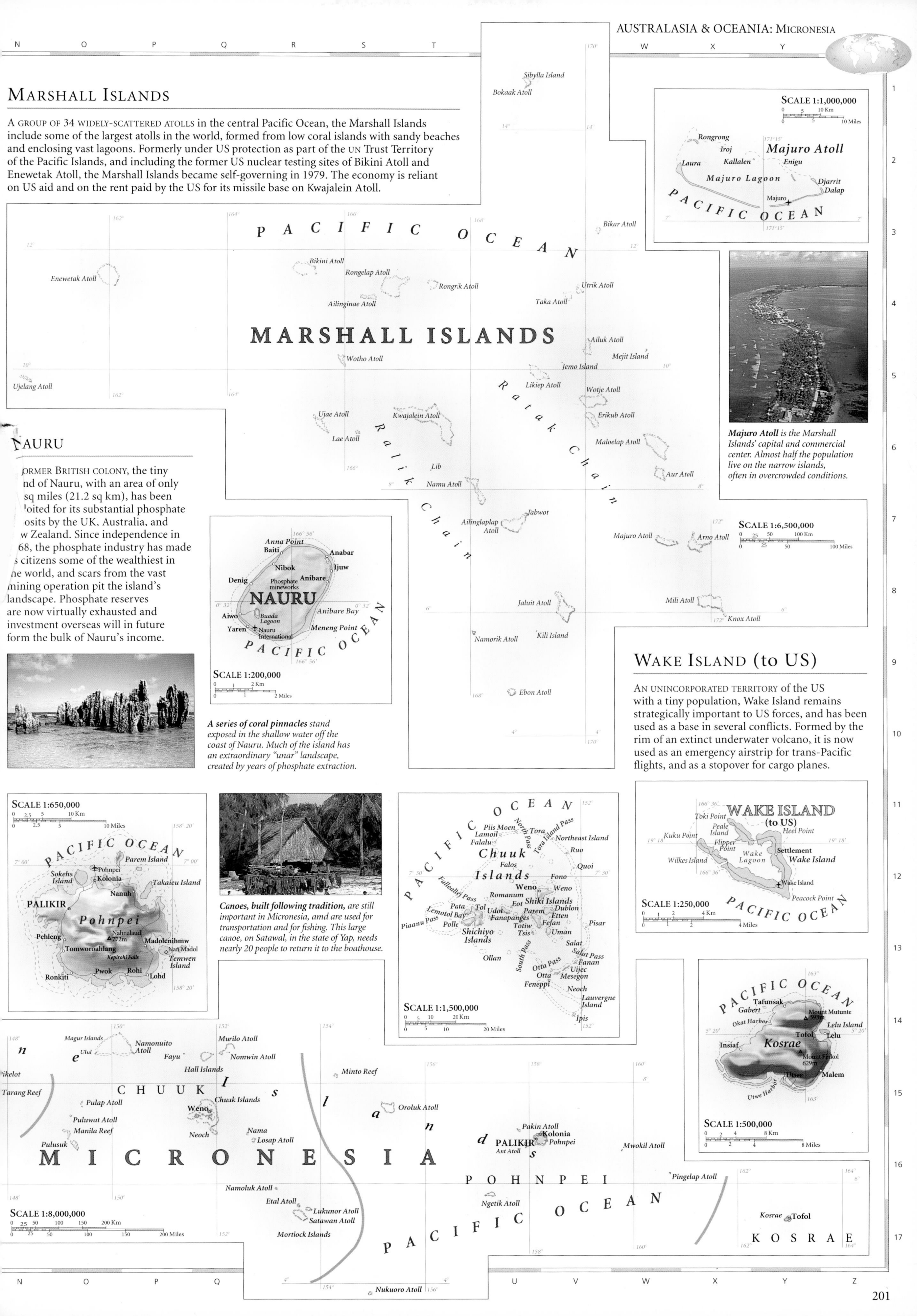

THE LANDSCAPE

ALTHOUGH IT IS STILL THE LARGEST OCEAN, the basin of the Pacific has been gradually decreasing in size due to the movement of the Indo-Australian Plate. The oldest parts are about 135 million years old. The eastern border of the Pacific is characterized by a continuous mountain chain running the length of the North and South American continents. The eastern basin has a low, uninterrupted relief, at depths averaging 15,000 ft (4570 m). In contrast, the western Pacific is scattered with island arcs and bounded by a series of deep ocean trenches. An almost continuous chain of volcanoes surrounds the ocean and an active mid-ocean ridge runs northeast–southwest.

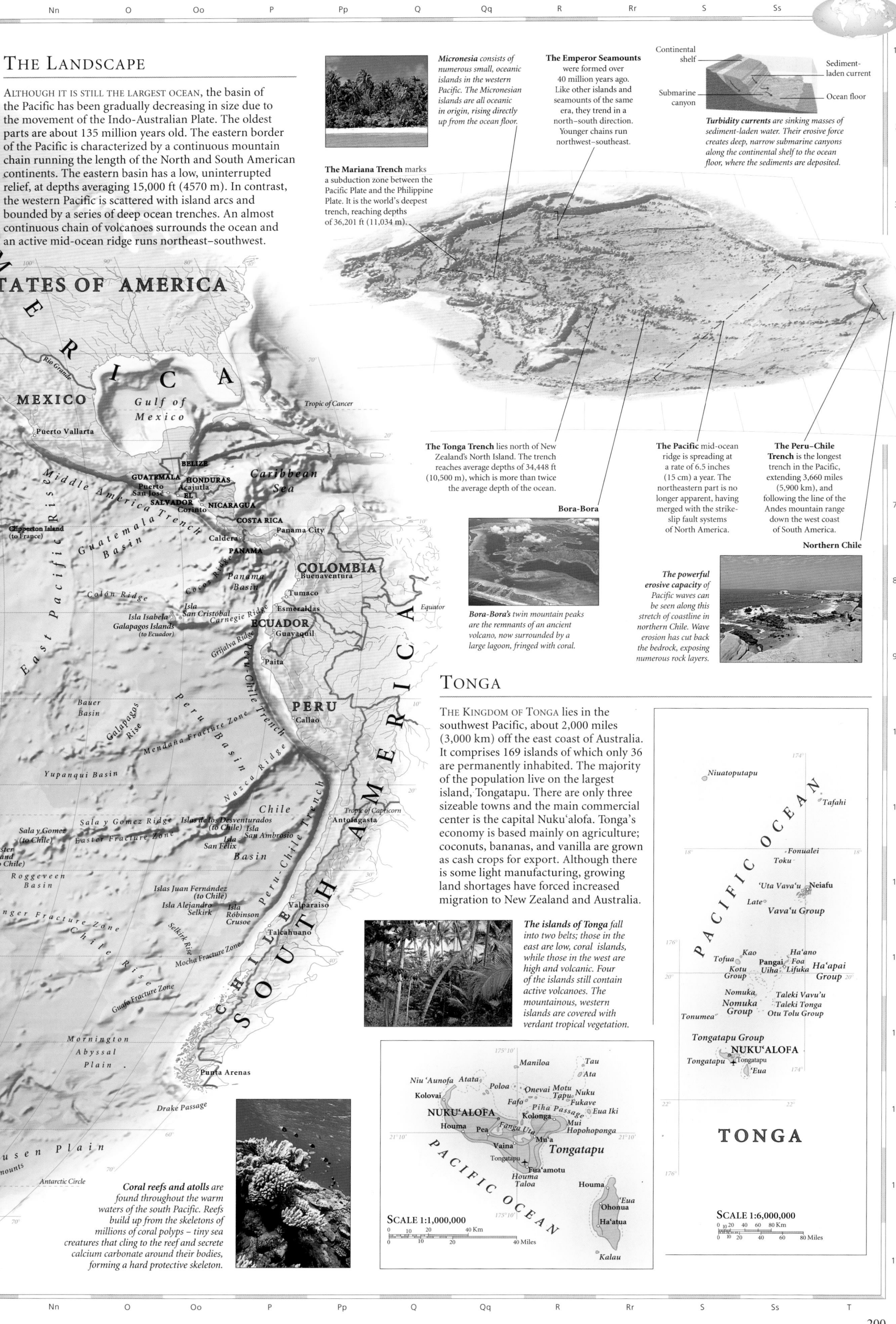

Micronesia consists of numerous small, oceanic islands in the western Pacific. The Micronesian islands are all oceanic in origin, rising directly up from the ocean floor.

The Emperor Seamounts were formed over 40 million years ago. Like other islands and seamounts of the same era, they trend in a north–south direction. Younger chains run northwest–southeast.

The Mariana Trench marks a subduction zone between the Pacific Plate and the Philippine Plate. It is the world's deepest trench, reaching depths of 36,201 ft (11,034 m).

Turbidity currents are sinking masses of sediment-laden water. Their erosive force creates deep, narrow submarine canyons along the continental shelf to the ocean floor, where the sediments are deposited.

Continental shelf — *Sediment-laden current* — *Submarine canyon* — *Ocean floor*

The Tonga Trench lies north of New Zealand's North Island. The trench reaches average depths of 34,448 ft (10,500 m), which is more than twice the average depth of the ocean.

The Pacific mid-ocean ridge is spreading at a rate of 6.5 inches (15 cm) a year. The northeastern part is no longer apparent, having merged with the strike-slip fault systems of North America.

The Peru–Chile Trench is the longest trench in the Pacific, extending 3,660 miles (5,900 km), and following the line of the Andes mountain range down the west coast of South America.

Bora-Bora

Bora-Bora's twin mountain peaks are the remnants of an ancient volcano, now surrounded by a large lagoon, fringed with coral.

Northern Chile

The powerful erosive capacity of Pacific waves can be seen along this stretch of coastline in northern Chile. Wave erosion has cut back the bedrock, exposing numerous rock layers.

TONGA

THE KINGDOM OF TONGA lies in the southwest Pacific, about 2,000 miles (3,000 km) off the east coast of Australia. It comprises 169 islands of which only 36 are permanently inhabited. The majority of the population live on the largest island, Tongatapu. There are only three sizeable towns and the main commercial center is the capital Nuku‘alofa. Tonga's economy is based mainly on agriculture; coconuts, bananas, and vanilla are grown as cash crops for export. Although there is some light manufacturing, growing land shortages have forced increased migration to New Zealand and Australia.

The islands of Tonga fall into two belts; those in the east are low, coral islands, while those in the west are high and volcanic. Four of the islands still contain active volcanoes. The mountainous, western islands are covered with verdant tropical vegetation.

Coral reefs and atolls are found throughout the warm waters of the south Pacific. Reefs build up from the skeletons of millions of coral polyps – tiny sea creatures that cling to the reef and secrete calcium carbonate around their bodies, forming a hard protective skeleton.

SCALE 1:1,000,000

SCALE 1:6,000,000

Wave action has eroded this shoreline near Port-Campbell in southeastern Australia leaving isolated pinnacles of rock cut off from the main coastline. They are known as the "Twelve Apostles."

POLYNESIA

KIRIBATI, TUVALU, Cook Islands, Easter Island, French Polynesia, Niue, Pitcairn Islands, Tokelau, Wallis & Futuna

THE NUMEROUS ISLAND GROUPS OF POLYNESIA lie to the east of Australia, scattered over a vast area in the south Pacific. The islands are a mixture of low-lying coral atolls, some of which enclose lagoons, and the tips of great underwater volcanoes. The populations on the islands are small, and most people are of Polynesian origin, as are the Maori of New Zealand. Local economies remain simple, relying mainly on subsistence crops, mineral deposits, many now exhausted, fishing, and tourism.

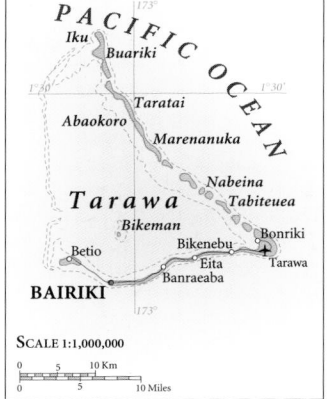

SCALE 1:1,000,000

With the exception of Banaba all the islands in Kiribati's three groups are low-lying, coral atolls. This aerial view shows the sparsely vegetated islands, intercut by many small lagoons.

KIRIBATI

A FORMER BRITISH COLONY, Kiribati became independent in 1979. Banaba's phosphate deposits ran out in 1980, following decades of exploitation by the British. Economic development remains slow and most agriculture is at a subsistence level, though coconuts provide export income, and underwater agriculture is being developed.

TUVALU

A CHAIN of nine coral atolls, 360 miles (579 km) long with a land area of just over 9 sq miles (23 sq km), Tuvalu is one of the world's smallest and most isolated states. As the Ellice Islands, Tuvalu was linked to the Gilbert Islands (now part of Kiribati) as a British colony until independence in 1978. Politically and socially conservative, Tuvaluans live by fishing and subsistence farming.

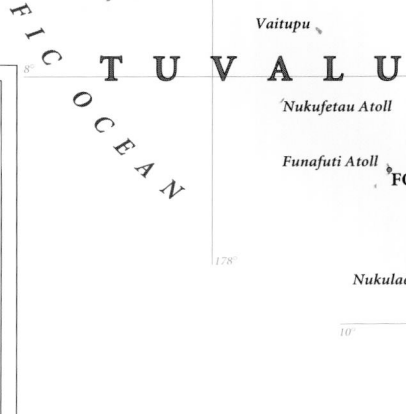

Funafuti Atoll contains more than 40% of Tuvalu's people, giving it an extremely high population density.

SCALE 1:500,000

SCALE 1:6,000,000

TOKELAU (to New Zealand)

A LOW-LYING CORAL ATOLL, Tokelau is a dependent territory of New Zealand with few natural resources. Although a 1990 cyclone destroyed crops and infrastructure, a tuna cannery and the sale of fishing licenses have raised revenue and a catamaran link between the islands has increased their tourism potential. Tokelau's small size and economic weakness makes independence from New Zealand unlikely.

Fishermen cast their nets to catch small fish in the shallow waters off Atafu Atoll, the most westerly island in Tokelau.

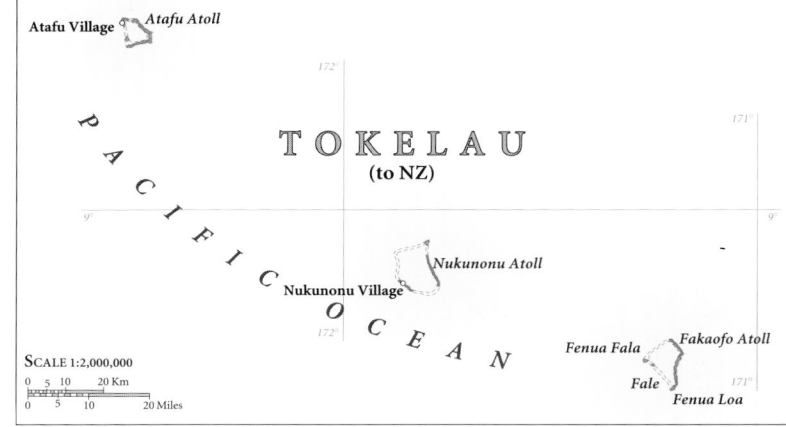

SCALE 1:2,000,000

WALLIS & FUTUNA (to France)

IN CONTRAST TO OTHER FRENCH overseas territories in the south Pacific, the inhabitants of Wallis and Futuna have shown little desire for greater autonomy. A subsistence economy produces a variety of tropical crops, while foreign currency remittances come from expatriates and from the sale of licenses to Japanese and Korean fishing fleets.

SCALE 1:1,000,000

SCALE 1:1,000,000

COOK ISLANDS (to New Zealand)

A MIXTURE OF CORAL ATOLLS and volcanic peaks, the Cook Islands achieved self-government in 1965 but exist in free association with New Zealand. A diverse economy includes pearl and giant clam farming, and an ostrich farm, plus tourism and banking. A 1991 friendship treaty with France provides for French surveillance of territorial waters.

Northern Cook Islands

Southern Cook Islands

COOK ISLANDS
(to NZ)

SCALE 1:20,000,000

NIUE (to New Zealand)

NIUE, the world's largest coral island, is self-governing but exists in free association with New Zealand. Tropical fruits are grown for local consumption; tourism and the sale of postage stamps provide foreign currency. The lack of local job prospects has led more than 10,000 Niueans to emigrate to New Zealand, which has now invested heavily in Niue's economy in the hope of reversing this trend.

Palm trees fringe the white sands of a beach on Aitutaki in the Southern Cook Islands, where tourism is of increasing economic importance.

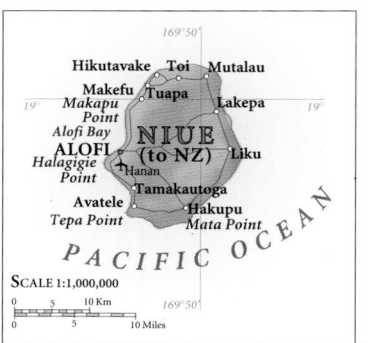

NIUE
(to NZ)

SCALE 1:1,000,000

Waves have cut back the original coastline, exposing a sandy beach, near Mutalau in the northeast corner of Niue.

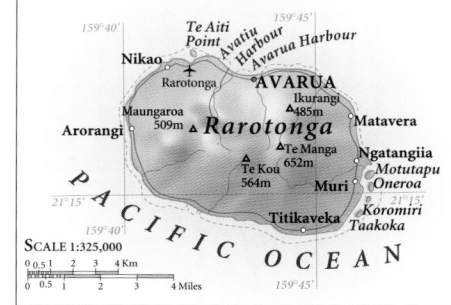

SCALE 1:325,000

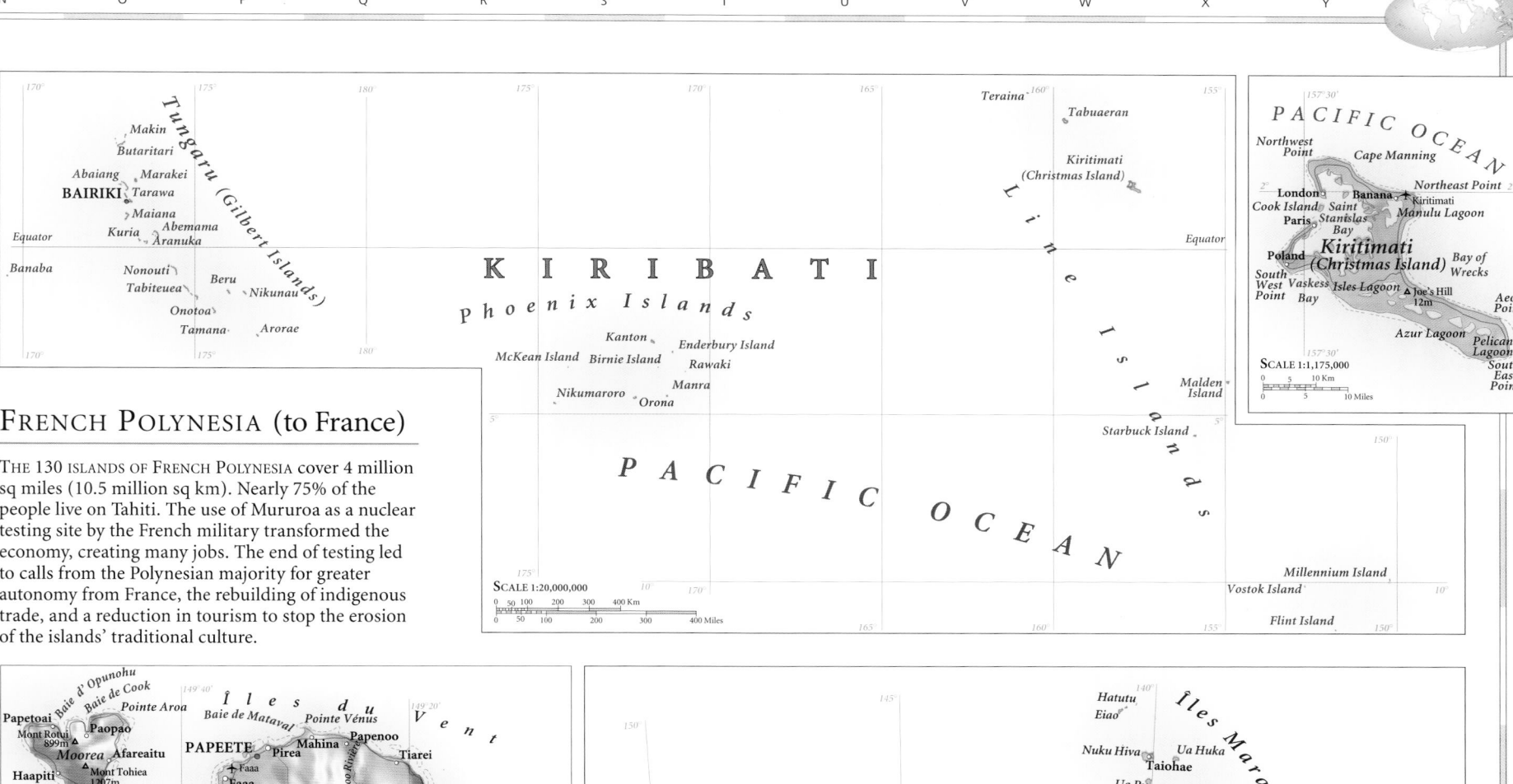

FRENCH POLYNESIA (to France)

THE 130 ISLANDS OF FRENCH POLYNESIA cover 4 million sq miles (10.5 million sq km). Nearly 75% of the people live on Tahiti. The use of Mururoa as a nuclear testing site by the French military transformed the economy, creating many jobs. The end of testing led to calls from the Polynesian majority for greater autonomy from France, the rebuilding of indigenous trade, and a reduction in tourism to stop the erosion of the islands' traditional culture.

The traditional Tahitian welcome for visitors, who are greeted by parties of canoes, has become a major tourist attraction.

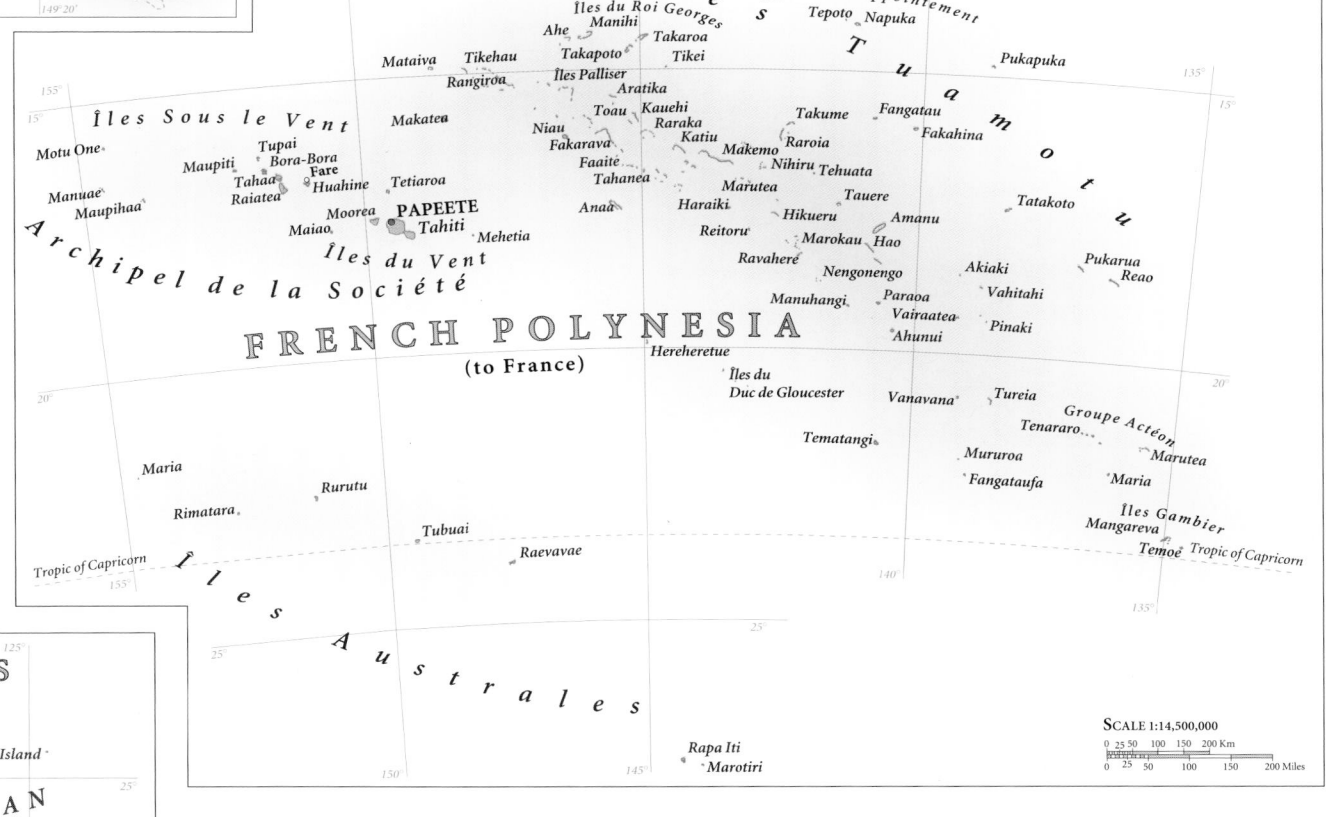

PITCAIRN ISLANDS (to UK)

BRITAIN'S MOST ISOLATED DEPENDENCY, Pitcairn Island was first populated by mutineers from the HMS *Bounty* in 1790. Emigration is further depleting the already limited gene pool of the island's inhabitants, with associated social and health problems. Barter, fishing, and subsistence farming form the basis of the economy although postage stamp sales provide foreign currency earnings, and offshore mineral exploitation may boost the economy in future.

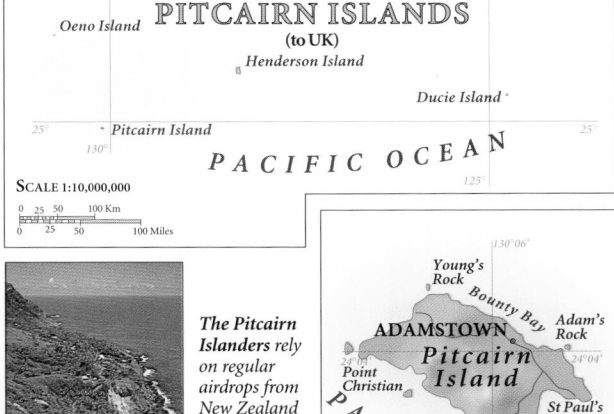

The Pitcairn Islanders rely on regular airdrops from New Zealand and periodic visits by supply vessels to provide them with basic commodities.

EASTER ISLAND (to Chile)

ONE OF THE MOST EASTERLY ISLANDS in Polynesia, Easter Island *(Isla de Pascua)* – also known as Rapa Nui, is part of Chile. The mainly Polynesian inhabitants support themselves by farming, which is mainly of a subsistence nature, and includes cattle rearing and crops such as sugar cane, bananas, corn, gourds, and potatoes. In recent years, tourism has become the most important source of income and the island sustains a small commercial airport.

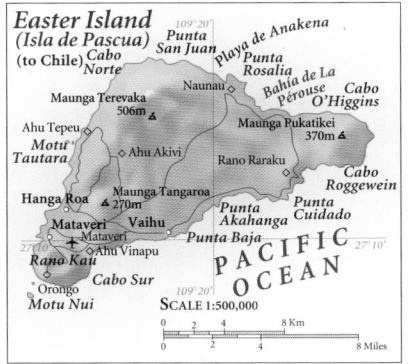

The Naunau, a series of huge stone statues overlook Playa de Anakena, on Easter Island. Carved from a soft volcanic rock, they were erected between 400 and 900 years ago.

ANTARCTICA

THE ICE-COVERED CONTINENT of Antarctica, which is the Earth's most southerly region, has drawn explorers and entrepreneurs seeking challenge and riches in its wintry lands for over 200 years. The extreme climate has deterred any large-scale settlement of the continent, and though commercial hunters built outposts in the past, habitation is now limited to scientific bases. The Antarctic Treaty, which came into force in 1961, provides for international governance and scientific cooperation in place of potential territorial conflict.

TERRITORIAL CLAIMS

Argentinian claim
Brazilian zone of interest
British claim
Norwegian undefined limit
Australian claim
Chilean claim
French claim
Australian claim
New Zealand claim

RESOURCES

MANY ORE MINERALS, including iron and gold, are found in the Antarctic, and there are also coal reserves in the Transantarctic Mountains. The severe conditions and environmental importance of the region mean that exploitation of potential mineral resources is both uneconomic and undesirable. The unique wildlife and landscape draw a small number of tourists annually.

Research Stations on King George Island

Arctowski (to Poland)
Artigas (to Uruguay)
Bellingshausen (to Russian Federation)
Comandante Ferraz (to Brazil)
Great Wall (to China)
Jubany (to Argentina)
King Sejong (to South Korea)
Teniente Rodolfo Marsh (to Chile)

Resources (including wildlife)

coal		seals	
fish		whales	
minerals		polar research base	
oil & gas			
penguins			

Most settlements in Antarctica are research bases such as this one at Rothera on Adelaide Island, although there is a small Chilean settlement on King George Island.

THE LANDSCAPE

THERE ARE TWO DISTINCT PARTS to Antarctica: Lesser Antarctica, a series of ice-covered, mountainous islands, joined together by the ice; and the high plateau of Greater Antarctica. The Ross Sea and the Weddell Sea are outliers of the Atlantic and Pacific oceans – deep bays partially covered by thick ice shelves.

Grease ice   Pancake ice   Sea-ice sheet   Ice floe

Pack ice forms out at sea in freezing temperatures. At the outer limits, grease ice congeals on the surface of the ocean. This is then spun around by wind and waves into irregular "pancakes," freezing and breaking up several times before bonding together again to form sea-ice sheets, which finally cement into enormous ice floes.

On Elephant Island, the coast is edged by glaciers, although the land is not permanently covered by ice.

During the winter the seas surrounding Antarctica freeze, increasing the size of the continent by 100%.

Elephant Island

Upper Wright Valley

Limit of winter pack ice

Limit of summer pack ice

High winds carrying snow form huge snowdrifts. The erosive power of the wind-borne snow can also sculpt the ice sheet to produce landforms known as sastrugi which align with the direction of the wind.

Many volcanoes, some of them still active, can be found in the mountains of the Antarctic Peninsula.

The Lambert Glacier is the largest glacier system in the world, up to 50 miles (80 km) wide at its seaward limit, and reaching 180 miles (300 km) into the interior by way of the Prince Charles Mountains.

Antarctica is the highest continent on Earth, because of the great thickness of ice which overlays the land. In places the ice alone can reach up to 15,700 ft (4,800 m) thick. Much of the basement rock of west Antarctica lies below sea level, pushed down by the weight of the ice.

The mountainous Antarctic Peninsula is formed of rocks 65–225 million years old, overlain by more recent rocks and glacial deposits. It is connected to the Andes in South America by a submarine ridge.

Nearly half – 44% – of the Antarctic coastline is bounded by ice shelves, like the Ronne Ice Shelf, which float on the Ocean. These are joined to the inland ice sheet by dome-shaped ice "rises."

More than 30% of Antarctic ice is contained in the Ross Ice Shelf.

The barren, flat-bottomed Upper Wright Valley was once filled by a glacier, but is now dry, strewn with boulders and pebbles. In some dry valleys, there has been no rain for over 2 million years.

Large colonies of seabirds live in the extremely harsh Antarctic climate. The Emperor penguins seen here, the smaller Adélie penguin, the Antarctic petrel and the South Polar skua are the only birds that breed exclusively on the continent.

Map labels

South Orkney Islands — Laurie Island, Orcadas (to Argentina), Coronation Island, Signy (to UK)

Scotia Sea

Clarence Island
Elephant Island
King George Island
Capitán Arturo Prat (to Chile)
Livingston Island
South Shetland Islands
Brabant Island
Anvers Island
Palmer (to US)
Biscoe Islands
Lavoisier Island
Cape Mascart
Adelaide Island
Rothera (to UK)
Marguerite Bay
Rothschild Island
Charcot Island
Latady Island

Drake Passage

Joinville Island
Dundee Island
General Bernardo O'Higgins (to Chile)
Esperanza (to Argentina)
Marambio (to Argentina)
Snowhill Island
James Ross Island
Robertson Island
Jason Peninsula
Churchill Peninsula
Larsen Ice Shelf
Cape Agassiz
Hearst Island
Ewing Island
Dolleman Island
Steele Island
Cape Bryant
Black Coast
Cape Knowles
Butler Island
Cape Mackintosh
Cape Deacon
Cape Fiske

Bransfield Strait
Danco Coast
Davis Coast
Graham Land
Fallières Coast
San Martín (to Argentina)
Antarctic Peninsula
Palmer Land
English Coast
Mount Jackson 3190m
Lassiter Coast

Weddell Sea

Ronne Ice Shelf
Korff Ice Rise
Henry Ice Rise

Douglas Range
Alexander Island
Wilkins Ice Shelf
Beethoven Peninsula
George VI Sound
Ronne Entrance
George VI Ice Shelf
Orville Coast
Zumberge Coast
Haag Nunataks
Spaatz Island
Case Island
Smyley Island
Rydberg Peninsula
Bryan Coast

Bellingshausen Sea

Rutford Ice Stream
Vinson Massif 4897m
Ellsworth Mountains
Ellsworth Land

Peter I Island (to Norway)
Dendtler Island
Farwell Island
Dustin Island
Thurston Island
Noville Peninsula
Eights Coast
Abbot Ice Shelf
Sherman Island
Pine Island Glacier
Walgreen Coast

Cape Flying Fish
King Peninsula
Canisteo Peninsula
Burke Island
Bear Peninsula
Martin Peninsula
Wright Island
Carney Island
Siple Island

Amundsen Sea

Bakutis Coast
Getz Ice Shelf
Mount Sidley 4181m
Executive Committee Range
Dean Island
Grant Island
Mount Siple 3100m
Hobbs Coast
Cape Burks
Russkaya (to Russian Federation)
Ruppert Coast
Newm... Isla...

PACIFIC OCEAN

Limit of winter pack ice
Limit of summer pack ice

Antarctic Circle

Small globe map

ATLANTIC OCEAN
INDIAN OCEAN
PACIFIC OCEAN
Weddell Sea
Dronning Maud Land
Palmer Land
Bellingshausen Sea
ANTARCTICA
Transantarctic Mountains
Amundsen Sea
Marie Byrd Land
Davis Sea
Wilkes Land
Ross Sea

198

204

ATLANTIC OCEAN

66

The sun sets over the Antarctic Peninsula for more than six months during the winter. However, there are more hours of sunshine during the brief Antarctic summer than most equatorial countries experience in a whole year.

▶ 180

Georg von Neumayer (to Germany)

Cape Norvegia

Fimbul Ice Shelf

Sanae (to South Africa)   Maitri (to India)

Novolazarevskaya (to Russian Federation)

Riiser-Larsen Sea

Riiser-Larsen Ice Shelf

Kronprinsesse Märtha Kyst

Borg Massif

Princess Astrid Kyst

Mühlig-Hofmann Mountains

Wohlthat Mountains

Prinsesse Ragnhild Kyst

Riiser-Larsen Peninsula

Lützow-Holmbukta

Prins Harald Kyst

INDIAN OCEAN

Idan Island

Brnt Ice Shelf

Maudheimvidda

Dronning Maud Land

Fimbulheimen

Asuka (to Japan)

Sør Rondane Mountains

Belgica Mountains

△ Mount Victor 2588m

Syowa (to Japan)   Thyer Glacier

Molodezhnaya (to Russian Federation)

Casey Bay

Amundsen Bay

Cape Batterbee

Nye Mountains

Napier Mountains

△ Mount Elkins 2300m

Kronprins Olav Kyst

Thorshavnheiane

Halley (to UK)   Caird Coast

Stancomb-Wills Glacier

Luitpold Coast

Coats Land

Belgrano II (to Argentina)

Theron Mountains

Slessor Glacier

Recovery Glacier

Filchner Ice Shelf

kner d

Enderby Land

Dismal Mountains

Edward VIII Gulf

Law Promontory

Mawson Coast

Mawson (to Australia)

Kemp Land

Hansen Mountains

Mac. Robertson Land

△ Mount Menzies 3355m

Prince Charles Mountains

Gustav Bull Mountains

Lars Christensen Coast

Cape Darnley

Pensacola Mountains

Support Force Glacier

Lambert Glacier

Amery Ice Shelf   Gillock Island

Mackenzie Bay

6

Princess Elizabeth Land

Ingrid Christensen Coast

Zhongshan (to China)   Prydz Bay

Davis (to Australia)

7

West Ice Shelf

Mikhaylov Island

ANTARCTICA

Greater Antarctica

King Leopold and Queen Astrid Land

Philippi Glacier

Davis Sea

Whitmore Mountains

Seelig

△ Amundsen-Scott (to US)

South Pole

Wilhelm II Land

Wilhelm II Coast

90°

Mirny (to Russian Federation)

Horlick Mountains

Gould Coast

Watson Escarpment

Transantarctic Mountains

Queen Maud Mountains

Amundsen Coast

Dufek Coast

Beardmore Glacier

Vostok (to Russian Federation)

+ South Geomagnetic Pole

Queen Mary Coast

Northcliffe Glacier

Denman Glacier

Scott Glacier

Masson Island

Mill Island

Bowman Island

Siple Coast

△ Mount Kirkpatrick 4528m

Mount Markham 4351m

Nimrod Glacier

Shackleton Coast

Knox Coast

Shackleton Ice Shelf

Land

Rockefeller Plateau

Byrd Glacier

Hillary Coast

△ Mount McClintock 3492m

Ross Ice Shelf

Roosevelt Island

Wilkes Land

Vincennes Bay

anders Coast

Shirase Coast

△ Mount Lister 4026m

Victoria Land

Casey (to Australia)

Budd Coast   Cape Poinsett

Sulzberger Bay

Edward VII Peninsula

Scott Base (to NZ)

McMurdo Base (to US)

Ross Island

Mount Erebus 3794m

Scott Coast

Sabrina Coast

Cape Waldron

13

Drygalski Ice Tongue

Dalton Iceberg Tongue

Ross Sea

Terre Adélie

Banzare Coast

Cape Goodenough

14

AN

Coulman Island

Borchgrevink Coast

George V Land

Porpoise Bay

Wilkes Coast

Cape Keltie

MAP KEY

ELEVATION

Oates Land

Rennick Glacier

Adélie Coast

Cape Adare

△ Mount Minto 4163m

George V Coast

Ninnis Glacier

Mertz Glacier

Dumont d'Urville (to France)

Cape Gray

ice cap

ice shelf

Immense, flat-topped icebergs are formed when blocks of ice break away from the main ice sheet. Though the exposed area is enormous, the volume of ice concealed beneath the water may be many times greater.

exposed land

Cape Cheetham

Leningradskaya (to Russian Federation)

Cape Freshfield

Cape Hudson

Dumont d'Urville Sea

SCALE 1:14,750,000
(projection: Lambert Azimuthal Equal Area)

Km
0 25 50   100   150   200   250   300   350   400   450   500

Miles
0   50   100   150   200   250   300   350   400   450   500

Ball'eny Islands

Antarctic Circle

Limit of summer pack ice

Scott Island

THE ARCTIC

THREE CONTINENTS, ASIA, NORTH AMERICA, AND EUROPE, reach into the Arctic Circle at their northernmost limits, almost entirely encircling the Arctic Ocean. Despite the region's extraordinarily harsh climate, it has been inhabited for thousands of years by peoples such as the European Lapps, the Russian Nenet, and the North American Inuit, who draw a living from fishing, herding, and hunting. More recently, particularly in the Russian Arctic, opportunities to exploit oil and other mineral reserves have encouraged immigration. Pollution of the Arctic's unique ecology and damage to the traditional lifestyles of many native peoples have been the unfortunate results of this activity, and international cooperation is needed to safeguard the future of the region.

MAP KEY

POPULATION

- ■ above 5 million
- ◼ 1 million to 5 million
- ◉ 500,000 to 1 million
- ◎ 100,000 to 500,000
- ⊕ 50,000 to 100,000
- ○ 10,000 to 50,000
- ○ below 10,000

SEA DEPTH

sea level
250m / 820ft
500m / 1640ft
1000m / 3281ft
2000m / 6562ft
3000m / 9843ft

SCALE 1:21,000,000
(projection: Lambert Azimuthal Equal Area)

Km 0 100 200 300 400 500 600
Miles 0 100 200 300 400 500 600

Windblown snow etches deep patterns in the ice sheet known as sastrugi. They align with the direction of the wind

RESOURCES

LARGE QUANTITIES of coal, oil, and natural gas are to be found in the basins of the Arctic Ocean, and in northern Canada, Alaska and the Russian Federation. The cost and difficulty of extraction and, more recently, awareness of damage to the environment, have limited exploitation to coastal regions. The unfrozen waters have stocks of fish including cod, flounder, and haddock. Quotas have now been put in place to restrict the number of fish caught annually. Reindeer are herded in large numbers by many of the native Arctic peoples. Most grain and vegetables are imported from elsewhere.

Bering Sea

NORTH AMERICA
Inuvik
ASIA
Tiksi
ARCTIC OCEAN
Qaanaaq
Noril'sk
Murmansk
Reykjavik
ATLANTIC OCEAN
EUROPE

Icebreakers are ships with specially strengthened hulls, designed to break a path through the ice. They are used to keep important routes open during the winter, when falling temperatures cause much of the Arctic Ocean to freeze over.

Resources
- coal
- fish
- mining
- oil & gas
- radioactive contamination
- major towns
- major ports

THE LANDSCAPE

THE ARCTIC OCEAN comprises two large ocean basins divided by three submarine ridges, the greatest of which, the Lomonosov Ridge, is a huge underwater mountain range which has an average height of more than 10,000 ft (3,000 m). The lands which encircle the Arctic Ocean are underlain by great shield areas of ancient rocks, which were heavily glaciated during the last Ice Age.

Icebergs are constantly broken up and reshaped by wind and the oceans. This flat-topped iceberg has been undercut, leaving a craggy ice cliff.

A complex and ancient mountain system, extending from the Queen Elizabeth Islands to eastern Greenland was formed more than 245 million years ago.

The Canadian Shield underlies almost all of the Canadian Arctic. It is a very stable plateau of ancient rock, now covered by glacial lakes and sediment, which supports tundra vegetation.

The Arctic Ocean is the world's smallest ocean with a total area of 5,440,000 sq miles (15,100,000 sq km).

At a latitude of more than 75° N, the Arctic Ocean is almost permanently covered by pack-ice, though high winds and the movement of the seas may cause the ice to crack and break up.

In the more southerly reaches of the Arctic, like Siberia, much of the land is covered by permafrost. In the summer, higher temperatures warm the frozen ground, causing a number of typical phenomena. These include solifluction, the fast downhill movement of top soil layers; freeze/thaw activity, which patterns the ground into regular polygonal shapes, and the formation of large domes with a frozen ice core, known as pingos.

Lomonosov Ridge

Lomonosov Ridge

Much of Greenland is covered by a massive ice sheet more than 650,000 sq miles (1,683,400 sq km) in extent. The weight of the ice has depressed the central land area to form a basin lying more than 1,000 ft (300 m) below sea level. Only at the edges of the island is bare rock visible.

Iceland has five major glaciers, sustained by heavy snowfall. Parts of the ice cap cover active volcanoes, such as Bárdharbunga, which periodically erupt causing the melted ice to form a great lake at the glacier margins.

Arctic ice shelf

Ice sheet
Iceberg
Crevasses occur at the edge of the ice sheet
Sea water melts the edge of the ice sheet

At the boundary of the Arctic ice shelves, sea water flows under the ice causing melting and forming crevasses on the surface. This eventually weakens blocks of ice which break away as icebergs. This process is known as calving.

Map labels

CANADA
NORTH AMERICA
ARCTIC
Great Bear Lake
Mackenzie
Great Slave Lake
Coppermine
Bathurst Inlet
Coronation Gulf
Cambridge Bay
Queen Maud Gulf
King William Island
Back
Boothia Peninsula
Nelson
Churchill
Southampton Island
Repulse Bay
Melville Peninsula
Hudson Bay
Coats Island
Mansel Island
Foxe Basin
Prince Charles Island
Ivujivik
Inukjuak
Hudson Strait
Foxe Peninsula
Baffin Island
Lake Harbour
Trobisher Bay
Cumberland Sound
Ungava Bay
Nain
Cape Chidley
Davis Strait
Maniitsoq
NUUK
Labrador Sea
Paamiut
Labrador Basin
Ivittuut
Qaqortoq
Nanortalik
Narsaq
Nunap Isua (Kap Farvel)
Eirik Ridge
ATLANTIC

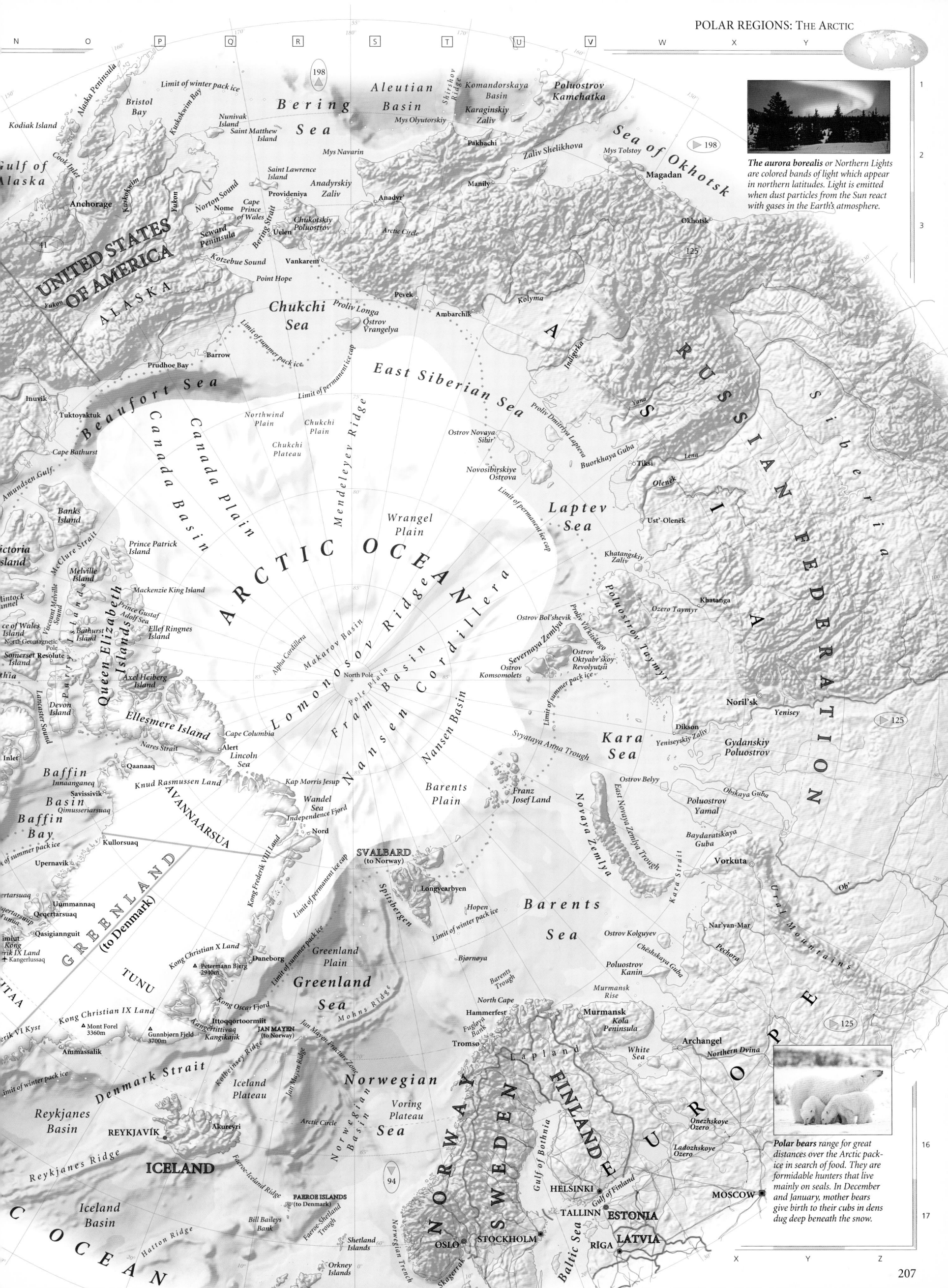

N   O   P   Q   R   S   T   U   V   W   X   Y

Bering Sea
Aleutian Basin
Komandorskaya Basin
Shirshov Ridge
Poluostrov Kamchatka
Karaginskiy Zaliv

Mys Olyutorskiy
Pakhachi
Zaliv Shelikhova
Mys Navarin
Mys Tolstoy

Alaska Peninsula
Bristol Bay
Kuskokwim Bay
Nunivak Island
Saint Matthew Island
Limit of winter pack ice
198

Kodiak Island
Cook Inlet
Gulf of Alaska
Anchorage
Yukon
Kuskokwim

Saint Lawrence Island
Anadyrskiy Zaliv
Anadyr'
Manily
Magadan
Okhotsk
Sea of Okhotsk
198
125

Norton Sound
Nome
Cape Prince of Wales
Providentiya
Chukotskiy Poluostrov

UNITED STATES OF AMERICA
41
Seward Peninsula
Bering Strait
Uelen
Arctic Circle

ALASKA
Kotzebue Sound
Vankarem

Point Hope
Chukchi Sea
Proliv Longa
Ostrov Vrangelya
Pevek
Kolyma
Ambarchik

Barrow
Limit of summer pack ice
Indigirka

Prudhoe Bay
Limit of permanent ice cap
East Siberian Sea
Yana

Beaufort Sea
Northwind Plain
Chukchi Plain
Mendeleyev Ridge
Wrangel Plain
Proliv Dmitriya Laptev

Inuvik
Tuktoyaktuk
Cape Bathurst
Canada Plain
Chukchi Plateau

R U S S I A N

Ostrov Novaya Sibir'
Buorkhaya Guba
Tiksi
Lena
Olenek

Amundsen Gulf
Banks Island
Canada Basin
Novosibirskiye Ostrova
Ust'-Olenek

Victoria Island
Prince Patrick Island
ARCTIC OCEAN
Laptev Sea
Khatangskiy Zaliv

McClure Strait
Melville Island
Mackenzie King Island
Limit of permanent ice cap
Khatanga

Viscount Melville
Prince Gustaf Adolf Sea
Ellef Ringnes Island
Ozero Taymyr

Queen Elizabeth Islands
Bathurst Island
Alpha Cordillera
Makarov Basin
Ostrov Bol'shevik
Poluostrov Taymyr

North Geomagnetic Pole
Axel Heiberg Island
Lomonosov Ridge
Severnaya Zemlya
Ostrov Komsomolets
Ostrov Oktyabr'skoy Revolyutsii

Somerset Island
Resolute Island
Devon Island
Ellesmere Island
Cape Columbia
North Pole
Pole Plain
Nansen Cordillera
Fram Basin

Lancaster Sound
Nares Strait
Alert
Lincoln Sea
Dikson
F E D E R A T I O N

Inlet
Baffin Basin
Qaanaaq
Kap Morris Jesup
Nansen Basin
Svyataya Anna Trough
Noril'sk
Yenisey

Innaanganeq
Savissivik
Qimusseriarsuaq
Wandel Sea
Independence Fjord
Franz Josef Land
Kara Sea
Yeniseyskiy Zaliv
Gydanskiy Poluostrov

Baffin Bay
Knud Rasmussen Land
Nord
Barents Plain
Ostrov Belyy
Poluostrov Yamal
Obskaya Guba

Kullorsuaq
Limit of summer pack ice
SVALBARD (to Norway)
Baydaratskaya Guba

Upernavik
AVANNAARSUA
Longyearbyen
Spitsbergen
Novaya Zemlya
East Novaya Zemlya Trough
Kara Strait
Vorkuta

Uummannaq
Qeqertarsuaq
Qasigiannguit
GREENLAND (to Denmark)
Hopen
Barents Sea
Nar'yan-Mar

Kong Frederik VIII Land
Limit of permanent ice cap
Limit of summer pack ice
Ostrov Kolguyev
Chëshskaya Guba
Pechora
Ob'

Kong Christian X Land
Daneborg
Greenland Plain
Bjørnøya
Poluostrov Kanin
U R A L

TUNU
Petermann Bjerg 2930m
Greenland Sea
Barents Trough

Kong Christian IX Land
Kong Oscar Fjord
Ostrov Kolguyev
North Cape
Murmansk Rise
Ural Mountains

ITAA
Mont Forel 3360m
Kangertittivaq
Ittoqqortoormiit
Mohns Ridge
Hammerfest
Fugløya Bank
Murmansk
Kola Peninsula
Archangel

Ammassalik
Gunnbjørn Fjeld 3700m
Kangikajik
JAN MAYEN (to Norway)
Jan Mayen Fracture Zone
Tromsø
White Sea
Northern Dvina

erik VI Kyst
Denmark Strait
Jan Mayen Ridge
Lapland
Onezhskoye Ozero

Reykjanes Basin
Iceland Plateau
Kolbeinsey Ridge
Norwegian Sea
Ladozhskoye Ozero

REYKJAVÍK
Akureyri
Voring Plateau
E U R O P E

Reykjanes Ridge
ICELAND
Arctic Circle
NORWAY
SWEDEN
FINLAND
HELSINKI
MOSCOW
125

Iceland Basin
Faroe-Iceland Ridge
94
Gulf of Bothnia
ESTONIA
TALLINN

A T L A N T I C
FAEROE ISLANDS (to Denmark)
Bill Baileys Bank
Norwegian Trench
STOCKHOLM
RIGA
LATVIA

O C E A N
Faeroe-Shetland Trough
Shetland Islands
OSLO
Baltic Sea
Skagerrak

Orkney Islands

1
2
3

16
17

X   Y   Z

207

GEOGRAPHICAL COMPARISONS

LARGEST COUNTRIES

Russian Federation	6,592,800 sq miles	(17,075,400 sq km)
Canada	3,851,788 sq miles	(9,976,140 sq km)
USA	3,681,760 sq miles	(9,372,610 sq km)
China	3,600,292 sq miles	(9,326,410 sq km)
Brazil	3,286,472 sq miles	(8,511,970 sq km)
Australia	2,967,893 sq miles	(7,686,850 sq km)
India	1,269,338 sq miles	(3,287,590 sq km)
Argentina	1,068,296 sq miles	(2,766,890 sq km)
Kazakhstan	1,049,150 sq miles	(2,717,300 sq km)
Sudan	967,493 sq miles	(2,505,815 sq km)

SMALLEST COUNTRIES

Vatican City	0.17 sq miles	(0.44 sq km)
Monaco	0.75 sq miles	(1.95 sq km)
Nauru	8.2 sq miles	(21.2 sq km)
Tuvalu	10 sq miles	(26 sq km)
San Marino	24 sq miles	(61 sq km)
Liechtenstein	62 sq miles	(160 sq km)
Marshall Islands	70 sq miles	(181 sq km)
Seychelles	108 sq miles	(280 sq km)
Maldives	116 sq miles	(300 sq km)
Malta	124 sq miles	(320 sq km)

LARGEST ISLANDS

(TO THE NEAREST 1,000 - OR 100,000 FOR THE LARGEST)

Greenland	849,400 sq miles	(2,200,000 sq km)
New Guinea	312,000 sq miles	(808,000 sq km)
Borneo	292,222 sq miles	(757,050 sq km)
Madagascar	229,300 sq miles	(594,000 sq km)
Sumatra	202,300 sq miles	(524,000 sq km)
Baffin Island	183,800 sq miles	(476,000 sq km)
Honshu	88,800 sq miles	(230,000 sq km)
Britain	88,700 sq miles	(229,800 sq km)
Victoria Island	81,900 sq miles	(212,000 sq km)
Ellesmere Island	75,700 sq miles	(196,000 sq km)

RICHEST COUNTRIES

(GNP PER CAPITA, IN US$)

Luxembourg	45,360
Switzerland	44,350
Japan	40,940
Liechtenstein	40,000
Norway	34,510
Denmark	32,100
Singapore	30,550
Germany	28,870
Austria	28,110
USA	28,020

POOREST COUNTRIES

(GNP PER CAPITA, IN US$)

Mozambique	80
Somalia	100
Ethiopia	100
Eritrea	100
Congo (Zaire)	130
Chad	160
Tanzania	170
Burundi	170
Malawi	180
Rwanda	190
Sierra Leone	200
Niger	200

MOST POPULOUS COUNTRIES

China	1,255,100,000
India	935,700,000
USA	263,300,000
Indonesia	197,600,000
Brazil	165,800,000
Russian Federation	147,000,000
Pakistan	140,500,000
Japan	125,100,000
Bangladesh	120,400,000
Nigeria	111,700,000

LEAST POPULOUS COUNTRIES

Vatican City	1,000
Tuvalu	9,000
Nauru	10,000
Palau	16,200
San Marino	24,000
Liechtenstein	30,630
Monaco	31,000
St Kitts & Nevis	44,000
Marshall Islands	52,000
Andorra	64,000
Dominica	71,000
Seychelles	73,000

MOST DENSELY POPULATED COUNTRIES

Monaco	41,332 people per sq mile	(15,897 per sq km)
Singapore	11,894 people per sq mile	(4,590 per sq km)
Vatican City	5,890 people per sq mile	(2,273 per sq km)
Malta	3,239 people per sq mile	(1,250 per sq km)
Maldives	2,591 people per sq mile	(1,000 per sq km)
Bangladesh	2,330 people per sq mile	(899 per sq km)
Bahrain	2,286 people per sq mile	(882 per sq km)
Barbados	1,809 people per sq mile	(698 per sq km)
Taiwan	1,682 people per sq mile	(649 per sq km)
Mauritius	1,542 people per sq mile	(595 per sq km)

MOST SPARSELY POPULATED COUNTRIES

Australia	5 people per sq mile	(2 per sq km)
Mauritania	5 people per sq mile	(2 per sq km)
Mongolia	5 people per sq mile	(2 per sq km)
Namibia	5 people per sq mile	(2 per sq km)
Suriname	5 people per sq mile	(2 per sq km)
Botswana	8 people per sq mile	(3 per sq km)
Canada	8 people per sq mile	(3 per sq km)
Iceland	8 people per sq mile	(3 per sq km)
Libya	8 people per sq mile	(3 per sq km)
Guyana	10 people per sq mile	(4 per sq km)

MOST WIDELY SPOKEN LANGUAGES

1. Chinese (Mandarin)	6. Arabic
2. English	7. Bengali
3. Hindi	8. Portuguese
4. Spanish	9. Malay-Indonesian
5. Russian	10. French

COUNTRIES WITH THE MOST LAND BORDERS

14: China (Afghanistan, Bhutan, Mynamar, India, Kazakhstan, Kyrgyzstan, Laos, Mongolia, Nepal, North Korea, Pakistan, Russian Federation, Tajikistan, Vietnam)

14: Russian Federation (Azerbaijan, Belarus, China, Estonia, Finland, Georgia, Kazakhstan, Latvia, Lithuania, Mongolia, North Korea, Norway, Poland, Ukraine)

10: Brazil (Argentina, Bolivia, Colombia, French Guiana, Guyana, Paraguay, Peru, Suriname, Uruguay, Venezuela)

9: Congo (Zaire) (Angola, Burundi, Central African Republic, Congo, Rwanda, Sudan, Tanzania, Uganda, Zambia)

9: Germany (Austria, Belgium, Czech Republic, Denmark, France, Luxembourg, Netherlands, Poland, Switzerland)

9: Sudan (Central African Republic, Chad, Congo (Zaire), Egypt, Eritrea, Ethiopia, Kenya, Libya, Uganda)

8: Austria *(Czech Republic, Germany, Hungary, Italy, Liechtenstein, Slovakia, Slovenia, Switzerland)*

8: France *(Andorra, Belgium, Germany, Italy, Luxembourg, Monaco, Spain, Switzerland)*

8: Tanzania *(Burundi, Congo (Zaire), Kenya, Malawi, Mozambique, Rwanda, Uganda, Zambia)*

8: Turkey *(Armenia, Azerbaijan, Bulgaria, Georgia, Greece, Iran, Iraq, Syria)*

8: Zambia *(Angola, Botswana, Congo (Zaire), Malawi, Mozambique, Namibia, Tanzania, Zimbabwe)*

LONGEST RIVERS

Nile (NE Africa) . . . . 4,160 miles . . . .(6,695 km)
Amazon (South America) . . . . 4,049 miles . . . .(6,516 km)
Yangtze (China) . . . . 3,915 miles . . . .(6,299 km)
Mississippi/Missouri (USA) . . . . 3,710 miles . . . .(5,969 km)
Ob'-Irtysh (Russian Federation) . . . . . . 3,461 miles . . . .(5,570 km)
Yellow River (China) . . . . 3,395 miles . . . .(5,464 km)
Congo (Central Africa) . . . . 2,900 miles . . . .(4,667 km)
Mekong (Southeast Asia) . . . . 2,749 miles . . . .(4,425 km)
Lena (Russian Federation) . . . . 2,734 miles . . . .(4,400 km)
Mackenzie (Canada) . . . . 2,640 miles . . . .(4,250 km)
Yenisey (Russian Federation) . . . . 2,541 miles . . . .(4,090km)

HIGHEST MOUNTAINS
(HEIGHT ABOVE SEA LEVEL)

Everest . . . . .29,030 ft . . . .(8,848 m)
K2 . . . . .28,253 ft . . . .(8,611 m)
Kanchenjunga I . . . . .28,210 ft . . . .(8,598 m)
Makalu I . . . . .27,767 ft . . . .(8,463 m)
Cho Oyu . . . . .26,907 ft . . . .(8,201 m)
Dhaulagiri I . . . . .26,796 ft . . . .(8,167 m)
Manaslu I . . . . .26,783 ft . . . .(8,163 m)
Nanga Parbat I . . . . .26,661 ft . . . .(8,126 m)
Annapurna I . . . . .26,547 ft . . . .(8,091 m)
Gasherbrum I . . . . .26,471 ft . . . .(8,068 m)

LARGEST BODIES OF INLAND WATER
(WITH AREA AND DEPTH)

Caspian Sea . . . . .143,243 sq miles (371,000 sq km) . . . . .3,215 ft (980 m)
Lake Superior . . . . .31,151 sq miles (83,270 sq km) . . . . .1,289 ft (393 m)
Lake Victoria . . . . .26,828 sq miles (69,484 sq km) . . . . . .328 ft (100 m)
Lake Huron . . . . . . .23,436 sq miles (60,700 sq km) . . . . . .751 ft (229 m)
Lake Michigan . . . .22,402 sq miles (58,020 sq km) . . . . . .922 ft (281 m)
Lake Tanganyika . .12,703 sq miles (32,900 sq km) . . . .4,700 ft (1435 m)
Great Bear Lake . . .12,274 sq miles (31,790 sq km) . . . . .1,047 ft (319 m)
Lake Baikal . . . . . . .11,776 sq miles (30,500 sq km) . . . .5,712 ft (1741 m)
Great Slave Lake . . .10,981 sq miles (28,440 sq km) . . . . . .459 ft (140 m)
Lake Erie . . . . .9,915 sq miles (25,680 sq km) . . . . . . .197 ft (60 m)

DEEPEST OCEAN FEATURES

Challenger Deep, Marianas Trench (Pacific) . . . . 36,201 ft . . . .(11,034 m)
Vityaz III Depth, Tonga Trench (Pacific) . . . . 35,704 ft . . . . .(10,882 m)
Vityaz Depth, Kurile-Kamchatka Trench (Pacific) . . . . 34,588 ft . . . . .(10,542 m)
Cape Johnson Deep, Philippine Trench (Pacific) . . . . . 34,441 ft . . . . .(10,497 m)
Kermadec Trench (Pacific) . . . . 32,964 ft . . . . .(10,047 m)
Ramapo Deep, Japan Trench (Pacific) . . . . 32,758 ft . . . . . .(9,984 m)
Milwaukee Deep, Puerto Rico Trench (Atlantic) . . . . . 30,185 ft . . . . . .(9,200 m)
Argo Deep, Torres Trench (Pacific) . . . . 30,070 ft . . . . . .(9,165 m)
Meteor Depth, South Sandwich Trench (Atlantic) . . . . . 30,000 ft . . . . . .(9,144 m)
Planet Deep, New Britain Trench (Pacific) . . . . 29,988 ft . . . . . .(9,140 m)

GREATEST WATERFALLS
(MEAN FLOW OF WATER)

Boyoma (Congo (Zaire)) . . . . .600,400 cu. ft/sec . .(17,000 cu.m/sec)
Khône (Laos/Cambodia) . . . . .410,000 cu. ft/sec . .(11,600 cu.m/sec)
Niagara (USA/Canada) . . . . .195,000 cu. ft/sec . . .(5,500 cu.m/sec)
Grande (Uruguay) . . . . .160,000 cu. ft/sec . . .(4,500 cu.m/sec)
Paulo Afonso (Brazil) . . . . .100,000 cu. ft/sec . . .(2,800 cu.m/sec)
Urubupunga (Brazil) .97,000 cu. ft/sec . . .(2,750 cu.m/sec)
Iguaçu (Argentina/Brazil) . . . . .62,000 cu. ft/sec . . .(1,700 cu.m/sec)
Maribondo (Brazil) . . . . .53,000 cu. ft/sec . . .(1,500 cu.m/sec)
Victoria (Zimbabwe) .39,000 cu. ft/sec . . .(1,100 cu.m/sec)
Kabalega (Uganda) . . . . .42,000 cu. ft/sec . . .(1,200 cu.m/sec)

Churchill (Canada) . . . . .35,000 cu. ft/sec . . .(1,000 cu.m/sec)
Cauvery (India) . . . . .33,000 cu. ft/sec . . . . .(900 cu.m/sec)

HIGHEST WATERFALLS

Angel (Venezuela) . . . . .3,212 ft . . . .(979 m)
Tugela (South Africa) . . . . .3,110 ft . . . .(948 m)
Utigard (Norway) . . . . .2,625 ft . . . .(800 m)
Mongefossen (Norway) . . . . .2,539 ft . . . .(774 m)
Mtarazi (Zimbabwe) . . . . .2,500 ft . . . .(762 m)
Yosemite (USA) . . . . .2,425 ft . . . .(739 m)
Ostre Mardola Foss (Norway) . . . . .2,156 ft . . . .(657 m)
Tyssestrengane (Norway) . . . . .2,119 ft . . . .(646 m)
***Cuquenan** (Venezuela) . . . . .2,001 ft . . . .(610 m)
Sutherland (New Zealand) . . . . .1,903 ft . . . .(580 m)
***Kjellfossen** (Norway) . . . . .1,841 ft . . . .(561 m)

* indicates that the total height is a single leap

LARGEST DESERTS

Sahara . . . . .3,450,000 sq miles . . . . . .(9,065,000 sq km)
Gobi .500,000 sq miles . . . . . .(1,295,000 sq km)
Ar Rub al Khali . . . . .289,600 sq miles . . . .(750,000 sq km)
Great Victorian . . . . .249,800 sq miles . . . .(647,000 sq km)
Sonoran . . . . .120,000 sq miles . . . .(311,000 sq km)
Kalahari . . . . .120,000 sq miles . . . .(310,800 sq km)
Kara Kum . . . . .115,800 sq miles . . . .(300,000 sq km)
Takla Makan . . . . .100,400 sq miles . . . .(260,000 sq km)
Namib .52,100 sq miles . . . .(135,000 sq km)
Thar . . . . .33,670 sq miles . . . .(130,000 sq km)

NB – Most of Antarctica is a polar desert, with only 50mm of precipitation annually

HOTTEST INHABITED PLACES

Djibouti (Djibouti)	86° F	(30 °C)
Timbouctou (Mali)	84.7° F	(29.3 °C)
Tirunelveli (India)		
Tuticorin (India)		
Nellore (India)	84.5° F	(29.2 °C)
Santa Marta (Colombia)		
Aden (Yemen)	84° F	(28.9 °C)
Madurai (India)		
Niamey (Niger)		
Hodeida (Yemen)	83.8° F	(28.8 °C)
Ouagadougou (Burkina)		
Thanjavur (India)		
Tiruchchirappalli (India)		

DRIEST INHABITED PLACES

Aswân (Egypt) . . . . .0.02 in . . . .(0.5 mm)
Luxor (Egypt) . . . . .0.03 in . . . .(0.7 mm)
Arica (Chile) . . . . .0.04 in . . . .(1.1 mm)
Ica (Peru) . . . . .0.1 in . . . .(2.3 mm)
Antofagasta (Chile) . . . . .0.2 in . . . .(4.9 mm)
El Minya (Egypt) . . . . .0.2 in . . . .(5.1 mm)
Asyût (Egypt) . . . . .0.2 in . . . .(5.2 mm)
Callao (Peru) . . . . .0.5 in . . . .(12.0 mm)
Trujillo (Peru) . . . . .0.55 in . . . .(14.0 mm)
El Faiyûm (Egypt) . . . . .0.8 in . . . .(19.0 mm)

WETTEST INHABITED PLACES

Buenaventura (Colombia) . . . . .265 in . . . .(6,743 mm)
Monrovia (Liberia) . . . . .202 in . . . .(5,131 mm)
Pago Pago (American Samoa) . . . . .196 in . . . .(4,990 mm)
Moulmein (Burma) . . . . .191 in . . . .(4,852 mm)
Lae (Papua New Guinea) . . . . .183 in . . . .(4,645 mm)
Baguio (Luzon Island, Philippines) . . . . .180 in . . . .(4,573 mm)
Sylhet (Bangladesh) . . . . .176 in . . . .(4,457 mm)
Padang (Sumatra, Indonesia) . . . . .166 in . . . .(4,225 mm)
Bogor (Java, Indonesia) . . . . .166 in . . . .(4,225 mm)
Conakry (Guinea) . . . . .171 in . . . .(4,341 mm)

THE TIME ZONES

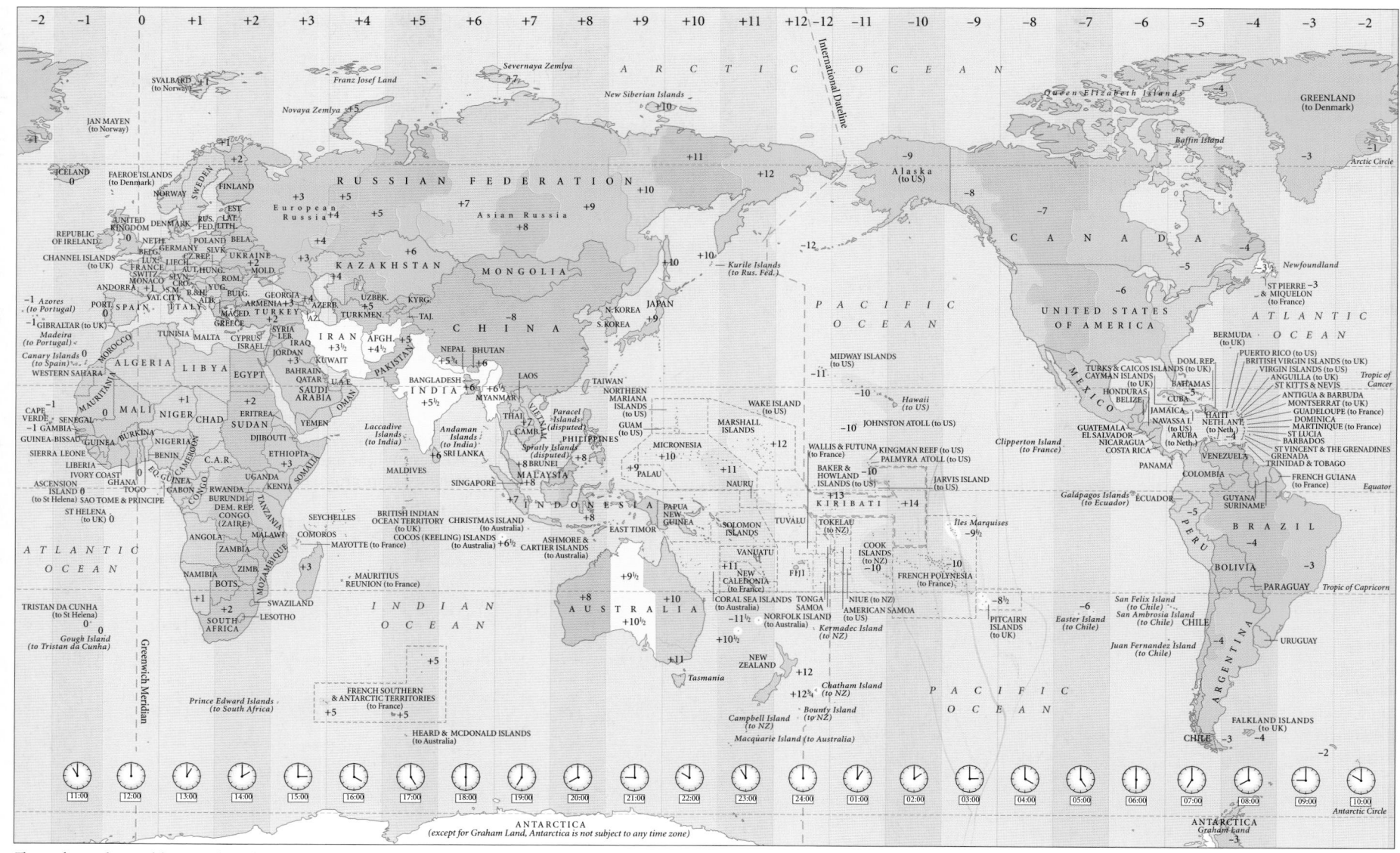

The numbers at the top of the map indicate the number of hours each time zone is ahead or behind Greenwich Mean Time (GMT). The clocks and 24-hour times given at the bottom of the map show the time in each time zone when it is 12:00 hours noon GMT.

TIME ZONES

The present system of international timekeeping divides the world into 24 time zones by means of 24 standard meridians of longitude, each 15° apart. Time is measured in each zone as so many hours ahead or behind the time at the Greenwich Meridian (GMT). Countries, or parts of countries, falling in the vicinity of each zone, adopt its time as shown on the map above. Therefore, using the map, when it is 12:00 noon GMT, it will be 2:00 pm in Zambia; similarly, when it is 4:30 pm GMT, it will be 11:30 am in Peru.

GREENWICH MEAN TIME (GMT)

Greenwich Mean Time (or Universal Time, as it is more correctly called) has been the internationally accepted basis for calculating solar time – measured in relation to the Earth's rotation around the Sun – since 1884. Greenwich Mean Time is specifically the solar time at the site of the former Royal Observatory in the London Borough of Greenwich, United Kingdom. The Greenwich Meridian is an imaginary line around the world that runs through the North and South poles. It corresponds to 0° of longitude, which lies on this site at Greenwich. Time is measured around the world in relation to the official time along the Meridian.

STANDARD TIME

Standard time is the official time, designated by law, in any specific country or region. Standard

time was initiated in 1884, after it became apparent that the practice of keeping various systems of local time was causing confusion – particularly in the US and Canada, where several railroad routes passed through scores of areas which calculated local time by different rules. The standard time of a particular region is calculated in reference to the longitudinal time zone in which it falls. In practice, these zones do not always match their longitudinal position; in some places the area of the zone has been altered in shape for the convenience of inhabitants, as can be seen in the map. For example, whilst Greenland occupies three time zones, the majority of the territory uses a standard time of -3 hours GMT. Similarly China, which spans five time zones, is standardized at -8 hours GMT.

THE INTERNATIONAL DATELINE

The International Dateline is an imaginary line that extends from pole to pole, and roughly corresponds to a line of 180° longitude for much of its length. This line is the arbitrary marker between calendar days. By moving from east to west across the line, a traveler will need to set their calendar back one day, whilst those traveling in the opposite direction will need to add a day. This is to compensate for the use of standard time around the world, which is based on the time at noon along the Greenwich Meridian, approximately halfway around the world. Wide deviations from 180° longitude occur through the

Bering Strait – to avoid dividing Siberia into two separate calendar days – and in the Pacific Ocean – to allow certain Pacific islands the same calendar day as New Zealand. Changes were made to the International Dateline in 1995 that made Millennium Island (formerly Caroline Island) in Kiribati the first land area to witness the beginning of the year 2000.

DAYLIGHT SAVING TIME

Also known as summer time, daylight saving is a system of advancing clocks in order to extend the waking day during periods of later daylight hours. This normally means advancing clocks by one hour in early spring, and reverting back to standard time in early autumn. The system of daylight saving is used throughout much of Europe, the US, Australia, and many other countries worldwide, although there are no standardized dates for the changeover to summer time due to the differences in hours of daylight at different latitudes. Daylight saving was first introduced in certain countries during the First World War, to decrease the need for artificial light and heat – the system stayed in place after the war, as it proved practical. During the Second World War, some countries went so far as to keep their clocks an hour ahead of standard time continuously, and the UK temporarily introduced "double summer time," which advanced clocks two hours ahead of standard time during the summer months.

Countries of the World

THERE ARE CURRENTLY 192 independent countries in the world – more than at any previous time – and 59 dependencies. Antarctica is the only land area on Earth that is not officially part of, and does not belong to, any single country.

In 1950, the world comprised 82 countries. In the decades following, many more states came into being as they achieved independence from their former colonial rulers. Most recently, the breakup of the former Soviet Union in 1991, and the former Yugoslavia in 1992, swelled the ranks of independent states.

Country Factfile Key

Formation Date of independence / date current borders were established
Population Total population / population density – based on total *land* area / percentage of urban-based population
Languages An asterisk (*) denotes the official language(s)
Calorie consumption Average number of calories consumed daily per person

Afghanistan
Central Asia

Official name Islamic State of Afghanistan
Formation 1919 / 1919
Capital Kabul
Population 23.4 million / 93 people per sq mile (36 people per sq km) / 20%
Total area 251,770 sq miles (652,090 sq km)
Languages Persian*, Pashtu*, Dari, Uzbek, Turkmen
Religions Sunni Muslim 84%, Shi'a Muslim 15%, other 1%
Ethnic mix Pashtun 38%, Tajik 25%, Hazara 19%, Uzbek 6%, other 12%
Government Mujahideen coalition
Currency Afghani = 100 puls
Literacy rate 31%
Calorie consumption 1,523 kilocalories

Albania
Southeastern Europe

Official name Republic of Albania
Formation 1912 / 1921
Capital Tiranë
Population 3.4 million / 321 people per sq mile (124 people per sq km) / 37%
Total area 11,100 sq miles (28,750 sq km)
Languages Albanian*, Greek, Macedonian
Religions Muslim 70%, Greek Orthodox 20%, Roman Catholic 10%
Ethnic mix Albanian 96%, Greek 2%, other (including Macedonian) 2%
Government Multiparty republic
Currency Lek = 100 qindars
Literacy rate 85%
Calorie consumption 2,605 kilocalories

Algeria
Northern Africa

Official name Democratic and Popular Republic of Algeria
Formation 1962 / 1962
Capital Algiers
Population 27.9 million / 33 people per sq mile (13 people per sq km) / 56%
Total area 919,590 sq miles (2,381,740 sq km)
Languages Arabic*, Berber, French
Religions Muslim 99%, Christian and Jewish 1%
Ethnic mix Arab and Berber 99%, European 1%
Government Multiparty republic
Currency Dinar = 100 centimes
Literacy rate 60%
Calorie consumption 2,897 kilocalories

Andorra
Southwestern Europe

Official name Principality of Andorra
Formation 1278 / 1278
Capital Andorra la Vella
Population 65,000 / 359 people per sq mile (139 people per sq km) / 63%
Total area 181 sq miles (468 sq km)
Languages Catalan*, Spanish, French, Portuguese
Religions Roman Catholic 94%, other 6%
Ethnic mix Catalan 61%, Spanish Castilian 30%, other 9%
Government Parliamentary democracy
Currency French franc, Spanish peseta
Literacy rate 100%
Calorie consumption 3,708 kilocalories

Angola
Southern Africa

Official name Republic of Angola
Formation 1975 / 1975
Capital Luanda
Population 12 million / 25 people per sq mile (10 people per sq km) / 32%
Total area 481,551 sq miles (1,246,700 sq km)
Languages Portuguese*, Umbundu, Kimbundu, Kongo
Religions Roman Catholic / Protestant 64%, traditional beliefs 34%, other 2%
Ethnic mix Ovimbundu 37%, Mbundu 25%, Bakongo 13%, other 25%
Government Multiparty republic
Currency Readjusted kwanza = 100 lwei
Literacy rate 45%
Calorie consumption 1,839 kilocalories

Antigua & Barbuda
West Indies

Official name Antigua and Barbuda
Formation 1981 / 1981
Capital St John's
Population 66,000 / 389 people per sq mile (150 people per sq km) / 36%
Total area 170 sq miles (440 sq km)
Languages English*, English patios
Religions Protestant 86%, Roman Catholic 10%, other 4%
Ethnic mix Black 98%, other 2%
Government Parliamentary democracy
Currency E. Caribbean dollar = 100 cents
Literacy rate 95%
Calorie consumption 2,458 kilocalories

Argentina
South America

Official name Republic of Argentina
Formation 1816 / 1816
Capital Buenos Aires
Population 36.1 million / 34 people per sq mile (13 people per sq km) / 87%
Total area 1,068,296 sq miles (2,766,890 sq km)
Languages Spanish*, Italian, English, German, French, Amerindian languages
Religions Roman Catholic 90%, Jewish 2%, other 8%
Ethnic mix European 85%, other (including *mestizo* and Indian) 15%
Government Multiparty republic
Currency Peso = 100 centavos
Literacy rate 96%
Calorie consumption 2,880 kilocalories

Armenia
Southwestern Asia

Official name Republic of Armenia
Formation 1991 / 1991
Capital Yerevan
Population 3.6 million / 313 people per sq mile (121 people per sq km) / 69%
Total area 11,505 sq miles (29,000 sq km)
Languages Armenian*, Azerbaijani, Russian, Kurdish
Religions Armenian Apostolic 90%, other Christian and Muslim 10%
Ethnic mix Armenian 93%, Azeri 3%, other 4%
Government Multiparty republic
Currency Dram = 100 louma
Literacy rate 99%
Calorie consumption NOT AVAILABLE

Australia
Australasia & Oceania

Official name Commonwealth of Australia
Formation 1901 / 1901
Capital Canberra
Population 18.5 million / 6 people per sq mile (2 people per sq km) / 85%
Total area 2,967,893 sq miles (7,686,850 sq km)
Languages English*, Greek, Italian, Vietnamese, Aboriginal languages
Religions Protestant 38%, Roman Catholic 26%, other 36%
Ethnic mix European 95%, Asian 4%, Aboriginal and other 1%
Government Parliamentary democracy
Currency Australian dollar = 100 cents
Literacy rate 99%
Calorie consumption 3,179 kilocalories

Austria
Central Europe

Official name Republic of Austria
Formation 1918 / 1919
Capital Vienna
Population 8.2 million / 257 people per sq mile (99 people per sq km) / 56%
Total area 32,375 sq miles (83,850 sq km)
Languages German*, Croat, Slovene
Religions Roman Catholic 78%, Protestant 5%, other (including Jewish and Muslim) 17%
Ethnic mix German 93%, Croat, Slovene, Hungarian 6%, other 1%
Government Multiparty republic
Currency Austrian Schilling = 100 groschen
Literacy rate 99%
Calorie consumption 3,497 kilocalories

Azerbaijan
Southwestern Asia

Official name Azerbaijani Republic
Formation 1991 / 1991
Capital Baku
Population 7.7 million / 230 people per sq mile (89 people per sq km) / 56%
Total area 33,436 sq miles (86,600 sq km)
Languages Azerbaijani*, Russian, Armenian
Religions Muslim 83%, Armenian Apostolic and Russian Orthodox 17%
Ethnic mix Azeri 83%, Armenian 6%, Russian 5%, Daghestani 3%, other 3%
Government Multiparty republic
Currency Manat = 100 gopik
Literacy rate 96%
Calorie consumption NOT AVAILABLE

Bahamas
West Indies

Official name Commonwealth of the Bahamas
Formation 1973 / 1973
Capital Nassau
Population 293,000 / 76 people per sq mile (29 people per sq km)/ 87%
Total area 5,359 sq miles (13,880 sq km)
Languages English*, English Creole,
Religions Protestant 64%, Roman Catholic 19%, other 17%
Ethnic mix Black 85%, White 15%
Government Parliamentary democracy
Currency Bahamian dollar = 100 cents
Literacy rate 96%
Calorie consumption 2,624 kilocalories

Bahrain
Southwestern Asia

Official name State of Bahrain
Formation 1971 / 1971
Capital Manama
Population 594,000 / 2,262 people per sq mile (874 people per sq km) / 90%
Total area 263 sq miles (680 sq km)
Languages Arabic*, English, Urdu
Religions Muslim (Shi'a majority) 85%, Christian 7%, other 8%
Ethnic mix Bahraini 70%, Iranian, Indian, Pakistani 24%, other Arab 4%, European 2%
Government Absolute monarchy (emirate)
Currency Bahrain dinar = 1,000 fils
Literacy rate 86%
Calorie consumption NOT AVAILABLE

Bangladesh
Southern Asia

Official name People's Republic of Bangladesh
Formation 1971 / 1971
Capital Dhaka
Population 124 million / 2,400 people per sq mile (926 people per sq km) / 18%
Total area 55,598 sq miles (143,998 sq km)
Languages Bengali*, Urdu, Chakma, Marma, Garo, Khasi, Santhali, Tripuri, Mro
Religions Muslim 87%, Hindu 12%, other 1%
Ethnic mix Bengali 98%, other 2%
Government Multiparty republic
Currency Taka = 100 paisa
Literacy rate 40%
Calorie consumption 2,019 kilocalories

Barbados
West Indies

Official name Barbados
Formation 1966 / 1966
Capital Bridgetown
Population 263,000 / 1,584 people per sq mile (612 people per sq km) / 47%
Total area 166 sq miles (430 sq km)
Languages English*, English Creole
Religions Protestant 55%, Roman Catholic 4%, other 41%
Ethnic mix Black 80%, mixed 15%, White 4%, other 1%
Government Parliamentary democracy
Currency Barbados dollar = 100 cents
Literacy rate 97%
Calorie consumption 3,207 kilocalories

Belarus
Eastern Europe

Official name Republic of Belarus
Formation 1991 / 1991
Capital Minsk
Population 10.3 million / 129 people per sq mile (50 people per sq km) / 68%
Total area 80,154 sq miles (207,600 sq km)
Languages Belarusian*, Russian*
Religions Russian Orthodox 60%, Roman Catholic 8%, other 32%
Ethnic mix Belarusian 78%, Russian 13%, Polish 4%, other 5%
Government Multiparty republic
Currency Belarusian rouble = 100 kopeks
Literacy rate 99%
Calorie consumption NOT AVAILABLE

Belgium
Northwestern Europe

Official name Kingdom of Belgium
Formation 1830 / 1919
Capital Brussels
Population 10.2 million / 805 people per sq mile (311 people per sq km) / 97%
Total area 12,780 sq miles (33,100 sq km)
Languages Flemish*, French*, German*
Religions Roman Catholic 88%, other 12%
Ethnic mix Fleming 58%, Walloon 33%, other 9%
Government Constitutional monarchy
Currency Belgian franc = 100 centimes
Literacy rate 99%
Calorie consumption: 3,681 kilocalories

Belize
Central America

Official name Belize
Formation 1981 / 1981
Capital Belmopan
Population 200,000 /23 people per sq mile (9 people per sq km) / 47%
Total area 8,865 sq miles (22,960 sq km)
Languages English*, English Creole, Spanish
Religions Christian 87%, other 13%
Ethnic mix *Mestizo* 44%, Creole 30%, Maya 11%, Asian Indian 4%, Garifuna 7%, other 4%
Government Parliamentary democracy
Currency Belizean dollar =100 cents
Literacy rate 75%
Calorie consumption 2,662 kilocalories

Benin
Western Africa

Official name Republic of Benin
Formation 1960 / 1960
Capital Porto-Novo
Population 5.9 million / 138 people per sq mile (53 people per sq km) / 31%
Total area 43,480 sq miles (112,620 sq km)
Languages French*, Fon, Bariba, Yoruba, Adja
Religions Traditional beliefs 70%, Muslim 15%, Christian 15%
Ethnic mix Fon 39%, Yoruba 12%, Adja 10%, other 39%
Government Multiparty republic
Currency CFA franc = 100 centimes
Literacy rate 34%
Calorie consumption 2,532 kilocalories

Bhutan
Southeastern Asia

Official name Kingdom of Bhutan
Formation 1656 / 1865
Capital Thimphu
Population 1.9 million / 105 people per sq mile (40 people per sq km) / 6%
Total area 18,147 sq miles (47,000 sq km)
Languages Dzongkha*, Nepali, Assamese
Religions Mahayana Buddhism 70%, Hindu 24%, Muslim 5%, other 1%
Ethnic mix Bhutia 61%, Gurung 15%, Assamese 13%, other 11%
Government Constitutional monarchy
Currency Ngultrum = 100 chetrum
Literacy rate 44%
Calorie consumption 2,553 kilocalories

Bolivia
South America

Official name Republic of Bolivia
Formation 1825 / 1938
Capitals Sucre (official)/La Paz (administrative)
Population 8 million / 19 people per sq mile (7 people per sq km) / 61%
Total area 424,162 sq miles (1,098,580 sq km)
Languages Spanish*, Quechua*, Aymará*
Religions Roman Catholic 93%, other 7%
Ethnic mix Quechua 37%, Aymará 32%, mixed 13%, European 10%, other 8%
Government Multiparty republic
Currency Boliviano = 100 centavos
Literacy rate 84%
Calorie consumption 2,094 kilocalories

Bosnia & Herzegovina
Southeastern Europe

Official name Republic of Bosnia and Herzegovina
Formation 1992 / 1992
Capital Sarajevo
Population 4 million / 203 people per sq mile (78 people per sq km) / 49%
Total area 19,741 sq miles (51,130 sq km)
Languages Serbian*, Croatian*
Religions Muslim 40%, Serbian Orthodox 31%, Roman Catholic 15%, other 14%
Ethnic mix Bosnian 44%, Serb 31%, Croat 17%, other 8%
Government Multiparty republic
Currency Maraka = 100 pfenniga
Literacy rate 93%
Calorie consumption NOT AVAILABLE

Botswana
Southern Africa

Official name Republic of Botswana
Formation 1966 / 1966
Capital Gaborone
Population 1.6 million / 7 people per sq mile (3 people per sq km) / 28%
Total area 224,600 sq miles (581,730 sq km)
Languages English*, Tswana, Shona, San
Religions Traditional beliefs 50%, Christian 50%
Ethnic mix Tswana 98%, other 2%
Government Multiparty republic
Currency Pula = 100 thebe
Literacy rate 74%
Calorie consumption 2,266 kilocalories

Brazil
South America

Official name Federative Republic of Brazil
Formation 1822 / 1889
Capital Brasília
Population 165.2 million / 51 people per sq mile (20 people per sq km) / 78%
Total area 3,286,472 sq miles (8,511,970 sq km)
Languages Portuguese*, German, Italian
Religions Roman Catholic 89%, other 11%
Ethnic mix White (Portuguese, Italian, German, Japanese) 66%, mixed 22%, Black 12%
Government Multiparty republic
Currency Real = 100 centavos
Literacy rate 84%
Calorie consumption 2,824 kilocalories

Brunei
Southeastern Asia

Official name Sultanate of Brunei
Formation 1984 / 1984
Capital Bandar Seri Begawan
Population 313,000 / 154 people per sq mile (57 people per sq km) / 59%
Total area 2,228 sq miles (5,770 sq km)
Languages Malay*, English, Chinese
Religions Muslim 63%, Buddhist 14%, Christian 10%, other 13%
Ethnic mix Malay 67%, Chinese 16%, other 17%
Government Absolute monarchy
Currency Brunei dollar = 100 cents
Literacy rate 90%
Calorie consumption 2,745 kilocalories

Bulgaria
Southeastern Europe

Official name Republic of Bulgaria
Formation 1908 / 1947
Capital Sofia
Population 8.4 million / 197 people per sq mile (76 people per sq km) / 71%
Total area 42,822 sq miles (110,910 sq km)
Languages Bulgarian*, Turkish, Macedonian, Romany, Armenian, Russian
Religions Christian 85%, Muslim 13%, Jewish 1%, other 1%
Ethnic mix Bulgarian 85%, Turkish 9%, Macedonian 3%, Romany 3%
Government Multiparty republic
Currency Lev = 100 stoninki
Literacy rate 98%
Calorie consumption 2,831 kilocalories

Burkina
Western Africa

Official name Burkina Faso
Formation 1960 / 1960
Capital Ouagadougou
Population 11.4 million / 108 people per sq mile (42 people per sq km) / 27%
Total area 105,870 sq miles (274,200 sq km)
Languages French*, Mossi, Fulani
Religions Indigenous beliefs 55%, Muslim 35%, Christian 10%
Ethnic mix Mossi 45%, Mande 10%, Fulani 10%, other 35%
Government Multiparty republic
Currency CFA franc = 100 centimes
Literacy rate 21%
Calorie consumption 2,387 kilocalories

Burundi
Central Africa

Official name Republic of Burundi
Formation 1962 / 1962
Capital Bujumbura
Population 6.6 million / 666 people per sq mile (257 people per sq km) / 7%
Total area 10,750 sq miles (27,830 sq km)
Languages Kirundi*, French*, Swahili
Religions Christian 68%, Traditional beliefs 32%
Ethnic mix Hutu 85%, Tutsi 14%, Twa pygmy 1%
Government Multiparty republic
Currency Burundi franc = 100 centimes
Literacy rate 45%
Calorie consumption 1,941 kilocalories

Cambodia
Southeastern Asia

Official name Kingdom of Cambodia
Formation 1953 / 1953
Capital Phnom Penh
Population 10.8 million / 158 people per sq mile (61 people per sq km) /21%
Total area 69,000 sq miles (181,040 sq km)
Languages Khmer*, French, Chinese, Vietnamese
Religions Theravada Buddhist 88%, Muslim 2%, other 10%
Ethnic mix Khmer 94%, Chinese 4%, other 2%
Government Constitutional monarchy
Currency Riel = 100 sen
Literacy rate 66%
Calorie consumption 2,021 kilocalories

Cameroon
Central Africa

Official name Republic of Cameroon
Formation 1960 / 1961
Capital Yaoundé
Population 14.3 million / 80 people per sq mile (31 people per sq km) / 45%
Total area 183,570 miles (475,440 sq km)
Languages English*, French*, Fang, Bulu, Yaundé, Duala
Religions Traditional beliefs 25%, Christian 53%, Muslim 22%
Ethnic mix Bamileke and Manum 20%, Fang 19%, other 61%
Government Multiparty republic
Currency CFA franc = 100 centimes
Literacy rate 72%
Calorie consumption 1,981 kilocalories

Canada
North America

Official name Canada
Formation 1867 / 1949
Capital Ottawa
Population 30.2 million / 8 people per sq mile (3 people per sq km) / 77%
Total area 3,851,788 sq miles (9,976,140 sq km)
Languages English*, French*, Chinese, Italian, German, Portuguese, Inuit
Religions Roman Catholic 47%, Protestant 41%, other 12%
Ethnic mix British origin 44%, French origin 25%, other European 20%, other 11%
Government Parliamentary democracy
Currency Canadian dollar = 100 cents
Literacy rate 99%
Calorie consumption 3,094 kilocalories

Cape Verde
Atlantic Ocean

Official Name Republic of Cape Verde
Formation 1975 / 1975
Capital Praia
Population 417,000 / 268 people per sq mile (103 people per sq km) / 54%
Total area 1,556 sq miles (4,030 sq km)
Languages Portuguese*, Creole
Religions Roman Catholic 98%, Protestant 2%
Ethnic mix Mestico 60%, African 30%, other 10%
Government Multiparty republic
Currency Cape Verde escudo = 100 centavos
Literacy rate 71%
Calorie consumption 2,805 kilocalories

Central African Republic
Central Africa

Official name Central African Republic
Formation 1960 / 1960
Capital Bangui
Population 3.5 million / 15 people per sq mile (6 people per sq km) / 39%
Total area 240,530 sq miles (622,980 sq km)
Languages French*, Sango, Banda, Gbaya
Religions Christian 50%, traditional beliefs 27%, Muslim 15%, other 8%
Ethnic mix Baya 34%, Banda 27%, Mandjia 21%, other 18%
Government Multiparty republic
Currency CFA franc = 100 centimes
Literacy rate 42%
Calorie consumption 1,690 kilocalories

Chad
Central Africa

Official name Republic of Chad
Formation 1960 / 1960
Capital N'Djamena
Population 6.9 million / 14 people per sq mile (5 people per sq km) / 21%
Total area 495,752 sq miles (1,284,000 sq km)
Languages French*, Sara, Maba
Religions Muslim 50%, Traditional beliefs 43%, Christian 7%
Ethnic mix Bagirmi, Sara and Kreish 31%, Sudanic Arab 26%, Teda 7%, other 36%
Government Multiparty republic
Currency CFA franc = 100 centimes
Literacy rate 50%
Calorie consumption 1,989 kilocalories

Chile
South America

Official name Republic of Chile
Formation 1818 / 1883
Capital Santiago
Population 14.8 million / 51 people per sq mile (20 people per sq km) / 84%
Total area 292,258 sq miles (756,950 sq km)
Languages Spanish*, Indian languages
Religions Roman Catholic 80%, Protestant and other 20%
Ethnic mix *Mestizo* and European 90%, Indian 10%
Government Multiparty republic
Currency Chilean peso = 100 centavos
Literacy rate 95%
Calorie consumption 2,582 kilocalories

China
Eastern Asia

Official name People's Republic of China
Formation 1949 / 1999
Capital Beijing
Population 1.3 billion / 349 people per sq mile (135 people per sq km) / 30%
Total area 3,628,166 sq miles (9,396,960 sq km)
Languages Mandarin*, Wu, Cantonese, Hsiang, Min, Hakka, Kan
Religions Nonreligious 59%, Traditional beliefs 20%, Buddhist 6%, other 15%
Ethnic mix Han 93%, Zhaung 1%, other 6%
Government Single-party republic
Currency Yuan = 10 jiao = 100 fen
Literacy rate 84%
Calorie consumption 2,727 kilocalories

Colombia
South America

Official name Republic of Colombia
Formation 1819 / 1922
Capital Bogotá
Population 37.7 million / 94 people per sq mile (36 people per sq km) / 73%
Total area 439,733 sq miles (1,138,910 sq km)
Languages Spanish*, Amerindian languages, English Creole
Religions Roman Catholic 95%, other 5%
Ethnic mix *Mestizo* 58%, White 20%, European-African 14%, other 8%
Government Multiparty republic
Currency Colombian peso = 100 centavos
Literacy rate 91%
Calorie consumption 2,677 kilocalories

Comoros
Indian Ocean

Official name Federal Islamic Republic of the Comoros
Formation 1975 / 1975
Capital Moroni
Population 672,000 / 780 people per sq mile (301 people per sq km) / 29%
Total area 861 sq miles (2,230 sq km)
Languages Arabic*, French*, Comoran
Religions Muslim 98%, Roman Catholic 1%, other 1%
Ethnic mix Comorian 96%, other 4%
Government Islamic republic
Currency Comoros franc = 100 centimes
Literacy rate 55%
Calorie consumption 1,897 kilocalories

Congo
Central Africa

Official name Republic of the Congo
Formation 1960 / 1960
Capital Brazzaville
Population 2.8 million / 21 people per sq km (8 people per sq km) / 59%
Total area 132,040 sq miles (342,000 sq km)
Languages French*, Kongo, Teke, Lingala
Religions Roman Catholic 50%, Traditional beliefs 48%, other 2%
Ethnic mix Bakongo 48%, Sangha 20%, Teke 17%, Mbochi 12%, other 3%
Government Multiparty republic
Currency CFA franc = 100 centimes
Literacy rate 77%
Calorie consumption 2,296 kilocalories

Congo, Dem. Rep. (Zaire)
Central Africa

Official name Democratic Republic of the Congo
Formation 1960 / 1960
Capital Kinshasa
Population 49.2 million / 56 people per sq mile (22 people per sq km) / 29%
Total area 905,563 sq miles (2,345,410 sq km)
Languages French*, Kiswahili, Tshiluba, Lingala
Religions Traditional beliefs 50%, Roman Catholic 37%, Protestant 13%
Ethnic mix Bantu 23%, Hamitic 23%, other 54%
Government Single-party republic
Currency Congolese franc = 100 centimes
Literacy rate 77%
Calorie consumption 2,060 kilocalories

Costa Rica
Central America

Official name Republic of Costa Rica
Formation 1821 / 1838
Capital San José
Population 3.7 million / 188 people per sq mile (72 people per sq km) / 50%
Total area 19,730 miles (51,100 sq km)
Languages Spanish*, English Creole, Bribri, Cabecar
Religions Roman Catholic 76%, other 24%
Ethnic mix White and *mestizo* 96%, Black 2%, Indian 2%
Government Multiparty republic
Currency Costa Rica colón = 100 centimos
Literacy rate 95%
Calorie consumption 2,883 kilocalories

Croatia
Southeastern Europe

Official name Republic of Croatia
Formation 1991 / 1991
Capital Zagreb
Population 4.5 million / 206 people per sq mile (80 people per sq km) / 64%
Total area 21,830 sq miles (56,540 sq km)
Languages Croatian*, Serbian, Hungarian (Magyar), Slovenian
Religions Roman Catholic 76%, Eastern Orthodox 11%, Protestant 1%, Muslim 1%, other 10%
Ethnic mix Croat 80%, Serb 12%, Hungarian, Slovenian, other 8%
Government Multiparty republic
Currency Kuna = 100 lipa
Literacy rate 98%
Calorie consumption NOT AVAILABLE

Cuba
West Indies

Official name Republic of Cuba
Formation 1902 / 1898
Capital Havana
Population 11.1 million / 259 people per sq mile (100 people per sq km) / 76%
Total area 42,803 sq miles (110,860 sq km)
Languages Spanish*, English, French
Religions Nonreligious **55%**, Roman Catholic 40%, other 5%
Ethnic mix White 66%, European-African 22%, Black 12%
Government Socialist republic
Currency Cuban peso = 100 centavos
Literacy rate 96%
Calorie consumption 2,833 kilocalories

Cyprus
Southeastern Europe

Official name Republic of Cyprus
Formation 1960 / 1983
Capital Nicosia
Population 766,000 / 218 people per sq mile (84 people per sq km) / 54%
Total area 3,572 sq miles (9,251 sq km)
Languages Greek*, Turkish, English
Religions Greek Orthodox 77%, Muslim 18%, other 5%
Ethnic mix Greek 77%, Turkish 18%, other (mainly British) 5%
Government Multiparty republic
Currency Cyprus pound / Turkish lira
Literacy rate 96%
Calorie consumption 3,779 kilocalories

Czech Republic
Central Europe

Official name Czech Republic
Formation 1993 / 1993
Capital Prague
Population 10.2 million / 335 people per sq mile (129 people per sq km) / 65%
Total area 30,260 sq miles (78,370 sq km)
Languages Czech*, Slovak, Romany, Hungarian (Magyar)
Religions Roman Catholic 39%, nonreligious 40%, other 21%
Ethnic mix Czech 85%, Moravian 13%, other 2%
Government Multiparty republic
Currency Czech koruna = 100 halura
Literacy rate 99%
Calorie consumption 3,156 kilocalories

Denmark
Northern Europe

Official name Kingdom of Denmark
Formation AD 960 / 1920
Capital Copenhagen
Population 5.3 million / 324 people per sq mile (125 people per sq km) / 85%
Total area 16,629 sq miles (43,069 sq km)
Languages Danish*, Faeroese, Inuit
Religions Evangelical Lutheran 89% other Christian 11%
Ethnic mix Danish 96%, Faeroese & Inuit 1%, other 3%
Government Constitutional monarchy
Currency Danish krone = 100 øre
Literacy rate 100%
Calorie consumption 3,664 kilocalories

Djibouti
Eastern Africa

Official name Republic of Djibouti
Formation 1977 / 1977
Capital Djibouti
Population 652,000 / 73 people per sq mile (28 people per sq km) /83%
Total area 8,958 sq miles (23,200 sq km)
Languages Arabic*, French*, Somali, Afar
Religions Christian 87%, other 13%
Ethnic mix Issa 60%, Afar 35%, other 5%
Government Multiparty republic
Currency Djibouti franc = 100 centimes
Literacy rate 48%
Calorie consumption 2,338 kilocalories

Dominica
West Indies

Official name Commonwealth of Dominica
Formation 1978 / 1978
Capital Roseau
Population 74,000 / 256 people per sq mile (99 people per sq km)/ 69%
Total area 290 sq miles (750 sq km)
Languages English*, French Creole, Carib, Cocoy
Religions Roman Catholic 77%, Protestant 15%, other 8%
Ethnic mix Black 98%, Indian 2%
Government Multiparty republic
Currency East Caribbean dollar = 100 cents
Literacy rate 94%
Calorie consumption 2,778 kilocalories

Dominican Republic
West Indies

Official name Dominican Republic
Formation 1865 / 1865
Capital Santo Domingo
Population 8.2 million / 439 people per sq mile (169 people per sq km) / 65%
Total area 18,815 sq miles (48,730 sq km)
Languages Spanish*, French Creole
Religions Roman Catholic 92%, other 8%
Ethnic mix European-African 73%, White 16%, Black 11%
Government Multiparty republic
Currency Dom. Republic peso = 100 centavos
Literacy rate 83%
Calorie consumption 2,286 kilocalories

Ecuador
South America

Official name Republic of Ecuador
Formation 1830 / 1941
Capital Quito
Population 12.2 million / 114 people per sq mile (44 people per sq km) / 58%
Total area 109,483 sq miles (283,560 sq km)
Languages Spanish*, Quechua, other Amerindian languages
Religions Roman Catholic 95%, other 5%
Ethnic mix *Mestizo* 55%, Indian 25%, Black 10%, White 10%
Government Multiparty republic
Currency Sucre = 100 centavos
Literacy rate 91%
Calorie consumption 2,583 kilocalories

Egypt
Northern Africa

Official name Arab Republic of Egypt
Formation 1936 / 1982
Capital Cairo
Population 65.7 million / 171 people per sq mile (66 people per sq km) / 45%
Total area 386,660 sq miles (1,001,450 sq km)
Languages Arabic*, French, English, Berber, Greek, Armenian
Religions Muslim 94%, other 6%
Ethnic mix Eastern Hamitic 90%, other (including Greek, Armenian) 10%
Government Multiparty republic
Currency Egyptian pound = 100 piastres
Literacy rate 53%
Calorie consumption 3,335 kilocalories

El Salvador
Central America

Official name Republic of El Salvador
Formation 1856 / 1838
Capital San Salvador
Population 6.1 million / 763 people per sq mile (294 people per sq km) /45%
Total area 8,124 sq miles (21,040 sq km)
Languages Spanish*, Nahua
Religions Roman Catholic 80%, other 20%
Ethnic mix *Mestizo* 89%, Indian 10%, White 1%
Government Multiparty republic
Currency Salvadorean colón = 100 centavos
Literacy rate 73%
Calorie consumption 2,663 kilocalories

Equatorial Guinea
Central Africa

Official name Republic of Equatorial Guinea
Formation 1968 / 1968
Capital Malabo
Population 430,000 / 40 people per sq mile (15 people per sq km) / 42%
Total area 10,830 sq miles (28,050 sq km)
Languages Spanish*, Fang, Bubi
Religions Christian 90%, other 10%
Ethnic mix Fang 72%, Bubi 14%, Duala 3%, other 11%
Government Multiparty republic
Currency CFA franc = 100 centimes
Literacy rate 80%
Calorie consumption NOT AVAILABLE

Eritrea
Eastern Africa

Official name State of Eritrea
Formation 1993 / 1993
Capital Asmara
Population 3.5 million / 97 people per sq mile (37 people per sq km) / 17%
Total area 36,170 sq miles (93,680 sq km)
Languages Tigrinya*, Arabic*, Tigre
Religions Christian 45%, Muslim 45%, other 10%
Ethnic mix Nine main ethnic groups
Government Provisional military government
Currency Nakfa = 100 cents
Literacy rate 25%
Calorie consumption 1,610 kilocalories

Estonia
Northeastern Europe

Official name Republic of Estonia
Formation 1991 / 1991
Capital Tallinn
Population 1.4 million / 80 people per sq mile (31 people per sq km) / 73%
Total area 17,423 sq miles (45,125 sq km)
Languages Estonian*, Russian
Religions Evangelical Lutheran 98%, Eastern Orthodox, Baptist 2%
Ethnic mix Russian 62%, Estonian 30%, Ukrainian 3%, other 5%
Government Multiparty republic
Currency Kroon = 100 cents
Literacy rate 99%
Calorie consumption NOT AVAILABLE

Ethiopia
Eastern Africa

Official name Federal Democratic Republic of Ethiopia
Formation 1896 / 1993
Capital Addis Ababa
Population 62.1 million / 146 people per sq mile (56 people per sq km) / 13%
Total area 435,605 sq miles (1,128,221 sq km)
Languages Amharic*, English, Arabic
Religions Muslim 40%, Christian 40%, Traditional beliefs 15%, other 5%
Ethnic mix Oromo 40%, Amhara 25%, Sidamo 9%, Somali 6%, other 20%
Government Multiparty republic
Currency Ethiopian birr = 100 cents
Literacy rate 35%
Calorie consumption 1,610 kilocalories

Fiji
Australasia & Oceania

Official name Sovereign Democratic Republic of Fiji
Formation 1970 / 1970
Capital Suva
Population 822,000 / 117 people per sq mile (45 people per sq km) / 40%
Total area 7,054 sq miles (18,270 sq km)
Languages English*, Fijian, Hindu, Urdu
Religions Christian 46%, Hindu 38%, Muslim 8%, other 8%
Ethnic mix Native Fijian 49%, Indo-Fijian 46%, other 5%
Government Multiparty republic
Currency Fiji dollar = 100 cents
Literacy rate 92%
Calorie consumption 3,089 kilocalories

Finland
Northern Europe

Official name Republic of Finland
Formation 1917 / 1947
Capital Helsinki
Population 5.2 million / 44 people per sq mile (17 people per sq km) / 63%
Total area 130,552 sq miles (338,130 sq km)
Languages Finnish*, Swedish*, Lappish
Religions Evangelical Lutheran 89%, Finnish Orthodox 1%, other 10%
Ethnic mix Finnish 93%, Swedish 6%, other (including Sami) 1%
Government Multiparty republic
Currency Markka = 100 pennia
Literacy rate 99%
Calorie consumption 3,018 kilocalories

France
Western Europe

Official name French Republic
Formation 486 / 1919
Capital Paris
Population 58.7 million / 276 people per sq mile (107 people per sq km) / 73%
Total area 212,930 sq miles (551,500 sq km)
Languages French*, Provençal, Breton, Catalan, Basque
Religions Roman Catholic 88%, Muslim 8%, other 4%
Ethnic mix French 90%, North African 6%, German 2%, Breton 1%, other 1%
Government Multiparty republic
Currency Franc = 100 centimes
Literacy rate 99%
Calorie consumption 3,633 kilocalories

Gabon
Central Africa

Official name Gabonese Republic
Formation 1960 / 1960
Capital Libreville
Population 1.2 million / 12 people per sq mile (5 people per sq km) / 50%
Total area 103,347 sq miles (267,670 sq km)
Languages French*, Fang, Punu, Sira, Nzebi, Mpongwe
Religions Christian 96%, Muslim 2%, other 2%
Ethnic mix Fang 3%, Eshira 25%, other Bantu 25%, European and other African 9%
Government Multiparty republic
Currency CFA franc = 100 centimes
Literacy rate 66%
Calorie consumption 2,500 kilocalories

Gambia
Western Africa

Official name Republic of the Gambia
Formation 1965 / 1965
Capital Banjul
Population 1.9 million / 309 people per sq mile (119 people per sq km) / 26%
Total area 4,363 sq miles (11,300 sq km)
Languages English*, Mandinka, Fulani, Wolof, Diola, Soninke
Religions Muslim 85%, Christian 9%, Traditional beliefs 6%
Ethnic mix Mandingo 42%, Fulani 18% Wolof 16%, Jola 10%, Serahull 9%, other 5%
Government Multiparty republic
Currency Dalasi = 100 butut
Literacy rate 33%
Calorie consumption 2,360 kilocalories

Georgia
Southwestern Asia

Official name Republic of Georgia
Formation 1991 / 1991
Capital Tbilisi
Population 5.4 million / 201 people per sq mile (77 people per sq km) / 58%
Total area 26,911 sq miles (69,700 sq km)
Languages Georgian*, Russian
Religions Georgian Orthodox 70%, Russian Orthodox 10%, other 20%
Ethnic mix Georgian 70%, Armenian 8%, Russian 6%, Azeri 6%, other 10%
Government Multiparty republic
Currency Lari = 100 tetri
Literacy rate 99%
Calorie consumption NOT AVAILABLE

Germany
Northern Europe

Official name Federal Republic of Germany
Formation 1871 / 1990
Capital Berlin
Population 82.4 million / 611 people per sq mile (236 people per sq km) / 87%
Total area 137,800 sq miles (356,910 sq km)
Languages German*, Sorbian, Turkish
Religions Protestant 36%, Roman Catholic 35%, Muslim 2%, other 27%
Ethnic mix German 92%, other 8%
Government Multiparty republic
Currency Deutsche Mark = 100 pfennigs
Literacy rate 99%
Calorie consumption 3,344 kilocalories

Ghana
Western Africa

Official name Republic of Ghana
Formation 1957 / 1957
Capital Accra
Population 18.9 million / 213 people per sq mile (82 people per sq km) / 36%
Total area 92,100 sq miles (238,540 sq km)
Languages English*, Akan, Mossi, Ewe
Religions Traditional beliefs 38%, Christian 43%, Muslim 11%, other 8%
Ethnic mix Akan 52%, Mossi 15%, Ewe 12%, Ga 8%, other 13%
Government Multiparty republic
Currency Cedi = 100 pesewas
Literacy rate 66%
Calorie consumption 2,199 kilocalories

Greece
Southeastern Europe

Official name Hellenic Republic
Formation 1829 / 1947
Capital Athens
Population 10.6 million / 210 people per sq mile (81 people per sq km) / 65%
Total area 50,961 sq miles (131,990 sq km)
Languages Greek*, Turkish, Albanian, Macedonian
Religions Greek Orthodox 98%, Muslim 1%, other 1%
Ethnic mix Greek 98%, other 2%
Government Multiparty republic
Currency Drachma = 100 lepta
Literacy rate 96%
Calorie consumption 3,815 kilocalories

Grenada
West Indies

Official name Grenada
Formation 1974 / 1974
Capital St George's
Population 98,600 / 751 people per sq mile (290 people per sq km) / 37%
Total area 131 sq miles (340 sq km)
Languages English*, English Creole
Religions Roman Catholic 68%, Anglican 17%, other 15%
Ethnic mix Black 84%, European-African 13%, South Asian 3%
Government Parliamentary democracy
Currency East Caribbean dollar = 100 cents
Literacy rate 98%
Calorie consumption 2,402 kilocalories

Guatemala
Central America

Official name Republic of Guatemala
Formation 1838 / 1838
Capital Guatemala City
Population 11.6 million / 277 people per sq mile (107 people per sq km) / 41%
Total area 42,043 sq miles (108,890 sq km)
Languages Spanish*, Quiché, Mam, Kekchí
Religions Christian 99%, other 1%
Ethnic mix Indian 60%, *mestizo* 30%, other 10%
Government Multiparty republic
Currency Quetzal = 100 centavos
Literacy rate 66%
Calorie consumption 2,255 kilocalories

Guinea
Western Africa

Official name Republic of Guinea
Formation 1958 / 1958
Capital Conakry
Population 7.7 million / 81 people per sq mile (31 people per sq km) / 30%
Total area 94,926 sq miles (245,860 sq km)
Languages French*, Fulani, Malinke, Soussou, Kissi
Religions Muslim 85%, Christian 8%, Traditional beliefs 7%
Ethnic mix Fila (Fulani) 30%, Malinke 30%, Soussou 15%, Kissi 10% other 20%
Government Multiparty republic
Currency Franc = 100 centimes
Literacy rate 37%
Calorie consumption 2,389 kilocalories

Guinea-Bissau
Western Africa

Official name Republic of Guinea-Bissau
Formation 1974 / 1974
Capital Bissau
Population 1.1 million / 101 people per sq mile (39 people per sq km) / 22%
Total area 13,940 sq miles (36,120 sq km)
Languages Portuguese*, Balante, Fulani, Malinke
Religions Traditional beliefs 52%, Muslim 40%, Christian 8%
Ethnic mix Balante 30%, Fila (Fulani) 22%, Malinke 12%, other 36%
Government Multiparty republic
Currency Guinea peso = 100 centavos
Literacy rate 33%
Calorie consumption 2,556 kilocalories

Guyana
South America

Official name Cooperative Republic of Guyana
Formation 1966 / 1966
Capital Georgetown
Population 856,000 / 11 people per sq mile (4 people per sq km) / 36%
Total area 83,000 sq miles (214,970 sq km)
Languages English*, English Creole, Hindi, Tamil, English
Religions Christian 57%, Hindu 33%, Muslim 9%, other 1%
Ethnic mix East Indian 52%, Black African 38%, Indian 4%, other 6%
Government Multiparty republic
Currency Guyana dollar =100 cents
Literacy rate 98%
Calorie consumption 2,384 kilocalories

Haiti
West Indies

Official name Republic of Haiti
Formation 1804 / 1844
Capital Port-au-Prince
Population 7.5 million / 705 people per sq mile (272 people per sq km) / 32%
Total area 10,714 sq miles (27,750 sq km)
Languages French*, French Creole*,
Religions Roman Catholic 80%, Protestant 16%, other 4%
Ethnic mix Black 95%, European-African 5%
Government Multiparty republic
Currency Gourde = 100 centimes
Literacy rate 45%
Calorie consumption 1,706 kilocalories

Honduras
Central America

Official name Republic of Honduras
Formation 1838 / 1838
Capital Tegucigalpa
Population 6.1 million / 141 people per sq mile (55 people per sq km) / 44%
Total area 43,278 sq miles (112,090 sq km)
Languages Spanish*, English Creole, Garifuna, Indian languages
Religions Roman Catholic 97%, other 3%
Ethnic mix *Mestizo* 90%, Black African 5%, Indian 4%, White 1%
Government Multiparty republic
Currency Lempira = 100 centavos
Literacy rate 70%
Calorie consumption 2,305 kilocalories

Hungary
Central Europe

Official name Republic of Hungary
Formation 1918 / 1947
Capital Budapest
Population 9.9 million / 278 people per sq mile (107 people per sq km) / 65%
Total area 35,919 sq miles (93,030 sq km)
Languages Hungarian (Magyar)*, German, Slovak
Religions Roman Catholic 64%, Protestant 27%, other 7%
Ethnic mix Hungarian (Magyar) 90%, German 2%, other 8%
Government Multiparty republic
Currency Forint = 100 filler
Literacy rate 99%
Calorie consumption 3,503 kilocalories

Iceland
Northwestern Europe

Official name Republic of Iceland
Formation 1944 / 1944
Capital Reykjavík
Population 277,000 / 7 people per sq mile (3 people per sq km) / 92%
Total area 39,770 sq miles (103,000 sq km)
Languages Icelandic*, English
Religions Evangelical Lutheran 93%, nonreligious 6%, other Christian 1%
Ethnic mix Icelandic (Norwegian-Celtic descent) 98%, other 2%
Government Constitutional republic
Currency New Icelandic króna = 100 aurar
Literacy rate 99%
Calorie consumption 3,058 kilocalories

India
Southern Asia

Official name Republic of India
Formation 1947 / 1947
Capital New Delhi
Population 976 million / 850 people per sq mile (328 people per sq km) / 27%
Total area 1,269,338 sq miles (3,287,590 sq km)
Languages Hindi*, English*, Urdu, Bengali, Marathi, Telugu, Tamil, Bihari
Religions Hindu 83%, Muslim 11%, Christian 2%, Sikh 2%, other 2%
Ethnic mix Indo-Aryan 72%, Dravidian 25%, Mongoloid and other 3%
Government Multiparty republic
Currency Rupee = 100 paisa
Literacy rate 53%
Calorie consumption 2,395 kilocalories

Indonesia
Southeastern Asia

Official name Republic of Indonesia
Formation 1949 / 1963
Capital Jakarta
Population 206.5 million / 295 people per sq mile (114 people per sq km) / 35%
Total area 735,555 sq miles (1,904,570 sq km)
Languages Bahasa Indonesia*, 250 (est.) languages or dialects
Religions Muslim 87%, Christian 9%, Hindu 2%, Buddhist 1%, other 1%
Ethnic mix Javanese 45%, Sundanese 14%, Madurese 8%, Coastal Malays 8%, other 25%
Government Multiparty republic
Currency Rupiah = 100 sen
Literacy rate 82%
Calorie consumption 2,752 kilocalories

Iran
Southwestern Asia

Official name Islamic Republic of Iran
Formation 1906 / 1906
Capital Tehran
Population 73.1 million / 116 people per sq mile (45 people per sq km) / 59%
Total area 636,293 sq miles (1,648,000 sq km)
Languages Farsi (Persian)*, Azerbaijani, Gilaki, Mazanderani, Kurdish, Baluchi, Arabic
Religions Shi'a Muslim 95%, Sunni Muslim 4%, other 1%
Ethnic mix Persian 50%, Azeri 20%, Lur and Bakhtiari 10%, Kurd 8%, Arab 2%, other 10%
Government Islamic Republic
Currency Iranian rial = 100 dinars
Literacy rate 73%
Calorie consumption 2,860 kilocalories

Iraq
Southwestern Asia

Official name Republic of Iraq
Formation 1932 / 1991
Capital Baghdad
Population 21.8 million / 129 people per sq mile (50 people per sq km) / 73%
Total area 169,235 sq miles (438,320 sq km)
Languages Arabic*, Kurdish, Armenian, Assyrian
Religions Shi'a ithna Muslim 62%, Sunni Muslim 33%, other 5%
Ethnic mix Arab 79%, Kurdish 16%, Persian 3%, Turkoman 2%
Government Single-party republic
Currency Iraqi dinar = 1000 fils
Literacy rate 58%
Calorie consumption 2,121 kilocalories

Ireland
Northwestern Europe

Official name Republic of Ireland
Formation 1922 / 1922
Capital Dublin
Population 3.6 million / 135 people per sq mile (52 people per sq km) / 57%
Total area 27,155 sq miles (70,280 sq km)
Languages English*, Irish Gaelic*
Religions Roman Catholic 88%, Protestant 3%, other 9%
Ethnic mix Irish 95%, other 5%
Government Multiparty republic
Currency Punt = 100 pence
Literacy rate 99%
Calorie consumption 3,847 kilocalories

Israel
Southwestern Asia

Official name State of Israel
Formation 1948 / 1994
Capital Jerusalem
Population 5.9 million / 752 people per sq mile (290 people per sq km) / 91%
Total area 7,992 sq miles (20,700 sq km)
Languages Hebrew*, Arabic*, Yiddish
Religions Jewish 82%, Muslim 14%, Christian 2%, Druze and other 2%
Ethnic mix Jewish 82%, Arab 18%
Government Multiparty republic
Currency New Israeli shekel = 100 agorat
Literacy rate 95%
Calorie consumption 3,050 kilocalories

Italy
Southern Europe

Official name Italian Republic
Formation 1871 / 1954
Capital Rome
Population 57.2 million / 504 people per sq mile (195 people per sq km) / 67%
Total area 116,320 sq miles (301,270 sq km)
Languages Italian*, German, French, Rhaeto-Romanic, Sardinian
Religions Roman Catholic 83%, other 17%
Ethnic mix Italian 94%, other 6%
Government Multiparty republic
Currency Lira = 100 centesimi
Literacy rate 98%
Calorie consumption 3,561 kilocalories

Ivory Coast
Western Africa

Official name Republic of the Ivory Coast
Formation 1960 / 1960
Capital Yamoussoukro
Population 14.6 million / 119 people per sq mile (45 people per sq km) / 42%
Total area 124,503 sq miles (322,463 sq km)
Languages French*, Akran, Kru, Voltaic
Religions Traditional beliefs 63%, Muslim 25%, Christian 12%
Ethnic mix Baoulé 23%, Bété 18%, Kru 17%, Malinke 15%, other 27%
Government Multiparty republic
Currency CFA franc = 100 centimes
Literacy rate 54%
Calorie consumption 2,491 kilocalories

Jamaica
West Indies

Official name Jamaica
Formation 1962 / 1962
Capital Kingston
Population 2.5 million / 598 people per sq mile (231 people per sq km) / 54%
Total area 4,243 sq miles (10,990 sq km)
Languages English*, English Creole
Religions Christian 55%, other 45%
Ethnic mix Black 75%, mixed 15%, South Asian 5%, other 5%
Government Parliamentary democracy
Currency Jamaican dollar = 100 cents
Literacy rate 85%
Calorie consumption 2,607 kilocalories

Japan
Eastern Asia

Official name Japan
Formation 1600 / 1972
Capital Tokyo
Population 125.9 million / 866 people per sq mile (334 people per sq km) / 78%
Total area 145,869 sq miles (377,800 sq km)
Languages Japanese*, Korean, Chinese
Religions Shinto and Buddhist 76%, Buddhist 16%, other 8%
Ethnic mix Japanese 99%, other 1%
Government Constitutional monarchy
Currency Yen = 100 sen
Literacy rate 99%
Calorie consumption 2,903 kilocalories

Jordan
Southwestern Asia

Official name Hashemite Kingdom of Jordan
Formation 1946 / 1976
Capital Amman
Population 6 million / 175 people per sq mile (67 people per sq km) / 71%
Total area 34,440 sq miles (89,210 sq km)
Languages Arabic*
Religions Muslim 95%, Christian 5%
Ethnic mix Arab 98%, (40% Palestinian), Armenian 1%, Circassian 1%
Government Constitutional monarchy
Currency Jordanian dinar = 1,000 fils
Literacy rate 87%
Calorie consumption 3,022 kilocalories

Kazakhstan
Central Asia

Official Name Republic of Kazakhstan
Formation 1991 / 1991
Capital Astana
Population 16.9 million / 16 people per sq mile (6 people per sq km) / 60%
Total area 1,049,150 sq miles (2,717,300 sq km)
Languages Kazakh*, Russian, German
Religions Muslim 47%, other 53% (mostly Russian Orthodox and Lutheran)
Ethnic mix Kazakh 44%, Russian 38%, Ukrainian 6%, German 2%, other 14%
Government Multiparty republic
Currency Tenge = 100 tein
Literacy rate 99%
Calorie consumption NOT AVAILABLE

Kenya
East Africa

Official name Republic of Kenya
Formation 1963 / 1963
Capital Nairobi
Population 29 million / 132 people per sq mile (51 people per sq km) / 28%
Total area 224,081 sq miles (580,370 sq km)
Languages Swahili*, English, Kikuyu, Luo, Kamba
Religions Christian 60%, Traditional beliefs 25%, Muslim 6%, other 9%
Ethnic mix Kikuyu 21%, Luhya 14%, Luo 13%, Kalenjin 11% other 41%
Government Multiparty republic
Currency Kenya shilling = 100 cents
Literacy rate 79%
Calorie consumption 2,075 kilocalories

Kiribati
Australasia & Oceania

Official Name Republic of Kiribati
Formation 1979 / 1979
Capital Bairiki
Population 78,000 / 284 people per sq mile (110 people per sq km) / 36%
Total area 274 sq miles (710 sq km)
Languages English*, Kiribati
Religions Roman Catholic 53%, Protestant 39%, other 8%
Ethnic mix I-Kiribati 98%, other 2%
Government Multiparty republic
Currency Australian dollar = 100 cents
Literacy rate 98%
Calorie consumption 2,651 kilocalories

Kuwait
Southwestern Asia

Official name State of Kuwait
Formation 1961 / 1961
Capital Kuwait
Population 1.8 million / 262 people per sq mile (101 people per sq km) / 97%
Total area 6,880 sq miles (17,820 sq km)
Languages Arabic*, English
Religions Muslim 92%, Christian 6%, other 2%
Ethnic mix Kuwaiti 45%, other Arab 35%, South Asian 9%, Iranian 4%, other 7%
Government Constitutional monarchy
Currency Dinar = 1,000 fils
Literacy rate 80%
Calorie consumption 2,523 kilocalories

Kyrgyzstan
Central Asia

Official name Kyrgyz Republic
Formation 1991 / 1991
Capital Bishkek
Population 4.5 million / 59 people per sq mile (23 people per sq km) / 39%
Total area 76,640 sq miles (198,500 sq km)
Languages Kyrgyz*, Russian*, Uzbek
Religions Muslim 65%, other (mostly Russian Orthodox) 35%
Ethnic mix Kyrgyz 57%, Russian 19%, Uzbek 13%, Tatar, Ukrainian, other 11%
Government Multiparty republic
Currency Som =100 teen
Literacy rate 97%
Calorie consumption NOT AVAILABLE

Laos
Southeastern Asia

Official name Lao People's Democratic Republic
Formation 1953 / 1953
Capital Vientiane
Population 5.4 million / 61 people per sq mile (23 people per sq km) / 22%
Total area 91,428 sq miles (236,800 sq km)
Languages Lao*, Miao, Yao
Religions Buddhist 85%, other (including Traditional beliefs) 15%
Ethnic mix Lao Loum 56%, Lao Theung 34%, Lao Soung 9%, other 1%
Government Single-party republic
Currency New kip = 100 cents
Literacy rate 58%
Calorie consumption 2,259 kilocalories

Latvia
Northeastern Europe

Official name Republic of Latvia
Formation 1991 / 1991
Capital Riga
Population 2.4 million / 96 people per sq mile (37 people per sq km) / 73%
Total area 24,938 sq miles (64,589 sq km)
Languages Latvian*, Russian
Religions Evangelical Lutheran 85%, other Christian 15%
Ethnic mix Latvian 52%, Russian 34%, Belarusian 5%, Ukrainian 4%, other 5%
Government Multiparty republic
Currency Lats = 100 santimi
Literacy rate 99%
Calorie consumption NOT AVAILABLE

Lebanon
Southwestern Asia

Official name Republic of Lebanon
Formation 1944 / 1944
Capital Beirut
Population 3.2 million / 810 people per sq mile (313 people per sq km) / 87%
Total area 4015 sq miles (10,400 sq km)
Languages Arabic*, French, Armenian,
Religions Muslim (mainly Shi'a) 70%, Christian (mainly Maronite) 30%
Ethnic mix Arab 93% (Lebanese 83%, Palestinian 10%), other 7%
Government Multiparty republic
Currency Lebanese pound = 100 piastres
Literacy rate 84%
Calorie consumption 3,317 kilocalories

Lesotho
Southern Africa

Official name Kingdom of Lesotho
Formation 1966 / 1966
Capital Maseru
Population 2.2 million / 188 people per sq mile (72 people per sq km) / 23%
Total area 11,718 sq miles (30,350 sq km)
Languages English*, Sesotho*, Zulu
Religions Christian 93%, other 7%
Ethnic mix Basotho 97%, European and Asian 3%
Government Constitutional monarchy
Currency Loti = 100 lisente
Literacy rate 82%
Calorie consumption 2,201 kilocalories

Liberia
Western Africa

Official name Republic of Liberia
Formation 1847 / 1947
Capital Monrovia
Population 2.7 million / 73 people per sq mile (28 people per sq km) / 45%
Total area 43,000 sq miles (111,370 sq km)
Languages English*, Kpelle, Bassa, Vai, Kru, Grebo, Kissi, Gola
Religions Traditional beliefs 70%, Muslim 20%, Christian 10%
Ethnic mix Indigenous tribes (16 main groups) 95%, Americo-Liberians 4%
Government Multiparty republic
Currency Liberian dollar = 100 cents
Literacy rate 38%
Calorie consumption 1,640 kilocalories

Libya
Northern Africa

Official name The Great Socialist People's Libyan Arab Jamahiriya
Formation 1951 / 1951
Capital Tripoli
Population 6 million / 9 people per sq mile (3 people per sq km) / 86%
Total area 679,358 sq miles (1,759,540 sq km)
Languages Arabic*, Tuareg
Religions Muslim (mainly Sunni) 97%, other 3%
Ethnic mix Arab and Berber 95%, other 5%
Government Single-party state
Currency Libyan dinar = 1,000 dirhams
Literacy rate 76%
Calorie consumption 3,308 kilocalories

Liechtenstein
Western Europe

Official name Principality of Liechtenstein
Formation 1719 / 1719
Capital Vaduz
Population 31,000 / 504 people per sq mile (195 people per sq km) / 87%
Total area 62 sq miles (160 sq km)
Languages German*, Alemannish, Italian
Religions Roman Catholic 81%, Protestant 7%, other 12%
Ethnic mix Liechtensteiner 63%, Swiss 15%, German 9%, other 13%
Government Constitutional monarchy
Currency Swiss franc = 100 centimes
Literacy rate 99%
Calorie consumption NOT AVAILABLE

Lithuania
Northeastern Europe

Official name Republic of Lithuania
Formation 1991 / 1991
Capital Vilnius
Population 3.7 million / 147 people per sq mile (57 people per sq km) / 72%
Total area 25,174 sq miles (65,200 sq km)
Languages Lithuanian*, Russian
Religions Roman Catholic 87%, Russian Orthodox 10%, other 3%
Ethnic mix Lithuanian 80%, Russian 9%, Polish 7%, Belarusian 2%, other 2%
Government Multiparty republic
Currency Litas = 100 centas
Literacy rate 98%
Calorie consumption NOT AVAILABLE

Luxembourg
Northwest Europe

Official name Grand Duchy of Luxembourg
Formation 1867 / 1867
Capital Luxembourg
Population 422,000 / 423 people per sq mile (163 people per sq km) / 89%
Total area 998 sq miles (2,586 sq km)
Languages Letzeburgish*, French*, German*
Religions Roman Catholic 97%, other 3%
Ethnic mix Luxemburger 72%, Portuguese 9%, Italian 5%, other 14%
Government Constitutional monarchy
Currency Franc = 100 centimes
Literacy rate 99%
Calorie consumption 3,681 kilocalories

Macedonia
Southeastern Europe

Official name Former Yugoslav Republic of Macedonia
Formation 1991 / 1991
Capital Skopje
Population 2.2 million / 222 people per sq mile (86 people per sq km) / 60%
Total area 9,929 sq miles (25,715 sq km)
Languages Macedonian*, Serbo-Croatian
Religions Christian 80%, Muslim 20%
Ethnic mix Macedonian 67%, Albanian 23%, Turkish 4%, Serb 2%, Romany 2%, other 2%
Government Multiparty republic
Currency Macedonian denar = 100 deni
Literacy rate 89%
Calorie consumption NOT AVAILABLE

Madagascar
Indian Ocean

Official name Democratic Republic of Madagascar
Formation 1960 / 1960
Capital Antananarivo
Population 16.3 million / 73 people per sq mile (28 people per sq km) / 27%
Total area 226,660 sq miles (587,040 sq km)
Languages Malagasy*, French*
Religions Traditional beliefs 52%, Christian 41%, Muslim 7%
Ethnic mix Merina 26%, Betsimisaraka 15%, Betsileo 12%, other 47%
Government Multiparty republic
Currency Franc = 100 centimes
Literacy rate 81%
Calorie consumption 2,135 kilocalories

Malawi
Southern Africa

Official name Republic of Malawi
Formation 1964 / 1964
Capital Lilongwe
Population 10.4 million / 286 people per sq mile (111 people per sq km) / 14%
Total area 45,745 sq miles (118,480 sq km)
Languages English*, Chewa, Lomwe, Yao
Religions Christian 75%, Muslim 20% traditional beliefs 5%
Ethnic mix Maravi 55%, Lomwe 17%, Yao 13%, other 15%
Government Multiparty republic
Currency Malawi kwacha = 100 tambala
Literacy rate 57%
Calorie consumption 1,825 kilocalories

Malaysia
Southeastern Asia

Official name Malaysia
Formation 1963 / 1965
Capital Kuala Lumpur
Population 21.5 million / 169 people per sq mile (65 people per sq km) / 54%
Total area 127,317 sq miles (329,750 sq km)
Languages Malay*, English*, Chinese, Tamil
Religions Muslim 53%, Buddhist 19%, Chinese faiths 12%, Christian 7%, other 9%
Ethnic mix Malay 47%, Chinese 32%, Indigenous tribes 12%, Indian 8%, other 1%
Government Federal constitutional monarchy
Currency Ringgit = 100 cents
Literacy rate 85%
Calorie consumption 2,888 kilocalories

Maldives
Indian Ocean

Official name Republic of Maldives
Formation 1965 / 1965
Capital Male'
Population 282,000 / 2,435 people per sq mile (940 people per sq km) / 26%
Total area 116 sq miles (300 sq km)
Languages Dhivehi (Maldivian)*, Sinhala, Tamil
Religions Sunni Muslim 100%
Ethnic mix Maldivian 99%, other 1%
Government Republic
Currency Rufiyaa = 100 laari
Literacy rate 91%
Calorie consumption 2,580 kilocalories

Mali
Western Africa

Official name Republic of Mali
Formation 1960 / 1960
Capital Bamako
Population 11.8 million / 25 people per sq mile (10 people per sq km) / 27%
Total area 478,837 sq miles (1,240,190 sq km)
Languages French*, Bambara, Fulani, Senufo, Soninké
Religions Muslim 80%, Traditional beliefs 18%, Christian 2%
Ethnic mix Bambara 31%, Fulani 13%, Senufo 12%, other 44%
Government Multiparty republic
Currency CFA franc = 100 centimes
Literacy rate 35%
Calorie consumption 2,278 kilocalories

Malta
Southern Europe

Official name Republic of Malta
Formation 1964 / 1964
Capital Valletta
Population 374,000 / 3,027 people per sq mile (1,169 people per sq km) / 89%
Total area 124 sq miles (320 sq km)
Languages Maltese*, English*
Religions Roman Catholic 98%, other (mostly Anglican) 2%
Ethnic mix Maltese (mixed Arab, Sicilian, Norman, Spanish, Italian, English) 98%, other 2%
Government Multiparty republic
Currency Maltese lira = 100 cents
Literacy rate 91%
Calorie consumption 3,486 kilocalories

Marshall Islands
Australasia & Oceania

Official name Republic of the Marshall Islands
Formation 1986 / 1986
Capital Majuro
Population 59,000 / 848 people per sq mile (327 people per sq km) / 69%
Total area 70 sq miles (181 sq km)
Languages English*, Marshallese*
Religions Protestant 80%, Roman Catholic 15%, other 5%
Ethnic mix Micronesian 97%, other 3%
Government Multiparty republic
Currency US dollar = 100 cents
Literacy rate 91%
Calorie consumption NOT AVAILABLE

Mauritania
West Africa

Official name Islamic Republic of Mauritania
Formation 1960 / 1960
Capital Nouakchott
Population 2.5 million / 6 people per sq mile (2 people per sq km) / 54%
Total area 395,953 sq miles (1,025,520 sq km)
Languages French*, Arabic*, Wolof
Religions Muslim 100%
Ethnic mix Maure 80%, Wolof 7%, Tukulor 5%, other 8%
Government Multiparty republic
Currency Ouguiya = 5 khoums
Literacy rate 38%
Calorie consumption 2,685 kilocalories

Mauritius
Indian Ocean

Official name Republic of Mauritius
Formation 1968 / 1968
Capital Port Louis
Population 1.2 million / 1,680 people per sq mile (649 people per sq km) / 41%
Total area 718 sq miles (1,860 sq km)
Languages English*, French Creole, Hindi, Urdu, Tamil, Chinese
Religions Hindu 52%, Roman Catholic, 26%, Muslim 17%, other 5%
Ethnic mix Creole 55%, South Asian 40%, Chinese 3%, other 2%
Government Multiparty republic
Currency Mauritian rupee = 100 cents
Literacy rate 83%
Calorie consumption 2,690 kilocalories

Mexico
North America

Official name United Mexican States
Formation 1836 / 1848
Capital Mexico City
Population 95.8 million / 130 people per sq mile (50 people per sq km) / 75%
Total area 756,061 sq miles (1,958,200 sq km)
Languages Spanish*, Mayan dialects
Religions Roman Catholic 95%, Protestant 1%, other 4%
Ethnic mix Indian *Mestizo* 55%, Indigenous Indian 20%, European 16%, other 9%
Government Multiparty republic
Currency Mexican peso = 100 centavos
Literacy rate 90%
Calorie consumption 3,146 kilocalories

Micronesia
Australasia & Oceania

Official name Federated States of Micronesia
Formation 1986 / 1986
Capital Palikir
Population 109,000 / 403 people per sq mile (156 people per sq km) / 28%
Total area 1,120 sq miles (2,900 sq km)
Languages English*, Trukese, Pohnpeian, Mortlockese, Kosrean
Religions Roman Catholic 50%, Protestant 48%, other 2%
Ethnic mix Micronesian 99%, other 1%
Government Republic
Currency US dollar = 100 cents
Literacy rate 90%
Calorie consumption NOT AVAILABLE

Moldova
Southeastern Europe

Official name Republic of Moldova
Formation 1991 / 1991
Capital Chişinău
Population 4.5 million / 346 people per sq mile (134 people per sq km) / 52%
Total area 13,000 sq miles (33,700 sq km)
Languages Romanian*, Moldovan, Russian
Religions Romanian Orthodox 98%, Jewish 1%, other 1%
Ethnic mix Moldovan 65%, Ukrainian 14%, Russian 13%, Gagauz 4%, other 4%
Government Multiparty republic
Currency Moldovan leu = 100 bani
Literacy rate 98%
Calorie consumption NOT AVAILABLE

Monaco
Southern Europe

Official name Principality of Monaco
Formation 1861 / 1861
Capital Monaco
Population 32,000 / 42,503 people per sq mile (16,410 people per sq km) / 100%
Total area 0.75 sq miles (1.95 sq km)
Languages French*, Italian, Monégasque, English
Religions Roman Catholic 89%, other 11%
Ethnic mix French 47%, Monégasque 17%, Italian 16%, other 20%
Government Constitutional monarchy
Currency French franc = 100 centimes
Literacy rate 99%
Calorie consumption NOT AVAILABLE

Mongolia
Eastern Asia

Official name Mongolia
Formation 1924 / 1924
Capital Ulan Bator
Population 2.6 million / 4 people per sq mile (2 people per sq km) / 61%
Total area 604,247 sq miles (1,565,000 sq km)
Languages Khalkha Mongol*, Turkic, Russian, Chinese
Religions Predominantly Tibetan Buddhist, with a Muslim minority
Ethnic mix Mongol 90%, Kazakh 4%, Chinese 2%, Russian 2%, other 2%
Government Multiparty republic
Currency Tughrik (togrog) = 100 möngös
Literacy rate 84%
Calorie consumption 1,899 kilocalories

Morocco
Northern Africa

Official name Kingdom of Morocco
Formation 1956 / 1956
Capital Rabat
Population 28 million / 162 people per sq mile (63 people per sq km) / 48%
Total area 269,757 sq miles (698,670 sq km)
Religions Muslim 98%, Jewish 1%, Christian 1%
Ethnic mix Arab and Berber 99%, European 1%
Government Constitutional monarchy
Currency Moroccan dirham = 100 centimes
Literacy rate 45%
Calorie consumption 2,984 kilocalories

Mozambique
Southern Africa

Official name Republic of Mozambique
Formation 1975 / 1975
Capital Maputo
Population 18.7 million / 62 people per sq mile (24 people per sq km) / 34%
Total area 309,493 sq miles (801,590 sq km)
Languages Portuguese*, Makua, Tsonga, Sena
Religions Traditional beliefs 60%, Christian 30%, Muslim 10%
Ethnic mix Makua-Lomwe 47%, Thonga 23%, Malawi 12%, Shona 11%, Yao 4%, other 3%
Government Multiparty republic
Currency Metical = 100 centavos
Literacy rate 40%
Calorie consumption 1,680 kilocalories

Myanmar
Southeastern Asia

Official name Union of Myanmar
Formation 1948 / 1948
Capital Rangoon
Population 47.6 million / 187 people per sq mile (72 people per sq km) / 26%
Total area 261,200 sq miles (676,550 sq km)
Languages Burmese*, Karen, Shan, Mon
Religions Buddhist 87%, Christian 6%, Muslim 4%, other 3%
Ethnic mix Burman 68%, Shan 9%, Karen 6%, Rakhine 4%, other 13%
Government Military regime
Currency Kyat = 100 pyas
Literacy rate 84%
Calorie consumption 2,598 kilocalories

Namibia
Southern Africa

Official name Republic of Namibia
Formation 1990 / 1994
Capital Windhoek
Population 1.7 million / 5 people per sq mile (2 people per sq km) / 37%
Total area 318,260 sq miles (824,290 sq km)
Languages English*, Afrikaans, Ovambo, Kavango, Bergdama
Religions Christian 90%, other 10%
Ethnic mix Ovambo 50%, Kavango 9%, Herero 8%, Damara 8%, other 25%
Government Multiparty republic
Currency Namibian dollar = 100 cents
Literacy rate 79%
Calorie consumption 2,134 kilocalories

Nauru
Australasia & Oceania

Official name Republic of Nauru
Formation 1968 / 1968
Capital No official capital
Population 11,000 / 1332 people per sq mile (514 people per sq km) / 100%
Total area 8.2 sq miles (21.2 sq km)
Languages Nauruan*, English, Kiribati, Chinese, Tuvaluan
Religions Christian 95%, other 5%
Ethnic mix Nauruan 62%, other Pacific islanders 25%, Chinese 8%, European 5%
Government Parliamentary democracy
Currency Australian dollar = 100 cents
Literacy rate 99%
Calorie consumption NOT AVAILABLE

Nepal
Southern Asia

Official name Kingdom of Nepal
Formation 1769 / 1769
Capital Kathmandu
Population 23.2 million / 439 people per sq mile (170 people per sq km) / 14%
Total area 54,363 sq miles (140,800 sq km)
Languages Nepali*, Maithilli, Bhojpuri
Religions Hindu 90%, Buddhist 4%, Muslim 3%, Christian 1%, other 2%
Ethnic mix Nepalese 58%, Bihari 19%, Tamang 6%, other 17%
Government Constitutional monarchy
Currency Nepalese rupee = 100 paisa
Literacy rate 38%
Calorie consumption 1,957 kilocalories

Netherlands
Northwest Europe

Official name Kingdom of the Netherlands
Formation 1815 / 1839
Capitals Amsterdam, The Hague
Population 15.7 million / 1,199 people per sq mile (463 people per sq km) / 89%
Total area 14,410 sq miles (37,330 sq km)
Languages Dutch*, Frisian
Religions Roman Catholic 36%, Protestant 27%, Muslim 3%, other 34%
Ethnic mix Dutch 96%, other 4%
Government Constitutional monarchy
Currency Netherland guilder = 100 cents
Literacy rate 99%
Calorie consumption 3,222 kilocalories

New Zealand
Australasia & Oceania

Official name New Zealand
Formation 1947 / 1947
Capital Wellington
Population 3.7 million / 36 people per sq mile (14 people per sq km) / 86%
Total area 103,730 sq miles (268,680 sq km)
Languages English*, Maori
Religions Protestant 47%, Roman Catholic 15%, other 38%
Ethnic mix European 82%, Maori 9%, Pacific Islanders 3%, other 6%
Government Constitutional monarchy
Currency NZ dollar = 100 cents
Literacy rate 99%
Calorie consumption 3,669 kilocalories

Nicaragua
Central America

Official name Republic of Nicaragua
Formation 1838 / 1838
Capital Managua
Population 4.5 million / 98 people per sq mile (38 people per sq km) / 63%
Total area 50,193 sq miles (130,000 sq km)
Languages Spanish*, English Creole, Miskito
Religions Roman Catholic 95%, other 5%
Ethnic mix *Mestizo* 69%, White 14%, Black 8%, Indigenous Indian 5%, Zambos 4%
Government Multiparty republic
Currency Córdoba ora = 100 pence
Literacy rate 63%
Calorie consumption 2,293 kilocalories

Niger
Western Africa

Official name Republic of Niger
Formation 1960 / 1960
Capital Niamey
Population 10.1 million / 21 people per sq mile (8 people per sq km) / 17%
Total area 489,188 sq miles (1,267,000 sq km)
Languages French*, Hausa, Djerma, Fulani, Tuareg, Teda
Religions Muslim 85%, traditional beliefs 14%, Christian 1%
Ethnic mix Hausa 54%, Djerma and Songhal 21%, Fulani 10%, Tuareg 9%, other 6%
Government Multiparty republic
Currency CFA franc = 100 centimes
Literacy rate 14%
Calorie consumption 2,257 kilocalories

Nigeria
Western Africa

Official name Federal Republic of Nigeria
Formation 1960 / 1961
Capital Abuja
Population 122 million / 346 people per sq mile (123 people per sq km) / 34%
Total area 356,668 sq miles (923,770 sq km)
Languages English*, Hausa, Yoruba, Ibo
Religions Muslim 50%, Christian 40%, Traditional beliefs 10%
Ethnic mix Hausa 21%, Yoruba 21%, Ibo 18%, Fulani 11%, other 29%
Government Multiparty republic
Currency Naira = 100 kobo
Literacy rate 59%
Calorie consumption 2,124 kilocalories

North Korea
Eastern Asia

Official name Democratic People's Republic of Korea
Formation 1948 / 1953
Capital Pyongyang
Population 23.2 million / 499 people per sq mile (193 people per sq km) / 61%
Total area 46,540 sq miles (120,540 sq km)
Languages Korean*, Chinese
Religions Traditional beliefs 16%, Ch'ondogyo 14%, Buddhist 2%, nonreligious 68%
Ethnic mix Korean 99%, other 1%
Government Single-party republic
Currency North Korean Won = 100 chon
Literacy rate 99%
Calorie consumption 2,833 kilocalories

Norway
Northern Europe

Official name Kingdom of Norway
Formation 1905 / 1905
Capital Oslo
Population 4.4 million / 37 people per sq mile (14 people per sq km) / 73%
Total area 125,060 sq miles (323,900 sq km)
Languages Norwegian* (Bokmal and *Nynorsk*), Lappish, Finnish
Religions Evangelical Lutheran 89%, Roman Catholic 1%, other and nonreligious 10%
Ethnic mix Norwegian 95%, Lapp 1%, other 4%
Government Constitutional monarchy
Currency Norwegian krone = 100 øre
Literacy rate 99%
Calorie consumption 3,244 kilocalories

Oman
Southwestern Asia

Official name Sultanate of Oman
Formation 1951 / 1951
Capital Muscat
Population 2.5 million / 30 people per sq mile (12 people per sq km) / 13%
Total area 82,030 sq miles (212,460 sq km)
Languages Arab*, Baluchi
Religions Ibadi Muslim 75%, other Muslim 11%, Hindu 14%
Ethnic mix Arab 75%, Baluchi 15%, other 15%
Government Monarchy with Consultative Council
Currency Omani rial = 1,000 baizas
Literacy rate 67%
Calorie consumption 3,013 kilocalories

Pakistan
Southern Asia

Official name Islamic Republic of Pakistan
Formation 1947 / 1947
Capital Islamabad
Population 147.8 million / 497 people per sq mile (192 people per sq km) / 35%
Total area 307,374 sq miles (796,100 sq km)
Main languages Urdu*, Punjabi, Sindhi, Pashtu, Baluchi
Religions Sunni Muslim 77%, Shi'a Muslim 20%, Hindu 2%, Christian 1%
Ethnic mix Punjabi 50%, Sindhi 15%, Pashtu 15%, Mohajir 8%, Baluch 5%, other 7%
Government Multiparty republic
Currency Pakistani rupee = 100 paisa
Literacy rate 40%
Calorie consumption 2,315 kilocalories

Palau
Australasia & Oceania

Official name Palau
Formation 1994 / 1994
Capital Oreor
Population 17,700 /90 people per sq mile (35 people per sq km) / 29%
Total area 192 sq miles (497 sq km)
Languages Palauan*, English*, Japanese
Religions Roman Catholic 66%, Modekngei 34%
Ethnic mix Palauan 99%, other 1%
Government Multiparty republic
Currency US dollar = 100 cents
Literacy rate 92%
Calorie consumption NOT AVAILABLE

Panama
Central America

Official name Republic of Panama
Formation 1903 / 1903
Capital Panama City
Population 2.8 million / 95 people per sq mile (37 people per sq km) / 53%
Total area 29,761 sq miles (77,080 sq km)
Languages Spanish*, English Creole, Indian languages
Religions Roman Catholic 93%, other 7%
Ethnic mix *Mestizo* 60%, White 14%, Black 12%, Indigenous Indian 8%, other 6%
Government Multiparty republic
Currency Balboa = 100 centesimos
Literacy rate 91%
Calorie consumption 2,242 kilocalories

Papua New Guinea
Australasia & Oceania

Official name Independent State of Papua New Guinea
Formation 1975 / 1975
Capital Port Moresby
Population 4.6 million / 26 people per sq mile (10 people per sq km) / 16%
Total area 178,700 sq miles (462, 840 sq km)
Languages English*, Pidgin English, Papuan, Motu, 750 (estimated) native languages
Religions Christian 62%, Traditional beliefs 38%
Ethnic mix Papuan 85%, other 15%
Government Parliamentary democracy
Currency Kina = 100 toea
Literacy rate 73%
Calorie consumption 2,613 kilocalories

Paraguay
South America

Official name Paraguay
Formation 1811 / 1938
Capital Asunción
Population 5.2 million / 34 people per sq mile (13 people per sq km) / 53%
Total area 157,046 sq miles (406,750 sq km)
Languages Spanish*, Guaraní
Religions Roman Catholic 90%, other 10%
Ethnic mix *Mestizo* 90%, Indigenous Indian 2%, other 8%
Government Multiparty republic
Literacy rate 92%
Calorie consumption 2,670 kilocalories

Peru
South America

Official name Republic of Peru
Formation 1824 / 1941
Capital Lima
Population 24.8 million / 50 people per sq mile (19 people per sq km) / 72%
Total area 496,223 sq miles (1,285,220 sq km)
Languages Spanish*, Quechua*, Aymará
Religions Roman Catholic 95%, other 5%
Ethnic mix Indigenous Indian 54%, *mestizo* 32%, White 12%, other 2%
Government Multiparty republic
Currency New sol = 100 centimos
Literacy rate 88%
Calorie consumption 1,882 kilocalories

Philippines
Southwestern Asia

Official name Republic of the Philippines
Formation 1946 / 1946
Capital Manila
Population 72.2 million / 627 people per sq mile (242 people per sq km) / 54%
Total area 115,831 sq miles (300,000 sq km)
Languages Filipino*, English*, Cebuano, Hiligaynon, Samaran, Bikol, Ilocano
Religions Roman Catholic 83%, Protestant 9%, Muslim 5%, other 3%
Ethnic mix Malay 50%, Indonesian and Polynesian 30%, Chinese 10%, other 10%
Government Multiparty republic
Currency Philippine peso = 100 centavos
Literacy rate 94%
Calorie consumption 2,257 kilocalories

Poland
Northern Europe

Official name Republic of Poland
Formation 1918 / 1945
Capital Warsaw
Population 38.7 million / 329 people per sq mile (127 people per sq km) / 65%
Total area 120,720 sq miles (312,680 sq km)
Languages Polish*, German
Religions Roman Catholic 93%, Eastern Orthodox 2%, other and nonreligious 5%
Ethnic mix Polish 98%, German 1%, other 1%
Government Multiparty republic
Currency Zloty = 100 groszy
Literacy rate 99%
Calorie consumption 3,301 kilocalories

Portugal
Southwestern Europe

Official name Republic of Portugal
Formation 1140 / 1640
Capital Lisbon
Population 9.8 million / 276 people per sq mile (107 people per sq km) / 36%
Total area 35,670 sq miles (92,390 sq km)
Languages Portuguese*
Religions Roman Catholic 97%, Protestant 1%, other 2%
Ethnic mix Portuguese 99%, African 1%
Government Multiparty republic
Currency Escudo = 100 centavos
Literacy rate 90%
Calorie consumption 3,634 kilocalories

Qatar
Southwestern Asia

Official name State of Qatar
Formation 1971 / 1971
Capital Doha
Population 600,000 / 136 people per sq mile (53 people per sq km) / 91%
Total area 4,247 sq miles (11,000 sq km)
Languages Arabic*, Farsi (Persian), Urdu, Hindi, English
Religions Sunni Muslim 86%, Hindu 10%, Christian 4%
Ethnic mix Arab 40%, Pakistani 18%, Iranian 10% Indian 18%, other 14%
Government Absolute monarchy
Currency Qatar riyal = 100 dirhams
Literacy rate 80%
Calorie consumption NOT AVAILABLE

Romania
Southeastern Europe

Official name Romania
Formation 1878 / 1947
Capital Bucharest
Population 22.6 million /254 people per sq mile (98 people per sq km) / 55%
Total area 91,700 sq miles (237,500 sq km)
Languages Romanian*, Hungarian,
Religions Romanian Orthodox 70%, Roman Catholic 5%, Protestant 4%, other 21%
Ethnic mix Romanian 89%, Magyar 9%, Romany 1%, other 1%
Government Multiparty republic
Currency Leu = 100 bani
Literacy rate 97%
Calorie consumption 3,051 kilocalories

Russian Federation
Europe / Asia

Official name Russian Federation
Formation 1991 / 1991
Capital Moscow
Population 147.2 million /22 people per sq mile (9 people per sq km) / 76%
Total area 6,592,800 sq miles (17,075,400 sq km)
Languages Russian*, Tatar, Ukrainian
Religions Russian Orthodox 75%, other (including Jewish, Muslim) 25%
Ethnic mix Russian 82%, Tatar 4%, Ukrainian 3%, Chuvash 1%, other 10%
Currency Rouble = 100 kopeks
Literacy rate 99%
Calorie consumption NOT AVAILABLE

Rwanda
Central Africa

Official name Rwandese Republic
Formation 1962 / 1962
Capital Kigali
Population 6.5 million / 675 people per sq mile (261 people per sq km) / 6%
Total area 10,170 sq miles (26,340 sq km)
Languages Kinyarwanda*, French*, Kiswahili, English
Religions Christian 74%, Traditional beliefs 25%, other 1%
Ethnic mix Hutu 90%, Tutsi 8%, Twa pygmy 2%
Government Multiparty republic
Currency Rwanda franc = 100 centimes
Literacy rate 63%
Calorie consumption 1,821 kilocalories

Saint Kitts & Nevis
West Indies

Official name Federation of Saint Christopher and Nevis
Formation 1983 / 1983
Capital Basseterre
Population 41,000 / 295 people per sq mile (114 people per sq km) / 42%
Total area 139 sq miles (360 sq km)
Languages English*, English Creole
Religions Protestant 71%, Roman Catholic 7%, other 22%
Ethnic mix Black 95%, mixed 5%
Government Parliamentary democracy
Currency E. Caribbean dollar = 100 cents
Literacy rate 90%
Calorie consumption 2,419 kilocalories

Saint Lucia
West Indies

Official name Saint Lucia
Formation 1979 / 1979
Capital Castries
Population 142,000 / 603 people per sq mile (233 people per sq km) / 48%
Total area 239 sq miles (620 sq km)
Languages English*, French Creole, Hindi, Urdu
Religions Roman Catholic 90%, other 10%
Ethnic mix Black 90%, African-European 6%, South Asian 4%
Government Parliamentary democracy
Currency E. Caribbean dollar = 100 cents
Literacy rate 93%
Calorie consumption 2,588 kilocalories

Saint Vincent & the Grenadines
West Indies

Official name Saint Vincent and the Grenadines
Formation 1979 / 1979
Capital Kingstown
Population 111,000 / 846 people per sq mile (327 people per sq km) / 46%
Total area 131 sq miles (340 sq km)
Languages English*, English Creole
Religions Protestant 62% Roman Catholic 19%, other 19%
Ethnic mix Black 82%, mixed 14%, White 3%, South Asian 1%
Government Parliamentary democracy
Currency E. Caribbean dollar = 100 cents
Literacy rate 82%
Calorie consumption 2,347 kilocalories

Samoa
Australasia & Oceania

Official name Independent State of Samoa
Formation 1962 / 1962
Capital Apia
Population 170,000 / 156 people per sq mile (60 people per sq km) / 21%
Total area 1,027 sq miles (2,840 sq km)
Languages Samoan*, English*
Religions Protestant 74%, Roman Catholic 26%
Ethnic mix Samoan 90%, other 10%
Government Parliamentary state
Currency Tala = 100 sene
Literacy rate 98%
Calorie consumption 2,828 kilocalories

San Marino
Southern Europe

Official name Republic of San Marino
Formation AD 301 / 301
Capital San Marino
Population 25,000 / 1,061 people per sq mile (410 people per sq km) / 94%
Total area 24 sq miles (61 sq km)
Languages *Italian
Religions Roman Catholic 93%, other and nonreligious 7%
Ethnic mix Sammarinese 95%, other 5%
Government Multiparty republic
Currency Lira = 100 centesimi
Literacy rate 96%
Calorie consumption 3,561 kilocalories

Sao Tome & Principe
Western Africa

Official name Democratic Republic of Sao Tome and Principe
Formation 1975 / 1975
Capital São Tomé
Population 131,000 / 354 people per sq mile (137 people per sq km) / 46%
Total area 372 sq miles (964 sq km)
Languages *Portuguese, Portuguese Creole
Religions Roman Catholic 90%, other Christian 10%
Ethnic mix Black 90%, Portuguese and Creole 10%
Government Multiparty republic
Currency Dobra = 100 centimos
Literacy rate 75%
Calorie consumption 2,129 kilocalories

Saudi Arabia
Southwestern Asia

Official name Kingdom of Saudi Arabia
Formation 1932 / 1935
Capital Riyadh
Population 20.2 million / 24 people per sq mile (8 people per sq km) / 80%
Total area 829,995 sq miles (2,149,690 sq km)
Languages Arabic*
Religions Sunni Muslim 85%, Shi'a Muslim 15%
Ethnic mix Arab 90%, Afroasian 10%
Government Absolute monarchy
Currency Saudi riyal = 100 malalah
Literacy rate 73%
Calorie consumption 2,735 kilocalories

Senegal
Western Africa

Official name Republic of Senegal
Formation 1960 / 1960
Capital Dakar
Population 9 million /121 people per sq mile (47 people per sq km) / 42%
Total area 75,950 sq miles (196,720 sq km)
Languages *French, Wolof, Fulani, Serer
Religions Muslim 90%, Traditional beliefs 5%, Christian 5%
Ethnic mix Wolof 46%, Fulani 25%, Serer 16%, other 13%
Government Multiparty republic
Currency CFA franc = 100 centimes
Literacy rate 35%
Calorie consumption 2,262 kilocalories

Seychelles
Indian Ocean

Official name Republic of Seychelles
Formation 1976 / 1976
Capital Victoria
Population 75,000 / 722 people per sq mile (279 people per sq km) / 54%
Total area 108 sq miles (280 sq km)
Languages *French Creole, French, English
Religions Roman Catholic 90%, other 10%
Ethnic mix Seychellois (mixed African, South Asian and European) 95%, Chinese and South Asian 5%
Government Multiparty republic
Currency Seychelles rupee = 100 cents
Literacy rate 84%
Calorie consumption 2,287 kilocalories

Sierra Leone
Western Africa

Official name Republic of Sierra Leone
Formation 1961 / 1961
Capital Freetown
Population 4.6 million / 166 people per sq mile (64 people per sq km) / 36%
Total area 27,699 sq miles (71,740 sq km)
Languages English*, Krio (Creole), Mende, Temne
Religions Traditional beliefs 52%, Muslim 40%, Christian 8%
Ethnic mix Mende 35%, Temne 32%, Limba 8%, Kuranko 4%, other 21%
Government Multiparty republic
Currency Leone = 100 cents
Literacy rate 33%
Calorie consumption 1,694 kilocalories

Singapore
Southeastern Asia

Official name Republic of Singapore
Formation 1965 / 1965
Capital Singapore
Population 3.5 million / 14,861 people per sq mile (5,738 people per sq km) / 100%
Total area 239 sq miles (620 sq km)
Languages Malay*, Chinese*, Tamil*, English*
Religions Buddhist 30%, Christian 20%, Muslim 17%, other 33%
Ethnic mix Chinese 78%, Malay 14%, Indian 6%, other 2%
Government Multiparty republic
Currency Singapore dollar = 100 cents
Literacy rate 91%
Calorie consumption 3,128 kilocalories

Slovakia
Central Europe

Official name Slovak Republic
Formation 1993 / 1993
Capital Bratislava
Population 5.4 million / 285 people per sq mile (110 people per sq km) / 59%
Total area 19,100 sq miles (49,500 sq km)
Languages Slovak*, Hungarian (Magyar), Romany, Czech
Religions Roman Catholic 60%, Atheist 10%, Protestant 8%, Orthodox 4%, other 18%
Ethnic mix Slovak 85%, Hungarian 9%, Czech 1%, other 5%
Government Multiparty republic
Currency Koruna = 100 halierov
Literacy rate 99%
Calorie consumption 3,156 kilocalories

Slovenia
Central Europe

Official name Republic of Slovenia
Formation 1991 / 1991
Capital Ljubljana
Population 1.9 million / 243 people per sq mile (94 people per sq km) / 64%
Total area 7,820 sq miles (20,250 sq km)
Languages Slovene*, Serbian, Croatian
Religions Roman Catholic 96%, Muslim 1%, other 3%
Ethnic mix Slovene 88%, Croat 3%, Serb 2%, Bosniak 1%, other 4%
Government Multiparty republic
Currency Tolar = 100 stotins
Literacy rate 99%
Calorie consumption NOT AVAILABLE

Solomon Islands
Australasia & Oceania

Official name Solomon Islands
Formation 1978 / 1978
Capital Honiara
Population 417,000 / 39 people per sq mile (15 people per sq km) / 17%
Total area 10,954 sq miles (28,370 sq km)
Languages English*, Pidgin English, Melanesian Pidgin
Religions Christian 91%, other 9%
Ethnic mix Melanesian 94%, other 6%
Government Parliamentary democracy
Currency Solomon Islands dollar = 100 cents
Literacy rate 62%
Calorie consumption 2,173 kilocalories

Somalia
Eastern Africa

Official name Somali Democratic Republic
Formation 1960 / 1960
Capital Mogadishu
Population 10.7 million / 39 people per sq mile (15 people per sq km) / 25%
Total area 246,200 sq miles (637,660 sq km)
Languages Somali*, Arabic*, English
Religions Sunni Muslim 98%, other (including Christian) 2%
Ethnic mix Somali 98%, Bantu, Arab and other 2%
Government Transitional
Currency Somaili shilling = 100 cents
Literacy rate 24%
Calorie consumption 1,499 kilocalories

South Africa
Southern Africa

Official name Republic of South Africa
Formation 1934 / 1994
Capitals Pretoria/Cape Town/Bloemfontein
Population 44.3 million / 94 people per sq mile (36 people per sq km) / 51%
Total area 471,443 sq miles (1,221,040 sq km)
Languages Afrikaans*, English*, 11 African languages
Religions Protestant 55%, Roman Catholic 9%, Hindu 1%, Muslim 1%, other 34%
Ethnic mix Other Black 38%, White 16%, Zulu 23%, mixed 10%, Xhosa 9%, other 4%
Government Multiparty republic
Currency Rand = 100 cents
Literacy rate 82%
Calorie consumption 2,695 kilocalories

South Korea
Eastern Asia

Official name Republic of Korea
Formation 1948 / 1953
Capital Seoul
Population 46.1 million / 1209 people per sq mile (467 people per sq km) / 81%
Total area 38,232 sq miles (99,020 sq km)
Languages Korean*, Chinese
Religions Mahayana Buddhist 47%, Protestant 38%, Roman Catholic 11%, Confucianist 3%, other 1%
Ethnic mix Korean 100%
Government Multiparty republic
Currency Won = 100 chon
Literacy rate 97%
Calorie consumption 3,285 kilocalories

Spain
Southwestern Europe

Official name Kingdom of Spain
Formation 1492 / 1713
Capital Madrid
Population 39.8 million / 206 people per sq mile (80 people per sq km) / 76%
Total area 194,900 sq miles (504,780 sq km)
Languages Castilian Spanish*, Catalan*, Galician*, Basque*
Religions Roman Catholic 96%, other 4%
Ethnic mix Castilian Spanish 72%, Catalan 17%, Galician 6%, other 5%
Government Constitutional monarchy
Currency Spanish Peseta = 100 céntimos
Literacy rate 95%
Calorie consumption 3,708 kilocalories

Sri Lanka
Southern Asia

Official name Democratic Socialist Republic of Sri Lanka
Formation 1948 / 1948
Capital Colombo
Population 18.5 million / 740 people per sq mile (286 people per sq km) / 22%
Total area 25,332 sq miles (65,610 sq km)
Languages Sinhala*, Tamil, English
Religions Buddhist 70%, Hindu 15%, Christian 8%, Muslim 7%
Ethnic mix Sinhalese 74%, Tamil 18%, other 8%
Government Multiparty republic
Currency Sri Lanka rupee = 100 cents
Literacy rate 90%
Calorie consumption 2,273 kilocalories

Sudan
Eastern Africa

Official name Republic of Sudan
Formation 1956 / 1956
Capital Khartoum
Population 28.5 million / 31 people per sq mile (12 people per sq km) / 25%
Total area 967,493 sq miles (2,505,815 sq km)
Languages Arabic*, Dinka, Nuer, Nubian, Beja, Zande, Bari, Fur
Religions Muslim 70%, Traditional beliefs 20%, Christian 9%, other 1%
Ethnic mix Arab 51%, Dinka 13%, Nuba 9%, Beja 7%, other 20%
Government Military regime
Currency Sudanese pound or dinar = 100 piastres
Literacy rate 53%
Calorie consumption 2,202 kilocalories

Suriname
South America

Official name Republic of Suriname
Formation 1975 / 1975
Capital Paramaribo
Population 442,000 / 7 people per sq mile (3 people per sq km) / 50%
Total area 63,039 sq miles (163,270 sq km)
Languages Dutch*, Pidgin English (Taki-Taki), Hindi, Javanese, Carib
Religions Christian 48%, Hindu 27%, Muslim 20%, other 5%
Ethnic mix Hindustani 34%, Creole 34%, Javanese 18%, Black 9%, other 5%
Government Multiparty republic
Currency Suriname guilder = 100 cents
Literacy rate 93%
Calorie consumption 2547 kilocalories

Swaziland
Southern Africa

Official name Kingdom of Swaziland
Formation 1968 / 1968
Capital Mbabane
Population 900,000 / 140 people per sq mile (54 people per sq km) / 29%
Total area 6,703 sq miles (17,360 sq km)
Languages Siswati*, English*, Zulu
Religions Christian 60%, Traditional beliefs 40%
Ethnic mix Swazi 95%, other 5%
Government Executive monarchy
Currency Lilangeni = 100 cents
Literacy rate 77%
Calorie consumption 2,706 kilocalories

Sweden
Northern Europe

Official name Kingdom of Sweden
Formation 1809 / 1905
Capital Stockholm
Population 8.9 million / 56 people per sq mile (22 people per sq km) / 83%
Total area 173,730 sq miles (449,960 sq km)
Languages Swedish*, Finnish, Lappish
Religions Evangelical Lutheran 89%, Roman Catholic 2%, Muslim 1%, other 8%
Ethnic mix Swedish 91%, other European 6%, Finnish and Lapp 3%
Government Constitutional monarchy
Currency Swedish krona = 100 öre
Literacy rate 99%
Calorie consumption 2,972 kilocalories

Switzerland
Central Europe

Official name Swiss Confederation
Formation 1291 / 1815
Capital Bern
Population 7.3 million / 475 people per sq mile (184 people per sq km) / 61%
Total area 15,940 sq miles (41,290 sq km)
Languages German*, French*, Italian*, Romansch
Religions Roman Catholic 48%, Protestant 44%, other 8%
Ethnic mix German 65%, French 18%, Italian 10%, other 7%
Government Federal republic
Currency Franc = 100 centimes
Literacy rate 99%
Calorie consumption 3,379 kilocalories

Syria
Southwest Asia

Official name Syrian Arab Republic
Formation 1946 / 1967
Capital Damascus
Population 15.3 million / 215 people per sq mile (83 people per sq km) / 52%
Total area 71,500 sq miles (185,180 sq km)
Languages Arabic*, French, Kurdish
Religions Sunni Muslim 74%, other Muslim 16%, Christian 10%
Ethnic mix Arab 89%, Kurdish 6%, Armenian, Turkmen, Circassian 2%, other 3%
Government Single-party republic
Currency Syrian pound = 100 piastres
Literacy rate 71%
Calorie consumption 3175 kilocalories

Taiwan
East Asia

Official name Republic of China
Formation 1949 / 1949
Capital Taipei
Population 21.5 million / 1724 people per sq mile (666 people per sq km) / 69%
Total area 13,969 sq miles (36,179 sq km)
Languages Mandarin Chinese*, Amoy Chinese, Hakka Chinese
Religions Buddhist, Confucianist, Taoist 93%, Christian 5%, other 2%
Ethnic mix Indigenous Chinese, Mainland Chinese 14%, Aborigine 2%
Government Multiparty republic
Currency Taiwan dollar = 100 cents
Literacy rate 94%
Calorie consumption NOT AVAILABLE

Tajikistan
Central Asia

Official name Republic of Tajikistan
Formation 1991 / 1991
Capital Dushanbe
Population 6.2 million / 112 people per sq mile (43 people per sq km) / 32%
Total area 55,251 sq miles (143,100 sq km)
Main languages Tajik*, Uzbek, Russian
Religions Sunni Muslim 80%, Shi'a Muslim 5%, other 15%
Ethnic mix Tajik 62%, Uzbek 24%, Russian 4%, Tatar 2%, other 8%
Government Multiparty republic
Currency Tajik rouble = 100 kopeks
Literacy rate 99%
Calorie consumption NOT AVAILABLE

Tanzania
East Africa

Official name United Republic of Tanzania
Formation 1961 / 1964
Capital Dodoma
Population 32.2 million / 94 people per sq mile (36 people per sq km) / 24%
Total area 364,900 sq miles (945,090 sq km)
Languages English*, Swahili*, Sukuma, Chagga, Nyamwezi, Hehe, Makonde
Religions Muslim 33%, Christian 33%, Traditional beliefs 30%, other 4%
Ethnic mix 120 small ethnic Bantu groups 99%, other 1%
Government Multiparty republic
Currency Tanzanian shilling = 100 cents
Literacy rate 71%
Calorie consumption 2018 kilocalories

Thailand
Southeastern Asia

Official name Kingdom of Thailand
Formation 1782 / 1907
Capital Bangkok
Population 59.6 million / 302 people per sq mile (117 people per sq km) / 20%
Total area 198,116 sq miles (513,120 sq km)
Languages Thai*, Chinese, Malay, Khmer, Mon, Karen
Religions Buddhist 95%, other 5%
Ethnic mix Thai 80%, Chinese 12%, Malay 4%, Khmer and other 4%
Government Constitutional monarchy
Currency Baht = 100 stangs
Literacy rate 94%
Calorie consumption 2432 kilocalories

Togo
Western Africa

Official name Togolese Republic
Formation 1960 / 1960
Capital Lomé
Population 4.4 million /210 people per sq mile (81 people per sq km) / 31%
Total area 21,927 sq miles (56,790 sq km)
Languages French*, Ewe, Kabye, Gurma
Religions Traditional beliefs 50%, Christian 35%, Muslim 15%
Ethnic mix Ewe 43%, Kabye 26%, Gurma 16%, other 15%
Government Multiparty republic
Currency CFA franc = 100 centimes
Literacy rate 53%
Calorie consumption 2242 kilocalories

Tonga
Australasia & Oceania

Official name Kingdom of Tonga
Formation 1970 / 1970
Capital Nuku'alofa
Population 97,000 / 351 people per sq mile (135 people per sq km) / 21%
Total area 290 sq miles (750 sq km)
Languages Tongan*, English*
Religions Protestant 64%, Roman Catholic 15%, other 21%
Ethnic mix Tongan 98%, other 2%
Government Constitutional monarchy
Currency Pa'anga = 100 seniti
Literacy rate 99%
Calorie consumption 2,946 kilocalories

Trinidad & Tobago
West Indies

Official name Republic of Trinidad and Tobago
Formation 1962 / 1962
Capital Port-of-Spain
Population 1.3 million / 656 people per sq mile (253 people per sq km) / 72%
Total area 1,981 sq miles (5,130 sq km)
Languages English*, English Creole, Hindi, French, Spanish
Religions Christian 58%, Hindu 30%, Muslim 8%, other 4%
Ethnic mix Black 43%, Asian 40%, mixed 19%, White and Chinese 1%
Government Multiparty republic
Currency Trinidad & Tobago dollar = 100 cents
Literacy rate 98%
Calorie consumption 2,585 kilocalories

Tunisia
Northern Africa

Official name Republic of Tunisia
Formation 1956 / 1956
Capital Tunis
Population 9.5 million / 158 people per sq mile (61 people per sq km) / 57%
Total area 63,170 sq miles (163,610 sq km)
Languages Arabic*, French
Religions Muslim 98%, Christian 1%, Jewish 1%
Ethnic mix Arab and Berber 98%, European 1%, other 1%
Government Multiparty republic
Currency Tunisian dinar = 1,000 millimes
Literacy rate 67%
Calorie consumption 3,330 kilocalories

Turkey
Asia / Europe

Official name Republic of Turkey
Formation 1923 / 1939
Capital Ankara
Population 63.8 million / 215 people per sq mile (83 people per sq km) / 69%
Total area 300,950 sq miles (779,450 sq km)
Languages Turkish*, Kurdish, Arabic, Circassian, Armenian
Religions Muslim 99%, other 1%
Ethnic mix Turkish 70%, Kurdish 20%, other 8%, Arab 2%
Government Multiparty republic
Currency Turkish lira = 100 krural
Literacy rate 83%
Calorie consumption 3,429 kilocalories

Turkmenistan
Central Asia

Official name Turkmenistan
Formation 1991 / 1991
Capital Ashgabat
Population 4.3 million / 23 people per sq mile (9 people per sq km) / 45%
Total area 188,455 sq miles (488,100 sq km)
Languages Turkmen*, Uzbek, Russian
Religions Muslim 87%, Eastern Orthodox 11%, other 2%
Ethnic mix Turkmen 72%, Russian 9%, Uzbek 9%, other 10%
Government Multiparty republic
Currency Manat = 100 tenge
Literacy rate 98%
Calorie consumption NOT AVAILABLE

Tuvalu
Australasia & Oceania

Official name Tuvalu
Formation 1978 / 1978
Capital Fongafale
Population 9000 / 976 people per sq mile (377 people per sq km) / 46%
Total area 10 sq miles (26 sq km)
Languages Tuvaluan*, Kiribati, English
Religions Protestant 97%, other 3%
Ethnic mix Polynesian 95% other 5%
Government Constitutional monarchy
Currency Australian dollar = 100 cents
Literacy rate 95%
Calorie consumption NOT AVAILABLE

UGANDA
Eastern Africa

Official name Republic of Uganda
Formation 1962 / 1962
Capital Kampala
Population 21.3 million / 276 people per sq mile (107 people per sq km) / 13%
Total area 91,073 sq miles (235,880 sq km)
Languages English*, Luganda, Nkole, Chiga, Lango, Acholi, Teso
Religions Christian 71%, Traditional beliefs 13%, Muslim 5%, other (including Hindu) 11%
Ethnic mix Buganda 18%, Banyoro 14%, Teso 9%, other 59%
Government Multiparty republic
Currency New Uganda shilling = 100 cents
Literacy rate 64%
Calorie consumption 2,159 kilocalories

UKRAINE
Eastern Europe

Official name Ukraine
Formation 1991 / 1991
Capital Kiev
Population 51.2 million / 220 people per sq mile (85 people per sq km) / 70%
Total area 223,090 sq miles (603,700 sq km)
Languages Ukrainian*, Russian, Tatar
Religions Mostly Ukrainian Orthodox, with Roman Catholic, Protestant and Jewish minorities
Ethnic mix Ukrainian 73%, Russian 22%, other (including Tatar) 5%
Government Multiparty republic
Currency Hryvna = 100 kopiykas
Literacy rate 98%
Calorie consumption NOT AVAILABLE

UNITED ARAB EMIRATES
Southwestern Asia

Official name United Arab Emirates
Formation 1971 / 1971
Capital Abu Dhabi
Population 2.4 million / 7 people per sq mile (29 people per sq km) / 84%
Total area 32,278 sq miles (83,600 sq km)
Languages Arabic*, Farsi (Persian), Urdu, Hindi, English
Religions Sunni Muslim 77%, Shi'a Muslim 19%, other 4%
Ethnic mix Asian 50%, Emirian 19%, other Arab 23%, other 8%
Government Federation of monarchs
Currency UAE dirham = 100 fils
Literacy rate 79%
Calorie consumption 3,384 kilocalories

UNITED KINGDOM
Northwestern Europe

Official name United Kingdom of Great Britain and Northern Ireland
Formation 1707 / 1922
Capital London
Population 58.2 million / 624 people per sq mile (241 people per sq mile) / 89%
Total area 94,550 sq miles (244,880 sq km)
Languages English*, Welsh*, Scottish, Gaelic
Religions Protestant 52%, Roman Catholic 9%, Muslim 3%, other 36%
Ethnic mix English 80%, Scottish 10%, Northern Irish 4%, Welsh 2%, West Indian, Asian 4%
Government Constitutional monarchy
Currency Pound sterling = 100 pence
Literacy rate 99%
Calorie consumption 3,317 kilocalories

UNITED STATES
North America

Official name United States of America
Formation 1787 / 1959
Capital Washington DC
Population 273.8 million / 77 people per sq mile (30 people per sq km) / 76%
Total area 3,681,760 sq miles (9,372,610 sq km)
Languages English*, Spanish, Italian, German, French, Polish, Chinese, Greek
Religions Protestant 61%, Roman Catholic 25%, Jewish 2%, other 12%
Ethnic mix White (including Hispanic) 84%, Black 12%, Chinese 1%, other 3%
Government Multiparty republic
Currency US dollar = 100 cents
Literacy rate 99%
Calorie consumption 3,732 kilocalories

URUGUAY
South America

Official name Oriental Republic of Uruguay
Formation 1828 / 1828
Capital Montevideo
Population 3.2 million / 47 people per sq mile (18 people per sq km) / 90%
Total area 67,494 sq miles (174,810 sq km)
Languages Spanish*
Religions Roman Catholic 66%, Protestant 2%, Jewish 2%, other 30%
Ethnic mix White 90%, *mestizo* 6% Black 4%
Government Multiparty republic
Currency Uruguayan peso = 100 centimes
Literacy rate 97%
Calorie consumption 2,750 kilocalories

UZBEKISTAN
Central Asia

Official name Republic of Uzbekistan
Formation 1991 / 1991
Capital Tashkent
Population 24.1 million / 140 people per sq mile (54 people per sq km) / 41%
Total area 172,741 sq miles (447,400 sq km)
Languages Uzbek*, Russian
Religions Muslim 88%, other (mostly Eastern Orthodox) 9%, other 3%
Ethnic mix Uzbek 71%, Russian 8%, Tajik 5%, Kazakh 4%, other 12%
Government Multiparty republic
Currency Sum = 100 teen
Literacy rate 99%
Calorie consumption NOT AVAILABLE

VANUATU
Australasia & Oceania

Official name Republic of Vanuatu
Formation 1980 / 1980
Capital Port-Vila
Population 200,000 / 42 people per sq mile (16 people per sq km) / 19%
Total area 4706 sq miles (12,190 sq km)
Languages Bislama*, English*, French*
Religions Protestant 77%, Roman Catholic 15%, Traditional beliefs 8%
Ethnic mix ni-Vanuatu 94%, other 6%
Government Multiparty republic
Currency Vatu = 100 centimes
Literacy rate 64%
Calorie consumption 2,739 kilocalories

VATICAN CITY
Southern Europe

Official name Vatican City State
Formation 1929 / 1929
Capital Not applicable
Population 1,000 / 5,886 people per sq mile (2,273 people per sq km) /100%
Total area 0.17 sq miles (0.44 sq km)
Languages Italian*, Latin*
Religions Roman Catholic 100%
Ethnic mix Italian 90%, Swiss 10% (including the Swiss Guard, which is responsible for papal security)
Government Papal Commission
Currency Italian lira = 100 centesimi
Literacy rate 99%
Calorie consumption 3,561 kilocalories

VENEZUELA
South America

Official name Republic of Venezuela
Formation 1821 / 1930
Capital Caracas
Population 23.2 million / 68 people per sq mile (26 people per sq km) / 93%
Total area 352,143 sq miles (912,050 sq km)
Languages Spanish*, Indian languages *
Religions Roman Catholic 89%, Protestant and other 11%
Ethnic mix *Mestizo* 69%, White 20%, Black 9%, Indian 2%
Government Multiparty republic
Currency Bolívar = 100 centimos
Literacy rate 92%
Calorie consumption 2,618 kilocalories

VIETNAM
Southeastern Asia

Official name Socialist Republic of Vietnam
Formation 1976 / 1976
Capital Hanoi
Population 77.9 million / 620 people per sq mile (239 people per sq km) / 21%
Total area 127,243 sq miles (329,560 sq km)
Languages Vietnamese*, Chinese, Thai, Khmer, Muong
Religions Buddhist 55%, Christian 7%, other and nonreligious 38%
Ethnic mix Vietnamese 88%, Chinese 4%, Thai 2%, other 6%
Government Single-party republic
Currency Dong = 10 hao = 100 xu
Literacy rate 91%
Calorie consumption 2,250 kilocalories

YEMEN
Southwestern Asia

Official name Republic of Yemen
Formation 1990 / 1990
Capital Sana
Population 16.9 million / 82 people per sq mile (32 people per sq km) / 34%
Total area 203,849 sq miles (527,970 sq km)
Languages Arabic*, Hindi, Tamil, Urdu
Religions Shi'a Muslim 55%, Sunni Muslim 42%, Christian, Hindu, Jewish 3%
Ethnic mix Arab 95%, Afro-Arab 3%, Indian, Somali, European 2%
Government Multiparty republic
Currency Rial (North), Dinar (South) – both are legal currency
Literacy rate 42%
Calorie consumption 2,203 kilocalories

YUGOSLAVIA (SERBIA & MONTENEGRO) *Europe*

Official name Federal Republic of Yugoslavia
Formation 1992 / 1992
Capital Belgrade
Population 10.4 million / 264 people per sq mile (102 people per sq km) / 57%
Total area 39,449 sq miles (102,173 sq km)
Languages Serbo-croat*, Albanian
Religions Roman Catholic, Eastern Orthodox 69%, Muslim 19%, Protestant 1%, other 11%
Ethnic mix Serb 62%, Albanian 17%, Montenegrin 5%, Magyar 3%, other 13%
Government Multiparty republic
Currency Dinar = 100 para
Literacy rate 93%
Calorie consumption NOT AVAILABLE

ZAMBIA
Southern Africa

Official name Republic of Zambia
Formation 1964 / 1964
Capital Lusaka
Population 8.7 million / 30 people per sq mile (12 people per sq km) / 45%
Total area 285,992 sq miles (740,720 sq km)
Languages English*, Bemba, Nyanja, Tonga, Kaonde, Lunda
Religions Christian 63%, Traditional beliefs 36%, other 1%
Ethnic mix Bemba 36%, Maravi 18%, Tonga 15%, other 31%
Government Multiparty republic
Currency Kwacha = 100 ngwee
Literacy rate 75%
Calorie consumption 1,931 kilocalories

ZIMBABWE
Southern Africa

Official name Republic of Zimbabwe
Formation 1980 / 1980
Capital Harare
Population 11.9 million / 80 people per sq mile (21 people per sq km) / 32%
Total area 150,800 sq miles (390,580 sq km)
Languages English*, Shona, Ndebele
Religions Syncretic (Christian and traditional beliefs) 50%, Christian 26%, Traditional beliefs 24%
Ethnic mix Shona 71%, Ndebele 16%, other African 11%, White, Asian 2%
Government Multiparty republic
Currency Zimbabwe dollar = 100 cents
Literacy rate 90%
Calorie consumption 1,985 kilocalories

GEOGRAPHICAL NAMES

THE FOLLOWING GLOSSARY lists all geographical terms occurring on the maps and in main-entry names in the Index-Gazetteer. These terms may precede, follow or be run together with the proper element of the name; where they precede it the term is reversed for indexing purposes – thus Poluostrov Yamal is indexed as Yamal, Poluostrov.

KEY
Geographical term *Language*, Term

A
Å *Danish, Norwegian*, River
Āb *Persian*, River
Adrar *Berber*, Mountains
Agía, Ágios *Greek*, Saint
Air *Indonesian*, River
Ákra *Greek*, Cape, point
Alpen *German*, Alps
Alt- *German*, Old
Altiplanicie *Spanish*, Plateau
Älve(en) *Swedish*, River
-ån *Swedish*, River
Anse *French*, Bay
'Aqabat *Arabic*, Pass
Archipiélago *Spanish*, Archipelago
Arcipelago *Italian*, Archipelago
Arquipélago *Portuguese*, Archipelago
Arrecife(s) *Spanish*, Reef(s)
Aru *Tamil*, River
Augstiene *Latvian*, Upland
Aukštuma *Lithuanian*, Upland
Aust- *Norwegian*, Eastern
Avtonomnyy Okrug *Russian*, Autonomous district
Āw *Kurdish*, River
'Ayn *Arabic*, Spring, well
'Ayoûn *Arabic*, Wells

B
Baelt *Danish*, Strait
Bahía *Spanish*, Bay
Baḥr *Arabic*, River
Baía *Portuguese*, Bay
Baie *French*, Bay
Bañado *Spanish*, Marshy land
Bandao *Chinese*, Peninsula
Banjaran *Malay*, Mountain range
Barajı *Turkish*, Dam
Barragem *Portuguese*, Reservoir
Bassin *French*, Basin
Batang *Malay*, Stream
Beinn, Ben *Gaelic*, Mountain
-berg *Afrikaans, Norwegian*, Mountain
Besar *Indonesian, Malay*, Big
Birkat, Birket *Arabic*, Lake, well, cistern
Boğazi *Turkish*, Lake
Boka *Serbo-Croatian*, Bay
Bol'sh-aya, -iye, -oy, -oye *Russian*, Big
Botigh(i) *Uzbek*, Depression basin
-bre(en) *Norwegian*, Glacier
Bredning *Danish*, Bay
Bucht *German*, Bay
Bugt(en) *Danish*, Bay
Buḥayrat *Arabic*, Lake, reservoir
Buḥeiret *Arabic*, Lake
Bukit *Malay*, Mountain
-bukta *Norwegian*, Bay
bukten *Swedish*, Bay
Bulag *Mongolian*, Spring
Bulak *Uighur*, Spring
Burnu *Turkish*, Cape, point
Buuraha *Somali*, Mountains

C
Cabo *Portuguese*, Cape
Caka *Tibetan*, Salt lake
Canal *Spanish*, Channel
Cap *French*, Cape
Capo *Italian*, Cape, headland
Cascada *Portuguese*, Waterfall
Cayo(s) *Spanish*, Islet(s), rock(s)
Cerro *Spanish*, Mountain
Chaîne *French*, Mountain range
Chapada *Portuguese*, Hills, upland
Chau *Cantonese*, Island
Chāy *Turkish*, River
Chhâk *Cambodian*, Bay
Chhu *Tibetan*, River
-chŏsuji *Korean*, Reservoir
Chott *Arabic*, Depression, salt lake
Chŭli *Uzbek*, Grassland, steppe
Ch'ün-tao *Chinese*, Island group
Chuŏr Phnum *Cambodian*, Mountains
Ciudad *Spanish*, City, town
Co *Tibetan*, Lake
Colline(s) *French*, Hill(s)
Cordillera *Spanish*, Mountain range
Costa *Spanish*, Coast
Côte *French*, Coast
Coxilha *Portuguese*, Mountains
Cuchilla *Spanish*, Mountains

D
Daban *Mongolian, Uighur*, Pass
Daği *Azerbaijani, Turkish*, Mountain
Dağlari *Azerbaijani, Turkish*, Mountains
-dake *Japanese*, Peak
-dal(en) *Norwegian*, Valley
Danau *Indonesian*, Lake
Dao *Chinese*, Island
Đao *Vietnamese*, Island
Daryā *Persian*, River
Daryācheh *Persian*, Lake
Dasht *Persian*, Desert, plain
Dawḥat *Arabic*, Bay
Denizi *Turkish*, Sea
Dere *Turkish*, Stream
Desierto *Spanish*, Desert
Dili *Azerbaijani*, Spit
-do *Korean*, Island
Dooxo *Somali*, Valley
Düzü *Azerbaijani*, Steppe
-dwīp *Bengali*, Island

E
-eilanden *Dutch*, Islands
Embalse *Spanish*, Reservoir
Ensenada *Spanish*, Bay
Erg *Arabic*, Dunes
Estany *Catalan*, Lake
Estero *Spanish*, Inlet
Estrecho *Spanish*, Strait
Étang *French*, Lagoon, lake
-ey *Icelandic*, Island
Ezero *Bulgarian, Macedonian*, Lake
Ezers *Latvian*, Lake

F
Feng *Chinese*, Peak
Fjord *Danish*, Fjord
-fjord(en) *Danish, Norwegian, Swedish*, fjord
-fjørdhur *Faeroese*, Fjord
Fleuve *French*, River
Fliegu *Maltese*, Channel
-fljór *Icelandic*, River
-flói *Icelandic*, Bay
Forêt *French*, Forest

G
-gan *Japanese*, Rock
-gang *Korean*, River
Ganga *Hindi, Nepali, Sinhala*, River
Gaoyuan *Chinese*, Plateau
Garagumy *Turkmen*, Sands
-gawa *Japanese*, River
Gebel *Arabic*, Mountain
-gebirge *German*, Mountain range
Ghadīr *Arabic*, Well
Ghubbat *Arabic*, Bay
Gjiri *Albanian*, Bay
Gol *Mongolian*, River
Golfe *French*, Gulf
Golfo *Italian, Spanish*, Gulf
Göl(ü) *Turkish*, Lake
Golyam, -a *Bulgarian*, Big
Gora *Russian, Serbo-Croatian*, Mountain
Góra *Polish*, Mountain
Gory *Russian*, Mountain
Gryada *Russian*, Ridge
Guba *Russian*, Bay
-gundo *Korean*, Island group
Gunung *Malay*, Mountain

H
Ḥadd *Arabic*, Spit
-haehyŏp *Korean*, Strait
Haff *German*, Lagoon
Hai *Chinese*, Bay, lake, sea
Haixia *Chinese*, Strait
Hamada *Arabic*, Plateau
Ḥammādat *Arabic*, Plateau
Hāmūn *Persian*, Lake
-hantō *Japanese*, Peninsula
Har, Haré *Hebrew*, Mountain
Ḥarrat *Arabic*, Lava-field
Hav(et) *Danish, Swedish*, Sea
Hawr *Arabic*, Lake
Hāyk' *Amharic*, Lake
He *Chinese*, River
-hegység *Hungarian*, Mountain range
Heide *German*, Heath, moorland
Helodrano *Malagasy*, Bay
Higashi- *Japanese*, East(ern)
Ḥiṣā' *Arabic*, Well
Hka *Burmese*, River
-ho *Korean*, Lake
Hô *Korean*, Reservoir
Ḥolot *Hebrew*, Dunes
Hora *Belorussian, Czech*, Mountain
Hrada *Belorussian*, Mountain, ridge
Hsi *Chinese*, River
Hu *Chinese*, Lake
Huk *Danish*, Point

I
Île(s) *French*, Island(s)
Ilha(s) *Portuguese*, Island(s)
Ilhéu(s) *Portuguese*, Islet(s)
Imeni *Russian*, In the name of
Inish- *Gaelic*, Island
Insel(n) *German*, Island(s)
Irmağı, Irmak *Turkish*, River
Isla(s) *Spanish*, Island(s)
Isola (Isole) *Italian*, Island(s)

J
Jabal *Arabic*, Mountain
Jāl *Arabic*, Ridge
-järv *Estonian*, Lake
-järvi *Finnish*, Lake
Jazā'ir *Arabic*, Islands
Jazīrat *Arabic*, Island
Jazīreh *Persian*, Island
Jebel *Arabic*, Mountain
Jezero *Serbo-Croatian*, Lake
Jezioro *Polish*, Lake
Jiang *Chinese*, River
-jima *Japanese*, Island
Jižní *Czech*, Southern
-jõgi *Estonian*, River
-joki *Finnish*, River
-jökull *Icelandic*, Glacier
Jūn *Arabic*, Bay
Juzur *Arabic*, Islands

K
Kaikyō *Japanese*, Strait
-kaise *Lappish*, Mountain
Kali *Nepali*, River
Kalnas *Lithuanian*, Mountain
Kalns *Latvian*, Mountain
Kang *Chinese*, Harbor
Kangri *Tibetan*, Mountain(s)
Kaôh *Cambodian*, Island
Kapp *Norwegian*, Cape
Káto *Greek*, Lower
Kavīr *Persian*, Desert
K'edi *Georgian*, Mountain range
Kediet *Arabic*, Mountain
Kepi *Albanian*, Cape, point
Kepulauan *Indonesian, Malay*, Island group
Khalīg, Khalīj *Arabic*, Gulf
Khawr *Arabic*, Inlet
Khola *Nepali*, River
Khrebet *Russian*, Mountain range
Ko *Thai*, Island
-ko *Japanese*, Inlet, lake
Kólpos *Greek*, Bay
-kopf *German*, Peak
Körfäzi *Azerbaijani*, Bay
Körfezi *Turkish*, Bay
Körgustik *Estonian*, Upland
Kosa *Russian, Ukrainian*, Spit
Koshi *Nepali*, River
Kou *Chinese*, River-mouth
Kowtal *Persian*, Pass
Kray *Russian*, Region, territory
Kryazh *Russian*, Ridge
Kuduk *Uighur*, Well
Kūh(hā) *Persian*, Mountain(s)
-kul' *Russian*, Lake
Kŭl(i) *Tajik, Uzbek*, Lake
-kundo *Korean*, Island group
-kysten *Norwegian*, Coast
Kyun *Burmese*, Island

L
Laaq *Somali*, Watercourse
Lac *French*, Lake
Lacul *Romanian*, Lake
Lagh *Somali*, Stream
Lago *Italian, Portuguese, Spanish*, Lake
Lagoa *Portuguese*, Lagoon
Laguna *Italian, Spanish*, Lagoon, lake
Laht *Estonian*, Bay
Laut *Indonesian*, Bay
Lembalemba *Malagasy*, Plateau
Lerr *Armenian*, Mountain
Lerrnashghťa *Armenian*, Mountain range
Les *Czech*, Forest
Lich *Armenian*, Lake
Liehtao *Chinese*, Island group
Liqeni *Albanian*, Lake
Límni *Greek*, Lake
Ling *Chinese*, Mountain range
Llano *Spanish*, Plain, prairie
Lumi *Albanian*, River
Lyman *Ukrainian*, Estuary

M
Madīnat *Arabic*, City, town
Mae Nam *Thai*, River
-mägi *Estonian*, Hill
Maja *Albanian*, Mountain
Mal *Albanian*, Mountains
Mal-aya, -oye, -yy, *Russian*, Small
-man *Korean*, Bay
Mar *Spanish*, Lake
Marios *Lithuanian*, Lake
Massif *French*, Mountains
Meer *German*, Lake
-meer *Dutch*, Lake
Melkosopochnik *Russian*, Plain
-meri *Estonian*, Sea
Mifraẕ *Hebrew*, Bay
Minami- *Japanese*, South(ern)
-misaki *Japanese*, Cape, point
Monkhafad *Arabic*, Depression
Montagne(s) *French*, Mountain(s)
Montañas *Spanish*, Mountains
Mont(s) *French*, Mountain(s)
Monte *Italian, Portuguese*, Mountain
More *Russian*, Sea
Mörön *Mongolian*, River
Mys *Russian*, Cape, point

N
-nada *Japanese*, Open stretch of water
Nagor'ye *Russian*, Upland
Naḥal *Hebrew*, River
Nahr *Arabic*, River
Nam *Laotian*, River
Namakzār *Persian*, Salt desert
Né-a, -on, -os *Greek*, New
Nedre- *Norwegian*, Lower
-neem *Estonian*, Cape, point
Nehri *Turkish*, River
-nes *Norwegian*, Cape, point
Nevado *Spanish*, Mountain (snow-capped)
Nieder- *German*, Lower
Nishi- *Japanese*, West(ern)
-nísi *Greek*, Island
Nisoi *Greek*, Islands
Nizhn-eye, -iy, -iye, -yaya *Russian*, Lower
Nizmennost' *Russian*, Lowland, plain
Nord *Danish, French, German*, North
Norte *Portuguese, Spanish*, North
Nos *Bulgarian*, Point, spit
Nosy *Malagasy*, Island
Nov-a, -i, *Bulgarian, Serbo-Croatian*, New
Nov-aya, -o, -oye, -yy, -yye *Russian*, New
Now-a, -e, -y *Polish*, New
Nur *Mongolian*, Lake
Nuruu *Mongolian*, Mountains
Nuur *Mongolian*, Lake
Nyzovyna *Ukrainian*, Lowland, plain

O
-ø *Danish*, Island
Ober- *German*, Upper
Oblast' *Russian*, Province
Órmos *Greek*, Bay
Orol(i) *Uzbek*, Island
Øster- *Norwegian*, Eastern
Ostrov(a) *Russian*, Island(s)
Otok *Serbo-Croatian*, Island
Oued *Arabic*, Watercourse
-oy *Faeroese*, Island
-øy(a) *Norwegian*, Island
Oya *Sinhala*, River
Ozero *Russian, Ukrainian*, Lake

P
Passo *Italian*, Pass
Pegunungan *Indonesian, Malay*, Mountain range
Pélagos *Greek*, Sea
Pendi *Chinese*, Basin
Penisola *Italian*, Peninsula
Pertuis *French*, Strait
Peski *Russian*, Sands
Phanom *Thai*, Mountain
Phou *Laotian*, Mountain
Pi *Chinese*, Point
Pic *Catalan, French*, Peak
Pico *Portuguese, Spanish*, Peak
-piggen *Danish*, Peak
Pik *Russian*, Peak
Pivostriv *Ukrainian*, Peninsula
Planalto *Portuguese*, Plateau
Planina, Planini *Bulgarian, Macedonian, Serbo-Croatian*, Mountain range
Plato *Russian*, Plateau
Ploskogor'ye *Russian*, Upland
Poluostrov *Russian*, Peninsula
Ponta *Portuguese*, Point
Porthmós *Greek*, Strait
Pótamos *Greek*, River
Presa *Spanish*, Dam
Prokhod *Bulgarian*, Pass
Proliv *Russian*, Strait
Pulau *Indonesian, Malay*, Island
Pulu *Malay*, Island
Punta *Spanish*, Point
Pushcha *Belarussian*, Forest
Puszcza *Polish*, Forest

Q
Qā' *Arabic*, Depression
Qalamat *Arabic*, Well
Qatorkŭh(i) *Tajik*, Mountain
Qiuling *Chinese*, Hills
Qolleh *Persian*, Mountain
Qu *Tibetan*, Stream
Quan *Chinese*, Well
Qulla(i) *Tajik*, Peak
Qundao *Chinese*, Island group

R
Raas *Somali*, Cape
-rags *Latvian*, Cape
Ramlat *Arabic*, Sands
Ra's *Arabic*, Cape, headland, point
Ravnina *Bulgarian, Russian*, Plain
Récif *French*, Reef
Recife *Portuguese*, Reef
Reka *Bulgarian*, River
Represa (Rep.) *Portuguese, Spanish*, Reservoir
Reshteh *Persian*, Mountain range
Respublika *Russian*, Republic, first-order administrative division
Respublika(si) *Uzbek*, Republic, first-order administrative division
-retsugan *Japanese*, Chain of rocks
-rettō *Japanese*, Island chain
Riacho *Spanish*, Stream
Riban' *Malagasy*, Mountains
Rio *Portuguese*, River
Río *Spanish*, River
Riu *Catalan*, River
Rivier *Dutch*, River
Rivière *French*, River
Rowd *Pashtu*, River
Rt *Serbo-Croatian*, Point
Rūd *Persian*, River
Rūdkhāneh *Persian*, River
Rudohorie *Slovak*, Mountains
Ruisseau *French*, Stream

S
-saar *Estonian*, Island
-saari *Finnish*, Island
Sabkhat *Arabic*, Salt marsh
Sāgar(a) *Hindi*, Lake, reservoir
Ṣaḥrā' *Arabic*, Desert
Saint, Sainte *French*, Saint
Salar *Spanish*, Salt-pan
Salto *Portuguese, Spanish*, Waterfall
Samudra *Sinhala*, Reservoir
-san *Japanese, Korean*, Mountain
-sanchi *Japanese*, Mountains
-sandur *Icelandic*, Beach
Sankt *German, Swedish*, Saint
-sanmaek *Korean*, Mountain range
-sanmyaku *Japanese*, Mountain range
San, Santa, Santo *Italian, Portuguese, Spanish*, Saint
São *Portuguese*, Saint
Sarīr *Arabic*, Desert
Sebkha, Sebkhet *Arabic*, Depression, salt marsh
Sedlo *Czech*, Pass
See *German*, Lake
Selat *Indonesian*, Strait
Selatan *Indonesian*, Southern
-selkä *Finnish*, Lake, ridge
Selseleh *Persian*, Mountain range
Serra *Portuguese*, Mountain
Serranía *Spanish*, Mountain
-seto *Japanese*, Channel, strait
Sever-naya, -noye, -nyy, -o *Russian*, Northern
Sha'īb *Arabic*, Watercourse
Shākh *Kurdish*, Mountain
Shamo *Chinese*, Desert
Shan *Chinese*, Mountain(s)
Shankou *Chinese*, Pass
Shanmo *Chinese*, Mountain range
Shaṭṭ *Arabic*, Distributary
Shet' *Amharic*, River
Shi *Chinese*, Municipality
-shima *Japanese*, Island
Shiqqat *Arabic*, Depression
-shotō *Japanese*, Group of islands
Shuiku *Chinese*, Reservoir
Shŭrkhog(i) *Uzbek*, Salt marsh
Sierra *Spanish*, Mountains
Sint *Dutch*, Saint
-sjø(en) *Norwegian*, Lake
-sjön *Swedish*, Lake
Solonchak *Russian*, Salt lake
Solonchakovyye Vpadiny *Russian*, Salt basin, wetlands
Sơn *Vietnamese*, Mountain
Sông *Vietnamese*, River
Sør- *Norwegian*, Southern
-spitze *German*, Peak
Star-á, -é *Czech*, Old
Star-aya, -oye, -yy, -yye *Russian*, Old
Stenó *Greek*, Strait
Step' *Russian*, Steppe
Štít *Slovak*, Peak
Stœng *Cambodian*, River
Stolovaya Strana *Russian*, Plateau
Stredné *Slovak*, Middle
Střední *Czech*, Middle
Stretto *Italian*, Strait
Su Anbari *Azerbaijani*, Reservoir
-suidō *Japanese*, Channel, strait
Sund *Swedish*, Sound, strait
Sungai *Indonesian, Malay*, River
Suu *Turkish*, River

T
Tal *Mongolian*, Plain
Tandavan' *Malagasy*, Mountain range
Tangorombohitr' *Malagasy*, Mountain massif
Tanjung *Indonesian, Malay*, Cape, point
Tao *Chinese*, Island
Ṭaraq *Arabic*, Hills
Tassili *Berber*, Mountain, plateau
Tau *Russian*, Mountain(s)
Taungdan *Burmese*, Mountain range
Techníti Límni *Greek*, Reservoir
Tekojärvi *Finnish*, Reservoir
Teluk *Indonesian, Malay*, Bay
Tengah *Indonesian*, Middle
Terara *Amharic*, Mountain
Timur *Indonesian*, Eastern
-tind(an) *Norwegian*, Peak
Tizma(si) *Uzbek*, Mountain range, ridge
-tō *Japanese*, Island
Tog *Somali*, Valley
-tōge *Japanese*, Pass
Togh(i) *Uzbek*, Mountain
Tônlé *Cambodian*, Lake
Top *Dutch*, Peak
-tunturi *Finnish*, Mountain
Ṭurāq *Arabic*, Hills
Tur'at *Arabic*, Channel

U
Udde(n) *Swedish*, Cape, point
'Uqlat *Arabic*, Well
Utara *Indonesian*, Northern
Uul *Mongolian*, Mountains

V
Väin *Estonian*, Strait
Vallée *French*, Valley
-vatn *Icelandic*, Lake
-vatnet *Norwegian*, Lake
Velayat *Turkmen*, Province
-vesi *Finnish*, Lake
Vestre- *Norwegian*, Western
-vidda *Norwegian*, Plateau
-vík *Icelandic*, Bay
-viken *Swedish*, Bay, inlet
Vinh *Vietnamese*, Bay
Víztárloló *Hungarian*, Reservoir
Vodaskhovishcha *Belarussian*, Reservoir
Vodokhranilishche (Vdkhr.) *Russian*, Reservoir
Vodoskhovyshche (Vdskh.) *Ukrainian*, Reservoir
Volcán *Spanish*, Volcano
Vostochn-o, yy *Russian*, Eastern
Vozvyshennost' *Russian*, Upland, plateau
Vozyera *Belarussian*, Lake
Vpadina *Russian*, Depression
Vrchovina *Czech*, Mountains
Vrha *Macedonian*, Peak
Vychodné *Slovak*, Eastern
Vysochyna *Ukrainian*, Upland
Vysočina *Czech*, Upland

W
Waadi *Somali*, Watercourse
Wādī *Arabic* Watercourse
Wāḥat, Wâhat *Arabic*, Oasis
Wald *German*, Forest
Wan *Chinese*, Bay
Way *Indonesian*, River
Webi *Somali*, River
Wenz *Amharic*, River
Wiloyat(i) *Uzbek*, Province
Wyżyna *Polish*, Upland
Wzgórza *Polish*, Upland
Wzvyshsha *Belarussian*, Upland

X
Xé *Laotian*, River
Xi *Chinese*, Stream

Y
-yama *Japanese*, Mountain
Yanchi *Chinese*, Salt lake
Yang *Chinese*, Bay
Yanhu *Chinese*, Salt lake
Yarımadası *Azerbaijani, Turkish*, Peninsula
Yaylası *Turkish*, Plateau
Yazovir *Bulgarian*, Reservoir
Yoma *Burmese*, Mountains
Ytre- *Norwegian*, Outer
Yü *Chinese*, Island
Yunhe *Chinese*, Canal
Yuzhn-o, -yy *Russian*, Southern

Z
-zaki *Japanese*, Cape, point
Zaliv *Bulgarian, Russian*, Bay
-zan *Japanese*, Mountain
Zangbo *Tibetan*, River
Zapadn-aya, -o, -yy *Russian*, Western
Západné *Slovak*, Western
Západní *Czech*, Western
Zatoka *Polish, Ukrainian*, Bay
-zee *Dutch*, Sea
Zemlya *Russian*, Earth, land
Zizhiqu *Chinese*, Autonomous region

Index

Glossary of Abbreviations

This glossary provides a comprehensive guide to the abbreviations used in this Atlas, and in the Index.

A

abbrev. abbreviated
AD Anno Domini
Afr. Afrikaans
Alb. Albanian
Amh. Amharic
anc. ancient
approx. approximately
Ar. Arabic
Arm. Armenian
ASEAN Association of South East Asian Nations
ASSR Autonomous Soviet Socialist Republic
Aust. Australian
Az. Azerbaijani
Azerb. Azerbaijan

B

Basq. Basque
BC before Christ
Bel. Belorussian
Ben. Bengali
Ber. Berber
B-H Bosnia-Herzegovina
bn billion (one thousand million)
BP British Petroleum
Bret. Breton
Brit. British
Bul. Bulgarian
Bur. Burmese

C

C central
C. Cape
°C degrees Centigrade
CACM Central America Common Market
Cam. Cambodian
Cant. Cantonese
CAR Central African Republic
Cast. Castilian
Cat. Catalan
CEEAC Central America Common Market
Chin. Chinese
CIS Commonwealth of Independent States
cm centimetre(s)
Cro. Croat
Cz. Czech
Czech Rep. Czech Republic

D

Dan. Danish
Div. Divehi
Dom. Rep. Dominican Republic
Dut. Dutch

E

E east
EC *see* EU
EEC *see* EU
ECOWAS Economic Community of West African States
ECU European Currency Unit
EMS European Monetary System
Eng. English
est estimated
Est. Estonian
EU European Union (previously European Community [EC], European Economic Community [EEC])

F

°F degrees Fahrenheit
Faer. Faeroese
Fij. Fijian
Fin. Finnish
Fr. French
Fris. Frisian
ft foot/feet
FYROM Former Yugoslav Republic of Macedonia

G

g gram(s)
Gael. Gaelic
Gal. Galician
GDP Gross Domestic Product (the total value of goods and services produced by a country excluding income from foreign countries)
Geor. Georgian
Ger. German
Gk Greek
GNP Gross National Product (the total value of goods and services produced by a country)

H

Heb. Hebrew
HEP hydro-electric power
Hind. Hindi
hist. historical
Hung. Hungarian

I

I. Island
Icel. Icelandic
in inch(es)
In. Inuit (Eskimo)
Ind. Indonesian
Intl International
Ir. Irish
Is Islands
It. Italian

J

Jap. Japanese

K

Kaz. Kazakh
kg kilogram(s)
Kir. Kirghiz
km kilometre(s)
km² square kilometre (singular)
Kor. Korean
Kurd. Kurdish

L

L. Lake
LAIA Latin American Integration Association
Lao. Laotian
Lapp. Lappish
Lat. Latin
Latv. Latvian
Liech. Liechtenstein
Lith. Lithuanian
Lux. Luxembourg

M

m million/metre(s)
Mac. Macedonian
Maced. Macedonia
Mal. Malay
Malg. Malagasy
Malt. Maltese
mi. mile(s)
Mong. Mongolian
Mt. Mountain
Mts Mountains

N

N north
NAFTA North American Free Trade Agreement
Nep. Nepali
Neth. Netherlands
Nic. Nicaraguan
Nor. Norwegian
NZ New Zealand

P

Pash. Pashtu
PNG Papua New Guinea
Pol. Polish
Poly. Polynesian
Port. Portuguese
prev. previously

R

Rep. Republic
Res. Reservoir
Rmsch Romansch
Rom. Romanian
Rus. Russian
Russ. Fed. Russian Federation

S

S south
SADC Southern Africa Development Community
SCr. Serbo-Croatian
Sinh. Sinhala
Slvk Slovak
Slvn. Slovene
Som. Somali
Sp. Spanish
St., St Saint
Strs Straits
Swa. Swahili
Swe. Swedish
Switz. Switzerland

T

Taj. Tajik
Th. Thai
Thai. Thailand
Tib. Tibetan
Turk. Turkish
Turkm. Turkmenistan

U

UAE United Arab Emirates
Uigh. Uighur
UK United Kingdom
Ukr. Ukrainian
UN United Nations
Urd. Urdu
US/USA United States of America
USSR Union of Soviet Socialist Republics
Uzb. Uzbek

V

var. variant
Vdkhr. Vodokhranilishche (Russian for reservoir)
Vdskh. Vodoskhovyshche (Ukrainian for reservoir)
Vtn. Vietnamese

W

W west
Wel. Welsh

Y

Yugo. Yugoslavia

This index lists all the placenames and features shown on the regional and continental maps in this Atlas. Placenames are referenced to the largest scale map on which they appear. The policy followed throughout the Atlas is to use the local spelling or local name at regional level; commonly-used English language names may occasionally be added (in parentheses) where this is an aid to identification e.g. Firenze (Florence). English names, where they exist, have been used for all international features e.g. oceans and country names; they are also used on the continental maps and in the introductory World Today section; these are then fully cross-referenced to the local names found on the regional maps. The index also contains commonly-found alternative names and variant spellings, which are also fully cross-referenced.

All main entry names are those of settlements unless otherwise indicated by the use of italicized definitions or representative symbols, which are keyed at the foot of each page.

1

25 de Mayo *see* Veinticinco de Mayo
143 *Y13* **26 Bakı Komissarı** ***Rus.*** Imeni 26 Bakinskikh Komissarov. SE Azerbaijan *39.18N 49.13E*
26 Baku Komissarlary Adyndaky *see* Imeni 26 Bakinskikh Komissarov
8 *M16* **100 Mile House** ***var.*** Hundred Mile House. British Columbia, SW Canada *51.39N 121.19W*

A

Aa *see* Gauja
Aabenraa *see* Åbenrå
Aabybro *see* Åbybro
103 *C16* **Aachen** ***Dut.*** Aken, ***Fr.*** Aix-la-Chapelle; ***anc.*** Aquae Grani, Aquisgranum. Nordrhein-Westfalen, W Germany *50.47N 6.06E*
Aaiún *see* Laâyoune
Aakirkeby *see* Åkirkeby
Aalborg *see* Ålborg
Aalborg Bugt *see* Ålborg Bugt
103 *J21* **Aalen** Baden-Württemberg, S Germany *48.49N 10.06E*
Aalestrup *see* Ålestrup
100 *I11* **Aalsmeer** Noord-Holland, C Netherlands *52.16N 4.43E*
101 *F18* **Aalst** ***Fr.*** Alost. Oost-Vlaanderen, C Belgium *50.57N 4.03E*
101 *K18* **Aalst** Noord-Brabant, S Netherlands *51.46N 5.07E*
100 *O12* **Aalten** Gelderland, E Netherlands *51.55N 6.34E*
101 *D17* **Aalter** Oost-Vlaanderen, NW Belgium *51.04N 3.28E*
Aanaar *see* Inari
Aanaarjävri *see* Inarijärvi
95 *M17* **Äänekoski** Länsi-Suomi, W Finland *62.33N 25.44E*
144 *H7* **Aanjar** ***var.*** 'Anjar. C Lebanon *33.45N 35.56E*
85 *G21* **Aansluit** Northern Cape, N South Africa *26.41S 22.24E*
Aar *see* Aare
110 *F7* **Aarau** Aargau, N Switzerland *47.22N 8.00E*
110 *D8* **Aarberg** Bern, W Switzerland *47.02N 7.15E*
101 *D16* **Aardenburg** Zeeland, SW Netherlands *51.16N 3.27E*
110 *D8* **Aare** ***var.*** Aar. ≈ W Switzerland
110 *F7* **Aargau** ***Fr.*** Argovie. ◆ *canton* N Switzerland
Aarhus *see* Århus
Aarlen *see* Arlon
Aars *see* Års
101 *I17* **Aarschot** Vlaams Brabant, C Belgium *50.58N 4.49E*
Aassi, Nahr el *see* Orontes
Aat *see* Ath
166 *G7* **Aba** ***prev.*** Ngawa. Sichuan, C China *32.51N 101.46E*
79 *V17* **Aba** Abia, S Nigeria *5.06N 7.22E*
81 *P16* **Aba** Orientale, NE Dem. Rep. Congo (Zaire) *3.52N 30.13E*
146 *J6* **Abā al Qazāz, Bi'r** *well* NW Saudi Arabia *26.37N 36.50E*
Abā as Su'ūd *see* Najrān
61 *G14* **Abacaxis, Rio** ≈ NW Brazil
Abaco Island *see* Great Abaco/Little Abaco
Abaco Island *see* Great Abaco, N Bahamas
148 *K10* **Ābādān** Khūzestān, SW Iran *30.24N 48.18E*
149 *O10* **Ābādeh** Fārs, C Iran *31.10N 52.39E*
76 *H8* **Abadla** W Algeria *31.04N 2.39W*
61 *M20* **Abaeté** Minas Gerais, SE Brazil *19.10S 45.24W*
169 *Q10* **Abag Qi** ***var.*** Xin Hot. Nei Mongol Zizhiqu, N China *43.58N 114.59E*
64 *P7* **Abaí** Caazapá, S Paraguay *25.58S 55.54W*
203 *O2* **Abaiang** ***var.*** Apia; ***prev.*** Charlotte Island. *atoll* Tungaru, W Kiribati
Abaj *see* Abay
79 *U15* **Abaji** Federal Capital District, C Nigeria *8.35N 6.54E*
39 *O7* **Abajo Peak** ▲ Utah, W USA *37.51N 109.28W*
79 *V16* **Abakaliki** Ebonyi, SE Nigeria *6.18N 8.07E*
126 *Hh15* **Abakan** Respublika Khakasiya, S Russian Federation *53.43N 91.25E*
126 *Hh15* **Abakan** ≈ S Russian Federation
79 *S11* **Abala** Tillabéri, SW Niger *14.55N 3.27E*
79 *U11* **Abalak** Tahoua, C Niger *15.28N 6.18E*
121 *N14* **Abalyanka** ***Rus.*** Obolyanka. ≈ N Belarus
126 *Ii14* **Aban** Krasnoyarskiy Kray, S Russian Federation *56.41N 96.04E*
149 *P9* **Āb Anbār-e Kān Sorkh** Yazd, C Iran *31.22N 53.37E*
59 *G16* **Abancay** Apurímac, SE Peru *13.37S 72.52W*
202 *H2* **Abaokoro** *atoll* Tungaru, W Kiribati
Abariringa *see* Kanton
149 *P10* **Abarkū** Yazd, C Iran
172 *Qq5* **Abashiri** ***var.*** Abasiri. Hokkaidō, NE Japan *44.00N 144.15E*
172 *Q6* **Abashiri-gawa** ≈ Hokkaidō, NE Japan
172 *Q5* **Abashiri-ko** ◎ Hokkaidō, NE Japan
Abasiri *see* Abashiri
43 *P10* **Abasolo** Tamaulipas, C Mexico *24.02N 98.18W*
194 *L16* **Abau** Central, S PNG *10.09S 148.40E*
151 *R10* **Abay** ***var.*** Abaj. Karaganda, C Kazakhstan *49.34N 72.54E*
83 *I15* **Ābaya Hāyk'** ***Eng.*** Lake Margherita, ***It.*** Abbaia. ◎ SW Ethiopia
Ābay Wenz *see* Blue Nile
126 *Hh15* **Abaza** Respublika Khakasiya, S Russian Federation *52.40N 89.58E*
Abbaia *see* Ābaya Hāyk'
149 *Q13* **Āb Bārīk** Fārs, S Iran
109 *C18* **Abbasanta** Sardegna, Italy, C Mediterranean Sea *40.08N 8.49E*
Abbatis Villa *see* Abbeville
32 *M3* **Abbaye, Point** *headland* Michigan, N USA *46.58N 88.08W*
Abbazia *see* Opatija
Abbé, Lake *see* Abhe, Lake
105 *N2* **Abbeville** ***anc.*** Abbatis Villa. Somme, N France *50.06N 1.50E*
25 *R7* **Abbeville** Alabama, S USA *31.34N 85.16W*
25 *U6* **Abbeville** Georgia, SE USA *31.58N 83.18W*
24 *I9* **Abbeville** Louisiana, S USA *29.58N 92.08W*
23 *P12* **Abbeville** South Carolina, SE USA *34.10N 82.22W*
99 *B20* **Abbeyfeale** ***Ir.*** Mainistir na Féile. SW Ireland *52.24N 9.21W*
108 *D8* **Abbiategrasso** Lombardia, NW Italy *45.24N 9.04E*
95 *I14* **Abborrträsk** Norrbotten, N Sweden *65.24N 19.33E*
204 *J9* **Abbot Ice Shelf** *ice shelf* Antarctica
8 *M17* **Abbotsford** British Columbia, SW Canada *49.01N 122.18W*
32 *K6* **Abbotsford** Wisconsin, N USA *44.57N 90.19W*
155 *U5* **Abbottābād** North-West Frontier Province, NW Pakistan *34.08N 73.10E*
121 *M14* **Abchuha** ***Rus.*** Obchuga. Minskaya Voblasts', NW Belarus *54.30N 29.23E*
100 *I10* **Abcoude** Utrecht, C Netherlands *52.16N 4.58E*
145 *N2* **'Abd al 'Azīz, Jabal** ▲▲ NE Syria
147 *U17* **'Abd al Kūrī** *island* SE Yemen
145 *Z13* **'Abd Allāh, Khawr** *bay* Iraq/Kuwait
131 *U6* **Abdulino** Orenburgskaya Oblast', W Russian Federation *53.37N 53.39E*
80 *J10* **Abéché** ***var.*** Abécher, Abeshr. Ouaddaï, SE Chad *13.49N 20.49E*
Abécher *see* Abéché
149 *S8* **Āb-e Garm va Sard** Khorāsān, E Iran
79 *R8* **Abeïbara** Kidal, NE Mali *19.07N 1.52E*
107 *P5* **Abejar** Castilla-León, N Spain *41.48N 2.46W*
56 *E9* **Abejorral** Antioquia, W Colombia *5.52N 75.34W*
Abela *see* Ávila
Abellinum *see* Avellino
94 *Q2* **Abeløya** *island* Kong Karls Land, E Svalbard
82 *I13* **Ābeltī** Oromo, C Ethiopia *8.09N 37.31E*
203 *O2* **Abemama** ***var.*** Apamama; ***prev.*** Roger Simpson Island. *atoll* Tungaru, W Kiribati
176 *Yy14* **Abemaree** ***var.*** Abemarre. Irian Jaya, E Indonesia *7.03S 140.10E*
79 *O16* **Abengourou** E Ivory Coast *6.42N 3.27W*
97 *G24* **Åbenrå** ***var.*** Aabenraa, ***Ger.*** Apenrade. Sønderjylland, SW Denmark *55.03N 9.25E*
103 *L22* **Abens** ≈ SE Germany
79 *S16* **Abeokuta** Ogun, SW Nigeria *7.07N 3.21E*
99 *I20* **Aberaeron** SW Wales, UK *52.15N 4.15W*
Aberbrothock *see* Arbroath
Abercorn *see* Mbala
31 *R6* **Abercrombie** North Dakota, N USA *46.25N 96.42W*
191 *T7* **Aberdeen** New South Wales, SE Australia *32.09S 150.55E*
11 *T15* **Aberdeen** Saskatchewan, S Canada *52.15N 106.19W*
85 *H25* **Aberdeen** Eastern Cape, S South Africa *32.30S 24.00E*
98 *L9* **Aberdeen** ***anc.*** Devana. NE Scotland, UK *57.10N 2.04W*
23 *X2* **Aberdeen** Maryland, NE USA *39.28N 76.09W*
25 *N3* **Aberdeen** Mississippi, S USA *33.49N 88.32W*
23 *T10* **Aberdeen** North Carolina, SE USA *35.07N 79.25W*
31 *P8* **Aberdeen** South Dakota, N USA *45.27N 98.29W*
34 *F8* **Aberdeen** Washington, NW USA *46.57N 123.48W*
98 *K9* **Aberdeen** *cultural region* NE Scotland, UK
15 *K6* **Aberdeen Lake** ◎ Nunavut, NE Canada
98 *J10* **Aberfeldy** C Scotland, UK *56.38N 3.48W*
99 *K21* **Abergavenny** ***anc.*** Gobannium. SE Wales, UK *51.50N 3.00W*
Abergwaun *see* Fishguard
Abermarre *see* Abemaree
27 *N5* **Abernathy** Texas, SW USA *33.49N 101.50W*
Abersee *see* Wolfgangsee
Abertawe *see* Swansea
Aberteifi *see* Cardigan
34 *I15* **Abert, Lake** ◎ Oregon, NW USA
99 *I20* **Aberystwyth** W Wales, UK *52.25N 4.04W*
Abeshr *see* Abéché
108 *F10* **Abetone** Toscana, C Italy *44.09N 10.42E*
129 *V5* **Abez'** Respublika Komi, NW Russian Federation *66.32N 61.41E*
148 *M5* **Āb Garm** Qazvin, N Iran
147 *N12* **Abhā** 'Asīr, SW Saudi Arabia *18.16N 42.31E*
148 *M5* **Abhar** Zanjān, NW Iran *36.07N 49.19E*
Abhé Bad/Ābhē Bid Hāyk' *see* Abhe, Lake
82 *K12* **Abhe, Lake** ***var.*** Lake Abbé, ***Amh.*** Ābhē Bid Hāyk', ***Som.*** Abhé Bad. ◎ Djibouti/Ethiopia
79 *V17* **Abia** ◆ *state* SE Nigeria
145 *V9* **'Abīd 'Alī** E Iraq *32.20N 45.58E*
121 *O17* **Abidavichy** ***Rus.*** Obidovichi. Mahilyowskaya Voblasts', E Belarus *53.21N 30.24E*
117 *L15* **Abide** Çanakkale, NW Turkey *40.04N 26.13E*
79 *N17* **Abidjan** S Ivory Coast *5.19N 4.01W*
Āb-i-Istāda *see* Īstādeh-ye Moqor, Āb-e-
29 *N4* **Abilene** Kansas, C USA *38.55N 97.12W*
27 *Q7* **Abilene** Texas, SW USA *32.27N 99.43W*
Abindonia *see* Abingdon
99 *M21* **Abingdon** ***anc.*** Abindonia. S England, UK *51.40N 1.16W*
32 *K12* **Abingdon** Illinois, N USA *40.48N 90.24W*
23 *P8* **Abingdon** Virginia, NE USA *36.42N 81.58W*
Abingdon *see* Pinta, Isla
20 *J15* **Abington** Pennsylvania, NE USA *40.06N 75.05W*
130 *K14* **Abinsk** Krasnodarskiy Kray, SW Russian Federation *44.51N 38.12E*
39 *R9* **Abiquiu Reservoir** ▣ New Mexico, SW USA
94 *I10* **Abisko** Norrbotten, N Sweden *68.21N 18.49E*
10 *G12* **Abitibi** ≈ Ontario, S Canada
10 *H12* **Abitibi, Lac** ◎ Ontario/Quebec, S Canada
82 *J10* **Ābīy Ādī** Tigray, N Ethiopia *13.40N 38.57E*
120 *H6* **Abja-Paluoja** Viljandimaa, S Estonia *58.07N 25.19E*
143 *Q8* **Abkhazia** ◆ *autonomous republic* NW Georgia
190 *F1* **Abminga** South Australia *26.07S 134.49E*
77 *W9* **Abnûb** C Egypt *27.18N 31.09E*
Åbo *see* Turku
158 *G9* **Abohar** Punjab, N India *30.10N 74.12E*
79 *O17* **Aboisso** SE Ivory Coast *5.27N 3.06W*
80 *H5* **Abo, Massif d'** ▲▲ NW Chad
79 *R16* **Abomey** S Benin *7.14N 2.00E*
81 *F16* **Abong Mbang** Est, SE Cameroon *3.58N 13.10E*
113 *L23* **Abony** Pest, C Hungary *47.10N 20.00E*
80 *J11* **Abou-Déïa** Salamat, SE Chad *11.30N 19.18E*
Aboudouhour *see* Abū aḍ Ḍuhūr
Abou Kémal *see* Abū Kamāl
Abou Simbel *see* Abu Simbel
143 *T12* **Abovyan** C Armenia *40.16N 44.33E*
179 *P8* **Abra** ≈ Luzon, N Philippines
147 *P15* **Abrād, Wādī** *seasonal river* W Yemen
Abraham Bay *see* The Carlton
106 *G10* **Abrantes** ***var.*** Abrántes. Santarém, C Portugal *39.28N 8.12W*
64 *J4* **Abra Pampa** Jujuy, N Argentina *22.46S 65.40W*
Abrashlare *see* Brezovo
56 *G7* **Abrego** Norte de Santander, N Colombia *8.07N 73.15W*
Abrene *see* Pytalovo
42 *C7* **Abreojos, Punta** *headland* W Mexico *26.43N 113.36W*
67 *J16* **Abrolhos Bank** *undersea feature* W Atlantic Ocean
121 *H19* **Abrova** ***Rus.*** Obrovo. Brestskaya Voblasts', SW Belarus *52.30N 25.31E*
118 *G11* **Abrud** ***Ger.*** Gross-Schlatten, ***Hung.*** Abrudbánya. Alba, SW Romania *46.15N 23.07E*
Abrudbánya *see* Abrud
120 *E6* **Abruka** *island* SW Estonia
109 *J15* **Abruzzese, Appennino** ▲▲ C Italy
109 *J14* **Abruzzo** ◆ *region* C Italy
147 *N14* **'Abs** ***var.*** Sūq 'Abs. W Yemen *16.42N 42.55E*
35 *T12* **Absaroka Range** ▲▲ Montana/Wyoming, NW USA
143 *Z11* **Abşeron Yarımadası** ***Rus.*** Apsheronskiy Poluostrov. *peninsula* E Azerbaijan
149 *N6* **Āb Shīrīn** Eşfahān, C Iran
145 *X10* **Abtān** SE Iraq *31.37N 47.06E*
111 *R6* **Abtenau** Salzburg, NW Austria *47.33N 13.21E*
170 *Dd12* **Abu** Yamaguchi, Honshū, SW Japan *34.30N 131.26E*
158 *E14* **Ābu** Rājasthān, N India *24.40N 72.49E*
144 *I4* **Abū aḍ Ḍuhūr** ***Fr.*** Aboudouhour. Idlib, NW Syria *35.30N 37.00E*
149 *P17* **Abū al Abyaḍ** *island* C UAE
144 *K10* **Abū al Ḥuṣayn, Khabrat** ◎ N Jordan
145 *R8* **Abū al Jīr** C Iraq *33.16N 42.55E*
145 *Y12* **Abū al Khaṣīb** ***var.*** Abul Khasib. SE Iraq *30.26N 48.00E*
145 *U12* **Abū at Tubrah, Thaqb** *well* S Iraq *30.54N 45.02E*
77 *V11* **Abu Balâs** ▲ SW Egypt *24.28N 27.36E*
Abu Dhabi *see* Abū Ẓaby
145 *R8* **Abū Farūkh** C Iraq *33.06N 43.18E*
82 *C12* **Abu Gabra** Southern Darfur, W Sudan *11.01N 26.49E*
145 *P10* **Abū Ghār, Sha'īb** *dry watercourse* S Iraq
82 *G7* **Abu Hamed** River Nile, N Sudan *19.31N 33.19E*
145 *O5* **Abū Ḥardān** ***var.*** Hajîne. Dayr az Zawr, E Syria *34.45N 40.49E*
145 *T7* **Abū Ḥāssawīyah** E Iraq *33.52N 44.47E*
144 *K10* **Abū Ḥifnah, Wādī** *dry watercourse* N Jordan
79 *V15* **Abuja** ● (Nigeria) Federal Capital District, C Nigeria *9.04N 7.28E*
145 *R9* **Abū Jahaf, Wādī** *dry watercourse* C Iraq
58 *F12* **Abujao, Río** ≈ E Peru
145 *U12* **Abū Jasrah** S Iraq *30.43N 44.50E*
145 *O6* **Abū Kamāl** ***Fr.*** Abou Kémal. Dayr az Zawr, E Syria *34.28N 40.55E*
175 *Q11* **Abuki, Pegunungan** ▲▲ Sulawesi, C Indonesia
171 *Ll14* **Abukuma-gawa** ≈ Honshū, C Japan
171 *Ll15* **Abukuma-sanchi** ▲▲ Honshū, C Japan
Abula *see* Ávila
Abul Khasib *see* Abū al Khaṣīb
81 *K16* **Abumombazi** ***var.*** Abumonbazi. Equateur, N Dem. Rep. Congo (Zaire) *3.43N 22.06E*
Abumonbazi *see* Abumombazi
61 *D15* **Abunã** Rondônia, W Brazil *9.40S 65.19W*
58 *K13* **Abunã, Rio** ***var.*** Río Abuná. ≈ Bolivia/Brazil
144 *G10* **Abū Nuṣayr** ***var.*** Abu Nuseir. 'Ammān, W Jordan *32.03N 35.58E*
Abu Nuseir *see* Abū Nuṣayr
145 *T12* **Abū Qabr** S Iraq *31.03N 44.34E*
144 *K5* **Abū Raḥbah, Jabal** ▲▲ C Syria
145 *S5* **Abū Rajāsh** N Iraq *34.47N 43.36E*
145 *W13* **Abū Raqrāq, Ghadīr** *well* S Iraq *30.15N 45.57E*
158 *E14* **Ābu Road** Rājasthān, N India *24.28N 72.46E*
82 *I6* **Abu Shagara, Ras** *headland* NE Sudan *18.04N 38.31E*
77 *W12* **Abu Simbel** ***var.*** Abou Simbel, Abū Sunbul. *ancient monument* S Egypt *22.25N 31.37E*
145 *U12* **Abū Sudayrah** S Iraq *30.55N 44.58E*
145 *T10* **Abū Şukhayr** S Iraq *31.54N 44.27E*
Abū Sunbul *see* Abu Simbel
172 *Nn6* **Abuta** Hokkaidō, NE Japan *42.34N 140.44E*
193 *E18* **Abut Head** *headland* South Island, NZ *43.06S 170.16E*
82 *E9* **Abu 'Urug** Northern Kordofan, C Sudan *15.52N 30.25E*
82 *K12* **Ābuyē Mēda** ▲ C Ethiopia *10.23N 39.46E*
179 *R13* **Abuyog** Leyte, C Philippines *10.45N 124.58E*
82 *D11* **Abu Zabad** Western Kordofan, C Sudan *12.21N 29.16E*
Abū Ẓabī *see* Abū Ẓaby
149 *P16* **Abū Ẓaby** ***var.*** Abū Ẓabī, ***Eng.*** Abu Dhabi. ● (UAE) Abū Ẓaby, C UAE *24.30N 54.20E*

◆ Country
● Country Capital
◇ Dependent Territory
○ Dependent Territory Capital
◆ Administrative Region
✕ International Airport
▲ Mountain
▲▲ Mountain Range
Volcano
≈ River
◎ Lake
▣ Reservoir

77 X8 **Abu Zenima** E Egypt 29.01N 33.08E
97 N17 **Åby** Östergötland, S Sweden 58.40N 16.19E
Abyad, Al Bahr al see White Nile
97 G20 **Åbybro** var. Aabybro. Nordjylland, N Denmark 57.09N 9.45E
82 K13 **Abyei** Western Kordofan, S Sudan 9.34N 28.26E
Abyla see Ávila
Abymes see les Abymes
Abyssinia see Ethiopia
Açâba see Assaba
56 F11 **Acacías** Meta, C Colombia 3.58N 73.46W
60 L13 **Açailándia** Maranhão, E Brazil 4.51S 47.25W
Acaill see Achill Island
44 E8 **Acajutla** Sonsonate, W El Salvador 13.35N 89.48W
81 D17 **Acalayong** SW Equatorial Guinea 1.05N 9.34E
43 N13 **Acámbaro** Guanajuato, C Mexico 20.01N 100.45W
56 C6 **Acandí** Chocó, NW Colombia 8.28N 77.18W
106 H4 **A Cañiza** var. La Cañiza. Galicia, NW Spain 42.13N 8.16W
42 J11 **Acaponeta** Nayarit, C Mexico 22.30N 105.21W
42 J11 **Acaponeta, Río de** ≈ C Mexico
43 O16 **Acapulco** var. Acapulco de Juárez. Guerrero, S Mexico 16.51N 99.53W
Acapulco de Juárez see Acapulco
57 T17 **Acaraí Mountains** Sp. Serra Acaraí. ▲ Brazil/Guyana
Acaraí, Serra see Acaraí Mountains
60 O13 **Acaraú** Ceará, NE Brazil 4.35S 37.37W
56 J6 **Acarigua** Portuguesa, N Venezuela 9.34N 69.12W
44 G6 **Acatenango, Volcán de** ℝ S Guatemala 14.30N 90.52W
42 Q15 **Acatlán** var. Acatlán de Osorio. Puebla, S Mexico 18.12N 98.01W
Acatlán de Osorio see Acatlán
43 S13 **Acayucan** var. Acayucán. Veracruz-Llave, E Mexico 17.58N 94.58W
Accho see 'Akko
23 Y5 **Accomac** Virginia, NE USA 37.43N 75.39W
79 Q17 **Accra** ● (Ghana) SE Ghana 5.33N 0.15W
99 L17 **Accrington** NW England, UK 53.46N 2.21W
63 B19 **Acebal** Santa Fe, C Argentina 33.13S 60.49W
103 Ee4 **Aceh** off. Daerah Istimewa Aceh, var. Acheen, Achin, Atchin, Atjeh. ◆ autonomous district NW Indonesia
109 M18 **Acerenza** Basilicata, S Italy 40.46N 15.51E
109 K17 **Acerra** anc. Acerrae. Campania, S Italy 40.55N 14.22E
Acerrae see Acerra
Ach'asar Lerr see Achkasar
59 J17 **Achacachi** La Paz, W Bolivia 16.04S 68.39W
56 K7 **Achaguas** Apure, C Venezuela 7.46N 68.13W
160 H12 **Achalpur** prev. Elichpur, Ellichpur. Mahārāshtra, C India 21.19N 77.30E
63 F18 **Achar** Tacuarembó, C Uruguay 32.26S 56.10W
117 H19 **Acharnés** var. Aharnes; prev. Akharnaí. Attikí, C Greece 38.09N 23.58E
Acheen see Aceh
101 K16 **Achel** Limburg, NE Belgium 51.15N 5.31E
117 D16 **Acheloós** var. Akhelóös, Aspropótamos; anc. Achelous. ≈ W Greece
Achelous see Acheloós
169 W8 **Acheng** Heilongjiang, NE China 45.31N 126.55E
111 N6 **Achenkirch** Tirol, W Austria 47.31N 11.42E
103 L24 **Achenpass** pass Austria/Germany 47.35N 11.39E
111 N7 **Achensee** ⊚ W Austria
103 F22 **Achern** Baden-Württemberg, SW Germany 48.37N 8.04E
117 C16 **Acherón** ≈ W Greece
79 W11 **Achétinamou** ≈ S Niger
158 J12 **Achhnera** Uttar Pradesh, N India 27.10N 77.45E
44 C7 **Achiguate, Río** ≈ S Guatemala
99 A16 **Achill Head** Ir. Ceann Acla. headland W Ireland 53.59N 10.14W
99 A16 **Achill Island** Ir. Acaill. island W Ireland
102 H14 **Achim** Niedersachsen, NW Germany 53.01N 9.01E
155 S15 **Achin** Nangarhār, E Afghanistan 34.04N 70.40E
Achin see Aceh
126 Hh14 **Achinsk** Krasnoyarskiy Kray, S Russian Federation 56.21N 90.25E
168 E5 **Achit Nuur** ⊚ NW Mongolia
143 T11 **Achkasar** Arm. Ach'asar Lerr. ▲ Armenia/Georgia 41.09N 43.55E
130 K13 **Achuyevo** Krasnodarskiy Kray, SW Russian Federation 46.00N 38.01E
83 J17 **Achwa** var. Aswa. ≈ N Uganda
142 E15 **Acigöl** salt lake SW Turkey
109 L24 **Acireale** Sicilia, Italy, C Mediterranean Sea 37.36N 15.10E
Aciris see Agri
27 N7 **Ackerly** Texas, SW USA 32.31N 101.43W
24 M4 **Ackerman** Mississippi, S USA 33.18N 89.10W
31 W13 **Ackley** Iowa, C USA 42.33N 93.03W
46 J5 **Acklins Island** island SE Bahamas
Acla, Ceann see Achill Head
64 G13 **Aconcagua, Cerro** ▲ W Argentina 32.36S 69.53W
106 G2 **A Coruña** Cast. La Coruña. ◆ province Galicia, NW Spain
106 H2 **A Coruña** Cast. La Coruña, Eng. Corunna; anc. Caronium. Galicia, NW Spain 43.22N 8.24W
44 L10 **Acoyapa** Chontales, S Nicaragua 12.01N 85.08W
108 I13 **Acquapendente** Lazio, C Italy 42.44N 11.52E

108 J13 **Acquasanta Terme** Marche, C Italy 42.46N 13.24E
108 I13 **Acquasparta** Lazio, C Italy 42.41N 12.31E
108 C8 **Acqui Terme** Piemonte, NW Italy 44.40N 8.28E
Acrae see Palazzolo Acreide
190 F7 **Acraman, Lake** salt lake South Australia
61 A15 **Acre** off. Estado do Acre. ◆ state W Brazil
Acre see 'Akko
61 C16 **Acre, Rio** ≈ W Brazil
109 N20 **Acri** Calabria, SW Italy 39.30N 16.22E
Acte see Ágion Óros
203 Y12 **Actéon, Groupe** island group Îles Tuamotu, SE French Polynesia
13 P12 **Acton-Vale** Quebec, SE Canada 45.39N 72.31W
43 P13 **Actopan** var. Actopán. Hidalgo, C Mexico 20.16N 98.57W
61 P14 **Açu** var. Assu. Rio Grande do Norte, E Brazil 5.33S 36.55W
Acunum Acusio see Montélimar
79 Q17 **Ada** SE Ghana 5.46N 0.37E
31 R5 **Ada** Minnesota, N USA 47.18N 96.31W
33 X12 **Ada** Ohio, N USA 40.46N 83.49W
29 O12 **Ada** Oklahoma, C USA 34.48N 96.38W
114 L8 **Ada** Serbia, N Yugoslavia 45.48N 20.08E
Ada Bazar see Adapazarı
42 D3 **Adair, Bahía de** bay NW Mexico
106 J3 **Adaja** ≈ N Spain
40 H17 **Adak Island** island Aleutian Islands, Alaska, USA
Adalia see Antalya
Adalia, Gulf of see Antalya Körfezi
147 X9 **Adam** N Oman 22.22N 57.30E
Adama see Nazrēt
62 I8 **Adamantina** São Paulo, S Brazil 21.40S 51.04W
81 E14 **Adamaoua** Eng. Adamawa. ◆ province C Cameroon
70 F11 **Adamaoua, Massif d'** Eng. Adamawa Highlands. plateau NW Cameroon
79 Y14 **Adamawa** ◆ state E Nigeria
Adamawa see Adamaoua
Adamawa Highlands see Adamaoua, Massif d'
108 F6 **Adamello** ▲ N Italy 46.09N 10.33E
83 J14 **Adami Tulu** Oromo, C Ethiopia 7.52N 38.39E
65 M23 **Adam, Mount** var. West Falkland, Falkland Islands 51.36S 60.00W
31 R16 **Adams** Nebraska, C USA 40.25N 96.30W
20 H8 **Adams** New York, NE USA 43.48N 75.57W
31 Q3 **Adams** North Dakota, N USA 48.23N 98.01W
161 I23 **Adam's Bridge** chain of shoals NW Sri Lanka
34 H10 **Adams, Mount** ▲ Washington, NW USA 46.12N 121.29W
Adam's Peak see Sri Pada
203 R16 **Adams's Rock** island Pitcairn Island, Pitcairn Islands
203 R16 **Adamstown** ○ (Pitcairn Islands) Pitcairn Island, Pitcairn Islands 25.04S 130.04W
22 G10 **Adamsville** Tennessee, S USA 35.14N 88.23W
27 S9 **Adamsville** Texas, SW USA 31.15N 98.09W
147 O17 **'Adan** Eng. Aden. SW Yemen 12.51N 45.04E
142 K16 **Adana** var. Seyhan. Adana, S Turkey 37.00N 35.19E
142 K16 **Adana** var. Seyhan. ◆ province S Turkey
Adâncata see Horlivka
175 Nn10 **Adang, Teluk** bay Borneo, C Indonesia
142 F11 **Adapazarı** prev. Ada Bazar. Sakarya, NW Turkey 40.48N 30.24E
82 H8 **Adarama** River Nile, NE Sudan 17.04N 34.57E
205 Q10 **Adare, Cape** headland Antarctica 71.24S 170.27E
108 E6 **Adda** anc. Addua. ≈ N Italy
82 A13 **Adda** ≈ W Sudan
149 Q17 **Ad Dab'iyah** Abū Ẓaby, C UAE 24.16N 54.07E
149 O17 **Ad Dafrah** desert S UAE
147 Q6 **Ad Dahna'** desert E Saudi Arabia
76 A11 **Ad Dakhla** var. Dakhla. SW Western Sahara 23.46N 15.56W
Ad Dalanj see Dilling
Ad Damar see Ed Damer
Ad Damazin see Ed Damazin
Ad Dāmir see Ed Damer
181 N2 **Ad Dammah** desert NE Saudi Arabia
147 R6 **Ad Dammām** var. Dammām. Ash Sharqīyah, NE Saudi Arabia 26.23N 50.04E
146 K5 **Ad Dār al Ḥamrā'** Tabūk, NW Saudi Arabia 27.21N 37.45E
146 M13 **Ad Darb** Jīzān, SW Saudi Arabia 17.45N 42.15E
147 O8 **Ad Dawādimī** Ar Riyāḍ, C Saudi Arabia 24.31N 44.21E
149 N16 **Ad Dawḥah** Eng. Doha. ● (Qatar) C Qatar 25.15N 51.36E
149 N16 **Ad Dawḥah** Eng. Doha. × C Qatar 25.15N 51.36E
145 Y11 **Ad Dawr** N Iraq 34.30N 43.49E
145 Y12 **Ad Dayr** S Iraq, Dayr, Shahbān. E Iraq 30.45N 47.36E
Addī Arkay see Ādī Ārk'ay
145 X13 **Ad Dibdibah** physical region Iraq/Kuwait
Aḍ Ḍiffah see Libyan Plateau
Addis Ababa see Ādīs Ābeba
145 U10 **Ad Dīwānīyah** var. Diwaniyah. C Iraq 32.00N 44.57E
Addua see Adda
145 U10 **Ad Dujayl** var. Ad Dujail. N Iraq 33.49N 44.16E
101 D16 **Adegem** Oost-Vlaanderen, NW Belgium 51.12N 3.31E
67 U7 **Adel** Georgia, SE USA 31.08N 83.25W

31 U14 **Adel** Iowa, C USA 41.36N 94.01W
190 H2 **Adelaide** state capital South Australia 34.55S 138.36E
190 I9 **Adelaide** × South Australia 34.55S 138.31E
46 H2 **Adelaide** New Providence, N Bahamas 24.59N 77.30W
190 I9 **Adelaide Peninsula** peninsula Nunavut, N Canada
189 P2 **Adelaide River** Northern Territory, N Australia 13.15S 131.06E
78 M10 **'Adel Bagrou** Hodh ech Chargui, SE Mauritania 15.33N 7.04W
194 I11 **Adelbert Range** ▲ N PNG
188 K3 **Adele Island** island Western Australia
109 O17 **Adelfia** Puglia, SE Italy 41.01N 16.52E
205 V16 **Adélie Coast** physical region Antarctica
205 V14 **Adélie, Terre** physical region Antarctica
Adelnau see Odolanów
Adelsberg see Postojna
Aden see 'Adan
79 Q17 **Aden, Gulf of** gulf SW Arabian Sea
79 Q17 **Aderbissinat** Agadez, C Niger 15.30N 7.57E
Adhaim see Al 'Uẓaym
149 N16 **Adh Dhayd** var. Al Dhaid. Ash Shāriqah, NE UAE 25.19N 55.51E
146 M4 **'Adhfa'** spring/well NW Saudi Arabia 29.15N 41.04E
144 J13 **'Ādhriyāt, Jabāl al** ▲ S Jordan
194 L12 **Adi, Pulau** island E Indonesia
82 I10 **Ādī Ārk'ay** var. Addi Arkay. Amhara, N Ethiopia 13.18N 37.56E
190 C7 **Adieu, Cape** headland South Australia 32.01S 132.12E
108 I8 **Adige** Ger. Etsch. ≈ N Italy
82 J10 **Ādīgrat** Tigray, N Ethiopia 14.17N 39.27E
160 I13 **Ādilābād** var. Ādilābād. Andhra Pradesh, C India 19.40N 78.31E
37 P2 **Adin** California, W USA 41.10N 120.57W
176 Vv12 **Adi, Pulau** island E Indonesia
20 A8 **Adirondack Mountains** ▲ New York, NE USA
82 J13 **Ādīs Ābeba** Eng. Addis Ababa. ● (Ethiopia) Ādīs Ābeba, C Ethiopia 8.59N 38.43E
82 J13 **Ādīs Ābeba** × Ādīs Ābeba, C Ethiopia 8.58N 38.53E
82 I11 **Ādīs Zemen** Amhara, N Ethiopia 12.00N 37.43E
101 **Ädi Ugri** see Mendefera
143 S16 **Adıyaman** Adıyaman, SE Turkey 37.46N 38.15E
143 N15 **Adıyaman** ◆ province S Turkey
118 L11 **Adjud** Vrancea, E Romania 46.06N 27.11E
47 T6 **Adjuntas** C Puerto Rico 18.10N 66.44W
Adjuntas, Presa de las see Vicente Guerrero, Presa
Ådkup see Erikub Atoll
130 L15 **Adler** Krasnodarskiy Kray, SW Russian Federation 43.25N 39.58E
Adler see Orlice
110 G7 **Adliswil** Zürich, NW Switzerland 47.20N 8.30E
15 L1 **Admiralty Inlet** fjord Baffin Island, Nunavut, NE Canada
34 G7 **Admiralty Inlet** inlet Washington, NW USA
41 X13 **Admiralty Island** island Alexander Archipelago, Alaska, USA
194 K8 **Admiralty Islands** island group N PNG
142 K16 **Adnan Menderes** × (İzmir) İzmir, W Turkey 38.16N 27.09E
39 V6 **Adobe Creek Reservoir** ☐ Colorado, C USA
79 T16 **Ado-Ekiti** Ekiti, SW Nigeria 7.42N 5.13E
83 I14 **Adola** see Kibre Mengist
63 C23 **Adolfo González Chaves** Buenos Aires, E Argentina 38.02N 60.06W
161 H17 **Ādoni** Andhra Pradesh, C India 15.37N 77.16E
104 K15 **Adour** ≈ SW France
107 O14 **Adowa** see Adua
107 O14 **Adra** Andalucía, S Spain 36.45N 3.01W
109 L24 **Adrano** Sicilia, Italy, C Mediterranean Sea 37.39N 14.49E
76 J9 **Adrar** C Algeria 27.55N 0.12W
78 I7 **Adrar** ◆ region C Mauritania
76 L11 **Adrar** ≈ SE Algeria
76 A12 **Adrar Soutouf** ▲ SW Western Sahara
Adrasman see Adrasmon
153 Q10 **Adrasman** Rus. Adrasmon. NW Tajikistan 40.38N 69.56E
108 I8 **Adria** anc. Atria, Hadria, Hatria. Veneto, NE Italy 45.03N 12.04E
33 R11 **Adrian** Michigan, N USA 41.54N 84.02W
31 S11 **Adrian** Minnesota, N USA 43.38N 95.55W
29 S8 **Adrian** Missouri, C USA 38.24N 94.21W
26 L2 **Adrian** Texas, SW USA 35.16N 102.39W
23 S4 **Adrian** West Virginia, NE USA 38.53N 80.14W
Adrianople/Adrianopolis see Edirne
123 Mm8 **Adriatic Basin** undersea feature N Mediterranean Sea
79 W8 **Adriatic Sea** Alb. Deti Adriatik, It. Mare Adriatico, SCr. Jadransko More, Slvn. Jadransko Morje. sea N Mediterranean Sea
Adriatico, Mare see Adriatic Sea
Adriatik, Deti see Adriatic Sea
44 L4 **Adua** prev. Adwa. E Honduras
Aduana del Sásabe see El Sásabe
81 O17 **Adusa** Orientale, NE Dem. Rep. Congo (Zaire) 1.28N 27.43E
120 J13 **Adutiškis** Švenčionys, E Lithuania 55.09N 26.34E
196 B17 **Adventure Sound** bay East Falkland, Falkland Islands
67 U7 **Adwa** var. Adowa, It. Aduwa. Tigray, N Ethiopia 14.08N 38.51E
83 I14 **Adwa** Oromo, C Ethiopia 7.52N 36.36E

126 M8 **Adycha** ≈ NE Russian Federation
130 L14 **Adygeya, Respublika** ◆ autonomous republic SW Russian Federation
152 C11 **Adzhikui** Turkm. Ajyguyy. Balkanskiy Velayat, W Turkmenistan 39.46N 53.57E
79 N17 **Adzopé** SE Ivory Coast 6.07N 3.54W
129 U4 **Adz'va** ≈ NW Russian Federation
129 U3 **Adz'vavom** Respublika Komi, NW Russian Federation 66.35N 59.13E
Ædua see Autun
117 K19 **Aegean Islands** island group Greece/Turkey
Aegean North see Vóreion Aigaíon
117 I17 **Aegean Sea** Gk. Aigaíon Pélagos, Aigaío Pélagos, Turk. Ege Denizi. sea NE Mediterranean Sea
Aegean South see Nótion Aigaíon
120 J3 **Aegviidu** Ger. Charlottenhof. Harjumaa, NW Estonia 59.16N 25.37E
Aegyptus see Egypt
Aelana see Al 'Aqabah
Aelok see Ailuk Atoll
Aelōninae see Ailinginae Atoll
Aelōnlaplap see Ailinglaplap Atoll
Æmilia see Emilia-Romagna
Æmilianum see Millau
Aemona see Ljubljana
Aenaria see Ischia
203 Z3 **Aeolian Islands** see Eolie, Isole
97 G24 **Ærø** Ger. Arrö. island C Denmark
97 H24 **Ærøskøbing** Fyn, C Denmark 54.52N 10.24E
Æsernia see Isernia
106 G3 **A Estrada** Galicia, NW Spain 42.40N 8.28W
117 C18 **Aetós** Itháki, Iónioi Nísoi, Greece, C Mediterranean Sea 38.21N 20.40E
203 Q8 **Afaahiti** Tahiti, W French Polynesia 17.43S 149.18W
145 U10 **'Afak** C Iraq 32.04N 45.16E
129 T14 **Afanasjevo** var. Afanas'yevo. Kirovskaya Oblast', NW Russian Federation 58.55N 53.13E
Afanas'yevo see Afanasjevo
117 F17 **Afántou** var. Afántou. Ródos, Dodekánisos, Greece, Aegean Sea 36.16N 28.10E
101 **Ädi Ugri** see Mendefera
82 K1 **Afar** ◆ region NE Ethiopia
203 O7 **Afareaitu** Moorea, W French Polynesia 17.33S 149.46W
146 L7 **'Afariyah, Bi'r al** well NW Saudi Arabia 29.15N 39.21E
85 D22 **Afars et des Issas, Territoire Français des** see Djibouti
Affenrücken Karas, SW Namibia 28.05S 15.49E
154 M6 **Afghānestān, Dowlat-e Eslāmī-ye** see Afghanistan
126 K16 **Afghanistan** off. Islamic State of Afghanistan. Pers. Dowlat-e Eslāmī-ye Afghānestān; prev. Republic of Afghanistan. ◆ Islamic state C Asia
Afgoi see Afgooye
83 N17 **Afgooye** It. Afgoi. Shabeellaha Hoose, S Somalia 2.09N 45.07E
147 N8 **'Afif** Ar Riyāḍ, C Saudi Arabia 23.57N 42.57E
79 V17 **Afikpo** Ebonyi, SE Nigeria 5.52N 7.58E
96 H7 **Afjord** Sør-Trøndelag, C Norway 63.57N 10.12E
111 V6 **Aflenz Kurort** Steiermark, E Austria 47.33N 15.14E
76 J6 **Aflou** N Algeria 34.09N 2.06E
83 L18 **Afmadow** Jubbada Hoose, S Somalia 0.24N 42.03E
41 V6 **Afognak Island** island Alaska, USA
106 J2 **A Fonsagrada** Galicia, NW Spain 43.09N 7.03W
194 L15 **Afore** Northern, S PNG 9.01S 148.22E
61 O15 **Afrânio** Pernambuco, E Brazil 8.31S 40.54W
68-69 **Africa** continent
70 L11 **Africa, Horn of** physical region Ethiopia/Somalia
180 K11 **Africana Seamount** undersea feature SW Indian Ocean 37.10S 29.10E
88 A14 **African Plate** tectonic feature
144 J12 **'Afrīn** Ḥalab, N Syria 36.31N 36.51E
142 M15 **'Afrin** Kahramanmaraş, C Turkey 38.13N 36.54E
100 J7 **Afsluitdijk** dam N Netherlands 53.00N 5.10E
31 U15 **Afton** Iowa, C USA 41.01N 94.12W
31 W9 **Afton** Minnesota, N USA 44.54N 92.46W
29 R8 **Afton** Oklahoma, C USA 36.41N 94.57W
142 F12 **Afyon** prev. Afyonkarahisar. Afyon, W Turkey 38.46N 30.31E
142 F12 **Afyon** var. Afiun Karahissar, Afyonkarahisar. ◆ province W Turkey
Afyonkarahisar see Afyon
79 V10 **Agadez** prev. Agadès. Agadez, C Niger 16.57N 7.55E
79 V9 **Agadez** ◆ department N Niger
76 G6 **Agadir** SW Morocco 30.26N 9.36W
76 H7 **Agadir Canyon** undersea feature SE Atlantic Ocean
151 R9 **Agadyr'** Zhezkazgan, C Kazakhstan 48.15N 72.54E
181 O7 **Agalega Islands** island group N Mauritius
44 K6 **Agalta, Sierra de** ▲ E Honduras
126 Gg10 **Agan** ≈ C Russian Federation
196 C16 **Agana/Agaña** see Hagåtña
196 B16 **Agana Bay** bay NW Guam
196 B17 **Agana Field** × (Agana) C Guam
160 G9 **Agar** Madhya Pradesh, C India 23.43N 76.01E
83 I14 **Agaro** Oromo, C Ethiopia 7.52N 36.36E

159 V15 **Agartala** Tripura, NE India 23.49N 91.15E
204 I5 **Agassiz, Cape** headland Antarctica 68.28S 62.58W
183 V13 **Agassiz Fracture Zone** tectonic feature S Pacific Ocean
196 B16 **Agat** W Guam 13.20N 144.38E
196 B16 **Agat Bay** bay W Guam
151 P13 **Agat, Gory** hill C Kazakhstan 46.55N 69.13E
Agatha see Agde
117 M20 **Agathónisi** island Dodekánisos, Greece, Aegean Sea
176 Y13 **Agats** Irian Jaya, E Indonesia 5.33S 138.07E
161 C21 **Agatti Island** island Lakshadweep, India, N Indian Ocean
40 D16 **Agattu Island** island Aleutian Islands, Alaska, USA
40 D16 **Agattu Strait** strait Aleutian Islands, Alaska, USA
12 B8 **Agawa** ≈ Ontario, S Canada
12 B8 **Agawa Bay** lake bay Ontario, S Canada
79 N17 **Agboville** SE Ivory Coast 5.56N 4.13W
143 V12 **Ağdam** Rus. Agdam. SW Azerbaijan 40.04N 46.00E
105 P16 **Agde** anc. Agatha. Hérault, S France 43.19N 3.28E
105 P16 **Agde, Cap d'** headland S France 43.17N 3.30E
Agedabia see Ajdābiyā
104 L14 **Agen** anc. Aginnum. Lot-et-Garonne, SW France 44.12N 0.37E
171 K16 **Ageo** Saitama, Honshū, S Japan 35.58N 139.36E
111 R5 **Ager** ≈ N Austria
44 **Agere Hiywet** see Hāgere Hiywet
110 G8 **Agerisee** ⊚ N Switzerland
148 M10 **Āghā Jārī** Khūzestān, SW Iran 30.48N 49.45E
41 O3 **Aghiyuk Island** island Alaska, USA
76 I9 **Aghouinit** SE Western Sahara 22.14N 13.10W
Aghri Dagh see Büyükağrı Dağı
76 A10 **Aghzoumal, Sebkhet** var. Sebjet Agsumal. salt lake W Western Sahara
117 F15 **Agiá** var. Ayiá. Thessalía, C Greece 39.43N 22.45E
Agialoúsa see Yenierenköy
117 M21 **Agía Marína** Dodekánisos, Greece, Aegean Sea
124 Nn4 **Agía Fýlaxis** var. Ayia Phyla. S Cyprus 34.43N 33.02E
117 L16 **Agía Nápa** var. Ayia Napa. E Cyprus 34.59N 34.00E
117 I18 **Agía Paraskeví** Lésvos, E Greece 39.13N 26.19E
117 J15 **Agiásos** var. Ayiásos, Ayiássos. Lésvos, E Greece 39.04N 26.22E
117 E21 **Agía Eírinis, Akrotírio** headland Límnos, E Greece 39.47N 25.21E
117 H14 **Agiasós** var. Áyios Óros, Eng. Mount Athos. ◆ monastic republic NE Greece
117 H14 **Ágion Óros** Eng. Akte, Aktí; anc. Acte. peninsula NE Greece
116 Z13 **Ágios Achílleios** religious building Dytikí Makedonía, N Greece 40.46N 21.04E
117 J16 **Ágios Efstrátios** var. Áyios Evstrátios, Hagios Evstrátios. island E Greece
117 Q23 **Ágios Geórgios** island SE Greece
117 E21 **Ágios Ilías** ▲ S Greece 37.09N 22.52E
117 K25 **Ágios Ioannis, Akrotírio** headland Kríti, Greece, E Mediterranean Sea 35.19N 25.46E
117 L20 **Ágios Kírykos** var. Áyios Kírikos. Ikaría, Dodekánisos, Greece, Aegean Sea 37.04N 26.16E
172 Oo14 **Aguni-jima** island Nansei-shotō, SW Japan
126 Mm12 **Aim** Khabarovskiy Kray, E Russian Federation 58.45N 134.08E
105 S13 **Agout** ≈ S France
117 H14 **Ágios Nikólaos** Thessalía, C Greece 39.33N 21.21E
180 K11 **Ágios Nikólaos** var. Áyios Nikólaos. Kríti, Greece, E Mediterranean Sea 35.12N 25.43E
117 E21 **Ágios Sérgios** see Yenibogaziçi
117 H14 **Agíou Órous, Kólpos** gulf N Greece
109 I24 **Agira** anc. Agyrium. Sicilia, Italy, C Mediterranean Sea 37.39N 14.31E
116 G12 **Agkístri** island S Greece
41 **Agkoro** see Angistro
105 O17 **Agly** ≈ S France
79 O16 **Agnibilékrou** E Ivory Coast 7.10N 3.10W
118 I11 **Agnita** Ger. Agnetheln, Hung. Szentágota. Sibiu, SW Romania 45.59N 24.39E
109 K15 **Agnone** Molise, C Italy 41.49N 14.21E
160 J6 **Agra** Uttar Pradesh, N India 27.09N 78.00E
Agra and Oudh, United Provinces of see Uttar Pradesh
Agram see Zagreb
143 S13 **Ağrı** var. Karaköse; prev. Karakılisse. Ağrı, E Turkey 39.43N 43.04E

143 S13 **Ağrı** ◆ province NE Turkey
Ağrı Dağı see Büyükağrı Dağı
109 J24 **Agrigento** Gk. Akragas; prev. Girgenti. Sicilia, Italy, C Mediterranean Sea 37.19N 13.33E
117 D18 **Agrínio** prev. Agrinion. Dytikí Ellás, W Greece 38.37N 21.25E
Agrinion see Agrínio
117 G17 **Agriovótano** Évvoia, C Greece 39.00N 23.18E
109 L18 **Agropoli** Campania, S Italy 40.21N 14.58E
131 T3 **Agryz** Udmurtskaya Respublika, NW Russian Federation 56.27N 52.58E
143 S13 **Ağstafa** Rus. Akstafa. NW Azerbaijan 41.06N 45.28E
Agsumal, Sebjet see Aghzoumal, Sebkhet
42 J11 **Agua Brava, Laguna** lagoon C Mexico
56 I7 **Aguachica** Cesar, N Colombia 8.16N 73.35E
61 I20 **Água Clara** Mato Grosso do Sul, SW Brazil 20.25S 52.58W
46 G9 **Aguada de Pasajeros** Cienfuegos, C Cuba 22.22N 80.50W
56 J5 **Aguada Grande** Lara, N Venezuela 10.34N 69.28W
47 S5 **Aguadilla** W Puerto Rico 18.24N 67.09W
45 S16 **Aguadulce** Coclé, S Panama 8.16N 80.31W
106 L14 **Aguadulce** Andalucía, S Spain 37.15N 4.58W
42 J9 **Aguanaval, Río** ≈ C Mexico
44 I5 **Aguán, Río** ≈ N Honduras
42 G8 **Agua Nueva** Texas, SW USA 26.57N 98.34W
62 J8 **Aguapeí, Rio** ≈ S Brazil
63 G14 **Aguapey, Río** ≈ NE Argentina
42 G3 **Agua Prieta** Sonora, NW Mexico 31.16N 109.33W
106 G5 **A Guardia** var. Laguardia, La Guardia. Galicia, NW Spain 41.54N 8.52W
56 G5 **Aguarico, Río** ≈ Ecuador/Peru
57 O6 **Aguasay** Monagas, NE Venezuela 9.25N 63.43W
42 M12 **Aguascalientes** Aguascalientes, C Mexico 21.53N 102.17W
42 M12 **Aguascalientes** ◆ state C Mexico
59 I18 **Aguas Calientes, Río** ≈ S Peru
107 R7 **Aguasvivas** ≈ NE Spain
62 J7 **Água Vermelha, Represa de** ☐ S Brazil
58 G12 **Aguaytía** Ucayali, C Peru 9.04S 75.32W
106 J5 **A Gudiña** var. La Gudiña. Galicia, NW Spain 42.04N 7.07W
106 G7 **Águeda** Aveiro, N Portugal 40.34N 8.28W
117 F18 **Águeda** ≈ Portugal/Spain 40.34N 8.28W
79 Q8 **Aguelhok** Kidal, NE Mali 19.26N 0.49E
105 U11 **Aiguille, Mont** ▲ E France 44.09N 3.34E
79 V12 **Aguié** Maradi, S Niger 13.28N 7.43E
196 K8 **Aguijan** island S Northern Mariana Islands
106 M14 **Aguilar de Campóo** Castilla-León, N Spain 42.46N 4.15W
106 N10 **Aguilar de la Frontera** Andalucía, S Spain 37.31N 4.40W
44 F7 **Aguilares** San Salvador, C El Salvador 13.58N 89.09W
107 Q14 **Águilas** Murcia, SE Spain 37.24N 1.36W
42 L15 **Aguililla** Michoacán de Ocampo, SW Mexico 18.43N 102.45W
180 J11 **Agulhas** var. L'Agulhas. Western Cape, SW South Africa 34.51S 19.59E
180 K11 **Agulhas Bank** undersea feature SW Indian Ocean
180 K11 **Agulhas Basin** undersea feature SW Indian Ocean
85 F26 **Agulhas, Cape** Afr. Kaap Agulhas. headland SW South Africa 34.51S 19.59E
62 O9 **Agulhas Negras, Pico das** ▲ SE Brazil 22.21S 44.39W
180 K11 **Agulhas Plateau** undersea feature SW Indian Ocean
Agurain see Salvatierra
179 R15 **Agusan** ≈ Mindanao, S Philippines
105 N10 **Agusan** ≈ Mindanao, S Philippines
56 G5 **Agustín Codazzi** var. Codazzi. Cesar, N Colombia 10.03N 73.15W
120 G7 **Agyrium** see Agira
Ahaggar see Hoggar
76 L2 **Ahaggar** high plateau region SE Algeria
Ahal Welayaty see Akhalskiy Velayat

Ahmedabad see Ahmadābād
Ahmednagar see Ahmadnagar
116 L9 **Ahmetbey** Kırklareli, NW Turkey 41.26N 27.35E
12 O11 **Ahmic Lake** ⊚ Ontario, S Canada
202 G12 **Ahoa** Île Uvea, E Wallis and Futuna 13.16S 176.12W
42 G8 **Ahome** Sinaloa, C Mexico 25.55N 109.10W
23 X8 **Ahoskie** North Carolina, SE USA 36.17N 76.59W
103 D17 **Ahr** ≈ W Germany
149 N12 **Ahram** var. Ahrom. Būshehr, S Iran 28.52N 51.18E
102 I9 **Ahrensburg** Schleswig-Holstein, N Germany 53.40N 10.13E
Ahrom see Ahram
95 L17 **Ähtäri** Länsi-Suomi, W Finland 62.31N 24.11E
42 K12 **Ahuacatlán** Nayarit, C Mexico 21.04N 104.32W
44 A9 **Ahuachapán** Ahuachapán, W El Salvador 13.55N 89.49W
44 A9 **Ahuachapán** ◆ department W El Salvador
203 V16 **Ahu Akivi** var. Siete Moai. ancient monument Easter Island, Chile, E Pacific Ocean
203 W11 **Ahunui** atoll Îles Tuamotu, C French Polynesia
193 E20 **Åhus** Skåne, S Sweden 55.55N 14.18E
47 S5 **Ahu Tahira** var. Ahu Vinapu. ancient monument Easter Island, Chile, E Pacific Ocean
203 V16 **Ahu Tepeu** ancient monument Easter Island, Chile, E Pacific Ocean
203 V17 **Ahu Vinapu** var. Ahu Tahira. ancient monument Easter Island, Chile, E Pacific Ocean
148 L9 **Ahvāz** var. Ahwāz; prev. Nāsiri. Khūzestān, SW Iran 31.19N 48.37E
Ahvenanmaa see Åland
147 X9 **Aḥwar** SW Yemen 13.34N 46.41E
Ahwāz see Ahvāz
Aibak see Āybak
103 K22 **Aichach** Bayern, SE Germany 48.26N 11.06E
171 I16 **Aichi** off. Aichi-ken, var. Aiti. ◆ prefecture Honshū, SW Japan
176 Ww12 **Aiduna** Irian Jaya, E Indonesia 4.20S 135.15E
Aidussina see Ajdovščina
194 F13 **Aiema** ≈ W PNG
Aifir, Clochán an see Giant's Causeway
Aigaíon Pélagos/Aigaío Pélagos see Aegean Sea
111 S3 **Aigen im Mülkreis** Oberösterreich, N Austria 48.39N 13.57E
117 G20 **Aígina** var. Aíyina, Egina. Aígina, C Greece 37.45N 23.25E
117 G20 **Aígina** island S Greece
117 E18 **Aígio** var. prev. Aíyion. Dytikí Ellás, S Greece 38.15N 22.04E
110 C10 **Aigle** Vaud, SW Switzerland 46.19N 6.58E
105 P14 **Aigrefeuille** ≈ W France 44.09N 3.34E
181 O16 **Aigrettes, Pointe des** headland W Réunion 21.01S 55.13E
63 G19 **Aiguá** var. Aigua. Maldonado, S Uruguay 34.13S 54.46W
105 S13 **Aigre** ≈ SE France
105 N10 **Aigurande** Indre, C France 46.26N 1.49E
Ai-hun see Heihe
171 K11 **Aikawa** Niigata, Sado, C Japan 38.04N 138.15E
23 Q13 **Aiken** South Carolina, SE USA 33.33N 81.43W
27 N4 **Aiken** Texas, SW USA 34.06N 101.31W
166 F13 **Ailao Shan** ▲ SW China
45 W14 **Ailigandí** San Blas, NE Panama 9.13N 78.04W
201 R4 **Ailinginae Atoll** var. Aelōninae. atoll Ralik Chain, SW Marshall Islands
201 T2 **Ailinglaplap Atoll** var. Aelōnlaplap. atoll Ralik Chain, S Marshall Islands
98 G7 **Ailion, Loch** see Allen, Lough
98 G11 **Ailsa Craig** island SW Scotland, UK
201 V5 **Ailuk Atoll** var. Aelok. atoll Ratak Chain, NE Marshall Islands
105 P14 **Ain** ◆ department E France
105 S10 **Ain** ≈ E France
120 G7 **Aïnaži** Est. Heinaste, Ger. Hainasch. Limbaži, N Latvia 57.51N 24.24E
76 L6 **Aïn Ben Tili** Tiris Zemmour, N Mauritania 25.58N 9.30W
76 J5 **'Aïn Defla** var. Aïn Eddefla. N Algeria 36.16N 1.58E
Aïn Eddefla see Aïn Defla
76 L5 **Aïn El Bey** × (Constantine) NE Algeria 36.16N 6.36E
117 C19 **Aínos** ▲ Kefalloniá, Iónioi Nísoi, Greece, C Mediterranean Sea 38.08N 20.39E
31 N13 **Ainsworth** Nebraska, C USA 42.33N 99.51W
76 I5 **Aïn Témouchent** N Algeria 35.18N 1.09W
Aintab see Gaziantep
194 J13 **Aiome** Madang, N PNG 5.04S 144.43E
76 **Aïoun el Atrous/Aïoun el Atroûss** see 'Ayoûn el 'Atroûs
56 F9 **Aipe** Huila, C Colombia 3.15N 75.16W
58 C11 **Aipena** ≈ N Peru
59 L19 **Aiquile** Cochabamba, C Bolivia 18.10S 65.10W
105 S15 **Air, Massif de l'**
110 F4 **Airai** Babeldaob, C Palau
196 E10 **Airai** × (Oreor) Babeldaob, C Palau
177 F8 **Airbangis** Sumatera, NW Indonesia 0.12S 99.23E
11 Q16 **Airdrie** Alberta, SW Canada 51.20N 114.00W
98 J12 **Airdrie** S Scotland, UK 55.52N 3.58W

◆ COUNTRY ● COUNTRY CAPITAL ◇ DEPENDENT TERRITORY ○ DEPENDENT TERRITORY CAPITAL ◆ ADMINISTRATIVE REGION × INTERNATIONAL AIRPORT ▲ MOUNTAIN ▲ MOUNTAIN RANGE ≈ RIVER ℝ VOLCANO ⊚ LAKE ☐ RESERVOIR

Air du Azbine see Aïr, Massif de l'
99 M17 **Aire** ✍ N England, UK
104 K15 **Aire-sur-l'Adour** Landes, SW France 43.43N 0.16W
105 O1 **Aire-sur-la-Lys** Pas-de-Calais, N France 50.39N 2.24E
16 N2 **Air Force Island** island Baffin Island, Nunavut, NE Canada
174 L11 **Airhitam, Teluk** bay Borneo, C Indonesia
175 R7 **Airmadidi** Sulawesi, N Indonesia 1.25N 124.58E
79 V8 **Aïr, Massif de l'** var. Air du Azbine, Asben. ▲ NC Niger
110 G10 **Airolo** Ticino, S Switzerland 46.32N 8.38E
104 K9 **Airvault** Deux-Sèvres, W France 46.51N 0.07W
103 K19 **Aisch** ✍ S Germany
65 G20 **Aisén** ◊ Región Aisén del General Carlos Ibáñez del Campo, var. Aisén. ◆ region S Chile
8 H7 **Aishihik Lake** ⊗ Yukon Territory, W Canada
105 P3 **Aisne** ◆ department N France
105 R4 **Aisne** ✍ N France
111 T4 **Aist** ✍ N Austria
116 K13 **Aísymi** Anatolikí Makedonía kai Thráki, NE Greece 41.00N 25.55E
107 S11 **Aitana** ▲ E Spain 38.39N 0.15W
194 F9 **Aitape** var. Eitape. Sandaun, NW PNG 3.07S 142.22E
Aiti see Aichi
31 V6 **Aitkin** Minnesota, N USA 46.31N 93.42W
117 D18 **Aitoликó** var. Etoliko; prev. Aitolikón. Dytikí Ellás, C Greece 38.25N 21.21E
Aitolikón see Aitoликó
202 L15 **Aitutaki** island S Cook Islands
118 H11 **Aiud** Ger. Strassburg, Hung. Nagyenyed; prev. Engeten. Alba, SW Romania 46.16N 23.42E
120 I9 **Aiviekste** ✍ C Latvia
201 Q8 **Aiwo** ✕ W Nauru 0.325 166.54E
196 E8 **Aiwokako Passage** passage Babeldaob, N Palau
Aix see Aix-en-Provence
105 S15 **Aix-en-Provence** var. Aix; anc. Aquae Sextiae. Bouches-du-Rhône, SE France 43.31N 5.27E
Aix-la-Chapelle see Aachen
105 T11 **Aix-les-Bains** Savoie, E France 45.40N 5.55E
194 E11 **Aiyang, Mount** ▲ NW PNG 5.03S 141.15E
Aíyina see Aígina
Aíyion see Aígio
159 W15 **Aīzawl** Mizoram, NE India 23.40N 92.45E
120 H9 **Aizkraukle** Aizkraukle, S Latvia 56.39N 25.07E
120 C9 **Aizpute** Liepāja, W Latvia 56.43N 21.32E
171 L14 **Aizu-Wakamatsu** var. Aizuwakamatu. Fukushima, Honshū, C Japan 37.27N 139.55E
Aizuwakamatu see Aizu-Wakamatsu
105 X15 **Ajaccio** Corse, France, C Mediterranean Sea 41.54N 8.43E
105 X15 **Ajaccio, Golfe d'** gulf Corse, France, C Mediterranean Sea
43 Q15 **Ajalpan** Puebla, S Mexico 18.25N 97.19W
160 F13 **Ajanta Range** ▲ C India
143 R10 **Ajaria** ◆ autonomous republic SW Georgia
Ajastan see Armenia
95 G14 **Ajaureforsen** Västerbotten, N Sweden 65.31N 15.43E
193 H17 **Ajax, Mount** ▲ South Island, NZ 42.34S 172.02E
168 F9 **Aj Bogd Uul** ▲ SW Mongolia 44.49N 95.01E
77 R8 **Ajdābiyā** var. Agedabia, Ajdābiyah. NE Libya 30.46N 20.13E
Ajdābiyah see Ajdābiyā
111 S12 **Ajdovščina** Ger. Haidenschaft. It. Aidussina. W Slovenia 45.52N 13.55E
171 Mm8 **Ajigasawa** Aomori, Honshū, C Japan 40.45N 140.11E
Ajjinena see El Geneina
113 H23 **Ajka** Veszprém, W Hungary 47.07N 17.31E
144 G9 **'Ajlūn** Irbid, N Jordan 32.19N 35.45E
144 H9 **'Ajlūn, Jabal** ▲ N Jordan
149 R15 **'Ajmān** var. Ajman, 'Ujmān. 'Ajmān, NE UAE 25.24N 55.27E
158 G12 **Ajmer** var. Ajmere. Rājasthān, N India 26.28N 74.40E
68 J15 **Ajo** Arizona, SW USA 32.22N 112.51W
107 N2 **Ajo, Cabo de** headland N Spain 43.31N 3.36W
68 J16 **Ajo Range** ▲ Arizona, SW USA
Ajyguyy see Adzhikui
172 P5 **Akabira** Hokkaidō, NE Japan 43.31N 142.03E
172 Q6 **Akan-ko** ⊗ Hokkaidō, NE Japan
Akanthoú see Tatlısu
155 J19 **Akaroa** Canterbury, South Island, NZ 43.48S 172.58E
82 E6 **Akasha** Northern, N Sudan 21.03N 30.45E
170 G14 **Akashi** var. Akasi. Ōsaka, Honshū, SW Japan 34.37N 134.59E
'**Akāsh, Wādī** var. Wādī 'Ukash. dry watercourse W Iraq
Akasi see Akashi
94 K13 **Äkäsjokisuu** Lappi, N Finland 67.24N 23.44E
131 S11 **Akbaba Dağı** ▲ Armenia/Turkey 41.04N 43.28E
142 B15 **Akbük Limanı** bay W Turkey
131 V8 **Akbulak** Orenburgskaya Oblast', W Russian Federation 51.01N 55.35E
143 O11 **Akçaabat** Trabzon, NE Turkey 41.00N 39.36E
143 N15 **Akçadağ** Malatya, C Turkey 38.21N 37.58E
142 G11 **Akçakoca** Bolu, NW Turkey 41.04N 31.07E
Akchakaya, Vpadina see Akdzhakaya, Vpadina
78 M7 **Akchâr** desert W Mauritania
151 L12 **Akchatau** Kaz. Aqshataū. Zhezkazgan, C Kazakhstan 47.58N 74.01E
142 L13 **Akdağlar** ▲ C Turkey
142 I13 **Ak Dağları** ▲ SW Turkey
142 K13 **Akdağmadeni** Yozgat, C Turkey 39.39N 35.48E
152 G8 **Akdepe** prev. Ak-Tepe, Leninsk, Turkm. Lenin. Dashkhovuzskiy Velayat, N Turkmenistan 42.10N 59.17E
124 O3 **Akdoğan** Gk. Lýsi. C Cyprus 35.06N 33.42E
126 Hh16 **Ak-Dovurak** Respublika Tyva, S Russian Federation 51.09N 90.36E
152 F9 **Akdzhakaya, Vpadina** var. Vpadina Akchakaya. depression N Turkmenistan
175 T17 **Akelamo** Pulau Halmahera, E Indonesia 1.12N 128.39E
Aken see Aachen
Akermanceaster see Bath
97 P15 **Åkersberga** Stockholm, C Sweden 59.28N 18.19E
97 H15 **Akershus** ◆ county S Norway
81 L16 **Aketi** Orientale, N Dem. Rep. Congo (Zaire) 2.46N 23.42E
Akgyr Erezi see Gryada Akkyr
152 E12 **Akhalskiy Velayat** Turkm. Ahal Welayaty. ◆ province C Turkmenistan
143 S10 **Akhalts'ikhe** SW Georgia
Akhangaran see Ohangaron
Akharnaí see Acharnés
77 R7 **Akhḍar, Al Jabal al** hill range NE Libya
Akhelóös see Acheloós
41 Q15 **Akhiok** Kodiak Island, Alaska, USA 56.57N 154.12W
142 C13 **Akhisar** Manisa, W Turkey 38.54N 27.49E
77 X10 **Akhmīm** anc. Panopolis. C Egypt 26.34N 31.50E
158 H6 **Akhnūr** Jammu and Kashmir, NW India 32.57N 74.43E
131 P11 **Akhtuba** ✍ SW Russian Federation
131 P11 **Akhtubinsk** Astrakhanskaya Oblast', SW Russian Federation 48.16N 46.13E
Akhtyrka see Okhtyrka
170 F15 **Aki** Kōchi, Shikoku, SW Japan 33.30N 133.54E
41 N12 **Akiachak** Alaska, USA 60.54N 161.25W
41 N12 **Akiak** Alaska, USA 60.54N 161.12W
203 X11 **Akiaki** atoll Îles Tuamotu, E French Polynesia
10 H9 **Akimiski Island** island Nunavut, C Canada
142 K17 **Akıncı Burnu** headland S Turkey 36.21N 35.47E
Akıncılar see Selçuk
119 U10 **Akinovka** Zaporiz'ka Oblast', S Ukraine
97 M24 **Åkirkeby** var. Aakirkeby. Bornholm, E Denmark 55.04N 14.55E
171 M10 **Akita** Akita, Honshū, C Japan 39.44N 140.06E
171 M10 **Akita** off. Akita-ken. ◆ prefecture Honshū, C Japan
78 H8 **Akjoujt** prev. Fort-Repoux. Inchiri, W Mauritania 19.42N 14.28W
94 H11 **Akkajaure** ⊗ N Sweden
Akkala see Oqqal'a
161 L25 **Akkaraipattu** Eastern Province, E Sri Lanka 7.13N 81.51E
94 H11 **Akkavare** ▲ N Sweden 67.33N 17.27E
151 P13 **Akkense** Zhezkazgan, C Kazakhstan 46.39N 68.06E
Akkerman see Bilhorod-Dnistrovs'kyy
131 U6 **Akkermanovka** Orenburgskaya Oblast', W Russian Federation 51.11N 58.03E
172 Qq7 **Akkeshi** Hokkaidō, NE Japan 43.03N 144.48E
172 Qq7 **Akkeshi-ko** ⊗ Hokkaidō, NE Japan
172 Qq8 **Akkeshi-wan** bay NW Pacific Ocean
144 F8 **'Akko** Eng. Acre, Fr. Saint-Jean-d'Acre; Bibl. Accho, Bibl. Ptolemais. Northern, N Israel 32.55N 35.04E
151 Q8 **Akkol'** Kaz. Aqköl; prev. Alekseyevka, Kaz. Alekseevka. Akmola, C Kazakhstan 51.58N 70.58E
151 T14 **Akkol'** Kaz. Aqköl. Almaty, SE Kazakhstan 45.01N 75.38E
151 Q16 **Akkol'** Kaz. Aqköl. Zhambyl, S Kazakhstan 43.25N 70.46E
150 M11 **Akkol', Ozero** prev. Ozero Zhaman-Akkol'. ⊗ C Kazakhstan
100 L6 **Akkrum** Friesland, N Netherlands 53.01N 5.52E
151 U8 **Akku** prev. Lebyazh'ye. Pavlodar, NE Kazakhstan 51.29N 77.48E
150 Qq12 **Akkystau** Kaz. Aqqystaū. Atyrau, W Kazakhstan 47.13N 51.01E
14 J3 **Aklavik** Northwest Territories, NW Canada 68.15N 135.01W
164 E8 **Akmenrags** headland W Latvia 56.49N 21.03E
47 V3 **Akmeqit** Xinjiang Uygur Zizhiqu, NW China 37.10N 76.59E
152 L12 **Akmeydan** Maryyskiy Velayat, C Turkmenistan 37.50N 62.08E
Akmola see Astana
151 P13 **Akmola** off. Akmolinskaya Oblast', Kaz. Aqmola Oblysy; prev. Tselinogradskaya Oblast'. ◆ province C Kazakhstan
Akmolinsk see Astana
Akmolinskaya Oblast' see Akmola
Akmyal Almanská Almanská
Aknavásár see Târgu Ocna
120 J11 **Akniste** Jēkabpils, S Latvia 56.09N 25.43E

---

170 G14 **Akō** Hyōgo, Honshū, SW Japan 34.44N 134.22E
83 G14 **Akobo** Jonglei, SE Sudan 7.49N 33.04E
83 G14 **Akobo** var. Ākobowenz. ✍ Ethiopia/Sudan
Ākobowenz see Akobo
160 H12 **Akola** Mahārāshtra, C India 20.44N 77.00E
160 H12 **Akot** Mahārāshtra, C India 20.45N 77.00E
79 N16 **Akoupé** SE Ivory Coast 6.19N 3.54W
10 M3 **Akpatok Island** island Nunavut, E Canada
126 G7 **Akqi** Xinjiang Uygur Zizhiqu, NW China 40.51N 78.20E
144 I2 **Akrad, Jabal al** ▲ N Syria
94 H3 **Akranes** Vesturland, W Iceland 64.19N 22.01W
145 S2 **Akra'** Ar. 'Aqrah. N Iraq 36.46N 43.52E
97 C16 **Åkrehamn** Rogaland, S Norway 59.15N 5.12E
79 V9 **Akrérèb** Agadez, C Niger 17.45N 9.01E
117 D22 **Akrítas, Akrotírio** headland S Greece 36.43N 21.52E
39 V3 **Akron** Colorado, C USA 40.09N 103.12W
31 R12 **Akron** Iowa, C USA 42.49N 96.33W
33 U12 **Akron** Ohio, N USA 41.04N 81.31W
124 N4 **Akrotíri** var. Akrotiri. UK air base S Cyprus 34.36N 32.57E
124 Nn4 **Akrotírion, Kólpos** var. Akrotiri Bay. bay S Cyprus
123 Mm4 **Akrotíri Sovereign Base Area** UK military installation S Cyprus 34.34N 32.58E
164 F11 **Aksai Chin** Chin. Aksayqin. disputed region China/India
25 V9 **Aksaj** see Aksay
142 I15 **Aksaray** Aksaray, C Turkey 38.23N 33.50E
142 I15 **Aksaray** ◆ province C Turkey
165 P8 **Aksay** var. Aksay Kazaku Zizhixian. Gansu, N China 39.25N 94.09E
150 G8 **Aksay** Kaz. Aksaj, Kaz. Aqsay. Zapadnyy Kazakhstan, NW Kazakhstan 51.10N 53.03E
131 O11 **Aksay** Volgogradskaya Oblast', SW Russian Federation 47.59N 43.54E
153 W10 **Aksay** var. Toxkan He. ✍ China/Kyrgyzstan
Aksay Kazakzu Zizhixian see Aksay
164 G11 **Aksayqin Hu** ⊗ NW China
142 G14 **Akşehir** Konya, W Turkey 38.22N 31.24E
142 G14 **Akşehir Gölü** ⊗ C Turkey
142 G16 **Akseki** Antalya, SW Turkey 37.03N 31.46E
126 L15 **Aksenovo-Zilovskoye** Chitinskaya Oblast', S Russian Federation 53.01N 117.26E
126 Kk16 **Aksha** Chitinskaya Oblast', S Russian Federation 50.16N 113.22E
151 V11 **Akshatau, Khrebet** ▲ E Kazakhstan
151 S8 **Ak-Shyyrak** Issyk-Kul'skaya Oblast', E Kyrgyzstan 41.46N 78.34E
151 M10 **Akshiganak** Kaz. Aqshyghanaq. Qostanay, NW Kazakhstan 48.16N 40.23E
151 Y11 **Aksu** Xinjiang Uygur Zizhiqu, NW China 41.16N 80.15E
151 R8 **Aksu** Kaz. Aqsū. Akmola, N Kazakhstan 52.31N 72.00E
151 T8 **Aksu** var. Jermak, Kaz. Ermak; prev. Yermak. Pavlodar, NE Kazakhstan 52.03N 76.55E
151 W13 **Aksu** Kaz. Aqsū. Almaty, SE Kazakhstan 45.11N 79.28E
151 X13 **Aksu** Kaz. Aqsū. ✍ SE Kazakhstan
151 X11 **Aksuat** Kaz. Aqsūat. Vostochnyy Kazakhstan, E Kazakhstan 47.46N 82.49E
151 Y11 **Aksuat** Kaz. Aqsūat. Vostochnyy Kazakhstan, SE Kazakhstan 48.16N 83.39E
131 S4 **Aksubayevo** Respublika Tatarstan, W Russian Federation 54.52N 50.58E
164 H7 **Aksu He** Rus. Sary-Dzhaz. ✍ China/Kyrgyzstan see also Sary-Dzhaz
82 J10 **Aksum** Tigray, N Ethiopia 14.06N 38.42E
151 O12 **Aktas** Kaz. Aqtas. Zhezkazgan, C Kazakhstan 48.03N 66.21E
68 J8 **Aktash** see Oqtosh
153 V9 **Ak-Tash, Gora** ▲ C Kyrgyzstan 40.53N 74.39E
151 R10 **Aktau** Kaz. Aqtaū. Karaganda, C Kazakhstan 50.13N 73.06E
150 E11 **Aktau** Kaz. Aqtaū; prev. Shevchenko. Mangistau, W Kazakhstan 43.37N 51.11E
26 K1 **Aktepe** see Akkystau
'**Aktau, Khrebet** see Oqtogh
SW Tajikistan
Aktau, Khrebet see Oqtow
Tizmasi, C Uzbekistan
Akte see Agion Oros
150 Tt2 **Ak-Tepe** see Akdepe
22 F9 **Akterek** Kaz. Aqterek. C Kazakhstan 43.46N 89.07W
153 X7 **Ak-Terek** Issyk-Kul'skaya Oblast', E Kyrgyzstan 42.14N 77.46E
68 J12 **Akti** see Ägion Óros
164 E8 **Akto** Xinjiang Uygur Zizhiqu, NW China 39.07N 75.43E
151 V12 **Aktogay** Kaz. Aqtoghay. Vostochnyy Kazakhstan, E Kazakhstan 46.56N 79.40E
151 T12 **Aktogay** Kaz. Aqtoghay. 47.46N 74.06E
121 M18 **Aktsyabrski** Rus. Oktyabr'skiy; prev. Karpilovka. Homyel'skaya Voblasts', SE Belarus 52.37N 28.52E
150 J10 **Aktumsyk** Kaz. Aqtöbe. Aktyubinsk, NW Kazakhstan 50.18N 57.09E
150 J11 **Aktyubinsk** off. Aktyubinskaya Oblast', Kaz. Aqtöbe Oblysy. ◆ province W Kazakhstan

---

153 W7 **Ak-Tyuz** var. Aktyuz. Chuyskaya Oblast', N Kyrgyzstan 42.50N 76.05E
81 J17 **Akula** Equateur, NW Dem. Rep. Congo (Zaire) 2.21N 20.13E
126 Bb15 **Akune** Kagoshima, Kyūshū, SW Japan 31.59N 130.11E
82 J9 **Akurdet** var. Agordat, Akordat. C Eritrea 15.33N 38.01E
T16 **Akure** Ondo, SW Nigeria 7.18N 5.12E
J2 **Akureyri** Nordhurland Eystra, N Iceland 65.40N 18.06W
A5 **Akutan** Akutan Island, Alaska, USA 54.08N 165.47W
V19 **Akwa Ibom** ◆ state SE Nigeria
Akyab see Sittwe
131 W7 **Ak'yar** Respublika Bashkortostan, W Russian Federation 51.51N 58.13E
151 Y11 **Akzhar** Kaz. Aqzhar. Vostochnyy Kazakhstan, E Kazakhstan 47.36N 83.37E
95 F13 **Ål** Buskerud, S Norway 60.37N 8.33E
22 H11 **Ala** Ola. ✍ SE Belarus
147 T6 **Al Araţāwiyah** Ar. Riyāḍ, N Saudi Arabia 26.33N 45.19E
175 O16 **Alas** Sumbawa, S Indonesia 8.27S 117.04E
142 D14 **Alaşehir** Manisa, W Turkey 38.19N 28.30E
145 N5 **Al 'Abd Allāh** var. Al Abdullah. S Iraq 32.06N 45.08E
145 W14 **Al 'Atshānah** var. Ashara. Dayr az Zawr, E Syria 34.51N 40.36E
145 N5 **Al Ashkharā** see Al Ashkharah
147 Z9 **Al Abdullah** see Al 'Abd Allāh
145 X8 **Al Ashkharah** var. Al Ashkhara. NE Oman 21.46N 59.30E
41 P8 **Alaska** off. State of Alaska; also known as Land of the Midnight Sun, The Last Frontier, Seward's Folly; prev. Russian America. ◆ state NW USA
41 T13 **Alaska, Gulf of** var. Golfo de Alasca. gulf Canada/USA
41 O15 **Alaska Peninsula** peninsula Alaska, USA
41 Q11 **Alaska Range** ▲ Alaska, USA
173 Ee4 **Alas, Lac** ✍ Chlef, NW Indonesia
175 O16 **Alas, Selat** strait Nusa Tenggara, C Indonesia
108 B10 **Alassio** Liguria, NW Italy 44.01N 8.12E
106 L8 **Alat** see Olot
175 Y12 **Älät** Rus. Alyaty; prev. Alyaty-Pristan'. SE Azerbaijan 39.57N 49.24E
145 S13 **Al 'Athāmīn** ▲ S Iraq 30.27N 43.40E
41 P8 **Alatna River** ✍ Alaska, USA
109 J15 **Alatri** Lazio, C Italy 41.43N 13.21E
106 J9 **Alatyr'** ✍ W Russian Federation
95 K16 **Alahärmä** Länsi-Suomi, W Finland 63.15N 22.49E
58 C7 **Alausí** Chimborazo, C Ecuador 2.07S 78.44W
107 O3 **Álava** Basq. Araba. ◆ province País Vasco, N Spain
143 T11 **Alaverdi** N Armenia 41.06N 44.37E
95 N14 **Ala-Vuokki** Oulu, E Finland 64.46N 29.29E
95 K17 **Alavus** Swe. Alavo. Länsi-Suomi, W Finland 62.33N 23.37E
90 K9 **Alawoona** South Australia 34.45S 140.28E
95 X9 **Alaykel'/Alay-Kuu** see Kek-Art
149 R17 **Al 'Ayn** var. Al Ain. Abū Ẓaby, E UAE 24.13N 55.44E
149 R17 **Al 'Ayn** var. Al Ain. ✕ Abū Ẓaby, E UAE 24.16N 55.31E
144 G12 **Al 'Aynā** Al Karak, W Jordan 30.59N 35.43E
Alayor see Alaior
Alayskiy Khrebet see Alai Range
127 N7 **Alazeya** ✍ NE Russian Federation
145 U8 **Al 'Azīzīyah** var. Aziziya. E Iraq 32.54N 45.05E
123 L14 **Al 'Azīzīyah** NW Libya 32.31N 13.01E
N11 **Al Azraq al Janūbī** Az Zarqā', N Jordan 31.49N 36.48E
108 B9 **Alba** anc. Alba Pompeia. Piemonte, NW Italy 44.40N 8.01E
145 X6 **Al 'Alamayn** see El Alamein
N15 **Alba** Texas, SW USA 32.47N 95.37W
W15 **Albia** Iowa, C USA 41.01N 92.48W
145 R1 **Al 'Amādīyah** N Iraq 37.09N 43.27E
196 K5 **Alamagan** island C Northern Mariana Islands
145 X10 **Al 'Amārah** var. Amara. E Iraq 31.51N 47.10E
82 J11 **Alamat'a** Tigray, N Ethiopia 12.22N 39.32E
39 R11 **Alameda** New Mexico, SW USA 35.09N 106.37W
124 Pp15 **'Alam el Rûm, Râs** headland N Egypt 31.21N 27.23E
42 M8 **Alamícamba** see Alamikamba
42 M8 **Alamikamba** var. Alamícamba. Región Autónoma Atlántico Norte, NE Nicaragua 13.29N 84.17W
26 K1 **Alamito Creek** ✍ Texas, SW USA
42 M8 **Alamitos, Sierra de los** ▲ NE Mexico 26.15N 102.14W
F10 **Alamo** Nevada, W USA 37.21N 115.07W
22 F9 **Alamo** Tennessee, S USA 35.46N 89.07W
39 Q12 **Alamo** Veracruz-Llave, C Mexico 20.55N 97.40W
39 S14 **Alamogordo** New Mexico, SW USA 32.52N 105.57W
68 F8 **Alamo Lake** ⊗ Arizona, SW USA
39 N7 **Alamos** Sonora, NW Mexico 26.59N 108.53W
39 S7 **Alamosa** Colorado, C USA 37.25N 105.51W
95 O13 **Åland** Fin. Ahvenanmaa. ◆ province SW Finland
95 L20 **Åland** var. Aland Islands, Fin. Ahvenanmaa. island group SW Finland
Åland Islands see Åland

---

115 L20 **Albania** off. Republic of Albania, Alb. Republika e Shqipërisë, Shqipëria; prev. People's Socialist Republic of Albania. ◆ republic SE Europe
Albania see Aubagne
109 H15 **Albania** Lazio, C Italy 41.43N 12.40E
28 J14 **Albany** Western Australia 35.03S 117.54E
25 S7 **Albany** Georgia, SE USA 31.34N 84.09W
23 P13 **Albany** Indiana, N USA 40.18N 85.14W
22 L8 **Albany** Kentucky, S USA 36.41N 85.07W
31 U7 **Albany** Minnesota, N USA 45.39N 94.33W
29 R2 **Albany** Missouri, C USA 40.15N 94.15W
20 L10 **Albany** state capital New York, NE USA 42.39N 73.45W
34 G12 **Albany** Oregon, NW USA 44.38N 123.06W
27 Q6 **Albany** Texas, SW USA 32.43N 99.18W
10 F10 **Albany** ✍ Ontario, S Canada
147 T14 **Al Buzūn** SE Yemen 15.40N 50.53E
95 G17 **Älä** Västernorrland, C Sweden 62.30N 15.25E
106 G14 **Albyn, Glen** Gen Mor, Glen C Scotland, UK
106 G12 **Alcácer do Sal** Setúbal, W Portugal 38.21N 8.29W
145 Q13 **Al Baṣrah** Eng. Basra; hist. Busra, Bussora. SE Iraq 30.31N 47.49E
145 V11 **Al Baṭḥā'** SE Iraq 31.06N 45.49E
147 X8 **Al Bāṭinah** var. Batinah. coastal region O Oman
1 H16 **Albatross Plateau** undersea feature E Pacific Ocean
Al Batrūn see Batroûn
124 Nn14 **Al Bayḍā'** var. Beida. NE Libya 32.46N 21.43E
147 P16 **Al Bayḍā'** var. Al Beida. SW Yemen 13.58N 45.38E
Al Bedei'ah see Al Badi'ah
42 S10 **Al Beida** see Al Bayḍā'
Albemarle var. Albemarle. North Carolina, SE USA 35.21N 80.12W
173 Ee4 **Albemarle Island** see Isabela, Isla
21 X10 **Albemarle Sound** inlet W Atlantic Ocean
108 B10 **Albenga** Liguria, NW Italy 44.04N 8.13E
106 L8 **Alberche** ✍ C Spain
28 J8 **Alberga Creek** seasonal river South Australia
106 G7 **Albergaria-a-Velha** Aveiro, N Portugal 40.42N 8.29W
107 S10 **Alberic** País Valenciano, E Spain 39.07N 0.31W
109 P18 **Alberobello** Puglia, SE Italy 40.47N 17.14E
110 J7 **Alberschwende** Vorarlberg, W Austria 47.28N 9.44E
11 O2 **Alberta** ◆ province SW Canada
194 K14 **Albert Edward, Mount** ▲ S PNG 8.23S 147.23E
63 C20 **Alberti** Buenos Aires, E Argentina 35.03S 60.15W
143 K23 **Albertirsa** Pest, C Hungary 47.15N 19.36E
149 R17 **Albert, Lake** var. Albert Nyanza, Lac Mobutu Sese Seko. ⊗ Uganda/Dem. Rep. Congo (Zaire)
31 V11 **Albert Lea** Minnesota, N USA 43.39N 93.22W
83 E14 **Albert Nile** ✍ NW Uganda
105 U8 **Albert Nyanza** see Albert, Lake
105 T11 **Albertville** Savoie, E France 45.41N 6.24E
25 Q2 **Albertville** Alabama, S USA 34.16N 86.12W
Albertville see Kalemie
105 N15 **Albi** anc. Albiga. Tarn, S France 43.55N 2.09E
31 W15 **Albia** Iowa, C USA 41.01N 92.48W
63 B19 **Albina** Marowijne, N Suriname 5.31N 54.04W
108 A8 **Albina, Ponta** headland SW Angola 15.52S 11.45E
32 M16 **Albion** Illinois, N USA 38.22N 88.03W
23 P11 **Albion** Indiana, N USA 41.22N 85.23W
31 S11 **Albion** Nebraska, C USA 41.41N 98.00W
20 D9 **Albion** New York, NE USA 43.13N 78.09W
18 D13 **Albion** Pennsylvania, NE USA 41.52N 80.18W
95 Q7 **Åland** ✍ see El Beqaa
2 F11 **Åland** Ontario, S Canada 46.07N 80.37W
90 K9 **Åland** var. Aland Islands, Fin. Ahvenanmaa. island group SW Finland
Åland Islands see Åland

---

149 O5 **Ålborg-Nørresundby** see Ålborg
97 Q14 **Alborz, Reshteh-ye Kūhhā-ye** Eng. Elburz Mountains. ▲ N Iran
37 O14 **Albox** Andalucía, S Spain 37.22N 2.08W
103 H23 **Albstadt** Baden-Württemberg, SW Germany 48.13N 9.01E
106 G14 **Albufeira** Beja, S Portugal 37.04N 8.15W
145 P5 **Âlbū Gharz, Sabkhat** ◊ W Iraq
39 V11 **Albuñol** Andalucía, S Spain 36.48N 3.10W
147 W8 **Al Buraymī** var. Buraimi. N Oman 24.16N 55.48E
149 R17 **Al Buraymī** var. Buraimi. spring/well Oman/UAE 24.27N 55.33E
Al Burayqah see Marsá al Burayqah
106 I10 **Alburquerque** Extremadura, W Spain 39.12N 7.00W
189 V14 **Albury** New South Wales, SE Australia 36.03S 146.52E
147 T14 **Al Buzūn** SE Yemen 15.40N 50.53E
95 G17 **Älä** Västernorrland, C Sweden 62.30N 15.25E
144 J6 **Alcácer do Sal** Setúbal, W Portugal
145 Q13 **Alcalá de Chisvert** var. Alcalá de Chisvert. País Valenciano, E Spain 40.19N 0.13E
107 T13 **Alcalá de Guadaira** Andalucía, S Spain 37.19N 5.49W
106 K14 **Alcalá de Henares** Ar. Alkal'a; anc. Complutum. Madrid, C Spain 40.28N 3.22W
107 O8 **Alcalá de los Gazules** Andalucía, S Spain 36.28N 5.43W
107 N14 **Alcalá La Real** Andalucía, S Spain 37.28N 3.55W
109 J23 **Alcamo** Sicilia, Italy, C Mediterranean Sea 37.58N 12.58E
107 T4 **Alcanadre** ✍ NE Spain
107 T8 **Alcanar** Cataluña, NE Spain 40.33N 0.28E
106 J5 **Alcañices** Castilla-León, N Spain 41.40N 6.21W
107 T7 **Alcañiz** Aragón, NE Spain 41.03N 0.09W
106 I9 **Alcántara** Extremadura, W Spain 39.42N 6.54W
106 J9 **Alcántara, Embalse de** ⊗ W Spain
107 R13 **Alcantarilla** Murcia, SE Spain 37.58N 1.12W
107 P11 **Alcaraz** Castilla-La Mancha, C Spain 38.40N 2.28W
190 F2 **Alcaraz, Sierra de** ▲ C Spain
106 G7 **Alcarrache** ✍ SW Spain
107 T6 **Alcarràs** Cataluña, NE Spain 41.34N 0.31E
107 N14 **Alcaudete** Andalucía, S Spain 37.34N 4.04W
Alcázar see Ksar-el-Kebir
107 O10 **Alcázar de San Juan** anc. Alce. Castilla-La Mancha, C Spain 39.24N 3.12W
Alcazarquivir see Ksar-el-Kebir
Alce see Alcázar de San Juan
59 B17 **Alcedo, Volcán** ▲ Galapagos Islands, Ecuador, E Pacific Ocean 0.25S 91.06W
145 X12 **Al Chabā'ish** var. Al Kaba'ish. SE Iraq 30.58N 47.01E
119 Y7 **Alchevs'k** prev. Kommunarsk, Voroshilovsk. Luhans'ka Oblast', E Ukraine 48.29N 38.52E
23 N9 **Alcira** see Alzira
32 H11 **Alcoa** Tennessee, S USA 35.47N 83.58W
106 F9 **Alcobaça** Leiria, C Portugal 39.31N 8.58W
107 N8 **Alcobendas** Madrid, C Spain 40.31N 3.37W
107 P7 **Alcocea del Pinar** Castilla-La Mancha, C Spain 41.01N 2.28W
106 I11 **Alconchel** Extremadura, W Spain 38.31N 7.04W
107 S9 **Alcora** País Valenciano, E Spain 40.04N 0.13W
107 R10 **Alcorcón** Madrid, C Spain 40.20N 3.50W
107 S9 **Alcorisa** Aragón, NE Spain 40.53N 0.24W
107 S10 **Alcoy** var. Alcoi. País Valenciano, E Spain 38.42N 0.30W
131 P11 **Alcúdia, Badia d'** bay Mallorca, Spain, W Mediterranean Sea
180 M7 **Aldabra Group** island group W Seychelles
145 U10 **Al Daghghārah** C Iraq 32.10N 44.57E
42 J3 **Aldama** Chihuahua, N Mexico 28.49N 105.52W
42 J5 **Aldama** Tamaulipas, C Mexico 22.55S 98.03W
126 Ll12 **Aldan** Respublika Sakha (Yakutiya), NE Russian Federation 58.31N 125.15E
168 Mm12 **Aldan** ✍ NE Russian Federation 47.43N 96.53W
109 M7 **Aldara** ▲ W Mongolia
al Dar al Baida see Rabat
99 Q14 **Aldeburgh** E England, UK 52.12N 1.35E
107 P5 **Aldehuela de Calatañazor** Castilla-León, N Spain 41.42N 2.46W
106 G13 **Aldeia Nova** see Aldeia Nova de São Bento
106 H13 **Aldeia Nova de São Bento var.** Aldeia Nova. Beja, S Portugal 37.55N 7.24W
31 V11 **Alden** Minnesota, N USA 43.40N 93.34W
192 H6 **Aldermen Islands, The** island group N NZ
42 G7 **Alderney** island Channel Islands
99 O22 **Aldershot** S England, UK 51.15N 0.46W
23 R6 **Alderson** West Virginia, NE USA 37.43N 80.38W

Al Dhaid see Adh Dhayd
32 J11 **Aledo** Illinois, N USA 41.12N 90.45W
78 H9 **Aleg** Brakna, SW Mauritania 17.03N 13.52W
66 Q10 **Alegranza** island Islas Canarias, Spain, NE Atlantic Ocean
39 P12 **Alegres Mountain** ▲ New Mexico, SW USA 34.09N 108.11W
63 F15 **Alegrete** Rio Grande do Sul, S Brazil 29.46S 55.46W
63 C16 **Alejandra** Santa Fe, C Argentina 29.54S 59.49W
200 Oo12 **Alejandro Selkirk, Isla** island Islas Juan Fernández, Chile, E Pacific Ocean
128 J12 **Alekhovshchina** Leningradskaya Oblast', NW Russian Federation 60.22N 33.57E
41 O13 **Aleknagik** Alaska, USA 59.16N 158.37W
Aleksandriya see Oleksandriya
Aleksandropol' see Gyumri
130 L3 **Aleksandrov** Vladimirskaya Oblast', W Russian Federation 56.24N 38.42E
115 N14 **Aleksandrovac** Serbia, C Yugoslavia 43.28N 21.05E
131 R9 **Aleksandrov Gay** Saratovskaya Oblast', W Russian Federation 50.08N 48.34E
131 U6 **Aleksandrovka** Orenburgskaya Oblast', W Russian Federation 52.47N 54.14E
Aleksandrovka see Oleksandrivka
116 J8 **Aleksandrovo** Lovech, N Bulgaria 43.16N 24.53E
129 V13 **Aleksandrovsk** Permskaya Oblast', NW Russian Federation 59.12N 57.27E
Aleksandrovsk see Zaporizhzhya
131 N14 **Aleksandrovskoye** Stavropol'skiy Kray, SW Russian Federation 44.43N 42.56E
127 O14 **Aleksandrovsk-Sakhalinskiy** Ostrov Sakhalin, Sakhalinskaya Oblast', SE Russian Federation 50.55N 142.12E
112 J10 **Aleksandrów Kujawski** Kujawsko-pomorskie, C Poland 52.51N 18.42E
112 K12 **Aleksandrów Łódzki** Łódzkie, C Poland 51.48N 19.18E
Alekseevka see Akkol'/Alekseyevka
151 P7 **Alekseyevka** Kaz. Alekseevka. Severnyy Kazakhstan, N Kazakhstan 53.31N 69.30E
151 Z10 **Alekseyevka** Kaz. Alekseevka. Vostochnyy Kazakhstan, E Kazakhstan 48.25N 85.38E
131 S7 **Alekseyevka** Samarskaya Oblast', W Russian Federation 52.37N 51.20E
Alekseyevka see Akkol'
126 Jj13 **Alekseyevsk** Irkutskaya Oblast', C Russian Federation 57.46N 108.07E
131 R4 **Alekseyevskoye** Respublika Tatarstan, W Russian Federation 55.18N 50.11E
130 K5 **Aleksin** Tul'skaya Oblast', W Russian Federation 54.30N 37.07E
115 O14 **Aleksinac** Serbia, SE Yugoslavia 43.33N 21.43E
202 G11 **Alele** Île Uvea, E Wallis and Futuna 13.13S 176.09W
97 N20 **Älem** Kalmar, S Sweden 56.57N 16.25E
104 L6 **Alençon** Orne, N France 48.25N 0.04E
60 I12 **Alenquer** Pará, NE Brazil 1.58S 54.45W
40 G10 **Alenuihaha Channel** channel Hawaii, USA, C Pacific Ocean
Alep/Aleppo see Ḩalab
105 N15 **Aléria** Corse, France, C Mediterranean Sea 42.06N 9.29E
207 Q11 **Alert** Ellesmere Island, Nunavut, N Canada 82.28N 62.13W
105 Q14 **Alès** prev. Alais. Gard, S France 44.07N 4.04E
118 G9 **Aleşd** Hung. Élesd. Bihor, SW Romania 47.03N 22.22E
108 C9 **Alessandria** Fr. Alexandrie. Piemonte, N Italy 44.54N 8.37E
97 J12 **Ålestrup** var. Aalestrup. Viborg, NW Denmark 56.42N 9.31E
96 D9 **Ålesund** Møre og Romsdal, S Norway 62.28N 6.10E
110 E10 **Aletschhorn** ▲ SW Switzerland 46.33N 8.01E
207 S1 **Aleutian Basin** undersea feature Bering Sea
40 H17 **Aleutian Islands** island group Alaska, USA
41 P14 **Aleutian Range** ▲ Alaska, USA
1 B5 **Aleutian Trench** undersea feature S Bering Sea
127 O10 **Alevina, Mys** headland E Russian Federation 58.52N 151.21E
13 Q6 **Alex** Quebec, SE Canada
30 J3 **Alexander** North Dakota, N USA 47.48N 103.38W
41 W14 **Alexander Archipelago** island group Alaska, USA
Alexanderbaai see Alexander Bay
85 D23 **Alexander Bay** Afr. Alexanderbaai. Northern Cape, W South Africa 28.35S 16.30E
25 Q5 **Alexander City** Alabama, S USA 32.56N 85.57W
204 J6 **Alexander Island** island Antarctica
Alexander Range see Kirghiz Range
191 O12 **Alexandra** Victoria, SE Australia 37.12S 145.43E
193 D22 **Alexandra** Otago, South Island, NZ 45.15S 169.24E
117 F14 **Alexándreia** var. Alexándria. Kentrikí Makedonía, N Greece 40.38N 22.27E
Alexandretta see İskenderun
Alexandretta, Gulf of see İskenderun Körfezi
5 N13 **Alexandria** Ontario, SE Canada 45.19N 74.37W
23 Q15 **Alexandria** Ar. Al Iskandarīyah. N Egypt 31.07N 29.51E
46 J12 **Alexandria** Jamaica 18.18N 77.21W
118 J15 **Alexandria** Teleorman, S Romania 43.58N 25.18E
28 P13 **Alexandria** Indiana, N USA 40.15N 85.40W

22 M4 **Alexandria** Kentucky, S USA 38.56N 84.21W
24 H7 **Alexandria** Louisiana, S USA 31.18N 92.27W
31 T7 **Alexandria** Minnesota, N USA 45.54N 95.22W
31 Q11 **Alexandria** South Dakota, N USA 43.39N 97.46W
23 W4 **Alexandria** Virginia, NE USA 38.48N 77.03W
Alexandria see Alexándreia
20 I7 **Alexandria Bay** New York, NE USA 44.20N 75.54W
Alexandrie see Alessandria
190 I10 **Alexandrina, Lake** ◎ South Australia
116 K13 **Alexandroúpoli** var. Alexandroúpolis, Turk. Dedeağaç, Dedeagach. Anatolikí Makedonía kai Thráki, NE Greece 40.51N 25.52E
Alexandroúpolis see Alexandroúpoli
8 L15 **Alexis Creek** British Columbia, SW Canada 52.06N 123.25W
126 Gg15 **Aleysk** Altayskiy Kray, S Russian Federation 52.32N 82.46E
145 S8 **Al Fallūjah** var. Falluja. C Iraq 33.21N 43.46E
107 R8 **Alfambra** ☊ E Spain
Al Faqa see Faq'
147 R15 **Al Farḍah** C Yemen 14.51N 48.33E
107 Q4 **Alfaro** La Rioja, N Spain 42.09N 1.46W
107 U5 **Alfarràs** Cataluña, NE Spain 41.49N 0.34E
Al Fāshir see El Fasher
Al Fashn see El Fashn
116 M7 **Alfatar** Silistra, NE Bulgaria 43.56N 27.17E
145 S3 **Al Fatḥah** C Iraq 35.06N 43.34E
145 Q3 **Al Fatsi** N Iraq 36.04N 42.39E
145 Z13 **Al Faw** var. Fao. SE Iraq 29.55N 48.25E
117 D20 **Alfeiós** prev. Alfiós, anc. Alpheus, Alpheus. ☊ S Greece
102 I13 **Alfeld** Niedersachsen, C Germany 51.58N 9.49E
Alfiós see Alfeiós
Alföld see Great Hungarian Plain
97 L15 **Ålfotbreen** glacier S Norway
21 P9 **Alfred** Maine, NE USA 43.28N 70.43W
20 F11 **Alfred** New York, NE USA 42.15N 77.47W
63 K14 **Alfredo Vagner** Santa Catarina, S Brazil 27.40S 49.22W
96 M12 **Alfta** Gävleborg, C Sweden 61.19N 16.04E
146 K12 **Al Fuḩayḩil** var. Fahaheel. SE Kuwait 29.01N 48.04E
145 U10 **Al Fuḩaymi** C Iraq 34.17N 42.09E
149 S16 **Al Fujayrah** Eng. Fujairah. NE UAE 25.09N 56.18E
149 S16 **Al Fujayrah** Eng. Fujairah. ✈ Al Fujayrah, NE UAE 25.04N 56.12E
Al Furāt see Euphrates
150 I10 **Alga** Kaz. Algha. Aktyubinsk, NW Kazakhstan 49.55N 57.19E
150 G9 **Algabas** Zapadnyy Kazakhstan, NW Kazakhstan 50.45N 52.07E
9 C17 **Ålgård** Rogaland, S Norway 58.45N 5.52E
106 G14 **Algarve** cultural region S Portugal
190 G3 **Algebuckina Bridge** South Australia 28.03S 135.48E
106 K16 **Algeciras** Andalucía, SW Spain 36.07N 5.27W
107 S10 **Algemesí** País Valenciano, E Spain 39.10N 0.27W
Al-Genain see El Geneina
123 I11 **Alger** var. Algiers, El Djazaïr, Al Jazair. ● (Algeria) N Algeria 36.47N 2.58E
123 T13 **Algeria** off. Democratic and Popular Republic of Algeria. ◆ republic N Africa
123 I9 **Algerian Basin** var. Balearic Plain undersea feature W Mediterranean Sea
Algha see Alga
144 I4 **Al Ghāb** ◎ NW Syria
147 X10 **Al Ghābah** var. Ghaba. C Oman 21.21N 57.13E
147 U13 **Al Ghaydah** E Yemen 16.15N 52.13E
146 M6 **Al Ghazālah** Ḩā'il, N Saudi Arabia 26.55N 41.23E
109 B17 **Alghero** Sardegna, Italy, C Mediterranean Sea 40.34N 8.19E
97 M20 **Älghult** Kronoberg, S Sweden 57.00N 15.34E
Al Ghurdaqah see Hurghada
Algiers see Alger
85 L26 **Algoa Bay** bay S South Africa
106 L4 **Algodonales** Andalucía, S Spain 36.54N 5.24W
107 N9 **Algodor** ☊ C Spain
33 N6 **Algoma** Wisconsin, N USA 44.41N 87.24W
31 U12 **Algona** Iowa, C USA 43.04N 94.13W
22 L8 **Algood** Tennessee, S USA 36.12N 85.27W
107 Q2 **Algorta** País Vasco, N Spain 43.20N 3.00W
63 E18 **Algorta** Río Negro, W Uruguay 32.21S 57.12W
158 J11 **Aligarh** Uttar Pradesh, N India 27.54N 78.04E
142 M7 **Alīgūdarz** Lorestān, W Iran 33.27N 49.33E
1 F12 **Alijos, Islas** island group California, SW USA

37 T15 **Alhambra** California, W USA 34.07N 118.06W
145 T12 **Al Ḩammām** S Iraq 31.09N 44.04E
147 X8 **Al Ḩamrā'** NE Oman 23.07N 57.22E
Al Ḩamrā' var. Al Ḩamadah al Ḩamrā'
147 O6 **Al Ḩamūdīyah** spring/well N Saudi Arabia 27.05N 44.24E
146 M7 **Al Ḩanākīyah** Al Madīnah, W Saudi Arabia 24.54N 40.31E
145 W14 **Al Ḩanīyah** escarpment Iraq/Saudi Arabia
145 Y12 **Al Ḩārithah** SE Iraq 30.43N 47.43E
146 L3 **Al Ḩarrah** desert NW Saudi Arabia
77 Q10 **Al Ḩarūj al Aswad** desert C Libya
145 N2 **Al Ḩasakah** var. Al Hasijah, El Haseke, Fr. Hassetché. Al Ḩasakah, NE Syria 36.22N 40.43E
145 O2 **Al Ḩasakah** var. Al Hasakah, Hasakah, Hassakeh. ◆ governorate NE Syria
Al Ḩasijah see Al Ḩasakah
106 M15 **Alhaurín el Grande** Andalucía, S Spain 36.39N 4.40W
147 Q16 **Al Ḩawrā** S Yemen 13.54N 47.36E
145 V10 **Al Ḩayy** var. Kut al Hai, Kūt al Ḩayy. E Iraq 32.10N 46.03E
147 U11 **Al Ḩibāk** desert E Saudi Arabia
144 H8 **Al Ḩijānah** var. Hejanah, Hijanah. Dimashq, W Syria 33.23N 36.34E
146 K7 **Al Ḩijāz** Eng. Hejaz. physical region NW Saudi Arabia
Al Hilbeh see 'Ulayyāniyah, Bi'r al
145 T9 **Al Ḩillah** var. Hilla. C Iraq 32.28N 44.28E
145 T9 **Al Ḩindīyah** var. Hindiya. C Iraq 32.31N 44.13E
144 G12 **Al Ḩisā** Āṭ Ṭafīlah, W Jordan 30.49N 35.58E
76 G5 **Al-Hoceïma** var. al Hoceima, Al-Hoceima, Alhucemas, prev. Villa Sanjurjo. N Morocco 35.13N 3.55W
107 N17 **Alhucemas, Peñón de** island group S Spain
147 N15 **Al Ḩudaydah** Eng. Hodeida. W Yemen 15.00N 42.50E
147 N15 **Al Ḩudaydah** Eng. Hodeida. ✈ W Yemen 14.45N 43.01E
146 M9 **Al Ḩudūd ash Shamālīyah** var. Minṭaqat al Ḩudūd ash Shamālīyah, Eng. Northern Border Region. ◆ province N Saudi Arabia
147 S7 **Al Ḩufūf** var. Hofuf. Ash Sharqīyah, NE Saudi Arabia 25.21N 49.33E
144 Q9 **Al Ḩuṣn** var. Husn. Irbid, N Jordan 32.28N 35.52E
145 U9 **'Alī** E Iraq 32.43N 45.21E
106 L10 **Alia** Extremadura, W Spain 39.25N 5.12W
149 P9 **'Aliābād** Yazd, C Iran 36.55N 54.33E
'Aliābād see Qā'emshahr
107 S7 **Aliaga** Aragón, NE Spain 40.40N 0.42W
142 B13 **Aliağa** İzmir, W Turkey 38.49N 26.58E
Aliákmon see Aliákmonas
117 F14 **Aliákmonas** prev. Aliákmon, anc. Haliacmon. ☊ N Greece
145 W9 **'Alī al Gharbi** E Iraq 32.28N 46.42E
145 U11 **'Alī al Ḩassānī** S Iraq 31.25N 44.50E
117 G13 **Aliártos** Stereá Ellás, C Greece 38.22N 23.06E
143 T12 **Äli-Bayramlı** Rus. Ali-Bayramly. SE Azerbaijan 39.57N 48.54E
Ali-Bayramly see Äli-Bayramlı
79 S13 **Alibey Baraji** ☐ NW Turkey
79 N9 **Alibori** ☊ N Benin
114 L10 **Alibunar** Serbia, NE Yugoslavia 45.06N 20.59E
107 S12 **Alicante** Cat. Alacant; Lat. Lucentum. País Valenciano, SE Spain 38.21N 0.28W
107 S12 **Alicante** ◆ province País Valenciano, SE Spain
107 S12 **Alicante** ✈ Murcia, E Spain 38.21N 0.28W
85 J25 **Alice** Eastern Cape, S South Africa 32.49S 26.47E
27 S14 **Alice** Texas, SW USA 27.45N 98.04W
85 I25 **Alicedale** Eastern Cape, S South Africa 33.19S 26.04E
47 R25 **Alice, Mount** hill West Falkland, Falkland Islands
109 P20 **Alice, Punta** headland S Italy 39.24N 17.09E
189 Q7 **Alice Springs** Northern Territory, C Australia 23.42S 133.52E
25 N4 **Aliceville** Alabama, S USA 33.07N 88.09W
153 U13 **Alichur** SE Tajikistan 37.49N 73.45E
153 U14 **Alichuri Janubí, Qatorkúhi** Rus. Yuzhno-Alichurskiy Khrebet. ▲ SE Tajikistan
153 U13 **Alichuri Shimolí, Qatorkúhi** Rus. Severo-Alichurskiy Khrebet. ▲ SE Tajikistan

57 V12 **Alimuni Piek** ▲ S Suriname 2.25N 55.46W
81 K15 **Alindao** Basse-Kotto, S Central African Republic 4.58N 21.16E
97 J23 **Alingsås** Västra Götaland, S Sweden 57.55N 12.30E
83 K18 **Alinjugul** spring/well E Kenya 0.03S 40.31E
155 S11 **Alipur** Punjab, E Pakistan 29.22N 70.58E
159 T12 **Alipur Duār** West Bengal, NE India 26.28N 89.25E
20 B14 **Aliquippa** Pennsylvania, NE USA 40.36N 80.15W
82 L12 **'Ali Sabieh** var. 'Ali Sabih. S Djibouti 11.07N 42.44E
'Ali Sabih see 'Ali Sabieh
146 K3 **Al 'Īsāwīyah** Al Jawf, NW Saudi Arabia 30.41N 37.58E
106 J10 **Aliseda** Extremadura, W Spain 39.25N 6.42W
145 T8 **Al Iskandarīyah** C Iraq 32.52N 44.22E
Al Iskandarīyah see Alexandria
127 Oo5 **Aliskerovo** Chukotskiy Avtonomnyy Okrug, NE Russian Federation 67.40N 167.37E
116 H13 **Alistráti** Kentrikí Makedonía, NE Greece 41.03N 23.58E
41 Q8 **Alitak Bay** bay Kodiak Island, Alaska, USA
117 N18 **Alivéri** var. Alivérion. Évvoia, C Greece 38.25N 24.02E
Alivérion see Alivéri
Aliwal-Noord see Aliwal North
85 I24 **Aliwal North** Afr. Aliwal-Noord. Eastern Cape, SE South Africa 30.39S 26.43E
124 Nn15 **Al Jabal al Akhḍar** ▲ NE Libya
144 H13 **Al Jafr** var. Jafr. S Jordan 30.18N 36.13E
77 T8 **Al Jaghbūb** NE Libya 29.45N 24.31E
148 K11 **Al Jahrā'** var. Al Jahra, Jahra. C Kuwait 29.17N 47.36E
Al Jahrah see Al Jahrā'
Al Jamāhīrīyah al 'Arabīyah al Lībīyah ash Sha'bīyah al Ishtirāk see Libya
146 K3 **Al Jarāwī** spring/well NW Saudi Arabia 30.12N 38.48E
147 X11 **Al Jaww** oasis E Oman 18.59N 57.16E
146 L3 **Al Jawf** var. Jauf. Al Jawf, NW Saudi Arabia 29.51N 39.49E
146 L4 **Al Jawf** off. Minṭaqat al Jawf. ◆ province N Saudi Arabia
Al Jawlān see Golan Heights
146 K9 **Al Jazair** see Alger
145 N4 **Al Jazirah** physical region Iraq/Syria
106 F14 **Aljezur** Faro, S Portugal 37.18N 8.49W
145 S13 **Al Jīl** S Iraq 30.28N 43.57E
144 G11 **Al Jīzah** var. Jiza. 'Ammān, N Jordan 31.42N 35.57E
Al Jīzah see El Gîza
147 S6 **Al Jubail** see Al Jubayl
147 S6 **Al Jubayl** var. Al Jubail. Ash Sharqīyah, NE Saudi Arabia 27.00N 49.36E
147 S7 **Al Juhaysh, Qalamat** well SE Saudi Arabia 20.35N 51.00E
149 N15 **Al Jumayliyah** N Qatar 25.37N 51.04E
Al Junaynah see El Geneina
106 G13 **Aljustrel** Beja, S Portugal 37.52N 8.10W
Al Kaba'ish see Al Chabā'ish
Al-Kadhimain see Al Kāẓimīyah
Al Kāf see El Kef
Alkal'a see Alcalá de Henares
37 W4 **Alkali Flat** salt flat Nevada, W USA
37 Q1 **Alkali Lake** ◎ Nevada, W USA
147 Q8 **Al Kāmil** NE Oman 22.15N 59.12E
144 G11 **Al Karak** var. El Kerak, Karak, Kerak; anc. Kir Moab, Kir of Moab. Al Karak, W Jordan 31.10N 35.42E
144 G12 **Al Karak** ◆ governorate W Jordan
145 W8 **Al Karmashīyah** E Iraq 32.57N 46.10E
Al-Kasr al-Kebir see Ksar-el-Kebir
144 H9 **Al Kāẓimīyah** var. Al-Kadhimain, Kadhimain. C Iraq 33.22N 44.19E
101 J18 **Alken** Limburg, NE Belgium 50.52N 5.19E
147 X8 **Al Khābūrah** var. Khabura. N Oman 23.56N 57.06E
Al Khalīl see Hebron
145 T7 **Al Khāliṣ** C Iraq 33.51N 44.33E
Al Khaluf see Khalūf
Al Khārijah see El Khârga
147 Q8 **Al Kharj** Ar Riyāḍ, C Saudi Arabia 24.12N 47.12E
147 N10 **Al Khaṣab** var. Khasab. N Oman 26.10N 56.18E
149 N15 **Al Khawr** var. Al Khor. N Qatar 25.40N 51.33E
148 K12 **Al Khīrān** var. Al Khiran. SE Kuwait 28.34N 48.21E
147 W9 **Al Khīrān** spring/well NW Oman 22.31N 55.42E
Al Khiyam see El Khiyam
Al-Khobar see Al Khubar
Al Khor see Al Khawr
147 S6 **Al Khubar** var. Al-Khobar. Ash Sharqīyah, NE Saudi Arabia 26.15N 50.10E
145 S6 **Al Khufrah** SE Libya 24.10N 23.19E
123 L14 **Al Khums** var. Homs, Khoms, Khums. NW Libya 32.39N 14.16E
147 R15 **Al Khuraybah** C Yemen 15.05N 48.16E
146 M9 **Al Khurmah** var. al-Hurma. Makkah, W Saudi Arabia 21.58N 42.00E
Al Kidan see Madinat ash Sha'b
100 H9 **Alkmaar** Noord-Holland, NW Netherlands 52.37N 4.45E
155 T10 **Al Kūfah** var. Kufa. S Iraq 32.01N 44.25E
145 V9 **Al Kūt** var. Kūt al 'Amārah, Kut al Imara. E Iraq 32.30N 45.51E
Al-Kuwait see Al Kuwayt
148 K11 **Al Kuwayt** var. Al-Kuwait, Eng. Kuwait, Kuwait City; prev. Qurein. ● (Kuwait) E Kuwait 29.23N 48.00E

148 K11 **Al Kuwayt** ✈ C Kuwait 29.13N 47.57E
117 G19 **Alykyonídon, Kólpos** gulf C Greece
147 N11 **Al Labbah** physical region N Saudi Arabia
144 G4 **Al Lādhiqīyah** Eng. Latakia, Fr. Lattaquié; anc. Laodicea, Laodicea ad Mare. Al Lādhiqīyah, W Syria 35.31N 35.46E
144 H4 **Al Lādhiqīyah** off. Muḩāfaẓat al Lādhiqīyah, var. Al Lathqiyah, Latakia, Lattakia. ◆ governorate W Syria
21 R2 **Allagash River** ☊ Maine, NE USA
158 M13 **Allahābād** Uttar Pradesh, N India 25.27N 81.49E
33 Q9 **Allāh Dāgh, Reshteh-ye** ▲ NE Iran
41 Q8 **Allakaket** Alaska, USA
126 Mm11 **Allakh-Yun'** ☊ NE Russian Federation
11 T15 **Allan** Saskatchewan, S Canada 51.50N 105.59W
177 Fj6 **Allanmyo** Magwe, C Myanmar 19.25N 95.13E
85 I22 **Allanridge** Free State, C South Africa 27.45S 26.43E
106 H4 **Allariz** Galicia, NW Spain 42.10N 7.48W
145 R13 **Al Laṣaf** var. Al Lussuf. S Iraq 31.38N 43.16E
25 S2 **Allatoona Lake** ☐ Georgia, SE USA
85 J19 **Alldays** Northern, NE South Africa 22.39S 29.04E
Alle see Lyna
33 P10 **Allegan** Michigan, N USA 42.31N 85.51W
19 Qq8 **Allegheny Mountains** ▲ New York/Pennsylvania, NE USA
20 E12 **Allegheny Plateau** ▲ New York/Pennsylvania, NE USA
20 D11 **Allegheny Reservoir** ☐ New York/Pennsylvania, NE USA
20 E12 **Allegheny River** ☊ New York/Pennsylvania, NE USA
24 K9 **Allemands, Lac des** ◎ Louisiana, S USA
27 N6 **Allen** Texas, SW USA 33.06N 96.40W
23 N6 **Allendale** South Carolina, SE USA 33.01N 81.19W
43 O9 **Allende** Coahuila de Zaragoza, NE Mexico 28.20N 100.47W
43 O9 **Allende** Nuevo León, NE Mexico 25.19N 100.01W
99 D16 **Allen, Lough** Ir. Loch Aillionn. ◎ NW Ireland
193 B26 **Allen, Mount** ▲ Stewart Island, Southland, NZ 47.05S 167.49E
111 V2 **Allensteig** Niederösterreich, N Austria 48.40N 15.24E
Allenstein see Olsztyn
20 I14 **Allentown** Pennsylvania, NE USA 40.37N 75.30W
161 G23 **Alleppey** var. Alappuzha; prev. Alleppi. Kerala, SW India 9.30N 76.22E
Alleppi see Alleppey
102 I12 **Aller** ☊ NW Germany
31 N16 **Allerton** Iowa, C USA 40.42N 93.22W
101 G18 **Alleur** Liège, E Belgium 50.40N 5.33E
103 J25 **Allgäuer Alpen** ▲ Austria/Germany
30 J13 **Alliance** Nebraska, C USA 42.05N 102.52W
21 U12 **Alliance** Ohio, N USA 40.55N 81.06W
105 O10 **Allier** ◆ department N France
145 R13 **Al Lifiyah** S Iraq 30.25N 43.03E
46 J13 **Alligator Pond** C Jamaica 17.52N 77.34W
23 V9 **Alligator River** ☊ North Carolina, SE USA
31 W12 **Allison** Iowa, C USA 42.45N 92.48W
12 G14 **Alliston** Ontario, S Canada 44.09N 79.51W
146 L11 **Al Lith** Makkah, SW Saudi Arabia 21.00N 41.00E
98 I12 **Alloa** C Scotland, UK 56.07N 3.49W
105 U14 **Allos** Alpes-de-Haute-Provence, SE France 44.16N 6.37E
111 U18 **Allschwil** Basel-Land, NW Switzerland 47.34N 7.32E
Al Lubnān see Lebanon
147 N14 **Al Luḩayyah** W Yemen 15.43N 42.45E
Al Lussuf see Al Laṣaf
12 K12 **Allumettes, Île des** island Quebec, SE Canada
Al Lussuf see Al Laṣaf
111 S5 **Alm** ☊ N Austria
13 Q7 **Alma** Quebec, SE Canada
29 S10 **Alma** Arkansas, C USA 35.28N 94.13W
23 V7 **Alma** Georgia, SE USA 31.32N 82.27W
29 T6 **Alma** Kansas, C USA 39.01N 96.17W
33 Q8 **Alma** Michigan, N USA 43.22N 84.39W
31 O17 **Alma** Nebraska, C USA 40.06N 99.21W
31 X9 **Alma** Wisconsin, N USA 44.21N 91.54W
Alma-Ata see Almaty
Alma-Atinskaya Oblast' see Almaty Oblysy
107 T5 **Almacellas** see Almacelles
107 T5 **Almacelles** var. Almacellas. Cataluña, NE Spain 41.43N 0.25E
106 F11 **Almada** Setúbal, W Portugal 38.40N 9.09W
107 N11 **Almadén** Castilla-La Mancha, C Spain 38.46N 4.49W
146 L6 **Al Madīnah** Eng. Medina. Al Madīnah, W Saudi Arabia 24.25N 39.29E
146 L7 **Al Madīnah** ◆ province W Saudi Arabia
144 I6 **Al Mafraq** var. Mafraq. Al Mafraq, N Jordan 32.20N 36.12E
144 I6 **Al Mafraq** off. Muḩāfaẓat al Mafraq. ◆ governorate NW Jordan

147 R15 **Al Maghārīm** C Yemen 15.00N 47.49E
107 N11 **Almagro** Castilla-La Mancha, C Spain 38.54N 3.43W
145 T9 **Al Maḩāwīl** C Iraq 32.39N 44.28E
145 T8 **Al Maḩmūdīyah** var. Mahmudiya. C Iraq 33.04N 44.22E
147 T14 **Al Mahrah** ▲ E Yemen
147 P7 **Al Majma'ah** Ar Riyāḍ, C Saudi Arabia 25.55N 45.18E
145 Q11 **Al Makmin** well S Iraq 31.38N 42.10E
145 Q11 **Al Mālikīyah** var. Malkiye. Al Ḩasakah, N Syria 37.12N 42.13E
Almalyk see Olmaliq
146 L3 **Al Manādir** var. Al Manadir. ◎ Oman/UAE
149 Q18 **Al Manāmah** Eng. Manama. ● (Bahrain) N Bahrain 26.13N 50.33E
145 O5 **Al Manāşif** ▲ E Syria
37 O4 **Almanor, Lake** ◎ California, W USA
107 R14 **Almansa** Castilla-La Mancha, C Spain 38.52N 1.06W
106 I6 **Al Manşūrah** see El Mansûra
106 L3 **Almanza** Castilla-León, N Spain 42.39N 5.01W
107 P14 **Almanzor** ▲ W Spain 40.13N 5.18W
107 Q11 **Almanzora** ☊ SE Spain
77 R7 **Al Mardah** C Iraq 32.35N 43.30E
Al-Mariyya see Almería
77 R7 **Al Marj** var. Barka, It. Barce. NE Libya 32.30N 20.54E
145 O5 **Al Mashrafah** Ar Raqqah, N Syria 36.25N 39.07E
147 X8 **Al Maşna'ah** var. Al Muşana'a. NE Oman 23.45N 57.37E
107 T9 **Almassora** País Valenciano, E Spain 39.55N 0.02W
151 U15 **Almaty** var. Alma-Ata. Almaty, SE Kazakhstan 43.19N 76.55E
151 S14 **Almaty** off. Almatinskaya Oblast', Kaz. Almaty Oblysy; prev. Alma-Atinskaya Oblast'. ◆ province SE Kazakhstan
151 U15 **Almaty** ✈ Almaty, SE Kazakhstan 43.15N 76.57E
Almaty Oblysy see Almaty
al-Mawaşilin see Al Muwayliḩ
145 R3 **Al Mawşil** Eng. Mosul. N Iraq 36.21N 43.07E
145 N5 **Al Mayādīn** var. Mayadin, Fr. Meyadine. Dayr az Zawr, E Syria 35.01N 40.28E
145 X10 **Al Maymūnah** var. Maimuna. SE Iraq 31.43N 46.55E
147 N5 **Al Mayyāh** Ḩā'il, N Saudi Arabia 27.56N 42.53E
107 P6 **Al Ma'zam** see Al Ma'zim
147 W8 **Al Ma'zim** var. Al Ma'zam. NW Oman 23.22N 56.16E
228 Kk11 **Al Mazra'ah** var. Al Mazra, Mazra'a. Al Karak, W Jordan 31.17N 35.31E
Al Mazra' see Al Mazra'ah
106 J6 **Almeirim** Santarém, C Portugal 39.12N 8.37W
100 O10 **Almelo** Overijssel, E Netherlands 52.22N 6.42E
107 S9 **Almenara** País Valenciano, E Spain 39.46N 0.13W
107 P12 **Almenaras** ▲ S Spain 38.31N 2.27W
107 P5 **Almenar de Soria** Castilla-León, N Spain 41.40N 2.12W
110 I15 **Almendra, Embalse de** ☐ Castilla-León, N Spain
106 J10 **Almendralejo** Extremadura, W Spain 38.40N 6.25W
100 J10 **Almere** var. Almere-stad. Flevoland, C Netherlands 52.22N 5.12E
100 J10 **Almere-Buiten** Flevoland, C Netherlands 52.24N 5.15E
100 I10 **Almere-Haven** Flevoland, C Netherlands 52.19N 5.13E
Almere-stad see Almere
106 M14 **Almería** Ar. Al-Mariyya; anc. Unci, Lat. Portus Magnus. Andalucía, S Spain 36.49N 2.25W
107 P15 **Almería** ◆ province Andalucía, S Spain
106 L15 **Almería, Golfo de** gulf S Spain
131 S1 **Al'met'yevsk** Respublika Tatarstan, W Russian Federation 54.52N 52.19E
97 L21 **Älmhult** Kronoberg, S Sweden 56.31N 14.10E
106 L17 **Almina, Punta** headland Ceuta, Spain, N Africa 35.54N 5.16W
32 I9 **Almira** Washington, NW USA 47.42N 118.57W
145 O17 **Almirante** Bocas del Toro, NW Panama 9.32N 82.25W
Almirós see Almyrós
146 M9 **Al Mislab** spring/well N Saudi Arabia 22.46N 40.47E
Almissa see Omiš
107 N13 **Almodóvar** var. Almodôvar. Beja, S Portugal 37.31N 8.03W
107 N11 **Almodóvar del Campo** Castilla-La Mancha, C Spain 38.46N 4.49W
106 M11 **Almodóvar del Pinar** Castilla-La Mancha, C Spain 39.43N 1.55W
33 S9 **Almont** Michigan, N USA 42.53N 83.02W
12 L13 **Almonte** Ontario, SE Canada 45.13N 76.12W
106 J12 **Almonte** Andalucía, S Spain 37.16N 6.31W
106 K9 **Almonte** ☊ W Spain
158 J14 **Almora** Uttar Pradesh, N India 29.36N 79.40E

106 M8 **Almorox** Castilla-La Mancha, C Spain 40.13N 4.22W
147 S7 **Al Mubarraz** Ash Sharqīyah, E Saudi Arabia 25.28N 49.34E
144 G15 **Al Mudawwarah** Ma'ān, SW Jordan 29.20N 36.00E
147 N19 **Al Muḑaybī** var. Al Muḑaybī. NE Oman 22.34N 58.07E
Al Muḑaybī see Al Muḑaybī
107 S5 **Almudébar** see Almudévar
107 S5 **Almudévar** Aragón, NE Spain 42.03N 0.34W
147 S15 **Al Mukallā** var. Mukalla. SE Yemen 14.36N 49.07E
147 N16 **Al Mukhā** Eng. Mocha. SW Yemen 13.18N 43.16E
107 N15 **Almuñécar** Andalucía, S Spain 36.43N 3.40W
145 Q7 **Al Muqdādīyah** C Iraq 33.58N 44.58E
146 L3 **Al Murayr** spring/well NW Saudi Arabia 30.06N 39.54E
142 M12 **Almus** Tokat, N Turkey 40.22N 36.54E
145 T9 **Al Muşana'a** see Al Maşna'ah
145 O5 **Al Muşayyib** var. Musaiyib. C Iraq 32.46N 44.19E
144 H10 **Al Muwaqqar** var. El Muwaqqar. 'Ammān, W Jordan 31.49N 36.06E
146 J5 **Al Muwayliḩ** var. al-Mawaşilin. Tabūk, NW Saudi Arabia 27.39N 35.33E
117 F17 **Almyrós** var. Almirós. Thessalía, C Greece 39.10N 22.45E
117 I24 **Almyroú, Órmos** bay Kríti, Greece, E Mediterranean Sea
98 L13 **Alnwick** N England, UK 55.26N 1.44W
Al Obayyid see El Obeid
Al Odaid see Al 'Udayd
202 L12 **Alofi** ● (Niue) W Niue 19.01S 169.55E
202 L12 **Alofi Bay** bay W Niue, C Pacific Ocean
202 L12 **Alofi, Île** island S Wallis and Futuna
Aloha State see Hawaii
120 G7 **Aloja** Limbaži, N Latvia 57.47N 24.53E
159 S10 **Along** Arunāchal Pradesh, NE India 28.15N 94.56E
186 M15 **Alónnisos** island Vóreioi Sporádes, Greece, Aegean Sea
106 M13 **Alora** Andalucía, S Spain 36.49N 4.43W
175 Rr15 **Alor, Kepulauan** island group E Indonesia
175 Rr16 **Alor, Pulau** prev. Ombai. island Kepulauan Alor, E Indonesia
175 R16 **Alor, Selat** strait Flores Sea/Savu Sea
173 G2 **Alor Setar** var. Alor Star, Alur Setar. Kedah, Peninsular Malaysia 6.06N 100.22E
Alost see Aalst
160 F9 **Álot** Madhya Pradesh, C India 23.58N 75.40E
195 N16 **Alotau** Milne Bay, SE PNG 10.18S 150.25E
176 Yy15 **Alotip** Irian Jaya, E Indonesia 8.07S 140.06E
Al Oued see El Oued
37 R12 **Alpaugh** California, W USA 35.52N 119.29W
33 R6 **Alpena** Michigan, N USA 45.04N 83.27W
105 S14 **Alpes-de-Haute-Provence** ◆ department SE France
105 U14 **Alpes-Maritimes** ◆ department SE France
189 W8 **Alpha** Queensland, E Australia 23.40S 146.38E
207 R9 **Alpha Cordillera** var. Alpha Ridge. undersea feature Arctic Ocean
Alpha Ridge see Alpha Cordillera
101 I15 **Alphen** Noord-Brabant, S Netherlands 51.29N 4.57E
100 H11 **Alphen aan den Rijn** var. Alphen. Zuid-Holland, C Netherlands 52.07N 4.40E
Alpheus see Alfeiós
106 G9 **Alpiarça** Santarém, C Portugal 39.15N 8.34W
26 L8 **Alpine** Texas, SW USA 30.22N 103.40W
110 F8 **Alpnach** Unterwalden, W Switzerland 46.56N 8.17E
110 D11 **Alps** Fr. Alpes, Ger. Alpen, It. Alpi. ▲ C Europe
147 O8 **Al Qābil** var. Qabil. N Oman 23.55N 55.49E
82 J5 **Al Qaddāḥīyah** N Libya 31.21N 15.16E
146 K4 **Al Qalībah** Tabūk, NW Saudi Arabia 28.28N 37.40E
145 O1 **Al Qāmishlī** var. Kamishli, Qamishly. Al Ḩasakah, NE Syria 37.00N 41.00E
144 I6 **Al Qaryatayn** var. Qaryatayn, Fr. Karietein. Ḩimş, C Syria 34.13N 37.13E
81 K11 **Al Qaşr** Al Kashaniya, NE Kuwait 29.59N 47.42E
146 J5 **Al Qaşr** ... 35.06N 37.39E
Al Qaşr see El Qaşr
Al Qaşrayn see Kasserine
147 S7 **Al Qaṭīf** Ash Sharqīyah, E Saudi Arabia 26.27N 50.01E
144 G11 **Al Qaţrānah** var. El Qaţrani. Al Karak, W Jordan 31.13N 36.03E
77 P11 **Al Qaţrūn** SW Libya 24.57N 14.40E
Al Qayrawān see Kairouan
Al-Qaşr al-Kebir see Ksar-el-Kebir
147 T9 **Al Qubbah** ...
Al Qubayyāt see Qoubaïyât
Al Quds/Al Quds ash Sharif see Jerusalem
146 G8 **Al Qunayṭirah** var. El Kuneitra, El Qunayṭirah, Kuneitra, Qunaytra. ◆ province SW Syria
144 G8 **Al Qunayṭirah** var. El Kuneitra, El Qunayṭirah, Kuneitra, Qunaytra. Al Qunayṭirah, SW Syria 33.07N 35.49E

◆ COUNTRY ◇ DEPENDENT TERRITORY ◆ ADMINISTRATIVE REGION ▲ MOUNTAIN ☒ VOLCANO ◎ LAKE
● COUNTRY CAPITAL ○ DEPENDENT TERRITORY CAPITAL ✈ INTERNATIONAL AIRPORT ▲ MOUNTAIN RANGE ☊ RIVER ☐ RESERVOIR

144 G8 **Al Qunayṭirah** off. Muḩāfaẓat al Qunayṭirah, var. El Q'unaytirah, Qunayṭirah, Fr. Kuneitra. ◆ governorate SW Syria

146 M11 **Al Qunfudhah** Makkah, SW Saudi Arabia 19.19N 41.02E

146 K2 **Al Qurayyāt** Al Jawf, NW Saudi Arabia 31.24N 37.25E

145 Y11 **Al Qurnah** var. Kurna. SE Iraq 31.01N 47.27E

145 V12 **Al Quşayr** S Iraq 30.36N 45.52E

144 I6 **Al Quşayr** var. El Quseir, Quşayr, Fr. Kousseir. Ḥimş, W Syria 34.36N 36.36E

Al Quşayr see Quseir

144 H7 **Al Quṭayfah** var. Quṭeifa, Qutayfe, Quteife, Fr. Kouteifé. Dimashq, W Syria 33.44N 36.33E

147 P8 **Al Quwayīyah** Ar Riyāḑ, C Saudi Arabia 24.06N 45.18E

Al Quwayr see Guwēr

144 F14 **Al Quwayrah** var. El Quweira. Ma'ān, SW Jordan 29.49N 35.19E

Al Rayyan see Ar Rayyān

Al Ruweis see Ar Ruways

97 G24 **Als** Ger. Alsen. island SW Denmark

105 U5 **Alsace** Fr. Elsass; anc. Alsatia. ◆ region NE France

11 R16 **Alsask** Saskatchewan, S Canada 51.24N 109.55W

Alsasua see Altsasu

Alsatia see Alsace

103 C16 **Alsdorf** Nordrhein-Westfalen, W Germany 50.52N 6.09E

8 G8 **Alsek** ≈ Canada/USA

Alsen see Als

103 F19 **Alsenz** ≈ W Germany

103 H17 **Alsfeld** Hessen, C Germany 50.45N 9.14E

121 K20 **Al'shany** Rus. Ol'shany. Brestskaya Voblasts', SW Belarus 52.04N 27.19E

Alsókubin see Dolný Kubín

120 C9 **Alsunga** Kuldīga, W Latvia 56.59N 21.31E

Alt see Olt

94 K9 **Alta** Fin. Alattio. Finnmark, N Norway 69.58N 23.16E

31 T12 **Alta** Iowa, C USA 42.40N 95.17W

110 I7 **Altach** Vorarlberg, W Austria 47.22N 9.39E

94 K9 **Altaelva** ≈ N Norway

94 J8 **Altafjorden** fjord NE Norwegian Sea

64 K10 **Alta Gracia** Córdoba, C Argentina 31.42S 64.25W

44 K11 **Altagracia** Zulia, NW Venezuela 10.43N 71.30W

56 H4 **Altagracia de Orituco** Guárico, N Venezuela 9.49N 66.22W

Altai see Altai Mountains

133 T7 **Altai Mountains** var. Altai, Chin. Altay Shan, Rus. Altay. ▲ Asia/Europe

25 V6 **Altamaha River** ≈ Georgia, SE USA

60 J13 **Altamira** Pará, NE Brazil 3.13S 52.15W

56 D12 **Altamira** Huila, S Colombia 2.02N 75.51W

44 M13 **Altamira** Alajuela, N Costa Rica 10.25N 84.21W

43 Q11 **Altamira** Tamaulipas, C Mexico 22.24N 97.57W

32 L15 **Altamont** Illinois, N USA 39.03N 88.45W

29 Q7 **Altamont** Kansas, C USA 37.11N 95.18W

34 H16 **Altamont** Oregon, NW USA 42.12N 121.44W

22 K10 **Altamont** Tennessee, S USA 35.25N 85.42W

25 X11 **Altamonte Springs** Florida, SE USA 28.39N 81.22W

109 O17 **Altamura** anc. Lupatia. Puglia, SE Italy 40.50N 16.33E

42 H9 **Altamura, Isla** island C Mexico

168 G7 **Altanbulag** Dzavhan, W Mongolia 48.05N 95.48E

168 G6 **Altanbulag** Dzavhan, N Mongolia 49.16N 96.22E

Altan Emel see Xin Barag Youqi

168 J8 **Altan-Ovoo** Arhangay, C Mongolia 47.24N 101.51E

168 E7 **Altanteel** Hovd, W Mongolia 47.05N 92.57E

42 F3 **Altar** Sonora, NW Mexico 30.44N 111.49W

42 D2 **Altar, Desierto de** var. Sonoran Desert. desert Mexico/USA see also Sonoran Desert

107 Q8 **Alta, Sierra** ▲ N Spain 40.29N 1.36W

42 H9 **Altata** Sinaloa, C Mexico 24.39N 107.55W

44 D4 **Alta Verapaz** off. Departamento de Alta Verapaz. ◆ department C Guatemala

109 L18 **Altavilla Silentia** Campania, S Italy 40.32N 15.06E

23 T7 **Altavista** Virginia, NE USA 37.06N 79.17W

164 L2 **Altay** Xinjiang Uygur Zizhiqu, NW China 47.51N 88.06E

168 G5 **Altay** Dzavhan, N Mongolia 49.40N 96.21E

168 G8 **Altay** Govĭ-Altay, W Mongolia 46.23N 96.16E

Altay see Altai Mountains

126 H16 **Altay, Respublika** var. Gornyy Altay; prev. Gorno-Altayskaya Respublika. ◆ autonomous republic S Russian Federation

Altay Shan see Altai Mountains

125 G15 **Altayskiy Kray** ◆ territory S Russian Federation

Altbetsche see Bečej

103 L20 **Altdorf** Bayern, SE Germany 49.23N 11.22E

110 G8 **Altdorf** var. Altorf. ◆ C Switzerland 46.52N 8.37E

107 T11 **Altea** Valenciano, E Spain 38.37N 0.03W

102 L10 **Alte Elde** ≈ N Germany

103 M16 **Altenburg** Thüringen, E Germany 50.59N 12.27E

Altenburg see Baia de Criş, Romania

102 P12 **Alte Oder** ≈ NE Germany

---

106 H10 **Alter do Chão** Portalegre, C Portugal 39.12N 7.40W

94 I10 **Altevatnet** ☉ N Norway

29 V12 **Altheimer** Arkansas, C USA 34.19N 91.51W

111 T9 **Althofen** Kärnten, S Austria 46.52N 14.27E

116 M7 **Altimir** Vratsa, NW Bulgaria 43.33N 23.48E

142 K11 **Altınkaya Barajı** ☒ N Turkey

145 S3 **Altın Köprü** var. Altun Kupri. N Iraq 35.45N 44.08E

142 E13 **Altıntaş** Kütahya, W Turkey 39.04N 30.07E

59 K18 **Altiplano** physical region W South America

Altkanischa see Kanjiža

105 U7 **Altkirch** Haut-Rhin, NE France 47.37N 7.14E

Altlublau see Stará Ľubovňa

102 L12 **Altmark** cultural region N Germany

Altmoldowa see Moldova Veche

27 W8 **Alto** Texas, SW USA 31.39N 95.04W

106 H11 **Alto Alentejo** physical region S Portugal

61 I19 **Alto Araguaia** Mato Grosso, C Brazil 17.19S 53.10W

60 L12 **Alto Bonito** Pará, NE Brazil 1.48S 46.18W

85 O15 **Alto Molócuè** Zambézia, NE Mozambique 15.41S 37.42E

32 K15 **Alton** Illinois, N USA 38.53N 90.10W

29 W8 **Alton** Missouri, C USA 36.41N 91.24W

11 X17 **Altona** Manitoba, S Canada 49.12N 97.38W

20 L14 **Altoona** Pennsylvania, NE USA 40.31N 78.22W

32 J6 **Altoona** Wisconsin, N USA 44.49N 91.22W

64 N3 **Alto Paraguay** off. Departamento del Alto Paraguay. ◆ department N Paraguay

61 L17 **Alto Paraíso de Goiás** Goiás, S Brazil 14.04S 47.15W

64 P6 **Alto Paraná** off. Departamento del Alto Paraná. ◆ department E Paraguay

Alto Paraná see Paraná

61 L15 **Alto Parnaíba** Maranhão, E Brazil 9.07S 45.55W

58 H13 **Alto Purús, Río** ≈ E Peru

Altorf see Altdorf

65 H19 **Alto Río Senguer** var. Alto Río Senguerr. Chubut, S Argentina 45.03S 70.48W

43 Q3 **Altotonga** Veracruz-Llave, E Mexico 19.43N 97.12W

103 N23 **Altötting** Bayern, SE Germany 48.12N 12.37E

Altpasua see Stara Pazova

168 I5 **Altraga** Hövsgöl, N Mongolia 50.08N 98.54E

Alt-Schwanenburg see Gulbene

107 P3 **Altsasu** var. Alsasua. Navarra, N Spain 42.54N 2.10W

Altsohl see Zvolen

110 I7 **Altstätten** Sankt Gallen, NE Switzerland 47.22N 9.33E

44 G1 **Altun Ha** ruins Belize, N Belize 17.46N 88.20W

Altun Kupri see Altın Köprü

164 D8 **Altun Shan** ▲ C China 39.19N 93.37E

164 L9 **Altun Shan** var. Altyn Tagh. ▲ NW China

37 P2 **Alturas** California, W USA 41.28N 120.32W

28 K12 **Altus** Oklahoma, C USA 34.38N 99.19W

28 K11 **Altus Lake** ☒ Oklahoma, C USA

Altvater see Praděd

Altyn Tagh see Altun Shan

Alu see Shortland Island

al-'Ubaila see Al 'Ubayla

145 O6 **Al 'Ubaydī** W Iraq 34.22N 41.15E

147 T9 **Al 'Ubaylah** var. al-'Ubaila. Ash Sharqīyah, E Saudi Arabia 22.01N 50.57E

147 T9 **Al 'Ubaylah** spring/well E Saudi Arabia 22.02N 50.56E

Al Ubayyiḑ see El Obeid

147 T7 **Al 'Udayd** var. Al Odaid. Abū Ẓaby, W UAE 24.34N 51.27E

120 J8 **Alūksne** Ger. Marienburg. Alūksne, NE Latvia 57.24N 27.02E

146 K6 **Al 'Ulā** Al Madīnah, NW Saudi Arabia 26.37N 37.55E

181 N4 **Alula-Fartak Trench** var. Illaue Fartak Trench. undersea feature W Indian Ocean

144 J11 **Al 'Umari** 'Ammān, E Jordan 31.30N 36.34E

108 J13 **Amandola** Marche, C Italy 42.58N 13.22E

109 N21 **Amantea** Calabria, SW Italy 39.06N 16.05E

203 W10 **Amanu** island Îles Tuamotu, C French Polynesia

60 I10 **Amapá** Amapá, NE Brazil 02.00N 50.50W

60 J11 **Amapá** off. Estado de Amapá; prev. Território do Amapá. ◆ state NE Brazil

119 T14 **Alupka** Respublika Krym, S Ukraine 44.24N 34.01E

77 P8 **Al 'Uqaylah** N Libya 30.13N 16.10E

Al Uqşur see Luxor

Al Urdunn see Jordan

173 G6 **Alur Panal** bay Sumatera, W Indonesia

147 V10 **Al 'Urūq al Mu'tariḑah** salt lake SE Saudi Arabia

145 Q7 **Alūs** C Iraq 34.04N 42.27E

119 T15 **Alushta** Respublika Krym, S Ukraine 44.40N 34.24E

77 N11 **Al 'Uwaynāt** var. Al Awaynāt. SW Libya 25.41N 10.33E

145 S9 **Al 'Uẓaym** var. Adhaim. E Iraq 34.12N 44.31E

28 L4 **Alva** Oklahoma, C USA 36.48N 98.40W

106 H8 **Alva** N Portugal

103 L20 **Alvdorf** Bayern, SE Germany 49.23N 11.22E

106 F13 **Alvaiázere** Leiria, C Portugal 39.49N 8.23W

60 J14 **Álvarães** Amazonas, NW Brazil 3.13S 64.53W

96 H11 **Alvdal** Hedmark, S Norway 62.07N 10.39E

---

96 K12 **Älvdalen** Dalarna, C Sweden 61.13N 14.04E

63 E15 **Alvear** Corrientes, NE Argentina 29.03S 56.30W

106 F10 **Alverca do Ribatejo** Lisboa, C Portugal 38.55N 9.01W

97 L20 **Alvesta** Kronoberg, S Sweden 56.52N 14.34E

96 D13 **Ålvik** Hordaland, S Norway 60.26N 6.27E

27 W12 **Alvin** Texas, SW USA 29.25N 95.14W

96 O13 **Älvkarleby** Uppsala, C Sweden 60.34N 17.30E

27 S5 **Alvord** Texas, SW USA 33.22N 97.39W

95 I18 **Älvros** Jämtland, C Sweden 62.04N 14.30E

94 J13 **Älvsbyn** Norrbotten, N Sweden 65.41N 21.00E

148 K8 **Al Wafrā'** SE Kuwait 28.37N 47.56E

146 J6 **Al Wajh** Tabūk, NW Saudi Arabia 26.15N 36.28E

149 N16 **Al Wakrah** var. Wakra. S Qatar 25.09N 51.36E

144 M8 **al Walaj, Sha'ib** dry watercourse W Iraq

158 I11 **Alwar** Rājasthān, N India 27.31N 76.34E

147 Q5 **Al Wari'ah** Ash Sharqīyah, N Saudi Arabia 27.54N 47.22E

161 G22 **Alwaye** Kerala, SW India 10.06N 76.22E

168 K14 **Alxa Zuoqi** var. Ehen Hudag. Nei Mongol Zizhiqu, N China 38.49N 105.40E

Al Yaman see Yemen

144 G9 **Al Yarmūk** Irbid, N Jordan 32.41N 35.55E

Alyat/Alyaty-Pristan' see Älät

117 I14 **Alykí** var. Aliki. Thásos, N Greece 40.36N 24.45E

121 F14 **Alytus** Pol. Olita. Alytus, S Lithuania 54.24N 24.02E

123 N23 **Alz** ≈ SE Germany

35 Y11 **Alzada** Montana, NW USA 45.00N 104.24W

26 Ii14 **Alzamay** Irkutskaya Oblast', S Russian Federation 55.33N 98.36E

101 M25 **Alzette** ≈ S Luxembourg

107 S10 **Alzira** var. Alcira; anc. Saetabicula, Suero. País Valenciano, E Spain 39.10N 0.27W

Al Zubair see Az Zubayr

189 O8 **Amadeus, Lake** seasonal lake Northern Territory, C Australia

83 E15 **Amadi** Western Equatoria, SW Sudan 31.39N 30.19E

16 Nn3 **Amadjuak Lake** ☉ Baffin Island, Nunavut, N Canada

97 J23 **Amager** island E Denmark

170 Cc13 **Amagi** Fukuoka, Kyūshū, SW Japan 33.24N 130.37E

171 J17 **Amagi-san** ▲ Honshū, S Japan 34.51N 138.57E

175 T11 **Amahai** var. Masohi. Pulau Seram, E Indonesia 3.19S 128.55E

78 M16 **Amak Island** island Alaska, USA

170 Bb14 **Amakusa-nada** gulf Kyūshū, SW Japan

97 J16 **Åmål** Västra Götaland, S Sweden 59.04N 12.40E

56 E8 **Amalfi** Antioquia, N Colombia 6.54N 75.04W

109 L18 **Amalfi** Campania, S Italy 40.37N 14.35E

117 D19 **Amaliáda** var. Amaliás. Dytikí Ellás, S Greece 37.48N 21.21E

Amaliás see Amaliáda

160 F12 **Amalner** Mahārāshtra, C India 21.03N 75.04E

176 X12 **Amamapare** Irian Jaya, E Indonesia 4.51S 136.43E

61 H21 **Amambaí, Serra de** var. Cordillera de Amambay, Serra de Amambay. ▲ Brazil/Paraguay see also Amambay, Cordillera de

64 P4 **Amambay** off. Departamento del Amambay. ◆ department E Paraguay

64 P5 **Amambay, Cordillera de** var. Serra de Amambaí, Serra de Amambay. ▲ Brazil/Paraguay see also Amambaí, Serra de

Amambay, Serra de/Amambay, Cordillera de see Amambaí, Serra de/Amambay, Cordillera de

172 U13 **Amami-guntō** island group SW Japan

172 Q13 **Amami-Ō-shima** island S Japan

194 E10 **Amanab** Sandaun, NW PNG 3.34S 141.10E

108 J13 **Amandola** Marche, C Italy 42.58N 13.22E

203 W10 **Amanu** island Îles Tuamotu, C French Polynesia

194 G10 **Amanu** Western California, W USA

37 V11 **Amargosa Range** ▲ California, W USA

27 O4 **Amarillo** Texas, SW USA 35.13N 101.49W

117 I21 **Amárynthos** var. Amarinthos. Évvoia, C Greece 38.24N 23.53E

Amasia see Amasya

142 K12 **Amasra** Bartın, N Turkey 40.40N 35.49E

176 Uu15 **Amasus** Pulau Yamdena, E Indonesia 7.48S 131.18E

129 N7 **Amderma** Nenetskiy Avtonomnyy Okrug, NW Russian Federation 69.45N 61.41E

165 N14 **Amdo** Xizang Zizhiqu, W China 32.15N 91.43E

160 H11 **Amla** prev. Amulla. Madhya Pradesh, C India 21.57N 78.06E

---

101 J20 **Amay** Liège, E Belgium 50.33N 5.19E

50 F7 **Amazon** Sp. Amazonas. ≈ Brazil/Peru

61 C14 **Amazonas** off. Estado do Amazonas. ◆ state N Brazil

56 G15 **Amazonas** off. Comisaría del Amazonas. ◆ province SE Colombia

58 C10 **Amazonas** off. Departamento de Amazonas. ◆ department N Peru

56 M12 **Amazonas** off. Territorio Amazonas. ◆ federal territory S Venezuela

50 F7 **Amazon Basin** basin N South America

49 V5 **Amazon Fan** undersea feature W Atlantic Ocean

60 K11 **Amazon, Mouths of the** delta NE Brazil

197 C12 **Ambae** var. Aoba, Omba. island C Vanuatu

158 I9 **Ambāla** Haryāna, NW India 30.22N 76.49E

161 J26 **Ambalangoda** Southern Province, SW Sri Lanka 6.13N 80.03E

161 K26 **Ambalantota** Southern Province, S Sri Lanka 6.07N 81.01E

180 I6 **Ambalavao** Fianarantsoa, C Madagascar 21.49S 46.55E

56 E10 **Ambalema** Tolima, C Colombia 4.49N 74.48W

81 E17 **Ambam** Sud, S Cameroon 2.22N 11.16E

180 J2 **Ambanja** Antsiranana, N Madagascar 13.40S 48.27E

127 O5 **Ambarchik** Respublika Sakha (Yakutiya), NE Russian Federation 69.33N 162.08E

64 K9 **Ambargasta, Salinas de** salt lake C Argentina

28 J6 **Ambarnyy** Respublika Kareliya, NW Russian Federation 65.53N 33.44E

56 C7 **Ambato** Tungurahua, C Ecuador 1.18S 78.39W

180 I5 **Ambatolampy** Antananarivo, C Madagascar 19.21S 47.27E

180 H4 **Ambatomainty** Mahajanga, W Madagascar 17.40S 45.39E

180 J4 **Ambatondrazaka** Toamasina, C Madagascar 17.49S 48.28E

175 Ss12 **Ambelau, Pulau** var. Ambelau, Pulau

103 L20 **Amberg** var. Amberg in der Oberpfalz. Bayern, SE Germany 49.27N 11.51E

Amberg in der Oberpfalz see Amberg

44 H1 **Ambergris Cay** island NE Belize

105 S11 **Ambérieu-en-Bugey** Ain, E France 45.57N 5.21E

193 J18 **Amberley** Canterbury, South Island, NZ 43.08S 172.51E

105 P11 **Ambert** Puy-de-Dôme, C France 45.34N 3.42E

Ambianum see Amiens

78 J11 **Ambidédi** Kayes, SW Mali 14.37N 11.49W

160 M10 **Ambikāpur** Madhya Pradesh, C India 23.09N 83.12E

180 J2 **Ambilobe** Antsiranana, N Madagascar 13.10S 49.03E

195 Q10 **Ambitle Island** island Feni Islands, NE PNG

41 O7 **Ambler** Alaska, USA 67.05N 157.51W

Amblève see Amel

Ambo see Hägere Hiywet

180 I8 **Amboasary** Toliara, S Madagascar 25.01S 46.22E

180 J4 **Ambodifototra** var. Ambodifototra. Toamasina, E Madagascar 16.58S 49.51E

Amboenten see Ambunten

180 J5 **Ambohidratrimo** Antananarivo, C Madagascar 18.48S 47.25E

180 I6 **Ambohimahasoa** Fianarantsoa, SE Madagascar 21.07S 47.13E

180 K3 **Ambohitralanana** Antsiraňana, NE Madagascar 15.13S 50.28E

176 X10 **Amboi, Kepulauan** island group E Indonesia

104 M8 **Amboise** Indre-et-Loire, C France 47.25N 1.00E

175 T11 **Ambon** prev. Amboina, Amboyna. Pulau Ambon, E Indonesia 3.40S 128.10E

175 T12 **Ambon, Pulau** island E Indonesia

141 I20 **Amboseli, Lake** ☉ Kenya/Tanzania

180 I6 **Ambositra** Fianarantsoa, SE Madagascar 20.31S 47.15E

180 J8 **Ambovombe** Toliara, S Madagascar 25.10S 46.06E

37 W10 **Amboy** California, W USA 34.33N 115.44W

32 L9 **Amboy** Illinois, N USA 41.42N 89.19W

Amboyna see Ambon

20 B14 **Ambridge** Pennsylvania, NE USA 40.33N 80.11W

Ambrim see Ambrym

84 A11 **Ambriz** Bengo, NW Angola 7.55S 13.11E

Ambrizete see N'Zeto

197 C13 **Ambrym** var. Ambrim. island C Vanuatu

174 Mm14 **Ambunten** prev. Amboenten. Pulau Madura, E Indonesia 6.55S 113.45E

194 G10 **Ambunti** East Sepik, NW PNG 4.06S 142.49E

161 I20 **Ambur** Tamil Nādu, SE India 12.48N 78.43E

40 E17 **Amchitka Island** island Aleutian Islands, Alaska, USA

40 E17 **Amchitka Pass** strait Aleutian Islands, Alaska, USA

176 Z12 **Amdam, Presa de la** ... Irian Jaya, E Indonesia 3.59S 140.35E

85 E20 **Aminuis** Omaheke, E Namibia 23.37S 19.21E

'Amiq, Wadi see 'Āmij, Wādī

148 J7 **Amīrābād** Īlām, NW Iran 33.19N 46.16E

153 O15 **Amu Darya** Rus. Amudar'ya, Taj. Dar''yoi Amu, Turkm. Amyderya, Uzb. Amudaryo; anc. Oxus. ≈ C Asia

Amudar'ya/Amudaryo/Amu, Dar''yoi see Amu Darya

130 J14 **Amisus** see Samsun

24 I3 **Amite** var. Amite City. Louisiana, S USA 30.40N 90.30W

Amite City see Amite

29 T12 **Amity** Arkansas, C USA 34.15N 93.43W

160 H11 **Amla** prev. Amulla. Madhya Pradesh, C India 21.57N 78.06E

---

43 P14 **Amecameca** var. Amecameca de Juárez. México, C Mexico 19.07N 98.45W

Amecameca de Juárez see Amecameca

63 A20 **Ameghino** Buenos Aires, E Argentina 34.51S 62.28W

101 M21 **Amel** Fr. Amblève. Liège, E Belgium 50.20N 6.13E

100 K4 **Ameland** Fris. It Amelân. island Waddeneilanden, N Netherlands

Amelân, It see Ameland

109 H14 **Amelia** Umbria, C Italy 42.33N 12.26E

23 V6 **Amelia Court House** Virginia, NE USA 37.19N 77.57W

25 W8 **Amelia Island** island Florida, SE USA

20 L12 **Amenia** New York, NE USA 41.51N 73.31W

America see United States of America

67 M21 **America-Antarctica Ridge** undersea feature S Atlantic Ocean

America in Miniature see Maryland

62 I9 **Americana** São Paulo, S Brazil 22.43S 47.19W

35 Q5 **American Falls** Idaho, NW USA 42.47N 112.51W

35 Q5 **American Falls Reservoir** ☒ Idaho, NW USA

68 L3 **American Fork** Utah, W USA 40.24N 111.47W

198 D8 **American Samoa** ◇ US unincorporated territory W Polynesia

25 S6 **Americus** Georgia, SE USA 32.04N 84.13W

100 J13 **Amerongen** Utrecht, C Netherlands 52.00N 5.30E

100 K11 **Amersfoort** Utrecht, C Netherlands 52.09N 5.23E

99 N21 **Amersham** SE England, UK 51.39N 0.37W

32 J5 **Amery** Wisconsin, N USA 45.18N 92.20W

205 W6 **Amery Ice Shelf** ice shelf Antarctica

31 V3 **Ames** Iowa, C USA 42.01N 93.37W

21 P10 **Amesbury** Massachusetts, NE USA 42.50N 70.55W

Amestratus see Mistretta

117 F18 **Amfíkleia** var. Amfíklia. Stereá Ellás, C Greece 38.37N 22.34E

Amfíklia see Amfíkleia

117 D17 **Amfilochía** var. Amfilokhía. Dytikí Ellás, C Greece 38.52N 21.09E

Amfilokhía see Amfilochía

116 H13 **Ámfissa** Stereá Ellás, C Greece 38.31N 22.22E

126 M11 **Amga** Respublika Sakha (Yakutiya), NE Russian Federation 60.55N 131.45E

126 M11 **Amga** ≈ NE Russian Federation

Amgalang see Xin Barag Zuoqi

195 N12 **Amgu** var. New Britain, C PNG

127 P4 **Amguema** ≈ NE Russian Federation

127 Nn14 **Amgun'** ≈ SE Russian Federation

82 J12 **Amhara** ◆ region N Ethiopia

11 P15 **Amherst** Nova Scotia, SE Canada 45.49N 64.13W

20 M11 **Amherst** Massachusetts, NE USA 42.22N 72.31W

20 D10 **Amherst** New York, NE USA 42.57N 78.47W

26 M4 **Amherst** Texas, SW USA 33.59N 102.24W

23 U6 **Amherst** Virginia, NE USA 37.35N 79.03W

Amherst see Kyaikkami

12 C18 **Amherstburg** Ontario, S Canada 42.05N 83.06W

23 Q6 **Amherstdale** West Virginia, NE USA 37.46N 81.46W

12 K15 **Amherst Island** island Ontario, SE Canada

30 J6 **Amidon** North Dakota, N USA 46.26N 103.18W

105 O3 **Amiens** anc. Ambianum, Samarobriva. Somme, N France 49.54N 2.18E

145 P8 **'Āmij, Wādī** var. Wadi 'Amiq. dry watercourse W Iraq

142 L17 **Amik Ovası** ◎ S Turkey

78 E9 **Amilcar Cabral** ✕ Sal, NE Cape Verde 16.45N 22.56E

148 J7 **Amīrābād** Īlām, NW Iran 33.19N 46.16E

Amirante Bank see Amirante Ridge

181 N6 **Amirante Basin** undersea feature W Indian Ocean

181 N6 **Amirante Islands** var. Amirantes Group. island group C Seychelles

181 N6 **Amirante Ridge** undersea feature W Indian Ocean

Amirante Trench see Amirante Group

181 N7 **Amirante Trench** undersea feature W Indian Ocean

11 U13 **Amisk Lake** ☉ Saskatchewan, C Canada

27 O12 **Amistad Reservoir** var. Presa de la Amistad. ☒ Mexico/USA

Amisus see Samsun

24 I3 **Amite** var. Amite City. Louisiana, S USA 30.40N 90.30W

Amite City see Amite

29 T12 **Amity** Arkansas, C USA 34.15N 93.43W

160 H11 **Amla** prev. Amulla. Madhya Pradesh, C India 21.57N 78.06E

---

40 I17 **Amlia Island** island Aleutian Islands, Alaska, USA

99 I18 **Amlwch** NW Wales, UK 53.25N 4.22W

144 H10 **'Ammān** var. Amman; anc. Philadelphia, Bibl. Rabbah Ammon, Rabbath Ammon. ● (Jordan) 'Ammān, NW Jordan 31.57N 35.55E

Amman see 'Ammān

144 H10 **'Ammān** off. Muḩāfaẓat 'Ammān. ◆ governorate NW Jordan

95 N14 **Ämmänsaari** Oulu, E Finland 64.51N 28.58E

94 M13 **Ammarnäs** Västerbotten, N Sweden 65.58N 16.10E

207 O15 **Ammassalik** var. Angmagssalik. Tunu, S Greenland 65.51N 37.30W

103 K24 **Ammer** ≈ SE Germany

100 L12 **Ammersee** ☉ SE Germany

100 J13 **Ammerzoden** Gelderland, C Netherlands 51.46N 5.10E

Ammóchostos see Gazimağusa

116 K13 **Ammóchostos, Kólpos** ≈ E Cyprus

Ammóchostos, Kólpos see Gazimağusa Körfezi

80 L10 **Amol** var. Āmul. Māzandarān, N Iran 36.30N 52.24E

117 K21 **Amorgós** Amorgós, Kykládes, Greece, Aegean Sea 36.49N 25.54E

117 K22 **Amorgós** island Kykládes, Greece, Aegean Sea

25 N3 **Amory** Mississippi, S USA 33.58N 88.29W

10 J13 **Amos** Quebec, SE Canada 48.34N 78.07W

97 G15 **Åmot** Buskerud, S Norway 59.52N 9.55E

97 E15 **Åmot** Telemark, S Norway 59.34N 7.59E

97 J15 **Åmotfors** Värmland, C Sweden 59.46N 12.24E

78 L10 **Amourj** Hodh ech Chargui, SE Mauritania 16.04N 7.12W

Amoy see Xiamen

180 H7 **Ampanihy** Toliara, SW Madagascar 24.40S 44.45E

161 L25 **Ampara** var. Amparai. Eastern Province, E Sri Lanka 7.16N 81.40E

180 J4 **Amparafaravola** Toamasina, E Madagascar 17.33S 48.13E

Amparai see Ampara

62 M9 **Amparo** São Paulo, S Brazil 22.40S 46.49W

180 J5 **Ampasimanolotra** Toamasina, E Madagascar 18.49S 49.04E

59 H17 **Ampato, Nevado** ▲ S Peru 15.52S 71.51W

103 L23 **Amper** ≈ SE Germany

66 M9 **Ampère Seamount** undersea feature E Atlantic Ocean 35.05N 13.00W

178 M10 **Amphitrite Group** island group N Paracel Islands

176 U15 **Amplawas** var. Emplawas. Pulau Babar, E Indonesia 8.01S 129.42E

107 O7 **Amposta** Cataluña, NE Spain 40.43N 0.34E

13 V7 **Amqui** Quebec, SE Canada 48.28N 67.27W

147 O14 **'Amrān** W Yemen 15.39N 43.59E

160 H12 **Amraoti** see Amrāvati

160 H12 **Amrāvati** prev. Amraoti. Mahārāshtra, C India 20.55N 77.45E

160 C11 **Amreli** Gujarāt, W India 21.36N 71.19E

144 H5 **'Amrit** ruins Ṭarṭūs, W Syria 34.48N 35.54E

158 H7 **Amritsar** Punjab, N India 31.38N 74.54E

158 J10 **Amroha** Uttar Pradesh, N India 28.54N 78.28E

102 G7 **Amrum** island NW Germany

95 J15 **Åmsele** Västerbotten, N Sweden 64.31N 19.24E

100 J10 **Amstelveen** Noord-Holland, C Netherlands 52.18N 4.49E

100 I10 **Amsterdam** ● (Netherlands) Noord-Holland, C Netherlands 52.22N 4.54E

20 K10 **Amsterdam** New York, NE USA 42.56N 74.11W

181 Q11 **Amsterdam Fracture Zone** tectonic feature S Indian Ocean

181 R11 **Amsterdam Island** island NE French Southern and Antarctic Territories

111 U4 **Amstetten** Niederösterreich, N Austria 48.07N 14.52E

80 J11 **Am Timan** Salamat, SE Chad 11.02N 20.17E

152 L12 **Amu-Bukhoro Kanali** var. Aral-Bukhorskiy Kanal. canal C Uzbekistan

145 O1 **'Āmūdah** var. Amude. Al Ḩasakah, N Syria 37.06N 40.56E

152 M14 **Amu-Dar'ya** Lebapskiy Velayat, NE Turkmenistan 37.58N 65.14E

153 O15 **Amu Darya** Rus. Amudar'ya, Taj. Dar''yoi Amu, Turkm. Amyderya, Uzb. Amudaryo; anc. Oxus. ≈ C Asia

Amudar'ya/Amudaryo/Amu, Dar''yoi see Amu Darya

Amude see 'Āmūdah

146 L3 **'Amūd, Jabal** ▲ NW Saudi Arabia 30.59N 39.17E

40 I17 **Amukta Island** island Aleutian Islands, Alaska, USA

40 I17 **Amukta Pass** strait Aleutian Islands, Alaska, USA

205 X3 **Amundsen Bay** bay Antarctica

205 P10 **Amundsen Coast** physical region Antarctica

15 M2 **Amundsen Gulf** gulf Northwest Territories, N Canada

199 Ll16 **Amundsen Plain** undersea feature S Pacific Ocean

205 Q14 **Amundsen-Scott** US research station Antarctica 89.59S 10.00E

204 I11 **Amundsen Sea** sea S Pacific Ocean

Amungen see Ancityum

175 N10 **Amuntai** prev. Amoentai. Borneo, C Indonesia 2.24S 115.13E

---

133 W6 **Amur** Chin. Heilong Jiang. ≈ China/Russian Federation

175 Rr7 **Amurang** prev. Amoerang. Sulawesi, C Indonesia 1.12N 124.37E

175 Rr7 **Amurang, Teluk** bay Sulawesi, C Indonesia

107 O3 **Amurrio** País Vasco, N Spain 43.03N 3.00W

127 Nn15 **Amursk** Khabarovskiy Kray, SE Russian Federation 50.13N 136.34E

126 M14 **Amurskaya Oblast'** ◆ province SE Russian Federation

82 X4 **'Amur, Wadi** ≈ NE Sudan

117 C17 **Amvrakikós Kólpos** gulf W Greece

Amvrosiyevka see Amvrosiyivka

119 X8 **Amvrosiyivka** Rus. Amvrosiyevka. Donets'ka Oblast', SE Ukraine 47.46N 38.31E

Amyderya see Amu Darya

116 E13 **Amýntaio** var. Amindeo; prev. Amíndaion. Dytikí Makedonía, N Greece 40.42N 21.42E

12 B6 **Amyot** Ontario, S Canada 48.28N 84.58W

203 U10 **Anaa** atoll Îles Tuamotu, C French Polynesia

Anaa see Portalegre

57 I6 **Anaco** Anzoátegui, NE Venezuela 9.30N 64.28W

35 Q10 **Anaconda** Montana, NW USA 46.09N 112.55W

34 H7 **Anacortes** Washington, NW USA 48.30N 122.36W

28 M11 **Anadarko** Oklahoma, C USA 35.04N 98.14W

116 N12 **Ana Dere** ≈ NW Turkey

106 G8 **Anadia** Aveiro, N Portugal 40.25N 8.27W

Anadolu Dağları see Doğu Karadeniz Dağları

127 Pp5 **Anadyr'** Chukotskiy Avtonomnyy Okrug, NE Russian Federation 64.40N 177.22E

127 P5 **Anadyr'** ≈ NE Russian Federation

Anadyr, Gulf of see Anadyrskiy Zaliv

133 X4 **Anadyrskiy Khrebet** var. Chukot Range. ▲ NE Russian Federation

127 Q4 **Anadyrskiy Zaliv** Eng. Gulf of Anadyr. gulf NE Russian Federation

117 K22 **Anáfi** anc. Anaphe. island Kykládes, Greece, Aegean Sea

109 I13 **Anagni** Lazio, C Italy 41.43N 13.12E

37 T15 **Anaheim** California, W USA 33.50N 117.54W

8 L13 **Anahim Lake** British Columbia, SW Canada 52.26N 125.13W

43 Q5 **Anáhuac** Nuevo León, NE Mexico 27.13N 100.09W

161 G22 **Anai Mudi** ▲ S India 10.16N 77.08E

Anaiza see 'Unayzah

161 M15 **Anakāpalle** Andhra Pradesh, E India 17.42N 83.06E

203 W15 **Anakena, Playa de** beach Easter Island, Chile, E Pacific Ocean

41 Q7 **Anaktuvuk Pass** Alaska, USA 68.08N 151.44W

41 Q7 **Anaktuvuk River** ≈ Alaska, USA

180 I4 **Analalava** Mahajanga, NW Madagascar 14.37S 47.46E

46 F6 **Ana Maria, Golfo de** gulf C Cuba

Anambas Islands see Anambas, Kepulauan

174 I5 **Anambas, Kepulauan** var. Anambas Islands. island group W Indonesia

79 U4 **Anambra** ◆ state SE Nigeria

31 Y3 **Anamoose** North Dakota, N USA 47.50N 100.14W

31 Y13 **Anamosa** Iowa, C USA 42.06N 91.17W

142 H17 **Anamur** İçel, S Turkey 36.06N 32.49E

142 H17 **Anamur Burnu** headland S Turkey 36.03N 32.49E

170 Ff16 **Anan** Tokushima, Shikoku, SW Japan 33.54N 134.40E

160 O12 **Ānandadur** Orissa, E India 21.33N 86.08E

161 H18 **Anantapur** Andhra Pradesh, S India 14.40N 77.36E

158 H5 **Anantnāg** var. Islamabad. Jammu and Kashmir, NW India 33.43N 75.10E

Ananyev see Anan'yiv

119 O9 **Anan'yiv** Rus. Ananyev. Odes'ka Oblast', SW Ukraine 47.43N 29.51E

130 J14 **Anapa** Krasnodarskiy Kray, SW Russian Federation 44.55N 37.20E

Anaphe see Anáfi

61 K18 **Anápolis** Goiás, C Brazil 16.19S 48.58W

149 R10 **Anār** Kermān, C Iran 30.48N 55.17E

149 P7 **Anār** Eşfahān, C Iran 33.21N 53.43E

Anar Dara see Anār Darreh

154 J7 **Anār Darreh** var. Anar Dara. Farāh, W Afghanistan 32.45N 61.37E

25 X9 **Anastasia Island** island Florida, SE USA

196 K7 **Anatahan** island N Mariana Islands

132 M6 **Anatolia** plateau C Turkey

88 F14 **Anatolian Plate** tectonic feature Asia/Europe

116 H13 **Anatoliki Makedonía kai Thráki** Eng. Macedonia East and Thrace. ◆ region NE Greece

Anatom see Aneityum

64 L8 **Añatuya** Santiago del Estero, C Argentina 28.27S 62.52W

An Baile Meánach see Ballymena

An Bhearú see Barrow
An Bhóinn see Boyne
An Blascaod Mór see Great Blasket Island
An Cabhán see Cavan
An Caisleán Nua see Newcastle
An Caisleán Riabhach see Castlereagh, Northern Ireland, UK
142 L16 An Caisleán Riabhach see Castlereagh, Ireland
58 C13 Ancash off. Departamento de Ancash. ◆ department W Peru
An Cathair see Caher
104 J8 Ancenis Loire-Atlantique, NW France 47.22N 1.10W
An Chanáil Ríoga see Royal Canal
An Cheacha see Caha Mountains
41 R11 Anchorage Alaska, USA 61.12N 149.52W
41 R12 Anchorage ✕ Alaska, USA 61.08N 150.00W
41 Q13 Anchor Point Alaska, USA 59.46N 151.49W
An Chorr Chríochach see Cookstown
67 M24 Anchorstack Point headland W Tristan da Cunha 37.07S 12.21W
An Clár see Clare
An Clochán see Clifden
An Clochán Liath see Dunglow
25 U12 Anclote Keys island group Florida, SE USA
An Cóbh see Cobh
59 J17 Anchuma, Nevado de ▲ W Bolivia 15.51S 68.33W
An Comar see Comber
59 D14 Ancón Lima, W Peru 11.47S 77.09W
108 J12 Ancona Marche, C Italy 43.37N 13.30E
Ancuabe see Ancuabi
84 Q13 Ancuabi var. Ancuabe. Cabo Delgado, NE Mozambique 12.57S 39.54E
65 H17 Ancud prev. San Carlos de Ancud. Los Lagos, S Chile 41.52S 73.49W
65 G17 Ancud, Golfo de gulf S Chile
Ancyra see Ankara
169 V8 Anda Heilongjiang, NE China 46.22N 125.15E
59 G16 Andahuaylas Apurímac, S Peru 13.38S 73.20W
An Daingean see Dingle
159 R15 Andal West Bengal, NE India 23.34N 87.13E
96 E9 Åndalsnes Møre og Romsdal, S Norway 62.33N 7.42E
106 K13 Andalucía Eng. Andalusia. ◆ autonomous community S Spain
25 P7 Andalusia Alabama, S USA 31.18N 86.29W
Andalusia see Andalucía
157 Q21 Andaman and Nicobar Islands var. Andamans and Nicobars. ◆ union territory India, NE Indian Ocean
181 N5 Andaman Basin undersea feature NE Indian Ocean
157 P19 Andaman Islands island group India, NE Indian Ocean
181 N4 Andaman Sea sea NE Indian Ocean
59 K19 Andamarca Oruro, C Bolivia 18.50S 67.24W
176 W10 Andamata Irian Jaya, E Indonesia 2.40S 132.30E
190 H5 Andamooka South Australia 30.26S 137.12E
147 Y9 'Andām, Wādī seasonal river NE Oman
180 J3 Andapa Antsiranana, NE Madagascar 14.39S 49.40E
155 R4 Andarāb var. Banow. Baghlān, NE Afghanistan 35.36N 69.18E
Andarbag see Andarbogh
153 S13 Andarbogh Rus. Andarbag, Andarbak. S Tajikistan 37.51N 71.45E
111 Z5 Andau Burgenland, E Austria 47.46N 17.03E
110 I10 Andeer Graubünden, S Switzerland 46.36N 9.24E
94 H9 Andenes Nordland, C Norway 69.18N 16.06E
101 J20 Andenne Namur, SE Belgium 50.28N 5.06E
79 S11 Andéramboukane Gao, E Mali 15.24N 3.03E
Anderbak see Andarbogh
101 G18 Anderlecht Brussels, C Belgium 50.50N 4.18E
101 G21 Anderlues Hainaut, S Belgium 50.24N 4.16E
110 G9 Andermatt Uri, C Switzerland 46.39N 8.36E
103 E17 Andernach anc. Antunnacum. Rheinland-Pfalz, SW Germany 50.25N 7.25E
196 L10 Andersen Air Force Base air base NE Guam 13.34N 144.55E
41 R9 Anderson Alaska, USA 64.20N 149.11W
37 N4 Anderson California, W USA 40.26N 122.21W
33 P12 Anderson Indiana, N USA 40.06N 85.40W
29 X8 Anderson Missouri, C USA 36.39N 94.26W
23 R4 Anderson South Carolina, SE USA 34.30N 82.39W
27 V10 Anderson Texas, SW USA
15 Gg3 Anderson ♏ Northwest Territories, NW Canada
97 K20 Anderstorp Jönköping, S Sweden 57.16N 13.46E
56 D9 Andes Antioquia, W Colombia 5.40N 75.55W
49 P7 Andes ▲ W South America
31 N2 Andes, Lake ◲ South Dakota, N USA
94 H9 Andfjorden fjord E Norwegian Sea
161 H16 Andhra Pradesh ◆ state E India
100 J8 Andijk Noord-Holland, NW Netherlands 52.38N 5.00E
153 S10 Andijon Rus. Andizhan. Andijon Wiloyati, E Uzbekistan 40.46N 72.19E
153 S10 Andijon Wiloyati Rus. Andizhanskaya Oblast'. ◆ province E Uzbekistan
Andikíthira see Antikýthira
180 J4 Andilamena Toamasina, C Madagascar 17.00S 48.35E
148 L8 Andimeshk var. Andimishk; prev. Salehābād. Khūzestān, SW Iran 32.28N 48.21E
Andimishk see Andimeshk
Andíparos see Antíparos
Andipaxi see Antípaxoi
Andípsara see Antípsara
142 L16 Andırın Kahramanmaraş, S Turkey 37.33N 36.18E
164 J8 Andirlangar Xinjiang Uygur Zizhiqu, NW China 37.38N 83.40E
Andírrion see Antírrio
Andissa see Ántissa
111 W7 Andizhan see Andijon
Andizhanskaya Oblast' see Andijon Wiloyati
155 N2 Andkhvoy Fāryāb, N Afghanistan 36.55N 65.07E
107 Q2 Andoain País Vasco, N Spain 43.13N 2.01W
176 W9 Andoi Irian Jaya, E Indonesia 0.53S 133.59E
169 Y15 Andong Jap. Antō. E South Korea 36.34N 128.42E
111 R4 Andorf Oberösterreich, N Austria 48.22N 13.33E
107 S7 Andorra Aragón, NE Spain 40.58N 0.27W
107 V4 Andorra off. Principality of Andorra, Cat. Valls d'Andorra, Fr. Vallée d'Andorre. ◆ monarchy SW Europe
Andorra see Andorra la Vella
107 V4 Andorra la Vella var. Andorra, Fr. Andorre la Vieille, Sp. Andorra la Vieja. ● (Andorra) C Andorra 42.30N 1.30E
Andorra la Vieja see Andorra la Vella
Andorra la Vella
Andorra, Valls d'/Andorre, Vallée d' see Andorra
Andorre la Vieille see Andorra la Vella
99 M22 Andover S England, UK 51.13N 1.28W
29 N6 Andover Kansas, C USA 37.42N 97.08W
94 H10 Andøya island C Norway
62 I8 Andradina São Paulo, S Brazil 20.54S 51.25W
41 N10 Andreafsky River ✍ Alaska, USA
40 H17 Andreanof Islands island group Aleutian Islands, Alaska, USA
128 H16 Andreas, Cape Cyp. Zafer Burnu, Gk. Akrotíri Apostólou Andréa. cape NE Cyprus
21 R14 Andreas, Cape cape Tverskaya Oblast', W Russian Federation 56.38N 32.17E
23 V6 Andrews North Carolina, SE USA 35.19N 84.01W
23 T13 Andrews South Carolina, SE USA 33.27N 79.33W
26 M7 Andrews Texas, SW USA 32.19N 102.33W
181 N5 Andrew Tablemount var. Gora Andryu. undersea feature W Indian Ocean 6.45N 50.30E
109 N17 Andreyevka see Kabanbay
115 K16 Andria Puglia, SE Italy 41.13N 16.16E
117 E20 Andrijevica Montenegro, SW Yugoslavia 42.45N 19.45E
Andritsaina Pelopónnisos, S Greece 37.29N 21.52E
An Droichead Nua see Newbridge
Andropov see Rybinsk
117 J20 Ándros Ándros, Kykládes, Greece, Aegean Sea 37.49N 24.55E
117 J20 Ándros island Kykládes, Greece, Aegean Sea
21 O14 Androscoggin River ✍ Maine/New Hampshire, NE USA
153 Q10 Andros Island island NW Bahamas
131 R7 Androsovka Samarskaya Oblast', W Russian Federation 52.41N 49.34E
46 G3 Andros Town Andros Island, NW Bahamas 24.40N 77.47W
161 D21 Ándrott Island island Lakshadweep, India, SW India
119 N5 Andrushivka Zhytomyrs'ka Oblast', N Ukraine 50.01N 29.02E
113 K16 Andrychów Małopolskie, S Poland 49.51N 19.23E
167 N1 Anduki Oriental, NE Dem. Rep. Congo (Zaire) 0.105 27.42E
94 I10 Andselv Troms, N Norway 69.05N 18.30E
81 I17 Andudu Orientale, NE Dem. Rep. Congo (Zaire) 2.25N 29.20E
107 N13 Andújar anc. Illiturgis. Andalucía, SW Spain 38.01N 4.03W
84 C12 Andulo Bié, W Angola 11.28S 16.43E
105 Q14 Anduze Gard, S France 44.03N 3.59E
An Eargail see Errigal Mountain
97 J16 Aneby Jönköping, S Sweden 57.49N 14.45E
79 U8 Anéfis Kidal, NE Mali 18.05N 0.38E
47 U8 Anegada island NE British Virgin Islands
41 O12 Anegada, Bahía bay E Argentina
201 R8 Anegada Passage passage Anguilla/British Virgin Islands
79 U16 Aného var. Anécho; prev. Petit-Popo. S Togo 6.13N 1.36E
197 D12 Aneityum var. Anatom; prev. Kéamu. island S Vanuatu
119 N10 Aненii Noi Rus. Novyye Aneny. C Moldova 46.52N 29.10E
194 L12 Anepmete New Britain, E PNG 5.47S 148.37E
131 V7 Anikhovka Orenburgskaya Oblast', W Russian Federation 51.27N 60.17E
79 T8 Aney Agadez, NE Niger 19.22N 13.00E
An Fheoir see Nore
Angara ✍ C Russian Federation
126 J15 Angarsk Irkutskaya Oblast', S Russian Federation 52.31N 103.55E
95 N14 Ånge Västernorrland, C Sweden 62.31N 15.40E
42 D4 Ángel de la Guarda, Isla island NW Mexico
171 O3 Angeles off. Angeles City. Luzon, N Philippines 15.16N 120.37E
Angeles City see Angeles
Angel Falls see Ángel, Salto
97 J22 Ängelholm Skåne, S Sweden 56.14N 12.52E
63 A17 Angélica Santa Fe, C Argentina 31.33S 61.33W
27 W8 Angelina River ✍ Texas, SW USA
57 Q9 Ángel, Salto Eng. Angel Falls. waterfall E Venezuela 5.52N 62.19W
'Anjar see Aanjar
97 M15 Ängelsberg Västmanland, C Sweden 59.57N 16.01E
37 P8 Angels Camp California, W USA 38.03N 120.31W
111 W7 Anger Steiermark, SE Austria 47.16N 15.41E
Angerapp see Ozersk
Angerburg see Węgorzewo
95 J15 Ångermanälven ✍ N Sweden
100 M5 Angermünde Brandenburg, NE Germany 53.01N 13.59E
104 K7 Angers anc. Juliomagus. Maine-et-Loire, NW France 47.30N 0.33W
13 W8 Angers ◲ Quebec, SE Canada
116 I13 Ängesön island N Sweden
178 J14 Ångk Tasaôm prev. Angtassom. Takev, S Cambodia 10.59N 104.39E
193 C25 Anglem, Mount ▲ Stewart Island, Southland, SW NZ 46.44S 167.56E
99 I18 Anglesey cultural region NW Wales, UK
99 I18 Anglesey island NW Wales, UK
104 I15 Anglet Pyrénées-Atlantiques, SW France 43.31N 1.30W
27 W12 Angleton Texas, SW USA 29.10N 95.25W
12 I9 Anglia see England
12 I9 Angliers Quebec, SE Canada 47.33N 79.17W
Anglo-Egyptian Sudan see Sudan
178 I18 Angmagssalik see Ammassalik
Nam Ngum ✍ C Laos
81 N16 Ango Orientale, N Dem. Rep. Congo (Zaire) 4.02N 25.49E
85 Q15 Angoche Nampula, E Mozambique 16.12S 39.55E
65 G14 Angol Araucanía, C Chile 37.48S 72.40W
33 Q11 Angola Indiana, N USA 41.37N 85.00W
84 A9 Angola off. Republic of Angola; prev. People's Republic of Angola, Portuguese West Africa. ◆ republic SW Africa
67 Q8 Angola Basin undersea feature E Atlantic Ocean
41 X13 Angoon Admiralty Island, Alaska, USA 57.33N 134.30W
153 O14 Angor Surkhondaryo Wiloyati, S Uzbekistan 37.30N 67.06E
95 I22 Angora see Ankara
130 M8 Angoram East Sepik, NW PNG 4.01S 144.03E
42 H8 Angostura Sinaloa, C Mexico 25.18N 108.10W
30 J11 Angostura see Ciudad Bolívar
104 K11 Angostura, Presa de la ◲ SE Mexico
66 Q2 Angostura Reservoir ◲ South Dakota, N USA
Angoulême anc. Iculisma. Charente, W France 45.39N 0.10E
111 T9 Angoumois cultural region W France
146 M5 Angra do Heroísmo Terceira, Azores, Portugal, NE Atlantic Ocean 38.40N 27.13W
145 P6 Angra dos Reis Rio de Janeiro, SE Brazil 22.58S 44.16W
62 D3 Angra Pequena see Lüderitz
117 J20 Angren Toshkent Wiloyati, E Uzbekistan 41.04N 70.17E
81 W16 Ang Thong var. Angthong. Ang Thong, C Thailand 14.34N 100.25E
Angthong see Ang Thong
46 G3 Anguang Oriental, N Dem. Rep. Congo (Zaire) 3.38N 24.14E
23 W4 Angués Aragón, NE Spain 42.07N 0.10W
47 U9 Anguilla ◇ UK dependent territory E West Indies
47 V9 Anguilla island E West Indies
46 F4 Anguilla Cays islets SW Bahamas
167 N1 Angul see Anugul
81 I21 Anguli Nur ◲ E China
91 W9 Angumu Orientale, E Dem. Rep. Congo (Zaire) 0.105 27.42E
52 G14 Angus ◆ Ontario, S Canada 44.19N 79.52W
98 J10 Angus cultural region E Scotland, UK
64 K19 Anhanguera Goiás, S Brazil 18.12S 48.19W
152 F13 Anhée Namur, S Belgium 50.18N 4.52E
84 C11 Anholt island C Denmark
167 P8 Anhua prev. Dongping. Hunan, S China 28.25N 111.10E
21 Ann, Cape headland Massachusetts, NE USA
79 U9 Anié S Togo 7.48N 1.12E
79 U9 Anié ✍ C Togo
104 I16 Anie, Pic d' ▲ SW France 42.56N 0.44W
81 A19 Anikhovka Orenburgskaya Oblast', W Russian Federation 51.27N 60.17E
188 I10 Annean, Lake ◲ Western Australia
39 R4 Animas New Mexico, SW USA 31.55N 108.49W
105 A19 Annobón island W Equatorial Guinea
39 R17 Animas Peak ▲ New Mexico, SW USA 31.34N 108.46W
39 R16 Animas Valley valley New Mexico, SW USA
167 R5 An Nu'ayriyah var. Nariya. Ash Sharqiyah, NE Saudi Arabia
190 M7 Anina Ger. Steierdorf, Hung. Stájerlakanina; prev. Steierdorf-Anina, Steierdorf-Anina, Steyerlak-Anina. Caraş-Severin, SW Romania 45.04N 21.51E
31 U14 Anita Iowa, C USA 41.27N 94.45W
127 Oo16 Aniva, Mys headland Ostrov Sakhalin, SE Russian Federation
127 Oo16 Aniva, Zaliv bay SE Russian Federation
197 E16 Aniwa island S Vanuatu
95 M19 Anjalankoski Etelä-Suomi, S Finland 60.39N 26.54E
45.15N 93.26W
12 B8 Anjigami Lake ◲ Ontario, S Canada
97 J22 Anjō var. Anzyō. Aichi, Honshū, SW Japan 34.58N 137.07E
104 J8 Anjou cultural region NW France
Anjouan var. Nzwani, Johanna Island. island SE Comoros
180 J4 Anjozorobe Antananarivo, C Madagascar 18.22S 47.52E
169 W13 Anju W North Korea 39.36N 125.44E
100 M5 Anjum Fris. Eanjum. Friesland, N Netherlands 53.22N 6.08E
180 G6 Ankaboa, Tanjona headland SW Madagascar
166 L7 Ankang prev. Xing'an. Shaanxi, C China 32.45N 109.00E
142 J12 Ankara prev. Angora, anc. Ancyra. ● (Turkey) Ankara, C Turkey 39.55N 32.49E
142 H12 Ankara ◆ province C Turkey
97 N19 Ankarsrum Kalmar, S Sweden 57.40N 16.19E
180 H6 Ankazoabo Toliara, SW Madagascar 22.18S 44.30E
180 I4 Ankazobe Antananarivo, C Madagascar 18.18S 47.07E
31 T15 Ankeny Iowa, C USA 41.43N 93.37W
178 Kk11 An Khê Gia Lai, C Vietnam 13.57N 108.39E
102 O9 Anklam Mecklenburg-Vorpommern, NE Germany 53.51N 13.42E
82 K13 Ankober Amhara, N Ethiopia 9.34N 39.44E
79 O17 Ankobra ✍ S Ghana
81 Q12 Ankoro Katanga, SE Dem. Rep. Congo (Zaire) 6.45S 26.58E
166 I13 Anlong Guizhou, S China 25.05N 105.29E
178 Ii11 Ánlong Vêng Siemréab, NW Cambodia 14.16N 104.07E
167 N8 Anlu Hubei, C China 31.15N 113.41E
An Mhí see Meath
An Muileann gCearr see Mullingar
95 J15 Ånn Jämtland, C Sweden 63.19N 12.34E
130 M8 Anna Voronezhskaya Oblast', W Russian Federation 51.31N 40.23E
32 L15 Anna Illinois, N USA 37.27N 89.15W
27 U5 Anna Texas, SW USA 33.21N 96.33W
76 L5 Annaba prev. Bône. NE Algeria 36.55N 7.46E
Annaberg-Buchholz Sachsen, E Germany 50.34N 13.26E
111 T9 Annabichl ✕ (Klagenfurt) Kärnten, S Austria 46.39N 14.21E
146 M5 An Nafūd desert NW Saudi Arabia
145 P6 'Annah var. 'Ānah. NW Iraq 34.25N 41.59E
145 T10 An Najaf var. Najaf. S Iraq
23 V5 Anna, Lake ◲ Virginia, NE USA
63.19 12.34E
99 F17 Annalee ✍ N Ireland
178 Jj9 Annamitique, Chaîne ▲ C Laos
99 J3 Annan S Scotland, UK 55.00N 3.19W
31 O8 Annandale Minnesota, N USA 45.15N 94.07W
23 V4 Annandale Virginia, NE USA 38.48N 77.10W
197 Q8 Anna Point headland N Nauru 0.30S 166.55E
33 X3 Annapolis state capital Maryland, NE USA 38.58N 76.29W
159 O10 Annapurna ▲ C Nepal 28.30N 83.49E
23 P11 Ann Arbor Michigan, N USA 42.16N 83.43W
145 W12 An Nāṣiriyah var. Nasiriya. SE Iraq 31.04N 46.16E
145 W11 An Naṣr E Iraq 31.34N 46.08E
152 F13 Annau Turkm. Änew. Akhalskiy Velayat, C Turkmenistan
123 Mm16 An Nawfaliyah var. Al Nūwfaliyah. N Libya
21 P11 Ann, Cape headland Massachusetts, NE USA
47 X10 Annobón island W Equatorial Guinea
105 T11 Annecy anc. Anneciacum. Haute-Savoie, E France 45.54N 6.07E
105 T11 Annecy, Lac d' ◲ E France
105 T11 Annemasse Haute-Savoie, E France 46.10N 6.15E
23 Z14 Annette Island island Alexander Archipelago, Alaska, USA
145 O14 An Nhon Binh Đinh
22 L6 An Nhon see Binh Đinh
23 Q3 Anniston Alabama, S USA 33.39N 85.49W
47 X10 Annobón island W Equatorial Guinea
105 S11 Annonay Ardèche, E France 45.15N 4.40E
46 C6 Annotto Bay C Jamaica 18.16N 76.45W
147 R5 An Nu'ayrīyah var. Nariya. Ash Sharqīyah, NE Saudi Arabia
145 O13 An Nu'mānīyah E Iraq 32.34N 45.22E
144 I7 Anti-Lebanon var. Jebel esh Sharqi, Ar. Al Jabal ash Sharqi, Fr. Anti-Liban. ▲ Lebanon/Syria
117 J22 Antimilos island Kykládes, Greece, Aegean Sea
68 L6 Antimony Utah, W USA 38.07N 112.00W
104 I10 Antioche, Pertuis d' inlet W France
56 D9 Antioquia Antioquia, C Colombia 6.24N 75.52W
56 E8 Antioquia ◆ province C Colombia
117 J21 Antíparos var. Andíparos. island Kykládes, Greece, Aegean Sea
117 H17 Antíparos island Kykládes, Greece, Aegean Sea
117 K16 Antípaxoi var. Andípaxi. island Iónioi Nísoi, Greece, C Mediterranean Sea
117 M18 Antíssa var. Ándissa. Lésvos, E Greece 39.15N 25.59E
193 Q7 Antipayuta Yamalo-Nenetskiy Avtonomnyy Okrug, N Russian Federation 69.08N 76.43E
199 J14 Antipodes Islands island group S NZ
117 J18 Antipolis see Antibes
117 J18 Antípsara var. Andípsara. island E Greece
Antiquaria see Antequera
11 E18 Antique ◆ province, SE Canada
117 E18 Antírrio var. Andírrion. Dytikí Ellás, C Greece 38.20N 21.46E
58 L8 Antisana ▲ N Ecuador 0.29S 78.08W
29 Q13 Antlers Oklahoma, C USA 34.13N 95.37W
95 J14 Antnäs Norrbotten, N Sweden 65.32N 21.53E
Antō see Andong
64 G5 Antofagasta Antofagasta, N Chile 23.40S 70.22W
64 G6 Antofagasta off. Región de Antofagasta. ◆ region N Chile
64 I7 Antofalla, Salar de salt lake NW Argentina
101 K16 Antoing Hainaut, SW Belgium 50.34N 3.26E
25 R5 Anton Texas, SW USA 33.48N 102.09W
39 U9 Antón Coclé, C Panama 8.22N 80.15W
39 T17 Anton Chico New Mexico, SW USA 35.12N 105.09W
62 K7 Antonina Paraná, S Brazil 25.28S 48.43W
105 O5 Antony Hauts-de-Seine, N France 48.45S 2.16E
119 Y8 Antratsit Rus. Antratsit. Luhans'ka Oblast', E Ukraine 48.07N 39.04E
99 G15 Antrim Ir. Aontroim. NE Northern Ireland, UK 54.43N 6.13W
99 G15 Antrim Ir. Aontroim. cultural region NE Northern Ireland, UK
99 G15 Antrim Mountains ▲ NE Northern Ireland, UK
180 H5 Antsalova Mahajanga, W Madagascar 18.40S 44.37E
180 J2 Antserana see Antsirañana
180 J2 Antsirañana prev. Antserana; prev. Diégo-Suarez. Antsirañana, N Madagascar 12.19S 49.16E
180 J2 Antsirañana ◆ province N Madagascar
180 I3 Antsirabe Antananarivo, C Madagascar 19.52S 47.00E
121 N7 Antsla Ger. Anzen. Võrumaa, SE Estonia 57.49N 26.34E
180 J3 Antsohihy Mahajanga, NW Madagascar 14.49S 47.58E
120 I7 Antu, Gunung ▲ Borneo, N Indonesia 0.57N 118.51E
179 P7 Antu, Gunung ▲ Borneo, N Indonesia
Antwerp see Antwerpen
Antwerpen Eng. Antwerp, Fr. Anvers. Antwerpen, N Belgium 51.13N 4.25E
101 H16 Antwerpen Eng. Antwerp. ◆ province N Belgium
An Uaimh see Navan
160 N4 Anugul var. Angul. Orissa, E India 20.51N 84.59E
158 F9 Anupshahr Uttar Pradesh, N India 29.10N 73.13E
160 N4 Anūppur Madhya Pradesh, C India 23.06N 81.45E
161 K24 Anuradhapura North Central Province, C Sri Lanka 8.19N 80.25E
195 X16 Anvik Alaska, USA 62.39N 160.12W
41 N11 Anvik River ✍ Alaska, USA
41 M7 Anvil Peak ▲ Semisopochnoi Island, Alaska, USA 51.59N 179.36E
167 N5 Anxi Fujian, SE China 25.04N 118.14E
170 H5 Anxi var. Anxi. Gansu, N China
159 P9 Anxious Bay bay South Australia
166 L4 Anyang Henan, C China 36.10N 114.18E
164 G4 A'nyêmaqên Shan ▲ C China
165 S11 Anykščiai Anykščiai, E Lithuania 55.30N 25.34E
167 O3 Anyuan Jiangxi, S China 25.10N 115.25E
127 R9 Anyuysky Chukotskiy Avtonomnyy Okrug, NE Russian Federation 68.22N 161.33E
120 O5 Anyuysky Khrebet ▲ NE Russian Federation
56 D8 Anza Antioquia, C Colombia 6.18N 75.54W
126 H13 Anzhero-Sudzhensk Kemerovskaya Oblast', S Russian Federation 56.08N 85.42E
109 H15 Anzio Lazio, C Italy 41.27N 12.37E
57 O6 Anzoátegui ◆ state NE Venezuela
153 P12 Anzob W Tajikistan 39.24N 68.55E
Anzyō see Anjō
Aoba see Ambae
172 Ss13 Aoga-shima island Izu-shotō, SE Japan
169 T12 Aohan Qi Nei Mongol Zizhiqu, N China 42.12N 119.57E
107 J8 Aoiz var. Aoiz-Agoitz. Navarra, N Spain 42.47N 1.23W
195 X16 Aola var. Tenaghau. Guadalcanal, C Solomon Islands 9.32S 160.28E
Aomen see Macao
172 N8 Aomori Aomori, Honshū, N Japan 40.49N 140.43E
171 Mm9 Aomori off. Aomori-ken. ◆ prefecture Honshū, C Japan
Aontroim see Antrim
117 C15 Aóos var. Vijosa, Vijosë, Alb. Lumi i Vjosës. ✍ Albania/Greece see also Vijosa; Lumi i
203 Q7 Aorai, Mont ▲ Tahiti, W French Polynesia 17.36S 149.28W
Aoraki see Cook, Mount
178 Ii13 Aôral, Phnum prev. Phnom Aural. ▲ W Cambodia 12.01N 104.10E
193 C13 Aorangi see Cook, Mount
193 C13 Aorangi Mountains ▲ North Island, NZ
107 U11 Aosta anc. Augusta Praetoria. Valle d'Aosta, NW Italy 45.43N 7.19E
79 O11 Aougoundou, Lac ◲ S Mali
78 K9 Aoukâr var. Aouker. plateau C Mauritania
80 J13 Aouk, Bahr ✍ Central African Republic/Chad
Aouker see Aoukâr
76 K11 Aousard SE Western Sahara 22.42N 14.22W
170 G12 Aoya Tottori, Honshū, SW Japan 35.31N 134.01E
80 H5 Aozou Borkou-Ennedi-Tibesti, N Chad 22.00N 17.11E
28 M11 Apache Oklahoma, C USA 34.57N 98.21W
64 G5 Apache Junction Arizona, SW USA 33.25N 111.33W
26 J9 Apache Mountains ▲ Texas, SW USA
68 M16 Apache Peak ▲ Arizona, SW USA 31.50N 110.25W
118 H10 Apahida Cluj, NW Romania 46.49N 23.45E
25 T9 Apalachee Bay bay Florida, SE USA
25 T3 Apalachee River ✍ Georgia, SE USA
25 S10 Apalachicola Florida, SE USA 29.43N 84.58W
25 S10 Apalachicola Bay bay Florida, SE USA
25 R9 Apalachicola River ✍ Florida, SE USA
Apam see Apan
43 P14 Apan var. Apam. Hidalgo, C Mexico 19.41N 98.24W
44 J8 Apanás, Lago de ◲ NW Nicaragua
56 I4 Apaporis, Río ✍ Brazil/Colombia
193 C23 Aparima ✍ South Island, NZ
179 P7 Aparri Luzon, N Philippines 18.16N 121.42E
114 J9 Apatin Vojvodina, NW Yugoslavia 45.40N 19.01E
128 J2 Apatity Murmanskaya Oblast', NW Russian Federation 67.33N 33.26E
57 X9 Apatou NW French Guiana 5.07N 54.20W
42 M14 Apatzingán var. Apatzingán de la Constitución. Michoacán de Ocampo, SW Mexico 19.04N 102.19W
176 W9 Apauwar Irian Jaya, E Indonesia 1.36S 138.10E
43 O15 Apaxtla de Castrejón var. Apaxtla. Guerrero, S Mexico 18.06N 99.55W
120 J7 Ape Alūksne, NE Latvia 57.32N 26.42E
100 I11 Apeldoorn Gelderland, E Netherlands 52.13N 5.57E
Apennines see Appennino
Apenrade see Åbenrå
59 J17 Apere, Río ✍ C Bolivia
57 V11 Apetina Sipaliwini, SE Suriname 3.30N 55.03W
23 U9 Apex North Carolina, SE USA 35.43N 78.51W
81 M16 Api ◇ Orientale, N Dem. Rep. Congo (Zaire) 3.42N 25.22E
158 M9 Api ▲ NW Nepal 30.07N 80.57E
Api ● see Abaiang
188 B8 Apia ● (Samoa) Upolu, SE Samoa 13.49S 171.46W
62 K11 Apiaí São Paulo, S Brazil 24.28S 48.51W
175 P16 Api, Gunung ▲ Pulau Sangeang, S Indonesia 8.09S 119.03E
195 Y16 Apio Maramasike Island, N Solomon Islands 9.36S 161.25E
43 P14 Apipilulco Guerrero, S Mexico 18.10N 99.40W
43 P14 Apizaco Tlaxcala, S Mexico 19.24N 98.10W
106 I4 A Pobla de Trives Cast. Puebla de Trives. Galicia, NW Spain 42.21N 7.16W
188 J3 Apoera Sipaliwini, NW Suriname 5.10N 57.08W
165 L16 Apolakkiá Ródos, Dodekánisos, Greece, Aegean Sea 36.02N 27.48E
103 L16 Apolda Thüringen, C Germany 51.01N 11.31E
198 B8 Apolima Strait strait C Pacific Ocean
190 M13 Apollo Bay Victoria, SE Australia 38.40S 143.44E
Apollonia see Sozopol
59 J16 Apolo La Paz, W Bolivia 14.40S 68.33W
59 J16 Apolobamba, Cordillera ▲ Bolivia/Peru
179 Re16 Apo, Mount ▲ Mindanao, S Philippines 6.54N 125.16E
25 W11 Apopka Florida, SE USA 28.40N 81.30W
25 W11 Apopka, Lake ◲ Florida, SE USA

61 J19 **Aporé, Rio** ≈ SW Brazil

32 K2 **Apostle Islands** *island group*
Wisconsin, N USA

Apostolas Andreas, Cape *see*
Zafer Burnu

63 F14 **Apóstoles** Misiones,
NE Argentina 27.54S 55.45W

Apostólou Andréa, Akrotíri
see Zafer Burnu

119 S9 **Apostolove** *Rus.* Apostolovo.
Dnipropetrovs'ka Oblast',
E Ukraine 47.40N 33.45E

Apostolovo *see* Apostolove

19 Q9 **Appalachian Mountains**
▲ E USA

97 K14 **Äppelbo** Dalarna, C Sweden
60.30N 14.00E

100 N7 **Appelscha** *Fris.* Appelskea.
Friesland, N Netherlands
52.57N 6.19E

Appelskea *see* Appelscha

108 G11 **Appennino** *Eng.* Apennines.
▲ Italy/San Marino

109 L17 **Appennino Campano**
▲ C Italy

110 F7 **Appenzell** Appenzell,
NW Switzerland 47.19N 9.25E

110 H7 **Appenzell** ◆ *canton*
NE Switzerland

57 V12 **Appikalo** Sipaliwini, S Suriname
2.07N 56.16W

100 O5 **Appingedam** Groningen,
NE Netherlands 53.18N 6.52E

27 X8 **Appleby** Texas, SW USA
31.43N 94.36W

99 L15 **Appleby-in-Westmorland**
NW England, UK 54.34N 2.26W

32 K0 **Apple River** ≈ Illinois, N USA

32 I5 **Apple River** ≈ Wisconsin,
N USA

27 W9 **Apple Springs** Texas, SW USA
31.13N 94.57W

31 S8 **Appleton** Minnesota, N USA
45.12N 96.01W

32 M7 **Appleton** Wisconsin, N USA
44.16N 88.24W

29 S5 **Appleton City** Missouri, C USA
38.11N 94.01W

37 U14 **Apple Valley** California, W USA
34.30N 117.11W

31 V9 **Apple Valley** Minnesota, N USA
44.43N 93.13W

23 U6 **Appomattox** Virginia, NE USA
37.21N 78.49W

196 B16 **Apra Harbour** *harbor* W Guam

196 B16 **Apra Heights** W Guam

108 F6 **Aprica, Passo dell'** *pass* N Italy
46.10N 10.08E

109 M15 **Apricena** *anc.* Hadria Picena.
Puglia, SE Italy 41.46N 15.27E

130 L14 **Apsheronsk** Krasnodarskiy Kray,
SW Russian Federation
44.27N 39.45E

Apsheronskiy Poluostrov *see*
Abşeron Yarımadası

105 S15 **Apt** *anc.* Apta Julia. Vaucluse,
SE France 43.54N 5.24E

Apta Julia *see* Apt

40 H12 **Apua Point** *headland* Hawaii,
USA, C Pacific Ocean
19.15N 155.13W

62 I10 **Apucarana** Paraná, S Brazil
23.34S 51.28W

109 L17 **Apulia** *see* Puglia

56 K8 **Apure** *off.* Estado Apure. ◆ *state*
C Venezuela

56 J7 **Apure, Río** ≈ W Venezuela

59 F16 **Apurímac** *off.* Departamento de
Apurímac. ◆ *department* C Peru

59 F15 **Apurímac, Río** ≈ S Peru

118 G10 **Apuseni, Munţii** ▲ W Romania

Aqaba/'Aqaba *see* Al 'Aqabah

144 F15 **Aqaba, Gulf of** *var.* Gulf of Elat,
Ar. Khalij al 'Aqabah; *anc.* Sinus
Aelaniticus. *gulf* NE Red Sea

145 R7 **'Aqabah** C Iraq 33.33N 42.55E

'Aqabah, Khalij al *see* Aqaba,
Gulf of

155 O2 **Āqchah** *var.* Āqcheh. Jowzjān,
N Afghanistan 36.59N 66.07E

Āqcheh *see* Āqchah

Aqköl *see* Akkol'

Aqmola *see* Astana

Aqmola Oblysy *see* Akmola

164 L10 **Aqqikkol Hu** ☞ NW China

'Aqrah *see* Ākrê

Aqsay *see* Aksay

Aqshataū *see* Akchatau

Aqsū *see* Aksu

Aqsüat *see* Aksuat

Aqtas *see* Aktas

Aqtaū *see* Aktau

Aqtöbe/Aqtöbe Oblysy *see*
Aktyubinsk

Aqtoghay *see* Aktogay

Aquae Augustae *see* Dax

Aquae Calidae *see* Bath

Aquae Flaviae *see* Chaves

Aquae Grani *see* Aachen

Aquae Panoniae *see* Baden

Aquae Sextiae *see* Aix-en-
Provence

Aquae Solis *see* Bath

Aquae Tarbelicae *see* Dax

68 J11 **Aquarius Mountains**
▲ Arizona, SW USA

64 O5 **Aquidabán, Río** ≈ E Paraguay

61 H20 **Aquidauana** Mato Grosso do Sul,
S Brazil 20.27S 55.45W

42 L13 **Aquila** Michoacán de Ocampo,
S Mexico 18.36N 103.32W

Aquila/Aquila degli Abruzzi
see L'Aquila

27 T8 **Aquila** Texas, SW USA
31.51N 97.13W

46 L3 **Aquin** S Haiti 18.16N 73.24W

Aquisgranum *see* Aachen

104 J13 **Aquitaine** ◆ *region* SW France

Aqzhar *see* Akzhar

159 P13 **Āra** *prev.* Arrah. Bihār, N India
25.34N 84.40E

107 S4 **Ara** ≈ NE Spain

25 P2 **Arab** Alabama, S USA
34.19N 86.30W

144 G12 **'Arab, Wādī al** *Heb.* Ha'Arava.
dry watercourse Israel/Jordan

119 U12 **Arabats'ka Strilka, Kosa** *spit*
S Ukraine

119 U12 **Arabats'ka Zatoka** *gulf*
S Ukraine

'Arab, Baḥr al *see* Arab, Bahr el

---

82 C12 **Arab, Bahr el** *var.* Baḥr al 'Arab.
≈ S Sudan

58 E7 **Arabela, Río** ≈ N Peru

181 O4 **Arabian Basin** *undersea feature*
N Arabian Sea

Arabian Desert *see* Sahara
el Sharqîya

147 N9 **Arabian Peninsula** *peninsula*
SW Asia

87 P15 **Arabian Plate** *tectonic feature*
Africa/Asia/Europe

147 W14 **Arabian Sea** *sea*
NW Indian Ocean

Arabicus, Sinus *see* Red Sea

'Arabī, Khalīj al *see* Persian Gulf

Arabistan *see* Khūzestān

**'Arabīyah as Su'ūdīyah,
Al Mamlakah al** *see* Saudi Arabia

**'Arabīyah Jumhūrīyah, Mişr
al** *see* Egypt

144 I9 **'Arab, Jabal al** ▲ S Syria

124 Pp14 **'Arab, Khalīg el** *Eng.* Arabs Gulf.
gulf N Egypt

Arab Republic of Egypt
see Egypt

145 Y12 **'Arab, Shaṭṭ al** *Eng.* Shatt al Arab,
Per. Arvand Rūd. ≈ Iran/Iraq

142 I11 **Araç** Kastamonu, N Turkey
41.13N 33.19E

61 O16 **Aracaju** *state capital* Sergipe,
E Brazil 10.45S 37.07W

56 F5 **Aracataca** Magdalena,
N Colombia 10.36N 74.13W

60 O13 **Aracati** Ceará, E Brazil
4.31S 37.45W

62 I8 **Araçatuba** São Paulo, S Brazil
21.12S 50.24W

106 J13 **Aracena** Andalucía, S Spain
37.54N 6.33W

115 F19 **Arachnaío** ▲ S Greece

117 D16 **Árachthos** *var.* Árta; *prev.*
Árakhthos, *anc.* Arachthus.
≈ W Greece

Arachthus *see* Árachthos

61 N18 **Araçuaí** Minas Gerais, SE Brazil
16.52S 42.03W

142 I11 **Araç Çayı** ≈ N Turkey

144 F11 **Arad** Southern, S Israel
31.16N 35.09E

118 F11 **Arad** Arad, W Romania
46.12N 21.20E

118 F11 **Arad** ◆ *county* W Romania

80 J9 **Arada** Biltine, N Chad
15.00N 20.38E

149 P18 **'Arādah** Abū Zaby, S UAE
22.57N 53.24E

124 O3 **Aradhippou** *see* Aradíppou

182 K6 **Arafura Sea** *Ind.* Laut Arafuru.
SE Cyprus 34.57N 33.37E

182 L6 **Arafura Sea** *Ind.* Laut Arafuru.
sea W Pacific Ocean

Arafura Shelf *undersea feature*
C Arafura Sea

Arafuru, Laut *see* Arafura Sea

143 T12 **Aragats, Gora** *see* Aragats Lerr

34 E14 **Arago, Cape** *headland* Oregon,
NW USA 43.17N 124.25W

107 R6 **Aragón** ◆ *autonomous community*
E Spain

107 O2 **Aragón** ≈ NE Spain

109 I24 **Aragona** Sicilia, Italy,
C Mediterranean Sea 37.25N 13.37E

107 O7 **Aragoncillo** ▲ C Spain
40.59N 2.01W

56 L5 **Aragua** *off.* Estado Aragua. ◆
state N Venezuela

57 N6 **Aragua de Barcelona**
Anzoátegui, NE Venezuela
9.30N 64.45W

61 K15 **Aragua de Maturín** Monagas,
NE Venezuela 9.58N 63.30W

Araguaia, Río *var.* Araguaya.
≈ C Brazil

61 J19 **Araguari** Minas Gerais, SE Brazil
18.37S 48.13W

60 J11 **Araguaya, Río** ≈ SW Brazil

Araguaya *see* Araguaia, Río

106 K14 **Arahal** Andalucía, S Spain
37.15N 5.33W

171 Jj13 **Arai** Niigata, Honshū, C Japan
36.58N 138.14E

Árainn *see* Inishmore

Árainn Mhór *see* Aran Island

Ara Jovis *see* Aranjuez

76 J11 **Arak** C Algeria 25.17N 3.45E

176 Yy15 **Arak** Irian Jaya, E Indonesia
7.14S 139.40E

148 M7 **Arāk** *prev.* Sultānābād. Markazī,
W Iran 34.07N 49.39E

196 D10 **Arakabesan** *island* Palau Islands,
N Palau

171 Kk12 **Arakawa** Niigata, Honshū,
C Japan 38.06N 139.25E

Árakhthos *see* Árachthos

Araks/Arak's *see* Aras

164 H7 **Aral** Xinjiang Uygur Zizhiqu,
NW China 40.40N 81.19E

Aral *see* Aral'sk, Kazakhstan

Aral *see* Vose', Tajikistan

Aral-Bukhorskiy Kanal *see*
Amu-Bukhoro Kanali

143 T12 **Aralık** Iğdır, E Turkey 39.54N 44.28E

152 H5 **Aral Sea** *Kaz.* Aral Tengizi, *Rus.*
Aral'skoye More, *Uzb.* Orol
Dengizi. *inland sea*
Kazakhstan/Uzbekistan

145 Kk1 **Aral Sea Kaz.** Kazylorda,
SW Kazakhstan 46.48N 61.40E

Aral'skoye More/Aral Tengizi
see Aral Sea

43 O10 **Aramberri** Nuevo León,
NE Mexico 24.05N 99.52W

194 F14 **Aramia** ≈ SW PNG

149 N6 **'Arān** *var.* Golārā. Eşfahān, C Iran
34.03N 51.30E

107 N5 **Aranda de Duero** Castilla-León,
N Spain 41.41N 3.41W

114 M12 **Aranđelovac** *prev.* Arandjelovac.
Serbia, C Yugoslavia 44.18N 20.32E

Arandjelovac *see* Aranđelovac

101 M15 **Aran, Emirate** *see* Arkhangel'sk

Aran Fawddwy ▲ NW Wales,
UK 52.48N 3.42W

99 D17 **Aran Island** *Ir.* Árainn Mhór.
island NW Ireland

---

99 A18 **Aran Islands** *island group*
W Ireland

107 N9 **Aranjuez** *anc.* Ara Jovis. Madrid,
C Spain 40.01N 3.37W

85 E20 **Aranos** Hardap, SE Namibia
24.11S 19.07E

27 U14 **Aransas Bay** *inlet* Texas, SW USA

27 T14 **Aransas Pass** Texas, SW USA
27.54N 97.09W

203 Q3 **Aranuka** *prev.* Nanouki. *atoll*
Tungaru, W Kiribati

178 I11 **Aranyaprathet** Prachin Buri,
S Thailand 13.42N 102.32E

Aranyasztal *see* Zlatý Stôl

Aranyosgyéres *see* Câmpia Turzii

Aranyosmarót *see* Zlaté Moravce

170 Cc14 **Arao** Kumamoto, Kyūshū,
SW Japan 33.16N 130.25E

79 O8 **Araouane** Tombouctou, N Mali
18.58N 3.39W

28 L10 **Arapaho** Oklahoma, C USA
35.34N 98.57W

29 T16 **Arapahoe** Nebraska, C USA
40.18N 99.54W

59 I18 **Arapa, Laguna** ☞ SE Peru

193 K14 **Arapawa Island** *island* C NZ

63 E17 **Arapey Grande, Río**
N Uruguay

146 M3 **'Ar'ar, Wādī** *dry watercourse*
Iraq/Saudi Arabia

133 N7 **Aras** *Arm.* Arak's, *Az.* Araz Nehri,
Per. Rūd-e Aras, *Rus.* Araks; *prev.*
Araxes. ≈ SW Asia

107 R9 **Aras de Alpuente** País
Valenciano, E Spain 39.55N 1.09W

143 S13 **Aras Güneyi Dağları**
▲ NE Turkey

Aras, Rūd-e *see* Aras

203 U9 **Aratika** *atoll* Îles Tuamotu,
C French Polynesia

Aratürük *see* Yiwu

56 I8 **Arauca** Arauca, N Colombia
7.03N 70.46W

56 I8 **Arauca** *off.* Intendencia de
Arauca. ◆ *province* NE Colombia

65 G15 **Araucanía** ◆ *región* C Chile

65 G14 **Arauco** Bío Bío, C Chile
37.16S 73.15W

56 H8 **Arauquita** Arauca, C Colombia
7.01N 71.20W

Arausio *see* Orange

158 F13 **Arāvali Range** ▲ N India

194 F14 **Arawa** Bougainville Island,
NE PNG 6.13S 155.37E

194 L12 **Arawe Islands** *island group*
E PNG

61 L20 **Araxá** Minas Gerais, SE Brazil
19.37S 46.49W

Araxes *see* Aras

107 P4 **Araya** Sucre, N Venezuela
10.34N 64.15W

83 N6 **Arba Minch'** Southern, S Ethiopia
6.02N 37.33E

145 U4 **Arbat** NE Iraq 35.26N 45.34E

109 D19 **Arbatax** Sardegna, Italy,
C Mediterranean Sea 39.57N 9.42E

Arbe *see* Rab

145 X4 **Arbīl** *var.* Erbil, Irbil, *Kurd.*
Hawlēr; *anc.* Arbela. N Iraq
36.12N 44.01E

97 M16 **Arboga** Västmanland, C Sweden
59.24N 15.49E

105 S9 **Arbois** Jura, E France 46.54N 5.45E

56 D6 **Arboletes** Antioquia,
NW Colombia 8.52N 76.25W

11 X15 **Arborg** Manitoba, S Canada
50.52N 97.20W

96 M13 **Arbrå** Gävleborg, C Sweden
61.27N 16.21E

98 K8 **Arbroath** *anc.* Aberbrothock. E
Scotland, UK 56.34N 2.34W

37 N6 **Arbuckle** California, W USA
39.00N 122.05W

28 M11 **Arbuckle Mountains**
▲ Oklahoma, C USA

119 O8 **Arbuzynka** Mykolayivs'ka
Oblast', S Ukraine 47.54N 31.19E

105 U12 **Arc** ≈ E France

104 J13 **Arcachon** Gironde, SW France
44.40N 1.10W

104 I13 **Arcachon, Bassin d'** *inlet*
SW France

25 W14 **Arcadia** Florida, SE USA
27.13N 81.51W

24 I4 **Arcadia** Louisiana, S USA
32.33N 92.55W

32 J7 **Arcadia** Wisconsin, N USA
44.15N 91.30W

121 O14 **Arcae Remorum** *see* Châlons-
en-Champagne

36 K4 **Arcata** California, W USA
40.51N 124.06W

42 L13 **Arcelia** Guerrero, S Mexico
18.19N 100.16W

100 M15 **Arcen** Limburg, SE Netherlands
51.28N 6.10E

47 O4 **Archangel** *see* Arkhangel'sk

Archangel Bay *see*
Chëshskaya Guba

---

117 O23 **Archángelos** *var.* Arhangelos,
Arkhángelos. Ródos, Dodekánisos,
Greece, Aegean Sea 36.13N 28.07E

116 F7 **Archar** ≈ NW Bulgaria

33 R3 **Archbold** Ohio, N USA
41.31N 84.18W

107 R12 **Archena** Murcia, SE Spain
38.07N 1.16W

27 R7 **Archer City** Texas, SW USA
33.36N 98.37W

106 M14 **Archidona** Andalucía, S Spain
37.06N 4.22W

67 S23 **Arch Islands** *island group*
SW Falkland Islands

108 G13 **Arcidosso** Toscana, C Italy
42.52N 11.30E

45 T15 **Arcenosa** Panamá, N Panama
9.02N 79.57W

190 F3 **Arckaringa Creek** *seasonal river*
South Australia

106 E12 **Arco** Trentino-Alto Adige, N Italy
45.53N 10.51E

35 Q14 **Arco** Idaho, NW USA
43.38N 113.18W

32 M4 **Arcola** Illinois, N USA
39.39N 88.19W

107 P6 **Arcos de Jalón** Castilla-León,
N Spain 41.12N 2.13W

106 K15 **Arcos de la Frontera** Andalucía,
S Spain 36.45N 5.49W

61 P15 **Arcoverde** Pernambuco, E Brazil
08.25S 37.00W

104 H5 **Arcvovest, Pointe de l'** *headland*
NW France 48.49N 2.58W

Arctic-Mid Oceanic Ridge *see*
Nansen Cordillera

207 R8 **Arctic Ocean** *ocean*

14 G4 **Arctic Red River** ≈ Northwest
Territories/Yukon Territory,
NW Canada

Arctic Red River *see* Tsiigehtchic

41 S6 **Arctic Village** Alaska, USA
68.07N 145.32W

204 H1 **Arctowski** *Polish research station*
South Shetland Islands, Antarctica
61.57S 58.23W

116 I12 **Arda** *var.* Ardhas, *Gk.* Ardas.
≈ Bulgaria/Greece *see also* Ardas

148 L2 **Ardabil** *var.* Ardebil. Ardabīl,
NW Iran 38.15N 48.18E

148 L2 **Ardabīl** *off.* Ostān-e Ardabīl. ◆
province NW Iran

143 N13 **Ardahan** Ardahan, NE Turkey
41.07N 42.40E

143 S11 **Ardahan** ◆ *province* NE Turkey

149 P8 **Ardakān** Yazd, C Iran 32.20N 54.02E

96 C12 **Ârdalstangen** Sogn og Fjordane,
S Norway 61.13N 7.43E

143 R11 **Ardanuç** Artvin, NE Turkey
41.07N 42.04E

116 L12 **Ardas** *var.* Ardhas, *Bul.* Arda.
≈ Bulgaria/Greece *see also* Arda

144 I13 **Arḍ aş Şawwān** *var.* Ardh es
Suwwān. *plain* S Jordan

131 P5 **Ardatov** Respublika Mordoviya,
W Russian Federation 54.49N 46.13E

12 G2 **Ardbeg** Ontario, S Canada
45.38N 80.05W

105 Q13 **Ardèche** ◆ *department* E France

105 Q13 **Ardèche** ≈ E France

99 F17 **Ardee** *Ir.* Baile Átha Fhirdhia.
NE Ireland 53.52N 6.33W

105 Q3 **Ardennes** ◆ *department*
NE France

101 J18 **Ardennes** *physical region*
Belgium/France

143 Q11 **Ardeşen** Rize, NE Turkey
41.12N 41.02E

149 O7 **Ardestān** *var.* Ardistan. Eşfahān,
C Iran 33.29N 52.16E

110 J9 **Ardez** Graubünden,
SE Switzerland 46.47N 10.09E

Ardhas *see* Arda/Ardas

Ardh es Suwwān *see*
Arḍ aş Şawwān

106 I12 **Ardila, Ribeira de** *Sp.* Ardilla.
≈ Portugal/Spain *see also* Ardilla

11 T17 **Ardill** Saskatchewan, S Canada
49.56N 105.49W

106 I12 **Ardilla** *Port.* Ribeira de Ardila.
≈ Portugal/Spain *see also* Ardila

82 K7 **Ardo** Northern, N Sudan
19.31N 30.25E

181 P7 **Argo Fracture Zone** *tectonic*
feature C Indian Ocean

116 I12 **Ardoğan** Kürdzhali, S Bulgaria
41.38N 25.22E

174 Mm15 **Argopuro, Gunung** ▲ Jawa,
S Indonesia 7.57S 113.32E

117 F22 **Árgos** Pelopónnisos, S Greece
37.38N 22.42E

145 S1 **Arbsh** S Iraq 37.07N 44.13E

117 D14 **Árgos Orestikó** Dytikí
Makedonía, N Greece 40.27N 21.15E

117 B19 **Argostóli** *var.* Argostólion.
Kefallinía, Iónioi Nísoi, Greece,
C Mediterranean Sea 38.10N 20.29E

Argostólion *see* Argostóli

104 J11 **Argovie** *see* Aargau

Arge *see* Arguello, Point

98 G12 **Argyll** *cultural region* W Scotland,
UK

168 A7 **Argyrokastron** *see* Gjirokastër

168 I7 **Arhangay** ◆ *province* C Mongolia

Arhangelos *see* Archángelos

97 P16 **Arholma** Stockholm, C Sweden
59.51N 19.01E

97 G15 **Ärhus** *var.* Aarhus. Århus,
C Denmark 56.09N 10.10E

97 G15 **Århus** ◆ *county* C Denmark

145 T1 **Arī** Iraq 37.07N 44.14E

194 M12 **Aria** ≈ New Britain, E PNG

Aria *see* Herāt

170 C13 **Ariake-kai** *bay* SW Japan

109 L17 **Ariano Irpino** Campania, S Italy
41.08N 15.00E

56 H6 **Ariari, Río** ≈ C Colombia

---

157 K19 **Ari Atoll** *atoll* C Maldives

64 G2 **Aribinda** N Burkina 14.12N 0.50W

64 G2 **Arica** *hist.* San Marcos de Arica.
Tarapacá, N Chile 18.30S 70.18W

56 H16 **Arica** Amazonas, S Colombia
2.09S 71.48W

64 G2 **Arica** ✈ Tarapacá, N Chile
18.30S 70.19W

170 G16 **Arida** Wakayama, Honshū,
SW Japan 34.05N 135.07E

116 E13 **Aridaía** *var.* Aridea, Aridhaía.
Dytikí Makedonía, N Greece
40.58N 22.04E

180 I15 **Aride, Île** *island* Inner Islands,
NE Seychelles

Aridhaía *see* Aridaía

105 N17 **Ariège** ◆ *department* S France

104 M16 **Ariège, var.** la Riege.
≈ Andorra/France

118 H11 **Arieş** ≈ W Romania

155 U10 **Arifwāla** Punjab, E Pakistan
30.14N 73.04E

Ariguani *see* El Difícil

144 G11 **Arīḥā** Al Karak, W Jordan
31.25N 35.46E

144 I3 **Arīḥā** *var.* Arīḥā. Idlib, S Syria
35.49N 36.36E

Arīḥā *see* Jericho

39 W4 **Arikaree River**
≈ Colorado/Nebraska, C USA

170 Bb12 **Arikawa** Nagasaki, Nakadōri-
jima, SW Japan 32.58N 129.06E

114 L13 **Arilje** Serbia, W Yugoslavia
43.45N 20.06E

47 U14 **Arima** Trinidad, Trinidad and
Tobago 10.38N 61.16W

Arime *see* Al 'Arīmah

42 M14 **Ario de Rosales** *var.* Ario de
Rosáles. Michoacán de Ocampo,
SW Mexico 19.12N 101.43W

120 F12 **Ariogala** Raseiniai, C Lithuania
55.16N 23.30E

49 T7 **Aripuanã** ≈ W Brazil

61 E15 **Ariquemes** Rondônia, W Brazil
9.55S 63.06W

124 Rr15 **'Arīsh, Wādī el** ≈ NE Egypt

56 K6 **Arismendi** Barinas, C Venezuela
8.28N 68.22W

8 J7 **Aristazabal Island** *island*
SW Canada

62 F13 **Aristóbulo del Valle** Misiones,
NE Argentina 27.09S 54.54W

10 I11 **Arizona** Entre Ríos, C Argentina
50.19N 89.01W

31 S16 **Armstrong** Texas, SW USA
26.55N 97.47W

119 S11 **Armyans'k** *Rus.* Armyansk.
Respublika Krym, S Ukraine
46.05S 33.43E

117 A14 **Arnaía** *var.* Arnea. Kentrikí
Makedonía, N Greece 40.30N 23.36E

123 Mm3 **Arnauti, Akrótiri** *var.*
Arnaoútis, Cape Arnaoútí.
headland W Cyprus. *see* West Cyprus

Arnaouti, Cape/Arnaoútis *see*
Arnaoúti, Akrótiri

10 L4 **Arnaud** ≈ Quebec, E Canada

105 Q8 **Arnay-le-Duc** Côte d'Or,
C France 47.08N 4.27E

Arnea *see* Arnaía

107 Q4 **Arnedo** La Rioja, N Spain
42.13N 2.04W

97 I14 **Ärnes** Akershus, S Norway
60.07N 11.28E

95 F25 **Ârnes** Sør-Trøndelag, S Norway
63.58N 10.12E

28 K9 **Arnett** Oklahoma, C USA
36.07N 99.46W

100 L12 **Arnhem** Gelderland,
SE Netherlands 51.58N 5.54E

189 Q2 **Arnhem Land** *physical region*
Northern Territory, N Australia

108 F11 **Arno** ≈ C Italy

201 W7 **Arno Atoll** *var.* Arpo. *atoll* Ratak
Chain, NE Marshall Islands

190 H8 **Arno Bay** South Australia
33.55S 136.31E

37 Q8 **Arnold** California, W USA
38.15N 120.19W

29 X3 **Arnold** Missouri, C USA
38.25N 90.22W

31 N3 **Arnold** Nebraska, C USA
41.25N 100.11W

111 I17 **Arnoldstein Slvn.** Pod Klošter.
Kärnten, S Austria 46.34N 13.43E

105 N9 **Arnon** ≈ C France

47 T4 **Arnos Vale** ✈ (Kingstown) Saint
Vincent, SE Saint Vincent and the
Grenadines 13.08N 61.13W

94 I8 **Arnøy** *island* N Norway

12 L12 **Arnprior** Ontario, SE Canada
45.26N 76.22W

101 G15 **Arnsberg** Nordrhein-Westfalen,
W Germany 51.24N 8.04E

101 H17 **Arnstadt** Thüringen, C Germany
50.49N 10.57E

Arnswalde *see* Choszczno

56 J4 **Aroa** Yaracuy, N Venezuela
10.25N 68.54W

85 E21 **Aroab** Karas, SE Namibia
26.47S 19.37E

203 O6 **Ároa, Pointe** *headland* Moorea,
W French Polynesia 17.27S 149.46W

147 M20 **Aroe Islands** *see* Aru, Kepulauan

103 H15 **Arolsen** Hessen, C Germany

24 E7 **Arona** Piemonte, NE Italy
45.45N 8.33E

Aroostook River *see*
Canada/USA

Arop Island *see* Long Island

40 N2 **Aropuk Lake** ☞ Alaska, USA

203 P4 **Arorae** *atoll* Tungaru, W Kiribati

202 G16 **Arorangi** Rarotonga, S Cook
Islands 21.13S 159.49W

110 I9 **Arosa** Graubünden, S Switzerland
46.48N 9.42E

106 F4 **Arousa, Ría de** *estuary* E Atlantic
Ocean

176 Uu16 **Aru, Tanjung** *headland*
Pulau Selaru, SE Indonesia
8.19S 130.45E

192 M7 **Arowhana** ▲ North Island, NZ
38.07S 177.52E

143 S11 **Arp'a** *Az.* Arpaçay.
≈ Armenia/Azerbaijan

143 S11 **Arpaçay** Kars, NE Turkey
40.51N 43.19E

Arpaçay see Arp'a
Arqalyq see Arkalyk
155 N14 Arra see Arrah SW Pakistan
Arrabona see Győr
Arrah see Ara
145 R9 Ar Raḥḥālīyah C Iraq 32.53N 43.21E
62 Q10 Arraial do Cabo Rio de Janeiro, SE Brazil 22.57S 42.00W
106 H11 Arraiolos Évora, S Portugal 38.43N 7.58W
145 R8 Ar Ramādī var. Ramadi, Rumadiya. SW Iraq 33.27N 43.19E
144 J6 Ar Rāmī Ḥimṣ, C Syria 34.32N 37.54E
Ar Rams see Rams
144 H9 Ar Ramtha var. Ramtha. Irbid, N Jordan 32.34N 36.00E
98 H13 Arran, Isle of island SW Scotland, UK
144 L3 Ar Raqqah see Rakka; anc. Nicephorium. Ar Raqqah, N Syria 35.57N 39.03E
144 L3 Ar Raqqah off. Muḥāfaẓat al Raqqah, var. Raqqah, Fr. Rakka. ◆ governorate N Syria
105 O2 Arras anc. Nemetocenna. Pas-de-Calais, N France 50.16N 2.46E
Arrasate see Mondragón
144 G12 Ar Rashādīyah Aṭ Ṭafīlah, W Jordan 30.42N 35.37E
144 I5 Ar Rastān var. Rastāne. Ḥimṣ, W Syria 34.57N 36.43E
145 X12 Ar Raṭāwī E Iraq 30.37N 47.12E
104 L15 Arrats ~ S France
147 N10 Ar Rawḍah Makkah, S Saudi Arabia 21.19N 42.48E
147 Q15 Ar Rawḍah S Yemen 14.26N 47.13E
148 K11 Ar Rawḍatayn var. Raudhatain. N Kuwait 29.52N 47.42E
149 N16 Ar Rayyan var. Al Rayyan. C Qatar 25.18N 51.24E
104 L17 Arreau Hautes-Pyrénées, S France 42.55N 0.21E
66 Q11 Arrecife var. Arrecife de Lanzarote, Puerto Arrecife. Lanzarote, Islas Canarias, NE Atlantic Ocean 28.57N 13.33W
Arrecife de Lanzarote see Arrecife
45 P6 Arrecife Edinburgh reef NE Nicaragua
63 C19 Arrecifes Buenos Aires, E Argentina 34.06S 60.09W
104 F6 Arrée, Monts d' ▲ NW France
Ar Refā'i see Ar Rifā'ī
Arretium see Arezzo
Arriaca see Guadalajara
111 S9 Arriach Kärnten, S Austria 46.43N 13.52E
43 T16 Arriaga Chiapas, SE Mexico 16.13N 93.54W
43 N12 Arriaga San Luis Potosí, C Mexico 21.55N 101.22W
145 W10 Ar Rifā'ī var. Ar Refā'ī. SE Iraq 31.46N 46.07E
145 V12 Ar Rihāb salt flat S Iraq
106 L2 Arriondas Asturias, N Spain 43.22N 5.10W
147 Q7 Ar Riyāḍ Eng. Riyadh. ● (Saudi Arabia) Ar Riyāḍ, C Saudi Arabia 24.49N 46.49E
147 O8 Ar Riyāḍ off. Minṭaqat ar Riyāḍ. ◆ province C Saudi Arabia
147 S15 Ar Riyān S Yemen 14.43N 49.18E
Arrō see Ærø
63 H18 Arroio Grande Rio Grande do Sul, S Brazil 32.15S 53.02W
104 K15 Arros ~ S France
105 Q9 Arroux ~ C France
27 R5 Arrowhead, Lake ⊞ Texas, SW USA
190 L5 Arrowsmith, Mount hill New South Wales, SE Australia 30.07S 141.37E
193 D21 Arrowtown Otago, South Island, NZ 44.57S 168.51E
63 D17 Arroyo Barú Entre Ríos, E Argentina 31.52S 58.25W
106 J10 Arroyo de la Luz Extremadura, W Spain 39.28N 6.36W
65 J16 Arroyo de la Ventana Río Negro, SE Argentina 41.41S 66.03W
37 P13 Arroyo Grande California, W USA 35.07N 120.35W
Ar Ru'ays see Ar Ruways
147 R11 Ar Rub' al Khālī Eng. Empty Quarter, Great Sandy Desert. desert SW Asia
145 V13 Ar Ruḍaymah S Iraq 30.19N 45.25E
63 A16 Arrufó Santa Fe, C Argentina 30.15S 61.45W
144 I7 Ar Ruḥaybah var. Ruhaybeh, Fr. Rouhaïbé. Dimashq, W Syria 33.45N 36.40E
145 V15 Ar Rukhaymiyah well S Iraq 29.22N 45.43E
145 U11 Ar Rumaythah var. Rumaitha. S Iraq 31.31N 45.15E
147 X8 Ar Rustāq var. Rostak, Rustaq. N Oman 23.34N 57.25E
145 N8 Ar Ruṭbah var. Rutba. SW Iraq 33.03N 40.16E
146 M3 Ar Rūthīyah spring/well NW Saudi Arabia 31.18N 41.23E
ar-Ruwaida see Ar Ruwaydah
147 O8 Ar Ruwaydah var. ar-Ruwaida. Jīzān, C Saudi Arabia 23.48N 44.44E
149 N15 Ar Ruways var. Al Ruweis, Ar Ru'ays, Ruwais. N Qatar 26.07N 51.13E
149 O17 Ar Ruways var. Ar Ru'ays, Ruwais. Abū Ẓaby, W UAE 24.09N 52.57E
97 G21 Års var. Aars. Nordjylland, N Denmark 56.49N 9.31E
Arsanias see Murat Nehri
157 N18 Arsen'yev Primorskiy Kray, SE Russian Federation 44.09N 133.26E
161 G19 Arsikere Karnātaka, W India 13.18N 76.15E
131 R3 Arsk Respublika Tatarstan, W Russian Federation 56.07N 49.54E
96 N10 Årskog Gävleborg, C Sweden 62.07N 17.19E
124 N3 Ársos C Cyprus 34.51N 32.46E
96 N13 Årsunda Gävleborg, C Sweden 60.31N 16.45E
96 C17 Árta var. Árachthos
96 C17 Árta anc. Ambracia. Ípeiros, W Greece 39.07N 20.59E
143 T16 Artashat S Armenia 39.57N 44.33E

42 M15 Arteaga Michoacán de Ocampo, SW Mexico 18.20N 102.18W
127 Nn18 Artem Primorskiy Kray, SE Russian Federation 43.24N 132.20E
46 C4 Artemisa La Habana, W Cuba 22.49N 82.46W
119 W7 Artemivs'k Donets'ka Oblast', E Ukraine 48.35N 37.58E
126 I14 Artemovsk Krasnoyarskiy Kray, S Russian Federation 54.22N 93.24E
126 Kk13 Artemovskiy Irkutskaya Oblast', C Russian Federation 58.15N 114.51E
125 Ee13 Artemovskiy Sverdlovskaya Oblast', C Russian Federation 57.22N 61.55E
107 U5 Artesa de Segre Cataluña, NE Spain 41.54N 1.03E
39 U14 Artesia New Mexico, SW USA 32.50N 104.24W
27 Q14 Artesia Wells Texas, SW USA 28.13N 99.18W
110 G8 Arth Schwyz, C Switzerland 47.05N 8.39E
12 F15 Arthur Ontario, S Canada
32 M14 Arthur Illinois, N USA 39.42N 88.28W
30 L14 Arthur Nebraska, C USA 41.32N 101.42W
31 Q5 Arthur North Dakota, N USA 47.07N 97.12W
193 B21 Arthur ~ South Island, NZ
20 B13 Arthur, Lake ⊞ Pennsylvania, NE USA
191 N15 Arthur River ~ Tasmania, SE Australia
193 G18 Arthur's Pass Canterbury, South Island, NZ 42.59S 171.33E
193 G17 Arthur's Pass pass South Island, NZ 42.57S 171.33E
46 I3 Arthur's Town Cat Island, C Bahamas 24.34N 75.39W
63 Q12 Artigas prev. San Eugenio, San Eugenio del Cuareim. Artigas, N Uruguay 30.25S 56.28W
63 E16 Artigas ◆ department N Uruguay
204 H1 Artigas Uruguayan research station Antarctica 61.57S 58.23W
67 M14 Art'ik W Armenia 40.38N 43.57E
197 G4 Art, Île island Îles Belep, W New Caledonia
105 O2 Artois cultural region N France
142 L12 Artova Tokat, N Turkey 40.06N 36.18E
107 Y9 Artrutx, Cap d' var. Cabo Dartuch. headland Menorca, Spain, W Mediterranean Sea 39.55N 3.49E
Artsiz see Artsyz
119 N11 Artsyz Rus. Artsiz. Odes'ka Oblast', SW Ukraine 45.59N 29.25E
164 E7 Artux Xinjiang Uygur Zizhiqu, NW China 39.45N 76.09E
143 R11 Artvin Artvin, NE Turkey 41.12N 41.48E
143 R11 Artvin ◆ province NE Turkey
152 G14 Artyk Akhalskiy Velayat, C Turkmenistan 37.32N 59.16E
81 Q16 Aru Dem. Rep. Congo (Zaire) 2.52N 30.49E
106 I4 A Rúa var. La Rúa. Galicia, NW Spain 42.27N 7.10W
83 E17 Arua NW Uganda 3.01N 30.55E
47 O15 Aruba ◇ Dutch autonomous region S West Indies
49 Q4 Aruba island Aruba, Lesser Antilles
Aru Islands see Aru, Kepulauan
171 W14 Aru, Kepulauan Eng. Aru Islands; prev. Aroe Islands. island group E Indonesia
159 W10 Arunāchal Pradesh prev. North East Frontier Agency, North East Frontier Agency of Assam. ◆ state NE India
169 U7 Arun Qi Nei Mongol Zizhiqu, N China 48.05N 123.28E
161 H23 Aruppukkottai Tamil Nādu, SE India 9.31N 78.03E
176 Ww9 Aruri, Selat strait Irian Jaya, E Indonesia
83 I21 Arusha Arusha, N Tanzania 3.22S 36.40E
83 I20 Arusha ◆ region E Tanzania
83 I20 Arusha ✕ Arusha, N Tanzania 3.26S 37.07E
56 C9 Arusí, Punta headland NW Colombia 5.36N 77.30W
174 L10 Arut, Sungai ~ Borneo, C Indonesia
161 J23 Aruvi Aru ~ NW Sri Lanka
81 M17 Aruwimi ~ NE Dem. Rep. Congo (Zaire)
Árva see Orava
39 T4 Arvada Colorado, C USA 39.48N 105.06W
168 J8 Arvayheer Övörhangay, C Mongolia 46.13N 102.47E
L8 Arviat prev. Eskimo Point. Nunavut, C Canada 61.10N 94.15W
95 I14 Arvidsjaur Norrbotten, N Sweden 65.34N 19.12E
94 J8 Ärviksand Troms, N Norway 70.10N 20.30E
37 S13 Arvin California, W USA 35.12N 118.52W
23 T9 Arvon, Mount ▲ Michigan, N USA 46.44N 88.09W
151 P17 Arys' Kaz. Arys. Yuzhnyy Kazakhstan, S Kazakhstan 42.25N 68.49E
Arys see Orzysz
Arys Köli see Arys, Ozero
151 O14 Arys, Ozero Kaz. Arys Köli. ⊗ C Kazakhstan
111 O16 Arzachena Sardegna, Italy, C Mediterranean Sea 41.05N 9.22E
129 D16 Arzamas Nizhegorodskaya Oblast', W Russian Federation 55.25N 43.51E
147 N13 Arzât S Oman 17.03N 54.19E
113 H3 Arzúa Galicia, NW Spain 42.55N 8.10W
113 A16 Aš Ger. Asch. Karlovarský Kraj, W Czech Republic 50.12N 12.12E

97 H15 Ås Akershus, S Norway 59.39N 10.48E
97 H20 Aså Nordjylland, N Denmark 57.07N 10.24E
85 E21 Asab Karas, S Namibia 25.28S 17.58E
79 U16 Asaba Delta, S Nigeria 6.10N 6.44E
155 S4 Asadābād var. Asadābād; prev. Chaghasarāy. Kunar, E Afghanistan 34.52N 71.09E
144 K3 Asad, Buḥayrat al ⊞ N Syria
65 H20 Asador, Pampa del plain S Argentina
171 K17 Asahi Chiba, Honshū, S Japan 36.08N 140.37E
171 J13 Asahi Toyama, Honshū, SW Japan 36.58N 137.34E
172 Pp5 Asahi-dake ▲ Hokkaidō, N Japan 43.42N 142.50E
170 Ff13 Asahi-gawa ~ Honshū, SW Japan
170 Q5 Asahikawa Hokkaidō, N Japan 43.46N 142.23E
153 S10 Asaka Rus. Assake; prev. Leninsk. Andijon Wiloyati, E Uzbekistan 40.39N 72.16E
79 P17 Asamankese SE Ghana 5.46N 0.41W
171 Jj15 Asama-yama ▲ Honshū, S Japan 36.25N 138.34E
196 B15 Asan W Guam 13.28N 144.43E
196 B15 Asan Point headland W Guam
159 F15 Asānsol West Bengal, NE India 23.40N 86.58E
82 K2 Āsayita Afar, NE Ethiopia 11.35N 41.23E
176 V9 Asbakin Irian Jaya, E Indonesia 0.45S 131.48E
Asben see Aïr, Massif de l'
125 Ee11 Asbest Sverdlovskaya Oblast', C Russian Federation 57.12N 61.18E
13 Q12 Asbestos Quebec, SE Canada 45.46N 71.55W
Y13 Asbury Iowa, C USA 42.30N 90.45W
20 K12 Asbury Park New Jersey, NE USA 40.13N 74.00W
43 Z12 Ascensión, Bahía de la bay NW Caribbean Sea
42 I3 Ascensión Chihuahua, N Mexico 31.07N 107.58W
67 M14 Ascension Fracture Zone tectonic feature C Atlantic Ocean
67 G14 Ascension Island ◇ dependency of St. Helena C Atlantic Ocean
67 N16 Ascension Island island C Atlantic Ocean
Asch see Aš
111 S3 Aschach an der Donau Oberösterreich, N Austria 48.22N 14.00E
103 H18 Aschaffenburg Bayern, SW Germany 49.58N 9.09E
103 F14 Ascheberg Nordrhein-Westfalen, W Germany 51.46N 7.36E
103 L14 Aschersleben Sachsen-Anhalt, C Germany 51.46N 11.28E
108 G12 Asciano Toscana, C Italy 43.15N 11.32E
108 J13 Ascoli Piceno anc. Asculum Picenum. Marche, C Italy 42.51N 13.34E
109 M17 Ascoli Satriano anc. Asculub, Ausculum. Puglia, SE Italy 41.13N 15.31E
110 D11 Ascona Ticino, S Switzerland 46.10N 8.45E
Asculub see Ascoli Satriano
Asculum Picenum see Ascoli Piceno
82 L11 Aseb var. Assab, Amh. Āseb. SE Eritrea 13.03N 42.36E
97 M20 Åseda Kronoberg, S Sweden 57.10N 15.19E
131 T6 Asekeyevo Orenburgskaya Oblast', W Russian Federation 53.36N 52.53E
194 J13 Aseki Morobe, C PNG 7.18S 156.16E
83 J14 Āsela var. Asella, Aselle, Asselle. Oromo, C Ethiopia 7.55N 39.08E
95 H15 Åsele Västerbotten, N Sweden 64.10N 17.19E
96 K12 Åsen Dalarna, C Sweden 61.18N 13.49E
116 J11 Asenovgrad prev. Stanimaka. Plovdiv, C Bulgaria 42.01N 24.54E
175 Q11 Asera Sulawesi, C Indonesia 3.24S 121.42E
97 E17 Åseral Vest-Agder, S Norway 58.37N 7.27E
120 J3 Aseri var. Asserien, Ger. Asserin. Ida-Virumaa, NE Estonia 59.28N 26.50E
42 J10 Aserradero Durango, W Mexico 23.47N 105.44W
97 H16 Åsgårdstrand Vestfold, S Norway 59.19N 10.28E
125 E11 Asha Chelyabinskaya Oblast', C Russian Federation 55.01N 57.11E
79 T4 Ashara var. Al 'Ashārah
141 O13 Ashburn Georgia, SE USA 31.42N 83.39W
193 G19 Ashburton Canterbury, South Island, NZ 43.55S 171.46E
193 G19 Ashburton ~ South Island, NZ
188 H8 Ashburton River ~ Western Australia
151 N10 Ashchysu ~ E Kazakhstan
8 M16 Ashcroft British Columbia, SW Canada 50.43N 121.20W
144 E10 Ashdod anc. Azotos, Lat. Azotus. Central, W Israel 31.48N 34.37E
29 S14 Ashdown Arkansas, C USA 33.40N 94.07W

29 T7 Ash Grove Missouri, C USA 37.19N 93.55W
171 Mm10 Ashikaga var. Asikaga. Tochigi, Honshū, S Japan 36.19N 139.26E
171 Mm10 Ashiro Iwate, Honshū, C Japan 40.04N 141.00E
170 E16 Ashizuri-misaki headland Shikoku, SW Japan 32.43N 132.59E
Ashkelon see Ashqelon
Ashkhabad see Ashgabat
25 Q4 Ashland Alabama, S USA 33.16N 85.50W
28 K7 Ashland Kansas, C USA 37.11N 99.46W
23 P5 Ashland Kentucky, S USA 38.28N 82.39W
21 S2 Ashland Maine, NE USA 46.36N 68.24W
24 M1 Ashland Mississippi, S USA 34.51N 89.10W
29 U4 Ashland Missouri, C USA 38.46N 92.15W
31 S15 Ashland Nebraska, C USA 41.01N 96.21W
33 T12 Ashland Ohio, N USA 40.52N 82.19W
34 G5 Ashland Oregon, NW USA 42.11N 122.42W
23 W4 Ashland Virginia, NE USA 37.45N 77.28W
32 K3 Ashland Wisconsin, N USA 46.34N 90.54W
27 I8 Ashland City Tennessee, S USA 36.16N 87.03W
191 S4 Ashley New South Wales, SE Australia 29.21S 149.49E
31 O7 Ashley North Dakota, N USA 46.00N 99.22W
181 W7 Ashmore and Cartier Islands ◇ Australian external territory E Indian Ocean
122 I14 Ashmyany Rus. Oshmyany. Hrodzyenskaya Voblasts', W Belarus 54.24N 25.57E
27 P6 Ashokan Reservoir ⊞ New York, NE USA
145 Y12 Ash Shāfī E Iraq 30.49N 47.30E
145 R4 Ash Shaddādah see Ash Shaddādī
Ash Shām/Ash Shām see Dimashq
145 T10 Ash Shāmīyah var. Shamiya. C Iraq 31.55N 44.37E
145 Y13 Ash Shāmīyah var. Al Bādiyah al Janūbīyah. desert S Iraq
145 T11 Ash Shanāfīyah var. Ash Shināfīyah. S Iraq 31.34N 44.38E
144 G13 Ash Sharā var. Esh Sharā. ▲ W Jordan
149 R16 Ash Shāriqah Eng. Sharjah. Ash Shāriqah, NE UAE 25.22N 55.28E
149 R16 Ash Shāriqah var. Sharjah. ✕ Ash Shāriqah, NE UAE 25.19N 55.37E
146 I4 Ash Sharmah var. Sharma. Tabūk, NW Saudi Arabia 28.01N 35.16E
145 R4 Ash Sharqāt NW Iraq 35.30N 43.15E
145 S10 Ash Sharqīyah off. Al Minṭaqah ash Sharqīyah, Eng. Eastern Region. ◆ province E Saudi Arabia
145 W11 Ash Shaṭrah var. Shatra. S Iraq 31.25N 46.10E
144 G13 Ash Shawbak Ma'ān, W Jordan 30.31N 35.34E
145 L5 Ash Shaykh Ibrāhīm Ḥimṣ, C Syria 35.03N 38.50E
147 O17 Ash Shaykh 'Uthmān SW Yemen 12.53N 45.00E
145 S15 Ash Shiḥr SE Yemen 14.45N 49.24E
Ash Shināfīyah see Ash Shanāfīyah
147 V12 Ash Shiṣar var. Shisur. SW Oman 18.13N 53.34E
145 S13 Ash Shubrum well S Iraq 30.09N 43.59E
147 R10 Ash Shuqqān desert E Saudi Arabia
145 O6 Ash Shuwayrif var. Ash Shwayrif. N Libya 29.54N 14.16E
Ash Shwayrif see Ash Shuwayrif
33 U10 Ashtabula Ohio, N USA 41.54N 80.46W
31 Q5 Ashtabula, Lake ⊞ North Dakota, N USA
143 T12 Ashtarak W Armenia 40.18N 44.22E
148 M6 Āshtiān var. Āshtiyān. Markazī, W Iran 34.24N 49.55E
Āshtiyān see Āshtiān
200 R15 Ata island Tongatapu Group, SW Tonga

143 P13 Aşkale Erzurum, NE Turkey 39.56N 40.39E
119 T11 Askaniya-Nova Khersons'ka Oblast', S Ukraine 46.27N 33.54E
97 H15 Asker Akershus, S Norway 59.49N 10.29E
97 L17 Askersund Örebro, C Sweden 58.55N 14.55E
Aski Kalak see Eski Kalak
97 J15 Askim Østfold, S Norway 59.34N 11.10E
131 V3 Askino Respublika Bashkortostan, W Russian Federation 56.07N 56.39E
117 D14 Åskio ▲ N Greece
158 L9 Askot Uttar Pradesh, N India 29.43N 80.19E
96 C12 Askvoll Sogn og Fjordane, S Norway 61.21N 5.04E
142 A13 Aslan Burnu headland W Turkey 38.44N 26.43E
142 L16 Aslantaş Barajı ⊞ S Turkey
155 S4 Asmār var. Bar Kunar. Kunar, E Afghanistan 34.58N 71.28E
82 I9 Asmara Amh. Āsmera. ● (Eritrea) C Eritrea 15.15N 38.57E
Åsmera see Asmara
97 L21 Åsnen ⊗ S Sweden
117 F19 Asopós ~ S Greece
176 X10 Asori Irian Jaya, E Indonesia 2.37S 136.06E
82 G12 Āsosa Benishangul, W Ethiopia 10.06N 34.27E
34 M10 Asotin Washington, NW USA 46.18N 117.03W
Aspadana see Eṣfahān
Aspang see Aspang Markt
111 X6 Aspang Markt var. Aspang. Niederösterreich, E Austria 47.34N 16.07E
107 S12 Aspe País Valenciano, E Spain 38.21N 0.43W
39 R5 Aspen Colorado, C USA 39.12N 106.49W
27 P6 Aspermont Texas, SW USA 33.07N 100.13W
Asphaltites, Lacus see Dead Sea
Aspinwall see Colón
193 C20 Aspiring, Mount ▲ South Island, NZ 44.21S 168.47E
117 B16 Asprókavos, Akrotírio headland Kérkyra, Iónioi Nísoi, Greece, C Mediterranean Sea 39.22N 20.07E
Aspropótamos see Acheloös
78 I10 Assaba var. Açaba. ◆ region S Mauritania
144 L4 Assabkah var. Sabkha. Ar Raqqah, N Syria 35.33N 39.07E
145 U6 As Sa'dīyah E Iraq 34.11N 45.09E
144 I10 Assad var. Assad, Buḥayrat al
148 J8 Aş Şafā ▲ S Syria 33.03N 37.07E
144 I10 Aş Şafāwī Al Mafraq, N Jordan 32.12N 32.30E
Aş Şaff see El Şaff
145 N2 Aş Şafīḥ Al Ḥasakah, N Syria 36.42N 40.12E
147 Q4 As Sālimī var. Salemy. SW Kuwait 29.07N 46.41E
69 W7 'Assal, Lac ⊗ C Djibouti
144 I5 As Sallūm see Salūm
145 G10 As Salṭ var. Salt. N Jordan 32.03N 35.43E
145 N2 As Salwá var. Salwa, Salwah. S Qatar 24.43N 50.52E
159 V12 Assam ◆ state NE India
155 V9 Assamaka var. Assamakka. Agadez, NW Niger 19.24N 5.52E
145 U11 As Samāwah var. Samawa. S Iraq 31.17N 45.05E
As Saqia al Hamra see Saguia al Hamra
145 J4 As Sa'rān Ḥamāh, C Syria 35.33N 37.28E
144 G9 As Sariḥ Irbid, N Jordan 32.31N 35.54E
23 Z5 Assateague Island island Maryland, NE USA
145 O6 As Sayyāl var. Sayyāl. Dayr az Zawr, E Syria 34.43N 40.52E
101 G18 Asse Vlaams Brabant, C Belgium 50.55N 4.12E
101 D16 Assebroek West-Vlaanderen, NW Belgium 51.12N 3.16E
Asselle see Āsela
109 C20 Assemini Sardegna, Italy, C Mediterranean Sea 39.16N 8.58E
100 N7 Assen Drenthe, NE Netherlands 53.00N 6.34E
101 E16 Assenede Oost-Vlaanderen, NW Belgium 51.15N 3.43E
97 G24 Assens Fyn, C Denmark 55.16N 9.54E
101 I21 Assesse Namur, SE Belgium 50.22N 5.01E
147 Y8 As Sib var. Seeb. NE Oman 23.40N 58.03E
145 Z13 As Sibah var. Sibah. SE Iraq 30.13N 47.24E
11 T7 Assiniboia Saskatchewan, S Canada 49.39N 105.58W
11 V5 Assiniboine ~ Manitoba, S Canada
1 P16 Assiniboine, Mount ▲ Alberta/British Columbia, SW Canada 50.54N 115.43W
76 K12 Assiout see Asyūṭ
62 J9 Assis São Paulo, S Brazil 22.37S 50.25W
108 I13 Assisi Umbria, C Italy 43.04N 12.36E
Assiut see Asyūṭ
79 R16 Assling see Jesenice
76 K11 Assouan see Aswān
59 Y8 Assu see Açu
Assuan see Aswān
145 K12 Aş Şukhnah var. Sukhne, Fr. Soukhné. Ḥimṣ, C Syria 34.55N 38.52E
147 X8 As Sufāl S Yemen 14.06N 48.42E
147 U4 As Sulaymānīyah, Kurd. Slēmānī. NE Iraq 35.31N 45.27E
37 T12 Atascadero California, W USA 35.28N 120.40W

147 P11 As Sulayyil Ar Riyāḍ, S Saudi Arabia 20.28N 45.33E
145 U5 As Sulṭān N Libya 31.01N 17.21E
147 Q5 Aş Şummān desert N Saudi Arabia
145 U8 As Şuwār var. Şuwär. Dayr az Zawr, E Syria 35.31N 40.37E
144 H9 As Suwaydā' var. El Suweida, Es Suweida, Suweida, Fr. Soueida. As Suwaydā', S Syria 32.43N 36.33E
144 H9 As Suwaydā' off. Muḥāfaẓat as Suwaydā', var. Soueida, Suwaydā, Suweida, Fr. Soueida. ◆ governorate S Syria
147 T9 As Suwayḥ NE Oman 22.07N 59.42E
147 X8 As Suwayq var. Suwaik. N Oman 23.51N 57.28E
145 T8 As Şuwayrah var. Suwaira. E Iraq 32.57N 44.46E
As Suways see Suez
Asta Colonia see Asti
117 M23 Astakída island SE Greece
151 Q9 Astana prev. Akmola, Akmolinsk, Tselinograd, Aqmola. ● (Kazakhstan) Akmola, N Kazakhstan 51.12N 71.25E
148 M3 Āstāneh Gīlān, NW Iran 37.16N 49.58E
Asta Pompeia see Asti
143 Y14 Astara S Azerbaijan 38.28N 48.51E
101 L15 Asten Noord-Brabant, SE Netherlands 51.24N 5.45E
108 C8 Asti anc. Asta Colonia, Asta Pompeia, Hasta Colonia, Hasta Pompeia. Piemonte, NW Italy 44.54N 8.10E
Astigi see Ecija
107 S12 Astipálaia see Astypálaia
154 L16 Astola Island island SW Pakistan
158 H4 Astor Jammu and Kashmir, NW India 35.21N 74.52E
106 K4 Astorga anc. Asturica Augusta. Castilla-León, N Spain 42.27N 6.04W
34 F10 Astoria Oregon, NW USA 46.12N 123.49W
1 F8 Astoria Fan undersea feature E Pacific Ocean
97 J22 Åstorp Skåne, S Sweden 56.09N 12.57E
Astrabad see Gorgān
131 Q13 Astrakhan' Astrakhanskaya Oblast', SW Russian Federation 46.20N 48.00E
Astrakhan-Bazar see Cälilabad
131 Q11 Astrakhanskaya Oblast' ◆ province SW Russian Federation
95 J15 Sträsk Västerbotten, N Sweden 64.38N 20.00E
Astrida see Butare
67 Q2 Astrid Ridge undersea feature S Atlantic Ocean
194 J11 Astrolabe Bay inlet N PNG
197 I4 Astrolabe, Récifs de l' reef C New Caledonia
124 Nn3 Astromerítis N Cyprus 35.09N 33.02E
117 F20 Ástros Pelopónnisos, S Greece 37.24N 22.43E
121 G16 Astryna Rus. Ostryna. Hrodzyenskaya Voblasts', W Belarus 53.44N 24.33E
Asturias ◆ autonomous community NW Spain
Asturias see Oviedo
Asturica Augusta see Astorga
117 L22 Astypálaia var. Astipálaia, It. Stampalia. island Kykládes, Greece, Aegean Sea
198 Aa8 Åsuisui, Cape headland Savai'i, W Samoa 13.43S 172.28W
205 S2 Asuka Japanese research station Antarctica 71.49S 23.43E
64 G8 Asunción ● (Paraguay) Central, S Paraguay 25.16S 57.36W
64 G8 Asunción ✕ Central, S Paraguay 25.15S 57.40W
196 K3 Asuncion Island island N Northern Mariana Islands
44 E6 Asunción Mita Jutiapa, SE Guatemala 14.19S 89.42W
Asunción Nochixtlán see Nochixtlán
43 P13 Asunción, Río ~ NW Mexico
97 M18 Åsunden ⊗ S Sweden
120 K11 Asveya Rus. Osveya. Vitsyebskaya Voblasts', N Belarus 56.00N 28.05E
Aswa see Achwa
77 X11 Aswān var. Assouan, Assuan; anc. Syene. SE Egypt 24.03N 32.58E
77 W9 Aswān High Dam dam SE Egypt 23.54N 32.51E
77 W8 Asyūṭ var. Assiout, Asiut, Siut; anc. Lycopolis. C Egypt 27.05N 31.10E
200 R15 Ata island Tongatapu Group, SW Tonga
202 H8 Atafu Atoll island NW Tokelau
202 H8 Atafu Village Atafu Atoll, NW Tokelau 8.40S 172.40W
76 K12 Atakor ▲ SE Algeria
79 R14 Atakora, Chaîne de l' var. Atakora Mountains. ▲ N Benin
Atakora Mountains see Atakora, Chaîne de l'
79 R16 Atakpamé C Togo 7.31N 1.07E
171 J13 Atami Shizuoka, Honshū, S Japan 35.04N 139.03E
171 X12 Atas Bogd ▲ SW Mongolia 43.17N 96.41E
37 S12 Atasu Sulaimaniya, Kurd. Slēmānī. NE Iraq 35.31N 45.27E
37 T12 Atascadero California, W USA 35.28N 120.40W

27 S13 Atascosa River ~ Texas, SW USA
151 R8 Atasu Zhezkazgan, C Kazakhstan 48.42N 71.37E
151 R12 Atasu ~ C Kazakhstan
200 Qq15 Atata island Tongatapu Group, S Tonga
142 H10 Atatürk ✕ (İstanbul) İstanbul, NW Turkey 40.58N 28.49E
143 N16 Atatürk Barajı ⊞ S Turkey
Atax see Aude
82 G8 Atbara var. 'Aṭbāra. River Nile, NE Sudan 17.42N 34.00E
82 H8 Atbara var. Nahr 'Aṭbārah. ~ Eritrea/Sudan
'Aṭbāra/'Aṭbarah, Nahr see Atbara
151 P9 Atbasar Akmola, N Kazakhstan 51.49N 68.18E
153 W3 At-Bashy var. At-Bashy. Narynskaya Oblast', C Kyrgyzstan 41.07N 75.48E
24 I8 Atchafalaya Bay bay Louisiana, S USA
24 I8 Atchafalaya River ~ Louisiana, S USA
Atchin see Aceh
29 Q3 Atchison Kansas, C USA 39.31N 95.07W
79 Q16 Atebubu C Ghana 7.47N 1.00W
107 Q6 Ateca Aragón, NE Spain 41.19N 1.49W
42 K11 Atengo, Río ~ C Mexico
Aternum see Pescara
109 K15 Atessa Abruzzo, C Italy 42.03N 14.25E
Ateste see Este
101 E19 Ath var. Aat. Hainaut, SW Belgium 50.37N 3.46E
11 Q13 Athabasca Alberta, SW Canada 54.43N 113.15W
11 Q12 Athabasca var. Athabaska. ~ Alberta, SW Canada
1 R10 Athabasca, Lake ⊗ Alberta/Saskatchewan, SW Canada
117 C16 Athamánon ▲ C Greece
97 F17 Athboy Ir. Baile Átha Buí. E Ireland 53.37N 6.54W
Athenae see Athens
99 C18 Athenry Ir. Baile Átha an Rí. W Ireland 53.19N 8.49W
25 S3 Athens Alabama, S USA 34.48N 86.58W
23 U12 Athens Georgia, SE USA 33.57N 83.24W
33 T14 Athens Ohio, N USA 39.19N 82.06W
22 M10 Athens Tennessee, S USA 35.26N 84.35W
27 V7 Athens Texas, SW USA 32.12N 95.51W
Athens see Athína
117 D19 Athéras, Akrotírio headland Kefallinía, Iónioi Nísoi, Greece, C Mediterranean Sea 38.20N 20.24E
189 W4 Atherton Queensland, NE Australia 17.18S 145.30E
83 J5 Athi ~ S Kenya
124 O3 Athiénou SE Cyprus 35.01N 33.31E
117 H19 Athína Eng. Athens; prev. Athínai, anc. Athenae. ● (Greece) Attikí, C Greece 37.58N 23.44E
Athínai see Athína
145 S10 Athiyah C Iraq 32.01N 44.04E
99 D18 Athlone Ir. Baile Átha Luain. C Ireland 53.25N 7.55W
161 F16 Athni Karnātaka, W India 16.43N 75.04E
193 C23 Athol Southland, South Island, NZ 45.30S 168.35E
21 N11 Athol Massachusetts, NE USA 42.35N 72.11W
117 I15 Áthos ▲ NE Greece 40.10N 24.21E
117 I15 Athos, Mount var. Ágion Óros, C Greece
147 P5 Ath Thumāmī spring/well N Saudi Arabia 27.56N 45.06E
101 L25 Athus Luxembourg, SE Belgium 49.34N 5.49E
99 E19 Athy Ir. Baile Átha Í. C Ireland 52.58N 6.58W
80 I10 Ati C Chad 13.13N 18.18E
83 E18 Atiak NW Uganda 3.13N 32.04E
59 G17 Atico Arequipa, SW Peru 16.13S 73.13W
107 O6 Atienza Castilla-La Mancha, C Spain 41.12N 2.52W
41 Q6 Atigun Pass Alaska, USA 68.01N 149.36W
10 B12 Atikokan Ontario, S Canada 48.45N 91.37W
11 O9 Atikonak Lac ⊗ Newfoundland, E Canada
44 C6 Atitlán, Lago de ⊗ W Guatemala
202 L16 Atiu island S Cook Islands
127 O9 Atka Magadanskaya Oblast', E Russian Federation 60.45N 151.34E
40 H17 Atka Atka Island, Alaska, USA 52.12N 174.13W
40 H17 Atka Island island Aleutian Islands, Alaska, USA
131 O7 Atkarsk Saratovskaya Oblast', W Russian Federation 52.15N 45.00E
29 U11 Atkins Arkansas, C USA 35.15N 92.56W
31 O13 Atkinson Nebraska, C USA 42.31N 98.57W
176 V13 Atkri Irian Jaya, E Indonesia 1.45S 130.04E
43 O13 Atlacomulco var. Atlacomulco de Fabela. México, C Mexico 19.48N 99.52W
Atlacomulco de Fabela see Atlacomulco
25 S5 Atlanta state capital Georgia, SE USA 33.45N 84.23W
33 R6 Atlanta Michigan, N USA 45.01N 84.07W
27 X6 Atlanta Texas, SW USA 33.06N 94.09W
31 T15 Atlantic Iowa, C USA 41.24N 95.00W
21 Y10 Atlantic North Carolina, SE USA 34.52N 76.20W
20 J17 Atlantic City New Jersey, NE USA 39.21N 74.25W
141 N11 Atlantic Beach Florida, SE USA 30.19N 81.24W
180 L14 Atlantic-Indian Basin undersea feature SW Indian Ocean

◆ COUNTRY ◇ DEPENDENT TERRITORY ◆ ADMINISTRATIVE REGION ▲ MOUNTAIN ▲ VOLCANO ⊗ LAKE
● COUNTRY CAPITAL ○ DEPENDENT TERRITORY CAPITAL ✕ INTERNATIONAL AIRPORT ▲ MOUNTAIN RANGE ~ RIVER ⊞ RESERVOIR

180 K13 **Atlantic-Indian Ridge** *undersea feature* SW Indian Ocean
56 E4 **Atlántico** *off.* Departamento del Atlántico. ◆ *province* NW Colombia
66-67 **Atlantic Ocean** *ocean*
44 K7 **Atlántico Norte, Región Autónoma** *prev.* Zelaya Norte. ◆ *autonomous region* NE Nicaragua
44 L10 **Atlántico Sur, Región Autónoma** *prev.* Zelaya Sur. ◆ *autonomous region* SE Nicaragua
44 I5 **Atlántida** ◆ *department* N Honduras
79 Y15 **Atlantika Mountains** ▲ E Nigeria
66 J10 **Atlantis Fracture Zone** *tectonic feature* N Atlantic Ocean
76 H7 **Atlas Mountains** ▲ NW Africa
127 Pp13 **Atlasova, Ostrov** *island* SE Russian Federation
127 Pp10 **Atlasovo** Kamchatskaya Oblast', E Russian Federation 55.42N 159.34E
123 H13 **Atlas Saharien** *var.* Saharan Atlas. ▲ Algeria/Morocco
123 Gg10 **Atlas, Tell** *see* Atlas Tellien
Atlas Tellien *Eng.* Tell Atlas. ▲ N Algeria
8 I9 **Atlin** British Columbia, W Canada 59.31N 133.40W
8 I9 **Atlin Lake** ◎ British Columbia, W Canada
43 P14 **Atlixco** Puebla, S Mexico 18.55N 98.25W
96 A11 **Atløyna** *island* S Norway
161 I17 **Ātmakūr** Andhra Pradesh, C India 15.52N 78.42E
25 O8 **Atmore** Alabama, S USA 31.01N 87.29W
103 J20 **Atmühl** ≈ S Germany
96 H11 **Atna** ≈ S Norway
170 E12 **Atō** Yamaguchi, Honshū, SW Japan 34.24N 131.42E
52 J21 **Atocha** Potosí, S Bolivia 20.55S 66.13W
29 P12 **Atoka** Oklahoma, C USA 34.23N 96.07W
29 O12 **Atoka Lake** *var.* Atoka Reservoir. ◎ Oklahoma, C USA
Atoka Reservoir *see* Atoka Lake
35 Q14 **Atomic City** Idaho, NW USA 43.26N 112.48W
42 L10 **Atotonilco** Zacatecas, C Mexico 24.12N 102.46W
42 M13 **Atotonilco el Alto** *var.* Atotonilco. Jalisco, SW Mexico 20.32N 102.27W
79 N7 **Atouila, 'Erg** *desert* N Mali
43 N16 **Atoyac** *var.* Atoyac de Alvarez. Guerrero, S Mexico 17.10N 100.27W
Atoyac de Alvarez *see* Atoyac
43 P15 **Atoyac, Río** ≈ S Mexico
41 O5 **Atqasuk** Alaska, USA 70.28N 157.24W
152 C13 **Atrak** *Per.* Rūd-e Atrak, *Rus.* Atrek, *Turkm.* Etrek.
Atrak, Rūd-e *see* Atrak ≈ Iran/Turkmenistan
97 J20 **Ätran** ≈ S Sweden
56 C7 **Atrato, Río** ≈ NW Colombia
Atrek *see* Atrak
109 K14 **Atri** Abruzzo, C Italy 42.34N 13.58E
Atria *see* Adria
171 Ij16 **Atsugi** *var.* Atugi. Kanagawa, Honshū, S Japan 35.27N 139.21E
171 L12 **Atsumi** Yamagata, Honshū, C Japan 38.38N 139.34E
172 Oo4 **Atsuta** Hokkaidō, NE Japan 43.28N 141.24E
176 Y13 **Atsy** Irian Jaya, E Indonesia 5.40S 138.19E
149 Q17 **Aṭ Ṭaff** *desert* C UAE
144 G12 **Aṭ Ṭafīlah** *var.* Eṭ Ṭafīla, Tafila. Aṭ Ṭafīlah, W Jordan 30.52N 35.36E
144 G12 **Aṭ Ṭafīlah** *off.* Muḥāfaẓat aṭ Ṭafīlah. ◆ *governorate* W Jordan
146 L10 **Aṭ Ṭā'if** Makkah, W Saudi Arabia 21.49N 40.49E
Attalea/Attalia *see* Antalya
25 Q3 **Attalla** Alabama, S USA 34.01N 86.05W
144 L2 **At Tall al Abyaḍ** *var.* Tall al Abyaḍ, Tell Abyad, *Fr.* Tell Abiad. Ar Raqqah, N Syria 36.36N 34.00E
144 L7 **Aṭ Ṭanf** Ḥimṣ, S Syria 33.29N 38.39E
Attapu *see* Samakhixai
145 S10 **Aṭ Ṭaqṭaqānah** C Iraq 32.03N 43.54E
117 O23 **Attávytos** ▲ Ródos, Dodekánisos, Greece, Aegean Sea 36.10N 27.47E
145 V15 **At Tawal** *desert* Iraq/Saudi Arabia
10 G9 **Attawapiskat** Ontario, C Canada 52.55N 82.25W
10 F9 **Attawapiskat** ≈ Ontario, S Canada
10 D9 **Attawapiskat Lake** ◎ Ontario, C Canada
At Taybé *see* Ṭayyibah
103 F16 **Attendorn** Nordrhein-Westfalen, W Germany 51.07N 7.54E
111 R5 **Attersee** Salzburg, NW Austria 47.55N 13.31E
111 R5 **Attersee** ◎ N Austria
101 L24 **Attert** Luxembourg, SE Belgium 49.45N 5.47E
155 U6 **Attock City** Punjab, E Pakistan 33.52N 72.19E
Attopeu *see* Samakhixai
X8 **Attoyac River** ≈ Texas, SW USA
40 C10 **Attu** Attu Island, Alaska, USA 52.53N 173.18E
145 Y12 **Aṭ Ṭubah** I Iraq 30.29N 47.28E
146 K4 **Aṭ Ṭubayq** *plain* Jordan/Saudi Arabia
40 C10 **Attu Island** *island* Aleutian Islands, Alaska, USA
Aṭ Ṭūr *see* El Ṭūr

161 I21 **Āttūr** Tamil Nādu, SE India 11.34N 78.39E
147 N17 **Aṭ Turbah** SW Yemen 12.42N 43.31E
64 I12 **Atuel, Río** ≈ C Argentina
203 X7 **Atuona** Hiva Oa, NE French Polynesia 9.46S 139.03W
Aturus *see* Adour
97 M18 **Åtvidaberg** Östergötland, S Sweden 58.12N 16.00E
37 P9 **Atwater** California, W USA 37.19N 120.33W
31 T8 **Atwater** Minnesota, N USA 45.08N 94.48W
28 I2 **Atwood** Kansas, C USA 39.48N 101.02W
33 U12 **Atwood Lake** ◎ Ohio, N USA
131 P5 **Atyashevo** Mordoviya, W Russian Federation 54.36N 46.04E
150 F12 **Atyrau** *prev.* Gur'yev. Atyrau, W Kazakhstan 47.07N 51.55E
150 E11 **Atyrau** *off.* Atyrauskaya Oblast', *var. Kaz.* Atyraū Oblysy; *prev.* Gur'yevskaya Oblast'. ◆ *province* W Kazakhstan
Atyrau Oblysy/Atyrauskaya Oblast' *see* Atyrau
110 J7 **Au** Vorarlberg, NW Austria 47.19N 10.01E
194 G8 **Aua Island** *island* NW PNG
105 S16 **Aubagne** *anc.* Albania. Bouches-du-Rhône, SE France 43.16N 5.34E
101 L25 **Aubange** Luxembourg, SE Belgium 49.34N 5.48E
105 Q6 **Aube** ◆ *department* N France
105 R6 **Aube** ≈ N France
101 L19 **Aubel** Liège, E Belgium 50.45N 5.49E
105 S18 **Aubenas** Ardèche, E France 44.37N 4.24E
105 O13 **Aubigny-sur-Nère** Cher, C France 47.30N 2.27E
105 O13 **Aubin** Aveyron, S France 44.30N 2.18E
105 Q13 **Aubrac, Monts d'** ▲ S France
68 J3 **Aubrey Cliffs** *cliff* Arizona, SW USA
25 R5 **Auburn** Alabama, S USA 32.37N 85.30W
37 P6 **Auburn** California, W USA 38.53N 121.03W
32 K14 **Auburn** Illinois, N USA 39.35N 89.45W
33 Q11 **Auburn** Indiana, N USA 41.22N 85.03W
21 J7 **Auburn** Kentucky, S USA 36.52N 86.42W
21 P8 **Auburn** Maine, NE USA 44.05N 70.15W
21 N11 **Auburn** Massachusetts, NE USA 42.11N 71.47W
31 S16 **Auburn** Nebraska, C USA 40.23N 95.50W
20 H10 **Auburn** New York, NE USA 42.55N 76.31W
34 H8 **Auburn** Washington, NW USA 47.18N 122.13W
105 N11 **Aubusson** Creuse, C France 45.58N 2.10E
123 J12 **Aurès, Massif de l'** ▲ NE Algeria
105 O10 **Auce** *Ger.* Autz. Dobele, SW Latvia 56.28N 22.54E
104 L15 **Auch** Lat. Augusta Auscorum, Elimberrum. Gers, S France 43.39N 0.37E
79 U16 **Auchi** Edo, S Nigeria 7.01N 6.17E
25 T9 **Aucilla River** ≈ Florida/Georgia, SE USA
192 L6 **Auckland** Auckland, North Island, NZ 36.53S 174.46E
192 K5 **Auckland** *off.* Auckland Region. ◆ *region* North Island, NZ
192 L6 **Auckland ✕** Auckland, North Island, NZ 37.01S 174.49E
199 Ii5 **Auckland Islands** *island group* S NZ
105 O16 **Aude** ◆ *department* S France
105 N16 **Aude** *anc.* Atax. ≈ S France
29 R16 **Audenarde** *see* Oudenaarde
104 E6 **Audern** *see* Audru
104 E6 **Audierne, Baie d'** *bay* NW France
104 E6 **Audincourt** Doubs, E France 47.28N 6.49E
120 G5 **Audru** *Ger.* Audern. Pärnumaa, SW Estonia 58.25N 24.21E
31 T14 **Audubon** Iowa, C USA 41.44N 94.56W
103 N17 **Aue** Sachsen, E Germany 50.34N 12.42E
102 F12 **Aue** ≈ N Germany
102 L9 **Auerbach** Bayern, SE Germany 49.41N 11.41E
103 M17 **Auerbach** Sachsen, E Germany 50.30N 12.24E
110 I10 **Auererrhein** ≈ SW Switzerland
81 E9 **Auersberg** ▲ E Germany 50.30N 12.42E
57 H16 **Ausangate, Nevado** ▲ C Peru 13.46S 71.13W
66 G12 **Aves Ridge** *undersea feature* SE Caribbean Sea
57 M14 **Avesta** Dalarna, C Sweden 60.09N 16.10E
105 O14 **Aveyron** ◆ *department* S France
105 N14 **Aveyron** ≈ S France
109 J15 **Avezzano** Abruzzo, C Italy 42.01N 13.25E
59 D16 **Avgó** ≈ C Greece 39.31N 21.24E
47 D16 **Austin** Texas, SW USA 31.09N 95.39W
33 P15 **Austin** Indiana, N USA 38.45N 85.48W
31 W11 **Austin** Minnesota, N USA 43.40N 92.58W
37 U5 **Austin** Nevada, W USA 39.28N 117.04W
193 Z12 **Austin** *state capital* Texas, SW USA 30.16N 97.44W
109 L25 **Augusta** *It.* Agosta. Sicília, Italy, C Mediterranean Sea 37.19N 15.13E
29 W11 **Augusta** Arkansas, C USA 35.16N 91.22W
23 V5 **Augusta** Georgia, SE USA 33.29N 81.58W
27 J6 **Augusta** Kansas, C USA 37.40N 96.59W
21 Q7 **Augusta** *state capital* Maine, NE USA 44.19N 69.44W
35 Q8 **Augusta** Montana, NW USA 47.28N 112.23W

Augusta Treverorum *see* Trier
Augusta Vangionum *see* Worms
Augusta Vindelicorum *see* Augsburg
97 G24 **Augustenborg** *Ger.* Augustenburg. Sønderjylland, SW Denmark 54.57N 9.52E
Augustenburg *see* Augustenborg
41 Q13 **Augustine Island** *island* Alaska, USA
12 L9 **Augustines, Lac des** ◎ Quebec, SE Canada
Augustobona Tricassium *see* Troyes
Augustodunum *see* Autun
Augustodurum *see* Bayeux
Augustoritum Lemovicensium *see* Limoges
112 O8 **Augustów** *Rus.* Avgustov. Podlaskie, NE Poland 53.51N 22.58E
Augustow, Kanał *see* Augustowski, Kanał
112 O8 **Augustowski, Kanał** *Eng.* Augustow Canal, *Rus.* Avgustovskiy Kanal. *canal* NE Poland
188 I9 **Augustus, Mount** ▲ Western Australia 24.42S 117.42E
195 X15 **Auki** Malaita, N Solomon Islands 8.48S 160.45E
23 W8 **Aulander** North Carolina, SE USA 36.15N 77.16W
188 L7 **Auld, Lake** *salt lake* Western Australia
Aulie Ata/Auliye-Ata *see* Taraz
104 I10 **Aulla** Toscana, C Italy 44.15N 10.00E
104 F6 **Aulne** ≈ NW France
39 T3 **Aulong** *see* Ulong
105 N3 **Ault** Colorado, C USA 40.34N 104.43W
105 N2 **Aumale** Seine-Maritime, N France 49.45N 1.45E
105 P9 **Aumont-Aubrac** *anc.* Auxumontis. Lozère, S France 44.43N 3.17E
42 K14 **Autlán** *var.* Autlán de Navarro. Jalisco, SW Mexico 19.46N 104.22W
Autlán de Navarro *see* Autlán
Autricum *see* Chartres
105 Q9 **Autun** *anc.* Ædua, Augustodunum. Saône-et-Loire, C France 46.57N 4.18E
Autz *see* Auce
29 U8 **Auvelais** Namur, S Belgium 50.27N 4.37E
105 P11 **Auvergne** ◆ *region* C France
105 M12 **Auvézère** ≈ N France
105 P7 **Auxerre** *anc.* Autesiodorum, Autissiodorum. Yonne, C France 47.48N 3.34E
105 N2 **Auxi-le-Château** Pas-de-Calais, N France 50.14N 2.06E
105 S8 **Auxonne** Côte d'Or, C France
57 P9 **Auyan Tepuy** ≈ SE Venezuela 5.48N 62.27W
202 H16 **Auzances** Creuse, C France 46.01N 2.29E
29 U8 **Ava** Missouri, C USA 36.57N 92.39W
97 C15 **Avaldsnes** Rogaland, S Norway 59.21N 5.16E
104 K6 **Avallon** Yonne, C France 47.30N 3.54E
37 S16 **Avalon** Santa Catalina Island, California, USA 33.20N 118.19W
20 J17 **Avalon** New Jersey, NE USA 39.04N 74.42W
11 V13 **Avalon Peninsula** *peninsula* Newfoundland, E Canada
207 Q11 **Avannaarsua** ◆ *province* N Greenland
62 K10 **Avaré** São Paulo, S Brazil 23.06S 48.57W
202 H16 **Avarua** ○ (Cook Islands) Rarotonga, S Cook Islands 21.12S 159.46W
202 H16 **Avarua Harbour** *harbor* Rarotonga, S Cook Islands 21.12S 159.46W
Avasfelsófalu *see* Negreşti-Oaş
54 L17 **Avatanak Island** *island* Aleutian Islands, Alaska, USA
202 B16 **Avatele** S Niue 19.06S 169.55E
202 H16 **Avatiu** Rarotonga, S Cook Islands 21.12S 159.46W
202 H15 **Avatiu Harbour** *harbor* Rarotonga, S Cook Islands 21.11S 159.46W
116 J13 **Avdeyevka** *see* Avdiyivka
116 J13 **Ávdira** Anatolikí Makedonía kai Thráki, NE Greece 40.58N 24.58E
119 X8 **Avdiyivka** *Rus.* Avdeyevka. Donets'ka Oblast', SE Ukraine 48.05N 37.45E
168 K7 **Avdzaga** C Mongolia 47.43N 103.30E
106 G6 **Ave** ≈ N Portugal
106 G7 **Aveiro** *anc.* Talabriga. Aveiro, W Portugal 40.37N 8.40W
106 G7 **Aveiro** ◆ *district* N Portugal
Avela *see* Ávila
63 D20 **Avellaneda** Buenos Aires, E Argentina 34.43S 58.23W
109 L17 **Avellino** *anc.* Abellinum. Campania, S Italy 40.54N 14.46E
37 Q12 **Avenal** California, W USA 36.00N 120.07W
Avenio *see* Avignon
98 I8 **Averøya** *island* S Norway
109 K17 **Aversa** Campania, S Italy 40.58N 14.10E
35 W9 **Avery** Idaho, NW USA 47.14N 115.48W
47 W5 **Avery** Texas, SW USA 33.33N 94.46W
Aves, Islas de *see* Las Aves, Islas
Avesnes *see* Avesnes-sur-Helpe
105 Q2 **Avesnes-sur-Helpe** *var.* Avesnes. Nord, N France 50.07N 3.57E

189 O7 **Australia** *off.* Commonwealth of Australia. ◆ *commonwealth republic*
182 M8 **Australia** *continent*
191 Q12 **Australian Alps** ▲ SE Australia
191 R11 **Australian Capital Territory** *prev.* Federal Capital Territory. ◆ *territory* SE Australia
Australie, Bassin Nord de l' *see* North Australian Basin
Austral Islands *see* Australes, Iles
Austrava *see* Ostrov
111 T6 **Austria** *off.* Republic of Austria, *Ger.* Österreich. ◆ *republic* C Europe
94 K3 **Austurland** ◆ *region* SE Iceland
94 G10 **Austvågøya** *island* C Norway
60 G13 **Autazes** Amazonas, N Brazil 3.37S 59.07W
104 M16 **Auterive** Haute-Garonne, S France 43.22N 1.29E
105 N2 **Authie** ≈ N France
Autissiodorum *see* Auxerre
147 X8 **Awābī** *var.* Al 'Awābī. NE Oman 23.19N 57.34E
170 G15 **Awaji-shima** *island* SW Japan
192 L9 **Awaküno** Waikato, North Island, NZ 38.40S 174.37E
148 J15 **'Awālī** C Bahrain 26.06N 50.33E
101 K19 **Awans** Liège, E Belgium 50.39N 5.30E
192 I2 **Awanui** Northland, North Island, NZ 35.03S 173.15E
154 M14 **Awārān** Baluchistan, SW Pakistan 26.31N 65.10E
83 K16 **Awara Plain** *plain* NE Kenya
82 M13 **Awarē** Somali, E Ethiopia 8.12N 44.09E
144 M6 **'Awārijī, Wādī** *dry watercourse* E Syria
83 J14 **Awasa** Southern, S Ethiopia 6.54N 38.26E
82 K13 **Awash** Afar, NE Ethiopia 8.59N 40.15E
82 K12 **Awash var.** Hawash. ≈ C Ethiopia
171 Kk13 **Awa-shima** *island* C Japan
82 L5 **Awaso** *see* Awaaso
164 H7 **Awat** Xinjiang Uygur Zizhiqu, NW China 40.36N 80.22E
193 J15 **Awatere** ≈ South Island, NZ
77 O10 **Awbārī** SW Libya 26.34N 12.42E
77 N9 **Awbārī, Idhān** *var.* Edeyen d'Oubari. *desert* SW Libya
82 C13 **Aweil** Northern Bahr el Ghazal, SW Sudan 8.44N 27.25E
98 H11 **Awe, Loch** ◎ W Scotland, UK
79 U16 **Awka** Anambra, SW Nigeria 6.12N 7.04E
41 O6 **Awuna River** ≈ Alaska, USA
Awwinorm *see* Avinurme
Ax *see* Dax
Axarfjördhur *see* Öxarfjördhur
105 N17 **Axat** Aude, S France 42.46N 2.13E
101 F16 **Axel** Zeeland, SW Netherlands 51.16N 3.55E
207 P9 **Axel Heiberg Island** *var.* Axel Heiburg. *island* Nunavut, N Canada
Axel Heiburg *see* Axel Heiberg Island
79 O17 **Axim** S Ghana 4.52N 2.13W
116 F13 **Axiós** *var.* Vardar. ≈ Greece/FYR Macedonia *see also* Vardar
105 N17 **Ax-les-Thermes** Ariège, S France 42.43N 1.49E
43 P16 **Axutla** *var.* Ayutla de los Libres. Guerrero, S Mexico 16.56N 99.22W
Ayutla de los Libres *see* Axutla
122 F13 **Ayachi, Jbel** ▲ C Morocco 32.30N 5.00W
63 D22 **Ayacucho** Buenos Aires, E Argentina 37.09S 58.30W
59 E16 **Ayacucho** Ayacucho, S Peru 13.05S 74.15W
59 E16 **Ayacucho** *off.* Departamento de Ayacucho. ◆ *department* S Peru
151 W11 **Ayagoz** *var.* Ayaguz, *Kaz.* Ayaköz; *prev.* Sergiopol. Vostochnyy Kazakhstan, E Kazakhstan 47.54N 80.25E
151 V12 **Ayagoz** *var.* Ayaguz, *Kaz.* Ayaköz. ≈ E Kazakhstan
Ayaguz *see* Ayagoz
Ayakagytma *see* Oyoqitghma
Ayakkuduk *see* Oyoqquduq
164 L10 **Ayakkum Hu** ◎ NW China
34 H6 **Ayakka** ≈ N Turkey
59 I16 **Ayaviri** Puno, S Peru 14.52S 70.34W
155 P3 **Āybak** *var.* Aibak, Haibak; *prev.* Samangān. Samangān, NE Afghanistan 36.16N 68.04E
153 N10 **Aydarkŭl** *Rus.* Ozero Aydarkul'. ◎ C Uzbekistan
Aydarkul', Ozero *see* Aydarkŭl
23 W13 **Ayden** North Carolina, SE USA 11.41N 10.00E
114 C15 **Aydın** Aydın, SW Turkey 37.51N 27.51E
114 C15 **Aydın** *var.* Aïdin; *anc.* Tralles, Tralles Aydın. ◆ *province* SW Turkey
142 L7 **Aydıncık** İçel, S Turkey 36.10N 33.16E
139 I2 **Aydın Dağları** ▲ W Turkey
164 I6 **Aydingkol Hu** ◎ NW China
131 X7 **Aydyrlinskiy** Orenburgskaya Oblast', W Russian Federation 52.03N 59.54E
79 W13 **Ayétoro** see Aiyétoro
131 M19 **Ayaguz** *see* Ayagoz

31 T14 **Avoca** Iowa, C USA 41.27N 95.20W
190 M11 **Avoca River** ≈ Victoria, SE Australia
31 P12 **Avon** South Dakota, N USA 43.00N 98.03W
20 F10 **Avon** New York, NE USA 42.53N 77.41W
99 M23 **Avon** ≈ S England, UK
99 L20 **Avon** ≈ C England, UK
68 K13 **Avondale** Arizona, SW USA 33.25N 112.20W
23 Z13 **Avon Park** Florida, SE USA 27.36N 81.30W
105 O4 **Avranches** Manche, N France 48.42N 1.21N
195 X16 **Avuavu** *var.* Kolotambu. Guadalcanal, C Solomon Islands 9.52S 160.25E
79 O17 **Awaaso** *var.* Awaso. SW Ghana 6.10N 2.18W
170 G15 **Awaji-shima** *island* SW Japan
147 X8 **Awābī** *var.* Al 'Awābī. NE Oman 23.19N 57.34E
145 V12 **'Ayn al 'Arab** Ḥalab, N Syria 36.55N 38.21E
145 V12 **'Ayn Ḥamūd** S Iraq 30.51N 45.37E
153 P12 **Ayni** *prev.* Varzimanor Ayni. W Tajikistan 39.24N 68.30E
148 M10 **'Aynin** var. Aynin. *spring/well* NW Saudi Arabia 20.52N 41.41E
23 U12 **Aynor** South Carolina, S USA 33.59N 79.11W
145 Q7 **'Ayn Zāzūn** Iraq 33.29N 42.34E
159 N12 **Ayodhya** Uttar Pradesh, N India 26.46N 82.12E
127 O5 **Ayon, Ostrov** *island* NE Russian Federation
107 R11 **Ayora** País Valenciano, E Spain 39.04N 1.04W
81 E16 **Ayorou** Tillabéri, W Niger 14.44N 0.54E
78 G8 **Ayos** Centre, S Cameroon 3.52N 12.31E
78 L5 **'Ayoûn 'Abd al Mâlek** *well* N Mauritania 24.51N 7.38W
78 K10 **'Ayoûn el 'Atroûs** *var.* Aïoun el Atrous, Aïoun el Atroûss. Hodh el Gharbi, SE Mauritania 16.37N 9.36W
103 U5 **Ayr** W Scotland, UK 55.28N 4.37W
98 I13 **Ayr** W Scotland, UK
98 I13 **Ayrshire** *cultural region* SW Scotland, UK
Aysen *see* Aisén
82 J13 **Aysha** Somali, E Ethiopia 10.36N 42.31E
150 L14 **Ayteke Bi** *Kaz.* Zhangaqazaly *prev.* Novokazalinsk. Kyzylorda, SW Kazakhstan 45.52N 62.09E
114 M12 **Aytos** Burgas, E Bulgaria 42.43N 27.13E
176 U7 **Ayu, Kepulauan** *island group* E Indonesia
175 O8 **Ayu, Tanjung** *headland* Borneo, N Indonesia 0.25N 117.34E
41 P16 **Ayutla** *var.* Ayutla de los Libres. Guerrero, S Mexico 16.56N 99.22W
178 H11 **Ayutthaya** *var.* Phra Nakhon Si Ayutthaya. Phra Nakhon Si Ayutthaya, C Thailand 14.19N 100.34E
147 N15 **Ayuttiyah** W Yemen 15.19N 43.03E
76 J11 **Azahar, Costa del** *coastal region* E Spain
106 F10 **Azambuja** Lisboa, C Portugal 39.04N 8.52W
159 N13 **Azamgarh** Uttar Pradesh, N India 26.03N 83.10E
79 O9 **Azaouâd** *desert* C Mali
79 S10 **Azaouagh, Vallée de l'** *var.* W Niger
Azaouak *see* Azaouagh, Vallée de l'
63 F14 **Azara** Misiones, NE Argentina 28.03S 55.42W
148 K3 **Azaran** Āẕarbāyjān-e Khāvarī, NI Iran 37.34N 47.10E
Āẕarbāyjān/Āẕarbāycan *see* Azerbaijan
Āẕarbāyjān-e Bākhtari *see* Āẕarbāyjān-e Gharbī
148 I4 **Āẕarbāyjān-e Gharbī** *off.* Ostān-e Āẕarbāyjān-e Gharbī *Eng.* West Azerbaijan *prev.* Āẕarbāyjān-e Bākhtari. ◆ *province* NW Iran
148 J3 **Āẕarbāyjān-e Khāvarī** *Eng.* East Azerbaijan, *Eng.* East Sharqī. ◆ *province* NW Iran
79 W13 **Azare** Bauchi, N Nigeria 11.41N 10.09E
131 M19 **Azarychy** *Rus.* Ozarichi. Homyel'skaya Voblasts', SE Belarus 52.39N 29.29E
104 L8 **Azay-le-Rideau** Indre-et-Loire, C France 47.16N 0.27E
147 X12 **A'zāz** Ḥalab, NW Syria 36.34N 37.03E
Azeffāl *see* Azaffal
80 F10 **Azerbaijan** *off.* Azerbaijani Republic, *Az.* Āẕarbāyjān Respublikası; *prev.* Azerbaijan SSR. ◆ *republic* SE Asia
107 S6 **Azhbulat, Ozero** ◎ NE Kazakhstan 41.16N 0.30W

21 O6 **Aziscohos Lake** ◎ Maine, NE USA
Azizbekov *see* Vayk'
Azizie *see* Telish
Aziziya *see* Al 'Azīzīyah
131 T4 **Aznakayevo** Respublika Tatarstan, W Russian Federation 54.55N 53.15E
58 C8 **Azogues** Cañar, S Ecuador 2.41S 78.54W
66 N2 **Azores** var. Açores, Ilhas dos Açores, *Port.* Arquipélago dos Açores. *island group* Portugal, NE Atlantic Ocean
66 L8 **Azores-Biscay Rise** *undersea feature* E Atlantic Ocean
80 K1 **Azotos/Azotus** *see* Ashdod
130 L12 **Azov** Rostovskaya Oblast', SW Russian Federation 47.06N 39.26E
130 J13 **Azov, Sea of** *Rus.* Azovskoye More, *Ukr.* Azovs'ke More. *sea* NE Black Sea
Azovs'ke More/Azovskoye More *see* Azov, Sea of
144 I10 **Azraq, Wāḥat al** *oasis* N Jordan 31.51N 36.51E
76 G6 **Azrou** C Morocco 33.30N 5.12W
155 R5 **Āzow var.** Āzro. Lowgar, E Afghanistan 34.10N 69.39E
39 P8 **Aztec** New Mexico, SW USA 36.49N 107.59W
68 M13 **Aztec Peak** ▲ Arizona, SW USA 33.48N 110.54W
47 N9 **Azua** var. Azua de Compostela. S Dominican Republic 18.25N 70.44W
Azua de Compostela *see* Azua
106 K12 **Azuaga** Extremadura, W Spain 38.16N 5.40W
58 B8 **Azuay** ◆ *province* W Ecuador
170 Bb11 **Azuchi-Ō-shima** *island* SW Japan
107 O11 **Azuer** ≈ C Spain
45 T17 **Azuero, Península de** *peninsula* S Panama
64 I6 **Azufre, Volcán** *var.* Volcán Lastarria. ▲ N Chile 25.16S 68.35W
118 J12 **Azuga** Prahova, SE Romania 45.27N 25.34E
63 C22 **Azul** Buenos Aires, E Argentina 36.46S 59.49W
64 I8 **Azul, Cerro** ▲ NW Argentina 28.28S 68.43W
171 Ll14 **Azul, Cordillera** ▲ C Peru
171 Kk15 **Azuma-san** ▲ Honshū, C Japan 37.44N 140.05E
105 V15 **Azur, Côte d'** *coastal region* SE France
203 Z3 **Azur Lagoon** ◎ Kiritimati, E Kiribati
'Azza *see* Gaza
144 H7 **Az Zāb al Kabir** *see* Great Zab
145 P7 **Az Zabdānī** *var.* Zabadani. Dimashq, W Syria 33.36N 36.07E
147 W8 **Aẓ Ẓāhirah** *desert* NW Oman
147 S6 **Aẓ Ẓahrān** *Eng.* Dhahran. Ash Sharqiyah, NE Saudi Arabia 26.18N 50.01E
147 R6 **Aẓ Ẓahrān al Khubar** *var.* Dhahran Al Khobar. ✕ Ash Sharqiyah, NE Saudi Arabia 26.28N 49.42E
Az Zaqāziq *see* Zagazig
144 H10 **Az Zarqā'** *var.* Zarqa, Az Zarqā', NW Jordan 32.04N 36.06E
144 I11 **Az Zarqā'** *off.* Muḥāfaẓat az Zarqā', *var.* Zarqa. ◆ *governorate* N Jordan
77 O7 **Az Zāwiyah** *var.* Zawia. NW Libya 32.45N 12.43E
146 M7 **Az Zilfī** Ar Riyāḍ, N Saudi Arabia 26.16N 44.48E
145 Y13 **Az Zubayr** *var.* Al Zubair. SE Iraq 30.24N 47.45E
155 P10 **Az Zuqur** *see* Jabal Zuuqar, Jazīrat

B

43 H14 **Ba** *prev.* Mba. Viti Levu, W Fiji 17.34S 177.40E
175 R18 **Baa** Pulau Rote, C Indonesia 10.43S 123.06E
197 G5 **Baaba, Île** *island* Îles Belep, N New Caledonia
144 H7 **Baalbek** var. Ba'labakk; *anc.* Heliopolis. E Lebanon 34.00N 36.15E
110 G8 **Baar** Zug, N Switzerland 47.12N 8.31E
82 Q12 **Baargaal** Bari, NE Somalia 11.12N 51.04E
101 I15 **Baarle-Hertog** Antwerpen, N Belgium 51.26N 4.56E
101 I15 **Baarle-Nassau** Noord-Brabant, S Netherlands 51.27N 4.56E
100 I11 **Baarn** Utrecht, C Netherlands 52.13N 5.16E
116 J13 **Baba** var. Buševa, *Gk.* Varnoús. ▲ FYR Macedonia/Greece 41.00N 21.42E
78 H10 **Babaçar** Brakna, W Mauritania 16.22N 13.57W
114 B12 **Baba Burnu** *headland* NW Turkey 41.18N 31.24E
119 N13 **Babadag** Tulcea, SE Romania 44.53N 28.46E
143 X10 **Babadağ Dağı** ▲ NE Azerbaijan 41.02N 48.04E
Babadaykhan *Turkm.* Babadaýhan; *prev.* Kirovsk. Ahalskaýy Velaýat, C Turkmenistan
152 H14 **Babadurmaz** Akhalskaya Velayat, C Turkmenistan 37.59 N 59.50E
116 M12 **Babaeski** Kırklareli, NW Turkey 41.26N 27.06E
145 S4 **Bāba Gurgur** Iraq 35.34N 44.18E
58 B7 **Babahoyo** *prev.* Bodegas. Los Ríos, C Ecuador 1.49S 79.33W
155 T5 **Bābā, Kūh-e** ▲ C Afghanistan
175 P10 **Babana** Sulawesi, C Indonesia 2.03S 119.13E

◆ COUNTRY ◇ DEPENDENT TERRITORY ◉ ADMINISTRATIVE REGION ▲ MOUNTAIN ▲ VOLCANO ◎ LAKE
● COUNTRY CAPITAL ○ DEPENDENT TERRITORY CAPITAL ✕ INTERNATIONAL AIRPORT ▲ MOUNTAIN RANGE ≈ RIVER ◎ RESERVOIR

227

176 U16 **Babar, Kepulauan** *island group* E Indonesia
176 U15 **Babar, Pulau** *island* Kepulauan Babar, E Indonesia
158 G4 **Bābāsar Pass** *pass* India/Pakistan
195 Q10 **Babase Island** *island* Feni Islands, NE PNG
152 C9 **Babashy** ▲▲ W Turkmenistan
174 Ll15 **Babat** Jawa, S Indonesia *7.07S 112.07E*
174 I10 **Babat** Sumatera, W Indonesia *2.45S 104.01E*
Babatag, Khrebet *see* Bobotogh, Qatorkŭhi
83 H21 **Babati** Arusha, NE Tanzania *4.12S 35.45E*
128 J13 **Babayevo** Vologodskaya Oblast', NW Russian Federation *59.22N 35.51E*
131 Q15 **Babayurt** Respublika Dagestan, SW Russian Federation *43.38N 46.49E*
35 P6 **Babb** Montana, NW USA *48.51N 113.26W*
31 X4 **Babbitt** Minnesota, N USA *47.42N 91.56W*
196 E9 **Babeldaob** *var.* Babeldaop, Babelthuap. *island* N Palau
Babeldaop *see* Babeldaob
147 N17 **Bab el Mandeb** *strait* Gulf of Aden/Red Sea
Babelthuap *see* Babeldaob
113 K17 **Babia Góra** *var.* Babia Hora. ▲ Czech Republic/Poland *49.33N 19.32E*
Babia Hora *see* Babia Góra
Babian Jiang *see* Black River
Babichi *see* Babichy
121 N19 **Babichy** *Rus.* Babichi. Homyel'skaya Voblasts', SE Belarus *52.17N 30.00E*
114 I10 **Babina Greda** Vukovar-Srijem, E Croatia *45.09N 18.33E*
8 K13 **Babine Lake** ◎ British Columbia, SW Canada
176 Vv10 **Babo** Irian Jaya, E Indonesia *2.29S 133.30E*
149 O4 **Bābol** *var.* Babul, Balfrush, Barfrush; *prev.* Barfurush. Māzandarān, N Iran *36.34N 52.39E*
149 O4 **Bābolsar** *var.* Babulsar; *prev.* Meshed-i-Sar. Māzandarān, N Iran *36.42N 52.37E*
68 L16 **Baboquivari Peak** ▲ Arizona, SW USA *31.46N 111.36W*
81 G15 **Baboua** Nana-Mambéré, W Central African Republic *5.46N 14.47E*
121 M17 **Babruysk** *Rus.* Bobruysk. Mahilyowskaya Voblasts', E Belarus *53.07N 29.13E*
Babu *see* Hexian
Babul *see* Bābol
Babulsar *see* Bābolsar
115 O19 **Babuna** ≈ C FYR Macedonia
115 O19 **Babuna** ▲▲ C FYR Macedonia
154 K7 **Bābūs, Dasht-e** *Pash.* Bebas, Dasht-i. ▲▲ W Afghanistan
126 Jj16 **Babushkin** Respublika Buryatiya, S Russian Federation *51.35N 105.49E*
179 P7 **Babuyan Channel** *channel* N Philippines
179 Pp7 **Babuyan Island** *island* N Philippines
145 T9 **Babylon** *site of ancient city* C Iraq *32.33N 44.25E*
114 J9 **Bač** *Ger.* Batsch. Serbia, NW Yugoslavia *45.24N 19.17E*
60 M13 **Bacabal** Maranhão, E Brazil *4.15S 44.45W*
43 Y14 **Bacalar** Quintana Roo, SE Mexico *18.38N 88.17W*
43 Y14 **Bacalar Chico, Boca** *strait* SE Mexico
175 Ss8 **Bacan, Kepulauan** *island group* E Indonesia
175 T8 **Bacan, Pulau** *prev.* Batjan. *island* Maluku, E Indonesia
118 L10 **Bacău** *Hung.* Bákó. Bacău, NE Romania *46.36N 26.55E*
118 K11 **Bacău** ◈ *county* E Romania
Băc Bô, Vinh *see* Tongking, Gulf of
178 Jj5 **Băc Can** Băc Thai, N Vietnam *22.07N 105.50E*
105 T5 **Baccarat** Meurthe-et-Moselle, NE France *48.27N 6.46E*
191 N12 **Bacchus Marsh** Victoria, SE Australia *37.46S 144.27E*
42 H4 **Bacerac** Sonora, NW Mexico *30.27N 108.55W*
118 L10 **Băceşti** Vaslui, E Romania *46.49N 27.13E*
178 Jj6 **Băc Giang** Ha Băc, N Vietnam *21.17N 106.12E*
56 I5 **Bachaquero** Zulia, NW Venezuela *9.57N 71.09W*
Bacher *see* Pohorje
120 M13 **Bacheykava** *Rus.* Bocheykovo. Vitsyebskaya Voblasts', N Belarus *55.01N 29.09E*
42 I5 **Bachíniva** Chihuahua, N Mexico *28.41N 107.13W*
164 G8 **Bachu** Xinjiang Uygur Zizhiqu, NW China *39.46N 78.30E*
15 J6 **Back** ≈ Nunavut, N Canada
114 K10 **Bačka Palanka** *prev.* Palanka. Serbia, NW Yugoslavia *44.22N 20.57E*
114 K8 **Bačka Topola** *Hung.* Topolya; *prev. Hung.* Bácstopolya. Serbia, N Yugoslavia *45.48N 19.39E*
97 J17 **Bäckefors** Västra Götaland, S Sweden *58.49N 12.07E*
Bäckermühle Schulzenmühle *see* Żywiec
97 L16 **Bäckhammar** Värmland, C Sweden *59.09N 14.13E*
114 K9 **Bački Petrovac** *Hung.* Petrőcz; *prev.* Petrovac, Petrovácz. Serbia, NW Yugoslavia *45.22N 19.34E*
103 I21 **Backnang** Baden-Württemberg, SW Germany *48.46N 9.25E*
178 J15 **Bac Liêu** *var.* Vinh Loi. Minh Hai, S Vietnam *9.19N 105.42E*
178 Jj6 **Băc Ninh** Ha Băc, N Vietnam *21.10N 106.04E*
42 G4 **Bacoachi** Sonora, NW Mexico *30.37N 109.57W*
179 Q13 **Bacolod** *off.* Bacolod City. Negros, C Philippines *10.43N 122.58E*
179 P12 **Baco, Mount** ▲ Mindoro, N Philippines *12.50N 121.08E*
113 K25 **Bácsalmás** Bács-Kiskun, S Hungary *46.09N 19.17E*
Bácsjózseffalva *see* Žednik
113 J24 **Bács-Kiskun** *off.* Bács-Kiskun Megye. ◈ *county* S Hungary
Bácsszenttamás *see* Srbobran
Bácstopolya *see* Bačka Topola
Bactra *see* Balkh
161 F21 **Badagara** Kerala, SW India *11.24N 75.45E*
103 M24 **Bad Aibling** Bayern, SE Germany *47.52N 12.00E*
168 I13 **Badain Jaran Shamo** *desert* N China
106 I11 **Badajoz** *anc.* Pax Augusta. Extremadura, W Spain *38.52N 6.58W*
106 J11 **Badajoz** ◈ *province* Extremadura, W Spain
155 S2 **Badakhshān** ◈ *province* NE Afghanistan
107 W6 **Badalona** *anc.* Baetulo. Cataluña, E Spain *41.27N 2.15E*
160 O11 **Bādāmapāhārh** Orissa, E India *22.04N 86.06E*
158 K8 **Badarīnāth** ▲ N India *30.43N 79.28E*
174 Jj8 **Badas, Kepulauan** *island group* W Indonesia
111 S6 **Bad Aussee** Salzburg, E Austria *47.35N 13.44E*
33 S8 **Bad Axe** Michigan, N USA *43.48N 83.00W*
103 G16 **Bad Berleburg** Nordrhein-Westfalen, W Germany *51.03N 8.24E*
103 L17 **Bad Blankenburg** Thüringen, C Germany *50.43N 11.19E*
Bad Borseck *see* Borsec
103 G18 **Bad Camberg** Hessen, W Germany *50.18N 8.15E*
102 L8 **Bad Doberan** Mecklenburg-Vorpommern, N Germany *54.06N 11.55E*
103 N14 **Bad Düben** Sachsen, E Germany *51.35N 12.34E*
111 X4 **Baden** *var.* Baden bei Wien; *anc.* Aquae Panoniae, Thermae Pannonicae. Niederösterreich, NE Austria *48.01N 16.13E*
110 F9 **Baden** Aargau, N Switzerland *47.28N 8.19E*
103 G21 **Baden-Baden** *anc.* Aurelia Aquensis. Baden-Württemberg, SW Germany *48.46N 8.13E*
Baden bei Wien *see* Baden
103 G22 **Baden-Württemberg** *Fr.* Bade-Wurtemberg. ◈ *state* SW Germany
114 A10 **Baderna** Istra, NW Croatia *45.12N 13.45E*
Bade-Wurtemberg *see* Baden-Württemberg
103 H20 **Bad Fredrichshall** Baden-Württemberg, S Germany *49.13N 9.15E*
102 P11 **Bad Freienwalde** Brandenburg, NE Germany *52.46N 14.03E*
111 Q8 **Badgastein** *var.* Gastein. Salzburg, NW Austria *47.07N 13.09E*
Badger State *see* Wisconsin
154 L4 **Bādghīs** ◈ *province* NW Afghanistan
111 T5 **Bad Hall** Oberösterreich, N Austria *48.03N 14.13E*
103 J14 **Bad Harzburg** Niedersachsen, C Germany *51.52N 10.34E*
103 I16 **Bad Hersfeld** Hessen, C Germany *50.52N 9.41E*
100 I10 **Badhoevedorp** Noord-Holland, C Netherlands *52.21N 4.46E*
111 Q8 **Bad Hofgastein** Salzburg, NW Austria *47.10N 13.07E*
Bad Homburg *see* Bad Homburg vor der Höhe
103 G18 **Bad Homburg vor der Höhe** *var.* Bad Homburg. Hessen, W Germany *50.13N 8.37E*
103 E17 **Bad Honnef** Nordrhein-Westfalen, W Germany *50.39N 7.13E*
155 Q17 **Badīn** Sind, SE Pakistan *24.40N 68.49E*
23 S10 **Badin Lake** ▣ North Carolina, SE USA
42 I8 **Badiraguato** Sinaloa, C Mexico *25.26N 107.33W*
111 R6 **Bad Ischl** Oberösterreich, N Austria *47.43N 13.35E*
Badjawa *see* Bajawa
103 J18 **Bad Kissingen** Bayern, SE Germany *50.12N 10.04E*
Bad Königswart *see* Lázně Kynžvart
103 F19 **Bad Kreuznach** Rheinland-Pfalz, SW Germany *49.49N 7.52E*
103 F24 **Bad Krozingen** Baden-Württemberg, SW Germany *47.55N 7.43E*
103 G16 **Bad Laasphe** Nordrhein-Westfalen, W Germany *50.57N 8.24E*
30 J6 **Badlands** *physical region* North Dakota, N USA
103 K16 **Bad Langensalza** Thüringen, C Germany *51.05N 10.40E*
111 T3 **Bad Leonfelden** Oberösterreich, N Austria *48.31N 14.17E*
103 I20 **Bad Mergentheim** Baden-Württemberg, SW Germany *49.30N 9.46E*
103 H17 **Bad Nauheim** Hessen, W Germany *50.22N 8.45E*
103 E17 **Bad Neuenahr-Arhweiler** Rheinland-Pfalz, W Germany *50.33N 7.07E*
Bad Neustadt *see* Bad Neustadt an der Saale
103 J18 **Bad Neustadt an der Saale** *var.* Bad Neustadt. Berlin, C Germany *50.21N 10.13E*
Badnur *see* Betūl
176 Yy15 **Bado** Irian Jaya, E Indonesia *7.06S 139.33E*
102 H13 **Bad Oeynhausen** Nordrhein-Westfalen, NW Germany *52.12N 8.48E*
102 J9 **Bad Oldesloe** Schleswig-Holstein, N Germany *53.49N 10.22E*
79 Q16 **Badou** C Togo *7.37N 0.37E*
Bad Polzin *see* Połczyn-Zdrój
102 H13 **Bad Pyrmont** Niedersachsen, C Germany *51.58N 9.16E*
111 X9 **Bad Radkersburg** Steiermark, SE Austria *46.40N 16.02E*
145 V8 **Badrah** E Iraq *33.06N 45.58E*
168 J6 **Badrah** Hövsgöl, N Mongolia *49.33N 101.58E*
103 N24 **Bad Reichenhall** Bayern, SE Germany *47.43N 12.52E*
146 K8 **Badr Ḩunayn** Al Madīnah, W Saudi Arabia *23.46N 38.45E*
30 M10 **Bad River** ≈ South Dakota, N USA
32 K4 **Bad River** ≈ Wisconsin, N USA
102 H13 **Bad Salzuflen** Nordrhein-Westfalen, NW Germany *52.04N 8.45E*
103 J16 **Bad Salzungen** Thüringen, C Germany *50.48N 10.15E*
111 V8 **Bad Sankt Leonhard im Lavanttal** Kärnten, S Austria *46.55N 14.51E*
102 K9 **Bad Schwartau** Schleswig-Holstein, N Germany *53.55N 10.42E*
103 L24 **Bad Tölz** Bayern, SE Germany *47.44N 11.34E*
189 U1 **Badu Island** *island* Queensland, NE Australia
161 K25 **Badulla** Uva Province, C Sri Lanka *6.58N 81.03E*
111 X5 **Bad Vöslau** Niederösterreich, NE Austria *47.58N 16.12E*
103 I24 **Bad Waldsee** Baden-Württemberg, S Germany *47.54N 9.44E*
37 U11 **Badwater Basin** *depression* California, W USA
103 J20 **Bad Windsheim** Bayern, C Germany *49.30N 10.25E*
103 J23 **Bad Wörishofen** Bayern, S Germany *48.00N 10.36E*
102 G10 **Bad Zwischenahn** Niedersachsen, NW Germany *53.10N 8.01E*
106 M13 **Baena** Andalucía, S Spain *37.37N 4.22W*
Baeterrae/Baeterrae Septimanorum *see* Béziers
Baetic Cordillera/Baetic Mountains *see* Béticos, Sistemas
Baetulo *see* Badalona
59 K18 **Baeza** Napo, NE Ecuador *0.30S 77.52W*
107 N13 **Baeza** Andalucía, S Spain *38.00N 3.28W*
81 D15 **Bafang** Ouest, W Cameroon *5.10N 10.10E*
78 H12 **Bafatá** C Guinea-Bissau *12.09N 14.37W*
155 U5 **Baffa** North-West Frontier Province, NW Pakistan *34.28N 73.14E*
207 O11 **Baffin Basin** *undersea feature* N Labrador Sea
207 N12 **Baffin Bay** *bay* Canada/Greenland
27 T15 **Baffin Bay** *inlet* Texas, SW USA
206 M12 **Baffin Island** *island* Nunavut, NE Canada
81 E15 **Bafia** Centre, C Cameroon *4.49N 11.13E*
79 R14 **Bafilo** NE Togo *9.22N 1.19E*
78 J12 **Bafing** ≈ W Africa
78 J12 **Bafoulabé** Kayes, W Mali *13.43N 10.49W*
81 D15 **Bafoussam** Ouest, W Cameroon *5.31N 10.25E*
149 R9 **Bāfq** Yazd, C Iran *31.34N 55.21E*
142 L10 **Bafra** Samsun, N Turkey *41.34N 35.55E*
142 L10 **Bafra Burnu** *headland* N Turkey *41.42N 36.02E*
149 S12 **Bāft** Kermān, S Iran *29.12N 56.36E*
81 N18 **Bafwabalinga** Orientale, NE Dem. Rep. Congo (Zaire) *0.52N 26.55E*
81 N18 **Bafwaboli** Orientale, NE Dem. Rep. Congo (Zaire) *0.42N 26.06E*
81 N17 **Bafwasende** Orientale, NE Dem. Rep. Congo (Zaire) *1.00N 27.09E*
194 J11 **Bagabag Island** *island* N PNG
44 K13 **Bagaces** Guanacaste, NW Costa Rica *10.29N 85.13W*
159 O12 **Bagaha** Bihār, N India *27.07N 84.04E*
161 F16 **Bāgalkot** Karnātaka, W India *16.10N 75.42E*
83 J22 **Bagamoyo** Pwani, E Tanzania *6.25S 38.55E*
174 Gg4 **Bagan Datuk** *var.* Bagan Datok. Perak, Peninsular Malaysia *3.58N 100.46E*
179 Rr15 **Baganga** Mindanao, S Philippines *7.31N 126.34E*
174 Gg6 **Bagansiapiapi** *var.* Pasirpangarayan. Sumatera, W Indonesia *2.09N 100.50E*
Bagaria *see* Bagheria
79 T11 **Bagaroua** Tahoua, W Niger *14.34N 4.24E*
81 I20 **Bagata** Bandundu, W Dem. Rep. Congo (Zaire) *3.46S 17.57E*
Bagdad *see* Baghdād
126 Kk15 **Bagdarin** Respublika Buryatiya, S Russian Federation *54.27N 113.34E*
63 G17 **Bagé** Rio Grande do Sul, S Brazil *31.22S 54.06W*
Bagenalstown *see* Muine Bheag
Bagerhat *see* Bagherhat
105 P16 **Bages et de Sigean, Étang de** ◎ S France
35 W17 **Baggs** Wyoming, C USA *41.02N 107.39W*
160 F11 **Bāgh** Madhya Pradesh, C India *22.22N 74.49E*
145 T8 **Baghdād** *var.* Bagdad, *Eng.* Baghdad. ● (Iraq) C Iraq *33.19N 44.25E*
145 T8 **Baghdād** ✕ C Iraq *33.19N 44.25E*
159 T16 **Bagherhat** *var.* Bagerhat. Khulna, S Bangladesh *22.40N 89.48E*
109 J23 **Bagheria** *var.* Bagaria. Sicilia, Italy, C Mediterranean Sea *38.04N 13.31E*
149 S10 **Bāghīn** Kermān, C Iran *30.50N 57.00E*
155 Q3 **Baghlān** Baghlān, NE Afghanistan *36.10N 68.43E*
155 Q3 **Baghlān** *var.* Bāghlān. ◈ *province* NE Afghanistan
154 M7 **Bāghrān** Helmand, S Afghanistan *32.55N 64.57E*
31 T4 **Bagley** Minnesota, N USA *47.31N 95.24W*
108 H10 **Bagnacavallo** Emilia-Romagna, C Italy *44.00N 12.59E*
104 K16 **Bagnères-de-Bigorre** Hautes-Pyrénées, S France *43.04N 0.09E*
104 L17 **Bagnères-de-Luchon** Hautes-Pyrénées, S France *42.46N 0.34E*
108 F11 **Bagni di Lucca** Toscana, C Italy *44.01N 10.38E*
108 H11 **Bagno di Romagna** Emilia-Romagna, C Italy *43.51N 11.57E*
105 R14 **Bagnols-sur-Cèze** Gard, S France *44.10N 4.37E*
168 M14 **Bag Nur** ◎ N China
179 Q13 **Bago** *off.* Bago City. Negros, C Philippines *10.30N 122.49E*
Bago *see* Pegu
78 M13 **Bagoé** ≈ Ivory Coast/Mali
Bagrāmē *see* Bagrāmī
155 R5 **Bagrāmī** *var.* Bagrāmē. Kābul, E Afghanistan *34.28N 69.16E*
121 B14 **Bagrationovsk** *Ger.* Preussisch Eylau. Kaliningradskaya Oblast', W Russian Federation *54.24N 20.39E*
Bagrax *see* Bohu
Bagrax Hu *see* Bosten Hu
58 C10 **Bagua** Amazonas, NE Peru *5.34S 78.24W*
179 P9 **Baguio** *off.* Baguio City. Luzon, N Philippines *16.25N 120.36E*
79 V9 **Bagzane, Monts** ▲ N Niger *17.48N 8.43E*
Bāḩah, Minţaqat al *see* Al Bāḩah
Bahama Islands *see* Bahamas
46 H3 **Bahamas** *off.* Commonwealth of the Bahamas. ◆ *commonwealth republic* N West Indies
1 L13 **Bahamas** *var.* Bahama Islands. *island group* N West Indies
159 S15 **Baharampur** *prev.* Berhampore. West Bengal, NE India *24.03N 88.16E*
175 Nn6 **Bahau, Sungai** ≈ Borneo, N Indonesia
155 U10 **Bahāwalnagar** Punjab, E Pakistan *30.00N 73.03E*
155 T11 **Bahāwalpur** Punjab, E Pakistan *29.24N 71.39E*
142 L16 **Bahçe** Osmaniye, S Turkey *37.11N 36.32E*
166 J8 **Ba He** ≈ C China
Bäherden *see* Bakharden
61 N16 **Bahia** *off.* Estado da Bahia. ◈ *state* E Brazil
63 B24 **Bahía Blanca** Buenos Aires, E Argentina *38.43S 62.19W*
42 L15 **Bahía Bufadero** Michoacán de Ocampo, SW Mexico
65 J19 **Bahía Bustamante** Chubut, SE Argentina *45.06S 66.30W*
42 D5 **Bahía de los Ángeles** Baja California, NW Mexico
42 C6 **Bahía de Tortugas** Baja California Sur, W Mexico *27.42N 114.54W*
44 J4 **Bahía, Islas de la** *Eng.* Bay Islands. *island group* N Honduras
42 E5 **Bahía Kino** Sonora, NW Mexico *28.48N 111.55W*
42 E9 **Bahía Magdalena** *var.* Puerto Magdalena. Baja California Sur, W Mexico *24.34N 112.07W*
56 C8 **Bahía Solano** *var.* Ciudad Mutis, Solano. Chocó, W Colombia *6.13N 77.27W*
82 I11 **Bahir Dar** *var.* Bahr Dar, Bahrdar Giyorgis. Amhara, N Ethiopia *11.33N 37.22E*
147 X8 **Bahlā'** *var.* Bahlah, Bahlat. NW Oman *22.55N 57.16E*
Bāhla *see* Bālān
Bahlah/Bahlat *see* Bahlā'
158 M11 **Bahraich** Uttar Pradesh, N India *27.34N 81.36E*
149 M14 **Bahrain** *off.* State of Bahrain, Dawlat al Bahrayn, *Ar.* Al Baḩrayn; *prev.* Bahrein, *anc.* Tylos or Tyros. ◆ *monarchy* SW Asia
148 M14 **Bahrain** ✕ C Bahrain *26.15N 50.39E*
148 M15 **Bahrain, Gulf of** *gulf* Persian Gulf, NW Arabian Sea
144 I7 **Baḩrat Mallāḩah** ◎ W Syria
Bahrayn, Dawlat al *see* Bahrain
Bahr Dar/Bahrdar Giyorgis *see* Bahir Dar
Bahrein *see* Bahrain
83 E16 **Bahr el Gabel** ◈ *state* S Sudan
82 E13 **Bahr ez Zaref** ≈ C Sudan
69 R8 **Bahr Kameur** ≈ N Central African Republic
Bahr Tabariya, Sea of *see* Tiberias, Lake
149 W15 **Bāhū Kalāt** Sīstān va Balūchestān, SE Iran *25.42N 61.28E*
120 N13 **Bahushewsk** *Rus.* Bogushëvsk. Vitsyebskaya Voblasts', NE Belarus *54.51N 30.13E*
Bai *see* Tagow Bāy
118 G13 **Baia de Aramă** Mehedinţi, SW Romania *45.00N 22.43E*
118 G11 **Baia de Criş** *Ger.* Altenburg, *Hung.* Körösbánya. Hunedoara, SW Romania *46.10N 22.40E*
85 A16 **Baia dos Tigres** Namibe, SW Angola *16.36S 11.44E*
84 A13 **Baia Farta** Benguela, W Angola *12.38S 13.12E*
118 H9 **Baia Mare** *Ger.* Frauenbach, *Hung.* Nagybánya; *prev.* Neustadt. Maramureş, NW Romania *47.39N 23.35E*
118 H8 **Baia Sprie** *Ger.* Mittelstadt, *Hung.* Felsőbánya. Maramureş, NW Romania *47.40N 23.42E*
80 G13 **Baïbokoum** Logone-Oriental, SW Chad *7.46N 15.43E*
166 F12 **Baicao Ling** ▲▲ SW China
169 U9 **Baicheng** *var.* Pai-ch'eng; *prev.* T'aon-an. Jilin, NE China *45.31N 122.50E*
164 I6 **Baicheng** *var.* Bay. Xinjiang Uygur Zizhiqu, NW China *41.49N 81.45E*
118 J13 **Băicoi** Prahova, SE Romania *45.01N 25.52E*
Baidoa *see* Baydhabo
13 U6 **Baie-Comeau** Quebec, SE Canada *49.12N 68.10W*
13 T7 **Baie-des-Bacon** Quebec, SE Canada *48.31N 69.17W*
13 S8 **Baie-des-Rochers** Quebec, SE Canada *47.57N 69.50W*
13 U6 **Baie-des-Sables** Quebec, SE Canada *48.41N 67.55W*
10 K11 **Baie-du-Poste** Quebec, SE Canada *50.19N 73.49W*
180 H17 **Baie Lazare** Mahé, NE Seychelles *4.45S 55.28E*
47 Y5 **Baie-Mahault** Basse Terre, C Guadeloupe *16.17N 61.34W*
13 R9 **Baie-St-Paul** Quebec, SE Canada *47.27N 70.30W*
13 V5 **Baie-Trinité** Quebec, SE Canada *49.25N 67.19W*
11 T11 **Baie Verte** Newfoundland, SE Canada *49.58N 56.06W*
169 X11 **Baihe** *prev.* Erdaobaihe. Jilin, NE China *42.24N 128.09E*
Baiguan *see* Shangyu
145 U11 **Bā'ij al Mahdī** S Iraq *31.21N 44.57E*
Baiji *see* Bayjī
Baikal, Lake *see* Baykal, Ozero
Bailādila *see* Kirandul
Baile an Chaistil *see* Ballycastle
Baile an Róba *see* Ballinrobe
Baile an tSratha *see* Ballintra
Baile Átha an Rí *see* Athenry
Baile Átha Buí *see* Athboy
Baile Átha Cliath *see* Dublin
Baile Átha Fhirdhia *see* Ardee
Baile Átha Í *see* Athy
Baile Átha Luain *see* Athlone
Baile Átha Troim *see* Trim
Baile Brigín *see* Balbriggan
Baile Easa Dara *see* Ballysadare
118 I13 **Băile Govora** Vâlcea, SW Romania *45.00N 24.08E*
118 F13 **Băile Herculane** *Ger.* Herkulesbad, *Hung.* Herkulesfürdő. Caraş-Severin, SW Romania *44.50N 22.23E*
Baile Locha Riach *see* Loughrea
Baile Mhistéala *see* Mitchelstown
Baile Monaidh *see* Ballymoney
107 N12 **Bailén** Andalucía, S Spain *38.06N 3.46W*
Baile na hInse *see* Ballynahinch
Baile na Lorgan *see* Castleblayney
Baile na Mainistreach *see* Newtownabbey
Baile Nua na hArda *see* Newtownards
118 I12 **Băile Olăneşti** Vâlcea, SW Romania *45.16N 24.13E*
118 H14 **Băileşti** Dolj, SW Romania *44.01N 23.20E*
Bailingmiao *see* Darhan Muminggan Lianheqi
60 K11 **Bailique, Ilha** *island* NE Brazil
105 O1 **Bailleul** Nord, N France *50.43N 2.43E*
80 H12 **Ba Illi** Chari-Baguirmi, SW Chad *10.31N 16.28E*
165 V12 **Bailong Jiang** ≈ C China
84 C13 **Bailundo** *Port.* Vila Teixeira da Silva. Huambo, C Angola *12.12S 15.52E*
165 T13 **Baima** *var.* Sêraitang. Qinghai, C China *32.55N 100.44E*
176 W14 **Baimun** Pulau Workai, E Indonesia *6.41S 134.40E*
194 H14 **Baimuru** Gulf, S PNG *7.31S 144.44E*
164 M16 **Bainang** Xizang Zizhiqu, W China *28.57N 89.31E*
25 S8 **Bainbridge** Georgia, SE USA *30.54N 84.33W*
175 Pp17 **Baing** Pulau Sumba, SE Indonesia *10.09S 120.34E*
164 M14 **Baingoin** Xizang Zizhiqu, W China *31.25N 90.01E*
106 G2 **Baio Grande** Galicia, NW Spain *43.08N 8.58W*
106 G4 **Baiona** Galicia, NW Spain *42.06N 8.49W*
169 V7 **Baiquan** Heilongjiang, NE China *47.37N 126.04E*
Bā'ir *see* Bāyir
164 I11 **Bairab Co** ◎ W China
27 Q7 **Baird** Texas, SW USA *32.23N 99.23W*
41 N7 **Baird Mountains** ▲▲ Alaska, USA
15 Mm2 **Baird Peninsula** *peninsula* Baffin Island, Nunavut, NE Canada
Baireuth *see* Bayreuth
202 H3 **Bairiki** ● (Kiribati) Tarawa, NW Kiribati *1.19N 173.01E*
169 S11 **Bairin Youqi** *var.* Daban. Nei Mongol Zizhiqu, N China *43.33N 118.40E*
169 S10 **Bairin Zuoqi** *var.* Lindong. Nei Mongol Zizhiqu, N China *43.59N 119.24E*
151 P17 **Bairkum** *Kaz.* Bayyrqum. Yuzhnyy Kazakhstan, S Kazakhstan *41.57N 68.05E*
191 P12 **Bairnsdale** Victoria, SE Australia *37.51S 147.37E*
179 Q14 **Bais** Negros, S Philippines *9.36N 123.07E*
104 L15 **Baïse** *var.* Baise. ≈ S France
169 W11 **Baishan** *prev.* Hunjiang. Jilin, NE China *41.57N 126.31E*
120 F12 **Baisogala** Radviliškis, C Lithuania *55.38N 23.44E*
201 Q7 **Baiti** N Nauru *0.30S 166.55E*
106 G13 **Baixo Alentejo** *physical region* S Portugal
66 P5 **Baixo, Ilhéu de** *island* Madeira, Portugal, NE Atlantic Ocean
85 E15 **Baixo Longa** Cuando Cubango, SE Angola *15.39S 18.39E*
165 V10 **Baiyin** Gansu, C China *36.33N 104.11E*
166 E8 **Baiyü** Sichuan, C China *30.37N 97.15E*
167 N14 **Baiyun** ✕ (Guangzhou) Guangdong, S China *23.12N 113.19E*
166 K4 **Baiyu Shan** ▲▲ C China
113 J25 **Baja** Bács-Kiskun, S Hungary *46.12N 18.56E*
42 C4 **Baja California** ◈ *state* NW Mexico
42 C4 **Baja California** *Eng.* Lower California. *peninsula* NW Mexico
42 E9 **Baja California Sur** ◈ *state* W Mexico
Bājah *see* Béja
Bajan *see* Bayan
203 V16 **Baja, Punta** *headland* Easter Island, Chile, E Pacific Ocean *27.10S 109.21W*
42 B4 **Baja, Punta** *headland* NW Mexico *29.57N 115.48W*
57 R5 **Baja, Punta** *headland* NE Venezuela
44 D5 **Baja Verapaz** *off.* Departamento de Baja Verapaz. ◈ *department* C Guatemala
175 Q16 **Bajawa** *prev.* Badjawa. Flores, S Indonesia *8.46S 120.58E*
159 S16 **Baj Baj** *prev.* Budge-Budge. West Bengal, E India *22.28N 88.10E*
147 N15 **Bājil** W Yemen *15.05N 43.16E*
191 U4 **Bajimba, Mount** ▲ New South Wales, SE Australia *29.19S 152.04E*
114 K13 **Bajina Bašta** Serbia, W Yugoslavia *43.58N 19.33E*
159 U14 **Bajitpur** Dhaka, E Bangladesh *24.12N 90.57E*
114 K8 **Bajmok** Serbia, NW Yugoslavia *45.59N 19.25E*
Bajo Boquete *see* Boquete
115 L17 **Bajram Curri** Kukës, N Albania *42.22N 20.06E*
81 J14 **Bakala** Ouaka, C Central African Republic *6.03N 20.31E*
131 T4 **Bakaly** Respublika Bashkortostan, W Russian Federation *55.10N 53.46E*
Bakan *see* Shimonoseki
151 U14 **Bakanas** *Kaz.* Baqanas. Almaty, SE Kazakhstan *44.50N 76.13E*
151 V12 **Bakanas** *Kaz.* Baqanas. ≈ E Kazakhstan
151 U14 **Bakbakty** *Kaz.* Baqbaqty. Almaty, SE Kazakhstan *44.36N 76.41E*
126 Gg13 **Bakchar** Tomskaya Oblast', C Russian Federation *56.58N 81.59E*
78 I11 **Bakel** E Senegal *13.43N 13.40W*
37 W13 **Baker** California, W USA *35.15N 116.04W*
24 J8 **Baker** Louisiana, S USA *30.35N 91.10W*
35 Y9 **Baker** Montana, NW USA *46.22N 104.16W*
34 L12 **Baker** Oregon, NW USA *44.46N 117.49W*
199 Jj8 **Baker and Howland Islands** ◇ *US unincorporated territory* W Polynesia
68 L12 **Baker Butte** ▲ Arizona, SW USA *34.24N 111.22W*
41 X15 **Baker Island** *island* Alexander Archipelago, Alaska, USA
15 Kk6 **Baker Lake** Nunavut, N Canada *64.19N 96.10W*
15 Kk6 **Baker Lake** ◎ Nunavut, N Canada
34 H6 **Baker, Mount** ▲ Washington, NW USA *48.46N 121.48W*
37 R13 **Bakersfield** California, W USA *35.22N 119.01W*
26 M9 **Bakersfield** Texas, SW USA *30.54N 102.21W*
23 P9 **Bakersville** North Carolina, SE USA *36.03N 82.10W*
Bakhābī *see* Bū Khābī
152 E12 **Bakharden** *Turkm.* Bäherden; *prev.* Bakherden. Akhalskiy Velayat, C Turkmenistan *38.30N 57.18E*
152 F12 **Bakhardok** *Turkm.* Bokurdak. Akhalskiy Velayat, C Turkmenistan *38.51N 58.34E*
149 U5 **Bākharz, Kuhhā-ye** ▲▲ NE Iran
158 D13 **Bākhāsar** Rājasthān, NW India *24.43N 71.09E*
Bakhchisaray *see* Bakhchysaray
119 T13 **Bakhchysaray** *Rus.* Bakhchisaray. Respublika Krym, S Ukraine *44.43N 33.53E*
Bakherden *see* Bakharden
119 R3 **Bakhmach** Chernihivs'ka Oblast', N Ukraine *51.10N 32.48E*
126 Hh11 **Bakhta** Krasnoyarskiy Kray, C Russian Federation *62.25N 88.56E*
148 K6 **Bākhtarān** *prev.* Kermānshāh, Qahremānshahr. Kermānshāh, W Iran *34.19N 47.04E*
Bākhtarān *see* Kermānshāh
149 Q11 **Bakhtegān, Daryācheh-ye** ◎ C Iran
151 X12 **Bakhty** Vostochnyy Kazakhstan, E Kazakhstan *46.41N 82.45E*
143 Z11 **Bakı** *Eng.* Baku. ● (Azerbaijan) E Azerbaijan *40.23N 49.51E*
143 Z11 **Bakı** ✕ E Azerbaijan *40.26N 49.55E*
142 C13 **Bakır Çayı** ≈ W Turkey
94 L1 **Bakkafjördhur** Austurland, NE Iceland *66.01N 14.49W*
94 L1 **Bakkaflói** *sea area* W Norwegian Sea
83 I15 **Bako** Southern, S Ethiopia *5.45N 36.39E*
78 L15 **Bako** NW Ivory Coast *9.06N 7.34W*
Bákó *see* Bacău
113 H23 **Bakony** *Eng.* Bakony Mountains, *Ger.* Bakonywald. ▲▲ W Hungary
Bakony Mountains/Bakonywald *see* Bakony
83 M16 **Bakool** *off.* Gobolka Bakool. ◈ *region* W Somalia
81 L15 **Bakouma** Mbomou, SE Central African Republic *5.42N 22.43E*
131 N15 **Baksan** Kabardino-Balkarskaya Respublika, SW Russian Federation *43.43N 43.31E*
121 I16 **Bakshty** Hrodzyenskaya Voblasts', W Belarus *53.56N 26.13E*
Baku *see* Bakı
204 K12 **Bakutis Coast** *physical region* Antarctica
Bakwanga *see* Mbuji-Mayi
151 O15 **Bakyrly** Yuzhnyy Kazakhstan, S Kazakhstan *44.30N 67.41E*
12 H13 **Bala** Ontario, S Canada *45.01N 79.37W*
99 J19 **Bala** NW Wales, UK *52.54N 3.31W*
142 I13 **Balâ** Ankara, C Turkey *39.34N 33.07E*
179 O16 **Balabac Island** *island* W Philippines
Balabac, Selat *see* Balabac Strait
175 O1 **Balabac Strait** *var.* Selat Balabac. *strait* Malaysia/Philippines
Ba'labakk *see* Baalbek
175 O10 **Balabalangan, Kepulauan** *island group* N Indonesia
197 H4 **Balabio, Île** *island* Province Nord, W New Caledonia
118 I14 **Balaci** Teleorman, S Romania *44.21N 24.55E*
145 S7 **Balad** N Iraq *34.00N 44.07E*
145 U7 **Balad Rūz** E Iraq *33.42N 45.04E*
126 J15 **Balagansk** Irkutskaya Oblast', S Russian Federation *54.02N 102.48E*
160 J11 **Bālāghāt** Madhya Pradesh, C India *21.48N 80.10E*
161 F14 **Bālāghāt Range** ▲▲ W India
105 X14 **Balagne** *physical region* Corse, France, C Mediterranean Sea
107 U5 **Balaguer** Cataluña, NE Spain *41.48N 0.48E*
107 S3 **Balaïtous** *var.* Pic de Balaitous, Pic de Balaïtous. ▲ France/Spain *42.51N 0.17W*
Balaitous, Pic de *see* Balaïtous
131 O3 **Balakhna** Nizhegorodskaya Oblast', W Russian Federation *56.26N 43.43E*
126 I14 **Balakhta** Krasnoyarskiy Kray, S Russian Federation *55.22N 91.24E*
190 I9 **Balaklava** South Australia *34.10S 138.22E*
Balakleya *see* Balakliya
119 V6 **Balakliya** *Rus.* Balakleya. Kharkivs'ka Oblast', E Ukraine *49.26N 36.51E*
131 Q7 **Balakovo** Saratovskaya Oblast', W Russian Federation *52.03N 47.47E*
85 P14 **Balama** Cabo Delgado, N Mozambique *13.18S 38.39E*
175 Nn1 **Balambangan, Pulau** *island* East Malaysia
154 L3 **Bālā Morghāb** Laghmān, NW Afghanistan *35.37N 63.21E*
158 E11 **Bālān** *prev.* Bāhla. Rājasthān, NW India *27.45N 71.31E*
118 J10 **Bălan** *Hung.* Balánbánya. Harghita, C Romania *46.39N 25.45E*
Balánbánya *see* Bălan
179 P10 **Balanga** Luzon, N Philippines *14.40N 120.32E*
160 M12 **Balāngīr** *prev.* Bolangir. Orissa, E India *20.46N 83.31E*
131 N8 **Balashov** Saratovskaya Oblast', W Russian Federation *51.31N 43.14E*
Balasore *see* Bāleshwar
113 K21 **Balassagyarmat** Nógrád, N Hungary *48.04N 19.16E*
31 S10 **Balaton** Minnesota, N USA *44.13N 95.52W*
113 H24 **Balaton** *var.* Lake Balaton, *Ger.* Plattensee. ◎ W Hungary
113 I23 **Balatonfüred** *var.* Füred. Veszprém, W Hungary *46.56N 17.51E*
Balaton, Lake *see* Balaton
118 I11 **Bălăuşeri** *Ger.* Bladenmarkt, *Hung.* Balavásár. Mureş, C Romania *46.24N 24.41E*
Balavásár *see* Bălăuşeri
107 Q11 **Balazote** Castilla-La Mancha, C Spain *38.54N 2.09W*
Balázsfalva *see* Blaj
121 F14 **Balbieriškis** Prienai, S Lithuania *54.29N 23.52E*
195 S12 **Balbi, Mount** ▲ Bougainville Island, NE PNG *5.51S 154.58E*
60 F11 **Balbina, Represa** ▣ NW Brazil
45 T15 **Balboa** Panamá, C Panama *8.55N 79.36W*
99 G17 **Balbriggan** *Ir.* Baile Brigín. E Ireland *53.37N 6.10W*
Balbunar *see* Kubrat
83 N17 **Balcad** Shabeellaha Dhexe, C Somalia *2.19N 45.19E*
63 D23 **Balcarce** Buenos Aires, E Argentina *37.51S 58.16W*
11 U16 **Balcarres** Saskatchewan, S Canada *50.49N 103.31W*
116 O8 **Balchik** Dobrich, NE Bulgaria *43.25N 28.11E*
193 E24 **Balclutha** Otago, South Island, NZ *46.15S 169.44E*
27 Q12 **Balcones Escarpment** *escarpment* Texas, SW USA
20 F14 **Bald Eagle Creek** ≈ Pennsylvania, NE USA
Baldenburg *see* Biały Bór
23 V12 **Bald Head Island** *island* North Carolina, SE USA
29 W10 **Bald Knob** Arkansas, C USA *35.18N 91.34W*
32 K17 **Bald Knob** *hill* Illinois, N USA *37.33N 89.21W*
Baldohn *see* Baldone
120 G9 **Baldone** *Ger.* Baldohn. Rīga, W Latvia *56.46N 24.18E*
24 I9 **Baldwin** Louisiana, S USA *29.50N 91.32W*
33 P7 **Baldwin** Michigan, N USA *43.54N 85.49W*
29 Q4 **Baldwin City** Kansas, C USA *38.43N 95.12W*
41 N8 **Baldwin Peninsula** *headland* Alaska, USA *66.45N 162.19W*
20 H9 **Baldwinsville** New York, NE USA *43.09N 76.19W*
25 N2 **Baldwyn** Mississippi, S USA *34.30N 88.38W*
11 W15 **Baldy Mountain** ▲ Manitoba, S Canada *51.29N 100.46W*
35 T7 **Baldy Mountain** ▲ Montana, NW USA *48.09N 109.39W*
39 O13 **Baldy Peak** ▲ Arizona, SW USA *33.56N 109.37W*
Bâle *see* Basel
107 X9 **Baleares** ◈ *autonomous community* E Spain
107 X11 **Baleares, Islas** *Eng.* Balearic Islands. *island group* Spain, W Mediterranean Sea
Baleares Major *see* Mallorca
Balearic Islands *see* Baleares, Islas
Balearic Plain *see* Algerian Basin
Balearis Minor *see* Menorca
174 M6 **Baleh, Batang** ≈ East Malaysia
10 J8 **Baleine, Grande Rivière de la** ≈ Quebec, E Canada
10 K7 **Baleine, Petite Rivière de la** ≈ Quebec, SE Canada
11 N6 **Baleine, Rivière à la** ≈ Quebec, E Canada
101 J16 **Balen** Antwerpen, N Belgium *51.11N 5.12E*
179 P9 **Baler** Luzon, N Philippines *15.47N 121.30E*
160 P11 **Bāleshwar** *prev.* Balasore. Orissa, E India *21.31N 86.58E*
126 L16 **Baley** Chitinskaya Oblast', S Russian Federation *51.30N 116.16E*
79 S12 **Baléyara** Tillabéri, W Niger *13.48N 2.57E*
131 T1 **Balezino** Udmurtskaya Respublika, NW Russian Federation *57.57N 53.03E*
44 J4 **Balfate** Colón, N Honduras *15.47N 86.24W*
11 O17 **Balfour** British Columbia, SW Canada *49.39N 116.57W*
31 N3 **Balfour** North Dakota, N USA *47.55N 100.34W*
Balfrush *see* Bābol
126 I16 **Balgazyn** Respublika Tyva, S Russian Federation *50.53N 95.12E*
11 U16 **Balgonie** Saskatchewan, S Canada *50.30N 104.12W*
Bălgrad *see* Alba Iulia
83 J19 **Balguda** *spring/well* S Kenya *1.28S 39.50E*
164 K6 **Balguntay** Xinjiang Uygur Zizhiqu, NW China *42.51N 86.19E*
147 R16 **Balḩāf** S Yemen *14.02N 48.15E*
158 F13 **Bāli** Rājasthān, N India *25.17N 73.16E*

◆ Country
● Country Capital
◇ Dependent Territory
○ Dependent Territory Capital
◈ Administrative Region
✕ International Airport
▲ Mountain
▲▲ Mountain Range
Volcano
≈ River
◎ Lake
▣ Reservoir

175 N15 **Bali** ◆ *province* S Indonesia
175 N16 **Bali** *island* C Indonesia
113 K16 **Balice ✈** (Kraków) Małopolskie, S Poland 49.57N 19.49E
176 Yy13 **Baliem, Sungai** ～ Irian Jaya, E Indonesia
142 C12 **Balıkesir** Balıkesir, W Turkey 39.38N 27.52E
144 L3 **Balıkesir** ◆ *province* NW Turkey
142 L3 **Balīkh, Nahr** ～ N Syria
175 O9 **Balikpapan** Borneo, C Indonesia 1.15S 116.49E
175 O9 **Balikpapan, Teluk** *bay* Borneo, C Indonesia
Bali, Laut *see* Bali Sea
195 O11 **Balima** ～ New Britain, C PNG
179 P17 **Balimbing** Tawitawi, SW Philippines 5.10N 120.00E
194 G14 **Balimo** Western, SW PNG 8.01S 142.52E
Bálinc *see* Balinţ
175 Qq9 **Balingara, Pegunungan** ▲ Sulawesi, N Indonesia
103 H23 **Balingen** Baden-Württemberg, SW Germany 48.16N 8.51E
118 F11 **Balinţ** *Hung.* Bálinc. Timiş, W Romania 45.52N 21.54E
179 Pp6 **Balintang Channel** *channel* N Philippines
144 K3 **Bālis** Ḩalab, N Syria 36.01N 38.03E
175 N15 **Bali Sea** *Ind.* Laut Bali. *sea* C Indonesia
175 N16 **Bali, Selat** *strait* C Indonesia
100 K7 **Balk** Friesland, N Netherlands 52.54N 5.34E
124 O7 **Balkan Mountains** *Bul./SCr.* Stara Planina. ▲ Bulgaria/Yugoslavia
152 B9 **Balkanskiy Velayat** *Turkm.* Balkan Welayaty. ◆ *province* W Turkmenistan
Balkan Welayaty *see* Balkanskiy Velayat
151 P8 **Balkashino** Akmola, N Kazakhstan 52.32N 68.43E
155 O2 **Balkh** *anc.* Bactra. Balkh, N Afghanistan 36.46N 66.54E
155 P2 **Balkh** ◆ *province* N Afghanistan
151 T13 **Balkhash** *Kaz.* Balqash. Zhezkazgan, SE Kazakhstan 46.52N 74.54E
Balkhash, Lake *see* Balkhash, Ozero
151 T13 **Balkhash, Ozero** *Eng.* Lake Balkhash, *Kaz.* Balqash. ⊜ SE Kazakhstan
Balla Balla *see* Mbalabala
98 H10 **Ballachulish** N Scotland, UK 56.40N 5.10W
188 M12 **Balladonia** Western Australia 32.21S 123.31E
99 C16 **Ballaghaderreen** *Ir.* Bealach an Doirín. C Ireland 53.51N 8.29W
94 H10 **Ballangen** Nordland, NW Norway 68.21N 16.48E
99 H14 **Ballantrae** W Scotland, UK 55.04N 5.00W
191 N12 **Ballarat** Victoria, SE Australia 37.36S 143.51E
188 K11 **Ballard, Lake** *salt lake* Western Australia
Ballari *see* Bellary
78 L11 **Ballé** Koulikoro, W Mali 15.18N 8.31W
42 D7 **Ballenas, Bahía de** *bay* W Mexico
42 D5 **Ballenas, Canal de** *channel* NW Mexico
205 R17 **Balleny Islands** *island group* Antarctica
42 J7 **Balleza** *var.* San Pablo Balleza. Chihuahua, N Mexico 26.55N 106.21W
116 M13 **Balı** Tekirdağ, NW Turkey 40.48N 27.03E
159 O13 **Ballia** Uttar Pradesh, N India 25.45N 84.09E
191 V4 **Ballina** New South Wales, SE Australia 28.49S 153.33E
99 C16 **Ballina** *Ir.* Béal an Átha. W Ireland 54.07N 9.09W
99 D16 **Ballinamore** *Ir.* Béal an Átha Móir. NW Ireland 54.03N 7.46W
99 D18 **Ballinasloe** *Ir.* Béal Átha na Sluaighe. W Ireland 53.19N 8.13W
27 P8 **Ballinger** Texas, SW USA 31.44N 99.57W
99 C17 **Ballinrobe** *Ir.* Baile an Róba. W Ireland 53.37N 9.14W
99 A21 **Ballinskelligs Bay** *Ir.* Bá na Scealg. *inlet* SW Ireland
99 D15 **Ballintra** *Ir.* Baile an tSratha. NW Ireland 54.34N 8.07W
105 T7 **Ballon d'Alsace** ▲ NE France 47.50N 6.54E
Ballon de Guebwiller *see* Grand Ballon
115 K21 **Ballsh** *var.* Ballshi. Fier, SW Albania 40.35N 19.45E
Ballshi *see* Ballsh
100 K4 **Ballum** Friesland, N Netherlands 53.27N 5.40E
99 F16 **Ballybay** *Ir.* Béal Átha Beithe. N Ireland 54.07N 6.54W
99 E14 **Ballybofey** *Ir.* Bealach Féich. NW Ireland 54.48N 7.46W
99 G14 **Ballycastle** *Ir.* Baile an Chaistil. N Northern Ireland, UK 55.12N 6.13W
99 G15 **Ballyclare** *Ir.* Bealach Cláir. E Northern Ireland, UK 54.45N 6.00W
99 E16 **Ballyconnell** *Ir.* Béal Átha Conaill. N Ireland 54.07N 7.34W
99 C17 **Ballyhaunis** *Ir.* Béal Átha hAmhnais. W Ireland 53.45N 8.45W
99 G14 **Ballymena** *Ir.* An Baile Meánach. NE Northern Ireland, UK 54.52N 6.16W
99 F14 **Ballymoney** *Ir.* Baile Monaidh. NE Northern Ireland, UK 55.10N 6.30W
99 G15 **Ballynahinch** *Ir.* Baile na hInse. SE Northern Ireland, UK 54.24N 5.54W
99 D16 **Ballysadare** *Ir.* Baile Easa Dara. NW Ireland 54.13N 8.30W
99 D15 **Ballyshannon** *Ir.* Béal Átha Seanaidh. NW Ireland 54.30N 8.10W
65 H19 **Balmaceda** Aisén, S Chile 51.27S 25.84W
65 G23 **Balmaceda, Cerro** ▲ S Chile 51.25 73.84W
111 N22 **Balmazújváros** Hajdú-Bihar, E Hungary 47.36N 21.18E

---

110 E10 **Balmhorn** ▲ SW Switzerland 46.27N 7.41E
190 L12 **Balmoral** Victoria, SE Australia 37.16S 141.38E
26 K9 **Balmorhea** Texas, SW USA 30.58N 103.44W
Balneario Claromecó *see* Claromecó
175 R9 **Balo** Sulawesi, N Indonesia 0.58S 123.19E
84 B13 **Balochistān** *see* Baluchistān
Balombo *Port.* Norton de Matos, Vila Norton de Matos. Benguela, W Angola 12.21S 14.46E
84 B13 **Balombo** ～ W Angola
189 X10 **Balonne River** ～ Queensland, E Australia
158 E13 **Bālotra** Rājasthān, N India 25.51N 72.18E
151 V14 **Balpyk Bi** *prev.* Kirovskiy *Kaz.* Kirov. Almaty, SE Kazakhstan 44.52N 78.10E
Balqá'/Balqā', Muḥāfaẓat al *see* Al Balqā'
Balqash *see* Balkhash/Balkhash, Ozero
158 M12 **Balrāmpur** Uttar Pradesh, N India 27.25N 82.10E
190 M9 **Balranald** New South Wales, SE Australia 34.39S 143.33E
118 H14 **Balş** Olt, S Romania 44.19N 24.06E
12 H11 **Balsam Creek** Ontario, S Canada 46.26N 79.10W
32 J5 **Balsam Lake** Wisconsin, N USA 45.27N 92.28W
12 J14 **Balsam Lake** ⊜ Ontario, SE Canada
61 M14 **Balsas** Maranhão, E Brazil 07.30S 46.00W
42 M15 **Balsas, Río** *var.* Río Mexcala. ～ S Mexico
45 W16 **Balsas, Río** ～ SE Panama
121 O18 **Bal'shavik** *Rus.* Bol'shevik. Homyel'skaya Voblasts', SE Belarus 52.34N 30.49E
97 O15 **Bålsta** Uppsala, C Sweden 59.33N 17.35E
110 E7 **Balsthal** Solothurn, NW Switzerland 47.20N 7.50E
119 O8 **Balta** Odes'ka Oblast', SW Ukraine 47.58N 29.39E
121 H14 **Baltaji Voke** Vilnius, SE Lithuania 54.35N 25.13E
107 N5 **Baltanás** Castilla-León, N Spain 41.56N 4.12W
63 E16 **Baltasar Brum** Artigas, N Uruguay 30.43S 57.19W
118 M9 **Bălţi** *Rus.* Bel'tsy. N Moldova 47.45N 27.57E
Baltic Port *see* Paldiski
120 B10 **Baltic Sea** *Ger.* Ostee, *Rus.* Baltiskoye More. *sea* N Europe
23 X3 **Baltimore** Maryland, NE USA 39.17N 76.36W
33 T13 **Baltimore** Ohio, N USA 39.48N 82.33W
23 X3 **Baltimore–Washington ✈** Maryland, E USA 39.10N 76.40W
Baltischport/Baltiski *see* Paldiski
Baltiskoye More *see* Baltic Sea
121 A14 **Baltiysk** *Ger.* Pillau. Kaliningradskaya Oblast', W Russian Federation 54.39N 19.54E
Baltkrievija *see* Belarus
194 K9 **Baluan Island** N PNG
Balúchestán va Sístán *see* Sístān va Balūchestān
154 M12 **Baluchistān** *var.* Balochistān, Beluchistan. ◆ *province* SW Pakistan
179 Q12 **Balud** Masbate, N Philippines 12.03N 123.12E
174 Mm6 **Balui, Batang** ～ East Malaysia
159 S13 **Bālurghāt** West Bengal, NE India 25.14N 88.43E
120 J8 **Balvi** Balvi, NE Latvia 57.07N 27.14E
194 H12 **Balyer River** Western Highlands, C PNG
153 W7 **Balykchy** *Kir.* Ysyk-Köl; *prev.* Issyk-Kul', Rybach'ye. Issyk-Kul'skaya Oblast', NE Kyrgyzstan 42.28N 76.08E
58 B7 **Balzar** Guayas, W Ecuador 1.25S 79.54W
110 I8 **Balzers** S Liechtenstein 47.04N 9.31E
149 T12 **Bam** Kermān, SE Iran 29.08N 58.27E
79 Y13 **Bama** Borno, NE Nigeria 11.28N 13.46E
78 L12 **Bamako ●** (Mali) Capital District, SW Mali 12.39N 8.01W
78 L12 **Bamba** Gao, C Mali 17.03N 1.19W
44 M8 **Bambana, Río** ～ NE Nicaragua
81 J15 **Bambari** Ouaka, C Central African Republic 5.45N 20.37E
189 W5 **Bambaroo** Queensland, NE Australia 19.05S 146.16E
103 K19 **Bamberg** Bayern, SE Germany 49.54N 10.52E
23 R14 **Bamberg** South Carolina, SE USA 33.18N 81.02W
81 M16 **Bambesa** Orientale, N Dem. Rep. Congo (Zaire) 3.25N 25.43E
78 G11 **Bambey** W Senegal 14.43N 16.26W
81 H16 **Bambio** Sangha-Mbaéré, SW Central African Republic 3.57N 16.54E
83 I24 **Bamboesberge** ▲ S South Africa 31.24S 26.10E
81 D14 **Bamenda** Nord-Ouest, W Cameroon 5.55N 10.09E
8 K17 **Bamfield** Vancouver Island, British Columbia, SW Canada 48.48N 125.05W
155 E12 **Bāmīān** *var.* Bāmiān. Bāmīān, E Afghanistan 34.50N 67.51E
155 O4 **Bāmīān** ◆ *province* N Afghanistan
81 J14 **Bamingui** Bamingui-Bangoran, C Central African Republic 7.38N 20.06E
80 D13 **Bamingui** ～ N Central African Republic
80 I13 **Bamingui-Bangoran** ◆ *prefecture* N Central African Republic
149 V13 **Bampūr** Sīstān va Balūchestān, SE Iran 27.13N 60.28E
194 G13 **Bamu** ～ SW PNG
Bamy *see* Bami
Bán *see* Bánovce nad Bebravou

---

83 N17 **Banaadir** *off.* Gobolka Banaadir. ◆ *region* S Somalia
203 N3 **Banaba** *var.* Ocean Island. *island* Tungaru, W Kiribati
61 O14 **Banabuiú, Açude** ⊟ NE Brazil
59 O19 **Bañados del Izozog** *salt lake* SE Bolivia
99 D18 **Banagher** *Ir.* Beannchar. C Ireland 53.12N 7.56W
83 L18 **Banalia** Orientale, N Dem. Rep. Congo (Zaire) 1.39N 25.19E
78 L12 **Banamba** Koulikoro, W Mali 13.33N 7.25W
42 G4 **Banámichi** Sonora, NW Mexico 30.01N 110.13W
178 I6 **Ban Donkon** Oudômxai, N Laos 20.20N 101.37E
81 I21 **Bandundu** ◆ *region* W Dem. Rep. Congo (Zaire)
157 Q22 **Banana** Andaman and Nicobar Islands, India, NE Indian Ocean 6.57N 93.54E
116 N13 **Banarlı** Tekirdağ, NW Turkey 41.04N 27.21E
158 H12 **Banās** ～ N India
57 Z11 **Banās, Râs** *headland* E Egypt 23.55N 35.47E
114 N10 **Banatski Karlovac** Serbia, NE Yugoslavia 45.03N 21.02E
147 P16 **Banā, Wādī** *dry watercourse* SW Yemen
142 E14 **Banaz** Uşak, W Turkey 38.46N 29.46E
142 E14 **Banaz Çayı** ～ W Turkey
165 P14 **Banbar** Xizang Zizhiqu, W China 31.01N 94.43E
99 G15 **Banbridge** *Ir.* Droichead na Banna. SE Northern Ireland, UK 54.21N 6.16W
97 M21 **Banbury** S England, UK 52.04N 1.19W
178 H7 **Ban Chiang Dao** Chiang Mai, NW Thailand 19.22N 98.59E
98 K9 **Banchory** NE Scotland, UK 58.04N 0.35W
12 J13 **Bancroft** Ontario, SE Canada 45.04N 77.49W
35 R15 **Bancroft** Idaho, NW USA 42.43N 111.54W
31 U11 **Bancroft** Iowa, C USA 43.17N 94.13W
160 I9 **Banda** Madhya Pradesh, C India 24.04N 78.57E
158 L13 **Bānda** Uttar Pradesh, N India 25.28N 80.19E
173 E3 **Bandaaceh** *var.* Banda Atjeh; *prev.* Koetaradja, Kutaradja, Kutaraja. Sumatera, W Indonesia 5.30N 95.19E
Banda Atjeh *see* Bandaaceh
176 U12 **Banda, Kepulauan** *island group* E Indonesia
Banda, Laut *see* Banda Sea
79 N17 **Bandama** *var.* Bandama Fleuve. ～ S Ivory Coast
79 N15 **Bandama Blanc** ～ C Ivory Coast
Bandama Fleuve *see* Bandama
Bandar 'Abbās *see* Bandar-e 'Abbās
159 W16 **Bandarban** Chittagong, SE Bangladesh 22.13N 92.13E
82 Q13 **Bandarbeyla** *var.* Bender Beila, Bender Beyla. Bari, NE Somalia 9.28N 50.48E
149 R14 **Bandar-e 'Abbās** *var.* Bandar 'Abbās; *prev.* Gombroon. Hormozgān, S Iran 27.10N 56.10E
148 M3 **Bandar-e Anzalī** Gīlān, NW Iran 37.25N 49.28E
149 N12 **Bandar-e Būshehr** *var.* Bûshehr, *Eng.* Bushire. Būshehr, S Iran 28.58N 50.49E
148 M11 **Bandar-e Gonāveh** *var.* Ganāveh; *prev.* Gonāveh. Būshehr, SW Iran 29.33N 50.39E
149 R14 **Bandar-e Khamīr** Hormozgān, S Iran 26.59N 55.30E
149 Q14 **Bandar-e Langeh** *var.* Bandar-e Lengeh, Lingeh. Hormozgān, S Iran 26.34N 54.52E
Bandar-e Lengeh *see* Bandar-e Langeh
148 L10 **Bandar-e Māhshahr** *var.* Māh-Shahr; *prev.* Bandar-e Ma'shūr. Khūzestān, SW Iran 30.33N 49.10E
Bandar-e Ma'shūr *see* Bandar-e Māhshahr
149 O14 **Bandar-e Nakhīlū** Hormozgān, S Iran
Bandar-e Shāh *see* Bandar-e Torkaman
149 N3 **Bandar-e Torkaman** *var.* Bandar-e Torkeman, Bandar-e Torkman; *prev.* Bandar-e Shāh. Golestān, N Iran 36.55N 54.04E
Bandar-e Torkeman/Bandar-e Torkman *see* Bandar-e Torkaman
Bandar Kassim *see* Boosaaso
174 I13 **Bandarlampung** *prev.* Tanjungkarang, Teloekbetoeng, Telukbetong. Sumatera, W Indonesia 5.28N 105.16E
Bandar Maharani *see* Muar
Bandar Masulipatnam *see* Machilipatnam
85 N16 **Bandar Penggaram** *see* Batu Pahat
175 N3 **Bandar Seri Begawan** *prev.* Brunei Town. ● (Brunei) N Brunei 4.53N 114.58E
175 Mm3 **Bandar Seri Begawan ✈** N Brunei 4.55N 114.58E
175 S13 **Banda Sea** *var.* Laut Banda. *sea* E Indonesia
106 H5 **Bande** Galicia, NW Spain 42.01N 7.58W
61 G15 **Bandeirantes** Mato Grosso, W Brazil 9.04S 57.53W
61 N20 **Bandeira, Pico da** ▲ SE Brazil 20.25S 41.45W
85 K19 **Bandelierkop** Northern, NE South Africa 23.25S 29.46E
64 L8 **Bandera** Santiago del Estero, N Argentina 28.52S 62.15W
27 Q11 **Bandera** Texas, SW USA 29.43N 99.07W
42 J12 **Banderas, Bahía de** *bay* W Mexico

---

79 U11 **Bandiagara** Mopti, C Mali 14.22N 3.42W
158 I12 **Bāndīkuī** Rājasthān, N India 27.07N 76.34E
142 C11 **Bandırma** *var.* Penderma. Balıkesir, NW Turkey 40.21N 27.58E
Bandjarmasin *see* Banjarmasin
79 C21 **Bandoeng** *see* Bandung
34 E14 **Bandon** Oregon, NW USA 43.07N 124.24W
178 J8 **Ban Dong Bang** Nong Khai, E Thailand 18.00N 104.08E
81 H20 **Bandundu** *prev.* Banningville. Bandundu, W Dem. Rep. Congo (Zaire) 3.18S 17.24E
174 J14 **Bandung** *prev.* Bandoeng. Jawa, C Indonesia 6.47S 107.28E
118 L15 **Băneasa** Constanţa, SW Romania 44.03N 27.42E
148 J4 **Bāneh** Kordestān, N Iran 35.58N 45.54E
46 I7 **Banes** Holguín, E Cuba 20.58N 75.43W
11 P16 **Banff** Alberta, SW Canada 51.10N 115.34W
98 K8 **Banff** NE Scotland, UK 57.39N 2.33W
98 K8 **Banffshire** *cultural region* NE Scotland, UK
79 N14 **Banfora** SW Burkina 10.36N 4.45W
161 H19 **Bangalore** Karnātaka, S India 12.58N 77.34E
159 S16 **Bangaon** West Bengal, NE India 23.01N 88.49E
179 P9 **Bangar** Luzon, N Philippines 16.51N 120.25E
81 L15 **Bangassou** Mbomou, SE Central African Republic 4.51N 22.55E
194 K12 **Bangeta, Mount** ▲ C PNG 6.11S 147.02E
175 Qq10 **Banggai, Kepulauan** *island group* C Indonesia
175 R9 **Banggai, Pulau** *island* Kepulauan Banggai, N Indonesia
176 X11 **Banggelapa** Irian Jaya, E Indonesia 3.47S 136.53E
175 O1 **Banggi, Pulau** *var.* Banggi, Pulau. *island* East Malaysia
82 N15 **Banghāzī** *Eng.* Bengazi, Benghazi, *It.* Bengasi. NE Libya 32.07N 20.04E
174 K8 **Bangkai, Tanjung** *var.* Bankai. *headland* Borneo, N Indonesia 0.21N 108.53E
175 M14 **Bangkalan** Pulau Madura, C Indonesia 7.04S 112.43E
175 S6 **Bangka, Pulau** *island* N Indonesia
174 J10 **Bangka, Pulau** *island* W Indonesia
174 Ii10 **Bangka, Selat** *strait* Sumatera, W Indonesia
175 Rr6 **Bangka, Selat** *var.* Selat Likupang. *strait* Sulawesi, N Indonesia
174 Gg8 **Bangkinang** Sumatera, W Indonesia 0.21N 100.56E
174 H10 **Bangko** Sumatera, W Indonesia 2.03S 102.15E
Bangkok *see* Krung Thep
159 T14 **Bangkok, Bight of** *see* Krung Thep, Ao
159 T14 **Bangladesh** *off.* People's Republic of Bangladesh; *prev.* East Pakistan. ◆ *republic* S Asia
178 Kk13 **Ba Ngôi** Khanh Hoa, S Vietnam 11.55N 109.07E
178 J10 **Bangong Co** *var.* Pangong Tso. ⊜ China/India *see also* Pangong Tso
9 W15 **Bangor** *Ir.* Beannchar. E Northern Ireland, UK 54.40N 5.40W
97 J18 **Bangor** NW Wales, UK 53.13N 4.07W
21 R6 **Bangor** Maine, NE USA 44.48N 68.46W
18 I12 **Bangor** Pennsylvania, NE USA 40.52N 75.12W
72 Q8 **Bangs, Mount** ▲ Arizona, SW USA 36.47N 113.54W
37 V15 **Banning** California, W USA 33.55N 116.52W
Banningville *see* Bandundu
178 Jj11 **Ban Nongsim** Champasak, S Laos 14.05N 106.00E
13 I6 **Bannu** *prev.* Edwardesabad. North-West Frontier Province, NW Pakistan 32.00N 70.36E
46 J7 **Banos** Tungurahua, C Ecuador 1.20S 78.24W
55 C19 **Baños** *see* Banyoles
191 R6 **Banora** New South Wales, SE Australia 28.13S 149.03E
Banow *see* Andarāb
159 W13 **Bang Saphan** *see* Bang Saphan Yai
82 I9 **Banska** California, W USA
160 N11 **Banra** ～ NE India

---

78 M12 **Banifing** *var.* Ngorolaka. ～ Burkina/Mali
79 R13 **Banikoara** N Benin 11.18N 2.25E
116 K8 **Baniski Lom** ～ N Bulgaria
23 U7 **Banister River** ～ Virginia, NE USA
77 08 **Bani Suwayf** *see* Beni Suef
77 W3 **Bani Walīd** NW Libya 31.46N 13.58E
144 H5 **Bāniyās** *var.* Banias, Baniyas, Paneas. Ţarţūs, W Syria 35.12N 35.57E
115 K14 **Banja** Serbia, W Yugoslavia 43.33N 19.35E
114 G11 **Banja Koviljača** Serbia, W Yugoslavia 44.31N 19.11E
114 G11 **Banja Luka** Republika Srpska, NW Bosnia and Herzegovina 44.46N 17.10E
175 O11 **Banjarmasin** *prev.* Bandjarmasin. Borneo, C Indonesia 3.22S 114.33E
79 F11 **Banjul** *prev.* Bathurst. ● (Gambia) W Gambia 13.25N 16.43W
78 F11 **Banjul ✈** W Gambia 13.18N 16.39W
Banjuwangi *see* Banyuwangi
143 Y13 **Bankā** *Rus.* Bank. SE Azerbaijan 39.25N 49.13E
178 Jj11 **Ban Kadian** *var.* Ban Kadiene. Champasak, S Laos 14.25N 105.42E
Ban Kadiene *see* Ban Kadian
178 I10 **Ban Kadian** *see* Bangkai, Tanjung
178 Gg15 **Ban Kam Phuam** Phangnga, SW Thailand 9.16N 98.24E
79 O11 **Ban Kantang** *var.* Kantang
178 H16 **Ban Yong Sata** Trang, SW Thailand 7.09N 99.42E
174 Mm16 **Banyuwangi** *var.* Banjuwangi; *prev.* Banjoewangi. Jawa, S Indonesia 8.12S 114.22E
205 X14 **Banzare Coast** *physical region* Antarctica
181 Q14 **Banzare Seamounts** *undersea feature* S Indian Ocean
Banzart *see* Bizerte
175 R9 **Baochang** *see* Taibus Qi
107 U8 **Ban Kadian var.** E Spain 40.34N 0.37E
81 E14 **Banyo** Adamaoua, NW Cameroon 6.46N 11.49E
107 X4 **Banyoles** *var.* Baños. Cataluña, NE Spain 42.07N 2.46E
176 X11 **Banyuwangi** *see* Banyuwangi
174 Mm16 **Banzare Coast** *physical region* Antarctica

---

113 J19 **Banská Bystrica** *Ger.* Neusohl, *Hung.* Besztercebánya. Banskobystrický Kraj, C Slovakia 48.45N 19.08E
113 K20 **Banskobystrický Kraj** ◆ *region* C Slovakia
178 J8 **Ban Sôppheung** Bolikhamxai, C Laos 18.33N 104.18E
158 G15 **Bānswāra** Rājasthān, N India 23.31N 74.28E
178 Gg15 **Ban Ta Khun** Surat Thani, SW Thailand 8.53N 98.52E
178 Jj9 **Ban Talak** Khammouan, C Laos 17.33N 105.40E
79 R5 **Bantè** N Benin 8.21N 1.55E
178 I8 **Ban Thabôk** Bolikhamxai, C Laos 18.27N 103.13E
178 Jj10 **Ban Tôp** Savannakhét, S Laos 16.07N 106.07E
99 B21 **Bantry** *Ir.* Beanntraí. SW Ireland 51.40N 9.27W
99 A21 **Bantry Bay** *Ir.* Bá Bheanntraí. *bay* SW Ireland
174 L15 **Bantul** *prev.* Bantoel. Jawa, C Indonesia 7.55S 110.21E
116 E11 **Banya** Burgas, E Bulgaria 42.46N 27.49E
116 N9 **Banya** Burgas, E Bulgaria
173 Ee6 **Banyak, Kepulauan** *prev.* Kepulauan Banjak. *island group* NW Indonesia
107 U8 **Ban La** *headland* E Spain 40.34N 0.37E
81 E14 **Banyo** Adamaoua, NW Cameroon 6.46N 11.49E
107 X4 **Banyoles** *var.* Baños. Cataluña, NE Spain 42.07N 2.46E

---

121 N14 **Baran'** Vitsyebskaya Voblasts', NE Belarus 54.28N 30.18E
158 I14 **Bārān** Rājasthān, N India 25.07N 76.31E
145 L4 **Bārānān, Shākh-i** ▲ E Iraq
121 I17 **Baranavichy Pol.** Baranowicze, *Rus.* Baranovichi. Brestskaya Voblasts', SW Belarus 53.07N 26.01E
122 Oo5 **Baranikha** Chukotskiy Avtonomnyy Okrug, NE Russian Federation 68.29N 168.13E
118 M4 **Baranivka** Zhytomyrs'ka Oblast', N Ukraine 50.16N 27.40E
41 W14 **Baranof Island** *island* Alexander Archipelago, Alaska, USA
Baranovichi/Baranowicze *see* Baranavichy
113 N15 **Baranów Sandomierski** Podkarpackie, SE Poland 50.28N 21.31E
113 I26 **Baranya** *off.* Baranya Megye. ◆ *county* S Hungary
159 R13 **Bārāri** Bihar, NE India 25.31N 87.22E
24 L10 **Barataria Bay** *bay* Louisiana, S USA
Barat Daya, Kepulauan *see* Damar, Kepulauan
120 L12 **Baravukha** *Rus.* Borovukha. Vitsyebskaya Voblasts', N Belarus 55.36N 28.33E
56 E11 **Baraya** Huila, C Colombia 3.10N 75.04W
56 M21 **Barbacena** Minas Gerais, SE Brazil 21.13S 43.46W
56 B13 **Barbacoas** Nariño, SW Colombia 1.37N 78.07W
56 L6 **Barbacoas** Aragua, N Venezuela 9.28N 66.58W
47 Z13 **Barbados** ◆ *commonwealth republic* SE West Indies
49 S3 **Barbados** *island* Barbados
107 U11 **Barbaria, Cap de** *var.* Cabo de Berbería. *headland* Formentera, E Spain 38.39N 1.24E
116 N13 **Barbaros** Tekirdağ, NW Turkey 40.55N 27.28E
76 A11 **Barbas, Cap** *headland* S Western Sahara 22.14N 16.45W
107 T5 **Barbastro** Aragón, NE Spain 42.02N 0.07E
106 K16 **Barbate** ～ SW Spain
106 K16 **Barbate de Franco** Andalucía, S Spain 36.11N 5.55W
85 K21 **Barberton** Mpumalanga, NE South Africa 25.45S 31.01E
33 T12 **Barberton** Ohio, N USA 41.02N 81.37W
104 K12 **Barbezieux-St-Hilaire** Charente, W France 45.28N 0.09W
56 G9 **Barbosa** Santander, C Colombia 5.57N 73.37W
23 N7 **Barbourville** Kentucky, S USA 36.52N 83.53W
47 W9 **Barbuda** *island* N Antigua and Barbuda
189 W8 **Barcaldine** Queensland, E Australia 23.33S 145.20E
106 H9 **Barcarrota** Extremadura, W Spain 38.31N 6.51W
Barcău *see* Berettyó
Barce *see* Al Marj
109 L23 **Barcelona** *anc.* Barcellona Pozzo di Gotto. Sicilia, Italy, C Mediterranean Sea 38.09N 15.15E
Barcellona Pozzo di Gotto *see* Barcellona
107 W6 **Barcelona** *anc.* Barcino, Barcinona. Cataluña, E Spain 41.25N 2.10E
57 S5 **Barcelona** Anzoátegui, NE Venezuela 10.07N 64.43W
107 S5 **Barcelona** ◆ *province* Cataluña, NE Spain
107 W6 **Barcelona ✈** Cataluña, E Spain 41.18N 2.05E
105 U14 **Barcelonnette** Alpes-de-Haute-Provence, SE France 44.24N 6.37E
60 I9 **Barcelos** Amazonas, N Brazil 0.58S 62.58W
106 G5 **Barcelos** Braga, N Portugal 41.31N 8.37W
112 I10 **Barcin** *Ger.* Bartschin. Kujawski-pomorskie, C Poland 52.51N 17.55E
Barcino/Barcinona *see* Barcelona
189 U7 **Barcoo** *see* Cooper Creek
113 I26 **Barcs** Somogy, SW Hungary 45.57N 17.26E
143 W11 **Bǎrdǎ** *Rus.* Barda. C Azerbaijan 40.28N 47.07E
80 H5 **Bardaï** Borkou-Ennedi-Tibesti, N Chad 21.21N 17.00E
145 R2 **Bardarash** N Iraq 36.32N 43.36E
145 Q7 **Bardasah** SW Iraq 34.02N 42.28E
159 S16 **Barddhamān** West Bengal, NE India 23.10N 88.03E
113 N18 **Bardejov** *Ger.* Bartfeld, *Hung.* Bártfa. Prešovský Kraj, E Slovakia 49.17N 21.18E
107 R4 **Bárdenas Reales** *physical region* N Spain
191 R6 **Baradine** New South Wales, SE Australia 30.57S 149.03E
94 K3 **Baraf Daja Islands** *see* Damar, Kepulauan
138 O9 **Baragarh** Orissa, E India 21.20N 83.36E
83 I17 **Baragoi** Rift Valley, W Kenya 1.39N 36.46E
47 J9 **Barahona** SW Dominican Republic 18.13N 71.07W
159 W13 **Barail Range** ▲ NE India
159 W13 **Barak** ～ NE India
82 G10 **Barakat** Gezira, C Sudan 14.18N 33.31E
155 Q6 **Barakī Barak** *var.* Baraki, Rajan. Lowgar, E Afghanistan 33.58N 68.58E
46 I7 **Baracoa** Guantánamo, E Cuba 20.19N 74.31W
203 N1 **Baraka** *var.* Barka, *Ar.* Khawr Barakah. *seasonal river* Eritrea/Sudan
159 N11 **Baram** ～ East Malaysia
202 J3 **Barama River** ～ N Guyana
57 S7 **Barama** *var.* Barka, Batang Baram, Batang. ～ East Malaysia
23 L6 **Bardstown** Kentucky, S USA 37.48N 85.28W
Barduli *see* Barletta
22 G7 **Bardwell** Kentucky, S USA 36.52N 89.01W
158 K11 **Bareilly** *var.* Bareli. Uttar Pradesh, N India 28.19N 79.24E
Bareli *see* Bareilly
100 H13 **Barendrecht** Zuid-Holland, SW Netherlands 51.52N 4.31E
104 M3 **Barentin** Seine-Maritime, N France 49.33N 0.57E
92 N3 **Barentsburg** Spitsbergen, W Svalbard 78.01N 14.19E
Barentsevo More/Barents Havet *see* Barents Sea
94 O3 **Barentsøya** *island* E Svalbard
207 T11 **Barents Plain** *undersea feature* N Barents Sea

◆ COUNTRY   ◇ DEPENDENT TERRITORY   ◈ ADMINISTRATIVE REGION   ▲ MOUNTAIN   ☒ VOLCANO   ⊜ LAKE
● COUNTRY CAPITAL   ○ DEPENDENT TERRITORY CAPITAL   ✕ INTERNATIONAL AIRPORT   ▲ MOUNTAIN RANGE   ～ RIVER   ⊟ RESERVOIR

229

129 P3 **Barents Sea** *Nor.* Barents Havet, *Rus.* Barentsevo More. *sea* Arctic Ocean

207 U14 **Barents Trough** *undersea feature* SW Barents Sea

82 J9 **Barentu** W Eritrea 15.08N 37.35E

104 J3 **Barfleur** Manche, N France 49.41N 1.18W

104 J3 **Barfleur, Pointe de** *headland* N France 49.46N 1.09W

Barfrush/Barfurush *see* Bābol

164 H14 **Barga** Xizang Zizhiqu, W China 30.51N 81.19E

107 N9 **Bargas** Castilla-La Mancha, C Spain 39.56N 4.00W

83 I15 **Bargē** Southern, S Ethiopia 6.11N 37.04E

108 A9 **Barge** Piemonte, NE Italy 44.49N 7.21E

159 U16 **Barguna** Khulna, S Bangladesh 22.09N 90.07E

Bärgusad *see* Vorotan

126 K15 **Barguzin** Respublika Buryatiya, S Russian Federation 53.37N 109.47E

159 O13 **Barhaj** Uttar Pradesh, N India 26.16N 83.43E

191 N10 **Barham** New South Wales, SE Australia 35.39S 144.09E

158 J12 **Barhan** Uttar Pradesh, N India 27.21N 78.10E

21 S7 **Bar Harbor** Mount Desert Island, Maine, NE USA 44.23N 68.14W

159 R14 **Barharwa** Bihār, NE India 24.52N 87.46E

159 P15 **Barhi** Bihār, N India 24.19N 85.25E

109 O17 **Bari** *var.* Bari delle Puglie; *anc.* Barium. Puglia, SE Italy 41.06N 16.52E

82 P12 **Bari** *off.* Gobolka Bari. ◆ *region* NE Somalia

178 K14 **Ba Ria** Ba Ria-Vung Tau, S Vietnam 10.30N 107.10E

Bāridah *see* Al Bāridah

Bari delle Puglie *see* Bari

Barikot *see* Barīkowṭ

155 T4 **Barīkowṭ** *var.* Barikot. Kunar, NE Afghanistan 35.18N 71.36E

44 C4 **Barillas** *var.* Santa Cruz Barillas. Huehuetenango, NW Guatemala 15.49N 91.19W

56 J6 **Barinas** Barinas, W Venezuela 8.36N 70.15W

56 I7 **Barinas** *off.* Estado Barinas; *prev.* Zamora. ◆ *state* C Venezuela

56 I6 **Barinitas** Barinas, NW Venezuela 8.47N 70.26W

160 P11 **Bāripada** Orissa, E India 21.58N 86.45E

62 K9 **Bariri** São Paulo, S Brazil 22.04S 48.46W

77 W11 **Bārīs** S Egypt 24.28N 30.39E

158 G14 **Bāri Sādri** Rājasthān, N India 24.25N 74.28E

159 U16 **Barisal** Khulna, S Bangladesh 22.40N 90.19E

173 G7 **Barisan, Pegunungan** ▲ Sumatera, W Indonesia

175 N10 **Barito, Sungai** ✍ Borneo, C Indonesia

Barium *see* Bari

Barka *see* Baraka

Barka *see* Al Marj

166 H8 **Barkam** Sichuan, C China 31.56N 102.22E

120 J9 **Barkava** Madona, C Latvia 56.43N 26.34E

8 M15 **Barkerville** British Columbia, SW Canada 53.06N 121.34W

12 J12 **Bark Lake** ◎ Ontario, SE Canada

22 H7 **Barkley, Lake** ◎ Kentucky/Tennessee, S USA

8 K17 **Barkley Sound** *inlet* British Columbia, W Canada

85 J24 **Barkly East** *Afr.* Barkly-Oos. Eastern Cape, SE South Africa 30.58S 27.34E

Barkly-Oos *see* Barkly East

189 S4 **Barkly Tableland** *plateau* Northern Territory/Queensland, N Australia

Barkly-Wes *see* Barkly West

85 H22 **Barkly West** *Afr.* Barkly-Wes. Northern Cape, N South Africa 28.31S 24.31E

165 O5 **Barkol** *var.* Barkol Kazak Zizhixian. Xinjiang Uygur Zizhiqu, NW China 43.37N 93.01E

165 O5 **Barkol Hu** ◎ NW China

Barkol Kazak Zizhixian *see* Barkol

32 J3 **Bark Point** *headland* Wisconsin, N USA 46.53N 91.11W

27 P11 **Barksdale** Texas, SW USA 29.43N 100.03W

Bar Kunar *see* Asmār

118 L11 **Bârlad** *prev.* Birlad. Vaslui, E Romania 46.12N 27.39E

118 M11 **Bârlad** *prev.* Birlad. ✍ E Romania

78 D9 **Barlavento, Ilhas de** *var.* Windward Islands. *island group* N Cape Verde

105 R5 **Bar-le-Duc** *var.* Bar-sur-Ornain. Meuse, NE France 48.46N 5.10E

188 K11 **Barlee, Lake** ◎ Western Australia

188 H8 **Barlee Range** ▲ Western Australia

109 N16 **Barletta** *anc.* Barduli. Puglia, SE Italy 41.19N 16.16E

112 E10 **Barlinek** *Ger.* Berlinchen. Zachodniopomorskie, NW Poland 53.00N 15.11E

29 S11 **Barling** Arkansas, C USA 35.19N 94.16W

176 Vv10 **Barma** Irian Jaya, E Indonesia 1.52S 132.57E

191 Q9 **Barmedman** New South Wales, SE Australia 34.09S 147.21E

Barmen-Elberfeld *see* Wuppertal

158 H12 **Bärmer** Rājasthān, NW India 25.46N 71.24E

190 K9 **Barmera** South Australia 34.14S 140.26E

99 J18 **Barmouth** NW Wales, UK 52.44N 4.03W

160 F10 **Barnagar** Madhya Pradesh, C India 23.01N 75.28E

158 H9 **Barnāla** Punjab, NW India 30.19N 75.33E

99 L15 **Barnard Castle** N England, UK 54.34N 1.55W

191 O6 **Barnato** New South Wales, SE Australia 31.39S 145.01E

126 H14 **Barnaul** Altayskiy Kray, C Russian Federation 53.21N 83.45E

---

111 V8 **Bärnbach** Steiermark, SE Austria 47.05N 15.07E

20 K16 **Barnegat** New Jersey, NE USA 39.43N 74.12W

25 S4 **Barnesville** Georgia, SE USA 33.03N 84.09W

31 R6 **Barnesville** Minnesota, N USA 46.39N 96.25W

33 U13 **Barnesville** Ohio, N USA 39.59N 81.10W

100 K11 **Barneveld** *var.* Barnveld. Gelderland, C Netherlands 52.10N 5.34E

27 P9 **Barnhart** Texas, SW USA 31.07N 101.09W

29 P2 **Barnsdall** Oklahoma, C USA 36.33N 96.09W

99 M17 **Barnsley** N England, UK 53.34N 1.28W

21 Q14 **Barnwell** South Carolina, SE USA 33.14N 81.21W

69 U8 **Baro** *var.* Baro Wenz. ✍ Ethiopia/Sudan

79 U15 **Baro** Niger, C Nigeria 8.35N 6.28E

Baro *see* Baro Wenz

155 U12 **Baroda** *see* Vadodara

155 U12 **Baroghil Pass** *var.* Kowtal-e Barowghil. *pass* Afghanistan/Pakistan 36.54N 73.22E

121 Q17 **Baron'ki** *Rus.* Boron'ki. Mahilyowskaya Voblasts', E Belarus 53.07N 32.09E

190 J9 **Barossa Valley** *valley* South Australia

Baroui *see* Salisbury

83 H14 **Baro Wenz** *var.* Baro, Nahr Barū. ✍ Ethiopia/Sudan

159 U12 **Barpeta** Assam, NE India 26.19N 91.05E

33 S7 **Barques, Pointe Aux** *headland* Michigan, N USA 44.04N 82.57W

56 L15 **Barquisimeto** Lara, NW Venezuela 10.03N 69.18W

61 N4 **Barra** Bahia, E Brazil 11.06S 43.15W

98 F8 **Barra** *island* NW Scotland, UK

191 T5 **Barraba** New South Wales, SE Australia 30.24S 150.37E

62 L9 **Barra Bonita** São Paulo, S Brazil 22.30S 48.34W

66 J12 **Barracuda Fracture Zone** *var.* Fifteen Twenty Fracture Zone. *tectonic feature* SW Atlantic Ocean

66 G11 **Barracuda Ridge** *undersea feature* N Atlantic Ocean

45 N12 **Barra del Colorado** Limón, NE Costa Rica 10.44N 83.35W

45 N9 **Barra de Río Grande** Región Autónoma Atlántico Sur, E Nicaragua 12.56N 83.30W

84 A11 **Barra do Cuanza** Luanda, NW Angola 9.12S 13.08E

62 O9 **Barra do Piraí** Rio de Janeiro, SE Brazil 22.36S 43.47W

63 D16 **Barra do Quaraí** Rio Grande do Sul, SE Brazil 31.03S 58.10W

61 G14 **Barra do São Manuel** Pará, N Brazil 7.12S 58.03W

85 N19 **Barra Falsa, Ponta da** *headland* S Mozambique 22.57S 35.36E

98 E10 **Barra Head** *headland* NW Scotland, UK 56.46N 7.37W

62 O9 **Barra Mansa** Rio de Janeiro, SE Brazil 22.25S 44.03W

59 D14 **Barranca** Lima, W Peru 10.46S 77.46W

56 E4 **Barrancabermeja** Santander, N Colombia 7.00N 73.51W

56 L6 **Barrancas** La Guajira, N Colombia 10.58N 72.46W

57 Q6 **Barrancas** Barinas, NW Venezuela 8.46N 70.07W

56 F6 **Barrancas** Monagas, NE Venezuela 8.45N 62.12W

106 I12 **Barranco de Loba** Bolívar, N Colombia 8.55N 74.07W

64 N7 **Barrancos** Beja, S Portugal 38.07N 6.58W

64 N7 **Barranqueras** Chaco, N Argentina 27.31S 58.53W

56 E4 **Barranquilla** Atlántico, N Colombia 10.58N 74.48W

85 N20 **Barra, Ponta da** *headland* S Mozambique 23.46S 35.33E

107 P11 **Barrax** Castilla-La Mancha, C Spain 39.04N 2.12W

21 N11 **Barre** Massachusetts, NE USA 42.24N 72.06W

20 M7 **Barre** Vermont, NE USA 44.09N 72.25W

61 M17 **Barreiras** Bahia, E Brazil 12.08S 44.58W

106 F11 **Barreiro** Setúbal, W Portugal 38.40N 9.04W

67 C26 **Barren Island** *island* S Falkland Islands

22 K7 **Barren River Lake** ◎ Kentucky, S USA

62 L7 **Barretos** São Paulo, S Brazil 20.33S 48.33W

11 P14 **Barrhead** Alberta, SW Canada 54.08N 114.28W

12 G14 **Barrie** Ontario, S Canada 44.24N 79.39W

11 N16 **Barrière** British Columbia, SW Canada 51.05N 120.07W

12 H8 **Barrière, Lac** ◎ Quebec, SE Canada

190 L6 **Barrier Range** *hill range* New South Wales, SE Australia

44 G3 **Barrier Reef** *reef* E Belize

196 C16 **Barrington Island** *see* Santa Fe, Isla

191 T7 **Barrington Tops** ▲ New South Wales, SE Australia 32.06S 151.18E

191 O4 **Barringun** New South Wales, SE Australia 29.02S 145.45E

61 K18 **Barro Alto** Goiás, S Brazil 15.07S 48.56W

61 N16 **Barro Duro** Piauí, NE Brazil 5.49S 42.30W

32 J5 **Barron** Wisconsin, N USA 45.24N 91.49W

63 H15 **Barros Cassal** Rio Grande do Sul, S Brazil 29.08S 52.37W

47 P14 **Barroualie** Saint Vincent, W Saint Vincent and the Grenadines 13.13N 61.16W

---

41 O4 **Barrow** Alaska, USA 71.17N 156.47W

99 E20 **Barrow** Ir. An Bhearú. ✍ SE Ireland

189 Q6 **Barrow Creek Roadhouse** Northern Territory, N Australia 21.30S 133.52E

99 I15 **Barrow-in-Furness** NW England, UK 54.07N 3.13W

188 G7 **Barrow Island** *island* Western Australia

41 O4 **Barrow, Point** *headland* Alaska, USA 71.23N 156.28W

11 V14 **Barrows** Manitoba, S Canada 52.49N 101.36W

99 J22 **Barry** S Wales, UK 51.24N 3.18W

12 J12 **Barry's Bay** Ontario, SE Canada 45.28N 77.40W

150 K14 **Barsakel'mes, Ostrov** *island* SW Kazakhstan

153 S14 **Barsem** S Tajikistan 37.36N 71.43E

151 V11 **Barshatas** Vostochnyy Kazakhstan, E Kazakhstan 48.04N 78.38E

161 F14 **Barsi** Mahārāshtra, W India 18.13N 75.42E

102 I13 **Barsinghausen** Niedersachsen, C Germany 53.19N 9.30E

153 X8 **Barskoon** Issyk-Kul'skaya Oblast', E Kyrgyzstan 42.07N 77.34E

102 F10 **Barssel** Niedersachsen, NW Germany 53.10N 7.46E

37 U14 **Barstow** California, W USA 34.52N 117.00W

26 L8 **Barstow** Texas, SW USA 31.27N 103.23W

105 R6 **Bar-sur-Aube** Aube, NE France 48.13N 4.43E

Bar-sur-Ornain *see* Bar-le-Duc

105 Q6 **Bar-sur-Seine** Aube, N France 48.06N 4.22E

153 S13 **Bartang** Tajikistan 38.06N 71.48E

153 T13 **Bartang** ✍ Tajikistan

Bärtfa/Bartfeld *see* Bardejov

102 N7 **Barth** Mecklenburg-Vorpommern, NE Germany 54.21N 12.43E

29 W13 **Bartholomew, Bayou** ✍ Arkansas/Louisiana, S USA

57 T8 **Bartica** N Guyana 6.24N 58.36W

142 H10 **Bartın** Bartın, NW Turkey 41.37N 32.19E

189 W4 **Bartle Frere** ▲ Queensland, E Australia 17.15S 145.43E

29 P8 **Bartlesville** Oklahoma, C USA 36.45N 95.58W

31 N14 **Bartlett** Nebraska, C USA 41.51N 98.32W

22 H9 **Bartlett** Tennessee, S USA 35.12N 89.52W

27 T9 **Bartlett** Texas, SW USA 30.47N 97.25W

68 L13 **Bartlett Reservoir** ◎ Arizona, SW USA

21 N6 **Barton** Vermont, NE USA 44.44N 72.09W

112 L7 **Bartoszyce** *Ger.* Bartenstein. Warmińsko-Mazurskie, NE Poland, 54.16N 20.49E

25 W12 **Bartow** Florida, SE USA 27.54N 81.50W

176 V10 **Barú** Irian Jaya, E Indonesia 1.44S 132.16E

173 G6 **Barumun, Sungai** ✍ Sumatera, W Indonesia

Barú, Nahr *see* Baro Wenz

174 M16 **Barung, Nusa** *island* S Indonesia

173 Fj6 **Barus** Sumatera, NW Indonesia 2.08N 98.15E

168 L10 **Baruunsuu** Ömnögovi, S Mongolia 43.46N 105.28E

169 P8 **Baruun-Urt** Sühbaatar, E Mongolia 46.39N 113.17E

45 P15 **Barú, Volcán** *var.* Volcán de Chiriquí. ▲ W Panama 8.49N 82.32W

101 K21 **Barvaux** Luxembourg, SE Belgium 50.21N 5.30E

44 M13 **Barva, Volcán** ▲ NW Costa Rica 10.07N 84.08W

119 W6 **Barvinkove** Kharkiv's'ka Oblast', E Ukraine 48.54N 37.03E

160 L10 **Barwāh** Madhya Pradesh, C India 22.17N 76.01E

Barwala Madhya Pradesh, C India 22.01N 74.55E

191 P5 **Barwon River** ✍ New South Wales, SE Australia

121 U13 **Barysaw** *Rus.* Borisov. Minskaya Voblasts', NE Belarus 54.14N 28.30E

131 Q4 **Barysh** Ul'yanovskaya Oblast', W Russian Federation 53.32N 47.06E

119 O2 **Baryshivka** Kyyivs'ka Oblast', N Ukraine 50.21N 31.21E

81 D17 **Basankusu** Equateur, NW Dem. Rep. Congo (Zaire) 1.12N 19.49E

119 N11 **Basarabeasca** *Rus.* Bessarabka. SE Moldova 46.22N 28.56E

118 M14 **Basarabi** Constanța, SW Romania 44.07N 28.27E

42 H6 **Basaseachic** Chihuahua, NW Mexico 28.18N 108.13W

107 T2 **Basauri** País Vasco, N Spain 43.13N 2.53W

63 D18 **Basavilbaso** Entre Ríos, E Argentina 32.23S 58.48W

81 F21 **Bas-Congo** *off.* Région du Bas-Congo; *prev.* Bas-Zaïre. ◆ *region* SW Dem. Rep. Congo (Zaire)

112 I7 **Basel** *Eng.* Basle, *Fr.* Bâle. Basel-Stadt, NW Switzerland 47.33N 7.36E

110 D7 **Basel** *Eng.* Basle, *Fr.* Bâle. ◆ *canton* NW Switzerland

149 T4 **Bashākerd, Kühhā-ye** ▲ SE Iran

11 Q15 **Bashaw** Alberta, SW Canada 52.40N 112.53W

147 R9 **Bashdedang** Maryyskiy Velayat, S Turkmenistan 37.34N 63.07E

167 T15 **Bashi Channel** *Chin.* Pa-shih Hai-hsia. *channel* Philippines/Taiwan

Bashkiria *see* Bashkortostan

125 Dd12 **Bashkortostan, Respublika** *prev.* Bashkiria. ◆ *autonomous republic* W Russian Federation

62 L8 **Bashmakovo** Penzenskaya Oblast', W Russian Federation 53.13N 43.00E

---

152 J10 **Bashsakarba** Lebapskiy Velayat, E Turkmenistan

119 R9 **Bashtanka** Mykolayivs'ka Oblast', S Ukraine 47.24N 32.27E

195 O17 **Basilaki Island** *island* SE PNG

24 H8 **Basile** Louisiana, S USA 30.28N 92.36W

109 M18 **Basilicata** ◆ *region* S Italy

35 V13 **Basin** Wyoming, C USA 44.22N 108.02W

99 N22 **Basingstoke** S England, UK 51.16N 1.08W

149 U8 **Basjān** Khorāsān, E Iran 31.57N 59.07E

114 B10 **Baška** *It.* Bescanuova. Primorje-Gorski Kotar, NW Croatia 45.00N 14.45E

12 L10 **Baskatong, Réservoir** ◎ Quebec, SE Canada

143 O14 **Baskil** Elâzığ, E Turkey 38.35N 38.52E

160 H9 **Bāsoda** Madhya Pradesh, C India 23.56N 77.58E

81 L17 **Basoko** Orientale, N Dem. Rep. Congo (Zaire) 1.13N 23.25E

Basque Country, The *see* País Vasco

Basra *see* Al Baṣrah

105 U5 **Bas-Rhin** ◆ *department* NE France

Bassam *see* Grand-Bassam

11 Q16 **Bassano** Alberta, SW Canada 50.48N 112.28W

108 H7 **Bassano del Grappa** Veneto, NE Italy 45.45N 11.45E

79 Q13 **Bassar** *var.* Bassari. NW Togo 9.15N 0.46E

Bassari *see* Bassar

180 L9 **Bassas da India** *island group* W Madagascar

110 D7 **Bassecourt** Jura, W Switzerland 47.20N 7.16E

177 F8 **Bassein** *var.* Pathein. Irrawaddy, SW Myanmar 16.46N 94.45E

81 J15 **Basse-Kotto** ◆ *prefecture* S Central African Republic

107 V5 **Bassella** Cataluña, NE Spain 42.01N 1.16E

104 I3 **Basse-Normandie** *Eng.* Lower Normandy. ◆ *region* N France

47 X6 **Basse-Pointe** N Martinique 14.52N 61.07W

78 H7 **Basse Santa Su** E Gambia 13.18N 14.10W

Basse-Saxe *see* Niedersachsen

47 X6 **Basse-Terre** ○ (Guadeloupe) Basse Terre, W Guadeloupe 16.07N 61.40W

47 X6 **Basse Terre** *island* W Guadeloupe

47 V10 **Basseterre** ● (Saint Kitts and Nevis) Saint Kitts, Saint Kitts and Nevis 17.15N 62.45W

31 Q13 **Bassett** Nebraska, C USA 42.34N 99.32W

23 V7 **Bassett** Virginia, NE USA 36.45N 79.59W

39 R14 **Bassett Peak** ▲ Arizona, SW USA 32.30N 110.16W

78 M10 **Bassikounou** Hodh ech Chargui, SE Mauritania 15.55N 5.58W

79 R15 **Bassila** W Benin 9.01N 1.46E

33 O11 **Bass Lake** Indiana, N USA 41.12N 86.35W

191 O14 **Bass Strait** *strait* SE Australia

102 H11 **Bassum** Niedersachsen, NW Germany 52.51N 8.43E

97 J21 **Bástad** S Iraq 36.20N 45.14E

159 U2 **Basti** Uttar Pradesh, N India 26.48N 82.43E

105 X14 **Bastia** Corse, France, C Mediterranean Sea 42.42N 9.27E

101 L22 **Bastogne** Luxembourg, SE Belgium 50.00N 5.43E

24 J5 **Bastrop** Louisiana, S USA 32.46N 91.54W

27 T11 **Bastrop** Texas, SW USA 30.06N 97.19W

95 J15 **Bastuträsk** Västerbotten, N Sweden 64.46N 20.05E

121 D19 **Bastyn'** *Rus.* Bostyn'. Brestskaya Voblasts', SW Belarus 52.23N 26.46E

Basuo *see* Dongfang

Basutoland *see* Lesotho

125 O5 **Basya** ✍ E Belarus

119 V8 **Basyl'kivka** Dnipropetrovs'ka Oblast', E Ukraine 48.12N 36.00E

81 D17 **Bata** NW Equatorial Guinea 1.50N 9.47E

81 D17 **Bata** ✕ S Equatorial Guinea 1.55N 9.48E

Batae Coritanorum *see* Leicester

81 N17 **Batama** Orientale, NE Dem. Rep. Congo (Zaire) 0.54N 26.25E

126 M8 **Batamay** Respublika Sakha (Yakutiya), NE Russian Federation 63.36N 134.44E

126 L8 **Batangay-Alyta** Respublika Sakha (Yakutiya), NE Russian Federation 67.36N 130.22E

114 L10 **Batajnica** Serbia, N Yugoslavia 44.55N 20.17E

142 H15 **Batakli Gölü** ◎ S Turkey

116 H11 **Batak, Yazovir** ◎ SW Bulgaria

158 H7 **Batāla** Punjab, N India 31.50N 75.10E

106 F9 **Batalha** Leiria, C Portugal 39.40N 8.49W

81 N17 **Batama** Orientale, NE Dem. Rep. Congo (Zaire) 0.54N 26.25E

171 Q8 **Batan** Respublika Sakha (Yakutiya), NE Russian Federation 63.28N 129.33E

173 F8 **Batam, Pulau** *island* Kepulauan Riau, W Indonesia

174 H7 **Batang** Sichuan, C China 30.04N 99.10E

173 G8 **Batang** Jawa, C Indonesia 6.55S 109.42E

81 I14 **Batangafo** Ouham, NW Central African Republic 7.19N 18.22E

171 P5 **Batangas** Luzon, N Philippines 13.47N 121.02E

179 P11 **Batanta, Pulau** *island* Maluku, E Indonesia

144 F10 **Bat Yam** Tel Aviv, C Israel 32.02N 34.44E

131 N6 **Batyrevo** Chuvashskaya Respublika, W Russian Federation 55.04N 47.34E

150 Q9 **Batys Qazaqstan Oblysy** *see* Zapadnyy Kazakhstan

---

181 T9 **Batavia** *see* Jakarta

181 T9 **Batavia Seamount** *undersea feature* E Indian Ocean 27.42S 100.36E

130 L12 **Bataysk** Rostovskaya Oblast', SW Russian Federation 47.10N 39.46E

12 B9 **Batchawana** Ontario, S Canada

12 B9 **Batchawana Bay** Ontario, S Canada 46.55N 84.36W

178 Ii12 **Bătdâmbâng** *prev.* Battambang. Bătdâmbâng, NW Cambodia 13.06N 103.13E

81 Q8 **Batéké, Plateaux** *plateau* S Congo

191 S11 **Batemans Bay** New South Wales, SE Australia 35.43S 150.09E

30 J4 **Bateland** South Dakota, N USA 43.05N 102.07W

29 T3 **Batesburg** South Carolina, SE USA 33.54N 81.33W

33 Q13 **Batesville** Arkansas, C USA 35.47N 91.37W

29 T4 **Batesville** Indiana, N USA 39.18N 85.13W

24 L2 **Batesville** Mississippi, S USA 34.18N 89.56W

27 Q13 **Batesville** Texas, SW USA 28.56N 99.38W

46 L13 **Bath** E Jamaica 17.56N 76.20W

99 L22 **Bath** *hist.* Akermanceaster, *anc.* Aquae Calidae, Aquae Solis. SW England, UK 51.22N 2.22W

21 Q8 **Bath** Maine, NE USA 43.54N 69.49W

20 F11 **Bath** New York, NE USA 42.20N 77.16W

Bath *see* Berkeley Springs

80 I10 **Batha** *seasonal river* C Chad

80 I10 **Batha** ◆ *prefecture* C Chad

147 Y8 **Baţhaʾ, Wādī al** *dry watercourse* NE Oman

158 H9 **Bathinda** Punjab, NW India 30.14N 74.54E

100 M11 **Bathmen** Overijssel, E Netherlands 52.15N 6.16E

47 X9 **Bathsheba** E Barbados 13.12N 59.31W

191 R8 **Bathurst** New South Wales, SE Australia 33.32S 149.34E

11 O13 **Bathurst** New Brunswick, SE Canada 47.37N 65.40W

Bathurst *see* Banjul

15 Q2 **Bathurst, Cape** *headland* Northwest Territories, N Canada 70.33N 128.00W

206 L8 **Bathurst Inlet** Nunavut, N Canada 66.23N 107.00W

15 J4 **Bathurst Inlet** *inlet* Nunavut, N Canada

189 N1 **Bathurst Island** *island* Northern Territory, N Australia

207 O9 **Bathurst Island** *island* Parry Islands, Nunavut, N Canada

79 O14 **Batié** SW Burkina 9.53N 2.57W

197 J14 **Batiki** *prev.* Mbatiki. *island* C Fiji

147 W9 **Bāţin, Wādī al** *dry watercourse* SW Asia

13 Q11 **Batiscan** ✍ Quebec, SE Canada

142 F16 **Batı Toroslar** ▲ SW Turkey

164 M8 **Batkorgan** Xinjiang Uygur Zizhiqu, NW China 39.05N 90.00E

25 V6 **Baxley** Georgia, SE USA 31.46N 82.20W

165 R15 **Baxoi** Xizang Zizhiqu, W China 30.01N 96.53E

31 U6 **Baxter** Minnesota, N USA 46.21N 94.19W

29 R8 **Baxter Springs** Kansas, C USA 37.01N 94.45W

83 M17 **Bay** off. Gobolka Bay. ◆ *region* SW Somalia

Bay *see* Baicheng

47 O5 **Bayamo** Granma, E Cuba 20.21N 76.38W

47 U5 **Bayamón** E Puerto Rico 18.24N 66.09W

169 W8 **Bayan** Heilongjiang, NE China 46.05N 127.24E

Bayan *see* Bajan. Pulau

175 Nn16 **Bayan** prev. Bajan. Pulau Lombok, C Indonesia 8.16S 116.28E

168 K6 **Bayan** Arhangay, C Mongolia 49.36N 99.36E

169 P7 **Bayan** Dornod, E Mongolia 47.56N 112.58E

169 N9 **Bayan** Dornogovi, SE Mongolia 46.15N 110.16E

168 F7 **Bayan** Govĭ-Altay, W Mongolia 45.15N 93.21E

161 L24 **Baticaloa** Eastern Province, E Sri Lanka 7.43N 81.43E

101 J18 **Battice** Liège, E Belgium 50.39N 5.50E

109 L18 **Battipaglia** Campania, S Italy 40.36N 15.00E

168 H8 **Bayanbulag** Bayanhongor, C Mongolia 46.46N 98.07E

169 N7 **Bayanbulag** Hentiy, C Mongolia 47.45N 107.20E

11 R15 **Battle** ✍ Alberta/Saskatchewan, SW Canada

9 O5 **Battle Born State** *see* Nevada

33 Q9 **Battle Creek** Michigan, N USA 42.19N 85.10W

29 T7 **Battlefield** Missouri, C USA 37.07N 93.22W

11 S15 **Battleford** Saskatchewan, S Canada 52.45N 108.19W

168 F5 **Bayan Gol** *see* Dengkou

37 S1 **Battle Mountain** Nevada, W USA 40.37N 116.55W

161 L24 **Batticaloa** Eastern Province, E Sri Lanka

158 H7 **Battonya** *Rom.* Bătania. Békés, SE Hungary 46.16N 21.00E

169 Q14 **Batuata, Pulau** *island* C Indonesia

173 N9 **Batu, Kepulauan** *prev.* Batoe. *island group* W Indonesia

173 F8 **Batam, Pulau** *island* Kepulauan Riau, W Indonesia

174 N6 **Batu Pahat** *prev.* Bandar Penggaram. Johor, Peninsular Malaysia 1.51N 102.55E

174 H4 **Batu, Tanjung** *headland* Borneo, N Indonesia

173 Q14 **Batudaka, Pulau** *island* N Indonesia

171 O13 **Baturebe** Sulawesi, N Indonesia 6.55S 109.42E

126 H13 **Baturino** Tomskaya Oblast', C Russian Federation 57.46N 85.08E

119 R3 **Baturyn** Chernihivs'ka Oblast', N Ukraine 51.20N 32.54E

171 O13 **Baubau** *var.* Baoebaoe. Pulau Buton, C Indonesia 5.30S 122.37E

79 W14 **Bauchi** Bauchi, NE Nigeria 10.18N 9.46E

79 W14 **Bauchi** ◆ *state* C Nigeria

104 H7 **Baud** Morbihan, NW France 47.52N 2.59W

31 T2 **Baudette** Minnesota, N USA 48.43N 94.35W

---

104 F5 **Batz, Île de** *island* NW France 48.57N 105.09E

174 L7 **Bau** Sarawak, East Malaysia 1.25N 110.10E

179 P9 **Bauang** Luzon, N Philippines 16.33N 120.19E

175 Qq13 **Baubau** *var.* Baoebaoe. Pulau Buton, C Indonesia 5.30S 122.37E

79 W14 **Bauchi** Bauchi, NE Nigeria 10.18N 9.46E

79 W14 **Bauchi** ◆ *state* C Nigeria

104 H7 **Baud** Morbihan, NW France 47.52N 2.59W

31 T2 **Baudette** Minnesota, N USA 48.43N 94.35W

200 Nn10 **Bauer Basin** *undersea feature* E Pacific Ocean

197 C14 **Bauer Field** *var.* Port Vila. ✕ (Port-Vila) Éfaté, C Vanuatu 17.42S 168.21E

11 W14 **Bauld, Cape** *headland* Newfoundland, E Canada 51.35N 55.22W

29 O7 **Bates City** Missouri, C USA

59 M15 **Baures, Río** ✍ N Bolivia

62 K9 **Baurú** São Paulo, S Brazil 22.19S 49.07W

120 G10 **Bauska** *Ger.* Bauske. Bauska, S Latvia 56.24N 24.11E

103 Q15 **Bautzen** *Lus.* Budyšin. Sachsen, E Germany 51.10N 14.28E

151 Q16 **Baūyrzhan Momysh-Uly** *Kaz.* Bauyrzhan Momyshuly; *prev.* Burnoye. Zhambyl, S Kazakhstan 42.36N 70.46E

158 N9 **Bavanum** *see* Bolzano

100 M11 **Bathmen** Overijssel, E Netherlands 52.15N 6.16E

111 N7 **Bavarian Alps** *Ger.* Bayrische Alpen. ▲ Austria/Germany

Bavière *see* Bayern

42 I5 **Bavispe, Río** ✍ NW Mexico

131 T5 **Bavly** Respublika Tatarstan, W Russian Federation 54.24N 53.21E

174 Kk10 **Bawal, Pulau** *island* N Indonesia

174 Mm10 **Bawan** Borneo, C Indonesia 1.36S 113.55E

191 O12 **Baw Baw, Mount** ▲ Victoria, SE Australia 37.49S 146.16E

176 W11 **Bawe** Irian Jaya, E Indonesia 2.56S 134.39E

174 L15 **Bawen** Jawa, S Indonesia 7.13S 110.25E

77 N8 **Bawiti** N Egypt 28.18N 28.52E

79 O12 **Bawku** N Ghana 11.00N 0.12W

178 Gg7 **Bawlakè** Kayah State, C Myanmar

143 R15 **Bawshan** Siirt, SE Turkey 38.07N 41.43E

173 Ff8 **Bawo Ofuloa** Pulau Tanahmasa, W Indonesia 0.10S 98.24E

167 N4 **Ba Xian** *see* Bazhou

190 L6 **Baxch山**

165 R12 **Bayan Har Shan** *var.* Bayan Khar. ▲ C China

168 C5 **Bayan-Ölgiy** ◆ *province* NW Mongolia

168 H6 **Bayan Mod** Nei Mongol Zizhiqu, N China 40.45N 104.29E

166 K13 **Bayan Nuru** Nei Mongol Zizhiqu, N China 40.09N 104.48E

168 J12 **Bayan Obo** Nei Mongol Zizhiqu, N China 41.45N 109.58E

175 O9 **Bayan, Pulau** *see* Bayan

169 Q14 **Bayan Shan** ▲ China

165 Q9 **Bayan Shan** ▲ China 37.36N 96.23E

168 J8 **Bayanteeg** Övörhangay, C Mongolia 45.39N 101.30E

---

168 L8 **Bayantöhöm** Töv, C Mongolia 46.57N 105.09E

168 H6 **Bayan-Uhaa** Dzavhan, C Mongolia 48.41N 98.46E

168 J8 **Bayan-Ulaan** Övörhangay, C Mongolia 45.38N 102.30E

30 I4 **Bayard** Nebraska, C USA 41.45N 103.19W

39 P11 **Bayard** New Mexico, SW USA 32.45N 108.07W

105 T13 **Bayard, Col** *pass* SE France

169 O8 **Bayasgalant** Sühbaatar, SE Mongolia

142 J12 **Bayat** Çorum, N Turkey 40.34N 34.07E

179 Q15 **Bayawan** Negros, C Philippines 9.22N 122.50E

149 R16 **Bayāż** Kermān, C Iran 30.40N 55.28E

179 Q15 **Baybay** Leyte, C Philippines 10.41N 124.48E

143 P11 **Bayburt** Bayburt, NE Turkey 40.15N 40.16E

143 P12 **Bayburt** ◆ *province* NE Turkey

33 R8 **Bay City** Michigan, N USA 43.34N 83.52W

27 W12 **Bay City** Texas, SW USA 28.59N 96.00W

125 U7 **Baydaratskaya Guba** *var.* Baydarata Bay. *bay* N Russian Federation

83 M16 **Baydhabo** *var.* Baydhowa, Isha Baydhabo, *It.* Baidoa. Bay, SW Somalia 3.07N 43.39E

Baydhowa *see* Baydhabo

103 N21 **Bayerischer Wald** ▲ SE Germany

103 K21 **Bayern** *Eng.* Bavaria, *Fr.* Bavière. ◆ *state* SE Germany

153 V9 **Bayetovo** Narynskaya Oblast', C Kyrgyzstan 41.14N 74.55E

104 K4 **Bayeux** *anc.* Augustodurum. Calvados, N France 49.16N 0.42W

12 E15 **Bayfield** Ontario, S Canada 43.33N 81.42W

32 J4 **Bayfield** Wisconsin, N USA 46.48N 90.48W

151 O15 **Baygakum** *Kaz.* Bäygequm. Kzylorda, S Kazakhstan 44.15N 66.34E

142 C14 **Bayındır** İzmir, SW Turkey 38.14N 27.37E

144 H12 **Bayir** *var.* Bā'ir. Ma'ān, S Jordan 30.46N 36.40E

145 X3 **Bayjī** *var.* Baiji. N Iraq 34.55N 43.28E

120 K15 **Baykal, Ozero** *Eng.* Lake Baikal. ◎ S Russian Federation

126 J16 **Baykal'sk** Irkutskaya Oblast', S Russian Federation 51.30N 104.03E

143 R15 **Baykan** Siirt, SE Turkey 38.07N 41.43E

127 N13 **Baykonur** *see* Baykonyr

150 M14 **Baykonyr** *var.* Baykonur. *Kaz.* Bayqongyr; *prev.* Leninsk. Kzylorda, S Kazakhstan 45.38N 63.20E

151 N12 **Baykonyr** *var.* Baykonur. Zhezkazgan, C Kazakhstan 47.50N 66.05E

164 T7 **Baykurt** Xinjiang Uygur Zizhiqu, W China 39.55N 75.33E

12 J9 **Bay, Laguna de** ◎ Luzon, N Philippines

131 W8 **Baymak** Respublika Bashkortostan, W Russian Federation 52.34N 58.20E

149 U17 **Bay Minette** Alabama, S USA

149 O17 **Baynūnah** *desert* W UAE

192 O8 **Bay of Plenty** off. Bay of Plenty Region. ◆ *region* North Island, NZ

203 P3 **Bay of Wrecks** *bay* Kiritimati, E Kiribati

179 P9 **Bayombong** Luzon, N Philippines 16.29N 121.08E

104 I15 **Bayonne** *anc.* Lapurdum. Pyrénées-Atlantiques, SW France 43.30N 1.28W

24 H7 **Bayou D'Arbonne Lake** ◎ Louisiana, S USA

25 N9 **Bayou La Batre** Alabama, S USA 30.24N 88.15W

Bayou State *see* Mississippi

Bayqadam *see* Saudakent

Bayqongyr *see* Baykonyr

Bayram-Ali *see* Bayramaly

152 J12 **Bayramaly** *prev.* Bayram-Ali. Maryyskiy Velayat, S Turkmenistan 37.33N 62.08E

103 J19 **Bayreuth** *var.* Baireuth. Bayern, SE Germany 49.57N 11.34E

Bayrische Alpen *see* Bavarian Alps

Bayrūt *see* Beyrouth

24 L9 **Bay Saint Louis** Mississippi, S USA 30.18N 89.19W

Baysān *see* Bet She'an

168 L8 **Bayshint** Töv, C Mongolia 47.22N 105.04E

12 H13 **Bays, Lake of** ◎ Ontario, S Canada

24 M7 **Bay Springs** Mississippi, S USA 31.58N 89.17W

Bay State *see* Massachusetts

Baysun *see* Boysun

12 S3 **Baysville** Ontario, S Canada 45.10N 79.03W

147 Q9 **Bayt al Faqīh** W Yemen 14.30N 43.20E

164 H4 **Baytik Shan** ▲ China/Mongolia

Bayt Laḥm *see* Bethlehem

27 W11 **Baytown** Texas, SW USA 29.44N 94.58W

175 O9 **Bayu** Borneo, N Indonesia 1.52N 115.00E

123 L16 **Bayy al Kabir, Wādī** *dry watercourse* NW Libya

64 M13 **Baza** Andalucía, S Spain 37.30N 2.45W

143 X10 **Bazardüzü Daği** *Rus.* Gora Bazardyuzyu. ▲ N Azerbaijan 41.13N 47.50E

Bazardüzü Daği

160 K16 **Bazaruto, Ilha do** *island* SE Mozambique

Bazardyuzyu, Gora *see* Bazardüzü Daği

Bazargic *see* Dobrich

85 N18 **Bazaruto, Ilha do** *island* SE Mozambique

104 K14 **Bazas** Gironde, SW France 44.27N 0.11W

107 O14 **Baza, Sierra de ▲** S Spain

166 J8 **Bazhong** Sichuan, C China 31.55N 106.64E

167 P3 **Bazhou** *prev.* Baxian, Ba Xian. Hebei, E China 39.04N 116.24E

12 M9 **Bazin ♒** Quebec, SE Canada

145 Q7 **Bāziyah** C Iraq 33.49N 42.41E

144 H6 **Bcharré** var. Bcharreh, Bsharrī, Bsherri. NE Lebanon 34.16N 36.01E

30 J5 **Beach** North Dakota, N USA 46.55N 104.00W

190 K12 **Beachport** South Australia 37.29S 140.03E

99 O23 **Beachy Head** *headland* SE England, UK 50.44N 0.16E

20 K13 **Beacon** New York, NE USA 41.30N 73.54W

65 J25 **Beagle Channel** *channel* Argentina/Chile

189 O1 **Beagle Gulf** *gulf* Northern Territory, N Australia

Bealach an Doirín *see* Ballaghaderreen

Bealach Cláir *see* Ballyclare

Bealach Féich *see* Ballybofey

180 J3 **Bealanana** Mahajanga, NE Madagascar 14.33S 48.43E

Béal an Átha *see* Ballina

Béal an Átha Móir *see* Ballinamore

Béal an Mhuirhead *see* Belmullet

Béal Átha Beithe *see* Ballybay

Béal Átha Conaill *see* Ballyconnell

Béal Átha hAmhnais *see* Ballyhaunis

Béal Átha na Sluaighe *see* Ballinasloe

Béal Átha Seanaidh *see* Ballyshannon

Bealdovuopmi *see* Peltovuoma

Béal Feirste *see* Belfast

Béal Tairbirt *see* Belturbet

Beanna Boirche *see* Mourne Mountains

Beannchar *see* Banagher, Ireland

Beannchar *see* Bangor, Northern Ireland, UK

Beanntraí *see* Bantry

25 N2 **Bear Creek ♒** Alabama/Mississippi, S USA

32 J13 **Bear Creek ♒** Illinois, N USA

29 U13 **Bearden** Arkansas, C USA 33.43N 92.37W

205 Q10 **Beardmore Glacier** *glacier* Antarctica

32 K13 **Beardstown** Illinois, N USA 40.01N 90.25W

30 L14 **Bear Hill ▲** Nebraska, C USA 41.24N 101.49W

Bear Island *see* Bjørnøya

12 H12 **Bear Lake** Ontario, S Canada 45.28N 79.31W

68 M1 **Bear Lake ☒** Idaho/Utah, NW USA

41 U11 **Bear, Mount ▲** Alaska, USA 61.16N 141.09W

104 J16 **Béarn** *cultural region* SW France

204 J11 **Bear Peninsula** *peninsula* Antarctica

158 I7 **Beās ♒** India/Pakistan

107 P3 **Beasain** País Vasco, N Spain 43.03N 2.10W

107 O12 **Beas de Segura** Andalucía, S Spain 38.16N 2.53W

47 N10 **Beata, Cabo** *headland* SW Dominican Republic 17.34N 71.25W

47 N10 **Beata, Isla** *island* SW Dominican Republic

66 F11 **Beata Ridge** *undersea feature* N Caribbean Sea

31 R17 **Beatrice** Nebraska, C USA 40.14N 96.43W

85 L16 **Beatrice** Mashonaland East, NE Zimbabwe 18.13S 30.52E

11 N11 **Beatton ♒** British Columbia, W Canada

11 N11 **Beatton River** British Columbia, W Canada 57.35N 121.10W

37 V10 **Beatty** Nevada, W USA 36.53N 116.44W

23 N6 **Beattyville** Kentucky, S USA 37.34N 83.39W

181 X16 **Beau Bassin** W Mauritius 20.13S 57.27E

105 R15 **Beaucaire** Gard, S France 43.49N 4.37E

12 I8 **Beauchastel, Lac** ☺ Quebec, SE Canada

12 I10 **Beauchêne, Lac** ☺ Quebec, SE Canada

191 V3 **Beaudesert** Queensland, E Australia 28.00S 152.27E

190 M12 **Beaufort** Victoria, SE Australia 37.27S 143.24E

23 X11 **Beaufort** North Carolina, SE USA 34.45N 76.50W

21 R15 **Beaufort** South Carolina, SE USA 32.25N 80.40W

40 M11 **Beaufort Sea** *sea* Arctic Ocean

Beaufort-West *Afr.* Beaufort-Wes. Western Cape, SW South Africa 32.21S 22.34E

105 N7 **Beaugency** Loiret, C France 47.46N 1.38E

21 R1 **Beau Lake** ☺ Maine, NE USA

98 I8 **Beauly** N Scotland, UK 57.28N 4.28W

101 G21 **Beaumont** Hainaut, S Belgium 50.12N 4.13E

193 E23 **Beaumont** Otago, South Island, NZ 45.48S 169.32E

24 M7 **Beaumont** Mississippi, S USA 31.10N 88.55W

27 X10 **Beaumont** Texas, SW USA 30.05N 94.06W

104 M15 **Beaumont-de-Lomagne** Tarnet-Garonne, S France 43.54N 1.00E

104 L6 **Beaumont-sur-Sarthe** Sarthe, NW France 48.15N 0.07E

105 R8 **Beaune** Côte d'Or, C France 47.01N 4.49E

13 R9 **Beaupré** Quebec, SE Canada 47.03N 70.52W

104 J8 **Beaupréau** Maine-et-Loire, NW France 47.13N 0.57W

101 I22 **Beauraing** Namur, SE Belgium 50.07N 4.57E

105 R12 **Beaurepaire** Isère, E France 45.20N 5.03E

11 Y16 **Beausejour** Manitoba, S Canada 50.04N 96.30W

105 N4 **Beauvais** *anc.* Bellovacum, Caesaromagus. Oise, N France 49.27N 2.04E

11 S13 **Beauval** Saskatchewan, C Canada 55.10N 107.37W

104 I9 **Beauvoir-sur-Mer** Vendée, NW France 46.54N 2.03W

41 R8 **Beaver** Alaska, USA 66.22N 147.31W

28 J8 **Beaver** Oklahoma, C USA 36.49N 100.31W

20 B14 **Beaver** Pennsylvania, NE USA 40.39N 80.19W

68 K6 **Beaver** Utah, W USA 38.16N 112.38W

8 L9 **Beaver ♒** British Columbia/Yukon Territory, W Canada

11 S13 **Beaver ♒** Saskatchewan, C Canada

31 N17 **Beaver City** Nebraska, C USA 40.08N 99.49W

8 G6 **Beaver Creek** Yukon Territory, W Canada 62.19N 140.45W

33 R14 **Beavercreek** Ohio, N USA 39.42N 83.58W

41 S8 **Beaver Creek ♒** Alaska, USA

28 H3 **Beaver Creek ♒** Kansas/Nebraska, C USA

30 J5 **Beaver Creek ♒** Montana/North Dakota, N USA

31 Q14 **Beaver Creek ♒** Nebraska, C USA

27 Q4 **Beaver Creek ♒** Texas, SW USA

32 M8 **Beaver Dam** Wisconsin, N USA 43.28N 88.49W

32 M8 **Beaver Dam Lake** ☺ Wisconsin, N USA

20 B14 **Beaver Falls** Pennsylvania, NE USA 40.45N 80.20W

35 P12 **Beaverhead Mountains** ▲ Idaho/Montana, NW USA

35 Q12 **Beaverhead River ♒** Montana, NW USA

67 A25 **Beaver Island** *island* W Falkland Islands

33 P5 **Beaver Island** *island* Michigan, N USA

29 S9 **Beaver Lake** ☒ Arkansas, C USA

10 I8 **Beaverlodge** Alberta, W Canada 55.10N 119.28W

28 J8 **Beaver River ♒** New York, NE USA

28 J8 **Beaver River ♒** Oklahoma, C USA

20 B13 **Beaver River ♒** Pennsylvania, NE USA

67 A25 **Beaver Settlement** Beaver Island, W Falkland Islands 51.30S 61.15W

Beaver State *see* Oregon

32 H14 **Beaverton** Ontario, S Canada 44.24N 79.07W

34 G11 **Beaverton** Oregon, NW USA 45.29N 122.48W

158 G12 **Beāwar** Rājasthān, N India 26.07N 74.21E

62 L8 **Bebedouro** São Paulo, S Brazil 20.58S 48.28W

103 I16 **Bebra** Hessen, C Germany 50.59N 9.46E

43 W12 **Becal** Campeche, SE Mexico 19.49N 90.28W

13 Q11 **Bécancour ♒** Quebec, SE Canada

99 Q19 **Beccles** E England, UK 52.27N 1.32E

114 L9 **Bečej** *Ger.* Altbecse, *Hung.* Óbecse, Rácz-Becse; *prev.* Magyar-Becse, Stari Bečej. Serbia, N Yugoslavia 45.36N 20.02E

106 I3 **Becerréa** Galicia, NW Spain 42.51N 7.10W

76 H7 **Béchar** *prev.* Colomb-Béchar. W Algeria 31.38N 2.10W

41 O14 **Becharof Lake** ☺ Alaska, USA

118 H15 **Bechet** var. Bechetu. Dolj, SW Romania 43.45N 23.57E

Bechetu *see* Bechet

23 R6 **Beckley** West Virginia, NE USA 37.46N 81.11W

103 G14 **Beckum** Nordrhein-Westfalen, W Germany 51.45N 8.03E

27 X7 **Beckville** Texas, SW USA 32.14N 94.27W

37 X4 **Becky Peak ▲** Nevada, W USA 39.59N 114.33W

118 I9 **Beclean** *Hung.* Bethlen; *prev.* Betlen. Bistriţa-Năsăud, N Romania 47.10N 24.10E

Bécs *see* Wien

113 H18 **Bečva** *Ger.* Betschau, *Pol.* Beczwa. ♒ E Czech Republic

Beczwa *see* Bečva

122 Dd12 **Beddouza, Cap** *headland* W Morocco 32.35N 9.16W

82 H13 **Bedelē** Oromo, C Ethiopia 8.27N 36.21E

153 Y8 **Bedel Pass** *Rus.* Pereval Bedel. *pass* China/Kyrgyzstan 41.22N 78.19E

Bedel, Pereval *see* Bedel Pass

97 H22 **Bedford** E England, UK 56.03N 10.13E

33 R14 **Bedford** Indiana, N USA 38.51N 86.29W

22 L4 **Bedford** Kentucky, S USA 40.40N 94.43W

20 D15 **Bedford** Pennsylvania, NE USA 40.00N 78.29W

23 T6 **Bedford** Virginia, NE USA 37.19N 79.31W

99 N20 **Bedfordshire** *cultural region* E England, UK

131 N5 **Bednodem'yanovsk** Penzenskaya Oblast', W Russian Federation 53.55N 43.14E

100 N5 **Bedum** Groningen, NE Netherlands 53.18N 6.36E

29 V11 **Beebe** Arkansas, C USA 35.04N 91.52W

47 T9 **Beef Island ✈** (Road Town) Tortola, E British Virgin Islands 18.25N 64.31W

Beehive State *see* Utah

101 L18 **Beek** Limburg, SE Netherlands 50.55N 5.46E

101 L18 **Beek ✈** (Maastricht) Limburg, SE Netherlands 50.55N 5.47E

101 K14 **Beek-en-Donk** Noord-Brabant, S Netherlands 51.31N 5.37E

144 F13 **Be'ér Menuḥa** var. Be'er Menukha. Southern, S Israel 30.21N 35.09E

101 D16 **Beernem** West-Vlaanderen, NW Belgium 51.09N 3.18E

101 I16 **Beerse** Antwerpen, N Belgium 51.20N 4.52E

172 Qq7 **Beersheba** *see* Be'ér Sheva'

144 E11 **Be'ér Sheva'** var. Beersheba, Ar. Bir es Saba. Southern, S Israel 31.15N 34.46E

11 S13 **Beese** Gelderland, C Netherlands 51.52N 5.12E

101 M16 **Beesel** Limburg, SE Netherlands 51.16N 6.01E

85 J21 **Beestekraal** North-West, N South Africa 25.21S 27.40E

204 J7 **Beethoven Peninsula** *peninsula* Alexander Island, Antarctica

Beetsterzweach *see* Beetsterzwaag

100 M6 **Beetsterzwaag** *Fris.* Beetstersweach. Friesland, N Netherlands 53.03N 6.04E

27 S13 **Beeville** Texas, SW USA 28.25N 97.46W

81 J18 **Befale** Equateur, NW Dem. Rep. Congo (Zaire) 0.25N 20.48E

180 I7 **Befandriana** *see* Befandriana Avaratra

180 J3 **Befandriana Avaratra** var. Befandriana, Befandriana Nord. Mahajanga, NW Madagascar 15.13S 48.33E

Befandriana Nord *see* Befandriana Avaratra

81 J18 **Befori** Equateur, N Dem. Rep. Congo (Zaire) 0.09N 22.18E

180 I7 **Befotaka** Fianarantsoa, S Madagascar 23.49S 47.00E

171 R11 **Bega** New South Wales, SE Australia 36.43S 149.49E

104 G5 **Bégard** Côtes d'Armor, NW France 48.37N 3.18W

114 M9 **Begejski Kanal** *canal* N Yugoslavia

151 V9 **Begen ♒** Vostochnyy Kazakhstan, E Kazakhstan 51.11N 79.03E

96 G13 **Begna ♒** S Norway

Begoml' *see* Byahoml'

159 Q13 **Begusarāi** Bihār, NE India 25.25N 86.07E

149 R9 **Behābād** Yazd, C Iran 32.22N 59.49E

79 V11 **Behagle** *see* Lai

57 Z10 **Béhague, Pointe** *headland* E French Guiana 4.37N 51.52W

148 M10 **Behar** *see* Bihār

148 M10 **Behbahān** var. Behbehan. Khūzestān, SW Iran 30.37N 50.07E

Behbehan *see* Behbahān

46 G3 **Behring Point** Andros Island, W Bahamas 24.28N 77.44W

149 P4 **Behshahr** *prev.* Ashraf. Māzandarān, N Iran 36.42N 53.36E

169 V6 **Bei'an** Heilongjiang, NE China 48.16N 126.28E

82 H13 **Beigi** Oromo, E Ethiopia 9.13N 34.48E

166 L16 **Beihai** Guangxi Zhuangzu Zizhiqu, S China 21.28N 109.10E

165 Q10 **Bei Hulsan Hu** ☺ C China

167 N13 **Bei Jiang ♒** S China

167 O2 **Beijing** var. Pei-ching, Eng. Peking; *prev.* Pei-p'ing. *country/municipality capital* (China) Beijing Shi, E China 39.58N 116.22E

167 P2 **Beijing ✈** Beijing Shi, E China 39.54N 116.22E

Beijing *see* Beijing Shi

167 O2 **Beijing Shi** var. Beijing, Jing, Pei-ching, Peking; *prev.* Pei-p'ing. *municipality* E China

78 G8 **Beïla** Trarza, W Mauritania 18.07N 15.55W

166 L15 **Beiliu** Guangxi Zhuangzu Zizhiqu, S China 22.43N 110.21E

165 O12 **Beilu He ♒** W China

99 H8 **Beinn Dearg ▲** N Scotland, UK 57.47N 4.52W

Beinn MacDuibh *see* Ben Macduí

162 I12 **Beipan Jiang ♒** S China

169 T12 **Beipiao** Liaoning, NE China 41.46N 120.51E

83 H4 **Beira** Sofala, C Mozambique 19.49S 34.52E

83 I4 **Beira ♒** Sofala, C Mozambique 19.39S 35.05E

106 I7 **Beira Alta** *former province* N Portugal

106 I8 **Beira Baixa** *former province* C Portugal

106 G8 **Beira Litoral** *former province* N Portugal

144 H6 **Beirut** *see* Beyrouth

Beisän *see* Bet She'an

11 Q16 **Beiseker** Alberta, SW Canada 51.20N 113.34W

85 K19 **Beitbridge** Matabeleland South, S Zimbabwe 22.10S 30.02E

118 G10 **Beiuş** *Hung.* Bihor, NW Romania 46.40N 22.18E

169 U12 **Beizhen** Liaoning, NE China 41.34N 121.51E

106 H12 **Beja** *anc.* Pax Julia. Beja, SE Portugal 38.01N 7.52W

106 G13 **Beja** *district* S Portugal

76 M5 **Béja** var. Bājah. N Tunisia 36.45N 9.04E

76 L4 **Bejaïa** *var.* Bejaïa, *Fr.* Bougie; *anc.* Saldae. NE Algeria 36.45N 5.02E

123 Iii1 **Béjar** Castilla-León, N Spain 40.24N 5.45W

152 A8 **Bek-Budi** *see* Qarshi

41 J14 **Bekasi** Jawa, C Indonesia 6.13S 106.59E

130 J8 **Bekabad** *see* Bekobod

174 J14 **Bekasi** Jawa, C Indonesia 6.13S 106.59E

Bekdash *see* Bekdaş

35 S11 **Bek-Budi** *see* Qarshi

152 A8 **Bekdash** Balkanskiy Velayat, NW Turkmenistan 41.33N 52.33E

153 T10 **Bek-Dzhar** Oshskaya Oblast', SW Kyrgyzstan 40.22N 73.08E

113 N24 **Békés** *Rom.* Bichiş. Békés, SE Hungary 46.47N 21.07E

113 M24 **Békés** *off.* Békés Megye. ♦ *county* SE Hungary

145 S2 **Bēkma** E Iraq 36.40N 44.15E

180 H7 **Bekitro** Toliara, S Madagascar 24.12S 45.19E

174 J14 **Bekkai** Hokkaidō, NE Japan 43.23N 145.01E

153 Q11 **Bekobod** *Rus.* Bekabad; *prev.* Begovat. Toshkent Wiloyati, E Uzbekistan 40.17N 69.10E

131 O7 **Bekovo** Penzenskaya Oblast', W Russian Federation 52.27N 43.41E

158 M13 **Bel** *see* Beliu

Bela Uttar Pradesh, N India 25.55N 82.00E

155 N15 **Bela** Baluchistān, SW Pakistan 26.12N 66.22E

81 F15 **Bélabo** Est, C Cameroon 4.54N 13.10E

114 N10 **Bela Crkva** *Ger.* Weisskirchen, *Hung.* Fehértemplom. Serbia, W Yugoslavia 44.54S 21.28E

181 Y16 **Bel Air** var. Rivière Sèche. E Mauritius

106 L12 **Belalcázar** Andalucía, S Spain 38.33N 5.07W

115 P15 **Bela Palanka** Serbia, SE Yugoslavia 43.13N 22.19E

121 H16 **Belarus** *off.* Republic of Belarus, var. Belorussia, *Latv.* Baltkrievija; *prev.* Belorussian SSR, *Rus.* Belorusskaya SSR. ♦ *republic* E Europe

Belau *see* Palau

61 H21 **Bela Vista** Mato Grosso do Sul, SW Brazil 22.04S 56.25W

85 L21 **Bela Vista** Maputo, S Mozambique 26.19S 32.40E

173 Ff4 **Belawan** Sumatera, W Indonesia 3.44N 98.39E

131 U4 **Běla Woda** *see* Weisswasser

127 N7 **Belaya ♒** W Russian Federation

127 N7 **Belaya Gora** Respublika Sakha (Yakutiya), NE Russian Federation 68.25N 146.12E

130 M11 **Belaya Kalitva** Rostovskaya Oblast', SW Russian Federation 48.09N 40.43E

129 R14 **Belaya Kholunitsa** Kirovskaya Oblast', NW Russian Federation 58.54N 50.52E

Belaya Tserkov' *see* Bila Tserkva

79 VI1 **Belbédji** Zinder, S Niger 14.35N 8.00E

112 K13 **Bełchatów** var. Belchatow. Łódzkie, C Poland 51.22N 19.19E

108 D6 **Belcher, Îles** *see* Belcher Islands

10 H7 **Belcher Islands** *Fr.* Îles Belcher. *island group* Nunavut, SE Canada

23 P6 **Belcher** Michigan, N USA 41.38N 85.12W

107 S6 **Belchite** Aragón, NE Spain 41.18N 0.45W

31 O2 **Belcourt** North Dakota, N USA 48.50N 99.44W

33 P9 **Belding** Michigan, N USA 43.06N 85.13W

131 U5 **Belebey** Respublika Bashkortostan, W Russian Federation 54.04N 54.13E

83 N16 **Beledweyne** var. Belet Huen, *It.* Beled Uen. Hiiraan, C Somalia 4.39N 45.12E

152 B10 **Belek** Balkanskiy Velayat, W Turkmenistan 39.57N 53.51E

60 L12 **Belém** var. Pará. *state capital* Pará, N Brazil 1.27S 48.29W

67 I14 **Belém Ridge** *undersea feature* C Atlantic Ocean

39 R12 **Belen** New Mexico, SW USA 34.37N 106.46W

64 I7 **Belén** Catamarca, NW Argentina 27.40S 67.01W

56 G9 **Belén** Boyacá, C Colombia 5.59N 72.55W

44 J11 **Belén** Rivas, SW Nicaragua 11.30N 85.55W

64 O5 **Belén** Concepción, C Paraguay 23.25S 57.13W

63 I20 **Belén** Salto, N Uruguay 30.46S 57.46W

63 D20 **Belén de Escobar** Buenos Aires, E Argentina 34.21S 58.46W

116 J7 **Belene** Pleven, N Bulgaria 43.39N 25.09E

116 J7 **Belene, Ostrov** *island* N Bulgaria

45 R13 **Belén, Río** ♒ C Panama

Belényes *see* Beiuş

197 G4 **Belep, Îles** *island group* W New Caledonia

106 H3 **Belesar, Embalse de** ☒ NW Spain

Belet Huen/Belet Uen *see* Beledweyne

130 J5 **Beleū Tul'skaya Oblast',** W Russian Federation 53.48N 36.07E

99 G15 **Belfast** *Ir.* Béal Feirste. ● E Northern Ireland, UK 54.34N 5.55W

21 R7 **Belfast** Maine, NE USA 44.25N 69.02W

99 G15 **Belfast Lough** *Ir.* Loch Lao *inlet* E Northern Ireland, UK 54.37N 6.11W

105 S10 **Belfort** North Dakota, N USA 46.53N 103.12W

105 U7 **Belfort** Territoire-de-Belfort, E France 47.37N 6.52E

161 E17 **Belgaum** Karnātaka, W India 15.52N 74.30E

Belgian Congo *see* Congo (Democratic Republic of the)

205 T3 **Belgica Mountains** ▲ Antarctica

101 F20 **België/Belgique** *see* Belgium

101 F20 **Belgium** *off.* Kingdom of Belgium, *Dut.* België, *Fr.* Belgique. ♦ *monarchy* NW Europe

130 J7 **Belgorod** Belgorodskaya Oblast', W Russian Federation 50.37N 36.37E

130 J8 **Belgorodskaya Oblast' ♦** *province* W Russian Federation

Belgrad *see* Beograd

31 T8 **Belgrade** Minnesota, N USA 45.27N 94.59W

35 S11 **Belgrade** Montana, NW USA 45.46N 111.10W

205 N5 **Belgrano II** *Argentinian research station* Antarctica 77.50S 35.25W

Belgrano, Cabo *see* Meredith, Cape

23 X9 **Belhaven** North Carolina, SE USA 35.36N 76.50W

109 I23 **Belice** var. Hypsas. ♒ Sicilia, Italy, C Mediterranean Sea

Belice *see* Belize/Belize City

115 M24 **Beli Drim** *Alb.* Drini i Bardhë. ♒ Albania/Yugoslavia

Beligrad *see* Berat

196 C8 **Beliliou** *prev.* Peleliu. *island* S Palau

116 L8 **Beli Lom, Yazovir** ☒ NE Bulgaria

114 I8 **Beli Manastir** *Hung.* Pélmonostor; *prev.* Monostor. Osijek-Baranja, NE Croatia 45.46N 18.38E

104 J13 **Bélin-Béliet** Gironde, SW France 44.30N 0.48W

81 F17 **Bélinga** Ogooué-Ivindo, NE Gabon 1.05N 13.12E

23 S4 **Belington** West Virginia, NE USA 39.01N 79.57W

131 O6 **Belinskiy** Penzenskaya Oblast', W Russian Federation 52.58N 43.25E

174 I9 **Belinyu** Pulau Bangka, W Indonesia 1.34S 105.45E

22 F9 **Belitung, Pulau** *island* W Indonesia

27 U5 **Bell** Texas, SW USA 33.36N 96.24W

94 N3 **Bell** *island* NW Svalbard

108 H6 **Belluno** Veneto, NE Italy 46.07N 12.06E

64 L11 **Bell Ville** Córdoba, C Argentina 32.42S 62.42W

85 E26 **Bellville** Western Cape, SW South Africa 33.49S 18.43E

27 U11 **Bellville** Texas, SW USA 29.57N 96.15W

31 V12 **Belmond** Iowa, C USA 42.51N 93.36W

18 K9 **Belmont** New York, NE USA 42.13N 78.01W

23 R10 **Belmont** North Carolina, SE USA 35.13N 81.01W

61 O18 **Belmonte** Bahia, E Brazil 15.52S 38.54W

106 I8 **Belmonte** Castelo Branco, C Portugal 40.21N 7.19W

107 P10 **Belmonte** Castilla-La Mancha, C Spain 39.34N 2.43W

44 G2 **Belmopan** ● (Belize) Cayo, C Belize 17.13N 88.48W

99 B16 **Belmullet** *Ir.* Béal an Mhuirhead. W Ireland 54.13N 9.58W

126 Mm15 **Belogorsk** Amurskaya Oblast', SE Russian Federation 50.53N 128.24E

116 F7 **Belogradchik** Vidin, NW Bulgaria 43.37N 22.42E

180 H8 **Beloha** Toliara, S Madagascar 25.09S 45.04E

61 M20 **Belo Horizonte** *prev.* Bello Horizonte. *state capital* Minas Gerais, SE Brazil 19.54S 43.54W

61 V11 **Beloit** Kansas, C USA 39.27N 98.06W

32 L9 **Beloit** Wisconsin, N USA 42.31N 89.01W

126 H15 **Belokorovichi** *see* Bilokorovychi

126 H15 **Belokurikha** Altayskiy Kray, S Russian Federation 51.57N 84.56E

128 E8 **Belomorsk** Respublika Kareliya, NW Russian Federation 64.30N 34.43E

Belomorsko-Baltiyskiy Kanal *Eng.* White Sea-Baltic Canal, White Sea. *canal* NW Russian Federation

159 V15 **Belonia** Tripura, NE India 23.15N 91.25E

29 V5 **Belopol'ye** *see* Bilopillya

107 O4 **Belorado** Castilla-León, N Spain 42.25N 3.10W

130 L14 **Belorechensk** Krasnodarskiy Kray, SW Russian Federation 44.46N 39.53E

Belorussia/Belorussian SSR *see* Belarus

Belorusskaya Gryada *see* Byelaruskaya Hrada

Belorusskaya SSR *see* Belarus

116 M8 **Beloslav** Varna, E Bulgaria 43.11N 27.42E

Belostok *see* Białystok

79 N12 **Belovo** *see* Belowe

131 W3 **Beloretsk** Respublika Bashkortostan, W Russian Federation 54.00N 58.22E

Belovezhskaya Pushcha *see* Białowieska, Puszcza/Byelavyezhskaya Pushcha

116 H10 **Belovo** *Bul.* Byalovo. C Bulgaria 42.12N 24.02E

126 H15 **Belovo** Kemerovskaya Oblast', S Russian Federation 54.25N 86.13E

31 W14 **Belle Plaine** Iowa, C USA 41.54N 92.16W

31 V9 **Belle Plaine** Minnesota, N USA 44.37N 93.44W

21 R7 **Belfast** Maine, NE USA 44.25N 69.02W

12 I5 **Belleterre** Quebec, SE Canada 47.24N 78.40W

12 J12 **Belleville** Ontario, SE Canada 44.10N 77.23W

105 R10 **Belleville** Rhône, E France 46.06N 4.44E

32 K15 **Belleville** Illinois, N USA 38.31N 89.58W

28 M3 **Belleville** Kansas, C USA 39.46N 97.37W

116 J10 **Belleville** Ohio, N USA 42.15N 90.25W

31 S15 **Bellevue** Iowa, C USA 41.08N 95.13W

33 S15 **Bellevue** Ohio, N USA 41.16N 82.50W

27 Z5 **Bellevue** Texas, SW USA 33.38N 98.00W

34 H7 **Bellevue** Washington, NW USA 47.36N 122.12W

109 U3 **Belpre** Ohio, N USA 39.14N 81.34W

100 M8 **Belterwijde** ☺ N Netherlands

29 R4 **Belton** Missouri, C USA 38.48N 94.31W

23 P11 **Belton** South Carolina, SE USA 34.31N 82.29W

27 T9 **Belton** Texas, SW USA 31.03N 97.27W

27 S9 **Belton Lake** ☒ Texas, SW USA

99 E16 **Belturbet** *Ir.* Béal Tairbirt. N Ireland 54.06N 7.25W

Beluchistan *see* Baluchistān

151 Z9 **Belukha, Gora** ▲ Kazakhstan/Russian Federation 49.42N 86.33E

109 M20 **Belvedere Marittimo** Calabria, SW Italy 39.37N 15.52E

32 L10 **Belvidere** Illinois, N USA 42.15N 88.50W

20 J14 **Belvidere** New Jersey, NE USA 40.49N 75.03W

Bely *see* Belyy

131 V8 **Belyayevka** Orenburgskaya Oblast', W Russian Federation 51.25N 56.26E

27 T8 **Belyy** var. Bely. Tverskaya Oblast', W Russian Federation 55.51N 32.57E

128 H17 **Belyy** var. Bely, Beyl. Tverskaya Oblast', W Russian Federation 55.51N 32.57E

130 I6 **Belyye Berega** Bryanskaya Oblast', W Russian Federation 53.11N 34.42E

126 H5 **Belyy, Ostrov** *island* N Russian Federation

126 H12 **Belyy Yar** Tomskaya Oblast', C Russian Federation 58.26N 84.57E

102 N13 **Belzig** Brandenburg, NE Germany 52.09N 12.37E

24 K4 **Belzoni** Mississippi, S USA 33.10N 90.29W

180 H4 **Bemaraha** ▲ W Madagascar

84 B10 **Bembe** Uíge, NW Angola 7.01S 14.18E

79 S14 **Bembèrèkè** var. Bimbéréké. N Benin 10.10N 2.40E

106 K12 **Bembézar ♒** SW Spain

106 J3 **Bembibre** Castilla-León, N Spain 42.37N 6.25W

31 T4 **Bemidji** Minnesota, N USA 47.27N 94.53W

100 L12 **Bemmel** Gelderland, SE Netherlands 51.52N 5.54E

176 U11 **Bemu** Pulau Seram, E Indonesia 3.21S 129.58E

107 T5 **Benabarre** var. Benavarn. Aragón, NE Spain 42.06N 0.28E

107 N3 **Benaco** *see* Garda, Lago di

81 L20 **Bena-Dibele** Kasai Oriental, C Dem. Rep. Congo (Zaire) 4.01S 22.50E

107 R9 **Benageber, Embalse de** ☒ E Spain

191 O11 **Benalla** Victoria, SE Australia 36.33S 146.00E

106 M14 **Benamejí** Andalucía, S Spain 37.16N 4.33W

Benares *see* Vārānasi

Benavarn *see* Benabarre

106 F10 **Benavente** Santarém, C Portugal 38.58N 8.49W

106 K5 **Benavente** Castilla-León, N Spain 42.00N 5.40W

27 S15 **Benavides** Texas, SW USA 27.36N 98.24W

98 F8 **Benbecula** *island* NW Scotland, UK

Bencovazzo *see* Benkovac

34 H13 **Bend** Oregon, NW USA 44.03N 121.18W

190 K7 **Benda Range** ▲ South Australia

191 T6 **Bendemeer** New South Wales, SE Australia 30.54S 151.12E

Bender *see* Tighina

Bender Beila/Bender Beyla *see* Bandarbeyla

Bender Cassim/Bender Qaasim *see* Boosaaso

Bendery *see* Tighina

191 N11 **Bendigo** Victoria, SE Australia 36.46S 144.18E

120 E10 **Bēne** Dobele, SW Latvia 56.30N 23.04E

100 K13 **Beneden-Leeuwen** Gelderland, C Netherlands 51.52N 5.32E

103 L24 **Benediktbeuern ▲** S Germany 47.39N 11.28E

Benemérita de San Cristóbal *see* San Cristóbal

79 N12 **Bénéna** Ségou, S Mali 13.04N 4.20W

180 I7 **Benenitra** Toliara, S Madagascar 23.25S 45.06E

Beneschau *see* Benešov

Beneški Zaliv *see* Venice, Gulf of

113 D17 **Benešov** *Ger.* Beneschau. Středočeský Kraj, W Czech Republic 49.48N 14.40E

126 LI3 **Benetta, Ostrov** *island* Novosibirskiye Ostrova, NE Russian Federation

109 L17 **Benevento** *anc.* Beneventum, Malventum. Campania, S Italy 41.07N 14.45E

181 S3 **Bengal, Bay of** *bay* N Indian Ocean

81 M17 **Bengamisa** Orientale, N Dem. Rep. Congo (Zaire) 0.58N 25.10E

Bengal, Sungai ♒ Jawa, S Indonesia

Bengasi *see* Banghāzī

167 P7 **Bengbu** var. Peng-pu. Anhui, E China 32.57N 117.17E

34 L9 **Benge** Washington, NW USA 46.55N 118.01W

Benghazi *see* Banghāzī

174 H7 **Bengkalis** Sumatera, W Indonesia 1.29N 102.07E

174 H6 **Bengkalis, Pulau** *island* W Indonesia

174 Kk7 **Bengkayang** Borneo, C Indonesia 0.45N 109.28E

174 H11 **Bengkoelen/Bengkoeloe** *see* Bengkulu

174 H11 **Bengkulu** *prev.* Bengkoeloe, Benkoelen, Benkulen. Sumatera, W Indonesia 3.46S 102.16E

174 H11 **Bengkulu** *off.* Propinsi Bengkulu; *prev.* Bengkoelen, Benkoelen, Benkulen. ♦ *province* W Indonesia

84 A11 **Bengo ♦** *province* W Angola

◆ COUNTRY ◇ DEPENDENT TERRITORY ◉ ADMINISTRATIVE REGION ▲ MOUNTAIN ☒ VOLCANO ☺ LAKE
● COUNTRY CAPITAL ○ DEPENDENT TERRITORY CAPITAL ✈ INTERNATIONAL AIRPORT ▲ MOUNTAIN RANGE ♒ RIVER ☒ RESERVOIR

97 J16 **Bengtsfors** Västra Götaland, S Sweden 59.03N 12.13E
84 B13 **Benguela** var. Benguella. Benguela, W Angola 12.34S 13.30E
85 A14 **Benguela** ◐ province W Angola
Benguella see Benguela
Bengweulu, Lake see Bangweulu, Lake
124 Qq15 **Benha** var. Banhā. N Egypt 30.22N 31.16E
198 G6 **Benham Seamount** undersea feature W Philippine Sea 15.48N 124.15E
98 H6 **Ben Hope** ▲ N Scotland, UK 58.25N 4.36W
81 P18 **Beni** Nord Kivu, NE Dem. Rep. Congo (Zaire) 0.31N 29.29E
59 L15 **Beni** var. El Beni. ◐ department N Bolivia
76 H8 **Beni Abbès** W Algeria 30.07N 2.09W
107 T8 **Benicarló** País Valenciano, E Spain 40.25N 0.25E
107 T9 **Benicasim** País Valenciano, E Spain 40.03N 0.03E
107 T12 **Benidorm** País Valenciano, SE Spain 38.33N 0.09W
77 W9 **Beni Mazâr** var. Banī Mazār. C Egypt 28.24N 30.38E
122 F12 **Beni-Mellal** C Morocco 32.20N 6.21W
79 R14 **Benin** off. Republic of Benin; prev. Dahomey. ◆ republic W Africa
79 S17 **Benin, Bight of** gulf W Africa
79 U16 **Benin City** Edo, SW Nigeria 6.22N 5.39E
59 K16 **Beni, Río** ≈ N Bolivia
123 Gg11 **Beni Saf** var. Beni-Saf. NW Algeria 35.16N 1.33W
82 H12 **Benishangul** ◐ region W Ethiopia
107 T11 **Benissa** País Valenciano, E Spain 38.43N 0.03E
124 Qq17 **Beni Suef** var. Banī Suwayf. N Egypt 29.09N 31.03E
11 V15 **Benito** Manitoba, S Canada 51.57N 101.24W
Benito see Uolo, Río
63 C23 **Benito Juárez** Buenos Aires, E Argentina 37.43S 59.48W
43 P14 **Benito Juaréz Internacional** ✈ (México) México, S Mexico 19.24N 99.02W
27 P5 **Benjamin** Texas, SW USA 33.34N 99.47W
60 B13 **Benjamin Constant** Amazonas, N Brazil 4.22S 70.01W
42 F4 **Benjamín Hill** Sonora, NW Mexico 30.13N 111.07W
65 F19 **Benjamín, Isla** island Archipiélago de los Chonos, S Chile
172 N5 **Benkei-misaki** headland Hokkaidō, NE Japan 42.49N 140.10E
30 L17 **Benkelman** Nebraska, C USA 40.04N 101.30W
98 I7 **Ben Klibreck** ▲ N Scotland, UK 58.15N 4.23W
Benkoelen see Bengkulu
114 D13 **Benkovac** It. Bencovazzo. Zadar, SW Croatia 44.02N 15.36E
Benkulen see Bengkulu
98 I11 **Ben Lawers** ▲ C Scotland, UK 56.33N 4.13W
98 J9 **Ben Macdui** var. Beinn MacDuibh. ▲ C Scotland, UK 57.02N 3.42W
98 G11 **Ben More** ▲ W Scotland, UK 56.26N 6.00W
98 I11 **Ben More** ▲ C Scotland, UK 56.22N 4.31W
98 H7 **Ben More Assynt** ▲ N Scotland, UK 58.09N 4.51W
193 E20 **Benmore, Lake** ◎ South Island, NZ
100 L12 **Bennekom** Gelderland, SE Netherlands 52.00N 5.40E
23 T11 **Bennettsville** South Carolina, SE USA 34.37N 79.41W
98 H10 **Ben Nevis** ▲ N Scotland, UK 56.46N 5.01W
192 M9 **Benneydale** Waikato, North Island, NZ 38.31S 175.22E
Bennichab see Bennichchâb
78 H8 **Bennichchâb** var. Bennichab. Inchiri, W Mauritania 19.25N 15.21W
20 L10 **Bennington** Vermont, NE USA 42.51N 73.09W
193 E20 **Ben Ohau Range** ▲▲ South Island, NZ
85 J21 **Benoni** Gauteng, NE South Africa 26.04S 28.18E
180 J2 **Be, Nosy** var. Nossi-Bé. island NW Madagascar
Bénoué see Benue
44 F2 **Benque Viejo del Carmen** Cayo, W Belize 17.04N 89.08W
103 G19 **Bensheim** Hessen, W Germany 49.40N 8.37E
39 N16 **Benson** Arizona, SW USA 31.55N 110.16W
31 S8 **Benson** Minnesota, N USA 45.19N 95.36W
23 U10 **Benson** North Carolina, SE USA 35.22N 78.33W
175 Pp14 **Benteng** Pulau Selayar, C Indonesia 6.07S 120.28E
85 A14 **Bentiaba** Namibe, SW Angola 14.18S 12.27E
189 T4 **Bentinck Island** island Wellesley Islands, Queensland, N Australia
82 E13 **Bentiu** Wahda, S Sudan 9.13N 29.49E
144 G8 **Bent Jbaïl** var. Bint Jubayl. S Lebanon 33.07N 35.25E
11 Q15 **Bentley** Alberta, SW Canada 52.27N 114.02W
63 I15 **Bento Gonçalves** Rio Grande do Sul, S Brazil 29.06S 51.29W
29 U12 **Benton** Arkansas, C USA 34.33N 92.35W
32 L16 **Benton** Illinois, N USA 38.00N 88.55W
22 H7 **Benton** Kentucky, S USA 36.51N 88.21W
24 G5 **Benton** Louisiana, S USA 32.41N 93.44W
29 Y7 **Benton** Missouri, C USA 37.05N 89.34W
22 M10 **Benton** Tennessee, S USA 35.10N 84.39W
33 O10 **Benton Harbor** Michigan, N USA 42.07N 86.27W
29 S9 **Bentonville** Arkansas, C USA 36.22N 94.12W
79 V16 **Benue** ◐ state SE Nigeria
80 F13 **Benue** Fr. Bénoué. ≈ Cameroon/Nigeria
174 Hh6 **Benut** Johor, Peninsular Malaysia 1.37N 103.15E
169 V12 **Benxi** prev. Pen-ch'i, Penhsihu, Penki. Liaoning, NE China 41.11N 123.46E
Benyakoni see Byenyakoni
114 K10 **Beočin** Serbia, N Yugoslavia 45.13N 19.43E
Beodericsworth see Bury St Edmunds
114 M11 **Beograd** Eng. Belgrade, Ger. Belgrad; anc. Singidunum. ● (Yugoslavia) Serbia, N Yugoslavia 44.48N 20.27E
114 L11 **Beograd** Eng. Belgrade. ✈ Serbia, N Yugoslavia 44.45N 20.21E
78 M16 **Béoumi** C Ivory Coast 7.40N 5.34W
37 V3 **Beowawe** Nevada, W USA 40.33N 116.31W
176 Ww8 **Bepondi, Pulau** see Bepondi, Pulau
170 D13 **Beppu** Ōita, Kyūshū, SW Japan 33.16N 131.28E
170 Dd14 **Beppu-wan** bay SW Japan
197 H15 **Beqa** prev. Mbengga. island W Fiji
Beqa Barrier Reef see Kavukavu Reef
47 Y14 **Bequia** island C Saint Vincent and the Grenadines
115 L16 **Berane** prev. Ivangrad. Montenegro, SW Yugoslavia 42.51N 19.51E
115 L21 **Berat** var. Berati, SCr. Beligrad. Berat, C Albania 40.42N 19.57E
115 L21 **Berat** ◐ district C Albania
Berătău see Berettyó
Berati see Berat
Beraun see Berounka, Czech Republic
Beraun see Beroun, Czech Republic
175 O6 **Berau, Sungai** ≈ Borneo, N Indonesia
176 V10 **Berau, Teluk** var. MacCluer Gulf. bay Irian Jaya, E Indonesia
82 G8 **Berber** River Nile, NE Sudan 18.01N 34.00E
82 N12 **Berbera** Woqooyi Galbeed, NW Somalia 10.24N 45.01E
81 H16 **Berbérati** Mambéré-Kadéï, SW Central African Republic 4.13N 15.49E
Berberia, Cabo de see Barbaria, Cap de
57 T9 **Berbice River** ≈ NE Guyana
Berchid see Berrechid
105 N2 **Berck-Plage** Pas-de-Calais, N France 50.24N 1.34E
27 T13 **Berclair** Texas, SW USA 28.33N 97.32W
119 W10 **Berda** ≈ SE Ukraine
Berdichev see Berdychiv
126 Ll11 **Berdigestyakh** Respublika Sakha (Yakutiya), NE Russian Federation 62.02N 127.03E
126 H14 **Berdsk** Novosibirskaya Oblast', C Russian Federation 54.42N 82.56E
119 W10 **Berdyans'k** Rus. Berdyansk; prev. Osipenko. Zaporiz'ka Oblast', SE Ukraine 46.46N 36.48E
119 W10 **Berdyans'ka Kosa** spit SE Ukraine
119 V10 **Berdyans'ka Zatoka** gulf S Ukraine
119 N5 **Berdychiv** Rus. Berdichev. Zhytomyrs'ka Oblast', N Ukraine 49.52N 28.39E
22 M6 **Berea** Kentucky, S USA 37.34N 84.18W
Beregovo/Beregszász see Berehove
118 G8 **Berehove** Cz. Berehovo, Hung. Beregszász, Rus. Beregovo. Zakarpats'ka Oblast', W Ukraine 48.13N 22.39E
Berehovo see Berehove
194 J15 **Bereina** Central, S PNG 8.33S 146.25E
47 O12 **Berekua** S Dominica 15.14N 61.19W
79 O16 **Berekum** W Ghana 7.27N 2.34W
77 Y11 **Berenice** var. Mînâ Baranîs. SE Egypt 23.58N 35.29E
15 L14 **Berens** ≈ Manitoba/Ontario, C Canada
11 X14 **Berens River** Manitoba, C Canada 52.22N 97.00W
31 R12 **Beresford** South Dakota, N USA 43.02N 96.45W
118 J4 **Berestechko** Volyns'ka Oblast', NW Ukraine 50.21N 25.06E
118 M11 **Bereşti** Galaţi, E Romania 46.04N 27.54E
119 U6 **Berestova** ≈ E Ukraine
Beretău see Berettyó
113 N23 **Berettyó** Rom. Barcău; prev. Berătău, Beretău. ≈ Hungary/Romania
113 N23 **Berettyóújfalu** Hajdú-Bihar, E Hungary 47.15N 21.33E
Berëza/Bereza Kartuska see Byaroza
119 Q4 **Berezan'** Kyyivs'ka Oblast', N Ukraine 50.18N 31.30E
119 Q10 **Berezanka** Mykolayivs'ka Oblast', S Ukraine 46.51N 31.24E
118 J6 **Berezhany** Pol. Brzeżany. Ternopil's'ka Oblast', W Ukraine 49.29N 25.00E
Berezina see Byerezino
Berezino see Byerazino
119 P10 **Berezivka** Rus. Berezovka. Odes'ka Oblast', SW Ukraine 47.12N 30.55E
119 Q2 **Berezna** Chernihivs'ka Oblast', NE Ukraine 51.35N 31.50E
118 L3 **Berezne** Rivnens'ka Oblast', NW Ukraine 51.00N 26.46E
119 R9 **Bereznehuvate** Mykolayivs'ka Oblast', S Ukraine 47.18N 32.52E
129 N10 **Bereznik** Arkhangel'skaya Oblast', NW Russian Federation 62.50N 42.40E
129 U13 **Berezniki** Permskaya Oblast', NW Russian Federation 59.25N 56.49E
Berezovka see Berezivka
125 Ff9 **Berezovo** Khanty-Mansiyskiy Avtonomnyy Okrug, N Russian Federation 63.48N 64.38E
131 O9 **Berezovskaya** Volgogradskaya Oblast', SW Russian Federation 50.17N 43.58E
126 H14 **Berezovskiy** Kemerovskaya Oblast', S Russian Federation 55.40N 86.06E
127 N14 **Berezovyy** Khabarovskiy Kray, E Russian Federation 51.42N 135.39E
85 E25 **Berg** ≈ W South Africa
Berg see Berg bei Rohrbach
107 V4 **Berga** Cataluña, NE Spain 42.06N 1.40E
97 N20 **Berga** Kalmar, S Sweden 57.13N 16.03E
142 B13 **Bergama** İzmir, W Turkey 39.07N 27.10E
108 E7 **Bergamo** anc. Bergomum. Lombardia, N Italy 45.42N 9.40E
107 P3 **Bergara** País Vasco, N Spain 43.05N 2.25W
111 S3 **Berg bei Rohrbach** var. Berg. Oberösterreich, N Austria 48.34N 14.02E
195 O11 **Bergberg** ≈ New Britain, C Papau New Guinea
102 O6 **Bergen** Mecklenburg-Vorpommern, NE Germany 54.25N 13.24E
103 I11 **Bergen** Niedersachsen, NW Germany 52.49N 9.57E
100 H8 **Bergen** Noord-Holland, NW Netherlands 52.40N 4.42E
96 C13 **Bergen** Hordaland, S Norway 60.24N 5.19E
Bergen see Mons
57 W9 **Berg en Dal** Brokopondo, C Suriname 5.15N 55.20W
101 G15 **Bergen op Zoom** Noord-Brabant, S Netherlands 51.30N 4.18E
104 L13 **Bergerac** Dordogne, SW France 44.51N 0.30E
101 J16 **Bergeyk** Noord-Brabant, S Netherlands 51.19N 5.21E
103 D16 **Bergheim** Nordrhein-Westfalen, W Germany 50.57N 6.39E
57 X10 **Bergi** Sipaliwini, E Suriname 4.36N 54.24W
103 E16 **Bergisch Gladbach** Nordrhein-Westfalen, W Germany 50.59N 7.07E
103 F14 **Bergkamen** Nordrhein-Westfalen, W Germany 51.36N 7.39E
97 N21 **Bergkvara** Kalmar, S Sweden 56.22N 16.04E
Bergomum see Bergamo
100 K13 **Bergse Maas** ≈ S Netherlands
97 P15 **Bergshamra** Stockholm, C Sweden 59.37N 18.40E
96 N10 **Bergsjö** Gävleborg, C Sweden 62.00N 17.10E
95 J14 **Bergsviken** Norrbotten, N Sweden 65.16N 21.24E
100 L6 **Bergum** Fris. Burgum. Friesland, N Netherlands 53.12N 5.58E
100 M6 **Bergumer Meer** ◎ N Netherlands
96 N12 **Bergviken** ◎ C Sweden
174 I9 **Berhala, Selat** strait Sumatera, W Indonesia
Berhampore see Baharampur
127 Q9 **Beringa, Ostrov** island E Russian Federation
101 J17 **Beringen** Limburg, NE Belgium 51.04N 5.13E
41 T12 **Bering Glacier** glacier Alaska, USA
Beringov Proliv see Bering Strait
127 Q5 **Beringovskiy** Chukotskiy Avtonomnyy Okrug, NE Russian Federation 63.04N 179.09E
199 K2 **Bering Sea** sea N Pacific Ocean
40 L9 **Bering Strait** Rus. Beringov Proliv. strait Bering Sea/Chukchi Sea
Berislav see Beryslav
107 O15 **Berja** Andalucía, S Spain 36.51N 2.55W
96 H9 **Berkåk** Sør-Trøndelag, S Norway 62.50N 10.01E
100 N11 **Berkel** ≈ Germany/Netherlands
37 N8 **Berkeley** California, W USA 37.52N 122.16W
67 E24 **Berkeley Sound** sound NE Falkland Islands
23 V2 **Berkeley Springs** var. Bath. West Virginia, NE USA 39.36N 78.12W
205 N6 **Berkner Island** island Antarctica
116 G8 **Berkovitsa** Montana, NW Bulgaria 43.16N 23.07E
99 M22 **Berkshire** cultural region S England, UK
101 H17 **Berlaar** Antwerpen, N Belgium 51.08N 4.39E
Berlanga see Berlanga de Duero
107 P6 **Berlanga de Duero** var. Berlanga. Castilla-León, N Spain 41.28N 2.51W
1 I16 **Berlanga Rise** undersea feature E Pacific Ocean
101 F17 **Berlare** Oost-Vlaanderen, NW Belgium 51.01N 4.01E
106 E9 **Berlenga, Ilha da** island C Portugal
94 M7 **Berlevåg** Finnmark, N Norway 70.51N 29.04E
102 O12 **Berlin** ● (Germany) Berlin, NE Germany 52.31N 13.26E
23 Z4 **Berlin** Maryland, NE USA 38.19N 75.13W
21 O7 **Berlin** New Hampshire, NE USA 44.27N 71.13W
20 D16 **Berlin** Pennsylvania, NE USA 39.54N 78.57W
32 L7 **Berlin** Wisconsin, N USA 43.57N 88.59W
102 O12 **Berlin** ◐ state NE Germany
Berlinchen see Barlinek
33 U12 **Berlin Lake** ⊠ Ohio, N USA
191 R11 **Bermagui** New South Wales, SE Australia 36.26S 150.01E
42 L8 **Bermejillo** Durango, C Mexico 25.55N 103.39W
64 M6 **Bermejo (viejo), Río** ≈ N Argentina
64 L5 **Bermejo, Río** ≈ N Argentina
64 I10 **Bermejo, Río** ≈ W Argentina
107 P2 **Bermeo** País Vasco, N Spain 43.25N 2.43W
106 K6 **Bermillo de Sayago** Castilla-León, N Spain 41.22N 6.07W
108 E6 **Bermina, Pizzo** Rmsch. Piz Bernina. ▲ Italy/Switzerland see also Bernina, Piz 46.22N 9.52E
66 A12 **Bermuda** var. Bermuda Islands, Bermudas; prev. Somers Islands. ◇ UK crown colony NW Atlantic Ocean
1 N11 **Bermuda** var. Great Bermuda, Long Island, Main Island. island Bermuda
Bermuda Islands see Bermuda
Bermuda-New England Seamount Arc see New England Seamounts
1 N11 **Bermuda Rise** undersea feature C Sargasso Sea
Bermudas see Bermuda
110 D8 **Bern** Fr. Berne. ● (Switzerland) Bern, W Switzerland 46.57N 7.25E
110 D9 **Bern** Fr. Berne. ◐ canton W Switzerland
39 R11 **Bernalillo** New Mexico, SW USA 35.18N 106.33W
12 H12 **Bernard Lake** ◎ Ontario, S Canada
63 B18 **Bernardo de Irigoyen** Santa Fe, NE Argentina 32.09S 61.06W
20 J14 **Bernardsville** New Jersey, NE USA 40.43N 74.34W
65 K14 **Bernasconi** La Pampa, C Argentina 37.55S 63.43W
102 O12 **Bernau** Brandenburg, NE Germany 52.40N 13.36E
104 L4 **Bernay** Eure, N France 49.04N 0.36E
103 L14 **Bernburg** Sachsen-Anhalt, C Germany 51.46N 11.45E
111 X5 **Berndorf** Niederösterreich, NE Austria 47.55N 16.10E
33 Q12 **Berne** Indiana, N USA 40.39N 84.57W
Berne see Bern
110 D10 **Berner Alpen** var. Berner Oberland, Eng. Bernese Oberland. ▲▲ SW Switzerland
Berner Oberland/Bernese Oberland see Berner Alpen
111 Y2 **Bernhardsthal** Niederösterreich, N Austria 48.40N 16.51E
24 H4 **Bernice** Louisiana, S USA 32.49N 92.39W
29 Y8 **Bernie** Missouri, C USA 36.40N 89.58W
188 G9 **Bernier Island** island Western Australia
Bernina Pass see Bernina, Passo del
110 J10 **Bernina, Passo del** Eng. Bernina Pass. pass SE Switzerland 46.23N 10.00E
110 J10 **Bernina, Piz** It. Pizzo Bermina. ▲ Italy/Switzerland see also Bermina, Pizzo 46.22N 9.55E
101 E20 **Bérnissart** Hainaut, SW Belgium 50.29N 3.37E
103 E18 **Bernkastel-Kues** Rheinland-Pfalz, W Germany 49.55N 7.04E
Beroea see Ḩalab
180 H6 **Beroroha** Toliara, SW Madagascar 21.40S 45.10E
Béroubouay see Gbéroubouè
113 C17 **Beroun** Ger. Beraun. Středočeský Kraj, W Czech Republic 49.58N 14.04E
113 C16 **Berounka** Ger. Beraun. ≈ W Czech Republic
115 Q18 **Berovo** E FYR Macedonia 41.45N 22.50E
76 F6 **Berrechid** var. Berchid. W Morocco 33.16N 7.32W
105 R15 **Berre, Étang de** ◎ SE France
105 S15 **Berre-l'Étang** Bouches-du-Rhône, SE France 43.28N 5.10E
190 K9 **Berri** South Australia 34.16S 140.35E
33 O10 **Berrien Springs** Michigan, N USA 41.57N 86.20W
191 O10 **Berrigan** New South Wales, SE Australia 35.41S 145.50E
105 N9 **Berry** cultural region C France
37 N7 **Berryessa, Lake** ◎ California, W USA
46 G2 **Berry Islands** island group N Bahamas
29 T9 **Berryville** Arkansas, C USA 36.21N 93.30W
23 V3 **Berryville** Virginia, NE USA 39.09N 77.58W
85 D21 **Berseba** Karas, S Namibia 26.00S 17.46E
119 O8 **Bershad'** Vinnyts'ka Oblast', C Ukraine 48.19N 29.28E
30 L3 **Berthold** North Dakota, N USA 48.16N 101.48W
39 T3 **Berthoud** Colorado, C USA 40.18N 105.04W
39 S4 **Berthoud Pass** pass Colorado, C USA 39.48N 105.46W
81 F15 **Bertoua** Est, E Cameroon 4.34N 13.42E
27 S10 **Bertram** Texas, SW USA 30.44N 98.03W
65 G22 **Bertrand, Cerro** ▲ S Argentina 50.00S 73.27W
101 J23 **Bertrix** Luxembourg, SE Belgium 49.52N 5.15E
203 P3 **Beru** var. Peru. atoll Tungaru, W Kiribati
Beruni see Beruniy
152 I9 **Beruniy** var. Biruni, Rus. Beruni. Qoraqalpoghiston Respublikasi, W Uzbekistan 41.48N 60.39E
60 F13 **Beruri** Amazonas, NW Brazil 3.44S 61.13W
20 H14 **Berwick** Pennsylvania, NE USA 41.03N 76.13W
98 K12 **Berwick** cultural region SE Scotland, UK
98 L12 **Berwick-upon-Tweed** N England, UK 55.46N 2.00W
119 S10 **Beryslav** Rus. Berislav. Khersons'ka Oblast', S Ukraine 46.51N 33.27E
Berytus see Beyrouth
180 H4 **Besalampy** Mahajanga, W Madagascar 16.43S 44.28E
105 T8 **Besançon** anc. Besontium, Vesontio. Doubs, E France 47.13N 6.01E
105 P10 **Besbre** ≈ C France
Bescanuova see Baška
Besdan see Bezdan
Besed' see Byesyedz'
153 R10 **Beshariq** Rus. Besharyk; prev. Kirovo. Farghona Wiloyati, E Uzbekistan 40.26N 70.33E
Besharyk see Beshariq
152 L9 **Beshbuloq** Rus. Beshulak. Nawoiy Wiloyati, N Uzbekistan 41.55N 64.13E
Beshenkovichi see Byeshankovichy
152 M13 **Beshkent** Qashqadaryo Wiloyati, S Uzbekistan 38.47N 65.42E
Beshulak see Beshbuloq
176 Uu8 **Besir** Irian Jaya, E Indonesia 0.25S 130.38E
114 L10 **Beška** Serbia, N Yugoslavia 45.09N 20.04E
Beskra see Biskra
131 O16 **Beslan** Respublika Severnaya Osetiya, SW Russian Federation 43.12N 44.33E
115 P16 **Besna Kobila** ▲ SE Yugoslavia 42.30N 22.16E
143 N16 **Besni** Adıyaman, S Turkey 37.42N 37.52E
Besontium see Besançon
124 Nn2 **Beşparmak Dağları** Eng. Kyrenia Mountains. ▲▲ N Cyprus
Bessarabka see Basarabeasca
94 O2 **Bessels, Kapp** headland C Svalbard 78.36N 21.43E
25 P4 **Bessemer** Alabama, S USA 33.24N 86.57W
32 K3 **Bessemer** Michigan, N USA 46.28N 90.03W
23 Q10 **Bessemer City** North Carolina, SE USA 35.16N 81.16W
104 M10 **Bessines-sur-Gartempe** Haute-Vienne, C France 46.06N 1.22E
101 K15 **Best** Noord-Brabant, S Netherlands 51.31N 5.24E
27 N9 **Best** Texas, SW USA 31.13N 101.34W
129 O11 **Bestuzhevo** Arkhangel'skaya Oblast', NW Russian Federation 61.36N 43.54E
126 M11 **Bestyakh** Respublika Sakha (Yakutiya), NE Russian Federation 61.25N 129.05E
Beszterce see Bistriţa
Besztercebánya see Banská Bystrica
180 I5 **Betafo** Antananarivo, C Madagascar 19.49S 46.49E
106 H2 **Betanzos** Galicia, NW Spain 43.16N 8.16W
106 G2 **Betanzos, Ría de** estuary NW Spain
81 G15 **Bétaré Oya** Est, E Cameroon 5.34N 14.09E
107 S9 **Bétera** País Valenciano, E Spain 39.34N 0.28W
79 R15 **Bétérou** C Benin 9.13N 2.18E
85 K21 **Bethal** Mpumalanga, NE South Africa 26.27S 29.28E
32 K15 **Bethalto** Illinois, N USA 38.54N 90.02W
85 D21 **Bethanie** var. Bethanien, Bethany. Karas, S Namibia 26.26S 17.06E
Bethanien see Bethanie
29 S2 **Bethany** Missouri, C USA 40.15N 94.03W
29 N10 **Bethany** Oklahoma, C USA 35.31N 97.37W
Bethany see Bethanie
41 N12 **Bethel** Alaska, USA 60.47N 161.45W
21 P7 **Bethel** Maine, NE USA 44.24N 70.47W
23 W9 **Bethel** North Carolina, SE USA 35.46N 77.21W
20 B15 **Bethel Park** Pennsylvania, NE USA 40.21N 80.03W
23 W3 **Bethesda** Maryland, NE USA 38.58N 77.06W
85 J22 **Bethlehem** Free State, C South Africa 28.11S 28.16E
20 I14 **Bethlehem** Pennsylvania, NE USA 40.36N 75.22W
144 F10 **Bethlehem** Ar. Bayt Laḩm, Heb. Bet Leḥem. C West Bank 31.43N 35.12E
Bethlen see Beclean
85 I24 **Bethulie** Free State, C South Africa 30.29S 25.54E
105 O1 **Béthune** Pas-de-Calais, N France 50.31N 2.37E
104 M3 **Béthune** ≈ N France
106 M14 **Béticos, Sistemas** var. Sistema Penibético, Eng. Baetic Cordillera, Baetic Mountains. ▲▲ S Spain
56 I6 **Betijoque** Trujillo, NW Venezuela 9.27N 70.47W
61 M20 **Betim** Minas Gerais, SE Brazil 19.55S 44.10W
202 H3 **Betio** Tarawa, W Kiribati 1.21N 172.55E
180 H7 **Betioky** Toliara, S Madagascar 23.42S 44.22E
Bet Leḥem see Bethlehem
Betlen see Beclean
178 H17 **Betong** Yala, SW Thailand 5.47N 101.04E
81 I16 **Bétou** La Likouala, N Congo 3.07N 18.30E
151 P14 **Betpak-Dala** Kaz. Betpaqdala. plateau S Kazakhstan
Betpaqdala see Betpak-Dala
180 H7 **Betroka** Toliara, S Madagascar 23.15S 46.07E
Betschau see Bečva
144 G9 **Bet She'an** Ar. Baysān, Beisān; anc. Scythopolis. Northern, N Israel 32.29N 35.25E
13 T6 **Betsiamites** Quebec, SE Canada 48.55N 68.40W
13 T6 **Betsiamites** ≈ Quebec, SE Canada
180 I4 **Betsiboka** ≈ N Madagascar
101 M25 **Bettembourg** Luxembourg, S Luxembourg 49.31N 6.06E
101 M23 **Bettendorf** Diekirch, NE Luxembourg 49.52N 6.13E
31 Z14 **Bettendorf** Iowa, C USA 41.31N 90.31W
77 R13 **Bette, Pic** var. Bikkū Bittī, It. Picco Bette. ▲ S Libya 22.02N 19.07E
Bette, Picco see Bette, Pic
159 P12 **Bettiah** Bihār, N India 26.49N 84.30E
41 Q7 **Bettles** Alaska, USA 66.54N 151.40W
97 N17 **Bettna** Södermanland, C Sweden 58.52N 16.40E
160 H11 **Betūl** prev. Badnur. Madhya Pradesh, C India 21.55N 77.54E
160 H9 **Betwa** ≈ C India
103 F16 **Betzdorf** Rheinland-Pfalz, W Germany 50.47N 7.50E
84 C9 **Béu** Uíge, NW Angola 6.16S 15.28E
33 P6 **Beulah** Michigan, N USA 44.37N 86.04W
30 L5 **Beulah** North Dakota, N USA 47.16N 101.48W
100 M8 **Beulakerwijde** ◎ N Netherlands
100 L13 **Beuningen** Gelderland, SE Netherlands 51.52N 5.47E
Beuthen see Bytom
105 N7 **Beuvron** ≈ C France
101 F16 **Beveren** Oost-Vlaanderen, N Belgium 51.13N 4.15E
23 T9 **B.Everett Jordan Reservoir** var. Jordan Lake. ⊠ North Carolina, SE USA
99 N17 **Beverley** E England, UK 53.51N 0.25W
Beverley see Beverly
101 J17 **Beverlo** Limburg, NE Belgium 51.06N 5.14E
21 P11 **Beverly** Massachusetts, NE USA 42.33N 70.49W
34 J9 **Beverly** var. Beverley. Washington, NW USA 46.50N 119.57W
37 S15 **Beverly Hills** California, W USA 34.02N 118.25W
103 I14 **Beverungen** Nordrhein-Westfalen, C Germany 51.39N 9.22E
100 H9 **Beverwijk** Noord-Holland, W Netherlands 52.28N 4.37E
110 C10 **Bex** Vaud, W Switzerland 46.15N 7.00E
99 P23 **Bexhill** var. Bexhill-on-Sea. SE England, UK 50.49N 0.28E
Bexhill-on-Sea see Bexhill
142 E17 **Bey Dağları** ▲▲ SW Turkey
Beyj see Belyy
142 E10 **Beykoz** İstanbul, NW Turkey 41.09N 29.06E
78 K15 **Beyla** Guinée-Forestière, SE Guinea 8.43N 8.41W
143 X12 **Beyläqan** prev. Zhdanov. SW Azerbaijan 39.43N 47.38E
82 L10 **Beylul** var. Beilul. SE Eritrea 13.10N 42.27E
150 H14 **Beyneu** Kaz. Beyneū. Mangistau, SW Kazakhstan 45.19N 55.11E
Beyneū see Beyneu
172 Ss14 **Beyonēsu-retsugan** Eng. Bayonnaise Rocks. island group SE Japan
142 G12 **Beypazarı** Ankara, NW Turkey 40.10N 31.55E
161 F21 **Beypore** Kerala, SW India 11.10N 75.49E
144 G7 **Beyrouth** var. Bayrūt, Eng. Beirut; anc. Berytus. ● (Lebanon) W Lebanon 33.54N 35.31E
144 G7 **Beyrouth** ✈ W Lebanon 33.52N 35.30E
142 G15 **Beyşehir** Konya, SW Turkey 37.39N 31.42E
142 G15 **Beyşehir Gölü** ◎ C Turkey
110 J7 **Bezau** Vorarlberg, NW Austria 47.24N 9.55E
114 J8 **Bezdan** Ger. Besdan, Hung. Bezdán. Serbia, NW Yugoslavia 45.51N 19.00E
Bezdezh see Byezdzyezh
128 G15 **Bezhanitsy** Pskovskaya Oblast', W Russian Federation 56.57N 29.53E
128 K15 **Bezhetsk** Tverskaya Oblast', W Russian Federation 57.47N 36.42E
105 P16 **Béziers** anc. Baeterrae, Baeterrae Septimanorum, Julia Beterrae. Hérault, S France 43.21N 3.13E
Bezmein see Byuzmeyin
Bezwada see Vijayawāda
160 P12 **Bhadrak** var. Bhadrakh. Orissa, E India 21.06N 86.31E
Bhadrakh see Bhadrak
161 F19 **Bhadra Reservoir** ⊠ SW India
161 F18 **Bhadrāvati** Karnātaka, SW India 13.52N 75.43E
159 R14 **Bhāgalpur** Bihār, NE India 25.13N 86.58E
Bhairab see Bhairab Bazar
159 U14 **Bhairab Bazar** var. Bhairab. Dhaka, C Bangladesh 24.04N 91.00E
159 O11 **Bhairahawa** Western, S Nepal 27.31N 83.27E
155 S8 **Bhakkar** Punjab, E Pakistan 31.41N 71.04E
159 P11 **Bhaktapur** Central, C Nepal 27.47N 85.21E
178 Gg3 **Bhamo** var. Banmo. Kachin State, N Myanmar 24.15N 97.15E
Bhāmragad see Bhāmragarh
160 K13 **Bhāmragarh** var. Bhāmragad. Mahārāshtra, C India 19.28N 80.39E
160 J12 **Bhandāra** Mahārāshtra, C India 21.10N 79.40E
Bhārat see India
158 J12 **Bharatpur** prev. Bhurtpore. Rājasthān, N India 27.13N 77.28E
160 D11 **Bharūch** Gujarāt, W India 21.48N 72.54E
161 E18 **Bhatkal** Karnātaka, W India 13.59N 74.34E
159 O13 **Bhatni** var. Bhatni Junction. Uttar Pradesh, N India 26.22N 83.55E
Bhatni Junction see Bhatni
159 S16 **Bhātpāra** West Bengal, NE India 22.55N 88.30E
155 U7 **Bhaun** Punjab, E Pakistan 32.53N 72.45E
Bhaunagar see Bhāvnagar
160 M13 **Bhavānīpātna** Orissa, E India 19.56N 83.09E
161 H21 **Bhavānīsāgar Reservoir** ⊠ S India
160 D11 **Bhāvnagar** prev. Bhaunagar. Gujarāt, W India 21.46N 72.13E
Bheanntraí, Bá see Bantry Bay
Bheara, Béal an see Gweebarra Bay
160 K12 **Bhilai** Madhya Pradesh, C India 21.13N 81.26E
158 G13 **Bhīlwāra** Rājasthān, N India 25.22N 74.39E
161 E14 **Bhīma** ≈ S India
161 K16 **Bhīmavaram** Andhra Pradesh, E India 16.34N 81.34E
160 I7 **Bhind** Madhya Pradesh, C India 26.33N 78.46E
158 E13 **Bhīnmāl** Rājasthān, N India 25.01N 72.22E
Bhir see Bīd
160 D13 **Bhiwandi** Mahārāshtra, W India 19.21N 73.07E
158 H10 **Bhiwāni** Haryāna, N India 28.49N 76.07E
158 L13 **Bhognīpur** Uttar Pradesh, N India 26.12N 79.48E
159 U16 **Bhola** Khulna, S Bangladesh 22.42N 90.43E
160 H10 **Bhopāl** Madhya Pradesh, C India 23.16N 77.24E
161 J14 **Bhopālpatnam** Madhya Pradesh, C India 18.51N 80.22E
161 E14 **Bhor** Mahārāshtra, W India 18.10N 73.55E
160 O12 **Bhubaneshwar** prev. Bhubaneswar, Bhuvaneshwar. Orissa, E India 20.16N 85.51E
Bhubaneswar see Bhubaneshwar
160 B9 **Bhuj** Gujarāt, W India 23.16N 69.40E
Bhuket see Phuket
Bhurtpore see Bharatpur
Bhusaval see Bhusāwal
160 G12 **Bhusāwal** prev. Bhusaval. Mahārāshtra, C India 21.01N 75.49E
159 T12 **Bhutan** off. Kingdom of Bhutan, var. Druk-yul. ◆ monarchy S Asia
Bhuvaneshwar see Bhubaneshwar
149 T15 **Bīābān, Kūh-e** ▲▲ S Iran
79 V18 **Biafra, Bight of** var. Bight of Bonny. bay W Africa
176 X9 **Biak** Irian Jaya, E Indonesia 1.10S 136.04E
176 Ww9 **Biak, Pulau** island E Indonesia
112 P12 **Biała Podlaska** Lubelskie, E Poland 52.03N 23.08E
112 F7 **Białogard** Ger. Belgard. Zachodniopomorskie, NW Poland 54.00N 15.58E
112 P10 **Białowieska, Puszcza** Bel. Byelavyezhskaya Pushcha, Rus. Belovezhskaya Pushcha. physical region Belarus/Poland see also Byelavyezhskaya Pushcha
112 G8 **Biały Bór** Ger. Baldenburg. Zachodniopomorskie, NW Poland 53.53N 16.49E
112 P9 **Białystok** Rus. Belostok, Bielostok. Podlaskie, NE Poland 53.08N 23.09E
109 L24 **Biancavilla** prev. Inessa. Sicilia, Italy, C Mediterranean Sea 37.37N 14.52E
Bianco, Monte see Blanc, Mont
78 L15 **Biankouma** W Ivory Coast 7.43N 7.37W
178 Ii7 **Bia, Phou** var. Pou Bia. ▲ C Laos 18.59N 103.09E
Bia, Pou see Bia, Phou
149 R5 **Biārjmand** Semnān, N Iran 36.04N 55.49E
107 P4 **Biarra** ≈ NE Spain
104 I15 **Biarritz** Pyrénées-Atlantiques, SW France 43.24N 1.39W
110 H10 **Biasca** Ticino, S Switzerland 46.22N 8.59E
63 E17 **Biassini** Salto, N Uruguay 31.18S 57.05W
172 Oo5 **Bibai** Hokkaidō, NE Japan 43.22N 141.52E
85 B15 **Bibala** Port. Vila Arriaga. Namibe, SW Angola 14.45S 13.18E
106 I4 **Bibéi** ≈ NW Spain
Biberach see Biberach an der Riss
103 I23 **Biberach an der Riss** var. Biberach, Ger. Biberach an der Riß. Baden-Württemberg, S Germany 48.06N 9.46E
110 E7 **Biberist** Solothurn, NW Switzerland 47.10N 7.34E
79 O16 **Bibiani** SW Ghana 6.28N 2.19W
114 C13 **Bibinje** Zadar, SW Croatia 44.04N 15.17E
Biblical Gebal see Jbaïl
118 I5 **Bibrka** Pol. Bóbrka, Rus. Bobrka. L'vivs'ka Oblast', NW Ukraine 49.39N 24.16E
119 N10 **Bîc** ≈ S Moldova
115 M18 **Bicaj** Kukës, NE Albania 42.00N 20.24E
118 K10 **Bicaz** Hung. Békás. Neamţ, NE Romania 46.53N 26.04E
191 Q16 **Bicheno** Tasmania, SE Australia 41.56S 148.15E
Bichiş see Békés
Bichiş-Ciaba see Békéscsaba
Bichitra see Phichit
143 P8 **Bichvint'a** Rus. Pitsunda. NW Georgia 43.12N 40.21E
13 T7 **Bic, Île du** island Quebec, SE Canada
34 J10 **Bickleton** Washington, NW USA 46.00N 120.16W
68 L6 **Bicknell** Utah, W USA 38.20N 111.32W
175 Tt7 **Bicoli** Pulau Halmahera, E Indonesia 0.34N 128.33E
113 J22 **Bicske** Fejér, C Hungary 47.28N 18.38E
161 F14 **Bīd** prev. Bhir. Mahārāshtra, W India 19.17N 75.22E
79 U15 **Bida** Niger, C Nigeria 9.06N 6.02E
161 H15 **Bīdar** Karnātaka, C India 17.55N 77.34E
147 Y8 **Bidbid** NE Oman 23.25N 58.07E
21 P9 **Biddeford** Maine, NE USA 43.28N 70.27W
100 L9 **Biddinghuizen** Flevoland, C Netherlands 52.28N 5.41E
35 X11 **Biddle** Montana, NW USA 45.04N 105.21W
99 J23 **Bideford** SW England, UK 51.01N 4.12W
84 D13 **Bié** ◐ province C Angola
37 O2 **Bieber** California, W USA 41.07N 121.09W
112 O9 **Biebrza** ≈ NE Poland
172 P5 **Biei** Hokkaidō, NE Japan 43.33N 142.28E
110 D8 **Biel** Fr. Bienne. Bern, W Switzerland 47.09N 7.16E
102 G13 **Bielefeld** Nordrhein-Westfalen, NW Germany 52.01N 8.31E
110 D8 **Bieler See** Fr. Lac de Bienne. ◎ W Switzerland
Bielitz/Bielitz-Biala see Bielsko-Biała
108 C7 **Biella** Piemonte, N Italy 45.33N 8.03E
Bielostok see Białystok
113 J17 **Bielsko-Biała** Ger. Bielitz, Bielitz-Biala. Śląskie, S Poland 49.48N 19.01E
112 P10 **Bielsk Podlaski** Białystok, E Poland 52.45N 23.11E
Bien Bien see Điên Biên
Biên Đông see South China Sea
11 V17 **Bienfait** Saskatchewan, S Canada 49.06N 102.47W
178 Jj14 **Biên Hoa** Đông Nai, S Vietnam 10.58N 106.49E
Bienne see Biel
Bienne, Lac de see Bieler See
10 K8 **Bienville, Lac** ◎ Quebec, C Canada
84 D13 **Bié, Planalto do** var. Bié Plateau. plateau C Angola
Bié Plateau see Bié, Planalto do
110 B9 **Bière** Vaud, W Switzerland 46.32N 6.19E

◆ Country ● Country Capital ◇ Dependent Territory ○ Dependent Territory Capital ◐ Administrative Region ✈ International Airport ▲ Mountain ▲▲ Mountain Range Volcano ≈ River ◎ Lake ⊠ Reservoir

100 O4 **Bierum** Groningen, NE Netherlands *53.25N 6.51E*
100 I13 **Biesbos** ***var.*** Biesbosch. *wetland* S Netherlands
Biesbosch *see* Biesbos
101 H21 **Biesme** Namur, S Belgium *50.19N 4.43E*
103 H21 **Bietigheim-Bissingen** Baden-Württemberg, SW Germany *48.57N 9.07E*
101 I23 **Bièvre** Namur, SE Belgium *49.56N 5.01E*
81 D18 **Bifoun** Moyen-Ogooué, NW Gabon *0.15S 10.24E*
172 Pp3 **Bifuka** Hokkaidō, NE Japan *44.28N 142.20E*
142 C11 **Biga** Çanakkale, NW Turkey *40.13N 27.13E*
142 C13 **Bigadiç** Balıkesir, W Turkey *39.24N 28.07E*
28 J7 **Big Basin** *basin* Kansas, C USA
193 B20 **Big Bay** *bay* South Island, NZ
197 B12 **Big Bay** *bay* C Vanuatu
33 O5 **Big Bay de Noc** ◎ Michigan, N USA
33 N3 **Big Bay Point** *headland* Michigan, N USA *46.51N 87.40W*
35 R10 **Big Belt Mountains** ⛰ Montana, NW USA
31 N10 **Big Bend Dam** *dam* South Dakota, N USA *44.03N 99.27W*
26 K12 **Big Bend National Park** *national park* Texas, S USA
24 K5 **Big Black River** ≈ Mississippi, S USA
29 O3 **Big Blue River** ≈ Kansas/Nebraska, C USA
26 M10 **Big Canyon** ≈ Texas, SW USA
35 N12 **Big Creek** Idaho, NW USA *45.05N 115.20W*
25 N8 **Big Creek Lake** ⊠ Alabama, S USA
25 X15 **Big Cypress Swamp** *wetland* Florida, SE USA
41 S9 **Big Delta** Alaska, USA *64.09N 145.50W*
32 K6 **Big Eau Pleine Reservoir** ⊠ Wisconsin, N USA
21 P5 **Bigelow Mountain** ▲ Maine, NE USA *45.09N 70.17W*
31 U3 **Big Falls** Minnesota, N USA *48.13N 93.48W*
35 P8 **Bigfork** Montana, NW USA *48.03N 114.04W*
31 U3 **Big Fork River** ≈ Minnesota, N USA
11 S15 **Biggar** Saskatchewan, S Canada *52.03N 107.58W*
188 L3 **Bigge Island** *island* Western Australia
37 O5 **Biggs** California, W USA *39.24N 121.44W*
34 I11 **Biggs** Oregon, NW USA *45.39N 120.49W*
12 K13 **Big Gull Lake** ◎ Ontario, SE Canada
39 P16 **Big Hachet Peak** ▲ New Mexico, SW USA *31.38N 108.24W*
35 P11 **Big Hole River** ≈ Montana, NW USA
35 V13 **Bighorn Basin** *basin* Wyoming, C USA
35 U11 **Bighorn Lake** ◎ Montana/Wyoming, N USA
35 W13 **Bighorn Mountains** ⛰ Wyoming, C USA
68 J13 **Big Horn Peak** ▲ Arizona, SW USA *33.40N 113.01W*
35 V11 **Bighorn River** ≈ Montana/Wyoming, NW USA
16 O5 **Big Island** *island* Nunavut, NE Canada
41 O16 **Big Koniuji Island** *island* Shumagin Islands, Alaska, USA
27 N9 **Big Lake** Texas, SW USA *31.11N 101.27W*
21 T5 **Big Lake** ◎ Maine, NE USA
32 I3 **Big Manitou Falls** *waterfall* Wisconsin, N USA *46.32N 92.07W*
37 R2 **Big Mountain** ▲ Nevada, W USA *41.18N 119.03W*
110 G10 **Bignasco** Ticino, S Switzerland *46.21N 8.37E*
31 R16 **Big Nemaha River** ≈ Nebraska, C USA
78 G12 **Bignona** SW Senegal *12.49N 16.16W*
Bigorra *see* Tarbes
Bigosovo *see* Bihosava
37 S10 **Big Pine** California, W USA *37.09N 118.18W*
37 Q14 **Big Pine Mountain** ▲ California, W USA *34.41N 119.37W*
29 V6 **Big Piney Creek** ≈ Missouri, C USA
67 M24 **Big Point** *headland* N Tristan da Cunha
33 P8 **Big Rapids** Michigan, N USA *43.42N 85.28W*
32 K6 **Big Rib River** ≈ Wisconsin, N USA
12 L14 **Big Rideau Lake** ◎ Ontario, SE Canada
11 T14 **Big River** Saskatchewan, C Canada *53.48N 106.55W*
29 X5 **Big River** ≈ Missouri, C USA
33 N7 **Big Sable Point** *headland* Michigan, N USA *44.03N 86.30W*
35 S7 **Big Sandy** Montana, NW USA *48.08N 110.09W*
27 W6 **Big Sandy** Texas, SW USA *32.34N 95.06W*
39 V5 **Big Sandy Creek** ≈ Colorado, C USA
31 Q16 **Big Sandy Creek** ≈ Nebraska, C USA
31 V5 **Big Sandy Lake** ◎ Minnesota, N USA
68 J11 **Big Sandy River** ≈ Arizona, SW USA
23 P5 **Big Sandy River** ≈ S USA
25 V6 **Big Satilla Creek** ≈ Georgia, SE USA
31 R12 **Big Sioux River** ≈ Iowa/South Dakota, N USA
37 U7 **Big Smoky Valley** *valley* Nevada, W USA
27 N7 **Big Spring** Texas, SW USA *32.15N 101.30W*
21 Q5 **Big Squaw Mountain** ▲ Maine, NE USA *45.28N 69.42W*
23 O7 **Big Stone Gap** Virginia, NE USA *36.52N 82.45W*
31 Q8 **Big Stone Lake** ◎ Minnesota/South Dakota, N USA
24 K4 **Big Sunflower River** ≈ Mississippi, S USA
35 T11 **Big Timber** Montana, NW USA *45.50N 109.57W*
10 D8 **Big Trout Lake** Ontario, C Canada *53.40N 90.00W*
12 I12 **Big Trout Lake** ◎ Ontario, SE Canada
37 O2 **Big Valley Mountains** ⛰ California, W USA
27 Q13 **Big Wells** Texas, SW USA *28.34N 99.34W*
12 F11 **Bigwood** Ontario, S Canada *46.03N 80.37W*
114 D11 **Bihać** Federacija Bosna I Hercegovina, NW Bosnia and Herzegovina *44.49N 15.53E*
159 P14 **Bihār** ***prev.*** Behar. ◐ *state* N India
Bihār *see* Bihār Sharīf
83 F20 **Biharamulo** Kagera, NW Tanzania *2.37S 31.19E*
159 R13 **Bihārīganj** Bihār, NE India *25.43N 86.58E*
159 P14 **Bihār Sharīf** ***var.*** Bihār. Bihār, N India *25.13N 85.31E*
118 F10 **Bihor** ◐ *county* NW Romania
172 Q6 **Bihoro** Hokkaidō, NE Japan *43.50N 144.05E*
120 K11 **Bihosava** ***Rus.*** Bigosovo. Vitsyebskaya Voblasts', NW Belarus *55.50N 27.45E*
Bijagos Archipelago *see* Bijagós, Arquipélago dos
78 G13 **Bijagós, Arquipélago dos** ***var.*** Bijagos Archipelago. *island group* W Guinea-Bissau
161 F16 **Bijāpur** Karnātaka, C India *16.49N 75.42E*
148 K5 **Bījār** Kordestān, W Iran *35.54N 47.36E*
114 J11 **Bijeljina** Republika Srpska, NE Bosnia and Herzegovina *44.46N 19.13E*
115 K15 **Bijelo Polje** Montenegro, SW Yugoslavia *43.03N 19.44E*
166 I11 **Bijie** Guizhou, S China *27.18N 105.15E*
158 J10 **Bijnor** Uttar Pradesh, N India *29.22N 78.09E*
158 F11 **Bīkāner** Rājasthān, NW India *28.01N 73.22E*
201 V3 **Bikar Atoll** ***var.*** Pikaar. *atoll* Ratak Chain, N Marshall Islands
202 H3 **Bikeman** *atoll* Tungaru, W Kiribati
202 I3 **Bikenebu** Tarawa, W Kiribati
127 Nn16 **Bikin** Khabarovskiy Kray, SE Russian Federation *46.45N 134.06E*
127 Nn16 **Bikin** ≈ SE Russian Federation
201 R3 **Bikini Atoll** ***var.*** Pikinni. *atoll* Ralik Chain, NW Marshall Islands
85 L17 **Bikita** Masvingo, E Zimbabwe *20.04S 31.38E*
Bīkkū Bīttī *see* Bette, Pic
81 I19 **Bikoro** Equateur, W Dem. Rep. Congo (Zaire) *0.45S 18.09E*
147 Z9 **Bilād Banī Bū 'Alī** NE Oman *22.01N 59.18E*
147 Z9 **Bilād Banī Bū Ḥasan** NE Oman *22.09N 59.13E*
147 X9 **Bilād Manaḥ** ***var.*** Manaḥ. NE Oman *22.37N 57.27E*
79 Q12 **Bilanga** C Burkina *12.35N 0.08W*
175 Q7 **Bilang, Teluk** *bay* Sulawesi, N Indonesia
158 F12 **Bilāra** Rājasthān, N India *26.14N 73.48E*
158 K10 **Bilāri** Uttar Pradesh, N India *28.37N 78.48E*
144 J5 **Bil'ās, Jabal al** ⛰ C Syria
158 I8 **Bilāspur** Himāchal Pradesh, N India *31.19N 76.46E*
160 L11 **Bilāspur** Madhya Pradesh, C India *22.06N 82.08E*
173 G6 **Bila, Sungai** ≈ Sumatera, W Indonesia
143 Y13 **Biläsuvar** ***Rus.*** Bilyasuvar; ***prev.*** Pushkino. SE Azerbaijan *39.26N 48.33E*
119 O5 **Bila Tserkva** ***Rus.*** Belaya Tserkov'. Kyyivs'ka Oblast', N Ukraine *49.48N 30.07E*
178 H11 **Bilauktaung Range** ***var.*** Thanintari Taungdan. ⛰ Myanmar/Thailand
107 O2 **Bilbao** ***Basq.*** Bilbo. País Vasco, N Spain *43.15N 2.55W*
Bilbo *see* Bilbao
94 H2 **Bíldudalur** Vestfirdhir, NW Iceland *65.40N 23.35W*
115 I16 **Bileća** Republika Srpska, S Bosnia and Herzegovina *42.53N 18.27E*
142 E12 **Bilecik** Bilecik, NW Turkey *39.59N 29.54E*
142 F12 **Bilecik** ◐ *province* NW Turkey
118 E11 **Biled** ***Ger.*** Billed, ***Hung.*** Billéd. Timiș, W Romania *45.55N 20.55E*
113 O15 **Biłgoraj** Lubelskie, E Poland *50.32N 22.42E*
119 P11 **Bilhorod-Dnistrovs'kyy** ***Rus.*** Belgorod-Dnestrovskiy, ***Rom.*** Cetatea Albă; ***prev.*** Akkerman, ***anc.*** Tyras. Odes'ka Oblast', SW Ukraine *46.10N 30.18E*
81 M16 **Bili** Orientale, N Dem. Rep. Congo (Zaire) *4.07N 25.09E*
127 Oo5 **Bilibino** Chukotskiy Avtonomnyy Okrug, NE Russian Federation *67.56N 166.45E*
178 Gg8 **Bilin** Mon State, S Myanmar *17.13N 97.12E*
179 Qq12 **Biliran Island** *island* C Philippines
115 N21 **Bilisht** ***var.*** Bilishti. Korçë, SE Albania *40.36N 21.00E*
Bilishti *see* Bilisht
191 N10 **Billabong Creek** ***var.*** Moulamein Creek. *seasonal river* New South Wales, SE Australia
190 G4 **Billa Kalina** South Australia *29.57S 136.13E*
207 Q17 **Bill Baileys Bank** *undersea feature* N Atlantic Ocean *60.34N 10.15W*
Billed/Billéd *see* Biled
159 N14 **Billi** Uttar Pradesh, N India *24.30N 82.58E*
99 M15 **Billingham** N England, UK *54.36N 1.16W*
35 U11 **Billings** Montana, NW USA *45.47N 108.32W*
97 J16 **Billingsfors** Västra Götaland, S Sweden *58.57N 12.14E*
Bill of Cape Clear, The *see* Clear, Cape
30 L9 **Billsburg** South Dakota, N USA *44.22N 101.40W*
97 F23 **Billund** Ribe, W Denmark *55.43N 9.07E*
68 L11 **Bill Williams Mountain** ▲ Arizona, SW USA *35.12N 112.12W*
68 I12 **Bill Williams River** ≈ Arizona, SW USA
79 Y8 **Bilma** Agadez, NE Niger *18.22N 13.01E*
79 Y8 **Bilma, Grand Erg de** *desert* NE Niger
189 Y9 **Biloela** Queensland, E Australia *24.27S 150.31E*
114 G8 **Bilo Gora** ⛰ N Croatia
119 U13 **Bilohirs'k** ***Rus.*** Belogorsk; ***prev.*** Karasubazar. Respublika Krym, S Ukraine *45.01N 34.45E*
118 M3 **Bilokorovychi** ***Rus.*** Belokorovichi. Zhytomyrs'ka Oblast', N Ukraine *51.07N 28.02E*
119 X5 **Bilokurakine** Luhans'ka Oblast', E Ukraine *49.32N 38.44E*
119 T3 **Bilopillya** ***Rus.*** Belopol'ye. Sums'ka Oblast', NE Ukraine *51.09N 34.16E*
119 Y6 **Bilovods'k** ***Rus.*** Belovodsk. Luhans'ka Oblast', E Ukraine *49.10N 39.34E*
24 M9 **Biloxi** Mississippi, S USA *30.24N 88.53W*
119 R10 **Bilozerka** Khersons'ka Oblast', S Ukraine *46.36N 32.23E*
119 W7 **Bilozers'ke** Donets'ka Oblast', E Ukraine *48.29N 37.03E*
100 J11 **Bilthoven** Utrecht, C Netherlands *52.07N 5.12E*
80 K9 **Biltine** Biltine, E Chad *14.30N 20.52E*
80 J9 **Biltine** ***off.*** Préfecture de Biltine. ◐ *prefecture* E Chad
168 D5 **Bilüü** Bayan-Ölgiy, W Mongolia *48.54N 89.40E*
Bilwi *see* Puerto Cabezas
Bilyasuvar *see* Biläsuvar
119 O11 **Bilyayivka** Odes'ka Oblast', SW Ukraine *46.28N 30.11E*
101 K18 **Bilzen** Limburg, NE Belgium *50.52N 5.31E*
Bimbéréké *see* Bembèrèkè
191 R10 **Bimberi Peak** ▲ New South Wales, SE Australia *35.42S 148.46E*
79 Q15 **Bimbila** E Ghana *8.54N 0.04E*
81 I15 **Bimbo** Ombella-Mpoko, SW Central African Republic *4.19N 18.27E*
46 F2 **Bimini Islands** *island group* W Bahamas
160 I9 **Bīna** Madhya Pradesh, C India *24.09N 78.10E*
149 T4 **Bīnālūd, Kūh-e** ⛰ NE Iran
101 F20 **Binche** Hainaut, S Belgium *50.25N 4.10E*
Bindloe Island *see* Marchena, Isla
85 L16 **Bindura** Mashonaland Central, NE Zimbabwe *17.18S 31.13E*
107 T5 **Binefar** Aragón, NE Spain *41.51N 0.16E*
85 J16 **Binga** Matabeleland North, W Zimbabwe *17.42S 27.21E*
191 T5 **Bingara** New South Wales, SE Australia *29.54S 150.36E*
103 F18 **Bingen am Rhein** Rheinland-Pfalz, SW Germany *49.58N 7.54E*
28 M11 **Binger** Oklahoma, C USA *35.19N 98.19W*
Bingerau *see* Węgrów
Bin Ghalfān, Jazā'ir *see* Ḥalāniyāt, Juzur al
21 Q6 **Bingham** Maine, NE USA *45.01N 69.51W*
20 H11 **Binghamton** New York, NE USA *42.06N 75.55W*
Bin Ghanīmah, Jabal *see* Bin Ghunaymah, Jabal
77 P11 **Bin Ghunaymah, Jabal** ***var.*** Jabal Bin Ghanīmah. ⛰ C Libya
145 U3 **Bingird** NE Iraq *36.03N 45.03E*
143 P14 **Bingöl** Bingöl, E Turkey *38.54N 40.28E*
143 P14 **Bingöl** ◐ *province* E Turkey
167 R6 **Binhai** ***var.*** Binhai Xian, Dongkan. Jiangsu, E China *34.03N 119.46E*
Binhai Xian *see* Binhai
178 Kk12 **Binh Đinh** ***var.*** An Nhon. Binh Đinh, C Vietnam *13.52N 109.07E*
178 Kk10 **Binh Sơn** ***var.*** Châu Ô. Quang Ngai, C Vietnam *15.18N 108.45E*
Binimani *see* Bintimani
173 Ff5 **Binjai** Sumatera, W Indonesia *3.37N 98.30E*
191 R6 **Binnaway** New South Wales, SE Australia *31.34S 149.24E*
110 E6 **Binningen** Basel-Land, NW Switzerland *47.31N 7.34E*
175 R13 **Binongko, Pulau** *island* Kepulauan Tukangbesi, C Indonesia
174 Gg3 **Bintang, Banjaran** ⛰ Peninsular Malaysia
174 I7 **Bintan, Pulau** *island* Kepulauan Riau, W Indonesia
78 J14 **Bintimani** ***var.*** Binimani. ▲ NE Sierra Leone *9.21N 11.09W*
Bint Jubayl *see* Bent Jbaïl
174 M5 **Bintulu** Sarawak, East Malaysia *3.12N 113.01E*
176 Vv10 **Bintuni** ***prev.*** Steenkool. Irian Jaya, E Indonesia *2.03S 133.45E*
176 Vv10 **Bintuni, Teluk** *bay* Irian Jaya, E Indonesia
169 W8 **Bin Xian** Heilongjiang, NE China *45.43N 127.24E*
166 K14 **Binyang** Guangxi Zhuangzu Zizhiqu, S China *23.15N 108.47E*
167 Q4 **Binzhou** Shandong, E China *37.22N 118.03E*
65 G14 **Bío Bío** ***off.*** Región del Bío Bío. ◐ *region* C Chile
65 G14 **Bío Bío, Río** ≈ C Chile
81 C16 **Bioco, Isla de** ***var.*** Bioko, ***Eng.*** Fernando Po, ***Sp.*** Fernando Póo; ***prev.*** Macías Nguema Biyogo. *island* NW Equatorial Guinea
114 D13 **Biograd na Moru** ***It.*** Zaravecchia. Zadar, SW Croatia *43.57N 15.27E*
Bioko *see* Bioco, Isla de
115 F14 **Biokovo** ⛰ S Croatia
Biorra *see* Birr
Bipontium *see* Zweibrücken
149 W13 **Bīrag, Kūh-e** ⛰ SE Iran
77 O10 **Birāk** ***var.*** Brak. C Libya *27.31N 14.16E*
145 S10 **Bi'r al Islām** C Iraq *32.15N 43.40E*
160 N11 **Biramitrapur** Orissa, E India *22.24N 84.42E*
145 T11 **Bi'r an Niṣf** S Iraq *31.22N 44.07E*
80 L12 **Birao** Vakaga, NE Central African Republic *10.14N 22.49E*
164 M6 **Biratar Bulak** *well* NW China *42.00N 90.26E*
159 R12 **Biratnagar** Eastern, SE Nepal *26.28N 87.16E*
172 Oo6 **Biratori** Hokkaidō, NE Japan *42.35N 142.07E*
41 S8 **Birch Creek** Alaska, USA *66.17N 145.54W*
40 M11 **Birch Creek** ≈ Alaska, USA
11 T14 **Birch Hills** Saskatchewan, S Canada *52.58N 105.22W*
190 M10 **Birchip** Victoria, SE Australia *36.01S 142.55E*
31 X4 **Birch Lake** ◎ Minnesota, N USA
11 Q11 **Birch Mountains** ⛰ Alberta, W Canada
11 V15 **Birch River** Manitoba, S Canada *52.22N 101.03W*
46 H12 **Birchs Hill** *hill* W Jamaica *18.22N 78.05W*
41 R11 **Birchwood** Alaska, USA *61.24N 149.28W*
196 I5 **Bird Island** *island* S Northern Mariana Islands
143 N16 **Birecik** Şanlıurfa, S Turkey *37.02N 38.01E*
158 M10 **Birendranagar** ***var.*** Surkhet. Mid Western, W Nepal *28.35N 81.36E*
Bir es Saba *see* Be'ér Sheva'
76 A12 **Bir-Gandouz** SW Western Sahara *21.35N 16.27W*
159 P12 **Birganj** Central, C Nepal *27.03N 84.53E*
83 B14 **Biri** ≈ W Sudan
Bi'r Ibn Hirmās *see* Al Bi'r
176 Yy10 **Biri, Sungai** ≈ Irian Jaya, E Indonesia
149 U8 **Bīrjand** Khorāsān, E Iran *32.54N 59.13E*
145 T11 **Birkat Ḥāmid** *well* S Iraq *31.16N 44.04E*
97 F18 **Birkeland** Aust-Agder, S Norway *58.18N 8.13E*
103 E19 **Birkenfeld** Rheinland-Pfalz, SW Germany *49.39N 7.09E*
99 K18 **Birkenhead** NW England, UK *53.24N 3.01W*
111 W7 **Birkfeld** Steiermark, SE Austria *47.21N 15.40E*
190 A2 **Birksgate Range** ⛰ South Australia
Bîrlad *see* Bârlad
99 K20 **Birmingham** C England, UK *52.30N 1.49W*
25 P4 **Birmingham** Alabama, S USA *33.30N 86.47W*
99 M20 **Birmingham** ✕ C England, UK *52.27N 1.46W*
Bir Moghrein *see* Bîr Mogreïn
78 J4 **Bîr Mogreïn** ***var.*** Bir Moghrein; ***prev.*** Fort-Trinquet. Tiris Zemmour, N Mauritania *25.10N 11.34W*
203 S4 **Birnie Island** *atoll* Phoenix Islands, C Kiribati
Birni-Ngaouré *see* Birnin Gaouré
79 S12 **Birnin Gaouré** ***var.*** Birni-Ngaouré. Dosso, SW Niger *12.59N 3.02E*
79 S12 **Birnin Kebbi** Kebbi, NW Nigeria *12.28N 4.08E*
Birni-Nkonni *see* Birnin Konni
79 T12 **Birnin Konni** ***var.*** Birni-Nkonni. Tahoua, SW Niger *13.50N 5.14E*
79 W13 **Birnin Kudu** Jigawa, N Nigeria *11.28N 9.29E*
127 N16 **Birobidzhan** Yevreyskaya Avtonomnaya Oblast', SE Russian Federation *48.41N 132.55E*
99 D18 **Birr** ***var.*** Parsonstown, ***Ir.*** Biorra. C Ireland *53.06N 7.54W*
191 P4 **Birrie River** ≈ New South Wales/Queensland, SE Australia
110 D7 **Birse** ≈ NW Switzerland
Birsen *see* Biržai
110 E6 **Birsfelden** Basel-Land, NW Switzerland *47.33N 7.37E*
131 U4 **Birsk** Respublika Bashkortostan, W Russian Federation *55.24N 55.33E*
121 F14 **Birštonas** Prienai, C Lithuania *54.37N 24.00E*
165 P14 **Biru** Xinjiang Uygur Zizhiqu, W China *31.30N 93.55E*
Biruni *see* Beruniy
126 Ii14 **Biryusa** ≈ C Russian Federation
126 Ii14 **Biryusinsk** Irkutskaya Oblast', C Russian Federation *55.52N 97.48E*
120 G10 **Biržai** ***Ger.*** Birsen. Biržai, NE Lithuania *56.12N 24.47E*
123 Jj17 **Birżebbuġa** SE Malta *35.50N 14.32E*
Bisanthe *see* Tekirdağ
175 T9 **Bisa, Pulau** *island* Maluku, E Indonesia
39 N17 **Bisbee** Arizona, SW USA *31.27N 109.55W*
31 O2 **Bisbee** North Dakota, N USA *48.36N 99.21W*
Biscaia, Baía de *see* Biscay, Bay of
104 I13 **Biscarrosse et de Parentis, Étang de** ◎ SW France
106 M1 **Biscay, Bay of** ***Sp.*** Golfo de Vizcaya, ***Port.*** Baía de Biscaia. *bay* France/Spain
25 Z16 **Biscayne Bay** *bay* Florida, SE USA
66 M7 **Biscay Plain** *undersea feature* SE Bay of Biscay
109 N17 **Bisceglie** Puglia, SE Italy *41.13N 16.31E*
Bischoflack *see* Škofja Loka
Bischofsburg *see* Biskupiec
111 Q7 **Bischofshofen** Salzburg, NW Austria *47.25N 13.13E*
103 P15 **Bischofswerda** Sachsen, E Germany *51.07N 14.13E*
105 V5 **Bischwiller** Bas-Rhin, NE France *48.46N 7.52E*
23 T10 **Biscoe** North Carolina, SE USA *35.20N 79.46W*
204 G5 **Biscoe Islands** *island group* Antarctica
12 E9 **Biscotasi Lake** ◎ Ontario, S Canada
12 E9 **Biscotasing** Ontario, S Canada *47.16N 82.04W*
56 J6 **Biscucuy** Portuguesa, NW Venezuela *9.22N 69.58W*
116 K11 **Biser** Khaskovo, S Bulgaria *41.52N 25.58E*
115 D15 **Biševo** ***It.*** Busi. *island* SW Croatia
147 N12 **Bīshah, Wādī** *dry watercourse* C Saudi Arabia
153 U7 **Bishkek** ***var.*** Pishpek; ***prev.*** Frunze. ● (Kyrgyzstan) Chuyskaya Oblast', N Kyrgyzstan *42.53N 74.26E*
153 U7 **Bishkek** ✕ Chuyskaya Oblast', N Kyrgyzstan *42.55N 74.37E*
159 R16 **Bishnupur** West Bengal, NE India *23.04N 87.19E*
85 J25 **Bisho** Eastern Cape, S South Africa *32.46S 27.21E*
37 S9 **Bishop** California, W USA *37.22N 118.24W*
27 S15 **Bishop** Texas, SW USA *27.36N 97.49W*
99 L15 **Bishop Auckland** N England, UK *54.40N 1.40W*
Bishop's Lynn *see* King's Lynn
99 O21 **Bishop's Stortford** E England, UK *51.45N 0.12E*
23 S12 **Bishopville** South Carolina, SE USA *34.13N 80.15W*
144 M5 **Bishrī, Jabal** ⛰ E Syria
169 U4 **Bishui** Heilongjiang, NE China *52.06N 123.42E*
83 G17 **Bisina, Lake** ***prev.*** Lake Salisbury. ◎ E Uganda
Biskara *see* Biskra
76 L6 **Biskra** ***var.*** Beskra, Biskara. NE Algeria *34.51N 5.44E*
112 M8 **Biskupiec** ***Ger.*** Bischofsburg. Warmińsko-Mazurskie, NE Poland *53.51N 20.56E*
179 Rr15 **Bislig** Mindanao, S Philippines *8.10N 126.18E*
29 X6 **Bismarck** Missouri, C USA *37.46N 90.37W*
30 M5 **Bismarck** *state capital* North Dakota, N USA *46.48N 100.46W*
194 K9 **Bismarck Archipelago** *island group* NE PNG
133 Z16 **Bismarck Plate** *tectonic feature* W Pacific Ocean
194 I12 **Bismarck Range** ⛰ N PNG
194 J10 **Bismarck Sea** *sea* W Pacific Ocean
143 P15 **Bismil** Dıyarbakır, SE Turkey *37.52N 40.37E*
45 N6 **Bismuna, Laguna** *lagoon* NE Nicaragua
Bisnulok *see* Phitsanulok
175 T6 **Bisoa, Tanjung** *headland* Pulau Halmahera, N Indonesia *2.15N 127.57E*
30 K7 **Bison** South Dakota, N USA *45.30N 102.25W*
95 H17 **Bispfors** Jämtland, C Sweden *63.00N 16.40E*
78 G13 **Bissau** ● (Guinea-Bissau) W Guinea-Bissau *11.52N 15.39W*
78 G13 **Bissau** ✕ W Guinea-Bissau *11.53N 15.41W*
101 M24 **Bissen** Luxembourg, C Luxembourg *49.46N 6.04E*
78 G12 **Bissorã** W Guinea-Bissau *12.16N 15.34W*
11 O10 **Bistcho Lake** ◎ Alberta, W Canada
24 G5 **Bistineau, Lake** ⊠ Louisiana, S USA
Bistrica *see* Ilirska Bistrica
118 I9 **Bistrița** ***Ger.*** Bistritz, ***Hung.*** Beszterce; ***prev.*** Nösen. Bistrița-Năsăud, N Romania *47.10N 24.30E*
118 K10 **Bistrița** ***Ger.*** Bistritz. ≈ NE Romania
118 I9 **Bistrița-Năsăud** ◐ *county* N Romania
Bistritz *see* Bistrița
Bistritz ober Pernstein *see* Bystřice nad Pernštejnem
158 L11 **Biswān** Uttar Pradesh, N India *27.30N 81.00E*
112 M7 **Bisztynek** Warmińsko-Mazurskie, NE Poland *54.05N 20.53E*
81 E17 **Bitam** Woleu-Ntem, N Gabon *2.04N 11.30E*
103 D18 **Bitburg** Rheinland-Pfalz, SW Germany *49.57N 6.31E*
105 U4 **Bitche** Moselle, NE France *49.01N 7.27E*
80 I11 **Bitkine** Guéra, C Chad *11.58N 18.13E*
143 R15 **Bitlis** Bitlis, SE Turkey *38.22N 42.04E*
143 R14 **Bitlis** ◐ *province* E Turkey
Bitoeng *see* Bitung
115 N20 **Bitola** ***Turk.*** Monastir; ***prev.*** Bitolj. S FYR Macedonia *41.01N 21.21E*
Bitolj *see* Bitola
109 O17 **Bitonto** ***anc.*** Butuntum. Puglia, SE Italy *41.07N 16.40E*
79 Q13 **Bitou** ***var.*** Bittou. SE Burkina *11.19N 0.16W*
161 C20 **Bitra Island** *island* Lakshadweep, India, N Indian Ocean
103 M14 **Bitterfeld** Sachsen-Anhalt, E Germany *51.36N 12.18E*
34 O9 **Bitterroot Range** ⛰ Idaho/Montana, NW USA
35 P10 **Bitterroot River** ≈ Montana, NW USA
109 D18 **Bitti** Sardegna, Italy, C Mediterranean Sea *40.30N 9.31E*
Bittou *see* Bitou
175 S7 **Bitung** ***prev.*** Bitoeng. Sulawesi, C Indonesia *1.28N 125.13E*
62 I12 **Bituruna** Paraná, S Brazil *26.11S 51.34W*
79 Y13 **Biu** Borno, E Nigeria *10.35N 12.13E*
Biumba *see* Byumba
171 H14 **Biwa-ko** ◎ Honshū, SW Japan
176 Y13 **Biwarlaut** Irian Jaya, E Indonesia *5.44S 138.14E*
29 P10 **Bixby** Oklahoma, C USA *35.56N 95.52W*
126 H15 **Biya** ≈ S Russian Federation
Biy-Khem *see* Bol'shoy Yenisey
126 H15 **Biysk** Altayskiy Kray, S Russian Federation *52.34N 85.09E*
170 Ff14 **Bizen** Okayama, Honshū, SW Japan *34.43N 134.10E*
Bizerta *see* Bizerte
123 K11 **Bizerte** ***Ar.*** Banzart, ***Eng.*** Bizerta. N Tunisia *37.18N 9.48E*
Bizkaia *see* Vizcaya
94 G2 **Bjargtangar** *headland* W Iceland *63.30N 24.28W*
Bjärnå *see* Perniö
97 K22 **Bjärnum** Skåne, S Sweden *56.15N 13.45E*
95 I16 **Bjästa** Västernorrland, C Sweden *63.12N 18.30E*
115 I14 **Bjelašnica** ▲ SE Bosnia and Herzegovina *43.13N 18.18E*
114 C10 **Bjelolasica** ▲ NW Croatia *45.13N 14.56E*
114 F8 **Bjelovar** ***Hung.*** Belovár. Bjelovar-Bilogora, N Croatia *45.54N 16.49E*
114 F8 **Bjelovar-Bilogora** ***off.*** Bjelovarsko-Bilogorska Županija. ◐ *province* NE Croatia
Bjelovarsko-Bilogorska Županija *see* Bjelovar-Bilogora
94 H10 **Bjerkvik** Nordland, C Norway *68.31N 16.08E*
97 G21 **Bjerringbro** Viborg, NW Denmark *56.22N 9.40E*
Bjeshkët e Namuna *see* North Albanian Alps
97 L14 **Björbo** Dalarna, C Sweden *60.28N 14.44E*
97 I15 **Bjørkelangen** Akershus, S Norway *59.52N 11.34E*
97 O14 **Björklinge** Uppsala, C Sweden *60.03N 17.33E*
95 I14 **Björksele** Västerbotten, N Sweden *64.58N 18.30E*
95 I16 **Björna** Västernorrland, C Sweden *63.34N 18.38E*
97 C14 **Bjørnafjorden** *fjord* S Norway
97 L16 **Björneborg** Värmland, C Sweden *59.13N 14.15E*
Björneborg *see* Pori
97 E14 **Bjornesfjorden** ◎ S Norway
94 M9 **Bjørnevatn** Finnmark, N Norway *69.40N 29.57E*
207 T13 **Bjørnøya** ***Eng.*** Bear Island. *island* N Norway
95 I15 **Bjurholm** Västerbotten, N Sweden *63.57N 19.16E*
97 J22 **Bjuv** Skåne, S Sweden *56.04N 12.57E*
78 M12 **Bla** Ségou, W Mali *12.58N 5.45W*
189 W8 **Blackall** Queensland, E Australia *24.25S 145.31E*
31 V2 **Black Bay** *lake bay* Minnesota, N USA
29 N9 **Black Bear Creek** ≈ Oklahoma, C USA
99 K17 **Blackburn** NW England, UK *53.45N 2.28W*
47 W10 **Blackburne** ✕ (Plymouth) E Montserrat *16.45N 62.09W*
41 T11 **Blackburn, Mount** ▲ Alaska, USA *61.43N 143.25W*
37 N5 **Black Butte Lake** ⊠ California, W USA
204 J5 **Black Coast** *physical region* Antarctica
11 Q16 **Black Diamond** Alberta, SW Canada *50.42N 114.09W*
20 K11 **Black Dome** ▲ New York, NE USA *42.16N 74.07W*
115 L18 **Black Drin** ***Alb.*** Lumi i Drinit të Zi, ***SCr.*** Crni Drim. ≈ Albania/FYR Macedonia
10 D6 **Black Duck** ≈ Ontario, C Canada
31 U4 **Blackduck** Minnesota, N USA *47.45N 94.33W*
35 R14 **Blackfoot** Idaho, NW USA *43.11N 112.20W*
35 P9 **Blackfoot River** ≈ Montana, NW USA
Black Forest *see* Schwarzwald
30 J10 **Blackhawk** South Dakota, N USA *44.09N 103.18W*
30 I10 **Black Hills** ⛰ South Dakota/Wyoming, N USA
11 T10 **Black Lake** ◎ Saskatchewan, C Canada
33 Q5 **Black Lake** ◎ Michigan, N USA
20 I7 **Black Lake** ◎ New York, NE USA
24 G6 **Black Lake** ⊠ Louisiana, S USA
28 F7 **Black Mesa** ▲ Oklahoma, C USA *37.00N 103.07W*
23 P10 **Black Mountain** North Carolina, SE USA *35.37N 82.19W*
37 P13 **Black Mountain** ▲ California, W USA *35.22N 120.21W*
39 Q2 **Black Mountain** ▲ Colorado, C USA *40.47N 107.23W*
23 O7 **Black Mountain** ▲ Kentucky, E USA *36.54N 82.53W*
98 K1 **Black Mountains** ⛰ SE Wales, UK
68 H10 **Black Mountains** ⛰ Arizona, SW USA
35 Q16 **Black Pine Peak** ▲ Idaho, NW USA *42.07N 113.07W*
99 K17 **Blackpool** NW England, UK *53.49N 3.03W*
39 Q14 **Black Range** ⛰ New Mexico, SW USA
46 I12 **Black River** W Jamaica *18.01N 77.52W*
12 J14 **Black River** ≈ Ontario, SE Canada
133 U12 **Black River** ***Chin.*** Babian Jiang, Lixian Jiang, ***Fr.*** Rivière Noire, ***Vtn.*** Sông Đa. ≈ China/Vietnam
46 I12 **Black River** ≈ W Jamaica
41 T7 **Black River** ≈ Alaska, USA
39 N13 **Black River** ≈ Arizona, SW USA
29 X7 **Black River** ≈ Arkansas/Missouri, C USA
24 I7 **Black River** ≈ Louisiana, S USA
33 S8 **Black River** ≈ Michigan, N USA
33 Q5 **Black River** ≈ Michigan, N USA
20 I8 **Black River** ≈ New York, NE USA
23 T13 **Black River** ≈ South Carolina, SE USA
32 J7 **Black River** ≈ Wisconsin, N USA
32 J7 **Black River Falls** Wisconsin, N USA *44.18N 90.51W*
37 R3 **Black Rock Desert** *desert* Nevada, W USA
Black Sand Desert *see* Garagumy
23 S7 **Blacksburg** Virginia, NE USA *37.16N 80.24W*
142 H10 **Black Sea** ***var.*** Euxine Sea, ***Bul.*** Cherno More, ***Rom.*** Marea Neagră, ***Rus.*** Chernoye More, ***Turk.*** Karadeniz, ***Ukr.*** Chorne More. *sea* Asia/Europe
119 Q10 **Black Sea Lowland** ***Ukr.*** Prychornomors'ka Nyzovyna. *depression* SE Europe
35 S17 **Blacks Fork** ≈ Wyoming, C USA
25 V7 **Blackshear** Georgia, SE USA *31.18N 82.14W*
25 S6 **Blackshear, Lake** ⊠ Georgia, SE USA
99 A16 **Blacksod Bay** ***Ir.*** Cuan an Fhóid Duibh. *inlet* W Ireland
23 V7 **Blackstone** Virginia, NE USA *37.04N 78.00W*
79 O14 **Black Volta** ***var.*** Borongo, Mouhoun, Moun Hou, ***Fr.*** Volta Noire. ≈ W Africa
25 O5 **Black Warrior River** ≈ Alabama, S USA
189 X8 **Blackwater** Queensland, E Australia *23.34S 148.51E*
99 D20 **Blackwater** ***Ir.*** An Abhainn Mhór. ≈ S Ireland
29 T4 **Blackwater River** ≈ Missouri, C USA
23 W7 **Blackwater River** ≈ Virginia, NE USA
Blackwater State *see* Nebraska
29 N8 **Blackwell** Oklahoma, C USA *36.48N 97.16W*
27 P7 **Blackwell** Texas, SW USA *32.05N 100.19W*
101 J15 **Bladel** Noord-Brabant, S Netherlands *51.22N 5.13E*
Bladenmarkt *see* Bălăușeri
116 G11 **Blagoevgrad** ***prev.*** Gorna Dzhumaya. Blagoevgrad, SW Bulgaria *42.01N 23.04E*
116 G11 **Blagoevgrad** ◐ *province* SW Bulgaria
126 Gg14 **Blagoveshchenka** Altayskiy Kray, S Russian Federation *52.49N 79.54E*
126 M16 **Blagoveshchensk** Amurskaya Oblast', SE Russian Federation *50.19N 127.30E*
131 V4 **Blagoveshchensk** Respublika Bashkortostan, W Russian Federation *55.03N 56.01E*
104 I7 **Blain** Loire-Atlantique, NW France *47.26N 1.47W*
31 V8 **Blaine** Minnesota, N USA *45.09N 93.13W*
34 H6 **Blaine** Washington, NW USA *48.59N 122.45W*
11 T15 **Blaine Lake** Saskatchewan, S Canada *52.49N 106.48W*
31 S14 **Blair** Nebraska, C USA *41.32N 96.07W*
98 J10 **Blairgowrie** C Scotland, UK *56.18N 3.24W*
20 C15 **Blairsville** Pennsylvania, NE USA *40.25N 79.12W*
118 H11 **Blaj** ***Ger.*** Blasendorf, ***Hung.*** Balázsfalva. Alba, SW Romania *46.10N 23.56E*
66 F9 **Blake-Bahama Ridge** *undersea feature* W Atlantic Ocean
25 S7 **Blakely** Georgia, SE USA *31.22N 84.55W*
66 E10 **Blake Plateau** ***var.*** Blake Terrace. *undersea feature* W Atlantic Ocean
32 M1 **Blake Point** *headland* Michigan, N USA *48.11N 88.25W*
Blake Terrace *see* Blake Plateau
63 B24 **Blanca, Bahía** *bay* E Argentina
58 C12 **Blanca, Cordillera** ⛰ W Peru
107 T12 **Blanca, Costa** *physical region* SE Spain
39 S7 **Blanca Peak** ▲ Colorado, C USA *37.34N 105.29W*
26 I9 **Blanca, Sierra** ▲ Texas, SW USA *31.15N 105.26W*
123 K11 **Blanc, Cap** *headland* N Tunisia *37.20N 9.41E*
Blanc, Cap *see* Nouâdhibou, Râs
33 R12 **Blanchard River** ≈ Ohio, N USA
190 E8 **Blanche, Cape** *headland* South Australia *33.03S 134.10E*
195 U15 **Blanche Channel** *channel* NW Solomon Islands
190 J4 **Blanche, Lake** ◎ South Australia
33 R14 **Blanchester** Ohio, N USA *39.17N 83.59W*
190 J9 **Blanchetown** South Australia *34.21S 139.36E*
47 U13 **Blanchisseuse** Trinidad, Trinidad and Tobago *10.47N 61.18W*
105 T11 **Blanc, Mont** ***It.*** Monte Bianco. ▲ France/Italy *45.45N 6.51E*
27 R11 **Blanco** Texas, SW USA *30.06N 98.25W*
44 K14 **Blanco, Cabo** *headland* NW Costa Rica *9.34N 85.06W*
34 D14 **Blanco, Cape** *headland* Oregon, NW USA *42.49N 124.33W*
64 H10 **Blanco, Río** ≈ W Argentina
58 F10 **Blanco, Río** ≈ NE Peru
13 O9 **Blanc, Réservoir** ⊠ Quebec, SE Canada
23 R7 **Bland** Virginia, NE USA *37.06N 81.07W*
94 I2 **Blanda** ≈ N Iceland
39 O7 **Blanding** Utah, W USA *37.37N 109.28W*
107 X5 **Blanes** Cataluña, NE Spain *41.40N 2.48E*
105 N3 **Blangy-sur-Bresle** Seine-Maritime, N France *49.55N 1.37E*
113 C18 **Blanice** ***Ger.*** Blanitz. ≈ SE Czech Republic
Blanitz *see* Blanice
101 C16 **Blankenberge** West-Vlaanderen, NW Belgium *51.19N 3.07E*
103 D17 **Blankenheim** Nordrhein-Westfalen, W Germany *50.25N 6.41E*
27 R8 **Blanket** Texas, SW USA *31.49N 98.47W*
57 O3 **Blanquilla, Isla** ***var.*** La Blanquilla. *island* N Venezuela
Blanquilla, La *see* Blanquilla, Isla
63 F18 **Blanquillo** Durazno, C Uruguay *32.52S 55.37W*
113 G18 **Blansko** ***Ger.*** Blanz. Brněnský Kraj, SE Czech Republic *49.22N 16.39E*
85 N15 **Blantyre** ***var.*** Blantyre-Limbe. Southern, S Malawi *15.45S 35.03E*
85 N15 **Blantyre** ✕ Southern, S Malawi *15.34S 35.03E*
Blantyre-Limbe *see* Blantyre
Blanz *see* Blansko
100 J10 **Blaricum** Noord-Holland, C Netherlands *52.16N 5.15E*
Blasendorf *see* Blaj
Blatnitsa *see* Durankulak
115 F15 **Blato** ***It.*** Blatta. Dubrovnik-Neretva, S Croatia *42.57N 16.47E*

◆ Country ● Country Capital ◇ Dependent Territory ○ Dependent Territory Capital ◐ Administrative Region ✕ International Airport ▲ Mountain ⛰ Mountain Range Volcano ≈ River ◎ Lake ⊠ Reservoir

Blatta see Blato
110 E10 **Blatten** Valais, SW Switzerland 46.22N 8.00E
103 J20 **Blaufelden** Baden-Württemberg, SW Germany 49.21N 10.01E
97 E23 **Blåvands Huk** headland W Denmark 55.33N 8.04E
104 G6 **Blavet** ♣ NW France
104 J12 **Blaye** Gironde, SW France 45.07N 0.36W
191 R8 **Blayney** New South Wales, SE Australia 33.33S 149.13E
67 D25 **Bleaker Island** island SE Falkland Islands
111 T10 **Bled** Ger. Veldes. NW Slovenia 46.23N 14.06E
101 D20 **Bléharies** Hainaut, SW Belgium 50.31N 3.25E
111 U9 **Bleiburg Slvn.** Pliberk. Kärnten, S Austria 46.36N 14.49E
103 L17 **Bleiloch-Stausee** ⊠ C Germany
100 H12 **Bleiswijk** Zuid-Holland, W Netherlands 52.01N 4.31E
97 L22 **Blekinge** ♦ county S Sweden
12 D17 **Blenheim** Ontario, S Canada 42.19N 81.58W
193 K15 **Blenheim** Marlborough, South Island, NZ 41.31S 174.00E
101 M15 **Blerick** Limburg, SE Netherlands 51.22N 6.10E
Blesae see Blois
27 V13 **Blessing** Texas, SW USA 28.52N 96.12W
12 I10 **Bleu, Lac** ⊗ Quebec, SE Canada
Blibba see Blitta
123 I11 **Blida** var. El Boulaida, El Boulaïda. N Algeria 36.32N 2.49E
97 P15 **Blidö** Stockholm, E Sweden 59.37N 18.55E
97 K18 **Blidsberg** Västra Götaland, S Sweden 57.55N 13.30E
193 A21 **Bligh Sound** sound South Island, NZ
197 H13 **Bligh Water** strait NW Fiji
12 D11 **Blind River** Ontario, S Canada 46.11N 82.55W
33 R11 **Blissfield** Michigan, N USA 41.49N 83.51W
174 Ll16 **Blitar** Jawa, C Indonesia 8.06S 112.12E
79 R15 **Blitta** prev. Blibba. C Togo 8.19N 0.58E
21 O13 **Block Island** island Rhode Island, NE USA
21 O13 **Block Island Sound** sound Rhode Island, NE USA
100 H10 **Bloemendaal** Noord-Holland, W Netherlands 52.23N 4.39E
85 H23 **Bloemfontein** var. Mangaung. ● (South Africa-judicial capital) Free State, C South Africa 29.07S 26.13E
85 I22 **Bloemhof** North-West, N South Africa 27.38S 25.33E
104 M7 **Blois** anc. Blesae. Loir-et-Cher, C France 47.36N 1.19E
100 L8 **Blokzijl** Overijssel, N Netherlands 52.46N 5.58E
97 N20 **Blomstermåla** Kalmar, S Sweden 56.58N 16.19E
94 I2 **Blönduós** Norðhurland Vestra, N Iceland 65.39N 20.15W
112 L11 **Blonie** Mazowieckie, C Poland 52.13N 20.36E
99 C14 **Bloody Foreland Ir.** Cnoc Fola. headland NW Ireland 9.18W
33 N15 **Bloomfield** Indiana, N USA 39.01N 86.58W
31 X16 **Bloomfield** Iowa, C USA 40.45N 92.24W
29 Y8 **Bloomfield** Missouri, C USA 36.53N 89.55W
39 P9 **Bloomfield** New Mexico, SW USA 36.42N 108.00W
27 U7 **Blooming Grove** Texas, SW USA 32.05N 96.43W
31 W10 **Blooming Prairie** Minnesota, N USA 43.52N 93.03W
32 L13 **Bloomington** Illinois, N USA 40.28N 88.59W
33 O15 **Bloomington** Indiana, N USA 39.10N 86.31W
31 V9 **Bloomington** Minnesota, N USA 44.50N 93.18W
27 T12 **Bloomington** Texas, SW USA 28.39N 96.53W
20 H14 **Bloomsburg** Pennsylvania, NE USA 40.58N 76.27W
189 X7 **Bloomsbury** Queensland, NE Australia 20.44S 148.34E
174 L14 **Blora** Jawa, C Indonesia 6.55S 111.28E
20 G12 **Blossburg** Pennsylvania, USA 41.38N 77.00W
27 V5 **Blossom** Texas, SW USA 33.39N 95.23W
127 Oo3 **Blossom, Mys** headland Ostrov Vrangelya, NE Russian Federation 70.49N 178.49E
25 R8 **Blountstown** Florida, SE USA 30.26N 85.03W
23 P8 **Blountville** Tennessee, SE USA 36.31N 82.19W
23 Q9 **Blowing Rock** North Carolina, SE USA 36.15N 81.53W
110 J8 **Bludenz** Vorarlberg, W Austria 47.10N 9.49E
68 L6 **Blue Bell Knoll** ▲ Utah, W USA 38.11N 111.31W
25 V11 **Blue Cypress Lake** ⊗ Florida, SE USA
31 U11 **Blue Earth** Minnesota, N USA 43.38N 94.06W
23 Q7 **Bluefield** Virginia, NE USA 37.15N 81.16W
23 R7 **Bluefield** West Virginia, NE USA 37.16N 81.13W
45 N10 **Bluefields** Región Autónoma Atlántico Sur, SE Nicaragua 12.01N 83.47W
45 N10 **Bluefields, Bahía de** bay W Caribbean Sea
31 X14 **Blue Grass** Iowa, C USA 41.30N 90.46W
Bluegrass State see Kentucky
Blue Hen State see Delaware
21 S7 **Blue Hill** Maine, NE USA
23 P16 **Blue Hill** Nebraska, C USA 40.19N 98.27W
32 J5 **Blue Hills** hill range Wisconsin, N USA
36 L3 **Blue Lake** California, W USA 40.52N 124.00W

12 K14 **Bobs Lake** ⊗ Ontario, SE Canada
56 I6 **Boquerón** Zulia, NW Venezuela 9.15N 71.10W
44 H1 **Boca Bacalar Chico** headland N Belize 25.50N 82.12W
114 G11 **Bočac** Republika Srpska, NW Bosnia and Herzegovina 44.32N 17.09E
43 R14 **Boca del Río** Veracruz-Llave, S Mexico 19.07N 96.07W
57 O4 **Boca de Pozo** Nueva Esparta, NE Venezuela 11.01N 64.21W
61 C15 **Boca do Acre** Amazonas, N Brazil 8.45S 67.22W
57 N12 **Boca Mavaca** Amazonas, S Venezuela 2.30N 65.10W
81 G14 **Bocaranga** Ouham-Pendé, W Central African Republic 7.07N 15.40E
25 Z15 **Boca Raton** Florida, SE USA 26.22N 80.05W
45 P14 **Bocas del Toro** Bocas del Toro, NW Panama 9.39N 96.38W
45 P15 **Bocas del Toro** off. Provincia de Bocas del Toro. ♦ province NW Panama
45 P15 **Bocas del Toro, Archipiélago de** island group NW Panama
44 L7 **Boceguillas** Castilla-León, N Spain 41.23N 3.37W
107 N6 **Bochechi** see Bacheykava
113 L17 **Bochnia** Małopolskie, SE Poland 49.58N 20.27E
101 K16 **Bocholt** Limburg, NE Belgium 51.10N 5.37E
103 D14 **Bocholt** Nordrhein-Westfalen, W Germany 51.49N 6.37E
103 E15 **Bochum** Nordrhein-Westfalen, W Germany 51.28N 7.13E
105 Y15 **Bocognano** Corse, France, C Mediterranean Sea 42.04N 9.03E
56 I6 **Boconó** Trujillo, NW Venezuela 9.12N 70.16W
118 F12 **Bocşa Ger.** Bokschen, Hung. Boksánbánya. Caras-Severin, SW Romania 45.24N 21.46E
81 H15 **Boda** Lobaye, SW Central African Republic 4.17N 17.25E
96 L12 **Boda** Dalarna, C Sweden 61.00N 15.15E
97 O20 **Böda** Kalmar, S Sweden 57.16N 17.04E
97 L19 **Bodafors** Jönköping, S Sweden 57.30N 14.40E
126 K14 **Bodaybo** Irkutskaya Oblast', E Russian Federation 57.52N 114.04E
24 G5 **Bodcau, Bayou** var. Bodcau Creek. ♣ Louisiana, S USA
46 D8 **Bodcau Creek** see Bodcau, Bayou
Boddam Town var. Boddentown. Grand Cayman, SW Cayman Islands 19.17N 81.10W
103 P14 **Bode** ♣ C Germany
36 L7 **Bodega Head** headland California, W USA 38.16N 123.04W
100 H11 **Bodegas** see Babahoyo
102 I9 **Bodegraven** Zuid-Holland, C Netherlands 52.05N 4.45E
94 H8 **Bodélé** depression W Chad
94 J13 **Boden** Norrbotten, N Sweden 65.49N 21.43E
Bodensee see Constance, Lake, C Europe
67 M15 **Bode Verde Fracture Zone** tectonic feature E Atlantic Ocean
161 N14 **Bodhan** Andhra Pradesh, C India 18.40N 77.51E
168 I9 **Bodi** Bayankhongor, C Mongolia 45.25N 100.33E
161 N12 **Bodinäyakkanür** Tamil Nādu, SE India 10.01N 77.18E
110 H10 **Bodio** Ticino, S Switzerland 46.23N 8.55E
99 G24 **Bodmin** SW England, UK 50.28N 4.43W
99 F24 **Bodmin Moor** moorland SW England, UK
94 C9 **Bodø** Nordland, C Norway 67.16N 14.22E
61 B16 **Bodoquena, Serra da** ▲ SW Brazil
142 B16 **Bodrum** Muğla, SW Turkey 37.03N 27.28E
Bodzaforduló see Intorsura Buzăului
101 I15 **Boekel** Noord-Brabant, SE Netherlands 51.35N 5.42E
79 J19 **Boende** Equateur, C Dem. Rep. Congo (Zaire) 0.12S 20.54E
27 R11 **Boerne** Texas, SW USA 29.47N 98.43W
Boeroe see Buru, Pulau
Boetoeng see Buton, Pulau
24 I5 **Boeuf River** ♣ Arkansas/Louisiana, S USA
35 N14 **Boffa** Guinée-Maritime, W Guinea 10.12N 14.01W
Bó Finne, Inis see Inishbofin
177 Ff9 **Boga** see Bogë
24 L8 **Bogalusa** Louisiana, S USA 30.47N 89.51W
79 O12 **Bogandé** C Burkina 13.01N 0.07W
81 I15 **Bogangolo** Ombella-Mpoko, C Central African Republic 5.36N 18.17E
191 Q7 **Bogan River** ♣ New South Wales, SE Australia
113 D14 **Bogatynia** Ger. Reichenau. Dolnośląskie, SW Poland 50.52N 14.54E
142 K13 **Boğazlıyan** Yozgat, C Turkey 39.13N 35.16E
81 J21 **Bogboma** Equateur, NW Dem. Rep. Congo (Zaire) 1.36N 19.24E
164 K12 **Bogdag Zangbo** ♣ NW China
164 L5 **Bogda Feng** ▲ NW China 43.51N 88.14E
116 I9 **Bogdan** ▲ C Bulgaria 42.37N 24.28E
115 I20 **Bogdanci** SE FYR Macedonia 41.12N 22.34E
164 M5 **Bogda Shan** var. Po-ko-to Shan. ▲ NW China
115 K17 **Bogë** var. Boga. N Albania 42.25N 19.38E
91 Q15 **Bogen** Fyn, C Denmark 55.34N 10.06E

191 T3 **Boggabilla** New South Wales, SE Australia 28.35S 150.21E
191 S6 **Boggabri** New South Wales, SE Australia 30.44S 150.00E
194 I10 **Bogia** Madang, N PNG 4.12S 144.55E
99 Qq13 **Bognor Regis** SE England, UK 50.46N 0.40W
179 Qq13 **Bogo** Cebu, C Philippines 11.04N 123.59E
Bogodukhov see Bohodukhiv
189 V15 **Bogong, Mount** ▲ Victoria, SE Australia 36.45S 147.19E
174 J14 **Bogor Dut.** Buitenzorg. Jawa, C Indonesia 6.34S 106.45E
130 L5 **Bogoroditsk** Tul'skaya Oblast', W Russian Federation 53.46N 38.09E
131 O3 **Bogorodsk** Nizhegorodskaya Oblast', W Russian Federation 56.06N 43.29E
127 Nn14 **Bogorodskoye** Khabarovskiy Kray, SE Russian Federation 52.22N 140.33E
129 R15 **Bogorodskoye** var. Bogorodskoje. Kirovskaya Oblast', NW Russian Federation 57.50N 50.41E
56 F10 **Bogotá** prev. Santa Fe de Bogotá. ● (Colombia) Cundinamarca, C Colombia 4.37N 74.04W
159 T14 **Bogra** Rajshahi, N Bangladesh 24.52N 89.28E
Bogschan see Boldu
126 Ii13 **Boguchany** Krasnoyarskiy Kray, C Russian Federation 58.20N 97.20E
130 M9 **Boguchar** Voronezhskaya Oblast', W Russian Federation 49.57N 40.34E
78 H10 **Bogué** Brakna, SW Mauritania 16.36N 14.15W
24 K8 **Bogue Chitto** ♣ Louisiana/Mississippi, S USA
81 J19 **Boko** Equateur, C Dem. Rep. Congo (Zaire) 1.27S 19.52E
113 E14 **Bolesławiec Ger.** Bunzlau. Dolnośląskie, SW Poland 51.16N 15.34E
46 H7 **Bog Walk** C Jamaica 18.06N 77.01W
167 Q3 **Bo Hai** var. Gulf of Chihli. gulf NE China
167 R3 **Bohai Haixia** strait NE China
167 Q3 **Bohai Wan** bay NE China
113 C17 **Bohemia** Cz. Čechy, Ger. Böhmen. cultural and historical region W Czech Republic
113 B18 **Bohemian Forest Cz.** Český Les, Šumava, Ger. Böhmerwald. ▲ C Europe
Bohemian-Moravian Highlands see Českomoravská Vrchovina
79 R16 **Bohicon** S Benin 7.08N 2.07E
111 S11 **Bohinjska Bistrica Ger.** Wocheiner Feistritz. NW Slovenia 46.16N 13.55E
179 Oo9 **Bohol** island C Philippines 16.22N 119.52E
29 T6 **Boligee** Alabama, S USA 37.36N 93.24W
22 F10 **Bolivar** Tennessee, S USA 35.15N 88.59W
56 C12 **Bolívar** Cauca, SW Colombia 1.49N 76.58W
56 F7 **Bolívar** Santander, N Colombia 5.51N 73.46W
57 N9 **Bolívar** off. Estado Bolívar. ♦ state SE Venezuela
27 X12 **Bolivar Peninsula** headland Texas, SW USA 29.26N 94.41W
56 I6 **Bolívar, Pico** ▲ W Venezuela 8.33N 71.05W
59 F19 **Bolivia** off. Republic of Bolivia. ◆ republic W South America
114 O13 **Boljevac** Serbia, E Yugoslavia 43.50N 21.57E
165 O10 **Bolkenhain** see Bolków
168 M9 **Bölkh** Dundgovi, C Mongolia 45.13N 108.12E
113 I17 **Bolków Ger.** Bolkenhain. Dolnośląskie, SW Poland 50.55N 15.49E
190 K3 **Bollards Lagoon** South Australia 28.58S 140.52E
105 R14 **Bollène** Vaucluse, SE France 44.16N 4.45E
96 H16 **Bollnäs** Gävleborg, C Sweden 61.21N 16.27E
189 W10 **Bollon** Queensland, C Australia 28.07S 147.28E
199 Jj14 **Bollons Tablemount** undersea feature S Pacific Ocean
C Sweden 63.00N 17.41E
178 Jj10 **Bolovens, Plateau des** plateau S Laos
108 H13 **Bolsena** Lazio, C Italy 42.39N 11.59E
130 B3 **Bol'shakovo** Ger. Kreuzingen; prev. Gross-Skaisgirren. Kaliningradskaya Oblast', W Russian Federation 54.53N 21.38E
108 G10 **Bologna** Emilia-Romagna, N Italy 44.30N 11.19E
128 I15 **Bologoye** Tverskaya Oblast', W Russian Federation 57.54N 34.04E
81 J18 **Bolomba** Equateur, NW Dem. Rep. Congo (Zaire) 0.24N 19.10E
102 K10 **Bolotnoye** Novosibirskaya Oblast', C Russian Federation 55.39N 84.19E
116 J13 **Boloústra, Akrotírio** headland NE Greece 40.56N 24.58E
174 Ll15 **Bolongongo** see Bodjonegoro. Jawa, C Indonesia 7.06S 111.49E
201 T1 **Bokaak Atoll** var. Bokak, Taongi. atoll Ratak Chain, NE Marshall Islands
109 Q16 **Bolsena, Lago di** ⊗ C Italy
Bokaro see Bokāro
159 Q15 **Bokāro** Bihār, N India 23.46N 85.55E
164 M5 **Bokatola** Equateur, NW Dem. Rep. Congo (Zaire) 0.37S 18.45E
78 H13 **Boké** Guinée-Maritime, W Guinea 10.57N 14.13W
Bokhara see Bukhoro
191 Q3 **Bokhara River** ♣ New South Wales/Queensland, SE Australia
116 J6 **Botoşani** ▲ N Russian Federation

97 CI6 **Boknafjorden** fjord S Norway
80 H11 **Bokoro** Chari-Baguirmi, W Chad 12.22N 17.03E
81 K19 **Bokota** Equateur, C Dem. Rep. Congo (Zaire) 0.56S 22.20E
178 Gg13 **Bokpyin** Tenasserim, S Myanmar 11.16N 98.47E
Boksánbánya/Bokschen see Bocşa
85 F21 **Bokspits** Kgalagadi, SW Botswana 26.50S 20.41E
81 K18 **Bokungu** Equateur, C Dem. Rep. Congo (Zaire) 0.41S 22.19E
Bokurdak see Bakhardok
80 G10 **Bol** Lac, W Chad 13.27N 14.40E
175 Qq9 **Bolaang** Sulawesi, N Indonesia 0.58S 122.10E
78 G13 **Bolama** SW Guinea-Bissau 11.34N 15.32W
Bolangir see Balāngīr
Bolanos see Bolanos, Mount, Guam
Bolaños see Bolaños de Calatrava, Spain
42 L12 **Bolaños, Río** ♣ C Mexico
196 B17 **Bolanos, Mount** var. Bolanos. ▲ S Guam 13.18N 144.41E
117 M14 **Bolayır** Çanakkale, NW Turkey 40.31N 26.46E
104 L3 **Bolbec** Seine-Maritime, N France 49.34N 0.31E
118 L13 **Boldu** var. Buzău. ♣ SE Romania 45.18N 27.15E
152 H8 **Boldumsaz** prev. Kalinin, Kalininsk, Porsy. Dashkhovuzskiy Velayat, N Turkmenistan 42.12N 59.33E
79 O15 **Bole** var. Bortala. Xinjiang Uygur Zizhiqu, NW China 44.52N 82.06E
79 O15 **Bole** NW Ghana 9.01N 2.28W
81 J19 **Boleko** Equateur, W Dem. Rep. Congo (Zaire) 1.27S 19.52E
113 E14 **Bolesławiec Ger.** Bunzlau. Dolnośląskie, SW Poland 51.16N 15.34E
131 W4 **Bolgar** prev. Kuybyshev. Respublika Tatarstan, W Russian Federation 54.58N 49.03E
79 P14 **Bolgatanga** N Ghana 10.45N 0.52W
119 N12 **Bolhrad Rus.** Bolgrad. Odes'ka Oblast', SW Ukraine 45.42N 28.34E
169 Y8 **Boli** Heilongjiang, NE China 45.45N 130.32E
81 J19 **Bolia** Bandundu, W Dem. Rep. Congo (Zaire) 1.34S 18.24E
95 J14 **Boliden** Västerbotten, N Sweden 64.52N 20.19E
176 Uu11 **Bolifar** Pulau Seram, E Indonesia 3.08S 130.34E
179 Oo9 **Bolinao** Luzon, N Philippines 16.22N 119.52E
100 K6 **Bolsward Fris.** Boalsert. Friesland, N Netherlands 53.04N 5.31E
12 G15 **Bolton** Ontario, S Canada 43.52N 79.45W
99 K17 **Bolton** prev. Bolton-le-Moors. NW England, UK 53.34N 2.25W
23 W8 **Bolton** North Carolina, SE USA 34.22N 78.26W
Bolton-le-Moors see Bolton
142 G11 **Bolu** NW Turkey 40.45N 31.37E
142 G11 **Bolu** ♦ province NW Turkey
94 H1 **Bolungarvík** Vestfirðir, NW Iceland 66.09N 23.16W
165 O10 **Boluntay** Qinghai, W China 36.30N 92.10E
142 F14 **Bolvadin** Afyon, W Turkey 38.43N 31.01E
116 M14 **Bolyarovo** prev. Pashkeni. Yambol, E Bulgaria 42.09N 26.49E
108 G6 **Bolzano Ger.** Bozen; anc. Bauzanum. Trentino-Alto Adige, N Italy 46.30N 11.22E
81 F22 **Boma** Bas-Congo, W Dem. Rep. Congo (Zaire) 5.42S 13.05E
191 R12 **Bombala** New South Wales, SE Australia 36.54S 149.15E
106 F10 **Bombarral** Leiria, C Portugal 39.15N 9.09W
Bombay see Mumbai
177 Vv11 **Bomberai** ♣ Irian Jaya, E Indonesia
177 Vv11 **Bomberai, Jazirah** peninsula Irian Jaya, E Indonesia
176 Vv11 **Bomberai, Semenanjung** headland Irian Jaya, E Indonesia 3.01S 133.25E
83 I17 **Bombo** S Uganda 0.38N 32.31E
81 I17 **Bomboma** Equateur, NW Dem. Rep. Congo (Zaire) 2.22N 19.03E
61 G14 **Bom Futuro** Pará, N Brazil 6.27S 54.44W
165 Q15 **Bomi** var. Bowo, Zhamo. Xizang Zizhiqu, W China 29.43N 96.12E
81 N17 **Bomili** Orientale, NE Dem. Rep. Congo (Zaire) 1.45N 27.01E
61 N17 **Bom Jesus da Lapa** Bahia, E Brazil 13.16S 43.22W
62 Q8 **Bom Jesus do Itabapoana** Rio de Janeiro, SE Brazil 21.07S 41.43W
97 B15 **Bømlafjorden** fjord S Norway
97 B15 **Bømlo** island S Norway
126 M14 **Bomnak** Amurskaya Oblast', SE Russian Federation 54.43N 128.50E
80 J13 **Bomongo** Equateur, NW Dem. Rep. Congo (Zaire) 1.22N 18.21E
83 G15 **Bom Retiro** Santa Catarina, S Brazil 27.52S 49.33W
81 M21 **Bombouki** ♣ NE Central African Republic/Dem. Rep. Congo (Zaire)
148 J2 **Bonāb var.** Benāb, Bunab. Āzarbāyjān-e Khāvarī, N Iran 37.24N 45.59E
47 Q16 **Bonaire** island E Netherlands Antilles
41 U11 **Bona, Mount** ▲ Alaska, USA 61.27N 141.45W
194 M16 **Bonando** ♣ SE Papua New Guinea
191 Q12 **Bonang** Victoria, SE Australia 37.13S 148.43E
44 L7 **Bonanza** Región Autónoma Atlántico Norte, NE Nicaragua 13.58N 84.37W

39 O4 **Bonanza** Utah, W USA 40.01N 109.12E
47 Q9 **Bonao** C Dominican Republic 18.55N 70.25W
188 E13 **Bonaparte Archipelago** island group Western Australia
34 K6 **Bonaparte, Mount** ▲ Washington, NW USA 48.45N 119.07W
41 N11 **Bonasila Dome** ▲ Alaska, USA 62.24N 160.28W
94 H11 **Bonåsjøen** Nordland, C Norway 67.35N 15.39E
47 T15 **Bonasse** Trinidad, Trinidad and Tobago 10.02N 61.50W
13 X7 **Bonaventure** Quebec, SE Canada 48.03N 65.30W
13 X7 **Bonaventure** ♣ Quebec, SE Canada
11 L14 **Bonavista** Newfoundland, SE Canada 48.38N 53.07W
11 L14 **Bonavista Bay** inlet NW Atlantic Ocean
123 Kk11 **Bon, Cap** headland N Tunisia 37.05N 11.04E
81 C16 **Bonda** Ogooué-Lolo, C Gabon 0.50S 12.28E
131 N6 **Bondari** Tambovskaya Oblast', W Russian Federation 52.58N 42.02E
108 G9 **Bondeno** Emilia-Romagna, C Italy 44.53N 11.24E
32 L4 **Bond Falls Flowage** ⊗ Michigan, N USA
81 L16 **Bondo** Orientale, N Dem. Rep. Congo (Zaire) 3.51N 23.41E
175 P17 **Bondokodi** Pulau Sumba, S Indonesia 9.36S 119.01E
79 O15 **Bondoukou** E Ivory Coast 8.03N 2.45W
Bondoukui/Bondoukuy see Boundoukui
174 Mm15 **Bondowoso** Jawa, C Indonesia 7.54S 113.49E
35 S14 **Bondurant** Wyoming, C USA 43.14N 110.26W
Bone see Watampone, Indonesia
Bône see Annaba, Algeria
32 I5 **Bone** ▲ Wisconsin, N USA
175 R12 **Bonelipu** Pulau Buton, C Indonesia 4.42S 123.09E
175 Q14 **Bonerate** Kepulauan var. Macan. island group C Indonesia
175 Pp15 **Bonerate, Pulau** island Kepulauan Bonerate, C Indonesia
31 O12 **Bonesteel** South Dakota, N USA 43.01N 98.55W
64 I8 **Bonete, Cerro** ▲ N Argentina 27.58S 68.22W
175 Pp11 **Bone, Teluk** bay Sulawesi, C Indonesia
110 D6 **Bonfol** Jura, NW Switzerland 47.28N 7.08E
159 U12 **Bongaigaon** Assam, NE India 26.30N 90.30E
81 K17 **Bongandanga** Equateur, NW Dem. Rep. Congo (Zaire) 1.30N 21.03E
80 G12 **Bongor** Mayo-Kébbi, SW Chad 10.18N 15.19E
79 N16 **Bongouanou** E Ivory Coast 6.39N 4.12W
178 Kk11 **Bồng Sơn var.** Hoai Nhon. Bình Đinh, C Vietnam 14.28N 109.00E
27 U5 **Bonham** Texas, SW USA 33.34N 96.10W
Bonhard see Bonyhád
105 U6 **Bonhomme, Col du** pass NE France 48.10N 7.07E
105 Y16 **Bonifacio** Corse, France, C Mediterranean Sea 41.23N 9.09E
Bonifacio, Bocche de/Bonifacio, Bouches de see Bonifacio, Strait of
105 Y16 **Bonifacio, Strait of Fr.** Bouches de Bonifacio, It. Bocche di Bonifacio. strait C Mediterranean Sea
25 Q8 **Bonifay** Florida, SE USA 30.49N 85.42W
Bonin Islands see Ogasawara-shotō
199 H6 **Bonin Trench** undersea feature NW Pacific Ocean
25 V13 **Bonita Springs** Florida, SE USA 26.19N 81.48W
44 I5 **Bonito, Pico** ▲ N Honduras 15.33N 86.55W
103 E17 **Bonn** Nordrhein-Westfalen, W Germany 50.43N 7.06E
12 I12 **Bonnechère** Ontario, SE Canada 45.39N 77.36W
35 N13 **Bonners Ferry** Idaho, NW USA 48.41N 116.19W
29 R4 **Bonner Springs** Kansas, C USA 39.03N 94.52W
104 L6 **Bonnétable** Sarthe, NW France 48.19N 0.24E
26 Xx **Bonne Terre** Missouri, C USA 37.55N 90.33W
8 J3 **Bonnet Plume** ♣ Yukon Territory, NW Canada
104 M6 **Bonneval** Eure-et-Loir, C France 48.11N 1.21E
105 T10 **Bonneville** Haute-Savoie, E France 46.04N 6.25E
68 J3 **Bonneville Salt Flats** salt flat Utah, W USA
79 O16 **Bonny** Rivers, S Nigeria 4.25N 7.13E
Bonny, Bight of see Biafra, Bight of
39 W4 **Bonny Reservoir** ⊠ Colorado, C USA
11 R14 **Bonnyville** Alberta, SW Canada 54.16N 110.46W
109 C18 **Bono** Sardegna, Italy, C Mediterranean Sea 40.24N 9.01E
176 Xx10 **Bonoi** Irian Jaya, E Indonesia 1.46S 137.45E
Bononia see Vidin, Bulgaria
Bononia see Boulogne-sur-Mer, France
109 H18 **Bonorva** Sardegna, Italy, C Mediterranean Sea 40.24N 8.46E
194 M16 **Bonua** SE Papua New Guinea
191 T4 **Bonshaw** New South Wales, SE Australia 29.06S 151.15E

◆ COUNTRY ◇ DEPENDENT TERRITORY ✱ ADMINISTRATIVE REGION ▲ MOUNTAIN ☆ VOLCANO ⊗ LAKE
● COUNTRY CAPITAL ○ DEPENDENT TERRITORY CAPITAL ✕ INTERNATIONAL AIRPORT ▲ MOUNTAIN RANGE ♣ RIVER ⊠ RESERVOIR

78 I16 **Bonthe** SW Sierra Leone 7.26N 12.32W
179 P8 **Bontoc** Luzon, N Philippines 17.04N 120.58E
194 M16 **Bonua** ⚡ S PNG
27 Y9 **Bon Wier** Texas, SW USA 30.43N 93.40W
113 J25 **Bonyhád** Ger. Bonhard. Tolna, S Hungary 46.17N 18.31E
Bonzabaai see Bonza Bay
85 J25 **Bonza Bay** Afr. Bonzabaai. Eastern Cape, S South Africa 32.58S 27.58E
190 D7 **Bookabie** South Australia 31.49S 132.41E
190 H6 **Bookaloo** South Australia 31.56S 137.21E
39 P5 **Book Cliffs** cliff Colorado/Utah, W USA
175 T19 **Boo, Kepulauan** island group E Indonesia
27 P1 **Booker** Texas, SW USA 36.27N 100.32W
78 K15 **Boola** Guinée-Forestière, SE Guinea 8.22N 8.40W
191 O8 **Booligal** New South Wales, SE Australia 33.56S 144.54E
101 G17 **Boom** Antwerpen, N Belgium 51.05N 4.24E
45 N6 **Boom** var. Boon. Región Autónoma Atlántico Norte, NE Nicaragua 14.52N 83.36W
191 S3 **Boomi** New South Wales, SE Australia 28.43S 149.35E
Boon see Boom
31 V13 **Boone** Iowa, C USA 42.04N 93.52W
23 Q8 **Boone** North Carolina, SE USA 36.13N 81.40W
29 S11 **Booneville** Arkansas, C USA 35.08N 93.55W
23 N6 **Booneville** Kentucky, S USA 37.27N 83.41W
25 N2 **Booneville** Mississippi, S USA 34.39N 88.34W
23 V3 **Boonsboro** Maryland, NE USA 39.30N 77.39W
168 H9 **Böön Tsagaan Nuur** ⊚ S Mongolia
36 L6 **Boonville** California, W USA 38.58N 123.21W
33 N16 **Boonville** Indiana, N USA 38.03N 87.16W
29 U4 **Boonville** Missouri, C USA 38.58N 92.44W
20 I9 **Boonville** New York, NE USA 43.28N 75.17W
82 M12 **Boorama** Woqooyi Galbeed, NW Somalia 9.58N 43.15E
191 O6 **Booroondarra, Mount** hill New South Wales, SE Australia 31.07S 145.20E
191 N9 **Booroorban** New South Wales, SE Australia 34.55S 144.45E
191 R9 **Boorowa** New South Wales, SE Australia 34.26S 148.42E
101 H17 **Boortmeerbeek** Vlaams Brabant, C Belgium 50.58N 4.27E
82 P11 **Boosaaso** var. Bandar Kassim, Bender Qaasim, Bosaso, It. Bender Cassim. Bari, N Somalia 11.26N 49.12E
21 Q8 **Boothbay Harbor** Maine, NE USA 43.52N 69.35W
Boothia Felix see Boothia Peninsula
15 K2 **Boothia, Gulf of** gulf Nunavut, NE Canada
15 K2 **Boothia Peninsula** prev. Boothia Felix. peninsula Nunavut, NE Canada
81 E18 **Booué** Ogooué-Ivindo, NE Gabon 0.03S 11.58E
103 J21 **Bopfingen** Baden-Württemberg, S Germany 48.51N 10.21E
103 F18 **Boppard** Rheinland-Pfalz, W Germany 50.13N 7.35E
64 M4 **Boquerón** off. Departamento de Boquerón. ◆ department W Paraguay
45 N13 **Boquete** var. Bajo Boquete. Chiriquí, W Panama 8.45N 82.26W
42 J6 **Boquilla, Presa de la** ⊞ N Mexico
42 L5 **Boquillas** var. Boquillas del Carmen. Coahuila de Zaragoza, NE Mexico 29.10N 102.55W
Boquillas del Carmen see Boquillas
126 I11 **Bor** Krasnoyarskiy Kray, C Russian Federation 61.28N 90.09E
83 E15 **Bor** Jonglei, S Sudan 6.12N 31.33E
97 L20 **Bor** Jönköping, S Sweden 57.04N 14.10E
114 P12 **Bor** Serbia, E Yugoslavia 44.05N 22.06E
203 S10 **Bora-Bora** island Îles Sous le Vent, W French Polynesia
178 I10 **Borabu** Maha Sarakham, E Thailand 16.01N 103.06E
35 J17 **Borah Peak** ▲ Idaho, NW USA 44.21N 113.53W
Boralday see Burunday
97 J19 **Borås** Västra Götaland, S Sweden 57.43N 12.55E
149 N11 **Borāzjān** var. Borazjān. Būshehr, S Iran 29.19N 51.12E
Borazjān see Borāzjān
58 G13 **Borba** Amazonas, N Brazil 4.39S 59.34W
106 H11 **Borba** Évora, S Portugal 38.48N 7.28W
Borbetomagus see Worms
57 O7 **Borbón** Bolívar, E Venezuela 7.55N 64.03W
61 Q15 **Borborema, Planalto da** plateau NE Brazil
118 M14 **Borcea, Braţul** ⚡ S Romania
Borchalo see Marneuli
205 R15 **Borchgrevink Coast** physical region Antarctica
143 Q12 **Borçka** Artvin, NE Turkey 41.24N 41.37E
100 I11 **Borculo** Gelderland, E Netherlands 52.07N 6.31E
190 K10 **Borda, Cape** headland South Australia 35.45S 136.34E
104 K13 **Bordeaux** anc. Burdigala. Gironde, SW France 44.49N 0.33W
11 T15 **Borden** Saskatchewan, S Canada 52.23N 107.10W
12 D8 **Borden Lake** ⊚ Ontario, S Canada

15 L1 **Borden Peninsula** peninsula Baffin Island, Nunavut, NE Canada
190 K11 **Bordertown** South Australia 36.21S 140.48E
94 H2 **Bordheyri** Vestfirdhir, NW Iceland 65.12N 21.09W
97 B18 **Bordhoy** Dan. Bordö Island Faeroe Islands 62.17N 6.30W
108 B11 **Bordighera** Liguria, NW Italy 43.48N 7.40E
76 K5 **Bordj-Bou-Arreridj** var. Bordj Bou Arreridj, Bordj Bou Arréridj. N Algeria 36.04N 4.45E
123 J10 **Bordj El Bahri, Cap de** headland N Algeria 36.52N 3.13E
76 L10 **Bordj Omar Driss** E Algeria 28.09N 6.52E
149 N13 **Bord Khūn** Hormozgān, S Iran
153 V7 **Bordunskiy** Chuyskaya Oblast', N Kyrgyzstan 42.37N 75.31E
97 M17 **Borensberg** Östergötland, S Sweden 58.33N 15.15E
Borgå see Porvoo
94 L2 **Borgarfjördhur** Austurland, NE Iceland 65.32N 13.46W
94 H3 **Borgarnes** Vesturland, W Iceland 64.33N 21.54W
95 G14 **Børgefjellet** ▲ C Norway
100 O7 **Borger** Drenthe, NE Netherlands 52.54N 6.48E
27 N2 **Borger** Texas, SW USA 35.40N 101.24W
97 N20 **Borgholm** Kalmar, S Sweden 56.50N 16.40E
109 N22 **Borgia** Calabria, SW Italy 38.48N 16.28E
101 J18 **Borgloon** Limburg, NE Belgium 50.48N 5.21E
205 P2 **Borg Massif** ▲ Antarctica
24 L9 **Borgne, Lake** ⊚ Louisiana, S USA
108 C7 **Borgomanero** Piemonte, NE Italy 45.42N 8.33E
108 G10 **Borgo Panigale** × (Bologna) Emilia-Romagna, N Italy 44.33N 11.16E
108 A9 **Borgo San Dalmazzo** Piemonte, N Italy 44.19N 7.28E
108 G11 **Borgo San Lorenzo** Toscana, C Italy 43.58N 11.22E
108 C7 **Borgosesia** Piemonte, NE Italy 45.41N 8.21E
108 E9 **Borgo Val di Taro** Emilia-Romagna, C Italy 44.29N 9.48E
108 G6 **Borgo Valsugana** Trentino-Alto Adige, N Italy 46.04N 11.31E
169 O11 **Borhoyn Tal** Dornogovĭ, SE Mongolia 43.43N 111.53E
178 I8 **Borikhan** var. Borikhane. Bolikhamxai, C Laos 18.36N 103.43E
Borikhane see Borikhan
Borislav see Boryslav
131 N8 **Borisoglebsk** Voronezhskaya Oblast', W Russian Federation 51.23N 42.00E
Borisov see Barysaw
Borisovgrad see Pŭrvomay
Borispol' see Boryspil'
180 I3 **Boriziny** Mahajanga, N Madagascar 15.31S 47.40E
107 Q5 **Borja** Aragón, NE Spain 41.49N 1.31W
Borjas Blancas see Les Borges Blanques
143 S10 **Borjomi** Rus. Borzhomi. C Georgia 41.50N 43.24E
120 L12 **Borkavichy** Rus. Borkovichi. Vitsyebskaya Voblasts', N Belarus 55.40N 28.18E
103 H16 **Borken** Hessen, C Germany 51.01N 9.16E
103 E14 **Borken** Nordrhein-Westfalen, W Germany 51.51N 6.51E
94 H10 **Borkenes** Troms, N Norway 68.46N 16.10E
80 H7 **Borkou-Ennedi-Tibesti** off. Préfecture du Borkou-Ennedi-Tibesti. ◆ prefecture N Chad
102 E9 **Borkum** island NW Germany
83 X7 **Bor, Lagh** var. Lak Bor. dry watercourse NE Kenya
Bor, Lak see Bor, Lagh
97 M14 **Borlänge** Dalarna, C Sweden 60.28N 15.25E
108 C9 **Bormida** ⚡ NW Italy
108 F6 **Bormio** Lombardia, N Italy 46.27N 10.24E
103 E14 **Born** Hessen, C Germany 51.07N 12.30E
100 O10 **Borne** Overijssel, E Netherlands 52.18N 6.45E
101 F17 **Bornem** Antwerpen, N Belgium 51.06N 4.13E
174 M6 **Borneo** island Brunei/Indonesia/Malaysia
103 E16 **Bornheim** Nordrhein-Westfalen, W Germany 50.46N 6.58E
97 L24 **Bornholm** ◆ county E Denmark
97 L24 **Bornholm** island E Denmark
79 Y13 **Borno** ◆ state NE Nigeria
106 K15 **Bornos** Andalucía, S Spain 36.49N 5.42W
168 L7 **Bornuur** Töv, C Mongolia 48.28N 106.15E
126 I14 **Borodino** Krasnoyarskiy Kray, S Russian Federation 55.54N 94.45E
119 O4 **Borodyanka** Kyyivs'ka Oblast', N Ukraine 50.40N 29.54E
126 M10 **Borogontsy** Respublika Sakha (Yakutiya), NE Russian Federation 62.42N 131.01E
164 I5 **Borohoro Shan** ▲ NW China
78 M15 **Boromo** SW Burkina 11.46N 2.54W
37 T13 **Boron** California, W USA 35.00N 117.42W
179 N12 **Borongan** Samar, C Philippines 11.26N 125.30E
Borongo see Black Volta
Boron'ki see Baron'ki
171 K17 **Borōshō-hantō** peninsula Honshū, S Japan
Borosjenő see Ineu
Borossebes see Sebiş
78 L15 **Borotou** NW Ivory Coast 8.46N 7.30W
119 X6 **Borova** Kharkivs'ka Oblast', E Ukraine 49.22N 37.39E
116 H8 **Borovan** Vratsa, NW Bulgaria 43.25N 23.45E
128 I14 **Borovichi** Novgorodskaya Oblast', W Russian Federation 58.23N 33.56E
Borovlje see Ferlach

114 J9 **Borovo** Vukovar-Srijem, E Croatia 45.22N 18.57E
151 Q7 **Borovoye** Kaz. Būrabay. Severnyy Kazakhstan, N Kazakhstan 53.07N 70.19E
130 K3 **Borovsk** Kaluzhskaya Oblast', W Russian Federation 55.12N 36.22E
125 F12 **Borovskiy** Tyumenskaya Oblast', C Russian Federation 57.04N 65.37E
151 N7 **Borovskoy** Kostanay, N Kazakhstan 53.49N 64.12E
Borovukha see Baravukha
97 L23 **Borrby** Skåne, S Sweden 55.27N 14.10E
189 R3 **Borroloola** Northern Territory, N Australia 16.09S 136.18E
118 F9 **Bors** Bihor, NW Romania 47.06N 21.47E
118 I9 **Borşa** Hung. Borsa. Maramureş, N Romania 47.40N 24.37E
118 J10 **Borsec** Ger. Bad Borseck, Hung. Borszék. Harghita, C Romania 46.57N 25.32E
94 K8 **Børselv** Finnmark, N Norway 70.18N 25.35E
115 L23 **Borsh** var. Borshi. Vlorë, S Albania 40.04N 19.51E
Borshchev see Borshchiv
118 K7 **Borshchiv** Pol. Borszczów, Rus. Borshchev. Ternopil's'ka Oblast', W Ukraine 48.48N 26.00E
Borshi see Borsh
118 L20 **Borsod-Abaúj-Zemplén** off. Borsod-Abaúj-Zemplén Megye. ◆ county NE Hungary
101 E15 **Borssele** Zeeland, SW Netherlands 51.26N 3.45E
Borszczów see Borshchiv
Borszék see Borsec
Bortala see Bole
105 O12 **Bort-les-Orgues** Corrèze, C France 45.26N 2.31E
168 E8 **Bor-Üdzüür** Hovd, W Mongolia 45.46N 92.13E
149 N9 **Borūjen** Chahār Maḥall va Bakhtīārī, C Iran 32.00N 51.08E
148 L7 **Borūjerd** var. Burujird. Lorestān, W Iran 33.55N 48.45E
118 H6 **Boryslav** Pol. Borysław, Rus. Borislav. L'vivs'ka Oblast', W Ukraine 49.18N 23.28E
Boryslaw see Boryslav
119 P4 **Boryspil'** Rus. Borispol'. Kyyivs'ka Oblast', N Ukraine 50.20N 30.58E
119 P4 **Boryspil'** Rus. Borispol'. × (Kyyiv) Kyyivs'ka Oblast', N Ukraine 50.21N 30.46E
116 L6 **Borzhomi** see Borjomi
119 R3 **Borzna** Chernihivs'ka Oblast', NE Ukraine 51.15N 32.25E
126 L16 **Borzya** Chitinskaya Oblast', S Russian Federation 50.18N 116.24E
108 B10 **Bosa** Sardegna, Italy, C Mediterranean Sea 40.18N 8.28E
114 F10 **Bosanska Dubica** var. Kozarska Dubica. Republika Srpska, NW Bosnia and Herzegovina 45.09N 16.47E
114 G10 **Bosanska Gradiška** var. Gradiška. Republika Srpska, N Bosnia and Herzegovina 45.09N 17.14E
114 F10 **Bosanska Kostajnica** var. Srpska Kostajnica. Republika Srpska, NW Bosnia and Herzegovina 45.12N 16.33E
114 G11 **Bosanska Krupa** var. Krupa, Krupa na Uni. Federacija Bosna I Hercegovina, NW Bosnia and Herzegovina 44.52N 16.09E
114 H10 **Bosanski Brod** var. Srpski Brod. Republika Srpska, N Bosnia and Herzegovina 45.07N 17.59E
114 G11 **Bosanski Novi** var. Novi Grad. Republika Srpska, NW Bosnia and Herzegovina 45.03N 16.22E
114 H11 **Bosanski Petrovac** var. Petrovac. Federacija Bosna I Hercegovina, NW Bosnia and Herzegovina 44.34N 16.21E
114 N12 **Bosanski Petrovac** Serbia, E Yugoslavia 44.22N 21.25E
114 G10 **Bosanski Šamac** var. Šamac. Republika Srpska, N Bosnia and Herzegovina 45.03N 18.27E
114 E12 **Bosansko Grahovo** var. Grahovo, Hrvatsko Grahovo. Federacija Bosna I Hercegovina, W Bosnia and Herzegovina 44.10N 16.22E
Bosaso see Boosaaso
194 Q13 **Bosavi, Mount** ▲ W PNG 6.33S 142.50E
166 J14 **Bose** Guangxi Zhuangzu Zizhiqu, S China 23.55N 106.31E
167 Q5 **Boshan** Shandong, E China 36.31N 117.46E
115 P16 **Bosilegrad** prev. Bosiljgrad. Serbia, SE Yugoslavia 42.30N 22.30E
Bosiljgrad see Bosilegrad
Bösing see Pezinok
100 H12 **Boskoop** Zuid-Holland, C Netherlands 52.04N 4.40E
113 G18 **Boskovice** Ger. Boskowitz. Brněnský Kraj, SE Czech Republic 49.30N 16.39E
Boskowitz see Boskovice
114 I10 **Bosna** ⚡ N Bosnia and Herzegovina
Bosna I Hercegovina, Federacija see Bosnia and Herzegovina
114 H12 **Bosnia and Herzegovina** off. Republic of Bosnia and Herzegovina. ◆ republic SE Europe
175 X9 **Bosnabaridi** Irian Jaya, E Indonesia 0.49S 136.00E
171 K17 **Bosobolo** var. Bosobolo, NW Dem. Rep. Congo (Zaire) 4.10N 19.55E
79 R12 **Bosó-hantō** peninsula Honshū, S Japan
Bosora see Busrá ash Shām
Bosphorus/Bosporus see Istanbul Boğazı
Bosporus Cimmerius see Kerch Strait
Bosporus Thracius see Istanbul Boğazı
Bosra see Busrá ash Shām
81 H17 **Bossangoa** Ouham, C Central African Republic 6.31N 17.24E
189 J7 **Bossé Bangou** see Bossey Bangou

81 I15 **Bossembélé** Ombella-Mpoko, C Central African Republic 5.13N 17.39E
81 H15 **Bossentélé** Ouham-Pendé, W Central African Republic 5.36N 16.37E
79 R12 **Bossey Bangou** var. Bossé Bangou. Tillabéri, SW Niger 13.22N 1.18E
24 G5 **Bossier City** Louisiana, S USA 32.31N 93.43W
85 D20 **Bossiesvlei** Hardap, S Namibia 25.01S 16.45E
79 Y11 **Bosso** Diffa, SE Niger 13.42N 13.18E
83 F15 **Bossoroca** Rio Grande do Sul, S Brazil 28.45S 54.54W
148 K3 **Bostān** Xinjiang Uygur Zizhiqu, W China 41.19N 83.15E
164 K6 **Bosten Hu** var. Bagrax Hu. ⊚ NW China
9 O18 **Boston** anc. St.Botolph's Town. E England, UK 52.58N 0.01W
21 O11 **Boston** state capital Massachusetts, NE USA 42.21N 71.03W
8 M17 **Boston Bar** British Columbia, SW Canada 49.54N 121.22W
29 T10 **Boston Mountains** ▲ Arkansas, C USA
13 P8 **Bostonnais** ⚡ Quebec, SE Canada
114 J10 **Bosut** ⚡ E Croatia
160 C11 **Botād** Gujarāt, W India 22.12N 71.43E
68 L2 **Botany Bay** inlet New South Wales, SE Australia
191 T9 **Botany Bay** inlet New South Wales, SE Australia
85 G18 **Boteti** var. Botletle. ⚡ N Botswana
116 J9 **Botev** ▲ C Bulgaria 42.45N 24.57E
116 H9 **Botevgrad** prev. Orkhaniye. Sofiya, W Bulgaria 42.55N 23.46E
95 J16 **Bothnia, Gulf of** Fin. Pohjanlahti, Swe. Bottniska Viken. gulf N Baltic Sea
191 P17 **Bothwell** Tasmania, SE Australia 42.24S 147.01E
106 H5 **Boticas** Vila Real, N Portugal 41.40N 7.40W
57 W10 **Boti-Pasi** Sipaliwini, C Suriname 4.15N 55.27W
85 I18 **Botletle** see Boteti
131 P16 **Botlikh** Chechenskaya Respublika, SW Russian Federation 42.39N 46.12E
119 N10 **Botna** ⚡ E Moldova
118 I9 **Botoşani** Hung. Botosány. Botoşani, NE Romania 47.43N 26.40E
118 K8 **Botoşani** ◆ county NE Romania
Botosány see Botoşani
167 P2 **Botou** prev. Bozhen. Hebei, E China 38.09N 116.37E
101 M20 **Botrange** ▲ E Belgium 50.30N 6.03E
109 O21 **Botricello** Calabria, SW Italy 38.56N 16.51E
85 I23 **Botshabelo** Free State, C South Africa 29.15S 26.45E
95 J15 **Botsmark** Västerbotten, N Sweden 64.15N 20.15E
85 G19 **Botswana** off. Republic of Botswana. ◆ republic S Africa
13 N2 **Bottineau** North Dakota, N USA 48.45N 100.28W
100 S11 **Bottniska Viken** see Bothnia, Gulf of
62 L9 **Botucatu** São Paulo, S Brazil 22.52S 48.30W
83 Q9 **Bouaflé** C Ivory Coast 6.58N 5.45W
79 N16 **Bouaké** var. Bwake. C Ivory Coast 7.39N 5.01W
81 H11 **Bouar** Nana-Mambéré, W Central African Republic 5.58N 15.38E
76 H7 **Bouarfa** NE Morocco 32.33N 1.54W
113 B19 **Boubín** ▲ SW Czech Republic 49.00N 13.51E
81 I14 **Bouca** Ouham, W Central African Republic 6.57N 18.18E
15 T5 **Boucher** ⚡ Quebec, SE Canada
105 S14 **Bouches-du-Rhône** ◆ department SE France
76 C9 **Bou Craa** var. Bu Craa. NW Western Sahara 26.31N 12.52W
79 O9 **Boû Djébéha** oasis C Mali 18.39N 3.45W
110 C8 **Boudry** Neuchâtel, W Switzerland 46.57N 6.46E
188 L2 **Bougainville, Cape** headland Western Australia 13.53S 126.01E
67 E24 **Bougainville, Cape** headland East Falkland, Falkland Islands 51.18S 58.28W
Bougainville, Détroit de see Bougainville Strait, Vanuatu
195 S13 **Bougainville Island** island NE PNG
195 S13 **Bougainville Strait** strait N Solomon Islands
197 R12 **Bougainville Strait** Fr. Détroit de Bougainville. strait C Vanuatu
176 U8 **Bougainville, Selat** strait Irian Jaya, E Indonesia
123 I12 **Bougaroun, Cap** headland NE Algeria 37.06N 6.18E
79 N13 **Boughessa** Kidal, NE Mali 20.05N 2.13E
Bougie see Béjaïa
78 L13 **Bougouni** Sikasso, SW Mali 11.22N 7.24W
101 J24 **Bouillon** Luxembourg, SE Belgium 49.47N 5.04E
76 I6 **Bouira** var. Bouïra. N Algeria 36.23N 3.54E
76 D8 **Bou-Izakarn** SW Morocco 29.12N 9.43W
76 G5 **Boujdour** var. Bojador. W Western Sahara 26.06N 14.28W
76 G5 **Boukhalef** × (Tanger) N Morocco 35.45N 5.53W
198 B4 **Boukombé** see Boukoumbé
78 J14 **Boukoumbé** var. Boukombé. C Benin 10.13N 1.09E
76 C10 **Boû Lanouâr** Dakhlet Nouâdhibou, W Mauritania 21.16N 16.28W
39 T7 **Boulder** Colorado, C USA 40.01N 105.18W
35 P13 **Boulder** Montana, NW USA 46.14N 112.07W
37 X12 **Boulder City** Nevada, W USA 35.58N 114.49W
189 T10 **Boulia** Queensland, C Australia 23.02S 139.58E

13 N10 **Boullé** ⚡ Quebec, SE Canada
104 J9 **Boulogne** ⚡ NW France
Boulogne see Boulogne-sur-Mer
104 L16 **Boulogne-sur-Gesse** Haute-Garonne, S France 43.18N 0.39E
105 N1 **Boulogne-sur-Mer** var. Boulogne; anc. Bononia, Gesoriacum, Gessoriacum. Pas-de-Calais, N France 50.43N 1.36E
197 I7 **Bouloupari** Province Sud, S New Caledonia 21.53S 165.29E
79 Q12 **Boulsa** C Burkina 12.40N 0.28W
79 W11 **Boultoum** Zinder, C Niger 14.43N 10.22E
197 K13 **Bouma** Taveuni, N Fiji 16.49S 179.53E
81 G16 **Bouma** ⚡ SE Cameroon
78 J9 **Boûmdeïd** var. Boumdeit. Assaba, S Mauritania 17.25N 11.21W
Boumdeït see Boûmdeïd
117 C17 **Boumistós** ▲ W Greece 38.48N 20.59E
79 O15 **Bouna** NE Ivory Coast 9.16N 3.00W
21 P4 **Boundary Bald Mountain** ▲ Maine, NE USA 45.45N 70.10W
37 S8 **Boundary Peak** ▲ Nevada, W USA 37.50N 118.21W
78 M14 **Boundiali** N Ivory Coast 9.31N 6.28W
81 E19 **Boundji** Cuvette, C Congo 1.04S 15.18E
79 O13 **Boundoukui** var. Bondoukui, Bondoukuy. W Burkina 11.51N 3.47W
68 L2 **Bountiful** Utah, W USA 40.53N 111.52W
Bounty Basin see Bounty Trough
203 Q16 **Bounty Bay** bay Pitcairn Island, C Pacific Ocean
192 J9 **Bounty Islands** island group S NZ
183 Q13 **Bounty Trough** var. Bounty Basin. undersea feature S Pacific Ocean
197 J6 **Bourail** Province Sud, C New Caledonia 21.33S 165.29E
29 V5 **Bourbeuse River** ⚡ Missouri, C USA
105 Q9 **Bourbon-Lancy** Saône-et-Loire, C France 46.39N 3.46E
33 N11 **Bourbonnais** Illinois, N USA 41.08N 87.52W
105 O10 **Bourbonnais** cultural region C France
105 S7 **Bourbonne-les-Bains** Haute-Marne, N France 48.00N 5.43E
105 O10 **Bourbon Vendée** see la Roche-sur-Yon
76 M8 **Bourdj Messaouda** E Algeria 30.18N 9.19E
79 Q10 **Bourem** Gao, C Mali 16.56N 0.21W
105 N11 **Bourg** see Bourg-en-Bresse
105 O11 **Bourganeuf** Creuse, C France 45.57N 1.47E
142 B12 **Bourgas** see Burgas
105 S10 **Bourg-en-Bresse** var. Bourg, Bourg-en-Bresse. Ain, E France 46.12N 5.13E
105 O8 **Bourges** anc. Avaricum. Cher, C France 47.06N 2.24E
105 T11 **Bourget, Lac du** ⊚ E France
105 P8 **Bourgogne** Eng. Burgundy. ◆ region E France
105 S11 **Bourgoin-Jallieu** Isère, E France 45.34N 5.16E
105 U11 **Bourg-St-Andéol** Ardèche, E France 44.24N 4.36E
105 U11 **Bourg-St-Maurice** Savoie, E France 45.37N 6.46E
110 C11 **Bourg St.Pierre** Valais, SW Switzerland 45.58N 7.12E
78 H8 **Boû Rjeimât** well W Mauritania 19.06N 15.16W
191 P5 **Bourke** New South Wales, SE Australia 30.07S 145.57E
9 N23 **Bournemouth** S England, UK 50.43N 1.54W
101 M23 **Bourscheid** Diekirch, NE Luxembourg 49.55N 6.04E
76 K6 **Bou Saâda** var. Bou Saada. N Algeria 35.13N 4.09E
68 I11 **Bouse Wash** ⚡ Arizona, SW USA
105 N10 **Boussac** Creuse, C France 46.20N 2.12E
104 M16 **Boussens** Haute-Garonne, S France 43.10N 0.58E
80 H11 **Bousso** prev. Fort-Bretonnet. Chari-Baguirmi, S Chad 10.31N 16.45E
78 H9 **Boutilimit** Trarza, SW Mauritania 17.33N 14.42W
67 D21 **Bouvet Island** ◇ Norwegian dependency S Atlantic Ocean
79 O13 **Bouza** Tahoua, SW Niger 14.25N 6.09E
111 R10 **Bovec** Ger. Flitsch, It. Plezzo. NW Slovenia 46.13N 13.33E
100 J8 **Bovenkarspel** Noord-Holland, NW Netherlands 52.33N 5.03E
31 V5 **Bovey** Minnesota, N USA 47.18N 93.25W
34 M9 **Bovill** Idaho, NW USA 46.50N 116.24W
26 L4 **Bovina** Texas, SW USA 34.30N 102.52W
109 M18 **Bovino** Puglia, SE Italy 41.15N 15.19E
63 C20 **Bovril** Entre Ríos, E Argentina 31.24S 59.25W
30 J2 **Bowbells** North Dakota, N USA 48.48N 102.15W
29 T9 **Bow City** Alberta, SW Canada 50.27N 112.14W
30 I2 **Bowdle** South Dakota, N USA 45.27N 99.39W
189 X6 **Bowen** Queensland, NE Australia 20.00S 148.10E
198 B4 **Bowers Ridge** undersea feature S Bering Sea
198 B4 **Bowes Point** headland Nunavut, N Canada 67.46N 101.51W
79 T7 **Bowie** Texas, SW USA 33.33N 97.51W
23 S5 **Bowie** Maryland, NE USA 39.00N 76.49W
22 J5 **Bowkān** see Būkān
29 Q4 **Bowling Green** Kentucky, S USA 36.59N 86.26W
29 V3 **Bowling Green** Missouri, C USA 39.20N 91.14W
33 Q11 **Bowling Green** Ohio, N USA 41.22N 83.40W

23 W5 **Bowling Green** Virginia, NE USA 38.01N 77.20W
30 J6 **Bowman** North Dakota, N USA 46.10N 103.25W
204 I5 **Bowman Coast** physical region Antarctica
204 J7 **Bowman-Haley Lake** ⊞ North Dakota, N USA
204 Z11 **Bowman Island** island Antarctica
191 S9 **Bowral** New South Wales, SE Australia 34.29S 150.28E
194 K14 **Bowutu Mountains** ▲ C PNG
85 I16 **Bowwood** Southern, S Zambia 17.09S 26.16E
30 I2 **Box Butte Reservoir** ⊞ Nebraska, C USA
30 I10 **Box Elder** South Dakota, N USA 44.06N 103.04W
97 M18 **Boxholm** Östergötland, S Sweden 58.11N 15.04E
Bo Xian/Boxian see Bozhou
101 L14 **Boxmeer** Noord-Brabant, SE Netherlands 51.39N 5.57E
101 J14 **Boxtel** Noord-Brabant, S Netherlands 51.36N 5.19E
142 H16 **Boyabat** Sinop, N Turkey 41.27N 34.45E
56 F9 **Boyacá** off. Departamento de Boyacá. ◆ province C Colombia
119 O4 **Boyarka** Kyyivs'ka Oblast', N Ukraine 50.19N 30.19E
24 H7 **Boyce** Louisiana, S USA 31.23N 92.40W
27 S6 **Boyd** Texas, SW USA 33.01N 97.33W
23 V8 **Boydton** Virginia, NE USA 36.40N 78.24W
79 F23 **Boyer River** ⚡ Iowa, C USA
12 G15 **Boyle** Alberta, SW Canada 54.38N 112.45W
99 D16 **Boyle** Ir. Mainistir na Búille. C Ireland 53.58N 8.18W
99 F17 **Boyne** Ir. An Bhóinn. ⚡ E Ireland
31 Q5 **Boyne City** Michigan, N USA 45.13N 85.00W
25 Z14 **Boynton Beach** Florida, SE USA 26.31N 80.04W
68 L2 **Boysen Reservoir** ⊞ Wyoming, C USA
151 O13 **Boysun** Rus. Baysun. Surkhondaryo Wiloyati, S Uzbekistan 38.13N 67.07E
142 B12 **Bozcaada** island Çanakkale, NW Turkey
142 C14 **Boz Dağları** ▲ W Turkey
35 S11 **Bozeman** Montana, NW USA 45.40N 111.02W
Bozen see Bolzano
81 J16 **Bozene** Equateur, NW Dem. Rep. Congo (Zaire) 2.55N 19.15E
157 P7 **Bozhou** var. Boxian, Bo Xian. Anhui, E China 33.49N 115.49E
22 K5 **Bozkır** Konya, S Turkey 37.10N 32.13E
81 H14 **Bozoum** Ouham-Pendé, W Central African Republic 6.17N 16.26E
143 N16 **Bozova** Şanlıurfa, S Turkey 37.22N 38.33E
110 C11 **Bozrah** see Buşrá ash Shām
99 N22 **Bozüyük** Bilecik, NW Turkey 39.55N 30.01E
108 J9 **Bra** Piemonte, NW Italy 44.42N 7.51E
114 D13 **Brač** var. Brach, It. Brazza; anc. Brattia. island S Croatia
115 F15 **Bracara Augusta** see Braga
12 H13 **Bracciano, Lago di** ⊚ C Italy
12 H13 **Bracebridge** Ontario, S Canada 45.01N 79.19W
Brach see Brač
95 G17 **Bräcke** Jämtland, C Sweden 62.42N 15.30E
27 P12 **Brackettville** Texas, SW USA 29.18N 100.25W
99 N22 **Bracknell** S England, UK 51.25N 0.46W
63 K13 **Braço do Norte** Santa Catarina, S Brazil 28.16S 49.11W
118 G11 **Brad** Hung. Brád. Hunedoara, SW Romania 46.07N 22.50E
109 N18 **Bradano** ⚡ S Italy
25 V12 **Bradenton** Florida, SE USA 27.30N 82.34W
12 H14 **Bradford** Ontario, S Canada 43.04N 80.21W
9 L16 **Bradford** N England, UK 53.48N 1.45W
29 V10 **Bradford** Arkansas, C USA 35.25N 91.27W
18 E12 **Bradford** Pennsylvania, NE USA 41.57N 78.38W
27 T5 **Bradley** Arkansas, C USA 33.06N 93.39W
27 Q9 **Brady** Texas, SW USA 31.08N 99.20W
27 Q9 **Brady Creek** ⚡ Texas, SW USA
98 I6 **Braemar** NE Scotland, UK 57.12N 2.52W
189 N2 **Brăeşti** Botoşani, NE Romania 47.50N 26.26E
106 G5 **Braga** anc. Bracara Augusta. Braga, NW Portugal 41.31N 8.25W
106 G5 **Braga** ◆ district N Portugal
118 J15 **Bragadiru** Teleorman, S Romania 43.44N 25.32E
63 C20 **Bragado** Buenos Aires, E Argentina 35.10S 60.28W
59 N14 **Bragança** Eng. Braganza; anc. Julio Briga. Bragança, NE Portugal 41.46N 6.46W
106 I5 **Bragança** ◆ district N Portugal
62 L8 **Bragança Paulista** São Paulo, S Brazil 22.55S 46.30W
Braganza see Bragança
Bragin see Brahin

31 V7 **Braham** Minnesota, N USA 45.43N 93.10W
Brahe see Brda
Brahestad see Raahe
121 O20 **Brahin** Rus. Bragin. Homyel'skaya Voblasts', SE Belarus 51.46N 30.16E
159 U15 **Brahmanbaria** Chittagong, E Bangladesh 23.58N 91.04E
160 O12 **Brāhmani** ⚡ E India
160 N13 **Brahmapur** Orissa, E India 19.21N 84.51E
133 S10 **Brahmaputra** var. Padma, Tsangpo, Ben. Jamuna, Chin. Yarlung Zangbo Jiang, Ind. Bramaputra, Dihang, Siang. ⚡ S Asia
99 H19 **Braich y Pwll** headland NW Wales, UK 52.47N 4.46W
191 R10 **Braidwood** New South Wales, SE Australia 35.26S 149.48E
32 M12 **Braidwood** Illinois, N USA 41.16N 88.12W
118 M13 **Brăila** Brăila, E Romania 45.17N 27.57E
118 L13 **Brăila** ◆ county E Romania
101 G19 **Braine-l'Alleud** Brabant Wallon, C Belgium 50.40N 4.22E
101 F19 **Braine-le-Comte** Hainaut, SW Belgium 50.37N 4.07E
31 U6 **Brainerd** Minnesota, N USA 46.20N 94.10W
101 J19 **Braives** Liège, E Belgium 50.37N 5.09E
85 H23 **Brak** ⚡ S South Africa
85 E18 **Brakel** Oost-Vlaanderen, SW Belgium 50.50N 3.48E
100 J13 **Brakel** Gelderland, C Netherlands 51.49N 5.05E
78 H9 **Brakna** ◆ region S Mauritania
97 J17 **Brålanda** Västra Götaland, S Sweden 58.32N 12.18E
97 F23 **Bramming** Ribe, W Denmark 55.28N 8.42E
12 G15 **Brampton** Ontario, S Canada 43.42N 79.46W
102 F12 **Bramsche** Niedersachsen, NW Germany 52.25N 7.58E
118 J12 **Bran** Ger. Törzburg, Hung. Törcsvár. Braşov, S Romania 45.31N 25.23E
31 W8 **Branch** Minnesota, N USA 45.29N 92.57W
23 R14 **Branchville** South Carolina, SE USA 33.15N 80.49W
59 Y6 **Branco, Cabo** headland E Brazil 7.07S 34.45W
60 F11 **Branco, Rio** ⚡ N Brazil
110 J8 **Brand** Vorarlberg, W Austria 47.07N 9.45E
85 B18 **Brandberg** ▲ NW Namibia 21.20S 14.22E
97 H14 **Brandbu** Oppland, S Norway 60.24N 10.30E
97 F22 **Brande** Ringkøbing, W Denmark 55.57N 9.07E
102 M12 **Brandenburg** var. Brandenburg an der Havel. Brandenburg, NE Germany 52.25N 12.34E
22 K5 **Brandenburg** Kentucky, S USA 37.58N 86.11W
102 N12 **Brandenburg** off. Freie und Hansestadt Hamburg, Fr. ◆ state NE Germany
Brandenburg an der Havel see Brandenburg
85 I23 **Brandfort** Free State, C South Africa 28.42S 26.28E
11 W16 **Brandon** Manitoba, S Canada 49.49N 99.57W
25 V12 **Brandon** Florida, SE USA 27.56N 82.17W
24 I6 **Brandon** Mississippi, S USA 32.16N 90.01W
99 A20 **Brandon Mountain** Ir. Cnoc Bréanainn. ▲ SW Ireland 52.13N 10.16W
97 I14 **Brandval** Hedmark, S Norway 60.18N 12.01E
85 F24 **Brandvlei** Northern Cape, W South Africa 30.19S 20.31E
25 U9 **Branford** Florida, SE USA 29.57N 82.54W
112 K7 **Braniewo** Ger. Braunsberg. Warmińsko-Mazurskie, NE Poland 54.24N 19.49E
204 H1 **Bransfield Strait** strait Antarctica
39 T8 **Branson** Colorado, C USA 37.01N 103.52W
29 T8 **Branson** Missouri, C USA 36.38N 93.13W
12 G16 **Brantford** Ontario, S Canada 43.04N 80.21W
104 L12 **Brantôme** Dordogne, SW France 45.21N 0.37E
190 M12 **Branxholme** Victoria, SE Australia 37.51S 141.48E
Brasil see Brazil
61 C16 **Brasiléia** Acre, W Brazil 10.58S 48.45W
61 K18 **Brasília** ● (Brazil) Distrito Federal, C Brazil 15.45S 47.57W
Braslav see Braslaw
120 J12 **Braslaw** Pol. Brasław, Rus. Braslav. Vitsyebskaya Voblasts', N Belarus 55.37N 27.01E
118 J13 **Braşov** Ger. Kronstadt, Hung. Brassó; prev. Oraşul Stalin. Braşov, C Romania 45.40N 25.34E
118 I12 **Braşov** ◆ county C Romania
79 U18 **Brass** Bayelsa, S Nigeria 4.19N 6.21E
101 H16 **Brasschaat** var. Brasschaet. Antwerpen, N Belgium 51.16N 4.30E
Brasschaet see Brasschaat
175 T4 **Brassey, Banjaran** var. Brassey Range. ▲ East Malaysia
Brassey Range see Brassey, Banjaran
Brassó see Braşov
25 T1 **Brasstown Bald** ▲ Georgia, SE USA 34.52N 83.48W
115 K22 **Brataj** Vlorë, SW Albania 40.18N 19.37E
116 J10 **Bratan** var. Morozov. ▲ C Bulgaria 42.31N 25.08E
113 F21 **Bratislava** Ger. Pressburg, Hung. Pozsony. ● (Slovakia) Bratislavský Kraj, SW Slovakia 48.10N 17.10E

113 *H21* **Bratislavský Kraj** ⬖ *region* W Slovakia
116 *H10* **Bratiya** ▲ C Bulgaria *42.36N 24.08E*
126 *J14* **Bratsk** Irkutskaya Oblast', C Russian Federation *56.19N 101.49E*
119 *Q8* **Brats'ke** Mykolayivs'ka Oblast', S Ukraine *47.52N 31.34E*
126 *J14* **Bratskoye Vodokhranilishche** ***Eng.*** Bratsk Reservoir. ▣ S Russian Federation
Bratsk Reservoir *see* Bratskoye Vodokhranilishche
Brattia *see* Brač
96 *D9* **Brattvåg** Møre og Romsdal, S Norway *62.36N 6.21E*
114 *K12* **Bratunac** Republika Srpska, E Bosnia and Herzegovina *44.10N 19.21E*
116 *J10* **Bratya Daskalovi** ***prev.*** Grozdovo. Stara Zagora, C Bulgaria *42.13N 25.21E*
111 *U2* **Braunau** ≈ N Austria
Braunau *see* Braunau am Inn
111 *Q4* **Braunau am Inn** ***var.*** Braunau. Oberösterreich, N Austria *48.16N 13.03E*
Braunsberg *see* Braniewo
102 *J13* **Braunschweig** ***Eng./Fr.*** Brunswick. Niedersachsen, N Germany *52.16N 10.31E*
Brava *see* Baraawe
107 *Y6* **Brava, Costa** *coastal region* NE Spain
45 *V16* **Brava, Punta** *headland* E Panama *8.21N 78.22W*
97 *N17* **Bråviken** *inlet* S Sweden
58 *B10* **Bravo, Cerro** ▲ N Peru *5.33S 79.10W*
Bravo del Norte, Río/Bravo, Río *see* Grande, Rio
37 *X17* **Brawley** California, W USA *32.58N 115.31W*
99 *G18* **Bray** ***Ir.*** Bré. E Ireland *53.12N 6.06W*
61 *G16* **Brazil** ***off.*** Federative Republic of Brazil, ***Port.*** Repùblica Federativa do Brasil, ***Sp.*** Brasil; ***prev.*** United States of Brazil. ◆ *federal republic* South America
67 *K15* **Brazil Basin** ***var.*** Brazilian Basin, Brazil'skaya Kotlovina. *undersea feature* W Atlantic Ocean
Brazilian Basin *see* Brazil Basin
Brazilian Highlands *see* Central, Planalto
Brazil'skaya Kotlovina *see* Brazil Basin
27 *U10* **Brazos River** ≈ Texas, SW USA
176 *Yy13* **Brazza** ≈ Irian Jaya, E Indonesia
Brazza *see* Brač
81 *G21* **Brazzaville** ● (Congo) Capital District, S Congo *4.13S 15.13E*
81 *G21* **Brazzaville** ✈ Le Pool, S Congo *4.15S 15.15E*
114 *J11* **Brčko** Republika Srpska, NE Bosnia and Herzegovina *44.52N 18.49E*
112 *H8* **Brda** ***Ger.*** Brahe. ≈ N Poland
Bré *see* Bray
193 *A23* **Breaksea Sound** *sound* South Island, NZ
192 *L4* **Bream Bay** *bay* North Island, NZ
192 *L4* **Bream Head** *headland* North Island, NZ *35.51S 174.35E*
Bréanainn, Cnoc *see* Brandon Mountain
47 *S6* **Brea, Punta** *headland* W Puerto Rico *17.56N 66.55W*
24 *I9* **Breaux Bridge** Louisiana, S USA *30.16N 91.54W*
118 *J13* **Breaza** Prahova, SE Romania *45.06N 25.44E*
174 *K14* **Brebes** Jawa, C Indonesia *6.54S 109.00E*
98 *K10* **Brechin** E Scotland, UK *56.44N 2.38W*
101 *H15* **Brecht** Antwerpen, N Belgium *51.21N 4.32E*
39 *S4* **Breckenridge** Colorado, C USA *39.28N 106.02W*
31 *R6* **Breckenridge** Minnesota, N USA *46.15N 96.35W*
27 *R6* **Breckenridge** Texas, SW USA *32.45N 98.54W*
99 *J21* **Brecknock** *cultural region* SE Wales, UK
65 *G25* **Brecknock, Península** *headland* S Chile *54.39S 71.48W*
113 *G19* **Břeclav** ***Ger.*** Lundenburg. Brněnský Kraj, SE Czech Republic *49.04N 16.51E*
99 *J21* **Brecon** E Wales, UK *51.57N 3.26W*
99 *J21* **Brecon Beacons** ⛰ S Wales, UK
101 *I14* **Breda** Noord-Brabant, S Netherlands *51.34N 4.46E*
97 *K20* **Bredaryd** Jönköping, S Sweden *57.10N 13.45E*
85 *F26* **Bredasdorp** Western Cape, SW South Africa *34.28S 20.03E*
95 *H16* **Bredbyn** Västernorrland, N Sweden *63.28N 18.04E*
125 *E13* **Bredy** Chelyabinskaya Oblast', C Russian Federation *52.23N 60.24E*
101 *K17* **Bree** Limburg, NE Belgium *51.07N 5.36E*
69 *T15* **Breede** ≈ S South Africa
100 *I7* **Breezand** Noord-Holland, NW Netherlands *52.52N 4.47E*
115 *P18* **Bregalnica** ≈ E FYR Macedonia
110 *I6* **Bregenz** ***anc.*** Brigantium. Vorarlberg, W Austria *47.31N 9.44E*
110 *J7* **Bregenzer Wald** ⛰ W Austria
116 *F6* **Bregovo** Vidin, NW Bulgaria *44.07N 22.40E*
104 *H5* **Bréhat, Île de** *island* NW France
94 *H2* **Breidhafjördhur** *bay* W Iceland
94 *L3* **Breidhdalsvík** Austurland, E Iceland *64.48N 14.02W*
110 *H9* **Breil** ***Ger.*** Brigels. Graubünden, S Switzerland *46.46N 9.04E*
94 *J8* **Breivikbotn** Finnmark, N Norway *70.36N 22.19E*
96 *I9* **Brekken** Sør-Trøndelag, S Norway *62.39N 11.49E*
96 *G7* **Brekstad** Sør-Trøndelag, S Norway *63.42N 9.40E*
96 *B10* **Bremangerlandet** *island* S Norway
Brême *see* Bremen
102 *H11* **Bremen** ***Fr.*** Brême. Bremen, NW Germany *53.05N 8.48E*
25 *R3* **Bremen** Georgia, SE USA *33.43N 85.09W*
33 *O11* **Bremen** Indiana, N USA *41.24N 86.07W*
102 *H10* **Bremen** ***off.*** Freie Hansestadt Bremen, ***Fr.*** Brême. ⬖ *state* N Germany
102 *G9* **Bremerhaven** Bremen, NW Germany *53.33N 8.34E*
Bremersdorp *see* Manzini
34 *G8* **Bremerton** Washington, NW USA *47.34N 122.37W*
102 *H10* **Bremervörde** Niedersachsen, NW Germany *53.29N 9.06E*
27 *U9* **Bremond** Texas, SW USA *31.10N 96.40W*
27 *U10* **Brenham** Texas, SW USA *30.10N 96.24W*
110 *M8* **Brenner** Tirol, W Austria *47.10N 11.51E*
Brenner, Col du/Brennero, Passo del *see* Brenner Pass
110 *M8* **Brenner Pass** ***var.*** Brenner Sattel, ***Fr.*** Col du Brenner, ***Ger.*** Brennerpass, ***It.*** Passo del Brennero. *pass* Austria/Italy *47.00N 11.29E*
Brenner Sattel *see* Brenner Pass
110 *G10* **Brenno** ≈ SW Switzerland
108 *F7* **Breno** Lombardia, N Italy *45.58N 10.18E*
25 *O5* **Brent** Alabama, S USA *32.54N 87.10W*
108 *H7* **Brenta** ≈ NE Italy
99 *P21* **Brentwood** E England, UK *51.38N 0.21E*
20 *L14* **Brentwood** Long Island, New York, NE USA *40.46N 73.12W*
108 *F7* **Brescia** ***anc.*** Brixia. Lombardia, N Italy *45.33N 10.13E*
101 *D15* **Breskens** Zeeland, SW Netherlands *51.24N 3.33E*
Breslau *see* Dolnośląskie
108 *H5* **Bressanone** ***Ger.*** Brixen. Trentino-Alto Adige, N Italy *46.43N 11.41E*
98 *M2* **Bressay** *island* NE Scotland, UK
104 *K9* **Bressuire** Deux-Sèvres, W France *46.50N 0.29W*
121 *F20* **Brest** ***Pol.*** Brześć nad Bugiem, ***Rus.*** Brest-Litovsk; ***prev.*** Brześć Litewski. Brestskaya Voblasts', SW Belarus *52.06N 23.42E*
104 *F5* **Brest** Finistère, NW France *48.24N 4.30W*
Brest-Litovsk *see* Brest
114 *A10* **Brestova** Istra, NW Croatia *45.09N 14.13E*
Brestskaya Oblast' *see* Brestskaya Voblasts'
121 *G19* **Brestskaya Voblasts'** ***prev. Rus.*** Brestskaya Oblast'. ⬖ *province* SW Belarus
104 *G6* **Bretagne** ***Eng.*** Brittany; ***Lat.*** Britannia Minor. ⬖ *region* NW France
118 *G12* **Bretea-Română** ***Hung.*** Oláhbrettye; ***prev.*** Bretea-Romînă. Hunedoara, W Romania *45.39N 23.00E*
Bretea-Romînă *see* Bretea-Română
105 *O3* **Breteuil** Oise, N France *49.37N 2.18E*
104 *I10* **Breton, Pertuis** *inlet* W France
24 *L10* **Breton Sound** *sound* Louisiana, S USA
192 *K2* **Brett, Cape** *headland* North Island, NZ *35.11S 174.21E*
103 *G21* **Bretten** Baden-Württemberg, SW Germany *49.01N 8.42E*
101 *K15* **Breugel** Noord-Brabant, S Netherlands *51.30N 5.30E*
108 *B6* **Breuil-Cervinia** ***It.*** Cervinia. Valle d'Aosta, NW Italy *45.57N 7.37E*
100 *I11* **Breukelen** Utrecht, C Netherlands *52.11N 5.01E*
23 *P10* **Brevard** North Carolina, SE USA *35.13N 82.43W*
40 *L9* **Brevig Mission** Alaska, USA *65.19N 166.29W*
97 *G16* **Brevik** Telemark, S Norway *59.03N 9.40E*
191 *P5* **Brewarrina** New South Wales, SE Australia *30.01S 146.50E*
21 *R6* **Brewer** Maine, NE USA *44.46N 68.44W*
31 *T11* **Brewster** Minnesota, N USA *43.43N 95.28W*
31 *N14* **Brewster** Nebraska, C USA *41.54N 99.52W*
33 *U12* **Brewster** Ohio, N USA *40.42N 81.36W*
191 *O8* **Brewster, Lake** ◎ New South Wales, SE Australia
25 *P7* **Brewton** Alabama, S USA *31.06N 87.04W*
Brezhnev *see* Naberezhnyye Chelny
111 *W12* **Brežice** ***Ger.*** Rann. E Slovenia *45.54N 15.35E*
116 *G9* **Breznik** Pernik, W Bulgaria *42.45N 22.54E*
113 *K19* **Brezno** ***Ger.*** Bries, Briesen, ***Hung.*** Breznóbánya; ***prev.*** Brezno nad Hronom. Banskobystrický Kraj, C Slovakia *48.49N 19.40E*
Breznóbánya/Brezno nad Hronom *see* Brezno
118 *I12* **Brezoi** Vâlcea, SW Romania *45.18N 24.15E*
116 *J10* **Brezovo** ***prev.*** Abrashlare. Plovdiv, C Bulgaria *42.19N 25.05E*
81 *K14* **Bria** Haute-Kotto, C Central African Republic *6.30N 22.00E*
105 *U13* **Briançon** ***anc.*** Brigantio. Hautes-Alpes, SE France *44.55N 6.37E*
68 *K7* **Brian Head** ▲ Utah, W USA *37.40N 112.49W*
105 *O7* **Briare** Loiret, C France *47.35N 2.46E*
191 *V2* **Bribie Island** *island* Queensland, E Australia
45 *O14* **Bribrí** Limón, E Costa Rica *9.37N 82.51W*
118 *L8* **Briceni** ***var.*** Brinceni, ***Rus.*** Brichany. N Moldova *48.21N 27.02E*
Bricgstow *see* Bristol
Brichany *see* Briceni
101 *M24* **Bridel** Luxembourg, C Luxembourg *49.40N 6.03E*
99 *J22* **Bridgend** S Wales, UK *51.30N 3.37W*
12 *I14* **Bridgenorth** Ontario, SE Canada *44.21N 78.22W*
25 *Q1* **Bridgeport** Alabama, S USA *34.57N 85.42W*
37 *R8* **Bridgeport** California, W USA *38.14N 119.13W*
20 *L13* **Bridgeport** Connecticut, NE USA *41.10N 73.12W*
33 *N15* **Bridgeport** Illinois, N USA *38.42N 87.45W*
30 *J14* **Bridgeport** Nebraska, C USA *41.37N 103.07W*
27 *S6* **Bridgeport** Texas, SW USA *33.12N 97.45W*
23 *S3* **Bridgeport** West Virginia, NE USA *39.17N 80.15W*
27 *S5* **Bridgeport, Lake** ▣ Texas, SW USA
35 *U11* **Bridger** Montana, NW USA *45.16N 108.55W*
20 *I17* **Bridgeton** New Jersey, NE USA *39.24N 75.10W*
188 *J14* **Bridgetown** Western Australia *34.01S 116.07E*
47 *Y14* **Bridgetown** ● (Barbados) SW Barbados *13.05N 59.36W*
191 *P17* **Bridgewater** Tasmania, SE Australia *42.47S 147.15E*
11 *P16* **Bridgewater** Nova Scotia, SE Canada *44.19N 64.30W*
21 *P12* **Bridgewater** Massachusetts, NE USA *41.59N 70.58W*
31 *Q11* **Bridgewater** South Dakota, N USA *43.33N 97.30W*
23 *U5* **Bridgewater** Virginia, NE USA *38.22N 78.58W*
21 *P8* **Bridgton** Maine, NE USA *44.04N 70.43W*
99 *K23* **Bridgwater** SW England, UK *51.08N 3.00W*
99 *K22* **Bridgwater Bay** *bay* SW England, UK
99 *O16* **Bridlington** E England, UK *54.04N 0.12W*
99 *O16* **Bridlington Bay** *bay* E England, UK
191 *P15* **Bridport** Tasmania, SE Australia *41.03S 147.26E*
99 *K24* **Bridport** S England, UK *50.43N 2.43W*
105 *O5* **Brie** *cultural region* N France
Brieg *see* Brzeg
Briel *see* Brielle
100 *G12* **Brielle** ***var.*** Briel, Bril, ***Eng.*** The Brill. Zuid-Holland, SW Netherlands *51.54N 4.10E*
110 *E9* **Brienz** Bern, C Switzerland *46.45N 8.00E*
110 *E9* **Brienzer See** ◎ SW Switzerland
Bries/Briesen *see* Brezno
Brietzig *see* Brzesko
105 *S4* **Briey** Meurthe-et-Moselle, NE France *49.15N 5.57E*
110 *E10* **Brig** ***Fr.*** Brigue, ***It.*** Briga. Valais, SW Switzerland *46.19N 8.00E*
Briga *see* Brig
103 *G24* **Brigach** ≈ S Germany
20 *K17* **Brigantine** New Jersey, NE USA *39.23N 74.21W*
Brigantio *see* Briançon
Brigantium *see* Bregenz
Brigels *see* Breil
27 *S9* **Briggs** Texas, SW USA *30.52N 97.55W*
68 *L1* **Brigham City** Utah, W USA *41.30N 112.00W*
12 *J15* **Brighton** Ontario, SE Canada *44.01N 77.44W*
99 *O23* **Brighton** SE England, UK *50.49N 0.10W*
39 *T4* **Brighton** Colorado, C USA *39.58N 104.46W*
32 *K15* **Brighton** Illinois, N USA *39.01N 90.09W*
105 *T16* **Brignoles** Var, W France *43.25N 6.03E*
Brigue *see* Brig
107 *O7* **Brihuega** Castilla-La Mancha, C Spain *40.45N 2.52W*
114 *A10* **Brijuni** ***It.*** Brioni. *island group* NW Croatia
78 *G12* **Brikama** W Gambia *13.13N 16.37W*
Bril *see* Brielle
Brill, The *see* Brielle
103 *G15* **Brilon** Nordrhein-Westfalen, W Germany *51.24N 8.34E*
Brinceni *see* Briceni
109 *Q18* **Brindisi** ***anc.*** Brundisium, Brundusium. Puglia, SE Italy *40.39N 17.55E*
29 *W11* **Brinkley** Arkansas, C USA *34.53N 91.11W*
Brioni *see* Brijuni
105 *P12* **Brioude** ***anc.*** Brivas. Haute-Loire, C France *45.18N 3.22E*
Briovera *see* St-Lô
191 *U2* **Brisbane** *state capital* Queensland, E Australia *27.30S 153.00E*
191 *V2* **Brisbane** ✈ Queensland, E Australia *27.30S 153.00E*
27 *P2* **Briscoe** Texas, SW USA *35.34N 100.17W*
108 *H10* **Brisighella** Emilia-Romagna, C Italy *44.12N 11.45E*
110 *G11* **Brissago** Ticino, S Switzerland *46.07N 8.40E*
99 *K22* **Bristol** ***anc.*** Bricgstow. SW England, UK *51.27N 2.34W*
20 *M12* **Bristol** Connecticut, NE USA *41.40N 72.56W*
25 *R9* **Bristol** Florida, SE USA *30.25N 84.58W*
21 *N9* **Bristol** New Hampshire, NE USA *43.34N 71.42W*
31 *Q8* **Bristol** South Dakota, N USA *45.18N 97.45W*
23 *P8* **Bristol** Tennessee, S USA *36.36N 82.11W*
20 *M8* **Bristol** Vermont, NE USA *44.07N 73.00W*
41 *N14* **Bristol Bay** *bay* Alaska, USA
99 *I22* **Bristol Channel** *inlet* England/Wales, UK
37 *W14* **Bristol Lake** ◎ California, W USA
29 *P10* **Bristow** Oklahoma, C USA *35.49N 96.23W*
88 *C10* **Britain** ***var.*** Great Britain. *island* UK
Britannia Minor *see* Bretagne
8 *L12* **British Columbia** ***Fr.*** Colombie-Britannique. ⬖ *province* SW Canada
British Guiana *see* Guyana
British Honduras *see* Belize
181 *Q7* **British Indian Ocean Territory** ◇ *UK dependent territory* C Indian Ocean
88 *B9* **British Isles** *island group* NW Europe
8 *I1* **British Mountains** ⛰ Yukon Territory, NW Canada
British North Borneo *see* Sabah
British Solomon Islands Protectorate *see* Solomon Islands
47 *S8* **British Virgin Islands** ***var.*** Virgin Islands. ◇ *UK dependent territory* E West Indies
85 *J21* **Brits** North-West, N South Africa *25.39S 27.46E*
85 *H24* **Britstown** Northern Cape, W South Africa *30.36S 23.30E*
12 *F12* **Britt** Ontario, S Canada *45.46N 80.34W*
31 *V12* **Britt** Iowa, C USA *43.06N 93.48W*
Brittany *see* Bretagne
31 *Q7* **Britton** South Dakota, N USA *45.47N 97.45W*
Briva Curretia *see* Brive-la-Gaillarde
Briva Isarae *see* Pontoise
Brivas *see* Brioude
Brive *see* Brive-la-Gaillarde
104 *M12* **Brive-la-Gaillarde** ***prev.*** Brive, ***anc.*** Briva Curretia. Corrèze, C France *45.09N 1.31E*
107 *O4* **Briviesca** Castilla-León, N Spain *42.33N 3.19W*
Brixen *see* Bressanone
Brixia *see* Brescia
151 *S15* **Brlik** ***prev.*** Novotroickoje, Novotroitskoye. Zhambyl, SE Kazakhstan *43.39N 73.45E*
113 *G19* **Brněnský Kraj** ⬖ *region* SE Czech republic
113 *G18* **Brno** ***Ger.*** Brünn. Brněnský Kraj, SE Czech Republic *49.10N 16.35E*
98 *G7* **Broad Bay** *bay* NW Scotland, UK
27 *X8* **Broaddus** Texas, SW USA *31.18N 94.16W*
191 *O12* **Broadford** Victoria, SE Australia *37.07S 145.04E*
98 *G9* **Broadford** N Scotland, UK *57.14N 5.54W*
98 *J13* **Broad Law** ▲ S Scotland, UK *55.30N 3.22W*
25 *U3* **Broad River** ≈ Georgia, SE USA
23 *N8* **Broad River** ≈ North Carolina/South Carolina, SE USA
189 *Y8* **Broadsound Range** ⛰ Queensland, E Australia
35 *X11* **Broadus** Montana, NW USA *45.28N 105.22W*
23 *U4* **Broadway** Virginia, NE USA *38.36N 78.48W*
120 *E9* **Brocēni** Saldus, SW Latvia *56.41N 22.31E*
11 *U11* **Brochet** Manitoba, C Canada *57.55N 101.40W*
11 *U10* **Brochet, Lac** ◎ Manitoba, C Canada
13 *S5* **Brochet, Lac au** ◎ Quebec, SE Canada
103 *K14* **Brocken** ▲ C Germany *51.48N 10.38E*
21 *O12* **Brockton** Massachusetts, NE USA *42.04N 71.01W*
12 *L14* **Brockville** Ontario, SE Canada *44.36N 75.42W*
20 *D13* **Brockway** Pennsylvania, NE USA *41.14N 78.45W*
Brod/Bród *see* Slavonski Brod
15 *Kk1* **Brodeur Peninsula** *peninsula* Baffin Island, Nunavut, NE Canada
98 *H13* **Brodick** W Scotland, UK *55.34N 5.09W*
Brod na Savi *see* Slavonski Brod
112 *K9* **Brodnica** ***Ger.*** Buddenbrock. Kujawski-pomorskie, C Poland *53.15N 19.22E*
114 *G10* **Brod-Posavina** ***off.*** Brodsko-Posavska Županija. ⬖ *province* NE Croatia
118 *J5* **Brody** L'vivs'ka Oblast', NW Ukraine *50.04N 25.07E*
97 *G22* **Brædstrup** Vejle, C Denmark *55.58N 9.37E*
100 *I10* **Broek-in-Waterland** Noord-Holland, C Netherlands *52.27N 4.58E*
34 *L13* **Brogan** Oregon, NW USA *44.15N 117.34W*
112 *N10* **Brok** Mazowieckie, C Poland *52.42N 21.53E*
29 *P9* **Broken Arrow** Oklahoma, C USA *36.03N 95.47W*
191 *T9* **Broken Bay** *bay* New South Wales, SE Australia
31 *N15* **Broken Bow** Nebraska, C USA *41.24N 99.38W*
29 *R13* **Broken Bow** Oklahoma, C USA *34.01N 94.44W*
29 *R12* **Broken Bow Lake** ▣ Oklahoma, C USA
190 *L6* **Broken Hill** New South Wales, SE Australia *31.58S 141.27E*
181 *S10* **Broken Ridge** *undersea feature* S Indian Ocean
194 *H10* **Broken Water Bay** *bay* W Bismarck Sea
57 *W10* **Brokopondo** Brokopondo, NE Suriname *05.04N 55.00W*
57 *W10* **Brokopondo** ⬖ *district* C Suriname
Bromberg *see* Bydgoszcz
97 *L22* **Bromölla** Skåne, S Sweden *56.04N 14.28E*
99 *L20* **Bromsgrove** W England, UK *52.19N 2.03W*
97 *G20* **Brønderslev** Nordjylland, N Denmark *57.16N 9.58E*
108 *D8* **Broni** Lombardia, N Italy *45.04N 9.18E*
8 *K11* **Bronlund Peak** ▲ British Columbia, W Canada *57.27N 126.43W*
95 *F14* **Brønnøysund** Nordland, C Norway *65.28N 12.13E*
25 *V10* **Bronson** Florida, SE USA *29.25N 82.38W*
33 *Q11* **Bronson** Michigan, N USA *41.52N 85.11W*
27 *X8* **Bronson** Texas, SW USA *31.20N 94.00W*
109 *L24* **Bronte** Sicilia, Italy, C Mediterranean Sea *37.46N 14.49E*
27 *P8* **Bronte** Texas, SW USA *31.53N 100.17W*
27 *Y9* **Brookeland** Texas, SW USA *31.05N 93.57W*
179 *O15* **Brooke's Point** Palawan, W Philippines *8.54N 117.54E*
29 *T3* **Brookfield** Missouri, C USA *39.46N 93.04W*
24 *K7* **Brookhaven** Mississippi, S USA *31.34N 90.26W*
34 *E16* **Brookings** Oregon, NW USA *42.03N 124.16W*
31 *R10* **Brookings** South Dakota, N USA *44.15N 96.46W*
31 *W14* **Brooklyn** Iowa, C USA *41.43N 92.27W*
31 *U8* **Brooklyn Park** Minnesota, N USA *45.06N 93.18W*
23 *U7* **Brookneal** Virginia, NE USA *37.03N 78.56W*
11 *R16* **Brooks** Alberta, SW Canada *50.34N 111.54W*
27 *V11* **Brookshire** Texas, SW USA *29.47N 95.57W*
40 *L8* **Brooks Mountain** ▲ Alaska, USA *65.31N 167.24W*
40 *M11* **Brooks Range** ⛰ Alaska, USA
33 *O12* **Brookston** Indiana, N USA *40.34N 86.53W*
25 *V11* **Brooksville** Florida, SE USA *28.33N 82.23W*
25 *N4* **Brooksville** Mississippi, S USA *33.13N 88.34W*
188 *J13* **Brookton** Western Australia *32.24S 117.04E*
33 *Q14* **Brookville** Indiana, N USA *39.25N 85.00W*
20 *D13* **Brookville** Pennsylvania, NE USA *41.07N 79.05W*
33 *Q14* **Brookville Lake** ▣ Indiana, N USA
188 *K5* **Broome** Western Australia *17.58S 122.15E*
39 *S4* **Broomfield** Colorado, C USA *39.55N 105.05W*
Broos *see* Orăștie
98 *J7* **Brora** N Scotland, UK *57.58N 4.00W*
98 *I7* **Brora** ≈ N Scotland, UK
97 *F23* **Brørup** Ribe, W Denmark *55.28N 9.01E*
97 *L23* **Brösarp** Skåne, S Sweden *55.43N 14.10E*
118 *J9* **Broșteni** Suceava, NE Romania *47.13N 25.43E*
104 *M6* **Brou** Eure-et-Loir, C France *48.12N 1.10E*
Broucsella *see* Brussel/Bruxelles
Broughton Bay *see* Tongjosŏn-man
16 *O1* **Broughton Island** Nunavut, NE Canada *67.34N 63.55W*
144 *G7* **Broummâna** C Lebanon *33.53N 35.39E*
24 *I9* **Broussard** Louisiana, S USA *30.09N 91.57W*
100 *E13* **Brouwersdam** *dam* SW Netherlands *51.46N 3.51E*
100 *E13* **Brouwershaven** Zeeland, SW Netherlands *51.44N 3.50E*
119 *P4* **Brovary** Kyyivs'ka Oblast', N Ukraine *50.30N 30.43E*
97 *G20* **Brovst** Nordjylland, N Denmark *57.06N 9.31E*
33 *S8* **Brown City** Michigan, N USA *43.12N 82.50W*
26 *M6* **Brownfield** Texas, SW USA *33.10N 102.16W*
35 *Q7* **Browning** Montana, NW USA *48.33N 113.00W*
35 *R6* **Brown, Mount** ▲ Montana, NW USA *48.52N 111.08W*
194 *K15* **Brown River** ≈ S PNG
1 *M9* **Browns Bank** *undersea feature* NW Atlantic Ocean
33 *O14* **Brownsburg** Indiana, N USA *39.50N 86.24W*
20 *J16* **Browns Mills** New Jersey, NE USA *39.58N 74.33W*
46 *J12* **Browns Town** C Jamaica *18.28N 77.22W*
33 *P15* **Brownstown** Indiana, N USA *38.52N 86.02W*
31 *R8* **Browns Valley** Minnesota, N USA *45.36N 96.49W*
22 *K7* **Brownsville** Kentucky, S USA *37.09N 86.13W*
22 *F9* **Brownsville** Tennessee, S USA *35.34N 89.15W*
27 *T17* **Brownsville** Texas, SW USA *25.55N 97.28W*
57 *W10* **Brownsweg** Brokopondo, C Suriname
31 *U9* **Brownton** Minnesota, N USA *44.43N 94.21W*
21 *R5* **Brownville Junction** Maine, NE USA *45.20N 69.04W*
27 *R8* **Brownwood** Texas, SW USA *31.41N 98.59W*
27 *R8* **Brownwood Lake** ▣ Texas, SW USA
106 *I9* **Brozas** Extremadura, W Spain *39.37N 6.48W*
121 *M18* **Brozha** Mahilyowskaya Voblasts', E Belarus *52.57N 29.07E*
105 *O2* **Bruay-en-Artois** Pas-de-Calais, N France *50.31N 2.30E*
105 *P2* **Bruay-sur-l'Escaut** Nord, N France *50.24N 3.33E*
12 *F13* **Bruce Peninsula** *peninsula* Ontario, S Canada
22 *H9* **Bruceton** Tennessee, S USA *36.02N 88.14W*
27 *T9* **Bruceville** Texas, SW USA *31.17N 97.15W*
103 *G21* **Bruchsal** Baden-Württemberg, SW Germany *49.07N 8.34E*
111 *Q7* **Bruck** Salzburg, NW Austria *47.18N 12.51E*
Bruck *see* Bruck an der Mur
111 *Y4* **Bruck an der Leitha** Niederösterreich, NE Austria *48.02N 16.47E*
111 *V7* **Bruck an der Mur** ***var.*** Bruck. Steiermark, C Austria *47.26N 15.13E*
103 *M24* **Bruckmühl** Bayern, SE Germany *47.52N 11.54E*
173 *Dd3* **Brueuh, Pulau** *island* NW Indonesia
Bruges *see* Brugge
110 *F6* **Brugg** Aargau, NW Switzerland *47.28N 8.13E*
101 *C16* **Brugge** ***Fr.*** Bruges. West-Vlaanderen, NW Belgium *51.13N 3.13E*
111 *R9* **Bruggen** Kärnten, S Austria *46.46N 13.13E*
103 *E16* **Brühl** Nordrhein-Westfalen, W Germany *50.49N 6.54E*
101 *F14* **Bruinisse** Zeeland, SW Netherlands *51.40N 4.04E*
174 *L5* **Bruit, Pulau** *island* East Malaysia
12 *K10* **Brûlé, Lac** ◎ Quebec, SE Canada
32 *M4* **Brule River** ≈ Michigan/Wisconsin, N USA
101 *H23* **Brûly** Namur, S Belgium *49.55N 4.31E*
61 *N17* **Brumado** Bahia, E Brazil *14.13S 41.37W*
100 *M11* **Brummen** Gelderland, E Netherlands *52.04N 6.10E*
96 *H13* **Brumunddal** Hedmark, S Norway *60.52N 10.55E*
25 *Q6* **Brundidge** Alabama, S USA *31.43N 85.49W*
Brundisium/Brundusium *see* Brindisi
35 *N15* **Bruneau River** ≈ Idaho, NW USA
Bruneck *see* Brunico
174 *Mm4* **Brunei** ***off.*** Sultanate of Brunei, ***Mal.*** Negara Brunei Darussalam. ◆ *monarchy* SE Asia
175 *N3* **Brunei Bay** ***var.*** Teluk Brunei. *bay* N Brunei
Brunei, Teluk *see* Brunei Bay
Brunei Town *see* Bandar Seri Begawan
108 *H5* **Brunico** ***Ger.*** Bruneck. Trentino-Alto Adige, N Italy *46.49N 11.57E*
Brünn *see* Brno
193 *G17* **Brunner, Lake** ◎ South Island, NZ
101 *M18* **Brunssum** Limburg, SE Netherlands *50.57N 5.58E*
25 *W7* **Brunswick** Georgia, SE USA *31.09N 81.30W*
21 *Q8* **Brunswick** Maine, NE USA *43.54N 69.58W*
23 *V3* **Brunswick** Maryland, NE USA *39.18N 77.37W*
29 *T3* **Brunswick** Missouri, C USA *39.25N 93.07W*
33 *T11* **Brunswick** Ohio, N USA *41.14N 81.50W*
Brunswick *see* Braunschweig
65 *H24* **Brunswick, Península** *headland* S Chile *53.30S 71.27W*
113 *H17* **Bruntál** ***Ger.*** Freudenthal. Ostravský Kraj, E Czech Republic *50.00N 17.27E*
205 *N3* **Brunt Ice Shelf** *ice shelf* Antarctica
Brusa *see* Bursa
39 *U3* **Brush** Colorado, C USA *40.15N 103.37W*
44 *M5* **Brus Laguna** Gracias a Dios, E Honduras *15.46N 84.31W*
62 *K13* **Brusque** Santa Catarina, S Brazil *27.07S 48.54W*
Brussa *see* Bursa
101 *E18* **Brussel** ***var.*** Brussels, ***Fr.*** Bruxelles, ***Ger.*** Brüssel; ***anc.*** Broucsella. ● (Belgium) Brussels, C Belgium ***see also*** Bruxelles *50.52N 4.21E*
Brüssel/Brussels *see* Brussel/Bruxelles
119 *O5* **Brusyliv** Zhytomyrs'ka Oblast', N Ukraine *50.16N 29.31E*
191 *Q12* **Bruthen** Victoria, SE Australia *37.43S 147.49E*
Bruttium *see* Calabria
Brüx *see* Most
101 *E18* **Bruxelles** ***var.*** Brussels, ***Dut.*** Brussel, ***Ger.*** Brüssel; ***anc.*** Broucsella. ● (Belgium) Brussels, C Belgium ***see also*** Brussel *50.52N 4.21E*
56 *J7* **Bruzual** Apure, W Venezuela *7.59N 69.18W*
33 *Q11* **Bryan** Ohio, N USA *41.28N 84.33W*
27 *U10* **Bryan** Texas, SW USA *30.40N 96.22W*
204 *J4* **Bryan Coast** *physical region* Antarctica
126 *I13* **Bryanka** Krasnoyarskiy Kray, C Russian Federation *59.01N 93.13E*
119 *Y7* **Bryanka** Luhans'ka Oblast', E Ukraine *48.30N 38.45E*
190 *J8* **Bryan, Mount** ▲ South Australia *33.25S 138.59E*
130 *I6* **Bryansk** Bryanskaya Oblast', W Russian Federation *53.15N 34.06E*
130 *H6* **Bryanskaya Oblast'** ⬖ *province* W Russian Federation
204 *J5* **Bryant, Cape** *headland* Antarctica
29 *U8* **Bryant Creek** ≈ Missouri, C USA
68 *K8* **Bryce Canyon** *canyon* Utah, W USA
121 *O15* **Bryli** ***Rus.*** Bryli. Mahilyowskaya Voblasts', E Belarus *53.55N 30.31E*
97 *C17* **Bryne** Rogaland, S Norway *58.43N 5.37E*
27 *R6* **Bryson** Texas, SW USA *33.09N 98.23W*
23 *N10* **Bryson City** North Carolina, SE USA *35.33N 83.39W*
12 *K11* **Bryson, Lac** ◎ Quebec, SE Canada
130 *K13* **Bryukhovetskaya** Krasnodarskiy Kray, SW Russian Federation *45.49N 38.01E*
113 *H15* **Brzeg** ***Ger.*** Brieg; ***anc.*** Civitas Altae Ripae. Opolskie, S Poland *50.52N 17.27E*
113 *G14* **Brzeg Dolny** ***Ger.*** Dyhernfurth. Dolnośląskie, SW Poland *51.15N 16.42E*
Brześć Litewski/Brześć nad Bugiem *see* Brest
113 *L17* **Brzesko** ***Ger.*** Brietzig. Małopolskie, S Poland *49.57N 20.35E*
Brzeżany *see* Berezhany
112 *K12* **Brzeziny** Łódzkie, C Poland *51.48N 19.42E*
Brzostowica Wielka *see* Vyalikaya Byerastavitsa
113 *O17* **Brzozów** Podkarpackie, SE Poland *49.38N 22.00E*
Bsharrī/Bsherri *see* Bcharré
197 *I13* **Bua** Vanua Levu, N Fiji *16.48S 178.36E*
97 *J20* **Bua** Halland, S Sweden *57.13N 12.07E*
84 *M13* **Bua** ≈ C Malawi
Bua *see* Čiovo
83 *L18* **Bu'aale** ***It.*** Buale. Jubbada Dhexe, SW Somalia *0.52N 42.37E*
201 *Q8* **Buada Lagoon** *lagoon* Nauru, C Pacific Ocean
195 *W14* **Buala** Santa Isabel, E Solomon Islands *8.06S 159.31E*
Buale *see* Bu'aale
202 *H1* **Buariki** *atoll* Tungaru, W Kiribati
178 *I10* **Bua Yai** ***var.*** Ban Bua Yai. Nakhon Ratchasima, E Thailand *15.34N 102.25E*
77 *P8* **Bu'ayrāt al Ḩasūn** ***var.*** Buwayrāt al Hasūn. C Libya *31.22N 15.41E*
78 *H13* **Buba** S Guinea-Bissau *11.36N 14.55W*
175 *Qq7* **Bubaa** Sulawesi, N Indonesia *0.32N 122.27E*
83 *D20* **Bubanza** NW Burundi *3.04S 29.22E*
85 *K18* **Bubi** ***prev.*** Bubye. ≈ S Zimbabwe
148 *L11* **Būbīyan, Jazīrat** *island* E Kuwait
Bublitz *see* Bobolice
Bubye *see* Bubi
197 *J13* **Buca** ***prev.*** Mbutha. Vanua Levu, N Fiji *16.39S 179.51E*
142 *F16* **Bucak** Burdur, SW Turkey *37.26N 30.32E*
56 *G8* **Bucaramanga** Santander, N Colombia *7.07N 73.10W*
109 *M18* **Buccino** Campania, S Italy *40.37N 15.25E*
118 *K9* **Bucecea** Botoșani, NE Romania *47.43N 26.24E*
118 *J6* **Buchach** ***Pol.*** Buczacz. Ternopil's'ka Oblast', W Ukraine *49.04N 25.22E*
191 *Q12* **Buchan** Victoria, SE Australia *37.26S 148.11E*
78 *J17* **Buchanan** ***prev.*** Grand Bassa. SW Liberia *5.52N 10.03W*
25 *R3* **Buchanan** Georgia, SE USA *33.48N 85.11W*
33 *O11* **Buchanan** Michigan, N USA *41.49N 86.21W*
23 *T6* **Buchanan** Virginia, NE USA *37.31N 79.40W*
27 *R10* **Buchanan Dam** Texas, SW USA *30.42N 98.24W*
27 *R10* **Buchanan, Lake** ▣ Texas, SW USA
98 *L8* **Buchan Ness** *headland* NE Scotland, UK *57.28N 1.46W*
11 *T12* **Buchans** Newfoundland, SE Canada *48.49N 56.44W*
Bucharest *see* București
103 *H20* **Buchen** Baden-Württemberg, SW Germany *49.31N 9.18E*
102 *I10* **Buchholz in der Nordheide** Niedersachsen, NW Germany *53.19N 9.52E*
110 *F7* **Buchs** Aargau, N Switzerland *47.24N 8.03E*
110 *I8* **Buchs** Sankt Gallen, NE Switzerland *47.10N 9.26E*
102 *H13* **Bückeburg** Niedersachsen, NW Germany *52.16N 9.03E*
68 *K14* **Buckeye** Arizona, SW USA *33.22N 112.34W*
Buckeye State *see* Ohio
23 *S4* **Buckhannon** West Virginia, NE USA *38.59N 80.13W*
27 *T9* **Buckholts** Texas, SW USA *30.52N 97.07W*
98 *K8* **Buckie** NE Scotland, UK *57.39N 2.55W*
12 *M12* **Buckingham** Quebec, SE Canada *45.34N 75.25W*
23 *U6* **Buckingham** Virginia, NE USA *37.33N 78.33W*
99 *N21* **Buckinghamshire** *cultural region* SE England, UK
41 *N8* **Buckland** Alaska, USA *65.58N 161.07W*
190 *G7* **Buckleboo** South Australia *32.55S 136.11E*
28 *K7* **Bucklin** Kansas, C USA *37.33N 99.37W*
29 *T3* **Bucklin** Missouri, C USA *39.46N 92.53W*
68 *I12* **Buckskin Mountains** ⛰ Arizona, SW USA
21 *R7* **Bucksport** Maine, NE USA *44.34N 68.46W*
84 *A9* **Buco Zau** Cabinda, NW Angola *4.47S 12.32E*
Bu Craa *see* Bou Craa
118 *K14* **București** ***Eng.*** Bucharest, ***Ger.*** Bukarest; ***prev.*** Altenburg, ***anc.*** Cetatea Damboviţei. ● (Romania) București, S Romania *44.27N 26.06E*
33 *S12* **Bucyrus** Ohio, N USA *40.48N 82.58W*
Buczacz *see* Buchach
96 *E9* **Bud** Møre og Romsdal, S Norway *62.55N 6.55E*
27 *S11* **Buda** Texas, SW USA *30.05N 97.50W*
121 *O18* **Buda-Kashalyova** ***Rus.*** Buda-Koshelëvo. Homyel'skaya Voblasts', SE Belarus *52.43N 30.34E*
Buda-Koshelëvo *see* Buda-Kashalyova
177 *G4* **Budalin** Sagaing, C Myanmar *22.24N 95.07E*
113 *J22* **Budapest** ***off.*** Budapest Főváros, ***SCr.*** Budimpešta. ● (Hungary) Pest, N Hungary *47.30N 19.03E*
158 *K11* **Budaun** Uttar Pradesh, N India *28.01N 79.07E*
147 *O9* **Budayyi'ah** *oasis* C Saudi Arabia *23.04N 43.29E*
205 *Y12* **Budd Coast** *physical region* Antarctica
Buddenbrock *see* Brodnica
109 *C17* **Budduso** Sardegna, Italy, C Mediterranean Sea *40.37N 9.19E*
99 *I23* **Bude** SW England, UK *50.49N 4.33W*
24 *J7* **Bude** Mississippi, S USA *31.27N 90.51W*
113 *C18* **Budějovický Kraj** ⬖ *region* S Czech Republic
101 *K16* **Budel** Noord-Brabant, SE Netherlands *51.16N 5.34E*
102 *I8* **Büdelsdorf** Schleswig-Holstein, N Germany *54.20N 9.40E*
131 *O14* **Budënnovsk** Stavropol'skiy Kray, SW Russian Federation *44.46N 44.07E*
118 *K14* **Budești** Călărași, SE Romania *44.13N 26.31E*
Budgewoi *see* Budgewoi Lake
191 *T8* **Budgewoi Lake** ***var.*** Budgewoi. New South Wales, SE Australia *33.13S 151.34E*
94 *I2* **Búdhardalur** Vesturland, W Iceland *65.07N 21.45W*
Budimpešta *see* Budapest
81 *J16* **Budjala** Equateur, NW Dem. Rep. Congo (Zaire) *2.39N 19.42E*
108 *G10* **Budrio** Emilia-Romagna, C Italy *44.33N 11.34E*
Budslav *see* Budslaw
121 *K14* **Budslaw** ***Rus.*** Budslav. Minskaya Voblasts', N Belarus *54.46N 27.26E*
Budua *see* Budva
174 *Ll5* **Budu, Tanjung** *headland* East Malaysia *2.51N 111.42E*

◆ Country
● Country Capital
◇ Dependent Territory
○ Dependent Territory Capital
⬖ Administrative Region
✈ International Airport
▲ Mountain
⛰ Mountain Range
Volcano
≈ River
◎ Lake
▣ Reservoir

115 *J17* **Budva** *It.* Budua. Montenegro, SW Yugoslavia 42.17N 18.49E
Budweis *see* České Budějovice
Budyšin *see* Bautzen

81 *D16* **Buea** Sud-Ouest, SW Cameroon 4.09N 9.13E
105 *S13* **Buech** ♣ SE France
20 *J17* **Buena** New Jersey, NE USA 39.30N 74.55W
64 *K12* **Buena Esperanza** San Luis, C Argentina 34.45S 65.15W
56 *C11* **Buenaventura** Valle del Cauca, W Colombia 3.54N 77.01W
42 *I4* **Buenaventura** Chihuahua, N Mexico 29.52N 107.25W
59 *M18* **Buena Vista** Santa Cruz, C Bolivia 17.27S 63.40W
42 *G10* **Buenavista** Baja California Sur, W Mexico 23.39N 109.40W
39 *S5* **Buena Vista** Colorado, C USA 38.50N 106.07W
25 *S5* **Buena Vista** Georgia, SE USA 32.19N 84.31W
23 *T6* **Buena Vista** Virginia, NE USA 37.43N 79.21W
46 *F5* **Buena Vista, Bahia de** *bay* N Cuba
37 *R13* **Buena Vista Lake Bed** ⊚ California, W USA
107 *P8* **Buendía, Embalse de** ⊡ C Spain
65 *F16* **Bueno, Río** ♣ S Chile
64 *N12* **Buenos Aires** *hist.* Santa Maria del Buen Aire. ● (Argentina) Buenos Aires, E Argentina 34.40S 58.30W
45 *O15* **Buenos Aires** Puntarenas, SE Costa Rica 9.09N 83.15W
63 *C20* **Buenos Aires** *off.* Provincia de Buenos Aires. ◆ *province* E Argentina
65 *H19* **Buenos Aires, Lago** *var.* Lago General Carrera. ⊚ Argentina/Chile
56 *C13* **Buesaco** Nariño, SW Colombia 1.22N 77.07W
31 *U8* **Buffalo** Minnesota, N USA 45.10N 93.49W
28 *T6* **Buffalo** Missouri, C USA 37.38N 93.05W
29 *D10* **Buffalo** New York, NE USA 42.53N 78.52W
29 *K8* **Buffalo** Oklahoma, C USA 36.50N 99.37W
30 *J7* **Buffalo** South Dakota, N USA 45.35N 103.32W
27 *V8* **Buffalo** Texas, SW USA 31.25N 96.04W
35 *W12* **Buffalo** Wyoming, C USA 44.21N 106.40W
31 *U11* **Buffalo Center** Iowa, C USA 43.23N 93.57W
26 *M3* **Buffalo Lake** ⊡ Texas, SW USA
32 *K7* **Buffalo Lake** ⊡ Wisconsin, N USA
11 *S12* **Buffalo Narrows** Saskatchewan, C Canada 55.52N 108.28W
29 *U9* **Buffalo River** ♣ Arkansas, C USA
31 *R5* **Buffalo River** ♣ Minnesota, N USA
22 *I10* **Buffalo River** ♣ Tennessee, S USA
32 *J6* **Buffalo River** ♣ Wisconsin, N USA
46 *L12* **Buff Bay** E Jamaica 18.18N 76.40W
25 *T3* **Buford** Georgia, SE USA 34.07N 84.00W
30 *J3* **Buford** North Dakota, N USA 48.00N 103.58W
35 *Y17* **Buford** Wyoming, C USA 41.05N 105.17W
118 *J14* **Buftea** București, S Romania 44.34N 25.57E
148 *I9* **Bug** *Bel.* Zakhodni Buh, *Eng.* Western Bug, *Rus.* Zapadnyy Bug, *Ukr.* Zakhidnyy Buh. ♣ E Europe
56 *D11* **Buga** Valle del Cauca, W Colombia 3.52N 76.16W
168 *F7* **Buga** Dzavhan, N Mongolia 47.42N 94.53E
105 *O17* **Bugarach, Pic du** ▲ S France 42.52N 2.23E
152 *B12* **Bugdaýly** Balkanskiy Velayat, W Turkmenistan 38.42N 54.14E
Buggs Island Lake *see* John H.Kerr Reservoir
Bughotu *see* Santa Isabel
175 *Q12* **Bungkalo** Sulawesi, C Indonesia 4.49S 121.42E
66 *P6* **Bugio** Island Madeira, Portugal, NE Atlantic Ocean
94 *M8* **Bugøynes** Finnmark, N Norway 69.57N 29.34E
129 *Q3* **Bugrino** Nenetskiy Avtonomnyy Okrug, NW Russian Federation 68.48N 49.12E
131 *T5* **Bugul'ma** Respublika Tatarstan, W Russian Federation 54.31N 52.45E
Bügür *see* Luntai
131 *T6* **Buguruslan** Orenburgskaya Oblast', W Russian Federation 53.37N 52.30E
165 *R9* **Buh He** ♣ C China
35 *O15* **Buhl** Idaho, NW USA 42.36N 114.45W
103 *F22* **Bühl** Baden-Württemberg, SW Germany 48.42N 8.07E
118 *K10* **Buhuși** Bacău, E Romania 46.34N 26.53E
Buie d'Istria *see* Buje
99 *I21* **Builth Wells** E Wales, UK 52.07N 3.27W
195 *S13* **Buin** Bougainville Island, NE PNG 6.50S 155.42E
110 *I9* **Buin, Piz** ▲ Austria/Switzerland 46.51N 10.07E
131 *Q4* **Buinsk** Chuvashskaya Respublika, W Russian Federation 55.09N 47.00E
131 *Q4* **Buinsk** Respublika Tatarstan, W Russian Federation 54.58N 48.16E
169 *R8* **Buir Nur** *Mong.* Buyr Nuur. ⊚ China/Mongolia *see also* Buyr Nuur
100 *M5* **Buitenpost** *Fris.* Bûtenpost. Friesland, N Netherlands 53.15N 6.09E
Buitenzorg *see* Bogor
85 *F19* **Buitepos** Omaheke, E Namibia 22.17S 19.59E
107 *N7* **Buitrago del Lozoya** Madrid, C Spain 41.00N 3.38W
Buj *see* Buy

106 *M13* **Bujalance** Andalucía, S Spain 37.54N 4.22W
115 *O17* **Bujanovac** Serbia, SE Yugoslavia 42.29N 21.43E
107 *S6* **Bujaraloz** Aragón, NE Spain 41.28N 0.10W
114 *A9* **Buje** *It.* Buie d'Istria. Istra, NW Croatia 45.23N 13.40E
Bujnurd *see* Bojnūrd
83 *D21* **Bujumbura** *prev.* Usumbura. ● (Burundi) W Burundi 3.25S 29.23E
83 *D20* **Bujumbura** ✕ W Burundi 3.21S 29.19E
126 *L15* **Bukachacha** Chitinskaya Oblast', S Russian Federation 52.49N 116.89E
165 *N11* **Bukadaban Feng** ▲ C China 36.09N 90.52E
195 *R11* **Buka Island** *island* NE PNG
83 *F18* **Bukata** S Uganda 0.18S 31.57E
81 *N24* **Bukama** Katanga, SE Dem. Rep. Congo (Zaire) 9.13S 25.52E
148 *I4* **Būkān** *var.* Bowkān. Āzarbāyjān-e Bākhtarī, NW Iran 36.31N 46.14E
Bukantau, Gory *see* Bukantau-Toghi
152 *K8* **Bukantow-Toghi** *Rus.* Gory Bukantau. ▲ N Uzbekistan
Bukarest *see* București
81 *O19* **Bukavu** *prev.* Costermansville. Sud Kivu, E Dem. Rep. Congo (Zaire) 2.18S 28.49E
83 *F21* **Bukene** Tabora, NW Tanzania 4.15S 32.51E
147 *W8* **Bū Khābī** *var.* Bakhābī. NW Oman 23.28N 56.06E
Bukhara *see* Bukhoro
Bukharskaya Oblast' *see* Bukhoro Wiloyati
152 *L11* **Bukhoro** *var.* Bokhara, *Rus.* Bukhara. Bukhoro Wiloyati, C Uzbekistan 39.50N 64.22E
152 *J11* **Bukhoro Wiloyati** *Rus.* Bukharskaya Oblast'. ◆ *province* C Uzbekistan
174 *I12* **Bukittkemuning** Sumatera, W Indonesia 4.43S 104.27E
173 *G8* **Bukittinggi** *prev.* Fort de Kock. Sumatera, W Indonesia 0.18S 100.19E
113 *L21* **Bükk** ▲ NE Hungary
83 *F19* **Bukoba** Kagera, NW Tanzania 1.19S 31.49E
115 *N20* **Bukovo** S FYR Macedonia 40.59N 21.20E
110 *G6* **Bülach** Zürich, NW Switzerland 47.31N 8.30E
179 *R16* **Bülaevo** see Bulayevo
168 *I6* **Bulag** Hövsgöl, N Mongolia 49.51N 100.41E
168 *M7* **Bulag** Töv, C Mongolia 48.09N 108.33E
168 *I8* **Bulagiyn Denj** Arhangay, C Mongolia 47.14N 100.56E
191 *U7* **Bulahdelah** New South Wales, SE Australia 32.24S 152.13E
176 *Yy15* **Bulaka, Sungai** ♣ Irian Jaya, E Indonesia
179 *Qq12* **Bulan** Luzon, N Philippines 12.40N 123.53E
143 *N11* **Bulancak** Giresun, N Turkey 40.57N 38.13E
158 *J10* **Bulandshahr** Uttar Pradesh, N India 28.30N 77.49E
143 *R14* **Bulanık** Muş, E Turkey 39.04N 42.16E
131 *V7* **Bulanovo** Orenburgskaya Oblast', W Russian Federation 52.27N 55.08E
85 *J17* **Bulawayo** *var.* Buluwayo. Matabeleland North, SW Zimbabwe 20.08S 28.36E
85 *J17* **Bulawayo** ✕ Matabeleland North, SW Zimbabwe 20.08S 28.36E
151 *Q6* **Bulayevo** *Kaz.* Bülaevo. Severnyy Kazakhstan, N Kazakhstan
142 *D15* **Buldan** Denizli, SW Turkey 38.03N 28.49E
160 *G12* **Buldāna** Mahārāshtra, C India 20.31N 76.18E
40 *E16* **Buldir Island** *island* Aleutian Islands, Alaska, USA
Buldur *see* Burdur
168 *H9* **Bulgan** Bayanhongor, C Mongolia 44.48N 98.39E
168 *K6* **Bulgan** Bulgan, N Mongolia 50.31N 101.30E
168 *F7* **Bulgan** Hovd, W Mongolia 46.57N 93.40E
168 *J5* **Bulgan** Hövsgöl, N Mongolia 50.30N 101.28E
168 *J10* **Bulgan** Ömnögovi, S Mongolia 44.07N 103.28E
168 *H7* **Bulgan** ◆ *province* N Mongolia
116 *H10* **Bulgaria** *off.* Republic of Bulgaria, *Bul.* Bŭlgariya; *prev.* People's Republic of Bulgaria, ◆ *republic* SE Europe
Bulgariya *see* Bulgaria
116 *L9* **Bŭlgarka** ♣ E Bulgaria
175 *T7* **Buli** Pulau Halmahera, E Indonesia 0.56N 128.17E
175 *T7* **Buli, Teluk** *bay* Pulau Halmahera, E Indonesia
166 *J13* **Buliu He** ♣ S China
106 *M11* **Bullaque** ♣ C Spain
108 *B13* **Bullas** Murcia, SE Spain 38.01N 1.40W
82 *M12* **Bullaxaar** Woqooyi Galbeed, NW Somalia 10.28N 44.15E
110 *C9* **Bulle** Fribourg, SW Switzerland 46.37N 7.04E
193 *G15* **Buller** ♣ South Island, NZ
191 *P12* **Buller, Mount** ▲ Victoria, SE Australia 37.10S 146.31E
68 *H11* **Bullhead City** Arizona, SW USA 35.07N 114.32W
101 *N21* **Büllingen** *Fr.* Bullange. Liège, E Belgium 50.23N 6.15E
Bullion State *see* Missouri
23 *Q14* **Bull Island** *island* South Carolina, SE USA
190 *M4* **Bulloo River Overflow** *wetland* New South Wales, SE Australia
192 *M13* **Bulls** Manawatu-Wanganui, North Island, NZ 40.10S 175.22E
23 *T14* **Bulls Bay** *bay* South Carolina, SE USA
29 *U9* **Bull Shoals Lake** ⊡ Arkansas/Missouri, C USA
189 *Q2* **Bulman** Northern Territory, N Australia 13.39S 134.21E
168 *I6* **Bulnayn Nuruu** ▲ N Mongolia

194 *J13* **Bulolo** Morobe, C PNG 7.11S 146.34E
175 *Qq7* **Bulowa, Gunung** ▲ Sulawesi, N Indonesia 0.33N 123.39E
115 *L19* **Bulqizë** *var.* Bulqizë. C Albania 41.30N 20.16E
Bulsar *see* Valsād
175 *R7* **Buludawa Keten, Pegunungan** ▲ Sulawesi, N Indonesia
175 *Pp13* **Bulukumba** *prev.* Boeloekoemba. Sulawesi, C Indonesia 5.34S 120.13E
153 *O11* **Bulunghur** Rus. Bulungur; *prev.* Krasnogvardeysk. Samarqand Wiloyati, C Uzbekistan 39.46N 67.18E
81 *I21* **Bulungu** Bandundu, SW Dem. Rep. Congo (Zaire) 4.34S 18.33E
81 *I21* **Bulungu** *see* Bulungur
81 *K17* **Bumba** Equateur, N Dem. Rep. Congo (Zaire) 2.14N 22.25E
124 *O15* **Bumbah, Khalij al** *gulf* N Libya
168 *K8* **Bumbat** Övörhangay, C Mongolia 46.30N 104.08E
83 *F19* **Bumbire Island** *island* N Tanzania
175 *Oo4* **Bum Bun, Pulau** *island* East Malaysia
83 *J17* **Buna** North Eastern, NE Kenya 2.40N 39.34E
27 *Y10* **Buna** Texas, SW USA 30.25N 94.00W
Bunab *see* Bonāb
Bunai *see* M'bunai
153 *S13* **Bunay** S Tajikistan 38.29N 71.41E
188 *I13* **Bunbury** Western Australia 33.24S 115.43E
99 *E14* **Buncrana** *Ir.* Bun Cranncha. NW Ireland 55.07N 7.27W
189 *Z9* **Bundaberg** Queensland, E Australia 24.49S 152.16E
191 *T5* **Bundarra** New South Wales, SE Australia 30.12S 151.06E
102 *G13* **Bünde** Nordrhein-Westfalen, NW Germany 52.12N 8.34E
158 *H13* **Bundi** Madhya, N India 25.28N 75.42E
194 *I12* **Bundi** Madang, N PNG 5.40S 145.10E
Bundoran *Ir.* Bun Dobhráin *see* Bundoran
115 *J18* **Bunë** *SCr.* Bojana. ♣ Albania/Yugoslavia *see also* Bojana
179 *R16* **Bunga** ♣ Mindanao, S Philippines
173 *Ff10* **Bungalaut, Selat** *strait* W Indonesia
178 *Ii8* **Bung Kan** Nong Khai, E Thailand 18.19N 103.39E
189 *N4* **Bungle Bungle Range** ▲ Western Australia
84 *C10* **Bungo** Uíge, NW Angola 7.30S 15.24E
83 *G18* **Bungoma** Western, W Kenya 0.34N 34.34E
171 *O12* **Bungo-suidō** *strait* SW Japan
170 *Dd13* **Bungo-Takada** Ōita, Kyūshū, SW Japan 33.36N 131.28E
102 *K8* **Bungsberg** *hill* N Germany 54.12N 10.45E
Bungur *see* Bunyu
81 *P17* **Bunia** Orientale, NE Dem. Rep. Congo (Zaire) 1.33N 30.16E
37 *U6* **Bunker Hill** ▲ Nevada, W USA 39.16N 117.06W
24 *I7* **Bunkie** Louisiana, S USA 30.58N 92.12W
25 *X10* **Bunnell** Florida, SE USA 29.28N 81.15W
107 *S10* **Buñol** País Valenciano, E Spain 39.25N 0.46W
100 *K11* **Bunschoten** Utrecht, C Netherlands 52.15N 5.22E
178 *Oo5* **Bünyan** Kayseri, C Turkey 38.51N 35.49E
178 *Oo5* **Bunyu** *var.* Bungur. Borneo, N Indonesia 3.33N 117.50E
178 *Oo5* **Bunyu, Pulau** *island* N Indonesia
103 *P23* **Buodobohki** *see* Patoniva
126 *L16* **Buorkhaya Guba** *bay* N Russian Federation
176 *Z15* **Bupul** Irian Jaya, E Indonesia 7.24S 140.57E
83 *K19* **Bura** Coast, SE Kenya 1.06S 40.01E
83 *P12* **Buraan** Sanaag, N Somalia 10.03N 49.08E
Bûrabay *see* Borovoye
Buraida *see* Buraydah
Buraimi *see* Al Buraymī
151 *Y12* **Buran** Vostochnyy Kazakhstan, E Kazakhstan 48.00N 85.09E
164 *G15* **Burang** Xizang Zizhiqu, W China 30.28N 81.13E
Burao *see* Burco
144 *H8* **Buraq** Dar'ā, S Syria 33.10N 36.28E
147 *O6* **Buraydah** *var.* Buraida. Al Qaşīm, N Saudi Arabia 26.20N 43.59E
37 *S13* **Burbank** California, W USA 34.10N 118.19W
33 *N11* **Burbank** Illinois, N USA 41.45N 87.48W
191 *U8* **Burcher** New South Wales, SE Australia 33.29S 147.30E
82 *N12* **Burco** *var.* Burao, Bur'o. Togdheer, NW Somalia 9.29N 45.30E
82 *H13* **Burē** Oromo, C Ethiopia 8.13N 35.09E
95 *J15* **Bure** Västerbotten, N Sweden 64.36N 21.15E
103 *G14* **Büren** Nordrhein-Westfalen, W Germany 51.34N 8.34E
34 *L15* **Burns Junction** Oregon, NW USA

8 *L13* **Burns Lake** British Columbia, SW Canada 54.13N 125.45W
31 *V9* **Burnsville** Minnesota, N USA 44.49N 93.14W
23 *P9* **Burnsville** North Carolina, SE USA 35.55N 82.18W
23 *R4* **Burnsville** West Virginia, NE USA 38.50N 80.39W
12 *I13* **Burnt River** ♣ Ontario, S Canada
12 *I11* **Burnt River** ♣ Ontario, S Canada
11 *W12* **Burntwood** ♣ Manitoba, C Canada
116 *N10* **Burgas** *Bul.* Burgas, Burgas. ♦ E Bulgaria 42.31N 27.30E
116 *N9* **Burgas** ✕ Burgas, E Bulgaria 42.35N 27.33E
116 *M10* **Burgas** ◆ *province* E Bulgaria
116 *M10* **Burgaski Zaliv** *gulf* E Bulgaria
116 *M10* **Burgasko Ezero** *lagoon* E Bulgaria
23 *V11* **Burgaw** North Carolina, SE USA 34.33N 77.54W
110 *E8* **Burg bei Magdeburg** *see* Burg
110 *E8* **Burgdorf** Bern, NW Switzerland 47.03N 7.37E
111 *IY7* **Burgenland** *off.* Land Burgenland. ◆ *state* SE Austria
11 *S13* **Burgeo** Newfoundland, SE Canada 47.42N 57.29W
85 *I24* **Burgersdorp** Eastern Cape, SE South Africa 31.00S 26.20E
85 *K20* **Burgersfort** Mpumalanga, NE South Africa 24.39S 30.18E
103 *N23* **Burghausen** Bayern, SE Germany 48.10N 12.48E
145 *O5* **Burghūth, Sabkhat al** ⊚ E Syria
103 *M20* **Burglengenfeld** Bayern, SE Germany 49.11N 12.01E
43 *P9* **Burgos** Tamaulipas, C Mexico 24.55N 98.46W
107 *N4* **Burgos** Castilla-León, N Spain 42.21N 3.40W
107 *N4* **Burgos** ◆ *province* Castilla-León, N Spain
113 *J18* **Burgstadlberg** *see* Hradiště
97 *P20* **Burgsvik** Gotland, SE Sweden 57.01N 18.18E
Burgum *see* Bergum
Burgundy *see* Bourgogne
165 *Q11* **Burhan Budai Shan** ▲ C China
142 *B12* **Burhaniye** Balikesir, W Turkey 39.28N 26.58E
160 *G12* **Burhānpur** Madhya Pradesh, C India 21.18N 76.13E
179 *Q11* **Burias Island** *island* C Philippines
131 *W7* **Buribay** Respublika Bashkortostan, W Russian Federation 51.57N 58.11E
45 *O17* **Burica, Punta** *headland* Costa Rica/Panama 8.02N 82.53W
45 *Q5* **Buriram** var. Buri Ram, Puriramya. Buri Ram, E Thailand 15.01N 103.06E
107 *S10* **Burjassot** País Valenciano, E Spain 39.30N 0.25W
151 *S14* **Burubaytal** *prev.* Burylbaytal. Zhambyl, SE Kazakhstan 45.01N 73.58E
152 *X8* **Burkan** ♣ E Kyrgyzstan
27 *R4* **Burkburnett** Texas, SW USA 34.06N 98.34W
31 *O12* **Burke** South Dakota, S USA 43.09N 99.18W
8 *K15* **Burke Channel** *channel* British Columbia, W Canada
22 *L7* **Burkesville** Kentucky, S USA 36.47N 85.22W
189 *X8* **Burketown** Queensland, NE Australia 17.48S 139.28E
27 *Q8* **Burkett** Texas, SW USA 32.01N 99.17W
27 *T9* **Burkeville** Texas, SW USA 30.58N 93.41W
23 *V7* **Burkeville** Virginia, NE USA 37.11N 78.12W
79 *O12* **Burkina** *off.* Burkina Faso; *prev.* Upper Volta. ◆ *republic* W Africa
Burkina Faso *see* Burkina
204 *L13* **Burks, Cape** *headland* Antarctica
12 *H7* **Burk's Falls** Ontario, S Canada 45.38N 79.25W
103 *P20* **Burladingen** Baden-Württemberg, S Germany 48.18N 9.05E
27 *T7* **Burleson** Texas, SW USA 32.32N 97.19W
35 *P15* **Burley** Idaho, NW USA 42.32N 113.47W
150 *G8* **Burlin** Zapadnyy Kazakhstan, NW Kazakhstan 51.25N 52.42E
33 *S13* **Burlington** Ontario, S Canada 43.21N 79.45W
39 *W4* **Burlington** Colorado, C USA 39.16N 102.16W
31 *Y15* **Burlington** Iowa, C USA 40.48N 91.05W
29 *P5* **Burlington** Kansas, C USA 38.11N 95.44W
23 *T9* **Burlington** North Carolina, SE USA 36.06N 79.26W
30 *M3* **Burlington** North Dakota, N USA 48.16N 101.25W
20 *L7* **Burlington** Vermont, NE USA 44.28N 73.13W
32 *M7* **Burlington** Wisconsin, N USA 42.38N 88.12W
29 *Q1* **Burlington Junction** Missouri, C USA 40.27N 95.04W
Burma *see* Myanmar
8 *L17* **Burnaby** British Columbia, SW Canada 49.16N 122.58W
191 *N16* **Burnie** Tasmania, SE Australia 41.06S 145.52E
99 *L17* **Burnley** NW England, UK 53.48N 2.13W
34 *L15* **Burney** California, W USA 40.52N 121.42W
191 *O16* **Burnie** Tasmania, SE Australia 41.06S 145.52E
144 *H9* **Burnie** NW England, UK
188 *I13* **Burney** see Burney
159 *R15* **Burnpur** West Bengal, NE India 23.39N 86.55E
116 *K14* **Burnas, Ozero** ⊚ SW Ukraine
27 *S10* **Burnet** Texas, SW USA 30.45N 98.13W
37 *O3* **Burney** California, W USA 40.52N 121.42W
191 *N16* **Burnie** Tasmania, SE Australia 41.06S 145.52E
113 *M15* **Burns-Zdrój** Świętokrzyskie, C Poland 50.52N 121.42W
191 *O16* **Busra** *see* Al Başrah
144 *H9* **Buşrá ash Shām** *var.* Bosora, Bosra, Bozrah, Buşrá. ♣ S Syria 32.31N 36.31E
191 *N16* **Burnye** see Bisevo
188 *I13* **Bustarek** Western Australia 33.43S 115.15E
83 *O14* **Buyo** SW Sudan
169 *R7* **Buyr Nuur** *Mong.* Buir Nur. ⊚ China/Mongolia *see also* Buir Nuur

8 *L13* **Bustan** *see* Büston
31 *J12* **Büsteni** Prahova, SE Romania 45.23N 25.31E
108 *D7* **Busto Arsizio** Lombardia, N Italy 45.37N 8.49E
153 *Q10* **Büston** *Rus.* Buston. Qoraqalpoghiston Respublikasi, W Uzbekistan 41.49N 60.51E
179 *P12* **Busuanga Island** *island* Calamian Group, W Philippines
102 *H8* **Büsum** Schleswig-Holstein, N Germany 54.08N 8.52E
81 *M16* **Buta** Orientale, N Dem. Rep. Congo (Zaire) 2.50N 24.41E
83 *E20* **Butare** *prev.* Astrida. S Rwanda 2.39S 29.44E
203 *O2* **Butaritari** *atoll* Tungaru, W Kiribati
Butawal *see* Butwal
98 *K13* **Bute** *cultural region* SW Scotland, UK
98 *H13* **Buteleyn Nuruu** ▲ N Mongolia
6 *L16* **Bute Inlet** *fjord* British Columbia, W Canada
83 *P18* **Butembo** Nord Kivu, NE Dem. Rep. Congo (Zaire) 0.09N 29.17E
109 *K25* **Butera** Sicilia, Italy, C Mediterranean Sea 37.12N 14.12E
101 *M20* **Bütgenbach** Liège, E Belgium 50.25N 6.12E
42 *M5* **Burro, Serranías del** ▲ NW Mexico
64 *K7* **Burruyacú** Tucumán, N Argentina 26.28S 64.30W
142 *E12* **Bursa** *var.* Brusa, *anc.* Prusa. Bursa, NW Turkey 40.12N 29.04E
142 *D12* **Bursa** *var.* Brusa, Brussa. ◆ *province* NW Turkey
77 *Y9* **Bûr Safâga** *var.* Bûr Safâjah. E Egypt 26.41N 33.58E
Bûr Safâjah *see* Bûr Safâga
Bûr Sa'id *see* Port Said
93 *O14* **Bûr Tînle** Mudug, C Somalia 7.50N 48.01E
25 *O5* **Burt Lake** ⊚ Michigan, N USA
120 *H7* **Burtnieks Ezers** *var.* Burtnieks. ⊚ N Latvia
25 *Q8* **Burton** Michigan, N USA 43.00N 84.16W
Burton on Trent *see* Burton upon Trent
99 *M19* **Burton upon Trent** *var.* Burton on Trent, Burton-upon-Trent. C England, UK 52.48N 1.36W
25 *N3* **Burträsk** Västerbotten, N Sweden 64.31N 20.40E
31 *Q2* **Butte** Montana, C USA 46.01N 112.33W
31 *Q2* **Butte** Nebraska, C USA 42.54N 98.51W
173 *G3* **Butterworth** Pinang, Peninsular Malaysia 5.24N 100.22E
85 *J25* **Butterworth** *var.* Gcuwa. Eastern Cape, SE South Africa 32.19S 28.09E
121 *I17* **Byelaruskaya Hrada** *Rus.* Belorusskaya Gryada. *ridge* C Belarus
121 *G18* **Byelavyezhskaya Pushcha** *Pol.* Puszcza Białowieska, *Rus.* Belovezhskaya Pushcha. *forest* Belarus/Poland *see also* Białowieża, Puszcza
121 *H15* **Byenyakoni** *Rus.* Benyakoni. Hrodzyenskaya Voblasts', W Belarus 54.12N 25.32E
121 *J14* **Byerazino** *Rus.* Berezino. Minskaya Voblasts', C Belarus 53.49N 28.58E
121 *L20* **Byerazino** *Rus.* Berezina. ♣ C Belarus
120 *M13* **Byeshankovichy** *Rus.* Beshenkovichi. Vitsyebskaya Voblasts', N Belarus 55.03N 29.28E
33 *U13* **Byesville** Ohio, N USA 39.58N 81.32W
121 *P18* **Byezdzyezh** *Rus.* Bezdezh. Brestskaya Voblasts', SW Belarus 52.18N 25.16E
95 *I15* **Bygdeå** Västerbotten, N Sweden 64.03N 20.49E
96 *F12* **Bygdøy** ◇ S Norway
95 *J15* **Bygdsiljum** Västerbotten, N Sweden 64.20N 20.30E
97 *D18* **Bygland** Aust-Agder, S Norway 58.46N 7.50E
97 *E18* **Byglandsfjord** Aust-Agder, S Norway 58.40N 7.50E
121 *M19* **Bykhaw** *Rus.* Bykhov. Mahilyowskaya Voblasts', E Belarus 53.31N 30.15E
Bykhov *see* Bykhaw
131 *Q9* **Bykovo** Volgogradskaya Oblast', SW Russian Federation 49.52N 45.24E
126 *L6* **Bykovskiy** Respublika Sakha (Yakutiya), NE Russian Federation 71.57N 129.07E
12 *K10* **Byrd Glacier** *glacier* Antarctica
205 *R12* **Byrd, Lac** ◇ Quebec, SE Canada
191 *S11* **Byrock** New South Wales, SE Australia 30.40S 146.24E
29 *O4* **Byron** Illinois, N USA
191 *V4* **Byron Bay** New South Wales, E Australia 28.39S 153.34E
191 *V4* **Byron, Cape** *headland* E Australia 28.37S 153.40E
65 *I23* **Byron, Isla** *island* S Chile
67 *B24* **Byron Sound** *sound* NW Falkland Islands
126 *J5* **Byrranga, Gora** ▲ N Russian Federation
95 *I14* **Byske** Västerbotten, N Sweden 64.58N 21.10E
113 *K18* **Bystrá** ▲ N Slovakia 49.10N 19.49E
113 *F18* **Bystrice nad Pernštejn** *Ger.* Bistritz ober Pernstein. Jihlavský Kraj, C Czech Republic 49.30N 16.16E
113 *J18* **Bytča** *Žilinský* Kraj, N Slovakia 49.15N 18.31E

121 L15 **Bytcha** *Rus.* Bytcha. Minskaya Voblasts', NE Belarus 54.19N 28.24E

Byteń/Byten' *see* Bytsyen'

113 J16 **Bytom** *Ger.* Beuthen. Śląskie, S Poland 50.21N 18.51E

112 H7 **Bytów** *Ger.* Bütow. Pomorskie, N Poland 54.09N 17.30E

121 H18 **Bytsyen'** *Pol.* Byteń, *Rus.* Byten'. Brestskaya Voblasts', SW Belarus 52.53N 25.32E

83 E19 **Byumba** *var.* Biumba. N Rwanda 1.37S 30.05E

152 F13 **Byuzmeyin** *Turkm.* Büzmeýin; *prev.* Bezmein. Akhalskiy Velayat, C Turkmenistan 38.07N 57.52E

121 O20 **Byval'ki** Homyel'skaya Voblasts', SE Belarus 51.51N 30.37E

97 O20 **Byxelkrok** Kalmar, S Sweden 57.18N 17.01E

Byzantium *see* İstanbul

Bzimah *see* Buzaymah

C

64 O6 **Caacupé** Cordillera, S Paraguay 25.22S 57.04W

64 P6 **Caaguazú** *off.* Departamento de Caaguazú. ◆ *department* C Paraguay

84 C13 **Caála** *var.* Kaala, Robert Williams, *Port.* Vila Robert Williams. Huambo, C Angola 12.51S 15.33E

64 P7 **Caazapá** Caazapá, S Paraguay 26.09S 56.21W

64 P7 **Caazapá** *off.* Departamento de Caazapá. ◆ *department* SE Paraguay

83 P15 **Cabaad, Raas** *headland* C Somalia 6.13N 49.01E

179 R14 **Cabadbaran** Mindanao, S Philippines 9.07N 125.34E

57 N10 **Cabadisocaña** Amazonas, S Venezuela 4.28N 64.45W

46 F5 **Cabaiguán** Sancti Spíritus, C Cuba 22.04N 79.31W

Caballeria, Cabo *see* Cavalleria, Cap de

39 Q14 **Caballo Reservoir** ☐ New Mexico, SW USA

42 L6 **Caballos Mesteños, Llano de los** *plain* N Mexico

106 G12 **Cabañaquinta** Asturias, N Spain 43.10N 5.37W

44 B9 **Cabañas** ◆ *department* E El Salvador

179 P10 **Cabanatuan** *off.* Cabanatuan City. Luzon, N Philippines 15.27N 120.57E

13 T8 **Cabano** Quebec, SE Canada 47.40N 68.55W

106 L11 **Cabeza del Buey** Extremadura, W Spain 38.43N 5.13W

47 V5 **Cabezas de San Juan** *headland* E Puerto Rico 18.23N 65.37W

107 N2 **Cabezón de la Sal** Cantabria, N Spain 43.19N 4.13W

Cabhán *see* Cavan

63 B23 **Cabildo** Buenos Aires, E Argentina 38.28S 61.49W

Cabillonum *see* Chalon-sur-Saône

56 H5 **Cabimas** Zulia, NW Venezuela 10.25N 71.27W

84 A9 **Cabinda** *var.* Kabinda. Cabinda, NW Angola 5.34S 12.12E

84 A9 **Cabinda** *var.* Kabinda. ◆ *province* NW Angola

35 N7 **Cabinet Mountains** ▲ Idaho/Montana, NW USA

84 B11 **Cabiri** Bengo, NW Angola 8.50S 13.42E

83 J20 **Cabo Blanco** Santa Cruz, SE Argentina 47.13S 65.43W

84 P13 **Cabo Delgado** *off.* Província de Cabo Delgado. ◆ *province* NE Mozambique

12 L9 **Cabonga, Réservoir** ☐ Quebec, SE Canada

29 V7 **Cabool** Missouri, C USA 37.07N 92.06W

191 V2 **Cabonture** Queensland, E Australia 27.05S 152.56E

Cabora Bassa, Lake *see* Cahora Bassa, Albufeira de

42 F3 **Caborca** Sonora, NW Mexico 30.44N 112.06W

Cabo San Lucas *see* San Lucas

29 V11 **Cabot** Arkansas, C USA 34.58N 92.01W

12 F12 **Cabot Head** *headland* Ontario, S Canada 45.13N 81.17W

16 S12 **Cabot Strait** *strait* E Canada

Cabo Verde, Ilhas do *see* Cape Verde

106 M14 **Cabra** Andalucía, S Spain 37.28N 4.28W

109 B19 **Cabras** Sardegna, Italy, C Mediterranean Sea 39.55N 8.30E

196 A15 **Cabras Island** *island* W Guam

47 O8 **Cabrera** N Dominican Republic 19.34N 69.55W

107 X10 **Cabrera** *anc.* Capraria. *island* Islas Baleares, W Mediterranean Sea

106 J4 **Cabrera** ≈ NW Spain

107 Q15 **Cabrera, Sierra** ▲ S Spain

11 S16 **Cabri** Saskatchewan, S Canada 50.37N 108.28W

107 R10 **Cabriel** ≈ E Spain

56 M7 **Cabruta** Guárico, C Venezuela 7.39N 66.19W

179 O08 **Cabugao** Luzon, N Philippines 17.55N 120.29E

56 G10 **Cabuyaro** Meta, C Colombia 4.16N 72.47W

62 I13 **Caçador** Santa Catarina, S Brazil 26.47S 51.00W

44 G8 **Cacaguatique, Cordillera** *var.* Cordillera. ▲ NE El Salvador

114 L13 **Čačak** Serbia, C Yugoslavia 43.52N 20.23E

57 Y10 **Cacao** NE French Guiana 4.33N 52.26W

83 H16 **Caçapava do Sul** Rio Grande do Sul, S Brazil 30.28S 53.28W

33 U3 **Capon River** ≈ West Virginia, NE USA

109 J23 **Caccamo** Sicilia, Italy, C Mediterranean Sea 37.55N 13.40E

109 A17 **Caccia, Capo** *headland* Sardegna, Italy, C Mediterranean Sea 40.34N 8.09E

61 G18 **Cáceres** Mato Grosso, W Brazil 16.04S 57.40W

106 J10 **Cáceres** *Ar.* Qazris. Extremadura, W Spain 39.28N 6.22W

106 J9 **Cáceres** ◆ *province* Extremadura, W Spain

Cachacrou *see* Scotts Head Village

63 C21 **Cacharí** Buenos Aires, E Argentina 36.24S 59.31W

28 L12 **Cache** Oklahoma, C USA 34.37N 98.37W

8 M16 **Cache Creek** British Columbia, SW Canada 50.49N 121.19W

37 N6 **Cache Creek** ≈ California, W USA

39 S3 **Cache La Poudre River** ≈ Colorado, C USA

Cacheo *see* Cacheu

29 W11 **Cache River** ≈ Arkansas, C USA

32 L17 **Cache River** ≈ Illinois, N USA

78 G12 **Cacheu, var.** Cacheo. W Guinea-Bissau 12.12N 16.10W

61 I15 **Cachimbo** Pará, NE Brazil 9.21S 54.58W

85 L15 **Cachina Bassa, Albufeira de var.** Lake Cabora Bassa. NW Mozambique

99 G20 **Cahore Point** *Ir.* Rinn Chathóir. *headland* SE Ireland 52.33N 6.11W

84 C11 **Cachingues** Bié, C Angola 13.05S 16.48E

56 L2 **Cáchira** Norte de Santander, N Colombia 7.46N 73.03W

63 H18 **Cachoeira do Sul** Rio Grande do Sul, S Brazil 29.58S 52.54W

61 O20 **Cachoeiro de Itapemirim** Espírito Santo, SE Brazil 20.51S 41.07W

84 E12 **Cacolo** Lunda Sul, NE Angola 10.09S 19.17E

85 C14 **Caconda** Huíla, C Angola 13.43S 15.03E

84 A9 **Cacongo** Cabinda, NW Angola 5.16S 12.10E

37 U9 **Cactus Peak** ▲ Nevada, W USA 37.42N 116.51W

84 A11 **Cacuaco** Luanda, NW Angola 8.49S 13.24E

85 B14 **Cacula** Huíla, SW Angola 14.31S 14.07E

69 R12 **Cacuchi** ≈ SW Angola

61 O19 **Caçumba, Ilha** *island* SE Brazil

57 N10 **Cacuri** Amazonas, S Venezuela

83 N17 **Cadale** Shabeellaha Dhexe, E Somalia 2.48N 46.19E

107 X4 **Cadaqués** Cataluña, NE Spain 42.16N 3.16E

113 J18 **Čadca** *Hung.* Csaca. Žilinský Kraj, N Slovakia 49.27N 18.46E

29 P13 **Caddo** Oklahoma, C USA 34.07N 96.15W

27 R6 **Caddo** Texas, SW USA 32.42N 98.40W

27 X6 **Caddo Lake** ☐ Louisiana/Texas, SW USA

29 S12 **Caddo Mountains** ▲ Arkansas, C USA

43 O8 **Cadereyta** Nuevo León, NE Mexico 25.35N 99.54W

12 I7 **Cadillac** Quebec, SE Canada 48.12N 78.23W

11 T17 **Cadillac** Saskatchewan, S Canada 49.43N 107.41W

104 K13 **Cadillac** Gironde, SW France 44.37N 0.16W

31 Q8 **Cadillac** Michigan, N USA 44.16N 85.22W

107 V4 **Cadí, Torre de** ▲ NE Spain 42.16N 1.38E

12 I7 **Cadiz** *off.* Cadiz City. Negros, C Philippines 10.58N 123.18E

22 H7 **Cadiz** Kentucky, S USA 36.53N 87.49W

33 U13 **Cadiz** Ohio, N USA 40.16N 81.00W

106 J15 **Cádiz** *anc.* Gades, Gadier, Gadir, Gadire. Andalucía, SW Spain 36.32N 6.18W

106 K15 **Cádiz** ◆ *province* Andalucía, SW Spain

106 I15 **Cadiz, Bahía de** *bay* SW Spain

106 I15 **Cádiz, Golfo de** *Eng.* Gulf of Cadiz. *gulf* Portugal/Spain

Cadiz, Gulf of *see* Cádiz, Golfo de

37 X14 **Cadiz Lake** ☐ California, W USA

190 E2 **Cadney Homestead** South Australia 27.52S 134.03E

Cadurcum *see* Cahors

85 F17 **Caecae** Ngamiland, NW Botswana 19.52S 21.04E

104 K4 **Caen** Calvados, N France 49.10N 0.19W

Caene/Caenepolis *see* Qena

Caerdydd *see* Cardiff

Caer Glou *see* Gloucester

Caer Gybi *see* Holyhead

Caerleon *see* Chester

Caer Luel *see* Carlisle

99 J18 **Caernarfon** *var.* Caernarvon, Carnarvon. NW Wales, UK 53.07N 4.16W

99 H18 **Caernarfon Bay** *bay* NW Wales, UK

99 J19 **Caernarvon** *cultural region* NW Wales, UK

Caernarvon *see* Caernarfon

Caesaraugusta *see* Zaragoza

Caesarea Mazaca *see* Kayseri

Caesarobriga *see* Talavera de la Reina

Caesarodunum *see* Tours

Caesaromagus *see* Beauvais

61 N17 **Caetité** Bahia, E Brazil 14.04S 42.28W

64 J6 **Cafayate** Salta, N Argentina 26.02S 66.00W

179 P9 **Cagayan** ≈ Luzon, N Philippines

179 R15 **Cagayan de Oro** *off.* Cagayan de Oro City. Mindanao, S Philippines 8.28N 124.38E

179 O017 **Cagayan de Tawi Tawi** *island* S Philippines

179 Pp14 **Cagayan Islands** *island group* C Philippines

33 O14 **Cagles Mill Lake** ☐ Indiana, N USA

118 G12 **Cagli** Marche, C Italy 43.33N 12.39E

109 C20 **Cagliari** *anc.* Caralis. Sardegna, Italy, C Mediterranean Sea 39.15N 9.06E

109 C20 **Cagliari, Golfo di** *gulf* Sardegna, Italy, C Mediterranean Sea

105 U15 **Cagnes-sur-Mer** Alpes-Maritimes, SE France 43.40N 7.09E

56 L5 **Cagua** Aragua, N Venezuela 10.09N 67.27W

179 Pp8 **Cagua, Mount** ▲ Luzon, N Philippines 18.10N 122.03E

56 F13 **Caguas** E Puerto Rico 18.13N 66.02W

47 U6 **Caguas** E Puerto Rico

44 A5 **Cahabón, Río** ≈ C Guatemala

85 B15 **Cahama** Cunene, SW Angola 16.16S 14.19E

99 B21 **Caha Mountains** *Ir.* An Cheacha. ▲ SW Ireland

99 D20 **Caher** *Ir.* An Cathair. S Ireland 52.21N 7.58W

99 A21 **Cahersiveen** *Ir.* Cathair Saidhbhín. SW Ireland 51.56N 10.12W

32 K15 **Cahokia** Illinois, N USA 38.34N 90.11W

85 L15 **Cahora Bassa, Albufeira de var.** Lake Cabora Bassa. NW Mozambique

99 G20 **Cahore Point** *Ir.* Rinn Chathóir. *headland* SE Ireland 52.33N 6.11W

104 M14 **Cahors** *anc.* Cadurcum. Lot, S France 44.26N 1.27E

58 D9 **Cahuapanas, Río** ≈ N Peru

118 M12 **Cahul, var.** Kagul. S Moldova 45.52N 28.13E

85 N16 **Caia** Sofala, C Mozambique 17.51S 35.22E

61 I13 **Caiapó, Serra do** ▲ C Brazil

46 F5 **Caibarién** Villa Clara, C Cuba 22.31N 79.28W

57 O5 **Caicara de Orinoco** Bolívar, C Venezuela 7.38N 66.10W

61 P14 **Caicó** Rio Grande do Norte, E Brazil 6.25S 37.04W

46 M6 **Caicos Islands** *island group* W Turks and Caicos Islands

46 L5 **Caicos Passage** *strait* Bahamas/Turks and Caicos Islands

167 O19 **Caidian** *prev.* Hanyang. Hubei, C China 30.33N 114.03E

Caiffa *see* Hefa

188 M12 **Caiguna** Western Australia 32.14S 125.33E

42 J1 **Caimanero, Laguna del var.** Laguna del Camaronero. *lagoon* E Pacific Ocean

119 N10 **Căinari** *Rus.* Kaynary. C Moldova 46.43N 29.00E

59 L19 **Caine, Río** ≈ C Bolivia

29 N8 **Caiphas** *see* Hefa

205 N4 **Caird Coast** *physical region* Antarctica

98 I11 **Cairn Gorm** ▲ C Scotland, UK 57.07N 3.38W

98 J9 **Cairngorm Mountains** ▲ C Scotland, UK

41 T12 **Cairn Mountain** ▲ Alaska, USA 61.07N 155.23W

189 W4 **Cairns** Queensland, NE Australia 16.51S 145.43E

124 Qq16 **Cairo** *Ar.* Al Qāhirah, *var.* El Qāhira. ● (Egypt) N Egypt 30.01N 31.18E

25 T8 **Cairo** Georgia, SE USA 30.52N 84.12W

32 L17 **Cairo** Illinois, N USA 37.00N 89.10W

77 W8 **Cairo** ✈ C Egypt 30.06N 31.36E

Caiseal *see* Cashel

Caisleán an Bharraigh *see* Castlebar

Caisleán na Finne *see* Castlefinn

98 J6 **Caithness** *cultural region* N Scotland, UK

85 D15 **Caiundo** Cuando Cubango, S Angola 15.41S 17.28E

106 K15 **Cádiz** ◆ *province* Andalucía, SW Spain

58 B11 **Cajamarca** *var.* Caxamarca. Cajamarca, NW Peru 7.09S 78.31W

58 B11 **Cajamarca** *off.* Departamento de Cajamarca. ◆ *department* N Peru

105 X14 **Cajarc** Lot, S France 44.28N 1.51E

179 Q12 **Cajidiocan** Sibuyan Island, C Philippines 12.20N 122.39E

44 G6 **Cajón, Represa El** ☐ NW Honduras

60 L10 **Caju, Ilha do** *island* NE Brazil

165 R10 **Caka Yanhu** ☐ C China

114 E7 **Čakovec** *Ger.* Csakathurn, *Hung.* Csáktornya; *prev.* Ger. Tschakathurn. Medimurje, N Croatia 46.24N 16.29E

79 V17 **Calabar** Cross River, S Nigeria 4.55N 8.25E

12 K13 **Calabogie** Ontario, SE Canada 45.18N 76.43W

56 M10 **Calabozo** Guárico, C Venezuela 8.53N 67.28W

119 N20 **Calabria** *anc.* Bruttium. ◆ *region* SW Italy

106 M16 **Calaburra, Punta de** *headland* S Spain 36.30N 4.38W

118 G14 **Calafat** Dolj, SW Romania 43.55N 23.01E

Calafate *see* El Calafate

73 Q4 **Calahorra** La Rioja, N Spain 42.19N 1.58W

105 N1 **Calais** Pas-de-Calais, N France 51.00N 1.53E

21 T8 **Calais** Maine, NE USA 45.09N 67.15W

Calais, Pas de *see* Dover, Strait of

Calalen *see* Kallalen

143 V13 **Cālilabad** *Rus.* Dzhalilabad; *prev.* Astrakhan-Bazar. S Azerbaijan 39.15N 48.30E

118 I12 **Călărași** *var.* Călăras, *Rus.* Kalarash. C Moldova 47.19N 28.13E

118 L14 **Călărași** Călărași, SE Romania 44.18N 26.52E

118 K14 **Călărași** ◆ *county* SE Romania

56 E10 **Calarca** Quindío, W Colombia 4.31N 75.37W

107 Q12 **Calasparra** Murcia, SE Spain 38.13N 1.40W

109 J24 **Calatafimi** Sicilia, Italy, C Mediterranean Sea 37.54N 12.52E

107 Q6 **Calatayud** Aragón, NE Spain 41.21N 1.37W

99 B21 **Caha Mountains** *Ir.* An Cheacha. ▲ SW Ireland

179 Pp11 **Calauag** Luzon, N Philippines 13.57N 122.18E

37 P8 **Calaveras River** ≈ California, W USA

179 Oo11 **Calavite, Cape** *headland* Mindoro, N Philippines 13.25N 120.16E

179 Qq12 **Calbayog** *off.* Calbayog City. Samar, C Philippines 12.07N 124.35E

179 R12 **Calbiga** Samar, C Philippines 11.37N 125.00E

24 G5 **Calcasieu Lake** ☐ Louisiana, S USA

24 H8 **Calcasieu River** ≈ Louisiana, S USA

58 B6 **Calceta** Manabí, W Ecuador 0.51S 80.09W

63 B18 **Calchaquí** Santa Fe, C Argentina 29.55S 60.13W

64 J6 **Calchaquí, Río** ≈ NW Argentina

60 I10 **Calçoene** Amapá, NE Brazil 2.29N 51.01W

159 S16 **Calcutta** West Bengal, NE India 22.30N 88.19E

159 S16 **Calcutta** ✈ West Bengal, N India 22.30N 88.19E

56 E9 **Caldas** *off.* Departamento de Caldas. ◆ *province* W Colombia

106 F10 **Caldas da Rainha** Leiria, W Portugal 39.24N 9.07W

106 G3 **Caldas de Reis** *var.* Caldas de Reyes. Galicia, NW Spain 42.36N 8.39W

Caldas de Reyes *see* Caldas de Reis

60 F11 **Caldeirão** Amazonas, NW Brazil 3.18S 60.22W

64 G7 **Caldera** Atacama, N Chile 27.04S 70.48W

44 L14 **Caldera** Puntarenas, W Costa Rica 9.55N 84.42W

107 N13 **Calderina** ▲ C Spain 39.18N 3.49W

143 T13 **Calderan** Van, E Turkey 39.10N 43.52E

34 M14 **Caldwell** Idaho, NW USA 43.39N 116.41W

29 N8 **Caldwell** Kansas, C USA 37.01N 97.36W

12 G16 **Caledon** Ontario, S Canada 43.51N 79.58W

85 I23 **Caledon** *var.* Mohokare. ≈ Lesotho/South Africa

44 G1 **Caledonia** Corozal, N Belize 18.13N 88.27W

12 G16 **Caledonia** Ontario, S Canada 43.04N 79.57W

31 X11 **Caledonia** Minnesota, N USA 43.37N 91.30W

107 X3 **Calella** *var.* Calella de la Costa. Cataluña, NE Spain 41.37N 2.40E

Calella de la Costa *see* Calella

25 P4 **Calera** Alabama, S USA 33.06N 86.45W

65 U19 **Caleta Olivia** Santa Cruz, SE Argentina 46.21S 67.37W

37 X17 **Calexico** California, W USA 32.39N 115.28W

99 H16 **Calf of Man** *island* SW Isle of Man

11 Q16 **Calgary** Alberta, SW Canada 51.04N 114.04W

11 Q16 **Calgary** ✈ Alberta, SW Canada 51.15N 114.03W

39 O5 **Calhan** Colorado, C USA 39.00N 104.18W

25 R4 **Calhoun** Georgia, SE USA 34.30N 84.57W

22 H6 **Calhoun** Kentucky, S USA 37.32N 87.10W

24 J6 **Calhoun City** Mississippi, S USA 33.51N 89.18W

25 R9 **Calhoun Falls** South Carolina, SE USA 34.05N 82.36W

56 D7 **Cali** Valle del Cauca, W Colombia 3.24N 76.30W

29 V9 **Calico Rock** Arkansas, C USA 36.07N 92.08W

161 F21 **Calicut** *var.* Kozhikode. Kerala, SW India 11.17N 75.49E

37 T9 **Caliente** Nevada, W USA 37.37N 114.30W

29 U5 **California** Missouri, C USA 38.38N 92.33W

18 B15 **California** Pennsylvania, NE USA 40.02N 79.52W

37 Q12 **California** *off.* State of California; *also known as* El Dorado, The Golden State. ◆ *state* W USA

37 P11 **California Aqueduct** *aqueduct* California, W USA

37 T13 **California City** California, W USA 35.06N 117.55W

42 F6 **California, Golfo de** *Eng.* Gulf of California; *prev.* Sea of Cortez. *gulf* W Mexico

California, Gulf of *see* California, Golfo de

143 V13 **Cālilabad** *Rus.* Dzhalilabad; *prev.* Astrakhan-Bazar. S Azerbaijan 39.15N 48.30E

118 I12 **Călărași** *var.* Călăras, *Rus.* Kalarash. C Moldova 47.19N 28.13E

111 J16 **Calitzdorp** Western Cape, SW South Africa 33.46N 21.59E

43 U12 **Calkiní** Campeche, E Mexico 20.21N 90.03W

190 K4 **Callabonna Creek** *var.* Tilcha Creek. *seasonal river* New South Wales/South Australia

190 J4 **Callabonna, Lake** ☐ South Australia

104 G5 **Callac** Côtes d'Armor, NW France 48.28N 3.22W

37 U5 **Callaghan, Mount** ▲ Nevada, W USA 39.38N 116.57W

Callan *see* Callan

12 I11 **Callan** *Ir.* Callainn. S Ireland 52.33N 7.22W

98 I11 **Callander** Ontario, S Canada 38.13N 1.40W

100 H7 **Callander** C Scotland, UK 56.14N 4.16W

179 P10 **Callantsoog** Noord-Holland, NW Netherlands 52.51N 4.41E

59 D14 **Callao** Callao, W Peru 12.03S 77.09W

59 D14 **Callao** *Departamento del Callao.* ◆ *constitutional province* W Peru

58 F11 **Callaria, Río** ≈ E Peru

11 Q13 **Calling Lake** Alberta, W Canada 55.12N 113.07W

107 T11 **Callosa de Ensarriá** *Callosa d'En Sarriá*

107 S12 **Callosa d'En Sarriá var.** Callosa de Ensarriá. País Valenciano, E Spain 38.40N 0.07W

107 S12 **Callosa de Segura** País Valenciano, E Spain 38.07N 0.52W

31 X11 **Calmar** Iowa, C USA 43.10N 91.51W

Calmar *see* Kalmar

45 R16 **Calobre** Veraguas, C Panama 8.18N 80.49W

179 P10 **Caloocan** *municipality* Luzon, N Philippines 14.38N 120.58E

25 X14 **Caloosahatchee River** ≈ Florida, SE USA

191 V2 **Caloundra** Queensland, E Australia 26.48S 153.07E

107 T11 **Calpe** País Valenciano, E Spain 38.39N 0.03E

43 P14 **Calpulalpan** Tlaxcala, S Mexico 19.36N 98.33W

109 K25 **Caltagirone** Sicilia, Italy, C Mediterranean Sea 37.13N 14.31E

109 J24 **Caltanissetta** Sicilia, Italy, C Mediterranean Sea 37.30N 14.00E

84 C11 **Caluango** Lunda Norte, NE Angola 8.16S 19.38E

84 C13 **Calucinga** Bié, W Angola 11.18S 16.10E

84 C12 **Calulo** Cuanza Sul, NW Angola 9.58S 14.56E

85 B14 **Caluquembe** Huíla, W Angola 13.46S 14.40E

82 Q9 **Caluula** Bari, NE Somalia 11.55N 50.51E

104 K4 **Calvados** ◆ *department* N France

195 P17 **Calvados Chain, The** *island group* SE PNG

27 V7 **Calvert** Texas, SW USA 30.58N 96.40W

22 H7 **Calvert City** Kentucky, S USA 37.01N 88.21W

105 X14 **Calvi** Corse, France, C Mediterranean Sea 42.34N 8.44E

42 I12 **Calvillo** Aguascalientes, C Mexico 21.51N 102.42W

85 E23 **Calvinia** Northern Cape, W South Africa 31.25S 19.45E

106 K13 **Calvitero** ▲ W Spain 40.16N 5.48W

103 G22 **Calw** Baden-Württemberg, SW Germany 48.43N 8.43E

Calydon *see* Kalydón

107 N13 **Calzada de Calatrava** Castilla-La Mancha, C Spain 38.42N 3.46W

84 C11 **Cama** *see* Kama

84 C11 **Camabatela** Cuanza Norte, NW Angola 8.13S 15.22E

24 D7 **Cameron** Louisiana, S USA 29.48N 93.19W

27 T9 **Cameron** Texas, SW USA 30.51N 96.58W

32 J5 **Cameron** Wisconsin, N USA 45.25N 91.42W

8 L7 **Cameron** ≈ British Columbia, W Canada

193 A24 **Cameron Mountains** ▲ South Island, NZ

81 D15 **Cameroon** *off.* Republic of Cameroon, Fr. Cameroun. ◆ *republic* W Africa

81 D15 **Cameroon Mountain** ▲ SW Cameroon 4.12N 9.00E

Cameroon Ridge *see* Cameroon

Cameroun *see* Cameroon

Camerounaise, Dorsale *Eng.* Cameroon Ridge. *ridge* NW Cameroon

Camellia State *see* Alabama

29 U6 **Camels Hump** ▲ Vermont, NE USA 44.18N 72.50W

119 N8 **Camenca** *Rus.* Kamenka. N Moldova 48.01N 28.43E

84 C11 **Camaxilo** Lunda Norte, NE Angola 8.19S 18.52E

106 G3 **Cambados** Galicia, NW Spain 42.31N 8.49W

62 L11 **Camamu** Bahia, E Brazil 13.55N 38.54W

24 D7 **Cameron** Louisiana, S USA

106 J8 **Cambundi-Catembo** *see* Nova Gaia

143 N11 **Çam Burnu** *headland* N Turkey 41.07N 37.48E

191 S9 **Camden** New South Wales, SE Australia 34.04S 150.40E

25 O6 **Camden** Alabama, S USA 31.59N 87.17W

29 U14 **Camden** Arkansas, C USA 33.34N 92.49W

23 Y3 **Camden** Delaware, NE USA 39.06N 75.30W

21 R7 **Camden** Maine, NE USA 44.12N 69.04W

20 I16 **Camden** New Jersey, NE USA 39.55N 75.07W

20 J9 **Camden** New York, NE USA 43.21N 75.45W

23 X3 **Camden** South Carolina, SE USA 34.15N 80.36W

22 H8 **Camden** Tennessee, S USA 36.03N 88.06W

27 X9 **Camden** Texas, SW USA 30.55N 94.43W

41 N5 **Camden Bay** *bay* S Beaufort Sea

29 U6 **Camdenton** Missouri, C USA 38.01N 92.44W

20 M7 **Camellia State** *see* Alabama

142 F3 **Çan Çanakkale, NW Turkey** 40.01N 26.59E

110 L12 **Canaan** Connecticut, NE USA 41.24N 73.34W

15 Kk13 **Canada** ◆ *commonwealth republic* N North America

207 N16 **Canada Basin** *undersea feature* Arctic Ocean

29 T4 **Cañada de Gómez** Santa Fe, C Argentina 32.49N 61.23W

207 P6 **Canada Plain** *undersea feature* Arctic Ocean

62 A18 **Cañada Rosquín** Santa Fe, C Argentina 32.04S 61.35W

27 P1 **Canadian** Texas, SW USA 35.54N 100.22W

15 K12 **Canadian River** ≈ SW USA

12 I18 **Canadian Shield** *physical region* Canada

61 I18 **Cañadón Grande, Sierra** ▲ S Argentina

57 P9 **Canaima** Bolívar, SE Venezuela 9.40N 72.33W
142 B11 **Çanakkale** *var.* Dardanelli; *prev.* Chanak, Kale Sultanie. Çanakkale, W Turkey 40.09N 26.25E
142 B12 **Çanakkale** ◆ *province* NW Turkey
142 B11 **Çanakkale Boğazı** *Eng.* Dardanelles. *strait* NW Turkey
197 I6 **Canala** Province Nord, C New Caledonia 21.31S 165.57E
61 A15 **Canamari** Amazonas, W Brazil 7.37S 72.33W
20 G10 **Canandaigua** New York, NE USA 42.52N 77.14W
20 F10 **Canandaigua Lake** ◎ New York, NE USA
42 G3 **Cananea** Sonora, NW Mexico 30.58N 110.19W
58 B8 **Cañar** ◆ *province* C Ecuador
66 N10 **Canarias, Islas** *Eng.* Canary Islands. ◆ *autonomous community* Spain, NE Atlantic Ocean
Canaries Basin *see* Canary Basin
46 C6 **Canarreos, Archipiélago de los** *island group* W Cuba
68 K3 **Canary Basin** *var.* Canaries Basin, Monaco Basin. *undersea feature* E Atlantic Ocean
Canary Islands *see* Canarias, Islas
44 L13 **Cañas** Guanacaste, NW Costa Rica 10.25N 85.07W
20 I10 **Canastota** New York, NE USA 43.04N 75.45W
42 K9 **Canatlán** Durango, C Mexico 24.33N 104.45W
106 J9 **Cañaveral** Extremadura, W Spain 39.46N 6.24W
25 Y11 **Canaveral, Cape** *headland* Florida, SE USA 28.27N 80.31W
61 O18 **Canavieiras** Bahia, E Brazil 15.43S 38.58W
45 R16 **Cañazas** Veraguas, W Panama 8.19N 81.09W
108 H6 **Canazei** Trentino-Alto Adige, N Italy 46.29N 11.50E
191 P6 **Canbelego** New South Wales, SE Australia 31.36S 146.20E
191 R10 **Canberra** ● (Australia) Australian Capital Territory, SE Australia 35.21S 149.08E
191 R10 **Canberra** ✕ Australian Capital Territory, SE Australia 35.19S 149.12E
37 P2 **Canby** California, W USA 41.27N 120.51W
31 S9 **Canby** Minnesota, N USA 44.42N 96.17W
105 N2 **Canche** ⫽ N France
104 L13 **Cancon** Lot-et-Garonne, SW France 44.33N 0.37E
43 Z11 **Cancún** Quintana Roo, SE Mexico 21.05N 86.48W
106 K2 **Candás** Asturias, N Spain 43.35N 5.45W
104 J7 **Cande** Maine-et-Loire, NW France 47.33N 1.03W
43 W14 **Candelaria** Campeche, SE Mexico 18.10N 91.00W
26 J11 **Candelaria** Texas, SW USA 30.05N 104.40W
43 W15 **Candelaria, Río** ⫽ Guatemala/Mexico
106 L8 **Candeleda** Castilla-León, N Spain 40.10N 5.13W
Candia *see* Irákleio
43 P8 **Cándido Aguilar** Tamaulipas, C Mexico 25.30N 97.57W
41 N8 **Candle** Alaska, USA 65.54N 161.55W
9 T14 **Candle Lake** Saskatchewan, C Canada 53.43N 105.09W
20 L13 **Candlewood, Lake** ▣ Connecticut, NE USA
31 O3 **Cando** North Dakota, N USA 48.29N 99.12W
179 P8 **Candon** Luzon, N Philippines 17.15N 120.25E
Canea *see* Chaniá
47 O12 **Canefield** ✕ (Roseau) SW Dominica 15.20N 61.24W
63 F20 **Canelones** *prev.* Guadalupe. Canelones, S Uruguay 34.31S 56.16W
63 E20 **Canelones** ◆ *department* S Uruguay
Canendiyú *see* Canindeyú
65 F14 **Cañete** Bío Bío, C Chile 37.48S 73.21W
107 Q9 **Cañete** Castilla-La Mancha, C Spain 40.03N 1.39W
Cañete *see* San Vicente de Cañete
29 P8 **Caney** Kansas, C USA 37.00N 95.56W
29 P8 **Caney River** ⫽ Kansas/Oklahoma, C USA
107 S3 **Canfranc-Estación** Aragón, NE Spain 42.42N 0.31W
85 E14 **Cangamba** *Port.* Vila de Aljustrel. Moxico, E Angola 13.39S 19.57E
84 C12 **Cangandala** Malanje, NW Angola 9.46S 16.27E
106 G4 **Cangas** Galicia, NW Spain 42.16N 8.46W
106 J2 **Cangas del Narcea** Asturias, N Spain 43.10N 6.31W
106 L2 **Cangas de Onís** Asturias, N Spain 43.21N 5.07W
167 S11 **Cangnan** *prev.* Lingxi. Zhejiang, SE China 27.29N 120.23E
84 C10 **Cangola** Uíge, NW Angola 7.54S 15.57E
85 E14 **Cangombe** Mexico, E Angola 14.27S 20.05E
65 H21 **Cangrejo, Cerro** ▲ S Argentina 49.19S 72.18W
63 H17 **Canguçu** Rio Grande do Sul, S Brazil 31.25S 52.37W
167 P3 **Cangzhou** Hebei, E China 38.19N 116.54E
10 M7 **Caniapiscau** ⫽ Quebec, E Canada
10 M8 **Caniapiscau, Réservoir de** ▣ Quebec, C Canada
109 J24 **Canicattì** Sicilia, Italy, C Mediterranean Sea 37.22N 13.51E
142 L11 **Canik Dağları** ▲ N Turkey
107 P14 **Caniles** Andalucía, S Spain 37.24N 2.41W
61 B16 **Canindé** Acre, W Brazil 10.55S 69.45W
64 P6 **Canindeyú** *var.* Canendiyú, Canindiyú. ◆ *department* E Paraguay
Canindiyú *see* Canindeyú
204 J10 **Canisteo Peninsula** *peninsula* Antarctica
20 F11 **Canisteo River** ⫽ New York, NE USA
42 M10 **Cañitas** *var.* Cañitas de Felipe Pescador. Zacatecas, C Mexico 23.35N 102.39W
Cañitas de Felipe Pescador *see* Cañitas
107 P15 **Canjáyar** Andalucía, S Spain 37.00N 2.45W
142 I12 **Çankırı** *var.* Chankiri; *anc.* Gangra, Germanicopolis. Çankın, N Turkey 40.36N 33.35E
142 I11 **Çankırı** *var.* Chankiri. ◆ *province* N Turkey
179 Qq13 **Canlaon Volcano** ⛰ Negros, C Philippines 10.24N 123.05E
9 P16 **Canmore** Alberta, SW Canada 51.07N 115.18W
98 F9 **Canna** *island* NW Scotland, UK
161 F20 **Cannanore** *var.* Kananur, Kannur. Kerala, SW India 11.52N 75.22E
33 O17 **Cannelton** Indiana, N USA 37.54N 86.44W
105 U15 **Cannes** Alpes-Maritimes, SE France 43.33N 6.58E
41 R5 **Canning River** ⫽ Alaska, USA
108 C6 **Cannobio** Piemonte, NE Italy 46.04N 8.39E
99 L19 **Cannock** C England, UK 52.40N 2.03W
30 M6 **Cannonball River** ⫽ North Dakota, N USA
31 W9 **Cannon Falls** Minnesota, N USA 44.30N 92.54W
20 I11 **Cannonsville Reservoir** ▣ New York, NE USA
191 R12 **Cann River** Victoria, SE Australia 37.34S 149.11E
63 I16 **Canoas** Rio Grande do Sul, S Brazil 29.42S 51.07W
63 I14 **Canoas, Rio** ⫽ S Brazil
12 I12 **Canoe Lake** ◎ Ontario, SE Canada
62 J12 **Canoinhas** Santa Catarina, S Brazil 26.12S 50.24W
39 T6 **Canon City** Colorado, C USA 38.25N 105.14W
57 P8 **Caño Negro** Bolívar, SE Venezuela
181 X15 **Canonniers Point** *headland* N Mauritius
25 W6 **Canoochee River** ⫽ Georgia, SE USA
9 V15 **Canora** Saskatchewan, S Canada 51.37N 102.28W
47 Y14 **Canouan** *island* S Saint Vincent and the Grenadines
11 R15 **Canso** Nova Scotia, SE Canada 45.20N 61.00W
106 M3 **Cantabria** ◆ *autonomous community* N Spain
106 K3 **Cantábrica, Cordillera** ▲ N Spain
Cantabrigia *see* Cambridge
105 O12 **Cantal** ◆ *department* C France
107 N6 **Cantalejo** Castilla-León, N Spain 41.15N 3.57W
105 O12 **Cantal, Monts du** ▲ C France
106 G8 **Cantanhede** Coimbra, C Portugal 40.21N 8.37W
Cantaño *see* Cataño
57 O6 **Cantaura** Anzoátegui, NE Venezuela 9.18N 64.21W
118 M11 **Cantemir** *Rus.* Kantemir. S Moldova 46.17N 28.12E
99 Q22 **Canterbury** *hist.* Cantwaraburh, *anc.* Durovernum, *Lat.* Cantuaria. SE England, UK 51.16N 1.04E
193 F19 **Canterbury** *off.* Canterbury Region. ◆ *region* South Island, NZ
193 H20 **Canterbury Bight** *bight* South Island, NZ
193 H19 **Canterbury Plains** *plain* South Island, NZ
178 Jj15 **Cần Thơ** Cần Thơ, S Vietnam 10.03N 105.46E
106 K13 **Cantillana** Andalucía, S Spain 37.34N 5.48W
61 N15 **Canto do Buriti** Piauí, NE Brazil 8.07S 43.00W
25 S2 **Canton** Georgia, SE USA 34.14N 84.29W
32 K12 **Canton** Illinois, N USA 40.33N 90.02W
24 L5 **Canton** Mississippi, S USA 32.36N 90.02W
29 V2 **Canton** Missouri, C USA 40.07N 91.31W
20 J7 **Canton** New York, NE USA 44.36N 75.10W
23 O10 **Canton** North Carolina, SE USA 35.31N 82.50W
33 U12 **Canton** Ohio, N USA 40.48N 81.22W
28 L9 **Canton** Oklahoma, C USA 36.03N 98.35W
20 G12 **Canton** Pennsylvania, NE USA 41.38N 76.49W
31 R11 **Canton** South Dakota, N USA 43.19N 96.33W
27 V7 **Canton** Texas, SW USA 32.34N 95.50W
Canton *see* Guangzhou
Canton Island *see* Kanton
28 L9 **Canton Lake** ▣ Oklahoma, C USA
108 D7 **Cantù** Lombardia, N Italy 45.43N 9.07E
Cantuaria/Cantwaraburh *see* Canterbury
41 R10 **Cantwell** Alaska, USA 63.23N 148.57W
61 O16 **Canudos** Bahia, E Brazil 9.51S 39.07W
49 T7 **Canumã, Rio** ⫽ N Brazil
Canusium *see* Puglia, Canosa di
26 G7 **Canutillo** Texas, SW USA 31.53N 106.34W
27 N3 **Canyon** Texas, SW USA 34.58N 101.55W
35 S12 **Canyon** Wyoming, C USA 44.44N 110.30W
34 K13 **Canyon City** Oregon, NW USA 44.22N 118.58W
35 R10 **Canyon Ferry Lake** ◎ Montana, NW USA
27 S11 **Canyon Lake** ▣ Texas, SW USA
178 Jj5 **Cao Bằng** *var.* Caobang. Cao Bằng, N Vietnam 22.40N 106.16E
166 J12 **Caodu He** ⫽ S China
178 J14 **Cao Lanh** Đồng Thap, S Vietnam 10.35N 105.25E
84 C11 **Caombo** Malanje, NW Angola 8.42S 16.33E
Caorach, Cuan na g *see* Sheep Haven
Caozhou *see* Heze
175 S10 **Capalulu** Pulau Mangole, E Indonesia 1.51S 125.53E
56 K8 **Capanaparo, Río** ⫽ Colombia/Venezuela
60 L12 **Capanema** Pará, NE Brazil 1.07S 47.07W
62 L10 **Capão Bonito** São Paulo, S Brazil 24.01S 48.22W
62 I13 **Capão Doce, Morro do** ▲ S Brazil 26.37S 51.28W
56 I4 **Capatárida** Falcón, N Venezuela 11.10N 70.38W
104 I15 **Capbreton** Landes, SW France 43.40N 1.25W
Cap-Breton, Île du *see* Cape Breton Island
13 W6 **Cap-Chat** Quebec, SE Canada 49.04N 66.43W
13 P11 **Cap-de-la-Madeleine** Quebec, SE Canada 46.22N 72.31W
105 N13 **Capdenac** Aveyron, S France 44.35N 2.06E
191 Q15 **Cape Barren Island** *island* Furneaux Group, Tasmania, SE Australia
67 O18 **Cape Basin** *undersea feature* S Atlantic Ocean
11 R14 **Cape Breton Island** *Fr.* Île du Cap-Breton. *island* Nova Scotia, SE Canada
25 Y11 **Cape Canaveral** Florida, SE USA 28.24N 80.36W
23 Y6 **Cape Charles** Virginia, NE USA 37.16N 76.01W
79 P17 **Cape Coast** *prev.* Cape Coast Castle. S Ghana 5.10N 1.13W
Cape Coast Castle *see* Cape Coast
21 Q12 **Cape Cod Bay** *bay* Massachusetts, NE USA
25 W15 **Cape Coral** Florida, SE USA 26.33N 81.57W
189 R4 **Cape Crawford Roadhouse** Northern Territory, N Australia 16.39S 135.44E
16 N4 **Cape Dorset** Baffin Island, Nunavut, NE Canada 64.12N 76.31W
23 N8 **Cape Fear River** ⫽ North Carolina, SE USA
29 Y7 **Cape Girardeau** Missouri, C USA 37.19N 89.31W
23 T14 **Cape Island** *island* South Carolina, SE USA
194 E11 **Capella** ▲ NW PNG 5.00S 141.09E
100 H12 **Capelle aan den IJssel** Zuid-Holland, SW Netherlands 51.55N 4.36E
85 C15 **Capelongo** Huíla, C Angola 14.45S 15.02E
20 J17 **Cape May** New Jersey, NE USA 38.54N 74.54W
20 J17 **Cape May Court House** New Jersey, NE USA 39.03N 74.46W
Cape Palmas *see* Harper
15 H2 **Cape Parry** Northwest Territories, N Canada 70.10N 124.33W
67 P19 **Cape Rise** *undersea feature* SW Indian Ocean
Cape Saint Jacques *see* Vung Tau
Capesterre *see* Capesterre-Belle-Eau
47 Y6 **Capesterre-Belle-Eau** *var.* Capesterre. Basse Terre, S Guadeloupe 16.03N 61.33W
85 D26 **Cape Town** *var.* Ekapa, *Afr.* Kaapstad, Kapstad. ● (South Africa-legislative capital) Western Cape, SW South Africa 33.55S 18.28E
85 E26 **Cape Town** ✕ Western Cape, SW South Africa 31.51S 21.06E
78 D9 **Cape Verde** *off.* Republic of Cape Verde, *Port.* Cabo Verde, Ilhas do Cabo Verde. ◆ *republic* E Atlantic Ocean
66 L11 **Cape Verde Basin** *undersea feature* E Atlantic Ocean
68 K5 **Cape Verde Islands** *island group* E Atlantic Ocean
66 L10 **Cape Verde Plain** *undersea feature* E Atlantic Ocean
Cape Verde Plateau/Cape Verde Rise *see* Cape Verde Terrace
66 L11 **Cape Verde Terrace** *var.* Cape Verde Plateau, Cape Verde Rise. *undersea feature* E Atlantic Ocean
189 V2 **Cape York Peninsula** *peninsula* Queensland, N Australia
46 M8 **Cap-Haïtien** *var.* Le Cap. N Haiti 19.43N 72.12W
45 T15 **Capira** Panamá, C Panama 8.45N 79.52W
12 K8 **Capitachouane** ⫽ Quebec, SE Canada
12 L8 **Capitachouane, Lac** ◎ Quebec, SE Canada
39 T13 **Capitan** New Mexico, SW USA 33.33N 105.34W
204 G3 **Capitán Arturo Prat** *Chilean research station* South Shetland Islands, Antarctica 62.24S 59.42W
39 S13 **Capitan Mountains** ▲ New Mexico, SW USA
64 M3 **Capitán Pablo Lagerenza** *var.* Mayor Pablo Lagerenza. Chaco, N Paraguay 19.55S 60.46W
39 T13 **Capitan Peak** ▲ New Mexico, SW USA 33.35N 105.15W
196 H5 **Capitol Hill** Saipan, S Northern Mariana Islands
62 I9 **Capivara, Represa** ▣ S Brazil
63 J16 **Capivari** Rio Grande do Sul, S Brazil 30.08S 50.32W
115 H15 **Čapljina** Federacija Bosna I Hercegovina, S Bosnia and Herzegovina 43.07N 17.42E
85 M15 **Capoche** *var.* Kapoche. ⫽ Mozambique/Zambia
Capo Delgado, Província de *see* Cabo Delgado
109 K17 **Capodichino** ✕ (Napoli) Campania, S Italy 40.53N 14.15E
Capodistria *see* Koper
108 E12 **Capraia, Isola** *island* Archipelago Toscano, C Italy
109 B16 **Caprara, Punta** *var.* Punta dello Scorno. *headland* Isola Asinara, W Italy 41.07N 8.19E
Capraria, Isola di to Capraia, Isola *see* Cabrera
12 F10 **Capreol** Ontario, S Canada 46.43N 80.55W
109 K18 **Capri** Campania, S Italy 40.33N 14.14E
183 S9 **Capricorn Tablemount** *undersea feature* W Pacific Ocean 18.34S 172.12W
109 J18 **Capri, Isola di** *island* S Italy
85 G16 **Caprivi** ◆ *district* NE Namibia
Caprivi Concession *see* Caprivi Strip
85 F16 **Caprivi Strip** *Ger.* Caprivizipfel; *prev.* Caprivi Concession. *cultural region* NE Namibia
Caprivizipfel *see* Caprivi Strip
27 O5 **Cap Rock Escarpment** *cliffs* Texas, SW USA
13 R10 **Cap-Rouge** Quebec, SE Canada 46.45N 71.18W
Cap Saint-Jacques *see* Vung Tau
40 F12 **Captain Cook** Hawaii, USA, C Pacific Ocean 19.30N 155.55W
191 R10 **Captains Flat** New South Wales, SE Australia 35.37S 149.28E
104 K14 **Captieux** Gironde, SW France 44.16N 0.15W
109 K17 **Capua** Campania, S Italy 41.06N 14.13E
56 F14 **Caquetá** *off.* Departamento del Caquetá. ◆ *province* S Colombia
56 E13 **Caquetá, Río** *var.* Rio Japurá, Yapurá. ⫽ Brazil/Colombia *see also* Japurá, Rio
CAR *see* Central African Republic
Cara *see* Kara
59 I16 **Carabaya, Cordillera** ▲ E Peru
56 K5 **Carabobo** *off.* Estado Carabobo. ◆ *state* N Venezuela
118 I14 **Caracal** Olt, S Romania 44.07N 24.18E
60 F10 **Caracaraí** Rondônia, W Brazil 1.46N 61.10W
56 L5 **Caracas** ● (Venezuela) Distrito Federal, N Venezuela 10.28N 66.53W
56 I5 **Carache** Trujillo, N Venezuela 9.40N 70.15W
62 N10 **Caraguatatuba** São Paulo, S Brazil 23.37S 45.24W
50 I7 **Carajás, Serra dos** ▲ N Brazil
Caralis *see* Cagliari
56 E9 **Caramanta** Antioquia, W Colombia 5.36N 75.37W
179 Q11 **Caramoan** Catanduanes Island, N Philippines 13.47N 123.49E
Caramurat *see* Mihail Kogălniceanu
118 F12 **Caransebeş** *Ger.* Karansebesch, *Hung.* Karánsebes. Caraş-Severin, SW Romania 45.23N 22.13E
Carapella *see* Carapelle
109 M16 **Carapelle** *var.* Carapella. ⫽ SE Italy
57 O9 **Carapo** Bolívar, SE Venezuela
11 P13 **Caraquet** New Brunswick, SE Canada 47.48N 64.58W
Caras *see* Caraz
118 F12 **Caraşova** *Hung.* Krassóvár. Caraş-Severin, SW Romania 45.10N 21.51E
118 F12 **Caraş-Severin** ◆ *county* SW Romania
44 M5 **Caratasca, Laguna de** *lagoon* NE Honduras
60 C13 **Carauari** Amazonas, NW Brazil 4.55S 66.57W
Caravaca *see* Caravaca de la Cruz
107 Q12 **Caravaca de la Cruz** *var.* Caravaca. Murcia, SE Spain 38.06N 1.51W
108 E7 **Caravaggio** Lombardia, N Italy 45.31N 9.39E
109 C18 **Caravai, Passo di** *pass* Sardegna, Italy, C Mediterranean Sea 40.06N 9.19E
61 O19 **Caravelas** Bahia, E Brazil 17.45S 39.15W
58 C12 **Caraz** *var.* Caras. Ancash, W Peru 9.01S 77.48W
63 H14 **Carazinho** Rio Grande do Sul, S Brazil 28.16S 52.46W
44 J11 **Carazo** ◆ *department* SW Nicaragua
Carballiño *see* O Carballiño
106 G2 **Carballo** Galicia, NW Spain 43.12N 8.42W
9 W16 **Carberry** Manitoba, S Canada 49.52N 99.19W
42 F4 **Carbó** Sonora, NW Mexico 29.40N 110.54W
109 C20 **Carbonara, Capo** *headland* Sardegna, Italy, C Mediterranean Sea 39.06N 9.31E
39 Q5 **Carbondale** Colorado, C USA 39.24N 107.12W
32 L17 **Carbondale** Illinois, N USA 37.43N 89.13W
29 Q4 **Carbondale** Kansas, C USA 38.49N 95.41W
20 I13 **Carbondale** Pennsylvania, NE USA 41.34N 75.30W
11 V12 **Carbonear** Newfoundland, SE Canada 47.45N 53.16W
107 Q9 **Carboneras de Guadazaón** *var.* Carboneras de Guadazón. Castilla-La Mancha, C Spain 39.54N 1.49W
Carboneras de Guadazón *see* Carboneras de Guadazaón
25 O3 **Carbon Hill** Alabama, S USA 33.53N 87.31W
109 B20 **Carbonia** *var.* Carbonia Centro. Sardegna, Italy, C Mediterranean Sea 39.10N 8.31E
Carbonia Centro *see* Carbonia
107 S10 **Carcaixent** País Valenciano, E Spain 39.07N 0.28W
Carcaso *see* Carcassonne
67 B24 **Carcass Island** *island* NW Falkland Islands
105 O16 **Carcassonne** *anc.* Carcaso. Aude, S France 43.13N 2.21E
107 R12 **Carche** ▲ S Spain 38.24N 1.11W
58 A13 **Carchi** ◆ *province* N Ecuador
8 I8 **Carcross** Yukon Territory, W Canada 60.11N 134.40W
Cardamomes, Chaîne des *see* Krâvanh, Chuŏr Phnum
161 G22 **Cardamom Hills** ▲ SW India
Cardamom Mountains *see* Krâvanh, Chuŏr Phnum
106 M12 **Cardeña** Andalucía, S Spain 38.16N 4.19W
46 D4 **Cárdenas** Matanzas, W Cuba 23.01N 81.12W
43 O11 **Cárdenas** San Luis Potosí, C Mexico 22.03N 99.30W
43 U15 **Cárdenas** Tabasco, SE Mexico 18.00N 93.21W
65 H21 **Cardiel, Lago** ◎ S Argentina
99 K22 **Cardiff** *Wel.* Caerdydd. ◐ S Wales, UK 51.30N 3.13W
99 J22 **Cardiff-Wales** ✕ S Wales, UK 51.24N 3.22W
99 I21 **Cardigan** *Wel.* Aberteifi. SW Wales, UK 52.06N 4.40W
99 I20 **Cardigan** *cultural region* W Wales, UK
99 I20 **Cardigan Bay** *bay* W Wales, UK
21 N8 **Cardigan, Mount** ▲ New Hampshire, NE USA 43.39N 71.52W
12 M13 **Cardinal** Ontario, SE Canada 44.48N 75.22W
107 V5 **Cardona** Cataluña, NE Spain 41.55N 1.40E
63 E19 **Cardona** Soriano, SW Uruguay 33.52S 57.18W
107 V4 **Cardoner** ⫽ NE Spain
9 Q17 **Cardston** Alberta, SW Canada 49.13N 113.19W
189 W5 **Cardwell** Queensland, NE Australia 18.24S 146.06E
118 G8 **Carei** *Ger.* Gross-Karol, Karol, *Hung.* Nagykároly; *prev.* Careii-Mari. Satu Mare, NW Romania 47.40N 22.27E
Careii-Mari *see* Carei
60 F13 **Careiro** Amazonas, NW Brazil 3.39S 60.22W
104 J4 **Carentan** Manche, N France 49.18N 1.15W
106 M2 **Cares** ⫽ N Spain
35 P14 **Carey** Idaho, NW USA 43.17N 113.58W
33 S12 **Carey** Ohio, N USA 40.57N 83.22W
27 P4 **Carey** Texas, SW USA 34.28N 100.18W
188 L11 **Carey, Lake** ◎ Western Australia
181 O8 **Cargados Carajos Bank** *undersea feature* C Indian Ocean
104 G6 **Carhaix-Plouguer** Finistère, NW France 48.16N 3.34W
63 A22 **Carhué** Buenos Aires, E Argentina 37.10S 62.45W
57 O5 **Cariaco** Sucre, NE Venezuela 10.28N 63.33W
109 O20 **Cariati** Calabria, SW Italy 39.30N 16.57E
2 H17 **Caribbean Plate** *tectonic feature*
46 I11 **Caribbean Sea** *sea* W Atlantic Ocean
9 N15 **Cariboo Mountains** ▲ British Columbia, SW Canada
9 W9 **Caribou** Manitoba, C Canada 59.27N 97.43W
21 S2 **Caribou** Maine, NE USA 46.51N 68.00W
9 P10 **Caribou Mountains** ▲ Alberta, SW Canada
Caribrod *see* Dimitrovgrad
42 I6 **Carichic** Chihuahua, N Mexico 27.57N 107.01W
179 Qq13 **Carigara** Leyte, C Philippines 11.15N 124.43E
105 R3 **Carignan** Ardennes, N France 49.38N 5.08E
191 Q5 **Carinda** New South Wales, SE Australia 30.26S 147.45E
107 R6 **Cariñena** Aragón, NE Spain 41.19N 1.13W
109 I23 **Carini** Sicilia, Italy, C Mediterranean Sea 38.06N 13.09E
109 K17 **Carinola** Campania, S Italy 41.14N 14.03E
Carinthi *see* Kärnten
57 O5 **Caripe** Monagas, NE Venezuela 10.06N 63.30W
57 P5 **Caripito** Monagas, NE Venezuela 10.03N 63.05W
13 W7 **Carleton** Quebec, SE Canada 48.07N 66.07W
33 S10 **Carleton** Michigan, N USA 42.03N 83.23W
11 O14 **Carleton, Mount** ▲ New Brunswick, SE Canada 47.10N 66.54W
12 L13 **Carleton Place** Ontario, SE Canada 45.09N 76.07W
37 V3 **Carlin** Nevada, W USA 40.40N 116.09W
32 K14 **Carlinville** Illinois, N USA 39.16N 89.52W
99 K14 **Carlisle** *anc.* Caer Luel, Luguvallium, Luguvallum. NW England, UK 54.54N 2.55W
29 V11 **Carlisle** Arkansas, C USA 34.46N 91.45W
33 N15 **Carlisle** Indiana, N USA 38.57N 87.23W
31 V14 **Carlisle** Iowa, C USA 41.30N 93.29W
23 N5 **Carlisle** Kentucky, S USA 38.19N 83.59W
20 F15 **Carlisle** Pennsylvania, NE USA 40.10N 77.10W
23 Q11 **Carlisle** South Carolina, SE USA 34.35N 81.27W
40 J17 **Carlisle Island** *island* Aleutian Islands, Alaska, USA
29 R7 **Carl Junction** Missouri, C USA 37.10N 94.34W
109 A20 **Carloforte** Sardegna, Italy, C Mediterranean Sea 39.10N 8.17E
Carlopago *see* Karlobag
63 B21 **Carlos Casares** Buenos Aires, E Argentina 35.39S 61.28W
63 E18 **Carlos Reyles** Durazno, C Uruguay 33.04S 56.28W
63 A21 **Carlos Tejedor** Buenos Aires, E Argentina 35.23S 62.31W
99 F19 **Carlow** *Ir.* Ceatharlach. SE Ireland 52.49N 6.55W
99 F19 **Carlow** *Ir.* Cheatharlach. *cultural region* SE Ireland
98 F7 **Carloway** NW Scotland, UK 58.16N 6.48W
Carlsbad *see* Karlovy Vary
37 U17 **Carlsbad** California, W USA 33.09N 117.21W
39 U15 **Carlsbad** New Mexico, SW USA 32.24N 104.14W
133 N13 **Carlsberg Ridge** *undersea feature* S Arabian Sea
Carlsruhe *see* Karlsruhe
31 W6 **Carlton** Minnesota, N USA 46.39N 92.25W
9 V17 **Carlyle** Saskatchewan, S Canada 49.39N 102.18W
32 L15 **Carlyle** Illinois, N USA 38.36N 89.22W
32 L15 **Carlyle Lake** ▣ Illinois, N USA
8 H7 **Carmacks** Yukon Territory, W Canada 62.04N 136.21W
108 B9 **Carmagnola** Piemonte, NW Italy 44.50N 7.43E
9 X16 **Carman** Manitoba, S Canada 49.31N 97.58W
Carmana/Carmania *see* Kermān
99 I21 **Carmarthen** SW Wales, UK 51.52N 4.19W
99 I21 **Carmarthen** *cultural region* SW Wales, UK
99 I22 **Carmarthen Bay** *inlet* SW Wales, UK
105 N14 **Carmaux** Tarn, S France 44.03N 2.09E
37 N11 **Carmel** California, W USA 36.32N 121.54W
33 O13 **Carmel** Indiana, N USA 39.58N 86.07W
20 L13 **Carmel** New York, NE USA 41.25N 73.40W
99 H18 **Carmel Head** *headland* NW Wales, UK 53.24N 4.35W
44 E2 **Carmelita** Petén, N Guatemala 17.33N 90.10W
63 D19 **Carmelo** Colonia, SW Uruguay 34.00S 58.20W
43 V14 **Carmen** *var.* Ciudad del Carmen. Campeche, SE Mexico 18.37N 91.49W
63 A25 **Carmen de Patagones** Buenos Aires, E Argentina 40.45S 63.00W
42 F8 **Carmen, Isla** *island* W Mexico
42 M5 **Carmen, Sierra del** ▲ NW Mexico
32 M16 **Carmi** Illinois, N USA 38.05N 88.09W
37 O7 **Carmichael** California, W USA 38.36N 121.21W
Carmiel *see* Karmi'él
27 U11 **Carmine** Texas, SW USA 30.07N 96.40W
106 K14 **Carmona** Andalucía, S Spain 37.28N 5.37W
Carmona *see* Uíge
Carnaro *see* Kvarner
12 I13 **Carnarvon** Ontario, SE Canada 45.03N 78.41W
85 G24 **Carnarvon** Northern Cape, W South Africa 30.58S 22.07E
Carnarvon *see* Caernarfon
188 K9 **Carnarvon Range** ▲ Western Australia
188 G9 **Carnarvon** Western Australia 24.57S 113.37E
Carn Domhnach *see* Carndonagh
98 E13 **Carndonagh** *Ir.* Carn Domhnach. NW Ireland 55.15N 7.15W
9 V17 **Carnduff** Saskatchewan, S Canada 49.10N 101.49W
28 L11 **Carnegie** Oklahoma, C USA 35.06N 98.36W
188 L9 **Carnegie, Lake** *salt lake* Western Australia
200 Oo8 **Carnegie Ridge** *undersea feature* E Pacific Ocean
98 H9 **Carn Eige** ▲ N Scotland, UK 57.18N 5.04W
190 F5 **Carnes** South Australia 30.12S 134.31E
204 J12 **Carney Island** *island* Antarctica
20 H16 **Carneys Point** New Jersey, NE USA 39.38N 75.29W
Carniche, Alpi *see* Karnische Alpen
157 Q21 **Car Nicobar** *island* Nicobar Islands, India, NE Indian Ocean
81 H15 **Carnot** Mambéré-Kadéï, W Central African Republic 4.58N 15.55E
190 F10 **Carnot, Cape** *headland* South Australia 34.57S 135.39E
98 K11 **Carnoustie** E Scotland, UK 56.29N 2.42W
99 F20 **Carnsore Point** *Ir.* Ceann an Chairn. *headland* SE Ireland 52.10N 6.22W
15 Gg4 **Carnwath** ⫽ Northwest Territories, NW Canada
33 R8 **Caro** Michigan, N USA 43.29N 83.24W
25 Z15 **Carol City** Florida, SE USA 25.56N 80.15W
61 L14 **Carolina** Maranhão, E Brazil 7.19S 47.25W
47 U5 **Carolina** E Puerto Rico 18.22N 65.57W
23 V12 **Carolina Beach** North Carolina, SE USA 34.02N 77.53W
Caroline Island *see* Millennium Island
201 N15 **Caroline Islands** *island group* C Micronesia
133 Z14 **Caroline Plate** *tectonic feature*
199 H7 **Caroline Ridge** *undersea feature* E Philippine Sea
Carolopois *see* Châlons-en-Champagne
47 V14 **Caroni Arena Dam** ▣ Trinidad, Trinidad and Tobago
Caronie, Monti *see* Nebrodi, Monti
57 P7 **Caroní, Río** ⫽ E Venezuela
47 U14 **Caroni River** ⫽ Trinidad, Trinidad and Tobago
Caronium *see* A Coruña
56 J5 **Carora** Lara, N Venezuela 10.09N 70.06W
88 F12 **Carpathian Mountains** *var.* Carpathians, *Cz./Pol.* Karpaty, *Ger.* Karpaten. ▲ E Europe
Carpathians *see* Carpathian Mountains
Carpathos/Carpathus *see* Kárpathos
118 H12 **Carpaţii Meridionali** *var.* Alpi Transilvaniei, Carpaţii Sudici, *Eng.* South Carpathians, Transylvanian Alps, *Ger.* Südkarpaten, Transsylvanische Alpen, *Hung.* Déli-Kárpátok, Erdélyi-Havasok. ▲ C Romania
Carpaţii Sudici *see* Carpaţii Meridionali
182 L7 **Carpentaria, Gulf of** *gulf* N Australia
Carpentoracte *see* Carpentras
105 R14 **Carpentras** *anc.* Carpentoracte. Vaucluse, SE France 44.03N 5.03E
108 F9 **Carpi** Emilia-Romagna, N Italy 44.46N 10.52E
118 E11 **Cărpiniş** *Hung.* Gyertyámos. Timiş, W Romania 45.46N 20.51E
37 R14 **Carpinteria** California, W USA 34.24N 119.30W
25 S9 **Carrabelle** Florida, SE USA 29.51N 84.39W
Carraig Aonair *see* Fastnet Rock
Carraig Fhearghais *see* Carrickfergus
Carraig Mhachaire Rois *see* Carrickmacross
Carraig na Siúire *see* Carrick-on-Suir
Carrantual *see* Carrauntoohil
108 E10 **Carrara** Toscana, C Italy 44.04N 10.07E
63 F20 **Carrasco** ✕ (Montevideo) Canelones, S Uruguay 34.51S 56.00W
107 P9 **Carrascosa del Campo** Castilla-La Mancha, C Spain 40.01N 2.34W
56 H4 **Carrasquero** Zulia, NW Venezuela 11.00N 72.01W
191 O9 **Carrathool** New South Wales, SE Australia 34.25S 145.30E
Carrauntohil *see* Carrauntoohil
99 B21 **Carrauntoohil** *Ir.* Carrantual, Carrauntohil, Corrán Tuathail. ▲ SW Ireland 51.59N 9.45W
47 Y15 **Carriacou** *island* N Grenada
99 G15 **Carrickfergus** *Ir.* Carraig Fhearghais. NE Northern Ireland, UK 54.43N 5.49W
99 F16 **Carrickmacross** *Ir.* Carraig Mhachaire Rois. N Ireland 53.58N 6.43W
99 D16 **Carrick-on-Shannon** *Ir.* Cora Droma Rúisc. NW Ireland 53.37N 8.04W
99 E20 **Carrick-on-Suir** *Ir.* Carraig na Siúire. S Ireland 52.21N 7.25W
190 I7 **Carrieton** South Australia 32.27S 138.33E
42 L7 **Carrillo** Chihuahua, N Mexico 25.53N 103.54W
31 O4 **Carrington** North Dakota, N USA 47.27N 99.07W
106 M4 **Carrión** ⫽ N Spain
106 M4 **Carrión de los Condes** Castilla-León, N Spain 42.19N 4.37W
27 P13 **Carrizo Springs** Texas, SW USA 28.31N 99.51W
39 S13 **Carrizozo** New Mexico, SW USA 33.38N 105.52W
31 T13 **Carroll** Iowa, C USA 42.04N 94.52W
25 N4 **Carrollton** Alabama, S USA 33.13N 88.05W
25 R3 **Carrollton** Georgia, SE USA 33.33N 85.04W
32 K14 **Carrollton** Illinois, N USA 39.18N 90.24W
22 L4 **Carrollton** Kentucky, S USA 38.40N 85.10W
33 R8 **Carrollton** Michigan, N USA 43.27N 83.55W
29 T3 **Carrollton** Missouri, C USA 39.21N 93.30W
33 U12 **Carrollton** Ohio, N USA 40.34N 81.05W
27 T6 **Carrollton** Texas, SW USA 32.57N 96.53W
9 U14 **Carrot** ⫽ Saskatchewan, S Canada
9 U14 **Carrot River** Saskatchewan, C Canada 53.18N 103.31W
20 J7 **Carry Falls Reservoir** ▣ New York, NE USA
142 L11 **Çarşamba** Samsun, N Turkey 41.13N 36.43E
30 L6 **Carson** North Dakota, N USA 46.21N 101.33W
37 Q6 **Carson City** *state capital* Nevada, W USA 39.10N 119.46W
37 R6 **Carson River** ⫽ Nevada, W USA
37 S5 **Carson Sink** *salt flat* Nevada, W USA
9 Q16 **Carstairs** Alberta, SW Canada 51.34N 114.01W
Carstensz, Puntjak *see* Jaya, Puncak
56 E5 **Cartagena** *var.* Cartagena de los Indes. Bolívar, NW Colombia 10.24N 75.33W
107 R13 **Cartagena** *anc.* Carthago Nova. Murcia, SE Spain 37.36N 0.58W
Cartagena de los Indes *see* Cartagena
56 D10 **Cartago** Valle del Cauca, W Colombia 4.45N 75.55W
45 N14 **Cartago** Cartago, C Costa Rica 9.49N 83.53W
44 M14 **Cartago** *off.* Provincia de Cartago. ◆ *province* C Costa Rica
27 O11 **Carta Valley** Texas, SW USA 29.46N 100.37W
106 F10 **Cartaxo** Santarém, C Portugal 39.10N 8.46W
106 I14 **Cartaya** Andalucía, S Spain 37.16N 7.09W
Carteret Islands *see* Tulun Islands
31 S15 **Carter Lake** Iowa, C USA 41.17N 95.55W
25 S3 **Cartersville** Georgia, SE USA 34.10N 84.48W
193 M14 **Carterton** Wellington, North Island, NZ 41.01S 175.32E
32 J13 **Carthage** Illinois, N USA 40.25N 91.09W
24 L5 **Carthage** Mississippi, S USA 32.43N 89.31W
29 R7 **Carthage** Missouri, C USA 37.10N 94.18W
20 I8 **Carthage** New York, NE USA 43.58N 75.36W
23 T10 **Carthage** North Carolina, SE USA 35.19N 79.24W
22 K8 **Carthage** Tennessee, S USA 36.16N 85.57W
27 X7 **Carthage** Texas, SW USA 32.09N 94.20W
76 M5 **Carthage** ✕ (Tunis) N Tunisia 36.51N 10.12E
Carthago Nova *see* Cartagena
12 E10 **Cartier** Ontario, S Canada 46.40N 81.31W
56 E13 **Cartogena de Chaira** Caquetá, S Colombia 1.19N 74.52W

11 S8 **Cartwright** Newfoundland, E Canada *53.40N 57.00W*
57 P9 **Caruana de Montaña** Bolívar, SE Venezuela *5.16N 63.12W*
61 Q15 **Caruaru** Pernambuco, E Brazil *8.15S 35.55W*
57 P5 **Carúpano** Sucre, NE Venezuela *10.39N 63.13W*
Carusbur *see* Cherbourg
60 M12 **Carutapera** Maranhão, E Brazil *1.12S 45.57W*
29 Y9 **Caruthersville** Missouri, C USA *36.07N 89.38W*
105 O1 **Carvin** Pas-de-Calais, N France *50.31N 3.00E*
60 E12 **Carvoeiro** Amazonas, NW Brazil *1.24S 61.59W*
106 E10 **Carvoeiro, Cabo** *headland* C Portugal *39.19N 9.27W*
23 U9 **Cary** North Carolina, SE USA *35.47N 78.46W*
190 M3 **Caryapundy Swamp** *wetland* New South Wales/Queensland, SE Australia
67 E24 **Carysfort, Cape** *headland* East Falkland, Falkland Islands *51.25S 57.49W*
76 F6 **Casablanca** *Ar.* Dar-el-Beida. NW Morocco *33.39N 7.30W*
62 M8 **Casa Branca** São Paulo, S Brazil *21.47S 47.05W*
38 L14 **Casa Grande** Arizona, SW USA *32.52N 111.45W*
108 C8 **Casale Monferrato** Piemonte, NW Italy *45.07N 8.28E*
108 E8 **Casalpusterlengo** Lombardia, N Italy *45.10N 9.37E*
56 H10 **Casanare** *off.* Intendencia de Casanare. ◆ *province* C Colombia
57 P5 **Casanay** Sucre, NE Venezuela *10.30N 63.25W*
26 K11 **Casa Piedra** Texas, SW USA *29.43N 104.03W*
109 Q19 **Casarano** Puglia, SE Italy *40.01N 18.10E*
44 J11 **Casares** Carazo, W Nicaragua *11.37N 86.19W*
107 R10 **Casas Ibáñez** Castilla-La Mancha, C Spain *39.16N 1.28W*
63 I14 **Casca** Rio Grande do Sul, S Brazil *28.39S 51.55W*
312 I17 **Cascade** Mahé, NE Seychelles *4.39S 55.28E*
35 N13 **Cascade** Idaho, NW USA *44.31N 116.02W*
31 Y13 **Cascade** Iowa, C USA *42.18N 91.00W*
35 R9 **Cascade** Montana, NW USA *47.15N 111.46W*
193 B20 **Cascade Point** *headland* South Island, NZ *44.00S 168.23E*
34 G13 **Cascade Range** ▲ Oregon/Washington, NW USA
35 N12 **Cascade Reservoir** ⊠ Idaho, NW USA
2 E8 **Cascadia Basin** *undersea feature* NE Pacific Ocean
106 E11 **Cascais** Lisboa, C Portugal *38.40N 9.25W*
13 W7 **Cascapédia** ∽ Québec, SE Canada
61 I22 **Cascavel** Ceará, E Brazil *4.10S 38.15W*
62 G11 **Cascavel** Paraná, S Brazil *24.55S 53.28W*
108 I13 **Cascia** Umbria, C Italy *42.45N 13.01E*
108 F11 **Cascina** Toscana, C Italy *43.40N 10.33E*
21 Q8 **Casco Bay** *bay* Maine, NE USA
204 J7 **Case Island** *island* Antarctica
108 B8 **Caselle** ✕ (Torino) Piemonte, NW Italy *45.06N 7.41E*
109 K17 **Caserta** Campania, S Italy *41.04N 14.19E*
13 N8 **Casey** Quebec, SE Canada *47.50N 74.09W*
32 M14 **Casey** Illinois, N USA *39.18N 87.59W*
205 Y12 **Casey** *Australian research station* Antarctica *65.58S 111.04E*
205 W3 **Casey Bay** *bay* Antarctica
82 Q11 **Caseyr, Raas** *headland* NE Somalia *11.51N 51.16E*
99 D20 **Cashel** *Ir.* Caiseal. S Ireland *52.31N 7.52W*
56 G6 **Casigua** Zulia, W Venezuela *8.46N 72.30W*
63 B19 **Casilda** Santa Fe, C Argentina *33.04S 61.10W*
Casim *see* General Toshevo
191 V4 **Casino** New South Wales, SE Australia *28.49S 153.01E*
Casinum *see* Cassino
113 E17 **Čáslav** *Ger.* Tschaslau. Střední Čechy, C Czech Republic *49.54N 15.22E*
58 C13 **Casma** Ancash, C Peru *9.27S 78.21W*
178 J7 **Ca, Sông** ∽ N Vietnam
109 K17 **Casoria** Campania, S Italy *40.54N 14.28E*
107 T6 **Caspe** Aragón, NE Spain *41.13N 0.03W*
35 X15 **Casper** Wyoming, C USA *42.48N 106.22W*
86 M10 **Caspian Depression** *Kaz.* Kaspiy Mangy Oypaty, *Rus.* Prikaspiyskaya Nizmennost'. *depression* Kazakhstan/Russian Federation
138 Kk9 **Caspian Sea** *Az.* Xäzär Dänizi, *Kaz.* Kaspiy Tengizi, *Per.* Baḩr-e Khazar, Daryā-ye Khazar, *Rus.* Kaspiyskoye More. *inland sea* Asia/Europe
85 L14 **Cassacatiza** Tete, NW Mozambique *14.20S 32.24E*
Cassai *see* Kasai
84 F13 **Cassamba** Moxico, E Angola *13.07S 20.22E*
109 N20 **Cassano allo Ionio** Calabria, SE Italy *39.46N 16.16E*
33 S8 **Cass City** Michigan, N USA *43.36N 83.10W*
Cassel *see* Kassel
12 M13 **Casselman** Ontario, SE Canada *45.19N 75.04W*
31 R5 **Casselton** North Dakota, N USA *46.53N 97.10W*
61 M16 **Cássia** *var.* Santa Rita de Cassia. Bahia, E Brazil *11.03S 44.16W*
8 J9 **Cassiar** British Columbia, W Canada *59.16N 129.40W*
8 K10 **Cassiar Mountains** ▲ British Columbia, W Canada
85 C15 **Cassinga** Huíla, SW Angola *15.06S 16.05E*
109 J16 **Cassino** *prev.* San Germano; *anc.* Casinum. Lazio, C Italy *41.28N 13.49E*
31 T4 **Cass Lake** Minnesota, N USA *47.22N 94.36W*
31 T4 **Cass Lake** ◎ Minnesota, N USA
33 P10 **Cassopolis** Michigan, N USA *41.56N 86.00W*
33 S8 **Cass River** ∽ Michigan, N USA
29 S8 **Cassville** Missouri, C USA *36.40N 93.52W*
Castamoni *see* Kastamonu
60 L12 **Castanhal** Pará, NE Brazil *1.16S 47.55W*
106 G8 **Castanheira de Pêra** Leiria, C Portugal *40.01N 8.12W*
43 N7 **Castaños** Coahuila de Zaragoza, NE Mexico *26.48N 101.25W*
110 I10 **Castasegna** Graubünden, SE Switzerland *46.21N 9.30E*
108 D8 **Casteggio** Lombardia, N Italy *45.01N 9.10E*
109 K23 **Castelbuono** Sicilia, Italy, C Mediterranean Sea *37.55N 14.04E*
109 K15 **Castel di Sangro** Abruzzo, C Italy *41.46N 14.03E*
108 H7 **Castelfranco Veneto** Veneto, NE Italy *45.40N 11.55E*
104 K14 **Casteljaloux** Lot-et-Garonne, SW France *44.19N 0.03E*
109 L18 **Castellabate** *var.* Santa Maria di Castellabate. Campania, S Italy *40.16N 14.57E*
109 I23 **Castellammare del Golfo** Sicilia, Italy, C Mediterranean Sea *38.01N 12.52E*
109 H22 **Castellammare, Golfo di** *gulf* Sicilia, Italy, C Mediterranean Sea
105 U15 **Castellane** Alpes-de-Haute-Provence, SE France *43.49N 6.34E*
109 O18 **Castellaneta** Puglia, SE Italy *40.37N 16.57E*
108 E9 **Castel l'Arquato** Emilia-Romagna, C Italy *44.52N 9.51E*
63 E21 **Castelli** Buenos Aires, E Argentina *36.07S 57.46W*
107 T9 **Castelló de la Plana** *var.* Castellón. País Valenciano, E Spain *39.58N 0.03W*
107 S8 **Castellón** ◆ *province* País Valenciano, E Spain
Castellón *see* Castelló de la Plana
107 S7 **Castellote** Aragón, NE Spain *40.46N 0.18W*
105 N16 **Castelnaudary** Aude, S France *43.18N 1.57E*
104 L16 **Castelnau-Magnoac** Hautes-Pyrénées, S France *43.18N 0.30E*
108 F10 **Castelnovo ne' Monti** Emilia-Romagna, C Italy *44.26N 10.24E*
Castelnuovo *see* Herceg-Novi
106 H9 **Castelo Branco** Castelo Branco, C Portugal *39.49N 7.30W*
106 H8 **Castelo Branco** ◆ *district* C Portugal
106 I10 **Castelo de Vide** Portalegre, C Portugal *39.25N 7.27W*
106 G9 **Castelo do Bode, Barragem do** ⊠ C Portugal
108 G10 **Castel San Pietro Terme** Emilia-Romagna, C Italy *44.22N 11.34E*
109 B17 **Castelsardo** Sardegna, Italy, C Mediterranean Sea *40.54N 8.42E*
104 M14 **Castelsarrasin** Tarn-et-Garonne, S France *44.01N 1.06E*
109 I24 **Casteltermini** Sicilia, Italy, C Mediterranean Sea *37.33N 13.37E*
109 H24 **Castelvetrano** Sicilia, Italy, C Mediterranean Sea *37.40N 12.46E*
190 L12 **Casterton** Victoria, SE Australia *37.37S 141.22E*
104 J15 **Castets** Landes, SW France *43.55N 1.08W*
108 H12 **Castiglione del Lago** Umbria, C Italy *43.07N 12.02E*
108 F13 **Castiglione della Pescaia** Toscana, C Italy *42.46N 10.53E*
108 F8 **Castiglione delle Stiviere** Lombardia, N Italy *45.24N 10.31E*
106 M9 **Castilla-La Mancha** ◆ *autonomous community* NE Spain
106 L5 **Castilla-León** *var.* Castillia y Leon. ◆ *autonomous community* NW Spain
107 N10 **Castilla Nueva** *cultural region* C Spain
107 N6 **Castilla Vieja** *cultural region* N Spain
Castillia y Leon *see* Castilla-León
Castillo de Locubim *see* Castillo de Locubín
107 N14 **Castillo de Locubín** *var.* Castillo de Locubim. Andalucía, S Spain *37.31N 3.55W*
104 K13 **Castillon-la-Bataille** Gironde, SW France *44.51N 0.01W*
65 I19 **Castillo, Pampa del** *plain* S Argentina
63 G19 **Castillos** Rocha, SE Uruguay *34.12S 53.52W*
99 B16 **Castlebar** *Ir.* Caisleán an Bharraigh. W Ireland *53.52N 9.16W*
99 F16 **Castleblayney** *Ir.* Baile na Lorgan. N Ireland *54.07N 6.43W*
47 O11 **Castle Bruce** E Dominica *15.25N 61.15W*
38 M5 **Castle Dale** Utah, W USA *39.10N 111.02W*
38 I14 **Castle Dome Peak** ▲ Arizona, SW USA *33.04N 114.08W*
99 J14 **Castle Douglas** S Scotland, UK *54.56N 3.55W*
99 E14 **Castlefinn** *Ir.* Caisleán na Finne. NW Ireland *54.48N 7.36W*
99 M17 **Castleford** N England, UK *53.43N 1.21W*
9 O17 **Castlegar** British Columbia, SW Canada *49.18N 117.48W*
66 B12 **Castle Harbour** *inlet* Bermuda, NW Atlantic Ocean
23 V12 **Castle Hayne** North Carolina, SE USA *34.23N 78.07W*
99 B20 **Castleisland** *Ir.* Oileán Ciarraí. SW Ireland *52.12N 9.30W*
191 N12 **Castlemaine** Victoria, SE Australia *37.06S 144.13E*
39 R5 **Castle Peak** ▲ Colorado, C USA *39.00N 106.51W*
35 O13 **Castle Peak** ▲ Idaho, NW USA *44.02N 114.42W*
192 N13 **Castlepoint** Wellington, North Island, NZ *40.54S 176.13E*
99 D17 **Castlerea** *Ir.* An Caisleán Riabhach. W Ireland *53.45N 8.31W*
99 G15 **Castlereagh** *Ir.* An Caisleán Riabhach. N Northern Ireland, UK *54.33N 5.53W*
191 R6 **Castlereagh River** ∽ New South Wales, SE Australia
39 T5 **Castle Rock** Colorado, C USA *39.22N 104.51W*
32 K7 **Castle Rock Lake** ⊠ Wisconsin, N USA
67 G25 **Castle Rock Point** *headland* S Saint Helena *16.01S 5.45W*
99 I16 **Castletown** SE Isle of Man *54.04N 4.39W*
31 R9 **Castlewood** South Dakota, N USA *44.43N 97.01W*
9 R15 **Castor** Alberta, SW Canada *52.13N 111.54W*
12 M13 **Castor** ∽ Ontario, SE Canada
29 X7 **Castor River** ∽ Missouri, C USA
Castra Albiensium *see* Castres
Castra Regina *see* Regensburg
105 N15 **Castres** *anc.* Castra Albiensium. Tarn, S France *43.36N 2.15E*
100 H9 **Castricum** Noord-Holland, W Netherlands *52.33N 4.40E*
47 S11 **Castries** ● (Saint Lucia) N Saint Lucia *14.01N 60.59W*
62 J11 **Castro** Paraná, S Brazil *24.45S 50.58W*
65 F17 **Castro** Los Lagos, W Chile *42.27S 73.48W*
106 H7 **Castro Daire** Viseu, N Portugal *40.54N 7.55W*
106 M13 **Castro del Río** Andalucía, S Spain *37.40N 4.28W*
Castrogiovanni *see* Enna
106 H14 **Castro Marim** Faro, S Portugal *37.13N 7.25W*
106 J2 **Castropol** Asturias, N Spain *43.30N 7.01W*
107 O2 **Castro-Urdiales** *var.* Castro Urdiales. Cantabria, N Spain *43.22N 3.10W*
106 G13 **Castro Verde** Beja, S Portugal *37.42N 8.04W*
109 N19 **Castrovillari** Calabria, SW Italy *39.48N 16.12E*
37 N10 **Castroville** California, W USA *36.46N 121.46W*
27 R12 **Castroville** Texas, SW USA *29.21N 98.52W*
106 K11 **Castuera** Extremadura, W Spain *38.43N 5.33W*
63 F19 **Casupá** Florida, S Uruguay *34.04S 55.39W*
193 A22 **Caswell Sound** *sound* South Island, NZ
143 Q13 **Çat** Erzurum, NE Turkey *39.40N 41.03E*
44 K6 **Catacamas** Olancho, C Honduras *14.50N 85.54W*
58 A10 **Catacaos** Piura, NW Peru *5.22S 80.40W*
24 I7 **Catahoula Lake** ◎ Louisiana, S USA
143 S15 **Çatak** Van, SE Turkey *38.01N 43.04E*
143 S15 **Çatak Çayi** ∽ SE Turkey
116 O12 **Çatalca** İstanbul, NW Turkey *41.09N 28.28E*
116 O12 **Çatalca Yarımadası** *physical region* NW Turkey
64 H6 **Catalina** Antofagasta, N Chile *25.19S 69.37W*
Catalonia *see* Cataluña
107 U5 **Cataluña** *Cat.* Catalunya; *Eng.* Catalonia. ◆ *autonomous community* N Spain
Catalunya *see* Cataluña
64 J7 **Catamarca** *off.* Provincia de Catamarca. ◆ *province* NW Argentina
Catamarca *see* San Fernando del Valle de Catamarca
179 Pp11 **Catanauan** Luzon, N Philippines *13.36N 122.19E*
85 M16 **Catandica** Manica, C Mozambique *18.04S 33.10E*
179 Qq11 **Catanduanes Island** *island* N Philippines
62 K8 **Catanduva** São Paulo, S Brazil *21.06S 48.57W*
109 L24 **Catania** Sicilia, Italy, C Mediterranean Sea *37.31N 15.04E*
109 M24 **Catania, Golfo di** *gulf* Sicilia, Italy, C Mediterranean Sea
47 U5 **Cataño** *var.* Cantaño. E Puerto Rico *18.26N 66.06W*
109 O21 **Catanzaro** Calabria, SW Italy *38.53N 16.36E*
109 O22 **Catanzaro Marina** *var.* Marina di Catanzaro. Calabria, S Italy *38.48N 16.33E*
27 Q14 **Catarina** Texas, SW USA *28.19N 99.36W*
179 Qq12 **Catarman** Samar, C Philippines *12.29N 124.34E*
107 S10 **Catarroja** País Valenciano, E Spain *39.24N 0.24W*
23 R11 **Catawba River** ∽ North Carolina/South Carolina, SE USA
179 R12 **Catbalogan** Samar, C Philippines *11.49N 124.55E*
12 I14 **Catchacoma** Ontario, SE Canada *44.43N 78.19W*
43 S15 **Catemaco** Veracruz-Llave, SE Mexico *18.28N 95.10W*
Cathair na Mart *see* Westport
Cathair Saidhbhín *see* Cahersiveen
33 P5 **Cat Head Point** *headland* Michigan, N USA *45.11N 85.37W*
25 Q2 **Cathedral Caverns** *cave* Alabama, S USA *34.36N 86.11W*
37 V16 **Cathedral City** California, W USA *33.45N 116.27W*
26 K10 **Cathedral Mountain** ▲ Texas, SW USA *30.10N 103.39W*
34 G10 **Cathlamet** Washington, NW USA *46.12N 123.24W*
78 G13 **Catió** S Guinea-Bissau *11.17N 15.16W*
57 O10 **Catisimiña** Bolívar, SE Venezuela *4.07N 63.40W*
46 J3 **Cat Island** *island* C Bahamas
10 B9 **Cat Lake** Ontario, S Canada *51.47N 91.52W*
23 P5 **Catlettsburg** Kentucky, S USA *38.24N 82.37W*
193 D24 **Catlins** ∽ South Island, NZ
37 R1 **Catnip Mountain** ▲ Nevada, W USA *41.53N 119.19W*
43 Z11 **Catoche, Cabo** *headland* SE Mexico *21.36N 87.04W*
29 P9 **Catoosa** Oklahoma, C USA *36.11N 95.45W*
43 N10 **Catorce** San Luis Potosí, C Mexico *23.42N 100.49W*
65 I14 **Catriel** Río Negro, C Argentina *37.54S 67.52W*
64 K13 **Catriló** La Pampa, C Argentina *36.24S 63.25W*
60 F11 **Catrimani** Roraima, N Brazil *0.24N 61.30W*
60 E10 **Catrimani, Rio** ∽ N Brazil
20 K11 **Catskill** New York, NE USA *42.13N 73.52W*
20 K11 **Catskill Creek** ∽ New York, NE USA
20 J11 **Catskill Mountains** ▲ New York, NE USA
20 D11 **Cattaraugus Creek** ∽ New York, NE USA
Cattaro *see* Kotor
Cattaro, Bocche di *see* Kotorska, Boka
109 I24 **Cattolica Eraclea** Sicilia, Italy, C Mediterranean Sea *37.27N 13.24E*
85 B14 **Catumbela** ∽ W Angola
85 N14 **Catur** Niassa, N Mozambique *13.50S 35.43E*
84 C10 **Cauale** ∽ NE Angola
179 Pp9 **Cauayan** Luzon, N Philippines *16.55N 121.46E*
56 C12 **Cauca** *off.* Departamento del Cauca. ◆ *province* SW Colombia
49 P5 **Cauca** ∽ SE Venezuela
60 P13 **Caucaia** Ceará, E Brazil *3.43S 38.45W*
56 E7 **Cauca, Río** ∽ N Colombia
56 E7 **Caucasia** Antioquia, NW Colombia *7.58N 75.13W*
143 Q8 **Caucasus** *Rus.* Kavkaz. ▲ Georgia/Russian Federation
64 I10 **Caucete** San Juan, W Argentina *31.37S 68.16W*
107 R11 **Caudete** Castilla-La Mancha, C Spain *38.42N 1.00W*
105 P2 **Caudry** Nord, N France *50.07N 3.24E*
84 D11 **Caungula** Lunda Norte, NE Angola *8.22S 18.37E*
64 G13 **Cauquenes** Maule, C Chile *35.58S 72.22W*
57 N8 **Caura, Río** ∽ C Venezuela
13 V7 **Causapscal** Quebec, SE Canada *48.22N 67.13W*
119 N10 **Căuşeni** *Rus.* Kaushany. E Moldova *46.37N 29.21E*
104 M14 **Caussade** Tarn-et-Garonne, S France *44.10N 1.31E*
104 K17 **Cauterets** Hautes-Pyrénées, S France *42.53N 0.08W*
8 J15 **Caution, Cape** *headland* British Columbia, SW Canada *51.10N 127.43W*
46 H7 **Cauto** ∽ E Cuba
Cauvery *see* Kāveri
104 L3 **Caux, Pays de** *physical region* N France
109 L18 **Cava dei Tirreni** Campania, S Italy *40.42N 14.42E*
106 G6 **Cávado** ∽ N Portugal
Cavaia *see* Kavajë
105 R15 **Cavaillon** Vaucluse, SE France *43.51N 5.01E*
105 U16 **Cavalaire-sur-Mer** Var, SE France *43.10N 6.31E*
108 G6 **Cavalese** *Ger.* Gablös. Trentino-Alto Adige, N Italy *46.18N 11.29E*
31 Q2 **Cavalier** North Dakota, N USA *48.47N 97.37W*
78 L17 **Cavalla** *var.* Cavally, Cavally Fleuve. ∽ Ivory Coast/Liberia
107 Y8 **Cavalleria, Cap de** *var.* Cabo Caballeria. *headland* Menorca, Spain, W Mediterranean Sea *40.04N 4.06E*
192 K2 **Cavalli Islands** *island group* N NZ
Cavally/Cavally Fleuve *see* Cavalla
99 E16 **Cavan** *Ir.* Cabhán. N Ireland *54.00N 7.21W*
99 E16 **Cavan** *Ir.* An Cabhán. *cultural region* N Ireland
108 H8 **Cavarzere** Veneto, NE Italy *45.07N 12.04E*
29 W9 **Cave City** Arkansas, C USA *35.56N 91.33W*
22 K7 **Cave City** Kentucky, S USA *37.08N 85.57W*
67 M25 **Cave Point** *headland* S Tristan da Cunha
23 N5 **Cave Run Lake** ⊠ Kentucky, S USA
60 K11 **Caviana de Fora, Ilha** *var.* Ilha Caviana. *island* N Brazil
Caviana, Ilha *see* Caviana de Fora, Ilha
115 I16 **Cavtat** *It.* Ragusavecchia. Dubrovnik-Neretva, SE Croatia *42.36N 18.13E*
Cawnpore *see* Kānpur
Caxamarca *see* Cajamarca
60 A13 **Caxias** Amazonas, W Brazil *4.27S 71.27W*
60 N13 **Caxias** Maranhão, E Brazil *4.52S 43.19W*
63 I15 **Caxias do Sul** Rio Grande do Sul, S Brazil *29.13S 51.10W*
44 J4 **Caxinas, Punta** *headland* N Honduras *16.01N 86.02W*
84 B11 **Caxito** Bengo, NW Angola *8.34S 13.37E*
142 F14 **Çay** Afyon, W Turkey *38.34N 31.01E*
42 L15 **Cayacal, Punta** *var.* Punta Mongrove. *headland* S Mexico *17.55N 102.09W*
58 C6 **Cayambe** Pichincha, N Ecuador *0.01N 78.10W*
58 C6 **Cayambe** ▲ N Ecuador *0.00N 77.58W*
23 R12 **Cayce** South Carolina, SE USA *33.58N 81.04W*
57 Y10 **Cayenne** ○ (French Guiana) NE French Guiana *4.55N 52.18W*
57 Y10 **Cayenne** ✕ NE French Guiana *4.55N 52.18W*
46 K10 **Cayes** *var.* Les Cayes. SW Haiti *18.10N 73.48W*
47 U6 **Cayey** C Puerto Rico *18.06N 66.09W*
47 U6 **Cayey, Sierra de** ▲ E Puerto Rico
105 N14 **Caylus** Tarn-et-Garonne, S France *44.13N 1.42E*
46 E8 **Cayman Brac** *island* E Cayman Islands
46 D8 **Cayman Islands** ◇ *UK dependent territory* W West Indies
66 D11 **Cayman Trench** *undersea feature* NW Caribbean Sea
49 O3 **Cayman Trough** *undersea feature* NW Caribbean Sea
82 O13 **Caynabo** Togdheer, N Somalia *8.55N 46.28E*
44 F3 **Cayo** ◆ *district* SW Belize
Cayo *see* San Ignacio
45 N9 **Cayos Guerrero** *reef* E Nicaragua
45 O9 **Cayos King** *reef* E Nicaragua
46 E4 **Cay Sal** *islet* SW Bahamas
12 G16 **Cayuga** Ontario, S Canada *42.57N 79.49W*
27 V8 **Cayuga** Texas, SW USA *31.55N 95.57W*
20 G10 **Cayuga Lake** ◎ New York, NE USA
106 K13 **Cazalla de la Sierra** Andalucía, S Spain *37.56N 5.46W*
118 L14 **Căzăneşti** Ialomiţa, SE Romania *44.36N 27.02E*
104 M16 **Cazères** Haute-Garonne, S France *43.15N 1.11E*
114 E10 **Cazin** Federacija Bosna I Hercegovina, NW Bosnia and Herzegovina *44.58N 15.58E*
84 G13 **Cazombo** Moxico, E Angola *11.53S 22.52E*
107 O13 **Cazorla** Andalucía, S Spain *37.55N 3.00W*
Cazza *see* Sušac
106 L4 **Cea** ∽ NW Spain
Ceadâr-Lunga *see* Ciadîr-Lunga
Ceanannas *see* Kells
Ceann Toirc *see* Kanturk
60 O13 **Ceará** *off.* Estado do Ceará. ◆ *state* C Brazil
Ceará *see* Fortaleza
Ceara Abyssal Plain *see* Ceará Plain
61 Q14 **Ceará Mirim** Rio Grande do Norte, E Brazil *5.30S 35.50W*
66 J13 **Ceará Plain** *var.* Ceara Abyssal Plain. *undersea feature* W Atlantic Ocean
66 I13 **Ceará Ridge** *undersea feature* C Atlantic Ocean
Ceatharlach *see* Carlow
45 Q17 **Cébaco, Isla** *island* SW Panama
42 K7 **Ceballos** Durango, C Mexico *26.33N 104.07W*
63 G19 **Cebollatí** Rocha, E Uruguay *33.15S 53.49W*
63 G19 **Cebollatí, Río** ∽ E Uruguay
107 P5 **Cebollera** ▲ N Spain *42.01N 2.40W*
106 M8 **Cebreros** Castilla-León, N Spain *40.27N 4.28W*
179 Qq14 **Cebu** *off.* Cebu City. Cebu, C Philippines *10.16N 123.45E*
179 Qq13 **Cebu** *island* C Philippines
109 J16 **Ceccano** Lazio, C Italy *41.34N 13.19E*
Čechy *see* Bohemia
108 F12 **Cecina** Toscana, C Italy *43.18N 10.31E*
28 K4 **Cedar Bluff Reservoir** ⊠ Kansas, C USA
32 M8 **Cedarburg** Wisconsin, N USA *43.18N 87.58W*
38 J7 **Cedar City** Utah, W USA *37.40N 113.03W*
27 T11 **Cedar Creek** Texas, SW USA *30.04N 97.30W*
30 L7 **Cedar Creek** ∽ North Dakota, N USA
27 U7 **Cedar Creek Reservoir** ⊠ Texas, SW USA
31 W13 **Cedar Falls** Iowa, C USA *42.31N 92.27W*
33 N8 **Cedar Grove** Wisconsin, N USA *43.31N 87.48W*
23 Y6 **Cedar Island** *island* Virginia, NE USA
25 U11 **Cedar Key** Cedar Keys, Florida, SE USA *29.08N 83.03W*
25 U11 **Cedar Keys** *island group* Florida, SE USA
9 V14 **Cedar Lake** ◎ Manitoba, C Canada
12 I11 **Cedar Lake** ◎ Ontario, SE Canada
26 M6 **Cedar Lake** ◎ Texas, SW USA
31 X13 **Cedar Rapids** Iowa, C USA *41.58N 91.39W*
31 X14 **Cedar River** ∽ Iowa/Minnesota, C USA
31 O14 **Cedar River** ∽ Nebraska, C USA
33 P8 **Cedar Springs** Michigan, N USA *43.13N 85.33W*
25 R3 **Cedartown** Georgia, SE USA *34.00N 85.16W*
29 O7 **Cedar Vale** Kansas, C USA *37.06N 96.30W*
37 Q2 **Cedarville** California, W USA *41.30N 120.10W*
106 H1 **Cedeira** Galicia, NW Spain *43.40N 8.03W*
44 H8 **Cedeño** Choluteca, S Honduras *13.10N 87.25W*
43 N10 **Cedral** San Luis Potosí, C Mexico *23.47N 100.40W*
44 I6 **Cedros** Francisco Morazán, C Honduras *14.38N 87.09W*
42 M9 **Cedros** Zacatecas, C Mexico *24.39N 101.47W*
42 B5 **Cedros, Isla** *island* W Mexico
199 Mm6 **Cedros Trench** *undersea feature* E Pacific Ocean
190 E7 **Ceduna** South Australia *32.09S 133.43E*
112 D10 **Cedynia** *Ger.* Zehden. Zachodniopomorskie, W Poland *52.54N 14.15E*
82 P12 **Ceelaayo** Sanaag, N Somalia *11.18N 49.20E*
83 O16 **Ceel Buur** *It.* El Bur; Galguduud, C Somalia *4.36N 46.33E*
83 N15 **Ceel Dheere** *var.* Ceel Dher, *It.* El Dere. Galguduud, C Somalia *5.18N 46.07E*
Ceel Dher *see* Ceel Dheere
83 P14 **Ceel Xamure** Mudug, E Somalia *7.15N 48.55E*
82 O12 **Ceerigaabo** *var.* Erigabo, Erigavo. Sanaag, N Somalia *10.34N 47.22E*
109 J23 **Cefalu** *anc.* Cephaloedium. Sicilia, Italy, C Mediterranean Sea *38.01N 14.01E*
107 N6 **Cega** ∽ N Spain
113 K23 **Cegléd** *prev.* Czegléd. Pest, C Hungary *47.08N 19.45E*
115 N18 **Čegrane** W FYR Macedonia *41.50N 20.59E*
107 Q13 **Cehegín** Murcia, SE Spain *38.04N 1.48W*
142 K12 **Çekerek** Yozgat, N Turkey *40.04N 35.30E*
109 J15 **Celano** Abruzzo, C Italy *42.06N 13.33E*
106 H4 **Celanova** Galicia, NW Spain *42.09N 7.58W*
44 F6 **Celaque, Cordillera de** ▲ W Honduras
43 N13 **Celaya** Guanajuato, C Mexico *20.31N 100.48W*
Celebes *see* Sulawesi
198 Ff8 **Celebes Basin** *undersea feature* SE South China Sea
175 Qq4 **Celebes Sea** *Ind.* Laut Sulawesi. *sea* Indonesia/Philippines
43 W12 **Celestún** Yucatán, E Mexico *20.49N 90.22W*
33 Q12 **Celina** Ohio, N USA *40.33N 84.34W*
22 L8 **Celina** Tennessee, S USA *36.30N 85.30W*
27 U5 **Celina** Texas, SW USA *33.19N 96.46W*
114 G11 **Čelinac Donji** Republika Srpska, N Bosnia and Herzegovina *44.43N 17.19E*
189 V10 **Celje** *Ger.* Cilli. C Slovenia *46.16N 15.14E*
113 G23 **Celldömölk** Vas, W Hungary *47.16N 17.10E*
102 J12 **Celle** *var.* Zelle. Niedersachsen, N Germany *52.37N 10.04E*
101 D19 **Celles** Hainaut, SW Belgium *50.42N 3.25E*
106 I7 **Celorico da Beira** Guarda, N Portugal *40.37N 7.24W*
Celovec *see* Klagenfurt
66 M7 **Celtic Sea** *Ir.* An Mhuir Cheilteach. *sea* SW British Isles
66 N7 **Celtic Shelf** *undersea feature* E Atlantic Ocean
116 L13 **Çeltik Gölü** ◎ NW Turkey
115 M14 **Čemerno** ▲ C Yugoslavia
175 Oo16 **Cempi, Teluk** *bay* Nusa Tenggara, S Indonesia
107 Q12 **Cenajo, Embalse del** ⊠ S Spain
176 Ww10 **Cenderawasih, Teluk** *var.* Teluk Irian, Teluk Sarera. *bay* W Pacific Ocean
107 P4 **Cenicero** La Rioja, N Spain *42.28N 2.37W*
108 E9 **Ceno** ∽ NW Italy
104 K13 **Cenon** Gironde, SW France *44.51N 0.33W*
12 K13 **Centennial Lake** ◎ Ontario, SE Canada
Centennial State *see* Colorado
39 S7 **Center** Colorado, C USA *37.45N 106.06W*
31 Q13 **Center** Nebraska, C USA *42.33N 97.51W*
30 M5 **Center** North Dakota, N USA *47.07N 101.18W*
27 X8 **Center** Texas, SW USA *31.46N 94.10W*
31 W8 **Center City** Minnesota, N USA *45.22N 92.46W*
38 L5 **Centerfield** Utah, W USA *39.07N 111.49W*
22 K9 **Center Hill Lake** ⊠ Tennessee, S USA
31 X13 **Center Point** Iowa, C USA *42.11N 91.47W*
27 R11 **Center Point** Texas, SW USA *29.56N 99.01W*
31 W16 **Centerville** Iowa, C USA *40.42N 92.49W*
29 W7 **Centerville** Missouri, C USA *37.27N 91.01W*
31 R12 **Centerville** South Dakota, N USA *43.07N 96.57W*
22 I9 **Centerville** Tennessee, S USA *35.43N 87.27W*
27 V9 **Centerville** Texas, SW USA *31.15N 95.58W*
42 M5 **Centinela, Picacho del** ▲ NE Mexico *29.07N 102.40W*
108 G9 **Cento** Emilia-Romagna, N Italy *44.43N 11.16E*
Centrafricaine, République *see* Central African Republic
41 S8 **Central** Alaska, USA *65.34N 144.48W*
39 P15 **Central** New Mexico, SW USA *32.46N 108.09W*
85 H18 **Central** ◆ *district* E Botswana
144 E10 **Central** ◆ *district* C Israel
83 I19 **Central** ◆ *province* C Kenya
84 M13 **Central** ◆ *region* C Malawi
159 P12 **Central** ◆ *zone* C Nepal
194 J15 **Central** ◆ *province* S PNG
65 I21 **Central** ◆ *department* C Paraguay
195 W15 **Central** *off.* Central Province. ◆ *province* S Solomon Islands
85 J14 **Central** ◆ *province* C Zambia
119 P11 **Central** ✕ (Odesa) Odes'ka Oblast', SW Ukraine *46.26N 30.41E*
Central *see* Centre
81 H14 **Central African Republic** *var.* République Centrafricaine, *abbrev.* CAR; *prev.* Ubangi-Shari, Oubangui-Chari, Territoire de l'Oubangui-Chari. ◆ *republic* C Africa
198 G6 **Central Basin Trough** *undersea feature* W Pacific Ocean
Central Borneo *see* Kalimantan Tengah
155 P12 **Central Brāhui Range** ▲ W Pakistan
Central Celebes *see* Sulawesi Tengah
31 Y13 **Central City** Iowa, C USA *42.12N 91.31W*
22 I6 **Central City** Kentucky, S USA *37.17N 87.07W*
31 P15 **Central City** Nebraska, C USA *41.04N 97.59W*
50 D6 **Central, Cordillera** ▲ W Bolivia
56 D11 **Central, Cordillera** ▲ W Colombia
44 M13 **Central, Cordillera** ▲ C Costa Rica
47 N9 **Central, Cordillera** ▲ C Dominican Republic
45 R16 **Central, Cordillera** ▲ C Panama
179 P8 **Central, Cordillera** ▲ Luzon, N Philippines
47 S6 **Central, Cordillera** ▲ Puerto Rico
44 H7 **Central District** *var.* Tegucigalpa. ◆ *district* C Honduras
32 L15 **Centralia** Illinois, N USA *38.31N 89.07W*
29 U4 **Centralia** Missouri, C USA *39.12N 92.08W*
34 G9 **Centralia** Washington, NW USA *46.43N 122.57W*
Central Indian Ridge *see* Mid-Indian Ridge
Central Java *see* Jawa Tengah
Central Kalimantan *see* Kalimantan Tengah
154 L14 **Central Makrān Range** ▲ W Pakistan
199 J8 **Central Pacific Basin** *undersea feature* C Pacific Ocean
61 M19 **Central, Planalto** *var.* Brazilian Highlands. ▲ E Brazil
34 F15 **Central Point** Oregon, NW USA *42.22N 122.55W*
161 K25 **Central Province** ◆ *province* C Sri Lanka
Central Provinces and Berar *see* Madhya Pradesh
194 G11 **Central Range** ▲ NW PNG
Central Russian Upland *see* Srednerusskaya Vozvyshennost'
Central Siberian Plateau/Central Siberian Uplands *see* Srednesibirskoye Ploskogor'ye
106 K8 **Central, Sistema** ▲ C Spain
Central Sulawesi *see* Sulawesi Tengah
37 N3 **Central Valley** California, W USA *40.39N 122.21W*
37 P8 **Central Valley** *valley* California, W USA
25 Q3 **Centre** Alabama, S USA *34.09N 85.40W*
81 E15 **Centre** *Eng.* Central. ◆ *province* C Cameroon
104 M8 **Centre** ◆ *region* N France
181 Y16 **Centre de Flacq** E Mauritius *20.12S 57.43E*
57 Y9 **Centre Spatial Guyanais** *space station* N French Guiana *5.11N 52.49W*
25 O5 **Centreville** Alabama, S USA *32.58N 87.08W*
23 X3 **Centreville** Maryland, NE USA *39.02N 76.04W*
24 J7 **Centreville** Mississippi, S USA *31.05N 91.04W*
Centum Cellae *see* Civitavecchia
166 M14 **Cenxi** Guangxi Zhuangzu Zizhiqu, S China *22.58N 111.00E*
Ceos *see* Kéa
Cephaloedium *see* Cefalu
114 I9 **Čepin** *Hung.* Csepén. Osijek-Baranja, E Croatia *45.32N 18.33E*
174 Ll15 **Cepu** *prev.* Tjepoe, Tjepu. Jawa, C Indonesia *7.07S 111.34E*
Ceram *see* Seram, Pulau
175 T10 **Ceram Sea** *Ind.* Laut Seram. *sea* E Indonesia
198 G9 **Ceram Trough** *undersea feature* W Pacific Ocean
Cerasus *see* Giresun
38 I10 **Cerbat Mountains** ▲ Arizona, SW USA
105 P17 **Cerbère, Cap** *headland* S France *42.28N 3.15E*
106 F13 **Cercal do Alentejo** Setúbal, S Portugal *37.48N 8.40W*
113 A18 **Čerchov** *Ger.* Czerkow. ▲ W Czech Republic *49.24N 12.47E*
105 O13 **Cère** ∽ C France
63 A16 **Ceres** Santa Fe, C Argentina *29.55S 61.55W*
61 K18 **Ceres** Goiás, C Brazil *15.21S 49.34W*
Ceresio *see* Lugano, Lago di
105 O17 **Céret** Pyrénées-Orientales, S France *42.30N 2.43E*
56 E6 **Cereté** Córdoba, NW Colombia *8.51N 75.48W*
312 I17 **Cerf, Île au** *island* Inner Islands, NE Seychelles
101 G22 **Cerfontaine** Namur, S Belgium *50.08N 4.25E*
Cergy-Pontoise *see* Pontoise
109 N16 **Cerignola** Puglia, SE Italy *41.16N 15.52E*
Cerigo *see* Kýthira
105 O9 **Cérilly** Allier, C France *46.38N 2.51E*
142 I11 **Çerkeş** Çankırı, N Turkey *40.51N 32.52E*
142 D10 **Çerkezköy** Tekirdağ, NW Turkey *41.18N 27.58E*
189 T12 **Cerknica** *Ger.* Zirknitz. SW Slovenia *45.48N 14.21E*
189 S11 **Cerkno** W Slovenia *46.07N 13.58E*
118 F10 **Cermei** *Hung.* Csermő. Arad, W Romania *46.33N 21.50E*
143 O15 **Çermik** Dıyarbakır, SE Turkey *38.09N 39.27E*
114 I10 **Cerna** Vukovar-Srijem, E Croatia *45.10N 18.36E*
Cernăuţi *see* Chernivtsi
118 M14 **Cernavodă** Constanţa, SW Romania *44.19N 28.01E*
105 U7 **Cernay** Haut-Rhin, NE France *47.49N 7.10E*
Černice *see* Schwarzach
43 O8 **Cerralvo** Nuevo León, NE Mexico *26.01N 99.37W*
42 G9 **Cerralvo, Isla** *island* W Mexico
109 L16 **Cerreto Sannita** Campania, S Italy *41.17N 14.39E*
115 L20 **Cërrik** *var.* Cerriku. Elbasan, C Albania *41.01N 19.55E*
Cerriku *see* Cërrik
43 O11 **Cerritos** San Luis Potosí, C Mexico *22.25N 100.16W*
62 K11 **Cerro Azul** Paraná, S Brazil *24.48S 49.13W*
63 F18 **Cerro Chato** Treinta y Tres, E Uruguay *33.08S 55.07W*

◆ Country ● Country Capital ◇ Dependent Territory ○ Dependent Territory Capital ◆ Administrative Region ✕ International Airport ▲ Mountain ▲ Mountain Range Volcano ∽ River ◎ Lake ⊠ Reservoir

63 F19 **Cerro Colorado** Florida, S Uruguay 33.52S 55.33W
58 E13 **Cerro de Pasco** Pasco, C Peru 10.43S 76.15W
63 G18 **Cerro Largo** ◆ *department* NE Uruguay
63 G14 **Cêrro Largo** Rio Grande do Sul, S Brazil 28.10S 54.43W
48 E7 **Cerrón Grande, Embalse** ⊠ N El Salvador
65 I14 **Cerros Colorados, Embalse** ⊠ W Argentina
107 V5 **Cervera** Cataluña, NE Spain 41.40N 1.16E
106 M3 **Cervera del Pisuerga** Castilla-León, N Spain 42.51N 4.30W
107 Q5 **Cervera del Río Alhama** La Rioja, N Spain 42.01N 1.58W
109 H15 **Cerveteri** Lazio, C Italy 42.00N 12.06E
108 H10 **Cesena** Emilia-Romagna, N Italy 44.14N 12.22E
108 J7 **Cervignano del Friuli** Friuli-Venezia Giulia, NE Italy 45.49N 13.18E
109 L17 **Cervinara** Campania, S Italy 41.01N 14.36E
108 B6 **Cervino, Monte** *var.* Matterhorn. ▲ Italy/Switzerland *see also* Matterhorn 46.00N 7.39E
105 Y14 **Cervione** Corse, France, C Mediterranean Sea 42.22N 9.28E
106 I1 **Cervo** Galicia, NW Spain 43.39N 7.25W
56 F5 **Cesar** *off.* Departamento del Cesar. ◆ *province* N Colombia
108 H10 **Cesena** Emilia-Romagna, N Italy
108 I10 **Cesenatico** Emilia-Romagna, N Italy 44.12N 12.24E
120 H8 **Cēsis** *Ger.* Wenden. Cēsis, C Latvia 57.19N 25.17E
113 D15 **Česká Lípa** *Ger.* Böhmisch-Leipa. Liberecký Kraj, N Czech Republic 50.40N 14.32E
Česká Republika *see* Czech Republic
113 F17 **Česká Třebová** *Ger.* Böhmisch-Trübau. Pardubický Kraj, C Czech Republic 49.54N 16.27E
113 D19 **České Budějovice** *Ger.* Budweis. Budějovický Kraj, S Czech Republic 48.58N 14.28E
113 D19 **České Velenice** Budějovický Kraj, S Czech Republic 48.49N 14.57E
113 E18 **Českomoravská Vrchovina** *var.* Českomoravská Vysočina, *Eng.* Bohemian-Moravian Highlands, *Ger.* Böhmisch-Mährische Höhe. ▲▲ S Czech Republic
Českomoravská Vysočina *see* Českomoravská Vrchovina
113 C19 **Český Krumlov** *Ger.* Böhmisch-Krumau, *Ger.* Krummau. Budějovický Kraj, S Czech Republic 48.48N 14.18E
114 F8 **Česma** ← N Croatia
142 A14 **Çeşme** İzmir, W Turkey 38.19N 26.19E
Cess *see* Cestos
191 T8 **Cessnock** New South Wales, SE Australia 32.51S 151.21E
78 K17 **Cestos** *var.* Cess. ← S Liberia
120 I9 **Cesvaine** Madona, E Latvia 56.58N 26.15E
118 G14 **Cetate** Dolj, SW Romania 44.06N 23.03E
Cetatea Albă *see* Bilhorod-Dnistrovs'kyy
115 J17 **Cetinje** *It.* Cettigne. Montenegro, SW Yugoslavia 42.23N 18.55E
109 N20 **Cetraro** Calabria, S Italy 39.30N 15.59E
Cette *see* Sète
196 A14 **Cetti Bay** *bay* SW Guam
Cettigne *see* Cetinje
106 L17 **Ceuta** *var.* Sebta. Ceuta, Spain, N Africa 35.52N 5.19W
90 C15 **Ceuta** *enclave* Spain, N Africa
108 B9 **Ceva** Piemonte, NW Italy 44.24N 8.01E
105 P14 **Cévennes** ▲ S France
110 G10 **Cevio** Ticino, S Switzerland 46.18N 8.36E
142 K16 **Ceyhan** Adana, S Turkey 37.01N 35.48E
142 K17 **Ceyhan Nehri** ← S Turkey
143 P17 **Ceylanpınar** Şanlıurfa, SE Turkey 36.53N 40.01E
Ceylon *see* Sri Lanka
181 R6 **Ceylon Plain** *undersea feature* N Indian Ocean
Ceyre to the Caribs *see* Marie-Galante
105 Q4 **Cèze** ← S France
152 H15 **Chaacha** *Turkm.* Chäche. Akhalskiy Velayat, S Turkmenistan 36.49N 60.33E
131 P6 **Chaadayevka** Penzenskaya Oblast', W Russian Federation 53.07N 45.55E
178 H12 **Cha-Am** Phetchaburi, SW Thailand 12.48N 99.58E
149 W15 **Chābahar** *var.* Chāh Bahār, Chahbar. Sīstān va Balūchestān, SE Iran 25.21N 60.38E
63 B19 **Chabas** Santa Fe, C Argentina 33.16S 61.22W
105 T10 **Chablais** *physical region* E France
63 B20 **Chacabuco** Buenos Aires, E Argentina 34.38S 60.31W
44 K8 **Chachagón, Cerro** ▲ N Nicaragua 13.18N 85.39W
58 C10 **Chachapoyas** Amazonas, NW Peru 6.13S 77.54W
Chàche *see* Chaacha
121 O18 **Chachersk** *Rus.* Chechersk. Homyel'skaya Voblasts', SE Belarus 52.54N 30.54E
121 N16 **Chachevichy** *Rus.* Chechevichi. Mahilyowskaya Voblasts', E Belarus 53.31N 29.49E
85 B14 **Chaco** *off.* Provincia de Chaco. ◆ *province* NE Argentina
Chaco *see* Gran Chaco
64 M6 **Chaco Austral** *physical region* N Argentina
64 M6 **Chaco Boreal** *physical region* N Argentina
64 M6 **Chaco Central** *physical region* N Argentina

41 Y15 **Chacon, Cape** *headland* Prince of Wales Island, Alaska, USA 54.41N 132.00W
80 H9 **Chad** *off.* Republic of Chad, *Fr.* Tchad. ◆ *republic* C Africa
126 Hh16 **Chadan** Respublika Tyva, S Russian Federation 51.16N 91.25E
23 U12 **Chadbourn** North Carolina, SE USA 34.19N 78.49W
85 L14 **Chadiza** Eastern, E Zambia 14.04S 32.27E
69 Q7 **Chad, Lake** *Fr.* Lac Tchad. ◎ C Africa
126 J13 **Chadobets** ← C Russian Federation
30 J12 **Chadron** Nebraska, C USA 42.48N 102.57W
Chadyr-Lunga *see* Ciadîr-Lunga
169 W14 **Chaeryŏng** SW North Korea 38.22N 125.35E
107 P17 **Chafarinas, Islas** *island group* S Spain
29 Y7 **Chaffee** Missouri, C USA 37.10N 89.39W
154 L12 **Chāgai Hills** *var.* Chāh Gay. ▲ Afghanistan/Pakistan
126 M12 **Chagda** Respublika Sakha (Yakutiya), NE Russian Federation 58.43N 130.38E
155 N5 **Chaghasarāy** *var.* Asadābād. Chakhcheran, Cheghcheran, Qala Ahangarān. Ghowr, C Afghanistan 34.28N 65.18E
105 R9 **Chagny** Saône-et-Loire, C France 46.54N 4.45E
181 Q7 **Chagos Archipelago** *var.* Oil Islands. *island group* British Indian Ocean Territory
133 O15 **Chagos Bank** *undersea feature* C Indian Ocean
133 O14 **Chagos-Laccadive Plateau** *undersea feature* N Indian Ocean
181 Q7 **Chagos Trench** *undersea feature* N Indian Ocean
45 T14 **Chagres, Río** ← C Panama
47 U14 **Chaguanas** Trinidad, Trinidad and Tobago 10.29N 61.24W
55 M6 **Chaguaramas** Guárico, N Venezuela 9.21N 66.15W
152 C9 **Chagyl** Balkanskiy Velayat, NW Turkmenistan 40.48N 55.21E
148 M9 **Chahār Maḥall and Bakhtīārī** *see* Chahār Maḥall va Bakhtīārī
148 M9 **Chahār Maḥall va Bakhtīārī** *off.* Ostān-e Chahār Maḥall va Bakhtīārī, *var.* Chahārmahāll and Bakhtīārī. ◆ *province* SW Iran
Chāh Bahār/Chahbar *see* Chābahār
149 V13 **Chāh Derāz** Sīstān va Balūchestān, SE Iran 26.45N 61.04E
118 Hh10 **Chai Badan** Lop Buri, C Thailand 15.07N 101.03E
159 Q16 **Chāībāsa** Bihār, N India 22.34N 85.48E
81 E19 **Chaillu, Massif du** ▲ C Gabon
178 Hh10 **Chai Nat** *var.* Chainat, Jainat, Jayanath. Chai Nat, C Thailand 15.12N 100.12E
67 M14 **Chain Fracture Zone** *tectonic feature* E Atlantic Ocean
181 N5 **Chain Ridge** *undersea feature* W Indian Ocean
Chairn, Ceann an *see* Carnsore Point
164 L5 **Chaiwopu** Xinjiang Uygur Zizhiqu, W China 43.31N 87.55E
178 I10 **Chaiyaphum** *var.* Jayabum. Chaiyaphum, C Thailand 15.49N 102.03E
64 N10 **Chajarí** Entre Ríos, E Argentina 30.45S 57.57W
44 C5 **Chajul** Quiché, W Guatemala 15.28N 91.02W
85 K16 **Chakari** Mashonaland West, N Zimbabwe 18.04S 29.49E
154 J9 **Chakhānsūr** Nīmrūz, SW Afghanistan 31.11N 62.06E
Chakhānsūr *see* Nīmrūz
Chakhcharān *see* Chaghcharān
155 V8 **Chak Jhumra** *var.* Jhumra. Punjab, E Pakistan 31.33N 73.13E
152 I16 **Chaknakdysonga** Akhalskiy Velayat, S Turkmenistan 35.39N 61.24E
159 P16 **Chakradharpur** Bihār, N India 22.37N 85.28E
158 J8 **Chakrāta** Uttar Pradesh, N India 30.42N 77.52E
155 U7 **Chakwāl** Punjab, NE Pakistan 32.56N 72.49E
59 F17 **Chala** Arequipa, SW Peru 15.52S 74.13W
104 K12 **Chalais** Charente, W France 45.16N 0.02E
110 D10 **Chalais** Valais, SW Switzerland 46.18N 7.37E
117 J20 **Chalándri** *var.* Halandri; *prev.* Khalándrion. *prehistoric site* Sýros, Kykládes, Greece, Aegean Sea 37.28N 24.56E
24 H6 **Chalan Kanoa** Saipan, S Northern Mariana Islands 15.07S 145.43E
196 C16 **Chalan Pago** C Guam
44 A6 **Chalap Dalam/Chalap Dalan** *see* Chehel Abdālān, Kūh-e
44 F7 **Chalatenango** Chalatenango, N El Salvador 14.03N 88.54W
44 A9 **Chalatenango** ◆ *department* NW El Salvador
85 P15 **Chalaua** Nampula, NE Mozambique 16.04S 39.08E
83 I16 **Chalbi Desert** *desert* N Kenya
44 D7 **Chalchuapa** Santa Ana, W El Salvador 13.58N 89.39W
Chalcidice *see* Chalkidikí
Chalcis *see* Chalkída
106 J11 **Châlette-sur-Loing** Loiret, C France 48.01N 2.45E
13 X8 **Chaleur Bay** *Fr.* Baie des Chaleurs. *bay* New Brunswick/Quebec, E Canada
104 M12 **Châlus** Haute-Vienne, C France 45.38N 1.00E
59 G16 **Chalhuanca** Apurímac, S Peru 14.21S 73.16W
160 F12 **Chālisgaon** Mahārāshtra, C India 20.28N 75.10E
117 N23 **Chálki** *island* Dodekánisos, Greece, Aegean Sea

117 F16 **Chalkiádes** Thessalía, C Greece 39.24N 22.25E
117 H18 **Chalkída** *var.* Halkida; *prev.* Khalkís, *anc.* Chalcis. Evvoia, E Greece 38.27N 23.37E
117 G14 **Chalkidikí** *var.* Khalkidhikí; *anc.* Chalcidice. *peninsula* NE Greece
193 A24 **Chalky Inlet** *inlet* South Island, NZ
41 S7 **Chalkyitsik** Alaska, USA 66.39N 143.43W
104 I9 **Challans** Vendée, NW France 46.51N 1.52W
59 K19 **Challapata** Oruro, SW Bolivia 19.02S 66.46W
199 H7 **Challenger Deep** *undersea feature* W Pacific Ocean
200 L12 **Challenger Fracture Zone** *tectonic feature* E Pacific Ocean
199 Ii13 **Challenger Plateau** *undersea feature* E Tasman Sea
35 P13 **Challis** Idaho, NW USA 44.31N 114.14W
24 M10 **Chalmette** Louisiana, S USA 29.56N 89.57W
128 J11 **Chalna** Respublika Kareliya, NW Russian Federation 61.53N 33.59E
105 Q5 **Châlons-en-Champagne** *var.* Châlons-sur-Marne, *hist.* Arcae Remorum, *anc.* Carolopois. Marne, NE France 48.58N 4.22E
105 Q5 **Châlons-sur-Marne** *see* Châlons-en-Champagne
105 R9 **Chalon-sur-Saône** *anc.* Cabillonum. Saône-et-Loire, C France 46.46N 4.51E
149 N4 **Chālūs** Māzandarān, N Iran 36.40N 51.25E
104 M11 **Châlus** Haute-Vienne, C France 45.38N 1.00E
103 N20 **Cham** Bayern, SE Germany 49.13N 12.40E
110 F7 **Cham** Zug, N Switzerland 47.10N 8.28E
39 R8 **Chama** New Mexico, SW USA 36.54N 106.34W
Cha Mai *see* Thung Song
85 E22 **Chamaites** Karas, S Namibia 27.13S 17.55E
155 O9 **Chaman** Baluchistān, SW Pakistan 30.55N 66.27E
39 R9 **Chama, Río** ← New Mexico, SW USA
158 J6 **Chamba** Himāchal Pradesh, N India 32.33N 76.10E
83 I25 **Chamba** Ruvuma, S Tanzania 11.33S 37.01E
156 H12 **Chambal** ← C India
9 U16 **Chamberlain** Saskatchewan, S Canada 50.49N 105.29W
31 O11 **Chamberlain** South Dakota, N USA 43.48N 99.19W
21 R3 **Chamberlain Lake** ◎ Maine, NE USA
41 S5 **Chamberlin, Mount** ▲ Alaska, USA 69.16N 144.54W
39 O11 **Chambers** Arizona, SW USA 35.11N 109.25W
20 F16 **Chambersburg** Pennsylvania, NE USA 39.54N 77.39W
33 N5 **Chambers Island** *island* Wisconsin, N USA
105 T11 **Chambéry** *anc.* Cambéria. Savoie, E France 45.34N 5.55E
84 L12 **Chambeshi** Northern, NE Zambia 10.55S 31.07E
84 L12 **Chambeshi** ← NE Zambia
76 M6 **Chambi, Jebel** *var.* Jabal ash Sha'nabī. ▲ W Tunisia 35.16N 8.39E
13 Q7 **Chambord** Quebec, SE Canada 48.25N 72.02W
194 G10 **Chambri Lake** ◎ W PNG
245 I11 **Chambri** ◆ Iraq 31.17N 45.05E
245 T4 **Chamchamāl** N Iraq 35.31N 44.49E
42 J14 **Chamela** Jalisco, SW Mexico 19.33N 105.04W
44 G5 **Chamelecón, Río** ← NW Honduras
64 J9 **Chamical** La Rioja, C Argentina 30.21S 66.19W
117 L23 **Chamilí** *island* Kykládes, Greece, Aegean Sea
178 I13 **Chamnar** Kaôh Kông, SW Cambodia 11.43N 103.32E
158 K9 **Chamoli** Uttar Pradesh, N India 30.22N 79.19E
105 U11 **Chamonix-Mont-Blanc** Haute-Savoie, E France 45.55N 6.52E
160 L11 **Chāmpa** Madhya Pradesh, C India 22.01N 82.42E
155 N9 **Champagne** Yukon Territory, W Canada 60.48N 136.22W
105 Q5 **Champagne** *cultural region* N France
Champagne *see* Campania
105 Q5 **Champagne-Ardenne** ◆ *region* N France
105 R5 **Champagnole** Jura, E France 46.43N 5.55E
32 M13 **Champaign** Illinois, N USA 40.07N 88.14W
178 Jj11 **Champasak** Champasak, S Laos 14.50N 105.51E
105 U6 **Champ de Feu** ▲ NE France 48.27N 7.15E
1 I07 **Champdoré, Lac** ◎ Quebec, NE Canada
43 A6 **Champerico** Retalhuleu, SW Guatemala 14.18N 91.54W
10 C11 **Champéry** Valais, SW Switzerland 46.12N 6.52E
20 L6 **Champlain** New York, NE USA 44.58N 73.25W
161 I10 **Champlain Canal** *canal* New York, NE USA
20 L9 **Champlain, Lac** ◎ Quebec, SE Canada
20 L7 **Champlain, Lake** ◎ Canada/USA
105 S7 **Champlitte** Haute-Saône, E France 47.36N 5.31E
43 X13 **Champotón** Campeche, SE Mexico 19.18N 90.43W
106 G10 **Chamusca** Santarém, C Portugal 39.21N 8.28W
121 O19 **Chamyarysy** *Rus.* Chemerisy. Homyel'skaya Voblasts', SE Belarus 51.42N 30.26E
26 M2 **Channing** Texas, SW USA 35.40N 102.19W
V 12 **Chantabun/Chantaburi** *see* Chanthaburi

Chanáil Mhór, An *see* Grand Canal
Chanak *see* Çanakkale
64 G7 **Chañaral** Atacama, N Chile 26.19S 70.34W
106 H13 **Chança, Rio** *var.* Chanza. ← Portugal/Spain
59 D14 **Chancay** Lima, W Peru 11.33S 77.16W
Chan-chiang/Chanchiang *see* Zhanjiang
64 G13 **Chanco** Maule, C Chile 35.43S 72.35W
41 R7 **Chandalar** Alaska, USA 67.30N 148.29W
41 R6 **Chandalar River** ← Alaska, USA
159 S16 **Chandannagar** *prev.* Chandernagore. West Bengal, E India 22.52N 88.21E
158 K10 **Chandausi** Uttar Pradesh, N India 28.27N 78.43E
24 M10 **Chandeleur Islands** *island group* Louisiana, S USA
24 M9 **Chandeleur Sound** *sound* N Gulf of Mexico
158 I8 **Chandigarh** Punjab, N India 30.41N 76.51E
159 Q16 **Chāndil** Bihār, NE India 22.58N 86.04E
190 D2 **Chandler** South Australia 26.59S 133.22E
13 Y7 **Chandler** Quebec, SE Canada 48.21N 64.40W
38 L14 **Chandler** Arizona, SW USA 33.18N 111.50W
25 O10 **Chandler** Oklahoma, C USA 35.42N 96.52W
27 V7 **Chandler** Texas, SW USA 32.18N 95.28W
41 Q6 **Chandler River** ← Alaska, USA
58 I13 **Chandless, Río** ← E Peru
169 N9 **Chandmanī** Dornogovĭ, SE Mongolia 45.36N 110.30E
12 J13 **Chandos Lake** ◎ Ontario, SE Canada
159 U15 **Chandpur** Chittagong, C Bangladesh 23.13N 90.43E
160 I13 **Chandrapur** Mahārāshtra, C India 19.58N 79.21E
85 J15 **Changa** Southern, S Zambia 16.24S 28.27E
Changan *see* Xi'an, Shaanxi, China
Chang'an *see* Rong'an, Guangxi Zhuangzu Zizhiqu, China
161 G23 **Changanācheri** Kerala, SW India 9.27N 76.34E
85 M19 **Changane** ← S Mozambique
85 M16 **Changara** Tete, NW Mozambique 16.54S 33.15E
169 X11 **Changbai** *var.* Changbai Chaoxianzu Zizhixian. Jilin, NE China 41.25N 128.08E
Changbai Chaoxianzu Zizhixian *see* Changbai
169 X11 **Changbai Shan** ▲ NE China
169 V10 **Changchun** *var.* Ch'angch'un, Ch'ang-ch'un; *prev.* Hsinking. Jilin, NE China 43.52N 125.18E
160 M10 **Changde** Hunan, S China 29.04N 111.42E
167 S13 **Changhua** *Jap.* Shōka. C Taiwan 24.06N 120.31E
174 I7 **Changi** ✈ (Singapore) E Singapore 1.22N 103.58E
165 U5 **Changji** Xinjiang Uygur Zizhiqu, NW China 44.02N 87.12E
150 O13 **Chang Jiang** *var.* Yangtze Kiang, *Eng.* Yangtze. ← C China
166 L17 **Changjiang Kou** *estuary* Shilu, Hainan, S China 19.16N 109.09E
178 Q7 **Changjiang Kou** *delta* E China
178 I6 **Chang, Ko** *island* S Thailand
167 Q2 **Changli** Hebei, E China 39.43N 119.13E
169 V10 **Changling** Jilin, NE China 44.15S 124.03E
167 N11 **Changning** *var.* Ch'angsha. Hunan, S China 28.10N 113.00E
167 Q10 **Changsha** Zhejiang, SE China 28.54N 118.26E
169 V11 **Changshan Qundao** *island group* NE China
167 S8 **Changshu** *var.* Ch'ang-shu. Jiangsu, E China 31.36N 120.42E
169 V11 **Changtu** Liaoning, NE China 42.49N 123.58E
45 P14 **Changuinola** Bocas del Toro, NW Panama 9.25S 82.31W
155 N9 **Changweiliang** Qinghai, W China 38.24N 92.07E
158 K6 **Changwu** Shaanxi, C China 35.12N 107.45E
166 M9 **Changyang** Hubei, C China 30.45N 111.13E
169 W14 **Changyŏn** SW North Korea 38.19N 125.14E
167 N5 **Changzhi** Shanxi, C China 36.09N 113.01E
167 R8 **Changzhou** Jiangsu, E China 31.35N 119.50E
117 H24 **Chaniá** *var.* Hania, Khaniá, *Eng.* Canea; *anc.* Cydonia. Kríti, Greece, E Mediterranean Sea 35.31N 24.01E
64 J5 **Chañi, Nevado de** ▲ NW Argentina 24.09S 65.44W
117 H24 **Chanión, Kólpos** *gulf* Kríti, Greece, E Mediterranean Sea
80 L10 **Chankiri** *see* Çankırı
23 M11 **Channahon** Illinois, N USA 41.25N 88.13W
161 H20 **Channapatna** Karnātaka, E India 12.43N 77.13E
99 K26 **Channel Islands** *Fr.* Îles Normandes. *island group* S English Channel
37 N16 **Channel Islands** *island group* California, W USA
37 T7 **Channel Islands National Park** *national park* California, W USA
13 S13 **Channel-Port aux Basques** Newfoundland, SE Canada 47.35N 59.02W
99 Q22 **Channel, The** *see* English Channel
99 Q23 **Channel Tunnel** *tunnel* France/UK

106 H3 **Chantada** Galicia, NW Spain 42.36N 7.46W
178 I12 **Chanthaburi** *var.* Chantabun, Chantaburi. Chantaburi, S Thailand 12.34N 102.07E
105 O4 **Chantilly** Oise, N France 49.12N 2.28E
15 Kk4 **Chantrey Inlet** *inlet* Nunavut, N Canada
245 V12 **Chanūn as Sa'ūdī** S Iraq 31.04N 46.00E
25 Q6 **Chanute** Kansas, C USA 37.40N 95.27W
125 G13 **Chany, Ozero** ◎ C Russian Federation
Chanza *see* Chança, Rio
167 P8 **Chao Hu** ◎ E China
178 Hh11 **Chao Phraya, Mae Nam** ← C Thailand
169 T8 **Chaor He** ← NE China
Chaouèn *see* Chefchaouen
169 S11 **Chaoyang** Guangdong, S China 23.16N 116.30E
169 T12 **Chaoyang** Liaoning, NE China 41.33N 120.28E
Chaoyang *see* Huinan, Jilin, China
Chaoyang *see* Jiayin, Heilongjiang, China
167 Q14 **Chaozhou** *var.* Chaoan, Chao'an, Ch'ao-an; *prev.* Chaochow. Guangdong, SE China 23.39N 116.34E
60 N13 **Chapadinha** Maranhão, E Brazil 3.45S 43.22W
10 K12 **Chapais** Quebec, SE Canada 49.46N 74.54W
42 L13 **Chapala** Jalisco, SW Mexico 20.17N 103.13W
42 L13 **Chapala, Lago de** ◎ C Mexico
152 F15 **Chapan, Gora** ▲ C Turkmenistan 37.48N 58.03E
59 M18 **Chapare** ← C Bolivia
56 E11 **Chaparral** Tolima, C Colombia 3.44N 75.33W
150 F9 **Chapayevo** Zapadnyy Kazakhstan, NW Kazakhstan 50.13S 51.05E
126 Kk12 **Chapayevo** Respublika Sakha (Yakutiya), NE Russian Federation 60.03N 117.19E
131 P5 **Chapayevsk** Samarskaya Oblast', W Russian Federation 52.57N 49.41E
62 H13 **Chapecó** Santa Catarina, S Brazil 27.06S 52.39W
62 I13 **Chapecó, Rio** ← S Brazil
22 J9 **Chapel Hill** Tennessee, S USA
46 J12 **Chapelton** C Jamaica 18.04N 77.16W
12 C8 **Chapleau** Ontario, S Canada 47.49N 83.24W
9 T16 **Chaplin** Saskatchewan, S Canada 50.27N 106.37W
130 M6 **Chaplygin** Lipetskaya Oblast', W Russian Federation 53.13N 39.58E
119 S11 **Chaplynka** Khersons'ka Oblast', S Ukraine 46.20N 33.34E
25 L3 **Chapman, Cape** *headland* Nunavut, NE Canada 69.15N 89.09W
27 T15 **Chapman Ranch** Texas, SW USA 27.32N 97.25W
22 P5 **Chapmanville** West Virginia, NE USA 37.58N 82.01W
30 K15 **Chappell** Nebraska, C USA 41.05N 102.28W
58 D9 **Chapra** Bihār, N India
78 I6 **Châr** *well* N Mauritania 21.31N 12.55W
126 Kk13 **Chara** Chitinskaya Oblast', S Russian Federation 56.57N 118.05E
126 Kk13 **Chara** ← C Russian Federation
56 G8 **Charala** Santander, C Colombia 6.16N 73.09W
8 N10 **Charcas** San Luis Potosí, C Mexico 23.09N 101.04W
27 T13 **Charco** Texas, SW USA 28.42N 97.35E
204 H7 **Charcot Island** *island* Antarctica
66 M8 **Charcot Seamounts** *undersea feature* E Atlantic Ocean
25 S9 **Chardara** *see* Shardara
151 P07 **Chardarinskoye Vodokhranilishche** ◎ S Kazakhstan
33 U11 **Chardon** Ohio, N USA 41.33N 81.10W
46 K9 **Chardonnières** SW Haiti 18.17N 74.09W
152 K12 **Chardzhev** *prev.* Chardzhou, Chardzou, Leninsk-Turkmenski, *Turkm.* Chärjew. Lebapskiy Velayat, E Turkmenistan 39.07N 63.30E
Chardzhevskaya Oblast' *see* Lebapskiy Velayat
Chardzou/Chardzhui *see* Chardzhev
104 I10 **Charente** ◆ *department* W France
104 J11 **Charente** ← W France
104 J10 **Charente-Maritime** ◆ *department* W France
143 Q12 **Ch'arents'avan** C Armenia 40.23N 44.41E
104 M4 **Chartres** *anc.* Autricum, Civitas Carnutum. Eure-et-Loir, C France 48.27N 1.27E
151 W15 **Charyn** *Kaz.* Sharyn. Almaty, SE Kazakhstan 43.45N 79.13E
33 Q4 **Charîkâr** *see* Chārīkār
121 F19 **Charnawchytsy** *Rus.* Chernavchitsy. Brestskaya Voblasts', SW Belarus 52.13N 23.43E
155 T5 **Chārsadda** North-West Frontier Province, NW Pakistan 34.12N 71.46E
152 M14 **Charshanga** *prev.* Charshangy, *Turkm.* Charshangngy. Lebapskiy Velayat, E Turkmenistan 37.31N 65.58E
Charshangngy/Charshangy *see* Charshanga
Charsk *see* Shar
189 W6 **Charters Towers** Queensland, NE Australia 20.01S 146.19E
13 P13 **Chartierville** Quebec, SE Canada 45.19N 71.13W
104 M4 **Chartres** *anc.* Autricum, Civitas Carnutum. Eure-et-Loir, C France 48.27N 1.27E
25 S4 **Charny** Quebec, SE Canada 46.43N 71.14W
155 T5 **Chärsadda** North-West Frontier Province, NW Pakistan 34.12N 71.46E
63 D21 **Chascomús** Buenos Aires, E Argentina 35.34S 58.01W
33 V15 **Chariton** Iowa, C USA 41.00N 93.18W
29 U3 **Chariton River** ← Missouri, C USA
9 S7 **Charity** NW Guyana 7.22N 58.34W
120 M13 **Chashniki** *Rus.* Chashniki. Vitsyebskaya Voblasts', N Belarus 54.52N 29.09E
Chärjew *see* Chardzhev
Chärjew Oblasty *see* Lebapskiy Velayat
117 D15 **Chásia** ▲ C Greece
13 V9 **Chaska** Minnesota, N USA 44.47N 93.36W
193 D25 **Chaslands Mistake** *headland* South Island, NZ 46.37S 169.21E
129 R11 **Chasovo** Respublika Komi, NW Russian Federation 61.58N 50.34E

Chasovo *see* Vazhgort
128 H14 **Chastova** Novgorodskaya Oblast', NW Russian Federation 58.37N 32.04E
149 R3 **Chāt Golestān**, N Iran 37.52N 55.27E
Chatak *see* Chhatak
41 R9 **Chatanika** Alaska, USA 65.06N 147.28W
41 R9 **Chatanika River** ← Alaska, USA
153 T8 **Chat-Bazar** Talasskaya Oblast', NW Kyrgyzstan 42.29N 72.37E
47 Y14 **Chateaubelair** Saint Vincent, W Saint Vincent and the Grenadines 13.16N 61.14W
104 J7 **Châteaubriant** Loire-Atlantique, NW France 47.43N 1.22W
105 Q8 **Château-Chinon** Nièvre, C France 47.04N 3.56E
110 C10 **Château d'Oex** Vaud, W Switzerland 46.28N 7.09E
104 M6 **Château-du-Loir** Sarthe, NW France 47.40N 0.25E
104 K7 **Châteaudun** Eure-et-Loir, C France 48.04N 1.19E
104 L6 **Château-Gontier** Mayenne, NW France 47.49N 0.42W
13 O13 **Châteauguay** Quebec, SE Canada 45.23N 73.46W
104 F6 **Châteaulin** Finistère, NW France 48.12N 4.07W
105 N9 **Châteaumeillant** Cher, C France 46.33N 2.10E
104 K11 **Châteauneuf-sur-Charente** Charente, W France 45.34N 0.33W
104 M7 **Château-Renault** Indre-et-Loire, C France 47.34N 0.52E
105 N9 **Châteauroux** *prev.* Indreville. Indre, C France 46.50N 1.42E
105 T5 **Château-Salins** Moselle, NE France 48.50N 6.29E
105 P4 **Château-Thierry** Aisne, N France 49.03N 3.24E
101 H21 **Châtelet** Hainaut, S Belgium 50.24N 4.31E
104 L9 **Châtellerault** *var.* Châtellerault. Vienne, W France 46.49N 0.33E
104 L9 **Châtellerault** *var.* Châtellerault. Vienne, W France 46.49N 0.33E
31 X10 **Chatfield** Minnesota, N USA 43.51N 92.11W
11 O14 **Chatham** New Brunswick, SE Canada 47.01N 65.30W
12 D17 **Chatham** Ontario, S Canada 42.24N 82.10W
99 P22 **Chatham** SE England, UK 51.22N 0.31E
32 K14 **Chatham** Illinois, N USA 39.40N 89.42W
23 T7 **Chatham** Virginia, NE USA 36.49N 79.24W
65 F22 **Chatham, Isla** *island* S Chile
183 R12 **Chatham Island** *island* Chatham Islands, NZ
Chatham Island *see* San Cristóbal, Isla
Chatham Island Rise *see* Chatham Rise
199 Jj14 **Chatham Islands** *island group* NZ, SW Pacific Ocean
183 Q12 **Chatham Rise** *var.* Chatham Island Rise. *undersea feature* S Pacific Ocean
41 X13 **Chatham Strait** *strait* Alaska, USA
Chathóir, Rinn *see* Cahore Point
104 M9 **Châtillon-sur-Indre** Indre, C France 46.58N 1.10E
105 Q7 **Châtillon-sur-Seine** Côte d'Or, C France 47.51N 4.30E
153 S8 **Chatkal** *Uzb.* Chotqol. ← Kyrgyzstan/Uzbekistan
153 R9 **Chatkal Range** *Rus.* Chatkal'skiy Khrebet. ▲ Kyrgyzstan/Uzbekistan
Chatkal'skiy Khrebet *see* Chatkal Range
25 N7 **Chatom** Alabama, S USA 31.28N 88.15W
149 S10 **Chatrapur** *see* Chhatrapur
149 S10 **Chatrūd** Kermān, C Iran 30.39N 56.57E
25 S2 **Chatsworth** Georgia, SE USA 34.46N 84.46W
23 S8 **Chattahoochee** Florida, SE USA 30.40N 84.51W
25 R8 **Chattahoochee River** ← SE USA
23 L10 **Chattanooga** Tennessee, S USA 35.05N 85.16W
153 V10 **Chatyr-Kël', Ozero** ◎ Kyrgyzstan
153 W9 **Chatyr-Tash** Narynskaya Oblast', C Kyrgyzstan 40.54N 76.22E
13 R10 **Chaudière** ← Quebec, SE Canada
178 Jj14 **Châu Độc** *var.* Chauphu, Chau Phu. An Giang, S Vietnam 10.52N 105.07E
117 Ff5 **Chauk** Magwe, W Myanmar 20.52N 94.49E
105 R6 **Chaumont** *prev.* Chaumont-en-Bassigny. Haute-Marne, N France 48.07N 5.07E
Chaumont-en-Bassigny *see* Chaumont
127 O14 **Chaunskaya Guba** *bay* NE Russian Federation
105 P3 **Chauny** Aisne, N France 49.37N 3.13E
Châu Độ *see* Binh Son
20 I1 **Chau Phu** *see* Châu Độc
199 I5 **Chausey, Îles** *island group* N France
Chausy *see* Chavusy
20 C11 **Chautauqua Lake** ◎ New York, NE USA
104 L9 **Chauvigny** Vienne, W France 46.35N 0.37E
128 L6 **Chavan'ga** Murmanskaya Oblast', NW Russian Federation 66.07N 37.44E
12 K10 **Chavannes, Lac** ◎ Quebec, SE Canada
63 D15 **Chavantes, Represa de** *see* Xavantes, Represa de
63 D15 **Chavarría** Corrientes, NE Argentina 28.57S 58.34W
Chavash Respubliki *see* Chuvashskaya Respublika
106 I5 **Chaves** *anc.* Aquae Flaviae. Vila Real, N Portugal 41.43N 7.28W

◆ COUNTRY   ◇ DEPENDENT TERRITORY   ◇ ADMINISTRATIVE REGION   ▲ MOUNTAIN   ▲ VOLCANO   ◎ LAKE
● COUNTRY CAPITAL   ○ DEPENDENT TERRITORY CAPITAL   ✈ INTERNATIONAL AIRPORT   ▲▲ MOUNTAIN RANGE   ← RIVER   ⊠ RESERVOIR

241

Chávez, Isla see Santa Cruz, Isla

84 G13 **Chavuma** North Western, NW Zambia 13.04S 22.43E

121 O16 **Chavusy** Rus. Chausy. Mahilyowskaya Voblasts', E Belarus 53.49N 30.59E

151 U6 **Chayan** Yuzhnyy Kazakhstan, S Kazakhstan 42.55N 69.32E

153 U8 **Chayek** Narynskaya Oblast', C Kyrgyzstan 41.54N 74.28E

245 T6 **Chäy Khänäh** E Iran 34.19N 44.33E

129 T16 **Chaykovskiy** Permskaya Oblast', NW Russian Federation 56.45N 54.09E

178 K12 **Chbar** Môndól Kiri, E Cambodia 12.46N 107.10E

25 Q4 **Cheaha Mountain** ▲ Alabama, S USA 33.29N 85.48W

Cheatharlach see Carlow

23 S2 **Cheat River** ✍ NE USA

113 A16 **Cheb** Ger. Eger. Karlovarský Kraj, W Czech Republic 50.04N 12.23E

131 Q3 **Cheboksary** Chuvashskaya Respublika, W Russian Federation 56.06N 47.14E

33 Q1 **Cheboygan** Michigan, N USA 45.40N 84.28W

Chechaouèn see Chefchaouen

Chechenia see Chechenskaya Respublika

131 O15 **Chechenskaya Respublika** Eng. Chechenia, Chechnia, Rus. Chechnya. ◆ autonomous republic SW Russian Federation

69 N4 **Chech, Erg** desert Algeria/Mali

Chechersk see Chachersk

Chechevichi see Chachevichy

Che-chiang see Zhejiang

Chechnia/Chechnya see Chechenskaya Respublika

169 Y15 **Chech'ŏn** Jap. Teisen. N South Korea 37.06N 128.15E

113 L15 **Chęciny** Świętokrzyskie, S Poland 50.51N 20.31E

29 Q10 **Checotah** Oklahoma, C USA 35.28N 95.31W

11 R15 **Chedabucto Bay** inlet Nova Scotia, E Canada

177 F7 **Cheduba Island** island W Myanmar

39 T5 **Cheesman Lake** ◉ Colorado, C USA

205 S16 **Cheetham, Cape** headland Antarctica 70.25S 162.40E

76 G5 **Chefchaouen** var. Chaouèn, Chechaouèn, Sp. Xauen. N Morocco 35.09N 5.16W

Chefoo see Yantai

40 M12 **Chefornak** Alaska, USA 60.09N 164.09W

126 Mm15 **Chegdomyn** Khabarovskiy Kray, SE Russian Federation 51.09N 132.58E

78 M4 **Chegga** Tiris Zemmour, NE Mauritania 25.27N 5.49W

Cheghcheran see Chaghcharān

34 G9 **Chehalis** Washington, NW USA 46.39N 122.57W

34 G9 **Chehalis River** ✍ Washington, NW USA

154 M6 **Chehel Abdālān, Kūh-e** var. Chalap Dalam, Pash. Chalap Dalan. ▲ C Afghanistan

117 D14 **Cheimaditis, Límni** ◉ N Greece

105 U15 **Cheiron, Mont** ▲ SE France 43.49N 7.00E

169 X17 **Cheju** Jap. Saishū. S South Korea 33.31N 126.34E

169 Y17 **Cheju** ✈ S South Korea 33.31N 126.28E

169 Y17 **Cheju-do** Jap. Saishū; prev. Quelpart. island S South Korea

169 X17 **Cheju-haehyŏp** strait S South Korea

Chekiang see Zhejiang

Chekichler see Chekishlyar

152 B13 **Chekishlyar** Turkm. Chekichler. Balkanskiy Velayat, W Turkmenistan 37.35N 53.52E

196 F8 **Chelab** Babeldaob, N Palau

153 N11 **Chelak** Rus. Chelek. Samarqand Wiloyati, C Uzbekistan 39.55N 66.45E

34 J7 **Chelan, Lake** ◉ Washington, NW USA

Chelek see Chelak

152 A11 **Cheleken** Balkanskiy Velayat, W Turkmenistan 39.25N 53.07E

Chélif/Chéliff see Chelif, Oued

76 J5 **Chelif, Oued** var. Chelif, Chéliff, Chellif, Sheliff. ✍ N Algeria

150 K12 **Chelkar** Aktyubinsk, W Kazakhstan 47.49N 59.28E

Chelkar, Ozero see Shalkar, Ozero

Chellif see Chelif, Oued

113 P14 **Chełm** Rus. Kholm. Lubelskie, SE Poland 51.07N 23.28E

112 I9 **Chełmno** Ger. Culm, Kulm. Kujawsko-pomorskie, C Poland 53.21N 18.27E

12 F10 **Chelmsford** Ontario, S Canada 46.33N 81.16W

99 P21 **Chelmsford** E England, UK 51.43N 0.28E

112 J9 **Chełmża** Ger. Culmsee, Kulmsee. Kujawsko-pomorskie, C Poland 53.12N 18.36E

29 Q3 **Chelsea** Oklahoma, C USA 36.32N 95.25W

20 M8 **Chelsea** Vermont, NE USA

99 L21 **Cheltenham** C England, UK 51.54N 2.04W

107 R9 **Chelva** País Valenciano, E Spain 39.45N 1.00W

125 Ee12 **Chelyabinsk** Chelyabinskaya Oblast', C Russian Federation 55.12N 61.25E

125 E12 **Chelyabinskaya Oblast'** ◆ province C Russian Federation

126 J1 **Chelyuskin, Mys** headland N Russian Federation 77.42N 104.13E

126 H15 **Chemal** Altayskiy Kray, S Russian Federation 51.23N 85.58E

43 V12 **Chemax** Yucatán, SE Mexico 20.41N 87.54W

85 N16 **Chemba** Sofala, C Mozambique 17.10S 34.52E

84 J13 **Chembe** Luapula, NE Zambia 11.58S 28.45E

152 L12 **Chemenibit** Maryyskiy Velayat, S Turkmenistan 35.27N 62.19E

Chemerisy see Chamyarysy

118 K7 **Chemeriwtsi** Khmel'nyts'ka Oblast', W Ukraine 49.00N 26.21E

104 J8 **Chemillé** Maine-et-Loire, NW France 47.15N 0.42W

181 X17 **Chemin Grenier** S Mauritius 20.28S 57.28E

103 N16 **Chemnitz** prev. Karl-Marx-Stadt. Sachsen, E Germany 50.49N 12.55E

Chemulpo see Inch'ŏn

34 H14 **Chemult** Oregon, NW USA 43.14N 121.48W

20 G12 **Chemung River** ✍ New York/Pennsylvania, NE USA

155 U8 **Chenāb** E India/Pakistan

41 S9 **Chena Hot Springs** Alaska, USA 65.06N 146.02W

20 I11 **Chenango River** ✍ New York, NE USA

174 Gg3 **Chenderoh, Tasik** ◉ Peninsular Malaysia

13 **Chêne, Rivière du** ✍ Quebec, SE Canada

34 L8 **Cheney** Washington, NW USA 47.29N 117.34W

28 M6 **Cheney Reservoir** ◼ Kansas, C USA

Chengchiatun see Liaoyuan

Ch'eng-chou/Chengchow see Zhengzhou

167 P1 **Chengde** var. Jehol. Hebei, E China 41.00N 117.57E

166 I9 **Chengdu** var. Chengtu, Ch'eng-tu. Sichuan, C China 30.40N 104.03E

167 Q14 **Chenghai** Guangdong, S China 23.33N 116.42E

166 L17 **Chenghsien** see Zhengzhou

Chengmai Hainan, S China 19.45N 109.56E

165 W12 **Chengxian** var. Cheng Xian. Gansu, C China 33.42N 105.45E

Chenkiang see Zhenjiang

161 J19 **Chennai** prev. Madras. Tamil Nādu, S India 13.04N 80.18E

161 J19 **Chennai** ✈ Tamil Nādu, S India 13.07N 80.13E

105 R8 **Chenôve** Côte d'Or, C France 47.16N 5.00E

166 L11 **Chenxi** Hunan, S China 28.01N 110.15E

Chen Xian/Chenxian/Chen Xiang see Chenzhou

167 N12 **Chenzhou** var. Chenxian, Chen Xian, Chen Xiang. Hunan, S China 25.51N 113.01E

178 Kk12 **Cheo Reo** var. A Yun Pa. Gia Lai, S Vietnam 13.19N 108.25E

116 I11 **Chepelare** Smolyan, S Bulgaria 41.43N 24.40E

116 I11 **Chepelarska Reka** ✍ S Bulgaria

58 B11 **Chepén** La Libertad, C Peru 7.12S 79.24W

64 J10 **Chepes** La Rioja, C Argentina 31.23S 66.34W

167 O15 **Chep Lap Kok** ✈ (Hong Kong) S China 22.23N 114.11E

45 U14 **Chepo** Panamá, C Panama 9.10N 79.05W

31 U2 **Chepping Wycombe** see High Wycombe

129 R14 **Cheptsa** ✍ NW Russian Federation

32 K3 **Chequamegon Point** headland Wisconsin, N USA 46.42N 90.45W

105 O8 **Cher** ◆ department C France

104 M8 **Cher** ✍ C France

83 H17 **Cherangani Hills** see Cherangany Hills

Cherangany Hills var. Cherangani Hills. ▲ W Kenya

23 S11 **Cheraw** South Carolina, SE USA 34.42N 79.52W

104 I3 **Cherbourg** anc. Carusbur. Manche, N France 49.39N 1.36W

131 R5 **Cherdakly** Ul'yanovskaya Oblast', W Russian Federation 54.21N 48.54E

129 U12 **Cherdyn'** Permskaya Oblast', NW Russian Federation 60.21N 56.39E

128 J14 **Cherekha** ✍ W Russian Federation

129 J15 **Cheremkhovo** Irkutskaya Oblast', S Russian Federation 53.16N 102.44E

126 Hh15 **Cheremushki** Respublika Khakasiya, S Russian Federation 52.48N 91.20E

128 L9 **Cheren** see Keren

Cherepovets Vologodskaya Oblast', NW Russian Federation 59.09N 37.49E

129 O11 **Cherevkovo** Arkhangel'skaya Oblast', NW Russian Federation 61.45N 45.16E

112 I9 **Chelmno** Ger. Culm, Kulm. Kujawsko-pomorskie, C Poland 53.21N 18.27E

76 I6 **Chergui, Chott ech** salt lake NW Algeria

119 P6 **Cherikov** see Cherykaw

Cherkas'ka Oblast' var. Cherkasy, Rus. Cherkasskaya Oblast'. ◆ province C Ukraine

Cherkasskaya Oblast' see Cherkas'ka Oblast'

119 Q6 **Cherkasy** Rus. Cherkassy. Cherkas'ka Oblast', C Ukraine 49.25N 32.04E

130 M15 **Cherkessk** Karachayevo-Cherkesskaya Respublika, SW Russian Federation 44.12N 42.06E

121 L16 **Cherlak** Omskaya Oblast', C Russian Federation 54.06N 74.59E

125 Ff14 **Cherlakskiy** Omskaya Oblast', C Russian Federation 54.34N 31.19E

125 U13 **Chermoz** Permskaya Oblast', NW Russian Federation 58.44N 56.07E

23 X5 **Chernavchitsy** see Charnawchytsy

99 K18 **Chernaya** Nenetskiy Avtonomnyy Okrug, NW Russian Federation 68.36N 56.34E

129 T4 **Chernaya** ✍ NW Russian Federation

Chernigov see Chernihiv

Chernigovskaya Oblast' see Chernihivs'ka Oblast'

119 Q2 **Chernihiv** Rus. Chernigov. Chernihivs'ka Oblast', NE Ukraine 48.18N 25.59E

119 V9 **Chernihivka** Zaporiz'ka Oblast', SE Ukraine 47.11N 36.10E

119 P2 **Chernihivs'ka Oblast'** var. Chernihiv, Rus. Chernigovskaya Oblast'. ◆ province NE Ukraine

116 I9 **Cherni Osŭm** ✍ N Bulgaria

116 I9 **Cherni Vit** ✍ N Bulgaria

116 G10 **Cherni Vrŭkh** ▲ W Bulgaria 42.33N 23.18E

118 K8 **Chernivtsi** Ger. Czernowitz, Rom. Cernăuţi, Rus. Chernovtsy. Chernivets'ka Oblast', W Ukraine 48.18N 25.55E

118 M7 **Chernivtsi** South Carolina, C Ukraine 48.33N 28.06E

Chernivtsi see Chernivets'ka Oblast'

126 Hh15 **Chernobyl'** see Chornobyl'

Chernogorsk Respublika Khakasiya, S Russian Federation 53.48N 91.03E

Cherno More see Black Sea

Chernomorskoye see Chornomors'ke

151 T7 **Chernoretskoye** Pavlodar, NE Kazakhstan 52.51N 76.37E

Chernovitskaya Oblast' see Chernivets'ka Oblast'

Chernovtsy see Chernivtsi

151 U8 **Chernoye** Pavlodar, NE Kazakhstan 51.40N 77.33E

Chernoye More see Black Sea

129 U16 **Chernushka** Permskaya Oblast', NW Russian Federation 56.30N 56.07E

119 N4 **Chernyakhiv** Rus. Chernyakhov. Zhytomyrs'ka Oblast', N Ukraine 50.31N 28.38E

Chernyakhov see Chernyakhiv

121 C14 **Chernyakhovsk** Ger. Insterburg. Kaliningradskaya Oblast', W Russian Federation 54.36N 21.49E

130 K8 **Chernyanka** Belgorodskaya Oblast', W Russian Federation 50.59N 37.54E

129 V5 **Chernysheva, Gryada** ▲ NW Russian Federation

150 J14 **Chernysheva, Zaliv** gulf SW Kazakhstan

126 L15 **Chernyshevsk** Chitinskaya Oblast', S Russian Federation 52.28N 116.52E

126 K11 **Chernyshevskiy** Respublika Sakha (Yakutiya), NE Russian Federation 62.57N 112.29E

131 P13 **Chernyye Zemli** plain SW Russian Federation

131 V7 **Chernyy Otrog** Orenburgskaya Oblast', W Russian Federation 52.03N 56.09E

31 T2 **Cherokee** Iowa, C USA 42.45N 95.33W

28 M8 **Cherokee** Oklahoma, C USA 36.45N 98.22W

27 R9 **Cherokee** Texas, SW USA 30.56N 98.42W

23 O8 **Cherokee Lake** ◼ Tennessee, S USA

Cherokees, Lake O' The see Grand Lake O' The Cherokees

46 H1 **Cherokee Sound** Great Abaco, N Bahamas 26.16N 77.03W

159 V13 **Cherrapunji** Meghālaya, NE India 25.16N 91.42E

30 L9 **Cherry Creek** ✍ South Dakota, N USA

20 J16 **Cherry Hill** New Jersey, NE USA 39.55N 75.01W

29 Q7 **Cherryvale** Kansas, C USA 37.16N 95.33W

23 Q10 **Cherryville** North Carolina, SE USA 35.22N 81.22W

127 O5 **Cherski Range** see Cherskogo, Khrebet

Cherskiy Respublika Sakha (Yakutiya), NE Russian Federation 68.45N 161.15E

126 Mm8 **Cherskogo, Khrebet** var. Cherski Range. ▲ NE Russian Federation

Cherso see Cres

130 L10 **Chertkovo** Rostovskaya Oblast', SW Russian Federation 49.22N 40.10E

178 I8 **Cherven Bryag** Pleven, N Bulgaria 43.17N 24.06E

116 H8 **Cherven'** see Chervyen'

118 M4 **Chervonoarmiys'k** Zhytomyrs'ka Oblast', N Ukraine 50.27N 28.15E

Chervonograd see Chervonohrad

118 I4 **Chervonohrad** Rus. Chervonograd. L'vivs'ka Oblast', W Ukraine 50.22N 24.11E

119 W6 **Chervonooskil's'ke Vodoskhovyshche** Rus. Krasnooskol'skoye Vodokhranilishche. ◼ NE Ukraine

119 Q6 **Chervonoye, Ozero** see Chyrvonaye, Vozyera

119 O4 **Chervonozavods'ke** Poltavs'ka Oblast', NE Ukraine 50.24N 33.22E

121 L16 **Chervyen'** Rus. Cherven'. Minskaya Voblasts', C Belarus 53.42N 28.23E

121 P16 **Cherykaw** Rus. Cherikov. Mahilyowskaya Voblasts', E Belarus 53.34N 31.19E

86 **Chesaning** Michigan, N USA 43.10N 84.07W

23 X5 **Chesapeake Bay** inlet NE USA

99 K18 **Cheshire** cultural region C England, UK

129 P5 **Chëshskaya Guba** var. Archangel Bay, Chesha Bay, Dvina Bay. bay NW Russian Federation

12 **Chesley** Ontario, S Canada

Chesne see Cheshire

23 Q10 **Chesnee** South Carolina, SE USA 35.09N 81.51W

99 K18 **Chester** Wel. Caerleon; hist. Legaceaster, Lat. Deva, Devana Castra. C England, UK 53.12N 2.54W

37 O4 **Chester** California, W USA 40.18N 121.14W

32 K16 **Chester** Illinois, S USA 37.54N 89.49W

20 I16 **Chester** Pennsylvania, NE USA 39.51N 75.21W

23 R1 **Chester** South Carolina, SE USA 34.42N 81.12W

27 X9 **Chester** Texas, SW USA 30.55N 94.36W

23 W4 **Chester** Virginia, NE USA 37.22N 77.27W

23 R4 **Chester** West Virginia, NE USA 40.34N 80.33W

99 M18 **Chesterfield** C England, UK 53.15N 1.25W

23 S11 **Chesterfield** South Carolina, SE USA 34.44N 80.05W

33 S10 **Chesterfield** Virginia, NE USA 37.22N 77.31W

199 I10 **Chesterfield, Îles** island group NW New Caledonia

15 L6 **Chesterfield Inlet** Nunavut, NW Canada 63.19N 90.57W

15 L6 **Chesterfield Inlet** inlet Nunavut, N Canada

23 Y3 **Chester River** ✍ Delaware/Maryland, NE USA

23 X3 **Chestertown** Maryland, NE USA 39.12N 76.04W

21 R4 **Chesuncook Lake** ◉ Maine, NE USA

32 J5 **Chetek** Wisconsin, N USA 45.19N 91.37W

11 R14 **Chéticamp** Nova Scotia, SE Canada 46.38N 61.19W

29 Q8 **Chetopa** Kansas, C USA 37.02N 95.05W

43 Y14 **Chetumal** var. Payo Obispo. Quintana Roo, SE Mexico 18.32N 88.15W

Chetumal, Bahia/Chetumal, Bahía de see Chetumal Bay

44 G1 **Chetumal, Bahía de** var. Bahía Chetumal, Bahía de Chetumal. bay Belize/Mexico

8 J4 **Chetwynd** British Columbia, W Canada 55.42N 121.36W

40 J10 **Chevak** Alaska, USA 61.31N 165.35W

38 M11 **Chevelon Creek** ✍ Arizona, SW USA

41 T9 **Chevak** Alaska, USA 64.04N 141.56W

193 J17 **Cheviot** Canterbury, South Island, NZ 42.48S 173.17E

98 L13 **Cheviot Hills** hill range England/Scotland, UK

98 L13 **Cheviot, The** ▲ NE England, UK 55.28N 2.10W

12 M11 **Chevreuil, Lac du** ◉ Quebec, SE Canada

83 I16 **Ch'ew Bahir** var. Lake Stefanie. ◉ Ethiopia/Kenya

34 L7 **Chewelah** Washington, NW USA 48.16N 117.42W

28 K10 **Cheyenne** Oklahoma, C USA 35.38N 99.40W

35 Z17 **Cheyenne** state capital Wyoming, C USA 41.08N 104.45W

28 L5 **Cheyenne Bottoms** ◉ Kansas, C USA

18 Kk6 **Cheyenne River** ✍ South Dakota/Wyoming, N USA

39 W5 **Cheyenne Wells** Colorado, C USA 38.49N 102.21W

110 C9 **Cheyres** Vaud, W Switzerland 46.48N 6.48E

159 S13 **Chhapra** prev. Chapra. Bihār, N India 25.49N 84.42E

159 V13 **Chhatak** var. Chatak. Chittagong, NE Bangladesh 25.02N 91.33E

160 J9 **Chhatarpur** Madhya Pradesh, C India 24.54N 79.34E

160 N13 **Chhatrapur** prev. Chatrapur; Orissa, E India 19.25N 85.01E

160 L12 **Chhattisgarh** plain C India

160 I11 **Chhindwāra** Madhya Pradesh, C India 22.04N 78.58E

159 T12 **Chhukha** SW Bhutan 27.01N 89.36E

167 S14 **Chiai** var. Chia-i, Chiayi, Kiayi, Jiayi, Jap. Kagi. C Taiwan 23.28N 120.27E

Chia-i see Chiai

85 B15 **Chiange** Port. Vila de Almoster. Huíla, SW Angola 15.49S 13.52E

Chiang-hsi see Jiangxi

167 S14 **Chiang Kai-shek** ✈ (T'aipei) N Taiwan 25.09N 121.20E

178 H8 **Chiang Khan** Loei, E Thailand 17.51N 101.43E

178 H7 **Chiang Mai** var. Chiangmai, Chiengmai, Kiangmai, Muang Chiang Mai. NW Thailand 18.48N 98.58E

178 H7 **Chiang Mai** ✈ Chiang Mai, NW Thailand 18.44N 98.53E

178 Hh6 **Chiang Rai** var. Chianpai, Chienrai, Muang Chiang Rai. Chiang Rai, NW Thailand 19.55N 99.51E

Chiang-su see Jiangsu

Chianning/Chian-ning see Nanjing

Chianpai see Chiang Rai

108 I8 **Chianti** cultural region C Italy

43 U16 **Chiapa** see Chiapa de Corzo

43 U16 **Chiapa de Corzo** var. Chiapa. Chiapas, SE Mexico 16.42N 92.58W

43 U16 **Chiapas** ◆ state SE Mexico

108 J12 **Chiaravalle** Marche, C Italy 43.36N 13.19E

109 N22 **Chiaravalle Centrale** Calabria, SW Italy 38.40N 16.25E

108 I7 **Chiari** Lombardia, N Italy 45.33N 10.00E

110 H12 **Chiasso** Ticino, S Switzerland 45.50N 9.02E

43 X5 **Chiautla de Tapia** var. Chiautla. Puebla, S Mexico 18.16N 98.31W

43 P15 **Chiautla de Tapia** see Chiautla de Tapia

108 E6 **Chiavari** Liguria, NW Italy 44.19N 9.22E

108 F6 **Chiavenna** Lombardia, N Italy 46.19N 9.25E

99 K18 **Chester** Wel. Caerleon

172 Ss12 **Chiba** var. Tiba. Chiba, Honshū, S Japan 35.37N 140.05E

171 K17 **Chiba off.** Chiba-ken, var. Tiba. ◆ prefecture Honshū, S Japan

85 M18 **Chibabava** Sofala, C Mozambique 20.16S 33.39E

85 B15 **Chibia** Port. João de Almeida, Vila João de Almeida. Huíla, SW Angola 15.09S 13.45E

85 M18 **Chiboma** Sofala, C Mozambique 20.06S 33.54E

84 J12 **Chibondo** Luapula, N Zambia 10.42S 28.42E

84 K11 **Chibote** Luapula, N Zambia 9.52S 29.33E

10 K12 **Chibougamau** Quebec, SE Canada 49.55N 74.24W

27 P4 **Childress** Texas, SW USA 34.25N 100.12W

65 G14 **Chile** off. Republic of Chile. ◆ republic SW South America

170 Ff11 **Chíburi-jima** island Oki-shotō, SW Japan

200 O13 **Chile Rise** undersea feature E Pacific Ocean

63 H20 **Chile Chico** Aisén, W Chile 46.34S 71.43W

64 H12 **Chilecito** La Rioja, NW Argentina 29.10S 67.30W

64 H12 **Chilecito** Mendoza, W Argentina 33.52S 69.03W

85 L14 **Chilembwe** Eastern, E Zambia 13.54S 31.38E

200 O13 **Chile Rise** undersea feature SE Pacific Ocean

119 N13 **Chilia Braţul** ✍ SE Romania

151 V15 **Chilik** Almaty, SE Kazakhstan 43.35N 78.17E

151 V15 **Chilik** ✍ SE Kazakhstan

160 O13 **Chilka Lake** var. Chilka Lake. ◉ E India

84 J13 **Chililabombwe** Copperbelt, C Zambia 12.19S 27.52E

Chi-lin see Jilin

160 O13 **Chilka Lake** var. Chilka Lake

8 H9 **Chilkoot Pass** pass British Columbia, W Canada 59.41N 135.15W

64 G13 **Chillán** Bío Bío, C Chile 36.37S 72.10W

63 C22 **Chillar** Buenos Aires, E Argentina 37.16S 59.58W

32 L10 **Chillicothe** Illinois, N USA 40.55N 89.29W

29 S3 **Chillicothe** Missouri, C USA 39.48N 93.33W

33 S14 **Chillicothe** Ohio, N USA 39.19N 82.58W

27 Q4 **Chillicothe** Texas, SW USA 34.15N 99.31W

8 M17 **Chilliwack** British Columbia, SW Canada 49.09N 121.54W

37 N5 **Chilo** California, W USA

85 L15 **Chiloé Tete**, NW Mozambique 15.45S 32.25E

85 M20 **Chioco** Tete, NW Mozambique 16.25S 32.49E

20 M11 **Chicopee** Massachusetts, NE USA 42.08N 72.34W

65 I19 **Chico, Río** ✍ SE Argentina

65 I20 **Chico, Río** ✍ S Argentina

29 W14 **Chicot, Lake** ◉ Arkansas, C USA

13 R7 **Chicoutimi** Quebec, SE Canada 48.24N 71.04W

13 Q8 **Chicoutimi** ✍ Quebec, SE Canada

85 L19 **Chicualacuala** Gaza, SW Mozambique 22.06S 31.42E

85 B14 **Chicuma** Benguela, C Angola 13.33S 14.41E

161 J21 **Chidambaram** Tamil Nādu, SE India 11.25N 79.42E

206 K13 **Chidley, Cape** headland Newfoundland, E Canada 60.25N 64.39W

103 N24 **Chiemsee** ◉ SE Germany

Chiengmai see Chiang Mai

Chienrai see Chiang Rai

108 B8 **Chieri** Piemonte, NW Italy 45.01N 7.49E

108 F8 **Chiese** ✍ N Italy

109 K14 **Chieti** var. Teate. Abruzzo, C Italy 42.21N 14.10E

170 E19 **Chièvres** Hainaut, SW Belgium 50.34N 3.49E

169 S12 **Chifeng** var. Ulanhad. Nei Mongol Zizhiqu, N China 42.16N 118.55E

84 J13 **Chifumage** ✍ E Angola

84 M13 **Chifunda** Eastern, NE Zambia 11.57S 32.36E

151 S14 **Chiganak** var. Čiganak. Zhambyl, SE Kazakhstan 45.10N 73.55E

41 P15 **Chiginagak, Mount** ▲ Alaska, USA 57.10N 157.00W

41 Q5 **Chigirin** see Chyhyryn

43 P13 **Chignahuapan** Puebla, S Mexico 19.52N 98.03W

41 O15 **Chignik** Alaska, USA 56.18N 158.24W

56 D7 **Chigorodó** Antioquia, NW Colombia 7.42N 76.45W

85 M19 **Chiguba** Gaza, S Mozambique

168 D6 **Chihertey** Bayan-Ölgiy, W Mongolia 48.10N 89.35E

152 H7 **Chihli** see Hebei

Chihli, Gulf of see Bo Hai

42 J6 **Chihuahua** Chihuahua, NW Mexico 28.40N 106.06W

42 J6 **Chihuahua** ◆ state N Mexico

151 O15 **Chiili** Kzylorda, S Kazakhstan 44.09N 66.44E

133 V7 **Chikaskia River** ✍ Kansas/Oklahoma, C USA

28 M7 **Chikaskia River** ✍ Kansas/Oklahoma, C USA

30 I14 **Chik Ballapur** Karnātaka, W India 13.28N 77.42E

85 M13 **Chikhachevo** Pskovskaya Oblast', W Russian Federation 57.17N 29.51E

84 K11 **Chikuma** ✍ Honshū, S Japan

43 O8 **Chikmagalūr** Karnātaka, W India 13.19N 75.46E

133 V7 **Chikoy** ✍ C Russian Federation

43 P15 **Chilapa** Guerrero, S Mexico 17.36N 99.09W

Chilapa see Chilapa de Alvarez

43 P16 **Chilapa de Alvarez** var. Chilapa. Guerrero, S Mexico 17.36N 99.09W

161 J25 **Chilaw** North Western Province, W Sri Lanka 7.34N 79.49E

59 D15 **Chilca** Lima, W Peru 12.33S 76.44W

25 Q4 **Childersburg** Alabama, S USA 33.16N 86.21W

27 P4 **Childress** Texas, SW USA

161 J16 **Chilakalūrupet** Andhra Pradesh, E India 16.09N 80.13E

152 L14 **Chilan** Lebapskiy Velayat, E Turkmenistan 32.57N 64.58E

43 P16 **Chilapa de Alvarez** var.

44 H9 **Chi-nan/Chinan** see Jinan

44 H9 **Chinandega** Chinandega, NW Nicaragua 12.37N 87.07W

44 H9 **Chinandega** ◆ department NW Nicaragua

China, People's Republic of see China

China, Republic of see Taiwan

26 J11 **Chinati Mountains** ▲ Texas, SW USA

Chinaz see Chinoz

59 E15 **Chincha Alta** Ica, SW Peru

9 N11 **Chinchaga** ✍ Alberta, SW Canada

Chin-chiang see Quanzhou

Chinchilla see Chinchilla de Monte Aragón

107 Q11 **Chinchilla de Monte Aragón** var. Chinchilla. Castilla-La Mancha, C Spain 38.55N 1.43W

56 D10 **Chinchiná** Caldas, W Colombia 4.58N 75.37W

107 O8 **Chinchón** Madrid, C Spain 40.07N 3.25W

43 Z14 **Chinchorro, Banco** island SE Mexico

23 Z5 **Chincoteague** Assateague Island, Virginia, NE USA 37.55N 75.22W

85 O17 **Chinde** Zambézia, NE Mozambique 18.34S 36.25E

169 X17 **Chin-do** Jap. Chin-tō. island SW South Korea

169 R13 **Chindu** Qinghai, C China 33.19N 97.08E

177 G2 **Chindwin** ✍ N Myanmar

Chinese Empire see China

152 L10 **Chingeldi** Rus. Chingildi. Nawoiy Wiloyati, N Uzbekistan 40.59N 64.13E

150 H9 **Chingirlau** Kaz. Shynghyrlaū. Zapadnyy Kazakhstan, W Kazakhstan 51.10N 53.44E

84 J13 **Chingola** Copperbelt, C Zambia 12.31S 27.52E

Ching-Tao/Ch'ing-tao see Qingdao

84 C13 **Chinguar** Huambo, C Angola 12.33S 16.22E

78 I7 **Chinguețți** var. Chinguetti. Adrar, C Mauritania 20.25N 12.24W

169 Z16 **Chinju** Jap. Chinkai. S South Korea 35.06N 128.48E

177 F4 **Chin Hills** ▲ W Myanmar

85 K16 **Chinhoyi** prev. Sinoia. Mashonaland West, N Zimbabwe 17.19S 30.06E

41 Q14 **Chiniak, Cape** headland Kodiak Island, Alaska, USA 57.37N 152.10W

12 H3 **Chiniguchi Lake** ◉ Ontario, S Canada

155 U8 **Chiniot** Punjab, NE Pakistan 31.40N 73.00E

169 Y16 **Chinju** Jap. Shinshū. S South Korea 35.11N 128.06E

80 M13 **Chinko** ✍ E Central African Republic

39 O9 **Chinle** Arizona, SW USA 36.09N 109.33W

167 W14 **Chinmen Tao** var. Jinmen Dao, Quemoy. island W Taiwan

Chinnchär see Shinshār

Chinnereth see Tiberias, Lake

171 J13 **Chino** var. Tino. Nagano, Honshū, S Japan 36.00N 138.10E

104 L8 **Chinon** Indre-et-Loire, C France 47.10N 0.15E

35 T7 **Chinook** Montana, NW USA 48.35N 109.13W

199 Jj3 **Chinook Trough** undersea feature N Pacific Ocean

38 L14 **Chino Valley** Arizona, SW USA 34.45N 112.27W

153 P10 **Chinoz** Rus. Chinaz. Toshkent Wiloyati, E Uzbekistan 40.58N 68.46E

84 L13 **Chinsali** Northern, NE Zambia 10.33S 32.04E

177 F4 **Chin State** ◆ state W Myanmar

Chinsura see Chunchura

56 E6 **Chinú** Córdoba, NW Colombia 9.07N 75.25W

101 K24 **Chiny, Forêt de** forest SE Belgium

85 M15 **Chioco** Tete, NW Mozambique 16.22S 32.50E

108 H8 **Chioggia** anc. Fossa Claudia. Veneto, NE Italy 45.13N 12.16E

116 J16 **Chionótrypa** ▲ NE Greece 41.16N 24.06E

117 L18 **Chíos** var. Híos, Khíos, It. Scio, Turk. Sakiz-Adasi. Chíos, E Greece 38.22N 26.07E

117 L18 **Chíos** var. Khíos. island E Greece

85 M14 **Chipata** prev. Fort Jameson. Eastern, E Zambia 13.40S 32.42E

85 C14 **Chipindo** Huíla, C Angola 13.53S 15.47E

58 R8 **Chipley** Florida, SE USA 30.46N 85.32W

161 D15 **Chiplun** Mahārāshtra, W India 17.31N 73.31E

83 H22 **Chipogolo** Dodoma, C Tanzania 6.52S 36.03E

25 R8 **Chipola River** ✍ Florida, SE USA

99 L22 **Chippenham** S England, UK 51.28N 2.07W

32 J4 **Chippewa Falls** Wisconsin, N USA 44.55N 91.25W

32 I4 **Chippewa, Lake** ◉ Wisconsin, N USA

33 Q4 **Chippewa River** ✍ Michigan, N USA

32 I6 **Chippewa River** ✍ Wisconsin, N USA

116 G8 **Chipping Wycombe** see High Wycombe

35 Z12 **Chiprovtsi** Montana, NW Bulgaria 43.23N 22.53E

21 Q7 **Chiputneticook Lakes** lakes Canada/USA

85 B15 **Chiquián** Ancash, W Peru 10.03S 77.11W

43 Y11 **Chiquilá** Quintana Roo, SE Mexico 21.25N 87.20W

44 E6 **Chiquimula** Chiquimula, SE Guatemala 14.46N 89.31W

44 A3 **Chiquimula** off. Departamento de Chiquimula. ◆ department SE Guatemala

44 D7 **Chiquimulilla** Santa Rosa, S Guatemala 14.06N 90.22W

56 F9 **Chiquinquirá** Boyacá, C Colombia 5.37N 73.51W

161 J17 **Chīrāla** Andhra Pradesh, E India 15.49N 80.21E

155 N4 **Chiras** Ghowr, N Afghanistan 35.51N 65.39S

158 H11 **Chirāwa** Rājasthān, N India 28.15N 75.42E

153 Q9 **Chirchiq** Rus. Chirchik. Toshkent Wiloyati, E Uzbekistan 41.30N 69.31E

153 P10 **Chirchiq** ♒ E Uzbekistan

Chire see Shire

85 L18 **Chiredzi** Masvingo, SE Zimbabwe 21.03S 31.40E

27 X8 **Chireno** Texas, SW USA 31.30N 94.21W

79 X7 **Chirfa** Agadez, NE Niger 21.01N 12.41E

39 O16 **Chiricahua Mountains** ▲ Arizona, SW USA

39 O16 **Chiricahua Peak** ▲ Arizona, SW USA 31.51N 109.17W

56 F6 **Chiriguaná** Cesar, N Colombia 9.24N 73.37W

41 P15 **Chirikof Island** island Alaska, USA

45 P16 **Chiriquí** off. Provincia de Chiriquí. ◆ province SW Panama

45 P17 **Chiriquí, Golfo de** Eng. Chiriqui Gulf. gulf SW Panama

45 P15 **Chiriquí Grande** Bocas del Toro, W Panama 8.55N 82.08W

Chiriquí Gulf see Chiriquí, Golfo de

45 P15 **Chiriquí, Laguna de** lagoon NW Panama

45 O16 **Chiriquí Viejo, Río** ♒ W Panama

Chiriquí, Volcán de see Barú, Volcán

85 N15 **Chiromo** Southern, S Malawi 16.32S 35.07E

116 J10 **Chirpan** Stara Zagora, C Bulgaria 42.13N 25.22E

45 N14 **Chirripó Atlántico, Río** ♒ E Costa Rica

Chirripó, Cerro see Chirripó Grande, Cerro

45 N14 **Chirripó Grande, Cerro** var. Cerro Chirripó. ▲ SE Costa Rica 9.31N 83.28W

45 N13 **Chirripó, Río** var. Río Chirripó del Pacífico. ♒ NE Costa Rica

Chirua, Lago see Chilwa, Lake

85 J15 **Chirundu** Southern, S Zambia 16.01S 28.52E

31 W8 **Chisago City** Minnesota, N USA 45.22N 92.53W

85 J14 **Chisamba** Central, C Zambia 14.58S 28.21E

41 T10 **Chisana** Alaska, USA 62.09N 142.07W

84 I13 **Chisasa** North Western, NW Zambia 12.09S 25.30E

10 I9 **Chisasibi** Quebec, C Canada 53.45N 79.01W

44 D4 **Chisec** Alta Verapaz, C Guatemala 15.47N 90.13W

131 U5 **Chishmy** Respublika Bashkortostan, W Russian Federation 54.33S 55.21E

31 V4 **Chisholm** Minnesota, N USA 47.29N 92.52W

166 I11 **Chishui He** ♒ C China

Chisimaio/Chisimayu see Kismaayo

119 N10 **Chişinău** Rus. Kishinev. ● (Moldova) C Moldova 47.00N 28.50E

119 N10 **Chişinău** ✈ S Moldova 46.54N 28.56E

Chişinău-Criş see Chişinău-Criş

118 F10 **Chişineu-Criş** Hung. Kisjenő; prev. Chişinău-Criş. Arad, W Romania 46.33N 21.29E

85 K14 **Chisomo** Central, C Zambia 13.30S 30.37E

108 A8 **Chisone** ♒ NW Italy

26 K12 **Chisos Mountains** ▲ Texas, SW USA

155 U10 **Chistian Mandi** Punjab, E Pakistan 29.52N 72.46E

41 T10 **Chistochina** Alaska, USA 62.34N 144.39W

131 R4 **Chistopol'** Respublika Tatarstan, W Russian Federation 55.20N 50.39E

151 O8 **Chistopol'ye** Severnyy Kazakhstan, N Kazakhstan 52.37N 67.13E

126 Kk16 **Chita** Chitinskaya Oblast', S Russian Federation 52.03N 113.34E

85 B16 **Chitado** Cunene, SW Angola 17.16S 13.54E

Chitaldroog/Chitaldrug see Chitradurga

85 C15 **Chitanda** ♒ S Angola

Chitangwiza see Chitungwiza

85 F10 **Chitato** Lunda Norte, NE Angola 7.23S 20.45E

85 C14 **Chitembo** Bié, C Angola 13.31S 16.44E

41 T11 **Chitina** Alaska, USA 61.31N 144.26W

41 T11 **Chitina River** ♒ Alaska, USA

126 Kk14 **Chitinskaya Oblast'** ◆ province S Russian Federation

84 M11 **Chitipa** Northern, N Malawi 9.40S 33.19E

172 O06 **Chitose** var. Titose. Hokkaidō, NE Japan 42.50N 141.39E

161 G18 **Chitradurga** prev. Chitaldroog, Chitaldrug. Karnātaka, W India 14.15N 76.24E

155 T3 **Chitrāl** North-West Frontier Province, NW Pakistan 35.51N 71.46E

45 S16 **Chitré** Herrera, S Panama 7.57N 80.25W

159 V16 **Chittagong** Ben. Chāttagām. Chittagong, SE Bangladesh 22.19N 91.48E

159 U16 **Chittagong** ◆ division E Bangladesh

159 Q15 **Chittaranjan** West Bengal, NE India 23.52N 86.40E

158 G14 **Chittaurgarh** Rājasthān, N India 24.54N 74.42E

161 I19 **Chittoor** Andhra Pradesh, E India 13.13N 79.06E

161 G21 **Chittūr** Kerala, SW India 10.42N 76.46E

85 K16 **Chitungwiza** prev. Chitangwiza. Mashonaland East, NE Zimbabwe 18.00S 31.06E

84 H4 **Chíuchíu** Antofagasta, N Chile 22.13S 68.34W

84 F12 **Chiumbe** var. Tshiumbe. ♒ Angola/Dem. Rep. Congo (Zaire)

85 F15 **Chiume** Moxico, E Angola 15.08S 21.09E

84 K13 **Chiundaponde** Northern, NE Zambia 12.13S 30.40E

108 H13 **Chiusi** Toscana, C Italy 43.00N 11.56E

56 J5 **Chivacoa** Yaracuy, N Venezuela 10.10N 68.54W

108 B8 **Chivasso** Piemonte, NW Italy 45.13S 7.54E

85 L17 **Chivhu** prev. Enkeldoorn. Midlands, C Zimbabwe 19.01S 30.54E

63 C20 **Chivilcoy** Buenos Aires, E Argentina 34.55S 60.00W

84 N12 **Chiweta** Northern, N Malawi 10.36S 34.09E

44 D4 **Chixoy, Río** var. Río Negro, Río Salinas. ♒ Guatemala/Mexico

84 H13 **Chizela** North Western, NW Zambia 13.11S 24.59E

129 O5 **Chizha** Nenetskiy Avtonomnyy Okrug, NW Russian Federation 67.04N 44.19E

170 G13 **Chizu** Tottori, Honshū, SW Japan 35.15N 134.14E

76 J5 **Chkalov** see Orenburg

117 G18 **Chlómo** ▲ C Greece 38.36N 22.57E

113 M15 **Chmielnik** Świętokrzyskie, C Poland 50.37N 20.43E

178 J11 **Chôâm Khsant** Preăh Vĭhéar, N Cambodia 14.13N 104.55E

64 G10 **Choapa, Río** var. Choapo. ♒ C Chile

Choapas see Las Choapas

Choarta see Chwārtā

85 H17 **Chobe** ◆ district NE Botswana

69 T13 **Chobe** ♒ N Botswana

12 K8 **Chochocouane** ♒ Quebec, SE Canada

112 E13 **Chocianów** Ger. Kotzenau. Dolnośląskie, SW Poland 51.25S 15.55E

56 C9 **Chocó** off. Departamento del Chocó. ◆ province W Colombia

37 X16 **Chocolate Mountains** ▲ California, W USA

23 W9 **Chocowinity** North Carolina, SE USA 35.33N 77.03W

29 N10 **Choctaw** Oklahoma, C USA 35.30N 97.16W

25 Q8 **Choctawhatchee Bay** bay Florida, SE USA

25 Q8 **Choctawhatchee River** ♒ Florida, SE USA

169 V14 **Chŏ-do** island SW North Korea

113 A16 **Chodov** Ger. Chodau. Karlovarský Kraj, W Czech Republic 50.15N 12.45E

112 G10 **Chodzież** Wielkopolskie, C Poland 53.00N 16.55E

65 J15 **Choele Choel** Río Negro, C Argentina 39.18S 65.42W

85 L14 **Chofombo** Tete, NW Mozambique 14.43S 31.48E

9 U14 **Choiceland** Saskatchewan, C Canada 53.28N 104.26W

195 U13 **Choiseul** var. Lauru. island NW Solomon Islands

65 M23 **Choiseul Sound** sound East Falkland, Falkland Islands

42 H7 **Choix** Sinaloa, C Mexico 26.43N 108.20W

112 D10 **Chojna** Zachodniopomorskie, W Poland 52.56N 14.25E

112 H8 **Chojnice** Ger. Konitz. Pomorskie, N Poland 53.41N 17.34E

113 F14 **Chojnów** Ger. Hainau, Haynau. Dolnośląskie, SW Poland 51.16N 15.55E

171 Ll11 **Chōkai-san** ▲ Honshū, C Japan 39.06N 140.02E

178 I11 **Chok Chai** Nakhon Ratchasima, C Thailand 14.43N 102.10E

82 I12 **Ch'ok'ē** var. Choke Mountains. ▲ NW Ethiopia

27 R13 **Choke Canyon Lake** ⊞ Texas, SW USA

Choke Mountains see Ch'ok'ē

151 T15 **Chokpar** Kaz. Shoqpar. Zhambyl, S Kazakhstan 43.49N 74.25E

153 W7 **Chok-Tal** var. Choktal. Issyk-Kul'skaya Oblast', E Kyrgyzstan 42.37N 76.45E

Chôkué see Chokwé

126 Mm6 **Chokurdakh** Respublika Sakha (Yakutiya), NE Russian Federation 70.38N 148.18E

85 L20 **Chokwé** var. Chókué. Gaza, S Mozambique 24.36S 33.06E

196 F8 **Chol** Babeldaob, N Palau

104 I8 **Cholet** Maine-et-Loire, NW France 47.03N 0.52W

85 H17 **Cholila** Chubut, W Argentina 42.33S 71.28W

3 V3 **Cholo** see Thyolo

153 V8 **Cholpon** Narynskaya Oblast', C Kyrgyzstan 42.07N 75.25E

153 X7 **Cholpon-Ata** Issyk-Kul'skaya Oblast', E Kyrgyzstan 42.39N 77.05E

43 P14 **Cholula** Puebla, S Mexico 19.03N 98.19W

44 G6 **Choluteca** Choluteca, S Honduras 13.16N 87.11W

44 H8 **Choluteca** ◆ department S Honduras

44 G6 **Choluteca, Río** ♒ SW Honduras

85 I15 **Choma** Southern, S Zambia 16.47S 26.58E

159 T11 **Chomo Lhari** ▲ NW Bhutan 27.59N 89.24E

178 H7 **Chom Thong** Chiang Mai, NW Thailand 18.28N 98.41E

113 B15 **Chomutov** Ger. Komotau. Ústecký Kraj, NW Czech Republic 50.28N 13.24E

126 K12 **Chona** ♒ C Russian Federation

169 X15 **Ch'ōnan** Jap. Tenan. W South Korea 36.51N 127.10E

178 Hh12 **Chon Buri** prev. Bang Pla Soi. Chon Buri, S Thailand 13.17N 100.58E

58 B6 **Chone** Manabí, W Ecuador 0.41S 80.06W

169 W13 **Ch'ŏngch'ŏn-gang** ♒ W North Korea

169 Y11 **Ch'ŏngjin** NE North Korea 41.48N 129.43E

169 W13 **Ch'ŏngju** W North Korea 39.43N 125.13E

167 S8 **Chongming Dao** island E China

166 J10 **Chongqing** var. Ch'ung-ching, Chungking, Pahsien, Tchongking, Yuzhou. Chongqing Shi, C China 29.34N 106.27E

167 O10 **Chongyang** Hubei, C China 29.34N 114.03E

169 Y16 **Chŏnju** prev. Chŏngup, Jap. Seiyu. SW South Korea 35.51N 127.08E

169 Y15 **Chŏnju** Jap. Zenshū. SW South Korea 35.51N 127.08E

169 Q9 **Chonnacht** see Connaught

65 Q9 **Chonogol** Sühbaatar, E Mongolia 45.55S 115.19E

65 F19 **Chonos, Archipiélago de los** island group S Chile

44 K10 **Chontales** ◆ department S Nicaragua

178 Jj14 **Chon Thanh** Sông Be, S Vietnam 11.25N 106.37E

164 K17 **Cho Oyu** var. Qowowuyag. ▲ China/Nepal 28.07N 86.37E

118 G7 **Chop** Cz. Čop, Hung. Csap. Zakarpats'ka Oblast', W Ukraine 48.25S 22.13E

Y3 **Choptank River** ♒ Maryland, NE USA

45 P15 **Chorcha, Cerro** ▲ W Panama 8.39N 82.07W

Chorku see Chorkŭh

153 R11 **Chorkŭh** Rus. Chorku. N Tajikistan 40.04N 70.30E

99 K17 **Chorley** NW England, UK 53.40N 2.37W

Chorne see Black Sea

119 R5 **Chornobay** Cherkas'ka Oblast', C Ukraine 49.40N 32.20E

119 O3 **Chornobyl'** Rus. Chernobyl'. Kyyivs'ka Oblast', N Ukraine 51.16N 30.15E

119 V5 **Chornomors'ke** Rus. Chernomorskoye. Respublika Krym, S Ukraine 45.29N 32.45E

119 R12 **Chornomors'ke** Rus. Chernomorskoye. Respublika Krym, S Ukraine 45.29N 32.45E

119 R4 **Chornukhy** Poltavs'ka Oblast', C Ukraine 50.15N 32.57E

Chorokh/Chorokhi see Çoruh Nehri

112 O9 **Choroszcz** Podlaskie, NE Poland 53.09N 22.59E

118 K6 **Chortkiv** Rus. Chortkov. Ternopil's'ka Oblast', W Ukraine 49.01N 25.45E

Chortkov see Chortkiv

Chorum see Çorum

112 M9 **Chorzele** Mazowieckie, C Poland 53.16N 20.53E

113 J16 **Chorzów** Ger. Königshütte; prev. Królewska Huta. Śląskie, S Poland 50.17N 18.57E

169 W12 **Ch'osan** N North Korea 40.45N 125.52E

171 Kk17 **Chōshi** var. Tyōsi. Chiba, Honshū, S Japan 35.43N 140.48E

65 H14 **Chos Malal** Neuquén, W Argentina 37.25S 70.16W

Chosŏn-minjujuŭi-inmin-kanghwaguk see North Korea

112 E9 **Choszczno** Ger. Arnswalde. Zachodniopomorskie, W Poland 53.10N 15.24E

159 O15 **Chota Nāgpur** plateau N India

35 R8 **Choteau** Montana, NW USA 47.48N 112.40W

12 M8 **Chouart** ♒ Quebec, SE Canada

78 I7 **Choûm** Adrar, C Mauritania 21.18N 12.58W

29 Q9 **Chouteau** Oklahoma, C USA 36.11N 95.20W

23 X8 **Chowan River** ♒ North Carolina, SE USA

37 Q10 **Chowchilla** California, W USA 37.06N 120.15W

169 P7 **Choybalsan** Dornod, E Mongolia 48.02N 114.31E

168 M9 **Choyr** Dornogovī, C Mongolia 46.20N 108.21E

193 I19 **Christchurch** Canterbury, South Island, NZ 43.33S 172.38E

99 M24 **Christchurch** S England, UK 50.43N 1.45W

193 I18 **Christchurch** ✈ Canterbury, South Island, NZ 43.28S 172.33E

83 H22 **Christiana** Free State, S Africa 27.55S 25.10E

117 J23 **Christiáni** island Kykládes, Greece, Aegean Sea

Christiana see Oslo

12 G13 **Christian Island** island Ontario, S Canada

203 P16 **Christian, Point** headland Pitcairn Island, Pitcairn Islands 25.04S 130.07E

Christian River see Kristiansand

Christiansand see Kristiansand

23 S7 **Christiansburg** Virginia, NE USA 37.07N 80.24W

Christiansfeld see Kristiansfeld

95 I24 **Christiansfeld** Sønderjylland, SW Denmark 55.21N 9.30E

97 G23 **Christianshāb** see Qasigiannguit

41 X14 **Christian Sound** inlet Alaska, USA

47 T9 **Christiansted** Saint Croix, S Virgin Islands (US) 17.43N 64.42W

Christiansund see Kristiansund

27 R13 **Christine** Texas, SW USA 28.47N 98.30W

181 U7 **Christmas Island** ◇ Australian external territory E Indian Ocean

133 T17 **Christmas Island** island E Indian Ocean

Christmas Island see Kiritimati

199 K8 **Christmas Ridge** undersea feature C Pacific Ocean

32 L16 **Christopher** Illinois, N USA 37.58N 89.03W

27 P9 **Christoval** Texas, SW USA 31.09N 100.30W

113 F17 **Chrudim** Pardubický Kraj, C Czech Republic 49.58N 15.49E

117 K25 **Chrýsi** island SE Greece

123 Mm3 **Chrysochoú, Kólpos** var. Khrysokhou Bay. bay E Mediterranean Sea

116 I13 **Chrysoúpoli** var. Hrisoupoli; prev. Khrisoúpolis. Anatolikí Makedonía kai Thráki, NE Greece 40.58N 24.42E

113 K16 **Chrzanów** var. Chrzanow, Ger. Zaumgarten. Śląskie, S Poland 50.09N 19.18E

133 Q7 **Chu** Kaz. Shū.

44 C5 **Chuacaús, Sierra de** ▲ W Guatemala

159 S15 **Chuadanga** Khulna, W Bangladesh 23.37N 88.52E

Chuan see Sichuan

Ch'uan-chou see Quanzhou

41 O11 **Chuathbaluk** Alaska, USA 61.36N 159.14W

194 J12 **Chuave** Chimbu, W PNG 6.06S 145.06E

65 J17 **Chubek** see Moskva

25 O8 **Chubut** off. Provincia de Chubut. ◆ province S Argentina

65 X9 **Chubut, Río** ♒ SE Argentina

45 O15 **Chucanti, Cerro** ▲ E Panama 8.39N 78.27W

Ch'u-chiang see Shaoguan

11 P9 **Chucunaque, Río** ♒ E Panama

9 Y9 **Chudin** see Chudzin

118 M5 **Chudniv** Zhytomyrs'ka Oblast', N Ukraine 50.02N 28.06E

128 H13 **Chudovo** Novgorodskaya Oblast', W Russian Federation 59.07N 31.42E

9 S12 **Chudskoye Ozero** see Peipus, Lake

121 J18 **Chudzin** Rus. Chudin. Brestskaya Voblasts', SW Belarus 52.43N 26.56E

41 Q13 **Chugach Islands** island group Alaska, USA

41 S11 **Chugach Mountains** ▲ Alaska, USA

31 O3 **Chugku-sanchi** ▲ Honshū, SW Japan

152 G12 **Chugoku-sanchi** ▲ Honshū, SW Japan

23 T5 **Churchs Ferry** North Dakota, N USA 48.15N 99.14W

158 G10 **Chūru** Rājasthān, NW India 28.18N 75.00E

56 J4 **Churuguara** Falcón, N Venezuela 10.48N 69.30W

28 M9 **Churchs** see Chucchwa

119 N11 **Cimişlia** Rus. Chimishliya. S Moldova 46.31N 28.45E

107 N3 **Cilleruelo de Bezana** Castilla-León, N Spain 42.58N 3.50W

111 T6 **Cilli** see Celje

11 K8 **Cill Mhantáin** see Wicklow

158 M10 **Cili** Hunan, S China 29.27N 111.03E

124 R12 **Cilicia Trough** undersea feature E Mediterranean Sea

119 V3 **Çildir Gölü** ⊞ NE Turkey

143 S11 **Çıldır** Ardahan, NE Turkey 41.07N 43.07E

21 Q3 **Cili** Hunan, S China

204 I5 **Churchill Peninsula** peninsula Antarctica

24 H8 **Church Point** Louisiana, S USA 30.24N 92.13W

31 N4 **Churubusco** Indiana, N USA 41.14N 85.19W

107 Q5 **Cintruénigo** Navarra, N Spain 42.05N 1.49W

105 X14 **Cinto, Monte** ▲ Corse, France, C Mediterranean Sea 42.22N 8.57E

28 J6 **Cimarron** Kansas, C USA 37.49N 100.20W

28 T9 **Cimarron** New Mexico, SW USA 36.30N 104.55W

64 K5 **Cimarron River** ♒ C USA

129 V14 **Chusovoy** Permskaya Oblast', NW Russian Federation 58.17N 57.54E

143 P15 **Çınar** Diyarbakır, SE Turkey 37.45N 40.22E

107 T5 **Cinca** ♒ NE Spain

114 G13 **Cincar** ▲ SW Bosnia and Herzegovina 43.55N 17.05E

33 Q15 **Cincinnati** Ohio, N USA 39.06N 84.31W

23 M4 **Cincinnati** ✕ Kentucky, USA 39.03N 84.39W

142 C15 **Çine** Aydın, SW Turkey 37.37N 28.03E

101 J21 **Ciney** Namur, SE Belgium 50.16N 5.06E

106 H6 **Cinfães** Viseu, N Portugal 41.04N 8.06W

108 J12 **Cingoli** Marche, C Italy 43.25N 13.09E

42 K14 **Cintalapa** var. Cintalapa de Figueroa. Chiapas, SE Mexico 16.42N 93.40W

Cintalapa de Figueroa see Cintalapa

105 X14 **Cintra** see Sintra

104 K14 **Ciron** ♒ SW France

Cirquenizza see Crikvenica

27 R7 **Cisco** Texas, SW USA 32.23N 98.58W

118 I12 **Cisnădie** Ger. Heltau, Hung. Nagydisznód. Sibiu, SW Romania 45.42N 24.08E

65 G18 **Cisnes, Río** ♒ S Chile

27 T11 **Cistern** Texas, SW USA 29.46N 97.12W

106 L3 **Cisterna** Castilla-León, N Spain 42.46N 5.07W

304 I14 **Citeureup** Jawa, S Indonesia 6.34S 105.41E

Citharista see la Ciotat

Citlaltépetl see Orizaba, Volcán Pico de

57 X10 **Citron** NW French Guiana 4.49N 53.56W

25 N7 **Citronelle** Alabama, S USA 31.05N 88.13W

37 O7 **Citrus Heights** California, W USA 38.42N 121.18W

108 H7 **Cittadella** Veneto, NE Italy 45.37N 11.46E

108 H13 **Città della Pieve** Umbria, C Italy 42.57N 12.01E

108 H12 **Città di Castello** Umbria, C Italy 43.27N 12.13E

109 I14 **Cittaducale** Lazio, C Italy

109 N22 **Cittanova** Calabria, SW Italy 38.21N 16.04E

Cittavecchia see Starigrad

118 G10 **Ciucea** Hung. Csucsa. Cluj, NW Romania 46.57N 22.51E

118 M13 **Ciucurova** Tulcea, SE Romania 44.57N 28.24E

43 N11 **Ciudad Acuña** see Villa Acuña

44 C6 **Ciudad Altamirano** Guerrero, S Mexico 18.20N 100.40W

44 K6 **Ciudad Barrios** San Miguel, NE El Salvador 13.46N 88.13W

57 N6 **Ciudad Bolívar** Barinas, NW Venezuela 8.23N 70.34W

57 N7 **Ciudad Bolívar** prev. Angostura. Bolívar, E Venezuela 8.07N 63.31W

42 K6 **Ciudad Camargo** Chihuahua, N Mexico 27.42N 105.10W

42 E8 **Ciudad Constitución** Baja California Sur, W Mexico 25.06N 111.42W

Ciudad Cortés see Cortés

43 V17 **Ciudad Cuauhtémoc** Chiapas, SE Mexico 15.25N 91.11W

44 J9 **Ciudad Darío** var. Darío. Matagalpa, W Nicaragua 12.42N 86.06W

Ciudad de Dolores Hidalgo see Dolores Hidalgo

44 C6 **Ciudad de Guatemala** Eng. Guatemala City; prev. Santiago de los Caballeros. ● (Guatemala) Guatemala, C Guatemala 14.37N 90.29W

44 C6 **Ciudad del Carmen** see Carmen

64 Q6 **Ciudad del Este** prev. Ciudad Presidente Stroessner, Presidente Stroessner, Puerto Presidente Stroessner. Alto Paraná, SE Paraguay 25.34S 54.40W

64 K5 **Ciudad de Libertador General San Martín** var. Libertador General San Martín. Jujuy, C Argentina 23.49S 64.45W

43 O11 **Ciudad Delicias** see Delicias

Ciudad del Maíz San Luis Potosí, C Mexico 22.25N 99.36W

Ciudad de México see México

56 J7 **Ciudad de Nutrias** Barinas, NW Venezuela 8.03N 69.17W

45 S7 **Ciudad de Panamá** see Panamá

56 F7 **Ciudad Guayana** prev. San Tomé de Guayana, Santo Tomé de Guayana. Bolívar, NE Venezuela 8.22N 62.37W

42 K14 **Ciudad Guzmán** Jalisco, SW Mexico 19.40N 103.30W

43 V17 **Ciudad Hidalgo** Chiapas, SE Mexico 14.40N 92.10W

43 N14 **Ciudad Hidalgo** Michoacán de Ocampo, SW Mexico 19.40N 100.34W

42 J3 **Ciudad Juárez** Chihuahua, N Mexico 31.39N 106.25W

43 Q8 **Ciudad Lerdo** Durango, C Mexico 25.34N 103.30W

43 Q11 **Ciudad Madero** var. Villa Cecilia. Tamaulipas, C Mexico 22.18N 97.55W

43 Q10 **Ciudad Mante** Tamaulipas, C Mexico 22.43N 99.01W

44 F2 **Ciudad Melchor de Mencos** var. Melchor de Mencos. Petén, NE Guatemala 17.03N 89.12W

43 P8 **Ciudad Miguel Alemán** Tamaulipas, C Mexico 26.19N 98.55W

42 G6 **Ciudad Mutis** see Bahía Solano

42 G6 **Ciudad Obregón** Sonora, NW Mexico 27.32N 109.52W

56 J5 **Ciudad Ojeda** Zulia, NW Venezuela 10.12N 71.17W

57 P7 **Ciudad Piar** Bolívar, E Venezuela 7.25N 63.19W

Ciudad Porfirio Díaz see Piedras Negras

Ciudad Quesada see Quesada

107 N11 **Ciudad Real** Castilla-La Mancha, C Spain 38.59N 3.55W

107 N11 **Ciudad Real** ◆ province Castilla-La Mancha, C Spain

106 J7 **Ciudad-Rodrigo** Castilla-León, N Spain

44 A6 **Ciudad Tecún Umán** San Marcos, SW Guatemala 14.40N 92.06W

Ciudad Trujillo see Santo Domingo

43 P12 **Ciudad Valles** San Luis Potosí, C Mexico 21.58N 99.00W

43 O10 **Ciudad Victoria** Tamaulipas, C Mexico 23.44N 99.07W

44 C6 **Ciudad Vieja** Suchitepéquez, S Guatemala 14.30N 90.46W

118 L8 **Ciuhuru** var. Reuțel. ♒ N Moldova

Ciutadella see Ciutadella de Menorca

107 Z8 **Ciutadella de Menorca** var. Ciutadella. Menorca, Spain, W Mediterranean Sea 40.00N 3.50E

142 L11 **Civa Burnu** headland N Turkey 41.22N 36.39E

◆ COUNTRY ◇ DEPENDENT TERRITORY ◆ ADMINISTRATIVE REGION ▲ MOUNTAIN ♒ VOLCANO ⊚ LAKE
● COUNTRY CAPITAL ○ DEPENDENT TERRITORY CAPITAL ✕ INTERNATIONAL AIRPORT ▲ MOUNTAIN RANGE ♒ RIVER ⊞ RESERVOIR

243

108 J7 **Cividale del Friuli** Friuli-
Venezia Giulia, NE Italy
46.06N 13.25E

109 H14 **Civita Castellana** Lazio, C Italy
42.16N 12.24E

108 J12 **Civitanova Marche** Marche,
C Italy 43.18N 13.40E
Civitas Altae Ripae see Brzeg
Civitas Carnutum see Chartres
Civitas Eburovicum see Évreux
Civitas Nemetum see Speyer

109 G15 **Civitavecchia** anc. Centum
Cellae, Trajani Portus. Lazio,
C Italy 42.04N 11.46E

104 L10 **Civray** Vienne, W France
46.10N 0.18E

142 E14 **Çivril** Denizli, W Turkey
38.18N 29.43E

167 O5 **Cixian** Hebei, E China
36.19N 114.22E

143 R16 **Cizre** Şırnak, SE Turkey
37.21N 42.10E
Clacton see Clacton-on-Sea

99 Q21 **Clacton-on-Sea** var. Clacton.
E England, UK 51.48N 1.09E

24 H5 **Claiborne, Lake** ◻ Louisiana,
S USA

104 L10 **Clain** ⚐ W France

9 Q11 **Claire, Lake** ◻ Alberta,
C Canada

27 O6 **Clairemont** Texas, SW USA
33.09N 100.45W

36 M3 **Clair Engle Lake** ◻ California,
W USA

20 B15 **Clairton** Pennsylvania, NE USA
40.17N 79.52W

34 F7 **Clallam Bay** Washington,
NW USA 48.13N 124.16W

105 P8 **Clamecy** Nièvre, C France
47.28N 3.30E

25 P5 **Clanton** Alabama, S USA
32.50N 86.37W

63 D17 **Clara** Entre Ríos, E Argentina
31.49S 58.48W

99 E18 **Clara** Ir. Clóirtheach. C Ireland
53.19N 7.36W

31 T9 **Clara City** Minnesota, N USA
44.57N 95.22W

63 D23 **Claraz** Buenos Aires, E Argentina
37.55S 59.18W
Clár Chlainne Mhuiris see
Claremorris

190 I8 **Clare** South Australia
33.49S 138.35E

99 C19 **Clare** Ir. An Clár. cultural region
W Ireland

99 C18 **Clare** ⚐ W Ireland

99 A16 **Clare Island** Ir. Cliara. island
W Ireland

46 J12 **Claremont** C Jamaica
18.22N 77.10W

31 W10 **Claremont** Minnesota, N USA
44.01N 93.00W

21 N9 **Claremont** New Hampshire,
NE USA 43.21N 72.18W

29 Q9 **Claremore** Oklahoma, C USA
36.18N 95.37W

99 C17 **Claremorris** Ir. Clár Chlainne
Mhuiris. W Ireland 53.47N 9.00W

193 J16 **Clarence** Canterbury, South
Island, NZ 42.07S 173.54E

193 J16 **Clarence** ⚐ South Island, NZ

67 F15 **Clarence Bay** bay Ascension
Island, C Atlantic Ocean

65 H25 **Clarence, Isla** island S Chile

204 H2 **Clarence Island** island South
Shetland Islands, Antarctica

191 V5 **Clarence River** ⚐ New South
Wales, SE Australia

46 J5 **Clarence Town** Long Island,
C Bahamas 23.03N 74.57W

29 W12 **Clarendon** Arkansas, C USA
34.41N 91.18W

27 O3 **Clarendon** Texas, SW USA
34.56N 100.53W

11 U12 **Clarenville** Newfoundland,
SE Canada 48.12N 54.01W

9 Q17 **Claresholm** Alberta, SW Canada
50.01N 113.33W

31 T16 **Clarinda** Iowa, C USA
40.44N 95.02W

57 N5 **Clarines** Anzoátegui,
NE Venezuela 9.55N 65.10W

31 V12 **Clarion** Iowa, C USA
42.43N 93.43W

20 C13 **Clarion** Pennsylvania, NE USA
41.11N 79.21W

199 L7 **Clarion Fracture Zone** tectonic
feature NE Pacific Ocean

20 D13 **Clarion River** ⚐ Pennsylvania,
C USA

25 Q9 **Clark** South Dakota, N USA
44.50N 97.44W

38 K11 **Clarkdale** Arizona, SW USA
34.46N 112.03W

13 W4 **Clarke City** Quebec, SE Canada
50.09N 66.36W

191 Q15 **Clarke Island** island Furneaux
Group, Tasmania, SE Australia

189 X6 **Clarke Range** ▲ Queensland,
E Australia

25 T2 **Clarkesville** Georgia, SE USA
34.36N 83.31W

31 S9 **Clarkfield** Minnesota, N USA
44.48N 95.49W

35 N7 **Clark Fork** Idaho, NW USA
48.06N 116.10W

35 N8 **Clark Fork** ⚐ Idaho/Montana,
NW USA

41 Q12 **Clark, Lake** ◻ Alaska, USA

37 W12 **Clark Mountain** ▲ California,
W USA 35.30N 115.34W

39 S3 **Clark Peak** ▲ Colorado, C USA
40.36N 105.57W

12 D14 **Clark, Point** headland Ontario,
S Canada 44.04N 81.45W

9 Q3 **Clarksburg** West Virginia,
NE USA

24 K2 **Clarksdale** Mississippi, S USA
34.12N 90.34W

35 U12 **Clarks Fork Yellowstone River**
⚐ Montana/Wyoming, NW USA

23 P13 **Clark Hill Lake** var. J. Storm
Thurmond Reservoir.
◻ Georgia/South Carolina,
SE USA

31 R14 **Clarkson** Nebraska, N USA
41.42N 97.07W

41 O13 **Clarks Point** Alaska, USA
58.50N 158.33W

21 I13 **Clarks Summit** Pennsylvania,
NE USA 41.29N 75.42W

34 M10 **Clarkston** Washington, NW USA
46.25N 117.02W

46 J12 **Clark's Town** C Jamaica
18.25N 77.32W

29 T10 **Clarksville** Arkansas, C USA
35.28N 93.28W

33 P13 **Clarksville** Indiana, N USA
40.01N 85.54W

22 I8 **Clarksville** Tennessee, S USA
36.31N 87.21W

27 W5 **Clarksville** Texas, SW USA
33.36N 95.03W

23 U8 **Clarksville** Virginia, NE USA
36.36N 78.36W

23 U11 **Clarkton** North Carolina,
SE USA 34.28N 78.39W

63 C24 **Claromecó** var. Balneario
Claromecó. Buenos Aires,
E Argentina 38.51S 60.01W

27 N3 **Claude** Texas, SW USA
35.06N 101.21W
Clausentum see Southampton

179 P7 **Claveria** Luzon, N Philippines
18.36N 121.04E

101 J20 **Clavier** Liège, E Belgium
50.27N 5.21E

25 W6 **Claxton** Georgia, SE USA
32.09N 81.54W

23 R4 **Clay** West Virginia, NE USA
38.28N 81.04W

29 N3 **Clay Center** Kansas, C USA
39.22N 97.08W

31 P16 **Clay Center** Nebraska, C USA
40.31N 98.03W

23 T2 **Claymont** Delaware, NE USA
39.48N 75.27W

38 M14 **Claypool** Arizona, SW USA
33.24N 110.50W

25 R6 **Clayton** Alabama, S USA
31.52N 85.27W

25 T1 **Clayton** Georgia, SE USA
34.52N 83.24W

24 J5 **Clayton** Louisiana, S USA
31.43N 91.32W

29 X5 **Clayton** Missouri, C USA
38.39N 90.21W

39 V9 **Clayton** New Mexico, SW USA
36.27N 103.12W

23 V9 **Clayton** North Carolina, SE USA
35.39N 78.27W

29 Q12 **Clayton** Oklahoma, C USA
34.35N 95.21W

190 I4 **Clayton River** seasonal river
South Australia

23 R7 **Claytor Lake** ◻ Virginia,
NE USA

29 P13 **Clear Boggy Creek**
⚐ Oklahoma, C USA

99 B22 **Clear, Cape** var. The Bill of Cape
Clear, Ir. Ceann Cléire. headland
SW Ireland 51.25N 9.31W

38 M12 **Clear Creek** ⚐ Arizona,
SW USA

41 S12 **Cleare, Cape** headland Montague
Island, Alaska, USA 59.46N 147.54W

20 E13 **Clearfield** Pennsylvania, NE USA
41.01N 78.27W

38 L2 **Clearfield** Utah, W USA
41.06N 112.03W

27 Q6 **Clear Fork Brazos River**
⚐ Texas, SW USA

33 T12 **Clear Fork Reservoir** ◻ Ohio,
N USA

9 N12 **Clear Hills** ▲ Alberta,
SW Canada

31 V12 **Clear Lake** Iowa, C USA
43.07N 93.27W

31 R9 **Clear Lake** South Dakota, N USA
44.45N 96.40W

36 M6 **Clear Lake** ◻ California, W USA

24 G6 **Clear Lake** ◻ Louisiana, S USA

36 M6 **Clearlake** California, W USA
38.57N 122.38W

37 P1 **Clear Lake Reservoir**
◻ California, W USA

9 N16 **Clearwater** British Columbia,
SW Canada 51.37N 120.01W

25 U12 **Clearwater** Florida, SE USA
27.58N 82.46W

9 R12 **Clearwater**
⚐ Alberta/Saskatchewan,
C Canada

29 W7 **Clearwater Lake** ◻ Missouri,
C USA

35 N10 **Clearwater Mountains**
▲ Idaho, NW USA

35 N10 **Clearwater River** ⚐ Idaho,
NW USA

31 S4 **Clearwater River** ⚐ Minnesota,
N USA

27 T7 **Cleburne** Texas, SW USA
32.21N 97.23W

34 J9 **Cle Elum** Washington, NW USA
47.12N 120.56W

99 O17 **Cleethorpes** E England, UK
53.34N 0.01W
Cléire, Ceann see Clear, Cape

23 O11 **Clemson** South Carolina, SE USA
34.40N 82.50W

23 Q4 **Clendenin** West Virginia,
NE USA 38.29N 81.21W

28 M9 **Cleo Springs** Oklahoma, C USA
36.25N 98.26W
Clerk Island see Onotoa

189 X8 **Clermont** Queensland,
E Australia 22.46S 147.40E

13 S8 **Clermont** Quebec, SE Canada
47.41N 70.15W

105 O4 **Clermont** Oise, N France
49.22N 2.25E

31 X12 **Clermont** Iowa, C USA
43.00N 91.39W

105 P11 **Clermont-Ferrand** Puy-de-
Dôme, C France 45.46N 3.04E

105 Q15 **Clermont-l'Hérault** Hérault,
S France 43.37N 3.25E

101 M22 **Clervaux** Diekirch,
N Luxembourg 50.03N 6.01E

108 G6 **Cles** Trentino-Alto Adige, N Italy
46.21N 11.04E

190 H8 **Cleve** South Australia
33.43S 136.30E
Cleve see Kleve

23 R4 **Cleveland** Georgia, SE USA
34.36N 83.45W

24 K3 **Cleveland** Mississippi, S USA
33.45N 90.43W

33 T11 **Cleveland** Ohio, N USA
41.30N 81.42W

29 O9 **Cleveland** Oklahoma, C USA
36.18N 96.27W

22 L10 **Cleveland** Tennessee, S USA
35.09N 84.52W

27 W10 **Cleveland** Texas, SW USA
30.19N 95.06W

33 N7 **Cleveland** Wisconsin, N USA
43.58N 87.45W

33 O4 **Cleveland Cliffs Basin**
◻ Michigan, N USA

33 U11 **Cleveland Heights** Ohio, N USA
41.31N 81.33W

35 P6 **Cleveland, Mount** ▲ Montana,
NW USA 48.55N 113.51W
Cleves see Kleve

99 B16 **Clew Bay** Ir. Cuan Mó. inlet
W Ireland

25 V13 **Clewiston** Florida, SE USA
26.45N 80.55W
Cliara see Clare Island

99 A17 **Clifden** Ir. An Clochán. W Ireland
53.30N 10.13W

39 O14 **Clifton** Arizona, SW USA
33.03N 109.18W

20 K14 **Clifton** New Jersey, NE USA
40.50N 74.28W

23 S6 **Clifton** Texas, SW USA
31.43N 97.36W

23 T6 **Clifton Forge** Virginia, NE USA
37.49N 79.49W

190 I1 **Clifton Hills** South Australia
27.03S 138.49E

9 S17 **Climax** Saskatchewan, S Canada
49.12N 108.22W

23 O8 **Clinch River**
⚐ Tennessee/Virginia, S USA

27 P12 **Cline** Texas, SW USA
29.14N 100.07W

23 N10 **Clingmans Dome** ▲ North
Carolina/Tennessee, SE USA
35.33N 83.30W

26 H8 **Clint** Texas, SW USA
31.35N 106.13W

8 L14 **Clinton** British Columbia,
SW Canada 51.06N 121.31W

12 E15 **Clinton** Ontario, S Canada
43.37N 81.31W

29 U10 **Clinton** Arkansas, C USA
35.36N 92.26W

32 L10 **Clinton** Illinois, N USA
40.09N 88.57W

31 Z14 **Clinton** Iowa, C USA
41.50N 90.11W

22 G7 **Clinton** Kentucky, S USA
36.39N 89.00W

24 J8 **Clinton** Louisiana, S USA
30.52N 91.01W

21 N11 **Clinton** Massachusetts, NE USA
42.25N 71.40W

33 R10 **Clinton** Michigan, N USA
42.04N 83.58W

24 K5 **Clinton** Mississippi, S USA
32.22N 90.22W

23 S5 **Clinton** Missouri, C USA
38.22N 93.51W

23 V10 **Clinton** North Carolina, SE USA
35.00N 78.19W

28 L10 **Clinton** Oklahoma, C USA
35.31N 98.58W

23 Q12 **Clinton** South Carolina, SE USA
34.28N 81.52W

23 M9 **Clinton** Tennessee, S USA
36.06N 84.07W

15 J7 **Clinton-Colden Lake**
◻ Northwest Territories,
NW Canada

8 H5 **Clinton Creek** Yukon Territory,
NW Canada 64.24N 140.35W

32 L13 **Clinton Lake** ◻ Illinois, N USA

29 Q4 **Clinton Lake** ◻ Kansas, C USA

23 T11 **Clio** South Carolina, SE USA
34.34N 79.33W

199 L7 **Clipperton Fracture Zone**
tectonic feature E Pacific Ocean

200 N7 **Clipperton Island** ◇ French
dependency of French Polynesia
E Pacific Ocean

48 K6 **Clipperton Island** island
E Pacific Ocean

2 F16 **Clipperton Seamounts**
undersea feature E Pacific Ocean

104 J8 **Clisson** Loire-Atlantique,
NW France 47.06N 1.19W

64 K7 **Clodomira** Santiago del Estero,
N Argentina 27.33S 64.07W
Cloich na Coillte see Clonakilty
Clóirtheach see Clara

99 C21 **Clonakilty** Ir. Cloich na Coillte.
SW Ireland 51.37N 8.54W

189 T6 **Cloncurry** Queensland,
C Australia 20.44S 140.34E

99 F18 **Clondalkin** Ir. Cluain Dolcáin.
E Ireland 53.19N 6.24W

99 E16 **Clones** Ir. Cluain Eois. N Ireland
54.10N 7.13W

189 P1 **Clonmel** Ir. Cluain Meala.
S Ireland 52.21N 7.42W

102 G11 **Cloppenburg** Niedersachsen,
NW Germany 52.51N 8.03E

31 W6 **Cloquet** Minnesota, N USA
46.43N 92.27W

39 S14 **Cloudcroft** New Mexico,
SW USA 32.57N 105.44W

35 W12 **Cloud Peak** ▲ Wyoming, C USA
44.22N 107.10W

193 K14 **Cloudy Bay** inlet South Island,
NZ

23 R10 **Clover** South Carolina, SE USA
35.06N 81.13W

36 M6 **Cloverdale** California, W USA
38.49N 123.03W

23 J5 **Cloverport** Kentucky, S USA
37.50N 86.37W

37 Q10 **Clovis** California, W USA
36.48N 119.43W

27 W12 **Clovis** New Mexico, SW USA
34.22N 103.12W

12 K13 **Cloyne** Ontario, SE Canada
44.48N 77.09W

99 C18 **Cluain Dolcáin** see Clondalkin
Cluain Eois see Clones
Cluainín see Manorhamilton
Cluain Meala see Clonmel

118 H10 **Cluj** ◆ county NW Romania
Cluj see Cluj-Napoca

118 H10 **Cluj-Napoca** Ger. Klausenburg,
Hung. Kolozsvár; prev. Cluj. Cluj,
NW Romania 46.47N 23.36E
Clunia see Feldkirch

105 R11 **Cluny** Saône-et-Loire, C France
46.25N 4.38E

105 T10 **Cluses** Haute-Savoie, E France
46.04N 6.34E

108 E7 **Clusone** Lombardia, N Italy
51.53N 10.00E

27 W12 **Clute** Texas, SW USA
29.01N 95.24W

99 J18 **Clwyd** cultural region NE Wales,
UK

193 D22 **Clyde** Otago, South Island, NZ
45.12S 169.21E

29 N3 **Clyde** Kansas, C USA
39.35N 97.24W

31 P2 **Clyde** North Dakota, N USA
48.44N 98.51W

33 S11 **Clyde** Ohio, N USA 41.18N 82.58W

27 Q7 **Clyde** Texas, SW USA
32.24N 99.29W

98 J13 **Clyde** ⚐ Ontario, SE Canada

98 H12 **Clyde** ⚐ W Scotland, UK
55.54N 4.24W

98 H13 **Clyde, Firth of** inlet S Scotland,
UK

35 S11 **Clyde Park** Montana, NW USA
45.56N 110.39W

106 I7 **Côa, Rio** ⚐ N Portugal

37 W16 **Coachella** California, SW USA
33.38N 116.10W

37 W16 **Coachella Canal** canal
California, W USA

23 I9 **Coacoyole** Durango, C Mexico
27.03S 138.49E

27 N7 **Coahoma** Texas, SW USA
32.18N 101.18W

42 L14 **Coalcomán** var. Coalcomán de
Matamoros. Michoacán de
Ocampo, S Mexico 18.49N 103.13W
Coalcomán de Matamoros see
Coalcomán

41 T8 **Coal Creek** Alaska, USA
65.21N 143.08W

9 Q17 **Coaldale** Alberta, SW Canada
49.42N 112.36W

29 P12 **Coalgate** Oklahoma, C USA
34.33N 96.14W

37 P11 **Coalinga** California, W USA
36.08N 120.21W

8 L9 **Coal River** British Columbia,
SW Canada 59.38N 126.45W

23 Q6 **Coal River** ⚐ West Virginia,
NE USA

38 M2 **Coalville** Utah, W USA
40.56N 111.22W

60 E13 **Coari** Amazonas, N Brazil
4.05S 63.07W

60 E14 **Coari, Rio** ⚐ NW Brazil

83 J20 **Coast** ◆ province SE Kenya
Coast see Pwani

14 F11 **Coast Mountains** Fr. Chaîne
Côtière. ▲ Canada/USA

17 FF3 **Coast Ranges** ▲ W USA

98 I12 **Coatbridge** S Scotland, UK
55.52N 4.01W

44 B6 **Coatepeque** Quezaltenango,
SW Guatemala 14.42N 91.49W

20 H16 **Coatesville** Pennsylvania, NE USA
39.58N 75.47W

13 Q13 **Coaticook** Quebec, SE Canada
45.07N 71.46W

15 Mm6 **Coats Island** island Nunavut,
NE Canada

205 O4 **Coats Land** physical region
Antarctica

43 T14 **Coatzacoalcos** var.
Quetzalcoalc; prev. Puerto
México. Veracruz-Llave, E Mexico
18.06N 94.25W

43 S14 **Coatzacoalcos, Río**
⚐ SE Mexico

118 M13 **Cobadin** Constanța,
SW Romania 44.02N 28.29E

12 H9 **Cobalt** Ontario, S Canada
47.23N 79.40W

44 D5 **Cobán** Alta Verapaz, C Guatemala
15.28N 90.19W

191 O6 **Cobar** New South Wales,
SE Australia 31.31S 145.50E

20 F12 **Cobb Hill** ▲ Pennsylvania,
NE USA 41.52N 77.52W

2 D8 **Cobb Seamount** undersea feature
E Pacific Ocean 47.00N 131.00W

12 K12 **Cobden** Ontario, SE Canada
45.36N 76.54W

99 D21 **Cobh** Ir. An Cóbh; prev. Cove of
Cork, Queenstown. SW Ireland
51.51N 8.16W

56 E10 **Cobija** Pando, NW Bolivia
11.04S 68.49W

59 J14 **Cobija** Tarapacá, N Chile

21 J10 **Cobleskill** New York, NE USA
42.40N 74.29W

12 I15 **Cobourg** Ontario, S Canada
43.57N 78.06W

189 P1 **Cobourg Peninsula** headland
Northern Territory, N Australia
11.27S 132.33E

191 O10 **Cobram** Victoria, SE Australia
35.56S 145.36E

84 N13 **Cóbuè** Niassa, N Mozambique
12.08S 34.46E

103 K18 **Coburg** Bayern, SE Germany
50.16N 10.58E

21 Q5 **Coburn Mountain** ▲ Maine,
NE USA 45.28N 70.07W
Coca see Puerto Francisco de
Orellana

57 H18 **Cocachacra** Arequipa, SW Peru
17.09S 71.46W

61 J17 **Cocalinho** Mato Grosso,
W Brazil 14.22S 51.00W
Cocanada see Kākināda

107 S11 **Cocentaina** País Valenciano,
E Spain 39.13N 1.04W

104 K11 **Cognac** anc. Compniacum.
Charente, W France 45.42N 0.19W
Cogoletto see Cogoleto

105 U16 **Cogolin** Var, SE France
43.15N 6.30E

107 O7 **Cogolludo** Castilla-La Mancha,
C Spain 40.58N 3.05W

39 R6 **Cohactepa Hills** ▲ Colorado,
C USA

F11 **Cohocton River** ⚐ New York,
NE USA

21 L10 **Cohoes** New York, NE USA
42.46N 73.42W

191 N10 **Cohuna** Victoria, SE Australia
35.48N 144.15E

45 P16 **Coiba, Isla de** island SW Panama

65 U5 **Coig, Río** ⚐ S Argentina

65 G19 **Coihaique** var. Coyhaique. Aisén,
S Chile 45.31S 72.00W

10 G12 **Cochrane** Ontario, S Canada
49.04N 81.01W

65 G20 **Cochrane** Aisén, S Chile
47.16S 72.33W

9 U10 **Cochrane**
⚐ Manitoba/Saskatchewan,
C Canada
Cochrane, Lago see Pueyrredón,
Lago

46 M6 **Cockburn Harbour** South
Caicos, S Turks and Caicos Islands
21.28N 71.30W

12 C11 **Cockburn Island** island Ontario,
S Canada

46 J3 **Cockburn Town** San Salvador,
E Bahamas 24.01N 74.30W

23 X2 **Cockeysville** Maryland, NE USA
39.29N 76.34W

189 N12 **Cocklebiddy** Western Australia
32.02S 125.54E

46 I12 **Cockpit Country, The** physical
region W Jamaica

45 S16 **Coclé** off. Provincia de Coclé. ◆
province C Panama

45 S15 **Coclé del Norte** Colón,
C Panama 9.04N 80.32W

25 Y12 **Cocoa** Florida, SE USA
28.23N 80.44W

25 Y12 **Cocoa Beach** Florida, SE USA
28.19N 80.36W

81 D17 **Cocobeach** Estuaire, NW Gabon
0.58N 9.34E

46 G5 **Coco, Cayo** island C Cuba

157 Q19 **Coco Channel** strait Andaman
Sea/Bay of Bengal

181 N6 **Coco-de-Mer Seamounts**
undersea feature W Indian Ocean

38 K10 **Coconino Plateau** plain Arizona,
SW USA

45 N6 **Coco, Río** var. Río Wanki,
Segovia o Wangkí.
⚐ Honduras/Nicaragua

181 T8 **Cocos (Keeling) Islands**
◇ Australian external territory
E Indian Ocean

181 T7 **Cocos Basin** undersea feature
E Indian Ocean

196 B17 **Cocos Island** island S Guam
Cocos Island Ridge see Cocos
Ridge

133 S17 **Cocos Islands** island group
E Indian Ocean

2 G15 **Cocos Plate** tectonic feature

200 Oo8 **Cocos Ridge** var. Cocos Island
Ridge. undersea feature E Pacific
Ocean

42 K13 **Cocula** Jalisco, SW Mexico
20.25N 103.52W

109 D17 **Coda Cavallo, Capo** headland
Sardegna, Italy, C Mediterranean
Sea 40.49N 9.43E

60 E13 **Codajás** Amazonas, N Brazil
3.49S 62.12W
Codazzi see Agustín Codazzi

21 Q12 **Cod, Cape** headland
Massachusetts, NE USA
41.50N 69.56W

193 B25 **Codfish Island** island SW NZ

108 H9 **Codigoro** Emilia-Romagna,
N Italy 44.49N 12.07E

11 P5 **Cod Island** island Newfoundland,
E Canada

118 J12 **Codlea** Ger. Zeiden, Hung.
Feketehalom. Brașov, C Romania
45.42N 25.27E

60 M13 **Codó** Maranhão, E Brazil
4.28S 43.51W

108 E8 **Codogno** Lombardia, N Italy
45.10N 9.42E

118 M10 **Codrii** hill range C Moldova

47 W9 **Codrington** Barbuda, Antigua
and Barbuda 17.37N 61.49W

108 J7 **Codroipo** Friuli-Venezia Giulia,
NE Italy 45.58N 13.00E

30 M12 **Cody** Nebraska, C USA
42.54N 101.13W

35 U12 **Cody** Wyoming, C USA
43.31N 109.04W

23 P7 **Coeburn** Virginia, NE USA
36.56N 82.27W

56 E10 **Coello** Tolima, W Colombia
4.16N 74.54W

189 V2 **Coen** Queensland, NE Australia
14.03S 143.16E

103 E14 **Coesfeld** Nordrhein-Westfalen,
W Germany 51.55N 7.10E

34 M8 **Coeur d'Alene** Idaho, NW USA
47.40N 116.46W

34 M8 **Coeur d'Alene Lake** ◻ Idaho,
NW USA

100 O8 **Coevorden** Drenthe,
NE Netherlands 52.39N 6.45E

8 H6 **Coffee Creek** Yukon Territory,
W Canada 62.39S 139.05W

32 L15 **Coffeen Lake** ◻ Illinois, N USA

24 L3 **Coffeeville** Mississippi, S USA
33.58N 89.40W

29 Q8 **Coffeyville** Kansas, C USA
37.02N 95.37W

190 G9 **Coffin Bay** South Australia
34.39S 135.30E

190 F9 **Coffin Bay Peninsula** peninsula
South Australia

191 V5 **Coffs Harbour** New South
Wales, SE Australia 30.18S 153.07E

107 R10 **Cofrentes** País Valenciano,
E Spain 39.13N 1.04W

191 P4 **Collarenebri** New South Wales,
SE Australia

188 I13 **Collie** Western Australia

188 L4 **Collier Bay** bay Western Australia

23 F10 **Collierville** Tennessee, S USA
35.02N 89.39W

108 F11 **Collina, Passo della** pass C Italy
44.02N 10.55E

12 G14 **Collingwood** Ontario, S Canada
44.28N 80.12W

192 I13 **Collingwood** Tasman, South
Island, NZ 40.41S 172.41E

194 M15 **Collingwood Bay** bay SE PNG

24 L7 **Collins** Mississippi, S USA
31.39N 89.33W

32 K15 **Collinsville** Illinois, S USA
38.40N 89.58W

29 P9 **Collinsville** Oklahoma, C USA
36.21N 95.50W

108 E7 **Collinwood** Tennessee, S USA
35.10N 87.44W
Collipo see Leiria

161 G21 **Coimbatore** Tamil Nādu, S India
11.00N 76.57E

106 G8 **Coimbra** anc. Conimbria,
Conimbriga. Coimbra, W Portugal
40.12N 8.25W

106 G8 **Coimbra** ◆ district N Portugal

106 L15 **Coín** Andalucía, S Spain
36.40N 4.45W

28 J3 **Colby** Kansas, C USA
39.24N 101.03W

55 H17 **Colca, Río** ⚐ SW Peru

99 P21 **Colchester** hist. Colneceaste, anc.
Camulodunum. E England, UK
51.54N 0.54E

21 N13 **Colchester** Connecticut, NE USA
41.34N 72.17W

40 M16 **Cold Bay** Alaska, USA
55.11N 162.43W

9 R14 **Cold Lake** Alberta, SW Canada
54.25N 110.16W

9 R13 **Cold Lake**
◻ Alberta/Saskatchewan, C Canada

31 U8 **Cold Spring** Minnesota, N USA
45.27N 94.25W

27 W10 **Coldspring** Texas, SW USA
30.35N 95.07W

9 N17 **Coldstream** British Columbia,
SW Canada 50.13N 119.09W

98 L13 **Coldstream** SE Scotland, UK
55.39N 2.19W

12 H13 **Coldwater** Ontario, S Canada
44.43N 79.36W

28 K7 **Coldwater** Kansas, C USA
37.13N 99.18W

33 Q10 **Coldwater** Michigan, N USA
41.56N 85.00W

27 N1 **Coldwater Creek**
⚐ Oklahoma/Texas, SW USA

24 K2 **Coldwater River** ⚐ Mississippi,
S USA

191 O9 **Coleambally** New South Wales,
SE Australia 34.48S 145.54E

21 O6 **Colebrook** New Hampshire,
NE USA 44.52N 71.27W

29 T5 **Cole Camp** Missouri, C USA
38.27N 93.12W

41 T6 **Coleen River** ⚐ Alaska, USA

9 P17 **Coleman** Alberta, SW Canada
49.36N 114.26W

27 Q8 **Coleman** Texas, SW USA
31.50N 99.26W

33 Q6 **Coleman** Michigan, N USA
43.45N 84.37W
Colemerik see Hakkâri

85 K22 **Colenso** KwaZulu/Natal, E South
Africa 28.39S 29.49E

190 L12 **Coleraine** Victoria, SE Australia
37.39S 141.42E

99 F14 **Coleraine** Ir. Cúil Raithin.
N Northern Ireland, UK
55.07N 6.40W

193 G18 **Coleridge, Lake** ◻ South Island,
NZ

85 H24 **Colesberg** Northern Cape,
C South Africa 30.41S 25.08E

23 L9 **Colfax** Louisiana, S USA
31.31N 92.43W

34 L9 **Colfax** Washington, NW USA
46.52N 117.21W

32 J6 **Colfax** Wisconsin, N USA
45.00N 91.44W

65 I19 **Colhué Huapí, Lago**
◻ S Argentina

47 Z6 **Colibris, Pointe des** headland
Grande Terre, E Guadeloupe
16.15N 61.10W

108 D6 **Colico** Lombardia, N Italy
46.08N 9.24E

101 E14 **Colijnsplaat** Zeeland,
SW Netherlands 51.36N 3.47E

42 L14 **Colima** Colima, S Mexico
19.12N 103.45W

42 L14 **Colima** ◆ state SW Mexico

42 L14 **Colima, Nevado de** ▲ C Mexico
19.36N 103.36W

61 M14 **Colinas** Maranhão, E Brazil
6.01S 44.15W

98 F10 **Coll** island W Scotland, UK

107 N7 **Collado Villalba** var. Villalba.
Madrid, C Spain 40.37N 3.58W

191 R4 **Collarenebri** New South Wales,
SE Australia 29.31S 148.33E

108 G12 **Colle di Val d'Elsa** Toscana,
C Italy 43.26N 11.06E

41 R9 **College** Alaska, USA
64.49N 148.06W

34 K10 **College Place** Washington,
NW USA 46.03N 118.23W

27 U10 **College Station** Texas, SW USA
30.36N 96.21W

191 P4 **Collerina** New South Wales,
SE Australia

Collipo see Leiria

65 G14 **Collipulli** Araucanía, C Chile
37.55S 72.30W

99 D16 **Collooney** Ir. Cúil Mhuine.
NW Ireland 54.10N 8.28W

31 R10 **Colman** South Dakota, N USA
43.58N 96.48W

105 U6 **Colmar** Ger. Kolmar. Haut-Rhin,
NE France 48.04N 7.21E

106 M15 **Colmenar** Andalucía, S Spain
36.54N 4.19W

107 O9 **Colmenar de Oreja** var.
Colmenar. Madrid, C Spain
40.06N 3.25W

107 N7 **Colmenar Viejo** Madrid,
C Spain 40.39N 3.46W

27 X9 **Colmesneil** Texas, SW USA
30.54N 94.25W
Cöln see Köln
Colneceaste see Colchester

43 C3 **Colnett, Cabo** headland Baja
California, NW Mexico

61 G15 **Colniza** Mato Grosso, W Brazil
9.16S 59.25W
Cologne see Köln

44 B6 **Colomba** Quezaltenango,
SW Guatemala 14.45N 91.39W
Colomb-Béchar see Béchar

56 E11 **Colombia** Huila, C Colombia
3.24N 74.49W

56 G10 **Colombia** off. Republic of
Colombia. ◆ republic N South
America

66 E12 **Colombian Basin** undersea
feature SW Caribbean Sea
Colombie-Britannique see
British Columbia

161 J25 **Colombo** ● (Sri Lanka) Western
Province, W Sri Lanka 6.55N 79.52E

161 J25 **Colombo** ✈ Western Province,
SW Sri Lanka 6.50N 79.59E

31 N11 **Colome** South Dakota, N USA
43.13N 99.42W

63 D18 **Colón** Entre Ríos, E Argentina
32.13S 58.15W

63 B19 **Colón** Buenos Aires, E Argentina
33.53S 61.06W

46 D5 **Colón** Matanzas, C Cuba
22.42N 80.54W

45 T14 **Colón** prev. Aspinwall. Colón,
C Panama 9.04N 80.32W

44 K5 **Colón** ◆ department NE Honduras

45 S15 **Colón** off. Provincia de Colón. ◆
province N Panama

59 A16 **Colón, Archipiélago de** var.
Islas de los Galápagos, Eng.
Galapagos Islands, Tortoise
Islands. island group Ecuador,
E Pacific Ocean

46 K5 **Colonel Hill** Crooked Island,
SE Bahamas 22.43N 74.12W

42 B3 **Colonet, Cabo** headland
NW Mexico 30.57N 116.19W

196 G14 **Colonia** Yap, W Micronesia
9.29N 138.06E

63 D19 **Colonia** ◆ department
SW Uruguay
Colonia see Kolonia, Micronesia
Colonia see Colonia del
Sacramento, Uruguay
Colonia Agrippina see Köln

63 D20 **Colonia del Sacramento** var.
Colonia. Colonia, SW Uruguay
34.28S 57.48W

64 L8 **Colonia Dora** Santiago del
Estero, N Argentina 28.34S 62.58W
Colonia Julia Fanestris see
Fano

23 W5 **Colonial Beach** Virginia,
NE USA 38.15N 76.57W

23 V6 **Colonial Heights** Virginia,
NE USA 37.15N 77.24W

200 Oo8 **Colón Ridge** undersea feature
E Pacific Ocean

98 F12 **Colonsay** island W Scotland, UK

59 K22 **Colorada, Laguna** ◻ SW Bolivia

39 R6 **Colorado** off. State of Colorado;
also known as Centennial State,
Silver State. ◆ state C USA

65 H22 **Colorado, Cerro** ▲ S Argentina
49.58S 71.38W

27 O7 **Colorado City** Texas, SW USA
32.23N 100.51W

38 M7 **Colorado Plateau** plateau
W USA

63 A24 **Colorado, Río** ⚐ E Argentina

45 N12 **Colorado, Río** ⚐ N Costa Rica
Colorado, Río see Colorado
River

18 Hh10 **Colorado River** var. Río
Colorado. ⚐ Mexico/USA

27 U11 **Colorado River** ⚐ Texas,
SW USA

37 W15 **Colorado River Aqueduct**
aqueduct California, W USA

39 T5 **Colorado Springs** Colorado,
C USA 38.49N 104.46W

42 L11 **Colotlán** Jalisco, SW Mexico
22.07N 103.15W

57 L19 **Colquechaca** Potosí, C Bolivia
18.39S 66.12W

25 S7 **Colquitt** Georgia, SE USA
31.10N 84.43W

31 R11 **Colton** South Dakota, N USA
43.47N 96.55W

37 U13 **Colton** California, W USA
34.04N 117.20W

31 Y9 **Columbia** Louisiana, S USA
31.51N 89.50W

29 X3 **Columbia** Mississippi, S USA
31.15N 89.50W

23 Y4 **Columbia** Maryland, NE USA
39.11N 76.52W

29 U4 **Columbia** Missouri, C USA
38.55N 92.19W

23 N9 **Columbia** North Carolina,
SE USA

23 T13 **Columbia** South Carolina,
SE USA 33.59N 81.00W

23 J9 **Columbia** Kentucky, S USA
37.06N 85.18W

32 K16 **Columbia** Illinois, N USA
38.26N 90.12W

24 L7 **Columbia** Mississippi, S USA
31.15N 89.50W

20 F15 **Columbia** Pennsylvania, NE USA
40.01N 76.30W

23 Q12 **Columbia** state capital South
Carolina, SE USA 34.00N 81.00W

22 I9 **Columbia** Tennessee, S USA *35.37N 87.02W*
2 F9 **Columbia** ≈ Canada/USA
34 K9 **Columbia Basin** *basin* Washington, NW USA
207 Q10 **Columbia, Cape** *headland* Ellesmere Island, Nunavut, NE Canada
33 Q12 **Columbia City** Indiana, N USA *41.09N 85.29W*
23 W3 **Columbia, District of** ◐ *federal district* NE USA
35 P7 **Columbia Falls** Montana, NW USA *48.22N 114.10W*
9 O15 **Columbia Icefield** *icefield* Alberta/British Columbia, S Canada
9 O15 **Columbia, Mount** ▲ Alberta/British Columbia, SW Canada *52.07N 117.30W*
9 N15 **Columbia Mountains** ▲▲ British Columbia, SW Canada
25 P4 **Columbiana** Alabama, S USA *33.10N 86.36W*
33 V12 **Columbiana** Ohio, N USA *40.53N 80.41W*
34 M14 **Columbia Plateau** *plateau* Idaho/Oregon, NW USA
31 P7 **Columbia Road Reservoir** ⊞ South Dakota, N USA
67 K16 **Columbia Seamount** *undersea feature* C Atlantic Ocean *20.30S 32.00W*
85 D25 **Columbine, Cape** *headland* SW South Africa *32.50S 17.39E*
107 U9 **Columbretes, Islas** *island group* E Spain
25 R5 **Columbus** Georgia, SE USA *32.28N 84.58W*
33 P14 **Columbus** Indiana, N USA *39.12N 85.55W*
29 R7 **Columbus** Kansas, C USA *37.10N 94.50W*
25 N4 **Columbus** Mississippi, S USA *33.30N 88.25W*
35 U11 **Columbus** Montana, NW USA *45.38N 109.15W*
31 Q15 **Columbus** Nebraska, C USA *41.25N 97.22W*
39 Q16 **Columbus** New Mexico, SW USA *31.49N 107.38W*
23 P10 **Columbus** North Carolina, SE USA *35.15N 82.09W*
30 K2 **Columbus** North Dakota, N USA *48.52N 102.47W*
33 S13 **Columbus** *state capital* Ohio, N USA *39.57N 83.00W*
27 U11 **Columbus** Texas, SW USA *29.42N 96.32W*
32 L8 **Columbus** Wisconsin, N USA *43.21N 89.00W*
33 R12 **Columbus Grove** Ohio, N USA *40.55N 84.03W*
31 Y15 **Columbus Junction** Iowa, C USA *41.16N 91.21W*
46 J3 **Columbus Point** *headland* Cat Island, C Bahamas *24.07N 75.19W*
37 T8 **Columbus Salt Marsh** *salt marsh* Nevada, W USA
37 N6 **Colusa** California, W USA *39.10N 122.03W*
34 L7 **Colville** Washington, NW USA *48.33N 117.54W*
192 M5 **Colville, Cape** *headland* North Island, NZ *36.28S 175.20E*
192 M5 **Colville Channel** *channel* North Island, NZ
41 P6 **Colville River** ≈ Alaska, USA
99 J18 **Colwyn Bay** N Wales, UK *53.18N 3.43W*
108 H9 **Comacchio** *var.* Commachio; *anc.* Comactium. Emilia-Romagna, N Italy *44.40N 12.10E*
108 H9 **Comacchio, Valli di** *lagoon* Adriatic Sea, N Mediterranean Sea
Comactium *see* Comacchio
43 V17 **Comalapa** Chiapas, SE Mexico *15.42N 92.06W*
43 U15 **Comalcalco** Tabasco, SE Mexico *18.16N 93.05W*
65 H16 **Comallo** Río Negro, SW Argentina *40.58S 70.13W*
28 M12 **Comanche** Oklahoma, C USA *34.22N 97.57W*
27 R8 **Comanche** Texas, SW USA *31.54N 98.36W*
204 H2 **Comandante Ferraz** *Brazilian research station* Antarctica *61.57S 58.23W*
64 N6 **Comandante Fontana** Formosa, N Argentina *25.19S 59.42W*
65 I22 **Comandante Luis Piedra Buena** Santa Cruz, S Argentina *50.04S 68.55W*
61 O18 **Comandatuba** Bahia, SE Brazil *15.13S 39.00W*
118 K11 **Comănești** *Hung.* Kománfalva. Bacău, SW Romania *46.24N 26.27E*
59 M19 **Comarapa** Santa Cruz, C Bolivia *17.52S 64.34W*
118 J13 **Comarnic** Prahova, SE Romania *45.13N 25.36E*
44 H6 **Comayagua** Comayagua, W Honduras *14.33N 87.37W*
44 H6 **Comayagua** ◐ *department* W Honduras
44 I6 **Comayagua, Montañas de** ▲▲ C Honduras
23 R15 **Combahee River** ≈ South Carolina, SE USA
64 G10 **Combarbalá** Coquimbo, C Chile *31.15S 71.03W*
105 S7 **Combeaufontaine** Haute-Saône, E France *47.43N 5.52E*
99 G15 **Comber** *Ir.* An Comar. E Northern Ireland, UK *54.33N 5.45W*
101 K20 **Comblain-au-Pont** Liège, E Belgium *50.29N 5.36E*
104 I6 **Combourg** Ille-et-Vilaine, NW France *48.21N 1.44W*
46 M9 **Comendador** *prev.* Elías Piña. W Dominican Republic *18.51N 71.40W*
Comer See *see* Como, Lago di
27 R11 **Comfort** Texas, SW USA *29.58N 98.54W*
159 V15 **Comilla** *Ben.* Kumillā. Chittagong, E Bangladesh *23.28N 91.10E*
101 B18 **Comines** Hainaut, W Belgium *50.46N 2.58E*
123 J16 **Comino** *Malt.* Kemmuna. *island* C Malta
109 D18 **Comino, Capo** *headland* Sardegna, Italy, C Mediterranean Sea *40.32N 9.49E*
109 K25 **Comiso** Sicilia, Italy, C Mediterranean Sea *36.57N 14.37E*
43 V16 **Comitán** *var.* Comitán de Domínguez. Chiapas, SE Mexico *16.14N 92.06W*
Comitán de Domínguez *see* Comitán
Commachio *see* Comacchio
Commander Islands *see* Komandorskiye Ostrova
105 O10 **Commentry** Allier, C France *46.18N 2.46E*
25 T2 **Commerce** Georgia, SE USA *34.12N 83.27W*
29 R8 **Commerce** Oklahoma, C USA *36.55N 94.52W*
27 V5 **Commerce** Texas, SW USA *33.16N 95.52W*
39 T4 **Commerce City** Colorado, C USA *39.45N 104.54W*
105 S5 **Commercy** Meuse, NE France *48.46N 5.36E*
57 W9 **Commewijne** *var.* Commewyne. ◐ *district* NE Suriname
Commewyne *see* Commewijne
13 P8 **Commissaires, Lac des** ◎ Quebec, SE Canada
66 A12 **Commissioner's Point** *headland* W Bermuda
15 Ll3 **Committee Bay** *bay* Nunavut, N Canada
108 D7 **Como** *anc.* Comum. Lombardia, N Italy *45.48N 9.04E*
65 J19 **Comodoro Rivadavia** Chubut, SE Argentina *45.49S 67.30W*
108 D6 **Como, Lago di** *var.* Lario, *Eng.* Lake Como, *Ger.* Comer See. ◎ N Italy
Como, Lake *see* Como, Lago di
42 E7 **Comondú** Baja California Sur, W Mexico *26.01N 111.50W*
118 F12 **Comorâște** *Hung.* Komornok. Caraș-Severin, SW Romania *45.13N 21.34E*
Comores, République Fédérale Islamique des *see* Comoros
161 G24 **Comorin, Cape** *headland* SE India *8.00N 77.10E*
312 M8 **Comoro Basin** *undersea feature* SW Indian Ocean
122 K14 **Comoro Islands** *island group* W Indian Ocean
312 H13 **Comoros** *off.* Federal Islamic Republic of the Comoros, *Fr.* République Fédérale Islamique des Comores. ◆ *republic* W Indian Ocean
8 L17 **Comox** Vancouver Island, British Columbia, SW Canada *49.40N 124.55W*
105 O4 **Compiègne** Oise, N France *49.25N 2.49E*
Complutum *see* Alcalá de Henares
Compniacum *see* Cognac
42 K12 **Compostela** Nayarit, C Mexico *21.14N 104.52W*
Compostella *see* Santiago
62 L11 **Comprida, Ilha** *island* S Brazil
119 N11 **Comrat** *Rus.* Komrat. S Moldova *46.18N 28.40E*
27 O11 **Comstock** Texas, SW USA *29.39N 101.10W*
33 P9 **Comstock Park** Michigan, N USA *43.00N 85.40W*
199 Kk3 **Comstock Seamount** *undersea feature* N Pacific Ocean *48.15N 156.55W*
Comum *see* Como
99 N17 **Cona** Xizang Zizhiqu, W China *27.58N 91.54E*
78 H14 **Conakry** ● (Guinea) Conakry, SW Guinea *9.31N 13.43W*
78 H14 **Conakry** ✕ Conakry, SW Guinea *9.37N 13.32W*
Conamara *see* Connemara
Conca *see* Cuenca
27 Q12 **Concan** Texas, SW USA *29.27N 99.43W*
104 F6 **Concarneau** Finistère, NW France *47.52N 3.55W*
85 O17 **Conceição** Sofala, C Mozambique *18.47S 36.18E*
61 K15 **Conceição do Araguaia** Pará, NE Brazil *8.15S 49.15W*
60 F10 **Conceição do Maú** Roraima, W Brazil *3.34N 59.52W*
63 D14 **Concepción** *var.* Concepcion. Corrientes, NE Argentina *28.25S 57.54W*
64 J8 **Concepción** Tucumán, N Argentina *27.19S 65.34W*
59 O17 **Concepción** Santa Cruz, E Bolivia *16.15S 62.07W*
64 G13 **Concepción** Bío Bío, C Chile *36.47S 73.01W*
56 E14 **Concepción** Putumayo, S Colombia *0.03N 75.39W*
64 O5 **Concepción** *var.* Villa Concepción. Concepción, C Paraguay *23.26S 57.23W*
64 O5 **Concepción** *off.* Departamento de Concepción. ◐ *department* E Paraguay
Concepción *see* La Concepción
Concepción de la Vega *see* La Vega
43 N9 **Concepción del Oro** Zacatecas, C Mexico *24.37N 101.25W*
63 D18 **Concepción del Uruguay** Entre Ríos, E Argentina *32.30S 58.15W*
44 K11 **Concepción, Volcán** ⛰ SW Nicaragua *11.31N 85.37W*
46 J4 **Conception Island** *island* C Bahamas
37 P14 **Conception, Point** *headland* California, W USA *34.27N 120.28W*
56 H6 **Concha** Zulia, W Venezuela *9.01N 71.45W*
62 L9 **Conchas** São Paulo, S Brazil *23.00S 47.58W*
39 U11 **Conchas Dam** New Mexico, SW USA *35.21N 104.11W*
39 U10 **Conchas Lake** ⊞ New Mexico, SW USA
104 M5 **Conches-en-Ouche** Eure, N France *49.00N 1.00E*
39 N12 **Concho** Arizona, SW USA *34.28N 109.33W*
42 J5 **Conchos, Río** ≈ NW Mexico
43 O8 **Conchos, Río** ≈ C Mexico
110 C8 **Concise** Vaud, W Switzerland *46.52N 6.40E*
37 N8 **Concord** California, W USA *37.58N 122.01W*
21 O9 **Concord** *state capital* New Hampshire, NE USA *43.10N 71.31W*
23 R10 **Concord** North Carolina, SE USA *35.30N 80.34W*
63 D17 **Concordia** Entre Ríos, E Argentina *31.25S 58.00W*
56 D9 **Concordia** Antioquia, W Colombia *6.03N 75.57W*
42 J10 **Concordia** Sinaloa, C Mexico *23.18N 106.03W*
59 I19 **Concordia** Tacna, SW Peru *18.12S 70.19W*
29 N3 **Concordia** Kansas, C USA *39.34N 97.39W*
29 S4 **Concordia** Missouri, C USA *38.58N 93.34W*
62 I13 **Concórdia** Santa Catarina, S Brazil *27.13S 52.01W*
178 J7 **Con Cuông** Nghê An, N Vietnam *19.02N 104.54E*
178 Jj16 **Côn Đao** *var.* Con Son. *island* S Vietnam
Condate *see* St-Claude, Jura, France
Condate *see* Rennes, Ille-et-Vilaine, France
Condate *see* Montereau-Faut-Yonne, Seine-St-Denis, France
31 P8 **Conde** South Dakota, N USA *45.08N 98.07W*
44 J8 **Condega** Estelí, NW Nicaragua *13.21N 86.23W*
105 P2 **Condé-sur-l'Escaut** Nord, N France *50.27N 3.36E*
104 K5 **Condé-sur-Noireau** Calvados, N France *48.52N 0.31W*
Condivincum *see* Nantes
191 P8 **Condobolin** New South Wales, SE Australia *33.04S 147.08E*
104 L15 **Condom** Gers, S France *43.56N 0.23E*
34 J11 **Condon** Oregon, NW USA *45.13N 120.11W*
56 D9 **Condoto** Chocó, W Colombia *5.06N 76.37W*
25 P7 **Conecuh River** ≈ Alabama/Florida, SE USA
108 H7 **Conegliano** Veneto, NE Italy *45.52N 12.18E*
63 C19 **Conesa** Buenos Aires, E Argentina *33.36S 60.21W*
12 F15 **Conestogo** ≈ Ontario, S Canada
195 O17 **Conflict Group** *island group* SE PNG
Confluentes *see* Koblenz
104 L10 **Confolens** Charente, W France *46.00N 0.40E*
38 J4 **Confusion Range** ▲▲ Utah, W USA
64 N6 **Confuso, Río** ≈ C Paraguay
23 R12 **Congaree River** ≈ South Carolina, SE USA
Công Hoa Xã Hôi Chu Nghia Viêt Nam *see* Vietnam
166 K12 **Congjiang** *prev.* Bingmei. Guizhou, S China *25.48N 108.55E*
81 K19 **Congo** *off.* Democratic Republic of Congo; *prev.* Zaire, Belgian Congo, Congo (Kinshasa). ◆ *republic* C Africa
81 G18 **Congo** *off.* Republic of the Congo, *Fr.* Moyen-Congo; *prev.* Middle Congo. ◆ *republic* C Africa
69 T11 **Congo** *var.* Kongo, *Fr.* Zaire. ≈ C Africa
Congo *see* Zaire (province, Angola)
Congo/Congo (Kinshasa) *see* Congo (Democratic Republic of)
122 G12 **Congo Basin** *drainage basin* W Dem. Rep. Congo (Zaire)
69 Q11 **Congo Canyon** *var.* Congo Seavalley, Congo Submarine Canyon. *undersea feature* E Atlantic Ocean
Congo Cone *see* Congo Fan
67 P15 **Congo Fan** *var.* Congo Cone. *undersea feature* E Atlantic Ocean
Coni *see* Cuneo
65 H18 **Cónico, Cerro** ▲ SW Argentina *43.12S 71.42W*
Conimbria/Conimbriga *see* Coimbra
Conjeeveram *see* Kānchipuram
9 R13 **Conklin** Alberta, C Canada *55.36N 111.06W*
26 M1 **Conlen** Texas, SW USA *36.16N 102.10W*
Con, Loch *see* Conn, Lough
Connacht *see* Connaught
99 B17 **Connaught** *var.* Connacht, *Ir.* Chonnacht, Cúige. *cultural region* W Ireland
33 V10 **Conneaut** Ohio, N USA *41.56N 80.32W*
20 L13 **Connecticut** *off.* State of Connecticut; also known as Blue Law State, Constitution State, Land of Steady Habits, Nutmeg State. ◐ *state* NE USA
21 N8 **Connecticut** ≈ Canada/USA
21 O6 **Connecticut Lakes** *lakes* New Hampshire, NE USA
34 K9 **Connell** Washington, NW USA *46.39N 118.51W*
99 B17 **Connemara** *Ir.* Conamara. *region* W Ireland
33 Q14 **Connersville** Indiana, N USA *39.38N 85.15W*
99 B16 **Conn, Lough** *Ir.* Loch Con. ◎ W Ireland
37 X6 **Connors Pass** *pass* Nevada, W USA *39.01N 114.37W*
189 X7 **Connors Range** ▲▲ Queensland, E Australia
58 E7 **Cononaco, Río** ≈ E Ecuador
31 W13 **Conrad** Iowa, C USA *42.13N 92.52W*
35 R7 **Conrad** Montana, NW USA *48.10N 111.58W*
27 W10 **Conroe** Texas, SW USA *30.18N 95.27W*
27 V10 **Conroe, Lake** ⊞ Texas, SW USA
63 C17 **Conscripto Bernardi** Entre Ríos, E Argentina *31.03S 59.04W*
61 M20 **Conselheiro Lafaiete** Minas Gerais, SE Brazil *20.40S 43.48W*
Consentia *see* Cosenza
99 L14 **Consett** N England, UK *54.49N 1.52W*
46 B5 **Consolación del Sur** Pinar del Río, W Cuba *22.29N 83.31W*
Con Son *see* Côn Đao
9 R15 **Consort** Alberta, SW Canada *51.58N 110.44W*
Constance *see* Konstanz
110 I6 **Constance, Lake** *Ger.* Bodensee. ◎ C Europe
106 G9 **Constância** Santarém, C Portugal *39.28N 8.22W*
119 N14 **Constanța** *var.* Küstendje, *Eng.* Constanza, *Ger.* Konstanza, *Turk.* Küstence. Constanța, SE Romania *44.09N 28.36E*
118 L14 **Constanța** ◐ *county* SE Romania
Constantia *see* Coutances, France
Constantia *see* Konstanz, Germany
106 K13 **Constantina** Andalucía, S Spain *37.54N 5.36W*
76 L5 **Constantine** *var.* Qacentina, *Ar.* Qoussantîna. NE Algeria *36.22N 6.43E*
41 O14 **Constantine, Cape** *headland* Alaska, USA *58.23N 158.53W*
Constantinople *see* İstanbul
Constantiola *see* Oltenița
Constanz *see* Konstanz
Constanza *see* Constanța
64 G13 **Constitución** Maule, C Chile *35.25S 72.19W*
63 D17 **Constitución** Salto, N Uruguay *31.04S 57.51W*
Constitution State *see* Connecticut
107 N10 **Consuegra** Castilla-La Mancha, C Spain *39.28N 3.36W*
189 X9 **Consuelo Peak** ▲ Queensland, E Australia *24.45S 148.01E*
58 E11 **Contamana** Loreto, N Peru *7.19S 75.04W*
Contrasto, Colle del *see* Contrasto, Portella del
109 K23 **Contrasto, Portella del** *var.* Colle del Contrasto. *pass* Sicilia, Italy, C Mediterranean Sea *37.51N 14.22E*
56 G8 **Contratación** Santander, C Colombia *6.18N 73.27W*
104 M8 **Contres** Loir-et-Cher, C France *47.24N 1.30E*
109 O17 **Conversano** Puglia, SE Italy *40.58N 17.07E*
29 U11 **Conway** Arkansas, C USA *35.05N 92.26W*
21 O8 **Conway** New Hampshire, NE USA *43.58N 71.05W*
23 U13 **Conway** South Carolina, SE USA *33.50N 79.03W*
27 N2 **Conway** Texas, SW USA *35.10N 101.23W*
29 U11 **Conway, Lake** ⊞ Arkansas, C USA
29 N7 **Conway Springs** Kansas, C USA *37.23N 97.38W*
99 J18 **Conwy** N Wales, UK *53.16N 3.51W*
25 T3 **Conyers** Georgia, SE USA *33.40N 84.01W*
Coo *see* Kos
190 F4 **Coober Pedy** South Australia *29.01S 134.46E*
189 P2 **Cooinda** Northern Territory, N Australia *12.54S 132.31E*
190 B6 **Cook** South Australia *30.37S 130.26E*
31 W4 **Cook** Minnesota, N USA *47.51N 92.41W*
203 N6 **Cook, Baie de** *bay* Moorea, W French Polynesia
8 J16 **Cook, Cape** *headland* Vancouver Island, British Columbia, SW Canada *50.04N 127.52W*
39 Q15 **Cookes Peak** ▲ New Mexico, SW USA *32.32N 107.43W*
22 L8 **Cookeville** Tennessee, S USA *36.09N 85.30W*
183 P9 **Cook Fracture Zone** *tectonic feature* S Pacific Ocean
Cook, Grand Récif de *see* Cook, Récif de
41 Q12 **Cook Inlet** *inlet* Alaska, USA
203 X2 **Cook Island** *island* Line Islands, E Kiribati
202 J14 **Cook Islands** ◇ *territory in free association with NZ* S Pacific Ocean
193 E19 **Cook, Mount** *prev.* Aoraki, Aorangi. ▲ South Island, NZ *43.38S 170.05E*
197 G4 **Cook, Récif de** *var.* Grand Récif de Cook. *reef* S New Caledonia
12 G14 **Cookstown** Ontario, S Canada *44.12N 79.39W*
99 F15 **Cookstown** *Ir.* An Chorr Chríochach. C Northern Ireland, UK *54.39N 6.45W*
193 K14 **Cook Strait** *var.* Raukawa. *strait* NZ
189 W3 **Cooktown** Queensland, NE Australia *15.28S 145.15E*
191 P6 **Coolabah** New South Wales, SE Australia *31.03S 146.42E*
190 J11 **Coola Coola Swamp** *wetland* South Australia
191 S7 **Coolah** New South Wales, SE Australia *31.49S 149.43E*
191 P9 **Coolamon** New South Wales, SE Australia *34.49S 147.13E*
191 T4 **Coolatai** New South Wales, SE Australia *29.16S 150.45E*
188 K12 **Coolgardie** Western Australia *31.00S 121.12E*
38 L14 **Coolidge** Arizona, SW USA *32.58N 111.29W*
27 U8 **Coolidge** Texas, SW USA *31.45N 96.39W*
191 Q11 **Cooma** New South Wales, SE Australia *36.16S 149.09E*
Coomassie *see* Kumasi
191 R6 **Coonabarabran** New South Wales, SE Australia *31.19S 149.18E*
190 J10 **Coonalpyn** South Australia *35.43S 139.50E*
191 R6 **Coonamble** New South Wales, SE Australia *30.56S 148.22E*
Coondapoor *see* Kundāpura
161 G21 **Coonoor** Tamil Nādu, SE India *11.21N 76.46E*
31 U14 **Coon Rapids** Iowa, C USA *41.52N 94.40W*
31 V8 **Coon Rapids** Minnesota, N USA *45.12N 93.18W*
27 V5 **Cooper** Texas, SW USA *33.22N 95.41W*
189 U9 **Cooper Creek** *var.* Barcoo, Cooper's Creek. *seasonal river* Queensland/South Australia
41 R12 **Cooper Landing** Alaska, USA *60.27N 149.59W*
23 T14 **Cooper River** ≈ South Carolina, SE USA
Cooper's Creek *see* Cooper Creek
46 H1 **Coopers Town** Great Abaco, N Bahamas *26.46N 77.27W*
20 J10 **Cooperstown** New York, NE USA *42.43N 74.55W*
31 P4 **Cooperstown** North Dakota, N USA *47.26N 98.07W*
33 P9 **Coopersville** Michigan, N USA *43.03N 85.55W*
190 D7 **Coorabie** South Australia *31.57S 132.18E*
25 Q3 **Coosa River** ≈ Alabama/Georgia, S USA
34 E14 **Coos Bay** Oregon, NW USA *43.22N 124.13W*
191 Q9 **Cootamundra** New South Wales, SE Australia *34.40S 148.03E*
99 E16 **Cootehill** *Ir.* Muinchille. N Ireland *54.04N 7.04W*
Çop *see* Chop
59 J17 **Copacabana** La Paz, W Bolivia *16.11S 69.02W*
65 H14 **Copahué, Volcán** ▲ C Chile *37.56S 71.04W*
43 U16 **Copainalá** Chiapas, SE Mexico *17.04N 93.13W*
34 F8 **Copalis Beach** Washington, NW USA *47.05N 124.11W*
44 F6 **Copán** ◐ *department* W Honduras
Copán *see* Copán Ruinas
27 T14 **Copano Bay** *bay* NW Gulf of Mexico
44 F6 **Copán Ruinas** *var.* Copán. Copán, W Honduras *14.51N 89.07W*
Copenhagen *see* København
109 Q19 **Copertino** Puglia, SE Italy *40.16N 18.03E*
64 H7 **Copiapó** Atacama, N Chile *27.17S 70.25W*
64 G8 **Copiapó, Bahía** *bay* N Chile
64 G7 **Copiapó, Río** ≈ N Chile
116 M12 **Çöpköy** Edirne, NW Turkey *41.14N 26.51E*
190 I5 **Copley** South Australia *30.36S 138.26E*
108 H9 **Copparo** Emilia-Romagna, C Italy *44.53N 11.53E*
57 V10 **Coppename Rivier** *var.* Koppename. ≈ C Suriname
27 S9 **Copperas Cove** Texas, SW USA *31.07N 97.54W*
84 J13 **Copperbelt** ◐ *province* C Zambia
41 S11 **Copper Center** Alaska, USA *61.57N 145.21W*
Coppermine *see* Kugluktuk
15 I5 **Coppermine** ≈ Northwest Territories/Nunavut, N Canada
41 T11 **Copper River** ≈ Alaska, USA
Copper State *see* Arizona
118 I11 **Copșa Mică** *Ger.* Kleinkopisch, *Hung.* Kiskapus. Sibiu, C Romania *45.48N 24.08E*
164 J14 **Coqên** Xizang Zizhiqu, W China *31.13N 85.12E*
Coquilhatville *see* Mbandaka
34 E14 **Coquille** Oregon, NW USA *43.10N 124.11W*
64 G9 **Coquimbo** Coquimbo, N Chile *29.59S 71.18W*
64 G9 **Coquimbo** *off.* Región de Coquimbo. ◐ *region* C Chile
118 I15 **Corabia** Olt, S Romania *43.46N 24.31E*
59 F17 **Coracora** Ayacucho, SW Peru *15.07S 73.45W*
Cora Droma Rúisc *see* Carrick-on-Shannon
46 K9 **Corail** SW Haiti *18.32N 73.54W*
191 V4 **Coraki** New South Wales, SE Australia *29.01S 153.15E*
188 G8 **Coral Bay** Western Australia *23.02S 113.51E*
25 Y16 **Coral Gables** Florida, SE USA *25.43N 80.16W*
15 M5 **Coral Harbour** Southampton Island, Northwest Territories, NE Canada *64.10N 83.15W*
199 I10 **Coral Sea** *sea* SW Pacific Ocean
182 M7 **Coral Sea Basin** *undersea feature* N Coral Sea
199 Hh10 **Coral Sea Islands** ◇ *Australian external territory* SW Pacific Ocean
190 M12 **Corangamite, Lake** ◎ Victoria, SE Australia
Corantijn Rivier *see* Courantyne River
20 B14 **Coraopolis** Pennsylvania, NE USA *40.28N 80.08W*
109 N17 **Corato** Puglia, SE Italy *41.09N 16.25E*
105 O17 **Corbières** ▲▲ S France
105 P8 **Corbigny** Nièvre, C France *47.15N 3.42E*
23 N7 **Corbin** Kentucky, S USA *36.57N 84.06W*
106 L14 **Corbones** ≈ SW Spain
Corcaigh *see* Cork
37 R11 **Corcoran** California, W USA *36.06N 119.33W*
49 T14 **Corcovado, Golfo** *gulf* S Chile
65 G18 **Corcovado, Volcán** ▲ S Chile *43.13S 72.45W*
106 F3 **Corcubión** Galicia, NW Spain *42.55N 9.12W*
Corcyra Nigra *see* Korčula
62 Q9 **Cordeiro** Rio de Janeiro, SE Brazil *22.01S 42.20W*
25 T6 **Cordele** Georgia, SE USA *31.58N 83.49W*
28 L11 **Cordell** Oklahoma, C USA *35.18N 98.58W*
105 N14 **Cordes** Tarn, S France *44.03N 1.57E*
64 O6 **Cordillera** *off.* Departamento de la Cordillera. ◐ *department* C Paraguay
190 K1 **Cordillo Downs** South Australia *26.44S 140.37E*
64 K10 **Córdoba** Córdoba, C Argentina *31.25S 64.10W*
43 R14 **Córdoba** Veracruz-Llave, E Mexico *18.52N 96.48W*
106 M13 **Córdoba** *var.* Cordoba, *Eng.* Cordova; *anc.* Corduba. Andalucía, SW Spain *37.52N 4.46W*
64 K11 **Córdoba** *off.* Provincia de Córdoba. ◐ *province* C Argentina
56 D7 **Córdoba** *off.* Departamento de Córdoba. ◐ *province* NW Colombia
106 L13 **Córdoba** ◐ *province* Andalucía, S Spain
64 K10 **Córdoba, Sierras de** ▲▲ C Argentina
25 O3 **Cordova** Alabama, S USA *33.45N 87.10W*
41 S12 **Cordova** Alaska, USA *60.32N 145.45W*
Cordova/Corduba *see* Córdoba
Corentyne River *see* Courantyne River
Corfu *see* Kérkyra
106 J9 **Coria** Extremadura, W Spain *39.58N 6.31W*
106 J14 **Coria del Río** Andalucía, S Spain *37.18N 6.04W*
191 S8 **Coricudgy, Mount** ▲ New South Wales, SE Australia *32.49S 150.28E*
109 N20 **Corigliano Calabro** Calabria, SW Italy *39.35N 16.30E*
Corinium/Corinium Dobunorum *see* Cirencester
25 N1 **Corinth** Mississippi, S USA *34.56N 88.29W*
Corinth *see* Kórinthos
Corinth Canal *see* Dióryga Korinthou
Corinth, Gulf of/Corinthiacus Sinus *see* Korinthiakós Kólpos
Corinthus *see* Kórinthos
44 I9 **Corinto** Chinandega, NW Nicaragua *12.28N 87.10W*
99 C21 **Cork** *Ir.* Corcaigh. S Ireland *51.54N 7.06W*
99 C21 **Cork** *Ir.* Corcaigh. *cultural region* SW Ireland
99 C21 **Cork** ✕ SW Ireland *51.52N 8.25W*
99 D21 **Cork Harbour** *Ir.* Cuan Chorcaí. *inlet* SW Ireland
109 I23 **Corleone** Sicilia, Italy, C Mediterranean Sea *37.49N 13.18E*
116 N13 **Çorlu** Tekirdağ, NW Turkey *41.10N 27.48E*
116 N12 **Çorlu Çayı** ≈ NW Turkey
Cormaiore *see* Courmayeur
9 V13 **Cormorant** Manitoba, C Canada *54.12N 100.33W*
25 T2 **Cornelia** Georgia, SE USA *34.30N 83.31W*
62 J10 **Cornélio Procópio** Paraná, S Brazil *23.07S 50.40W*
57 V9 **Corneliskondre** Sipaliwini, N Suriname *5.21N 56.10W*
32 J5 **Cornell** Wisconsin, N USA *45.09N 91.10W*
11 S12 **Corner Brook** Newfoundland, E Canada *48.58N 57.58W*
Corner Rise Seamounts *see* Corner Seamounts
66 I9 **Corner Seamounts** *var.* Corner Rise Seamounts. *undersea feature* NW Atlantic Ocean
118 M9 **Cornești** *Rus.* Korneshty. C Moldova *47.23N 28.00E*
Corneto *see* Tarquinia
Cornhusker State *see* Nebraska
29 X8 **Corning** Arkansas, C USA *36.25N 90.35W*
37 N5 **Corning** California, W USA *39.54N 122.12W*
31 U15 **Corning** Iowa, C USA *40.58N 94.46W*
20 G11 **Corning** New York, NE USA *42.08N 77.03W*
Corn Islands *see* Maíz, Islas del
109 J14 **Corno Grande** ▲ C Italy *42.26N 13.29E*
13 N13 **Cornwall** Ontario, SE Canada *45.01N 74.45W*
99 H25 **Cornwall** *cultural region* SW England, UK
99 G25 **Cornwall, Cape** *headland* SW England, UK *50.11N 5.39W*
56 J4 **Coro** *prev.* Santa Ana de Coro. Falcón, NW Venezuela *11.27N 69.40W*
59 J18 **Corocoro** La Paz, W Bolivia *17.10S 68.28W*
59 K17 **Coroico** La Paz, W Bolivia *16.09S 67.41W*
192 M5 **Coromandel** Waikato, North Island, NZ *36.47S 175.30E*
161 K20 **Coromandel Coast** *coast* E India
192 M5 **Coromandel Peninsula** *peninsula* North Island, NZ
192 M6 **Coromandel Range** ▲▲ North Island, NZ
179 P12 **Coron** Busuanga Island, W Philippines *12.02N 120.10E*
37 T15 **Corona** California, W USA *33.52N 117.34W*
39 T12 **Corona** New Mexico, SW USA *34.15N 105.36W*
9 U17 **Coronach** Saskatchewan, S Canada *49.07N 105.33W*
37 U17 **Coronado** California, W USA *32.41N 117.10W*
45 N15 **Coronado, Bahía de** *bay* S Costa Rica
9 R15 **Coronation** Alberta, SW Canada *52.06N 111.25W*
15 Ii4 **Coronation Gulf** *gulf* Nunavut, N Canada
204 I1 **Coronation Island** *island* Antarctica
41 X14 **Coronation Island** *island* Alexander Archipelago, Alaska, USA
63 B18 **Coronda** Santa Fe, C Argentina *31.58S 60.55W*
65 F14 **Coronel** Bío Bío, C Chile *37.01S 73.07W*
63 D20 **Coronel Brandsen** *var.* Brandsen. Buenos Aires, E Argentina *35.07S 58.15W*
64 K4 **Coronel Cornejo** Salta, N Argentina *22.46S 63.49W*
63 B24 **Coronel Dorrego** Buenos Aires, E Argentina *38.38S 61.15W*
64 P6 **Coronel Oviedo** Caaguazú, SE Paraguay *25.30S 56.27W*
63 B23 **Coronel Pringles** Buenos Aires, E Argentina *37.58S 61.26W*
63 B23 **Coronel Suárez** Buenos Aires, E Argentina *37.27S 61.57W*
63 E22 **Coronel Vidal** Buenos Aires, E Argentina *37.28S 57.45W*
57 V9 **Coronie** ◐ *district* NW Suriname
59 G17 **Coropuna, Nevado** ▲ S Peru *15.31S 72.31W*
Çorovoda *see* Çorovodë
115 L22 **Çorovodë** *var.* Çorovoda. Berat, S Albania *40.29N 20.15E*
191 P11 **Corowa** New South Wales, SE Australia *36.01S 146.22E*
44 G1 **Corozal** Corozal, N Belize *18.22N 88.22W*
56 E6 **Corozal** Sucre, NW Colombia *9.18N 75.19W*
44 G1 **Corozal** ◐ *district* N Belize
27 T14 **Corpus Christi** Texas, SW USA *27.48N 97.24W*
27 T14 **Corpus Christi Bay** *inlet* Texas, SW USA
27 R14 **Corpus Christi, Lake** ⊞ Texas, SW USA
65 F16 **Corral** Los Lagos, C Chile *39.55S 73.30W*
107 O9 **Corral de Almaguer** Castilla-La Mancha, C Spain *39.45N 3.10W*
106 K6 **Corrales** Castilla-León, N Spain *41.22N 5.43W*
39 R11 **Corrales** New Mexico, SW USA *35.11N 106.37W*
Corrán Tuathail *see* Carrauntoohil
108 F9 **Correggio** Emilia-Romagna, C Italy *44.47N 10.46E*
179 P11 **Corregidor Island** *island* NW Philippines
61 M16 **Corrente** Piauí, E Brazil *10.28S 45.10W*
61 I19 **Correntes, Rio** ≈ SW Brazil
105 N12 **Corrèze** ◐ *department* C France
99 C17 **Corrib, Lough** *Ir.* Loch Coirib. ◎ W Ireland
63 C14 **Corrientes** Corrientes, NE Argentina *27.28S 58.42W*
63 D15 **Corrientes** *off.* Provincia de Corrientes. ◐ *province* NE Argentina
46 A5 **Corrientes, Cabo** *headland* W Cuba *21.48N 84.30W*
42 I13 **Corrientes, Cabo** *headland* SW Mexico *20.25N 105.42W*
Corrientes, Provincia de *see* Corrientes
63 C16 **Corrientes, Río** ≈ NE Argentina
58 E8 **Corrientes, Río** ≈ Ecuador/Peru
27 W9 **Corrigan** Texas, SW USA *31.00N 94.49W*
57 U9 **Corriverton** E Guyana *5.55N 57.09W*
Corriza *see* Korçë
191 Q11 **Corryong** Victoria, SE Australia *36.14S 147.54E*
105 Y12 **Corse** *Eng.* Corsica. ◐ *region* France, C Mediterranean Sea
105 X13 **Corse** *Eng.* Corsica. *island* France, C Mediterranean Sea
105 Y13 **Corse, Cap** *headland* Corse, France, C Mediterranean Sea *43.01N 9.25E*
105 X15 **Corse-du-Sud** ◐ *department* Corse, France, C Mediterranean Sea
31 P11 **Corsica** South Dakota, N USA *43.25N 98.24W*
Corsica *see* Corse
27 U7 **Corsicana** Texas, SW USA *32.04N 96.27W*
105 Y15 **Corte** Corse, France, C Mediterranean Sea *42.19N 9.09E*
65 G16 **Corte Alto** Los Lagos, S Chile *40.58S 73.04W*
106 I13 **Cortegana** Andalucía, S Spain *37.55N 6.49W*
45 N15 **Cortés** *var.* Ciudad Cortés. Puntarenas, SE Costa Rica *8.59N 83.32W*
44 G5 **Cortés** ◐ *department* NW Honduras
39 P8 **Cortez** Colorado, C USA *37.22N 108.36W*
Cortez, Sea of *see* California, Golfo de
108 H6 **Cortina d'Ampezzo** Veneto, NE Italy *46.33N 12.09E*
20 H11 **Cortland** New York, NE USA *42.34N 76.09W*
33 V11 **Cortland** Ohio, N USA *41.19N 80.43W*
108 H12 **Cortona** Toscana, C Italy *43.15N 12.01E*
78 H13 **Corubal, Rio** ≈ E Guinea-Bissau
106 G10 **Coruche** Santarém, C Portugal *38.58N 8.31W*
Çoruh *see* Rize
143 R11 **Çoruh Nehri** *Geor.* Chorokhi, *Rus.* Chorokh. ≈ Georgia/Turkey
142 K12 **Çorum** *var.* Chorum. Çorum, N Turkey *40.31N 34.57E*
142 J12 **Çorum** *var.* Chorum. ◐ *province* N Turkey
61 H19 **Corumbá** Mato Grosso do Sul, S Brazil *19.00S 57.35W*
12 D16 **Corunna** Ontario, S Canada *42.49N 82.25W*
Corunna *see* A Coruña
34 F12 **Corvallis** Oregon, NW USA *44.34N 122.36W*
66 M1 **Corvo** *var.* Ilha do Corvo. *island* Azores, Portugal, NE Atlantic Ocean
Corvo, Ilha do *see* Corvo
33 O16 **Corydon** Indiana, N USA *38.12N 86.07W*
31 V16 **Corydon** Iowa, C USA *40.45N 93.19W*
Cos *see* Kos
42 I9 **Cosalá** Sinaloa, C Mexico *24.25N 106.39W*
43 R15 **Cosamaloapan** *var.* Cosamaloapan de Carpio. Veracruz-Llave, E Mexico *18.21N 95.50W*

◆ Country
● Country Capital
◇ Dependent Territory
○ Dependent Territory Capital
◐ Administrative Region
✕ International Airport
▲ Mountain
▲▲ Mountain Range
⛰ Volcano
≈ River
◎ Lake
⊞ Reservoir

Cosamaloapan de Carpio see Cosamaloapan
109 N21 Cosenza anc. Consentia. Calabria, SW Italy 39.16N 16.15E
33 T13 Coshocton Ohio, N USA 40.16N 81.51W
44 H9 Cosigüina, Punta headland NW Nicaragua 12.53N 87.42W
31 T9 Cosmos Minnesota, N USA 44.56N 94.42W
105 O8 Cosne-sur-Loire Nièvre, C France 47.25N 2.56E
110 B9 Cossonay Vaud, W Switzerland 46.37N 6.28E
Cossyra see Pantelleria
49 R4 Costa, Cordillera de la var. Cordillera de Venezuela. ▲ N Venezuela
44 K13 Costa Rica off. Republic of Costa Rica. ◆ republic Central America
45 N15 Costeña, Fila ▲ S Costa Rica
Costermansville see Bukavu
39 I14 Costeşti Argeş, SW Romania 44.38N 24.52E
39 S8 Costilla New Mexico, SW USA 36.58N 105.31W
37 O7 Cosumnes River ≈ California, W USA
103 O16 Coswig Sachsen, E Germany 51.07N 13.36E
103 M14 Coswig Sachsen-Anhalt, E Germany 51.53N 12.26E
Cosyra see Pantelleria
179 R16 Cotabato Mindanao, S Philippines 7.13N 124.12E
58 C5 Cotacachi ▲ N Ecuador 0.29N 78.17W
52 L21 Cotagaita Potosí, S Bolivia 20.46S 65.40W
105 V15 Côte d'Azur prev. Nice. ✈ (Nice) Alpes-Maritimes, SE France 43.40N 7.12E
Côte d'Ivoire see Ivory Coast
105 R8 Côte d'Or cultural region C France
105 R7 Côte d'Or ◆ department E France
Côte Française des Somalis see Djibouti
104 J4 Cotentin peninsula N France
104 G6 Côtes d'Armor prev. Côtes-du-Nord. ◆ department NW France
Côtes-du Nord see Côtes d'Armor
Côthen see Köthen
Côtière, Chaîne see Coast Mountains
42 M13 Cotija var. Cotija de la Paz. Michoacán de Ocampo, SW Mexico 19.49N 102.39W
Cotija de la Paz see Cotija
79 R16 Cotonou var. Kotonu. S Benin 6.24N 2.25E
79 R16 Cotonou ✈ S Benin 6.31N 2.18E
58 B6 Cotopaxi prev. León. ◆ province C Ecuador
58 C6 Cotopaxi ▲ N Ecuador 0.42S 78.24W
Cotrone see Crotone
99 L21 Cotswold Hills var. Cotswolds. hill range S England, UK
Cotswolds see Cotswold Hills
34 F13 Cottage Grove Oregon, NW USA 43.48N 123.03W
23 S14 Cottageville South Carolina, SE USA 32.55N 80.28W
103 P14 Cottbus prev. Kottbus. Brandenburg, E Germany 51.42N 14.22E
29 U9 Cotter Arkansas, C USA 36.16N 92.30W
108 A9 Cottian Alps Fr. Alpes Cottiennes, It. Alpi Cozie. ▲ France/Italy
Cottiennes, Alpes see Cottian Alps
Cotton State, The see Alabama
24 G4 Cotton Valley Louisiana, S USA 32.49N 93.25W
38 L12 Cottonwood Arizona, SW USA 34.43N 112.00W
34 M10 Cottonwood Idaho, NW USA 46.01N 116.20W
31 S9 Cottonwood Minnesota, N USA 44.37N 95.41W
27 Q7 Cottonwood Texas, SW USA 32.12N 99.14W
29 O5 Cottonwood Falls Kansas, C USA 38.22N 96.32W
38 L3 Cottonwood Heights Utah, W USA 40.37N 111.48W
31 S10 Cottonwood River ≈ Minnesota, N USA
47 N9 Cotuí C Dominican Republic 19.04N 70.10W
27 Q12 Cotulla Texas, SW USA 28.26N 99.13W
Cotyora see Ordu
104 I11 Coubre, Pointe de la headland W France 45.39N 1.23W
20 E12 Coudersport Pennsylvania, NE USA 41.45N 78.00W
13 S9 Coudres, Île aux island Quebec, SE Canada
190 E12 Couedic, Cape de headland South Australia 36.04S 136.43E
104 I6 Couentrey see Coventry
34 H10 Cougar Washington, NW USA 46.03N 122.18W
104 L10 Couhé Vienne, W France
34 K8 Coulee City Washington, NW USA 47.36N 119.18W
205 Q15 Coulman Island island Antarctica
105 P5 Coulommiers Seine-et-Marne, N France 48.49N 3.04E
12 K11 Coulonge ≈ Quebec, SE Canada
12 K11 Coulonge Est ≈ Quebec, SE Canada
37 O9 Coulterville California, W USA 37.41N 120.10W
40 M9 Council Alaska, USA 64.54N 163.40W
34 M12 Council Idaho, NW USA 44.45N 116.26W
31 S15 Council Bluffs Iowa, C USA 41.15N 95.51W
29 O5 Council Grove Kansas, C USA 38.37N 96.27W
29 O5 Council Grove Lake ◙ Kansas, C USA

34 G7 Coupeville Washington, NW USA 48.13N 122.41W
57 U12 Courantyne River var. Corantijn Rivier, Corentyne River. ≈ Guyana/Suriname
101 G21 Courcelles Hainaut, S Belgium 50.28N 4.22E
110 D7 Courgenay Jura, NW Switzerland 47.24N 7.09E
130 B2 Courland Lagoon Ger. Kurisches Haff, Rus. Kurskiy Zaliv. lagoon Lithuania/Russian Federation
120 B12 Courland Spit Lith. Kuršių Nerija, Rus. Kurshskaya Kosa. spit Lithuania/Russian Federation
108 A6 Courmayeur prev. Cormaiore. Valle d'Aosta, NW Italy 45.48N 7.00E
110 D7 Courroux Jura, NW Switzerland
8 K17 Courtenay Vancouver Island, British Columbia, SW Canada 49.40N 124.58W
23 W7 Courtland Virginia, NE USA 36.41N 77.01W
27 V10 Courtney Texas, SW USA 30.16N 96.04W
32 J4 Court Oreilles, Lac ◙ Wisconsin, N USA
Courtrai see Kortrijk
101 H19 Court-Saint-Étienne Wallon Brabant, C Belgium 50.38N 4.34E
24 G6 Coushatta Louisiana, S USA 32.00N 93.20W
180 I16 Cousin island Inner Islands, NE Seychelles
180 I16 Cousine island Inner Islands, NE Seychelles
104 J4 Coutances anc. Constantia. Manche, N France 49.04N 1.27W
104 K12 Coutras Gironde, SW France 45.01N 0.07W
47 U14 Couva Trinidad, Trinidad and Tobago 10.25N 61.27W
110 B8 Couvet Neuchâtel, W Switzerland 46.57N 6.41E
101 H22 Couvin Namur, S Belgium 50.03N 4.30E
118 K12 Covasna Ger. Kovasana, Hung. Kovászna. Covasna, E Romania 45.51N 26.09E
118 J11 Covasna ◆ county E Romania
12 E12 Cove Island island Ontario, S Canada
36 M5 Covelo California, W USA 39.46N 123.16W
99 O22 Coventry anc. Couentrey. C England, UK 52.25N 1.30W
26 I8 Coventry Lake ◙ Connecticut, NE USA
23 U5 Covesville Virginia, NE USA 37.52N 78.41W
84 B13 Covilhã Castelo Branco, E Portugal 40.16N 7.30W
25 T3 Covington Georgia, SE USA 33.34N 83.52W
33 N13 Covington Indiana, N USA 40.08N 87.23W
20 M3 Covington Kentucky, S USA 39.04N 84.30W
24 K8 Covington Louisiana, S USA 30.28N 90.06W
33 Q13 Covington Ohio, N USA 40.07N 84.21W
22 F9 Covington Tennessee, S USA 35.33N 89.39W
23 S6 Covington Virginia, NE USA 37.47N 79.59W
191 Q8 Cowal, Lake seasonal lake New South Wales, SE Australia
9 W15 Cowan Manitoba, S Canada 51.59N 100.36W
20 F12 Cowanesque River ≈ New York/Pennsylvania, NE USA
188 L12 Cowan, Lake ◙ Western Australia
13 P13 Cowansville Quebec, SE Canada 45.13N 72.43W
190 H8 Cowell South Australia 33.43S 136.53E
99 M23 Cowes S England, UK 50.45N 1.19W
29 Q10 Coweta Oklahoma, C USA 35.57N 95.39W
2 H6 Cowie Seamount undersea feature NE Pacific Ocean 54.15N 149.30W
34 G10 Cowlitz River ≈ Washington, NW USA
23 Q11 Cowpens South Carolina, SE USA 35.01N 81.48W
191 R8 Cowra New South Wales, SE Australia 33.52S 148.36E
178 M10 Coxen Hole see Roatán
65 H14 Coxim Mato Grosso do Sul, S Brazil 18.28S 54.45W
65 H14 Coxim, Rio ≈ SW Brazil
Coxin Hole see Roatán
159 V17 Cox's Bazar Chittagong, S Bangladesh 21.25N 92.01E
78 H14 Coyah Conakry, W Guinea 9.45N 13.25W
42 K5 Coyame Chihuahua, N Mexico 29.28N 105.01W
26 L9 Coyanosa Draw ≈ Texas, SW USA
Coyhaique see Coihaique
44 C7 Coyolate, Río ≈ S Guatemala
30 J6 Coyote State see South Dakota
42 L10 Coyotitán Sinaloa, C Mexico 23.46N 106.34W
43 N15 Coyuca de Benítez Guerrero, S Mexico 18.21N 100.39W
43 O16 Coyuca de Benítez, Río ≈ Guerrero, S Mexico 16.57N 100.01W
Coyuca de Benítez/Coyuca de Catalán see Coyuca
31 N15 Cozad Nebraska, C USA 40.52N 99.58W
108 A9 Cozie, Alpi see Cottian Alps
Cozmeni see Kitsman'
42 E3 Cozón, Cerro ▲ NW Mexico 31.26N 112.29W
43 Z12 Cozumel Quintana Roo, E Mexico 20.30N 86.54W
43 X8 Cozumel, Isla island SE Mexico
107 X4 Crab Creek ≈ Washington, NW USA
46 H12 Crab Pond Point headland W Jamaica 18.07N 78.01W
118 L8 Cracovia/Cracow see Kraków
85 I25 Cradock Eastern Cape, S South Africa 32.06S 25.37E

41 Y14 Craig Prince of Wales Island, Alaska, USA 55.29N 133.04W
23 V7 Craig Virginia, NE USA 37.10N 78.07W
45 Q15 Crexa see Cres
63 K14 Criciúma Santa Catarina, S Brazil 28.39S 49.23W
98 J11 Crieff C Scotland, UK 56.22N 3.49W
114 B10 Crikvenica It. Cirquenizza; prev. Cirkvenica, Crjkvenica. Primorje-Gorski Kotar, NW Croatia 45.12N 14.40E
Crimea/Crimean Oblast see Krym, Respublika
103 M16 Crimmitschau var. Krimmitschau. Sachsen, E Germany 50.48N 12.22E
118 G10 Crişcior Hung. Kristyor. Hunedoara, W Romania 46.09N 22.54E
23 Y5 Crisfield Maryland, NE USA 37.58N 75.51W
33 P3 Crisp Point headland Michigan, N USA 46.45N 85.15W
61 L19 Cristalina Goiás, C Brazil 16.43S 47.37W
46 J7 Cristal, Sierra del ▲ E Cuba
45 T14 Cristóbal Colón, C Panama 9.18N 79.52W
56 F7 Cristóbal Colón, Pico ▲ N Colombia 10.52N 73.46W
Cristur/Cristuru Săcuiesc see Cristuru Secuiesc
118 I11 Cristuru Secuiesc prev. Cristur, Cristuru Săcuiesc, Sitaş Cristuru, Ger. Kreutz, Hung. Székelykeresztúr, Szitás-Keresztúr. Mureş, C Romania 46.16N 25.01E
118 F10 Crişul Alb var. Weisse Kreisch, Ger. Weisse Körös, Hung. Fehér-Körös. ≈ Hungary/Romania
118 F10 Crişul Negru Ger. Schwarze Körös, Hung. Fekete-Körös. ≈ Hungary/Romania
118 G10 Crişul Repede var. Schnelle Kreisch, Ger. Schnelle Körös, Hung. Sebes-Körös. ≈ Hungary/Romania
119 N10 Criuleni Rus. Kriulyany. C Moldova 47.12N 29.09E
181 P11 Crivadia Vulcanului see Vulcan
115 O17 Crna Gora ▲ FYR Macedonia/Yugoslavia 6.17N 70.15W
115 O20 Crna Gora see Montenegro
115 O20 Crna Reka ≈ S FYR Macedonia
189 V10 Crni Drim see Black Drin
189 V13 Crni Vrh ▲ NE Slovenia 46.28N 15.14E
189 V13 Črnomelj Ger. Tschernembl. SE Slovenia 45.33N 15.10E
99 A17 Croagh Patrick Ir. Cruach Phádraig. ▲ W Ireland 53.45N 9.39W
114 D10 Croatia off. Republic of Croatia, Ger. Kroatien, SCr. Hrvatska. ◆ republic SE Europe
13 P8 Croce, Picco di see Wilde Kreuzspitze
175 Nn3 Croche ≈ Quebec, SE Canada
Crocker, Banjaran var. Crocker Range. ▲ East Malaysia
Crocker Range see Crocker, Banjaran
27 V9 Crockett Texas, SW USA 31.19N 95.27W
69 V14 Crocodile var. Krokodil. ≈ N South Africa
Crocodile see Limpopo
22 I7 Crofton Kentucky, S USA 37.01N 87.25W
31 Q12 Crofton Nebraska, C USA 42.43N 97.30W
Croia see Krujë
105 R16 Croisette, Cap headland SE France 43.12N 5.21E
104 G8 Croisic, Pointe du headland W France 47.16N 2.42W
105 S13 Croix Haute, Col de la pass E France 44.43N 5.39E
13 U5 Croix, Pointe à la headland Quebec, SE Canada 49.16N 67.46W
12 F13 Croker, Cape headland Ontario, S Canada 44.56N 80.57W
189 P1 Croker Island island Northern Territory, N Australia
98 I8 Cromarty N Scotland, UK 57.40N 4.01W
101 M4 Crombach Liège, E Belgium 50.14N 6.07E
99 Q18 Cromer E England, UK 52.55N 1.06E
193 D22 Cromwell Otago, South Island, NZ 45.03S 169.13E
193 H16 Cronadun West Coast, South Island, NZ 42.03S 171.52E
41 O11 Crooked Creek Alaska, USA 61.52N 158.06W
46 K5 Crooked Island island SE Bahamas
46 J5 Crooked Island Passage channel SE Bahamas
34 I13 Crooked River ≈ Oregon, NW USA
31 R4 Crookston Minnesota, N USA 47.46N 96.36W
30 I10 Crooks Tower ▲ South Dakota, N USA 44.09N 103.55W
33 T14 Crooksville Ohio, N USA 39.46N 82.05W
191 R9 Crookwell New South Wales, SE Australia 34.28S 149.27E
99 K17 Crosby var. Great Crosby. NW England, UK 53.30N 3.01W
31 U6 Crosby Minnesota, N USA 46.30N 93.58W
30 J2 Crosby North Dakota, N USA 48.54N 103.17W
27 Q5 Crosbyton Texas, SW USA 33.39N 101.14W
79 V14 Cross ≈ Cameroon/Nigeria
31 U10 Cross City Florida, SE USA 29.37N 83.08W
25 T6 Crossen see Krosno Odrzańskie
29 V11 Crossett Arkansas, S USA 33.07N 91.57W
99 K15 Cross Fell ▲ N England, UK 54.42N 2.30W
9 P16 Crossfield Alberta, SW Canada 51.24N 114.03W
23 Q12 Cross Hill South Carolina, SE USA 34.18N 81.58W

99 L18 Crewe C England, UK 53.04N 2.27W
23 V7 Crewe Virginia, NE USA 37.10N 78.07W
9 X13 Cross Lake Manitoba, C Canada 54.37N 97.34W
24 F5 Cross Lake ◙ Louisiana, S USA
38 I12 Crossman Peak ▲ Arizona, SW USA 34.33N 114.09W
27 Q7 Cross Plains Texas, SW USA 32.08N 99.10W
79 V17 Cross River ◆ state SE Nigeria
23 L9 Crossville Tennessee, S USA 35.57N 85.01W
33 S8 Croswell Michigan, N USA 43.16N 82.37W
12 K13 Crotch Lake ◙ Ontario, SE Canada
109 O21 Croton/Crotona see Crotone
109 O21 Crotone var. Cotrone; anc. Croton, Crotona. Calabria, SW Italy 39.04N 17.07E
35 V11 Crow Agency Montana, NW USA 45.35N 107.28W
191 U7 Crowdy Head headland New South Wales, SE Australia 31.52S 152.45E
27 Q4 Crowell Texas, SW USA 33.58N 99.43W
191 O6 Crowl Creek seasonal river New South Wales, SE Australia
24 H9 Crowley, Lake ◙ California, W USA 30.11N 92.21W
37 S9 Crowley, Lake ◙ California, W USA
29 X10 Crowleys Ridge hill range Arkansas, C USA
194 J11 Crown Island island N Papua New Guinea
33 N11 Crown Point Indiana, N USA 41.24N 87.21W
39 P10 Crownpoint New Mexico, SW USA 35.40N 108.09W
35 R10 Crow Peak ▲ Montana, NW USA 46.17N 111.54W
9 P17 Crowsnest Pass pass Alberta/British Columbia, SW Canada 49.38N 114.43W
31 T6 Crow Wing River ≈ Minnesota, N USA
99 O22 Croydon SE England, UK 51.21N 0.06W
181 P11 Crozet Basin undersea feature S Indian Ocean
181 O12 Crozet Islands island group French Southern and Antarctic Territories
181 O12 Crozet Plateau var. Crozet Plateaus. undersea feature SW Indian Ocean
Crozet Plateaus see Crozet Plateau
104 E6 Crozon Finistère, NW France 48.14N 4.31W
Cruacha Dubha, Na see Macgillycuddy's Reeks
Cruach Phádraig see Croagh Patrick
118 M14 Crucea Constanţa, SE Romania 44.30N 28.18E
46 E5 Cruces Cienfuegos, C Cuba 22.18N 80.18W
109 O20 Crucoli Torretta Calabria, SW Italy 39.26N 17.03E
43 P9 Cruillas Tamaulipas, C Mexico 24.43N 98.26W
63 G14 Cruz Alta Rio Grande do Sul, S Brazil 28.35S 53.37W
46 G8 Cruz, Cabo headland S Cuba 19.50N 77.43W
62 N9 Cruzeiro São Paulo, S Brazil 22.33S 44.55W
62 H10 Cruzeiro do Oeste Paraná, S Brazil 23.45S 53.03W
61 A15 Cruzeiro do Sul Acre, W Brazil 7.40S 72.39W
25 U11 Crystal Bay bay Florida, SE USA
190 I8 Crystal Brook South Australia 33.24S 138.10E
9 X17 Crystal City Manitoba, S Canada 49.07N 98.54W
29 X5 Crystal City Missouri, C USA 38.13N 90.22W
27 P13 Crystal City Texas, SW USA 28.40N 99.49W
32 M4 Crystal Falls Michigan, N USA 46.06N 88.19W
25 Q8 Crystal Lake Florida, SE USA 30.26N 85.41W
23 O6 Crystal Lake ◙ Michigan, N USA
25 V11 Crystal River Florida, SE USA 28.54N 82.35W
39 Q5 Crystal River ≈ Colorado, C USA
24 K6 Crystal Springs Mississippi, S USA 31.59N 90.21W
114 H22 Csaca see Čadca
Csakathurn/Csáktornya see Čakovec
34 I13 Csap see Chop
Cserépalja see Crepaja
Csermő see Cermei
Csíkszereda see Miercurea-Ciuc
113 L24 Csongrád Csongrád, SE Hungary 46.42N 20.05E
113 L24 Csongrád off. Csongrád Megye. ◆ county SE Hungary
191 R9 Csorna Győr-Moson-Sopron, NW Hungary 47.37N 17.13E
113 H22 Csorna ≈ Croatia, SE Canada 44.39N 76.13W
113 G25 Csurgó Somogy, SW Hungary 46.16N 17.09E
Csurog see Čurug
191 U6 Cúa Miranda, N Venezuela 10.07N 66.53W
110 Q10 Cuale Malanje, NW Angola 8.13S 16.11E
69 T12 Cuando var. Kwando. ≈ S Africa 33.39N 101.14W
85 E15 Cuando Cubango var. Kuando-Kubango. ◆ province SE Angola
85 D16 Cuangar Cuando Cubango, S Angola 17.34S 18.39E
85 E16 Cuango Lunda Norte, NE Angola 9.09S 18.01E
83 D23 Cuango Uíge, NW Angola 6.17S 16.41E
Cuango var. Kwango. ≈ Angola/Dem. Rep. Congo (Zaire)
see also Kwango

21 U6 Cross Island island Maine, NE USA
54.37N 97.34W
84 C12 Cuanza var. Kwanza. ≈ C Angola
84 B11 Cuanza Norte var. Kuanza Norte. ◆ province NW Angola
84 B12 Cuanza Sul var. Kuanza Sul. ◆ province NE Angola
63 K14 Cuareim, Río var. Rio Quaraí. ≈ Brazil/Uruguay see also Quaraí, Rio
Cuareim, Río see Quaraí, Rio
85 D15 Cuatir ◊ S Angola
42 M7 Cuatro Ciénegas var. Cuatro Ciénegas de Carranza. Coahuila de Zaragoza, NE Mexico 26.59N 102.04W
Cuatro Ciénegas de Carranza see Cuatro Ciénegas
42 L9 Cuauhtémoc Chihuahua, N Mexico 28.25N 106.51W
43 P13 Cuautla Morelos, S Mexico 18.47N 98.56W
106 H12 Cuba Beja, S Portugal 38.10N 7.54W
29 W6 Cuba Missouri, C USA 38.03N 91.24W
39 R10 Cuba New Mexico, SW USA 36.01N 106.57W
46 E6 Cuba off. Republic of Cuba. ◆ republic W West Indies
46 F5 Cuba island W West Indies
84 B13 Cubal Benguela, W Angola 12.58S 14.16E
84 D13 Cubango var. Kuvango, Port. Vila Artur de Paiva, Vila da Ponte. Huíla, SW Angola 14.27S 16.17E
85 D16 Cubango var. Kavango, Kavengo, Kubango, Okavango, Okavanggo. ≈ S Africa see also Okavango
Cubango see Okavango
56 H8 Cubará Boyacá, N Colombia 7.01N 72.07W
142 I12 Çubuk Ankara, N Turkey 40.13N 33.01E
85 C15 Cuchi Cuando Cubango, C Angola 14.40N 16.58E
56 G7 Cuchumatanes, Sierra de los ▲ W Guatemala
84 E12 Cúcuta var. San José de Cúcuta. Norte de Santander, N Colombia 7.55N 72.31W
33 N9 Cudahy Wisconsin, N USA 42.57N 87.51W
161 J21 Cuddalore Tamil Nādu, SE India 11.43N 79.46E
161 I18 Cuddapah Andhra Pradesh, S India 14.30N 78.49E
106 M6 Cuéllar Castilla-León, N Spain 41.24N 4.19W
84 D13 Cuemba var. Coemba. Bié, C Angola 12.09S 18.07E
58 B8 Cuenca Azuay, S Ecuador 2.54S 79.00W
107 Q9 Cuenca anc. Conca. Castilla-La Mancha, C Spain 40.04N 2.07W
107 P9 Cuenca ◆ province Castilla-La Mancha, C Spain
42 L9 Cuencamé var. Cuencamé de Ceniceros. Durango, C Mexico 24.51N 103.42W
Cuencamé de Ceniceros see Cuencamé
107 Q8 Cuenca, Serranía de ▲ C Spain
107 P5 Cuera see Chur
Cummin in Pommern see Kamień Pomorski
43 O14 Cuernavaca Morelos, S Mexico 18.57N 99.15W
27 T12 Cuero Texas, SW USA 29.04N 97.16W
46 I7 Cueto Holguín, E Cuba 20.39N 75.55W
43 Q13 Cuetzalán var. Cuetzalán del Progreso. Puebla, S Mexico 20.00N 97.27W
Cuetzalán del Progreso see Cuetzalán
Ciudad Presidente Stroessner see Ciudad del Este
118 H12 Cugir Hung. Kudzsir. Alba, SW Romania 45.48N 23.24E
61 H18 Cuiabá prev. Cuyabá. state capital Mato Grosso, SW Brazil 15.31S 56.04W
61 H19 Cuiabá, Rio ≈ SW Brazil
100 L13 Cuijck Noord-Brabant, SE Netherlands 51.40N 5.55E
Cúil an tSúdaire see Portarlington
Cúil Mhuine see Collooney
Cúil Raithin see Coleraine
84 C14 Cuima Huambo, C Angola 13.16S 15.39E
84 C15 Cuito var. Kwito. ≈ SE Angola
Cuito Cuanavale Cuando Cubango, S Angola 15.01S 19.07E
113 G25 Csurgó Somogy, SW Hungary 46.16N 17.09E
29 W4 Cuivre River ≈ Missouri, C USA
304 Hh4 Cukai var. Chukai, Kemaman. Terengganu, Peninsular Malaysia 4.15N 103.25E
115 L23 Çukë var. Cukë. Vlorë, S Albania 39.50N 20.01E
Cularo see Grenoble
179 Pp13 Culasi Panay Island, C Philippines 11.22N 122.05E
35 Y7 Culbertson Montana, NW USA 48.08N 104.30W
30 M16 Culbertson Nebraska, C USA 40.14N 100.38W
191 P10 Culcairn New South Wales, SE Australia 35.41S 147.01E
47 W5 Culebra, Isla de island E Puerto Rico
39 T8 Culebra Peak ▲ Colorado, C USA 37.07N 105.11W
106 J5 Culebra, Sierra de la ▲ NW Spain
100 J12 Culemborg Gelderland, C Netherlands 51.57N 5.17E
143 V14 Culfa Rus. Dzhul'fa. SW Azerbaijan 38.58N 45.37E
191 P4 Culgoa River ≈ New South Wales/Queensland, SE Australia
42 J9 Culiacán var. Culiacán-Rosales, Culiacán-Rosales. Sinaloa, C Mexico 24.48N 107.24W
Culiacán-Rosales/Culiacán Rosales see Culiacán
179 P13 Culion Island island Calamian Group, W Philippines
107 P14 Cúllar-Baza Andalucía, S Spain 37.34N 2.34W
107 S10 Cullera País Valenciano, E Spain 39.10N 0.15W
23 O3 Cullman Alabama, S USA 34.10N 86.50W
110 B10 Cully Vaud, W Switzerland 46.58N 6.46E
Culm see Chełmno
Culmsee see Chełmża
23 V4 Culpeper Virginia, NE USA 38.28N 78.00W
193 I17 Culverden Canterbury, South Island, NZ 42.46S 172.51E
57 N5 Cumaná Sucre, NE Venezuela 10.28N 64.12W
57 O5 Cumanacoa Sucre, NE Venezuela 10.16N 63.58W
56 C13 Cumbal, Nevado de elevation S Colombia 0.51N 77.58W
23 O7 Cumberland Kentucky, S USA 36.55N 83.00W
23 U2 Cumberland Maryland, NE USA 39.39N 78.45W
23 V6 Cumberland Virginia, NE USA 37.30N 78.13W
197 A11 Cumberland, Cape var. Cape Nahoi. headland Espíritu Santo, N Vanuatu 14.39S 166.35E
9 V14 Cumberland House Saskatchewan, S Canada 53.57N 102.21W
25 W8 Cumberland Island island Georgia, SE USA
L7 · Cumberland, Lake ◙ Kentucky, S USA
16 O1 Cumberland Peninsula peninsula Baffin Island, Nunavut, NE Canada
2 N9 Cumberland Plateau plateau E USA
32 L1 Cumberland Point headland Michigan, N USA 47.51N 89.14W
23 O7 Cumberland River ≈ Kentucky/Tennessee, S USA
16 O2 Cumberland Sound inlet Baffin Island, Nunavut, NE Canada
98 I12 Cumbernauld S Scotland, UK 55.57N 4.00W
99 K15 Cumbria cultural region NW England, UK
99 K15 Cumbrian Mountains ▲ NW England, UK
25 S2 Cumming Georgia, SE USA 34.12N 84.08W
190 G9 Cummins South Australia 34.17S 135.43E
98 I13 Cumnock W Scotland, UK 55.31N 4.28W
42 G4 Cumpas Sonora, NW Mexico 30.00N 109.48W
142 H16 Çumra Konya, C Turkey 37.32N 32.52E
65 G15 Cunco Araucanía, C Chile 38.57S 72.13W
56 C8 Cundinamarca off. Departamento de Cundinamarca. ◆ province C Colombia
43 U15 Cunduacán Tabasco, SE Mexico 18.04N 93.17W
85 A16 Cunene var. Kunene. ◆ province S Angola
Angola/Namibia see also Kunene
108 A9 Cuneo Fr. Coni. Piemonte, NW Italy 44.22N 7.31E
43 R15 Cunjamba Cuando Cubango, E Angola 15.22S 20.07E
189 V10 Cunnamulla Queensland, E Australia 28.09S 145.43E
108 I7 Cuorgne Piemonte, NE Italy 45.23N 7.34E
98 I13 Cupar E Scotland, UK 56.19N 3.01W
118 L8 Cupcina Rus. Kupchino; prev. Calinisc, Kalinisk. N Moldova 48.07N 27.22E
56 C8 Cupica Chocó, W Colombia 6.43N 77.31W
56 C8 Cupica, Golfo de gulf W Colombia
114 N13 Ćuprija Serbia, E Yugoslavia 43.57N 21.21E
Cura see Villa de Cura
47 P16 Curaçao island Netherlands Antilles
58 I13 Curanja, Río ≈ E Peru
59 P14 Curaray, Río ≈ Ecuador/Peru
118 K14 Curcani Călăraşi, SE Romania 44.04N 26.39E
190 H4 Curdimurka South Australia 29.27S 136.56E
105 P7 Cure ≈ C France
181 Y16 Curepipe C Mauritius 20.19S 57.31E
84 D12 Curiapo Delta Amacuro, NE Venezuela 10.03N 62.18W
Curia Rhaetorum see Chur
64 G12 Curicó Maule, C Chile 35.00S 71.15W
56 G15 Curieta see Krk
180 I15 Curieuse island Inner Islands, NE Seychelles
61 H16 Curitiba prev. Curytyba. state capital Paraná, S Brazil 25.25S 49.25W
62 J12 Curitibanos Santa Catarina, S Brazil 27.18S 50.34W
191 S6 Curlewis New South Wales, SE Australia 31.09S 150.18E

246

190 J6 Curnamona South Australia 31.39S 139.35E
85 A15 Curoca ∿ SW Angola
191 T6 Currabubula New South Wales, SE Australia 31.17S 150.43E
61 Q14 Currais Novos Rio Grande do Norte, E Brazil 6.12S 36.30W
37 W7 Currant Nevada, W USA 38.43N 115.27W
37 W6 Currant Mountain ▲ Nevada, W USA 38.56N 115.19W
46 H2 Current Eleuthera Island, C Bahamas 25.24N 76.44W
29 W8 Current River ∿ Arkansas/Missouri, C USA
23 Y8 Currituck North Carolina, SE USA 36.27N 76.02W
23 Y8 Currituck Sound sound North Carolina, SE USA
41 R11 Curry Alaska, USA 62.36N 150.00W
Curtbunar see Tervel
118 I13 Curtea de Arges var. Curtea-de-Arges. Arges, S Romania 45.06N 24.40E
118 E10 Curtici Ger. Kurtitsch, Hung. Kürtös. Arad, W Romania 46.21N 21.17E
30 M16 Curtis Nebraska, C USA 40.36N 100.27W
106 H2 Curtis-Estación Galicia, NW Spain 43.09N 8.10W
191 O14 Curtis Group island group Tasmania, SE Australia
189 Y8 Curtis Island island Queensland, SE Australia
60 K11 Curuá, Ilha do island NE Brazil
49 U7 Curuá, Rio ∿ N Brazil
61 A14 Curuçá, Rio ∿ NW Brazil
114 L9 Čurug Hung. Csurog. Serbia, N Yugoslavia 45.30N 20.02E
63 D16 Curuzú Cuatiá Corrientes, NE Argentina 29.45S 58.01W
61 M19 Curvelo Minas Gerais, SE Brazil 18.45S 44.27W
20 E14 Curwensville Pennsylvania, NE USA 40.57N 78.29W
32 M3 Curwood, Mount ▲ Michigan, N USA 46.42N 88.14W
Curytiba see Curitiba
Curzola see Korčula
44 A10 Cuscatlán ◆ department C El Salvador
59 H15 Cuzco var. Cuzco. Cuzco, C Peru 13.34S 72.01W
59 H15 Cusco off. Departamento de Cusco; var. Cuzco. ◆ department C Peru
25 S6 Cushing Oklahoma, C USA 36.01N 96.46W
27 W8 Cushing Texas, SW USA 31.48N 94.50W
42 I6 Cusihuiriáchic Chihuahua, N Mexico 28.16N 106.46W
105 P10 Cusset Allier, C France 46.07N 3.27E
25 S6 Cusseta Georgia, SE USA 32.18N 84.46W
30 J10 Custer South Dakota, N USA 43.46N 103.36W
Cüstrin see Kostrzyn
35 Q7 Cut Bank Montana, NW USA 48.37N 112.19W
Cutch, Gulf of see Kachchh, Gulf of
25 S6 Cuthbert Georgia, SE USA 31.46N 84.47W
9 S15 Cut Knife Saskatchewan, S Canada 52.40N 108.54W
25 Y16 Cutler Ridge Florida, SE USA 25.34N 80.21W
24 K10 Cut Off Louisiana, S USA 29.32N 90.20W
65 I15 Cutral-Có Neuquén, C Argentina 38.55S 69.13W
109 O21 Cutro Calabria, SW Italy 39.01N 16.59E
191 O4 Cuttaburra Channels seasonal river New South Wales, SE Australia
160 O12 Cuttack Orissa, E India 20.28N 85.52E
85 C15 Cuvelai Cunene, SW Angola 15.40S 15.48E
81 G18 Cuvette var. Région de la Cuvette. ◆ province C Congo
181 V9 Cuvier Basin undersea feature E Indian Ocean
181 U9 Cuvier Plateau undersea feature E Indian Ocean
84 B12 Cuvo ∿ W Angola
102 H9 Cuxhaven Niedersachsen, NW Germany 53.51N 8.42E
Cuyabá see Cuiabá
179 Pp13 Cuyo East Pass passage C Philippines
179 P13 Cuyo West Pass passage C Philippines
57 S8 Cuyuni, Río var. Río Cuyuni. ∿ Guyana/Venezuela
Cuzco see Cusco
99 K22 Cwmbran Wel. Cwmbrân. SW Wales, UK 51.39N 3.00W
30 K15 C.W.McConaughy, Lake ☒ Nebraska, C USA
83 D20 Cyangugu SW Rwanda 2.27S 29.00E
112 D11 Cybinka Ger. Ziebingen. Lubuskie, W Poland 52.11N 14.46E
Cyclades see Kykládes
Cydonia see Chaniá
Cymru see Wales
23 M5 Cynthiana Kentucky, S USA 38.23N 84.17W
9 S17 Cypress Hills ▲ Alberta/Saskatchewan, SW Canada
Cypro-Syrian Basin see Cyprus Basin
123 Mm1 Cyprus off. Republic of Cyprus, Gk. Kypros, Turk. Kıbrıs Cumhuriyeti. ◆ republic E Mediterranean Sea
86 L14 Cyprus Gk. Kýpros, Turk. Kıbrıs. island E Mediterranean Sea
123 Gg10 Cyprus Basin var. Cypro-Syrian Basin. undersea feature E Mediterranean Sea
Cythera see Kýthira
Cythnos see Kýthnos
112 F9 Czaplinek Ger. Tempelburg. Zachodniopomorskie, NW Poland 53.33N 16.14E

Czarna Woda see Wda
112 G8 Czarne Pomorskie, N Poland 53.40N 17.00E
112 G10 Czarnków Wielkopolskie, C Poland 52.52N 16.31E
113 E17 Czech Republic Cz. Česká Republika. ◆ republic C Europe
112 G12 Czempiń Wielkopolskie, C Poland 52.10N 16.46E
Czenstochau see Częstochowa
112 I8 Czersk Pomorskie, N Poland 53.48N 17.58E
113 J15 Częstochowa Ger. Czenstochau, Tschenstochau, Rus. Chenstokhov. Śląskie, S Poland 50.51N 19.09E
112 F10 Człopa Ger. Schloppe. Zachodniopomorskie, NW Poland 53.04N 16.04E
112 H8 Człuchów Ger. Schlochau. Pomorskie, NW Poland 53.40N 17.19E

D

169 V9 Da'an var. Dalai. Jilin, NE China 45.28N 124.18E
13 S10 Daaquam Quebec, SE Canada 46.36N 70.03W
56 I4 Dabajuro Falcón, NW Venezuela 11.00N 70.41W
79 N15 Dabakala NE Ivory Coast 8.19N 4.24W
Daban see Bairin Youqi
113 K23 Dabas Pest, C Hungary 47.13N 19.18E
166 L8 Daba Shan ▲ C China
146 J5 Dabbāgh, Jabal ▲ NW Saudi Arabia 27.52N 35.48E
56 D8 Dabeiba Antioquia, NW Colombia 6.57N 76.13W
160 E11 Dabhoi Gujarāt, W India 22.07N 73.28E
167 P8 Dabie Shan ▲ C China
78 J13 Dabola Haute-Guinée, C Guinea 10.48N 11.01W
79 N17 Dabou S Ivory Coast 5.19N 4.22W
112 P8 Dabrowa Białostocka Podlaskie, NE Poland 53.38N 23.18E
113 M16 Dabrowa Tarnowska Małopolskie, S Poland 50.10N 21.00E
121 M20 Dabryn' Rus. Dobryn'. Homyel'skaya Voblasts', SE Belarus 51.46N 29.12E
165 P10 Dabsan Hu ☒ C China
167 Q13 Dabu prev. Huliao. Guangdong, S China 24.19N 116.07E
118 H15 Dăbuleni Dolj, SW Romania 43.47N 24.05E
Dacca see Dhaka
103 L23 Dachau Bayern, SE Germany 48.16N 11.25E
166 K8 Dachuan prev. Daxian, Da Xian. Sichuan, C China 31.16N 107.31E
66 M10 Dacia Bank see Dacia Seamount
Dacia Seamount var. Dacia Bank. undersea feature E Atlantic Ocean 31.10N 13.42W
39 T3 Dacono Colorado, C USA 40.04N 104.56W
25 W12 Dade City Florida, SE USA 28.21N 82.12W
158 L10 Dadeldhura var. Dandeldhura. Far Western, W Nepal 29.12N 80.31E
25 Q5 Dadeville Alabama, S USA 32.49N 85.45W
105 N15 Dadou ∿ S France
160 D12 Dādra and Nagar Haveli ◆ union territory W India
155 P14 Dādu Sind, SE Pakistan
178 K11 Da Du Boloc Kon Tum, C Vietnam 14.06N 107.40E
166 G9 Dadu He ∿ C China
Daegu see Taegu
Daerah Istimewa Aceh see Aceh
179 Q11 Daet Luzon, N Philippines 14.06N 122.57E
166 I11 Dafang Guizhou, S China 27.07N 105.40E
159 W11 Dafla Hills ▲ NE India
9 U15 Dafoe Saskatchewan, S Canada 51.46N 104.11W
78 G10 Dagana N Senegal 16.28N 15.35W
Dagana see Dahana, Tajikistan
Dagana see Massakory, Chad
120 K11 Dagda Krāslava, SE Latvia 56.06N 27.36E
Dagden see Hiiumaa
Dagden-Sund see Soela Väin
131 P16 Dagestan, Respublika prev. Dagestanskaya ASSR, Eng. Daghestan. ◆ autonomous republic SW Russian Federation
Dagestanskaya ASSR see Dagestan, Respublika
131 R17 Dagestanskiye Ogni Respublika Dagestan, SW Russian Federation 42.09N 48.08E
193 A23 Dagg Sound sound South Island, NZ
Daghestan see Dagestan, Respublika
147 Y8 Daghmar NE Oman 23.09N 59.10E
Dağlıq Qarabağ see Nagornyy Karabakh
Dago see Hiiumaa
56 C11 Dagua Valle del Cauca, W Colombia 3.37N 76.42W
179 P9 Dagupan off. Dagupan City. Luzon, N Philippines 16.04N 120.21E
165 N16 Dagzê Xizang Zizhiqu, W China 29.38N 91.15E
153 Q13 Dahana Rus. Dagana, Dakhana. SW Tajikistan 38.03N 69.51E
169 Y10 Dahei var. Dahai. NE China
169 T7 Da Hinggan Ling Eng. Great Khingan Range. ▲ NE China
82 K9 Dahlak Archipelago var. Dahlak Archipelago. island group E Eritrea
25 T2 Dahlonega Georgia, SE USA 34.31N 83.59W

103 O14 Dahme Brandenburg, E Germany 52.10N 13.47E
102 O13 Dahme ∿ E Germany
147 O14 Dahm, Ramlat desert NW Yemen
160 E10 Dāhod prev. Dohad. Gujarāt, W India 22.48N 74.18E
Dahomey see Benin
164 G10 Dahongliutan Xinjiang Uygur Zizhiqu, NW China 35.59N 79.12E
Dahra see Dara
245 R2 Dahūk var. Dohuk, Kurd. Dihôk. ◆ N Iraq 36.52N 43.01E
118 J15 Daia Giurgiu, S Romania 44.00N 25.59E
171 L15 Daigo Ibaraki, Honshū, S Japan 36.43N 140.22E
169 O13 Dai Hai ☒ N China
Daihoku see T'aipei
195 X14 Dai Island island N Solomon Islands
177 G8 Daik-u Pegu, SW Myanmar
Dali see Idálion
144 H9 Dā'il Dar'ā, S Syria 32.45N 36.07E
178 Kk12 Dai Lanh Khanh Hoa, S Vietnam 12.49N 109.20E
167 Q13 Daimao Shan ▲ SE China
207 N11 Daimiel Castilla-La Mancha, C Spain 39.04N 3.37W
117 F22 Daimoniá Pelopónnisos, S Greece 36.38N 22.54E
Dainan see T'ainan
27 W6 Daingerfield Texas, SW USA 33.01N 94.43W
165 R13 Dainkognubma Xizang Zizhiqu, W China 32.25N 97.58E
171 Hh17 Daiō-zaki headland Honshū, SW Japan 34.15N 136.50E
63 B22 Daireaux Buenos Aires, E Argentina 36.36S 61.42W
Dairbhre see Valencia Island
Dairen see Dalian
77 W9 Dairût var. Dayrūt. C Egypt 27.31N 30.46E
170 Ff12 Dai-sen ▲ Kyūshū, SW Japan 35.22N 133.33E
27 X10 Daisetta Texas, SW USA 30.06N 94.38W
199 Gg5 Daitō-jima island group SW Japan
199 Gg5 Daitō Ridge undersea feature N Philippine Sea
167 N3 Daixian var. Dai Xian. Shanxi, C China 39.07N 112.54E
46 M8 Dajabón NW Dominican Republic 19.29N 71.40W
166 G8 Dajin Chuan ∿ C China
154 J6 Dak ◆ W Afghanistan
78 F11 Dakar ● (Senegal) W Senegal 14.43N 17.27W
78 F11 Dakar × W Senegal 14.42N 17.27W
178 K11 Đak Glây Kon Tum, C Vietnam 15.05N 107.42E
Dakhana see Dahana
Dakhla see Ad Dakhla
78 F7 Dakhlet Nouâdhibou ◆ region NW Mauritania
178 K13 Đak Nông Đắc Lắc, S Vietnam 11.58N 107.42E
79 U11 Dakoro S Niger 14.28N 6.45E
31 U12 Dakota City Iowa, C USA 42.42N 94.13W
31 R13 Dakota City Nebraska, C USA 42.25N 96.25W
115 M17 Dakovica var. Djakovica, Alb. Gjakovë. Serbia, S Yugoslavia 42.22N 20.30E
114 I10 Dakovo var. Djakovo, Hung. Diakovár. Osijek-Baranja, E Croatia 45.18N 18.24E
Dakshin see Deccan
178 K11 Đak Tô var. Đăc Tô. Kon Tum, C Vietnam 14.35N 107.55E
45 N7 Dakura Región Autónoma Atlántico Norte, NE Nicaragua 14.22N 83.13W
84 E12 Dala Lunda Sul, S Angola 11.03S 20.12E
110 J8 Dalaas Vorarlberg, W Austria 47.08N 10.03E
78 T13 Dalaba Moyenne-Guinée, W Guinea 10.46N 12.12W
Dalai see Da'an
169 Q11 Dalai Nur lake ☒ N China
Dala-Jarna see Järna
97 M14 Dalälven ∿ C Sweden
142 C16 Dalaman Muğla, SW Turkey 36.46N 28.46E
142 C16 Dalaman Çayı ∿ SW Turkey 36.37N 28.51E
168 K11 Dalandzadgad Ömnögovĭ, S Mongolia 43.35N 104.23E
197 D17 Dalane physical region S Norway
201 Z2 Dalap-Uliga-Djarrit var. Delap-Uliga-Darrit, D-U-D. island group Ratak Chain, SE Marshall Islands
96 J12 Dalarna prev. Kopparberg ◆ county C Sweden
96 L13 Dalarna Eng. Dalecarlia. cultural region C Sweden
97 P16 Dalarö Stockholm, C Sweden 59.08N 18.24E
178 Kk13 Đa Lat Lâm Đông, S Vietnam 11.55N 108.25E
168 J11 Dalay Ömnögovĭ, S Mongolia 43.22N 103.56E
154 L12 Dālbandīn var. Dāl Bandin. Baluchistān, SW Pakistan 28.48N 64.08E
97 J17 Dalbosjön lake bay S Sweden
189 Y10 Dalby Queensland, E Australia 27.14S 151.16E
96 D13 Dale Hordaland, S Norway 60.34N 5.48E
96 C12 Dale Sogn og Fjordane, S Norway 61.22S 5.24E
34 K12 Dale Oregon, NW USA 44.58N 118.56W
Dalecarlia see Dalarna
34 T11 Dale City Virginia, NE USA 38.38N 77.18W

22 L8 Dale Hollow Lake ☒ Kentucky/Tennessee, S USA
100 O8 Dalen Drenthe, NE Netherlands 52.42N 6.45E
97 E15 Dalen Telemark, S Norway 59.25N 7.58E
177 F4 Daletme Chin State, W Myanmar 21.44N 92.48E
25 Q7 Daleville Alabama, S USA 31.18N 85.42W
100 M9 Dalfsen Overijssel, E Netherlands 52.31N 6.16E
26 M1 Dalhart Texas, SW USA 36.04N 102.31W
11 O13 Dalhousie New Brunswick, SE Canada 48.03N 66.22W
158 I6 Dalhousie Himāchal Pradesh, N India 32.31N 76.01E
166 F12 Dali var. Xiaguan. Yunnan, SW China 25.33N 100.10E
Dali see Idálion
169 U14 Dalian var. Dairen, Dalien, Lüda, Ta-lien, Rus. Dalny. Liaoning, NE China 38.53N 121.36E
107 O15 Dalías Andalucía, S Spain 36.49N 2.50W
Dalien see Dalian
Dalijan see Delījān
114 J9 Dalj Hung. Dalja. Osijek-Baranja, E Croatia 45.29N 19.00E
Dalja see Dalj
34 F12 Dallas Oregon, NW USA 44.55N 123.19W
27 T7 Dallas Texas, SW USA 32.46N 96.48W
27 T7 Dallas-Fort Worth × Texas, SW USA 32.37N 97.16W
160 K12 Dallī Rājhara Madhya Pradesh, C India 20.33N 81.06E
41 X15 Dall Island island Alexander Archipelago, Alaska, USA
40 M12 Dall Lake ☒ Alaska, USA
79 S12 Dallol Bosso seasonal river W Niger
115 E14 Dalmacija Eng. Dalmatia, Ger. Dalmatien, It. Dalmazia. cultural region S Croatia
Dalmatia/Dalmatien/Dalmazia see Dalmacija
127 Nn17 Dal'negorsk Primorskiy Kray, SE Russian Federation 44.33N 135.35E
127 Nn17 Dal'nerechensk Primorskiy Kray, SE Russian Federation 45.57N 133.42E
Dalny see Dalian
78 M16 Daloa C Ivory Coast 6.52N 6.28W
166 J11 Dalou Shan ▲ S China
189 X7 Dalrymple Lake ☒ Queensland, E Australia
12 L12 Dalrymple Lake ☒ Ontario, S Canada
189 X7 Dalrymple, Mount ▲ Queensland, E Australia 21.01S 148.34E
95 K20 Dalsbruk Fin. Taalintehdas. Länsi-Suomi, SW Finland 60.01N 22.33E
97 K19 Dalsjöfors Västra Götaland, S Sweden 57.43N 13.04E
97 I17 Dals Långed var. Långed. Västra Götaland, S Sweden 58.54N 12.20E
159 O15 Dāltenganj prev. Daltonganj. Bihār, N India 23.59N 84.07E
Dāltenganj see Daltonganj
25 R2 Dalton Georgia, SE USA 34.46N 84.58W
205 X14 Dalton Iceberg Tongue ice tongue Antarctica
94 J1 Dalvík Norðhurland Eystra, N Iceland 65.58N 18.31W
37 M8 Daly City California, W USA 37.44N 122.27W
189 P2 Daly River ∿ Northern Territory, N Australia
189 Q3 Daly Waters Northern Territory, N Australia 16.21S 133.21E
121 F20 Damachava var. Damachova, Pol. Domaczewo, Rus. Domachëvo. Brestskaya Voblasts', SW Belarus 51.45N 23.36E
79 W11 Damagaram Takaya Zinder, S Niger 14.02N 9.28E
160 D12 Damān Damān and Diu, W India 20.25N 72.58E
160 B12 Damān and Diu ◆ union territory W India
77 V7 Damanhûr anc. Hermopolis Parva. N Egypt 31.02N 30.34E
175 T15 Damar, Kepulauan var. Baraf Daja Islands, Kepulauan Barat Daja. island group C Indonesia
175 T15 Damar, Pulau island Maluku, E Indonesia
81 I15 Damara S Central African Republic 5.00N 18.45E
85 D18 Damaraland physical region C Namibia
Damas see Dimashq
79 Y12 Damasak Borno, NE Nigeria 13.10N 12.40E
23 Q8 Damascus Virginia, NE USA 36.37N 81.46W
Damascus see Dimashq
79 N12 Damaturu Yobe, NE Nigeria 11.44N 11.58E
149 O5 Damāvand, Qolleh-ye ▲ N Iran 35.59N 52.06E
85 B10 Damba Uíge, NW Angola 6.42S 15.07E
116 M12 Dambaslar Tekirdağ, NW Turkey 41.13N 27.13E
118 J13 Dâmboviţa prev. Dîmboviţa. ◆ county SE Romania
118 J13 Dâmboviţa var. Dîmboviţa. ∿ S Romania
Y15 D'Ambre, Île island NE Mauritius
K24 Dambulla Central Province, C Sri Lanka 7.51N 80.40E

46 J9 Dame Marie, Cap headland SW Haiti 18.37N 74.24W
149 Q4 Dāmghān Semnān, N Iran 36.13N 54.22E
144 G10 Dāmiyā Al Balqā', NW Jordan 32.07N 35.33E
152 G11 Damla Dashkhovuzskiy Velayat, N Turkmenistan 40.05N 59.15E
102 G12 Damme Niedersachsen, NW Germany 52.31N 8.12E
159 R15 Dāmodar ∿ NE India
160 J9 Damoh Madhya Pradesh, C India 23.52N 79.24E
79 P15 Damongo NW Ghana 9.06N 1.46W
144 G7 Damoûr var. Ad Dāmūr. W Lebanon 33.34N 35.30E
175 Pp7 Dampal, Teluk bay Sulawesi, C Indonesia
188 H7 Dampier Western Australia 20.40S 116.40E
188 H6 Dampier Archipelago island group Western Australia
176 Uu9 Dampier, Selat strait Irian Jaya, E Indonesia
194 L12 Dampier Strait strait NE PNG
147 U14 Damqawt var. Damqut. E Yemen 16.35N 52.39E
165 O13 Dam Qu ∿ C China
Damqut see Damqawt
178 Ii13 Dāmrei, Chuŏr Phnum Fr. Chaîne de l'Éléphant. ▲ SW Cambodia
110 C7 Damvant Jura, NW Switzerland 47.22N 6.55E
Damwâld see Damwoude
100 L5 Damwoude Fris. Damwâld. Friesland, N Netherlands 53.18N 5.59E
165 N15 Damxung Xizang Zizhiqu, W China 30.28N 91.01E
82 K11 Danakil Desert var. Afar Depression, Danakil Plain. desert E Africa
Danakil Plain see Danakil Desert
37 R8 Dana, Mount ▲ California, W USA 37.54N 119.13W
78 L16 Danané W Ivory Coast 7.16N 8.09W
178 Kk10 Đa Nang prev. Tourane. Quang Nam-Đà Nàng, C Vietnam 16.04N 108.13E
179 Qq15 Danao var. Danao City. Cebu, C Philippines 10.34N 124.60E
166 G9 Danba Sichuan, C China 30.54N 101.49E
20 L13 Danbury Connecticut, NE USA 41.21N 73.27W
27 W12 Danbury Texas, SW USA 29.13N 95.20W
37 X15 Danby Lake ☒ California, W USA
204 H4 Danco Coast physical region Antarctica
84 B11 Dande ∿ NW Angola
Dandeldhura see Dadeldhura
161 E17 Dandeli Karnātaka, W India 15.18N 74.42E
191 O12 Dandenong Victoria, SE Australia 38.01S 145.13E
169 V8 Dandong var. Tan-tung; prev. An-tung. Liaoning, NE China 40.09N 124.23E
207 Q14 Daneborg var. Danborg. Tunu, N Greenland 74.34N 19.51W
27 V12 Danevang Texas, SW USA 29.03N 96.11W
165 V12 Dangara Rus. Danghara. SW Tajikistan 38.04N 69.14E
166 G9 Dangchang Gansu, C China 34.01N 104.19E
165 N8 Dangchengwan see Subei
Dang Raek, Phnom/Dangrek, Chaine des see Dângrêk, Chuŏr Phnum
178 Ii13 Dângrêk, Chuŏr Phnum var. Phanom Dang Raek, Phanom Dong Rak, Fr. Chaîne des Dangrek. ▲ Cambodia/Thailand
44 G3 Dangriga prev. Stann Creek. Stann Creek, E Belize 16.58N 88.13W
167 P6 Dangshan Anhui, E China 34.28N 116.24E
142 M14 Darende Malatya, C Turkey 38.33N 37.31E
35 T15 Daniel Wyoming, C USA 42.49N 110.04W
85 H22 Daniëlskuil Northern Cape, N South Africa 28.11N 23.33E
21 N12 Danielson Connecticut, NE USA 41.48N 71.53W
129 M15 Danilov Yaroslavskaya Oblast', W Russian Federation 58.11N 40.11E
131 O9 Danilovka Volgogradskaya Oblast', SW Russian Federation 50.21N 44.03E
166 L7 Dan Jiang ∿ C China
167 M7 Danjiangkou Shuiku ☒ C China
147 W8 Dank var. Dhank. NW Oman 23.34N 56.16E
158 J7 Dankhar Himāchal Pradesh, N India 32.07N 78.12E
131 L5 Dankov Lipetskaya Oblast', W Russian Federation 53.17N 39.07E
44 H9 Danlí El Paraíso, S Honduras 14.02N 86.34W
168 K7 Danli Bulgan, C Mongolia 48.07N 103.64E

169 N8 Darhan Hentiy, C Mongolia 46.38N 109.25E
168 L6 Darhan Selenge, N Mongolia 49.24N 105.57E
169 N12 Darhan Muminggan Lianheqi var. Bailingmiao. Nei Mongol Zizhiqu, N China 41.41N 110.25E
102 K11 Dannenberg Niedersachsen, N Germany 53.05N 11.06E
192 N12 Dannevirke Manawatu-Wanganui, North Island, NZ 40.13S 176.04E
23 U8 Dan River ∿ Virginia, SE USA
20 P10 Dannemora New York, NE USA 44.42N 73.42W
97 O14 Dannemora Uppsala, C Sweden 60.13N 17.49E
Danmark see Denmark
Danmarksstraedet see Denmark Strait
159 S9 Dāmodar see NE India
160 J9 Damoh Madhya Pradesh, C India 23.52N 79.24E
178 Hh9 Dan Sai Loei, C Thailand 17.15N 101.04E
178 P10 Dansville New York, NE USA 42.34N 77.40W
44 K9 Dariense, Cordillera ▲ C Nicaragua
25 W7 Darien Georgia, SE USA 31.22N 81.25W
45 W16 Darién off. Provincia del Darién. ◆ province SE Panama
Darién, Golfo del see Darién, Gulf of
45 X14 Darién, Gulf of Sp. Golfo del Darién. gulf S Caribbean Sea
Darién, Istmo de see Panamá, Istmo de
45 W15 Darién, Serranía del ▲ Colombia/Panama
Dario see Ciudad Darío
Dariorigum see Vannes
Dariv see Darvi
Darj see Dirj
Darjeeling see Darjiling
159 S12 Darjiling prev. Darjeeling. West Bengal, NE India 27.00N 88.13E
Darkehnen see Ozersk
165 S12 Darlag Qinghai, C China 33.43N 99.42E
191 T3 Darling Downs hill range Queensland, E Australia
30 M2 Darling, Lake ☒ North Dakota, N USA
188 I12 Darling Range ▲ Western Australia
190 L8 Darling River ∿ New South Wales, SE Australia
99 M15 Darlington N England, UK 54.31N 1.34W
23 T12 Darlington South Carolina, SE USA 34.18N 79.52W
32 K9 Darlington Wisconsin, N USA 42.40N 90.07W
112 G7 Darłowo Zachodniopomorskie, NW Poland 54.24N 16.21E
103 G18 Darmstadt Hessen, SW Germany 49.52N 8.39E
77 S7 Darnah var. Dérna. NE Libya 32.46N 22.39E
105 S6 Darney Vosges, NE France 48.06N 5.58E
190 M7 Darnick New South Wales, SE Australia 32.52S 143.38E
205 Y6 Darnley, Cape headland Antarctica 67.36S 70.04E
107 R7 Daroca Aragón, NE Spain 41.07N 1.25W
153 S11 Daroot-Korgon var. Daraut-Kurgan. Oshskaya Oblast', SW Kyrgyzstan 39.34N 72.13E
6 A23 Darregueira Buenos Aires, E Argentina 37.40S 63.12W
Darregueira see Darragueira
Darreh Gaz see Dargaz
148 K7 Darreh Shahr var. Darreh-ye Shahr. Īlām, W Iran 33.10N 47.18E
Darreh-ye Shahr see Darreh Shahr
27 P1 Darrouzett Texas, SW USA 36.27N 100.19W
S15 Darsana var. Darshana. Khulna, N Bangladesh 23.33N 88.49E
Darshana see Darsana
102 M7 Darss peninsula NE Germany
102 M7 Darsser Ort headland NE Germany 54.28N 12.31E
99 I24 Dart ∿ SW England, UK
99 P22 Dartford SE England, UK 51.27N 0.13E
190 L12 Dartmoor Victoria, SE Australia 37.56S 141.18E
99 I24 Dartmoor moorland SW England, UK
11 Q15 Dartmouth Nova Scotia, SE Canada 44.40N 63.34W
99 J24 Dartmouth SW England, UK 50.20N 3.34W
13 Y6 Dartmouth Quebec, SE Canada
191 Q11 Dartmouth Reservoir ☒ Victoria, SE Australia
Dartuch, Cabo see Artrutx, Cap d'
194 G15 Daru Western, SW PNG 9.04S 143.12E
114 G9 Daruvar Hung. Daruvár. Bjelovar-Bilogora, NE Croatia 45.34N 17.12E
152 T6 Darvaza Turkm. Derweze. Akhalskiy Velayat, C Turkmenistan 40.10N 58.27E
Darvaza see Darwoza
Darvazskiy Khrebet see Darvoz, Qatorkūhi
114 J9 Darvi var. Dariv. Govĭ-Altay, W Mongolia 46.20N 94.11E
152 L9 Darvīshān var. Darweshan, Garmser. Helmand, S Afghanistan 31.01N 64.12E
153 R13 Darvoz, Qatorkūhi Rus. Darvazskiy Khrebet. ▲ C Tajikistan
Darwazeh see Darvāshin
57 R10 Darwin Río Negro, S Argentina 39.13S 65.41W
67 D24 Darwin prev. Palmerston, Port Darwin. territory capital Northern Territory, N Australia 12.27S 130.52E
67 D24 Darwin var. Darwin Settlement. East Falkland, Falkland Islands 51.51S 58.55W
64 B17 Darwin, Cordillera ▲ S Chile
58 B17 Darwin, Volcán ⏣ Galápagos Islands, Ecuador, E Pacific Ocean 0.12S 91.17W
153 O10 Darwoza Rus. Darvaza. Jizzakh Wiloyati, C Uzbekistan 40.59N 67.19E
155 S8 Darya Khān Punjab, E Pakistan 31.48N 71.05E
Dar'yalyktakyr, Ravnina plain S Kazakhstan
149 T11 Dārzīn Kermān, S Iran 29.10N 58.09E
166 L8 Dashennongjia ▲ C China 31.24N 110.16E

Dashhowuz see Dashkhovuz
Dashhowuz Welayaty see
Dashhowuzskiy Welayat
121 O16 **Dashkawka** Rus. Dashkovka.
Mahilyowskaya Voblasts', E Belarus
53.42N 30.17E
152 H8 **Dashkhovuz** Turkm.
Dashhowuz; prev. Tashauz.
Dashkhovuzskiy Velayat,
N Turkmenistan 41.51N 59.52E
Dashkhovuz see Dashkhovuzskiy
Velayat
152 E9 **Dashkhovuzskiy Velayat** var.
Dashhowuz, Turkm.
Dashhowuz Welayaty. ◆ province
N Turkmenistan
Dashköpri see Tashkepri
Dashkovka see Dashkawka
154 J15 **Dasht** ॐ SW Pakistan
Dashtidzhum see Dashtijum
153 R13 **Dashtijum** Rus. Dashtidzhum.
SW Tajikistan 38.06N 70.11E
155 W7 **Daska** Punjab, NE Pakistan
32.21N 74.20E
Đa, Sông see Black River
79 R15 **Dassa** var. Dassa-Zoumé. S Benin
7.46N 2.15E
Dassa-Zoumé see Dassa
31 U4 **Dassel** Minnesota, N USA
45.06N 94.18W
158 H3 **Dastegil Sar** var. Disteghil Sār.
▲ N India
142 C16 **Datça** Muğla, SW Turkey
36.46N 27.40E
172 Nn6 **Date** Hokkaidō, NE Japan
42.28N 140.51E
160 I8 **Datia** prev. Duttia. Madhya
Pradesh, C India 25.40N 78.28E
165 T10 **Datong** Qinghai, C China
37.01N 101.33E
167 N2 **Datong** var. Tatung, Ta-t'ung.
Shanxi, C China 40.09N 113.16E
165 S9 **Datong He** ॐ C China
165 S9 **Datong Shan** ▲ C China
304 Kk6 **Datu, Tanjung** headland
Indonesia/Malaysia 2.01N 109.37E
Daua see Dawa Wenz
180 H16 **Dauban, Mount** ▲ Silhouette,
NE Seychelles
155 T7 **Dāūd Khel** Punjab, E Pakistan
32.52N 71.34E
121 G15 **Daugai** Alytus, S Lithuania
54.22N 24.20E
Daugava see Western Dvina
120 J11 **Daugavpils** Ger. Dünaburg; prev.
Rus. Dvinsk. municipality
Daugvapils, SE Latvia 55.53N 26.33E
Dauka see Dawkah
Daulatabad see Malāyer
103 D18 **Daund** prev. Dhond.
Mahārāshtra, W India 18.28N 74.37E
178 Gg12 **Daung Kyun** island S Myanmar
9 W15 **Dauphin** Manitoba, S Canada
51.09N 100.04W
105 S13 **Dauphiné** cultural region E France
25 N9 **Dauphin Island** island Alabama,
S USA
9 X15 **Dauphin River** Manitoba,
S Canada 51.55N 98.03W
79 V12 **Daura** Katsina, N Nigeria
13.03N 8.18E
158 H12 **Dausa** prev. Daosa. Rājasthān,
N India 26.54N 76.18E
Dauwa see Dawwah
143 Y10 **Däväçi** Rus. Divichi.
NE Azerbaijan 41.15N 48.58E
161 F18 **Dāvangere** Karnātaka, W India
14.30N 75.52E
179 Rr16 **Davao** off. Davao City. Mindanao,
S Philippines 7.06N 125.35E
179 Rr16 **Davao Gulf** gulf Mindanao,
S Philippines
13 Q11 **Daveluyville** Québec, SE Canada
46.12N 72.07W
31 Z14 **Davenport** Iowa, C USA
41.31N 90.34W
34 L8 **Davenport** Washington,
NW USA 47.39N 118.09W
45 P6 **David** Chiriquí, W Panama
8.25N 82.25W
13 O11 **David** ॐ Québec, SE Canada
31 R15 **David City** Nebraska, C USA
41.15N 97.07W
David-Gorodok see Davyd-
Haradok
9 T16 **Davidson** Saskatchewan,
S Canada 51.15N 105.58W
23 R10 **Davidson** North Carolina,
SE USA 35.29N 80.49W
28 K12 **Davidson** Oklahoma, C USA
34.15N 99.06W
41 S6 **Davidson Mountains**
▲ Alaska, USA
180 M8 **Davie Ridge** undersea feature
W Indian Ocean
190 A1 **Davies, Mount** ▲ South
Australia 26.14S 129.14E
37 O7 **Davis** California, W USA
38.31N 121.46W
29 N14 **Davis** Oklahoma, C USA
34.30N 97.07W
205 Y13 **Davis** Australian research station
Antarctica 68.30S 78.15E
204 P13 **Davis Coast** physical region
Antarctica
20 C16 **Davis, Mount** ▲ Pennsylvania,
C USA 39.47N 79.10W
26 K9 **Davis Mountains** ▲ Texas,
SW USA
205 Z9 **Davis Sea** sea Antarctica
67 O20 **Davis Seamounts** undersea
feature S Atlantic Ocean
206 M13 **Davis Strait** strait Baffin
Bay/Labrador Sea
131 O13 **Davlekanovo** Respublika
Bashkortostan, W Russian
Federation 54.13N 55.06E
110 J9 **Davos** Rmsch. Tavau.
Graubünden, E Switzerland
46.48N 9.50E
121 J20 **Davyd-Haradok** Pol.
Dawidgródek, Rus. David-
Gorodok. Brestskaya Voblasts',
SW Belarus 52.03N 27.13E
169 U12 **Dawa** Liaoning, NE China
40.55N 122.02E
147 O11 **Dawāsir, Wādī ad** dry watercourse
S Saudi Arabia
83 K15 **Dawa Wenz** var. Daua, Webi
Daawo. ॐ E Africa

Dawaymah, Birkat ad see Umm
al Baqar, Hawr
Dawei see Tavoy
121 K14 **Dawhinava** Rus. Dolginovo.
Minskaya Voblasts', N Belarus
54.39N 27.28E
Dawidgródek see Davyd-
Haradok
147 V12 **Dawkah** var. Dauka. SW Oman
18.32N 54.03E
Dawlat Qatar see Qatar
26 M11 **Daws** Al Bāḩah, SW Saudi Arabia
20.19N 41.12E
8 H5 **Dawson** var. Dawson City. Yukon
Territory, NW Canada
64.04N 139.24W
25 T4 **Dawson** Georgia, SE USA
31.46N 84.27W
31 S9 **Dawson** Minnesota, N USA
44.55N 96.03W
9 N13 **Dawson Creek** British Columbia,
W Canada 55.48N 120.18W
8 H7 **Dawson Range** ▲ Yukon
Territory, W Canada
189 Y9 **Dawson River** ॐ Queensland,
E Australia
8 J15 **Dawsons Landing** British
Columbia, W Canada
51.33N 127.38W
22 I7 **Dawson Springs** Kentucky,
S USA 37.10N 87.41W
25 S2 **Dawsonville** Georgia, SE USA
34.23N 84.07W
166 G8 **Dawu** Sichuan, C China
30.55N 101.08E
Dawu see Maqên
147 Y10 **Dawwah** var. Dauwa. W Oman
20.36N 58.52E
104 J15 **Dax** var. Ax; anc. Aquae Augustae,
Aquae Tarbelicae. Landes,
SW France 43.43N 1.03W
Da Xian/Daxian see Dachuan
166 G9 **Daxue Shan** ▲ C China
166 G12 **Dayao** Yunnan, SW China
25.41N 101.23E
Dayishan see Gaoyou
191 N12 **Daylesford** Victoria, SE Australia
37.24S 144.07E
37 U10 **Daylight Pass** pass California,
W USA 36.44N 116.55W
63 D17 **Daymán, Río** ॐ N Uruguay
Dayr see Ad Dayr
144 G10 **Dayr 'Allā** var. Deir 'Alla.
Al Balqā', N Jordan 32.39N 36.06E
144 M5 **Dayr az Zawr** var. Deir ez Zor.
Dayr az Zawr, E Syria 35.12N 40.12E
144 M5 **Dayr az Zawr** off. Muḩāfaẓat
Dayr az Zawr, var. Dayr Az-Zor. ◆
governorate E Syria
Dayr Az-Zor see Dayr az Zawr
Dayrūṭ see Dairūṭ
9 Q15 **Daysland** Alberta, SW Canada
52.53N 112.19W
33 R14 **Dayton** Ohio, N USA
39.45N 84.11W
22 L10 **Dayton** Tennessee, S USA
35.30N 85.01W
27 W11 **Dayton** Texas, SW USA
30.03N 94.53W
34 L8 **Dayton** Washington, NW USA
46.19N 117.58W
175 O10 **Daytona Beach** Florida, SE USA
29.12N 81.03W
175 O9 **Dayu** Borneo, C Indonesia
1.58S 115.04E
167 O11 **Dayu Ling** ▲ S China
167 R7 **Da Yunhe** Eng. Grand Canal.
canal E China
167 S11 **Dayu Shan** island SE China
166 J9 **Dazhu** Sichuan, C China
30.45N 107.10E
166 J9 **Dazu** Chongqing Shi, C China
29.41N 105.46E
85 H24 **De Aar** Northern Cape, C South
Africa 30.40S 24.01E
204 K5 **Deacon, Cape** headland
Antarctica
41 R5 **Deadhorse** Alaska, USA
70.15N 148.28W
35 T12 **Dead Indian Peak** ▲ Wyoming,
C USA 44.36N 109.45W
46 J4 **Deadman's Cay** Long Island,
C Bahamas 23.09N 75.06W
144 G14 **Dead Sea** var. Bahret Lut, Lacus
Asphaltites, Ar. Al Baḩr al Mayyit,
Baḩrat Lūṭ, Heb. Yam HaMelah. salt
lake Israel/Jordan
30 J9 **Deadwood** South Dakota, N USA
44.22N 103.43W
99 Q22 **Deal** SE England, UK 51.14N 1.22E
85 I22 **Dealesville** Free State, C South
Africa 28.40S 25.46E
Dealnu see Tana/Teno
167 P10 **De'an** Jiangxi, S China
29.24N 115.46E
64 K9 **Deán Funes** Córdoba,
C Argentina 30.25S 64.22W
204 L12 **Dean Island** island Antarctica
33 S10 **Dearborn** Michigan, N USA
42.16N 83.13W
29 R3 **Dearborn** Missouri, C USA
39.31N 94.46W
34 X9 **Deary** Idaho, NW USA
46.46N 116.33W
8 J10 **Dease Lake** British Columbia,
W Canada
8 J10 **Dease Lake** British Columbia,
W Canada 58.28N 130.04W
34 U11 **Death Valley** California, W USA
35.25N 116.50W
37 U11 **Death Valley** valley California,
W USA
Dées see Dej
104 L4 **Deauville** Calvados, N France
49.21N 0.06E
119 X7 **Debal'tseve** Rus. Debal'tsevo.
Donets'ka Oblast', SE Ukraine
48.21N 38.25E
115 M19 **Debar** Ger. Dibra, Turk. Debre.
W FYR Macedonia 41.31N 20.32E
41 O9 **Debauch Mountain** ▲ Alaska,
USA 64.31N 159.52W
De Behagle see Laï
34 X7 **De Berry** Texas, SW USA

131 T2 **Debesy** Udmurtskaya Respublika,
NW Russian Federation
57.41N 53.56E
200 N16 **De Gerlache Seamounts**
undersea feature SE Pacific Ocean
113 N16 **Dębica** Podkarpackie, SE Poland
50.03N 21.24E
100 J11 **De Bilt** var. De Bilt. Utrecht,
C Netherlands 52.06N 5.10E
127 O9 **Debin** Ros. Dauka. SW Oman
112 N13 **Dęblin** Rus. Ivangorod. Lubelskie,
E Poland 51.34N 21.49E
112 D10 **Dębno** Zachodniopomorskie,
NW Poland 52.44N 14.42E
41 S10 **Deborah, Mount** ▲ Alaska, USA
63.38N 147.13W
35 N8 **De Borgia** Montana, NW USA
47.23N 115.24W
82 J13 **Debre Birhan** see Debre Birhan.
Debra Marcos see Debre Mark'os
Debra Tabor see Debre Tabor
Debre see Debar
82 J13 **Debre Birhan** var. Debre Birhan.
Amhara, N Ethiopia 9.45N 39.40E
113 N22 **Debrecen** Ger. Debreczin, Rom.
Debreṭin; prev. Debreczen. Hajdú-
Bihar, E Hungary 47.31N 21.37E
Debreczen/Debreczin see
Debrecen
82 I12 **Debre Mark'os** var. Debra
Marcos. Amhara, N Ethiopia
10.18N 37.48E
115 N19 **Debrešte** SW FYR Macedonia
41.29N 21.20E
82 J11 **Debre Tabor** var. Debra Tabor.
Amhara, N Ethiopia 11.46N 38.06E
Debreṭin see Debrecen
82 J13 **Debre Zeyt** Oromo, C Ethiopia
8.41N 39.00E
115 L16 **Dečani** Serbia, S Yugoslavia
42.34N 20.18E
25 P2 **Decatur** Alabama, S USA
34.36N 86.58W
25 S3 **Decatur** Georgia, SE USA
33.46N 84.18W
32 L13 **Decatur** Illinois, N USA
39.50N 88.57W
33 Q12 **Decatur** Indiana, N USA
40.48N 84.55W
24 M5 **Decatur** Mississippi, S USA
32.26N 89.06W
31 S14 **Decatur** Nebraska, C USA
42.00N 96.19W
27 W5 **Decatur** Texas, SW USA
33.14N 97.32W
22 H9 **Decaturville** Tennessee, S USA
35.34N 88.05W
105 O13 **Decazeville** Aveyron, S France
44.34N 2.18E
161 H17 **Deccan** Hind. Dakshin. plateau
C India
12 J8 **Decelles, Réservoir** ☐ Québec,
SE Canada
10 K2 **Déception** Québec, NE Canada
62.06N 74.36W
166 G11 **Dechang** Sichuan, C China
27.24N 102.09E
113 C15 **Děčín** Ger. Tetschen. Ústecký Kraj,
NW Czech Republic 50.48N 14.15E
105 P9 **Decize** Nièvre, C France
46.51N 3.25E
100 I6 **De Cocksdorp** Noord-Holland,
NW Netherlands 53.09N 4.52E
31 X11 **Decorah** Iowa, C USA
43.18N 91.47W
Dedeagac/Dedeagach see
Alexandroúpoli
196 G15 **Dededo** Guam 13.30N 144.51E
100 N9 **Dedemsvaart** Overijssel,
E Netherlands 52.36N 6.28E
21 O11 **Dedham** Massachusetts, NE USA
42.14N 71.10W
65 H19 **Dedo, Cerro** ▲ SW Argentina
46.51N 71.48W
79 O13 **Dédougou** W Burkina
12.28N 3.27W
128 G15 **Dedovichi** Pskovskaya Oblast',
W Russian Federation 57.31N 29.53E
169 V6 **Dedu** var. Qingshan.
Heilongjiang, NE China
48.30N 126.17E
161 J24 **Deduru Oya** ॐ W Sri Lanka
85 N14 **Dedza** Central, S Malawi
14.24S 34.15E
85 N14 **Dedza Mountain** ▲ C Malawi
14.23S 34.16E
99 J19 **Dee** Wel. Afon Dyfrdwy.
☐ England/Wales, UK
98 K9 **Dee** ॐ NE Scotland, UK
23 T3 **Deep Bay** see Chilumba
T3 **Deep Creek Lake** ☐ Maryland,
NE USA
38 J4 **Deep Creek Range** ▲ Utah,
W USA
29 P10 **Deep Fork** ॐ Oklahoma, C USA
12 J11 **Deep River** Ontario, SE Canada
46.06N 77.28W
23 T10 **Deep River** ॐ North Carolina,
SE USA
191 U4 **Deepwater** New South Wales,
SE Australia 29.27S 151.52E
33 S14 **Deer Creek Lake** ☐ Ohio,
N USA
175 Z15 **Deerfield Beach** Florida,
SE USA 26.19N 80.06W
41 N8 **Deering** Alaska, USA
66.04N 162.43W
40 M16 **Deer Island** island Alaska, USA
21 S7 **Deer Isle** island Maine, NE USA
11 S11 **Deer Lake** Newfoundland,
E Canada
101 D18 **Deerlijk** West-Vlaanderen,
W Belgium 50.52N 3.21E
35 Q10 **Deer Lodge** Montana, N USA
46.23N 112.43W
34 L8 **Deer Park** Washington, NW USA
47.55N 117.28W
161 J23 **Delft** island NW Sri Lanka
31 W3 **Deer River** Minnesota, N USA
47.19N 93.47W
33 R11 **Defiance** Ohio, N USA
41.16N 84.21W
25 Q8 **De Funiak Springs** Florida,
SW USA 30.43N 86.07W
97 L23 **Degeberga** Skåne, S Sweden
55.48N 14.06E
106 H12 **Degebe, Ribeira** ॐ S Portugal
82 M13 **Degeh Bur** Somali, E Ethiopia
8.08N 43.35E
79 U17 **Dégélis** Rivers, S Nigeria
4.46N 6.47E

97 L16 **Degerfors** Örebro, C Sweden
59.13N 14.26E
103 N21 **Deggendorf** Bayern, SE Germany
48.49N 12.58E
124 Nn2 **Değirmenlik** Gk. Kythréa.
N Cyprus 35.14N 33.28E
22 J11 **Degoña** Amhara, N Ethiopia
12.22N 37.36E
42 J6 **Délicias** var. Ciudad Delicias.
Chihuahua, N Mexico
28.08N 105.22W
188 J6 **De Grey River** ॐ Western
Australia
130 M10 **Degtevo** Rostovskaya Oblast',
SW Russian Federation
49.12N 40.39E
149 X13 **Dehak** Sīstān va Balūchestān,
SE Iran 27.10N 62.34E
149 R9 **Deh 'Alī** Kermān, C Iran
31.40N 56.10E
149 S13 **Dehbārez** var. Rūdān.
Hormozgān, S Iran 27.30N 57.10E
149 P10 **Deh Bid** Fārs, C Iran 30.37N 53.11E
148 M10 **Deh Dasht** Kohkilūyeh va Būyer
Aḩmadī, SW Iran 30.49N 50.36E
77 N8 **Dehibat** SE Tunisia 31.58N 10.43E
Dehli see Delhi
148 K8 **Dehlorān** Īlām, W Iran
32.40N 47.18E
153 N13 **Dehqonobod** Rus.
Dekhkanabad. Qashqadaryo
Wiloyati, S Uzbekistan 38.44N 66.31E
158 I8 **Dehra Dūn** Uttar Pradesh,
N India 30.18N 78.03E
159 O14 **Dehri** Bihār, N India 24.55N 84.10E
154 K10 **Deh Shū** var. Deshu. Helmand,
S Afghanistan 30.28N 63.21E
101 D17 **Deinze** Oost-Vlaanderen,
NW Belgium 50.58N 3.31E
Deir 'Alla see Dayr 'Allā
Deir ez Zor see Dayr az Zawr
Deirgeirt, Loch see Derg, Lough
118 H9 **Dej** Hung. Dés; prev. Deés. Cluj,
NW Romania 47.08N 23.55E
97 K15 **Deje** Värmland, C Sweden
59.34N 13.28E
176 Y14 **De Jongs, Tanjung** headland
Irian Jaya, SE Indonesia
6.55S 138.31E
9 W17 **Deloraine** Manitoba, S Canada
49.12N 100.28W
32 M9 **Delavan** Wisconsin, N USA
43.03N 88.22W
63 C23 **De La Garma** Buenos Aires,
E Argentina 37.58S 60.25W
12 K10 **Delahey** Québec,
SE Canada
82 E11 **Delami** Southern Kordofan,
C Sudan 11.51N 30.30E
41 S9 **Delta Junction** Alaska, USA
64.02N 145.43W
175 Y10 **Deltona** Florida, SE USA
28.54N 81.15W
191 S3 **Delungra** New South Wales,
SE Australia 29.40S 150.49E
160 C12 **Delvāda** Gujarāt, W India
20.46N 71.01E
63 B17 **Del Valle** Buenos Aires,
E Argentina 35.55S 60.42W
117 C15 **Delvináki** var. Dhelvinákion;
prev. Pogónion. Ípeiros, W Greece
39.56N 20.27E
115 L23 **Delvinë** var. Delvina, It. Delvino.
Vlorë, S Albania 39.56N 20.07E
Delvino see Delvinë
118 H7 **Delyatyn** Ivano-Frankivs'ka
Oblast', W Ukraine 48.32N 24.38E
131 O11 **Dëma** ॐ W Russian Federation
107 O5 **Demanda, Sierra de la**
▲ N Spain
81 K21 **Dema** Kasai Occidental, C Dem.
Rep. Congo 5.24S 22.16E
180 H13 **Dembeni** Grande Comore,
NW Comoros 11.49S 43.25E
81 M15 **Dembia** Mbomou, SE Central
African Republic 5.08N 24.25E
H13 **Dembi Dolo** var. Dembidollo.
Oromo, C Ethiopia 8.33N 34.49E
158 K6 **Dêmchok** var. Dêmqog.
China/India see also Dêmqog
32.30N 79.42E
158 I6 **Demchok** var. Dêmqog. disputed
region China/India see also
Dêmqog
100 I12 **De Meern** Utrecht, C Netherlands
52.06N 5.00E
101 I17 **Demer** ॐ C Belgium
66 H12 **Demerara Plain** undersea
feature W Atlantic Ocean
66 H12 **Demerara Plateau** undersea
feature W Atlantic Ocean
57 T9 **Demerara River** ॐ NE Guyana
130 H3 **Demidov** Smolenskaya Oblast',
W Russian Federation 55.15N 31.30E
39 Q15 **Deming** New Mexico, SW USA
32.13N 107.46W
34 H4 **Deming** Washington, NW USA
48.50N 122.13W
60 E10 **Demini, Rio** ॐ NW Brazil
33 N11 **Demmin** Mecklenburg-
Vorpommern, NE Germany
53.53N 13.03E
25 O5 **Demopolis** Alabama, S USA
32.30N 87.50W
33 N11 **Demotte** Indiana, N USA
41.13N 87.07W
164 F2 **Dêmqog** var. Dêmchok.
China/India see also Dêmchok
32.36N 79.28E

158 L6 **Dêmqog** var. Demchok. disputed
region China/India see also
Demchok
176 Yy10 **Den** Irian Jaya, E Indonesia
2.19S 140.06E
125 G11 **Dem'yanka** ॐ C Russian
Federation
128 H15 **Demyansk** Novgorodskaya
Oblast', W Russian Federation
57.39N 32.31E
125 FJ11 **Dem'yanskoye** Tyumenskaya
Oblast', C Russian Federation
59.39N 69.15E
105 P2 **Denain** Nord, N France
50.19N 3.24E
41 S10 **Denali** Alaska, USA 63.08N 147.33W
Denali see McKinley, Mount
83 M14 **Denan** Somali, E Ethiopia
6.40N 43.31E
99 J18 **Denbigh** Wel. Dinbych.
NE Wales, UK 53.10N 3.25W
99 J18 **Denbigh** cultural region N Wales,
UK
100 I8 **Den Burg** Noord-Holland,
NW Netherlands 53.03N 4.46E
101 F18 **Dender** Fr. Dendre.
ॐ W Belgium
101 F18 **Denderleeuw** Oost-Vlaanderen,
NW Belgium 50.54N 4.04E
101 F17 **Dendermonde** Fr. Termonde.
Oost-Vlaanderen, NW Belgium
51.01N 4.07E
Dendre see Dender
99 C19 **Derg, Lough** Ir. Loch Deirgeirt.
◎ W Ireland
100 P10 **Denekamp** Overijssel,
E Netherlands 52.23N 7.00E
79 W12 **Denga** Zinder, S Niger
13.15N 9.43E
168 L13 **Dengkagoin** see Têwo
168 L13 **Dengkou** var. Bayan Gol. Nei
Mongol Zizhiqu, N China
40.15N 106.58E
165 Q14 **Dêngqên** Xizang Zizhiqu,
W China 31.28N 95.28E
166 M7 **Deng Xian** see Dengzhou
169 O13 **Dengzhou** prev. Deng Xian.
Henan, C China 32.43N 112.02E
100 N9 **Den Ham** Overijssel,
E Netherlands 52.30N 6.30E
188 H10 **Denham** Western Australia
25.56S 113.35E
46 J12 **Denham, Mount** ▲ C Jamaica
18.13N 77.33W
24 J8 **Denham Springs** Louisiana,
S USA 30.29N 90.57W
100 I7 **Den Helder** Noord-Holland,
NW Netherlands 52.54N 4.45E
107 T11 **Denia** País Valenciano, E Spain
38.51N 0.07E
191 S7 **Deniliquin** New South Wales,
SE Australia 35.33S 144.58E
31 Y14 **Denison** Iowa, C USA
42.00N 95.22W
27 U5 **Denison** Texas, SW USA
33.45N 96.32W
142 D15 **Denizli** Denizli, SW Turkey
37.46N 29.04E
142 D15 **Denizli** ◆ province SW Turkey
Denjong see Sikkim
191 S7 **Denman** New South Wales,
SE Australia 32.24S 150.43E
205 Y10 **Denman Glacier** glacier
Antarctica
23 R14 **Denmark** South Carolina,
SE USA 33.19N 81.08W
97 G23 **Denmark** off. Kingdom of
Denmark, Dan. Danmark;
anc. Hafnia. ◆ monarchy N Europe
94 H1 **Denmark Strait** var.
Danmarksstraedet. strait
Greenland/Iceland
47 T11 **Dennery** E Saint Lucia
13.55N 60.53W
100 I7 **Den Oever** Noord-Holland,
NW Netherlands 52.56N 5.01E
153 O13 **Denow** Rus. Denau.
Surkhondaryo Wiloyati,
S Uzbekistan 38.19N 67.48E
175 N16 **Denpasar** prev. Paloe. Bali,
C Indonesia 8.40S 115.13E
118 E12 **Denta** Timiş, W Romania
45.18N 21.14E
23 Y3 **Denton** Maryland, NE USA
38.52N 75.49W
27 T6 **Denton** Texas, SW USA
33.10N 97.08W
195 O15 **D'Entrecasteaux Islands** island
group SE PNG
39 T4 **Denver** state capital Colorado,
C USA 39.44N 105.00W
18 K8 **Denver** ✈ Colorado, C USA
39.57N 104.38W
26 L6 **Denver City** Texas, SW USA
32.57N 102.49W
158 J2 **Deoband** Uttar Pradesh, N India
29.40N 77.40E
Deoghar see Devghar
160 E13 **Deolāli** Mahārāshtra, W India
19.57N 73.50E
160 I10 **Deori** Madhya Pradesh, C India
23.09N 78.39E
159 O12 **Deoria** Uttar Pradesh, N India
26.31N 83.48E
101 A17 **De Panne** West-Vlaanderen,
W Belgium 51.06N 2.34E
27 V5 **Deport** Texas, SW USA
33.31N 95.19W
126 M7 **Deputatskiy** Respublika S
akha (Yakutiya), NE Russian
Federation 69.18N 139.48E
31 V14 **Des Moines** state capital Iowa,
C USA 41.36N 93.36W
31 V14 **Des Moines** ✈ Iowa, C USA
34.02N 94.20W
19 N8 **Des Moines River** ॐ C USA
129 P4 **Desna** ॐ C Russian
Federation/Ukraine
65 F24 **Desolación, Isla** island S Chile

24 J5 **Delhi** Louisiana, S USA
32.28N 91.29W
20 J1 **Delhi** New York, NE USA
42.16N 74.55W
158 I10 **Delhi** ◆ union territory NW India
142 J17 **Deli Burnu** headland S Turkey
57 X10 **Délices** C French Guiana
4.46N 53.43W
142 J12 **Delice Çayı** ॐ C Turkey
149 N7 **Delijan** var. Dalijan, Dilijan.
Markazi, W Iran 34.01N 50.39E
114 P12 **Deli Jovan** ▲ E Yugoslavia
Déli-Kárpátok see Carpaţii
Meridionali
15 H6 **Déline** prev. Fort Franklin.
Northwest Territories, NW Canada
65.10N 123.30W
13 Q7 **Delisle** Québec, SE Canada
48.39N 71.42W
9 T15 **Delisle** Saskatchewan, S Canada
51.54N 107.01W
103 M15 **Delitzsch** Sachsen, E Germany
51.31N 12.19E
35 Q12 **Dell** Montana, NW USA
44.41N 112.42W
20 K7 **Dell City** Texas, SW USA
31.56N 105.12W
105 U7 **Delle** Territoire-de-Belfort,
E France 47.30N 7.00E
30 J8 **Dellenbaugh, Mount**
▲ Arizona, SW USA 36.06N 113.32W
31 R11 **Dell Rapids** South Dakota,
N USA 43.50N 96.42W
23 Y4 **Delmar** Maryland, NE USA
38.26N 75.32W
20 K1 **Delmar** New York, NE USA
42.37N 73.49W
102 G11 **Delmenhorst** Niedersachsen,
NW Germany 53.03N 8.37E
114 C9 **Delnice** Primorje-Gorski Kotar,
NW Croatia 45.24N 14.49E
39 R7 **Del Norte** Colorado, C USA
37.40N 106.21W
41 N6 **De Long Mountains** ▲ Alaska,
USA
191 P16 **Deloraine** Tasmania, SE Australia
41.34S 146.43E
38 K6 **Delta** Utah, W USA 39.21N 112.34W
79 T17 **Delta** ◆ state S Nigeria
57 Q6 **Delta Amacuro** off. Territorio
Delta Amacuro. ◆ federal district
NE Venezuela
41 S9 **Delta Junction** Alaska, USA

158 L6 **Der'a/Derá/Déraa**
see Dar'ā
155 S5 **Dera Ghāzi Khān** var. Dera
Ghāzikhan. Punjab, C Pakistan
30.01N 70.37E
155 S8 **Dera Ismāïl Khān**
North-West Frontier Province,
C Pakistan 31.51N 70.55E
115 L16 **Đeravica** ▲ S Yugoslavia
42.32N 20.08E
118 L6 **Derazhnya** Khmel'nyts'ka
Oblast', W Ukraine 49.16N 27.24E
131 R17 **Derbent** Respublika Dagestan,
SW Russian Federation
42.01N 48.16E
41 M14 **Derbent** Surkhondaryo Wiloyati,
S Uzbekistan 38.15N 66.54E
83 M15 **Derbissaka** Mbomou, SE Central
African Republic 5.43N 24.48E
188 L4 **Derby** Western Australia
17.18S 123.36E
99 M19 **Derby** C England, UK 52.55N 1.30W
29 N7 **Derby** Kansas, C USA
37.33N 97.16W
99 L18 **Derbyshire** cultural region
C England, UK
114 O11 **Derdap** physical region E Yugoslavia
Dereli see Gónnoi
176 X11 **Derew** ॐ Irian Jaya, E Indonesia
131 R8 **Dergachi** Saratovskaya Oblast',
W Russian Federation 51.15N 48.58E
Dergachi see Derhachi
99 C19 **Derg, Lough** Ir. Loch Deirgeirt.
◎ W Ireland
119 V5 **Derhachi** Rus. Dergachi.
Kharkivs'ka Oblast', E Ukraine
50.08N 36.10E
24 G8 **De Ridder** Louisiana, S USA
30.51N 93.18W
143 P16 **Derik** Mardin, SE Turkey
37.22N 40.16E
85 E20 **Derm** Hardap, C Namibia
23.38S 18.12E
150 M14 **Dermentobe** prev.
Dyurmen'tyube. Kzylorda,
S Kazakhstan 45.46N 63.42E
29 V4 **Dermott** Arkansas, C USA
33.31N 91.26W
Dérna see Darnah
Dernberg, Cape see Dolphin
Head
24 J11 **Dernieres, Isles** island group
Louisiana, S USA
Dernis see Drniš
104 I3 **Déroute, Passage de la** strait
Channel Islands/France
Derrá see Dar'ā
Derry see Londonderry
82 G12 **Dertona** see Tortona
Dertosa see Tortosa
107 T11 **Derudeb** Red Sea, NE Sudan
17.31N 36.07E
114 H10 **Derventa** Republika Srpska,
N Bosnia and Herzegovina
44.57N 17.55E
191 O16 **Derwent Bridge** Tasmania,
SE Australia 42.10S 146.13E
191 O17 **Derwent, River** ॐ Tasmania,
SE Australia
Derweze see Darvaza
Deržavinsk see Derzhavinsk
151 O9 **Derzhavinsk** var. Deržavinsk.
Akmola, C Kazakhstan
51.04N 66.19E
59 J18 **Desaguadero** Puno, S Peru
16.35S 69.01W
59 J18 **Desaguadero, Río**
ॐ Bolivia/Peru
203 W9 **Désappointement, Îles du**
island group Îles Tuamotu, C French
Polynesia
29 X5 **Des Arc** Arkansas, C USA
34.58N 91.30W
12 C10 **Desbarats** Ontario, S Canada
46.20N 83.52W
43 H13 **Descabezado Grande, Volcán**
▲ C Chile 35.34S 70.40W
42 B2 **Descanso** Baja California,
NW Mexico 32.08N 116.51W
104 I9 **Descartes** Indre-et-Loire,
C France 46.58N 0.40E
9 T13 **Deschambault Lake**
◎ Saskatchewan, C Canada
118 E12 **Deschnaer Koppe** see Velká
Deštná
34 J11 **Deschutes River** ॐ Oregon,
NW USA
82 J11 **Desē** var. Desse, It. Dessie.
Amhara, N Ethiopia 11.01N 39.39E
65 J16 **Deseado, Río** ॐ S Argentina
108 F8 **Desenzano del Garda**
Lombardia, N Italy 45.28N 10.31E
38 K3 **Deseret Peak** ▲ Utah, W USA
40.27N 112.37W
91 C13 **Deserta Grande** island Madeira,
Portugal, NE Atlantic Ocean
91 C13 **Desertas, Ilhas** island group
Madeira, Portugal,
NE Atlantic Ocean
37 X16 **Desert Center** California,
W USA 33.42N 115.22W
37 X16 **Desert Hot Springs** California,
W USA 33.57N 116.33W
12 K10 **Désert, Lac** ◎ Québec,
SE Canada
38 J2 **Desert Peak** ▲ Utah, W USA
41.03N 113.22W
33 Y14 **Deshler** Ohio, N USA
41.12N 83.54W
Deshu see Deh Shū
117 S10 **Desiderii Fanum** see St-Dizier
108 D7 **Desio** Lombardia, N Italy
45.37N 9.12E
117 L15 **Deskáti** var. Dheskáti. Dytikí
Makedonía, N Greece 39.55N 21.49E
30 L2 **Des Lacs River** ॐ North
Dakota, N USA
29 X6 **Desloge** Missouri, C USA
37.52N 90.31W
9 T13 **Desmarais** Alberta, W Canada
55.58N 113.55W
31 Q10 **De Smet** South Dakota, N USA
44.23N 97.33W

◆ COUNTRY ◇ DEPENDENT TERRITORY ▲ ADMINISTRATIVE REGION ▲ MOUNTAIN ☒ VOLCANO ◎ LAKE
● COUNTRY CAPITAL ○ DEPENDENT TERRITORY CAPITAL ✈ INTERNATIONAL AIRPORT ▲ MOUNTAIN RANGE ॐ RIVER ☐ RESERVOIR

31 V14 De Soto Iowa, C USA 41.31N 94.00W
25 Q4 De Soto Falls waterfall Alabama, S USA 33.22N 86.12W
85 I25 Despatch Eastern Cape, S South Africa 33.48S 25.28E
107 N12 Despeñaperros, Desfiladero de pass S Spain 38.25N 3.26W
33 N10 Des Plaines Illinois, N USA 42.01N 87.52W
117 J21 Despotikó island Kykládes, Greece, Aegean Sea
114 N12 Despotovac Serbia, E Yugoslavia 44.06N 21.25E
103 M14 Dessau Sachsen-Anhalt, E Germany 51.51N 12.15E
Desse see Desē
101 J16 Dessel Antwerpen, N Belgium 51.15N 5.07E
Dessie see Desē
Destêrro see Florianópolis
25 P9 Destin Florida, SE USA 30.23N 86.30W
Deštná see Velká Deštná
200 Oo11 Desventurados, Islas de los island group W Chile
105 N1 Desvres Pas-de-Calais, N France 50.41N 1.48E
118 E12 Deta Ger. Detta. Timiş, W Romania 45.22N 21.13E
103 H14 Detmold Nordrhein-Westfalen, W Germany 51.55N 8.52E
33 S10 Detroit Michigan, N USA 42.19N 83.03W
27 W5 Detroit Texas, SW USA 33.39N 95.16W
33 S10 Detroit ≈ Canada/USA
31 S6 Detroit Lakes Minnesota, N USA 46.49N 95.49W
33 S10 Detroit Metropolitan ✕ Michigan, N USA 42.12N 83.16W
Detta see Deta
178 J11 Det Udom Ubon Ratchathani, E Thailand 14.54N 105.03E
113 K20 Detva Hung. Gyeva. Banskobystrický Kraj, C Slovakia 48.34N 19.25E
160 G13 Deūlgaon Rāja Mahārāshtra, C India 20.04N 76.08E
101 L15 Deurne Noord-Brabant, SE Netherlands 51.28N 5.46E
101 H16 Deurne ✕ (Antwerpen) Antwerpen, N Belgium 51.10N 4.28E
Deutsch-Brod see Havlíčkův Brod
Deutschendorf see Poprad
Deutsch-Eylau see Iława
189 Y6 Deutschkreutz Burgenland, E Austria 47.36N 16.38E
Deutsch Krone see Wałcz
Deutschland/Deutschland, Bundesrepublik see Germany
189 V9 Deutschlandsberg Steiermark, SE Austria 46.52N 15.13E
Deutsch-Südwestafrika see Namibia
189 Y3 Deutsch-Wagram Niederösterreich, E Austria 48.19N 16.33E
Deux-Ponts see Zweibrücken
12 I11 Deux Rivieres Ontario, SE Canada 46.13N 78.16W
104 K9 Deux-Sèvres ◐ department W France
118 G11 Deva Ger. Diemrich, Hung. Déva. Hunedoara, W Romania 45.55N 22.54E
Deva see Chester
Devana see Aberdeen
Devana Castra see Chester
Devdelija see Gevgelija
142 L12 Deveci Dağları ▲▲ N Turkey
143 P15 Devegeçidi Barajı ⊠ SE Turkey
142 K15 Develi Kayseri, C Turkey 38.22N 35.28E
100 M11 Deventer Overijssel, E Netherlands 52.15N 6.10E
12 O10 Devenyns, Lac ◎ Quebec, SE Canada
98 K8 Deveron ≈ NE Scotland, UK
159 R14 Devghar prev. Deoghar. Bihār, NE India
29 R10 Devil's Den plateau Arkansas, C USA
37 R7 Devils Gate pass California, W USA 38.20N 119.23W
32 J2 Devils Island island Apostle Islands, Wisconsin, N USA
Devil's Island see Diable, Île du
31 P3 Devils Lake North Dakota, N USA 48.07N 98.49W
33 R10 Devils Lake ◎ Michigan, N USA
31 O3 Devils Lake ◎ North Dakota, N USA
37 W13 Devils Playground desert California, W USA
27 O11 Devils River ≈ Texas, SW USA
35 Y12 Devils Tower ▲ Wyoming, C USA 44.33N 104.45W
116 I11 Devin prev. Dovlen. Smolyan, SW Bulgaria 41.45N 24.24E
27 R12 Devine Texas, SW USA 29.08N 98.54W
158 H13 Devli Rājasthān, N India 25.46N 75.22E
Devne see Devnya
116 N8 Devnya prev. Devne. Varna, E Bulgaria 43.13N 27.36E
33 U14 Devola Ohio, N USA 39.28N 81.28W
Devoll see Devollit, Lumi i
115 M21 Devollit, Lumi i var. Devoll. ≈ SE Albania
9 Q14 Devon Alberta, SW Canada 53.21N 113.47W
99 I23 Devon cultural region SW England, UK
207 N10 Devon Island prev. North Devon Island. island Parry Islands, Nunavut, NE Canada
191 O16 Devonport Tasmania, SE Australia 41.14S 146.20E
142 H11 Devrek Zonguldak, N Turkey 41.13N 31.57E
160 G10 Dewās Madhya Pradesh, C India 22.58N 76.03E
De Westerein see Zwaagwesteinde
29 P8 Dewey Oklahoma, C USA 36.48N 95.56W
Dewey see Culebra
100 M8 De Wijk Drenthe, NE Netherlands 52.41N 6.13E
29 W12 De Witt Arkansas, C USA 34.17N 91.20W
31 Z14 De Witt Iowa, C USA 41.49N 90.32W
31 R16 De Witt Nebraska, C USA 40.23N 96.55W
99 M17 Dewsbury N England, UK 53.42N 1.37W
167 Q10 Dexing Jiangxi, S China 28.49N 117.37E
29 Y8 Dexter Missouri, C USA 36.48N 89.57W
39 U14 Dexter New Mexico, SW USA 33.12N 104.25W
166 I8 Deyang Sichuan, C China 31.07N 104.22E
190 C4 Dey-Dey, Lake salt lake South Australia
149 S7 Deyhūk Khorāsān, E Iran 33.18N 57.30E
152 K12 Deynau var. Dyanev, Turkm. Dänew. Lebapskiy Velayat, NE Turkmenistan 39.16N 63.09E
148 L8 Dezfūl var. Dizful. Khūzestān, SW Iran 32.22N 48.28E
133 X4 Dezhneva, Mys headland NE Russian Federation 66.07N 69.40W
167 P4 Dezhou Shandong, E China 37.28N 116.18E
Dezh Shāhpūr see Marīvān
Dhahran see Az̧ Z̧ahrān
Dhahran Al Khobar see Az̧ Z̧ahrān al Khubar
159 U14 Dhaka prev. Dacca. ● (Bangladesh) Dhaka, C Bangladesh 23.42N 90.22E
159 T15 Dhaka ◐ division C Bangladesh
Dhali see Idálion
147 O15 Dhamār W Yemen 14.31N 44.25E
Dhambul see Taraz
160 K12 Dhamtari Madhya Pradesh, C India 20.43N 81.36E
159 Q15 Dhanbād Bihār, NE India 23.48N 86.27E
158 L10 Dhangadhi var. Dhangarhi. Far Western, W Nepal 28.45N 80.38E
Dhangarhi see Dhangadhi
Dhank see Ḑank
159 R12 Dhankuta Eastern, E Nepal 27.06N 87.21E
158 I6 Dhaola Dhār ▲▲ NE India
160 F10 Dhār Madhya Pradesh, C India 22.36N 75.23E
159 R12 Dharan var. Dharan Bazar. Eastern, E Nepal 26.51N 87.18E
161 H21 Dhārāpuram Tamil Nādu, SE India 10.45N 77.33E
161 H20 Dharmapuri Tamil Nādu, SE India 12.10N 78.07E
161 H18 Dharmavaram Andhra Pradesh, E India 14.27N 77.43E
160 M11 Dharmjaygarh Madhya Pradesh, C India 22.27N 83.16E
Dharmsāla see Dharmshāla
158 I7 Dharmshāla prev. Dharmsāla. Himāchal Pradesh, N India 32.13N 76.24E
161 F17 Dhārwād prev. Dharwar. Karnātaka, SW India 15.30N 75.04E
Dharwar see Dhārwād
159 O10 Dhaulāgiri ▲ C Nepal 28.45N 83.27E
83 L18 Dheere Laaq var. Lak Dera, It. Lach Dera. seasonal river Kenya/Somalia
124 O3 Dhekéleia Sovereign Base Area UK military installation E Cyprus 34.59N 33.45E
124 O3 Dhekélia Eng. Dhekelia. Gk. Dekéleia. UK air base SE Cyprus 35.00N 33.45E
Dhelvinákion see Delvinaki
115 M22 Dhëmbelit, Majae ▲ S Albania 40.10N 20.22E
160 O12 Dhenkānāl Orissa, E India 20.40N 85.36E
Dheskáti see Deskáti
144 G11 Dhībān 'Ammān, NW Jordan 31.30N 35.46E
Dhidhimótikhon see Didymóteicho
Dhíkti Ori see Díkti
144 I12 Dhirwah, Wādī adh dry watercourse C Jordan
Dhístomon see Dístomo
Dhodhekánisos see Dodekánisos
Dhodhóni see Dodóni
Dhofar see Z̧ufār
Dhomokós see Domokós
Dhond see Daund
161 H17 Dhone Andhra Pradesh, C India 15.25N 77.52E
160 B11 Dhorāji Gujarāt, W India 21.43N 70.27E
Dhráma see Dráma
160 C10 Dhrāngadhra Gujarāt, W India 22.58N 71.31E
Dhrepanon, Akrotírio see Drépano, Akrotírio
159 T13 Dhuburi Assam, NE India 26.06N 89.55E
160 F12 Dhule prev. Dhulia. Mahārāshtra, C India 20.54N 74.46E
Dhulia see Dhule
Dhún Dealgan, Cuan see Dundalk Bay
Dhún Droma, Cuan see Dundrum Bay
Dhún na nGall, Bá see Donegal Bay
Dhū Shaykh see Qazānīyah
82 Q13 Dhuudo Bari, NE Somalia 9.21N 50.19E
83 N15 Dhuusa Marreeb var. Dusa Marreb, It. Dusa Mareb. Galguduud, C Somalia 5.33N 46.24E
117 J24 Día island SE Greece
57 Y9 Diable, Île du var. Devil's Island. island N French Guiana
13 N12 Diable, Rivière du ≈ Quebec, SE Canada
37 N8 Diablo, Mount ▲ California, W USA 37.52N 121.57W
37 O9 Diablo Range ▲▲ California, W USA
26 I8 Diablo, Sierra ▲▲ Texas, SW USA
47 O11 Diablotins, Morne ▲ N Dominica 15.30N 61.23W
79 N11 Diafarabé Mopti, C Mali 14.12N 5.01W
79 N11 Diaka ≈ SW Mali
Diakovár see Đakovo
78 I12 Dialakoto S Senegal 13.21N 13.19W
63 B18 Diamante Entre Ríos, E Argentina 32.04S 60.40W
64 I12 Diamante, Río ≈ C Argentina
61 M19 Diamantina Minas Gerais, SE Brazil 18.16S 43.37W
61 N17 Diamantina, Chapada ▲▲ E Brazil
181 U11 Diamantina Fracture Zone tectonic feature E Indian Ocean
189 T8 Diamantina River ≈ Queensland/South Australia
40 D9 Diamond Head headland Oahu, Hawaii, USA, C Pacific Ocean 21.15N 157.48W
39 P2 Diamond Peak ▲ Colorado, C USA 40.56N 108.56W
37 W5 Diamond Peak ▲ Nevada, W USA 39.34N 115.46W
Diamond State see Delaware
78 J11 Diamou Kayes, SW Mali 14.04N 11.16W
97 I23 Dianalund Vestsjælland, C Denmark 55.31N 11.30E
67 G25 Diana's Peak ▲ C Saint Helena
166 M16 Dianbai Guangdong, S China 21.33N 110.58E
166 G13 Dian Chi ◎ SW China
108 B10 Diano Marina Liguria, NW Italy 43.55N 8.06E
79 R13 Diapaga E Burkina 12.04N 1.47E
Diarbekr see Diyarbakır
109 J15 Diavolo, Passo del pass C Italy 41.55N 13.42E
63 B18 Díaz Santa Fe, C Argentina 32.22S 61.04W
147 W6 Dibā al Ḩiṣn var. Dibāh, Dibba. Ash Shāriqah, NE UAE 25.34N 56.16E
245 S3 Dibaga N Iraq 35.51N 43.49E
Dibāh see Dibā al Ḩiṣn
81 L22 Dibaya Kasai Occidental, S Dem. Rep. Congo (Zaire) 6.31S 22.57E
Dibba see Dibā al Ḩiṣn
205 W15 Dibble Iceberg Tongue ice feature Antarctica
115 L19 Dibër ◐ district E Albania
85 I20 Dibete Central, SE Botswana 23.45S 26.29E
27 W9 Diboll Texas, SW USA 31.11N 94.46W
Dibra see Debar
159 X11 Dibrugarh Assam, NE India 27.29N 94.49E
56 G4 Dibulla La Guajira, N Colombia 11.14N 73.22W
27 O5 Dickens Texas, SW USA 33.37N 100.50W
21 R2 Dickey Maine, NE USA 47.04N 69.05W
32 K9 Dickeyville Wisconsin, N USA 42.37N 90.36W
30 K5 Dickinson North Dakota, N USA 46.54N 102.48W
2 E6 Dickins Seamount undersea feature NE Pacific Ocean
29 O13 Dickson Oklahoma, C USA 34.11N 96.58W
22 I9 Dickson Tennessee, S USA 36.04N 87.23W
Dicle see Tigris
Dicsőszentmárton see Târnăveni
100 M12 Didam Gelderland, E Netherlands 51.55N 6.07E
169 Y8 Didao Heilongjiang, NE China 45.20N 130.54E
78 L12 Didiéni Koulikoro, W Mali 13.48N 8.01W
Didimo see Dídymo
Didimotiho see Didymóteicho
83 K17 Didimtu spring/well NE Kenya 2.58N 40.07E
69 U9 Didinga Hills ▲▲ S Sudan
9 Q16 Didsbury Alberta, SW Canada 51.39N 114.09W
158 G11 Didwāna Rājasthān, N India 27.22N 74.36E
117 G20 Dídymo var. Didimo. ▲ S Greece 37.28N 23.12E
116 L12 Didymóteicho var. Dhidhimótikhon, Didimotiho. Anatolikí Makedonía kai Thráki, NE Greece 41.22N 26.28E
105 S13 Die Drôme, E France 44.46N 5.21E
79 O13 Diébougou SW Burkina 11.00N 3.12W
Diedenhofen see Thionville
9 S16 Diefenbaker, Lake ⊠ Saskatchewan, S Canada
64 H7 Diego de Almagro Atacama, N Chile 26.24S 70.10W
65 F23 Diego de Almagro, Isla island S Chile
63 A20 Diego de Alvear Santa Fe, C Argentina 34.24S 62.04W
181 Q7 Diego Garcia island S British Indian Ocean Territory
Diégo-Suarez see Antsirañana
101 M23 Diekirch Diekirch, C Luxembourg 49.52N 6.10E
101 L23 Diekirch ◐ district N Luxembourg
78 K11 Diéma Kayes, W Mali 14.28N 9.10W
103 H15 Diemel ≈ W Germany
100 I10 Diemen Noord-Holland, C Netherlands 52.21N 4.58E
Diemrich see Deva
178 I6 Điện Biên var. Bien Bien, Dien Bien Phu. Lai Châu, N Vietnam 21.22N 103.01E
Dien Bien Phu see Điện Biên
178 J8 Diễn Châu Nghê An, N Vietnam 18.54N 105.35E
101 K18 Diepenbeek Limburg, NE Belgium 50.54N 5.25E
100 N11 Diepenheim Overijssel, E Netherlands 52.10N 6.37E
100 M10 Diepenveen Overijssel, E Netherlands 52.18N 6.09E
102 G12 Diepholz Niedersachsen, NW Germany 52.36N 8.22E
104 M3 Dieppe Seine-Maritime, N France 49.55N 1.04E
100 M12 Dieren Gelderland, E Netherlands 52.03N 6.06E
29 S13 Dierks Arkansas, C USA 34.07N 94.01W
101 J17 Diest Vlaams Brabant, C Belgium 50.58N 5.03E
110 F7 Dietikon Zürich, NW Switzerland 47.24N 8.25E
105 R13 Dieulefit Drôme, E France 44.30N 5.01E
105 T5 Dieuze Moselle, NE France 48.49N 6.41E
121 H15 Dieveniškės Šalčininkai, SE Lithuania 54.12N 25.38E
100 N7 Diever Drenthe, NE Netherlands 52.49N 6.19E
103 F17 Diez Rheinland-Pfalz, W Germany 50.22N 8.01E
79 Y12 Diffa Diffa, SE Niger 13.20N 12.39E
79 Y10 Diffa ◐ department SE Niger
101 L25 Differdange Luxembourg, SW Luxembourg 49.31N 5.52E
11 O16 Digby Nova Scotia, SE Canada 44.37N 65.46W
28 J5 Dighton Kansas, C USA 38.28N 100.28W
Dignano d'Istria see Vodnjan
105 T14 Digne var. Digne-les-Bains. Alpes-de-Haute-Provence, SE France 44.04N 6.13E
Digne-les-Bains see Digne
Digoel see Digul, Sungai
105 Q10 Digoin Saône-et-Loire, C France 46.30N 3.59E
179 Rr16 Digos Mindanao, S Philippines 6.46N 125.21E
155 Q16 Digri Sind, SE Pakistan 25.10N 69.10E
176 Z13 Digul Barat, Sungai ≈ Irian Jaya, E Indonesia
176 Z14 Digul, Sungai prev. Digoel. ≈ Irian Jaya, E Indonesia
176 Z13 Digul Timur, Sungai ≈ Irian Jaya, E Indonesia
Dihang see Brahmaputra
159 X10 Dihang ≈ NE India
Dihōk see Dahūk
83 L17 Diinsoor Bay, S Somalia 2.28N 43.00E
Dijlah see Tigris
101 H17 Dijle ≈ C Belgium
105 R8 Dijon anc. Dibio. Côte d'Or, C France 47.21N 5.03E
95 H14 Dikanäs Västerbotten, N Sweden 65.15N 16.00E
82 L12 Dikhil SW Djibouti 11.07N 42.18E
142 B13 Dikili İzmir, W Turkey 39.04N 26.52E
101 B17 Diksmuide var. Dixmuide, Fr. Dixmude. West-Vlaanderen, W Belgium 51.01N 2.52E
126 Hh6 Dikson Taymyrskiy (Dolgano-Nenetskiy) Avtonomnyy Okrug, N Russian Federation 73.30N 80.35E
117 K25 Díkti var. Dhíkti Ori. ▲ Kríti, Greece, E Mediterranean Sea
79 Z13 Dikwa Borno, NE Nigeria 12.00N 13.57E
83 J15 Dila Southern, S Ethiopia 6.19N 38.16E
101 G18 Dilbeek Vlaams Brabant, C Belgium 50.51N 4.16E
175 S16 Dili var. Dilli, Dilly. ○ (East Timor) N East Timor 8.33S 125.34E
79 Y11 Dilia var. Dillia. ≈ SE Niger
Dilijan see Delijan
178 K13 Di Linh Lâm Đông, S Vietnam 11.34N 108.04E
103 G16 Dillenburg Hessen, W Germany 50.45N 8.16E
27 Q13 Dilley Texas, SW USA 28.40N 99.10W
Dilli see Delhi, India
Dilli see Dili, East Timor
Dillia see Dilia
82 E11 Dilling var. Ad Dalanj. Southern Kordofan, C Sudan 12.01N 29.40E
103 D20 Dillingen Saarland, SW Germany 49.20N 6.43E
Dillingen see Dillingen an der Donau
103 J22 Dillingen an der Donau var. Dillingen. Bayern, S Germany 48.34N 10.29E
41 O13 Dillingham Alaska, USA 59.03N 158.30W
35 Q12 Dillon Montana, NW USA 45.13N 112.37W
23 T12 Dillon South Carolina, SE USA 34.25N 79.22W
33 T13 Dillon Lake ⊠ Ohio, N USA
Dilly see Dili
Dilman see Salmās
81 K24 Dilolo Katanga, S Dem. Rep. Congo (Zaire) 10.42S 22.21E
117 J20 Dílos island Kykládes, Greece, Aegean Sea
147 Y11 Dil', Ra's ad headland E Oman 19.12N 57.53E
31 R5 Dilworth Minnesota, N USA 46.53N 96.38W
144 H7 Dimashq var. Ash Shām, Esh Sham, Eng. Damascus, Fr. Damas, It. Damasco. ● (Syria) Dimashq, SW Syria 33.30N 36.19E
144 I8 Dimashq off. Muḩāfaz̧at Dimashq, var. Damascus, Ar. Ash Sham, Ash Shām, Damasco, Esh Sham, Fr. Damas. ◐ governorate S Syria
144 I7 Dimashq ✕ Dimashq, S Syria 33.30N 36.19E
81 L21 Dimbelenge Kasai Occidental, C Dem. Rep. Congo (Zaire) 5.36S 23.04E
79 N16 Dimbokro E Ivory Coast 6.39N 4.43W
190 L11 Dimboola Victoria, SE Australia 36.29S 142.03E
Dimbovița see Dâmbovița
Dimitrov see Dymytrov
116 K11 Dimitrovgrad Khaskovo, S Bulgaria 42.03N 25.36E
131 R5 Dimitrovgrad Ul'yanovskaya Oblast', W Russian Federation 54.14N 49.37E
115 Q15 Dimitrovgrad prev. Caribrod. Serbia, SE Yugoslavia 43.01N 22.46E
Dimitrovo see Pernik
Dimlang see Vogel Peak
26 M3 Dimmitt Texas, SW USA 34.33N 102.18W
116 F7 Dimovo Vidin, NW Bulgaria 43.46N 22.46E
61 A16 Dimpolis Acre, W Brazil 9.52S 71.51W
117 O23 Dimyliá Ródos, Dodekánisos, Greece, Aegean Sea 36.17N 27.59E
179 R13 Dinagat Dinagat Island, S Philippines 10.00N 125.36E
179 Rr13 Dinagat Island island S Philippines
159 S13 Dinajpur Rajshahi, NW Bangladesh 25.37N 88.39E
104 I6 Dinan Côtes d'Armor, NW France 48.27N 2.01W
101 I21 Dinant Namur, S Belgium 50.16N 4.55E
142 E15 Dinar Afyon, SW Turkey 38.04N 30.09E
114 F13 Dinara ▲ W Croatia 43.49N 16.42E
Dinara see Dinaric Alps
104 I5 Dinard Ille-et-Vilaine, NW France 48.38N 2.04W
114 F13 Dinaric Alps var. Dinara. ▲▲ Bosnia and Herzegovina/Croatia
149 N10 Dīnār, Kūh-e ▲ C Iran 30.51N 51.36E
Dinbych see Denbigh
161 H22 Dindigul Tamil Nādu, SE India 10.23N 78.00E
85 M19 Dindiza Gaza, S Mozambique 23.22S 33.28E
155 V7 Dinga Punjab, E Pakistan 32.37N 73.45E
81 H21 Dinga Bandundu, SW Dem. Rep. Congo (Zaire) 5.00S 16.29E
164 L16 Dinggyê Xizang Zizhiqu, W China 28.18N 88.06E
99 A20 Dingle Ir. An Daingean. SW Ireland 52.08N 10.16W
99 A20 Dingle Bay Ir. Bá an Daingin. bay SW Ireland
20 I13 Dingmans Ferry Pennsylvania, NE USA 41.12N 74.51W
103 N22 Dingolfing Bayern, SE Germany 48.37N 12.28E
179 P8 Dingras Luzon, N Philippines 18.06N 120.43E
78 J13 Dinguiraye Haute-Guinée, N Guinea 11.19N 10.49W
98 I8 Dingwall N Scotland, UK 57.36N 4.25W
165 V10 Dingxi Gansu, C China 35.36N 104.33E
167 Q7 Dingyuan Anhui, E China 32.30N 117.40E
167 O3 Dingzhou prev. Ding Xian. Hebei, E China 38.31N 114.52E
178 K6 Đinh Lập Lang Sơn, N Vietnam 21.33N 107.03E
178 K14 Đinh Quan Đông Nai, S Vietnam 11.11N 107.20E
102 E13 Dinkel ≈ Germany/Netherlands
103 J21 Dinkelsbühl Bayern, S Germany 49.06N 10.18E
103 D14 Dinslaken Nordrhein-Westfalen, W Germany 51.34N 6.43E
37 R11 Dinuba California, W USA 36.32N 119.23W
23 W7 Dinwiddie Virginia, NE USA 37.04N 77.34W
100 N13 Dinxperlo Gelderland, E Netherlands 51.51N 6.30E
117 F14 Dió anc. Dium. site of ancient city Kentrikí Makedonía, N Greece 40.13N 22.30E
Diófás see Nucet
78 M12 Dioïla Koulikoro, W Mali 12.28N 6.43W
117 G19 Dióryga Korinthou Eng. Corinth Canal. canal S Greece
78 G12 Diouloulou SW Senegal 13.00N 16.34W
79 N11 Dioura Mopti, W Mali 14.48N 5.20W
78 G11 Diourbel W Senegal 14.38N 16.12W
158 L10 Dipayal Far Western, W Nepal 29.09N 80.46E
124 Oo2 Dipkarpaz Gk. Rizokarpaso, Rizokárpason. NE Cyprus 35.36N 34.23E
155 R17 Diplo Sind, SE Pakistan 24.29N 69.36E
179 Qq15 Dipolog var. Dipolog City. Mindanao, S Philippines 8.31N 123.20E
193 C23 Dipton Southland, South Island, NZ 45.55S 168.21E
79 O10 Diré Tombouctou, C Mali 16.12N 3.31W
82 L13 Dirē Dawa Dirē Dawa, E Ethiopia 9.34N 41.53E
Dirfis see Dírfys
117 H18 Dírfys var. Dirfis. ▲▲ Évvoia, C Greece
77 N9 Dirj var. Daraj, Darj. W Libya 30.09N 10.25E
188 G10 Dirk Hartog Island island Western Australia
79 Y8 Dirkou Agadez, NE Niger 18.45N 13.00E
189 X11 Dirranbandi Queensland, E Australia 28.37S 148.13E
83 O16 Dirri Galguduud, C Somalia 4.15N 46.31E
Dirschau see Tczew
39 N6 Dirty Devil River ≈ Utah, W USA
34 E10 Disappointment, Cape headland Washington, NW USA 46.16N 124.06W
188 L8 Disappointment, Lake salt lake Western Australia
191 R12 Disaster Bay bay New South Wales, SE Australia
46 J11 Discovery Bay C Jamaica 18.27N 77.23W
190 K13 Discovery Bay inlet SE Australia
69 Y15 Discovery II Fracture Zone tectonic feature SW Indian Ocean
Discovery Seamount/Discovery Seamounts see Discovery Tablemount
67 O19 Discovery Tablemount var. Discovery Seamount, Discovery Seamounts. undersea feature SW Indian Ocean 42.00S 0.10E
110 G9 Disentis Rmsch. Mustér. Graubünden, S Switzerland 46.43N 8.52E
41 O10 Dishna River ≈ Alaska, USA
205 X4 Dismal Mountains ▲▲ Antarctica
30 M14 Dismal River ≈ Nebraska, C USA
Disna see Dzisna
101 I19 Dison Liège, E Belgium 50.37N 5.52E
159 V12 Dispur Assam, NE India 26.03N 91.52E
13 R11 Disraeli Quebec, SE Canada 45.58N 71.21W
117 F18 Dístomo prev. Dhístomon. Stereá Ellás, C Greece 38.25N 22.40E
117 H18 Dístos, Límni ◎ Évvoia, C Greece
61 L18 Distrito Federal Eng. Federal District. ◐ federal district C Brazil
43 P14 Distrito Federal ◐ federal district S Mexico
56 L4 Distrito Federal off. Territorio Distrito Federal. ◐ federal district N Venezuela
Distrito Federal, Territorio see Distrito Federal
118 J10 Ditrău Hung. Ditró. Harghita, C Romania 46.47N 25.30E
Ditró see Ditrău
160 B12 Diu Damān and Diu, W India 20.42N 70.58E
179 Rr14 Diuata Mountains ▲▲ Mindanao, S Philippines
Dium see Dió
189 S13 Divača SW Slovenia 45.40N 13.58E
104 K5 Dives ≈ N France
Divichi see Däväçi
35 Q11 Divide Montana, NW USA 45.44N 112.47W
Divin see Dzivin
85 N18 Divinhe Sofala, E Mozambique 20.41S 34.46E
61 L20 Divinópolis Minas Gerais, SE Brazil 20.07S 44.55W
126 I14 Divnogorsk Krasnoyarskiy Kray, S Russian Federation 55.53N 92.15E
131 N13 Divnoye Stavropol'skiy Kray, SW Russian Federation 45.54N 43.18E
78 M17 Divo S Ivory Coast 5.49N 5.22W
Divodurum Mediomatricum see Metz
143 N13 Divriği Sivas, C Turkey 39.22N 38.06E
Dīwaniyah see Ad Dīwānīyah
12 J10 Dix Milles, Lac ◎ Quebec, SE Canada
12 M8 Dix Milles, Lac des ◎ Quebec, SE Canada
Dixmude/Dixmuide see Diksmuide
37 N7 Dixon California, W USA 38.19N 121.49W
32 L10 Dixon Illinois, N USA 41.51N 89.26W
23 I6 Dixon Kentucky, S USA 37.30N 87.39W
29 V6 Dixon Missouri, C USA 37.59N 92.05W
39 S9 Dixon New Mexico, SW USA 36.10N 105.49W
41 Y15 Dixon Entrance strait Canada/USA
20 D14 Dixonville Pennsylvania, NE USA 40.43N 79.01W
143 T13 Diyadin Ağrı, E Turkey 39.33N 43.40E
245 V5 Diyālá, Nahr var. Rudkhaneh-ye Sirvān, Sirwan. ≈ Iran/Iraq see also Sirvān, Rudkhaneh-ye
143 P15 Diyarbakır var. Diarbekr; anc. Amida. Dıyarbakır, SE Turkey 37.55N 40.13E
143 P15 Diyarbakır var. Diarbekr. ◐ province SE Turkey
Dizful see Dezfūl
81 F16 Dja ≈ SE Cameroon
Djadié see Zadié
79 X7 Djado Agadez, NE Niger 21.00N 12.11E
79 X6 Djado, Plateau du ▲▲ NE Niger
Djailolo see Halmahera, Pulau
Djajapura see Jayapura
Djakarta see Jakarta
Djakovica see Đakovica
Djakovo see Đakovo
81 G20 Djambala Plateaux, C Congo 2.31S 14.43E
Djambi see Jambi
Djambi see Hari, Batang, Sumatera, W Indonesia
76 M9 Djanet E Algeria 28.43N 8.57E
76 M11 Djanet prev. Fort Charlet. SE Algeria 24.34N 9.33E
Djatiwangi see Jatiwangi
Djaul see Dyaul Island
Djawa see Jawa
Djéblé see Jablah
80 I10 Djédaa Batha, C Chad 13.31N 18.34E
76 J6 Djelfa var. El Djelfa. N Algeria 34.42N 3.16E
81 M14 Djéma Haut-Mbomou, E Central African Republic 6.03N 25.19E
Djeneponto see Jeneponto
79 N12 Djenné var. Jenné. Mopti, C Mali 13.55N 4.31W
Djérablous see Jarābulus
Djerba see Jerba, Île de
81 F15 Djérem ≈ C Cameroon
Djevdjelija see Gevgelija
79 P11 Djibo N Burkina 14.09N 1.37W
82 L12 Djibouti var. Jibuti. ● (Djibouti) E Djibouti 11.32N 42.55E
82 L12 Djibouti off. Republic of Djibouti, var. Jibuti; prev. French Somaliland, French Territory of the Afars and Issas, Fr. Côte Française des Somalis, Territoire Français des Afars et des Issas. ◆ republic E Africa
82 L12 Djibouti ✕ C Djibouti 11.29N 42.54E
Djidjel/Djidjelli see Jijel
57 W10 Djoemoe Sipaliwini, C Suriname 4.00N 55.27W
Djokjakarta see Yogyakarta
81 K21 Djoku-Punda Kasai Occidental, S Dem. Rep. Congo (Zaire) 5.27S 20.58E
81 K18 Djolu Equateur, N Dem. Rep. Congo (Zaire) 0.42N 22.23E
Djorče Petrov see Đorče Petrov
81 F17 Djoua ≈ Congo/Gabon
79 R14 Djougou W Benin 9.42N 1.38E
81 F16 Djoum Sud, S Cameroon 2.38N 12.51E
80 I8 Djourab, Erg du dunes N Chad
81 P17 Djugu Orientale, NE Dem. Rep. Congo (Zaire) 1.55N 30.31E
Djumbir see Ďumbier
94 L3 Djúpivogur Austurland, SE Iceland 64.39N 14.18W
96 L13 Djura Dalarna, C Sweden 60.37N 15.00E
Djurdjevac see Đurđevac
85 G18 D'Kar Ghanzi, NW Botswana 21.31S 21.55E
207 U6 Dmitriya Lapteva, Proliv strait N Russian Federation
130 J7 Dmitriyev-L'govskiy Kurskaya Oblast', W Russian Federation 52.08N 35.09E
Dmitriyevsk see Makiyivka
130 K3 Dmitrov Moskovskaya Oblast', W Russian Federation 56.23N 37.30E
Dmitrovichi see Dzmitravichy
130 J6 Dmitrovsk-Orlovskiy Orlovskaya Oblast', W Russian Federation 52.28N 35.01E
119 R3 Dmytrivka Chernihivs'ka Oblast', N Ukraine 50.56N 32.57E
Dnepr see Dnieper
Dneprodzerzhinsk see Dniprodzerzhyns'k
Dneprodzerzhinskoye Vodokhranilishche see Dniprodzerzhyns'ke Vodoskhovyshche
Dnepropetrovsk see Dnipropetrovs'k
Dnepropetrovskaya Oblast' see Dnipropetrovs'ka Oblast'
Dneprorudnoye see Dniprorudne
Dneprovskiy Liman see Dniprovs'kyy Lyman
Dneprovsko-Bugskiy Kanal see Dnyaprowska-Buhski, Kanal
Dnestr see Dniester
Dnestrovskiy Liman see Dnistrovs'kyy Lyman
88 H11 Dnieper Bel. Dnyapro, Rus. Dnepr, Ukr. Dnipro. ≈ E Europe
119 P3 Dnieper Lowland Bel. Prydnyaprowskaya Nizina, Ukr. Prydniprovs'ka Nyzovyna. lowlands Belarus/Ukraine
118 M8 Dniester Rom. Nistru, Rus. Dnestr, Ukr. Dnister; anc. Tyras. ≈ Moldova/Ukraine
Dnipro see Dnieper
119 T7 Dniprodzerzhyns'k Rus. Dneprodzerzhinsk; prev. Kamenskoye. Dnipropetrovs'ka Oblast', E Ukraine 48.30N 34.35E
119 T7 Dniprodzerzhyns'ke Vodoskhovyshche Rus. Dneprodzerzhinskoye Vodokhranilishche. ⊠ C Ukraine
119 U7 Dnipropetrovs'k Rus. Dnepropetrovsk; prev. Yekaterinoslav. Dnipropetrovs'ka Oblast', E Ukraine 48.28N 34.59E
119 U8 Dnipropetrovs'k ✕ Dnipropetrovs'ka Oblast', S Ukraine 48.20N 35.04E
Dnipropetrovs'k see Dnipropetrovs'ka Oblast'
119 T7 Dnipropetrovs'ka Oblast' var. Dnipropetrovs'k, Rus. Dnepropetrovskaya Oblast'. ◐ province E Ukraine
119 U9 Dniprorudne Rus. Dneprorudnoye. Zaporiz'ka Oblast', SE Ukraine 47.21N 35.00E
119 Q11 Dniprovs'kyy Lyman Rus. Dneprovskiy Liman. bay S Ukraine
Dnister see Dniester
119 O11 Dnistrovs'kyy Lyman Rus. Dnestrovskiy Liman. inlet S Ukraine
128 G14 Dno Pskovskaya Oblast', W Russian Federation 57.48N 29.58E
Dnyapro see Dnieper
121 H20 Dnyaprowska-Buhski, Kanal Rus. Dneprovsko-Bugskiy Kanal. canal SW Belarus
11 O14 Doaktown New Brunswick, SE Canada 46.34N 66.06W
80 H13 Doba Logone-Oriental, S Chad 8.40N 16.49E
120 E9 Dobele Ger. Doblen. Dobele, W Latvia 56.36N 23.14E
103 N16 Döbeln Sachsen, E Germany 51.07N 13.07E
176 Vv9 Doberai, Jazirah Dut. Vogelkop. peninsula Irian Jaya, E Indonesia
112 F10 Dobiegniew Ger. Lubuskie, W Poland 52.58N 15.43E
Doblen see Dobele
83 K18 Dobli spring/well SW Somalia 0.24N 41.18E
176 W13 Dobo Pulau Wamar, E Indonesia 5.45S 134.12E
114 H11 Doboj Republika Srpska, N Bosnia and Herzegovina 44.45N 18.03E
112 L8 Dobre Miasto Ger. Guttstadt. Warmińsko-Mazurskie, NE Poland 53.59N 20.25E
116 N7 Dobrich Rom. Bazargic; prev. Tolbukhin. Dobrich, NE Bulgaria 43.34N 27.49E
116 N7 Dobrich ◐ province NE Bulgaria
130 M8 Dobrinka Lipetskaya Oblast', W Russian Federation 52.10N 40.30E
130 M7 Dobrinka Volgogradskaya Oblast', SW Russian Federation 50.52N 41.48E
Dobrla Vas see Eberndorf
113 I15 Dobrodzień Ger. Guttentag. Opolskie, S Poland 50.43N 18.24E
Dobrogea see Dobruja
119 W7 Dobropillya Rus. Dobropol'ye. Donets'ka Oblast', SE Ukraine 48.29N 37.06E
Dobropol'ye see Dobropillya
119 P8 Dobrovelychkivka Kirovohrads'ka Oblast', C Ukraine 48.22N 31.12E
Dobrudja/Dobrudzha see Dobruja
116 O7 Dobruja var. Dobrudja, Bul. Dobrudzha, Rom. Dobrogea. physical region Bulgaria/Romania
121 P19 Dobrush Homyel'skaya Voblasts', SE Belarus 52.25N 31.21E
129 U14 Dobryanka Permskaya Oblast', NW Russian Federation 58.28N 56.27E
119 P2 Dobryanka Chernihivs'ka Oblast', N Ukraine 52.03N 31.09E
Dobryn' see Dabryn'
23 R8 Dobson North Carolina, SE USA 36.30N 80.54W
61 N20 Doce, Rio ≈ SE Brazil

◆ Country ● Country Capital ◇ Dependent Territory ○ Dependent Territory Capital ◐ Administrative Region ✕ International Airport ▲ Mountain ▲▲ Mountain Range ⛰ Volcano ≈ River ◎ Lake ⊠ Reservoir

95 I16 **Docksta** Västernorrland, C Sweden 63.06N 18.22E
43 N10 **Doctor Arroyo** Nuevo León, NE Mexico 23.40N 100.09W
64 L4 **Doctor Pedro P. Peña** Boquerón, W Paraguay 22.25 62.22W
175 T7 **Dodaga** Pulau Halmahera, E Indonesia 1.06N 128.10E
161 G21 **Dodda Betta** ▲ S India 11.28N 76.44E
Dodecanese see Dodékánisos
117 M22 **Dodekánisos** var. Nóties Sporádes, Eng. Dodecanese; prev. Dhodhekánisos. island group SE Greece
28 J6 **Dodge City** Kansas, C USA 37.45N 100.01W
32 K9 **Dodgeville** Wisconsin, N USA 42.57N 90.07W
99 H25 **Dodman Point** headland SW England, UK 50.13N 4.47W
83 J14 **Dodola** Oromo, C Ethiopia 7.00N 39.15E
83 H22 **Dodoma** ● (Tanzania) Dodoma, C Tanzania 6.10S 35.45E
83 H22 **Dodoma** ◆ region C Tanzania
117 C16 **Dodóni** var. Dhodhóni. site of ancient city Ípeiros, W Greece 39.33N 20.47E
35 U7 **Dodson** Montana, NW USA 48.25N 108.18W
27 P3 **Dodson** Texas, SW USA 34.46N 100.01W
100 M12 **Doesburg** Gelderland, E Netherlands 52.01N 6.07E
100 N12 **Doetinchem** Gelderland, E Netherlands 51.58N 6.16E
164 L12 **Dogai Coring** var. Lake Montcalm. ◎ W China
143 N15 **Doğanşehir** Malatya, C Turkey 38.06N 37.52E
86 E9 **Dogger Bank** undersea feature C North Sea
25 S10 **Dog Lake** island Florida, SE USA
12 C7 **Dog Lake** ◎ Ontario, S Canada
108 B9 **Dogliani** Piemonte, NE Italy 44.33N 7.55E
170 G11 **Dōgo** island Oki-shotō, SW Japan
Do Gonbadān see Do Gonbadān
79 S12 **Dogondoutchi** Dosso, SW Niger 13.37N 4.03E
170 F13 **Dōgo-yama** var. Dōgo-yama. ▲ Kyūshū, SW Japan 35.03N 133.12E
Dogrular see Pravda
143 T13 **Doğubayazıt** Ağrı, E Turkey 39.33N 44.07E
143 P12 **Doğu Karadeniz Dağları** var. Anadolu Dağları. ▲ NE Turkey
Doha see Ad Dawḩah
Dohad see Dāhod
Dohuk see Dahūk
165 N16 **Doilungdêqên** Xizang Zizhiqu, W China 29.41N 90.58E
116 F12 **Doïráni, Límnis** Bul. Ezero Doyransko. ◎ N Greece
Doire see Londonderry
101 H22 **Doische** Namur, S Belgium 50.09N 4.43E
61 P13 **Dois de Julho** × (Salvador) Bahia, NE Brazil 12.54 38.58W
62 H12 **Dois Vizinhos** Paraná, S Brazil 25.47S 53.03W
82 H10 **Doka** Gedaref, E Sudan 13.30N 35.46E
Doka see Kéita, Bahr
245 T3 **Dokan** var. Dūkān. E Iraq 35.55N 44.58E
96 H13 **Dokka** Oppland, S Norway 60.49N 10.04E
100 L5 **Dokkum** Friesland, N Netherlands 53.20N 6.00E
100 L5 **Dokkumer Ee** ≈ N Netherlands
78 K13 **Doko** Haute-Guinée, NE Guinea 11.46N 8.58W
Dokshitsy see Dokshytsy
120 K13 **Dokshytsy** Rus. Dokshitsy. Vitsyebskaya Voblasts', N Belarus 54.54N 27.46E
119 X8 **Dokuchayevs'k** var. Dokuchayevsk. Donets'ka Oblast', SE Ukraine 47.43N 37.40E
Dolak, Pulau see Yos Sudarso, Pulau
31 N9 **Doland** South Dakota, N USA 44.51N 98.06W
65 J13 **Dolavón** Chaco, S Argentina 43.20S 65.42W
13 P6 **Dolbeau** Québec, SE Canada 48.52N 72.15W
104 I5 **Dol-de-Bretagne** Ille-et-Vilaine, NW France 48.33N 1.45W
66 J13 **Doldrums Fracture Zone** tectonic feature W Atlantic Ocean
108 K6 **Dôle** Jura, E France 47.04N 5.30E
99 J19 **Dolgellau** NW Wales, UK 52.44N 3.54W
Dolginovo see Dawhinava
Dolgi, Ostrov see Dolgiy, Ostrov
129 U2 **Dolgiy, Ostrov** var. Ostrov Dolgi. island NW Russian Federation
168 J9 **Dölgöön** Övörhangay, C Mongolia 45.57N 103.14E
109 C20 **Dolianova** Sardegna, Italy, C Mediterranean Sea 39.23N 9.08E
Dolina see Dolyna
127 Oo15 **Dolinsk** Ostrov Sakhalin, Sakhalinskaya Oblast', SE Russian Federation 47.20N 142.52E
Dolinskaya see Dolyns'ka
81 F21 **Dolisie** prev. Loubomo. Le Niari, S Congo 4.12S 12.40E
118 G14 **Dolj** ◆ county SW Romania
100 P5 **Dollard** bay NW Germany
204 J5 **Dolleman Island** island Antarctica
116 I8 **Dolni Dŭbnik** Pleven, N Bulgaria 43.24N 24.25E
116 F8 **Dolni Lom** Vidin, NW Bulgaria 43.31N 22.46E
Dolnja Lendava see Lendava
116 K9 **Dolno Panicherevo** var. Panicherevo. Sliven, C Bulgaria 42.36N 25.51E
113 F14 **Dolnośląskie** ◆ province SW Poland
113 K18 **Dolný Kubín** Hung. Alsókubin. Žilinský Kraj, N Slovakia 49.13N 19.16E

108 H8 **Dolo** Veneto, NE Italy 45.25N 12.06E
Dolomites/Dolomiti see Dolomitiche, Alpi
108 H6 **Dolomitiche, Alpi** var. Dolomiti, Eng. Dolomites. ▲ NE Italy
Dolonnur see Duolun
168 K10 **Doloon** Ömnögovi, S Mongolia 44.28N 105.22E
63 J23 **Dolores** Buenos Aires, E Argentina 36.21S 57.39W
44 E3 **Dolores** Petén, N Guatemala 16.33N 89.25W
179 R12 **Dolores** Samar, C Philippines 12.01N 125.27E
107 S12 **Dolores** País Valenciano, E Spain 38.09N 0.45W
63 D19 **Dolores** Soriano, SW Uruguay 33.34S 58.15W
43 N12 **Dolores Hidalgo** var. Ciudad de Dolores Hidalgo. Guanajuato, C Mexico 21.10N 100.55W
15 Hh3 **Dolphin and Union Strait** strait Northwest Territories / Nunavut, N Canada
67 D23 **Dolphin, Cape** headland East Falkland, Falkland Islands 51.15S 58.57W
46 H12 **Dolphin Head** hill W Jamaica 18.21N 78.08W
85 B21 **Dolphin Head** var. Cape Dernberg. headland SW Namibia 25.33S 14.36E
112 G12 **Dolsk** Ger. Dolzig. Wielkopolskie, C Poland 51.59N 17.03E
178 J8 **Đô Lương** Nghệ An, N Vietnam 18.51N 105.19E
118 I6 **Dolyna** Rus. Dolina. Ivano-Frankivs'ka Oblast', W Ukraine 48.58N 24.01E
119 R8 **Dolyns'ka** Rus. Dolinskaya. Kirovohrads'ka Oblast', S Ukraine 48.06N 32.46E
Dolzig see Dolsk
Domachëvo/Domaczewo see Damachava
119 P9 **Domanivka** Mykolayivs'ka Oblast', S Ukraine 47.40N 30.56E
159 S13 **Domar** Rajshahi, N Bangladesh 26.09N 88.49E
110 I9 **Domat/Ems** Graubünden, SE Switzerland 46.49N 9.28E
113 A18 **Domažlice** Ger. Taus. Plzeňský Kraj, W Czech Republic 49.26N 12.56E
131 X8 **Dombarovskiy** Orenburgskaya Oblast', W Russian Federation 50.53N 59.18E
96 G10 **Dombås** Oppland, S Norway 62.04N 9.07E
85 M17 **Dombe** Manica, C Mozambique 19.59S 33.24E
84 A13 **Dombe Grande** Benguela, C Angola 12.57S 13.07E
105 R10 **Dombes** physical region E France
176 Xx10 **Dombo** Irian Jaya, E Indonesia 1.52S 137.09E
113 I25 **Dombóvár** Tolna, S Hungary 46.24N 18.09E
101 D14 **Domburg** Zeeland, SW Netherlands 51.34N 3.30E
60 L13 **Dom Eliseu** Pará, NE Brazil 4.02S 47.31W
Domel Island see Letsôk-aw Kyun
105 O11 **Dôme, Puy de** ▲ C France 45.46N 3.00E
38 H13 **Dome Rock Mountains** ▲ Arizona, SW USA
Domesnes, Cape see Kolkasrags
64 G8 **Domeyko** Atacama, N Chile 28.57S 70.54W
64 N5 **Domeyko, Cordillera** ▲ N Chile
104 K5 **Domfront** Orne, N France 48.35N 0.39W
176 Xx10 **Dom, Gunung** ▲ Irian Jaya, E Indonesia 2.41S 137.00E
47 X11 **Dominica** off. Commonwealth of Dominica. ◆ republic E West Indies
49 S3 **Dominica** island Dominica
Dominica Channel see Martinique Passage
45 N15 **Dominical** Puntarenas, SE Costa Rica 9.16N 83.52W
47 X11 **Dominican Republic** ◆ republic C West Indies
47 X11 **Dominica Passage** passage E Caribbean Sea
101 K14 **Dommel** ≈ S Netherlands
83 O14 **Domo** Somali, E Ethiopia 7.53N 46.55E
130 L4 **Domodedovo** × (Moskva) Moskovskaya Oblast', W Russian Federation 55.19N 37.55E
108 C6 **Domodossola** Piemonte, NE Italy 46.07N 8.19E
117 F17 **Domokós** var. Dhomokós. Stereá Ellás, C Greece 39.07N 22.18E
180 I14 **Domoni** Anjouan, SE Comoros 12.15S 44.39E
63 G16 **Dom Pedrito** Rio Grande do Sul, S Brazil 30.55S 54.39W
Dompoe see Dompu
175 Oo16 **Dompu** prev. Dompoe. Sumbawa, C Indonesia 8.30S 118.28E
178 Hh11 **Don Muang** × (Krung Thep) Nonthaburi, C Thailand 13.51N 100.40E
64 H13 **Domuyo, Volcán** ▲ W Argentina 36.36S 70.22W
189 U11 **Domžale** Ger. Domschale. C Slovenia 46.09N 14.33E
131 O10 **Don** ≈ NW Russian Federation
98 K9 **Don** ≈ NE Scotland, UK
190 M11 **Donald** Victoria, SE Australia 36.25S 143.03E
24 J9 **Donaldsonville** Louisiana, S USA 30.06N 90.59W
25 S8 **Donalsonville** Georgia, SE USA 31.02N 84.52W
Donau see Danube
103 G23 **Donaueschingen** Baden-Württemberg, SW Germany 47.57N 8.30E
103 K22 **Donauwörth** Bayern, S Germany 48.43N 10.46E
189 U7 **Donawitz** Steiermark, SE Austria 47.23N 15.00E
119 X7 **Donbass** industrial region Russian Federation/Ukraine

106 K11 **Don Benito** Extremadura, W Spain 38.57N 5.52W
99 M17 **Doncaster** anc. Danum. N England, UK 53.31N 1.07W
46 K12 **Don Christophers Point** headland C Jamaica 18.19N 76.48W
57 V9 **Donderkamp** Sipaliwini, NW Suriname 5.18N 56.22W
84 B12 **Dondo** Cuanza Norte, NW Angola 9.40S 14.24E
175 Q9 **Dondo** Sulawesi, N Indonesia 0.54S 121.33E
85 N17 **Dondo** Sofala, C Mozambique 19.36S 34.46E
161 K26 **Dondra Head** headland S Sri Lanka 5.57N 80.33E
Donduşani var. Donduşeni
118 M8 **Donduşeni** var. Donduşani, Rus. Dondyushany. N Moldova 48.13N 27.38E
Dondyushany see Donduşeni
99 D15 **Donegal** Ir. Dún na nGall. NW Ireland 54.39N 8.06W
99 D14 **Donegal** Ir. Dún na nGall. cultural region NW Ireland
99 C15 **Donegal Bay** Ir. Bá Dhún na nGall. bay NW Ireland
86 K10 **Donets** ≈ Russian Federation/Ukraine
119 X8 **Donets'k** Donetsk; prev. Stalino. Donets'ka Oblast', E Ukraine 48.03N 37.44E
119 W8 **Donets'k** × Donets'ka Oblast', E Ukraine 48.03N 37.44E
Donets'k see Donets'ka Oblast'
119 W8 **Donets'ka Oblast'** var. Donets'k, Rus. Donetskaya Oblast'; prev. Rus. Stalinskaya Oblast'. ◆ province SE Ukraine
Donetskaya Oblast' see Donets'ka Oblast'
69 P8 **Donga** ≈ Cameroon/Nigeria
163 O13 **Dongchuan** Yunnan, SW China 26.09N 103.10E
101 I14 **Dongen** Noord-Brabant, S Netherlands 51.37N 4.55E
166 K17 **Dongfang** var. Basuo. Hainan, S China 19.05N 108.40E
169 Z7 **Dongfanghong** Heilongjiang, NE China 46.13N 133.13E
169 W11 **Dongfeng** Jilin, NE China 42.39N 125.33E
175 P9 **Donggala** Sulawesi, C Indonesia 0.40S 119.43E
169 V13 **Donggou** Liaoning, NE China 39.52N 124.07E
167 O14 **Dongguan** Guangdong, S China 23.03N 113.43E
178 K9 **Đông Ha** Quang Tri, C Vietnam 16.45N 107.10E
166 M16 **Dong Hai** see East China Sea
178 Jj9 **Đông Hôi** Quang Binh, C Vietnam 17.30N 106.34E
110 J10 **Dongio** Ticino, S Switzerland 46.27N 8.58E
166 L11 **Dongkan** see Binhai
Dongkou Hunan, S China 27.06N 110.34E
Dongliao see Liaoyuan
Dong-nai see Đông Nai, Sông
178 K13 **Đông Nai, Sông** var. Dong-nai, Dong Noi, Donnai. ≈ S Vietnam
167 O14 **Dongnan Qiuling** plateau SE China
169 Y9 **Dongning** Heilongjiang, NE China 44.01N 131.03E
85 C14 **Dong Noi** see Đông Nai, Sông
82 E7 **Dongola** var. Donqola. Dunqulah. Northern, N Sudan 19.10N 30.27E
81 I17 **Dongou** La Likouala, NE Congo 2.04N 18.40E
178 Jj13 **Đông Phu** Sông Be, S Vietnam 11.31N 106.55E
Dong Rak, Phanom see Dângrêk, Chuôr Phnum
167 Q14 **Dongshan Dao** island SE China
169 N14 **Dongsheng** Nei Mongol Zizhiqu, C China 39.51N 110.00E
167 R7 **Dongtai** Jiangsu, E China 32.52N 120.13E
167 N10 **Dongting Hu** var. Tung-t'ing Hu. ◎ S China
167 P10 **Dongxiang** Jiangxi, S China 28.11N 116.36E
167 Q9 **Dongying** Shandong, E China 37.27N 118.01E
29 X8 **Doniphan** Missouri, C USA 36.37N 90.49W
8 G7 **Donjek** ≈ Yukon Territory, W Canada
114 E11 **Donji Lapac** Lika-Senj, W Croatia 44.33N 15.58E
114 H8 **Donji Miholjac** Osijek-Baranja, E Croatia 45.45N 18.10E
114 P12 **Donji Milanovac** Serbia, E Yugoslavia 44.27N 22.06E
114 G12 **Donji Vakuf** var. Srbobran, Federacija Bosna I Hercegovina, C Bosnia & Herzegovina 44.08N 17.23E
100 M6 **Donkerbroek** Friesland, N Netherlands 53.00N 6.05E
178 Hh11 **Don Muang** × (Krung Thep)
191 U5 **Dorrigo** New South Wales, SE Australia 30.22S 152.43E
37 N1 **Dorris** California, W USA 41.58N 121.54W
12 J13 **Dorset** Ontario, SE Canada 45.12N 78.52W
99 K23 **Dorset** cultural region S England, UK
103 E14 **Dorsten** Nordrhein-Westfalen, W Germany 51.40N 6.58E
Dort see Dordrecht
103 F15 **Dortmund** Nordrhein-Westfalen, W Germany 51.31N 7.28E
102 F12 **Dortmund-Ems-Kanal** canal W Germany
142 L17 **Dörtyol** Hatay, S Turkey 36.51N 36.10E
148 L7 **Do Rūd** var. Dow Rūd, Durud. Lorestān, W Iran 33.31N 49.03E
81 O15 **Doruma** Orientale, N Dem. Rep. Congo (Zaire) 4.35N 27.43E
13 O2 **Dorval** × (Montréal) Québec, SE Canada 45.27N 73.46W
47 T5 **Dos Bocas, Lago** ◎ C Puerto Rico
106 K16 **Dos Hermanas** Andalucía, S Spain 37.16N 5.55W
27 O9 **Dos Palos** California, W USA 37.00N 120.39W
116 I11 **Dospat** Smolyan, S Bulgaria 41.39N 24.10E
116 H11 **Dospat, Yazovir** ◙ SW Bulgaria
102 M11 **Dosse** ≈ NE Germany
31 V12 **Dostuk** Narynskaya Oblast', C Kyrgyzstan 41.19N 75.40E
25 I15 **Dothan** Alabama, S USA 31.13N 85.23W
41 T9 **Dot Lake** Alaska, USA 63.39N 144.10W
170 F11 **Dōzen** var. Oki-shotō, SW Japan
25 K9 **Dozois, Réservoir** ◎ Québec, SE Canada
76 D7 **Drâa** seasonal river S Morocco
Drâa, Hammada du see Dra, Hamada du

179 Q11 **Donsol** Luzon, N Philippines 12.56N 123.34E
83 L16 **Doolow** Somali, E Ethiopia 4.10N 42.04E
41 Q7 **Doonerak, Mount** ▲ Alaska, USA 67.54N 150.33W
100 J12 **Doorn** Utrecht, C Netherlands 52.01N 5.21E
33 N9 **Door Peninsula** peninsula Wisconsin, N USA
82 F13 **Dooxo Nugaaleed** var. Nogal Valley. valley E Somalia
166 G3 **Do Qu** ≈ C China
108 B7 **Dora Baltea** anc. Duria Major. ≈ NW Italy
188 K7 **Dora, Lake** salt lake Western Australia
108 A8 **Dora Riparia** anc. Duria Minor. ≈ NW Italy
Dorbiljin see Emin
169 V8 **Dorbod** var. Dorbod Mongolzu Zizhixian, Talkang. Heilongjiang, NE China 46.50N 124.25E
Dorbod Mongolzu Zizhixian see Dorbod
115 N18 **Đorče Petrov** var. Djorče Petrov, Gorče Petrov. N FYR Macedonia 42.01N 21.21E
12 F16 **Dorchester** Ontario, S Canada 43.00N 81.04W
99 L24 **Dorchester** anc. Durnovaria. S England, UK 50.43N 2.25W
15 Mm4 **Dorchester, Cape** headland Baffin Island, Nunavut, NE Canada 65.25N 77.25W
85 D19 **Dordabis** Khomas, C Namibia 22.55S 17.39E
104 L12 **Dordogne** ◆ department SW France
105 O9 **Dordogne** ≈ W France
100 H13 **Dordrecht** var. Dordt, Dort. Zuid-Holland, SW Netherlands 51.48N 4.40E
8 J14 **Dore** see Dordrecht
105 P11 **Dore** ≈ C France
9 S13 **Doré Lake** Saskatchewan, C Canada 54.37N 107.36W
105 O12 **Dore, Monts** ▲ C France
103 M23 **Dorfen** Bayern, SE Germany 48.16N 12.06E
41 Q13 **Dorgali** Sardegna, Italy, C Mediterranean Sea 40.18N 9.34E
168 F7 **Dörgön Nuur** ◎ NW Mongolia
79 Q12 **Dori** N Burkina 14.03N 0.01W
103 E16 **Doring** ≈ S South Africa
105 P4 **Dormagen** Nordrhein-Westfalen, W Germany 51.06N 6.49E
110 E6 **Dornach** Solothurn, NW Switzerland 47.28N 7.37E
101 E21 **Dornbirn** Vorarlberg, W Austria 47.25N 9.46E
98 I7 **Dornoch** N Scotland, UK 57.52N 4.00W
98 J7 **Dornoch Firth** inlet N Scotland, UK
169 P7 **Dornod** ◆ province E Mongolia
169 N10 **Dornogovĭ** ◆ province SE Mongolia
79 P10 **Doro** Tombouctou, S Mali 16.07N 0.57W
118 L14 **Dorohoi** Botoşani, NE Romania 47.57N 26.24E
191 P17 **Dorotea** Västerbotten, N Sweden 64.16N 16.30E

12 L9 **Douaire, Lac** ◎ Québec, SE Canada
81 D16 **Douala** var. Duala. Littoral, W Cameroon 4.04N 9.43E
81 D16 **Douala** × Littoral, W Cameroon 3.57N 9.48E
104 F6 **Douarnenez** Finistère, NW France 48.04N 4.19W
104 F6 **Douarnenez, Baie de** bay NW France
Douay see Douai
27 O6 **Double Mountain Fork Brazos River** ≈ Texas, SW USA
25 O3 **Double Springs** Alabama, USA 34.09N 87.24W
105 T8 **Doubs** ◆ department E France
105 T8 **Doubs** ≈ France/Switzerland
193 A22 **Doubtful Sound** sound South Island, NZ
192 I7 **Doubtless Bay** bay North Island, NZ
27 X9 **Douglas** Texas, SW USA
104 K8 **Doué-la-Fontaine** Maine-et-Loire, NW France 47.12N 0.16W
79 O11 **Douentza** Mopti, S Mali 14.59N 2.57W
67 D24 **Douglas** East Falkland, Falkland Islands
99 I16 **Douglas** ○ (Isle of Man) E Isle of Man 54.09N 4.28W
85 H23 **Douglas** Northern Cape, C South Africa 29.03S 23.46E
41 X13 **Douglas** Alexander Archipelago, Alaska, USA 58.12N 134.18W
39 O17 **Douglas** Arizona, SW USA 31.20N 109.32W
25 U7 **Douglas** Georgia, SE USA 31.30N 82.51W
35 Y15 **Douglas** Wyoming, C USA 42.48N 105.22W
23 O7 **Douglas, Cape** headland SW Alaska USA 64.59N 166.41W
8 J14 **Douglas Channel** channel British Columbia, W Canada
190 G3 **Douglas Creek** seasonal river South Australia
33 P5 **Douglas Lake** ◎ Michigan, N USA
23 O9 **Douglas Lake** ◎ Tennessee, S USA
41 Q13 **Douglas, Mount** ▲ Alaska, USA 58.51N 153.31W
204 I6 **Douglas Range** ▲ Alexander Island, Antarctica
124 N10 **Doukáto, Akrotírio** headland Lefkáda, W Greece 38.35N 20.33E
105 P4 **Doullens** Somme, N France 50.09N 2.21E
81 F15 **Doumé** Est, E Cameroon 4.13N 13.27E
Douma see Dūmā
106 I7 **Douro** Sp. Duero. ≈ Portugal/Spain see also Duero
106 G6 **Douro Litoral** former province N Portugal
104 K15 **Douze** ≈ SW France
191 P17 **Dover** Tasmania, SE Australia 43.19S 147.01E
99 Q22 **Dover** Fr. Douvres; Lat. Dubris Portus. SE England, UK 51.08N 1.19E
19 Rr8 **Dover** state capital Delaware, NE USA 39.09N 75.31W
21 P9 **Dover** New Hampshire, NE USA 43.10N 70.50W
21 N13 **Dover** New Jersey, NE USA 40.51N 74.33W
33 U12 **Dover** Ohio, N USA 40.31N 81.28W
22 H8 **Dover** Tennessee, S USA 36.29N 87.50W
99 Q23 **Dover, Strait of** var. Straits of Dover, Fr. Pas de Calais. strait England, UK/France
Dover, Straits of see Dover, Strait of
Dovlen see Devin
96 G13 **Dovre** Oppland, S Norway 61.59N 9.16E
96 G13 **Dovrefjell** plateau S Norway
85 M14 **Dowa** Central, C Malawi 13.42S 33.55E
32 H10 **Dowagiac** Michigan, N USA 41.58N 86.06W
149 N10 **Dow Gonbadān** var. Do Gonbadān, Gonbadān. Kohkīlūyeh va Būyer Aḩmadī, SW Iran 30.24N 50.45E
154 M2 **Dowlatābād** Fāryāb, N Afghanistan 36.30N 64.51E
35 R16 **Downey** Idaho, NW USA 42.25N 112.06W
37 P5 **Downieville** California, W USA 39.33N 120.49W
99 G16 **Downpatrick** Ir. Dún Pádraig. SE Northern Ireland, UK 54.19N 5.43W
37 P10 **Downs** Kansas, C USA 39.30N 98.33W
20 J12 **Downsville** New York, NE USA 42.03N 74.59W
35 S14 **Driggs** Idaho, NW USA 43.44N 111.06W
119 N9 **Drin** Rus. Drin, Lumi i and Herzegovina/Yugoslavia
114 K12 **Drin, Gulf of** see Drinit, Gjiri i
115 O17 **Drinit, Gjiri i** var. Pelg i Drinit, Eng. Gulf of Drin. gulf NW Albania
115 O17 **Drinit, Lumi i** ≈ NW Albania
Drinit, Pellg i see Drinit, Gjiri i
Drin i Zi see Black Drin
115 L22 **Drinos, Lumi i/Drínos Pótamos** see Dríno
Dríno var. Drino, Alb. Lumi i Drinos. ≈ S Albania/Greece
27 S11 **Dripping Springs** Texas, SW USA 30.11N 98.04W
27 S15 **Dripping Springs** Texas

24 H5 **Driskill Mountain** ▲ Louisiana, S USA 32.25N 92.54W
Drissa see Drysa
96 G13 **Driva** ≈ S Norway
114 E13 **Drniš** It. Šibenik-Knin, S Croatia 43.51N 16.10E
97 H15 **Drøbak** Akershus, S Norway 59.40N 10.37E
118 G13 **Drobeta-Turnu Severin** prev. Turnu Severin. Mehedinţi, SW Romania 44.39N 22.39E
118 M8 **Drochia** Ir. Drokiya. N Moldova 48.02N 27.46E
99 F17 **Drogheda** Ir. Droichead Átha. NE Ireland 53.43N 6.21W
Drogichin see Drahichyn
Drogobych see Drohobych
118 H6 **Drohiczyn Poleski** see Drahichyn
Drohobych Pol. Drohobycz, Rus. Drogobych. L'vivs'ka Oblast', NW Ukraine 49.22N 23.30E
Drohobycz see Drohobych
Droichead Átha see Drogheda
Droicheadna Bandan see Bandon
Droichead na Banna see Banbridge
Droim Mór see Dromore
Drokiya see Drochia
105 R13 **Drôme** ◆ department E France
105 S13 **Drôme** ≈ E France
99 G15 **Dromore** Ir. Droim Mór. SE Northern Ireland, UK
108 A9 **Dronero** Piemonte, NE Italy 44.28N 7.25E
104 L13 **Dronne** ≈ SW France
205 Q3 **Dronning Maud Land** physical region Antarctica
100 K6 **Dronrijp** Fris. Dronryp. Friesland, N Netherlands 53.12S 5.37E
100 L9 **Dronryp** see Dronrijp
100 L9 **Dronten** Flevoland, C Netherlands 52.31N 5.40E
Dronthem see Trondheim
105 L13 **Dropt** ≈ SW France
155 T4 **Drosh** North-West Frontier Province, NW Pakistan 35.33N 71.48E
Drossen see Ośno Lubuskie
Drug see Durg
152 I9 **Drujba** Rus. Druzhba. Khorazm Wiloyati, N Uzbekistan 41.14N 61.13E
120 L17 **Drūkšiai** ◎ NE Lithuania
9 Q16 **Drumheller** Alberta, SW Canada 51.28N 112.42W
35 U6 **Drummond** Montana, NW USA 46.39N 113.12W
33 R4 **Drummond Island** island Michigan, N USA
Drummond Island see Tabiteuea
23 X7 **Drummond, Lake** ◎ Virginia, NE USA
13 P12 **Drummondville** Québec, SE Canada 45.52N 72.28W
41 Q13 **Drum, Mount** ▲ Alaska, USA 62.11N 144.37W
29 O9 **Drumright** Oklahoma, C USA 35.59N 96.36W
101 B18 **Drunen** Noord-Brabant, S Netherlands 51.40N 5.07E
120 M11 **Druskininkai** Pol. Druskieniki. Druskininkai, S Lithuania 54.00N 24.00E
Druskienniki see Druskininkai
100 N7 **Druten** Gelderland, SE Netherlands 51.52N 5.37E
120 M11 **Druya** Vitsyebskaya Voblasts', NW Belarus 55.48N 27.27E
119 S2 **Druzhba** Sums'ka Oblast', NE Ukraine 52.01N 33.56E
Druzhba see Dostyk, Kazakhstan
Druzhba see Druzhba, Uzbekistan
126 Mm7 **Druzhina** Respublika Sakha (Yakutiya), NE Russian Federation 68.01N 144.58E
119 X7 **Druzhkivka** Donets'ka Oblast', E Ukraine 48.38N 37.31E
114 E12 **Drvar** Federacija Bosna I Hercegovina, Bosnia and Herzegovina 44.21N 16.24E
115 G15 **Drvenik** Split-Dalmacija, SE Croatia 43.10N 17.13E
116 K9 **Dryanovo** Gabrovo, N Bulgaria 42.59N 25.28E
28 G7 **Dry Cimarron River** ≈ Kansas/Oklahoma, C USA
12 B11 **Dryden** Ontario, C Canada 49.48N 92.48W
26 M11 **Dryden** Texas, SW USA 30.01N 102.06W
205 Q14 **Drygalski Ice Tongue** ice feature Antarctica
120 L13 **Drysa** Rus. Drissa. ≈ N Belarus
25 V17 **Dry Tortugas** island Florida, SE USA
81 D15 **Dschang** Ouest, W Cameroon 5.28N 10.01E
56 J5 **Duaca** Lara, N Venezuela 10.16N 69.12W
Duacum see Douai
Duala see Douala
47 N9 **Duarte, Pico** ▲ C Dominican Republic 19.02N 70.57W
146 J5 **Đubā** Tabūk, NW Saudi Arabia 27.25N 35.42E
191 N9 **Dubbo** New South Wales, SE Australia 32.16S 148.40E
110 G8 **Dübendorf** Zürich, NW Switzerland 47.24N 8.36E
99 F18 **Dublin** Ir. Baile Átha Cliath; anc. Eblana. ● (Ireland), E Ireland 53.19N 6.15W

188 G10 **Dorre Island** island Western Australia
103 G23 **Dow Rūd** see Do Rūd
102 I5 **Dowsk** see Dowsk
33 O10 **Dowsk** Rus. Dowsk. Homyel'skaya Voblasts', SE Belarus 53.09N 30.27E
57 Q4 **Doyle** California, W USA 40.00N 120.06W
20 I15 **Doylestown** Pennsylvania, NE USA 40.18N 75.07W
104 M5 **Dreux** anc. Drocae, Durocasses. Eure-et-Loir, C France 48.43N 1.22E
96 I11 **Drevsjø** Hedmark, S Norway 61.52N 12.01E
24 K3 **Drew** Mississippi, S USA 33.49N 90.31W
120 L11 **Drisa** Rus. Drissa. ≈ N Belarus
25 V17 **Dry Tortugas** island Florida, SE USA
81 D15 **Dschang** Ouest, W Cameroon 5.28N 10.01E
147 U7 **Dubayy** Eng. Dubai. Dubayy, NE UAE 25.10N 55.18E
147 W7 **Dubayy** Eng. Dubai. × Dubayy, NE UAE 25.15N 55.22E

116 G9 **Dragoman** Sofiya, W Bulgaria 42.57N 22.53E
117 L25 **Dragonáda** island SE Greece
Dragonera, Isla see Sa Dragonera
47 T14 **Dragon's Mouths, The** strait Trinidad and Tobago/Venezuela
97 J23 **Dragør** København, E Denmark 55.36N 12.42E
116 F10 **Dragovishtitsa** Kyustendil, W Bulgaria 42.22N 22.39E
105 U15 **Draguignan** Var, SE France 43.31N 6.31E
76 E9 **Dra, Hamada du** var. Hammada du Drâa, Haut Plateau du Dra. plateau W Algeria
76 E9 **Dra, Haut Plateau du** see Dra, Hamada du
121 H19 **Drahichyn** Pol. Drohiczyn Poleski, Rus. Drogichin. Brestskaya Voblasts', SW Belarus 52.10N 25.10E
31 N4 **Drake** North Dakota, N USA 47.54N 100.23W
85 K23 **Drakensberg** ▲ Lesotho/South Africa
204 F3 **Drake Passage** passage Atlantic Ocean/Pacific Ocean
116 L8 **Dralfa** Tŭrgovishte, N Bulgaria 43.17N 26.25E
116 I12 **Dráma** var. Dhráma. Anatolikí Makedonía kai Thráki, NE Greece 41.09N 24.10E
Dramburg see Drawsko Pomorskie
97 H15 **Drammen** Buskerud, S Norway 59.43N 10.12E
97 H15 **Drammensfjorden** fjord S Norway
94 H1 **Drangajökull** ▲ NW Iceland
97 F16 **Drangedal** Telemark, S Norway 59.04N 9.01E
156 I2 **Drangsnes** Vestfirðir, NW Iceland 65.42N 21.27W
189 W9 **Drann** see Dravinja
189 W9 **Drau** var. Drava, Eng. Drave, Hung. Dráva. ≈ C Europe see also Drava
86 I11 **Drava** var. Drau, Eng. Drave, Hung. Dráva. ≈ C Europe see also Drau
Dráva see Drau/Drava
Drave see Drau/Drava
189 W10 **Dravinja** Ger. Drann. ≈ NE Slovenia
189 V9 **Dravograd** Ger. Unterdrauburg; prev. Spodnji Dravograd. N Slovenia 46.36N 15.00E
112 F10 **Drawa** ≈ NW Poland
112 F9 **Drawno** NW Poland 53.12N 15.44E
112 F10 **Drawsko Pomorskie** Ger. Dramburg. Zachodniopomorskie, NW Poland 53.31N 15.48E
99 Q22 **Drayton** North Dakota, N USA 48.34N 97.10W
9 P15 **Drayton Valley** Alberta, SW Canada 53.15N 115.00W
194 F10 **Dreikikir** East Sepik, NW PNG 3.34S 142.41E
100 N7 **Dreikirchen** see Teiuş
100 N7 **Drenthe** ◆ province NE Netherlands
117 H15 **Drépano, Akrotírio** var. Akra Dhrepanon. headland N Greece 39.56N 23.57E
Drepanum see Trapani
119 V3 **Dresden** Ontario, S Canada 42.35N 82.11W
103 O16 **Dresden** Sachsen, E Germany 51.02N 13.43E
22 G8 **Dresden** Tennessee, S USA 36.17N 88.42W
104 M5 **Dretun'** Rus. Dretun'. Vitsyebskaya Voblasts', N Belarus 55.42N 29.12E
Dreux anc. Drocae, Durocasses.
105 S13 **Drable** see José Enrique Rodó
Drac ≈ E France
Drač/Draç see Durrës
62 I8 **Dracena** São Paulo, S Brazil 21.27S 51.30W
100 M6 **Drachten** Friesland, N Netherlands 53.07N 6.06E
94 H11 **Drag** Nordland, C Norway
118 L14 **Drăgăneşti-Vlaşca** Teleorman, S Romania 44.05N 25.39E
118 J14 **Drăgăşani** Vâlcea, SW Romania 44.40N 24.16E
99 Q22 **Driffield** E England, UK 54.00N 0.28W
67 D25 **Driftwood Point** headland East Falkland, Falkland Islands 52.15S 59.00W

◆ COUNTRY ◇ DEPENDENT TERRITORY ◆ ADMINISTRATIVE REGION ▲ MOUNTAIN ▲ VOLCANO ◎ LAKE
● COUNTRY CAPITAL ○ DEPENDENT TERRITORY CAPITAL × INTERNATIONAL AIRPORT ▲ MOUNTAIN RANGE ≈ RIVER ◙ RESERVOIR

25 U5 **Dublin** Georgia, SE USA 32.32N 82.54W

27 R7 **Dublin** Texas, SW USA 32.05N 98.20W

99 G18 **Dublin** Ir. Baile Átha Cliath; anc. Eblana. cultural region E Ireland

99 G18 **Dublin Airport** ✈ E Ireland 53.25N 6.18W

201 V12 **Dublon** var. Tonoas. island Chuuk Islands, C Micronesia

130 K2 **Dubna** Moskovskaya Oblast', W Russian Federation 56.45N 37.09E

113 G19 **Dubňany** Ger. Dubnian. Brněnský Kraj, SE Czech Republic 48.54N 17.00E

113 I19 **Dubnian** see Dubňany

113 I19 **Dubnica nad Váhom** Hung. Máriatölgyes; prev. Dubnicz. Trenčiansky Kraj, W Slovakia 48.58N 18.10E

118 K4 **Dubno** Rivnens'ka Oblast', NW Ukraine 50.27N 25.39E

20 D13 **Du Bois** Pennsylvania, NE USA 41.07N 78.45W

35 R13 **Dubois** Idaho, NW USA 44.10N 112.13W

35 T14 **Dubois** Wyoming, C USA 43.31N 109.37W

Dubossary see Dubăsari

131 O10 **Dubovka** Volgogradskaya Oblast', SW Russian Federation 49.10N 48.49E

78 H14 **Dubréka** Guinée-Maritime, SW Guinea 9.48N 13.31W

12 B7 **Dubreuilville** Ontario, S Canada 48.21N 84.31W

Dubris Portus see Dover

121 L20 **Dubrova** Rus. Dubrova. Homyel'skaya Voblasts', SE Belarus 51.46N 28.13E

130 I5 **Dubrovka** Bryanskaya Oblast', W Russian Federation 53.44N 33.27E

115 H16 **Dubrovnik** It. Ragusa. Dubrovnik-Neretva, SE Croatia 42.39N 18.06E

115 I16 **Dubrovnik** ✈ Dubrovnik-Neretva, SE Croatia 42.34N 18.17E

115 F16 **Dubrovnik-Neretva** off. Dubrovačko-Neretvanska Županija. ◆ province SE Croatia

Dubrovno see Dubrowna

118 L2 **Dubrovytsya** Rivnens'ka Oblast', NW Ukraine 51.34N 26.37E

121 O14 **Dubrowna** Rus. Dubrovno. Vitsyebskaya Voblasts', N Belarus 54.34N 30.40E

31 Z13 **Dubuque** Iowa, C USA 42.30N 90.39W

120 E12 **Dubysa** ∿ C Lithuania

178 K12 **Đưc Cơ** Gia Lai, C Vietnam 13.48N 107.41E

203 V12 **Duc de Gloucester, Îles du** Eng. Duke of Gloucester Islands. island group C French Polynesia

113 C15 **Duchcov** Ger. Dux. Ústecký Kraj, NW Czech Republic 50.37N 13.40E

39 N3 **Duchesne** Utah, W USA 40.09N 110.24W

203 P17 **Ducie Island** atoll E Pitcairn Islands

9 W15 **Duck Bay** Manitoba, S Canada 52.11N 100.08W

25 X17 **Duck Key** island Florida Keys, Florida, SE USA

9 T14 **Duck Lake** Saskatchewan, S Canada 52.52N 106.12W

9 V15 **Duck Mountain** ▲ Manitoba, S Canada

22 I9 **Duck River** ∿ Tennessee, S USA

23 M10 **Ducktown** Tennessee, S USA 35.01N 84.24W

178 Kk11 **Đưc Phô** Quang Ngai, C Vietnam 14.55N 108.55E

178 Jj8 **Đưc Thọ** Ha Tinh, N Vietnam 18.30N 105.36E

178 Kk13 **Đưc Trong** var. Liên Nghia. Lâm Dông, S Vietnam 11.45N 108.24E

D-U-D see Dalap-Uliga-Djarrit

101 M25 **Dudelange** var. Forge du Sud, Ger. Dudelingen. Luxembourg, S Luxembourg 49.29N 6.04E

Dudelingen see Dudelange

103 J15 **Duderstadt** Niedersachsen, C Germany 51.31N 10.16E

159 N15 **Dudhi** Uttar Pradesh, N India 24.13N 83.18E

156 I8 **Dudinka** Taymyrskiy (Dolgano-Nenetskiy) Avtonomnyy Okrug, N Russian Federation 69.27N 86.13E

99 L20 **Dudley** C England, UK 52.30N 2.04W

160 G13 **Dudna** ∿ C India

78 L16 **Duékoué** W Ivory Coast 6.45N 7.21W

106 M5 **Dueñas** Castilla-León, N Spain 41.52N 4.33W

106 K4 **Duerna** ∿ NW Spain

107 O6 **Duero** Port. Douro.

— **Portugal/Spain see also** Douro

Duesseldorf see Düsseldorf

23 P12 **Due West** South Carolina, SE USA 34.19N 82.23W

201 H17 **Dufek Coast** physical region Antarctica

99 L18 **Duffel** Antwerpen, C Belgium 51.06N 4.30E

37 S2 **Duffer Peak** ▲ Nevada, W USA 41.40N 118.45W

195 X7 **Duff Islands** island group E Solomon Islands

Dufour, Pizzo/Dufour, Punta see Dufour Spitze

110 E12 **Dufour Spitze** It. Pizzo Dufour, Punta Dufour. ▲ Italy/Switzerland 45.54N 7.56E

114 D9 **Duga Resa** Karlovac, C Croatia 45.25N 15.30E

24 H5 **Dugdemona River** ∿ Louisiana, S USA

160 J12 **Duggipar** Mahārāshtra, C India 21.06N 80.10E

114 B13 **Dugi Otok** var. Isola Grossa, It. Isola Lunga. island W Croatia

115 F14 **Dugopolje** Split-Dalmacija, S Croatia 43.35N 16.35E

156 L8 **Du He** ∿ C China

56 M11 **Duida, Cerro** ▲ S Venezuela 3.21N 65.45W

Duinekerke see Dunkerque

103 E15 **Duisburg** prev. Duisburg-Hamborn. Nordrhein-Westfalen, W Germany 51.24N 6.47E

Duisburg-Hamborn see Duisburg

101 F14 **Duiveland** island SW Netherlands

101 M12 **Duiven** Gelderland, E Netherlands 51.57N 6.02E

245 W10 **Dujaylah, Hawr ad** ⊕ S Iraq

83 L18 **Dujuuma** Shabeellaha Hoose, S Somalia 1.04N 42.37E

Dūkān see Dokan

41 Z14 **Duke Island** island Alexander Archipelago, Alaska, USA

Dukelský Priesmy/Dukelský Průsmyk see Dukla Pass

83 F14 **Duk Faiwil** Jonglei, SE Sudan 7.30N 31.27E

147 T7 **Dukhān** C Qatar 25.29N 50.48E

Dukhan Heights see Dukhān, Jabal

149 N16 **Dukhān, Jabal** var. Dukhan Heights. hill range S Qatar

131 Q7 **Dukhovnitskoye** Saratovskaya Oblast', W Russian Federation 52.31N 48.32E

130 H4 **Dukhovshchina** Smolenskaya Oblast', W Russian Federation 55.15N 32.22E

120 I12 **Dūkštas** Ignalina, E Lithuania 55.32N 26.21E

113 N17 **Dukla** Podkarpackie, SE Poland 49.33N 21.40E

113 N18 **Duklai Hág** see Dukla Pass

113 N18 **Dukla Pass** Cz. Dukelský Průsmyk, Ger. Dukla-Pass, Hung. Duklai Hág, Pol. Przełęcz Dukielska, Slvk. Dukelský Priesmy. pass Poland/Slovakia 49.25N 21.42E

Dukou see Panzhihua

168 M8 **Dulaan** Hentiy, C Mongolia 47.09N 108.48E

165 R10 **Dulan** var. Qagan Us. Qinghai, C China 36.11N 97.51E

39 R8 **Dulce** New Mexico, SW USA 36.55N 107.00W

45 N16 **Dulce, Golfo** gulf S Costa Rica

44 K6 **Dulce Nombre de Culmí** Olancho, C Honduras 15.04N 85.35W

64 L9 **Dulce, Río** ∿ C Argentina

126 M9 **Dulgalakh** ∿ NE Russian Federation

116 M8 **Dŭlgopol** Varna, E Bulgaria 43.05N 27.24E

159 V14 **Dullabchara** Assam, NE India 24.25N 92.22E

22 D3 **Dulles** ✈ (Washington DC) Virginia, NE USA 39.00N 77.27W

103 E14 **Dülmen** Nordrhein-Westfalen, W Germany 51.51N 7.17E

116 M7 **Dulovo** Silistra, NE Bulgaria 43.50N 27.10E

31 W5 **Duluth** Minnesota, N USA 46.46N 92.06W

144 H7 **Dūmā** Fr. Douma. Dimashq, SW Syria 33.33N 36.24E

176 Y11 **Dumai** Sumatera, W Indonesia 1.41N 101.27E

9 T15 **Dumaresq River** ∿ New South Wales/Queensland, SE Australia

27 N1 **Dumas** Arkansas, C USA 33.53N 91.29W

27 N1 **Dumas** Texas, SW USA 35.51N 101.57W

144 I7 **Dumayr** Dimashq, W Syria 33.36N 36.28E

98 I12 **Dumbarton** W Scotland, UK 55.57N 4.34W

98 I12 **Dumbarton** cultural region C Scotland, UK

197 J7 **Dumbéa** Province Sud, S New Caledonia 22.11S 166.27E

113 K19 **Dumbier** Ger. Djumbir, Hung. Gyömbér. ▲ C Slovakia 48.54N 19.36E

118 I11 **Dumbrăveni** Ger. Elisabethstadt, Hung. Erzsébetváros; prev. Ebesfalva, Eppeschdorf, Ibaşfalău. Sibiu, C Romania 46.13N 24.38E

118 L12 **Dumbrăveni** Vrancea, E Romania 45.30N 27.08E

99 J14 **Dumfries** S Scotland, UK 55.04N 3.37W

99 J14 **Dumfries** cultural region SW Scotland, UK

159 R15 **Dumka** Bihār, NE India 24.16N 87.15E

Dümmer see Dümmersee

102 G12 **Dümmersee** var. Dümmer. ⊕ NW Germany

12 J11 **Dumoine** ∿ Quebec, SE Canada

12 J10 **Dumoine, Lac** ⊕ Quebec, SE Canada

205 V16 **Dumont d'Urville** French research station Antarctica 66.24S 139.38E

205 W15 **Dumont d'Urville Sea** S Pacific Ocean

12 K11 **Dumont, Lac** ⊕ Quebec, SE Canada

77 W7 **Dumyât** Eng. Damietta. N Egypt 31.25N 31.48E

Duna see Don, Russian Federation

Duna see Danube, C Europe

Dûnas see Western Dvina

Dünaburg see Daugavpils

123 I24 **Dunaföldvár** Tolna, C Hungary 46.16N 18.54E

Dunaj see Wien, Austria

Dunaj see Danube, C Europe

113 I18 **Dunajec** ∿ S Poland

113 H21 **Dunajská Streda** Hung. Dunaszerdahely. Trnavský Kraj, W Slovakia 48.00N 17.27E

Dunapentele see Dunaújváros

118 M13 **Dunărea Veche, Brațul** ∿ SE Romania

119 N13 **Dunării, Delta** delta SE Romania

Dunaszerdahely see Dunajská Streda

113 J13 **Dunaújváros** prev. Dunapentele, Sztálinváros. Fejér, C Hungary 47.00N 18.55E

Dunav see Danube

116 J8 **Dunavska Ravnina** Eng. Danubian Plain. plain N Bulgaria

116 G7 **Dunavtsi** Vidin, NW Bulgaria 43.56N 22.49E

Dunayevtsy see Dunayivtsi

118 L7 **Dunayivtsi** Rus. Dunayevtsy. Khmel'nyts'ka Oblast', NW Ukraine 48.54N 26.51E

193 F22 **Dunback** Otago, South Island, NZ 45.22S 170.37E

8 L17 **Duncan** Vancouver Island, British Columbia, SW Canada 48.46N 123.10W

39 O15 **Duncan** Arizona, SW USA 32.43N 109.06W

28 M12 **Duncan** Oklahoma, C USA 34.30N 97.57W

Duncan Island see Pinzón, Isla

157 Q20 **Duncan Passage** strait Andaman Sea/Bay of Bengal

98 K6 **Duncansby Head** headland N Scotland, UK 58.37N 3.01W

12 G12 **Dunchurch** Ontario, S Canada 45.36N 79.54W

120 D7 **Dundaga** Talsi, NW Latvia 57.29N 22.19E

12 G14 **Dundalk** Ontario, S Canada 44.11N 80.22W

99 F16 **Dundalk** Ir. Dún Dealgan. NE Ireland 54.01N 6.25W

25 X3 **Dundalk** Maryland, NE USA 39.15N 76.31W

99 F16 **Dundalk Bay** Ir. Cuan Dhún Dealgan. bay NE Ireland

12 G16 **Dundas** Ontario, S Canada 43.16N 79.55W

188 L12 **Dundas, Lake** salt lake Western Australia

169 O7 **Dundbürd** Hentiy, E Mongolia 47.55N 111.37E

13 N13 **Dún Dealgan** see Dundalk

85 K22 **Dundee** KwaZulu/Natal, E South Africa 28.08S 30.12E

98 K11 **Dundee** E Scotland, UK 56.28N 3.00W

31 R10 **Dundee** Michigan, N USA 41.57N 83.39W

27 R5 **Dundee** Texas, SW USA 33.43N 98.52W

204 H3 **Dundee Island** island Antarctica

168 L9 **Dundgovĭ** ◆ province C Mongolia

176 Y11 **Dundu** var. Irian Jaya, E Indonesia

9 T15 **Dundurn** Saskatchewan, S Canada 51.43S 106.27W

118 E6 **Dund-Us** Hovd, W Mongolia 48.06N 91.22E

116 I07 **Dunduklak** Rom. Răcari; prev. Blatnitsa, Dunavtsi. Dobrich, NE Bulgaria 43.41N 28.31E

32 L4 **Dunedin** Otago, South Island, NZ 45.51S 170.31E

29 P13 **Dunedin** Oklahoma, C USA 33.59N 96.22W

107 N5 **Dunfermline** C Scotland, UK 56.04N 3.28W

98 J12 **Dún Fhionnachaidh** see Dunfanaghy

99 F15 **Dungannon** Ir. Dún Geanainn, C Northern Ireland, UK 54.31N 6.46W

158 F15 **Dungarpur** Rājasthān, N India 23.53N 73.39E

99 E21 **Dungarvan** Ir. Dún Garbháin. S Ireland 52.04N 7.37W

103 N21 **Dungau** cultural region SE Germany

Dún Geanainn see Dungannon

99 P23 **Dungeness, Punta** headland S Argentina 52.25S 68.25W

65 I23 **Dungeness** headland SE England, UK 50.55N 0.58E

Dungloe see Dunglow

99 D14 **Dunglow** var. Dungloe, Ir. An Clochán Liath. NW Ireland 54.57N 8.22W

191 T7 **Dungog** New South Wales, SE Australia 32.24S 151.45E

81 O16 **Dungu** Orientale, NE Dem. Rep. Congo (Zaire) 3.40N 28.31E

304 Hh3 **Dungun** var. Kuala Dungun. Terengganu, Peninsular Malaysia 4.46N 103.25E

82 I6 **Dungúnab** Red Sea, NE Sudan 21.06N 37.06E

21 P13 **Dunham** Quebec, SE Canada 45.08N 72.48W

159 U13 **Dungarpur** Dhaka, N Bangladesh 25.10N 90.41E

159 R15 **Durgāpur** West Bengal, NE India 23.30N 87.19E

169 X10 **Dunhua** Jilin, NE China 43.23N 128.12E

165 P8 **Dunhuang** Gansu, N China 40.10N 94.43E

190 L12 **Dunkeld** Victoria, SE Australia 37.41S 142.19E

105 O1 **Dunkerque** Eng. Dunkirk, Flem. Duinekerke; prev. Dunquerque. Nord, N France 51.06N 2.34E

99 K23 **Dunkery Beacon** ▲ SW England, UK 51.10N 3.36W

20 C11 **Dunkirk** New York, NE USA 42.28N 79.19W

79 P17 **Dunkwa** SW Ghana 5.58N 1.45W

99 G18 **Dún Laoghaire** Eng. Dunleary; prev. Kingstown. E Ireland 53.16N 6.07W

Dunleary see Dún Laoghaire

39 H6 **Dunn** N Scotland, UK 58.34N 4.45W

189 Y3 **Dürnkrut** Niederösterreich, NE Austria 48.28N 16.50E

113 A16 **Dyleň** Ger. Tillenberg. ▲ NW Czech Republic 49.58N 12.31E

112 K9 **Dylewska Góra** ▲ N Poland 53.33N 19.57E

119 O4 **Dymer** Kyyivs'ka Oblast', N Ukraine 50.50N 30.21E

119 W7 **Dymytrov** Rus. Dimitrov. Donets'ka Oblast', SE Ukraine 48.18N 37.19E

115 K19 **Dýrrachium** see Durrës

31 N14 **Dunning** Nebraska, C USA 41.49N 100.04W

67 B24 **Dunnose Head Settlement** West Falkland, Falkland Islands 51.24S 60.28W

12 G17 **Dunnville** Ontario, S Canada 42.54N 79.36W

144 P9 **Dún Pádraig** see Downpatrick

144 P9 **Dunqul** see Dunkerque

Dunqulah see Dongola

98 L12 **Duns** SE Scotland, UK 55.46N 2.13W

31 N2 **Dunseith** North Dakota, N USA 48.48N 100.03W

37 N2 **Dunsmuir** California, W USA 41.12N 122.19W

99 N21 **Dunstable** Lat. Durocobrivae. E England, UK 51.52N 0.31W

193 D21 **Dunstan Mountains** ▲ South Island, NZ

105 Q5 **Dun-sur-Auron** Cher, C France 46.52N 2.40E

193 F21 **Duntroon** Canterbury, South Island, NZ

155 T10 **Dunyāpur** Punjab, E Pakistan 29.48N 71.48E

178 Ii4 **Dương Đông** Kiên Giang, S Vietnam 10.15N 103.58E

116 G10 **Dupnitsa** prev. Marek, Stanke Dimitrov. Kyustendil, W Bulgaria 42.15S 23.09E

30 L8 **Dupree** South Dakota, N USA 45.03N 101.36W

35 Q7 **Dupuyer** Montana, NW USA 48.13N 112.34W

147 Y11 **Duqm** var. Daqm. E Oman 19.42N 57.39E

65 F23 **Duque de York, Isla** island S Chile

189 N4 **Durack Range** ▲ Western Australia

K10 **Durağan** Sinop, N Turkey 41.52N 35.03E

105 U5 **Durance** ∿ SE France

33 R9 **Durand** Michigan, N USA 42.54N 83.58W

31 V6 **Durand** Wisconsin, N USA 44.37N 91.55W

197 L7 **Durand, Récif** reef SE New Caledonia

42 K10 **Durango** var. Victoria de Durango. Durango, W Mexico 24.03N 104.37W

107 P3 **Durango** País Vasco, N Spain 43.10N 2.37W

39 Q8 **Durango** Colorado, C USA 37.13N 107.51W

42 J9 **Durango** ◆ state C Mexico

116 I07 **Durankulak** Rom. Răcari; prev. Blatnitsa, Dunavtsi. Dobrich, NE Bulgaria 43.41N 28.31E

28 L4 **Durant** Mississippi, S USA 33.04N 89.51W

29 P13 **Durant** Oklahoma, C USA 33.59N 96.22W

Duranulac see Durankulak

107 N6 **Duratón** ∿ N Spain

63 E19 **Durazno** var. San Pedro de Durazno. Durazno, S Uruguay 33.24S 56.28W

63 E19 **Durazno** ◆ department C Uruguay

Durazzo see Durrës

85 K23 **Durban** var. Port Natal. KwaZulu/Natal, E South Africa 29.51S 31.00E

85 K23 **Durban** ✈ KwaZulu/Natal, E South Africa 29.55S 31.01E

120 C9 **Durbe** Ger. Durben. Liepāja, W Latvia 56.34N 21.22E

Durben see Durbe

101 K21 **Durbuy** Luxembourg, SE Belgium 50.21N 5.27E

107 N15 **Dúrcal** Andalucía, S Spain 37.00N 3.24W

114 F8 **Đurđevac** Ger. Sankt Georgen, Hung. Szentgyörgy; prev. Djurdjevac, Gjurgjevac, Koprivnica-Križevci, N Croatia 46.02N 17.03E

115 K15 **Đurđevica Tara** Montenegro, SW Yugoslavia 43.09N 19.18E

99 L24 **Durdle Door** natural arch S England, UK

164 L3 **Düre** Xinjiang Uygur Zizhiqu, W China 46.30N 88.52E

103 D16 **Düren** anc. Marcodurum. Nordrhein-Westfalen, W Germany 50.48N 6.28E

160 K12 **Durg** prev. Drug. Madhya Pradesh, C India 21.12N 81.19E

159 U13 **Durgapur** Dhaka, N Bangladesh 25.10N 90.41E

12 F14 **Durham** Ontario, S Canada 44.10N 80.48W

99 M14 **Durham** hist. Dunholme. N England, UK 54.46N 1.34W

23 U9 **Durham** North Carolina, SE USA 35.59N 78.54W

99 M14 **Durham** cultural region N England, UK

22 F8 **Dyersburg** Tennessee, S USA 36.01N 89.23W

31 Y13 **Dyersville** Iowa, C USA 42.29N 91.07W

99 I21 **Dyfed** cultural region SW Wales, UK

43 X11 **Dzilam de Bravo** Yucatán, E Mexico 21.24N 88.52W

113 I19 **Dzisna** Rus. Disna. Vitsyebskaya Voblasts', N Belarus 55.33N 28.13E

120 K12 **Dzisna** Lith. Dysna, Rus. Disna. ∿ Belarus/Lithuania

116 G20 **Dzivin** Rus. Divin. Brestskaya Voblasts', SW Belarus 51.58N 24.33E

121 M15 **Dzmitravichy** Rus. Dmitrovichi. Minskaya Voblasts', C Belarus 53.58N 29.14E

133 S8 **Dzogsool** Töv, C Mongolia 46.46N 107.18E

113 S8 **Dzungaria** var. Sungaria, Zungaria. physical region NW China

168 G5 **Dzungarian Basin** see Junggar Pendi

168 G5 **Dzüünbulag** Dornod, E Mongolia 46.48N 115.21E

169 O8 **Dzüünbulag** Sühbaatar, E Mongolia 46.30N 114.22E

99 A21 **Dursey Island** Ir. Oileán Baoi. island SW Ireland

Dursi see Durrës

116 P12 **Durusu** İstanbul, NW Turkey 41.18N 28.41E

116 O12 **Durusu Gölü** ⊕ NW Turkey

144 H9 **Durūz, Jabal ad** ▲ SW Syria 32.42N 36.36W

192 K13 **D'Urville Island** island C NZ

176 Xx9 **D'Urville, Tanjung** headland Irian Jaya, E Indonesia 1.26S 137.52E

120 I11 **Dusetos** Zarasai, NE Lithuania 55.44N 25.49E

152 H14 **Dushak** Akhalskiy Velayat, S Turkmenistan 37.15N 59.57E

166 K12 **Dushan** Guizhou, S China 25.45N 107.33E

153 P13 **Dushanbe** var. Dyushambe; prev. Stalinabad, Taj. Stalinobod. ● (Tajikistan) W Tajikistan 38.35N 68.43E

153 P13 **Dushanbe** ✈ W Tajikistan 38.33N 68.49E

143 T9 **Dushet'i** E Georgia 42.07N 44.44E

20 H13 **Dushore** Pennsylvania, NE USA 41.30N 76.23W

193 A23 **Dusky Sound** sound South Island, NZ

103 E15 **Düsseldorf** var. Duesseldorf. Nordrhein-Westfalen, W Germany 51.13N 6.49E

153 P14 **Dūstí** Rus. Dusti. SW Tajikistan 37.22N 68.41E

153 O10 **Dŭstlik** Jizzakh Wiloyati, C Uzbekistan 40.37N 67.59E

118 M5 **Dusti** var. Dusi. SW Ukraine

40 L17 **Dutch Harbor** Unalaska Island, Alaska, USA 53.50N 166.33W

38 J3 **Dutch Mount** ▲ Utah, W USA 40.16N 113.56W

Dutch New Guinea see Irian Jaya

Dutch West Indies see Netherlands Antilles

85 H20 **Dutlwe** Kweneng, S Botswana 23.56S 23.53E

129 U8 **Du Toit Fracture Zone** tectonic feature SW Indian Ocean

79 V13 **Dutovo** Respublika Komi, NW Russian Federation 63.45N 56.38E

79 V13 **Dutsan Wai** var. Victoria de Kaduna, C Nigeria 10.49N 8.15E

79 W13 **Dutse** Jigawa, N Nigeria 11.43N 9.25E

Dutsen Wai see Dutsan Wai

Duttia see Datia

12 E17 **Dutton** Ontario, S Canada 42.40N 81.28W

38 L7 **Dutton, Mount** ▲ Utah, W USA 38.00N 112.10W

168 E7 **Duut** Hovd, W Mongolia 47.28N 91.52E

12 K11 **Duval, Lac** ⊕ Quebec, SE Canada

131 W3 **Duvan** Respublika Bashkortostan, W Russian Federation 55.42N 57.56E

144 L9 **Duwayhilat Satih ar Ruwayshid** seasonal river SE Jordan

Dux see Duchcov

166 J13 **Duyang Shan** ▲ S China

178 Jj15 **Duyên Hai** Tra Vinh, S Vietnam 9.39N 106.28E

166 K12 **Duyun** Guizhou, S China 26.16N 107.28E

142 G11 **Düzce** Bolu, NW Turkey 40.51N 31.09E

152 B11 **Duzdab** see Zāhedān

Duzkyr, Khrebet see Duzkyr, Khrebet

152 I16 **Duzkyr, Khrebet** prev. Khrebet Duzenkyr. ▲ S Turkmenistan 39.42N 54.10E

153 Y7 **Dvina Bay** see Chëshskaya Guba

Dvinsk see Daugavpils

128 L7 **Dvinskaya Guba** bay NW Russian Federation

114 E10 **Dvor** Sisak-Moslavina, C Croatia 45.05N 16.22E

119 W5 **Dvorichna** Kharkivs'ka Oblast', E Ukraine 49.52N 37.43E

113 F16 **Dvůr Králové nad Labem** Ger. Königinhof an der Elbe. Hradecký Kraj, NE Czech Republic 50.25N 15.48E

160 A10 **Dwārka** Gujarāt, W India

32 M12 **Dwight** Illinois, N USA 41.05N 88.25W

100 N8 **Dwingeloo** Drenthe, NE Netherlands 52.49N 6.20E

35 N10 **Dworshak Reservoir** ⊠ Idaho, NW USA

113 K17 **Dyal** see Dyaul Island

113 M15 **Dyanev** see Deynau

113 L16 **Dyatlovo** see Dzyatlava

195 N9 **Dyaul Island** var. Djaul, Dyal. island NE PNG

22 G8 **Dyer** Tennessee, S USA 36.04N 88.59W

13 O1 **Dyer, Cape** headland Baffin Island, Nunavut, NE Canada 66.37N 61.13W

31 X13 **Dysart** Iowa, C USA 42.10N 92.18W

Dysna see Dzisna

117 D18 **Dytiki Ellás** Eng. Greece West. ◆ region C Greece

117 C14 **Dytiki Makedonía** Eng. Macedonia West. ◆ region N Greece

131 U4 **Dyurtyuli** Respublika Bashkortostan, W Russian Federation 55.20N 54.49E

168 I7 **Dyushambe** see Dushanbe

168 I7 **Dzaanhushuu** Arhangay, C Mongolia 47.30N 101.06E

168 I8 **Dzadgay** Bayanhongor, C Mongolia 46.54N 99.11E

168 H10 **Dzag** Bayanhongor, C Mongolia 44.29N 99.19E

180 J14 **Dzaoudzi** E Mayotte 12.48S 45.18E

152 H14 **Dzaudzhikau** see Vladikavkaz

168 G7 **Dzavhan** ◆ province NW Mongolia

168 G7 **Dzavhan Gol** ∿ NW Mongolia

168 J7 **Dzegstey** Arhangay, C Mongolia 47.38N 102.31E

131 O3 **Dzerzhinsk** Nizhegorodskaya Oblast', W Russian Federation 56.20N 43.22E

Dzerzhinsk see Dzyarzhynsk, Belarus

Dzerzhinsk see Dzerzhyns'k, Ukraine

Dzerzhinskiy see Nar'yan-Mar

151 W13 **Dzerzhinskoye** Almaty, SE Kazakhstan 45.49N 81.04E

119 X7 **Dzerzhyns'k** Rus. Dzerzhinsk. Donets'ka Oblast', SE Ukraine 48.21N 37.50E

118 M5 **Dzerzhyns'k** Zhytomyrs'ka Oblast', N Ukraine 50.07N 27.56E

Dzhailgan see Jayilgan

153 T10 **Dzhalal-Abad** Kir. Jalal-Abad. Jalal-Abadskaya Oblast', W Kyrgyzstan 40.55N 73.00E

153 S9 **Dzhalal-Abadskaya Oblast'** Kir. Jalal-Abad Oblasty. ◆ province W Kyrgyzstan

Dzhalilabad see Cälilabad

150 G9 **Dzhambeyty** Kaz. Zhympity. Zapadnyy Kazakhstan, W Kazakhstan 50.16N 52.34E

Dzhambul see Zhambyl

34 G15 **Dzhankel'dy** see Jangeldi

151 T12 **Dzhankoy** Respublika Krym, S Ukraine 45.42N 34.24E

151 V14 **Dzhansugurov** Kaz. Zhansúgirov. Almaty, SE Kazakhstan 45.25N 79.23E

153 R9 **Dzhany-Bazar** var. Yangibazar. W Kyrgyzstan 41.40N 70.49E

150 D9 **Dzhanybek** Kaz. Zhánibek. Zapadnyy Kazakhstan, W Kazakhstan 49.27N 46.51E

126 L8 **Dzhardzhan** Respublika Sakha (Yakutiya), NE Russian Federation 68.47N 123.51E

119 S11 **Dzharylhats'ka Zatoka** gulf S Ukraine

Dzhayilgan see Jayilgan

112 S14 **Dzhelandy** SE Tajikistan 37.34N 72.34E

153 Y7 **Dzhergalan** Kir. Jyrgalan. Issyk-Kul'skaya Oblast', NE Kyrgyzstan 42.37N 78.55E

150 L8 **Dzhetygara** Kaz. Zhetiqara. Kostanay, NW Kazakhstan 52.10N 61.12E

Dzhetysay see Zhetysay

Dzhezkazgan see Zhezkazgan

152 J10 **Dzhigirbent** Turkm. Jigerbent. Lebapskiy Velayat, NE Turkmenistan 40.44N 61.56E 50.25N 15.48E

160 A10 **Dzhirgatal'** see Jirgatol

Dzhizak see Jizzakh

Dzhizakskaya Oblast' see Jizzakh Wiloyati

127 N12 **Dzhugdzhur, Khrebet** ▲ E Russian Federation

Dzhul'fa see Culfa

32 M12 **Dzhuma** see Juma

112 L9 **Dzialdowo** Warmińsko-Mazurskie, C Poland 53.13N 20.11E

113 L16 **Działoszyce** Świętokrzyskie, C Poland 50.21N 20.19E

113 G15 **Dzierżoniów** Ger. Reichenbach. Dolnośląskie, SW Poland 50.43N 16.40E

43 X11 **Dzidzantún** Yucatán, E Mexico 21.15N 89.01W

168 M8 **Dzogsool** Töv, C Mongolia 46.46N 107.18E

133 S8 **Dzungaria** var. Sungaria, Zungaria. physical region NW China

168 G5 **Dzungarian Basin** see Junggar Pendi

168 H7 **Dzuunmod** Dzavhan, C Mongolia 48.09N 97.22E

168 L8 **Dzuunmod** Töv, C Mongolia 47.45N 107.00E

Dzüün Soyonî Nuruu see Eastern Sayans

168 F8 **Dzyel** Govĭ-Altay, SW Mongolia 46.09N 93.55E

Dzvina see Western Dvina

121 J16 **Dzyarzhynsk** Rus. Dzerzhinsk; prev. Kaydanovo. Minskaya Voblasts', C Belarus 53.41N 27.09E

121 H17 **Dzyatlava** Pol. Zdzięcioł, Rus. Dyatlovo. Hrodzyenskaya Voblasts', W Belarus 53.27N 25.23E

E

E see Hubei

Éadan Doire see Edenderry

39 W6 **Eads** Colorado, C USA 38.28N 102.46 W

39 O13 **Eagar** Arizona, SW USA 34.05N 109.17 W

41 T8 **Eagle** Alaska, USA 64.47N 141.12W

11 S8 **Eagle** ∿ Newfoundland, E Canada

8 I3 **Eagle** ∿ Yukon Territory, NW Canada

31 T7 **Eagle Bend** Minnesota, N USA 46.10N 95.02W

30 M8 **Eagle Butte** South Dakota, N USA 44.58N 101.13W

31 V12 **Eagle Grove** Iowa, C USA 42.39N 93.54W

21 R2 **Eagle Lake** Maine, NE USA 47.01N 68.35W

27 U11 **Eagle Lake** Texas, SW USA 29.35N 96.19W

10 A11 **Eagle Lake** ⊕ Ontario, S Canada

37 P3 **Eagle Lake** ⊕ California, W USA

21 R3 **Eagle Lake** ⊕ Maine, NE USA

31 Y3 **Eagle Mountain** ▲ Minnesota, N USA 47.90N 90.33W

27 T6 **Eagle Mountain Lake** ⊠ Texas, SW USA

39 S9 **Eagle Nest Lake** ⊠ New Mexico, SW USA

27 P13 **Eagle Pass** Texas, SW USA 28.43N 100.31W

67 C25 **Eagle Passage** passage SW Atlantic Ocean

37 R8 **Eagle Peak** ▲ California, W USA 38.11N 119.22W

37 Q2 **Eagle Peak** ▲ California, W USA 41.16N 120.12W

39 P13 **Eagle Peak** ▲ New Mexico, SW USA 33.39N 109.36W

8 I4 **Eagle Plain** Yukon Territory, NW Canada 66.23N 136.42W

34 G15 **Eagle Point** Oregon, NW USA 42.28N 122.48W

195 N17 **Eagle Point** headland SE PNG 10.31S 149.53E

41 R11 **Eagle River** Alaska, USA 61.18N 149.38W

32 M2 **Eagle River** Michigan, N USA 47.24N 88.18W

32 L4 **Eagle River** Wisconsin, N USA 45.55N 89.15W

23 S6 **Eagle Rock** Virginia, NE USA 37.40N 79.46W

38 J13 **Eagletail Mountains** ▲ Arizona, SW USA

178 Kk2 **Ea Hleo** Đắc Lắc, S Vietnam 13.09N 108.14E

178 Kk12 **Ea Kar** Đắc Lắc, S Vietnam 12.47N 108.26E

— **Eanjum** see Anjum

— **Eanodat** see Enontekiö

10 B10 **Ear Falls** Ontario, C Canada 50.37N 93.13W

29 X10 **Earle** Arkansas, C USA 35.16N 90.28W

37 R12 **Earlimart** California, W USA 35.53N 119.17W

22 I6 **Earlington** Kentucky, S USA 37.16N 87.30W

12 H8 **Earlton** Ontario, S Canada 47.41N 79.46W

31 T13 **Early** Iowa, C USA 42.27N 95.09W

98 I11 **Earn** ∿ N Scotland, UK

193 C21 **Earnslaw, Mount** ▲ South Island, NZ 44.34S 168.26E

26 M4 **Earth** Texas, SW USA 34.13N 102.24W

23 P11 **Easley** South Carolina, SE USA 34.49N 82.36W

East see Est

East Açores Fracture Zone see East Azores Fracture Zone

99 P19 **East Anglia** physical region E England, UK

13 Q12 **East Angus** Quebec, SE Canada 45.29N 71.39W

East Antarctica see Greater Antarctica

20 E10 **East Aurora** New York, NE USA 42.44N 78.34W

East Australian Basin see Tasman Basin

East Azerbaijan see Äzarbāyān-e Sharqī

66 L9 **East Azores Fracture Zone** var. East Açores Fracture Zone. tectonic feature E Atlantic Ocean

24 M11 **East Bay** bay Louisiana, S USA

27 V11 **East Bernard** Texas, SW USA 29.31N 96.04W

31 V8 **East Bethel** Minnesota, N USA

East Borneo see Kalimantan Timur

99 P23 **Eastbourne** SE England, UK 50.46N 0.16E

13 R11 **East-Broughton** Quebec, SE Canada 45.29N 71.05W

46 M6 **East Caicos** island E Turks and Caicos Islands

192 R7 **East Cape** headland North Island, NZ 37.40S 178.31E

182 M4 **East Caroline Basin** undersea feature SW Pacific Ocean

East China Sea Chin. Dong Hai. sea NW Pacific Ocean

99 P19 **East Dereham** E England, UK 52.40N 0.55E

32 I9 **East Dubuque** Illinois, N USA 42.29N 90.38W

9 S17 **Eastend** Saskatchewan, S Canada 49.29N 108.48W

◆ COUNTRY ◇ DEPENDENT TERRITORY ◆ ADMINISTRATIVE REGION ▲ MOUNTAIN ✖ VOLCANO ⊕ LAKE
● COUNTRY CAPITAL ○ DEPENDENT TERRITORY CAPITAL ✈ INTERNATIONAL AIRPORT ▲ MOUNTAIN RANGE ∿ RIVER ⊠ RESERVOIR

251

200 Nn11 **Easter Fracture Zone** tectonic feature E Pacific Ocean
Easter Island see Pascua, Isla de
83 J18 **Eastern** ◆ province Kenya
159 Q12 **Eastern** ◆ zone E Nepal
84 L13 **Eastern** ◆ province E Zambia
85 H24 **Eastern Cape** off. Eastern Cape Province, Afr. Oos-Kaap. ◆ province SE South Africa
Eastern Desert see Sahara el Sharqiya
83 F15 **Eastern Equatoria** ◆ state SE Sudan
Eastern Euphrates see Murat Nehri
161 J17 **Eastern Ghats** ▲ SE India
194 I13 **Eastern Highlands** ◆ province C PNG
161 K25 **Eastern Province** ◆ province E Sri Lanka
Eastern Region see Ash Sharqiyah
126 I15 **Eastern Sayans** Mong. Dzüün Soyoni Nuruu, Rus. Vostochnyy Sayan. ▲ Mongolia/Russian Federation
Eastern Scheldt see Oosterschelde
Eastern Sierra Madre see Madre Oriental, Sierra
Eastern Transvaal see Mpumalanga
9 W14 **Easterville** Manitoba, C Canada 53.06N 99.52W
Easterwälde see Oosterwolde
65 M23 **East Falkland** var. Isla Soledad. island E Falkland Islands
21 P12 **East Falmouth** Massachusetts, NE USA 41.34N 70.31W
East Fayu see Fayu
East Flanders see Oost Vlaanderen
41 S6 **East Fork Chandalar River** ↔ Alaska, USA
31 U12 **East Fork Des Moines River** ↔ Iowa/Minnesota, C USA
East Frisian Islands see Ostfriesische Inseln
20 K10 **East Glenville** New York, NE USA 42.53N 73.55W
31 R4 **East Grand Forks** Minnesota, N USA 47.54N 97.59W
99 O23 **East Grinstead** SE England, UK 51.07N 0.00W
20 M12 **East Hartford** Connecticut, NE USA 41.45N 72.36W
20 M13 **East Haven** Connecticut, NE USA 41.16N 72.52W
181 T9 **East Indiaman Ridge** undersea feature E Indian Ocean
133 V16 **East Indies** island group SE Asia
East Java see Jawa Timur
33 Q6 **East Jordan** Michigan, N USA 45.09N 85.07W
East Kalimantan see Kalimantan Timur
East Kazakhstan see Vostochnyy Kazakhstan
98 I12 **East Kilbride** S Scotland, UK 55.46N 4.10W
27 R7 **Eastland** Texas, SW USA 32.24N 98.49W
33 Q9 **East Lansing** Michigan, C USA 42.44N 84.28W
37 X11 **East Las Vegas** Nevada, W USA 36.05N 115.02W
99 M23 **Eastleigh** S England, UK 50.58N 1.22W
33 V12 **East Liverpool** Ohio, N USA 40.37N 80.34W
85 J25 **East London** Afr. Oos-Londen; prev. Emonti, Port Rex. Eastern Cape, S South Africa 33.00S 27.54E
98 K12 **East Lothian** cultural region SE Scotland, UK
10 I10 **Eastmain** Quebec, E Canada 52.11N 78.27W
10 J10 **Eastmain** ↔ Quebec, C Canada
13 P13 **Eastmain** Quebec, SE Canada 45.19N 72.18W
23 U6 **Eastman** Georgia, SE USA 32.12N 83.10W
183 O3 **East Mariana Basin** undersea feature W Pacific Ocean
32 K11 **East Moline** Illinois, N USA 41.30N 90.26W
195 O12 **East New Britain** ◆ province E PNG
31 T15 **East Nishnabotna River** ↔ Iowa, C USA
207 V12 **East Novaya Zemlya Trough** var. Novaya Zemlya Trough. undersea feature W Kara Sea
East Nusa Tenggara see Nusa Tenggara Timur
23 X4 **Easton** Maryland, NE USA 38.46N 76.04W
20 I14 **Easton** Pennsylvania, NE USA 40.40N 75.13W
200 N8 **East Pacific Rise** undersea feature E Pacific Ocean
East Pakistan see Bangladesh
33 V12 **East Palestine** Ohio, N USA 40.49N 80.32W
32 L12 **East Peoria** Illinois, N USA 40.39N 89.34W
25 S3 **East Point** Georgia, SE USA 33.40N 84.26W
21 U6 **Eastport** Maine, NE USA 44.54N 66.59W
29 Z8 **East Prairie** Missouri, C USA 36.46N 89.23W
21 O9 **East Providence** Rhode Island, NE USA 41.48N 71.20W
22 L7 **East Ridge** Tennessee, S USA 35.00N 85.15W
99 N16 **East Riding** cultural region N England, UK
20 F9 **East Rochester** New York, NE USA 43.06N 77.29W
32 K15 **East Saint Louis** Illinois, N USA 38.35N 90.07W
67 K21 **East Scotia Basin** undersea feature E Scotia Sea
East Sea see Japan, Sea of
194 F11 **East Sepik** ◆ province NW PNG
181 N4 **East Sheba Ridge** undersea feature W Arabian Sea
East Siberian Sea see Vostochno-Sibirskoye More
20 I14 **East Stroudsburg** Pennsylvania, NE USA 40.59N 75.10W
East Tasmanian Rise/East

Tasmania Plateau/East Tasmania Rise see East Tasman Plateau
199 Hh13 **East Tasman Plateau** var. East Tasmanian Plateau, East Tasmania Plateau, East Tasmania Rise. undersea feature SW Tasman Sea
66 L7 **East Thulean Rise** undersea feature N Atlantic Ocean
175 S16 **East Timor** var. Loro Sae prev. Portuguese Timor, Timor Timur ◇ disputed territory SE Asia
23 V4 **Eastville** Virginia, NE USA 37.18N 75.57W
37 T3 **East Walker River** ↔ California/Nevada, W USA
190 D1 **Eateringinna Creek** ↔ South Australia
39 T3 **Eaton** Colorado, C USA 40.31N 104.42W
12 Q12 **Eaton** Quebec, SE Canada
9 S16 **Eatonia** Saskatchewan, S Canada 51.13N 109.22W
33 Q10 **Eaton Rapids** Michigan, N USA 42.30N 84.39W
25 U4 **Eatonton** Georgia, SE USA 33.19N 83.23W
34 H9 **Eatonville** Washington, NW USA 46.51N 122.19W
32 K6 **Eau Claire** Wisconsin, N USA 44.49N 91.30W
Eau Claire, Lac à L' See St.Clair, Lake
10 I7 **Eau Claire, Lac à l'** ❍ Quebec, SE Canada
32 K6 **Eau Claire River** ↔ Wisconsin, N USA
196 J16 **Eauripik Atoll** atoll Caroline Islands, C Micronesia
199 I8 **Eauripik Rise** undersea feature W Pacific Ocean
104 K15 **Eauze** Gers, S France 43.52N 0.06E
43 P11 **Ébano** San Luis Potosí, C Mexico 22.13N 98.22W
99 J23 **Ebbw Vale** S Wales, UK 51.48N 3.13W
81 E17 **Ebebiyin** NE Equatorial Guinea 2.08N 11.15E
97 H22 **Ebeltoft** Århus, C Denmark 56.12N 10.40E
189 X5 **Ebenfurth** Niederösterreich, E Austria 47.51N 16.21E
20 D14 **Ebensburg** Pennsylvania, NE USA 40.28N 78.42W
189 S5 **Ebensee** Oberösterreich, N Austria 47.48N 13.45E
103 H20 **Eberbach** Baden-Württemberg, SW Germany 49.28N 8.58E
124 Q9 **Eber Gölü** salt lake C Turkey
189 U9 **Eberndorf** Slvn. Dobrla Vas. Kärnten, S Austria 46.33N 14.35E
189 R4 **Eberschwang** Oberösterreich, N Austria 48.09N 13.37E
102 O11 **Eberswalde-Finow** Brandenburg, E Germany 52.49N 13.48E
172 Oo5 **Ebetsu** var. Ebetu. Hokkaidō, NE Japan 43.08N 141.37E
Ebetu see Ebetsu
164 I4 **Ebinur Hu** ❍ NW China
144 I3 **Ebla** Ar. Tell Mardikh. site of ancient city Idlib, NW Syria
Eblana see Dublin
110 H7 **Ebnat** Sankt Gallen, NE Switzerland 47.16N 9.07E
109 L18 **Eboli** Campania, S Italy 40.39N 15.01E
81 E16 **Ebolowa** Sud, S Cameroon 2.55N 11.10E
81 N21 **Ebombo** Kasai Oriental, C Dem. Rep. Congo (Zaire) 5.42S 26.07E
201 T9 **Ebon Atoll** var. Epoon. atoll Ralik Chain, S Marshall Islands
Ebora see Évora
Eboracum see York
Eborodunum see Yverdon
103 J19 **Ebrach** Bayern, C Germany 49.49N 10.30E
189 X5 **Ebreichsdorf** Niederösterreich, E Austria 47.58N 16.24E
107 S6 **Ebro** ↔ NE Spain
107 N3 **Ebro, Embalse del** ❑ N Spain
123 Hh8 **Ebro Fan** undersea feature W Mediterranean Sea
Eburacum see York
Ebusus see Eivissa
101 F20 **Écaussinnes-d'Enghien** Hainaut, SW Belgium 50.34N 4.10E
Ecbatana see Hamadān
23 Q6 **Eccles** West Virginia, NE USA 37.47N 81.22W
117 L14 **Eceabat** Çanakkale, NW Turkey 40.11N 26.23E
179 P9 **Echague** Luzon, N Philippines 16.42N 121.37E
Ech Cheliff/Ech Chleff see Chlef
Echeng see Ezhou
171 Kk13 **Echigo-sanmyaku** ▲ Honshū, C Japan
117 C18 **Echinádes** island group W Greece
116 J12 **Echínos** var. Ehinos, Echínos. Anatolikí Makedonía kai Thráki, NE Greece 41.16N 25.00E
142 B9 **Echizen-misaki** headland Honshū, SW Japan 35.59N 135.57E
Echmiadzin see Ejmiatsin
15 Hh5 **Echo Bay** Northwest Territories, NW Canada
7 Y11 **Echo Bay** Nevada, W USA
38 L9 **Echo Cliffs** cliff Arizona, SW USA
37 Q10 **Echo Summit** ▲ California, W USA 38.47N 120.06W
120 L12 **Échouani, Lac** ❍ Quebec, SE Canada
12 L8 **Echt** Limburg, SE Netherlands 51.07N 5.52E
101 M16 **Echterdingen** ✈ (Stuttgart) Baden-Württemberg, SW Germany 48.40N 9.13E
103 K21 **Echternach** Grevenmacher, E Luxembourg 49.49N 6.25E
101 N24 **Echuca** Victoria, SE Australia 36.10S 144.45E
191 N11 **Écija** anc. Astigi. Andalucía, SW Spain 37.33N 5.04W
106 L14 **Eckengraf** see Viesite
Eckernförde Schleswig-Holstein,
100 I7 ...N Germany 54.28N 9.49E

102 J7 **Eckernförder Bucht** inlet N Germany
104 L7 **Écommoy** Sarthe, NW France 47.51N 0.15E
12 L10 **Écorce, Lac de l'** ❍ Quebec, SE Canada
13 Q8 **Écorces, Rivière aux** ↔ Quebec, SE Canada
58 C7 **Ecuador** off. Republic of Ecuador. ◆ republic NW South America
82 L10 **Ed** var. Edd. SE Eritrea 13.54N 41.39E
97 I17 **Eda** Västra Götaland, S Sweden 58.55N 11.55E
100 I9 **Edam** Noord-Holland, C Netherlands 52.31N 5.03E
98 K4 **Eday** island NE Scotland, UK
27 S17 **Edcouch** Texas, SW USA 26.17N 97.57W
Edd see Ed
82 I7 **Ed Da'ein** Southern Darfur, W Sudan 11.25N 26.07E
82 G8 **Ed Damazin** var. Ad Damazin. Blue Nile, E Sudan 11.45N 34.20E
82 E8 **Ed Damer** var. Ad Damar, Ad Dāmir. River Nile, NE Sudan 17.37N 33.58E
82 E8 **Ed Debba** Northern, N Sudan 18.01N 30.55E
82 F10 **Ed Dueim** var. Ad Duwaym, Ad Duwēm. White Nile, C Sudan 14.00N 32.19E
24 K5 **Eddceville** Mississippi, S USA 31.29N 90.36W
191 Q16 **Eddystone Point** headland Tasmania, SE Australia 41.01S 148.18E
99 I25 **Eddystone Rocks** rocks SW England, UK
31 W15 **Eddyville** Iowa, C USA 41.09N 92.37W
22 N7 **Eddyville** Kentucky, S USA 37.05N 88.04W
100 I13 **Ede** Gelderland, C Netherlands 52.03N 5.40E
79 T16 **Ede** Osun, SW Nigeria 7.40N 4.21E
81 D16 **Edéa** Littoral, SW Cameroon 3.46N 10.07E
113 M20 **Edelény** Borsod-Abaúj-Zemplén, NE Hungary 48.19N 20.44E
191 R12 **Eden** New South Wales, SE Australia 37.04S 149.51E
23 T8 **Eden** North Carolina, SE USA 36.29N 79.46W
27 P9 **Eden** Texas, SW USA 31.13N 99.51W
99 K14 **Eden** ↔ NW England, UK
85 L23 **Edenburg** Free State, C South Africa 29.43S 25.54E
193 D24 **Edendale** Southland, South Island, NZ 46.18S 168.48E
99 E18 **Edenderry** Ir. Éadan Doire. C Ireland 53.21N 7.03W
190 L11 **Edenhope** Victoria, SE Australia 37.04S 141.15E
23 X8 **Edenton** North Carolina, SE USA 36.06N 76.46W
103 G16 **Eder** ↔ NW Germany
176 Yy14 **Ederah** ↔ Irian Jaya, E Indonesia
103 H15 **Edersee** ❍ W Germany
Edessa see Şanlıurfa
116 E13 **Édessa** var. Édhessa. Kentrikí Makedonía, N Greece 40.48N 22.03E
Edfu see Idfu
31 P16 **Edgar** Nebraska, C USA 40.22N 97.58W
21 P13 **Edgartown** Martha's Vineyard, Massachusetts, NE USA 41.22N 70.30W
41 X13 **Edgecumbe, Mount** ▲ Baranof Island, Alaska, USA 57.03N 135.45W
23 Q13 **Edgefield** South Carolina, SE USA 33.47N 81.55W
31 P6 **Edgeley** North Dakota, N USA 46.19N 98.42W
30 J11 **Edgemont** South Dakota, N USA 43.18N 103.49W
156 O3 **Edgeøya** island S Svalbard
29 Q4 **Edgerton** Kansas, C USA 38.45N 95.00W
31 S10 **Edgerton** Minnesota, N USA 43.52N 96.07W
23 X3 **Edgewood** Maryland, NE USA 39.20N 76.21W
27 V6 **Edgewood** Texas, SW USA 32.42N 95.53W
31 V9 **Edina** Minnesota, N USA 44.53N 93.21W
29 U2 **Edina** Missouri, C USA 40.10N 92.10W
27 S17 **Edinburg** Texas, SW USA 26.18N 98.09W
87 M24 **Edinburgh** var. Settlement of Edinburgh. O (Tristan da Cunha) NW Tristan da Cunha 37.03S 12.18W
98 J12 **Edinburgh** O S Scotland, UK 55.57N 3.13W
33 P14 **Edinburgh** Indiana, N USA 39.19N 86.00W
98 J12 **Edinburgh** ✈ S Scotland, UK 55.57N 3.22W
118 L8 **Edineţ** var. Edineţi, Rus. Yedintsy. NW Moldova 48.10N 27.18E
Edineţi see Edineţ
Edingen see Enghien
142 B9 **Edirne** Eng. Adrianople; anc. Adrianopolis, Hadrianopolis. Edirne, NW Turkey 41.40N 26.34E
117 H13 **Edirne** ◆ province NW Turkey
20 K15 **Edison** New Jersey, NE USA 40.31N 74.24W
23 S15 **Edisto Island** South Carolina, SE USA 32.34N 80.17W
23 S15 **Edisto River** ↔ South Carolina, SE USA
35 S10 **Edith, Mount** ▲ Montana, NW USA 46.25N 111.10W
35 V10 **Edmond** Oklahoma, C USA 35.40N 97.28W
34 H8 **Edmonds** Washington, NW USA 47.48N 122.22W
9 Q16 **Edmonton** Alberta, SW Canada 53.34N 113.25W
22 K7 **Edmonton** Kentucky, S USA 36.57N 85.37W
8 Q4 **Edmonton** ✈ Alberta, SW Canada 53.22N 113.43W
77 V9 **Edmore** North Dakota, N USA 48.22N 98.26W
11 N13 **Edmundston** New Brunswick, SE Canada 47.22N 68.20W
27 U12 **Edna** Texas, SW USA 28.58N 96.39W
41 X14 **Edna Bay** Kosciusko Island, Alaska, USA 55.54N 133.40W

79 U16 **Edo** ◆ state S Nigeria
108 F6 **Edolo** Lombardia, N Italy 46.12N 10.19E
66 L6 **Edoras Bank** undersea feature C Atlantic Ocean
98 G3 **Edrachillis Bay** bay NW Scotland, UK
142 H12 **Edremit** Balıkesir, NW Turkey 39.34N 27.01E
142 H12 **Edremit Körfezi** gulf NW Turkey
97 P14 **Edsbro** Stockholm, C Sweden 59.54N 18.30E
97 N18 **Edsbruk** Kalmar, S Sweden 58.01N 16.30E
96 M12 **Edsbyn** Gävleborg, C Sweden 61.22N 15.45E
9 O14 **Edson** Alberta, SW Canada 53.36N 116.28W
64 K13 **Eduardo Castex** La Pampa, C Argentina 35.52S 64.15W
60 F12 **Eduardo Gomes** ✈ (Manaus) Amazonas, NW Brazil 3.05S 55.15W
69 U9 **Edward, Lake** var. Albert Edward Nyanza, Edward Nyanza, Lac Idi Amin, Lake Rutanzige. ❍ Uganda/Dem. Rep. Congo (Zaire)
Edward Nyanza see Edward, Lake
27 O10 **Edwards Plateau** plain Texas, SW USA
32 J11 **Edwards River** ↔ Illinois, N USA
32 K15 **Edwardsville** Illinois, N USA 38.48N 89.57W
205 O13 **Edward VII Peninsula** peninsula Antarctica
205 X4 **Edward VIII Gulf** bay Antarctica
8 J11 **Edziza, Mount** ▲ British Columbia, W Canada 57.43N 130.39W
25 H17 **Edzo** prev. Rae-Edzo. Northwest Territories, NW Canada 62.43N 115.55W
41 N12 **Eek** Alaska, USA 60.13N 162.01W
101 D16 **Eeklo** var. Eekloo. Oost-Vlaanderen, NW Belgium 51.10N 3.34E
Eekloo see Eeklo
41 N12 **Eek River** ↔ Alaska, USA
100 N6 **Eelde** Drenthe, NE Netherlands 53.07N 6.30E
37 N12 **Eel River** ↔ California, W USA
33 P12 **Eel River** ↔ Indiana, N USA
Eems see Ems
100 O4 **Eemshaven** Groningen, NE Netherlands 53.27N 6.50E
100 M11 **Eems Kanaal** canal NE Netherlands
100 O5 **Eerbeek** Gelderland, E Netherlands 52.06N 6.04E
101 C17 **Eernegem** West-Vlaanderen, W Belgium 51.08N 3.03E
101 J15 **Eersel** Noord-Brabant, S Netherlands 51.21N 5.19E
Eesti Vabariik see Estonia
189 S4 **Eferding** Oberösterreich, N Austria 48.18N 14.00E
32 M15 **Effingham** Illinois, N USA 39.07N 88.32W
119 N15 **Eforie-Nord** Constanţa, SE Romania 44.04N 28.37E
119 N15 **Eforie Sud** Constanţa, E Romania 44.00N 28.38E
85 F18 **Efyrnwy, Afon** see Vyrnwy
169 N7 **Eg** Hentiy, N Mongolia 48.42N 110.07E
109 G23 **Egadi, Isole** island group S Italy
37 X6 **Egan Range** ▲ Nevada, W USA
12 J12 **Eganville** Ontario, SE Canada 45.33N 77.03W
Ege Denizi see Aegean Sea
41 O14 **Egegik** Alaska, USA 58.13N 157.22W
113 L21 **Éger** Ger. Erlau. Heves, NE Hungary 47.54N 20.21E
Éger see Cheb, Czech Republic
Eger see Ohre, Czech Republic
Egeria see Égina
181 P8 **Egeria Fracture Zone** tectonic feature W Indian Ocean
97 C17 **Egersund** Rogaland, S Norway 58.27N 6.01E
110 J7 **Egg** Vorarlberg, NW Austria 47.27N 9.55E
103 L15 **Egge-gebirge** ▲ C Germany
189 Q4 **Eggelsberg** Niederösterreich, N Austria 48.04N 13.00E
189 W2 **Eggenburg** Niederösterreich, NE Austria 48.36N 15.49E
103 N22 **Eggenfelden** Bayern, SE Germany 48.24N 12.45E
20 J17 **Egg Harbor City** New Jersey, NE USA 39.31N 74.39W
67 G25 **Egg Island** island W Saint Helena
191 N14 **Egg Lagoon** Tasmania, SE Australia 39.42S 143.57E
42 E8 **Egido Insurgentes** Baja California Sur, W Mexico 25.14N 111.45W
143 T12 **Egilsstaðir** Austurland, E Iceland 65.14N 14.21W
Egina see Aígina
105 N12 **Egletons** Corrèze, C France 45.24N 2.01E
100 H9 **Egmond aan Zee** Noord-Holland, NW Netherlands 52.37N 4.37E
Egmont see Taranaki, Mount
192 I10 **Egmont, Cape** headland North Island, NZ 39.18S 173.44E
35 Y10 **Egoli** see Johannesburg
Eğri Palanka see Kriva Palanka
127 Pp4 **Egvekinot** Chukotskiy Avtonomnyy Okrug, NE Russian Federation 66.19N 179.10E
77 V9 **Egypt** off. Arab Republic of Egypt, Ar. Jumhūrīyah Miṣr al 'Arabīyah, prev. United Arab Republic, anc. Aegyptus. ◆ republic NE Africa
151 T8 **Egypt, Lake Of** ❍ Illinois, N USA
127 N14 **Ehen Hudag** see Alxa Zuoqi

170 E15 **Ehime** off. Ehime-ken. ◆ prefecture SW Japan
103 J21 **Ehingen** Baden-Württemberg, S Germany 48.16N 9.43E
23 R14 **Ehrhardt** South Carolina, SE USA 33.06N 81.00W
110 L7 **Ehrwald** Tirol, W Austria 47.24N 10.52E
97 P14 **Eibar** País Vasco, N Spain 43.10N 2.28W
100 O11 **Eibergen** Gelderland, E Netherlands 52.06N 6.39E
189 X10 **Eibiswald** Steiermark, SE Austria 46.41N 15.14E
189 P8 **Eichham** ▲ SW Austria
103 J15 **Eichsfeld** hill range C Germany
103 K21 **Eichstätt** Bayern, SE Germany 48.53N 11.11E
102 H8 **Eider** ↔ N Germany
96 E13 **Eidfjord** Hordaland, S Norway 60.26N 7.05E
96 D13 **Eidfjorden** fjord S Norway
96 E13 **Eidsvåg** Møre og Romsdal, S Norway 62.46N 8.00E
97 I14 **Eidsvoll** Akershus, S Norway 60.19N 11.16E
156 N2 **Eidsvollfjellet** ▲ NW Svalbard 79.13N 13.23E
Eier-Berg see Suur Munamägi
110 E9 **Eiger** ▲ C Switzerland 46.33N 8.02E
98 G10 **Eigg** island W Scotland, UK
161 D24 **Eight Degree Channel** channel India/Maldives
46 G1 **Eight Mile Rock** Grand Bahama Island, N Bahamas 26.28N 78.43W
204 J9 **Eights Coast** physical region Antarctica
188 K6 **Eighty Mile Beach** beach Western Australia
101 L18 **Eijsden** Limburg, SE Netherlands 50.47N 5.41E
97 G15 **Eikeren** ❍ S Norway
Eil see Eyl
Eilat see Elat
191 O12 **Eildon** Victoria, SE Australia 37.17S 145.57E
191 O12 **Eildon, Lake** ❍ Victoria, SE Australia
82 E8 **Eilei** Northern Kordofan, C Sudan 16.33N 30.54E
103 N15 **Eilenburg** Sachsen, E Germany 51.28N 12.37E
96 H13 **Eina** Oppland, S Norway 60.37N 10.37E
103 I14 **Einbeck** Niedersachsen, C Germany 51.50N 9.52E
101 K15 **Eindhoven** Noord-Brabant, S Netherlands 51.25N 5.30E
110 G8 **Einsiedeln** Schwyz, NE Switzerland 47.06N 8.44E
Eipel see Ipel'
Éire see Ireland, Republic of
Éireann, Muir see Irish Sea
Eirik Outer Ridge see Eirik Ridge
66 I6 **Eirik Ridge** var. Eirik Outer Ridge. undersea feature E Labrador Sea
156 I3 **Eiriksjökull** ▲ C Iceland 64.45N 20.23W
61 B14 **Eirunepé** Amazonas, N Brazil 6.37S 69.52W
101 L17 **Eisden** Limburg, NE Belgium 51.05N 5.42E
103 J16 **Eisenach** Thüringen, C Germany 50.58N 10.19E
189 U6 **Eisenerz** Steiermark, SE Austria 47.33N 14.52E
102 L13 **Eisenhüttenstadt** Brandenburg, E Germany 52.09N 14.36E
189 U10 **Eisenkappel** Slvn. Železna Kapela. Kärnten, S Austria 46.27N 14.33E
189 Y5 **Eisenstadt** Burgenland, E Austria 47.49N 16.31E
Eishü see Yŏngju
121 H15 **Eišiškes** Šalčininkai, SE Lithuania 54.10N 24.59E
103 L15 **Eisleben** Sachsen-Anhalt, C Germany 51.31N 11.33E
202 I3 **Eita** Tarawa, W Kiribati 1.21N 173.04E
Eitape see Aitape
107 V11 **Eivissa** var. Iviza, Cast. Ibiza; anc. Ebusus. Eivissa, Spain, W Mediterranean Sea 38.54N 1.25E
103 N22 **Eivissa** var. Iviza, Cast. Ibiza; anc. Ebusus. island Islas Baleares, Spain, W Mediterranean Sea
45 N10 **Ejea de los Caballeros** Aragón, NE Spain 42.07N 1.09W
191 N14 **Eja Lagoon** Tasmania, SE Australia 39.42S 143.57E
42 E8 **Ejido Insurgentes** Baja California Sur, W Mexico 25.14N 111.45W
101 G20 **Éghezée** Namur, C Belgium 50.36N 4.55E
156 L2 **Egilsstaðir** Austurland, E Iceland 65.14N 14.21W
168 I12 **Ejin Qi** var. Dalain Hob. Nei Mongol Zizhiqu, N China 41.59N 101.04E
143 T12 **Ejmiadzin** var. Ejmiatsin, Rus. Echmiadzin. W Armenia 40.10N 44.17E
Ejmiatsin see Ejmiatsin
42 E8 **Ejutla** see Ejutla de Crespo
35 Y10 **Ejutla de Crespo** var. Ejutla. Oaxaca, SE Mexico 16.30N 96.40W
131 N16 **Ekalaka** Montana, NW USA 45.52N 104.32W
192 J10 **Ekapa** see Cape Town
127 Pp4 **Ekaterinodar** see Krasnodar
130 M15 **Ekenäs** Fin. Tammisaari. Etelä-Suomi, SW Finland 60.00N 23.30E
83 H18 **Ekerem** see Okarem
192 M13 **Eketahuna** Manawatu-Wanganui, North Island, NZ 40.41S 175.40E
Ekhinos see Echínos
127 O6 **Ekiatapskiy Khrebet** ▲ NE Russian Federation
151 T8 **Ekibastuz** Pavlodar, NE Kazakhstan 51.45N 75.22E
127 N14 **Ekimchan** Amurskaya Oblast', SE Russian Federation 53.04N 132.56E

97 O15 **Ekoln** ❍ C Sweden
82 I7 **Ekowit** Red Sea, NE Sudan 18.46N 37.07E
97 L19 **Eksjö** Jönköping, S Sweden 57.40N 15.00E
95 I15 **Ekträsk** Västerbotten, N Sweden 64.28N 19.49E
41 O13 **Ekuk** Alaska, USA 58.48N 158.25W
10 F9 **Ekwan** ↔ Ontario, C Canada
41 O13 **Ekwok** Alaska, USA 59.21N 157.28W
177 G6 **Ela** Mandalay, C Myanmar 19.37N 96.15E
83 N15 **El Ábrēd** Somali, E Ethiopia
117 F22 **Elafónisos** island S Greece
117 F22 **Elafónisou, Porthmós** strait S Greece
El-Aïoun see El Aaiún
77 U8 **'Alamein** see Al 'Alamayn
103 O18 **El Alazán** Veracruz-Llave, C Mexico 21.06N 97.43W
59 J18 **El Alto** var. La Paz. ✈ (La Paz) La Paz, W Bolivia 16.31S 68.07W
56 J8 **El Amparo** see El Amparo de Apure
56 J8 **El Amparo de Apure** var. El Amparo. Apure, C Venezuela 7.03N 70.46W
175 Ss12 **Elara** Pulau Ambelau, E Indonesia 3.49S 127.10E
175 Ss12 **Araïch/El Araïche** see Larache
43 Q12 **El Arco** Baja California, NW Mexico 28.03N 113.25W
77 X7 **El 'Arish** var. Al 'Arīsh. NE Egypt 33.00N 31.00E
117 L25 **Elása** island S Greece
117 E15 **Elassón** var. Elassóna. Thessalía, C Greece 39.54N 22.11E
107 N2 **El Astillero** Cantabria, N Spain 43.23N 3.45W
144 F14 **Elat** var. Eilat, Elath. Southern, S Israel 29.33N 34.57E
Elath see Elat, Israel
Elath see Al 'Aqabah, Jordan
117 D18 **Eláti** ▲ Lefkáda, Iónioi Nísoi, C Greece, C Mediterranean Sea 38.43N 20.38E
196 L16 **Elato Atoll** atoll Caroline Islands, C Micronesia
82 C7 **El'Atrun** Northern Darfur, NW Sudan 18.10N 26.40E
76 H6 **El Ayoun** var. El Aaiun, El-Aïoun, La Youne. NE Morocco 34.38N 2.30W
143 N14 **Elâziğ** var. Elâziz. Elâziğ, E Turkey 38.41N 39.13E
143 O14 **Elâziğ** var. Elâziz. ◆ province C Turkey
Elâziz see Elâziğ
31 W13 **Eldora** Iowa, C USA 42.21N 93.06W
62 G12 **Eldorado** Misiones, NE Argentina 26.22S 54.33W
42 I9 **El Dorado** Sinaloa, C Mexico 24.19N 107.22W
29 U14 **El Dorado** Arkansas, C USA 33.12N 92.40W
32 M17 **Eldorado** Illinois, N USA 37.48N 88.26W
29 O6 **El Dorado** Kansas, C USA 37.49N 96.51W
28 K12 **Eldorado** Oklahoma, C USA 34.28N 99.38W
27 O9 **Eldorado** Texas, SW USA 30.51N 100.36W
57 Q8 **El Dorado** Bolívar, E Venezuela 6.45N 61.37W
56 F10 **El Dorado** ✈ (Bogotá) Cundinamarca, C Colombia 1.15N 71.52W
29 S8 **El Dorado Lake** ❍ Kansas, C USA
29 U9 **El Dorado Springs** Missouri, C USA 37.53N 94.01W
83 H18 **Eldoret** Rift Valley, W Kenya 0.31N 35.16E
31 Z14 **Eldridge** Iowa, C USA 41.39N 90.34W
97 J23 **Eldsberga** Halland, S Sweden 56.36N 13.00E
97 Q8 **Electra** Texas, SW USA 34.01N 98.55W
39 R8 **Electra, Lake** ❍ Colorado, C USA
7 O7 **Electra Lake** ❍ Colorado, C USA
40 B8 **Eleele** Haw. 'Ele'ele. Kauai, Hawaii, USA, C Pacific Ocean 21.54N 159.35W
Elefantes see Olifants
117 H19 **Elefsína** prev. Elevsís. Attikí, C Greece 38.02N 23.33E
117 G19 **Elefthérai** anc. Eleutherae. site of ancient city Attikí/Sterá Ellás, C Greece 38.12N 23.12E
116 I13 **Eleftheroúpoli** prev. Elevtheroúpolis. Anatolikí Makedonía kai Thráki, NE Greece 40.56N 24.16E
45 N10 **El Eglab** see Al Eglab
76 F10 **El Eglab** ▲ SW Algeria
62 F10 **Eleja** Jelgava, C Latvia 56.24N 23.41E
121 E16 **Elek** see Elk
83 H15 **Elektrėnai** Kaišiadorys, SE Lithuania 54.47N 24.39E
130 L13 **Elektrostal'** Moskovskaya Oblast', W Russian Federation 55.46N 38.23E
83 H15 **Elemi Triangle** disputed region Kenya/Sudan
56 G16 **El Encanto** Amazonas, S Colombia 1.33N 73.12W
39 R14 **Elephant Butte Reservoir** ❑ New Mexico, SW USA
El Éphant, Chaine de l' see Dâmrei, Chuŏr Phnum
204 G2 **Elephant Island** island South Shetland Islands, Antarctica
Elephant River see Olifants
El Escorial see San Lorenzo de El Escorial
Élesd see Aleşd
116 F11 **Eleshnitsa** ▲ W Bulgaria
143 S13 **Eleşkirt** Ağrı, E Turkey 39.22N 42.48E
44 F5 **El Estor** Izabal, E Guatemala 15.31N 89.19W
Eleutherae see Eléftheres
46 I2 **Eleuthera Island** island N Bahamas

65 H22 **El Calafate** var. Calafate. Santa Cruz, S Argentina 50.19S 72.12W
97 O8 **El Callao** Bolívar, E Venezuela 7.18N 61.48W
27 U12 **El Campo** Texas, SW USA 29.12N 96.16W
56 I7 **El Cantón** Barinas, W Venezuela 7.25N 71.16W
37 Q8 **El Cajon** California, W USA 32.46N 119.39W
56 H5 **El Carmelo** Zulia, NW Venezuela 10.21N 71.46W
64 J5 **El Carmen** Jujuy, NW Argentina 24.24S 65.16W
56 E5 **El Carmen de Bolívar** Bolívar, NW Colombia 9.45N 75.11W
97 O8 **El Carmen de Bolívar** Bolívar, SE Colombia 6.25N 63.34W
44 M12 **El Castillo de La Concepción** Río San Juan, S Nicaragua 10.58N 84.24W
El Cayo see San Ignacio
37 X17 **El Centro** California, ✈ USA 32.47N 115.33W
57 N6 **El Chaparro** Anzoátegui, NE Venezuela 9.08N 65.01W
107 Q12 **Elche** var. Elx-Elche; anc. Ilici, Lat. Illicis. País Valenciano, E Spain 38.16N 0.40W
107 Q12 **Elche de la Sierra** Castilla-La Mancha, C Spain 38.27N 2.03W
43 U15 **El Chichonal, Volcán** ▲ SE Mexico 17.20N 93.12W
42 C2 **El Chinero** Baja California, NW Mexico
189 X15 **Elcho Island** island Wessel Islands, Northern Territory, N Australia
65 H18 **El Corcovado** Chubut, SW Argentina 43.31S 71.30W
107 R12 **Elda** País Valenciano, E Spain 38.28N 0.46W
102 M10 **Elde** ↔ NE Germany
100 L12 **Elden** Gelderland, E Netherlands 51.57N 5.53E
83 J16 **El Der** spring/well S Ethiopia 3.55N 39.48E
El Dere see Ceel Dheere
42 J3 **El Desemboque** Sonora, NW Mexico 30.33N 112.58W
56 F5 **El Difícil** var. Ariguaní. Magdalena, N Colombia 9.49N 74.16W
126 Mm11 **El'dikan** Respublika Sakha (Yakutiya), NE Russian Federation 60.46N 135.04E
El Djazair see Alger
El Djelfa see Djelfa
31 X15 **Eldon** Iowa, C USA 40.55N 92.13W
29 U5 **Eldon** Missouri, C USA 38.21N 92.34W
56 E13 **El Doncello** Caquetá, S Colombia 1.43N 75.16W
El Eglab see Al Eglab
76 F10 **El Eglab** ▲ SW Algeria
62 F10 **Eleja** Jelgava, C Latvia 56.24N 23.41E
Elek see Elk
83 H15 **Elektrėnai** Kaišiadorys, SE Lithuania 54.47N 24.39E
130 L13 **Elektrostal'** Moskovskaya Oblast', W Russian Federation 55.46N 38.23E
116 F11 **Eleshnitsa** ▲ W Bulgaria
143 S13 **Eleşkirt** Ağrı, E Turkey 39.22N 42.48E
44 F5 **El Estor** Izabal, E Guatemala 15.31N 89.19W
Eleutherae see Eléftheres
46 I2 **Eleuthera Island** island N Bahamas

252

◆ COUNTRY   ◇ DEPENDENT TERRITORY   ◆ ADMINISTRATIVE REGION   ▲ MOUNTAIN   ▲ VOLCANO   ❍ LAKE
● COUNTRY CAPITAL   O DEPENDENT TERRITORY CAPITAL   ✈ INTERNATIONAL AIRPORT   ▲ MOUNTAIN RANGE   ↔ RIVER   ❑ RESERVOIR

39 S5 **Elevenmile Canyon Reservoir** ☒ Colorado, C USA
29 W8 **Eleven Point River** ↭ Arkansas/Missouri, C USA
Elevsís see Elefsína
Elevtheroúpoli see Eleftheroúpoli
77 W8 **El Faiyûm** var. Al Fayyûm. N Egypt 29.24N 30.52E
82 B10 **El Fasher** var. Al Fâshir. Northern Darfur, W Sudan 13.37N 25.22E
77 W8 **El Fashn** var. Al Fashn. C Egypt 28.49N 30.54E
El Ferrol/El Ferrol del Caudillo see Ferrol
41 W13 **Elfin Cove** Chichagof Island, Alaska, USA 58.09N 136.16W
107 N4 **El Fluvià** ↭ NE Spain
42 H7 **El Fuerte** Sinaloa, W Mexico 26.28N 108.34W
82 D11 **El Geili** Western Kordofan, C Sudan 11.43N 28.19E
El Gedaref see Gedaref
82 A10 **El Geneina** var. Ajjinena, Al-Genain, Al Junaynah. Western Darfur, W Sudan 13.27N 22.30E
98 J8 **Elgin** NE Scotland, UK 57.39N 3.19W
32 M10 **Elgin** Illinois, N USA 42.02N 88.16W
31 P14 **Elgin** Nebraska, C USA 41.58N 98.04W
37 Y9 **Elgin** Nevada, W USA 37.19N 114.30W
30 L6 **Elgin** North Dakota, N USA 46.24N 101.51W
28 M12 **Elgin** Oklahoma, C USA 34.46N 98.17W
27 T10 **Elgin** Texas, SW USA 30.21N 97.22W
127 N9 **El'ginskiy** Respublika Sakha (Yakutiya), NE Russian Federation 64.27N 141.57E
77 W8 **El Giza** var. Al Jizah, Gîza, Gizeh. N Egypt 30.01N 31.13E
76 J8 **El Goléa** var. Al Golea. C Algeria 30.35N 2.58E
42 D2 **El Golfo de Santa Clara** Sonora, NW Mexico 31.44N 114.34W
83 G18 **Elgon, Mount** ▲ E Uganda 1.07N 34.29E
96 I10 **Elgpiggen** ▲ S Norway 62.13N 11.18E
107 T4 **El Grado** Aragón, NE Spain 42.09N 0.13E
42 L6 **El Guaje, Laguna** ☒ NE Mexico
56 H6 **El Guayabo** Zulia, W Venezuela 8.37N 72.19W
79 O6 **El Guettâra** oasis N Mali 22.01N 3.00W
78 J6 **El Hammâmi** desert N Mauritania
78 M5 **El Hank** cliff N Mauritania
El Haseke see Al Ḥasakah
82 H10 **El Hawata** Gedaref, E Sudan 13.25N 34.42E
El Higo see Higos
176 Uu16 **Eliase** Pulau Selaru, E Indonesia 8.16S 130.49E
Elías Piña see Comendador
27 R6 **Eliasville** Texas, SW USA 32.55N 98.46W
Elichpur see Achalpur
39 V13 **Elida** New Mexico, SW USA 33.57N 103.39W
117 F18 **Elikónas** ▲ C Greece
69 T10 **Elila** ↭ W Dem. Rep. Congo (Zaire)
41 N9 **Elim** Alaska, USA 64.37N 162.15W
Elimberrum see Auch
Eliocroca see Lorca
63 B16 **Elisa** Santa Fe, C Argentina 30.42S 61.04W
Elisabethstedt see Dumbrăveni
Élisabethville see Lubumbashi
131 O13 **Elista** Respublika Kalmykiya, SW Russian Federation 46.17N 44.09E
190 I9 **Elizabeth** South Australia 34.44S 138.39E
23 Q3 **Elizabeth** West Virginia, NE USA 39.03N 81.22W
21 Q9 **Elizabeth, Cape** headland Maine, NE USA 43.34N 70.12W
23 Y8 **Elizabeth City** North Carolina, SE USA 36.18N 76.13W
23 P8 **Elizabethton** Tennessee, S USA 36.21N 82.12W
32 M17 **Elizabethtown** Illinois, N USA 37.24N 88.21W
22 K6 **Elizabethtown** Kentucky, S USA 37.41N 85.51W
20 L7 **Elizabethtown** New York, NE USA 44.13N 73.37W
23 U11 **Elizabethtown** North Carolina, SE USA 34.37N 78.36W
21 G15 **Elizabethtown** Pennsylvania, NE USA 40.08N 76.36W
76 E6 **El-Jadida** prev. Mazagan. W Morocco 33.15N 8.27W
82 F11 **El Jebelein** White Nile, E Sudan 12.37N 32.51E
112 N8 **Ełk** Ger. Lyck. Warmińsko-Mazurskie, NE Poland 53.51N 22.19E
112 O8 **Ełk** ↭ NE Poland
31 Y12 **Elkader** Iowa, C USA 42.51N 91.24W
82 C11 **El Kamlin** Gezira, C Sudan 15.01N 33.02E
35 N11 **Elk City** Idaho, NW USA 45.50N 115.28W
28 K10 **Elk City** Oklahoma, C USA 35.24N 99.24W
29 P7 **Elk City Lake** ☒ Kansas, C USA
36 M5 **Elk Creek** California, W USA 39.34N 122.34W
25 J10 **Elk Creek** ↭ South Dakota, N USA
76 M5 **El Kef** var. Al Kāf, Le Kef. NW Tunisia 36.13N 8.44E
76 F7 **El Kelâa Srarhna** var. Kal al Sraghna. C Morocco 32.05N 7.20W
9 P17 **Elkford** British Columbia, SW Canada 49.58N 114.57W
El Khalil see Hebron
82 E7 **El Khandaq** Northern, N Sudan 18.34N 30.34E
77 W10 **El Khârga** var. Al Khārijah. C Egypt 25.31N 30.36E
31 P11 **Elkhart** Indiana, N USA 41.40N 85.58W
28 H7 **Elkhart** Kansas, C USA 37.00N 101.51W

27 V8 **Elkhart** Texas, SW USA 31.37N 95.34W
32 M7 **Elkhart Lake** ☒ Wisconsin, N USA
39 Q3 **Elkhead Mountains** ▲ Colorado, C USA
20 I12 **Elk Hill** ▲ Pennsylvania, NE USA 41.42N 75.33W
144 G8 **El Khiyam** var. Al Khiyām, Khiam. S Lebanon 33.12N 35.42E
31 S15 **Elkhorn** Nebraska, C USA 41.17N 96.13W
32 M9 **Elkhorn** Wisconsin, N USA 42.40N 88.34W
31 R14 **Elkhorn River** ↭ Nebraska, C USA
131 O16 **El'khotovo** Respublika Severnaya Osetiya, SW Russian Federation 43.18N 44.17E
39 R8 **Elkin** North Carolina, SE USA 36.14N 80.51W
23 S4 **Elkins** West Virginia, NE USA 38.55N 79.51W
205 X3 **Elkins, Mount** ▲ Antarctica 66.25S 53.54E
12 G8 **Elk Lake** Ontario, S Canada 47.44N 80.19W
33 P6 **Elk Lake** ☒ Michigan, N USA
20 F12 **Elkland** Pennsylvania, NE USA 41.59N 77.16W
37 W3 **Elko** Nevada, W USA 40.48N 115.46W
31 R12 **Elk Point** Alberta, SW Canada 53.52N 110.49W
31 N14 **Elk Point** South Dakota, N USA 42.42N 96.37W
31 W8 **Elk River** Minnesota, N USA 45.18N 93.34W
22 J10 **Elk River** ↭ Alabama/Tennessee, S USA
23 R4 **Elk River** ↭ West Virginia, NE USA
22 I7 **Elkton** Kentucky, S USA 36.48N 87.07W
21 Y2 **Elkton** Maryland, NE USA 39.36N 75.49W
31 R10 **Elkton** South Dakota, N USA 44.14N 96.28W
22 I10 **Elkton** Tennessee, S USA 35.01N 86.51W
23 U5 **Elkton** Virginia, NE USA 38.22N 78.35W
83 L15 **El Kure** Somali, E Ethiopia 5.37N 42.05E
82 D12 **El Lagowa** Western Kordofan, C Sudan 11.22N 29.10E
41 S12 **Ellamar** Alaska, USA 60.54N 146.37W
Ellás see Greece
23 T9 **Ellaville** Georgia, SE USA 32.14N 84.18W
207 P9 **Ellef Ringnes Island** island Nunavut, N Canada
31 V10 **Ellendale** Minnesota, N USA 43.53N 93.19W
31 P7 **Ellendale** North Dakota, N USA 45.57N 98.33W
38 M6 **Ellen, Mount** ▲ Utah, W USA 38.06N 110.48W
32 J9 **Ellensburg** Washington, NW USA 47.00N 124.34W
20 K12 **Ellenville** New York, NE USA 41.43N 74.24W
23 T10 **Ellerbe** North Carolina, SE USA 35.04N 79.45W
207 P10 **Ellesmere Island** island Queen Elizabeth Islands, Nunavut, N Canada
193 H19 **Ellesmere, Lake** ☒ South Island, NZ
99 K18 **Ellesmere Port** C England, UK 53.16N 2.54W
35 O14 **Elletsville** Indiana, N USA 39.13N 86.37W
101 F19 **Ellezelles** Hainaut, SW Belgium 50.44N 3.40E
15 Jj4 **Ellice** ↭ Nunavut, NE Canada
Ellice Islands see Tuvalu
Ellichpur see Achalpur
23 W3 **Ellicott City** Maryland, NE USA 39.16N 76.48W
25 S2 **Ellijay** Georgia, SE USA 34.42N 84.28W
29 W7 **Ellington** Missouri, C USA 37.14N 90.58W
28 L5 **Ellinwood** Kansas, C USA 38.21N 98.34W
85 J24 **Elliot** Eastern Cape, SE South Africa 31.19S 27.51E
12 D10 **Elliot Lake** Ontario, S Canada 46.23N 82.39W
189 X6 **Elliot, Mount** ▲ Queensland, E Australia 19.29S 146.58W
23 T5 **Elliott Knob** ▲ Virginia, NE USA 38.10N 79.18W
28 K4 **Ellis** Kansas, C USA 38.56N 99.33W
190 F8 **Elliston** South Australia 33.39S 134.56E
24 M7 **Ellisville** Mississippi, S USA 31.36N 89.12W
107 V5 **Elló** NE Spain
98 L9 **Ellon** NE Scotland, UK 57.22N 2.06W
Ellore see Elūru
63 S13 **Elloree** South Carolina, SE USA 33.30N 80.37W
28 M4 **Ellsworth** Kansas, C USA 38.43N 98.13W
21 S7 **Ellsworth** Maine, NE USA 44.32N 68.25W
30 J6 **Ellsworth** Wisconsin, N USA 44.43N 92.28W
28 M11 **Ellsworth, Lake** ☒ Oklahoma, C USA
204 K9 **Ellsworth Land** physical region Antarctica
204 L9 **Ellsworth Mountains** ▲ Antarctica
103 J21 **Ellwangen** Baden-Württemberg, S Germany 48.58N 10.07E
20 B14 **Ellwood City** Pennsylvania, NE USA 40.49N 80.15W
110 H8 **Elm** Glarus, NE Switzerland 46.55N 9.09E
32 G9 **Elma** Washington, NW USA 47.00N 123.24W

124 Qq15 **El Maḥalla el Kubra** var. Al Maḥallah al Kubrá, Mahalla el Kubra. N Egypt 30.58N 31.10E
76 E9 **El Mahbas** var. Mahbés. SW Western Sahara 27.25N 9.09W
65 H17 **El Maitén** Chubut, W Argentina 42.03S 71.10W
142 E16 **Elmalı** Antalya, SW Turkey 36.40N 29.54E
82 G10 **El Manaqil** Gezira, C Sudan 14.12N 33.01E
56 M12 **El Mango** Amazonas, S Venezuela 1.55N 66.34W
77 W7 **El Mansûra** var. Al Manṣūrah, Mansûra. N Egypt 31.02N 31.30E
57 P8 **El Manteco** Bolívar, E Venezuela 7.17N 62.31W
31 O16 **Elm Creek** Nebraska, C USA 40.43N 99.22W
79 V9 **Elméki** Agadez, C Niger 17.52N 8.07E
110 K7 **Elmen** Tirol, W Austria 47.22N 10.34E
20 I16 **Elmer** New Jersey, NE USA 39.34N 75.09W
144 G6 **El Mina** var. Al Mînâ'. N Lebanon 34.28N 35.49E
77 W9 **El Minya** var. Al Minyâ, Minya. C Egypt 28.06N 30.40E
20 G11 **Elmira** New York, NE USA 42.06N 76.49W
38 L7 **El Mirage** Arizona, SW USA 33.36N 112.19W
31 O7 **Elm Lake** ☒ South Dakota, N USA
47 P9 **El Mojàn** see San Rafael
107 N7 **El Molar** Madrid, C Spain 40.43N 3.34W
78 L7 **El Mrâyer** well C Mauritania 21.40N 7.50W
78 L8 **El Mreïti** well N Mauritania 23.40N 7.23W
78 L8 **El Mreyyé** desert E Mauritania
31 P8 **Elm River** ↭ North Dakota/South Dakota, N USA
102 I9 **Elmshorn** Schleswig-Holstein, N Germany 53.45N 9.39E
82 D12 **El Muglad** Western Kordofan, C Sudan 11.01N 27.43E
El Muwaqqar see Al Muwaqqar
12 G14 **Elmvale** Ontario, S Canada 44.34N 79.53W
32 K12 **Elmwood** Illinois, N USA 40.46N 89.58W
28 J8 **Elmwood** Oklahoma, C USA 36.37N 100.31W
105 P17 **Elne** anc. Illiberis. Pyrénées-Orientales, S France 42.36N 2.58E
56 F11 **El Nevado, Cerro** elevation C Colombia 3.56N 74.20W
179 Oo13 **El Nido** Palawan, W Philippines 11.10N 119.25E
64 I12 **El Nihuil** Mendoza, W Argentina 35.05S 68.38W
77 W7 **El Nouzha** ✕ (Alexandria) N Egypt 31.06N 29.58E
82 E10 **El Obeid** var. Al Obayyid, Al Ubayyiḍ. Northern Kordofan, C Sudan 13.10N 30.10E
43 O13 **El Oro** México, S Mexico 19.51N 100.07W
58 B9 **El Oro** ◆ province SW Ecuador
63 B19 **Elortondo** Santa Fe, C Argentina 33.42S 61.37W
56 J8 **Elorza** Apure, C Venezuela 7.01N 69.30W
131 Q10 **El'ton** Volgogradskaya Oblast', SW Russian Federation
76 L7 **El Oued** var. Al Oued, El Ouâdi, El Wad. NE Algeria 33.19N 6.52E
34 K10 **Eloy** Arizona, SW USA 32.47N 111.33W
57 Q7 **El Palmar** Bolívar, E Venezuela 8.03N 61.51W
42 K8 **El Palmito** Durango, C Mexico 25.40N 104.58W
57 P7 **El Pao** Bolívar, E Venezuela 8.02N 62.38W
56 K5 **El Pao** Cojedes, N Venezuela 9.42N 68.12W
44 J7 **El Paraíso** El Paraíso, S Honduras 13.52N 86.32W
44 G7 **El Paraíso** ◆ department SE Honduras
32 L12 **El Paso** Illinois, N USA 40.44N 89.01W
26 G8 **El Paso** Texas, SW USA 31.45N 106.30W
26 G8 **El Paso** ✕ Texas, SW USA 31.48N 106.24W
107 V6 **El Perelló** Cataluña, NE Spain 40.52N 0.43E
57 P5 **El Pilar** Sucre, NE Venezuela 10.33N 63.13W
56 J8 **El Pital, Cerro** ▲ El Salvador/Honduras 14.19N 89.06W
37 Q9 **El Portal** California, W USA 37.40N 119.46W
42 J3 **El Porvenir** Chihuahua, N Mexico 31.13N 105.51W
45 U14 **El Porvenir** San Blas, N Panama 9.33N 78.55W
107 W6 **El Prat de Llobregat** Cataluña, NE Spain 41.19N 2.04E
44 H5 **El Progreso** Yoro, NW Honduras 15.25N 87.49W
44 A2 **El Progreso** off. Departamento de El Progreso. ◆ department C Guatemala
El Progreso see Guastatoya
35 P13 **El Puente del Arzobispo** Castilla-La Mancha, C Spain 39.48N 5.10W
106 J15 **El Puerto de Santa María** Andalucía, S Spain 36.36N 6.13W
64 I8 **El Puesto** Catamarca, NW Argentina 27.55S 67.37W
El Qáhira see Cairo
77 V10 **El Qaṣr** var. Al Qaṣr. C Egypt 25.39N 28.54E
El Qaṭrani see Al Qaṭrānah
42 G9 **El Quelite** Sinaloa, C Mexico 23.37N 106.26W
64 G9 **Elqui, Río** ↭ N Chile
El Quneitra see Al Qunayṭirah
El Quseir see Al Quṣayr
147 O15 **El-Rahaba** ✕ (Şan'ā') W Yemen 15.28N 44.12E

44 M10 **El Rama** Región Autónoma Atlántico Sur, SE Nicaragua 12.12N 84.13W
44 W16 **El Real** var. El Real de Santa María. Darién, SE Panama 8.07N 77.42W
El Real de Santa María see El Real
28 M10 **El Reno** Oklahoma, C USA 35.31N 97.57W
42 K9 **El Rodeo** Durango, C Mexico 25.08N 104.34W
106 J13 **El Ronquillo** Andalucía, S Spain 37.43N 6.09W
9 S16 **Elrose** Saskatchewan, S Canada 51.07N 107.59W
32 K8 **Elroy** Wisconsin, N USA 43.43N 90.16W
27 S17 **Elsa** Texas, SW USA 26.17N 97.59W
77 W8 **El Şaff** var. Aş Şaff. N Egypt 29.26N 31.19E
42 J10 **El Salto** Durango, C Mexico 23.46N 105.22W
44 D8 **El Salvador** off. Republica de El Salvador. ◆ republic Central America
56 K7 **El Samán de Apure** Apure, C Venezuela 7.51N 68.47W
12 D7 **Elsas** Ontario, S Canada 48.31N 82.53W
42 F3 **El Sásabe** var. Aduana del Sásabe. Sonora, NW Mexico 31.27N 111.31W
Elsass see Alsace
42 J5 **El Sáuz** Chihuahua, N Mexico 29.03N 106.15W
56 I5 **Elsberry** Missouri, C USA 39.10N 90.46W
47 P9 **El Seibo** var. Santa Cruz de El Seibo, Santa Cruz del Seibo. E Dominican Republic 18.45N 69.04W
44 B7 **El Semillero Barra Nahualate** Escuintla, SW Guatemala 14.01N 91.28W
Elsene see Ixelles
165 N11 **Elsen Nur** ☒ C China
38 L6 **Elsinore** Utah, W USA 38.40N 112.09W
Elsinore see Helsingør
101 L18 **Elsloo** Limburg, SE Netherlands 50.57N 5.46E
62 G13 **El Soberbio** Misiones, NE Argentina 27.15S 54.04W
57 N6 **El Socorro** Guárico, C Venezuela 9.00N 65.42W
56 L6 **El Sombrero** Guárico, N Venezuela 9.25N 67.06W
100 L10 **Elspeet** Gelderland, E Netherlands 52.19N 5.47E
100 L12 **Elst** Gelderland, E Netherlands 51.55N 5.51E
103 O15 **Elsterwerda** Brandenburg, E Germany 51.27N 13.32E
42 J4 **El Sueco** Chihuahua, N Mexico 29.52N 106.23W
El Suweida see As Suwaydā'
El Suweis see Suez
56 D12 **El Tambo** Cauca, SW Colombia 2.25N 76.49W
183 T13 **Eltanin Fracture Zone** tectonic feature SE Pacific Ocean
107 X5 **El Ter** ↭ NE Spain
192 K11 **Eltham** Taranaki, North Island, NZ 39.25S 174.17E
57 O6 **El Tigre** Anzoátegui, NE Venezuela 8.55N 64.15W
El Tigrito see San José de Guanipa
56 J5 **El Tocuyo** Lara, N Venezuela 9.47N 69.48W
131 Q10 **El'ton** Volgogradskaya Oblast', SW Russian Federation
34 K10 **Eltopia** Washington, NW USA 46.33N 118.59W
63 A18 **El Trébol** Santa Fe, C Argentina 32.12S 61.40W
42 J13 **El Tuito** Jalisco, SW Mexico 20.20N 105.19W
77 X8 **El Tûr** var. Aţ Ţūr. NE Egypt 28.18N 33.37E
181 K16 **Elūru** prev. Ellore. Andhra Pradesh, E India 16.45N 81.10E
120 H13 **Elva** Ger. Elwa. Tartumaa, SE Estonia 58.13N 26.27E
57 R9 **El Vado Reservoir** ☒ New Mexico, SW USA
45 S15 **El Valle** Coclé, C Panama 8.39N 80.07W
106 H11 **Elvas** Portalegre, C Portugal 38.52N 7.10W
56 C14 **El Venado** Apure, C Venezuela 7.25N 68.46W
107 V6 **El Vendrell** Cataluña, NE Spain 41.13N 1.31E
94 H13 **Elverum** Hedmark, S Norway 60.52N 11.34E
57 R4 **El Viejo** Chinandega, NW Nicaragua 12.37N 87.09W
56 G7 **El Viejo, Cerro** ▲ C Colombia 7.31N 72.56W
56 H6 **El Vigía** Mérida, NW Venezuela 8.37N 71.39W
107 Q4 **El Villar de Arnedo** La Rioja, N Spain 42.19N 2.05W
64 A14 **Elvira** Amazonas, W Brazil 7.12S 69.56W
Elwa see Elva
83 K17 **El Wad** see El Oued
83 K17 **El Wak** North Eastern, NE Kenya 2.46N 40.57E
35 R7 **Elwell, Lake** ☒ Montana, NW USA
33 P13 **Elwood** Indiana, N USA 40.16N 85.50W
29 R3 **Elwood** Kansas, C USA 39.43N 94.52W
31 O16 **Elwood** Nebraska, C USA 40.35N 99.51W
102 F10 **Elx-Elche** see Elche
25 X4 **Ely** Minnesota, N USA 47.54N 91.52W
37 X4 **Ely** Nevada, W USA 39.15N 114.53W
99 P17 **Ely** E England, UK 52.23N 0.16E
El Yopal see Yopal
31 T11 **Elyria** Ohio, N USA 41.22N 82.06W
47 S9 **El Yunque** ▲ E Puerto Rico 18.15N 65.46W
103 F23 **Elz** ↭ SW Germany

197 C14 **Emae** island Shepherd Islands, C Vanuatu
120 I5 **Emajõgi** Ger. Embach. ↭ SE Estonia
155 Q2 **Emämrüd** see Shährüd
155 Q2 **Emäm Şäḥeb** var. Emam Saheb, Hazarat Imam. Kunduz, NE Afghanistan 37.10N 68.55E
Emämshahr see Shährüd
57 M20 **Emän** ↭ S Sweden
197 D14 **Emao** island C Vanuatu
150 J17 **Emba** Kaz. Embi. Aktyubinsk, W Kazakhstan 48.49N 58.10E
150 H12 **Emba** Kaz. Zhem. ↭ W Kazakhstan
64 K5 **Embarcación** Salta, N Argentina 23.15S 64.04W
32 M15 **Embarras River** ↭ Illinois, N USA
Embi see Emba
83 J19 **Embu** Eastern, C Kenya 0.30N 37.30E
102 E10 **Emden** Niedersachsen, NW Germany 53.22N 7.12E
31 Q4 **Emerado** North Dakota, N USA 47.55N 97.21W
189 X8 **Emerald** Queensland, E Australia 23.33S 148.10E
Emerald Isle see Montserrat
56 H6 **Emerald** Zulia, NW Venezuela 9.01N 72.16W
9 Y17 **Emerson** Manitoba, S Canada 49.01N 97.07W
31 T15 **Emerson** Iowa, C USA 41.00N 95.22W
31 R13 **Emerson** Nebraska, C USA 42.16N 96.43W
38 M5 **Emery** Utah, W USA 38.54N 111.16W
142 E13 **Emet** Kütahya, W Turkey 39.21N 29.15E
194 G14 **Emeti** Western, SW PNG 7.51S 143.14E
37 V3 **Emigrant Pass** pass Nevada, W USA 40.39N 116.15W
80 I6 **Emi Koussi** ▲ N Chad 19.52N 18.34E
45 V13 **Emiliano Zapata** Chiapas, SE Mexico 17.43N 91.45W
108 E9 **Emilia-Romagna** prev. Emilia, anc. Æmilia. ◆ region N Italy
164 J3 **Emin** var. Dorbiljin. Xinjiang Uygur Zizhiqu, NW China 46.31N 83.35E
155 W4 **Emïnäbäd** Punjab, E Pakistan 32.01N 73.51E
22 L5 **Eminence** Kentucky, S USA 38.22N 85.10W
29 W7 **Eminence** Missouri, C USA 37.09N 91.21W
116 N9 **Emine, Nos** headland E Bulgaria 42.43N 27.53E
164 I3 **Emin He** ↭ NW China
195 M8 **Emirau Island** island N PNG
142 F13 **Emirdağ** Afyon, W Turkey 39.01N 31.09E
57 M21 **Emmaboda** Kalmar, S Sweden 56.36N 15.30E
120 E5 **Emmaste** Hiiumaa, W Estonia 58.43N 22.36E
20 I15 **Emmaus** Pennsylvania, NE USA 40.32N 75.28W
191 U4 **Emmaville** New South Wales, SE Australia 29.26S 151.38E
110 E9 **Emme** ↭ W Switzerland
100 L8 **Emmeloord** Flevoland, C Netherlands 52.43N 5.46E
100 O8 **Emmen** Drenthe, NE Netherlands 52.48N 6.57E
110 F8 **Emmen** Luzern, C Switzerland 47.03N 8.14E
103 F23 **Emmendingen** Baden-Württemberg, SW Germany 48.07N 7.51E
100 P8 **Emmer-Compascuum** Drenthe, NE Netherlands 52.47N 7.03E
103 D14 **Emmerich** Nordrhein-Westfalen, W Germany 51.49N 6.16E
31 U12 **Emmetsburg** Iowa, C USA 43.06N 94.40W
34 M14 **Emmett** Idaho, NW USA 43.52N 116.30W
40 M10 **Emmonak** Alaska, USA 62.46N 164.31W
26 L12 **Emory Peak** ▲ Texas, SW USA 29.15N 103.18W
42 F6 **Empalme** Sonora, NW Mexico 27.57N 110.49W
85 L23 **Empangeni** KwaZulu/Natal, E South Africa 28.45S 31.57E
62 N8 **Empedrado** Corrientes, NE Argentina 27.57S 58.46W
199 I63 **Emperor Seamounts** undersea feature NW Pacific Ocean
199 J3 **Emperor Trough** undersea feature N Pacific Ocean
33 O6 **Empire** Nevada, W USA 40.26N 119.21W
Empire State of the South see Georgia
Emplawas see Amplawas
108 F9 **Empoli** Toscana, C Italy 43.43N 10.57E
28 M5 **Emporia** Kansas, C USA 38.24N 96.10W
23 W7 **Emporia** Virginia, SE USA 36.41N 77.32W
20 E13 **Emporium** Pennsylvania, NE USA 41.31N 78.13W
195 S12 **Empress Augusta Bay** inlet Bougainville Island, PNG
Empty Quarter see Rub' al Khāli
102 G11 **Ems** var. Eems. ↭ NW Germany
102 I13 **Emsdetten** Nordrhein-Westfalen, NW Germany 52.10N 7.31E
102 F10 **Emsland** cultural region NW Germany
102 H13 **Ems-Hunte Canal** canal Küstenkanal
102 G10 **Ems-Jade-Kanal** canal NW Germany
96 N11 **Enâfors** Jämtland, C Sweden 63.16N 12.24E
96 N11 **Enånger** Gävleborg, C Sweden 61.30N 17.10E
98 G7 **Enârd Bay** bay NW Scotland, UK

97 N15 **Enköping** Uppsala, C Sweden 59.39N 17.07E
109 K24 **Enna** var. Castrogiovanni, Henna. Sicilia, Italy, C Mediterranean Sea 37.34N 14.18E
82 D11 **En Nahud** Western Kordofan, C Sudan 12.40N 28.28E
144 F8 **En Nâqoûra** var. An Nāqūrah. SW Lebanon 33.06N 33.30E
80 K8 **En Nazira** see Nazaret
103 E15 **Ennepetal** Nordrhein-Westfalen, W Germany 51.18N 7.22E
191 P4 **Enngonia** New South Wales, SE Australia 29.19S 145.52E
99 C19 **Ennis** Ir. Inis. W Ireland 52.49N 8.58W
35 R11 **Ennis** Montana, NW USA 45.21N 111.45W
27 U7 **Ennis** Texas, SW USA 32.20N 96.37W
99 F20 **Enniscorthy** Ir. Inis Córthaidh. SE Ireland 52.30N 6.34W
99 E15 **Enniskillen** var. Inniskilling. SW Northern Ireland, UK 54.21N 7.37W
99 B19 **Ennistimon** Ir. Inis Díomáin. W Ireland 52.56N 9.17W
111 T4 **Enns** Oberösterreich, N Austria 48.12N 14.29E
111 T4 **Enns** ↭ C Austria
95 O16 **Eno** Itä-Suomi, E Finland 62.45N 30.15E
26 M5 **Enochs** Texas, SW USA 33.51N 102.46W
95 N17 **Enonkoski** Isä-Suomi, E Finland 62.04N 28.53E
156 K10 **Enontekiö** Lapp. Eanodat. Lappi, N Finland 68.25N 23.40E
23 Q11 **Enoree** South Carolina, SE USA 34.39N 81.58W
23 P11 **Enoree River** ↭ South Carolina, SE USA
189 T4 **Enosburg Falls** Vermont, NE USA 44.54N 72.47W
175 P17 **Enrekang** Sulawesi, C Indonesia 3.33S 119.46E
47 N9 **Enriquillo** SW Dominican Republic 17.53N 71.13W
47 N9 **Enriquillo, Lago** ☒ SW Dominican Republic
100 I13 **Ens** Flevoland, N Netherlands 52.39N 5.49E
100 P11 **Enschede** Overijssel, E Netherlands 52.13N 6.55E
42 B2 **Ensenada** Baja California, NW Mexico 31.52N 116.31W
103 E20 **Ensheim** ✕ (Saarbrücken) Saarland, SW Germany 49.13N 7.09E
166 L9 **Enshi** Hubei, C China 30.16N 109.25E
171 H21 **Enshū-nada** gulf SW Japan
25 O8 **Ensley** Florida, SE USA 30.31N 87.16W
Enso see Svetogorsk
83 F18 **Entebbe** var. Entebe. ◉ S Uganda 0.07N 32.29E
83 F18 **Entebbe** ✕ C Uganda 0.04N 32.29E
103 M18 **Entenbühl** ▲ Czech Republic/Germany 50.09N 12.10E
100 N10 **Enter** Overijssel, E Netherlands 52.19N 6.34E
25 Q7 **Enterprise** Alabama, S USA 31.19N 85.50W
34 L11 **Enterprise** Oregon, NW USA 45.25N 117.16W
38 J7 **Enterprise** Utah, W USA 37.33N 113.42W
34 J8 **Entiat** Washington, NW USA 47.40N 120.12W
107 P15 **Entinas, Punta de las** headland S Spain 36.40N 2.44W
110 F9 **Entlebuch** Luzern, W Switzerland 47.02N 8.04E
65 I22 **Entrada, Punta** headland S Argentina
105 O13 **Entraygues-sur-Truyère** Aveyron, S France 44.39N 2.36E
197 G3 **Entrecasteaux, Récifs d'** reef N New Caledonia
63 C17 **Entre Ríos** off. Provincia de Entre Ríos. ◆ province NE Argentina
44 K7 **Entre Ríos, Cordillera** ▲ Honduras/Nicaragua
106 G9 **Entroncamento** Santarém, C Portugal 39.28N 8.28W
176 Z10 **Entrop** Irian Jaya, E Indonesia 2.37S 140.43E
79 V16 **Enugu** Enugu, S Nigeria 6.24N 7.24E
79 U16 **Enugu** ◆ state SE Nigeria
127 Pp3 **Enurmino** Chukotskiy Avtonomnyy Okrug, NE Russian Federation 66.46N 171.40W
56 E9 **Envigado** Antioquia, W Colombia 6.09N 75.37W
61 B15 **Envira** Amazonas, W Brazil 7.12S 69.59W
Enyélé see Enyélé
81 I16 **Enyélé** var. Enyellé. La Likouala, NE Congo 2.48N 18.01E
103 H21 **Enz** ↭ SW Germany
171 I16 **Enzan** Yamanashi, Honshū, S Japan 35.42N 138.43E
106 I2 **Eo** ↭ NW Spain
Eochaill see Youghal
Eochaille, Cuan see Youghal Bay
109 K22 **Eolie, Isole** var. Isole Lipari, Eng. Aeolian Islands, Lipari Islands. island group S Italy
117 J25 **Eot** island Chuuk, C Micronesia
203 Q4 **Epáno Archánes** var. Áno Arkhánai; prev. Epáno Arkhánai. Kríti, Greece, E Mediterranean Sea 35.12N 25.10E
Epáno Arkhánai see Epáno Archánes
171 G14 **Epanomí** Kentrikí Makedonía, N Greece 40.25N 22.55E
100 M10 **Epe** Gelderland, E Netherlands 52.21N 5.58E
79 S16 **Epe** Lagos, S Nigeria 6.37N 4.01E
81 H17 **Epéna** La Likouala, NE Congo 1.27N 17.28E
105 Q4 **Épernay** anc. Sparnacum. Marne, N France 49.01N 3.58E
Eperies/Eperjes see Prešov
38 L5 **Ephraim** Utah, W USA 39.21N 111.35W
21 H15 **Ephrata** Pennsylvania, NE USA 40.09N 76.08W
34 J8 **Ephrata** Washington, NW USA 47.19N 119.33W

◆ COUNTRY   ◇ DEPENDENT TERRITORY   ⬡ ADMINISTRATIVE REGION   ▲ MOUNTAIN   ☒ VOLCANO   ☒ LAKE
● COUNTRY CAPITAL   ◎ DEPENDENT TERRITORY CAPITAL   ✕ INTERNATIONAL AIRPORT   ▲ MOUNTAIN RANGE   ↭ RIVER   ☒ RESERVOIR

197 C13 **Epi** *var.* Épi *island* C Vanuatu
107 R6 **Épila** Aragón, NE Spain 41.34N 1.19W
105 T6 **Épinal** Vosges, NE France 48.10N 6.28E
Epiphania *see* Ḩamāh
Epirus *see* Ípeiros
124 N4 **Episkopí** SW Cyprus 34.37N 32.53E
Episkopi Bay *see* Episkopí, Kólpos
124 N4 **Episkopí, Kólpos** *var.* Episkopi Bay. *bay* SE Cyprus
Epitoli *see* Pretoria
Epoon *see* Ebon Atoll
Eporedia *see* Ivrea
Eppeschdorf *see* Dumbrăveni
103 H21 **Eppingen** Baden-Württemberg, SW Germany 49.09N 8.54E
85 E18 **Epukiro** Omaheke, E Namibia 21.40S 19.09E
31 Y13 **Epworth** Iowa, C USA 42.27N 90.55W
149 O10 **Eqlīd** *var.* Iqlid. Fārs, C Iran 30.54N 52.43E
Equality State *see* Wyoming
81 J18 **Equateur** *off.* Région de l' Equateur. ◐ *region* N Dem. Rep. Congo (Zaire)
157 K22 **Equatorial Channel** *channel* S Maldives
81 B17 **Equatorial Guinea** *off.* Republic of Equatorial Guinea. ◆ *republic* C Africa
194 H14 **Era** ≈ S PNG
124 R13 **Eratosthenes Tablemount** *undersea feature* E Mediterranean Sea 33.48N 32.53E
Erautini *see* Johannesburg
194 H13 **Erave** Southern Highlands, W PNG 6.36S 143.55E
142 L12 **Erbaa** Tokat, N Turkey 40.39N 36.37E
103 E19 **Erbeskopf** ▲ W Germany 49.43N 7.04E
Erbil *see* Arbīl
124 Nn3 **Ercan** ✈ (Nicosia) N Cyprus 35.07N 33.30E
Ercegnovi *see* Herceg-Novi
143 T14 **Erçek Gölü** ◎ E Turkey
143 S14 **Erciş** Van, E Turkey 39.02N 43.21E
142 K14 **Erciyes Dağı** *anc.* Argaeus. ▲ C Turkey 38.35N 35.27E
113 J22 **Érd** *Ger.* Hanselbeck. Pest, C Hungary 47.22N 18.55E
Erdaobaihe *see* Baihe
165 O12 **Erdaogou** Qinghai, W China 34.30N 92.49E
169 X11 **Erdao Jiang** ▲▲ NE China
Erdăt-Sângeorz *see* Sângeorgiu de Pădure
142 C11 **Erdek** Balıkesir, NW Turkey 40.25N 27.49E
Erdély *see* Transylvania
Erdélyi-Havasok *see* Carpaţii Meridionali
142 J17 **Erdemli** İçel, S Turkey 36.39N 34.18E
168 K6 **Erdenet** Bulgan, N Mongolia 49.01N 104.06E
168 I8 **Erdenetsogt** Bayanhongor, C Mongolia 46.27N 100.53E
80 K7 **Erdi** *plateau* NE Chad
80 L7 **Erdi Ma** *desert* NE Chad
103 M23 **Erding** Bayern, SE Germany 48.18N 11.54E
Erdőszáda *see* Ardusat
Erdőszentgyörgy *see* Sângeorgiu de Pădure
104 I7 **Erdre** ≈ NW France
205 R13 **Erebus, Mount** ▲ Ross Island, Antarctica 78.11S 165.09E
63 H14 **Erechim** Rio Grande do Sul, S Brazil 27.34S 52.15W
169 O7 **Ereen Davaanï Nuruu** ▲▲ NE Mongolia
169 Q6 **Ereentsav** Dornod, NE Mongolia 49.51N 115.41E
142 I16 **Ereğli** Konya, S Turkey 37.30N 34.01E
142 I15 **Ereğli Gölü** ◎ W Turkey
117 A15 **Ereíkoussa** *island* Iónioi Nísoi, Greece, C Mediterranean Sea
169 O11 **Erenhot** *var.* Erlian. Nei Mongol Zizhiqu, NE China 43.35N 112.00E
106 M6 **Eresma** ≈ N Spain
117 K17 **Eresós** *var.* Eressós. Lésvos, E Greece 39.10N 25.57E
Eressós *see* Eresós
Erevan *see* Yerevan
Ereymentaū *see* Yereymentau
101 K21 **Érezée** Luxembourg, SE Belgium 50.16N 5.34E
76 G7 **Erfoud** SE Morocco 31.29N 4.18W
103 D16 **Erft** ≈ W Germany
103 K16 **Erfurt** Thüringen, C Germany 50.59N 11.02E
143 P15 **Ergani** Dıyarbakır, SE Turkey 38.16N 39.43E
169 N11 **Ergel** Dornogovĭ, SE Mongolia 43.10N 109.13E
Ergene Irmağı *see* Ergene Çayı
168 L11 **Ergenetsogt** Ömnögovĭ, S Mongolia 42.54N 106.16E
142 C10 **Ergene Çayı** *var.* Ergene Irmağı. ≈ NW Turkey
120 I9 **Ērgļi** Madona, C Latvia 56.54N 25.37E
80 H11 **Erguig, Bahr** ≈ SW Chad
Ergun He *see* Argun
169 S5 **Ergun Youqi** Nei Mongol Zizhiqu, N China 50.13N 120.09E
169 T5 **Ergun Zuoqi** Nei Mongol Zizhiqu, N China 50.48N 121.30E
166 F12 **Er Hai** ▣ SW China
104 G4 **Er, Îles d'** *island group* NW France
106 K4 **Ería** ≈ NW Spain
82 H8 **Eriba** Kassala, NE Sudan 16.37N 36.04E
98 I6 **Eriboll, Loch** *inlet* NW Scotland, UK
67 Q18 **Erica Seamount** *undersea feature* SW Indian Ocean 38.15S 14.30E
109 H23 **Erice** Sicilia, Italy, C Mediterranean Sea 38.02N 12.35E
106 E10 **Ericeira** Lisboa, C Portugal 38.58N 9.25W
98 H10 **Ericht, Loch** ◎ C Scotland, UK
28 J11 **Erick** Oklahoma, C USA 35.13N 99.52W
20 B11 **Erie** Pennsylvania, NE USA 42.06N 80.03W
20 E9 **Erie Canal** *canal* New York, NE USA

Érié, Lac *see* Erie, Lake
33 T10 **Erie, Lake** *Fr.* Lac Érié. ◎ Canada/USA
Erigabo *see* Ceerigaabo
79 N8 **'Erîgât** *desert* N Mali
Erigavo *see* Ceerigaabo
156 P2 **Erik Eriksenstretet** *strait* E Svalbard
9 X15 **Eriksdale** Manitoba, S Canada 50.52N 98.07W
201 V6 **Erikub Atoll** *var.* Ādkup. *atoll* Ratak Chain, C Marshall Islands
Erimanthos *see* Erýmanthos
172 P8 **Erimo** Hokkaidō, NE Japan 42.01N 143.07E
172 Oo8 **Erimo-misaki** *headland* Hokkaidō, NE Japan 41.57N 143.12E
22 H8 **Erin** Tennessee, S USA 36.19N 87.41W
98 E9 **Eriskay** *island* NW Scotland, UK
Erithraí *see* Erythrés
82 I9 **Eritrea** *off.* State of Eritrea, *Tig.* Ērtra. ◆ *transitional government* E Africa
Erivan *see* Yerevan
103 D16 **Erkelenz** Nordrhein-Westfalen, W Germany 51.04N 6.19E
97 P15 **Erken** ◎ C Sweden
103 K19 **Erlangen** Bayern, S Germany 49.35N 11.00E
166 G9 **Erlang Shan** ▲ C China 29.56N 102.24E
Erlau *see* Eger
189 V5 **Erlauf** ≈ NE Austria
189 Q8 **Erldunda Roadhouse** Northern Territory, N Australia 25.13S 133.13E
Erlian *see* Erenhot
29 T15 **Erling, Lake** ▣ Arkansas, USA
189 O8 **Erlsbach** Tirol, W Austria 46.54N 12.15E
Ermak *see* Aksu
100 K10 **Ermelo** Gelderland, C Netherlands 52.18N 5.37E
85 K21 **Ermelo** Mpumalanga, NE South Africa 26.31S 29.58E
142 H17 **Ermenek** Karaman, S Turkey 36.37N 32.55E
Érmihályfalva *see* Valea lui Mihai
117 G20 **Ermióni** Pelopónnisos, S Greece 37.24N 23.15E
117 J20 **Ermoúpoli** *var.* Hermoupolis; *prev.* Ermoúpolis. Sýros, Kykládes, Greece, Aegean Sea 37.26N 24.55E
Ermoúpolis *see* Ermoúpoli
Ernabella *see* Pukatja
161 G22 **Ernākulam** Kerala, SW India 10.04N 76.18E
104 J6 **Ernée** Mayenne, NW France 48.18N 0.54W
63 H14 **Ernestina, Barragem** ▣ S Brazil
56 E4 **Ernesto Cortissoz** ✈ (Barranquilla) Atlántico, N Colombia
161 H21 **Erode** Tamil Nādu, SE India 11.21N 77.43E
Eroj *see* Iroj
85 C19 **Erongo** ◐ *district* W Namibia
101 F21 **Erquelinnes** Hainaut, S Belgium 50.18N 4.07E
76 G7 **Er-Rachidia** *var.* Ksar al Soule. E Morocco 31.58N 4.22W
82 E11 **Er Rahad** *var.* Ar Rahad. Northern Kordofan, C Sudan 12.42N 30.33E
Er Ramle *see* Ramla
85 O15 **Errego** Zambézia, NE Mozambique 16.02S 37.12E
Errenteria *see* Rentería
Er Rif/Er Riff *see* Rif
99 D14 **Errigal Mountain** *Ir.* An Earagail. ▲ N Ireland 55.03N 8.09W
99 A15 **Erris Head** *Ir.* Ceann Iorrais. *headland* W Ireland 54.18N 10.01W
197 D15 **Erromango** *island* S Vanuatu
Error Guyot *see* Error Tablemount
181 O4 **Error Tablemount** *var.* Error Guyot. *undersea feature* W Indian Ocean 10.19N 56.04E
82 G11 **Er Roseires** Blue Nile, E Sudan 11.52N 34.22E
Erseka *see* Ersekë
115 M22 **Ersekë** *var.* Erseka, Kolonjë. Korçë, SE Albania 40.19N 20.39E
Érsekújvár *see* Nové Zámky
31 S4 **Erskine** Minnesota, N USA 47.42N 96.00W
105 V6 **Erstein** Bas-Rhin, NE France 48.24N 7.39E
110 G9 **Erstfeld** Uri, C Switzerland 46.49N 8.41E
164 M3 **Ertai** Xinjiang Uygur Zizhiqu, NW China 46.04N 90.06E
130 M7 **Ertil'** Voronezhskaya Oblast', W Russian Federation 51.51N 40.46E
Ertis *see* Irtysh, C Asia
Ertis *see* Irtyshsk, Kazakhstan
164 K2 **Ertix He** *Rus.* Chërnyy Irtysh. ≈ China/Kazakhstan
Ērtra *see* Eritrea
23 P9 **Erwin** North Carolina, SE USA 35.19N 78.40W
116 L12 **Erydropótamos** *Bul.* Byala Reka. ≈ Bulgaria/Greece
117 E19 **Erýmanthos** *var.* Erimanthos. ▲ S Greece 37.57N 21.51E
117 G19 **Erythrés** *prev.* Erithraí. Stereá Ellás, C Greece 38.19N 23.20E
166 F12 **Eryuan** Yunnan, SW China 26.09N 100.01E
189 U6 **Erzbach** ≈ W Austria
Erzerum *see* Erzurum
103 N17 **Erzgebirge** *Cz.* Krušné Hory, *Eng.* Ore Mountains. ▲▲ Czech Republic/Germany *see also* Krušné Hory
126 I16 **Erzin** Respublika Tyva, S Russian Federation 50.17N 95.03E
143 O13 **Erzincan** *var.* Erzinjan. Erzincan, E Turkey 39.43N 39.30E
143 N13 **Erzincan** *var.* Erzinjan. ◐ *province* NE Turkey
Erzinjan *see* Erzincan
Erzsébetváros *see* Dumbrăveni
143 Q13 **Erzurum** *prev.* Erzerum. Erzurum, NE Turkey 39.57N 41.16E
143 Q12 **Erzurum** *prev.* Erzerum. ◐ *province* NE Turkey
195 N16 **Esa'ala** Normanby Island, SE PNG 9.45S 150.47E
172 Nn7 **Esan-misaki** *headland* Hokkaidō, N Japan 41.49N 141.12E

172 Pp3 **Esashi** Hokkaidō, NE Japan 44.57N 142.32E
171 M11 **Esashi** *var.* Esasi. Iwate, Honshū, C Japan 39.13N 141.08E
171 Mm6 **Esasho** Hokkaidō, N Japan
Esasi *see* Esashi
97 F23 **Esbjerg** Ribe, W Denmark 55.28N 8.28E
Esbo *see* Espoo
38 L7 **Escalante** Utah, W USA 37.46N 111.36W
38 M7 **Escalante River** ≈ Utah, W USA
12 L12 **Escalier, Réservoir l'** ▣ Quebec, SE Canada
42 K7 **Escalón** Chihuahua, N Mexico 26.43N 104.20W
106 M8 **Escalona** Castilla-La Mancha, C Spain 40.10N 4.24W
25 O8 **Escambia River** ≈ Florida, SE USA
33 N5 **Escanaba** Michigan, N USA 45.45N 87.03W
33 N4 **Escanaba River** ≈ Michigan, N USA
107 R8 **Escandón, Puerto de** *pass* E Spain 40.17N 0.57W
43 W14 **Escárcega** Campeche, SE Mexico 18.33N 90.41W
179 Pp7 **Escarpada Point** *headland* Luzon, N Philippines 18.28N 122.10E
25 N8 **Escatawpa River** ≈ Alabama/Mississippi, S USA
105 P2 **Escaut** ≈ N France
Escaut *see* Scheldt
101 M25 **Esch-sur-Alzette** Luxembourg, S Luxembourg 49.30N 5.58E
103 J15 **Eschwege** Hessen, C Germany 51.10N 10.03E
103 D16 **Eschweiler** Nordrhein-Westfalen, W Germany 50.48N 6.15E
Esclaves, Grand Lac des *see* Great Slave Lake
47 O8 **Escocesa, Bahía** *bay* N Dominican Republic
45 W15 **Escocés, Punta** *headland* NE Panama 8.50N 77.37W
37 U17 **Escondido** California, W USA 33.07N 117.05W
44 M10 **Escondido, Río** ≈ SE Nicaragua
13 S7 **Escoumins, Rivière des** ≈ Quebec, SE Canada
39 O13 **Escudilla Mountain** ▲ Arizona, SW USA 33.57N 109.07W
42 J11 **Escuinapa** *var.* Escuinapa de Hidalgo. Sinaloa, C Mexico 22.49N 105.46W
Escuinapa de Hidalgo *see* Escuinapa
44 C6 **Escuintla** Escuintla, S Guatemala 14.16N 90.46W
43 V17 **Escuintla** Chiapas, SE Mexico 15.15N 92.39W
44 A2 **Escuintla** *off.* Departamento de Escuintla. ◐ *department* S Guatemala
13 W7 **Escuminac** ≈ Quebec, SE Canada
81 D16 **Eséka** Centre, SW Cameroon 3.40N 10.48E
142 I12 **Esenboğa** ✈ (Ankara) Ankara, C Turkey 40.05N 33.01E
Esenguly *see* Gasan-Kuli
142 D17 **Eşen Çayı** ≈ SW Turkey
107 T4 **Ésera** ≈ NE Spain
149 N8 **Eşfahān** *Eng.* Isfahan; *anc.* Aspadana. Eşfahān, C Iran 32.40N 51.40E
149 O7 **Eşfahān** *off.* Ostān-e Eşfahān. ◐ *province* C Iran
107 N5 **Esgueva** ≈ N Spain
155 Q3 **Eshkamesh** Takhār, NE Afghanistan 36.25N 69.10E
155 T2 **Eshkāshem** Badakhshān, NE Afghanistan 36.43N 71.34E
85 L23 **Eshowe** KwaZulu/Natal, E South Africa 28.49S 31.29E
149 T5 **'Eshqābād** Khorāsān, NE Iran 36.00N 59.01E
Esh Sham *see* Dimashq
Esh Sharā *see* Ash Sharāh
Esik *see* Yesik
Esil *see* Ishim, Kazakhstan/Russian Federation
Esil *see* Yesil', Kazakhstan
191 V2 **Esk** Queensland, E Australia 27.15S 152.22E
192 O11 **Eskdale** Hawke's Bay, North Island, NZ 39.24S 176.51E
Eski Dzhumaya *see* Tŭrgovishte
156 L2 **Eskifjördhur** Austurland, E Iceland 65.04N 14.01W
245 S3 **Eski Kaļak** *var.* Askī Kalak, Kalak. N Iraq 36.16N 43.40E
97 N16 **Eskilstuna** Södermanland, C Sweden 59.22N 16.31E
14 G2 **Eskimo Lakes** *lakes* Northwest Territories, NW Canada
15 L8 **Eskimo Point** *headland* Nunavut, E Canada 61.19N 93.49W
Eskimo Point *see* Arviat
245 Q2 **Eski Mosul** N Iraq 36.31N 42.45E
153 T10 **Eski-Nookat** *var.* Iski-Nauket. Oshskaya Oblast', SW Kyrgyzstan 40.18N 72.29E
142 F12 **Eskişehir** *var.* Eskishehr. Eskişehir, W Turkey 39.46N 30.30E
142 F13 **Eskişehir** *var.* Eski shehr. ◐ *province* NW Turkey
Eskishehr *see* Eskişehir
106 K5 **Esla** ≈ NW Spain
148 J6 **Eslāmābād** *var.* Eslāmābād-e Gharb; *prev.* Harunabad, Shāhābād. Kermānshāhān, W Iran 34.07N 46.34E
Eslāmābād-e Gharb *see* Eslāmābād
154 J4 **Eslām Qal'eh** *Pash.* Islam Qala. Herāt, W Afghanistan 34.41N 61.03E
97 K23 **Eslöv** Skåne, S Sweden 55.49N 13.19E
149 S12 **Esmā'īlābād** Kermān, S Iran 28.48N 56.39E
149 U8 **Esmā'īlābād** Khorāsān, E Iran 35.19N 60.30E
142 D14 **Eşme** Uşak, W Turkey 38.25N 28.58E
46 G6 **Esmeralda** Camagüey, E Cuba 21.51N 78.10W
65 F21 **Esmeralda, Isla** *island* S Chile
58 B5 **Esmeraldas** Esmeraldas, N Ecuador 0.55N 79.40W
58 B5 **Esmeraldas** ◐ *province* NW Ecuador

Esna *see* Isna
12 B6 **Esnagi Lake** ◎ Ontario, S Canada
149 V14 **Espakeh** Sīstān va Balūchestān, SE Iran 26.54N 60.09E
105 O13 **Espalion** Aveyron, S France 44.31N 2.45E
España *see* Spain
12 E11 **Espanola** Ontario, S Canada 46.15N 81.46W
39 S10 **Espanola** New Mexico, SW USA 35.59N 106.04W
59 C18 **Española, Isla** *var.* Hood Island. *island* Galapagos Islands, Ecuador, E Pacific Ocean
106 M13 **Espejo** Andalucía, S Spain 37.40N 4.34W
96 C13 **Espeland** Hordaland, S Norway 60.22N 5.27E
102 G12 **Espelkamp** Nordrhein-Westfalen, NW Germany 52.22N 8.37E
40 M8 **Espenberg, Cape** *headland* Alaska, USA 66.33N 163.36W
188 L13 **Esperance** Western Australia 33.49S 121.52E
195 W15 **Esperance, Cape** *headland* Guadacanal, C Solomon Islands 9.09S 159.38E
59 P18 **Esperancita** Santa Cruz, E Bolivia
63 B17 **Esperanza** Santa Fe, C Argentina 31.25S 60.59W
42 G6 **Esperanza** Sonora, NW Mexico 27.37N 109.51W
26 H9 **Esperanza** Texas, SW USA 31.09N 105.40W
204 H3 **Esperanza** *Argentinian research station* Antarctica 63.29S 56.53W
106 E12 **Espichel, Cabo** *headland* S Portugal 38.25N 9.15W
56 E10 **Espinal** Tolima, C Colombia 4.09N 74.54W
50 K10 **Espinhaço, Serra do** ▲▲ SE Brazil
106 G6 **Espinho** Aveiro, N Portugal 41.01N 8.37W
61 N18 **Espinosa** Minas Gerais, SE Brazil 14.58S 42.49W
105 O15 **Espinouse** ▲▲ S France
62 Q8 **Espírito Santo** *off.* Estado do Espírito Santo. ◐ *state* E Brazil
197 B12 **Espiritu Santo** *var.* Santo. *island* W Vanuatu
43 Z13 **Espíritu Santo, Bahía del** *bay* SE Mexico
42 F9 **Espíritu Santo, Isla del** *island* W Mexico
43 Y12 **Espita** Yucatán, SE Mexico 21.00N 88.17W
13 Y7 **Espoir, Cap d'** *headland* Quebec, SE Canada 48.24N 64.21W
Esponsede/Esponsende *see* Esposende
95 L20 **Espoo** *Swe.* Esbo. Etelä-Suomi, S Finland 60.10N 24.42E
106 G5 **Esposende** *var.* Esponsede, Esponsende. Braga, N Portugal 41.30N 8.46W
85 M18 **Espungabera** Manica, SW Mozambique 20.27S 32.48E
65 H17 **Esquel** Chubut, SW Argentina 42.55S 71.19W
8 L17 **Esquimalt** Vancouver Island, British Columbia, SW Canada 48.25N 123.27W
63 C16 **Esquina** Corrientes, NE Argentina 30.00S 59.30W
44 E6 **Esquipulas** Chiquimula, SE Guatemala 14.34N 89.20W
44 K9 **Esquipulas** Matagalpa, C Nicaragua 12.39N 85.48W
96 I8 **Essandjøen** ◎ S Norway
76 E7 **Essaouira** *prev.* Mogador. W Morocco 31.33N 9.40W
Esseg *see* Osijek
Es Semara *see* Smara
101 G15 **Essen** Antwerpen, N Belgium 51.28N 4.28E
103 E15 **Essen** *var.* Essen an der Ruhr. Nordrhein-Westfalen, W Germany 51.28N 7.01E
Essen an der Ruhr *see* Essen
76 I5 **Es Senia** ✈ (Oran) NW Algeria 35.34N 0.42W
57 T8 **Essequibo Islands** *island group* N Guyana
57 T11 **Essequibo River** ≈ C Guyana
12 C18 **Essex** Ontario, S Canada 42.06N 82.52W
31 T16 **Essex** Iowa, C USA 40.49N 95.18W
99 P21 **Essex** *cultural region* E England, UK
33 R8 **Essexville** Michigan, N USA 43.37N 83.50W
103 H22 **Esslingen** *var.* Esslingen am Neckar. Baden-Württemberg, SW Germany 48.45N 9.18E
Esslingen am Neckar *see* Esslingen
105 N6 **Essonne** ◐ *department* N France
Es Suweida *see* As Suwaydā'
81 F16 **Est** *Eng.* East. ◐ *province* SE Cameroon
106 I1 **Estaca de Bares, Punta da** *point* NW Spain 43.46N 7.40W
26 M5 **Estacado, Llano** *plain* New Mexico/Texas, SW USA
65 K25 **Estados, Isla de los** *prev. Eng.* Staten Island. *island* S Argentina
149 P12 **Eşţahbān** Fārs, S Iran
12 F11 **Estaire** Ontario, S Canada 46.19N 80.47W
39 S12 **Estancia** New Mexico, SW USA 34.45N 106.03W
61 P16 **Estância** Sergipe, E Brazil 11.15S 37.28W
106 G7 **Estarreja** Aveiro, N Portugal 40.45N 8.34W
104 M17 **Estats, Pic d'** *Sp.* Pico d'Estats. ▲ France/Spain 42.39N 1.24E
Estats, Pico d' *see* Estats, Pic d'
85 K23 **Estcourt** KwaZulu/Natal, E South Africa 28.58S 29.54E
108 H8 **Este** *anc.* Ateste. Veneto, NE Italy 45.14N 11.40E
44 J9 **Estelí** Estelí, NW Nicaragua 13.04N 86.21W
44 J9 **Estelí** ◐ *department* NW Nicaragua
Estella *see* Estella-Lizarra
107 Q4 **Estella-Lizarra** *Bas.* Lizarra *var* Estella Navarra, N Spain 42.40N 2.01W

31 R9 **Estelline** South Dakota, N USA 44.34N 96.54W
27 P4 **Estelline** Texas, SW USA 34.33N 100.26W
106 L14 **Estepa** Andalucía, S Spain 37.16N 4.52W
106 L16 **Estepona** Andalucía, S Spain 36.25N 5.09W
41 R9 **Ester** Alaska, USA 64.49N 148.03W
9 V16 **Esterhazy** Saskatchewan, S Canada 50.40N 102.01W
39 S3 **Estes Park** Colorado, C USA 40.22N 105.31W
9 V17 **Estevan** Saskatchewan, S Canada 49.07N 103.04W
31 T11 **Estherville** Iowa, C USA 43.24N 94.49W
23 R15 **Estill** South Carolina, SE USA 32.45N 81.14W
105 Q6 **Estissac** Aube, N France 48.17N 3.51E
13 T9 **Est, Lac de l'** ◎ Quebec, SE Canada
Estland *see* Estonia
9 S16 **Eston** Saskatchewan, S Canada 51.09N 108.41W
120 G5 **Estonia** *off.* Republic of Estonia, *Est.* Eesti Vabariik, *Ger.* Estland, *Latv.* Igaunija; *prev.* Estonian SSR, *Rus.* Estonskaya SSR. ◆ *republic* NE Europe
Estonskaya SSR *see* Estonia
106 E11 **Estoril** Lisboa, W Portugal 38.42N 9.22W
61 L14 **Estreito** Maranhão, E Brazil 6.34S 47.22W
106 I8 **Estrela, Serra da** ▲▲ C Portugal
42 D3 **Estrella, Punta** *headland* NW Mexico 30.53N 114.45W
Estremadura *see* Extremadura
106 F10 **Estremadura** *cultural and historical region* W Portugal
106 H11 **Estremoz** Évora, S Portugal 38.49N 7.34W
81 D18 **Estuaire** *off.* Province de l'Estuaire, *var.* L'Estuaire. ◐ *province* NW Gabon
Eszék *see* Osijek
113 I22 **Esztergom** *Ger.* Gran; *anc.* Strigonium. Komárom-Esztergom, N Hungary 47.46N 18.44E
158 K11 **Etah** Uttar Pradesh, N India 27.33N 78.39E
201 R17 **Etal Atoll** *atoll* Mortlock Islands, C Micronesia
101 K24 **Étalle** Luxembourg, SE Belgium 49.40N 5.36E
105 N6 **Étampes** Essonne, N France 48.25N 2.10E
190 J1 **Etamunbanie, Lake** *salt lake* South Australia
105 N1 **Étaples** Pas-de-Calais, N France 50.31N 1.37E
158 K12 **Etāwah** Uttar Pradesh, N India 26.46N 79.01E
13 R10 **Etchemin** ≈ Quebec, SE Canada
Etchmiadzin *see* Ejmiatsin
42 G7 **Etchojoa** Sonora, NW Mexico 26.54N 109.37W
95 L19 **Etelä-Suomi** ◐ *province* S Finland
85 B16 **Etengua** Kunene, NW Namibia 17.24S 13.05E
101 K25 **Éthe** Luxembourg, SE Belgium 49.34N 5.32E
9 W15 **Ethelbert** Manitoba, S Canada 51.30N 100.22W
82 H12 **Ethiopia** *off.* Federal Democratic Republic of Ethiopia; *prev.* Abyssinia, People's Democratic Republic of Ethiopia. ◆ *republic* E Africa
82 I13 **Ethiopian Highlands** *var.* Ethiopian Plateau. *plateau* N Ethiopia
Ethiopian Plateau *see* Ethiopian Highlands
36 M2 **Etna** California, W USA 41.25N 122.53W
20 B14 **Etna** Pennsylvania, NE USA 40.29N 79.55W
96 G12 **Etna** ≈ S Norway
109 L24 **Etna, Monte** *Eng.* Mount Etna. 🌋 Sicilia, Italy, C Mediterranean Sea 37.46N 15.00E
Etna, Mount *see* Etna, Monte
97 C15 **Etne** Hordaland, S Norway 59.40N 5.55E
Etoliko *see* Aitolikó
41 Y14 **Etolin Island** *island* Alexander Archipelago, Alaska, USA
40 L12 **Etolin Strait** *strait* Alaska, USA
85 C17 **Etosha Pan** *salt lake* N Namibia
81 G18 **Etoumbi** Cuvette, NW Congo 0.01N 14.57E
22 M10 **Etowah** Tennessee, S USA 35.19N 84.31W
25 S2 **Etowah River** ≈ Georgia, SE USA
Etrek *see* Atrak
104 L3 **Étretat** Seine-Maritime, N France 49.46N 0.23E
116 H9 **Etropole** Sofiya, W Bulgaria 42.50N 24.00E
Etsch *see* Adige
Et Tafila *see* Aţ Ţafīlah
101 M23 **Ettelbrück** Diekirch, C Luxembourg 49.51N 6.06E
201 V12 **Etten** *atoll* Chuuk Islands, C Micronesia
101 H14 **Etten-Leur** Noord-Brabant, S Netherlands 51.34N 4.37E
78 G7 **Et Tîdra** *var.* Ile Tîdra. *island* Dakhlet Nouâdhibou, NW Mauritania
103 G21 **Ettlingen** Baden-Württemberg, SW Germany 48.57N 8.25E
104 M2 **Eu** Seine-Maritime, N France 50.01N 1.24E
200 Rr16 **'Eua** *prev.* Middleburg Island. *island* Tongatapu Group, SE Tonga
200 R15 **Eua Iki** *island* Tongatapu Group, S Tonga
Euboea *see* Évvoia
189 O12 **Eucla** Western Australia 31.40S 128.50E
33 U11 **Euclid** Ohio, N USA 41.35N 81.31W
29 W14 **Eudora** Arkansas, C USA 33.06N 91.15W
29 Q4 **Eudora** Kansas, C USA 38.56N 95.06W
190 J9 **Eudunda** South Australia 34.11S 139.03E

25 R6 **Eufaula** Alabama, S USA 31.53N 85.09W
29 Q11 **Eufaula** Oklahoma, C USA 35.17N 95.34W
29 Q11 **Eufaula Lake** *var.* Eufaula Reservoir. ▣ Oklahoma, C USA
Eufaula Reservoir *see* Eufaula Lake
34 F13 **Eugene** Oregon, NW USA 44.03N 123.05W
42 B6 **Eugenia, Punta** *headland* W Mexico 27.48N 115.03W
191 Q8 **Eugowra** New South Wales, SE Australia 33.28S 148.21E
106 I2 **Eume** ≈ NW Spain
106 H2 **Eume, Embalse do** ▣ NW Spain
Eumolpias *see* Plovdiv
61 O18 **Eunápolis** Bahia, SE Brazil 16.19S 39.36W
24 H8 **Eunice** Louisiana, S USA 30.29N 92.25W
39 W15 **Eunice** New Mexico, SW USA 32.26N 103.09W
101 M19 **Eupen** Liège, E Belgium 50.37N 6.01E
138 Jj9 **Euphrates** *Ar.* Al Furāt, *Turk.* Firat Nehri. ≈ SW Asia
144 L3 **Euphrates Dam** *dam* N Syria 35.52N 38.34E
24 M4 **Eupora** Mississippi, S USA 33.32N 89.16W
95 K19 **Eura** Länsi-Suomi, W Finland 61.06N 22.12E
95 K19 **Eurajoki** Länsi-Suomi, W Finland 61.13N 21.44E
2 **Eurasian Plate** *tectonic feature*
104 L4 **Eure** ◐ *department* N France
104 M4 **Eure** ≈ N France
104 M6 **Eure-et-Loir** ◐ *department* C France
36 K3 **Eureka** California, W USA 40.47N 124.12W
29 P6 **Eureka** Kansas, C USA 37.47N 96.15W
35 O6 **Eureka** Montana, NW USA 48.52N 115.03W
37 V5 **Eureka** Nevada, W USA 39.30N 115.58W
31 O7 **Eureka** South Dakota, N USA 45.46N 99.37W
38 L4 **Eureka** Utah, W USA 39.57N 112.07W
34 K10 **Eureka** Washington, NW USA 46.21N 118.41W
29 S9 **Eureka Springs** Arkansas, C USA 36.25N 93.43W
190 K6 **Eurinilla Creek** *seasonal river* South Australia
191 O11 **Euroa** Victoria, SE Australia 36.46S 145.35E
180 M9 **Europa** *island* W Madagascar
106 L3 **Europa, Picos de** ▲▲ N Spain
106 L16 **Europa Point** *headland* S Gibraltar 36.07N 5.20W
86-87 **Europe** *continent*
100 F12 **Europoort** Zuid-Holland, W Netherlands 51.59N 4.08E
Euskadi *see* País Vasco
103 D17 **Euskirchen** Nordrhein-Westfalen, W Germany 50.40N 6.47E
25 W11 **Eustis** Florida, SE USA 28.51N 81.41W
190 M9 **Euston** New South Wales, SE Australia 34.34S 142.45E
25 N5 **Eutaw** Alabama, S USA 32.50N 87.53W
102 K8 **Eutin** Schleswig-Holstein, N Germany 54.07N 10.37E
8 K14 **Eutsuk Lake** ▣ British Columbia, SW Canada
Euxine Sea *see* Black Sea
85 C16 **Evale** Cunene, SW Angola 16.36S 15.46E
39 T3 **Evans** Colorado, C USA 40.22N 104.41W
9 P14 **Evansburg** Alberta, SW Canada 53.34N 114.57W
31 X13 **Evansdale** Iowa, C USA 42.28N 92.16W
191 V4 **Evans Head** New South Wales, SE Australia 29.07S 153.27E
10 J11 **Evans, Lac** ◎ Quebec, SE Canada
39 S5 **Evans, Mount** ▲ Colorado, C USA 39.15N 106.10W
15 Mm6 **Evans Strait** *strait* Nunavut, N Canada
33 N10 **Evanston** Illinois, N USA 42.02N 87.41W
35 S17 **Evanston** Wyoming, C USA 41.16N 110.57W
12 D11 **Evansville** Manitoulin Island, Ontario, S Canada 45.48N 82.34W
33 N16 **Evansville** Indiana, N USA 37.58N 87.33W
32 L9 **Evansville** Wisconsin, N USA 42.46N 89.16W
27 S8 **Evant** Texas, SW USA 31.28N 98.09W
149 P13 **Evaz** Fārs, S Iran 27.48N 53.58E
31 W4 **Eveleth** Minnesota, N USA 47.27N 92.32W
190 E3 **Evelyn Creek** *seasonal river* South Australia
189 Q2 **Evelyn, Mount** ▲ Northern Territory, N Australia 13.28S 132.50E
126 I11 **Evenkiyskiy Avtonomnyy Okrug** ◐ *autonomous district* N Russian Federation
191 R13 **Everard, Cape** *headland* Victoria, SE Australia 37.48S 149.21E
190 F6 **Everard, Lake** *salt lake* South Australia
190 C2 **Everard Ranges** ▲▲ South Australia
159 R11 **Everest, Mount** *Chin.* Qomolangma Feng, *Nep.* Sagarmatha. ▲ China/Nepal 27.58N 86.57E
20 E15 **Everett** Pennsylvania, NE USA 40.00N 78.22W
34 H7 **Everett** Washington, NW USA 47.58N 122.12W
101 E17 **Evergem** Oost-Vlaanderen, NW Belgium 51.07N 3.43E
25 X16 **Everglades City** Florida, SE USA 25.51N 81.22W
25 Y16 **Everglades, The** *wetland* Florida, SE USA
25 P7 **Evergreen** Alabama, S USA 31.25N 86.55W
39 T4 **Evergreen** Colorado, C USA 39.37N 105.19W

Evergreen State *see* Washington
99 L21 **Evesham** C England, UK 52.06N 1.57W
105 T10 **Évian-les-Bains** Haute-Savoie, E France 46.22N 6.34E
95 K16 **Evijärvi** Länsi-Suomi, W Finland 63.22N 23.30E
81 D17 **Evinayong** *var.* Ebinayon, Evinayoung. C Equatorial Guinea 1.28N 10.17E
Evinayoung *see* Evinayong
117 E18 **Évinos** ≈ C Greece
97 E17 **Evje** Aust-Agder, S Norway 58.34N 7.49E
Evmolpia *see* Plovdiv
106 H11 **Évora** *anc.* Ebora, *Lat.* Liberalitas Julia. Évora, C Portugal 38.34N 7.54W
106 G11 **Évora** ◐ *district* S Portugal
104 M4 **Évreux** *anc.* Civitas Eburovicum. Eure, N France 49.01N 1.09E
104 K6 **Évron** Mayenne, NW France 48.10N 0.24W
116 L13 **Évros** *Bul.* Maritsa, *Turk.* Meriç; *anc.* Hebrus. ≈ SE Europe *see also* Maritsa/Meriç
117 F21 **Evrótas** ≈ S Greece
105 O5 **Évry** Essonne, N France 48.38N 2.27E
27 O8 **E.V.Spence Reservoir** ▣ Texas, SW USA
117 I18 **Évvoia** *Lat.* Euboea. *island* C Greece
40 D9 **Ewa Beach** Oahu, Hawaii, USA, C Pacific Ocean 21.19N 158.00W
34 L9 **Ewan** Washington, NW USA 47.06N 117.46W
46 K12 **Ewarton** C Jamaica 18.10N 77.04W
83 J18 **Ewaso Ng'iro** *var.* Nyiro. ≈ C Kenya
194 E13 **Ewe** ≈ W PNG
31 P13 **Ewing** Nebraska, C USA 42.13N 98.20W
204 J5 **Ewing Island** *island* Antarctica
67 P17 **Ewing Seamount** *undersea feature* E Atlantic Ocean 23.19S 8.45E
164 L6 **Ewirgol** Xinjiang Uygur Zizhiqu, W China 42.55N 87.39E
81 G19 **Ewo** Cuvette, W Congo 0.55S 14.49E
29 S3 **Excelsior Springs** Missouri, C USA 39.20N 94.13W
99 J23 **Exe** ≈ SW England, UK
204 L12 **Executive Committee Range** ▲▲ Antarctica
12 E16 **Exeter** Ontario, S Canada 43.19N 81.26W
99 J24 **Exeter** *anc.* Isca Damnoniorum. SW England, UK 50.43N 3.31W
37 R11 **Exeter** California, W USA 36.17N 119.08W
21 P10 **Exeter** New Hampshire, NE USA 42.57N 70.55W
Exin *see* Kcynia
31 T14 **Exira** Iowa, C USA 41.36N 94.55W
99 J23 **Exmoor** *moorland* SW England, UK
23 Y6 **Exmore** Virginia, NE USA 37.31N 75.48W
188 G8 **Exmouth** Western Australia 22.00S 114.06E
99 J24 **Exmouth** SW England, UK 50.36N 3.24W
188 G8 **Exmouth Gulf** *gulf* Western Australia
181 V8 **Exmouth Plateau** *undersea feature* E Indian Ocean
117 J20 **Exompourgo** *ancient monument* Tínos, Kykládes, Greece, Aegean Sea 37.34N 25.12E
106 I10 **Extremadura** *var.* Estremadura. ◐ *autonomous community* W Spain
80 F12 **Extrême-Nord** *Eng.* Extreme North. ◐ *province* N Cameroon
Extreme North *see* Extrême-Nord
46 I3 **Exuma Cays** *islets* C Bahamas
46 I3 **Exuma Sound** *sound* C Bahamas
83 H20 **Eyasi, Lake** ◎ N Tanzania
97 F17 **Eydehavn** Aust-Agder, S Norway 58.31N 8.52E
98 L12 **Eyemouth** SE Scotland, UK 55.51N 2.06W
98 G7 **Eye Peninsula** *peninsula* NW Scotland, UK
82 Q13 **Eyl** *It.* Eil. Nugaal, E Somalia 8.03N 49.49E
105 N11 **Eymoutiers** Haute-Vienne, C France 45.45N 1.43E
Eyo (lower course) *see* Uolo, Río
31 X10 **Eyota** Minnesota, N USA 44.00N 92.13W
190 H2 **Eyre Basin, Lake** *salt lake* South Australia
190 I1 **Eyre Creek** *seasonal river* Northern Territory/South Australia
182 L9 **Eyre, Lake** *salt lake* South Australia
193 C22 **Eyre Mountains** ▲▲ South Island, NZ
190 H3 **Eyre North, Lake** *salt lake* South Australia
190 G7 **Eyre Peninsula** *peninsula* South Australia
190 H4 **Eyre South, Lake** *salt lake* South Australia
97 B18 **Eysturoy** *Dan.* Østerø *Island* Faeroe Islands 62.15N 6.55W
63 D20 **Ezeiza** ✈ (Buenos Aires) Buenos Aires, E Argentina 34.49S 58.30W
Ezeres *see* Ezeriş
118 F12 **Ezeriş** *Hung.* Ezeres. Caraş-Severin, W Romania 45.21N 21.55E
167 O9 **Ezhou** *prev.* Echeng. Hubei, C China 30.22N 114.52E
129 R11 **Ezhva** Respublika Komi, NW Russian Federation 61.45N 50.43E
142 B12 **Ezine** Çanakkale, NW Turkey 39.46N 26.19E
Ezo *see* Hokkaidō
Ezra/Ezraa *see* Izra'

F

203 P7 **Faaa** Tahiti, W French Polynesia 17.31S 149.36W
203 P7 **Faaa** ✈ (Papeete) Tahiti, W French Polynesia 17.31S 149.36W
Faaborg *see* Fåborg
157 K19 **Faadhippolhu Atoll** *var.* Fadiffolu, Lhaviyani Atoll. *atoll* N Maldives

◆ Country ● Country Capital ◇ Dependent Territory ○ Dependent Territory Capital ◐ Administrative Region ✈ International Airport ▲ Mountain ▲▲ Mountain Range 🌋 Volcano ≈ River ◎ Lake ▣ Reservoir

203 U10 **Faaite** atoll Îles Tuamotu, C French Polynesia

203 Q8 **Faaone** Tahiti, W French Polynesia 17.39S 149.18W

26 H8 **Fabens** Texas, SW USA 31.30N 106.09W

96 H12 **Fåberg** Oppland, S Norway 61.15N 10.21E

97 H24 **Fåborg** var. Faaborg. Fyn, C Denmark 55.06N 10.15E

108 I12 **Fabriano** Marche, C Italy 43.19N 15.52E

151 U16 **Fabrichnyy** Almaty, SE Kazakhstan 43.12N 76.19E

56 F10 **Facatativá** Cundinamarca, C Colombia 4.52N 74.27W

79 X9 **Fachi** Agadez, C Niger 18.01N 11.36E

196 B16 **Facpi Point** headland W Guam

20 I13 **Factoryville** Pennsylvania, NE USA 41.34N 75.45W

80 K8 **Fada** Borkou-Ennedi-Tibesti, E Chad 17.13N 21.31E

79 Q13 **Fada-Ngourma** E Burkina 12.07N 0.20E

126 J4 **Faddeya, Zaliv** bay N Russian Federation

126 M4 **Faddeyevskiy, Ostrov** island Novosibirskiye Ostrova, NE Russian Federation

147 W12 **Fadhi** S Oman 17.54N 55.30E

108 H10 **Faenza** anc. Faventia. Emilia-Romagna, N Italy 44.16N 11.52E

66 M5 **Faeroe-Iceland Ridge** undersea feature N Norwegian Sea

66 M5 **Faeroe Islands** Dan. Færøerne, Faer. Føroyar. ◇ Danish external territory N Atlantic Ocean

88 C8 **Faeroe Islands** island group N Atlantic Ocean

Færøerne see Faeroe Islands

66 N6 **Faeroe-Shetland Trough** undersea feature NE Atlantic Ocean

106 H6 **Fafe** Braga, N Portugal 41.27N 8.10W

82 K13 **Fafen Shet'** �◈ E Ethiopia

200 Qq15 **Fafo** island Tongatapu Group, S Tonga

198 Bb8 **Fagaloa Bay** bay Upolu, C Samoa

198 B7 **Fagamālo** Savai'i, N Samoa 13.27S 172.22W

118 I12 **Făgăraş** Ger. Fogarasch, Hung. Fogaras. Braşov, C Romania 45.49N 24.58E

97 M20 **Fagerhult** Kalmar, S Sweden 57.07N 15.40E

96 G13 **Fagernes** Oppland, S Norway 60.58N 9.13E

156 I9 **Fagernes** Troms, N Norway 69.31N 19.16E

97 M14 **Fagersta** Västmanland, C Sweden 59.58N 15.49E

79 W13 **Faggo** var. Foggo. Bauchi, N Nigeria 11.22N 9.55E

Faghman see Fughmah

Fagibina, Lac see Faguibine, Lac

65 J25 **Fagnano, Lago** ◎ S Argentina

101 G22 **Fagne** hill range S Belgium

79 N10 **Faguibine, Lac** var. Lake Fagibina. ◎ NW Mali

Fahaheel see Al Fuḩayḩīl

Fahlun see Falun

149 U12 **Fahraj** Kermān, SE Iran 29.03N 58.54E

66 P5 **Faial** Madeira, Portugal, NE Atlantic Ocean 32.46N 16.52W

66 N2 **Faial** var. Ilha do Faial. island Azores, Portugal, NE Atlantic Ocean

Faial, Ilha do see Faial

110 G10 **Faido** Ticino, S Switzerland 46.30N 8.48E

Faifo see Hôi An

Failaka Island see Faylakah

202 G12 **Faioa, Île** island N Wallis and Futuna

189 W8 **Fairbairn Reservoir** ◎ Queensland, E Australia

41 R9 **Fairbanks** Alaska, USA 64.48N 147.47W

23 U12 **Fair Bluff** North Carolina, SE USA 34.18N 79.02W

33 R14 **Fairborn** Ohio, N USA 39.49N 84.01W

25 S3 **Fairburn** Georgia, SE USA 33.34N 84.34W

32 M12 **Fairbury** Illinois, N USA 40.45N 88.30W

31 Q17 **Fairbury** Nebraska, C USA 40.08N 97.10W

31 T9 **Fairfax** Minnesota, N USA 44.31N 94.43W

29 O8 **Fairfax** Oklahoma, C USA 36.34N 96.42W

23 R14 **Fairfax** South Carolina, SE USA 32.57N 81.14W

21 W4 **Fairfax** Virginia, NE USA 38.50N 77.18W

35 N8 **Fairfield** California, W USA 38.14N 122.03W

35 O14 **Fairfield** Idaho, NW USA 43.20N 114.45W

32 M16 **Fairfield** Illinois, N USA 38.22N 88.22W

31 X15 **Fairfield** Iowa, C USA 41.00N 91.57W

33 R8 **Fairfield** Ohio, N USA 39.21N 84.33W

27 U8 **Fairfield** Texas, SW USA 31.43N 96.10W

29 T7 **Fair Grove** Missouri, C USA 37.22N 93.09W

21 P12 **Fairhaven** Massachusetts, NE USA 41.38N 70.51W

23 P4 **Fairhope** Alabama, S USA 30.31N 87.54W

94 E9 **Fair Isle** island NE Scotland, UK

193 F20 **Fairlie** Canterbury, South Island, NZ 44.05S 170.50E

31 U11 **Fairmont** Minnesota, N USA 43.40N 94.27W

31 Q16 **Fairmont** Nebraska, C USA 40.37N 97.36W

21 R4 **Fairmont** West Virginia, NE USA 39.29N 80.08W

33 P12 **Fairmount** Indiana, N USA 40.25N 85.39W

20 I7 **Fairmount** New York, NE USA 43.03N 76.14W

31 R7 **Fairmount** North Dakota, N USA 46.02N 96.36W

39 S5 **Fairplay** Colorado, C USA 39.13N 106.00W

20 F9 **Fairport** New York, NE USA 43.06N 77.26W

9 O12 **Fairview** Alberta, W Canada 56.03N 118.28W

28 L9 **Fairview** Oklahoma, C USA 36.16N 98.28W

38 L4 **Fairview** Utah, W USA 39.37N 111.26W

37 T6 **Fairview Peak** ▲ Nevada, W USA 39.13N 118.09W

196 H14 **Fais** atoll Caroline Islands, W Micronesia

155 U8 **Faisalābād** prev. Lyallpur. Punjab, NE Pakistan 31.25N 73.06E

Faisaliya var. Fayşalīyah

30 L8 **Faith** South Dakota, N USA 45.01N 102.02W

159 N12 **Faizābād** Uttar Pradesh, N India 26.46N 82.07E

47 S9 **Fajardo** E Puerto Rico 18.20N 65.39W

245 R9 **Fajj, Wādī al** dry watercourse S Iraq

146 K4 **Fajr, Bi'r** well NW Saudi Arabia 28.46N 37.51E

203 W10 **Fakahina** atoll Îles Tuamotu, C French Polynesia

202 L10 **Fakaofo Atoll** island SE Tokelau

203 U10 **Fakarava** atoll Îles Tuamotu, C French Polynesia

161 K17 **Fakel** Afr. Valsbaai. bay SW South Africa

131 T2 **Fakel** Udmurtskaya Respublika, NW Russian Federation 57.35N 53.00E

97 I19 **Fakenham** E England, UK 52.50N 0.51E

176 V11 **Fakfak** Irian Jaya, E Indonesia 2.55S 132.16E

176 V11 **Fakfak, Pegunungan** ▲ Irian Jaya, E Indonesia

159 T12 **Fakiragram** Assam, NE India 26.22N 90.15E

116 M10 **Fakiyska Reka** ☆ SE Bulgaria

97 J24 **Fakse** Storstrøm, SE Denmark 55.16N 12.07E

97 J24 **Fakse Bugt** bay SE Denmark

97 J24 **Fakse Ladeplads** Storstrøm, SE Denmark 55.13N 12.10E

169 V11 **Faku** Liaoning, NE China 42.30N 123.27E

78 J14 **Falaba** N Sierra Leone 9.54N 11.22W

104 K5 **Falaise** Calvados, N France 48.52N 0.12W

116 H12 **Falakró** ▲ NE Greece

201 T12 **Falalu** island Chuuk, C Micronesia

177 F14 **Falam** Chin State, W Myanmar 22.58N 93.45E

149 N8 **Falāvarjān** Eşfahān, C Iran 32.33N 51.28E

118 M11 **Fălciu** Vaslui, E Romania 46.19N 28.10E

56 I4 **Falcón** off. Estado Falcón. ◇ state NW Venezuela

117 L20 **Falconara Marittima** Marche, C Italy 43.37N 13.22E

109 A16 **Falcone, Capo del** Falcone, Punta del

109 A16 **Falcone, Punta del** var. Capo del Falcone. headland Sardegna, Italy, C Mediterranean Sea 40.57N 8.12E

9 Y16 **Falcon Lake** Manitoba, S Canada 49.43N 95.18W

Falcon Lake see Falcón, Presa/Falcon Reservoir

43 O7 **Falcón, Presa** var. Falcon Lake, Falcon Reservoir. ◎ Mexico/USA

see also Falcon Reservoir

27 Q16 **Falcon Reservoir** var. Falcon Lake, Presa Falcón. ◎ Mexico/USA **see also** Falcón, Presa

202 L10 **Fale** island Fakaofo Atoll, SE Tokelau

198 Aa7 **Faleā!upo** Savai'i, NW Samoa 13.30S 172.46W

202 B10 **Falefatu** island Funafuti Atoll, C Tuvalu

198 Aa7 **Falelima** Savai'i, NW Samoa 13.30S 172.40W

97 N18 **Falerum** Östergötland, S Sweden 58.07N 16.15E

Faleshty see Fălești

118 M9 **Fălești** Rus. Faleshty. NW Moldova 47.33N 27.43E

27 S15 **Falfurrias** Texas, SW USA 27.13N 98.08W

9 O13 **Falher** Alberta, W Canada 55.45N 117.18W

Falkenau an der Eger see Sokolov

97 J21 **Falkenberg** Halland, S Sweden 56.55N 12.30E

Falkenberg see Niemodlin

Falkenberg in Pommern see Złocieniec

102 N12 **Falkensee** Brandenburg, NE Germany 52.34N 13.04E

98 J12 **Falkirk** C Scotland, UK 56.00N 3.48W

67 I20 **Falkland Escarpment** undersea feature SW Atlantic Ocean

65 K24 **Falkland Islands** var. Falklands, Islas Malvinas. ◇ UK dependent territory SW Atlantic Ocean

65 I20 **Falkland Islands** island group SW Atlantic Ocean

65 I20 **Falkland Plateau** var. Argentine Rise. undersea feature SW Atlantic Ocean

Falklands see Falkland Islands

65 M23 **Falkland Sound** var. Estrecho de San Carlos. strait C Falkland Islands

117 J23 **Falkonéra** island S Greece

97 K18 **Falköping** Västra Götaland, S Sweden 58.10N 13.31E

245 U6 **Fallāḩ** E Iraq 32.58N 45.09E

37 S10 **Fallbrook** California, W USA 33.22N 117.15W

201 U12 **Falleallej Pass** passage Chuuk Islands, C Micronesia

95 J14 **Fällfors** Västerbotten, N Sweden 65.07N 20.46E

204 I6 **Fallières Coast** physical region Antarctica

102 I11 **Fallingbostel** Niedersachsen, NW Germany 52.52N 9.42E

35 X9 **Fallon** Montana, NW USA 46.49N 105.07W

37 S5 **Fallon** Nevada, W USA 39.28N 118.46W

21 O12 **Fall River** Massachusetts, NE USA 41.42N 71.09W

29 P6 **Fall River Lake** ◎ Kansas, C USA

37 O3 **Fall River Mills** California, W USA 41.00N 121.28W

23 W4 **Falls Church** Virginia, NE USA 38.53N 77.11W

31 S17 **Falls City** Nebraska, C USA 40.03N 95.36W

27 S12 **Falls City** Texas, SW USA 28.58N 98.01W

79 U12 **Falluja** see Al Fallūjah

79 S12 **Falmey** Dosso, SW Niger 12.29N 2.58E

47 W10 **Falmouth** Antigua, Antigua and Barbuda 16.59N 61.48W

46 J11 **Falmouth** W Jamaica 18.28N 77.39W

99 H25 **Falmouth** SW England, UK 50.07N 5.04W

22 M4 **Falmouth** Kentucky, S USA 38.40N 84.19W

21 P13 **Falmouth** Massachusetts, NE USA 41.31N 70.36W

23 W5 **Falmouth** Virginia, NE USA 38.19N 77.28W

85 L26 **False Bay** Afr. Valsbaai. bay SW South Africa

161 K17 **False Divi Point** headland E India 15.46N 80.43E

40 M16 **False Pass** Unimak Island, Alaska, USA 54.52N 163.15W

160 P12 **False Point** headland E India 20.23N 86.52E

107 U6 **Falset** Cataluña, NE Spain 41.07N 0.49E

97 J25 **Falster** island SE Denmark

97 J23 **Falsterbo** Skåne, S Sweden 55.22N 12.49E

118 K9 **Fălticeni** Hung. Falticsén. Suceava, NE Romania 47.27N 26.18E

Falticsén see Fălticeni

96 M13 **Falun** var. Fahlun. Dalarna, C Sweden 60.36N 15.36E

64 I8 **Famatina** La Rioja, NW Argentina 28.53S 67.31W

101 J21 **Famenne** physical region SE Belgium

115 D22 **Fan** var. Fani. ☆ N Albania

79 X15 **Fan** ☆ E Nigeria

78 M12 **Fana** Koulikoro, SW Mali 12.45N 6.55W

117 K19 **Fána** ancient harbor Chíos, SE Greece 38.11N 25.35E

201 T13 **Fanan** island Chuuk, C Micronesia

201 U12 **Fanapanges** island Chuuk, C Micronesia

117 L20 **Fanári, Akrotírio** headland Ikaría, Dodekánisos, Greece, Aegean Sea 37.40N 26.21E

47 Q13 **Fancy** Saint Vincent, Saint Vincent and the Grenadines 13.22N 61.10W

181 O11 **Fandriana** Fianarantsoa, SE Madagascar 20.13S 47.21E

178 H6 **Fang** Chiang Mai, NW Thailand 19.55N 99.13E

82 E13 **Fangak** Jonglei, SE Sudan 9.04N 30.52E

203 W10 **Fangatau** atoll Îles Tuamotu, C French Polynesia

203 U13 **Fangataufa** island Îles Tuamotu, SE French Polynesia

200 Qq5 **Fanga Uta** bay S Tonga

167 N7 **Fangcheng** Henan, C China 33.18N 113.03E

166 K15 **Fangchenggang** var. Fangcheng Gezu Zizhixian; prev. Fangcheng. Guangxi Zhuangzu Zizhiqu, S China 21.49N 108.21E

167 S15 **Fangshan** S Taiwan 22.19N 120.41E

169 X8 **Fangzheng** Heilongjiang, NE China 45.49N 128.49E

Fani see Fan

121 K16 **Fanipal'** Rus. Fanipol'. Minskaya Voblasts', C Belarus 53.43N 27.17E

Fanipol' see Fanipal'

27 T13 **Fannin** Texas, SW USA 28.41N 97.13W

Fanning Island see Tabuaeran

96 G8 **Fannrem** Sør-Trøndelag, S Norway 63.16N 9.48E

108 I11 **Fano** anc. Colonia Iulia Fanestris, Fanum Fortunae. Marche, C Italy 43.51N 13.01E

97 E23 **Fanø** island W Denmark

178 I15 **Fan Si Pan** ▲ N Vietnam 22.18N 103.36E

147 W7 **Faq'** var. Al Faqa. Dubayy, E UAE 24.42N 55.37E

152 X12 **Farab** Turkm. Farap. Lebapskiy Velayat, NE Turkmenistan 39.15N 63.52E

204 H5 **Faraday** UK research station Antarctica 65.35S 64.09W

193 G16 **Faraday, Mount** ▲ South Island, NZ 42.01S 171.37E

81 P16 **Faradje** Orientale, NE Dem. Rep. Congo (Zaire) 3.45N 29.43E

181 O16 **Faradofay** see Tôlañaro

181 O16 **Farafangana** Fianarantsoa, SE Madagascar 22.49S 47.49E

154 J6 **Farāh** var. Farah, Fararud. Farāh, W Afghanistan 32.22N 62.07E

154 K7 **Farāh** ◇ province W Afghanistan

154 J6 **Farāh Rūd** ☆ W Afghanistan

196 K7 **Farallon de Medinilla** island C Northern Mariana Islands

196 J2 **Farallon de Pajaros** var. Uracas. island N Northern Mariana Islands

78 H12 **Faranah** Haute-Guinée, S Guinea 10.01N 10.43W

Farap see Farab

154 J7 **Fararud** see Farāh

146 M6 **Farasān, Jazā'ir** island group SW Saudi Arabia

181 O15 **Faratsiho** Antananarivo, C Madagascar 19.24S 46.57E

196 M13 **Faraulep Atoll** atoll Caroline Islands, C Micronesia

101 H20 **Farciennes** Hainaut, S Belgium 50.25N 4.33E

107 O14 **Fardes** ☆ S Spain

203 S10 **Fare** Huahine, W French Polynesia 16.42S 151.01W

99 M23 **Fareham** S England, UK 50.51N 1.10W

41 P11 **Farewell** Alaska, USA 62.35N 153.59W

192 H13 **Farewell, Cape** headland South Island, NZ 40.30S 172.39E

192 H13 **Farewell, Cape** see Nunap Isua

192 H13 **Farewell Spit** spit South Island, NZ

97 J17 **Färgelanda** Västra Götaland, S Sweden 58.34N 11.58E

Farghana Rus. Fergana; prev. Novyy Margilan. Farghona Wiloyati, E Uzbekistan 40.27N 71.43E

Farghona Valley see Fergana Valley

153 R10 **Farghona Wiloyati** Rus. Ferganskaya Oblast'. ◇ province E Uzbekistan

128 L13 **Fáurei** prev. Filimon Sirbu. Brăila, SE Romania 45.04N 27.15E

195 T13 **Fauro** island Shortland Islands, NW Solomon Islands

156 G12 **Fauske** Nordland, C Norway 67.15N 15.27E

9 P13 **Faust** Alberta, W Canada 55.19N 115.33W

101 L23 **Fauvillers** Luxembourg, SE Belgium 49.52N 5.40E

109 J24 **Favara** Sicilia, Italy, C Mediterranean Sea 37.19N 13.39E

109 G23 **Faventia** see Faenza

109 G23 **Favignana, Isola** island Isole Egadi, S Italy

10 D8 **Fawn** ☆ Ontario, S Canada

156 H3 **Faxa Bay** see Faxaflói

156 H3 **Faxaflói** Eng. Faxa Bay. bay W Iceland

80 I7 **Faya** prev. Faya-Largeau, Largeau. Borkou-Ennedi-Tibesti, N Chad 17.58N 19.06E

Faya-Largeau see Faya

197 J5 **Fayaoué** Province des Îles Loyauté, C New Caledonia 20.41S 166.31E

144 M5 **Faydāt** hill range E Syria

25 O3 **Fayette** Alabama, S USA 33.40N 87.49W

31 X12 **Fayette** Iowa, C USA 42.50N 91.48W

24 J6 **Fayette** Mississippi, S USA 31.42N 91.03W

29 U4 **Fayette** Missouri, C USA 39.09N 92.40W

29 S9 **Fayetteville** Arkansas, C USA 36.03N 94.09W

23 U10 **Fayetteville** North Carolina, SE USA 35.03N 78.52W

22 J10 **Fayetteville** Tennessee, S USA 35.09N 86.34W

27 U11 **Fayetteville** Texas, SW USA 29.52N 96.40W

23 R5 **Fayetteville** West Virginia, NE USA 38.03N 81.06W

147 R4 **Faylakah** var. Failaka Island. island E Kuwait

245 T10 **Fayşalīyah** var. Faisaliya. S Iraq 31.48N 44.36E

201 P15 **Fayu** var. East Fayu. island Hall Islands, C Micronesia

158 G8 **Fāzilka** Punjab, NW India 30.25N 74.04E

78 I6 **Fdérik** var. Fdérick, Fr. Fort Gouraud. Tiris Zemmour, NW Mauritania 22.40N 12.40W

99 B20 **Feale** ☆ SW Ireland

23 V12 **Fear, Cape** headland Bald Head Island, North Carolina, SE USA 33.50N 77.57W

37 O6 **Feather River** ☆ California, W USA

193 H14 **Featherston** Wellington, North Island, NZ 41.07S 175.19E

104 L3 **Fécamp** Seine-Maritime, N France 49.45N 0.22E

63 D17 **Fedala** see Mohammedia

63 D17 **Federación** Entre Ríos, E Argentina 30.57S 58.04W

63 D17 **Federal** Entre Ríos, E Argentina 30.53S 58.45W

79 T15 **Federal Capital District** ◇ capital territory C Nigeria

Federal Capital Territory see Australian Capital Territory

Federal District see Distrito Federal

23 Y4 **Federalsburg** Maryland, NE USA 38.41N 75.46W

79 M6 **Fedjaj, Chott el** var. Chott el Fejaj, Shaṭṭ al Fijāj. salt lake C Tunisia

96 B13 **Fedje** island S Norway

150 M7 **Fedorovka** Kostanay, N Kazakhstan 51.12N 62.00E

131 U6 **Fedorovka** Respublika Bashkortostan, W Russian Federation 53.09N 55.07E

119 U13 **Fedotova Kosa** spit SE Ukraine

96 H10 **Feragen** ◎ S Norway

76 L5 **Fer, Cap de** headland NE Algeria 37.05N 7.10E

113 O21 **Ferdinand** Indiana, N USA 38.13N 86.51W

Ferdinand see Mihail Kogălniceanu, Romania

Ferdinand see Montana, Bulgaria

Ferdinandsberg see Oţelu Roşu

149 T7 **Ferdows** var. Firdaus; prev. Tūn. Khorāsān, E Iran 34.00N 58.09E

Fère-Champenoise Marne, N France 48.45N 3.59E

Ferencz-József Csúcs see Gerlachovský štit

109 J16 **Ferentino** Lazio, C Italy 41.40N 13.16E

116 I13 **Féres** Anatolikí Makedonía kai Thráki, NE Greece 40.54N 26.12E

189 S9 **Fergana** see Farghona

Fergana Valley var. Farghona Valley, Rus. Ferganskaya Dolina, Taj. Wodii Farghona, Uzb. Farghona Wodiysi. basin Tajikistan/Uzbekistan

Ferganskaya Dolina see Fergana Valley

Ferganskaya Oblast' see Farghona Wiloyati

Ferganskiy Khrebet ▲ C Kyrgyzstan

130 J7 **Fatezh** Kurskaya Oblast', W Russian Federation 52.01N 35.51E

78 K11 **Fatick** W Senegal 14.20N 16.29W

106 G9 **Fátima** Santarém, W Portugal 39.37N 8.39W

142 M11 **Fatsa** Ordu, N Turkey 41.01N 37.31E

202 D12 **Fatshan** see Foshan

202 D12 **Fatua, Pointe** var. Pointe Nord. headland Île Futuna, S Wallis and Futuna

203 X7 **Fatu Hiva** island Îles Marquises, NE French Polynesia

81 N7 **Fatunda** var. Fatunda. Bandundu, W Dem. Rep. Congo (Zaire) 4.07S 17.13E

31 O8 **Faulkton** South Dakota, N USA 45.01N 99.07W

157 A20 **Fáurei** prev. Filimon Sirbu. Brăila, SE Romania

(see column 4)

103 F24 **Feldberg** ▲ SW Germany 47.52N 8.01E

118 J12 **Feldioara** Ger. Marienburg, Hung. Földvár. Braşov, C Romania 45.48N 25.37E

110 I7 **Feldkirch** anc. Clunia. Vorarlberg, W Austria 47.15N 9.37E

189 S9 **Feldkirchen in Kärnten** Slvn. Trg. Kärnten, S Austria 46.42N 14.01E

198 B8 **Feleolo** K. (Ápia) Upolu, C Samoa 13.49S 171.59W

106 H6 **Felgueiras** Porto, N Portugal 41.22N 8.12W

99 E16 **Fermanagh** cultural region SW Northern Ireland, UK

180 J16 **Félicité** island Inner Islands, NE Seychelles

108 J12 **Fermo** anc. Firmum Picenum. Marche, C Italy 43.09N 13.43E

157 A20 **Felidhu Atoll** atoll C Maldives

106 J6 **Fermoselle** Castilla-León, N Spain 41.19N 6.24W

43 Y13 **Felipe Carrillo Puerto** Quintana Roo, SE Mexico 19.33N 88.01W

99 D20 **Fermoy Ir.** Mainistir Fhear Maí. SW Ireland 52.07N 8.16W

97 Q21 **Felixstowe** E England, UK 51.58N 1.19E

25 W8 **Fernandina Beach** Amelia Island, Florida, SE USA 30.40N 81.27W

105 N11 **Felletin** Creuse, C France 45.53N 2.12E

59 A17 **Fernandina, Isla** var. Narborough Island. island Galapagos Islands, Ecuador, E Pacific Ocean

157 N10 **Fellin** see Viljandi

49 Y5 **Fernando de Noronha** island E Brazil

37 N10 **Felton** California, W USA 37.03N 122.04W

62 J7 **Fernandópolis** São Paulo, S Brazil 20.12S 50.14W

106 M13 **Feltre** Veneto, NE Italy 46.01N 11.53E

97 H25 **Femerbelt** Ger. Fehmarnbelt. strait Denmark/Germany

85 Q14 **Fernão Veloso, Baia de** bay NE Mozambique

97 J24 **Femø** island SE Denmark

36 K3 **Ferndale** California, W USA 40.34N 124.16W

96 H6 **Femunden** ◎ S Norway

34 H6 **Ferndale** Washington, NW USA 48.51N 122.35W

106 H2 **Fene** Galicia, NW Spain 43.28N 8.10W

9 P17 **Fernie** British Columbia, SW Canada 49.30N 115.00W

12 I14 **Fenelon Falls** Ontario, SE Canada 44.34N 78.43W

37 R5 **Fernley** Nevada, W USA 39.35N 119.15W

143 O11 **Fener Burnu** headland N Turkey 41.07N 39.26E

Ferozepore see Firozpur

117 J14 **Fengári** ▲ Samothráki, E Greece 40.27N 25.37E

109 N18 **Ferrandina** Basilicata, S Italy 40.30N 16.25E

169 V13 **Fengcheng** var. Feng-cheng, Fenghwangcheng. Liaoning, NE China 40.28N 124.01E

108 G9 **Ferrara** anc. Forum Alieni. Emilia-Romagna, N Italy 44.49N 11.36E

156 K11 **Fenggang** prev. Longquan. Guizhou, S China 27.52N 107.42E

123 H11 **Ferrat, Cap** headland NW Algeria 35.52N 0.24W

156 L9 **Fenghua** Zhejiang, SE China 29.38N 121.23E

109 D20 **Ferrato, Capo** headland Sardegna, Italy, C Mediterranean Sea 39.18N 9.37E

166 M14 **Fengkai** prev. Jiankou. Guangdong, S China 23.26N 111.28E

106 G8 **Ferreira do Alentejo** Beja, S Portugal 38.04N 8.06W

167 U11 **Fenggang** see Fengcheng

58 B11 **Ferreñafe** Lambayeque, W Peru 6.37S 79.45W

167 T13 **Fenglin** Jap. Hōrin. C Taiwan 23.52N 121.30E

110 C12 **Ferret, Cap** headland W France 44.37N 1.15W

167 P1 **Fengning** prev. Dagezhen. Hebei, E China 41.12N 116.37E

110 D9 **Ferret, Val** valley W Switzerland 45.57N 7.04E

147 R4 **Fengqing** Yunnan, SW China 24.38N 99.54E

24 I6 **Ferriday** Louisiana, S USA 31.37N 91.33W

158 E13 **Fengqiu** Henan, C China 35.01N 114.24E

Ferro see Hierro

167 O6 **Fengrun** Hebei, E China

109 D16 **Ferro, Capo** headland Sardegna, Italy, C Mediterranean Sea 41.09N 9.31E

167 Q2 **Fengshui Shan** ▲ NE China 52.20N 123.22E

106 H2 **Ferrol** var. El Ferrol; prev. El Ferrol del Caudillo. Galicia, NW Spain 43.28N 8.13W

167 T4 **Fengshun** Guangdong, S China 23.51N 116.11E

58 B12 **Ferrol, Península de** peninsula W Peru

167 P14 **Fengtien** see Liaoning, China

38 M5 **Ferron** Utah, W USA 39.05N 111.07W

Fengtien see Shenyang, China

167 P9 **Fengxian** prev. Feng Xian; prev. Shuangshuyu. Shaanxi, C China 33.50N 106.33E

14 R4 **Ferrum** Virginia, NE USA 36.54N 80.01W

169 T13 **Fengxian** var. Feng Xian; prev. Shuangshuyu. Shaanxi, C China

25 O8 **Ferry Pass** Florida, SE USA 30.29N 87.09W

169 P13 **Fengxiang** see Luobei

Fengzhen Nei Mongol Zizhiqu, N China 40.25N 113.09E

166 M6 **Fen He** ☆ C China

31 S4 **Fertile** Minnesota, N USA 47.32N 96.16W

159 V15 **Feni** Chittagong, E Bangladesh 23.00N 91.24E

Fertő see Neusiedler See

195 Q10 **Feni Islands** island group NE PNG

100 L5 **Ferwerd Fris.** Ferwert. Friesland, N Netherlands 53.21N 5.46E

40 H7 **Fenimore Pass** strait Aleutian Islands, Alaska, USA

Ferwert see Ferwerd

86 B9 **Feni Ridge** undersea feature N Atlantic Ocean

76 G6 **Fès Eng.** Fez. N Morocco 34.06N 4.57W

202 B10 **Fennern** see Vändra

81 I22 **Feshi** Bandundu, SW Dem. Rep. Congo (Zaire) 6.07S 18.12E

32 J9 **Fennimore** Wisconsin, N USA 42.58N 90.39W

31 O4 **Fessenden** North Dakota, N USA 47.39N 99.37W

180 I4 **Fenoarivo** Toamasina, E Madagascar 20.52S 46.52E

29 X5 **Festus** Missouri, C USA 38.13N 90.24W

97 J24 **Fensmark** Storstrøm, SE Denmark 55.16N 11.41E

118 M14 **Fetești** Ialomiţa, SE Romania 44.22N 27.51E

99 O19 **Fens, The** wetland E England, UK

142 D17 **Fethiye** Muğla, SW Turkey 36.37N 29.07E

33 R9 **Fenton** Michigan, N USA 42.48N 83.43W

94 M1 **Fetlar** island NE Scotland, UK

166 F6 **Fenyang** Shanxi, C China 37.14N 111.40E

97 I15 **Fetsund** Akershus, S Norway 59.55N 11.03E

119 U13 **Feodosiya** var. Kefe, It. Kaffa; anc. Theodosia. Respublika Krym, S Ukraine 45.03N 35.23E

90 K3 **Ffestiniog** NW Wales, UK

Fêdory see Fyadory

118 F15 **Fès** see Fès

149 T8 **Feragen** ◎ S Norway

Fhóid Duibh, Cuan an see Blacksod Bay

64 I8 **Fiambalá** Catamarca, NW Argentina 27.45S 67.37W

181 I7 **Fianarantsoa** Fianarantsoa, C Madagascar 21.27S 47.05E

181 N16 **Fianarantsoa** ◇ province SE Madagascar

80 K12 **Fianga** Mayo-Kébbi, SW Chad 9.57N 15.09E

82 I13 **Ficce** see Fichē

82 J12 **Fichē** It. Ficce. Oromo, C Ethiopia 9.48N 38.43E

103 P24 **Fichtelberg** ▲ Czech Republic/Germany 50.26N 12.57E

103 M19 **Fichtelgebirge** ▲ SE Germany

115 K21 **Fier** var. Fieri. Fier, SW Albania 40.44N 19.34E

103 N19 **Fichtelnaab** ☆ SE Germany

115 K21 **Fier** ◇ district SW Albania

115 K21 **Fier** ◆ *district* W Albania
Fieri *see* Fier
Fierza *see* Fierzë
115 L17 **Fierzë** *var.* Fierza. Shkodër, N Albania *42.15N 20.02E*
115 L17 **Fierzës, Liqeni i** ☒ N Albania
110 F10 **Fiesch** Valais, SW Switzerland *46.25N 8.09E*
108 G11 **Fiesole** Toscana, C Italy *43.50N 11.18E*
144 K12 **Fifah** Aţ Ţafilah, W Jordan *30.55N 35.25E*
98 K11 **Fife** *var.* Kingdom of Fife. *cultural region* E Scotland, UK
98 K11 **Fife Ness** *headland* E Scotland, UK *56.16N 2.35W*
Fifteen Twenty Fracture Zone *see* Barracuda Fracture Zone
105 N13 **Figeac** Lot, S France *44.37N 2.01E*
97 N19 **Figeholm** Kalmar, SE Sweden *57.12N 16.34E*
Figig *see* Figuig
85 J18 **Figtree** Matabeleland South, SW Zimbabwe *20.20S 28.20E*
106 F8 **Figueira da Foz** Coimbra, W Portugal *40.09N 8.51W*
107 X4 **Figueres** Cataluña, E Spain *42.16N 2.57E*
76 H7 **Figuig** *var.* Figig. E Morocco *32.09N 1.13W*
Fijāj, Shaţţ al *see* Fedjaj, Chott el
197 J14 **Fiji** *off.* Sovereign Democratic Republic of Fiji, *Fij.* Viti. ◆ *republic* SW Pacific Ocean
197 I13 **Fiji** *island group* SW Pacific Ocean
183 Q8 **Fiji Plate** *tectonic feature*
107 P14 **Filabres, Sierra de los** ▲ SE Spain
85 K18 **Filabusi** Matabeleland South, S Zimbabwe *20.31S 29.16E*
44 K13 **Filadelfia** Guanacaste, W Costa Rica *10.24N 85.33W*
113 K20 **Fiľakovo** *Hung.* Fülek. Banskobystrický Kraj, C Slovakia *48.15N 19.53E*
205 N5 **Filchner Ice Shelf** *ice shelf* Antarctica
12 J11 **Fildegrand** ◆ Quebec, SE Canada
35 O15 **Filer** Idaho, NW USA *42.34N 114.36W*
Filevo *see* Vŭrbitsa
118 H14 **Filiaşi** Dolj, SW Romania *44.32N 23.30E*
117 B16 **Filiátes** Ípeiros, W Greece *39.36N 20.19E*
117 D21 **Filiatrá** Pelopónnisos, S Greece *37.10N 21.35E*
109 K22 **Filicudi, Isola** *island* Isole Eolie, S Italy
147 Y10 **Filim** E Oman *20.37N 58.11E*
Filimon Sîrbu *see* Făurei
79 S11 **Filingué** Tillabéri, W Niger *14.12N 3.16E*
116 K13 **Filiouri** ﹏ NE Greece
116 I13 **Filippoi** *anc.* Philippi. *site of ancient city* Anatolikí Makedonía kai Thráki, NE Greece *41.01N 24.15E*
97 L15 **Filipstad** Värmland, C Sweden *59.43N 14.10E*
110 I9 **Filisur** Graubünden, S Switzerland *46.40N 9.43E*
96 E12 **Fillefjell** ﹏ S Norway
37 R14 **Fillmore** California, W USA *34.23N 118.56W*
38 K5 **Fillmore** Utah, W USA *38.57N 112.19W*
12 J10 **Fils, Lac du** ◎ Quebec, SE Canada
142 H11 **Filyos Çayı** ﹏ N Turkey
205 Q2 **Fimbulheimen** *physical region* Antarctica
205 Q1 **Fimbul Ice Shelf** *ice shelf* Antarctica
108 G9 **Finale Emilia** Emilia-Romagna, C Italy *44.50N 11.17E*
108 C10 **Finale Ligure** Liguria, NW Italy *44.11N 8.21E*
107 P14 **Fiñana** Andalucía, S Spain *37.09N 2.47W*
180 I6 **Finandrahana** Fianarantsoa, SE Madagascar
23 S6 **Fincastle** Virginia, NE USA *37.29N 79.51W*
101 M25 **Findel** ✈ (Luxembourg) Luxembourg, C Luxembourg *49.38N 6.16E*
98 J9 **Findhorn** ﹏ N Scotland, UK
33 R12 **Findlay** Ohio, N USA *41.02N 83.39W*
20 G11 **Finger Lakes** *lakes* New York, NE USA
85 L14 **Fíngoè** Tete, NW Mozambique *15.01S 31.52E*
142 E17 **Finike** Antalya, SW Turkey *36.18N 30.07E*
104 F6 **Finistère** ◆ *department* NW France
194 J12 **Finisterre, Mount** ▲ C PNG *5.58S 146.30E*
194 J12 **Finisterre Range** ▲ N PNG
189 Q8 **Finke** Northern Territory, N Australia *25.37S 134.35E*
189 S10 **Finkenstein** Kärnten, S Austria *46.34N 13.53E*
201 Y15 **Finkol, Mount** *var.* Mount Crozer. ▲ Kosrae, E Micronesia *5.18N 163.00E*
95 L17 **Finland** *off.* Republic of Finland, *Fin.* Suomen Tasavalta, Suomi. ◆ *republic* N Europe
128 F12 **Finland, Gulf of** *Est.* Soome Laht, *Fin.* Suomenlahti, *Ger.* Finnischer Meerbusen, *Rus.* Finskiy Zaliv, *Swe.* Finska Viken. *gulf* E Baltic Sea
8 L11 **Finlay** ﹏ British Columbia, W Canada
191 O10 **Finley** New South Wales, SE Australia *35.41S 145.33E*
31 Q4 **Finley** North Dakota, N USA *47.30N 97.50W*
Finnischer Meerbusen *see* Finland, Gulf of
156 K9 **Finnmark** ◆ *county* N Norway
156 K9 **Finnmarksvidda** *physical region* N Norway
156 I9 **Finnsnes** Troms, N Norway *69.13N 17.58E*
194 K13 **Finschhafen** Morobe, C PNG *6.38S 147.49E*

96 E13 **Finse** Hordaland, S Norway *60.35N 7.33E*
181 Y15 **Finska Viken/Finskiy Zaliv** *see* Finland, Gulf of
97 M17 **Finspång** Östergötland, S Sweden *58.42N 15.45E*
110 F10 **Finsteraarhorn** ▲ Switzerland *46.33N 8.07E*
103 O14 **Finsterwalde** Brandenburg, E Germany *51.37N 13.43E*
193 A23 **Finstrøm** *physical region* South Island, NZ
108 E9 **Fiorenzuola d'Arda** Emilia-Romagna, C Italy *44.57N 9.53E*
Firat Nehri *see* Euphrates
Firdaus *see* Ferdows
20 M14 **Fire Island** *island* New York, NE USA
108 G11 **Firenze** *Eng.* Florence; *anc.* Florentia. Toscana, C Italy *43.46N 11.15E*
108 G10 **Firenzuola** Toscana, C Italy *44.07N 11.22E*
12 C6 **Firesteel** ﹏ Ontario, S Canada *48.46N 83.34W*
Firliug *see* Fârliug
63 B19 **Firmat** Santa Fe, C Argentina *33.28S 61.28W*
105 Q22 **Firminy** Loire, E France *45.22N 4.18E*
Firmum Picenum *see* Fermo
158 J12 **Firozabad** Uttar Pradesh, N India *27.09N 78.24E*
158 G8 **Firozpur** *var.* Ferozepore. Punjab, NW India *30.55N 74.37E*
First State *see* Delaware
149 O12 **Firūzābād** Fārs, S Iran *28.51N 52.34E*
Fischamend *see* Fischamend Markt
189 Y4 **Fischamend Markt** *var.* Fischamend. Niederösterreich, E Austria *48.08N 16.37E*
189 W6 **Fischbacher Alpen** ▲ E Austria
Fischhausen *see* Primorsk
85 J13 **Fish** *var.* Vis. ﹏ S Namibia
85 F24 **Fish** *Afr.* Vis. ﹏ SW South Africa
9 X15 **Fisher Branch** Manitoba, S Canada *51.09N 97.34W*
21 N13 **Fisher River** Manitoba, S Canada *51.25N 97.23W*
21 N13 **Fishers Island** *island* New York, NE USA
39 U6 **Fishers Peak** ▲ Colorado, C USA *37.06N 104.27W*
15 M6 **Fisher Strait** *strait* Nunavut, N Canada
99 I22 **Fishguard** *Wel.* Abergwaun. SW Wales, UK *51.58N 4.49W*
21 R2 **Fish River Lake** ◎ Maine, NE USA
204 K6 **Fiske, Cape** *headland* Antarctica *74.27S 60.28W*
105 P4 **Fismes** Marne, N France *49.19N 3.41E*
106 F3 **Fisterra, Cabo** *headland* NW Spain *42.53N 9.16W*
21 N11 **Fitchburg** Massachusetts, NE USA *42.34N 71.48W*
98 L3 **Fitful Head** *headland* NE Scotland, UK *59.57N 1.24W*
97 C14 **Fitjar** Hordaland, S Norway *59.55N 5.19E*
198 B68 **Fito** ▲ Upolu, C Samoa *13.57S 171.42W*
25 U6 **Fitzgerald** Georgia, SE USA *31.42N 83.15W*
188 M5 **Fitzroy Crossing** Western Australia *18.10S 125.40E*
65 G21 **Fitzroy, Monte** *var.* Cerro Chaltel. ▲ S Argentina *49.18S 73.06W*
189 Y8 **Fitzroy River** ﹏ Queensland, E Australia
188 L5 **Fitzroy River** ﹏ Western Australia
12 E12 **Fitzwilliam Island** *island* S Canada
109 J15 **Fiuggi** Lazio, C Italy *41.47N 13.16E*
Fiume *see* Rijeka
109 H15 **Fiumicino** Lazio, C Italy *41.46N 12.13E*
Fiumicino *see* Leonardo da Vinci
108 E10 **Fivizzano** Toscana, C Italy *44.13N 10.06E*
81 O21 **Fizi** Sud Kivu, E Dem. Rep. Congo (Zaire) *4.15S 28.57E*
Fizuli *see* Füzuli
156 I11 **Fjällåsen** Norrbotten, N Sweden *67.31N 20.07E*
95 G22 **Fjerritslev** Nordjylland, N Denmark *57.06N 9.16E*
97 L16 **Fjugesta** Örebro, C Sweden *59.10N 14.50E*
39 V5 **Flagler** Colorado, C USA *39.17N 103.04W*
25 X10 **Flagler Beach** Florida, SE USA *29.28N 81.07W*
38 L11 **Flagstaff** Arizona, SW USA *35.12N 111.39W*
67 H24 **Flagstaff Bay** *bay* Saint Helena, C Atlantic Ocean
21 P5 **Flagstaff Lake** ◎ Maine, NE USA
96 E13 **Flåm** Sogn og Fjordane, S Norway *60.51N 7.06E*
13 O8 **Flamand** ﹏ Quebec, SE Canada
32 J5 **Flambeau River** ﹏ Wisconsin, N USA
99 O16 **Flamborough Head** *headland* E England, UK *54.06N 0.03W*
102 N13 **Fläming** *hill range* NE Germany
18 I7 **Flaming Gorge Reservoir** ☒ Utah/Wyoming, NW USA
176 X13 **Flamingo, Teluk** *bay* N Arafura Sea
101 B18 **Flanders** *Dut.* Vlaanderen, *Fr.* Flandre. *cultural region* Belgium/France
Flandre *see* Flanders
31 N10 **Flandreau** South Dakota, N USA *44.03N 96.36W*
98 I6 **Flannan Isles** *island group* NW Scotland, UK
30 M6 **Flasher** North Dakota, N USA *46.25N 101.12W*
95 G15 **Flåsjön** ◎ N Sweden
41 O11 **Flat** Alaska, USA *62.27N 158.00W*
14 G7 **Flat** ﹏ Northwest Territories, NW Canada
156 H1 **Flateyri** Vestfirðir, Nw Iceland *66.03N 23.28W*

35 P8 **Flathead Lake** ◎ Montana, NW USA
181 Y15 **Flat Island** *Fr.* Île Plate. *island* N Mauritius
179 N14 **Flat Island** *island* NE Spratly Islands
27 T11 **Flatonia** Texas, SW USA *29.41N 97.06W*
193 M14 **Flat Point** *headland* North Island, NZ *41.12S 176.03E*
29 X6 **Flat Rock** Missouri, C USA *37.51N 90.31W*
33 P8 **Flat Rock** Michigan, N USA
33 P14 **Flatrock River** ﹏ Indiana, N USA
34 E6 **Flattery, Cape** *headland* Washington, NW USA *48.22N 124.43W*
66 M3 **Flatts Village** *var.* The Flatts Village. C Bermuda *32.19N 64.43W*
110 H7 **Flawil** Sankt Gallen, NE Switzerland *47.25N 9.12E*
99 N22 **Fleet** S England, UK *51.16N 0.49W*
99 K16 **Fleetwood** NW England, UK *53.55N 3.01W*
20 H15 **Flemington** Pennsylvania, NE USA *40.27N 75.49W*
97 D18 **Flekkefjord** Vest-Agder, S Norway *58.16N 6.40E*
23 N5 **Flemingsburg** Kentucky, C USA *38.24N 83.43W*
20 J15 **Flemington** New Jersey, NE USA *40.30N 74.51W*
66 I7 **Flemish Cap** *undersea feature* NW Atlantic Ocean
97 N16 **Flen** Södermanland, C Sweden *59.03N 16.37E*
102 J8 **Flensburg** Schleswig-Holstein, N Germany *54.46N 9.25E*
102 I8 **Flensburger Förde** *inlet* Denmark/Germany
104 K15 **Flers** Orne, N France *48.45N 0.34W*
97 C14 **Flesland** ✈ (Bergen) Hordaland, S Norway *60.18N 5.15E*
Flessingue *see* Vlissingen
23 S4 **Fletcher** North Carolina, SE USA *35.24N 82.29W*
33 R6 **Fletcher Pond** ◎ Michigan, N USA
104 L15 **Fleurance** Gers, S France *43.50N 0.39E*
110 B8 **Fleurier** Neuchâtel, W Switzerland *46.55N 6.37E*
101 H20 **Fleurus** Hainaut, S Belgium *50.28N 4.33E*
105 N7 **Fleury-les-Aubrais** Loiret, C France *47.55N 1.55E*
100 K10 **Flevoland** ◆ *province* C Netherlands
Flickertail State *see* North Dakota
110 H9 **Flims** Glarus, NE Switzerland *46.50N 9.16E*
190 F8 **Flinders Island** *island* Investigator Group, South Australia
191 P14 **Flinders Island** *island* Furneaux Group, Tasmania, SE Australia
190 I6 **Flinders Ranges** ▲ South Australia
189 U5 **Flinders River** ﹏ Queensland, NE Australia
9 V13 **Flin Flon** Manitoba, C Canada *54.46N 101.51W*
99 K18 **Flint** NE Wales, UK *53.15N 3.09W*
33 R9 **Flint** Michigan, N USA *43.00N 83.41W*
99 J18 **Flint** *cultural region* NE Wales, UK
29 O7 **Flint Hills** *hill range* Kansas, C USA
203 Y6 **Flint Island** *island* Line Islands, E Kiribati
25 S4 **Flint River** ﹏ Georgia, SE USA
33 R9 **Flint River** ﹏ Michigan, N USA
201 X12 **Flipper Point** *headland* C Wake Island *19.18N 166.37E*
96 I13 **Flisa** Hedmark, S Norway
96 J13 **Flisa** ﹏ S Norway
126 Hb4 **Flissingskiy, Mys** *headland* Novaya Zemlya, NW Russian Federation *76.43N 69.01E*
Flitsch *see* Bovec
107 U6 **Flix** Cataluña, NE Spain *41.13N 0.32E*
117 J19 **Flóka** ﹏ E Germany
27 O4 **Flomot** Texas, SW USA *34.13N 100.58W*
31 V9 **Floodwood** Minnesota, N USA *46.55N 92.55W*
32 M15 **Flora** Illinois, N USA *38.40N 88.29W*
105 P14 **Florac** Lozère, S France *44.18N 3.35E*
25 U2 **Florala** Alabama, S USA *31.00N 86.19W*
105 S4 **Florange** Moselle, NE France *49.21N 6.06E*
25 N3 **Florence** Alabama, S USA *34.48N 87.40W*
38 L11 **Florence** Arizona, SW USA *33.01N 111.23W*
39 U7 **Florence** Colorado, C USA *38.20N 105.06W*
29 O5 **Florence** Kansas, C USA *38.13N 96.54W*
23 N5 **Florence** Kentucky, SW USA *39.00N 84.37W*
34 E14 **Florence** Oregon, NW USA *43.58N 124.06W*
21 T12 **Florence** South Carolina, SW USA *34.12N 79.45W*
32 L7 **Florence** Wisconsin, N USA *45.55N 88.14W*
Florence *see* Firenze
54 E11 **Florencia** Caquetá, S Colombia *1.37N 75.37W*
101 I21 **Florennes** Namur, S Belgium *50.15N 4.36E*
Florentia *see* Firenze
65 J18 **Florentino Ameghino, Embalse** ☒ S Argentina
101 J24 **Florenville** Luxembourg, SE Belgium *49.42N 5.18E*
25 S9 **Flores** Alabama, S USA *30.50N 97.47W*
44 B4 **Flores** Petén, N Guatemala *16.54N 89.55W*
63 G19 **Flores** ◆ *department* S Uruguay
175 Pp16 **Flores** *island* Nusa Tenggara, C Indonesia

66 M1 **Flores** *island* Azores, Portugal, NE Atlantic Ocean
44 H8 **Flores, Golfo de** *see* Fonseca, Gulf of
105 O6 **Flores, Lago** *see* Petén Itzá, Lago
175 P15 **Flores Sea** *Ind.* Laut Flores. *sea* C Indonesia
118 M8 **Floreşti** *Rus.* Floreshty. N Moldova *47.52N 28.19E*
27 S12 **Floresville** Texas, SW USA *29.07N 98.09W*
61 N14 **Floriano** Piauí, E Brazil *6.45S 43.00W*
63 K14 **Florianópolis** *prev.* Desterro. *state capital* Santa Catarina, S Brazil *27.34S 48.31W*
46 G6 **Florida** Camagüey, C Cuba *21.31N 78.13W*
63 F19 **Florida** Florida, S Uruguay *34.04S 56.13W*
63 F19 **Florida** ◆ *department* S Uruguay
25 U9 **Florida** *off.* State of Florida; also known as Peninsular State, Sunshine State. ◆ *state* SE USA
25 Y17 **Florida Bay** *bay* Florida, SE USA
56 G8 **Floridablanca** Santander, N Colombia *7.04N 73.06W*
195 X15 **Florida Islands** *island group* C Solomon Islands
25 Y17 **Florida Keys** *island group* Florida, SE USA
39 Q16 **Florida Mountains** ▲ New Mexico, SW USA
66 D10 **Florida, Straits of** *strait* Atlantic Ocean/Gulf of Mexico
116 J13 **Flórina** *var.* Phlórina. Dytikí Makedonía, N Greece *40.48N 21.25E*
29 X4 **Florissant** Missouri, C USA *38.47N 90.19W*
96 C11 **Florø** Sogn og Fjordane, S Norway *61.34N 5.01E*
37 R13 **Flóyd, Akrotírio** *headland* Astypálaia, Kykládes, Greece, Aegean Sea *36.38N 26.23E*
33 N4 **Floyd** Virginia, NE USA *36.53N 80.21W*
191 O4 **Floyd River** ﹏ Michigan, N USA
27 N4 **Floydada** Texas, SW USA *33.58N 101.20W*
Flüela Wisshorn *see* Weisshorn
29 U13 **Fluessen** ◎ N Netherlands
107 S5 **Flúmen** ﹏ NE Spain
109 C20 **Flumendosa** ﹏ Sardegna, Italy, C Mediterranean Sea
207 O14 **Forel, Mont** ▲ SE Greenland
33 R9 **Flushing** Michigan, N USA *43.03N 83.51W*
Flushing *see* Vlissingen
27 R17 **Fluvanna** Texas, SW USA *32.54N 101.06W*
194 F14 **Fly** ﹏ Indonesia/PNG
204 I10 **Flying Fish, Cape** *headland* Thurston Island, Antarctica *72.00S 102.25W*
Flylân *see* Vlieland
200 Ss13 **Foa** *island* Ha'apai Group, C Tonga
9 U15 **Foam Lake** Saskatchewan, S Canada *51.37N 103.31W*
34 F6 **Focia** *var.* Srbinje, Republika Srpska, Bosnia and Herzegovina *43.32N 18.45E*
118 J10 **Focşani** Vrancea, E Romania *45.43N 27.13E*
109 M16 **Foggia** Puglia, SE Italy *41.28N 15.31E*
25 S3 **Foggo** *see* Fuga
78 D10 **Fogo** *island* Ilhas de Sotavento, SW Cape Verde
11 U11 **Fogo Island** *island* Newfoundland, E Canada *49.45N 69.04W*
189 U7 **Fohnsdorf** Steiermark, SE Austria *47.13N 14.40E*
102 I7 **Föhr** *island* NW Germany *54.50N 8.30E*
106 F14 **Fóia** ▲ S Portugal *37.19N 8.39W*
12 I10 **Foins, Lac aux** ◎ Quebec, SE Canada
105 N17 **Foix** Ariège, S France *42.58N 1.39E*
130 I5 **Fokino** Bryanskaya Oblast', W Russian Federation *53.22N 34.22E*
96 E13 **Fola, Cnoc** *see* Bloody Foreland
96 E13 **Folarskardnuten** ▲ S Norway *60.34N 7.18E*
15 O10 **Foldafjorden** *fjord* C Norway
95 E14 **Foldafjorden** *fjord* C Norway
189 U7 **Føldvær** *see* Feldioara
117 J19 **Foldereid** Nord-Trøndelag, C Norway *64.58N 12.09E*
117 J22 **Folégandros** *island* Kykládes, Greece, Aegean Sea
99 K17 **Folkestone** SE England, UK *51.04N 1.10E*
25 W8 **Folkston** Georgia, SE USA *30.49N 82.00W*
96 I10 **Folldal** Hedmark, S Norway *62.08N 10.00E*
25 V9 **Follett** Texas, SW USA *36.25N 100.08W*
108 F13 **Follonica** Toscana, C Italy *42.55N 10.45E*
23 U12 **Folly Beach** South Carolina, SE USA *32.39N 79.56W*
37 O7 **Folsom** California, W USA *38.40N 121.11W*
118 M12 **Folteşti** Galaţi, E Romania *45.45N 28.00E*
97 I22 **Fomboni** Mohéli, S Comoros *12.18S 43.46E*
108 K6 **Fonda** New York, NE USA *42.57N 74.24W*
9 S10 **Fond-du-Lac** Saskatchewan, C Canada *59.19N 107.09W*
32 M8 **Fond du Lac** Wisconsin, N USA *43.48N 88.27W*
9 T10 **Fond-du-Lac** ﹏ Saskatchewan, C Canada
202 C9 **Fongafale** *var.* Funafuti. ● *country capital* (Tuvalu) Funafuti Atoll, C Tuvalu *8.31N 179.11E*
109 C18 **Fonni** Sardegna, Italy, C Mediterranean Sea *40.07N 9.17E*
201 V12 **Fono** *island* Chuuk, C Micronesia
56 G4 **Fonseca** La Guajira, N Colombia *10.54N 72.54W*

97 K15 **Fonseca, Golfo de** *see* Fonseca, Gulf of
44 H8 **Fonseca, Gulf of** *Sp.* Golfo de Fonseca. *gulf* Central America
105 O6 **Fontainebleau** Seine-et-Marne, N France *48.24N 2.42E*
23 N10 **Fontana Lake** ☒ North Carolina, SE USA
109 L24 **Fontanarossa** ✈ (Catania) Sicilia, Italy, C Mediterranean Sea *37.28N 15.04E*
9 N11 **Fontas** ﹏ British Columbia, W Canada
60 D12 **Fonte Boa** Amazonas, N Brazil *2.31S 66.01W*
104 J10 **Fontenay-le-Comte** Vendée, NW France *46.28N 0.48W*
35 T16 **Fontenelle Reservoir** ☒ Wyoming, C USA
200 Ss12 **Fonualei** *island* Vava'u Group, N Tonga
113 H24 **Fonyód** Somogy, W Hungary *46.43N 17.31E*
Foochow *see* Fuzhou
41 Q10 **Foraker, Mount** ▲ Alaska, USA *62.57N 151.24W*
197 O14 **Forari** Éfaté, C Vanuatu *17.42S 168.33E*
105 U4 **Forbach** Moselle, NE France *49.10N 6.54E*
191 Q8 **Forbes** New South Wales, SE Australia *33.24S 148.00E*
79 T17 **Forcados** Delta, S Nigeria *5.16N 5.25E*
105 S14 **Forcalquier** Alpes-de-Haute-Provence, SE France *43.57N 5.46E*
103 K19 **Forchheim** Bayern, SE Germany *49.43N 11.07E*
176 V14 **Fordate, Pulau** *island* Kepulauan Tanimbar, E Indonesia
37 R13 **Ford City** California, W USA *35.08N 119.27W*
33 N4 **Ford River** ﹏ Michigan, N USA
191 O4 **Fords Bridge** New South Wales, SE Australia *29.46S 145.25E*
22 J6 **Fordsville** Kentucky, S USA *37.36N 86.39W*
29 X9 **Fordyce** Arkansas, C USA *33.49N 92.24W*
78 I14 **Forécariah** Guinée-Maritime, SW Guinea *9.28N 13.06W*
191 O4 **Foremost** Alberta, SW Canada *49.30N 111.34W*
12 D16 **Forest** Ontario, S Canada *43.05N 82.00W*
25 S3 **Forest** Mississippi, S USA *32.22N 89.30W*
33 S12 **Forest** Ohio, N USA *40.47N 83.26W*
31 V11 **Forest City** Iowa, C USA *43.15N 93.38W*
23 Q10 **Forest City** North Carolina, SE USA *35.19N 81.52W*
34 G4 **Forest Grove** Oregon, NW USA *45.31N 123.06W*
191 P17 **Forestier Peninsula** *peninsula* Tasmania, SE Australia
31 V8 **Forest Lake** Minnesota, N USA *45.16N 92.59W*
25 S3 **Forest Park** Georgia, SE USA *33.37N 84.22W*
31 Q3 **Forest River** ﹏ North Dakota, N USA
13 T6 **Forestville** Quebec, SE Canada *48.45S 69.04W*
105 Q11 **Forez, Monts du** ▲ C France
98 K10 **Forfar** E Scotland, UK *56.37N 2.54W*
27 P2 **Forgan** Oklahoma, C USA *36.54N 100.32W*
103 J23 **Forggensee** ◎ S Germany
153 N10 **Forīsh** *Rus.* Farish. Jizzakh Wiloyati, C Uzbekistan *40.33N 66.52E*
22 J6 **Forked Deer River** ﹏ Tennessee, s USA
34 F7 **Forks** Washington, NW USA *47.57N 124.22W*
156 K12 **Forlandsundet** *sound* W Svalbard
108 H10 **Forlì** *anc.* Forum Livii. Emilia-Romagna, N Italy *44.13N 12.01E*
31 Q7 **Forman** North Dakota, N USA *46.07N 97.39W*
99 K17 **Formby** NW England, UK *53.34N 3.04W*
107 V11 **Formentera** *anc.* Ophiusa, *Lat.* Frumentum. *island* Islas Baleares, Spain, W Mediterranean Sea *38.41N 1.28E*
107 Y9 **Formentor, Cabo de** *var.* Cabo de Formentor, Cape Formentor. *headland* Mallorca, Spain, W Mediterranean Sea *39.57N 3.12E*
107 Y9 **Formentor, Cape** *see* Formentor, Cap de
109 J16 **Formia** Lazio, C Italy *41.16N 13.37E*
64 O7 **Formosa** Formosa, NE Argentina *26.07S 58.13W*
64 O7 **Formosa** *off.* Provincia de Formosa. ◆ *province* NE Argentina
61 I17 **Formosa** Goiás, S Brazil *15.30S 47.20W*
Formosa/Formo'sa *see* Taiwan
Formosa, Serra ▲ C Brazil
Formosa Strait *see* Taiwan Strait
97 H15 **Fornebu** ✈ (Oslo) Akershus, S Norway *59.54N 10.37E*
95 Z15 **Fornæs** *headland* C Denmark *56.26N 10.57E*
108 E9 **Fornovo di Taro** Emilia-Romagna, C Italy *44.41N 10.04E*
119 T24 **Foros** Respublika Krym, S Ukraine *44.24N 33.47E*
46 M8 **Forssa** Etelä-Suomi, S Finland *19.37N 71.51W*
46 M8 **Fort-Liberté** NE Haiti *19.37N 71.51W*
20 K13 **Forsythe** ﹏ Quebec, SE Canada
Fort Loudoun Lake ☒ Tennessee, S USA
25 X11 **Forrest City** Arkansas, C USA *35.01N 90.48W*
202 C9 **Forrester Island** *island* Alexander Archipelago, Alaska, USA
27 N7 **Forsan** Texas, SW USA *32.06N 101.22W*
109 C18 **Fonni** Sardegna, Italy, C Mediterranean Sea
189 V5 **Forsayth** Queensland, NE Australia *18.31S 143.37E*
97 L19 **Forserum** Jönköping, S Sweden *57.42N 14.28E*

97 K15 **Forshaga** Värmland, C Sweden *59.33N 13.28E*
103 Q14 **Forst** *Lus.* Barść Łużyca. Brandenburg, E Germany *51.43N 14.38E*
191 S10 **Forster-Tuncurry** New South Wales, SE Australia *32.13S 152.31E*
25 T4 **Forsyth** Georgia, SE USA *33.00N 83.57W*
29 T8 **Forsyth** Missouri, C USA *36.41N 93.07W*
35 W10 **Forsyth** Montana, NW USA *46.15N 106.40W*
155 U11 **Fort Abbās** Punjab, E Pakistan *29.11N 72.54E*
10 D12 **Fort Albany** Ontario, C Canada
58 L13 **Fortaleza** Pando, N Bolivia *9.48S 65.28W*
60 P13 **Fortaleza** *prev.* Ceará. *state capital* Ceará, NE Brazil *3.45S 38.34W*
61 D16 **Fortaleza** Rondônia, W Brazil *8.45S 64.06W*
58 C13 **Fortaleza, Río** ﹏ W Peru
23 O3 **Fort Ashby** West Virginia, NE USA *39.30N 78.46W*
98 I9 **Fort Augustus** N Scotland, UK *57.13N 4.37W*
Fort-Bayard *see* Zhanjiang
35 S8 **Fort Benton** Montana, NW USA *47.49N 110.40W*
37 O7 **Fort Bidwell** California, W USA *41.50N 120.07W*
36 L5 **Fort Bragg** California, W USA *39.25N 123.48W*
33 N16 **Fort Branch** Indiana, N USA *38.15N 87.34W*
Fort-Bretonnet *see* Bousso
35 T17 **Fort Bridger** Wyoming, C USA *41.18N 110.19W*
Fort-Cappolani *see* Tidjikja
Fort Charlet *see* Djanet
Fort-Chimo *see* Kuujjuaq
14 J7 **Fort Chipewyan** Alberta, C Canada *58.42N 111.07W*
39 U3 **Fort Collins** Colorado, C USA *40.35N 105.04W*
79 S13 **Fort-Crampel** *see* Kaga Bandoro
Fort-Dauphin *see* Tôlanaro
26 K10 **Fort Davis** Texas, SW USA *30.34N 103.55W*
39 O10 **Fort Defiance** Arizona, SW USA *35.44N 109.04W*
27 P12 **Fort-de-France** *prev.* Fort-Royal. ● (Martinique) W Martinique *14.36N 61.04W*
27 P12 **Fort-de-France, Baie de** *bay* W Martinique
25 P6 **Fort Deposit** Alabama, S USA *31.58N 86.34W*
31 V8 **Fort Dodge** Iowa, C USA *42.30N 94.10W*
11 V3 **Forteau** Quebec, E Canada
108 E11 **Forte dei Marmi** Toscana, C Italy *43.59N 10.10E*
12 H17 **Fort Erie** Ontario, S Canada *42.55N 78.55W*
188 H7 **Fortescue River** ﹏ Western Australia
15 H8 **Fort Simpson** *var.* Simpson. Northwest Territories, W Canada *61.52N 121.22W*
15 H8 **Fort Smith** *district capital* Northwest Territories, W Canada *60.01N 111.55W*
29 R10 **Fort Smith** Arkansas, C USA *35.23N 94.24W*
39 T9 **Fort Stanton** New Mexico, SW USA *33.28N 105.31W*
26 L9 **Fort Stockton** Texas, SW USA *30.54N 102.54W*
39 T9 **Fort Sumner** New Mexico, SW USA *34.28N 104.15W*
27 N2 **Fort Supply** Oklahoma, C USA *36.34N 99.34W*
27 N2 **Fort Supply Lake** ☒ Oklahoma, C USA
31 O4 **Fort Thompson** South Dakota, N USA *44.01N 99.22W*
Fort-Trinquet *see* Bir Mogreïn**
107 R12 **Fortuna** Murcia, SE Spain *38.10N 1.07W*
36 K3 **Fortuna** California, W USA *40.35N 124.07W*
30 J2 **Fortuna** North Dakota, N USA *48.53N 103.46W*
25 T7 **Fort Valley** Georgia, SE USA *32.33N 83.53W*
8 I5 **Fort Vermilion** Alberta, W Canada *58.22N 115.58W*
25 U2 **Fort Walton Beach** Florida, SE USA *30.24N 86.37W*
33 P9 **Fort Wayne** Indiana, N USA *41.07N 85.07W*
98 H10 **Fort William** N Scotland, UK *56.49N 5.07W*
27 T6 **Fort Worth** Texas, SW USA *32.43N 97.19W*
30 M7 **Fort Yates** North Dakota, N USA *46.05N 100.37W*
41 S7 **Fort Yukon** Alaska, USA *66.35N 145.05W*
Forum Alieni *see* Ferrara
Forum Julii *see* Fréjus
Forum Livii *see* Forlì
149 G5 **Forūr, Jazīreh-ye** *island* S Iran
96 H7 **Fosen** *physical region* S Norway
167 N14 **Foshan** *var.* Fatshan, Fo-shan, Namhoi. Guangdong, S China *23.03N 113.05E*
Fossa Claudia *see* Chioggia
2 Savona Piemonte, NW Italy *44.33N 7.43E*
101 H21 **Fosses-la-Ville** Namur, S Belgium *50.24N 4.42E*
34 J12 **Fossil** Oregon, NW USA *44.58N 120.15W*
110 J3 **Foss Lake** *see* Foss Reservoir
108 I11 **Fossombrone** Marche, C Italy *43.42N 12.48E*
27 K10 **Foss Reservoir** *var.* Foss Lake. ☒ Oklahoma, C USA
31 S4 **Fosston** Minnesota, N USA *47.34N 95.45W*
191 O13 **Foster** Victoria, SE Australia *38.40S 146.15E*
9 T12 **Foster Lakes** ◎ Saskatchewan, C Canada
33 S12 **Fostoria** Ohio, N USA *41.09N 83.24W*
81 D19 **Fougamou** Ngounié, C Gabon *1.15S 10.37E*
104 J8 **Fougères** Ille-et-Vilaine, NW France *48.21N 1.12W*
Fou-hsin *see* Fuxin

29 S14 **Fouke** Arkansas, C USA 33.15N 93.53W
98 K2 **Foula** island NE Scotland, UK
67 D24 **Foul Bay** bay East Falkland, Falkland Islands
99 P21 **Foulness Island** island SE England, UK
193 F15 **Foulwind, Cape** headland South Island, NZ 41.45S 171.28E
81 E15 **Foumban** Ouest, NW Cameroon 5.43N 10.49E
180 H13 **Foumbouni** Grande Comore, NW Comoros 11.49S 43.30E
205 N8 **Foundation Ice Stream** glacier Antarctica
37 T6 **Fountain** Colorado, C USA 38.40N 104.42W
38 L4 **Fountain Green** Utah, W USA 39.37N 111.37W
23 P11 **Fountain Inn** South Carolina, SE USA 34.41N 82.12W
29 S11 **Fourche LaFave River** Arkansas, C USA
35 Z13 **Four Corners** Wyoming, C USA 44.04N 104.08W
105 Q2 **Fourmies** Nord, N France 50.01N 4.03E
40 J17 **Four Mountains, Islands of** island group Aleutian Islands, Alaska, USA
181 P17 **Fournaise, Piton de la** ▲ SE Réunion 21.13S 55.43E
12 J8 **Fournière, Lac** ◎ Quebec, SE Canada
117 L20 **Foúrnoi** island Dodekánisos, Greece, Aegean Sea
66 K13 **Four North Fracture Zone** tectonic feature N Atlantic Ocean
Fouron-Saint-Martin see Sint-Martens-Voeren
32 L3 **Fourteen Mile Point** headland Michigan, N USA 46.59N 89.07W
78 I13 **Fouta Djallon** var. Futa Jallon. ▲ W Guinea
193 L13 **Foveaux Strait** strait S NZ
37 Q11 **Fowler** California, W USA 36.35N 119.40W
39 U6 **Fowler** Colorado, C USA 38.07N 104.01W
33 N12 **Fowler** Indiana, N USA 40.36N 87.20W
190 D7 **Fowlers Bay** bay South Australia
27 R13 **Fowlerton** Texas, SW USA 28.27N 98.48W
148 M2 **Fowman** var. Fuman, Fumen. Gilān, NW Iran 37.15N 49.19E
67 C25 **Fox Bay East** West Falkland, Falkland Islands
67 C25 **Fox Bay West** West Falkland, Falkland Islands
12 J14 **Foxboro** Ontario, SE Canada 44.16N 77.23W
9 O14 **Fox Creek** Alberta, W Canada 54.25N 116.57W
66 G5 **Foxe Basin** sea Nunavut, N Canada
66 G5 **Foxe Channel** channel Nunavut, N Canada
97 I16 **Foxen** ◎ C Sweden
16 N4 **Foxe Peninsula** peninsula Baffin Island, Nunavut, NE Canada
193 E19 **Fox Glacier** West Coast, South Island, NZ 43.28S 170.00E
40 L17 **Fox Islands** island Aleutian Islands, Alaska, USA
32 M10 **Fox Lake** Illinois, N USA 42.24N 88.10W
9 V12 **Fox Mine** Manitoba, C Canada 56.36N 101.48W
37 R3 **Fox Mountain** ▲ Nevada, W USA 41.01N 119.30W
67 E25 **Fox Point** headland East Falkland, Falkland Islands 51.55S 58.24W
32 M11 **Fox River** ✍ Illinois/Wisconsin, N USA
32 L7 **Fox River** ✍ Wisconsin, N USA
192 L13 **Foxton** Manawatu-Wanganui, North Island, NZ 40.30S 175.17E
9 S16 **Fox Valley** Saskatchewan, S Canada 50.28N 109.28W
9 W16 **Foxwarren** Manitoba, S Canada 50.30N 101.09W
99 C14 **Foyle, Lough** Ir. Loch Feabhail. inlet N Ireland
204 H5 **Foyn Coast** physical region Antarctica
106 I2 **Foz** Galicia, NW Spain 43.33N 7.16W
62 I12 **Foz do Areia, Represa de** □ S Brazil
61 N4 **Foz do Breu** Acre, Brazil 9.21S 72.40W
85 A16 **Foz do Cunene** Namibe, SW Angola 17.11S 11.52E
62 G12 **Foz do Iguaçu** Paraná, S Brazil 25.33S 54.31W
60 D2 **Foz do Mamoriá** Amazonas, NW Brazil 2.28S 66.06W
107 T6 **Fraga** Aragón, NE Spain 41.31N 0.21E
46 F5 **Fragoso, Cayo** island C Cuba
63 H14 **Fraile Muerto** Cerro Largo, NE Uruguay 32.30S 54.30W
101 H21 **Fraire** Namur, S Belgium 50.16N 4.30E
101 L21 **Fraiture, Baraque de** hill SE Belgium 50.22N 5.50E
Frakitát see Hlohovec
207 S10 **Fram Basin** var. Amundsen Basin. undersea feature Arctic Ocean
101 F20 **Frameries** Hainaut, S Belgium 50.25N 3.40E
21 O11 **Framingham** Massachusetts, NE USA 42.17N 71.24W
62 L7 **Franca** São Paulo, S Brazil 20.33S 47.27W
197 G4 **Français, Récif des** reef N New Caledonia
109 K14 **Francavilla al Mare** Abruzzo, C Italy 42.25N 14.16E
109 P18 **Francavilla Fontana** Puglia, SE Italy 40.31N 17.34E
104 M8 **France** off. French Republic. It./Sp. Francia; prev. Gaul, Gaule, Lat. Gallia. ◆ republic W Europe
47 O8 **Francés Viejo, Cabo** headland NE Dominican Republic 19.39N 69.57W
81 F19 **Franceville** var. Massoukou, Masuku. Haut-Ogooué, E Gabon 1.40S 13.31E
81 F19 **Franceville** × Haut-Ogooué, E Gabon 1.38S 13.24E
Francfort see Frankfurt am Main

105 T8 **Franche-Comté** ◆ region E France
Francia see France
31 O11 **Francis Case, Lake** □ South Dakota, N USA
62 H12 **Francisco Beltrão** Paraná, S Brazil 26.04S 53.04W
Francisco I. Madero see Villa Madero
63 N23 **Francisco Madero** Buenos Aires, E Argentina 35.52S 62.03W
44 H6 **Francisco Morazán** prev. Tegucigalpa. ◆ department C Honduras
85 J18 **Francistown** North East, NE Botswana 21.08S 27.31E
Franconian Forest see Frankenwald
Franconian Jura see Fränkische Alb
100 K6 **Franeker** Fris. Frjentsjer. Friesland, N Netherlands 53.10N 5.33E
Frankenalb see Fränkische Alb
103 H16 **Frankenberg** Hessen, C Germany 51.04N 8.49E
103 J20 **Frankenhöhe** hill range C Germany
33 R8 **Frankenmuth** Michigan, N USA 43.33N 83.44W
103 F20 **Frankenstein** hill W Germany 49.24N 8.04E
Frankenstein/Frankenstein in Schlesien see Ząbkowice Śląskie
103 G20 **Frankenthal** Rheinland-Pfalz, W Germany 49.33N 8.21E
103 I18 **Frankenwald** Eng. Franconian Forest. ▲ C Germany
46 J12 **Frankfield** C Jamaica 18.07N 77.22W
12 J14 **Frankford** Ontario, SE Canada 44.12N 77.36W
33 O13 **Frankfort** Indiana, N USA 40.16N 86.30W
29 O3 **Frankfort** Kansas, C USA 39.42N 96.25W
22 L5 **Frankfort** state capital Kentucky, C USA 38.12N 84.52W
Frankfort on the Main see Frankfurt am Main
Frankfurt see Słubice, Poland
Frankfurt see Frankfurt am Main, Germany
103 G18 **Frankfurt am Main** var. Frankfurt, Fr. Francfort; prev. Eng. Frankfort on the Main. Hessen, SW Germany 50.07N 8.40E
102 Q12 **Frankfurt an der Oder** Brandenburg, E Germany 52.19N 14.31E
103 L21 **Fränkische Alb** var. Frankenalb, Eng. Franconian Jura. ▲ S Germany
103 J18 **Fränkische Saale** ✍ C Germany
103 I19 **Fränkische Schweiz** hill range C Germany
25 R4 **Franklin** Georgia, SE USA 33.15N 85.06W
33 P14 **Franklin** Indiana, N USA 39.28N 86.01W
22 J7 **Franklin** Kentucky, S USA 36.43N 86.34W
24 I9 **Franklin** Louisiana, S USA 29.48N 91.30W
31 O17 **Franklin** Nebraska, C USA 40.06N 98.57W
23 N10 **Franklin** North Carolina, SE USA 35.07N 83.22W
20 C13 **Franklin** Pennsylvania, NE USA 41.24N 79.49W
22 J9 **Franklin** Tennessee, S USA 35.55N 86.52W
27 V9 **Franklin** Texas, SW USA 31.01N 96.29W
23 X3 **Franklin** Virginia, NE USA 36.40N 76.55W
23 T4 **Franklin** West Virginia, NE USA 38.39N 79.19W
32 M9 **Franklin** Wisconsin, N USA 42.53N 88.00W
15 N4 **Franklin Bay** inlet Northwest Territories, N Canada
34 K7 **Franklin D.Roosevelt Lake** □ Washington, NW USA
37 W4 **Franklin Lake** ◎ Nevada, W USA
193 B22 **Franklin Mountains** ▲ South Island, NZ
41 R5 **Franklin Mountains** ▲ Alaska, USA
15 N4 **Franklin, Point** headland Alaska, USA 70.54N 158.48W
191 O17 **Franklin River** ✍ Tasmania, SE Australia
15 K2 **Franklin Strait** strait Nunavut, N Canada
24 K8 **Franklinton** Louisiana, S USA 30.51N 90.09W
23 V9 **Franklinton** North Carolina, SE USA 36.06N 78.27W
27 V7 **Frankston** Texas, SW USA 32.03N 95.30W
35 U2 **Frannie** Wyoming, C USA 44.57N 108.37W
13 O13 **Franquelin** Quebec, SE Canada 49.17N 67.52W
13 S5 **Franquelin** ✍ Quebec, SE Canada
126 H11 **Frantsa-Iosifa, Zemlya** Eng. Franz Josef Land. island group N Russian Federation
193 H18 **Franz Josef Glacier** West Coast, South Island, NZ 43.22S 170.11E
Franz Josef Land see Frantsa-Iosifa, Zemlya
Franz-Josef Spitze see Gerlachovský štít
103 D23 **Franz Strauss** abbrev. F.J.S. × (München) Bayern, SE Germany 48.07N 11.43E
109 A19 **Frasca, Capo della** headland Sardegna, Italy, C Mediterranean Sea 39.46N 8.27E
103 I15 **Frascati** Lazio, C Italy 41.48N 12.40E

9 N14 **Fraser** ✍ British Columbia, SW Canada
85 G24 **Fraserburg** Western Cape, SW South Africa 31.49S 21.29E
98 L8 **Fraserburgh** NE Scotland, UK 57.41N 2.19W
189 Z9 **Fraser Island** var. Great Sandy Island. island Queensland, E Australia
8 L14 **Fraser Lake** British Columbia, SW Canada 54.00N 124.45W
8 L15 **Fraser Plateau** plateau British Columbia, SW Canada
192 P10 **Frasertown** Hawke's Bay, North Island, NZ 38.58S 177.25E
101 E19 **Frasnes-lez-Buissenal** Hainaut, SW Belgium 50.40N 3.37E
110 I7 **Frastanz** Vorarlberg, NW Austria 47.13N 9.37E
12 B8 **Frater** Ontario, S Canada 47.19N 84.28W
Frauenbach see Baia Mare
Frauenburg see Saldus, Latvia
Frauenburg see Frombork, Poland
110 H6 **Frauenfeld** Thurgau, NE Switzerland 47.34N 8.54E
189 Z5 **Frauenkirchen** Burgenland, E Austria 47.49N 16.57E
83 D19 **Fray Bentos** Río Negro, W Uruguay 33.09S 58.14W
63 F19 **Fray Marcos** Florida, S Uruguay 34.13S 55.43W
31 S6 **Frazee** Minnesota, N USA 46.42N 95.40W
97 G23 **Fredericia** Vejle, C Denmark 55.34N 9.46E
22 J14 **Frederick** Maryland, NE USA 39.24N 77.24W
28 L12 **Frederick** Oklahoma, C USA 34.23N 99.01W
31 P7 **Frederick** South Dakota, N USA 45.49N 98.31W
23 N5 **Fredericksburg** Iowa, C USA 42.58N 92.12W
27 N10 **Fredericksburg** Texas, SW USA 30.16N 98.52W
23 W5 **Fredericksburg** Virginia, NE USA 38.16N 77.27W
41 X13 **Frederick Sound** sound Alaska, USA
29 X6 **Fredericktown** Missouri, C USA 37.33N 90.17W
62 H13 **Frederico Westphalen** Rio Grande do Sul, S Brazil 27.22S 53.20W
11 O15 **Fredericton** New Brunswick, SE Canada 45.57N 66.40W
97 J23 **Frederiksborg** off. Frederiksborg Amt. ◆ county E Denmark
Frederikshåb see Paamiut
97 H19 **Frederikshavn** prev. Fladstrand. Nordjylland, N Denmark 57.28N 10.33E
97 J22 **Frederikssund** Frederiksborg, E Denmark 55.51N 12.04E
47 T9 **Frederiksted** Saint Croix, S Virgin Islands (US) 17.41N 64.51W
97 J22 **Frederiksværk** off. Frederiksværk og Hanehoved. Frederiksborg, E Denmark 55.58N 12.01E
Frederiksværk og Hanehoved see Frederiksværk
56 I9 **Fredonia** Antioquia, W Colombia 5.57N 75.42W
38 M6 **Fredonia** Arizona, SW USA 36.57N 112.31W
29 P5 **Fredonia** Kansas, C USA 37.31N 95.49W
20 C11 **Fredonia** New York, NE USA 42.26N 79.19W
37 M9 **Fredonyer Pass** pass California, W USA 40.21N 120.52W
95 I15 **Fredrika** Västerbotten, N Sweden 64.03N 18.25E
107 T6 **Fredriksberg** Dalarna, C Sweden 60.07N 14.22E
97 H16 **Fredrikstad** Østfold, S Norway 59.12N 10.57E
32 K16 **Freeburg** Illinois, N USA 38.25N 89.54W
20 J16 **Freeland** Pennsylvania, NE USA 41.01N 75.54W
190 J7 **Freeling Heights** ▲ South Australia 30.09S 139.24E
37 W2 **Freel Peak** ▲ California, W USA 38.52N 119.52W
16 T7 **Freels, Cape** headland Newfoundland, E Canada 49.16N 53.30W
31 Q11 **Freeman** South Dakota, N USA 43.21N 97.26W
46 J2 **Freeport** Grand Bahama Island, N Bahamas 26.28N 78.43W
32 L10 **Freeport** Illinois, N USA 42.18N 89.37W
27 W12 **Freeport** Texas, SW USA 28.57N 95.21W
46 J1 **Freeport** × Grand Bahama Island, N Bahamas 26.31N 78.48W
85 R14 **Free State** off. Free State Province; prev. Orange Free State, Afr. Oranje Vrystaat. ◆ province C South Africa
Free State see Maryland
78 I15 **Freetown** ● (Sierra Leone) W Sierra Leone 8.27N 13.16W
180 I16 **Frégate** island Inner Islands, NE Seychelles
106 I12 **Fregenal de la Sierra** Extremadura, W Spain 38.10N 6.39W
190 G2 **Fregon** South Australia 26.44S 132.03E
104 I4 **Fréhel, Cap** headland NW France 48.41N 2.21W
96 F7 **Frei** Møre og Romsdal, S Norway 63.02N 7.47E
103 O16 **Freiberg** Sachsen, E Germany 50.55N 13.21E
103 O16 **Freiberger Mulde** ✍ E Germany

Freiburg see Fribourg, Switzerland
Freiburg see Freiburg im Breisgau, Germany
103 F23 **Freiburg im Breisgau** var. Freiburg, Fr. Fribourg-en-Brisgau. Baden-Württemberg, SW Germany 48.00N 7.52E
Freiburg in Schlesien see Świebodzice
Freie Hansestadt Bremen see Bremen
Freie und Hansestadt Hamburg see Hamburg
103 L22 **Freising** Bayern, SE Germany 48.24N 11.45E
189 T3 **Freistadt** Oberösterreich, N Austria 48.30N 14.42E
Freistadt see Hlohovec
103 O16 **Freital** Sachsen, E Germany 51.00N 13.40E
Freiwaldau see Jeseník
106 J6 **Freixo de Espada à Cinta** Bragança, N Portugal 41.04N 6.49W
105 U15 **Fréjus** anc. Forum Julii. Var, SE France 43.25N 6.43E
188 I13 **Fremantle** Western Australia 32.07S 115.43E
37 N9 **Fremont** California, W USA 37.32N 121.56W
33 Q11 **Fremont** Indiana, N USA 41.43N 84.54W
31 W15 **Fremont** Iowa, C USA 41.12N 92.26W
33 P8 **Fremont** Michigan, N USA 43.28N 85.56W
31 R15 **Fremont** Nebraska, C USA 41.25N 96.30W
33 S11 **Fremont** Ohio, N USA 41.23N 83.07W
35 T4 **Fremont Peak** ▲ Wyoming, C USA 43.07N 109.37W
38 M6 **Fremont River** ✍ Utah, W USA
23 O9 **French Broad River** ✍ Tennessee, S USA
20 C12 **French Creek** ✍ Pennsylvania, NE USA
34 K15 **Frenchglen** Oregon, NW USA 42.49N 118.55W
57 Y10 **French Guiana** var. Guiana, Guyane. ◇ French overseas department N South America
33 O15 **French Lick** Indiana, N USA 38.33N 86.37W
193 J14 **French Pass** Marlborough, South Island, NZ 40.57S 173.49E
203 T11 **French Polynesia** ◇ French overseas territory C Polynesia
12 F11 **French River** ✍ Ontario, S Canada
French Somaliland see Djibouti
181 P12 **French Southern and Antarctic Territories** Fr. Terres Australes et Antarctiques Françaises. ◇ French overseas territory S Indian Ocean
French Sudan see Mali
French Territory of the Afars and Issas see Djibouti
French Togoland see Togo
76 J6 **Frenda** NW Algeria 35.06N 1.03E
113 J18 **Frenštát pod Radhoštěm** Ger. Frankstadt. Ostravský Kraj, E Czech Republic 49.33N 18.10E
205 U6 **Freshfield, Cape** headland Antarctica
42 L10 **Fresnillo** var. Fresnillo de González Echeverría. Zacatecas, C Mexico 23.10N 102.52W
Fresnillo de González Echeverría see Fresnillo
37 Q10 **Fresno** California, W USA 36.44N 119.48W
34 G4 **Friday Harbor** San Juan Islands, Washington, NW USA 48.31N 123.01W
194 F11 **Frieda** ✍ NW PNG
Friedau see Ormož
103 K23 **Friedberg** Bayern, SE Germany 48.21N 10.58E
103 H18 **Friedberg** Hessen, W Germany 50.19N 8.46E
Friedeberg Neumark see Strzelce Krajeńskie
Friedek-Mistek see Frýdek-Místek
Friedland see Pravdinsk
166 M13 **Friedrichshafen** Baden-Württemberg, S Germany 47.39N 9.28E
167 S11 **Friedrichstadt** see Jaunjelgava

103 F22 **Friesenheim** Baden-Württemberg, SW Germany 48.27N 7.56E
Friesische Inseln see Frisian Islands
100 K6 **Friesland** ◆ province N Netherlands
26 M3 **Friona** Texas, SW USA 34.38N 102.43W
44 L2 **Frío, Río** ✍ N Costa Rica
27 R13 **Frío River** ✍ Texas, SW USA
101 M25 **Frisange** Luxembourg, S Luxembourg 49.31N 6.12E
38 J6 **Frisco Peak** ▲ Utah, W USA 38.31N 111.17W
86 F9 **Frisian Islands** Dut. Friesche Eilanden, Ger. Friesische Inseln. island group N Europe
20 L12 **Frissell, Mount** ▲ Connecticut, NE USA 42.01N 73.25W
97 J19 **Fristad** Västra Götaland, S Sweden 57.49N 13.01E
27 N2 **Fritch** Texas, SW USA 35.38N 101.36W
97 J19 **Fritsla** Västra Götaland, S Sweden 57.33N 12.46E
103 H16 **Fritzlar** Hessen, C Germany 51.09N 9.16E
108 H6 **Friuli-Venezia Giulia** ◆ region NE Italy
Frjentsjer see Franeker
206 L13 **Frobisher Bay** inlet Baffin Island, Nunavut, NE Canada
Frobisher Bay see Iqaluit
9 S12 **Frobisher Lake** ◎ Saskatchewan, C Canada
96 G7 **Frohavet** sound C Norway
Frohenbruck see Veselí nad Lužnicí
189 X7 **Frohnleiten** Steiermark, SE Austria 47.16N 15.19E
101 O9 **Froidchapelle** Hainaut, S Belgium 50.10N 4.18E
131 O9 **Frolovo** Volgogradskaya Oblast', SW Russian Federation 49.46N 43.38E
112 K7 **Frombork** Ger. Frauenburg. Warmińsko-Mazurskie, NE Poland 54.21N 19.40E
99 L21 **Frome** S West England, UK 51.15N 2.21W
190 I4 **Frome Creek** seasonal river South Australia
190 J6 **Frome Downs** South Australia 31.17S 139.48E
190 J5 **Frome, Lake** salt lake South Australia
Frome, Lake see Wronki
106 H10 **Fronteira** Portalegre, C Portugal 39.03N 7.39W
42 M7 **Frontera** Coahuila de Zaragoza, NE Mexico 26.53N 101.27W
43 U14 **Frontera** Tabasco, SE Mexico 18.32N 92.35W
42 G3 **Fronteras** Sonora, NW Mexico 30.51N 109.36W
105 Q16 **Frontignan** Hérault, S France 43.27N 3.45E
21 X7 **Frontino** Antioquia, NW Colombia 6.46N 76.10W
23 V4 **Front Royal** Virginia, NE USA 38.52N 78.09W
109 I16 **Frosinone** anc. Frusino. Lazio, C Italy 41.37N 13.19E
109 K16 **Frosolone** Molise, C Italy 41.34N 14.25E
27 U7 **Frost** Texas, SW USA 32.04N 96.48W
23 V2 **Frostburg** Maryland, NE USA 39.39N 78.55W
25 X13 **Frostproof** Florida, SE USA 27.45N 81.31W
Frostviken see Kvarnbergsvattnet
95 M15 **Frövi** Örebro, C Sweden 59.31N 15.24E
96 F7 **Frøya** island W Norway
39 P5 **Fruita** Colorado, C USA 39.10N 108.43W
30 J9 **Fruitdale** South Dakota, N USA 44.39N 103.38W
25 W11 **Fruitland Park** Florida, SE USA 28.51N 81.54W
Frumentum see Formentera
153 S11 **Frunze** Oshskaya Oblast', SW Kyrgyzstan 40.07N 71.40E
Frunze see Bishkek
119 O9 **Frunzivka** Odes'ka Oblast', SW Ukraine 47.19N 29.46E
Frusino see Frosinone
110 E9 **Frutigen** Bern, W Switzerland 46.35N 7.38E
113 I17 **Frýdek-Místek** Ger. Friedek-Mistek. Ostravský Kraj, E Czech Republic 49.40N 18.22E
200 Q16 **Fua'amotu** Tongatapu, S Tonga 21.15S 175.08W
202 A9 **Fuafatu** island Funafuti Atoll, C Tuvalu
202 A9 **Fuagea** island Funafuti Atoll, C Tuvalu
202 B8 **Fualifeke** atoll C Tuvalu
202 A8 **Fualopa** island Funafuti Atoll, C Tuvalu
157 A22 **Fuammulah** var. Gnaviyani Atoll. atoll S Maldives
95 R14 **Funäsdalen** Jämtland, C Sweden 62.33N 12.33E
106 G14 **Fuengirola** Andalucía, S Spain 36.31N 4.37W
106 I12 **Fuente de Cantos** Extremadura, W Spain 38.15N 6.16W
66 O6 **Fuente del Maestre** Extremadura, W Spain 38.31N 6.26W
106 J6 **Fuente Obejuna** Andalucía, S Spain 38.15N 5.25W
106 I11 **Fuentesaúco** Castilla-León, N Spain 41.13N 5.30W

64 O3 **Fuerte Olimpo** var. Olimpo. Alto Paraguay, NE Paraguay 21.01S 57.51W
42 H8 **Fuerte, Río** ✍ C Mexico
66 Q11 **Fuerteventura** island Islas Canarias, Spain, NE Atlantic Ocean
147 S14 **Fughmah** var. Faghman, Fugma. C Yemen 16.01N 47.19E
156 M2 **Fuglehuken** headland W Svalbard 78.54N 10.30E
97 B18 **Fugløy** Dan. Fuglø. island Faeroe Islands 62.21N 6.15W
207 T5 **Fugløya Bank** undersea feature E Norwegian Sea
166 E11 **Fugong** Yunnan, SW China
83 K6 **Fugugo** spring/well NE Kenya 3.19N 39.39E
164 L2 **Fuhai** var. Burultokay. Xinjiang Uygur Zizhiqu, NW China 47.15N 87.39E
157 P10 **Fu He** ✍ S China
Fuhien see Fujian
102 J9 **Fuhlsbüttel** × (Hamburg) Hamburg, N Germany 53.37N 9.57E
103 L22 **Fuhne** ✍ C Germany
Fu-hsin see Fuxin
171 J17 **Fujairah** var. Al Fujayrah
171 J17 **Fuji** var. Huzi. Shizuoka, Honshū, S Japan 35.08N 138.39E
167 Q12 **Fujian** var. Fu-chien, Fuhkien, Fujian Sheng, Fukien, Min. ◆ province SE China
166 I9 **Fu Jiang** ✍ C China
Fujian Sheng see Fujian
171 Ii7 **Fujieda** var. Huzieda. Shizuoka, Honshū, S Japan 34.52N 138.15E
171 J16 **Fujin** Heilongjiang, NE China 47.12N 132.01E
171 Ji7 **Fujinomiya** var. Huzinomiya. Shizuoka, Honshū, S Japan 35.19N 138.34E
171 J16 **Fuji-san** var. Fujiyama, Eng. Mount Fuji. ▲ Honshū, SE Japan 35.23N 138.44E
171 Jii7 **Fujisawa** var. Huzisawa. Kanagawa, Honshū, S Japan 35.22N 139.28E
171 Jii7 **Fuji-Yoshida** var. Huziyosida. Yamanashi, Honshū, S Japan 35.30N 138.46E
171 Ii7 **Fukagawa** var. Hukagawa. Hokkaidō, NE Japan 43.44N 142.01E
164 L5 **Fukang** Xinjiang Uygur Zizhiqu, W China 44.07N 87.55E
171 M8 **Fukaura** Aomori, Honshū, C Japan 40.38N 139.55E
200 R5 **Fukave** island Tongatapu Group, S Tonga
Fukien see Fujian
171 Gg14 **Fukuchiyama** var. Hukutiyama. Kyōto, Honshū, SW Japan 35.16N 135.07E
170 Bf2 **Fukue** var. Hukue. Nagasaki, Fukue-jima, SW Japan 32.40N 128.46E
170 Bf2 **Fukue-jima** island Gotō-rettō, SW Japan
171 Hh13 **Fukui** var. Hukui. Fukui, Honshū, SW Japan 36.03N 136.12E
171 Hh14 **Fukui** off. Fukui-ken, var. Hukui. ◆ prefecture Honshū, SW Japan
170 C12 **Fukuoka** var. Hukuoka; hist. Najima. Fukuoka, Kyūshū, SW Japan 33.36N 130.24E
170 C13 **Fukuoka** off. Fukuoka-ken, var. Hukuoka. ◆ prefecture Kyūshū, SW Japan
171 L13 **Fukushima** var. Hukusima. Fukushima, Honshū, C Japan 37.44N 140.27E
171 Mm7 **Fukushima** Hokkaidō, NE Japan 41.27N 140.14E
171 Kk14 **Fukushima** off. Fukushima-ken, var. Hukusima. ◆ prefecture Honshū, C Japan
170 F14 **Fukuyama** var. Hukuyama. Hiroshima, Honshū, SW Japan 34.28N 133.23E
78 G13 **Fulacunda** South Guinea-Bissau 11.46N 15.09W
133 P8 **Fūlādī, Kūh-e** ▲ E Afghanistan 34.37N 67.31E
197 K16 **Fulaga** island Lau Group, E Fiji
103 I17 **Fulda** Hessen, C Germany 50.33N 9.40E
31 Q4 **Fulda** Minnesota, N USA 43.52N 95.36W
103 I16 **Fulda** ✍ C Germany

11 O16 **Fundy, Bay of** bay Canada/USA
Fünen see Fyn
56 C13 **Fúnes** Nariño, SW Colombia 0.58N 77.27W
85 M19 **Funhalouro** Inhambane, S Mozambique 23.04S 34.01E
167 R6 **Funing** Jiangsu, E China 33.43N 119.47E
166 I14 **Funing** Yunnan, SW China
166 M7 **Funiu Shan** ▲ C China
79 U13 **Funtua** Katsina, N Nigeria 11.31N 7.19E
167 R12 **Fuqing** Fujian, SE China 25.40N 119.23E
85 M14 **Furancungo** Tete, NW Mozambique 14.49S 33.32E
172 P5 **Furano** var. Hurano. Hokkaidō, NE Japan 43.20N 142.23E
118 I15 **Furculești** Teleorman, S Romania 43.51N 25.05E
Füred see Balatonfüred
172 Q7 **Füren-ko** ◎ Hokkaidō, NE Japan
149 R12 **Fürg** Fārs, S Iran 28.16N 55.13E
Furluk see Fârliug
103 L22 **Fürstenau** Niedersachsen, NW Germany 52.30N 7.40E
189 X8 **Fürstenfeld** Steiermark, SE Austria 47.03N 16.01E
103 L23 **Fürstenfeldbruck** Bayern, S Germany 48.10N 11.16E
102 P12 **Fürstenwalde** Brandenburg, NE Germany 52.22N 14.04E
103 K20 **Fürth** Bayern, S Germany 49.28N 10.58E
189 W3 **Furth bei Göttweig** Niederösterreich, NW Austria 48.22N 15.33E
172 O4 **Furubira** Hokkaidō, NE Japan 43.14N 140.38E
96 L12 **Furudal** Dalarna, C Sweden 61.10N 15.07E
171 H14 **Furukawa** Gifu, Honshū, SW Japan 36.13N 137.11E
171 M12 **Furukawa** var. Hurukawa. Miyagi, Honshū, C Japan 38.36N 140.56E
56 F10 **Fusagasugá** Cundinamarca, C Colombia 4.22N 74.21W
Fusan see Pusan
115 L18 **Fushë-Arëzi/Fushë-Arrësi** var. Fushë-Arëzi, Fushë-Arrësi. Shkodër, N Albania 42.05N 20.01E
Fushë-Arrësi see Fushë-Arëzi
115 K19 **Fushë-Kruja** var. Fushë-Krujë. Durrës, C Albania 41.30N 19.43E
Fushë-Krujë see Fushë-Krujë
169 V12 **Fushun** var. Fou-shan, Fu-shun. Liaoning, NE China 41.49N 123.54E
Fusin see Fuxin
110 G10 **Fusio** Ticino, S Switzerland 46.27N 8.39E
169 X11 **Fusong** Jilin, NE China 42.19N 127.16E
103 K24 **Füssen** Bayern, S Germany 47.34N 10.42E
166 K15 **Fusui** prev. Funan. Guangxi Zhuangzu Zizhiqu, S China 22.39N 107.49E
Futa Jallon see Fouta Djallon
65 G18 **Futaleufú** Los Lagos, S Chile 43.12S 71.53W
114 K10 **Futog** Serbia, NW Yugoslavia 45.15N 19.43E
171 K17 **Futtsu** var. Huttu. Chiba, Honshū, S Japan 35.11N 139.52E
197 E16 **Futuna** island S Vanuatu
202 D12 **Futuna, Île** island S Wallis and Futuna
167 Q12 **Fuxian Xi** ✍ SE China
166 L5 **Fuxian** var. Fu Xian. Shaanxi, C China 36.03N 109.19E
Fuxian see Wafangdian
166 G13 **Fuxian Hu** ◎ SW China
169 U12 **Fuxin** var. Fou-hsin, Fu-hsin, Fusin. Liaoning, NE China 41.59N 121.39E
167 P7 **Fuyang** Anhui, E China 32.54N 115.47E
167 O4 **Fuyang He** ✍ E China
169 W3 **Fuyu** Heilongjiang, NE China 47.48N 124.25E
Fuyu/Fu-yü see Songyuan
169 Z6 **Fuyuan** Heilongjiang, NE China 48.20N 134.22E
164 M3 **Fuyun** var. Koktokay. Xinjiang Uygur Zizhiqu, NW China 46.57N 89.29E
113 L22 **Füzesabony** Heves, E Hungary 47.44N 20.23E
137 W13 **Füzuli** Rus. Fizuli. SW Azerbaijan 39.33N 47.09E
121 Q20 **Fyadory** Rus. Fëdory. Brestskaya Voblasts', SW Belarus 51.56N 26.21E
97 G24 **Fyn** Fyns Amt, var. Fünen. ◆ county C Denmark
97 G23 **Fyn** Ger. Fünen. ◆ island C Denmark
98 H11 **Fyne, Loch** inlet W Scotland, UK
97 E16 **Fyresvatnet** ◎ S Norway
FYR Macedonia/FYROM see Macedonia, FYR
Fyzabad see Feyzābād

G

83 O14 **Gaalkacyo** var. Galka'yo, It. Galcaio. Mudug, C Somalia 6.49N 47.23E
Gabakly see Kabakly
118 K7 **Gabare** Vratsa, NW Bulgaria 43.20N 23.57E
104 K15 **Gabas** ✍ SW France
37 T7 **Gabbs** Nevada, W USA 38.51N 117.55W

◆ COUNTRY   ◇ DEPENDENT TERRITORY   ◆ ADMINISTRATIVE REGION   ▲ MOUNTAIN   ▲ VOLCANO   ◎ LAKE
● COUNTRY CAPITAL   ○ DEPENDENT TERRITORY CAPITAL   × INTERNATIONAL AIRPORT   ▲ MOUNTAIN RANGE   ✍ RIVER   □ RESERVOIR

257

84 B12 **Gabela** Cuanza Sul, W Angola
10.49S 14.21E
Gaberones see Gaborone
201 X14 **Gabert** island Caroline Islands,
E Micronesia
76 M7 **Gabès** var. Qābis. E Tunisia
33.53N 10.03E
76 M6 **Gabès, Golfe de** Ar. Khalīj
Qābis. gulf E Tunisia
Gablonz an der Neisse see
Jablonec nad Nisou
Gablös see Cavalese
81 E18 **Gabon** off. Gabonese Republic.
◆ republic C Africa
85 I20 **Gaborone** prev. Gaberones.
● (Botswana) South East,
SE Botswana 24.42S 25.49E
85 I20 **Gaborone ✈** South East,
SE Botswana 24.45S 25.49E
106 K8 **Gabriel y Galán, Embalse de**
☒ W Spain
149 U15 **Gābrīk, Rūd-e ↔** SE Iran
116 J9 **Gabrovo** Gabrovo, N Bulgaria
42.54N 25.19E
66 J9 **Gabrovo ◆** province N Bulgaria
78 H12 **Gabú** prev. Nova Lamego.
E Guinea-Bissau 12.16N 14.09W
31 O6 **Gackle** North Dakota, N USA
46.34N 99.07W
115 I15 **Gacko** Republika Srpska, Bosnia
and Herzegovina 43.08N 18.29E
161 F17 **Gadag** Karnātaka, W India
15.25N 75.37E
95 G15 **Gäddede** Jämtland, C Sweden
64.30N 14.15E
165 S12 **Gadê** Qinghai, C China
33.56N 99.49E
Gades/Gadier/Gadir/Gadire
see Cádiz
107 P15 **Gádor, Sierra de ▲** S Spain
155 S15 **Gadra** Sind, SE Pakistan
25.39N 70.28E
25 Q3 **Gadsden** Alabama, S USA
34.00N 86.00W
38 H15 **Gadsden** Arizona, SW USA
32.33N 114.45W
Gadyach see Hadyach
81 H15 **Gadzi** Mambéré-Kadéï,
SW Central African Republic
4.46N 16.42E
118 J13 **Găeşti** Dâmboviţa, S Romania
44.41N 25.18E
109 J17 **Gaeta** Lazio, C Italy 41.12N 13.34E
109 I17 **Gaeta, Golfo di** var. Gulf of
Gaeta. gulf C Italy
196 L14 **Gaferut** atoll Caroline Islands,
W Micronesia
23 Q10 **Gaffney** South Carolina, SE USA
35.04N 81.39W
Gâfle see Gävle
Gäfleborg see Gävleborg
76 M6 **Gafsa** var. Qafşah. W Tunisia
34.24N 8.51E
Gafurov see Ghafurov
130 J3 **Gagarin** Smolenskaya Oblast',
W Russian Federation 55.33N 35.00E
153 O10 **Gagarin** Jizzakh Wiloyati,
C Uzbekistan 40.40N 68.04E
103 G21 **Gaggenau** Baden-Württemberg,
SW Germany 48.48N 8.19E
196 F16 **Gagil Tamil** var. Gagil-Tomil.
island Caroline Islands,
W Micronesia
Gagil-Tomil see Gagil Tamil
131 O4 **Gagino** Nizhegorodskaya Oblast',
W Russian Federation 55.18N 45.01E
109 O19 **Gagliano del Capo** Puglia,
SE Italy 39.49N 18.22E
96 L13 **Gagnef** Dalarna, C Sweden
60.34N 15.04E
78 M7 **Gagnoa** C Ivory Coast 6.10N 5.56W
11 N10 **Gagnon** Quebec, E Canada
51.55N 68.16W
Gago Coutinho see Lumbala
N'Guimbo
175 T8 **Gag, Pulau** island E Indonesia
1 P8 **Gagra** NW Georgia 43.17N 40.17E
33 S13 **Gahanna** Ohio, N USA
40.01N 82.52W
149 R13 **Gahkom** Hormozgān, S Iran
28.14N 55.48E
Gahnpa see Ganta
75 Q19 **Gaíba, Laguna ☒** E Bolivia
159 T13 **Gaibanda** var. Gaibandah.
Rajshahi, N Bangladesh
25.15N 89.32E
Gaibandah see Gaibanda
Gaibhlte, Cnoc Mór na n see
Galtymore Mountain
189 R9 **Gail ↔** S Austria
103 J21 **Gaildorf** Baden-Württemberg,
S Germany 48.41N 10.08E
105 N15 **Gaillac** var. Gaillac-sur-Tarn.
Tarn, S France 43.54N 1.54E
Gaillac-sur-Tarn see Gaillac
Gaillimh see Galway
Gaillimhe, Cuan na see Galway
Bay
Gailtaler Alpen ▲ S Austria
189 Q9 **Gaiman** Chaco, S Argentina
43.15S 65.30W
22 K8 **Gainesboro** Tennessee, S USA
36.21N 85.39W
25 V10 **Gainesville** Florida, SE USA
29.39N 82.19W
25 T2 **Gainesville** Georgia, SE USA
34.18N 83.49W
29 U8 **Gainesville** Missouri, C USA
36.36N 92.25W
27 T5 **Gainesville** Texas, SW USA
33.37N 97.09W
189 X5 **Gainfarn** Niederösterreich,
NE Austria 47.59N 16.11E
99 N18 **Gainsborough** E England, UK
53.24N 0.48W
190 G6 **Gairdner, Lake** salt lake South
Australia
23 **Gaissane** see Gáissát
94 L8 **Gáissát** var. Gaissane
45 T15 **Gaital, Cerro ▲** C Panama
8.37N 80.04W
23 W3 **Gaithersburg** Maryland,
NE USA 39.07N 77.07W
169 U13 **Gaizhou** Liaoning, NE China
40.24N 122.16E
120 I9 **Gaizina Kalns** var. Gaiziņ.
▲ Latvia 56.51N 25.58E
120 I9 **Gaiziņa Kalns** var. Gaiziņ.
304 L15 **Gajahmungkur, Danau ☒** Jawa,
S Indonesia

41 S10 **Gakona** Alaska, USA
62.21N 145.16W
Galaassiya see Galaosiye
Galăgil see Jalājil
64 J6 **Galán, Cerro ▲** NW Argentina
25.66 66.45W
113 H21 **Galanta** Hung. Galánta. Trnavský
Kraj, W Slovakia 48.11N 17.45E
152 L11 **Galaosiye** Rus. Galaassiya.
Bukhoro Wiloyati, C Uzbekistan
39.53N 64.25E
59 B17 **Galápagos** off. Provincia de
Galápagos. ◆ province Ecuador,
E Pacific Ocean
199 M9 **Galapagos Fracture Zone**
tectonic feature E Pacific Ocean
200 O10 **Galapagos Rise** undersea feature
E Pacific Ocean
98 X13 **Galashiels** SE Scotland, UK
55.37N 2.49W
118 M12 **Galaţi** Ger. Galatz. Galaţi,
E Romania 45.27N 28.00E
118 L12 **Galaţi ◆** county E Romania
109 Q19 **Galatina** Puglia, SE Italy
40.10N 18.10E
109 Q19 **Galatone** Puglia, SE Italy
40.09N 18.04E
Galatz see Galaţi
23 R8 **Galax** Virginia, NE USA
36.39N 80.55W
Galaymoor see Kala-i-Mor
66 P11 **Gáldar** Gran Canaria, Islas
Canarias, NE Atlantic Ocean
28.09N 15.40W
96 H11 **Galdhøpiggen ▲** S Norway
61.30N 8.18E
42 I4 **Galeana** Chihuahua, N Mexico
30.07N 107.35W
43 O9 **Galeana** Nuevo León, NE Mexico
24.45N 99.59W
62 P9 **Galeão ✈** (Rio de Janeiro) Rio de
Janeiro, SE Brazil 22.48S 43.16W
175 T6 **Galela** Pulau Halmahera,
E Indonesia 1.52N 127.48E
41 O9 **Galena** Alaska, USA
64.43N 156.55W
32 K10 **Galena** Illinois, N USA
42.25N 90.25W
29 R7 **Galena** Kansas, C USA
37.04N 94.38W
29 T8 **Galena** Missouri, C USA
36.45N 93.30W
47 V15 **Galeota Point** headland Trinidad,
Trinidad and Tobago 10.07N 60.59W
107 P13 **Galera** Andalucía, S Spain
37.45N 2.33W
47 Y16 **Galera Point** headland Trinidad,
Trinidad and Tobago 10.49N 60.54W
58 A5 **Galera, Punta** headland
NW Ecuador 0.49N 80.03W
32 K12 **Galesburg** Illinois, N USA
40.57N 90.22W
32 J7 **Galesville** Wisconsin, N USA
44.04N 91.21W
20 F12 **Galeton** Pennsylvania, NE USA
41.43N 77.38W
118 H9 **Gâlgău** Hung. Galgó; prev. Gilgău.
Sălaj, NW Romania 47.15N 23.44E
Galgó see Gâlgău
Galgóc see Hlohovec
81 O16 **Galguduud** off. Gobolka
Galguduud. ◆ region E Somalia
143 Q9 **Galich** W Georgia 42.40N 41.39E
129 N14 **Galich** Kostromskaya Oblast',
NW Russian Federation
58.21N 4.21E
116 H7 **Galiche** Vratsa, NW Bulgaria
43.36N 23.53E
106 H3 **Galicia** anc. Gallaecia. ◆
autonomous community NW Spain
66 M8 **Galicia Bank** undersea feature
E Atlantic Ocean
Galilee see HaGalil
189 W7 **Galilee, Lake** ☒ Queensland,
NE Australia
Galilee, Sea of see Tiberias, Lake
108 E11 **Galileo Galilei ✈** (Pisa) Toscana,
C Italy 43.40N 10.22E
33 S12 **Galion** Ohio, N USA 40.43N 82.47W
Galka'yo see Gaalkacyo
82 H11 **Gallabat** Gedaref, E Sudan
12.56N 36.08E
Gallaecia see Galicia
108 C7 **Gallarate** Lombardia, NW Italy
45.39N 8.46E
29 S2 **Gallatin** Missouri, C USA
39.54N 93.57W
22 J8 **Gallatin** Tennessee, S USA
36.23N 86.27W
33 R11 **Gallatin Peak ▲** Montana,
NW USA 45.22N 111.21W
33 R12 **Gallatin River ↔**
Montana/Wyoming, NW USA
161 J26 **Galle** prev. Point de Galle.
Southern Province, SW Sri Lanka
6.04N 80.11E
107 S5 **Gállego ↔** NE Spain
200 N9 **Gallego Rise** undersea feature
E Pacific Ocean
Gallegos see Río Gallegos
65 H23 **Gallegos, Río ↔**
Argentina/Chile
Gallia see France
24 K10 **Galliano** Louisiana, S USA
29.26N 90.18W
116 G13 **Gallikós ↔** N Greece
39 S12 **Gallina Peak ▲** New Mexico,
SW USA 34.14N 105.47W
37 P10 **Gallinas, Punta** headland
NE Colombia 12.27N 71.43W
39 T11 **Gallinas River ↔** New Mexico,
SW USA 41.03N 0.25E
109 Q19 **Gallipoli** Puglia, SE Italy
40.07N 18.00E
Gallipoli see Gelibolu
Gallipoli Peninsula see Gelibolu
Yarımadası
33 T15 **Gallipolis** Ohio, N USA
38.48N 82.13W
156 J12 **Gällivare** Norrbotten, N Sweden
67.08N 20.39E
189 T4 **Gallneukirchen** Niederösterreich,
N Austria 48.21N 14.22E
107 Q7 **Gallo ↔** C Spain
95 G17 **Gällö** Jämtland, C Sweden
62.57N 15.15E
109 I23 **Gallo, Capo** headland Sicilia, Italy,
C Mediterranean Sea 38.13N 13.18E
39 P13 **Gallo Mountains ▲** New
Mexico, SW USA

20 G8 **Galloo Island** island New York,
NE USA
99 H15 **Galloway, Mull of** headland
S Scotland, UK 54.37N 4.54W
39 P10 **Gallup** New Mexico, SW USA
35.31N 108.45W
107 R5 **Gallur** Aragón, NE Spain
41.51N 1.21W
Gâlma see Guelma
76 C10 **Galtat-Zemmour** C Western
Sahara 25.07N 12.21W
97 Q22 **Galten** Århus, C Denmark
56.09N 9.54E
99 D20 **Galtymore Mountain** Ir. Cnoc
Mór na nGaibhlte. ▲ S Ireland
52.21N 8.09W
99 D20 **Galty Mountains** Ir. Na
Gaibhlte. ▲ S Ireland
32 K11 **Galva** Illinois, N USA
41.10N 90.02W
27 X12 **Galveston** Texas, SW USA
29.16N 94.48W
27 W11 **Galveston Bay** inlet Texas,
SW USA
27 W12 **Galveston Island** island Texas,
SW USA
63 B18 **Gálvez** Santa Fe, C Argentina
31.57S 61.13W
99 C18 **Galway** Ir. Gaillimh. W Ireland
53.16N 9.03W
99 B18 **Galway** Ir. Gaillimh. cultural region
W Ireland
99 B18 **Galway Bay** Ir. Cuan na
Gaillimhe. bay W Ireland
85 F18 **Gam** Otjozondjupa, N Namibia
20.10S 20.51E
194 G14 **Gam ↔** SW PNG
171 Hh16 **Gamagōri** Aichi, Honshū,
SW Japan 34.49N 137.12E
56 F7 **Gámara** Cesar, N Colombia
8.21N 73.46W
Gámas see Kaamanen
164 L17 **Gamba** Xizang Zizhiqu, W China
28.13N 88.31E
79 P14 **Gambaga** NE Ghana 10.32N 0.28W
82 G13 **Gambēla** Gambēla, W Ethiopia
8.09N 34.15E
83 H14 **Gambēla ◆** region, W Ethiopia
8.09N 34.15E
40 E12 **Gambell** Saint Lawrence Island,
Alaska, USA 63.43N 171.40W
78 E12 **Gambia** off. Republic of The
Gambia, The Gambia. ◆ republic
W Africa
78 D12 **Gambia** Fr. Gambie. ↔ W Africa
66 K12 **Gambia Plain** undersea feature
E Atlantic Ocean
33 T13 **Gambier** Ohio, N USA
40.22N 82.24W
203 Y13 **Gambier, Îles** island group
E French Polynesia
190 G10 **Gambier Islands** island group
South Australia
81 H19 **Gambona** Plateaux, E Congo
1.52S 15.51E
81 G16 **Gamboula** Mambéré-Kadéï,
SW Central African Republic
4.09N 15.12E
39 P10 **Gamerco** New Mexico, SW USA
35.34N 108.45W
143 V12 **Gamiş Dağı ▲** W Azerbaijan
40.18N 46.15E
97 N18 **Gamleby** Kalmar, S Sweden
57.54N 16.25E
95 J14 **Gammelstad** var. Gammelstaden.
Gammelstad. Norrbotten,
N Sweden 65.37N 22.04E
Gammelstaden see Gammelstad
Gammouda see Sidi Bouzid
161 J25 **Gampaha** Western Province,
W Sri Lanka 07.05N 80.00E
161 K25 **Gampola** Central Province, C Sri
Lanka 7.10N 80.34E
176 Uu8 **Gam, Pulau** island E Indonesia
178 J25 **Gâm, Sông ↔** N Vietnam
156 L7 **Gamvik** Finnmark, N Norway
71.04N 28.08E
156 H13 **Gan** Addu Atoll, C Maldives
Gan see Gansu, China
Gan see Jiangxi, China
Ganaane see Juba
83 O10 **Ganado** Arizona, SW USA
35.42N 109.31W
27 U12 **Ganado** Texas, SW USA
29.02N 96.30W
14 L14 **Gananoque** Ontario, SE Canada
44.19N 76.10W
66 O11 **Ganāveh** see Bandar-e Ganāveh
143 V11 **Gäncä** Rus. Gyandzha; prev.
Kirovabad, Yelisavetpol.
W Azerbaijan 40.41N 46.22E
161 J26 **Ganchi** see Ghonchi
Gand see Gent
84 B13 **Ganda** var. Mariano Machado,
Port. Vila Mariano Machado.
Benguela, W Angola 12.59S 14.37E
81 L22 **Gandajika** Kasai Oriental,
S Dem. Rep. Congo (Zaire)
6.42S 24.00E
159 O12 **Gandak** Nep. Nārāyāni.
↔ India/Nepal
11 U11 **Gander** Newfoundland,
SE Canada 48.58N 54.33W
11 U11 **Gander ✈** Newfoundland,
E Canada 49.03N 54.49W
102 N5 **Ganderkesee** Niedersachsen,
NW Germany 53.01N 8.33E
107 T7 **Gandesa** Cataluña, NE Spain
41.03N 0.25E
158 B10 **Gändhidhäm** Gujarāt, W India
23.07N 70.05E
158 D10 **Gändhinagar** Gujarāt, W India
23.12N 72.37E
107 T11 **Gandía** País Valenciano, E Spain
38.58N 0.10W
165 O10 **Gang** Qinghai, W China
55 G9 **Gangānagar** Rājasthān, N India
158 G9 **Gangāpur** Rājasthān, N India
26.30N 76.49E
159 S17 **Ganga Sāgar** West Bengal,
NE India 21.39N 88.04E
160 K4 **Gangāvati** var. Gangāwati.
Karnātaka, S India 15.26N 76.35E
Gangāwati see Gangāvati
159 S9 **Gangca** var. Shaliuhe. Qinghai,
C China 37.21N 100.09E

164 H14 **Gangdisê Shan** Eng. Kailas
Range. ▲ W China
105 Q15 **Ganges** Hérault, S France
43.57N 3.42E
159 P13 **Ganges** Ben. Padma.
↔ Bangladesh/India see also
Padma
Ganges Cone see Ganges Fan
181 S3 **Ganges Fan** var. Ganges Cone.
undersea feature N Bay of Bengal
159 U17 **Ganges, Mouths of the** delta
Bangladesh/India
109 K23 **Gangi** anc. Engyum. Sicilia, Italy,
C Mediterranean Sea 37.48N 14.13E
158 K8 **Gangotri** Uttar Pradesh, N India
30.55N 79.01E
159 S11 **Gangtok** Sikkim, N India
27.19N 88.39E
165 W11 **Gangu** Gansu, C China
34.46N 105.21E
169 T9 **Gan He ↔** NE China
175 T9 **Gani** Pulau Halmahera,
E Indonesia 0.45S 128.13E
167 O12 **Gan Jiang ↔** S China
152 H15 **Gannaly** Akhalskiy Velayat,
S Turkmenistan 37.02N 60.43E
169 U17 **Gannan** Heilongjiang, NE China
47.58N 123.36E
105 P10 **Gannat** Allier, C France
46.06N 3.12E
35 T9 **Gannett Peak ▲** Wyoming,
C USA 43.10N 109.39W
31 O10 **Gannvalley** South Dakota,
N USA 44.01N 98.59W
189 Y3 **Gänserndorf** Niederösterreich,
NE Austria 48.22N 16.43E
165 T9 **Gansos, Lago dos** see Goose
Lake
165 T9 **Gansu** var. Gan, Gansu Sheng,
Kansu. ◆ province N China
Gansu Sheng see Gansu
78 K6 **Ganta** var. Gahnpa. NE Liberia
7.15N 8.58W
190 I11 **Gantheaume, Cape** headland
South Australia 36.04S 137.28E
167 O12 **Gantsevichi** see Hantsavichy
150 O12 **Ganyu** var. Qingkou, Jiangsu,
E China 34.55N 119.06E
167 O12 **Ganyushkino** Atyrau,
SW Kazakhstan 46.35N 49.15E
79 O10 **Gao** Gao, E Mali 16.15N 0.03E
79 R10 **Gao ◆** region SE Mali
167 O10 **Gao'an** Jiangxi, S China
28.24N 115.27E
167 N5 **Gaoping** Shanxi, C China
35.51N 112.55E
165 S8 **Gaotai** Gansu, N China
39.22N 99.44E
79 O4 **Gaoua** SW Burkina 10.18N 3.12W
78 I13 **Gaoual** Moyenne-Guinée,
N Guinea 11.43N 13.13W
167 R7 **Gaoxiong** see Kaohsiung
165 R7 **Gaoyou** var. Dayishan. Jiangsu,
E China 32.45N 119.30E
166 M15 **Gaozhou** Guangdong, S China
21.56N 110.49E
105 T13 **Gap** anc. Vapincum. Hautes-
Alpes, SE France 44.33N 6.04E
164 G13 **Gar** var. Gar Xincun. Xizang
Zizhiqu, W China 32.04N 80.01E
Garabekevyul/Garabekewül
see Karabekaul
Garabogazköl see Kara-Bogaz-
Gol
83 K19 **Garissa** Coast, E Kenya
0.27S 39.39E
3 V11 **Garland** North Carolina, SE USA
34.45N 78.25W
27 T6 **Garland** Texas, SW USA
32.54N 96.36W
36 L1 **Garland** Utah, W USA
41.43N 112.07W
108 D8 **Garlasco** Lombardia, N Italy
45.12N 8.59E
121 F14 **Garliava** Kaunas, S Lithuania
54.49N 23.52E
148 M9 **Garm, &c-var.** Rüd-e Khersän.
↔ SW Iran
103 K25 **Garmisch-Partenkirchen**
Bayern, S Germany 47.30N 11.06E
149 O5 **Garmsār** prev. Qishlaq. Semnān,
N Iran 35.18N 52.21E
Garmser see Darvishän
31 V12 **Garner** Iowa, C USA 43.06N 93.36W
23 U9 **Garner** North Carolina, SE USA
35.42N 78.36W
29 Q5 **Garnett** Kansas, C USA
38.16N 95.14W
101 M25 **Garnich** Luxembourg,
SW Luxembourg 49.37N 5.57E
190 M8 **Garnpung, Lake** salt lake New
South Wales, SE Australia
Garoe see Garoowe
Garoet see Garut
159 U13 **Gāro Hills** hill range NE India
104 K13 **Garonne ↔** S France
104 J14 **Garonne** anc. Garumna.
↔ S France
79 O13 **Garango** S Burkina 11.45N 0.30W
61 Q15 **Garanhuns** Pernambuco, E Brazil
8.52S 36.28W
196 H5 **Garapan** Saipan, S Northern
Mariana Islands 15.12S 145.43E
80 F12 **Garoua** var. Garua. Nord,
N Cameroon 9.16N 13.22E
80 J13 **Garoua Boulaï** Est, E Cameroon
5.54N 14.33E
79 O10 **Garou, Lac ☒** C Mali
194 M11 **Garove Island** island Witu
Islands, C PNG
97 K16 **Garphyttan** Örebro, C Sweden
59.18N 14.54E
31 R11 **Garretson** South Dakota, N USA
43.43N 96.30W
33 Q11 **Garrett** Indiana, N USA
41.21N 85.08W
31 N14 **Garrison** Iowa, C USA
42.08N 92.10W
33 M4 **Garrison** North Dakota, N USA
47.36N 101.25W
27 X8 **Garrison** Texas, SW USA
31.49N 94.29W
30 L4 **Garrison Dam** dam North
Dakota, N USA 47.29N 101.24W
106 J9 **Garrovillas** Extremadura,
W Spain 39.43N 6.33W
58 D9 **Garupá** Sikasso, SW Mali
10.58N 7.26W
106 L10 **Garrucha** Andalucía, S Spain
37.10N 1.49W
105 Q14 **Gard ◆** department S France
105 Q14 **Gard ↔** S France
108 F7 **Garda, Lago di** var. Benaco, Eng.
Lake Garda, Ger. Gardasee.
☒ NE Italy
Garda, Lake see Garda, Lago di

155 Q5 **Gardan Dīvāl** see Gardan Dīwāl.
155 Q5 **Gardan Dīwāl** var. Gardan Dīvāl.
Wardag, C Afghanistan
34.30N 68.15E
28 L12 **Gardanne** Bouches-du-Rhône,
SE France 43.27N 5.28E
102 L12 **Gardasee** see Garda, Lago di
12 M10 **Garden ↔** Ontario, S Canada
46.33N 80.51W
25 X6 **Garden City** Georgia, SE USA
32.06N 81.09W
28 I6 **Garden City** Kansas, C USA
37.58N 100.52W
29 S5 **Garden City** Missouri, C USA
38.34N 94.12W
27 N8 **Garden City** Texas, SW USA
31.50N 101.29W
25 P3 **Gardendale** Alabama, S USA
33.39N 86.48W
33 P5 **Garden Island** island Michigan,
N USA
24 M11 **Garden Island Bay** bay
Louisiana, S USA
33 O3 **Garden Peninsula** peninsula
Michigan, N USA
33 N1 **Gary** Indiana, N USA
41.35N 87.21W
27 X7 **Gary** Texas, SW USA 32.00N 94.21W
164 G13 **Garyū-san ▲** Kyūshū, SW Japan
34.40N 132.12E
164 G13 **Gar Zangbo ↔** W China
166 F8 **Garzê** Sichuan, C China
31.40N 99.58E
56 E24 **Garzón** Huila, S Colombia
2.13N 75.37W
152 R13 **Gasan-Kuli** var. Esenguly.
Balkanskiy Velayat,
W Turkmenistan 37.29N 53.56E
33 T7 **Gas City** Indiana, N USA
40.29N 85.36W
104 K15 **Gascogne** Eng. Gascony. cultural
region S France
28 V3 **Gasconade River ↔** Missouri,
C USA
Gascony see Gascogne
188 H9 **Gascoyne Junction** Western
Australia 25.06S 115.10E
181 V8 **Gascoyne Plain** undersea feature
E Indian Ocean
188 H9 **Gascoyne River ↔** Western
Australia
199 U12 **Gascoyne Tablemount** undersea
feature N Tasman Sea 36.30S 156.30E
69 U6 **Gash** var. Nahr al Qāsh.
↔ W Sudan
155 X3 **Gasherbrum ▲** NE Pakistan
35.39N 76.34E
165 N9 **Gas Hu ☒** C China
79 X12 **Gashua** Yobe, NE Nigeria
12.50N 11.03E
176 Uu9 **Gasim** Irian Jaya, E Indonesia
1.21S 131.27E
195 N22 **Gasmata** New Britain, E PNG
6.12S 150.25E
25 V14 **Gasparilla Island** island Florida,
SE USA
304 Jj11 **Gaspar, Selat** strait W Indonesia
13 Y6 **Gaspé** Quebec, SE Canada
48.50N 64.33W
13 X6 **Gaspé, Cap de** headland Quebec,
SE Canada 48.45N 64.10W
13 X6 **Gaspé, Péninsule de** var.
Péninsule de la Gaspésie. peninsula
Quebec, SE Canada
Gaspésie, Péninsule de la see
Gaspé, Péninsule de
171 Ll2 **Gas-san ▲** Honshū, C Japan
38.33N 140.02E
79 W15 **Gassol** Taraba, E Nigeria
8.28N 10.24E
23 R10 **Gastonia** North Carolina, SE USA
35.15N 81.11W
23 V8 **Gaston, Lake ☒** North
Carolina/Virginia, SE USA
117 D19 **Gastoúni** Dytikí Ellás, S Greece
37.51N 21.15E
65 I17 **Gastre** Chubut, S Argentina
42.20S 69.10W
107 P15 **Gata, Cabo de** headland S Spain
36.43N 2.11E
107 N11 **Gata, Cape** see Gátas, Akrotíri
118 E12 **Gătaia** Ger. Gataja, Hung. Gátalja;
prev. Gáttája. Timiş, W Romania
45.24N 21.25E
Gataja/Gátalja see Gátaia
152 K12 **Gâtas, Akrotíri** var. Cape Gata.
headland S Cyprus 34.34N 33.03E
106 J8 **Gata, Sierra de ▲** W Spain
128 G13 **Gatchina** Leningradskaya Oblast',
NW Russian Federation
59.33N 30.06E
23 P8 **Gate City** Virginia, NE USA
36.38N 82.34W
99 M14 **Gateshead** NE England, UK
54.57N 1.37W
15 Jj2 **Gateshead Island** island
Nunavut, N Canada
23 X8 **Gatesville** North Carolina,
SE USA 36.23N 76.43W
27 S8 **Gatesville** Texas, SW USA
31.26N 97.44W
12 L12 **Gatineau** Quebec, SE Canada
45.28N 75.40W
12 L11 **Gatineau ↔** Ontario/Quebec,
SE Canada
22 N9 **Gatlinburg** Tennessee, S USA
35.43N 83.30W
45 T14 **Gatún, Lago ☒** C Panama
45 S13 **Gatún, Lago ☒** C Panama
108 F13 **Gdov** Pskovskaya Oblast',
W Russian Federation 58.43N 27.51E
112 I6 **Gdynia** Ger. Gdingen.
N Poland 54.31N 18.30E
28 M10 **Geary** Oklahoma, C USA
35.37N 98.19W
Geavvú see Kevo
78 H2 **Gêba, Rio ↔** C Guinea-Bissau
175 Tt8 **Gebe, Pulau** island E Indonesia
142 E11 **Gebze** Kocaeli, NW Turkey
40.48N 29.25E
82 H10 **Gedaref** var. Al Qadārif,
El Gedaref. Gedaref, E Sudan
14.03N 35.24E
82 H10 **Gedaref ◆** state E Sudan
82 B11 **Gedid Ras el Fil** Southern
Darfur, W Sudan 12.48N 25.42E

111 W3 **Gars am Kamp** var. Gars.
Niederösterreich, NE Austria
48.35N 15.40E
83 K20 **Garsen** Coast, S Kenya 2.16S 40.07E
Garshy see Karshi
12 F10 **Garson** Ontario, S Canada
46.33N 80.51W
102 L12 **Gardelegen** Sachsen-Anhalt,
C Germany 52.31N 11.25E
104 M10 **Gartempe ↔** C France
Gartog see Markam
85 D21 **Garub** Karas, SW Namibia
26.33S 16.00E
304 Jj15 **Garut** prev. Garoet. Jawa,
C Indonesia 7.15S 107.55E
193 C20 **Garvie Mountains ▲** South
Island, NZ
112 N12 **Garwolin** Mazowieckie, E Poland
51.53N 21.36E
27 U12 **Garwood** Texas, SW USA
29.25N 96.25W
33 N1 **Gary** Indiana, N USA
41.35N 87.21W

176 Y10 **Gauttier, Pegunungan ▲** Irian
Jaya, E Indonesia
149 P14 **Gāvbandī** Hormozgān, S Iran
117 H25 **Gavdopoúla** island SE Greece
117 H26 **Gávdos** island SE Greece
104 K16 **Gave de Pau** var. Gave-de-Pay.
↔ SW France
Gave-de-Pay see Gave de Pau
104 J16 **Gave d'Oloron ↔** SW France
101 E18 **Gavere** Oost-Vlaanderen,
NW Belgium 50.56N 3.40E
96 N13 **Gävle** var. Gäfle; prev. Gefle.
Gävleborg, C Sweden 60.40N 17.09E
96 M11 **Gävleborg** var. Gäfleborg,
Gefleborg. ◆ county C Sweden
96 O13 **Gävlebukten** bay C Sweden
128 L16 **Gavrilov-Yam** Yaroslavskaya
Oblast', W Russian Federation
57.19N 39.52E
190 I9 **Gawler** South Australia
34.37S 138.43E
190 G7 **Gawler Ranges** hill range South
Australia
168 H11 **Gaxun Nur ☒** N China
159 P14 **Gaya** Bihār, N India 24.48N 85.00E
79 S13 **Gaya** Dosso, SW Niger 11.54N 3.25E
Gaya see Kyjov
33 Q6 **Gaylord** Michigan, N USA
45.01N 84.40W
31 U9 **Gaylord** Minnesota, N USA
44.33N 94.13W
189 Y9 **Gayndah** Queensland, E Australia
25.37S 151.38E
129 T9 **Gayny** Komi-Permyatskiy
Avtonomnyy Okrug, NW Russian
Federation 60.19N 54.15E
Gaysin see Haysyn
Gayvoron see Hayvoron
144 E11 **Gaza** Ar. Ghazzah, Heb. 'Azza.
NE Gaza Strip 31.30N 34.00E
85 L20 **Gaza** off. Província de Gaza. ◆
province SW Mozambique
152 I9 **Gaz-Achak** Turkm. Gazojak.
Lebapskiy Velayat,
NE Turkmenistan 41.12N 61.24E
Gazalkent see Ghazalkent
152 C11 **Gazandzhyk** Turkm. Gazanjyk;
prev. Kazandzhik. Balkanskiy
Velayat, W Turkmenistan
39.16N 55.27E
Gazanjyk see Gazandzhyk
79 V12 **Gazaoua** Maradi, S Niger
13.28N 7.54E
144 E11 **Gaza Strip** Ar. Qitā' Ghazzah.
disputed region SW Asia
195 P11 **Gazelle Peninsula** headland New
Britain, E PNG 4.32S 151.56E
197 J5 **Gazelle, Récif de la** reef C New
Caledonia
Gazgan see Ghozghon
Gazi Antep see Gaziantep
142 M16 **Gaziantep** var. Gazi Antep; prev.
Aintab, Antep. Gaziantep, S Turkey
37.04N 37.21E
142 M17 **Gaziantep** var. Gazi Antep. ◆
province S Turkey
116 M13 **Gaziköy** Tekirdağ, NW Turkey
40.45N 27.18E
124 O3 **Gazimağusa** var. Famagusta, Gk.
Ammóchostos. E Cyprus
35.06N 33.57E
124 Nn2 **Gazimağusa Körfezi** var.
Famagusta Bay, Gk. Kólpos
Ammóchostos. bay E Cyprus
152 K11 **Gazli** Bukhoro Wiloyati,
C Uzbekistan 40.09N 63.28E
Gazojak see Gaz-Achak
81 K15 **Gbadolite** Equateur, NW Dem.
Rep. Congo (Zaire) 4.18N 20.55E
78 K16 **Gbanga** var. Gbarnga. N Liberia
7.01N 9.30W
Gbarnga see Gbanga
79 S14 **Gbéroubouè** var. Béroubouay.
N Benin 10.35N 2.47E
79 W16 **Gboko** Benue, S Nigeria
7.21N 8.57E
Gcuwa see Butterworth
112 J7 **Gdańsk** Fr. Dantzig, Ger. Danzig.
Pomorskie, N Poland 54.21N 18.35E
**Gdan'skaya Bukhta/Gdańsk,
Gulf of** see Danzig, Gulf of
Gdańska, Zakota see Danzig,
Gulf
Gdingen see Gdynia

101 I23 **Gedinne** Namur, SE Belgium *49.57N 4.55E*
142 E13 **Gediz** Kütahya, W Turkey *39.04N 29.25E*
142 C14 **Gediz Nehri** ≈ W Turkey
83 N14 **Gedlegubē** Somali, E Ethiopia *6.53N 45.08E*
83 L17 **Gedo** *off.* Gobolka Gedo. ◈ *region* SW Somalia
97 I25 **Gedser** Storstrøm, SE Denmark *54.34N 11.57E*
101 I16 **Geel** *var.* Gheel. Antwerpen, N Belgium *51.10N 4.58E*
191 N13 **Geelong** Victoria, SE Australia *38.09S 144.20E*
Ge'e'mu *see* Golmud
101 I14 **Geertruidenberg** Noord-Brabant, S Netherlands *51.43N 4.52E*
102 H10 **Geeste** ≈ NW Germany
102 J10 **Geesthacht** Schleswig-Holstein, N Germany *53.25N 10.22E*
191 P17 **Geeveston** Tasmania, SE Australia *43.12S 146.54E*
Gefle *see* Gävle
Gefleborg *see* Gävleborg
164 G13 **Gê'gyai** Xizang Zizhiqu, W China *32.28N 81.03E*
79 X12 **Geidam** Yobe, NE Nigeria *12.52N 11.55E*
9 T11 **Geikie** ≈ Saskatchewan, C Canada
96 F13 **Geilo** Buskerud, S Norway *60.31N 8.13E*
96 E10 **Geiranger** Møre og Romsdal, S Norway *62.07N 7.12E*
103 I22 **Geislingen** *var.* Geislingen an der Steige. Baden-Württemberg, SW Germany *48.37N 9.52E*
Geislingen an der Steige *see* Geislingen
83 F20 **Geita** Mwanza, NW Tanzania *2.52S 32.12E*
97 G15 **Geithus** Buskerud, S Norway *59.55N 9.57E*
166 H14 **Gejiu** *var.* Kochiu. Yunnan, S China *23.21N 103.07E*
Gëkdepe *see* Geok-Tepe
152 E9 **Geklengkui, Solonchak** *var.* Solonchak Goklenkuy. *salt marsh* NW Turkmenistan
83 D14 **Gel** ≈ W Sudan
109 K25 **Gela** *prev.* Terranova di Sicilia. Sicilia, Italy, C Mediterranean Sea *37.04N 14.15E*
Gêladaindong *see* Gladaindong
83 N14 **Geladī** Somali, E Ethiopia *6.58N 46.24E*
174 Kk11 **Gelam, Pulau** *var.* Pulau Galam. *island* N Indonesia
100 L11 **Gelderland** *prev. Eng.* Guelders. ◈ *province* E Netherlands
100 J13 **Geldermalsen** Gelderland, C Netherlands *51.52N 5.16E*
103 D14 **Geldern** Nordrhein-Westfalen, W Germany *51.31N 6.19E*
101 K15 **Geldrop** Noord-Brabant, SE Netherlands *51.25N 5.31E*
101 L17 **Geleen** Limburg, SE Netherlands *50.57N 5.49E*
130 K14 **Gelendzhik** Krasnodarskiy Kray, SW Russian Federation *44.34N 38.06E*
Gelib *see* Jilib
142 B11 **Gelibolu** *Eng.* Gallipoli. Çanakkale, NW Turkey *40.25N 26.40E*
117 L14 **Gelibolu Yarımadası** *Eng.* Gallipoli Peninsula. *peninsula* NW Turkey
Geliting, Teluk *see* Geliting, Teluk
175 Qq16 **Geliting, Teluk** *var.* Teluk Gelinting. *bay* Nusa Tenggara, S Indonesia
83 O14 **Gellinsor** Mudug, C Somalia *6.25N 46.44E*
103 H18 **Gelnhausen** Hessen, C Germany *50.12N 9.12E*
103 E14 **Gelsenkirchen** Nordrhein-Westfalen, W Germany *51.33N 7.06E*
85 C20 **Geluk** Hardap, SW Namibia *24.35S 15.48E*
101 H20 **Gembloux** Namur, Belgium *50.34N 4.42E*
194 I12 **Gembogl** Chimbu, C PNG *5.52S 145.06E*
81 J16 **Gemena** Equateur, NW Dem. Rep. Congo (Zaire) *3.13N 19.49E*
101 L14 **Gemert** Noord-Brabant, SE Netherlands *51.33N 5.40E*
142 E11 **Gemlik** Bursa, NW Turkey *40.25N 29.10E*
Gem of the Mountains *see* Idaho
108 J6 **Gemona del Friuli** Friuli-Venezia Giulia, NE Italy *46.18N 13.11E*
Gem State *see* Idaho
Genalē Wenz *see* Juba
174 Ll7 **Genali, Danau** ◎ Borneo, N Indonesia
101 G19 **Genappe** Wallon Brabant, C Belgium *50.39N 4.27E*
143 P14 **Genç** Bingöl, E Turkey *38.45N 40.31E*
Genck *see* Genk
100 M9 **Genemuiden** Overijssel, E Netherlands *52.38N 6.03E*
65 K14 **General Acha** La Pampa, C Argentina *37.24S 64.34W*
63 C21 **General Alvear** Buenos Aires, E Argentina *36.03S 60.01W*
64 I12 **General Alvear** Mendoza, W Argentina *34.58S 67.40W*
63 B20 **General Arenales** Buenos Aires, E Argentina *34.21S 61.19W*
63 D21 **General Belgrano** Buenos Aires, E Argentina *35.46S 58.30W*
204 H3 **General Bernardo O'Higgins** *Chilean research station* Antarctica *63.09S 57.13W*
43 O8 **General Bravo** Nuevo León, NE Mexico *25.47N 99.04W*
64 M7 **General Capdevila** Chaco, N Argentina *27.25S 61.30W*
General Carrera, Lago *see* Buenos Aires, Lago
43 N9 **General Cepeda** Coahuila de Zaragoza, NE Mexico *25.18N 101.24W*
65 K15 **General Conesa** Río Negro, E Argentina *40.07S 64.32W*
63 G18 **General Enrique Martínez** Treinta y Tres, E Uruguay *33.13S 53.46W*
64 L3 **General Eugenio A. Garay** *var.* Fortín General Eugenio Garay; *prev.* Yrendagüé. Nueva Asunción, NW Paraguay *20.31S 62.09W*
63 C18 **General Galarza** Entre Ríos, E Argentina *32.43S 59.24W*
63 E22 **General Guido** Buenos Aires, E Argentina *36.36S 57.45W*
General José F.Uriburu *see* Zárate
63 E22 **General Juan Madariaga** Buenos Aires, E Argentina *37.02S 57.06W*
43 O16 **General Juan N Alvarez** ✕ (Acapulco) Guerrero, S Mexico *16.47N 99.47W*
63 B22 **General La Madrid** Buenos Aires, E Argentina *37.13S 61.10W*
63 E21 **General Lavalle** Buenos Aires, E Argentina *36.25S 56.55W*
General Machado *see* Camacupa
64 I8 **General Manuel Belgrano, Cerro** ▲ W Argentina *29.05S 67.05W*
43 O8 **General Mariano Escobero** ✕ (Monterrey) Nuevo León, NE Mexico *25.47N 100.00W*
63 B20 **General O'Brien** Buenos Aires, E Argentina *34.54S 60.45W*
64 K13 **General Pico** La Pampa, C Argentina *35.40S 63.44W*
64 M7 **General Pinedo** Chaco, N Argentina *27.16S 61.19W*
63 B20 **General Pinto** Buenos Aires, E Argentina *34.45S 61.49W*
63 E22 **General Pirán** Buenos Aires, E Argentina *37.16S 57.46W*
45 N15 **General, Río** ≈ S Costa Rica
65 I15 **General Roca** Río Negro, C Argentina *39.00S 67.35W*
179 Rr17 **General Santos** *off.* General Santos City. Mindanao, S Philippines *6.09N 125.10E*
43 O9 **General Terán** Nuevo León, NE Mexico *25.17N 99.37W*
116 N7 **General Toshevo** *Rom.* I.G.Duca, *prev.* Casim, Kasimköj. Dobrich, NE Bulgaria *43.43N 28.04E*
63 B20 **General Viamonte** Buenos Aires, E Argentina *35.01S 61.00W*
63 A20 **General Villegas** Buenos Aires, E Argentina *35.01S 63.01W*
Gênes *see* Genova
20 E11 **Genesee River** ≈ New York/Pennsylvania, NE USA
32 K11 **Geneseo** Illinois, N USA *41.27N 90.08W*
20 F10 **Geneseo** New York, NE USA *42.48N 77.46W*
59 L14 **Geneshuaya, Río** ≈ N Bolivia
25 Q8 **Geneva** Alabama, S USA *31.01N 85.51W*
32 M10 **Geneva** Illinois, N USA *41.53N 88.18W*
31 Q16 **Geneva** Nebraska, C USA *40.31N 97.36W*
20 G10 **Geneva** New York, NE USA *42.52N 76.58W*
33 U10 **Geneva** Ohio, NE USA *41.48N 80.53W*
Geneva *see* Genève
110 B10 **Geneva, Lake** *Fr.* Lac de Genève, Lac Léman, *le* Léman, *Ger.* Genfer See. ◎ France/Switzerland
110 A10 **Genève** *Eng.* Geneva, *Ger.* Genf, *It.* Ginevra. Genève, SW Switzerland *46.13N 6.09E*
110 A11 **Genève** *Eng.* Geneva, *Ger.* Genf, *It.* Ginevra. ◈ *canton* SW Switzerland
110 A10 **Genève** *var.* Geneva. ✕ Vaud, SW Switzerland *46.13N 6.06E*
Genève, Lac de *see* Geneva, Lake
Genf *see* Genève
Genfer See *see* Geneva, Lake
169 S5 **Gen He** ≈ NE China
Genichesk *see* Heniches'k
106 L14 **Genil** ≈ S Spain
101 K18 **Genk** *var.* Genck. Limburg, NE Belgium *50.58N 5.30E*
170 Cc12 **Genkai-nada** *gulf* Kyūshū, SW Japan
109 C19 **Gennargentu, Monti del** ▲ Sardegna, Italy, C Mediterranean Sea *40.01N 9.14E*
101 M14 **Gennep** Limburg, SE Netherlands *51.43N 5.58E*
32 M10 **Genoa** Illinois, N USA *42.06N 88.41W*
31 Q15 **Genoa** Nebraska, C USA *41.27N 97.43W*
Genoa *see* Genova
Genoa, Gulf of *see* Genova, Golfo di
108 D10 **Genova** *Eng.* Genoa, *Fr.* Gênes; *anc.* Genua. Liguria, NW Italy *44.28N 9.00E*
108 D10 **Genova, Golfo di** *Eng.* Gulf of Genoa. *gulf* NW Italy
59 C17 **Genovesa, Isla** *var.* Tower Island. *island* Galapagos Islands, Ecuador, E Pacific Ocean
Genshū *see* Wŏnju
101 E17 **Gent** *Eng.* Ghent, *Fr.* Gand. Oost-Vlaanderen, NW Belgium *51.01N 3.42E*
174 J15 **Genteng** Jawa, C Indonesia *7.21S 106.19E*
102 M12 **Genthin** Sachsen-Anhalt, E Germany *52.24N 12.10E*
29 R9 **Gentry** Arkansas, C USA *36.16N 94.28W*
Genua *see* Genova
109 I15 **Genzano di Roma** Lazio, C Italy *41.42N 12.42E*
Geokchay *see* Göyçay
152 F13 **Geok-Tepe** *var.* Gëkdepe, *Turkm.* Gökdepe. Akhalskiy Velayat, C Turkmenistan *38.05N 58.07E*
126 Gg1 **Georga, Zemlya** *Eng.* George Land. *island* Zemlya Frantsa-Iosifa, N Russian Federation
85 G26 **George** Western Cape, S South Africa *33.51S 22.28E*
31 S11 **George** Iowa, C USA *43.20N 96.00W*
11 O5 **George** ≈ Newfoundland/Quebec, E Canada
67 C25 **George Island** *island* S Falkland Islands
191 R10 **George, Lake** ◎ New South Wales, SE Australia
83 E18 **George, Lake** ◎ SW Uganda
25 W10 **George, Lake** ◎ Florida, SE USA
20 L8 **George, Lake** ⊠ New York, NE USA
George Land *see* Georga, Zemlya
Georgenburg *see* Jurbarkas
George River *see* Kangiqsualujjuaq
66 G8 **Georges Bank** *undersea feature* W Atlantic Ocean
193 A21 **George Sound** *sound* South Island, NZ
67 F15 **Georgetown** ○ (Ascension Island) NW Ascension Island *17.55S 14.25W*
189 V5 **Georgetown** Queensland, NE Australia *18.17S 143.37E*
191 P15 **George Town** Tasmania, SE Australia *41.07S 146.50E*
46 I4 **George Town** Great Exuma Island, C Bahamas *23.28N 75.47W*
46 D8 **George Town** *var.* Georgetown. ○ (Cayman Islands) Grand Cayman, SW Cayman Islands *19.15N 81.22W*
78 H12 **Georgetown** E Gambia *13.33N 14.49W*
57 T8 **Georgetown** ● (Guyana) N Guyana *6.46N 58.10W*
173 Ff3 **George Town** *var.* Penang, Pinang. Pinang, Peninsular Malaysia *5.28N 100.19E*
47 Y14 **Georgetown** Saint Vincent, Saint Vincent and the Grenadines *13.14N 61.07W*
23 Y4 **Georgetown** Delaware, NE USA *38.39N 75.22W*
25 R6 **Georgetown** Georgia, SE USA *31.52N 85.04W*
22 M5 **Georgetown** Kentucky, S USA *38.13N 84.33W*
23 T13 **Georgetown** South Carolina, SE USA *33.22N 79.17W*
45 S10 **Georgetown** Texas, SW USA *30.37N 97.40W*
57 T8 **Georgetown** ✕ N Guyana *6.46N 58.10W*
205 U16 **George V Coast** *physical region* Antarctica
205 T15 **George V Land** *physical region* Antarctica
204 J7 **George VI Ice Shelf** *ice shelf* Antarctica
204 J6 **George VI Sound** *sound* Antarctica
45 S14 **George West** Texas, SW USA *28.19N 98.07W*
143 R9 **Georgia** *off.* Republic of Georgia, *Geor.* Sak'art'velo, *Rus.* Gruzinskaya SSR, Gruziya; *prev.* Georgian SSR. ◆ *republic* SW Asia
25 S5 **Georgia** *off.* State of Georgia; *also known as* Empire State of the South, Peach State. ◈ *state* SE USA
12 F12 **Georgian Bay** *lake bay* Ontario, S Canada
8 L17 **Georgia, Strait of** *strait* British Columbia, W Canada
Georgi Dimitrov *see* Kostenets
Georgi Dimitrov, Yazovir *see* Koprinka, Yazovir
116 M9 **Georgi Traykov, Yazovir** ⊠ NE Bulgaria
Georgiu-Dezh *see* Liski
151 W10 **Georgiyevka** Vostochnyy Kazakhstan, E Kazakhstan *49.19N 81.34E*
Georgiyevka *see* Korday
131 N15 **Georgiyevsk** Stavropol'skiy Kray, SW Russian Federation *44.07N 43.22E*
102 G13 **Georgsmarienhütte** Niedersachsen, NW Germany *52.12N 8.04E*
205 O1 **Georg von Neumayer** *German research station* Antarctica *70.41S 8.18W*
103 M16 **Gera** Thüringen, E Germany *50.51N 12.13E*
103 K16 **Gera** ≈ C Germany
101 E19 **Geraardsbergen** Oost-Vlaanderen, SW Belgium *50.46N 3.52E*
117 F21 **Geráki** Pelopónnisos, S Greece *36.56N 22.46E*
29 W5 **Gerald** Missouri, C USA *38.24N 91.20W*
49 V8 **Geral de Goiás, Serra** ▲ E Brazil
193 G20 **Geraldine** Canterbury, South Island, NZ *44.06S 171.13E*
188 H11 **Geraldton** Western Australia *28.47S 114.39E*
10 E11 **Geraldton** Ontario, S Canada *49.43N 86.58W*
62 J12 **Geral, Serra** ▲▲ S Brazil
175 P16 **Gerampi** Sumbawa, S Indonesia *8.47S 118.51E*
105 U6 **Gérardmer** Vosges, NE France *48.05N 6.54E*
Gerasa *see* Jarash
Gerdauen *see* Zheleznodorozhnyy
41 Q11 **Gerdine, Mount** ▲ Alaska, USA *61.40N 152.21W*
142 H11 **Gerede** Bolu, N Turkey *40.48N 32.13E*
142 H11 **Gerede Çayı** ≈ N Turkey
154 M8 **Gereshk** Helmand, SW Afghanistan *31.49N 64.31E*
103 L24 **Geretsried** Bayern, S Germany *47.51N 11.28E*
107 P14 **Gérgal** Andalucía, S Spain *37.07N 2.34W*
194 K13 **Gerhards, Cape** *headland* C PNG *6.43S 147.31E*
30 I14 **Gering** Nebraska, C USA *41.49N 103.39W*
37 R3 **Gerlach** Nevada, W USA *40.38N 119.21W*
Gerlachfalvi Csúcs/Gerlachovka *see* Gerlachovský štít
113 L18 **Gerlachovský štít** *var.* Gerlachovka, *Ger.* Gerlsdorfer Spitze, *Hung.* Gerlachfalvi Csúcs; *prev.* Stalinov Štít, *Ger.* Franz-Josef Spitze, *Hung.* Ferencz-József Csúcs. ▲ N Slovakia *49.12N 20.09E*
110 E8 **Gerlafingen** Solothurn, NW Switzerland *47.10N 7.34E*
Gerlsdorfer Spitze *see* Gerlachovský štít
145 V3 **Germak** E Iraq *35.49N 46.09E*
German East Africa *see* Tanzania
Germanicopolis *see* Çankırı
Germanicum, Mare/German Ocean *see* North Sea
Germanovichi *see* Hyermanavichy
German Southwest Africa *see* Namibia
22 E10 **Germantown** Tennessee, S USA *35.06N 89.51W*
103 I15 **Germany** *off.* Federal Republic of Germany, *Ger.* Bundesrepublik Deutschland, Deutschland. ◆ *federal republic* N Europe
103 L23 **Germering** Bayern, SE Germany *48.07N 11.22E*
85 J21 **Germiston** *var.* Gauteng. Gauteng, NE South Africa *26.15S 28.10E*
Gernika *see* Gernika-Lumo
107 P2 **Gernika-Lumo** *var.* Gernika, Guernica, Guernica y Lumo. País Vasco, N Spain *43.19N 2.40W*
171 Ii15 **Gero** Gifu, Honshū, SW Japan *35.48N 137.15E*
117 F22 **Geroliménas** Pelopónnisos, S Greece *36.28N 22.25E*
Gerona *see* Girona
101 H21 **Gerpinnes** Hainaut, S Belgium *50.20N 4.37E*
104 L15 **Gers** ◈ *department* S France
104 L14 **Gers** ≈ S France
Gerunda *see* Girona
142 K10 **Gerze** Sinop, N Turkey *41.48N 35.13E*
164 I13 **Gêrzê** Xizang Zizhiqu, W China *32.19N 84.05E*
Gesoriacum/Gessoriacum *see* Boulogne-sur-Mer
101 J21 **Gesves** Namur, SE Belgium *50.24N 5.04E*
95 J20 **Geta** Åland, SW Finland *60.22N 19.49E*
107 N8 **Getafe** Madrid, C Spain *40.18N 3.43W*
97 J21 **Getinge** Halland, S Sweden *56.46N 12.42E*
20 F16 **Gettysburg** Pennsylvania, NE USA *39.49N 77.13W*
31 N8 **Gettysburg** South Dakota, N USA *45.00N 99.57W*
204 K12 **Getz Ice Shelf** *ice shelf* Antarctica
143 S15 **Gevaş** Van, SE Turkey *38.16N 43.04E*
Gevgeli *see* Gevgelija
115 Q20 **Gevgelija** *var.* Đevđelija, Djevdjelija, *Turk.* Gevgeli. SE FYR Macedonia *41.09N 22.30E*
105 T10 **Gex** Ain, E France *46.21N 6.02E*
94 I3 **Geysir** *physical region* SW Iceland
142 F11 **Geyve** Sakarya, NW Turkey *40.31N 30.18E*
82 G10 **Gezira** ◈ *state* E Sudan
111 V3 **Gföhl** Niederösterreich, N Austria *48.30N 15.27E*
85 H22 **Ghaap Plateau** *Afr.* Ghaapplato. *plateau* C South Africa
Ghaapplato *see* Ghaap Plateau
Ghaba *see* Al Ghābah
144 J8 **Ghāb, Tall** ▲ SE Syria *33.09N 37.48E*
145 Q9 **Ghadaf, Wādī al** *dry watercourse* C Iraq
Ghadamès *see* Ghadāmis
76 M9 **Ghadāmis** *var.* Ghadamès, Rhadames. W Libya *30.07N 9.30E*
147 Y10 **Ghadan** E Oman *20.20N 57.58E*
77 O10 **Ghaddūwah** C Libya *26.36N 14.26E*
153 Q11 **Ghafurov** *Rus.* Gafurov; *prev.* Sovetabad. NW Tajikistan *40.13N 69.42E*
159 N12 **Ghāghara** ≈ S Asia
155 P13 **Ghaibi Dero** Sind, SE Pakistan *27.34N 67.42E*
147 Y10 **Ghalat** E Oman *21.06N 58.51E*
153 O11 **Ghallaorol** Jizzakh Wiloyati, C Uzbekistan *40.01N 67.30E*
145 W11 **Ghamūkah, Hawr** ◎ S Iraq
79 P15 **Ghana** *off.* Republic of Ghana. ◆ *republic* W Africa
147 X12 **Ghānah** *spring/well* S Oman *18.35N 56.34E*
Ghanongga *see* Ranongga
Ghansi/Ghansiland *see* Ghanzi
85 F18 **Ghanzi** *var.* Khanzi. Ghanzi, W Botswana *21.39S 21.38E*
85 G19 **Ghanzi** *var.* Ghansi, Ghansiland, Khanzi. ◈ *district* C Botswana
69 T14 **Ghanzi** *var.* Knanzi. ≈ Botswana/South Africa
Ghap'an *see* Kapan
144 F13 **Gharandal** Ma'ān, SW Jordan *30.12N 35.18E*
Gharbt, Jabal al *see* Liban, Jebel
76 K7 **Ghardaïa** N Algeria *32.29N 3.44E*
145 U14 **Gharībiyah, Sha'īb al** ≈ S Iraq
153 R12 **Gharm** *Rus.* Garm. C Tajikistan *39.03N 70.25E*
155 P17 **Gharo** Sind, SE Pakistan *24.44N 67.35E*
145 W10 **Gharrāf, Shaṭṭ al** ≈ S Iraq
Gharvān *see* Gharyān
77 O7 **Gharyān** *var.* Gharvān. NW Libya *32.10N 13.01E*
76 M11 **Ghāt** *var.* Gat. SW Libya *24.58N 10.10E*
Ghawdex *see* Gozo
147 U8 **Ghayathi** Abū Ẓaby, W UAE *23.51N 53.01E*
Ghazāl, Baḥr al *see* Ghazal, Bahr el
80 H9 **Ghazal, Bahr el** *var.* Soro. *seasonal river* C Chad
82 E13 **Ghazal, Bahr el** *var.* Baḥr al Ghazāl. ≈ S Sudan
153 Q9 **Ghazalkent** *Rus.* Gazalkent. Toshkent Wiloyati, E Uzbekistan *41.30N 69.46E*
76 H6 **Ghazaouet** NW Algeria *35.05N 1.52W*
158 J10 **Ghāziābād** Uttar Pradesh, N India *28.42N 77.28E*
159 O13 **Ghāzīpur** Uttar Pradesh, N India *25.38N 83.33E*
155 Q6 **Ghaznī** *var.* Ghazni. Ghaznī, E Afghanistan *33.31N 68.24E*
155 P7 **Ghaznī** ◈ *province* SE Afghanistan
Ghazzah *see* Gaza
Gheel *see* Geel
Ghelīzāne *see* Relizane
Ghent *see* Gent
Gheorghe Brațul *see* Sfântu Gheorghe, Brațul
Gheorghe Gheorghiu-Dej *see* Onești
118 J10 **Gheorgheni** *prev.* Gheorghieni, Sînt-Miclăuș, *Ger.* Niklasmarkt, *Hung.* Gyergyószentmiklós. Harghita, C Romania *46.43N 25.34E*
Gheorghieni *see* Gheorgheni
118 H10 **Gherla** *Ger.* Neuschliss, *Hung.* Szamosújvár; *prev.* Armenierstadt. Cluj, NW Romania *47.02N 23.55E*
Gheweifat *see* Ghuwayfāt
152 L11 **Ghijduwon** *Rus.* Gizhduvan. Bukhoro Wiloyati, C Uzbekistan *40.06N 64.38E*
Ghilan *see* Gīlān
109 C18 **Ghilarza** Sardegna, Italy, C Mediterranean Sea *40.09N 8.50E*
Ghilizane *see* Relizane
Ghimbi *see* Gīmbī
Ghiriş *see* Câmpia Turzii
105 Y15 **Ghisonaccia** Corse, France, C Mediterranean Sea *42.00N 9.25E*
Ghizo *see* Gizo
153 Q11 **Ghonchí** *Rus.* Ganchi. NW Tajikistan *39.57N 69.10E*
Ghor *see* Ghowr
159 T13 **Ghoraghat** Rajshahi, NW Bangladesh *25.17N 89.16E*
155 R13 **Ghotki** Sind, SE Pakistan *28.00N 69.21E*
154 M5 **Ghowr** *var.* Ghor. ◈ *province* C Afghanistan
152 M10 **Ghozghon** *Rus.* Gazgan. Nawoiy Wiloyati, C Uzbekistan *40.36N 65.29E*
153 T13 **Ghūdara** *var.* Gudara, *Rus.* Kudara. SE Tajikistan *38.28N 72.39E*
159 R13 **Ghugri** ≈ N India
153 S14 **Ghund** *Rus.* Gunt. ≈ SE Tajikistan
Ghurdaqah *see* Hurghada
154 J5 **Ghūrīān** Herāt, W Afghanistan *34.19N 61.25E*
147 T8 **Ghuwayfāt** *var.* Gheweifat. Abū Ẓaby, W UAE *24.06N 51.40E*
123 Mm17 **Ghuzayyil, Sabkhat** *salt lake* N Libya
153 N13 **Ghuzor** *Rus.* Guzar. Qashqadaryo Wiloyati, S Uzbekistan *38.41N 66.12E*
117 G17 **Giáltra** Évvoia, C Greece *38.21N 22.58E*
Giamame *see* Jamaame
116 F13 **Giannitsá** *var.* Yiannitsá. Kentrikí Makedonía, N Greece *40.48N 22.24E*
109 F14 **Giannutri, Isola di** *island* Archipelago Toscano, C Italy
98 F13 **Giant's Causeway** *Ir.* Clochán an Aifir. *lava flow* N Northern Ireland, UK
178 J15 **Gia Rai** Minh Hai, S Vietnam *9.16N 105.25E*
109 L24 **Giarre** Sicilia, Italy, C Mediterranean Sea *37.43N 15.12E*
46 I7 **Gibara** Holguín, E Cuba *21.05N 76.08W*
31 O16 **Gibbon** Nebraska, C USA *40.45N 98.50W*
34 K11 **Gibbon** Oregon, NW USA *45.40N 118.22W*
35 P11 **Gibbonsville** Idaho, NW USA *45.33N 113.55W*
66 A13 **Gibb's Hill** *hill* S Bermuda *32.15N 64.51W*
94 I9 **Gibostad** Troms, N Norway *69.21N 18.01E*
106 I14 **Gibraleón** Andalucía, S Spain *37.22N 6.58W*
106 L16 **Gibraltar** ○ (Gibraltar) S Gibraltar *36.08N 5.21W*
106 L16 **Gibraltar** ◇ *UK dependent territory* SW Europe
Gibraltar, Détroit de/Gibraltar, Estrecho de *see* Gibraltar, Strait of
106 J17 **Gibraltar, Strait of** *Fr.* Détroit de Gibraltar, *Sp.* Estrecho de Gibraltar. *strait* Atlantic Ocean/Mediterranean Sea
33 S11 **Gibsonburg** Ohio, N USA *41.22N 83.19W*
32 M13 **Gibson City** Illinois, N USA *40.27N 88.24W*
188 L8 **Gibson Desert** *desert* Western Australia
8 L17 **Gibsons** British Columbia, SW Canada *49.24N 123.31W*
155 N12 **Gīdār** Baluchistān, SW Pakistan *28.16N 66.00E*
161 I17 **Giddalūr** Andhra Pradesh, E India *15.24N 78.54E*
45 U10 **Giddings** Texas, SW USA *30.10N 96.56W*
29 Y8 **Gideon** Missouri, C USA *36.27N 89.55W*
83 I15 **Gīdolē** Southern, S Ethiopia *5.31N 37.26E*
120 H13 **Giedraičiai** Molėtai, E Lithuania *55.05N 25.16E*
105 O7 **Gien** Loiret, C France *47.40N 2.37E*
103 G17 **Giessen** Hessen, W Germany *50.34N 8.40E*
100 O6 **Gieten** Drenthe, NE Netherlands *53.00N 6.43E*
25 Y13 **Gifford** Florida, SE USA *27.40N 80.24W*
15 Ll1 **Gifford** ≈ Baffin Island, Nunavut, NE Canada
102 J12 **Gifhorn** Niedersachsen, N Germany *52.28N 10.33E*
9 P13 **Gift Lake** Alberta, W Canada *55.49N 115.57W*
171 Hh15 **Gifu** *var.* Gihu. Gifu, Honshū, SW Japan *35.23N 136.46E*
171 I14 **Gifu** *off.* Gifu-ken, *var.* Gihu. ◈ *prefecture* Honshū, SW Japan
130 M13 **Gigant** Rostovskaya Oblast', SW Russian Federation *46.29N 41.18E*
42 E8 **Giganta, Sierra de la** ▲▲ W Mexico
56 E12 **Gigante** Huila, S Colombia *2.24N 75.34W*
116 I7 **Gigen** Pleven, N Bulgaria *43.40N 24.31E*
Giggiga *see* Jijiga
98 G12 **Gigha Island** *island* SW Scotland, UK
109 E14 **Giglio, Isola del** *island* Archipelago Toscano, C Italy
Gihu *see* Gifu
106 L2 **Gijón** *var.* Xixón. Asturias, NW Spain *43.31N 5.40W*
83 D20 **Gikongoro** SW Rwanda *2.30S 29.32E*
38 K14 **Gila Bend** Arizona, SW USA *32.57N 112.43W*
38 J14 **Gila Bend Mountains** ▲▲ Arizona, SW USA
39 N14 **Gila Mountains** ▲▲ Arizona, SW USA
38 I15 **Gila Mountains** ▲▲ Arizona, SW USA
148 M4 **Gīlān** *off.* Ostān-e Gīlān; *var.* Ghilan, Guilan. ◈ *province* NW Iran
Gilani *see* Gnjilane
38 L14 **Gila River** ≈ Arizona, SW USA
31 W4 **Gilbert** Minnesota, N USA *47.29N 92.27W*
Gilbert Islands *see* Tungaru
8 L16 **Gilbert, Mount** ▲ British Columbia, SW Canada *50.49N 124.03W*
189 U4 **Gilbert River** ≈ Queensland, NE Australia
(0) C6 **Gilbert Seamounts** *undersea feature* NE Pacific Ocean
35 S7 **Gildford** Montana, NW USA *48.34N 110.21W*
85 P15 **Gilé** Zambézia, NE Mozambique *16.04S 38.16E*
32 K4 **Gile Flowage** ◎ Wisconsin, N USA
190 G7 **Giles, Lake** *salt lake* South Australia
77 U12 **Gilf Kebir Plateau** *Ar.* Haḍabat al Jilf al Kabīr. *plateau* SW Egypt
191 R6 **Gilgandra** New South Wales, SE Australia *31.43S 148.39E*
Gilgău *see* Gâlgău
83 I19 **Gilgil** Rift Valley, SW Kenya *0.33S 36.18E*
191 S4 **Gil Gil Creek** ≈ New South Wales, SE Australia
155 V3 **Gilgit** Jammu and Kashmir, NE Pakistan *35.54N 74.19E*
155 V3 **Gilgit** ≈ N Pakistan
9 X11 **Gillam** Manitoba, C Canada *56.25N 94.45W*
97 J22 **Gilleleje** Frederiksborg, E Denmark *56.05N 12.17E*
32 K14 **Gillespie** Illinois, N USA *39.07N 89.49W*
29 W13 **Gillett** Arkansas, C USA *34.07N 91.22W*
35 X12 **Gillette** Wyoming, C USA *44.17N 105.30W*
99 P22 **Gillingham** SE England, UK *51.24N 0.33E*
205 X6 **Gillock Island** *island* Antarctica
181 O16 **Gillot** ✕ (St-Denis) N Réunion *20.52S 55.31E*
67 H25 **Gill Point** *headland* E Saint Helena *15.58S 5.37W*
32 M12 **Gilman** Illinois, N USA *40.44N 87.58W*
45 W6 **Gilmer** Texas, SW USA *32.43N 94.56W*
Gilolo *see* Halmahera, Pulau
83 G14 **Gilo Wenz** ≈ SW Ethiopia
37 O10 **Gilroy** California, W USA *37.00N 121.34W*
194 H12 **Giluwe, Mount** ▲ W PNG *6.03S 143.52E*
126 M14 **Gilyuy** ≈ SE Russian Federation
101 I14 **Gilze** Noord-Brabant, S Netherlands *51.33N 4.55E*
172 O14 **Gima** Okinawa, Kume-jima, SW Japan
82 H13 **Gīmbī** *It.* Ghimbi. Oromo, C Ethiopia *9.13N 35.39E*
47 T12 **Gimie, Mount** ▲ C Saint Lucia *13.51N 61.00W*
9 X16 **Gimli** Manitoba, S Canada *50.39N 97.00W*
Gimma *see* Jīma
97 O14 **Gimo** Uppsala, C Sweden *60.10N 18.12E*
104 L15 **Gimone** ≈ S France
Gimpoe *see* Gimpu
175 Pp9 **Gimpu** *prev.* Gimpoe. Sulawesi, C Indonesia *1.38S 120.00E*
190 F5 **Gina** South Australia *29.56S 134.33E*
Ginevra *see* Genève
101 J19 **Gingelom** Limburg, NE Belgium *50.46N 5.09E*
188 I12 **Gingin** Western Australia *31.22S 115.51E*
179 R14 **Gingoog** Mindanao, S Philippines *8.47N 125.05E*
83 K14 **Ginir** Oromo, C Ethiopia *7.12N 40.43E*
Giohar *see* Jawhar
109 O17 **Gioia del Colle** Puglia, SE Italy *40.46N 16.55E*
109 M22 **Gioia, Golfo di** *gulf* S Italy
Giona *see* Gkióna
117 I16 **Gioúra** *island* Vóreioi Sporádes, Greece, Aegean Sea
109 O17 **Giovinazzo** Puglia, SE Italy *41.10N 16.40E*
Gipeswic *see* Ipswich
Gipuzkoa *see* Guipúzcoa
Giran *see* Ilan
32 K14 **Girard** Illinois, N USA *39.27N 89.46W*
29 R7 **Girard** Kansas, C USA *37.30N 94.50W*
45 O6 **Girard** Texas, SW USA *33.18N 100.38W*
56 E10 **Girardot** Cundinamarca, C Colombia *4.19N 74.46W*
180 M7 **Giraud Seamount** *undersea feature* SW Indian Ocean *9.57S 46.55E*
85 A15 **Giraul** ≈ SW Angola
98 L9 **Girdle Ness** *headland* NE Scotland, UK *57.09N 2.04W*
143 N11 **Giresun** *var.* Kerasunt; *anc.* Cerasus, Pharnacia. Giresun, NE Turkey *40.55N 38.34E*
143 N12 **Giresun** *var.* Kerasunt. ◈ *province* NE Turkey
143 N12 **Giresun Dağları** ▲▲ N Turkey
77 X10 **Girga** *var.* Girgeh, Jirjā. C Egypt *26.19N 31.49E*
Girgeh *see* Girga
Girgenti *see* Agrigento
194 H10 **Girgir, Cape** *headland* NW PNG *3.48S 144.29E*
159 Q15 **Girīdīh** Bihār, NE India *24.10N 86.19E*
191 P6 **Girilambone** New South Wales, SE Australia *31.19S 146.57E*
Girin *see* Jilin
124 R12 **Girne** *Gk.* Kerýneia, Kyrenia. N Cyprus *35.19N 33.19E*
107 X5 **Girona** *var.* Gerona; *anc.* Gerunda. Cataluña, NE Spain *41.58N 2.49E*
107 W5 **Girona** *var.* Gerona ◈ *province* Cataluña, NE Spain
104 J12 **Gironde** ◈ *department* SW France
104 J11 **Gironde** *estuary* SW France
107 V5 **Gironella** Cataluña, NE Spain *42.01N 1.52E*
105 N15 **Girou** ≈ S France
99 H14 **Girvan** W Scotland, UK *55.14N 4.53W*
44 M9 **Girvin** Texas, SW USA *31.05N 102.24W*
192 Q9 **Gisborne** Gisborne, North Island, NZ *38.41S 178.01E*
192 P9 **Gisborne** *off.* Gisborne District. ◈ *unitary authority* North Island, NZ
Giseifu *see* Ŭijŏngbu
Gisenye *see* Gisenyi
83 D19 **Gisenyi** *var.* Gisenye. NW Rwanda *1.42S 29.18E*
97 K20 **Gislaved** Jönköping, S Sweden *57.19N 13.30E*
105 N4 **Gisors** Eure, N France *49.18N 1.46E*
Gissar *see* Hisor
153 P12 **Gissar Range** *Rus.* Gissarskiy Khrebet. ▲▲ Tajikistan/Uzbekistan
Gissarskiy Khrebet *see* Gissar Range
101 B16 **Gistel** West-Vlaanderen, W Belgium *51.09N 2.58E*
110 F9 **Giswil** Unterwalden, C Switzerland *46.49N 8.11E*
117 B16 **Gitánes** *ancient monument* Ípeiros, W Greece *39.34N 20.19E*
83 E20 **Gitarama** C Rwanda *2.05S 29.45E*
83 E20 **Gitega** C Burundi *3.20S 29.56E*
Githio *see* Gýtheio
110 H11 **Giubiasco** Ticino, S Switzerland *46.10N 9.01E*
108 K13 **Giulianova** Abruzzo, C Italy *42.45N 13.58E*
Giulie, Alpi *see* Julian Alps
Giumri *see* Gyumri
118 M13 **Giurgeni** Ialomița, SE Romania *44.46N 27.51E*
118 J15 **Giurgiu** Giurgiu, S Romania *43.54N 25.58E*
118 J14 **Giurgiu** ◈ *county* SE Romania
97 F22 **Give** Vejle, C Denmark *55.51N 9.15E*
105 R2 **Givet** Ardennes, N France *50.08N 4.50E*
105 R11 **Givors** Rhône, E France *45.36N 4.46E*
85 K19 **Giyani** Northern, NE South Africa *23.19S 30.37E*
82 I13 **Giyon** Oromo, C Ethiopia *8.31N 37.56E*
Giza/Gizeh *see* El Gîza
77 V8 **Giza, Pyramids of** *ancient monument* N Egypt *29.46N 31.03E*
Gizhduvan *see* Ghijduwon
127 Oo8 **Gizhiga** Magadanskaya Oblast', E Russian Federation *61.57N 160.16E*
127 Oo8 **Gizhiginskaya Guba** *bay* E Russian Federation
195 T14 **Gizo** Gizo, NW Solomon Islands *8.03S 156.49E*
195 T14 **Gizo** *var.* Ghizo. *island* NW Solomon Islands
112 N7 **Giżycko** *Ger.* Warmińsko-Mazurskie, NE Poland *54.03N 21.48E*
Gîzymałów *see* Hrymayliv
Gjakovë *see* Đakovica
96 F12 **Gjende** ◎ S Norway
97 F17 **Gjerstad** Aust-Agder, S Norway *58.54N 9.03E*
Gjilan *see* Gnjilane
Gjinokastër *see* Gjirokastër
115 L23 **Gjirokastër** *var.* Gjirokastra; *prev.* Gjinokastër, *Gk.* Argyrokastron, *It.* Argirocastro. Gjirokastër, S Albania *40.04N 20.09E*
115 L22 **Gjirokastër** ◈ *district* S Albania
15 K3 **Gjoa Haven** King William Island, Nunavut, NW Canada *68.37N 95.57W*
96 H13 **Gjøvik** Oppland, S Norway *60.46N 10.40E*
115 J22 **Gjuhëzës, Kepi i** *headland* SW Albania *40.25N 19.19E*
Gjurgjevac *see* Đurđevac
117 E18 **Gkióna** *var.* Giona. ▲ C Greece
124 Oo3 **Gkréko, Akrotíri** *var.* Cape Greco, Pidálion. *headland* E Cyprus *34.57N 34.06E*
101 I18 **Glabbeek-Zuurbemde** Vlaams Brabant, C Belgium *50.54N 4.58E*
11 R14 **Glace Bay** Cape Breton Island, Nova Scotia, SE Canada *46.12N 59.57W*
9 O16 **Glacier** British Columbia, SW Canada *51.12N 117.33W*
41 W12 **Glacier Bay** *inlet* Alaska, USA
34 I7 **Glacier Peak** ▲ Washington, NW USA *48.06N 121.06W*
165 N13 **Gladaindong** *var.* Gêladaindong. ▲ C China *33.24N 91.00E*
23 Q7 **Glade Spring** Virginia, NE USA *36.47N 81.46W*
45 W7 **Gladewater** Texas, SW USA *32.32N 94.57W*
189 Y8 **Gladstone** Queensland, E Australia *23.52S 151.16E*
190 I8 **Gladstone** South Australia *33.16S 138.21E*
9 X16 **Gladstone** Manitoba, S Canada *50.12N 98.56W*
33 O5 **Gladstone** Michigan, N USA *45.51N 87.01W*
29 R4 **Gladstone** Missouri, C USA *39.12N 94.33W*
33 Q7 **Gladwin** Michigan, N USA *43.58N 84.29W*
97 J15 **Glafsfjorden** ◎ C Sweden
94 H2 **Gláma** *physical region* NW Iceland
96 I12 **Glåma** *var.* Glommen, Glomma. ≈ S Norway
114 F13 **Glamoč** Federacija Bosna I Hercegovina, NE Bosnia and Herzegovina *44.01N 16.51E*
99 J22 **Glamorgan** *cultural region* S Wales, UK
97 G24 **Glamsbjerg** Fyn, C Denmark *55.16N 10.07E*
179 Rr17 **Glan** Mindanao, S Philippines *5.49N 125.11E*
97 M17 **Glan** ◎ S Sweden
111 T9 **Glan** ≈ SE Austria
103 F19 **Glan** ≈ W Germany
Glaris *see* Glarus
110 H9 **Glarner Alpen** *Eng.* Glarus Alps. ▲▲ E Switzerland
110 H8 **Glarus** Glarus, E Switzerland *47.03N 9.04E*

◆ Country
● Country Capital
◇ Dependent Territory
○ Dependent Territory Capital
◈ Administrative Region
✕ International Airport
▲ Mountain
▲▲ Mountain Range
Volcano
≈ River
◎ Lake
⊠ Reservoir

110 H9 **Glarus** *Fr.* Glaris, ◆ *canton* C Switzerland
Glarus Alps *see* Glarner Alpen
29 N3 **Glasco** Kansas, C USA 39.21N 97.50W
98 I12 **Glasgow** S Scotland, UK 55.52N 4.15W
22 K7 **Glasgow** Kentucky, S USA 37.00N 85.54W
29 T4 **Glasgow** Missouri, C USA 39.13N 92.51W
35 W7 **Glasgow** Montana, NW USA 48.12N 106.37W
23 T6 **Glasgow** Virginia, NE USA 37.37N 79.27W
98 I12 **Glasgow** ✕ W Scotland, UK 55.52N 4.27W
9 S14 **Glaslyn** Saskatchewan, S Canada 53.20N 108.18W
20 I10 **Glassboro** New Jersey, NE USA 39.40N 75.05W
44 L10 **Glass Mountains** ▲ Texas, SW USA
99 K23 **Glastonbury** SW England, UK 51.09N 2.43W
Glatz *see* Kłodzko
103 N16 **Glauchau** Sachsen, E Germany 50.48N 12.31E
Glavn'a Morava *see* Velika Morava
115 N16 **Glavnik** Serbia, S Yugoslavia 42.53N 21.10E
131 T1 **Glazov** Udmurtskaya Respublika, NW Russian Federation 58.05N 52.38E
Glda *see* Gwda
111 U8 **Gleinalpe** ▲ SE Austria
111 W8 **Gleisdorf** Steiermark, SE Austria 47.07N 15.43E
Gleiwitz *see* Gliwice
41 S11 **Glenallen** Alaska, USA 62.06N 145.33W
104 F7 **Glénan, Îles** *island group* NW France
193 G21 **Glenavy** Canterbury, South Island, NZ 44.53S 171.04E
8 H5 **Glenboyle** Yukon Territory, NW Canada 63.55N 138.43W
23 X3 **Glen Burnie** Maryland, NE USA 39.09N 76.37W
38 L8 **Glen Canyon** *canyon* Utah, W USA
38 L8 **Glen Canyon Dam** *dam* Arizona, SW USA 36.56N 111.28W
32 K15 **Glen Carbon** Illinois, N USA 38.45N 89.58W
12 E17 **Glencoe** Ontario, S Canada 42.44N 81.42W
85 K22 **Glencoe** KwaZulu/Natal, E South Africa 28.09S 30.12E
31 U9 **Glencoe** Minnesota, N USA 44.46N 94.09W
98 H10 **Glen Coe** *valley* N Scotland, UK
38 K13 **Glendale** Arizona, SW USA 33.32N 112.11W
37 S15 **Glendale** California, W USA 34.09N 118.17W
190 G5 **Glendambo** South Australia 30.59S 135.45E
35 Y8 **Glendive** Montana, NW USA 47.07N 104.42W
35 Y15 **Glendo** Wyoming, C USA 42.27N 105.01W
57 S10 **Glendor Mountians** ▲ C Guyana
190 K12 **Glenelg River** ✍ South Australia/Victoria, SE Australia
31 P4 **Glenfield** North Dakota, N USA 47.25N 98.33W
45 V12 **Glen Flora** Texas, SW USA 29.22N 96.12W
189 P7 **Glen Helen** Northern Territory, N Australia 23.45S 132.46E
191 U5 **Glen Innes** New South Wales, SE Australia 29.42S 151.45E
33 P6 **Glen Lake** ◎ Michigan, N USA
8 I7 **Glenlyon Peak** ▲ Yukon Territory, W Canada 62.32N 134.51W
39 N16 **Glenn, Mount** ▲ Arizona, SW USA 31.55N 110.00W
35 N15 **Glenns Ferry** Idaho, NW USA 42.57N 115.18W
25 W6 **Glennville** Georgia, SE USA 31.56N 81.55W
8 L10 **Glenora** British Columbia, W Canada 57.52N 131.16W
190 M11 **Glenorchy** Victoria, SE Australia 36.56S 142.39E
191 V5 **Glenreagh** New South Wales, SE Australia 30.04S 153.00E
35 X15 **Glenrock** Wyoming, C USA 42.51N 105.52W
98 K11 **Glenrothes** E Scotland, UK 56.11N 3.09W
20 L9 **Glens Falls** New York, NE USA 43.18N 73.38W
99 D14 **Glenties** *Ir.* Na Gleannta. NW Ireland 54.46N 8.16W
30 L11 **Glen Ullin** North Dakota, N USA 46.49N 101.49W
23 R4 **Glenville** West Virginia, NE USA 38.55N 80.50W
27 T12 **Glenwood** Arkansas, C USA 34.19N 93.33W
31 S9 **Glenwood** Iowa, C USA 41.03N 95.44W
31 T7 **Glenwood** Minnesota, N USA 45.39N 95.23W
38 L5 **Glenwood** Utah, W USA 38.45N 111.59W
32 I5 **Glenwood City** Wisconsin, N USA 45.04N 92.11W
39 Q4 **Glenwood Springs** Colorado, C USA 39.33N 107.21W
110 F10 **Gletsch** Valais, S Switzerland 46.34N 8.21E
Glevum *see* Gloucester
31 U14 **Glidden** Iowa, C USA 42.03N 94.43W
112 D9 **Glina** Sisak-Moslavina, NE Croatia 45.19N 16.07E
96 F11 **Glittertind** ▲ S Norway 61.40N 8.19E
113 J16 **Gliwice** *Ger.* Gleiwitz. Śląskie, S Poland 50.19N 18.49E
38 M17 **Globe** Arizona, SW USA 33.24N 110.47W
110 D7 **Globino** *Rus.* Globino.
118 I2 **Glodeni** *Rus.* Glodyany. N Moldova 47.47N 27.33E

111 S9 **Glödnitz** Kärnten, S Austria 46.57N 14.03E
Glodyany *see* Glodeni
Glogau *see* Głogów
111 W6 **Gloggnitz** Niederösterreich, E Austria 47.41N 15.57E
112 F13 **Głogów** *Ger.* Glogau, Glogow. Dolnośląskie, SW Poland 51.39N 16.04E
94 G12 **Glomfjord** Nordland, C Norway 66.48N 13.57E
Glommen *see* Glåma
95 I14 **Glommersträsk** Norrbotten, N Sweden 65.16N 19.40E
180 I1 **Glorieuses, Nosy** *island group* N Madagascar
67 C25 **Glorious Hill** *hill* East Falkland, Falkland Islands
40 J12 **Glory of Russia Cape** *headland* Saint Matthew Island, Alaska, USA 60.36N 172.57W
24 J7 **Gloster** Mississippi, S USA 31.12N 91.01W
191 U7 **Gloucester** New South Wales, SE Australia 32.01S 152.00E
194 L12 **Gloucester** New Britain, E PNG 5.28S 148.28E
99 L21 **Gloucester** *hist.* Caer Glou. *Lat.* Glevum. C England, UK 51.52N 2.13W
21 P11 **Gloucester** Massachusetts, NE USA 42.36N 70.36W
23 X6 **Gloucester** Virginia, NE USA 37.23N 76.30W
99 K22 **Gloucestershire** *cultural region* C England, UK
33 T14 **Glouster** Ohio, N USA 39.30N 82.04W
44 H3 **Glovers Reef** *reef* E Belize
20 K10 **Gloversville** New York, NE USA 43.03N 74.20W
112 K12 **Głowno** Łódź, C Poland 51.59N 19.43E
113 H16 **Głubczyce** *Ger.* Leobschütz. Opolskie, S Poland 50.13N 17.49E
130 L11 **Glubokiy** Rostovskaya Oblast', SW Russian Federation 48.34N 40.16E
151 W9 **Glubokoye** Vostochnyy Kazakhstan, E Kazakhstan 50.10N 82.16E
Glubokoye *see* Hlybokaye
113 H16 **Głuchołazy** *Ger.* Ziegenhals. Opolskie, S Poland 50.18N 17.22E
102 I9 **Glückstadt** Schleswig-Holstein, N Germany 53.47N 9.25E
Glukhov *see* Hlukhiv
Glushkevichi *see* Hlushkavichy
Glusk/Glussk *see* Hlusk
Glybokaya *see* Hlyboka
97 E21 **Glyngøre** Viborg, NW Denmark 56.46N 8.51E
131 Q9 **Gmelinka** Volgogradskaya Oblast', SW Russian Federation 50.50N 46.51E
111 R8 **Gmünd** Kärnten, S Austria 46.56N 13.32E
111 U2 **Gmünd** Niederösterreich, N Austria 48.45N 14.57E
Gmünd *see* Schwäbisch Gmünd
111 S5 **Gmunden** Oberösterreich, N Austria 47.54N 13.46E
Gmundner See *see* Traunsee
96 N10 **Gnarp** Gävleborg, C Sweden 62.03N 17.19E
111 W8 **Gnas** Steiermark, SE Austria 46.53N 15.48E
Gnesen *see* Gniezno
97 O16 **Gnesta** Södermanland, C Sweden 59.02N 17.19E
112 H11 **Gniezno** *Ger.* Gnesen. Wielkopolskie, C Poland 52.33N 17.35E
115 L15 **Gnjilane** *var.* Gilani, *Alb.* Gjilan. Serbia, S Yugoslavia 42.27N 21.28E
97 K20 **Gnosjö** Jönköping, S Sweden 57.22N 13.43E
161 L12 **Goa** *prev.* Old Goa, Vela Goa, Velha Goa. Goa, W India 15.31N 73.55E
161 L16 **Goa** *var.* Old Goa. ◆ *state* W India
44 H7 **Goascorán, Río** ✍ El Salvador/Honduras
79 O16 **Goaso** *var.* Gawso. W Ghana 6.49N 2.27W
83 K14 **Goba** *It.* Oromo, S Ethiopia 7.02N 39.58E
85 C20 **Gobabeb** Erongo, W Namibia 23.36S 15.03E
85 C20 **Gobabis** Omaheke, E Namibia 22.24S 18.58E
Gobannium *see* Abergavenny
66 M7 **Goban Spur** *undersea feature* NW Atlantic Ocean
Gobbà *see* Goba
65 H21 **Gobernador Gregores** Santa Cruz, S Argentina 48.43S 70.21W
63 F14 **Gobernador Ingeniero Virasoro** Corrientes, NE Argentina
168 L13 **Gobi** *desert* China/Mongolia
170 G16 **Gōbō** Wakayama, Honshū, SW Japan 33.52N 135.09E
103 D14 **Goch** Nordrhein-Westfalen, W Germany 51.41N 6.10E
85 E20 **Gochas** Hardap, S Namibia 24.54S 18.43E
161 I14 **Godāvari** *var.* Godavari.
161 L16 **Godāvari, Mouths of the** *delta* E India
13 V3 **Godbout** Quebec, SE Canada 49.19N 67.37W
13 V3 **Godbout** ✍ Quebec, SE Canada
13 V3 **Godbout Est** ✍ Quebec, SE Canada
29 N6 **Goddard** Kansas, C USA 37.39N 97.34W
12 E15 **Goderich** Ontario, S Canada 43.45N 81.42W
Godhavn *see* Qeqertarsuaq
160 G13 **Godhra** Gujarāt, W India 22.49N 73.40E
Goding *see* Hodonín
113 N22 **Gödöllő** Pest, N Hungary 47.36N 19.19E
64 I10 **Godoy Cruz** Mendoza, W Argentina 32.58S 68.49W
9 Y13 **Gods** ✍ Manitoba, C Canada
9 Y13 **Gods Lake** ◎ Manitoba, C Canada

9 X13 **Gods Lake** ◎ Manitoba, C Canada
Godthaab/Godthåb *see* Nuuk
Godwin Austen, Mount *see* K2
116 H9 **Goede Hoop, Kaap de** *see* Good Hope, Cape of
Goedgegun *see* Nhlangano
116 F9 **Goeie Hoop, Kaap die** *see* Good Hope, Cape of
Goelands, Lac aux *see* Goéland, Lac aux
11 O7 **Goéland, Lac aux** ◎ Quebec, SE Canada
100 E13 **Goeree** *island* SW Netherlands
101 F15 **Goes** Zeeland, SW Netherlands 51.30N 3.55E
21 L14 **Goffstown** New Hampshire, NE USA 43.01N 71.34W
12 L8 **Gogama** Ontario, S Canada 47.35N 81.35W
32 L3 **Gogebic, Lake** ◎ Michigan, N USA
32 K3 **Gogebic Range** *hill range* Michigan/Wisconsin, N USA
143 V13 **Gogi, Mount** *Arm.* Gogi Lerr, *Az.* Kükü dağ. ▲ Armenia/Azerbaijan 39.33N 45.35E
128 F12 **Gogland, Ostrov** *island* NW Russian Federation
113 I15 **Gogolin** Opolskie, S Poland 50.28N 18.04E
79 S14 **Gogounou** *var.* Gogonou. N Benin 10.49N 2.49E
158 I10 **Gohāna** Haryāna, N India 29.06N 76.43E
61 L14 **Goianésia** Goiás, C Brazil 15.21S 49.01W
61 K18 **Goiânia** *prev.* Goyania. *state capital* Goiás, C Brazil 16.43S 49.18W
61 K18 **Goiás** Goiás, C Brazil 15.57S 50.07W
61 J18 **Goiás** *off.* Estado de Goiás; *prev.* Goiaz, Goyaz. ◆ *state* C Brazil
Goiaz *see* Goiás
165 R14 **Goinsargoin** Xizang Zizhiqu, W China 31.55N 98.04E
62 H10 **Goio-Erê** Paraná, SW Brazil 24.08S 53.07W
101 I15 **Goirle** Noord-Brabant, S Netherlands 51.31N 5.04E
106 H8 **Góis** Coimbra, N Portugal 40.10N 8.06W
171 Gg16 **Gojō** *var.* Gozyō. Nara, Honshū, SW Japan 34.21N 135.42E
171 M10 **Gojōme** Akita, Honshū, NW Japan 39.55N 140.07E
155 U9 **Gojra** Punjab, E Pakistan 31.09N 72.39E
170 D15 **Gokase-gawa** ✍ Kyūshū, SW Japan
142 M11 **Gökçeada** *var.* Imroz Adası, *Gk.* Imbros. *island* NW Turkey
Gökçeada *see* Imroz
Gökdepe *see* Geok-Tepe
142 M12 **Gökırmak** ✍ N Turkey
79 X14 **Gokwe** Midlands, NW Zimbabwe 18.10S 28.54E
96 F13 **Gol** Buskerud, S Norway 60.42N 8.57E
159 X12 **Golāghāt** Assam, NE India 26.31N 93.54E
112 H10 **Golańcz** Wielkopolskie, C Poland 52.59N 17.17E
144 G8 **Golan Heights** *Ar.* Al Jawlān, *Heb.* HaGolan. ▲ SW Syria
Golārā *see* Ārān
Golaya Pristan *see* Hola Prystan'
149 T11 **Golbāf** Kermān, C Iran 29.51N 57.43E
142 M15 **Gölbaşı** Adıyaman, S Turkey 37.46N 37.40E
111 P9 **Gölbner** ▲ SW Austria 46.51N 12.31E
32 M17 **Golconda** Illinois, N USA 37.23N 88.30W
23 U9 **Golconda** Nevada, W USA 40.56N 117.29W
142 E11 **Gölcük** Kocaeli, NW Turkey 40.42N 29.50E
110 I7 **Goldach** Sankt Gallen, NE Switzerland 47.28N 9.28E
112 N7 **Goldap** *Ger.* Goldap. Warmińsko-Mazurskie, NE Poland 54.18N 22.23E
34 G5 **Gold Beach** Oregon, NW USA 42.24N 124.25W
112 E13 **Goldberg** *see* Złotoryja
70 D11 **Gold Coast** *coastal region* S Ghana
191 V3 **Gold Coast** *cultural region* Queensland, E Australia
41 N10 **Gold Creek** Alaska, USA 62.48N 149.40W
9 O14 **Golden** British Columbia, SW Canada 51.19N 116.58W
39 T4 **Golden** Colorado, C USA 39.40N 105.12W
192 I13 **Golden Bay** *bay* South Island, NZ
29 R7 **Golden City** Missouri, C USA 37.23N 94.05W
34 I11 **Goldendale** Washington, NW USA 45.49N 120.49W
165 N16 **Gonggar** Xizang Zizhiqu, W China 29.18N 90.56E
166 G9 **Gongga Shan** ▲ C China 29.49N 101.55E
Goldener Tisch *see* Zlatý Stôl
46 L13 **Golden Grove** E Jamaica 17.55N 76.16W
40 L13 **Golden Lake** ◎ Ontario, SE Canada
24 K10 **Golden Meadow** Louisiana, S USA 29.22N 90.15W
23 V10 **Golden Rock** ✕ (Basseterre) Saint Kitts, Saint Kitts and Nevis 17.16N 62.43W
35 K16 **Golden State, The** *see* California
85 K16 **Golden Valley** Mashonaland West, N Zimbabwe 18.15S 29.46E
37 O15 **Goldfield** Nevada, W USA 37.40N 117.13W
8 K17 **Gold River** Vancouver Island, British Columbia, SW Canada 49.48N 126.01W
23 V9 **Goldsboro** North Carolina, SE USA 35.22N 77.59W
45 U8 **Goldsmith** Texas, SW USA 31.58N 102.36W
9 Y13 **Goldthwaite** Texas, SW USA 31.27N 98.34W

143 R11 **Göle** Ardahan, NE Turkey 40.46N 42.36E
Golema Ada *see* Ostrovo
116 H9 **Golema Planina** ▲ W Bulgaria
116 F9 **Golemi Vrŭkh** ▲ W Bulgaria 42.41N 22.38E
112 D8 **Goleniów** *Ger.* Gollnow. Zachodniopomorskie, NW Poland 53.33N 14.48E
149 N15 **Golestān** ◆ *province* N Iran
37 Q14 **Goleta** California, W USA 30.14N 90.55W
45 O16 **Golfito** Puntarenas, SE Costa Rica 8.37N 83.07W
45 T13 **Goliad** Texas, SW USA 28.40N 97.23W
115 L14 **Golija** ▲ SW Yugoslavia
115 O16 **Goljak** ▲ SE Yugoslavia
142 M12 **Gölköy** Ordu, N Turkey 40.42N 37.37E
Gollel *see* Lavumisa
Gollnow *see* Goleniów
111 X3 **Göllersbach** ✍ NE Austria
Golmo *see* Golmud
165 P10 **Golmud** *var.* Ge'e'mu, Golmo, *Chin.* Ko-erh-mu. Qinghai, C China 36.22N 94.56E
105 Y14 **Golo** ✍ Corse, France, C Mediterranean Sea
41 N9 **Golovin** Alaska, USA 64.33N 162.54W
148 M7 **Golpāyegān** *var.* Gulpaigan. Eşfahān, W Iran 33.22N 50.18E
Golshan *see* Tabas
Gol'shany *see* Hal'shany
98 J7 **Golspie** N Scotland, UK 57.58N 3.55W
114 O11 **Golubac** Serbia, NE Yugoslavia 44.39N 21.37E
112 J9 **Golub-Dobrzyń** Kujawski-pomorskie, C Poland 53.06N 19.03E
151 S7 **Golubovka** Pavlodar, N Kazakhstan 53.07N 74.11E
84 N1 **Golungo Alto** Cuanza Norte, NW Angola 9.10S 14.45E
116 M8 **Golyama Kamchiya** ✍ E Bulgaria
116 L8 **Golyama Reka** ✍ N Bulgaria
116 H11 **Golyama Syutkya** ▲ SW Bulgaria 41.55N 24.03E
116 I12 **Golyam Perelik** ▲ S Bulgaria 41.37N 24.34E
116 I11 **Golyam Persenk** ▲ S Bulgaria 41.50N 24.33E
125 F12 **Golyshmanovo** Tyumenskaya Oblast', C Russian Federation 56.22N 68.25E
81 P19 **Goma** Nord Kivu, NE Dem. Rep. Congo (Zaïre) 1.36S 29.07E
171 Gg16 **Gomadan-zan** ▲ Honshū, SW Japan 34.03N 135.34E
79 X14 **Gombe** Gombe, E Nigeria 10.19N 11.02E
69 U10 **Gombe** *var.* Igombe. ✍ E Tanzania
79 Y14 **Gombi** Adamawa, E Nigeria 10.07N 12.45E
Gombroon *see* Bandar-e 'Abbās
Gomel' *see* Homyel'
Gomel'skaya Oblast' *see* Homyel'skaya Voblasts'
66 N11 **Gomera** *island* Islas Canarias, Spain, NE Atlantic Ocean
42 I5 **Gómez Farías** Chihuahua, N Mexico 29.25N 107.46W
42 L8 **Gómez Palacio** Durango, C Mexico 25.39N 103.30W
164 J13 **Gomo** Xizang Zizhiqu, W China 33.37N 86.40E
175 T11 **Gomumu, Pulau** *see* Gomumu, Pulau
149 T6 **Gonābād** *var.* Gunabad. Khorāsān, NE Iran 36.21N 58.58E
Gonaïves *see* Les Gonaïves
126 M13 **Gonam** ✍ NE Russian Federation
46 L9 **Gonâve, Canal de la** *var.* Canal de Sud. *channel* N Caribbean Sea
46 K9 **Gonâve, Golfe de la** *gulf* N Caribbean Sea
46 K9 **Gonâve, Île de la** *island* C Haiti
Gonbadān *see* Dow Gonbadān
149 Q3 **Gonbad-e Kāvūs** *var.* Gonbad-i-Qawus. Golestān, N Iran 37.15N 55.10E
158 M12 **Gonda** Uttar Pradesh, N India 27.07N 81.58E
Gondar *see* Gonder
82 I11 **Gonder** *var.* Gondar. Amhara, N Ethiopia 12.35N 37.27E
160 I12 **Gondiā** Mahārāshtra, C India 21.27N 80.12E
106 G6 **Gondomar** Porto, NW Portugal 41.10N 8.34W
142 C12 **Gönen** Balıkesir, W Turkey 40.06N 27.39E
142 C12 **Gönen Çayı** ✍ NW Turkey
165 O15 **Gongbo'gyamda** Xizang Zizhiqu, W China 30.03S 93.10E
165 N16 **Gonggar** Xizang Zizhiqu, W China 29.18N 90.56E
166 G9 **Gongga Shan** ▲ C China 29.49N 101.55E
165 T10 **Gonghe** Qinghai, C China 36.22N 100.44E
165 X9 **Gongliu** *var.* Tokkuztara. Xinjiang Uygur Zizhiqu, NW China 43.29N 82.16E
79 V10 **Gongola** ✍ E Nigeria
191 P5 **Gongolgon** New South Wales, SE Australia 30.18S 146.57E
165 Q8 **Gongpoquan** Gansu, N China 41.45N 100.27E
166 I4 **Gongxian** *var.* Gong Xian. Sichuan, C China 28.25N 104.51E
162 Q9 **Gongzhuling** *prev.* Huaide. Jilin, NE China 43.30N 124.48E
165 S14 **Gonjo** Xizang Zizhiqu, W China 30.51N 98.16E
109 B20 **Gonnesa** Sardegna, Italy, C Mediterranean Sea 39.15N 8.27E
Gonnì/Gónnos *see* Gónnoi
117 I7 **Gónnoi** *var.* Gonni, Gónnos; *prev.* Derelí. Thessalía, C Greece 39.52N 22.27E

172 N9 **Gonohe** Aomori, Honshū, C Japan 40.34N 141.18E
170 Cc11 **Gōnoura** Nagasaki, Iki, SW Japan 33.44N 129.41E
37 O11 **Gonzales** California, W USA 36.30N 121.26W
24 J9 **Gonzales** Louisiana, S USA 30.14N 90.55W
45 T12 **Gonzales** Texas, SW USA 29.30N 97.27W
43 P11 **González** Tamaulipas, C Mexico 22.52N 98.25W
23 V6 **Goochland** Virginia, NE USA 37.40N 77.53W
195 N16 **Goodenough Bay** *inlet* SE PNG
205 X14 **Goodenough, Cape** *headland* Antarctica 66.15S 126.34E
195 N15 **Goodenough Island** *var.* Morata. *island* SE PNG
Good Hope *see* Fort Good Hope
41 N8 **Goodhope Bay** *bay* Alaska, USA
85 D26 **Goodhope, Kaap de Goede Hoop, Kaap die Goeie Hoop.** *headland* SW South Africa 34.19S 18.25E
8 K10 **Good Hope Lake** British Columbia, W Canada 59.15N 129.18W
85 E23 **Goodhouse** Northern Cape, W South Africa 28.54S 18.13E
35 O5 **Gooding** Idaho, NW USA 42.56N 114.42W
28 H3 **Goodland** Kansas, C USA 39.21N 101.42W
181 Y15 **Goodlands** NW Mauritius 20.01S 57.39E
22 J8 **Goodlettsville** Tennessee, S USA 36.19N 86.42W
41 N13 **Goodnews** Alaska, USA 59.07N 161.35W
45 O3 **Goodnight** Texas, SW USA 35.00N 101.07W
191 Q4 **Goodooga** New South Wales, SE Australia 29.09S 147.30E
31 N4 **Goodrich** North Dakota, N USA 47.24N 100.07W
45 W10 **Goodrich** Texas, SW USA 30.36N 94.57W
31 X10 **Goodview** Minnesota, N USA
28 H8 **Goodwell** Oklahoma, C USA 36.35N 101.38W
99 N17 **Goole** E England, UK 53.43N 0.46W
191 O8 **Goodoogow** New South Wales, SE Australia 34.00S 145.43E
190 I10 **Goolwa** South Australia 35.31S 138.43E
189 Y11 **Goondiwindi** Queensland, E Australia 28.33S 150.22E
100 O11 **Goor** Overijssel, E Netherlands 52.13N 6.33E
Goose Bay *see* Happy Valley-Goose Bay
35 V4 **Gooseberry Creek** ✍ Wyoming, C USA
23 S14 **Goose Creek** South Carolina, SE USA 32.58N 80.01W
65 M24 **Goose Green** *var.* Prado del Ganso. East Falkland, Falkland Islands 51.52S 59.00W
17 G6 **Goose Lake** *var.* Lago dos Gansos. ◎ California/Oregon, W USA
31 Q4 **Goose River** ✍ North Dakota, N USA
159 T16 **Gopalganj** Dhaka, S Bangladesh 23.03N 89.52E
159 O12 **Gopalganj** Bihār, N India 26.28N 84.25E
Gopher State *see* Minnesota
103 O22 **Göppingen** Baden-Württemberg, SW Germany 48.42N 9.39E
112 G13 **Góra** *Ger.* Guhrau. Dolnośląskie, SW Poland 51.40N 16.30E
112 M12 **Góra Kalwaria** Mazowieckie, C Poland 52.00N 21.14E
159 O12 **Gorakhpur** Uttar Pradesh, N India 26.45N 83.22E
Gorany *see* Harany
115 J14 **Goražde** Federacija Bosna I Hercegovina, Bosnia and Herzegovina 43.39N 18.58E
Gorbovichi *see* Harbavichy
Gorče Petrov *see* Đorče Petrov
(0) (E) **Gorda Ridges** *undersea feature* NE Pacific Ocean
Gordiaz *see* Gardēz
80 K12 **Gordil** Vakaga, N Central African Republic 9.37N 21.42E
25 U5 **Gordon** Georgia, SE USA 32.52N 83.19W
30 M7 **Gordon** Nebraska, C USA 42.48N 102.12W
45 R7 **Gordon** Texas, SW USA 32.32N 98.34W
30 L13 **Gordon Creek** ✍ Nebraska, C USA
65 K25 **Gordon, Isla** *island* S Chile
191 O17 **Gordon, Lake** ◎ Tasmania, SE Australia
191 O17 **Gordon River** ✍ Tasmania, SE Australia
23 V5 **Gordonsville** Virginia, NE USA 38.08N 78.11W
80 I13 **Goré** Oromo, E Ethiopia 8.08N 35.33E
193 D24 **Gore** Southland, South Island, NZ 46.06S 168.58E
80 H13 **Goré** Logone-Oriental, S Chad 7.55N 16.37E
12 I7 **Gore Bay** Manitoulin Island, Ontario, S Canada 45.54N 82.28W
45 O3 **Goree** Texas, SW USA 33.28N 99.31W
143 N11 **Görele** Giresun, NE Turkey 41.00N 39.00E
21 Q11 **Gore Mountain** ▲ Vermont, NE USA 44.55N 71.47W
41 O9 **Gore Point** *headland* Alaska, USA 59.12N 150.57W
39 P9 **Gore Range** ▲ Colorado, C USA
99 F19 **Gorey** *Ir.* Guaire. SE Ireland 52.40N 6.18W
Görz *see* Gorizia

21 P8 **Gorham** Maine, NE USA 43.41N 70.27W
100 I13 **Gorinchem** *var.* Gorkum. Zuid-Holland, C Netherlands 51.49N 4.58E
143 V13 **Goris** SE Armenia 39.31N 46.20E
128 K16 **Goritsy** Tverskaya Oblast', W Russian Federation 57.09N 36.44E
108 J7 **Gorizia** *Ger.* Friuli-Venezia Giulia, NE Italy 45.55N 13.37E
111 W12 **Gorjanci** *var.* Uskočke Planine, Žumberak, Žumberačko Gorje, *Ger.* Uskokengebirge; *prev.* Sichelburger Gebirge. ▲ Croatia/Slovenia
see also Žumberačko Gorje
Gorki *see* Jirkov
Gorki *see* Horki
Gor'kiy *see* Nizhniy Novgorod
Gor'kiy Reservoir *see* Gor'kovskoye Vodokhranilishche
125 D9 **Gor'kovskoye Vodokhranilishche** *Eng.* Gor'kiy Reservoir. ▨ W Russian Federation
Gorkum *see* Gorinchem
97 I23 **Gørlev** Vestsjælland, E Denmark 55.33N 11.13E
113 M17 **Gorlice** Małopolskie, S Poland 49.37N 21.08E
103 Q15 **Görlitz** Sachsen, E Germany 51.09N 14.57E
Görlitz *see* Zgorzelec
Gorlovka *see* Horlivka
45 R7 **Gorman** Texas, SW USA 32.12N 98.40W
23 T3 **Gormania** West Virginia, NE USA 39.16N 79.18W
Gorna Dzhumaya *see* Blagoevgrad
116 K8 **Gorna Oryakhovitsa** Veliko Tŭrnovo, N Bulgaria 43.12N 25.38E
116 J8 **Gorna Studena** Veliko Tŭrnovo, N Bulgaria 43.26N 25.21E
111 X9 **Gornja Radgona** *Ger.* Oberradkersburg. NE Slovenia 46.39N 16.00E
114 M13 **Gornji Milanovac** Serbia, C Yugoslavia 44.01N 20.26E
114 G13 **Gornji Vakuf** *var.* Uskoplje. Federacija Bosna I Hercegovina, C Bosnia and Herzegovina 43.55N 17.34E
126 H15 **Gorno-Altaysk** Respublika Altay, S Russian Federation 51.58N 85.55E
Gorno-Altayskaya Respublika *see* Altay, Respublika
126 K13 **Gorno-Chuyskiy** Irkutskaya Oblast', C Russian Federation 57.33N 111.38E
129 V14 **Gornozavodsk** Permskaya Oblast', NW Russian Federation 58.21N 58.24E
127 O16 **Gornozavodsk** Ostrov Sakhalin, Sakhalinskaya Oblast', SE Russian Federation 46.34N 141.52E
126 Gg15 **Gornyak** Altayskiy Kray, S Russian Federation 50.58N 81.24E
131 R8 **Gornyy** Saratovskaya Oblast', W Russian Federation 51.42N 48.26E
Gornyy Altay *see* Altay, Respublika
131 O10 **Gornyy Balykley** Volgogradskaya Oblast', SW Russian Federation 49.37N 45.03E
100 J10 **Gorredijk** *Fris.* De Gordyk. Friesland, N Netherlands 53.00N 6.04E
100 M7 **Gorssel** Gelderland, E Netherlands 52.12N 6.13E
86 C14 **Gorringe Ridge** *undersea feature* E Atlantic Ocean
Gorssel *see* Gorssel
20 I7 **Gouverneur** New York, NE USA 44.20N 75.27W
101 L21 **Gouvy** Luxembourg, E Belgium 50.10N 5.55E
47 R14 **Gouyave** *var.* Charlotte Town. NW Grenada 12.10N 61.43W
79 U12 **Goverla, Gora** *see* Hoverla, Hora
61 N20 **Governador Valadares** Minas Gerais, SE Brazil 18.51S 41.57W
179 R16 **Governor Generoso** Mindanao, S Philippines 6.36N 126.06E
46 F2 **Governor's Harbour** Eleuthera Island, C Bahamas 25.11N 76.15W
168 I10 **Govĭ-Altay** ◆ *province* SW Mongolia
168 I10 **Govĭ Altayn Nuruu** ▲ S Mongolia
160 L9 **Govind Ballabh Pant Sāgar** ◎ C India

33 P11 **Goshen** Indiana, N USA 41.34N 85.49W
20 I12 **Goshen** New York, NE USA 41.24N 74.17W
171 Mm8 **Goshogawara** *var.* Gosyogawara. Aomori, Honshū, C Japan 40.46N 140.24E
152 I8 **Goshquduq Qum** *var.* Tosqudug Qumlari, *Rus.* Peski Taskuduk. *desert* W Uzbekistan
103 J14 **Goslar** Niedersachsen, C Germany 51.55N 10.25E
29 Y9 **Gosnell** Arkansas, C USA 35.57N 89.58W
114 C11 **Gospić** Lika-Senj, C Croatia 44.33N 15.22E
99 N23 **Gosport** S England, UK 50.48N 1.07W
96 H7 **Gossa** *island* S Norway
110 H7 **Gossau** Sankt Gallen, NE Switzerland 47.25N 9.16E
101 G20 **Gossélies** *var.* Goss'lies. Hainaut, S Belgium 50.28N 4.25E
79 P10 **Gossi** Tombouctou, C Mali 15.44N 1.19W
Goss'lies *see* Gossélies
115 N18 **Gostivar** W FYR Macedonia 41.48N 20.55E
Gostomel' *see* Hostomel'
112 G12 **Gostyń** *var.* Gostyn. Wielkopolskie, C Poland 51.53N 16.59E
112 K11 **Gostynin** Mazowieckie, C Poland 52.25N 19.27E
97 J18 **Göta Älv** ✍ S Sweden
97 N17 **Göta kanal** *canal* S Sweden
97 K18 **Götaland** *cultural region* S Sweden
97 I18 **Göteborg** *Eng.* Gothenburg. Västra Götaland, S Sweden
Göteborg Eng. Gothenburg.
79 X16 **Gotel Mountains** ▲ E Nigeria
97 K17 **Götene** Västra Götaland, S Sweden 58.31N 13.28E
Gotera *see* San Francisco
103 K16 **Gotha** Thüringen, C Germany 50.57N 10.43E
31 N15 **Gothenburg** Nebraska, C USA 40.57N 100.09W
Gothenburg *see* Göteborg
79 X12 **Gothèye** Tillabéri, SW Niger 13.52N 1.27E
Gothland *see* Gotland
97 P19 **Gotland** *island* SE Sweden
97 P19 **Gotland** *var.* Gothland, Gottland. ◆ *county* SE Sweden
170 B12 **Gotō-rettō** *island group* SW Japan
116 H12 **Gotse Delchev** *prev.* Nevrokop. Blagoevgrad, SW Bulgaria 41.35N 23.43E
97 J18 **Gotska Sandön** *island* SE Sweden
170 Ee12 **Gotsu** *var.* Gōtsu. Shimane, Honshū, SW Japan 35.00N 132.13E
103 I15 **Göttingen** *var.* Goettingen. Niedersachsen, C Germany 51.33N 9.55E
Gottland *see* Gotland
95 I16 **Gottne** Västernorrland, C Sweden 63.27N 18.25E
Gottschee *see* Kočevje
Gottwaldov *see* Zlín
Gōtu *see* Gōtsu
110 I7 **Götzis** Vorarlberg, NW Austria 47.21N 9.40E
100 H12 **Gouda** Zuid-Holland, C Netherlands 52.01N 4.42E
78 I13 **Goudiri** *var.* Goudiry. E Senegal 14.12N 12.40W
Goudiry *see* Goudiri
79 X12 **Goudoumaria** Diffa, S Niger 13.42N 11.09E
13 O12 **Gouffre, Rivière du** ✍ Quebec, SE Canada
67 M19 **Gough Fracture Zone** *tectonic feature* S Atlantic Ocean
67 M19 **Gough Island** *island* Tristan da Cunha, S Atlantic Ocean
13 N8 **Gouin, Réservoir** ▨ Quebec, S Canada
12 B10 **Goulais River** Ontario, S Canada 46.41N 84.22W
191 R9 **Goulburn** New South Wales, SE Australia 34.45S 149.43E
191 O10 **Goulburn River** ✍ Victoria, SE Australia
205 Q16 **Gould Coast** *physical region* Antarctica
Goulimime *see* Guelmime
116 F13 **Goumenissa** Kentrikí Makedonía, N Greece 40.55N 22.27E
79 O10 **Goundam** Tombouctou, NW Mali 16.25N 3.41W
80 H12 **Goundi** Moyen-Chari, S Chad 9.18N 17.21E
80 G12 **Gounou-Gaya** Mayo-Kébbi, SW Chad 9.37N 15.30E
79 O12 **Gourci** *var.* Gourcy. NW Burkina 13.14N 2.22W
Gourcy *see* Gourci
104 L14 **Gourdon** Lot, S France 44.45N 1.22E
79 W11 **Gouré** Zinder, SE Niger 13.58N 10.16E
104 G6 **Gourin** Morbihan, NW France 48.07N 3.37W
79 Q4 **Gourma-Rharous** Tombouctou, C Mali 16.54N 1.55W
105 N4 **Gournay-en-Bray** Seine-Maritime, N France 49.29N 1.42E
80 J2 **Gouro** Borkou-Ennedi-Tibesti, N Chad 19.26N 19.36E
106 H8 **Gouveia** Guarda, N Portugal 40.28N 7.34W

| ◆ COUNTRY | ◇ DEPENDENT TERRITORY | ◈ ADMINISTRATIVE REGION | ▲ MOUNTAIN | ▰ VOLCANO | ◎ LAKE |
| ● COUNTRY CAPITAL | ○ DEPENDENT TERRITORY CAPITAL | ✕ INTERNATIONAL AIRPORT | ▲ MOUNTAIN RANGE | ✍ RIVER | ▨ RESERVOIR |

158 I7 **Govind Sāgar** ⊠ NE India
153 N14 **Govurdak** Turkm. Gowurdak; prev. Guardak. Lebapskiy Velayat, E Turkmenistan 37.50N 66.06E
20 D11 **Gowanda** New York, NE USA 42.25N 78.55W
154 J10 **Gowd-e Zereh, Dasht-e** var. Guad-i-Zirreh. marsh SW Afghanistan
12 F8 **Gowganda** Ontario, S Canada 47.39N 80.43W
12 G8 **Gowganda Lake** ◎ Ontario, S Canada
31 U13 **Gowrie** Iowa, C USA 42.16N 94.17W
Gowurdak see Govurdak
63 C15 **Goya** Corrientes, NE Argentina 29.10S 59.15W
Goyania see Goiânia
143 X11 **Göyçay** Rus. Geokchay. C Azerbaijan 40.38N 47.44E
Goymat see Koymat
Goymatdag see Koymatdag, Gory
142 F12 **Göynük** Bolu, NW Turkey 40.24N 30.45E
172 N12 **Goyō-san** ▲ Honshū, C Japan 39.12N 141.40E
80 K11 **Goz Beïda** Ouaddaï, SE Chad 12.06N 21.22E
164 H11 **Gozha Co** ◎ W China
123 J15 **Gozo** Malt. Ghawdex. island N Malta
82 H9 **Gōz Regeb** Kassala, NE Sudan 16.03N 35.33E
Gozyō see Gojō
85 H25 **Graaff-Reinet** Eastern Cape, S South Africa 32.16S 24.31E
Graasten see Gråsten
78 L17 **Grabo** SW Ivory Coast 4.57N 7.30W
114 P11 **Grabovica** Serbia, E Yugoslavia 44.30N 22.29E
112 I13 **Grabów nad Prosną** Wielkopolskie, C Poland 51.30N 18.06E
110 I8 **Grabs** Sankt Gallen, NE Switzerland 47.10N 9.27E
114 D12 **Gračac** Zadar, C Croatia 44.18N 15.52E
114 I11 **Gračanica** Federacija Bosna I Hercegovina, NE Bosnia and Herzegovina 44.41N 18.20E
12 L11 **Gracefield** Quebec, SE Canada 46.06N 76.03W
101 K19 **Grâce-Hollogne** Liège, E Belgium 50.38N 5.30E
25 R8 **Graceville** Florida, SE USA 30.57N 85.31W
31 R8 **Graceville** Minnesota, N USA 45.34N 96.25W
44 G6 **Gracias** Lempira, W Honduras 14.34N 88.34W
Gracias see Lempira
44 L5 **Gracias a Dios** ◐ department E Honduras
45 O6 **Gracias a Dios, Cabo de** headland Honduras/Nicaragua 15.00N 83.10W
66 O2 **Graciosa** var. Ilha Graciosa. island Azores, Portugal, NE Atlantic Ocean
66 Q11 **Graciosa** island Islas Canarias, Spain, NE Atlantic Ocean
Graciosa, Ilha see Graciosa
114 I11 **Gradačac** Federacija Bosna I Hercegovina, N Bosnia and Herzegovina 44.51N 18.24E
61 J15 **Gradaús, Serra dos** ▲▲ C Brazil
106 L3 **Gradefes** Castilla-León, N Spain 42.37N 5.13W
Gradiška see Bosanska Gradiška
Gradizhsk see Hradyz'k
108 J7 **Grado** Friuli-Venezia Giulia, NE Italy 45.41N 13.24E
106 K2 **Grado** Asturias, N Spain 43.22N 6.04W
115 P19 **Gradsko** C FYR Macedonia 41.34N 21.56E
39 V11 **Grady** New Mexico, SW USA 34.49N 103.19W
31 T12 **Graettinger** Iowa, C USA 43.14N 94.45W
103 M23 **Grafing** Bayern, SE Germany 48.01N 11.57E
45 S6 **Graford** Texas, SW USA 32.56N 98.15W
191 V5 **Grafton** New South Wales, SE Australia 29.41S 152.55E
31 Q3 **Grafton** North Dakota, N USA 48.24N 97.24W
23 S3 **Grafton** West Virginia, NE USA 39.20N 80.01W
23 T9 **Graham** North Carolina, SE USA 36.04N 79.24W
45 R6 **Graham** Texas, SW USA 33.06N 98.34W
Graham Bell Island see Greem-Bell, Ostrov
8 I13 **Graham Island** island Queen Charlotte Islands, British Columbia, SW Canada
21 S6 **Graham Lake** ◎ Maine, NE USA
204 H4 **Graham Land** physical region Antarctica
39 N15 **Graham, Mount** ▲ Arizona, SW USA 32.42N 109.52W
Grahamstad see Grahamstown
85 I25 **Grahamstown** Afr. Grahamstad. Eastern Cape, S South Africa 33.18S 26.31E
Grahovo see Bosansko Grahovo
70 C11 **Grain Coast** coastal region S Liberia
174 Mm16 **Grajagan** Jawa, S Indonesia 8.33S 114.13E
174 Mm16 **Grajagan, Teluk** bay Jawa, S Indonesia
61 L14 **Grajaú** Maranhão, E Brazil 5.49S 45.12W
60 M13 **Grajaú, Rio** ≈ NE Brazil
112 O8 **Grajewo** Podlaskie, NE Poland 53.38N 22.25E
97 F24 **Gram** Sønderjylland, SW Denmark 55.18N 9.03E
105 N13 **Gramat** Lot, S France 44.45N 1.45E
24 H5 **Grambling** Louisiana, S USA 32.31N 92.43W
117 C14 **Grámmos** ▲▲ Albania/Greece
98 I9 **Grampian Mountains** ▲▲ C Scotland, UK
190 L12 **Grampians, The** ▲▲ Victoria, SE Australia
100 O9 **Gramsbergen** Overijssel, E Netherlands 52.37N 6.39E
115 L21 **Gramsh** var. Gramshi. Elbasan, C Albania 40.52N 20.12E
Gramshi see Gramsh
Gran see Hron, Slovakia
Gran see Esztergom, N Hungary
56 F11 **Granada** Meta, C Colombia 3.36N 73.44W
44 J10 **Granada** Granada, SW Nicaragua 11.55N 85.58W
107 N14 **Granada** Andalucía, S Spain 37.13N 3.40W
39 W6 **Granada** Colorado, C USA 38.00N 102.18W
44 J11 **Granada** ◐ department SW Nicaragua
107 N14 **Granada** ◐ province Andalucía, S Spain
65 I21 **Gran Altiplanicie Central** plain S Argentina
99 E17 **Granard** Ir. Gránard. C Ireland 53.46N 7.30W
65 J20 **Gran Bajo** basin S Argentina
65 J15 **Gran Bajo del Gualicho** basin E Argentina
65 I21 **Gran Bajo de San Julián** basin SE Argentina
45 S7 **Granbury** Texas, SW USA 32.26N 97.47W
13 P12 **Granby** Quebec, SE Canada 45.24N 72.40W
29 S8 **Granby** Missouri, C USA 36.55N 94.14W
39 S3 **Granby, Lake** ⊠ Colorado, C USA
66 O12 **Gran Canaria** var. Grand Canary. island Islas Canarias, Spain, NE Atlantic Ocean
49 T11 **Gran Chaco** var. Chaco. lowland plain South America
47 R14 **Grand Anse** SW Grenada 12.01N 61.45W
Grand-Anse see Portsmouth
46 G1 **Grand Bahama Island** island N Bahamas
Grand Balé see Tui
105 U7 **Grand Ballon** Ger. Ballon de Guebwiller. ▲ NE France 47.53N 7.06E
11 T13 **Grand Bank** Newfoundland, SE Canada 47.04N 55.46W
66 I7 **Grand Banks of Newfoundland** undersea feature NW Atlantic Ocean
Grand Bassa see Buchanan
79 N17 **Grand-Bassam** var. Bassam. SE Ivory Coast 5.13N 3.46W
12 E16 **Grand Bend** Ontario, S Canada 43.17N 81.46W
78 L17 **Grand-Bérébi** var. Grand-Béréby. SW Ivory Coast 4.37N 6.55W
Grand-Béréby see Grand-Bérébi
47 X11 **Grand-Bourg** Marie-Galante, SE Guadeloupe 15.53N 61.18W
46 M6 **Grand Caicos** var. Middle Caicos. island C Turks and Caicos Islands
12 K12 **Grand Calumet, Île du** island Quebec, SE Canada
99 E18 **Grand Canal** Ir. An Chanáil Mhór. canal C Ireland
Grand Canary see Gran Canaria
38 K10 **Grand Canyon** Arizona, SW USA 36.01N 112.10W
38 J9 **Grand Canyon** canyon Arizona, SW USA
Grand Canyon State see Arizona
46 D8 **Grand Cayman** island SW Cayman Islands
9 R14 **Grand Centre** Alberta, SW Canada 54.25N 110.13W
78 L17 **Grand Cess** SE Liberia 4.36N 8.12W
110 D12 **Grand Combin** ▲ S Switzerland 45.58N 7.27E
34 K8 **Grand Coulee** Washington, NW USA 47.56N 119.00W
34 J8 **Grand Coulee** valley Washington, NW USA
47 X5 **Grand Cul-de-Sac Marin** bay N Guadeloupe
Grand Duchy of Luxembourg see Luxembourg
65 J22 **Grande, Bahía** bay S Argentina
9 N14 **Grande Cache** Alberta, W Canada 53.52N 119.07W
105 U12 **Grande Casse** ▲ E France 45.22N 6.50E
180 G12 **Grande Comore** var. Njazidja, Great Comoro. island NW Comoros
63 G18 **Grande, Cuchilla** hill range E Uruguay
47 S5 **Grande de Añasco, Río** ≈ W Puerto Rico
Grande de Chiloé, Isla see Chiloé, Isla de
60 J12 **Grande de Gurupá, Ilha** river island NE Brazil
59 K21 **Grande de Lipez, Río** ≈ SW Bolivia
47 U6 **Grande de Loíza, Río** ≈ E Puerto Rico
47 T5 **Grande de Manatí, Río** ≈ C Puerto Rico
44 L9 **Grande de Matagalpa, Río** ≈ C Nicaragua
42 K12 **Grande de Santiago, Río** var. Santiago. ≈ C Mexico
45 O15 **Grande de Térraba, Río** var. Río Térraba. ≈ SE Costa Rica
10 J9 **Grande Deux, Réservoir la** ⊠ Quebec, E Canada
62 O10 **Grande, Ilha** island SE Brazil
9 O13 **Grande Prairie** Alberta, W Canada 55.10N 118.52W
76 I8 **Grand Erg Occidental** desert W Algeria
76 L9 **Grand Erg Oriental** desert Algeria/Tunisia
61 J20 **Grande, Rio** ≈ S Brazil
(0) F15 **Grande, Rio** var. Río Bravo, Sp. Río Bravo del Norte, Bravo del Norte. ≈ Mexico/USA
59 M18 **Grande, Río** ≈ C Bolivia
13 Y7 **Grande-Rivière** Quebec, SE Canada 48.27N 64.37W
13 Y6 **Grande Rivière** ≈ Quebec, SE Canada
46 M8 **Grande-Rivière-du-Nord** N Haiti 19.28N 72.07W
64 K9 **Grande, Salina** var. Gran Salitral. salt lake C Argentina
13 S7 **Grandes-Bergeronnes** Quebec, SE Canada 48.16N 69.32W
49 W6 **Grande, Serra** ▲▲ W Brazil
42 K4 **Grande, Sierra** ▲▲ N Mexico
105 S12 **Grandes Rousses** ▲▲ E France
65 K17 **Grandes, Salinas** salt lake E Argentina
47 Y5 **Grande Terre** island E West Indies
13 X5 **Grande-Vallée** Quebec, SE Canada 49.12N 65.08W
47 Y5 **Grande Vigie, Pointe de la** headland Grande Terre, N Guadeloupe 16.31N 61.27W
11 N14 **Grand Falls** New Brunswick, SE Canada 47.01N 67.46W
11 T11 **Grand Falls** Newfoundland, SE Canada 48.57N 55.48W
44 L9 **Grandfalls** Texas, SW USA 31.20N 102.51W
23 P9 **Grandfather Mountain** ▲ North Carolina, SE USA 36.06N 81.48W
28 L13 **Grandfield** Oklahoma, C USA 34.15N 98.40W
9 N17 **Grand Forks** British Columbia, SW Canada 49.01N 118.30W
31 R4 **Grand Forks** North Dakota, N USA 47.54N 97.02W
33 O9 **Grand Haven** Michigan, N USA 43.03N 86.13W
Grandichi see Hrandzichy
31 P15 **Grand Island** Nebraska, C USA 40.55N 98.20W
33 O3 **Grand Island** island Michigan, N USA
24 K10 **Grand Isle** Louisiana, S USA 29.12N 90.00W
67 A23 **Grand Jason** island Jason Islands, NW Falkland Islands
39 P5 **Grand Junction** Colorado, C USA 39.03N 108.33W
22 F10 **Grand Junction** Tennessee, S USA 35.03N 89.11W
12 J9 **Grand-Lac-Victoria** Quebec, SE Canada 47.33N 77.28W
12 J9 **Grand lac Victoria** ◎ Quebec, SE Canada
79 N17 **Grand-Lahou** var. Grand Lahu. S Ivory Coast 5.09N 5.01W
Grand Lahu see Grand-Lahou
39 S3 **Grand Lake** Colorado, C USA 40.15N 105.49W
11 S11 **Grand Lake** ◎ Newfoundland, E Canada
24 G9 **Grand Lake** ◎ Louisiana, S USA
33 R5 **Grand Lake** ◎ Michigan, N USA
33 Q13 **Grand Lake** ⊠ Ohio, N USA
29 R9 **Grand Lake O' The Cherokees** var. Lake O' The Cherokees. ⊠ Oklahoma, C USA
33 Q9 **Grand Ledge** Michigan, N USA 42.45N 84.45W
104 I8 **Grand-Lieu, Lac de** ◎ NW France
21 U6 **Grand Manan Channel** channel Canada/USA
11 O15 **Grand Manan Island** island New Brunswick, SE Canada
31 Y4 **Grand Marais** Minnesota, N USA 47.45N 90.19W
13 P10 **Grand-Mère** Quebec, SE Canada 46.36N 72.41W
39 P5 **Grand Mesa** ▲▲ Colorado, C USA
110 C10 **Grand Muveran** ▲ W Switzerland 46.16N 7.12E
106 G12 **Grândola** Setúbal, S Portugal 38.10N 8.34W
Grand Paradis see Gran Paradiso
197 G4 **Grand Passage** passage N New Caledonia
79 R16 **Grand-Popo** S Benin 6.19N 1.49E
31 Z3 **Grand Portage** Minnesota, N USA 48.00N 89.36W
45 T6 **Grand Prairie** Texas, SW USA 32.45N 97.00W
9 W14 **Grand Rapids** Manitoba, C Canada 53.12N 99.19W
33 P9 **Grand Rapids** Michigan, N USA 42.57N 86.40W
31 V5 **Grand Rapids** Minnesota, N USA 47.14N 93.31W
197 G5 **Grand Récif de Koumac** reef W New Caledonia
197 J8 **Grand Récif Sud** reef S New Caledonia
12 L10 **Grand-Remous** Quebec, SE Canada 46.36N 75.53W
12 F15 **Grand River** ≈ Ontario, S Canada
33 P9 **Grand River** ≈ Michigan, N USA
29 T3 **Grand River** ≈ Missouri, C USA
30 M7 **Grand River** ≈ South Dakota, N USA
47 Q11 **Grand' Rivière** N Martinique 14.51N 61.12W
34 F11 **Grand Ronde** Oregon, NW USA 45.03N 123.43W
34 L11 **Grand Ronde River** ≈ Oregon/Washington, NW USA
Grand-Saint-Bernard, Col du see Great Saint Bernard Pass
45 V6 **Grand Saline** Texas, SW USA 32.40N 95.42W
57 X10 **Grand-Santi** W French Guiana 4.19N 54.24W
Grandsee see Grandson
110 B9 **Grandson** prev. Grandsee. Vaud, W Switzerland 46.49N 6.39E
180 J16 **Grand Sœur** island Les Sœurs, NE Seychelles
35 S14 **Grand Teton** ▲ Wyoming, C USA 43.44N 110.48W
33 P5 **Grand Traverse Bay** lake bay Michigan, N USA
47 N6 **Grand Turk** ○ (Turks and Caicos Islands) Grand Turk Island, S Turks and Caicos Islands 21.24N 71.08W
47 N6 **Grand Turk Island** island SE Turks and Caicos Islands
105 S13 **Grand Veymont** ▲ E France 44.51N 5.32E
9 W15 **Grandview** Manitoba, S Canada 51.10N 100.40W
29 R4 **Grandview** Missouri, C USA 38.53N 94.31W
38 I10 **Grand Wash Cliffs** cliff Arizona, SW USA
12 J8 **Granet, Lac** ◎ Quebec, SE Canada
97 L14 **Grangärde** Dalarna, C Sweden 60.15N 15.00E
46 H12 **Grange Hill** W Jamaica 18.18N 78.10W
98 J12 **Grangemouth** C Scotland, UK 56.01N 3.43W
45 T10 **Granger** Texas, SW USA 30.43N 97.26W
34 J10 **Granger** Washington, NW USA 46.20N 120.11W
35 T17 **Granger** Wyoming, C USA 41.37N 109.58W
Granges see Grenchen
97 L14 **Grängesberg** Dalarna, C Sweden 60.06N 15.00E
35 N11 **Grangeville** Idaho, NW USA 45.55N 116.07W
8 K13 **Granisle** British Columbia, SW Canada 54.55N 126.13W
32 K15 **Granite City** Illinois, N USA 38.42N 90.09W
31 S9 **Granite Falls** Minnesota, N USA 44.48N 95.33W
23 Q9 **Granite Falls** North Carolina, SE USA 35.48N 81.25W
38 K12 **Granite Mountain** ▲ Arizona, SW USA 34.38N 112.34W
35 T12 **Granite Peak** ▲ Montana, NW USA 45.09N 109.48W
37 T2 **Granite Peak** ▲ Nevada, W USA 41.40N 117.35W
38 J3 **Granite Peak** ▲ Utah, W USA 40.09N 113.18W
Granite State see New Hampshire
109 H24 **Granitola, Capo** headland Sicilia, Italy, C Mediterranean Sea 37.33N 12.39E
193 H15 **Granity** West Coast, South Island, NZ 41.37S 171.53E
Gran Lago see Nicaragua, Lago de
65 J18 **Gran Laguna Salada** ◎ S Argentina
Gran Malvina, Isla see West Falkland
97 L18 **Gränna** Jönköping, S Sweden 58.01N 14.30E
107 W5 **Granollers** var. Granollérs. Cataluña, NE Spain 41.37N 2.18E
108 A7 **Gran Paradiso** Fr. Grand Paradis. ▲ NW Italy 45.31N 7.13E
Gran Pilastro see Hochfeiler
Gran Salitral see Grande, Salina
Gran San Bernardo, Passo di see Great Saint Bernard Pass
Gran Santiago see Santiago
109 J14 **Gran Sasso d'Italia** ▲▲ C Italy
102 N11 **Gransee** Brandenburg, NE Germany 53.00N 13.10E
30 L15 **Grant** Nebraska, C USA 40.50N 101.43W
29 R1 **Grant City** Missouri, C USA 40.29N 94.24W
99 N19 **Grantham** E England, UK 52.55N 0.39W
67 D24 **Grantham Sound** sound East Falkland, Falkland Islands
204 K13 **Grant Island** island Antarctica
47 Z14 **Grantley Adams** ✕ (Bridgetown) SE Barbados 13.04N 59.29W
37 S7 **Grant, Mount** ▲ Nevada, W USA 38.34N 118.47W
98 J9 **Grantown-on-Spey** N Scotland, UK 57.11N 3.52W
37 W8 **Grant Range** ▲▲ Nevada, W USA
39 Q11 **Grants** New Mexico, SW USA 35.09N 107.50W
32 I4 **Grantsburg** Wisconsin, N USA 45.46N 92.40W
34 F15 **Grants Pass** Oregon, NW USA 42.26N 123.19W
38 K3 **Grantsville** Utah, W USA 40.36N 112.27W
23 R4 **Grantsville** West Virginia, NE USA 38.54N 81.04W
104 I5 **Granville** Manche, N France 48.49N 1.34W
9 V12 **Granville Lake** ◎ Manitoba, C Canada
45 V8 **Grapeland** Texas, SW USA 31.29N 95.28W
45 T6 **Grapevine** Texas, SW USA 32.55N 97.04W
85 K20 **Graskop** Mpumalanga, NE South Africa 24.58S 30.49E
97 P14 **Gräsö** Uppsala, C Sweden 60.22N 18.30E
95 I19 **Gräsö** island C Sweden
105 U15 **Grasse** Alpes-Maritimes, SE France 43.42N 6.52E
20 E14 **Grassflat** Pennsylvania, NE USA 41.00N 78.04W
35 U9 **Grassrange** Montana, NW USA 47.02N 108.48W
20 J6 **Grass River** ≈ New York, NE USA
37 P6 **Grass Valley** California, W USA 39.12N 121.04W
191 N14 **Grassy** Tasmania, SE Australia 40.03S 144.04E
30 K4 **Grassy Butte** North Dakota, N USA 47.20N 103.13W
23 R5 **Grassy Knob** ▲ West Virginia, NE USA 38.04N 80.31W
97 G24 **Gråsten** var. Graasten. Sønderjylland, SW Denmark 54.55N 9.37E
97 J18 **Grästorp** Västra Götaland, S Sweden 58.19N 12.45E
Gratianopolis see Grenoble
111 V8 **Gratwein** Steiermark, SE Austria 47.06N 15.18E
Gratz see Graz
110 I9 **Graubünden** Fr. Grisons, It. Grigioni. ◐ canton SE Switzerland
Graudenz see Grudziądz
105 N15 **Graulhet** Tarn, S France 43.45N 1.58E
107 T4 **Graus** Aragón, NE Spain 42.11N 0.21E
63 I16 **Gravataí** Rio Grande do Sul, S Brazil 29.55S 51.00W
100 L13 **Grave** Noord-Brabant, SE Netherlands 51.45N 5.45E
9 T17 **Gravelbourg** Saskatchewan, S Canada 49.52N 106.33W
105 N1 **Gravelines** Nord, N France 51.00N 2.07E
Graven see Grez-Doiceau
12 H13 **Gravenhurst** Ontario, S Canada 44.55N 79.22W
35 O10 **Grave Peak** ▲ Idaho, NW USA 46.24N 114.43W
104 I11 **Grave, Pointe de** headland W France 45.33N 1.24W
191 S4 **Gravesend** New South Wales, SE Australia 29.37S 150.15E
99 P22 **Gravesend** SE England, UK 51.27N 0.24E
109 N17 **Gravina di Puglia** Eng. Gravina in Puglia. Puglia, SE Italy 40.48N 16.25E
Gravina in Puglia see Gravina di Puglia
105 S8 **Gray** Haute-Saône, E France 47.28N 5.34E
25 T4 **Gray** Georgia, SE USA 33.00N 83.31W
205 V16 **Gray, Cape** headland Antarctica 67.30S 143.30E
34 F9 **Grayland** Washington, NW USA 46.46N 124.07W
41 N10 **Grayling** Alaska, USA 62.55N 160.07W
33 Q6 **Grayling** Michigan, N USA 44.40N 84.43W
34 F9 **Grays Harbor** inlet Washington, NW USA
23 O5 **Grayson** Kentucky, S USA 38.19N 82.57W
39 S4 **Grays Peak** ▲ Colorado, C USA 39.37N 105.49W
32 M16 **Grayville** Illinois, N USA 38.15N 87.59W
111 V8 **Graz** prev. Gratz. Steiermark, SE Austria 47.04N 15.23E
106 L15 **Grazalema** Andalucía, S Spain 36.46N 5.22W
115 P15 **Grdelica** Serbia, SE Yugoslavia 42.54N 22.05E
46 H1 **Great Abaco** var. Abaco Island. island N Bahamas
Great Admiralty Island see Manus Island
Great Alfold see Great Hungarian Plain
Great Ararat see Büyükağrı Dağı
189 U8 **Great Artesian Basin** lowlands Queensland, C Australia
197 I15 **Great Astrolabe Reef** reef Kadavu, SW Fiji
189 O12 **Great Australian Bight** bight S Australia
66 E11 **Great Bahama Bank** undersea feature E Gulf of Mexico
192 M4 **Great Barrier Island** island N NZ
189 X4 **Great Barrier Reef** reef Queensland, NE Australia
20 L11 **Great Barrington** Massachusetts, NE USA 42.11N 73.20W
(0) F10 **Great Basin** basin W USA
15 H5 **Great Bear Lake** Fr. Grand Lac de l'Ours. ◎ Northwest Territories, NW Canada
Great Belt see Storebælt
28 L5 **Great Bend** Kansas, C USA 38.21N 98.45W
Great Bermuda see Bermuda
99 A20 **Great Blasket Island** Ir. An Blascaod Mór. island SW Ireland
Great Britain see Britain
157 Q23 **Great Channel** channel Andaman Sea/Indian Ocean
177 F10 **Great Coco Island** island SW Myanmar
Great Crosby see Crosby
23 X7 **Great Dismal Swamp** wetland North Carolina/Virginia, SE USA
35 V16 **Great Divide Basin** basin Wyoming, C USA
189 W7 **Great Dividing Range** ▲▲ NE Australia
12 D12 **Great Duck Island** island Ontario, S Canada
Great Elder Reservoir see Waconda Lake
205 V8 **Greater Antarctica** var. East Antarctica. physical region Antarctica
46 G8 **Greater Antilles** island group West Indies
133 V16 **Greater Sunda Islands** var. Sunda Islands. island group Indonesia
192 I1 **Great Exhibition Bay** inlet North Island, NZ
46 H4 **Great Exuma Island** island C Bahamas
35 R8 **Great Falls** Montana, NW USA 47.30N 111.18W
23 R11 **Great Falls** South Carolina, SE USA 34.34N 80.54W
86 F9 **Great Fisher Bank** undersea feature C North Sea
Great Glen see Mor, Glen
Great Grimsby see Grimsby
46 I4 **Great Guana Cay** island C Bahamas
66 I5 **Great Hellefiske Bank** undersea feature N Atlantic Ocean
113 L24 **Great Hungarian Plain** var. Great Alfold, Plain of Hungary, Hung. Alföld. plain SE Europe
46 L7 **Great Inagua** var. Inagua Islands. island S Bahamas
Great Indian Desert see Thar Desert
85 G25 **Great Karoo** var. Great Karroo, High Veld, Afr. Groot Karoo, Hoë Karoo. plateau region S South Africa
Great Karroo see Great Karoo
Great Kei see Groot-Kei
Great Khingan Range see Da Hinggan Ling
12 E11 **Great La Cloche Island** island Ontario, S Canada
191 P16 **Great Lake** ◎ Tasmania, SE Australia
Great Lake see Tônlé Sap
16 O16 **Great Lakes** lakes Ontario, Canada/USA
Great Lakes State see Michigan
99 L20 **Great Malvern** W England, UK 52.07N 2.19W
192 M5 **Great Mercury Island** island N NZ
Great Meteor Seamount see Great Meteor Tablemount
66 K10 **Great Meteor Tablemount** var. Great Meteor Seamount. undersea feature E Atlantic Ocean 30.00N 28.30W
33 Q14 **Great Miami River** ≈ Ohio, N USA
157 Q24 **Great Nicobar** island Nicobar Islands, India, NE Indian Ocean
99 O19 **Great Ouse** var. Ouse. ≈ E England, UK
191 Q17 **Great Oyster Bay** bay Tasmania, SE Australia
46 I13 **Great Pedro Bluff** headland W Jamaica 17.51N 77.44W
23 T12 **Great Pee Dee River** ≈ North Carolina/South Carolina, SE USA
133 W9 **Great Plain of China** plain E China
(0) F12 **Great Plains** var. High Plains. plains Canada/USA
39 W6 **Great Plains Reservoirs** ⊠ Colorado, C USA
21 Q13 **Great Point** headland Nantucket Island, Massachusetts, NE USA 41.23N 70.03W
70 I13 **Great Rift Valley** var. Rift Valley. depression Asia/Africa
83 I23 **Great Ruaha** ≈ S Tanzania
20 K10 **Great Sacandaga Lake** ⊠ New York, NE USA
110 C12 **Great Saint Bernard Pass** Fr. Col du Grand-Saint-Bernard, It. Passo di Gran San Bernardo. pass Italy/Switzerland 45.52N 7.10E
46 F1 **Great Sale Cay** island N Bahamas
Great Salt Desert see Kavīr, Dasht-e
38 K1 **Great Salt Lake** salt lake Utah, W USA
38 J3 **Great Salt Lake Desert** plain Utah, W USA
28 M8 **Great Salt Plains Lake** ⊠ Oklahoma, C USA
77 T9 **Great Sand Sea** desert Egypt/Libya
188 L6 **Great Sandy Desert** desert Western Australia
Great Sandy Desert see Ar Rub' al Khālī
Great Sandy Island see Fraser Island
197 I13 **Great Sea Reef** reef Vanua Levu, N Fiji
40 H17 **Great Sitkin Island** island Aleutian Islands, Alaska, USA
15 I8 **Great Slave Lake** Fr. Grand Lac des Esclaves. ◎ Northwest Territories, NW Canada
23 O10 **Great Smoky Mountains** ▲▲ North Carolina/Tennessee, SE USA
8 L11 **Great Snow Mountain** ▲ British Columbia, W Canada 57.22N 124.08W
66 A12 **Great Sound** bay Bermuda, NW Atlantic Ocean
188 M10 **Great Victoria Desert** desert South Australia/Western Australia
204 H2 **Great Wall** Chinese research station South Shetland Islands, Antarctica 61.57S 58.23W
21 T7 **Great Wass Island** island Maine, NE USA
99 Q19 **Great Yarmouth** var. Yarmouth. E England, UK 52.37N 1.43E
145 S1 **Great Zab** Ar. Az Zāb al Kabīr, Kurd. Zē-i Bādīnān, Turk. Büyükzap Suyu. ≈ Iraq/Turkey
97 I17 **Grebbestad** Västra Götaland, S Sweden 58.42N 11.15E
Grebenka see Hrebinka
44 M13 **Grecia** Alajuela, C Costa Rica 10.04N 84.19W
63 E18 **Greco** Río Negro, W Uruguay 32.49S 57.03W
Greco, Cape see Gkréko, Akrotíri
106 L8 **Gredos, Sierra de** ▲▲ W Spain
20 F9 **Greece** New York, NE USA 43.12N 77.41W
117 E17 **Greece** off. Hellenic Republic, Gk. Ellás; anc. Hellas. ◆ republic SE Europe
Greece Central see Stereá Ellás
Greece West see Dytikí Ellás
39 T3 **Greeley** Colorado, C USA 40.21N 104.41W
31 P14 **Greeley** Nebraska, C USA 41.33N 98.31W
126 Hh1 **Greem-Bell, Ostrov** Eng. Graham Bell Island. island Zemlya Frantsa-Iosifa, N Russian Federation
32 M6 **Green Bay** Wisconsin, N USA 44.32N 88.00W
33 N6 **Green Bay** lake bay Michigan/Wisconsin, N USA
23 S5 **Greenbrier River** ≈ West Virginia, NE USA
31 S2 **Greenbush** Minnesota, N USA 48.42N 96.10W
191 R12 **Green Cape** headland New South Wales, SE Australia 37.15S 150.03E
33 O14 **Greencastle** Indiana, N USA 39.38N 86.51W
20 F16 **Greencastle** Pennsylvania, NE USA 39.47N 77.43W
29 T2 **Green City** Missouri, C USA 40.16N 92.57W
23 O9 **Greeneville** Tennessee, S USA 36.09N 82.49W
37 O11 **Greenfield** California, W USA 36.19N 121.15W
33 P14 **Greenfield** Indiana, N USA 39.47N 85.46W
31 U15 **Greenfield** Iowa, C USA 41.18N 94.27W
20 M11 **Greenfield** Massachusetts, NE USA 42.33N 72.34W
29 S7 **Greenfield** Missouri, C USA 37.25N 93.50W
33 S14 **Greenfield** Ohio, N USA 39.21N 83.22W
22 G8 **Greenfield** Tennessee, S USA 36.09N 88.48W
32 M9 **Greenfield** Wisconsin, N USA 42.55N 87.59W
29 T9 **Green Forest** Arkansas, C USA 36.19N 93.24W
39 T7 **Greenhorn Mountain** ▲ Colorado, C USA 37.50N 104.59W
Green Island see Lü Tao
195 R10 **Green Islands** var. Nissan Islands. island group NE PNG
9 S14 **Green Lake** Saskatchewan, C Canada 54.15N 107.51W
32 L8 **Green Lake** ◎ Wisconsin, N USA
207 O14 **Greenland** Dan. Grønland, Inuit Kalaallit Nunaat. ◇ Danish external territory NE North America
86 D4 **Greenland** island NE North America
207 R13 **Greenland Plain** undersea feature N Greenland Sea
207 R14 **Greenland Sea** sea Arctic Ocean
39 R4 **Green Mountain Reservoir** ⊠ Colorado, C USA
20 M8 **Green Mountains** ▲▲ Vermont, NE USA
Green Mountain State see Vermont
98 I12 **Greenock** W Scotland, UK 55.57N 4.45W
41 T5 **Greenough, Mount** ▲ Alaska, USA 69.15N 141.37W
194 E10 **Green River** Sandaun, NW PNG 3.46S 141.10E
39 N5 **Green River** Utah, W USA 39.00N 110.07W
35 U17 **Green River** Wyoming, C USA 41.33N 109.27W
28 I7 **Green River** ≈ W USA
32 K11 **Green River** ≈ Illinois, N USA
22 J7 **Green River** ≈ Kentucky, S USA
30 K5 **Green River** ≈ North Dakota, N USA
39 N6 **Green River** ≈ Utah, W USA
35 T16 **Green River** ≈ Wyoming, C USA
22 L7 **Green River Lake** ⊠ Kentucky, S USA
25 O5 **Greensboro** Alabama, S USA 32.42N 87.36W
25 U3 **Greensboro** Georgia, SE USA 33.34N 83.10W
23 T9 **Greensboro** North Carolina, SE USA 36.04N 79.47W
33 P14 **Greensburg** Indiana, N USA 39.20N 85.28W
28 K6 **Greensburg** Kansas, C USA 37.36N 99.17W
22 L7 **Greensburg** Kentucky, S USA 37.15N 85.28W
20 C15 **Greensburg** Pennsylvania, NE USA 40.18N 79.32W
39 O13 **Greens Peak** ▲ Arizona, SW USA 34.06N 109.34W
23 V12 **Green Swamp** wetland North Carolina, SE USA
23 O4 **Greenup** Kentucky, S USA 38.34N 82.49W
38 M16 **Green Valley** Arizona, SW USA 31.49N 111.00W
78 K17 **Greenville** var. Sino, Sinoe. SE Liberia 5.01N 9.03W
25 P6 **Greenville** Alabama, S USA 31.49N 86.37W
25 T8 **Greenville** Florida, SE USA 30.28N 83.37W
25 S4 **Greenville** Georgia, SE USA 33.03N 84.42W
32 L15 **Greenville** Illinois, N USA 38.53N 89.24W
22 I7 **Greenville** Kentucky, S USA 37.12N 87.10W
21 Q5 **Greenville** Maine, NE USA 45.26N 69.36W
33 P9 **Greenville** Michigan, N USA 43.10N 85.15W
24 J4 **Greenville** Mississippi, S USA 33.24N 91.03W
23 W9 **Greenville** North Carolina, SE USA 35.36N 77.22W
33 Q13 **Greenville** Ohio, N USA 40.06N 84.37W
21 O12 **Greenville** Rhode Island, NE USA 41.52N 71.33W
23 P11 **Greenville** South Carolina, SE USA 34.51N 82.23W
45 U6 **Greenville** Texas, SW USA 33.09N 96.06W
33 T12 **Greenwich** Ohio, N USA 41.01N 82.31W
29 S11 **Greenwood** Arkansas, C USA 35.13N 94.15W
33 O14 **Greenwood** Indiana, N USA 39.38N 86.06W
24 K4 **Greenwood** Mississippi, S USA 33.30N 90.11W
23 P12 **Greenwood** South Carolina, SE USA 34.12N 82.09W
23 Q12 **Greenwood, Lake** ⊠ South Carolina, SE USA
23 P11 **Greer** South Carolina, SE USA 34.56N 82.13W
29 V10 **Greers Ferry Lake** ⊠ Arkansas, C USA
29 S13 **Greeson, Lake** ⊠ Arkansas, C USA
31 O12 **Gregory** South Dakota, N USA 43.11N 99.26W
190 J3 **Gregory, Lake** salt lake South Australia
188 J9 **Gregory Lake** ◎ Western Australia
189 V5 **Gregory Range** ▲▲ Queensland, E Australia
Greifenberg/Greifenberg in Pommern see Gryfice
Greifenhagen see Gryfino
102 O8 **Greifswald** Mecklenburg-Vorpommern, NE Germany 54.04N 13.23E
102 O8 **Greifswalder Bodden** bay NE Germany
111 U4 **Grein** Oberösterreich, N Austria 48.14N 14.50E
103 M17 **Greiz** Thüringen, C Germany 50.40N 12.10E
Gremicha/Gremiha see Gremikha
128 M4 **Gremikha** var. Gremicha, Gremiha. Murmanskaya Oblast', NW Russian Federation 68.01N 39.31E
129 V14 **Gremyachinsk** Permskaya Oblast', NW Russian Federation 58.33N 57.52E
97 H21 **Grenå** var. Grenaa. Århus, C Denmark 56.25N 10.52E
Grenaa see Grenå
24 L3 **Grenada** Mississippi, S USA 33.46N 89.48W
47 W15 **Grenada** ◆ commonwealth republic SE West Indies
49 S4 **Grenada** island Grenada
49 R4 **Grenada Basin** undersea feature W Atlantic Ocean
24 L3 **Grenada Lake** ⊠ Mississippi, S USA
47 Y14 **Grenadines, The** island group Grenada/St Vincent and the Grenadines
110 D7 **Grenchen** Fr. Granges. Solothurn, NW Switzerland 47.12N 7.30E

◆ Country ● Country Capital ◇ Dependent Territory ○ Dependent Territory Capital ◐ Administrative Region ✕ International Airport ▲ Mountain ▲▲ Mountain Range Ж Volcano ≈ River ◎ Lake ⊠ Reservoir

191 Q9 **Grenfell** New South Wales, SE Australia 33.54S 148.09E

9 V16 **Grenfell** Saskatchewan, S Canada 50.24N 102.55W

94 J7 **Grenivík** Nordhurland Eystra, N Iceland 65.57N 18.10W

105 S12 **Grenoble** anc. Cularo, Gratianopolis. Isère, E France 45.10N 5.41E

30 J2 **Grenora** North Dakota, N USA 48.36N 103.57W

94 N8 **Grense-Jakobselv** Finnmark, N Norway 69.46N 30.39E

47 S14 **Grenville** E Grenada 12.07N 61.37W

34 G11 **Gresham** Oregon, NW USA 45.30N 122.25W

Gresk see Hresk

108 B7 **Gressoney-St-Jean** Valle d'Aosta, NW Italy 45.48N 7.49E

24 K9 **Gretna** Louisiana, S USA 29.54N 90.03W

23 T7 **Gretna** Virginia, NE USA 36.57N 79.21W

100 F13 **Grevelingen** inlet S North Sea

102 F13 **Greven** Nordrhein-Westfalen, NW Germany 52.06N 7.37E

117 D15 **Grevena** Dytikí Makedonía, N Greece 40.04N 21.26E

103 D16 **Grevenbroich** Nordrhein-Westfalen, W Germany 51.06N 6.34E

101 N24 **Grevenmacher** Grevenmacher, E Luxembourg 49.40N 6.27E

101 M24 **Grevenmacher** ◆ district E Luxembourg

102 K9 **Grevesmühlen** Mecklenburg-Vorpommern, N Germany 53.52N 11.12E

193 H16 **Grey ◆** South Island, NZ

35 V12 **Greybull** Wyoming, C USA 44.29N 108.03W

35 U13 **Greybull River ◆** Wyoming, C USA

67 A24 **Grey Channel** sound Falkland Islands

Greyerzer See see Gruyère, Lac de la

11 T10 **Grey Islands** island group Newfoundland, E Canada

20 L10 **Greylock, Mount ▲** Massachusetts, NE USA 42.38N 73.09W

193 G17 **Greymouth** West Coast, South Island, NZ 42.28S 171.13E

189 U10 **Grey Range ▲** New South Wales/Queensland, E Australia

99 G18 **Greystones** Ir. Na Clocha Liatha. E Ireland 53.07N 6.04W

193 M14 **Greytown** Wellington, North Island, NZ 41.05S 175.25E

85 K23 **Greytown** KwaZulu/Natal, E South Africa 29.04S 30.34E

Greytown see San Juan del Norte

101 N19 **Grez-Doiceau** Dut. Graven. Wallon Brabant, C Belgium 50.43N 4.41E

117 J19 **Griá, Akrotírio** headland Andros, Kykládes, Greece, Aegean Sea 37.54N 24.57E

131 N8 **Gribanovskiy** Voronezhskaya Oblast', W Russian Federation 51.27N 41.53E

80 I13 **Gribingui ◆** N Central African Republic

37 O6 **Gridley** California, W USA 39.21N 121.41W

85 G23 **Griekwastad** Northern Cape, C South Africa 28.50S 23.16E

25 S4 **Griffin** Georgia, SE USA 33.15N 84.16W

191 O9 **Griffith** New South Wales, SE Australia 34.16S 146.01E

12 F13 **Griffith Island** island Ontario, S Canada

23 W10 **Grifton** North Carolina, SE USA 35.22N 77.26W

Grigioni see Graubünden

121 H14 **Grigiškes** Trakai, SE Lithuania 54.42N 25.00E

119 N10 **Grigoriopol** C Moldova 47.09N 29.18E

153 X7 **Grigor'yevka** Issyk-Kul'skaya Oblast', E Kyrgyzstan 42.43N 77.27E

200 Oo9 **Grijalva Ridge** undersea feature E Pacific Ocean

43 U15 **Grijalva, Río** var. Tabasco. ◆ Guatemala/Mexico

100 N5 **Grijpskerk** Groningen, NE Netherlands 53.15N 6.18E

85 C22 **Grillenthal** Karas, SW Namibia 26.75S 15.24E

81 J15 **Grimari** Ouaka, C Central African Republic 5.44N 20.02E

Grimaylov see Hrymayliv

101 G18 **Grimbergen** Vlaams Brabant, C Belgium 50.55N 4.22E

191 N15 **Grim, Cape** headland Tasmania, SE Australia 40.42S 144.42E

102 N8 **Grimmen** Mecklenburg-Vorpommern, NE Germany 54.06N 13.03E

12 G16 **Grimsby** Ontario, S Canada 43.10N 79.34W

99 O17 **Grimsby** prev. Great Grimsby. E England, UK 53.34N 0.04W

94 J1 **Grímsey** var. Grimsey. island N Iceland

9 O12 **Grimshaw** Alberta, W Canada 56.12N 117.37W

97 F18 **Grimstad** Aust-Agder, S Norway 58.19N 8.34E

94 H4 **Grindavík** Reykjanes, W Iceland 63.57N 18.10W

110 F9 **Grindelwald** Bern, S Switzerland

97 F23 **Grindsted** Ribe, W Denmark 55.46N 8.55E

31 W4 **Grinnell** Iowa, C USA 41.44N 92.43W

111 U10 **Grintavec ▲** N Slovenia 46.21N 14.31E

190 J11 **Griselda, Lake** salt lake South Australia

Grisons see Graubünden

97 P14 **Grisslehamn** Stockholm, C Sweden 60.04N 18.49E

31 Y5 **Griswold** Iowa, C USA 41.14N 95.08W

104 M3 **Griz Nez, Cap** headland N France 50.51N 1.34E

114 P13 **Grljan** Serbia, E Yugoslavia 43.52N 22.18E

114 L11 **Grmeč ▲** NW Bosnia and Herzegovina

101 H16 **Grobbendonk** Antwerpen, N Belgium 51.12N 4.41E

120 C10 **Grobiņa** Ger. Grobin. Liepāja, W Latvia 56.32N 21.12E

85 K20 **Groblersdal** Mpumalanga, NE South Africa 25.15S 29.25E

85 G23 **Groblershoop** Northern Cape, W South Africa 28.51S 22.01E

111 V3 **Gross-Siegharts** Niederösterreich, N Austria 48.48N 15.25E

Gross-Skaisgirren see Bol'shakovo

113 H18 **Grodków** Opolskie, S Poland 50.42N 17.23E

Grodnenskaya Oblast' see Hrodzyenskaya Voblasts'

Grodno see Hrodna

112 L12 **Grodzisk Mazowiecki** Mazowieckie, C Poland 52.07N 20.40E

112 F12 **Grodzisk Wielkopolski** Wielkopolskie, C Poland 52.13N 16.21E

100 O12 **Groenlo** Gelderland, E Netherlands 52.01N 6.36E

85 E22 **Groenrivier** Karas, SE Namibia 27.27S 18.52E

45 U8 **Groesbeck** Texas, SW USA 31.31N 96.34W

100 L13 **Groesbeek** Gelderland, SE Netherlands 51.46N 5.55E

104 G7 **Groix, Îles de** island group NW France

112 M12 **Grójec** Mazowieckie, C Poland 51.51N 20.52E

67 A24 **Gröll Seamount** undersea feature C Atlantic Ocean 12.54S 33.24W

102 E12 **Gronau** var. Gronau in Westfalen. Nordrhein-Westfalen, NW Germany 52.12N 7.01E

Gronau in Westfalen see Gronau

95 F15 **Grong** Nord-Trøndelag, C Norway 64.29N 12.19E

95 N22 **Grönhögen** Kalmar, S Sweden 56.16N 16.09E

100 N5 **Groningen** Groningen, NE Netherlands 53.13N 6.34E

57 W9 **Groningen** Saramacca, N Suriname 5.45N 55.31W

100 N5 **Groningen ◆** province NE Netherlands

110 H11 **Grono** Graubünden, S Switzerland 46.15N 9.07E

97 M20 **Grönskåra** Kalmar, S Sweden 57.04N 15.45E

45 U2 **Groom** Texas, SW USA 35.12N 101.06W

37 W9 **Groom Lake** ◎ Nevada, W USA

85 H25 **Groot ◆** S South Africa

189 S2 **Groote Eylandt** island Northern Territory, N Australia

100 M6 **Grootegast** Groningen, NE Netherlands 53.11N 6.12E

85 D17 **Grootfontein** Otjozondjupa, N Namibia 19.31S 18.04E

85 E22 **Groot Karasberg ▲** S Namibia

85 J25 **Groot Karoo** var. Great Karoo. S South Africa

Groot-Kei Eng. Great Kei. ◆ S South Africa

47 T10 **Gros Islet** N Saint Lucia 14.04N 60.57W

46 L8 **Gros-Morne** NW Haiti 19.37N 72.39W

11 S11 **Gros Morne ▲** Newfoundland, E Canada 49.38N 57.45W

105 R9 **Grosne ◆** C France

47 S12 **Gros Piton ▲** SW Saint Lucia 13.48N 61.04W

Grossa, Isola see Dugi Otok

Grossbetschkerek see Zrenjanin

Gross Isper see Grosse Ysper

Grosse Kokel see Târnava Mare

103 M21 **Grosse Laaber** var. Grosse Laber. ◆ SE Germany

Grosse Laber see Grosse Laaber

Grosse Morava see Velika Morava

103 O15 **Grossenhain** Sachsen, E Germany 51.18N 13.31E

111 Y4 **Grossenzersdorf** Niederösterreich, NE Austria 48.11N 16.34E

103 O21 **Grosser Arber ▲** SE Germany 49.07N 13.10E

103 K23 **Grosser Beerberg ▲** C Germany 49.09N 10.45E

103 G18 **Grosser Feldberg ▲** W Germany 50.13N 8.28E

111 U8 **Grosser Löffler** It. Monte Lovello. ▲ Austria/Italy 47.02N 11.56E

111 N8 **Grosser Möseler** var. Mesule. ▲ Austria/Italy 47.01N 11.52E

102 J8 **Grosser Plöner See** ◎ N Germany

103 O21 **Grosser Rachel ▲** SE Germany 48.59N 13.23E

13 V6 **Grosses-Roches** Quebec, SE Canada 48.55N 67.06W

111 P8 **Grosses Weiesbachhorn** var. Wiesbachhorn. ▲ W Austria 47.09N 12.44E

108 E7 **Grosseto** Toscana, C Italy 42.45N 11.07E

103 N22 **Grosse Vils** ◆ SE Germany

111 U4 **Grosse Ysper** var. Grosse Isper. ◆ N Austria

103 G19 **Gross-Gerau** Hessen, W Germany 49.58N 8.28E

111 U3 **Gross Gerungs** Niederösterreich, N Austria 48.33N 14.58E

112 D9 **Gryfino** Ger. Greifenhagen. Zachodniopomorskie, NW Poland 53.15N 14.30E

111 P8 **Grossglockner ▲** W Austria 47.05N 12.94E

Grosskanizsa see Nagykanizsa

Gross-Karol see Carei

Grosskikinda see Kikinda

111 W4 **Grossklein** Steiermark, SE Austria 46.43N 15.26E

Grosskoppe see Velká Deštná

Grossmeseritsch see Velké Meziříčí

Grossmichel see Michalovce

103 N23 **Grosspetersdorf** Burgenland, SE Austria 47.15N 16.19E

111 T5 **Grossraming** Oberösterreich, C Austria 47.54N 14.00E

103 P14 **Grossräschen** Brandenburg, E Germany 51.34N 14.00E

Grossräschenbach see Revúca

Gross-Sankt-Johannis see Suure-Jaani

Gross-Schlatten see Abrud

Gross-Steffelsdorf see Rimavská Sobota

Gross Strehlitz see Strzelce Opolskie

111 O8 **Grossvenediger ▲** W Austria 47.07N 12.19E

Grosswardein see Oradea

Gross Wartenberg see Syców

117 S3 **Grouplje ◆** S Slovenia 46.00N 14.36E

101 H17 **Grote Nete ◆** N Belgium

96 E10 **Grotli** Oppland, S Norway 62.02N 7.36E

21 N13 **Groton** Connecticut, NE USA 41.20N 72.03W

31 P8 **Groton** South Dakota, N USA 45.27N 98.06W

109 P18 **Grottaglie** Puglia, SE Italy 40.31N 17.25E

109 L17 **Grottaminarda** Campania, S Italy 41.04N 15.02E

108 K13 **Grottammare** Marche, C Italy 43.00N 13.52E

23 U5 **Grottoes** Virginia, NE USA 38.16N 78.49W

Grou see Grouw

11 N10 **Groulx, Monts ▲** Quebec, E Canada

12 C7 **Groundhog ◆** Ontario, S Canada

38 J1 **Grouse Creek** Utah, W USA 41.41N 113.52W

38 J1 **Grouse Creek Mountains ▲** Utah, W USA

100 L6 **Grouw** Fris. Grou. Friesland, N Netherlands 53.07N 5.51E

29 R8 **Grove** Oklahoma, C USA 36.35N 94.46W

33 S13 **Grove City** Ohio, N USA 39.52N 83.05W

20 B13 **Grove City** Pennsylvania, NE USA 41.09N 80.02W

25 O5 **Grove Hill** Alabama, S USA 31.42N 87.46W

35 S15 **Grover** Wyoming, C USA 42.48N 110.57W

37 P13 **Grover City** California, W USA 35.08N 120.37W

45 Y11 **Groves** Texas, SW USA 29.57N 93.55W

21 O7 **Groveton** New Hampshire, NE USA 44.35N 71.28W

45 W8 **Groveton** Texas, SW USA 31.03N 95.07W

38 M6 **Growler Mountains ▲** Arizona, SW USA

Grozdovo see Bratya Daskalovi

131 P16 **Groznyy** Chechenskaya Respublika, SW Russian Federation 43.20N 45.42E

114 G9 **Grubišno Polje** Bjelovar-Bilogora, NE Croatia 45.42N 17.09E

Grudovo see Sredets

112 J7 **Grudziądz** Ger. Graudenz. Kujawsko-pomorskie, C Poland 53.28N 18.45E

45 R17 **Grulla** var. La Grulla. Texas, SW USA 26.15N 98.37W

42 K14 **Grullo** Jalisco, SW Mexico 19.45N 104.15W

69 V10 **Grumeti ◆** N Tanzania

97 K16 **Grums** Värmland, C Sweden 59.22N 13.10E

111 S5 **Grünau im Almtal** Oberösterreich, N Austria 47.51N 13.56E

103 H17 **Grünberg** Hessen, W Germany 50.36N 8.57E

Grünberg/Grünberg in Schlesien see Zielona Góra

Grünberg in Schlesien see Zielona Góra

94 H3 **Grundarfjördhur** Vestfirdhir, W Iceland 64.55N 23.15W

23 P7 **Grundy** Virginia, NE USA 37.16N 82.06W

31 W13 **Grundy Center** Iowa, C USA 42.21N 92.46W

Grüneberg see Zielona Góra

45 W1 **Gruver** Texas, SW USA 36.16N 101.24W

110 C9 **Gruyère, Lac de la** Ger. Greyerzer See. ◎ SW Switzerland

110 C9 **Gruyères** Fribourg, W Switzerland 46.34N 7.04E

120 E11 **Gruzdžiai** Šiauliai, N Lithuania 56.06N 23.15E

Gruzinskaya SSR/Gruziya see Georgia

152 C10 **Gryada Akkyr** Turkm. Akgyr Erezi. hill range NW Turkmenistan

196 B15 **Gryfice** Ger. Greifenberg. Zachodniopomorskie, NW Poland 53.55N 15.30E

112 E8 **Gryfice** Ger. Greifenberg. Greifenberg in Pommern. Zachodniopomorskie, NW Poland 53.55N 15.12E

96 E11 **Gryllefjord** Troms, N Norway 69.21N 17.07E

95 L15 **Grythyttan** Örebro, C Sweden 59.52N 14.31E

110 D10 **Gstaad** Bern, W Switzerland 46.30N 7.16E

45 P14 **Guabito** Bocas del Toro, NW Panama 9.30N 82.35W

42 N12 **Guacamayas, Golfo de** gulf S Cuba

22 J7 **Guachochi** Chihuahua, N Mexico

42 O8 **Guadajira ◆** SW Spain

106 M13 **Guadajoz ◆** S Spain

42 L13 **Guadalajara** Jalisco, C Mexico 20.43N 103.23W

107 O4 **Guadalajara** Ar. Wad Al-Hajarah; anc. Arriaca. Castilla-La Mancha, C Spain 40.37N 3.10W

107 O7 **Guadalajara ◆** province Castilla-La Mancha, C Spain

106 G12 **Guadalcanal** Andalucía, S Spain 38.06N 5.49W

195 W16 **Guadalcanal** off. Guadalcanal Province. ◆ province S Solomon Islands

195 W16 **Guadalcanal** island S Solomon Islands

107 O12 **Guadalén ◆** S Spain

107 R13 **Guadalentín ◆** SE Spain

105 K15 **Guadalete ◆** SW Spain

107 O13 **Guadalimar ◆** S Spain

107 P12 **Guadalmena ◆** S Spain

106 L11 **Guadalmez ◆** W Spain

107 S7 **Guadalope ◆** E Spain

106 K13 **Guadalquivir ◆** W Spain

106 J14 **Guadalquivir, Marismas del** var. Las Marismas. wetland SW Spain

42 M11 **Guadalupe** Zacatecas, C Mexico 22.44N 102.27W

59 E16 **Guadalupe** Ica, W Peru 13.59S 75.49W

106 L10 **Guadalupe** Extremadura, W Spain 39.26N 5.18W

38 L14 **Guadalupe** Arizona, SW USA 33.20N 111.57W

37 P13 **Guadalupe** California, W USA 34.55N 120.34W

199 Mm5 **Guadalupe** island NW Mexico

Guadalupe see Canelones

42 J5 **Guadalupe Bravos** Chihuahua, N Mexico 21.32N 106.04W

42 A4 **Guadalupe, Isla** island NW Mexico

39 U15 **Guadalupe Mountains ▲** New Mexico/Texas, SW USA

46 J9 **Guantánamo** Guantánamo, SE Cuba 20.06N 75.16W

44 J8 **Guadalupe Peak ▲** Texas, SW USA 31.53N 104.51W

45 R11 **Guadalupe River ◆** SW USA

106 K10 **Guadalupe, Sierra de ▲** W Spain

42 I3 **Guadalupe Victoria** Durango, C Mexico 24.30N 104.03W

42 J5 **Guadalupe y Calvo** Chihuahua, N Mexico 26.04N 106.58W

107 N7 **Guadarrama** Madrid, C Spain 40.40N 4.06W

107 M7 **Guadarrama, Puerto de** pass C Spain 40.41N 4.14W

107 N9 **Guadarrama, Sierra de ▲** C Spain

107 Q9 **Guadazaón ◆** C Spain

47 X10 **Guadeloupe** ◇ French overseas department E West Indies

49 S3 **Guadeloupe** island group E West Indies

47 W10 **Guadeloupe Passage** passage E Caribbean Sea

106 H13 **Guadiana** ◆ Portugal/Spain

107 O12 **Guadiana Menor ◆** S Spain

107 Q8 **Guadiela ◆** C Spain

107 O14 **Guadix** Andalucía, S Spain 37.19N 3.07W

Guad-i-Zirreh see Gowd-e Zereh, Dasht-e

200 O13 **Guafo Fracture Zone** tectonic feature SE Pacific Ocean

65 F18 **Guafo, Isla** island S Chile

44 J6 **Guaimaca** Francisco Morazán, C Honduras 14.33N 86.49W

56 L7 **Guainía** off. Comisaría del Guainía. ◆ province C Colombia

56 L7 **Guainía, Río** var. Colombia/Venezuela

57 O9 **Guaiquinima, Cerro** elevation SE Venezuela 5.45N 63.46W

64 O7 **Guaíra** off. Departamento del Guairá. ◆ department S Paraguay

62 G10 **Guaíra** Paraná, S Brazil 24.04S 54.15W

62 L7 **Guaíra** São Paulo, S Brazil 20.17S 48.21W

Guaira see Gorey

65 F18 **Guaitecas, Isla** island S Chile

46 G6 **Guajaba, Cayo** headland C Cuba 21.50N 77.33W

61 D16 **Guajará-Mirim** Rondônia, W Brazil 10.49S 65.21W

44 H4 **Guajira, Península de la** peninsula N Colombia

56 J6 **Guajira, Río ◆** C Venezuela

36 J4 **Guala** California, W USA 38.45N 123.33W

44 D5 **Gualán** Zacapa, C Guatemala 15.06N 89.20W

63 D13 **Gualeguay** Entre Ríos, E Argentina 33.09S 59.19W

63 D18 **Gualeguaychú** Entre Ríos, E Argentina 33.05S 58.30W

63 C18 **Gualicho, Salina del** salt lake E Argentina

65 K16 **Gualicho, Salina del** salt lake E Argentina

200 J10 **Guam** ◇ US unincorporated territory W Pacific Ocean

196 B15 **Guam** ◇ US unincorporated territory W Pacific Ocean

63 I14 **Guaminí** Buenos Aires, E Argentina 37.04S 62.22W

42 H6 **Guamúchil** Sinaloa, C Mexico 25.23N 108.00W

56 J6 **Guana** var. Misión de Guana. Zulia, NW Venezuela 11.07N 72.17W

46 C4 **Guanabacoa** La Habana, W Cuba 23.01N 82.12W

44 K13 **Guanacaste** off. Provincia de Guanacaste. ◆ province NW Costa Rica

44 J7 **Guanacaste, Cordillera de ▲** NW Costa Rica

46 C4 **Guanahacabibes, Golfo de** gulf W Cuba

44 K4 **Guanaja, Isla de** island Islas de la Bahía, N Honduras

44 J6 **Guanajay** Guanajuato, C Mexico 22.52N 82.39W

42 M12 **Guanajuato** Guanajuato, C Mexico 21.00N 101.16W

42 L12 **Guanajuato ◆** state C Mexico

56 J6 **Guanare** Portuguesa, N Venezuela 9.04N 69.45W

56 K7 **Guanare, Río ◆** W Venezuela

56 J6 **Guanarito** Portuguesa, NW Venezuela 8.39N 69.12W

166 M3 **Guancen Shan ▲** C China

46 I9 **Guandacol** La Rioja, W Argentina 29.31S 68.30W

46 A5 **Guane** Pinar del Río, W Cuba 22.12N 84.05W

167 N14 **Guangdong** var. Guangdong Sheng, Kuang-tung, Kwangtung, Yue. ◆ province S China

Guangdong Sheng see Guangdong

Guanghua see Laohekou

Guangiu see Kwangju

166 I13 **Guangnan** Yunnan, SW China

167 N8 **Guangshui** prev. Yingshan. Hubei, C China 31.41N 113.53E

166 L11 **Guangxi** var. Guangxi Zhuangzu Zizhiqu

166 J4 **Guangxi Zhuangzu Zizhiqu** var. Guangxi, Gui, Kuang-hsi, Kwangsi, Eng. Kwangsi Chuang Autonomous Region. ◆ autonomous region S China

166 J8 **Guangyuan** var. Kuang-yuan, Kwangyuan. Sichuan, C China 32.27N 105.49E

167 N14 **Guangzhou** var. Kuang-chou, Kwangchow, Eng. Canton. Guangdong, S China 23.10N 113.19E

61 N19 **Guanhães** Minas Gerais, SE Brazil 18.46S 42.58W

166 I12 **Guanling** var. Guanling Bouyeizu Miaozu Zizhixian. Guizhou, S China 25.56N 105.36E

Guanling Bouyeizu Miaozu Zizhixian see Guanling

57 N5 **Guanta** Anzoátegui, NE Venezuela 10.15N 64.37W

46 J9 **Guantánamo** Guantánamo, SE Cuba 20.06N 75.16W

166 H9 **Guanxian** var. Guan Xian. Sichuan, C China 31.01N 103.40E

167 Q6 **Guanyun** Jiangsu, E China 34.19N 119.16E

56 C12 **Guapi** Cauca, SW Colombia 2.36N 77.54W

45 V11 **Guápiles** Limón, NE Costa Rica 10.11N 83.45W

63 G17 **Guaporé** Rio Grande do Sul, S Brazil 28.55S 51.53W

61 D14 **Guaporé, Río ◆** var. Río Iténez, Río

43 S8 **Guaporé, Río ◆** Bolivia/Brazil see also Iténez, Río

56 B8 **Guaranda** Bolívar, C Ecuador 1.34S 78.58W

62 H11 **Guaraniaçu** Paraná, S Brazil 25.05S 52.52W

61 O20 **Guarapari** Espírito Santo, SE Brazil 20.39S 40.31W

62 I12 **Guarapuava** Paraná, S Brazil 25.22S 51.28W

62 N8 **Guararapes** São Paulo, S Brazil 21.16S 50.37W

107 S4 **Guara, Sierra de ▲** NE Spain

62 N10 **Guaratinguetá** São Paulo, S Brazil 22.44S 45.16W

106 I7 **Guarda** Guarda, N Portugal 40.31N 7.16W

106 I7 **Guarda ◆** district N Portugal

106 M3 **Guardo** Castilla-León, N Spain 42.48N 4.49W

106 K11 **Guareña** Extremadura, W Spain 38.51N 6.06W

61 E15 **Guaria** Guarda, N Portugal

42 M8 **Guaricana, Pico ▲** S Brazil 25.13S 48.56W

56 J8 **Guárico** off. Estado Guárico. ◆ state N Venezuela

46 J7 **Guárico, Punta** headland E Cuba 20.36N 74.43W

62 M10 **Guárico, Río ◆** C Venezuela

63 L22 **Guarujá** São Paulo, S Brazil 23.50S 46.27W

45 R17 **Guarumal** Veraguas, S Panama 7.48N 81.15W

63 L22 **Guarulhos ✈** (São Paulo) São Paulo, S Brazil 23.23S 46.32W

42 D6 **Guasave** Sinaloa, C Mexico 25.33N 108.28W

56 I8 **Guasdualito** Apure, C Venezuela 7.13N 70.45W

57 O7 **Guasipati** Bolívar, E Venezuela 7.29N 61.54W

108 F9 **Guastalla** Emilia-Romagna, C Italy 44.54N 10.38E

44 D6 **Guastatoya** var. El Progreso. El Progreso, C Guatemala 14.51N 90.02W

44 B4 **Guatemala** off. Republic of Guatemala. ◆ republic Central America

44 A2 **Guatemala** off. Departamento de Guatemala. ◆ department S Guatemala

Guatemala City see Ciudad de Guatemala

200 O10 **Guatemala Basin** undersea feature E Pacific Ocean

Guatemala City see Ciudad de Guatemala

47 Y17 **Guatuaro Point** headland Trinidad, Trinidad and Tobago 10.19N 60.58W

56 G13 **Guaviare** off. Comisaría Guaviare. ◆ province S Colombia

56 E15 **Guaviare, Río ◆** E Colombia

56 I5 **Guavi ◆** SW PNG

194 G14 **Guavi ◆** SW PNG

85 G13 **Guaviare** off. Comisaría Guaviare. ◆ province S Colombia

47 U6 **Guayama** E Puerto Rico 17.58N 66.07W

44 J7 **Guayambre, Río ◆** S Honduras

45 S13 **Guayanas, Macizo de las ▲** NW Costa Rica

44 J7 **Guayapes, Punta** headland E Puerto Rico 18.08N 65.48W

57 O9 **Guayaquil** var. Santiago de Guayaquil. Guayas, SW Ecuador 2.13S 79.54W

56 A8 **Guayaquil** var. Santiago de Guayaquil. Guayas, SW Ecuador 2.13S 79.54W

56 A7 **Guayaquil, Golfo de** var. Gulf of Guayaquil. gulf SW Ecuador

Guayaquil, Gulf of see Guayaquil, Golfo de

58 A7 **Guayas ◆** province W Ecuador

64 N7 **Guaycurú, Río ◆** NE Argentina

42 F6 **Guaymas** Sonora, NW Mexico 27.56N 110.54W

47 U9 **Guaynabo** E Puerto Rico 18.19N 66.05W

82 H12 **Guba** Benishangul, W Ethiopia 11.11N 35.21E

152 K8 **Gubadag** Turkm. Tel'man; prev. Tel'mansk. Dashkhovuzskiy Velayat, N Turkmenistan 42.07N 59.55E

129 T1 **Guba Dolgaya** Nenetskiy Avtonomnyy Okrug, NW Russian Federation 70.16N 59.04E

129 X13 **Gubakha** Permskaya Oblast', NW Russian Federation 58.52N 57.35E

108 I12 **Gubbio** Umbria, C Italy 43.21N 12.34E

112 D12 **Guben** var. Wilhelm-Pieck-Stadt. Brandenburg, E Germany 51.58N 14.42E

Gubin Ger. Guben. Lubuskie, W Poland 51.58N 14.43E

130 K8 **Gubkin** Belgorodskaya Oblast', W Russian Federation 51.16N 37.32E

107 S8 **Gúdar, Sierra de ▲** E Spain

143 P8 **Gudaut'a** NW Georgia 43.06N 40.35E

96 G12 **Gudbrandsdalen** valley S Norway

97 G21 **Gudenå** var. Gudenaa. ◆ C Denmark

Gudenaa see Gudenå

131 N16 **Gudermes** Chechenskaya Respublika, SW Russian Federation 43.21N 46.06E

161 I13 **Gūdūr** Andhra Pradesh, E India 22.50N 82.02W

152 B13 **Gudurolum** Balkanskiy Velayat, W Turkmenistan 37.28N 54.30E

96 D13 **Gudvangen** Sogn og Fjordane, S Norway 60.54N 6.49E

105 V12 **Guebwiller** Haut-Rhin, NE France 47.55N 7.13E

12 K8 **Guéckédou** var. Guékédou. ◆ Quebec, SE Canada

78 J15 **Guékédou** var. Guéckédou. Guinée-Forestière, S Guinea 8.33N 10.08W

43 R16 **Guelatao** Oaxaca, SE Mexico

Guelders see Gelderland

80 G11 **Guélengdeng** Mayo-Kébbi, W Chad 10.55N 15.31E

76 L5 **Guelma** var. Gâlma. NE Algeria 36.28N 7.25E

12 G15 **Guelph** Ontario, S Canada 43.33N 80.12W

104 I7 **Guémené-Penfao** Loire-Atlantique, NW France 47.37N 1.49W

104 I7 **Guer** Morbihan, NW France 47.54N 2.07W

80 I11 **Guéra** off. Préfecture du Guéra. ◆ prefecture S Chad

104 H8 **Guérande** Loire-Atlantique, NW France 47.19N 2.25W

80 K9 **Guéréda** Biltine, E Chad 14.30N 22.04E

105 N10 **Guéret** Creuse, C France 46.10N 1.52E

35 T9 **Guernsey** Wyoming, C USA 42.16N 104.44W

99 Q25 **Guernsey** island Channel Islands, NW Europe

161 G15 **Gurgaon** Bāgpha Karnātaka, C India 17.22N 76.46E

80 G11 **Guéra, off.** Préfecture du Guéra. ◆ prefecture S Chad

105 L7 **Guérou** Assaba, S Mauritania 16.48N 11.40W

45 O13 **Guerra** Texas, SW USA 26.54N 98.53W

42 I13 **Guerrero ◆** state S Mexico

42 D6 **Guerrero Negro** Baja California Sur, NW Mexico 27.55N 114.04W

105 P9 **Gueugnon** Saône-et-Loire, C France 46.36N 4.03E

78 M17 **Guéyo** S Ivory Coast 5.25N 6.04W

109 L19 **Guglionesi** Molise, C Italy 41.54N 14.54E

196 K5 **Guguan** island C Northern Mariana Islands

67 H2 **Guiana Basin** undersea feature W Atlantic Ocean

49 P4 **Guiana Highlands** var. Macizo de las Guayanas. ▲ South America

Guiba see Juba

104 I7 **Guichen** Ille-et-Vilaine, NW France 47.57N 1.47W

63 E18 **Guichón** Paysandú, N Uruguay 32.21S 57.12W

63 E18 **Guichón** Paysandú, N Uruguay 32.21S 57.12W

79 U12 **Guidan-Roumji** Maradi, S Niger 13.40N 6.41E

Guidder see Guider

80 G12 **Guide** Qinghai, C China

80 F12 **Guider** var. Guidder. Nord, N Cameroon 9.55N 13.58E

78 M11 **Guidimouni** Zinder, S Niger 13.40N 9.31E

78 M10 **Guier, Lac de** var. Lac de Guiers. ◎ N Senegal

56 E15 **Guaviare, Río ◆** E Colombia

56 G12 **Guayabero, Río ◆** SW Colombia

47 U6 **Guayama** E Puerto Rico 17.58N 66.07W

78 L16 **Guiglo** W Ivory Coast 6.33N 7.28W

56 L5 **Güigüe** Carabobo, N Venezuela 10.04N 67.48W

85 M20 **Guija, Lago de** ◎ El Salvador/Guatemala

166 L13 **Guilin** var. Kuei-lin, Kweilin. Guangxi Zhuangzu Zizhiqu, S China 25.15N 110.16E

10 J6 **Guillaume-Delisle, Lac** ◎ Quebec, NE Canada

105 U13 **Guillestre** Hautes-Alpes, SE France 44.41N 6.41E

106 H6 **Guimarães** var. Guimaráes. Braga, N Portugal 41.25N 8.19W

60 D11 **Guimarães Rosas, Pico ▲** NW Brazil

25 J3 **Guin** Alabama, S USA 33.58N 87.54W

Gûina see Wina

78 I14 **Guinea** off. Republic of Guinea, var. Guinée; prev. French Guinea, People's Revolutionary Republic of Guinea. ◆ republic W Africa

66 N13 **Guinea Basin** undersea feature E Atlantic Ocean

78 E12 **Guinea-Bissau** off. Republic of Guinea-Bissau, Fr. Guinée-Bissau, Port. Guiné-Bissau; prev. Portuguese Guinea. ◆ republic W Africa

68 K7 **Guinea Fracture Zone** tectonic feature E Atlantic Ocean

66 O13 **Guinea, Gulf of** Fr. Golfe de Guinée. gulf E Atlantic Ocean

Guiné-Bissau see Guinea-Bissau

Guinée see Guinea

Guinée-Bissau see Guinea-Bissau

78 H13 **Guinée-Forestière ◆** state SE Guinea

Guinée, Golfe de see Guinea, Gulf of

78 H13 **Guinée-Maritime ◆** state W Guinea

46 C4 **Güines** La Habana, W Cuba 22.50N 82.02W

104 G5 **Guingamp** Côtes d'Armor, NW France 48.34N 3.09W

107 P3 **Guipúzcoa** Basq. Gipuzkoa. ◆ province País Vasco, N Spain

46 C5 **Güira de Melena** La Habana, W Cuba 22.43N 82.31W

76 G8 **Guir, Hamada du** desert Algeria/Morocco

57 P7 **Güiria** Sucre, NE Venezuela 10.37N 62.21W

166 L5 **Gui Shui ◆** S China

106 H2 **Guitiriz** Galicia, NW Spain 43.10N 7.52W

79 N17 **Guitri** S Ivory Coast 5.31N 5.13W

179 Q13 **Guiuan** Samar, C Philippines 11.02N 125.45E

166 J12 **Guiyang** var. Kuei-Yang, Kuei-yang, Kueyang, Kweiyang; prev. Kweichu. Guizhou, S China 26.33N 106.44E

166 J12 **Guizhou** var. Guizhou Sheng, Kuei-chou, Kweichow, Qian. ◆ province S China

Guizhou Sheng see Guizhou

104 I13 **Guîtres** Gironde, SW France 45.02N 0.11W

104 J13 **Guîtres** Gironde, SW France 43.39N 1.04W

160 B10 **Gujar Khan** Punjab, E Pakistan 33.15N 73.18E

160 G5 **Gujarāt** var. Gujerat. ◆ state W India

155 V6 **Gujar Khan** Punjab, E Pakistan 33.15N 73.18E

155 V7 **Gujrānwāla** Punjab, NE Pakistan 32.11N 74.08E

155 V7 **Gujrāt** Punjab, E Pakistan 32.33N 74.03E

165 V9 **Gulang** Gansu, C China 37.31N 102.55E

191 R6 **Gulargambone** New South Wales, SE Australia 31.19S 148.31E

161 G15 **Gulbarga ▲** Karnātaka, C India 17.22N 76.46E

120 J8 **Gulbene** Ger. Alt-Schwanenburg. Gulbene, NE Latvia 57.09N 26.44E

153 U10 **Gul'cha** Kir. Gülchö. Oshskaya Oblast', SW Kyrgyzstan 40.16N 73.27E

Gülchö see Gul'cha

181 T10 **Gulden Draak Seamount** undersea feature E Indian Ocean 33.45S 101.00E

142 J16 **Gülek Boğazı** var. Cilician Gates. pass S Turkey 37.19N 34.49E

194 I14 **Gulf ◆** province S PNG

25 O9 **Gulf Breeze** Florida, SE USA 30.21N 87.09W

25 Q9 **Gulfport** Florida, SE USA 27.45N 82.42W

24 M8 **Gulfport** Mississippi, S USA 30.22N 89.05W

25 O9 **Gulf Shores** Alabama, S USA 30.15N 87.40W

Gulf, The see Persian Gulf

191 R7 **Gulgong** New South Wales, SE Australia 32.22S 149.31E

166 I11 **Gulin** Sichuan, C China 28.06N 105.47E

176 V12 **Gulir** Pulau Kasiui, E Indonesia 4.27S 131.41E

153 P10 **Gulistan** Rus. Gulistan. Sirdaryo Wiloyati, E Uzbekistan 40.28N 68.45E

169 T6 **Guliya Shan ▲** NE China 49.42N 122.22E

41 S4 **Gulja** see Yining

41 U13 **Gulkana** Alaska, USA 62.17N 145.25W

9 S16 **Gull Lake** Saskatchewan, S Canada 50.04N 108.30W

33 O6 **Gull Lake** ◎ Michigan, C USA

31 T6 **Gull Lake** ◎ Minnesota, N USA

95 I16 **Gullspång** Västra Götaland, S Sweden 58.58N 14.04E

158 F7 **Gulmarg** Jammu and Kashmir, NW India 34.04N 74.25E

101 L18 **Gulpen** Limburg, SE Netherlands 50.48N 5.53E

151 S13 **Gul'shad** Kaz. Gulshat. Zhezkazgan, E Kazakhstan 46.37N 74.21E

Gulshat see Gul'shad

83 F17 **Gulu** N Uganda 2.46N 32.21E

116 K10 **Gŭlŭbovo** Stara Zagora, C Bulgaria 42.10N 25.52E

116 I7 **Gulyantsi** Pleven, N Bulgaria 43.37N 24.40E

Gulyaypole see Hulyaypole

Guma see Pishan

81 K16 **Gumba** Equateur, NW Dem. Rep. Congo (Zaire) 2.58N 23.21E

Gumbinnen see Gusev

83 H24 **Gumbiro** Ruvuma, S Tanzania 10.19S 35.40E
152 B11 **Gumdag** *prev.* Kum-Dag. Balkanskiy Velayat, W Turkmenistan 39.13N 54.35E
79 W12 **Gumel** Jigawa, N Nigeria 12.37N 9.23E
107 N5 **Gumiel de Hizán** Castilla-León, N Spain 41.46N 3.42W
194 I12 **Gumine** *var.* Gumire. Chimbu, C PNG 6.12S 144.53E
Gumire *see* Gumine
159 P16 **Gumla** Bihār, N India 23.03N 84.36E
Gumma *see* Gunma
103 F16 **Gummersbach** Nordrhein-Westfalen, W Germany 51.01N 7.34E
79 T13 **Gummi** Zamfara, NW Nigeria 12.07N 5.07E
Gumpolds *see* Humpolec
159 N13 **Gumti** *var.* Gomati. ⁓ N India
Gümülcine/Gümüljina *see* Komotiní
Gümüşane *see* Gümüşhane
143 O12 **Gümüşhane** *var.* Gümüşane, Gumushkhane. Gümüşhane, NE Turkey 40.30N 39.27E
143 O12 **Gümüşhane** *var.* Gümüşane, Gumushkhane. ◐ *province* NE Turkey
Gumushkhane *see* Gümüşhane
176 W13 **Gumzai** Pulau Kola, E Indonesia 5.27S 134.38E
160 H9 **Guna** Madhya Pradesh, C India 24.39N 77.21E
Gunabad *see* Gonābād
Gunbad-i-Qawus *see* Gonbad-e Kāvūs
191 O9 **Gunbar** New South Wales, SE Australia 34.03S 145.32E
191 O9 **Gun Creek** *seasonal river* New South Wales, SE Australia
191 Q10 **Gundagai** New South Wales, SE Australia 35.06S 148.03E
81 K17 **Gundji** Equateur, N Dem. Rep. Congo (Zaire) 2.13N 21.31E
161 G20 **Gundlupet** Karnātaka, W India 11.48N 76.42E
142 G16 **Gündoğmuş** Antalya, S Turkey 36.52N 32.01E
143 O14 **Güney Doğu Toroslar** ▲▲ SE Turkey
81 J21 **Gungu** Bandundu, SW Dem. Rep. Congo (Zaire) 5.43S 19.19E
131 P17 **Gunib** Respublika Dagestan, SW Russian Federation 42.24N 46.55E
114 J11 **Gunja** Vukovar-Srijem, E Croatia 44.53N 18.51E
33 P9 **Gun Lake** ◎ Michigan, N USA
171 Jj15 **Gunma** *off.* Gunma-ken, *var.* Gumma. ◐ *prefecture* Honshū, S Japan
207 P15 **Gunnbjørn Fjeld** *var.* Gunnbjörns Bjerge. ▲ C Greenland 69.03N 29.36W
191 S6 **Gunnedah** New South Wales, SE Australia 30.58S 150.15E
181 Y15 **Gunner's Quoin** *var.* Coin de Mire. *island* N Mauritius
39 R6 **Gunnison** Colorado, C USA 38.33N 106.55W
38 L5 **Gunnison** Utah, W USA 39.09N 111.49W
39 P5 **Gunnison River** ⁓ Colorado, C USA
23 X2 **Gunpowder River** ⁓ Maryland, NE USA
Güns *see* Kőszeg
Gunsan *see* Kunsan
111 S4 **Gunskirchen** Oberösterreich, N Austria 48.07N 13.54E
Gunt *see* Ghund
161 H17 **Guntakal** Andhra Pradesh, C India 15.10N 77.24E
25 Q2 **Guntersville** Alabama, S USA 34.21N 86.17W
25 Q2 **Guntersville Lake** ⊠ Alabama, S USA
111 X4 **Guntramsdorf** Niederösterreich, E Austria 48.03N 16.19E
161 J16 **Guntūr** *var.* Guntur. Andhra Pradesh, SE India 16.19N 80.27E
173 F7 **Gunungsitoli** Pulau Nias, W Indonesia 1.11N 97.35E
161 M14 **Gunupur** Orissa, E India 19.04N 83.52E
103 J23 **Günz** ⁓ S Germany
Gunzan *see* Kunsan
103 J22 **Günzburg** Bayern, S Germany 48.26N 10.18E
103 K21 **Gunzenhausen** Bayern, S Germany 49.07N 10.45E
167 P7 **Guoyang** Anhui, E China 33.29N 116.14E
118 G11 **Gurahonț** *Hung.* Honctő. Arad, W Romania 46.16N 22.20E
Gurahumora *see* Gura Humorului
118 K9 **Gura Humorului** *Ger.* Gurahumora. Suceava, NE Romania 47.31N 26.00E
164 K4 **Gurbantünggüt Shamo** *desert* W China
158 H7 **Gurdāspur** Punjab, N India 32.02N 75.23E
29 T13 **Gurdon** Arkansas, C USA 33.55N 93.09W
Gurdzhaani *see* Gurjaani
Gurgan *see* Gorgān
158 I10 **Gurgaon** Haryāna, N India 28.24N 76.59E
61 M15 **Gurguéia, Rio** ⁓ NE Brazil
57 Q7 **Guri, Embalse de** ⊠ E Venezuela
143 V10 **Gurjaani** *Rus.* Gurdzhaani. E Georgia 41.42N 45.47E
111 T8 **Gurk** Kärnten, S Austria 46.52N 14.17E
111 T9 **Gurk** *Slvn.* Krka. ⁓ S Austria
Gurkfeld *see* Krško
116 K9 **Gurkovo** *prev.* Kolupchii. Stara Zagora, C Bulgaria 42.42N 25.46E
111 S9 **Gurktaler Alpen** ▲▲ S Austria
152 H8 **Gurlan** *Rus.* Gurlen. Khorazm Wiloyati, W Uzbekistan 41.54N 60.18E
Gurlen *see* Gurlan
85 M16 **Guro** Manica, C Mozambique 17.25S 33.23E
142 M14 **Gürün** Sivas, C Turkey 38.43N 37.15E
61 K16 **Gurupi** Tocantins, C Brazil 11.43S 49.01W
60 L12 **Gurupi, Rio** ⁓ NE Brazil
158 E14 **Guru Sikhar** ▲ NW India 24.45N 72.51E
Gur'yev/Gur'yevskaya Oblast' *see* Atyrau
79 U13 **Gusau** Zamfara, NW Nigeria 12.18N 6.27E
130 C3 **Gusev** *Ger.* Gumbinnen. Kaliningradskaya Oblast', W Russian Federation 54.36N 22.13E
152 J17 **Gushgy** *prev.* Kushka. Maryyskiy Velayat, S Turkmenistan 35.18N 62.17E
Gushiago *see* Gushiegu
79 Q14 **Gushiegu** *var.* Gushiago. NE Ghana 9.54N 0.12W
172 P15 **Gushikawa** Okinawa, Okinawa, SW Japan 26.21N 127.49E
115 L16 **Gusinje** Montenegro, SW Yugoslavia 42.34N 19.51E
126 Jj16 **Gusinoozersk** Respublika Buryatiya, S Russian Federation 51.18N 106.28E
130 M4 **Gus'-Khrustal'nyy** Vladimirskaya Oblast', W Russian Federation 55.39N 40.42E
109 B19 **Guspini** Sardegna, Italy, C Mediterranean Sea 39.33N 8.39E
111 X8 **Güssing** Burgenland, SE Austria 47.04N 16.18E
111 V6 **Gusswerk** Steiermark, E Austria 47.43N 15.18E
94 O2 **Gustav Adolf Land** *physical region* NE Svalbard
205 X5 **Gustav Bull Mountains** ▲▲ Antarctica
41 W13 **Gustavus** Alaska, USA 58.24N 135.44W
94 O1 **Gustav V Land** *physical region* NE Svalbard
37 P9 **Gustine** California, W USA 37.14N 121.00W
45 R8 **Gustine** Texas, SW USA 31.51N 98.24W
102 M9 **Güstrow** Mecklenburg-Vorpommern, NE Germany 53.48N 12.11E
97 N18 **Gusum** Östergötland, S Sweden 58.15N 16.30E
Guta/Gúta *see* Kolárovo
Gutenstein *see* Ravne na Koroškem
103 G14 **Gütersloh** Nordrhein-Westfalen, W Germany 51.54N 8.22E
29 N10 **Guthrie** Oklahoma, C USA 35.52N 97.25W
45 P5 **Guthrie** Texas, SW USA 33.37N 100.21W
31 U14 **Guthrie Center** Iowa, C USA 41.40N 94.30W
43 Q13 **Gutiérrez Zamora** Veracruz-Llave, E Mexico 20.29N 97.07W
Gutta *see* Kolárovo
31 Y12 **Guttenberg** Iowa, C USA 42.47N 91.06W
Guttentag *see* Dobrodzień
Guttstadt *see* Dobre Miasto
168 G8 **Guulin** Govĭ-Altay, C Mongolia 46.33N 97.21E
159 V12 **Guwāhāti** *prev.* Gauhāti. Assam, NE India 26.09N 91.42E
145 R3 **Guwēr** *var.* Al Kuwayr, Al Quwayr, Quwair. N Iraq 36.03N 43.30E
Guwlumayak *see* Kuuli-Mayak
57 R9 **Guyana** *off.* Cooperative Republic of Guyana; *prev.* British Guiana. ◆ *republic* N South America
23 P5 **Guyandotte River** ⁓ West Virginia, NE USA
Guyane *see* French Guiana
Guyi *see* Sanjiang
28 H8 **Guymon** Oklahoma, C USA 36.40N 101.28W
152 K12 **Guynuk** Lebapskiy Velayat, NE Turkmenistan 39.18N 63.00E
23 O9 **Guyot, Mount** ▲ North Carolina/Tennessee, SE USA 35.42N 83.15W
191 U5 **Guyra** New South Wales, SE Australia 30.13S 151.42E
165 W10 **Guyuan** Ningxia, N China 35.57N 106.13E
Guzar *see* Ghuzor
124 N3 **Güzelyurt** *Gk.* Mórfou, Morphou. W Cyprus 35.11N 33.00E
124 N2 **Güzelyurt Körfezi** *var.* Morfou Bay, Morphou Bay, *Gk.* Kólpos Mórfou. *bay* W Cyprus
42 I3 **Guzmán** Chihuahua, N Mexico 31.13N 107.27W
121 B14 **Gvardeysk** *Ger.* Tapaiu. Kaliningradskaya Oblast', W Russian Federation 54.39N 21.02E
Gvardeyskoye *see* Hvardiys'ke
191 R5 **Gwabegar** New South Wales, SE Australia 30.34S 148.58E
154 J16 **Gwādar** *var.* Gwadur. Baluchistān, SW Pakistan 25.09N 62.21E
154 J16 **Gwādar East Bay** *bay* SW Pakistan
154 J16 **Gwādar West Bay** *bay* SW Pakistan
Gwadur *see* Gwādar
85 J17 **Gwai** Matabeleland North, W Zimbabwe 19.17S 27.37E
160 I7 **Gwalior** Madhya Pradesh, C India 26.15N 78.12E
85 J18 **Gwanda** Matabeleland South, SW Zimbabwe 20.56S 29.00E
81 N15 **Gwane** Orientale, N Dem. Rep. Congo (Zaire) 4.40N 25.51E
85 I17 **Gwayi** ⁓ W Zimbabwe
112 G8 **Gwda** *var.* Głda, *Ger.* Küddow. ⁓ NW Poland
99 C14 **Gweebarra Bay** *Ir.* Béal an Bheara. *inlet* W Ireland
99 D14 **Gweedore** *Ir.* Gaoth Dobhair. NW Ireland 55.03N 8.13W
Gwelo *see* Gweru
99 K21 **Gwent** *cultural region* S Wales, UK
85 K17 **Gweru** *prev.* Gwelo. Midlands, C Zimbabwe 19.27S 29.49E
31 Q7 **Gwinner** North Dakota, N USA 46.10N 97.42W
79 Y13 **Gwoza** Borno, NE Nigeria 11.07N 13.40E
Gwy *see* Wye
191 R4 **Gwydir River** ⁓ New South Wales, SE Australia
99 I19 **Gwynedd** *var.* Gwyneth. *cultural region* NW Wales, UK
Gwyneth *see* Gwynedd
165 O16 **Gyaca** Xizang Zizhiqu, W China 29.06N 92.37E
Gya'gya *see* Saga
117 M22 **Gyalí** *var.* Yialí. *island* Dodekánisos, Greece, Aegean Sea
Gyandzha *see* Gäncä
164 M16 **Gyangzê** Xizang Zizhiqu, W China 28.49N 89.37E
164 L14 **Gyaring Co** ◎ W China
165 Q12 **Gyaring Hu** ◎ C China
117 I20 **Gyáros** *var.* Yioúra. *island* Kykládes, Greece, Aegean Sea
126 Hh7 **Gyda** Yamalo-Nenetskiy Avtonomnyy Okrug, N Russian Federation 70.55N 78.34E
126 H7 **Gydanskiy Poluostrov** *Eng.* Gyda Peninsula. *peninsula* N Russian Federation
Gyda Peninsula *see* Gydanskiy Poluostrov
Gyéres *see* Câmpia Turzii
Gyergyószentmiklós *see* Gheorgheni
Gyergyótölgyes *see* Tulgheș
Gyertyámos *see* Cărpiniș
Gyeva *see* Detva
Gyigang *see* Zayü
97 I23 **Gyldenløves Høy** *hill range* C Denmark
189 Z10 **Gympie** Queensland, E Australia 26.04S 152.40E
177 Ff7 **Gyobingauk** Pegu, SW Myanmar 18.13N 95.39E
113 M23 **Gyomaendrőd** Békés, SE Hungary 46.55N 20.49E
Gyömbér *see* Ďumbier
113 L22 **Gyöngyös** Heves, NE Hungary 47.43N 19.48E
113 H22 **Győr** *Ger.* Raab; *Lat.* Arrabona. Győr-Moson-Sopron, NW Hungary 47.40N 17.40E
113 G22 **Győr-Moson-Sopron** *off.* Győr-Moson-Sopron Megye. ◐ *county* NW Hungary
9 X15 **Gypsumville** Manitoba, S Canada 51.46N 98.37W
10 M4 **Gyrfalcon Islands** *island group* Nunavut, NE Canada
97 N14 **Gysinge** Gävleborg, C Sweden 60.16N 16.55E
117 F22 **Gýtheio** *var.* Githio; *prev.* Yíthion. Pelopónnisos, S Greece 36.46N 22.34E
152 L13 **Gyuichbirleshik** Lebapskiy Velayat, E Turkmenistan 38.10N 64.33E
113 N24 **Gyula** *Rom.* Jula. Békés, SE Hungary 46.37N 21.19E
Gyulafehérvár *see* Alba Iulia
Gyulovo *see* Roza
143 T11 **Gyumri** *var.* Giumri, *Rus.* Kumayri; *prev.* Aleksandropol', Leninakan. W Armenia 40.48N 43.51E
152 D13 **Gyunuzyndag, Gora** ▲ W Turkmenistan 38.15N 56.25E
152 D12 **Gyzylarbat** *prev.* Kizyl-Arvat. Balkanskiy Velayat, W Turkmenistan 39.01N 56.14E
Gyzylbaydak *see* Krasnoye Znamya
Gyzyletrek *see* Kizyl-Atrek
Gyzylgaya *see* Kizyl-Kaya
Gyzylsu *see* Kizyl-Su

H

159 T12 **Ha** W Bhutan 27.16N 89.22E
Haabai *see* Ha'apai Group
101 H17 **Haacht** Vlaams Brabant, C Belgium 50.58N 4.37E
111 T4 **Haag** Niederösterreich, NE Austria 48.07N 14.32E
204 L8 **Haag Nunataks** ▲▲ Antarctica
94 N2 **Haakon VII Land** *physical region* NW Svalbard
100 O11 **Haaksbergen** Overijssel, E Netherlands 52.09N 6.45E
101 E14 **Haamstede** Zeeland, SW Netherlands 51.43N 3.45E
200 Ss13 **Ha'ano** *island* Ha'apai Group, C Tonga
200 Ss13 **Ha'apai Group** *var.* Haabai. *island group* C Tonga
95 L15 **Haapajärvi** Oulu, C Finland 63.45N 25.19E
95 L17 **Haapamäki** Länsi-Suomi, W Finland 62.11N 24.32E
95 L15 **Haapavesi** Oulu, C Finland 64.09N 25.25E
203 N7 **Haapiti** Moorea, W French Polynesia 17.33S 149.52W
120 F4 **Haapsalu** *Ger.* Hapsal. Läänemaa, W Estonia 58.57N 23.32E
Ha'Arava *see* 'Arabah, Wādī al
Haarby *see* Hårby
100 H10 **Haarlem** *prev.* Harlem. Noord-Holland, W Netherlands 52.22N 4.39E
193 D19 **Haast** West Coast, South Island, NZ 43.53S 169.01E
193 C20 **Haast** ⁓ South Island, NZ
193 D20 **Haast Pass** *pass* South Island, NZ 44.07S 169.18E
200 R16 **Ha'atua** 'Eau, E Tonga 21.23S 174.57W
155 P15 **Hab** ⁓ SW Pakistan
147 W7 **Haba** *var.* Al Haba. Dubayy, NE UAE 25.01N 55.37E
164 K2 **Habahe** *var.* Kaba. Xinjiang Uygur Zizhiqu, NW China 48.04N 86.20E
147 U13 **Ḩabarūt** *var.* Habrut. SW Oman 17.19N 52.45E
83 J18 **Habaswein** North Eastern, NE Kenya 1.01N 39.27E
101 L24 **Habay-la-Neuve** Luxembourg, SE Belgium 49.43N 5.38E
145 S8 **Ḩabbānīyah, Buḩayrat** ◎ C Iraq
Habelschwerdt *see* Bystrzyca Kłodzka
159 V14 **Habiganj** Chittagong, NE Bangladesh 24.22N 91.25E
169 Q12 **Habirag** Nei Mongol Zizhiqu, N China 42.18N 115.40E
97 L19 **Habo** Västra Götaland, S Sweden 57.55N 14.04E
127 P16 **Habomai Islands** *island group* Kuril'skiye Ostrova, SE Russian Federation
172 P3 **Haboro** Hokkaidō, NE Japan 44.19N 141.42E
159 S16 **Habra** West Bengal, NE India 22.39N 88.17E
Habrut *see* Ḩabarūt
149 P17 **Ḩabshān** Abū Ẓaby, C UAE 23.51N 53.34E
56 E14 **Hacha** Putumayo, S Colombia 00.02S 75.30W
172 Ss13 **Hachijō** Tōkyō, Hachijō-jima, SE Japan 35.40N 139.19E
172 Ss13 **Hachijō-jima** *var.* Hatizyô Zima. *island* Izu-shotō, SE Japan
171 I14 **Hachiman** Gifu, Honshū, SW Japan 35.46N 136.57E
171 M9 **Hachimori** Akita, Honshū, C Japan 40.22N 139.59E
172 N10 **Hachinohe** Aomori, Honshū, C Japan 40.30N 141.28E
171 Jj16 **Hachiōji** *var.* Hatiôzi. Tōkyō, Honshū, S Japan 35.39N 139.17E
95 G17 **Hackås** Jämtland, C Sweden 62.55N 14.31E
20 K14 **Hackensack** New Jersey, NE USA 40.51N 73.57W
Hadama *see* Nazrēt
171 J16 **Hadano** Kanagawa, Honshū, S Japan 35.23N 139.13E
147 W13 **Ḩadbaram** S Oman 17.27N 55.13E
145 U13 **Ḩaddānīyah** *well* S Iraq 30.27N 44.40E
98 K12 **Haddington** SE Scotland, UK 55.59N 2.45W
147 Z8 **Ḩadd, Ra's al** *headland* NE Oman 22.28N 59.58E
Haded *see* Xadeed
79 W12 **Hadejia** Jigawa, N Nigeria 12.22N 10.02E
79 W12 **Hadejia** ⁓ N Nigeria
144 F9 **Hadera** *var.* Khadera. Haifa, C Israel 32.25N 34.55E
Hadersleben *see* Haderslev
97 G24 **Haderslev** *Ger.* Hadersleben. Sønderjylland, SW Denmark 55.15N 9.30E
157 J21 **Hadhdhunmathi Atoll** *var.* Haddummati Atoll, Laamu Atoll. *atoll* S Maldives
Hadhramaut *see* Ḩaḑramawt
147 W17 **Hadīboh** Suquṭrā, SE Yemen 12.37N 54.04E
164 K9 **Hadilik** Xinjiang Uygur Zizhiqu, W China 37.51N 86.10E
142 H16 **Hadım** Konya, S Turkey 36.58N 32.27E
146 K7 **Hadīyah** Al Madīnah, W Saudi Arabia 25.36N 38.31E
15 J1 **Hadley Bay** *bay* Victoria Island, Nunavut, N Canada
178 Jj6 **Ha Đông** *var.* Hadong. Ha Tây, N Vietnam 20.58N 105.46E
147 R15 **Ḩaḑramawt** *Eng.* Hadhramaut. ▲▲ S Yemen
Hadria *see* Adria
Hadrianopolis *see* Edirne
Hadria Picena *see* Apricena
97 G22 **Hadsten** Århus, C Denmark 56.19N 10.03E
97 G21 **Hadsund** Nordjylland, N Denmark 56.43N 10.07E
119 S4 **Hadyach** *Rus.* Gadyach. Poltavs'ka Oblast', NE Ukraine 50.21N 34.00E
114 I13 **Hadžići** Federacija Bosna I Hercegovina, SE Bosnia and Herzegovina 43.49N 18.12E
169 W14 **Haeju** S North Korea 38.04N 125.40E
Haerbin/Haerhpin/Ha-erh-pin *see* Harbin
147 P5 **Ḩafar al Bāṭin** Ash Sharqīyah, N Saudi Arabia 28.25N 45.58E
9 T15 **Hafford** Saskatchewan, S Canada 52.43N 107.19W
142 M13 **Hafik** Sivas, N Turkey 39.52N 37.24E
155 V8 **Hāfizābād** Punjab, E Pakistan 32.05N 73.37E
94 H4 **Hafnarfjördhur** Reykjanes, W Iceland 64.03N 21.57W
Hafnia *see* København, Denmark
Hafnia *see* Denmark
Hafren *see* Severn
Hafun *see* Xaafuun
Hafun, Ras *see* Xaafuun, Raas
82 G10 **Hag 'Abdullah** Sinnar, E Sudan 13.58N 33.34E
83 K18 **Hagadera** North Eastern, E Kenya 0.06N 40.23E
144 G8 **HaGalil** *Eng.* Galilee. ▲▲ N Israel
12 G10 **Hagar** Ontario, S Canada 46.27N 80.22W
161 G18 **Hagari** *var.* Vedāvati. ⁓ W India
196 B16 **Hagåtña** *var.* Agana, Agaña. ○ (Guam) NW Guam 13.27N 144.45E
102 M13 **Hagelberg** *hill* NE Germany 52.03N 12.33E
41 N14 **Hagemeister Island** *island* Alaska, USA
103 F15 **Hagen** Nordrhein-Westfalen, W Germany 51.21N 7.27E
102 K10 **Hagenow** Mecklenburg-Vorpommern, N Germany 53.27N 11.10E
8 K15 **Hagensborg** British Columbia, SW Canada 52.24N 126.24W
82 I13 **Hāgere Hiywet** *var.* Agere Hiywet, Ambo. Oromo, C Ethiopia 9.00N 37.55E
35 O15 **Hagerman** Idaho, NW USA 42.48N 114.53W
39 U14 **Hagerman** New Mexico, SW USA 33.07N 104.19W
23 V2 **Hagerstown** Maryland, NE USA 39.38N 77.43W
12 G16 **Hagersville** Ontario, S Canada 42.58N 80.03W
104 J15 **Hagetmau** Landes, SW France 43.40N 0.36W
97 K14 **Hagfors** Värmland, C Sweden 60.03N 13.45E
95 G16 **Häggenås** Jämtland, C Sweden 63.24N 14.53E
170 Dd12 **Hagi** Yamaguchi, Honshū, SW Japan 34.24N 131.22E
178 J5 **Ha Giang** Ha Giang, N Vietnam 22.49N 104.58E
Hagios Evstrátios *see* Ágios Efstrátios
HaGolan *see* Golan Heights
105 T4 **Hagondange** Moselle, NE France 49.15N 6.06E
99 B18 **Hag's Head** *Ir.* Ceann Caillí. *headland* W Ireland 52.56N 9.29W
104 I3 **Hague, Cap de la** *headland* N France 49.43N 1.56W
105 V5 **Haguenau** Bas-Rhin, NE France 48.49N 7.46E
172 T16 **Hahajima-rettō** *island group* SE Japan
13 R8 **Há Há , Lac** ◎ Quebec, SE Canada
180 H13 **Hahaya** ✈ (Moroni) Grande Comore, NW Comoros
24 K9 **Hahnville** Louisiana, S USA 29.58N 90.24W
85 E22 **Haib** Karas, S Namibia 28.12S 18.19E
Haibak *see* Āybak
155 N15 **Haibo** ⁓ SW Pakistan
169 U12 **Haicheng** Liaoning, NE China 40.52N 122.45E
Haida *see* Nový Bor
Haidarabad *see* Hyderābād
Haidenschaft *see* Ajdovščina
178 Jj6 **Hai Dương** Hai Hưng, N Vietnam 20.55N 106.21E
144 F9 **Haifa** ◐ *district* NW Israel
Haifa *see* Hefa
Haifa, Bay of *see* Hefa, Mifraz
167 P14 **Haifeng** Guangdong, S China 22.58N 115.16E
Haifong *see* Hai Phong
167 P3 **Hai He** ⁓ E China
Haikang *see* Leizhou
166 L17 **Haikou** *var.* Hai-k'ou, Hoihow, *Fr.* Hoï-Hao. Hainan, S China 20.00N 110.16E
146 M6 **Ḩā'il** Ḩā'il, NW Saudi Arabia 27.00N 42.50E
147 N5 **Ḩā'il** *off.* Minṭaqah Ḩā'il. ◐ *province* N Saudi Arabia
Hai-la-erh *see* Hailar
169 S6 **Hailar** *var.* Hai-la-erh; *prev.* Hulun. Nei Mongol Zizhiqu, N China 49.15N 119.40E
169 S6 **Hailar He** ⁓ NE China
35 P14 **Hailey** Idaho, NW USA 43.31N 114.18W
12 H9 **Haileybury** Ontario, S Canada 47.27N 79.39W
169 X9 **Hailin** Heilongjiang, NE China 44.37N 129.24E
Ḩā'il, Minṭaqah *see* Ḩā'il
Hailong *see* Meihekou
95 K14 **Hailuoto** *Swe.* Karlö. *island* W Finland
Haima *see* Haymā'
Haimen *see* Taizhou
166 M17 **Hainan** *var.* Hainan Sheng, Qiong. ◐ *province* S China
166 K17 **Hainan Dao** *island* S China
Hainan Sheng *see* Hainan
Hainan Strait *see* Qiongzhou Haixia
Hainasch *see* Ainaži
Hainau *see* Chojnów
101 E20 **Hainaut** ◐ *province* SW Belgium
Hainburg *see* Hainburg an der Donau
111 Z4 **Hainburg an der Donau** *var.* Hainburg. Niederösterreich, NE Austria 48.08N 16.57E
41 W12 **Haines** Alaska, USA 59.13N 135.27W
34 L12 **Haines** Oregon, NW USA 44.53N 117.56W
25 W12 **Haines City** Florida, SE USA 28.06N 81.37W
8 H8 **Haines Junction** Yukon Territory, W Canada 60.45N 137.30W
111 W4 **Hainfeld** Niederösterreich, NE Austria 48.01N 15.45E
103 N16 **Hainichen** Sachsen, E Germany 50.58N 13.07E
178 K6 **Hai Phong** *var.* Haifong, Haiphong. N Vietnam 20.49N 106.40E
167 S12 **Haitan Dao** *island* SE China
46 K8 **Haiti** *off.* Republic of Haiti. ◆ *republic* C West Indies
37 T11 **Haiwee Reservoir** ⊠ California, W USA
82 I7 **Haiya** Red Sea, NE Sudan 18.16N 36.21E
165 T10 **Haiyan** Qinghai, W China 36.55N 100.54E
166 M13 **Haiyang Shan** ▲▲ S China
165 V10 **Haiyuan** Ningxia, N China 36.32N 105.31E
Hajda *see* Nový Bor
113 M22 **Hajdú-Bihar** *off.* Hajdú-Bihar Megye. ◐ *county* E Hungary
113 N22 **Hajdúböszörmény** Hajdú-Bihar, E Hungary 47.39N 21.32E
113 N22 **Hajdúhadház** Hajdú-Bihar, E Hungary 47.40N 21.40E
113 N21 **Hajdúnánás** Hajdú-Bihar, E Hungary 47.49N 21.25E
113 N22 **Hajdúszoboszló** Hajdú-Bihar, E Hungary 47.27N 21.23E
148 I3 **Ḩājī Ebrāhīm, Kūh-e** ▲ Iran/Iraq 36.53N 44.56E
171 Kk11 **Hajiki-zaki** *headland* Sado, C Japan 38.19N 138.28E
Hajîne *see* Abū Ḩardān
159 P13 **Hājīpur** Bihār, N India 25.40N 85.13E
147 N14 **Ḩajjah** W Yemen 15.43N 43.33E
145 U11 **Hajjama** S Iraq 31.24N 45.20E
149 R12 **Ḩājjīābād** Hormozgān, C Iran
145 U14 **Ḩājj, Thaqb al** *well* S Iraq 29.58N 44.32E
115 L16 **Hajla** ▲▲ SW Yugoslavia
112 P10 **Hajnówka** *Ger.* Hermhausen. Podlaskie, NE Poland 52.43N 23.37E
177 Ff4 **Haka** Chin State, W Myanmar 22.42N 93.40E
Hakapehi *see* Punaauia
Hakâri *see* Hakkâri
143 T16 **Hakkâri** *var.* Çölemerik, Hakâri. Hakkâri, SE Turkey 37.36N 43.45E
143 T16 **Hakkâri** *var.* Hakkari. ◐ *province* SE Turkey
94 J12 **Hakkas** Norrbotten, N Sweden 66.52N 21.36E
171 Gg16 **Hakken-zan** ▲ Honshū, SW Japan 34.11N 135.57E
172 N9 **Hakkōda-san** ▲ Honshū, C Japan 40.40N 140.49E
172 Pp3 **Hako-dake** ▲ Hokkaidō, NE Japan 44.40N 142.22E
172 N7 **Hakodate** Hokkaidō, NE Japan 41.46N 140.42E
171 Ii12 **Hakui** Ishikawa, Honshū, SW Japan 36.52N 136.45E
202 B16 **Hakupu** SE Niue 19.06S 169.49E
171 I14 **Haku-san** ▲ Honshū, SW Japan 36.07N 136.45E
Hal *see* Halle
155 Q15 **Hāla** Sind, SE Pakistan 25.46N 68.28E
144 J3 **Ḩalab** *Eng.* Aleppo, *Fr.* Alep; *anc.* Beroea. Ḩalab, NW Syria 36.13N 37.10E
144 J3 **Ḩalab** *off.* Muḩāfaẓat Ḩalab, *var.* Aleppo, Halab. ◐ *governorate* NW Syria
144 J3 **Ḩalab** ✈ Ḩalab, NW Syria 36.12N 37.10E
147 O8 **Ḩalabān** *var.* Halibān. Ar Riyāḑ, C Saudi Arabia 23.28N 44.19E
145 V4 **Ḩalabja** NE Iraq 35.10N 45.58E
202 A16 **Halagigie Point** *headland* W Niue
77 Z11 **Halaib** SE Egypt 22.10N 36.33E
202 G12 **Halalo** Île Uvea, N Wallis and Futuna 13.21S 176.10W
Halandri *see* Chalándri
147 X13 **Ḩalānīyāt, Juzur al** *var.* Jazā'ir Bin Ghalfān, *Eng.* Kuria Muria Islands. *island group* S Oman
147 W13 **Ḩalānīyāt, Khalīj al** *Eng.* Kuria Muria Bay. *bay* S Oman
Halas *see* Kiskunhalas
40 G11 **Halawa** *Haw.* Hālawa. Hawaii, USA, C Pacific Ocean 20.13N 155.46W
40 F9 **Halawa, Cape** *headland* Molokai, Hawaii, USA, C Pacific Ocean 21.09N 156.43W
168 H6 **Halban** Hövsgöl, N Mongolia 49.30N 97.33E
103 K14 **Halberstadt** Sachsen-Anhalt, C Germany 51.54N 11.04E
192 M12 **Halcombe** Manawatu-Wanganui, North Island, NZ 40.09S 175.30E
97 I16 **Halden** *prev.* Fredrikshald. Østfold, S Norway 59.07N 11.19E
102 L13 **Haldensleben** Sachsen-Anhalt, C Germany 52.18N 11.25E
Háldi *see* Halti
159 S17 **Haldia** West Bengal, NE India 22.07N 88.06E
158 K10 **Haldwāni** Uttar Pradesh, N India 29.13N 79.31E
40 F10 **Haleakala** *crater* Maui, Hawaii, USA, C Pacific Ocean 20.45N 156.12W
45 N4 **Hale Center** Texas, SW USA 34.03N 101.50W
101 J18 **Halen** Limburg, NE Belgium 50.35N 5.08E
25 O2 **Haleyville** Alabama, S USA 34.13N 87.37W
79 O17 **Half Assini** SW Ghana 5.03N 2.57W
37 R8 **Half Dome** ▲ California, W USA 37.46N 119.27W
193 C25 **Halfmoon Bay** *var.* Oban. Stewart Island, Southland, NZ 46.52S 168.08E
190 E5 **Half Moon Lake** *salt lake* South Australia
169 R7 **Halhgol** Dornod, E Mongolia 47.57N 118.07E
Haliacmon *see* Aliákmonas
Halibān *see* Ḩalabān
12 I13 **Haliburton** Ontario, SE Canada 45.03N 78.32W
12 I12 **Haliburton Highlands** *var.* Madawaska Highlands. *hill range* Ontario, SE Canada
11 Q15 **Halifax** Nova Scotia, SE Canada 44.37N 63.34W
99 L17 **Halifax** N England, UK 53.43N 1.52W
23 W8 **Halifax** North Carolina, SE USA 36.18N 77.35W
23 U7 **Halifax** Virginia, NE USA 36.46N 78.55W
11 Q15 **Halifax** ✈ Nova Scotia, SE Canada 44.33N 63.48W
149 T13 **Halīl Rūd** *seasonal river* SE Iran
144 I6 **Ḩalīmah** ▲ Lebanon/Syria 34.12N 36.37E
168 G8 **Haliun** Govĭ-Altay, W Mongolia 45.55N 96.06E
120 I3 **Haljala** *Ger.* Halljal. Lääne-Virumaa, N Estonia 59.25N 26.18E
41 Q4 **Halkett, Cape** *headland* Alaska, USA 70.48N 152.11W
Halkida *see* Chalkída
98 J6 **Halkirk** N Scotland, UK 58.30N 3.29W
13 X7 **Hall** ⁓ Quebec, SE Canada
Hall *see* Schwäbisch Hall
95 H15 **Hälla** Västerbotten, N Sweden 63.55N 17.19E
98 J6 **Halladale** ⁓ N Scotland, UK
97 J21 **Halland** ◐ *county* S Sweden
25 Z15 **Hallandale** Florida, SE USA 25.58N 80.09W
97 K22 **Hallandsås** *physical region* S Sweden
15 M2 **Hall Beach** Nunavut, N Canada 68.10N 81.55W
101 G19 **Halle** *Fr.* Hal. Vlaams Brabant, C Belgium 50.43N 4.13E
103 M15 **Halle** *var.* Halle an der Saale. Sachsen-Anhalt, C Germany 51.28N 11.58E
Halle an der Saale *see* Halle
37 W3 **Halleck** Nevada, W USA 40.57N 115.27W
97 L15 **Hällefors** Örebro, C Sweden 59.48N 14.27E
97 N16 **Hälleforsnäs** Södermanland, C Sweden 59.10N 16.30E
111 Q6 **Hallein** Salzburg, N Austria 47.40N 13.06E
103 L15 **Halle-Neustadt** Sachsen-Anhalt, C Germany 51.28N 11.54E
45 U12 **Hallettsville** Texas, SW USA 29.27N 96.57W
205 N4 **Halley** *UK research station* Antarctica 75.42S 26.30W
30 L4 **Halliday** North Dakota, N USA 47.19N 102.19W
39 S2 **Halligan Reservoir** ⊠ Colorado, C USA
102 G7 **Halligen** *island group* N Germany
96 G13 **Hallingdal** *valley* S Norway
40 J12 **Hall Island** *island* Alaska, USA
Hall Island *see* Maiana
201 P15 **Hall Islands** *island group* C Micronesia
120 H6 **Halliste** ⁓ S Estonia
Halljal *see* Haljala
95 I15 **Hällnäs** Västerbotten, N Sweden 64.19N 19.41E
31 R2 **Hallock** Minnesota, N USA 48.47N 96.56W
16 Oo3 **Hall Peninsula** *peninsula* Baffin Island, Nunavut, NE Canada
22 F9 **Halls** Tennessee, S USA 35.52N 89.24W
97 M16 **Hallsberg** Örebro, C Sweden 59.04N 15.07E
189 N5 **Halls Creek** Western Australia 18.17S 127.39E
190 L12 **Halls Gap** Victoria, SE Australia 37.09S 142.30E
97 N15 **Hallstahammar** Västmanland, C Sweden 59.25N 16.16E
111 R6 **Hallstatt** Salzburg, W Austria 47.32N 13.39E
111 R6 **Hallstatter See** ◎ C Austria
97 P14 **Hallstavik** Stockholm, C Sweden 60.12N 18.45E
45 X7 **Hallsville** Texas, SW USA 32.31N 94.30W
105 P1 **Halluin** Nord, N France 50.46N 3.07E
Halmahera, Laut *see* Halmahera Sea
175 T7 **Halmahera, Pulau** *prev.* Djailolo, Gilolo, Jailolo. *island* E Indonesia
175 Tt8 **Halmahera Sea** *Ind.* Laut Halmahera. *sea* E Indonesia
97 J21 **Halmstad** Halland, S Sweden 56.41N 12.48E
175 T11 **Halong** Pulau Ambon, E Indonesia 3.39S 128.13E
121 N15 **Halowchyn** *Rus.* Golovchin. Mahilyowskaya Voblasts', E Belarus 54.03N 29.52E
97 H20 **Hals** Nordjylland, N Denmark 57.00N 10.19E
96 F8 **Halsa** Møre og Romsdal, S Norway 63.04N 8.13E
121 I15 **Hal'shany** *Rus.* Gol'shany. Hrodzyenskaya Voblasts', W Belarus 54.15N 26.01E
Hälsingborg *see* Helsingborg
31 R5 **Halstad** Minnesota, N USA 47.21N 96.49W
29 N6 **Halstead** Kansas, C USA 38.00N 97.30W
101 G15 **Halsteren** Noord-Brabant, S Netherlands 51.31N 4.16E
95 L16 **Halsua** Länsi-Suomi, W Finland 63.28N 24.10E
103 E14 **Haltern** Nordrhein-Westfalen, W Germany 51.45N 7.10E
94 J9 **Halti** *var.* Haltiatunturi, *Lapp.* Háldi. ▲ Finland/Norway 69.18N 21.19E
Haltiatunturi *see* Halti
118 J6 **Halych** Ivano-Frankivs'ka Oblast', W Ukraine 49.08N 24.44E
Halycus *see* Platani
105 P3 **Ham** Somme, N France 49.46N 3.03E
Hama *see* Ḩamāh
170 Ee12 **Hamada** Shimane, Honshū, SW Japan 34.54N 132.07E
148 L6 **Hamadān** *anc.* Ecbatana. Hamadān, W Iran 34.50N 48.31E
148 L6 **Hamadān** *off.* Ostān-e Hamadān. ◐ *province* W Iran
144 I5 **Ḩamāh** *var.* Hama; *anc.* Epiphania, *Bibl.* Hamath. Ḩamāh, W Syria 35.09N 36.43E
144 I5 **Ḩamāh** *off.* Muḩāfaẓat Ḩamāh, *var.* Hama. ◐ *governorate* C Syria
171 I17 **Hamakita** Shizuoka, Honshū, S Japan 34.46N 137.46E
172 O4 **Hamamasu** Hokkaidō, NE Japan 43.37N 141.24E
171 I17 **Hamamatsu** *var.* Hamamatu. Shizuoka, Honshū, S Japan 34.43N 137.45E
Hamamatu *see* Hamamatsu
172 Qq7 **Hamanaka** Hokkaidō, NE Japan 43.05N 145.05E
171 I17 **Hamana-ko** ◎ Honshū, S Japan
96 I13 **Hamar** *prev.* Storhammer. Hedmark, S Norway 60.48N 11.04E
147 U10 **Ḩamārīr al Kidan, Qalamat** *well* E Saudi Arabia 21.40N 53.13E
170 G12 **Hamasaka** Hyōgo, Honshū, SW Japan 35.37N 134.27E
Hamath *see* Ḩamāh
172 Pp2 **Hamatonbetsu** Hokkaidō, NE Japan 45.07N 142.21E
161 K26 **Hambantota** Southern Province, SE Sri Lanka 6.11N 81.10E
194 G10 **Hambili** ⁓ NW PNG
Hambourg *see* Hamburg
102 J9 **Hamburg** Hamburg, N Germany 53.33N 10.02E
29 V14 **Hamburg** Arkansas, C USA 33.13N 91.48W
31 S16 **Hamburg** Iowa, C USA 40.36N 95.39W
20 D10 **Hamburg** New York, NE USA 42.40N 78.49W
102 I10 **Hamburg** *Fr.* Hambourg. ◐ *state* N Germany
154 K5 **Hamdam Āb, Dasht-e** *Pash.* Dasht-i Hamdamab. ▲▲ W Afghanistan
Hamdamab, Dasht-i *see* Hamdam Āb, Dasht-e
20 M13 **Hamden** Connecticut, NE USA 41.22N 72.55W
146 K6 **Ḩamḑ, Wādī al** *dry watercourse* W Saudi Arabia
95 K18 **Hämeenkyrö** Länsi-Suomi, W Finland 61.39N 23.10E
95 L19 **Hämeenlinna** *Swe.* Tavastehus. Etelä-Suomi, S Finland 61.00N 24.25E
HaMelaḩ, Yam *see* Dead Sea
Hamelin *see* Hameln
102 I13 **Hameln** *Eng.* Hamelin. Niedersachsen, N Germany 52.05N 9.21E
188 I8 **Hamersley Range** ▲▲ Western Australia
169 Y12 **Hamgyŏng-sanmaek** ▲▲ N North Korea
169 X13 **Hamhŭng** C North Korea 39.53N 127.31E
165 O6 **Hami** *var.* Ha-mi, *Uigh.* Kumul, Qomul. Xinjiang Uygur Zizhiqu, NW China 42.48N 93.27E
145 X10 **Ḩāmid Amīn** E Iraq 32.06N 46.53E
147 W11 **Hamīdān, Khawr** *oasis* SE Saudi Arabia 20.25N 54.43E

◆ Country
● Country Capital
◇ Dependent Territory
○ Dependent Territory Capital
◐ Administrative Region
✈ International Airport
▲ Mountain
▲▲ Mountain Range
Volcano
⁓ River
◎ Lake
⊠ Reservoir

144 H5 **Ḥamīdīyah** var. Hamīdīyé. Ṭarṭūs, W Syria 34.43N 35.58E

116 L12 **Hamidiye** Edirne, NW Turkey 41.09N 26.40E

Hamidiyé see Ḥamīdīyah

190 L12 **Hamilton** Victoria, SE Australia 37.45S 142.04E

66 B12 **Hamilton** ○ (Bermuda) C Bermuda 32.18N 64.48W

12 G16 **Hamilton** Ontario, S Canada 43.15N 79.49W

192 M7 **Hamilton** Waikato, North Island, NZ 37.48S 175.15E

98 I12 **Hamilton** S Scotland, UK 55.46N 4.03W

25 N3 **Hamilton** Alabama, S USA 34.08N 87.59W

40 M10 **Hamilton** Alaska, USA 62.54N 163.53W

32 J13 **Hamilton** Illinois, N USA 40.24N 91.20W

29 S3 **Hamilton** Missouri, C USA 39.44N 94.00W

35 R8 **Hamilton** Montana, NW USA 46.15N 114.09W

45 S8 **Hamilton** Texas, SW USA 31.42N 98.07W

12 G16 **Hamilton** × Ontario, SE Canada 43.12N 79.54W

66 I6 **Hamilton Bank** undersea feature SE Labrador Sea

190 E1 **Hamilton Creek** seasonal river South Australia

11 R8 **Hamilton Inlet** inlet Newfoundland, E Canada

29 T12 **Hamilton, Lake** ☒ Arkansas, C USA

37 W6 **Hamilton, Mount** ▲ Nevada, W USA 39.15N 115.30W

77 S8 **Ḥamīm, Wādī al** ♂ NE Libya

95 N19 **Hamina** Swe. Fredrikshamn. Etelä-Suomi, S Finland 60.33N 27.15E

9 W16 **Hamiota** Manitoba, S Canada 50.13N 100.37W

158 L13 **Hamirpur** Uttar Pradesh, N India 25.57N 80.07E

Ḥamīs Musaiṭ see Khamis Mushayt

23 T11 **Hamlet** North Carolina, SE USA 34.52N 79.41W

45 P6 **Hamlin** Texas, SW USA 32.52N 100.07W

23 P5 **Hamlin** West Virginia, NE USA 38.16N 82.06W

33 O7 **Hamlin Lake** ☒ Michigan, N USA

103 F14 **Hamm** var. Hamm in Westfalen. Nordrhein-Westfalen, W Germany 51.39N 7.49E

Ḥammāmāt, Khalīj al see Hammamet, Golfe de

77 N1 **Hammamet, Golfe de** Ar. Khalīj al Ḥammāmāt. gulf NE Tunisia

145 R3 **Ḥammām al 'Alīl** N Iraq 36.07N 43.15E

145 X12 **Ḥammār, Hawr al** ♂ SE Iraq

95 J20 **Hammarland** Åland, SW Finland 60.13N 19.45E

95 H16 **Hammarstrand** Jämtland, C Sweden 63.07N 16.27E

95 O17 **Hammaslahti** Itä-Suomi, E Finland 62.26N 29.58E

101 F17 **Hamme** Oost-Vlaanderen, NW Belgium 51.06N 4.07E

102 H10 **Hamme** ♂ NW Germany

97 G22 **Hammel** Århus, C Denmark 56.15N 9.52E

103 I18 **Hammelburg** Bayern, C Germany 50.06N 9.50E

101 H18 **Hamme-Mille** Wallon Brabant, C Belgium 50.48N 4.42E

102 H10 **Hamme-Oste-Kanal** canal NW Germany

95 G16 **Hammerdal** Jämtland, C Sweden 63.34N 15.19E

94 K8 **Hammerfest** Finnmark, N Norway 70.40N 23.40E

103 D14 **Hamminkeln** Nordrhein-Westfalen, W Germany 51.43N 6.36E

Hamm in Westfalen see Hamm

28 K10 **Hammon** Oklahoma, C USA 35.37N 99.22W

33 N11 **Hammond** Indiana, N USA 41.35N 87.30W

24 K8 **Hammond** Louisiana, S USA 30.30N 90.27W

101 K20 **Hamoir** Liège, E Belgium 50.28N 5.35E

101 J21 **Hamois** Namur, SE Belgium 50.21N 5.09E

101 K16 **Hamont** Limburg, NE Belgium 51.15N 5.33E

193 F22 **Hampden** Otago, South Island, NZ 45.18S 170.49E

21 R6 **Hampden** Maine, NE USA 44.44N 68.51W

99 M23 **Hampshire** cultural region S England, UK

11 O15 **Hampton** New Brunswick, SE Canada 45.30N 65.49W

29 U14 **Hampton** Arkansas, C USA 33.33N 92.28W

31 V12 **Hampton** Iowa, C USA 42.44N 93.12W

21 P10 **Hampton** New Hampshire, NE USA 42.55N 70.48W

23 R14 **Hampton** South Carolina, SE USA 32.52N 81.06W

23 X7 **Hampton** Tennessee, S USA 36.16N 82.10W

23 X7 **Hampton** Virginia, NE USA 37.01N 76.21W

96 L11 **Hamra** Gävleborg, C Sweden 61.40N 15.00E

82 D10 **Ḥamra esh Sheikh** Northern Kordofan, C Sudan 14.37N 27.55E

145 S5 **Ḥamrīn, Jabal** ▲ N Iraq

123 J16 **Ḥamrun** C Malta 35.53N 14.28E

178 K14 **Ham Thuận Nam** Bình Thuận, S Vietnam 10.49N 107.49E

176 Ww11 **Hamūkti** var. Hamuku. Irian Jaya, E Indonesia 3.18S 135.00E

Hāmūn, Daryācheh-ye see Ṣāberī, Hāmūn-e/Sīstān, Daryācheh-ye

Hamwih see Southampton

40 G10 **Hana** Haw. Hāna. Maui, Hawaii, USA, C Pacific Ocean 20.45N 155.59W

23 S14 **Hanahan** South Carolina, SE USA 32.55N 80.01W

40 F10 **Hanalei** Kauai, Hawaii, USA, C Pacific Ocean 22.12N 159.30W

178 Kk10 **Ha Nam** Quang Nam–Đà Nẵng, C Vietnam 15.42N 108.24E

171 Mm11 **Hanamaki** Iwate, Honshū, C Japan 39.25N 141.04E

40 F10 **Hanamanioa, Cape** headland Maui, Hawaii, USA, C Pacific Ocean 20.34N 156.22W

202 B16 **Hanan** × (Alofi) SW Niue

103 H18 **Hanau** Hessen, W Germany 50.06N 8.56E

15 J7 **Hanbury** ♂ Northwest Territories, NW Canada

Hânceşti see Hînceşti

8 L7 **Hanceville** British Columbia, SW Canada 51.54N 122.56W

25 P3 **Hanceville** Alabama, S USA 34.03N 86.46W

166 L6 **Hancheng** Shaanxi, C China 35.22N 110.27E

23 V2 **Hancock** Maryland, NE USA 39.42N 78.10W

32 M3 **Hancock** Michigan, N USA 47.07N 88.34W

31 S8 **Hancock** Minnesota, N USA 45.30N 95.47W

20 I12 **Hancock** New York, NE USA 41.57N 75.15W

82 Q12 **Handa** Bari, NE Somalia 10.35N 51.09E

167 O13 **Handan** var. Han-tan. Hebei, E China 36.34N 114.28E

97 P16 **Handen** Stockholm, C Sweden 59.12N 18.09E

83 J22 **Handeni** Tanga, E Tanzania 5.25S 38.04E

39 Q7 **Handies Peak** ▲ Colorado, C USA 37.54N 107.30W

113 J19 **Handlová** Ger. Krickerhäu. Hung. Nyitrabánya; prev. Ger. Kriegerhaj. Trenčiansky Kraj, W Slovakia 48.45N 18.45E

171 K17 **Haneda** × (Tōkyō) Tōkyō, Honshū, S Japan 35.33N 139.45E

144 F13 **HaNegev** Eng. Negev. desert S Israel

37 Q11 **Hanford** California, W USA 36.19N 119.39W

203 V16 **Hanga Roa** Easter Island, Chile, E Pacific Ocean 27.09S 109.25W

168 H7 **Hangayn Nuruu** ▲ C Mongolia

Hang-chou/Hangchow see Hangzhou

97 N20 **Hänger** Jönköping, S Sweden 57.06N 13.58E

Hangō see Hanko

167 R9 **Hangzhou** var. Hang-chou, Hangchow. Zhejiang, SE China 30.18N 120.07E

168 F5 **Hanhöhiy Uul** ▲ NW Mongolia

143 P15 **Hani** Diyarbakır, SE Turkey 38.25N 40.22E

Hanía see Chaniá

147 R11 **Ḥanīsh al Kabīr, Jazīrat al** island SW Yemen

Hanka, Lake see Khanka, Lake

95 M17 **Hankasalmi** Länsi-Suomi, W Finland 62.25N 26.27E

31 R7 **Hankinson** North Dakota, N USA 46.04N 96.54W

95 K20 **Hanko** Swe. Hangö. Etelä-Suomi, SW Finland 59.50N 23.00E

Han-kou/Han-k'ou/Hankow see Wuhan

38 M6 **Hanksville** Utah, W USA 38.21N 110.43W

158 K6 **Hanle** Jammu and Kashmir, NW India 32.46N 79.01E

193 I17 **Hanmer Springs** Canterbury, South Island, NZ 42.30S 172.48E

9 R16 **Hanna** Alberta, SW Canada 51.37N 111.55W

29 V3 **Hannibal** Missouri, C USA 39.42N 91.23W

188 M3 **Hann, Mount** ▲ Western Australia 15.53S 125.48E

102 H12 **Hannover** Eng. Hanover. Niedersachsen, NW Germany 52.23N 9.43E

101 J19 **Hannut** Liège, E Belgium 50.40N 5.04E

122 J17 **Hanöbukten** bay S Sweden

178 Jj6 **Ha Nôi** Eng. Hanoi, Fr. Ha noi. ● (Vietnam) N Vietnam 21.01N 105.52E

12 F14 **Hanover** Ontario, S Canada 44.22N 81.01W

33 P15 **Hanover** Indiana, N USA 38.42N 85.28W

20 G16 **Hanover** Pennsylvania, NE USA 39.46N 76.57W

23 W6 **Hanover** Virginia, NE USA 37.44N 77.21W

Hanover see Hannover

65 G20 **Hanover, Isla** island S Chile

29 W9 **Hardy** Arkansas, C USA 36.19N 91.28W

96 D10 **Hareid** Møre og Romsdal, S Norway 62.21N 6.01E

15 G5 **Hare Indian** ♂ Northwest Territories, NW Canada

101 D18 **Harelbeke** var. Harlebeke. West-Vlaanderen, W Belgium 50.51N 3.19E

164 F3 **Hantengri Feng** var. Pik Khan-Tengri. ▲ China/Kazakhstan see also Khan-Tengri, Pik 42.17N 80.11E

121 I18 **Hantsavichy** Pol. Hancewicze, Rus. Gantsevichi. Brestskaya Voblasts', SW Belarus 52.45N 26.27E

16 N2 **Hantzsch** ♂ Baffin Island, Nunavut, NE Canada

158 G9 **Hanumāngarh** Rājasthān, NW India 29.17N 74.21E

191 O10 **Hanwood** New South Wales, SE Australia 34.19S 146.03E

167 N8 **Hanyang** see Caidian

Hanyang see Wuhan

166 H10 **Hanyuan** var. Fulin. Sichuan, C China 29.22N 102.39E

167 J7 **Hanzhong** Shaanxi, C China 33.12N 106.59E

203 W11 **Hao** atoll Îles Tuamotu, C French Polynesia

159 T10 **Hāora** prev. Howrah. West Bengal, NE India 22.34N 88.18E

174 V7 **Hari, Batang** prev. Djambi. ♂ Sumatera, W Indonesia

158 J9 **Haridwār** prev. Hardwar. Uttar Pradesh, N India 29.58N 78.09E

161 L13 **Harihar** Karnātaka, W India 14.33N 75.43E

193 F18 **Harihari** West Coast, South Island, NZ 43.09S 170.35E

36 M1 **Happy Camp** California, W USA 41.48N 123.24W

11 O13 **Happy Valley-Goose Bay** prev. Goose Bay. Newfoundland, E Canada 53.19N 60.24W

Hapsal see Haapsalu

158 J10 **Hāpur** Uttar Pradesh, N India 28.43N 77.46E

144 F12 **HaQatan, HaMakhtesh** ▲ S Israel

146 I1 **Ḥaql** Tabūk, NW Saudi Arabia 29.18N 34.58E

176 Vv13 **Har** Pulau Kai Besar, E Indonesia 5.21S 133.09E

168 M8 **Haraat** Dundgovi, C Mongolia 46.30N 107.39E

147 R8 **Ḥaraḍ** var. Haradh. Ash Sharqīyah, E Saudi Arabia 24.08N 49.01E

Ḥarad see Ḥaraḍ

120 N12 **Haradok** Rus. Gorodok. Vitsyebskaya Voblasts', N Belarus 55.27N 30.00E

94 J13 **Harads** Norrbotten, N Sweden 66.04N 21.04E

121 G19 **Haradzyets** Rus. Gorodets. Brestskaya Voblasts', SW Belarus 52.11N 24.41E

121 J17 **Haradzyeya** Rus. Gorodeya. Minskaya Voblasts', C Belarus 53.18N 26.33E

203 V10 **Haraiki** atoll Îles Tuamotu, C French Polynesia

171 Ll14 **Haramachi** Fukushima, Honshū, E Japan 37.39N 140.55E

120 M12 **Harany** Rus. Gorany. Vitsyebskaya Voblasts', N Belarus 55.25N 29.03E

85 L16 **Harare** prev. Salisbury. ● (Zimbabwe) Mashonaland East, NE Zimbabwe 17.47S 31.03E

85 L16 **Harare** × Mashonaland East, NE Zimbabwe 17.51S 31.06E

80 J10 **Haraz-Djombo** Batha, C Chad 14.10N 19.35E

121 O16 **Harbavichy** Rus. Gorbovichi. Mahilyowskaya Voblasts', E Belarus 53.51N 30.42E

78 J16 **Harbel** W Liberia 6.19N 10.19W

169 W8 **Harbin** var. Haerbin, Ha-erh-pin, Kharbin; prev. Haerhpin, Pingkiang, Pinkiang. Heilongjiang, NE China 45.45N 126.40E

33 S5 **Harbor Beach** Michigan, N USA 43.50N 82.39W

11 T13 **Harbour Breton** Newfoundland, E Canada 47.28N 55.49W

67 D25 **Harbours, Bay of** bay East Falkland, Falkland Islands

97 G24 **Hårby** var. Haarby. Fyn, C Denmark 55.13N 10.07E

38 I13 **Harcuvar Mountains** ▲ Arizona, SW USA

110 I7 **Hard** Vorarlberg, NW Austria 47.28N 9.41E

160 H11 **Harda Khäs** Madhya Pradesh, C India 22.22N 77.06E

96 D13 **Hardanger** physical region S Norway

96 D9 **Hardangerfjorden** fjord S Norway

96 D13 **Hardangerjøkulen** glacier S Norway

96 E13 **Hardangervidda** plateau S Norway

85 D20 **Hardap** ♦ district S Namibia

23 R15 **Hardeeville** South Carolina, SE USA 32.18N 81.04W

100 L5 **Hardegarijp** Fris. Hurdegaryp. Friesland, N Netherlands 53.13N 5.57E

100 O9 **Hardenberg** Overijssel, E Netherlands 52.34N 6.37E

100 K10 **Harderwijk** Gelderland, C Netherlands 52.21N 5.36E

32 J14 **Hardin** Illinois, N USA 39.10N 90.37W

35 V11 **Hardin** Montana, NW USA 45.43N 107.34W

11 L13 **Harding, Lake** ☒ Alabama/Georgia, SE USA

22 J6 **Hardinsburg** Kentucky, S USA 37.46N 86.27W

100 L13 **Hardinxveld-Giessendam** Zuid-Holland, C Netherlands 51.52N 4.49E

9 R15 **Hardisty** Alberta, SW Canada 52.42N 111.22W

158 L12 **Hardoi** Uttar Pradesh, N India 27.22N 80.06E

30 I14 **Hardwick** Georgia, SE USA 33.03N 83.13W

21 N7 **Hardwick** Vermont, NE USA 44.30N 72.21W

36 L11 **Hardy** ♂ S England, UK 60.13N 18.25E

82 M13 **Hargeysa** var. Hargeisa. Woqooyi Galbeed, NW Somalia 9.31N 44.06E

118 H9 **Harghita** ♦ county NE Romania

45 S17 **Hargill** Texas, SW USA 26.26N 98.00W

168 J8 **Harhorin** Övörhangay, C Mongolia 47.13N 102.48E

165 Q9 **Har Hu** ☒ C China

Hariana see Haryāna

98 F7 **Harris, Sound of** strait NW Scotland, UK

29 W7 **Harrisonville** Missouri, C USA 38.39N 94.21W

116 L6 **Harris Ridge** see Lomonosov Ridge

199 J3 **Harris Seamount** undersea feature N Pacific Ocean 46.09N 161.25W

98 F7 **Harris** physical region NW Scotland, UK

29 X10 **Harrisburg** Arkansas, C USA 35.33N 90.43W

32 M17 **Harrisburg** Illinois, N USA 37.44N 88.32W

30 I14 **Harrisburg** Nebraska, C USA 41.31N 103.43W

34 F12 **Harrisburg** Oregon, NW USA 44.15N 123.10W

155 R6 **Harrisburg** state capital Pennsylvania, NE USA 40.16N 76.52W

33 R9 **Harrison** Michigan, N USA 44.01N 84.49W

30 J14 **Harrison** Nebraska, C USA 42.39N 103.53W

45 T12 **Harrison** Arkansas, C USA 36.13N 93.06W

22 M9 **Harriman** Tennessee, S USA 35.57N 84.33W

31 N14 **Harrington** North Dakota, N USA 47.43N 99.55W

11 R11 **Harrington Harbour** Quebec, E Canada 50.34N 59.29W

66 H6 **Harrington Sound** bay Bermuda, NW Atlantic Ocean

99 M21 **Harrow** SE England, UK

147 N17 **Ḥaryān, Ṭawī al** spring/well NE Oman 21.56N 58.33E

103 J14 **Harz** ▲ C Germany

Ḥasakah see Al Ḥasakah

171 P9 **Hasama** Miyagi, Honshū, C Japan 38.42N 141.09E

142 J15 **Hasan Dağı** ▲ C Turkey 38.09N 34.15E

145 Y9 **Ḥasan Ibn Ḥassūn** C Iraq 32.24N 44.13E

155 R6 **Ḥasan Khēl** var. Ahmad Khel. Paktiā, SE Afghanistan 33.46N 69.37E

102 F12 **Hase** ♂ NW Germany

102 F12 **Haselberg** see Krasnoznamensk

102 F12 **Haseldorf** Niedersachsen, NW Germany 52.40N 7.28E

168 K9 **Hashaat** Dundgovi, C Mongolia 45.09N 104.51E

Hashemite Kingdom of Jordan see Jordan

171 G16 **Hashima** var. Hasimoto. Wakayama, Honshū, SW Japan 34.18N 135.34E

147 X7 **Ḥāsik** Oman 17.22N 55.18E

95 P14 **Hāsilpur** Punjab, E Pakistan 29.44N 72.33E

95 N15 **Haskell** Oklahoma, C USA 35.49N 95.40W

45 Q6 **Haskell** Texas, SW USA 33.09N 99.43W

116 M16 **Hasköy** Edirne, NW Turkey 41.37N 26.51E

97 E18 **Hasle** Bornholm, E Denmark 55.12N 14.43E

99 P22 **Haslemere** SE England, UK 51.06N 0.45W

104 L4 **Hasparren** Pyrénées-Atlantiques, SW France 43.22N 1.18W

144 F5 **Ḥasbaiya** var. Hasbaya. ♂ S Lebanon

144 J15 **Ḥāṣ, Jabal al** ▲ NW Syria

96 N10 **Hassela** Gävleborg, C Sweden 62.06N 16.45E

101 K18 **Hasselt** Limburg, NE Belgium 50.55N 5.19E

100 M9 **Hasselt** Overijssel, E Netherlands 52.36N 6.06E

Hassetché see Al Ḥasakah

103 J18 **Hassfurt** Bayern, C Germany 50.02N 10.32E

76 I3 **Hassi Bel Guebbour** E Algeria 28.41N 6.29E

76 L8 **Hassi Messaoud** E Algeria 31.41N 6.10E

97 K22 **Hässleholm** Skåne, S Sweden 56.09N 13.45E

191 O13 **Hastings** Victoria, SE Australia 38.18S 145.12E

192 O11 **Hastings** Hawke's Bay, North Island, NZ 39.39S 176.51E

99 P23 **Hastings** SE England, UK 50.51N 0.36E

33 R9 **Hastings** Michigan, N USA 42.37N 85.16W

31 W9 **Hastings** Minnesota, N USA 44.44N 92.51W

31 P16 **Hastings** Nebraska, C USA 40.35N 98.23W

97 K22 **Hästveda** Skåne, S Sweden 56.16N 13.55E

94 J8 **Hasvik** Finnmark, N Norway 70.29N 22.08E

39 V6 **Haswell** Colorado, C USA 38.27N 103.09W

168 I10 **Hatansuudal** Bayankhongor, C Mongolia 44.34N 100.41E

169 P9 **Hatavch** Sühbaatar, E Mongolia 46.10N 112.57E

142 K17 **Hatay** ♦ province S Turkey

39 R15 **Hatch** New Mexico, USA 32.40N 107.10W

38 K7 **Hatch** Utah, W USA 37.39N 112.25W

22 J6 **Hatchie River** ♂ Tennessee, S USA

118 L9 **Haţeg** Ger. Wallenthal, Hung. Hátszeg; prev. Hatzeg, Hötzing. Hunedoara, SW Romania

172 Oo17 **Hateruma-jima** island Yaeyama-shotō, SW Japan

168 I5 **Hatgal** Hövsgöl, N Mongolia 50.24N 100.12E

159 V16 **Hathazari** Chittagong, SE Bangladesh 22.30N 91.46E

147 T13 **Ḥaṭībah, Ḥiṣá'** oasis NE Yemen 17.46N 51.14E

178 Ii14 **Ha Tiên** Kiên Giang, S Vietnam 10.24N 104.30E

178 Jj8 **Ha Tinh** Ha Tinh, N Vietnam 18.21N 105.55E

172 M5 **Hatizyo** see Hachijō

144 F13 **Hatira, Haré** hill range S Israel

178 J6 **Hat Lot** Son La, N Vietnam 21.07N 104.10E

102 H12 **Havelländ Grosse** var. Hauptkanal. canal NE Germany

12 J15 **Havelock** Ontario, SE Canada 44.22N 77.57W

193 H16 **Havelock** Marlborough, South Island, NZ 41.17S 173.46E

23 X11 **Havelock** North Carolina, SE USA 34.53N 76.55W

192 O11 **Havelock North** Hawke's Bay, North Island, NZ 39.40S 176.53E

100 H10 **Havelte** Drenthe, NE Netherlands 52.46N 6.14E

29 R6 **Haven** Kansas, C USA 37.54N 97.46W

99 J20 **Haverfordwest** SW Wales, UK 51.49N 4.57W

99 P20 **Haverhill** E England, UK 52.04N 0.26E

21 O10 **Haverhill** Massachusetts, NE USA 42.46N 71.02W

95 G17 **Haverö** Västernorrland, C Sweden 62.25N 15.04E

113 I17 **Havířov** Ostravský Kraj, E Czech Republic 49.47N 18.30E

113 E17 **Havlíčkův Brod** Ger. Deutsch-Brod; prev. Německý Brod. Jihlavský Kraj, C Czech Republic 49.41N 15.47E

94 I7 **Havøysund** Finnmark, N Norway 70.59N 24.39E

35 T7 **Havre** Montana, NW USA 48.33N 109.40W

Havre see Le Havre

101 H20 **Havré** Hainaut, S Belgium 50.28N 4.03E

11 P11 **Havre-St-Pierre** Quebec, E Canada 50.16N 63.36W

116 M12 **Havsa** Edirne, NW Turkey 41.33N 26.49E

40 D8 **Hawaii** off. State of Hawaii; also known as Aloha State, Paradise of the Pacific. ♦ state USA, C Pacific Ocean

40 G12 **Hawaii** Haw. Hawai'i. island Hawaiian Islands, USA, C Pacific Ocean

199 I3 **Hawaiian Islands** prev. Sandwich Islands. island group Hawaii, USA, C Pacific Ocean

199 J6 **Hawaiian Ridge** undersea feature N Pacific Ocean

199 K6 **Hawaiian Trough** undersea feature N Pacific Ocean

31 R12 **Hawarden** Iowa, C USA 43.00N 96.29W

Hawash see Āwash

145 P6 **Hawbayn al Gharbīyah** C Iraq 34.24N 42.06E

193 D21 **Hawea, Lake** ☒ South Island, NZ

192 K11 **Hawera** Taranaki, North Island, NZ 39.35S 174.14E

22 J5 **Hawesville** Kentucky, S USA 37.53N 86.44W

40 G11 **Hawi** Haw. Hāwī. Hawaii, USA, C Pacific Ocean 20.13N 155.49W

98 J13 **Hawick** SE Scotland, UK 55.24N 2.49W

145 X4 **Ḥawījah** C Iraq 35.15N 43.54E

145 Y9 **Ḥawr al** see Ḥammār, Hawr al

193 E21 **Hawkdun Range** ▲ South Island, NZ

192 P10 **Hawke Bay** bay North Island, NZ

190 O6 **Hawker** South Australia 31.54S 138.25E

192 N11 **Hawke's Bay** off. Hawkes Bay Region. ♦ region North Island, NZ

155 O16 **Hawkes Bay** bay SE Pakistan

13 N12 **Hawkesbury** Ontario, SE Canada 45.35N 74.37W

Hawkeye State *see* Iowa

25 T5 **Hawkinsville** Georgia, SE USA 32.16N 83.28W

12 B7 **Hawk Junction** Ontario, S Canada 48.05N 84.34W

23 N10 **Haw Knob** ▲ North Carolina/Tennessee, SE USA 35.18N 84.01W

23 Q9 **Hawksbill Mountain** ▲ North Carolina, USA 35.54N 81.53W

35 Z16 **Haw Springs** Wyoming, C USA 41.48N 104.17W

Hawlêr *see* Arbil

31 S5 **Hawley** Minnesota, N USA 46.53N 96.18W

45 P7 **Hawley** Texas, SW USA 32.36N 99.47W

147 R14 **Hawrā'** C Yemen 15.39N 48.20E

145 P7 **Hawrān, Wadi** *dry watercourse* W Iraq

23 T9 **Haw River** �❦ North Carolina, SE USA

145 U5 **Hawshqūrah** E Iraq 34.34N 45.33E

37 S7 **Hawthorne** Nevada, W USA 38.30N 118.38W

39 W3 **Haxtun** Colorado, C USA 40.36N 102.38W

191 N9 **Hay** New South Wales, SE Australia 34.31S 144.50E

9 O10 **Hay** ☆ W Canada

176 U11 **Haya** Pulau Seram, E Indonesia 3.22S 129.31E

172 N11 **Hayachine-san** ▲ Honshū, C Japan 39.31N 141.28E

105 S4 **Hayange** Moselle, NE France 49.20N 6.04E

HaYarden *see* Jordan

Hayastani Hanrapetut'yun *see* Armenia

41 N9 **Haycock** Alaska, USA 65.12N 161.10W

38 M14 **Hayden** Arizona, SW USA 33.00N 110.46W

39 Q3 **Hayden** Colorado, C USA 40.29N 107.15W

30 M10 **Hayes** South Dakota, N USA 44.20N 101.01W

9 X13 **Hayes** ☆ Manitoba, C Canada

15 Kk11 **Hayes** ☆ Nunavut, NE Canada

30 M16 **Hayes Center** Nebraska, USA 40.28N 101.01W

41 S10 **Hayes, Mount** ▲ Alaska, USA 63.37N 146.43W

23 N11 **Hayesville** North Carolina, SE USA 35.15N 84.15W

37 N10 **Hayford Peak** ▲ Nevada, W USA 36.40N 115.10W

36 M3 **Hayfork** California, W USA 40.33N 123.10W

Hayir, Qasr al *see* Hayr al Gharbī, Qasr al

169 P8 **Haylaastay** Sühbaatar, E Mongolia 46.14N 113.51E

12 I12 **Hay Lake** ☉ Ontario, SE Canada

147 X11 **Haymā'** *var.* Haima. E Oman 19.58N 56.20E

142 H13 **Haymana** Ankara, C Turkey 39.25N 32.30E

144 J7 **Haymūr, Jabal** ▲ W Syria

Haynau *see* Chojnów

24 G4 **Haynesville** Louisiana, S USA

25 P6 **Hayneville** Alabama, S USA 32.13N 86.34W

116 M12 **Hayrabolu** Tekirdağ, NW Turkey 41.12N 27.08E

142 C10 **Hayrabolu Deresi** ☆ NW Turkey

144 J6 **Hayr al Gharbī, Qasr al** *var.* Qasr al Hayir, Qasr al Hir al Gharbī. *ruins* Ḥimṣ, C Syria 34.23N 37.40E

144 L5 **Hayr ash Sharqī, Qasr al** *var.* Qasr al Hir Ash Sharqī. *ruins* Ḥimṣ, C Syria 35.07N 39.06E

15 Hh9 **Hay River** Northwest Territories, W Canada 60.51N 115.42W

28 K4 **Hays** Kansas, C USA 38.52N 99.19W

30 K12 **Hay Springs** Nebraska, USA 42.40N 102.41W

67 H25 **Haystack, The** ▲ NE Saint Helena 15.55S 5.40W

29 N7 **Haysville** Kansas, C USA 37.34N 97.21W

119 O7 **Haysyn** *Rus.* Gaysin. Vinnyts'ka Oblast', C Ukraine 48.49N 29.29E

29 Y9 **Hayti** Missouri, C USA 36.13N 89.45W

31 Q9 **Hayti** South Dakota, N USA 44.39N 97.11W

119 O8 **Hayvoron** *Rus.* Gayvoron. Kirovohrads'ka Oblast', C Ukraine 48.19N 29.54E

37 N9 **Hayward** California, W USA 37.40N 122.07W

32 J4 **Hayward** Wisconsin, N USA 46.01N 91.25W

99 I23 **Haywards Heath** SE England, UK 51.00N 0.06W

149 U11 **Hazārān, Kūh-e** *var.* Kūh-e à Hazr. ▲ SE Iran 29.26N 57.15E

Hazarat Imam *see* Emām Şāḥeb

23 O7 **Hazard** Kentucky, S USA 37.15N 83.11W

143 O15 **Hazar Gölü** ☉ C Turkey

159 P15 **Hazārībāg** *var.* Hazārībāgh. Bihār, N India 24.00N 85.23E

Hazārībāgh *see* Hazārībāg

105 O1 **Hazebrouck** Nord, N France 50.43N 2.33E

32 K9 **Hazel Green** Wisconsin, N USA 42.33N 90.26W

199 Ii10 **Hazel Holme Bank** *undersea feature* S Pacific Ocean 12.49S 174.30E

8 K13 **Hazelton** British Columbia, SW Canada 55.15N 127.37W

31 N6 **Hazelton** North Dakota, N USA 46.27N 100.17W

37 S6 **Hazen** Nevada, W USA 39.33N 119.02W

30 L5 **Hazen** North Dakota, N USA 47.18N 101.37W

40 L12 **Hazen Bay** *bay* Sea of Bering Sea

145 S2 **Hazim, Bi'r** *well* C Iraq 34.50N 43.25E

25 V6 **Hazlehurst** Georgia, SE USA 31.51N 82.35W

24 K2 **Hazlehurst** Mississippi, S USA 31.51N 90.24W

20 K15 **Hazlet** New Jersey, NE USA 40.24N 74.10W

152 I9 **Hazorasp** *Rus.* Khazarasp. Khorazm Wiloyati, W Uzbekistan 41.21N 61.01E

153 R13 **Hazratishoh, Qatorkŭhi** *var.* Khrebet Khazretishi, *Rus.* Khrebet Khozretishi. ▲ S Tajikistan

155 U6 **Hazro** Punjab, E Pakistan 33.55N 72.33E

25 R7 **Headland** Alabama, S USA 31.21N 85.20W

190 C6 **Head of Bight** *headland* South Australia 31.33S 131.05E

35 N10 **Headquarters** Idaho, NW USA 46.38N 115.52W

36 M7 **Healdsburg** California, W USA 38.36N 122.52W

29 N13 **Healdton** Oklahoma, C USA 34.13N 97.29W

191 O12 **Healesville** Victoria, SE Australia 37.41S 145.31E

41 Q10 **Healy** Alaska, USA 63.51N 148.58W

181 R13 **Heard and McDonald Islands** ◇ *Australian external territory* S Indian Ocean

181 R13 **Heard Island** *island* Heard and McDonald Islands, S Indian Ocean

45 U9 **Hearne** Texas, SW USA 30.52N 96.35W

10 F12 **Hearst** Ontario, S Canada 49.42N 83.40W

204 J5 **Hearst Island** *island* Antarctica

30 L5 **Heart River** ☆ North Dakota, N USA

T13 **Heath** Ohio, N USA 40.01N 82.26W

191 N11 **Heathcote** Victoria, SE Australia 36.57S 144.43E

99 O22 **Heathrow** ✈ (London) SE England, UK 51.28N 0.27W

23 X5 **Heathsville** Virginia, NE USA 37.54N 76.25W

29 R11 **Heavener** Oklahoma, C USA 34.53N 94.36W

45 R15 **Hebbronville** Texas, SW USA 27.18N 98.40W

169 Q13 **Hebei** *var.* Hebei Sheng, Hopeh, Hopei, Ji; *prev.* Chihli. ◆ *province* E China

Hebei Sheng *see* Hebei

176 U9 **Hebera** Irian Jaya, E Indonesia 1.08S 129.54E

38 M3 **Heber City** Utah, W USA 40.29N 111.24W

29 V10 **Heber Springs** Arkansas, C USA 35.30N 91.58W

161 N5 **Hebi** Henan, C China 35.57N 114.07E

34 F11 **Hebo** Oregon, NW USA 45.10N 123.55W

98 F9 **Hebrides, Sea of the** *sea* NW Scotland, UK

11 P5 **Hebron** Newfoundland, E Canada 58.15N 62.45W

33 N11 **Hebron** Indiana, N USA 41.19N 87.12W

31 Q17 **Hebron** Nebraska, USA 40.10N 97.35W

30 L5 **Hebron** North Dakota, N USA 46.54N 102.03W

144 F11 **Hebron** *var.* Al Khalīl, El Khalīl, *Heb.* Hevron; *anc.* Kiriath-Arba. S West Bank 31.30N 35.00E

Hebrus *see* Évros/Maritsa/Meriç

97 N14 **Heby** Västmanland, C Sweden 59.55N 16.52E

8 I14 **Hecate Strait** *strait* British Columbia, W Canada

43 W14 **Hecelchakán** Campeche, SE Mexico 20.09N 90.04W

166 K13 **Hechi** *var.* Jinchengjiang. Guangxi Zhuangzu Zizhiqu, S China 24.40N 108.05E

103 H23 **Hechingen** Baden-Württemberg, S Germany 48.20N 8.58E

101 K17 **Hechtel** Limburg, NE Belgium 51.07N 5.24E

166 J9 **Hechuan** Chongqing Shi, C China 30.01N 106.15E

31 T9 **Hecla** South Dakota, N USA 45.52N 98.09W

31 T9 **Hector** Minnesota, N USA 44.44N 94.43W

95 F17 **Hede** Jämtland, C Sweden 62.25N 13.30E

97 M14 **Hedemora** Dalarna, C Sweden 60.18N 15.58E

94 K13 **Hedenäset** Norrbotten, N Sweden 66.12N 23.40E

97 G23 **Hedensted** Vejle, C Denmark 55.46N 9.43E

97 N15 **Hedesunda** Gävleborg, C Sweden 60.25N 17.00E

45 U3 **Hedley** Texas, SW USA 34.52N 100.39W

96 I12 **Hedmark** ◆ *county* S Norway

172 Pp14 **Hedo-misaki** *headland* Okinawa, SW Japan 26.55N 128.15E

31 X15 **Hedrick** Iowa, C USA 41.10N 92.18W

101 L16 **Heel** Limburg, SE Netherlands 51.13N 5.59E

102 G8 **Heel Point** *point* Wake Island 19.18N 166.39E

100 J10 **Heemskerk** Noord-Holland, W Netherlands 52.31N 4.40E

100 M10 **Heerde** Gelderland, E Netherlands 52.24N 6.01E

100 L7 **Heerenveen** *Fris.* It Hearrenfean. Friesland, N Netherlands 52.57N 5.55E

100 J8 **Heerhugowaard** Noord-Holland, NW Netherlands 52.40N 4.49E

94 G13 **Heer Land** *physical region* S Svalbard

101 M18 **Heerlen** Limburg, SE Netherlands 50.55N 6.00E

101 J19 **Heers** Limburg, NE Belgium 50.46N 5.17E

Heerwegen *see* Polkowice

100 K13 **Heesch** Noord-Brabant, SE Netherlands 51.43N 5.31E

100 K15 **Heeze** Noord-Brabant, SE Netherlands 51.22N 5.34E

144 F6 **Hefa** *var.* Haifa; *anc.* Caiffa, Caiphas, *anc.* Sycaminum. Haifa, N Israel 32.49N 34.58E

144 F6 **Hefa, Mifraz** *Eng.* Bay of Haifa. *bay* N Israel

167 Q8 **Hefei** *var.* Hofei; *hist.* Luchow. Anhui, E China 31.51N 117.20E

25 R3 **Heflin** Alabama, S USA 33.39N 85.35W

169 X7 **Hegang** Heilongjiang, NE China 47.18N 130.15E

171 Ii11 **Hegura-jima** *island* SW Japan

102 H4 **Heide** Schleswig-Holstein, N Germany 54.12N 9.06E

103 G20 **Heidelberg** Baden-Württemberg, SW Germany 49.24N 8.40E

85 J21 **Heidelberg** Gauteng, NE South Africa 26.27S 28.21E

24 M6 **Heidelberg** Mississippi, S USA 31.53N 88.58W

Heidenheim *see* Heidenheim an der Brenz

103 J22 **Heidenheim an der Brenz** *var.* Heidenheim. Baden-Württemberg, S Germany 48.40N 10.09E

111 U2 **Heidenreichstein** Niederösterreich, N Austria 48.53N 15.07E

170 E14 **Heigun-tō** *var.* Heguri-jima. *island* SW Japan

169 W5 **Heihe** *prev.* Ai-hun. Heilongjiang, NE China 50.13N 127.29E

Hei-ho *see* Nagqu

85 J22 **Heilbron** Free State, N South Africa 27.16S 27.58E

103 H21 **Heilbronn** Baden-Württemberg, SW Germany 49.09N 9.13E

102 K7 **Heiligenhafen** Schleswig-Holstein, N Germany 54.22N 10.57E

Heiligenkreuz *see* Žiar nad Hronom

103 J15 **Heiligenstadt** Thüringen, C Germany 51.22N 10.09E

169 W8 **Heilong Jiang** *see* Amur

169 W8 **Heilongjiang** *var.* Hei, Heilongjiang Sheng, Hei-lung-chiang, Heilungkiang. ◆ *province* NE China

Heilongjiang Sheng *see* Heilongjiang

100 H9 **Heiloo** Noord-Holland, NW Netherlands 52.36N 4.43E

Heilsberg *see* Lidzbark Warmiński

Hei-lung-chiang/Heilungkiang *see* Heilongjiang

94 I4 **Heimaey** *var.* Heimaey. *island* S Iceland

96 H8 **Heimdal** Sør-Trøndelag, S Norway 63.21N 10.22E

95 N17 **Heinävesi** Itä-Suomi, E Finland 62.22N 28.42E

101 M22 **Heinerscheid** Diekirch, N Luxembourg 50.06N 6.04E

100 M10 **Heino** Overijssel, E Netherlands 52.26N 6.13E

95 M18 **Heinola** Etelä-Suomi, S Finland 61.13N 26.04E

103 C16 **Heinsberg** Nordrhein-Westfalen, W Germany 51.02N 6.01E

169 U12 **Heishan** Liaoning, NE China 41.43N 122.12E

166 H8 **Heishui** Sichuan, C China 32.08N 102.54E

101 H17 **Heist-op-den-Berg** Antwerpen, C Belgium 51.04N 4.43E

Heitō *see* P'ingtung

176 T14 **Heitske** Irian Jaya, E Indonesia 7.02S 138.45E

Hejanah *see* Al Hijānah

Hejaz *see* Al Ḥijāz

166 M14 **He Jiang** ☆ S China

174 K6 **Hejing** Xinjiang Uygur Zizhiqu, NW China 42.21N 86.19E

165 S12 **Héjjasfalva** *var.* Vânători. Hunedoara, W China 35.49N 99.49E

143 N14 **Hekimhan** Malatya, C Turkey 38.49N 37.55E

94 I4 **Hekla** ▲ S Iceland 63.56N 19.42W

112 J16 **Hel** *Ger.* Hela. Pomorskie, N Poland 54.35N 18.48E

112 J16 **Hela** *see* Hel

95 F17 **Helagsfjället** ▲ C Sweden 62.57N 12.31E

165 W8 **Helan** *var.* Xigang. Ningxia, N China 38.33N 106.21E

168 K4 **Helan Shan** ▲ N China

101 M16 **Helden** Limburg, SE Netherlands 51.20N 6.00E

29 V12 **Helena** Arkansas, C USA 34.32N 90.34W

35 R9 **Helena** *state capital* Montana, NW USA 46.35N 112.02W

98 H12 **Helensburgh** W Scotland, UK 56.00N 4.45W

192 K5 **Helensville** Auckland, North Island, NZ 36.42S 174.25E

97 L20 **Helgasjön** ☉ S Sweden

102 G8 **Helgoland** *Eng.* Heligoland. *island* NW Germany

Helgoland Bay *see* Helgoländer Bucht

102 G8 **Helgoländer Bucht** *var.* Helgoland Bay, Heligoland Bight. *bay* NW Germany

Heligoland *see* Helgoland

Heligoland Bight *see* Helgoländer Bucht

94 I4 **Hella** Sudhurland, SW Iceland 63.51N 20.24W

115 — **Hellas** *see* Greece

149 N11 **Helleh, Rūd-e** ☆ S Iran

100 N10 **Hellendoorn** Overijssel, E Netherlands 52.22N 6.27E

123 Gg10 **Hellenic Republic** *see* Greece

123 Gg10 **Hellenic Trough** *undersea feature* Aegean Sea, C Mediterranean Sea

96 E10 **Hellesylt** Møre og Romsdal, S Norway 62.05N 6.51E

100 F13 **Hellevoetsluis** Zuid-Holland, SW Netherlands 51.48N 4.08E

107 Q12 **Hellín** Castilla-La Mancha, C Spain 38.31N 1.43W

117 H19 **Hellinikon** ✈ (Athína) Attikí, C Greece 37.53N 23.43E

34 M7 **Hells Canyon** *valley* Idaho/Oregon, NW USA

154 L9 **Helmand** ◇ *province* S Afghanistan

154 K10 **Helmand, Daryā-ye** *Rus.* Gil'mend, *Rud-e* Hirmand. ☆ Afghanistan/Iran *see also* Hirmand, Rūd-e

Helmantica *see* Salamanca

103 N15 **Helme** ☆ C Germany

101 L15 **Helmond** Noord-Brabant, S Netherlands 51.28N 5.40E

98 K7 **Helmsdale** N Scotland, UK 58.06N 3.36W

103 K15 **Helmstedt** Niedersachsen, N Germany 52.13N 11.01E

169 Y10 **Helong** Jilin, NE China 42.35N 129.01E

38 M4 **Helper** Utah, W USA 39.40N 110.52W

102 O10 **Helpter Berge** *hill* NE Germany 53.29N 13.37E

96 J15 **Helsingborg** *prev.* Hälsingborg. Skåne, S Sweden 55.59N 12.48E

95 — **Helsingfors** *see* Helsinki

97 I22 **Helsingør** *Eng.* Elsinore. Frederiksborg, E Denmark 56.03N 12.37E

95 M20 **Helsinki** *Swe.* Helsingfors. ● (Finland) Etelä-Suomi, S Finland 60.18N 24.58E

99 H25 **Helston** SW England, UK 50.04N 5.16W

Heltau *see* Cisnădie

63 C17 **Helvecia** Santa Fe, C Argentina 31.09S 60.09W

97 K15 **Helvellyn** ▲ NW England, UK 54.31N 3.00W

77 W8 **Helwân** *var.* Hilwān, Hulwan, Hulwân. N Egypt 29.51N 31.20E

37 U16 **Hemet** California, W USA 33.45N 116.58W

30 J13 **Hemingford** Nebraska, USA 42.18N 103.02W

24 T13 **Hemingway** South Carolina, SE USA 33.45N 79.25W

74 G13 **Hemnesberget** Nordland, C Norway 66.13N 13.33E

45 Y8 **Hemphill** Texas, SW USA 31.20N 93.51W

45 V11 **Hempstead** Texas, SW USA 30.06N 96.04W

97 P20 **Hemse** Gotland, SE Sweden 57.12N 18.22E

96 F13 **Hemsedal** Sør-Trøndelag, S Norway

165 T11 **Henan** *var.* Henan Mongolzu Zizhixian, Yéganmyin. Qinghai, C China 34.42N 101.36E

92 T13 **Henan** *var.* Henan Sheng, Honan, Yu. ◆ *province* C China

Henan Mongolzu Zizhixian/Henan Sheng *see* Henan

107 O7 **Henares** ☆ C Spain

171 M8 **Henashi-zaki** *headland* Honshū, C Japan 40.37N 139.51E

104 I16 **Hendaye** Pyrénées-Atlantiques, SW France 43.22N 1.46W

63 B21 **Henderson** Buenos Aires, E Argentina 36.18S 61.43W

22 I5 **Henderson** Kentucky, S USA 37.50N 87.35W

37 X11 **Henderson** Nevada, W USA 36.02N 114.58W

23 V8 **Henderson** North Carolina, SE USA 36.19N 78.24W

22 M7 **Henderson** Tennessee, S USA 35.25N 88.37W

45 W7 **Henderson** Texas, SW USA 32.09N 94.48W

32 M8 **Henderson Creek** ☆ Illinois, N USA

195 X16 **Henderson Field** ✈ (Honiara) Guadalcanal, C Solomon Islands 9.28S 160.02E

203 O17 **Henderson Island** *atoll* N Pitcairn Islands

24 P11 **Hendersonville** North Carolina, SE USA 35.19N 82.27W

23 J8 **Hendersonville** Tennessee, S USA 36.18N 86.36W

149 O14 **Hendorābi, Jazireh-ye** *island* S Iran

57 W16 **Hendrik Top** *var.* Hendriktop. *elevation* C Suriname 4.14N 56.07W

99 L20 **Hendū Kosh** *see* Hindu Kush

113 L22 **Heney, Lac** ☉ Quebec, SE Canada

194 J12 **Henganofi** Eastern Highlands, C PNG 6.13S 145.31E

167 S15 **Hengchow** *see* Hengyang

167 S15 **Hengchun** ✈ Taiwan 22.09N 120.43E

100 M12 **Hengduan Shan** ▲ SW China

100 N12 **Hengelo** Gelderland, E Netherlands 52.15N 6.48E

100 O11 **Hengelo** Overijssel, E Netherlands 52.16N 6.48E

167 N11 **Hengnan** *see* Hengyang

166 L4 **Hengshan** Shaanxi, C China 37.57N 109.17E

167 O4 **Hengshui** Hebei, E China 37.42N 115.39E

167 N12 **Hengyang** *var.* Hengnan, Heng-yang; *prev.* Hengchow. Hunan, S China 26.54N 112.33E

119 U11 **Henichesk** *Rus.* Genichesk. Khersons'ka Oblast', S Ukraine 46.10N 34.49E

23 Z4 **Henlopen, Cape** *headland* Delaware, NE USA 38.48N 75.06W

96 J10 **Henna** *see* Enna

96 M10 **Hennan** Gävleborg, C Sweden 62.01N 15.55E

104 G7 **Hennebont** Morbihan, NW France 47.48N 3.16W

28 M9 **Hennepin** Illinois, USA 41.13N 89.21W

167 R2 **Hennessey** Oklahoma, C USA 36.06N 97.54W

102 M12 **Hennigsdorf** *var.* Hennigsdorf bei Berlin. Brandenburg, NE Germany 52.58N 110.53W

102 N11 **Hennigsdorf bei Berlin** *see* Hennigsdorf

21 M9 **Henniker** New Hampshire, NE USA 43.10N 71.47W

45 V11 **Henrietta** Texas, SW USA 33.49N 98.12W

32 L12 **Henry** Illinois, N USA 41.06N 89.21W

23 Y7 **Henry, Cape** *headland* Virginia, NE USA 36.55N 76.01W

29 P10 **Henryetta** Oklahoma, C USA 35.26N 95.58W

204 M7 **Henry Ice Rise** *ice cap* Antarctica

16 N1 **Henry Kater, Cape** *headland* Baffin Island, Nunavut, NE Canada 69.00N 66.45W

35 R13 **Henrys Fork** ☆ Idaho, NW USA

12 E15 **Hensall** Ontario, S Canada 43.25N 81.28W

102 J9 **Henstedt-Ulzburg** Schleswig-Holstein, N Germany 53.43N 9.59E

169 N7 **Hentiy** ◆ *province* N Mongolia

169 N7 **Hentiyn Nuruu** ▲ N Mongolia

191 N7 **Henty** New South Wales, SE Australia 35.33S 147.03E

177 N9 **Henzada** Irrawaddy, SW Myanmar 17.36N 95.25E

103 G19 **Heppenheim** Hessen, W Germany 49.39N 8.38E

34 J11 **Heppner** Oregon, NW USA 45.21N 119.33W

166 L15 **Hepu** *prev.* Lianzhou. Guangxi Zhuangzu Zizhiqu, S China 21.40N 109.12E

94 J2 **Heradhsvötn** ☆ C Iceland

154 K5 **Herakleion** *see* Irákleio

154 L5 **Herāt** *var.* Herat; *anc.* Aria. Herāt, W Afghanistan 34.22N 62.11E

154 J5 **Herāt** ◆ *province* W Afghanistan

105 P14 **Hérault** ◆ *department* S France

105 P15 **Hérault** ☆ S France

9 T16 **Herbert** Saskatchewan, S Canada 50.27N 107.09W

193 F22 **Herbert** Otago, South Island, NZ 45.14S 170.48E

41 J17 **Herbert Island** *island* Aleutian Islands, Alaska, USA

194 I12 **Herbert, Mount** ▲ C PNG 5.44S 145.00E

13 Q7 **Herbertshöhe** *see* Kokopo

9 Q7 **Hérbertville** Quebec, SE Canada 48.22N 71.42W

103 G17 **Herborn** Hessen, W Germany 50.40N 8.18E

115 L17 **Herceg-Novi** *It.* Castelnuovo; *prev.* Ercegnovi. Montenegro, SW Yugoslavia 42.28N 18.35E

99 X10 **Herchmer** Manitoba, C Canada 57.25N 94.12W

99 K21 **Herdhubreidh** ▲ C Iceland 65.12N 16.26W

44 M13 **Heredia** Heredia, C Costa Rica 10.00N 84.06W

44 M13 **Heredia** *off.* Provincia de Heredia. ◆ *province* N Costa Rica

99 K21 **Hereford** W England, UK 52.04N 2.43W

44 M3 **Hereford** Texas, SW USA 34.49N 102.25W

13 Q7 **Hereford, Mont** ▲ Quebec, SE Canada 45.04N 71.38W

99 K21 **Herefordshire** *cultural region* W England, UK

203 U11 **Hereheretue** *atoll* Îles Tuamotu, C French Polynesia

107 N10 **Herencia** Castilla-La Mancha, C Spain 39.19N 3.07W

101 H13 **Herent** Vlaams Brabant, C Belgium 50.54N 4.40E

101 I16 **Herentals** *var.* Herenthals. Antwerpen, N Belgium 51.10N 4.49E

Herenthals *see* Herentals

101 J18 **Herenthout** Antwerpen, N Belgium 51.09N 4.45E

97 J23 **Herfølge** Roskilde, E Denmark 55.25N 12.09E

102 G13 **Herford** Nordrhein-Westfalen, NW Germany 52.07N 8.40E

29 O5 **Herington** Kansas, C USA 38.37N 96.51W

110 H7 **Herisau** *Fr.* Hérisau. Appenzell Ausser Rhoden, NE Switzerland 47.22N 9.16E

Héristal *see* Herstal

101 J18 **Herk-de-Stad** Limburg, NE Belgium 50.57N 5.10E

Herkulesbad/Herkulesfürdő *see* Băile Herculane

Herlen Gol/Herlen He *see* Kerulen

37 Q4 **Herlong** California, W USA 40.07N 120.06W

99 L23 **Herm** *island* Channel Islands

111 R9 **Hermagor** *Slvn.* Šmohor. Kärnten, S Austria 46.37N 13.24E

31 S7 **Herman** Minnesota, N USA 45.49N 96.08W

1 — **Herma Ness** *headland* NE Scotland, UK 60.51N 0.55W

29 V4 **Hermann** Missouri, C USA 38.40N 91.25W

189 Q8 **Hermannsburg** Northern Territory, N Australia 23.59S 132.55E

Hermannstadt *see* Sibiu

12 — **Hermansverk** Sogn og Fjordane, S Norway 61.10N 6.52E

144 H4 **Hermel** *var.* Hirmil. NE Lebanon 34.23N 36.19E

191 P6 **Hermidale** New South Wales, SE Australia 31.36S 146.42E

34 J8 **Hermiston** Oregon, NW USA 45.50N 119.17W

29 T6 **Hermitage** Missouri, C USA 37.57N 93.21W

194 I8 **Hermit Islands** *island group* N PNG

45 O7 **Hermleigh** Texas, SW USA 32.37N 100.44W

144 G7 **Hermon, Mount** *Ar.* Jabal ash Shaykh. ▲ S Syria 33.30N 33.30E

28 M9 **Hermon** Illinois, USA 41.13N 89.21W

30 M13 **Hermosa** South Dakota, N USA 43.49N 103.12W

42 F5 **Hermosillo** Sonora, NW Mexico 28.58N 110.53W

113 N20 **Hernád** *var.* Hornád, *Ger.* Kundert. ☆ Hungary/Slovakia

63 C18 **Hernández** Entre Ríos, E Argentina 32.21S 60.01W

25 V11 **Hernando** Florida, SE USA 28.54N 82.22W

24 L1 **Hernando** Mississippi, S USA 34.49N 89.59W

107 Q2 **Hernani** País Vasco, N Spain 43.16N 1.58W

101 F19 **Herne** Vlaams Brabant, C Belgium 50.45N 4.03E

103 E14 **Herne** Nordrhein-Westfalen, W Germany 51.33N 7.13E

97 F22 **Herning** Ringkøbing, W Denmark 56.07N 8.58E

124 Q13 **Herodotus Basin** *undersea feature* E Mediterranean Sea

124 Nn14 **Herodotus Trough** *undersea feature* C Mediterranean Sea

31 T11 **Heron Lake** Minnesota, N USA 43.48N 95.18W

Herowābad *see* Khalkhāl

97 G16 **Herre** Telemark, S Norway 59.06N 9.34E

31 N7 **Herreid** South Dakota, N USA 45.49N 100.04W

103 H22 **Herrenberg** Baden-Württemberg, S Germany 48.36N 8.52E

106 L14 **Herrera** Andalucía, S Spain 37.22N 4.49W

45 R17 **Herrera** *off.* Provincia de Herrera. ◆ *province* S Panama

106 L10 **Herrera del Duque** Extremadura, W Spain 39.10N 5.03W

106 M4 **Herrera de Pisuerga** Castilla-León, N Spain 42.35N 4.20W

43 Z13 **Herrero, Punta** *headland* SE Mexico 19.16N 87.28W

191 P16 **Herrick** Tasmania, SE Australia 41.07S 147.53E

32 L12 **Herrin** Illinois, N USA 37.48N 89.01W

22 M6 **Herrington Lake** ☉ Kentucky, S USA

97 K18 **Herrljunga** Västra Götaland, S Sweden 58.04N 13.01E

105 N16 **Hers** ☆ S France

8 I11 **Herschel Island** *island* Yukon Territory, NW Canada

101 I17 **Herselt** Antwerpen, C Belgium 51.03N 4.52E

20 G15 **Hershey** Pennsylvania, NE USA 76.39W

101 K19 **Herstal** *Fr.* Héristal. Liège, E Belgium 50.40N 5.37E

99 O21 **Hertford** E England, UK 51.48N 0.04W

23 X8 **Hertford** North Carolina, SE USA 36.11N 76.28W

99 O21 **Hertfordshire** *cultural region* E England, UK

101 E18 **Hertsberge** Brabant, NW Belgium 50.57N 3.26E

103 K20 **Herzogenaurach** Bayern, SE Germany 49.34N 10.52E

111 W4 **Herzogenburg** Niederösterreich, NE Austria 48.18N 15.43E

— **'s-Hertogenbosch** *see* 's-Hertogenbosch

105 N2 **Hesdin** Pas-de-Calais, N France 50.21N 2.00E

166 K14 **Heshan** Guangxi Zhuangzu Zizhiqu, S China 23.45N 108.58E

165 X10 **Heshui** *var.* Xihuachi. Gansu, C China 35.42N 108.06E

37 U14 **Hesperia** California, W USA 34.25N 117.17W

39 P7 **Hesperus Mountain** ▲ Colorado, C USA 37.27N 108.05W

8 J6 **Hess** ☆ Yukon Territory, NW Canada

Hesse *see* Hessen

103 J21 **Hesselberg** ▲ S Germany 49.04N 10.32E

97 I22 **Hesselø** *island* E Denmark

103 G18 **Hessen** *Eng./Fr.* Hesse. ◆ *state* C Germany

199 Jj6 **Hess Tablemount** *undersea feature* C Pacific Ocean 17.49N 174.15W

29 N6 **Hesston** Kansas, C USA 38.08N 97.25W

20 J13 **Hesston** *hill* New Jersey, NE USA 43.14N 73.57E

9 P13 **Hetch Hetchy** NW England, UK 53.19N 3.06W

159 Pf2 **Hetauda** Central, C Nepal

30 K7 **Hettinger** North Dakota, N USA 46.00N 102.38W

103 L14 **Hettstedt** Sachsen-Anhalt, C Germany 51.39N 11.31E

34 P3 **Heuglin, Kapp** *headland* SE Svalbard 78.15N 22.49E

195 Y16 **Heuru** San Cristobal, SE Solomon Islands 10.13S 161.25E

101 I17 **Heusden** Limburg, NE Belgium 51.01N 5.16E

100 J13 **Heusden** Noord-Brabant, S Netherlands 51.43N 5.05E

104 K3 **Hève, Cap de la** *headland* N France 49.28N 0.11E

101 H18 **Heverlee** Vlaams Brabant, C Belgium 50.52N 4.41E

113 L22 **Heves** Heves, C Hungary 47.37N 20.17E

113 L22 **Heves** *off.* Heves Megye. ◆ *county* NE Hungary

144 — **Hevron** *see* Hebron

47 Y13 **Hewanorra** ✈ (Saint Lucia) S Saint Lucia 13.44N 60.57W

194 I8 **Hermit Islands** *island group* N PNG

96 M13 **Hexian** *var.* Babu, He Xian. Guangxi Zhuangzu Zizhiqu, S China 24.25N 111.31E

166 I16 **Heyang** Shaanxi, C China 35.03N 109.55E

97 K16 **Heysham** NW England, UK 54.02N 2.54W

167 P9 **Heyuan** Guangdong, S China 23.50N 114.43E

190 L12 **Heywood** Victoria, SE Australia 38.09S 141.38E

188 K3 **Heywood Islands** *island group* Western Australia

167 O6 **Heze** *var.* Caozhou. Shandong, E China 35.16N 115.27E

165 U11 **Hezheng** Gansu, C China 35.24N 103.21E

165 U11 **Hezuozhen** Gansu, C China 34.55N 102.49E

25 Z6 **Hialeah** Florida, SE USA 25.51N 80.16W

29 Q3 **Hiawatha** Kansas, C USA 39.48N 95.31W

38 M4 **Hiawatha** Utah, W USA 39.28N 111.00W

31 V4 **Hibbing** Minnesota, N USA 47.24N 92.55W

191 N17 **Hibbs, Point** *headland* Tasmania, SE Australia 42.37S 145.15E

Hibernia *see* Ireland

170 D12 **Hibiki-nada** *inlet* SW Japan

22 F8 **Hickman** Kentucky, S USA 36.34N 89.11W

23 Q9 **Hickory** North Carolina, SE USA 35.41N 81.20W

23 Q9 **Hickory, Lake** ☉ North Carolina, SE USA

192 Q7 **Hicks Bay** Gisborne, North Island, NZ 37.36S 178.18E

45 S8 **Hico** Texas, SW USA 31.58N 98.01W

170 D4 **Hiodaka** Hokkaidō, NE Japan 42.53N 142.24E

171 Gg13 **Hidaka** Hyōgo, Honshū, SW Japan 35.27N 134.43E

172 P7 **Hidaka** Hokkaidō, NE Japan

Hidaka-sanmyaku ▲ Hokkaidō, NE Japan

43 O6 **Hidalgo** *var.* Villa Hidalgo. Coahuila de Zaragoza, NE Mexico 27.46N 99.54W

43 N8 **Hidalgo** Nuevo León, NE Mexico 29.58N 100.27W

43 O10 **Hidalgo** Tamaulipas, C Mexico 24.17N 99.21W

43 O13 **Hidalgo** ◆ *state* C Mexico

42 J7 **Hidalgo del Parral** *var.* Parral. Chihuahua, N Mexico 26.58N 105.40W

171 J14 **Hida-sanmyaku** ▲ Honshū, S Japan

102 N3 **Hiddensee** *island* NE Germany

82 G6 **Hidiglib, Wadi** ☆ NE Sudan

111 H6 **Hieflau** Salzburg, E Austria 47.36N 14.34E

117 H5 **Hienghène** Province Nord, C New Caledonia 20.43S 164.54E

Hierosolyma *see* Jerusalem

66 N12 **Hierro** *var.* Ferro. *island* Islas Canarias, Spain, NE Atlantic Ocean

170 Ee13 **Higashi-Hiroshima** *var.* Higashihirosima. Hiroshima, Honshū, SW Japan 34.25N 132.45E

171 J18 **Higashi-Izu** Shizuoka, Honshū, S Japan 34.43N 138.58E

170 Ll12 **Higashine** *var.* Higasine. Yamagata, Honshū, C Japan 38.26N 140.23E

170 C11 **Higashi-suidō** *strait* SW Japan

Higashihirosima *see* Higashi-Hiroshima

Higasine *see* Higashine

45 P7 **Higgins** Texas, SW USA 36.06N 100.01W

33 P7 **Higgins Lake** ☉ Michigan, N USA

29 S4 **Higginsville** Missouri, C USA 39.04N 93.43W

— **High Atlas** *see* Haut Atlas

32 M5 **High Falls Reservoir** ☉ Wisconsin, N USA

46 K12 **Highgate** C Jamaica 18.15N 76.53W

45 X11 **High Island** Texas, SW USA 29.35N 94.24W

33 O5 **High Island** *island* Michigan, N USA

33 N13 **Highland** Illinois, N USA 38.44N 89.40W

33 N10 **Highland Park** Illinois, N USA 42.10N 87.48W

23 Q11 **Highlands** North Carolina, SE USA 35.04N 83.10W

9 O11 **High Level** Alberta, W Canada 58.31N 117.07W

31 Q9 **Highmore** South Dakota, N USA 44.29N 99.26W

179 Oo10 **High Peak** ▲ Luzon, N Philippines 15.28N 120.07E

— **High Plains** *see* Great Plains

23 S9 **High Point** North Carolina, SE USA 35.58N 80.00W

20 J13 **High Point** *hill* New Jersey, NE USA 41.18N 74.39E

9 P13 **High Prairie** Alberta, W Canada 55.27N 116.28W

9 Q16 **High River** Alberta, SW Canada 50.34N 113.49W

23 S9 **High Rock Lake** ☉ North Carolina, SE USA

25 V10 **High Springs** Florida, SE USA 29.49N 82.36W

— **High Veld** *see* Great Karoo

99 J24 **High Willhays** ▲ SW England, UK 50.39N 3.58W

99 N22 **High Wycombe** *prev.* Chepping Wycombe, Chipping Wycombe. SE England, UK 51.37N 0.46W

43 V12 **Higos** Veracruz-Llave, E Mexico 21.47N 98.28W

104 H7 **Higuer, Cap** *headland* NE Spain 43.23N 1.46W

47 P9 **Higüero, Punta** *headland* W Puerto Rico 18.21N 67.15W

47 P9 **Higüey** *var.* Salvaleón de Higüey. E Dominican Republic 18.34N 68.43W

202 O11 **Hihifo** ✈ (Matā'utu) Uvea, N Wallis and Futuna

83 N14 **Hiiraan** *off.* Gobolka Hiiraan. ◆ *region* C Somalia

120 F4 **Hiiumaa** *off.* Hiiumaa Maakond. ◆ *province* W Estonia

120 D4 **Hiiumaa** *Ger.* Dagden, *Swe.* Dagö. *island* W Estonia

107 S6 **Híjar** Aragón, NE Spain 41.10N 0.27W

170 E13 **Hikari** Yamaguchi, Honshū, SW Japan 33.55N 131.58E

170 F15 **Hiketa** Kagawa, Shikoku, SW Japan 35.15N 134.14E

171 Hh15 **Hikone** Shiga, Honshū, SW Japan 35.15N 136.14E

170 D3 **Hiko-san** ▲ Kyūshū, SW Japan 38.09S 141.38E

203 V10 **Hikueru** *atoll* Îles Tuamotu, C French Polynesia

192 K3 **Hikurangi** Northland, North Island, NZ 35.37S 174.16E

◆ COUNTRY ● COUNTRY CAPITAL ◇ DEPENDENT TERRITORY ○ DEPENDENT TERRITORY CAPITAL ◆ ADMINISTRATIVE REGION ✈ INTERNATIONAL AIRPORT ▲ MOUNTAIN ▲ MOUNTAIN RANGE ☆ RIVER ☉ LAKE ☉ RESERVOIR ☒ VOLCANO

192 Q8 **Hikurangi** ▲ North Island, NZ 37.55S 177.59E

199 J13 **Hikurangi Trench** var. Hikurangi Trough. undersea feature SW Pacific Ocean
Hikurangi Trough see Hikurangi Trench

202 B15 **Hikutavake** NW Niue

124 Nn14 **Hilāl, Ra's al** headland N Libya 32.55N 22.09E

63 A24 **Hilario Ascasubi** Buenos Aires, E Argentina 39.22S 62.39W

103 K17 **Hildburghausen** Thüringen, C Germany 50.26N 10.44E

103 E15 **Hilden** Nordrhein-Westfalen, W Germany 51.10N 6.55E

102 I13 **Hildesheim** Niedersachsen, N Germany 52.09N 9.57E

35 T9 **Hilger** Montana, NW USA 47.15N 109.18W
Hili see Hilli
Hilla see Al Ḥillah

47 O14 **Hillaby, Mount** ▲ N Barbados 13.12N 59.34W

97 K19 **Hillared** Västra Götaland, S Sweden 57.37N 13.10E

205 R12 **Hillary Coast** physical region Antarctica

44 G2 **Hill Bank** Orange Walk, N Belize 17.36N 88.43W

35 O14 **Hill City** Idaho, NW USA 43.18N 115.03W

25 W6 **Hillsville** Georgia, SE USA 31.51N 81.36W

28 K3 **Hill City** Kansas, C USA 39.21N 99.51W

31 V5 **Hill City** Minnesota, N USA 46.59N 93.36W

30 J10 **Hill City** South Dakota, N USA 43.54N 103.33W

67 C24 **Hill Cove Settlement** West Falkland, Falkland Islands

100 H10 **Hillegom** Zuid-Holland, W Netherlands 52.18N 4.34E

97 J22 **Hillerød** Frederiksborg, E Denmark 55.55N 12.19E

38 M7 **Hillers, Mount** ▲ Utah, W USA 37.53N 110.42W

159 S13 **Hilli** var. Hili. Rajshahi, NW Bangladesh 25.17N 89.02E

31 R11 **Hillsboro** Illinois, N USA 43.31N 96.21W

32 L14 **Hillsboro** Illinois, N USA 39.09N 89.29W

29 N5 **Hillsboro** Kansas, C USA 38.21N 97.12W

29 X5 **Hillsboro** Missouri, C USA 38.13N 90.33W

21 N10 **Hillsboro** New Hampshire, NE USA 43.06N 71.52W

39 Q14 **Hillsboro** New Mexico, SW USA 32.55N 107.33W

31 R4 **Hillsboro** North Dakota, N USA 47.25N 97.03W

33 R14 **Hillsboro** Ohio, N USA 39.12N 83.36W

34 G11 **Hillsboro** Oregon, NW USA 45.31N 122.59W

45 T8 **Hillsboro** Texas, SW USA 32.01N 97.08W

32 K8 **Hillsboro** Wisconsin, N USA 43.40N 90.21W

25 Y14 **Hillsboro Canal** canal Florida, SE USA

47 Y15 **Hillsborough** Carriacou, N Grenada 12.28N 61.28W

99 G15 **Hillsborough** E Northern Ireland, UK 54.27N 6.06W

23 U9 **Hillsborough** North Carolina, SE USA 36.04N 79.06W

33 Q10 **Hillsdale** Michigan, N USA 41.55N 84.37W

191 O8 **Hillston** New South Wales, SE Australia 33.30S 145.33E

23 R7 **Hillsville** Virginia, NE USA 36.45N 80.44W

98 L2 **Hillswick** NE Scotland, UK 60.28N 1.37W
Hill Tippera see Tripura

40 H11 **Hilo** Hawaii, USA, C Pacific Ocean 19.42N 155.04W

20 F9 **Hilton** New York, NE USA 43.17N 77.47W

12 C10 **Hilton Beach** Ontario, S Canada 46.14N 83.51W

23 R16 **Hilton Head Island** South Carolina, SE USA 32.13N 80.45W

23 R16 **Hilton Head Island** island South Carolina, SE USA 32.13N 80.45W

101 J15 **Hilvarenbeek** Noord-Brabant, S Netherlands 51.28N 5.07E

100 J11 **Hilversum** Noord-Holland, C Netherlands 52.13N 5.10E
Hilwān see Helwân

158 J11 **Himāchal Pradesh** ◆ state NW India

158 M9 **Himalaya/Himalaya Shan** see Himalayas

158 M9 **Himalayas** var. Himalaya, Chin. Himalaya Shan. ▲ S Asia

179 Q14 **Himamaylan** Negros, C Philippines 10.04N 122.52E

95 K15 **Himanka** Länsi-Suomi, W Finland 64.03N 24.40E
Himara see Himarë

115 L23 **Himarë** var. Himara. Vlorë, S Albania 40.06N 19.45E

144 M2 **Ḩimār, Wādī al** dry watercourse N Syria

160 P9 **Himatnagar** Gujarāt, W India 23.37N 73.01E

111 H4 **Himberg** Niederösterreich, E Austria 48.03N 16.27E

171 I14 **Hime-gawa** ♦ Honshū, S Japan

170 G14 **Himeji** var. Himezi. Hyōgo, SW Japan 34.49N 134.42E
Himeji-shima island SW Japan

170 Dd13 **Hime-jima** island SW Japan
Himezi see Himeji

171 Ii13 **Himi** Toyama, Honshū, SW Japan 36.52N 136.59E

111 I9 **Himmelberg** Kärnten, S Austria 46.45N 14.01E

144 I5 **Ḩimṣ** var. Homs; anc. Emesa. Ḩimṣ, C Syria 34.43N 36.43E

144 K6 **Ḩimṣ** off. Muḩāfaz̧at Ḩimṣ, var. Homs. ◆ governorate C Syria

144 J5 **Ḩimṣ, Buḩayrat** var. Buḩayrat Qaṭṭīnah. ☒ W Syria

179 R15 **Hinatuan** Mindanao, S Philippines 8.21N 126.19E

116 N10 **Hînceşti** prev. Hânceşti; prev. Kotovsk. C Moldova 46.48N 28.33E

46 M9 **Hindelang** see Chíos

189 X5 **Hinchinbrook Island** island Queensland, NE Australia

41 S12 **Hinchinbrook Island** island Alaska, USA

99 M19 **Hinckley** C England, UK 52.33N 1.21W

31 N1 **Hinckley** Minnesota, N USA 46.01N 92.57W

38 K5 **Hinckley** Utah, W USA 39.21N 112.39W

20 J9 **Hinckley Reservoir** ☒ New York, NE USA

158 I12 **Hindaun** Rājasthān, N India 26.43N 77.01E
Hindenburg/Hindenburg in Oberschlesien see Zabrze
Hindiya see Al Hindīyah

23 O6 **Hindman** Kentucky, S USA 37.20N 82.58W

190 L10 **Hindmarsh, Lake** ☒ Victoria, SE Australia

193 G19 **Hinds** Canterbury, South Island, NZ 44.00S 171.33E

193 G19 **Hinds** ♦ South Island, NZ

97 H23 **Hindsholm** island C Denmark

155 S4 **Hindu Kush** Per. Hendū Kosh. ▲ Afghanistan/Pakistan

161 H19 **Hindupur** Andhra Pradesh, E India 13.46N 77.33E

9 O12 **Hines Creek** Alberta, W Canada 56.14N 118.36W

25 W6 **Hinesville** Georgia, SE USA 31.51N 81.36W

160 I12 **Hinganghāt** Mahārāshtra, C India 20.31N 78.52E

155 N15 **Hingol** ♦ SW Pakistan

160 H13 **Hingoli** Mahārāshtra, C India 19.45N 77.08E

143 R13 **Hınıs** Erzurum, E Turkey 39.22N 41.43E

94 O12 **Hinlopenstretet** strait N Svalbard

94 G10 **Hinnøya** island C Norway

170 D15 **Hinokage** Miyazaki, Kyūshū, SW Japan 32.39N 131.20E

170 F11 **Hino-misaki** headland Honshū, SW Japan 35.25N 132.37E

110 H10 **Hinterrhein** ♦ SW Switzerland

9 O14 **Hinton** Alberta, SW Canada 53.24N 117.34W

28 M10 **Hinton** Oklahoma, C USA 35.28N 98.21W

23 R6 **Hinton** West Virginia, NE USA 37.40N 80.53W
Hios see Chíos

43 N4 **Hipólito** Coahuila de Zaragoza, NE Mexico 25.42N 101.22W
Hipponium see Vibo Valentia

170 C12 **Hirado** Nagasaki, Hirado-shima, SW Japan 33.22N 129.32E

170 C12 **Hirado-shima** island SW Japan

171 Gg15 **Hirakata** Ōsaka, Honshū, SW Japan 34.48N 135.37E

172 P17 **Hirakubo-saki** headland Ishigaki-jima, SW Japan 24.36N 124.19E

160 H7 **Hirakud Reservoir** ☒ E India
Hir al Gharbi, Qasr al see Ḩayr al Gharbi, Qasr al

172 N9 **Hiranai** Aomori, Honshū, N Japan 40.56N 140.55E

172 Pp16 **Hirara** Okinawa, Miyako-jima, SW Japan 24.48N 125.16E
Qasr al Hair Ash Sharqi see Ḩayr ash Sharqī, Qaṣr al

170 F12 **Hirata** Shimane, Honshū, SW Japan 35.26N 132.50E

171 Jj17 **Hiratsuka** var. Hiratuka. Kanagawa, Honshū, S Japan 35.20N 139.20E
Hiratuka see Hiratsuka

142 I13 **Hirfanlı Baraji** ☒ C Turkey

161 G18 **Hiriyūr** Karnātaka, W India 13.58N 76.33E
Hirlău see Hârlău

154 K10 **Hirmand, Rūd-e** var. Daryā-ye Helmand. ♦ Afghanistan/Iran see also Helmand, Daryā-ye
Hirmil see Hermel

172 P8 **Hiroo** Hokkaidō, NE Japan 42.16N 143.16E

171 Mm9 **Hirosaki** Aomori, Honshū, C Japan 40.34N 140.28E

170 D13 **Hiroshima** var. Hirosima. Hiroshima, Honshū, SW Japan 34.22N 132.25E

170 Ee13 **Hiroshima** off. Hiroshima-ken, var. Hirosima. ♦ prefecture Honshū, SW Japan
Hirosima see Hiroshima

105 Q3 **Hirson** Aisne, N France 49.55N 4.04E
Hîrşova see Hârşova

97 G19 **Hirtshals** Nordjylland, N Denmark 57.34N 9.58E

171 H16 **Hisai** Mie, Honshū, SW Japan 34.38N 136.27E

158 H10 **Hisār** Haryāna, NW India 29.10N 75.45E

194 J15 **Hisiu** Central, SW PNG 9.01S 146.49E

153 P13 **Hisor** Rus. Gissar. W Tajikistan 38.34N 68.29E
Hispalis see Sevilla

Hispana/Hispania see Spain

Hispaniola island Dominican Republic/Haiti

66 F11 **Hispaniola Basin** var. Hispaniola Trough. undersea feature SW Atlantic Ocean
Hispaniola Trough see Hispaniola Basin
Histonium see Vasto

145 R7 **Hīt** var. Heet. SW Iraq 33.38N 42.50E

170 D14 **Hita** Ōita, Kyūshū, SW Japan 33.20N 130.55E

171 L16 **Hitachi** var. Hitati. Ibaraki, Honshū, S Japan 36.40N 140.42E

171 L16 **Hitachi-Ōta** var. Hitatiōta. Ibaraki, Honshū, S Japan 36.31N 140.31E
Hitati see Hitachi

171 L16 **Hitatiōta** see Hitachi-Ōta

99 O21 **Hitchin** E England, UK 51.57N 0.16W

203 Q7 **Hitiaa** Tahiti, W French Polynesia 17.34S 149.16W

170 G13 **Hitoyoshi** var. Hitoyosi. Kumamoto, Kyūshū, SW Japan 32.12N 130.45E
Hitoyosi see Hitoyoshi

96 F7 **Hitra** prev. Hitteren. island S Norway
Hitteren see Hitra

197 B10 **Hiu** island Torres Islands, N Vanuatu

171 K14 **Hiuchiga-take** ▲ Honshū, C Japan 36.57N 139.18E

170 Ee14 **Hiuchi-nada** gulf S Japan

203 X7 **Hiva Oa** island Îles Marquises, N French Polynesia

76 J6 **Hiwassee** Lake ☒ North Carolina, SE USA

22 M10 **Hiwassee River** ♦ SE USA

97 H20 **Hjallerup** Nordjylland, N Denmark 57.10N 10.10E

97 I14 **Hjälmaren** Eng. Lake Hjalmar. ☒ C Sweden
Hjalmar, Lake see Hjälmaren

97 C14 **Hjellestad** Hordaland, S Norway 60.15N 5.13E

97 D16 **Hjelmeland** Rogaland, S Norway 59.12N 6.07E

96 G10 **Hjerkinn** Oppland, S Norway 62.13N 9.37E

97 L18 **Hjo** Västra Götaland, S Sweden 58.16N 14.07E

97 G19 **Hjørring** Nordjylland, N Denmark 57.28N 9.58E

178 M14 **Hkakabo Razi** ▲ Myanmar/China 28.17N 97.28E

178 M1 **Hkring Bum** ▲ N Myanmar 27.05N 97.16E

85 L21 **Hlathikulu** var. Hlatikulu. S Swaziland 26.57S 31.19E
Hlatikulu see Hlathikulu
Hiboka see Hlyboka

113 F17 **Hlinsko** var. Hlinsko v Čechách. Pardubický Kraj, C Czech Republic 49.46N 15.54E
Hlinsko v Čechách see Hlinsko

119 S6 **Hlobyne** Rus. Globino. Poltavs'ka Oblast', C Ukraine 49.24N 33.16E

113 H20 **Hlohovec** Ger. Freistadtl, Hung. Galgóc; prev. Frakštát. Trnavský Kraj, W Slovakia 48.27N 17.47E

85 J23 **Hlotse** var. Leribe. NW Lesotho 28.55S 28.01E

113 I17 **Hlučín** Ger. Hultschin, Pol. Hulczyn. Ostravský Kraj, E Czech Republic 49.54N 18.10E

119 S2 **Hlukhiv** Rus. Glukhov. Sums'ka Oblast', NE Ukraine 51.39N 33.52E

121 K21 **Hlushkavichy** Rus. Glushkevichi. Homyel'skaya Voblasts', SE Belarus 51.33N 27.48E

121 L18 **Hlusk** Rus. Glusk, Glussk. Mahilyowskaya Voblasts', E Belarus 52.54N 28.40E

118 K8 **Hlyboka** Ger. Hliboka, Rus. Glybokaya. Chernivets'ka Oblast', W Ukraine 48.04N 25.55E

121 K13 **Hlybokaye** Rus. Glubokoye. Vitsyebskaya Voblasts', N Belarus 55.08N 27.40E

79 Q16 **Ho** SE Ghana 6.36N 0.28E

178 Jj6 **Hoa Binh** Hoa Binh, N Vietnam 20.49N 105.19E

85 E20 **Hoachanas** Hardap, C Namibia 23.52S 18.02E

178 Jj8 **Hoa Lac** Quang Binh, C Vietnam 17.54N 106.24E

178 J5 **Hoang Liên Sơn** ▲ N Vietnam

85 B17 **Hoanib** ♦ NW Namibia

35 S15 **Hoback Peak** ▲ Wyoming, C USA 43.04N 110.34W

191 P17 **Hobart** prev. Hobarton, Hobart Town. state capital Tasmania, SE Australia 42.54S 147.18E

28 L11 **Hobart** Oklahoma, C USA 35.01N 99.05W

191 P17 **Hobart** × Tasmania, SE Australia 42.52S 147.28E
Hobarton/Hobart Town see Hobart

39 W4 **Hobbs** New Mexico, SW USA 32.42N 103.08W

204 L12 **Hobbs Coast** physical region Antarctica

25 Y12 **Hobe Sound** Florida, SE USA 27.03N 80.08W
Hobicaurikány see Uricani

56 E12 **Hobo** Huila, S Colombia 2.34N 75.28W

101 L14 **Hoboken** Antwerpen, N Belgium 51.12N 4.22E

164 K13 **Hoboksar** var. Hoboksar Mongol Zizhixian. Xinjiang Uygur Zizhiqu, NW China 46.48N 85.42E
Hoboksar Mongol Zizhixian see Hoboksar

165 O15 **Hobucken** North Carolina, SE USA 35.15N 76.31W

97 O20 **Hobro** Nordjylland, N Denmark 56.39N 9.51E

83 P15 **Hobyo** It. Obbia. Mudug, E Somalia 5.16N 48.24E

111 Q4 **Hochburg** Oberösterreich, N Austria 48.10N 12.57E

110 F8 **Hochdorf** Luzern, N Switzerland 47.10N 8.16E

111 N18 **Hochfeiler** It. Gran Pilastro. ▲ Austria/Italy 46.59N 11.42E

178 M14 **Hồ Chí Minh** var. Ho Chi Minh City; prev. Saigon. S Vietnam 10.46N 106.43E
Ho Chi Minh City see Hồ Chí Minh

110 H7 **Höchst** Vorarlberg, NW Austria 47.28N 9.40E
Höchstadt an der Aisch see Aisch

103 K19 **Höchstadt an der Aisch** var. Höchstadt. Bayern, C Germany 49.43N 10.48E

170 E13 **Hōchu** var. Hooshu.

178 K10 **Hô Chí Minh** ...

78 L9 **Hodh ech Chargui** ♦ region E Mauritania
Hodh el Garbi see Hodh el Gharbi

78 J10 **Hodh el Gharbi** var. Hodh el Garbi. ♦ region S Mauritania

113 G25 **Hódmezővásárhely** Csongrád, SE Hungary 46.27N 20.17E

76 J6 **Hodna, Chott El** var. Chott el-Hodna, Ar. Shatt al-Hodna. salt lake N Algeria
Hodna, Shatt al- see Hodna, Chott El

113 G19 **Hodonín** Ger. Göding. Brněnský Kraj, SE Czech Republic 48.51N 17.07E

168 G6 **Hödrögö** Dzavhan, N Mongolia 48.51N 96.48E
Hodság/Hodschag see Odžaci
Hoei see Huy

101 F12 **Hoeilaart** Vlaams Brabant, C Belgium 50.46N 4.28E
Hoë Karoo see Great Karoo

100 F12 **Hoek van Holland** Eng. Hook of Holland. Zuid-Holland, W Netherlands 52.00N 4.07E

100 L11 **Hoenderloo** Gelderland, E Netherlands 52.08N 5.46E

101 L18 **Hoensbroek** Limburg, SE Netherlands 50.55N 5.55E

169 Y11 **Hoeryŏng** NE North Korea 42.23N 129.46E

101 K18 **Hoeselt** Limburg, NE Belgium 50.49N 5.30E

100 K11 **Hoevelaken** Gelderland, C Netherlands 52.10N 5.27E
Hoey see Huy

103 M18 **Hof** Bayern, SE Germany 50.19N 11.55E
Höfdhakaupstadhur see Skagaströnd
Hofei see Hefei

103 G18 **Hofheim am Taunus** Hessen, W Germany 50.04N 8.27E
Hofmarkt see Odorheiu Secuiesc

94 L3 **Höfn** Austurland, SE Iceland 64.14N 15.17W

96 N13 **Hofors** Gävleborg, C Sweden 60.33N 16.21E

94 J6 **Hofsjökull** glacier C Iceland

94 J1 **Hofsós** Nordhurland Vestra, N Iceland 65.54N 19.25E

170 Dd13 **Hōfu** Yamaguchi, Honshū, SW Japan 34.01N 131.34E
Hofuf see Al Hufūf

97 J22 **Höganäs** Skåne, S Sweden 56.11N 12.39E

191 P14 **Hogan Group** island group Tasmania, SE Australia

25 R4 **Hogansville** Georgia, SE USA 33.10N 84.55W

30 I14 **Hogback Mountain** ▲ Nebraska, C USA 41.40N 103.44W

97 G14 **Høgevarde** ▲ S Norway 60.19N 9.27E

33 S5 **Høgfors** see Karkkila

23 Y6 **Hog Island** island Michigan, C USA

191 P14 **Hog Island** island Virginia, NE USA
Hogoley Islands see Chuuk Islands

97 N19 **Högsby** Kalmar, S Sweden 57.10N 16.03E

38 L7 **Hogup Mountains** ▲ Utah, W USA

103 E17 **Hohe Acht** ▲ W Germany 50.23N 7.00E
Hohenelbe see Vrchlabí

110 I7 **Hohenems** Vorarlberg, W Austria 47.22N 9.43E
Hohenmauth see Vysoké Mýto
Hohensalza see Inowrocław
Hohenstadt see Zábřeh
Hohenstein in Ostpreussen see Olsztynek

22 I9 **Hohenwald** Tennessee, S USA 35.33N 87.33W

103 L17 **Hohenwarte-Stausee** ☒ C Germany
Hohes Venn see Hautes Fagnes

110 I8 **Hohe Tauern** ▲ W Austria

169 O13 **Hohhot** var. Huhehot, Huhuohaote, Mong. Kukukhoto; prev. Kweisui, Kweisui. Nei Mongol Zizhiqu, N China 40.49N 111.37E

105 U6 **Hohneck** ▲ NE France 48.04N 7.01E

75 D16 **Hohoe** E Ghana 7.07N 0.31E

170 D12 **Hōhoku** Yamaguchi, Honshū, SW Japan 34.15N 130.56E

165 H15 **Hoh Sai Hu** ☒ C China

165 L11 **Hoh Xil Hu** ☒ C China

164 L11 **Hoh Xil Shan** ▲ W China

178 Kk10 **Hôi An** prev. Faifo. Quang Nam-Đa Nẵng, C Vietnam 15.54N 108.19E

83 F17 **Hoima** W Uganda 1.25N 31.22E

28 L5 **Hoisington** Kansas, C USA 38.31N 98.46W
Hojagala see Khodzhakala
Hojambaz see Khodzhambas

97 H23 **Højby** Fyn, C Denmark 55.19N 10.27E

97 F23 **Højer** Sønderjylland, SW Denmark 54.57N 8.43E

170 Ee14 **Hōjō** var. Hōzyō. Ehime, Shikoku, SW Japan 33.58N 132.47E

192 I3 **Hokianga Harbour** inlet SE Tasman Sea

193 F17 **Hokitika** West Coast, South Island, NZ 42.43S 170.58E

172 P5 **Hokkai-dō** ◆ territory Hokkaidō, NE Japan

172 Oo5 **Hokkaidō** prev. Ezo, Yeso, Yezo. island NE Japan

37 X17 **Holbrook** California, W USA 32.48N 115.22W

100 L5 **Holwerd** Fris. Holwert. Friesland, N Netherlands 53.22N 5.51E
Holwert see Holwerd

149 S4 **Hokmābād** Khorāsān, N Iran 36.37N 57.34E

170 Ee15 **Hokō** see P'enghu
Hoko-guntō/Hoko-shotō see P'enghu Liehtao

143 T4 **Hoktemberyan** Rus. Oktemberyan. SW Armenia 40.09N 43.58E

143 X12 **Hoctún** var. Hoctun. E Mexico 20.48N 89.13W
Hoctun see Hoctún

23 T4 **Hocking River** ♦ Ohio, N USA

96 M1 **Hof Buskerud, S Norway** 60.36N 8.18E

119 R11 **Hola Prystan'** Rus. Golaya Pristan. Khersons'ka Oblast', S Ukraine 46.31N 32.31E

97 I23 **Holbæk** Vestsjælland, E Denmark 55.42N 11.42E

168 L6 **Holboo** Dzavhan, W Mongolia

191 P10 **Holbrook** New South Wales, SE Australia 35.43S 147.18E

39 N11 **Holbrook** Arizona, SW USA 34.54N 110.09W

29 S5 **Holden** Missouri, C USA 38.42N 93.59W

38 K5 **Holden** Utah, W USA 39.06N 112.16W

29 O11 **Holdenville** Oklahoma, C USA 35.04N 96.24W

31 O16 **Holdrege** Nebraska, C USA 40.28N 99.28W

37 X3 **Hole in the Mountain Peak** ▲ Nevada, W USA 40.54N 115.06W

161 G20 **Hole Narsipur** Karnātaka, W India 12.46N 76.13E

113 H18 **Holešov** Ger. Holleschau. Zlínský Kraj, E Czech Republic 49.19N 17.34E

47 N14 **Holetown** prev. Jamestown. W Barbados 13.09N 59.37W

33 Q12 **Holgate** Ohio, N USA 41.12N 84.06W

44 H7 **Holguín** Holguín, SE Cuba 20.51N 76.16W

25 V7 **Holiday** Florida, SE USA 28.11N 82.44W

41 Q13 **Holitna River** ♦ Alaska, USA

96 J13 **Höljes** Värmland, C Sweden 60.54N 12.34E

111 X3 **Hollabrunn** Niederösterreich, NE Austria 48.33N 16.06E

38 L3 **Holladay** Utah, W USA 40.39N 111.49W

9 X16 **Holland** Manitoba, C Canada 49.36N 98.52W

33 O9 **Holland** Michigan, N USA 42.47N 86.06W

45 V8 **Holland** Texas, SW USA 30.52N 97.24W
Holland see Netherlands

24 K4 **Hollandale** Mississippi, S USA 33.10N 90.51W
Hollandia see Jayapura
Hollandsch Diep see Hollands Diep

101 H14 **Hollands Diep** var. Hollandsch Diep. channel SW Netherlands
Holleschau see Holešov

80 R5 **Holliday** Texas, SW USA 33.49N 98.41W

20 E15 **Hollidaysburg** Pennsylvania, NE USA 40.24N 78.22W

23 S6 **Hollins** Virginia, NE USA 37.20N 79.56W

28 I2 **Hollis** Oklahoma, C USA 34.42N 99.54W

37 O10 **Hollister** California, W USA 36.51N 121.25W

29 V11 **Hollister** Missouri, C USA 36.37N 93.13W

95 M19 **Hollola** Etelä-Suomi, S Finland 60.59N 25.31E

100 N4 **Hollum** Friesland, N Netherlands 53.27N 5.38E

97 K23 **Höllviksnäs** Skåne, S Sweden 55.25N 12.57E

39 W6 **Holly** Colorado, C USA 38.03N 102.07W

33 R9 **Holly** Michigan, N USA 42.47N 83.37W

23 S14 **Holly Hill** South Carolina, SE USA 33.19N 80.24W

23 W11 **Holly Ridge** North Carolina, SE USA 34.31N 77.31W

24 L1 **Holly Springs** Mississippi, S USA 34.46N 89.25W

25 Z15 **Hollywood** Florida, SE USA 26.00N 80.09W

15 I2 **Holman** Victoria Island, Northwest Territories, N Canada 70.42N 117.45W

94 I2 **Hólmavík** Vestfirðhir, NW Iceland 65.42N 21.43W

32 J7 **Holmen** Wisconsin, N USA 43.57N 91.14W

25 R8 **Holmes Creek** ♦ Alabama/Florida, SE USA

97 H16 **Holmestrand** Vestfold, S Norway 59.28N 10.19E

95 J16 **Holmön** island N Sweden

97 E22 **Holmsland Klit** beach W Denmark

95 J16 **Holmsund** Västerbotten, N Sweden 63.42N 20.25E

97 Q18 **Holmudden** headland SE Sweden 57.59N 19.04E

194 L13 **Holnicote Bay** headland SW PNG 8.30S 148.18E

144 J7 **Holon** var. Kholon. Tel Aviv, C Israel 32.01N 34.46E

32 J7 **Holstebro** Wisconsin, N USA 43.57N 91.14W

119 P8 **Holovanivs'k** Rus. Golovanevsk. Kirovohrads'ka Oblast', C Ukraine 48.21N 30.26E

97 F21 **Holstebro** Ringkøbing, W Denmark 56.22N 8.37E
Holsted Ribe, W Denmark 55.30N 8.54E

31 T13 **Holstein** Iowa, C USA 42.29N 95.32W
Holsteinborg/Holstenborg/Holstensborg see Sisimiut

22 M8 **Holston River** ♦ Tennessee, S USA

33 Q9 **Holt** Michigan, N USA 42.38N 84.31W

100 H10 **Holten** Overijssel, E Netherlands 52.16N 6.25E

28 K5 **Holton** Kansas, C USA 39.28N 95.44W

31 N6 **Holstein** Overijssel, N Netherlands

143 Q11 **Hopa** Artvin, NE Turkey 41.23N 41.27E

20 J14 **Hopatcong** New Jersey, NE USA 40.55N 74.39W

8 M17 **Hope** British Columbia, SW Canada 49.21N 121.28W

27 V14 **Hope** Alaska, USA 60.55N 149.39W

29 T14 **Hope** Arkansas, C USA 33.40N 93.35W

33 O13 **Hope** Indiana, N USA 39.18N 85.46W

31 Q5 **Hope** North Dakota, N USA 47.18N 97.42W

11 O7 **Hopedale** Newfoundland, NE Canada 55.25N 60.14W
Hopeh/Hopei see Hebei

188 K13 **Hope, Lake** salt lake Western Australia

43 Y14 **Hopelchén** Campeche, SE Mexico 19.44N 89.52W

23 U8 **Hope Mills** North Carolina, SE USA 34.58N 78.57W

191 O7 **Hope, Mount** New South Wales, SE Australia 32.49S 145.55E

94 P6 **Hopen** island SE Svalbard

207 Q4 **Hope, Point** headland Alaska, USA

10 M3 **Hopes Advance, Cap** headland Quebec, NE Canada 61.07N 69.30W

190 L10 **Hopetoun** Victoria, SE Australia 35.46S 142.23E

85 H23 **Hopetown** Northern Cape, W South Africa 29.38S 24.06E

23 W6 **Hopewell** Virginia, NE USA 37.16N 77.15W

111 I7 **Hopfgarten-im-Brixental** Tirol, W Austria

189 N8 **Hopkins Lake** salt lake Western Australia

190 M12 **Hopkins River** ♦ Victoria, SE Australia

22 I7 **Hopkinsville** Kentucky, S USA

36 M6 **Hopland** California, W USA 38.58N 123.09W

34 F9 **Hoquiam** Washington, NW USA 46.58N 123.53W

143 Q11 **Horasan** Erzurum, NE Turkey 40.03N 42.10E

103 G20 **Horb am Neckar** Baden-Württemberg, S Germany 48.27N 8.42E

97 K23 **Hörby** Skåne, S Sweden 55.50N 13.42E

45 N12 **Horconcitos** Chiriquí, W Panama 8.17N 82.10W

97 C14 **Hordaland** ◆ county S Norway

118 N11 **Horezu** prev. Hurezu. Vâlcea, SW Romania 45.06N 24.00E

110 F7 **Horgen** Zürich, N Switzerland 47.16N 8.36E

168 J7 **Horgo** Arhangay, C Mongolia 48.06N 99.52E
Hōrin see Fenglin

169 U11 **Horinger** Nei Mongol Zizhiqu, N China 40.23N 111.48E

168 J9 **Horiult** Bayanhongor, C Mongolia 45.09N 100.50E

9 U17 **Horizon** Saskatchewan, S Canada 49.33N 105.05W

207 J19 **Hönö** Västra Götaland, S Sweden 57.42N 11.39E

40 G11 **Honokaa** Haw. Honoka'a. Hawaii, USA, C Pacific Ocean 20.04N 155.27W

40 G11 **Honokohau** Haw. Honokōhau. Hawaii, USA, C Pacific Ocean 19.40N 156.01W

40 D9 **Honolulu** ● Oahu, Hawaii, USA, C Pacific Ocean 21.18N 157.51W

40 H11 **Honomu** Haw. Honōmū. Hawaii, USA, C Pacific Ocean 19.51N 155.06W

107 N9 **Honrubia** Castilla-La Mancha, C Spain 39.36N 2.16W

171 L15 **Honshū** var. Hondo, Honsyū. island New Japan
Honsyū see Honshū
Honte see Westerschelde
Honzyō see Honjō

15 I5 **Hood** ♦ Nunavut, NW Canada

34 H11 **Hood, Mount** ▲ Oregon, NW USA 45.22N 121.41W

34 H11 **Hood River** Oregon, NW USA 45.42N 121.31W

100 H10 **Hoofddorp** Noord-Holland, W Netherlands 52.18N 4.42E

101 G15 **Hoogerheide** Noord-Brabant, S Netherlands 51.25N 4.19E

100 N8 **Hoogeveen** Drenthe, NE Netherlands 52.43N 6.30E

100 O6 **Hoogezand-Sappemeer** Groningen, NE Netherlands 53.10N 6.46E

100 J8 **Hoogkarspel** Noord-Holland, NW Netherlands 52.42N 4.59E

100 N5 **Hoogkerk** Groningen, NE Netherlands 53.13N 6.30E

100 O13 **Hoogvliet** Zuid-Holland, SW Netherlands 51.51N 4.23E

28 I8 **Hooker** Oklahoma, C USA 36.51N 101.12W

99 E21 **Hook Head Ir.** Rinn Duáin. headland SE Ireland 52.07N 6.55W

121 L19 **Hook of Holland** see Hoek van Holland

168 J9 **Hoolt** Övörhangay, C Mongolia 45.31N 103.06E

100 J8 **Hoonah** Chichagof Island, Alaska, USA 58.05N 135.21W

41 Q11 **Hooper Bay** Alaska, USA 61.31N 166.06W

33 N3 **Hoopeston** Illinois, N USA 40.28N 87.40W

97 K23 **Höör** Skåne, S Sweden 55.55N 13.33E

100 I9 **Hoorn** Noord-Holland, NW Netherlands 52.37N 5.04E

20 L10 **Hoosic River** ♦ New York, NE USA
Hoosier State see Indiana

37 Y11 **Hoover Dam** dam Arizona/Nevada, W USA 36.01N 114.44W

168 J9 **Höövör** Övörhangay, C Mongolia 45.10N 101.19E

168 J9 **Höövör** Övörhangay, C Mongolia

44 G6 **Honduras** off. Republic of Honduras. ♦ republic Central America
Honduras, Golfo de see Honduras, Gulf of

44 H4 **Honduras, Gulf of** Sp. Golfo de Honduras. gulf W Caribbean Sea

9 V12 **Hone** Manitoba, C Canada 56.13N 101.12W

23 Q12 **Honea Path** South Carolina, SE USA 34.26N 82.23W

97 H14 **Hønefoss** Buskerud, S Norway 60.10N 10.15E

33 S12 **Honey Creek** ♦ Ohio, N USA

45 V5 **Honey Grove** Texas, SW USA 33.34N 95.54W

37 Q4 **Honey Lake** ☒ California, W USA

104 L4 **Honfleur** Calvados, N France 49.25N 0.13E

96 M5 **Hommelvik** Sør-Trøndelag, S Norway 63.24N 10.46E

97 C16 **Hommersåk** Rogaland, S Norway 58.55N 5.51E

161 H15 **Honnābād** Karnātaka, C India 17.46N 77.08E

24 J7 **Homochitto River** ♦ Mississippi, S USA

85 N20 **Homoine** Inhambane, SE Mozambique 23.51S 35.04E

114 O12 **Homoljske Planine** ▲ E Yugoslavia
Homonna see Humenné

79 X13 **Hongwon** E North Korea 40.03N 127.54E

166 H7 **Hongyuan** prev. Hurama. Sichuan, C China 32.49N 102.40E

167 Q10 **Hongze Hu** var. Hung-tse Hu. ☒ E China

195 W16 **Honiara** ● (Solomon Islands) Guadalcanal, C Solomon Islands 9.27S 159.55E

99 K24 **Honiton** SW England, UK 47.16N 8.36E

172 O9 **Honjō** var. Honzyō. Akita, Honshū, C Japan 39.23N 140.03E

95 K18 **Honkajoki** Länsi-Suomi, SW Finland 62.00N 22.15E

169 O13 **Honningsvåg** Finnmark, N Norway 70.58N 25.58E

199 J10 **Horizon Bank** *undersea feature* S Pacific Ocean
199 Jj11 **Horizon Deep** *undersea feature* W Pacific Ocean
97 L14 **Hörken** Örebro, S Sweden *60.03N 14.55E*
121 O15 **Horki** ***Rus.*** Gorki. Mahilyowskaya Voblasts', E Belarus *54.17N 30.59E*
205 O10 **Horlick Mountains** ▲ Antarctica
119 X7 **Horlivka** ***Rom.*** Adâncata, ***Rus.*** Gorlovka. Donets'ka Oblast', E Ukraine *48.19N 38.04E*
149 V11 **Ḩormak** Sīstān va Balūchestān, SE Iran *30.00N 60.50E*
149 R13 **Hormozgān** ***off.*** Ostān-e Hormozgān. ◈ *province* S Iran
Hormoz, Tangeh-ye *see* Hormuz, Strait of
147 W6 **Hormuz, Strait of** ***var.*** Strait of Ormuz, ***Per.*** Tangeh-ye Hormoz. *strait* Iran/Oman
111 W2 **Horn** Niederösterreich, NE Austria *48.39N 15.37E*
97 M18 **Horn** Östergötland, S Sweden *57.54N 15.49E*
15 Hh8 **Horn** ↝ Northwest Territories, NW Canada
Hornád *see* Hernád
15 H3 **Hornaday** ↝ Northwest Territories, NW Canada
94 H13 **Hornavan** ◎ N Sweden
67 C24 **Hornby Mountains** *hill range* West Falkland, Falkland Islands
Horn, Cape *see* Hornos, Cabo de
99 O18 **Horncastle** E England, UK *53.12N 0.07W*
97 N14 **Horndal** Dalarna, C Sweden *60.16N 16.25E*
95 I16 **Hörnefors** Västerbotten, N Sweden *63.37N 19.54E*
20 F11 **Hornell** New York, NE USA *42.19N 77.38W*
Horné Nové Mesto *see* Kysucké Nové Mesto
10 F12 **Hornepayne** Ontario, S Canada *49.13N 84.48W*
96 D10 **Hornindalsvatnet** ◎ S Norway
103 G22 **Hornisgrinde** ▲ SW Germany *48.37N 8.13E*
24 M9 **Horn Island** *island* Mississippi, S USA
65 J26 **Hornos, Cabo de** ***Eng.*** Cape Horn. *headland* S Chile *55.52S 67.00W*
119 S10 **Hornostayivka** Khersons'ka Oblast', S Ukraine *47.00N 33.42E*
191 T9 **Hornsby** New South Wales, SE Australia *33.44S 151.08E*
99 O16 **Hornsea** E England, UK *53.54N 0.09W*
96 O11 **Hornslandet** *peninsula* C Sweden
97 H22 **Hornslet** Århus, C Denmark *56.19N 10.19E*
94 O4 **Hornsundtind** ▲ S Svalbard *76.54N 16.07E*
Horochów *see* Horokhiv
118 J7 **Horodenka** ***Rus.*** Gorodenka. Ivano-Frankivs'ka Oblast', W Ukraine *48.41N 25.28E*
119 Q2 **Horodnya** ***Rus.*** Gorodnya. Chernihivs'ka Oblast', NE Ukraine *51.54N 31.30E*
118 K6 **Horodok** Khmel'nyts'ka Oblast', W Ukraine *49.10N 26.34E*
118 H5 **Horodok** ***Pol.*** Gródek Jagielloński, ***Rus.*** Gorodok, Gorodok Yagellonski. L'vivs'ka Oblast', NW Ukraine *49.48N 23.39E*
119 Q6 **Horodyshche** ***Rus.*** Gorodishche. Cherkas'ka Oblast', C Ukraine *49.18N 31.27E*
172 P4 **Horokanai** Hokkaidō, NE Japan *44.02N 142.08E*
118 J4 **Horokhiv** ***Pol.*** Horochów, ***Rus.*** Gorokhov. Volyns'ka Oblast', NW Ukraine *50.31N 24.50E*
172 P7 **Horoshiri-dake** ***var.*** Horosiri Dake. ▲ Hokkaidō, N Japan *42.43N 142.41E*
Horosiri Dake *see* Horoshiri-dake
113 C17 **Hořovice** ***Ger.*** Horowitz. Středočeský Kraj, W Czech Republic *49.49N 13.53E*
Horowitz *see* Hořovice
169 T9 **Horqin Youyi Zhongqi** Nei Mongol Zizhiqu, NE China *45.02N 121.33E*
169 U11 **Horqin Zuoyi Houqi** Nei Mongol Zizhiqu, N China *42.53N 122.22E*
169 T9 **Horqin Zuoyi Zhongqi** Nei Mongol Zizhiqu, N China *45.02N 121.28E*
64 O5 **Horqueta** Concepción, C Paraguay *23.23S 57.04W*
57 O12 **Horqueta Minas** Amazonas, S Venezuela *2.19N 63.31W*
97 J20 **Horred** Västra Götaland, S Sweden *57.22N 12.25E*
157 J19 **Horsburgh Atoll** *atoll* N Maldives
22 K7 **Horse Cave** Kentucky, S USA *37.10N 85.54W*
39 V6 **Horse Creek** ↝ Colorado, C USA
29 S6 **Horse Creek** ↝ Missouri, C USA
20 G11 **Horseheads** New York, NE USA *42.10N 76.49W*
39 P13 **Horse Mount** ▲ New Mexico, SW USA *33.58N 108.10W*
97 G22 **Horsens** Vejle, C Denmark *55.52N 9.52E*
67 F25 **Horse Pasture Point** *headland* W Saint Helena *15.57S 5.46W*
35 N13 **Horseshoe Bend** Idaho, NW USA *43.55N 116.11W*
38 L13 **Horseshoe Reservoir** ▣ Arizona, SW USA
66 M9 **Horseshoe Seamounts** *undersea feature* E Atlantic Ocean
190 L11 **Horsham** Victoria, SE Australia *36.44S 142.13E*
99 O23 **Horsham** SE England, UK *51.01N 0.21W*
101 M15 **Horst** Limburg, SE Netherlands *51.29N 6.04E*
66 N2 **Horta** Faial, Azores, Portugal, NE Atlantic Ocean *38.31N 28.39W*
97 H16 **Horten** Vestfold, S Norway *59.25N 10.24E*
113 M23 **Hortobágy-Berettyó** ↝ E Hungary
29 Q3 **Horton** Kansas, C USA *39.39N 95.31W*
15 H3 **Horton** ↝ Northwest Territories, NW Canada
97 I23 **Hørve** Vestsjælland, E Denmark *55.46N 11.28E*
97 L22 **Hörvik** Blekinge, S Sweden *56.01N 14.45E*
144 E11 **Horvot Haluza** ***var.*** Khorvot Khalutsa. *ruins* Southern, S Israel *30.49N 34.50E*
12 E7 **Horwood Lake** ◎ Ontario, S Canada
118 K4 **Horyn'** ***Rus.*** Goryn. ↝ NW Ukraine
83 I14 **Hosa'ina** ***var.*** Hosseina, ***It.*** Hosanna. Southern, S Ethiopia *7.38N 37.58E*
Hosanna *see* Hosa'ina
103 H18 **Hösbach** Bayern, C Germany *50.00N 9.12E*
Hose Mountains *see* Hose, Pegunungan
174 Mm6 **Hose, Pegunungan** ***var.*** Hose Mountains. ▲ East Malaysia
154 L15 **Hoshāb** Baluchistān, SW Pakistan *26.01N 63.51E*
160 H10 **Hoshangābād** Madhya Pradesh, C India *22.43N 77.45E*
118 L4 **Hoshcha** Rivnens'ka Oblast', NW Ukraine *50.37N 26.38E*
158 I7 **Hoshiārpur** Punjab, NW India *31.35N 75.57E*
168 J7 **Höshööt** Arhangay, C Mongolia *48.06N 102.34E*
101 M23 **Hosingen** Diekirch, NE Luxembourg *50.01N 6.04E*
195 N12 **Hoskins** New Britain, E PNG *5.28S 150.25E*
161 G17 **Hospet** Karnātaka, C India *15.16N 76.19E*
106 K4 **Hospital de Orbigo** Castilla-León, N Spain *42.27N 5.52W*
Hospitalet *see* L'Hospitalet de Llobregat
94 N13 **Hossa** Oulu, E Finland *65.28N 29.36E*
Hosseina *see* Hosa'ina
Hosszúmezjő *see* Câmpulung Moldovenesc
65 I25 **Hoste, Isla** *island* S Chile
119 O4 **Hostomel'** ***Rus.*** Gostomel'. Kyyivs'ka Oblast', N Ukraine *50.40N 30.15E*
161 H20 **Hosūr** Tamil Nādu, SE India *12.45N 77.51E*
178 H8 **Hot** Chiang Mai, NW Thailand *18.14N 98.35E*
164 G10 **Hotan** ***var.*** Khotan, ***Chin.*** Ho-t'ien. Xinjiang Uygur Zizhiqu, NW China *37.10N 79.51E*
164 H9 **Hotan He** ↝ NW China
85 G22 **Hotazel** Northern Cape, N South Africa *27.12S 22.58E*
39 Q5 **Hotchkiss** Colorado, C USA *38.47N 107.43W*
37 V7 **Hot Creek Range** ▲ Nevada, W USA
Hote *see* Hoti
176 U11 **Hoti** ***var.*** Hote. Pulau Seram, E Indonesia *2.58S 130.19E*
Ho-t'ien *see* Hotan
Hotin *see* Khotyn
95 H15 **Hoting** Jämtland, C Sweden *64.07N 16.14E*
168 L14 **Hotong Qagan Nur** ◎ N China
168 J8 **Hotont** Arhangay, C Mongolia *47.21N 102.27E*
29 T12 **Hot Springs** Arkansas, C USA *34.30N 93.03W*
30 J11 **Hot Springs** South Dakota, N USA *43.25N 103.28W*
23 S5 **Hot Springs** Virginia, NE USA *38.00N 79.50W*
37 Q4 **Hot Springs Peak** ▲ California, W USA *40.23N 120.06W*
29 T12 **Hot Springs Village** Arkansas, C USA *34.39N 93.03W*
Hotspur Bank *see* Hotspur Seamount
67 J16 **Hotspur Seamount** ***var.*** Hotspur Bank. *undersea feature* C Atlantic Ocean *18.00S 35.00W*
15 Hh6 **Hottah Lake** ◎ Northwest Territories, NW Canada
46 K9 **Hotte, Massif de la** ▲ SW Haiti
101 K21 **Hotton** Luxembourg, SE Belgium *50.18N 5.25E*
Hötzing *see* Haţeg
197 I6 **Houaïlou** Province Nord, C New Caledonia *21.17S 165.37E*
76 K5 **Houari Boumediène** ✕ (Alger) N Algeria *36.38N 3.15E*
178 Hh6 **Houayxay** ***var.*** Ban Houayxay, Ban Houei Sai. Bokèo, N Laos *20.16N 100.27E*
105 N5 **Houdan** Yvelines, N France *48.48N 1.36E*
101 F20 **Houdeng-Goegnies** ***var.*** Houdeng-Gœgnies. Hainaut, S Belgium *50.28N 4.10E*
104 K14 **Houeillès** Lot-et-Garonne, SW France *44.15N 0.02E*
101 L22 **Houffalize** Luxembourg, SE Belgium *50.08N 5.47E*
32 M3 **Houghton** Michigan, N USA *41.55N 84.37W*
33 Q7 **Houghton Lake** Michigan, N USA *44.18N 84.45W*
33 Q7 **Houghton Lake** ◎ Michigan, N USA
21 T3 **Houlton** Maine, NE USA *46.09N 67.49W*
166 M5 **Houma** Shanxi, C China *35.33N 111.19E*
200 Q15 **Houma** 'Eua, C Tonga *21.10S 175.17W*
200 R16 **Houma** Tongatapu, S Tonga *21.18S 174.55W*
24 J10 **Houma** Louisiana, S USA *29.34N 90.43W*
200 Qq16 **Houma Taloa** *headland* Tongatapu, S Tonga *21.16S 175.07W*
79 O13 **Houndé** SW Burkina *11.34N 3.31W*
104 J12 **Hourtin-Carcans, Lac d'** ◎ SW France
38 J5 **House Range** ▲ Utah, W USA
8 K13 **Houston** British Columbia, SW Canada *54.24N 126.39W*
41 R11 **Houston** Alaska, USA *61.37N 149.50W*
31 X10 **Houston** Minnesota, N USA *43.45N 91.34W*
24 M3 **Houston** Mississippi, S USA *33.54N 89.00W*
29 V7 **Houston** Missouri, C USA *37.19N 91.57W*
45 W11 **Houston** Texas, SW USA *29.45N 95.21W*
45 W11 **Houston** ✕ Texas, SW USA *30.03N 95.18W*
100 J12 **Houten** Utrecht, C Netherlands *52.01N 5.10E*
101 K17 **Houthalen** Limburg, NE Belgium *51.01N 5.22E*
101 I22 **Houyet** Namur, SE Belgium *50.10N 5.00E*
97 H22 **Hov** Århus, C Denmark *55.54N 10.13E*
97 L17 **Hova** Västra Götaland, S Sweden *58.52N 14.13E*
168 E6 **Hovd** ***var.*** Khovd. Hovd, W Mongolia *47.58N 91.40E*
168 J10 **Hovd** Övörhangay, C Mongolia *44.43N 102.08E*
168 E7 **Hovd** ◈ *province* W Mongolia
168 C5 **Hovd Gol** ↝ NW Mongolia
99 O23 **Hove** SE England, UK *50.49N 0.10W*
31 N8 **Hoven** South Dakota, N USA *45.12N 99.47W*
118 I8 **Hoverla, Hora** ***Rus.*** Gora Goverla. ▲ W Ukraine *48.09N 24.30E*
168 H8 **Höviyn Am** Bayanhongor, C Mongolia *47.08N 98.41E*
97 M21 **Hovmantorp** Kronoberg, S Sweden *56.46N 15.07E*
169 N11 **Hövsgöl** Dornogovĭ, SE Mongolia *43.35N 109.40E*
168 I5 **Hövsgöl** ◈ *province* N Mongolia
Hovsgol, Lake *see* Hövsgöl Nuur
168 J5 **Hövsgöl Nuur** ***var.*** Lake Hovsgol. ◎ N Mongolia
80 L9 **Howa, Ouadi** ***var.*** Wādi Howar. ↝ Chad/Sudan ***see also*** Howar, Wādi
29 P7 **Howard** Kansas, C USA *37.28N 96.15W*
31 Q10 **Howard** South Dakota, N USA *43.58N 97.31W*
45 N10 **Howard Draw** *valley* Texas, SW USA
31 U8 **Howard Lake** Minnesota, N USA *45.03N 94.03W*
82 B8 **Howar, Wādi** ***var.*** Ouadi Howa. ↝ Chad/Sudan ***see also*** Howa, Ouadi
45 U5 **Howe** Texas, SW USA *33.29N 96.38W*
191 R12 **Howe, Cape** *headland* New South Wales/Victoria, SE Australia *37.30S 149.58E*
33 R9 **Howell** Michigan, N USA *42.36N 83.55W*
30 L9 **Howes** South Dakota, N USA *44.34N 102.03W*
85 K23 **Howick** KwaZulu/Natal, E South Africa *29.29S 30.13E*
Howrah *see* Hāora
29 W9 **Hoxie** Arkansas, C USA *36.03N 90.58W*
28 J3 **Hoxie** Kansas, C USA *39.21N 100.26W*
103 I14 **Höxter** Nordrhein-Westfalen, W Germany *51.46N 9.22E*
164 K6 **Hoxud** Xinjiang Uygur Zizhiqu, NW China *42.18N 86.51E*
98 J5 **Hoy** *island* N Scotland, UK
45 S17 **Hoya, Cerro** ▲ S Panama *7.22N 80.38W*
96 D12 **Høyanger** Sogn og Fjordane, S Norway *61.13N 6.04E*
103 P15 **Hoyerswerda** Sachsen, E Germany *51.27N 14.17E*
170 Dd15 **Hōyo-kaikyō** ***var.*** Hayasui-seto. *strait* SW Japan
106 J8 **Hoyos** Extremadura, W Spain *40.10N 6.43W*
31 W4 **Hoyt Lakes** Minnesota, N USA *47.31N 92.08W*
89 V2 **Hoyvík** Streymoy, N Faeroe Islands
143 O14 **Hozat** Tunceli, E Turkey *39.09N 39.13E*
Hôzyô *see* Hōjō
113 F16 **Hradec Králové** ***Ger.*** Königgrätz. Hradecký Kraj, N Czech Republic *50.13N 15.49E*
113 E16 **Hradecký Kraj** ◈ *region* N Czech Republic
113 B16 **Hradiště** ***Ger.*** Burgstadlberg. ▲ NW Czech Republic *50.12N 13.04E*
119 R6 **Hradyz'k** ***Rus.*** Gradizhsk. Poltavs'ka Oblast', NE Ukraine *49.12N 33.08E*
121 M16 **Hradzyanka** ***Rus.*** Grodzyanka. Mahilyowskaya Voblasts', E Belarus *53.36N 28.47E*
121 F16 **Hrandzichy** ***Rus.*** Grandichi. Hrodzyenskaya Voblasts', W Belarus *53.43N 23.50E*
113 H18 **Hranice** ***Ger.*** Mährisch-Weisskirchen. Olomoucký Kraj, E Czech Republic *49.34N 17.45E*
114 I13 **Hrasnica** Federacija Bosna I Hercegovina, SE Bosnia and Herzegovina *43.48N 18.19E*
111 V11 **Hrastnik** C Slovenia *46.09N 15.08E*
143 U12 **Hrazdan** ***Rus.*** Razdan. C Armenia *40.30N 44.50E*
143 T12 **Hrazdan** ***var.*** Zanga, ***Rus.*** Razdan. ↝ C Armenia
119 R5 **Hrebinka** ***Rus.*** Grebenka. Poltavs'ka Oblast', NE Ukraine *50.08N 32.27E*
121 K17 **Hresk** ***Rus.*** Gresk. Minskaya Voblasts', C Belarus *53.10N 27.28E*
Hrisoupoli *see* Chrysoúpoli
121 F16 **Hrodna** ***Pol.*** Grodno. Hrodzyenskaya Voblasts', W Belarus *53.40N 23.50E*
121 F16 **Hrodzyenskaya Voblasts'** ***prev. Rus.*** Grodnenskaya Oblast'. ◈ *province* W Belarus
113 J21 **Hron** ***Ger.*** Gran, ***Hung.*** Garam. ↝ C Slovakia
113 Q14 **Hrubieszów** ***Rus.*** Grubeshov. Lubelskie, E Poland *50.48N 23.54E*
114 F13 **Hrvace** Split-Dalmacija, SE Croatia *43.46N 16.35E*
Hrvatska *see* Croatia
114 F10 **Hrvatska Kostajnica** ***var.*** Kostajnica. Sisak-Moslavina, C Croatia *45.14N 16.35E*
Hrvatsko Grahovo *see* Bosansko Grahovo
118 K6 **Hrymayliv** ***Pol.*** Gízymałów, ***Rus.*** Grimaylov. Ternopil's'ka Oblast', W Ukraine *49.18N 26.02E*
178 H4 **Hsenwi** Shan State, E Myanmar *23.20N 97.59E*
Hsia-men *see* Xiamen
Hsiang-t'an *see* Xiangtan
Hsi Chiang *see* Xi Jiang
178 Gg6 **Hsihseng** Shan State, C Myanmar *20.07N 97.16E*
167 S13 **Hsinchu** *municipality* N Taiwan *24.51N 121.01E*
Hsing-k'ai Hu *see* Khanka, Lake
Hsi-ning/Hsining *see* Xining
Hsinking *see* Changchun
Hsin-yang *see* Xinyang
167 S14 **Hsinying** ***var.*** Sinying, ***Jap.*** Shinei. C Taiwan *23.12N 120.15E*
178 Gg4 **Hsipaw** Shan State, C Myanmar *22.36N 97.16E*
Hsu-chou *see* Xuzhou
167 S13 **Hsüeh Shan** ▲ N Taiwan
Hu *see* Shanghai Shi
85 B18 **Huab** ↝ W Namibia
59 M21 **Huacaya** Chuquisaca, S Bolivia *20.55S 63.24W*
59 J19 **Huachacalla** Oruro, SW Bolivia *19.01S 68.22W*
165 X9 **Huachi** ***var.*** Rouyuanchengzi. Gansu, C China *36.24N 107.58E*
59 N16 **Huachi, Laguna** ◎ N Bolivia
59 D14 **Huacho** Lima, W Peru *11.09S 77.37W*
169 Y8 **Huachuan** Heilongjiang, NE China *46.21N 130.43E*
169 P12 **Huade** Nei Mongol Zizhiqu, N China *41.52N 113.58E*
169 W10 **Huadian** Jilin, NE China *42.58N 126.37E*
58 E13 **Huagaruncho, Cordillera** ▲ C Peru
Hua Hin *see* Ban Hua Hin
203 S10 **Huahine** *island* Îles Sous le Vent, W French Polynesia
Huahua, Rio *see* Wawa, Río
178 Ii9 **Huai** ↝ E Thailand
167 P6 **Huaibei** Anhui, E China *34.00N 116.48E*
Huaide *see* Gongzhuling
163 T10 **Huai He** ↝ C China
166 L11 **Huaihua** Hunan, S China *27.36N 109.56E*
167 N14 **Huaiji** Guangdong, S China *23.54N 112.12E*
167 O2 **Huailai** ***prev.*** Shacheng. Hebei, E China *40.22N 115.34E*
167 P7 **Huainan** ***var.*** Huai-nan, Hwainan. Anhui, E China *32.36N 116.56E*
167 N2 **Huairen** Shanxi, C China *35.28N 110.29E*
167 O7 **Huaiyang** Henan, C China *33.39N 114.34E*
167 Q7 **Huaiyin** ***var.*** Qingjiang. Jiangsu, E China *33.33N 119.03E*
178 Gg16 **Huai Yot** Trang, SW Thailand *7.45N 99.36E*
43 Q15 **Huajuapan** ***var.*** Huajuapan de León. Oaxaca, SE Mexico *17.49N 97.48W*
Huajuapan de León *see* Huajuapan
43 O9 **Hualahuises** Nuevo León, NE Mexico *24.55N 99.44W*
38 I11 **Hualapai Mountains** ▲ Arizona, SW USA
38 I11 **Hualapai Peak** ▲ Arizona, SW USA *35.04N 113.54W*
64 J7 **Hualfin** Catamarca, N Argentina *27.15S 66.52W*
167 T13 **Hualien** ***var.*** Hwalien, ***Jap.*** Karen. C Taiwan *23.58N 121.34E*
58 E10 **Huallaga, Río** ↝ N Peru
58 C11 **Huamachuco** La Libertad, C Peru *7.50S 78.03W*
43 Q14 **Huamantla** Tlaxcala, S Mexico *19.18N 97.57W*
84 C13 **Huambo** ***Port.*** Nova Lisboa. Huambo, C Angola *12.48S 15.45E*
84 B13 **Huambo** ◈ *province* C Angola
43 P15 **Huamuxtitlán** Guerrero, S Mexico *17.49N 98.34W*
65 H17 **Huancache, Sierra** ▲ SW Argentina
59 I17 **Huancané** Puno, SE Peru *15.15S 69.47W*
59 F16 **Huancapi** Ayacucho, C Peru *13.36S 74.09W*
59 E15 **Huancavelica** Huancavelica, SW Peru *12.45S 75.03W*
59 E15 **Huancavelica** ***off.*** Departamento de Huancavelica. ◈ *department* W Peru
59 E14 **Huancayo** Junín, C Peru *12.03S 75.13W*
59 K20 **Huanchaca, Cerro** ▲ S Bolivia *20.12S 66.35W*
58 C12 **Huandoy, Nevado** ▲ W Peru *8.48S 77.33W*
167 O8 **Huangchuan** Henan, C China *32.08N 115.03E*
Huang Hai *see* Yellow Sea
163 Q8 **Huang He** ***var.*** Yellow River. ↝ C China
167 Q4 **Huanghe Kou** *delta* E China
166 L5 **Huangling** Shaanxi, C China *35.34N 109.12E*
167 O9 **Huangpi** Hubei, C China *30.53N 114.16E*
169 P13 **Huangqi Hai** ◎ N China
167 Q9 **Huang Shan** ▲ Anhui, E China *29.43N 118.19E*
167 Q9 **Huangshan** ***var.*** Tunxi. Anhui, E China *29.48N 118.17E*
167 O9 **Huangshi** ***var.*** Huang-shih, Hwangshih. Hubei, C China *30.14N 115.00E*
Huang-shih *see* Huangshi
166 L5 **Huangtu Gaoyuan** *plateau* C China
63 B22 **Huanguelén** Buenos Aires, E Argentina *37.01S 61.57W*
167 S10 **Huangyan** Zhejiang, SE China *28.42N 121.13E*
165 T10 **Huangyuan** Qinghai, C China *36.40N 101.12E*
165 T10 **Huangzhong** Qinghai, C China *36.31N 101.32E*
169 W12 **Huanren** Liaoning, NE China *41.16N 125.25E*
59 F15 **Huanta** Ayacucho, C Peru *12.54S 74.13W*
58 E13 **Huánuco** Huánuco, C Peru *9.57S 76.15W*
58 D13 **Huánuco** ***off.*** Departamento de Huánuco. ◈ *department* C Peru
59 K19 **Huanuni** Oruro, W Bolivia *18.15S 66.54W*
165 X9 **Huan Xian** Gansu, C China *36.30N 107.20E*
167 S12 **Huap'ing Yu** *island* N Taiwan
64 H3 **Huara** Tarapacá, N Chile *19.59S 69.42W*
59 D14 **Huaral** Lima, W Peru *11.28S 77.12W*
Huarás *see* Huaraz
58 D13 **Huaraz** ***var.*** Huarás. Ancash, W Peru *9.30S 77.31W*
59 I16 **Huari Huari, Río** ↝ S Peru
58 C13 **Huarmey** Ancash, W Peru *10.03S 78.09W*
42 H4 **Huásabas** Sonora, NW Mexico *29.46N 109.18W*
58 D8 **Huasaga, Río** ↝ Ecuador/Peru
178 H16 **Hua Sai** Nakhon Si Thammarat, SW Thailand *8.01N 100.18E*
58 D12 **Huascarán, Nevado** ▲ W Peru *9.01S 77.27W*
64 G8 **Huasco** Atacama, N Chile *28.28S 71.12W*
64 G8 **Huasco, Río** ↝ C Chile
165 S11 **Huashikia** Qinghai, W China
42 G7 **Huatabampo** Sonora, NW Mexico *26.49N 109.40W*
165 W10 **Huating** Gansu, C China *35.13N 106.39E*
178 Jj7 **Huatt, Phou** ▲ N Vietnam *19.45N 104.48E*
43 Q14 **Huatusco** ***var.*** Huatusco de Chicuellar. Veracruz-Llave, C Mexico *19.13N 96.57W*
Huatusco de Chicuellar *see* Huatusco
43 P13 **Huauchinango** Puebla, S Mexico *20.12N 98.03W*
Huaunta *see* Wounta
43 R15 **Huautla** ***var.*** Huautla de Jiménez. Oaxaca, SE Mexico *18.10N 96.51W*
Huautla de Jiménez *see* Huautla
167 O5 **Huaxian** ***var.*** Daokou, Hua Xian. Henan, C China *35.33N 114.30E*
31 V13 **Hubbard** Iowa, C USA *42.18N 93.18W*
45 U8 **Hubbard** Texas, SW USA *31.52N 96.43W*
45 Q6 **Hubbard Creek Lake** ◎ Texas, SW USA
33 R6 **Hubbard Lake** ◎ Michigan, N USA
166 M9 **Hubei** ***var.*** E, Hubei Sheng, Hupeh, Hupei. ◈ *province* C China
Hubei Sheng *see* Hubei
111 P8 **Huben** Tirol, W Austria *46.55N 12.35E*
33 R13 **Huber Heights** Ohio, N USA *39.50N 84.07W*
161 F17 **Hubli** Karnātaka, SW India *15.19N 75.13E*
169 X12 **Huch'ang** N North Korea *41.24N 127.04E*
99 M18 **Hucknall** C England, UK *53.01N 1.10W*
99 L17 **Huddersfield** N England, UK *53.39N 1.46W*
97 O16 **Huddinge** Stockholm, C Sweden *59.15N 17.57E*
96 N11 **Hudiksvall** Gävleborg, C Sweden *61.45N 17.12E*
31 W13 **Hudson** Iowa, C USA *42.24N 92.27W*
21 O11 **Hudson** Massachusetts, NE USA *42.24N 71.34W*
33 Q11 **Hudson** Michigan, N USA *41.51N 84.21W*
32 H6 **Hudson** Wisconsin, N USA *44.58N 92.43W*
9 V14 **Hudson Bay** Saskatchewan, S Canada *58.13N 90.48W*
10 G6 **Hudson Bay** *bay* NE Canada
205 T16 **Hudson, Cape** *headland* Antarctica *68.15S 154.00E*
Hudson, Détroit d' *see* Hudson Strait
29 Q9 **Hudson, Lake** ▣ Oklahoma, C USA
20 K9 **Hudson River** ↝ New Jersey/New York, NE USA
8 M12 **Hudson's Hope** British Columbia, W Canada *56.03N 121.58W*
10 L2 **Hudson Strait** ***Fr.*** Détroit d'Hudson. *strait* Nunavut/Quebec, NE Canada
Hudūd ash Shamālīyah, Minṭaqat al *see* Al Ḥudūd ash Shamālīyah
Hudur *see* Xuddur
178 K9 **Huê** Thưa Thiên-Huê, C Vietnam *16.28N 107.34E*
106 J7 **Huebra** ↝ W Spain
44 H8 **Hueco Mountains** ▲ Texas, SW USA
118 G10 **Huedin** ***Hung.*** Bánffyhunyad. Cluj, NW Romania *46.51N 23.01E*
42 J10 **Huehuento, Cerro** ▲ C Mexico *24.04N 105.42W*
44 B5 **Huehuetenango** Huehuetenango, W Guatemala *15.19N 91.25W*
44 B4 **Huehuetenango** ***off.*** Departamento de Huehuetenango. ◈ *department* W Guatemala
42 L11 **Huejuquilla** Jalisco, SW Mexico *22.40N 103.52W*
43 P12 **Huejutla** ***var.*** Huejutla de Reyes. Hidalgo, C Mexico *21.08N 98.16W*
Huejutla de Reyes *see* Huejutla
104 G6 **Huelgoat** Finistère, NW France *48.22N 3.45W*
107 O13 **Huelma** Andalucía, S Spain *37.39N 3.28W*
106 I14 **Huelva** ***anc.*** Onuba. Andalucía, SW Spain *37.15N 6.55W*
106 I13 **Huelva** ◈ *province* Andalucía, SW Spain
106 J13 **Huelva** ↝ SW Spain
107 Q14 **Huercal-Overa** Andalucía, S Spain *37.22N 1.55W*
39 Q9 **Huerfano Mountain** ▲ New Mexico, SW USA *36.25N 107.50W*
39 T7 **Huerfano River** ↝ Colorado, C USA
107 S12 **Huertas, Cabo** *headland* SE Spain *38.21N 0.25W*
107 R6 **Huerva** ↝ N Spain
107 S4 **Huesca** ***anc.*** Osca. Aragón, NE Spain *42.07N 0.25W*
107 T4 **Huesca** ◈ *province* Aragón, NE Spain
107 P13 **Huéscar** Andalucía, S Spain *37.39N 2.31W*
43 N15 **Huetamo** ***var.*** Huetamo de Núñez. Michoacán de Ocampo, SW Mexico *18.37N 100.53W*
Huetamo de Núñez *see* Huetamo
107 P8 **Huete** Castilla-La Mancha, C Spain *40.07N 2.40W*
25 P4 **Hueytown** Alabama, S USA *33.27N 87.00W*
30 L16 **Hugh Butler Lake** ▣ Nebraska, C USA
189 V6 **Hughenden** Queensland, NE Australia *20.56S 144.15E*
190 A6 **Hughes** South Australia *30.41S 129.31E*
41 P8 **Hughes** Alaska, USA *66.03N 154.15W*
29 X11 **Hughes** Arkansas, C USA *34.57N 90.28W*
45 W6 **Hughes Springs** Texas, SW USA *33.00N 94.37W*
39 V5 **Hugo** Colorado, C USA *39.08N 103.28W*
29 Q13 **Hugo** Oklahoma, C USA *34.01N 95.31W*
29 Q13 **Hugo Lake** ▣ Oklahoma, C USA
28 H7 **Hugoton** Kansas, C USA *37.10N 101.21W*
Huhehot/Huhuohaote *see* Hohhot
167 R13 **Hui'an** Fujian, SE China *25.06N 118.45E*
192 O9 **Huiarau Range** ▲ North Island, NZ
85 D22 **Huib-Hoch Plateau** *plateau* S Namibia
43 O13 **Huichapán** Hidalgo, C Mexico *20.22N 99.42W*
Huicheng *see* Shexian
169 W13 **Hŭich'ŏn** C North Korea *40.09N 126.17E*
56 E12 **Huila** ***off.*** Departamento del Huila. ◈ *province* S Colombia
85 B15 **Huíla** ◈ *province* SW Angola
56 D11 **Huila, Nevado del** *elevation* C Colombia *2.56N 75.59W*
85 B15 **Huíla Plateau** *plateau* S Angola
166 G12 **Huili** Sichuan, C China *26.39N 102.13E*
167 P4 **Huimin** Shandong, E China *37.28N 117.30E*
169 W11 **Huinan** ***var.*** Chaoyang. Jilin, NE China *42.40N 126.03E*
64 K12 **Huinca Renancó** Córdoba, C Argentina *34.51S 64.22W*
165 V10 **Huining** Gansu, C China *35.42N 105.01E*
166 J12 **Huishui** Guizhou, S China *26.07N 106.39E*
104 L6 **Huisne** ↝ NW France
100 L12 **Huissen** Gelderland, SE Netherlands *51.57N 5.57E*
165 N11 **Huiten Nur** ◎ C China
95 K19 **Huittinen** Länsi-Suomi, W Finland *61.10N 22.40E*
43 O15 **Huitzuco** ***var.*** Huitzuco de los Figueroa. Guerrero, S Mexico *18.18N 99.22W*
Huitzuco de los Figueroa *see* Huitzuco
165 W11 **Huixian** ***var.*** Hui Xian. Gansu, C China *33.48N 106.02E*
43 V17 **Huixtla** Chiapas, SE Mexico *15.07N 92.29W*
166 H12 **Huize** Yunnan, SW China *26.28N 103.18E*
100 J10 **Huizen** Noord-Holland, C Netherlands *52.16N 5.15E*
167 O14 **Huizhou** Guangdong, S China *23.05N 114.22E*
168 J6 **Hujirt** Arhangay, C Mongolia *48.49N 101.20E*
168 J8 **Hujirt** Övörhangay, C Mongolia *46.50N 102.38E*
168 K8 **Hujrt** Töv, C Mongolia *46.41N 104.40E*
Hukagawa *see* Fukagawa
Hŭksan-chedo *see* Hŭksan-gundo
169 W17 **Hŭksan-gundo** ***var.*** Hŭksan-chedo. *island group* SW South Korea
Hukue *see* Fukue
Hukui *see* Fukui
85 G20 **Hukuntsi** Kgalagadi, SW Botswana *23.58S 21.43E*
Hukuoka *see* Fukuoka
Hukusima *see* Fukushima
Hukutiyama *see* Fukuchiyama
Hukuyama *see* Fukuyama
169 W8 **Hulan** Heilongjiang, NE China *45.58N 126.37E*
169 W8 **Hulan He** ↝ NE China
33 Q4 **Hulbert Lake** ◎ Michigan, N USA
Hulczyn *see* Hlučín
169 Z8 **Hulin** Heilongjiang, NE China *45.48N 133.06E*
169 S9 **Hulingol** ***prev.*** Huolin Gol. Nei Mongol Zizhiqu, N China *45.35N 119.53E*
12 L12 **Hull** Quebec, SE Canada *45.25N 75.45W*
31 S12 **Hull** Iowa, C USA *43.11N 96.07W*
Hull *see* Kingston upon Hull
Hull Island *see* Orona
101 F16 **Hulst** Zeeland, SW Netherlands *51.16N 4.03E*
169 Q7 **Hulstay** Dornod, NE Mongolia *48.25N 114.56E*
Hultschin *see* Hlučín
97 M19 **Hultsfred** Kalmar, S Sweden *57.30N 15.49E*
Hulun *see* Hailar
Hu-lun Ch'ih *see* Hulun Nur
169 Q6 **Hulun Nur** ***var.*** Hu-lun Ch'ih; ***prev.*** Dalai ***Nor.*** ◎ NE China
Hulwan/Hulwân *see* Helwân
119 V8 **Hulyaypole** ***Rus.*** Gulyaypole. Zaporiz'ka Oblast', SE Ukraine *47.41N 36.10E*
169 V4 **Huma** Heilongjiang, NE China *51.40N 126.38E*
47 V6 **Humacao** E Puerto Rico *18.09N 65.49W*
169 U4 **Huma He** ↝ NE China
64 J5 **Humahuaca** Jujuy, N Argentina *23.13S 65.19W*
61 E14 **Humaitá** Amazonas, N Brazil *7.33S 63.01W*
64 N7 **Humaitá** Ñeembucú, S Paraguay *27.06S 58.28W*
85 H26 **Humansdorp** Eastern Cape, S South Africa *34.01S 24.45E*
29 S6 **Humansville** Missouri, C USA *37.47N 93.34W*
42 I8 **Humaya, Río** ↝ C Mexico
85 C16 **Humbe** Cunene, SW Angola *16.37S 14.52E*
99 N17 **Humber** *estuary* E England, UK
99 N17 **Humberside** *cultural region* E England, UK
Humberto *see* Umberto
45 W11 **Humble** Texas, SW USA *29.58N 95.15W*
9 U15 **Humboldt** Saskatchewan, S Canada *52.13N 105.09W*
31 U12 **Humboldt** Iowa, C USA *42.42N 94.13W*
29 Q6 **Humboldt** Kansas, C USA *37.48N 95.26W*
31 S17 **Humboldt** Nebraska, C USA *40.09N 95.56W*
37 S3 **Humboldt** Nevada, W USA *40.36N 118.15W*
22 G9 **Humboldt** Tennessee, S USA *35.49N 88.55W*
36 K3 **Humboldt Bay** *bay* California, W USA
37 S4 **Humboldt Lake** ◎ Nevada, W USA
197 J7 **Humboldt, Mont** ▲ S New Caledonia *21.57S 166.24E*
37 S4 **Humboldt River** ↝ Nevada, W USA
37 T5 **Humboldt Salt Marsh** *wetland* Nevada, W USA
191 P11 **Hume, Lake** ◎ New South Wales/Victoria, SE Australia
113 N19 **Humenné** ***Ger.*** Homenau, ***Hung.*** Homonna. Prešovský Kraj, E Slovakia *48.57N 21.54E*
31 V15 **Humeston** Iowa, C USA *40.51N 93.30W*
56 J5 **Humocaro Bajo** Lara, N Venezuela *9.42N 70.02W*
31 Q14 **Humphrey** Nebraska, C USA *41.38N 97.29W*
37 S9 **Humphreys, Mount** ▲ California, W USA *37.11N 118.39W*
38 L11 **Humphreys Peak** ▲ Arizona, SW USA *35.18N 111.40W*
113 E17 **Humpolec** ***Ger.*** Gumpolds, Humpoletz. Jihlavský Kraj, C Czech Republic *49.33N 15.22E*
Humpoletz *see* Humpolec
95 K19 **Humppila** Etelä-Suomi, S Finland *60.54N 23.21E*
34 F8 **Humptulips** Washington, NW USA *47.13N 123.57W*
44 H7 **Humuya, Río** ↝ W Honduras
77 P9 **Hūn** N Libya *29.06N 15.56E*
Hunabasi *see* Funabashi
94 I1 **Húnaflói** *bay* NW Iceland
166 M11 **Hunan** ***var.*** Hunan Sheng, Xiang. ◈ *province* S China
Hunan Sheng *see* Hunan
169 Y10 **Hunchun** Jilin, NE China *42.51N 130.21E*
97 I22 **Hundested** Frederiksborg, E Denmark *55.58N 11.52E*
Hundred Mile House *see* 100 Mile House
118 G12 **Hunedoara** ***Ger.*** Eisenmarkt, ***Hung.*** Vajdahunyad. Hunedoara, SW Romania *45.45N 22.54E*
118 G12 **Hunedoara** ◈ *county* W Romania
103 I17 **Hünfeld** Hessen, C Germany *50.40N 9.46E*
113 H23 **Hungary** ***off.*** Republic of Hungary, ***Ger.*** Ungarn, ***Hung.*** Magyarország, ***Rom.*** Ungaria, ***SCr.*** Mađarska, ***Ukr.*** Uhorshchyna; ***prev.*** Hungarian People's Republic. ◆ *republic* C Europe
Hungary, Plain of *see* Great Hungarian Plain
168 F6 **Hungiy** Dzavhan, W Mongolia *48.31N 94.15E*
169 X13 **Hŭngnam** E North Korea *39.50N 127.36E*
35 P8 **Hungry Horse Reservoir** ▣ Montana, NW USA
Hungt'ou *see* Lan Yü
Hung-tse Hu *see* Hongze Hu
178 Jj6 **Hưng Yên** Hai Hưng, N Vietnam *20.37N 106.04E*
Hunjiang *see* Baishan
97 I18 **Hunnebostrand** Västra Götaland, S Sweden *58.26N 11.19E*
103 E19 **Hunsrück** ▲ W Germany
99 P18 **Hunstanton** E England, UK *52.57N 0.28E*
161 G20 **Hunsūr** Karnātaka, E India *12.18N 76.15E*
168 I7 **Hunt** Arhangay, C Mongolia *47.49N 99.24E*
102 G12 **Hunte** ↝ NW Germany
31 Q5 **Hunter** North Dakota, N USA *47.10N 97.11W*
45 S11 **Hunter** Texas, SW USA *29.47N 98.01W*
193 D20 **Hunter** ↝ South Island, NZ
191 N15 **Hunter Island** *island* Tasmania, SE Australia
20 K11 **Hunter Mountain** ▲ New York, NE USA *42.10N 74.13W*
193 B23 **Hunter Mountains** ▲ South Island, NZ
191 S7 **Hunter River** ↝ New South Wales, SE Australia
34 L7 **Hunters** Washington, NW USA *48.07N 118.13W*
193 F20 **Hunters Hills, The** *hill range* South Island, NZ
192 M12 **Hunterville** Manawatu-Wanganui, North Island, NZ *39.55S 175.34E*
33 N16 **Huntingburg** Indiana, N USA *38.18N 86.57W*
99 O20 **Huntingdon** E England, UK *52.19N 0.12W*
20 E15 **Huntingdon** Pennsylvania, NE USA *40.28N 78.00W*
22 G9 **Huntingdon** Tennessee, S USA *36.00N 88.25W*
99 O20 **Huntingdonshire** *cultural region* C England, UK

◆ Country
● Country Capital
◇ Dependent Territory
○ Dependent Territory Capital
◈ Administrative Region
✕ International Airport
▲ Mountain
▲ Mountain Range
Volcano
↝ River
◎ Lake
▣ Reservoir

33 P12 **Huntington** Indiana, N USA 40.52N 85.30W
34 L13 **Huntington** Oregon, NW USA 44.22N 117.18W
45 X9 **Huntington** Texas, SW USA 31.16N 94.34W
38 M5 **Huntington** Utah, W USA 39.19N 110.57W
23 P5 **Huntington** West Virginia, NE USA 38.24N 82.27W
37 T16 **Huntington Beach** California, W USA 33.39N 118.00W
37 W4 **Huntington Creek** ≈ Nevada, W USA
192 L7 **Huntly** Waikato, North Island, NZ 37.33S 175.09E
98 K8 **Huntly** NE Scotland, UK 57.25N 2.48W
8 K8 **Hunt, Mount** ▲ Yukon Territory, NW Canada 61.29N 129.10W
12 H12 **Huntsville** Ontario, S Canada 45.18N 79.12W
25 P2 **Huntsville** Alabama, S USA 34.43N 86.35W
29 S9 **Huntsville** Arkansas, C USA 36.05N 93.43W
29 U3 **Huntsville** Missouri, C USA 39.26N 92.33W
22 M8 **Huntsville** Tennessee, S USA 36.25N 84.30W
45 V10 **Huntsville** Texas, SW USA 30.43N 95.34W
38 L2 **Huntsville** Utah, W USA 41.16N 111.47W
43 W12 **Hunucmá** Yucatán, SE Mexico 20.59N 89.55W
155 W3 **Hunza** var. Karīmābād. Jammu and Kashmir, NE Pakistan 36.22N 74.43E
155 W3 **Hunza** ≈ NE Pakistan
Hunze see Oostermoers Vaart
164 H4 **Huocheng** var. Shuiding. Xinjiang Uygur Zizhiqu, NW China 44.03N 80.49E
167 N6 **Huojia** Henan, C China 35.13N 113.37E
Huolin Gol see Hulingol
197 P3 **Huon** reef N New Caledonia
194 K13 **Huon Gulf** gulf E PNG
194 K13 **Huon Peninsula** headland C PNG 6.24S 147.50E
Huoshao Dao see Lü Tao
Huoshao Tao see Lan Yü
Hupeh/Hupei see Hubei
Hurano see Furano
97 H14 **Hurdalssjøen** ≈ S Norway
12 E13 **Hurd, Cape** headland Ontario, S Canada 45.12N 81.43W
Hurdegaryp see Hardegarijp
31 N4 **Hurdsfield** North Dakota, N USA 47.24N 99.55W
168 J7 **Hüremt** Bulgan, C Mongolia 48.40N 102.33E
168 J8 **Hüremt** Övörhangay, C Mongolia 46.18N 102.27E
77 X9 **Hurghada** var. Al Ghurdaqah, Ghurdaqah. E Egypt 27.16N 33.46E
69 V9 **Huri Hills** ▲ NW Kenya
39 P15 **Hurley** New Mexico, SW USA 32.42N 108.07W
32 K4 **Hurley** Wisconsin, N USA 46.25N 90.15W
23 V4 **Hurlock** Maryland, NE USA 38.37N 75.51W
31 P10 **Huron** South Dakota, N USA 44.19N 98.13W
33 S6 **Huron, Lake** ⊗ Canada/USA
33 N3 **Huron Mountains** hill range Michigan, N USA
38 J8 **Hurricane** Utah, W USA 37.10N 113.18W
23 P5 **Hurricane** West Virginia, NE USA 38.28S 82.01W
38 J8 **Hurricane Cliffs** cliff Arizona, SW USA
25 V6 **Hurricane Creek** ≈ Georgia, SE USA
96 E12 **Hurrungane** ▲ S Norway 61.25N 7.48E
103 E16 **Hürth** Nordrhein-Westfalen, W Germany 50.52N 6.52E
Hurukawa see Furukawa
193 I17 **Hurunui** ≈ South Island, NZ
97 F21 **Hurup** Viborg, NW Denmark 56.46N 8.25E
119 T14 **Hurzuf** Respublika Krym, S Ukraine 44.33N 34.18E
Huş see Huşi
97 B19 **Húsavík** Dan. Husevig. Faeroe Islands 61.19N 6.41W
94 K1 **Húsavík** Nordhurland Eystra, NE Iceland 66.03N 17.19W
118 M10 **Huşi** var. Huş. Vaslui, E Romania 46.40N 28.05E
97 L19 **Huskvarna** Jönköping, S Sweden 57.46N 14.15E
41 P8 **Huslia** Alaska, USA 65.42N 156.24W
Husn see Al Ḩuşn
97 C15 **Husnes** Hordaland, S Norway 59.52N 5.46E
96 D8 **Hustadvika** sea area S Norway
Husté see Khust
102 H7 **Husum** Schleswig-Holstein, N Germany 54.28N 9.04E
95 I16 **Husum** Västernorrland, C Sweden 63.21N 17.54E
118 K6 **Husyatyn** Ternopil's'ka Oblast', W Ukraine 49.04N 26.10E
Huszt see Khust
168 K6 **Hutag** Bulgan, N Mongolia 49.22N 102.50E
28 M6 **Hutchinson** Kansas, C USA 38.03N 97.55W
31 U9 **Hutchinson** Minnesota, N USA 44.88N 94.22W
25 Y13 **Hutchinson Island** island Florida, SE USA
38 L4 **Hutch Mountain** ▲ Arizona, SW USA 34.49N 111.22W
147 O14 **Ḩuth** NW Yemen 16.13N 44.00E
195 R11 **Hutjena** Buka Island, NE PNG 5.19S 154.40E
111 T8 **Hüttenberg** Kärnten, S Austria 46.58N 14.33E
45 T10 **Hutto** Texas, SW USA 30.32N 97.33W
110 E8 **Huttwil** Bern, W Switzerland 47.06N 7.48E
164 K5 **Hutubi** Xinjiang Uygur Zizhiqu, NW China 44.10N 86.51E
167 N4 **Hutuo He** ≈ C China
Hutyú see Fuchū

193 E20 **Huxley, Mount** ▲ South Island, NZ 44.02S 169.42E
101 J20 **Huy** Dut. Hoei, Hoey. Liège, E Belgium 50.31N 5.13E
167 R8 **Huzhou** var. Wuxing. Zhejiang, SE China 30.54N 120.04E
Huzi see Fuji
Huzieda see Fujieda
Huzinomiya see Fujinomiya
Huziyosida see Fuji-Yoshida
94 I2 **Hvammstangi** Nordhurland Vestra, N Iceland 65.22N 20.54W
94 K4 **Hvannadalshnúkur** ▲ S Iceland 64.01N 16.39W
115 E15 **Hvar** It. Lesina. Split-Dalmacija, S Croatia 43.10N 16.27E
115 F15 **Hvar** It. Lesina; anc. Pharus. island S Croatia
119 T13 **Hvardiys'ke** Rus. Gvardeyskoye. Respublika Krym, S Ukraine 45.07N 34.01E
94 I4 **Hveragerdhi** Sudhurland, SW Iceland 64.00N 21.13W
97 E22 **Hvide Sande** Ringkøbing, W Denmark 56.00N 8.08E
94 I3 **Hvítá** ≈ C Iceland
97 G15 **Hvittingfoss** Buskerud, S Norway 59.28N 10.00E
94 I4 **Hvolsvöllur** Sudhurland, SW Iceland 63.44N 20.12W
Hwach'ŏn-chŏsuji see P'aro-ho
Hwainan see Huainan
Hwang-Hae see Yellow Sea
Hwangshih see Huangshi
85 L17 **Hwedza** Mashonaland East, E Zimbabwe 18.15S 29.48E
65 Q20 **Hyades, Cerro** ▲ S Chile 46.57S 73.09W
21 Q12 **Hyannis** Massachusetts, NE USA 41.38N 70.15W
30 L13 **Hyannis** Nebraska, C USA 41.58N 101.45W
168 F6 **Hyargas Nuur** ⊗ NW Mongolia
41 Y14 **Hydaburg** Prince of Wales Island, Alaska, USA 55.13N 132.44W
193 F22 **Hyde** Otago, South Island, NZ 45.17S 170.17E
23 O7 **Hyde** Kentucky, S USA 37.07N 83.22W
20 K12 **Hyde Park** New York, USA 41.46N 73.52W
161 I15 **Hyderābād** var. Haidarabad. Andhra Pradesh, C India 17.22N 78.25E
155 Q16 **Hyderābād** var. Haidarabad. Sind, SE Pakistan 25.25N 68.21E
105 T16 **Hyères** Var, SE France 43.07N 6.07E
105 T16 **Hyères, Îles d'** island group S France
120 K12 **Hyermanavichy** Rus. Germanovichi. Vitsyebskaya Voblasts', N Belarus 55.25N 27.43E
169 X12 **Hyesan** NE North Korea 41.17N 128.13E
8 K8 **Hyland** ≈ Yukon Territory, NW Canada
97 K18 **Hyltebruk** Halland, S Sweden 57.00N 13.14E
20 L10 **Hyndman** Pennsylvania, NE USA 39.49N 78.42W
35 Q4 **Hyndman Peak** ▲ Idaho, NW USA 43.45N 114.07W
170 G13 **Hyōgo** off. Hyōgo-ken. ◆ prefecture Honshū, SW Japan
170 G13 **Hyōno-sen** ▲ Kyūshū, SW Japan 35.21N 134.30E
Hypanis see Kuban'
38 L1 **Hyrum** Utah, W USA 41.37N 111.51W
95 N14 **Hyrynsalmi** Oulu, C Finland 64.40N 28.30E
30 J3 **Hysham** Montana, NW USA 46.16N 107.14W
99 Q23 **Hythe** SE England, UK 51.04N 1.04E
170 D15 **Hyūga** Miyazaki, Kyūshū, SW Japan 32.24N 131.34E
Hyvinge see Hyvinkää
95 L19 **Hyvinkää** Swe. Hyvinge. Etelä-Suomi, S Finland 60.37N 24.49E

─── **I** ───

118 J9 **Iacobeni** Ger. Jakobeny. Suceava, NE Romania 47.24N 25.19E
Iader see Zadar
180 I7 **Iakora** Fianarantsoa, SE Madagascar 23.04S 46.40E
194 H12 **Ialibu** Southern Highlands, W PNG 6.15S 143.55E
118 K14 **Ialomiţa** var. Jalomitsa. ◆ county SE Romania
118 K14 **Ialomiţa** ≈ SE Romania
119 N10 **Ialoveni** Rus. Yaloveny. C Moldova 46.57N 28.47E
119 N11 **Ialpug** var. Ialpugul Mare, Rus. Yalpug. ≈ Moldova/Ukraine
119 N11 **Ialpugul Mare** see Ialpug
25 T8 **Iamonia, Lake** ⊗ Florida, SE USA
118 L13 **Iana** Brăila, SE Romania 45.06N 27.29E
118 M10 **Iana** Ger. Jassy. Iaşi, NE Romania 47.08N 27.38E
81 L9 **Ianga** Ger. Jassy, Yassy. ◆ county NE Romania
116 J13 **Iasmos** Anatolikí Makedonía kai Thráki, NE Greece 41.07N 25.12E
81 C15 **Iatt, Lake** ⊗ Louisiana, S USA
60 B11 **Iauaretê** Amazonas, NW Brazil 0.37N 69.10W
179 O10 **Iba** Luzon, N Philippines 15.25N 119.55E
79 S16 **Ibadan** Oyo, SW Nigeria 7.21N 4.01E
58 B12 **Ibagué** Tolima, C Colombia 4.27N 75.13W
62 J10 **Ibaiti** Paraná, S Brazil 23.58S 50.09W

179 Pp12 **Ibajay** Panay Island, C Philippines 11.42N 122.17E
38 I4 **Ibapah Peak** ▲ Utah, W USA 39.51N 113.55W
81 J21 **Ibar** Alb. Ibër. ≈ C Yugoslavia
170 F14 **Ibara** Okayama, Honshū, SW Japan 34.36N 133.27E
171 Kk16 **Ibaraki** off. Ibaraki-ken. ◆ prefecture Honshū, S Japan
58 C5 **Ibarra** var. San Miguel de Ibarra. Imbabura, N Ecuador 0.22S 78.07W
Ibasfalău see Dumbrăveni
147 O16 **Ibb** W Yemen 13.55N 44.10E
102 F13 **Ibbenbüren** Nordrhein-Westfalen, NW Germany 52.17N 7.43E
81 H16 **Ibenga** ≈ N Congo
59 I14 **Iberia** Madre de Dios, E Peru 11.21S 69.36W
68 M1 **Iberia** see Spain
Iberian Basin see Iberia Plain
Iberian Mountains see Ibérico, Sistema
86 D12 **Iberian Peninsula** physical region Portugal/Spain
66 M8 **Iberian Plain** undersea feature E Atlantic Ocean
Ibérica, Cordillera see Ibérico, Sistema
107 P6 **Ibérico, Sistema** var. Cordillera Ibérica, Eng. Iberian Mountains. ▲ NE Spain
10 M7 **Iberville, Lac d'** ⊗ Quebec, NE Canada
79 T14 **Ibeto** Niger, W Nigeria 10.30N 5.07E
79 W15 **Ibi** Taraba, C Nigeria 8.13N 9.46E
107 S11 **Ibi** País Valenciano, E Spain 38.37N 0.34W
63 F15 **Ibiá** Minas Gerais, SE Brazil 19.30S 46.31W
63 C19 **Ibicuy** Entre Ríos, E Argentina 33.46S 59.07W
63 F15 **Ibirapuitã** ≈ S Brazil
63 I20 **Ibiza** see Eivissa
107 Q11 **Ibiza** var. Iviza. island Islas Baleares, Spain, W Mediterranean Sea
144 I4 **Ibn Wardān, Qaşr** ruins Ḩamāh, C Syria 35.19N 37.13E
79 T16 **Ibo** see Sassandra
196 Vv11 **Ibobang** Babeldaob, N Palau
61 N17 **Ibotirama** Bahia, E Brazil 12.13S 43.12W
147 Y8 **Ibrā** NE Oman 22.45N 58.30E
131 Q4 **Ibresi** Chuvashskaya Respublika, W Russian Federation 55.22N 47.04E
147 X8 **'Ibrī** NW Oman 23.12N 56.28E
170 B16 **Ibusuki** Kagoshima, Kyūshū, SW Japan 31.13N 130.37E
59 E16 **Ica** var. Ica. Ica, SW Peru 14.01S 75.48W
59 E16 **Ica** off. Departamento de Ica. ◆ department SW Peru
60 C11 **Içana** Amazonas, NW Brazil 0.22N 67.25W
Icaria see Ikaría
60 B13 **Içá, Río** var. Río Putumayo. ≈ N South America see also Putumayo, Río
142 I17 **İçel** var. Ichili. ◆ province S Turkey
94 I3 **Iceland** off. Republic of Iceland, Dan. Island, Icel. Ísland. ◆ republic N Atlantic Ocean
88 B7 **Iceland** island N Atlantic Ocean
66 L5 **Iceland Basin** undersea feature N Atlantic Ocean
Icelandic Plateau see Iceland Plateau
207 Q15 **Iceland Plateau** var. Icelandic Plateau. undersea feature S Greenland Sea
161 E16 **Ichalkaranji** Mahārāshtra, W India 16.42N 74.28E
170 Cc15 **Ichifusa-yama** ▲ Kyūshū, SW Japan 32.18N 131.05E
171 K17 **Ichihara** var. Itihara. Chiba, Honshū, S Japan 35.30N 140.08E
Ichili see İçel
171 I15 **Ichinomiya** var. Itinomiya. Aichi, Honshū, SW Japan 35.19N 136.47E
171 Mm12 **Ichinoseki** var. Itinoseki. Iwate, Honshū, C Japan 38.55N 141.16E
119 R3 **Ichnya** Chernihivs'ka Oblast', NE Ukraine 50.52N 32.24E
59 E17 **Ichoa, Río** ≈ C Bolivia
I-ch'un see Yichun
Iconium see Konya
124 P7 **Iculisma** headland NW Turkey 41.54N 28.03E
41 N5 **Icy Bay** inlet Alaska, USA
41 N5 **Icy Cape** headland Alaska, USA 70.19N 161.52W
41 W13 **Icy Strait** strait Alaska, USA
29 R13 **Idabel** Oklahoma, S USA 33.54N 94.49W
31 T13 **Ida Grove** Iowa, C USA 42.21N 95.28W
79 S13 **Idah** Kogi, S Nigeria 7.06N 6.45E
35 N13 **Idaho** off. State of Idaho; also known as Gem of the Mountains, Gem State. ◆ state NW USA
35 N14 **Idaho City** Idaho, NW USA 43.48N 115.51W
35 R14 **Idaho Falls** Idaho, NW USA 43.28N 112.01W
124 Nn3 **Idálion** var. Dali, Dhali. C Cyprus 35.01N 33.25E
45 N5 **Idalou** Texas, SW USA 33.40N 101.40W
106 I9 **Idanha-a-Nova** Castelo Branco, C Portugal 39.55N 7.15W
103 E19 **Idar-Oberstein** Rheinland-Pfalz, SW Germany 49.43N 7.19E
120 J3 **Ida-Virumaa** off. Ida-Viru Maakond. ◆ province NE Estonia
64 Q6 **Idel'** Respublika Kareliya, NW Russian Federation 64.08N 34.12E
81 C15 **Idenao** Sud-Ouest, SW Cameroon 4.04N 9.01E
Idenburg-rivier see Taritatu, Sungai
Idensalmi see Iisalmi
180 I6 **Ider** Hövsgöl, C Mongolia 48.45N 99.52E
77 N9 **Idfu** var. Edfu. SE Egypt 24.57N 32.51E
157 K18 **Idhi Óros** see Idi
Idhra see Ýdra
173 F3 **Idi** prev. Idhi Óros. ▲ Kríti, Greece 35.13N 24.55E
62 J10 **Idi Amin, Lac** see Edward, Lake

108 G10 **Idice** ≈ N Italy
78 G9 **Idini** Trarza, W Mauritania 17.58N 15.40W
81 J21 **Idiofa** Bandundu, SW Dem. Rep. Congo (Zaire) 5.00S 19.38E
41 O9 **Iditarod River** ≈ Alaska, USA
97 M14 **Idkerberget** Dalarna, C Sweden 35.30N 137.48E
144 I3 **Idlib** Idlib, N Syria 35.57N 36.37E
144 I4 **Idlib** off. Muḩāfaẓat Idlib. ◆ governorate NW Syria
96 J11 **Idra** see Ýdra
120 J4 **Idre** Dalarna, C Sweden 61.52N 12.45E
Idria see Idrija
111 S11 **Idrija** It. Idria. W Slovenia 46.00N 14.10E
103 G18 **Idstein** Hessen, W Germany 50.13N 8.16E
85 J25 **Idutywa** Eastern Cape, SE South Africa 32.06N 28.18E
120 G9 **Iecava** Bauska, S Latvia 56.36N 24.10E
172 P14 **Ie-jima** var. Ii-shima. island Nansei-shotō, SW Japan
101 B18 **Ieper** Fr. Ypres. West-Vlaanderen, W Belgium 50.51N 2.52E
117 K25 **Ierápetra** Kríti, Greece, E Mediterranean Sea 35.00N 25.45E
117 G22 **Iérax, Akrotírio** headland S Greece 36.45N 23.06E
Ierissós see Ierissós
117 H14 **Ierissós** var. Ierissós. Kentrikí Makedonía, N Greece 40.24N 23.52E
118 I11 **Iernut** Hung. Radnót. Mureş, C Romania 46.27N 24.18E
201 R8 **Ijuw** NE Nauru 0.30S 166.57E
101 I13 **Izendijke** Zeeland, SW Netherlands 51.20N 3.36E
94 K9 **Iešjávri** ≈ N Norway
95 K18 **Ikaalinen** Länsi-Suomi, W Finland 61.46N 23.04E
116 H10 **Ikalamavony** Fianarantsoa, SE Madagascar 21.10S 46.34E
170 Cc12 **Iki-suidō** strait SW Japan
131 O13 **Iki Burul** Respublika Kalmykiya, SW Russian Federation 45.58N 44.44E
170 Bb12 **Ikitsuki-shima** island SW Japan
143 P11 **Ikizdere** Rize, NE Turkey 40.47N 40.33E
41 P14 **Ikolik, Cape** headland Kodiak Island, Alaska, USA 57.12N 154.46W
79 V17 **Ikom** Cross River, SE Nigeria 5.57N 8.43E
180 I6 **Ikongo** prev. Fort-Carnot. Fianarantsoa, SE Madagascar 21.52S 47.27E
41 P9 **Ikpikpuk River** ≈ Alaska, USA
202 H1 **Iku** prev. Lone Tree Islet. atoll Tungaru, W Kiribati
171 Gg14 **Ikuno** Hyōgo, Honshū, SW Japan 35.10N 134.48E
202 H16 **Ikurangi** ▲ Rarotonga, S Cook Islands 21.12S 159.45W
176 Xx12 **Ilaga** Irian Jaya, E Indonesia 3.54S 137.30E
179 P5 **Ilagan** Luzon, N Philippines 17.07N 121.54E
148 J7 **Ilam** Eastern, E Nepal 26.52N 87.58E
148 J7 **Īlām** var. Elam. Īlām, W Iran 33.40N 46.24E
148 J8 **Īlām** off. Ostān-e Īlām. ◆ province W Iran
167 T13 **Ilan** Jap. Giran. N Taiwan 24.46N 121.46E
152 G9 **Ilanly Obvodnitel'nyy Kanal** canal N Turkmenistan
126 I14 **Ilanskiy** Krasnoyarskiy Kray, S Russian Federation 56.16N 95.59E
110 H9 **Ilanz** Graubünden, S Switzerland 46.46N 9.10E
131 O10 **Ilovlya** Volgogradskaya Oblast', SW Russian Federation 49.45N 44.18E
131 O10 **Ilovlya** ≈ SW Russian Federation
127 P8 **Il'pyrskiy** Koryakskiy Avtonomnyy Okrug, E Russian Federation 60.00N 164.16E
130 K10 **Il's'kiy** Krasnodarskiy Kray, SW Russian Federation 44.52N 38.26E
9 S13 **Ile-à-la-Crosse** Saskatchewan, C Canada 55.29N 108.00W
81 J21 **Ilebo** prev. Port-Francqui. Kasai Occidental, W Dem. Rep. Congo (Zaire) 4.19S 20.31E
120 I11 **Ilūkste** Daugavpils, SE Latvia 55.58N 26.21E
176 Uu12 **Ilur** Pulau Gorong, E Indonesia 4.03S 131.25E
34 F10 **Ilwaco** Washington, NW USA 46.18N 124.03W
176 Y11 **Ilugwa** Irian Jaya, E Indonesia 3.42S 139.09E
3.42S 139.09E

168 L8 **Ihhayrhan** Töv, C Mongolia 46.57N 105.51E
180 I6 **Ihosy** Fianarantsoa, S Madagascar 22.25S 46.09E
168 L7 **Ihsüüj** Töv, C Mongolia 48.12N 106.23E
191 L14 **Iida** Nagano, Honshū, S Japan 35.30N 137.48E
171 Bb16 **Iide-san** ▲ Honshū, C Japan
95 M14 **Iijoki** ≈ C Finland
120 J4 **Iisaku** NE Estonia 59.07N 27.18E
95 M16 **Iisalmi** var. Idensalmi. Itä-Suomi, C Finland 63.31N 27.10E
164 I5 **Ili** ≈ China/Kazakhstan
25 N18 **Iitti** Etelä-Suomi, S Finland 61.13N 28.49E
171 H14 **Iizuka** Fukuoka, Kyūshū, SW Japan 35.25N 136.00E
170 D13 **Iizuka** Fukuoka, Kyūshū, SW Japan 33.38N 130.40E
79 S16 **Ijebu-Ode** Ogun, SW Nigeria 6.46N 3.57E
100 H9 **IJmuiden** Noord-Holland, W Netherlands 52.28N 4.34E
100 M12 **IJssel** var. Yssel. ≈ Netherlands/Germany
100 I8 **IJsselmeer** prev. Zuider Zee. ⊗ N Netherlands
100 J8 **IJsselmuiden** Overijssel, E Netherlands 52.34N 5.55E
100 I12 **IJsselstein** Utrecht, C Netherlands 52.01N 5.01E
61 G14 **Ijuí** Rio Grande do Sul, S Brazil 28.22S 53.55W
61 G15 **Ijuí, Rio** ≈ S Brazil
104 M6 **Illiers-Combray** Eure-et-Loir, C France 48.18N 1.15E
32 K12 **Illinois** off. State of Illinois; also known as Prairie State, Sucker State. ◆ state C USA
32 J13 **Illinois River** ≈ Illinois, N USA
119 N6 **Illintsi** Vinnyts'ka Oblast', C Ukraine 49.07N 29.13E
76 M10 **Illizi** SE Algeria 26.30N 8.28E
29 Y7 **Illmo** Missouri, C USA 37.13N 89.30W
Illur co see Lorca
Illuro see Mataró
Illyrisch-Feistritz see Ilirska Bistrica
103 K17 **Ilmenau** Thüringen, C Germany 50.40N 10.54E
59 I16 **Ilo** Moquegua, SW Peru 17.39S 71.22W
179 O11 **Iloilo** off. Iloilo City. Panay Island, C Philippines 10.42N 122.34E
114 K8 **Ilok** Hung. Újlak. Serbia, NW Yugoslavia 45.12N 19.22E
95 O11 **Ilomantsi** Itä-Suomi, E Finland 62.40N 30.55E
79 S16 **Ilorin** Kwara, W Nigeria 8.30N 4.34E
119 X8 **Ilovays'k** Rus. Ilovaysk. Donets'ka Oblast', SE Ukraine 47.54N 38.13E
42 F10 **Imuris** Sonora, NW Mexico 30.48N 110.52W
179 P10 **Imus** Luzon, N Philippines 14.27N 120.55E
171 J15 **Ina** Nagano, Honshū, S Japan
67 M18 **Inaccessible Island** island W Tristan da Cunha
117 D17 **Ínachos** ≈ S Greece
196 H6 **I Naftan, Puntan** headland Saipan, S Northern Mariana Islands
Inagua Islands see Great Inagua/Little Inagua
193 H15 **Inangahua** West Coast, South Island, NZ 41.51S 171.58E
176 V10 **Inanwatan** Irian Jaya, E Indonesia 2.06S 132.07E
59 H14 **Iñapari** Madre de Dios, E Peru 11.00S 69.34W
196 H1 **Inarajan** W Guam 13.16N 144.45E
94 L10 **Inari** Lapp. Anár, Aanaar. Lappi, N Finland 68.54N 27.01E
94 L10 **Inarijärvi** Lapp. Aanaarjävri, Swe. Enareträsk. ⊗ N Finland
94 L10 **Inarijoki** Lapp. Anárjohka. ≈ Finland/Norway
Inău see Ineu

145 T11 **Imān 'Abd Allāh** S Iraq 31.36N 44.34E
128 J4 **Imandra, Ozero** ⊗ NW Russian Federation
170 E16 **Imano-yama** ▲ Shikoku, SW Japan 32.51N 132.48E
170 C13 **Imari** Saga, Kyūshū, SW Japan 33.16N 129.51E
66 J6 **Imarssuak Mid-Ocean Seachannel** see Imarssuak Seachannel
66 J6 **Imarssuak Seachannel** var. Imarssuak Mid-Ocean Seachannel. channel N Atlantic Ocean
89 N18 **Imatra** Etelä-Suomi, S Finland 61.13N 28.49E
171 H14 **Imazu** Shiga, Honshū, SW Japan 35.25N 136.00E
58 C6 **Imbabura** ◆ province N Ecuador
57 R9 **Imbituba** Santa Catarina, S Brazil 5.44N 60.23W
63 K14 **Imbituva** Santa Catarina, S Brazil 28.15S 48.43W
29 W8 **Imboden** Arkansas, C USA 36.12N 91.10W
Imbros see Gökçeada
152 B11 **Imeni 26 Bakinskikh Komissarov** Turkm. 26 Baku Komissarlary Adyndaky. Balkanskiy Velayat, W Turkmenistan 39.24N 54.04E
152 I14 **Imeni 26 Bakinskikh Komissarov** see 26 Bakı
129 N13 **Imeni Babushkina** Vologodskaya Oblast', NW Russian Federation 59.40N 43.04E
130 J7 **Imeni Karla Libknekhta** Kurskaya Oblast', W Russian Federation 51.36N 35.28E
152 I14 **Imeni Mollanepesa** Maryyskiy Velayat, S Turkmenistan
127 N14 **Imeni Poliny Osipenko** Khabarovskiy Kray, SE Russian Federation 52.21N 136.17E
152 J15 **Imeni S.A.Niyazova** Maryyskiy Velayat, S Turkmenistan 36.44N 62.23E
Imeni Sverdlova Rudnik see Sverdlovs'k
196 B9 **Imeong** Babeldaob, N Palau
83 I14 **Imi** Somali, E Ethiopia 6.27N 42.10E
117 M21 **Imia** Turk. Kardak. island Dodekánisos, Greece, Aegean Sea
Imishli see Imişli
143 X12 **Imişli** Rus. Imishli. C Azerbaijan 39.54N 48.04E
169 X14 **Imjin-gang** ≈ North Korea/South Korea
37 S3 **Imlay** Nevada, USA 40.39N 118.10W
33 S9 **Imlay City** Michigan, N USA 43.01N 83.04W
25 S3 **Immokalee** Florida, SE USA 26.24N 81.25W
79 S16 **Imo** ◆ state SE Nigeria
108 G10 **Imola** Emilia-Romagna, N Italy 44.22N 11.43E
194 E9 **Imonda** Sandaun, NW PNG 3.19S 141.10E
Imoschi see Imotski
115 G14 **Imotski** var. Imoschi. Split-Dalmacija, SE Croatia 43.28N 17.13E
61 L13 **Imperatriz** Maranhão, NE Brazil 5.31S 47.28W
108 B10 **Imperia** Liguria, NW Italy 43.52N 8.03E
59 E14 **Imperial** Lima, W Peru 13.04S 76.20W
37 V17 **Imperial** California, W USA 32.51N 115.34W
30 L16 **Imperial** Nebraska, C USA 40.30N 101.37W
44 M9 **Imperial** Texas, SW USA 31.15N 102.40W
37 Y17 **Imperial Dam** dam California, W USA 32.52N 114.27W
81 I17 **Impfondo** La Likouala, NE Congo 1.40N 18.02E
159 X14 **Imphal** Manipur, NE India 24.46N 93.55E
105 P9 **Imphy** Nièvre, C France 46.55N 3.16E
108 G11 **Impruneta** Toscana, C Italy 43.42N 11.16E
117 K15 **Imroz** var. Gökçeada. Çanakkale, NW Turkey 40.11N 25.53E
42 I7 **Imroz Adası** see Gökçeada
78 G7 **Imuris** NW Mauritania
179 O15 **Inch'ŏn** off. Inch'ŏn-gwangyŏksi, Jap. Jinsen; prev. Chemulpo. NW South Korea 37.27N 126.40E

◆ COUNTRY ◇ DEPENDENT TERRITORY ✪ ADMINISTRATIVE REGION ▲ MOUNTAIN ⊠ VOLCANO ⊗ LAKE
◆ COUNTRY CAPITAL ◇ DEPENDENT TERRITORY CAPITAL ✕ INTERNATIONAL AIRPORT ▲ MOUNTAIN RANGE ≈ RIVER ⊡ RESERVOIR

85 M17 **Inchope** Manica, C Mozambique 19.09S 33.54E
Incoronata see Kornat
105 Y15 **Incudine, Monte** ▲ Corse, France, C Mediterranean Sea 41.52N 9.13E
62 M10 **Indaiatuba** São Paulo, S Brazil 23.03S 47.11W
95 H17 **Indal** Västernorrland, C Sweden 62.36N 17.06E
95 H17 **Indalsälven** ✍ C Sweden
42 K8 **Inde** Durango, C Mexico 25.55N 105.10W
Indefatigable Island see Santa Cruz, Isla
37 S10 **Independence** California, W USA 36.48N 118.12W
31 X13 **Independence** Iowa, C USA 42.28N 91.42W
29 P7 **Independence** Kansas, C USA 37.13N 95.42W
22 M4 **Independence** Kentucky, S USA 38.56N 84.32W
29 R4 **Independence** Missouri, C USA 39.05N 94.25W
23 R8 **Independence** Virginia, NE USA 36.37N 81.09W
32 J7 **Independence** Wisconsin, N USA 44.21N 91.25W
207 R12 **Independence Fjord** fjord N Greenland
Independence Island see Malden Island
37 W2 **Independence Mountains** ▲ Nevada, W USA
59 K18 **Independencia** Cochabamba, C Bolivia 17.07S 66.52W
59 E16 **Independencia, Bahía de la** bay W Peru
Independencia, Monte see Adam, Mount
118 M12 **Independenţa** Galaţi, SE Romania 45.27N 27.45E
Inderagiri see Indragiri, Sungai
Inderbor see Inderborskiy
150 F11 **Inderborskiy** Kaz. Inderbor. Atyrau, W Kazakhstan 48.35N 51.45E
157 I14 **India** off. Republic of India, var. Indian Union, Union of India, Hind. Bhārat. ◆ republic S Asia
India see Indija
20 D14 **Indiana** Pennsylvania, NE USA 40.37N 79.09W
33 N13 **Indiana** off. State of Indiana; also known as The Hoosier State. ◆ state N USA
33 O14 **Indianapolis** state capital Indiana, N USA 39.46N 86.09W
9 O10 **Indian Cabins** Alberta, W Canada 59.51N 117.06W
44 G1 **Indian Church** Orange Walk, N Belize 17.47N 88.39W
Indian Desert see Thar Desert
9 U16 **Indian Head** Saskatchewan, S Canada 50.31N 103.40W
33 O4 **Indian Lake** ⊗ Michigan, N USA
20 K9 **Indian Lake** ⊗ New York, NE USA
33 R13 **Indian Lake** ⊗ Ohio, N USA
180-181 **Indian Ocean** ocean
31 V15 **Indianola** Iowa, C USA 41.21N 93.33W
24 K4 **Indianola** Mississippi, S USA 33.27N 90.39W
38 J6 **Indian Peak** ▲ Utah, W USA 38.18N 113.52W
25 Y13 **Indian River** lagoon Florida, SE USA
37 W10 **Indian Springs** Nevada, W USA 36.33N 115.40W
25 Y14 **Indiantown** Florida, SE USA 27.01N 80.29W
61 K19 **Indiara** Goiás, S Brazil 17.12S 50.09W
129 Q4 **Indiga** Nenetskiy Avtonomnyy Okrug, NW Russian Federation 67.40N 49.01E
126 Mm6 **Indigirka** ✍ NE Russian Federation
114 L10 **Indija** Hung. India; prev. Indjija. Serbia, N Yugoslavia 45.03N 20.04E
37 V16 **Indio** California, W USA 33.42N 116.13W
44 M2 **Indio, Río** ✍ SE Nicaragua
158 I10 **Indira Gandhi International** ✕ (Delhi) Delhi, N India
157 Q23 **Indira Point** headland Andaman and Nicobar Islands, India, NE Indian Ocean 6.54N 93.54E
195 X15 **Indispensable Strait** strait C Solomon Islands
Indjija see Indija
133 Q13 **Indo-Australian Plate** tectonic feature
181 N11 **Indomed Fracture Zone** tectonic feature SW Indian Ocean
175 Nn12 **Indonesia** off. Republic of Indonesia, Ind. Republik Indonesia; prev. Dutch East Indies, Netherlands East Indies, United States of Indonesia. ◆ republic SE Asia
Indonesian Borneo see Kalimantan
160 G10 **Indore** Madhya Pradesh, C India 22.42N 75.50E
174 Hh8 **Indragiri, Sungai** var. Batang Kuantan, Inderagiri. ✍ Sumatera, W Indonesia
Indramajoe/Indramaju see Indramayu
174 K14 **Indramayu** prev. Indramajoe, Indramaju. Jawa, C Indonesia 6.22S 108.19E
161 K14 **Indrāvati** ✍ S India
105 N9 **Indre** ◆ department C France
104 M8 **Indre** ✍ C France
104 L8 **Indre-et-Loire** ◆ department C France
Indreville see Châteauroux
158 G3 **Indus** Chin. Yindu He; prev. Yin-tu Ho. ✍ S Asia
Indus Cone see Indus Fan
181 P3 **Indus Fan** var. Indus Cone. undersea feature N Arabian Sea
155 P17 **Indus, Mouths of the** delta S Pakistan
85 L21 **Indwe** Eastern Cape, SE South Africa 31.28S 27.21E
142 H10 **İnebolu** Kastamonu, N Turkey 41.57N 33.45E
79 P8 **I-n-Échaï** oasis C Mali 20.04N 2.00W

116 M13 **İnecik** Tekirdağ, NW Turkey 40.55N 27.16E
142 E12 **İnegöl** Bursa, NW Turkey 40.06N 29.31E
Inessa see Biancavilla
118 F10 **Ineu** Hung. Borosjenő; prev. Inău. Arad, W Romania 46.25N 21.50E
Ineul/Ineu, Vîrful see Ineu, Vârful
118 J9 **Ineu, Vârful** var. Ineul; prev. Vîrful Ineu. ▲ N Romania 47.31N 24.52E
23 P6 **Inez** Kentucky, S USA 37.53N 82.33W
76 E8 **Inezgane** ✕ (Agadir) W Morocco 30.35N 9.27W
43 T17 **Inferior, Laguna** lagoon S Mexico
42 M15 **Infiernillo, Presa del** ⊠ S Mexico
106 L2 **Infiesto** Asturias, N Spain
95 L20 **Ingå** Fin. Inkoo. Etelä-Suomi, S Finland 60.03N 24.01E
79 U10 **Ingal** var. I-n-Gall. Agadez, C Niger 16.52N 6.57E
I-n-Gall see Ingal
101 C18 **Ingelmunster** West-Vlaanderen, W Belgium 50.12N 3.15E
81 I18 **Ingende** Equateur, W Dem. Rep. Congo (Zaire) 0.15S 18.58E
64 L5 **Ingeniero Guillermo Nueva Juárez** Formosa, N Argentina 23.55S 61.49W
65 H16 **Ingeniero Jacobacci** Río Negro, C Argentina 41.21S 69.46W
12 F16 **Ingersoll** Ontario, S Canada 43.03N 80.52W
168 K6 **Ingettolgoy** Bulgan, N Mongolia 49.27N 103.59E
189 W5 **Ingham** Queensland, NE Australia 18.34S 146.12E
152 M11 **Ingichka** Samarqand Wiloyati, C Uzbekistan 39.46N 65.56E
99 L16 **Ingleborough** ▲ N England, UK 54.07N 2.22W
45 T14 **Ingleside** Texas, SW USA 27.52N 97.12W
192 K10 **Inglewood** Taranaki, North Island, NZ 39.10S 174.12E
37 S15 **Inglewood** California, W USA 33.57N 118.21W
126 Kk16 **Ingoda** ✍ S Russian Federation
103 I21 **Ingólfshöfði** headland S Iceland
118 K12 **Ingolstadt** Bayern, S Germany 48.46N 11.25E
35 V9 **Ingomar** Montana, NW USA 46.34N 107.21W
11 R14 **Ingonish Beach** Cape Breton Island, Nova Scotia, SE Canada 46.41N 60.22W
159 S14 **Ingrāj Bāzār** prev. English Bazar. West Bengal, NE India 25.00N 88.10E
45 Q11 **Ingram** Texas, SW USA 30.04N 99.14W
205 X7 **Ingrid Christensen Coast** physical region Antarctica
76 K14 **I-n-Guezzam** S Algeria 19.35N 5.49E
Inguletz see Inhulets'
Inguri see Enguri
Ingushetia/Ingushetiya, Respublika see Ingushskaya Respublika
131 O15 **Ingushskaya Respublika** var. Respublika Ingushetiya, Eng. Ingushetia. ◆ autonomous republic SW Russian Federation
85 N20 **Inhambane** Inhambane, SE Mozambique 23.51S 35.31E
85 M20 **Inhambane** off. Província de Inhambane. ◆ province S Mozambique
85 N17 **Inhaminga** Sofala, C Mozambique 18.22S 35.02E
85 N20 **Inharrime** Inhambane, SE Mozambique 24.28S 35.01E
85 M18 **Inhassoro** Inhambane, E Mozambique 21.31S 35.13E
119 S9 **Inhulets'** Rus. Ingulets. Dnipropetrovs'ka Oblast', E Ukraine 47.40N 33.51E
119 R10 **Inhulets'** Rus. Ingulets. ✍ S Ukraine
107 Q10 **Iniesta** Castilla-La Mancha, C Spain 39.27N 1.45W
I-ning see Yining
56 K11 **Inírida, Río** ✍ E Colombia
Inis see Ennis
Inis Ceithleann see Enniskillen
Inis Córthaidh see Enniscorthy
Inis Díomáin see Ennistimon
99 A17 **Inishbofin** Ir. Inis Bó Finne. island W Ireland
97 B18 **Inisheer** var. Inishere A, Ir. Inis Oírr. island W Ireland
Inishere see Inisheer
97 B18 **Inishmaan** Ir. Inis Meáin. island W Ireland
97 A18 **Inishmore** Ir. Árainn. island W Ireland
98 E13 **Inishtrahull** Ir. Inis Trá Tholl. island NW Ireland
97 A17 **Inishturk** Ir. Inis Toirc. island W Ireland
Inkoo see Ingå
193 J16 **Inland Kaikoura Range** ▲ South Island, NZ
Inland Sea see Seto-naikai
23 P11 **Inman** South Carolina, SE USA 35.03N 82.05W
110 L7 **Inn** ✍ C Europe
207 O11 **Innaanganeq** var. Kap York. headland NW Greenland 75.54N 66.21W
190 K2 **Innamincka** South Australia 27.47S 140.45E
113 J12 **Innbygda** ✍ Hedmark, S Norway 61.18N 12.16E
94 G12 **Inndyr** Nordland, C Norway 67.01N 14.00E
98 F11 **Inner Channel** inlet SE Belize
180 H15 **Inner Hebrides** island group W Scotland, UK
180 H15 **Inner Islands** var. Central Group. island group N Seychelles
Inner Mongolia/Inner Mongolian Autonomous Region see Nei Mongol Zizhiqu
98 G8 **Inner Sound** strait N Scotland, UK
112 J13 **Innerste** ✍ C Germany
189 W5 **Innisfail** Queensland, NE Australia 17.29S 146.03E

9 Q15 **Innisfail** Alberta, SW Canada 52.01N 113.58W
Inniskilling see Enniskillen
41 O11 **Innoko River** ✍ Alaska, USA
170 F14 **Innoshima** var. Innosima. Hiroshima, SW Japan 34.18N 133.09E
Innosima see Innoshima
110 M7 **Innsbruck** Ger. Innsbruck. Tirol, W Austria 47.16N 11.25E
81 J19 **Inongo** Bandundu, W Dem. Rep. Congo (Zaire) 1.55S 18.19E
Inoucdjouac see Inukjuak
112 I10 **Inowrocław** Ger. Hohensalza; prev. Inowrazlaw. Kujawski-pomorskie, C Poland 52.47N 18.15E
Inowrazlaw see Inowrocław
59 K18 **Inquisivi** La Paz, W Bolivia 17.01S 67.03W
79 O8 **In-Sâkâne, 'Erg** desert N Mali
76 J10 **I-n-Salah** var. In Salah. C Algeria 27.10N 2.31E
131 O5 **Insar** Respublika Mordoviya, W Russian Federation 53.52N 44.26E
201 X15 **Insiaf** Kosrae, E Micronesia
96 L13 **Insjön** Dalarna, C Sweden 60.40N 15.05E
Insterburg see Chernyakhovsk
Insula see Lille
118 L13 **Însurăţei** Brăila, SE Romania 44.55N 27.40E
129 V6 **Inta** Respublika Komi, NW Russian Federation 66.00N 60.09E
79 R9 **I-n-Tebezas** Kidal, E Mali 17.58N 1.51E
Interamna see Teramo
Interamna Nahars see Terni
30 L11 **Interior** South Dakota, N USA 43.42N 101.57W
110 E9 **Interlaken** Bern, W Switzerland 46.40N 7.51E
31 V2 **International Falls** Minnesota, N USA 48.37N 93.25W
178 Gg7 **Inthanon, Doi** ▲ NW Thailand 18.33N 98.29E
44 G7 **Intibucá** ◆ department SW Honduras
44 G8 **Intibucá** La Unión, SE El Salvador 13.10N 88.03W
63 B15 **Intiyaco** Santa Fe, C Argentina 28.43S 60.04W
118 K12 **Întorsura Buzăului** Ger. Bozau, Hung. Bodzaforduló. Covasna, E Romania 45.49N 26.10E
24 H9 **Intracoastal Waterway** inland waterway system Louisiana, S USA
45 V13 **Intracoastal Waterway** inland waterway system Texas, SW USA
110 G11 **Intragna** Ticino, S Switzerland 46.11N 8.42E
171 Kk17 **Inubō-zaki** headland Honshū, S Japan 35.42N 140.51E
170 D14 **Inukai** Ōita, Kyūshū, SW Japan 33.05N 131.37E
10 I5 **Inukjuak** var. Inoucdjouac; prev. Port Harrison. Quebec, NE Canada 58.28N 77.58W
65 I24 **Inútil, Bahía** bay S Chile
14 G3 **Inuvik** var. Inuuvik. Northwest Territories, NW Canada 68.25N 133.34W
171 I15 **Inuyama** Aichi, Honshū, SW Japan 35.22N 136.55E
58 G13 **Inuya, Río** ✍ E Peru
129 U13 **In'va** ✍ NW Russian Federation
98 H11 **Inveraray** W Scotland, UK 56.13N 5.04W
193 C24 **Invercargill** Southland, South Island, NZ 46.25S 168.22E
191 T5 **Inverell** New South Wales, SE Australia 29.49S 151.07E
98 I8 **Invergordon** N Scotland, UK 57.42N 4.01W
9 P16 **Invermere** British Columbia, SW Canada 50.30N 116.00W
11 R14 **Inverness** Cape Breton Island, Nova Scotia, SE Canada 46.13N 61.19W
98 I8 **Inverness** N Scotland, UK 57.27N 4.15W
25 V11 **Inverness** Florida, SE USA 28.50N 82.19W
98 I8 **Inverness** cultural region NW Scotland, UK
98 K9 **Inverurie** NE Scotland, UK 57.13N 2.13W
190 F8 **Investigator Group** island group South Australia
181 T7 **Investigator Ridge** undersea feature E Indian Ocean
190 H10 **Investigator Strait** strait South Australia
32 R11 **Inwood** Iowa, C USA 43.16N 96.25W
126 H16 **Inya** Respublika Altay, S Russian Federation 50.27N 86.45E
126 N10 **Inya** ✍ E Russian Federation
Inyanga see Nyanga
85 M16 **Inyangani** ▲ NE Zimbabwe 18.22S 32.57E
85 J17 **Inyathi** Matabeleland North, SW Zimbabwe 19.36S 28.52E
37 T12 **Inyokern** California, W USA 35.37N 117.48W
37 T10 **Inyo Mountains** ▲ California, W USA
131 P6 **Inza** Ul'yanovskaya Oblast', W Russian Federation 53.51N 46.21E
131 W3 **Inzer** Respublika Bashkortostan, W Russian Federation 54.11N 57.37E
131 N7 **Inzhavino** Tambovskaya Oblast', W Russian Federation 52.18N 42.28E
115 C16 **Ioánnina** var. Janina, Yannina. Ípeiros, W Greece 39.39N 20.52E
170 B17 **Iō-jima** var. Iwojima. Nansei-shotō, SW Japan
128 L4 **Iokan'ga** ✍ NW Russian Federation
23 Q6 **Iola** Kansas, C USA 37.55N 95.24W
Iolcos see Iolkós
127 G16 **Iolkós** anc. Iolcus. site of ancient city Thessalía, C Greece 39.24N 22.56E
85 A16 **Iona** Namibe, SW Angola 16.54S 12.39E
98 F11 **Iona** island W Scotland, UK
118 M15 **Ion Corvin** Constanţa, SE Romania 44.06N 27.49E

37 P7 **Ione** California, W USA 38.21N 120.55W
118 I13 **Ioneşti** Vâlcea, SW Romania 44.51N 24.12E
32 Q9 **Ionia** Michigan, N USA 42.59N 85.04W
123 Mm12 **Ionian Basin** var. Ionia Basin. undersea feature C Mediterranean Sea
Ionian Islands see Iónioi Nísoi
123 Mm11 **Ionian Sea** Gk. Iónio Pélagos, It. Mar Ionio. sea C Mediterranean Sea
117 B17 **Iónioi Nísoi** Eng. Ionian Islands. island group W Greece
117 B17 **Iónioi Nísoi** Eng. Ionian Islands. ◆ region W Greece
Ionio, Mar/Iónio Pélagos see Ionian Sea
143 U10 **Iori** var. Qaburi. ✍ Azerbaijan/Georgia
Iorrais, Ceann see Erris Head
117 J22 **Íos** Íos, Kykládes, Greece, Aegean Sea 36.43N 25.16E
117 J22 **Íos** var. Nio. island Kykládes, Greece, Aegean Sea
24 G9 **Iowa** Louisiana, S USA 30.12N 93.00W
31 V13 **Iowa** off. State of Iowa; also known as The Hawkeye State. ◆ state C USA
31 Y14 **Iowa City** Iowa, C USA 41.39N 91.31W
31 X14 **Iowa Falls** Iowa, C USA 42.31N 93.15W
45 R4 **Iowa Park** Texas, SW USA 33.57N 98.40W
31 Y14 **Iowa River** ✍ Iowa, C USA
121 M19 **Ipa** Rus. Ipa. ✍ SE Belarus
61 N20 **Ipatinga** Minas Gerais, SE Brazil 19.31S 42.30W
131 N13 **Ipatovo** Stavropol'skiy Kray, SW Russian Federation 45.43N 42.54E
117 C16 **Ípeiros** Eng. Epirus. ◆ region W Greece
113 J21 **Ipel'** var. Ipoly, Ger. Eipel. ✍ Hungary/Slovakia
56 C13 **Ipiales** Nariño, SW Colombia 0.50N 77.42W
201 V14 **Ipis** atoll Chuuk Islands, C Micronesia
61 A14 **Ipixuna** Amazonas, W Brazil 6.57S 71.42W
174 Gg4 **Ipoh** Perak, Peninsular Malaysia 4.36N 101.01E
Ipoly see Ipel'
197 D15 **Ipota** Erromango, S Vanuatu 18.54S 169.19E
81 K14 **Ippy** Ouaka, C Central African Republic 6.17N 21.13E
116 L13 **Ipsala** Edirne, NW Turkey 40.55N 26.24E
Ipsario see Ypsário
191 V3 **Ipswich** Queensland, E Australia 27.36S 152.49E
99 Q20 **Ipswich** hist. Gipeswic. E England, UK 52.05N 1.08E
31 O8 **Ipswich** South Dakota, N USA 45.24N 99.00W
121 P18 **Iputs'** Rus. Iputs'. ✍ Belarus/Russian Federation
16 J3 **Iqaluit** prev. Frobisher Bay. Baffin Island, Nunavut, NE Canada 63.43N 68.28W
165 P9 **Iqe** Qinghai, W China 38.03N 94.45E
165 P9 **Iqe He** ✍ C China
Iqlid see Eqlid
63 G3 **Iquique** Tarapacá, N Chile 20.15S 70.07W
58 C8 **Iquitos** Loreto, N Peru 3.51S 73.13W
45 N9 **Iraan** Texas, SW USA 30.52N 101.52W
81 K14 **Ira Banda** Haute-Kotto, E Central African Republic 5.57N 22.05E
172 Pp16 **Irabu-jima** island Miyako-shotō, SW Japan
99 Y9 **Iracoubo** N French Guiana 5.28N 53.15W
171 Hh17 **Irago-misaki** headland Honshū, SW Japan 34.35N 137.00E
62 H13 **Iraí** Rio Grande do Sul, S Brazil 27.15S 53.16W
116 G12 **Irákleia** Kentrikí Makedonía, N Greece 41.09N 23.16E
117 J21 **Irákleia** island Kykládes, Greece, Aegean Sea
117 J25 **Irákleio** var. Herakleion, Eng. Candia; prev. Iráklion. Kríti, Greece, C Mediterranean Sea 35.19N 25.07E
117 J25 **Irákleio** ✕ Kríti, Greece, E Mediterranean Sea 35.20N 25.10E
117 F15 **Irákleio** anc. Heracleum. castle Kentrikí Makedonía, N Greece 40.02N 22.34E
Iráklion see Irákleio
149 O7 **Iran** off. Islamic Republic of Iran; prev. Persia. ◆ republic SW Asia
175 W3 **Iran, Pegunungan** var. Iran Mountains. ▲ Indonesia/Malaysia
Iran, Plateau of see Iranian Plateau
Iranian Plate tectonic feature
Iranian Plateau var. Plateau of Iran. plateau N Iran
149 W13 **Īrānshahr** Sīstān va Balūchestān, SE Iran 27.14N 60.40E
57 P5 **Irapa** Sucre, NE Venezuela 10.33N 62.37E
43 N13 **Irapuato** Guanajuato, C Mexico 20.40N 101.22W
60 I13 **Iranduba** Amazonas, NW Brazil 3.19S 60.09W
145 R7 **Iraq** off. Republic of Iraq, Ar. 'Irāq. ◆ republic SW Asia
62 J12 **Irati** Paraná, S Brazil 25.25S 50.37W
107 R3 **Irati** ✍ N Spain
129 T8 **Irayėl'** Respublika Komi, NW Russian Federation 64.28N 55.20E
44 I13 **Irazú, Volcán** ▲ C Costa Rica 9.57N 83.52W
144 H7 **Irbid** Irbid, N Jordan 32.33N 35.51E
Irbenskiy Zaliv/Irbes Šaurums see Irbe Strait

120 D7 **Irbe Strait** Est. Kura Kurk, Latv. Irbes Šaurums, Rus. Irbenskiy Zaliv; prev. Est. Irbe Väin. strait Estonia/Latvia
Irbe Väin see Irbe Strait
144 G9 **Irbid** off. Muḥāfaẓat Irbid. ◆ governorate N Jordan
Irbīl see Arbīl
125 F11 **Irbit** Sverdlovskaya Oblast', C Russian Federation 57.37N 63.10E
81 I18 **Irebu** Equateur, W Dem. Rep. Congo (Zaire) 0.32S 17.44E
86 C9 **Ireland** Lat. Hibernia. island Ireland/UK
66 A12 **Ireland Island North** island W Bermuda
66 A12 **Ireland Island South** island W Bermuda
99 D17 **Ireland, Republic of** off. Republic of Ireland, var. Ireland, Ir. Éire. ◆ republic NW Europe
129 V15 **Iren'** ✍ NW Russian Federation
193 A22 **Irene** var. ▲ South Island, NZ 45.04S 167.24E
24 G9 **Irgalem** see Yirga 'Alem
150 L11 **Irgiz** Aktyubinsk, C Kazakhstan 48.37N 61.12E
Irian see New Guinea
176 Y13 **Irian Barat** see Irian Jaya
176 Y13 **Irian Jaya** var. Irian Barat, West Irian, West New Guinea, West Papua; prev. Dutch New Guinea, Netherlands New Guinea. ◆ province E Indonesia
Irian, Teluk see Cenderawasih, Teluk
80 K9 **Iriba** Biltine, NE Chad 15.10N 22.10E
172 O17 **Iriomote-jima** island Sakishima-shotō, SW Japan
44 L4 **Iriona** Colón, N Honduras 15.53N 85.08W
81 N18 **Iringa** Iringa, C Tanzania 7.49S 35.39E
83 H23 **Iringa** ◆ region S Tanzania
60 I13 **Iri, Rio** ✍ C Brazil
37 W9 **Irish, Mount** ▲ Nevada, W USA 37.39N 115.22W
99 H17 **Irish Sea** Ir. Muir Éireann. sea C British Isles
145 U12 **Irjal ash Shaykhiyah** ✍ Iraq 30.49N 44.58E
153 U11 **Irkeshtam** Oshskaya Oblast', SW Kyrgyzstan 39.39N 73.49E
126 J16 **Irkut** ✍ S Russian Federation
126 Jj16 **Irkutsk** Irkutskaya Oblast', S Russian Federation 52.18N 104.15E
126 Jj13 **Irkutskaya Oblast'** ◆ province S Russian Federation
152 K8 **Irlir, Gora** ▲ N Uzbekistan 42.43N 63.24E
Irlir Toghi var. Gora Irlir. ▲ N Uzbekistan
23 R12 **Irmo** South Carolina, SE USA 34.05N 81.10W
104 L6 **Iroise** sea NW France
201 X2 **Iroj** var. Eroj. island Ratak Chain, SE Marshall Islands
190 I7 **Iron Baron** South Australia 33.01S 137.13E
195 X15 **Iron Bottom Sound** sound C Solomon Islands
12 C10 **Iron Bridge** Ontario, S Canada 46.16N 83.12W
12 I13 **Iron City** Tennessee, S USA 35.01N 87.34W
Irondale ✍ Ontario, SE Canada
190 I7 **Iron Knob** South Australia 32.46S 137.08E
32 M5 **Iron Mountain** Michigan, N USA 45.51N 88.03W
34 M3 **Iron River** Michigan, N USA 46.05N 88.38W
32 J3 **Iron River** Wisconsin, N USA 46.34N 91.22W
29 X6 **Ironton** Missouri, C USA 37.36N 90.37W
33 S15 **Ironton** Ohio, N USA 38.32N 82.40W
32 K4 **Ironwood** Michigan, N USA 46.27N 90.10W
12 H12 **Iroquois Falls** Ontario, S Canada 48.46N 80.40W
171 Ii8 **Irō-zaki** headland Honshū, S Japan 34.36N 138.49E
Irpen' see Irpin'
119 O4 **Irpin'** Rus. Irpen'. Kyyivs'ka Oblast', N Ukraine 50.31N 30.15E
119 O4 **Irpin'** Rus. Irpen'. ✍ N Ukraine
147 O7 **'Irqah** SW Yemen 13.42N 47.21E
177 M13 **Irrawaddy** var. Ayeyarwady. ◆ division SW Myanmar
177 G6 **Irrawaddy** var. Ayeyarwady. ✍ W Myanmar
177 F9 **Irrawaddy, Mouths of the** delta SW Myanmar
119 N4 **Irsha** ✍ N Ukraine
118 H7 **Irshava** Zakarpats'ka Oblast', W Ukraine 48.19N 23.03E
109 N18 **Irsina** Basilicata, S Italy 40.42N 16.18E
Irtish see Irtysh
123 R5 **Irtysh** var. Irtish, Kaz. Ertis. ✍ C Asia
151 S7 **Irtyshsk** Kaz. Ertis. Pavlodar, NE Kazakhstan 53.21N 75.27E
151 S11 **Irtyshskiy** ✍ N Kazakhstan
34 N3 **Irumu** Orientale, NE Dem. Rep. Congo (Zaire) 1.27N 29.52E
107 Q2 **Irún** País Vasco, N Spain 43.19N 1.48W
Iruña see Pamplona
107 Q3 **Irurtzun** Navarra, N Spain 42.55N 1.49W
T15 **Irvine** W Scotland, UK 55.37N 4.40W
22 M5 **Irvine** Kentucky, S USA 37.42N 83.58W

45 T6 **Irving** Texas, SW USA 32.48N 96.57W
22 K5 **Irvington** Kentucky, S USA 37.52N 86.16W
Isaak see Iisaku
195 W14 **Isabel** off. Isabel Province. ◆ province N Solomon Islands
179 Q17 **Isabela** Basilan Island, SW Philippines 6.41N 121.58E
47 S5 **Isabela** W Puerto Rico 18.30N 67.01W
47 N8 **Isabela, Cabo** headland N Dominican Republic 19.54N 71.03W
44 I12 **Isabela, Isla** var. Albemarle Island. island Galapagos Islands, Ecuador, E Pacific Ocean
44 K9 **Isabelia, Cordillera** ▲ NW Nicaragua
42 S12 **Isabella Lake** ⊠ California, W USA
33 N2 **Isabelle, Point** headland Michigan, N USA 47.20N 87.56W
Isabel Segunda see Vieques
118 M13 **Isaccea** Tulcea, E Romania 45.16N 28.28E
94 H1 **Ísafjarðardjúp** inlet NW Iceland
94 H1 **Ísafjörður** Vestfirðir, NW Iceland 66.04N 23.09W
170 C13 **Isahaya** Nagasaki, Kyūshū, SW Japan 32.51N 130.04E
126 Gg14 **Iskitim** Novosibirskaya Oblast', C Russian Federation 54.36N 83.05E
166 J11 **Iskra** prev. Popovo. Kürdzhali, S Bulgaria 41.53N 25.12E
180 H7 **Isalo, Massif de l'** var. Massif de L'Isalo. ▲ SW Madagascar
Isalo, Massif de L' see Isalo
81 K20 **Isangel** Tanna, S Vanuatu 19.34S 169.17E
81 M18 **Isangi** Orientale, N Dem. Rep. Congo (Zaire) 0.46N 24.15E
103 L24 **Isar** ✍ Austria/Germany
103 M23 **Isar-Kanal** canal SE Germany
Isbarta see Isparta
Isca Damnoniorum see Exeter
109 K18 **Ischia, Isola d'** island S Italy
109 K18 **Ischia, Isola d'** anc. Aenaria. Campania, S Italy 40.43N 13.57E
171 J18 **Ise** Mie, Honshū, SW Japan 34.29N 136.42E
102 J12 **Ise** ✍ N Germany
97 I23 **Isefjord** fjord E Denmark
Iseghem see Izegem
199 Jj17 **Iselin Seamount** undersea feature S Pacific Ocean
Isenhof see Püssi
108 E7 **Iseo** Lombardia, N Italy 45.40N 10.03E
105 U12 **Iseran, Col de l'** pass E France 45.26N 7.06E
105 S12 **Isère** ◆ department E France
105 S12 **Isère** ✍ E France
112 F15 **Iserlohn** Nordrhein-Westfalen, W Germany 51.22N 7.42E
109 K16 **Isernia** var. Æsernia. Molise, C Italy 41.34N 14.13E
171 Ji15 **Isesaki** Gunma, Honshū, S Japan 36.19N 139.16E
133 Q5 **Ise-wan** bay S Japan
79 S15 **Iseyin** Oyo, W Nigeria 7.56N 3.33E
153 Q11 **Isfana** Oshskaya Oblast', SW Kyrgyzstan 39.51N 69.31E
153 R11 **Isfara** N Tajikistan 40.06N 70.34E
155 O4 **Isfi Maidān** Ghowr, N Afghanistan 35.05N 66.16E
94 O3 **Isfjorden** fjord W Svalbard
Isha Baydhabo see Baydhabo
129 V11 **Isherim, Gora** ▲ NW Russian Federation 61.06N 59.09E
131 Q5 **Isheyevka** Ul'yanovskaya Oblast', W Russian Federation 54.27N 48.18E
170 O17 **Ishigaki** Okinawa, Ishigaki-jima, SW Japan 24.19N 124.09E
172 P17 **Ishigaki-jima** var. Isigaki Zima. island Sakishima-shotō, SW Japan
172 O5 **Ishikari-gawa** var. Isikari Gawa. ✍ Hokkaidō, NE Japan
170 O05 **Ishikari-wan** bay Hokkaidō, NE Japan
171 L14 **Ishikawa** Fukushima, Honshū, C Japan 37.08N 140.26E
172 Oo14 **Ishikawa** var. Isikawa. Okinawa, SW Japan 26.25N 127.46E
171 I13 **Ishikawa** off. Ishikawa-ken, var. Isikawa. ◆ prefecture Honshū, SW Japan
123 R6 **Ishim** var. Esil. ✍ Kazakhstan/Russian Federation
131 V6 **Ishimbay** Respublika Bashkortostan, W Russian Federation 53.28N 56.03E
151 O9 **Ishimskoye** Akmola, C Kazakhstan 51.22N 67.07E
171 M13 **Ishinomaki** var. Isinomaki. Miyagi, Honshū, C Japan 38.25N 141.16E
171 Kk16 **Ishioka** var. Isioka. Ibaraki, Honshū, S Japan 36.12N 140.18E
153 N11 **Ishkashim** Rus. Ishkashim. ✍ C Tajikistan
Ishkoshim see Ishkashim
Ishkoshimskiy Khrebet see Ishkashimskiy Khrebet
109 N18 **Ishkoshim, Qatorkŭhi** Rus. Ishkashimskiy Khrebet. ▲ SE Tajikistan
34 N3 **Ishpeming** Michigan, N USA 46.29N 87.40W
153 N11 **Ishtikhon** Rus. Ishtykhan. Samarqand Wiloyati, C Uzbekistan 39.59N 66.28E
Ishtykhan see Ishtikhon
155 T15 **Ishurdi** var. Iswardi. Rajshahi, W Bangladesh 24.08N 89.04E
171 Ee15 **Ishizuchi-san** ▲ Shikoku, SW Japan 33.44N 133.07E

142 C11 **Işıklar Dağı** ▲ NW Turkey
109 C19 **Isili** Sardegna, Italy, C Mediterranean Sea 39.46N 9.06E
125 Ff13 **Isil'kul'** Omskaya Oblast', C Russian Federation 54.51N 71.07E
83 I18 **Isinomaki** see Ishinomaki
Isioka see Ishioka
81 O16 **Isiro** Orientale, NE Dem. Rep. Congo (Zaire) 2.46N 27.37E
81 O16 **Isiolo** Eastern, C Kenya 0.21N 37.33E
94 P2 **Isispynten** headland NE Svalbard 79.51N 26.44E
126 Lll1 **Isit** Respublika Sakha (Yakutiya), NE Russian Federation 60.53N 125.32E
153 Q9 **Iskabad Canal** canal N Afghanistan
Iskander see Iskandar
153 Q9 **Iskandar** Rus. Iskander. Toshkent Wiloyati, E Uzbekistan 41.32N 69.46E
Iskander see Iskandar
Iskär see Iskŭr
124 O2 **İskele** var. Trikomo, Gk. Trikomon. E Cyprus 35.16N 33.54E
142 K17 **İskenderun** Eng. Alexandretta. Hatay, S Turkey 36.34N 36.08E
144 H2 **İskenderun Körfezi** Eng. Gulf of Alexandretta. gulf S Turkey
142 J11 **İskilip** Çorum, N Turkey 40.45N 34.28E
Iski-Nauket see Eski-Nookat
126 Gg14 **Iskitim** Novosibirskaya Oblast', C Russian Federation 54.36N 83.05E
166 J11 **Iskra** prev. Popovo. Kürdzhali, S Bulgaria 41.53N 25.12E
116 G10 **Iskŭr** var. Iskar. ✍ NW Bulgaria
116 H10 **Iskŭr, Yazovir** prev. Yazovir Stalin. ⊠ W Bulgaria
43 S15 **Isla** Veracruz-Llave, SE Mexico 18.01N 95.30W
121 J15 **Islach** Rus. Isloch'. ✍ C Belarus
106 H14 **Isla Cristina** Andalucía, S Spain 37.12N 7.20W
Isla de León see San Fernando
109 K18 **Islāmābād** ● (Pakistan) Federal Capital Territory Islāmābād, NE Pakistan 33.40N 73.07E
155 V6 **Islāmābād** ✕ Federal Capital Territory Islāmābād, NE Pakistan 33.40N 73.01E
Islamabad see Anantnag
155 F17 **Islāmkot** Sind, SE Pakistan 24.37N 70.04E
25 Y17 **Islamorada** Florida Keys, Florida, SE USA 24.55N 80.37W
159 P14 **Islāmpur** Bihār, N India 25.09N 85.13E
20 E7 **Island Beach** spit New Jersey, NE USA
21 S4 **Island Falls** Maine, NE USA 45.59N 68.16W
190 H13 **Island Lagoon** ⊗ South Australia
9 Y13 **Island Lake** ⊗ Manitoba, C Canada
31 W5 **Island Lake Reservoir** ⊠ Minnesota, N USA
35 R13 **Island Park** Idaho, NW USA 44.27N 111.21W
21 N6 **Island Pond** Vermont, NE USA 44.48N 71.51W
192 K2 **Islands, Bay of** inlet North Island, NZ
105 R7 **Is-sur-Tille** Côte d'Or, C France 47.34N 5.03E
44 J3 **Islas de la Bahía** ◆ department N Honduras
67 L20 **Islas Orcadas Rise** undersea feature S Atlantic Ocean
98 F12 **Islay** island SW Scotland, UK
118 I15 **Islaz** Teleorman, S Romania 43.43N 24.52E
104 I7 **Isle** ✍ W France
104 M12 **Isle** ✍ W France
99 I16 **Isle of Man** ◆ UK crown dependency NW Europe
23 X7 **Isle of Wight** Virginia, NE USA 36.54N 76.41W
99 M24 **Isle of Wight** cultural region S England, UK
203 Y3 **Isles Lagoon** ⊗ Kiritimati, E Kiribati
39 R11 **Isleta Pueblo** New Mexico, SW USA 34.54N 106.40W
Isloch' see Islach
63 E9 **Ismael Cortinas** Flores, S Uruguay 33.57S 57.04W
Ismailia see Ismâ'ilîya
77 W7 **Ismâ'ilîya** var. Ismailia. N Egypt 30.31N 32.13E
Ismid see İzmit
77 X10 **Isna** var. Esna. SE Egypt 25.16N 32.24E
84 M12 **Isoka** Muchinga, NE Zambia 10.07S 32.42E
Isola d'Ischia see Ischia
Isola d'Istria see Izola
Isonzo see Soča
53 U4 **Isoukstouc** ✍ Quebec, SE Canada
142 F15 **Isparta** var. Isbarta. Isparta, SW Turkey 37.46N 30.31E
142 F15 **Isparta** var. Isbarta. Isparta. ◆ province SW Turkey
116 J7 **Isperih** prev. Kemanlar. Razgrad, N Bulgaria 43.43N 26.49E
109 L26 **Ispica** Sicilia, Italy, C Mediterranean Sea 36.46N 14.55E
155 X9 **Ispikān** Baluchistan, SW Pakistan 26.21N 62.15E
142 Q12 **İspir** Erzurum, NE Turkey 40.28N 41.01E
144 E12 **Israel** off. State of Israel, var. Medinat Israel, Heb. Yisra'el, Yisra'el. ◆ republic SW Asia
Issa see Vis
59 S9 **Issano** Cuyuni-Mazaruni, C Guyana 5.49N 59.28W
78 M14 **Issia** SW Ivory Coast 6.33N 6.33W
Issiq Köl see Issyk-Kul', Ozero
105 P11 **Issoire** Puy-de-Dôme, C France 45.33N 3.15E
105 N9 **Issoudun** anc. Uxellodunum. Indre, C France 46.57N 1.58E
83 H22 **Issuna** Singida, C Tanzania 5.24S 34.48E
Issyk see Yesik
Issyk-Kul' see Balykchy

Symbol	Meaning	Symbol	Meaning
◆	COUNTRY	◆	ADMINISTRATIVE REGION
●	COUNTRY CAPITAL	✕	INTERNATIONAL AIRPORT
◇	DEPENDENT TERRITORY	▲	MOUNTAIN
○	DEPENDENT TERRITORY CAPITAL	▲	MOUNTAIN RANGE
✕	VOLCANO	⊗	LAKE
✍	RIVER	⊠	RESERVOIR

153 X7 **Issyk-Kul', Ozero** var. Issiq Köl, Kir. Ysyk-Köl. © E Kyrgyzstan
153 X7 **Issyk-Kul'skaya Oblast'** Kir. Ysyk-Köl Oblasty. ♦ province E Kyrgyzstan
155 Q7 **Istädeh-ye Moqor, Āb-e-** var. Āb-i-Istāda. ⊘ SE Afghanistan
142 D11 **Istanbul** Bul. Tsarigrad, Eng. Istanbul; prev. Constantinople, anc. Byzantium. Istanbul, NW Turkey 41.01N 28.57E
116 P12 **İstanbul** ♦ province NW Turkey
116 P12 **İstanbul Boğazı** var. Bosporus Thracius, Eng. Bosphorus, Bosporus, Turk. Karadeniz Boğazı. strait NW Turkey
Istarska Županija see Istra
117 G19 **Isthmía** Pelopónnisos, S Greece 37.55N 23.02E
117 G17 **Istiaía** Évvoia, C Greece 38.57N 23.09E
56 D9 **Istmina** Chocó, W Colombia 5.09N 76.42W
25 W3 **Istokpoga, Lake** ⊘ Florida, SE USA
114 A9 **Istra** off. Istarska županija. ♦ province NW Croatia
114 I10 **Istra** Eng. Istria, Ger. Istrien. cultural region NW Croatia
105 R15 **Istres** Bouches-du-Rhône, SE France 43.30N 4.58E
Istria/Istrien see Istra
179 R16 **Isulan** Mindanao, S Philippines 6.36N 124.36E
194 I11 **Isumrud Strait** strait NE PNG
194 I11 **Iswardi** see Ishurdi
131 V7 **Isyangulovo** Respublika Bashkortostan, W Russian Federation 52.10N 56.38E
64 O6 **Itá** Central, S Paraguay 25.28S 57.21W
61 O17 **Itaberaba** Bahia, E Brazil 12.34S 40.21W
61 M20 **Itabira** prev. Presidente Vargas. Minas Gerais, SE Brazil 19.39S 43.13W
61 O18 **Itabuna** Bahia, E Brazil 14.48S 39.18W
61 J18 **Itacaiu** Mato Grosso, S Brazil 15.49S 51.21W
60 G12 **Itacoatiara** Amazonas, N Brazil 3.06S 58.22W
56 D9 **Itagüí** Antioquia, W Colombia 6.12N 75.40W
62 D13 **Itá Ibaté** Corrientes, NE Argentina 27.27S 57.24W
62 G11 **Itaipú, Represa de** ⊡ Brazil/Paraguay
60 H11 **Itaituba** Pará, NE Brazil 4.15S 55.55W
62 K13 **Itajaí** Santa Catarina, S Brazil 26.49S 48.39W
Italia/Italiana, Republica/Italian Republic, The see Italy
Italian Somaliland see Somalia
45 T7 **Italy** Texas, SW USA 32.10N 96.52W
108 G12 **Italy** off. The Italian Republic, It. Italia, Republica Italiana. ♦ republic S Europe
61 O19 **Itamaraju** Bahia, E Brazil 16.58S 39.31W
61 C14 **Itamarati** Amazonas, W Brazil 6.12S 68.16W
61 M19 **Itambé, Pico de** ▲ SE Brazil 18.22S 43.21W
117 H15 **Ítamos** ▲ N Greece 40.06N 23.51E
159 W11 **Itānagar** Arunāchal Pradesh, NE India 27.09N 93.35E
Itany see Litani
61 N19 **Itaobim** Minas Gerais, SE Brazil 16.34S 41.27W
61 P15 **Itaparica, Represa de** ⊡ E Brazil
60 M13 **Itapecuru-Mirim** Maranhão, E Brazil 3.24S 44.19W
62 Q8 **Itaperuna** Rio de Janeiro, SE Brazil 21.13S 41.51W
61 O18 **Itapetinga** Bahia, E Brazil 15.16S 40.16W
62 L10 **Itapetininga** São Paulo, S Brazil 23.33S 48.03W
62 K10 **Itapeva** São Paulo, S Brazil 23.58S 48.54W
49 W6 **Itapicuru, Rio** ⊘ NE Brazil
60 O13 **Itapipoca** Ceará, E Brazil 3.28S 39.34W
62 M9 **Itapira** São Paulo, S Brazil 22.25S 46.46W
62 K8 **Itápolis** São Paulo, S Brazil 21.36S 48.43W
62 L9 **Itaporanga** São Paulo, S Brazil 23.43S 49.28W
64 P7 **Itapúa** off. Departamento de Itapúa. ♦ department SE Paraguay
61 E15 **Itapuã do Oeste** Rondônia, W Brazil 9.21S 63.07W
63 G15 **Itaqui** Rio Grande do Sul, S Brazil 29.10S 56.28W
62 K10 **Itararé** São Paulo, S Brazil 24.07S 49.16W
160 H11 **Itārsi** Madhya Pradesh, C India 22.42N 77.55E
45 T7 **Itasca** Texas, SW USA 32.09N 97.09W
Itassi see Vieille Case
95 N17 **Itä-Suomi** ♦ province E Finland
62 D13 **Itatí** Corrientes, NE Argentina 27.16S 58.15W
62 L9 **Itatinga** São Paulo, S Brazil 23.08S 48.36W
117 F18 **Itéas, Kólpos** gulf C Greece
59 N15 **Iténez, Río** var. Rio Guaporé. ⊘ Bolivia/Brazil see also Guaporé, Rio
56 H11 **Itéviate, Río** ⊘ C Colombia
102 I13 **Ith** hill range C Germany
33 Q8 **Ithaca** Michigan, N USA 43.17N 84.36W
20 H11 **Ithaca** New York, NE USA 42.25N 76.30W
117 C18 **Itháki** Itháki, Iónioi Nísoi, Greece, C Mediterranean Sea
117 C18 **Itháki** island Iónioi Nísoi, Greece, C Mediterranean Sea
It Hearrenfean see Heerenveen
Itihara see Ichihara
81 L17 **Itimbiri** ⊘ N Dem. Rep. Congo (Zaire)

Itinomiya see Ichinomiya
Itinoseki see Ichinoseki
41 Q5 **Itkilik River** ⊘ Alaska, USA
171 J17 **Itō** Shizuoka, Honshū, S Japan 34.59N 139.03E
171 J13 **Itoigawa** Niigata, Honshū, S Japan 37.01N 137.52E
13 R6 **Itomamo, Lac** ⊘ Quebec, SE Canada
172 Oo15 **Itoman** Okinawa, SW Japan 26.04N 127.40E
104 M5 **Iton** ⊘ N France
59 M16 **Itonamas Río** ⊘ NE Bolivia
Itoupé, Mont see Sommet Tabulaire
Itseqqortoormiit see Ittoqqortoormiit
24 K4 **Itta Bena** Mississippi, S USA 33.30N 90.19W
109 B17 **Ittiri** Sardegna, Italy, C Mediterranean Sea 40.36N 8.34E
207 Q14 **Ittoqqortoormiit** var. Itseqqortoormiit, Dan. Scoresbysund, Eng. Scoresby Sound. Tunu, C Greenland 70.33N 21.52W
62 M10 **Itu** São Paulo, S Brazil 23.17S 47.16W
178 Mm14 **Itu Aba Island** island W Spratly Islands
56 D8 **Ituango** Antioquia, NW Colombia 7.06N 75.51W
56 C7 **Ituí, Río** ⊘ W Brazil
81 O20 **Itula** Sud Kivu, E Dem. Rep. Congo (Zaire) 3.30S 27.49E
61 K19 **Itumbiara** Goiás, C Brazil 18.25S 49.15W
57 T9 **Ituni** E Guyana 5.24N 58.18W
43 X13 **Iturbide** Campeche, SE Mexico 19.59N 89.40W
Ituri see Aruwimi
127 Pp16 **Iturup, Ostrov** island Kuril'skiye Ostrova, SE Russian Federation
62 L7 **Ituverava** São Paulo, S Brazil 20.22S 47.48W
61 C15 **Ituxi, Río** ⊘ W Brazil
63 E14 **Ituzaingó** Corrientes, NE Argentina 27.34S 56.43W
103 K18 **Itz** ⊘ C Germany
102 J9 **Itzehoe** Schleswig-Holstein, N Germany 53.55N 9.31E
25 N2 **Iuka** Mississippi, S USA 34.48N 88.11W
62 I11 **Ivaiporã** Paraná, S Brazil 24.16S 51.46W
94 L10 **Ivalo** Lapp. Avveel, Avvil. Lappi, N Finland 68.34N 27.29E
94 L10 **Ivalojoki** Lapp. Avreel. ⊘ N Finland
121 N6 **Ivanava** Pol. Janów, Janów Poleski, Rus. Ivanovo. Brestskaya Voblasts', SW Belarus 52.07N 25.31E
Ivangorod see Dęblin
191 N7 **Ivanhoe** New South Wales, SE Australia 32.54S 144.20E
31 S9 **Ivanhoe** Minnesota, N USA 44.46N 96.15W
12 D8 **Ivanhoe** ⊘ Ontario, S Canada
114 E8 **Ivanić-Grad** Sisak-Moslavina, N Croatia 45.43N 16.23E
119 T10 **Ivanivka** Khersons'ka Oblast', S Ukraine 46.43N 34.28E
119 P10 **Ivanivka** Odes'ka Oblast', SW Ukraine 46.57N 30.26E
115 L14 **Ivanjica** Serbia, C Yugoslavia 43.36N 20.14E
114 G11 **Ivanjska** var. Potkozarje. Republika Srpska, NW Bosnia & Herzegovina 44.54N 17.04E
113 H17 **Ivanka** × (Bratislava) Bratislavský Kraj, W Slovakia 48.10N 17.13E
119 O3 **Ivankiv** Rus. Ivankov. Kyyivs'ka Oblast', N Ukraine 50.55N 29.53E
Ivankov see Ivankiv
41 O15 **Ivanof Bay** Alaska, USA 55.55N 159.28W
118 J7 **Ivano-Frankivs'k** Ger. Stanislau, Pol. Stanisławów, Rus. Ivano-Frankovsk; prev. Stanisław. Ivano-Frankivs'ka Oblast', W Ukraine 48.55N 24.45E
118 J7 **Ivano-Frankivs'k** Ger. Stanislau, Pol. Stanisławów, Rus. Ivano-Frankovsk; prev. Stanisław. Ivano-Frankivs'ka Oblast', Rus. Ivano-Frankovskaya Oblast', prev. Stanislavskaya Oblast'. ♦ province W Ukraine
Ivano-Frankovsk see Ivano-Frankivs'k
128 M16 **Ivanovo** Ivanovskaya Oblast', W Russian Federation 57.01N 40.58E
Ivanovo see Ivanava
125 A16 **Ivanovskaya Oblast'** ♦ province W Russian Federation
37 X12 **Ivanpah Lake** ⊘ California, W USA
114 F7 **Ivanščica** ▲ NE Croatia
116 M8 **Ivanski** Shumen, NE Bulgaria 43.09N 27.02E
131 R7 **Ivanteyevka** Saratovskaya Oblast', W Russian Federation 52.13N 49.06E
Ivantsevichi/Ivantsevichy see Ivatsevichy
118 I4 **Ivanychi** Volyns'ka Oblast', NW Ukraine 50.37N 24.22E
121 H16 **Ivatsevichy** Pol. Iwacewicze, Rus. Ivantsevichi, Ivantsevichy. Brestskaya Voblasts', SW Belarus 52.43N 25.21E
143 L12 **Ivaylovgrad** Khaskovo, S Bulgaria 41.32N 26.06E
116 K11 **Ivaylovgrad, Yazovir** ⊡ S Bulgaria
125 F10 **Ivdel'** Sverdlovskaya Oblast', C Russian Federation 60.42N 60.07E
Ivenets see Ivyanets
118 L12 **Iveşti** Galaţi, E Romania 45.27N 28.00E
81 F18 **Ivindo** ⊘ Gabon/Congo
61 I21 **Ivinheima** Mato Grosso do Sul, SW Brazil 22.16S 53.52W
206 M15 **Ivittuut** var. Ivigtut. Kitaa, S Greenland 61.28N 48.13W
180 I6 **Ivohibe** Fianarantsoa, SE Madagascar 22.28S 46.52E
Ivoire, Côte d' see Ivory Coast

194 I14 **Ivori** ⊘ S PNG
78 L15 **Ivory Coast** off. Republic of the Ivory Coast, Fr. Côte d'Ivoire, République de la Côte d'Ivoire. ♦ republic W Africa
70 C11 **Ivory Coast** Fr. Côte d'Ivoire. coastal region S Ivory Coast
97 L22 **Ivösjön** ⊘ S Sweden
108 B7 **Ivrea** anc. Eporedia. Piemonte, NW Italy 45.28N 7.52E
10 J2 **Ivujivik** Quebec, NE Canada 62.25N 77.49W
121 J16 **Ivyanets** Rus. Ivenets. Minskaya Voblasts', C Belarus 53.52N 26.45E
Iv'ye see Iwye
172 N11 **Iwaizumi** Iwate, Honshū, NE Japan 39.48N 141.46E
171 L15 **Iwaki** Fukushima, Honshū, N Japan 37.01N 140.52E
171 Mm9 **Iwaki-san** ▲ Honshū, C Japan
170 E13 **Iwakuni** Yamaguchi, Honshū, SW Japan 34.07N 132.06E
172 Nn5 **Iwanai** Hokkaidō, NE Japan 42.51N 140.21E
171 Ll13 **Iwanuma** Miyagi, Honshū, C Japan 38.07N 140.49E
171 J17 **Iwata** Shizuoka, Honshū, S Japan 34.42N 137.49E
172 N10 **Iwate** Iwate, Honshū, N Japan 40.02N 141.12E
171 Mm11 **Iwate** ♦ prefecture Honshū, C Japan
171 Mm10 **Iwate-san** ▲ Honshū, C Japan 39.52N 140.59E
79 S16 **Iwo** Oyo, SW Nigeria 7.21N 3.58E
121 I16 **Iwojima** see Iō-jima
121 I16 **Iwye** Pol. Iwje, Rus. Iv'ye. Hrodzyenskaya Voblasts', W Belarus 53.55N 25.46E
C4 C4 **Ixcán, Río** ⊘ Guatemala/Mexico
101 G18 **Ixelles** Dut. Elsene. Brussels, C Belgium 50.49N 4.21E
59 I14 **Ixiamas** La Paz, NW Bolivia 13.43S 68.10W
4 O13 **Ixmiquilpan** var. Ixmiquilpán. Hidalgo, C Mexico 20.28N 99.11W
85 K23 **Ixopo** KwaZulu/Natal, E South Africa 30.07S 30.03E
42 M16 **Ixtapa** Guerrero, S Mexico 17.37N 101.29W
43 S16 **Ixtepec** Oaxaca, SE Mexico 16.32N 95.03W
42 K12 **Ixtlán** var. Ixtlán del Río. Nayarit, C Mexico 21.03N 104.23W
42 K12 **Ixtlán del Río** see Ixtlán
125 F12 **Iyevlevo** Tyumenskaya Oblast', C Russian Federation 57.36N 67.20E
170 E14 **Iyo** Ehime, Shikoku, SW Japan 33.44N 132.42E
170 F15 **Iyomishima** var. Iyomisima. Ehime, Shikoku, SW Japan 33.58N 133.31E
Iyomisima see Iyomishima
170 Dd14 **Iyo-nada** sea S Japan
44 I4 **Izabal** off. Departamento de Izabal. ♦ department E Guatemala
44 F5 **Izabal, Lago de** prev. Golfo Dulce. ⊘ E Guatemala
115 L14 **Izacía** Serbia, C Yugoslavia 43.36N 20.14E
143 G11 **Izad Khvāst** Fārs, C Iran 31.31N 52.08E
43 X12 **Izamal** Yucatán, SE Mexico 20.58N 89.00W
131 T2 **Izberbash** Respublika Dagestan, SW Russian Federation 42.32N 47.51E
101 C18 **Izegem** prev. Iseghem. West-Vlaanderen, W Belgium 50.55N 3.13E
Izena see Ivankiv
148 M9 **Ïzeh** Khūzestān, SW Iran 55.55N 159.28W
118 J7 **Izena-jima** island Nansei-shotō, SW Japan
116 N10 **Izgrev** Burgas, E Bulgaria 42.09N 27.49E
131 T2 **Izhevsk** prev. Ustinov. Udmurtskaya Respublika, NW Russian Federation 56.48N 53.12E
129 S7 **Izhma** Respublika Komi, NW Russian Federation 64.56N 53.52E
129 S7 **Izhma** ⊘ NW Russian Federation
119 V12 **Izmail** see Izmayil
118 H4 **Izmayil** Rus. Izmail. Odes'ka Oblast', SW Ukraine 45.19N 28.48E
142 C14 **İzmir** prev. Smyrna. İzmir, W Turkey 38.25N 27.10E
142 C14 **İzmir** prev. Smyrna. ♦ province W Turkey
142 E11 **İzmit** var. Ismid; anc. Astacus. Kocaeli, NW Turkey 40.46N 29.55E
106 M14 **Iznajar, Embalse de** ⊡ S Spain 37.17N 4.16W
107 N14 **Iznalloz** Andalucía, S Spain 37.23N 3.31W
142 E11 **İznik** Bursa, NW Turkey 40.27N 29.43E
142 E12 **İznik Gölü** ⊘ NW Turkey
130 M14 **Izobil'nyy** Stavropol'skiy Kray, SW Russian Federation 45.22N 41.40E
111 S13 **Izola** It. Isola d'Istria. SW Slovenia 45.31N 13.40E
193 C19 **Jackson Bay** bay South Island, NZ
144 H9 **'Izra'** var. Ezra, Ezraa. Dar'ā, S Syria 32.52N 36.15E
43 P14 **Iztaccíhuatl, Volcán** var. Volcán Ixtaccíhuatl. ▲ S Mexico 19.07N 98.37W
43 P14 **Iztapa** Escuintla, SE Guatemala 13.55S 90.45W
Izúcar de Matamoros see Matamoros
171 J17 **Izu-hantō** peninsula Honshū, S Japan
170 F15 **Izuhara** Nagasaki, Tsushima, SW Japan 34.11N 129.16E
170 C14 **Izumi** Kagoshima, Kyūshū, SW Japan 32.05N 130.22E
171 Gg15 **Izumi-Sano** Ōsaka, Honshū, SW Japan 34.29N 135.25E
171 Gg15 **Izumi-Sano** Ōsaka, Honshū, SW Japan 34.24N 135.19E

170 F12 **Izumo** Shimane, Honshū, SW Japan 35.22N 132.45E
Izu Shichito see Izu-shotō
172 Ss13 **Izu-shotō** var. Izu Shichito. island group S Japan
199 H4 **Izu Trench** undersea feature NW Pacific Ocean
126 I4 **Izvestiy TsIK, Ostrova** island N Russian Federation
116 G10 **Izvor** Pernik, W Bulgaria 42.27N 22.53E
118 L5 **Izyaslav** Khmel'nyts'ka Oblast', W Ukraine 50.08N 26.53E
119 W6 **Izyum** Kharkivs'ka Oblast', E Ukraine 49.12N 37.18E

J

95 M18 **Jaala** Etelä-Suomi, S Finland 61.04N 26.30E
146 J5 **Jabal ash Shifā** desert NW Saudi Arabia
147 U8 **Jabal az Zannah** var. Jebel Dhanna. Abū Ẓaby, W UAE 24.10N 52.36E
144 E11 **Jabaliya** var. Jabālīyah. NE Gaza Strip 31.30N 34.25E
107 N11 **Jabalón** ⊘ C Spain
160 J10 **Jabalpur** prev. Jubbulpore. Madhya Pradesh, C India 23.10N 79.58E
147 N15 **Jabal Zuqar, Jazirat** var. Az Zuqur. island SW Yemen
144 J3 **Jabbul, Sabkhat al** salt flat NW Syria
189 P1 **Jabiru** Northern Territory, N Australia 12.44S 132.48E
144 H4 **Jablah** var. Jeble, Fr. Djéblé. Al Lādhiqīyah, W Syria 35.00N 36.00E
114 C11 **Jablanac** Lika-Senj, W Croatia 44.43N 14.54E
115 H14 **Jablanica** Federacija Bosna I Hercegovina, SW Bosnia and Herzegovina 43.39N 17.43E
115 M20 **Jablanica** Alb. Mali i Jabllanicës, var. Malet e Jabllanicës. ▲ Albania/FYR Macedonia see also Jabllanicës, Malet e
115 M20 **Jabllanicës, Mali i** var. Malet e Jabllanicës, Mac. Jablanica. ▲ Albania/FYR Macedonia see also Jablanica
113 E15 **Jablonec nad Nisou** Ger. Gablonz an der Neisse. Liberecký Kraj, N Czech Republic 50.43N 15.10E
Jablonków/Jablunkau see Jablunkov
112 J9 **Jablonowo Pomorskie** Kujawsko-pomorskie, C Poland 53.24N 19.08E
113 J17 **Jablunkov** Ger. Jablunkau, Pol. Jablonków. Ostravský Kraj, E Czech Republic 49.34N 18.45E
61 L16 **Jaboatão** Pernambuco, E Brazil 08.05S 35.00W
62 L8 **Jaboticabal** São Paulo, S Brazil 21.15S 48.16W
201 U7 **Jabwot** var. Jabat, Jebat, Jōwat. island Ralik Chain, S Marshall Islands
107 S4 **Jaca** Aragón, NE Spain 42.34N 0.33W
44 B4 **Jacaltenango** Huehuetenango, W Guatemala 15.39N 91.46W
61 G14 **Jacaré-a-Canga** Pará, NE Brazil 5.58S 57.31W
62 N10 **Jacareí** São Paulo, S Brazil 23.18S 45.55W
61 E15 **Jaciparaná** Rondônia, W Brazil 9.20S 64.27W
21 P5 **Jackman** Maine, NE USA 45.35N 70.14W
37 X5 **Jackpot** Nevada, W USA 41.57N 114.41W
37 T2 **Jacksboro** Tennessee, S USA 36.19N 84.10W
45 S5 **Jacksboro** Texas, SW USA 33.13N 98.10W
23 R4 **Jackson** Alabama, S USA 31.30N 87.53W
37 P7 **Jackson** California, W USA 38.19N 120.46W
23 O6 **Jackson** Georgia, SE USA 33.17N 83.58W
24 J8 **Jackson** Kentucky, S USA 37.30N 83.22W
31 T4 **Jackson** Louisiana, S USA 30.50N 91.13W
33 Q10 **Jackson** Michigan, C USA 42.15N 84.24W
31 T11 **Jackson** Minnesota, N USA 43.38N 95.00W
24 K5 **Jackson** state capital Mississippi, S USA 32.19N 90.12W
29 Y7 **Jackson** Missouri, C USA 37.22N 89.40W
23 W8 **Jackson** North Carolina, SE USA 36.24N 77.22W
33 T15 **Jackson** Ohio, NE USA 39.03N 82.40W
22 G9 **Jackson** Tennessee, S USA 35.37N 88.46W
35 S14 **Jackson** Wyoming, C USA 43.28N 110.46W
204 J6 **Jackson, Mount** ▲ Antarctica 71.43S 63.45W
35 V12 **Jackson Reservoir** ⊡ Colorado, C USA
23 T5 **Jacksonville** Alabama, S USA 33.48N 85.45W
29 V11 **Jacksonville** Arkansas, C USA 34.52N 92.06W
25 V9 **Jacksonville** Florida, SE USA 30.19N 81.39W

32 K14 **Jacksonville** Illinois, N USA 39.43N 90.13W
23 W11 **Jacksonville** North Carolina, SE USA 34.45N 77.25W
45 W7 **Jacksonville** Texas, SW USA 31.57N 95.16W
25 X9 **Jacksonville Beach** Florida, SE USA 30.17N 81.23W
46 L9 **Jacmel** var. Jaquemel. S Haiti 18.13N 72.33W
155 Q12 **Jacobābād** Sind, SE Pakistan 28.16N 68.16E
57 T11 **Jacobs Ladder Falls** waterfall S Guyana 2.57N 58.06W
47 O11 **Jaco, Pointe** headland N Dominica 15.38N 61.25W
13 Q9 **Jacques-Cartier** ⊘ Quebec, SE Canada
11 P11 **Jacques-Cartier, Détroit de** var. Jacques-Cartier Passage. strait Gulf of St. Lawrence/St. Lawrence River
13 W6 **Jacques-Cartier, Mont** ▲ Quebec, SE Canada 48.58N 66.00W
Jacques-Cartier Passage see Jacques-Cartier, Détroit de
195 O12 **Jacquinot Bay** inlet New Britain, PNG
60 L11 **Jacuí, Rio** ⊘ S Brazil
62 L11 **Jacupiranga** São Paulo, S Brazil 24.42S 48.00W
102 G10 **Jade** ⊘ NW Germany
102 G10 **Jadebusen** bay NW Germany
69 Z2 **Jado** island N Tunisia
77 S9 **Jādū** var. Jūlā. NE Libya 29.01N 21.33E
201 U8 **Jaluit Atoll** var. Jālwōj. atoll Ralik Chain, S Marshall Islands
201 U8 **Jālwōj** see Jaluit Atoll
58 C10 **Jaén** Cajamarca, N Peru 5.43S 78.46W
107 N13 **Jaén** Andalucía, SW Spain 37.46N 3.48W
107 N13 **Jaén** ♦ province Andalucía, S Spain
161 J23 **Jaffna** Northern Province, N Sri Lanka 9.42N 80.03E
161 K23 **Jaffna Lagoon** lagoon N Sri Lanka
21 N10 **Jaffrey** New Hampshire, NE USA 42.46N 72.00W
144 H13 **Jafr, Qā' al** var. El Jafr. salt pan S Jordan
158 J9 **Jagādhri** Haryāna, N India 30.10N 77.18E
120 H4 **Jägala** var. Jägala Jõgi, Ger. Jaggowal. ⊘ NW Estonia
120 H4 **Jägala Jõgi** see Jägala
Jaggowal see Jägala
161 L12 **Jagdalpur** Madhya Pradesh, C India 19.07N 82.04E
169 U5 **Jagdaqi** Nei Mongol Zizhiqu, N China 50.25N 124.05E
Jägerndorf see Krnov
115 L14 **Jagodina** prev. Svetozarevo. Serbia, C Yugoslavia 43.59N 21.15E
114 K12 **Jagodnja** ▲ W Yugoslavia
103 J18 **Jagst** ⊘ SW Germany
161 I14 **Jagtiāl** Andhra Pradesh, C India 18.47N 78.55E
63 H18 **Jaguarão** Rio Grande do Sul, S Brazil 32.32S 53.20W
63 H18 **Jaguarão, Rio** var. Río Yaguarón. ⊘ Brazil/Uruguay
62 J8 **Jaguariaíva** Paraná, S Brazil 24.15S 49.43W
46 J6 **Jagüey Grande** Matanzas, W Cuba 22.31N 81.09W
159 P12 **Jahānābād** Bihār, N India 25.13N 84.58E
149 P12 **Jahrom** var. Jahrum. Fārs, S Iran 28.34N 53.32E
Jahrum see Jahrom
Jahra see Al Jahrā'
149 P12 **Jailolo, Selat** strait E Indonesia
175 Tt8 **Jailolo** var. Halmahera, Pulau
Jainat see Chai Nat
Jainti see Jayanti
158 H12 **Jaipur** Rājasthān, N India 26.54N 75.46E
159 T14 **Jaipur Hat** Rajshahi, NW Bangladesh 25.04N 89.03E
158 D11 **Jaisalmer** Rājasthān, NW India 26.55N 70.56E
160 O12 **Jājapur** Orissa, E India 18.54N 82.36E
149 R4 **Jājarm** Khorāsān, NE Iran 36.58N 56.25E
114 G12 **Jajce** Federacija Bosna I Hercegovina, N Bosnia and Herzegovina 44.20N 17.16E
Jajị see 'Alī Kheyl
85 D17 **Jakalsberg** Otjozondjupa, N Namibia 19.23S 17.28E
174 I14 **Jakarta** prev. Djakarta, Dut. Batavia. ● (Indonesia) Jawa, C Indonesia 6.07S 106.45E
8 I8 **Jakes Corner** Yukon Territory, W Canada 60.18N 134.00W
18 H4 **Jakin** Georgia, SE USA 31.06N 84.55W
115 J20 **Jakupica** ▲ C FYR Macedonia
38 W15 **Jal** New Mexico, SW USA 32.07N 103.10W
147 P7 **Jalājil** var. Galājil. Ar Riyāḍ, C Saudi Arabia 25.42N 45.22E
149 R5 **Jalāl-Abad** var. Dzhalal-Abad, Dzhalal-Abadskaya Oblast'. W Kyrgyzstan
155 S5 **Jalālābād** var. Jalalabad, Jelalabad. Nangarhār, E Afghanistan 34.25N 70.28E
Jalal-Abad Oblasty see Dzhalal-Abadskaya Oblast'
158 H8 **Jalandhar** prev. Jullundur. Punjab, N India 31.19N 75.36E
44 G7 **Jalapa** Nueva Segovia, NW Nicaragua 13.57N 86.09W

44 A3 **Jalapa** off. Departamento de Jalapa. ♦ department SE Guatemala
44 E6 **Jalapa, Río** ⊘ SE Guatemala
149 X13 **Jālaq** Sīstān va Balūchestān, SE Iran
95 K17 **Jalasjärvi** Länsi-Suomi, W Finland 62.30N 22.49E
155 O8 **Jaldak** Zābul, SE Afghanistan 32.00N 66.45E
62 J7 **Jales** São Paulo, S Brazil 20.15S 50.34W
160 P11 **Jaleshwar** var. Jaleswar. Orissa, NE India 21.51N 87.15E
160 F12 **Jaleswar** see Jaleshwar
79 X15 **Jalingo** Taraba, E Nigeria 8.54N 11.22E
42 K13 **Jalisco** ♦ state W Mexico
158 E13 **Jālor** Rājasthān, N India 25.21N 72.43E
114 K11 **Jalovik** Serbia, W Yugoslavia 44.37N 19.48E
42 L12 **Jalpa** Zacatecas, C Mexico 21.40N 103.00W
159 S12 **Jalpāiguri** West Bengal, NE India 26.43N 88.24E
43 O12 **Jalpan** var. Jalpan. Querétaro de Arteaga, C Mexico 21.13N 99.28W
69 Z2 **Jalta** island N Tunisia
77 S9 **Jālū** var. Jūlā. NE Libya 29.01N 21.33E
201 O15 **Jālwōj** see Jaluit Atoll
113 O15 **Janów Lubelski** Lubelskie, E Poland 50.43N 22.24E
112 H10 **Janów Poleski** see Ivanava
207 Q15 **Jan Mayen** ◇ Norwegian dependency N Atlantic Ocean
86 D5 **Jan Mayen** island N Atlantic Ocean
207 R15 **Jan Mayen Fracture Zone** tectonic feature Greenland Sea/Norwegian Sea
207 R13 **Jan Mayen Ridge** undersea feature Greenland Sea/Norwegian Sea
42 H3 **Janos** Chihuahua, N Mexico 30.53N 108.21W
113 K25 **Jánoshalma** SCr. Jankovac. Bács-Kiskun, S Hungary 46.18N 19.16E
112 H11 **Janow/Janów** see Jonava, Lithuania
112 H10 **Janowiec Wielkopolski** Ger. Janowitz. Kujawski-pomorskie, C Poland 52.47N 17.30E
Janowitz see Janowiec Wielkopolski
Janów Lubelski see Janowiec
113 O15 **Janów Lubelski** Lubelskie, E Poland 50.43N 22.24E
Janów Poleski see Ivanava
176 W12 **Jantan** Irian Jaya, E Indonesia 3.53S 134.20E
61 M18 **Januária** Minas Gerais, SE Brazil 15.28S 44.22W
Janūbīyah, Al Bādiyah al see Ash Shāmiyah
104 I7 **Janzé** Ille-et-Vilaine, NW France 47.55N 1.28W
160 H13 **Jaora** Madhya Pradesh, C India 23.40N 75.10E
171 H12 **Japan** var. Nippon, Jap. Nihon. ♦ monarchy E Asia
133 Y9 **Japan** island group E Asia
199 H3 **Japan Basin** undersea feature N Sea of Japan
133 Y8 **Japan, Sea of** var. East Sea, Rus. Yapanskoye More. sea NW Pacific Ocean
199 H4 **Japan Trench** undersea feature Sea of Japan
61 A15 **Japiim** var. Máncio Lima. Acre, W Brazil 8.03S 73.39W
60 D12 **Japurá** Amazonas, N Brazil 1.43S 66.14W
60 C12 **Japurá, Rio** var. Río Caquetá, Yapurá. ⊘ Brazil/Colombia see also Caquetá, Río
45 W17 **Jaqué** Darién, SE Panama 7.30N 78.09W
Jaquemel see Jacmel
Jarablos see Jarābulus
144 K2 **Jarābulus** var. Jarablos, Jerablus, Fr. Djérablous. Ḥalab, N Syria 36.51N 38.02E
62 K13 **Jaraguá do Sul** Santa Catarina, S Brazil 26.28S 49.07W
106 K9 **Jaraicejo** Extremadura, W Spain 39.40N 5.49W
107 O7 **Jarafuel** Valencia, E Spain 39.08N 1.05W
106 K9 **Jaráiz de la Vera** Extremadura, W Spain 40.04N 5.45W
107 O7 **Jarandilla de la Vega** see Jarandilla de la Vera
106 K8 **Jarandilla de la Vera** var. Jarandilla de la Vega. Extremadura, W Spain 40.07N 5.39W
155 V9 **Jaränwäla** Punjab, E Pakistan 31.19N 73.25E
144 G9 **Jarash** var. Jerash; anc. Gerasa. Irbid, NW Jordan 32.16N 35.54E
96 F7 **Jarbah, Jazirat** see Jerba, Île de
97 O16 **Järbo** Gävleborg, C Sweden 60.43N 16.40E
46 F7 **Jardines de la Reina, Archipiélago de los** island group C Cuba
168 J7 **Jargalant** Arhangay, C Mongolia 47.46N 101.56E
168 J8 **Jargalant** Bayanhongor, C Mongolia 47.14N 99.43E
168 D7 **Jargalant** Bulgan, N Mongolia 49.09N 104.19E
168 G9 **Jargalant** Govĭ-Altay, W Mongolia 45.57N 97.10E
60 L13 **Jarí, Rio** var. Jary. ⊘ N Brazil
147 N7 **Jarīr, Wādī** dry watercourse C Saudi Arabia
Jarja see Yur'ya
96 L13 **Järna** var. Dala-Järna. Dalarna, C Sweden 60.31N 14.22E
104 K11 **Jarnac** Charente, W France 45.41N 0.10W
112 H12 **Jarocin** Wielkopolskie, C Poland 51.59N 17.31E
113 F16 **Jaroměř** Ger. Jermer. Hradecký Kraj, N Czech Republic 50.22N 15.55E
113 O16 **Jarosław** Ger. Jaroslau, Rus. Yaroslav. Podkarpackie, SE Poland 50.01N 22.41E
60 L13 **Jarí, Rio** see Jarí, Rio
153 S14 **Jarqŭrghon** Rus. Dzharkurgan. Surkhondaryo Wiloyati, S Uzbekistan 37.31N 67.17E
145 J7 **Jarrāḥ, Wadi** dry watercourse NE Syria
168 K8 **Jartai Yanchi** ⊘ N China

♦ COUNTRY   ◇ DEPENDENT TERRITORY   ♦ ADMINISTRATIVE REGION   ▲ MOUNTAIN   ⊘ VOLCANO   ⊘ LAKE
● COUNTRY CAPITAL   ○ DEPENDENT TERRITORY CAPITAL   × INTERNATIONAL AIRPORT   ▲ MOUNTAIN RANGE   ⊘ RIVER   ⊡ RESERVOIR

61 *E16* **Jaru** Rondônia, W Brazil
10.24S 62.45W

169 *T10* **Jarud Qi** Nei Mongol Zizhiqu,
N China 44.25N 121.12E

120 *I4* **Järva-Jaani** *Ger.* Sankt-Johannis.
58.47N 24.49E

120 *G5* **Järvakandi** *Ger.* Jerwakant.
Raplamaa, NW Estonia
58.47N 24.49E

120 *H4* **Järvamaa** *off.* Järva Maakond. ◆
province N Estonia

95 *L19* **Järvenpää** Etelä-Suomi, S Finland
60.28N 25.06E

12 *G17* **Jarvis** Ontario, S Canada
42.53N 80.06W

185 *R8* **Jarvis Island** ◇ *US unincorporated
territory* C Pacific Ocean

96 *M11* **Järvsö** Gävleborg, C Sweden
61.43N 16.25E

Jary *see* Jari, Rio

114 *M9* **Jaša Tomić** Serbia, NE Yugoslavia
45.26N 20.51E

114 *D12* **Jasenice** Zadar, SW Croatia
44.15N 15.33E

144 *I11* **Jashshat al 'Adlah, Wādī al** *dry
watercourse* C Jordan

79 *Q16* **Jasikan** E Ghana 7.24N 0.28E

149 *T15* **Jāsk** Hormozgān, SE Iran
25.35N 58.06E

152 *F6* **Jasliq** *Rus.* Zhaslyk.
Qoraqalpoghiston Respublikasi,
NW Uzbekistan 43.57N 57.30E

113 *N17* **Jasło** Podkarpackie, SE Poland
49.45N 21.28E

9 *U16* **Jasmin** Saskatchewan, S Canada
51.11N 103.34W

67 *A23* **Jason Islands** *island group*
NW Falkland Islands

204 *I4* **Jason Peninsula** *peninsula*
Antarctica

33 *N15* **Jasonville** Indiana, N USA
39.09N 87.12W

9 *O15* **Jasper** Alberta, S Canada
52.55N 118.04W

12 *L13* **Jasper** Ontario, SE Canada
44.50N 75.57W

25 *O3* **Jasper** Alabama, S USA
33.49N 87.16W

29 *T9* **Jasper** Arkansas, C USA
36.00N 93.11W

25 *U8* **Jasper** Florida, SE USA
30.31N 82.57W

33 *N16* **Jasper** Indiana, N USA
38.22N 86.57W

31 *R11* **Jasper** Minnesota, N USA
43.51N 96.24W

29 *S7* **Jasper** Missouri, C USA
37.20N 94.18W

22 *K10* **Jasper** Tennessee, S USA
35.04N 85.37W

45 *Y9* **Jasper** Texas, SW USA
30.55N 94.00W

9 *O15* **Jasper National Park** *national
park* Alberta/British Columbia,
SW Canada

Jassy *see* Iași

115 *N14* **Jastrebac** ▲ SE Yugoslavia

114 *D9* **Jastrebarsko** Zagreb, N Croatia
45.40N 15.40E

Jastrow *see* Jastrowie

112 *G9* **Jastrowie** *Ger.* Jastrow.
Wielkopolskie, C Poland
53.25N 16.48E

113 *J17* **Jastrzębie-Zdrój** Śląskie,
S Poland 49.58N 18.34E

113 *L22* **Jászapáti** Jász-Nagykun-Szolnok,
E Hungary 47.31N 20.09E

113 *L22* **Jászberény** Jász-Nagykun-
Szolnok, E Hungary 47.30N 19.54E

113 *L23* **Jász-Nagykun-Szolnok** *off.*
Jász-Nagykun-Szolnok Megye. ◆
county E Hungary

61 *J19* **Jataí** Goiás, C Brazil 17.58S 51.45W

60 *G12* **Jatapu, Serra do** ▲ N Brazil

43 *N16* **Jatate, Río** ← SE Mexico

155 *P17* **Jāti** Sind, SE Pakistan 24.19N 68.18E

46 *F6* **Jatibonico** Sancti Spíritus,
C Cuba 21.55N 79.12W

174 *Ji14* **Jatiluhur, Danau** ◎ Jawa,
S Indonesia

Jativa *see* Xátiva

174 *K14* **Jatiwangi** *prev.* Djatiwangi. Jawa,
C Indonesia 6.45S 108.12E

155 *S11* **Jattoi** Punjab, E Pakistan
29.20N 70.55E

62 *L9* **Jaú** São Paulo, S Brazil
22.17S 48.32W

60 *F11* **Jauaperi, Rio** ← N Brazil

101 *I19* **Jauche** Wallon Brabant,
C Belgium 50.42N 4.55E

Jauer *see* Jawor

Jauf *see* Al Jawf

155 *U7* **Jauharābād** Punjab, E Pakistan
32.19N 72.15E

59 *E14* **Jauja** Junín, C Peru 11.44S 75.30W

43 *O10* **Jaumave** Tamaulipas, C Mexico
23.28N 99.22W

120 *H10* **Jaunjelgava** *Ger.* Friedrichstadt.
Aizkraukle, S Latvia 56.36N 25.06E

120 *I8* **Jaunlatgale** *see* Pytalovo

120 *E9* **Jaunpiebalga** Gulbene,
NE Latvia 57.10N 26.02E

120 *E9* **Jaunpils** Tukums, C Latvia
56.45N 23.03E

159 *N13* **Jaunpur** Uttar Pradesh, N India
25.43N 82.40E

31 *N8* **Java** South Dakota, N USA
45.29N 99.54W

Java *see* Jawa

107 *R9* **Javalambre** ▲ E Spain
40.02N 1.06W

181 *V7* **Java Ridge** *undersea feature*
E Indian Ocean

61 *A14* **Javari, Rio** *var.* Yavarí.
← Brazil/Peru

159 *S11* **Javarthushuu** Dornod,
NE Mongolia 49.05N 112.40E

174 *Kk13* **Java Sea** *Ind.* Laut Jawa. *sea*
W Indonesia

181 *U7* **Java Trench** *var.* Sunda Trench.
undersea feature E Indian Ocean

114 *I8* **Jávea** *var.* Xábia. País Valenciano,
E Spain 38.48N 0.10E

169 *O7* **Javhlant** Hentiy, E Mongolia
47.46N 112.06E

65 *G20* **Javier, Isla** *island* S Chile

115 *L14* **Javor** ▲ Bosnia and
Herzegovina/Yugoslavia

113 *K20* **Javorie** *Hung.* Jávoros.
▲ S Slovakia 48.26N 19.16E

Jávoros *see* Javorie

95 *J15* **Jävre** Norrbotten, N Sweden
65.10N 21.31E

---

174 *K14* **Jawa** *Eng.* Java; *prev.* Djawa. *island*
C Indonesia

174 *J15* **Jawa Barat** *off.* Propinsi Jawa
Barat, *Eng.* West Java. ◆ *province*
S Indonesia

174 *Kk15* **Jawa, Laut** *see* Java Sea

145 *R3* **Jawān** NW Iraq 35.57N 43.03E

174 *Jj15* **Jawa Tengah** *off.* Propinsi Jawa
Tengah, *Eng.* Central Java. ◆
province S Indonesia

174 *Ll15* **Jawa Timur** *off.* Propinsi Jawa
Timur, *Eng.* East Java. ◆ *province*
S Indonesia

83 *N17* **Jawhar** *var.* Jowhar, *It.* Giohar.
Shabeellaha Dhexe, S Somalia
2.36N 45.30E

113 *F14* **Jawor** *Ger.* Jauer. Dolnośląskie,
SW Poland 51.01N 16.10E

113 *J12* **Jaworów** *see* Yavoriv

113 *I12* **Jaworzno** Śląskie, S Poland
50.13N 19.07E

29 *R9* **Jay** Oklahoma, C USA
36.25N 94.48W

111 *X8* **Jayabum** *see* Chaiyaphum

159 *T12* **Jayanti** *prev.* Jainti. West Bengal,
NE India 26.43N 89.43E

176 *Xx12* **Jaya, Puncak** *prev.* Puntjak
Carstensz, Puntjak Sukarno.
▲ Irian Jaya, E Indonesia
4.00S 137.10E

176 *Z10* **Jayapura** *var.* Djajapura, *Dut.*
Hollandia; *prev.* Kotabaru,
Sukarnapura. Irian Jaya,
E Indonesia 2.37S 140.39E

176 *Yi12* **Jayawijaya, Pegunungan**
▲ Irian Jaya, E Indonesia

Jay Dairen *see* Dalian

Jayhawker State *see* Kansas

153 *Si2* **Jayilgan** *Rus.* Dzhailgan,
Dzhayilgan. C Tajikistan
39.17N 71.32E

161 *L14* **Jaypur** *var.* Jeypore, Jeypur.
Orissa, E India 18.44 82.36E

45 *O6* **Jayton** Texas, SW USA
33.15N 100.34W

149 *U13* **Jaz Murian, Hāmūn-e** ◎ SE Iran

144 *M4* **Jazrah** Ar Raqqah, C Syria
36.56N 39.02E

144 *G6* **Jbail** *var.* Jebeil, Jubayl, Jubeil;
anc. Biblical Gebal, Byblos.
W Lebanon 34.08N 35.45E

45 *O7* **J.B.Thomas, Lake** ◎ Texas,
SW USA

Jdaïdé *see* Judaydah

37 *X12* **Jean**, Nevada, W USA
35.45N 115.20W

24 *I9* **Jeanerette** Louisiana, S USA
29.54N 91.39W

46 *L8* **Jean-Rabel** NW Haiti
19.49N 73.12W

149 *T12* **Jebāl Bārez, Kūh-e** ▲ SE Iran

79 *T15* **Jebba** Kwara, W Nigeria 9.04N 4.50E

Jebeil *see* Jbail

118 *E12* **Jebel** *Hung.* Széphely; *prev.* Hung.
Zsebely. Timiș, W Romania
45.33N 21.13E

Jebel *see* Dzhebel

Jebel, Bahr el *see* White Nile

Jebel Dhanna *see* Jabal aż
Zannah

Jeble *see* Jablah

98 *K13* **Jedburgh** SE Scotland, UK
55.28N 2.34W

Jedda *see* Jiddah

113 *L15* **Jędrzejów** *Ger.* Endersdorf.
Świętokrzyskie, C Poland
50.39N 20.18E

102 *K12* **Jeetze** *var.* Jeetzel. ← C Germany

Jeetzel *see* Jeetze

31 *U14* **Jefferson** Iowa, C USA
42.01N 94.22W

23 *Q8* **Jefferson** North Carolina,
SE USA 36.24N 81.33W

45 *X6* **Jefferson** Texas, SW USA
32.45N 94.21W

32 *M9* **Jefferson** Wisconsin, N USA
43.01N 88.48W

32 *M9* **Jefferson City** *state capital*
Missouri, C USA 38.33N 92.12W

35 *R10* **Jefferson City** Montana,
NW USA 46.24N 112.01W

23 *N9* **Jefferson City** Tennessee, S USA
36.07N 83.29W

37 *U7* **Jefferson, Mount** ▲ Nevada,
W USA 38.49N 116.54W

34 *H12* **Jefferson, Mount** ▲ Oregon,
NW USA 44.40N 121.48W

22 *L5* **Jeffersontown** Kentucky, S USA
38.11N 85.33W

33 *P16* **Jeffersonville** Indiana, N USA
38.16N 85.45W

35 *V15* **Jeffrey City** Wyoming, C USA
42.29N 107.49W

79 *T13* **Jega** Kebbi, NW Nigeria
12.15N 4.21E

79 *P14* **Jejui-Guazú, Río** ← E Paraguay

120 *I10* **Jēkabpils** *Ger.* Jakobstadt.
Jēkabpils, S Latvia 56.30N 25.56E

25 *W7* **Jekyll Island** *island* Georgia,
SE USA

174 *L11* **Jelai, Sungai** ← Borneo,
C Indonesia

113 *H14* **Jelcz-Laskowice** Dolnośląskie,
SW Poland 51.01N 17.24E

113 *E14* **Jelenia Góra** *Ger.* Hirschberg,
Hirschberg im Riesengebirge,
Hirschberg in Riesengebirge,
Hirschberg in Schlesien.
Dolnośląskie, SW Poland
50.54N 15.48E

159 *S11* **Jelep La** *pass* N India 27.24N 88.51E

120 *F9* **Jelgava** *Ger.* Mitau. Jelgava,
C Latvia 56.38N 23.47E

114 *L13* **Jelica** ▲ C Yugoslavia

22 *M8* **Jellico** Tennessee, S USA
36.33N 84.06W

95 *G23* **Jelling** Vejle, C Denmark
55.45N 9.24E

174 *Ii5* **Jemaja, Pulau** *island* W Indonesia

Jemaluang *see* Jemaluang

101 *E20* **Jemappes** Hainaut, S Belgium
50.27N 3.52E

174 *M16* **Jember** *prev.* Djember. Jawa,
C Indonesia 8.07S 113.45E

101 *I20* **Jemeppe-sur-Sambre** Namur,
S Belgium 50.27N 4.41E

39 *R10* **Jemez Pueblo** New Mexico,
SW USA 35.36N 106.43W

---

164 *K2* **Jeminay** Xinjiang Uygur Zizhiqu,
NW China 47.28N 85.49E

201 *U5* **Jemo Island** *atoll* Ratak Chain,
C Marshall Islands

175 *Nn8* **Jempang, Danau** ◎ Borneo,
N Indonesia

103 *L16* **Jena** Thüringen, C Germany
50.55N 11.34E

24 *I6* **Jena** Louisiana, S USA
31.40N 92.07W

110 *I8* **Jenaz** Graubünden,
SE Switzerland 46.56N 9.43E

111 *N7* **Jenbach** Tirol, W Austria
47.23N 11.47E

175 *P13* **Jeneponto** *prev.* Djeneponto.
Sulawesi, C Indonesia 5.40S 119.42E

144 *P9* **Jenin** N West Bank 32.28N 35.17E

23 *P7* **Jenkins** Kentucky, S USA
37.10N 82.37W

29 *P1* **Jenks** Oklahoma, C USA
36.01N 95.58W

Jenné *see* Djenné

24 *H9* **Jennings** Louisiana, S USA
30.13N 92.39W

15 *Jj4* **Jenny Lind Island** *island*
Nunavut, N Canada

25 *Y13* **Jensen Beach** Florida, SE USA
27.15N 80.13W

15 *M2* **Jens Munk Island** *island*
Nunavut, N Canada

61 *O17* **Jequié** Bahia, E Brazil 13.52S 40.06W

61 *O18* **Jequitinhonha, Rio** ← E Brazil

76 *H6* **Jerablus** *see* Jarābulus

76 *M7* **Jerada** NE Morocco 34.16N 2.07W

77 *N7* **Jerba, Île de** *var.* Djerba, Jazīrat
Jarbah. *island* E Tunisia

46 *K9* **Jérémie** SW Haiti 18.38N 74.10W

Jerez *see* Jeréz de la Frontera,
Spain

Jeréz *see* Jerez de García Salinas,
Mexico

42 *L11* **Jerez de García Salinas** *var.*
Jeréz. Zacatecas, C Mexico
22.38N 102.58W

106 *J15* **Jeréz de la Frontera** *var.* Jerez;
prev. Xeres. Andalucía, SW Spain
36.40N 6.07W

106 *I12* **Jerez de los Caballeros**
Extremadura, W Spain 38.19N 6.45W

Jergucati *see* Jorgucat

144 *G10* **Jericho** *Ar.* Arīḥā, *Heb.* Yeriho.
E West Bank 31.51N 35.27E

76 *M7* **Jerid, Chott el** *var.* Shaṭṭ al Jarīd.
salt lake SW Tunisia

191 *O10* **Jerilderie** New South Wales,
SE Australia 35.09S 145.43E

Jerischmarkt *see* Câmpia Turzii

94 *K11* **Jerisjärvi** ◎ NW Finland

Jermentau *see* Yereymentau

Jermer *see* Jaroměř

38 *K11* **Jerome** Arizona, SW USA
34.45N 112.06W

35 *O15* **Jerome** Idaho, NW USA
42.43N 114.31W

99 *L26* **Jersey** *island* Channel Islands,
NW Europe

20 *K14* **Jersey City** New Jersey, NE USA
40.44N 74.01W

20 *F13* **Jersey Shore** Pennsylvania,
NE USA 41.12N 77.13W

32 *K14* **Jerseyville** Illinois, N USA
39.07N 90.19W

106 *K8* **Jerte** ← W Spain

144 *F10* **Jerusalem** *Ar.* Al Quds, Al Quds
ash Sharif, *Heb.* Yerushalayim; *anc.*
Hierosolyma. ● (Israel) Jerusalem,
NE Israel 31.46N 35.13E

144 *G10* **Jerusalem** *district* E Israel

191 *S10* **Jervis Bay** New South Wales,
SE Australia 35.09S 150.42E

191 *S10* **Jervis Bay Territory** ◇ *territory*
SE Australia

Jerwakant *see* Järvakandi

111 *S10* **Jesenice** *Ger.* Assling.
NW Slovenia 46.26N 14.00E

113 *H16* **Jeseník** *Ger.* Freiwaldau.
Olomoucký Kraj, E Czech Republic
50.14N 17.12E

Jesi *see* Iesi

108 *I8* **Jesolo** *var.* Iesolo. Veneto, NE Italy
45.32N 12.37E

94 *I14* **Jessheim** Akershus, S Norway
60.07N 11.10E

159 *T15* **Jessore** Khulna, W Bangladesh
23.10N 89.12E

25 *W6* **Jesup** Georgia, SE USA
31.36N 81.54W

43 *S15* **Jesús Carranza** Veracruz-Llave,
SE Mexico 17.30N 95.01W

64 *K10* **Jesús María** Córdoba,
C Argentina 30.58S 64.04W

28 *K6* **Jesus, Point** *var.* Iesolo.
38.04N 99.53W

105 *Q2* **Jeumont** Nord, N France
50.18N 4.06E

97 *H14* **Jevnaker** Oppland, S Norway
60.13N 10.22E

45 *V9* **Jewett** Texas, SW USA
31.21N 96.08W

21 *N12* **Jewett City** Connecticut, NE USA
41.36N 71.58W

Jewish Autonomous Oblast *see*
Yevreyskaya Avtonomnaya Oblast'

Jeypore/Jeypur *see* Jaypur,
Orissa, India

Jeypur *see* Jaipur, Rājasthān,
India

115 *L17* **Jezercës, Maja e** ▲ N Albania
42.27N 19.49E

113 *B18* **Jezerní Hora** ▲ SW Czech
Republic 49.10N 13.11E

160 *F10* **Jhābua** Madhya Pradesh, C India
22.47N 74.36E

158 *H14* **Jhālāwār** Rājasthān, N India
24.33N 76.10E

155 *U9* **Jhang Sadar** *var.* Jhang, Jhang
Sadar. Punjab, NE Pakistan
31.16N 72.19E

158 *J13* **Jhānsi** Uttar Pradesh, N India
25.27N 78.34E

161 *M11* **Jhārsuguda** Orissa, E India
21.54N 84.09E

155 *V7* **Jhelum** Punjab, NE Pakistan
32.55N 73.42E

155 *V7* **Jhelum** ← E Pakistan

Jhenaidaha *see* Jhenida

---

159 *T15* **Jhenida** *var.* Jhenaidaha. Dhaka,
W Bangladesh 23.34N 89.39E

155 *P16* **Jhimpir** Sind, SE Pakistan
25.01N 67.59E

Jhind *see* Jind

155 *R16* **Jhudo** Sind, SE Pakistan
24.58N 69.23E

Jhumra *see* Chak Jhumra

158 *H11* **Jhunjhunūn** Rājasthān, N India
28.07N 75.21E

Ji *see* Hebei, China

Ji *see* Jilin, China

159 *S14* **Jiāganj** West Bengal, NE India
24.18N 88.07E

166 *J7* **Jialing Jiang** ← C China

169 *Y7* **Jiamusi** *var.* Chia-mu-ssu,
Kiamusze. Heilongjiang, NE China
46.45N 130.19E

167 *O11* **Ji'an** Jiangxi, S China 27.06N 114.57E

169 *W12* **Ji'an** Jilin, NE China 41.04N 126.07E

169 *T13* **Jianchang** Liaoning, NE China

166 *F11* **Jianchuan** Yunnan, SW China
26.28N 99.49E

164 *M4* **Jiangjunmiao** Xinjiang Uygur
Zizhiqu, W China 44.42N 90.06E

166 *K11* **Jiangkou** Guizhou, S China
27.46N 108.53E

167 *Q12* **Jiangle** Fujian, SE China
26.44N 117.26E

167 *N15* **Jiangmen** Guangdong, S China
22.34N 113.01E

167 *Q10* **Jiangshan** Zhejiang, S China
28.41N 118.33E

167 *Q7* **Jiangsu** *var.* Chiang-su, Jiangsu
Sheng, Kiangsu, Su. ◆ *province*
E China

167 *Q10* **Jiangshan Sheng** *see* Jiangsu

167 *O11* **Jiangxi** *var.* Chiang-hsi, Gan,
Jiangxi Sheng, Kiangsi. ◆ *province*
S China

166 *M9* **Jiangyou** *prev.* Zhongba. Sichuan,
C China 31.52N 104.52E

167 *N9* **Jianli** Hubei, C China
29.48N 112.45E

167 *Q11* **Jian'ou** Fujian, SE China
27.04N 118.19E

166 *I9* **Jianshi** Hubei, C China
30.37N 109.42E

133 *V11* **Jian Xi** ← SE China

167 *Q11* **Jianyang** Fujian, SE China
27.20N 118.03E

166 *I9* **Jianyang** Sichuan, C China
30.25N 104.33E

165 *X10* **Jiaohe** Jilin, NE China
43.41N 127.20E

167 *N6* **Jiaozuo** Henan, C China
35.13N 113.13E

164 *F8* **Jiashi** *var.* Payzawat. Xinjiang
Uygur Zizhiqu, NW China
39.27N 76.45E

167 *R10* **Jiashan** *see* Mingguang

160 *L9* **Jiāwān** Madhya Pradesh, C India

167 *S9* **Jiaxing** Zhejiang, S China
30.43N 120.46E

167 *X6* **Jiayi** *see* Chiai

167 *R13* **Jiayin** *var.* Chaoyang.
Heilongjiang, NE China

164 *I4* **Jiayuguan** Gansu, N China
39.47N 98.14E

176 *W14* **Jin, Kepulauan** *island group*
E Indonesia

166 *I8* **Jibou** *Hung.* Zsibó. Sălaj,
NW Romania 47.13N 23.17E

147 *Z9* **Jibsh, Ra's al** *headland* E Oman
21.20N 59.23E

113 *E15* **Jičín** *Ger.* Jitschin. Hradecký Kraj,
N Czech Republic 50.27N 15.20E

146 *K10* **Jiddah** *Eng.* Jedda. Makkah,
W Saudi Arabia 21.33N 39.13E

147 *W11* **Jiddat al Ḥarāsīs** *desert* C Oman

116 *M4* **Jiexiu** Shanxi, C China
37.00N 111.50E

167 *P14* **Jieyang** Guangdong, S China
23.33S 116.21E

121 *F14* **Jieznas** Prienai, S Lithuania
60.07N 11.10E

147 *P15* **Jif'īyah, Bi'r** *see* Jif'īyah, Bi'r
Y Yemen 14.48N 46.00E

79 *W13* **Jigawa** ◆ *state* N Nigeria

46 *I7* **Jiguaní** Granma, E Cuba
20.24N 76.25W

165 *T12* **Jigzhi** Qinghai, C China
33.23N 101.25E

118 *I14* **Jieznas** *var.* Iesolo.
38.04N 99.53W

113 *E18* **Jihlava** *var.* Iglau, *Pol.* Iglawa.
Jihlavský Kraj, C Czech Republic
49.22N 15.36E

113 *E18* **Jihlava** *var.* Igel, *Ger.* Iglawa.
← S Czech Republic

113 *E18* **Jihlavský Kraj** ◆ *region*
C Czech Republic

116 *L9* **Jijel** *var.* Djidjel; *prev.* Djidjelli.
NE Algeria 36.49N 5.43E

118 *L9* **Jijia** ← N Romania

118 *L13* **Jijiga** *It.* Giggiga. Somali,
E Ethiopia 9.21N 42.53E

107 *S12* **Jijona** *var.* Xixona. País
Valenciano, E Spain 38.34N 0.29W

83 *L18* **Jilib** *It.* Gelib. Jubbada Dhexe,
S Somalia 0.17N 42.54E

169 *W10* **Jilin** *var.* Chi-lin, Girin, Kirin;
prev. Yungki, Yunki. Jilin,
NE China 43.46N 126.31E

169 *W11* **Jilin Hada Ling** ▲ NE China

169 *W11* **Jilin Sheng** *see* Jilin

154 *G9* **Jiliu He** ← NE China

83 *J14* **Jima** *var.* Jimma, *It.* Gimma.
Oromo, C Ethiopia 7.41N 36.51E

46 *M9* **Jimaní** W Dominican Republic
18.27N 71.51E

118 *E11* **Jimbolia** *Ger.* Hatzfeld, *Hung.*
Zsombolya. Timiș, W Romania
45.47N 20.43E

115 *V7* **Jimena de la Frontera**
Andalucía, S Spain 36.27N 5.28W

---

42 *K7* **Jiménez** Chihuahua, N Mexico
27.09N 104.54W

43 *N7* **Jiménez** Coahuila de Zaragoza,
NE Mexico 29.04N 100.43W

43 *P9* **Jiménez** *var.* Santander Jiménez.
Tamaulipas, C Mexico 24.11N 98.29W

42 *L10* **Jiménez del Teul** Zacatecas,
C Mexico 23.13N 103.46W

79 *Y14* **Jimeta** Adamawa, E Nigeria
9.16N 12.25E

164 *M5* **Jimsar** Xinjiang Uygur Zizhiqu,
NW China 44.05N 88.48E

20 *I14* **Jim Thorpe** Pennsylvania,
NE USA 40.51N 75.43W

167 *P5* **Jinan** *var.* Chinan, Chi-nan,
Tsinan. Shandong, E China
36.42N 116.57E

165 *T8* **Jinchang** Gansu, N China
38.31N 102.07E

167 *N5* **Jincheng** Shanxi, C China
35.33N 112.51E

Jinchengjiang *see* Hechi

158 *I9* **Jind** *prev.* Jhind. Haryāna,
N India 29.25N 76.14E

166 *K11* **Jing** Guizhou, S China

191 *Q11* **Jingcheng** ← SE China

113 *O18* **Jindřichův Hradec** *Ger.*
Neuhaus. Budějovický Kraj, S
Czech Republic 49.09N 15.01E

165 *X10* **Jingchuan** Gansu, C China
35.19N 107.23E

167 *Q10* **Jingdezhen** Jiangxi, S China
29.18N 117.18E

167 *O12* **Jinggangshan** Jiangxi, S China
26.36N 114.11E

167 *P3* **Jinghai** Tianjin Shi, E China
38.53N 116.45E

166 *K6* **Jing He** ← C China

166 *I4* **Jinghe** *var.* Jing. Xinjiang Uygur
Zizhiqu, NW China 44.35N 82.55E

166 *F15* **Jinghong** *var.* Yunjinghong.
Yunnan, SW China 22.03N 100.55E

166 *M9* **Jingmen** Hubei, C China
30.58N 112.09E

165 *V9* **Jingtai** *var.* Yitiaoshan. Gansu,
C China 37.12N 104.06E

166 *J14* **Jingxi** Guangxi Zhuangzu
Zizhiqu, S China 23.10N 106.22E

169 *W11* **Jingyu** Jilin, NE China
42.22N 126.48E

165 *V10* **Jingyuan** Gansu, C China
36.34N 104.43E

166 *L12* **Jingzhou** *var.* Jing Xian. Hunan,
C China 26.36N 109.43E

166 *M9* **Jingzhou** *prev.* Shashi, Sha-shih,
Shasi. Hubei, C China 30.21N 112.09E

167 *R10* **Jinhua** Zhejiang, S China
29.15N 119.36E

169 *P13* **Jining** Nei Mongol Zizhiqu,
N China 40.58N 113.08E

167 *P5* **Jining** Shandong, E China
35.25N 116.35E

81 *G18* **Jinja** S Uganda 0.27N 33.13E

167 *O11* **Jin Jiang** ← S China

167 *R13* **Jinjiang** *var.* Qingyang. Fujian,
SE China 24.45N 118.35E

176 *W14* **Jin, Kepulauan** *island group*
E Indonesia

167 *O10* **Jinmen Dao** *see* Chinmen Tao

144 *M4* **Jibli** Ar Raqqah, C Syria
35.49N 39.23E

44 *K7* **Jinotega** Jinotega, NW Nicaragua
13.03N 85.59W

44 *K7* **Jinotega** ◆ *department*
N Nicaragua

44 *J9* **Jinotepe** Carazo, SW Nicaragua
11.49N 86.11W

42 *M9* **Jinping** *prev.* Sanjiang. Guizhou,
S China 26.42N 109.13E

166 *H14* **Jinping** Yunnan, SW China
22.47N 103.12E

166 *J11* **Jinsha** Guizhou, S China
27.24N 106.16E

163 *N12* **Jinsha Jiang** ← SW China

166 *M10* **Jinshi** Hunan, S China
29.42N 111.46E

165 *R7* **Jinta** Gansu, N China 40.01N 98.57E

179 *Q12* **Jintotolo Channel** *channel*
C Philippines

167 *Q12* **Jin Xi** *see* SE China

167 *O10* **Jinxi** *see* Lianshan

167 *O11* **Jin Jiang** *see* Jinzhou

167 *P6* **Jinxiang** Shandong, E China
35.07N 116.19E

167 *P8* **Jinzhai** *prev.* Meishan. Anhui,
E China 31.42N 115.47E

169 *U14* **Jinzhou** *prev.* Jinxian. Liaoning,
NE China 39.04N 121.45E

169 *T12* **Jinzhou** *var.* Chin-chou,
Chinchow; *prev.* Chinhsien.
Liaoning, NE China 41.07N 121.06E

169 *W11* **Jinz, Qā' al** ◎ C Jordan

49 *S8* **Jiparaná, Rio** ← W Brazil

58 *A7* **Jipijapa** Manabí, W Ecuador
1.22S 80.34W

44 *F8* **Jiquilisco** Usulután, S El Salvador
13.19N 88.34W

153 *S12* **Jirgatol** *Rus.* Dzhirgatal'.
C Tajikistan 39.13N 71.09E

113 *B15* **Jirkov** *Ger.* Görkau. Ústecký Kraj,
NW Czech Republic 50.30N 13.28E

149 *Q9* **Jiroft** *var.* Sabzvārān.

166 *L11* **Jishou** Hunan, S China
28.20N 109.43E

166 *M7* **Jisr ash Shadadi** *see* Ash
Shaddādah

---

166 *K13* **Jiuwan Dashan** ▲ S China

154 *I16* **Jiwani** Baluchistan, SW Pakistan
25.05N 61.46E

169 *Y8* **Jixi** Heilongjiang, NE China
45.16N 131.01E

169 *Y7* **Jixian** Heilongjiang, NE China

166 *M5* **Jixian** *var.* Ji Xian. Shanxi,
C China 36.15N 110.41E

Jiza *see* Al Jīzah

147 *N13* **Jīzān** *var.* Qīzān. Jīzān, SW Saudi
Arabia 17.49N 42.49E

147 *N13* **Jīzān** *var.* Minṭaqat Jīzān. ◆
province SW Saudi Arabia

146 *K6* **Jizl, Wādī al** *dry watercourse*
W Saudi Arabia

170 *FJ12* **Jizō-zaki** *headland* Honshū,
SW Japan 35.34N 133.16E

147 *U14* **Jiz', Wādī al** *dry watercourse*
E Yemen

153 *O11* **Jizzakh** *Rus.* Dzhizak. Jizzakh
Wiloyati, C Uzbekistan
40.07N 67.47E

153 *N10* **Jizzakh Wiloyati** *Rus.*
Dzhizakskaya Oblast'. ◆ *province*
C Uzbekistan

62 *I13* **Joaçaba** Santa Catarina, S Brazil
27.08S 51.30W

78 *F11* **Joal** *see* Joal-Fadiout

78 *F11* **Joal-Fadiout** *var.* Joal.
W Senegal 14.16N 16.51W

78 *E10* **João Barrosa** Boa Vista, E Cape
Verde 16.01N 22.44W

João Belo *see* Xai-Xai

João de Almeida *see* Chibia

61 *Q15* **João Pessoa** *prev.* Paraíba. *state*
capital Paraíba, E Brazil 7.06S 34.52W

45 *X7* **Joaquín** Texas, SW USA
31.58N 94.03W

64 *K6* **Joaquín V.González** Salta,
N Argentina 25.03S 64.06W

Joazeiro *see* Juazeiro

Jo'burg *see* Johannesburg

111 *O7* **Jochberger Ache** ← W Austria

120 *G13* **Jonava** *Ger.* Janow, *Pol.* Janów.
E Lithuania 55.04N 24.19E

152 *L11* **Jondor** *Rus.* Zhondor. Bukhoro
Wiloyati, C Uzbekistan
39.46N 64.11E

165 *V11* **Jonê** Gansu, C China 34.36N 103.39E

29 *X9* **Jonesboro** Arkansas, C USA
35.50N 90.42W

25 *S4* **Jonesboro** Georgia, SE USA
33.31N 84.21W

32 *L17* **Jonesboro** Illinois, N USA
37.25N 89.19W

24 *H5* **Jonesboro** Louisiana, S USA
32.14N 92.43W

23 *P8* **Jonesboro** Tennessee, S USA

21 *T6* **Jonesport** Maine, NE USA
44.33N 67.35W

(0) *J4* **Jones Sound** *channel* Nunavut,
N Canada

24 *I6* **Jonesville** Louisiana, S USA
31.37N 91.49W

33 *Q10* **Jonesville** Michigan, N USA
41.58N 84.39W

23 *Q11* **Jonesville** South Carolina,
SE USA 34.49N 81.38W

83 *F14* **Jongli** Jonglei, SE Sudan
6.54N 31.19E

83 *F14* **Jonglei** *var.* Gongoleh State. ◆
state SE Sudan

81 *E18* **Jong Falls** *waterfall* Kiritimati,
W India 14.16N 74.44E

161 *E18* **Jog Falls** *waterfall* Karnātaka,
W India 14.16N 74.44E

149 *S4* **Joghatāy** Khorāsān, NE Iran
36.34N 57.01E

159 *U12* **Jogighopa** Assam, NE India
26.13N 90.34E

158 *I7* **Jogindarnagar** Himāchal
Pradesh, N India 31.55N 76.55E

Jogjakarta *see* Yogyakarta

171 *Jj13* **Jōhana** Toyama, Honshū,
SW Japan 36.30N 136.53E

97 *I22* **Jöhana** *see* Yogyakarta

120 *I5* **Jõgeva** *Ger.* Laisholm. Jõgevamaa,
E Estonia 58.46N 26.23E

120 *I4* **Jõgevamaa** *off.* Jõgeva Maakond.
◆ *province* E Estonia

85 *J21* **Johannesburg** *var.* Egoli,
Erautini, Gauteng, *abbrev.* Jo'burg.
Gauteng, NE South Africa
26.10S 28.01E

166 *J11* **Johannesburg** California,
W USA 35.20N 117.37W

85 *J21* **Johannesburg** × Gauteng,
NE South Africa 26.08S 28.01E

155 *P14* **Johi** Sind, SE Pakistan 26.46N 67.28E

57 *T13* **Johi Village** S Guyana
1.48N 58.33W

34 *K13* **John Day** Oregon, NW USA
44.25N 118.57W

34 *J11* **John Day River** ← Oregon,
NW USA

144 *H12* **John F Kennedy** × (New York)
Long Island, New York, NE USA
40.39N 73.45W

169 *U14* **Jinzhou** *prev.* Jinxian. Liaoning,
NE China 39.04N 121.45E

23 *V8* **John H.Kerr Reservoir** *var.*
Buggs Island, Kerr Lake.
❒ North Carolina/Virginia,
SE USA

9 *V8* **John Martin Reservoir**
❒ Colorado, C USA

96 *K6* **John o'Groats** N Scotland, UK
58.37N 3.03W

29 *P5* **John Redmond Reservoir**
❒ Kansas, C USA

39 *O4* **John River** ← Alaska, USA

26 *H6* **Johnson** Kansas, C USA
37.33N 101.46W

21 *M7* **Johnson** Vermont, NE USA
44.39N 72.40W

20 *H11* **Johnson City** New York, NE USA
42.06N 75.54W

23 *P8* **Johnson City** Tennessee, S USA
36.18N 82.21W

45 *R10* **Johnson City** Texas, SW USA
30.16N 98.24W

23 *S12* **Johnsondale** California, W USA
35.58N 118.32W

Johnsons Crossing Yukon
Territory, W Canada 60.30N 133.15W

23 *T13* **Johnsonville** South Carolina,
SE USA 33.50N 79.26W

20 *D13* **Johnsonburg** Pennsylvania,
NE USA 41.28N 78.47W

199 *K6* **Johnston Atoll** ◇ *US
unincorporated territory*
C Pacific Ocean

199 *K6* **Johnston Atoll** *atoll*
C Pacific Ocean

32 *L17* **Johnston City** Illinois, N USA
37.49N 88.55W

---

188 *K12* **Johnston, Lake** *salt lake*
Western Australia

33 *S13* **Johnstown** Ohio, N USA
40.08N 82.39W

20 *D15* **Johnstown** Pennsylvania,
NE USA 40.19N 78.55W

174 *Hh6* **Johor** *var.* Johore. ◆ *state*
Peninsular Malaysia

Johor Baharu *see* Johor Bahru

174 *I6* **Johor Bahru** *var.* Johor Baharu,
Johore Bahru. Johor, Peninsular
Malaysia 1.28N 103.45E

Johore *see* Johor

Johore Bahru *see* Johor Bahru

120 *K3* **Jõhvi** *Ger.* Jewe. Ida-Virumaa,
NE Estonia 59.21N 27.25E

105 *P7* **Joigny** Yonne, C France
47.59N 3.24E

62 *K12* **Joinville** *var.* Joinvile. Santa
Catarina, S Brazil 26.19S 48.55W

105 *R6* **Joinville** Haute-Marne, N France
48.26N 5.07E

204 *H3* **Joinville Island** *island* Antarctica

43 *O15* **Jojutla** *var.* Jojutla de Juárez.
Morelos, C Mexico 18.36N 99.11W

94 *I12* **Jokkmokk** Norrbotten, N Sweden
66.35N 19.56E

94 *L2* **Jökulsá á Dal** ← E Iceland

94 *K2* **Jökulsá á Fjöllum** ←
NE Iceland

Jokyakarta *see* Yogyakarta

32 *M11* **Joliet** Illinois, N USA 41.33N 88.04W

13 *O11* **Joliette** Quebec, SE Canada
46.02N 73.27W

179 *Pp17* **Jolo** Jolo Island, SW Philippines
6.02N 121.00E

179 *Pp17* **Jolo Island** *island* SW Philippines

174 *Ll15* **Jombang** *prev.* Djombang. Jawa,
S Indonesia 7.33S 112.13E

165 *R14* **Jomda** Xizang Zizhiqu, W China
31.26N 98.09E

58 *A6* **Jome, Punta de** *headland*
W Ecuador 0.57S 80.49W

111 *O7* **Jochberger Ache** ← W Austria

94 *K12* **Jock** Norrbotten, N Sweden

44 *I5* **Jocón** Yoro, N Honduras
15.17N 86.55W

107 *O13* **Jódar** Andalucía, S Spain
37.51N 3.18W

158 *F12* **Jodhpur** Rājasthān, NW India
26.16N 73.01E

101 *I19* **Jodoigne** Wallon Brabant,
C Belgium 50.43N 4.52E

97 *I22* **Jægerspris** Frederiksborg,
E Denmark 55.52N 11.58E

95 *O16* **Joensuu** Itä-Suomi, E Finland
62.36N 29.45E

39 *C17* **Jæren** *physical region* S Norway
39.36N 102.40W

94 *I13* **Joes** Colorado, C USA
39.36N 102.40W

203 *Z3* **Joe's Hill** *hill* Kiritimati,
NE Kiribati 1.48N 157.19W

171 *Ji13* **Jōetsu** *var.* Zyoetu. Niigata,
Honshū, C Japan 37.09N 138.13E

85 *M18* **Jofane** Inhambane,
S Mozambique 21.16S 34.12E

159 *N13* **Jogbani** Bihār, NE India
26.21N 87.16E

83 *F14* **Jongli** Jonglei, SE Sudan
6.54N 31.19E

159 *X12* **Jorhāt** Assam, NE India
26.45N 94.09E

95 *J14* **Jörn** Västerbotten, N Sweden
65.02N 20.04E

39 *R14* **Jornada Del Muerto** *valley* New
Mexico, SW USA

95 *N17* **Joroinen** Isä-Suomi, E Finland
62.06N 25.54E

97 *C17* **Jørpeland** Rogaland, S Norway
59.01N 6.01E

79 *W14* **Jos** Plateau, C Nigeria 9.58N 8.57E

179 *Rr17* **Jose Abad Santos** *var.* Trinidad.
Mindanao, SW Philippines
5.51N 125.35E

63 *F19* **José Battle y Ordóñez** *var.*
Battle y Ordóñez. Florida,
C Uruguay 33.28S 55.07W

65 *H18* **José de San Martín** Chubut,
S Argentina 44.03S 70.27W

63 *E19* **José Enrique Rodó** *var.* Rodó,
F.E.Rodó; *prev.* Drabble, Drable.
Soriano, SW Uruguay 33.43S 57.33W

José E.Rodo *see* José Enrique
Rodó

Josefsdorf *see* Žabalj

46 *C4* **José Martí** × (La Habana)
Cuidad de La Habana, N Cuba
23.03N 82.22W

◆ COUNTRY   ◇ DEPENDENT TERRITORY   ◆ ADMINISTRATIVE REGION   ▲ MOUNTAIN   ▲ VOLCANO   ◎ LAKE
● COUNTRY CAPITAL   ○ DEPENDENT TERRITORY CAPITAL   × INTERNATIONAL AIRPORT   ▲ MOUNTAIN RANGE   ← RIVER   ❒ RESERVOIR

271

63 F19 **José Pedro Varela** var. José P.Varela. Lavalleja, S Uruguay 33.30S 54.28W

189 N2 **Joseph Bonaparte Gulf** gulf N Australia

39 N11 **Joseph City** Arizona, SW USA 34.56N 110.18W

11 O9 **Joseph, Lake** ◎ Newfoundland, E Canada

12 G13 **Joseph, Lake** ◎ Ontario, S Canada

194 I11 **Josephstaal** Madang, N PNG 4.42S 144.54E

José P.Varela see José Pedro Varela

61 J14 **José Rodrigues** Pará, N Brazil 5.45S 51.19W

158 K9 **Joshimath** Uttar Pradesh, N India 30.33N 79.34E

45 T7 **Joshua** Texas, SW USA 32.27N 97.23W

37 V15 **Joshua Tree** California, W USA 34.07N 116.18W

79 V14 **Jos Plateau** plateau C Nigeria

104 H6 **Josselin** Morbihan, NW France 47.57N 2.35W

Jos Sudarso, Pulau see Yos Sudarso, Pulau

96 E11 **Jostedalsbreen** glacier S Norway

96 F12 **Jotunheimen** ▲ S Norway

144 G7 **Joûnié** var. Juníyah. W Lebanon 33.54N 33.36E

45 R13 **Jourdanton** Texas, SW USA 28.55N 98.33W

100 L7 **Joure** Fris. De Jouwer. Friesland, N Netherlands 52.58N 5.48E

95 M18 **Joutsa** Länsi-Suomi, W Finland 61.46N 26.09E

95 N18 **Joutseno** Etelä-Suomi, S Finland 61.06N 28.30E

94 M12 **Joutsijärvi** Lappi, NE Finland 66.40N 28.00E

110 A9 **Joux, Lac de** ◎ W Switzerland

46 D5 **Jovellanos** Matanzas, W Cuba 22.49N 81.14W

159 V13 **Jowai** Meghálaya, NE India 25.25N 92.21E

Jöwat see Jabwot

Jowhar see Jawhar

149 O12 **Jowkän** Färs, S Iran

149 Q10 **Jowzam** Kermän, C Iran

155 N2 **Jowzjän** ◆ province N Afghanistan

Józseffalva see Žabalj

J.Storm Thurmond Reservoir see Clark Hill Lake

47 T6 **Juana Díaz** C Puerto Rico 18.03N 66.30W

42 L9 **Juan Aldama** Zacatecas, C Mexico 24.18N 103.23W

(0) E9 **Juan de Fuca Plate** tectonic feature

34 F7 **Juan de Fuca, Strait of** strait Canada/USA

Juan Fernandez Islands see Juan Fernández, Islas

200 Oo12 **Juan Fernández, Islas** Eng. Juan Fernandez Islands. island group W Chile

57 O4 **Juangriego** Nueva Esparta, NE Venezuela 11.03N 63.58W

58 D11 **Juanjuí** var. Juanjuy. San Martín, N Peru 7.12S 76.45W

Juanjuy see Juanjuí

95 N16 **Juankoski** Itä-Suomi, C Finland 63.01N 28.24E

Juan Lacaze see Juan L.Lacaze

63 F20 **Juan L.Lacaze** var. Juan Lacaze, Puerto Sauce; prev. Sauce. Colonia, SW Uruguay 34.25S 57.25W

64 L5 **Juan Solá** Salta, N Argentina 23.30S 62.42W

65 F21 **Juan Stuven, Isla** island S Chile

61 H16 **Juará** Mato Grosso, W Brazil 11.16S 57.28W

43 N7 **Juárez** var. Villa Juárez. Coahuila de Zaragoza, NE Mexico 27.39N 100.43W

42 C2 **Juárez, Sierra de** ▲ NW Mexico

61 O15 **Juazeiro** prev. Joazeiro. Bahia, E Brazil 9.25S 40.30W

61 P14 **Juazeiro do Norte** Ceará, E Brazil 7.10S 39.18W

83 F15 **Juba** var. Jūbā. Bahr el Gabel, S Sudan 4.49N 31.34E

83 L17 **Juba** Amh. Genalē Wenz, It. Guiba, Som. Ganaane, Webi Jubba. ♒ Ethiopia/Somalia

204 H2 **Jubany** Argentinian research station Antarctica 61.57S 58.23W

Jubayl see Jbail

83 L18 **Jubbada Dhexe** off. Gobolka Jubbada Dhexe. ◆ region SW Somalia

83 K18 **Jubbada Hoose** ◆ region SW Somalia

Jubba, Webi see Juba

Jubbulpore see Jabalpur

Jubeil see Jbail

76 B9 **Juby, Cap** headland SW Morocco 27.58N 12.56W

107 R10 **Júcar** var. Jucar. ♒ C Spain

42 L12 **Juchipila** Zacatecas, C Mexico 21.25N 103.06W

43 S16 **Juchitán** var. Juchitán de Zaragoza. Oaxaca, SE Mexico 16.27N 95.00W

Juchitán de Zaragoza see Juchitán

144 G11 **Judaea** cultural region Israel/West Bank

144 F11 **Judaean Hills** Heb. Harē Yehuda. hill range E Israel

144 H8 **Judayda'at Hämir** I Syria 31.29N 41.25E

111 U8 **Judenburg** Steiermark, C Austria 47.09N 14.42E

35 T8 **Judith River** ♒ Montana, NW USA

29 V11 **Judsonia** Arkansas, C USA 35.16N 91.38W

147 P14 **Jufrah, Wädi al** dry watercourse NW Yemen

Jugoslavija/Jugoslavija, Savezna Republika see Yugoslavia

44 N10 **Juigalpa** Chontales, S Nicaragua 12.04N 85.21W

167 T13 **Juishui** C Taiwan 23.43N 121.28E

102 E9 **Juist** island NW Germany

61 M21 **Juiz de Fora** Minas Gerais, SE Brazil 21.46S 43.22W

---

64 J5 **Jujuy** off. Provincia de Jujuy. ◆ province N Argentina

Jujuy see San Salvador de Jujuy

94 J11 **Jukkasjärvi** Norrbotten, N Sweden 67.52N 20.39E

Jula see Gyula, Hungary

39 W2 **Julesburg** Colorado, C USA 40.59N 102.15W

59 J17 **Juli** Puno, SE Peru 15.32S 70.10W

189 U6 **Julia Creek** Queensland, C Australia 20.40S 141.49E

37 V17 **Julian** California, W USA 33.04N 116.36W

100 H7 **Julianadorp** Noord-Holland, NW Netherlands 52.53N 4.43E

111 S11 **Julian Alps** Ger. Julische Alpen, It. Alpi Giulie, Slvn. Julijske Alpe. ▲ Italy/Slovenia

57 V11 **Juliana Top** ▲ C Suriname 3.39N 56.36W

Julianehåb see Qaqortoq

Julijske Alpe see Julian Alps

42 J6 **Julimes** Chihuahua, N Mexico 28.29N 105.21W

63 G15 **Júlio Briga** see Bragança, Portugal

Júlio de Castilhos Rio Grande do Sul, S Brazil 29.14S 53.42W

Juliomagus see Angers

Julische Alpen see Julian Alps

Jullundur see Jalandhar

153 N11 **Juma** Rus. Dzhuma. Samarqand Wiloyati, C Uzbekistan 39.46N 66.37E

167 O3 **Juma He** ♒ E China

83 L18 **Jumboo** Jubbada Hoose, S Somalia 0.12S 42.34E

37 Y11 **Jumbo Peak** ▲ Nevada, W USA 36.12N 114.09W

107 R12 **Jumilla** Murcia, SE Spain 38.28N 1.19W

159 N10 **Jumla** Mid Western, NW Nepal 29.18N 82.13E

Jummoo see Jammu

Jumna see Yamuna

Jumporn see Chumphon

32 K5 **Jump River** ♒ Wisconsin, N USA

160 B11 **Jūnāgadh** var. Junagarh. Gujarāt, W India 21.31N 70.31E

Junagarh see Jūnāgadh

167 Q6 **Junan** prev. Shizilu. Shandong, E China 35.11N 118.47E

64 G11 **Juncal, Cerro** ▲ C Chile 33.03S 70.02W

45 Q10 **Junction** Texas, SW USA 30.29N 99.46W

38 K6 **Junction** Utah, W USA 38.14N 112.13W

29 Q4 **Junction City** Kansas, C USA 39.01N 96.49W

34 F13 **Junction City** Oregon, NW USA 44.13N 123.12W

62 H10 **Jundiaí** São Paulo, S Brazil 23.10S 46.54W

194 E13 **June** ♒ W PNG

41 X12 **Juneau** state capital Alaska, USA 58.13N 134.11W

32 M8 **Juneau** Wisconsin, N USA 43.22N 88.42W

107 U6 **Juneda** Cataluña, NE Spain 41.33N 0.50E

191 Q9 **Junee** New South Wales, SE Australia 34.51S 147.33E

37 R8 **June Lake** California, W USA 37.46N 119.04W

Jungbunzlau see Mladá Boleslav

164 L4 **Junggar Pendi** Eng. Dzungarian Basin. basin NW China

101 N24 **Junglinster** Grevenmacher, C Luxembourg 49.43N 6.15E

20 F14 **Juniata River** ♒ Pennsylvania, NE USA

63 B20 **Junín** Buenos Aires, E Argentina 34.36S 61.01W

59 E14 **Junín** Junín, C Peru 11.13S 76.01W

59 E14 **Junín** off. Departamento de Junín. ◆ department C Peru

65 H15 **Junín de los Andes** Neuquén, W Argentina 39.57S 71.04W

59 D14 **Junín, Lago de** ◎ C Peru

Juniyah see Joûnié

171 I11 **Junkseylon** see Phuket

166 I11 **Junlian** Sichuan, C China 28.11N 104.31E

45 Q11 **Juno** Texas, SW USA 30.09N 101.07W

94 J11 **Junosuando** Norrbotten, N Sweden 67.25N 22.28E

95 H15 **Junsele** Västernorrland, C Sweden 63.42N 16.54E

34 L14 **Juntura** Oregon, NW USA 43.43N 118.05W

95 N14 **Juntusranta** Oulu, E Finland 65.12N 29.30E

120 H11 **Juodupė** Rokiškis, NE Lithuania

121 H14 **Juozapinės Kalnas** ▲ SE Lithuania 54.29N 25.27E

101 K19 **Juprelle** Liège, E Belgium 50.43N 5.31E

82 D13 **Jur** ♒ C Sudan

105 S9 **Jura** ◆ department E France

110 C7 **Jura** ◆ canton NW Switzerland

110 B8 **Jura** var. Jura Mountains. ▲ France/Switzerland

98 G12 **Jura** island SW Scotland, UK

98 G12 **Jura, Sound of** strait W Scotland, UK

145 V15 **Juraybiyät, Bi'r** well S Iraq 31.39N 45.28E

120 E13 **Jurbarkas** Ger. Georgenburg, Jurburg. Jurbarkas, W Lithuania 55.04N 22.45E

Jurburg see Jurbarkas

120 F9 **Jūrmala** Rīga, C Latvia 56.56N 23.42E

176 Ww13 **Jursian, Pulau** island E Indonesia

60 D13 **Juruá** Amazonas, NW Brazil 3.08S 65.59W

61 E14 **Juruá, Rio** var. Río Yuruá. ♒ Brazil/Peru

---

61 G16 **Juruena** Mato Grosso, W Brazil 10.32S 58.38W

61 G16 **Juruena** ♒ W Brazil

171 Mm8 **Jūsan-ko** ◎ Honshū, C Japan

45 O6 **Justiceburg** Texas, SW USA 32.57N 101.07W

Justinianopolis see Kırşehir

64 K11 **Justo Daract** San Luis, C Argentina 33.52S 65.12W

61 C14 **Jutaí** Amazonas, W Brazil 5.10S 68.45W

60 C13 **Jutaí, Rio** ♒ NW Brazil

102 N13 **Jüterbog** Brandenburg, E Germany 51.58N 13.06E

44 E6 **Jutiapa** Jutiapa, S Guatemala 14.18N 89.52W

44 E6 **Jutiapa** off. Departamento de Jutiapa. ◆ department SE Guatemala

44 J6 **Juticalpa** Olancho, C Honduras 14.39N 86.12W

84 I13 **Jutila** North Western, NW Zambia 12.35S 26.09E

86 F8 **Jutland Bank** undersea feature SE North Sea

Jutland see Jylland

95 N16 **Juuka** Itä-Suomi, E Finland 63.12N 29.16E

95 N17 **Juva** Itä-Suomi, SE Finland 61.55N 27.54E

Juvavum see Salzburg

45 A6 **Juventud, Isla de la** var. Isla de Pinos, Eng. Isle of Youth; prev. The Isle of the Pines. island W Cuba

Ju Xian see Juxian

167 Q5 **Juxian** var. Ju Xian. Shandong, E China 35.33N 118.45E

167 P6 **Juye** Shandong, E China 35.25N 116.04E

115 O15 **Južna Morava** Ger. Südliche Morava. ♒ SE Yugoslavia

97 I23 **Jyderup** Vestsjælland, E Denmark 55.40N 11.25E

97 F22 **Jylland** Eng. Jutland. peninsula W Denmark

95 M17 **Jyväskylä** Länsi-Suomi, W Finland 62.07N 25.47E

K

155 X3 **K2** Chin. Qogir Feng, Eng. Mount Godwin Austen. ▲ China/Pakistan 35.53N 76.35E

40 D9 **Kaaawa** Haw. Ka'a'awa. Oahu, Hawaii, USA, C Pacific Ocean 21.33N 157.51W

83 G16 **Kaabong** NE Uganda 3.30N 34.07E

57 V9 **Kaaden** see Kadaň

57 V9 **Kaaimanston** Sipaliwini, N Suriname 5.06N 56.04W

152 G14 **Kaakhka** var. Kaka. Akhalskiy Velayat, S Turkmenistan 37.19N 59.36E

Kaala see Caála

197 H5 **Kaala-Gomen** Province Nord, W New Caledonia 20.40S 164.24E

94 L9 **Kaamanen** Lapp. Gámas. Lappi, N Finland 69.04N 27.16E

Kaapstad see Cape Town

Kaarasjoki see Karasjok

94 J10 **Kaaresuanto** Lapp. Gárassavon. Lappi, N Finland 68.28N 22.29E

95 K19 **Kaarina** Länsi-Suomi, W Finland 60.24N 22.25E

101 I14 **Kaatsheuvel** Noord-Brabant, S Netherlands 51.39N 5.01E

95 N16 **Kaavi** Itä-Suomi, C Finland 62.58N 28.30E

Ka'a'wa see Kaaawa

78 M14 **Kaba** Irian Jaya, E Indonesia 7.34S 138.27E

Kaba see Habahe

175 Q13 **Kabaena, Pulau** island C Indonesia

175 Q13 **Kabaena, Selat** strait Sulawesi, C Indonesia

152 J11 **Kabakly** Turkm. Gabakly. Lebapskiy Velayat, NE Turkmenistan 39.45N 62.30E

78 I14 **Kabala** N Sierra Leone 9.40N 11.36W

83 E19 **Kabale** SW Uganda 1.15S 29.58E

57 U10 **Kabalebo Rivier** ♒ W Suriname

81 M22 **Kabalo** Katanga, SE Dem. Rep. Congo (Zaire) 6.02S 26.55E

151 W13 **Kabanbay** Kaz. Qabanbay prev. Andreyevka, Kaz. Andreevka. Almaty, SE Kazakhstan 45.49N 80.34E

81 O21 **Kabambare** Maniema, E Dem. Rep. Congo (Zaire) 2.10S 27.40E

197 K15 **Kabara** prev. Kambara. island Lau Group, E Fiji

Kabardino-Balkaria see Kabardino-Balkarskaya Respublika

130 M15 **Kabardino-Balkarskaya Respublika** Eng. Kabardino-Balkaria. ◆ autonomous republic SW Russian Federation

81 O19 **Kabare** Sud Kivu, E Dem. Rep. Congo (Zaire) 2.13S 28.40E

176 Uu8 **Kabarei** Irian Jaya, E Indonesia 0.01S 130.58E

179 Q16 **Kabasalan** Mindanao, S Philippines 7.46N 122.49E

79 U15 **Kabba** Kogi, S Nigeria 7.48N 6.07E

94 I13 **Kåbdalis** Norrbotten, N Sweden

144 M6 **Kabd as Sārim** hill range E Syria

12 B7 **Kaboni** Ontario, S Canada

192 J2 **Kaeo** Northland, North Island, NZ 35.03S 173.40E

169 X14 **Kaesŏng** var. Kaesŏng-si. N North Korea 37.57N 126.30E

Kaesŏng-si see Kaesŏng

169 X14 **Kabin, Pulau** var. Pulau Kabia. island W Indonesia

175 P13 **Kabin, Pulau** var. Pulau Kabia. island W Indonesia

175 Rr16 **Kabir Pulau Pantar, S Indonesia 8.15S 124.12E

155 T10 **Kabīrwäla** Punjab, E Pakistan 30.24N 71.51E

176 U8 **Kable Bet** Irian Jaya, E Indonesia 0.24S 129.54E

80 I13 **Kabo** Ouham, NW Central African Republic 7.40N 18.38E

Kăbol see Kåbul

61 A14 **Kaburé, Rio** var. Río Yuruá. ♒ Brazil/Peru

---

85 H14 **Kabompo** ♒ W Zambia

81 M22 **Kabongo** Katanga, S Dem. Rep. Congo (Zaire) 7.19S 25.34E

123 Kk12 **Kaboudia, Rass** headland E Tunisia 35.13N 11.10E

128 J14 **Kabozha** Novgorodskaya Oblast', W Russian Federation 58.48N 35.00E

149 U4 **Kabūd Gonbad** Khorāsān, NE Iran 37.01N 59.46E

148 L5 **Kabūd Rāhang** Hamadān, W Iran 35.12N 48.43E

84 L12 **Kabuko** Northern, NE Zambia

155 Q5 **Kābul** var. Kabul, Per. Kābol. ● (Afghanistan) Kābul, E Afghanistan 34.34N 69.07E

155 Q5 **Kābul** Eng. Kabul, Per. Kābol. ◆ province E Afghanistan

155 Q5 **Kābul** ♒ Kābul, E Afghanistan 34.31N 69.10E

155 R5 **Kābul, Daryā-ye** see Kābul, Daryā-ye

155 S5 **Kābul, Daryā-ye** var. Kabul. ♒ Afghanistan/Pakistan see also Kabul

81 O25 **Kabunda** Katanga, SE Dem. Rep. Congo (Zaire) 12.21S 29.14E

175 Ss4 **Kaburuang, Pulau** island Kepulauan Talaud, N Indonesia

82 G8 **Kabushiya** River Nile, NE Sudan 16.54N 33.40E

83 E19 **Kabwe** Central, C Zambia 14.28S 28.25E

194 K12 **Kabwum** Morobe, C PNG 6.07S 147.11E

115 N17 **Kačanik** Serbia, S Yugoslavia 42.13N 21.16E

120 F13 **Kacergine** Kaunas, C Lithuania 54.55N 23.40E

119 S13 **Kacha** Respublika Krym, S Ukraine 44.46N 33.33E

160 A10 **Kachchh, Gulf of** var. Gulf of Cutch, Gulf of Kutch. gulf W India

160 I11 **Kachchhidhāna** Madhya Pradesh, C India 21.33N 78.54E

155 Q11 **Kachchh, Rann of** var. Rann of Kachh, Rann of Kutch. salt marsh India/Pakistan

41 Q13 **Kachemak Bay** bay Alaska, USA

79 V14 **Kachia** Kaduna, C Nigeria 9.52N 8.00E

166 K6 **Kachin State** ◆ state N Myanmar

151 T7 **Kachiry** Pavlodar, NE Kazakhstan 53.04N 76.05E

126 Ji15 **Kachug** Irkutskaya Oblast', S Russian Federation 53.52N 105.54E

143 Q11 **Kaçkar Dağları** ▲ NE Turkey

161 C21 **Kadamatt Island** island Lakshadweep, India, N Indian Ocean

113 B15 **Kadaň** Ger. Kaaden. Ústecký Kraj, NW Czech Republic 50.22N 13.14E

178 Gg12 **Kadan Kyun** prev. King Island. island Mergui Archipelago, S Myanmar

197 I16 **Kadavu** prev. Kandavu. island S Fiji

197 I15 **Kadavu Passage** channel S Fiji

81 G16 **Kadéï** ♒ Cameroon/Central African Republic

Kadhimain see Al Kāẓimīyah

116 M13 **Kadıköy Baraji** ⚒ NW Turkey

142 H15 **Kadınhanı** Konya, C Turkey 38.15N 32.13E

78 M14 **Kadiolo** Sikasso, S Mali 10.30N 5.43W

142 L16 **Kadirli** Osmaniye, S Turkey 37.22N 36.64E

30 L10 **Kadoka** South Dakota, N USA 43.49N 101.30W

131 N5 **Kadom** Ryazanskaya Oblast', W Russian Federation 54.35N 42.27E

85 K16 **Kadoma** prev. Gatooma. Mashonaland West, C Zimbabwe 18.18S 29.55E

82 E12 **Kadugli** Southern Kordofan, S Sudan 11.00N 29.44E

79 V14 **Kaduna** Kaduna, C Nigeria 10.32N 7.25E

79 V13 **Kaduna** ◆ state C Nigeria

79 V15 **Kaduna** ♒ C Nigeria

128 I14 **Kaduy** Vologodskaya Oblast', NW Russian Federation 59.10N 37.11E

118 M12 **Kahul, Ozero** var. Lacul Cahul, Rus. Ozero Kagul. ◎ Moldova/Ukraine

160 E13 **Kadwa** ♒ W India

127 Nn9 **Kadykchan** Magadanskaya Oblast', E Russian Federation 62.54N 146.53E

127 T7 **Kadzherom** Respublika Komi, NW Russian Federation 64.42N 55.51E

153 X8 **Kadzhi-Say** Kir. Kajisay. Issyk-Kul'skaya Oblast', NE Kyrgyzstan 42.07N 77.10E

81 I10 **Kaédi** Gorgol, S Mauritania 16.12N 13.31W

80 G12 **Kaélé** Extrême-Nord, N Cameroon 10.09N 14.25E

40 C9 **Kaena Point** headland Oahu, Hawaii, USA, C Pacific Ocean 21.34N 158.16W

192 H2 **Kaeo** Northland, North Island, NZ 35.03S 173.40E

169 X14 **Kaesŏng** var. Kaesŏng-si. N North Korea 37.57N 126.30E

164 K6 **Kaesŏng-si** see Kaesŏng

164 K6 **Kāf** NW China

79 U14 **Kafanchan** Kaduna, C Nigeria 9.32N 8.18E

176 V13 **Kai Kecil, Pulau** island Kepulauan Kai, E Indonesia 5.35S 132.39E

78 G11 **Kaffrine** C Senegal 14.07N 15.27W

176 V14 **Kai, Kepulauan** prev. Kei Islands. island group Maluku, SE Indonesia

117 I19 **Kafiréas, Akrotírio** headland Évvoia, C Greece 38.04N 24.32E

117 I19 **Kafiréos, Stenó** strait Évvoia/Kykládes, C Greece, Aegean Sea

---

193 J16 **Kaikoura Peninsula** peninsula South Island, NZ

Kailas Range see Gangdisê Shan

166 K12 **Kaili** Guizhou, S China 26.34N 107.58E

40 F10 **Kailua** Maui, Hawaii, USA, C Pacific Ocean 20.53N 156.13W

40 G11 **Kailua** var. Kailua-Kona. Kona, Hawaii, USA, C Pacific Ocean 19.43N 155.58W

Kailua-Kona see Kailua

194 E13 **Kaim** ♒ W PNG

176 Y13 **Kaima** Irian Jaya, E Indonesia 5.36S 138.39E

192 M7 **Kaimai Range** ▲ North Island, NZ

193 E18 **Kaimana** Irian Jaya, E Indonesia 3.39S 133.44E

117 E14 **Kaïmaktsalán** var. Kajmakčalan. ▲ Greece/FYR Macedonia 40.57N 21.48E

193 C20 **Kaimanawa Mountains** ▲ North Island, NZ

120 E4 **Kaina** Ger. Keinis; prev. Keina. Hiiumaa, W Estonia 58.49N 22.45E

111 V7 **Kainach** ♒ SE Austria

170 G16 **Kainan** Tokushima, Shikoku, SW Japan 33.36N 134.20E

170 Fj16 **Kainan** Wakayama, Honshū, SW Japan 34.10N 135.11E

194 J12 **Kainantu** Eastern Highlands, C PNG 6.16S 145.49E

153 U7 **Kaindy** Kir. Kayyngdy. Chuyskaya Oblast', N Kyrgyzstan 42.48N 73.39E

79 T14 **Kainji Dam** dam W Nigeria 9.52N 4.36E

79 T14 **Kainji Lake** see Kainji Reservoir

79 T14 **Kainji Reservoir** var. Kainji Lake. ⚒ W Nigeria

194 J14 **Kaintiba** var. Kamina. Gulf, S PNG 7.29S 146.04E

94 K12 **Kainulaisjärvi** Norrbotten, N Sweden 67.00N 22.13E

192 K5 **Kaipara Harbour** harbor North Island, NZ

158 I10 **Kairāna** Uttar Pradesh, N India 29.24N 77.10E

194 G9 **Kairiru Island** island NW PNG

76 M6 **Kairouan** var. Al Qayrawān. E Tunisia 35.45N 10.11E

103 F20 **Kaiserslautern** Rheinland-Pfalz, SW Germany 49.27N 7.46E

120 G13 **Kaišiadorys** Kaišiadorys, S Lithuania 54.51N 24.27E

192 I2 **Kaitaia** Northland, North Island, NZ 35.07S 173.13E

193 E24 **Kaitangata** Otago, South Island, NZ 46.15S 169.49E

158 I9 **Kaithal** Haryāna, NW India 29.46N 76.26E

174 J11 **Kait, Tanjung** headland Sumatera, W Indonesia 3.13S 106.03E

40 G12 **Kaiwi Channel** channel Hawaii, USA, C Pacific Ocean

166 K9 **Kaixian** var. Kai Xian. Sichuan, C China 31.13N 108.25E

169 V11 **Kaiyuan** var. K'ai-yüan. Liaoning, NE China 42.36N 124.03E

166 H14 **Kaiyuan** Yunnan, SW China 23.42N 103.13E

41 O9 **Kaiyuh Mountains** ▲ Alaska, USA

95 M15 **Kajaani** Swe. Kajana. Oulu, C Finland 64.16N 27.46E

155 N7 **Kajaki, Band-e** ⚒ S Afghanistan

155 N5 **Kajan** see Kayan, Sungai

119 S11 **Kajang** Selangor, Peninsular Malaysia 2.59N 101.48E

175 Pp15 **Kajang, Pulau** island Kepulauan Bonerate, S Indonesia

175 Q15 **Kajaotoa, Pulau** island W Indonesia

161 I24 **Kala Oya** ♒ NW Sri Lanka

Kalarash see Călăraşi

149 I9 **Kālårne** Jämtland, C Sweden 63.00N 16.10E

149 V15 **Kalar Rüd** ♒ SE Iran

118 I9 **Kalasin** var. Muang Kalasin. Kalasin, E Thailand 16.28N 103.31E

155 O8 **Kalåt** Per. Qalåt. Zābul, S Afghanistan 32.10N 66.54E

10 C12 **Kalabaka Falls** Ontario, S Canada 48.24N 89.40W

85 F23 **Kakamas** Northern Cape, W South Africa 28.45S 20.33E

83 H18 **Kakamega** Western, W Kenya 0.13N 34.43E

114 H13 **Kakanj** Federacija Bosna I Hercegovina, Bosnia and Herzegovina 44.06N 18.07E

193 F22 **Kakanui Mountains** ▲ South Island, NZ

192 K11 **Kakaramea** Taranaki, North Island, NZ 39.42S 174.27E

78 J16 **Kakata** C Liberia 6.34N 10.19W

192 M11 **Kakatahi** Manawatu-Wanganui, North Island, NZ 39.40S 175.20E

115 M23 **Kakavi** Gjirokastër, S Albania 39.55N 20.19E

153 O14 **Kakaydi** Surkhondaryo Wiloyati, S Uzbekistan 37.37N 67.30E

170 Ee13 **Kake** Hiroshima, Honshū, SW Japan 34.36N 132.17E

41 X13 **Kake** Kupreanof Island, Alaska, USA 56.48N 133.57E

172 Qq13 **Kakeromajima** Kagoshima, SW Japan

149 T6 **Kākhak** var. Kåkhk. Khorāsān, E Iran

120 L11 **Kakhanovichy** Rus. Kokhanovichi. Vitsyebskaya Voblasts', N Belarus 55.57N 28.06E

119 S10 **Kakhovka** Khersons'ka Oblast', S Ukraine 46.48N 33.30E

119 U9 **Kakhovs'ka Vodoskhovyshche** Rus. Kakhovskoye Vodokhranilishche. ⚒ SE Ukraine

119 U9 **Kakhovskoye Vodokhranilishche** see Kakhovs'ka Vodoskhovyshche

119 T11 **Kakhovs'kyy Kanal** canal S Ukraine

119 S12 **Kakia** see Khakhea

161 L16 **Kakināda** prev. Cocanada. Andhra Pradesh, E India 16.55N 82.13E

116 O8 **Kaliakra, Nos** headland NE Bulgaria 43.22N 28.28E

---

193 J16 **Kakshaal-Too, Khrebet** see Kokshaal-Tau

41 S5 **Kaktovik** Alaska, USA 70.07N 143.37W

171 Ll13 **Kakuda** Miyagi, Honshū, C Japan 37.59N 140.47E

171 M11 **Kakunodate** Akita, Honshū, C Japan 39.36N 140.38E

155 T7 **Kalaallit Nunaat** see Greenland

175 Rr16 **Kalabahi** Pulau Alor, S Indonesia 8.13S 124.31E

196 I5 **Kalabera** Saipan, S Northern Mariana Islands

85 G14 **Kalabo** Western, W Zambia 14.52S 22.33E

130 M9 **Kalach** Voronezhskaya Oblast', W Russian Federation 50.24N 41.00E

125 G13 **Kalachinsk** Omskaya Oblast', C Russian Federation 55.03N 74.30E

131 N10 **Kalach-na-Donu** Volgogradskaya Oblast', SW Russian Federation 48.45N 43.29E

177 F5 **Kaladan** ♒ W Myanmar

81 T18 **Kaladar** Ontario, SE Canada 44.37N 77.06W

40 G13 **Ka Lae** var. South Cape, South Point. headland Hawaii, USA, C Pacific Ocean 18.54N 155.40W

85 G19 **Kalahari Desert** desert Southern Africa

40 B8 **Kalaheo** Haw. Kalāheo. Kauai, Hawaii, USA, C Pacific Ocean 21.55N 159.31W

Kalaï-Mor see Qal'aikhum

152 J11 **Kala-i-Mor** Turkm. Galaymor. Maryyskiy Velayat, S Turkmenistan 35.40N 62.28E

95 K15 **Kalajoki** Oulu, W Finland 64.15N 24.00E

Kalak see Eski Kaļak

Kal al Sraghna see El Kelâa Srarhna

34 G10 **Kalama** Washington, NW USA 46.00N 122.50W

117 D20 **Kalámai** see Kalámata

117 G14 **Kalamariá** Kentrikí Makedonía, N Greece 40.36N 22.58E

117 E21 **Kalámata** prev. Kalámai. Peloponnísos, S Greece 37.01N 22.07E

33 P10 **Kalamazoo** Michigan, N USA 42.17N 85.35W

33 P9 **Kalamazoo River** ♒ Michigan, N USA

119 S13 **Kalamitska Zatoka** Rus. Kalamitskiy Zaliv. gulf S Ukraine

119 S13 **Kalamitskiy Zaliv** see Kalamitska Zatoka

117 H18 **Kálamos** Attikí, C Greece 38.16N 23.51E

117 C18 **Kálamos** island Iónioi Nísoi, Greece, C Mediterranean Sea

117 D15 **Kalampáka** var. Kalambaka. Thessalía, C Greece 39.43N 21.36E

Kalan see Călan, Romania

Kalan see Tunceli, Turkey

119 S11 **Kalanchak** Khersons'ka Oblast', S Ukraine 46.14N 33.19E

175 Pp15 **Kalao, Pulau** island Kepulauan Bonerate, S Indonesia

175 Q15 **Kalaotoa, Pulau** island W Indonesia

161 I24 **Kala Oya** ♒ NW Sri Lanka

95 H17 **Kälarne** Jämtland, C Sweden 63.00N 16.10E

149 V15 **Kalar Rüd** ♒ SE Iran

118 I9 **Kalasin** var. Muang Kalasin. Kalasin, E Thailand 16.28N 103.31E

155 O8 **Kalåt** Per. Qalåt. Zābul, S Afghanistan 32.10N 66.54E

155 O11 **Kalåt** var. Kelat, Khelat. Baluchistān, SW Pakistan 29.02N 66.34E

117 J14 **Kalathriá, Akrotírio** headland Samothráki, NE Greece 40.24N 25.34E

117 G14 **Kalávrita** see Kalávryta

117 E18 **Kalávryta** var. Kalávrita. Dytikí Ellás, S Greece 38.01N 22.06E

147 V10 **Kalbān** W Oman 20.19N 58.40E

188 H1 **Kalbarri** Western Australia 27.43S 114.08E

151 X10 **Kalbinskiy Khrebet** Kaz. Qalba Zhotasy. ▲ E Kazakhstan

150 I7 **Kaldygayty** ♒ W Kazakhstan

142 I12 **Kalecik** Ankara, N Turkey 40.08N 33.24E

81 O19 **Kalehe** Sud Kivu, E Dem. Rep. Congo (Zaire) 2.04S 28.52E

127 N24 **Kalí Límni** ▲ Kárpathos, SE Greece 35.34N 27.08E

◆ COUNTRY   ◇ DEPENDENT TERRITORY   ◆ ADMINISTRATIVE REGION   ▲ MOUNTAIN   ⛰ VOLCANO   ◎ LAKE
● COUNTRY CAPITAL   ○ DEPENDENT TERRITORY CAPITAL   ✕ INTERNATIONAL AIRPORT   ▲ MOUNTAIN RANGE   ♒ RIVER   ⚒ RESERVOIR

81 N20 **Kalima** Maniema, E Dem. Rep. Congo (Zaire) 2.33S 26.27E

174 M8 **Kalimantan** Eng. Indonesian Borneo. *geopolitical region* Borneo, C Indonesia

174 L8 **Kalimantan Barat** off. Propinsi Kalimantan Barat, Eng. West Borneo, West Kalimantan. ◆ *province* N Indonesia

174 Mm11 **Kalimantan Selatan** off. Propinsi Kalimantan Selatan, Eng. South Borneo, South Kalimantan. ◆ *province* N Indonesia

174 M9 **Kalimantan Tengah** off. Propinsi Kalimantan Tengah, Eng. Central Borneo, Central Kalimantan. ◆ *province* N Indonesia

175 N7 **Kalimantan Timur** off. Propinsi Kalimantan Timur, Eng. East Borneo, East Kalimantan. ◆ *province* N Indonesia

Kálimnos see Kálymnos

159 S12 **Kálimpang** West Bengal, N India 27.05N 88.25E

Kalinin see Tver', Russian Federation

Kalinin see Boldumsaz, Turkmenistan

Kalininabad see Kalininobod

130 B3 **Kaliningrad** Kaliningradskaya Oblast', W Russian Federation 54.48N 21.33E

Kaliningrad see Kaliningradskaya Oblast'

130 A3 **Kaliningradskaya Oblast'** var. Kaliningrad. ◆ *province and enclave* W Russian Federation

Kalinino see Tashir

153 P14 **Kalininobod** Rus. Kalininabad. SW Tajikistan 37.49N 68.55E

131 08 **Kalininsk** Saratovskaya Oblast', W Russian Federation 51.31N 44.25E

Kalininsk see Boldumsaz

121 M19 **Kalinkavichy** Rus. Kalinkovichi. Homyel'skaya Voblasts', SE Belarus 52.07N 29.19E

Kalinkovichi see Kalinkavichy

83 G18 **Kaliro** SE Uganda 0.54N 33.30E

Kalisch/Kalish see Kalisz

35 O7 **Kalispell** Montana, NW USA 48.12N 114.18W

112 I13 **Kalisz** Ger. Kalisch, Rus. Kalish; anc. Calisia. Wielkopolskie, C Poland 51.46N 18.04E

112 F9 **Kalisz Pomorski** Ger. Kallies. Zachodniopomorskie, NW Poland 53.55N 15.55E

130 M10 **Kalitva** ♂ SW Russian Federation

83 F21 **Kaliua** Tabora, C Tanzania 5.03S 31.48E

94 K13 **Kalix** Norrbotten, N Sweden 65.51N 23.13E

94 K12 **Kalixälven** ♂ N Sweden

94 J11 **Kalixfors** Norrbotten, N Sweden 67.45N 20.20E

151 T8 **Kalkaman** Pavlodar, NE Kazakhstan 51.57N 75.58E

Kalkandelen see Tetovo

189 O4 **Kalkarindji** Northern Territory, N Australia 17.31S 130.40E

33 P6 **Kaska** Michigan, N USA 44.43N 85.12W

95 F16 **Kall** Jämtland, C Sweden 63.30N 13.16E

201 X2 **Kallalen** var. Calalen. *island* Ratak Chain, SE Marshall Islands

120 J5 **Kallaste** Ger. Krasnogor. Tartumaa, SE Estonia 58.37N 27.12E

95 N16 **Kallavesi** ◎ SE Finland

117 F17 **Kallidromo** ▲ C Greece

Kallies see Kalisz Pomorski

97 M22 **Kallinge** Blekinge, S Sweden 56.13N 15.16E

117 L16 **Kalloní** Lésvos, E Greece 39.14N 26.15E

95 F16 **Kallsjön** ◎ C Sweden

97 N21 **Kalmar** var. Calmar. Kalmar, S Sweden 56.40N 16.22E

97 M19 **Kalmar** var. Calmar. ◆ *county* S Sweden

97 N21 **Kalmarsund** *strait* S Sweden

154 L16 **Kalmat, Khor** Eng. Kalmat Lagoon. *lagoon* SW Pakistan

Kalmat Lagoon see Kalmat, Khor

119 X9 **Kal'mius** ♂ E Ukraine

101 H15 **Kalmthout** Antwerpen, N Belgium 51.24N 4.27E

Kalmykia/Kalmykiya-Khal'mg Tangch, Respublika see Kalmykiya, Respublika

131 O12 **Kalmykiya, Respublika** var. Respublika Kalmykiya-Khal'mg Tangch, Eng. Kalmykia; prev. Kalmytskaya ASSR. ◆ *autonomous republic* SW Russian Federation

Kalmytskaya ASSR see Kalmykiya, Respublika

120 F9 **Kalnciems** Jelgava, C Latvia 56.46N 23.37E

116 L10 **Kalnitsa** ♂ SE Bulgaria

113 J24 **Kalocsa** Bács-Kiskun, S Hungary 46.31N 19.00E

116 J9 **Kalofer** Plovdiv, C Bulgaria 42.35N 25.00E

40 E10 **Kalohi Channel** *channel* C Pacific Ocean

85 I16 **Kalomo** Southern, S Zambia 17.04S 26.27E

31 X14 **Kalona** Iowa, C USA 41.28N 91.42W

117 K22 **Kalotási, Akrotírio** *headland* Amorgós, Kykládes, Greece, Aegean Sea 36.47N 25.45E

158 J8 **Kalpa** Himáchal Pradesh, N India 31.33N 78.16E

117 C15 **Kalpáki** Ípeiros, W Greece 39.53N 20.38E

161 C22 **Kalpeni Island** *island* Lakshadweep, India, N Indian Ocean

158 K13 **Kálpi** Uttar Pradesh, N India 26.07N 79.43E

164 G7 **Kalpin** Xinjiang Uygur Zizhiqu, NW China 40.35N 78.52E

152 K8 **Kalqudug** Rus. Kulkuduk. Nawoiy Wiloyati, N Uzbekistan 42.36N 63.24E

195 P16 **Kalri Lake** ◎ SE Pakistan

49 R5 **Kál Shūr** ♂ N Iran

41 N11 **Kalskag** Alaska, USA 61.32N 160.15W

97 B18 **Kalsoy** Dan. Kalsø *Island* Faeroe Islands 62.20N 6.46W

41 Q9 **Kaltag** Alaska, USA 64.19N 158.43W

110 H7 **Kaltbrunn** Sankt Gallen, NE Switzerland 47.11N 9.00E

Kaltdorf see Pruszków

79 X14 **Kaltungo** Gombe, E Nigeria 9.49N 11.22E

130 K4 **Kaluga** Kaluzhskaya Oblast', W Russian Federation 54.31N 36.16E

161 J26 **Kalu Ganga** ♂ S Sri Lanka

84 J13 **Kalulushi** Copperbelt, C Zambia 12.52S 28.06E

188 M2 **Kalumburu** Western Australia 14.11S 126.40E

97 H23 **Kalundborg** Vestsjælland, E Denmark 55.42N 11.06E

84 K11 **Kalungwishi** ♂ N Zambia

118 I6 **Kalūr Kot** Punjab, E Pakistan 32.07N 71.19E

118 I6 **Kalush** Pol. Kałusz. Ivano-Frankivs'ka Oblast', W Ukraine 49.01N 24.21E

Kalusz see Kalush

112 N11 **Kałuszyn** Mazowieckie, C Poland 52.12N 21.43E

161 J26 **Kalutara** Western Province, SW Sri Lanka 6.34N 79.58E

130 I5 **Kaluwawa** see Fergusson Island

Kaluzhskaya Oblast' ◆ *province* W Russian Federation

121 E14 **Kalvarija** Pol. Kalwaria. Marijampolė, S Lithuania 54.25N 23.13E

95 K15 **Kälviä** Länsi-Suomi, W Finland 63.50N 23.31E

111 U6 **Kalwang** Steiermark, E Austria 47.25N 14.48E

Kalwaria see Kalvarija

160 D13 **Kalyān** Mahārāshtra, W India 19.16N 73.10E

128 K16 **Kalyazin** Tverskaya Oblast', W Russian Federation 57.15N 37.53E

117 D18 **Kalydón** anc. Calydon. *site of ancient city* Dytikí Ellás, C Greece 38.24N 21.31E

117 M21 **Kálymnos** var. Kálimnos. Kálymnos, Dodekánisos, Greece, Aegean Sea 36.57N 26.58E

117 M21 **Kálymnos** var. Kálimnos. *island* Dodekánisos, Greece, Aegean Sea

119 O5 **Kalynivka** Kyyivs'ka Oblast', N Ukraine 50.14N 30.16E

119 N6 **Kalynivka** Vinnyts'ka Oblast', C Ukraine 49.27N 28.32E

44 M10 **Kama** var. Cama. Región Autónoma Atlántico Sur, SE Nicaragua 12.06N 83.55W

125 E9 **Kama** ♂ NW Russian Federation

172 N12 **Kamaishi** var. Kamaisi. Iwate, Honshū, C Japan 39.17N 141.51E

Kamaisi see Kamaishi

120 H13 **Kamajai** Molėtai, E Lithuania 55.49N 25.30E

120 H11 **Kamajai** Rokiškis, NE Lithuania 55.16N 25.30E

171 Jj17 **Kamakura** Kanagawa, Honshū, S Japan 35.17N 139.31E

155 U9 **Kamália** Punjab, NE Pakistan 30.43N 72.39E

85 I14 **Kamalondo** North Western, NW Zambia 13.42S 25.38E

142 I13 **Kaman** Kırşehir, C Turkey 39.22N 33.43E

81 O20 **Kamanyola** Sud Kivu, E Dem. Rep. Congo (Zaire) 2.54S 29.04E

147 N14 **Kamarán** *island* W Yemen

57 R9 **Kamarang** N Guyana 5.49N 60.38W

Kāmāreddi/Kamareddy see Rāmāreddi

154 K13 **Kamarod** Baluchistān, SW Pakistan 27.34N 63.36E

175 R13 **Kamaru** Pulau Buton, C Indonesia 5.10S 123.03E

41 Q13 **Kamishak Bay** *bay* Alaska, USA

172 Pp6 **Kami-Shihoro** Hokkaidō, NE Japan 43.14N 143.18E

Kamishli see Al Qāmishlī

Kamissar see Kamsar

170 Cc10 **Kami-Tsushima** Nagasaki, Tsushima, SW Japan 34.40N 129.27E

81 O20 **Kamituga** Sud Kivu, E Dem. Rep. Congo (Zaire) 3.05 28.10E

170 B17 **Kamiyaku** Kagoshima, Yaku-shima, SW Japan 30.24N 130.32E

9 N16 **Kamloops** British Columbia, SW Canada 50.39N 120.24W

109 G25 **Kamma** Sicilia, Italy, C Mediterranean Sea 36.46N 12.03E

199 Ii4 **Kammu Seamount** *undersea feature* N Pacific Ocean 32.09N 173.00E

111 U11 **Kamnik** Ger. Stein. C Slovenia 46.13N 14.34E

111 I15 **Kamniše Alpe** see Kamniško-Savinjske Alpe

111 T10 **Kamniško-Savinjske Alpe** var. Kamniše Alpe, Sanntaler Alpen, Ger. Steiner Alpen. ▲ N Slovenia

171 K17 **Kamogawa** Chiba, Honshū, S Japan 35.05N 140.04E

155 W8 **Kāmoke** Punjab, E Pakistan 31.58N 74.13E

84 L13 **Kamoto** Eastern, E Zambia 13.16S 32.04E

83 V3 **Kamp** ♂ N Austria

83 F19 **Kampala** ● (Uganda) S Uganda 0.21N 32.28E

161 J19 **Kampar, Sungai** ♂ Sumatera, W Indonesia

174 Ii10 **Kampa, Teluk** *bay* Pulau Bangka, W Indonesia

100 L9 **Kampen** Overijssel, E Netherlands 52.33N 5.55E

81 N20 **Kampene** Maniema, E Dem. Rep. Congo (Zaire) 3.34S 26.40E

31 Q9 **Kampeska, Lake** ◎ South Dakota, N USA

178 Gg9 **Kamphaeng Phet** var. Kambaeng Petch. Kamphaeng Phet, W Thailand 16.28N 99.31E

Kampo see Campo, Cameroon

178 J13 **Kâmpóng Cham** prev. Kompong Cham. Kâmpóng Cham, C Cambodia 12.00N 105.27E

178 J13 **Kâmpóng Chhnăng** prev. Kompong, Kâmpóng Chhnăng, C Cambodia 12.15N 104.40E

178 Ii12 **Kâmpóng Khleäng** prev. Kompong Kleang. Siěmréap, NW Cambodia 13.04N 104.07E

131 L8 **Kâmpóng Saôm** prev. Kompong Som, Sihanoukville. Kâmpóng Saôm, SW Cambodia 10.37N 103.30E

178 J13 **Kâmpóng Spoe** prev. Kompong Speu. Kâmpóng Spœ, S Cambodia 11.28N 104.29E

124 N3 **Kámpos** see Kambos

178 Ii14 **Kâmpôt** prev. Kâmpôt, SW Cambodia 10.37N 104.10E

Kamptee see Kāmthi

79 O14 **Kampti** SW Burkina 10.07N 3.22W

Kampuchea see Cambodia

174 Li5 **Kampung Sirik** Sarawak, East Malaysia 2.42N 111.28E

176 Y13 **Kampung, Sungai** ♂ Irian Jaya, E Indonesia

176 Vv12 **Kamrau, Teluk** *bay* Irian Jaya, E Indonesia

9 V15 **Kamsack** Saskatchewan, S Canada 51.34N 101.51W

78 H13 **Kamsar** var. Kamissar. Guinée-Maritime, W Guinea 10.36N 14.34W

130 L11 **Kamskoye Ust'ye** Respublika Tatarstan, W Russian Federation 55.13N 49.11E

129 U14 **Kamskoye Vodokhranilishche** var. Kama Reservoir. ◎ NW Russian Federation

160 I12 **Kamthi** prev. Kamptee. Mahārāshtra, C India 21.19N 79.11E

Kamuela see Waimea

172 Nn5 **Kamuenai** Hokkaidō, NE Japan 43.07N 140.25E

172 P7 **Kamui-dake** ▲ Hokkaidō, NE Japan 42.24N 142.57E

172 Nn4 **Kamui-misaki** *headland* Hokkaidō, NE Japan 43.20N 140.20E

45 O15 **Kamuk, Cerro** ▲ SE Costa Rica 9.15N 83.01W

176 Vv9 **Kamura, Sungai** ♂ Irian Jaya, E Indonesia

176 X12 **Kamura, Sungai** ♂ Irian Jaya, E Indonesia

118 K7 **Kam"yanets'-Podil's'kyy** Rus. Khmel'nyts'ka Oblast', W Ukraine 48.42N 26.36E

119 Q6 **Kam"yanka** Rus. Kamenka. Cherkas'ka Oblast', C Ukraine 49.03N 32.06E

119 T9 **Kam"yanka-Buz'ka** Rus. Kamenka-Bugskaya. L'vivs'ka Oblast', W Ukraine 50.03N 24.20E

171 Ll12 **Kam"yanka-Dniprovs'ka** Rus. Kamenka Dneprovskaya. Zaporiz'ka Oblast', SE Ukraine 47.28N 34.24E

85 G20 **Kang** Kgalagadi, C Botswana 23.40S 22.49E

78 L13 **Kangaba** Koulikoro, SW Mali 11.57N 8.24W

142 M13 **Kangal** Sivas, C Turkey 39.15N 37.22E

149 O13 **Kangān** Büshehr, S Iran 25.50N 57.30E

173 G2 **Kangar** Perlis, Peninsular Malaysia 6.28N 100.10E

190 F10 **Kangaroo Island** *island* South Australia

95 M17 **Kangasniemi** Itä-Suomi, E Finland 61.58N 26.36E

148 K6 **Kangāvar** var. Kangāvar. Kermānshāh, W Iran 34.30N 47.53E

197 J13 **Kanacea** prev. Kanathea. Taveuni, N Fiji 16.59S 179.57E

197 K14 **Kanacea** *island* Lau Group, E Fiji 17.36N 88.06E

40 G17 **Kanaga Island** *island* Aleutian Islands, Alaska, USA

40 G17 **Kanaga Volcano** ▲ Kanaga Island, Alaska, USA 51.55N 177.09W

171 J17 **Kanagawa** off. Kanagawa-ken. ◆ *prefecture* Honshū, S Japan

11 Q8 **Kanairiktok** ♂ Newfoundland, E Canada

Kanaky see New Caledonia

81 K22 **Kananga** prev. Luluabourg. Kasai Occidental, S Dem. Rep. Congo (Zaire) 5.51S 22.22E

Kananur see Cannanore

131 Q4 **Kanash** Chuvashskaya Respublika, W Russian Federation 55.30N 47.27E

207 P15 **Kangikajik** var. Kap Brewster. *headland* E Greenland 70.10N 22.00W

11 N5 **Kangiqsualujjuaq** prev. George River, Port-Nouveau-Quebec, Quebec, E Canada 58.34N 65.58W

10 L2 **Kangiqsujuaq** prev. Maricourt, Wakeham Bay. Quebec, NE Canada 61.35N 72.00W

169 X12 **Kangirsuk** prev. Bellin, Payne. Quebec, E Canada 60.00N 70.01W

207 P15 **Kangmar** Xizang Zizhiqu, W China 30.45N 85.43E

164 M16 **Kangmar** Xizang Zizhiqu, W China 28.30N 89.40E

Kängnüng see Kangnung

163 V15 **Kangnüng** Jap. Kōryo. NE South Korea 37.47N 128.52E

171 G7 **Kango** Estuaire, NW Gabon 0.17N 10.00E

165 O17 **Kangto** ▲ China/India 27.54N 92.33E

165 S9 **Kangping** Liaoning, NE China 42.45N 123.22E

177 G4 **Kanbalu** Sagaing, C Myanmar 23.10N 95.31E

177 F8 **Kangyidaung** Irrawaddy, SW Myanmar 16.40N 96.01E

178 H11 **Kanchanaburi** Kanchanaburi, W Thailand 14.01N 99.31E

Känchenjunga see Kangchenjunga

151 V11 **Kanchingiz, Khrebet** ▲ E Kazakhstan

161 I19 **Känchipuram** Tamil Nādu, SE India 12.49N 79.43E

175 N8 **Kandahār** Per. Qandahār. Kandahār, S Afghanistan 31.36N 65.48E

175 N9 **Kandahār Per.** Qandahār. ◆ *province* SE Afghanistan

194 I5 **Kandalaksha** var. Kandalaksa, Fin. Kantalahti. Murmanskaya Oblast', NW Russian Federation 67.09N 32.13E

Kandalakshskaya Guba/Kandalakshskaya Guba see Kandalakshskiy Zaliv

128 K6 **Kandalakshskiy Zaliv** var. Kandalakshskaya Guba, Eng. Kandalaksha Gulf. *bay* NW Russian Federation

178 J13 **Kâmpóng Cham** prev. Kompong Cham. Kâmpóng Cham, C Cambodia 12.00N 105.27E

85 G17 **Kandalangodi** see Kandalengoti

Kandalengoti var. Kandalangodi. Ngamiland, NW Botswana 19.25S 22.12E

175 N10 **Kandangan** Borneo, C Indonesia 2.49S 115.15E

120 E8 **Kandava** Ger. Kandau. Tukums, W Latvia 57.02N 22.46E

Kandavu see Kadavu

79 R14 **Kandé** var. Kanté. NE Togo 9.55N 1.01E

103 F23 **Kandel** ▲ SW Germany 48.03N 8.00E

194 G12 **Kandep** Enga, W PNG 5.10S 143.26E

78 J18 **Kandh Kot** Sind, SE Pakistan 28.15N 69.18E

79 S13 **Kandi** N Benin 11.04N 2.58E

155 S15 **Kandiáro** Sind, SE Pakistan 27.01N 68.16E

142 F13 **Kandra** Kocaeli, NW Turkey 41.04N 30.07E

154 M16 **Kandrách** var. Kanrach. Baluchistān, SW Pakistan 25.25N 65.28E

180 I4 **Kandreho** Mahajanga, C Madagascar 17.27S 46.06E

194 M12 **Kandrian** New Britain, E PNG 6.10S 149.33E

Kandukur see Kondukūr

161 K25 **Kandy** Central Province, C Sri Lanka 7.16N 80.40E

155 I10 **Kandyagash** Kaz. Qandyaghash; prev. Oktyabr'sk. Aktyubinsk, W Kazakhstan 49.25N 57.24E

20 D12 **Kane** Pennsylvania, NE USA 41.39N 78.47W

66 I11 **Kane Fracture Zone** *tectonic feature* NW Atlantic Ocean

Kaneka see Kanëvka

80 G9 **Kanem** off. Préfecture du Kanem. ◆ *prefecture* W Chad

Kaneohe Haw. Kāne'ohe. Oahu, Hawaii, USA, C Pacific Ocean 21.25N 157.48W

Kanestron, Akrotírio see Palioúri, Akrotírio

78 L13 **Kanëvka** var. Kanëka. Murmanskaya Oblast', NW Russian Federation 67.07N 39.43E

130 K13 **Kanevskaya** Krasnodarskiy Kray, SW Russian Federation 46.07N 38.57E

Kanevskoye Vodokhranilishche see Kaniv's'ke Vodoskhovyshche

171 Ll12 **Kaneyama** Yamagata, Honshū, C Japan 38.54N 140.22E

85 G20 **Kang** Kgalagadi, C Botswana 23.40S 22.49E

38 K8 **Kanab** Utah, W USA 37.03N 112.31W

38 K9 **Kanab Creek** ♂ Arizona/Utah, SW USA 36.24N 112.38W

197 J13 **Kanacea** prev. Kanathea. Taveuni, N Fiji 16.59S 179.57E

166 G9 **Kangding** Sichuan, C China 30.03N 101.56E

175 Nn14 **Kangean, Kepulauan** *island group* S Indonesia

175 N14 **Kangean, Pulau** *island* Kepulauan Kangean, S Indonesia

69 U8 **Kangen** var. Kengen. ♂ E Sudan

207 N14 **Kangerlussuaq** Dan. Søndre Strømfjord ● Kitaa, W Greenland 66.59N 50.38W

207 Q15 **Kangertittivaq** Dan. Scoresby Sund. *fjord* E Greenland

178 H2 **Kangfang** Kachin State, N Myanmar 26.09N 98.36E

176 Z13 **Kangi** Irian Jaya, E Indonesia 5.56S 140.49E

169 X12 **Kangiqsualujjuaq** prev. George River

193 G17 **Kaniere, Lake** ◎ South Island, NZ

196 E17 **Kanifaay** Yap, W Micronesia

129 O4 **Kanin Kamen'** ▲ NW Russian Federation

129 N3 **Kanin Nos** Nenetskiy Avtonomnyy Okrug, NW Russian Federation 68.38N 43.19E

120 E8 **Kanin Nos, Mys** *headland* NW Russian Federation 68.39N 43.14E

129 O5 **Kanin, Poluostrov** *peninsula* NW Russian Federation

145 V8 **Käni Sakht** E Iraq 33.19N 46.04E

145 T3 **Käni Sulaymän** N Iraq 35.54N 44.35E

172 N8 **Kanita** Aomori, Honshū, C Japan 41.04N 140.36E

119 Q5 **Kaniv** Rus. Kanëv. Cherkas'ka Oblast', C Ukraine 49.46N 31.28E

190 K11 **Kaniva** Victoria, SE Australia 36.25S 141.13E

119 Q5 **Kaniv's'ke Vodoskhovyshche** Eng. Kaniv Reservoir, Rus. Kanevskoye Vodokhranilishche. ◎ C Ukraine

114 L8 **Kanjiža** Ger. Altkanizsa, Hung. Magyarkanizsa, Ókanizsa; prev. Stara Kanjiža. Serbia, N Yugoslavia 46.03N 20.03E

95 K18 **Kankaanpää** Länsi-Suomi, W Finland 61.46N 22.25E

32 M12 **Kankakee** Illinois, N USA 41.07N 87.51W

33 O11 **Kankakee River** ♂ Illinois/Indiana, N USA

78 K14 **Kankan** Haute-Guinée, E Guinea 10.25N 9.19W

160 K13 **Känker** Madhya Pradesh, C India 20.19N 81.29E

78 J10 **Kankossa** Assaba, S Mauritania 15.54N 11.31W

178 Gg13 **Kanmaw Kyun** var. Kisseraing, Kitharang. *island* Mergui Archipelago, S Myanmar

170 E13 **Kanmuri-yama** ▲ Kyūshū, SW Japan 34.28N 132.03E

23 R10 **Kannapolis** North Carolina, SE USA 35.29N 80.37W

95 L16 **Kannonkoski** Länsi-Suomi, W Finland 62.58N 25.19E

95 K15 **Kannus** Länsi-Suomi, W Finland 63.51N 23.55E

79 V13 **Kano** Kano, N Nigeria 11.56N 8.30E

79 V13 **Kano** ◆ *state* N Nigeria

79 V13 **Kano** ✕ Kano, N Nigeria 11.56N 8.26E

170 F14 **Kan'onji** var. Kanonzi. Kagawa, Shikoku, SW Japan 34.10N 133.38E

28 M5 **Kanopolis Lake** ◎ Kansas, C USA

38 K5 **Kanosh** Utah, W USA 38.48N 112.26W

174 Ll6 **Kanowit** Sarawak, East Malaysia 2.03N 112.15E

176 Yy10 **Kanoya** Kagoshima, Kyūshū, SW Japan 31.23N 130.50E

158 L13 **Känpur** Eng. Cawnpore. Uttar Pradesh, N India 26.80N 80.21E

Kanrach see Kandrách

29 R9 **Kansas** Oklahoma, C USA 36.14N 94.46W

28 L5 **Kansas** off. State of Kansas; also known as Jayhawker State, Sunflower State. ◆ *state* C USA

29 S4 **Kansas City** Kansas, C USA 39.06N 94.37W

29 R4 **Kansas City** Missouri, C USA 39.06N 94.34W

29 R4 **Kansas City** ✕ Missouri, C USA 39.18N 94.45W

29 P4 **Kansas River** ♂ Kansas, C USA

126 I14 **Kansk** Krasnoyarskiy Kray, S Russian Federation 56.11N 95.32E

Kansu see Gansu

153 V7 **Kant** Chuyskaya Oblast', N Kyrgyzstan 42.54N 74.47E

12 D6 **Kantalaksha** ♂ Ontario, S Canada

Kantalahti see Kandalaksha

79 T25 **Kantang** var. Ban Kantang. Trang, SW Thailand 7.25N 99.30E

117 H25 **Kántanos** Kríti, Greece, E Mediterranean Sea 35.20N 23.42E

79 R12 **Kantchari** E Burkina 12.28N 1.31E

Kanté see Kandé

130 L9 **Kantemirovka** Voronezhskaya Oblast', W Russian Federation 49.43N 39.53E

179 J11 **Kantharalak** Si Sa Ket, E Thailand 14.32N 104.37E

Kantipur see Kathmandu

41 Q9 **Kantishna River** ♂ Alaska, USA

79 R14 **Kara** var. Lama-Kara. NE Togo 9.36N 1.12E

79 Q14 **Kara** ♂ N Togo

153 U7 **Kara-Balta** Chuyskaya Oblast', N Kyrgyzstan 42.50N 73.51E

150 G11 **Kara-Balau** Atyrau, W Kazakhstan 48.29N 53.05E

99 C20 **Kanturk Ir.** Ceann Toirc. SW Ireland 52.12N 8.54W

152 E7 **Kanu** var. Cara. Región Autónoma Atlántico Sur, E Nicaragua 12.52N 83.35W

79 R14 **Kara** var. Lama-Kara. NE Togo 9.36N 1.12E

121 J17 **Kapyl'** Rus. Kopyl'. Minskaya Voblasts', C Belarus 53.09N 27.05E

54 N9 **Kara** var. Cara. Región Autónoma Atlántico Sur, E Nicaragua 12.52N 83.35W

152 E7 **Karabekaul** var. Garabekevyul, Turkm. Garabekewül. Lebapskiy Velayat, E Turkmenistan

152 K15 **Karabil', Vozvyshennost'** ▲ S Turkmenistan

152 A9 **Kara-Bogaz-Gol** Turkm. Garabogazköl. Balkanskiy Velayat, W Turkmenistan 41.03N 52.52E

152 B9 **Kara-Bogaz-Gol, Zaliv** *bay* W Turkmenistan

151 R15 **Karaboget** Kaz. Qaraböget. Zhambyl, S Kazakhstan 44.36N 72.03E

142 H11 **Karabük** Karabük, NW Turkey 41.12N 32.36E

143 H11 **Karabük** ◆ *province* NW Turkey

126 li13 **Karabula** Krasnoyarskiy Kray, C Russian Federation 58.01N 97.17E

151 V14 **Karabulak** Kaz. Qarabulaq. Vostochnyy Kazakhstan, E Kazakhstan 44.54N 78.28E

151 Y11 **Karabulak** Kaz. Qarabulaq. Vostochnyy Kazakhstan, E Kazakhstan 47.33N 84.38E

◆ COUNTRY ● COUNTRY CAPITAL ◇ DEPENDENT TERRITORY ◉ DEPENDENT TERRITORY CAPITAL ◆ ADMINISTRATIVE REGION ✕ INTERNATIONAL AIRPORT ▲ MOUNTAIN ▲ MOUNTAIN RANGE ♂ RIVER ◎ LAKE ◈ VOLCANO ◈ RESERVOIR

273

151 Q17 **Karabulak** *Kaz.* Qarabulaq. Yuzhnyy Kazakhstan, S Kazakhstan 42.51N 69.46E

142 C17 **Kara Burnu** *headland* SW Turkey 36.34N 28.00E

150 K10 **Karabutak** *Kaz.* Qarabutaq. Aktyubinsk, W Kazakhstan 49.58N 60.06E

142 D12 **Karacabey** Bursa, NW Turkey 40.13N 28.22E

116 O12 **Karaköy** İstanbul, NW Turkey 41.24N 28.21E

116 M12 **Karacaoğlan** Kırklareli, NW Turkey 41.30N 27.06E

Karachay-Cherkessia *see* Karachayevo-Cherkesskaya Respublika

130 L15 **Karachayevo-Cherkesskaya Respublika** *Eng.* Karachay-Cherkessia. ◆ *autonomous republic* SW Russian Federation

130 M15 **Karachayevsk** Karachayevo-Cherkesskaya Respublika, SW Russian Federation 43.43N 41.53E

130 J6 **Karachev** Bryanskaya Oblast', W Russian Federation 53.07N 35.56E

155 O16 **Karāchi** Sind, SE Pakistan 24.51N 67.01E

155 O16 **Karāchi** ✕ Sind, S Pakistan 24.51N 67.01E

Karácsonkő *see* Piatra-Neamţ

161 E15 **Karād** Mahārāshtra, W India 17.19N 74.15E

142 H16 **Karadağ** ▲ S Turkey 37.00N 33.00E

153 T10 **Karadar'ya** *Uzb.* Qoradaryo. ♦ Kyrgyzstan/Uzbekistan

Karadeniz *see* Black Sea

Karadeniz Boğazı *see* İstanbul Boğazı

152 B13 **Karadepe** Balkanskiy Velayat, W Turkmenistan 38.04N 54.01E

Karadzhar *see* Qorajar

Karaferiye *see* Véroia

152 E13 **Karagan** *Turkm.* Garagan. Akhalskiy Velayat, C Turkmenistan 38.16N 57.34E

151 R10 **Karaganda** *Kaz.* Qaraghandy. Karaganda, C Kazakhstan 49.52N 73.07E

151 R10 **Karaganda** *off.* Karagandinskaya Oblast', *Kaz.* Qaraghandy Oblysy. ♦ *province* C Kazakhstan

Karagandinskaya Oblast' *see* Karaganda

151 T10 **Karagayly** *Kaz.* Qaraghayly. Karaganda, C Kazakhstan 49.25N 75.31E

152 A11 **Karagel'** *Turkm.* Garagöl. Balkanskiy Velayat, W Turkmenistan 39.24N 53.13E

127 Pp8 **Karaginskiy, Ostrov** *island* E Russian Federation

207 T1 **Karaginskiy Zaliv** *bay* E Russian Federation

143 P13 **Karagöl Dağları** ▲ NE Turkey

116 L13 **Karahisar** Edirne, NW Turkey 40.47N 26.34E

131 V3 **Karaidel'** Respublika Bashkortostan, W Russian Federation 55.50N 56.55E

131 V3 **Karaidel'skiy** Respublika Bashkortostan, W Russian Federation 55.51N 57.09E

116 L13 **Karaidemir Baraji** ◙ NW Turkey

161 J21 **Kāraikāl** Pondicherry, SE India 10.58N 79.49E

161 I22 **Kāraikkudi** Tamil Nādu, SE India 10.04N 78.46E

151 Y11 **Kara Irtysh** *Rus.* Chërnyy Irtysh. ♦ NE Kazakhstan

149 N5 **Karaj** Tehrān, N Iran 35.43N 51.25E

174 H5 **Karak** Pahang, Peninsular Malaysia 3.24N 101.58E

Karak *see* Al Karak

153 T11 **Kara-Kabak** Oshskaya Oblast', SW Kyrgyzstan 39.40N 72.45E

152 D12 **Kara-Kala** *var.* Garrygala. Balkanskiy Velayat, W Turkmenistan 38.27N 56.15E

Karakala *see* Oqqal'a

Karakalpakstan, Respublika *see* Qoraqalpoghiston Respublikasi

Karakalpakya *see* Qoraqalpoghiston

Karakax *see* Moyu

164 G10 **Karakax He** ♦ NW China

124 S9 **Karakaya Baraji** ◙ C Turkey

175 Ss4 **Karakelang, Pulau** *island* N Indonesia

Karakılısse *see* Ağrı

Karak, Muḥāfaẓat al *see* Al Karak

Kara-Köl *see* Kara-Kul'

153 Y7 **Karakol** *prev.* Przheval'sk. Issyk-Kul'skaya Oblast', NE Kyrgyzstan 42.31N 78.20E

153 X8 **Karakol** *var.* Karakolka. Issyk-Kul'skaya Oblast', NE Kyrgyzstan 41.30N 77.18E

Karakolka *see* Karakol

155 W2 **Karakoram Highway** *road* China/Pakistan

155 Z3 **Karakoram Pass** *Chin.* Karakoram Shankou. *pass* C Asia 35.23N 77.45E

158 I3 **Karakoram Range** ▲ C Asia

Karakoram Shankou *see* Karakoram Pass

Karaköse *see* Ağrı

151 P14 **Karakoyyn, Ozero** *Kaz.* Qaraqoyyn. ◙ C Kazakhstan

85 F19 **Karakubis** Ghanzi, W Botswana 22.03S 20.36E

153 T9 **Kara-Kul'** *Kir.* Kara-Köl. Dzhalal-Abadskaya Oblast', W Kyrgyzstan 40.55N 73.36E

Karakul' *see* Qarokŭl, Tajikistan

Karakul' *see* Qarokŭl, Uzbekistan

153 U10 **Kara-Kul'dzha** Oshskaya Oblast', SW Kyrgyzstan 40.34N 73.33E

131 T3 **Karakulino** Udmurtskaya Respublika, NW Russian Federation 56.02N 53.45E

Karakul', Ozero *see* Qarokŭl

Kara Kum *see* Garagumy

Kara Kum Canal/Karakumskiy Kanal *see* Garagumskiy Kanal

---

85 E17 **Karakuwisa** Okavango, NE Namibia 18.55S 19.40E

126 Jj14 **Karam** Irkutskaya Oblast', S Russian Federation 55.07N 107.21E

Karamai *see* Karamay

175 N13 **Karamain, Pulau** *island* N Indonesia

142 H16 **Karaman** Karaman, S Turkey 37.10N 33.13E

116 M8 **Karaman** ♦ *province* S Turkey

164 J4 **Karamay** *var.* Karamai, Kelamayi, *prev. Chin.* K'o-la-ma-i. Xinjiang Uygur Zizhiqu, NW China 45.33N 84.45E

175 Nn11 **Karambu** Borneo, N Indonesia 3.48S 116.06E

193 H14 **Karamea** West Coast, South Island, NZ 41.15S 172.07E

193 H14 **Karamea** ♦ South Island, NZ

193 G15 **Karamea Bight** *gulf* South Island, NZ

152 L14 **Karamet-Niyaz** *Turkm.* Garamätnyyaz. Lebapskiy Velayat, E Turkmenistan 37.45N 64.28E

164 K10 **Karamiran He** ♦ NW China

176 Yy11 **Karamor, Pengunungan** ▲ Irian Jaya, E Indonesia

153 S11 **Karamyk** Oshskaya Oblast', SW Kyrgyzstan 39.28N 71.45E

175 Nn16 **Karangasem** Bali, S Indonesia 8.24S 115.40E

160 H12 **Kāranja** Mahārāshtra, C India 20.30N 77.26E

Karanpur *see* Karanpura

158 F9 **Karanpura** *var.* Karanpur. Rājasthān, NW India 29.46N 73.30E

Karánsebes/Karansebesch *see* Caransebeş

151 T14 **Karaoy** *Kaz.* Qaraoy. Almaty, SE Kazakhstan 45.52N 74.44E

116 N7 **Karapelit** *Rom.* Stejarul. Dobrich, NE Bulgaria 43.40N 27.33E

142 I15 **Karapınar** Konya, C Turkey 37.43N 33.34E

85 D22 **Karas** ♦ *district* S Namibia

153 Y8 **Kara-Say** Issyk-Kul'skaya Oblast', NE Kyrgyzstan 41.34N 77.55E

85 E22 **Karasburg** Karas, S Namibia 27.59S 18.45E

94 K9 **Kárášjohka** ♦ N Norway

94 L9 **Karasjok** *Fin.* Kaarasjoki. Finnmark, N Norway 69.27N 25.28S

Kárášjohka *see* Kárášjohka

Kara Strait *see* Karskiye Vorota, Proliv

151 N8 **Karasu** *Kaz.* Qarasü. Kostanay, N Kazakhstan 52.43N 65.28E

142 F11 **Karasu** Sakarya, NW Turkey 41.03N 30.39E

125 G14 **Karasubazar** *see* Bilohirs'k

Karasuk Novosibirskaya Oblast', C Russian Federation 53.41N 78.04E

151 U13 **Karatal** *Kaz.* Qaratal. ♦ SE Kazakhstan

142 K17 **Karataş** Adana, S Turkey 36.37N 35.24E

151 Q16 **Karatau** *Kaz.* Qarataū. Zhambyl, S Kazakhstan 43.09N 70.28E

Karatau *see* Karatau, Khrebet

151 P16 **Karatau, Khrebet** *var.* Karatau, *Kaz.* Qarataū. ▲ S Kazakhstan

150 G13 **Karaton** *Kaz.* Qaraton. Atyrau, W Kazakhstan 46.33N 53.31E

170 C12 **Karatsu** *var.* Karatu. Saga, Kyūshū, SW Japan 33.27N 129.55E

Karatu *see* Karatsu

116 Hh7 **Karaul** Taymyrskiy (Dolgano-Nenetskiy) Avtonomnyy Okrug, N Russian Federation 70.07N 83.12E

Karaulbazar *see* Qorowulbozor

Karauzyak *see* Qoraŭzak

177 D16 **Karáva** ▲ C Greece 39.19N 21.33E

117 F22 **Karavás** Kýthira, S Greece 36.21N 22.57E

115 J20 **Karavastasë, Laguna e** *var.* Kënet' e Karavastas, Kravasta Lagoon. *lagoon* W Albania

Karavastas, Kënet' e *see* Karavastasë, Laguna e

220 I5 **Karavere** Tartumaa, E Estonia 58.25S 26.29E

117 L23 **Karavonísia** *island* Kykládes, Greece, Aegean Sea

174 Jj14 **Karawang** *prev.* Krawang. Jawa, C Indonesia 6.13S 107.16E

111 T10 **Karawanken** *Slvn.* Karavanke. ▲ Austria/Yugoslavia

Karaxahar *see* Kaidu He

143 R13 **Karayazı** Erzurum, NE Turkey 39.40N 42.09E

151 Q12 **Karazhal** Zhezkazgan, C Kazakhstan 48.02N 70.52E

145 S9 **Karbalā'** *var.* Kerbala, Kerbela. S Iraq 32.37N 44.03E

96 L11 **Kärböle** Gävleborg, C Sweden 61.59N 15.16E

113 M23 **Karcag** Jász-Nagykun-Szolnok, E Hungary 47.21N 20.51E

116 N7 **Kardam** Dobrich, NE Bulgaria 43.45N 28.06E

Kardámaina *see* Kardámyla

117 M22 **Kardámaina** Kós, Dodekánisos, Greece, Aegean Sea 36.46N 27.08E

Kardamíla *see* Kardámyla

117 L18 **Kardámyla** *var.* Kardamíla, Kardamíla. Chíos, E Greece 38.33N 26.04E

Kardeljevo *see* Ploče

118 G5 **Kardh** *see* Qardho

Kardhámila *see* Kardámyla

Kardhítsa *var.* Kardhitsa

117 E16 **Kardítsa** *var.* Kardhitsa. Thessalía, C Greece 39.22N 21.55E

120 E4 **Kärdla** *Ger.* Kertel. Hiiumaa, W Estonia 59.00N 22.42E

121 I16 **Karelia** *see* Kareliya, Respublika

Karelichy *Pol.* Korelicze, *Rus.* Korelichi. Hrodzyenskaya Voblasts', W Belarus 53.54N 26.07E

128 L10 **Kareliya, Respublika** *prev.* Karel'skaya ASSR, *Eng.* Karelia. ♦ *autonomous republic* NW Russian Federation

Karel'skaya ASSR *see* Kareliya, Respublika

83 E22 **Karema** Rukwa, W Tanzania 6.49S 30.25E

I14 **Karen** *see* Hualien

I14 **Karenda** Central, C Zambia 14.42S 26.52E

Gg8 **Karen State** *var.* Kawthule State, Kayin State. ♦ *state* S Myanmar

J10 **Karesuando** *Lapp.* Kaaresuanto. Norrbotten, N Sweden 68.25N 22.28E

Karet *see* Kâghet

Kareýz-e-Elýãs/Kärez Iliäs *see* Kärïz-e Elyãs

Gg12 **Kargasok** Tomskaya Oblast', C Russian Federation 59.01N 80.34E

Gg14 **Kargat** Novosibirskaya Oblast', C Russian Federation 55.07N 80.19E

J11 **Karg** Çorum, N Turkey 41.09N 34.31E

I5 **Kargil** Jammu and Kashmir, NW India 34.34N 76.06E

Kargilik *see* Yecheng

L11 **Kargopol'** Arkhangel'skaya Oblast', NW Russian Federation 61.30N 38.53E

F12 **Kargowa** *Ger.* Unruhstadt. Lubuskie, W Poland 52.05N 15.50E

X13 **Kari** Bauchi, E Nigeria 11.13N 10.34E

H14 **Kariba** Mashonaland West, N Zimbabwe 16.28S 28.47E

J16 **Kariba, Lake** ◙ Zambia/Zimbabwe

Nn5 **Kariba-yama** ▲ Hokkaidō, NE Japan 42.36N 139.55E

C19 **Karibib** Erongo, C Namibia 21.56S 15.51E

L9 **Karies** *see* Karyés

Karigasniemi *Lapp.* Garegegasnjárga. Lappi, N Finland 69.24N 25.52E

P6 **Karikachi-tōge** *pass* Hokkaidō, NE Japan 43.05N 142.46E

J2 **Karikari, Cape** *headland* North Island, NZ 34.47S 173.24E

Karīmābād *see* Hunza

K10 **Karimata, Kepulauan** *island group* N Indonesia

K9 **Karimata, Pulau** *island* N Indonesia

K10 **Karimata, Selat** *strait* W Indonesia

I14 **Karīmnagar** Andhra Pradesh, C India 18.28N 79.09E

I13 **Karimui** Chimbu, C PNG 6.19S 144.48E

L13 **Karimunjawa, Pulau** *island* S Indonesia

N12 **Karin** Woqooyi Galbeed, N Somalia 10.48N 45.46E

L20 **Kariot** *see* Ikaría

Karis *Fin.* Karjaa. Etelä-Suomi, SW Finland 60.05N 23.39E

J4 **Kärīz-e Elyãs** *var.* Kareýz-e-Elýãs, Kärez Iliäs. Herāt, NW Afghanistan 35.26N 61.24E

Karjaa *see* Karis

T10 **Karkaralinsk** *Kaz.* Qarqaraly. Karaganda, E Kazakhstan 49.31N 75.53E

J11 **Karkar Island** *island* N PNG

N7 **Karkas, Kūh-e** ▲ C Iran

K8 **Karkheh, Rūd-e** ♦ SW Iran

L20 **Karkínagrio** Ikaría, Dodekánisos, Greece, Aegean Sea 37.31N 26.01E

R12 **Karkinits'ka Zatoka** *Rus.* Karkinitskiy Zaliv. *gulf* S Ukraine

Karkinitskiy Zaliv *see* Karkinits'ka Zatoka

L19 **Karkkila** *Swe.* Högfors. Etelä-Suomi, S Finland 60.31N 24.10E

M19 **Kärkölä** Etelä-Suomi, S Finland 60.52N 25.17E

G9 **Karkoo** South Australia 34.03S 135.45E

D5 **Kärkü** *see* Kirkūk

Kärla *Ger.* Kergel. Saaremaa, W Estonia 58.19N 22.14E

F7 **Karleby** *see* Kokkola

Karlino *Ger.* Körlin an der Persante. Zachodniopomorskie, NW Poland 54.02N 15.52E

Q13 **Karlova** Bíngöl, E Turkey 39.16N 41.01E

U6 **Karlivka** Poltavs'ka Oblast', C Ukraine 49.27N 35.08E

I5 **Karl-Marx-Stadt** *see* Chemnitz

Karló *see* Hailuoto

Karlobag *It.* Carlopago. Lika-Senj, W Croatia 44.31N 15.06E

D9 **Karlovac** *Ger.* Karlstadt, *Hung.* Károlyváros. Karlovac, C Croatia 45.28N 15.31E

C10 **Karlovac** *off.* Karlovačka Županija. ♦ *province* C Croatia

Karlovačka Županija *see* Karlovac

A16 **Karlovarský Kraj** ♦ W Czech Republic

J9 **Karlovo** *prev.* Levskigrad. Plovdiv, C Bulgaria 42.39N 24.49E

A16 **Karlovy Vary** *Ger.* Karlsbad; *prev. Eng.* Carlsbad. Karlovarský Kraj, W Czech Republic 50.13N 12.51E

Karlsbad *see* Karlovy Vary

L17 **Karlsborg** Västra Götaland, S Sweden 58.31N 14.31E

Karlsburg *see* Alba Iulia

K17 **Karlshamn** Blekinge, S Sweden 56.10N 14.49E

L16 **Karlskoga** Örebro, C Sweden 59.19N 14.33E

M22 **Karlskrona** Blekinge, S Sweden 56.11N 15.38E

G21 **Karlsruhe** *var.* Carlsruhe. Baden-Württemberg, SW Germany 49.01N 8.24E

K16 **Karlstad** Värmland, C Sweden 59.22N 13.36E

R3 **Karlstad** Minnesota, N USA 48.34N 96.31E

I16 **Karlstadt** Bayern, C Germany 49.56N 9.46E

Q14 **Karluk** Kodiak Island, Alaska, USA 57.34N 154.27W

Y14 **Karluk** *see* Qarluq

G17 **Karma** *Rus.* Korma. Homyel'skaya Voblasts', SE Belarus 53.07N 30.48E

F14 **Karmala** Mahārāshtra, W India 18.26N 75.08E

---

144 G8 **Karmi'él** *var.* Carmiel. Northern, N Israel 32.55N 35.21E

97 B16 **Karmøy** *island* S Norway

158 I9 **Karnāl** Haryāna, N India 29.40N 76.58E

159 W15 **Karnaphuli Reservoir** ◙ NE India

161 F17 **Karnātaka** *var.* Kanara; *prev.* Maisur, Mysore. ♦ *state* W India

45 S13 **Karnes City** Texas, SW USA 28.52N 97.54W

111 P9 **Karnische Alpen** *It.* Alpi Carniche. ▲ Austria/Italy

116 M9 **Karnobat** Burgas, E Bulgaria 42.39N 26.58E

111 Q9 **Kärnten** *off.* Land Kärnten, *Eng.* Carinthi, *Slvn.* Koroška. ♦ *state* S Austria

Karnul *see* Kurnool

85 K16 **Karoi** Mashonaland West, N Zimbabwe 16.49N 29.40E

Karol *see* Carei

Károly-Fehérvár *see* Alba Iulia

Károlyváros *see* Karlovac

161 E20 **Kāroli** *var.* Kasari Jõgi, *Ger.* Kasargen. ♦ W Estonia

179 Qq15 **Karomatan** Mindanao, S Philippines 7.47N 123.48E

Karool-Tëbë Narynskaya Oblast', C Kyrgyzstan 40.33N 75.52E

84 M12 **Karonga** Northern, N Malawi 9.56S 33.54E

85 J14 **Karool-Tëbë** Narynskaya Oblast', C Kyrgyzstan 40.33N 75.52E

190 J9 **Karoonda** South Australia 35.04S 139.58E

155 S9 **Karor Lāl Esan** Punjab, E Pakistan 31.15N 70.54E

155 T11 **Karor Pacca** *var.* Kahror, Kahror Pakka. Punjab, E Pakistan 29.37N 71.58E

175 P10 **Karosa** Sulawesi, C Indonesia 1.38S 119.21E

Karpas/Karpas Peninsula *see* Kírpaşa

Karpaten *see* Carpathian Mountains

117 L22 **Karpáthio Pélagos** *sea* Dodekánisos, Greece, Aegean Sea

117 N24 **Kárpathos** Kárpathos, SE Greece 35.30N 27.13E

117 N24 **Kárpathos** *It.* Scarpanto; *anc.* Carpathos, Carpathus. *island* SE Greece

Karpathos Strait *see* Karpathou, Stenó

117 N24 **Karpathou, Stenó** *var.* Karpathos Strait, Scarpanto Strait. *strait* Dodekánisos, Greece, Aegean Sea

Karpenísi *prev.* Karpenísion. Stereá Ellás, C Greece 38.55N 21.45E

117 E17 **Karpenísi** *prev.* Karpenísion. Stereá Ellás, C Greece 38.55N 21.45E

Karpenísion *see* Karpenísi

129 O8 **Karpogory** Arkhangel'skaya Oblast', NW Russian Federation 64.01N 44.22E

188 I7 **Karratha** Western Australia 20.43S 116.52E

143 S12 **Kars** *var.* Qars. Kars, NE Turkey 40.34N 43.04E

143 S12 **Kars** *var.* Qars. ♦ *province* NE Turkey

155 O12 **Karsakpay** *Kaz.* Qarsaqbay. Zhezkazgan, C Kazakhstan 47.51N 66.42E

95 L15 **Kärsämäki** Oulu, C Finland 63.58N 25.49E

95 L17 **Kārsava** *Ger.* Karsau; *prev. Rus.* Korsovka. Ludza, E Latvia 56.46N 27.39E

152 A9 **Karshi** *Turkm.* Garshy. Balkanskiy Velayat, NW Turkmenistan 40.45N 52.50E

Karshi *see* Qarshi

Karshinskaya Step *see* Qarshi Chŭli

Karshinskiy Kanal *see* Qarshi Kanali

86 I5 **Karskiye Vorota, Proliv** *Eng.* Kara Strait. *strait* N Russian Federation

126 Gg5 **Karskoye More** *Eng.* Kara Sea. *sea* Arctic Ocean

95 L17 **Kārstula** Länsi-Suomi, W Finland 62.52N 24.48E

131 Q5 **Karsun** Ul'yanovskaya Oblast', W Russian Federation 54.12N 47.00E

125 E12 **Kartaly** Chelyabinskaya Oblast', C Russian Federation 53.02N 60.42E

94 K13 **Kartung** Norrbotten, N Sweden 66.03N 23.55E

94 K13 **Karunki** Lappi, N Finland 66.01N 24.06E

167 G6 **Kārūn, Rūd-e** *see* Kārūn

161 H21 **Kārūr** Tamil Nādu, SE India 10.58N 78.03E

95 K17 **Karvia** Länsi-suomi, W Finland 62.07N 22.34E

113 J17 **Karviná** *Ger.* Karwin, *Pol.* Karwina; *prev.* Nová Karvinná. Ostravský Kraj, E Czech Republic 49.51N 18.33E

161 H22 **Kārwār** Karnātaka, W India 14.49N 74.09E

117 M24 **Kásos** *island* S Greece

117 M25 **Kasos Strait** *see* Kasou, Stenó

117 I14 **Karyés** *var.* Karyés. Ágion Óros, N Greece 40.15N 24.15E

126 Kk16 **Karymskoye** Chitinskaya Oblast', S Russian Federation 51.36N 114.02E

116 I19 **Kárystos** *var.* Káristos. Évvoia, C Greece 38.01N 24.25E

142 E17 **Kaş** Antalya, SW Turkey 36.12N 29.31E

131 Q16 **Kasaan** Prince of Wales Island, Alaska, USA 55.32N 132.24W

170 G14 **Kasai** *Hyōgo*, Honshū, SW Japan 34.56N 134.49E

81 K21 **Kasai** *var.* Cassai, Kassai. ♦ Angola/Dem. Rep. Congo (Zaire)

152 M11 **Karmana** Nawoiy Wiloyati, C Uzbekistan 40.09N 65.18E

---

82 I9 **Kassala** Kassala, E Sudan 15.24N 36.25E

82 H9 **Kassala** ♦ *state* NE Sudan

117 G15 **Kassándra** *prev.* Pallíni; *anc.* Pallene. *peninsula* N Greece

117 G15 **Kassándra, Akrotírio** *headland* N Greece 39.58N 23.22E

117 H15 **Kassándras, Kólpos** *var.* Kólpos Toronaíos. *gulf* N Greece

111 H15 **Kassel** *prev.* Cassel. Hessen, C Germany 51.19N 9.30E

76 M6 **Kasserine** *var.* Al Qasrayn. W Tunisia 35.15N 8.52E

12 I14 **Kasshabog Lake** ◙ Ontario, SE Canada

145 O5 **Kassir, Sabkhat al** ◙ E Syria

5 W10 **Kasson** Minnesota, N USA 44.00N 92.42W

117 C17 **Kassópi** *site of ancient city* Ípeiros, W Greece 39.09N 20.38E

117 N24 **Kastállou, Akrotírio** *headland* Kárpathos, SE Greece 35.24N 27.08E

142 J11 **Kastamonu** *var.* Castamoni, Kastamuni. Kastamonu, N Turkey 41.22N 33.46E

142 J10 **Kastamonu** *var.* Kastamoni, Kastamuni. ♦ *province* N Turkey

117 E18 **Kastaneá** Kentrikí Makedonía, N Greece 40.25N 22.09E

117 E17 **Kastélli** Kríti, Greece, E Mediterranean Sea 35.30N 23.39E

117 I21 **Kastellórizon** *see* Megísti

117 N21 **Kastlösa** Kalmar, S Sweden 56.25N 16.25E

117 D14 **Kastória** Dytikí Makedonía, N Greece 40.30N 21.16E

130 K7 **Kastornoye** Kurskaya Oblast', W Russian Federation 51.49N 38.07E

117 I21 **Kástro** Sífnos, Kykládes, Greece, Aegean Sea 36.58N 24.45E

97 J23 **Kastrup** ✕ (København). København, E Denmark 55.36N 12.39E

121 Q17 **Kastsyukovichy** *Rus.* Kostyukovichi. Mahilyowskaya Voblasts', E Belarus 53.19N 32.03E

121 O18 **Kastsyukowka** *Rus.* Kostyukovka. Homyel'skaya Voblasts', SE Belarus 52.32N 30.54E

170 Cc12 **Kasuga** Fukuoka, Kyūshū, SW Japan 33.31N 130.27E

171 I15 **Kasugai** Aichi, Honshū, SW Japan 35.15N 136.57E

171 I15 **Kasukabe** Saitama, Honshū, S Japan 35.59N 139.45E

171 Kk16 **Kasumiga-ura** ◙ Honshū, S Japan

131 R17 **Kasumkent** Respublika Dagestan, SW Russian Federation 41.39N 48.09E

84 M13 **Kasungu** Central, C Malawi 13.01S 33.30E

155 W9 **Kasūr** Punjab, E Pakistan 31.07N 74.30E

85 G15 **Kataba** Western, W Zambia 15.28S 23.25E

21 R4 **Katahdin, Mount** ▲ Maine, NE USA 45.55N 68.52W

81 M20 **Katako-Kombe** Kasai Oriental, C Dem. Rep. Congo (Zaire) 3.24S 24.25E

41 T12 **Katalla** Alaska, USA 60.12N 144.31W

Katana *see* Qaţanā

81 L24 **Katanga** *off.* Région du Katanga; *prev.* Shaba. ♦ *region* SE Dem. Rep. Congo (Zaire)

126 J12 **Katanga** ♦ S Russian Federation

160 J11 **Katangi** Madhya Pradesh, C India 21.46N 79.49E

188 I13 **Katanning** Western Australia 33.44S 117.33E

189 P8 **Kata Tjuta** *var.* Mount Olga ▲ Northern Territory, C Australia 25.20S 130.47E

79 U12 **Katchall Island** *island* Nicobar Islands, NE Indian Ocean

117 F14 **Katerini** Kentrikí Makedonía, N Greece 40.17N 22.30E

119 P7 **Kateryno pil'** Cherkas'ka Oblast', C Ukraine 49.00N 30.59E

178 Gg3 **Katha** Sagaing, N Myanmar 24.10N 96.19E

189 P2 **Katherine** Northern Territory, N Australia 14.28S 132.19E

159 P11 **Kathmandu** *prev.* Kantipur. ● (Nepal) Central, C Nepal 27.46N 85.16E

158 H7 **Kathua** Jammu and Kashmir, NW India 32.24N 75.33E

78 L12 **Kati** Koulikoro, SW Mali 12.45N 8.06W

159 P13 **Katihār** Bihār, NE India 25.33N 87.34E

192 N7 **Katikati** Bay of Plenty, North Island, NZ 37.33S 175.55E

85 H16 **Katima Mulilo** Caprivi, NE Namibia 17.31S 24.19E

78 N15 **Katiola** Côte d'Ivoire 8.12N 5.04W

203 V10 **Katiu** *atoll* Îles Tuamotu, C French Polynesia

161 I15 **Katol** Mahārāshtra, C India 21.16N 78.35E

81 N22 **Katompi** Katanga, SE Dem. Rep. Congo (Zaire) 6.10S 26.19E

85 K14 **Katondwe** Lusaka, C Zambia 15.08S 30.10E

41 P14 **Katmai, Mount** ▲ Alaska, USA 58.16N 154.57W

160 J9 **Katni** Madhya Pradesh, C India 23.46N 80.28E

78 L11 **Kéti** *see* Kati

17 D19 **Káto Acháïa** *var.* Kato Ahaia, Káto Akhaía. Dytikí Ellás, S Greece 38.08N 21.33E

Kato Ahaia/Káto Akhaía *see* Káto Acháïa

161 G17 **Kato Lakatámeia** *var.* Kato Lakatamia. C Cyprus 35.07N 33.20E

Kato Lakatamia *see* Kato Lakatámeia

---

83 E18 **Katonga** ♦ S Uganda

117 F15 **Káto Ólympos** ▲ C Greece

117 D17 **Katoúna** Dytikí Ellás, C Greece 38.46N 21.07E

117 E19 **Káto Vlasiá** Dytikí Makedonía, S Greece 38.02N 21.54E

113 J16 **Katowice** *Ger.* Kattowitz. Śląskie, S Poland 50.14N 19.00E

159 S15 **Kátoya** West Bengal, NE India 23.39N 88.10E

142 E16 **Katrançik Dağı** ▲ SW Turkey

97 N16 **Katrineholm** Södermanland, C Sweden 58.58N 16.15E

98 I11 **Katrine, Loch** ◙ C Scotland, UK

79 V12 **Katsina** Katsina, N Nigeria 12.58N 7.33E

79 U12 **Katsina** ♦ *state* N Nigeria

69 P8 **Katsina Ala** ♦ S Nigeria

170 C11 **Katsumoto** Nagasaki, Iki, SW Japan 33.49N 129.42E

171 L16 **Katsuta** *var.* Katuta. Ibaraki, Honshū, S Japan 36.24N 140.31E

171 K17 **Katsuura** *var.* Katuura. Chiba, Honshū, SW Japan 35.09N 140.16E

171 I14 **Katsuyama** *var.* Katuyama. Fukui, Honshū, SW Japan 36.03N 136.28E

170 FfJ3 **Katsuyama** Okayama, Honshū, SW Japan 35.06N 133.43E

117 E14 **Kattakurgan** *see* Kattaqŭrghon

153 N11 **Kattaqŭrghon** *Rus.* Kattakurgan. Samarqand Wiloyati, C Uzbekistan 39.55N 66.11E

117 O23 **Kattavía** Ródos, Dodekánisos, Greece, Aegean Sea 35.56N 27.47E

97 I21 **Kattegat** *Dan.* Kattegatt. *strait* N Europe

Kattegatt *see* Kattegat

97 P19 **Katthammarsvik** Gotland, SE Sweden 57.27N 18.54E

Kattowitz *see* Katowice

9 N17 **Katun'** ♦ S Russian Federation

Katuta *see* Katsuta

Katuura *see* Katsuura

Katuyama *see* Katsuyama

100 G11 **Katwijk aan Zee** *var.* Katwijk. Zuid-Holland, W Netherlands 52.12N 4.24E

40 B8 **Kauai** *Haw.* Kaua'i. *island* Hawaiian Islands, Hawaii, USA, C Pacific Ocean

40 C8 **Kauai Channel** *channel* Hawaii, USA, C Pacific Ocean

175 Ss11 **Kaubalatmada, Gunung** *var.* Kaplamada. ▲ Pulau Buru, E Indonesia 3.16S 126.14E

203 U10 **Kauehi** *atoll* Îles Tuamotu, C French Polynesia

Kauen *see* Kaunas

103 K24 **Kaufbeuren** Bayern, S Germany 47.52N 10.37E

45 U7 **Kaufman** Texas, SW USA 32.35N 96.18W

103 I15 **Kaufungen** Hessen, C Germany 51.16N 9.39E

95 K17 **Kauhajoki** Länsi-Suomi, W Finland 62.24N 22.12E

95 K16 **Kauhava** Länsi-Suomi, W Finland 63.06N 23.07E

32 M7 **Kaukauna** Wisconsin, N USA 44.18N 88.18W

94 L11 **Kaukonen** Lappi, N Finland 67.28N 24.49E

40 A8 **Kaulakahi Channel** *channel* Hawaii, USA, C Pacific Ocean

40 E9 **Kaunakakai** Molokai, Hawaii, USA, C Pacific Ocean 21.05N 157.01W

40 F12 **Kauna Point** *headland* Hawaii, USA, C Pacific Ocean 19.02N 155.52W

120 F13 **Kaunas** *Ger.* Kauen, *Pol.* Kowno; *prev. Rus.* Kovno. Kaunas, C Lithuania 54.54N 23.57E

194 H10 **Kaup** East Sepik, NW PNG 3.48S 143.56E

79 U12 **Kaura Namoda** Zamfara, NW Nigeria 12.43N 6.17E

95 K16 **Kaustinen** Länsi-Suomi, W Finland 63.33N 23.40E

175 T7 **Kau, Teluk** *bay* Pulau Halmahera, E Indonesia

101 M23 **Kautenbach** Diekirch, NE Luxembourg 49.57N 6.01E

94 K10 **Kautokeino** Finnmark, N Norway 69.00N 23.01E

115 J16 **Kavadarci** *Turk.* Kavadar. C FYR Macedonia 41.25N 22.00E

116 M13 **Kavak Çayı** ♦ NW Turkey

116 I13 **Kavála** *prev.* Kaválla. Anatolikí Makedonía kai Thráki, NE Greece 40.57N 24.25E

116 I13 **Kavála, Kólpos** *gulf* Aegean Sea, NE Mediterranean Sea

127 Nn17 **Kavalerovo** Primorskiy Kray, SE Russian Federation 44.17N 135.04E

161 I21 **Kāvali** Andhra Pradesh, E India 15.04N 80.02E

161 C21 **Kavarna** Dobrich, NE Bulgaria 43.26N 28.21E

120 G12 **Kavarskas** Anykščiai, E Lithuania 55.27N 24.55E

78 I14 **Kavendou** ▲ C Guinea 10.49N 12.14W

115 F20 **Kavi** Gujarāt, W India 22.12N 72.40E

195 N9 **Kaviéng** *var.* Kaewieng. NE PNG 2.34S 150.48E

124 Nn3 **Kavīr, Dasht-e** *var.* Great Salt Desert. *salt pan* N Iran

149 Q6 **Kavīr, Dasht-e** *var.* Great Salt Desert. *salt pan* N Iran

197 I15 **Kavukavu Reef** *reef* Beqa Barrier Reef, Cakaubalavu Reef. *reef* Viti Levu, SW Fiji

84 G12 **Kavungo** Moxico, E Angola
11.31S 22.59E
Kazankteken see Qizqetkan
Kazanlık see Kazanlŭk
116 J9 **Kazanlŭk** prev. Kazanlik. Stara
Zagora, C Bulgaria 42.38N 25.24E
172 T17 **Kazan-rettō** Eng. Volcano
Islands. island group SE Japan
125 F12 **Kazanskoye** Tyumenskaya
Oblast', C Russian Federation
55.39N 69.06E
119 V12 **Kazantip, Mys** headland
S Ukraine 45.27N 35.50E
153 U9 **Kazarman** Narynskaya Oblast',
C Kyrgyzstan 41.21N 74.03E
Kazatin see Kozyatyn
Kazbegi see Kazbek
Kazbek see Kazbegi, Geor.
143 T9 **Kazbek** var. Kazbegi, Geor.
Mqinvartsveri. ▲ N Georgia
42.43N 44.28E
84 M13 **Kazembe** Eastern, NE Zambia
12.06S 32.45E
149 N11 **Käzerün** Färs, S Iran 29.40N 51.38E
129 R12 **Kazhym** Respublika Komi,
NW Russian Federation
60.19N 51.26E
Kazi Ahmad see Qāzi Ahmad
Kazi Magomed see
Qazimämmäd
142 H16 **Kazımkarabekir** Karaman,
S Turkey 37.13N 32.68E
113 M20 **Kazincbarcika** Borsod-Abaúj-
Zemplén, NE Hungary 48.15N 20.40E
121 H17 **Kazlowshchyna** Pol.
Kozlowszczyzna, Rus.
Kozlovshchina. Hrodzyenskaya
Voblasts', W Belarus 53.19N 25.18E
121 E14 **Kazlų Rūda** Marijampolė,
S Lithuania 54.45N 23.28E
150 E9 **Kaztalovka** Zapadnyy
Kazakhstan, W Kazakhstan
49.47N 48.40E
81 K22 **Kazumba** Kasai Occidental,
S Dem. Rep. Congo (Zaire)
6.19S 21.57E
171 Mm10 **Kazuno** Akita, Honshū, C Japan
40.08N 140.47E
Kazvin see Qazvin
120 J12 **Kaz'yany** Rus. Koz'yany.
Vitsyebskaya Voblasts',
NW Belarus 55.19N 26.52E
125 F19 **Kazym** ♂ N Russian Federation
112 H10 **Kcynia** Ger. Exin. Kujawsko-
pomorskie, C Poland 53.00N 17.29E
117 I20 **Kéa** Kéa, Kykládes, Greece,
Aegean Sea 37.22N 24.21E
117 I20 **Kéa** prev. Kéos, anc. Ceos. island
Kykládes, Greece, Aegean Sea
40 H11 **Keaau** Haw. Kea'au. Hawaii, USA,
C Pacific Ocean 19.36N 155.01W
40 F11 **Keahole Point** headland Hawaii,
USA, C Pacific Ocean
19.43S 156.03W
40 G12 **Kealakekua** Hawaii, USA,
C Pacific Ocean 19.31N 155.55W
40 H11 **Kea, Mauna** ▲ Hawaii, USA,
C Pacific Ocean 19.50N 155.30W
39 N10 **Keams** Arizona, SW USA
35.47N 110.09W
31 O16 **Kearney** Nebraska, C USA
40.42N 99.06W
38 L3 **Kearns** Utah, W USA
40.39N 112.00W
117 H26 **Kéas, Stenó** strait SE Greece
143 O14 **Keban Barajı** dam C Turkey
38.49N 38.46E
143 O14 **Keban Barajı** ☐ C Turkey
79 S13 **Kebbi** ◆ state NW Nigeria
78 G10 **Kébémèr** NW Senegal
15.24N 16.25W
76 M7 **Kebili** var. Qibili. C Tunisia
33.42N 9.06E
144 H4 **Kebir, Nahr el** ♂ NW Syria
82 A10 **Kebkabiya** Northern Darfur,
W Sudan 13.39N 24.04E
94 I11 **Kebnekaise** ▲ N Sweden
83 M14 **K'ebrī Dehar** Somali, E Ethiopia
6.43N 44.15E
154 K15 **Kech** ♂ SW Pakistan
8 K10 **Kechika** ♂ British Columbia,
W Canada
113 K23 **Kecskemét** Bács-Kiskun,
C Hungary 46.54N 19.41E
174 Gg7 **Kedah** ◆ state Peninsular Malaysia
120 F12 **Kėdainiai** Kėdainiai, C Lithuania
55.19N 24.00E
11 N13 **Kedgwick** New Brunswick,
SE Canada 47.37N 67.21W
174 Ll15 **Kediri** Jawa, C Indonesia
7.45S 112.01E
176 Y10 **Kedir Sarmi** Irian Jaya,
E Indonesia 2.00S 139.01E
169 V7 **Kedong** Heilongjiang, NE China
48.00N 126.15E
78 I12 **Kédougou** SE Senegal
12.34N 12.09W
126 Gg13 **Kedrovyy** Tomskaya Oblast',
C Russian Federation 57.31N 79.45E
113 H16 **Kędzierzyn-Kozle** Ger.
Heydebrech. Opolskie, S Poland
50.20N 18.12E
14 G6 **Keele** ♂ Northwest Territories,
NW Canada
8 K6 **Keele Peak** ▲ Yukon Territory,
NW Canada 63.31N 130.21W
Keelung see Chilung
21 N10 **Keene** New Hampshire, NE USA
42.56N 72.14W
174 H3 **Keerbergen** Vlaams Brabant,
C Belgium 51.01N 4.39E
85 E21 **Keetmanshoop** Karas,
S Namibia 26.36S 18.07E
12 C12 **Keewatin** Ontario, S Canada
49.46N 94.30W
31 V4 **Keewatin** Minnesota, N USA
47.24N 93.04W
117 B18 **Kefallinía** var. Kefallonía. island
Iónioi Nísoi, Greece,
C Mediterranean Sea
Kefallonía see Kefallinía
117 M22 **Kéfalos** Kós, Dodekánisos,
Greece, Aegean Sea 36.44N 26.58E
175 Rr17 **Kefamenanu** Timor, C Indonesia
9.31S 124.28E
144 F10 **Kefar Sava** var. Kfar Saba.
Central, C Israel 32.11N 34.54E
Kefe see Feodosiya
79 V15 **Keffi** Nassarawa, C Nigeria
8.52N 7.54E
94 H4 **Keflavík** ✈ (Reykjavík)
Reykjanes, W Iceland 63.58N 22.37W

94 H4 **Keflavík** Reykjanes, W Iceland
64.01N 22.35W
161 J25 **Kegalee** see Kegalla
Kegalla var. Kegalee. Kegalle.
Sabaragamuwa Province, C Sri
Lanka 7.13N 80.21E
Kegalle see Kegalla
152 H7 **Kegayli** Rus. Kegeyli.
Qoraqalpoghiston Respublikasi,
W Uzbekistan 42.46N 59.49E
Kegel see Keila
151 W16 **Kegen** Almaty, SE Kazakhstan
42.57N 79.15E
103 F22 **Kehl** Baden-Württemberg,
SW Germany 48.34N 7.49E
120 H3 **Kehra** Ger. Kedder. Harjumaa,
NW Estonia 59.19N 25.22E
119 U6 **Kehychivka** Kharkiv'ska Oblast',
E Ukraine 49.51N 35.12E
97 L17 **Keighley** N England, UK
53.51N 1.53W
Kei Islands see Kai, Kepulauan
Keijō see Sŏul
120 G3 **Keila** Ger. Kegel. Harjumaa,
NW Estonia 59.19N 24.28E
Keilberg see Klínovec
85 F23 **Keimoes** Northern Cape,
W South Africa 28.41S 20.57E
Keina/Keinis see Käina
176 Yy14 **Keisak** Irian Jaya, E Indonesia
7.01S 140.02E
Keishū see Kyŏngju
79 T11 **Keïta** Tahoua, C Niger 14.43N 5.45E
80 J12 **Kéita, Bahr** var. Doka.
♂ S Chad
95 M16 **Keitele** ◎ C Finland
190 K10 **Keith** South Australia
36.01S 140.22E
98 K8 **Keith** NE Scotland, UK
57.33N 2.57W
28 K3 **Keith Sebelius Lake** ☐ Kansas,
C USA
34 G11 **Keizer** Oregon, NW USA
44.59N 123.01W
40 A8 **Kekaha** Kauai, Hawaii, USA,
C Pacific Ocean 21.58N 159.43W
153 U10 **Kēk-Art** prev. Alaykel', Alay-Ku.
Oshskaya Oblast', SW Kyrgyzstan
40.15N 74.21E
153 W10 **Kēk-Aygyr** var. Keyaygyr.
Narynskaya Oblast', C Kyrgyzstan
42.40N 75.37E
153 V9 **Kēk-Dzhar** Narynskaya Oblast',
C Kyrgyzstan 41.28N 74.48E
12 L8 **Kekek** ♂ Quebec, SE Canada
193 K15 **Kekerengu** Canterbury, South
Island, NZ 41.55S 174.05E
113 L21 **Kékes** ▲ N Hungary 47.53N 19.59E
175 Rr17 **Kekneno, Gunung** ▲ Timor,
S Indonesia
153 S9 **Kēk-Tash** Kir. Kök-Tash. Dzhalal-
Abadskaya Oblast', W Kyrgyzstan
40.15N 72.59E
83 M15 **K'elafo** Somali, E Ethiopia
5.36N 44.12E
175 O6 **Kelai, Sungai** ♂ Borneo,
N Indonesia
174 L10 **Kelamayi** see Karamay
Kelang see Klang
174 H3 **Kelantan** ◆ state Peninsular
Malaysia
Kelantan see Kelantan, Sungai
174 H3 **Kelantan, Sungai** var. Kelantan.
♂ Peninsular Malaysia
Kelat see Kālat
Kelcyra see Këlcyrë
115 L22 **Këlcyrë** var. Këlcyra. Gjirokastër,
S Albania 40.19N 20.12E
152 L14 **Kelifskiy Uzboy** salt marsh
E Turkmenistan
143 O12 **Kelkit** Gümüşhane, NE Turkey
40.05N 39.25E
143 O12 **Kelkit Çayı** ♂ N Turkey
79 V11 **Kéllé** Zinder, S Niger 14.10N 10.10E
81 G18 **Kéllé** Cuvette, W Congo
0.04S 14.33E
151 P7 **Kellerovka** Severnyy Kazakhstan,
N Kazakhstan 53.51N 69.15E
15 H1 **Kellett, Cape** headland Banks
Island, Northwest Territories,
NW Canada 71.57N 125.55W
33 S11 **Kelleys Island** island Ohio, N
USA
35 N8 **Kellogg** Idaho, NW USA
47.30N 116.07W
94 M12 **Kelloselkä** Lappi, N Finland
66.55N 28.52E
99 F17 **Kells** Ir. Ceanannas. E Ireland
120 E12 **Kelmė** Kelmė, C Lithuania
55.39N 22.57E
101 M19 **Kelmis** var. La Calamine. Liège,
E Belgium 50.43N 6.01E
80 H12 **Kélo** Tandjilé, SW Chad
9.21N 15.49E
85 I14 **Kelongwa** North Western,
NW Zambia 13.41S 26.19E
9 X12 **Kelowna** British Columbia,
SW Canada 49.49N 119.28W
35 M6 **Kelseyville** California, W USA
38.58N 122.51W
98 K13 **Kelso** SE Scotland, UK
55.36N 2.27W
34 G10 **Kelso** Washington, NW USA
46.09N 122.54W
205 W15 **Keltie, Cape** headland Antarctica
174 Jj8 **Keluang** var. Kluang. Johor,
Peninsular Malaysia 2.01N 103.18E
174 L9 **Kelume** Pulau Lingga,
W Indonesia 0.12S 104.22E
9 U15 **Kelvington** Saskatchewan,
S Canada 52.10N 103.30W
21 R2 **Kem'** Respublika Kareliya,
NW Russian Federation
64.58N 34.39E
128 I7 **Kem'** ♂ NW Russian Federation
143 O13 **Kemah** Erzincan, E Turkey
39.34N 39.01E
143 N13 **Kemaliye** Erzincan, E Turkey
39.17N 38.30E
Kemaman see Cukai
Kemanlar see Isperikh
76 F6 **Kénitra** prev. Port-Lyautey.
NW Morocco 34.19N 6.29W
23 V9 **Kenly** North Carolina, SE USA
35.59N 78.16W
99 B21 **Kenmare** Ir. Neidín. S Ireland
51.52N 9.34W
29 N3 **Kenmare** North Dakota, N USA
48.40N 102.04W
99 A21 **Kenmare River** Ir. An Ribhéar.
inlet NE Atlantic Ocean
20 D10 **Kenmore** New York, NE USA
42.58N 78.52W
29 N7 **Kennebec** South Dakota, N USA
43.53N 99.52W
19 Q6 **Kennebec River** ♂ Maine,
NE USA
19 R7 **Kennebunk** Maine, NE USA
43.22N 70.33W
8 F9 **Kennedy Entrance** strait Alaska,
USA
177 Ff3 **Kennedy Peak** ▲ W Myanmar
23.29N 93.52E
24 K9 **Kenner** Louisiana, S USA
29.57N 90.15W
29 V9 **Kennett** Missouri, C USA
36.14N 90.03W
20 I16 **Kennett Square** Pennsylvania,
NE USA 39.50N 75.40W
34 K10 **Kennewick** Washington,
NW USA 46.11N 119.07W

12 G8 **Kenogami Lake** Ontario,
S Canada 48.04N 80.10W
12 F7 **Kenogamissi Lake** ◎ Ontario,
S Canada
8 I6 **Keno Hill** Yukon Territory,
NW Canada 63.54N 135.18W
10 A11 **Kenora** Ontario, S Canada
49.46N 94.25W
33 N9 **Kenosha** Wisconsin, N USA
42.34N 87.49W
28 L3 **Kensington** Prince Edward
Island, SE Canada 46.25N 63.39W
28 L3 **Kensington** Kansas, C USA
39.46N 99.01W
34 I11 **Kent** Oregon, NW USA
45.14N 120.43W
24 H8 **Kent** Texas, SW USA
31.03N 104.13W
34 H8 **Kent** Washington, NW USA
47.22N 122.13W
99 P22 **Kent** cultural region SE England,
UK
35 S16 **Kenton** Ohio, N USA
40.39N 83.36W
15 J4 **Kent Peninsula** peninsula
Nunavut, N Canada
117 F14 **Kentrikí Makedonía** Eng.
Macedonia Central. ◆ region
N Greece
22 J6 **Kentucky** off. Commonwealth of
Kentucky; also known as The
Bluegrass State. ◆ state C USA
22 H8 **Kentucky Lake** ◎
Kentucky/Tennessee, S USA
11 P15 **Kentville** Nova Scotia, SE Canada
45.04N 64.30W
24 K8 **Kentwood** Louisiana, S USA
30.56N 90.30W
33 P9 **Kentwood** Michigan, N USA
42.52N 85.33W
83 H7 **Kenya** off. Republic of Kenya.
◆ republic E Africa
41 R12 **Kenai, Mount** see Kirinyaga
(0) D5 **Kenai Mountains** ▲ Alaska,
USA
41 R12 **Kenai Peninsula** peninsula
Alaska, USA
23 V11 **Kenansville** North Carolina,
SE USA 34.57N 77.54W
124 Pp15 **Kenâyis, Râs el-** headland
N Egypt 31.13N 27.53E
39 W16 **Kendal** NW England, UK
54.19N 2.45W
25 Y16 **Kendall** Florida, SE USA
25.39N 80.18W
15 Ll6 **Kendall, Cape** headland Nunavut,
E Canada 63.31N 87.09W
20 J15 **Kendall Park** New Jersey,
NE USA 40.25N 74.33W
33 Q11 **Kendallville** Indiana, N USA
41.24N 85.10W
175 Qq12 **Kendari** Sulawesi, C Indonesia
3.57S 122.36E
174 L10 **Kendawangan** Borneo,
C Indonesia 2.31S 110.13E
160 O12 **Kendräparha** see Kendrapara
Kendräparha prev. Kendräpara.
Orissa, E India 20.29N 86.25E
160 O11 **Kendujhargarh** prev.
Keonjhargarh. Orissa, E India
21.42N 85.36E
45 S13 **Kenedy** Texas, SW USA
28.49N 97.51W
152 E13 **Kënekesir** Turkm. Könekesir.
Balkanskiy Velayat,
W Turkmenistan 38.16N 56.51E
161 F22 **Kerala** ◆ state S India
194 H10 **Keram** ♂ N PNG
172 O14 **Kerama-rettō** island group
SW Japan
191 N10 **Kerang** Victoria, SE Australia
35.46S 144.01E
152 G8 **Këneurgench** Turkm.
Köneürgench; prev. Kunya-
Urgench. Dashkhovuzskiy Velayat,
N Turkmenistan 42.20N 59.09E
117 H19 **Keratéa** var. Keratia. Attikí,
C Greece 37.48N 23.58E
81 M19 **Kerava** Swe. Kervo. Etelä-Suomi,
S Finland 60.22N 25.01E
82 Q11 **Kerba/Kerbela** see Karbalā'
34 F15 **Kerby** Oregon, NW USA
42.10N 123.39W
10 H11 **Kerch** Rus. Kerch'. Respublika
Krym, SE Ukraine 45.22N 36.30E
10 H13 **Kerch** Rus. Kerch'. Respublika
**Kerchens'ka
Protska/Kerchenskiy Proliv**
see Kerch Strait
119 V13 **Kerchens'kyy Pivostriv**
peninsula S Ukraine
124 R4 **Kerch Strait** var. Bosporus
Cimmerius, Enikale Strait, Rus.
Kerchenskiy Proliv, Ukr.
Kerchens'ka Protska. strait Black
Sea/Sea of Azov
194 J14 **Kerema** Gulf, S PNG 7.58S 145.46E
82 J9 **Keremitlik** see Lyulyakovo
142 J9 **Kerempe Burnu** headland
N Turkey 42.01N 33.20E
83 H19 **Kericho** Rift Valley, W Kenya
0.21S 35.16E
192 K2 **Kerikeri** Northland, North
Island, NZ 35.13S 173.57E
95 K15 **Kerimäki** Itä-Suomi, E Finland
54.30N 3.03W
82 Q7 **Keren** var. Cheren. C Eritrea
15.45N 38.22E
192 M6 **Kerepehi** Waikato, North Island,
NZ 37.18S 175.33E
95 P10 **Kerey, Ozero** ◎ C Kazakhstan
177 Ff3 **Kergel** ▲
181 O13 **Kerguelen** var. C French
Southern and Antarctic Territories
181 O13 **Kerguelen Plateau** undersea
feature S Indian Ocean
41 Y14 **Ketchikan** Revillagigedo Island,
Alaska, USA 55.20N 131.39W
35 O14 **Ketchum** Idaho, NW USA
43.40N 114.21W
174 L10 **Ketapang** Borneo, C Indonesia
1.49S 109.58E
79 R17 **Ket** SE Ghana 5.54N 1.02E
174 Kk10 **Ketapang** Jawa, C Indonesia
1.49S 109.58E
131 O12 **Ketchenery** Respublika
Kalmykiya, SW Russian Federation
47.18N 44.31E

151 W16 **Ketmen', Khrebet**
▲ SE Kazakhstan
79 S16 **Kétou** SE Benin 7.25N 2.36E
112 M7 **Kętrzyn** Ger. Rastenburg.
Warmińsko-Mazurskie, NE
Poland, 54.03N 21.22E
99 N20 **Kettering** C England, UK
52.24N 0.43W
33 R14 **Kettering** Ohio, N USA
39.41N 84.10W
20 F13 **Kettle Creek** ♂ Pennsylvania,
NE USA
34 L7 **Kettle Falls** Washington,
NW USA 48.36N 118.03W
21 D16 **Kettle Point** headland Ontario,
S Canada 43.12N 82.01W
31 V6 **Kettle River** ♂ Minnesota,
N USA
194 E12 **Ketu** ♂ W PNG
20 G10 **Keuka Lake** ◎ New York,
NE USA
Keupriya see Primorsko
95 L17 **Keuruu** Länsi-Suomi, W Finland
62.15N 24.34E
Kevevára see Kovin
94 L9 **Kevo** Lapp. Geavvú. Lappi,
N Finland 69.42N 27.00E
95 M6 **Kew** North Caicos, N Turks and
Caicos Islands 21.52N 71.57W
32 K11 **Kewanee** Illinois, N USA
41.15N 89.55W
33 N7 **Kewaunee** Wisconsin, N USA
44.27N 87.31W
33 M3 **Keweenaw Bay** ◎ Michigan,
N USA
33 N2 **Keweenaw Peninsula** peninsula
Michigan, N USA 47.15N 88.19W
33 N2 **Keweenaw Point** headland
Michigan, N USA 47.24N 87.42W
31 N12 **Keya Paha River**
♂ Nebraska/South Dakota,
N USA
25 Z16 **Key Biscayne** Florida, SE USA
25.41N 80.09W
28 G8 **Keyes** Oklahoma, C USA
36.48N 102.15W
25 Y17 **Key Largo** Key Largo, Florida,
SE USA 25.06N 80.24W
23 U3 **Keyser** West Virginia, NE USA
39.26N 78.58W
39 O9 **Keystone Lake** ◎ Oklahoma,
C USA
38 L16 **Keystone Peak** ▲ Arizona,
SW USA 31.52N 111.12W
23 U7 **Keystone State** see Pennsylvania
23 T3 **Keysville** Virginia, NE USA
37.02N 78.28W
29 T3 **Keytesville** Missouri, C USA
39.25N 92.56W
25 W17 **Key West** Florida Keys, Florida,
SE USA 24.33N 81.47W
131 T1 **Kez** Udmurtskaya Respublika,
NW Russian Federation
57.55S 53.42E
Kezdivásárhely see Târgu
Secuiesc
126 J13 **Kezhma** Krasnoyarskiy Kray,
C Russian Federation 58.57N 101.00E
113 L18 **Kéžmárk** Prešovský Kraj,
E Slovakia 49.09N 20.25E
85 F20 **Kgalagadi** ◆ district
SW Botswana
85 I20 **Kgatleng** ◆ district
S Botswana
196 F8 **Kgkeklau** Babeldaob, N Palau
129 R6 **Khabarikha** var. Chabarikha.
Respublika Komi, NW Russian
Federation 65.52N 52.33E
127 N16 **Khabarovsk** Khabarovskiy Kray,
SE Russian Federation
48.31N 135.07E
126 Mm12 **Khabarovskiy Kray** ◆ territory
E Russian Federation
147 W7 **Khabb** Abū Ẓaby, E UAE
24.39N 55.43E
Khabour, Nahr al see Khābūr,
Nahr al
145 N2 **Khabura** see Al Khābūrah
Khābūr, Nahr al var. Nahr
al Khabour. ♂ Syria/Turkey
55.27N 10.40E
82 B12 **Khadari** ♂ W Sudan
Khadera see Hadera
147 X12 **Khadra** var. Khudal. SE Oman
18.48N 56.48E
161 E14 **Khadki** prev. Kirkee.
Mahārāshtra, W India 18.34N 73.52E
130 L14 **Khadyzhensk** Krasnodarskiy
Kray, SW Russian Federation
44.26N 39.31E
116 N9 **Khadzhiyska Reka**
♂ E Bulgaria
119 P10 **Khadzhybeys'kyy Lyman**
♂ SW Ukraine
144 K3 **Khafsah** Ḥalab, N Syria
36.16N 38.03E
158 M13 **Khāga** Uttar Pradesh, N India
25.48N 81.04E
159 Q13 **Khagaria** Bihār, NE India
25.31N 86.27E
155 Q13 **Khairpur** Sind, SE Pakistan
27.30N 68.49E
126 Hh5 **Khakasiya, Respublika** prev.
Khakasskaya Avtonomnaya
Oblast', Eng. Khakassia. ◆
autonomous republic C Russian
Federation
**Khakassia/Khakasskaya
Avtonomnaya Oblast'** see
Khakasiya, Respublika
178 H9 **Kha Khaeng, Khao**
▲ W Thailand 16.13N 99.03E
85 G20 **Khakhea** var. Kakia. Southern,
S Botswana 24.40S 23.28E
152 L13 **Khalach** Lebapskiy Velayat,
E Turkmenistan 38.00N 64.46E
131 W7 **Khalándrion** see Chalándri
Khalílovo Orenburgskaya
Oblast', W Russian Federation
51.25N 58.13E
Khalkabad see Khalqobod
148 L3 **Khalkhāl** prev. Herowābād.
Ardabīl, NW Iran 37.40N 48.34E
Khalkidhikí see Chalkidikí
Khalkís see Chalkída
131 W3 **Khal'mer-Yu** Respublika Komi,
NW Russian Federation
68.00N 64.40E
121 M14 **Khalopyenichy** Rus.
Kholopenichi. Minskaya Voblasts',
NE Belarus 54.31N 28.58E

◆ COUNTRY ◇ DEPENDENT TERRITORY ◆ ADMINISTRATIVE REGION ▲ MOUNTAIN ✗ VOLCANO ◎ LAKE
● COUNTRY CAPITAL ○ DEPENDENT TERRITORY CAPITAL ✗ INTERNATIONAL AIRPORT ▲ MOUNTAIN RANGE ♂ RIVER ☐ RESERVOIR

275

152 H7 **Khalqobod** *Rus.* Khalkabad.
Qoraqalpoghiston Respublikasi,
W Uzbekistan 42.42N 59.46E
Khalturin *see* Orlov
147 Y10 **Khalūf** *var.* Al Khalūf. E Oman
20.27N 57.58E
160 K10 **Khamaria** Madhya Pradesh,
C India 23.07N 80.54E
160 D11 **Khambhāt** Gujarāt, W India
22.19N 72.39E
160 C12 **Khambhāt, Gulf of** *Eng.* Gulf of
Cambay. *gulf* W India
178 K10 **Khâm Đưc** Quang Nam–Đa
Năng, C Vietnam 15.28N 107.49E
160 G12 **Khāmgaon** Mahārāshtra, C India
20.40N 76.34E
147 O14 **Khamir** *var.* Khamr. W Yemen
15.56N 43.56E
147 N12 **Khamis Mushayt** *var.* Hamis
Musait. 'Asir, SW Saudi Arabia
18.19N 42.41E
126 L10 **Khampa** Respublika Sakha
(Yakutiya), NE Russian Federation
63.43N 123.02E
Khamr *see* Khamir
85 C19 **Khan** *≈* W Namibia
155 Q2 **Khānābād** Kunduz,
NE Afghanistan 36.42N 69.07E
**Khān Abou Châmâte/Khan
Abou Ech Cham** *see* Khān Abū
Shāmāt
144 I7 **Khān Abū Shāmāt** *var.* Khān
Abou Châmâte, Khan Abou Ech
Cham. Dimashq, W Syria
33.43N 36.56E
Khān al Baghdādī *see*
Al Baghdādī
145 T7 **Khān al Maḥāwīl** *see* Al Maḥāwīl
145 T10 **Khān al Mashāhidah** C Iraq
33.40N 44.15E
145 T10 **Khān al Muṣallá** S Iraq
32.09N 44.19E
145 U6 **Khānaqin** E Iraq 34.22N 45.22E
145 T11 **Khān ar Ruḥbah** S Iraq
31.42N 44.18E
145 S2 **Khān as Sūr** N Iraq 36.28N 41.36E
145 T8 **Khān Āzād** C Iraq 33.07N 44.22E
160 N13 **Khandaparha** *prev.* Khandpara.
Orissa, E India 20.15N 85.10E
Khandpara *see* Khandaparha
155 T2 **Khandūd** *var.* Khandud, Wakhan.
Badakhshān, NE Afghanistan
36.57N 72.19E
160 G11 **Khandwa** Madhya Pradesh,
C India 21.49N 76.22E
126 Mm10 **Khandyga** Respublika Sakha
(Yakutiya), NE Russian Federation
62.39N 135.30E
155 T10 **Khānewāl** Punjab, NE Pakistan
30.18N 71.55E
155 S10 **Khāngarh** Punjab, E Pakistan
29.56N 71.10E
Khanh Hung *see* Soc Trăng
Khaniá *see* Chaniá
Khanka *see* Khonqa
169 Z8 **Khanka, Lake** *var.* Hsing-k'ai
Hu, Lake Hanka, *Chin.* Xingkai
Hu, *Rus.* Ozero Khanka.
☺ China/Russian Federation
Khanka, Ozero *see* Khanka, Lake
Khankendi *see* Xankändi
Khanlar *see* Xanlar
126 Kk10 **Khannya** *≈* NE Russian
Federation
155 S12 **Khānpur** Punjab, SE Pakistan
23.37N 70.40E
155 S12 **Khānpur** Punjab, E Pakistan
28.31N 70.30E
144 I4 **Khān Shaykhūn** *var.* Khan
Sheikhun. Idlib, NW Syria
35.27N 36.37E
Khan Sheikhun *see* Khān
Shaykhūn
151 S15 **Khantau** Zhambyl, S Kazakhstan
44.13N 73.47E
151 W16 **Khan Tengri, Pik**
▲ SE Kazakhstan 42.13N 80.13E
178 J9 **Khanthabouli** *prev.*
Savannakhét. Savannakhét, S Laos
16.37N 104.48E
125 Ff10 **Khanty-Mansiysk** *prev.*
Ostyako-Voguls'k. Khanty-
Mansiyskiy Avtonomnyy Okrug,
C Russian Federation 61.01N 69.00E
129 V8 **Khanty-Mansiyskiy
Avtonomnyy Okrug** *◇*
autonomous district C Russian
Federation
145 R4 **Khānūqah** C Iraq 35.25N 43.15E
144 E11 **Khān Yūnis** *var.* Khān Yūnus.
S Gaza Strip 31.23N 34.19E
Khān Yūnus *see* Khān Yūnis
Khanzi *see* Ghanzi
145 U5 **Khān Zūr** E Iraq 35.03N 45.08E
178 H10 **Khao Laem Reservoir** *☐*
☐ W Thailand
126 Kk17 **Khapcheranga** Chitinskaya
Oblast', S Russian Federation
49.46N 112.21E
131 Q12 **Kharabali** Astrakhanskaya
Oblast', SW Russian Federation
47.28N 47.14E
159 R16 **Kharagpur** West Bengal,
NE India 22.30N 87.19E
145 V11 **Kharā'ib 'Abd al Karīm** S Iraq
31.07N 45.33E
149 Q8 **Kharānaq** Yazd, C Iran
31.54N 54.21E
Kharbin *see* Harbin
152 H13 **Khardzhagaz** Akhalskiy Velayat,
C Turkmenistan 37.54N 60.10E
160 F11 **Khargon** Madhya Pradesh,
C India 21.49N 75.39E
155 V7 **Khāriān** Punjab, NE Pakistan
32.52N 73.52E
119 V5 **Kharisyz'k** Donets'ka Oblast',
E Ukraine 48.01N 38.10E
119 U5 **Kharkiv** *Rus.* Khar'kov.
Kharkivs'ka Oblast', NE Ukraine
50.00N 36.14E
119 V5 **Kharkiv** *≈* Kharkivs'ka Oblast',
NE Ukraine 49.54N 36.20E
Kharkiv, Rus. Khar'kovskaya
Oblast'. *◇ province* E Ukraine
Khar'kov *see* Kharkiv
Khar'kovskaya Oblast' *see*
Kharkivs'ka Oblast'

128 L3 **Kharlovka** Murmanskaya
Oblast', NW Russian Federation
68.47N 37.09E
116 K11 **Kharmanli** Khaskovo, S Bulgaria
41.55N 25.54E
116 K11 **Kharmanliyska Reka**
≈ S Bulgaria
128 M13 **Kharovsk** Vologodskaya Oblast',
NW Russian Federation
59.57N 40.05E
82 F9 **Khartoum** *var.* El Khartûm,
Khartum. *●* (Sudan) Khartoum,
C Sudan 15.33N 32.31E
82 F9 **Khartoum** *☓* state NE Sudan
82 F9 **Khartoum** *☓* Khartoum,
C Sudan 15.35N 32.33E
82 F9 **Khartoum North** Khartoum,
C Sudan 15.37N 32.33E
119 X8 **Khartsyz'k** Khartsyzsk.
Donets'ka Oblast', SE Ukraine
48.01N 38.10E
Khartsyzsk *see* Khartsyz'k
Khartum *see* Khartoum
127 N18 **Khasan** Primorskiy Kray,
SE Russian Federation
42.24N 130.45E
131 P16 **Khasavyurt** Respublika
Dagestan, SW Russian Federation
43.16N 46.33E
149 W12 **Khash** *prev.* Vāsht. Sīstān va
Balūchestān, SE Iran 28.15N 61.11E
154 K8 **Khāsh, Dasht-e** *Eng.* Khash
Desert. *desert* SW Afghanistan
Khash Desert *see* Khāsh,
Dasht-e
82 H9 **Khashm el Girba** *var.* Khashm
al Qirba, Khashm al Qirbah.
Kassala, E Sudan 15.00N 35.59E
144 G14 **Khashsh, Jabal al** *▲* S Jordan
159 V13 **Khāsi Hills** *hill range* NE India
116 K11 **Khaskovo** Khaskovo, S Bulgaria
41.56N 25.34E
116 K11 **Khaskovo** *◇ province* S Bulgaria
126 J7 **Khatanga** Taymyrskiy (Dolgano-
Nenetskiy) Avtonomnyy Okrug,
N Russian Federation 71.55N 102.17E
126 J7 **Khatanga** *≈* N Russian
Federation
126 J6 **Khatanga, Gulf of** *see*
Khatangskiy Zaliv
126 J6 **Khatangskiy Zaliv** *var.* Gulf of
Khatanga. *bay* N Russian
Federation
147 W7 **Khatmat al Malāḥah** N Oman
24.56N 56.22E
149 S16 **Khatmat al Malāḥah** Ash
Shāriqah, E UAE
127 Q6 **Khatyrka** Chukotskiy
Avtonomnyy Okrug, NE Russian
Federation 62.03N 175.09E
152 I14 **Khauz-Khan** *Turkm.* Hanhowuz.
Akhalskiy Velayat, S Turkmenistan
37.15N 61.12E
152 I14 **Khauzkhanskoye
Vodokhranilishche**
☐ S Turkmenistan
85 D19 **Khavaling** *see* Khovaling
85 D19 **Khavast** *see* Khowos
145 W10 **Khawr, Nahr al** *≈* S Iraq
Khawr Barakah *see* Baraka
147 W7 **Khawr Fakkān** *var.* Khor
Fakkan. Ash Shāriqah, NE UAE
25.21N 56.19E
146 L6 **Khaybar** Al Madīnah, NW Saudi
Arabia 25.52N 39.15E
Khaybar, Kowtal-e *see* Khyber
Pass
153 S11 **Khaydarkan** *var.* Khaydarken.
Oshskaya Oblast', SW Kyrgyzstan
39.56N 71.16E
Khaydarken *see* Khaydarkan
129 U2 **Khaypudyrskaya Guba** *bay*
NW Russian Federation
145 S1 **Khazar, Baḥr-e/Khazar, Daryā-
ye** *see* Caspian Sea
Khazarasp *see* Hazorasp
Khazretishi, Khrebet *see*
Hazratishoh, Qatorkühi
Khelat *see* Kālat
76 M7 **Khemisset** NW Morocco
33.52N 6.04W
178 J10 **Khemmarat** *var.* Kemarat. Ubon
Ratchathani, E Thailand
16.03N 105.10E
75 L6 **Khenchela** *var.* Khenchla.
NE Algeria 35.22N 7.09E
76 G7 **Khenifra** C Morocco 32.59N 5.37W
119 R10 **Khersân, Rūd-e** *see* Garm, Āb-e
119 R10 **Kherson** Khersons'ka Oblast',
S Ukraine 46.39N 32.37E
119 S14 **Khersones, Mys** *Rus.* Mys
Khersonesskiy. *headland* S Ukraine
44.34N 33.24E
Khersonesskiy, Mys *see*
Khersones, Mys
119 R10 **Khersons'ka Oblast'** *var.*
Kherson, *Rus.* Khersonskaya
Oblast'. *◇ province* S Ukraine
Khersonskaya Oblast' *see*
Khersons'ka Oblast'
126 J7 **Kheta** Taymyrskiy (Dolgano-
Nenetskiy) Avtonomnyy Okrug,
N Russian Federation 71.33N 99.40E
126 J7 **Kheta** *≈* N Russian Federation
85 C18 **Khiam** *see* El Khiyam
128 K12 **Khibiny** *▲* NW Russian
Federation
126 K16 **Khilok** Chitinskaya Oblast',
S Russian Federation 51.26N 110.25E
126 K16 **Khilok** *≈* S Russian Federation
130 K3 **Khimki** Moskovskaya Oblast',
W Russian Federation 55.57N 37.48E
153 S12 **Khingov** *Rus.* Obi-Khingou.
≈ C Tajikistan
155 R15 **Khipro** Sind, SE Pakistan
25.50N 69.21E
145 S10 **Khirr, Wādī al** *dry watercourse*
S Iraq
116 I10 **Khisarya** Plovdiv, C Bulgaria
42.33N 24.43E

552 H9 **Khiva** *see* Khiwa
152 H9 **Khiwa** *Rus.* Khiva. Khorazm
Wiloyati, W Uzbekistan
41.22N 60.21E
178 H9 **Khlong Khlung** Kamphaeng
Phet, W Thailand 16.15N 99.41E
178 Gg16 **Khlong Thom** Krabi,
SW Thailand 7.55N 99.09E
178 I12 **Khlung** Chantaburi, S Thailand
12.25N 102.12E
76 I7 **Khmel'nik** *see* Khmil'nyk
Khmel'nitskaya Oblast' *see*
Khmel'nyts'ka Oblast'
Khmel'nyts'kiy *see* Khmel
'nyts'kyy
118 K5 **Khmel'nyts'ka Oblast'** *var.*
Khmel'nyts'kyy, *Rus.*
Khmel'nitskaya Oblast'; *prev.*
Kamenets-Podol'skaya Oblast'. *◇*
province NW Ukraine
118 L6 **Khmel 'nyts'kyy** *Rus.*
Khmel'nitskiy; *prev.* Proskurov.
Khmel'nyts'ka Oblast', W Ukraine
49.24N 26.59E
Khmel'nyts'kyy *see*
Khmel'nyts'ka Oblast'
118 M6 **Khmil'nyk** *Rus.* Khmel'nik.
Vinnyts'ka Oblast', C Ukraine
49.36N 27.59E
143 R9 **Khobi** W Georgia 42.20N 41.54E
121 P15 **Khodasy** *Rus.* Khodosy.
Mahilyowskaya Voblasts', E Belarus
53.56N 31.28E
118 I6 **Khodorov** *Pol.* Chodorów, *Rus.*
Khodorov. L'vivs'ka Oblast',
NW Ukraine 49.19N 24.19E
Khodorov *see* Khodoriv
Khodosy *see* Khodasy
152 D12 **Khodzhakala** *Turkm.* Hojagala.
Balkanskiy Velayat,
W Turkmenistan 38.46N 56.14E
152 M13 **Khodzhambas** *Turkm.*
Hojambaz. Lebapskiy Velayat,
E Turkmenistan 38.11N 64.33E
Khodzhent *see* Khüjand
Khodzheyli *see* Khüjayli
Khoi *see* Khvoy
130 L8 **Khokhol'skiy** Voronezhskaya
Oblast', W Russian Federation
51.33N 38.43E
178 Hh10 **Khok Samrong** Lop Buri,
C Thailand 15.03N 100.43E
155 P2 **Kholm** *var.* Tashqurghan, *Pash.*
Khulm. Balkh, N Afghanistan
36.42N 67.40E
128 H15 **Kholm** Novgorodskaya Oblast',
W Russian Federation 57.10N 31.06E
Kholm *see* Chelm
159 T16 **Kholmech** *see* Kholmyech
159 T16 **Kholmsk** Ostrov Sakhalin,
Sakhalinskaya Oblast', SE Russian
Federation 46.57N 142.10E
121 O19 **Kholmyech** *Rus.* Kholmech'.
Homyel'skaya Voblasts', SE Belarus
52.09N 30.37E
159 P16 **Kholon** *see* Holon
85 D19 **Kholopenichi** *see* Khalopyenichy
85 D19 **Khomas** *◇ district* C Namibia
Khomas Hochland *var.*
Khomasplato. *plateau* C Namibia
Khomasplato *see* Khomas
Hochland
148 M7 **Khomein** *see* Khomeyn
147 R7 **Khomeyn** *var.* Khomein,
Khumain. Markazi, W Iran
33.37N 50.03E
149 N8 **Khomeynīshahr** *prev.*
Homāyūnshahr. Esfahān, C Iran
32.39N 51.34E
Khoms *see* Al Khums
178 I9 **Khong Sedone** *see* Muang
Khôngxédôn
178 I9 **Khon Kaen** *var.* Muang Khon
Kaen. Khon Kaen, E Thailand
16.25N 102.49E
152 I9 **Khonqa** *Rus.* Khanka. Khorazm
Wiloyati, W Uzbekistan
41.31N 60.39E
178 I9 **Khon San** Khon Kaen,
E Thailand 16.40N 101.51E
127 N16 **Khonuu** Respublika Sakha
(Yakutiya), NE Russian Federation
66.24N 143.15E
131 N8 **Khopër** *var.* Khoper.
≈ SW Russian Federation
127 Nn16 **Khor** Khabarovskiy Kray,
SE Russian Federation
47.43N 134.48E
127 Nn16 **Khor** *≈* SE Russian Federation
149 S6 **Khorāsān** *off.* Ostān-e Khorāsān,
var. Khorassan, Khurasan. *◇*
province NE Iran
Khorassan *see* Khorāsān
Khorat *see* Nakhon Ratchasima
152 H9 **Khorazm Wiloyati** *Rus.*
Khorezmskaya Oblast'. *◇ province*
W Uzbekistan
160 O13 **Khordha** *prev.* Khurda. Orissa,
E India 20.13N 85.39E
129 U4 **Khorey-Ver** Nenetskiy
Avtonomnyy Okrug, NW Russian
Federation 67.25N 58.05E
Khorezmskaya Oblast' *see*
Khorazm Wiloyati
151 W15 **Khorgos** Almaty, SE Kazakhstan
44.13N 80.22E
126 K16 **Khorinsk** Respublika Buryatiya,
S Russian Federation 52.13N 109.52E
85 C18 **Khorixas** Kunene, NW Namibia
20.22S 14.55E
119 S5 **Khorol** Poltavs'ka Oblast',
NE Ukraine 49.49N 33.16E
148 L7 **Khorramābād** *var.*
Khurramabad. Lorestān, W Iran
33.28N 48.21E
148 K10 **Khorramshahr** *var.*
Khurramshahr, Muhammerah;
prev. Mohammerah. Khūzestān,
SW Iran 30.29N 48.09E
153 S14 **Khorugh** *Rus.* Khorog.
S Tajikistan 37.29N 71.31E
131 Q12 **Khosheutovo** Astrakhanskaya
Oblast', SW Russian Federation
47.04N 47.49E

794 M8 **Khotan** *see* Hotan
Khorvot Khalutsa *see* Horvot
Haluza
97 F22 **Khotsimsk** *Rus.* Khotsimsk.
Mahilyowskaya Voblasts', E Belarus
53.24N 32.34E
121 R16 **Khotsimsk** *Rus.* Khotsimsk.
Mahilyowskaya Voblasts', E Belarus
53.24N 32.34E
118 K7 **Khotyn** *Rom.* Hotin, *Rus.* Khotin.
Chernivets'ka Oblast', W Ukraine
48.29N 26.30E
76 I7 **Khouribga** C Morocco
32.54N 6.51W
153 Q13 **Khovaling** *Rus.* Khavaling.
SW Tajikistan 38.22N 69.54E
Khovd *see* Hovd
153 P11 **Khowos** *var.* Ursat'yevskaya, *Rus.*
Khavast. Sirdaryo Wiloyati,
E Uzbekistan 40.14N 68.46E
155 R2 **Khowst** Paktiā, E Afghanistan
33.22N 69.57E
Khoy *see* Khvoy
121 N20 **Khoyniki** *Rus.* Khoyniki.
Homyel'skaya Voblasts', SE Belarus
51.53N 29.58E
32 J8 **Kickapoo River** *≈* Wisconsin,
N USA
9 P16 **Kicking Horse Pass** *pass*
Alberta/British Columbia,
SW Canada 51.27N 116.13W
79 R9 **Kidal** Kidal, C Mali 18.22N 1.21E
79 R8 **Kidal** *◇ region* NE Mali
179 R16 **Kidapawan** Mindanao,
S Philippines 7.02N 125.04E
99 L20 **Kidderminster** C England, UK
52.22N 2.13W
191 U3 **Kidira** E Senegal 14.27N 12.18W
192 O11 **Kidnappers, Cape** *headland*
North Island, NZ 41.13S 175.15E
102 J8 **Kiel** Schleswig-Holstein,
N Germany 54.21N 10.04E
99 B20 **Kilbeo** Ir. Cill Airne.
SW Ireland 52.03N 9.30W
30 K4 **Kielder Bucht** *bay* N Germany
30 J4 **Kieler Förde** *inlet* N Germany
47 V15 **Kielce** *Rus.* Keltsy.
Świętokrzyskie, C Poland
50.52N 20.39E
Kiev *see* Kyyiv
117 C19 **Kiev Reservoir** *see* Kyyivs'ke
Vodoskhovyshche
78 J10 **Kiffa** Assaba, S Mauritania
16.37N 11.22W
117 H19 **Kifisiá** Attikí, C Greece
38.04N 23.49E
117 E19 **Kifisós** *≈* C Greece
145 U5 **Kifri** N Iraq 34.43N 44.58E
83 D20 **Kigali** *●* (Rwanda) C Rwanda
1.58S 30.02E
83 E21 **Kiği** Bingöl, E Turkey 39.19N 40.19E
83 E21 **Kigoma** Kigoma, W Tanzania
4.52S 29.36E
83 E21 **Kigoma** *◇ region* W Tanzania
40 F10 **Kihei** *Haw.* Kīhei. Maui, Hawaii,
USA, C Pacific Ocean
20.47N 156.28W
95 K17 **Kihniö** Länsi-Suomi, W Finland
62.10N 23.10E
120 F6 **Kihnu** *var.* Kihnu Saar, *Ger.*
Kühnö. *island* SW Estonia
Kihnu Saar *see* Kihnu
40 A8 **Kiʻi Landing** Niʻihau, Hawaii,
USA, C Pacific Ocean
21.58N 160.03W
95 J16 **Kiiminki** Oulu, C Finland
65.05N 25.46E
171 Gg16 **Kii-sanchi** *▲* Honshū, SW Japan
171 Gg16 **Kiistala** Lappi, N Finland
67.52N 25.19E
94 L11 **Kiistala** Lappi, N Finland
67.52N 25.19E
172 R14 **Kii-suidō** *strait* S Japan
172 R14 **Kikai-shima** *var.* Kikaiga-shima.
island Nansei-shotō, SW Japan
114 M8 **Kikinda** *Ger.* Grosskikinda,
Hung. Nagykikinda; *prev.* Velika
Kikinda. Serbia, N Yugoslavia
45.48N 20.29E
172 N7 **Kikladhes** *see* Kykládes
194 G13 **Kikori** Gulf, S PNG 7.31S 144.16E
194 G13 **Kikori** *≈* W PNG
170 Cc14 **Kikuchi** *var.* Kikuti. Kumamoto,
Kyūshū, SW Japan 33.00N 130.49E
Kikuti *see* Kikuchi
131 Q7 **Kikvidze** Volgogradskaya Oblast',
SW Russian Federation
50.47N 42.58E
149 N12 **Khvormūj** *var.* Khormuj.
Büshehr, S Iran 28.32N 51.22E
148 I2 **Khvoy** *var.* Khoi, Khoy.
Āzarbāyjān-e Bākhtarī, NW Iran
38.36N 45.03E
97 K15 **Kil** Värmland, C Sweden
59.30N 13.19E
96 N12 **Kilafors** Gävleborg, C Sweden
61.13N 16.34E
155 S5 **Khyber Pass** *var.* Kowtal-e
Khaybar. *pass*
Afghanistan/Pakistan 34.07N 71.05E
195 V14 **Kia** Santa Isabel, N Solomon
Islands 7.34S 158.31E
191 S10 **Kiama** New South Wales,
SE Australia 34.40S 150.49E
179 R17 **Kiamba** Mindanao, S Philippines
5.59N 124.36E
81 O22 **Kiambi** Katanga, SE Dem. Rep.
Congo (Zaire) 7.15S 28.01E
85 C18 **Kiamichi River** *≈* Oklahoma,
C USA
9 Q12 **Kiana** Alaska, USA 66.58N 160.25W
41 N7 **Kiangmai** *see* Chiang Mai
Kiang-ning *see* Nanjing
Kiangsi *see* Jiangxi
Kiangsu *see* Jiangsu
95 M14 **Kiantajärvi** *☺* E Finland
117 F19 **Kiáto** *prev.* Kiáton. Peloponnísos,
S Greece 38.01N 22.45E
Kiáton *see* Kiáto
116 K9 **Kiayi** *see* Chiai
81 E20 **Kibali** *var.* Uele (upper course).
≈ NE Dem. Rep. Congo (Zaire)
201 U9 **Kili Island** *var.* Köle. island Ralik
Chain, S Marshall Islands
5.27S 12.21E

155 V2 **Kilik Pass** *pass*
Afghanistan/China 37.03N 74.31E
Kilimane *see* Quelimane
83 I21 **Kilimanjaro** *◇ region* E Tanzania
83 I20 **Kilimanjaro** *▲* Uhuru Peak.
▲ NE Tanzania 3.01S 37.14E
Kilimbangara *see* Kolombangara
Kilinailau Islands *see* Tulun
Islands
83 K23 **Kilindoni** Pwani, E Tanzania
7.55S 39.40E
120 H6 **Kilingi-Nõmme** *Ger.* Kurkund.
Pärnumaa, SW Estonia
58.07N 24.00E
142 M17 **Kilis** Kilis, S Turkey 36.43N 37.07E
142 M16 **Kilis** *◇ province* S Turkey
119 N12 **Kiliya** *Rom.* Chilia-Nouă. Odes'ka
Oblast', SW Ukraine 45.29N 29.16E
99 B19 **Kilkee** *Ir.* Cill Chaoi. W Ireland
52.40N 9.37W
99 E19 **Kilkenny** *Ir.* Cill Chainnigh.
S Ireland 52.39N 7.15W
99 E19 **Kilkenny** *Ir.* Cill Chainnigh.
cultural region S Ireland
9 S16 **Kindersley** Saskatchewan,
S Canada 51.28N 109.08W
78 I14 **Kinda** Guinée-Maritime,
SW Guinea 9.28N 12.06W
30 B18 **Kilkieran Bay** *Ir.* Cuan Chill
Chiaráin. *bay* W Ireland
116 G13 **Kilkís** Kentrikí Makedonía,
N Greece 40.59N 22.54E
9 R15 **Killam** Alberta, SW Canada
52.45N 111.46W
191 U3 **Killarney** Queensland,
E Australia 28.18S 152.15E
9 W17 **Killarney** Manitoba, S Canada
49.12N 99.40W
12 E11 **Killarney** Ontario, S Canada
45.58N 81.27W
99 B20 **Killarney** *Ir.* Cill Airne.
SW Ireland 52.03N 9.30W
30 K4 **Killdeer** North Dakota, N USA
47.21N 102.45W
30 J4 **Killdeer Mountains** *▲* North
Dakota, N USA
47 V15 **Killdeer River** *≈* Trinidad,
Trinidad and Tobago
45 S9 **Killeen** Texas, SW USA
31.07N 97.43W
41 P8 **Killik River** *≈* Alaska, USA
16 P4 **Killini** *island* Nunavut,
NE Canada
117 C19 **Killíni, Akrotírio** *headland*
S Greece 37.55N 21.07E
99 D15 **Killybegs** *Ir.* Na Cealla Beaga.
NW Ireland 54.37N 8.27W
99 B19 **Kilmain** *see* Quelimane
98 I13 **Kilmarnock** W Scotland, UK
55.37N 4.30W
23 X6 **Kilmarnock** Virginia, NE USA
37.42N 76.22W
129 S16 **Kil'mez'** Kirovskaya Oblast',
NW Russian Federation
56.55N 51.03E
131 S2 **Kil'mez'** Udmurtskaya
Respublika, NW Russian
Federation 57.04N 51.22E
129 R16 **Kil'mez'** *≈* NW Russian
Federation
69 V11 **Kilombero** *≈* S Tanzania
94 J10 **Kilpisjärvi** Lappi, N Finland
69.03N 20.49E
99 B19 **Kilrush** *Ir.* Cill Rois. W Ireland
52.39N 9.28W
81 O24 **Kilwa** Katanga, SE Dem. Rep.
Congo (Zaire) 9.22S 28.19E
Kilwa *see* Kilwa Kivinje
83 K24 **Kilwa Kivinje** *var.* Kilwa. Lindi,
SE Tanzania 8.45S 39.21E
83 K24 **Kilwa Masoko** Lindi,
SE Tanzania 8.55S 39.31E
176 Uu11 **Kilwo** Pulau Seram, E Indonesia
3.36S 130.48E
116 P12 **Kilyos** Istanbul, NW Turkey
41.15N 29.01E
39 V8 **Kim** Colorado, C USA
37.12N 103.22W
175 N3 **Kimanis, Teluk** *bay* Sabah, East
Malaysia
190 H8 **Kimba** South Australia
33.09S 136.26E
30 I5 **Kimball** Nebraska, C USA
41.16N 103.40W
31 O11 **Kimball** South Dakota, N USA
43.45N 98.57W
81 N20 **Kimbao** Bandundu, SW Dem.
Rep. Congo (Zaire) 5.27S 17.40E
195 N12 **Kimbe** New Britain, E PNG
5.36S 150.10E
195 N11 **Kimbe Bay** *inlet* New Britain,
E PNG
9 P17 **Kimberley** British Columbia,
SW Canada 49.40N 115.58W
85 H23 **Kimberley** Northern Cape,
C South Africa 28.45S 24.46E
188 M4 **Kimberley Plateau** *plateau*
Western Australia
35 P15 **Kimberly** Idaho, NW USA
42.31N 114.21W
169 Y12 **Kimch'aek** *prev.* Sŏngjin. E North
Korea 40.42N 129.13E
169 Y15 **Kimch'ŏn** S South Korea
36.08N 128.06E
169 Z16 **Kim Hae** *var.* Pusan. *☓* (Pusan)
SE South Korea 35.10N 128.57E
95 K20 **Kími** *var.* Kými
81 K20 **Kimito** Swe. Kemiö. Länsi-Suomi,
SW Finland 60.10N 22.65E
172 O6 **Kimobetsu** Hokkaidō, NE Japan
42.47N 140.55E
117 I21 **Kímolos** island Kykládes, Greece,
Aegean Sea
117 I21 **Kímolou Sífnou, Stenó** *strait*
Kykládes, Greece, Aegean Sea
130 L5 **Kimovsk** Tul'skaya Oblast',
W Russian Federation 53.59N 38.34E
169 X15 **Kimpo** *☓* (Sŏul) NW South Korea
37.37N 126.42E
23 O8 **Kimpolung** *see* Câmpulung
Moldovenesc
128 K16 **Kimry** Tverskaya Oblast',
W Russian Federation 56.52N 37.21E
81 H21 **Kimvula** Bas-Congo, SW Dem.
Rep. Congo (Zaire) 5.45S 15.58E
175 Nn2 **Kinabalu, Gunung** *▲* East
Malaysia 5.52N 116.08E
146 K13 **Kinabatangan** *see* Kinabatangan,
Sungai
175 Oo3 **Kinabatangan, Sungai** *var.*
Kinabatangan. *≈* East Malaysia

98 I7 **Kinbrace** N Scotland, UK
58.16N 2.59W
12 E14 **Kincardine** Ontario, S Canada
44.10N 81.35W
98 K10 **Kincardine** *cultural region*
E Scotland, UK
81 K21 **Kinda** Kasai Occidental, SE Dem.
Rep. Congo (Zaire) 4.48S 21.49E
81 M24 **Kinda** Katanga, SE Dem. Rep.
Congo (Zaire) 9.19S 25.06E
177 FF3 **Kindat** Sagaing, N Myanmar
23.42N 94.28E
111 V6 **Kindberg** Steiermark, C Austria
47.31N 15.27E
24 H8 **Kinder** Louisiana, S USA
30.29N 92.51W
100 H13 **Kinderdijk** Zuid-Holland,
SW Netherlands 51.52N 4.37E
99 M17 **Kinder Scout** *▲* C England, UK
53.25N 1.52W
81 K21 **Kindu** *prev.* Kindu-Port-Empain.
Maniema, C Dem. Rep. Congo
(Zaire) 2.57S 25.54E
Kindu-Port-Empain *see* Kindu
131 S6 **Kinel'** Samarskaya Oblast',
W Russian Federation 53.14N 50.40E
129 N15 **Kineshma** Ivanovskaya Oblast',
W Russian Federation 57.28N 42.07E
146 K10 **King Abdul Aziz** *☓* (Makkah)
Makkah, W Saudi Arabia
21.44N 39.08E
23 X6 **King and Queen Court House**
Virginia, NE USA 37.40N 76.49W
King Charles Islands *see* Kong
Karls Land
King Christian IX Land *see*
Kong Frederik IX Land
King Christian X Land *see*
Kong Frederik X Land
King City California, W USA
36.12N 121.09W
29 X2 **King City** Missouri, C USA
40.03N 94.31W
41 M16 **King Cove** Alaska, USA
28 M10 **Kingfisher** Oklahoma, C USA
35.49N 97.56W
King Frederik VI Coast *see*
Kong Frederik VI Kyst
King Frederik VIII Land *see*
Kong Frederik VIII Land
129 S16 **Kil'mez'** Kirovskaya Oblast',
NW Russian Federation
204 G3 **King George Island** *var.* King
George Land. *island* South Shetland
Islands, Antarctica
10 I6 **King George Islands** *island*
group Nunavut, C Canada
King George Land *see* King
128 G13 **Kingisepp** Leningradskaya
Oblast', NW Russian Federation
59.23N 28.37E
191 N14 **King Island** *island* Tasmania,
SE Australia
8 J15 **King Island** *island* British
Columbia, SW Canada
King Island *see* Kadan Kyun
Kingissepp *see* Kuressaare
147 Q7 **King Khalid** *☓* (Ar Riyāḑ) Ar
Riyāḑ, C Saudi Arabia 25.00N 46.40E
37 S2 **King Lear Peak** *▲* Nevada,
W USA 41.13N 118.30W
205 Y8 **King Leopold and Queen
Astrid Land** *physical region*
Antarctica
188 M4 **King Leopold Ranges**
▲ Western Australia
19 M9 **Kingman** Arizona, SW USA
35.12N 114.02W
27 N5 **Kingman** Kansas, C USA
37.39N 98.06W
195 K7 **Kingman Reef** *◇ US territory*
C Pacific Ocean
80 E8 **Kingombe** Maniema, E Dem.
Rep. Congo (Zaire) 2.37S 26.59E
190 F5 **Kingoonya** South Australia
30.56S 135.20E
204 H2 **King Peninsula** *peninsula*
Antarctica
41 P13 **King Salmon** Alaska, USA
58.41N 156.39W
7 Q6 **Kings Beach** California, W USA
39.13N 120.02W
37 R11 **Kingsburg** California, W USA
36.30N 119.33W
190 I10 **Kingscote** South Australia
35.41S 137.36E
King's County *see* Offaly
204 H2 **King Sejong** South Korean research
station Antarctica 61.57S 58.23W
191 T9 **Kingsford Smith** *☓* (Sydney)
New South Wales, SE Australia
33.58S 151.09E
9 P17 **Kingsgate** British Columbia,
SW Canada 48.58N 116.09W
23 W8 **Kingsland** Georgia, SE USA
30.48N 81.41W
9 O19 **King's Lynn** *var.* Bishop's Lynn,
Kings Lynn, Lynn Regis.
E England, UK 52.45N 0.24E
115 Mm3 **Kings Mountain** North
Carolina, USA 35.15N 81.20W
188 K4 **King Sound** *sound* Western
Australia
39 N2 **Kings Peak** *▲* Utah, W USA
40.43N 110.27W
23 O8 **Kingsport** Tennessee, S USA
36.32N 82.31W
37 R12 **Kings River** *≈* California,
W USA
191 P17 **Kingston** Tasmania, SE Australia
42.58S 147.18E
175 K14 **Kingston** Ontario, SE Canada
44.13N 76.30W
Kingston *●* (Jamaica) E Jamaica
17.58N 76.48W
193 C22 **Kingston** Otago, South Island,
NZ 45.20S 168.45E
21 P12 **Kingston** Massachusetts,
NE USA 41.59N 70.43W
29 S3 **Kingston** Missouri, C USA
39.36N 94.02W

20 K12 **Kingston** New York, NE USA 41.55N 74.00W
33 S14 **Kingston** Ohio, N USA 39.28N 82.54W
21 O13 **Kingston** Rhode Island, NE USA 41.28N 71.31W
22 M9 **Kingston** Tennessee, S USA 35.52N 84.30W
37 W12 **Kingston Peak** ▲ California, W USA 35.43N 115.54W
190 J11 **Kingston Southeast** South Australia 36.51S 139.53E
99 N17 **Kingston upon Hull** *var.* Hull. E England, UK 53.45N 0.19W
99 N22 **Kingston upon Thames** SE England, UK 51.25N 0.18W
47 P14 **Kingstown** ● (Saint Vincent and the Grenadines) Saint Vincent, Saint Vincent and the Grenadines 13.09N 61.13W
Kingstown *see* Dún Laoghaire
23 T13 **Kingstree** South Carolina, SE USA 33.40N 79.49W
66 L8 **Kings Trough** *undersea feature* E Atlantic Ocean
12 C18 **Kingsville** Ontario, S Canada 42.03N 82.43W
45 S15 **Kingsville** Texas, SW USA 27.31N 97.52W
23 W6 **King William** Virginia, NE USA 37.42N 77.03W
15 K3 **King William Island** *island* Nunavut, N Canada Arctic Ocean
85 I25 **King William's Town** *var.* King, Kingwilliamstown. Eastern Cape, S South Africa 32.51S 27.20E
23 T3 **Kingwood** West Virginia, NE USA 39.28N 79.40W
142 C13 **Kınık** İzmir, W Turkey 39.04N 27.25E
81 G21 **Kinkala** Le Pool, S Congo 4.18S 14.49E
171 Mm14 **Kinka-san** *headland* Honshū, C Japan 38.17N 141.34E
192 M8 **Kinleith** Waikato, North Island, NZ 38.16S 175.53E
97 J19 **Kinna** Västra Götaland, S Sweden 57.31N 12.42E
98 L8 **Kinnaird Head** *var.* Kinnairds Head. *headland* NE Scotland, UK 58.39N 3.22W
97 K20 **Kinnared** Halland, S Sweden 57.01N 13.04E
Kinneret, Yam *see* Tiberias, Lake
161 K24 **Kinniyai** Eastern Province, NE Sri Lanka 8.30N 81.10E
95 L16 **Kinnula** Länsi-Suomi, W Finland 63.24N 25.00E
12 I8 **Kinojévis** ∿ Quebec, SE Canada
170 G16 **Kino-kawa** ∿ Honshū, SW Japan
9 U11 **Kinoosao** Saskatchewan, C Canada 57.06N 101.01W
101 L17 **Kinrooi** Limburg, NE Belgium 51.09N 5.47E
98 J11 **Kinross** C Scotland, UK 56.13N 3.26W
98 J11 **Kinross** *cultural region* C Scotland, UK
99 C21 **Kinsale** *Ir.* Cionn tSáile. SW Ireland 51.42N 8.31W
97 D14 **Kinsarvik** Hordaland, S Norway 60.22N 6.43E
81 G21 **Kinshasa** *prev.* Léopoldville. ● (Zaire) Kinshasa, W Dem. Rep. Congo (Zaire) 4.21S 15.16E
81 G21 **Kinshasa** *off.* Ville de Kinshasa, *var.* Kinshasa City. ◈ *region* SW Dem. Rep. Congo (Zaire)
81 G21 **Kinshasa** ✕ Kinshasa, SW Dem. Rep. Congo (Zaire) 4.23S 15.30E
Kinshasa City *see* Kinshasa
119 U9 **Kins'ka** ∿ SE Ukraine
28 K6 **Kinsley** Kansas, C USA 37.52N 99.25W
23 W10 **Kinston** North Carolina, SE USA 35.15N 77.34W
79 P15 **Kintampo** W Ghana 6.36N 0.28E
190 B1 **Kintore, Mount** ▲ South Australia 26.30S 130.24E
98 G13 **Kintyre** *peninsula* W Scotland, UK
98 G13 **Kintyre, Mull of** *headland* W Scotland, UK 55.16N 5.46W
177 G4 **Kin-u** Sagaing, C Myanmar 22.46N 95.36E
10 G8 **Kinushseo** ∿ Ontario, C Canada
9 P13 **Kinuso** Alberta, W Canada 55.19N 115.23W
160 I13 **Kinwat** Mahārāshtra, C India 19.37N 78.12E
83 F16 **Kinyeti** ▲ S Sudan 3.56N 32.52E
103 I17 **Kinzig** ∿ SW Germany
197 J13 **Kioa** *island* N Fiji
Kioga, Lake *see* Kyoga, Lake
28 M8 **Kiowa** Kansas, C USA 37.01N 98.29W
29 P12 **Kiowa** Oklahoma, C USA 34.43N 95.54W
Kiparissía *see* Kyparissía
12 H10 **Kipawa, Lac** ◎ Quebec, SE Canada
83 G24 **Kipengere Range** ⛰ SW Tanzania
83 E23 **Kipili** Rukwa, W Tanzania 7.30S 30.39E
83 K20 **Kipini** Coast, SE Kenya 2.30S 40.30E
9 V16 **Kipling** Saskatchewan, S Canada 50.04N 102.45W
40 M13 **Kipnuk** Alaska, USA 59.56N 164.02W
99 F18 **Kippure** *Ir.* Cipiúr. ▲ E Ireland 53.10N 6.22W
81 N25 **Kipushi** Katanga, SE Dem. Rep. Congo (Zaire) 11.45S 27.19E
195 Y17 **Kirakira** *var.* Kaokaona. San Cristobal, SE Solomon Islands 10.28S 161.54E
161 K14 **Kirandul** *var.* Bailādila. Madhya Pradesh, C India 18.46N 81.18E
161 I21 **Kīranūr** Tamil Nādu, SE India 11.37N 79.10E
121 N21 **Kiraw** *Rus.* Kirovo. Homyel'skaya Voblasts', SE Belarus 51.30N 29.25E
121 M17 **Kirawsk** *Rus.* Kirovsk; *prev.* Startsy. Mahilyowskaya Voblasts', E Belarus 53.16N 29.28E
120 F5 **Kirbla** Läänemaa, W Estonia 58.45N 23.57E
45 Y9 **Kirbyville** Texas, SW USA 30.39N 93.53W
116 M12 **Kırcasalih** Edirne, NW Turkey 41.24N 26.48E
111 W8 **Kirchbach** *var.* Kirchbach in Steiermark. Steiermark, SE Austria 46.55N 15.40E
Kirchbach in Steiermark *see* Kirchbach
110 H7 **Kirchberg** Sankt Gallen, NE Switzerland 47.24N 9.03E
111 S5 **Kirchdorf an der Krems** Oberösterreich, N Austria 47.54N 14.06E
Kirchheim *see* Kirchheim unter Teck
103 I22 **Kirchheim unter Teck** *var.* Kirchheim. Baden-Württemberg, SW Germany 48.39N 9.27E
Kirdzhali *see* Kŭrdzhali
126 Jj14 **Kirenga** ∿ S Russian Federation
126 Jj13 **Kirensk** Irkutskaya Oblast', C Russian Federation 57.37N 107.54E
Kirghizia *see* Kyrgyzstan
151 S16 **Kirghiz Range** *Rus.* Kirgizskiy Khrebet; *prev.* Alexander Range. ⛰ Kazakhstan/Kyrgyzstan
Kirghiz SSR *see* Kyrgyzstan
Kirghiz Steppe *see* Kazakhskiy Melkosopochnik
Kirgizskaya SSR *see* Kyrgyzstan
Kirgizskiy Khrebet *see* Kirghiz Range
81 I19 **Kiri** Bandundu, W Dem. Rep. Congo (Zaire) 1.29S 19.00E
Kiriath-Arba *see* Hebron
203 R3 **Kiribati** *off.* Republic of Kiribati. ◆ *republic* C Pacific Ocean
142 L17 **Kırıkhan** Hatay, S Turkey 36.30N 36.19E
142 I13 **Kırıkkale** Kırıkkale, C Turkey 39.50N 33.31E
142 C10 **Kırıkkale** ◈ *province* C Turkey
128 L13 **Kirillov** Vologodskaya Oblast', NW Russian Federation 59.52N 38.24E
Kirin *see* Jilin
83 I18 **Kirinyaga** *prev.* Mount Kenya. ✶ C Kenya 0.02S 37.19E
128 H13 **Kirishi** *var.* Kirisi. Leningradskaya Oblast', NW Russian Federation 59.28N 32.02E
170 C16 **Kirishima-yama** ▲ Kyūshū, SW Japan 31.58N 130.51E
Kirisi *see* Kirishi
203 Y2 **Kiritimati** ✕ Kiritimati, E Kiribati 2.00N 157.30W
203 Y2 **Kiritimati** *prev.* Christmas Island. *atoll* Line Islands, E Kiribati
195 O15 **Kiriwina Island** *Eng.* Trobriand Island. *island* SE PNG
195 O15 **Kiriwina Islands** *var.* Trobriand Islands. *island group* S PNG
98 K12 **Kirkcaldy** E Scotland, UK 56.07N 3.10W
99 I14 **Kirkcudbright** S Scotland, UK 54.49N 4.03W
99 I14 **Kirkcudbright** *cultural region* S Scotland, UK
Kirkee *see* Khadki
94 M8 **Kirkenes** *var.* Kirkkoniemi. Finnmark, N Norway 69.43N 30.01E
97 I14 **Kirkenær** Hedmark, S Norway 60.27N 12.04E
94 J4 **Kirkjubæjarklaustur** Sudhurland, S Iceland 63.46N 18.03W
Kirk-Kilissa *see* Kırklareli
Kirkkoniemi *see* Kirkenes
95 L20 **Kirkkonummi** *Swe.* Kyrkslätt. Etelä-Suomi, S Finland 60.06N 24.25E
12 G7 **Kirkland Lake** Ontario, S Canada 48.10N 80.01W
142 C9 **Kırklareli** *prev.* Kirk-Kilissa. Kırklareli, NW Turkey 41.45N 27.12E
142 I13 **Kırklareli** ◈ *province* NW Turkey
193 F20 **Kirkliston Range** ⛰ South Island, NZ
12 D10 **Kirkpatrick Lake** ◎ Ontario, S Canada
205 Q11 **Kirkpatrick, Mount** ▲ Antarctica 84.37S 164.36E
29 U2 **Kirksville** Missouri, C USA 40.11N 92.34W
145 T4 **Kirkūk** *var.* Karkūk, Kerkuk. N Iraq 35.28N 44.25E
145 U7 **Kir Kush** E Iraq 33.42N 45.15E
98 K5 **Kirkwall** NE Scotland, UK
85 H25 **Kirkwood** Eastern Cape, S South Africa 33.23S 25.19E
29 X5 **Kirkwood** Missouri, C USA 38.34N 90.24W
Kirman *see* Kermān
Kir Moab/Kir of Moab *see* Al Karak
130 I5 **Kirov** Kaluzhskaya Oblast', W Russian Federation 54.01N 34.16E
129 R14 **Kirov** *prev.* Vyatka. Kirovskaya Oblast', NW Russian Federation 58.34N 49.38E
Kīrov *see* Balpyk Bi, Kazakhstan
Kīrov *see* Kirova, Almaty, Kazakhstan
151 U13 **Kirova** *Kaz.* Kīrov. Almaty, SE Kazakhstan 46.24N 77.16E
Kirovabad *see* Gäncä, Azerbaijan
Kirovabad *see* Panj, Tajikistan
Kirovakan *see* Vanadzor
Kirovo *see* Kiraw, Belarus
Kirovo/Kirovograd *see* Kirovohrad, Ukraine
Kirovo *see* Beshariq, Uzbekistan
129 R14 **Kirovo-Chepetsk** Kirovskaya Oblast', NW Russian Federation 58.33N 50.06E
Kirovogradskaya Oblast'/Kirovohrad *see* Kirovohrads'ka Oblast'
119 R7 **Kirovohrad** *Rus.* Kirovograd; *prev.* Kirovo, Yelizavetgrad, Zinov'yevsk. Kirovohrads'ka Oblast', C Ukraine 48.30N 31.17E
119 P7 **Kirovohrads'ka Oblast'** *var.* Kirovohrad, *Rus.* Kirovogradskaya Oblast'. ◈ *province* C Ukraine
128 J4 **Kirovsk** Murmanskaya Oblast', NW Russian Federation 67.37N 33.38E
Kirovsk *see* Babadaykhan, Turkmenistan
Kirovsk *see* Kirawsk, Belarus
119 X7 **Kirovs'k** Luhans'ka Oblast', E Ukraine 48.40N 38.39E
125 Dd9 **Kirovskaya Oblast'** ◈ *province* NW Russian Federation
119 X8 **Kirov's'ke** Donets'ka Oblast', E Ukraine 48.12N 38.19E
119 U13 **Kirovs'ke** *Rus.* Kirovskoye. Respublika Krym, S Ukraine 45.13N 35.12E
127 P12 **Kirovskiy** Kamchatskaya Oblast', E Russian Federation 54.06N 155.48E
Kirovskiy *see* Balpyk Bi
Kirovskoye *see* Kyzyl-Adyr
Kirovskoye *see* Kirovs'ke
124 Oo2 **Kırpaşa** *var.* Karpas Peninsula, *Gk.* Karpasía. *peninsula* NE Cyprus
152 E11 **Kirpili** Akhalskiy Velayat, C Turkmenistan 39.31N 57.13E
98 K10 **Kirriemuir** E Scotland, UK 56.37N 3.00W
129 S13 **Kirs** Kirovskaya Oblast', NW Russian Federation 59.22N 52.20E
131 N7 **Kirsanov** Tambovskaya Oblast', W Russian Federation 52.40N 42.48E
142 J14 **Kırşehir** *anc.* Justinianopolis. Kırşehir, C Turkey 39.09N 34.07E
142 I13 **Kırşehir** ◈ *province* C Turkey
155 P4 **Kīrthar Range** ⛰ S Pakistan
39 P9 **Kirtland** New Mexico, SW USA 36.43N 108.21W
Kirun/Kirun' *see* Chilung
94 J11 **Kiruna** Norrbotten, N Sweden 67.50N 20.16E
81 M18 **Kirundu** Orientale, NE Dem. Rep. Congo (Zaire) 0.45S 25.28E
28 L3 **Kirwin Reservoir** ⊟ Kansas, C USA
131 Q4 **Kirya** Chuvashskaya Respublika, W Russian Federation 55.04N 46.50E
Kiryat Gat *see* Qiryat Gat
171 K15 **Kiryū** Gunma, Honshū, S Japan 36.24N 139.19E
97 M18 **Kisa** Östergötland, S Sweden 58.00N 15.39E
171 Ll11 **Kisakata** Akita, Honshū, C Japan 39.12N 139.55E
Kisalföld *see* Little Alföld
81 L18 **Kisangani** *prev.* Stanleyville. Orientale, NE Dem. Rep. Congo (Zaire) 0.30N 25.13E
41 N12 **Kisaralik River** ∿ Alaska, USA
171 K17 **Kisarazu** Chiba, Honshū, S Japan 35.19N 139.51E
113 I22 **Kisbér** Komárom-Esztergom, NW Hungary 47.30N 18.00E
9 V17 **Kisbey** Saskatchewan, S Canada 49.41N 102.39W
126 H14 **Kiselevsk** Kemerovskaya Oblast', S Russian Federation 54.00N 86.38E
159 R13 **Kishanganj** Bihār, NE India 26.06N 87.57E
158 G12 **Kishangarh** Rājasthān, N India 26.33N 74.52E
Kishegyes *see* Mali Iđoš
79 S15 **Kishi** Oyo, W Nigeria 9.01N 3.53E
Kishinev *see* Chişinău
Kishiözen *see* Malyy Uzen'
171 Gg15 **Kishiwada** *var.* Kisiwada. Ōsaka, Honshū, SW Japan 34.28N 135.22E
149 P14 **Kīsh, Jazīreh-ye** *var.* Qeys. *island* S Iran
151 R7 **Kishkenekol'** *prev.* Kzyltu. *Kaz.* Qyzyltu; Severnyy Kazakhstan, N Kazakhstan 53.39N 72.22E
158 I6 **Kishtwār** Jammu and Kashmir, NW India 33.19N 75.49E
83 H19 **Kisii** Nyanza, SW Kenya 0.40S 34.46E
83 J23 **Kisiju** Pwani, E Tanzania 7.25S 39.19E
Kisiwada *see* Kishiwada
Kisjenő *see* Chişineu-Criş
40 E17 **Kiska Island** *island* Aleutian Islands, Alaska, USA
Kiskapus *see* Copşa Mică
113 M22 **Kiskörei-víztároló** ⊟ E Hungary
Kis-Küküllo *see* Târnava Mică
113 L24 **Kiskunfélegyháza** *var.* Félegyháza. Bács-Kiskun, C Hungary 46.42N 19.52E
113 K25 **Kiskunhalas** *var.* Halas. Bács-Kiskun, S Hungary 46.25N 19.28E
113 K24 **Kiskunmajsa** Bács-Kiskun, S Hungary 46.31N 19.45E
131 N15 **Kislovodsk** Stavropol'skiy Kray, SW Russian Federation 43.55N 42.44E
83 L18 **Kismaayo** *var.* Chisimayu, Kismayu, *It.* Chisimaio. Jubbada Hoose, S Somalia 0.04S 42.34E
Kismayu *see* Kismaayo
171 Ii15 **Kiso-sanmyaku** ⛰ Honshū, S Japan
Kisseraing *see* Kanmaw Kyun
78 K14 **Kissidougou** Guinée-Forestière, S Guinea 9.15N 10.07W
25 X12 **Kissimmee** Florida, SE USA 28.17N 81.24W
25 X12 **Kissimmee, Lake** ◎ Florida, SE USA
25 X13 **Kissimmee River** ∿ Florida, SE USA
9 V13 **Kississing Lake** ◎ Manitoba, C Canada
113 L24 **Kistelek** Csongrád, SE Hungary 46.27N 19.58E
Kistna *see* Krishna
113 M23 **Kisújszállás** Jász-Nagykun-Szolnok, E Hungary 47.13N 20.43E
170 F12 **Kisuki** Shimane, Honshū, SW Japan 35.25N 133.15E
83 H18 **Kisumu** *prev.* Port Florence. Nyanza, W Kenya 0.02N 34.42E
Kisutzaneustadtl *see* Kysucké Nové Mesto
113 O20 **Kisvárda** *Ger.* Kleinwardein. Szabolcs-Szatmár-Bereg, E Hungary 48.13N 22.03E
83 J24 **Kiswere** Lindi, SE Tanzania 9.24S 39.37E
Kiszucaújhely *see* Kysucké Nové Mesto
78 K12 **Kita** Kayes, W Mali 13.04N 9.29W
207 N14 **Kitaa** ◈ *province* W Greenland
Kitab *see* Kitob
172 N5 **Kitahiyama** Hokkaidō, NE Japan 42.25N 139.55E
171 L15 **Kita-Ibaraki** Ibaraki, Honshū, S Japan 36.48N 140.43E
172 Ss17 **Kita-Iō-jima** *Eng.* San Alessandro. *island* SE Japan
171 Mm11 **Kitakami** Iwate, Honshū, C Japan 39.16N 141.06E
171 M13 **Kitakami-gawa** ∿ Honshū, C Japan
172 N11 **Kitakami-sanchi** ⛰ Honshū, C Japan
171 L13 **Kitakata** Fukushima, Honshū, C Japan 37.38N 139.51E
170 D12 **Kitakyūshū** *var.* Kitakyûsyû. Fukuoka, Kyūshū, SW Japan 33.51N 130.49E
Kitakyûsyû *see* Kitakyūshū
83 H18 **Kitale** Rift Valley, W Kenya 1.01N 35.01E
172 Q5 **Kitami** Hokkaidō, NE Japan 43.51N 143.50E
172 Pp4 **Kitami-sanchi** ⛰ Hokkaidō, NE Japan
171 Kk17 **Kita-ura** ◎ Honshū, S Japan
195 O15 **Kitava Island** *island* Kiriwina Islands, SE PNG
39 W5 **Kit Carson** Colorado, C USA 38.45N 102.47W
188 M12 **Kitchener** Western Australia 31.03S 124.00E
12 F16 **Kitchener** Ontario, S Canada 43.28N 80.27W
95 O17 **Kitee** Itä-Suomi, E Finland 62.06N 30.09E
83 G16 **Kitgum** N Uganda 3.16N 32.54E
Kithareng *see* Kanmaw Kyun
Kíthira *see* Kýthira
Kíthnos *see* Kýthnos
8 J13 **Kitimat** British Columbia, SW Canada 54.04N 128.37W
94 L11 **Kitinen** ∿ N Finland
153 N12 **Kitob** *Rus.* Kitab. Qashqadaryo Wiloyati, S Uzbekistan 39.06N 66.46E
118 K7 **Kitsman'** *Ger.* Kotzman, *Rom.* Cozmeni, *Rus.* Kitsman. Chernivets'ka Oblast', W Ukraine 48.27N 25.46E
170 Dd14 **Kitsuki** *var.* Kituki. Ōita, Kyūshū, SW Japan 33.25N 131.37E
20 C14 **Kittanning** Pennsylvania, NE USA 40.48N 79.28W
21 P10 **Kittery** Maine, NE USA 43.05N 70.44W
94 L11 **Kittilä** Lappi, N Finland 67.39N 24.52E
111 Z4 **Kittsee** Burgenland, E Austria 48.05N 17.05E
83 J19 **Kitui** Eastern, S Kenya 1.25S 38.00E
Kituki *see* Kitsuki
83 G22 **Kitunda** Tabora, C Tanzania 6.47S 33.13E
8 K13 **Kitwanga** British Columbia, SW Canada 55.07N 128.03W
84 J13 **Kitwe** *var.* Kitwe-Nkana. Copperbelt, C Zambia 12.48S 28.13E
Kitwe-Nkana *see* Kitwe
111 O7 **Kitzbühel** Tirol, W Austria 47.27N 12.22E
111 O7 **Kitzbüheler Alpen** ⛰ W Austria
103 J19 **Kitzingen** Bayern, SE Germany 49.43N 10.10E
159 Q14 **Kiul** Bihār, NE India 25.10N 86.06E
194 E12 **Kiunga** Western, SW PNG 6.06S 141.12E
95 M16 **Kiuruvesi** Itä-Suomi, C Finland 63.37N 26.40E
40 M7 **Kivalina** Alaska, USA 67.43N 164.31W
94 L13 **Kivalo** *ridge* C Finland
118 J3 **Kivertsi** *Pol.* Kiwerce, *Rus.* Kivertsy. Volyns'ka Oblast', NW Ukraine 50.49N 25.31E
Kivertsy *see* Kivertsi
95 L16 **Kivijärvi** Länsi-Suomi, W Finland 63.09N 25.04E
97 L23 **Kivik** Skåne, S Sweden 55.40N 14.15E
120 J3 **Kiviõli** Ida-Virumaa, NE Estonia 59.20N 27.00E
Kivu, Lac *see* Kivu, Lake
69 U10 **Kivu, Lake** *Fr.* Lac Kivu. ◎ Rwanda/Dem. Rep. Congo (Zaire)
194 G15 **Kiwai Island** *island* SW PNG
41 N8 **Kiwalik** Alaska, USA 66.01N 161.50W
Kiwerce *see* Kivertsi
Kiyev *see* Kyyiv
151 R10 **Kiyevka** Karaganda, C Kazakhstan 50.15N 71.33E
Kiyevskaya Oblast' *see* Kyyivs'ka Oblast'
Kiyevskoye Vodokhranilishche *see* Kyyivs'ke Vodoskhovyshche
142 D10 **Kıyıköy** Kırklareli, NW Turkey 41.37N 28.07E
151 O9 **Kiyma** Akmola, C Kazakhstan 51.37N 67.31E
129 V13 **Kizel** Permskaya Oblast', NW Russian Federation 58.59N 57.37E
129 O12 **Kizema** *var.* Kizëma. Arkhangel'skaya Oblast', NW Russian Federation 61.06N 44.51E
Kizilagach *see* Elkhovo
142 H12 **Kızılcahamam** Ankara, N Turkey 40.28N 32.37E
142 J10 **Kızıl Irmak** ∿ C Turkey
Kızılkoca *see* Şefaatli
Kizil Kum *see* Kyzyl Kum
143 P16 **Kızıltepe** Mardin, SE Turkey 37.12N 40.36E
Ki Zil Uzen *see* Qezel Owzan
131 Q16 **Kizilyurt** Respublika Dagestan, SW Russian Federation 43.13N 46.54E
131 Q15 **Kizlyar** Respublika Dagestan, SW Russian Federation 43.51N 46.39E
131 S3 **Kizner** Udmurtskaya Respublika, NW Russian Federation 56.19N 51.37E
Kizyl-Arvat *see* Gyzylarbat
152 B13 **Kizyl-Atrek** *Turkm.* Gyzyletrek. Balkanskiy Velayat, W Turkmenistan 37.40N 54.44E
152 D10 **Kizyl-Kaya** *Turkm.* Gyzylgaya. Balkanskiy Velayat, NW Turkmenistan 40.37N 55.15E
152 A10 **Kizyl-Su** *Turkm.* Gyzylsu. Balkanskiy Velayat, W Turkmenistan 39.49N 53.00E
97 H16 **Kjerkøy** *island* S Norway
Kjølen *see* Kölen
94 L7 **Kjøllefjord** Finnmark, N Norway 70.55N 27.19E
94 H11 **Kjøpsvik** Nordland, C Norway 68.07N 16.22E
174 Ii9 **Klabat, Teluk** *bay* Pulau Bangka, W Indonesia
114 I12 **Kladanj** Federacija Bosan I Hercegovina, C Bosnia and Herzegovina 44.14N 18.42E
176 Xx16 **Kladar** Irian Jaya, E Indonesia 8.14S 137.46E
113 C16 **Kladno** Středočeský Kraj, NW Czech Republic 50.10N 14.04E
114 P11 **Kladovo** Serbia, E Yugoslavia 44.37N 22.36E
178 Hh12 **Klaeng** Rayong, S Thailand 12.48N 101.41E
111 T9 **Klagenfurt** *Slvn.* Celovec. Kärnten, S Austria 46.37N 14.19E
120 B11 **Klaipėda** *Ger.* Memel. Klaipėda, NW Lithuania 55.42N 21.09E
174 M15 **Klakah** Jawa, C Indonesia 7.55S 113.12E
97 B18 **Klaksvík** *Dan.* Klaksvig Faeroe Islands 62.13N 6.43W
36 L2 **Klamath** California, W USA 41.31N 124.02W
34 H16 **Klamath Falls** Oregon, NW USA 42.13N 121.46W
36 M1 **Klamath Mountains** ⛰ California/Oregon, W USA
36 L2 **Klamath River** ∿ California/Oregon, W USA
174 Gg5 **Klang** *var.* Kelang; *prev.* Port Swettenham. Selangor, Peninsular Malaysia 3.01N 101.27E
96 J13 **Klarälven** ∿ Norway/Sweden
113 B15 **Klášterec nad Ohří** *Ger.* Klösterle an der Eger. Ústecký Kraj, NW Czech Republic 50.24N 13.10E
174 L15 **Klaten** Jawa, C Indonesia 7.40S 110.31E
113 B18 **Klatovy** *Ger.* Klattau. Plzeňský Kraj, W Czech Republic 49.24N 13.16E
Klattau *see* Klatovy
Klausenburg *see* Cluj-Napoca
41 Y14 **Klawock** Prince of Wales Island, Alaska, USA 55.33N 133.06W
100 P8 **Klazienaveen** Drenthe, NE Netherlands 52.43N 7.00E
Kleck *see* Klyetsk
112 H11 **Kłecko** Wielkolpolskie, C Poland 52.37N 17.27E
112 I11 **Kleczew** Wielkopolskie, C Poland 52.22N 18.12E
8 L15 **Kleena Kleene** British Columbia, SW Canada 51.55N 124.54W
85 D20 **Klein Aub** Hardap, C Namibia 23.48S 16.39E
Kleine Donau *see* Mosoni-Duna
103 O14 **Kleine Elster** ∿ E Germany
Kleine Kokel *see* Târnava Mică
101 I16 **Kleine Nete** ∿ N Belgium
Kleines Ungarisches Tiefland *see* Little Alföld
85 E22 **Klein Karas** Karas, S Namibia 27.37S 18.05E
Kleinkopisch *see* Copşa Mică
Klein-Marien *see* Väike-Maarja
Kleinschlatten *see* Zlatna
85 D23 **Kleinsee** Northern Cape, W South Africa 29.43S 17.03E
Kleinwardein *see* Kisvárda
117 C16 **Kleisoúra** Ípeiros, W Greece 39.21N 20.52E
97 C17 **Klepp** Rogaland, S Norway 58.46N 5.39E
85 I22 **Klerksdorp** North-West, N South Africa 26.52S 26.39E
130 I5 **Kletnya** Bryanskaya Oblast', W Russian Federation 53.25N 32.58E
Kletsk *see* Klyetsk
103 D14 **Kleve** *Eng.* Cleves, *Fr.* Clèves; *prev.* Cleve. Nordrhein-Westfalen, W Germany 51.46N 6.07E
115 J16 **Kličevo** Montenegro, SW Yugoslavia 42.45N 18.58E
121 M16 **Klichaw** *Rus.* Klichev. Mahilyowskaya Voblasts', E Belarus 53.28N 29.21E
Klichev *see* Klichaw
121 Q16 **Klimavichy** *Rus.* Klimovichi. Mahilyowskaya Voblasts', E Belarus 53.37N 31.58E
116 M7 **Kliment** Shumen, NE Bulgaria 43.37N 27.00E
Klimovichi *see* Klimavichy
95 G14 **Klimpfjäll** Västerbotten, N Sweden 65.04N 14.49E
130 K3 **Klin** Moskovskaya Oblast', W Russian Federation 56.19N 36.45E
115 M16 **Klina** Serbia, S Yugoslavia 42.38N 20.35E
113 B15 **Klínovec** *Ger.* Keilberg. ▲ NW Czech Republic 50.23N 12.57E
97 P19 **Klintehamn** Gotland, SE Sweden 57.22N 18.15E
131 R8 **Klintsovka** Saratovskaya Oblast', W Russian Federation 51.42N 49.17E
130 H6 **Klintsy** Bryanskaya Oblast', W Russian Federation 52.46N 32.20E
97 K22 **Klippan** Skåne, S Sweden 56.07N 13.10E
94 G13 **Klippen** Västerbotten, N Sweden 65.50N 15.07E
124 Nn3 **Klírou** W Cyprus 35.01N 33.11E
116 I9 **Klisura** Plovdiv, C Bulgaria 42.42N 24.28E
97 F20 **Klitmøller** Viborg, NW Denmark 57.01N 8.29E
114 F11 **Ključ** Federacija Bosna I Hercegovina, NW Bosnia and Herzegovina 44.32N 16.46E
113 J14 **Kłobuck** Śląskie, S Poland 50.55N 18.54E
112 J11 **Kłodawa** Wielkopolskie, C Poland 52.14N 18.55E
113 G16 **Kłodzko** *Ger.* Glatz. Dolnośląskie, SW Poland 50.27N 16.37E
97 I14 **Kløfta** Akershus, S Norway 60.04N 11.09E
114 P12 **Klokočevac** Serbia, E Yugoslavia 44.19N 22.11E
120 G3 **Klooga** *Ger.* Lodensee. Harjumaa, NW Estonia 59.18N 24.15E
101 F15 **Kloosterzande** Zeeland, SW Netherlands 51.22N 4.01E
115 L19 **Klos** *var.* Klosi. Dibër, C Albania 41.30N 20.07E
Klosi *see* Klos
Klösterle an der Eger *see* Klášterec nad Ohří
111 X3 **Klosterneuburg** Niederösterreich, NE Austria 48.19N 16.19E
110 J9 **Klosters** Graubünden, SE Switzerland 46.54N 9.52E
110 G7 **Kloten** Zürich, N Switzerland 47.27N 8.34E
110 G7 **Kloten** ✕ (Zürich) Zürich, N Switzerland 47.25N 8.36E
102 K12 **Klötze** Sachsen-Anhalt, C Germany 52.37N 11.09E
10 K3 **Klotz, Lac** ◎ Quebec, NE Canada
103 O15 **Klotzsche** ✕ (Dresden) Sachsen, E Germany 51.06N 13.44E
8 H7 **Kluane Lake** ◎ Yukon Territory, W Canada
Kluang *see* Keluang
113 I14 **Kluczbork** *Ger.* Kreuzburg, Kreuzburg in Oberschlesien. Opolskie, S Poland 50.59N 18.13E
41 W12 **Klukwan** Alaska, USA 59.24N 135.49W
Klyastitsy *see* Klyastsitsy
120 L11 **Klyastsitsy** *Rus.* Klyastitsy. Vitsyebskaya Voblasts', N Belarus 55.54N 28.38E
131 T5 **Klyavlino** Samarskaya Oblast', W Russian Federation 54.21N 52.12E
86 K9 **Klyaz'in** ∿ W Russian Federation
131 N3 **Klyaz'ma** ∿ W Russian Federation
121 J17 **Klyetsk** *Pol.* Kleck, *Rus.* Kletsk. Minskaya Voblasts', SW Belarus 53.04N 26.38E
153 S8 **Klyuchevka** Talasskaya Oblast', NW Kyrgyzstan 42.33N 71.45E
127 Pp10 **Klyuchevskaya Sopka, Vulkan** ✶ E Russian Federation 56.03N 160.37E
127 Pp10 **Klyuchi** Kamchatskaya Oblast', E Russian Federation 56.18N 160.44E
97 D17 **Knaben** Vest-Agder, S Norway 58.46N 7.04E
Knanzi *see* Ghanzi
97 K21 **Knäred** Halland, S Sweden 56.30N 13.21E
99 M16 **Knaresborough** N England, UK 54.01N 1.35W
116 H8 **Knezha** Vratsa , NW Bulgaria 43.29N 24.04E
45 O9 **Knickerbocker** Texas, SW USA 31.18N 100.35W
30 K5 **Knife River** ∿ North Dakota, N USA
8 K16 **Knight Inlet** *inlet* British Columbia, W Canada
41 S12 **Knight Island** *island* Alaska, USA
99 K20 **Knighton** E Wales, UK 52.20N 3.00W
37 O7 **Knights Landing** California, W USA 38.47N 121.43W
114 E13 **Knin** Šibenik-Knin, S Croatia 44.03N 16.12E
45 Q12 **Knippal** Texas, SW USA
111 U7 **Knittelfeld** Steiermark, C Austria 47.13N 14.51E
97 O15 **Knivsta** Uppsala, C Sweden 59.43N 17.49E
115 P14 **Knjaževac** Serbia, E Yugoslavia 43.34N 22.16E
29 S4 **Knob Noster** Missouri, C USA 38.47N 93.33W
101 D15 **Knokke-Heist** West-Vlaanderen, NW Belgium 51.21N 3.19E
97 H20 **Knøsen** *hill* N Denmark 57.09N 10.15E
Knosós *see* Knossos
117 J25 **Knossos** *Gk.* Knosós. *prehistoric site* Kríti, Greece, E Mediterranean Sea 35.17N 25.10E
45 N7 **Knott** Texas, SW USA 32.21N 101.35W
204 K5 **Knowles, Cape** *headland* Antarctica 71.45S 60.19W
33 O11 **Knox** Indiana, N USA 41.16N 86.37W
31 O3 **Knox** North Dakota, N USA 48.19N 99.43W
20 C13 **Knox** Pennsylvania, NE USA 41.13N 79.33W
201 X8 **Knox Atoll** *var.* Ņadikdik, Narikrik. *atoll* Ratak Chain, SE Marshall Islands
8 H13 **Knox, Cape** *headland* Graham Island, British Columbia, SW Canada 54.05N 133.02W
45 P5 **Knox City** Texas, SW USA 33.25N 99.49W
205 Y11 **Knox Coast** *physical region* Antarctica
33 T12 **Knox Lake** ⊟ Ohio, N USA
25 T5 **Knoxville** Georgia, SE USA 32.44N 83.58W
32 K12 **Knoxville** Illinois, N USA 40.54N 90.16W
31 W15 **Knoxville** Iowa, C USA 41.19N 93.06W
23 N9 **Knoxville** Tennessee, S USA 35.57N 83.55W
207 P11 **Knud Rasmussen Land** *physical region* N Greenland
Knüll *see* Knüllgebirge
103 I16 **Knüllgebirge** *var.* Knüll. ⛰ C Germany
Knyazhevo *see* Sredishte
Knyazhitsy *see* Knyazhytsy
121 O15 **Knyazhytsy** *Rus.* Knyazhitsy. Mahilyowskaya Voblasts', E Belarus 54.10N 30.27E
85 G26 **Knysna** Western Cape, SW South Africa 34.01S 23.05E
176 V10 **Koagas** Irian Jaya, E Indonesia 2.40S 132.16E
Koartac *see* Quaqtaq
174 J10 **Koba** Pulau Bangka, W Indonesia 2.29S 106.22E
170 C16 **Kobayashi** *var.* Kobayasi. Miyazaki, Kyūshū, SW Japan 32.01N 130.55E
Kobayasi *see* Kobayashi
171 Gg14 **Kōbe** Hyōgo, Honshū, SW Japan 34.39N 135.10E
Kobelyaki *see* Kobelyaky
119 T6 **Kobelyaky** *Rus.* Kobelyaki. Poltavs'ka Oblast', NE Ukraine 49.10N 34.13E
97 J22 **København** *Eng.* Copenhagen; *anc.* Hafnia. ● (Denmark) Sjælland, København, E Denmark 55.43N 12.34E
97 J23 **København** *off.* Københavns Amt. ◈ *county* E Denmark
78 K10 **Kobenni** Hodh el Gharbi, S Mauritania 15.58N 9.24W
176 U11 **Kobi** Pulau Seram, E Indonesia 2.56S 129.53E
103 F17 **Koblenz** *prev.* Coblenz, *Fr.* Coblence, *anc.* Confluentes. Rheinland-Pfalz, W Germany 50.21N 7.36E
110 F6 **Koblenz** Aargau, N Switzerland 47.34N 8.16E
176 Ww11 **Kobowre, Pegunungan** ⛰ Irian Jaya, E Indonesia
Kobrin *see* Kobryn
176 W14 **Kobroor, Pulau** *island* Kepulauan Aru, E Indonesia
121 G19 **Kobryn** *Pol.* Kobryn, *Rus.* Kobrin. Brestskaya Voblasts', SW Belarus 52.13N 24.21E
41 O7 **Kobuk** Alaska, USA 66.54N 156.52W
41 O7 **Kobuk River** ∿ Alaska, USA
143 Q10 **K'obulet'i** W Georgia 41.47N 41.46E
126 Ll10 **Kobyay** Respublika Sakha (Yakutiya), NE Russian Federation 63.36N 126.33E
142 E11 **Kocaeli** ◈ *province* NW Turkey
115 P18 **Kočani** NE FYR Macedonia 41.55N 22.25E
114 K12 **Koceljevo** Serbia, W Yugoslavia 44.28N 19.49E
111 U12 **Kočevje** *Ger.* Gottschee. S Slovenia 45.41N 14.47E
159 T12 **Koch Bihār** West Bengal, NE India 26.19N 89.25E
126 J10 **Kochechum** ∿ N Russian Federation
103 I20 **Kocher** ∿ SW Germany
129 T13 **Kochevo** Komi-Permyatskiy Avtonomnyy Okrug, NW Russian Federation 59.37N 54.16E
170 Ee15 **Kōchi** *var.* Kôti. Kōchi, Shikoku, SW Japan 33.31N 133.30E
170 Ee15 **Kōchi** *off.* Kōchi-ken, *var.* Kôti. ◈ *prefecture* Shikoku, SW Japan
Kochi *see* Cochin
Kochiu *see* Gejiu
Kochkor *see* Kochkorka
153 V8 **Kochkorka** *Kir.* Kochkor. Narynskaya Oblast', C Kyrgyzstan 42.09N 75.42E
129 V5 **Kochmes** Respublika Komi, NW Russian Federation 66.10N 60.46E
131 P15 **Kochubey** Respublika Dagestan, SW Russian Federation 44.25N 46.33E
117 I17 **Kochýlas** ▲ Skýros, Vóreioi Sporádes, Greece, Aegean Sea 38.50N 24.35E
112 O13 **Kock** Lubelskie, E Poland 51.39N 22.26E
83 J19 **Kodacho** *spring/well* S Kenya 1.52S 39.22E
161 K24 **Koddiyar Bay** *bay* NE Sri Lanka
41 Q14 **Kodiak** Kodiak Island, Alaska, USA 57.47N 152.24W
41 Q14 **Kodiak Island** *island* Alaska, USA
160 B12 **Kodīnār** Gujarāt, W India 20.43N 70.46E
128 M9 **Kodino** Arkhangel'skaya Oblast', NW Russian Federation 63.36N 39.54E
126 Ii13 **Kodinsk** Krasnoyarskiy Kray, C Russian Federation 58.37N 99.18E
82 F12 **Kodok** Upper Nile, SE Sudan 9.51N 32.07E
119 N8 **Kodyma** Odes'ka Oblast', SW Ukraine 48.05N 29.09E
101 B17 **Koekelare** West-Vlaanderen, W Belgium 51.07N 2.58E
Koeln *see* Köln
Koepang *see* Kupang
Ko-erh-mu *see* Golmud
101 J17 **Koersel** Limburg, NE Belgium 51.04N 5.17E
85 E21 **Koës** Karas, SE Namibia 25.57S 19.04E
Koetai *see* Mahakam, Sungai
Koetaradja *see* Bandaaceh
38 I14 **Kofa Mountains** ⛰ Arizona, SW USA
176 Z15 **Kofarau** Irian Jaya, E Indonesia 7.29S 140.28E
153 P13 **Kofarnihon** *Rus.* Kofarnikhon; *prev.* Ordzhonikidzeabad, *Taj.* Orjonikidzeobod, Yangi-Bazar. W Tajikistan 38.32N 68.56E
153 P14 **Kofarnihon** *Rus.* Kafirnigan. ∿ SW Tajikistan
Kofarnikhon *see* Kofarnihon
116 M11 **Kofçaz** Kırklareli, NW Turkey 41.57N 27.07E
176 U9 **Kofiau, Pulau** *var.* Kafiau. *island* Kepulauan Raja Ampat, E Indonesia
117 J25 **Kófinas** ▲ Kríti, Greece, E Mediterranean Sea 34.58N 25.03E
124 Nn4 **Kofínou** *var.* Kophinou. S Cyprus 34.49N 33.24E
111 V8 **Köflach** Steiermark, SE Austria 47.04N 15.04E
79 Q17 **Koforidua** SE Ghana 6.04N 0.17W
170 Ff12 **Kōfu** Tottori, Honshū, SW Japan 35.16N 133.31E
171 J15 **Kōfu** *var.* Kôhu. Yamanashi, Honshū, S Japan 35.40N 138.33E
171 K16 **Koga** Ibaraki, Honshū, S Japan 36.12N 139.42E
83 F22 **Koga** Tabora, C Tanzania 6.08S 32.20E
Kogălniceanu *see* Mihail Kogălniceanu
11 P6 **Kogaluk** ∿ Newfoundland, E Canada
10 J4 **Kogaluk** ∿ Quebec, NE Canada
126 Gg10 **Kogalym** Khanty-Mansiyskiy Avtonomnyy Okrug, C Russian Federation 62.13N 74.34E
97 J23 **Køge** Roskilde, E Denmark 55.28N 12.12E
97 J23 **Køge Bugt** *bay* E Denmark
79 U16 **Kogi** ◈ *state* C Nigeria
152 L11 **Kogon** *Rus.* Kagan. Bukhoro Wiloyati, C Uzbekistan 39.46N 64.28E
169 Y17 **Kŏgŭm-do** *island* S South Korea
Kőhalom *see* Rupea
155 T6 **Kohāt** North-West Frontier Province, NW Pakistan 33.37N 71.30E
120 G4 **Kohila** *Ger.* Koil. Raplamaa, NW Estonia 59.07N 24.46E

◆ Country
● Country Capital
◇ Dependent Territory
○ Dependent Territory Capital
◈ Administrative Region
✕ International Airport
▲ Mountain
⛰ Mountain Range
✶ Volcano
∿ River
◎ Lake
⊟ Reservoir

159 X13 **Kohīma** Nāgāland, E India 25.40N 94.07E
Koh I Noh see Büyükağrı Dağı
148 L10 **Kohkīlūyeh va Būyer Aḥmadī** *off.* Ostān-e Kohkīlūyeh va Būyer Aḥmadī, *var.* Boyer Ahmadī va Kohkīlūyeh. ◐ *province* SW Iran
Kohsān see Kūhestān
120 J3 **Kohtla-Järve** Ida-Virumaa, NE Estonia 59.22N 27.21E
Kôhu see Kōfu
119 N10 **Kohyl'nyk** *Rom.* Cogîlnic. ≈ Moldova/Ukraine
171 K13 **Koide** Niigata, Honshū, C Japan 37.13N 138.58E
8 G7 **Koidern** Yukon Territory, W Canada 61.55N 140.22W
78 J15 **Koidu** E Sierra Leone 8.39N 11.01W
120 I4 **Koigi** Järvamaa, C Estonia 58.51N 25.45E
Koil see Kohila
180 H13 **Koimbani** Grande Comore, NW Comoros 11.37S 43.22E
145 T3 **Koi Sanjaq** *var.* Koysanjaq, Küysanjaq. N Iraq 36.04N 44.37E
95 O16 **Koitere** ◎ E Finland
Koivisto see Primorsk
169 Z16 **Kōje-do** *Jap.* Kyōsai-tō. *island* S South Korea
82 J13 **K'ok'a Hāyk'** ◎ C Ethiopia
Kokand see Qŭqon
190 F6 **Kokatha** South Australia 31.17S 135.16E
Kokcha see Kŭkcha
Kokchetav see Kokshetau
95 K18 **Kokemäenjoki** ≈ SW Finland
176 Z12 **Kokenau** *var.* Kokonau. Irian Jaya, E Indonesia 4.38S 136.24E
85 E22 **Kokerboom** Karas, SE Namibia 28.10S 19.25E
121 N14 **Kokhanava** *Rus.* Kokhanovo. Vitsyebskaya Voblasts', NE Belarus 54.28N 29.58E
Kokhanovichi see Kakhanavichy
Kokhanovo see Kokhanava
Kök-Janggak see Kok-Yangak
95 K16 **Kokkola** *Swe.* Karleby; *prev. Swe.* Gamlakarleby. Länsi-Suomi, W Finland 63.49N 23.10E
164 L3 **Kok Kuduk** *well* N China 46.03N 87.34E
120 H9 **Koknese** Aizkraukle, C Latvia 56.38N 25.27E
79 T13 **Koko** Kebbi, W Nigeria 11.25N 4.33E
194 K15 **Kokoda** Northern, S PNG 8.51S 147.37E
78 K12 **Kokofata** Kayes, W Mali 12.48N 9.56W
41 N6 **Kokolik River** ≈ Alaska, USA
33 O13 **Kokomo** Indiana, N USA 40.29N 86.07W
Kokonau see Kokenau
Koko Nor see Qinghai Hu, China
Koko Nor see Qinghai, China
195 P10 **Kokopo** *var.* Kopopo; *prev.* Herbertshöhe. New Britain, E PNG 4.19S 152.13E
151 X10 **Kokpekti** *Kaz.* Kökpekti. Vostochnyy Kazakhstan, E Kazakhstan 48.45N 82.24E
151 X11 **Kokpekti** ≈ E Kazakhstan
41 P9 **Kokrines** Alaska, USA 64.57N 154.42W
41 P9 **Kokrines Hills** ▲▲ Alaska, USA
151 P17 **Koksaray** Yuzhnyy Kazakhstan, S Kazakhstan 42.40N 68.09E
153 X9 **Kokshaal-Tau** *Rus.* Khrebet Kakshaal-Too. ▲▲ China/Kyrgyzstan
151 P7 **Kokshetau** *Kaz.* Kökshetaū; *prev.* Kokchetav. Severnyy Kazakhstan, N Kazakhstan 53.18N 69.25E
101 A17 **Koksijde** West-Vlaanderen, W Belgium 51.07N 2.39E
10 M5 **Koksoak** ≈ Quebec, E Canada
85 K24 **Kokstad** KwaZulu/Natal, E South Africa 30.23S 29.22E
151 W15 **Koktal** *Kaz.* Köktal. Almaty, SE Kazakhstan 44.04N 79.43E
151 Q12 **Koktas** ≈ C Kazakhstan
Kök-Tash see Kěk-Tash
Koktokay see Fuyun
170 C16 **Kokubu** Kagoshima, Kyūshū, SW Japan 31.44N 130.44E
126 L15 **Kokuy** Chitinskaya Oblast', S Russian Federation 52.13N 117.18E
153 T9 **Kok-Yangak** *Kir.* Kök-Janggak. Dzhalal-Abadskaya Oblast', W Kyrgyzstan 41.02N 73.11E
164 F9 **Kokyar** Xinjiang Uygur Zizhiqu, W China 37.24N 77.15E
155 O13 **Kolāchi** *var.* Kulachi. ≈ SW Pakistan
78 J15 **Kolahun** N Liberia 8.24N 10.01W
175 Q12 **Kolaka** Sulawesi, C Indonesia 4.04S 121.37E
Kolam see Quilon
K'o-la-ma-i see Karamay
Kola Peninsula see Kol'skiy Poluostrov
161 H19 **Kolār** Karnātaka, E India 13.10N 78.10E
161 H19 **Kolār Gold Fields** Karnātaka, E India 12.56N 78.16E
94 K11 **Kolari** Lappi, NW Finland 67.19N 23.51E
113 I21 **Kolárovo** *Ger.* Gutta; *prev.* Guta, *Hung.* Gúta. Nitriansky Kraj, SW Slovakia 47.54N 18.00E
115 K16 **Kolašin** Montenegro, SW Yugoslavia 42.49N 19.32E
158 F11 **Kolāyat** Rājasthān, NW India 27.55N 73.01E
97 N15 **Kolbäck** Västmanland, C Sweden 59.33N 16.15E
Kolbcha see Kowlbcha
207 Q15 **Kolbeinsey Ridge** *undersea feature* Denmark Strait/Norwegian Sea
Kolberg see Kołobrzeg
97 H15 **Kolbotn** Akershus, S Norway 62.15N 10.24E
113 N16 **Kolbuszowa** Podkarpackie, SE Poland 50.12N 22.07E
130 L3 **Kol'chugino** Vladimirskaya Oblast', W Russian Federation 56.19N 39.24E
78 H12 **Kolda** S Senegal 12.58N 14.58W
97 G23 **Kolding** Vejle, C Denmark 55.28N 9.30E
81 M17 **Kole** Orientale, N Dem. Rep. Congo (Zaire) 2.09N 25.17E
81 K20 **Kole** Kasai Oriental, SW Dem. Rep. Congo (Zaire) 3.27S 22.28E
Kōle see Kili Island
86 F6 **Kölen** *Nor.* Kjølen. ▲▲ Norway/Sweden
Kolepom, Pulau see Yos Sudarso, Pulau
120 H3 **Kolga Laht** *Ger.* Kolko-Wiek. *bay* N Estonia
129 Q3 **Kolguyev, Ostrov** *island* NW Russian Federation
161 E16 **Kolhāpur** Mahārāshtra, SW India 16.42N 74.13E
157 K21 **Kolhumadulu Atoll** *var.* Kolumadulu Atoll, Thaa Atoll. *atoll* S Maldives
95 O16 **Koli** *var.* Kolinkylä. Itä-Suomi, E Finland 63.06N 29.45E
41 O13 **Koliganek** Alaska, USA 59.43N 157.16W
113 E16 **Kolín** *Ger.* Kolin. Středočeský Kraj, C Czech Republic 50.01N 15.10E
Kolinkylä see Koli
202 E12 **Koliu** Île Futuna, W Wallis and Futuna
120 E7 **Kolka** Talsi, NW Latvia 57.43N 22.33E
120 E7 **Kolkasrags** *prev. Eng.* Cape Domesnes. *headland* NW Latvia 57.45N 22.35E
Kolkhozabad see Kolkhozobod
153 P14 **Kolkhozobod** *Rus.* Kolkhozabad; *prev.* Kaganovichabad, Tugalan. SW Tajikistan 37.33N 68.34E
Kolki/Kołki see Kolky
Kolko-Wiek see Kolga Laht
118 K3 **Kolky** *Pol.* Kołki, *Rus.* Kolki. Volyns'ka Oblast', NW Ukraine 51.05N 25.40E
Kollam see Quilon
161 G20 **Kollegāl** Karnātaka, W India 12.07N 77.06E
100 M5 **Kollum** Friesland, N Netherlands 53.16N 6.09E
Kolmar see Colmar
103 E16 **Köln** *var.* Koeln, *Eng./Fr.* Cologne; *prev.* Cöln, *anc.* Colonia Agrippina, Oppidum Ubiorum. Nordrhein-Westfalen, W Germany 50.57N 6.57E
112 N9 **Kolno** Podlaskie, NE Poland 53.24N 21.57E
112 J12 **Koło** Wielkopolskie, C Poland 52.10N 18.39E
40 B8 **Koloa** *Haw.* Kōloa. Kauai, Hawaii, USA, C Pacific Ocean 21.54N 159.28W
112 E7 **Kołobrzeg** *Ger.* Kolberg. Zachodniopomorskie, NW Poland 54.10N 15.33E
130 H4 **Kolodnya** Smolenskaya Oblast', W Russian Federation 54.57N 32.22E
202 E13 **Kolofau, Mont** ⛰ Île Alofi, S Wallis and Futuna 14.21S 178.01W
129 O14 **Kologriv** Kostromskaya Oblast', NW Russian Federation 58.49N 44.22E
78 L12 **Kolokani** Koulikoro, W Mali 13.34N 8.01W
79 N13 **Koloko** W Burkina 11.06N 5.18W
195 U14 **Kolombangara** *var.* Kilimbangara, Nduke. *island* New Georgia Islands, NW Solomon Islands
Kolomea see Kolomyya
130 L4 **Kolomna** Moskovskaya Oblast', W Russian Federation 55.02N 38.52E
118 J7 **Kolomyya** *Ger.* Kolomea. Ivano-Frankivs'ka Oblast', W Ukraine 48.31N 25.00E
78 M13 **Kolondiéba** Sikasso, SW Mali 11.04N 6.55W
200 R15 **Kolonga** Tongatapu, S Tonga 21.07S 175.04W
201 U16 **Kolonia** *var.* Colonia. Pohnpei, E Micronesia 6.57N 158.12E
Kolonja see Kolonjë
115 K21 **Kolonjë** *var.* Kolonja. Fier, C Albania 40.49N 19.37E
Kolonjë see Ersekë
Kolotambu see Avuavu
200 Q15 **Kolovai** Tongatapu, S Tonga 21.05S 175.20W
175 R13 **Kolowanawatobo, Teluk** *bay* Pulau Buton, C Indonesia
Kolozsvár see Cluj-Napoca
114 C9 **Kolpa** *Ger.* Kulpa, *SCr.* Kupa. ≈ Croatia/Slovenia
126 H12 **Kolpashevo** Tomskaya Oblast', C Russian Federation 58.21N 82.44E
128 H13 **Kolpino** Leningradskaya Oblast', NW Russian Federation 59.43N 30.39E
102 M10 **Kölpinsee** ◎ NE Germany
128 K5 **Kol'skiy Poluostrov** *Eng.* Kola Peninsula. *peninsula* NW Russian Federation
131 T6 **Koltubanovskiy** Orenburgskaya Oblast', W Russian Federation 53.00N 52.00E
114 L11 **Kolubara** ≈ C Yugoslavia
Kolupchii see Gurkovo
112 K13 **Koluszki** Łodzkie, C Poland 51.43N 19.49E
129 T6 **Kolva** ≈ NW Russian Federation
95 E14 **Kolvereid** Nord-Trøndelag, W Norway 64.47N 11.22E
154 L15 **Kolwa** Baluchistān, SW Pakistan 26.03N 64.00E
81 M24 **Kolwezi** Katanga, S Dem. Rep. Congo (Zaire) 10.43S 25.29E
127 Nn7 **Kolyma** ≈ NE Russian Federation
Kolyma Lowland see Kolymskaya Nizmennost'
Kolyma Range/Kolymskiy, Khrebet see Kolymskoye Nagor'ye
127 Nn6 **Kolymskaya Nizmennost'** *Eng.* Kolyma Lowland. *lowlands* NE Russian Federation
127 Nn6 **Kolymskoye** Respublika Sakha (Yakutiya), NE Russian Federation 68.42N 158.46E
127 N17 **Kolymskoye Nagor'ye** *var.* Khrebet Kolymskiy, *Eng.* Kolyma Range. ▲▲ E Russian Federation
127 N17 **Kolyuchinskaya Guba** *bay* NE Russian Federation
151 W15 **Kol'zhat** Almaty, SE Kazakhstan 43.30N 80.37E
116 G8 **Kom** ▲ NW Bulgaria 43.10N 23.02E
82 I13 **Koma** Oromo, C Ethiopia 8.19N 36.48E
79 X12 **Komadugu Gana** ≈ NE Nigeria
171 Ii15 **Komagane** Nagano, Honshū, S Japan 35.46N 137.56E
81 P17 **Komanda** Orientale, NE Dem. Rep. Congo (Zaire) 1.23N 29.44E
207 U1 **Komandorskaya Basin** *var.* Kamchatka Basin. *undersea feature* SW Bering Sea
129 Pp9 **Komandorskiye Ostrova** *Eng.* Commander Islands. *island group* E Russian Federation
Kománfalva see Comănești
113 I22 **Komárno** *Ger.* Komorn, *Hung.* Komárom. Nitriansky Kraj, SW Slovakia 47.46N 18.07E
113 I22 **Komárom** Komárom-Esztergom, NW Hungary 47.44N 18.06E
Komárom see Komárno
113 I22 **Komárom-Esztergom** *off.* Komárom-Esztergom Megye. ◐ *county* N Hungary
171 I13 **Komatsu** *var.* Komatu. Ishikawa, Honshū, SW Japan 36.24N 136.27E
170 Ff15 **Komatsushima** Tokushima, Shikoku, SW Japan 34.00N 134.36E
Komatu see Komatsu
85 D17 **Kombat** Otjozondjupa, N Namibia 19.42S 17.45E
Kombissiguiri see Kombissiri
79 P13 **Kombissiri** *var.* Kombissiguiri. C Burkina 12.03N 1.14W
196 E10 **Komebail Lagoon** *lagoon* N Palau
83 F20 **Kome Island** *island* N Tanzania
Komeyo see Wandai
176 W13 **Komfane** Pulau Wokam, E Indonesia 5.36S 134.42E
119 P10 **Kominternivs'ke** Odes'ka Oblast', SW Ukraine 46.52N 30.56E
129 R12 **Komi-Permyatskiy Avtonomnyy Okrug** ◐ *autonomous district* W Russian Federation
129 R8 **Komi, Respublika** ◐ *autonomous republic* NW Russian Federation
113 I25 **Komló** Baranya, SW Hungary 46.11N 18.19E
Kommunarsk see Alchevs'k
153 S12 **Kommunizm, Qullai** ▲ E Tajikistan
194 G12 **Komo** Southern Highlands, W PNG 6.06S 142.52E
175 P16 **Komodo** Pulau Komodo, S Indonesia 8.35S 119.27E
175 P16 **Komodo, Pulau** *island* Nusa Tenggara, S Indonesia
79 N15 **Komoé** *var.* Komoé Fleuve. ≈ E Ivory Coast
Komoé Fleuve see Komoé
77 X11 **Kôm Ombo** *var.* Kawm Umbū. SE Egypt 24.23N 32.58E
81 F20 **Komono** La Lékoumou, SW Congo 3.15S 13.14E
176 Y16 **Komoran** Irian Jaya, E Indonesia 8.14S 138.51E
176 Y16 **Komoran, Pulau** *island* E Indonesia
Komorn see Komárno
Komornok see Comorâște
171 J14 **Komoro** Nagano, Honshū, S Japan 36.22N 138.25E
Komosolabad see Komsomolobod
Komotau see Chomutov
116 K13 **Komotiní** *var.* Gümüljina, *Turk.* Gümülcine. Anatolikí Makedonía kai Thráki, NE Greece 41.06N 25.27E
115 K16 **Komovi** ▲ SW Yugoslavia
119 R8 **Kompaniyivka** Kirovohrads'ka Oblast', C Ukraine 48.16N 32.12E
194 H12 **Kompiam** Enga, W PNG 5.23S 143.54E
Kompong see Kâmpóng Chhnăng
Kompong Cham see Kâmpóng Cham
Kompong Kleang see Kâmpóng Khleăng
Kompong Som see Kâmpóng Saôm
Kompong Speu see Kâmpóng Spoe
Komrat see Comrat
Komsomol see Komsomol'skiy, Atyrau, Kazakhstan
Komsomol see Komsomolets, Kostanay, Kazakhstan
150 L7 **Komsomolets** *Kaz.* Komsomol. Kostanay, N Kazakhstan 53.48N 61.58E
126 I2 **Komsomolets, Ostrov** *island* Severnaya Zemlya, N Russian Federation
150 F13 **Komsomolets, Zaliv** *lake gulf* SW Kazakhstan
153 Q12 **Komsomolobod** *Rus.* Komosolabad. C Tajikistan 38.51N 69.54E
128 M16 **Komsomol'sk** Ivanovskaya Oblast', W Russian Federation 56.58N 40.15E
119 S6 **Komsomol's'k** Poltavs'ka Oblast', C Ukraine 49.01N 33.37E
152 M11 **Komsomol'sk** Nawoiy Wiloyati, N Uzbekistan 40.14N 65.10E
150 G12 **Komsomol'skiy** *Kaz.* Komsomol. Atyrau, W Kazakhstan 47.18N 53.37E
129 W4 **Komsomol'skiy** Respublika Komi, NW Russian Federation 67.33N 64.00E
127 Nn15 **Komsomol'sk-na-Amure** Khabarovskiy Kray, SE Russian Federation 50.31N 136.58E
Komsomol'sk-na-Ustyurte see Komsomol'sk-Ustyurt
150 K10 **Komsomol'skoye** Aktyubinsk, NW Kazakhstan
131 Q8 **Komsomol'skoye** Saratovskaya Oblast', W Russian Federation 50.45N 47.00E
152 G6 **Komsomol'sk-Ustyurt** *Rus.* Komsomol'sk-na-Ustyurte. Qoraqalpoghiston Respublikasi, NW Uzbekistan 44.06N 58.14E
151 P10 **Kon** ≈ C Kazakhstan
Kona see Kailua
128 K16 **Konakovo** Tverskaya Oblast', W Russian Federation 56.42N 36.44E
149 V15 **Konārak** Sīstān va Balūchestān, SE Iran 25.26N 60.22E
Konarhā see Kunar
29 O11 **Konawa** Oklahoma, C USA 34.57N 96.45W
176 V9 **Konda** Irian Jaya, E Indonesia 1.34S 131.58E
125 Ff11 **Konda** ≈ C Russian Federation
160 L13 **Kondagaon** Madhya Pradesh, C India 19.38N 81.41E
12 K10 **Kondiaronk, Lac** ◎ Quebec, SE Canada
188 J13 **Kondinin** Western Australia 32.31S 118.15E
83 H21 **Kondoa** Dodoma, C Tanzania 4.54S 35.46E
131 P6 **Kondol'** Penzenskaya Oblast', W Russian Federation 52.49N 45.03E
116 N10 **Kondolovo** Burgas, E Bulgaria 42.07N 27.43E
176 Z16 **Kondomirat** Irian Jaya, E Indonesia 8.57S 140.55E
128 J10 **Kondopoga** Respublika Kareliya, NW Russian Federation 62.12N 34.16E
Kondoz see Kunduz
161 J17 **Kondukūr** *var.* Kandukur. Andhra Pradesh, E India 15.17N 79.49E
Kondūz see Kunduz
197 H6 **Koné** Province Nord, W New Caledonia 21.04S 164.51E
Könekesir see Këneksir
Köneürgench see Këneurgench
79 N16 **Kong** N Ivory Coast 9.06N 4.34W
41 S5 **Kongakut River** ≈ Alaska, USA
207 O14 **Kong Christian IX Land** *Eng.* King Christian IX Land. *physical region* SE Greenland
207 P13 **Kong Christian X Land** *Eng.* King Christian X Land. *physical region* E Greenland
207 N13 **Kong Frederik IX Land** *Eng.* King Frederik IX Land. *physical region* SW Greenland
207 Q12 **Kong Frederik VIII Land** *Eng.* King Frederik VIII Land. *physical region* NE Greenland
207 N15 **Kong Frederik VI Kyst** *Eng.* King Frederik VI Coast. *physical region* SE Greenland
178 I13 **Kong, Kaôh** *prev.* Kas Kong. *island* SW Cambodia
94 P2 **Kong Karls Land** *Eng.* King Charles Islands. *island group* SE Svalbard
83 G14 **Kong Kong** ≈ SE Sudan
Kongo see Congo (river)
85 G16 **Kongola** Caprivi, NE Namibia 17.47S 23.24E
81 N21 **Kongolo** Katanga, E Dem. Rep. Congo (Zaire) 5.20S 26.57E
83 F14 **Kongor** Jonglei, SE Sudan 7.09N 31.44E
207 Q14 **Kong Oscar Fjord** *fjord* E Greenland
79 P12 **Kongoussi** N Burkina 13.19N 1.31W
97 G15 **Kongsberg** Buskerud, S Norway 59.39N 9.38E
94 Q2 **Kongsøya** *island* Kong Karls Land, E Svalbard
97 I14 **Kongsvinger** Hedmark, S Norway 60.10N 12.00E
178 Jj11 **Kông, Tônle** *Lao.* Xê Kong. ≈ Cambodia/Laos
164 E8 **Kongur Shan** ▲ NW China 38.39N 75.21E
83 I22 **Kongwa** Dodoma, C Tanzania 6.13S 36.28E
Kong, Xê see Kông, Tônle
Konia see Konya
153 R11 **Konibodom** *Rus.* Kanibadam. N Tajikistan 40.16N 70.20E
113 K15 **Koniecpol** Śląskie, S Poland 50.47N 19.45E
Konieh see Konya
Königgrätz see Hradec Králové
Königinhof an der Elbe see Dvůr Králové nad Labem
103 K23 **Königsbrunn** Bayern, S Germany 48.16N 10.52E
Königshütte see Chorzów
103 O24 **Königssee** ◎ SE Germany
111 S8 **Königstuhl** ▲ S Austria 46.57N 13.47E
111 U3 **Königswiesen** Oberösterreich, N Austria 48.25N 14.48E
103 E17 **Königswinter** Nordrhein-Westfalen, W Germany 50.40N 7.12E
152 M11 **Konimekh** *Rus.* Kenimekh. Nawoiy Wiloyati, N Uzbekistan 40.14N 65.10E
112 I12 **Konin** *Ger.* Kuhnau. Wielkopolskie, C Poland 52.13N 18.16E
Koninkrijk der Nederlanden see Netherlands
115 L24 **Konispol** *var.* Konispoli. Vlorë, S Albania 39.40N 20.10E
Konispoli see Konispol
117 C15 **Kónitsa** Ípeiros, W Greece 40.04N 20.48E
Konitz see Chojnice
110 D8 **Köniz** Bern, W Switzerland 46.55N 7.26E
115 H14 **Konjic** Federacija Bosna I Hercegovina, C Bosnia and Herzegovina 43.39N 17.55E
94 J10 **Könkämäälven** ≈ Finland/Sweden
161 D14 **Konkan** W India
85 D22 **Konkiep** ≈ S Namibia
78 I14 **Konkouré** ≈ W Guinea
79 O11 **Konna** Mopti, S Mali 14.58N 3.49W
195 P10 **Konogaiang, Mount** ▲ New Ireland, NE PNG 4.05S 152.43E
195 P10 **Konogogo** New Ireland, NE PNG 3.25S 152.09E
110 E9 **Konolfingen** Bern, W Switzerland 46.53N 7.36E
79 P16 **Konongo** C Ghana 6.39N 1.06W
195 O9 **Konos** New Ireland, NE PNG 3.07S 151.43E
128 M12 **Konosha** Arkhangel'skaya Oblast', NW Russian Federation 60.58N 40.09E
119 R3 **Konotop** Sums'ka Oblast', NE Ukraine 51.15N 33.13E
164 L7 **Konqi He** ≈ NW China
113 L14 **Końskie** Świętokrzyskie, C Poland 51.12N 20.26E
Konstantinovka see Kostyantynivka
130 M11 **Konstantinovsk** Rostovskaya Oblast', SW Russian Federation 47.37N 41.07E
103 H24 **Konstanz** *var.* Constanz, *Eng.* Constance; *hist.* Kostnitz, *anc.* Constantia. Baden-Württemberg, S Germany 47.40N 9.10E
Konstanza see Constanța
79 T14 **Kontagora** Niger, W Nigeria 10.25N 5.29E
80 E13 **Kontcha** Nord, N Cameroon 8.00N 12.13E
101 G17 **Kontich** Antwerpen, N Belgium 51.07N 4.27E
95 O16 **Kontiolahti** Itä-Suomi, E Finland 62.46N 29.51E
95 M15 **Kontiomäki** Oulu, C Finland 64.20N 28.09E
178 K11 **Kon Tum** *var.* Kontum. Kon Tum, C Vietnam 14.23N 108.00E
Konur see Sulakyurt
142 H15 **Konya** *var.* Konieh; *prev.* Konia, *anc.* Iconium. Konya, C Turkey 37.51N 32.30E
142 H15 **Konya** *var.* Konia, Konieh. ◐ *province* C Turkey
151 T13 **Konyrat** *var.* Kounradskiy, *Kaz.* Qongyrat. Karaganda, C Kazakhstan 46.58N 74.54E
151 W15 **Konyrolen** Almaty, SE Kazakhstan 44.16N 79.18E
83 I19 **Konza** Eastern, S Kenya 1.44S 37.07E
100 I9 **Koog aan den Zaan** Noord-Holland, C Netherlands 52.28N 4.49E
190 E7 **Koonibba** South Australia 31.55S 133.23E
33 O11 **Koontz Lake** Indiana, N USA 41.25N 86.24W
176 V8 **Koor** Irian Jaya, E Indonesia 0.21S 132.28E
191 R9 **Koorawatha** New South Wales, SE Australia 34.03S 148.33E
120 J5 **Koosa** Tartumaa, E Estonia 58.31N 27.06E
35 N7 **Kootenai** *var.* Kootenay. ≈ Canada/USA *see also* Kootenay
9 P17 **Kootenay** *var.* Kootenai. ≈ Canada/USA *see also* Kootenai
85 F24 **Kootjieskolk** Northern Cape, W South Africa 31.16S 20.21E
115 M15 **Kopaonik** ▲▲ S Yugoslavia
Kopar see Koper
94 K1 **Kópasker** Nordhurland Eystra, N Iceland 66.15N 16.23W
94 H4 **Kópavogur** Reykjanes, W Iceland 64.06N 21.47W
111 S13 **Koper** *It.* Capodistria; *prev.* Kopar. SW Slovenia 45.32N 13.42E
97 C16 **Kopervik** Rogaland, S Norway 59.16N 5.18E
Kopetdag, Khrebet see Koppeh Dāgh
125 Ee12 **Kopeysk** Kurganskaya Oblast', C Russian Federation 55.06N 61.31E
Kophinou see Kofinou
190 G8 **Kopi** South Australia 33.24S 135.40E
Kopiago see Lake Copiago
159 W12 **Kopili** ≈ NE India
97 M15 **Köping** Västmanland, C Sweden 59.31N 16.00E
115 K17 **Koplik** *var.* Kopliku. Shkodër, NW Albania 42.12N 19.26E
Kopliku see Koplik
Kopopo see Kokopo
96 I11 **Koppang** Hedmark, S Norway 61.34N 11.01E
Kopparberg see Dalarna
149 S3 **Koppeh Dāgh** *var.* Khrebet Kopetdag. ▲▲ Iran/Turkmenistan
Koppename see Coppename Rivier
97 J15 **Koppom** Värmland, C Sweden 59.42N 12.07E
Kopreinitz see Koprivnica
116 K9 **Koprinka, Yazovir** *prev.* Yazovir Georgi Dimitrov. ⊠ C Bulgaria 42.36N 25.19E
114 F7 **Koprivnica** *Ger.* Kopreinitz, *Hung.* Kaproncza. Koprivnica-Križevci, N Croatia 46.10N 16.49E
114 F8 **Koprivnica-Križevci** *off.* Koprivničko-Križevačka Županija. ◐ *province* N Croatia
113 I17 **Kopřivnice** *Ger.* Nesselsdorf. Ostravský Kraj, E Czech Republic 49.36N 18.09E
Köprülü see Veles
Koptsevichi see Kaptsevichy
Kopyl' see Kapyl'
121 O14 **Kopys'** *Rus.* Kopys'. Vitsyebskaya Voblasts', NE Belarus 54.21N 30.21E
115 M18 **Korab** ▲ Albania/FYR Macedonia 41.48N 20.33E
Korabavur Pastligi see Karabaur', Uval
83 M14 **K'orahē** Somali, E Ethiopia 6.36N 44.21E
117 L16 **Kórakas, Akrotírio** *headland* Lésvos, E Greece 39.20N 26.20E
114 D9 **Korana** ≈ C Croatia
161 L14 **Korāput** Orissa, E India 18.49N 82.43E
Korat see Nakhon Ratchasima
178 I9 **Korat Plateau** *plateau* E Thailand
145 T1 **Korāwa, Sar-i** ▲ NE Iraq 37.07N 44.39E
160 L11 **Korba** Madhya Pradesh, C India 22.25N 82.43E
103 H15 **Korbach** Hessen, C Germany 51.16N 8.52E
Korça see Korçë
115 M21 **Korçë** *var.* Korça, *Gk.* Korytsa, *It.* Corriza; *prev.* Koritsa. Korçë, SE Albania 40.37N 20.46E
115 M21 **Korçë** ◐ *district* SE Albania
115 G15 **Korčula** *It.* Curzola. Dubrovnik-Neretva, S Croatia 42.57N 17.08E
115 F15 **Korčula** *It.* Curzola; *anc.* Corcyra Nigra. *island* S Croatia
115 F15 **Korčulanski Kanal** *channel* S Croatia
151 T6 **Korday** *prev.* Georgiyevka. Zhambyl, SE Kazakhstan 43.06N 74.42E
148 J5 **Kordestān** *off.* Ostān-e Kordestān, *var.* Kurdestan. ◐ *province* W Iran
149 P4 **Kord Kūy** *var.* Kurd Kui. Golestān, N Iran 36.49N 54.04E
169 V13 **Korea Bay** *bay* China/North Korea
Korea, Democratic People's Republic of see North Korea
176 Uu15 **Koreare** Pulau Yamdena, E Indonesia 7.33S 131.13E
Korea, Republic of see South Korea
169 Z17 **Korea Strait** *Jap.* Chōsen-kaikyō, *Kor.* Taehan-haehyŏp. *channel* Japan/South Korea
Korelichi/Korelicze see Karelichy
82 J11 **Korem** Tigray, N Ethiopia 12.32N 39.29E
79 U11 **Korén Adoua** ≈ C Niger
130 I7 **Korenevo** Kurskaya Oblast', W Russian Federation 51.21N 34.53E
130 L13 **Korenovsk** Krasnodarskiy Kray, SW Russian Federation 45.28N 39.25E
118 L4 **Korets'** *Pol.* Korzec, *Rus.* Korets. Rivnens'ka Oblast', NW Ukraine 50.38N 27.12E
127 Pp8 **Korf** Koryakskiy Avtonomnyy Okrug, E Russian Federation 60.20N 165.37E
204 L7 **Korff Ice Rise** *ice cap* Antarctica
94 G13 **Korgen** Troms, N Norway 66.04N 13.51E
153 R9 **Korgon-Dĕbĕ** Dzhalal-Abadskaya Oblast', W Kyrgyzstan 41.51N 70.52E
78 M14 **Korhogo** N Ivory Coast 9.28N 5.38W
117 F19 **Korinthiakós Kólpos** *Eng.* Gulf of Corinth; *anc.* Corinthiacus Sinus. *gulf* C Greece
117 F19 **Kórinthos** *Eng.* Corinth; *anc.* Corinthus. Pelopónnisos, S Greece 37.55N 22.55E
115 M18 **Koritnik** ▲ S Yugoslavia 42.06N 20.34E
Koritsa see Korçë
171 L14 **Kōriyama** Fukushima, Honshū, C Japan 37.25N 140.20E
142 E16 **Korkuteli** Antalya, SW Turkey 37.04N 30.12E
164 K6 **Korla** *Chin.* K'u-erh-lo. Xinjiang Uygur Zizhiqu, NW China 41.48N 86.10E
126 H11 **Korliki** Khanty-Mansiyskiy Avtonomnyy Okrug, C Russian Federation 61.28N 82.12E
Körlin an der Persante see Karlino
Korma see Karma
12 D8 **Kormak** Ontario, S Canada 47.38N 83.00W
Kormakíti, Akrotíri/Kormakiti, Cape/Kormakítis see Koruçam Burnu
113 G23 **Körmend** Vas, W Hungary 47.01N 16.34E
145 T5 **Körmör** E Iraq 35.06N 44.47E
114 C13 **Kornat** *It.* Incoronata. *island* W Croatia
Korneshty see Cornești
111 X3 **Korneuburg** Niederösterreich, NE Austria 48.22N 16.20E
151 P7 **Korneyevka** Severnyy Kazakhstan, N Kazakhstan 54.01N 68.30E
97 I17 **Kornsjø** Østfold, S Norway 58.55N 11.40E
79 O11 **Koro** Mopti, S Mali 14.05N 3.06W
197 J14 **Koro** *island* C Fiji
194 F12 **Koroba** Southern Highlands, W PNG 5.46S 142.48E
130 K8 **Korocha** Belgorodskaya Oblast', W Russian Federation 50.49N 37.08E
142 H12 **Köroğlu Dağları** ▲▲ C Turkey
191 V6 **Korogoro Point** *headland* New South Wales, SE Australia 31.03S 153.04E
83 J21 **Korogwe** Tanga, E Tanzania 5.12S 38.26E
190 L13 **Koroit** Victoria, SE Australia 38.17S 142.22E
197 H15 **Korolevu** Viti Levu, W Fiji 18.12S 177.44E
202 I17 **Koromiri** *island* S Cook Islands
179 R16 **Koronadal** Mindanao, S Philippines 6.23N 124.54E
117 E22 **Koróni** Pelopónnisos, S Greece 36.46N 21.57E
116 G13 **Korónia, Límni** ◎ N Greece
112 I9 **Koronowo** *Ger.* Krone an der Brahe. Kujawski-pomorskie, C Poland 53.18N 17.56E
119 R2 **Korop** Chernihivs'ka Oblast', N Ukraine 51.35N 32.57E
117 H19 **Koropí** Attikí, C Greece 37.54N 23.52E
Koror see Oreor
Kőrös see Križevci
113 L23 **Körös** ≈ E Hungary
Körösbánya see Baia de Criş
197 J14 **Koro Sea** *sea* C Fiji
Koroška see Kärnten
119 N3 **Korosten'** Zhytomyrs'ka Oblast', NW Ukraine 50.56N 28.39E
Korostyshev see Korostyshiv
119 N4 **Korostyshiv** *Rus.* Korostyshev. Zhytomyrs'ka Oblast', N Ukraine 50.18N 29.04E
129 V3 **Korotaikha** ≈ NW Russian Federation
126 H9 **Korotchayevo** Yamalo-Nenetskiy Avtonomnyy Okrug, N Russian Federation 66.00N 78.11E
80 I8 **Koro Toro** Borkou-Ennedi-Tibesti, N Chad 16.01N 18.27E
41 N16 **Korovin Island** *island* Shumagin Islands, Alaska, USA
197 I14 **Korovou** Viti Levu, W Fiji 17.48S 178.32E
95 M17 **Korpilahti** Länsi-Suomi, W Finland 62.01N 25.34E
94 K12 **Korpilombolo** Norrbotten, N Sweden 66.51N 23.00E
127 Oo16 **Korsakov** Ostrov Sakhalin, Sakhalinskaya Oblast', SE Russian Federation 46.41N 142.45E
95 J16 **Korsholm** *Fin.* Mustasaari. Länsi-Suomi, W Finland 63.07N 21.45E
97 I23 **Korsør** Vestsjælland, E Denmark 55.19N 11.09E
Korsovka see Kārsava
119 P6 **Korsun'-Shevchenkivs'kyy** *Rus.* Korsun'-Shevchenkovskiy. Cherkas'ka Oblast', C Ukraine 49.25N 31.15E
Korsun'-Shevchenkovskiy see Korsun'-Shevchenkivs'kyy
101 C17 **Kortemark** West-Vlaanderen, W Belgium 51.03N 3.03E
101 H18 **Kortenberg** Vlaams Brabant, C Belgium 50.52N 4.33E
101 K18 **Kortessem** Limburg, NE Belgium 50.52N 5.22E
101 E14 **Kortgene** Zeeland, SW Netherlands 51.34N 3.48E
82 F8 **Korti** Northern, N Sudan 18.06N 31.33E
101 C18 **Kortrijk** *Fr.* Courtrai. West-Vlaanderen, W Belgium 50.49N 3.16E
124 N2 **Koruçam Burnu** *var.* Cape Kormakiti, Kormakítis, *Gk.* Akrotíri Kormakíti. *headland* N Cyprus 35.24N 32.55E
191 O13 **Korumburra** Victoria, SE Australia 38.27S 145.48E
Koryak Range see Koryakskoye Nagor'ye
127 P8 **Koryakskiy Avtonomnyy Okrug** ◐ *autonomous district* E Russian Federation
Koryakskiy Khrebet see Koryakskoye Nagor'ye
127 Pp7 **Koryakskoye Nagor'ye** *var.* Koryakskiy Khrebet, *Eng.* Koryak Range. ▲▲ NE Russian Federation
129 P11 **Koryazhma** Arkhangel'skaya Oblast', NW Russian Federation 61.16N 47.06E
Kōryō see Kangnŭng
Korytsa see Korçë
119 Q2 **Koryukivka** Chernihivs'ka Oblast', N Ukraine 51.45N 32.16E
Korzec see Korets'
117 N21 **Kos** Kos, Dodekánisos, Greece, Aegean Sea 36.53N 27.18E
117 M21 **Kos** *It.* Coo; *anc.* Cos. *island* Dodekánisos, Greece, Aegean Sea
129 T12 **Kosa** Komi-Permyatskiy Avtonomnyy Okrug, NW Russian Federation 59.55N 54.54E
129 T13 **Kosa** ≈ NW Russian Federation
170 C11 **Kō-saki** *headland* Nagasaki, Tsushima, SW Japan 34.06N 129.13E
169 X13 **Kosan** SE North Korea 38.50N 127.26E
121 H18 **Kosava** *Rus.* Kosovo. Brestskaya Voblasts', SW Belarus 52.45N 25.16E
Kosch see Kose
150 G12 **Koschagyl** *Kaz.* Qosshaghyl. Atyrau, W Kazakhstan 46.52N 53.46E
112 G12 **Kościan** *Ger.* Kosten. Wielkopolskie, C Poland 52.04N 16.37E
112 I7 **Kościerzyna** Pomorskie, NW Poland 54.06N 17.55E
24 L4 **Kosciusko** Mississippi, S USA 33.03N 89.35W
Kosciusko, Mount see Kosciuszko, Mount
191 R11 **Kosciuszko, Mount** *prev.* Mount Kosciusko ▲ New South Wales, SE Australia 36.28S 148.15E
120 H4 **Kose** *Ger.* Kosch. Harjumaa, NW Estonia 59.10N 25.10E
116 G6 **Koshava** Vidin, NW Bulgaria 44.03N 23.00E
153 U9 **Kosh-Dĕbĕ** *var.* Koshtebĕ. Narynskaya Oblast', C Kyrgyzstan 41.03N 74.08E
171 K16 **Koshigaya** *var.* Kosigaya. Saitama, Honshū, S Japan 35.54N 139.46E
K'o-shih see Kashi
170 B15 **Koshikijima-rettō** *var.* Kosikizima Rettô. *island group* SW Japan
151 W13 **Koshkarkol', Ozero** ◎ SE Kazakhstan
32 L9 **Koshkonong, Lake** ◎ Wisconsin, N USA
152 B10 **Koshoba** *Turkm.* Goshoba. Balkanskiy Velayat, NW Turkmenistan 40.28N 54.11E
171 J14 **Kōshoku** *var.* Kôsyoku. Nagano, Honshū, S Japan 36.31N 138.07E
Koshtebĕ see Kosh-Dĕbĕ
Kōshū see Kwangju
113 N19 **Košice** *Ger.* Kaschau, *Hung.* Kassa. Košický Kraj, E Slovakia 48.43N 21.15E
113 M20 **Košický Kraj** ◐ *region* E Slovakia
Kosigaya see Koshigaya
Kosikizima Rettô see Koshikijima-rettō
159 R12 **Kosi Reservoir** ⊠ E Nepal
J8 **Kosiv** Ivano-Frankivs'ka Oblast', W Ukraine 48.19N 25.04E
151 O11 **Koskol'** Zhezkazgan, C Kazakhstan 49.34N 67.03E
129 Q9 **Koslan** Respublika Komi, NW Russian Federation 63.27N 48.52E
Köslin see Koszalin
152 M12 **Koson** *Rus.* Kasan. Qashqadaryo Wiloyati, S Uzbekistan 39.03N 65.34E
169 Y13 **Kosŏng** SE North Korea 38.40N 128.13E
153 S9 **Kosonsoy** *Rus.* Kassansay. Namangan Wiloyati, E Uzbekistan 41.15N 71.28E
115 M16 **Kosovo** *prev.* Autonomous Province of Kosovo and Metohija. *region* S Yugoslavia
Kosovo see Kosava
Kosovo and Metohija, Autonomous Province of see Kosovo

◆ Country ● Country Capital ◇ Dependent Territory ○ Dependent Territory Capital ◐ Administrative Region ✈ International Airport ▲ Mountain ▲▲ Mountain Range ⛰ Volcano ≈ River ◎ Lake ⊠ Reservoir

115 N16 Kosovo Polje Serbia, S Yugoslavia 42.40N 21.07E
115 O16 Kosovska Kamenica Serbia, SE Yugoslavia 42.37N 21.33E
115 M16 Kosovska Mitrovica Alb. Mitrovicë; prev. Mitrovica, Titova Mitrovica. Serbia, S Yugoslavia 42.54N 20.52E
201 X17 Kosrae ◆ state E Micronesia
201 Y14 Kosrae prev. Kusaie. island Caroline Islands, E Micronesia
27 U9 Kosse Texas, SW USA 31.16N 96.38W
111 P6 Kössen Tirol, W Austria 47.40N 12.24E
78 M16 Kossou, Lac de ◎ C Ivory Coast
Kossukavak see Krumovgrad
Kostajnica see Hrvatska Kostajnica
150 M7 Kostanay var. Kustanay, Kaz. Qostanay. N Kazakhstan 53.15N 63.34E
150 L8 Kostanay var. Kostanayskaya Oblast, Kaz. Qostanay Oblysy. ◆ province N Kazakhstan
Kostanayskaya Oblast see Kostanay
Kostamus see Kostomuksha
Kosten see Kościan
116 H10 Kostenets prev. Georgi Dimitrov. Sofiya, W Bulgaria 42.17N 23.52E
82 F10 Kosti White Nile, C Sudan 13.10N 32.37E
Kostnitz see Konstanz
128 H7 Kostomuksha Fin. Kostamus. Respublika Kareliya, NW Russian Federation 64.33N 30.28E
118 K3 Kostopil' Rus. Kostopol'. Rivnens'ka Oblast', NW Ukraine 50.20N 26.28E
Kostopol' see Kostopil'
128 M15 Kostroma Kostromskaya Oblast', NW Russian Federation 57.46N 40.59E
129 N14 Kostroma ⩳ NW Russian Federation
129 N14 Kostromskaya Oblast' ◆ province NW Russian Federation
112 D11 Kostrzyn Ger. Cüstrin, Küstrin. Lubuskie, W Poland 52.35N 14.39E
112 H11 Kostrzyn Wielkolpolskie, C Poland 52.23N 17.13E
119 X7 Kostyantynivka Rus. Konstantinovka. Donets'ka Oblast', SE Ukraine 48.30N 37.45E
Kostyukovichi see Kastsyukovichy
Kostyukovka see Kastsyukowka
Kôsyoku see Kôshoku
129 U6 Kos'yu Respublika Komi, NW Russian Federation 65.39N 59.01E
129 U6 Kos'yu ⩳ NW Russian Federation
112 F7 Koszalin Ger. Köslin. Koszalin, NW Poland 54.11N 16.10E
113 F22 Kőszeg Ger. Güns. Vas, W Hungary 47.24N 16.33E
158 H13 Kota prev. Kotah. Rājasthān, N India 25.13N 75.51E
174 H9 Kota Baru Sumatera, W Indonesia 1.07S 101.43E
175 Nn11 Kotabaru Pulau Laut, C Indonesia 3.15S 116.15E
Kotabaru see Jayapura
174 H2 Kota Bharu var. Kota Baharu, Kota Bahru. Kelantan, Peninsular Malaysia 6.07N 102.15E
Kotaboemi see Kotabumi
174 Ii12 Kotabumi prev. Kotaboemi. Sumatera, W Indonesia 4.49S 104.54E
155 S10 Kot Addu Punjab, E Pakistan 30.25N 70.54E
Kotah see Kota
175 Nn2 Kota Kinabalu prev. Jesselton. Sabah, East Malaysia 5.58N 116.04E
175 Nn2 Kota Kinabalu ✕ Sabah, East Malaysia 5.58N 116.04E
94 M12 Kotala Lappi, N Finland 67.01N 29.00E
Kotamobagoe see Kotamobagu
175 Rr7 Kotamobagu prev. Kotamobagoe. Sulawesi, C Indonesia 0.46N 124.21E
161 L14 Kotapad var. Kotapārh. Orissa, E India 19.10N 82.23E
Kotapārh see Kotapad
178 Gg17 Ko Ta Ru Tao island SW Thailand
174 L11 Kotawaringin, Teluk bay Borneo, C Indonesia
155 Q13 Kot Diji Sind, SE Pakistan 27.16N 68.43E
158 K9 Kotdwāra Uttar Pradesh, N India 29.43N 78.33E
129 Q14 Kotel'nich Kirovskaya Oblast', NW Russian Federation 58.19N 48.12E
131 N12 Kotel'nikovo Volgogradskaya Oblast', SW Russian Federation 47.37N 43.07E
126 Ll4 Kotel'nyy, Ostrov island Novosibirskiye Ostrova, N Russian Federation
119 T5 Kotel'va Poltavs'ka Oblast', C Ukraine 50.04N 34.46E
103 M14 Köthen var. Cöthen. Sachsen-Anhalt, C Germany 51.46N 11.58E
Kôti see Kôchi
83 G17 Kotido NE Uganda 3.03N 34.07E
95 N19 Kotka Etelä-Suomi, S Finland 60.28N 26.54E
129 P11 Kotlas Arkhangel'skaya Oblast', NW Russian Federation 61.13N 46.43E
40 M10 Kotlik Alaska, USA 63.01N 163.33W
79 Q17 Kotoka ✕ (Accra) S Ghana 5.41N 0.10W
Kotonu see Cotonou
115 J17 Kotor It. Cattaro. Montenegro, SW Yugoslavia 42.25N 18.47E
Kotor see Kotoriba
114 F7 Kotoriba Hung. Kotor. Medimurje, N Croatia 46.20N 16.47E
115 I17 Kotorska, Boka It. Bocche di Cattaro. bay Montenegro, SW Yugoslavia
114 H11 Kotorsko Republika Srpska, N Bosnia and Herzegovina 44.50N 18.03E
114 G11 Kotor Varoš Republika Srpska, N Bosnia and Herzegovina 44.37N 17.24E
Koto Sho/Kotosho see Lan Yü
130 M7 Kotovsk Tambovskaya Oblast', W Russian Federation 52.39N 41.31E
119 O9 Kotovs'k Rus. Kotovsk. Odes'ka Oblast', SW Ukraine 47.42N 29.30E
Kotovsk see Hînceşti
121 G16 Kotra Rus. Kotra. ⩳ W Belarus
155 P16 Kotri Sind, SE Pakistan 25.22N 68.16E
111 Q9 Kötschach Kärnten, S Austria 46.41N 12.57E
161 K15 Kottagūdem Andhra Pradesh, E India 17.36N 80.40E
161 F21 Kottappadi Kerala, SW India 11.38N 76.03E
161 G23 Kottayam Kerala, SW India 9.37N 76.31E
Kottbus see Cottbus
Kotte see Sri Jayawardanapura
81 K15 Kotto ⩳ Central African Republic/Dem. Rep. Congo (Zaire)
200 S13 Kotu Group island group W Tonga
152 B11 Koturdepe Turkm. Goturdepe. Balkanskiy Velayat, W Turkmenistan 39.32N 53.39E
126 Jj9 Kotuy ⩳ N Russian Federation
85 M16 Kotwa Mashonaland East, NE Zimbabwe 16.58S 32.46E
41 N7 Kotzebue Alaska, USA 66.54N 162.36W
40 M7 Kotzebue Sound inlet Alaska, USA
Kotzenan see Chocianów
Kotzman see Kitsman'
79 R14 Kouandé NW Benin 10.19N 1.42E
81 J15 Kouango Ouaka, S Central African Republic 5.00N 20.01E
79 O13 Koudougou C Burkina 12.15N 2.22W
100 K7 Koudum Friesland, N Netherlands 52.55N 5.26E
117 L25 Koufonísi island SE Greece
117 K21 Koufonísi island Kykládes, Greece, Aegean Sea
40 M8 Kougarok Mountain ▲ Alaska, USA 65.41N 165.29W
81 E20 Kouilou ⩳ S Congo
16 N3 Koukdjuak ⩳ Baffin Island, Nunavut, NE Canada
124 N4 Koúklia SW Cyprus 34.42N 32.35E
81 E19 Koulamoutou Ogooué-Lolo, C Gabon 1.06S 12.26E
78 L12 Koulikoro Koulikoro, SW Mali 12.55N 7.35W
78 L11 Koulikoro ◆ region SW Mali
197 H5 Koumac Province Nord, W New Caledonia 20.34S 164.18E
171 J15 Koumi Nagano, Honshū, S Japan 36.06N 138.27E
80 I13 Kouma Moyen-Chari, S Chad 8.55N 17.31E
Kounadougou see Koundougou
78 M15 Kounahiri C Ivory Coast 7.47N 5.51W
78 I12 Koundâra Moyenne-Guinée, NW Guinea 12.28N 13.15W
79 N13 Koundougou var. Kounadougou. C Burkina 11.43N 4.40W
78 H11 Koungheul C Senegal 14.00N 14.48W
Kounradskiy see Konyrat
27 X10 Kountze Texas, SW USA 30.22N 94.18W
79 Q13 Koupéla C Burkina 12.09N 0.23W
79 N13 Kouri Sikasso, SW Mali 12.09N 4.46W
57 Y9 Kourou N French Guiana 5.07N 52.37W
116 J12 Kourou ⩳ NE Greece
78 K14 Kouroussa Haute-Guinée, C Guinea 10.40N 9.49W
Kousseir see Al Quşayr
80 G11 Kousséri prev. Fort-Foureau. Extrême-Nord, NE Cameroon 12.01N 15.03E
Kouteifé see Al Quţayfah
78 M13 Koutiala Sikasso, S Mali 12.24N 5.30W
78 M14 Kouto NW Ivory Coast 9.51N 6.25W
95 M19 Kouvola Etelä-Suomi, S Finland 60.50N 26.48E
81 G18 Kouyou ⩳ C Congo
114 M10 Kovačica Hung. Antafalva; prev. Kovacsicza. Serbia, N Yugoslavia 45.08N 20.36E
Kovacsicza see Kovačica
Kővárhosszúfalu see Satulung
Kovászna see Covasna
128 I4 Kovdor Murmanskaya Oblast', NW Russian Federation 67.32N 30.27E
128 I5 Kovdozero, Ozero ◎ NW Russian Federation
118 J3 Kovel' Pol. Kowel. Volyns'ka Oblast', NW Ukraine 51.13N 24.42E
114 M11 Kovin Hung. Kevevára; prev. Temes-Kubin. Serbia, NE Yugoslavia 44.45N 20.59E
Kovno see Kaunas
131 N3 Kovrov Vladimirskaya Oblast', W Russian Federation 56.24N 41.21E
131 O5 Kovylkino Respublika Mordoviya, W Russian Federation 54.03N 43.52E
112 J11 Kowal Kujawsko-pomorskie, C Poland 52.31N 19.08E
112 J9 Kowalewo Pomorskie Ger. Schönsee. Kujawsko-pomorskie, C Poland 53.07N 18.48E
Kowasna see Covasna
121 M16 Kowbcha Rus. Kolbcha. Mahilyowskaya Voblasts', E Belarus 53.40N 29.13E
Koweit see Kuwait
Kowel see Kovel'
193 F17 Kowhitirangi West Coast, South Island, NZ 42.54S 171.01E
167 O15 Kowloon Chin. Jiulong. Hong Kong, S China
Kowno see Kaunas
165 N7 Kox Kuduk well NW China 40.32N 92.30E
142 D16 Köyceğiz Muğla, SW Turkey 36.58N 28.38E
129 N6 Koyda Arkhangel'skaya Oblast', NW Russian Federation 66.22N 42.42E
152 D10 Koymat Turkm. Goymat. Balkanskiy Velayat, NW Turkmenistan 40.23N 55.45E
152 D10 Koymatdag, Gory Turkm. Goymatdag. hill range NW Turkmenistan
Koyna Reservoir see Shivāji Sāgar
171 M11 Koyoshi-gawa ⩳ Honshū, C Japan
Koysanjaq see Koi Sanjaq
Koytash see Qŭytosh
41 N9 Koyuk Alaska, USA 64.55N 161.09W
41 N9 Koyuk River ⩳ Alaska, USA
41 O9 Koyukuk Alaska, USA 64.52N 157.42W
41 O9 Koyukuk River ⩳ Alaska, USA
142 J13 Kozaklı Nevşehir, C Turkey 39.13N 34.51E
170 F13 Kōzan Hiroshima, Honshū, SW Japan 34.35N 133.02E
142 K16 Kozan Adana, S Turkey 37.27N 35.46E
117 E14 Kozáni Dytikí Makedonía, N Greece 40.18N 21.48E
114 F10 Kozara ▲ NW Bosnia and Herzegovina
Kozarska Dubica see Bosanska Dubica
119 P3 Kozelets' Rus. Kozelets. Chernihivs'ka Oblast', NE Ukraine 50.54N 31.09E
119 S6 Kozel'shchyna Poltavs'ka Oblast', C Ukraine 49.13N 33.49E
130 J5 Kozel'sk Kaluzhskaya Oblast', W Russian Federation 54.04N 35.51E
Kozhikode see Calicut
129 V9 Kozhimiz, Gora ▲ NW Russian Federation 63.13N 58.54E
128 L9 Kozhozero, Ozero ◎ NW Russian Federation
129 T7 Kozhva var. Kozya. Respublika Komi, NW Russian Federation 65.06N 57.00E
129 T7 Kozhva ⩳ NW Russian Federation
129 U6 Kozhym Respublika Komi, NW Russian Federation 65.43N 59.25E
112 N13 Kozienice Mazowieckie, C Poland 51.37N 21.30E
111 S13 Kozina SW Slovenia 45.36N 13.56E
116 H7 Kozloduy Vratsa, NW Bulgaria 43.47N 23.42E
131 Q3 Kozlovka Chuvashskaya Respublika, W Russian Federation 55.53N 48.07E
Kozlovshchina/Kozlowszczyzna see Kazlowshchyna
131 P3 Koz'modem'yansk Respublika Mariy El, W Russian Federation 56.19N 46.33E
118 J6 Kozova Ternopil's'ka Oblast', W Ukraine 49.25N 25.09E
115 P20 Kožuf ▲ S FYR Macedonia 41.10N 22.14E
172 S13 Kōzu-shima island E Japan
Kozya see Kozhva
Koz'yany see Kaz'yany
119 N5 Kozyatyn Rus. Kazatin. Vinnyts'ka Oblast', C Ukraine 49.40N 28.48E
79 Q16 Kpalimé var. Palimé. SW Togo 6.54N 0.37E
79 Q16 Kpandu E Ghana 7.00N 0.18E
101 F15 Krabbendijke Zeeland, SW Netherlands 51.25N 4.07E
178 Gg16 Krabi var. Muang Krabi. Krabi, SW Thailand 8.04N 98.52E
178 Gg14 Kra Buri Ranong, SW Thailand 10.25N 98.48E
178 Jj13 Krâchéh prev. Kratie. Krâchéh, E Cambodia 12.28N 106.01E
97 G17 Kragerø Telemark, S Norway 58.53N 9.22E
114 M13 Kragujevac Serbia, C Yugoslavia 44.01N 20.54E
Krainburg see Kranj
178 Gg14 Kra, Isthmus of isthmus Malaysia/Thailand
114 D12 Krajina cultural region SW Croatia
Krakatau, Pulau see Rakata, Pulau
Krakau see Małopolskie
113 L16 Kraków Eng. Cracow, Ger. Krakau; anc. Cracovia. Małopolskie, S Poland 50.03N 19.57E
102 L9 Krakower See ◎ NE Germany
178 Ii12 Krâlănh Siĕmréab, NW Cambodia 13.35N 103.27E
47 Q16 Kralendijk Bonaire, E Netherlands Antilles 12.07N 68.13W
114 B10 Kraljevica It. Porto Re. Primorje-Gorski Kotar, NW Croatia 45.15N 14.36E
114 M13 Kraljevo prev. Rankovićevo. Serbia, C Yugoslavia 43.44N 20.40E
Kralup an der Moldau see Kralupy nad Vltavou
113 C16 Kralupy nad Vltavou Ger. Kralup an der Moldau. Středočeský Kraj, NW Czech Republic 50.13N 14.17E
119 W7 Kramators'k Rus. Kramatorsk. Donets'ka Oblast', SE Ukraine 48.43N 37.34E
Kramatorsk see Kramators'k
95 H17 Kramfors Västernorrland, C Sweden 62.55N 17.49E
117 D15 Kranéa Dytikí Makedonía, N Greece 39.54N 21.21E
110 M7 Kranebitten ✕ (Innsbruck) Tirol, W Austria 47.18N 11.21E
117 G20 Kranídi Pelopónnisos, S Greece 37.21N 23.09E
111 T11 Kranj Ger. Krainburg. NW Slovenia 46.16N 14.16E
117 F16 Krannón battleground Thessalía, C Greece 39.32N 22.20E
Kranz see Zelenogradsk
114 D7 Krapina Krapina-Zagorje, N Croatia 46.12N 15.52E
114 E8 Krapina ⩳ N Croatia
114 D8 Krapina-Zagorje off. Krapinsko-Zagorska Županija. ◆ province N Croatia
116 L7 Krapinets ⩳ NE Bulgaria
113 I15 Krapkowice Ger. Krappitz. Opolskie, S Poland 50.28N 17.55E
Krappitz see Krapkowice
129 O12 Krasavino Vologodskaya Oblast', NW Russian Federation 60.56N 46.27E
125 Ff5 Krasino Novaya Zemlya, Arkhangel'skaya Oblast', N Russian Federation 70.45N 54.16E
127 N18 Kraskino Primorskiy Kray, SE Russian Federation 42.40N 130.51E
120 J11 Krāslava Krāslava, SE Latvia 55.56N 27.08E
121 M14 Krasnaluki Rus. Krasnoluki. Vitsyebskaya Voblasts', N Belarus 54.37N 28.49E
121 P17 Krasnapollye Rus. Krasnopol'ye. Mahilyowskaya Voblasts', E Belarus 53.19N 31.24E
130 L15 Krasnaya Polyana Krasnodarskiy Kray, SW Russian Federation 43.40N 40.13E
121 J18 Krasnaya Slabada var. Chyrvonaya Slabada, Rus. Krasnaya Sloboda. Minskaya Voblasts', S Belarus 52.51N 27.10E
Krasnaya Sloboda see Krasnaya Slabada
121 J15 Krasnaye Rus. Krasnoye. Minskaya Voblasts', C Belarus 54.13N 27.04E
113 O14 Kraśnik Ger. Kratznick. Lubelskie, E Poland 50.55N 22.13E
113 O14 Kraśnik Fabryczny Lubelskie, SE Poland 50.57N 22.07E
119 O9 Krasni Okny Odes'ka Oblast', SW Ukraine 47.33N 29.28E
151 P7 Krasnoarmeysk Severnyy Kazakhstan, N Kazakhstan 53.52N 69.51E
131 P8 Krasnoarmeysk Saratovskaya Oblast', W Russian Federation 51.01N 45.42E
Krasnoarmeysk see Krasnoarmiys'k/Tayynsha
127 Oo4 Krasnoarmeyskiy Chukotskiy Avtonomnyy Okrug, NE Russian Federation 69.30N 171.44E
119 W7 Krasnoarmiys'k Rus. Krasnoarmeysk. Donets'ka Oblast', SE Ukraine 48.16N 37.13E
129 P11 Krasnoborsk Arkhangel'skaya Oblast', NW Russian Federation 61.31N 45.57E
130 K14 Krasnodar prev. Ekaterinodar, Yekaterinodar. Krasnodarskiy Kray, SW Russian Federation 45.02N 39.00E
130 K13 Krasnodarskiy Kray ◆ territory SW Russian Federation
119 Z7 Krasnodon Luhans'ka Oblast', E Ukraine 48.16N 39.45E
Krasnogor see Kallaste
131 T2 Krasnogorskoye Latv. Sarkaņi. Udmurtskaya Respublika, NW Russian Federation 57.42N 52.29E
Krasnograd see Krasnohrad
Krasnogvardeysk see Bulunghur
130 M13 Krasnogvardeyskoye Stavropol'skiy Kray, SW Russian Federation 45.49N 41.31E
Krasnogvardeyskoye see Krasnohvardiys'ke
119 U6 Krasnohrad Rus. Krasnograd. Kharkivs'ka Oblast', E Ukraine 49.22N 35.27E
119 S12 Krasnohvardiys'ke Rus. Krasnogvardeyskoye. Respublika Krym, S Ukraine 45.30N 34.19E
126 L16 Krasnokamensk Chitinskaya Oblast', S Russian Federation 50.03N 118.01E
129 U14 Krasnokamsk Permskaya Oblast', W Russian Federation 58.07N 55.48E
131 U8 Krasnokholm Orenburgskaya Oblast', W Russian Federation 51.34N 54.11E
119 U5 Krasnokuts'k Rus. Krasnokutsk. Kharkivs'ka Oblast', E Ukraine 50.01N 35.03E
130 L7 Krasnolesnyy Voronezhskaya Oblast', W Russian Federation 51.53N 39.37E
Krasnoluki see Krasnaluki
Krasnoosol'skoye Vodokhranilishche see Chervonooskil's'ke Vodoskhovyshche
119 S11 Krasnoperekops'k Rus. Krasnoperekopsk. Respublika Krym, S Ukraine 45.56N 33.46E
119 U4 Krasnopillya Sums'ka Oblast', NE Ukraine 50.46N 35.17E
Krasnopol'ye see Krasnapollye
126 H9 Krasnosel'kup Yamalo-Nenetskiy Avtonomnyy Okrug, N Russian Federation 65.46N 82.11E
128 L5 Krasnoshchel'ye Murmanskaya Oblast', NW Russian Federation 67.22N 37.03E
131 O5 Krasnoslobodsk Respublika Mordoviya, W Russian Federation 54.24N 43.51E
131 T2 Krasnoslobodsk Volgogradskaya Oblast', SW Russian Federation 48.41N 44.34E
Krasnostav see Krasnystaw
125 F10 Krasnotur'insk Sverdlovskaya Oblast', C Russian Federation 59.45N 60.19E
125 E11 Krasnoufimsk Sverdlovskaya Oblast', C Russian Federation 56.43N 57.39E
125 Ee10 Krasnoural'sk Sverdlovskaya Oblast', C Russian Federation 58.24N 59.46E
131 V5 Krasnousol'skiy Respublika Bashkortostan, W Russian Federation 53.55N 56.22E
129 U12 Krasnovishersk Permskaya Oblast', NW Russian Federation 60.22N 57.04E
Krasnovodsk see Turkmenbashi
152 A10 Krasnovodskiy Zaliv Turkm. Krasnowodsk Aylagy. lake gulf W Turkmenistan
152 B10 Krasnovodskoye Plato Turkm. Krasnowodsk Platosy. plateau NW Turkmenistan
Krasnowodsk Aylagy see Krasnovodskiy Zaliv
Krasnowodsk Platosy see Krasnovodskoye Plato
126 Hh14 Krasnoyarsk Krasnoyarskiy Kray, S Russian Federation 56.04N 92.46E
131 X7 Krasnoyarskiy Orenburgskaya Oblast', W Russian Federation 51.56N 59.54E
126 I12 Krasnoyarskiy Kray ◆ territory C Russian Federation
126 I14 Krasnoyarskoye Vodokhranilishche ⊡ S Russian Federation
Krasnoye see Krasnaye
152 J15 Krasnoye Znamya Turkm. Gyzylbaydak. Maryyskiy Velayat, S Turkmenistan 36.51N 62.24E
129 R11 Krasnozatonskiy Respublika Komi, NW Russian Federation 61.39N 51.00E
120 D13 Krasnoznamensk prev. Lasdehnen, Ger. Haselberg. Kaliningradskaya Oblast', W Russian Federation 54.57N 22.28E
119 R11 Krasnoznam"yans'kyy Kanal canal S Ukraine
113 P14 Krasnystaw Rus. Krasnostav. Lubelskie, SE Poland 51.00N 23.10E
130 H4 Krasnyy Smolenskaya Oblast', W Russian Federation 54.36N 31.27E
131 P2 Krasnyye Baki Nizhegorodskaya Oblast', W Russian Federation 57.07N 45.12E
131 Q13 Krasnyye Barrikady Astrakhanskaya Oblast', SW Russian Federation 46.14N 47.48E
128 K15 Krasnyy Kholm Tverskaya Oblast', W Russian Federation 58.04N 37.05E
131 Q8 Krasnyy Kut Saratovskaya Oblast', W Russian Federation 50.54N 46.58E
Krasnyy Liman see Krasnyy Lyman
119 Y7 Krasnyy Luch prev. Krindachevka. Luhans'ka Oblast', E Ukraine 48.08N 38.52E
119 X6 Krasnyy Lyman Rus. Krasnyy Liman. Donets'ka Oblast', SE Ukraine 49.00N 37.50E
131 R3 Krasnyy Steklovar Respublika Mariy El, W Russian Federation 56.14N 48.49E
131 P8 Krasnyy Tekstil'shchik Saratovskaya Oblast', W Russian Federation 51.35N 45.49E
131 R13 Krasnyy Yar Astrakhanskaya Oblast', SW Russian Federation 46.33N 48.21E
Krassóvár see Caraşova
118 L5 Krasyliv Khmel'nyts'ka Oblast', W Ukraine 49.38N 26.59E
113 O21 Kraszna Rom. Crasna. ⩳ Hungary/Romania
Kratie see Krâchéh
194 I13 Kratke Range ▲ C PNG
115 P17 Kratovo NE FYR Macedonia 42.04N 22.08E
Kratznick see Kraśnik
176 Yy11 Krau Irian Jaya, E Indonesia 3.15S 140.07E
178 Ii13 Krâvanh, Chuŏr Phnum Eng. Cardamom Mountains, Fr. Chaîne des Cardamomes. ▲ W Cambodia
Kravasta Lagoon see Karavastasë, Laguna e
Krawang see Karawang
Kraxatau see Rakata, Pulau
131 Q15 Kraynovka Respublika Dagestan, SW Russian Federation 43.58N 47.24E
120 D12 Kražiai Kelmė, C Lithuania 55.36N 22.41E
29 P11 Krebs Oklahoma, C USA 34.55N 95.43W
103 D15 Krefeld Nordrhein-Westfalen, W Germany 51.19N 6.34E
Kreisstadt see Krosno Odrzańskie
117 D17 Kremastón, Technití Límni ⊡ C Greece
Kremenchug see Kremenchuk
Kremenchugskoye Vodokhranilishche/Kremenchuk Reservoir see Kremenchuts'ke Vodoskhovyshche
119 S6 Kremenchuk Rus. Kremenchug. Poltavs'ka Oblast', NE Ukraine 49.03N 33.27E
119 R6 Kremenchuts'ke Vodoskhovyshche Eng. Kremenchuk Reservoir, Rus. Kremenchugskoye Vodokhranilishche. ⊡ C Ukraine
118 K5 Kremenets' Pol. Krzemieniec, Rus. Kremenets. Ternopil's'ka Oblast', W Ukraine 50.05N 25.43E
Kremennaya see Kreminna
119 X6 Kreminna Rus. Kremennaya. Luhans'ka Oblast', E Ukraine 49.03N 38.14E
39 R4 Kremmling Colorado, C USA 40.03N 106.23W
111 V3 Krems ⩳ NE Austria
Krems see Krems an der Donau
111 W3 Krems an der Donau var. Krems. Niederösterreich, N Austria 48.25N 15.34E
Kremsier see Kroměříž
111 S4 Kremsmünster Oberösterreich, N Austria 48.04N 14.08E
40 M17 Krenitzin Islands island Aleutian Islands, Alaska, USA
Kresena see Kresna
116 G11 Kresna var. Kresena. Blagoevgrad, SW Bulgaria 41.43N 23.12E
114 O12 Krespoljin Serbia, E Yugoslavia 44.37N 21.36E
27 N4 Kress Texas, SW USA 34.21N 101.43W
127 Pp4 Kresta, Zaliv bay E Russian Federation
117 D20 Kréstena prev. Selinoús. Dytikí Ellás, S Greece 37.36N 21.36E
128 H14 Krestty Novgorodskaya Oblast', W Russian Federation 58.15N 32.28E
126 Kk11 Krestyakh Respublika Sakha (Yakutiya), NE Russian Federation 62.10N 116.24E
Kretikon Delagos see Kritikó Pélagos
120 C11 Kretinga Ger. Krottingen. Kretinga, NW Lithuania 55.53N 21.13E
Kreutz see Cristuru Secuiesc
Kreuz see Križevci, Croatia
Kreuz see Risti, Estonia
Kreuzburg/Kreuzburg in Oberschlesien see Kluczbork
Kreuzingen see Bol'shakovo
110 H6 Kreuzlingen Thurgau, NE Switzerland 47.37N 9.10E
103 K25 Kreuzspitze ▲ S Germany 47.30N 10.55E
103 F16 Kreuztal Nordrhein-Westfalen, W Germany 50.58N 8.00E
121 I15 Kreva Rus. Krevo. Hrodzyenskaya Voblasts', W Belarus 54.19N 26.16E
Krevo see Kreva
Kría Vrísi see Krýa Vrýsi
81 D16 Kribi Sud, SW Cameroon 2.53N 9.57E
Krichëv see Krychaw
Krickerhäu/Kriegerhaj see Handlová
111 W6 Krieglach Steiermark, E Austria 47.33N 15.37E
110 F8 Kriens Luzern, W Switzerland 47.01N 8.16E
Krimmitschau see Crimmitschau
100 H12 Krimpen aan den IJssel Zuid-Holland, SW Netherlands 51.56N 4.39E
Krindachevka see Krasnyy Luch
117 G25 Kríos, Akrotírio headland Kríti, Greece, E Mediterranean Sea 35.17N 23.31E
161 J16 Krishna prev. Kistna. ⩳ C India
161 H20 Krishnagiri Tamil Nādu, SE India 12.33N 78.10E
161 K17 Krishna, Mouths of the delta SE India
159 S15 Krishnanagar West Bengal, N India 23.22N 88.31E
161 G20 Krishnarājāsāgara Reservoir ⊡ W India
97 N19 Kristdala Kalmar, S Sweden 57.24N 16.12E
Kristiania see Oslo
97 E18 Kristiansand var. Christiansand. Vest-Agder, S Norway 58.07N 7.52E
97 L22 Kristianstad Skåne, S Sweden 56.01N 14.10E
96 F8 Kristiansund var. Christiansund. Møre og Romsdal, S Norway 63.07N 7.45E
Kristiinankaupunki see Kristinestad
95 I14 Kristineberg Västerbotten, N Sweden 65.07N 18.36E
97 L16 Kristinehamn Värmland, C Sweden 59.16N 14.09E
95 J17 Kristinestad Fin. Kristiinankaupunki. Länsi-Suomi, W Finland 62.15N 21.24E
Kristyor see Crişcior
117 J25 Kríti Eng. Crete. ◆ region Greece, Aegean Sea
117 J24 Kríti Eng. Crete. island Greece, Aegean Sea
117 J23 Kritikó Pélagos var. Kretikon Delagos, Eng. Sea of Crete; anc. Mare Creticum. sea Greece, Aegean Sea
Kriulyany see Criuleni
114 I12 Krivaja ⩳ NE Bosnia and Herzegovina
Krivaja see Mali Idoš
115 P17 Kriva Palanka Turk. Eğri Palanka. NE FYR Macedonia 42.13N 22.19E
Krivichi see Kryvychy
116 H8 Krivodol Vratsa, NW Bulgaria 43.23N 23.30E
130 M10 Krivorozh'ye Rostovskaya Oblast', SW Russian Federation 48.51N 40.49E
Krivoshin see Kryvoshyn
Krivoy Rog see Kryvyy Rih
114 F7 Križevci Ger. Kreuz, Hung. Kőrös. Varaždin, NE Croatia 46.02N 16.32E
114 B10 Krk It. Veglia. Primorje-Gorski Kotar, NW Croatia 45.01N 14.36E
114 B10 Krk It. Veglia; anc. Curieta. island NW Croatia
111 V12 Krka ⩳ SE Slovenia
Krka see Gurk
111 R11 Krn ▲ NW Slovenia 46.15N 13.37E
113 H16 Krnov Ger. Jägerndorf. Ostravský Kraj, E Czech Republic 50.05N 17.42E
Kroatien see Croatia
97 G14 Krøderen Buskerud, S Norway 60.06N 9.48E
97 G14 Krøderen ◎ S Norway
Kroi see Krui
97 N17 Krokek Östergötland, S Sweden 58.40N 16.25E
Krokodil see Crocodile
95 G16 Krokom Jämtland, C Sweden 63.19N 14.30E
119 S2 Krolevets' Rus. Krolevets. Sums'ka Oblast', NE Ukraine 51.34N 33.24E
Królewska Huta see Chorzów
113 H18 Kroměříž Ger. Kremsier. Zlínský Kraj, E Czech Republic 49.18N 17.24E
100 I9 Krommenie Noord-Holland, C Netherlands 52.30N 4.46E
130 J6 Kromy Orlovskaya Oblast', W Russian Federation 52.41N 35.45E
103 L18 Kronach Bayern, E Germany 50.14N 11.19E
Krone an der Brahe see Koronowo
178 I13 Krŏng Kaôh Kŏng Kaôh Kŏng, SW Cambodia 11.37N 102.58E
97 K21 Kronoberg ◆ county S Sweden
127 Pp11 Kronotskiy Zaliv bay E Russian Federation
205 Q2 Kronprinsesse Märtha Kyst physical region Antarctica
205 V3 Kronprins Olav Kyst physical region Antarctica
128 G12 Kronshtadt Leningradskaya Oblast', NW Russian Federation 60.01N 29.42E
Kronstadt see Braşov
85 I22 Kroonstad Free State, C South Africa 27.40S 27.15E
126 Kk13 Kropotkin Irkutskaya Oblast', C Russian Federation 58.30N 115.21E
130 L14 Kropotkin Krasnodarskiy Kray, SW Russian Federation 45.28N 40.30E
112 J11 Krośniewice Łódzkie, C Poland 52.14N 19.10E
113 N17 Krosno Ger. Krossen. Podkarpackie, SE Poland 49.40N 21.46E
112 E12 Krosno Odrzańskie Ger. Crossen, Kreisstadt. Lubuskie, W Poland 52.02N 15.06E
Krossen see Krosno
Krotoschin see Krotoszyn
112 H13 Krotoszyn Ger. Krotoschin. Wielkopolskie, C Poland 51.43N 17.24E
Krottingen see Kretinga
Krousón see Krousónas
117 J25 Krousónas prev. Krousón, Kroussón. Kríti, Greece, E Mediterranean Sea 35.13N 24.58E
Kroussón see Krousónas
Kraba see Krrabë
115 L20 Krrabë var. Kraba. Tiranë, C Albania 41.15N 19.56E
115 L17 Krrabit, Mali i ▲ N Albania
111 W12 Krško Ger. Gurkfeld; prev. Videm-Krško. E Slovenia 45.57N 15.31E
85 K19 Kruger National Park national park Northern, N South Africa
85 J21 Krugersdorp Gauteng, NE South Africa 26.04S 27.46E
40 D16 Krugloi Point headland Agattu Island, Alaska, USA 52.30N 173.46E
Krugloye see Kruhlaye
121 N15 Kruhlaye Rus. Krugloye. Mahilyowskaya Voblasts', E Belarus 54.15N 29.48E
174 I13 Krui var. Kroi. Sumatera, SW Indonesia 5.11S 103.55E
101 G16 Kruibeke Oost-Vlaanderen, N Belgium 51.10N 4.18E
85 G25 Kruidfontein Western Cape, SW South Africa 32.50S 21.59E
101 F15 Kruiningen Zeeland, SW Netherlands 51.28N 4.01E
Kruja see Krujë
115 L19 Krujë var. Kruja, It. Croia. Durrës, C Albania 41.30N 19.48E
Krulevshchina see Krulewshchyna
120 K13 Krulewshchyna Rus. Krulevshchina. Vitsyebskaya Voblasts', N Belarus 55.01N 27.46E
27 T6 Krum Texas, SW USA 33.15N 97.14W
103 J23 Krumbach Bayern, S Germany 48.12N 10.21E
115 M17 Krumë Kukës, NE Albania 42.11N 20.25E
Krummau see Český Krumlov
116 K12 Krumovgrad prev. Kossukavak. Kŭrdzhali, S Bulgaria 41.27N 25.40E
116 K12 Krumovitsa ⩳ S Bulgaria
116 L10 Krumovo Yambol, E Bulgaria 42.16N 26.25E
178 Hh11 Krung Thep var. Krung Thep Mahanakhon, Eng. Bangkok. ● (Thailand) Bangkok, C Thailand 13.43N 100.30E
178 Hh12 Krung Thep, Ao var. Bight of Bangkok. bay S Thailand
Krung Thep Mahanakhon see Krung Thep
Krupa/Krupa na Uni see Bosanska Krupa
121 M15 Krupki Rus. Krupki. Minskaya Voblasts', C Belarus 54.19N 29.07E
97 G24 Kruså var. Krusaa. Sønderjylland, SW Denmark 54.49N 9.25E
Krusaa see Kruså
15 I4 Krusenstern, Cape headland Nunavut, NW Canada 68.17N 114.00W
115 N14 Kruševac Serbia, C Yugoslavia 43.36N 21.19E
115 N19 Kruševo SW FYR Macedonia 41.22N 21.15E
113 A15 Krušné Hory Eng. Ore Mountains, Ger. Erzgebirge. ▲ Czech Republic/Germany see also Erzgebirge
41 W13 Kruzof Island island Alexander Archipelago, Alaska, USA
116 F13 Krýa Vrýsi var. Kría Vrísi. Kentrikí Makedonía, N Greece 40.40N 22.18E
121 P16 Krychaw Rus. Krichëv. Mahilyowskaya Voblasts', E Belarus 53.42N 31.43E
66 K11 Krylov Seamount undersea feature E Atlantic Ocean 17.34N 30.07W
Krym see Krym, Respublika
119 S13 Krym, Respublika var. Krym, Eng. Crimea, Crimean Oblast; prev. Rus. Krymskaya ASSR, Krymskaya Oblast'. ◆ province SE Ukraine
130 K14 Krymsk Krasnodarskiy Kray, SW Russian Federation 44.56N 38.02E
Krymskaya ASSR/Krymskaya Oblast' see Krym, Respublika
119 T13 Kryms'ki Hory ▲ S Ukraine
119 T13 Kryms'kyy Pivostriv peninsula S Ukraine
113 M18 Krynica Ger. Tannenhof. Małopolskie, S Poland 49.26N 20.57E
119 P8 Kryve Ozero Odes'ka Oblast', SW Ukraine 47.54N 30.19E
121 I18 Kryvoshyn Rus. Krivoshin. Brestskaya Voblasts', SW Belarus 52.52N 26.07E
121 K14 Kryvychy Rus. Krivichi. Minskaya Voblasts', C Belarus 54.43N 27.16E
119 S8 Kryvyy Rih Rus. Krivoy Rog. Dnipropetrovs'ka Oblast', SE Ukraine 47.53N 33.24E
119 N8 Kryzhopil' Vinnyts'ka Oblast', C Ukraine 48.22N 28.51E
Krzemieniec see Kremenets'
113 J14 Krzepice Śląskie, S Poland 50.58N 18.42E
112 F10 Krzyż Wielkopolski Wielkopolskie, C Poland 52.52N 16.03E
Ksar al Kabir see Ksar-el-Kebir
Ksar al Soule see Er-Rachidia
76 J5 Ksar El Boukhari N Algeria 35.57N 2.49E
76 G5 Ksar-el-Kebir var. Alcázar, Ksar al Kabir, Ksar-el-Kébir, Ar. Al-Kasr al-Kebir, Al-Qsar al-Kbir, Sp. Alcazarquivir. NW Morocco 35.04N 5.55W
112 H12 Książ Wielkopolski Ger. Xions. Wielkopolskie, C Poland 52.03N 17.10E
131 O3 Kstovo Nizhegorodskaya Oblast', W Russian Federation 56.07N 44.12E
174 Mm4 Kuala Belait W Brunei 4.48N 114.12E
Kuala Dungun see Dungun
174 M7 Kualakerian Borneo, C Indonesia
174 M10 Kualakuayan Borneo, C Indonesia 2.01S 112.34E
174 H4 Kuala Lipis Pahang, Peninsular Malaysia 04.11N 102.00E
174 H5 Kuala Lumpur ● (Malaysia) Kuala Lumpur, Peninsular Malaysia 3.07N 101.42E

◆ Country ● Country Capital ◇ Dependent Territory ○ Dependent Territory Capital ◈ Administrative Region ✕ International Airport ▲ Mountain ▲ Mountain Range ⩯ Volcano ⩳ River ◎ Lake ⊡ Reservoir

279

Kuala Pelabohan Kelang see Pelabuhan Klang
175 Nn3 **Kuala Penyu** Sabah, East Malaysia 5.37N 115.36E
40 E9 **Kualapuu** *Haw.* Kualapu'u. Molokai, Hawaii, USA, C Pacific Ocean 21.09N 157.02W
173 G6 **Kuala, Sungai** ♒ Sumatera, W Indonesia
174 Hh3 **Kuala Terengganu** *var.* Kuala Trengganu. Terengganu, Peninsular Malaysia 5.19N 103.07E
174 Hh9 **Kualatungkal** Sumatera, W Indonesia 0.49S 103.22E
175 O3 **Kuamut, Sungai** ♒ East Malaysia
175 Qq7 **Kuandang** Sulawesi, N Indonesia 0.50N 122.55E
175 Qq7 **Kuandang, Teluk** *bay* Sulawesi, N Indonesia
169 V12 **Kuandian** Liaoning, NE China 40.41N 124.46E
Kuando-Kubango see Cuando Cubango
Kuang-chou see Guangzhou
Kuang-hsi see Guangxi Zhuangzu Zizhiqu
Kuang-tung see Guangdong
Kuang-yuan see Guangyuan
174 Hh4 **Kuantan** Pahang, Peninsular Malaysia 3.49N 103.19E
Kuantan, Batang see Indragiri, Sungai
Kuanza Norte see Cuanza Norte
Kuanza Sul see Cuanza Sul
Kuba see Quba
125 Aa12 **Kuban'** *var.* Hypanis.
Kubango see Cubango/Okavango
147 X8 **Kubārah** NW Oman 23.03N 56.52E
95 H16 **Kubbe** Västernorrland, C Sweden 63.31N 18.04E
82 A11 **Kubbum** Southern Darfur, W Sudan 11.46N 23.46E
128 L13 **Kubenskoye, Ozero** ⊚ NW Russian Federation
170 Ee16 **Kubokawa** Kōchi, Shikoku, SW Japan 33.22N 133.14E
116 L7 **Kubrat** *prev.* Balbunar. Razgrad, N Bulgaria 43.48N 26.31E
175 Oo15 **Kubu** Sumbawa, S Indonesia 8.15S 115.30E
114 O13 **Kučajske Planine** ▲ E Yugoslavia
172 Pp2 **Kuccharo-ko** ⊚ Hokkaidō, N Japan
114 O11 **Kučevo** Serbia, NE Yugoslavia 44.29N 21.42E
Kuchan see Qūchān
174 L6 **Kuching** *prev.* Sarawak. Sarawak, East Malaysia 1.31N 110.19E
174 L7 **Kuching** ✈ Sarawak, East Malaysia 1.31N 110.19E
170 Aa17 **Kuchinoerabu-jima** *island* Nansei-shotō, SW Japan
170 Cl3 **Kuchinotsu** Nagasaki, Kyūshū, SW Japan 32.36N 130.11E
111 Q6 **Kuchl** Salzburg, NW Austria 47.37N 13.22E
154 L9 **Küchnay Darwēyshān** Helmand, S Afghanistan 31.01N 64.09E
Kuchurgan see Kuchurhan
119 O9 **Kuchurhan** *Rus.* Kuchurgan. ♒ NE Ukraine
Kuçova see Kuçovë
115 L21 **Kuçovë** *var.* Kuçova; *prev.* Qyteti Stalin. Berat, C Albania 40.48N 19.55E
142 D11 **Küçük Çekmece** İstanbul, NW Turkey 41.01N 28.46E
170 Dd13 **Kudamatsu** *var.* Kudamatu. Yamaguchi, Honshū, SW Japan 34.00N 131.53E
Kudamatu see Kudamatsu
Kudara see Ghūdara
175 O1 **Kudat** Sabah, East Malaysia 6.54N 116.46E
161 G17 **Kūdligi** Karnātaka, W India 14.58N 76.24E
Kudowa see Kudowa-Zdrój
113 F16 **Kudowa-Zdrój** *Ger.* Kudowa. Wałbrzych, SW Poland 50.27N 16.13E
119 P9 **Kudryavtsivka** Mykolayivs'ka Oblast', S Ukraine 47.18N 31.02E
174 Ll4 **Kudus** *prev.* Koedoes. Jawa, C Indonesia 6.46S 110.48E
129 T13 **Kudymkar** Komi-Permyatskiy Avtonomnyy Okrug, NW Russian Federation 59.01N 54.40E
Kudzsir see Cugir
Kuei-chou see Guizhou
Kuei-lin see Guilin
Kuei-yang see Guiyang
K'u-erh-lo see Korla
Kueyang see Guiyang
Kufa see Al Kūfah
142 E14 **Küfiçayı** ♒ C Turkey
111 O6 **Kufstein** Tirol, W Austria 47.36N 12.10E
151 V14 **Kugaly** Kaz. Qoghaly. Almaty, SE Kazakhstan 44.30N 78.40E
25 I4 **Kugluktuk** *var.* Qurlurtuuq; *prev.* Coppermine. Nunavut, N Canada 67.49N 115.12W
149 Y13 **Kūhak** Sīstān va Balūchestān, SE Iran 27.10N 63.15E
149 R9 **Kūhbonān** Kermān, C Iran 31.22N 56.16E
154 J5 **Kūhestān** *var.* Kohsān. Herāt, W Afghanistan 34.40N 61.10E
95 N15 **Kuhmo** Oulu, E Finland 64.04N 29.34E
95 L18 **Kuhmoinen** Länsi-Suomi, W Finland 61.32N 25.09E
Kuhnau see Konin
Kühnö see Kihnu
149 O8 **Kührpäyeh** Eşfahān, C Iran 32.42N 52.25E
178 H13 **Kui Buri** *var.* Ban Kui Nua. Prachuap Khiri Khan, SW Thailand 12.10N 99.49E
Kuibyshev see Kuybyshevskoye Vodokhranilishche
84 D13 **Kuito** *Port.* Silva Porto. Bié, C Angola 12.21S 16.54E
41 X14 **Kuiu Island** *island* Alexander Archipelago, Alaska, USA
94 L13 **Kuivaniemi** Oulu, C Finland 65.34N 25.13E
79 V14 **Kukawa** Kaduna, C Nigeria 10.27N 7.39E
112 J10 **Kujawsko-pomorskie** ♦ *province*, C Poland

172 N10 **Kuji** *var.* Kuzi. Iwate, Honshū, C Japan 40.12N 141.47E
Kujto, Ozero see Kuyto, Ozero
170 D14 **Kujū-renzan** *var.* Kujū-san.
45 N7 ▲ Kyūshū, SW Japan 33.07N 131.13E
Kujū-san *var.* Kujū-renzan.
115 O16 **Kukalaya, Rio** *var.* Rio Cuculaya, Rio Kukulaya. ♒ NE Nicaragua
152 M10 **Kükcha** Rus. Kokcha. Bukhoro Wiloyati, C Uzbekistan 40.30N 64.58E
115 M18 **Kukës** *var.* Kukësi. Kukës, NE Albania 42.03N 20.25E
115 L18 **Kukësi** see Kukës
Kukësi ♦ *district* NE Albania
194 J14 **Kukipi** Gulf, S PNG 8.10S 146.09E
131 S3 **Kukmor** Respublika Tatarstan, W Russian Federation 56.11N 50.56E
161 J21 **Kukobakonam** Tamil Nādu, SE India 10.58N 79.24E
176 Z16 **Kumbe, Sungai** ♒ Irian Jaya, E Indonesia
Kum-Dag see Gumdag
172 O14 **Kume-jima** *island* Nansei-shotō, SW Japan
131 V6 **Kumertau** Respublika Bashkortostan, W Russian Federation 52.48N 55.48E
125 F11 **Kuminskiy** Khanty-Mansiyskiy Avtonomnyy Okrug, C Russian Federation 58.42N 65.56E
37 R4 **Kumiva Peak** ▲ Nevada, W USA 40.24N 119.16W
165 N8 **Kum Kuduk** *well* NW China 40.21N 91.43E
165 N7 **Kumkuduk** Xinjiang Uygur Zizhiqu, W China 40.15N 91.55E
97 M16 **Kumla** Örebro, C Sweden 59.08N 15.08E
142 E17 **Kumluca** Antalya, SW Turkey 36.22N 30.16E
102 N9 **Kummerower See** ⊚ NE Germany
79 X14 **Kumo** Gombe, E Nigeria 10.03N 11.13E
151 O13 **Kumola** ♒ C Kazakhstan
178 N1 **Kumon Range** ▲ N Myanmar
126 K14 **Kumora** Respublika Buryatiya, S Russian Federation 55.43N 110.47E
85 F22 **Kums** Karas, SE Namibia 28.07S 19.40E
161 E18 **Kumta** Karnātaka, W India 14.25N 74.24E
164 L6 **Kümüx** Xinjiang Uygur Zizhiqu, W China
40 H12 **Kumukahi, Cape** *headland* Hawaii, USA, C Pacific Ocean 19.31N 154.48W
131 Q17 **Kumukh** Respublika Dagestan, SW Russian Federation 42.10N 47.07E
131 N9 **Kumul** see Hami
131 N9 **Kumylzhenskaya** Volgogradskaya Oblast', SW Russian Federation 55.25N 42.31E
114 E11 **Kumzār** N Oman 26.19N 56.26E
147 W6 **Kunar** *Per.* Konarhā. ♦ *province* E Afghanistan
127 P16 **Kunashir, Ostrov** *var.* Kunashir. *island* Kuril'skiye Ostrova, SE Russian Federation
120 I3 **Kunda** Lääne-Virumaa, NE Estonia 59.31N 26.32E
158 M13 **Kunda** Uttar Pradesh, N India 25.43N 81.31E
161 E19 **Kundāpura** *var.* Coondapoor. Karnātaka, W India 13.39N 74.41E
81 O24 **Kundelungu, Monts** ▲ S Dem. Rep. Congo (Zaire)
31 O6 **Kulm** North Dakota. N USA 46.18N 98.57W
152 D12 **Kundiawa** Chimbu, W PNG 06.05S 144.57E
174 Hh7 **Kundla** see Sāvarkundla
194 I12 **Kundu** var. Kondoz, Kondūz, Qondūz, *Per.* Kondūz, Kundūz, NE Afghanistan 36.48N 68.50E
Kunduz see Kunduz
155 Q2 **Kunduz** *var.* Kondoz, Kundūz, Qondūz, *Per.* Kundūz, ♦ *province* NE Afghanistan
85 B18 **Kuneitra** see Al Qunayṭirah
145 S1 **Kunene** *var.* Cunene.
161 F15 **Kurdish** see Kordestän
145 S1 **Kurd Kui** see Kord Kūy
116 J11 **Kürdzhali** *var.* Kirdzhali.
116 K11 **Kürdzhali** ♦ *province* S Bulgaria
170 Ee13 **Kürdzhali, Yazovir** ⊚ S Bulgaria
97 J19 **Küngälv** Västra Götaland,
164 J5 **Künes He** ♒ NW China
97 J19 **Küngälv** Västra Götaland, C Sweden 57.54N 12.00E
149 R13 **Kūl, Rūd-e** *var.* Kūl. ♒ S Iran
150 G12 **Kul'sary** *Kaz.* Qulsary. Atyrau, W Kazakhstan 46.58N 53.58E
159 R15 **Kulti** West Bengal, NE India 23.45N 86.49E
95 G14 **Kultsjön** ⊚ N Sweden
142 I14 **Kulu** Konya, W Turkey 39.06N 33.01E
127 Nn10 **Kulu** ♒ E Russian Federation
125 G14 **Kulunda** Altayskiy Kray, S Russian Federation 52.33N 79.04E
151 T7 **Kulunda Steppe** *Kaz.* Qulyndy Zhazyghy, Rus. Kulundinskaya Ravnina. *grassland* Kazakhstan/Russian Federation
Kulundinskaya Ravnina see Kulunda Steppe
190 M9 **Kulwin** Victoria, SE Australia 35.04S 142.37E
Kulyab see Külöb
111 Q3 **Kulykivka** Chernihivs'ka Oblast', N Ukraine 51.23N 31.39E
Kum see Qom
170 Ee15 **Kuma** Ehime, Shikoku, SW Japan 33.36N 132.53E
131 P14 **Kuma** ♒ SW Russian Federation
Kumafa see Kumawa, Pegunungan
171 K15 **Kumagaya** Saitama, Honshū, S Japan 36.10N 139.22E
172 N6 **Kumaishi** Hokkaidō, NE Japan 42.08N 139.57E
174 Ll11 **Kumai, Teluk** *bay* Borneo, C Indonesia
131 X7 **Kumak** Orenburgskaya Oblast', W Russian Federation 51.16N 60.06E
176 Y9 **Kumamba, Kepulauan** *island group* E Indonesia
170 Cc14 **Kumamoto** Kumamoto, Kyūshū, SW Japan 32.49N 130.40E
170 C14 **Kumamoto** *off.* Kumamoto-ken. ♦ *prefecture* Kyūshū, SW Japan

171 Gg17 **Kumano** Mie, Honshū, SW Japan 33.54N 136.03E
169 X16 **Kumanova** see Kumanovo
115 O17 **Kumanovo** *Turk.* Kumanova. N FYR Macedonia 42.08N 21.42E
193 G17 **Kumara** West Coast, South Island, New Zealand
188 J8 **Kumarina Roadhouse** Western Australia 24.46S 119.39E
159 T15 **Kumarkhali** Khulna, W Bangladesh 23.52N 89.13E
79 P16 **Kumasi** *prev.* Coomassie. C Ghana 6.40N 1.39W
176 Vv17 **Kumawa, Pegunungan** *var.* Kumafa. ▲ Irian Jaya, E Indonesia
Kumayri see Gyumri
81 D15 **Kumba** Sud-Ouest, W Cameroon 4.39N 9.25E
116 N13 **Kumbag** Tekirdağ, NW Turkey 40.51N 27.26E
Kumboka see Cuando
169 V12 **Kumdian** see Kuandian
131 M12 **Kurikoma-yama** ▲ Honshū, C Japan 38.57N 140.44E
199 Hh3 **Kurile Basin** *undersea feature* NW Pacific Ocean
Kurile Islands see Kuril'skiye Ostrova
Kurile-Kamchatka Depression see Kurile Trench
199 Hh3 **Kurile Trench** *var.* Kurile-Kamchatka Depression. *undersea feature* NW Pacific Ocean
131 Q9 **Kurilovka** Saratovskaya Oblast', W Russian Federation 50.39N 48.02E
127 P15 **Kuril'sk** Kuril'skiye Ostrova, Sakhalinskaya Oblast', SE Russian Federation 45.10N 147.51E
127 Pp15 **Kuril'skiye Ostrova** *Eng.* Kurile Islands. *island group* SE Russian Federation
44 M9 **Kurinwas, Río** ♒ E Nicaragua
Kurisches Haff *var.* Courland Lagoon.
130 M4 **Kurkund** see Kilingi-Nõmme
Kurlovskiy Vladimirskaya Oblast', W Russian Federation 55.25N 40.39E
82 G12 **Kurmuk** Blue Nile, SE Sudan 10.36N 34.16E
Kurna see Al Qurnah
161 H17 **Kurnool** *var.* Karnul. Andhra Pradesh, S India 15.51N 78.01E
171 J13 **Kurobe** Toyama, Honshū, SW Japan 36.52N 137.26E
170 Cc13 **Kurogi** Fukuoka, Kyūshū, SW Japan 33.09N 130.45E
171 Mm9 **Kuroishi** *var.* Kuroisi. Aomori, Honshū, C Japan 40.40N 140.34E
Kuroisi see Kuroishi
171 Kk14 **Kuroiso** Tochigi, Honshū, S Japan 36.58N 140.01E
172 N5 **Kuromatsunai** Hokkaidō, NE Japan 42.40N 140.18E
172 Oo17 **Kuro-shima** *island* SW Japan
171 H16 **Kuroso-yama** ▲ Honshū, SW Japan 34.31N 136.10E
193 F21 **Kurow** Canterbury, South Island, NZ 44.44S 170.29E
131 N15 **Kursavka** Stavropol'skiy Kray, SW Russian Federation 44.28N 42.31E
120 E11 **Kuršėnai** Šiauliai, N Lithuania 56.00N 22.56E
119 W5 **Kürshim** see Kurchum
Kurshskaya Kosa/Kuršių Nerija see Courland Spit
130 J7 **Kursk** Kurskaya Oblast', W Russian Federation 51.43N 36.46E
130 J7 **Kurskaya Oblast'** ♦ *province* W Russian Federation
115 N15 **Kurskiy Zaliv** see Courland Lagoon
143 R15 **Kuršumlija** Serbia, S Yugoslavia 43.09N 21.16E
170 Ee13 **Kuşadası** Siirt, SE Turkey 37.56N 41.43E
125 Ee12 **Kurtamysh** Kurganskaya Oblast', C Russian Federation 54.51N 64.46E
Kurtbunar see Tervel
Kurt-Dere see Vŭlchidol
Kurtitsch/Kürtös see Curtici
151 U15 **Kurtty** ♒ SE Kazakhstan
95 L18 **Kuru** Länsi-Suomi, W Finland 61.51N 23.46E
82 B13 **Kuru** ♒ W Sudan
143 M13 **Kuru Dağı** ▲ NW Turkey
164 L7 **Kuruktag** ▲ NW China
85 Q22 **Kuruman** Northern Cape, N South Africa 27.28S 23.27E
69 T14 **Kuruman** ♒ W South Africa
170 Cc13 **Kurume** Fukuoka, Kyūshū, SW Japan 33.15N 130.27E
155 J25 **Kurunegala** North Western Province, C Sri Lanka 7.28N 80.22E
57 T10 **Kurupukari** ♒ C Guyana 4.39N 58.39W
169 X6 **Kurbin He** ♒ NE China
151 X10 **Kurchum** *Kaz.* Kürshim. Vostochnyy Kazakhstan, E Kazakhstan 48.35N 83.37E
174 Hh7 **Kundur, Pulau** *island* W Indonesia
155 Q2 **Kundūz** *var.* Kondoz, Kundūz, Qondūz, *Per.* Kondūz, ♦ *province* NE Afghanistan
170 Fj14 **Kurashiki** *var.* Kurasiki. Okayama, Honshū, SW Japan 34.35N 133.44E
160 L10 **Kurasia** Madhya Pradesh, C India 23.11N 82.16E
Kurasiki see Kurashiki
170 G12 **Kurayoshi** *var.* Kurayosi. Tottori, Honshū, SW Japan 35.25N 133.51E
Kurayosi see Kurayoshi
Kurchum *Kaz.* Kürshim. ♒ E Kazakhstan
143 X11 **Kürdämir** *Rus.* Kyurdamir. C Azerbaijan 40.21N 48.08E
161 P14 **Kurduvādi** Mahārāshtra, W India 18.06N 75.31E
145 T10 **Kurdzhali** *var.* Kirdzhali.
Kure Hiroshima, Honshū, SW Japan 34.13N 132.33E
199 J5 **Kure Atoll** *var.* Ocean Island. *atoll* Hawaiian Islands, Hawaii, USA, C Pacific Ocean
142 J10 **Küre Dağları** ▲ N Turkey
Kurenets see Kuranyets
120 E6 **Kuressaare** *Ger.* Arensburg; *prev.* Kingissepp. Saaremaa, W Estonia 58.15N 22.25E
126 I9 **Kureyka** ♒ N Russian Federation
151 P10 **Kurgal'dzhin, Ozero** ⊚ C Kazakhstan
151 Q10 **Kurgal'dzhinskiy** see Kurgal'dzhino
172 Q7 **Kurgan** Permskaya Oblast', NW Russian Federation
Kurgan-Tyube see Qürghonteppa
203 O2 **Kuria** Kiribati
164 H10 **Kunlun Shan** *Eng.* Kunlun Mountains. ▲ NW China
165 P11 **Kunlun Shankou** *pass* C China 35.45N 93.59E
166 Y16 **Kunming** var. K'un-ming; *prev.* Yunnan. Yunnan, SW China 25.04N 102.40E
172 N6 **Kunnui** Hokkaidō, NE Japan 42.26N 140.17E

97 B18 **Kunoy Dan.** Kunø Island Faeroe Islands 62.18N 6.40W
169 X16 **Kunsan** *var.* Gunsan, *Jap.* Gunzan. W South Korea
113 L24 **Kunszentmárton** Jász-Nagykun-Szolnok, E Hungary 46.49N 20.15E
113 J23 **Kunszentmiklós** Bács-Kiskun, C Hungary 47.02N 19.08E
189 X3 **Kununurra** Western Australia 15.49S 128.43E
174 Mm8 **Kunya** Borneo, C Indonesia
103 I20 **Künzelsau** Baden-Württemberg, S Germany 49.22N 9.43E
167 S10 **Kuocang Shan** ▲ SE China
128 H5 **Kuolayarvi** var. Luolajarvi. Murmanskaya Oblast', NW Russian Federation
95 N16 **Kuopio** Itä-Suomi, C Finland 62.42N 27.30E
95 K17 **Kuortane** Länsi-Suomi, W Finland 62.48N 23.30E
95 K17 **Kuortti** Itä-Suomi, E Finland 61.25N 26.25E
Kupa see Kolpa
175 R17 **Kupang** *prev.* Koepang. Timor, C Indonesia 10.13S 123.37E
37 R4 **Kuparuk River** ♒ Alaska, USA
Kupcina see Cupcina
194 L16 **Kupiano** Central, S PNG 10.04S 148.16E
188 M4 **Kupingarri** Western Australia 16.46S 125.57E
125 G14 **Kupino** Novosibirskaya Oblast', C Russian Federation 54.22N 77.09E
120 H11 **Kupiškis** Kupiškis, NE Lithuania 55.51N 24.58E
116 L13 **Küplü** Edirne, NW Turkey 41.06N 26.23E
41 X13 **Kupreanof Island** *island* Alexander Archipelago, Alaska, USA
41 O16 **Kupreanof Point** *headland* Alaska, USA 55.54N 159.36W
114 G13 **Kupres** Federacija Bosna I Hercegovina, SW Bosnia and Herzegovina 44.00N 17.15E
119 W5 **Kup"yans'k** *Rus.* Kupyansk. Kharkivs'ka Oblast', E Ukraine 49.40N 37.41E
119 W5 **Kup"yans'k-Vuzlovyy** Kharkivs'ka Oblast', E Ukraine 49.40N 37.41E
143 X11 **Kuqa** Xinjiang Uygur Zizhiqu, NW China 41.43N 82.58E
57 R8 **Kura Az.** Kür, *Geor.* Mtkvari, *Turk.* Kura Nehri. ♒ SW Asia
170 Ee13 **Kuracki** SW Japan
170 Ee13 **Kurahashi-jima** *island* SW Japan
Kura Kurk see Irbe Strait
153 Q10 **Kurama Range** ▲ Rus. Kuraminskiy Khrebet. ▲ Tajikistan/Uzbekistan
Kuraminskiy Khrebet see Kurama Range
151 U5 **Kuran, Kepulauan** *island group* E Indonesia
121 J14 **Kuranyets** *Rus.* Kurenets. Minskaya Voblasts', C Belarus 54.34N 26.58E

41 P11 **Kuskokwim Mountains** ▲ Alaska, USA
41 N12 **Kuskokwim River** ♒ Alaska, USA
110 G7 **Küsnacht** Zürich, N Switzerland 47.21N 8.32E
172 Qq6 **Kussharo-ko** *var.* Kussyaro. ⊚ Hokkaidō, NE Japan
Küssnacht see Küssnacht am Rigi
110 F8 **Küssnacht am Rigi** *var.* Küssnacht. Schwyz, C Switzerland 47.03N 8.25E
Kussyaro see Kussharo-ko
Kustanay see Kostanay
Küstence/Küstendje see Constanța
102 F11 **Küstenkanal** *var.* Ems-Hunte Canal. *canal* NW Germany
Küstrin see Kostrzyn
175 T7 **Kusu** Pulau Halmahera, E Indonesia
175 Nn16 **Kuta** Pulau Lombok, S Indonesia 8.52S 116.15E
145 T4 **Kūtabān** N Iraq 35.21N 44.45E
142 E13 **Kütahya** *prev.* Kutaia. Kütahya, W Turkey 39.25N 29.55E
142 E13 **Kütahya** *var.* Kutaia. ♦ *province* W Turkey
Kutai see Mahakam, Sungai
143 R9 **K'ut'aisi** W Georgia 42.15N 42.42E
Kūt al 'Amārah see Al Kūt
Kūt al Hai/Kūt al Ḩayy see Al Ḩayy
Kūt al Imara see Al Kūt
126 M12 **Kutana** Respublika Sakha (Yakutiya), NE Russian Federation 59.05N 131.43E
201 S6 **Kutaradja/Kutaraja** see Bandaaceh
172 Nn5 **Kutchan** Hokkaidō, NE Japan 42.54N 140.46E
114 F9 **Kutina** Sisak-Moslavina, NE Croatia 45.29N 16.45E
114 H9 **Kutjevo** Požega-Slavonija, NE Croatia 45.26N 17.54E
113 E17 **Kutná Hora** *Ger.* Kuttenberg. Středočeský Kraj, C Czech Republic 49.57N 15.15E
112 K12 **Kutno** Łódzkie, C Poland 52.13N 19.23E
81 I20 **Kutu** Bandundu, W Dem. Rep. Congo (Zaire) 2.42S 18.07E
82 B10 **Kutum** Northern Darfur, W Sudan
153 Y7 **Kuturgu** Issyk-Kul'skaya Oblast', C Kyrgyzstan 42.45N 78.04E
10 M5 **Kuujjuaq** *prev.* Fort-Chimo. Quebec, E Canada 58.10N 68.15W
10 I7 **Kuujjuarapik** Quebec, C Canada 55.20N 77.45W
152 A10 **Kuuli-Mayak** *Turkm.* Guwlumayak. Balkanskiy Velayat, NW Turkmenistan 40.14N 52.43E
120 I4 **Kuulse magi** ▲ S Estonia
94 N13 **Kuusamo** Oulu, E Finland 65.57N 29.15E
95 M19 **Kuusankoski** Etelä-Suomi, S Finland 60.51N 26.40E
131 W7 **Kuvandyk** Orenburgskaya Oblast', W Russian Federation 51.27N 57.18E
Kuvango see Cubango
Kuvasay see Quwasoy
116 I16 **Kuvshinovo** Tverskaya Oblast', W Russian Federation 57.03N 34.09E
147 Q4 **Kuwait** *off.* State of Kuwait, *var.* Dawlat al Kuwait, Koweit, Kuweit. ♦ *monarchy* SW Asia
Kuwait see Al Kuwayt
Kuwait City see Al Kuwayt
147 R9 **Kuwait, Dawlat al** see Kuwait
Kuwajleen see Kwajalein Atoll
171 H15 **Kuwana** Mie, Honshū, SW Japan 35.03N 136.40E
176 V9 **Kuwawin** Irian Jaya, E Indonesia 1.10S 132.40E
145 T12 **Kusak** ♒ C Kazakhstan
Kusary see Qusar
178 Hh8 **Ku Sathan, Doi** ▲ NW Thailand 18.22N 100.31E
171 I25 **Kusatsu** Shiga, Honshū, SW Japan 35.02N 135.58E
144 F11 **Kuseifa** Southern, C Israel 31.15N 35.01E
142 C12 **Kuş Gölü** ⊚ NW Turkey
131 O12 **Kushchevskaya** Krasnodarskiy Kray, SW Russian Federation 46.35N 39.40E
171 H16 **Kushida-gawa** ♒ Honshū, SW Japan
170 Bb15 **Kushikino** var. Kusikino. Kagoshima, Kyūshū, SW Japan 31.42N 130.13E
170 C17 **Kushima** var. Kusima. Miyazaki, Kyūshū, SW Japan 31.27N 131.11E
170 G17 **Kushimoto** Wakayama, Honshū, SW Japan 33.28N 135.46E
172 Q7 **Kushiro** *var.* Kusiro. Hokkaidō, NE Japan 42.58N 144.24E
Kushk see Gushgy
154 K4 **Kūshk** Herāt, W Afghanistan 34.54N 62.09E
151 R4 **Kushmurun** *Kaz.* Qusmuryn. Kostanay, N Kazakhstan 52.27N 64.31E
151 N8 **Kushmurun, Ozero** *Kaz.* Qusmuryn Köli. ⊚ N Kazakhstan
131 U4 **Kushnarenkovo** Respublika Bashkortostan, W Russian Federation 55.05N 55.19E
159 T15 **Kushtia** Rupt. Kustia. Khulna, W Bangladesh 23.54N 89.07E
125 Ee10 **Kushva** Sverdlovskaya Oblast', C Russian Federation 58.14N 59.36E
Kusikino see Kushikino
Kusima see Kushima
Kusiro see Kushiro
95 K17 **Kuskivesi** Länsi-Suomi, W Finland 62.36N 22.25E

57 S12 **Kuyuwini Landing** S Guyana 2.06N 59.14W
40 M9 **Kuzitrin River** ♒ Alaska, USA
131 P6 **Kuznetsk** Penzenskaya Oblast', W Russian Federation 53.06N 46.27E
118 K3 **Kuznetsovs'k** Rivnens'ka Oblast', NW Russian Federation 51.12N 25.51E
128 K6 **Kuzomen'** Murmanskaya Oblast', NW Russian Federation 66.16N 36.47E
172 N10 **Kuzumaki** Iwate, Honshū, C Japan 40.04N 141.26E
94 H9 **Kvaløya** ▲ N Norway
94 K8 **Kvalsund** Finnmark, N Norway 70.30N 23.56E
96 G11 **Kvam** Oppland, S Norway 61.42N 9.43E
131 X7 **Kvarkeno** Orenburgskaya Oblast', W Russian Federation 52.09N 59.44E
95 C15 **Kvarnbergsvattnet** var. Frostviken. ⊚ N Sweden
114 A11 **Kvarner** var. Carnaro, *It.* Quarnero. *gulf* W Croatia
114 B11 **Kvarnerić** *strait* W Croatia
41 O14 **Kvichak Bay** *bay* Alaska, USA
94 H12 **Kvikkjokk** Norrbotten, N Sweden 66.58N 17.45E
94 D13 **Kvina** ♒ S Norway
94 Q1 **Kvitøya** *island* NE Svalbard
97 F16 **Kvitseid** Telemark, S Norway 59.23N 8.31E
97 H24 **Kvivndrup** Fyn, C Denmark 55.10N 10.31E
81 H20 **Kwa** ♒ W Dem. Rep. Congo (Zaire)
79 Q15 **Kwadwokurom** C Ghana 7.49N 0.15W
195 X14 **Kwailibesi** Malaita, N Solomon Islands 8.25S 160.48E
201 S6 **Kwajalein Atoll** *var.* Kuwajleen. *atoll* Ralik Chain, C Marshall Islands
57 W9 **Kwakoegron** Brokopondo, N Suriname 5.13N 55.19W
83 J21 **Kwale** Coast, S Kenya 4.11S 39.30E
79 U17 **Kwale** Delta, S Nigeria 5.51N 6.29E
81 H20 **Kwamouth** Bandundu, W Dem. Rep. Congo (Zaire) 3.10S 16.16E
Kwando see Cuando
Kwangchow see Guangzhou
Kwangchu see Guangzhou
Kwangju see Kwangju
169 X16 **Kwangju** *off.* Kwangju-gwangyŏksi, *var.* Guangju, Kwangchu, *Jap.* Kōshū. SW South Korea 35.09N 126.52E
81 H20 **Kwango** *Port.* Cuango. ♒ Angola/Dem. Rep. Congo (Zaire) *see also* Cuango
Kwangsi/Kwangsi Chuang Autonomous Region see Guangxi Zhuangzu Zizhiqu
Kwangtung see Guangdong
Kwangyuan see Guangyuan
83 F17 **Kwania, Lake** ⊚ C Uganda
Kwanza see Cuanza
85 X17 **Kwara** ♦ *state* SW Nigeria
176 Ww11 **Kwatisore** Irian Jaya, E Indonesia 3.14S 134.57E
85 K22 **KwaZulu/Natal** *off.* KwaZulu-Natal Province; *prev.* Natal. ♦ *province* E South Africa
Kweichow see Guizhou
Kweichu see Guiyang
Kweilin see Guilin
Kweiyang see Guiyang
85 K17 **Kwekwe** *prev.* Que Que. Midlands, C Zimbabwe 18.55S 29.48E
85 G20 **Kweneng** ♦ *district* S Botswana
Kwesui see Hohhot
41 N12 **Kwethluk** Alaska, USA 60.48N 161.26W
112 J8 **Kwidzyn** *Ger.* Marienwerder. Pomorskie, N Poland 53.44N 18.55E
40 M13 **Kwigillingok** Alaska, USA 59.51N 163.07W
194 K16 **Kwikila** Central, S PNG 9.48S 147.37E
81 I20 **Kwilu** ♒ W Dem. Rep. Congo (Zaire)
Kwito see Cuito
175 V8 **Kwoka, Gunung** ▲ Irian Jaya, E Indonesia 0.34S 132.25E
81 G17 **Kyabé** Moyen-Chari, S Chad 9.28N 18.54E
191 O11 **Kyabram** Victoria, SE Australia 36.21S 145.04E
178 Gg9 **Kyaikkami** *prev.* Amherst. Mon State, S Myanmar 16.02N 97.36E
177 F9 **Kyaiklat** Irrawaddy, SW Myanmar 16.25N 95.42E
177 G8 **Kyaikto** Mon State, S Myanmar 17.16N 97.01E
126 J16 **Kyakhta** Respublika Buryatiya, S Russian Federation 50.24N 106.12E
190 G2 **Kyancutta** South Australia 33.10S 135.33E
77 F7 **Kyangin** Irrawaddy
178 J19 **Kyangin** Irrawaddy, SW Myanmar 18.19N 95.15E
177 F6 **Kyaukpadaung** Mandalay, C Myanmar 20.49N 95.07E
177 V9 **Kyaukpyu** Arakan State, W Myanmar 19.27N 93.33E
177 G7 **Kyaukse** Mandalay, C Myanmar 21.34N 96.12E
177 F5 **Kyaunggon** Irrawaddy, SW Myanmar 17.04N 95.12E
120 I8 **Kybartai** *Pol.* Kibarty. Vilkaviškis, S Lithuania 54.37N 22.44E
113 G19 **Kyjov** *Ger.* Gaya. Brněnský Kraj, SE Czech Republic 49.00N 17.07E
Kyklades see Kikládhes. *Eng.* Cyclades. *island group* SE Greece
27 S11 **Kyle** Texas, SW USA 29.59N 97.52W
98 G5 **Kyle of Lochalsh** N Scotland, UK 57.17N 5.39W
103 N8 **Kyll** ♒ W Germany
117 F19 **Kyllíni** var. Killini. ▲ S Greece
95 M19 **Kymijoki** ♒ S Finland
117 H18 **Kými, Akrotírio** *headland* Évvoia, C Greece 38.39N 24.08E
129 W14 **Kyn** Permskaya Oblast', NW Russian Federation 57.48N 58.58E
126 J15 **Kyutyn** Irkutskaya Oblast', S Russian Federation 54.18N 101.28E
126 Il12 **Kyuta** Evenkiyskiy Avtonomnyy Okrug, C Russian Federation 61.00N 97.07E
191 N12 **Kyneton** Victoria, SE Australia 37.14S 144.26E
83 G17 **Kyoga, Lake** *var.* Lake Kioga. ⊚ C Uganda

◆ COUNTRY   ◇ DEPENDENT TERRITORY   ◆ ADMINISTRATIVE REGION   ▲ MOUNTAIN   ▲ VOLCANO   ⊚ LAKE
● COUNTRY CAPITAL   ○ DEPENDENT TERRITORY CAPITAL   ✕ INTERNATIONAL AIRPORT   ▲ MOUNTAIN RANGE   ♒ RIVER   ▣ RESERVOIR

171 H13 Kyōga-misaki *headland* Honshū, SW Japan 35.46N 135.13E
191 V4 Kyogle New South Wales, SE Australia 28.37S 153.00E
169 W15 Kyŏnggi-man *bay* NW South Korea
169 Z16 Kyŏngju *Jap.* Keishū. SE South Korea 35.49N 129.09E
Kyŏngsŏng *see* Sŏul
Kyōsai-tō *see* Kōje-do
83 F19 Kyotera S Uganda 0.37S 31.34E
171 H15 Kyōto Kyōto, Honshū, SW Japan 35.01N 135.46E
171 H14 Kyōto *off.* Kyōto-fu, *var.* Kyōto Hu. ◆ *urban prefecture* Honshū, SW Japan
Kyōto-fu/Kyōto Hu *see* Kyōto
117 D21 Kyparissía *var.* Kiparissía. Pelopónnisos, S Greece 37.13N 21.39E
117 D20 Kyparissiakós Kólpos *gulf* S Greece
124 N3 Kyperoúnda *var.* Kyperounda. C Cyprus 34.57N 33.02E
Kypros *see* Cyprus
117 H16 Kyrá Panagía *island* Vóreioi Sporádes, Greece, Aegean Sea
Kyrenia *see* Girne
Kyrenia Mountains *see* Beşparmak Dağları
Kyrgyz Republic *see* Kyrgyzstan
153 U9 Kyrgyzstan *off.* Kyrgyz Republic, *var.* Kirghizia; *prev.* Kirgizskaya SSR, Kirghiz SSR, Republic of Kyrgyzstan. ◆ *republic* C Asia
102 M11 Kyritz Brandenburg, NE Germany 52.56N 12.24E
Kyrkslätt *see* Kirkkonummi
96 G8 Kyrksæterøra Sør-Trøndelag, S Norway 63.16N 9.04E
129 U8 Kyrta Respublika Komi, NW Russian Federation 64.03N 57.41E
125 Ee12 Kyshtym Chelyabinskaya Oblast', S Russian Federation 55.39N 60.31E
113 J18 Kysucké Nové Mesto *prev.* Horné Nové Mesto, *Ger.* Kisutzaneustadtl, Oberneustadtl, *Hung.* Kiszucaújhely. Žilinský Kraj, N Slovakia 49.19N 18.47E
119 N12 Kytay, Ozero ⊘ SW Ukraine
117 F23 Kýthira *var.* Kíthira, *It.* Cerigo; *Lat.* Cythera. Kýthira, S Greece 36.09N 22.58E
117 F23 Kýthira *var.* Kíthira, *It.* Cerigo; *Lat.* Cythera. *island* S Greece
117 I20 Kýthnos Kýthnos, Kykládes, Greece, Aegean Sea 37.24N 24.28E
117 I20 Kýthnos *var.* Kíthnos, Thermiá, *It.* Termia; *anc.* Cythnos. *island* Kykládes, Greece, Aegean Sea
117 I20 Kýthnou, Stenó *strait* Kykládes, Greece, Aegean Sea
Kythréa *see* Değirmenlik
Kyungёy Ala-Too, Khrebet *see* Kungei Ala-Tau
152 C11 Kyuren, Gora ▲ W Turkmenistan 39.05N 55.09E
172 C15 Kyūshū *var.* Kyûsyû. *island* SW Japan
199 Gg6 Kyushu-Palau Ridge *var.* Kyusyu-Palau Ridge. *undersea feature* W Pacific Ocean
170 Cc15 Kyūshū-sanchi ▲ Kyūshū, SW Japan
116 F10 Kyustendil *anc.* Pautalia. Kyustendil, W Bulgaria 42.17N 22.42E
116 G11 Kyustendil ◆ *province* W Bulgaria
Kyûsyû *see* Kyūshū
Kyusyu-Palau Ridge *see* Kyushu-Palau Ridge
126 L7 Kyusyur Respublika Sakha (Yakutiya), NE Russian Federation 70.36N 127.19E
191 P10 Kywong New South Wales, SE Australia 34.59S 146.42E
119 P4 Kyyiv *Eng.* Kiev, *Rus.* Kiyev. ● (Ukraine) Kyyivs'ka Oblast', N Ukraine 50.26N 30.31E
Kyyiv *see* Kyyivs'ka Oblast'
119 O4 Kyyivs'ka Oblast' *var.* Kyyiv, *Rus.* Kiyevskaya Oblast'. ◆ *province*
119 P3 Kyyivs'ke Vodoskhovyshche *Eng.* Kiev Reservoir, *Rus.* Kiyevskoye Vodokhranilishche. ⊠ N Ukraine
95 L16 Kyyjärvi Länsi-Suomi, W Finland 63.01N 24.34E
126 I16 Kyzyl Respublika Tyva, C Russian Federation 51.45N 94.28E
153 S8 Kyzyl-Adyr *prev.* Kirovskoye. Talasskaya Oblast', NW Kyrgyzstan 42.37N 71.34E
151 V14 Kyzylagash Almaty, SE Kazakhstan 45.19N 78.45E
152 C13 Kyzylbair Balkanskiy Velayat, W Turkmenistan 38.13N 55.38E
Kyzyl-Dzhiik, Pereval *see* Uzbel Shankou
151 S7 Kyzylkak, Ozero ⊘ NE Kazakhstan
151 X11 Kyzylkesek Vostochnyy Kazakhstan, E Kazakhstan 47.55N 82.01E
153 S10 Kyzyl-Kiya *Kir.* Kyzyl-Kyya. Oshskaya Oblast', SW Kyrgyzstan 40.15N 72.07E
150 L11 Kyzylkol', Ozero ⊘ C Kazakhstan
151 N15 Kyzylorda *var.* Kzyl-Orda, Qyzyl Orda *Kaz.* Qyzylorda; *prev.* Perovsk, S Kazakhstan 44.54N 65.30E
150 L14 Kyzylorda *off.* Kyzylordinskaya Oblast' *Kaz.* Qyzylorda Oblysy. ◆ *province* S Kazakhstan
138 L9 Kyzyl Kum *var.* Kizil Kum, Qizil Qum, *Uzb.* Qizilqum. *desert* Kazakhstan/Uzbekistan
Kyzyl-Kyya *see* Kyzyl-Kiya
Kyzylrabat *see* Qizilrabot
Kyzylsu *see* Kyzyl-Suu
153 X7 Kyzyl-Suu *prev.* Pokrovka. Issyk-Kul'skaya Oblast', NE Kyrgyzstan 42.19N 77.55E
153 S12 Kyzyl-Suu *var.* Kyzylsu. ⊠ Kyrgyzstan/Tajikistan

153 X8 Kyzyl-Tuu Issyk-Kul'skaya Oblast', E Kyrgyzstan 42.06N 76.54E
151 Q12 Kyzylzhar *Kaz.* Qyzylzhar. Zhezkazgan, C Kazakhstan 48.22N 70.00E
Kzyl-Orda *see* Kyzylorda
Kyzylordinskaya Oblast' *see* Kyzylorda
Kzyltu *see* Kishkenekol'

L

111 X2 Laa an der Thaya Niederösterreich, NE Austria 48.42N 16.22E
65 N15 La Adela La Pampa, SE Argentina 38.57S 64.02W
Laagen *see* Numedalslågen
111 S5 Laakirchen Oberösterreich, N Austria 47.59N 13.49E
Laaland *see* Lolland
106 I11 La Albuera Extremadura, W Spain 38.43N 6.49W
107 O7 La Alcarria *physical region* C Spain
106 K14 La Algaba Andalucía, S Spain 37.27N 6.01W
107 P9 La Almarcha Castilla-La Mancha, C Spain 39.40N 2.22W
107 R6 La Almunia de Doña Godina Aragón, NE Spain 41.28N 1.22W
43 N5 La Amistad, Presa ⊠ NE Mexico
120 F4 Läänemaa *off.* Lääne Maakond.
120 I3 Lääne-Virumaa *off.* Lääne-Viru Maakond. ◆ *province* NE Estonia
64 J9 La Antigua, Salina *salt lake* W Argentina
101 E17 Laarne Oost-Vlaanderen, NW Belgium 51.03N 3.49E
82 O13 Laas Caanood Nugaal, N Somalia 8.33N 47.44E
43 O9 La Ascensión Nuevo León, NE Mexico 24.15N 99.53W
82 N12 Laas Dhaareed Woqooyi Galbeed, N Somalia 10.12N 46.09E
57 O4 La Asunción Nueva Esparta, NE Venezuela 11.06N 63.53W
Laatokka *see* Ladozhskoye Ozero
102 I13 Laatzen Niedersachsen, NW Germany 51.19N 9.46E
40 E9 Laau Point *headland* Molokai, Hawaii, USA, C Pacific Ocean 21.06N 157.18W
44 D6 La Aurora ✕ (Ciudad de Guatemala) Guatemala, C Guatemala 14.33N 90.30W
76 C9 Laâyoune *var.* Aaiún. ○ (Western Sahara) NW Western Sahara 27.10N 13.10W
130 L14 Laba ⊠ SW Russian Federation
42 M6 La Babia Coahuila de Zaragoza, NE Mexico 28.36N 102.04W
13 R7 La Baie Quebec, SE Canada 48.20N 70.54W
175 R16 Labala Pulau Lomblen, S Indonesia 8.30S 123.27E
64 K8 La Banda Santiago del Estero, N Argentina 27.43S 64.13W
106 K4 La Bañeza Castilla-León, N Spain 42.18N 5.54W
42 M13 La Barca Jalisco, SW Mexico 20.18N 102.30W
42 K14 La Barra de Navidad Jalisco, C Mexico 19.12N 104.38W
197 J13 Labasa *prev.* Lambasa. Vanua Levu, N Fiji 16.25S 179.24E
179 Q15 Labason Mindanao, S Philippines 8.03N 122.31E
104 H8 La Baule-Escoublac Loire-Atlantique, NW France 47.16N 2.24W
Labe *see* Elbe
78 I13 Labé Moyenne-Guinée, NW Guinea 11.19N 12.16W
78 S14 La Belle Florida, SE USA 26.45N 81.26W
13 N11 Labelle Quebec, SE Canada 46.15N 74.43W
8 H7 Laberge, Lake ⊘ Yukon Territory, W Canada
Labes *see* Łobez
Labiau *see* Polessk
114 A10 Labin *It.* Albona. Istra, NW Croatia 45.05N 14.07E
130 L14 Labinsk Krasnodarskiy Kray, SW Russian Federation 44.30N 40.43E
107 X5 La Bisbal d'Empordà Cataluña, NE Spain 41.58N 3.01E
121 P16 Labkovichy *Rus.* Lobkovichi. Mahilyowskaya Voblasts', E Belarus 53.49N 31.43E
13 S4 La Blache, Lac de ⊘ Quebec, SE Canada
179 Q11 Laboc Luzon, N Philippines 14.10N 122.47E
Labohanbadjo *see* Labuhanbadjo
Laborca *see* Laborec
113 N18 Laborec *Hung.* Laborca. ⊠ E Slovakia
110 D11 La Borgne ⊠ S Switzerland
47 T12 La Borie *see* Saint Louis
44 C5 La Bouenza ◆ *province* S Congo
104 J14 Labouheyre Landes, SW France 44.12N 0.55W
64 L12 Laboulaye Córdoba, C Argentina 34.07S 63.23W
1 Q7 Labrador *cultural region* Newfoundland, SE Canada
11 N9 Labrador City Newfoundland, E Canada 52.55N 66.52W
11 Q5 Labrador Sea *sea* NW Atlantic Ocean
Labrador Sea Basin *see* Labrador Basin
56 D9 Labranzagrande Boyacá, C Colombia 5.30N 72.33W
47 U15 La Brea Trinidad, Trinidad and Tobago 10.13N 61.36W
61 D14 Labrea Amazonas, N Brazil 7.19S 64.46W
13 S6 Labrieville Quebec, SE Canada 49.15N 69.31W
104 K14 Labrit Landes, SW France 44.03N 0.29W

110 C9 La Broye ⊠ SW Switzerland
104 M15 Labruguière Tarn, S France 43.31N 2.15E
174 I8 Labu Pulau Singkep, W Indonesia 0.34S 104.24E
175 N3 Labuan *var.* Victoria. Labuan, East Malaysia 5.19N 115.13E
175 N3 Labuan ◆ *federal territory* East Malaysia
175 N3 Labuan *see* Labuan, Pulau
175 Pp16 Labuhanbajo *prev.* Labohanbadjo. Flores, S Indonesia 8.29S 119.54E
173 G6 Labuhanbilik Sumatera, N Indonesia 2.33N 100.09E
173 Ee5 Labuhanhaji Sumatera, W Indonesia 3.31N 97.00E
175 Oo2 Labuk *var.* Labuk, Sungai
175 O2 Labuk, Teluk *var.* Labuk Bay, Telukan Labuk. *bay* S Sulu Sea
175 O2 Labuk, Telukan *see* Labuk, Teluk
175 Ff9 Labutta Irrawaddy, SW Myanmar 16.07N 94.45E
125 G8 Labytnangi Yamalo-Nenetskiy Avtonomnyy Okrug, N Russian Federation 66.39N 66.26E
80 F10 Lac *off.* Préfecture du Lac. ◆ *prefecture* W Chad
115 K19 Laç *var.* Laci. Lezhë, C Albania 41.37N 19.37E
64 G11 La Calera Valparaíso, C Chile 32.48S 71.13W
11 P11 Lac-Allard Quebec, E Canada 50.37N 63.26W
106 L13 La Campana Andalucía, S Spain 37.35N 5.24W
104 J12 Lacanau Gironde, SW France 44.59N 1.04W
44 C2 Lacandón, Sierra del ▲ Guatemala/Mexico
La Cañiza *see* A Cañiza
43 W16 Lacantún, Río ⊠ SE Mexico
105 Q3 La Capelle Aisne, N France 49.58N 3.55E
114 K10 Lačarak Serbia, NW Yugoslavia 45.00N 19.34E
64 L11 La Carlota Córdoba, C Argentina 33.27S 63.16W
179 Q13 La Carlota Negros, S Philippines 10.21N 122.55E
106 L13 La Carlota Andalucía, S Spain 37.40N 4.55W
107 N12 La Carolina Andalucía, S Spain 38.15N 3.37W
105 O15 Lacaune Tarn, S France 43.42N 2.42E
13 P7 Lac-Bouchette Quebec, SE Canada 48.14N 72.11W
Laccadive Islands/Laccadive Minicoy and Amindivi Islands, the *see* Lakshadweep
175 P9 Lacepede Bay *bay* South Australia
34 K8 Lacey Washington, NW USA 47.01N 122.49W
105 P12 la Chaise-Dieu Haute-Loire, C France 45.19N 3.41E
116 G13 Lachanás Kentrikí Makedonía, N Greece 40.57N 23.15E
128 L11 Lacha, Ozero ⊘ NW Russian Federation
105 O8 la Charité-sur-Loire Nièvre, C France 47.10N 2.59E
105 N9 la Châtre Indre, C France 46.34N 1.58E
110 C8 La Chaux-de-Fonds Neuchâtel, W Switzerland 47.07N 6.51E
110 G8 Lachen Schwyz, C Switzerland 47.11N 8.49E
191 Q8 Lachlan River ⊠ New South Wales, SE Australia
45 T15 La Chorrera Panamá, C Panama 8.51N 79.46W
13 V7 Lac-Humqui Quebec, SE Canada 48.35N 0.21N
13 N12 Lachute Quebec, SE Canada 45.39N 74.21W
143 W13 Laçın *Rus.* Lachyn. SW Azerbaijan 39.36N 46.34E
105 S16 La Ciotat *anc.* Citharista. Bouches-du-Rhône, SE France 43.10N 5.36E
20 D10 Lackawanna New York, NE USA 41.01N 89.24W
32 L12 Lac La Biche Alberta, SW Canada 54.46N 111.58W
111 X7 Lac La Martre *see* Wha Ti
31 R12 Lac-Mégantic *var.* Mégantic. Quebec, SE Canada 45.34N 70.52W
24 M8 La Follette Tennessee, S USA 36.22N 84.07W
31 N12 Lafontaine Quebec, SE Canada
24 K10 Lafourche, Bayou ⊠ Louisiana, S USA

13 P9 La Croche Quebec, SE Canada 47.38N 72.42W
31 X3 La Croix, Lac ⊘ Canada/USA
28 K5 La Crosse Kansas, C USA 38.31N 99.18W
23 V7 La Crosse Virginia, NE USA 36.41N 78.03W
34 L9 La Crosse Washington, NW USA 46.48N 117.51W
32 J7 La Crosse Wisconsin, N USA 43.48N 91.12W
56 J13 La Cruz Nariño, SW Colombia 1.31N 77.01W
44 K12 La Cruz Guanacaste, NW Costa Rica 11.04N 85.37W
42 I10 La Cruz Sinaloa, W Mexico 23.52N 106.52W
63 F19 La Cruz Florida, S Uruguay 33.54S 56.10W
44 M9 La Cruz de Río Grande Región Autónoma Atlántico Sur, E Nicaragua 13.04N 84.07W
56 J4 La Cruz de Taratara Falcón, N Venezuela 11.03N 69.43W
42 M6 La Cuesta Coahuila de Zaragoza, NE Mexico 28.43N 102.33W
95 A17 La Cumbra, Volcán ⊻ Galapagos Islands, Ecuador, E Pacific Ocean 0.21S 91.30W
158 J5 Ladákh Range ▲ NE India
28 I5 Ladder Creek ⊠ Kansas, C USA
47 X10 la Désirade *atoll* E Guadeloupe
85 F25 Ladismith Western Cape, SW South Africa 33.27S 21.15E
158 G11 Lādnūn Rājasthān, NW India 27.36N 74.25E
Ladoga, Lake *see* Ladozhskoye Ozero
117 E19 Ládon ⊠ S Greece
56 E9 La Dorada Caldas, C Colombia 5.28N 74.40W
128 H11 Ladozhskoye Ozero *Eng.* Lake Ladoga, *Fin.* Laatokka. ⊘ NW Russian Federation
39 R12 Ladron Peak ▲ New Mexico, SW USA 34.25N 107.04W
128 J11 Ladva-Vetka Respublika Kareliya, NW Russian Federation 61.18N 34.24E
191 Q15 Lady Barron Tasmania, SE Australia 40.12S 148.12E
12 G9 Lady Evelyn Lake ⊘ Ontario, S Canada
25 W11 Lady Lake Florida, SE USA 28.55N 81.55W
8 L17 Ladysmith Vancouver Island, British Columbia, SW Canada 48.55S 123.45E
85 J22 Ladysmith KwaZulu/Natal, E South Africa 28.34S 29.46E
32 J5 Ladysmith Wisconsin, N USA 45.27N 91.07W
151 P9 Ladyzhenka Akmola, C Kazakhstan 50.58N 68.44E
201 R6 Lae Atoll *atoll* Ralik Chain, W Marshall Islands
186 E12 Lae Morobe, W PNG 6.45S 146.55E
42 C3 La Encantada, Cerro de ▲ NW Mexico 31.03N 115.25W
176 E11 Lærdalsøyri Sogn og Fjordane, S Norway 61.04N 7.24E
57 N11 La Esmeralda Amazonas, S Venezuela 3.16N 65.33W
44 G7 La Esperanza Intibucá, SW Honduras 14.18N 88.10W
32 K8 La Farge Wisconsin, N USA 43.36N 90.39W
25 R5 Lafayette Alabama, S USA 32.54N 85.24W
39 T4 Lafayette Colorado, C USA 39.59N 105.06W
25 R8 Lafayette Georgia, SE USA 34.42N 85.16W
33 O13 Lafayette Indiana, N USA 40.25N 86.52W
23 O4 Lafayette Louisiana, S USA 30.13N 92.01W
21 X7 Lafayette Tennessee, S USA 36.31N 86.01W
20 K14 Lafayette, Mount ▲ New Hampshire, NE USA 44.09N 71.37W
105 P3 la Fère Aisne, N France 49.41N 3.20E
104 L6 la Ferté-Bernard Sarthe, C France 48.13N 0.40E
104 K5 la Ferté-Macé Orne, N France 48.35N 0.21N
105 N7 la Ferté-St-Aubin Loiret, C France 47.42N 1.57E
105 P5 la Ferté-sous-Jouarre Seine-et-Marne, N France 48.57N 3.07E
79 V15 Lafia Nassarawa, C Nigeria 8.29N 8.34E
79 T15 Lafiagi Kwara, W Nigeria 8.52N 5.25E
17 T17 Laflèche Saskatchewan, S Canada 49.40N 106.27W
104 K7 la Flèche Sarthe, NW France 47.42N 0.04W
111 X7 Lafnitz *Hung.* Lapines. ⊠ Austria/Hungary
197 I6 La Foa Province Sud, S New Caledonia 21.46S 165.49E

39 R7 La Garita Mountains ▲ Colorado, C USA
179 P9 Lagawe Luzon, N Philippines 16.46N 121.06E
80 F13 Lagdo Nord, N Cameroon 9.12N 13.43E
80 F13 Lagdo, Lac de ⊘ N Cameroon
102 H13 Lage Nordrhein-Westfalen, W Germany 52.00N 8.48E
96 H12 Lågen ⊠ S Norway
63 J14 Lages Santa Catarina, S Brazil 27.44S 50.16W
155 R4 Laghmān ◆ *province* E Afghanistan
76 J6 Laghouat N Algeria 33.49N 2.59E
107 Q10 La Gineta Castilla-La Mancha, C Spain 39.08N 2.00W
117 E21 Lagkáda *var.* Langada. Pelopónnisos, S Greece 36.49N 22.19E
116 G13 Lagkadás *var.* Langades. Langadhás. Kentrikí Makedonía, N Greece 40.45N 23.04E
117 E20 Lagkádia *var.* Langadia. Langadia. Pelopónnisos, S Greece 37.39N 22.03E
56 F6 La Gloria Cesar, N Colombia 8.37N 73.49W
43 O7 La Gloria Nuevo León, NE Mexico
94 X3 La Goagira *see* La Guajira
106 G14 Lagoa Faro, S Portugal 37.07N 8.27W
63 J14 Lagoa Vermelha Rio Grande do Sul, S Brazil 28.13S 51.32W
143 V10 Lagodekhi SE Georgia 41.49N 46.15E
85 F25 La Gomera Escuintla, S Guatemala 14.04N 91.03W
109 M19 Lagonegro Basilicata, S Italy 40.06N 15.42E
65 G16 Lago Ranco Los Lagos, C Chile 40.21S 72.29W
79 S16 Lagos Lagos, SW Nigeria 6.24N 3.16E
106 F14 Lagos *anc.* Lacobriga. Faro, S Portugal 37.04N 8.40W
79 S16 Lagos ◆ *state* SW Nigeria
42 M12 Lagos de Moreno Jalisco, SW Mexico 21.21N 101.55W
79 S16 Lagosta *see* Lastovo
76 A12 Lagouira SW Western Sahara 20.55N 17.04W
34 L11 La Grande Oregon, NW USA 45.21N 118.04W
105 Q14 la Grande-Combe Gard, S France 44.13N 4.01E
10 K9 la Grande Rivière *var.* Fort George. ⊠ Quebec, SE Canada
25 R4 La Grange Georgia, SE USA 33.01N 85.01W
33 P11 Lagrange Indiana, N USA 41.38N 85.25W
22 L5 La Grange Kentucky, C USA 38.22N 85.07W
29 X10 La Grange Missouri, C USA 40.00N 91.31W
23 R4 La Grange North Carolina, S USA 35.18N 77.47W
27 U11 La Grange Texas, SW USA 29.54N 96.52W
107 N7 La Granja Castilla-León, C Spain 40.53N 4.01W
57 O9 La Gran Sabana *grassland* E Venezuela
57 Q9 La Grita Táchira, NW Venezuela 8.09N 71.58W
13 R11 La Guadeloupe Quebec, SE Canada 45.57N 70.56W
66 F12 La Guaira Distrito Federal, N Venezuela 10.35N 66.52W
56 G4 La Guajira ◆ Departamento de La Guajira, *var.* Guajira, La Goagira. ◆ *province* NE Colombia
196 I4 Lagua Lichan, Punta *headland* Saipan, S Northern Mariana Islands
29 V7 La Guardia ✕ (New York) Long Island, New York, NE USA 40.44N 73.51W
La Guardia/Laguardia *see* A Guardia
107 P4 Laguardia País Vasco, N Spain 42.32N 2.31W
28 H9 La Guerche-sur-l'Aubois Cher, C France 46.55N 3.00E
105 O13 Laguiole Aveyron, S France 44.49N 2.50E
63 V9 Laguna Santa Catarina, S Brazil 28.29S 48.47W
39 U7 Laguna New Mexico, SW USA 35.03N 107.38W
37 T16 Laguna Beach California, W USA 33.32N 117.46W
28 F5 Laguna Dam *dam* Arizona/California, SW USA 32.49N 114.30W
37 Y17 Laguna Mountains ▲ California, W USA
42 L7 Laguna El Rey Coahuila de Zaragoza, N Mexico
59 V9 Laguna Paiva Santa Fe, C Argentina 31.01S 60.40W
42 H3 Lagunas Loreto, N Peru 5.16S 75.40W
58 E9 Lagunillas Santa Cruz, SE Bolivia 19.37S 63.39W
56 H6 Lagunillas Mérida, W Venezuela 8.31N 71.24W
42 M8 Lagunillas Zacatecas, C Mexico 36.13N 84.09W
46 C4 La Habana ● (Cuba) Ciudad de La Habana, W Cuba 23.07N 82.25W
175 Oo3 Lahad Datu, Teluk *var.* Telukan Lahad Datu, Teluk Darvel, Teluk Datu; Darvel Bay. *bay* Sabah, East Malaysia
97 L20 Lagan Kronoberg, S Sweden 56.55N 14.01E
97 K21 Lågan ⊠ S Sweden
174 Hh12 Lahat Sumatera, W Indonesia 3.46S 103.31E

La Haye *see* 's-Gravenhage
Lahej *see* Laḥij
64 G9 La Higuera Coquimbo, N Chile 29.33S 71.15W
147 S13 Laḥj, Ḥiṣā' al *spring/well* NE Yemen 17.28N 50.05E
147 O16 Laḥij *var.* Laḥj, *Eng.* Lahej, SW Yemen 13.03N 44.55E
148 M3 Lāhījān Gīlān, NW Iran 37.15N 50.03E
121 I19 Lahishyn *Pol.* Lohiszyn, *Rus.* Logishin. Brestskaya Voblasts', SW Belarus 52.19N 25.59E
103 F18 Lahn ⊠ W Germany
Lähn *see* Wleń
97 J21 Laholm Halland, S Sweden 56.30N 13.04E
97 J21 Laholmsbukten *bay* S Sweden
155 W8 Lahore Punjab, NE Pakistan 31.35N 74.18E
155 W8 Lahore ✕ Punjab, E Pakistan 31.34N 74.22E
57 Q6 La Horqueta Delta Amacuro, NE Venezuela 9.13N 62.02W
121 K15 Lahoysk *Rus.* Logoysk. Minskaya Voblasts', C Belarus 54.12N 27.53E
103 F22 Lahr Baden-Württemberg, S Germany 48.21N 7.51E
95 M19 Lahti *Swe.* Lahtis. Etelä-Suomi, S Finland 61.00N 25.40E
Lahtis *see* Lahti
42 M14 La Huacana Michoacán de Ocampo, SW Mexico 18.56N 101.52W
42 N9 La Huerta Jalisco, SW Mexico 19.28N 104.40W
80 H12 Laï *var.* Behagle, De Behagle. Tandjilé, S Chad 9.22N 16.13E
194 G12 Laiagam Enga, W PNG 5.31S 143.28E
Laibach *see* Ljubljana
178 I5 Lai Châu Lai Châu, N Vietnam 22.04N 103.10E
40 D9 Laie *Haw.* Lā'ie. Oahu, Hawaii, USA, C Pacific Ocean 21.39N 157.55W
104 L5 l'Aigle Orne, N France 48.46N 0.37E
105 P2 Laignes Côte d'Or, C France 47.51N 4.24E
95 K17 Laihia Länsi-Suomi, W Finland 62.58N 22.00E
85 K17 Laila *see* Laylā
85 F25 Laingsburg Western Cape, SW South Africa 33.09S 20.48E
111 U2 Lainsitz *Cz.* Lužnice. ⊠ Austria/Czech Republic
98 I7 Lairg N Scotland, UK 58.02N 4.22W
81 I17 Laisamis Eastern, N Kenya 1.35N 37.49E
Laisberg *see* Leisi
131 R4 Laishevo Respublika Tatarstan, W Russian Federation 55.26N 49.27E
94 H13 Laisholm *see* Jõgeva
94 J17 Laitila Länsi-Suomi, W Finland 60.52N 21.40E
167 P5 Laixi *var.* Shuiji. Shandong, E China 36.53N 120.33E
167 R4 Laiyang Shandong, E China 37.03N 120.48E
167 O3 Laiyuan Hebei, E China 39.19N 114.43E
167 R4 Laizhou *var.* Ye Xian. Shandong, E China 37.12N 120.01E
167 Q4 Laizhou Wan *bay* Laichow Bay. *bay* E China
39 S8 La Jara Colorado, C USA 37.16N 105.57W
42 I6 La Junta Chihuahua, N Mexico 28.27N 107.21W
39 T7 La Junta Colorado, C USA 37.58N 103.34W
94 J13 Laiträsk Norrbotten, N Sweden 66.16N 21.16E
Lak Dera *see* Lakeamu
Lakeamu *see* Lakekamu
31 P12 Lake Andes South Dakota, N USA 43.08N 98.33W
158 H4 Lake Arthur Louisiana, S USA 30.04N 92.40W
128 H11 Lakhdenpokh'ya Respublika Kareliya, NW Russian Federation
197 L14 Lakeba *prev.* Lakemba. *island* Lau Group, E Fiji
197 L15 Lakeba Passage *channel* E Fiji
158 L11 Lakhimpur Uttar Pradesh, N India 27.57N 80.46E
160 I11 Lakhnādon Madhya Pradesh, C India 22.36N 79.36E
Lakhnau *see* Lucknow
160 A9 Lakhpat Gujarāt, W India
121 K19 Lakhva *Rus.* Lakhva. Brestskaya Voblasts', SW Belarus 52.13N 27.15E
28 I6 Lakin Kansas, C USA 37.56N 101.18W
155 S7 Lakki Marwat North-West Frontier Province, E Pakistan 32.36N 70.55E
117 F21 Lakonía *historical region* S Greece
117 F21 Lakonikós Kólpos *gulf* S Greece
78 L13 Lakota Ivory Coast 5.50N 5.40W
31 Q5 Lakota North Dakota, N USA 48.02N 98.20W
Lak Sao *see* Ban Lakxao

20 D10 Lake Erie Beach New York, NE USA 42.37N 79.04W
3 T11 Lakefield Minnesota, N USA 43.40N 95.10W
27 V6 Lake Fork Reservoir ⊠ Texas, SW USA
32 M9 Lake Geneva Wisconsin, N USA 42.36N 88.25W
20 L9 Lake George New York, NE USA 43.25N 73.45W
16 O4 Lake Harbour Baffin Island, Nunavut, NE Canada 69.49W
38 I12 Lake Havasu City Arizona, SW USA 34.26N 114.20W
27 W12 Lake Jackson Texas, SW USA 29.01N 95.25W
194 J14 Lakekamu *var.* Lakeamu. ⊠ S PNG
188 K13 Lake King Western Australia 33.09S 119.46E
194 F12 Lake Kutubu ⊠ W PNG
25 V12 Lakeland Florida, SE USA 28.03N 81.57W
25 U7 Lakeland Georgia, SE USA 31.02N 83.04W
189 W4 Lakeland Downs Queensland, NE Australia 15.54S 144.54E
9 P16 Lake Louise Alberta, SW Canada 51.25N 116.10W
Lakemba *see* Lakeba
31 V11 Lake Mills Iowa, C USA 43.25N 93.31W
41 Q10 Lake Minchumina Alaska, USA 63.55N 152.25W
194 E13 Lake Murray Western, SW PNG 6.45S 141.25E
82 F5 Lake Nasser *var.* Buhayrat Nāşir, Buḩayrat Nāşir, Buḩeiret Nāşir. ☒ Egypt/Sudan
33 R9 Lake Orion Michigan, N USA 42.46N 83.14W
202 B14 Lakepa Niue 18.58S 169.48E
31 T11 Lake Park Iowa, C USA 43.27N 95.19W
20 K7 Lake Placid New York, NE USA 44.16N 73.57W
20 K9 Lake Pleasant New York, NE USA 43.27N 74.24W
36 M6 Lakeport California, W USA 39.03N 122.55W
31 Q10 Lake Preston South Dakota, N USA 44.21N 97.22W
24 J5 Lake Providence Louisiana, S USA 32.48N 91.10W
193 E20 Lake Pukaki Canterbury, South Island, NZ 44.12S 170.10E
191 Q12 Lakes Entrance Victoria, SE Australia 37.52S 147.58E
39 N12 Lakeside Arizona, SW USA 34.09N 109.58W
37 V17 Lakeside California, W USA 32.50N 116.55W
25 S9 Lakeside Florida, SE USA 30.22N 84.18W
30 K13 Lakeside Nebraska, C USA 42.01N 102.27W
34 E13 Lakeside Oregon, NW USA 43.34N 124.10W
23 W6 Lakeside Virginia, NE USA 37.37N 77.28W
Lakes State *see* El Buhayrat
Lake State *see* Michigan
193 F20 Lake Tekapo Canterbury, South Island, NZ 44.01S 170.29E
23 Q10 Lake Toxaway North Carolina, SE USA 35.06N 82.57W
T13 Lake View Iowa, C USA 42.18N 95.04W
34 L9 Lakeview Oregon, NW USA 42.13N 120.21W
27 O3 Lakeview Texas, SW USA 34.38N 100.36W
24 L8 Lake Village Arkansas, S USA 33.19N 91.16W
25 W12 Lake Wales Florida, SE USA 27.54N 81.35W
31 T4 Lakewood Colorado, C USA 39.38N 105.07W
29 O15 Lakewood New Jersey, NE USA 40.04N 74.11W
20 C11 Lakewood Ohio, N USA 41.28N 81.48W
25 Y13 Lakewood Park Florida, SE USA 27.32N 80.24W
27 Z14 Lake Worth Florida, SE USA 26.37N 80.03W
158 H4 Lake Wular ⊠ NE India
128 H11 Lakhdenpokh'ya Respublika Kareliya, NW Russian Federation
158 L11 Lakhimpur Uttar Pradesh, N India 27.57N 80.46E
160 I11 Lakhnādon Madhya Pradesh, C India 22.36N 79.36E
160 A9 Lakhpat Gujarāt, W India
121 K19 Lakhva *Rus.* Lakhva. Brestskaya Voblasts', SW Belarus 52.13N 27.15E
28 I6 Lakin Kansas, C USA 37.56N 101.18W
155 S7 Lakki Marwat North-West Frontier Province, E Pakistan 32.36N 70.55E
94 L8 Laksefjorden *fjord* N Norway
94 K8 Lakselv Finnmark, N Norway 70.01N 24.57E
161 B21 Lakshadweep *prev.* the Laccadive, Minicoy and Amindivi Islands. ◆ *union territory* India, N Indian Ocean
161 C22 Lakshadweep *Eng.* Laccadive Islands. *island group* India, N Indian Ocean
159 S17 Lakshmikantapur West Bengal, NE India 22.04N 88.19E
114 G11 Laktaši Republika Srpska, N Bosnia and Herzegovina 44.54N 17.18E
155 V7 Lāla Mūsa Punjab, NE Pakistan 32.40N 73.57E
la Laon *see* Laon

◆ COUNTRY ◇ DEPENDENT TERRITORY ◈ ADMINISTRATIVE REGION ▲ MOUNTAIN ⊻ VOLCANO ⊘ LAKE
● COUNTRY CAPITAL ○ DEPENDENT TERRITORY CAPITAL ✕ INTERNATIONAL AIRPORT ▲ MOUNTAIN RANGE ⊠ RIVER ⊟ RESERVOIR

281

116 M11 **Lalapaşa** Edirne, NW Turkey 41.52N 26.43E

85 P14 **Lalaua** Nampula, N Mozambique 14.21S 38.16E

107 S10 **L'Alcúdia** var. L'Alcudia. País Valenciano, E Spain 39.10N 0.30W

82 J11 **Lalibela** Amhara, N Ethiopia, 12.01N 39.05E

159 T12 **Lalmanirhat** Rajshahi, N Bangladesh 25.51N 89.34E

81 F20 **Le Lkoumou ◆** province SW Congo

44 E8 **La Libertad** SW El Salvador 13.28N 89.17W

44 E3 **La Libertad** Petén, N Guatemala 16.46N 90.07W

44 H6 **La Libertad** Comayagua, SW Honduras 14.44N 87.37W

42 E4 **La Libertad** var. Puerto Libertad. Sonora, NW Mexico 29.52N 112.39W

44 K10 **La Libertad** Chontales, S Nicaragua 12.14N 85.15W

44 A9 **La Libertad ◆** department SW El Salvador

58 B11 **La Libertad** off. Departamento de La Libertad. ◆ department W Peru

64 G11 **La Ligua** Valparaíso, C Chile 32.23S 71.16W

145 U5 **La'li Khān** E Iraq 34.58N 45.36E

81 H16 **La Likouala ◆** province NE Congo

106 H3 **Lalín** Galicia, NW Spain 42.40N 8.06W

104 L13 **Lalinde** Dordogne, SW France 44.52N 0.42E

106 K16 **La Línea** var. La Línea de la Concepción. Andalucía, S Spain 36.10N 5.21W

La Línea de la Concepción see La Línea

158 J14 **Lalitpur** Uttar Pradesh, N India 24.42N 78.24E

159 P11 **Lalitpur** Central, C Nepal 27.45N 85.17E

158 K10 **Lālkua** Uttar Pradesh, N India 29.04N 79.31E

9 R12 **La Loche** Saskatchewan, C Canada 56.31N 109.27W

104 M6 **la Loupe** Eure-et-Loir, C France 48.30N 1.04E

101 G20 **La Louvière** Hainaut, S Belgium 50.28N 4.15E

L'Altissima see Hochwilde

106 L14 **La Luisiana** Andalucía, S Spain 37.30N 5.14W

39 S14 **La Luz** New Mexico, SW USA 32.58N 105.56W

109 D16 **La Maddalena** Sardegna, Italy, C Mediterranean Sea 41.13N 9.25E

64 J7 **La Madrid** Tucumán, N Argentina 27.37S 65.16W

Lama-Kara see Kara

175 R16 **Lamakera, Selat** strait Nusa Tenggara, S Indonesia

13 S8 **La Malbaie** Quebec, SE Canada 47.39N 70.10W

178 Jj10 **Lamam** Xékong, S Laos 15.20N 106.40E

107 P10 **La Mancha** physical region C Spain

la Manche see English Channel

197 C13 **Lamar** Malekula, C Vanuatu 16.26S 167.47E

39 W6 **Lamar** Colorado, C USA 38.03N 102.36W

29 S7 **Lamar** Missouri, C USA 37.30N 94.16W

23 S12 **Lamar** South Carolina, SE USA 34.10N 80.03W

109 C19 **La Marmora, Punta ▲** Sardegna, Italy, C Mediterranean Sea 39.58N 9.20E

15 N7 **La Martre, Lac ⊚** Northwest Territories, NW Canada

58 D10 **Lamas** San Martín, N Peru 6.27S 76.32W

44 I5 **La Masica** Atlántida, NW Honduras 15.37N 87.04W

105 R12 **Lamastre** Ardèche, E France 44.59N 4.32E

La Matepec see Santa Ana, Volcán de

46 I7 **La Maya** Santiago de Cuba, E Cuba 20.09N 75.40W

111 S5 **Lambach** Oberösterreich, N Austria 48.06N 13.52E

173 Fj8 **Lambale** Pulau Pini, W Indonesia 0.08N 98.36E

104 H5 **Lamballe** Côtes d'Armor, NW France 48.28N 2.31W

81 D18 **Lambaréné** Moyen-Ogooué, W Gabon 0.40S 10.13E

Lambasa see Labasa

175 Q12 **Lambasina Besar, Pulau** island C Indonesia

58 B11 **Lambayeque** Lambayeque, W Peru 6.39S 79.54W

58 A10 **Lambayeque** off. Departamento de Lambayeque. ◆ department NW Peru

99 G17 **Lambay Island** Ir. Reachrainn. island E Ireland

195 O10 **Lambert, Cape** headland New Britain, E PNG 4.15S 151.31E

205 N6 **Lambert Glacier** glacier Antarctica

31 T10 **Lamberton** Minnesota, N USA 44.14N 95.15W

29 X4 **Lambert-Saint Louis ✕** Missouri, C USA 38.43N 90.19W

33 R11 **Lambertville** Michigan, N USA 41.46N 83.37W

20 I15 **Lambertville** New Jersey, NE USA 40.20N 74.55W

175 Pp9 **Lamboya** N Indonesia 0.57S 120.23E

108 D8 **Lambro ⊿** N Italy

15 H2 **Lambton, Cape** headland Banks Island, Northwest Territories, NW Canada 71.04N 123.07W

35 W11 **Lame Deer** Montana, NW USA 45.37N 106.37W

106 H6 **Lamego** Viseu, N Portugal 41.04N 7.49W

197 C13 **Lamen Bay** Épi, C Vanuatu 16.36S 168.10E

47 X6 **Lamentin** Basse Terre, N Guadeloupe 16.16N 61.37W

Lamentin see Le Lamentin

190 K10 **Lameroo** South Australia 35.22S 140.30E

56 F10 **La Mesa** Cundinamarca, C Colombia 4.39N 74.24W

37 U17 **La Mesa** California, W USA 32.44N 117.00W

39 R16 **La Mesa** New Mexico, SW USA 32.03N 106.41W

27 N6 **Lamesa** Texas, SW USA 32.43N 101.57W

197 I15 **Lami** Viti Levu, C Fiji 18.07S 178.25E

117 F17 **Lamía** Stereá Ellás, C Greece 38.54N 22.26E

179 Q17 **Lamitan** Basilan Island, SW Philippines 6.40N 122.07E

197 J14 **Lamiti** Gau, C Fiji 18.00S 179.20E

176 Uu8 **Lamlam** Irian Jaya, E Indonesia 0.03S 130.46E

196 B16 **Lamlam, Mount ▲** SW Guam 13.19N 144.40E

111 Q6 **Lammer ⊿** E Austria

193 E23 **Lammerlaw Range ▲** South Island, NZ

97 G20 **Lammhult** Kronoberg, S Sweden 57.09N 14.34E

95 L18 **Lammi** Etelä-Suomi, S Finland 61.06N 25.00E

201 U11 **Lamoil** island Chuuk, C Micronesia

37 W3 **Lamoille** Nevada, W USA 40.47N 115.37W

20 M7 **Lamoille River ⊿** Vermont, NE USA

32 J13 **La Moine River ⊿** Illinois, N USA

179 Pp10 **Lamon Bay** bay Luzon, N Philippines

31 V16 **Lamoni** Iowa, C USA 40.37N 93.56W

37 R13 **Lamont** California, W USA 35.15N 118.54W

29 N8 **Lamont** Oklahoma, C USA 36.41N 97.33W

56 E13 **La Montañita** var. Montañita. Caquetá, S Colombia 1.23N 75.28W

45 N8 **La Mosquitia** var. Miskito Coast, Eng. Mosquito Coast. coastal region E Nicaragua

104 I9 **la Mothe-Achard** Vendée, NW France 46.37N 1.37W

196 L15 **Lamotrek Atoll** atoll Caroline Islands, C Micronesia

31 P6 **La Moure** North Dakota, N USA 46.21N 98.17W

178 H8 **Lampang** var. Muang Lampang. Lampang, NW Thailand 18.16N 99.30E

178 Ii9 **Lam Pao Reservoir ⊡** E Thailand

27 S9 **Lampasas** Texas, SW USA 31.03N 98.10W

27 S9 **Lampasas River ⊿** Texas, SW USA

43 N7 **Lampazos** var. Lampazos de Naranjo. Nuevo León, NE Mexico 27.00N 100.28W

Lampazos de Naranjo see Lampazos

117 E19 **Lámpeia** Dytikí Ellás, S Greece 37.51N 21.48E

103 G19 **Lampertheim** Hessen, W Germany 49.36N 8.28E

99 I20 **Lampeter** SW Wales, UK 52.07N 4.03W

178 H7 **Lamphun** var. Lampun, Muang Lamphun. Lamphun, NW Thailand 18.36N 99.01E

5 X10 **Lamplugh** Manitoba, C Canada 58.18N 94.06W

Lampun see Lamphun

174 Ii13 **Lampung** off. Propinsi Lampung. ◆ province SW Indonesia

174 Ii13 **Lampung, Teluk** bay Sumatera, SW Indonesia

130 K6 **Lamskoye** Lipetskaya Oblast', W Russian Federation 52.57N 38.06E

83 K20 **Lamu** Coast, SE Kenya 2.17S 40.49E

45 N14 **La Muerte, Cerro ▲** C Costa Rica 9.33N 83.47W

105 S13 **la Mure** Isère, E France 44.54N 5.48E

39 S10 **Lamy** New Mexico, SW USA 35.28N 105.52W

121 J18 **Lan' Rus.** Lan'. ⊿ C Belarus

40 E10 **Lanai** Haw. Lāna'i. island Hawaii, USA, C Pacific Ocean

40 E10 **Lanai City** Lanai, Hawaii, USA, C Pacific Ocean 20.49N 156.55W

101 L18 **Lanaken** Limburg, NE Belgium 50.53N 5.39E

179 R15 **Lanao, Lake** var. Lake Sultan Alonto. ⊚ Mindanao, S Philippines

98 J12 **Lanark** S Scotland, UK 55.38N 4.24W

98 I13 **Lanark** cultural region C Scotland, UK

106 L9 **La Nava de Ricomalillo** Castilla-La Mancha, C Spain 39.40N 4.58W

178 Gg13 **Lanbi Kyun** prev. Sullivan Island. island Mergui Archipelago, S Myanmar

Lancang Jiang see Mekong

99 K17 **Lancaster** cultural region NW England, UK

13 **Lancaster** Ontario, SE Canada 45.10N 74.31W

37 R13 **Lancaster** California, W USA 34.42N 118.08W

37 T14 **Lancaster** Kentucky, S USA 37.35N 84.34W

33 U11 **Lancaster** Missouri, C USA 40.46N 83.37W

21 O7 **Lancaster** New Hampshire, NE USA 44.29N 71.34W

20 D10 **Lancaster** New York, NE USA 42.54N 78.40W

33 T14 **Lancaster** Ohio, N USA 39.42N 82.36W

18 F15 **Lancaster** Pennsylvania, NE USA 40.03N 76.18W

23 P10 **Lancaster** South Carolina, SE USA 34.43N 80.46W

27 V8 **Lancaster** Texas, SW USA 32.35N 96.45W

23 X5 **Lancaster** Virginia, NE USA 37.45N 76.25W

32 L8 **Lancaster** Wisconsin, N USA 42.52N 90.43W

207 N10 **Lancaster Sound** sound Nunavut, N Canada

109 K14 **Lanciano** Abruzzo, C Italy 42.15N 14.22E

113 O16 **Lańcut** Podkarpackie, SE Poland 50.04N 22.13E

174 Kk8 **Landak, Sungai ⊿** Borneo, N Indonesia

Landao see Lantau Island

Landau see Landau an der Isar, Bayern, Germany

Landau see Landau in der Pfalz, Rheinland-Pfalz, Germany

103 O22 **Landau an der Isar** var. Landau. Bayern, SE Germany 48.40N 12.41E

103 F20 **Landau in der Pfalz** var. Landau. Rheinland-Pfalz, SW Germany 49.12N 8.07E

Land Burgenland see Burgenland

110 K8 **Landeck** Tirol, W Austria 47.09N 10.34E

101 J19 **Landen** Vlaams Brabant, C Belgium 50.45N 5.04E

35 U15 **Lander** Wyoming, C USA 42.49N 108.43W

104 F5 **Landerneau** Finistère, NW France 48.27N 4.16W

97 J20 **Landeryd** Halland, S Sweden 57.05N 13.13E

104 J15 **Landes ◆** department SW France

Landeshut/Landeshut in Schlesien see Kamienna Góra

107 R9 **Landete** Castilla-La Mancha, C Spain 39.54N 1.22W

101 M18 **Landgraaf** Limburg, SE Netherlands 50.54N 6.04E

104 F5 **Landivisiau** Finistère, NW France 48.31N 4.03W

Land of Enchantment see New Mexico

Land of Opportunity see Arkansas

Land of Steady Habits see Connecticut

Land of the Midnight Sun see Alaska

110 I8 **Landquart** Graubünden, SE Switzerland 46.58N 9.35E

110 I9 **Landquart ⊿** Austria/Switzerland

23 P10 **Landrum** South Carolina, SE USA 35.10N 82.11W

178 H8 **Landsberg** var. Gorzów Iławeckie, Warmińsko-Mazurskie, NE Poland

Landsberg see Gorzów Wielkopolski, Gorzów, Poland

103 K23 **Landsberg am Lech** Bayern, S Germany 48.04N 10.53E

Landsberg an der Warthe see Gorzów Wielkopolski

99 G25 **Land's End** headland SW England, UK 50.02N 5.41W

103 M22 **Landshut** Bayern, SE Germany 48.31N 12.09E

Landskron see Lanškroun

97 J24 **Landskrona** Skåne, S Sweden 55.52N 12.52E

100 I10 **Landsmeer** Noord-Holland, C Netherlands 52.25N 4.55E

97 J19 **Landvetter ✕** (Göteborg) Västra Götaland, S Sweden 57.39N 12.22E

Landwarów see Lentvaris

25 R5 **L'Anguille River ⊿** Arkansas, C USA

95 I16 **Långviksmon** Västernorrland, N Sweden 63.39N 18.45E

103 K22 **Langweid** Bayern, S Germany 48.29N 10.50E

166 J8 **Langzhong** Sichuan, C China 31.46N 105.55E

9 U15 **Lan Hsü** see Lan Yü

9 U15 **Lanigan** Saskatchewan, S Canada 51.49N 105.01W

118 K5 **Lanivtsi** Ternopil's'ka Oblast', W Ukraine 49.52N 26.05E

143 Y13 **Länkäran Rus.** Lenkoran'. S Azerbaijan 38.46N 48.50E

104 L16 **Lannemezan** Hautes-Pyrénées, S France 43.07N 0.22E

104 G5 **Lannion** Côtes d'Armor, NW France 48.43N 3.27W

12 M11 **L'Annonciation** Quebec, SE Canada 46.22N 74.51W

107 V5 **L'Anoia ⊿** NE Spain

20 I15 **Lansdale** Pennsylvania, NE USA 40.14N 75.13W

12 L14 **Lansdowne** Ontario, SE Canada 44.25N 76.00W

158 K9 **Lansdowne** Uttar Pradesh, N India 29.49N 78.42E

32 M3 **L'Anse** Michigan, N USA 46.45N 88.27W

13 S7 **L'Anse-St-Jean** Quebec, SE Canada 48.14N 70.13W

95 K18 **Länsi-Suomi ◆** province W Finland

31 Y11 **Lansing** Iowa, C USA 43.22N 91.11W

29 R4 **Lansing** Kansas, C USA 39.15N 94.54W

33 Q9 **Lansing** state capital Michigan, N USA 42.43N 84.33W

95 M16 **Lapinlahti** Itä-Suomi, C Finland 63.21N 27.25E

24 K9 **Laplace** Louisiana, S USA 30.04N 90.28W

113 G17 **La Plaine** SE Dominica 15.19N 61.15W

178 Gg10 **Lanta, Ko** island S Thailand

167 O15 **Lantau Island** Chin. Landao. island Hong Kong, S China

Lan-ts'ang Chiang see Mekong

175 Q7 **Lanu** Sulawesi, N Indonesia 1.00N 121.33E

109 D19 **Lanusei** Sardegna, Italy, C Mediterranean Sea 39.55N 9.31E

104 H7 **Lanvaux, Landes de** physical region NW France

169 W8 **Lanxi** Heilongjiang, NE China 46.06N 126.10E

167 R10 **Lanxi** Zhejiang, SE China 29.13N 119.30E

175 T15 **Lan Yü** var. Huoshao Tao, var. Hungt'ou, Lan Hsü, Lanyü, Eng. Orchid Island, prev. Kotosho, Koto Sho. island SE Taiwan

105 U4 **Lanzarote** island Islas Canarias, Spain, NE Atlantic Ocean

165 V10 **Lanzhou** var. Lanchou, Lanchow, Lan-chow; prev. Kaolan. Gansu, C China 36.01N 103.52E

108 H13 **Lanzo Torinese** Piemonte, NE Italy 45.18N 7.26E

179 P8 **Laoag** Luzon, N Philippines 18.11N 120.34E

179 R12 **Laoang** Samar, C Philippines 12.29N 125.00E

96 D10 **Langevågen** Møre og Romsdal, S Norway 62.25N 6.13E

167 P3 **Langfang** Hebei, E China 39.30N 116.39E

96 E9 **Langfjorden** fjord S Norway

31 Q8 **Langford** South Dakota, N USA 45.36N 97.48W

173 G6 **Langgapayung** Sumatera, W Indonesia 1.42N 99.57E

108 E9 **Langhirano** Emilia-Romagna, C Italy 44.37N 10.16E

99 K14 **Langholm** S Scotland, UK 55.13N 3.11W

173 Fj2 **Langjökull** glacier C Iceland

175 R13 **Langkawi, Pulau** island Peninsular Malaysia

178 Gg15 **Langka, Kepulauan** island group C Indonesia

12 L8 **Langlade** Quebec, SE Canada 48.13N 75.58W

178 Jj7 **Lang Mô** Thanh Hoa, N Vietnam 19.36N 105.30E

110 E8 **Langnau** see Langnau im Emmental

110 E8 **Langnau im Emmental** var. Langnau. Bern, W Switzerland 46.57N 7.46E

105 Q13 **Langogne** Lozère, S France 44.40N 3.52E

164 K16 **Langol Kangri ▲** W China 28.51N 87.28E

104 K13 **Langon** Gironde, SW France 47.43N 1.49W

94 G4 **La Ngounié** see Ngounié

164 G14 **Langoya** island C Norway

Langqên Zangbo see China/India

106 K2 **Langreo** var. Sama de Langreo. Asturias, N Spain 43.18N 5.40W

105 S7 **Langres** Haute-Marne, C France 8.24N 78.09W

105 R8 **Langres, Plateau de** plateau C France

173 F4 **Langsa** Sumatera, W Indonesia 4.29N 97.53E

95 H15 **Långsele** Västernorrland, C Sweden 63.10N 17.04E

97 M14 **Långshyttan** Dalarna, C Sweden 60.27N 16.02E

178 K5 **Lang Son** var. Langson. Lang Son, N Vietnam 21.49N 106.45E

178 Gg14 **Lang Suan** Chumphon, SW Thailand 9.59N 99.03E

95 J14 **Långträsk** Norrbotten, N Sweden 65.21N 20.16E

27 N11 **Langtry** Texas, SW USA 29.46N 101.25W

105 P16 **Languedoc** cultural region S France

105 P15 **Languedoc-Roussillon ◆** region S France

29 X10 **L'Anguille River ⊿** Arkansas, C USA

172 Pp1 **La Perouse Strait** Jap. Sōya-kaikyō, Rus. Proliv Laperuza. strait Japan/Russian Federation

32 M3 **L'Anse** Michigan, N USA 46.45N 88.27W

65 I14 **La Perra, Salitral de** salt lake C Argentina

13 S7 **L'Anse-St-Jean** Quebec, SE Canada 48.14N 70.13W

43 Q10 **La Pesca** Tamaulipas, C Mexico 23.49N 97.45W

42 M13 **La Piedad Cavadas** Michoacán de Ocampo, C Mexico 20.19N 102.01W

95 M16 **Lapines** see Lafnitz

95 M16 **Lapinlahti** Itä-Suomi, C Finland 63.21N 27.25E

Lápithos see Lapta

24 K9 **Laplace** Louisiana, S USA 30.04N 90.28W

25 V12 **Largo** Florida, SE USA 27.54N 82.47W

39 V12 **Largo, Canon** valley New Mexico, SW USA

48 D6 **Largo, Cayo** island W Cuba

25 V12 **Largo, Key** island Florida Keys, Florida, SE USA

98 H12 **Largs** W Scotland, UK 55.47N 4.50W

104 I16 **la Rhune** var. Larrún. ▲ France/Spain see also Larrún 43.19N 1.34W

la Riege see Ariège

31 O3 **Larimore** North Dakota, N USA 47.54N 97.37W

109 L15 **Larino** Molise, C Italy 41.46N 14.50E

65 J9 **La Rioja** La Rioja, NW Argentina 29.25S 66.49W

64 J9 **La Rioja** off. Provincia de La Rioja. ◆ province NW Argentina

107 O4 **La Rioja ◆** autonomous community N Spain

117 F16 **Lárisa** var. Larissa. Thessalía, C Greece 39.38N 22.27E

Larissa see Lárisa

155 Q13 **Lārkāna** var. Larkhana. Sind, SE Pakistan 27.31N 68.18E

Larkhana see Lārkāna

124 Nn3 **Lárnaca** var. Larnaca, Larnax. SE Cyprus 34.54N 33.38E

124 Nn3 **Lárnaca ✕** SE Cyprus 34.52N 33.38E

Larnax see Lárnaca

99 G14 **Larne Ir.** Latharna. E Northern Ireland, UK 54.51N 5.49W

106 L3 **La Robla** Castilla-León, N Spain 42.48N 5.37W

106 J10 **La Roca de la Sierra** Extremadura, W Spain 39.06N 6.41W

101 K22 **La Roche-en-Ardenne** Luxembourg, SE Belgium 50.11N 5.35E

104 L11 **la Rochefoucauld** Charente, W France 45.43N 0.23E

104 J10 **la Rochelle** anc. Rupella. Charente-Maritime, W France 46.09N 1.07W

104 I9 **la Roche-sur-Yon** prev. Bourbon Vendée, Napoléon-Vendée. Vendée, NW France 46.40N 1.25W

107 Q10 **La Roda** Castilla-La Mancha, C Spain 39.13N 2.10W

106 L14 **La Roda de Andalucía** Andalucía, S Spain 37.12N 4.45W

47 P9 **La Romana** E Dominican Republic 18.25N 69.00W

9 T13 **La Ronge** Saskatchewan, C Canada 55.07N 105.18W

9 U13 **La Ronge, Lac ⊚** Saskatchewan, C Canada

24 K10 **Larose** Louisiana, S USA 29.34N 90.22W

44 M9 **La Rosita** Región Autónoma Atlántico Norte, NE Nicaragua 13.55N 84.23W

189 Q3 **Larrimah** Northern Territory, N Australia 15.30S 133.12E

64 N11 **Larroque** Entre Ríos, E Argentina 33.03S 59.06W

107 Q2 **Larrún Fr.** la Rhune. ▲ France/Spain see also la Rhune 43.18N 1.35W

205 X6 **Lars Christensen Coast** physical region Antarctica

41 Q14 **Larsen Bay** Kodiak Island, Alaska, USA 57.32N 153.58W

204 I5 **Larsen Ice Shelf** ice shelf Antarctica

15 K3 **Larsen Sound** sound Nunavut, N Canada

106 K8 **La Rúa** see A Rúa

104 K16 **Laruns** Pyrénées-Atlantiques, SW France 43.00N 0.25W

97 G16 **Larvik** Vestfold, S Norway 59.03N 10.01E

126 H11 **Lar'yak** Khanty-Mansiyskiy Avtonomnyy Okrug, C Russian Federation 61.09N 80.01E

La-sa see Lhasa

175 Tt11 **Lasahata** Pulau Seram, E Indonesia 2.52S 128.27E

Lasahau see Lasahau

39 O6 **La Sal** Utah, W USA 38.19N 109.14W

12 C17 **La Salle** Ontario, S Canada 42.13N 83.05W

32 L11 **La Salle** Illinois, N USA 41.19N 89.06W

47 O9 **Las Americas ✕** (Santo Domingo) S Dominican Republic 18.24N 69.38W

81 G17 **La Sangha ◆** province N Congo

39 V6 **Las Animas** Colorado, C USA 38.04N 103.13W

110 D10 **La Sarine** var. Sarine. ⊿ SW Switzerland

110 B9 **La Sarraz** Vaud, W Switzerland 46.40N 6.32E

10 H12 **La Sarre** Quebec, SE Canada 48.49N 79.12W

56 L3 **Las Aves, Islas** var. Islas de Aves. island group N Venezuela

57 N7 **Las Bonitas** Bolívar, C Venezuela 7.50N 65.40W

106 K15 **Las Cabezas de San Juan** Andalucía, S Spain 36.58N 5.55W

65 G19 **Lascano** Rocha, E Uruguay 33.40S 54.12W

64 I3 **Lascar, Volcán ▲** N Chile 23.22S 67.33W

57 T15 **Las Choapas var.** Choapas. Veracruz-Llave, SE Mexico 17.51N 94.00W

39 R14 **Las Cruces** New Mexico, SW USA 32.19N 106.49W

Lasdehnen see Krasnoznamensk

107 V4 **La Seu d'Urgel** var. La Seu d'Urgell, Seo de Urgel. Cataluña, NE Spain 42.22N 1.27E

La Selle Selle, Pic de la

64 G9 **La Serena** Coquimbo, C Chile 29.54S 71.18W

106 K11 **La Serena** physical region W Spain

La Seu d'Urgell see La Seu d'Urgell

105 T16 **La Seyne-sur-Mer** Var, SE France 43.07N 5.52E

154 M8 **Lashkar Gāh var.** Lash-Kar-Gar'. Helmand, S Afghanistan 31.34N 64.21E

175 Qq13 **Lasihao var.** Lasahau. Pulau Muna, C Indonesia 5.01S 122.23E

109 N21 **La Sila ▲** SW Italy

115 H23 **La Silueta, Cerro ▲** S Chile 52.22S 72.09W

112 J13 **Lask** Łódzkie, C Poland

111 V11 **Laško Ger.** Tüffer. C Slovenia 46.09N 15.11E

65 H14 **Las Lajas** Neuquén, W Argentina 38.30S 70.22W

65 H15 **Las Lajas, Cerro ▲** W Argentina 38.46S 69.56W

65 N6 **Las Lomitas** Formosa, N Argentina 24.44S 60.34W

42 E8 **Las Margaritas** Chiapas, SE Mexico 16.19N 91.58W

Las Marismas see Guadalquivir, Marismas del

56 M6 **Las Mercedes** Guárico, N Venezuela 9.06N 66.22W

44 F6 **Las Minas, Cerro ▲** W Honduras 14.33N 88.41W

◆ COUNTRY ◇ DEPENDENT TERRITORY ◆ ADMINISTRATIVE REGION ▲ MOUNTAIN ▲ VOLCANO ⊚ LAKE
◆ COUNTRY CAPITAL ○ DEPENDENT TERRITORY CAPITAL ✕ INTERNATIONAL AIRPORT ▲ MOUNTAIN RANGE ⊿ RIVER ⊡ RESERVOIR

107 O11 **La Solana** Castilla-La Mancha, C Spain 38.55N 3.13W
47 Q14 **La Soufrière** ☼ Saint Vincent, Saint Vincent and the Grenadines 13.20N 61.11W
104 M10 **la Souterraine** Creuse, C France 46.15N 1.28E
64 N7 **Las Palmas** Chaco, N Argentina 27.07S 58.45W
45 Q16 **Las Palmas** Veraguas, W Panama 8.09N 81.28W
66 P12 **Las Palmas** var. Las Palmas de Gran Canaria. Gran Canaria, Islas Canarias, Spain, NE Atlantic Ocean 28.07N 15.27W
66 P12 **Las Palmas** ◆ province Islas Canarias, Spain, NE Atlantic Ocean
66 Q12 **Las Palmas** ✈ Gran Canaria, Islas Canarias, Spain, NE Atlantic Ocean
Las Palmas de Gran Canaria see Las Palmas
42 D6 **Las Palomas** Baja California Sur, W Mexico 31.43N 107.37W
107 P10 **Las Pedroñeras** Castilla-La Mancha, C Spain 39.27N 2.40W
108 E10 **La Spezia** Liguria, NW Italy 44.07N 9.49E
63 F20 **Las Piedras** Canelones, S Uruguay 34.42S 56.13W
65 J18 **Las Plumas** Chubut, S Argentina 43.46S 67.15W
63 B18 **Las Rosas** Santa Fe, C Argentina 32.27S 61.30W
Lassa see Lhasa
37 O4 **Lassen Peak** ▲ California, W USA 40.27N 121.28W
204 K6 **Lassiter Coast** physical region Antarctica
111 V9 **Lassnitz** ✍ SE Austria
13 O12 **L'Assomption** Quebec, SE Canada 45.48N 73.27W
13 N11 **L'Assomption** ✍ Quebec, SE Canada
45 S17 **Las Tablas** Los Santos, S Panama 7.45N 80.17W
Lastarria, Volcán see Azufre, Volcán
39 V4 **Last Chance** Colorado, C USA 39.41N 103.34W
Last Frontier, The see Alaska
9 U16 **Last Mountain Lake** ⊚ Saskatchewan, S Canada
64 H9 **Las Tórtolas, Cerro** ▲ W Argentina 29.57S 69.49W
63 C14 **Las Toscas** Santa Fe, C Argentina 28.22S 59.19W
81 F19 **Lastoursville** Ogooué-Lolo, E Gabon 0.49S 12.43E
115 F16 **Lastovo** It. Lagosta. island SW Croatia
115 F16 **Lastovski Kanal** channel SW Croatia
42 E6 **Las Tres Vírgenes, Volcán** ☼ W Mexico 27.27N 112.34W
42 F4 **Las Trincheras** Sonora, NW Mexico 30.21N 111.27W
57 N8 **Las Trincheras** Bolívar, E Venezuela 6.57N 64.49W
46 H7 **Las Tunas** var. Victoria de las Tunas. Las Tunas, E Cuba 20.58N 76.58W
La Suisse see Switzerland
42 I5 **Las Varas** Chihuahua, N Mexico 29.35S 108.01W
42 J12 **Las Varas** Nayarit, C Mexico 21.11N 105.09W
64 L10 **Las Varillas** Córdoba, C Argentina 31.54S 62.45W
37 X11 **Las Vegas** Nevada, W USA 36.09N 115.10W
39 T10 **Las Vegas** New Mexico, SW USA 35.35N 105.15W
195 W8 **Lata** Nendö, Solomon Islands 10.45S 165.43E
11 R10 **La Tabatière** Quebec, E Canada 50.51N 58.58W
58 C6 **Latacunga** Cotopaxi, C Ecuador 0.58S 78.36W
204 I7 **Latady Island** island Antarctica
56 E14 **La Tagua** Putumayo, S Colombia 0.04S 74.39W
Latakia see Al Lādhiqīyah
94 J10 **Lätäseno** ✍ NW Finland
12 H9 **Latchford** Ontario, S Canada 47.20N 79.45W
12 J13 **Latchford Bridge** Ontario, SE Canada 47.20N 79.25W
200 Ss12 **Late** island Vava'u Group, N Tonga
159 P15 **Lätehär** Bihār, N India 23.48N 84.28E
13 R7 **Laterrière** Quebec, SE Canada 48.17N 71.10W
104 J13 **la Teste** Gironde, SW France 44.35N 1.04W
27 V8 **Latexo** Texas, SW USA 31.24N 95.28W
20 L10 **Latham** New York, NE USA 42.45N 73.45W
Latharna see Larne
110 B9 **La Thiele** var. Thièle. ✍ W Switzerland
29 R3 **Lathrop** Missouri, C USA 39.33N 94.19W
109 I16 **Latina** prev. Littoria. Lazio, C Italy 41.28N 12.52E
43 R14 **La Tinaja** Veracruz-Llave, S Mexico
108 J7 **Latisana** Friuli-Venezia Giulia, NE Italy 45.47N 13.01E
Latium see Lazio
117 K25 **Lató** site of ancient city Kríti, Greece, E Mediterranean Sea 35.09N 25.46E
197 J7 **La Tontouta** ✈ (Nouméa) Province Sud, S New Caledonia 22.06S 166.12E
57 N4 **La Tortuga, Isla** var. Isla Tortuga. island N Venezuela
110 C10 **La Tour-de-Peilz** var. La Tour de Peilz. Vaud, SW Switzerland
105 S11 **la Tour-du-Pin** Isère, E France
104 J11 **la Tremblade** Charente-Maritime, W France 45.45N 1.07W
104 L10 **la Trimouille** Vienne, W France 46.27N 1.02E
114 J9 **La Trinidad** Estelí, NW Nicaragua 12.57N 86.13W
179 P9 **La Trinidad** Luzon, N Philippines 16.30N 120.39E

43 V16 **La Trinitaria** Chiapas, SE Mexico 16.02N 92.00W
47 Q11 **la Trinité** E Martinique 14.43N 60.57W
13 U7 **La Trinité-des-Monts** Quebec, SE Canada 48.07N 68.31W
20 C15 **Latrobe** Pennsylvania, NE USA 40.18N 79.19W
191 P13 **La Trobe River** ✍ Victoria, SE Australia
Lattakia/Lattaquié see Al Lādhiqīyah
175 T11 **Latu** Pulau Seram, E Indonesia 3.24S 128.37E
13 P9 **La Tuque** Quebec, SE Canada 47.25N 72.46W
161 G14 **Lātūr** Mahārāshtra, C India 18.24N 76.34E
120 G8 **Latvia** off. Republic of Latvia, Ger. Lettland, Latv. Latvija, Latvijas Republika; prev. Latvian SSR, Rus. Latviyskaya SSR. ◆ republic NE Europe
Latvian SSR/Latvija/Latvijas Republika/Latvijskaya SSR see Latvia
195 O12 **Lau** New Britain, E PNG 5.46S 151.21E
183 R9 **Lau Basin** undersea feature S Pacific Ocean
103 O15 **Lauchhammer** Brandenburg, E Germany 51.27N 13.32E
Laudunum see Laon
Laudus see St-Lô
Lauenburg/Lauenburg in Pommern see Lauenburg
103 L20 **Lauf an der Pegnitz** Bayern, SE Germany 49.31N 11.16E
110 D7 **Laufen** Basel, NW Switzerland 47.25N 7.31E
111 P5 **Lauffen** Salzburg, NW Austria 47.54N 12.57E
94 I2 **Laugarbakki** Nordhurland Vestra, N Iceland 65.18N 20.51W
94 I4 **Laugarvatn** Sudhurland, SW Iceland 64.09N 20.43W
33 O3 **Laughing Fish Point** headland Michigan, N USA 46.31N 87.01W
197 L14 **Lau Group** island group E Fiji
Lauis see Lugano
95 M17 **Laukaa** Länsi-Suomi, W Finland 62.27N 25.58E
120 D12 **Laukuva** Šilalė, W Lithuania 55.37N 22.12E
Laun see Louny
191 P16 **Launceston** Tasmania, SE Australia 41.25S 147.07E
99 I24 **Launceston** anc. Dunheved. SW England, UK 50.37N 4.21W
56 C13 **La Unión** Nariño, SW Colombia 1.34N 77.09W
44 H8 **La Unión**, SE El Salvador 13.19N 87.52W
44 I6 **La Unión** Olancho, C Honduras 15.02N 86.40W
42 M15 **La Unión** Guerrero, S Mexico 17.59N 101.48W
43 Y14 **La Unión** Quintana Roo, E Mexico 18.00N 101.48W
107 S13 **La Unión** Murcia, SE Spain 37.37N 0.53W
56 L7 **La Unión** Barinas, C Venezuela 8.12N 67.46W
44 B10 **La Unión** ◆ department E El Salvador
40 H11 **Laupahoehoe** Haw. Laupāhoehoe. Hawaii, USA, C Pacific Ocean 20.00N 155.15W
103 I23 **Laupheim** Baden-Württemberg, S Germany 48.13N 9.54E
189 W3 **Laura** Queensland, NE Australia 15.37S 144.32E
201 X2 **Laura** atoll Majuro Atoll, SE Marshall Islands
Laurana see Lovran
56 L8 **La Urbana** Bolívar, C Venezuela 7.05N 66.58W
25 V14 **Laurel** Delaware, NE USA 38.33N 75.34W
23 W3 **Laurel** Maryland, NE USA 39.06N 76.51W
24 M6 **Laurel** Mississippi, S USA 31.41N 89.10W
35 U11 **Laurel** Montana, NW USA 45.40N 108.46W
31 R13 **Laurel** Nebraska, C USA 42.25N 97.04W
20 H15 **Laureldale** Pennsylvania, NE USA 40.24N 75.52W
20 C16 **Laurel Hill** ridge Pennsylvania, NE USA
31 T12 **Laurens** Iowa, C USA 42.51N 94.51W
23 P11 **Laurens** South Carolina, SE USA 34.29N 82.01W
Laurentian Highlands see Laurentian Mountains
13 P10 **Laurentian Mountains** var. Laurentian Highlands, Fr. Les Laurentides. plateau Newfoundland/Quebec, Canada
13 O12 **Laurentides** Quebec, SE Canada 45.51N 73.49W
Laurentides, Les see Laurentian Mountains
109 M19 **Lauria** Basilicata, S Italy 40.03S 15.49E
204 I1 **Laurie Island** island Antarctica
23 T11 **Laurinburg** North Carolina, SE USA 34.51N 79.40W
32 M2 **Laurium** Michigan, N USA 47.14N 88.26W
110 B9 **Lausanne** It. Losanna. Vaud, SW Switzerland 46.31N 6.39E
103 Q16 **Lausche** ▲ Czech Republic/Germany see also Luže 50.52N 14.39E
103 Q16 **Lausitzer Bergland** var. Lausitzer Gebirge. ▲ E Germany see also Luže
Lausitzer Gebirge see Lausitzer Bergland
Lausitzer Neisse see Neisse
105 T12 **Lautaret, Col du** pass SE France 45.03N 6.23E
65 G15 **Lautaro** Araucanía, C Chile 38.31S 72.27W
114 L12 **Lautawsar** Serbia, E Yugoslavia 44.25N 20.17E
103 F21 **Lauter** ✍ W Germany

110 I7 **Lauterach** Vorarlberg, NW Austria 47.28N 9.43E
103 I17 **Lauterbach** Hessen, C Germany 50.37N 9.24E
110 E9 **Lauterbrunnen** Bern, C Switzerland 46.36N 7.52E
175 Nn12 **Laut Kecil, Kepulauan** island group N Indonesia
197 H14 **Lautoka** Viti Levu, W Fiji 17.40S 177.25E
175 Nn11 **Laut, Pulau** prev. Laoet. island Borneo, C Indonesia
174 J14 **Laut, Pulau** island Kepulauan Natuna, W Indonesia
175 Nn11 **Laut, Selat** strait Borneo, C Indonesia
173 F4 **Laut Tawar, Danau** ⊚ Sumatera, NW Indonesia
201 V14 **Lauvergne Island** island Chuuk, C Micronesia
100 M5 **Lauwers Meer** ⊚ N Netherlands
100 M4 **Lauwersoog** Groningen, NE Netherlands 53.25N 6.14E
104 M14 **Lauzerte** Tarn-et-Garonne, S France 44.15N 1.08E
27 U13 **Lavaca Bay** ✍ Texas, SW USA
27 U12 **Lavaca River** ✍ Texas, SW USA
13 O12 **Laval** Quebec, SE Canada 45.32N 73.44W
104 J6 **Laval** Mayenne, NW France 48.04N 0.46W
13 T6 **Laval** ✍ Quebec, SE Canada
13 O12 **Lavalleja** ◆ department S Uruguay
13 O12 **Lavaltrie** Quebec, SE Canada 45.56N 73.14W
195 X17 **Lavanggu** Rennell, S Solomon Islands 11.39S 160.13E
149 O14 **Lāvān, Jazīreh-ye** island S Iran
111 U8 **Lavant** ✍ S Austria
120 G5 **Lavassaare** Ger. Lawassaar. Pärnumaa, SW Estonia 58.31N 24.22E
106 L3 **La Vecilla de Curueño** Castilla-León, N Spain 42.51N 5.24W
47 N8 **La Vega** var. Concepción de la Vega. C Dominican Republic 19.15N 70.32W
56 J4 **La Vela de Coro** var. La Vela. Falcón, N Venezuela 11.26N 69.35W
55 N17 **Lavelanet** Ariège, S France 42.55N 1.49E
109 M17 **Lavello** Basilicata, S Italy 41.03N 15.48E
29 R3 **La Verkin** Utah, W USA 37.12N 113.16W
34 I8 **Laverne** Oklahoma, C USA 36.42N 99.53W
27 S12 **La Vernia** Texas, SW USA 29.19N 98.07W
95 K18 **Lavia** Länsi-Suomi, W Finland 61.36N 22.34E
12 I12 **Lavieille, Lake** ⊚ Ontario, SE Canada
96 C12 **Lavik** Sogn og Fjordane, S Norway 61.06N 5.25E
35 U10 **Lavina** Montana, NW USA 46.18N 108.55W
204 H5 **Lavoisier Island** island Antarctica
175 Oo11 **Lavonia** Georgia, SE USA 34.26N 83.06W
105 R13 **la Voulte-sur-Rhône** Ardèche, E France 44.49N 4.46E
127 Q3 **Lavrentiya** Chukotskiy Avtonomnyy Okrug, NE Russian Federation 65.33N 171.12W
117 H20 **Lávrio** prev. Lávrion. Attikí, C Greece 37.43N 24.03E
Lávrion see Lávrio
85 L22 **Lavumisa** prev. Gollel. SE Swaziland 27.20S 31.51E
155 T4 **Lawarai Pass** pass N Pakistan 35.22N 71.48E
Lawassaare see Lavassaare
21 P7 **Lawdar** SW Yemen 13.49N 45.54E
27 Q7 **Lawn** Texas, SW USA 32.07N 99.45W
205 Y4 **Law Promontory** headland Antarctica
79 O14 **Lawra** NW Ghana 10.40N 2.55W
193 E23 **Lawrence** Otago, South Island, NZ 45.53S 169.43E
33 P14 **Lawrence** Indiana, N USA 39.49N 86.01W
29 Q4 **Lawrence** Kansas, C USA 38.58N 95.14W
21 O10 **Lawrence** Massachusetts, NE USA 42.42N 71.09W
22 L5 **Lawrenceburg** Kentucky, S USA 38.02N 84.54W
22 I10 **Lawrenceburg** Tennessee, S USA 35.14N 87.19W
25 T3 **Lawrenceville** Georgia, SE USA 33.57N 83.59W
33 N15 **Lawrenceville** Illinois, N USA 38.43N 87.40W
23 V7 **Lawrenceville** Virginia, NE USA 36.45N 77.51W
29 S3 **Lawson** Missouri, C USA 39.26N 94.12W
28 L12 **Lawton** Oklahoma, C USA 34.35N 98.19W
146 I4 **Lawz, Jabal al** ▲ NW Saudi Arabia 28.43N 35.20E
97 L16 **Laxå** Örebro, S Sweden 59.00N 14.37E
129 T5 **Laya** ✍ NW Russian Federation
59 I19 **La Yarada** Tacna, SW Peru 18.14S 70.30W
147 S15 **Layjūn** C Yemen 15.27N 49.16E
147 Q9 **Laylá** var. Laila. Ar Riyāḍ, C Saudi Arabia 22.13N 46.39E
25 P4 **Lay Lake** ⊠ Alabama, S USA
112 H6 **Layou** Saint Vincent, Saint Vincent and the Grenadines 13.11N 61.16W
94 L9 **la Younne** see El Ayoun
105 Ji5 **Laysan Island** island Hawaiian Islands, Hawaii, USA, C Pacific Ocean
38 L2 **Layton** Utah, W USA 41.03N 112.00W
36 L5 **Laytonville** California, W USA 39.40N 123.30W
180 H17 **Lazare, Pointe** headland Mahé, NE Seychelles 4.46S 55.28E
127 O14 **Lazarev** Khabarovskiy Kray, SE Russian Federation 52.11N 141.18E
114 I12 **Lazarevac** Serbia, W Yugoslavia 44.25N 20.17E
105 U15 **le Cannet** Alpes-Maritimes, SE France 43.33N 7.00E

67 N22 **Lazarev Sea** sea Antarctica
42 M15 **Lázaro Cárdenas** Michoacán de Ocampo, SW Mexico 17.55N 102.12W
121 F15 **Lazdijai** Lazdijai, S Lithuania 54.13N 23.33E
109 H15 **Lazio** anc. Latium. ◆ region C Italy
113 A16 **Lázně Kynžvart** Ger. Bad Königswart. Karlovarský Kraj, W Czech Republic 50.00N 12.40E
Lazovsk see Singerei
178 Ii13 **Leach** Pôŭthisăt, W Cambodia 12.19N 103.45E
33 X9 **Leachville** Arkansas, C USA 35.56N 90.15W
31 O6 **Lead** South Dakota, N USA 44.21N 103.45W
9 S16 **Leader** Saskatchewan, S Canada 50.55N 109.31W
21 S6 **Lead Mountain** ▲ Maine, NE USA 44.53N 68.07W
39 R5 **Leadville** Colorado, C USA 39.15N 106.17W
9 V12 **Leaf Rapids** Manitoba, C Canada 56.30N 100.01W
24 M7 **Leaf River** ✍ Mississippi, S USA
25 N7 **League City** Texas, SW USA 29.30N 95.05W
27 Q11 **Leakesville** Mississippi, S USA 31.09N 88.33W
27 Q11 **Leakey** Texas, SW USA 29.43N 99.45W
Leal see Lihula
85 G15 **Lealui** Western, W Zambia 15.12S 22.58E
12 C18 **Leamington** Ontario, S Canada 42.03N 82.34W
Leamington/Leamington Spa see Royal Leamington Spa
Leammi see Lemmenjoki
27 S10 **Leander** Texas, SW USA 30.34N 97.51W
62 F13 **Leandro N.Alem** Misiones, NE Argentina 27.34S 55.15W
99 A20 **Leane, Lough** Ir. Loch Léin. ⊚ SW Ireland
188 G8 **Learmonth** Western Australia 22.17S 114.03E
Leau see Zoutleeuw
L'Eau d'Heure see Plate Taille, Lac de la
202 D12 **Leava** Île Futuna, S Wallis and Futuna
29 R3 **Leavenworth** Kansas, C USA 39.17N 94.55W
34 I8 **Leavenworth** Washington, NW USA 47.36N 120.39W
94 L8 **Leavvajohka** var. Levajok. Finnmark, N Norway 69.57N 26.18E
29 R4 **Leawood** Kansas, C USA 38.57N 94.37W
112 H6 **Leba** Ger. Leba. Pomorskie, N Poland 54.45N 17.31E
112 I6 **Leba** Ger. Leba. ✍ N Poland
103 D20 **Lebach** Saarland, SW Germany 49.25N 6.54E
179 R17 **Lebak** Mindanao, S Philippines 6.28N 124.03E
175 Oo11 **Lebani,Teluk** bay Sulawesi, C Indonesia
33 O13 **Lebanon** Indiana, N USA 40.03N 86.28W
22 L6 **Lebanon** Kentucky, S USA 37.34N 85.15W
29 U6 **Lebanon** Missouri, C USA 37.40N 92.39W
21 N9 **Lebanon** New Hampshire, NE USA 43.40N 72.15W
34 G2 **Lebanon** Oregon, NW USA 44.32N 122.54W
20 H15 **Lebanon** Pennsylvania, NE USA 40.20N 76.24W
22 J8 **Lebanon** Tennessee, S USA 36.13N 86.16W
23 P7 **Lebanon** Virginia, NE USA 36.54N 82.04W
144 G6 **Lebanon** off. Republic of Lebanon, Ar. Al Lubnān, Fr. Liban. ◆ republic SW Asia
188 I14 **Lebanon Junction** Kentucky, S USA 37.49N 85.43W
78 R8 **Lebanon, Mount** see Liban, Jebel
152 J10 **Lebap** Lebapskiy Velayat, NE Turkmenistan 41.04N 61.49E
152 H11 **Lebap** ◆ province E Turkmenistan
Lebapskiy Velayat Turkm. Lebap Welayaty; prev. Rus. Chardzhevskaya Oblast', Turkm. Chärjew Oblasty. ◆ province E Turkmenistan
81 G20 **Lebap Welayaty** see Lebapskiy Velayat
Lebasee see Lebsko, Jezioro
117 C17 **Lebbeke** Oost-Vlaanderen, NW Belgium 51.00N 4.08E
117 B17 **Lebec** California, W USA 34.51N 118.52W
126 Ll12 **Lebedinyy** Respublika Sakha (Yakutiya), NE Russian Federation 53.00N 39.11E
119 T4 **Lebedyn** Rus. Lebedin. Sums'ka Oblast', NE Ukraine 50.36N 34.30E
10 I12 **Lebel-sur-Quévillon** Quebec, SE Canada 49.01N 76.55W
94 L8 **Lebesby** Finnmark, N Norway 70.31N 27.00E
104 M9 **le Blanc** Indre, C France 46.38N 1.04E
28 L6 **Lebo** Kansas, C USA 38.22N 95.50W
81 L15 **Lebo** Orientale, N Dem. Rep. Congo (Zaire) 4.30N 23.58E
112 H6 **Lębork** var. Ľbork, Ger. Lauenburg, Lauenburg in Pommern. Pomorskie, N Poland 54.33N 17.44E
107 N8 **Lebrija** Andalucía, S Spain 36.55N 6.04W
101 K24 **Léglise** Luxembourg, SE Belgium 49.48N 5.31E
108 G8 **Legnago** Lombardia, NE Italy 45.13N 11.18E
108 D7 **Legnano** Veneto, NE Italy 45.36N 8.54E
110 F6 **Leça da Palmeira** Porto, N Portugal 41.12N 8.43W
113 F14 **Legnica** Ger. Liegnitz. Dolnośląskie, SW Poland 51.12N 16.11E

105 N9 **Le Cap** see Cap-Haïtien
105 P2 **le Cateau-Cambrésis** Nord, N France 50.06N 3.33E
109 Q18 **Lecce** Puglia, SE Italy 40.22N 18.10E
108 D7 **Lecco** Lombardia, N Italy 45.49N 9.27E
31 V10 **Le Center** Minnesota, N USA 44.23N 93.43W
110 J7 **Lech** Vorarlberg, W Austria 47.14N 10.10E
103 K22 **Lech** ✍ Austria/Germany
117 D19 **Lechainá** var. Lehena, Lekhainá. Dytikí Ellás, S Greece 37.56N 21.16E
104 J11 **le Château d'Oléron** Charente-Maritime, W France 45.53N 1.12W
103 R3 **le Chesne** Ardennes, N France 49.33N 4.42E
105 R13 **le Cheylard** Ardèche, E France 44.55N 4.27E
110 K7 **Lechtaler Alpen** ▲ W Austria
102 H6 **Leck** Schleswig-Holstein, N Germany 54.50N 8.58E
2 L9 **Lecointre, Lac** ⊚ Quebec, SE Canada
24 H7 **Lecompte** Louisiana, S USA 31.05N 92.24W
105 Q9 **le Creusot** Saône-et-Loire, C France 46.48N 4.25E
Lecumberri see Lekunberri
112 P13 **Łęczna** Lubelskie, E Poland 51.18N 22.51E
112 J12 **Łęczyca** Ger. Lentschiza, Rus. Lenchitsa. Łódzkie, C Poland 52.04N 19.10E
102 F10 **Leda** ✍ NW Germany
101 F17 **Lede** Oost-Vlaanderen, NW Belgium 50.58N 3.58E
100 K6 **Ledesma** Castilla-León, N Spain 41.05N 5.59W
47 Q12 **le Diamant** SW Martinique 14.28N 61.02W
99 K17 **Ledge** New England, UK 53.30N 2.33W
190 I5 **Leigh Creek** South Australia 30.25S 138.23E
25 O2 **Leighton** Alabama, S USA 34.42N 87.31W
99 M21 **Leighton Buzzard** E England, UK 51.55N 0.40W
9 Q14 **Leduc** Alberta, SW Canada 53.16N 113.30W
99 F19 **Leinster, Mount** Ir. Pisa. Laighean, SE Ireland 52.36N 6.45W
121 F15 **Leipalingis** Lazdijai, S Lithuania 54.05N 23.52E
94 J12 **Leipojärvi** Norrbotten, N Sweden 67.03N 21.15E
33 R12 **Leipsic** Ohio, N USA 41.06N 83.58W
Leipsic see Leipzig
117 M20 **Leipsoí** island Dodekánisos, Greece, Aegean Sea
103 M15 **Leipzig** Pol. Lipsk; hist. Leipsic, anc. Lipsia. Sachsen, E Germany 51.19N 12.24E
103 M15 **Leipzig Halle** ✈ Sachsen, E Germany 51.26N 12.14E
106 G9 **Leiria** var. Collipo. Leiria, C Portugal 39.45N 8.49W
106 F9 **Leiria** ◆ district C Portugal
97 C15 **Leirvik** Hordaland, S Norway 59.48N 5.26E
120 E5 **Leisi** Ger. Laisberg. Saaremaa, W Estonia 58.33N 22.42E
22 J8 **Leitchfield** Kentucky, S USA 37.28N 86.17W
111 Y5 **Leitha** Hung. Lajta. ✍ Austria/Hungary
Leitir Ceanainn see Letterkenny
Leitmeritz see Litoměřice
99 D16 **Leitrim** Ir. Liatroim. cultural region NW Ireland
117 F18 **Leivádia** prev. Levádhia. Stereá Ellás, C Greece 38.24N 22.51E
Leix see Laois
99 F18 **Leixlip** Eng. Salmon Leap, Ir. Léim an Bhradáin. E Ireland 53.22N 6.31W
81 G20 **Léfini** ✍ SE Congo
Léfka see Lefke
117 C17 **Lefkáda** prev. Levkás. Lefkáda, Iónioi Nísoi, Greece, C Mediterranean Sea 38.50N 20.43E
117 B17 **Lefkáda** It. Santa Maura; prev. Levkás, anc. Leucas. island Iónioi Nísoi, Greece, C Mediterranean Sea
117 H25 **Lefká Óri** ▲ Kríti, Greece, E Mediterranean Sea
124 N3 **Lefke** Gk. Léfka. W Cyprus 35.06N 32.52E
117 B16 **Lefkímmi** var. Levkímmi. Kérkyra, Iónioi Nísoi, Greece, C Mediterranean Sea 39.25N 20.03E
Lefkosía/Lefkosa see Nicosia
80 A3 **Le Kef** see El Kef
106 G11 **Lékéti, Monts de la** ▲ S Congo
Lekhainá see Lechainá
116 H8 **Lekhchevo** Montana, NW Bulgaria 43.32N 23.31E
94 I4 **Lekhovo** see Nicosia
81 E20 **Lekoui** ◆ province SW Congo
96 L3 **Leksand** Dalarna, C Sweden 60.44N 14.50E
188 L12 **Leksula** see Leksula
128 H8 **Leksozero, Ozero** ⊚ NW Russian Federation
107 N8 **Leganés** Madrid, C Spain 40.19N 3.46W
107 Q3 **Lekunberri** var. Lecumberri. Navarra, N Spain 43.00N 1.54W
179 T16 **Lelai, Tanjung** headland Pulau Halmahera, N Indonesia 1.31N 128.43E
112 M11 **Legionowo** Mazowieckie, C Poland 52.23N 20.53E
101 K24 **Léglise** Luxembourg, SE Belgium 49.48N 5.31E
108 G8 **Legnago** Lombardia, NE Italy 45.13N 11.18E
108 D7 **Legnano** Veneto, NE Italy 45.36N 8.54E
113 F14 **Legnica** Ger. Lignitz. Dolnośląskie, SW Poland 51.12N 16.11E
105 U15 **Le Cannet** Alpes-Maritimes, SE France 43.33N 7.00E

Le Léman see Geneva, Lake
27 Q9 **Lelia Lake** Texas, SW USA 34.52N 100.42W
115 I14 **Lelija** ▲ SE Bosnia and Herzegovina 43.25N 18.31E
110 C8 **Le Locle** Neuchâtel, W Switzerland 47.04N 6.45E
201 T13 **Lelu** Kosrae, E Micronesia
201 T13 **Lelu** see Lelu Island
201 V14 **Lelu Island** var. Lelu. island Kosrae, E Micronesia
57 W9 **Lelydorp** Wanica, N Suriname 5.36N 55.04W
100 K9 **Lelystad** Flevoland, C Netherlands 52.30N 5.25E
65 K25 **Le Maire, Estrecho de** strait S Argentina
174 Hh7 **Lemang** Pulau Rangsang, W Indonesia 1.04N 102.44E
195 R11 **Lemankoa** Buka Island, NE PNG 5.04S 154.37E
Léman, Lac see Geneva, Lake
104 L6 **Le Mans** Sarthe, NW France 48.00N 0.12E
31 S12 **Le Mars** Iowa, C USA 42.47N 96.10W
174 I11 **Lematan, Air** ✍ Sumatera, W Indonesia
111 S3 **Lembach Im Mühlkreis** Oberösterreich, N Austria 48.28N 13.53E
103 G23 **Lemberg** ▲ SW Germany 48.09N 8.47E
Lemberg see L'viv
Lemdiyya see Médéa
124 Qq12 **Lemesós** var. Limassol. SW Cyprus 34.40N 33.02E
102 H13 **Lemgo** Nordrhein-Westfalen, W Germany 52.01N 8.54E
35 P5 **Lemhi Range** ▲ Idaho, NW USA
10 Oo2 **Lemieux Islands** island group Nunavut, NE Canada
175 Q7 **Lemito** Sulawesi, N Indonesia 0.34N 121.31E
94 L10 **Lemmenjoki** Lapp. Leammi. ✍ NE Finland
100 L7 **Lemmer** Fris. De Lemmer. Friesland, N Netherlands 52.49N 5.43E
30 L7 **Lemmon** South Dakota, N USA 45.54N 102.09W
38 M15 **Lemmon, Mount** ▲ Arizona, SW USA 32.26N 110.47W
Lemnos see Límnos
33 O14 **Lemoore** California, W USA 36.16N 119.48W
104 J5 **le Mont St-Michel** castle Manche, N France 48.37N 1.31N
37 Q11 **Lemoore** California, W USA 36.16N 119.48W
201 T13 **Lemotol Bay** bay Chuuk Islands, C Micronesia
47 Y5 **le Moule** var. Moule. Grande Terre, NE Guadeloupe 16.20N 61.20W
Lemovices see Limoges
Le Moyen-Ogooué see Moyen-Ogooué
10 M6 **le Moyne, Lac** ⊚ Quebec, E Canada
95 L18 **Lempäälä** Länsi-Suomi, W Finland 61.33N 23.46E
44 E7 **Lempa, Río** ✍ Central America
44 F7 **Lempira** prev. Gracias. ◆ department SW Honduras
Lemsalu see Limbaži
129 V6 **Lemva** ✍ NW Russian Federation
129 N17 **Le Murge** ▲ SE Italy
97 F21 **Lemvig** Ringkøbing, W Denmark 56.31N 8.19E
177 F8 **Lemyethna** Irrawaddy, SW Myanmar 17.36N 95.07E
32 K10 **Lena** Illinois, N USA 42.22N 89.49W
133 V4 **Lena** ✍ NE Russian Federation
181 N13 **Lena Tablemount** undersea feature S Pacific Ocean 51.06S 56.54E
Lenchitsa see Łęczyca
61 N17 **Lençóis** Bahia, E Brazil 12.36S 41.24W
62 K9 **Lençóis Paulista** São Paulo, S Brazil 22.35S 48.51W
111 Y9 **Lendava** Hung. Lendva, Ger. Unterlimbach; prev. Dolnja Lendava. NE Slovenia 46.33N 16.27E
85 F20 **Lendeba** Hardap, SE Namibia 24.41S 19.58E
128 H9 **Lendery** Respublika Kareliya, NW Russian Federation 63.20N 31.18E
29 R4 **Lenexa** Kansas, C USA 38.57N 94.43W
111 Q5 **Lengau** Oberösterreich, N Austria 48.01N 13.17E
165 Q17 **Lenger** Yuzhnyy Kazakhstan, S Kazakhstan 42.10N 69.54E
165 O9 **Lenghu** Qinghai, C China 38.50N 93.25E
165 T9 **Lenglong Ling** ▲ N China 37.40N 102.13E
110 D7 **Lengnau** Bern, N Switzerland 47.11N 7.22E
166 M12 **Lengshuijiang** Hunan, S China 26.31N 111.38E
97 M20 **Lenhovda** Kronoberg, S Sweden 57.00N 15.16E
81 E20 **Le Niari** ◆ province SW Congo
Lenin see Akdepe, Turkmenistan
Lenin see Leningradskiy
Leninakan see Gyumri
119 V12 **Lenina, Pik** ✍ C Asia
Lenine Rus. Lenino. Respublika Krym, S Ukraine 45.18N 35.47E
Leningor see Leninogorsk
153 Q13 **Leningrad** Rus. Leningradskiy; prev. Mŭ'minobod, Rus. Muminabad. SW Tajikistan 38.03N 69.50E
Leningrad see Sankt-Peterburg
130 L13 **Leningradskaya** Krasnodarskiy Kray, SW Russian Federation 46.19N 39.23E
205 S16 **Leningradskaya** Russian research station Antarctica 69.30S 159.51E
128 H12 **Leningradskaya Oblast'** ◆ province NW Russian Federation
Leningradskiy see Leningrad, Tajikistan
Leningradskiy see Lenine, Ukraine
Lenino see Lyenina, Belarus

◆ COUNTRY ● COUNTRY CAPITAL ◇ DEPENDENT TERRITORY ○ DEPENDENT TERRITORY CAPITAL ◈ ADMINISTRATIVE REGION ✕ INTERNATIONAL AIRPORT ▲ MOUNTAIN ▲ MOUNTAIN RANGE ▼ VOLCANO ✍ RIVER ⊚ LAKE ⊠ RESERVOIR

283

151 X9 **Leninobod** see Khŭjand
Leninogorsk Kaz. Leningor. Vostochnyy Kazakhstan, E Kazakhstan 50.20N 83.33E
131 T5 **Leninogorsk** Respublika Tatarstan, W Russian Federation 54.34N 52.27E
153 T12 **Lenin Peak** Rus. Pik Lenina, Taj. Qullai Lenin. ▲ Kyrgyzstan/Tajikistan 39.20N 72.50E
153 S8 **Leninpol'** Talasskaya Oblast', NW Kyrgyzstan 42.29N 71.54E
Lenin, Qullai see Lenin Peak
131 P11 **Leninsk** Volgogradskaya Oblast', SW Russian Federation 48.41N 45.18E
Leninsk see Akdepe, Turkmenistan
Leninsk see Asaka, Uzbekistan
Leninsk see Baykonyr, Kazakhstan
151 T8 **Leninskiy** Pavlodar, E Kazakhstan 52.18N 76.48E
126 H14 **Leninsk-Kuznetskiy** Kemerovskaya Oblast', S Russian Federation 54.42N 86.16E
151 N7 **Leninskoye** Kaz. Lenin. Kostanay, N Kazakhstan 54.04N 65.22E
129 P15 **Leninskoye** Kirovskaya Oblast', NW Russian Federation 58.19N 47.03E
Leninsk-Turkmenski see Chardzhev
Leninváros see Tiszaújváros
Lenkoran' see Länkäran
103 F15 **Lenne** ☑ W Germany
103 G16 **Lennestadt** Nordrhein-Westfalen, W Germany 51.07N 8.04E
31 R11 **Lennox** South Dakota, N USA 43.21N 96.53W
65 J25 **Lennox, Isla** Eng. Lennox Island. island S Chile
Lennox Island see Lennox, Isla
23 O9 **Lenoir** North Carolina, SE USA 35.54N 81.32W
22 M9 **Lenoir City** Tennessee, S USA 35.48N 84.15W
110 C7 **Le Noirmont** Jura, NW Switzerland 47.14N 6.57E
12 L9 **Lenôtre, Lac** ◎ Quebec, SE Canada
31 U15 **Lenox** Iowa, C USA 40.52N 94.33W
105 O2 **Lens** anc. Lendum, Lentium. Pas-de-Calais, N France 50.25N 2.49E
126 Kk12 **Lensk** Respublika Sakha (Yakutiya), NE Russian Federation 60.43N 115.16E
113 F24 **Lenti** S Hungary 46.38N 16.30E
Lentia see Linz
95 N14 **Lentiira** Oulu, E Finland 64.22N 29.52E
109 L25 **Lentini** anc. Leontini. Sicilia, Italy, C Mediterranean Sea 37.17N 15.00E
Lentium see Lens
Lentschiza see Łęczyca
95 N15 **Lentua** ◎ E Finland
113 H14 **Lentvaris** Pol. Landwarów. Trakai, SE Lithuania 24.39N 24.58E
110 F7 **Lenzburg** Aargau, N Switzerland 47.24N 8.09E
111 R5 **Lenzing** Oberösterreich, N Austria 47.58N 13.14E
79 P13 **Léo** SW Burkina 11.09N 2.04W
111 V7 **Leoben** Steiermark, C Austria 47.22N 15.06E
Leobschütz see Głubczyce
46 L9 **Léogáne** N Haiti 18.28N 72.39W
175 Q7 **Leok** Sulawesi, N Indonesia 1.10N 121.20E
31 O7 **Leola** South Dakota, N USA 45.41N 98.58W
99 K20 **Leominster** W England, UK 52.09N 2.18W
21 N11 **Leominster** Massachusetts, NE USA 42.29N 71.43W
31 V16 **Leon** Iowa, C USA 40.44N 93.45W
42 M12 **León** var. León de los Aldamas. Guanajuato, C Mexico 21.05N 101.43W
44 I10 **León** León, NW Nicaragua 12.24N 86.52W
106 L4 **León** Castilla-León, NW Spain 42.34N 5.33W
44 I9 **León** ◆ department W Nicaragua
106 K4 **León** ◆ province Castilla-León, NW Spain
León see Cotopaxi
104 I15 **Léon** Landes, SW France 43.54N 1.17W
27 V9 **Leona** Texas, SW USA 31.09N 95.58W
188 K11 **Leonora** Western Australia 28.52S 121.16E
27 U10 **Leonard** Texas, SW USA 33.22N 96.15W
Leonard Murray Mountains see Murray Range
109 H15 **Leonardo da Vinci** prev. Fiumicino. ✈ (Roma) Lazio, C Italy 41.48N 12.13E
23 X5 **Leonardtown** Maryland, NE USA 38.17N 76.35W
27 Q13 **Leona River** ☑ Texas, SW USA
43 Z11 **Leona Vicario** Quintana Roo, SE Mexico 20.57N 87.06W
103 H21 **Leonberg** Baden-Württemberg, SW Germany 48.48N 9.01E
64 M3 **León, Cerro** ▲ NW Paraguay 20.21S 60.16W
León de los Aldamas see León
111 T4 **Leonding** Oberösterreich, N Austria 48.17N 14.15E
109 I14 **Leonessa** Lazio, C Italy 42.36N 12.56E
109 K24 **Leonforte** Sicilia, Italy, C Mediterranean Sea 37.37N 14.22E
191 O13 **Leongatha** Victoria, SE Australia 38.30S 145.56E
117 F21 **Leonídi** Pelopónnisos, S Greece 37.10N 22.50E
106 L4 **Leónidas, Montes de** ▲ NW Spain
27 S8 **Leon River** ☑ Texas, SW USA
101 J17 **Leopoldsburg** Limburg, NE Belgium 51.07N 5.16E
Léopold II, Lac see Mai-Ndombe, Lac
Léopoldville see Kinshasa

28 I5 **Leoti** Kansas, C USA 38.28N 101.21W
118 M11 **Leova** Rus. Leovo. SW Moldova 46.31N 28.16E
Leovo see Leova
104 G8 **le Palais** Morbihan, NW France 47.20N 3.08W
29 X10 **Lepanto** Arkansas, C USA 35.34N 90.21W
174 J11 **Lepar, Pulau** island W Indonesia
106 I14 **Lepe** Andalucía, S Spain 37.15N 7.12W
Lepel' see Lyepyel'
85 I19 **Lephepe** Kweneng, SE Botswana 23.15S 25.48E
167 Q10 **Leping** Jiangxi, S China 29.01N 117.07E
Lépontiennes, Alpes/Lepontine, Alpi see Lepontine Alps
110 G10 **Lepontine Alps** Fr. Alpes Lépontiennes, It. Alpi Lepontine. ▲ Italy/Switzerland
81 Q8 **le Port** NW Réunion
181 O16 **le Port** NW Réunion
105 N1 **le Portel** Pas-de-Calais, N France 50.42N 1.34E
95 N17 **Leppävirta** Itä-Suomi, C Finland 62.30N 27.49E
47 Q11 **le Prêcheur** NW Martinique 14.48N 61.13W
Lepsi see Lepsy
151 V13 **Lepsy** Kaz. Lepsi. Almaty, SE Kazakhstan 46.13N 78.55E
151 V13 **Lepsy** Kaz. Lepsi. ☑ SE Kazakhstan
Le Puglie see Puglia
105 Q12 **Le Puy** prev. le Puy-en-Velay, hist. Anicium, Podium Anicensis. Haute-Loire, C France 45.03N 3.52E
Le Puy-en-Velay see le Puy
47 X1 **Le Raizet** var. Le Raizet. ✈ (Pointe-à-Pitre) Grande Terre, C Guadeloupe 16.16N 61.31W
109 J24 **Lercara Friddi** Sicilia, Italy, C Mediterranean Sea 37.45N 13.21E
80 G12 **Léré** Mayo-Kébbi, SW Chad 9.40N 14.16E
Leribe see Hlotse
108 E10 **Lerici** Liguria, NW Italy 46.06N 9.53E
56 I14 **Lérida** Vaupés, S Colombia 0.01S 70.28W
107 N5 **Lerma** Castilla-León, N Spain 42.01N 3.46W
42 M13 **Lerma, Río** ☑ C Mexico
117 F20 **Lérna** prehistoric site Pelopónnisos, S Greece 37.31N 22.43E
47 R11 **le Robert** E Martinique 14.40N 60.56W
117 M21 **Léros** island Dodekánisos, Greece, Aegean Sea
32 L13 **Le Roy** Illinois, N USA 40.21N 88.45W
29 Q6 **Le Roy** Kansas, C USA 38.04N 95.37W
31 W11 **Le Roy** Minnesota, N USA 43.30N 92.30W
20 L8 **Le Roy** New York, NE USA 42.58N 77.58W
Lerrnayin Gharabakh see Nagornyy Karabakh
97 J19 **Lerum** Västra Götaland, S Sweden 57.46N 12.12E
98 M2 **Lerwick** NE Scotland, UK 60.09N 1.09W
47 Y6 **les Abymes** var. Abymes. Grande Terre, C Guadeloupe 16.16N 61.30W
les Albères, Chaine des
104 M4 **les Andelys** Eure, N France 49.15N 1.27E
47 Q12 **les Anses-d'Arlets** SW Martinique 14.29N 61.05W
107 U6 **Les Borges Blanques** var. Borjas Blancas. Cataluña, NE Spain 41.31N 0.52E
Lesbos see Lésvos
55 Y6 **Les Cayes** see Cayes
33 Q4 **Les Cheneaux Islands** island Michigan, N USA
105 T12 **Les Écrins** ▲ E France 44.54N 6.25E
110 C10 **Le Sépey** Vaud, W Switzerland 46.21N 7.04E
13 T7 **Les Escoumins** Quebec, SE Canada 48.21N 69.25W
Les Gonaïves see Gonaïves
166 H9 **Leshan** Sichuan, C China 29.42N 103.43E
110 D11 **Les Haudères** Valais, SW Switzerland 46.02N 7.27E
104 J9 **les Herbiers** Vendée, NW France 46.52N 1.01W
129 O8 **Leshukonskoye** Arkhangel'skaya Oblast', NW Russian Federation 64.54N 45.48E
Lesina see Hvar
109 M15 **Lesina, Lago di** ◎ SE Italy
96 G10 **Lesja** Oppland, S Norway 62.07N 8.56E
97 L15 **Lesjöfors** Värmland, C Sweden 59.57N 14.12E
113 O18 **Lesko** Podkarpackie, SE Poland 49.28N 22.19E
115 O15 **Leskovac** Serbia, SE Yugoslavia 43.00N 21.56E
115 M22 **Leskoviku** var. Leskovik. Korçë, S Albania 40.09N 20.39E
Leskovik see Leskoviku
38 L4 **Leslie** Idaho, NW USA 43.51N 113.28W
33 Q10 **Leslie** Michigan, N USA 42.27N 84.25W
Leśna/Lesnaya see Lyasnaya
104 F5 **Lesneven** Finistère, NW France 48.35N 4.19W
113 I14 **Lešnica** Serbia, W Yugoslavia 44.40N 19.18E
131 Q17 **Lesnoy** Kirovskaya Oblast', NW Russian Federation 59.49N 52.07E
26 M5 **Lesnoy** Respublika Dagestan, SW Russian Federation
127 Nn17 **Lesozavodsk** Primorskiy Kray, SE Russian Federation 45.23N 133.15E

104 J12 **Lesparre-Médoc** Gironde, SW France 45.18N 0.57W
110 C8 **Les Ponts-de-Martel** Neuchâtel, W Switzerland 47.00N 6.45E
104 I9 **les Sables-d'Olonne** Vendée, NW France 46.30N 1.46W
105 P1 **Lesquin** ✈ Nord, N France 50.34N 3.07E
111 S7 **Lessach** var. Lessachbach. ☑ E Austria
Lessachbach see Lessach
47 W11 **les Saintes** var. Iles des Saintes. island group S Guadeloupe
76 L5 **Les Salines** ✈ (Annaba) NE Algeria 36.45N 7.57E
97 M21 **Lessebo** Kronoberg, S Sweden 56.45N 15.19E
204 M10 **Lesser Antarctica** var. West Antarctica. physical region Antarctica
47 P15 **Lesser Antilles** island group W Indies
143 T4 **Lesser Caucasus** Rus. Malyy Kavkaz. ▲ SW Asia
Lesser Khingan Range see Xiao Hinggan Ling
9 **Lesser Slave Lake** ◎ Alberta, W Canada
Lesser Sunda Islands see Nusa Tenggara
101 E19 **Lessines** Hainaut, SW Belgium 50.43N 3.49E
105 R16 **les Stes-Maries-de-la-Mer** Bouches-du-Rhône, SE France 43.27N 4.26E
12 **Lester B. Pearson** var. Toronto. ✈ (Toronto) Ontario, S Canada 43.59N 81.30W
31 U9 **Lester Prairie** Minnesota, N USA 44.52N 94.02W
95 L16 **Lestijärvi** Länsi-Suomi, W Finland 63.29N 24.41E
31 U9 **Le Sueur** Minnesota, N USA 44.27N 93.53W
110 B8 **Les Verrières** Neuchâtel, W Switzerland 46.54N 6.29E
117 L17 **Lésvos** anc. Lesbos. island E Greece
112 G12 **Leszno** Ger. Lissa. Wielkopolskie, C Poland 51.51N 16.34E
181 P17 **le Tampon** S Réunion
99 O21 **Letchworth** E England, UK 51.58N 0.13W
113 G25 **Letenye** Zala, SW Hungary 46.25N 16.42E
57 S11 **Lethem** S Guyana 3.24N 59.45W
85 H18 **Letiahau** ☑ W Botswana
56 J18 **Leticia** Amazonas, S Colombia 4.09S 69.57W
175 T15 **Leti, Kepulauan** island group E Indonesia
85 I18 **Letlhakane** Central, C Botswana 21.28S 25.39E
85 H20 **Letlhakeng** Kweneng, SE Botswana 24.04S 25.03E
116 J8 **Letnitsa** Lovech, N Bulgaria 43.19N 25.02E
105 N1 **le Touquet-Paris-Plage** Pas-de-Calais, N France 50.31N 1.34E
177 G8 **Letpadan** Pegu, SW Myanmar 17.22N 94.10E
177 Fj6 **Letpan** Arakan State, W Myanmar 19.22N 94.11E
104 M2 **le Tréport** Seine-Maritime, N France 50.03N 1.21E
178 Gg13 **Letsôk-aw Kyun** var. Letsutan Island; prev. Domel Island. island Mergui Archipelago, S Myanmar
Letsutan Island see Letsôk-aw Kyun
99 E14 **Letterkenny** Ir. Leitir Ceanainn. NW Ireland 54.57N 7.43W
Lettland see Latvia
118 M4 **Letychiv** Khmel'nyts'ka Oblast', W Ukraine 49.24N 27.39E
118 H14 **Lëtzebuerg** see Luxembourg
Letzlingen see Lefkáda
105 P17 **Leucate** Aude, S France 42.55N 3.03E
105 P17 **Leucate, Étang de** ◎ S France
110 E10 **Leuk** Valais, SW Switzerland 46.18N 7.46E
110 E10 **Leukerbad** Valais, SW Switzerland 46.02N 7.47E
Leusden see Leusden-Centrum
100 K11 **Leusden-Centrum** var. Leusden. Utrecht, C Netherlands 52.07N 5.25E
Leutensdorf see Litvínov
Leuthen see Lutynia
101 H18 **Leuven** Fr. Louvain, Ger. Löwen. Vlaams Brabant, C Belgium 50.52N 4.42E
120 I20 **Leuze** Namur, C Belgium 50.33N 4.55E
Leuze see Leuze-en-Hainaut
115 K18 **Leuze-en-Hainaut** var. Leuze. Hainaut, SW Belgium 50.36N 3.37E
Léva see Levice
Levádhia see Leivádia
38 L4 **Levan** Utah, W USA 39.33N 111.52W
95 E16 **Levanger** Nord-Trøndelag, C Norway 63.51N 11.18E
124 P14 **Levantine Basin** undersea feature E Mediterranean Sea
108 D10 **Levanto** Liguria, NW Italy 44.12N 9.33E
109 I24 **Levanzo, Isola di** island Isole Egadi, S Italy
131 Q17 **Levashi** Respublika Dagestan, SW Russian Federation 42.27N 47.19E
26 M5 **Levelland** Texas, SW USA 33.35N 102.23W
41 P13 **Levelock** Alaska, USA 59.07N 156.51W
103 E16 **Leverkusen** Nordrhein-Westfalen, W Germany 51.01N 7.00E
113 I22 **Levice** Ger. Lewentz, Lewenz, Hung. Léva. Nitriansky Kraj, SW Slovakia 48.13N 18.37E

108 G6 **Levico Terme** Trentino-Alto Adige, N Italy 46.02N 11.19E
117 E20 **Levídi** Pelopónnisos, S Greece 37.39N 22.13E
105 P14 **le Vigan** Gard, S France 43.00N 3.36E
13 R10 **Lévis** var. Levis. Quebec, SE Canada 46.46N 71.10W
23 P6 **Levisa Fork** ☑ Kentucky/Virginia, S USA
117 L21 **Levítha** island Kykládes, Greece, Aegean Sea
20 L14 **Levittown** Long Island, New York, NE USA 40.42N 73.29W
20 J15 **Levittown** Pennsylvania, NE USA 40.09N 74.50W
Levkás see Lefkáda
Levká Óri see Lefká Óri
Levkímmi see Lefkímmi
Lévkosia/Levkosía see Nicosia
113 L19 **Levoča** Ger. Leutschau, Hung. Locse. Prešovský Kraj, E Slovakia 49.01N 20.34E
166 M11 **Levroux** Indre, C France 47.00N 1.37E
116 J8 **Levski** Pleven, N Bulgaria 43.22N 25.10E
130 L6 **Lev Tolstoy** Lipetskaya Oblast', W Russian Federation 53.12N 39.28E
197 I14 **Levuka** Ovalau, C Fiji 17.42S 178.49E
177 G6 **Lewe** Mandalay, C Myanmar 19.40N 96.04E
Lewentz/Lewenz see Levice
99 O19 **Lewes** SE England, UK 50.52N 0.01E
21 Z4 **Lewes** Delaware, NE USA 38.46N 75.08W
31 Q3 **Lewis And Clark Lake** ◎ Nebraska/South Dakota, N USA
20 G14 **Lewisburg** Pennsylvania, NE USA 40.57N 76.52W
22 J10 **Lewisburg** Tennessee, S USA 35.27N 86.47W
23 S6 **Lewisburg** West Virginia, NE USA 37.48N 80.27W
98 F6 **Lewis, Butt of** headland N Scotland, UK 58.31N 6.18W
98 F7 **Lewis, Isle of** island NW Scotland, UK
37 U4 **Lewis, Mount** ▲ Nevada, W USA 40.22N 116.50W
193 H16 **Lewis Pass** pass South Island, NZ 42.23S 172.21E
35 P7 **Lewis Range** ▲ Montana, NW USA
25 O3 **Lewis Smith Lake** ◎ Alabama, S USA
34 M10 **Lewiston** Idaho, NW USA 46.25N 117.01W
21 P7 **Lewiston** Maine, NE USA 44.07N 70.13W
31 X10 **Lewiston** Minnesota, N USA 43.58N 91.52W
20 D9 **Lewiston** New York, NE USA 43.10N 79.02W
38 L1 **Lewiston** Utah, W USA 41.58N 111.52W
32 K13 **Lewistown** Illinois, N USA 40.23N 90.09W
35 T9 **Lewistown** Montana, NW USA 47.04N 109.25W
29 T4 **Lewisville** Arkansas, C USA 33.21N 93.34W
27 T2 **Lewisville** Texas, SW USA 33.06N 96.57W
27 T6 **Lewisville, Lake** ◎ Texas, SW USA
25 U3 **Lexington** Georgia, SE USA 33.51N 83.04W
22 M5 **Lexington** Kentucky, S USA 38.03N 84.30W
29 S3 **Lexington** Mississippi, S USA 33.06N 90.03W
29 S4 **Lexington** Missouri, C USA 39.10N 93.52W
31 N16 **Lexington** Nebraska, C USA 40.46N 99.44W
22 Q9 **Lexington** North Carolina, SE USA 35.49N 80.15W
29 N11 **Lexington** Oklahoma, C USA 35.00N 97.20W
22 G9 **Lexington** Tennessee, S USA 35.39N 88.23W
27 T10 **Lexington** Texas, SW USA 30.25N 97.00W
23 S6 **Lexington** Virginia, NE USA 37.47N 79.26W
23 X5 **Lexington Park** Maryland, NE USA 38.16N 76.27W
Leyden see Leiden
29 R4 **Leyre** ☑ SW France
179 R13 **Leyte** island C Philippines
179 R13 **Leyte Gulf** gulf E Philippines
113 O16 **Leżajsk** Podkarpackie, SE Poland 50.15N 22.24E
Lezha see Lezhë
115 K18 **Lezha** var. Lezha; prev. Lesh, Leshi. Lezhë, N Albania 41.46N 19.40E
115 K18 **Lezhë** ◆ district NW Albania
105 O16 **Lézignan-Corbières** Aude, S France 43.12N 2.46E
130 J7 **L'gov** Kurskaya Oblast', W Russian Federation 51.38N 35.17E
165 N13 **Lhari** Xizang Zizhiqu, W China 30.40N 93.40E
165 N16 **Lhasa** Xizang Zizhiqu, W China 29.40N 91.07E
165 N16 **Lhasa He** ☑ W China
164 K16 **Lhazê** var. Quxar. Xizang Zizhiqu, W China 29.07N 87.32E
165 J8 **Lhazhong** Xizang Zizhiqu, W China 31.58N 86.43E
173 F3 **Lhoksukon** Sumatera, W Indonesia 5.04N 97.19E
165 Q15 **Lhorong** Xizang Zizhiqu, W China 30.51N 95.41E
159 R11 **Lhotse** ▲ China/Nepal 27.57N 156.51W
165 N17 **Lhozhag** Xizang Zizhiqu, W China 28.21N 90.47E
165 O16 **Lhünzê** Xizang Zizhiqu, W China 28.25N 92.30E

165 N15 **Lhünzhub** var. Poindo. Xizang Zizhiqu, W China 30.14N 91.20E
178 N8 **Li** Lamphun, NW Thailand 17.46N 98.54E
179 Rr14 **Lianga** Mindanao, S Philippines 8.36N 126.04E
167 P22 **Liangcheng** Fujian, SE China 25.47N 116.42E
166 K9 **Liangping** Chongqing Shi, C China 30.40N 107.46E
167 O9 **Liangzi Hu** ◎ C China
Liangzhou see Wuwei
166 I14 **Lianjiang** Fujian, SE China 26.12N 119.33E
166 L15 **Lianjiang** Guangdong, S China 21.37N 110.18E
167 O13 **Lianping** Guangdong, S China 24.22N 114.23E
167 T13 **Lianshan** prev. Jinxi. Liaoning, NE China 40.42N 120.52E
166 M11 **Lianyuan** prev. Lantian. Hunan, S China 27.51N 111.44E
167 Q6 **Lianyungang** var. Xinpu. Jiangsu, E China 34.37N 119.12E
167 N13 **Lianzhou** var. Linxian; prev. Lian Xian. Guangdong, S China 24.48N 112.20E
Liao see Liaoning
167 P5 **Liaocheng** Shandong, E China 36.31N 115.59E
169 U13 **Liaodong Bandao** var. Liaotung Peninsula. peninsula NE China
169 U13 **Liaodong Wan** Eng. Gulf of Lantung, Gulf of Liaotung. gulf NE China
169 U11 **Liao He** ☑ NE China
169 W12 **Liao Ling** ▲ NE China
169 U12 **Liaoning** var. Liao, Liaoning Sheng, Shengking; hist. Fengtien, Shenking. ◆ province NE China
Liaoning Sheng see Liaoning
Liaotung, Gulf of see Liaodong Wan
Liaotung Peninsula see Liaodong Bandao
169 V12 **Liaoyang** var. Liao-yang. Liaoning, NE China 41.16N 123.12E
169 V11 **Liaoyuan** var. Dongliao, Shuang-liao, Jap. Chengchiatun. Jilin, NE China 42.51N 125.10E
169 U12 **Liaozhong** Liaoning, NE China 41.33N 122.54E
Liaqatabad see Piplän
8 M10 **Liard** ☑ W Canada
Liard see Fort Liard
8 L10 **Liard River** British Columbia, W Canada 59.22N 126.04W
155 O15 **Liäri** Baluchistän, SW Pakistan 25.43N 66.28E
Liatroim see Leitrim
201 S6 **Lib** var. Ellep. island Ralik Chain, C Marshall Islands
144 H6 **Liban** see Lebanon
144 H6 **Liban, Jebel** Ar. Jabal al Gharbī, Jabal Lubnän, Eng. Mount Lebanon. ▲ C Lebanon
Libau see Liepāja
35 N7 **Libby** Montana, NW USA 48.25N 115.33W
81 I18 **Libenge** Equateur, NW Dem. Rep. Congo (Zaire) 3.39N 18.39E
28 I7 **Liberal** Kansas, C USA 37.01N 100.55W
29 R7 **Liberal** Missouri, C USA 37.33N 94.31W
Liberalitas Julia see Évora
113 D15 **Liberec** Ger. Reichenberg. Liberecký Kraj, N Czech Republic 50.44N 15.04E
113 D15 **Liberecký Kraj** ◆ region N Czech Republic
44 K12 **Liberia** Guanacaste, NW Costa Rica 10.36N 85.26W
78 K17 **Liberia** off. Republic of Liberia. ◆ republic W Africa
63 D16 **Libertad** Corrientes, NE Argentina 30.01S 57.51W
56 I7 **Libertad** San José, S Uruguay 34.37S 56.39W
56 I7 **Libertad** Barinas, NW Venezuela 8.21N 69.39W
56 K6 **Libertad** Cojedes, N Venezuela 9.19N 68.43W
64 G12 **Libertador** off. Región del Libertador General Bernardo O'Higgins. ◆ region C Chile
Libertador General San Martín see Ciudad de Libertador General San Martín
22 L6 **Liberty** Kentucky, S USA 37.19N 84.54W
24 J7 **Liberty** Mississippi, S USA 31.09N 90.49W
29 R4 **Liberty** Missouri, C USA 39.14N 94.22W
20 J12 **Liberty** New York, NE USA 41.48N 74.45W
23 R9 **Liberty** North Carolina, SE USA 35.49N 79.34W
99 E14 **Lifford** Ir. Leifear. NW Ireland 54.49N 7.34W
101 J23 **Libian Desert** see Libyan Desert
115 K18 **Libiaz** Małopolskie, S Poland 50.10N 19.25E
101 J23 **Libin** Luxembourg, SE Belgium 50.00N 5.13E
166 K13 **Libo** Guizhou, S China 25.28N 107.52E
115 L20 **Libohova** see Libohovë
115 L20 **Libohovë** var. Libohova. Gjirokastër, S Albania 40.03N 20.13E
104 K13 **Libourne** Gironde, SW France 44.55N 0.13W
101 L21 **Libramont** Luxembourg, SE Belgium 49.55N 5.21E
115 M20 **Librazhd** var. Librazhdi. Elbasan, E Albania 41.10N 20.22E
81 C18 **Libreville** ● (Gabon) Estuaire, NW Gabon 0.23N 9.27E
179 Rr15 **Libuganon** ☑ Mindanao, S Philippines 10.51S 37.10E
77 P10 **Libya** off. Socialist People's Libyan Arab Jamahiriya, Ar. Al Jamāhīrīyah al 'Arabīyah al Lībīyah ash Sha'bīyah al Ishtirākīyah; prev. Libyan Arab Republic. ◆ Islamic state N Africa
77 T11 **Libyan Desert** var. Libian Desert, Ar. Aṣ Ṣaḥrā' al Lībīyah. desert N Africa

77 T8 **Libyan Plateau** var. Aḍ Ḍiffah. plateau Egypt/Libya
Libyah, Aṣ Ṣaḥrā' al see Libyan Desert
64 G12 **Licantén** Maule, C Chile 35.00S 72.00W
109 J25 **Licata** anc. Phintias. Sicilia, Italy, C Mediterranean Sea 37.07N 13.56E
143 P14 **Lice** Diyarbakır, SE Turkey 38.28N 40.39E
99 L19 **Lichfield** C England, UK 52.42N 1.48W
85 N14 **Lichinga** Niassa, N Mozambique 13.19S 35.15E
111 V3 **Lichtenau** Niederösterreich, N Austria 48.29N 15.24E
85 I21 **Lichtenburg** North-West, N South Africa 26.06S 26.08E
103 K18 **Lichtenfels** Bayern, SE Germany 50.09N 11.03E
100 O12 **Lichtenvoorde** Gelderland, E Netherlands 51.58N 6.34E
101 C17 **Lichtervelde** West-Vlaanderen, W Belgium 51.01N 3.09E
166 L9 **Lichuan** Hubei, C China 30.19N 108.55E
29 V7 **Licking** Missouri, C USA 37.30N 91.51W
22 M4 **Licking River** ☑ Kentucky, S USA
114 C11 **Lički Osik** Lika-Senj, C Croatia 44.36N 15.24E
114 C11 **Ličko-Senjska Županija** ◆ province W Croatia
81 N25 **Licosa, Punta** headland S Italy
121 H16 **Lida** Rus. Lida. Hrodzyenskaya Voblasts', W Belarus 53.53N 25.19E
95 H17 **Liden** Västernorrland, C Sweden 62.43N 16.49E
31 R7 **Lidgerwood** North Dakota, N USA 46.04N 97.09W
105 X14 **l'Île-Rousse** Corse, France, C Mediterranean Sea 42.39N 8.59E
Lidhorikíon see Lidoríki
167 N11 **Liling** Hunan, S China 27.42N 113.49E
97 K17 **Lidköping** Västra Götaland, S Sweden 58.30N 13.10E
108 I8 **Lido di Jesolo** var. Lido di Jesolo. Veneto, NE Italy 45.30N 12.37E
109 H15 **Lido di Ostia** Lazio, C Italy 41.42N 12.19E
117 E18 **Lidoríki** prev. Lidhorikíon, Lidokhorikon. Stereá Ellás, C Greece 38.31N 22.12E
96 H12 **Lidzbark** Warmińsko-Mazurskie, NE Poland 53.15N 19.49E
112 L7 **Lidzbark Warmiński** Ger. Heilsberg. Warmińsko-Mazurskie, NE Poland 54.07N 20.34E
111 U3 **Liebenau** Oberösterreich, N Austria 48.33N 14.48E
189 P7 **Liebig, Mount** ▲ Northern Territory, C Australia 23.19S 131.30E
111 V8 **Lieboch** Steiermark, SE Austria 47.00N 15.21E
110 J8 **Liechtenstein** off. Principality of Liechtenstein. ◆ principality C Europe
101 F18 **Liedekerke** Vlaams Brabant, C Belgium 50.51N 4.05E
101 K19 **Liège** Dut. Luik, Ger. Lüttich. Liège, E Belgium 50.37N 5.34E
101 K20 **Liège** Dut. Luik. ◆ province E Belgium
Liegnitz see Legnica
95 O16 **Lieksa** Itä-Suomi, E Finland 63.20N 30.00E
120 F10 **Lielupe** ☑ Latvia/Lithuania
120 G9 **Lielvārde** Ogre, C Latvia 56.45N 24.48E
178 Kk14 **Liên Hương** var. Tuy Phong. Bình Thuận, S Vietnam 11.13N 108.40E
111 P9 **Lienz** Tirol, W Austria 46.49N 12.45E
120 B10 **Liepāja** Ger. Libau. Liepāja, W Latvia 56.31N 21.02E
101 H17 **Lier** Fr. Lierre. Antwerpen, N Belgium 51.07N 4.34E
111 T6 **Liezen** Steiermark, SE Austria 47.34N 14.12E
99 E14 **Lifford** Ir. Leifear. NW Ireland 54.49N 7.34W
197 K5 **Lifou** var. Lifu. island Îles Loyauté, E New Caledonia
Lifu see Lifou
166 K13 **Ligao** Luzon, N Philippines 13.16N 123.30E
83 K18 **Ligoni** North Eastern, E Kenya 0.23N 40.55E
83 K18 **Liger** see Loire
91 N1 **Lighthouse Reef** reef E Belize
191 Q4 **Lightning Ridge** New South Wales, SE Australia 29.25S 148.00E
105 N9 **Lignières** Cher, C France 46.46N 2.12E
105 S5 **Ligny-en-Barrois** Meuse, NE France 48.42N 5.22E
85 P15 **Ligonha** ☑ NE Mozambique
33 I11 **Ligonier** Indiana, N USA 41.25N 85.33W
108 C9 **Ligure, Mar** see Ligurian Sea
108 C9 **Liguria** ◆ region NW Italy
Ligurian Mountains see Ligure, Appennino
108 C9 **Ligurian Sea** Fr. Mer Ligurienne, It. Mar Ligure. sea N Mediterranean Sea

195 P9 **Lihir Group** island group NE PNG
195 P9 **Lihir Island** island Lihir Group, N PNG
40 B8 **Lihue** Haw. Lihu'e. Kauai, Hawaii, USA, C Pacific Ocean 21.58N 159.22W
120 F5 **Lihula** Ger. Leal. Läänemaa, W Estonia 58.43N 23.52E
128 I2 **Liinakhamari** var. Linacmamari. Murmanskaya Oblast', NW Russian Federation 69.40N 31.27E
Liivi Laht see Riga, Gulf of
166 F11 **Lijiang** var. Dayan, Lijiang Naxizu Zizhixian. Yunnan, SW China 26.52N 100.10E
114 C11 **Lika-Senj** off. Ličko-Senjska Županija. ◆ province W Croatia
81 N25 **Likasi** prev. Jadotville. Katanga, SE Dem. Rep. Congo (Zaire) 11.01S 26.51E
81 L16 **Likati** Orientale, N Dem. Rep. Congo (Zaire) 3.28N 23.45E
8 M15 **Likely** British Columbia, SW Canada 52.40N 121.34W
159 Y11 **Likhapäni** Assam, NE India 27.24N 95.51E
128 J16 **Likhoslavl'** Tverskaya Oblast', W Russian Federation 57.08N 35.27E
201 U5 **Likiep Atoll** atoll Ratak Chain, C Marshall Islands
97 D18 **Liknes** Vest-Agder, S Norway 58.19N 6.58E
81 H18 **Likouala** ◆ province W Congo
81 H18 **Likouala aux Herbes** ☑ E Congo
202 B16 **Liku** E Niue 19.01S 169.46E
Likupang, Selat see Bangka, Selat
29 Y8 **Lilbourn** Missouri, C USA 36.35N 89.37W
105 X14 **l'Île-Rousse** Corse, France, C Mediterranean Sea 42.39N 8.59E
111 W5 **Lilienfeld** Niederösterreich, NE Austria 48.01N 15.36E
167 N11 **Liling** Hunan, S China 27.42N 113.49E
97 J18 **Lilla Edet** Västra Götaland, S Sweden 58.07N 12.08E
105 P1 **Lille** var. l'Isle, Dut. Rijssel, Flem. Ryssel; prev. Lisle, anc. Insula. Nord, N France 50.37N 3.04E
97 G24 **Lille Bælt** var. Lille Bælt, Eng. Little Belt. strait S Denmark
96 H12 **Lillehammer** Oppland, S Norway 61.07N 10.27E
105 O1 **Lillers** Pas-de-Calais, N France 50.34N 2.26E
97 E18 **Lillesand** Aust-Agder, S Norway 58.13N 8.22E
95 F17 **Lillestrøm** Akershus, S Norway 59.58N 11.04E
95 I18 **Lillhärdal** Jämtland, C Sweden 61.51N 14.04E
23 U10 **Lillington** North Carolina, SE USA 35.24N 78.49W
107 O9 **Lillo** Castilla-La Mancha, C Spain 39.43N 3.19W
8 M16 **Lillooet** British Columbia, SW Canada 50.40N 121.58W
85 M14 **Lilongwe** ● (Malawi) Central, W Malawi 13.58S 33.48E
85 M14 **Lilongwe** ✈ Central, W Malawi 13.46S 33.44E
85 M14 **Lilongwe** ☑ W Malawi
179 Q15 **Liloy** Mindanao, S Philippines 8.04N 122.42E
190 J7 **Lilydale** South Australia 32.57S 140.00E
191 P16 **Lilydale** Tasmania, SE Australia 41.15S 147.13E
115 J14 **Lim** ☑ Bosnia and Herzegovina/Yugoslavia
59 D15 **Lima** ● (Peru) Lima, W Peru 12.06S 77.03W
96 K13 **Lima** Dalarna, C Sweden 60.55N 13.19E
33 R12 **Lima** Ohio, NE USA 40.43N 84.06W
59 D14 **Lima** ◆ department W Peru
Lima see Jorge Chávez International
106 G5 **Lima, Rio** Sp. Limia. ☑ Portugal/Spain see also Limia
113 L17 **Limanowa** Małopolskie, S Poland 49.43N 20.25E
174 I4 **Limas** Pulau Sebangka, W Indonesia 0.09N 104.31E
99 F14 **Limavady** Ir. Léim an Mhadaidh. NW Northern Ireland, UK 55.03N 6.57W
65 J14 **Limay Mahuida** La Pampa, C Argentina 37.09S 66.40W
61 B16 **Limay, Río** ☑ W Argentina
103 N16 **Limbach-Oberfrohna** Sachsen, E Germany 50.52N 12.46E
83 F22 **Limba Limba** ☑ C Tanzania
109 C17 **Limbara, Monte** ▲ Sardegna, Italy, C Mediterranean Sea 40.52N 9.09E
46 M8 **Limbé** N Haiti 19.40N 72.25W
175 O7 **Limboto, Danau** ◎ Sulawesi, N Indonesia
101 L19 **Limbourg** Liège, E Belgium 50.37N 5.55E
101 K17 **Limburg** ◆ province NE Belgium
101 L16 **Limburg** ◆ province SE Netherlands
103 F17 **Limburg an der Lahn** Hessen, W Germany 50.22N 8.04E
96 H12 **Limedsforsen** Dalarna, C Sweden 60.52N 13.25E
62 I13 **Limeira** São Paulo, S Brazil 22.34S 47.25W
21 S2 **Limerick** Ir. Luimneach. SW Ireland 52.40N 8.37W
99 C21 **Limerick** Ir. Luimneach. cultural region SW Ireland
21 S2 **Limestone** Maine, NE USA 46.52N 67.49W
27 U9 **Limestone, Lake** ◎ Texas, SW USA
41 P12 **Lime Village** Alaska, USA 61.21N 155.28W
97 F20 **Limfjorden** fjord N Denmark
97 J23 **Limhamn** Skåne, S Sweden 55.34N 12.57E

◆ COUNTRY ◇ DEPENDENT TERRITORY ◆ ADMINISTRATIVE REGION ▲ MOUNTAIN ▲ VOLCANO ◎ LAKE
● COUNTRY CAPITAL ○ DEPENDENT TERRITORY CAPITAL ✈ INTERNATIONAL AIRPORT ▲ MOUNTAIN RANGE ☑ RIVER ◎ RESERVOIR

106 H5 **Limia** *Port.* Rio Lima ⩫ Portugal/ Spain *see also* Lima, Rio
95 L14 **Liminka** Oulu, C Finland *64.48N 25.19E*
Limín Vathéos *see* Sámos
117 G17 **Límni** Évvoia, C Greece *38.46N 23.20E*
117 J15 **Límnos** *anc.* Lemnos. *island* E Greece
104 M11 **Limoges** *anc.* Augustoritum Lemovicensium, Lemovices. Haute-Vienne, C France *45.50N 1.16E*
39 U5 **Limon** Colorado, C USA *39.15N 103.41W*
45 O13 **Limón** *var.* Puerto Limón. Limón, E Costa Rica *9.59N 83.02W*
44 K4 **Limón** Colón, NE Honduras *15.51N 85.30W*
45 N13 **Limón** *off.* Provincia de Limón. ◐ *province* E Costa Rica
108 A10 **Limone Piemonte** Piemonte, NE Italy *44.12N 7.37E*
Limones *see* Valdéz
Limonum *see* Poitiers
105 N11 **Limousin** ◐ *region* C France
105 N16 **Limoux** Aude, S France *43.03N 2.13E*
85 L19 **Limpopo** *var.* Crocodile. ⩫ S Africa
166 K17 **Limu Ling** ▲ S China
115 M20 **Lin** *var.* Lini. Elbasan, E Albania *41.03N 20.37E*
Linacmamari *see* Liinakhamari
179 P13 **Linapacan Island** *island* W Philippines
64 G13 **Linares** Maule, C Chile *35.49S 71.37W*
56 C13 **Linares** Nariño, SW Colombia *1.23N 77.33W*
43 O9 **Linares** Nuevo León, NE Mexico *24.50N 99.33W*
107 N12 **Linares** Andalucía, S Spain *38.04N 3.37W*
109 G15 **Linaro, Capo** *headland* C Italy *42.01N 11.49E*
108 D8 **Linate** ✕ (Milano) Lombardia, N Italy *45.27N 9.18E*
166 F13 **Lincang** Yunnan, SW China *23.55N 100.03E*
167 P11 **Linchuan** *var.* Fuzhou. Jiangxi, S China *27.58N 116.19E*
63 B20 **Lincoln** Buenos Aires, E Argentina *34.50S 61.32W*
193 H19 **Lincoln** Canterbury, South Island, NZ *43.37S 172.30E*
99 N18 **Lincoln** *anc.* Lindum, Lindum Colonia. E England, UK *53.13N 0.33W*
37 O6 **Lincoln** California, W USA *38.52N 121.18W*
32 L13 **Lincoln** Illinois, N USA *40.09N 89.21W*
28 M4 **Lincoln** Kansas, C USA *39.03N 98.09W*
21 S5 **Lincoln** Maine, NE USA *45.22N 68.30W*
29 T5 **Lincoln** Missouri, C USA *38.23N 93.19W*
31 R16 **Lincoln** *state capital* Nebraska, C USA *40.46N 96.42W*
34 F11 **Lincoln City** Oregon, NW USA *44.57N 124.01W*
178 M10 **Lincoln Island** *island* E Paracel Islands
207 Q11 **Lincoln Sea** *sea* Arctic Ocean
99 N18 **Lincolnshire** *cultural region* E England, UK
23 R10 **Lincolnton** North Carolina, SE USA *35.28N 81.15W*
27 V7 **Lindale** Texas, SW USA *32.31N 95.24W*
103 I25 **Lindau** *var.* Lindau am Bodensee. Bayern, S Germany *47.33N 9.40E*
Lindau am Bodensee *see* Lindau
126 L9 **Linde** ⩫ NE Russian Federation
57 T9 **Linden** E Guyana *5.58N 58.11W*
25 O6 **Linden** Alabama, S USA *32.18N 87.48W*
22 H9 **Linden** Tennessee, S USA *35.37N 87.50W*
27 X6 **Linden** Texas, SW USA *33.01N 94.22W*
20 J16 **Lindenwold** New Jersey, NE USA *39.47N 74.58W*
97 M15 **Lindesberg** Örebro, C Sweden *59.36N 15.15E*
97 D18 **Lindesnes** *headland* S Norway *57.58N 7.03E*
Líndhos *see* Líndos
83 K24 **Lindi** Lindi, SE Tanzania *10.00S 39.41E*
83 J24 **Lindi** ◐ *region* SE Tanzania
81 N17 **Lindi** ⩫ NE Dem. Rep. Congo (Zaire)
169 V7 **Lindian** Heilongjiang, NE China *47.10N 124.51E*
193 E21 **Lindis Pass** *pass* South Island, NZ *44.33S 169.40E*
85 J22 **Lindley** Free State, C South Africa *27.48S 27.57E*
97 J19 **Lindome** Västra Götaland, S Sweden *57.34N 12.04E*
Lindong *see* Bairin Zuoqi
117 O23 **Líndos** *var.* Líndhos. Ródos, Dodekánisos, Greece, Aegean Sea *36.04N 28.04E*
12 I14 **Lindsay** Ontario, SE Canada *44.21N 78.43W*
37 R11 **Lindsay** California, W USA *36.11N 119.06W*
35 X8 **Lindsay** Montana, NW USA *47.13N 105.10W*
29 N11 **Lindsay** Oklahoma, C USA *34.50N 97.37W*
29 N5 **Lindsborg** Kansas, C USA *38.34N 97.39W*
97 N21 **Lindsdal** Kalmar, S Sweden *56.43N 16.18E*
175 Pp9 **Lindu, Danau** ◎ Sulawesi, N Indonesia
Lindum/Lindum Colonia *see* Lincoln
203 W3 **Line Islands** *island group* E Kiribati
Linëvo *see* Linova
166 M5 **Linfen** *var.* Lin-fen. Shanxi, C China *36.07N 111.34E*
161 F18 **Linganamakki Reservoir** ▣ SW India
166 L17 **Lingao** Hainan, S China *19.44N 109.23E*
179 Oo9 **Lingayen** Luzon, N Philippines *16.00N 120.12E*
179 Oo9 **Lingayen Gulf** *gulf* Luzon, N Philippines
166 M6 **Lingbao** *var.* Guolüezhen. Henan, C China *34.34N 110.50E*
96 N12 **Lingbo** Gävleborg, C Sweden *61.04N 16.45E*
Lingeh *see* Bandar-e Langeh
102 E12 **Lingen** *var.* Lingen an der Ems. Niedersachsen, NW Germany *52.31N 7.19E*
Lingen an der Ems *see* Lingen
174 Ii8 **Lingga, Kepulauan** *island group* W Indonesia
174 I8 **Lingga, Pulau** *island* Kepulauan Lingga, W Indonesia
12 J14 **Lingham Lake** ◎ Ontario, SE Canada
96 M13 **Linghed** Dalarna, C Sweden *60.48N 15.55E*
35 Z15 **Lingle** Wyoming, C USA *42.07N 104.21W*
20 G15 **Linglestown** Pennsylvania, NE USA *40.20N 76.46W*
81 K18 **Lingomo II** Equateur, NW Dem. Rep. Congo (Zaire) *0.42N 21.59E*
166 L15 **Lingshan** Guangxi Zhuangzu Zizhiqu, S China *22.28N 109.19E*
166 L17 **Lingshui** Hainan, S China *18.35N 110.03E*
161 G16 **Lingsugūr** Karnātaka, C India *16.13N 76.33E*
109 L23 **Linguaglossa** Sicilia, Italy, C Mediterranean Sea *37.51N 15.06E*
78 H10 **Linguère** N Senegal *15.24N 15.06W*
165 W8 **Lingwu** Ningxia, N China *38.04N 106.21E*
Lingxi *see* Yongshun
167 O12 **Lingxian** *var.* Ling Xian. Hunan, S China *26.32N 113.48E*
169 S12 **Lingyuan** Liaoning, NE China *41.09N 119.24E*
169 U4 **Linhai** Heilongjiang, NE China *51.30N 124.18E*
167 S10 **Linhai** *var.* Taizhou. Zhejiang, SE China *28.53N 121.08E*
61 O20 **Linhares** Espírito Santo, SE Brazil *19.22S 40.04W*
168 M13 **Linhe** Nei Mongol Zizhiqu, N China *40.46N 107.27E*
Lini *see* Lin
145 S1 **Linik, Chiyā-ē** ▲ N Iraq
97 M18 **Linköping** Östergötland, S Sweden *58.25N 15.37E*
169 Y8 **Linkou** Heilongjiang, NE China *45.18N 130.16E*
120 F11 **Linkuva** Pakruojis, N Lithuania *56.06N 23.58E*
29 V5 **Linn** Missouri, C USA *38.29N 91.51W*
27 S16 **Linn** Texas, SW USA *26.32N 98.06W*
29 T2 **Linneus** Missouri, C USA *39.53N 93.10W*
98 H10 **Linnhe, Loch** *inlet* W Scotland, UK
121 G19 **Linova** *Rus.* Linëvo. Brestskaya Voblasts', SW Belarus *52.28N 24.33E*
167 O5 **Linqing** Shandong, E China *36.49N 115.39E*
167 N6 **Linruzhen** Henan, C China *34.10N 112.51E*
62 K8 **Lins** São Paulo, S Brazil *21.40S 49.43W*
95 F17 **Linsell** Jämtland, C Sweden *62.10N 14.00E*
166 J9 **Linshui** Sichuan, C China *30.24N 106.54E*
46 K12 **Linstead** C Jamaica *18.07N 77.01W*
165 U11 **Lintan** Gansu, N China *34.43N 101.27E*
165 V11 **Lintao** Gansu, C China *35.23N 103.54E*
13 S12 **Lintère** ⩫ Quebec, SE Canada
110 H8 **Linth** ⩫ NW Switzerland
110 H8 **Linthal** Glarus, NE Switzerland *46.59N 8.57E*
33 N15 **Linton** Indiana, N USA *39.01N 87.10W*
31 N6 **Linton** North Dakota, N USA *46.16N 100.13W*
169 R11 **Linxi** Nei Mongol Zizhiqu, N China *43.29N 117.59E*
165 U11 **Linxia** *var.* Linxia Huizu Zizhizhou. Gansu, C China *35.33N 103.08E*
Linxia Huizu Zizhizhou *see* Linxia
Linxian *see* Lianzhou
167 Q6 **Linyi** Shandong, S China *37.12N 116.54E*
167 P4 **Linyi** Shandong, E China *37.13N 116.50E*
166 M6 **Linyi** Shanxi, C China *35.10N 110.45E*
111 T4 **Linz** *anc.* Lentia. Oberösterreich, N Austria *48.19N 14.18E*
165 S8 **Linze** *var.* Shahepu. Gansu, N China *39.06N 100.03E*
46 J13 **Lionel Town** C Jamaica *17.49N 77.13W*
105 Q16 **Lion, Golfe du** *Eng.* Gulf of Lion, Gulf of Lions; *anc.* Sinus Gallicus. *gulf* S France
Lion, Gulf of/Lions, Gulf of *see* Lion, Golfe du
85 K16 **Lions Den** Mashonaland West, N Zimbabwe *17.16S 30.00E*
12 F13 **Lion's Head** Ontario, S Canada *44.59N 81.16W*
Lios Ceannúir, Bá *see* Liscannor Bay
Lios Mór *see* Lismore
Lios na gCearrbhach *see* Lisburn
Lios Tuathail *see* Listowel
81 G17 **Liouesso** La Sangha, N Congo *1.01N 15.43E*
Liozno *see* Lyozna
179 P11 **Lipa** *off.* Lipa City. Luzon, N Philippines *13.57N 121.10E*
27 S7 **Lipan** Texas, SW USA *32.31N 98.03W*
Lipari Islands/Lipari, Isole *see* Eolie, Isole
109 L22 **Lipari, Isola** *island* Isole Eolie, S Italy
118 L8 **Lipcani** *Rus.* Lipkany. N Moldova *48.16N 26.47E*
95 N17 **Liperi** Itä-Suomi, E Finland *62.30N 29.25E*
130 L7 **Lipetsk** Lipetskaya Oblast', W Russian Federation *52.37N 39.37E*
130 K6 **Lipetskaya Oblast'** ◐ *province* W Russian Federation
59 K22 **Lipez, Cordillera de** ▲ SW Bolivia
112 E10 **Lipiany** *Ger.* Lippehne. Zachodniopomorskie, W Poland *53.00N 14.58E*
114 G9 **Lipik** Požega-Slavonija, NE Croatia *45.24N 17.08E*
128 L12 **Lipin Bor** Vologodskaya Oblast', NW Russian Federation *60.12N 38.04E*
166 L12 **Liping** Guizhou, S China *26.16N 109.07E*
Lipkany *see* Lipcani
121 H15 **Lipnishki** *Rus.* Lipnishki. Hrodzyenskaya Voblasts', W Belarus *54.01N 25.39E*
112 J10 **Lipno** Kujawsko-pomorskie, C Poland *52.51N 19.11E*
118 F11 **Lipova** *Hung.* Lippa. Arad, W Romania *46.06N 21.40E*
Lipovets *see* Lypovets'
Lippa *see* Lipova
103 E14 **Lippe** ⩫ W Germany
Lippehne *see* Lipiany
103 G14 **Lippstadt** Nordrhein-Westfalen, W Germany *51.40N 8.21E*
27 P1 **Lipscomb** Texas, SW USA *36.12N 100.13W*
Lipsia/Lipsk *see* Leipzig
Liptau-Sankt-Nikolaus/Liptószentmiklós *see* Liptovský Mikuláš
113 K19 **Liptovský Mikuláš** *Ger.* Liptau-Sankt-Nikolaus, *Hung.* Liptószentmiklós. Žilinský Kraj, N Slovakia *49.06N 19.36E*
191 O13 **Liptrap, Cape** *headland* Victoria, SE Australia *38.55S 145.58E*
166 L13 **Lipu** Guangxi Zhuangzu Zizhiqu, S China *24.29N 110.24E*
147 X12 **Liqbi** S Oman *18.27N 56.37E*
83 G17 **Lira** N Uganda *2.15N 32.55E*
59 F15 **Lircay** Huancavelica, C Peru *12.58S 74.43W*
109 J15 **Liri** ⩫ C Italy
150 M8 **Lisakovsk** Kostanay, NW Kazakhstan *52.37N 62.34E*
81 K17 **Lisala** Equateur, N Dem. Rep. Congo (Zaire) *2.10N 21.28E*
106 F11 **Lisboa** *Eng.* Lisbon; *anc.* Felicitas Julia, Olisipo. ● (Portugal) Lisboa, W Portugal *38.43N 9.07W*
106 F10 **Lisboa** ◐ *Eng.* Lisbon. *district* C Portugal
21 N7 **Lisbon** New Hampshire, NE USA *44.11N 71.52W*
31 Q6 **Lisbon** North Dakota, N USA *46.27N 97.42W*
Lisbon *see* Lisboa
21 Q8 **Lisbon Falls** Maine, NE USA *44.00N 70.03W*
99 G15 **Lisburn** *Ir.* Lios na gCearrbhach. E Northern Ireland, UK *54.31N 6.03W*
40 L6 **Lisburne, Cape** *headland* Alaska, USA *68.52N 166.13W*
99 B19 **Liscannor Bay** *Ir.* Bá Lios Ceannúir. *inlet* W Ireland
115 Q18 **Lisec** ▲ E FYR Macedonia *41.46N 22.30E*
166 F13 **Lishe Jiang** ⩫ SW China
166 M4 **Lishi** Shanxi, C China *37.27N 111.05E*
169 V10 **Lishu** Jilin, NE China *43.25N 124.19E*
167 R10 **Lishui** Zhejiang, SE China *28.27N 119.25E*
199 Jj5 **Lisianski Island** *island* Hawaiian Islands, Hawaii, USA, C Pacific Ocean
Lisichansk *see* Lysychans'k
104 L4 **Lisieux** *anc.* Noviomagus. Calvados, N France *49.09N 0.13E*
130 L8 **Liski** *prev.* Georgiu-Dezh. Voronezhskaya Oblast', W Russian Federation *51.00N 39.36E*
Lisle/l'Isle *see* Lille
105 N4 **l'Isle-Adam** Val-d'Oise, N France *49.07N 2.13E*
105 R15 **l'Isle-sur-la-Sorgue** Vaucluse, SE France *43.55N 5.03E*
13 S9 **L'Islet** Quebec, SE Canada *47.07N 70.18W*
190 M12 **Lismore** Victoria, SE Australia *37.59S 143.18E*
99 D20 **Lismore** *Ir.* Lios Mór. S Ireland *52.10N 7.10W*
Lissa *see* Vis, Croatia
Lissa *see* Leszno, Poland
100 H11 **Lisse** Zuid-Holland, W Netherlands *52.15N 4.33E*
97 D18 **Lista** *peninsula* S Norway
97 D18 **Listafjorden** *fjord* S Norway
205 R13 **Lister, Mount** ▲ Antarctica *78.12S 161.46E*
130 M8 **Listopadovka** Voronezhskaya Oblast', W Russian Federation *51.54N 41.08E*
12 F15 **Listowel** Ontario, S Canada *43.44N 80.57W*
99 B20 **Listowel** *Ir.* Lios Tuathail. SW Ireland *52.27N 9.28W*
166 L14 **Litang** Guangxi Zhuangzu Zizhiqu, S China *23.09N 109.07E*
166 F9 **Litang** Sichuan, C China *30.03N 100.12E*
166 F10 **Litang Qu** ⩫ C China
57 X12 **Litani** *var.* Itany. ⩫ French Guiana/Suriname
144 G8 **Lītani, Nahr el** *var.* Nahr al Litant. ⩫ C Lebanon
Litant, Nahr al *see* Lītani, Nahr el
Litauen *see* Lithuania
32 K14 **Litchfield** Illinois, N USA *39.17N 89.52W*
31 U8 **Litchfield** Minnesota, N USA *45.09N 94.31W*
38 K13 **Litchfield Park** Arizona, SW USA *33.29N 112.21W*
191 S8 **Lithgow** New South Wales, SE Australia *33.30S 150.09E*
117 I26 **Líthino, Akrotírio** *headland* Kríti, Greece, E Mediterranean Sea *34.55N 24.43E*
120 D12 **Lithuania** *off.* Republic of Lithuania, *Ger.* Litauen, *Lith.* Lietuva, *Pol.* Litwa, *Rus.* Litva; *prev.* Lithuanian SSR, *Rus.* Litovskaya SSR. ◆ *republic* NE Europe
Lithuanian SSR *see* Lithuania
111 U11 **Litija** *Ger.* Littai. C Slovenia *46.03N 14.50E*
20 H15 **Lititz** Pennsylvania, NE USA *40.09N 76.18W*
117 F15 **Litóchoro** *var.* Litohoro, Litókhoron. Kentrikí Makedonía, N Greece *40.06N 22.30E*
Litohoro/Litókhoron *see* Litóchoro
113 C15 **Litoměřice** *Ger.* Leitmeritz. Ústecký Kraj, NW Czech Republic *50.32N 14.09E*
113 F17 **Litomyšl** *Ger.* Leitomischl. Pardubický Kraj, C Czech Republic *49.52N 16.16E*
113 G17 **Litovel** *Ger.* Littau. Olomoucký Kraj, E Czech Republic *49.42N 17.04E*
127 Nn15 **Litovko** Khabarovskiy Kray, SE Russian Federation *49.22N 135.10E*
Litovskaya SSR *see* Lithuania
Littai *see* Litija
Littau *see* Litovel
46 G1 **Little Abaco** *var.* Abaco Island. *island* N Bahamas
113 I21 **Little Alföld** *Ger.* Kleines Ungarisches Tiefland, *Hung.* Kisalföld, *Slvk.* Podunajská Rovina. *plain* Hungary/Slovakia
157 Q20 **Little Andaman** *island* Andaman Islands, India, NE Indian Ocean
28 M5 **Little Arkansas River** ⩫ Kansas, C USA
192 L4 **Little Barrier Island** *island* N NZ
Little Belt *see* Lillebælt
40 M11 **Little Black River** ⩫ Alaska, USA
29 O2 **Little Blue River** ⩫ Kansas/Nebraska, C USA
46 D8 **Little Cayman** *island* E Cayman Islands
9 X11 **Little Churchill** ⩫ Manitoba, C Canada
177 Ee10 **Little Coco Island** *island* SW Myanmar
38 L10 **Little Colorado River** ⩫ Arizona, SW USA
12 E11 **Little Current** Manitoulin Island, Ontario, S Canada *45.57N 81.55W*
10 E11 **Little Current** ⩫ Ontario, S Canada
40 L8 **Little Diomede Island** *island* Alaska, USA
46 I4 **Little Exuma** *island* C Bahamas
31 U7 **Little Falls** Minnesota, N USA *45.59N 94.21W*
20 J10 **Little Falls** New York, NE USA *43.02N 74.51W*
26 M5 **Littlefield** Texas, SW USA *33.55N 102.19W*
31 V3 **Littlefork** Minnesota, N USA *48.24N 93.33W*
31 V3 **Little Fork River** ⩫ Minnesota, N USA
9 N16 **Little Fort** British Columbia, SW Canada *51.27N 120.15W*
9 Y14 **Little Grand Rapids** Manitoba, C Canada *52.06N 95.29W*
99 N23 **Littlehampton** SE England, UK *50.48N 0.33W*
37 T2 **Little Humboldt River** ⩫ Nevada, W USA
46 K6 **Little Inagua** *var.* Inagua Islands. *island* S Bahamas
23 Q4 **Little Kanawha River** ⩫ West Virginia, NE USA
85 F25 **Little Karoo** *plateau* S South Africa
41 O16 **Little Koniuji Island** *island* Shumagin Islands, Alaska, USA
46 H12 **Little London** W Jamaica *18.14N 78.13W*
11 R10 **Little Mecatina** *Fr.* Rivière du Petit Mécatina. ⩫ Newfoundland/Quebec, E Canada
98 F8 **Little Minch, The** *strait* NW Scotland, UK
29 T13 **Little Missouri River** ⩫ Arkansas, USA
30 J7 **Little Missouri River** ⩫ NW USA
30 J3 **Little Muddy River** ⩫ North Dakota, N USA
157 Q22 **Little Nicobar** *island* Nicobar Islands, India, NE Indian Ocean
29 R6 **Little Osage River** ⩫ Missouri, C USA
99 P20 **Little Ouse** ⩫ E England, UK
155 V2 **Little Pamir** *Pash.* Pāmīr-e Khord, *Rus.* Malyy Pamir. ▲ Afghanistan/Tajikistan
23 U12 **Little Pee Dee River** ⩫ North Carolina/South Carolina, SE USA
29 V10 **Little Red River** ⩫ Arkansas, C USA
Little Rhody *see* Rhode Island
193 I19 **Little River** Canterbury, South Island, NZ *43.45S 172.49E*
23 U12 **Little River** South Carolina, SE USA *33.52N 78.36W*
29 Y9 **Little River** ⩫ Arkansas/Missouri, C USA
29 R13 **Little River** ⩫ Arkansas/Oklahoma, USA
25 T7 **Little River** ⩫ Georgia, SE USA
24 H6 **Little River** ⩫ Louisiana, S USA
27 T10 **Little River** ⩫ Texas, SW USA
29 V12 **Little Rock** *state capital* Arkansas, C USA *34.45N 92.17W*
33 N8 **Little Sable Point** *headland* Michigan, N USA *43.38N 86.32W*
105 U11 **Little Saint Bernard Pass** *Fr.* Col du Petit St-Bernard, *It.* Colle di Piccolo San Bernardo. *pass* France/Italy *45.41N 6.54E*
38 K7 **Little Salt Lake** ◎ Utah, W USA
188 K8 **Little Sandy Desert** *desert* Western Australia
31 S13 **Little Sioux River** ⩫ Iowa, C USA
40 E17 **Little Sitkin Island** *island* Aleutian Islands, Alaska, USA
9 O13 **Little Smoky** Alberta, W Canada *54.35N 117.06W*
9 O14 **Little Smoky** ⩫ Alberta, W Canada
39 P3 **Little Snake River** ⩫ Colorado, C USA
66 A12 **Little Sound** *bay* Bermuda, NW Atlantic Ocean
39 T4 **Littleton** Colorado, C USA *39.36N 105.01W*
21 N7 **Littleton** New Hampshire, NE USA *44.18N 71.46W*
20 D11 **Little Valley** New York, NE USA *42.15N 78.46W*
32 M15 **Little Wabash River** ⩫ Illinois, N USA
12 D10 **Little White River** ⩫ Ontario, S Canada
30 M12 **Little White River** ⩫ South Dakota, N USA
27 R5 **Little Wichita River** ⩫ Texas, SW USA
148 I4 **Little Zab** *Ar.* Nahraz Zāb aş Şaghīr, *Kurd.* Zē-i Kōya, *Per.* Rūdkhāneh-ye Zāb-e Kūchek. ⩫ Iran/Iraq
81 D15 **Littoral** ◐ *province* W Cameroon
Littoria *see* Latina
Litva/Litwa *see* Lithuania
113 B15 **Litvínov** *Ger.* Leutensdorf. Ústecký Kraj, NW Czech Republic *50.37N 13.37E*
118 M6 **Lityn** Vinnyts'ka Oblast', C Ukraine *49.19N 28.06E*
Liu-chou/Liuchow *see* Liuzhou
169 W11 **Liuhe** Jilin, NE China *42.15N 125.49E*
Liukang Tenggaya, Kepulauan *see* Sabalana, Kepulauan
85 Q15 **Liúpo** Nampula, NE Mozambique *15.36S 39.57E*
85 G14 **Liuwa Plain** *plain* W Zambia
166 L13 **Liuzhou** *var.* Liu-chou, Liuchow. Guangxi Zhuangzu Zizhiqu, S China *24.09N 108.55E*
118 H8 **Livada** *Hung.* Sárköz. Satu Mare, NW Romania *47.52N 23.03E*
117 J20 **Liváda, Akrotírio** *headland* Tínos, Kykládes, Greece, Aegean Sea *37.36N 25.15E*
117 L21 **Livádi** *island* Kykládes, Greece, Aegean Sea
Livanátai *see* Livanátes
117 G18 **Livanátes** *prev.* Livanátai. Stereá Ellás, C Greece *38.43N 23.01E*
120 I10 **Līvāni** *Ger.* Lievenhof. Preiļi, SE Latvia *56.22N 26.12E*
67 E25 **Lively Island** *island* SE Falkland Islands
67 D25 **Lively Sound** *sound* SE Falkland Islands
41 R8 **Livengood** Alaska, USA *65.31N 148.32W*
108 I7 **Livenza** ⩫ NE Italy
37 O6 **Live Oak** California, W USA *39.17N 121.41W*
25 U9 **Live Oak** Florida, SE USA *30.18N 82.59W*
37 O9 **Livermore** California, W USA *37.40N 121.46W*
22 I6 **Livermore** Kentucky, S USA *37.31N 87.08W*
21 Q7 **Livermore Falls** Maine, NE USA *44.30N 70.09W*
26 J10 **Livermore, Mount** ▲ Texas, SW USA *30.37N 104.10W*
11 P16 **Liverpool** Nova Scotia, SE Canada *44.03N 64.43W*
99 K17 **Liverpool** NW England, UK *53.25N 2.55W*
191 S7 **Liverpool Range** ▲ New South Wales, SE Australia
98 J12 **Livingston** C Scotland, UK *55.51N 3.31W*
25 N5 **Livingston** Alabama, S USA *32.34N 88.12W*
37 P9 **Livingston** California, W USA *37.22N 120.45W*
24 J8 **Livingston** Louisiana, S USA *30.30N 90.45W*
35 S11 **Livingston** Montana, NW USA *45.40N 110.33W*
22 L8 **Livingston** Tennessee, S USA *36.22N 85.19W*
27 W9 **Livingston** Texas, SW USA *30.42N 94.55W*
44 F4 **Livingston** Izabal, E Guatemala *15.49N 88.46W*
85 I16 **Livingstone** *var.* Maramba. Southern, S Zambia *17.51S 25.48E*
193 B22 **Livingstone Mountains** ▲ South Island, NZ
82 K13 **Livingstone Mountains** ▲ S Tanzania
84 N12 **Livingstonia** Northern, N Malawi *10.29S 34.06E*
204 G4 **Livingston Island** *island* Antarctica
27 W9 **Livingston, Lake** ▣ Texas, SW USA
114 F13 **Livno** Federacija Bosna I Hercegovina, SW Bosnia and Herzegovina *43.49N 17.00E*
130 K7 **Livny** Orlovskaya Oblast', W Russian Federation *52.25N 37.42E*
95 M14 **Livojoki** ⩫ C Finland
33 R10 **Livonia** Michigan, N USA *42.22N 83.22W*
108 E11 **Livorno** *Eng.* Leghorn. Toscana, C Italy *43.31N 10.18E*
Livramento *see* Santana do Livramento
147 U8 **Liwā** *var.* Al Liwā'. *oasis region* S UAE
83 I24 **Liwale** Lindi, SE Tanzania *9.46S 37.56E*
165 W9 **Liwangbu** Ningxia, N China *36.42N 106.04E*
85 N15 **Liwonde** Southern, S Malawi *15.04S 35.12E*
165 V11 **Lixian** *var.* Li Xian. Gansu, C China *34.15N 105.07E*
166 H8 **Lixian** *var.* Li Xian; *prev.* Zagunao. Sichuan, C China *31.27N 103.06E*
Lixian Jiang *see* Black River
117 B18 **Lixoúri** *prev.* Lixoúrion. Kefallinía, Iónioi Nísoi, Greece, C Mediterranean Sea *38.12N 20.25E*
Lixoúrion *see* Lixoúri
Lixus *see* Larache
Lizarra *see* Estella-Lizarra
35 U15 **Lizard Head Peak** ▲ Wyoming, C USA *42.47N 109.12W*
99 H25 **Lizard Point** *headland* SW England, UK *49.57N 5.12W*
114 L12 **Ljig** Serbia, C Yugoslavia *44.14N 20.16E*
Ljouwert *see* Leeuwarden
Ljubelj *see* Loibl Pass
111 U11 **Ljubljana** *Ger.* Laibach, *It.* Lubiana; *anc.* Aemona, Emona. ● (Slovenia) C Slovenia *46.03N 14.28E*
111 T11 **Ljubljana** ✕ C Slovenia *46.14N 14.26E*
115 N17 **Ljuboten** ▲ S Yugoslavia *42.12N 21.06E*
97 P19 **Ljugarn** Gotland, SE Sweden *57.23N 18.45E*
86 G7 **Ljungan** ⩫ N Sweden
95 F17 **Ljungan** ⩫ C Sweden
97 K21 **Ljungby** Kronoberg, S Sweden *56.49N 13.55E*
97 M17 **Ljungsbro** Östergötland, S Sweden *58.31N 15.30E*
97 I18 **Ljungskile** Västra Götaland, S Sweden *58.13N 11.55E*
96 M11 **Ljusdal** Gävleborg, C Sweden *61.49N 16.10E*
96 M11 **Ljusnan** ⩫ C Sweden
96 N12 **Ljusne** Gävleborg, C Sweden *61.11N 17.07E*
97 P15 **Ljusterö** Stockholm, C Sweden *59.31N 18.40E*
111 X9 **Ljutomer** *Ger.* Luttenberg. NE Slovenia *46.31N 16.12E*
65 G15 **Llaima, Volcán** ▲ S Chile *39.01S 71.38W*
107 X4 **Llançà** *var.* Llansá. Cataluña, NE Spain *42.23N 3.08E*
99 J21 **Llandovery** C Wales, UK *52.01N 3.47W*
99 J20 **Llandrindod Wells** E Wales, UK *52.15N 3.22W*
99 J18 **Llandudno** N Wales, UK *53.19N 3.49W*
99 I21 **Llanelli** *prev.* Llanelly. SW Wales, UK *51.41N 4.11W*
Llanelly *see* Llanelli
106 M2 **Llanes** Asturias, N Spain *43.24N 4.46W*
99 K19 **Llangollen** NE Wales, UK *52.58N 3.10W*
27 R10 **Llano** Texas, SW USA *30.45N 98.40W*
27 Q10 **Llano River** ⩫ Texas, SW USA
56 I9 **Llanos** *physical region* Colombia/Venezuela
65 G16 **Llanquihue, Lago** ◎ S Chile
Llansá *see* Llançà
107 U5 **Lleida** *Cast.* Lérida; *anc.* Ilerda. Cataluña, NE Spain *41.37N 0.36E*
107 U5 **Lleida** *Cast.* Lérida ◐ *province* Cataluña, NE Spain
106 K12 **Llerena** Extremadura, W Spain *38.13N 6.00W*
107 S9 **Lliria** País Valenciano, E Spain *39.37N 0.36W*
107 W4 **Llivia** Cataluña, NE Spain *42.27N 2.00E*
107 O3 **Llodio** País Vasco, N Spain *43.07N 2.58W*
107 X5 **Lloret de Mar** Cataluña, NE Spain *41.42N 2.51E*
Llorri *see* Tossal de l'Orri
8 L11 **Lloyd George, Mount** ▲ British Columbia, W Canada *57.46N 124.57W*
9 R14 **Lloydminster** Alberta/Saskatchewan, SW Canada *53.18N 110.00W*
38 L6 **Loa** Utah, W USA *38.24N 111.38W*
174 Mm4 **Loagan Bunut** ◎ East Malaysia
178 Mm14 **Loaita Island** *island* W Spratly Islands
40 G12 **Loa, Mauna** ▲ Hawaii, USA, C Pacific Ocean *19.28N 155.39W*
Loanda *see* Luanda
81 J22 **Loange** ⩫ S Dem. Rep. Congo (Zaire)
81 E21 **Loango** Le Kouilou, S Congo *4.37S 11.49E*
108 B10 **Loano** Liguria, NW Italy *44.07N 8.15E*
64 H4 **Loa, Río** ⩫ N Chile
85 I20 **Lobatse** *var.* Lobatsi. Kgatleng, SE Botswana *25.10S 25.40E*
Lobatsi *see* Lobatse
103 Q15 **Löbau** Sachsen, E Germany *51.06N 14.39E*
81 H16 **Lobaye** ◐ *prefecture* SW Central African Republic
81 I16 **Lobaye** ⩫ SW Central African Republic
101 G21 **Lobbes** Hainaut, S Belgium *50.21N 4.16E*
63 D23 **Lobería** Buenos Aires, E Argentina *38.07S 58.48W*
112 F8 **Łobez** *Ger.* Labes. Zachodniopomorskie, NW Poland *53.29N 15.39E*
84 A13 **Lobito** Benguela, W Angola *12.19S 13.34E*
Lobkovichi *see* Labkovichy
Lob Nor *see* Lop Nur
176 W11 **Lobo** Irian Jaya, E Indonesia *3.41S 134.06E*
106 J11 **Lobón** Extremadura, W Spain *38.51N 6.37W*
63 D20 **Lobos** Buenos Aires, E Argentina *35.10S 59.07W*
42 E4 **Lobos, Cabo** *headland* NW Mexico *29.53N 112.43W*
42 F6 **Lobos, Isla** *island* NW Mexico
Lobositz *see* Lovosice
Lobsens *see* Łobżenica
Loburi *see* Lop Buri
112 H9 **Łobżenica** *Ger.* Lobsens. Wielkopolskie, C Poland *53.19N 17.11E*
110 G11 **Locarno** *Ger.* Luggarus. Ticino, S Switzerland *46.10N 8.47E*
98 E9 **Lochboisdale** NW Scotland, UK *57.08N 7.17W*
100 N11 **Lochem** Gelderland, E Netherlands *52.10N 6.25E*
104 M8 **Loches** Indre-et-Loire, C France *47.08N 1.00E*
Loch Garman *see* Wexford
98 H12 **Lochgilphead** W Scotland, UK *56.02N 5.27W*
98 H7 **Lochinver** N Scotland, UK *58.10N 5.14W*
98 F8 **Lochmaddy** NW Scotland, UK *57.35N 7.10W*
98 J10 **Lochnagar** ▲ C Scotland, UK *56.58N 3.09W*
101 E17 **Lochristi** Oost-Vlaanderen, NW Belgium *51.07N 3.49E*
98 H9 **Lochy, Loch** ◎ N Scotland, UK
190 G8 **Lock** South Australia *33.37S 135.45E*
99 J14 **Lockerbie** S Scotland, UK *55.10N 3.27W*
29 S13 **Lockesburg** Arkansas, C USA *33.58N 94.10W*
191 P10 **Lockhart** New South Wales, SE Australia *35.15S 146.43E*
27 S11 **Lockhart** Texas, SW USA *29.52N 97.40W*
20 F13 **Lock Haven** Pennsylvania, NE USA *41.08N 77.27W*
27 N4 **Lockney** Texas, SW USA *34.06N 101.27W*
102 O12 **Löcknitz** ⩫ NE Germany
20 E9 **Lockport** New York, NE USA *43.09N 78.40W*
178 Jj13 **Lôc Ninh** Sông Be, S Vietnam *11.51N 106.34E*
109 N23 **Locri** Calabria, SW Italy *38.16N 16.16E*
Locse *see* Levoča
29 T2 **Locust Creek** ⩫ Missouri, C USA
25 P3 **Locust Fork** ⩫ Alabama, S USA
29 Q9 **Locust Grove** Oklahoma, C USA *36.12N 95.10W*
96 E11 **Lodalskåpa** ▲ S Norway *61.47N 7.10E*
191 N10 **Loddon River** ⩫ Victoria, SE Australia
Lodensee *see* Klooga
105 P15 **Lodève** *anc.* Luteva. Hérault, S France *43.43N 3.19E*
128 I12 **Lodeynoye Pole** Leningradskaya Oblast', NW Russian Federation *60.41N 33.29E*
35 V11 **Lodge Grass** Montana, NW USA *45.19N 107.20W*
30 J15 **Lodgepole Creek** ⩫ Nebraska/Wyoming, C USA
155 T11 **Lodhrān** Punjab, E Pakistan *29.36N 71.34E*
108 D8 **Lodi** Lombardia, NW Italy *45.15N 9.36E*
37 O8 **Lodi** California, W USA *38.07N 121.17W*
33 T12 **Lodi** Ohio, N USA *41.00N 82.01W*
94 H10 **Lødingen** Nordland, C Norway *68.24N 15.55E*
81 L20 **Lodja** Kasai Oriental, C Dem. Rep. Congo (Zaire) *3.28S 23.24E*
39 O3 **Lodore, Canyon of** *canyon* Colorado, C USA
107 Q4 **Lodosa** Navarra, N Spain *42.25N 2.04W*
83 H16 **Lodwar** Rift Valley, NW Kenya *3.06N 35.37E*
112 K13 **Łódź** *Rus.* Lodz. Łódź, C Poland *51.51N 19.26E*
112 J13 **Łódzkie** ◐ *province* C Poland *51.51N 19.26E*
178 I8 **Loei** *var.* Loey, Muang Loei. Loei, C Thailand *17.28N 101.42E*
100 I11 **Loenen** Utrecht, C Netherlands *52.13N 5.01E*
178 J9 **Loeng Nok Tha** Yasothon, E Thailand *16.12N 104.31E*
85 F24 **Loeriesfontein** Northern Cape, W South Africa *30.53S 19.28E*
97 H20 **Læsø** *island* N Denmark
Loewoek *see* Luwuk
Loey *see* Loei
78 J16 **Lofa** ⩫ N Liberia
111 P6 **Lofer** Salzburg, C Austria *47.37N 12.42E*
94 F11 **Lofoten** *var.* Lofoten Islands. *island group* C Norway
Lofoten Islands *see* Lofoten
97 N18 **Loftahammar** Kalmar, S Sweden *57.55N 16.45E*
131 O10 **Log** Volgogradskaya Oblast', SW Russian Federation *49.32N 43.52E*
79 S12 **Loga** Dosso, SW Niger *13.33N 3.18E*
31 S14 **Logan** Iowa, C USA *41.38N 95.47W*
28 K3 **Logan** Kansas, C USA *39.39N 99.34W*
33 T14 **Logan** Ohio, N USA *39.32N 82.24W*
38 L1 **Logan** Utah, W USA *41.45N 111.50W*
23 P6 **Logan** West Virginia, NE USA *37.51N 81.59W*
37 Y10 **Logandale** Nevada, W USA *36.36N 114.28W*
21 O11 **Logan International** ✕ (Boston) Massachusetts, NE USA *42.22N 71.00W*
9 N16 **Logan Lake** British Columbia, SW Canada *50.28N 120.42W*
25 Q4 **Logan Martin Lake** ▣ Alabama, S USA
8 G8 **Logan, Mount** ▲ Yukon Territory, W Canada *60.32N 140.34W*
34 I7 **Logan, Mount** ▲ Washington, NW USA *48.32N 120.57W*
35 P7 **Logan Pass** *pass* Montana, NW USA *48.43N 113.44W*
33 O12 **Logansport** Indiana, N USA *40.44N 86.25W*
24 F6 **Logansport** Louisiana, S USA *31.58N 94.00W*
Logar *see* Lowgar
69 R11 **Loge** ⩫ NW Angola
Logishin *see* Lahishyn
Log na Coille *see* Lugnaquillia Mountain
80 G11 **Logone** *var.* Logone. ⩫ Cameroon/Chad
80 G13 **Logone-Occidental** *off.* Préfecture du Logone-Occidental. ◐ *prefecture* SW Chad
80 H13 **Logone Occidental** ⩫ SW Chad
80 G13 **Logone-Oriental** *off.* Préfecture du Logone-Oriental. ◐ *prefecture* SW Chad
80 H13 **Logone Oriental** ⩫ SW Chad
Logone Oriental *see* Pendé
L'Ogooué-Ivindo *see* Ogooué-Ivindo
L'Ogooué-Lolo *see* Ogooué-Lolo
L'Ogooué-Maritime *see* Ogooué-Maritime
Logoysk *see* Lahoysk
107 P4 **Logroño** *anc.* Vareia, *Lat.* Juliobriga. La Rioja, N Spain *42.28N 2.25W*
106 L10 **Logrosán** Extremadura, W Spain *39.21N 5.28W*
97 G20 **Løgstør** Nordjylland, N Denmark *56.57N 9.19E*
97 H22 **Løgten** Århus, C Denmark *56.16N 10.19E*

◆ Country ● Country Capital ◇ Dependent Territory ○ Dependent Territory Capital ◐ Administrative Region ✕ International Airport ▲ Mountain ▲ Mountain Range ⛰ Volcano ⩫ River ◎ Lake ▣ Reservoir

97 F24 **Løgumkloster** Sønderjylland, SW Denmark 55.04N 8.58E
Lögurinn *see* Lagarfljót
197 B10 **Loh** *island* Torres Islands, N Vanuatu
159 P15 **Lohārdaga** Bihār, N India 23.27N 84.42E
158 H10 **Lohāru** Haryāna, N India 28.27N 75.53E
103 D15 **Lohausen** ✕ (Düsseldorf) Nordrhein-Westfalen, W Germany 51.18N 6.51E
201 O14 **Lohd** Pohnpei, E Micronesia
194 I14 **Lohiki** ≈ S PNG
94 L12 **Lohiniva** Lappi, N Finland 67.09N 25.04E
Lohiszyn *see* Lahishyn
95 L20 **Lohja** *var.* Lojo. Etelä-Suomi, S Finland 60.14N 24.07E
175 O8 **Lohjanan** Borneo, C Indonesia
27 Q9 **Lohn** Texas, SW USA 31.15N 99.22W
102 G12 **Lohne** Niedersachsen, NW Germany 52.40N 8.13E
Lohr *see* Lohr am Main
103 I18 **Lohr am Main** *var.* Lohr. Bayern, C Germany 50.00N 9.30E
111 T10 **Loibl Pass** *Ger.* Loiblpass, *Slvn.* Ljubelj. *pass* Austria/Slovenia 46.25N 14.15E
178 Gg6 **Loi-Kaw** Kayah State, C Myanmar 19.40N 97.12E
95 K19 **Loimaa** Länsi-Suomi, W Finland 60.51N 23.03E
105 O6 **Loing** ≈ C France
178 Ii6 **Loi, Phou** ▲ N Laos 20.18N 103.14E
104 L7 **Loir** ≈ C France
105 Q11 **Loire** ◐ *department* E France
104 M7 **Loire** *var.* Liger. ≈ C France
104 I7 **Loire-Atlantique** ◐ *department* NW France
105 O7 **Loiret** ◐ *department* C France
104 M8 **Loir-et-Cher** ◐ *department* C France
103 L24 **Loisach** ≈ SE Germany
58 B9 **Loja** Loja, S Ecuador 3.58S 79.16W
106 M14 **Loja** Andalucía, S Spain 37.10N 4.09W
58 B9 **Loja** ◐ *province* S Ecuador
Lojo *see* Lohja
118 J4 **Lokachi** Volyns'ka Oblast', NW Ukraine 50.44N 24.39E
81 M20 **Lokandu** Maniema, C Dem. Rep. Congo (Zaire) 2.34S 25.43E
94 M11 **Lokan Tekojärvi** ◎ NE Finland
143 Z11 **Lökbatan** *Rus.* Lokbatan. E Azerbaijan 40.21N 49.43E
101 F17 **Lokeren** Oost-Vlaanderen, NW Belgium 51.06N 3.58E
Lokhvitsa *see* Lokhvytsya
119 S4 **Lokhvytsya** *Rus.* Lokhvitsa. Poltavs'ka Oblast', NE Ukraine 50.21N 33.16E
83 H17 **Lokichar** Rift Valley, NW Kenya 2.22N 35.40E
83 G16 **Lokichokio** Rift Valley, NW Kenya 4.16N 34.22E
83 H16 **Lokitaung** Rift Valley, NW Kenya 4.15N 35.45E
94 M11 **Lokka** Lappi, N Finland 67.47N 27.40E
96 G8 **Løkken Verk** Sør-Trøndelag, S Norway 63.07N 9.40E
128 G16 **Loknya** Pskovskaya Oblast', W Russian Federation 56.48N 30.08E
79 V15 **Loko** Nassarawa, C Nigeria 8.00N 7.48E
79 U15 **Lokoja** Kogi, C Nigeria 7.47N 6.44E
83 H17 **Lokori** Rift Valley, W Kenya 1.55N 36.03E
79 R16 **Lokossa** S Benin 6.37N 1.43E
120 I3 **Loksa** *Ger.* Loxa. Harjumaa, NW Estonia 59.36N 25.43E
16 P3 **Loks Land** *island* Nunavut, NE Canada
82 C13 **Lol** ≈ S Sudan
78 K15 **Lola** Guinée-Forestière, SE Guinea 7.52N 8.28W
37 Q5 **Lola, Mount** ▲ California, W USA 39.27N 120.20W
83 H20 **Loliondo** Arusha, NE Tanzania 2.03S 35.46E
97 H25 **Lolland** *prev.* Laaland. *island* S Denmark
195 O11 **Lolobau Island** *island* E PNG
175 T6 **Loloda Utara, Kepulauan** *island group* E Indonesia
81 E16 **Lolodorf** Sud, SW Cameroon 3.16N 10.49E
116 G7 **Lom** *prev.* Lom-Palanka. Oblast Montana, NW Bulgaria 43.48N 23.16E
116 G7 **Lom** ≈ Montana, NW Bulgaria
81 M19 **Lomami** ≈ C Dem. Rep. Congo (Zaire)
59 F17 **Lomas** Arequipa, SW Peru 15.29S 74.54W
65 I23 **Lomas, Bahía** *bay* S Chile
63 D20 **Lomas de Zamora** Buenos Aires, E Argentina 34.52S 58.26W
63 D20 **Loma Verde** Buenos Aires, E Argentina 35.16S 58.24W
188 K4 **Lombadina** Western Australia 16.39S 122.54E
108 E6 **Lombardia** *Eng.* Lombardy. ◐ *region* N Italy
Lombardy *see* Lombardia
104 M15 **Lombez** Gers, S France 43.28N 0.54E
175 R15 **Lomblen, Pulau** *island* Nusa Tenggara, S Indonesia
181 W7 **Lombok Basin** *undersea feature* E Indian Ocean
175 Nn16 **Lombok, Pulau** *island* Nusa Tenggara, C Indonesia
175 Nn16 **Lombok, Selat** *strait* S Indonesia
79 Q16 **Lomé** ● (Togo) S Togo 6.08N 1.13E
79 Q16 **Lomé** ✕ S Togo 6.08N 1.13E
81 L19 **Lomela** Kasai Oriental, C Dem. Rep. Congo (Zaire) 2.19S 23.15E
27 R9 **Lometa** Texas, SW USA 31.13N 98.23W
81 F16 **Lomié** Est, SE Cameroon 3.09N 13.34E
32 M8 **Lomira** Wisconsin, N USA 43.36N 88.26W
97 K23 **Lomma** Skåne, S Sweden 55.40N 13.04E
101 J16 **Lommel** Limburg, N Belgium 51.13N 5.19E
98 I11 **Lomond, Loch** ◎ C Scotland, UK
207 R9 **Lomonosov Ridge** *var.* Harris Ridge, *Rus.* Khrebet Lomonsova. *undersea feature* Arctic Ocean
Lomonsova, Khrebet *see* Lomonosov Ridge
Lom-Palanka *see* Lom
Lomphat *see* Lumphăt
37 P14 **Lompoc** California, W USA 34.39N 120.29W
178 Hh9 **Lom Sak** *var.* Muang Lom Sak. Phetchabun, C Thailand 16.45N 101.12E
112 N9 **Łomża** *Rus.* Lomzha. Podlaskie, NE Poland 53.10N 22.04E
Lomzha *see* Łomża
161 D14 **Lonāvale** *prev.* Lonaula. Mahārāshtra, W India 18.45N 73.27E
65 G15 **Loncoche** Araucanía, C Chile 39.21S 72.34W
65 H14 **Loncopue** Neuquén, W Argentina 38.06S 70.36W
101 G17 **Londerzeel** Vlaams Brabant, C Belgium 51.00N 4.19E
Londinium *see* London
12 E16 **London** Ontario, S Canada 42.59N 81.12W
203 Y2 **London** Kiritimati, E Kiribati 02.00N 157.28W
99 O22 **London** *anc.* Augusta, *Lat.* Londinium. ● (UK) SE England, UK 51.30N 0.10W
23 N7 **London** Kentucky, S USA 37.06N 84.03W
33 S13 **London** Ohio, NE USA 39.52N 83.27W
27 Q10 **London** Texas, SW USA 30.40N 99.33W
99 O22 **London City** ✕ SE England, UK 51.31N 0.07E
99 E14 **Londonderry** *var.* Derry, *Ir.* Doire. NW Northern Ireland, UK 55.00N 7.19W
99 F14 **Londonderry** *cultural region* NW Northern Ireland, UK
188 M2 **Londonderry, Cape** *headland* Western Australia 13.46S 126.56E
65 H25 **Londonderry, Isla** *island* S Chile
45 O7 **Londres, Cayos** *reef* NE Nicaragua
62 I10 **Londrina** Paraná, S Brazil 23.18S 51.13W
29 N13 **Lone Grove** Oklahoma, C USA 34.11N 97.15W
12 E12 **Lonely Island** *island* Ontario, S Canada
37 T8 **Lone Mountain** ▲ Nevada, W USA 38.01N 117.28W
27 V6 **Lone Oak** Texas, SW USA 33.02N 95.58W
37 T11 **Lone Pine** California, W USA 36.36N 118.04W
Lone Star State *see* Texas
85 D14 **Longa** Cuando Cubango, C Angola 14.37S 18.27E
84 B12 **Longa** ≈ W Angola
85 E15 **Longa** ≈ SE Angola
169 W11 **Longang Shan** ▲▲ NE China
207 S4 **Longa, Proliv** *Eng.* Long Strait. *strait* NE Russian Federation
46 J13 **Long Bay** *bay* W Jamaica
23 V13 **Long Bay** *bay* North Carolina/South Carolina, E USA
37 T16 **Long Beach** California, W USA 33.46N 118.11W
24 M9 **Long Beach** Mississippi, S USA 30.21N 89.09W
20 L14 **Long Beach** Long Island, New York, NE USA 40.34N 73.38W
34 F9 **Long Beach** Washington, NW USA 46.21N 124.03W
20 K16 **Long Beach Island** *island* New Jersey, NE USA
67 M25 **Longbluff** *headland* SW Tristan da Cunha
25 U13 **Longboat Key** *island* Florida, SE USA
20 K15 **Long Branch** New Jersey, NE USA 40.18N 73.59W
46 J5 **Long Cay** *island* SE Bahamas
167 P14 **Longchuan** *prev.* Laolong. Guangdong, S China 24.07N 115.10E
Longchuan Jiang *see* Shweli
34 K12 **Long Creek** Oregon, NW USA 44.40N 119.07W
165 W10 **Longde** Ningxia, N China 35.37N 106.07E
191 P16 **Longford** Tasmania, SE Australia 41.41S 147.03E
99 D17 **Longford** *Ir.* An Longfort. C Ireland 53.44N 7.49W
99 E17 **Longford** *Ir.* An Longfort. *cultural region* C Ireland
167 P1 **Longhua** Hebei, E China 41.18N 117.43E
175 Nn8 **Longiram** Borneo, C Indonesia 0.01S 115.36E
46 J4 **Long Island** *island* C Bahamas
10 H8 **Long Island** *island* Nunavut, C Canada
194 K11 **Long Island** *var.* Arop Island. *island* N PNG
20 L14 **Long Island** *island* New York, NE USA
Long Island *see* Bermuda
20 M14 **Long Island Sound** *sound* NE USA
166 K13 **Long Jiang** ≈ S China
169 U7 **Longjiang** Heilongjiang, NE China 47.20N 123.09E
169 Y10 **Longjing** *var.* Yanji. Jilin, NE China 42.48N 129.26E
167 R4 **Longkou** Shandong, E China 37.40N 120.21E
10 E11 **Longlac** Ontario, S Canada 49.46N 86.34W
21 S1 **Long Lake** ◎ Maine, NE USA
33 Q6 **Long Lake** ◎ Michigan, N USA
33 R5 **Long Lake** ◎ Michigan, N USA
31 N6 **Long Lake** ◎ North Dakota, N USA
32 J4 **Long Lake** ⊞ Wisconsin, N USA
101 K23 **Longlier** Luxembourg, SE Belgium 49.51N 5.27E
166 I13 **Longlin** *var.* Longlin Gezu Zizhixian. Guangxi Zhuangzu Zizhiqu, S China 24.46N 105.19E
39 T3 **Longmont** Colorado, C USA 40.09N 105.07W
31 N13 **Long Pine** Nebraska, C USA 42.32N 99.42W
Longping *see* Luodian
12 F17 **Long Point** *headland* Ontario, S Canada 42.33N 80.15W
12 K15 **Long Point** *headland* Ontario, SE Canada 43.56N 76.53W
192 P10 **Long Point** *headland* North Island, NZ 39.07S 177.41E
32 L2 **Long Point** *headland* Michigan, N USA 47.50N 89.09W
12 G17 **Long Point Bay** *lake bay* Ontario, S Canada
31 T7 **Long Prairie** Minnesota, N USA 45.58N 94.52W
11 S11 **Long Range Mountains** *hill range* Newfoundland, E Canada
67 H25 **Long Range Point** *headland* SE Saint Helena 16.00S 05.41W
189 V8 **Longreach** Queensland, E Australia 23.31S 144.18E
166 H7 **Longriba** Sichuan, C China 32.32N 102.20E
166 L10 **Longshan** Hunan, S China 29.25N 109.28E
39 S3 **Longs Peak** ▲ Colorado, C USA 40.15N 105.37W
Long Strait *see* Longa, Proliv
104 K8 **Longué** Maine-et-Loire, NW France 47.23N 0.07W
11 P11 **Longue-Pointe** Quebec, E Canada 50.20N 64.13W
105 S4 **Longuyon** Meurthe-et-Moselle, NE France 49.25N 5.37E
27 W7 **Longview** Texas, SW USA 32.30N 94.44W
34 G10 **Longview** Washington, NW USA 46.08N 122.56W
67 H25 **Longwood** C Saint Helena
27 P7 **Longworth** Texas, SW USA 32.37N 100.20W
105 S3 **Longwy** Meurthe-et-Moselle, NE France 49.31N 5.46E
165 V11 **Longxi** Gansu, C China 35.00N 104.34E
178 J14 **Long Xuyên** *var.* Longxuyen. An Giang, S Vietnam 10.22N 105.25E
167 Q13 **Longyan** Fujian, SE China 25.06N 117.01E
94 O3 **Longyearbyen** ○ (Svalbard) Spitsbergen, W Svalbard 78.12N 15.39E
166 J15 **Longzhou** Guangxi Zhuangzu Zizhiqu, S China 22.22N 106.46E
102 F12 **Löningen** Niedersachsen, NW Germany 52.43N 7.42E
29 V11 **Lonoke** Arkansas, C USA 34.46N 91.54W
97 L21 **Lönsboda** Skåne, S Sweden 56.24N 14.19E
105 S9 **Lons-le-Saunier** *anc.* Ledo Salinarius. Jura, E France 46.40N 5.31E
33 O15 **Loogootee** Indiana, N USA 38.40N 86.54W
33 Q9 **Looking Glass River** ≈ Michigan, N USA
23 X11 **Lookout, Cape** *headland* North Carolina, SE USA 34.36N 76.31W
41 O6 **Lookout Ridge** *ridge* Alaska, USA
189 N11 **Loongana** Western Australia 30.53S 127.15E
101 I14 **Loon op Zand** Noord-Brabant, S Netherlands 51.37N 5.04E
99 A19 **Loop Head** *Ir.* Ceann Léime. *headland* W Ireland 52.55N 10.33W
111 V4 **Loosdorf** Niederösterreich, NE Austria 48.13N 15.25E
164 G10 **Lop** Xinjiang Uygur Zizhiqu, NW China 37.06N 80.12E
114 J11 **Lopare** Republika Srpska, NE Bosnia and Herzegovina 44.39N 18.49E
Lopatichi *see* Lapatsichy
131 Q15 **Lopatin** Respublika Dagestan, SW Russian Federation 43.52N 47.40E
131 P7 **Lopatino** Penzenskaya Oblast', W Russian Federation 52.38N 45.46E
178 Hh10 **Lop Buri** *var.* Loburi. Lop Buri, C Thailand 14.46N 100.40E
27 R16 **Lopeno** Texas, SW USA 26.42N 99.06W
197 D13 **Lopevi** *var.* Ulveah. *island* C Vanuatu
81 C18 **Lopez, Cap** *headland* W Gabon 0.39S 8.44E
100 I12 **Lopik** Utrecht, C Netherlands 51.58N 4.57E
Lop Nor *see* Lop Nur
164 M7 **Lop Nur** *var.* Lob Nor, Lop Nor, Lo-pu Po. *seasonal lake* NW China
Lopnur *see* Yuli
81 K17 **Lopori** ≈ NW Dem. Rep. Congo (Zaire)
100 O5 **Loppersum** Groningen, NE Netherlands 53.19N 6.45E
94 I8 **Lopphavet** *sound* N Norway
Lo-pu Po *see* Lop Nur
Lora *see* Lowrah
190 F3 **Lora Creek** *seasonal river* South Australia
106 K13 **Lora del Río** Andalucía, S Spain 37.39N 5.31W
154 M11 **Lora, Hāmūn-i** *wetland* SW Pakistan
33 T11 **Lorain** Ohio, N USA 41.27N 82.10W
27 O7 **Loraine** Texas, SW USA 32.24N 100.42W
33 R13 **Loramie, Lake** ⊞ Ohio, N USA
107 Q13 **Lorca** *Ar.* Lurka; *anc.* Eliocroca, *Lat.* Illur co. Murcia, S Spain 37.40N 1.40W
199 I12 **Lord Howe Island** *island* E Australia
Lord Howe Island *see* Ontong Java Atoll
183 O10 **Lord Howe Rise** *undersea feature* SW Pacific Ocean
199 I12 **Lord Howe Seamounts** *undersea feature* W Pacific Ocean
39 P15 **Lordsburg** New Mexico, SW USA 32.19N 108.42W
194 K8 **Lorengau** *var.* Lorungau. Manus Island, N PNG 2.03S 147.16E
27 N5 **Lorenzo** Texas, SW USA 33.40N 101.31W
148 K7 **Lorestān** *off.* Ostān-e Lorestān, *var.* Luristan. ◐ *province* W Iran
59 M17 **Loreto** Beni, N Bolivia 15.19S 64.40W
108 J12 **Loreto** Marche, C Italy 43.25N 13.37E
42 F8 **Loreto** Baja California Sur, W Mexico 25.59N 111.21W
42 M11 **Loreto** Zacatecas, C Mexico 22.15N 102.00W
58 E9 **Loreto** *off.* Departamento de Loreto. ◐ *department* NE Peru
83 K18 **Lorian Swamp** *swamp* E Kenya
56 E6 **Lorica** Córdoba, NW Colombia 9.13N 75.49W
104 G7 **Lorient** *prev.* l'Orient. Morbihan, NW France 47.45N 3.22W
113 K22 **Lőrinci** Heves, NE Hungary 47.43N 19.39E
12 G11 **Loring** Ontario, S Canada 45.55N 79.59W
35 V6 **Loring** Montana, NW USA 48.49N 107.48W
105 R13 **Loriol-sur-Drôme** Drôme, E France 44.46N 4.51E
23 U12 **Loris** South Carolina, SE USA 34.03N 78.53W
59 I18 **Loriscota, Laguna** ◎ S Peru
191 N13 **Lorne** Victoria, SE Australia 38.33S 145.57E
98 G11 **Lorn, Firth of** *inlet* W Scotland, UK
Loro Sae *see* East Timor
103 F24 **Lörrach** Baden-Württemberg, S Germany 47.37N 7.40E
105 T5 **Lorraine** ◐ *region* NE France
Lorungau *see* Lorengau
96 L11 **Los** Gävleborg, C Sweden 61.42N 15.15E
37 P14 **Los Alamos** California, W USA 34.44N 120.16W
39 S10 **Los Alamos** New Mexico, SW USA 35.52N 106.17W
44 F5 **Los Amates** Izabal, E Guatemala 15.16N 89.07W
37 S15 **Los Angeles** California, W USA 34.03N 118.14W
37 S15 **Los Angeles** ✕ California, W USA 33.54N 118.24W
65 G14 **Los Ángeles** Bío Bío, C Chile 37.29S 72.18W
37 T13 **Los Angeles Aqueduct** *aqueduct* California, W USA
Losanna *see* Lausanne
65 H20 **Los Antiguos** Santa Cruz, SW Argentina 46.36S 71.31W
201 Q16 **Losap Atoll** *atoll* C Micronesia
37 P10 **Los Banos** California, W USA 37.00N 120.39W
106 K16 **Los Barrios** Andalucía, S Spain 36.10N 5.30W
64 L5 **Los Blancos** Salta, N Argentina 23.39S 62.36W
44 L12 **Los Chiles** Alajuela, NW Costa Rica 11.00N 84.42W
107 O2 **Los Corrales de Buelna** Cantabria, N Spain 43.15N 4.04W
27 T17 **Los Fresnos** Texas, SW USA 26.03N 97.28W
37 N9 **Los Gatos** California, W USA 37.13N 121.58W
112 O11 **Łosice** Mazowieckie, E Poland 52.13N 22.42E
114 B11 **Lošinj** *Ger.* Lussin, *It.* Lussino. *island* W Croatia
Los Jardines *see* Ngetik Atoll
65 G15 **Los Lagos** Los Lagos, C Chile 39.52S 72.52W
65 F17 **Los Lagos** *off.* Región de los Lagos. ◐ *region* C Chile
Loslau *see* Wodzisław Śląski
66 N11 **Los Llanos** *var.* Los Llanos de Aridane. La Palma, Islas Canarias, Spain, NE Atlantic Ocean 28.39N 17.54W
Los Llanos de Aridane *see* Los Llanos
39 R11 **Los Lunas** New Mexico, SW USA 34.48N 106.43W
65 I16 **Los Menucos** Río Negro, C Argentina 40.52S 68.07W
42 H8 **Los Mochis** Sinaloa, C Mexico 25.48N 108.57W
37 N4 **Los Molinos** California, W USA 40.00N 122.05W
106 M9 **Los Navalmorales** Castilla-La Mancha, C Spain 39.43N 4.37W
27 S15 **Los Olmos Creek** ≈ Texas, SW USA
Losonc/Losontz *see* Lučenec
178 Jj5 **Lô, Sông** *Chin.* Panlong Jiang. ≈ China/Vietnam
46 B5 **Los Palacios** Pinar del Río, W Cuba 22.30N 83.19W
106 K14 **Los Palacios y Villafranca** Andalucía, S Spain 37.10N 5.55W
175 Ss16 **Lospalos** E East Timor 8.28S 126.56E
39 R12 **Los Pinos Mountains** ▲▲ New Mexico, SW USA
39 R11 **Los Ranchos De Albuquerque** New Mexico, SW USA 35.09N 106.37W
42 M14 **Los Reyes** Michoacán de Ocampo, SW Mexico 19.36N 102.29W
58 B7 **Los Ríos** ◐ *province* C Ecuador
66 O11 **Los Rodeos** ✕ (Santa Cruz de Tenerife) Tenerife, Islas Canarias, Spain, NE Atlantic Ocean 28.27N 16.19W
56 L4 **Los Roques, Islas** *island group* N Venezuela
45 S17 **Los Santos** Los Santos, S Panama 7.55N 80.25W
45 S17 **Los Santos** *off.* Provincia de Los Santos. ◐ *province* S Panama
Los Santos *see* Los Santos de Maimona
106 J12 **Los Santos de Maimona** *var.* Los Santos. Extremadura, W Spain 38.27N 6.22W
100 P10 **Losser** Overijssel, E Netherlands 52.16N 7.01E
98 J8 **Lossiemouth** NE Scotland, UK 57.43N 3.18W
63 B14 **Los Tábanos** Santa Fe, C Argentina 28.27S 59.57W
56 J4 **Los Taques** Falcón, N Venezuela 11.49N 70.16W
12 G11 **Lost Channel** Ontario, S Canada 45.54N 80.20W
56 L5 **Los Teques** Miranda, N Venezuela 10.23N 67.01W
37 Q12 **Lost Hills** California, W USA 35.35N 119.40W
38 I7 **Lost Peak** ▲ Utah, W USA 37.30N 113.57W
35 P11 **Lost Trail Pass** *pass* Montana, NW USA 45.40N 113.58W
195 N15 **Losuia** Kiriwina Island, SE PNG 8.30S 151.04E
64 G10 **Los Vilos** Coquimbo, C Chile 31.52S 71.28W
107 N10 **Los Yébenes** Castilla-La Mancha, C Spain 39.34N 3.52W
105 N13 **Lot** ◐ *department* S France
105 N13 **Lot** ≈ S France
65 F14 **Lota** Bío Bío, C Chile 37.08S 73.07W
83 G15 **Lotagipi Swamp** *wetland* Kenya/Sudan
104 K14 **Lot-et-Garonne** ◐ *department* SW France
85 K21 **Lothair** Mpumalanga, NE South Africa 26.22S 30.25E
35 R7 **Lothair** Montana, NW USA 48.28N 111.15W
81 L20 **Loto** Kasai Oriental, C Dem. Rep. Congo (Zaire) 2.48S 22.30E
198 B8 **Lotofagā** Upolu, SE Samoa 13.57S 171.51W
110 E10 **Lötschbergtunnel** *tunnel* Valais, SW Switzerland
27 T9 **Lott** Texas, SW USA 31.12N 97.02W
128 H3 **Lotta** *var.* Lutto. ≈ Finland/Russian Federation
192 Q7 **Lottin Point** *headland* North Island, NZ 37.26S 178.07E
Lötzen *see* Giżycko
Loualaba *see* Lualaba
178 I6 **Louangnamtha** *var.* Luong Nam Tha. Louang Namtha, N Laos 20.55N 101.24E
178 I7 **Louangphabang** *var.* Louangphrabang, Luang Prabang. Louangphabang, N Laos 19.51N 102.08E
Louangphrabang *see* Louangphabang
204 H5 **Loubet Coast** *physical region* Antarctica
Loubomo *see* Dolisie
Louch *see* Loukhi
104 H6 **Loudéac** Côtes d'Armor, NW France 48.10N 2.45W
166 M11 **Loudi** Hunan, S China 27.51N 111.58E
81 F21 **Loudima** La Bouenza, S Congo 4.06S 13.04E
22 M9 **Loudon** Tennessee, S USA 35.43N 84.19W
33 T12 **Loudonville** Ohio, N USA 40.38N 82.13W
104 L8 **Loudun** Vienne, W France 47.01N 0.04E
104 K7 **Loué** Sarthe, NW France 48.00N 0.14W
78 G10 **Louga** NW Senegal 15.36N 16.14W
99 M19 **Loughborough** C England, UK 52.46N 1.10W
99 C18 **Loughrea** *Ir.* Baile Locha Riach. W Ireland 53.12N 8.34W
105 S9 **Louhans** Saône-et-Loire, C France 46.38N 5.12E
23 P5 **Louisa** Kentucky, S USA 38.06N 82.40W
23 V5 **Louisa** Virginia, NE USA 38.02N 78.00W
23 V9 **Louisburg** North Carolina, SE USA 36.05N 78.18W
27 U12 **Louise** Texas, SW USA 29.07N 96.22W
13 P11 **Louiseville** Quebec, SE Canada 46.15N 72.54W
195 Q17 **Louisiade Archipelago** *island group* SE PNG
29 W3 **Louisiana** Missouri, C USA 39.25N 91.03W
24 G8 **Louisiana** *off.* State of Louisiana; *also known as* Creole State, Pelican State. ◐ *state* S USA
194 K9 **Lou Island** *island* N PNG
85 K19 **Louis Trichardt** Northern, NE South Africa 23.06S 29.55E
25 V4 **Louisville** Georgia, SE USA 33.00N 82.24W
32 M15 **Louisville** Illinois, N USA 38.46N 88.32W
22 K5 **Louisville** Kentucky, S USA 38.15N 85.45W
24 M4 **Louisville** Mississippi, S USA 33.07N 89.03W
31 S15 **Louisville** Nebraska, C USA 41.00N 96.09W
199 Jj12 **Louisville Ridge** *undersea feature* S Pacific Ocean
128 J6 **Loukhi** *var.* Louch. Respublika Kareliya, NW Russian Federation 66.05N 33.04E
81 H19 **Loukoléla** Cuvette, E Congo 1.04S 17.10E
106 G14 **Loulé** Faro, S Portugal 37.07N 8.01W
113 C16 **Louny** *Ger.* Laun. Ústecký Kraj NW Czech Republic 50.22N 13.49E
31 O15 **Loup City** Nebraska, C USA 41.16N 98.58W
31 P15 **Loup River** ≈ Nebraska, C USA
13 S9 **Loup, Rivière du** ≈ Quebec, SE Canada
10 K7 **Loups Marins, Lacs des** *lakes* Quebec, NE Canada
104 K16 **Lourdes** Hautes-Pyrénées, S France 43.06N 0.03W
Lourenço Marques *see* Maputo
106 F11 **Loures** Lisboa, C Portugal 38.49N 9.10W
106 F10 **Lourinhã** Lisboa, C Portugal 39.13N 9.19W
117 C16 **Loúros** ≈ W Greece
106 G8 **Lousã** Coimbra, N Portugal 40.07N 8.15W
166 M10 **Lou Shui** ≈ C China
191 O5 **Louth** New South Wales, SE Australia 30.34S 145.07E
99 O18 **Louth** E England, UK 53.18N 0.00W
99 F17 **Louth** *Ir.* Lú. *cultural region* NE Ireland
117 H15 **Loutrá** Kentrikí Makedonía, N Greece 39.55N 23.37E
117 G19 **Loutráki** Pelopónnisos, S Greece 37.55N 22.55E
Louvain *see* Leuven
101 H19 **Louvain-la Neuve** Wallon Brabant, C Belgium 50.39N 4.36E
12 J8 **Louvicourt** Quebec, SE Canada 48.04N 77.22W
104 M4 **Louviers** Eure, N France 49.13N 1.11E
32 K14 **Lou Yaeger, Lake** ⊞ Illinois, N USA
95 J15 **Lövånger** Västerbotten, N Sweden 64.22N 21.19E
128 J14 **Lovat'** ≈ NW Russian Federation
115 J17 **Lovćen** ▲ S Yugoslavia 42.22N 18.49E
116 I8 **Lovech** Lovech, N Bulgaria 43.09N 24.42E
116 I9 **Lovech** ◐ *province* N Bulgaria
27 V9 **Lovelady** Texas, SW USA 31.07N 95.27W
39 T3 **Loveland** Colorado, C USA 40.24N 105.04W
35 U12 **Lovell** Wyoming, C USA 44.50N 108.23W
Lovello, Monte *see* Grosser Löffler
37 S4 **Lovelock** Nevada, W USA 40.11N 118.30W
108 E7 **Lovere** Lombardia, N Italy 45.51N 10.06E
32 L10 **Loves Park** Illinois, N USA 42.19N 89.03W
28 M2 **Lovewell Reservoir** ⊞ Kansas, C USA
95 M19 **Loviisa** *Swe.* Lovisa. Etelä-Suomi, S Finland 60.27N 26.15E
39 V15 **Loving** New Mexico, SW USA 32.17N 104.06W
23 U6 **Lovingston** Virginia, NE USA 37.45N 78.47W
39 V14 **Lovington** New Mexico, SW USA 32.56N 103.21W
Lovisa *see* Loviisa
113 C15 **Lovosice** *Ger.* Lobositz. Ústecký Kraj, NW Czech Republic 50.29N 14.01E
128 K4 **Lovozero** Murmanskaya Oblast', NW Russian Federation 68.00N 35.03E
128 K4 **Lovozero, Ozero** ◎ NW Russian Federation
114 B9 **Lovran** *It.* Laurana. Primorje-Gorski Kotar, NW Croatia 45.16N 14.15E
118 E11 **Lovrin** *Ger.* Lowrin. Timiş, W Romania 45.58N 20.48E
84 E10 **Lóvua** Lunda Norte, NE Angola 7.21S 20.09E
84 G12 **Lóvua** Moxico, E Angola 11.33S 23.35E
67 D25 **Low Bay** *bay* East Falkland, Falkland Islands
15 M6 **Low, Cape** *headland* Nunavut, E Canada 63.05N 85.27W
35 N10 **Lowell** Idaho, NW USA 46.07N 115.36W
21 O10 **Lowell** Massachusetts, NE USA 42.37N 71.19W
Löwen *see* Leuven
Löwenberg in Schlesien *see* Lwówek Śląski
Lower Austria *see* Niederösterreich
Lower Bann *see* Bann
Lower California *see* Baja California
Lower Danube *see* Niederösterreich
193 L14 **Lower Hutt** Wellington, North Island, NZ 41.13S 174.51E
41 N11 **Lower Kalskag** Alaska, USA 61.30N 160.28W
37 O1 **Lower Klamath Lake** ◎ California, W USA
37 Q2 **Lower Lake** ◎ California/Nevada, W USA
99 E15 **Lower Lough Erne** ◎ SW Northern Ireland, UK
Lower Lusatia *see* Niederlausitz
Lower Normandy *see* Basse-Normandie, France
8 K9 **Lower Post** British Columbia, W Canada 59.53N 128.19W
31 T4 **Lower Red Lake** ◎ Minnesota, N USA
Lower Rhine *see* Neder Rijn
Lower Saxony *see* Niedersachsen
Lower Tunguska *see* Nizhnyaya Tunguska
99 Q19 **Lowestoft** E England, UK 52.28N 1.45E
155 Q5 **Lowgar** *var.* Logar. ◐ *province* E Afghanistan
190 H7 **Low Hill** South Australia 32.17S 136.46E
112 K12 **Łowicz** Łódzkie, C Poland 52.06N 19.55E
35 N13 **Lowman** Idaho, NW USA 44.04N 115.37W
155 P8 **Lowrah** *var.* Lora. ≈ SE Afghanistan
Lowrin *see* Lovrin
191 N17 **Low Rocky Point** *headland* Tasmania, SE Australia 42.59S 145.28E
20 I8 **Lowville** New York, NE USA 43.47N 75.29W
Loxa *see* Loksa
190 K9 **Loxton** South Australia 34.30S 140.36E
83 G21 **Loya** Tabora, C Tanzania 4.57S 33.53E
32 K6 **Loyal** Wisconsin, N USA 44.45N 90.30W
20 G13 **Loyalsock Creek** ≈ Pennsylvania, NE USA
37 Q5 **Loyalton** California, W USA 39.39N 120.16W
Lo-yang *see* Luoyang
197 J6 **Loyauté, Îles** *island group* S New Caledonia
Loyev *see* Loyew
121 O20 **Loyew** *Rus.* Loyev. Homyel'skaya Voblasts', SE Belarus 51.55N 30.48E
129 S13 **Loyno** Kirovskaya Oblast', NW Russian Federation 59.44N 52.42E
105 P13 **Lozère** ◐ *department* S France
105 Q14 **Lozère, Mont** ▲ S France 44.27N 3.44E
114 J11 **Loznica** Serbia, W Yugoslavia 44.32N 19.13E
119 V7 **Lozova** *Rus.* Lozovaya. Kharkivs'ka Oblast', E Ukraine 48.54N 36.22E
Lozovaya *see* Lozova
107 N7 **Lozoyuela** Madrid, C Spain 40.55N 3.36W
Lœvvajok *see* Leavvajohka
Lu *see* Shandong, China
Lú *see* Louth, Ireland
84 F12 **Luacano** Moxico, E Angola 11.19S 21.30E
81 N21 **Lualaba** *Fr.* Loualaba. ≈ SE Dem. Rep. Congo (Zaire)
85 H14 **Luampa** Western, NW Zambia 15.02S 24.27E
85 H15 **Luampa Kuta** Western, W Zambia 15.22S 24.40E
167 P8 **Lu'an** Anhui, E China 31.46N 116.31E
106 K2 **Luanco** Asturias, N Spain 43.36N 5.48W
84 A11 **Luanda** *var.* Loanda, *Port.* São Paulo de Loanda. ● (Angola) Luanda, NW Angola 8.48S 13.17E
84 A11 **Luanda** ◐ *province* NW Angola
84 A11 **Luanda** ✕ Luanda, NW Angola 8.49S 13.16E
84 D12 **Luando** ≈ C Angola
Luang *see* Tapi, Mae Nam
85 G14 **Luanginga** *var.* Luanguinga. ≈ Angola/Zambia
178 Gg15 **Luang, Khao** ▲ SW Thailand 8.21N 99.46E
Luang Prabang *see* Louangphabang
178 I8 **Luang Prabang Range** *Th.* Thiukhaoluang Phrahang. ▲▲ Laos/Thailand
178 H16 **Luang, Thale** *lagoon* S Thailand
Luangua, Rio *see* Luangwa
84 E11 **Luangue** ≈ NE Angola
Luanguinga *see* Luanginga
85 K15 **Luangwa** *var.* Aruângua. Lusaka, C Zambia 15.34S 30.23E
85 K14 **Luangwa** *var.* Aruângua, Rio Luangua. ≈ Mozambique/Zambia
167 Q2 **Luan He** ≈ E China
202 G11 **Luaniva, Île** *island* E Wallis and Futuna
167 P2 **Luanping** *var.* Anjiangying. Hebei, E China 40.55N 117.19E
84 J13 **Luanshya** Copperbelt, C Zambia 13.09S 28.24E
64 K13 **Luan Toro** La Pampa, C Argentina 36.14S 65.08W
167 Q2 **Luanxian** *var.* Luan Xian. Hebei, E China 39.47N 118.46E
84 J12 **Luapula** ◐ *province* N Zambia
81 O25 **Luapula** ≈ Dem. Rep. Congo (Zaire)/Zambia
106 J2 **Luarca** Asturias, N Spain 43.33N 6.31W
174 Ll7 **Luar, Danau** ◎ Borneo, N Indonesia
81 L25 **Luashi** Katanga, S Dem. Rep. Congo (Zaire) 10.54S 23.55E
84 G12 **Luau** *Port.* Vila Teixeira de Sousa. Moxico, NE Angola 10.43S 22.07E
81 C16 **Luba** *prev.* San Carlos. Isla de Bioco, NW Equatorial Guinea 3.26N 8.36E
44 F4 **Lubaantun** *ruins* Toledo, S Belize 16.18N 88.57W
113 P16 **Lubaczów** *var.* Lúbaczów. Podkarpackie, SE Poland 50.09N 23.08E
Lubale *see* Lubalo
84 E11 **Lubalo** Lunda Norte, NE Angola 9.02S 19.11E
84 E11 **Lubalo** *var.* Lubale. ≈ Angola/Zaire
120 J9 **Lubāna** Madona, E Latvia 56.55N 29.43E
Lubānas Ezers *see* Lubāns
179 P11 **Lubang Island** *island* N Philippines
85 B15 **Lubango** *Port.* Sá da Bandeira. Huíla, SW Angola 14.54S 13.33E
120 J9 **Lubāns** *var.* Lubānas Ezers. ◎ E Latvia
81 M21 **Lubao** Kasai Oriental, C Dem. Rep. Congo (Zaire) 5.21S 25.42E
112 O13 **Lubartów** *Ger.* Qumälisch. Lubelskie, E Poland 51.26N 22.36E
102 G13 **Lübbecke** Nordrhein-Westfalen, NW Germany 52.18N 8.37E
102 O13 **Lübben** Brandenburg, E Germany 51.55N 13.51E
103 P14 **Lübbenau** Brandenburg, E Germany 51.52N 13.57E
27 N5 **Lubbock** Texas, SW USA 33.34N 101.51W
Lubcz *see* Lyubcha
21 U6 **Lubec** Maine, NE USA 44.49N 67.00W
102 K9 **Lübeck** Schleswig-Holstein, N Germany 53.52N 10.40E
102 K8 **Lübecker Bucht** *bay* N Germany
81 M21 **Lubefu** Kasai Oriental, C Dem. Rep. Congo (Zaire) 4.43S 24.25E
113 O14 **Lubelska, Wyżyna** *plateau* SE Poland
113 O14 **Lubelskie** ◐ *province* E Poland
Lubembe *see* Luembe
Lüben *see* Lubin
150 H9 **Lubenka** Zapadnyy Kazakhstan, W Kazakhstan 50.27N 54.07E
81 P18 **Lubero** Nord Kivu, E Dem. Rep. Congo (Zaire) 0.10S 29.12E
81 L22 **Lubi** ≈ S Dem. Rep. Congo (Zaire)
Lubiana *see* Ljubljana
112 J11 **Lubień Kujawski** Kujawsko-pomorskie, C Poland 52.25N 19.10E
69 T11 **Lubilandji** ≈ S Dem. Rep. Congo (Zaire)
112 F13 **Lubin** *Ger.* Lüben. Dolnośląskie, SW Poland 51.22N 16.12E
113 O14 **Lublin** *Rus.* Lyublin. Lubelskie, E Poland 51.15N 22.33E
113 J15 **Lubliniec** Śląskie, S Poland 50.40N 18.40E
Lubnān, Jabal *see* Liban, Jebel
119 R5 **Lubny** Poltavs'ka Oblast', NE Ukraine 50.00N 33.00E
Lubombo *see* Lyubomi'
112 G11 **Luboń** *Ger.* Peterhof. Wielkolpolskie, C Poland 52.22N 16.54E
112 D12 **Lubsko** *Ger.* Sommerfeld. Lubuskie, W Poland 51.46N 14.56E
81 N24 **Lubudi** Katanga, SE Dem. Rep. Congo (Zaire) 9.57S 25.58E
174 Hh11 **Lubuklinggau** Sumatera, W Indonesia 3.15S 102.51E
81 N25 **Lubumbashi** *prev.* Élisabethville. Katanga, SE Dem. Rep. Congo (Zaire) 11.39S 27.31E
85 I14 **Lubungu** Central, C Zambia 14.28S 26.30E
112 E12 **Lubuskie** ◐ *province* W Poland
81 N18 **Lubutu** Maniema, E Dem. Rep. Congo (Zaire) 0.42S 26.31E
Luca *see* Lucca
84 C11 **Lucala** ≈ W Angola
12 E16 **Lucan** Ontario, S Canada 43.10N 81.22W
99 F18 **Lucan** *Ir.* Leamhcán. E Ireland 53.22N 6.27W

◆ Country
● Country Capital
◇ Dependent Territory
○ Dependent Territory Capital
◐ Administrative Region
✕ International Airport
▲ Mountain
▲▲ Mountain Range
Volcano
≈ River
◎ Lake
⊞ Reservoir

Lucanian Mountains see Lucano, Appennino
109 M18 **Lucano, Appennino** Eng. Lucanian Mountains. ▲ S Italy
84 F11 **Lucapa** var. Lukapa. Lunda Norte, NE Angola 8.23S 20.42E
31 V15 **Lucas** Iowa, C USA 41.01N 93.26W
63 C18 **Lucas González** Entre Ríos, E Argentina 32.25S 59.33W
67 C25 **Lucas Point** headland West Falkland, Falkland Islands 52.10S 60.22W
33 S15 **Lucasville** Ohio, N USA 38.52N 83.00W
108 F11 **Lucca** anc. Luca. Toscana, C Italy 43.49N 10.30E
46 H12 **Lucea** W Jamaica 18.26N 78.10W
99 H15 **Luce Bay** inlet SW Scotland, UK
24 M8 **Lucedale** Mississippi, S USA 30.55N 88.35W
179 Pp11 **Lucena** off. Lucena City. Luzon, N Philippines 13.57N 121.38E
106 M14 **Lucena** Andalucía, S Spain 37.25N 4.28W
107 S8 **Lucena del Cid** País Valenciano, E Spain 40.07N 0.15W
113 D15 **Lučenec** Ger. Losontz, Hung. Losonc. Banskobystrický Kraj, C Slovakia 48.21N 19.36E
Lucentum see Alicante
109 M16 **Lucera** Puglia, SE Italy 41.30N 15.19E
Lucerna/Lucerne see Luzern
Lucerne, Lake of see Vierwaldstätter See
42 J4 **Lucero** Chihuahua, N Mexico 30.51N 106.27W
127 Nn17 **Luchegorsk** Primorskiy Kray, SE Russian Federation 46.26N 134.10E
107 Q13 **Luchena** ≈ SE Spain
84 N13 **Lucheringo** var. Luchulingo. ≈ N Mozambique
Luchesa see Luchosa
Luchin see Luchyn
120 N13 **Luchosa** Rus. Luchesa. ≈ N Belarus
Luchow see Hefei
102 K11 **Lüchow** Mecklenburg-Vorpommern, N Germany 52.57N 11.10E
Luchulingo see Lucheringo
121 N17 **Luchyn** Rus. Luchin. Homyel'skaya Voblasts', SE Belarus 53.01N 30.01E
57 U11 **Lucie Rivier** ≈ W Suriname
190 K11 **Lucindale** South Australia 36.57S 140.20E
175 T13 **Lucipara, Kepulauan** island group E Indonesia
85 A14 **Lucira** Namibe, SW Angola 13.51S 12.35E
Łuck see Luts'k
103 O14 **Luckau** Brandenburg, E Germany 51.50N 13.42E
102 N13 **Luckenwalde** Brandenburg, E Germany 52.06N 13.11E
12 E15 **Lucknow** Ontario, S Canada 43.58N 81.30W
158 L12 **Lucknow** var. Lakhnau. Uttar Pradesh, N India 26.49N 80.54E
104 J10 **Luçon** Vendée, NW France 46.27N 1.10W
46 I7 **Lucrecia, Cabo** headland E Cuba 21.00N 75.34W
84 F13 **Lucusse** Moxico, E Angola 12.32S 20.46E
Lüda see Dalian
116 M9 **Luda Kamchiya** ≈ E Bulgaria
Ludasch see Luduş
116 I10 **Luda Yana** ≈ C Bulgaria
114 F7 **Ludbreg** Varaždin, N Croatia 46.15N 16.36E
31 P7 **Ludden** North Dakota, N USA 45.58N 98.07W
103 F15 **Lüdenscheid** Nordrhein-Westfalen, W Germany 51.13N 7.37E
85 C21 **Lüderitz** prev. Angra Pequena. Karas, SW Namibia 26.37S 15.10E
158 H8 **Ludhiāna** Punjab, N India 30.55N 75.52E
33 O7 **Ludington** Michigan, N USA 43.58N 86.27W
99 K20 **Ludlow** W England, UK 52.19N 2.27W
37 W14 **Ludlow** California, W USA 34.43N 116.07W
30 J7 **Ludlow** South Dakota, N USA 45.48N 103.21W
20 M9 **Ludlow** Vermont, NE USA 43.24N 72.39W
116 L7 **Ludogorie** physical region NE Bulgaria
25 W6 **Ludowici** Georgia, SE USA 31.42N 81.44W
Ludsan see Ludza
118 I10 **Luduş** Ger. Ludasch, Hung. Marosludas. Mureş, C Romania 46.27N 24.04E
97 M14 **Ludvika** Dalarna, C Sweden 60.07N 15.13E
103 H21 **Ludwigsburg** Baden-Württemberg, SW Germany 48.54N 9.12E
102 O13 **Lüdwigsfelde** Brandenburg, NE Germany 52.17N 13.15E
103 G20 **Ludwigshafen** var. Ludwigshafen am Rhein. Rheinland-Pfalz, W Germany 49.28N 8.24E
Ludwigshafen am Rhein see Ludwigshafen
103 L20 **Ludwigskanal** canal SE Germany
102 L10 **Ludwigslust** Mecklenburg-Vorpommern, N Germany 53.19N 11.28E
120 K10 **Ludza** Ger. Ludsan. Ludza, E Latvia 56.32N 27.41E
81 K21 **Luebo** Kasai Occidental, SW Dem. Rep. Congo (Zaire) 5.19S 21.21E
27 Q6 **Lueders** Texas, SW USA 32.46N 99.38W
81 N20 **Lueki** Maniema, C Dem. Rep. Congo (Zaire) 3.25S 25.49E
84 F10 **Luembe** var. Lubembe. ≈ Angola/Dem. Rep. Congo (Zaire)
84 E13 **Luena** var. Lwena, Port. Luso. Moxico, E Angola 11.46S 19.52E
81 M24 **Luena** Katanga, SE Dem. Rep. Congo (Zaire) 9.28S 25.45E
84 K12 **Luena** Northern, NE Zambia 10.31S 30.12E
84 F13 **Luena** ≈ E Angola
85 F16 **Luengue** ≈ SE Angola
69 V13 **Luenha** ≈ W Mozambique
85 G15 **Lueti** ≈ Angola/Zambia
166 J7 **Lüeyang** Shaanxi, C China 33.12N 106.31E
167 P14 **Lufeng** Guangdong, S China 22.58N 115.36E
81 N24 **Lufira** ≈ SE Dem. Rep. Congo (Zaire)
81 N25 **Lufira, Lac de Retenue de la** var. Lac Tshangalele. ◎ SE Dem. Rep. Congo (Zaire)
27 W8 **Lufkin** Texas, SW USA 31.20N 94.43W
84 L11 **Lufubu** ≈ N Zambia
128 G14 **Luga** Leningradskaya Oblast', NW Russian Federation 58.43N 29.46E
128 G13 **Luga** ≈ NW Russian Federation
Luganer See see Lugano, Lago di
110 H11 **Lugano** Ger. Lauis. Ticino, S Switzerland 46.01N 8.57E
110 H12 **Lugano, Lago di** var. Ceresio, Ger. Luganer See. ◎ S Switzerland
Lugansk see Luhans'k
197 B12 **Luganville** Espiritu Santo, C Vanuatu 15.31S 167.12E
Lugdunum see Lyon
Lugdunum Batavorum see Leiden
85 O15 **Lugela** Zambézia, NE Mozambique 16.27S 36.47E
85 O16 **Lugela** ≈ C Mozambique
84 P13 **Lugenda, Rio** ≈ N Mozambique
Luggarus see Locarno
Lugh Ganana see Luuq
99 G19 **Lugnaquillia Mountain** Ir. Log na Coille. ▲ E Ireland 52.58N 6.27W
108 H10 **Lugo** Emilia-Romagna, N Italy 44.25N 11.52E
106 I3 **Lugo** anc. Lugus Augusti. Galicia, NW Spain 43.00N 7.33W
106 I3 **Lugo** ◈ province Galicia, NW Spain
23 R12 **Lugoff** South Carolina, SE USA 34.13N 80.41W
118 F12 **Lugoj** Ger. Lugosch, Hung. Lugos. Timiş, W Romania 45.40N 21.56E
Lugos/Lugosch see Lugoj
Lugovoy/Lugovoye see Kulan
164 I13 **Lugu** Xizang Zizhiqu, W China 33.26N 84.10E
Lugus Augusti see Lugo
Luguvallium/Luguvallum see Carlisle
119 Y7 **Luhans'k** Rus. Lugansk; prev. Voroshilovgrad. Luhans'ka Oblast', E Ukraine 48.32N 39.21E
119 Y7 **Luhans'k** ✕ Luhans'ka Oblast', E Ukraine 48.25N 39.24E
119 X6 **Luhans'ka Oblast'** var. Luhans'k; prev. Voroshilovgrad, Rus. Voroshilovgradskaya Oblast'. ◈ province E Ukraine
167 Q7 **Luhe** Jiangsu, E China 32.22N 118.51E
175 T11 **Luhu** Pulau Seram, E Indonesia 3.20S 127.58E
166 G8 **Luhuo** var. Zhaggo. Sichuan, C China 31.25N 100.39E
118 M3 **Luhyny** Zhytomyrs'ka Oblast', N Ukraine 51.06N 28.24E
85 G15 **Lui** ≈ W Zambia
85 G16 **Luiana** ≈ SE Angola
85 L15 **Luia, Rio** var. Ruya. ≈ Mozambique/Zimbabwe
Luichow Peninsula see Leizhou Bandao
Luik see Liège
84 C13 **Luimbale** Huambo, C Angola 12.15S 15.19E
Luimneach see Limerick
108 D6 **Luino** Lombardia, N Italy 46.00N 8.45E
84 E13 **Luio** ≈ E Angola
94 L11 **Luiro** ≈ NE Finland
81 N25 **Luishia** Katanga, SE Dem. Rep. Congo (Zaire) 11.18S 27.08E
61 M19 **Luislândia do Oeste** Minas Gerais, SE Brazil 17.59S 45.35W
42 K5 **Luis L.León, Presa** ⊠ N Mexico
Luis Muñoz Marin see San Juan
205 N5 **Luitpold Coast** physical region Antarctica
81 K22 **Luiza** Kasai Occidental, S Dem. Rep. Congo (Zaire) 7.10S 22.27E
63 D20 **Luján** Buenos Aires, E Argentina 34.34S 59.07W
81 N24 **Lukafu** Katanga, SE Dem. Rep. Congo (Zaire) 10.28S 27.31E
Lukapa see Lucapa
114 I11 **Lukavac** Federacija Bosna I Hercegovina, NE Bosnia and Herzegovina 44.33N 18.31E
81 I20 **Lukenie** ≈ C Dem. Rep. Congo (Zaire)
81 H19 **Lukolela** Equateur, W Dem. Rep. Congo (Zaire) 1.03S 17.07E
121 M14 **Lukoml'skaye, Vozyera** Rus. Ozero Lukoml'skoye. ◎ N Belarus
Lukoml'skoye, Ozero see Lukoml'skaye, Vozyera
116 I8 **Lukovit** Lovech, N Bulgaria 43.13N 24.10E
112 O12 **Łuków** Ger. Bogendorf. Lubelskie, E Poland 51.57N 22.22E
131 O4 **Lukoyanov** Nizhegorodskaya Oblast', W Russian Federation 55.02N 44.26E
81 N22 **Lukuga** ≈ SE Dem. Rep. Congo (Zaire)
81 F21 **Lukula** Bas-Congo, SW Dem. Rep. Congo (Zaire) 5.22S 12.57E
85 G14 **Lukulu** Western, NW Zambia 14.24S 23.12E
201 R17 **Lukunor Atoll** atoll Mortlock Islands, C Micronesia
84 J12 **Lukwesa** Luapula, NE Zambia 10.03S 28.42E
95 K14 **Luleå** Norrbotten, N Sweden 65.34N 22.10E
94 J13 **Luleälven** ≈ N Sweden
142 C10 **Lüleburgaz** Kırklareli, NW Turkey 41.25N 27.22E
166 M4 **Lüliang Shan** ▲ C China
81 O21 **Lulimba** Maniema, E Dem. Rep. Congo (Zaire) 4.42S 28.37E
24 K9 **Luling** Louisiana, S USA 29.55N 90.22W
27 T11 **Luling** Texas, SW USA 29.40N 97.39W
81 I18 **Lulonga** ≈ NW Dem. Rep. Congo (Zaire)
81 K22 **Lulua** ≈ S Dem. Rep. Congo (Zaire)
Luluabourg see Kananga
198 Dd8 **Luma** Ta'ū, E American Samoa 14.15S 169.30W
174 M16 **Lumajang** Jawa, C Indonesia 8.06S 113.13E
164 G12 **Lumajangdong Co** ◎ W China
84 G13 **Lumbala Kaquengue** Moxico, E Angola 12.40S 22.34E
85 F14 **Lumbala N'Guimbo** var. Nguimbo, Port. Gago Coutinho, Vila Gago Coutinho. Moxico, E Angola 14.04S 21.25E
23 T11 **Lumber River** ≈ North Carolina/South Carolina, SE USA
Lumber State see Maine
24 L8 **Lumberton** Mississippi, S USA 31.00N 89.27W
23 U11 **Lumberton** North Carolina, SE USA 34.37N 79.00W
107 R4 **Lumbier** Navarra, N Spain 42.39N 1.19W
85 Q15 **Lumbo** Nampula, NE Mozambique 15.00S 40.40E
128 M4 **Lumbovka** Murmanskaya Oblast', NW Russian Federation 67.41N 40.31E
106 J7 **Lumbrales** Castilla-León, N Spain 40.57N 6.43W
159 W13 **Lumding** Assam, NE India 25.46N 93.10E
84 F12 **Lumege** var. Lumeje. Moxico, E Angola 11.30S 20.57E
Lumeje see Lumege
194 F10 **Lumi** Sandaun, NW PNG 3.30S 142.04E
101 J17 **Lummen** Limburg, NE Belgium 50.58N 5.12E
95 J20 **Lumparland** Åland, SW Finland 60.06N 20.15E
178 K12 **Lumphăt** prev. Lomphat. Rôtânôkiri, NE Cambodia 13.32N 106.57E
9 U16 **Lumsden** Saskatchewan, S Canada 50.39N 104.52W
193 C23 **Lumsden** Southland, South Island, NZ 45.43S 168.26E
174 J11 **Lumut, Tanjung** headland Sumatera, W Indonesia 3.47S 105.55E
163 P4 **Lün** Töv, C Mongolia 47.51N 105.11E
166 H13 **Lunan** var. Lunan Yizu Zizhixian. Yunnan, SW China 24.46N 103.12E
Lunan Yizu Zizhixian see Lunan
118 I13 **Lunca Corbului** Argeş, S Romania 44.41N 24.46E
97 K23 **Lund** Skåne, S Sweden 55.42N 13.10E
37 X6 **Lund** Nevada, W USA 38.50N 115.00W
84 D11 **Lunda Norte** ◈ province NE Angola
84 E12 **Lunda Sul** ◈ province NE Angola
84 M13 **Lundazi** Eastern, NE Zambia 12.19S 33.10E
97 G16 **Lunde** Telemark, S Norway 61.31N 6.37E
Lundenburg see Břeclav
97 C17 **Lundevatnet** ◎ S Norway
Lundi see Runde
99 I23 **Lundy** island SW England, UK
102 J10 **Lüneburg** Niedersachsen, N Germany 53.15N 10.25E
102 J11 **Lüneburger Heide** heathland NW Germany
105 Q15 **Lunel** Hérault, S France 43.40N 4.08E
103 F14 **Lünen** Nordrhein-Westfalen, W Germany 51.37N 7.31E
11 P16 **Lunenburg** Nova Scotia, SE Canada 44.22N 64.21W
23 V7 **Lunenburg** Virginia, NE USA 36.56N 78.15W
105 T5 **Lunéville** Meurthe-et-Moselle, NE France 48.34N 6.30E
85 I14 **Lunga** ≈ C Zambia
Lunga, Isola see Dugi Otok
164 H12 **Lungdo** Xizang Zizhiqu, W China 33.45N 82.09E
164 I14 **Lunggar** Xizang Zizhiqu, W China 31.10N 84.01E
78 I15 **Lungi** ✕ (Freetown) W Sierra Leone 8.36N 13.10W
Lungkiang see Qiqihar
Lungleh see Lunglei
159 W15 **Lunglei** prev. Lungleh. Mizoram, NE India 22.55N 92.49E
164 L15 **Lungsang** Xizang Zizhiqu, W China 29.49N 88.27E
84 E13 **Lungué-Bungo** var. Lungwebungu. ≈ Angola/Zambia see also Lungwebungu
85 G14 **Lungwebungu** var. Lungué-Bungo. ≈ Angola/Zambia see also Lungué-Bungo
158 F12 **Lūni** Rājasthān, N India 26.03N 73.00E
158 F12 **Lūni** ≈ N India
Luninets see Luninyets
37 S7 **Luning** Nevada, W USA 38.29N 118.10W
Łuniniec see Luninyets
131 P6 **Lunino** Penzenskaya Oblast', W Russian Federation 53.35N 45.12E
121 J19 **Luninyets** Pol. Łuniniec, Rus. Luninets. Brestskaya Voblasts', SW Belarus 52.15N 26.49E
158 F10 **Lūnkaransar** Rājasthān, NW India 28.31N 73.49E
121 G17 **Lunna** Pol. Łunna, Rus. Lunna. Hrodzyenskaya Voblasts', W Belarus 53.27N 24.16E
78 I15 **Lunsar** W Sierra Leone 8.40N 12.31W
85 K14 **Lunsemfwa** ≈ C Zambia
164 J6 **Luntai** var. Bügür. Xinjiang Uygur Zizhiqu, NW China 41.48N 84.14E
100 K11 **Lunteren** Gelderland, C Netherlands 52.04N 5.37E
175 O16 **Lunyuk** Sumbawa, S Indonesia 8.56S 117.15E
111 U5 **Lunz am See** Niederösterreich, C Austria 47.54N 15.01E
169 Y7 **Luobei** var. Fengxiang. Heilongjiang, NE China 47.35N 130.51E
166 J13 **Luodian** var. Longping. Guizhou, S China 25.25N 106.49E
166 M15 **Luoding** Guangdong, S China 22.44N 111.28E
166 M6 **Luo He** ≈ C China
166 L5 **Luo He** ≈ C China
167 N7 **Luohe** Henan, C China 33.37N 114.00E
Luolajarvi see Kuoloyarvi
Luong Nam Tha see Louangnamtha
166 L13 **Luoqing Jiang** ≈ S China
167 O8 **Luoshan** Henan, C China 32.12N 114.30E
167 O12 **Luoxiao Shan** ▲ S China
167 N6 **Luoyang** var. Honan, Lo-yang. Henan, C China 34.40N 112.25E
167 R12 **Luoyuan** Fujian, SE China 26.29N 119.32E
81 F21 **Luozi** Bas-Congo, W Dem. Rep. Congo (Zaire) 4.57S 14.07E
85 J17 **Lupane** Matabeleland North, W Zimbabwe 18.46S 27.47E
166 I12 **Lupanshui** prev. Shuicheng. Guizhou, S China 26.38N 104.49E
174 Ll7 **Lupar, Batang** ≈ East Malaysia
Lupatia see Altamura
118 G12 **Lupeni** Hung. Lupény. Hunedoara, SW Romania 45.20N 23.07E
Lupény see Lupeni
84 N13 **Lupiliche** Niassa, N Mozambique 11.36S 35.15E
85 E14 **Lupire** Cuando Cubango, E Angola 14.39S 19.39E
179 Rr16 **Lupon** Mindanao, S Philippines 6.53N 126.00E
81 L22 **Luputa** Kasai Oriental, S Dem. Rep. Congo (Zaire) 7.07S 23.43E
123 Jj17 **Luqa** ✕ (Valletta) S Malta 35.53N 14.27E
165 U11 **Luqu** Gansu, C China 34.34N 102.27E
47 U5 **Luquillo, Sierra de** ▲ E Puerto Rico
28 L4 **Luray** Kansas, C USA 39.06N 98.41W
23 U4 **Luray** Virginia, NE USA 38.40N 78.27W
105 T7 **Lure** Haute-Saône, E France 47.42N 6.30E
84 D11 **Luremo** Lunda Norte, NE Angola 8.32S 17.55E
99 F15 **Lurgan** Ir. An Lorgain. S Northern Ireland, UK 54.28N 6.19W
59 K18 **Luribay** La Paz, W Bolivia 17.09S 67.39W
85 Q14 **Lúrio** Nampula, NE Mozambique 13.32S 40.33E
85 P14 **Lúrio, Rio** ≈ NE Mozambique
Luristan see Lorestān
Lurka see Lorca
85 J15 **Lusaka** ● (Zambia) Lusaka, SE Zambia 15.23S 28.16E
85 J15 **Lusaka** ◈ province C Zambia
85 J15 **Lusaka** ✕ Lusaka, C Zambia 15.10S 28.22E
81 L21 **Lusambo** Kasai Oriental, C Dem. Rep. Congo (Zaire) 4.54S 23.25E
195 N14 **Lusancay Islands and Reefs** island group SE PNG
81 I21 **Lusanga** Bandundu, SW Dem. Rep. Congo (Zaire) 4.55S 18.40E
81 N21 **Lusangi** Maniema, E Dem. Rep. Congo (Zaire) 4.39S 27.10E
Lusatian Mountains see Lausitzer Bergland
Lushnja see Lushnjë
115 K21 **Lushnjë** var. Lushnja. Fier, C Albania 40.54N 19.43E
83 J21 **Lushoto** Tanga, E Tanzania 4.48S 38.19E
104 L10 **Lusignan** Vienne, W France 46.25N 0.06E
35 Z15 **Lusk** Wyoming, C USA 42.45N 104.27W
Luso see Luena
104 L10 **Lussac-les-Châteaux** Vienne, W France 46.23N 0.44E
Lussin/Lussino see Lošinj
Lussinpiccolo see Mali Lošinj
110 I7 **Lustenau** Vorarlberg, W Austria 47.26N 9.39E
167 T14 **Lü Tao** var. Huoshao Dao, Lütao, Eng. Green Island. island SE Taiwan
Lūt, Baḩrat/Lut, Bahret see Dead Sea
24 K9 **Lutcher** Louisiana, S USA 30.02N 90.42W
149 T9 **Lūt, Dasht-e** var. Kavīr-e Lūt. desert E Iran
85 F14 **Lutembo** Moxico, E Angola 13.30S 21.21E
Lutetia/Lutetia Parisiorum see Paris
Luteva see Lodève
12 G15 **Luther Lake** ◎ Ontario, S Canada
195 U13 **Luti** Choiseul Island, NW Solomon Islands 7.13S 157.01E
Lūt, Kavīr-e see Lūt, Dasht-e
99 N21 **Luton** SE England, UK 51.52N 0.25W
99 N21 **Luton** ✕ (London) SE England, UK 51.54N 0.24W
110 B10 **Lutry** Vaud, SW Switzerland 46.31N 6.31E
15 Ii8 **Łutselk'e** prev. Snowdrift. Northwest Territories, W Canada 62.24N 110.42W
31 Y4 **Lutsen** Minnesota, N USA 47.39N 90.37W
118 J4 **Luts'k** Pol. Łuck, Rus. Lutsk. Volyns'ka Oblast', NW Ukraine 50.45N 25.22E
Luttenberg see Ljutomer
Lüttich see Liège
85 G25 **Luttig** Western Cape, SW South Africa 32.33S 22.13E
Lutto see Lotta
84 E13 **Lutuai** Moxico, E Angola 12.38S 20.06E
119 Y7 **Lutuhyne** Luhans'ka Oblast', E Ukraine 48.24N 39.12E
176 Ww13 **Lutur, Pulau** island Kepulauan Aru, E Indonesia
25 V12 **Lutz** Florida, SE USA 28.09N 82.27W
Lutzow-Holm Bay see Lützow Holmbukta
205 V2 **Lützow Holmbukta** var. Lutzow-Holm Bay. bay Antarctica
83 L16 **Luuq** It. Lugh Ganana. Gedo, SW Somalia 3.42N 42.34E
94 M12 **Luusua** Lappi, NE Finland 66.28N 27.16E
25 Q6 **Luverne** Alabama, S USA 31.43N 86.15W
31 S11 **Luverne** Minnesota, N USA 43.39N 96.12W
81 O22 **Luvua** ≈ SE Dem. Rep. Congo (Zaire)
84 F13 **Luvuei** Moxico, E Angola 13.08S 21.09E
83 H24 **Luwego** ≈ S Tanzania
84 K12 **Luwingu** Northern, NE Zambia 10.13S 29.55E
175 Qq9 **Luwuk** prev. Loewoek. Sulawesi, C Indonesia 0.55S 122.46E
25 N3 **Luxapallila Creek** ≈ Alabama/Mississippi, S USA
101 M25 **Luxembourg** ● (Luxembourg) Luxembourg, S Luxembourg 49.37N 6.07E
101 M25 **Luxembourg** off. Grand Duchy of Luxembourg, var. Lëtzebuerg, Luxemburg. ◆ monarchy NW Europe
101 J23 **Luxembourg** ◈ province SE Belgium
101 L24 **Luxembourg** ◈ district S Luxembourg
33 N6 **Luxemburg** Wisconsin, N USA 44.32N 87.42W
Luxemburg see Luxembourg
105 U7 **Luxeuil-les-Bains** Haute-Saône, E France 47.49N 6.22E
166 E13 **Luxi** prev. Mangshi. Yunnan, SW China 24.27N 98.31E
84 E10 **Luxico** ≈ Angola/Dem. Rep. Congo (Zaire)
77 X10 **Luxor** Ar. Al Uqşur. E Egypt 25.39N 32.39E
77 X10 **Luxor** ✕ C Egypt 25.39N 32.48E
166 M4 **Luya Shan** ▲ C China
104 J15 **Luy de Béarn** ≈ SW France
104 J15 **Luy de France** ≈ SW France
129 P12 **Luza** Kirovskaya Oblast', NW Russian Federation 60.37N 47.13E
129 Q12 **Luza** ≈ NW Russian Federation
106 I16 **Luz, Costa de la** coastal region SW Spain
113 K20 **Luže** var. Lausche. ▲ Czech Republic/Germany see also Lausche 50.51N 14.40E
110 F8 **Luzern** Fr. Lucerne, It. Lucerna. Luzern, C Switzerland 47.03N 8.16E
110 E8 **Luzern** Fr. Lucerne. ◈ canton C Switzerland
166 L13 **Luzhai** Guangxi Zhuangzu Zizhiqu, S China 24.33N 109.46E
120 K12 **Luzhki** Rus. Luzhki. Vitsyebskaya Voblasts', N Belarus 55.20N 27.54E
166 I10 **Luzhou** Sichuan, C China 28.55N 105.28E
Lužická Nisa see Neisse
Lužické Hory see Lausitzer Bergland
Lužnice see Lainsitz
179 Pp9 **Luzon** island N Philippines
179 Oo6 **Luzon Strait** strait Philippines/Taiwan
118 I5 **L'viv** Ger. Lemberg, Pol. Lwów, Rus. L'vov. L'vivs'ka Oblast', W Ukraine 49.48N 24.04E
L'viv see L'vivs'ka Oblast'
118 I4 **L'vivs'ka Oblast'** var. L'viv, Rus. L'vovskaya Oblast'. ◈ province NW Ukraine
L'vov see L'viv
L'vovskaya Oblast' see L'vivs'ka Oblast'
Lwena see Luena
Lwów see L'viv
112 F11 **Lwówek** Ger. Neustadt bei Pinne. Wielkolpolskie, C Poland 52.27N 16.10E
113 E14 **Lwówek Śląski** Ger. Löwenberg in Schlesien. Dolnośląskie, SW Poland 51.06N 15.35E
121 I18 **Lyakhavichy** Rus. Lyakhovichi. Brestskaya Voblasts', SW Belarus 53.01N 26.15E
Lyakhovichi see Lyakhavichy
193 B22 **Lyall, Mount** ▲ South Island, NZ 45.14S 167.31E
Lyallpur see Faisalābād
125 G10 **Lyamin** ≈ C Russian Federation
Lyangar see Langar
125 G10 **Lyantor** Khanty-Mansiyskiy Avtonomnyy Okrug, C Russian Federation 61.40N 72.21E
128 H11 **Lyaskelya** Respublika Kareliya, NW Russian Federation 61.42N 31.06E
121 I18 **Lyasnaya** Rus. Lesnaya. Brestskaya Voblasts', SW Belarus 52.58N 25.46E
121 F19 **Lyasnaya** Pol. Leśna, Rus. Lesnaya. ≈ SW Belarus
128 H15 **Lychkovo** Novgorodskaya Oblast', W Russian Federation 57.55N 32.24E
Lyck see Ełk
95 I15 **Lycksele** Västerbotten, N Sweden 64.34N 18.40E
20 G13 **Lycoming Creek** ≈ Pennsylvania, NE USA
Lycopolis see Asyût
205 N3 **Lyddan Island** island Antarctica
85 K20 **Lydenburg** Mpumalanga, NE South Africa 25.01S 30.27E
121 L20 **Lyel'chytsy** Rus. Lel'chitsy. Homyel'skaya Voblasts', SE Belarus 51.46N 28.19E
121 P14 **Lyenina** Rus. Lenino. Mahilyowskaya Voblasts', E Belarus 54.26N 31.07E
120 L13 **Lyepyel'** Rus. Lepel'. Vitsyebskaya Voblasts', N Belarus 54.54N 28.43E
27 S17 **Lyford** Texas, SW USA 26.24N 97.47W
97 E17 **Lygna** ≈ S Norway
20 G14 **Lykens** Pennsylvania, NE USA 40.33N 76.42W
117 E21 **Lykódimo** ▲ S Greece 36.56N 21.49E
99 K24 **Lyme Bay** bay S England, UK
99 K24 **Lyme Regis** S England, UK 50.44N 2.55W
112 L7 **Łyna** Ger. Alle. ≈ N Poland
31 P12 **Lynch** Nebraska, C USA 42.49N 98.27W
22 J10 **Lynchburg** Tennessee, S USA 35.15N 86.22W
23 T6 **Lynchburg** Virginia, NE USA 37.24N 79.08W
23 T12 **Lynches River** ≈ South Carolina, SE USA
34 H6 **Lynden** Washington, NW USA 48.57N 122.27W
190 I5 **Lyndhurst** South Australia 30.19S 138.20E
29 Q5 **Lyndon** Kansas, C USA 38.37N 95.40W
21 N7 **Lyndonville** Vermont, NE USA 44.31N 71.58W
97 D18 **Lyngdal** Vest-Agder, S Norway 58.07N 7.04E
94 I9 **Lyngen** inlet Arctic Ocean
97 G17 **Lyngør** Aust-Agder, S Norway 58.38N 9.05E
94 I9 **Lyngseidet** Troms, N Norway 69.36N 20.07E
21 P11 **Lynn** Massachusetts, NE USA 42.28N 70.57W
Lynn see King's Lynn
25 R9 **Lynn Haven** Florida, SE USA 30.15N 85.39W
9 V11 **Lynn Lake** Manitoba, C Canada 56.51N 101.01W
Lynn Regis see King's Lynn
120 I13 **Lyntupy** Rus. Lyntupy. Vitsyebskaya Voblasts', NW Belarus 55.03N 26.19E
105 R11 **Lyon** Eng. Lyons; anc. Lugdunum. Rhône, E France 45.46N 4.49E
15 H3 **Lyon, Cape** headland Northwest Territories, NW Canada 69.47N 123.10W
20 K6 **Lyon Mountain** ▲ New York, NE USA 44.42N 73.52W
105 Q11 **Lyonnais, Monts du** ▲ C France
67 N25 **Lyon Point** headland SE Tristan da Cunha 37.06S 12.13W
190 E5 **Lyons** South Australia 30.40S 133.50E
39 T3 **Lyons** Colorado, C USA 40.13N 105.16W
25 V6 **Lyons** Georgia, SE USA 32.12N 82.19W
28 M5 **Lyons** Kansas, C USA 38.21N 98.12W
31 R14 **Lyons** Nebraska, C USA 41.56N 96.28W
20 G10 **Lyons** New York, NE USA 43.03N 76.58W
Lyons see Lyon
120 O13 **Lyozna** Rus. Liozno. Vitsyebskaya Voblasts', NE Belarus 55.01N 30.48E
119 S4 **Lypova Dolyna** Sums'ka Oblast', NE Ukraine 50.36N 33.50E
119 N6 **Lypovets'** Rus. Lipovets. Vinnyts'ka Oblast', C Ukraine 49.13N 29.06E
Lys see Leie
113 I18 **Lysá Hora** ▲ E Czech Republic 49.31N 18.27E
97 D16 **Lysefjorden** fjord S Norway
97 I18 **Lysekil** Västra Götaland, S Sweden 58.16N 11.25E
Lýsi see Akdoğan
35 V14 **Lysite** Wyoming, C USA 43.16N 107.42W
131 P3 **Lyskovo** Nizhegorodskaya Oblast', W Russian Federation 56.04N 45.01E
110 D8 **Lyss** Bern, W Switzerland 47.04N 7.19E
97 H22 **Lystrup** Århus, C Denmark 56.13N 10.13E
129 V14 **Lys'va** Permskaya Oblast', NW Russian Federation 58.04N 57.48E
119 P6 **Lysyanka** Cherkas'ka Oblast', C Ukraine 49.15N 30.50E
119 X6 **Lysychans'k** Rus. Lisichansk. Luhans'ka Oblast', E Ukraine 48.52N 38.27E
99 K17 **Lytham St Anne's** NW England, UK 53.45N 3.01W
193 I19 **Lyttelton** Canterbury, South Island, NZ 43.35S 172.44E
8 M17 **Lytton** British Columbia, SW Canada 50.12N 121.34W
121 L18 **Lyuban'** Rus. Lyuban'. Minskaya Voblasts', S Belarus 52.48N 28.00E
121 L18 **Lyubanskaye Vodaskhovishcha** ⊠ C Belarus
118 M5 **Lyubar** Zhytomyrs'ka Oblast', N Ukraine 49.54N 27.48E
Lyubashëvka see Lyubashivka
119 O8 **Lyubashivka** Rus. Lyubashëvka. Odes'ka Oblast', SW Ukraine 47.49N 30.18E
121 I16 **Lyubcha** Pol. Lubcz, Rus. Lyubcha. Hrodzyenskaya Voblasts', W Belarus 53.46N 26.04E
130 L4 **Lyubertsy** Moskovskaya Oblast', W Russian Federation 55.37N 38.02E
118 K2 **Lyubeshiv** Volyns'ka Oblast', NW Ukraine 51.46N 25.33E
128 M14 **Lyubim** Yaroslavskaya Oblast', NW Russian Federation 58.21N 40.46E
116 K11 **Lyubimets** Khaskovo, S Bulgaria 41.51N 26.03E
Lyublin see Lublin
118 I3 **Lyuboml'** Pol. Luboml. Volyns'ka Oblast', NW Ukraine 51.12N 24.01E
Lyubotin see Lyubotyn
119 U5 **Lyubotyn** Rus. Lyubotin. Kharkivs'ka Oblast', E Ukraine 49.57N 35.57E
130 I5 **Lyudinovo** Kaluzhskaya Oblast', W Russian Federation 53.52N 34.28E
131 T2 **Lyuk** Udmurtskaya Respublika, NW Russian Federation 56.55N 52.45E
116 M9 **Lyulyakovo** prev. Keremitlik. Burgas, E Bulgaria 42.53N 27.05E
121 I18 **Lyusina** Rus. Lyusino. Brestskaya Voblasts', SW Belarus 52.37N 26.31E
Lyusino see Lyusina

M

144 G9 **Ma'ād** Irbid, N Jordan 32.37N 35.36E
Maalahti see Malax
157 **Maale** see Male'
144 G13 **Ma'ān** Ma'ān, SW Jordan 30.10N 35.45E
144 H13 **Ma'ān** off. Muḩāfaz̧at Ma'ān, var. Ma'an, Ma'ān. ◈ governorate S Jordan
95 M16 **Maaninka** Itä-Suomi, C Finland 63.10N 27.19E
168 K7 **Maanĭt** Bulgan, C Mongolia 48.17N 103.29E
168 M8 **Maanĭt** Töv, C Mongolia 47.14N 107.34E
95 N15 **Maanselkä** Oulu, C Finland 63.53N 28.27E
167 Q8 **Ma'anshan** Anhui, E China 31.45N 118.31E
196 F16 **Maap** island Caroline Islands, W Micronesia
120 H3 **Maardu** Ger. Maart. Harjumaa, NW Estonia 59.28N 25.01E
Ma'aret-en-Nu'man see Ma'arrat an Nu'mān
101 K16 **Maarheeze** Noord-Brabant, SE Netherlands 51.19N 5.37E
Maarianhamina see Mariehamn
144 I4 **Ma'arrat an Nu'mān** var. Ma'aret-en-Nu'man, Fr. Maarret enn Naamâne. Idlib, NW Syria 35.40N 36.40E
Maarret enn Naamâne see Ma'arrat an Nu'mān
100 I11 **Maarssen** Utrecht, C Netherlands 52.07N 5.03E
Maart see Maardu
101 L17 **Maas** Fr. Meuse. ≈ W Europe see also Meuse
101 M15 **Maasbree** Limburg, SE Netherlands 51.22N 6.03E
101 L17 **Maaseik** prev. Maeseyck. Limburg, NE Belgium 51.04N 5.48E
179 R13 **Maasin** Leyte, C Philippines 10.10N 124.55E
101 L17 **Maasmechelen** Limburg, NE Belgium 50.58N 5.42E
100 G12 **Maassluis** Zuid-Holland, SW Netherlands 51.55N 4.15E
101 L18 **Maastricht** var. Maestricht; anc. Traietum ad Mosam, Traiectum Tungorum. Limburg, SE Netherlands 50.51N 5.42E
191 N18 **Maatsuyker Group** island group Tasmania, SE Australia
Maba see Qujiang
85 L20 **Mabalane** Gaza, S Mozambique 23.43S 32.37E
27 V7 **Mabank** Texas, SW USA 32.22N 96.06W
172 N10 **Mabechi-gawa** var. Mabuchi-gawa. ≈ Honshū, C Japan
99 O18 **Mablethorpe** E England, UK 53.20N 0.14E
176 W9 **Maboi** Irian Jaya, E Indonesia 1.00S 134.02E
85 M19 **Mabote** Inhambane, S Mozambique 22.03S 34.09E
34 J10 **Mabton** Washington, NW USA 46.13N 120.00W
Mabuchi-gawa see Mabechi-gawa
85 H20 **Mabutsane** Southern, S Botswana 24.22S 23.34E
65 G19 **Macá, Cerro** ▲ S Chile 45.07S 73.11W
62 Q9 **Macaé** Rio de Janeiro, SE Brazil 22.21S 41.48W
84 N13 **Macaloge** Niassa, N Mozambique 12.30S 35.25E
Macan see Bonerate, Kepulauan
167 N15 **Macao** Chin. Aomen, Port. Macau. S China
106 H9 **Mação** Santarém, C Portugal 39.33N 8.00W
60 J11 **Macapá** state capital Amapá, N Brazil 0.04N 51.04W
45 S17 **Macaracas** Los Santos, S Panama 7.43N 80.33W
57 P6 **Macare, Caño** ≈ NE Venezuela
57 Q6 **Macareo, Caño** ≈ NE Venezuela
Macarsca see Makarska
MacArthur see Ormoc
190 L12 **Macarthur** Victoria, SE Australia 38.04S 142.02E
58 C7 **Macas** Morona Santiago, SE Ecuador 2.22S 78.07W
Macassar see Ujungpandang
61 Q14 **Macau** Rio Grande do Norte, E Brazil 5.04S 36.37W
Macau see Macao
Macău see Makó, Hungary
67 E24 **Macbride Head** headland East Falkland, Falkland Islands 51.25S 57.55W
25 V9 **Macclenny** Florida, SE USA 30.16N 82.07W
99 L18 **Macclesfield** C England, UK 53.16N 2.07W
198 F6 **Macclesfield Bank** undersea feature N South China Sea
MacCluer Gulf see Berau, Teluk
189 N7 **Macdonald, Lake** salt lake Western Australia
189 Q7 **Macdonnell Ranges** ▲ Northern Territory, C Australia
98 K8 **Macduff** NE Scotland, UK 57.39N 2.28W
106 I6 **Macedo de Cavaleiros** Bragança, N Portugal 41.31N 6.57W
Macedonia Central see Kentrikí Makedonía
Macedonia East and Thrace see Anatolikí Makedonía kai Thráki
115 O19 **Macedonia, FYR** off. the Former Yugoslav Republic of Macedonia, var. Macedonia, Mac. Makedonija, abbrev. FYR Macedonia, FYROM. ◆ republic SE Europe
Macedonia West see Dytikí Makedonía
61 Q16 **Maceió** state capital Alagoas, E Brazil 9.40S 35.43W
78 K15 **Macenta** Guinée-Forestière, SE Guinea 8.31N 9.31W
108 J12 **Macerata** Marche, C Italy 43.19N 13.28E
9 S11 **MacFarlane** ≈ Saskatchewan, C Canada
190 H7 **Macfarlane, Lake** var. Lake Mcfarlane. ◎ South Australia
Macgillicuddy's Reeks Mountains see Macgillycuddy's Reeks
99 B21 **Macgillycuddy's Reeks** var. Macgillicuddy's Reeks Mountains, Ir. Na Cruacha Dubha. ▲ SW Ireland
9 X16 **MacGregor** Manitoba, S Canada 49.58N 98.49W
155 O10 **Mach** Baluchistān, SW Pakistan 29.52N 67.19E

◆ Country
● Country Capital
◇ Dependent Territory
○ Dependent Territory Capital
◈ Administrative Region
✕ International Airport
▲ Mountain
▲ Mountain Range
Volcano
≈ River
◎ Lake
⊠ Reservoir

58 C6 **Machachi** Pichincha, C Ecuador 0.33S 78.34W

85 M19 **Machaila** Gaza, S Mozambique 22.15S 32.57E

Machaire Fíolta see Magherafelt

Machaire Rátha see Maghera

83 I19 **Machakos** Eastern, S Kenya 1.33S 37.17E

58 B8 **Machala** El Oro, SW Ecuador 3.19S 79.57W

85 J19 **Machaneng** Central, SE Botswana 23.12S 27.28E

85 M18 **Machanga** Sofala, E Mozambique 20.55S 35.03E

82 G13 **Machar Marshes** wetland SE Sudan

104 I8 **Machecoul** Loire-Atlantique, NW France 46.59N 1.51W

167 O8 **Macheng** Hubei, C China 31.10N 115.00E

161 J16 **Mācherla** Andhra Pradesh, C India 16.28N 79.25E

159 O11 **Machhapuchhre** ▲ C Nepal 28.30N 83.57E

21 T6 **Machias** Maine, NE USA 44.43N 67.28W

21 T6 **Machias River** ✎ Maine, NE USA

21 T6 **Machias River** ✎ Maine, NE USA

66 P5 **Machico** Madeira, Portugal, NE Atlantic Ocean 32.43N 16.46W

161 K16 **Machilipatnam** var. Bandar Masulipatnam. Andhra Pradesh, E India 16.12N 81.10E

56 G5 **Machiques** Zulia, NW Venezuela 10.01N 72.40W

59 G15 **Machupicchu** Cusco, C Peru 13.07S 72.30W

85 M20 **Macia** var. Vila de Macia. Gaza, S Mozambique 25.01S 33.05E

Macías Nguema Biyogo see Bioco, Isla de

118 M13 **Măcin** Tulcea, SE Romania 45.15N 28.09E

191 N14 **Macintyre River** ✎ New South Wales/Queensland, SE Australia

189 Y7 **Mackay** Queensland, NE Australia 21.10S 149.10E

189 O7 **Mackay, Lake** salt lake Northern Territory/Western Australia

8 M13 **Mackenzie** British Columbia, W Canada 55.18N 123.09W

15 G6 **Mackenzie** ✎ Northwest Territories, NW Canada

205 T6 **Mackenzie Bay** bay Antarctica

8 J1 **Mackenzie Bay** bay NW Canada

2 D9 **Mackenzie Delta** delta Northwest Territories, NW Canada

207 P8 **Mackenzie King Island** island Queen Elizabeth Islands, Northwest Territories, N Canada

14 G5 **Mackenzie Mountains** ▲ Northwest Territories, NW Canada

33 Q5 **Mackinac, Straits Of** ◎ Michigan, N USA

204 K5 **Mackintosh, Cape** headland Antarctica 72.52S 60.00W

9 R15 **Macklin** Saskatchewan, S Canada 52.19N 109.51W

191 V6 **Macksville** New South Wales, SE Australia 30.39S 152.54E

191 V5 **Maclean** New South Wales, SE Australia 29.30S 153.15E

85 J24 **Maclear** Eastern Cape, SE South Africa 31.04S 28.22E

191 U6 **Macleay River** ✎ New South Wales, SE Australia

MacLeod see Fort Macleod

188 G9 **Macleod, Lake** ◎ Western Australia

8 I6 **Macmillan** ✎ Yukon Territory, NW Canada

32 J12 **Macomb** Illinois, N USA 40.27N 90.40W

109 B18 **Macomer** Sardegna, Italy, C Mediterranean Sea 40.14N 8.46E

84 Q13 **Macomia** Cabo Delgado, NE Mozambique 12.15S 40.06E

25 T5 **Macon** Georgia, SE USA 32.48N 83.41W

25 N4 **Macon** Mississippi, S USA 33.06N 88.33W

29 U3 **Macon** Missouri, C USA 39.44N 92.28W

105 R10 **Mâcon** anc. Matisco, Matisco Ædourum. Saône-et-Loire, C France 46.19N 4.48E

24 J2 **Macon, Bayou** ✎ Arkansas/Louisiana, S USA

84 G13 **Macondo** Moxico, E Angola 12.31S 23.45E

85 M16 **Macossa** Manica, C Mozambique 17.51S 33.54E

9 T12 **Macoun Lake** ◎ Saskatchewan, C Canada

32 K14 **Macoupin Creek** ✎ Illinois, N USA

Macouria see Tonate

85 N18 **Macovane** Inhambane, SE Mozambique 21.30S 35.07E

191 N17 **Macquarie Harbour** inlet Tasmania, SE Australia

199 I15 **Macquarie Island** island NZ, SW Pacific Ocean

191 T8 **Macquarie, Lake** lagoon New South Wales, SE Australia

191 Q6 **Macquarie Marshes** wetland New South Wales, SE Australia

183 O13 **Macquarie Ridge** undersea feature SW Pacific Ocean

191 Q6 **Macquarie River** ✎ New South Wales, SE Australia

191 P17 **Macquarie River** ✎ Tasmania, SE Australia

205 V3 **Mac. Robertson Land** physical region Antarctica

99 C21 **Macroom** Ir. Maigh Chromtha. SW Ireland 51.54N 8.57W

44 G5 **Macuelizo** Santa Bárbara, NW Honduras 15.21N 88.31W

190 G2 **Macumba River** ✎ South Australia

59 I16 **Macusani** Puno, S Peru 14.05N 70.27W

58 E8 **Macuspana** Tabasco, SE Mexico 17.43N 92.36W

144 G10 **Ma'daba** var. Mādabā, Madeba; anc. Medeba. 'Ammān, NW Jordan 31.43N 35.48E

---

180 G2 **Madagascar** off. Democratic Republic of Madagascar, Malg. Madagasikara; prev. Malagasy Republic. ◆ republic W Indian Ocean

180 I5 **Madagascar** island W Indian Ocean

132 L17 **Madagascar Basin** undersea feature W Indian Ocean

132 L16 **Madagascar Plain** undersea feature W Indian Ocean

69 Y14 **Madagascar Plateau** var. Madagascar Ridge, Madagascar Rise, Rus. Madagaskarskiy Khrebet. undersea feature W Indian Ocean

Madagascar Ridge/Madagascar Rise see Madagascar Plateau

Madagaskara see Madagascar

Madagaskarskiy Khrebet see Madagascar Plateau

66 N2 **Madalena** Pico, Azores, Portugal, NE Atlantic Ocean 38.31N 28.15W

79 Y6 **Madama** Agadez, NE Niger 21.54N 13.43E

116 J12 **Madan** Smolyan, S Bulgaria 41.30N 24.58E

161 I19 **Madanapalle** Andhra Pradesh, E India 13.33N 78.31E

194 V11 **Madang** Madang, N PNG 5.09S 145.48E

194 I11 **Madang** ◆ province N PNG

152 G7 **Madaniyat** Rus. Madeniyet. Qoraqalpoghiston Respublikasi, W Uzbekistan 42.48N 59.00E

Madaniyin see Médenine

79 U12 **Madaoua** Tahoua, SW Niger 14.06N 6.01E

Madaras see Vtáčnik

159 U15 **Madaripur** Dhaka, C Bangladesh 23.09N 90.10E

79 U12 **Madarounfa** Maradi, S Niger 13.16N 7.07E

Madarska see Hungary

152 B13 **Madau** Turkm. Madaw. Balkanskiy Velayat, N Turkmenistan 38.11N 54.46E

195 P15 **Madau Island** island SE PNG

Madaw see Madau

21 S1 **Madawaska** Maine, NE USA 47.19N 68.19W

12 J13 **Madawaska** ✎ Ontario, SE Canada

Madawaska Highlands see Haliburton Highlands

177 M14 **Madaya** Mandalay, C Myanmar 22.12N 96.04E

109 L17 **Maddaloni** Campania, S Italy 41.03N 14.22E

31 O3 **Maddock** North Dakota, N USA 47.57N 99.31W

101 I14 **Made** Noord-Brabant, S Netherlands 51.40N 4.48E

Madeba see Ma'daba

66 L9 **Madeira** var. Ilha de Madeira. island Madeira, Portugal, NE Atlantic Ocean

66 O5 **Madeira, Ilha de** see Madeira

66 L9 **Madeira Islands** Port. Região Autónoma da Madeira. ◆ autonomous region Madeira, Portugal, NE Atlantic Ocean

66 V9 **Madeira Plain** undersea feature E Atlantic Ocean

66 V9 **Madeira Ridge** undersea feature E Atlantic Ocean

61 F14 **Madeira, Rio** Sp. Río Madera. ✎ Bolivia/Brazil see also Madera, Río

13 X6 **Madeleine** ✎ Quebec, SE Canada

13 X5 **Madeleine, Cap de la** headland Quebec, SE Canada 49.13N 65.20W

11 Q13 **Madeleine, Îles de la** Eng. Magdalen Islands. island group Quebec, E Canada

189 N12 **Madeley** Western Australia 31.52S 127.01E

31 U10 **Madelia** Minnesota, N USA 44.03N 94.26W

37 P3 **Madeline** California, W USA 41.02N 120.28W

32 K3 **Madeline Island** island Apostle Islands, Wisconsin, N USA

143 O15 **Maden** Elazığ, SE Turkey 38.24N 39.42E

151 V12 **Madeniyet** Vostochnyy Kazakhstan, E Kazakhstan 47.51N 78.37E

Madeniyet see Madaniyat

42 H5 **Madera** Chihuahua, N Mexico 29.10N 108.10W

37 Q10 **Madera** California, W USA 36.57N 120.02W

58 L13 **Madera, Río** Port. Rio Madeira. ✎ Bolivia/Brazil see also Madeira, Rio

108 D6 **Madesimo** Lombardia, N Italy 46.09N 9.26E

147 O14 **Madhāb, Wādī** dry watercourse NW Yemen

159 R13 **Madhepura** var. Madhipure. Bihār, NE India 25.55S 86.48E

Madhipure see Madhepura

159 Q13 **Madhubani** Bihār, N India 26.21N 86.04E

159 Q13 **Madhupur** Bihār, NE India 24.16N 86.37E

158 K15 **Madhya Pradesh** prev. Central Provinces and Berar. ◆ state C India

59 K14 **Madidi, Río** ✎ W Bolivia

161 F20 **Madikeri** prev. Mercara. Karnātaka, W India 12.28N 75.40E

29 O13 **Madill** Oklahoma, C USA 34.05N 96.46W

81 G21 **Madimba** Bas-Congo, SW Dem. Rep. Congo (Zaire) 4.58S 15.07E

144 M4 **Mā'din** Ar Raqqah, N Syria 35.45N 39.36E

78 M14 **Madinani** NW Ivory Coast 9.37N 6.57W

147 O14 **Madīnat ash Sha'b** prev. Al Ittiḥād. SW Yemen 12.52N 44.55E

144 K3 **Madīnat ath Thawrah** var. Ath Thawrah. Ar Raqqah, N Syria Asia 35.36N 39.00E

181 O6 **Madingley Rise** undersea feature W Indian Ocean

---

81 E21 **Madingo-Kayes** Le Kouilou, S Congo 4.22S 11.40E

81 F21 **Madingou** La Bouenza, S Congo 4.10S 13.33E

25 U4 **Madioen** see Madiun

25 T3 **Madison** Florida, SE USA 30.27N 83.24W

33 T3 **Madison** Georgia, SE USA 33.37N 83.28W

33 P15 **Madison** Indiana, N USA 38.44N 85.22W

29 P6 **Madison** Kansas, C USA 38.08N 96.08W

21 Q6 **Madison** Maine, NE USA 44.48N 69.52W

31 S9 **Madison** Minnesota, N USA 45.00N 96.12W

24 K5 **Madison** Mississippi, S USA 32.27N 90.07W

31 Q14 **Madison** Nebraska, C USA 41.49N 97.27W

31 R10 **Madison** South Dakota, N USA 44.00N 97.06W

23 V5 **Madison** Virginia, NE USA 38.24N 78.12W

23 Q5 **Madison** West Virginia, NE USA 38.04N 81.49W

32 L9 **Madison** state capital Wisconsin, N USA 43.04N 89.22W

23 T6 **Madison Heights** Virginia, NE USA 37.25N 79.07W

22 I6 **Madisonville** Kentucky, S USA 37.19N 87.30W

22 M10 **Madisonville** Tennessee, S USA 35.31N 84.21W

27 V9 **Madisonville** Texas, SW USA 30.57N 95.54W

174 L15 **Madiun** prev. Madioen. Jawa, C Indonesia 7.37S 111.33E

12 J14 **Madjene** see Majene

83 J18 **Mado Gashi** North Eastern, E Kenya 0.40N 39.09E

165 R11 **Madoi** Qinghai, C China 34.53N 98.07E

201 O13 **Madolenihmw** Pohnpei, E Micronesia

120 I9 **Madona** Ger. Modohn. Madona, E Latvia 56.51N 26.10E

109 J23 **Madonie** ▲ Sicilia, Italy, C Mediterranean Sea

147 Y11 **Madrakah, Ra's** headland E Oman 18.56N 57.54E

34 I12 **Madras** Oregon, NW USA 44.37N 121.07W

59 J14 **Madras** see Chennai

65 F20 **Madre de Dios** off. Departamento de Madre de Dios. ◆ department E Peru

65 J14 **Madre de Dios, Isla** island S Chile

59 J14 **Madre de Dios, Río** ✎ Bolivia/Peru

27 T16 **Madre, Laguna** ◎ Texas, SW USA

43 N9 **Madre, Laguna** lagoon NE Mexico

39 W13 **Madre Mount** ▲ New Mexico, SW USA 34.45N 108.10W

107 N8 **Madrid** ● (Spain) Madrid, C Spain 40.25N 3.43W

31 V4 **Madrid** Iowa, C USA 41.52N 93.49W

107 N7 **Madrid** ◆ autonomous community C Spain

107 N10 **Madridejos** Castilla-La Mancha, C Spain 39.28N 3.31W

106 L7 **Madrigal de las Altas Torres** Castilla-León, N Spain 41.05N 5.00W

106 K10 **Madrigalejo** Extremadura, W Spain 39.08N 5.36W

36 L3 **Mad River** ✎ California, W USA

44 J8 **Madriz** ◆ department W Nicaragua

106 K10 **Madroñera** Extremadura, W Spain 39.25N 5.46W

189 N12 **Madura** Western Australia 31.52S 127.01E

174 M15 **Madura** see Madurai

161 H21 **Madurai** prev. Madura, Mathurai. Tamil Nādu, S India 9.55N 78.07E

174 M15 **Madura, Pulau** prev. Madoera. island C Indonesia

174 Mm15 **Madura, Selat** strait C Indonesia

131 Q17 **Madzhalis** Respublika Dagestan, SW Russian Federation 42.12N 47.46E

85 M14 **Madzimoyo** Eastern, E Zambia 13.39S 32.31E

171 K15 **Mae** var. Maebasi, Mayebashi. Gunma, Honshū, S Japan 36.24N 139.01E

178 H4 **Mae Chan** Chiang Rai, NW Thailand 20.13N 99.52E

178 Sg7 **Mae Hong Son** var. Maehongson, Muai To. Mae Hong Son, NW Thailand 19.16N 97.55E

178 H6 **Mae Nam Khong** see Mekong

178 H10 **Mae Nam Tha Chin** ✎ W Thailand

178 H7 **Mae Nam Yom** ✎ W Thailand

178 H7 **Mae Sariang** Mae Hong Son, NW Thailand 18.07N 97.57E

39 Q3 **Maeser** Utah, W USA 40.28N 109.35W

178 Sg9 **Mae Sot** var. Ban Mae Sot. Tak, W Thailand 16.43N 98.31E

39 V10 **Maestre de Campo** see Magdalena

97 J15 **Magnor** Hedmark, S Norway 59.57N 12.14E

197 X14 **Mago** prev. Mango. island Lau Group, E Fiji

85 L17 **Mágoé** Tete, NW Mozambique 15.51S 31.49E

13 Q8 **Magog** Quebec, SE Canada 45.16N 72.09W

85 J23 **Magogong** Southern, S Zambia 16.01S 27.37E

43 Q12 **Magozal** Veracruz-Llave, C Mexico 21.33N 97.57W

11 O7 **Magpie** ✎ Ontario, S Canada

9 P15 **Magrath** Alberta, SW Canada 49.27N 112.52W

107 R10 **Magro** ✎ E Spain

---

83 K23 **Mafia** island E Tanzania

83 J23 **Mafia Channel** sea waterway E Tanzania

85 L21 **Mafikeng** North-West, N South Africa 25.52S 25.39E

62 P13 **Mafra** Santa Catarina, S Brazil 26.07S 49.46W

106 F10 **Mafra** Lisboa, C Portugal 38.57N 9.19W

149 Q17 **Mafraq** Abū Ẓaby, C UAE 24.21N 54.33E

Mafraq/Mafraq, Muḥāfaẓat al see Al Mafraq

127 L10 **Magadan** Magadanskaya Oblast', E Russian Federation 59.37N 150.49E

127 Nn8 **Magadanskaya Oblast'** ◆ province E Russian Federation

110 O11 **Magadino** Ticino, S Switzerland 46.09N 8.50E

65 G23 **Magallanes** off. Región de Magallanes y de la Antártica Chilena. ◆ region S Chile

Magallanes see Punta Arenas

Magallanes, Estrecho de see Magellan, Strait of

12 I10 **Maganasipi, Lac** ◎ Quebec, SE Canada

56 F6 **Magangué** Bolívar, N Colombia 9.13N 74.46W

79 V12 **Magaria** Zinder, S Niger 13.00N 8.55E

194 M16 **Magarida** Central, SW PNG 10.13S 149.17E

179 Pp9 **Magat** ✎ Luzon, N Philippines

29 T11 **Magazine Mountain** ▲ Arkansas, C USA 35.10N 93.38W

78 I15 **Magburaka** C Sierra Leone 8.43N 11.57W

126 M14 **Magdagachi** Amurskaya Oblast', SE Russian Federation 53.25N 125.41E

64 O12 **Magdalena** Buenos Aires, E Argentina 35.04S 57.30W

59 M15 **Magdalena** Beni, N Bolivia 13.22S 64.07W

42 F4 **Magdalena** Sonora, NW Mexico 30.37N 110.58W

39 Q13 **Magdalena** New Mexico, SW USA 34.07N 107.14W

56 F5 **Magdalena** off. Departamento del Magdalena. ◆ province N Colombia

42 F9 **Magdalena, Bahía** bay W Mexico

65 G19 **Magdalena, Isla** island Archipiélago de los Chonos, S Chile

42 E9 **Magdalena, Isla** island W Mexico

49 P6 **Magdalena, Río** ✎ C Colombia

42 F5 **Magdalena, Río** ✎ NW Mexico

Magdalen Islands see Madeleine, Îles de la

102 L13 **Magdeburg** Sachsen-Anhalt, C Germany 52.07N 11.39E

24 L6 **Magee** Mississippi, S USA 31.52N 89.43W

174 Kk15 **Magelang** Jawa, C Indonesia 7.28S 110.10E

199 J7 **Magellan Rise** undersea feature C Pacific Ocean

65 **Magellan, Strait of** Sp. Estrecho de Magallanes. strait Argentina/Chile

108 D7 **Magenta** Lombardia, NW Italy 45.28N 8.52E

94 K7 **Mageroya** see Magerøya

170 B17 **Mage-shima** island Nansei-shotō, SW Japan

110 G11 **Maggia** Ticino, S Switzerland 46.15N 8.42E

110 G12 **Maggia** ✎ SW Switzerland

Maggiore, Lago see Maggiore, Lake

108 C6 **Maggiore, Lake** It. Lago Maggiore. ◎ Italy/Switzerland

49 W2 **Maggotty** W Jamaica 18.09N 77.46W

78 I10 **Magham** Gorgol, S Mauritania 15.31N 12.49W

99 F14 **Maghera** Ir. Machaire Rátha. C Northern Ireland, UK 54.51N 6.40W

99 F15 **Magherafelt** Ir. Machaire Fíolta. C Northern Ireland, UK 54.45N 6.36W

196 H6 **Magicienne Bay** bay Saipan, S Northern Mariana Islands

107 O13 **Magina** ▲ S Spain 37.43N 3.24W

83 H24 **Magingo** Ruvuma, S Tanzania 9.57S 35.23E

126 Jj14 **Magistral'nyy** Irkutskaya Oblast', S Russian Federation 56.18N 107.27E

114 H11 **Maglaj** Federacija Bosna I Hercegovina, N Bosnia and Herzegovina 44.32N 18.03E

109 Q19 **Maglie** Puglia, SE Italy 40.07N 18.18E

111 X8 **Magndorf** Steiermark, SE Austria 46.54N 15.55E

Magna Utah, W USA 40.42N 112.06W

12 G12 **Magnetawan** ✎ Ontario, S Canada

Magnesia see Manisa

125 Dd12 **Magnitogorsk** Chelyabinskaya Oblast', C Russian Federation 53.28N 59.06E

27 T4 **Magnolia** Arkansas, C USA 33.16N 93.14W

24 K7 **Magnolia** Mississippi, S USA 31.08N 90.27W

Magnolia State see Mississippi

47 J15 **Mahón** Cat. Maó, Eng. Port Mahon; anc. Portus Magonis. Menorca, Spain, W Mediterranean Sea 39.54N 4.15E

120 H13 **Mahiliaŭ** see Mahilyow

---

78 I9 **Magta' Lahjar** var. Magta Lahjar, Magta' Lahjar, Magtá Lajar. Brakna, SW Mauritania 17.27N 13.07W

85 L20 **Magude** Maputo, S Mozambique 25.01S 32.40E

79 Y12 **Magumeri** Borno, NE Nigeria 12.07N 12.48E

201 O14 **Magur Islands** island group Caroline Islands, C Micronesia

177 F16 **Magwe** var. Magway. Magwe, W Myanmar 20.07N 94.59E

177 F16 **Magwe** var. Magway. ◆ division C Myanmar

Magyar-Becse see Bečej

Magyarkanizsa see Kanjiža

Magyarország see Hungary

Magyarszombor see Zimbor

148 J4 **Mahābād** prev. Mehabad; prev. Sāūjbulāgh. Āzarbāyjān-e Bākhtarī, NW Iran 36.43N 45.43E

180 I13 **Mahabo** Toliara, W Madagascar 20.22S 44.39E

161 D14 **Maha Chai** see Samut Sakhon

83 N17 **Mahadday Weyne** Shabeellaha Dhexe, C Somalia 2.55N 45.30E

81 Q17 **Mahagi** Orientale, NE Dem. Rep. Congo (Zaire) 2.16N 30.58E

180 I4 **Mahāil** see Muhāyil

158 G10 **Mahajamba** seasonal river NW Madagascar

158 G10 **Mahajan** Rājasthān, NW India 28.46N 73.49E

180 I3 **Mahajanga** var. Majunga. Mahajanga, NW Madagascar 15.40S 46.19E

180 I3 **Mahajanga** ◆ province W Madagascar

180 I3 **Mahajanga** ✕ Mahajanga, NW Madagascar 15.40S 46.19E

171 X16 **Mahakam, Sungai** var. Koetai, Kutai. ✎ Borneo, C Indonesia

85 I19 **Mahalapye** var. Mahalatswe. Central, SE Botswana 23.01S 26.52E

Mahalatswe see Mahalapye

175 Q10 **Mahalona** Sulawesi, C Indonesia 2.37S 121.36E

149 S11 **Mahān** Kermān, E Iran 30.07N 57.15E

149 O17 **Mahanadi** ✎ E India

180 J5 **Mahanoro** Toamasina, E Madagascar 19.52S 48.48E

159 O15 **Mahārājganj** Bihār, N India 26.07N 84.31E

160 J13 **Mahārāshtra** ◆ state W India

161 K24 **Mahaweli Ganga** ✎ C Sri Lanka

161 J15 **Mahbés** see El Mahbas

161 H16 **Mahbūbābād** Andhra Pradesh, E India 17.35N 80.00E

161 H16 **Mahbūbnagar** Andhra Pradesh, C India 16.45N 78.01E

146 M8 **Mahd adh Dhahab** Al Madīnah, W Saudi Arabia 23.33N 40.56E

180 H3 **Mahdia** C Guyana 5.16N 59.08W

77 N6 **Mahdia** var. Al Mahdīyah, Mehdia. NE Tunisia 35.14N 11.06E

161 F20 **Mahe** Fr. Mahé; prev. Mayyali. Pondicherry, SW India 11.44N 75.33E

180 I16 **Mahé** ✕ Mahé, NE Seychelles 4.37S 55.27E

181 Y17 **Mahé** island Inner Islands, NE Seychelles

181 Y17 **Mahebourg** SE Mauritius 20.24S 57.42E

158 L10 **Mahendranagar** Far Western, W Nepal 28.58N 80.13E

83 J23 **Mahenge** Morogoro, SE Tanzania 8.40S 36.40E

193 F22 **Maheno** Otago, South Island, NZ 45.10S 170.51E

160 D9 **Mahesāna** Gujarāt, W India 23.37N 72.28E

160 F11 **Maheshwar** Madhya Pradesh, C India 22.12N 75.40E

157 F14 **Mahi** ✎ N India

192 O10 **Mahia Peninsula** peninsula North Island, NZ

105 N5 **Maintenon** Eure-et-Loir, C France 48.35N 1.34E

180 H4 **Maintirano** Mahajanga, W Madagascar 18.01S 44.03E

95 M15 **Mainua** Oulu, C Finland 64.05N 27.28E

103 G18 **Mainz** Fr. Mayence. Rheinland-Pfalz, SW Germany 50.00N 8.16E

203 P7 **Maiana** Tahiti, W French Polynesia 17.28S 149.27W

193 E23 **Maungerangi, Lake** ◎ South Island, NZ

57 N9 **Maiquetía** Distrito Federal, N Venezuela 10.36N 66.57W

110 I10 **Maira** It. Mera. ✎ Italy/Switzerland

108 A9 **Maira** ✎ NW Italy

177 H13 **Maịtarị** Assam, NE India 25.18N 79.52E

Māh-Shahr see Bandar-e Māhshahr

81 N19 **Mahulu** Maniema, E Dem. Rep. Congo (Zaire) 1.04S 27.10E

160 C12 **Mahya** Gujarāt, W India 21.06N 71.46E

116 N12 **Mahya Dağı** ▲ NW Turkey 41.47N 27.34E

107 T6 **Maials** var. Mayals. Cataluña, NE Spain 41.22N 0.30E

203 O2 **Maiana** Prev. Hall Island. atoll Tungaru, W Kiribati

203 S11 **Maiana** var. Tapuaemanu, Tubuai-Manu. island Îles du Vent, W French Polynesia

56 M7 **Maicao** La Guajira, N Colombia 11.25N 72.15W

105 U8 **Maiche** Doubs, E France 47.15N 6.43E

99 P22 **Maidenhead** S England, UK 51.31N 0.43W

99 Q22 **Maidstone** Saskatchewan, S Canada 53.06N 109.21W

99 P22 **Maidstone** SE England, UK 51.16N 0.31E

79 Y13 **Maiduguri** Borno, NE Nigeria 11.51N 13.09E

110 I8 **Maienfeld** Sankt Gallen, NE Switzerland 47.01N 9.30E

118 J22 **Maïeru** Hung. Szászmagyarós. Braşov, C Romania 45.55N 25.30E

160 K9 **Maihar** Madhya Pradesh, C India 24.18N 80.46E

69 T10 **Maiko** ✎ W Dem. Rep. Congo (Zaire)

158 L11 **Maila** Uttar Pradesh, N India 28.16N 80.19E

155 U10 **Mailsi** Punjab, E Pakistan 29.46N 72.15E

153 R8 **Maimak** Talasskaya Oblast', NW Kyrgyzstan 42.40N 71.12E

Maimāna see Meymaneh

Maimansingh see Mymensingh

176 Vv11 **Maimawa** Irian Jaya, E Indonesia 3.21S 133.36E

44 M8 **Maïn** ancient monument Peloponnisos, S Greece 36.24N 22.28E

117 F20 **Maïnalo** ▲ S Greece

103 L22 **Mainburg** Bayern, SE Germany 48.40N 11.48E

Main Camp see Banana

12 E12 **Main Channel** lake channel Ontario, S Canada

81 I20 **Mai-Ndombe, Lac** prev. Lac Léopold II. ◎ W Dem. Rep. Congo (Zaire)

21 R6 **Maine** off. State of Maine; also known as Lumber State, Pine Tree State. ◆ state NE USA

104 J8 **Maine** cultural region NW France

104 J7 **Maine-et-Loire** ◆ department NW France

21 N5 **Maine, Gulf of** gulf NE USA

79 X12 **Maïné-Soroa** Diffa, SE Niger 13.13N 12.05E

178 O2 **Maingkwan** var. Mungkawn. Kachin State, N Myanmar 26.19N 96.37E

Main Island see Bermuda

Mainistir Fhear Maí see Fermoy

Mainistir na Corann see Midleton

Mainistir na Féile see Abbeyfeale

98 J5 **Mainland** island Orkney, N Scotland, UK

98 L2 **Mainland** island Shetland, NE Scotland, UK

165 P16 **Maining** Xizang Zizhiqu, W China 29.12N 94.06E

158 K12 **Mainpuri** Uttar Pradesh, N India 27.13N 79.01E

180 H4 **Mintirano** Mahajanga, W Madagascar 18.01S 44.03E

---

45 O10 **Maíz, Islas del** var. Corn Islands. island group SE Nicaragua

171 H14 **Maizuru** Kyōto, Honshū, SW Japan 35.28N 135.21E

56 F6 **Majagual** Sucre, N Colombia 8.36N 74.30W

45 Z13 **Majahual** Quintana Roo, E Mexico 18.43N 87.43W

203 O2 **Maiana** Prev. Hall Island. atoll Tungaru, W Kiribati

175 F11 **Majene** prev. Madjene. Sulawesi, C Indonesia 3.33S 118.59E

45 U9 **Majé, Serranía de** ▲ E Panama

114 I11 **Majevica** ▲ NE Bosnia and Herzegovina

83 H15 **Maji** Southern, S Ethiopia 6.11N 35.32E

147 X7 **Majis** NW Oman 24.25N 56.34E

Majorca see Mallorca

Mājro see Majuro Atoll

Majunga see Mahajanga

201 T3 **Majuro** ✕ Majuro Atoll, SE Marshall Islands 7.04N 171.07E

201 V2 **Majuro Atoll** var. Mājro. atoll Ratak Chain, SE Marshall Islands

201 X2 **Majuro Lagoon** lagoon Majuro Atoll, SE Marshall Islands

78 C11 **Maka** C Senegal 13.39N 14.25W

81 F20 **Makaba** Le Niari, SW Congo 3.28S 12.36E

40 D9 **Makaha** Haw. Makaha. Oahu, Hawaii, USA, C Pacific Ocean 21.28N 158.13W

40 B8 **Makahuena Point** headland Kauai, Hawaii, USA, C Pacific Ocean 21.52N 159.28W

40 D9 **Makakilo City** Oahu, Hawaii, USA, C Pacific Ocean 21.21N 158.05W

85 H18 **Makalamabedi** Central, C Botswana 20.18S 23.52E

Makale see Mek'elē

164 K17 **Makalu** Chin. Makaru Shan. ▲ China/Nepal 27.53N 87.09E

83 G23 **Makampi** Mbeya, S Tanzania 8.00S 33.17E

151 Q13 **Makanchi** Kaz. Maqanshy. Vostochnyy Kazakhstan, E Kazakhstan 46.47N 82.00E

44 M8 **Makantaka** Región Autónoma Atlántico Norte, NE Nicaragua 13.13N 84.04W

202 B16 **Makatea** island headland W Niue 18.58S 169.55E

193 C24 **Makarewa** Southland, South Island, NZ 46.17S 168.16E

119 O4 **Makariv** Kyyivs'ka Oblast', N Ukraine 50.28N 29.49E

193 D20 **Makarora** ✎ South Island, NZ

127 Oo15 **Makarov** Ostrov Sakhalin, Sakhalinskaya Oblast', SE Russian Federation 48.24N 142.37E

207 R9 **Makarov Basin** undersea feature Arctic Ocean

199 Hh4 **Makarov Seamount** undersea feature W Pacific Ocean 29.30N 153.30E

115 F15 **Makarska** It. Macarsca. Split-Dalmacija, SE Croatia 43.18N 17.00E

125 W13 **Makar'yev** Kostromskaya Oblast', NW Russian Federation 57.52N 43.46E

84 L11 **Makasa** Northern, NE Zambia 9.42S 31.54E

Makasar see Ujungpandang

Makasar, Selat see Makassar Straits

Makassar see Ujungpandang

198 Ff8 **Makassar Straits** Ind. Selat Makasar. strait C Indonesia

150 G12 **Mākat** Kaz. Maqat. Atyrau, SW Kazakhstan 47.41N 53.28E

203 T10 **Makatea** island Îles Tuamotu, C French Polynesia

145 X3 **Makātū** E Iraq 33.55N 45.25E

180 H6 **Makay** var. Massif du Makay. ▲ SW Madagascar

116 J12 **Makaza** pass Bulgaria/Greece 41.16N 25.26E

176 Uu9 **Makbon** Irian Jaya, E Indonesia 0.43S 131.30E

Makedonija see Macedonia, FYR

202 B16 **Makefu** W Niue 18.58S 169.55W

203 V10 **Makemo** atoll Îles Tuamotu, C French Polynesia

78 I15 **Makeni** C Sierra Leone 8.57N 12.01W

Makenzen see Orlyak

Makeyevka see Makiyivka

131 Q16 **Makhachkala** prev. Petrovsk-Port. Respublika Dagestan, SW Russian Federation 42.59N 47.30E

150 F11 **Makhambet** Atyrau, W Kazakhstan 47.35N 51.35E

145 X3 **Makharadze** see Ozurget'i

9 S5 **Makhfar al Buşayyah** S Iraq 30.09N 46.09E

145 R4 **Makhmūr** N Iraq 35.46N 43.31E

144 I11 **Makhrūq, Wadi al** dry watercourse E Jordan

145 X4 **Makhūl, Jabal** ▲ C Iraq

147 R13 **Makhyah, Wādī** dry watercourse N Yemen

176 W11 **Maki** Irian Jaya, E Indonesia 3.00S 134.10E

175 S8 **Makian, Pulau** island Maluku, E Indonesia

193 G23 **Makikihi** Canterbury, South Island, NZ 44.36S 171.09E

203 O2 **Makin** prev. Pitt Island. atoll Tungaru, W Kiribati

83 K18 **Makindu** Eastern, S Kenya 2.15S 37.49E

151 Q8 **Makinsk** Akmola, N Kazakhstan 52.37N 70.26E

195 V17 **Makira** off. Makira Province. ◆ province SE Solomon Islands

Makira see San Cristobal

119 X8 **Makiyivka** Rus. Makeyevka; prev. Dmitriyevsk. Donets'ka Oblast', E Ukraine 47.57N 37.47E

146 L10 **Makkah** Eng. Mecca. Makkah, W Saudi Arabia 21.27N 39.50E

146 M10 **Makkah** var. Minṭaqat Makkah. ◆ province W Saudi Arabia

11 R7 **Makkovik** Newfoundland, NE Canada 55.06N 59.06W

100 K6 **Makkum** Friesland, N Netherlands 53.03N 5.25E

◆ COUNTRY ◇ DEPENDENT TERRITORY ◆ ADMINISTRATIVE REGION ▲ MOUNTAIN ☒ VOLCANO ◎ LAKE
● COUNTRY CAPITAL ○ DEPENDENT TERRITORY CAPITAL ✕ INTERNATIONAL AIRPORT ▲ MOUNTAIN RANGE ✎ RIVER ▨ RESERVOIR

Mako see Makung
113 M25 **Makó** *Rom.* Macău. Csongrád, SE Hungary 46.14N 20.28E
12 G9 **Makobe Lake** ◎ Ontario, S Canada
197 J14 **Makogai** *island* C Fiji
81 F18 **Makokou** Ogooué-Ivindo, NE Gabon 0.37N 12.46E
83 G23 **Makongolosi** Mbeya, S Tanzania 8.24S 33.09E
83 G19 **Makota** SW Uganda 0.37S 30.12E
81 G18 **Makoua** Cuvette, C Congo 0.01S 15.40E
112 M10 **Maków Mazowiecki** Mazowieckie, C Poland 52.51N 21.06E
113 K17 **Maków Podhalański** Małopolskie, S Poland 49.43N 19.40E
149 V14 **Makran** *cultural region* Iran/Pakistan
158 G12 **Makrāna** Rājasthān, N India 27.01N 74.43E
149 U15 **Makran Coast** *coastal region* SE Iran
121 F20 **Makrany** *Rus.* Mokrany. Brestskaya Voblasts', SW Belarus 51.49N 24.15E
Makrinoros see Makrynóros
117 H20 **Makrónisos** *island* Kykládes, Greece, Aegean Sea
117 D17 **Makrynóros** *var.* Makrinoros. ▲ C Greece
117 G19 **Makryplági** ▲ C Greece 38.00N 23.06E
Maksamaa see Maxmo
128 J15 **Maksatikha** *var.* Maksatha, Maksaticha. Tverskaya Oblast', W Russian Federation 57.49N 35.46E
160 G10 **Maksi** Madhya Pradesh, C India 23.18N 76.09E
148 I1 **Mākū** Āžarbāyjān-e Bākhtarī, NW Iran 39.16N 44.33E
159 Y11 **Mākum** Assam, NE India 27.28N 95.28E
Makun see Makung
167 R14 **Makung** *prev.* Mako, Makun. W Taiwan 23.34N 119.34E
170 Bb16 **Makurazaki** Kagoshima, Kyūshū, SW Japan 31.15N 130.15E
79 V15 **Makurdi** Benue, C Nigeria 7.41N 8.35E
125 F12 **Makushino** Kurganskaya Oblast', C Russian Federation 55.11N 67.16E
40 L17 **Makushin Volcano** ▲ Unalaska Island, Alaska, USA 53.53N 166.55W
85 K16 **Makwiro** Mashonaland West, N Zimbabwe 17.52S 30.24E
59 D15 **Mala** Lima, W Peru 12.45S 76.38W
Mala see Mallow, Ireland
Mala see Malaita, Solomon Islands
95 J14 **Malå** Västerbotten, N Sweden 65.12N 18.45E
202 G12 **Mala'atoli** Île Uvea, E Wallis and Futuna
79 Qq15 **Malabang** see Malabang
161 E21 **Malabar Coast** *coast* SW India
81 C16 **Malabo** *prev.* Santa Isabel. ● (Equatorial Guinea) Isla de Bioco, NW Equatorial Guinea 3.43N 8.51E
81 C16 **Malabo** × Isla de Bioco, N Equatorial Guinea 3.44N 8.51E
Malaca see Malaga
Malacca see Melaka
173 G4 **Malacca, Strait of** *Ind.* Selat Malaka. *strait* Indonesia/Malaysia
Malacka see Malacky
113 G20 **Malacky** *Hung.* Malacka. Bratislavský Kraj, W Slovakia 48.25N 17.01E
35 R16 **Malad City** Idaho, NW USA 42.10N 112.16W
119 Q4 **Mala Divytsya** Chernihivs'ka Oblast', N Ukraine 50.40N 32.13E
121 J15 **Maladzyechna** *Pol.* Molodeczno, *Rus.* Molodechno. Minskaya Voblasts', C Belarus 54.19N 26.51E
202 D12 **Malaee** Île Futuna, N Wallis and Futuna
39 V15 **Malaga** New Mexico, SW USA 32.10N 104.04W
56 G8 **Málaga** Santander, C Colombia 6.42N 72.43W
106 M15 **Málaga** *anc.* Malaca. Andalucía, S Spain 36.43N 4.25W
106 L15 **Málaga** ◆ *province* Andalucía, S Spain
106 M15 **Málaga** × Andalucía, S Spain 36.38N 4.36W
Malagasy Republic see Madagascar
107 N10 **Malagón** Castilla-La Mancha, C Spain 39.10N 3.51W
99 G18 **Malahide** *Ir.* Mullach Íde. E Ireland 53.27N 6.09W
195 Y14 **Malaita** *off.* Malaita Province. ◆ *province* N Solomon Islands
195 Y15 **Malaita** *var.* Mala. *island* N Solomon Islands
82 F13 **Malakal** Upper Nile, S Sudan 9.31N 31.40E
114 C10 **Mala Kapela** ▲ NW Croatia
27 V7 **Malakoff** Texas, SW USA 32.10N 96.00W
Malakula see Malekula
155 V7 **Malakwāl** *var.* Mālikwāla. Punjab, E Pakistan 32.31N 73.18E
121 J12 **Malalamai** Madang, W PNG 5.47S 146.40E
73 J14 **Malalaua** Gulf, S PNG 8.04S 146.09E
175 Q11 **Malamala** Sulawesi, C Indonesia 3.21S 120.58E
174 M15 **Malang** Jawa, C Indonesia 7.58S 112.45E
82 O14 **Malanje** *var.* Malange. Niassa, N Mozambique 13.27S 36.05E
Malange see Malanje
94 J9 **Malanje** *var.* Malange. Malanje, NW Angola 9.33S 16.25E
84 C11 **Malanje** *var.* Malange. ◆ *province* N Angola
154 M16 **Malān, Rās** *headland* SW Pakistan 25.13N 63.31E
79 S13 **Malanville** NE Benin 11.49N 3.22E
45 T17 **Mala, Punta** *headland* S Panama 7.28N 79.58W
97 N16 **Mälaren** ◎ C Sweden

64 H13 **Malargüe** Mendoza, W Argentina 35.31S 69.34W
12 J8 **Malartic** Quebec, SE Canada 48.09N 78.09W
121 F20 **Malaryta** *Pol.* Maloryta, *Rus.* Malorita. Brestskaya Voblasts', SW Belarus 51.46N 24.04E
65 J19 **Malaspina** Chubut, SE Argentina 44.56S 66.52W
41 U12 **Malaspina Glacier** *glacier* Alaska, USA
143 N15 **Malatya** *anc.* Melitene. Malatya, SE Turkey 38.22N 38.18E
142 M4 **Malatya** ◆ *province* C Turkey
119 Q7 **Mala Vyska** *Rus.* Malaya Viska. Kirovohrads'ka Oblast', S Ukraine 48.37N 31.36E
85 M14 **Malaŵi** *off.* Republic of Malawi; *prev.* Nyasaland, Nyasaland Protectorate. ◆ *republic* S Africa
Malawi, Lake see Nyasa, Lake
95 J17 **Malax** *Fin.* Maalahti. Länsi-Suomi, W Finland 62.55N 21.30E
128 H14 **Malaya Vishera** Novgorodskaya Oblast', W Russian Federation 58.52N 32.12E
179 R15 **Malaybalay** Mindanao, S Philippines 8.10N 125.08E
148 L6 **Malāyer** *prev.* Daulatabad. Hamadān, W Iran 34.19N 48.46E
Malaya Viska see Mala Vyska
174 I3 **Malay Peninsula** *peninsula* Malaysia/Thailand
174 I3 **Malaysia** *var.* Federation of Malaysia; *prev.* the separate territories of Federation of Malaya, Sarawak and Sabah (North Borneo) and Singapore. ◆ *monarchy* SE Asia
143 R14 **Malazgirt** Muş, E Turkey 39.09N 42.30E
13 M8 **Malbaie** ℘ Quebec, SE Canada
79 T12 **Malbaza** Tahoua, S Niger 13.57N 5.32E
112 J7 **Malbork** *Ger.* Marienburg, Marienburg in Westpreussen. Pomorskie, N Poland 54.01N 19.02E
102 N9 **Malchin** Mecklenburg-Vorpommern, N Germany 53.43N 12.46E
100 L13 **Malchin** Gelderland, SE Netherlands 51.54N 5.51E
18 O11 **Malden** Massachusetts, NE USA 42.25N 71.04W
29 Y8 **Malden** Missouri, C USA 36.33N 89.58W
203 X4 **Malden Island** *prev.* Independence Island. *atoll* E Kiribati
181 Q6 **Maldives** *off.* Maldivian Divehi, Republic of Maldives. ◆ *republic* N Indian Ocean
99 P21 **Maldon** E England, UK 51.43N 0.40E
63 G20 **Maldonado** Maldonado, S Uruguay 34.57S 54.58W
63 G20 **Maldonado** ◆ *department* S Uruguay
43 P17 **Maldonado, Punta** *headland* S Mexico 16.18N 98.31W
157 K19 **Male' Div.** Maale ● (Maldives) Male' Atoll, C Maldives 4.10N 73.29E
108 G6 **Male** Trentino-Alto Adige, N Italy 46.21N 10.51E
78 K13 **Maléa** *var.* Maléya. Haute-Guinée, NE Guinea 11.46N 9.43W
117 G22 **Maléas, Akrotírio** *headland* S Greece 36.25N 23.11E
117 L17 **Maléas, Akrotírio** *headland* Lésvos, E Greece 39.01N 26.36E
157 K19 **Male' Atoll** *var.* Kaafu Atoll. *atoll* C Maldives
Malebo, Pool see Stanley Pool
160 E12 **Mālegaon** Mahārāshtra, W India 20.33N 74.31E
83 F15 **Malek** Jonglei, S Sudan 6.04N 31.36E
197 B13 **Malekula** *var.* Malakula, *prev.* Mallicolo. *island* W Vanuatu
201 Y15 **Malem** Kosrae, E Micronesia 5.16N 163.01E
81 O15 **Malema** Nampula, N Mozambique 14.57S 37.28E
81 N23 **Malemba-Nkulu** Katanga, SE Dem. Rep. Congo (Zaire) 8.01S 26.48E
195 Q10 **Malendok Island** *island* Tanga Islands, NE PNG
128 K9 **Malen'ga** Respublika Kareliya, NW Russian Federation 63.50N 36.21E
13 L15 **Małopolska** *plateau* S Poland
113 K17 **Małopolskie** ◆ *province* S Poland
97 M20 **Mālerås** Kalmar, S Sweden 56.55N 15.34E
105 O6 **Malesherbes** Loiret, C France 48.18N 2.25E
117 G18 **Malesína** Stereá Ellás, E Greece 38.37N 23.15E
151 V15 **Malovodnoye** Almaty, SE Kazakhstan 43.31N 77.42E
96 C10 **Måløy** Sogn og Fjordane, S Norway 61.55S 5.06E
130 K4 **Maloyaroslavets** Kaluzhskaya Oblast', W Russian Federation 55.03N 36.31E
125 F6 **Malozemel'skaya Tundra** *physical region* NW Russian Federation
86 K9 **Malpartida de Cáceres** Extremadura, W Spain 39.25N 6.30W
106 K9 **Malpartida de Plasencia** Extremadura, W Spain 39.58N 6.03W
108 C7 **Malpensa** × (Milano) Lombardia, N Italy 45.41N 8.40E
120 J10 **Malqat'er** *desert* N Mauritania
35 V7 **Malta** Montana, NW USA 48.21N 107.52W
118 J3 **Malta** *off.* Republic of Malta. ◆ *republic* C Mediterranean Sea
123 Jj14 **Malta** *island* Malta, C Mediterranean Sea
123 L11 **Malta** Sulawesi, N Indonesia 0.36S 123.13E
Malik, Wadi al see Milk, Wadi el
Malta, Canale di see Malta Channel

178 Gg12 **Mali Kyun** *var.* Tavoy Island. *island* Mergui Archipelago, S Myanmar
97 M19 **Målilla** Kalmar, S Sweden 57.24N 15.48E
114 B11 **Mali Lošinj** *It.* Lussinpiccolo. Primorje-Gorski Kotar, W Croatia 44.31N 14.28E
Malin see Malyn
179 Q15 **Malindang, Mount** ▲ Mindanao, S Philippines 8.12N 123.37E
83 A20 **Malindi** Coast, SE Kenya 3.13S 40.04E
Malines see Mechelen
98 E13 **Malin Head** *Ir.* Cionn Mhálanna. *headland* NW Ireland 55.37N 7.37W
175 Pp7 **Malino, Gunung** ▲ Sulawesi, N Indonesia 0.44N 120.45E
115 M21 **Maliq** *var.* Maliqi. Korçë, SE Albania 40.45N 20.45E
Maliqi see Maliq
179 Rr16 **Malita** Mindanao, S Philippines 6.13N 125.39E
160 G12 **Malkāpur** Mahārāshtra, C India 20.52N 76.18E
142 B10 **Malkara** Tekirdağ, NW Turkey 40.55N 26.56E
121 J19 **Mal'kavichy** *Rus.* Mal'kovichi. Brestskaya Voblasts', SW Belarus 52.28N 26.39E
Malkiye see Al Mālikiyah
116 L11 **Malko Sharkovo, Yazovir** ◎ SE Bulgaria
116 N11 **Malko Tŭrnovo** Burgas, E Bulgaria 42.00N 27.31E
Mal'kovichi see Mal'kavichy
191 R12 **Mallacoota** Victoria, SE Australia 37.34S 149.45E
98 G10 **Mallaig** N Scotland, UK 57.03N 5.48W
190 I9 **Mallala** South Australia 34.29S 138.30E
77 W9 **Mallawī** C Egypt 27.49N 30.43E
107 R5 **Mallén** Aragón, NE Spain 41.52N 1.25W
108 F5 **Malles Venosta** Trentino-Alto Adige, N Italy 46.40N 10.37E
Mallicolo see Malekula
111 Q8 **Mallnitz** Salzburg, S Austria 46.58N 13.09E
107 W9 **Mallorca** *Eng.* Majorca; *anc.* Baliares Major. *island* Islas Baleares, Spain, W Mediterranean Sea
99 C20 **Mallow** *Ir.* Mala. SW Ireland 52.07N 8.39W
95 E15 **Malm** Nord-Trøndelag, C Norway 64.04N 11.12E
97 L19 **Malmbäck** Jönköping, S Sweden 57.34N 14.30E
94 J12 **Malmberget** Norrbotten, N Sweden 67.09N 20.39E
101 M20 **Malmédy** Liège, E Belgium 50.25N 6.01E
83 E25 **Malmesbury** Western Cape, SW South Africa 33.28S 18.43E
97 N16 **Malmköping** Södermanland, C Sweden 59.07N 16.49E
97 K23 **Malmö** Skåne, S Sweden 55.35N 13.00E
97 K23 **Malmö** × Skåne, S Sweden 55.33N 13.22E
94 Q16 **Malmok** *headland* Bonaire, S Netherlands Antilles 12.16N 68.21W
97 M18 **Malmslätt** Östergötland, S Sweden 58.25N 15.30E
129 R16 **Malmyzh** Kirovskaya Oblast', NW Russian Federation 56.30N 50.37E
197 B12 **Malo** *island* W Vanuatu
130 J7 **Maloarkhangel'sk** Orlovskaya Oblast', W Russian Federation 52.25N 36.37E
Maloelap see Maloelap Atoll
201 V6 **Maloelap Atoll** *var.* Maļoeļap. *atoll* E Marshall Islands
Maloenda see Malunda
110 I10 **Maloja** Graubünden, S Switzerland 46.25N 9.42E
84 L12 **Malole** Northern, NE Zambia 10.05S 31.37E
197 H13 **Malolo** *island* Mamanuca Group, W Fiji
197 H13 **Malolo Barrier Reef** *var.* Ro Ro Reef. *reef* W Fiji
179 P10 **Malolos** Luzon, N Philippines 14.51N 120.49E
20 K6 **Malone** New York, NE USA 44.51N 74.18W
81 K25 **Malonga** Katanga, S Dem. Rep. Congo (Zaire) 10.30S 23.06E
13 L15 **Małopolska** *plateau* S Poland
113 L15 **Małopolskie** ◆ *province* S Poland
113 K17 **Małopolskie** ◆ *province* S Poland
128 K9 **Maloshuyka** Arkhangel'skaya Oblast', NW Russian Federation 63.50N 37.20E
116 G10 **Mal'ovitsa** ▲ W Bulgaria 42.12N 23.19E
151 V15 **Malovodnoye** Almaty, SE Kazakhstan 43.31N 77.42E
96 C10 **Måløy** Sogn og Fjordane, S Norway 61.55S 5.06E
130 K4 **Maloyaroslavets** Kaluzhskaya Oblast', W Russian Federation 55.03N 36.31E
125 F6 **Malozemel'skaya Tundra** *physical region* NW Russian Federation
86 K9 **Malpartida de Cáceres** Extremadura, W Spain 39.25N 6.30W
106 K9 **Malpartida de Plasencia** Extremadura, W Spain 39.58N 6.03W
108 C7 **Malpensa** × (Milano) Lombardia, N Italy 45.41N 8.40E
120 J10 **Malqat'er** *desert* N Mauritania
57 X9 **Mana** NW French Guiana 5.40N 53.49W
58 A6 **Manabí** ◆ *province* W Ecuador
44 G4 **Manabique, Punta** *var.* Cabo Tres Puntas. *headland* E Guatemala 15.57N 88.37W
56 G11 **Manacacías, Río** ℘ C Colombia
60 F13 **Manacapuru** Amazonas, N Brazil 3.16S 60.37W
123 L11 **Manado** *prev.* Menado. Sulawesi, C Indonesia 1.31N 124.55E
196 H5 **Managaha** *island* S Northern Mariana Islands

123 L12 **Malta Channel** *It.* Canale di Malta. *strait* Italy/Malta
85 D20 **Maltahöhe** Hardap, SW Namibia 24.49S 16.58E
99 N16 **Malton** N England, UK 54.07N 0.49W
175 T11 **Maluku** *off.* Propinsi Maluku, *Dut.* Molukken, *Eng.* Moluccas. ◆ *province* E Indonesia
175 Ss9 **Maluku** *Dut.* Molukken, *Eng.* Moluccas; *prev.* Spice Islands. *island group* E Indonesia
175 Ss9 **Maluku, Laut** see Molucca Sea
79 V13 **Malumfashi** Katsina, N Nigeria 11.51N 7.39E
96 K13 **Malung** Dalarna, C Sweden 60.40N 13.45E
96 K13 **Malungsfors** Dalarna, C Sweden 60.43N 13.34E
195 X14 **Maluu** *var.* Malu'u. Malaita, N Solomon Islands 8.22S 160.39E
161 D16 **Mālvan** Mahārāshtra, W India 16.05N 73.28E
29 U12 **Malvern** Arkansas, C USA 34.21N 92.48W
31 S15 **Malvern** Iowa, C USA 40.59N 95.36W
46 I13 **Malvern** ▲ W Jamaica 17.59N 77.42W
Malvinas, Islas see Falkland Islands
119 N4 **Malyn** *Rus.* Malin. Zhytomyrs'ka Oblast', N Ukraine 50.46N 29.14E
127 O5 **Malyy Anyuy** ℘ NE Russian Federation
131 O11 **Malyye Derbety** Respublika Kalmykiya, SW Russian Federation 47.57N 44.39E
159 U12 **Manās** *var.* Dangme Chu. ℘ Bhutan/India
125 R8 **Manas, Gora** ▲ Kyrgyzstan/Uzbekistan 42.17N 71.04E
164 K3 **Manas Hu** ◎ NW China
159 U12 **Manās** *var.* Dangme Chu.
126 M5 **Malyy Lyakhovskiy, Ostrov** *island* NE Russian Federation
126 Jj4 **Malyy Taymyr, Ostrov** *island* Severnaya Zemlya, N Russian Federation
150 E10 **Malyy Uzen'** *Kaz.* Kishiözen. ℘ Kazakhstan/Russian Federation
126 I16 **Malyy Yenisey** *Ka.* Ka-Krem. ℘ S Russian Federation
126 K13 **Mama** Irkutskaya Oblast', C Russian Federation 58.13N 112.45E
131 S3 **Mamadysh** Respublika Tatarstan, W Russian Federation 55.46N 51.22E
119 N14 **Mamaia** Constanţa, E Romania 44.13N 28.37E
197 G14 **Mamanuca Group** *island group* Yasawa Group, W Fiji
152 L13 **Mamash** Lebapskiy Velayat, E Turkmenistan 38.24N 64.12E
176 W11 **Mamasiware** Irian Jaya, E Indonesia 2.46S 134.26E
175 Q10 **Mambasa** Orientale, NE Dem. Rep. Congo (Zaire) 1.22N 29.02E
176 Xx10 **Mamberamo, Sungai** ℘ Irian Jaya, E Indonesia
81 I21 **Mambéré** ℘ SW Central African Republic
81 H18 **Mambéré-Kadéï** ◆ *prefecture* SW Central African Republic
176 X9 **Mambetaloi** Irian Jaya, E Indonesia 1.38S 136.12E
81 H18 **Mambili** ℘ W Congo
85 N18 **Mambone** *var.* Nova Mambone. Inhambane, E Mozambique 20.59S 35.04E
179 P11 **Mamburao** Mindoro, N Philippines 13.16N 120.36E
180 I16 **Mamelles** *island* Inner Islands, NE Seychelles
101 M25 **Mamer** Luxembourg, SW Luxembourg 49.37N 6.01E
104 L6 **Mamers** Sarthe, NW France 48.21N 0.22E
81 D15 **Mamfé** Sud-Ouest, W Cameroon 5.46N 9.18E
151 P6 **Mamlyutka** Severnyy Kazakhstan, N Kazakhstan 54.55N 68.31E
133 X7 **Manchurian Plain** *plain* NE China
Máncio Lima see Japiim
Mancunium see Manchester
154 J15 **Mand** Baluchistān, SW Pakistan 26.06N 61.58E
Mand see Mand, Rūd-e
83 H25 **Manda** Iringa, SW Tanzania 10.25S 34.38E
180 H6 **Mandabe** Toliara, W Madagascar 21.01S 44.55E
81 I15 **Mandal** Hövsgöl, N Mongolia 49.55N 99.21E
168 I7 **Mandal** Töv, C Mongolia 48.24N 106.47E
96 E13 **Mandal** Vest-Agder, S Norway 58.01N 7.27E
177 G5 **Mandalay** Mandalay, C Myanmar 21.57N 96.04E
178 G4 **Mandalay** ◆ *division* C Myanmar
168 I9 **Mandalgovi** Dundgovĭ, C Mongolia 45.47N 106.18E
145 Q6 **Mandalī** E Iraq 33.43N 45.33E

101 G20 **Manage** Hainaut, S Belgium 50.30N 4.13E
44 J10 **Managua** ● (Nicaragua) Managua, W Nicaragua 12.07N 86.15W
44 J10 **Managua** ◆ *department* W Nicaragua
44 J10 **Managua** × Managua, W Nicaragua 12.07N 86.11W
20 K16 **Manahawkin** New Jersey, NE USA 39.39N 74.12W
79 V13 **Manaia** Taranaki, North Island, NZ 39.32S 174.04E
192 K11 **Manakara** Fianarantsoa, SE Madagascar 22.09S 48.00E
180 J6 **Manāli** Himāchal Pradesh, NW India 32.18N 77.12E
133 U12 **Ma, Nam** *Vtn.* Sông Mã. ℘ Laos/Vietnam
Manama see Al Manāmah
194 J10 **Manam Island** *island* N PNG
69 Y13 **Manananara** ℘ SE Madagascar
190 M9 **Manangatang** Victoria, SE Australia 35.04S 142.53E
180 J6 **Mananjary** Fianarantsoa, SE Madagascar 21.13S 48.19E
78 L14 **Manankoro** Sikasso, SW Mali 10.33N 7.25W
78 J12 **Manantali, Lac de** ◎ W Mali
Manáos see Manaus
193 B23 **Manapouri** Southland, South Island, NZ 45.33S 167.38E
193 B23 **Manapouri, Lake** ◎ South Island, NZ
60 F12 **Manaquiri** Amazonas, NW Brazil 3.27S 60.37W
Manar see Mannar
164 K5 **Manas** Xinjiang Uygur Zizhiqu, NW China 44.16N 86.12E
159 P18 **Manaslu** ▲ C Nepal 28.33N 84.33E
39 S8 **Manassa** Colorado, C USA 37.10N 105.56W
21 W7 **Manassas** Virginia, NE USA 38.45N 77.28W
47 T5 **Manatí** C Puerto Rico 18.25N 66.29W
175 S16 **Manatuto** N East Timor 8.31S 126.00E
194 L14 **Manau** Northern, S PNG 8.05S 147.57E
56 H7 **Manaure** La Guajira, N Colombia 11.46N 72.28W
60 F12 **Manaus** *prev.* Manáos. *state capital* Amazonas, NW Brazil 3.06S 60.00W
142 G17 **Manavgat** Antalya, SW Turkey 36.46N 31.28E
192 M13 **Manawatu** ℘ North Island, NZ
192 L11 **Manawatu-Wanganui** *off.* Manawatu-Wanganui Region. ◆ *region* North Island, NZ
176 Uu12 **Manawoka, Pulau** *island* Kepulauan Gorong, E Indonesia
179 Rr16 **Manay** Mindanao, S Philippines 7.12N 126.29E
144 K2 **Manbij** *var.* Mambij, *Fr.* Membidj. Ḥalab, N Syria 36.31N 37.55E
107 X9 **Mancha Real** Andalucía, S Spain 37.46N 3.37W
104 I4 **Manche** ◆ *department* N France
99 L17 **Manchester** *Lat.* Mancunium. NW England, UK 53.30N 2.15W
23 S5 **Manchester** Georgia, SE USA 32.51N 84.37W
31 Y13 **Manchester** Iowa, C USA 42.28N 91.27W
23 N7 **Manchester** Kentucky, S USA 37.10N 83.40W
21 O10 **Manchester** New Hampshire, NE USA 42.58N 71.25W
22 K10 **Manchester** Tennessee, S USA 35.28N 86.05W
20 M9 **Manchester** Vermont, NE USA 43.09N 73.03W
99 L18 **Manchester** × NW England, UK 53.21N 2.16W
154 J15 **Manchhar Lake** ◎ SE Pakistan
Man-chou-li see Manzhouli
133 X7 **Manchurian Plain** *plain* NE China

155 U10 **Mandi Bürewāla** *var.* Būrewāla. Punjab, E Pakistan 30.04N 72.46E
158 G9 **Mandi Dabwāli** Haryāna, NW India 29.55N 74.40E
Mandidzudzure see Chimanimani
85 M15 **Mandié** Manica, NW Mozambique 16.27S 33.28E
85 N14 **Mandimba** Niassa, N Mozambique 14.21S 35.40E
175 Ss9 **Mandioli, Pulau** *island* Kepulauan Bacan, E Indonesia
59 G9 **Mandioré, Laguna** ◎ E Bolivia
160 J10 **Mandla** Madhya Pradesh, C India 22.39N 80.21E
180 J6 **Mandlakazi** *var.* Manjacaze. Gaza, S Mozambique 24.43S 33.57E
97 E24 **Mandø** *var.* Manø. *island* W Denmark
117 G19 **Mándra** Attikí, C Greece 38.04N 23.31E
180 I7 **Mandrare** ℘ S Madagascar
116 M10 **Mandra, Yazovir** *salt lake* SE Bulgaria
109 L23 **Mandrazzi, Portella** *pass* Sicilia, Italy, C Mediterranean Sea
180 J3 **Mandritsara** Mahajanga, N Madagascar 15.49S 48.49E
149 O13 **Mand, Rūd-e** *var.* Mand. ℘ S Iran
160 F9 **Mandsaur** *prev.* Mandasor. Madhya Pradesh, C India 24.05N 75.04E
160 F11 **Māndu** Madhya Pradesh, C India 22.20N 75.24E
175 Oo5 **Mandul, Pulau** *island* N Indonesia
85 G15 **Mandundu** Western, W Zambia 16.34S 22.18E
188 I13 **Mandurah** Western Australia 32.31S 115.40E
109 P18 **Manduria** Puglia, SE Italy 40.24N 17.37E
161 G20 **Mandya** Karnātaka, C India 12.34N 76.55E
79 P12 **Mané** C Burkina 12.59N 1.21W
100 E8 **Manerbio** Lombardia, N Italy 45.22N 10.09E
118 K3 **Manevychi** *var.* Manevychi. *Rus.* Manevichi. Volyns'ka Oblast', NW Ukraine 51.18N 25.29E
109 N16 **Manfredonia** Puglia, SE Italy 41.38N 15.54E
109 N16 **Manfredonia, Golfo di** *gulf* Adriatic Sea, S Mediterranean Sea
79 P13 **Manga** C Burkina 11.40N 1.04W
79 F14 **Mangabeiras, Chapada das** ▲ E Brazil
81 M20 **Mangai** Bandundu, W Dem. Rep. Congo (Zaire) 3.57S 19.32E
202 L17 **Mangaia** *island group* S Cook Islands
192 M9 **Mangakino** Waikato, North Island, NZ 38.22S 175.45E
118 M15 **Mangalia** *anc.* Callatis. Constanţa, SE Romania 43.46N 28.34E
81 P17 **Mangbwalu** Orientale, NE Dem. Rep. Congo (Zaire) 2.06N 30.04E
173 R1 **Mangfall** ℘ SE Germany
174 K1 **Manggar** Pulau Belitung, W Indonesia 2.51S 108.14E
176 Vv12 **Manggawitu** Irian Jaya, E Indonesia 4.11S 133.28E
152 H8 **Manghit** *Rus.* Mangit. Qoraqalpoghiston Respublikasi, W Uzbekistan 42.06N 60.02E
177 N9 **Mangin Range** ▲ N Myanmar
145 R1 **Mangish** N Iraq 37.03N 43.04E
150 F15 **Mangistau** *Kaz.* Mangghystaū Oblïsy; *prev.* Mangyshlaska. ◆ *province* SW Kazakhstan
Mangit see Manghit
85 A13 **Manglares, Cabo** *headland* SW Colombia 1.36N 79.01W
155 V6 **Mangla Reservoir** ◎ NE Pakistan
165 N9 **Mangnai** *var.* Lao Mangnai. Qinghai, C China 37.52N 91.39E
Mango see Mago, Fiji
Mango see Sansanné-Mango, Togo
Mangoche see Mangochi
85 N14 **Mangochi** *var.* Mangoche; *prev.* Fort Johnston. Southern, SE Malawi 14.27S 35.15E
180 G7 **Mangoky** ℘ W Madagascar
175 S10 **Mangole, Pulau** *island*
192 J2 **Mangonui** Northland, North Island, NZ 35.00S 173.31E
Mangqystaū Oblysy see Mangistau
Mangqystaū Shyghanaghy see Mangyshlakskiy Zaliv
106 H7 **Mangualde** Viseu, N Portugal 40.36N 7.46W
61 F15 **Mangueira, Lagoa** ◎ S Brazil
79 X6 **Mangueni, Plateau du** ▲ NE Niger
26 K11 **Mangum** Oklahoma, C USA 34.52N 99.30W
81 O18 **Manguredjipa** Nord Kivu, E Dem. Rep. Congo (Zaire) 0.28N 28.33E
85 L16 **Mangwendi** Mashonaland East, E Zimbabwe 18.22S 31.24E
158 K10 **Māngāpur** Uttar Pradesh, N India 27.03N 82.12E
150 F15 **Mangyshlak, Plato** *plateau* SW Kazakhstan

150 E14 **Mangyshlakskiy Zaliv** *Kaz.* Mangqystaū Shyghanaghy. *gulf* SW Kazakhstan
Mangyshlaskaya see Mangistau
168 I5 **Manhan** *var.* Tögrög. Hövsgöl, N Mongolia 50.05N 100.01E
29 O4 **Manhattan** Kansas, C USA 39.11N 96.33W
101 L21 **Manhay** Luxembourg, SE Belgium 50.13N 5.43E
85 L21 **Manhiça** *prev.* Vila de Manhiça. Maputo, S Mozambique 25.20S 32.49E
85 L21 **Manhoca** Maputo, S Mozambique 26.47S 32.37E
61 N20 **Manhuaçu** Minas Gerais, SE Brazil 20.16S 42.01W
149 R11 **Māni** Kermān, C Iran
56 H10 **Maní** Casanare, C Colombia 4.49N 72.15W
48 S6 **Mania, Bahía de** *bay* W Ecuador
85 M17 **Manica** *var.* Vila de Manica. Manica, W Mozambique 18.51S 32.52E
85 M17 **Manica** *off.* Província de Manica. ◇ *province* W Mozambique
85 L17 **Manicaland** ◆ *province* E Zimbabwe
13 U5 **Manic Deux, Réservoir** ▨ Quebec, SE Canada
Manich see Manych
61 F14 **Manicoré** Amazonas, N Brazil 5.48S 61.16W
11 N11 **Manicouagan** Quebec, SE Canada
11 N11 **Manicouagan** ℘ Quebec, SE Canada
13 U6 **Manicouagan, Péninsule de** *peninsula* Quebec, SE Canada
11 N11 **Manicouagan, Réservoir** ▨ Quebec, E Canada
13 T4 **Manic Trois, Réservoir** ▨ Quebec, SE Canada
81 M20 **Maniema** *off.* Région du Maniema. ◇ *region* E Dem. Rep. Congo (Zaire)
Maniewicze see Manevychi
166 F8 **Maniganggo** Sichuan, C China 32.01N 99.04E
9 Y15 **Manigotagan** Manitoba, S Canada 51.06N 96.18W
159 R13 **Manihāri** Bihār, N India 25.21N 87.37E
203 O9 **Manihi** *island* Îles Tuamotu, C French Polynesia
202 L13 **Manihiki** *atoll* N Cook Islands
183 U8 **Manihiki Plateau** *undersea feature* C Pacific Ocean
206 M14 **Maniitsoq** *var.* Manītsoq, *Dan.* Sukkertoppen. Kita, S Greenland 65.12N 52.05W
159 T15 **Manikganj** Dhaka, C Bangladesh 23.52N 90.00E
158 M14 **Mānikpur** Uttar Pradesh, N India 25.04N 81.06E
179 P11 **Manila** *off.* City of Manila. ● (Philippines) Luzon, N Philippines 14.34N 120.58E
29 Y9 **Manila** Arkansas, C USA 35.52N 90.10W
201 N16 **Manila Reef** *reef* W Micronesia
191 R8 **Manilla** New South Wales, SE Australia 30.44S 150.43E
200 Qq14 **Maniloa** *island* Tongatapu Group, S Tonga
127 Z **Manily** Koryakskiy Avtonomnyy Okrug, E Russian Federation 62.33N 165.03E
176 Ww9 **Maninian, Plateau** *island* E Indonesia
Maninjau, Danau ◎ Sumatera, W Indonesia
159 W13 **Manipur** ◆ *state* NE India
159 V13 **Manipur Hills** *hill range* E India
142 C14 **Manisa** *var.* Manissa; *prev.* Saruhan, *anc.* Magnesia. Manisa, W Turkey 38.36N 27.28E
142 C13 **Manisa** *var.* Manissa. ◆ *province* W Turkey
Manissa see Manisa
18 I7 **Manistee** Michigan, N USA 44.14N 86.19W
18 I6 **Manistee River** ℘ Michigan, N USA
18 I5 **Manistique** Michigan, N USA 45.57N 86.15W
18 I4 **Manistique Lake** ◎ Michigan, N USA
9 X16 **Manitoba** ◆ *province* S Canada
9 X16 **Manitoba, Lake** ◎ Manitoba, S Canada
10 I5 **Manitou** Manitoba, S Canada 49.12N 98.28W
18 I5 **Manitou Island** *island* Michigan, N USA
10 G5 **Manitou Lake** ◎ Ontario, SE Canada
10 C11 **Manitoulin Island** *island* Ontario, S Canada
39 T5 **Manitou Springs** Colorado, C USA 38.51N 104.56W
10 G12 **Manitouwabing Lake** ◎ Ontario, S Canada
10 E12 **Manitowaning** Manitoulin Island, Ontario, S Canada 45.43N 81.49W
2 B7 **Manitowik Lake** ◎ Ontario, S Canada
33 N7 **Manitowoc** Wisconsin, N USA 44.04N 87.40W
145 S10 **Mangole, Pulau** *island*
10 J14 **Maniwaki** Quebec, SE Canada 46.24N 75.58W
176 X11 **Maniwori** Irian Jaya, E Indonesia 2.49S 136.00E
56 E10 **Manizales** Caldas, W Colombia 5.03N 75.32W
114 F11 **Manjača** ▲ NW Bosnia and Herzegovina
Manjacaze see Mandlakazi
188 J14 **Manjimup** Western Australia 34.18S 116.14E
111 V4 **Mank** Niederösterreich, C Austria 48.06N 15.13E
159 N12 **Mankāpur** Uttar Pradesh, N India 27.03N 82.12E
28 M3 **Mankato** Kansas, C USA 39.45N 98.10W

◆ COUNTRY   ◇ DEPENDENT TERRITORY   ◇ ADMINISTRATIVE REGION   ▲ MOUNTAIN   ✕ VOLCANO   ◎ LAKE
● COUNTRY CAPITAL   ○ DEPENDENT TERRITORY CAPITAL   ✕ INTERNATIONAL AIRPORT   ▲ MOUNTAIN RANGE   ℘ RIVER   ▨ RESERVOIR

289

31 U10 **Mankato** Minnesota, N USA 44.10N 94.00W
119 O7 **Man'kivka** Cherkas'ka Oblast', C Ukraine 48.58N 30.10E
78 M15 **Mankono** C Ivory Coast 8.06N 6.07W
9 T17 **Mankota** Saskatchewan, S Canada 49.25N 107.04W
161 K23 **Mankulam** Northern Province, N Sri Lanka 9.09N 80.27E
41 Q9 **Manley Hot Springs** Alaska, USA 65.00N 150.37W
20 H10 **Manlius** New York, NE USA 43.00N 75.58W
107 W5 **Manlleu** Cataluña, NE Spain 41.58N 2.16E
31 V11 **Manly** Iowa, C USA 43.17N 93.12W
160 E13 **Manmād** Mahārāshtra, W India 20.15N 74.28E
190 J7 **Mannahill** South Australia 32.29S 139.58E
161 J23 **Mannar** var. Manar. Northern Province, NW Sri Lanka 9.01N 79.53E
161 I24 **Mannar, Gulf of** gulf India/Sri Lanka
161 J23 **Mannar Island** island N Sri Lanka
Mannersdorf see Mannersdorf am Leithagebirge
111 Y5 **Mannersdorf am Leithagebirge** var. Mannersdorf. Niederösterreich, E Austria 47.58N 16.36E
111 Y6 **Mannersdorf an der Rabnitz** Burgenland, E Austria 47.25N 16.32E
103 G20 **Mannheim** Baden-Württemberg, SW Germany 49.28N 8.29E
9 O12 **Manning** Alberta, W Canada 56.52N 117.39W
31 T14 **Manning** Iowa, C USA 41.54N 95.03W
30 K5 **Manning** North Dakota, N USA 47.13N 102.46W
23 S13 **Manning** South Carolina, SE USA 33.42N 80.12W
203 Y2 **Manning, Cape** headland Kiritimati, NE Kiribati 2.01N 157.25W
195 V13 **Manning Strait** strait NW Solomon Islands
23 S3 **Mannington** West Virginia, NE USA 39.31N 80.20W
190 A1 **Mann Ranges** ▲ South Australia
109 C19 **Mannu** ≈ Sardegna, Italy, C Mediterranean Sea
9 R14 **Mannville** Alberta, SW Canada 53.19N 111.08W
78 I11 **Mano** ≈ Liberia/Sierra Leone
Manø see Mandø
41 O13 **Manokotak** Alaska, USA 59.00N 158.58W
176 W9 **Manokwari** Irian Jaya, E Indonesia 0.49S 134.04E
81 N22 **Manono** Shaba, SE Dem. Rep. Congo (Zaire) 7.18S 27.25E
27 T10 **Manor** Texas, SW USA
99 D16 **Manorhamilton** Ir. Cluainín. NW Ireland 54.18N 8.10W
105 S15 **Manosque** Alpes-de-Haute-Provence, SE France 43.49N 5.46E
10 L11 **Manouane, Lac** ⊚ Québec, SE Canada
169 W12 **Manp'o** var. Manp'ojin. NW North Korea 41.10N 126.24E
Manp'ojin see Manp'o
203 T4 **Manra** prev. Sydney Island. atoll Phoenix Islands, C Kiribati
107 V5 **Manresa** Cataluña, NE Spain 41.44N 1.52E
158 H9 **Mānsa** Punjab, NW India 30.00N 75.25E
84 J12 **Mansa** prev. Fort Rosebery. Luapula, N Zambia 11.13S 28.55E
78 G12 **Mansa Konko** C Gambia 13.26N 15.29W
13 Q11 **Manseau** Quebec, SE Canada 46.23N 71.59W
155 U15 **Mānsehra** North-West Frontier Province, NW Pakistan 34.22N 73.18E
15 Mm6 **Mansel Island** island Nunavut, NE Canada
191 O12 **Mansfield** Victoria, SE Australia 37.04S 146.06E
99 M18 **Mansfield** C England, UK 53.09N 1.10W
29 U11 **Mansfield** Arkansas, C USA 35.03N 94.15W
24 G6 **Mansfield** Louisiana, S USA 32.02N 93.42W
21 Q12 **Mansfield** Massachusetts, NE USA 42.00N 71.11W
33 T12 **Mansfield** Ohio, N USA 40.45N 82.31W
20 I12 **Mansfield** Pennsylvania, NE USA 41.46N 77.02W
20 M7 **Mansfield, Mount** ▲ Vermont, NE USA 44.31N 72.49W
61 N16 **Mansidão** Bahia, E Brazil 10.46S 44.03W
104 L11 **Mansle** Charente, W France 45.52N 0.11E
78 G12 **Mansôa** C Guinea-Bissau 12.07N 15.18W
49 V8 **Manso, Rio** ≈ C Brazil
Mansûra see El Mansûra
Mansurabad see Mehrān, Rūd-e
58 A6 **Manta** Manabí, W Ecuador 0.57S 80.39W
59 F14 **Mantaro, Río** ≈ C Peru
37 O8 **Manteca** California, W USA 37.48N 121.13W
56 J7 **Mantecal** Apure, C Venezuela 7.34N 69.07W
33 N10 **Manteno** Illinois, C USA 41.15N 87.49W
23 Y9 **Manteo** Roanoke Island, North Carolina, SE USA 35.55N 75.39W
Mantes-Gassicourt see Mantes-la-Jolie
105 N5 **Mantes-la-Jolie** prev. Mantes-Gassicourt, Mantes sur Seine; anc. Medunta. Yvelines, N France 48.58N 1.42E
Mantes-sur-Seine see Mantes-la-Jolie
38 L5 **Manti** Utah, W USA 39.16N 111.38W
Mantinea see Mantíneia

117 F20 **Mantíneia** anc. Mantinea. site of ancient city Pelopónnisos, S Greece 37.36N 22.22E
61 M21 **Mantiqueira, Serra da** ▲ S Brazil
31 W10 **Mantorville** Minnesota, N USA 44.04N 92.45W
117 G22 **Mantoúdi** var. Mandoudi; prev. Mandoúdhion. Évvoia, C Greece 38.46N 23.28E
Mantoue see Mantova
108 F8 **Mantova** Eng. Mantua, Fr. Mantoue. Lombardia, NW Italy 45.10N 10.46E
95 M19 **Mäntsälä** Etelä-Suomi, S Finland 60.38S 25.21E
95 L17 **Mänttä** Länsi-Suomi, W Finland 62.00N 24.36E
Mantua see Mantova
129 O14 **Manturovo** Kostromskaya Oblast', NW Russian Federation 58.19N 44.42E
95 M18 **Mäntyharju** Ita-Suomi, SE Finland 61.25N 26.52E
94 M13 **Mäntyjärvi** Lappi, N Finland 66.00N 27.55E
202 L16 **Manuae** island S Cook Islands
203 Q10 **Manuae** atoll Îles Sous le Vent, W French Polynesia
198 Dd8 **Manu'a Islands** island group E American Samoa
42 L5 **Manuel Benavides** Chihuahua, N Mexico 29.07N 103.52W
63 D21 **Manuel J.Cobo** Buenos Aires, E Argentina 35.49S 57.54W
60 M12 **Manuel Luís, Recife** reef E Brazil
63 F15 **Manuel Viana** Rio Grande do Sul, S Brazil 29.33S 55.28W
61 I14 **Manuel Zinho** Pará, N Brazil 7.21S 54.47W
203 V11 **Manuhangi** atoll Îles Tuamotu, C French Polynesia
193 E22 **Manuherikia** ≈ South Island, NZ
175 R11 **Manui, Pulau** island N Indonesia
Manukau see Manurewa
192 L6 **Manukau Harbour** harbor North Island, NZ
174 K14 **Manuk, Ci** ≈ Jawa, S Indonesia
176 U12 **Manuk, Pulau** island Maluku, E Indonesia
203 Z2 **Manulu Lagoon** ⊚ Kiritimati, E Kiribati
190 J7 **Manunda Creek** seasonal river South Australia
59 K15 **Manupari, Río** ≈ N Bolivia
192 L6 **Manurewa** var. Manukau. Auckland, North Island, NZ 37.03S 174.55E
59 K15 **Manurimi, Río** ≈ NW Bolivia
194 I8 **Manus** ◆ province N PNG
194 J8 **Manus Island** var. Great Admiralty Island. island N PNG
176 U15 **Manuwui** Pulau Babar, E Indonesia 7.47S 129.39E
31 Q3 **Manvel** North Dakota, N USA 48.07N 97.15W
35 Z14 **Manville** Wyoming, C USA 42.45N 104.38W
24 Q6 **Many** Louisiana, S USA 31.34N 93.28W
83 H21 **Manyara, Lake** ⊚ NE Tanzania
130 L12 **Manych** var. Manich. ≈ SW Russian Federation
131 N13 **Manych-Gudilo, Ozero** salt lake SW Russian Federation
85 H14 **Manyinga** North Western, NW Zambia 13.28S 24.18E
107 O11 **Manzanares** Castilla-La Mancha, C Spain 39.00N 3.23W
46 M7 **Manzanillo** Granma, E Cuba 20.21N 77.07W
44 D8 **Manzanillo** Colima, SW Mexico 19.00N 104.18W
45 X5 **Manzanillo, Bahía** bay SW Mexico
39 S11 **Manzano Mountains** ▲ New Mexico, SW USA
39 R12 **Manzano Peak** ▲ New Mexico, SW USA 34.35N 106.27W
169 R6 **Manzhouli** var. Man-chou-li. Nei Mongol Zizhiqu, N China 49.36N 117.28E
Manzil Bū Ruqaybah see Menzel Bourguiba
145 X9 **Manziliyah** E Iraq 32.26N 47.01E
85 L21 **Manzini** prev. Bremersdorp. C Swaziland 26.30S 31.33E
85 L21 **Manzini** ✕ (Mbabane) C Swaziland 26.36S 31.25E
80 G10 **Mao** Kanem, C Chad 14.06N 15.16E
47 N9 **Mao** NW Dominican Republic 19.33N 71.09W
Maó see Mahón
165 W9 **Maojing** Gansu, N China 36.25N 106.36E
176 Xx12 **Maoke, Pegunungan** Dut. Sneeuw-gebergte, Eng. Snow Mountains. ▲ Irian Jaya, E Indonesia
Maol Réidh, Caoc see Mweelrea
166 M15 **Maoming** Guangdong, S China 21.45N 110.50E
166 H8 **Maoxian** var. Mao Xian; prev. Fengyizhen. Sichuan, C China 31.42N 103.48E
85 L19 **Mapai** Gaza, SW Mozambique 22.52S 32.00E
164 H15 **Mapam Yumco** ⊚ W China
85 I15 **Mapanza** Southern, S Zambia 16.16S 26.54E
56 J4 **Maparari** Falcón, N Venezuela 10.47N 69.26W
108 J7 **Mapello** Lombardia, N Italy 45.44N 9.33E
175 O5 **Mapi, Pulau** island N Indonesia
176 Yy14 **Mapi** Irian Jaya, E Indonesia 7.02S 139.24E
176 Vv7 **Mapia, Kepulauan** island group E Indonesia
42 L8 **Mapimí** Durango, C Mexico 25.50N 103.50W
85 N19 **Mapinhane** Inhambane, SE Mozambique 22.14S 35.07E
57 N7 **Mapire** Monagas, NE Venezuela 7.48N 64.40W
9 T17 **Maple Creek** Saskatchewan, S Canada 49.55N 109.28W
33 O7 **Maple River** ≈ Michigan, N USA
31 P7 **Maple River** ≈ North Dakota/South Dakota, N USA

31 S13 **Mapleton** Iowa, C USA 42.10N 95.47W
31 U10 **Mapleton** Minnesota, N USA 43.55N 93.57W
31 R5 **Mapleton** North Dakota, C USA 46.51N 97.04W
34 F13 **Mapleton** Oregon, NW USA 44.01N 123.56W
38 L5 **Mapleton** Utah, W USA 40.07N 111.37W
199 I15 **Mapmaker Seamounts** undersea feature N Pacific Ocean
194 G10 **Maprik** East Sepik, NW PNG 3.35S 143.03E
85 L21 **Maputo** prev. Lourenço Marques. ● (Mozambique) Maputo, S Mozambique 25.58S 32.34E
85 L21 **Maputo** ✕ Maputo, S Mozambique 25.58S 32.36E
69 U17 **Maputo** ◆ province S Mozambique
85 L21 **Maputo** ≈ S Mozambique
Maqanshy see Makanchi
Maqat see Makat
115 K19 **Maqellarë** Dibër, C Albania 41.36N 20.29E
115 M19 **Maqellarë** Dibër, C Albania 41.36N 20.29E
165 S12 **Maqên** var. Dawu. Qinghai, C China 34.32N 100.17E
165 S11 **Maqên Gangri** ▲ C China 34.44N 99.25E
165 U12 **Maqu** Gansu, C China 34.02N 102.00E
106 M9 **Maqueda** Castilla-La Mancha, C Spain 40.04N 4.22W
84 J9 **Maquela do Zombo** Uíge, NW Angola 6.03S 15.05E
65 I16 **Maquinchao** Río Negro, C Argentina 41.19S 68.46W
31 Z13 **Maquoketa** Iowa, C USA 42.03N 90.42W
31 Y13 **Maquoketa River** ≈ Iowa, C USA
12 F13 **Mar** Ontario, S Canada 44.48N 81.12W
97 F14 **Mår** ≈ S Norway
203 P8 **Maraa** Tahiti, W French Polynesia 17.45S 149.34W
60 D12 **Maraã** Amazonas, NW Brazil 1.48S 65.21W
203 O8 **Maraa, Pointe** headland Tahiti, W French Polynesia 17.43S 149.34W
61 N14 **Marabá** Pará, NE Brazil 5.23S 49.10W
56 H5 **Maracaibo** Zulia, NW Venezuela 10.39N 71.39W
56 H5 **Maracaibo, Gulf of** see Venezuela, Golfo de
56 H5 **Maracaibo, Lago de** var. Lake Maracaibo. inlet NW Venezuela
Maracaibo, Lake see Maracaibo, Lago de
60 K10 **Maracá, Ilha de** island NE Brazil
61 H20 **Maracaju, Serra de** ▲ S Brazil
60 I11 **Maracanaquará, Planalto** ▲ NE Brazil
56 L5 **Maracay** Aragua, N Venezuela 10.15N 67.36W
Marada see Marādah
77 R9 **Marādah** var. Marada. N Libya 29.15N 19.28E
79 R12 **Maradi** Maradi, S Niger 13.30N 7.05E
79 U11 **Maradi** ◆ department S Niger
83 E21 **Maragarazi** var. Muragarazi. ≈ Burundi/Tanzania
Maragha see Marāgheh
147 R3 **Maragha** var. Marāgheh. Āzarbāyjān-e Khāvarī, NW Iran 37.21N 46.13E
147 P7 **Marāh** var. Marrāt. Ar Riyāḍ, C Saudi Arabia 25.04N 45.30E
57 N11 **Marahuaca, Cerro** ▲ S Venezuela 3.30N 65.30W
29 R9 **Marais des Cygnes River** ≈ Kansas/Missouri, C USA
60 L9 **Marajó, Baía de** bay N Brazil
60 K12 **Marajó, Ilha de** island N Brazil
203 O2 **Marakei** atoll Tungaru, W Kiribati
Marakesh see Marrakech
83 J18 **Maralal** Rift Valley, C Kenya 1.04N 36.42E
85 H24 **Maralaleng** Kgalagadi, S Botswana 25.42S 22.39E
151 L18 **Maraldy, Ozero** ⊚ NE Kazakhstan
190 C5 **Maralinga** South Australia 30.16S 131.35E
179 R15 **Maramag** Mindanao, S Philippines 7.45N 124.58E
Máramarossziget see Sighetu Marmaţiei
195 Y16 **Maramasike** var. Small Malaita. island N Solomon Islands
Maramba see Livingstone
204 R3 **Marambio** Argentinian research station Antarctica 64.22S 57.18W
118 H9 **Maramureş** ◆ county NW Romania
38 L12 **Marana** Arizona, SW USA 32.24N 111.12W
107 P7 **Maranchón** Castilla-La Mancha, C Spain 41.03N 2.10W
148 J2 **Marand** var. Merend. Āzarbāyjān-e Khāvarī, NW Iran 38.23N 45.48E
202 I2 **Maranhão** off. Estado do Maranhão. ≈ state E Brazil
61 H14 **Maranhão, Barragem do** ☐ C Portugal
155 U11 **Mārān, Koh-i** ▲ SW Pakistan 29.24N 66.56E
108 J7 **Marano, Laguna di** lagoon NE Italy
58 E9 **Marañón, Río** ≈ N Peru
104 I10 **Marans** Charente-Maritime, W France 46.19N 0.58W
85 N19 **Marão** Inhambane, S Mozambique 24.15S 34.09E
193 B23 **Mararoa** ≈ South Island, NZ
Maras/Marash see Kahramanmaraş
106 G10 **Marateca** Setúbal, S Portugal
117 B20 **Marathiá, Akrotírio** headland Zákynthos, Iónioi Nísoi, Greece, C Mediterranean Sea 37.39N 20.49E
31 P7 **Maple River** ≈ North Dakota/South Dakota, N USA

25 Y17 **Marathon** Florida Keys, Florida, SE USA 24.42N 81.05W
26 L10 **Marathon** Texas, SW USA 30.10N 103.14W
Marathón see Marathónas
117 N16 **Marathónas** prev. Marathón. Attikí, C Greece 38.09N 23.58E
175 Oo6 **Maratua, Pulau** island N Indonesia
61 O8 **Maraú** Bahia, SE Brazil 14.05S 39.02W
149 R13 **Marāveh Tappeh** Golestān, N Iran 37.52S 55.57E
26 L11 **Maravillas Creek** ≈ Texas, SW USA
194 J13 **Marawaka** Eastern Highlands, C PNG 6.56S 145.54E
179 R15 **Marawi** Mindanao, S Philippines 7.58N 124.16E
106 L16 **Marbella** Andalucía, S Spain 36.31N 4.49W
188 J7 **Marble Bar** Western Australia 21.13S 119.48E
38 L9 **Marble Canyon** canyon Arizona, SW USA
27 S9 **Marble Falls** Texas, SW USA 30.34N 98.16W
29 Y7 **Marble Hill** Missouri, C USA 37.18N 89.58W
35 T15 **Marbleton** Wyoming, C USA 45.31N 110.06W
Marburg see Maribor
Marburg see Marburg an der Lahn, Germany
103 H16 **Marburg an der Lahn** hist. Marburg. Hessen, W Germany 50.49N 8.46E
113 X23 **Marcal** ≈ W Hungary
44 G7 **Marcala** La Paz, SW Honduras 14.08N 88.02W
113 H24 **Marcali** Somogy, SW Hungary 46.33N 17.24E
85 A16 **Marca, Ponta da** headland SW Angola 16.31S 11.42E
61 I16 **Marcelândia** Mato Grosso, W Brazil 11.18S 54.49W
29 T3 **Marceline** Missouri, C USA 39.42N 92.57W
63 C22 **Marcelino Ramos** Rio Grande do Sul, S Brazil 27.31S 51.57W
57 V12 **Marcel, Mont** ▲ S French Guiana 2.32N 53.00W
99 O15 **March** E England, UK 52.37N 0.13E
111 Z3 **March** var. Morava. ≈ C Europe see also Morava
105 N11 **Marche-en-Famenne** Luxembourg, SE Belgium 50.13N 5.21E
108 I12 **Marche** Eng. Marches. ◆ region W Italy
108 I12 **Marche** cultural region C France
106 K4 **Marchena** Andalucía, S Spain 37.19N 5.24W
59 B17 **Marchena, Isla** var. Bindloe Island. island Galápagos Islands, Ecuador, E Pacific Ocean
Marches see Marche
101 J20 **Marchin** Liège, E Belgium 50.28N 5.14E
189 S1 **Marchinbar Island** island Wessel Islands, Northern Territory, N Australia
64 J9 **Mar Chiquita, Laguna** ⊚ C Argentina
105 Q10 **Marcigny** Saône-et-Loire, C France 46.16N 4.04E
25 W16 **Marco** Florida, SE USA 25.56N 81.43W
61 O15 **Marcolândia** Pernambuco, E Brazil 7.21S 40.40W
108 I7 **Marco Polo** ✕ (Venezia) Veneto, NE Italy 45.30N 12.21E
7 Q5 **Marcus Baker, Mount** ▲ Alaska, USA 61.26N 147.45W
199 Hh5 **Marcus Island** var. Minami Tori Shima. island E Japan
20 K9 **Marcy, Mount** ▲ New York, NE USA 44.06N 73.55W
155 T5 **Mardān** North-West Frontier Province, N Pakistan 34.13N 71.59E
65 N6 **Mar del Plata** Buenos Aires, E Argentina 38.10S 57.38W
143 Q16 **Mardin** Mardin, SE Turkey 37.19N 40.43E
143 Q16 **Mardin** ◆ province SE Turkey
143 Q16 **Mardin Dağları** ▲ SE Turkey
168 J9 **Mardzad** Övörhangay, C Mongolia 45.22N 101.30E
197 L6 **Maré** island Îles Loyauté, E New Caledonia
Marea Neagră see Black Sea
107 Z8 **Mare de Déu del Toro** ▲ Menorca, Spain 39.59N 4.06E
189 W4 **Mareeba** Queensland, NE Australia 17.03S 145.30E
98 G8 **Maree, Loch** ⊚ N Scotland, UK
Mareeq see Mereeg
Marek see Dupnitsa
78 J11 **Maréna** Kayes, W Mali 14.36N 10.57W
202 I2 **Marenanuka** atoll Tungaru, W Kiribati
31 X11 **Marengo** Iowa, C USA 41.48N 92.04W
104 J11 **Marennes** Charente-Maritime, W France 45.49N 1.04W
20 D13 **Marenisco** Michigan, N USA 46.22N 89.41W
101 M24 **Marennwille** Pennsylvania, NE USA 37.04N 79.07W
101 G18 **Marienne** ≈ SE Belgium
26 K5 **Marfa** Texas, SW USA 30.18N 104.01W
59 I17 **Marfil, Laguna** ⊚ E Bolivia
60 F10 **Margai Caka** ⊚ W China

105 P13 **Margeride, Montagnes de la** ▲ C France
Margherita see Jamaame
109 N16 **Margherita di Savoia** Puglia, SE Italy 41.22N 16.09E
Margherita, Lake see Ābaya Hāyk'
81 E18 **Margherita Peak** Fr. Pic Marguerite. ▲ Uganda/Dem. Rep. Congo (Zaire) 0.28N 29.58E
155 O4 **Marghi** Bāmiān, N Afghanistan 35.10N 66.26E
153 S10 **Marghilon** var. Margelan, Rus. Margilan. Farghona Wiloyati, E Uzbekistan 40.29N 71.43E
118 G9 **Marghita** Hung. Margitta. Bihor, NW Romania 47.20N 22.19E
118 K8 **Margilan** see Marghilon
154 K9 **Märgow, Dasht-e** desert SW Afghanistan
101 H18 **Margraten** Limburg, SE Netherlands 50.49N 5.49E
8 M15 **Marguerite** British Columbia, SW Canada 52.17N 122.18W
204 I6 **Marguerite Bay** bay Antarctica
Marguerite, Pic see Margherita Peak
119 T9 **Marhanets'** Rus. Marganets. Dnipropetrovs'ka Oblast', E Ukraine 47.34N 34.37E
194 K13 **Mari** Western, SW PNG 9.10S 141.39E
203 R13 **Maria** island Îles Australes, SW French Polynesia
203 V12 **Maria** atoll Groupe Actéon, SE French Polynesia
42 I12 **María Cleofas, Isla** island C Mexico
64 H4 **María Elena** var. Oficina María Elena. Antofagasta, N Chile 22.18S 69.40W
97 I22 **Mariager** Århus, C Denmark 56.39N 9.58E
63 G22 **María Ignacia** Buenos Aires, E Argentina 37.24S 59.30W
191 P17 **Maria Island** island Tasmania, SE Australia
42 I12 **María Madre, Isla** island C Mexico
42 I12 **María Magdalena, Isla** island C Mexico
199 N16 **Mariana Islands** island group Guam/Northern Mariana Islands
183 N3 **Mariana Trench** var. Challenger Deep. undersea feature W Pacific Ocean
159 X12 **Mariāni** Assam, NE India 26.39N 94.18E
29 X11 **Marianna** Arkansas, C USA 34.46N 90.45W
25 R8 **Marianna** Florida, SE USA 30.46N 85.13W
180 J16 **Marianne** island Inner Islands, NE Seychelles
97 X11 **Mariannelund** Jönköping, S Sweden 57.37N 15.32E
63 G19 **Mariano I.Loza** Corrientes, NE Argentina 29.22S 58.12W
Mariano Machado see Ganda
113 A16 **Mariánské Lázně** Ger. Marienbad. Karlovarský Kraj, W Czech Republic 49.57N 12.42E
Máriaradna see Radna
35 S7 **Marias River** ≈ Montana, NW USA
Maria-Theresiopel see Subotica
Máriatölgyes see Dubnica nad Váhom
192 H1 **Maria van Diemen, Cape** headland North Island, NZ 34.27S 172.38E
114 G12 **Mariazell** Steiermark, E Austria 47.45N 15.17E
147 P13 **Ma'rib** var. Marib. NW Yemen 15.28N 45.25E
97 J22 **Maribo** Storstrøm, S Denmark 54.46N 11.30E
111 W9 **Maribor** Ger. Marburg. NE Slovenia 46.33N 15.40E
Marica see Maritsa
37 O13 **Maricopa** California, W USA 35.03N 119.24W
Maricourt see Kangiqsujuaq
83 B17 **Maridi** Western Equatoria, SW Sudan 4.55N 29.30E
204 M11 **Marie Byrd Land** physical region Antarctica
199 Ll16 **Marie Byrd Seamount** undersea feature N Amundsen Sea 70.00S 118.00W
47 X11 **Marie-Galante** var. Ceyre to the Caribs. island SE Guadeloupe
47 Y6 **Marie-Galante, Canal de** channel S Guadeloupe
95 J20 **Mariehamn** Fin. Maarianhamina. Åland, SW Finland 60.04N 19.55E
46 C4 **Mariel** La Habana, W Cuba 22.58N 82.49W
101 I21 **Mariembourg** Namur, S Belgium 50.07N 4.30E
Marienberg see Mariánské Lázně
Marienburg see Alūksne, Latvia
Marienburg see Malbork, Poland
Marienburg see Feldioara, Romania
Marienburg in Westpreussen see Malbork
Marienhausen see Viļaka
Marienwerder see Kwidzyń
60 G10 **Marié, Rio** ≈ NW Brazil
97 K17 **Mariestad** Västra Götaland, S Sweden 58.42N 13.49E
23 T2 **Marietta** Georgia, SE USA 33.57N 84.33W
33 V14 **Marietta** Ohio, N USA 39.25N 81.27W
27 O13 **Marietta** Oklahoma, C USA 33.56N 97.07W
57 N7 **Margarita, Isla de** island N Venezuela
83 J17 **Marigat** Rift Valley, W Kenya
105 S16 **Marignane** Bouches-du-Rhône, SE France 43.25N 5.12E
Marignano see Melegnano
47 O11 **Marigot** NE Dominica 15.31N 61.17W

126 Hh14 **Mariinsk** Kemerovskaya Oblast', S Russian Federation 56.13N 87.27E
131 Q3 **Mariinskiy Posad** Respublika Mariy El, W Russian Federation 56.07N 47.44E
121 G12 **Marikostenovo** Blagoevgrad, SW Bulgaria 41.32N 23.21E
62 J9 **Marília** São Paulo, S Brazil 22.13S 49.58W
84 D11 **Marimba** Malanje, NW Angola 8.18S 16.58E
145 T1 **Mari Milā** E Iraq 36.58N 44.42E
106 G4 **Marín** Galicia, NW Spain 42.22N 8.42W
37 N10 **Marina** California, W USA 36.40N 121.48W
Marina di Catanzaro see Catanzaro Marina
Mar'ina Gorka see Mar''ina Horka
119 L17 **Mar''ina Horka** Rus. Mar'ina Gorka. Minskaya Voblasts', C Belarus 53.30N 28.09E
179 Pp11 **Marinduque** island C Philippines
33 S9 **Marine City** Michigan, N USA 42.43N 82.29W
33 N6 **Marinette** Wisconsin, N USA 45.06N 87.37W
62 I10 **Maringá** Paraná, S Brazil 23.25S 51.55W
85 N16 **Maríngue** Sofala, C Mozambique 17.57S 34.23E
106 F9 **Marinha Grande** Leiria, C Portugal 39.45N 8.55W
109 N16 **Marino** Lazio, C Italy 41.46N 12.40E
61 A15 **Mário Lobão** Acre, W Brazil 8.25N 72.58W
23 O5 **Marion** Alabama, S USA 32.37N 87.19W
29 Y11 **Marion** Arkansas, C USA 35.12N 90.12W
32 L17 **Marion** Illinois, N USA 37.43N 88.55W
33 P13 **Marion** Indiana, N USA 40.31N 85.40W
31 X13 **Marion** Iowa, C USA 42.01N 91.36W
29 O5 **Marion** Kansas, C USA
22 H6 **Marion** Kentucky, S USA 37.19N 88.04W
23 P9 **Marion** North Carolina, S USA 35.43N 82.00W
33 S12 **Marion** Ohio, N USA 40.34N 83.07W
23 T12 **Marion** South Carolina, SE USA 34.10N 79.24W
23 Q7 **Marion** Virginia, NE USA 36.49N 81.31W
29 O5 **Marion Lake** ⊚ Kansas, C USA
23 U13 **Marion, Lake** ⊚ South Carolina, SE USA
57 N7 **Maripa** Bolívar, E Venezuela 7.27N 65.10W
57 X11 **Maripasoula** W French Guiana 3.43N 54.04W
37 Q9 **Mariposa** California, W USA 37.28N 119.59W
63 G19 **Mariscala** Lavalleja, S Uruguay 34.05S 54.46W
64 M4 **Mariscal Estigarribia** Boquerón, NW Paraguay 22.02S 60.39W
58 C12 **Mariscal Sucre** var. Quito. ✕ (Quito) Pichincha, C Ecuador 0.21S 78.37W
32 K16 **Marissa** Illinois, N USA 38.15N 89.45W
105 U14 **Maritime Alps** Fr. Alpes Maritimes, It. Alpi Marittime. ▲ France/Italy
Maritimes, Alpes see Maritime Alps
Maritime Territory see Primorskiy Kray
116 K11 **Maritsa** var. Marica, Gk. Évros, Turk. Meriç; anc. Hebrus. ≈ SW Europe see also Évros/Meriç
Maritsa see Simeonovgrad
Marittime, Alpi see Maritime Alps
Maritzburg see Pietermaritzburg
119 X9 **Mariupol'** prev. Zhdanov. Donets'ka Oblast', SE Ukraine 47.06N 37.33E
47 Q6 **Marival, Caño** ≈ NE Venezuela
148 J5 **Marivan** prev. Dezh Shāhpūr. Kordestān, W Iran 35.32N 46.09E
125 R3 **Mariyets** Respublika Mariy El, W Russian Federation 56.31N 49.48E
175 R3 **Mariyskaya ASSR** see Mariy El, Respublika
120 G4 **Märjamaa** Ger. Merjama. Raplamaa, NW Estonia 58.53S 24.24E
101 M23 **Mark** Fr. Marcq. ≈ Belgium/Netherlands
83 N17 **Marka** var. Merca. Shabeellaha Hoose, S Somalia 1.43N 44.45E
78 M11 **Markala** Ségou, W Mali 13.38N 6.07W
165 S15 **Markam** var. Gartog. Xizang Zizhiqu, W China 29.40N 98.33E
97 K21 **Markaryd** Kronoberg, S Sweden 56.25N 13.34E
148 L2 **Markazī** off. Ostān-e Markazī. ◆ province NW Iran
12 F14 **Markdale** Ontario, S Canada 44.19N 80.37W
81 D20 **Marke** Rwanda
203 V14 **Marotiri** var. Îlots de Bass, Morotiri. island group Îles Australes, SW French Polynesia
80 J7 **Maroua** Extrême-Nord, N Cameroon 10.34N 14.19E
180 J4 **Marovoay** Mahajanga, NW Madagascar 16.06S 46.40E
57 W7 **Marowijne** ◆ district NE Suriname
57 W8 **Marowijne** see Maroni
199 M20 **Marquesas Fracture Zone** tectonic feature E Pacific Ocean
199 L17 **Marquesas Islands** island group Marquises, Îles
25 W17 **Marquesas Keys** island group Florida, SE USA
31 Y12 **Marquette** Iowa, C USA 43.03N 91.10W

27 V12 **Markham** Texas, SW USA 28.57N 96.04W
194 J13 **Markham** ≈ C PNG
205 Q11 **Markham, Mount** ▲ Antarctica 82.58S 163.30E
112 M11 **Marki** Mazowieckie, C Poland 52.19N 21.07E
164 F8 **Markit** Xinjiang Uygur Zizhiqu, NW China 38.55N 77.40E
119 Y5 **Markivka** Rus. Markovka. Luhans'ka Oblast', E Ukraine 49.34N 39.35E
37 S9 **Markleeville** California, W USA 38.41N 119.47W
100 I4 **Marknesse** Flevoland, N Netherlands 52.44N 5.54E
81 H14 **Markounda** var. Marcounda. Ouham, NW Central African Republic 7.37N 17.00E
127 P6 **Markovo** Chukotskiy Avtonomnyy Okrug, NE Russian Federation 64.35N 170.13E
131 L17 **Marks** Saratovskaya Oblast', W Russian Federation 51.40N 46.44E
24 K4 **Marks** Mississippi, S USA 34.15N 90.16W
24 I7 **Marksville** Louisiana, S USA 31.07N 92.04W
103 I19 **Marktheidenfeld** Bayern, C Germany 49.50N 9.36E
103 J24 **Marktoberdorf** Bayern, S Germany 47.45N 10.36E
103 M18 **Marktredwitz** Bayern, E Germany 50.00N 12.04E
29 V3 **Mark Twain Lake** ☐ Missouri, C USA
Markuleşti see Mărculeşti
103 E14 **Marl** Nordrhein-Westfalen, W Germany 51.40N 7.06E
190 L2 **Marla** South Australia 27.19S 133.35E
189 I7 **Marlborough** Queensland, E Australia 22.55S 150.07E
99 M22 **Marlborough** S England, UK 51.25N 1.44W
193 I15 **Marlborough** off. Marlborough District. ◇ unitary authority South Island, NZ
105 P3 **Marle** Aisne, N France 49.54N 3.48E
33 S8 **Marlette** Michigan, N USA 43.20N 83.05W
27 T9 **Marlin** Texas, SW USA 31.18N 96.54W
23 S5 **Marlinton** West Virginia, NE USA 38.13N 80.05W
28 M12 **Marlow** Oklahoma, C USA 34.39N 97.57W
161 E17 **Marmagao** Goa, W India 15.22N 73.53E
Marmanda see Marmande
104 L13 **Marmande** anc. Marmanda. Lot-et-Garonne, SW France 44.30N 0.10E
142 C11 **Marmara** Balikesir, NW Turkey 40.36N 27.34E
142 D11 **Marmara Denizi** Eng. Sea of Marmara. sea NW Turkey
116 N13 **Marmaraereğlisi** Tekirdağ, NW Turkey 40.58N 27.57E
Marmara, Sea of see Marmara Denizi
142 C16 **Marmaris** Muğla, SW Turkey 36.52N 28.16E
30 M4 **Marmarth** North Dakota, N USA 46.17N 103.55W
23 Q5 **Marmet** West Virginia, NE USA 38.12N 81.31W
108 H5 **Marmolada, Monte** ▲ N Italy 46.26N 11.58E
106 M13 **Marmolejo** Andalucía, S Spain 38.03N 4.10W
12 J4 **Marmora** Ontario, SE Canada 44.28N 77.40W
41 Q14 **Marmot Bay** bay Alaska, USA
105 Q4 **Marne** ◆ department N France
105 Q5 **Marne** ≈ N France
143 U10 **Marneuli** prev. Borchalo, Sarvani. S Georgia 41.28N 44.45E
80 J3 **Maro** Moyen-Chari, S Chad 8.25N 18.46E
56 L12 **Maroa** Amazonas, S Venezuela 2.40N 67.33W
180 J3 **Maroantsetra** Toamasina, NE Madagascar 15.22S 49.43E
203 W11 **Marokau** atoll Îles Tuamotu, C French Polynesia
180 J5 **Marolambo** Toamasina, E Madagascar 20.03S 48.07E
180 J2 **Maromokotro** ▲ N Madagascar
85 L16 **Marondera** prev. Marandellas. Mashonaland East, NE Zimbabwe 18.10S 31.33E
57 V9 **Maroni** Dut. Marowijne. ≈ French Guiana/Suriname
191 R3 **Maroochydore-Mooloolaba** Queensland, E Australia 26.36S 153.04E
175 R3 **Maros** Sulawesi, C Indonesia 4.58S 119.34E
118 H11 **Maros** var. Mureş, Mureşul, Ger. Marosch, Mieresch. ≈ Hungary/Romania see also Mureş
Marosch see Maros/Mureş
Maroshévíz see Topliţa
Marosillye see Ilia
Marosludas see Luduş
Marosújvár/Marosújvárakna see Ocna Mureş
Marosvásárhely see Târgu Mureş

◆ COUNTRY ● COUNTRY CAPITAL ◇ DEPENDENT TERRITORY ○ DEPENDENT TERRITORY CAPITAL ◆ ADMINISTRATIVE REGION ✕ INTERNATIONAL AIRPORT ▲ MOUNTAIN ▲ MOUNTAIN RANGE ▲ VOLCANO ≈ RIVER ⊚ LAKE ☐ RESERVOIR

33 N3 **Marquette** Michigan, N USA
46.32N 87.24W

105 N1 **Marquise** Pas-de-Calais,
N France 50.49N 1.42E

203 X7 **Marquises, Îles** Eng. Marquesas
Islands. *island group* N French
Polynesia

191 Q6 **Marra Creek** ॐ New South
Wales, SE Australia

82 B10 **Marra Hills** *plateau* W Sudan

82 B11 **Marra, Jebel** ▲ W Sudan
12.59N 24.16E

76 E7 **Marrakech** *var.* Marakesh, *Eng.*
Marrakesh; *prev.* Morocco.
W Morocco 31.39N 7.57W

Marrakesh *see* Marrakech
Marrät *see* Marāh

191 N15 **Marrawah** Tasmania,
SE Australia 40.55S 144.41E

190 I4 **Marree** South Australia
29.39S 138.06E

83 L17 **Marrehan** ◆ SW Somalia

85 N17 **Marromeu** Sofala,
C Mozambique 18.18S 35.58E

106 J17 **Marroquí, Punta** *headland*
SW Spain 36.01N 5.39W

191 N8 **Marrowie Creek** *seasonal river*
New South Wales, SE Australia

85 O14 **Marrupa** Niassa, N Mozambique
13.13S 37.30E

190 D1 **Marryat** South Australia
26.12S 133.22E

77 Y10 **Marsa 'Alam** SE Egypt
25.01N 34.52E

77 R8 **Marsá al Burayqah** *var.*
Al Burayqah. N Libya 30.21N 19.37E

83 J17 **Marsabit** Eastern, N Kenya
2.19N 37.58E

109 H23 **Marsala** *anc.* Lilybaeum. Sicilia,
Italy, C Mediterranean Sea
37.48N 12.26E

123 J17 **Marsaxlokk Bay** *bay* SE Malta

67 G15 **Mars Bay** *bay* Ascension Island,
C Atlantic Ocean

103 H15 **Marsberg** Nordrhein-Westfalen,
W Germany 51.28N 8.51E

9 R15 **Marsden** Saskatchewan, S Canada
52.50N 109.45W

100 H7 **Marsdiep** *strait* NW Netherlands

105 R16 **Marseille** *Eng.* Marseilles; *anc.*
Massilia. Bouches-du-Rhône,
SE France 43.19N 5.21E

Marseille-Marignane *see*
Provence

32 M11 **Marseilles** Illinois, N USA
41.19N 88.42W

Marseilles *see* Marseille

78 J16 **Marshall** W Liberia 6.10N 10.22W

41 N11 **Marshall** Alaska, USA
61.52N 162.04W

29 U9 **Marshall** Arkansas, C USA
35.54N 92.37W

33 N14 **Marshall** Illinois, N USA
39.23N 87.41W

33 Q10 **Marshall** Michigan, S USA
42.16N 84.57W

31 S9 **Marshall** Minnesota, N USA
44.26N 95.48W

29 T4 **Marshall** Missouri, C USA
39.07N 93.12W

23 O9 **Marshall** North Carolina, SE USA
35.49N 82.41W

27 X6 **Marshall** Texas, SW USA
32.32N 94.22W

201 S4 **Marshall Islands** *off.* Republic of
the Marshall Islands. ◆ *republic*
W Pacific Ocean

183 Q3 **Marshall Islands** *island group*
W Pacific Ocean

199 I7 **Marshall Seamounts** *undersea
feature* SW Pacific Ocean

31 W13 **Marshalltown** Iowa, C USA
42.01N 92.54W

21 P12 **Marshfield** Massachusetts,
NE USA 42.04N 70.40W

29 T7 **Marshfield** Missouri, C USA
37.20N 92.54W

32 K6 **Marshfield** Wisconsin, N USA
44.41N 90.12W

46 H1 **Marsh Harbour** Great Abaco,
W Bahamas 26.31N 77.03W

21 S3 **Mars Hill** Maine, NE USA
46.31N 67.51W

23 P9 **Mars Hill** North Carolina,
SE USA 35.49N 82.33W

24 H10 **Marsh Island** Louisiana,
S USA

23 S11 **Marshville** North Carolina,
SE USA 34.59N 80.22W

13 W5 **Marsoui** Quebec, SE Canada
49.12N 65.58W

13 R8 **Mars, Rivière à** ॐ Quebec,
SE Canada

97 O15 **Märsta** Stockholm, C Sweden
59.37N 17.52E

97 H24 **Marstal** Fyn, C Denmark
54.52N 10.31E

97 I19 **Marstrand** Västra Götaland,
S Sweden 57.54N 11.31E

27 U8 **Mart** Texas, SW USA 31.32N 96.49W

178 G9 **Martaban, Gulf of** *see* Mottama,
Gulf of

177 G9 **Martaban, Gulf of** *gulf*
S Myanmar

109 Q19 **Martano** Puglia, SE Italy
40.12N 18.19E

175 N11 **Martapoera** *see* Martapura

175 N11 **Martapura** *prev.* Martapoera.
Borneo, C Indonesia 3.25S 114.51E

101 L23 **Martelange** Luxembourg,
SE Belgium 49.50N 5.43E

116 L7 **Marten** Ruse, N Bulgaria
43.57N 26.08E

12 H10 **Marten River** Ontario, S Canada
46.43N 79.45W

9 T15 **Martensville** Saskatchewan,
S Canada 52.15N 106.42W

Marteskirch *see* Tārnāveni
Martes Tolosane *see* Martres-
Tolosane

117 K25 **Mártha** Kriti, Greece,
E Mediterranean Sea 35.03N 25.22E

191 Q6 **Marthaguy Creek** ॐ New
South Wales, SE Australia

21 P13 **Martha's Vineyard** *island*
Massachusetts, NE USA

110 C11 **Martigny** Valais, SW Switzerland
46.04N 7.03E

105 R16 **Martigues** Bouches-du-Rhône,
SE France 43.24N 5.03E

---

113 J19 **Martin** *Ger.* Sankt Martin, *Hung.*
Turócszentmárton; *prev.*
Turčianský Svätý Martin. Žilinský
Kraj, N Slovakia 49.03N 18.54E

30 L11 **Martin** South Dakota, N USA
43.10N 101.43W

22 G8 **Martin** Tennessee, S USA
36.20N 88.51W

107 S7 **Martín** ॐ E Spain

109 P18 **Martina Franca** Puglia, SE Italy
40.42N 17.19E

193 M14 **Martinborough** Wellington,
North Island, NZ 41.15S 175.28E

27 S11 **Martindale** Texas, SW USA
29.49N 97.49W

37 N8 **Martinez** California, W USA
38.00N 122.12W

25 V3 **Martinez** Georgia, SE USA
33.31N 82.04W

43 Q13 **Martínez De La Torre** Veracruz-
Llave, E Mexico 20.06N 97.03W

47 Y12 **Martinique** ◇ *French overseas
department* E West Indies

1 O15 **Martinique** *island* E West Indies

Martinique Channel *see*
Martinique Passage

47 X12 **Martinique Passage** *var.*
Dominica Channel, Martinique
Channel. *channel*
Dominica/Martinique

25 Q5 **Martin Lake** ⊟ Alabama, S USA

117 G18 **Martíno** *prev.* Martíno. Stereá
Ellás, C Greece 38.34N 23.13E

204 J11 **Martin Peninsula** *peninsula*
Antarctica

41 S5 **Martin Point** *headland* Alaska,
USA 70.06N 143.04W

111 V3 **Martinsberg** Niederösterreich,
NE Austria 48.23N 15.09E

23 V3 **Martinsburg** West Virginia,
NE USA 39.25N 77.55W

33 V13 **Martins Ferry** Ohio, N USA
40.06N 80.43W

33 O14 **Martinsville** Indiana, N USA
39.25N 86.25W

23 S8 **Martinsville** Virginia, NE USA
36.41N 79.52W

67 K16 **Martin Vaz, Ilhas** *island group*
E Brazil

Martók *see* Martuk

192 M12 **Marton** Manawatu-Wanganui,
North Island, NZ 40.05S 175.22E

107 N13 **Martos** Andalucía, S Spain
37.43N 3.58W

104 M16 **Martres-Tolosane** *var.* Martes
Tolosane. Haute-Garonne, S France
43.13N 1.00E

94 M11 **Martti** Lappi, NE Finland
67.28N 28.19E

150 I9 **Martuk** *Kaz.* Martók.
Aktyubinsk, NW Kazakhstan
50.45N 56.30E

143 U12 **Martuni** ॐ Armenia 40.07N 45.20E

60 L11 **Marudá** Pará, E Brazil 5.25S 49.04W

175 O2 **Marudu, Teluk** *bay* East Malaysia

155 O8 **Ma'ruf** Kandahār, SE Afghanistan
31.37N 67.08E

170 F14 **Marugame** Kagawa, Shikoku,
SW Japan 34.16N 133.46E

193 H16 **Maruia** ॐ South Island, NZ

100 M6 **Marum** Groningen,
NE Netherlands 53.07N 6.16E

197 C13 **Marum, Mount** ▲ Ambrym,
C Vanuatu 16.15S 168.07E

81 P23 **Marungu** ▲ SE Dem. Rep.
Congo (Zaire)

203 Y12 **Marutea** *atoll* Groupe Actéon,
C French Polynesia

149 O11 **Marv Dasht** *var.* Mervdasht.
Fārs, S Iran 29.51N 52.44E

105 P13 **Marvejols** Lozère, S France
44.35N 3.16E

29 X12 **Marvell** Arkansas, C USA
34.33N 90.52W

38 L6 **Marvine, Mount** ▲ Utah,
W USA 38.40N 111.38W

145 Q7 **Marwānīyah** ॐ C Iraq 33.58N 42.31E

158 F13 **Mārwār** *var.* Marwar Junction.
Rājasthān, N India 25.43N 73.39E

9 R14 **Marwayne** Alberta, SW Canada
53.30N 110.25W

152 I14 **Mary** *prev.* Merv. Maryyskiy
Velayat, S Turkmenistan
37.24N 61.48E

Mary *see* Maryyskiy Velayat

189 Z7 **Maryborough** Queensland,
E Australia 25.31S 152.36E

190 M11 **Maryborough** Victoria,
SE Australia 37.04S 143.43E

Maryborough *see* Port Laoise

85 G23 **Marydale** Northern Cape,
W South Africa 29.25S 22.06E

119 W8 **Mar"yinka** Donets'ka Oblast',
E Ukraine 47.57N 37.27E

Mary Island *see* Kanton

23 W4 **Maryland** *off.* State of Maryland;
also known as America in
Miniature, Cockade State, Free
State, Old Line State. ◆ *state*
NE USA

27 P7 **Maryneal** Texas, SW USA
32.12N 100.25W

99 J15 **Maryport** NW England, UK
54.44N 3.28W

11 U13 **Marystown** Newfoundland,
SE Canada 47.10N 55.10W

38 K6 **Marysvale** Utah, W USA
38.26N 112.14W

37 O6 **Marysville** California, W USA
39.08N 121.35W

23 R2 **Marysville** Missouri, C USA
40.20N 94.52W

33 N9 **Marysville** Michigan, S USA
42.54N 82.29W

33 S9 **Marysville** Ohio, NE USA
40.13N 83.22W

34 H7 **Marysville** Washington,
NW USA 48.03N 122.10W

31 R8 **Maryville** Missouri, C USA
40.20N 94.51W

22 H9 **Maryville** Tennessee, S USA
35.45N 83.58W

Mary Welayaty *see* Maryyskiy
Velayat

152 I15 **Maryyskiy Velayat** *var.* Mary,
Turkm. Mary Welayaty. ◆ *province*
S Turkmenistan

Marzūq *see* Murzuq

99 V11 **Mas** Irian Jaya, E Indonesia
3.28S 132.40E

---

44 J11 **Masachapa** *var.* Puerto
Masachapa. Managua,
W Nicaragua 11.47N 86.31W

83 G19 **Masai Mara National Reserve**
reserve C Kenya

83 I21 **Masai Steppe** *grassland*
NW Tanzania

83 F19 **Masaka** SW Uganda 0.19S 31.46E

175 N12 **Masalembo Besar, Pulau** *island*
S Indonesia

143 Y13 **Masallı** *Rus.* Masally.
S Azerbaijan 39.03N 48.39E
Masally *see* Masallı

175 Pp10 **Masamba** Sulawesi, C Indonesia
2.33S 120.19E

Masampo *see* Masan

169 Y16 **Masan** *prev.* Masampo. S South
Korea 35.10N 128.36E

Masandam Peninsula *see*
Musandam Peninsula

83 J25 **Masasi** Mtwara, SE Tanzania
10.43S 38.48E

44 J10 **Masawa** *see* Massawa

44 J10 **Masaya** Masaya, W Nicaragua
11.58N 86.06W

44 J10 **Masaya** ◆ *department*
W Nicaragua

179 Q12 **Masbate** Masbate, N Philippines
12.21N 123.34E

76 I6 **Mascara** *var.* Mouaskar.
NW Algeria 35.25N 0.10E

181 O7 **Mascarene Basin** *undersea feature*
W Indian Ocean

181 O9 **Mascarene Islands** *island group*
W Indian Ocean

181 N9 **Mascarene Plain** *undersea feature*
W Indian Ocean

181 O7 **Mascarene Plateau** *undersea
feature* W Indian Ocean

204 H5 **Mascart, Cape** *headland* Adelaide
Island, Antarctica

64 J10 **Mascasín, Salinas de** *salt lake*
C Argentina

42 K13 **Mascota** Jalisco, C Mexico
20.31N 104.46W

13 O12 **Mascouche** Quebec, SE Canada
45.46N 73.37W

128 J9 **Masel'gskaya** Respublika
Kareliya, NW Russian Federation
63.09N 34.22E

85 J23 **Maseru** ● (Lesotho) W Lesotho
29.21S 27.34E

85 J23 **Maseru** ✈ W Lesotho 29.27S 27.37E

166 K14 **Mashan** Guangxi Zhuangzu
Zizhiqu, S China 23.40N 108.10E

85 K17 **Mashava** *prev.* Mashaba.
Masvingo, SE Zimbabwe
20.03S 30.28E

149 U4 **Mashhad** *var.* Meshed, Khorāsān,
NE Iran 36.16N 59.34E

172 Oo4 **Mashike** Hokkaidō, NE Japan
43.51N 141.30E

155 N14 **Māshkel** var. SW Pakistan

149 X13 **Māshkel** *var.* Rūd-i Māshkel,
Rūd-e Māshkīd. ॐ Iran/Pakistan

154 K12 **Māshkel, Hāmūn-i** *salt marsh*
SW Pakistan

**Māshkel, Rūd-i/Māshkīd,
Rūd-e** *see* Māshkel

85 K17 **Mashonaland Central** ◆
province N Zimbabwe

85 N16 **Mashonaland East** ◆ *province*
NE Zimbabwe

85 J16 **Mashonaland West** ◆ *province*
NW Zimbabwe

Mashtagi *see* Maştağa

172 Qq6 **Mashū-ko** *var.* Masyū Ko.
☉ Hokkaidō, NE Japan

147 S14 **Masilah, Wādī al** *dry watercourse*
SE Yemen

79 I19 **Masi-Manimba** Bandundu,
SW Dem. Rep. Congo (Zaire)
4.44S 17.56E

172 E9 **Masjed-e Soleymān** *see* Masjed
Soleymān

148 L9 **Masjed Soleymān** *var.* Masjed-e
Soleymān, Masjid-i Sulaiman.
Khūzestān, SW Iran 31.58N 49.17E

Masjid-i Sulaiman *see* Masjed
Soleymān

Maskat *see* Masqaţ

145 Q7 **Maskānah** C Iraq 33.41N 42.46E

147 X8 **Maskin** *var.* Miskin. NW Oman

99 B17 **Mask, Lough** *Ir.* Loch Measca.
☉ W Ireland

116 N10 **Maslen Nos** *headland* E Bulgaria
42.19N 27.47E

180 K3 **Masoala, Tanjona** *headland*
NE Madagascar 15.58N 50.13E

Masohi *see* Amahai

33 Q9 **Mason** Michigan, N USA
42.33N 84.25W

33 R14 **Mason** Ohio, N USA 39.21N 84.18W

27 Q10 **Mason** Texas, SW USA
30.44N 99.15W

23 P4 **Mason** West Virginia, NE USA
39.01N 82.01W

193 B25 **Mason Bay** *bay* Stewart Island,
NZ

32 K13 **Mason City** Illinois, N USA
40.12N 89.42W

31 V11 **Mason City** Iowa, C USA
43.09N 93.12W

Ma, Sông *see* Ma, Nam

32 B16 **Masontown** Pennsylvania,
NE USA 39.49N 79.53W

147 Y8 **Masqaţ** *var.* Maskat, *Eng.* Muscat.
● (Oman) NE Oman 23.34N 58.36E

21 P15 **Massachusetts** *off.*
Commonwealth of Massachusetts;
also known as Bay State, Old Bay
State, Old Colony State. ◆ *state*
NE USA

21 P11 **Massachusetts Bay** *bay*
Massachusetts, NE USA

37 R2 **Massacre Lake** ☉ Nevada,
W USA

109 L18 **Massafra** Puglia, SE Italy
40.35N 17.05E

110 G11 **Massagno** Ticino, S Switzerland
46.01N 8.55E

80 G11 **Massaguet** Chari-Baguirmi,
W Chad 12.28N 15.25E

80 G10 **Massakori** *see* Massakory

80 G10 **Massakory** *var.* Massakori; *prev.*
Dagana. Chari-Baguirmi, W Chad
13.01N 15.43E

80 H11 **Massalassef** Chari-Baguirmi,
SW Chad 11.37N 17.09E

75 S8 **Massa Marittima** Toscana,
C Italy 43.03N 10.55E

85 B11 **Massangano** Cuanza Norte,
NW Angola 9.36S 14.19E

85 M18 **Massangena** Gaza,
S Mozambique 21.34S 32.57E

82 J9 **Massawa** *var.* Masawa, *Amh.*
Mits'iwa. E Eritrea 15.37N 39.27E

82 K9 **Massawa Channel** *channel*
E Eritrea

20 J6 **Massena** New York, NE USA
44.55N 74.53W

80 H11 **Massenya** Chari-Baguirmi,
SW Chad 11.21N 16.09E

8 I13 **Masset** Graham Island, British
Columbia, SW Canada
54.00N 132.09W

104 L16 **Masseube** Gers, S France
43.26N 0.33E

12 E11 **Massey** Ontario, S Canada
46.13N 82.06W

105 P12 **Massiac** Cantal, C France
45.16N 3.13E

105 P12 **Massif Central** *plateau* C France

13 S12 **Massilia** *see* Marseille

33 U12 **Massillon** Ohio, N USA
40.48N 81.31W

79 N12 **Massina** Ségou, W Mali
13.58N 5.24W

85 N19 **Massinga** Inhambane,
SE Mozambique 23.16S 35.23E

85 L20 **Massingir** Gaza,
SW Mozambique 23.57S 32.12E

205 Z10 **Masson Island** *island* Antarctica

45 W11 **Massoukou** *see* Franceville

143 Z11 **Maştağa** *Rus.* Mashtagi, Mastaga.
E Azerbaijan 40.31N 50.01E

192 M13 **Masterton** Wellington, North
Island, NZ 40.56S 175.39E

20 M14 **Mastic** Long Island, New York,
NE USA 40.48N 72.50W

155 O10 **Mastung** Baluchistān,
SW Pakistan 29.46N 66.48E

121 J20 **Mastva** *Rus.* Mostva.
ॐ SW Belarus

G17 **Masty** *Rus.* Mosty.
Hrodzyenskaya Voblasts',
W Belarus 53.25N 24.30E

170 E12 **Masuda** Shimane, Honshū,
SW Japan 34.40N 131.50E

94 J11 **Masugnsbyn** Norrbotten,
N Sweden 67.28N 22.01E

85 K17 **Masuku** *see* Franceville

85 M15 **Masvingo** *prev.* Fort Victoria,
Nyanda, Victoria. Masvingo,
SE Zimbabwe 20.04S 30.49E

85 K18 **Masvingo** *prev.* Victoria. ◆
province SE Zimbabwe

176 W10 **Maswaar, Pulau** *island*
East Indies

144 H5 **Maşyāf** *Fr.* Misiaf. Ḥamāh, C Syria
35.04N 36.21E

Masyū Ko *see* Mashū-ko

83 H7 **Maszewo** Zachodniopomorskie,
NW Poland 53.29N 15.01E

85 J17 **Matabeleland North** ◆ *province*
W Zimbabwe

85 J18 **Matabeleland South** ◆ *province*
S Zimbabwe

84 O13 **Mataca** Niassa, N Mozambique
12.27S 36.13E

197 J13 **Matacawa Levu** *island* Yasawa
Group, NW Fiji

12 G8 **Matachewan** Ontario, S Canada
47.57N 80.39W

81 F22 **Matadi** Bas-Congo, W Dem. Rep.
Congo (Zaire) 5.49S 13.31E

27 O4 **Matador** Texas, SW USA
34.01N 100.50W

44 J9 **Matagalpa** Matagalpa,
C Nicaragua 12.54N 85.57W

44 K9 **Matagalpa** ◆ *department*
W Nicaragua

10 I12 **Matagami** Quebec, S Canada
49.46N 77.37W

27 U13 **Matagorda** Texas, SW USA
28.40N 96.57W

27 U13 **Matagorda Bay** *inlet* Texas,
SW USA

27 U14 **Matagorda Island** *island* Texas,
SW USA

27 V13 **Matagorda Peninsula** *headland*
Texas, SW USA 28.34N 96.01W

203 Q8 **Mataiea** Tahiti, W French
Polynesia 17.46S 149.25W

203 T9 **Mataiva** *atoll* Îles Tuamotu,
C French Polynesia

191 O7 **Matakana** New South Wales,
SE Australia 32.59S 145.53E

192 N7 **Matakana Island** *island* NE NZ

85 C15 **Matala** Huila, SW Angola
14.45S 15.01E

202 O12 **Matala'a Pointe** *headland* Île
Uvea, N Wallis and Futuna
13.19S 176.07W

202 E12 **Matalesina, Pointe** *headland* Île
Alofi, W Wallis and Futuna

83 J18 **Matale** Central Province, C Sri
Lanka 7.28N 80.37E

78 I10 **Matam** NE Senegal 15.40N 13.18W

192 M8 **Matamata** Waikato, North Island,
NZ 37.49S 175.45E

85 C15 **Matamba** Huíla, SW Angola

202 E12 **Matala'a Pointe** *headland* Île
Uvea, N Wallis and Futuna
13.19S 176.07W

202 D12 **Matapu, Pointe** *headland* Île
Futuna, W Wallis and Futuna
13.21N 129.40E

202 B17 **Mata Point** *headland* SE Niue
19.07S 169.51E

198 B8 **Mataiva** *atoll* see of Bay Île Uvea,
13.57S 171.55W

---

43 Q8 **Matamoros** Tamaulipas,
C Mexico 25.49N 97.31W

175 Q10 **Matana, Danau** ☉ Sulawesi,
C Indonesia

77 S13 **Ma'ţan as Sārah** SE Libya
21.45N 21.55E

84 J12 **Matanda** Luapula, N Zambia
11.24S 28.25E

83 J24 **Matandu** ॐ Tanzania

13 V6 **Matane** Quebec, SE Canada
48.48N 67.31W

13 V6 **Matane** ॐ Quebec, SE Canada

79 S12 **Matankari** Dosso, SW Niger
13.39N 4.03E

56 G7 **Matanza** Santander, N Colombia
7.22N 73.01W

44 D4 **Matanzas** Matanzas, NW Cuba
23.00N 81.32W

13 V7 **Matapédia** ॐ Quebec,
SE Canada

13 V6 **Matapédia, Lac** ☉ Quebec,
SE Canada

202 B17 **Mata Point** *headland* SE Niue
19.07S 169.51E

167 S12 **Matsu Tao** Chin. Mazu Dao.
island NW Taiwan

170 C12 **Matsue** *var.* Mattō
Shimane, Honshū, SW Japan
35.27N 133.03E

171 Mm7 **Matsumae** Hokkaidō, NE Japan
41.27N 140.04E

171 J14 **Matsumoto** *var.* Matumoto.
Nagano, Honshū, S Japan
36.18N 137.58E

171 H16 **Matsusaka** *var.* Matsuzaka.
Mie, Honshū, SW Japan
34.34N 136.30E

110 C12 **Matsuura** *var.* Matuura.
Nagasaki, Kyūshū, SW Japan
33.21N 129.40E

170 Ee14 **Matsuyama** *var.* Matuyama.
Ehime, Shikoku, SW Japan
33.49N 132.46E

Matsuye *see* Matsue
Matsuzaka *see* Matsusaka

12 F8 **Mattagami** ॐ Ontario, S Canada

12 F8 **Mattagami Lake** ☉ Ontario,
S Canada

64 K12 **Mataldi** Córdoba, C Argentina
34.32S 64.18W

33 Y9 **Mattamuskeet, Lake** ☉ North
Carolina, SE USA

23 W6 **Mattaponi River** ॐ Virginia,
NE USA

12 I11 **Mattawa** Ontario, SE Canada
46.19N 78.42W

12 I11 **Mattawa** ॐ Ontario, SE Canada

21 S5 **Mattawamkeag** Maine, NE USA
45.30N 68.20W

21 S4 **Mattawamkeag Lake** ☉ Maine,
NE USA

110 D7 **Matterhorn** *It.* Monte Cervino.
▲ Italy/Switzerland *see also*
Cervino, Monte 45.58N 7.36E

37 W1 **Matterhorn** ▲ Nevada, W USA
41.48N 115.22W

34 L12 **Matterhorn Peak** ▲ California,
W USA 38.06N 119.19W

111 Y5 **Mattersburg** Burgenland,
E Austria 47.44N 16.23E

111 S7 **Matter Vispa** ॐ S Switzerland

46 H7 **Matthew Town** Great Inagua,
S Bahamas 20.56N 73.40W

111 Q4 **Mattighofen** Oberösterreich,
NW Austria 48.07N 13.09E

109 N16 **Mattinata** Puglia, SE Italy
41.41N 16.01E

32 L13 **Mattoon** Illinois, N USA
39.28N 88.22W

20 M14 **Mattituck** Long Island, New
York, NE USA 40.59N 72.31W

171 I13 **Mattō** *var.* Matsutō. Ishikawa,
Honshū, SW Japan 36.31N 136.34E

57 P5 **Matto Grosso** *see* Mato Grosso

85 M17 **Mavita** Manica, W Mozambique
19.31S 33.09E

117 K22 **Mavrópetra, Akrotírío**
headland Thíra, Kykládes, Greece,
Aegean Sea 36.28N 25.22E

117 F16 **Mavrovoúni** ▲ C Greece
39.37N 22.45E

192 Q8 **Mawhai Point** *headland* North
Island, NZ 38.08S 178.24E

177 F3 **Mawlaik** Sagaing, C Myanmar
23.40N 94.25E

Mawlamyine *see* Moulmein

147 N14 **Mawr, Wādī** *dry watercourse*
NW Yemen

205 X5 **Mawson** *Australian research station*
Antarctica 67.43S 63.16E

205 X5 **Mawson Coast** *physical region*
Antarctica

30 M4 **Max** North Dakota, N USA
47.48N 101.18E

43 W12 **Maxcanú** Yucatán, SE Mexico
20.35N 90.00W

Maxesibebi *see* Mount Ayliff

111 Q5 **Maxglan** ✈ (Salzburg) Salzburg,
W Austria 47.46N 13.00E

95 K16 **Maxmo** *Fin.* Maksamaa. Länsi-
Suomi, W Finland 63.13N 22.04E

23 T11 **Maxton** North Carolina, SE USA
34.47N 79.34W

28 T8 **May** Texas, SW USA 31.58N 98.54W

194 E10 **May** ॐ NW PNG

127 N17 **Maya** ॐ E Russian Federation

157 O19 **Māyābandar** Andaman and
Nicobar Islands, India, E Indian
Ocean 12.43N 92.52E

Mayadin *see* Al Mayādīn

46 L5 **Mayaguana** *island* SE Bahamas

46 L5 **Mayaguana Passage** *passage*
SE Bahamas

47 S6 **Mayagüez** W Puerto Rico
18.12N 67.08W

47 R6 **Mayagüez, Bahía de** *bay*
W Puerto Rico

Mayals *see* Maials

81 G20 **Mayama** Le Pool, SE Congo
3.49S 14.52E

149 R4 **Maya, Mesa De** ▲ Colorado,
C USA 37.06N 103.30W

149 R4 **Mayamey** Semnān, N Iran
36.26N 55.49E

44 F3 **Maya Mountains** *Sp.* Montañas
Mayas. ▲ Belize/Guatemala

46 I7 **Mayarí** Holguín, E Cuba
20.40N 75.42W

20 J11 **May, Cape** *headland* New Jersey,
NE USA 38.55N 74.57W

82 I9 **Maych'ew** *var.* Mai Chio, *It.*
Mai Ceu, Tigray, N Ethiopia
12.55N 39.30E

144 I2 **Maydān Ikbiz** Ḥalab, N Syria
36.51N 36.46E

155 Q5 **Maydān Shahr** Wardag,
E Afghanistan 34.27N 68.48E

83 O12 **Maydh** Sanaag, N Somalia
10.57N 47.07E

---

104 J16 **Mauléon-Licharre** Pyrénées-
Atlantiques, SW France
43.14N 0.51W

64 G13 **Maule, Río** ॐ C Chile

65 G17 **Maullín** Los Lagos, S Chile
41.37S 73.34W

33 R11 **Maumee** Ohio, N USA
41.34N 83.40W

33 Q12 **Maumee River** ॐ
Indiana/Ohio, N USA

29 U1 **Maumelle** Arkansas, C USA
34.51N 92.24W

29 T11 **Maumelle, Lake** ⊟ Arkansas,
C USA

175 Q4i6 **Maumere** *prev.* Maoemere.
Flores, S Indonesia 8.34S 122.13E

85 G17 **Maun** Ngamiland, C Botswana
20.00S 23.25E

Maunāth Bhanjan *see* Mau
Maunawai *see* Waimea

202 H16 **Maungaroa** ▲ Rarotonga, S Cook
Islands 21.13S 159.48W

192 K3 **Maungatapere** Northland, North
Island, NZ 35.46S 174.10E

192 K4 **Maungaturoto** Northland, North
Island, NZ 36.06S 174.21E

203 R10 **Maupiti** *var.* Maurua. *island* Îles
Sous le Vent, W French Polynesia

158 K14 **Mau Rānīpur** Uttar Pradesh,
N India 25.13N 79.07E

24 K9 **Maurepas, Lake** ☉ Louisiana,
S USA

105 T16 **Maures** ▲ SE France

105 O12 **Mauriac** Cantal, C France
45.13N 2.21E

Maurice *see* Mauritius

67 J20 **Maurice Ewing Bank** *undersea
feature* SW Atlantic Ocean

190 C4 **Maurice** *var.* Lake *salt lake* South
Australia

20 I7 **Maurice River** ॐ New Jersey,
NE USA

27 Y10 **Mauriceville** Texas, SW USA
30.13S 93.52W

100 K12 **Maurik** Gelderland,
C Netherlands 51.57N 5.25E

78 H8 **Mauritania** *off.* Islamic Republic
of Mauritania, *Ar.* Mūrītānīyah.
◆ *republic* W Africa

181 W15 **Mauritius** *off.* Republic of
Mauritius, *Fr.* Maurice. ◆ *republic*
W Indian Ocean

181 N9 **Mauritius** *island* W Indian Ocean

181 N9 **Mauritius Trench** *undersea
feature* W Indian Ocean

104 H6 **Mauron** Morbihan, NW France
48.06N 2.16W

105 N13 **Maurs** Cantal, C France

Maurua *see* Maupiti

Maury Mid-Ocean Channel
see Maury Seachannel

66 L6 **Maury Seachannel** *var.* Maury
Mid-Ocean Channel. *undersea
feature* N Atlantic Ocean

32 K8 **Mauston** Wisconsin, N USA
43.45N 90.01W

111 R8 **Mauterndorf** Salzburg,
NW Austria 47.09N 13.39E

111 T4 **Mauthausen** Oberösterreich,
N Austria 48.13N 14.30E

111 Q9 **Mauthen** Kärnten, S Austria
46.39N 12.58E

85 F15 **Mavinga** Cuando Cubango,
SE Angola 15.49S 20.23E

Maydī see Midi
Mayebashi see Maebashi
Mayence see Mainz
104 K6 Mayenne Mayenne, NW France 48.18N 0.37W
104 J6 Mayenne ◆ department NW France
104 J7 Mayenne ♒ N France
38 K12 Mayer Arizona, SW USA 34.25N 112.15W
24 J4 Mayersville Mississippi, S USA 32.54N 91.04W
9 P14 Mayerthorpe Alberta, SW Canada 53.58N 115.06W
23 S12 Mayesville South Carolina, SE USA 34.00N 80.10W
193 G19 Mayfield Canterbury, South Island, NZ 43.50S 171.24E
35 N14 Mayfield Idaho, NW USA 43.24N 115.57W
22 G7 Mayfield Kentucky, S USA 36.44N 88.38W
38 L5 Mayfield Utah, W USA 39.06N 111.42W
168 K9 Mayhan Övörhangay, C Mongolia 46.02N 104.00E
39 T14 Mayhill New Mexico, SW USA 32.52N 105.28W
151 T9 Maykain Kaz. Mayqayyng. Pavlodar, NE Kazakhstan 51.24N 75.46E
130 L14 Maykop Respublika Adygeya, SW Russian Federation 44.36N 40.06E
Maylibash see Maylybas
Mayli-Say see Mayluu-Suu
153 T9 Maylu-Suu prev. Mayli-Say, Kir. Mayly-Say. Dzhalal-Abadskaya Oblast', W Kyrgyzstan 41.16N 72.27E
150 L14 Maylybas prev. Maylibash. Kzylorda, S Kazakhstan 45.51N 62.37E
Mayly-Say see Mayluu-Suu
Maymana see Meymaneh
178 Gg5 Maymyo Mandalay, C Myanmar 22.03N 96.30E
127 P6 Mayn ♒ NE Russian Federation
131 Q5 Mayna Ul'yanovskaya Oblast', W Russian Federation 54.04N 47.20E
23 N8 Maynardville Tennessee, S USA 36.15N 83.48W
12 J13 Maynooth Ontario, SE Canada 45.14N 77.54W
8 I6 Mayo Yukon Territory, NW Canada 63.37N 135.48W
25 U9 Mayo Florida, SE USA 30.03N 83.10W
99 B16 Mayo Ir. Maigh Eo. cultural region W Ireland
Mayo see Maio
80 G12 Mayo-Kébbi off. Préfecture du Mayo-Kébbu, var. Mayo-Kébi. ◆ prefecture SW Chad
Mayo-Kébi see Mayo-Kébbi
81 F19 Mayoko Le Niari, SW Congo 2.18S 12.45E
179 Q11 Mayon Volcano ☷ Luzon, N Philippines 13.15N 123.40E
63 A24 Mayor Buratovich Buenos Aires, E Argentina 39.12S 62.41W
106 L4 Mayorga Castilla-León, N Spain 42.10N 5.16W
192 N6 Mayor Island island NE NZ
Mayor Pablo Lagerenza see Capitán Pablo Lagerenza
181 I14 Mayotte ◇ French territorial collectivity E Africa
Mayoumba see Mayumba
46 J13 May Pen C Jamaica 17.58N 77.15W
179 P7 Mayraira Point headland Luzon, N Philippines 18.36N 120.47E
111 N8 Mayrhofen Tirol, W Austria 47.09N 11.52E
194 F10 May River East Sepik, NW PNG 4.10S 141.51E
126 Mm15 Mayskiy Amurskaya Oblast', SE Russian Federation 52.13N 129.30E
131 O15 Mayskiy Kabardino-Balkarskaya Respublika, SW Russian Federation 43.37N 44.04E
151 U9 Mayskoye Pavlodar, NE Kazakhstan 50.55N 78.11E
20 J17 Mays Landing New Jersey, NE USA 39.27N 74.43W
23 N4 Maysville Kentucky, S USA 38.39N 83.44W
29 R2 Maysville Missouri, C USA 39.53N 94.21W
176 Y14 Mayu channel Irian Jaya, E Indonesia
81 D20 Mayuma var. Mayoumba. Nyanga, S Gabon 3.22S 10.37E
175 Ss7 Mayu, Pulau island Maluku, E Indonesia
33 S8 Mayville Michigan, N USA 43.18N 83.16W
20 C11 Mayville New York, NE USA 42.15N 79.31W
31 Q4 Mayville North Dakota, N USA 47.27N 97.17W
126 M11 Mayya Respublika Sakha (Yakutiya), NE Russian Federation 61.45N 130.16E
Mayyali see Mahe
Mayyit, Al Baḩr al see Dead Sea
85 L15 Mazabuka Southern, S Zambia 15.52S 27.46E
Mazaca see Kayseri
34 J7 Mazagan see El-Jadida
Mazagão Amazonas, N Brazil 48.34N 120.26W
105 O15 Mazamet Tarn, S France 43.30N 2.21E
149 O4 Māzandarān off. Ostān-i Māzandarān. ◆ province N Iran
162 F7 Mazar Xinjiang Uygur Zizhiqu, NW China 36.31N 76.59E
109 H24 Mazara del Vallo Sicilia, Italy, C Mediterranean Sea 37.39N 12.36E
155 O2 Mazār-e Sharīf var. Mazār-i Sharif. Balkh, N Afghanistan 36.44N 67.06E
Mazār-i Sharif see Mazār-e Sharīf
107 R13 Mazarrón Murcia, SE Spain 37.36N 1.19W
107 R14 Mazarrón, Golfo de gulf SE Spain
57 S9 Mazaruni River ♒ N Guyana
44 B6 Mazatenango Suchitepéquez, SW Guatemala 14.31N 91.28W

42 I10 Mazatlán Sinaloa, C Mexico 23.15N 106.24W
38 L12 Mazatzal Mountains ▲ Arizona, SW USA
120 D10 Mažeikiai Mažeikiai, NW Lithuania 56.19N 22.21E
120 D7 Mazirbe Talsi, NW Latvia 57.39N 22.16E
42 G5 Mazocahui Sonora, NW Mexico 29.34N 110.07W
59 I18 Mazocruz Puno, S Peru 16.41S 69.42W
Mazoe, Rio see Mazowe
81 I16 Mazomeno Maniema, E Dem. Rep. Congo (Zaire) 4.54S 27.13E
165 Q6 Mazong Shan ▲ N China 41.40N 97.10E
85 L16 Mazowe var. Rio Mazoe. ♒ Mozambique/Zimbabwe
112 L11 Mazowieckie ◆ province C Poland
Mazra'a see Al Mazra'ah
144 G6 Mazraat Kfar Debiâne C Lebanon 34.00N 35.51E
120 H7 Mazsalaca Est. Väike-Salatsi, Ger. Salisburg. Valmiera, N Latvia 57.52N 25.03E
112 L9 Mazury physical region NE Poland
121 M20 Mazyr Rus. Mozyr'. Homyel'skaya Voblasts', SE Belarus 52.03N 29.14E
109 K25 Mazzarino Sicilia, Italy, C Mediterranean Sea 37.18N 14.13E
Mba see Ba
85 L21 Mbabane ● (Swaziland) NW Swaziland 26.24S 31.13E
Mbacké see Mbaké
79 N16 Mbahiakro E Ivory Coast 7.25N 4.18W
81 I16 Mbaïki var. M'Baïki. Lobaye, SW Central African Republic 3.52N 17.58E
81 F21 Mbakaou, Lac de ☒ C Cameroon
78 G11 Mbaké var. Mbacké. W Senegal 14.50N 15.52W
84 L11 Mbala prev. Abercorn. Northern, NE Zambia 8.49S 31.22E
85 J18 Mbalabala prev. Balla Balla. Matabeleland South, SW Zimbabwe 20.27S 29.03E
81 E16 Mbale E Uganda 1.04N 34.12E
81 E16 Mbalmayo var. M'Balmayo. Centre, S Cameroon 3.30N 11.31E
83 I23 Mbamba Bay Ruvuma, S Tanzania 11.15S 34.44E
81 I18 Mbandaka prev. Coquilhatville. Equateur, NW Dem. Rep. Congo (Zaire) 0.07N 18.11E
84 B9 M'Banza Congo var. Mbanza Congo; prev. São Salvador, São Salvador do Congo. Zaire, NW Angola 6.18S 14.16E
81 G21 Mbanza-Ngungu Bas-Congo, W Dem. Rep. Congo (Zaire) 5.19S 14.45E
69 V11 Mbarangandu ♒ E Tanzania
83 E19 Mbarara SW Uganda 0.36S 30.40E
81 L15 Mbari ♒ SE Central African Republic
83 I24 Mbarika Mountains ▲ S Tanzania
85 J24 Mbashe ♒ S South Africa
80 F13 Mbatiki var. Batiki 7.51N 13.36E
81 I21 Mbemkuru var. Mbwemkuru. ♒ S Tanzania
Mbengga see Beqa
180 H13 Mbéni Grande Comore, NW Comoros
85 K18 Mberengwa Midlands, S Zimbabwe 20.25S 29.57E
83 G24 Mbeya Mbeya, SW Tanzania 8.54S 33.28E
83 G23 Mbeya ◆ region S Tanzania
81 E19 Mbigou Ngounié, C Gabon 01.54S 12.00E
81 L15 Mbilua see Vella Lavella
81 F19 Mbinda Le Niari, SW Congo 2.07S 12.52E
81 D17 Mbini W Equatorial Guinea 1.34N 9.39E
85 L18 Mbinyi Masvingo, SE Zimbabwe 21.21S 30.58E
83 G23 Mbogo Mbeya, W Tanzania 7.24S 33.26E
81 L15 Mboki Haut-Mbomou, SE Central African Republic 5.18N 25.52E
81 G18 Mbomo Cuvette, NW Congo 0.25N 14.42E
81 L15 Mbomou ◆ prefecture SE Central African Republic
Mbomou/M'Bomu/Mbomu see Bomu
78 F13 Mbour W Senegal 14.24N 16.58W
78 I10 Mbout Gorgol, S Mauritania 16.01N 12.37W
81 J14 Mbrès var. Mbrés. Nana-Grébizi, C Central African Republic 6.40N 19.46E
81 J22 Mbuji-Mayi prev. Bakwanga. Kasai Oriental, S Dem. Rep. Congo (Zaire) 6.08S 23.30E
194 J9 M'buke Islands island group N PNG
83 H21 Mbulu Arusha, N Tanzania 3.45S 35.33E
194 K8 M'bunai var. Bunai. Manus Island, N PNG 2.09S 147.11E
64 N9 Mburucuyá Corrientes, NE Argentina 28.03S 58.15W
83 J22 Mbutha see Buca
83 I23 Mbwemkuru var. Mbuthku ♒ S Tanzania
83 J23 Mbwikwe Singida, C Tanzania 5.19S 34.09E
11 O16 McAdam New Brunswick, SE Canada 45.34N 67.19W
27 O14 McAdoo Texas, SW USA 33.41N 100.58W
37 V3 McAfee Peak ▲ Nevada, W USA 41.31N 115.57W
27 S17 McAllen Texas, SW USA 26.12N 98.13W
23 S11 McBee South Carolina, SE USA 34.30N 80.12W
9 N14 McBride British Columbia, SW Canada 53.21N 120.19W
27 S5 McCamey Texas, SW USA 31.08N 102.13W

35 R15 McCammon Idaho, NW USA 42.38N 112.10W
37 X11 McCarran ✈ (Las Vegas) Nevada, W USA 36.04N 115.07W
41 T11 McCarthy Alaska, USA 61.25N 142.55W
32 M5 McCaslin Mountain hill Wisconsin, N USA 45.24N 88.24W
27 O2 McClellan Creek ♒ Texas, SW USA
23 T14 McClellanville South Carolina, SE USA 33.07N 79.27W
15 J2 McClintock Channel channel Nunavut, N Canada
205 R12 McClintock, Mount ▲ Antarctica 80.09S 156.42E
37 N2 McCloud California, W USA 41.15N 122.09W
37 N3 McCloud River ♒ California, W USA
37 S3 McClure, Lake ☒ California, W USA
207 O8 McClure Strait strait Northwest Territories, N Canada
23 T11 McColl South Carolina, SE USA 34.40N 79.33W
24 K7 McComb Mississippi, S USA 31.14N 90.27W
21 E16 McConnellsburg Pennsylvania, NE USA 39.56N 78.00W
21 U10 McConnelsville Ohio, N USA 39.39N 81.51W
30 M17 McCook Nebraska, C USA 40.12N 100.37W
23 P13 McCormick South Carolina, SE USA 33.54N 82.17W
29 W11 McCrory Arkansas, C USA 35.15N 91.12W
27 T10 McDade Texas, SW USA 30.15N 97.15W
25 O8 McDavid Florida, SE USA 30.51N 87.18W
37 T1 McDermitt Nevada, W USA 41.57N 117.43W
25 S4 McDonough Georgia, SE USA 33.27N 84.09W
38 L12 McDowell Mountains ▲ Arizona, SW USA
22 H8 McEwen Tennessee, S USA 36.06N 87.37W
37 R12 McFarland California, W USA 35.41N 119.14W
Mcfarlane, Lake see Macfarlane, Lake
29 P12 McGee Creek Lake ☒ Oklahoma, C USA
29 W13 McGehee Arkansas, C USA 33.37N 91.22W
37 X5 Mcgill Nevada, W USA 39.24N 114.46W
12 K12 McGillivray, Lac ☒ Quebec, SE Canada
41 O10 Mcgrath Alaska, USA 62.57N 155.36W
27 T8 McGregor Texas, SW USA 31.26N 97.24W
35 O12 McGuire, Mount ▲ Idaho, NW USA 45.10N 114.36W
85 K18 Mchinji prev. Fort Manning. Central, W Malawi 13.47S 32.51E
30 M7 McIntosh South Dakota, N USA 45.55N 101.19W
16 O3 McKeand ♒ Baffin Island, Nunavut, NE Canada
203 R4 McKean island Phoenix Islands, C Kiribati
21 O15 McKee Creek ♒ Illinois, N USA
20 C15 Mckeesport Pennsylvania, NE USA 40.18N 79.48W
23 V7 McKenney Virginia, NE USA 36.57N 77.42W
22 G8 McKenzie Tennessee, S USA 36.07N 88.31W
193 B20 McKerrow, Lake ☒ South Island, NZ
41 Q10 McKinley, Mount var. Denali. ▲ Alaska, USA 63.04N 151.00W
41 R10 McKinley Park Alaska, USA 63.42N 149.01W
36 K3 McKinleyville California, W USA 40.56N 124.06W
28 J6 McKinney Texas, SW USA 33.12N 96.37W
28 J5 McKinney, Lake ☒ Kansas, C USA
30 M7 McLaughlin South Dakota, N USA 45.48N 100.48W
27 O2 McLean Texas, SW USA 35.13N 100.36W
32 M16 Mcleansboro Illinois, N USA 38.05N 88.32W
9 O13 McLennan Alberta, W Canada 55.42N 116.49W
12 L9 McLennan, Lac ☒ Quebec, SE Canada
8 M13 McLeod Lake British Columbia, SW Canada 55.03N 123.02W
29 N10 McLoud Oklahoma, C USA 35.26N 97.05W
34 G15 McLoughlin, Mount ▲ Oregon, NW USA 42.27N 122.18W
39 U15 McMillan, Lake ☒ New Mexico, SW USA
34 G11 McMinnville Oregon, NW USA 45.13N 123.12W
22 M9 McMinnville Tennessee, S USA 35.42N 85.46W
205 R15 McMurdo US research station Antarctica 77.40S 167.16E
39 P4 McNary Texas, SW USA 31.15N 105.46W
25 Q4 Mcnary Arizona, SW USA 34.04N 109.51W
41 S13 McPherson Kansas, C USA 38.22N 97.39W
McPherson see Fort McPherson
25 U6 McRae Georgia, SE USA 32.04N 82.54W
37 N13 McVille North Dakota, N USA 47.45N 98.10W
85 J25 Mdantsane Eastern Cape, SE South Africa 32.54S 27.71.09W
178 Jj6 Me Ninh Binh, N Vietnam 20.21N 105.49E
28 J7 Meade Kansas, C USA 37.14N 100.20W
41 O5 Meade River ♒ Alaska, USA 70.14N 98.33W
37 Y11 Mead, Lake ☒ Arizona/Nevada, W USA 38.00N 102.13E

26 M5 Meadow Texas, SW USA 33.20N 102.12W
9 S14 Meadow Lake Saskatchewan, C Canada 55.08N 108.25W
37 Y10 Meadow Valley Wash ♒ Nevada, W USA
24 J7 Meadville Mississippi, S USA 31.28N 90.51W
20 B12 Meadville Pennsylvania, NE USA 41.38N 80.09W
12 F14 Meaford Ontario, S Canada 44.35N 80.35W
106 G8 Mealhada Aveiro, N Portugal 40.22N 8.27W
11 R8 Mealy Mountains ▲ Newfoundland, E Canada
9 O10 Meander River Alberta, SW Canada 59.01N 117.42W
34 E11 Meares, Cape headland Oregon, NW USA 45.29N 123.59W
49 V6 Mearim, Rio ♒ NE Brazil
99 F17 Measca, Loch see Mask, Lough
9 T14 Meath Ir. An Mhí. cultural region E Ireland
9 T14 Meath Park Saskatchewan, S Canada 53.25N 105.18W
105 O5 Meaux Seine-et-Marne, N France 48.58N 2.54E
23 T9 Mebane North Carolina, SE USA 36.06N 79.16W
176 W9 Mebo, Gunung ▲ Irian Jaya, E Indonesia 1.10S 133.53E
96 I8 Mebonden Sør-Trøndelag, S Norway 63.13N 11.00E
84 A10 Mebridege ♒ NW Angola
37 W16 Mecca California, W USA 33.34N 116.04W
31 N7 Mecca see Makkah
20 L10 Mechanicsville Iowa, C USA 41.54N 91.15W
101 H17 Mechelen Eng. Mechlin, Fr. Malines. Antwerpen, C Belgium 51.01N 4.28E
196 C8 Mecherchar var. Eil Malk. island Palau Islands, Palau
103 D17 Mechernich Nordrhein-Westfalen, W Germany 50.36N 6.39E
130 I24 Mechetinskaya Rostovskaya Oblast', SW Russian Federation 46.46N 40.30E
116 J11 Mechka ♒ S Bulgaria
63 D23 Mechongué Buenos Aires, E Argentina 38.09S 58.13W
117 L14 Mecidiye Edirne, NW Turkey
103 I24 Meckenbeuren Baden-Württemberg, S Germany 47.42N 9.34E
102 L8 Mecklenburger Bucht bay N Germany
102 M10 Mecklenburgische Seenplatte wetland NE Germany
102 L9 Mecklenburg-Vorpommern ◆ state NE Germany
85 Q15 Meconta Nampula, NE Mozambique 15.01S 39.52E
85 J25 Mecsek ▲ SW Hungary
85 Q14 Mecubúri ♒ N Mozambique
85 Q14 Mecúfi Cabo Delgado, NE Mozambique 13.18S 40.33E
84 O13 Mecula Niassa, N Mozambique 12.03S 37.37E
173 Ff5 Medan Sumatera, E Indonesia 3.34N 98.39E
63 A24 Médanos var. Medanos. Buenos Aires, E Argentina 38.51S 62.44W
63 C19 Médanos Entre Ríos, C Argentina 33.25S 59.03W
161 K24 Medawachchiya North Central Province, N Sri Lanka 8.32N 80.30E
108 C8 Mede Lombardia, N Italy 45.06N 8.43E
76 J5 Médéa var. El Mediyya, Lemdiyya. N Algeria 36.24N 2.42E
56 F4 Medeba see Ma'dabã
Medellín Antioquia, NW Colombia 6.15N 75.36W
102 H9 Medem ♒ NW Germany
100 J8 Medemblik Noord-Holland, NW Netherlands 52.46N 5.06E
77 N7 Médenine var. Madaniyin. SE Tunisia 33.23N 10.30E
78 G9 Mederdra Trarza, SW Mauritania 16.55N 15.40W
F4 Medesto Méndez Izabal, NE Guatemala 15.54N 89.13W
21 O11 Medford Massachusetts, NE USA 42.25N 71.08W
29 N8 Medford Oklahoma, C USA 36.49N 97.45W
34 G15 Medford Oregon, NW USA 42.19N 122.52W
32 K6 Medford Wisconsin, N USA 45.09N 90.21W
41 P10 Medfra Alaska, USA 63.06N 154.40W
118 M14 Medgidia Constanța, SE Romania 44.16N 28.13E
Medgyes see Mediaș
116 G11 Media Luna, Arrecifes de la reef E Honduras
49 O5 Mediapolis Iowa, C USA 41.00N 91.09W
118 I11 Mediaș Ger. Mediasch, Hung. Medgyes. Sibiu, C Romania 46.09N 24.20E
45 S15 Medias Aguas Veracruz-Llave, SE Mexico 17.40N 95.01W
Mediasch see Mediaș
108 G10 Medicina Emilia-Romagna, C Italy 44.29N 11.38E
37 X16 Medicine Bow Wyoming, C USA 41.52N 106.11W
39 V9 Medicine Bow Mountains ▲ Colorado/Wyoming, C USA 0.55S 131.46E
37 X16 Medicine Bow River ♒ Wyoming, C USA
9 V17 Medicine Hat Alberta, SW Canada 50.03N 110.40W
38 L7 Medicine Lodge Kansas, C USA 37.14N 98.33W
28 L7 Medicine Lodge River ♒ Kansas/Oklahoma, C USA

114 E7 Međimurje off. Međimurska Županija. ◆ province N Croatia
Međimurska Županija see Međimurje
56 G10 Medina Cundinamarca, C Colombia 4.31N 73.21W
20 E9 Medina New York, NE USA 43.13N 78.23W
31 O5 Medina North Dakota, N USA 46.53N 99.18W
33 T11 Medina Ohio, N USA 41.08N 81.51W
27 Q11 Medina Texas, SW USA 29.46N 99.14W
Medina see Al Madīnah
123 LI12 Medina Bank undersea feature C Mediterranean Sea
107 P6 Medinaceli Castilla-León, N Spain 41.10N 2.25W
106 L6 Medina del Campo Castilla-León, N Spain 41.18N 4.55W
106 L5 Medina de Ríoseco Castilla-León, N Spain 41.52N 5.03W
Médina Gonassé see Médina Gounas
78 H12 Médina Gounas var. Médina Gonassé. S Senegal 13.06N 13.49W
27 S12 Medina River ♒ Texas, SW USA
106 K16 Medina Sidonia Andalucía, S Spain 36.26N 5.55W
121 H14 Medininkai Vilnius, SE Lithuania 54.31N 25.39E
159 R16 Medinipur West Bengal, NE India 22.27N 87.19E
124 O13 Mediterranean Ridge undersea feature C Mediterranean Sea
123 L11 Mediterranean Sea Fr. Mer Méditerranée, var. Africa/Asia/Europe
123 M11 Méditerranée, Mer see Mediterranean Sea
81 N17 Medje Orientale, NE Dem. Rep. Congo (Zaire) 2.27N 27.14E
123 K11 Medjerda, Oued var. Mejerda, Wâdi Majardah. ♒ Algeria/Tunisia see also Mejerda
116 C8 Medkovets Montana, NW Bulgaria 43.39N 23.22E
95 J15 Medle Västerbotten, N Sweden 64.45N 20.45E
131 W7 Mednogorsk Orenburgskaya Oblast', W Russian Federation 51.24N 57.37E
127 Qq9 Mednyy, Ostrov island E Russian Federation
165 Q16 Mêdog Xizang Zizhiqu, W China 29.25N 95.25E
30 J5 Medora North Dakota, N USA 46.52N 103.32W
81 E17 Médouneu Woleu-Ntem, N Gabon 0.58N 10.49E
108 I7 Medua see Italy
Medunta see Mantes-la-Jolie
128 J16 Medvedevo ♒ W Russian Federation
131 O9 Medvedsta Ger. Zwischenwässern. ♒ SW Russian Federation
114 E8 Medvednica ▲ NE Croatia
129 R15 Medvezh'i, Ostrova island group NW Russian Federation
127 Nn5 Medvezh'yegorsk Respublika Kareliya, NW Russian Federation 62.53N 34.29E
44 F4 Mejicanos San Salvador, C El Salvador 13.50N 89.13W
Méjico see Mexico
64 G5 Mejillones Antofagasta, N Chile 23.03S 70.25W
130 J4 Medyn' Kaluzhskaya Oblast', W Russian Federation 54.59N 35.52E
188 J10 Meekatharra Western Australia 26.36S 118.34E
39 Q4 Meeker Colorado, C USA 40.02N 107.54W
11 T12 Meelpaeg Lake ☒ Newfoundland, E Canada
101 H14 Meenen see Menen
103 M16 Meerane Sachsen, E Germany 50.49N 12.28E
103 D15 Meerbusch Nordrhein-Westfalen, W Germany 51.19N 6.43E
100 I12 Meerkerk Zuid-Holland, C Netherlands 51.55N 5.00E
101 L18 Meerssen var. Mersen. Limburg, SE Netherlands 50.52N 5.45E
158 J10 Meerut Uttar Pradesh, N India 29.01N 77.40E
35 U13 Meeteetse Wyoming, C USA 44.10N 108.53W
101 KV12 Meeuwen Limburg, NE Belgium 51.04N 5.36E
83 K16 Mēga Oromo, C Ethiopia 4.03N 38.15E
83 J16 Mēga Escarpment escarpment S Ethiopia
117 I14 Megála Kalívia see Megála Kalývia
117 I14 Megála Kalývia var. Megála Kalívia. Thessalía, C Greece 39.30N 21.48E
117 H14 Megáli Panagiá see Megáli Panagía
117 I14 Megáli Panagía var. Megáli Panagía. Kentrikí Makedonía, N Greece 40.24N 23.41E
117 K20 Megáli Préspa, Límni see Prespa, Lake
116 K12 Megálo Livádi ▲ Bulgaria/Greece 41.18N 25.51E
117 E20 Megalópoli prev. Megalópolis. Pelopónnisos, S Greece 37.23N 22.08E
117 C18 Meganísi island Iónioi Nísoi, Greece, C Mediterranean Sea
117 C18 Meganom, Mys see Mehanom, Mys
117 E19 Mégara Attikí, C Greece 38.00N 23.20E

27 R5 Megargel Texas, SW USA 33.27N 98.55W
100 K13 Megen Noord-Brabant, S Netherlands 51.49N 5.34E
159 U16 Meghna ♒ S Bangladesh
159 U16 Meghna ◆ state NE India
143 V14 Meghri Rus. Megri. SE Armenia 38.57N 46.15E
126 Gg11 Megion Khanty-Mansiyskiy Avtonomnyy Okrug, C Russian Federation 61.01N 76.15E
117 O22 Megísti var. Kastellórizon. island SE Greece
Megri see Meghri
Mehabad see Mahābād
118 F13 Mehadia Hung. Mehádia. Caraș-Severin, SW Romania 44.55N 22.22E
94 U13 Mehamn Finnmark, N Norway 71.01N 27.46E
119 U13 Mehanom, Mys Rus. Mys Meganom. headland S Ukraine
155 P14 Mehar Sind, SE Pakistan
188 J8 Meharry, Mount ▲ Western Australia 23.17S 118.48E
118 G14 Mehdia ◆ county SW Romania
159 S15 Meherpur Khulna, C Bangladesh 23.46N 88.40E
23 W8 Meherrin River ♒ North Carolina/Virginia, SE USA
203 T11 Mehetia island Îles du Vent, W French Polynesia
120 K6 Mehikoorma Tartumaa, E Estonia 58.14N 27.29E
149 T4 Mehrabad ✈ (Tehrān) Tehrān, N Iran 35.46N 51.17E
148 J7 Mehrān Īlam, W Iran 33.07N 46.10E
149 Q14 Mehrān, Rûd-e prev. Mansurabad. ♒ W Iran
149 Q9 Mehriz Yazd, C Iran 31.31N 54.28E
155 R5 Mehtarlām var. Mehtar Lām, Meterlam, Methariam, Metharlam. Laghmān, E Afghanistan 34.39N 70.10E
73 N1 Mehdia enclave Spain, N Africa
105 N8 Mehun-sur-Yèvre Cher, C France 47.09N 2.15E
81 G14 Meiganga Adamaoua, NE Cameroon 6.31N 14.07E
166 H10 Meigu Sichuan, C China 28.22N 103.07E
169 W11 Meihekou var. Hailong. Jilin, NE China 42.31N 125.40E
101 L15 Meijel Limburg, SE Netherlands 51.22N 5.52E
177 G5 Meiktila Mandalay, C Myanmar 20.52N 95.54E
110 G7 Meilen Zürich, N Switzerland 47.16N 8.39E
Meilu see Wuchuan
167 O3 Meinhua Yu island N Taiwan
103 J17 Meiningen Thüringen, C Germany 50.34N 10.25E
110 F9 Meiringen Bern, S Switzerland 46.42N 8.13E
103 O15 Meissen var. Meißen. Sachsen, E Germany 51.10N 13.28E
103 I15 Meißner ▲ C Germany 51.13N 9.52E
97 K15 Mellan-Fryken ☒ C Sweden
101 E17 Melle Oost-Vlaanderen, NW Belgium 51.00N 3.48E
102 G13 Melle Niedersachsen, NW Germany 52.12N 8.19E
97 K18 Mellerud Västra Götaland, S Sweden 58.42N 12.28E
104 H4 Melle-sur-Bretonne Deux-Sèvres, W France 46.13N 0.07W
31 N4 Mellette South Dakota, N USA 45.08N 98.30W
123 J16 Mellieha E Malta 35.58N 14.21E
82 B10 Melit Northern Darfur, W Sudan 14.07N 25.34E
77 N6 Mellita ✈ SE Tunisia 33.47N 10.51E
65 G21 Mellizo Sur, Cerro ▲ S Chile 48.25S 73.10W
10 N1 Melmoth KwaZulu/Natal, E South Africa 28.36S 31.23E
113 D16 Mělník Ger. Melnik. Středočeský Kraj, NW Czech Republic 50.21N 14.30E
126 H13 Mel'nikovo Tomskaya Oblast', C Russian Federation 56.35N 84.11E
63 G18 Melo Cerro Largo, NE Uruguay 32.22S 54.10W
81 G18 Melo see Melun
191 P7 Melrose New South Wales, SE Australia 32.41S 146.58E
190 I7 Melrose South Australia 32.52S 138.16E
31 T7 Melrose Minnesota, N USA 45.40N 94.46W
35 Q11 Melrose Montana, NW USA 45.33N 112.41W
39 V9 Melrose New Mexico, SW USA 34.25N 103.37W
110 I6 Mels Sankt Gallen, NE Switzerland 47.03N 9.25E
Melsetter see Chimanimani
35 V5 Melstone Montana, NW USA 46.37N 107.49W
103 I16 Melsungen Hessen, C Germany 51.07N 9.33E
94 L12 Meltaus Lappi, NW Finland 66.54N 25.18E
97 N19 Melton Mowbray C England, UK 52.46N 0.53W
84 Q13 Meluco Cabo Delgado, NE Mozambique 12.52S 39.55E
105 O5 Melun anc. Melodunum. Seine-et-Marne, N France 48.32N 2.40E
82 E7 Melut Upper Nile, SE Sudan 10.27N 32.13E
29 P5 Melvern Lake ☒ Kansas, C USA
9 V16 Melville Saskatchewan, S Canada 50.57N 102.49W
Melville Bay/Melville Bugt see Qimusseriarsuaq
47 O11 Melville Hall ✈ (Dominica) NE Dominica 15.33N 61.19W
189 O1 Melville Island island Northern Territory, N Australia
207 O8 Melville Island island Parry Islands, Northwest Territories, NW Canada
11 R7 Melville, Lake ☒ Newfoundland, E Canada
15 L12 Melville Peninsula peninsula Nunavut, NE Canada
Melville Sound see Viscount Melville Sound

27 Q9 **Melvin** Texas, SW USA 31.12N 99.34W
99 D15 **Melvin, Lough** *Ir.* Loch Meilbhe. ◎ S Northern Ireland, UK/Ireland
174 M9 **Memala** Borneo, C Indonesia 1.43S 112.36E
115 L22 **Memaliaj** Gjirokastër, S Albania 40.21N 19.56E
85 Q14 **Memba** Nampula, NE Mozambique 14.07S 40.33E
85 Q14 **Memba, Baia de** *inlet* NE Mozambique
Membidj *see* Manbij
Memel *see* Neman, NE Europe
Memel *see* Klaipėda, Lithuania
103 J23 **Memmingen** Bayern, S Germany 47.58N 10.10E
29 U1 **Memphis** Missouri, C USA 40.27N 92.10W
22 E10 **Memphis** Tennessee, S USA 35.09N 90.03W
27 P3 **Memphis** Texas, SW USA 34.43N 100.31W
22 E10 **Memphis** ✕ Tennessee, S USA 35.02N 89.57W
13 Q13 **Memphrémagog, Lac** *var.* Lake Memphremagog. ◎ Canada/USA *see also* Memphremagog, Lake
21 N6 **Memphremagog, Lake** *var.* Lac Memphrémagog. ◎ Canada/USA *see also* Memphrémagog, Lac
119 Q2 **Mena** Chernihivs'ka Oblast', NE Ukraine 51.30N 32.15E
29 S12 **Mena** Arkansas, C USA 34.35N 94.14W
Menaam *see* Menaldum
Menado *see* Manado
108 D6 **Menaggio** Lombardia, N Italy 46.03N 9.14E
31 T6 **Menahga** Minnesota, N USA 46.45N 95.06W
79 R10 **Ménaka** Goa, E Mali 15.54N 2.25E
100 K5 **Menaldum** *Fris.* Menaam. Friesland, N Netherlands 53.13N 5.38E
Mènam Khong *see* Mekong
76 E7 **Menara** ✕ (Marrakech) C Morocco 31.36N 8.00W
27 Q9 **Menard** Texas, SW USA 30.55N 99.47W
199 M14 **Menard Fracture Zone** *tectonic feature* E Pacific Ocean
32 M7 **Menasha** Wisconsin, N USA 44.13N 88.25W
Mencezi Garagum *see* Tsentral'nyye Nizmennyye Garagumy
200 O10 **Mendaña Fracture Zone** *tectonic feature* E Pacific Ocean
174 M10 **Mendawai, Sungai** ≈ Borneo, C Indonesia
105 P13 **Mende** *anc.* Mimatum. Lozère, S France 44.31N 3.30E
83 J14 **Mendebo** ▲▲ C Ethiopia
82 J9 **Mendefera** *prev.* Adi Ugri. S Eritrea 14.53N 38.51E
207 S7 **Mendeleyev Ridge** *undersea feature* Arctic Ocean
131 T3 **Mendeleyevsk** Respublika Tatarstan, W Russian Federation 55.54N 52.19E
103 F15 **Menden** Nordrhein-Westfalen, W Germany 51.25N 7.48E
24 L6 **Mendenhall** Mississippi, S USA 31.57N 89.52W
40 L13 **Mendenhall, Cape** *headland* Nunivak Island, Alaska, USA 59.45N 166.10W
43 P9 **Méndez** *var.* Villa de Méndez. Tamaulipas, C Mexico 25.06N 98.32W
82 H13 **Mendī** Oromo, C Ethiopia 9.43N 35.07E
194 G12 **Mendi** Southern Highlands, W PNG 6.07S 143.39E
99 K22 **Mendip Hills** *var.* Mendips. *hill range* S England, UK
Mendips *see* Mendip Hills
36 L6 **Mendocino** California, W USA 39.18N 123.48W
36 J3 **Mendocino, Cape** *headland* California, W USA 40.26N 124.24W
1 B8 **Mendocino Fracture Zone** *tectonic feature* NE Pacific Ocean
37 P10 **Mendota** California, W USA 36.44N 120.24W
32 L11 **Mendota** Illinois, N USA 41.32N 89.04W
32 K8 **Mendota, Lake** ◎ Wisconsin, N USA
64 I11 **Mendoza** Mendoza, W Argentina 33.00S 68.47W
64 I12 **Mendoza** *off.* Provincia de Mendoza. ◐ *province* W Argentina
110 H12 **Mendrisio** Ticino, S Switzerland 45.52N 8.58E
174 Hh7 **Mendung** Pulau Mendol, W Indonesia 0.33N 103.09E
56 I5 **Mene de Mauroa** Falcón, NW Venezuela 10.39N 71.04W
56 I5 **Mene Grande** Zulia, NW Venezuela 9.51N 70.57W
142 B14 **Menemen** İzmir, W Turkey 38.34N 27.03E
101 C18 **Menen** *var.* Meenen, *Fr.* Menin. West-Vlaanderen, W Belgium 50.48N 3.07E
169 Q8 **Menengiyn Tal** *plain* E Mongolia
201 R9 **Meneng Point** *headland* SW Nauru 0.33S 166.57E
94 L10 **Menesjärvi** *Lapp.* Menešjávri. Lappi, N Finland 68.39N 26.22E
Menešjávri *see* Menesjärvi
109 I24 **Menfi** Sicilia, Italy, C Mediterranean Sea 37.34N 12.58E
167 P7 **Mengcheng** Anhui, E China 33.17N 116.31E
166 F15 **Menghai** Yunnan, SW China 22.02N 100.18E
175 Q11 **Mengkoka, Pegunungan** *var.* Pegunungan Mekongga. ▲▲ Sulawesi, C Indonesia
166 F15 **Mengla** Yunnan, SW China 21.30N 101.33E
67 F24 **Menguera Point** *headland* East Falkland, Falkland Islands
166 M13 **Mengzhu Ling** ▲▲ S China
166 H14 **Mengzi** Yunnan, SW China 23.20N 103.32E
Menin *see* Menen
190 L7 **Menindee** New South Wales, SE Australia 32.24S 142.25E
190 L7 **Menindee Lake** ◎ New South Wales, SE Australia
190 J10 **Meningie** South Australia 35.43S 139.20E
105 O5 **Mennecy** Essonne, N France 48.34N 2.25E
31 Q12 **Menno** South Dakota, N USA 43.14N 97.34W
116 H13 **Menoíkio** ▲▲ NE Greece
33 N5 **Menominee** Michigan, N USA 45.06N 87.36W
32 M5 **Menominee River** ≈ Michigan/Wisconsin, N USA
32 M8 **Menomonee Falls** Wisconsin, N USA 43.11N 88.09W
32 I6 **Menomonie** Wisconsin, N USA 44.52N 91.55W
85 D14 **Menongue** *var.* Vila Serpa Pinto, *Port.* Serpa Pinto. Cuando Cubango, C Angola 14.38S 17.38E
123 I18 **Menorca** *Eng.* Minorca; *anc.* Balearis Minor. *island* Islas Baleares, Spain, W Mediterranean Sea
107 S13 **Menor, Mar** *lagoon* SE Spain
41 S10 **Mentasta Lake** ◎ Alaska, USA
41 S10 **Mentasta Mountains** ▲▲ Alaska, USA
173 Ff10 **Mentawai, Kepulauan** *island group* W Indonesia
173 G10 **Mentawai, Selat** *strait* W Indonesia
174 Ii10 **Mentok** Pulau Bangka, W Indonesia 2.01S 105.10E
105 V15 **Menton** *It.* Mentone. Alpes-Maritimes, SE France 43.46N 7.30E
26 K8 **Mentone** Texas, SW USA 31.42N 103.36W
Mentone *see* Menton
33 U11 **Mentor** Ohio, N USA 41.40N 81.20W
175 Nn7 **Menyapa, Gunung** ▲ Borneo, N Indonesia 1.04N 116.01E
165 T9 **Menyuan** *var.* Menyuan Huizu Zizhixian. Qinghai, C China 37.27N 101.33E
Menyuan Huizu Zizhixian *see* Menyuan
76 M5 **Menzel Bourguiba** *var.* Manzil Bū Ruqaybah; *prev.* Ferryville. N Tunisia 37.09N 9.51E
142 M15 **Menzelet Barajı** ⊠ C Turkey
131 T4 **Menzelinsk** Respublika Tatarstan, W Russian Federation 55.44N 53.00E
188 K11 **Menzies** Western Australia 29.42S 121.04E
205 V6 **Menzies, Mount** ▲ Antarctica 73.32S 61.02E
42 J6 **Meoqui** Chihuahua, N Mexico 28.19N 105.30W
85 N14 **Meponda** Niassa, NE Mozambique 13.19S 34.52E
100 M8 **Meppel** Drenthe, NE Netherlands 52.42N 6.12E
102 E12 **Meppen** Niedersachsen, NW Germany 52.42N 7.18E
Meqerghane, Sebkha *see* Mekerrhane, Sebkha
107 T6 **Mequinenza, Embalse de** ⊠ NE Spain
32 M8 **Mequon** Wisconsin, N USA 43.13N 87.57W
Mera *see* Maira
190 D3 **Meramangye, Lake** *salt lake* South Australia
29 W5 **Meramec River** ≈ Missouri, C USA
Meran *see* Merano
174 H10 **Merangin** ≈ Sumatera, W Indonesia
108 G5 **Merano** *Ger.* Meran. Trentino-Alto Adige, N Italy 46.40N 11.10E
174 H4 **Meraputh Lama** Pahang, Peninsular Malaysia 4.37N 101.58E
108 D7 **Merate** Lombardia, N Italy 45.42N 9.26E
175 Nn11 **Meratus, Pegunungan** ▲▲ Borneo, N Indonesia
176 Z16 **Merauke** Irian Jaya, E Indonesia 8.28S 140.28E
176 Z16 **Merauke, Sungai** ≈ Irian Jaya, E Indonesia
190 L9 **Merbein** Victoria, SE Australia 34.11S 142.03E
101 F21 **Merbes-le-Château** Hainaut, S Belgium 50.19N 4.09E
Merca *see* Marka
56 C13 **Mercaderes** Cauca, SW Colombia 1.47N 77.10W
Mercara *see* Madikeri
37 P9 **Merced** California, W USA 37.17N 120.30W
63 C20 **Mercedes** Buenos Aires, E Argentina 34.42S 59.30W
63 D15 **Mercedes** Corrientes, NE Argentina 29.09S 58.04W
64 J11 **Mercedes** *prev.* Villa Mercedes. San Luis, C Argentina 33.40S 65.24W
63 D19 **Mercedes** Soriano, SW Uruguay 33.16S 58.01W
27 S17 **Mercedes** Texas, SW USA 26.09N 97.54W
37 R9 **Merced Peak** ▲ California, W USA 37.34N 119.30W
37 P9 **Merced River** ≈ California, W USA
20 B13 **Mercer** Pennsylvania, NE USA 41.13N 80.13W
101 G18 **Merchtem** Vlaams Brabant, C Belgium 50.57N 4.13E
13 O13 **Mercier** Quebec, SE Canada 45.15N 73.45W
27 Q9 **Mercury** Texas, SW USA 31.23N 99.09W
192 M5 **Mercury Islands** *island group* N NZ
21 O9 **Meredith** New Hampshire, NE USA 43.36N 71.28W
67 B25 **Meredith, Cape** *var.* Cabo Belgrano *headland* West Falkland, Falkland Islands 52.15S 60.40W
39 V6 **Meredith, Lake** ◎ Colorado, C USA
27 N2 **Meredith, Lake** ⊠ Texas, SW USA
83 O16 **Mereeg** *var.* Mareeq, *It.* Meregh. Galguduud, E Somalia 3.47N 47.19E
119 V5 **Merefa** Kharkivs'ka Oblast', E Ukraine 49.48N 36.04E
Meregh *see* Mereeg
197 C11 **Mere Lava** *island* Banks Islands, N Vanuatu
101 E17 **Merelbeke** Oost-Vlaanderen, NW Belgium 51.00N 3.45E
Merend *see* Marand
127 Oo9 **Merenga** Magadanskaya Oblast', E Russian Federation 61.43N 156.02E
178 K12 **Mereuch** Mōndól Kiri, E Cambodia 13.01N 107.26E
Mergate *see* Margate
150 F9 **Mergenevo** Zapadnyy Kazakhstan, NW Kazakhstan 49.58N 51.19E
178 Gg12 **Mergui** Tenasserim, S Myanmar 12.25N 98.34E
177 G12 **Mergui Archipelago** *island group* S Myanmar
116 L12 **Meriç** Edirne, NW Turkey 41.12N 26.24E
116 L12 **Meriç** *Bul.* Maritsa, *Gk.* Évros; *anc.* Hebrus. ≈ SE Europe *see also* Évros/Maritsa
43 X12 **Mérida** Yucatán, SW Mexico 20.58N 89.35W
106 J11 **Mérida** *anc.* Augusta Emerita. Extremadura, W Spain 38.55N 6.19W
56 I6 **Mérida** Mérida, W Venezuela 8.36N 71.07W
56 H7 **Mérida** *off.* Estado Mérida. ◐ *state* W Venezuela
20 M13 **Meriden** Connecticut, NE USA 41.32N 72.48W
24 M5 **Meridian** Mississippi, S USA 32.24N 88.43W
27 S8 **Meridian** Texas, SW USA 31.55N 97.39W
104 J13 **Mérignac** Gironde, SW France 44.50N 0.39W
104 J13 **Mérignac** ✕ (Bordeaux) Gironde, SW France 44.51N 0.44W
95 J18 **Merikarvia** Länsi-Suomi, W Finland 61.51N 21.30E
191 R12 **Merimbula** New South Wales, SE Australia 36.52S 149.51E
190 L9 **Meringur** Victoria, SE Australia 34.26S 141.19E
Merín, Laguna *see* Mirim Lagoon
99 I19 **Merioneth** *cultural region* W Wales, UK
196 A11 **Merir** *island* Palau Islands, N Palau
196 B17 **Merizo** SW Guam 13.15N 144.40E
Merjama *see* Märjamaa
151 S16 **Merke** Zhambyl, S Kazakhstan 42.52N 73.09E
27 P7 **Merkel** Texas, SW USA 32.28N 100.00W
121 F15 **Merkinė** Varėna, S Lithuania 54.09N 24.11E
101 G16 **Merksem** Antwerpen, N Belgium 51.17N 4.26E
101 I15 **Merksplas** Antwerpen, N Belgium 51.22N 4.54E
Merkulovichi *see* Myerkulavichy
121 G15 **Merkys** ≈ S Lithuania
34 F15 **Merlin** Oregon, NW USA 42.34N 123.23W
63 C20 **Merlo** Buenos Aires, E Argentina 34.39S 58.45W
144 G8 **Meron, Haré** ▲ N Israel 35.06N 33.00E
76 K6 **Merouane, Chott** *salt lake* NE Algeria
82 F7 **Merowe** Northern, N Sudan 18.28N 31.49E
188 J12 **Merredin** Western Australia 31.31S 118.18E
99 I14 **Merrick** ▲ S Scotland, UK 55.09N 4.28W
34 H16 **Merrill** Oregon, NW USA 42.00N 121.37W
32 L5 **Merrill** Wisconsin, N USA 45.12N 89.43W
33 N11 **Merrillville** Indiana, N USA 41.28N 87.19W
21 O10 **Merrimack River** ≈ Massachusetts/New Hampshire, NE USA
30 L12 **Merriman** Nebraska, C USA 42.54N 101.42W
9 N17 **Merritt** British Columbia, SW Canada 50.09N 120.49W
25 Y12 **Merritt Island** Florida, SE USA 28.21N 80.42W
25 Y11 **Merritt Island** *island* Florida, SE USA
30 M12 **Merritt Reservoir** ⊠ Nebraska, C USA
191 S7 **Merriwa** New South Wales, SE Australia 32.09S 150.24E
191 O8 **Merriwagga** New South Wales, SE Australia 33.51S 145.38E
24 G8 **Merryville** Louisiana, S USA 30.45N 93.32W
82 K9 **Mersa Fatma** E Eritrea 14.52N 40.16E
104 M7 **Mer St-Aubin** Loir-et-Cher, C France 47.42N 1.31E
Mersa Maṭrûḥ *see* Maṭrûḥ
101 M24 **Mersch** Luxembourg, C Luxembourg 49.45N 6.06E
103 M15 **Merseburg** Sachsen-Anhalt, C Germany 51.22N 12.00E
Mersen *see* Meerssen
99 K18 **Mersey** ≈ NW England, UK
142 J17 **Mersin** İçel, S Turkey 36.49N 34.39E
174 I6 **Mersing** Johor, Peninsular Malaysia 2.25N 103.49E
120 E8 **Mērsrags** Talsi, NW Latvia 57.21N 23.05E
158 G12 **Merta** *var.* Merta City. Rājasthān, N India 26.40N 74.04E
Merta City *see* Merta
158 F12 **Merta Road** Rājasthān, N India 26.45N 73.59E
99 J21 **Merthyr Tydfil** S Wales, UK 51.46N 3.22W
106 H13 **Mértola** Beja, S Portugal 37.37N 7.40W
205 V16 **Mertz Glacier** *glacier* Antarctica
101 M24 **Mertzig** Diekirch, C Luxembourg 49.50N 6.00E
27 O9 **Mertzon** Texas, SW USA 31.15N 100.49W
83 I18 **Meru** Eastern, C Kenya 0.03N 37.37E
105 N4 **Méru** Oise, N France 49.15N 2.07E
83 I20 **Meru, Mount** ▲ NE Tanzania 3.12S 36.45E
Merv *see* Mary
Mervdasht *see* Marv Dasht
142 K11 **Merzifon** Amasya, N Turkey 40.52N 35.28E
103 D20 **Merzig** Saarland, SW Germany 49.27N 6.39E
38 L14 **Mesa** Arizona, SW USA 33.25N 111.49W
31 V4 **Mesabi Range** ▲▲ Minnesota, N USA
56 H6 **Mesa Bolívar** Mérida, NW Venezuela 8.30N 71.37W
109 Q18 **Mesagne** Puglia, SE Italy 40.33N 17.49E
41 P12 **Mesa Mountain** ▲ Alaska, USA 60.26N 155.14W
117 J25 **Mesará** *lowland* Kríti, Greece, E Mediterranean Sea
39 S14 **Mescalero** New Mexico, SW USA 33.09N 105.46W
103 G15 **Meschede** Nordrhein-Westfalen, W Germany 51.21N 8.16E
143 Q12 **Mescit Dağları** ▲▲ NE Turkey
201 V13 **Mesegon** *island* Chuuk, C Micronesia
Meseritz *see* Międzyrzecz
56 F11 **Mesetas** Meta, C Colombia 3.14N 74.09W
Meshchera Lowland *see* Meshcherskaya Nizina
130 M4 **Meshcherskaya Nizina** *Eng.* Meshchera Lowland. *basin* W Russian Federation
130 J5 **Meshchovsk** Kaluzhskaya Oblast', W Russian Federation 54.21N 35.23E
129 R9 **Meshchura** Respublika Komi, NW Russian Federation 63.18N 50.56E
Meshed *see* Mashhad
Meshed-i-Sar *see* Bābolsar
82 E13 **Meshra'er Req** Warab, S Sudan 8.30N 29.27E
39 R15 **Mesilla** New Mexico, SW USA 32.15N 106.49W
110 H10 **Mesocco** *Ger.* Misox. Ticino, S Switzerland 46.18N 9.13E
117 D18 **Mesolóngi** *prev.* Mesolóngion. Dytikí Ellás, W Greece 38.22N 21.26E
Mesolóngion *see* Mesolóngi
12 E8 **Mesomikenda Lake** ◎ Ontario, S Canada
63 D15 **Mesopotamia** *var.* Mesopotamia Argentina. *physical region* NE Argentina
Mesopotamia Argentina *see* Mesopotamia
37 Y10 **Mesquite** Nevada, W USA 36.47N 114.04W
84 Q13 **Messalo, Rio** *var.* Mualo. ≈ NE Mozambique
Messana/Messene *see* Messina
101 L25 **Messancy** Luxembourg, SE Belgium 49.36N 5.49E
109 M23 **Messina** *var.* Messana, Messene; *anc.* Zancle. Sicilia, Italy, C Mediterranean Sea 38.11N 15.33E
85 K19 **Messina** Northern, NE South Africa 22.18S 30.02E
Messina, Strait of *see* Messina, Stretto di
109 M23 **Messina, Stretto di** *Eng.* Strait of Messina. *strait* SW Italy
117 E21 **Messíni** Pelopónnisos, S Greece 37.03N 22.00E
117 E21 **Messinía** *peninsula* S Greece
117 E22 **Messiniakós Kólpos** *gulf* S Greece
126 H8 **Messoyakha** ≈ N Russian Federation
116 H11 **Mesta** *Gk.* Néstos, *Turk.* Kara Su. ≈ Bulgaria/Greece *see also* Néstos
Mestghanem *see* Mostaganem
143 R8 **Mestia** *var.* Mestiya. N Georgia 43.03N 42.49E
Mestiya *see* Mestia
117 K18 **Mestón, Akrotírio** *headland* Chíos, E Greece 38.15N 25.52E
108 H8 **Mestre** Veneto, NE Italy 45.30N 12.13E
61 M16 **Mestre, Espigão** ▲ E Brazil
174 Ii11 **Mesuji** ≈ Sumatera, W Indonesia
Mesule *see* Grosser Möseler
8 J10 **Meszah Peak** ▲ British Columbia, W Canada 58.31N 131.28W
56 G11 **Meta** *off.* Departamento del Meta. ◐ *province* C Colombia
13 Q8 **Metabetchouane** ≈ Quebec, SE Canada
16 O4 **Meta Incognita Peninsula** *peninsula* Baffin Island, Nunavut, NE Canada
24 K9 **Metairie** Louisiana, S USA 29.58N 90.09W
34 M6 **Metaline Falls** Washington, NW USA 48.51N 117.21W
64 K6 **Metán** Salta, N Argentina 25.25S 64.52W
84 N13 **Metangula** Niassa, N Mozambique 12.40S 34.49E
44 E7 **Metapán** Santa Ana, NW El Salvador 14.19N 89.30W
56 K9 **Meta, Río** ≈ Colombia/Venezuela
108 I11 **Metauro** ≈ C Italy
82 H11 **Metema** Amhara, N Ethiopia 12.53N 36.10E
117 D15 **Metéora** *religious building* Thessalía, C Greece 39.45N 21.37E
67 O20 **Meteor Rise** *undersea feature* SW Indian Ocean
195 N9 **Meteran** New Hanover, NE PNG 2.36S 150.09E
Meterlam/Methariam/Metharlam *see* Mehtarlām
117 G20 **Methanon** *peninsula* S Greece
34 J6 **Methow River** ≈ Washington, NW USA
21 O10 **Methuen** Massachusetts, NE USA 42.43N 71.10W
193 G19 **Methven** Canterbury, South Island, NZ 43.37S 171.38E
Metis *see* Metz
115 G15 **Metković** Dubrovnik-Neretva, SE Croatia 43.02N 17.37E
41 Y14 **Metlakatla** Annette Island, Alaska, USA 55.07N 131.34W
111 V13 **Metlika** *Ger.* Möttling. SE Slovenia 45.38N 15.18E
111 T8 **Metnitz** Kärnten, S Austria 46.58N 14.09E
29 W12 **Meto, Bayou** ≈ Arkansas, C USA
174 I13 **Metro** Sumatera, W Indonesia 5.05S 105.17E
32 M17 **Metropolis** Illinois, N USA 37.09N 88.43W
Metropolitan *see* Santiago
37 N8 **Metropolitan Oakland** ✕ California, W USA 37.42N 122.13W
117 D15 **Métsovo** *prev.* Métsovon. Ípeiros, C Greece 39.47N 21.12E
Métsovon *see* Métsovo
25 V5 **Metter** Georgia, SE USA 32.24N 82.03W
101 H21 **Mettet** Namur, S Belgium 50.19N 4.43E
103 D20 **Mettlach** Saarland, SW Germany 49.28N 6.37E
Mettu *see* Metu
82 H13 **Metu** *var.* Mattu, Mettu. Oromo, C Ethiopia 8.18N 35.39E
175 N7 **Metulang** Borneo, N Indonesia 1.28N 114.40E
144 G8 **Metulla** Northern, N Israel 33.16N 35.35E
150 G14 **Metvyy Kultuk, Sor** *salt flat* SW Kazakhstan
105 T4 **Metz** *anc.* Divodurum Mediomatricum, Mediomatrica, Metis. Moselle, NE France 49.07N 6.09E
103 H22 **Metzingen** Baden-Württemberg, S Germany 48.31N 9.16E
173 E4 **Meulaboh** Sumatera, W Indonesia 4.10N 96.09E
101 D18 **Meulebeke** West-Vlaanderen, W Belgium 50.57N 3.18E
105 U6 **Meurthe** ≈ NE France
105 S5 **Meurthe-et-Moselle** ◐ *department* NE France
105 S4 **Meuse** ◐ *department* NE France
86 F10 **Meuse** *Dut.* Maas. ≈ W Europe *see also* Maas
195 O11 **Mevelo** ≈ New Britain, C Papau New Guinea
Mexcala, Río *see* Balsas, Río
27 U8 **Mexia** Texas, SW USA 31.40N 96.28W
60 K11 **Mexiana, Ilha** *island* NE Brazil
42 C1 **Mexicali** Baja California, NW Mexico 32.34N 115.26W
29 V4 **Mexico** Missouri, C USA 39.10N 91.52W
20 H9 **Mexico** New York, NE USA 43.27N 76.14W
42 L7 **Mexico** *off.* United Mexican States, *var.* Méjico, México, *Sp.* Estados Unidos Mexicanos. ◆ *federal republic* N Central America
43 O14 **México** *var.* Ciudad de México, *Eng.* Mexico City. ● (Mexico) México, C Mexico 19.24N 99.04W
43 O13 **México** ◐ *state* S Mexico
1 J13 **Mexico Basin** *var.* Sigsbee Deep. *undersea feature* C Gulf of Mexico
Mexico City *see* México
México, Golfo de *see* Mexico, Gulf of
46 B4 **Mexico, Gulf of** *Sp.* Golfo de México. *gulf* W Atlantic Ocean
Meyadine *see* Al Mayādīn
41 Y14 **Meyers Chuck** Etolin Island, Alaska, USA 55.44N 132.15W
154 M3 **Meymaneh** *var.* Maimāna, Maymana. Fāryāb, NW Afghanistan 35.57N 64.48E
149 N7 **Meymeh** Eşfahān, C Iran 33.28N 51.09E
127 Pp6 **Meynypil'gyno** Chukotskiy Avtonomnyy Okrug, NE Russian Federation 62.33N 177.00E
110 A10 **Meyrin** Genève, SW Switzerland 46.13N 6.04E
177 Ff8 **Mezaligon** Irrawaddy, SW Myanmar 17.53N 95.12E
43 O15 **Mezcala** Guerrero, S Mexico 17.55N 99.34W
116 H8 **Mezdra** Vratsa, NW Bulgaria 43.09N 23.44E
105 P16 **Mèze** Hérault, S France 43.25N 3.37E
129 O6 **Mezen'** Arkhangel'skaya Oblast', NW Russian Federation 65.54N 44.10E
129 P8 **Mezen'** ≈ NW Russian Federation
Mezen, Bay of *see* Mezenskaya Guba
105 Q13 **Mézenc, Mont** ▲ C France 44.57N 4.15E
129 O8 **Mezenskaya Guba** *var.* Bay of Mezen. *bay* NW Russian Federation
125 Bb7 **Mezha** ≈ W Russian Federation
Mezha *see* Myazha
126 Hh15 **Mezhdurechensk** Kemerovskaya Oblast', S Russian Federation 53.37N 87.59E
125 F4 **Mezhdusharskiy, Ostrov** *island* Novaya Zemlya, N Russian Federation
Mezhëvo *see* Myezhava
Mezhgor'ye *see* Mizhhir"ya
119 V8 **Mezhova** Dnipropetrovs'ka Oblast', E Ukraine 48.15N 36.44E
8 J12 **Meziadin Junction** British Columbia, W Canada 56.06N 129.15W
113 G16 **Mezileské Sedlo** *var.* Przełęcz Międzyleska. *pass* Czech Republic/Poland
104 L14 **Mézin** Lot-et-Garonne, SW France 44.03N 0.16E
113 M24 **Mezőberény** Békés, SE Hungary 46.49N 21.00E
113 M25 **Mezőhegyes** Békés, SE Hungary 46.19N 20.51E
113 M25 **Mezőkovácsháza** Békés, SE Hungary 46.25N 20.52E
113 M21 **Mezőkövesd** Borsod-Abaúj-Zemplén, NE Hungary 47.48N 20.34E
Mezőtelegd *see* Tileagd
113 M23 **Mezőtúr** Jász-Nagykun-Szolnok, E Hungary 47.00N 20.37E
42 K10 **Mezquital** Durango, C Mexico 23.29N 104.24W
108 G6 **Mezzolombardo** Trentino-Alto Adige, N Italy 46.13N 11.08E
84 L13 **Mfuwe** Northern, N Zambia 13.00S 31.45E
123 J16 **Mġarr** Gozo, N Malta 36.01N 14.18E
130 H6 **Mglin** Bryanskaya Oblast', W Russian Federation 53.01N 32.54E
Mhálanna, Cionn *see* Malin Head
160 G10 **Mhow** Madhya Pradesh, C India 22.36N 75.47E
Miadziol Nowy *see* Myadzyel
179 Q13 **Miagao** Panay Island, C Philippines 10.40N 122.15E
43 R17 **Miahuatlán** *var.* Miahuatlán de Porfirio Díaz. Oaxaca, SE Mexico 16.21N 96.36W
Miahuatlán de Porfirio Díaz *see* Miahuatlán
106 K10 **Miajadas** Extremadura, W Spain 39.10N 5.54W
Miajlar *see* Myājlār
38 M14 **Miami** Arizona, SW USA 33.23N 110.53W
25 Z16 **Miami** Florida, SE USA 25.46N 80.11W
29 R8 **Miami** Oklahoma, C USA 36.52N 94.52W
27 O2 **Miami** Texas, SW USA 35.41N 100.38W
25 Z16 **Miami** ✕ Florida, SE USA 25.47N 80.16W
25 Z16 **Miami Beach** Florida, SE USA 25.47N 80.07W
25 Y15 **Miami Canal** *canal* Florida, SE USA
33 R14 **Miamisburg** Ohio, N USA 39.38N 84.17W
155 U10 **Miān Channūn** Punjab, E Pakistan 30.27N 72.24E
148 J4 **Miāndowāb** *var.* Mianduab, Miyāndoāb. Āzarbāyjān-e Bākhtarī, NW Iran 36.58N 46.06E
180 H5 **Miandrivazo** Toliara, C Madagascar 19.31S 45.28E
Mianduab *see* Miāndowāb
148 K3 **Miāneh** *var.* Miyāneh. Āzarbāyjān-e Khāvarī, NW Iran 37.25N 47.43E
155 O16 **Miāni Hōr** *lagoon* S Pakistan
166 G10 **Mianning** Sichuan, C China 28.34N 102.12E
155 T7 **Miānwāli** Punjab, NE Pakistan 32.31N 71.33E
166 J7 **Mianxian** *var.* Mian Xian. Shaanxi, C China 33.12N 106.36E
166 I8 **Mianyang** Sichuan, C China 31.28N 104.43E
Mianyang *see* Xiantao
167 R3 **Miaodao Qundao** *island group* E China
167 S13 **Miaoli** N Taiwan 24.33N 120.48E
125 E12 **Miass** Chelyabinskaya Oblast', C Russian Federation 55.00N 59.55E
112 G8 **Miastko** *Ger.* Rummelsburg in Pommern. Pomorskie, N Poland 54.00N 16.58E
Miava *see* Myjava
9 O15 **Mica Creek** British Columbia, SW Canada 51.58N 118.29W
166 J7 **Micang Shan** ▲▲ C China
194 I12 **Michael, Mount** ▲ C PNG 6.24S 145.18E
Mi Chai *see* Nong Khai
113 O19 **Michalovce** *Ger.* Grossmichel, *Hung.* Nagymihály. Košický Kraj, E Slovakia 48.46N 21.54E
101 M20 **Michel, Baraque** *hill* E Belgium 50.38N 6.09E
41 S5 **Michelson, Mount** ▲ Alaska, USA 69.19N 144.16W
47 P9 **Miches** E Dominican Republic 18.56N 69.04W
32 M4 **Michigamme, Lake** ◎ Michigan, N USA
32 M4 **Michigamme Reservoir** ⊠ Michigan, N USA
33 N4 **Michigamme River** ≈ Michigan, N USA
33 O7 **Michigan** *off.* State of Michigan; also known as Great Lakes State, Lake State, Wolverine State. ◐ *state* N USA
33 O11 **Michigan City** Indiana, N USA 41.43N 86.52W
33 O8 **Michigan, Lake** ◎ N USA
33 P2 **Michipicoten Bay** *lake bay* Ontario, N Canada
12 A8 **Michipicoten Island** *island* Ontario, S Canada
12 B7 **Michipicoten River** Ontario, S Canada 47.56N 84.48W
Michurin *see* Tsarevo
130 M6 **Michurinsk** Tambovskaya Oblast', W Russian Federation 52.56N 40.30E
Mico, Punta/Mico, Punto *see* Monkey Point
44 L10 **Mico, Río** ≈ SE Nicaragua
47 T12 **Micoud** SE Saint Lucia 13.49N 60.54W
201 N16 **Micronesia** *off.* Federated States of Micronesia. ◆ *federation* W Pacific Ocean
183 P4 **Micronesia** *island group* W Pacific Ocean
174 Jj5 **Midai, Pulau** *island* Kepulauan Natuna, W Indonesia
Mid-Atlantic Cordillera *see* Mid-Atlantic Ridge
67 M17 **Mid-Atlantic Ridge** *var.* Mid-Atlantic Cordillera, Mid-Atlantic Rise, Mid-Atlantic Swell. *undersea feature* Atlantic Ocean
Mid-Atlantic Rise/Mid-Atlantic Swell *see* Mid-Atlantic Ridge
101 E15 **Middelburg** Zeeland, SW Netherlands 51.30N 3.36E
85 H24 **Middelburg** Eastern Cape, S South Africa 31.28S 25.01E
85 K21 **Middelburg** Mpumalanga, NE South Africa 25.46S 29.28E
97 G23 **Middelfart** Fyn, C Denmark 55.30N 9.43E
100 G13 **Middelharnis** Zuid-Holland, SW Netherlands 51.45N 4.10E
101 B16 **Middelkerke** West-Vlaanderen, W Belgium 51.11N 2.51E
100 I9 **Middenbeemster** Noord-Holland, C Netherlands 52.33N 4.55E
100 I8 **Middenmeer** Noord-Holland, NW Netherlands 52.48N 4.58E
37 Q2 **Middle Alkali Lake** ◎ California, W USA
200 Nn6 **Middle America Trench** *undersea feature* E Pacific Ocean
157 P19 **Middle Andaman** *island* Andaman Islands, India, NE Indian Ocean
Middle Atlas *see* Moyen Atlas
23 R3 **Middlebourne** West Virginia, NE USA 39.29N 80.54W
25 W9 **Middleburg** Florida, SE USA 30.03N 81.55W
Middleburg Island *see* 'Eua
Middle Caicos *see* Grand Caicos
27 N8 **Middle Concho River** ≈ Texas, SW USA
Middle Congo *see* Congo (Republic of)
41 R6 **Middle Fork Chandalar River** ≈ Alaska, USA
41 Q7 **Middle Fork Koyukuk River** ≈ Alaska, USA
35 O12 **Middle Fork Salmon River** ≈ Idaho, NW USA
9 T15 **Middle Lake** Saskatchewan, S Canada 52.31N 105.16W
30 L13 **Middle Loup River** ≈ Nebraska, C USA
193 E22 **Middlemarch** Otago, South Island, NZ 45.30S 170.07E
33 T15 **Middleport** Ohio, N USA 39.00N 82.03W
31 U14 **Middle Raccoon River** ≈ Iowa, C USA
31 R3 **Middle River** ≈ Minnesota, N USA
23 N8 **Middlesboro** Kentucky, S USA 36.37N 83.42W
99 M15 **Middlesbrough** N England, UK 54.34N 1.13W
44 G3 **Middlesex** Stann Creek, C Belize 17.00N 88.31W
99 N22 **Middlesex** *cultural region* SE England, UK
11 P15 **Middleton** Nova Scotia, SE Canada 44.54N 65.01W
22 F10 **Middleton** Tennessee, S USA 35.05N 88.57W
32 L9 **Middleton** Wisconsin, N USA 43.06N 89.30W
41 S13 **Middleton Island** *island* Alaska, USA
36 M7 **Middletown** California, W USA 38.44N 122.39W
23 Y2 **Middletown** Delaware, NE USA 39.25N 75.39W
20 K15 **Middletown** New Jersey, NE USA 40.22N 74.07W
20 K13 **Middletown** New York, NE USA 41.27N 74.25W
33 R14 **Middletown** Ohio, N USA 39.33N 84.19W
20 G15 **Middletown** Pennsylvania, NE USA 40.11N 76.42W
147 N14 **Mīdī** *var.* Maydī. NW Yemen 16.18N 42.51E
105 O16 **Midi, Canal du** *canal* S France
104 K17 **Midi de Bigorre, Pic du** ▲ S France 42.57N 0.08E
104 K17 **Midi d'Ossau, Pic du** ▲ SW France 42.51N 0.27W
181 R7 **Mid-Indian Basin** *undersea feature* N Indian Ocean
181 P7 **Mid-Indian Ridge** *var.* Central Indian Ridge. *undersea feature* C Indian Ocean
105 N14 **Midi-Pyrénées** ◐ *region* S France
27 N8 **Midkiff** Texas, SW USA 31.35N 101.51W
12 G13 **Midland** Ontario, S Canada 44.43N 79.51W
33 R8 **Midland** Michigan, N USA 43.37N 84.15W
30 M10 **Midland** South Dakota, N USA 44.04N 101.07W
26 M8 **Midland** Texas, SW USA 32.00N 102.04W
85 K17 **Midlands** ◐ *province* C Zimbabwe
99 D21 **Midleton** *Ir.* Mainistir na Corann. SW Ireland 51.55N 8.10W
27 T7 **Midlothian** Texas, SW USA 32.28N 96.59W
98 K12 **Midlothian** *cultural region* S Scotland, UK
180 I7 **Midongy** Fianarantsoa, S Madagascar 21.58S 47.46E
104 K15 **Midou** ≈ SW France
199 Ii6 **Mid-Pacific Mountains** *var.* Mid-Pacific Seamounts. *undersea feature* NW Pacific Ocean
Mid-Pacific Seamounts *see* Mid-Pacific Mountains
179 R16 **Midsayap** Mindanao, S Philippines 7.12N 124.31E
38 L3 **Midway** Utah, W USA 40.30N 111.28W
199 Jj5 **Midway Islands** ◇ *US territory* C Pacific Ocean
35 X14 **Midwest** Wyoming, C USA 43.24N 106.15W
29 N10 **Midwest City** Oklahoma, C USA 35.27N 97.24W
158 M10 **Mid Western** ◐ *zone* W Nepal
100 P5 **Midwolda** Groningen, NE Netherlands 53.12N 7.00E
143 Q16 **Midyat** Mardin, SE Turkey 37.25N 41.19E
116 F8 **Midzhur** *SCr.* Midžor. ▲ Bulgaria/Yugoslavia *see also* Midžor 43.24N 22.41E
115 Q14 **Midžor** *Bul.* Midzhur. ▲ Bulgaria/Yugoslavia *see also* Midzhur 43.24N 22.40E
171 H16 **Mie** *off.* Mie-ken. ◐ *prefecture* Honshū, SW Japan
113 L16 **Miechów** Małopolskie, S Poland 50.20N 20.00E
112 F11 **Międzychód** *Ger.* Mitteldorf. Wielkopolskie, C Poland 52.36N 15.52E
Międzyleska, Przełęcz *see* Mezileské Sedlo
112 O12 **Międzyrzec Podlaski** Lubelskie, E Poland 52.00N 22.47E
112 E11 **Międzyrzecz** *Ger.* Meseritz. Lubelskie, W Poland 52.26N 15.33E
Mie-ken *see* Mie
104 L16 **Miélan** Gers, S France 43.25N 0.18E
113 N16 **Mielec** Podkarpackie, SE Poland 50.18N 21.27E
97 L21 **Mien** ◎ S Sweden
43 O8 **Mier** Tamaulipas, C Mexico 26.25N 99.10W
118 J11 **Miercurea-Ciuc** *Ger.* Szeklerburg, *Hung.* Csíkszereda. Harghita, C Romania 46.23N 25.47E
Mieresch *see* Maros/Mureş
Mieres del Camín *see* Mieres del Camino
106 K2 **Mieres del Camino** *var.* Mieres del Camín. Asturias, NW Spain 43.15N 5.46W
101 K15 **Mierlo** Noord-Brabant, SE Netherlands 51.27N 5.37E
43 O10 **Mier y Noriega** Nuevo León, NE Mexico 23.24N 100.06W

◆ Country ● Country Capital ◇ Dependent Territory ○ Dependent Territory Capital ◐ Administrative Region ✕ International Airport ▲ Mountain ▲▲ Mountain Range Volcano ≈ River ◎ Lake ⊠ Reservoir

Mies see Stříbro
82 K13 **Mi'ěso** var. Meheso, Oromo. C Ethiopia 9.13N 40.47E
Miesso see Mi'ěso
112 D10 **Mieszkowice** Ger. Bärwalde Neumark. Zachodniopomorskie, W Poland 52.45N 14.24E
20 G14 **Mifflinburg** Pennsylvania, NE USA 40.55N 77.03W
20 F14 **Mifflintown** Pennsylvania, NE USA 40.34N 77.24W
43 R15 **Miguel Alemán, Presa** ⊠ SE Mexico
42 I9 **Miguel Asua** var. Miguel Auza. Zacatecas, C Mexico 24.16N 103.28W
Miguel Auza see Miguel Asua
45 S15 **Miguel de la Borda** var. Donoso. Colón, C Panama 9.06N 80.19W
43 R15 **Miguel Hidalgo** ✈ (Guadalajara) Jalisco, SW Mexico 20.52N 101.09W
42 H7 **Miguel Hidalgo, Presa** ⊠ W Mexico
118 J14 **Mihăileşti** Giurgiu, S Romania 44.19N 25.54E
118 M14 **Mihail Kogălniceanu** var. Kogălniceanu; prev. Caramurat, Ferdinand. Constanţa, SE Romania 44.23N 28.24E
119 N14 **Mihai Viteazu** Constanţa, SE Romania 44.37N 28.41E
142 G12 **Mihalıçcık** Eskişehir, NW Turkey 39.52N 31.30E
170 Ec13 **Mihara** Hiroshima, Honshū, SW Japan 34.24N 133.03E
171 Ji7 **Mihara-yama** ⚱ Miyako-jima, SE Japan 34.43N 139.22E
107 S8 **Mijares** ♒ E Spain
100 I11 **Mijdrecht** Utrecht, C Netherlands 52.12N 4.52E
172 Oo5 **Mikasa** Hokkaidō, NE Japan 43.19N 141.54E
Mikashevichi see Mikashevichy
121 K19 **Mikashevichy** Pol. Mikaszewicze, Rus. Mikashevichi. Brestskaya Voblasts', SW Belarus 52.13N 27.28E
Mikaszewicze see Mikashevichy
171 Hh16 **Mikawa-wan** bay E Japan
130 L5 **Mikhaylov** Ryazanskaya Oblast', W Russian Federation 54.12N 39.03E
Mikhaylovgrad see Montana
205 Z8 **Mikhaylov Island** island Antarctica
151 T6 **Mikhaylovka** Pavlodar, N Kazakhstan 53.49N 76.31E
131 N9 **Mikhaylovka** Volgogradskaya Oblast', SW Russian Federation 50.06N 43.17E
Mikhaylovka see Mykhaylivka
170 G14 **Miki** Hyōgo, Honshū, SW Japan 34.46N 135.04E
83 K24 **Mikindani** Mtwara, SE Tanzania 10.16S 40.04E
95 N18 **Mikkeli** Swe. Sankt Michel. Itä-Suomi, E Finland 61.41N 27.14E
112 M8 **Mikołajki** Ger. Nikolaiken. Warmińsko-Mazurskie, NE Poland 53.49N 21.31E
Mikonos see Mýkonos
116 I9 **Mikre** Lovech, N Bulgaria 43.02N 24.32E
116 C13 **Mikrí Préspa, Límni** ⊜ N Greece
129 P4 **Mikulkin, Mys** headland NW Russian Federation 67.50N 46.36E
83 I23 **Mikumi** Morogoro, SE Tanzania 7.22S 37.00E
129 R10 **Mikun'** Respublika Komi, NW Russian Federation 62.20N 50.02E
171 Hh13 **Mikuni** Fukui, Honshū, SW Japan 36.12N 136.09E
171 Jj14 **Mikuni-tōge** pass Honshū, C Japan 36.48N 138.47E
152 Si3 **Mikura-jima** island E Japan
31 V7 **Milaca** Minnesota, N USA 45.45N 93.40W
64 J10 **Milagro** La Rioja, C Argentina 31.00S 66.01W
58 B7 **Milagro** Guayas, SW Ecuador 2.08S 79.34W
33 P4 **Milakokia Lake** ⊜ Michigan, N USA
32 I1 **Milan** Illinois, N USA 41.27N 90.33W
33 R10 **Milan** Michigan, N USA 42.05N 83.40W
29 T2 **Milan** Missouri, C USA 40.12N 93.07W
39 Q11 **Milan** New Mexico, SW USA 35.10N 107.53W
22 G9 **Milan** Tennessee, S USA 35.55N 88.45W
Milan see Milano
97 F15 **Miland** Telemark, S Norway 59.57N 8.48E
85 L15 **Milange** Zambézia, NE Mozambique 16.08S 35.51E
108 D8 **Milano** Eng. Milan, Ger. Mailand; anc. Mediolanum. Lombardia, N Italy 45.28N 9.10E
27 U10 **Milano** Texas, SW USA 30.42N 96.51W
142 C15 **Milas** Muğla, SW Turkey 37.16N 27.46E
121 K21 **Milashavichy** Rus. Milashevichi. Homyel'skaya Voblasts', SE Belarus 51.38N 27.54E
Milashevichi see Milashavichy
121 I18 **Milavidy** Rus. Milovidy. Brestskaya Voblasts', SW Belarus 52.54N 25.51E
109 L23 **Milazzo** anc. Mylae. Sicilia, Italy, C Mediterranean Sea 38.13N 15.15E
31 R8 **Milbank** South Dakota, N USA 45.12N 96.36W
21 T7 **Milbridge** Maine, NE USA 44.31N 67.55W
102 L11 **Milde** ♒ C Germany
12 F14 **Mildmay** Ontario, S Canada 44.03N 81.07W
190 L10 **Mildura** Victoria, SE Australia 34.13S 142.09E
143 X12 **Mil Düzü** Rus. Mil'skaya Ravnina, Mil'skaya Step'; physical region C Azerbaijan
166 H13 **Mile** Yunnan, SW China 24.28N 103.25E
Mile see Mili Atoll
189 Y10 **Miles** Queensland, E Australia 26.41S 150.15E

27 P8 **Miles** Texas, SW USA 31.36N 100.10W
119 Z5 **Milove** Luhans'ka Oblast', E Ukraine 49.22N 40.09E
Milovidy see Milavidy
35 X9 **Miles City** Montana, NW USA 46.24N 105.48W
9 U17 **Milestone** Saskatchewan, S Canada 50.00N 104.24W
109 N22 **Mileto** Calabria, SW Italy 38.35N 16.03E
109 K16 **Miletto, Monte** ⚠ C Italy 41.28N 14.21E
20 M13 **Milford** Connecticut, NE USA 41.12N 73.01W
23 Y3 **Milford** var. Milford City. Delaware, NE USA 38.54N 75.25W
31 T11 **Milford** Iowa, C USA 43.19N 95.09W
21 S6 **Milford** Maine, NE USA 44.57N 68.37W
31 R16 **Milford** Nebraska, C USA 40.46N 97.03W
21 O10 **Milford** New Hampshire, NE USA 42.49N 71.38W
20 J13 **Milford** Pennsylvania, NE USA 41.19N 74.48W
27 T7 **Milford** Texas, SW USA 32.07N 96.57W
38 K6 **Milford** Utah, W USA 38.22N 112.57W
Milford see Millau
99 H21 **Milford Haven** prev. Milford. SW Wales, UK 51.43N 5.01W
Milford City see Milford
29 O4 **Milford Lake** ⊠ Kansas, C USA
193 B21 **Milford Sound** Southland, South Island, NZ 44.40S 167.57E
193 B21 **Milford Sound** inlet South Island, NZ
Milhau see Millau
201 W8 **Mili Atoll** var. Mile. atoll Ratak Chain, SE Marshall Islands
112 H13 **Milicz** Dolnośląskie, SW Poland 51.31N 17.18E
109 L25 **Militello in Val di Catania** Sicilia, Italy, C Mediterranean Sea 37.16N 14.46E
127 Pp11 **Mil'kovo** Kamchatskaya Oblast', E Russian Federation 54.12N 158.35E
9 R17 **Milk River** Alberta, SW Canada 49.10N 112.06W
46 J13 **Milk River** ♒ C Jamaica
35 W7 **Milk River** ♒ Montana, NW USA
82 D9 **Milk, Wadi el** var. Wadi al Malik. ♒ C Sudan
101 L14 **Mill** Noord-Brabant, SE Netherlands 51.42N 5.46E
105 P24 **Millau** var. Milhau; anc. Æmilianum. Aveyron, S France 44.06N 3.04E
12 I14 **Millbrook** Ontario, SE Canada 44.09N 78.26W
25 U4 **Milledgeville** Georgia, SE USA 33.04N 83.13W
10 C12 **Mille Lacs, Lac des** ⊠ Ontario, S Canada
31 V6 **Mille Lacs Lake** ⊜ Minnesota, C USA
25 V4 **Millen** Georgia, SE USA 32.50N 81.56W
203 T9 **Millennium Island** prev. Caroline Island, Thornton Island. atoll Line Islands, E Kiribati
31 O9 **Miller** South Dakota, N USA 44.31N 98.59W
32 K5 **Miller Dam Flowage** ⊠ Wisconsin, N USA
41 U12 **Miller, Mount** ⚠ Alaska, USA 60.29N 142.16W
130 L10 **Millerovo** Rostovskaya Oblast', SW Russian Federation 48.57N 40.25E
39 N17 **Miller Peak** ⚠ Arizona, SW USA 31.23N 110.17W
33 T12 **Millersburg** Ohio, N USA 40.33N 81.55W
20 G15 **Millersburg** Pennsylvania, NE USA 40.31N 76.56W
193 D23 **Millers Flat** Otago, South Island, NZ 45.42S 169.25E
27 Q8 **Millersview** Texas, SW USA 31.26N 99.44W
108 B10 **Millesimo** Piemonte, NE Italy 44.24N 8.09E
15 Mm15 **Milles Lacs, Lac des** ⊠ Ontario, S Canada
27 Q13 **Millett** Texas, SW USA 28.33N 99.10W
105 N11 **Millevaches, Plateau de** plateau C France
190 K12 **Millicent** South Australia 37.37S 140.21E
100 M13 **Millingen aan den Rijn** Gelderland, SE Netherlands 51.52N 6.02E
22 E10 **Millington** Tennessee, S USA 35.20N 89.54W
21 R4 **Millinocket** Maine, NE USA 45.38N 68.45W
21 R4 **Millinocket Lake** ⊜ Maine, NE USA
205 L11 **Mill Island** island Antarctica
191 T3 **Millmerran** Queensland, E Australia 27.52S 151.15E
111 R9 **Millstatt** Kärnten, S Austria 46.45N 13.36E
99 B19 **Milltown Malbay** Ir. Sráid na Cathrach. W Ireland 52.51N 9.23W
20 J17 **Millville** New Jersey, NE USA 39.24N 75.01W
29 S13 **Millwood Lake** ⊠ Arkansas, C USA
105 S16 **Milne Bank** see Milne Seamounts
195 O17 **Milne Bay** ✧ province SE PNG
195 Q17 **Milne Bay** bay SE PNG
66 J8 **Milne Seamounts** var. Milne Bank. undersea feature N Atlantic Ocean
30 J4 **Milnor** North Dakota, N USA 46.15N 97.27W
21 R5 **Milo** Maine, NE USA 45.15N 69.01W
117 I22 **Mílos** Mílos, Kykládes, Greece, Aegean Sea 36.45N 24.26E
117 I22 **Mílos** island Kykládes, Greece, Aegean Sea
112 H11 **Milosław** Wielkopolskie, C Poland 52.13N 17.28E
115 K19 **Milot** var. Miloti. Lezhë, C Albania 41.42N 19.43E

Miloti see Milot
119 Z5 **Milove** Luhans'ka Oblast', E Ukraine 49.22N 40.09E
190 L4 **Milparinka** New South Wales, SE Australia 29.48S 141.57E
37 N9 **Milpitas** California, W USA 37.25N 121.54W
12 G15 **Milton** Ontario, S Canada 43.31N 79.52W
193 E24 **Milton** Otago, South Island, NZ 46.07S 169.59E
23 Y4 **Milton** Delaware, NE USA 38.48N 75.21W
25 P8 **Milton** Florida, SE USA 30.37N 87.02W
20 G14 **Milton** Pennsylvania, NE USA 41.01N 76.49W
20 L7 **Milton** Vermont, NE USA 44.37N 73.04W
34 K11 **Milton-Freewater** Oregon, NW USA 45.54N 118.24W
99 N21 **Milton Keynes** SE England, UK 52.00N 0.43W
29 N3 **Miltonvale** Kansas, C USA 39.21N 97.27W
167 N10 **Miluo** Hunan, S China 29.13N 113.00E
32 M9 **Milwaukee** Wisconsin, N USA 43.03N 87.55W
Milyang see Miryang
Mimatum see Mende
39 G16 **Mimbres Mountains** ⛰ New Mexico, SW USA
190 D2 **Mimili** South Australia
104 J14 **Mimizan** Landes, SW France 44.12N 1.12W
81 E19 **Mimmaya** see Minmaya
Mimongo Ngounié, C Gabon 1.36S 11.43E
37 T7 **Mina** Nevada, W USA 38.23N 118.07W
149 S14 **Mīnāb** Hormozgān, SE Iran 27.08N 57.02E
84 **Mina Baranis** see Berenice
155 R9 **Mina Bāzār** Baluchistān, SW Pakistan 31.00N 67.48E
170 C15 **Minamata** Kumamoto, Kyūshū, SW Japan 32.12N 130.23E
172 Si7 **Minami-Iō-jima** Eng. San Augustine. island SE Japan
172 Nn7 **Minami-Kayabe** Hokkaidō, NE Japan 41.54N 140.58E
170 B17 **Minamitane** Kagoshima, Tanega-shima, SW Japan 30.23N 130.54E
172 Tori **Minami Tori Shima** see Marcus Island
64 J4 **Mina Pirquitas** Jujuy, NW Argentina 22.48S 66.24W
181 O3 **Minā' Qābūs** NE Oman
63 F19 **Minas** Lavalleja, S Uruguay 34.19S 55.15W
11 P15 **Minas Basin** bay Nova Scotia, SE Canada
63 C17 **Minas de Corrales** Rivera, NE Uruguay 31.34S 55.19W
46 A5 **Minas de Matahambre** Pinar del Río, W Cuba 22.34N 83.57W
106 J13 **Minas de Ríotinto** Andalucía, S Spain 37.40N 6.36W
62 K7 **Minas Gerais** off. Estado de Minas Gerais. ✧ state E Brazil
44 E5 **Minas, Sierra de las** ⛰ E Guatemala
43 T14 **Minatitlán** Veracruz-Llave, E Mexico 17.58N 94.31W
177 Ff6 **Minbu** Magwe, W Myanmar 20.09N 94.52E
155 V10 **Minchinābād** Punjab, E Pakistan 30.10N 73.40E
65 G17 **Minchinmávida, Volcán** ⚱ S Chile 42.51S 72.23W
98 C5 **Minch, The** var. North Minch. strait NW Scotland, UK
108 F8 **Mincio** anc. Mincius. ♒ N Italy
28 M11 **Minco** Oklahoma, C USA 35.18N 97.56W
179 Rr16 **Mindanao** island S Philippines
Mindanao Sea see Bohol Sea
103 J23 **Mindel** ♒ S Germany
103 J23 **Mindelheim** Bayern, S Germany 48.03N 10.29E
Mindello see Mindelo
78 C9 **Mindelo** var. Mindello; prev. Porto Grande. São Vicente, N Cape Verde 16.54N 25.01W
12 I13 **Minden** Ontario, SE Canada 44.54N 78.41W
102 H13 **Minden** anc. Minthun. Nordrhein-Westfalen, NW Germany 52.18N 8.55E
24 G7 **Minden** Louisiana, S USA 32.37N 93.17W
31 O16 **Minden** Nebraska, C USA 40.30N 98.57W
37 O6 **Minden** Nevada, W USA 38.58N 119.46W
190 I4 **Mindona Lake** seasonal lake New South Wales, SE Australia
179 Pp12 **Mindoro** island N Philippines
179 P12 **Mindoro Strait** strait W Philippines
165 S9 **Mine** Gansu, N China
170 Dd12 **Mine** Yamaguchi, Honshū, SW Japan 34.10N 131.12E
99 E21 **Mine Head** Ir. Mionn Ard. headland S Ireland 51.58N 7.36W
99 J23 **Minehead** SW England, UK 51.13N 3.28W
61 I16 **Minéiros** Goiás, C Brazil 17.34S 52.33W
17 V6 **Mineola** New York, NE USA 33.39N 95.29W
27 S13 **Mineral** Texas, SW USA 28.32N 97.54W
131 N15 **Mineral'nye Vody** Stavropol'skiy Kray, SW Russian Federation 44.13N 43.06E
32 K8 **Mineral Point** Wisconsin, N USA 42.54N 90.09W
27 S5 **Mineral Wells** Texas, SW USA 32.48N 98.06W
38 M5 **Minersville** Utah, W USA 38.12N 112.56W
33 U12 **Minerva** Ohio, N USA 40.43N 81.06W
109 N17 **Minervino Murge** Puglia, SE Italy 41.06N 16.04E

105 O16 **Minervois** physical region S France
164 I10 **Minfeng** var. Niya. Xinjiang Uygur Zizhiqu, NW China 37.07N 82.43E
81 O25 **Minga** Katanga, SE Dem. Rep. Congo (Zaire) 11.06S 27.57E
143 W11 **Mingäçevir** var. Mingechaur, Mingechevir. C Azerbaijan 40.46N 47.02E
143 W11 **Mingäçevir Su Anbarı** Rus. Mingechaurskoye Vodokhranilishche, Mingechevirskoye Vodokhranilishche. ⊠ NW Azerbaijan
9 U17 **Minton** Saskatchewan, S Canada 49.12N 104.33W
Mingaora see Mingora
177 G8 **Mingaladon** ✈ (Yangon) Yangon, SW Myanmar 16.55N 96.11E
11 P11 **Mingan** Quebec, E Canada 50.19N 64.01W
155 U5 **Mingāora** var. Mingora, Mingaora. North-West Frontier Province, N Pakistan 34.46N 72.22E
152 K9 **Mingbuloq** Rus. Mynbulak. Nawoiy Wiloyati, N Uzbekistan 42.18N 62.53E
152 K9 **Mingbuloq Botighi** Rus. Vpadina Mynbulak. depression N Uzbekistan
Mingechaur/Mingechevir see Mingäçevir
Mingechaurskoye Vodokhranilishche/Mingechevirskoye Vodokhranilishche see Mingäçevir Su Anbarı
167 Q7 **Mingguang** prev. Jiashan. Anhui, S China 32.45N 117.58E
177 Ff4 **Mingin** Sagaing, C Myanmar 22.51N 94.30E
107 Q10 **Minglanilla** Castilla-La Mancha, C Spain 39.31N 1.36W
33 W12 **Mingo Junction** Ohio, N USA 40.19N 80.36W
Mingora see Mingāora
169 V7 **Mingshui** Heilongjiang, NE China 47.10N 125.52E
Mingteke Daban see Mintaka Pass
85 Q14 **Mínguri** Nampula, NE Mozambique 14.30S 40.37E
165 U10 **Minhe** var. Shangchuankou. Qinghai, C China 36.23N 102.40E
177 Ff6 **Minhla** Magwe, W Myanmar 19.57N 94.58E
178 J15 **Minh Lương** Kiên Giang, S Vietnam 9.52N 105.10E
106 G6 **Minho, Rio** Sp. Miño. ♒ Portugal/Spain see also Miño
106 G5 **Minho** former province N Portugal
161 G24 **Minicoy Island** island SW India
35 P15 **Minidoka** Idaho, NW USA 42.45N 113.29W
120 C11 **Minija** ♒ W Lithuania
188 G9 **Minilya** Western Australia 23.45S 114.03E
47 T12 **Minitte Point** headland S Saint Lucia 13.42N 60.57W
9 V15 **Minitonas** Manitoba, S Canada 52.07N 101.02W
Minius see Miño
167 R12 **Min Jiang** ♒ SE China
166 H10 **Min Jiang** ♒ C China
190 H9 **Minlaton** South Australia 34.52S 137.33E
172 N8 **Minmaya** var. Mimmaya. Aomori, Honshū, C Japan 41.10N 140.24E
79 U16 **Minna** Niger, C Nigeria 9.33N 6.33E
172 Pp16 **Minna-jima** island Sakishima-shotō, SW Japan
29 N4 **Minneapolis** Kansas, C USA 39.07N 97.42W
31 V8 **Minneapolis** Minnesota, N USA 44.58N 93.15W
31 V8 **Minneapolis-Saint Paul** ✈ Minnesota, N USA 44.53N 93.13W
15 Kk15 **Minnedosa** Manitoba, S Canada 50.13N 99.49W
28 J7 **Minneola** Kansas, C USA 37.26N 100.00W
31 S7 **Minnesota** off. State of Minnesota; also known as Gopher State, New England of the West, North Star State. ✧ state N USA
31 U8 **Minnesota River** ♒ Minnesota/South Dakota, N USA
31 V9 **Minnetonka** Minnesota, N USA 44.55N 93.28W
31 Q4 **Minnewaukan** North Dakota, N USA 48.05N 99.14W
190 F2 **Minnipa** South Australia 32.52S 135.07E
106 G5 **Miño** Galicia, NW Spain 43.21N 8.12W
106 G5 **Miño** var. Mino, Minius, Port. Rio Minho. ♒ Portugal/Spain see also Minho, Rio
171 Ei6 **Minobu** Yamanashi, Honshū, S Japan 35.22N 138.30E
32 L4 **Minocqua** Wisconsin, N USA 45.53N 89.42W
171 Ii10 **Minokamo** Gifu, Honshū, SW Japan 35.24N 136.57E
32 L12 **Minonk** Illinois, N USA 40.54N 89.01W
Minorca see Menorca
30 M3 **Minot** North Dakota, N USA 48.15N 101.19W
165 U8 **Minqin** Gansu, N China 38.35N 103.07E
121 J16 **Minsk** ● (Belarus) Minskaya Voblasts', C Belarus 53.52N 27.34E
121 J16 **Minsk** ✈ Minskaya Voblasts', C Belarus 53.52N 27.58E
Minskaya Oblast' see Minskaya Voblasts'
121 J16 **Minskaya Voblasts'** prev. Rus. Minskaya Oblast'. ✧ province C Belarus
121 J16 **Minskaya Wzvyshsha** ⛰ C Belarus
112 N12 **Mińsk Mazowiecki** var. Nowo-Mińsk. Mazowieckie, C Poland 52.11N 21.33E
33 Q13 **Minster** Ohio, N USA 40.23N 84.22W
81 F15 **Minta** Centre, C Cameroon 4.34N 12.54E

155 W2 **Mintaka Pass** Chin. Mingteke Daban. pass China/Pakistan 36.59N 75.04E
117 D20 **Mínthi** ⚠ S Greece
Minthun see Minden
11 C14 **Minto** New Brunswick, SE Canada 46.04N 66.05W
8 H6 **Minto** Yukon Territory, W Canada 62.33N 136.45W
41 Q5 **Minto** Alaska, USA 65.07N 149.22W
31 Q3 **Minto** North Dakota, N USA 48.17N 97.22W
10 K6 **Minto, Lac** ⊜ Quebec, C Canada 57.12S 104.33W
205 R16 **Minto, Mount** ⚠ Antarctica 71.38S 169.11E
201 R15 **Minto Reef** atoll Caroline Islands, C Micronesia
39 R4 **Minturn** Colorado, C USA 39.35N 106.21W
109 K16 **Minturno** Lazio, C Italy 41.15N 13.47E
126 Hh15 **Minusinsk** Krasnoyarskiy Kray, S Russian Federation 53.37N 91.49E
110 G11 **Minusio** Ticino, S Switzerland 46.11N 8.47E
81 E17 **Minvoul** Woleu-Ntem, N Gabon 2.07N 12.12E
147 N13 **Minwakh** N Yemen 16.54N 48.04E
165 V11 **Minxian** var. Min Xian. Gansu, C China 34.22N 104.02E
Minya see El Minya
33 R6 **Mio** Michigan, N USA 44.40N 84.09W
Mionn Ard see Mine Head
176 W9 **Mios Num, Selat** strait Irian Jaya, E Indonesia
177 Ff4 **Mingin** Sagaing, C Myanmar
164 L5 **Miquan** Xinjiang Uygur Zizhiqu, NW China 44.04N 87.40E
121 I17 **Mir** Hrodzyenskaya Voblasts', W Belarus 53.25N 26.28E
108 H8 **Mira** Veneto, NE Italy 45.25N 12.07E
106 G13 **Mira, Rio** ♒ S Portugal
10 K15 **Mira** ✈ (Montréal) Quebec, SE Canada 71.38S 169.11E
62 Q8 **Miracema** Rio de Janeiro, SE Brazil 21.24S 42.10W
56 Q9 **Miraflores** Boyacá, C Colombia 5.07N 73.09W
42 G10 **Miraflores** Baja California Sur, W Mexico 23.24N 109.45W
46 L9 **Miragoâne** S Haiti 18.25N 73.07W
161 E16 **Miraj** Mahārāshtra, W India 16.51N 74.42E
63 E23 **Miramar** Buenos Aires, E Argentina 38.15S 57.50W
105 R15 **Miramas** Bouches-du-Rhône, SE France 43.34N 4.58E
104 K12 **Mirambeau** Charente-Maritime, W France 45.23N 0.33W
104 L13 **Miramont-de-Guyenne** Lot-et-Garonne, SW France 44.34N 0.20E
117 L25 **Mirampéllou Kólpos** gulf Kriti, Greece, E Mediterranean Sea
164 L3 **Miran** Xinjiang Uygur Zizhiqu, NW China 39.13N 88.58E
56 M5 **Miranda** off. Estado Miranda. ✧ state N Venezuela
Miranda de Corvo see Miranda do Corvo
107 O3 **Miranda de Ebro** La Rioja, N Spain 42.40N 2.57W
106 G8 **Miranda do Corvo** var. Miranda de Corvo. Coimbra, N Portugal 40.04N 8.19W
106 J6 **Miranda do Douro** Bragança, N Portugal 41.30N 6.16W
104 L15 **Mirande** Gers, S France 43.31N 0.25E
106 I6 **Mirandela** Bragança, N Portugal 41.28N 7.10W
27 R15 **Mirando City** Texas, SW USA 27.24N 99.00W
108 G9 **Mirandola** Emilia-Romagna, N Italy 44.52N 11.04E
62 K13 **Mirandópolis** São Paulo, S Brazil 21.10S 51.03W
62 K8 **Mirassol** São Paulo, S Brazil 20.50S 49.30W
106 J3 **Miravalles** ⚠ NW Spain 42.52N 6.45W
44 L12 **Miravalles, Volcán** ⚱ NW Costa Rica 10.43S 85.07W
147 W13 **Mirbāţ** var. Marbat. S Oman 17.03N 54.44E
46 M9 **Mirebalais** C Haiti 18.46N 72.03W
105 T6 **Mirecourt** Vosges, NE France 48.19N 6.04E
105 N16 **Mirepoix** Ariège, S France 44.55N 93.28W
Mirgorod see Myrhorod
145 W10 **Mīr Ḩājī Khalīl** Iraq 32.11N 46.19E
27 V11 **Missouri City** Texas, SW USA 29.37N 95.32W
174 Mm4 **Miri** Sarawak, East Malaysia 4.22N 113.58E
79 W2 **Miria** Zinder, S Niger 13.39N 9.15E
190 F3 **Mirikata** South Australia 29.56S 135.13E
56 K4 **Mirimire** Falcón, N Venezuela 11.07N 68.36W
63 I16 **Mirim Lagoon** var. Lake Mirim, Sp. Laguna Merín. lagoon Brazil/Uruguay
Mirim, Lake see Mirim Lagoon
Mírina see Mýrina
180 H14 **Miringoni** Mohéli, S Comoros 12.16S 43.39E
149 W11 **Mīrjāveh** Sīstān va Balūchestān, SE Iran 29.01N 61.23E
205 X2 **Mirny** Russian research station Antarctica 66.25S 93.09E
126 Kk11 **Mirnyy** Respublika Sakha (Yakutiya), NE Russian Federation 62.30N 113.58E
113 H22 **Mironovka** see Myronivka
112 F9 **Mirosławiec** Zachodniopomorskie, NW Poland 53.21N 16.04E
131 V5 **Mirovo** see Vrattsa
102 L11 **Mirow** Mecklenburg-Vorpommern, N Germany 53.16N 12.48E
158 F6 **Mirpur** Jammu and Kashmir, NW India 33.10N 73.49E
Mirpur see New Mirpur
112 N12 **Mirpur Batoro** Sind, SE Pakistan 24.40N 68.15E
155 P17 **Mírpur Khás** Sind, SE Pakistan 25.33N 69.00E
155 P17 **Mírpur Sakro** Sind, SE Pakistan 24.31N 67.37E

149 T14 **Mīr Shahdād** Hormozgān, S Iran 26.15N 58.28E
Mirtoan Sea see Mirtóo Pélagos
117 G21 **Mirtóo Pélagos** Eng. Mirtoan Sea; anc. Myrtoum Mare. sea S Greece
169 Z16 **Miryang** var. Milyang, Eng. Mirinae; Jap. Mitsuō. SE South Korea 35.30N 128.46E
170 Di4 **Misaki** Ehime, Shikoku, SW Japan 33.22N 132.04E
170 Lac **Misawa** Aomori, Honshū, C Japan 40.41N 141.22E
59 Q14 **Misha** Heilongjiang, NE China 45.30N 131.53E
33 O11 **Mishawaka** Indiana, N USA 41.40N 86.10W
171 Ji17 **Mishima** var. Misima. Shizuoka, Honshū, S Japan 35.07N 138.55E
170 Dd11 **Mi-shima** island SW Japan
131 V4 **Mishkino** Respublika Bashkortostan, W Russian Federation 55.31N 55.57E
159 V6 **Mishmi Hills** hill range NE India
167 N11 **Mi Shui** ♒ S China
84 **Misiaf** see Maşyāf
107 S4 **Misilmeri** Sicilia, Italy, C Mediterranean Sea 38.03N 13.27E
Misima see Mishima
195 P17 **Misima Island** island SE PNG
Misión de Guana see Guana
62 F13 **Misiones** off. Provincia de Misiones. ✧ province NE Argentina
64 P8 **Misiones** off. department de las Misiones. ✧ department de S Paraguay
Misión San Fernando see San Fernando
Miskin see Maskin
Miskito Coast see La Mosquitia
44 M6 **Miskitos, Cayos** island group NE Nicaragua
113 M21 **Miskolc** Borsod-Abaúj-Zemplén, NE Hungary 48.04N 20.46E
175 T10 **Misoöl, Pulau** island Maluku, E Indonesia
Misox see Mesocco
180 H12 **Mitsamiouli** Grande Comore, NW Comoros 11.22S 43.19E
180 I3 **Mitsinjo** Mahajanga, NW Madagascar 16.00S 45.52E
180 H12 **Mitsoudjé** Grande Comore, NW Comoros
Mitspe Ramon see Mizpé Ramon
172 Oo7 **Mitsuishi** Hokkaidō, NE Japan 42.12N 142.40E
171 K13 **Mitsuke** var. Mituke. Niigata, Honshū, C Japan 37.33N 138.57E
Mitsuō see Miryang
170 Cc10 **Mitsushima** Nagasaki, Tsushima, SW Japan 34.16N 129.16E
102 J12 **Mittelandkanal** canal NW Germany
110 I7 **Mittelberg** Vorarlberg, NW Austria 47.19N 10.09E
Mitteldorf see Międzychód
Mittelstadt see Baia Sprie
Mitterburg see Pazin
111 P7 **Mittersill** Salzburg, NW Austria 47.16N 12.27E
103 N16 **Mittweida** Sachsen, E Germany 50.59N 12.57E
56 I13 **Mitú** Vaupés, SE Colombia 1.07N 70.04W
81 N23 **Mitumba, Monts** var. Chaîne des Mitumba, Mitumba Range. ⛰ E Dem. Rep. Congo (Zaire)
81 N23 **Mitumba, Chaine des/Mitumba Range** see Mitumba, Monts
81 O22 **Mitwaba** Katanga, SE Dem. Rep. Congo (Zaire) 8.37S 27.19E
81 E18 **Mitzic** Woleu-Ntem, N Gabon 0.48N 11.30E
84 K11 **Miueru Wantipa, Lake** ⊜ N Zambia
171 Ji17 **Miura** Kanagawa, Honshū, S Japan 35.07N 139.37E
171 M13 **Miyagi** off. Miyagi-ken. ♦ prefecture Honshū, C Japan
144 M7 **Miyāh, Wādī al** dry watercourse E Syria
172 Si3 **Miyake** Tōkyō, Miyako-jima, SE Japan 34.34N 135.33E
172 N11 **Miyako** Iwate, Honshū, C Japan 39.39N 141.57E
172 Q16 **Miyako-jima** island Sakishima-shotō, SW Japan
170 C16 **Miyakonojō** var. Miyakonzyō. Miyazaki, Kyūshū, SW Japan 31.42N 131.03E
Miyakonzyō see Miyakonojō
172 Pp16 **Miyako-shotō** island group SW Japan
150 H4 **Miyaly** Atyrau, W Kazakhstan 48.52S 53.55E
170 C16 **Miyazaki** Kagoshima, Kyūshū, SW Japan 31.55N 130.29E
170 Cc16 **Miyazaki** Miyazaki, Kyūshū, SW Japan 31.55N 131.23E
170 C15 **Miyazaki** off. Miyazaki-ken. ♦ prefecture Kyūshū, SW Japan
171 H13 **Miyazu** Kyōto, Honshū, SW Japan 35.28N 135.21E
Miyory see Myory
170 F13 **Miyoshi** var. Miyosi. Hiroshima, Honshū, SW Japan 34.48N 132.51E
Miyosi see Miyoshi
Miza see Mizë
83 H17 **Mizan Teferi** Southern, S Ethiopia 6.57N 35.30E
Mizda see Mizdah
77 O8 **Mizdah** var. Mizda. NW Libya 31.25N 12.58E
115 K20 **Mizë** var. Miza. Fier, W Albania 40.58N 19.29E
99 A22 **Mizen Head** Ir. Carn Uí Néid. headland SW Ireland 51.26N 9.50W
118 I7 **Mizhhir"ya** Rus. Mezhgor'ye. Zakarpats'ka Oblast', W Ukraine 48.28N 23.31E

◆ COUNTRY   ◇ DEPENDENT TERRITORY   ◈ ADMINISTRATIVE REGION   ⚠ MOUNTAIN   ⚱ VOLCANO   ⊜ LAKE
● COUNTRY CAPITAL   ○ DEPENDENT TERRITORY CAPITAL   ✈ INTERNATIONAL AIRPORT   ⛰ MOUNTAIN RANGE   ♒ RIVER   ⊠ RESERVOIR

166 L4 **Mizhi** Shaanxi, C China 37.43N 110.13E
118 K13 **Mizil** Prahova, SE Romania 45.00N 26.29E
116 H7 **Miziya** Vratsa, NW Bulgaria 43.42N 23.52E
159 W15 **Mizo Hills** hill range E India
159 W15 **Mizoram** ◈ state NE India
144 F12 **Mizpé Ramon** var. Mitspe Ramon. Southern, S Israel 30.37N 34.46E
59 L19 **Mizque** Cochabamba, C Bolivia 17.58S 65.18W
59 M19 **Mizque, Río** ≈ C Bolivia
171 I15 **Mizunami** Gifu, Honshū, SW Japan 35.19N 137.12E
171 Mm12 **Mizusawa** Iwate, Honshū, C Japan 39.09N 141.07E
97 M18 **Mjölby** Östergötland, S Sweden 58.19N 15.10E
97 G15 **Mjøndalen** Buskerud, S Norway 59.45N 9.58E
97 J19 **Mjörn** ◎ S Sweden
96 I13 **Mjøsa** var. Mjøsen. ◎ S Norway
Mjøsen see Mjøsa
83 G21 **Mkalama** Singida, C Tanzania 4.09S 34.34E
82 K13 **Mkata** ≈ C Tanzania
85 K14 **Mkushi** Central, C Zambia 13.37S 29.27E
85 L22 **Mkuze** KwaZulu/Natal, E South Africa 27.40S 32.05E
83 J22 **Mkwaja** Tanga, E Tanzania 5.42S 38.48E
113 D16 **Mladá Boleslav** Ger. Jungbunzlau. Středočeský Kraj, N Czech Republic 50.24N 14.55E
114 M12 **Mladenovac** Serbia, C Yugoslavia 44.27N 20.42E
116 L11 **Mladinovo** Khaskovo, S Bulgaria 41.57N 26.13E
115 O17 **Mlado Nagoričane** N FYR Macedonia 42.11N 21.49E
Mlanje see Mulanje
114 N12 **Mlava** ≈ E Yugoslavia
112 L9 **Mława** Mazowieckie, C Poland 53.07N 20.23E
115 G16 **Mljet** It. Meleda; anc. Melita. island S Croatia
118 K4 **Mlyniv** Rivnens'ka Oblast', NW Ukraine 50.31N 25.36E
85 I21 **Mmabatho** North-West, N South Africa 25.51S 25.37E
85 I19 **Mmashoro** Central, E Botswana 21.56S 26.39E
46 J7 **Moa** Holguín, E Cuba 20.38N 74.36W
78 J15 **Moa** ≈ Guinea/Sierra Leone
39 O6 **Moab** Utah, W USA 38.34N 109.34W
189 V1 **Moa Island** island Queensland, NE Australia
197 J15 **Moala** island S Fiji
85 L21 **Moamba** Maputo, SW Mozambique 25.33S 32.15E
81 F19 **Moanda** var. Mouanda. Haut-Ogooué, SE Gabon 1.31S 13.07E
175 T16 **Moa, Pulau** island Kepulauan Leti, E Indonesia
85 M15 **Moatize** Tete, NW Mozambique 16.03S 33.49E
81 P22 **Moba** Katanga, E Dem. Rep. Congo (Zaire) 7.03S 29.51E
171 K17 **Mobara** Chiba, Honshū, S Japan 35.25N 140.19E
Mobay see Montego Bay
81 K15 **Mobaye** Basse-Kotto, S Central African Republic 4.19N 21.17E
81 K15 **Mobayi-Mbongo** Equateur, NW Dem. Rep. Congo (Zaire) 4.19N 21.18E
27 P2 **Mobeetie** Texas, SW USA 35.33N 100.25W
29 U3 **Moberly** Missouri, C USA 39.25N 92.26W
25 N8 **Mobile** Alabama, S USA 30.41N 88.02W
25 N9 **Mobile Bay** bay Alabama, S USA
25 N8 **Mobile River** ≈ Alabama, S USA
31 N8 **Mobridge** South Dakota, N USA 45.32N 100.25W
Mobutu Sese Seko, Lac see Albert, Lake
47 N8 **Moca** N Dominican Republic 19.23N 70.31W
Moçâmedes see Namibe
178 J6 **Môc Châu** Sơn La, N Vietnam 20.52N 104.38E
197 L15 **Moce** island Lau Group, E Fij
85 Q15 **Moçambique** Nampula, NE Mozambique 15.00S 40.44E
Mocha see Al Mukhā
200 Oo13 **Mocha Fracture Zone** tectonic feature SE Pacific Ocean
65 F14 **Mocha, Isla** island C Chile
58 C12 **Moche, Río** ≈ W Peru
178 J14 **Môc Hoa** Long An, S Vietnam 10.46N 105.55E
85 I20 **Mochudi** Kgatleng, SE Botswana 24.25S 26.07E
84 Q13 **Mocímboa da Praia** var. Vila de Mocímboa da Praia. Cabo Delgado, N Mozambique 11.16S 40.21E
96 L13 **Mockfjärd** Dalarna, C Sweden 60.30N 14.57E
23 R9 **Mocksville** North Carolina, SE USA 35.53N 80.33W
34 F8 **Moclips** Washington, NW USA 47.11N 124.13W
84 C13 **Môco** var. Morro de Môco. ▲ W Angola 12.36S 15.09E
56 D13 **Mocoa** Putumayo, SW Colombia 1.07N 76.37W
62 M8 **Mococa** São Paulo, S Brazil 21.30S 47.00W
Môco, Morro de see Môco
42 H8 **Mocorito** Sinaloa, C Mexico 25.24N 107.55W
42 J4 **Moctezuma** Chihuahua, N Mexico 30.10N 106.24W
43 N11 **Moctezuma** San Luis Potosí, C Mexico 22.44N 101.04W
42 G4 **Moctezuma** Sonora, NW Mexico 29.49N 109.40W
43 P12 **Moctezuma, Río** ≈ C Mexico
Mó, Cuan see Clew Bay
85 O16 **Mocuba** Zambézia, NE Mozambique 16.49S 37.01E
105 U12 **Modane** Savoie, E France 45.14N 6.41E
108 F9 **Modena** anc. Mutina. Emilia-Romagna, N Italy 44.39N 10.55E
38 I7 **Modena** Utah, W USA 37.46N 113.54W
37 O9 **Modesto** California, W USA 37.38N 121.01W
109 L25 **Modica** anc. Motyca. Sicilia, Italy, C Mediterranean Sea 36.52N 14.45E
81 K17 **Modjamboli** Equateur, N Dem. Rep. Congo (Zaire) 2.27N 22.03E
111 X4 **Mödling** Niederösterreich, NE Austria 48.07N 16.15E
Modohn see Madona
169 N8 **Modot** Hentiy, C Mongolia 47.45N 109.03E
176 W12 **Modowi** Irian Jaya, E Indonesia 4.05S 134.39E
114 I12 **Modračko Jezero** ⊠ NE Bosnia and Herzegovina
114 I10 **Modriča** Republika Srpska, N Bosnia and Herzegovina 44.57N 18.17E
191 O13 **Moe** Victoria, SE Australia 38.10S 146.18E
Moearatewe see Muaratewe
Moei, Mae Nam see Thaungyin
96 H13 **Moelv** Hedmark, S Norway 60.55N 10.47E
94 I10 **Moen** Troms, N Norway 69.08N 18.35E
Moen see Weno, Micronesia
Möen see Møn, Denmark
Moena see Muna, Pulau
38 M10 **Moenkopi Wash** ≈ Arizona, SW USA
193 F22 **Moeraki Point** headland South Island, NZ 45.23S 170.52E
101 F16 **Moerbeke** Oost-Vlaanderen, NW Belgium 51.11N 3.57E
101 H14 **Moerdijk** Noord-Brabant, S Netherlands 51.42N 4.37E
Moero, Lac see Mweru, Lake
103 D15 **Moers** var. Mörs. Nordrhein-Westfalen, W Germany 51.27N 6.37E
Moesi see Musi, Air
Moeskroen see Mouscron
98 J13 **Moffat** S Scotland, UK 55.28N 3.36W
193 C22 **Moffat Peak** ▲ South Island, NZ 44.57S 168.10E
158 H8 **Moga** Punjab, N India 30.49N 75.13E
81 N19 **Moga** Sud Kivu, E Dem. Rep. Congo (Zaire) 2.16S 26.54E
Mogadiscio/Mogadishu see Muqdisho
Mogador see Essaouira
106 J6 **Mogadouro** Bragança, N Portugal 41.19N 6.43W
171 Ll12 **Mogami-gawa** ≈ Honshū, C Japan
178 Gg2 **Mogaung** Kachin State, N Myanmar 25.19N 96.54E
112 L13 **Mogielnica** Mazowieckie, C Poland 51.40N 20.42E
Mogilëv see Mahilyow
Mogilev-Podol'skiy see Mohyliv-Podil's'kyy
Mogilëvskaya Oblast' see Mahilyowskaya Voblasts'
112 I11 **Mogilno** Kujawsko-pomorskie, C Poland 52.39N 17.58E
62 L9 **Mogi-Mirim** var. Moji-Mirim. São Paulo, S Brazil 22.26S 46.55W
85 Q15 **Mogincual** Nampula, NE Mozambique 15.33S 40.28E
116 E13 **Moglenítsas** ≈ N Greece
108 H8 **Mogliano Veneto** Veneto, NE Italy 45.34N 12.13E
115 M21 **Moglicë** Korçë, SE Albania 40.43N 20.22E
126 L15 **Mogocha** Chitinskaya Oblast', S Russian Federation 53.39N 119.47E
126 H13 **Mogochin** Tomskaya Oblast', C Russian Federation 57.42N 83.24E
82 F13 **Mogogh** Jonglei, SE Sudan 8.25N 31.19E
176 Vv10 **Mogoi** Irian Jaya, E Indonesia 1.44S 133.13E
178 Gg4 **Mogok** Mandalay, C Myanmar 22.55N 96.28E
39 P14 **Mogollon Mountains** ▲▲ New Mexico, SW USA
38 M12 **Mogollon Rim** cliff Arizona, SW USA
63 E23 **Mogotes, Punta** headland E Argentina 38.03S 57.31W
44 J8 **Mogotón** ▲ NW Nicaragua 13.45N 86.22W
106 I14 **Moguer** Andalucía, S Spain 37.15N 6.52W
113 J26 **Mohács** Baranya, SW Hungary 46.00N 18.40E
193 C20 **Mohaka** ≈ North Island, NZ
30 M2 **Mohall** North Dakota, N USA 48.45N 101.30W
Moḥammadābād see Dargaz
76 F6 **Mohammedia** prev. Fédala. NW Morocco 33.46N 7.16W
76 F6 **Mohammed V** ✈ (Casablanca) W Morocco 33.07N 8.28W
Mohammerah see Khorramshahr
38 H10 **Mohave, Lake** ⊠ Arizona/Nevada, W USA
38 I12 **Mohave Mountains** ▲▲ Arizona, SW USA
38 I15 **Mohawk Mountains** ▲▲ Arizona, SW USA
20 J10 **Mohawk River** ≈ New York, NE USA
169 T3 **Mohe** Heilongjiang, NE China 53.00N 122.33E
97 L20 **Moheda** Kronoberg, S Sweden 57.00N 14.34E
180 H13 **Mohéli** var. Mwali, Mohilla, Mohila, Fr. Moili. island S Comoros
158 I11 **Mohendergarh** Haryāna, N India 28.16N 76.13E
40 K12 **Mohican, Cape** headland Nunivak Island, Alaska, USA 60.12N 167.25W
Mohn see Muhu
103 G15 **Möhne** ≈ W Germany
103 G15 **Möhne-Stausee** ⊠ W Germany
94 P2 **Mohn, Kapp** headland NW Svalbard 79.26N 25.44E
207 S14 **Mohns Ridge** undersea feature Greenland Sea/Norwegian Sea
59 I17 **Moho** Puno, SW Peru 15.21S 69.32W
Mohokare see Caledon
97 L17 **Moholm** Västra Götaland, S Sweden 58.37N 14.04E
38 J11 **Mohon Peak** ▲ Arizona, SW USA 34.55N 113.07W
83 J23 **Mohoro** Pwani, E Tanzania 8.09S 39.10E
Mohra see Moravice
Mohrungen see Morąg
118 M7 **Mohyliv-Podil's'kyy** Rus. Mogilev-Podol'skiy. Vinnyts'ka Oblast', C Ukraine 48.28N 27.49E
97 D17 **Moi** Rogaland, S Norway 58.27N 6.31E
197 I6 **Moindou** Province Sud, C New Caledonia 21.42S 165.40E
118 K11 **Moinești** Hung. Mojnest. Bacău, E Romania 46.27N 26.31E
Móinteach Mílic see Mountmellick
12 J14 **Moira** ≈ Ontario, SE Canada
94 G13 **Mo i Rana** Nordland, C Norway 66.19N 14.10E
159 X14 **Moirāng** Manipur, NE India 24.28N 93.45E
117 J25 **Moíres** Kríti, Greece, E Mediterranean Sea 35.03N 24.51E
120 H6 **Mõisaküla** Ger. Moiseküll. Viljandimaa, S Estonia 58.05N 25.11E
Moiseküll see Mõisaküla
13 W4 **Moisie** Quebec, E Canada 50.12N 66.06W
13 W3 **Moisie** ≈ Quebec, SE Canada
104 M14 **Moissac** Tarn-et-Garonne, S France 44.07N 1.04E
80 I13 **Moïssala** Moyen-Chari, S Chad 8.21N 17.46E
57 O7 **Moitaco** Bolívar, E Venezuela 8.00N 64.22W
97 P15 **Möja** Stockholm, C Sweden 59.25N 18.55E
107 Q14 **Mojácar** Andalucía, S Spain 37.09N 1.49W
37 T13 **Mojave** California, W USA 35.03N 118.10W
37 V13 **Mojave Desert** plain California, W USA
37 V13 **Mojave River** ≈ California, W USA
Moji-Mirim see Mogi-Mirim
115 K15 **Mojkovac** Montenegro, SW Yugoslavia 42.57N 19.34E
Mojnest see Moinești
174 Ll15 **Mojokerto** prev. Modjokerto. Jawa, C Indonesia 7.25S 112.31E
Mōka see Mooka
159 Q13 **Mokāma** prev. Mokameh, Mukama. Bihār, N India 25.24N 85.55E
81 O25 **Mokambo** Katanga, SE Dem. Rep. Congo (Zaire) 12.23S 28.21E
Mokameh see Mokāma
40 D9 **Mokapu Point** headland Oahu, Hawaii, USA, C Pacific Ocean 21.27N 157.43W
192 L9 **Mokau** Waikato, North Island, NZ 38.42S 174.37E
192 L9 **Mokau** ≈ North Island, NZ
37 P7 **Mokelumne River** ≈ California, W USA
85 J23 **Mokhotlong** NE Lesotho 29.19S 29.06E
Mokil Atoll see Mwokil Atoll
97 N14 **Möklinta** Västmanland, C Sweden 60.04N 16.34E
192 L4 **Mokohinau Islands** island group N NZ
159 X12 **Mokokchūng** Nāgāland, NE India 26.19N 94.30E
80 F12 **Mokolo** Extrême-Nord, N Cameroon 10.49N 13.54E
193 D24 **Mokoreta** ≈ South Island, NZ
169 X17 **Mokp'o** Jap. Moppo. SW South Korea 34.49N 126.26E
115 L16 **Mokra Gora** ▲▲ S Yugoslavia
Mokrany see Makrany
131 O5 **Moksha** ≈ W Russian Federation
Moktama see Martaban
79 T14 **Mokwa** Niger, W Nigeria 9.19N 5.01E
101 J16 **Mol** prev. Moll. Antwerpen, N Belgium 51.10N 5.07E
109 O17 **Mola di Bari** Puglia, SE Italy 41.03N 17.04E
Molai see Moláoi
43 P13 **Molango** Hidalgo, C Mexico 20.48N 98.43W
117 F22 **Moláoi** var. Molai. Pelopónnisos, S Greece 36.47N 22.50E
43 Z12 **Molas del Norte, Punta** var. Punta Molas. headland SE Mexico 20.34N 86.43W
Molas, Punta see Molas del Norte, Punta
107 R11 **Molatón** ▲ C Spain 38.58N 1.19W
99 K18 **Mold** NE Wales, UK 53.10N 3.07W
Moldau see Moldova
Moldau see Vltava, Czech Republic
Moldavia see Moldova
Moldavian SSR/Moldavskaya SSR see Moldova
96 E9 **Molde** Møre og Romsdal, S Norway 62.43N 7.07E
Moldotau, Khrebet see Moldo-Too, Khrebet
153 V9 **Moldo-Too, Khrebet** prev. Khrebet Moldotau. ▲▲ C Kyrgyzstan
118 K9 **Moldova** ≈ N Romania
118 K9 **Moldova** Eng. Moldavia, Ger. Moldau. former province NE Romania
118 L9 **Moldova** off. Republic of Moldova, var. Moldavia; prev. Moldavian SSR, Rus. Moldavskaya SSR. ◆ republic SE Europe
118 F13 **Moldova Nouă** Ger. Neumoldowa, Hung. Újmoldova. Caraș-Severin, SW Romania 44.45N 21.39E
118 F13 **Moldova Veche** Ger. Altmoldowa, Hung. Ómoldova. Caraș-Severin, SW Romania 44.45N 21.13E
Moldoveanul see Vârful Moldoveanu
85 I20 **Molepolole** Kweneng, SE Botswana 24.25S 25.30E
46 L8 **Môle-St-Nicolas** NW Haiti 19.46N 73.19W
120 H13 **Molėtai** Molėtai, E Lithuania 55.14N 25.25E
109 O17 **Molfetta** Puglia, SE Italy 41.12N 16.34E
175 R8 **Molibagu** Sulawesi, N Indonesia 0.25N 123.57E
64 G12 **Molina** Maule, C Chile 35.06S 71.18W
107 Q7 **Molina de Aragón** Castilla-La Mancha, C Spain 40.49N 1.54W
107 R13 **Molina de Segura** Murcia, SE Spain 38.03N 1.10W
32 J11 **Moline** Illinois, N USA 41.30N 90.31W
29 P7 **Moline** Kansas, C USA 37.21N 96.18W
81 P23 **Moliro** Katanga, SE Dem. Rep. Congo (Zaire) 8.10S 30.31E
109 K16 **Molise** ◈ region S Italy
97 K15 **Molkom** Värmland, C Sweden 59.36N 13.43E
Moll see Mol
111 Q9 **Möll** ≈ S Austria
97 J22 **Mölle** Skåne, S Sweden 56.15N 12.19E
59 H18 **Mollendo** Arequipa, SW Peru 17.01S 72.01W
107 U5 **Mollerussa** Cataluña, NE Spain 41.37N 0.52E
110 H8 **Mollis** Glarus, NE Switzerland 47.05N 9.03E
97 J19 **Mölndal** Västra Götaland, S Sweden 57.39N 12.05E
97 J19 **Mölnlycke** Västra Götaland, S Sweden 57.42N 12.19E
119 U9 **Molochans'k** Rus. Molochansk. Zaporiz'ka Oblast', SE Ukraine 47.10N 35.38E
119 U10 **Molochna** Rus. Molochnaya. ≈ S Ukraine
Molochnaya see Molochna
119 U10 **Molochnyy Lyman** bay N Black Sea
Molodechno/Molodeczno see Maladzyechna
205 V3 **Molodezhnaya** Russian research station Antarctica 67.33S 46.12E
128 J14 **Mologa** ≈ NW Russian Federation
40 E9 **Molokai** Haw. Moloka'i. island Hawaii, USA, C Pacific Ocean
183 X3 **Molokai Fracture Zone** tectonic feature NE Pacific Ocean
128 K15 **Molokovo** Tverskaya Oblast', W Russian Federation 58.10N 36.43E
129 Q14 **Moloma** ≈ NW Russian Federation
191 R8 **Molong** New South Wales, SE Australia 33.07S 148.52E
85 H21 **Molopo** seasonal river Botswana/South Africa
117 F17 **Mólos** Stereá Ellás, C Greece 38.48N 22.40E
175 Q7 **Molosipat** Sulawesi, N Indonesia 0.28N 121.08E
Molotov see Severodvinsk, Arkhangel'skaya Oblast', Russian Federation
Molotov see Perm', Permskaya Oblast', Russian Federation
81 G17 **Moloundou** Est, SE Cameroon 2.03N 15.13E
105 U5 **Molsheim** Bas-Rhin, NE France 48.33N 7.30E
15 L12 **Molson Lake** ◎ Manitoba, C Canada
Moluccas see Maluku
175 Rr8 **Molucca Sea** Ind. Laut Maluku. sea E Indonesia
Molukken see Maluku
85 O15 **Molumbo** Zambézia, N Mozambique 15.33S 36.19E
176 Uu14 **Molu, Pulau** island Maluku, E Indonesia
85 P16 **Moma** Nampula, NE Mozambique 16.42S 39.12E
127 N17 **Moma** ≈ NE Russian Federation
176 Xx13 **Momats** ≈ Irian Jaya, E Indonesia
44 J11 **Mombacho, Volcán** ⛰ SW Nicaragua 11.49N 85.58W
83 K21 **Mombasa** Coast, SE Kenya 4.04N 39.40E
83 J21 **Mombasa** ✈ Coast, SE Kenya 4.01S 39.31E
Mombetsu see Monbetsu
176 Y16 **Mombum** Irian Jaya, E Indonesia 8.16S 138.51E
116 J12 **Momchilgrad** prev. Mastanli. Kŭrdzhali, S Bulgaria 41.33N 25.25E
101 F23 **Momignies** Hainaut, S Belgium 50.02N 4.10E
56 E6 **Momil** Córdoba, NW Colombia 9.15N 75.40W
44 I10 **Momotombo, Volcán** ⛰ W Nicaragua 12.25N 86.33W
58 B5 **Mompiche, Ensenada de** bay NW Ecuador
81 K18 **Mompono** Equateur, NW Dem. Rep. Congo (Zaire) 0.11N 21.31E
56 F6 **Mompós** Bolívar, NW Colombia 9.10N 74.21W
97 J24 **Møn** prev. Möen. island SE Denmark
38 L4 **Mona** Utah, W USA 39.49N 111.52W
Mona, Canal de la see Mona Passage
98 E8 **Monach Islands** island group NW Scotland, UK
105 V14 **Monaco** var. Monaco-Ville; anc. Monoecus. ● (Monaco) S Monaco 43.46N 7.22E
105 V14 **Monaco** off. Principality of Monaco. ◆ monarchy W Europe
Monaco see München
Monaco Basin see Canary Basin
Monaco-Ville see Monaco
98 I9 **Monadhliath Mountains** ▲▲ N Scotland, UK
57 O6 **Monagas** off. Estado Monagas. ◈ state NE Venezuela
99 F16 **Monaghan** Ir. Muineachán. N Ireland 54.15N 6.58W
99 E16 **Monaghan** Ir. Muineachán. cultural region N Ireland
45 S16 **Monagrillo** Herrera, S Panama 7.58N 80.23W
26 L8 **Monahans** Texas, SW USA 31.33N 102.52W
47 Q9 **Mona, Isla** island W Puerto Rico
47 Q9 **Mona Passage** Sp. Canal de la Mona. channel Dominican Republic/Puerto Rico
45 O14 **Mona, Punta** headland E Costa Rica 9.44N 82.48W
161 K25 **Monaragala** Uva Province, SE Sri Lanka 6.52N 81.22E
35 S9 **Monarch** Montana, NW USA 47.04N 110.51W
14 Ff14 **Monarch Mountain** ▲ British Columbia, SW Canada 51.59N 125.56W
Monasterio see Monesterio
Monasterzyska see Monastyrys'ka
Monastir see Bitola
Monastyriska see Monastyrys'ka
119 O7 **Monastyryshche** Cherkas'ka Oblast', C Ukraine 48.59N 29.47E
118 J6 **Monastyrys'ka** Pol. Monasterzyska, Rus. Monastyriska. Ternopil's'ka Oblast', W Ukraine 49.04N 25.10E
81 E15 **Monatélé** Centre, SW Cameroon 4.16N 11.12E
172 Q4 **Monbetsu** var. Mombetsu, Monbetu. Hokkaidō, NE Japan 44.22N 143.22E
Monbetu see Monbetsu
108 B8 **Moncalieri** Piemonte, NW Italy 45.00N 7.41E
106 G4 **Monção** Viana do Castelo, N Portugal 42.03N 8.29W
107 Q5 **Moncayo** ▲ N Spain 41.43N 1.51W
107 Q5 **Moncayo, Sierra del** ▲▲ N Spain
128 J4 **Monchegorsk** Murmanskaya Oblast', NW Russian Federation 67.55N 32.46E
103 D15 **Mönchengladbach** prev. München-Gladbach. Nordrhein-Westfalen, W Germany 51.12N 6.25E
106 F14 **Monchique** Faro, S Portugal 37.19N 8.33W
106 G14 **Monchique, Serra de** ▲▲ S Portugal
23 S14 **Moncks Corner** South Carolina, SE USA 33.12N 80.00W
43 N7 **Monclova** Coahuila de Zaragoza, NE Mexico 26.55N 101.25W
Moncorvo see Torre de Moncorvo
11 P14 **Moncton** New Brunswick, SE Canada 46.04N 64.49W
106 F8 **Mondego, Cabo** headland N Portugal 40.10N 8.58W
106 G8 **Mondego, Rio** ≈ N Portugal
106 I2 **Mondoñedo** Galicia, NW Spain 43.25N 7.22W
101 N25 **Mondorf-les-Bains** Grevenmacher, SE Luxembourg 49.30N 6.16E
104 M7 **Mondoubleau** Loir-et-Cher, C France 48.00N 0.49E
32 J6 **Mondovi** Wisconsin, N USA 44.34N 91.40W
108 B9 **Mondovì** Piemonte, NW Italy 44.22N 7.55E
107 P3 **Mondragón** var. Arrasate. País Vasco, N Spain 43.04N 2.30W
109 J17 **Mondragone** Campania, S Italy 41.07N 13.52E
111 R5 **Mondsee** ◎ N Austria
126 J16 **Mondy** Respublika Buryatiya, S Russian Federation 51.41N 101.03E
117 G22 **Monemvasía** Pelopónnisos, S Greece 36.22N 23.03E
20 B15 **Monessen** Pennsylvania, NE USA 40.07N 79.51W
106 J12 **Monesterio** var. Monasterio. Extremadura, W Spain 38.04N 6.16W
12 L8 **Monet** Quebec, SE Canada 48.09N 75.37W
29 S8 **Monett** Missouri, C USA 36.55N 93.55W
29 X9 **Monette** Arkansas, C USA 35.53N 90.20W
12 G11 **Monetville** Ontario, S Canada 46.08N 80.24W
108 J7 **Monfalcone** Friuli-Venezia Giulia, NE Italy 45.49N 13.31E
106 H10 **Monforte** Portalegre, C Portugal 39.03N 7.25W
106 I4 **Monforte** Galicia, NW Spain 42.31N 7.30W
83 I24 **Monga** Lindi, SE Tanzania 9.05S 37.51E
81 L16 **Monga** Orientale, N Dem. Rep. Congo (Zaire) 4.12N 22.49E
83 F15 **Mongalla** Bahr el Gabel, S Sudan 5.12N 31.42E
159 U11 **Mongar** E Bhutan 27.16N 91.07E
178 K6 **Mong Cai** Quang Ninh, N Vietnam 21.33N 107.56E
188 I11 **Mongers Lake** salt lake Western Australia
195 U14 **Mongga** Kolombangara, NW Solomon Islands 7.51S 157.00E
178 Hh6 **Möng Hpayak** Shan State, E Myanmar 20.56N 100.00E
Monghyr see Munger
108 B10 **Mongioie** ▲ NW Italy 44.13N 7.46E
178 Gg5 **Möng Küng** Shan State, E Myanmar 21.39N 97.31E
Mongla see Mungla
196 C15 **Mongmong** C Guam
178 Gg6 **Möng Nai** Shan State, E Myanmar 20.28N 97.51E
80 I11 **Mongo** Guéra, C Chad 12.11N 18.39E
78 I14 **Mongo** ≈ N Sierra Leone
169 I8 **Mongolia** Mong. Mongol Uls. ◆ republic E Asia
133 V8 **Mongolia, Plateau of** plateau E Mongolia
Mongolküre see Zhaosu
Mongol Uls see Mongolia
81 E17 **Mongomo** E Equatorial Guinea 1.39N 11.18E
79 Y12 **Mongonu** var. Monguno. Borno, NE Nigeria 12.42N 13.37E
Mongora see Mingāora
80 K11 **Mongororo** Ouaddaï, SE Chad 12.03N 22.26E
Mongos, Chaîne des see Bongo, Massif des
81 I16 **Mongoumba** Lobaye, SW Central African Republic 3.39N 18.30E
Mongrove, Punta see Cayacal, Punta
85 G15 **Mongu** Western, W Zambia 15.13S 23.09E
78 I10 **Mônguel** Gorgol, SW Mauritania 16.25N 13.07W
Monguno see Mongonu
178 H4 **Möng Yai** Shan State, E Myanmar 22.25N 98.02E
178 Hh5 **Möng Yang** Shan State, E Myanmar 21.52N 99.31E
178 H3 **Möng Yu** Shan State, E Myanmar 24.00N 97.57E
168 K8 **Mönhbulag** Övörhangay, C Mongolia 46.48N 103.25E
Mönh Sarïdag see Munku-Sardyk, Gora
194 L15 **Moni** ≈ S Papau New Guinea
117 I15 **Moní Megístis Lávras** monastery Kentrikí Makedonía, N Greece 40.10N 24.22E
117 F18 **Moní Osíou Loúkas** monastery Stereá Ellás, C Greece 38.22N 22.42E
56 F9 **Moniquirá** Boyacá, C Colombia 5.57N 73.35W
105 Q12 **Monistrol-sur-Loire** Haute-Loire, C France 45.19N 4.12E
37 V7 **Monitor Range** ▲▲ Nevada, W USA
117 I14 **Moní Vatopedíou** monastery Kentrikí Makedonía, N Greece 40.19N 24.13E
179 Rr15 **Monkayo** Mindanao, S Philippines 7.45N 125.58E
Monkchester see Newcastle upon Tyne
85 N14 **Monkey Bay** Southern, SE Malawi 14.09S 34.53E
45 N11 **Monkey Point** var. Punta Mico, Punte Mono, Punto Mico. headland SE Nicaragua 11.37N 83.39W
Monkey River see Monkey River Town
44 G3 **Monkey River Town** var. Monkey River. Toledo, SE Belize 16.22N 88.28W
12 M13 **Monkland** Ontario, SE Canada 45.11N 74.51W
81 J19 **Monkoto** Equateur, NW Dem. Rep. Congo (Zaire) 1.35S 20.43E
99 K21 **Monmouth** Wel. Trefynwy. SE Wales, UK 51.49N 2.43W
32 J12 **Monmouth** Illinois, N USA 40.54N 90.39W
34 F12 **Monmouth** Oregon, NW USA 44.51N 123.13W
99 K21 **Monmouth** cultural region SE Wales, UK
100 I10 **Monnickendam** Noord-Holland, C Netherlands 52.28N 5.01E
79 R15 **Mono** ≈ C Togo
Monoecus see Monaco
37 R8 **Mono Lake** ◎ California, W USA
117 O23 **Monólithos** Ródos, Dodekánisos, Greece, Aegean Sea 36.08N 27.45E
21 Q12 **Monomoy Island** island Massachusetts, NE USA
33 O12 **Monon** Indiana, N USA 40.52N 86.54W
31 Y12 **Monona** Iowa, C USA 43.03N 91.23W
32 L9 **Monona** Wisconsin, N USA 43.03N 89.18W
20 B15 **Monongahela** Pennsylvania, NE USA 40.10N 79.54W
20 B16 **Monongahela River** ≈ NE USA
109 P17 **Monopoli** Puglia, SE Italy 40.57N 17.18E
Mono, Punte see Monkey Point
113 K23 **Monor** Pest, C Hungary 47.19N 19.28E
Monostor see Beli Manastir
80 K8 **Monou** Borkou-Ennedi-Tibesti, NE Chad 16.22N 22.15E
107 S12 **Monóvar** País Valenciano, E Spain 38.25N 0.49W
107 R7 **Monreal del Campo** Aragón, NE Spain 40.46N 1.19W
109 I23 **Monreale** Sicilia, Italy, C Mediterranean Sea 38.04N 13.16E
25 T3 **Monroe** Georgia, SE USA 33.47N 83.42W
31 W14 **Monroe** Iowa, C USA 41.31N 93.06W
24 I5 **Monroe** Louisiana, S USA 32.31N 92.06W
33 S10 **Monroe** Michigan, N USA 41.55N 83.24W
20 K13 **Monroe** New York, NE USA 41.18N 74.09W
23 S11 **Monroe** North Carolina, SE USA 34.59N 80.33W
38 L6 **Monroe** Utah, W USA 38.37N 112.07W
34 H7 **Monroe** Washington, NW USA 47.51N 121.58W
32 L9 **Monroe** Wisconsin, N USA 42.34N 89.39W
29 V3 **Monroe City** Missouri, C USA 39.39N 91.43W
33 O15 **Monroe Lake** ⊠ Indiana, N USA
25 O7 **Monroeville** Alabama, S USA 31.31N 87.19W
20 C15 **Monroeville** Pennsylvania, NE USA 40.24N 79.44W
78 J16 **Monrovia** ● (Liberia) W Liberia 6.18N 10.48W
78 J16 **Monrovia** ✈ W Liberia 6.22N 10.50W
107 T7 **Monroyo** Aragón, NE Spain 40.46N 0.03W
101 F20 **Mons** Dut. Bergen. Hainaut, S Belgium 50.28N 3.58E
106 I8 **Monsanto** Castelo Branco, C Portugal 40.01N 7.07W
108 H8 **Monselice** Veneto, NE Italy 45.15N 11.47E
178 Gg9 **Mon State** ◈ state S Myanmar
100 G12 **Monster** Zuid-Holland, W Netherlands 52.01N 4.10E
97 N20 **Mönsterås** Kalmar, S Sweden 57.03N 16.27E
103 F17 **Montabaur** Rheinland-Pfalz, W Germany 50.25N 7.48E
108 G8 **Montagnana** Veneto, NE Italy 45.14N 11.31E
37 N1 **Montague** California, W USA 41.43N 122.31W
27 S5 **Montague** Texas, SW USA 33.39N 97.41W
191 S11 **Montague Island** island New South Wales, SE Australia
41 S12 **Montague Island** island Alaska, USA
41 S13 **Montague Strait** strait N Gulf of Alaska
104 J8 **Montaigu** Vendée, NW France 46.58N 1.18W
Montaigu see Scherpenheuvel
107 S7 **Montalbán** Aragón, NE Spain 40.49N 0.48W
108 G13 **Montalcino** Toscana, C Italy 43.01N 11.34E
106 H5 **Montalegre** Vila Real, N Portugal 41.49N 7.48W
116 G8 **Montana** prev. Ferdinand, Mikhaylovgrad. Montana, NW Bulgaria 43.25N 23.14E
110 D10 **Montana** Valais, SW Switzerland 46.23N 7.29E
41 R11 **Montana** Alaska, USA 62.06N 150.03W
116 G8 **Montana** ◈ province NW Bulgaria
35 T9 **Montana** off. State of Montana; also known as Mountain State, Treasure State. ◈ state NW USA
106 J10 **Montánchez** Extremadura, W Spain 39.15N 6.07W
Montañita see La Montañita
13 Q8 **Mont-Apica** Quebec, SE Canada 47.57N 71.24W
106 G10 **Montargil** Portalegre, C Portugal 39.04N 8.10W
106 G10 **Montargil, Barragem de** ⊠ C Portugal
105 O7 **Montargis** Loiret, C France 48.00N 2.43E
105 O4 **Montataire** Oise, N France 49.16N 2.24E
104 M14 **Montauban** Tarn-et-Garonne, S France 44.01N 1.19E
21 N14 **Montauk** Long Island, New York, NE USA 41.01N 71.58W
21 N14 **Montauk Point** headland Long Island, New York, NE USA 41.04N 71.51W
105 Q7 **Montbard** Côte d'Or, C France 47.35N 4.25E
105 U7 **Montbéliard** Doubs, E France 47.31N 6.49E
27 W11 **Mont Belvieu** Texas, SW USA 29.51N 94.53W
107 U6 **Montblanc** var. Montblanch. Cataluña, NE Spain 41.22N 1.10E
Montblanch see Montblanc
105 Q11 **Montbrison** Loire, E France 45.37N 4.04E
Montcalm, Lake see Dogai Coring
105 Q9 **Montceau-les-Mines** Saône-et-Loire, C France 46.40N 4.19E
105 U12 **Mont Cenis, Col du** pass E France 45.16N 6.54E
104 K15 **Mont-de-Marsan** Landes, SW France 43.54N 0.30W
105 O3 **Montdidier** Somme, N France 49.39N 2.34E
197 J7 **Mont-Dore** Province Sud, S New Caledonia 22.18S 166.34E
22 K10 **Monteagle** Tennessee, S USA 35.15N 85.47W
59 M20 **Monteagudo** Chuquisaca, S Bolivia 19.48S 63.57W
43 R16 **Monte Albán** ruins Oaxaca, S Mexico 17.01N 96.46W
107 R11 **Montealegre del Castillo** Castilla-La Mancha, C Spain 38.48N 1.18W
61 N18 **Monte Azul** Minas Gerais, SE Brazil 15.13S 42.52W
12 M12 **Montebello** Quebec, SE Canada 45.40N 74.55W
108 H7 **Montebelluna** Veneto, NE Italy 45.46N 12.03E
62 G13 **Montecarlo** Misiones, NE Argentina 26.37S 54.45W
63 D16 **Monte Caseros** Corrientes, NE Argentina 30.15S 57.39W
62 J13 **Monte Castelo** Santa Catarina, S Brazil 26.34S 50.12W
108 F11 **Montecatini Terme** Toscana, C Italy 43.52N 10.46E
44 H7 **Montecillos, Cordillera de** ▲▲ W Honduras
64 I12 **Monte Comén** Mendoza, W Argentina 34.34S 67.52W
46 M8 **Monte Cristi** var. San Fernando de Monte Cristi. NW Dominican Republic 19.52N 71.39W
60 C13 **Monte Cristo** Amazonas, W Brazil 3.13S 68.00W
109 E14 **Montecristo, Isola di** island Archipelago Toscano, C Italy
Monte Croce Carnico, Passo di see Plöcken Pass
60 J12 **Monte Dourado** Pará, NE Brazil 0.48S 52.32W
42 L11 **Monte Escobedo** Zacatecas, C Mexico 22.19N 103.30W
108 I13 **Montefalco** Umbria, C Italy 42.54N 12.40E
109 H14 **Montefiascone** Lazio, C Italy 42.33N 12.01E
107 N14 **Montefrío** Andalucía, S Spain 37.19N 4.00W
46 I11 **Montego Bay** var. Mobay. W Jamaica 18.28N 77.55W
Montego Bay see Sangster
106 J8 **Montehermoso** Extremadura, W Spain 40.04N 6.19W
106 F10 **Montejunto, Serra de** ▲ C Portugal 39.10N 9.01W
Monteleone di Calabria see Vibo Valentia
56 E7 **Montelíbano** Córdoba, NW Colombia 7.58N 75.24W
105 R13 **Montélimar** anc. Acunum Acusio, Montilium Adhemari. Drôme, E France 44.33N 4.45E
106 K15 **Montellano** Andalucía, S Spain 37.00N 5.34W
37 Y2 **Montello** Nevada, W USA 41.18N 114.10W
32 L8 **Montello** Wisconsin, N USA 43.46N 89.19W
65 J18 **Montemayor, Meseta de** plain SE Argentina
43 O9 **Montemorelos** Nuevo León, NE Mexico 25.10N 99.51W
106 G11 **Montemor-o-Novo** Évora, S Portugal 38.37N 8.13W
106 G8 **Montemor-o-Velho** var. Montemor-o-Vélho. Coimbra, N Portugal 40.10N 8.40W
106 H7 **Montemuro, Serra de** ▲ N Portugal 40.59N 7.59W
104 K12 **Montendre** Charente-Maritime, W France 45.17N 0.24W
63 I15 **Montenegro** Rio Grande do Sul, S Brazil 29.40S 51.32W
115 J16 **Montenegro** Serb. Crna Gora. ◈ republic SW Yugoslavia

◆ Country
● Country Capital
◇ Dependent Territory
○ Dependent Territory Capital
◈ Administrative Region
✈ International Airport
▲ Mountain
▲▲ Mountain Range
⛰ Volcano
≈ River
◎ Lake
⊠ Reservoir

64 G10 **Monte Patria** Coquimbo, N Chile 30.40S 71.00W

47 O9 **Monte Plata** E Dominican Republic 18.46N 69.43W

85 P14 **Montepuez** Cabo Delgado, N Mozambique 13.11S 38.59E

85 P14 **Montepuez** ♨ N Mozambique

108 G13 **Montepulciano** Toscana, C Italy 43.02N 11.51E

64 L6 **Monte Quemado** Santiago del Estero, N Argentina 25.46S 62.51W

105 O6 **Montereau-Faut-Yonne** anc. Condate. Seine-St-Denis, N France 48.22N 2.57E

37 N11 **Monterey** California, W USA 36.36N 121.53W

22 L9 **Monterey** Tennessee, S USA 36.09N 85.16W

23 T5 **Monterey** Virginia, NE USA 38.24N 79.33W

Monterey see Monterrey

37 N10 **Monterey Bay** bay California, W USA

56 D6 **Montería** Córdoba, NW Colombia 8.45N 75.54W

59 N18 **Montero** Santa Cruz, C Bolivia 17.19S 63.15W

64 J7 **Monteros** Tucumán, C Argentina 27.12S 65.30W

106 I5 **Monterrei** Galicia, NW Spain 41.56N 7.27W

43 O8 **Monterrey** var. Monterey. Nuevo León, NE Mexico 25.40N 100.16W

34 F9 **Montesano** Washington, NW USA 46.58N 123.37W

109 M19 **Montesano sulla Marcellana** Campania, S Italy 40.15N 15.41E

109 N16 **Monte Sant' Angelo** Puglia, SE Italy 41.43N 15.58E

61 O16 **Monte Santo** Bahia, E Brazil 10.25S 39.18W

109 D18 **Monte Santu, Capo di** headland Sardegna, Italy, C Mediterranean Sea 40.05N 9.43E

61 M19 **Montes Claros** Minas Gerais, SE Brazil 16.45S 43.52W

109 K14 **Montesilvano Marina** Abruzzo, C Italy 42.28N 14.07E

25 P4 **Montevallo** Alabama, S USA 33.06N 86.51W

108 G12 **Montevarchi** Toscana, C Italy 43.31N 11.34E

31 S9 **Montevideo** Minnesota, N USA 44.56N 95.43W

63 F20 **Montevideo ●** (Uruguay) ● Uruguay S Uruguay 34.55S 56.10W

39 S7 **Monte Vista** Colorado, C USA 37.33N 106.08W

25 T5 **Montezuma** Georgia, SE USA 32.18N 84.01W

31 W14 **Montezuma** Iowa, C USA 41.35N 92.31W

28 J6 **Montezuma** Kansas, C USA 37.33N 100.25W

105 U12 **Montgenèvre, Col de** pass France/Italy 44.56N 6.45E

99 K20 **Montgomery** E Wales, UK 52.37N 3.05W

25 Q5 **Montgomery** state capital Alabama, S USA 32.22N 86.18W

31 V9 **Montgomery** Minnesota, N USA 44.26N 93.34W

20 G13 **Montgomery** Pennsylvania, NE USA 41.08N 76.52W

23 Q5 **Montgomery** West Virginia, NE USA 38.07N 81.19W

99 K19 **Montgomery** cultural region E Wales, UK

Montgomery see Sahiwal

29 V4 **Montgomery City** Missouri, C USA 38.58N 91.30W

37 S8 **Montgomery Pass** pass Nevada, W USA 37.57N 118.21W

104 K12 **Montguyon** Charente-Maritime, W France 45.12N 0.13W

110 C10 **Monthey** Valais, SW Switzerland 46.15N 6.55E

29 V13 **Monticello** Arkansas, C USA 33.37N 91.44W

25 T4 **Monticello** Florida, SE USA 30.33N 83.52W

25 S8 **Monticello** Georgia, SE USA 33.18N 83.40W

32 M13 **Monticello** Illinois, N USA 40.01N 88.34W

33 O12 **Monticello** Indiana, N USA 40.45N 86.46W

31 Y13 **Monticello** Iowa, C USA 42.14N 91.11W

22 L7 **Monticello** Kentucky, S USA 36.51N 84.51W

31 V8 **Monticello** Minnesota, N USA 45.19N 93.45W

24 K7 **Monticello** Mississippi, S USA 31.33N 90.06W

29 V2 **Monticello** Missouri, C USA 40.07N 91.42W

20 J12 **Monticello** New York, NE USA 41.39N 74.41W

39 O7 **Monticello** Utah, W USA 37.52N 109.20W

108 F8 **Montichiari** Lombardia, N Italy 45.24N 10.27E

104 M12 **Montignac** Dordogne, SW France 45.24N 0.54E

101 G22 **Montignies-le-Tilleul** var. Montigny-le-Tilleul. Hainaut, S Belgium 50.22N 4.22E

12 J8 **Montigny, Lac de** ☺ Quebec, SE Canada

105 S6 **Montigny-le-Roi** Haute-Marne, N France 48.02N 5.26E

Montigny-le-Tilleul see Montignies-le-Tilleul

45 R16 **Montijo** Veraguas, S Panama 7.58N 81.12W

106 I7 **Montijo** Setúbal, W Portugal 38.42N 8.58W

106 J11 **Montijo** Extremadura, W Spain 38.55N 6.37W

Montilium Adhemari see Montélimar

106 M13 **Montilla** Andalucía, S Spain 37.35N 4.38W

104 L3 **Montivilliers** Seine-Maritime, N France 49.31N 0.10E

13 O11 **Mont-Joli** Quebec, SE Canada 48.33N 68.12W

12 M10 **Mont-Laurier** Quebec, SE Canada 46.33N 75.31W

13 X5 **Mont-Louis** Quebec, SE Canada 49.15N 65.44W

105 N17 **Mont-Louis** var. Mont Louis. Pyrénées-Orientales, S France 42.30N 2.07E

105 O10 **Montluçon** Allier, C France 46.21N 2.37E

13 R9 **Montmagny** Quebec, SE Canada 47.00N 70.31W

105 S3 **Montmédy** Meuse, NE France 49.31N 5.21E

105 P5 **Montmirail** Marne, N France 48.53N 3.31E

13 R9 **Montmorency** ♨ Quebec, SE Canada

104 M10 **Montmorillon** Vienne, W France 46.25N 0.52E

109 J14 **Montorio al Vomano** Abruzzo, C Italy 42.31N 13.39E

106 M13 **Montoro** Andalucía, S Spain 38.00N 4.21W

35 S16 **Montpelier** Idaho, NW USA 42.19N 111.18W

31 P6 **Montpelier** North Dakota, N USA 46.40N 98.34W

20 M7 **Montpelier** state capital Vermont, NE USA 44.15N 72.32W

105 Q15 **Montpellier** Hérault, S France 43.37N 3.52E

104 L12 **Montpon-Ménestérol** Dordogne, SW France 45.01N 0.10E

12 L5 **Montreal** ♨ Ontario, S Canada

12 C8 **Montreal** ♨ Ontario, S Canada

Montreal see Mirabel

10 K15 **Montréal** Eng. Montreal. Quebec, SE Canada 45.30N 73.34W

9 T14 **Montreal Lake** ☺ Saskatchewan, C Canada

12 B9 **Montreal River** Ontario, S Canada 47.13N 84.36W

105 N2 **Montreuil** Pas-de-Calais, N France 50.28N 1.46E

104 K8 **Montreuil-Bellay** Maine-et-Loire, NW France 47.07N 0.10W

110 C10 **Montreux** Vaud, SW Switzerland 46.27N 6.55E

110 B9 **Montricher** Vaud, W Switzerland 46.37N 6.24E

98 K10 **Montrose** E Scotland, UK 56.43N 2.28W

29 W14 **Montrose** Arkansas, C USA 33.18N 91.29W

39 Q6 **Montrose** Colorado, C USA 38.28N 107.52W

31 Y16 **Montrose** Iowa, C USA 40.31N 91.24W

20 H12 **Montrose** Pennsylvania, NE USA 41.49N 75.52W

23 X5 **Montross** Virginia, NE USA 38.04N 76.50W

13 O12 **Mont-St-Hilaire** Quebec, SE Canada 45.34N 73.10W

105 S3 **Mont-St-Martin** Meurthe-et-Moselle, NE France 49.31N 5.51E

47 V10 **Montserrat** var. Emerald Isle. ◇ UK dependent territory E West Indies

107 V5 **Montserrat** ▲ NE Spain 41.39N 1.44E

106 M7 **Montuenga** Castilla-León, N Spain 41.04N 4.37W

101 M19 **Montzen** Liège, E Belgium 50.42N 5.59E

39 N8 **Monument Valley** valley Arizona/Utah, SW USA

177 G4 **Monywa** Sagaing, C Myanmar 22.04N 95.12E

108 D7 **Monza** Lombardia, N Italy 45.34N 9.16E

85 J15 **Monze** Southern, S Zambia 16.15S 27.29E

107 T5 **Monzón** Aragón, NE Spain 41.54N 0.12E

27 T9 **Moody** Texas, SW USA 31.18N 97.21W

100 L13 **Mook** Limburg, SE Netherlands 51.45N 5.52E

110 C8 **Mont, Lac de** Ger. Murtensee. ☺ W Switzerland

86 I11 **Morava** var. March. ♨ C Europe see also March

Morava see Moravia, Czech Republic

Morava see Velika Morava, Yugoslavia

31 W15 **Moravia** Iowa, C USA 40.53N 92.49W

113 F18 **Moravia** Cz. Morava, Ger. Mähren. cultural region E Czech Republic

113 H17 **Moravice** Ger. Mohra. ♨ NE Czech Republic

116 J12 **Moraviţa** Timiş, SW Romania 45.15N 21.17E

113 G17 **Moravská Třebová** Ger. Mährisch-Trübau. Pardubický Kraj, C Czech Republic 49.45N 16.40E

113 E19 **Moravské Budějovice** Ger. Mährisch-Budwitz. Jihlavský Kraj, C Czech Republic 49.03N 15.48E

113 F19 **Moravský Krumlov** Ger. Mährisch-Kromau. Brněnský Kraj, SE Czech Republic 48.58N 16.30E

98 J8 **Moray** cultural region N Scotland, UK

98 J8 **Moray Firth** N Scotland, UK

44 B10 **Morazán** ◇ department NE El Salvador

160 C10 **Morbi** Gujarāt, W India 22.51N 70.49E

105 I7 **Morbihan** ◇ department NW France

Mörbisch am See see Mörbisch.

111 Y5 **Mörbisch am See** var. Mörbisch. Burgenland, E Austria 47.43N 16.40E

97 N21 **Mörbylånga** Kalmar, S Sweden 56.30N 16.23E

104 J14 **Morcenx** Landes, SW France 44.04N 0.55W

Morchen Khort see Mürcheh Khvort

15 C19 **Morden** Manitoba, S Canada 49.12N 98.04W

200 O14 **Mornington Abyssal Plain** undersea feature SE Pacific Ocean

189 T4 **Mornington, Isla** island S Chile

189 T4 **Mornington Island** island Wellesley Islands, Queensland, N Australia

21 Q4 **Moosehead Lake** ☺ Maine, NE USA

9 U16 **Moose Jaw** Saskatchewan, S Canada 50.25N 105.29W

9 V14 **Moose Lake** Manitoba, C Canada 53.42N 100.22W

31 W6 **Moose Lake** Minnesota, N USA 46.28N 92.46W

21 P6 **Mooselookmeguntic Lake** ☺ Maine, NE USA

41 R12 **Moose Pass** Alaska, USA 60.28N 149.21W

21 P5 **Moose River** ♨ Maine, NE USA

20 J9 **Moose River** ♨ New York, NE USA

9 V16 **Moosomin** Saskatchewan, S Canada 50.09N 101.40W

10 H10 **Moosonee** Ontario, SE Canada 51.18N 80.40W

21 N12 **Moosup** Connecticut, NE USA 41.42N 71.51W

85 N16 **Mopeia** Zambézia, NE Mozambique 17.58S 35.43E

85 H18 **Mopipi** Central, C Botswana 21.05S 24.54E

Moppo see Mokp'o

79 N15 **Mopti** Mopti, C Mali 14.30N 4.15W

79 O11 **Mopti** ◇ region S Mali

59 H18 **Moquegua** Moquegua, SE Peru 17.12S 70.55W

59 H18 **Moquegua** off. Departamento de Moquegua. ◇ department S Peru

113 J23 **Mór** Ger. Moor. Fejér, C Hungary 47.21N 18.13E

80 G11 **Mora** Extrême-Nord, N Cameroon 11.01N 14.07E

106 G11 **Mora** Évora, S Portugal 38.55N 8.10W

107 N9 **Mora** Castilla-La Mancha, C Spain 39.40N 3.46W

96 L12 **Mora** Dalarna, C Sweden 61.00N 14.30E

31 V7 **Mora** Minnesota, N USA 45.52N 93.18W

39 T10 **Mora** New Mexico, SW USA 35.56N 105.16W

158 K10 **Morādābād** Uttar Pradesh, N India 28.49N 78.45E

107 N9 **Móra d'Ebre** var. Mora de Ebro. Cataluña, NE Spain 41.04N 0.37E

Mora de Ebro see Móra d'Ebre

107 S8 **Mora de Rubielos** Aragón, NE Spain 40.14N 0.45W

112 K8 **Morag** Ger. Mohrungen. Warmińsko-Mazurskie, NE Poland 53.55N 19.55E

113 L25 **Mórahalom** Csongrád, S Hungary 46.13N 19.51E

107 N11 **Moral de Calatrava** Castilla-La Mancha, C Spain 38.49N 3.34W

65 G19 **Moraleda, Canal** strait SE Pacific Ocean

56 J3 **Morales** Bolívar, N Colombia 8.16N 73.52W

56 D12 **Morales** Cauca, SW Colombia 2.43N 76.36W

44 F5 **Morales** Izabal, E Guatemala 15.29N 88.46W

180 J5 **Moramanga** Toamasina, E Madagascar 18.57S 48.13E

29 Q6 **Moran** Kansas, C USA 37.55N 95.10W

27 T5 **Moran** Texas, SW USA 32.33N 99.10W

189 X7 **Moranbah** Queensland, NE Australia 22.01S 148.07E

46 L13 **Morant Bay** E Jamaica 17.52N 76.24W

98 G9 **Morar, Loch** ◇ N Scotland, UK

105 T5 **Morata** see Goodenough Island

107 Q12 **Moratalla** Murcia, SE Spain 38.10N 1.52W

110 C8 **Morat, Lac de** Ger. Murtensee. ◇ W Switzerland

172 Nm6 **Mori** Hokkaidō, NE Japan

37 Y6 **Moriah, Mount** ▲ Nevada, W USA 39.16N 114.10W

39 S11 **Moriarty** New Mexico, SW USA 34.59N 106.03W

56 J12 **Morichal** Guainía, E Colombia 2.18N 69.54W

194 H14 **Moriguá Island** island S PNG

169 U7 **Morin Dawa** var. Morin Dawa Daurzu Zhiqi. Nei Mongol Zizhiqu, N China 48.21N 124.32E

Morin Dawa Daurzu Zizhiqi see Morin Dawa

15 C15 **Morinville** Alberta, SW Canada 53.48N 113.37W

171 Mm1 **Morioka** Iwate, Honshū, C Japan 39.42N 141.08E

191 T8 **Morisset** New South Wales, SE Australia 33.05S 151.32E

171 Mm10 **Moriyoshi-yama** ▲ Honshū, C Japan 39.58N 140.32E

94 K13 **Morjärv** Norrbotten, N Sweden 66.03N 22.45E

131 R3 **Morki** Respublika Mariy El, W Russian Federation 56.27N 49.01E

126 K10 **Morkoka** ♨ NE Russian Federation

104 F5 **Morlaix** Finistère, NW France 48.35N 3.49W

97 M20 **Mörlunda** Kalmar, S Sweden 57.19N 15.52E

109 N19 **Mormanno** Calabria, SW Italy 39.55N 15.58E

38 L11 **Mormon Lake** ◇ Arizona, SW USA

37 Y10 **Mormon Peak** ▲ Nevada, W USA 36.59N 114.25W

Mormon State see Utah

81 Y5 **Morne-à-l'Eau** Grande Terre, N Guadeloupe 16.20N 61.28W

193 Y15 **Morning Sun** Iowa, C USA 41.06N 91.15W

200 O14 **Mornington Abyssal Plain** undersea feature SE Pacific Ocean

Morea see Pelopónnisos

30 K8 **Moreau River** ♨ South Dakota, N USA

99 K16 **Morecambe** NW England, UK 54.04N 2.52W

99 K16 **Morecambe Bay** inlet NW England, UK

191 S4 **Moree** New South Wales, SE Australia 29.28S 149.52E

194 E15 **Morehead** Western, SW PNG 8.42S 141.37E

23 Q3 **Morehead** Kentucky, S USA 38.13N 83.25W

194 E15 **Morehead** ♨ SW PNG

23 X11 **Morehead City** North Carolina, SE USA 34.43N 76.43W

29 Y8 **Morehouse** Missouri, C USA 36.51N 89.41W

110 E10 **Mörel** Valais, SW Switzerland 46.22N 8.03E

56 K5 **Morón** Ciego de Ávila, C Cuba 22.04N 78.39W

56 K5 **Morón** Carabobo, N Venezuela 10.28N 68.10W

Morón see Morón de la Frontera

106 K14 **Morón de la Frontera** var. Morón. Andalucía, S Spain 37.07N 5.27W

180 I6 **Morondava** Toliara, W Madagascar 20.19S 44.16E

180 H5 **Morondo** Toliara, W Madagascar 20.15S 44.17E

175 T6 **Morotai, Pulau** island E Indonesia

175 T6 **Morotai, Selat** strait Maluku, E Indonesia

83 H17 **Moroto** NE Uganda 2.31N 34.40E

Morozov see Bratan

130 H3 **Morozovsk** Rostovskaya Oblast', SW Russian Federation 48.21N 41.54E

99 L14 **Morpeth** N England, UK 55.10N 1.40W

Morphou see Güzelyurt

Morphou Bay see Güzelyurt Körfezi

129 U4 **More-Yu** ♨ NW Russian Federation

105 T9 **Morez** Jura, E France 46.33N 6.01E

30 M11 **Morrill** Nebraska, C USA 41.57N 103.55W

29 U11 **Morrilton** Arkansas, C USA 35.09N 92.44W

14 N9 **Morrin** Alberta, SW Canada 51.40N 112.45W

192 M7 **Morrinsville** Waikato, North Island, NZ 37.40S 175.32E

31 S8 **Morris** Manitoba, S Canada 49.21N 97.21W

32 M11 **Morris** Illinois, N USA 41.21N 88.25W

31 S8 **Morris** Minnesota, N USA 45.31N 95.52W

56 B12 **Morrison** Nariño, SW Colombia 2.31N 78.24W

207 R11 **Morris Jesup, Kap** headland N Greenland 83.33N 32.40W

190 B1 **Morris, Mount** ▲ South Australia 26.04S 131.03E

38 K10 **Morrison** Illinois, N USA 41.48N 89.58W

20 J14 **Morristown** New Jersey, NE USA 40.48N 74.28W

23 O8 **Morristown** Tennessee, S USA 36.12N 83.18W

45 S8 **Morrito** Río San Juan, SW Nicaragua 11.37N 85.03W

45 W14 **Morrocoy, Punta** headland E Nicaragua 12.18N 83.38W

45 Q15 **Morro Bay** California, W USA 35.21N 120.51W

45 N5 **Morro, Punta** headland NE Panama 9.06N 77.52W

37 N7 **Mörrum** Blekinge, S Sweden 56.10N 14.45E

85 N16 **Morrumbala** Zambézia, NE Mozambique 17.16S 35.34E

85 N20 **Morrumbene** Inhambane, SE Mozambique 23.38S 35.22E

27 N1 **Morse** Texas, SW USA 36.03N 101.28W

131 N6 **Morshansk** Tambovskaya Oblast', W Russian Federation 53.25N 41.48E

104 I5 **Mortagne-au-Perche** Orne, N France 48.32N 0.31E

104 J8 **Mortagne-sur-Sèvre** Vendée, NW France 47.00N 0.57W

106 G8 **Mortágua** Viseu, N Portugal 40.24N 8.13W

104 J5 **Mortain** Manche, N France 48.39N 0.51W

108 D8 **Mortara** Lombardia, N Italy 45.15N 8.43E

61 P14 **Mossoró** Rio Grande do Norte, NE Brazil 5.10S 37.19W

25 N9 **Moss Point** Mississippi, S USA 30.24N 88.31W

34 G8 **Mossyrock** Washington, NW USA 46.32N 122.28W

113 B15 **Most** Ger. Brüx. Ústecký Kraj, NW Czech Republic 50.30N 13.37E

121 Ji16 **Mosta** var. Musta. C Malta 35.54N 14.25E

92 L7 **Mostar** Federacija Bosna i Hercegovina, S Bosnia and Herzegovina 43.20N 17.47E

63 J17 **Mostardas** Rio Grande do Sul, S Brazil 31.06S 50.52W

118 K14 **Mostiștea** ♨ S Romania 44.20N 26.56E

103 H20 **Mosbach** Baden-Württemberg, SW Germany 49.21N 9.06E

97 E16 **Mosby** Vest-Agder, S Norway 58.12N 7.55E

35 V9 **Mosby** Montana, NW USA 46.58N 107.53W

34 M9 **Moscow** Idaho, NW USA 46.43N 117.00W

22 F10 **Moscow** Tennessee, S USA 35.04N 89.27W

Moscow see Moskva

103 D19 **Mosel** Fr. Moselle. ♨ W Europe see also Moselle

105 T4 **Moselle** ◇ department NE France

105 T6 **Moselle** Ger. Mosel. ♨ W Europe see also Mosel

34 K9 **Moses Lake** ◇ Washington, NW USA

85 I18 **Mosetse** Central, E Botswana 20.40S 26.37E

94 H4 **Mosfellsbær** Suðurland, SW Iceland 64.10N 21.42W

193 F23 **Mosgiel** Otago, South Island, NZ 45.51S 170.21E

128 M11 **Mosha** ♨ NW Russian Federation

83 I20 **Moshi** Kilimanjaro, NE Tanzania 3.21S 37.19E

112 I12 **Mosina** Wielkopolskie, C Poland 52.13N 16.48E

32 L6 **Mosinee** Wisconsin, N USA 44.45N 89.39W

94 F13 **Mosjøen** Nordland, C Norway 65.49N 13.12E

127 Nn13 **Moskal'vo** Ostrov Sakhalin, Sakhalinskaya Oblast', SE Russian Federation 53.36N 142.31E

94 I13 **Moskosel** Norrbotten, N Sweden 65.52N 19.30E

130 K4 **Moskovskaya Oblast'** ◇ province W Russian Federation

Moskovskiy see Moskva

130 J3 **Moskva** Eng. Moscow. ● (Russian Federation) Gorod Moskva, W Russian Federation 55.45N 37.42E

153 Q14 **Moskva** Rus. Moskovskiy; prev. Chubek. SW Tajikistan 37.41N 69.33E

130 L4 **Moskva** ♨ W Russian Federation

85 I20 **Mosomane** Kgatleng, SE Botswana 24.05S 26.16E

Moson and Magyaróvár see Mosonmagyaróvár

113 H21 **Mosoni-Duna** Ger. Kleine Donau. ♨ NW Hungary

113 H21 **Mosonmagyaróvár** Ger. Wieselburg-Ungarisch-Altenburg; prev. Moson and Magyaróvár; Ger. Wieselburg und Ungarisch-Altenburg. Győr-Moson-Sopron, NW Hungary 47.51N 17.15E

119 X8 **Mospyne** Rus. Mospino. Donets'ka Oblast', E Ukraine 47.53N 38.03E

56 B12 **Mosquera** Nariño, SW Colombia 2.31N 78.24W

39 U10 **Mosquero** New Mexico, SW USA 35.46N 103.57W

Mosquito Coast see La Mosquitia

33 U11 **Mosquito Creek Lake** ◇ Ohio, N USA

Mosquito Gulf see Mosquitos, Golfo de los

25 X11 **Mosquito Lagoon** wetland Florida, SE USA

45 N5 **Mosquito, Punta** headland E Nicaragua

45 W14 **Mosquito, Punta** headland NE Panama 9.06N 77.52W

45 Q15 **Mosquitos, Golfo de los Eng.** Mosquito Gulf. gulf N Panama

97 S13 **Moss** Østfold, S Norway 59.25N 10.40E

Mossâmedes see Namibe

24 G8 **Moss Bluff** Louisiana, S USA 30.18N 93.11W

85 N16 **Mossel Bay** var. Western Cape, SW South Africa 34.10S 22.07E

85 G25 **Mosselbaai/Mossel Bay** see Mosselbaai

81 F20 **Mossendjo** Le Niari, SW Congo 2.57S 12.59E

191 N8 **Mossgiel** New South Wales, SE Australia 33.16S 144.34E

103 H22 **Mössingen** Baden-Württemberg, S Germany 48.22N 9.01E

189 W4 **Mossman** Queensland, NE Australia 16.34S 145.27E

97 L17 **Motala** Östergötland, S Sweden 58.33N 15.05E

197 C10 **Mota Lava** island Banks Islands, N Vanuatu

203 X7 **Motane** var. Mohotani. island Îles Marquises, NE French Polynesia

158 K13 **Motihāri** var. Mottihari. Uttar Pradesh, N India 25.43N 78.55E

Mother of Presidents/Mother of States see Virginia

98 I12 **Motherwell** C Scotland, UK 55.48N 4.00W

159 P12 **Motīhāri** Bihār, N India 26.40N 84.55E

107 Q10 **Motilla del Palancar** Castilla-La Mancha, C Spain 39.34N 1.55W

158 K13 **Motihari Island** island NE NZ

67 E25 **Motley Island** island SE Falkland Islands

85 J19 **Motloutse** ♨ E Botswana

43 V17 **Motozintla de Mendoza** Chiapas, SE Mexico 15.22N 92.11W

107 N15 **Motril** Andalucía, S Spain 36.45N 3.29W

118 G13 **Motru** Gorj, SW Romania 44.49N 22.55E

171 Mm5 **Motsuta-misaki** headland Hokkaidō, NE Japan 42.36N 139.48E

30 L6 **Mott** North Dakota, N USA 46.21N 102.17W

109 O18 **Möttling** var. Metlika

192 P8 **Motu** ♨ North Island, NZ

193 I14 **Motueka** Tasman, South Island, NZ 41.08S 173.01E

193 I14 **Motueka** ♨ South Island, NZ

Motu Iti see Tupai

43 S12 **Motul** var. Motul de Felipe Carrillo Puerto. Yucatán, SE Mexico 21.06N 89.16W

Motul de Felipe Carrillo Puerto see Motul

203 U17 **Motu Nui** island Easter Island, Chile, E Pacific Ocean

203 Q10 **Motu One** var. Bellingshausen. atoll Îles Sous le Vent, W French Polynesia

200 R15 **Motu Tapu** island Tongatapu Group, S Tonga

192 L3 **Motutapu Island** island N NZ

Motyca see Modica

126 K14 **Motygino** Krasnoyarskiy Kray, C Russian Federation 58.09N 94.35E

107 U3 **Mouaскar** see Mascara

Moubermé, Tuc de Fr. Pic de Maubermé, Sp. Pico Maubermé; prev. Tuc de Maubermé. ▲ France/Spain see also Maubermé, Pic de 42.48N 0.57E

47 N7 **Mouchoir Passage** passage SE Turks and Caicos Islands

78 I9 **Moudjéria** Tagant, SW Mauritania 17.52N 12.19W

110 C9 **Moudon** Vaud, W Switzerland 46.41N 6.49E

81 E19 **Mouhoun** see Black Volta

81 E19 **Mouila** Ngounié, C Gabon 1.49S 11.01E

81 K14 **Mouka** Haute-Kotto, C Central African Republic 7.12N 21.52E

Moukden see Shenyang

191 N10 **Moulamein** New South Wales, SE Australia 35.06S 144.03E

Moulamein Creek see Billabong Creek

76 F6 **Moulay-Bousselham** NW Morocco 34.54N 6.15W

82 M11 **Moulhoulé** N Djibouti 12.34N 43.06E

105 P9 **Moulins** Allier, C France 46.34N 3.20E

178 G9 **Moulmein** var. Maulmain, Mawlamyine. Mon State, S Myanmar 16.30N 97.39E

177 F9 **Moulmeingyun** Irrawaddy, SW Myanmar 16.24N 95.15E

76 G6 **Moulouya** var. Mulucha, Muluya, Mulwiya. seasonal river NE Morocco

25 O2 **Moulton** Alabama, S USA 34.28N 87.16W

31 W16 **Moulton** Iowa, C USA 40.41N 92.40W

27 T11 **Moulton** Texas, SW USA 29.34N 97.08W

25 T7 **Moultrie** Georgia, SE USA 31.10N 83.47W

23 S14 **Moultrie, Lake** ◇ South Carolina, SE USA

24 K3 **Mound Bayou** Mississippi, S USA 33.52N 90.43W

32 L17 **Mound City** Illinois, N USA 37.06N 89.09W

29 R3 **Mound City** Kansas, C USA 38.08N 94.48W

29 S2 **Mound City** Missouri, C USA 40.07N 95.13W

31 N7 **Mound City** South Dakota, N USA 45.42N 100.04W

80 B13 **Moundou** Logone-Occidental, SW Chad 8.34N 16.01E

29 Q4 **Mounds** Oklahoma, C USA 35.52N 96.03W

23 R3 **Moundsville** West Virginia, NE USA 39.55N 80.44W

178 M9 **Moüng Roessei** Bătdâmbâng, W Cambodia 12.46N 103.28E

Moun Hou see Black Volta

14 G5 **Mountain** ♨ Northwest Territories, NW Canada

39 S12 **Mountainair** New Mexico, SW USA 34.31N 106.15W

37 T7 **Mountain City** Nevada, W USA 41.48N 115.58W

21 S4 **Mountain City** Tennessee, S USA 36.28N 81.48W

29 U9 **Mountain Grove** Missouri, C USA 37.07N 92.15W

29 U8 **Mountain Home** Arkansas, C USA 36.19N 92.24W

35 N15 **Mountain Home** Idaho, NW USA 43.07N 115.42W

27 T4 **Mountain Home** Texas, SW USA 30.11N 99.19W

31 T10 **Mountain Iron** Minnesota, N USA 47.31N 92.37W

31 T10 **Mountain Lake** Minnesota, N USA 43.57N 94.54W

25 S3 **Mountain Park** Georgia, SE USA 34.04N 84.24W

Legend (bottom):

◆ COUNTRY ◇ DEPENDENT TERRITORY ▲ ADMINISTRATIVE REGION ▲ MOUNTAIN ♨ VOLCANO ☺ LAKE
● COUNTRY CAPITAL ○ DEPENDENT TERRITORY CAPITAL ✕ INTERNATIONAL AIRPORT ▲ MOUNTAIN RANGE ♨ RIVER ☺ RESERVOIR

37 W12 **Mountain Pass** pass California,
W USA 35.28N 115.31W

29 T12 **Mountain Pine** Arkansas,
C USA 34.34N 93.10W

41 Y14 **Mountain Point** Annette Island,
Alaska, USA 55.17N 131.31W
Mountain State see Montana,
USA
Mountain State see West
Virginia, USA

29 V7 **Mountain View** Arkansas,
C USA 35.52N 92.07W

40 H12 **Mountain View** Hawaii, USA,
C Pacific Ocean 19.31N 155.03W

29 V10 **Mountain View** Missouri,
C USA 37.00N 91.42W

40 M11 **Mountain Village** Alaska, USA
62.06N 163.42W

23 R8 **Mount Airy** North Carolina,
SE USA 36.30N 80.36W

85 K24 **Mount Ayliff** Xh. Maxesibebi.
Eastern Cape, SE South Africa
30.48S 29.22E

31 U16 **Mount Ayr** Iowa, C USA
40.42N 94.14W

190 J9 **Mount Barker** South Australia
35.06S 138.52E

188 J14 **Mount Barker** Western Australia
34.42S 117.40E

191 P11 **Mount Beauty** Victoria,
SE Australia 36.47S 147.12E

12 E16 **Mount Brydges** Ontario,
S Canada 42.54N 81.29W

33 N16 **Mount Carmel** Illinois, N USA
38.23N 87.46W

32 K10 **Mount Carroll** Illinois, N USA
42.04N 89.58W

33 S9 **Mount Clemens** Michigan,
N USA 42.36N 82.52W

193 E19 **Mount Cook** Canterbury, South
Island, NZ 43.46S 170.06E

85 L16 **Mount Darwin** Mashonaland
Central, NE Zimbabwe 16.45S 31.32E

21 S7 **Mount Desert Island** island
Maine, NE USA

25 W11 **Mount Dora** Florida, SE USA
28.48N 81.38W

190 G5 **Mount Eba** South Australia
30.11S 135.40E

27 W8 **Mount Enterprise** Texas,
SW USA 31.53N 94.40W

190 J4 **Mount Fitton** South Australia
29.55S 139.26E

85 J24 **Mount Fletcher** Eastern Cape,
SE South Africa 30.40S 28.30E

12 F15 **Mount Forest** Ontario, S Canada
43.58N 80.43W

190 K12 **Mount Gambier** South Australia
37.47S 140.48E

189 W5 **Mount Garnet** Queensland,
NE Australia 17.41S 145.07E

23 P6 **Mount Gay** West Virginia,
NE USA 37.49N 82.00W

35 S12 **Mount Gilead** Ohio, N USA
40.33N 82.49W

194 H12 **Mount Hagen** Western
Highlands, C PNG 5.53S 144.12E

20 J16 **Mount Holly** New Jersey,
NE USA 39.59N 74.46W

23 R10 **Mount Holly** North Carolina,
SE USA 35.18N 81.01W

29 T12 **Mount Ida** Arkansas, C USA
34.33N 93.37W

189 T6 **Mount Isa** Queensland,
C Australia 20.48S 139.32E

23 U4 **Mount Jackson** Virginia,
NE USA 38.45N 78.38W

20 D12 **Mount Jewett** Pennsylvania,
NE USA 41.43N 78.37W

20 L13 **Mount Kisco** New York, NE USA
41.12N 73.42W

20 B15 **Mount Lebanon** Pennsylvania,
NE USA 40.21N 80.03W

190 J8 **Mount Lofty Ranges** ▲ South
Australia

188 J10 **Mount Magnet** Western
Australia 28.09S 117.52E

192 N7 **Mount Maunganui** Bay of
Plenty, North Island, NZ
37.39S 176.11E

99 E18 **Mountmellick** Ir. Móinteach
Mílic. C Ireland 53.07N 7.19W

32 L10 **Mount Morris** Illinois, N USA
42.03N 89.25W

33 R9 **Mount Morris** Michigan, N USA
43.07N 83.42W

20 F10 **Mount Morris** New York,
NE USA 42.43N 77.51W

20 B16 **Mount Morris** Pennsylvania,
NE USA 39.43N 80.06W

32 K15 **Mount Olive** Illinois, N USA
39.04N 89.43W

23 V10 **Mount Olive** North Carolina,
SE USA 35.12N 78.03W

23 N4 **Mount Olivet** Kentucky, S USA
38.32N 84.01W

31 Y15 **Mount Pleasant** Iowa, C USA
40.57N 91.33W

33 Q8 **Mount Pleasant** Michigan,
N USA 43.36N 84.46W

20 C15 **Mount Pleasant** Pennsylvania,
NE USA 40.07N 79.33W

23 T14 **Mount Pleasant** South Carolina,
SE USA 32.47N 79.51W

22 J9 **Mount Pleasant** Tennessee,
S USA 35.32N 87.11W

27 W6 **Mount Pleasant** Texas,
SW USA 33.10N 94.49W

35 L4 **Mount Pleasant** Utah, W USA
39.33N 111.27W

65 N23 **Mount Pleasant ✕** (Stanley) East
Falkland, Falkland Islands

99 G25 **Mount's Bay** inlet SW England,
UK

37 N2 **Mount Shasta** California,
W USA 41.18N 122.19W

32 J13 **Mount Sterling** Illinois, N USA
39.59N 90.45W

23 N5 **Mount Sterling** Kentucky, S USA
38.03N 83.56W

20 E15 **Mount Union** Pennsylvania,
NE USA 40.23N 77.51W

25 V6 **Mount Vernon** Georgia, SE USA
32.10N 82.35W

32 L12 **Mount Vernon** Illinois, N USA
38.19N 88.54W

22 M6 **Mount Vernon** Kentucky, S USA
37.22N 84.22W

29 S7 **Mount Vernon** Missouri, C USA
37.06N 93.49W

35 T13 **Mount Vernon** Ohio, N USA
40.23N 82.29W

34 K13 **Mount Vernon** Oregon,
NW USA 44.22N 119.07W

27 W6 **Mount Vernon** Texas, SW USA
33.11N 95.13W

34 H7 **Mount Vernon** Washington,
NW USA 48.25N 122.19W

22 L5 **Mount Washington** Kentucky,
S USA 38.03N 85.33W

190 F8 **Mount Wedge** South Australia
33.29S 135.08E

32 L14 **Mount Zion** Illinois, N USA
39.46N 88.52W

189 Y9 **Moura** Queensland, NE Australia
24.34S 149.57E

60 F12 **Moura** Amazonas, NW Brazil
1.32S 61.43W

106 H12 **Moura** Beja, S Portugal
38.07N 7.27W

106 I12 **Mourão** Évora, S Portugal
38.22N 7.19W

78 L11 **Mourdiah** Koulikoro, W Mali
14.28N 7.31W

80 K7 **Mourdi, Dépression du** desert
lowland Chad/Sudan

104 J16 **Mourenx** Pyrénées-Atlantiques,
SW France 43.24N 0.37W
Mourgana see Mourgána

117 C15 **Mourgána**
▲ Albania/Greece 39.48N 20.24E

99 G16 **Mourne Mountains** Ir. Beanna
Boirche. ▲ SE Northern Ireland,
UK

117 I15 **Moúrtzeflos, Akrotírio**
headland Límnos, E Greece
40.00N 25.02E

101 C19 **Mouscron** Dut. Moeskroen.
Hainaut, W Belgium 50.43N 3.13E
Mouse River see Souris River

80 H10 **Moussoro** Kanem, W Chad
13.40N 16.31E

105 T11 **Moûtiers** Savoie, E France
45.28N 6.31E

180 J14 **Moutsamoudou** var.
Mutsamudu. Anjouan,
SE Comoros 12.05 44.25E

76 K1 **Mouydir, Monts de** ▲ S Algeria

81 F20 **Mouyondzi** La Bouenza, S Congo
3.58S 13.57E

117 E16 **Mouzáki** prev. Mouzákion.
Thessalía, C Greece 39.25N 21.40E
Mouzákion see Mouzáki

84 E13 **Moville** Iowa, C USA 42.30N 96.04W

180 I14 **Moya** Anjouan, SE Comoros
12.18S 44.27E

42 L12 **Moyahua** Zacatecas, C Mexico
21.19N 103.10W

83 J16 **Moyale** Oromo, C Ethiopia
3.34N 38.58E

78 I15 **Moyamba** W Sierra Leone
8.04N 12.30W

76 G7 **Moyen Atlas** Eng. Middle Atlas.
▲ N Morocco

80 H13 **Moyen-Chari** off. Préfecture du
Moyen-Chari. ◆ prefecture S Chad
Moyen-Chari see Moyen-Chari
Moyen-Congo see Congo
(Republic of)

85 J24 **Moyeni** var. Quthing. SW Lesotho
30.25S 27.43E

78 H13 **Moyenne-Guinée** ◆ state
NW Guinea

81 D18 **Moyen-Ogooué** off. Province du
Moyen-Ogooué, var. Le Moyen-
Ogooué. ◆ province C Gabon

105 S4 **Moyeuvre-Grande** Moselle,
NE France 49.15N 6.03E

35 N7 **Moyie Springs** Idaho, NW USA
48.43N 116.15W

151 S15 **Moynkum** prev. Fumanovka,
Kaz. Fürmanov. Zhambyl, S
Kazakhstan
44.15N 72.55E

83 J9 **Moyo** NW Uganda 3.37N 31.43E

58 D10 **Moyobamba** San Martín,
NW Peru 6.04S 76.56W

175 O16 **Moyo, Pulau** island S Indonesia

80 H10 **Moyto** Chari-Baguirmi, W Chad
12.34N 16.33E

164 G9 **Moyu** var. Karakax. Xinjiang
Uygur Zizhiqu, NW China
37.16N 79.39E

126 J19 **Moyyero** ⚟ N Russian
Federation

151 Q15 **Moyynkum, Peski** Kaz.
Moyynqum. desert S Kazakhstan
Moyynqum see Moyynkum, Peski

151 S12 **Moyynty** Zhezkazgan,
C Kazakhstan 47.10N 73.24E
Moyynty ⚟ C Kazakhstan

85 M18 **Mozambique** off. Republic of
Mozambique; prev. People's
Republic of Mozambique,
Portuguese East Africa. ◆ republic
S Africa
Mozambique Basin see Natal
Basin

85 P17 **Mozambique, Canal de** see
Mozambique Channel
Mozambique Channel Fr.
Canal de Mozambique, Mal.
Lakandranon' i Mozambika. strait
W Indian Ocean

180 L11 **Mozambique Escarpment** var.
Mozambique Scarp. undersea
feature SW Indian Ocean

180 L10 **Mozambique Plateau** var.
Mozambique Rise. undersea feature
SW Indian Ocean
Mozambique Rise see
Mozambique Plateau
Mozambique Scarp see
Mozambique Escarpment

131 O15 **Mozdok** Respublika Severnaya
Osetiya, SW Russian Federation
43.48N 44.42E

59 K17 **Mozetenes, Serranías de**
▲ C Bolivia

130 J4 **Mozhaysk** Moskovskaya Oblast',
W Russian Federation 55.31N 36.01E

131 T3 **Mozhga** Udmurtskaya
Respublika, NW Russian
Federation 56.24N 52.13E
Mozyr' see Mazyr

81 P22 **Mpala** Katanga, E Dem. Rep.
Congo (Zaire) 6.45S 29.28E

81 G19 **Mpama** ⚟ C Congo

83 E22 **Mpanda** Rukwa, W Tanzania
6.21S 31.01E

85 L11 **Mpande** Northern, NE Zambia
9.13S 31.42E

85 J18 **Mphoengs** Matabeleland South,
SW Zimbabwe 21.04S 27.56E

83 F18 **Mpigi** S Uganda 0.13N 32.19E

84 L13 **Mpika** Northern, NE Zambia
11.49S 31.27E

85 J14 **Mpongwe** Copperbelt, C Zambia
13.25S 28.13E

84 K11 **Mporokoso** Northern, N Zambia
9.22S 30.06E

81 H20 **Mpouya** Plateaux, SE Congo
2.39S 16.12E

79 P16 **Mpraeso** C Ghana 6.46N 0.41W

84 L11 **Mpulungu** Northern, N Zambia
8.47S 31.09E

85 K21 **Mpumalanga** prev. Eastern
Transvaal, Afr. Oos-Transvaal.
◆ province NE South Africa

85 D16 **Mpungu** Okavango, N Namibia
17.36S 18.16E

83 I22 **Mpwapwa** Dodoma, C Tanzania
6.21S 36.30E

112 M8 **Mqinvartsveri** see Kazbek

85 O16 **Mragowo** Ger. Sensburg.
Warmińsko-Mazurskie, NE
Poland, 53.52N 21.19E

131 V6 **Mrakovo** Respublika
Bashkortostan, W Russian
Federation 52.43N 56.37E

180 J13 **Mramani** Anjouan, E Comoros
12.17S 44.31E

114 F12 **Mrkonjić Grad** Republika
Srpska, W Bosnia and Herzegovina
44.25N 17.04E

112 H9 **Mrocza** Kujawsko-pomorskie,
NW Poland 53.15N 17.38E

128 I14 **Msta** ⚟ NW Russian Federation
Mtkvari see Kura

130 K6 **Mtoko** see Mutoko

130 L12 **Mtsensk** Orlovskaya Oblast',
W Russian Federation 53.17N 36.34E

83 K24 **Mtwara** Mtwara, SE Tanzania
10.16S 40.10E

83 J25 **Mtwara** ◆ region SE Tanzania

106 G14 **Mu** ▲ S Portugal 37.24N 8.04W

200 Qq15 **Mu'a** Tongatapu, S Tonga
21.11S 175.07W
Muai To see Mae Hong Son

85 P16 **Mualama** Zambézia,
NE Mozambique 16.51S 38.21E
Mualo see Messalo, Rio

81 E22 **Muanda** Bas-Congo, SW Dem.
Rep. Congo (Zaire) 5.53S 12.17E
Muang Chiang Rai see Chiang
Rai

178 J6 **Muang Ham** Houaphan, N Laos
20.19N 104.00E

178 J9 **Muang Hinboun** Khammouan,
C Laos 17.37N 104.37E
Muang Kalasin see Kalasin
Muang Khammouan see
Thakhek

178 Jj11 **Muang Khôngxédôn** var.
Khong Sedone. Salavan, S Laos
15.34N 105.46E
Muang Khon Kaen see Khon
Kaen

178 Ii6 **Muang Khoua** Phôngsali, N Laos
21.07N 102.31E
Muang Krabi see Krabi
Muang Lampang see Lampang
Muang Lamphun see Lamphun
Muang Loei see Loei
Muang Lom Sak see Lom Sak
Muang Nakhon Sawan see
Nakhon Sawan
Muang Namo Oudômxai,
N Laos 20.58N 101.46E
Muang Nan see Nan

178 Ii6 **Muang Ngoy** Louangphabang,
N Laos 20.43N 102.42E

178 I5 **Muang Ou Tai** Phôngsali, N Laos
22.06N 101.59E
Muang Pak Lay see Pak Lay
Muang Pakxan see Pakxan

178 Jj10 **Muang Pakxong** Champasak,
S Laos 15.10N 106.17E

178 Jj9 **Muang Phalan** var. Muang
Phalane. Savannakhét, S Laos
16.40N 105.33E
Muang Phalane see Muang
Phalan
Muang Phan see Phan
Muang Phayao see Phayao
Muang Phichit see Phichit

178 Jj9 **Muang Phin** Savannakhét,
S Laos 16.31N 106.01E
Muang Phitsanulok see
Phitsanulok
Muang Phrae see Phrae
Muang Roi Et see Roi Et
Muang Sakon Nakhon see
Sakon Nakhon
Muang Samut Prakan see
Samut Prakan

178 I6 **Muang Sing** Louang Namtha,
N Laos 21.12N 101.09E
Muang Ubon see Ubon
Ratchathani
Muang Uthai Thani see
Uthai Thani

178 I7 **Muang Vangviang** Viangchan,
C Laos 18.53N 102.27E
Muang Xaignabouri see
Xaignabouri
Muang Xay see Xai

178 Jj9 **Muang Xépôn** var. Sepone.
Savannakhét, S Laos 16.40N 106.15E

174 H6 **Muar** var. Bandar Maharani.
Johor, Peninsular Malaysia
2.01N 102.34E

173 Fj6 **Muara** Sumatera, W Indonesia
2.18N 98.54E

174 Hh11 **Muarabeliti** Sumatera,
W Indonesia 3.13S 103.00E

174 H9 **Muarabungo** Sumatera,
W Indonesia 1.36S 103.37E

174 I11 **Muaraenim** Sumatera,
W Indonesia 3.46S 103.43E

174 Mm8 **Muarajuloi** Borneo, C Indonesia
0.12S 114.43E

175 Nn9 **Muarakaman** Borneo,
C Indonesia 0.09S 116.43E

173 Ff9 **Muarasigep** Pulau Siberut,
W Indonesia 1.01S 98.48E

174 Hh9 **Muaratembesi** Sumatera,
W Indonesia 1.42S 103.07E

175 N9 **Muaratewe** var. Muaratewe;
prev. Muarateweh. Borneo,
C Indonesia 0.58S 114.52E

175 O7 **Muaraweweh** see Muaratewe
Muarawahau Borneo,
N Indonesia 1.03N 116.48E

144 G13 **Mubārak, Jabal** ▲ S Jordan
29.19N 35.13E

159 N13 **Mubārakpur** Uttar Pradesh,
N India 26.05N 83.19E
Mubarek see Muborak

83 F19 **Mubende** SW Uganda 0.34N 31.24E

79 Y14 **Mubi** Adamawa, NE Nigeria
10.15N 13.18E

152 M12 **Mubrak** Rus. Mubarek.
Qashqadaryo Wiloyati,
S Uzbekistan 39.17N 65.10E

176 Vv9 **Mubrani** Irian Jaya, E Indonesia
0.42S 133.25E

69 U12 **Muchinga Escarpment**
escarpment NE Zambia

131 N7 **Muchkapskiy** Tambovskaya
Oblast', W Russian Federation
51.51N 42.25E

58 B5 **Muisne** Esmeraldas, NW Ecuador
0.34N 79.58W

95 P14 **Muite** Nampula, NE Mozambique
14.15S 38.27E

43 Z11 **Mujeres, Isla** island E Mexico

118 G7 **Mukacheve** Hung. Munkács, Rus.
Mukachevo. Zakarpats'ka Oblast',
W Ukraine 48.26N 22.44E
Mukachevo see Mukacheve

44 J5 **Mukalla** see Al Mukallā

152 J14 **Mucur** Kırşehir, C Turkey
39.04N 34.25E

149 U8 **Mūd** Khorāsān, E Iran

169 J7 **Mudanjiang** var. Mu-tan-chiang.
Heilongjiang, NE China
44.33N 129.40E

169 Y9 **Mudan Jiang** ⚟ NE China

142 D11 **Mudanya** Bursa, NW Turkey
40.22N 28.52E

30 K8 **Mud Butte** South Dakota, N USA
45.00N 102.51W

161 G16 **Muddebihāl** Karnātaka, C India
16.26N 76.07E

29 P12 **Muddy Boggy Creek**
⚟ Oklahoma, C USA

38 M6 **Muddy Creek** ⚟ Utah, W USA

39 V7 **Muddy Creek Reservoir**
☒ Colorado, C USA

35 W15 **Muddy Gap** Wyoming, C USA
42.21N 107.27W

37 Y11 **Muddy Peak** ▲ Nevada, W USA
36.17N 114.40W

191 R7 **Mudgee** New South Wales,
SE Australia 32.37S 149.34E

31 S3 **Mud Lake** ☒ Minnesota, N USA

31 P7 **Mud Lake Reservoir** ☒ South
Dakota, N USA

178 Gg9 **Mudon** Mon State, S Myanmar
16.14N 97.46E

83 O14 **Mudug** off. Gobolka Mudug. ◆
region NE Somalia

83 O14 **Mudug** var. Mudugh. plain
N Somalia
Mudugh see Mudug

85 Q15 **Muecate** Nampula,
NE Mozambique 14.56S 39.38E

84 Q13 **Mueda** Cabo Delgado,
NE Mozambique 11.40S 39.36E

44 L10 **Muelle de los Bueyes** Región
Autónoma Atlántico Sur,
SE Nicaragua 12.03N 84.34W

85 M14 **Muende** Tete, NW Mozambique
14.22S 33.00E

27 P9 **Muenster** Texas, SW USA
33.39N 97.22W

29 R10 **Muenster** see Münster
Muerto, Cayo reef NE Nicaragua

42 E7 **Muerto, Mar** lagoon SE Mexico

66 F11 **Muertos Trough** undersea feature
N Caribbean Sea

85 H14 **Mufaya Kuta** Western,
NW Zambia 14.30S 24.18E

84 J13 **Mufulira** Copperbelt, C Zambia
12.33S 28.15E

167 O10 **Mufu Shan** ▲ C China

169 I6 **Mugalzhary, Gory Kaz.**
Mugalzhar Taūlary.
Mugalzhary, Gory see
Mugalzhar Taūlary

143 Y12 **Mūgan Düzü** Rus. Muganskaya
Ravnina, Muganskaya Step'.
physical region S Azerbaijan
Muganskaya
Ravnina/Muganskaya Step' see
Mūgan Düzü

108 K8 **Múggia** Friuli-Venezia Giulia,
NE Italy 45.36N 13.48E

159 N14 **Mughal Sarāi** Uttar Pradesh,
N India 25.18N 83.07E
Mughla see Muğla

147 W11 **Mughshin** var. Muqshin. S Oman
19.25N 54.38E

153 S12 **Mughsu** Rus. Muksu.
⚟ Tajikistan

170 Ff16 **Mugi** Tokushima, Shikoku,
SW Japan 33.39N 134.24E

142 C16 **Muğla** var. Mughla. Muğla,
SW Turkey 37.13N 28.22E

142 C16 **Muğla** var. Mughla. ◆ province
SW Turkey

150 J11 **Mugodzhary, Gory Kaz.**
Mugalzhar Taūlary.

85 O15 **Mugulama** Zambézia,
NE Mozambique 16.15S 37.33E

145 U9 **Muhammad** E Iraq 32.46N 45.14E

145 R8 **Muhammadīyah** C Iraq
33.22N 42.48E

82 I6 **Muhammad Qol** Red Sea,
NE Sudan 20.52N 37.09E

77 Y9 **Muhammad, Rās** headland
E Egypt 27.45N 34.18E

146 M12 **Muhāyil** Mah. 'Asir,
SW Saudi Arabia 18.34N 42.01E

145 O7 **Muhaywir** W Iraq 34.41N 41.06E

103 H21 **Mühlacker** Baden-Württemberg,
SW Germany 48.57N 8.51E
Mühlbach see Sebeş

101 M17 **Mühldorf** see Mühldorf am Inn

103 N23 **Mühldorf am Inn** var.
Mühldorf. Bayern, SE Germany
48.14N 12.32E

101 K17 **Mühlhausen** var. Mühlhausen in
Thüringen. Thüringen, C Germany
51.13N 10.28E
Mühlhausen in Thüringen see
Mühlhausen

205 Q2 **Mühlig-Hofmann Mountains**
▲ Antarctica

95 L14 **Muhos** Oulu, C Finland
64.48N 26.00E

120 E5 **Muhu** Ger. Mohn, Moon. island
W Estonia

83 F19 **Muhutwe** Kagera, NW Tanzania
1.31S 31.40E

100 J10 **Muiden** Noord-Holland,
C Netherlands 52.19N 5.04E

200 R15 **Mui Hopohopoonga** headland
Tongatapu, S Tonga 21.09S 175.01W

171 K14 **Muika** var. Muikamachi. Niigata,
Honshū, C Japan 37.04N 138.53E
Muikamachi see Muika
Muinchille see Cootehill
Muineachán see Monaghan

99 F19 **Muine Bheag** Eng.
Bagenalstown. SE Ireland
52.42N 6.57W
Muinchille see Mouluoya

85 K14 **Mulungushi** Central, C Zambia
14.15S 28.27E

85 K14 **Mulungwe** Central, C Zambia
13.57S 29.51E

29 N7 **Mulvane** Kansas, C USA
37.28N 97.14W

191 O10 **Mulwala** New South Wales,
SE Australia 35.59S 146.00E
Mulwiya see Mouluoya

190 K6 **Mulyah** South Australia
31.29S 140.45E

171 K14 **Mumbai** prev. Bombay.
Mahārāshtra, W India 18.55N 72.51E

160 D13 **Mumbai** ★ Mahārāshtra, W India
19.10N 72.51E

85 D14 **Mumbué** Bié, C Angola
13.52S 17.15E

85 H15 **Mumeng** Morobe, C PNG
6.57S 146.37E

194 J13 **Mumeng** Morobe, C PNG
6.57S 146.37E

176 W9 **Mumi** Irian Jaya, E Indonesia
1.33S 134.09E

172 Oo6 **Mu-kawa** ⚟ Hokkaidō,
NE Japan

145 S6 **Mukayshifah** var. Mukāshafa,
Mukashshafah. N Iraq 34.24N 43.44E

178 J9 **Mukdahan** Mukdahan,
E Thailand 16.31N 104.43E
Mukden see Shenyang

172 Ss16 **Mukojima-rettō** Eng. Parry
group. island group SE Japan

152 M14 **Mukry** Lebapskiy Velayat,
E Turkmenistan 37.39N 65.37E
Muksu see Mughsu

159 U14 **Muktagacha** var. Muktagachha
Dhaka, N Bangladesh 24.46N 90.16E
Muktagachha see Muktagacha

84 K13 **Mukuku** Central, C Zambia
12.10S 29.50E

84 K11 **Mukupa Kaoma** Northern,
NE Zambia 9.55S 30.19E

83 I18 **Mukutan** Rift Valley, W Kenya
1.06N 36.16E

85 F16 **Mukwe** Caprivi, NE Namibia
18.01S 21.24E

157 K20 **Mulaku Atoll** var. Meemu Atoll.
atoll C Maldives

85 J15 **Mulalika** Lusaka, C Zambia
15.37S 28.48E

169 X8 **Mulan** Heilongjiang, NE China
45.57N 128.00E

85 O15 **Mulanje** var. Mlanje. Southern,
S Malawi 16.04S 35.35E

42 H5 **Mulatos** Sonora, NW Mexico
28.42N 108.44W

25 P3 **Mulberry Fork** ⚟ Alabama,
S USA

41 P17 **Mulchatna River** ⚟ Alaska,
USA

129 W4 **Mul'da** Respublika Komi,
NW Russian Federation
67.29N 63.55E

101 E18 **Mulde** ⚟ E Germany

29 R10 **Muldrow** Oklahoma, C USA
35.25N 94.34W

31 N10 **Mundelein** Illinois, N USA
42.15N 88.00W

103 I15 **Münden** Niedersachsen,
C Germany 51.26N 9.54E

110 I10 **Mulegns** Graubünden,
S Switzerland 46.30N 9.36E

85 M21 **Mulenda** Kasai Oriental, C Dem.
Rep. Congo (Zaire) 4.19S 24.55E

26 M4 **Muleshoe** Texas, SW USA
34.13N 102.43W

85 O15 **Mulevala** Zambézia,
NE Mozambique 16.18S 37.40E

191 P5 **Mulgoa Creek** seasonal river New
South Wales, SE Australia

107 O15 **Mulhacén** var. Cerro de
Mulhacén. ▲ S Spain 37.07N 3.11W
Mulhacén, Cerro de see
Mulhacén

103 E24 **Mülheim** Baden-Württemberg,
SW Germany 47.45N 7.16E

101 D16 **Mülheim** var. Mülheim an der
Ruhr. Nordrhein-Westfalen,
W Germany 51.25N 6.52E
Mülheim an der Ruhr see
Mülheim

105 T13 **Mulhouse** Ger. Mülhausen. Haut-
Rhin, NE France 47.45N 7.19E

166 G11 **Muli** var. Bowa, Muli Zangzu
Zizhixian. Sichuan, C China
27.49N 101.10E

176 Y15 **Muli** channel Irian Jaya,
E Indonesia
Muli Zangzu Zizhixian see Muli

169 Y9 **Muling** Heilongjiang, NE China
44.54N 130.35E

169 T16 **Mungla** var. Mongla. Khulna,
S Bangladesh 22.18N 89.34E

84 C13 **Mungo** Huambo, W Angola
11.46S 16.13E

196 F16 **Munguuy Bay** bay Yap,
W Micronesia

85 N8 **Mullan** Idaho, NW USA
47.28N 115.48W

30 M13 **Mullen** Nebraska, C USA
42.02N 101.01W

191 Q6 **Mullengudgery** New South
Wales, SE Australia 31.42S 147.24E

25 Q5 **Mullens** West Virginia, NE USA
37.34N 81.22W
Müller-gerbirge see Muller,
Pegunungan

174 Mm7 **Muller, Pegunungan** Dut.
Müller-gerbirge. ▲
C Indonesia

33 O5 **Mullett Lake** ☒ Michigan,
N USA

20 G13 **Mullica River** ⚟ New Jersey,
NE USA

27 P8 **Mullin** Texas, SW USA
31.33N 98.40W

99 E17 **Mullingar** Ir. An Muileann
gCearr. C Ireland 53.31N 7.19W

23 T12 **Mullins** South Carolina, SE USA
34.12N 79.15W

110 D8 **Münsingen** Bern, W Switzerland
46.52N 7.36E

105 U6 **Munster** Haut-Rhin, NE France
48.03N 7.09E

102 I11 **Munster** Niedersachsen,
NW Germany 52.58N 10.05E

99 B20 **Munster** Ir. Cúige Mumhan.
cultural region S Ireland

102 F13 **Münster** var. Muenster, Münster
in Westfalen. Nordrhein-Westfalen,
W Germany 51.58N 7.37E

110 F10 **Münster** Valais, S Switzerland
46.31N 8.18E
Münsterberg in Schlesien see
Ziębice
Münster in Westfalen see
Münster

102 E13 **Münsterland** cultural region
NW Germany

102 F13 **Münster-Osnabrück**
★ Nordrhein-Westfalen,
NW Germany 52.08N 7.41E

33 R4 **Munuscong Lake** ☒ Michigan,
N USA

85 K17 **Munyati** ⚟ C Zimbabwe

111 R3 **Münzkirchen** Oberösterreich,
N Austria 48.29N 13.37E

94 K11 **Muodoslompolo** Norrbotten,
N Sweden 67.57N 23.31E

94 M13 **Muojärvi** ☒ NE Finland

178 J6 **Mường Khên** Hoa Binh,
N Vietnam 20.34N 105.18E
Muong Sai see Xai

178 I7 **Muong Xiang Ngeun** var. Xieng
Ngeun. Louangphabang, N Laos
19.43N 102.09E

94 K11 **Muonio** Lappi, N Finland
67.58N 23.40E

94 K11 **Muonioälv/Muoniojoki** see
Muoniojoki

94 K11 **Muoniojoki** var. Muonioälv,
Swe. Muoniojoki. ⚟
Finland/Sweden

85 N17 **Mupa** ⚟ C Mozambique

85 E16 **Mupini** Okavango, NE Namibia
17.55S 19.34E

82 F8 **Muqaddam, Wadi** ⚟ N Sudan

144 K9 **Muqāṭ** Al Mafraq, E Jordan
32.28N 38.04E

147 X7 **Muqaz** N Oman 24.13N 56.48E

83 N17 **Muqdisho** Eng. Mogadishu, It.
Mogadiscio. ● (Somalia) Banaadir,
S Somalia 2.06N 45.22E

158 C12 **Munābāo** Rājasthān, NW India
25.46N 70.19E
Munamägi see Suur Munamägi

1 T8 **Mur** SCr. Mura. ⚟ C Europe
Mura see Mur

143 T14 **Muradiye** Van, E Turkey
39.00N 43.44E
Muragarazi see Maragarazi

171 L12 **Murakami** Niigata, Honshū,
C Japan 38.13N 139.28E

65 G22 **Murallón, Cerro** ▲ S Argentina
49.49S 73.25W

83 E20 **Muramvya** C Burundi 3.18S 29.41E

83 I19 **Murang'a** prev. Fort Hall.
Central, SW Kenya 0.43S 37.10E

83 H16 **Murangering** Rift Valley,
NW Kenya 3.48N 35.29E
Murapara see Murupara

146 M5 **Murār, Bi'r al** well NW Saudi
Arabia 27.40N 40.21E

129 Q13 **Murashi** Kirovskaya Oblast',
NW Russian Federation
59.27N 48.02E

105 O12 **Murat** C France 45.07N 2.52E

116 N12 **Muratlı** Tekirdağ, NW Turkey
41.12N 27.30E

143 R14 **Murat Nehri** var. Eastern
Euphrates; anc. Arsanias.
⚟ NE Turkey

109 D20 **Muravera** Sardegna, Italy,
C Mediterranean Sea 39.24N 9.34E

171 L12 **Murayama** Yamagata, Honshū,
C Japan 38.30N 140.25E

124 Oo15 **Murayash, Ra's al** headland
N Libya 31.58N 25.00E

106 I6 **Murça** Vila Real, N Portugal
41.25N 7.27W

82 Q11 **Murcanyo** Bari, NE Somalia
11.39N 50.27E

149 N8 **Mürcheh Khvort** var. Morcheh
Khort. Eṣfahān, C Iran 33.07N 51.26E

193 H15 **Murchison** Tasman, South
Island, NZ 41.48S 172.19E

193 B22 **Murchison Mountains**
▲ South Island, NZ

188 I10 **Murchison River** ⚟ Western
Australia

107 R13 **Murcia** Murcia, SE Spain
37.58N 1.07W

107 Q13 **Murcia** ◆ autonomous community
SE Spain

105 O13 **Mur-de-Barrez** Aveyron,
S France 44.48N 2.39E

190 G8 **Murdinga** South Australia
33.46S 135.46E

30 M10 **Murdo** South Dakota, N USA
43.53N 100.42W

13 X6 **Murdochville** Quebec,
SE Canada 48.57N 65.30W

111 W9 **Mureck** Steiermark, SE Austria
46.43N 15.46E

116 M13 **Müreftë** Tekirdağ, NW Turkey
40.40N 27.16E

118 H10 **Mureş** ◆ county N Romania

86 J11 **Mureş** var. Maros, Mureşul, Ger.
Marosch, Mieresch.
⚟ Hungary/Romania see also
Maros
Mureşul see Maros/Mureş

104 M15 **Muret** Haute-Garonne, S France
43.28N 1.19E

29 T13 **Murfreesboro** Arkansas, C USA
34.03N 93.41W

23 X8 **Murfreesboro** North Carolina,
SE USA 36.26N 77.06W

22 J9 **Murfreesboro** Tennessee, S USA
35.51N 86.23W

154 J14 **Murgab** prev. Murgap see also
Morghāb, Daryā-ye. Maryyskiy
Velayat, S Turkmenistan
37.19N 61.48E
Murgab see Murghāb, Pash.
Daryā-ye Morghāb, Turkm.
Murgap Deryasy
Murgab/Murghab see also
Morghāb, Daryā-ye

152 J16 **Murgap** see Murghāb, Pash.
Daryā-ye Morghāb, Turkm.
Murgap Deryasy
Murgap Deryasy see
Afghanistan/Turkmenistan see also
Morghāb, Daryā-ye
Murgap see Murghob
Murgap Deryasy see Murgap
Murgap Deryasy var. Murgap,
Daryā-ye/Murgab

297

116 H9 **Murgash** ▲ W Bulgaria
42.51N 23.58E

Murghab see Morghāb,
Daryā-ye /Murgab

153 U13 **Murghob** Rus. Murgab.
SE Tajikistan 38.11N 73.59E

153 U13 **Murghob** Rus. Murgab.
◆ SE Tajikistan

189 Z10 **Murgon** Queensland, E Australia
26.075 152.03E

202 I16 **Muri** Rarotonga, S Cook Islands
21.155 159.43W

110 F7 **Muri** Aargau, W Switzerland
47.16N 8.21E

110 D8 **Muri** var. Muri bei Bern. Bern,
W Switzerland 46.55N 7.30E

106 K3 **Murias de Paredes** Castilla-
León, N Spain 42.51N 6.11W

Muri bei Bern see Muri

84 F11 **Muriege** Lunda Sul, NE Angola
9.555 21.12E

201 P14 **Murilo Atoll** atoll Hall Islands,
C Micronesia

Mūritānīyah see Mauritania

102 N10 **Müritz** var. Müritzee.
◎ NE Germany

Müritzee see Müritz

102 L10 **Müritz-Elde-Kanal** canal
N Germany

192 K6 **Muriwai Beach** Auckland, North
Island, NZ 36.565 174.28E

94 J13 **Murjek** Norrbotten, N Sweden
66.27N 20.54E

128 J3 **Murmansk** Murmanskaya
Oblast', NW Russian Federation
68.58N 33.07E

128 I4 **Murmanskaya Oblast'** ◆
province NW Russian Federation

207 V14 **Murmansk Rise** undersea feature
SW Barents Sea

128 J3 **Murmashi** Murmanskaya
Oblast', NW Russian Federation
68.49N 32.42E

130 M5 **Murmino** Ryazanskaya Oblast',
W Russian Federation 54.31N 40.01E

103 K24 **Murnau** Bayern, SE Germany
47.41N 11.12E

105 X16 **Muro, Capo di** headland Corse,
France, C Mediterranean Sea
41.45N 8.40E

109 M18 **Muro Lucano** Basilicata, S Italy
40.48N 15.33E

131 N4 **Murom** Vladimirskaya Oblast',
W Russian Federation 55.33N 42.03E

125 G13 **Muromtsevo** Omskaya Oblast',
C Russian Federation 56.18N 75.15E

172 Nn6 **Muroran** Hokkaidō, NE Japan
42.19N 140.58E

106 G3 **Muros** Galicia, NW Spain
42.46N 9.03W

106 F3 **Muros e Noia, Ría de** estuary
NW Spain

170 F16 **Muroto** Kōchi, Shikoku,
SW Japan 33.18N 134.07E

170 F16 **Muroto-zaki** headland Shikoku,
SW Japan 33.15N 134.09E

118 L7 **Murovani Kurylivtsi** Vinnyts'ka
Oblast', C Ukraine 48.43N 27.31E

112 GII **Murowana Goślina**
Wielkopolskie, C Poland
52.33N 16.59E

34 M14 **Murphy** Idaho, NW USA
43.14N 116.36W

23 N10 **Murphy** North Carolina, SE USA
35.05N 84.01W

37 P8 **Murphys** California, W USA
38.07N 120.27W

32 L17 **Murphysboro** Illinois, N USA
37.45N 89.20W

31 V15 **Murray** Iowa, C USA 41.03N 93.56W

22 H8 **Murray** Kentucky, S USA
36.36N 88.18W

190 J10 **Murray Bridge** South Australia
35.065 139.15E

183 X2 **Murray Fracture Zone** tectonic
feature NE Pacific Ocean

194 E13 **Murray, Lake** ◎ SW PNG

23 P12 **Murray, Lake** ◎ South Carolina,
SE USA

8 K8 **Murray, Mount** ▲ Yukon
Territory, NW Canada
60.49N 128.57W

194 H13 **Murray Range** var. Leonard
Murray Mountains. ▲ W PNG

Murray Range see Murray Ridge

181 O3 **Murray Ridge** var. Murray
Range. undersea feature N Arabian
Sea

191 N10 **Murray River** ↔ SE Australia

190 K10 **Murrayville** Victoria,
SE Australia 35.175 141.12E

155 U5 **Murree** Punjab, E Pakistan
33.55N 73.25E

103 I21 **Murrhardt** Baden-Württemberg,
S Germany 49.00N 9.34E

191 O9 **Murrumbidgee River** ↔ New
South Wales, SE Australia

85 P15 **Murrupula** Nampula,
NE Mozambique 15.265 38.46E

191 T7 **Murrurundi** New South Wales,
SE Australia 31.475 150.51E

111 X9 **Murska Sobota** Ger. Olsnitz.
NE Slovenia 46.40N 16.09E

160 G12 **Murtajāpur** prev. Murtazapur.
Mahārāshtra, C India 20.43N 77.28E

79 S16 **Murtala Muhammed** ✈ (Lagos)
Ogun, SW Nigeria 6.31N 3.12E

Murtazapur see Murtajāpur

110 C8 **Murten** Neuchâtel, W Switzerland
46.55N 7.06E

Murtensee see Morat, Lac de

190 L11 **Murtoa** Victoria, SE Australia
36.40N 142.27E

94 N13 **Murtovaara** Oulu, E Finland
65.40N 29.25E

161 D14 **Murud** Mahārāshtra, W India
18.27N 72.56E

192 O9 **Murupara** var. Murupara. Bay of
Plenty, North Island, NZ

203 X12 **Mururoa** var. Moruroa. atoll Îles
Tuamotu, SE French Polynesia

Murviedro see Sagunto

160 J9 **Murwāra** Madhya Pradesh,
N India 23.50N 80.23E

191 V4 **Murwillumbah** New South
Wales, SE Australia 28.195 153.24E

152 H11 **Murzechirla** prev. Mirzachila.
Akhalskiy Velayat, C Turkmenistan
39.33N 60.02E

---

85 L16 **Mutoko** prev. Mtoko.
Mashonaland East, NE Zimbabwe
17.245 32.13E

83 J20 **Mutomo** Eastern, S Kenya
1.495 38.13E

126 J12 **Mutoray** Evenkiyskiy
Avtonomnyy Okrug, C Russian
Federation 61.30N 101.00E

Mutrah see Maţraḩ

81 M24 **Mutshatsha** Katanga, S Dem.
Rep. Congo (Zaire) 10.405 24.25E

172 Nn8 **Mutsu** var. Mutu. Aomori,
Honshū, N Japan 41.18N 141.11E

172 N8 **Mutsu-wan** bay N Japan

110 E6 **Muttenz** Basel-Land,
NW Switzerland 47.31N 7.39E

193 A26 **Muttonbird Islands** island group
SW NZ

Mutu see Mutsu

85 O15 **Mutuáli** Nampula,
N Mozambique 14.515 37.01E

84 D13 **Mutumbo** Bié, C Angola
11.055 17.22E

201 Y14 **Mutunte, Mount** var. Mount
Buache. ▲ Kosrae, E Micronesia
5.21N 163.00E

161 K24 **Mutur** Eastern Province, E Sri
Lanka 8.27N 81.15E

94 L13 **Muurola** Lappi, NW Finland
66.19N 25.00E

168 M14 **Mu Us Shamo** var. Ordos Desert.
desert N China

84 B11 **Muxima** Bengo, NW Angola

128 I8 **Muyezerskiy** Respublika
Kareliya, NW Russian Federation
63.54N 32.00E

83 E20 **Muyinga** NE Burundi 2.545 30.19E

44 K9 **Muy Muy** Matagalpa,
C Nicaragua 12.43N 85.37W

Muynak see Mŭynoq

152 G6 **Mŭynoq** Rus. Muynak.
Qoraqalpoghiston Respublikasi,
NW Uzbekistan 43.45N 59.03E

81 N22 **Muyumba** Katanga, SE Dem.
Rep. Congo (Zaire) 7.135 27.02E

155 V5 **Muzaffarābād** Jammu and
Kashmir, NE Pakistan 34.24N 73.30E

155 S10 **Muzaffargarh** Punjab,
E Pakistan 30.04N 71.10E

158 J9 **Muzaffarnagar** Uttar Pradesh,
N India 29.28N 77.42E

159 P13 **Muzaffarpur** Bihār, N India
26.07N 85.22E

164 H6 **Muzat He** ↔ W China

85 L15 **Muze** Tete, NW Mozambique
15.055 31.16E

125 Fj8 **Muzhi** Yamalo-Nenetskiy
Avtonomnyy Okrug, N Russian
Federation 65.25N 64.40E

104 H7 **Muzillac** Morbihan, NW France
47.34N 2.30W

114 L9 **Mužlja** Hung. Felsőmuzslya; prev.
Gornja Mužlja. Serbia,
N Yugoslavia 45.21N 20.25E

56 F9 **Muzo** Boyacá, C Colombia
5.34N 74.07W

85 J15 **Muzoka** Southern, S Zambia
16.395 27.21E

41 Y15 **Muzon, Cape** headland Dall
Island, Alaska, USA 54.39N 132.41W

42 M4 **Múzquiz** Coahuila de Zaragoza,
NE Mexico 27.52N 101.31W

153 U13 **Muzqŭl, Qatorkŭhi** Rus.
Khrebet Muzkol. ▲ SE Tajikistan

164 G10 **Muztag** ▲ NW China 36.02N 80.13E

164 D8 **Muztag** ▲ W China
38.16N 75.03E

164 K10 **Muztag Feng** var. Ulugh Muztag.
▲ W China 36.26N 87.15E

85 K17 **Mvuma** prev. Umvuma.
Midlands, C Zimbabwe 19.165 30.31E

84 L13 **Mwanza** Eastern, Z Zambia
12.405 32.15E

83 G20 **Mwanza** Mwanza, NW Tanzania
2.315 32.55E

81 N23 **Mwanza** Katanga, SE Dem. Rep.
Congo (Zaire) 7.495 26.49E

83 F20 **Mwanza** ◆ region N Tanzania

84 M13 **Mwase Lundazi** Eastern,
E Zambia 12.265 33.20E

99 B17 **Mweelrea** Ir. Caoc Maol Réidh.
▲ W Ireland 53.37N 9.47W

84 K21 **Mweka** Kasai Occidental, C Dem.
Rep. Congo (Zaire) 4.515 21.37E

84 L22 **Mwene-Ditu** Kasai Oriental,
S Dem. Rep. Congo (Zaire)
7.055 23.33E

85 J18 **Mwenezi** S Zimbabwe

81 O20 **Mwenga** Sud Kivu, E Dem.
Congo (Zaire) 3.00S 28.28E

84 K11 **Mweru, Lake** var. Lac Moero.
◎ Dem. Rep. Congo
(Zaire)/Zambia

84 H13 **Mwinilunga** North Western,
NW Zambia 11.435 24.24E

201 V16 **Mwokil Atoll** var. Mokil Atoll.
atoll Caroline Islands, E Micronesia

120 J13 **Myadzyel** Pol. Miadzioł Nowy,
Rus. Myadel'. Minskaya Voblasts',
N Belarus 54.51N 26.51E

84 I13 **Myaksa** Vologodskaya Oblast',
NW Russian Federation
58.54N 38.15E

191 U8 **Myall Lake** ◎ New South Wales,
SE Australia

177 Ff7 **Myanaung** Irrawaddy,
SW Myanmar 18.16N 95.19E

178 Gg4 **Myanmar** off. Union of
Myanmar, var. Burma. ◆ military
dictatorship SE Asia

177 Ff9 **Myaungmya** Irrawaddy,
SW Myanmar 16.33N 94.55E

120 M13 **Myazha** Rus. Mezha. Vitsyebskaya
Voblasts', NE Belarus 55.40N 30.25E

177 Ff5 **Myerkulavichy** Rus.
Merkulovichi. Homyel'skaya
Voblasts', SE Belarus 52.57N 30.33E

121 N14 **Myezhava** Rus. Mezhëvo.
Vitsyebskaya Voblasts', NE Belarus
54.39N 30.18E

---

177 Ff5 **Myingyan** Mandalay, C Myanmar
21.25N 95.19E

177 G5 **Myinmu** Sagaing, C Myanmar
21.25N 95.07E

178 Gg2 **Myitkyina** Kachin State,
N Myanmar 25.24N 97.25E

177 G5 **Myittha** Mandalay, C Myanmar
21.21N 96.06E

113 H19 **Myjava** Hung. Miava. Trenčiansky
Kraj, W Slovakia 48.48N 17.31E

119 V9 **Myjeldino** see Myyëldino

119 V9 **Mykhaylivka** Rus. Mikhaylovka.
Zaporiz'ka Oblast', SE Ukraine
47.16N 35.14E

97 A18 **Mykines Dan.** Myggenaes Island
Faeroe Islands 62.07N 7.38W

118 I5 **Mykolayiv** L'vivs'ka Oblast',
W Ukraine 49.31N 23.59E

119 Q10 **Mykolayiv** Rus. Nikolayev.
Mykolayivs'ka Oblast', S Ukraine
46.57N 31.58E

119 Q10 **Mykolayiv ✕** Mykolayivs'ka
Oblast', S Ukraine 47.02N 31.54E

119 S13 **Mykolayivka** Respublika Krym,
S Ukraine 44.58N 33.37E

119 O9 **Mykolayiv, Res'ka Oblast'** var.
Mykolayiv, Rus. Nikolayevskaya
Oblast'. ◆ province S Ukraine

117 J20 **Mýkonos** Mýkonos, Kykládes,
Greece, Aegean Sea 37.27N 25.20E

117 K20 **Mýkonos** var. Míkonos. island
Kykládes, Greece, Aegean Sea

129 R7 **Myla** Respublika Komi,
NW Russian Federation
65.24N 50.51E

Mylae see Milazzo

95 L19 **Myllykoski** Etelä-Suomi,
S Finland 60.45N 26.52E

159 U14 **Mymensing** see Mymensingh

159 U14 **Mymensingh** var. Maimansingh,
Mymensing; prev. Nasirābād.
Dhaka, N Bangladesh 24.45N 90.22E

95 K19 **Mynämäki** Länsi-Suomi, W
Finland 60.41N 22.00E

151 S14 **Mynaral Kaz.** Myngaral.
Zhambyl, S Kazakhstan
45.25N 73.37E

Mynbulak see Mingbuloq

85 Q14 **Mynbulak, Vpadina** see
Mingbuloq Botighi

Myngaral see Mynaral

169 W13 **Myohaung** Arakan State,
W Myanmar 20.34N 93.12E

171 Jj13 **Myohyang-sanmaek** ▲ C North
Korea

85 I15 **Myōkō-san** ▲ Honshū, S Japan
36.54N 138.05E

85 I15 **Myooye** Central, C Zambia
15.05 27.24E

120 K12 **Myory** var. Miyory. Vitsyebskaya
Voblasts', N Belarus 55.39N 27.39E

94 J4 **Mýrdalsjökull** glacier S Iceland

94 G10 **Myre** Nordland, C Norway
68.54N 15.04E

119 J15 **Myrhorod** Rus. Mirgorod.
Poltavs'ka Oblast', NE Ukraine
49.57N 33.36E

117 J15 **Mýrina** var. Mírina. Límnos,
SE Greece 39.52N 25.04E

119 P5 **Myronivka** Rus. Mironovka.
Kyyivs'ka Oblast', N Ukraine

23 U13 **Myrtle Beach** South Carolina,
SE USA 33.41N 78.52W

34 F14 **Myrtle Creek** Oregon, NW USA
43.01N 123.19W

191 P11 **Myrtleford** Victoria, SE Australia
36.345 146.45E

34 E14 **Myrtle Point** Oregon, NW USA
43.04N 124.08W

117 K25 **Mýrtos** Kriti, Greece,
E Mediterranean Sea 35.00N 25.34E

Myrtoum Mare see Mirtóo
Pélagos

95 G17 **Myrvik** Jämtland, C Sweden
62.59N 14.19E

97 J15 **Mysen** Østfold, S Norway
59.33N 11.19E

128 L15 **Myshkin** Yaroslavskaya Oblast',
NW Russian Federation
57.47N 38.28E

113 H17 **Myślenice** Małopolskie, S Poland
49.49N 19.55E

112 D10 **Myślibórz** Zachodniopomorskie,
NW Poland 52.55N 14.51E

161 G20 **Mysore** var. Maisur. Karnātaka,
W India 12.18N 76.37E

123 J16 **Mysore** see Karnātaka

117 F21 **Mystrás** var. Mistras.
Pelopónnisos, S Greece
57.03N 22.22E

118 I7 **Myszyna Pol.** Komi-Permyatskiy
Nadvornaya. Ivano-Frankivs'ka
Oblast', W Ukraine 48.27N 24.30E

113 K15 **Myszków** Śląskie, S Poland
63.52N 34.17E

178 Jj14 **My Tho** var. Mi Tho. Tiên Giang,
S Vietnam 10.21N 106.21E

117 L17 **Mytilene** var. Mitilíni; anc.
Mytilene. Lésvos, E Greece
39.05N 26.33E

130 K3 **Mytishchi** Moskovskaya Oblast',
W Russian Federation 56.00N 37.51E

39 N3 **Myton** Utah, W USA
40.11N 110.03W

94 K2 **Mývatn** ◎ C Iceland

129 T11 **Myyëldino** var. Myjeldino.
Respublika Komi, NW Russian
Federation 61.46N 54.48E

95 K19 **Mäntäli Swe.** Nådendal. Länsi-
Suomi, SW Finland 60.29N 22.10E

— N —

100 J10 **Naarden** Noord-Holland,
C Netherlands 52.18N 5.10E

111 U4 **Naarn** ↔ N Austria

99 F18 **Naas Ir.** An Nás, Nás na Ríogh.
C Ireland 53.13N 6.39W

94 M9 **Näätämöjoki Lapp.** Njávdám.
↔ NE Finland

85 E23 **Nababeep** var. Nababiep.
Northern Cape, W South Africa
29.365 17.46E

Nababiep see Nababeep

171 H16 **Nabari** Mie, Honshū, SW Japan
34.37N 136.06E

144 G8 **Nabatié var.** An Nabatīyah at
Taḩtā, Nabatié, Nabatiyet et Tahta.
SW Lebanon 33.18N 35.36E

197 I13 **Nabavatu** Vanua Levu, N Fiji
16.35S 178.55E

202 I2 **Nabeina** island Tungaru,
W Kiribati

131 T4 **Naberezhnyye Chelny** prev.
Brezhnev. Respublika Tatarstan,
W Russian Federation 55.43N 52.21E

44 I10 **Nabarote** León, SW Nicaragua

41 T10 **Nabesna** Alaska, USA

41 T10 **Nabesna River** ↔ Alaska, USA

77 N5 **Nabeul** var. Nābul. NE Tunisia
36.32N 10.45E

158 I9 **Nabha** Punjab, NW India
30.22N 76.12E

176 Ww11 **Nabire** Irian Jaya, E Indonesia
3.225 135.31E

147 O15 **Nabi Shu'ayb, Jabal an**
▲ W Yemen 15.24N 44.04E

197 I13 **Nabiti** Vanua Levu, N Fiji
16.375 178.54E

144 F10 **Nablus** var. Nābulus, Heb.
Shekhem; anc. Neapolis, Bibl.
Shechem. N West Bank

100 I3 **Nagele** Flevoland, N Netherlands
52.39N 5.43E

161 H24 **Nabouwalu** Vanua Levu, N Fiji
17.005 178.43E

159 X12 **Nāginimāra** Nāgāland, NE India
26.43N 94.51E

Nābul see Nabeul

Nābulus see Nablus

197 J13 **Nabuna** Vanua Levu, N Fiji

151 S14 **Nabunturan** Mindanao,
S Philippines 7.34N 125.54E

85 Q14 **Nacala** Nampula,
NE Mozambique 14.305 40.37E

44 H8 **Nacaome** Valle, S Honduras
13.30N 87.31W

Na Cealla Beaga see Killybegs

Na-ch'ii see Nagqu

171 Gg17 **Nachikatsuura var.** Nachi-
Katsuura. Wakayama, Honshū,
SW Japan 33.37N 135.54E

83 J24 **Nachingwea** Lindi, SE Tanzania
10.205 38.46E

113 F16 **Náchod** Hradecký Kraj, N Czech
Republic 50.25N 16.09E

Na Clocha Liatha see Greystones

42 G3 **Naco** Sonora, NW Mexico
31.16N 109.56W

27 X8 **Nacogdoches** Texas, SW USA
31.36N 94.40W

42 G4 **Nacozari de García** Sonora,
NW Mexico 30.27N 109.41W

197 H13 **Nacula** prev. Nathula. island
Yasawa Group, NW Fiji

42 J6 **Nada** see Danzhou

79 Q11 **Nadawli** NW Ghana 10.30N 2.40W

106 I3 **Nadela** Galicia, NW Spain
42.58N 7.33W

150 M7 **Nadendah** see Naantali

47 O8 **Nadezhdinka** prev.
Nadezhdinskiy. Kostanay,
N Kazakhstan 53.46N 63.43E

Nadezhdinskiy var. Nadezhdinka

197 H14 **Nadi** prev. Nandi. Viti Levu,
W Fiji 17.48S 177.25E

197 H14 **Nadi** prev. Nandi. ✕ Viti Levu,
W Fiji 17.46S 177.28E

160 D10 **Nadiād** Gujarāt, W India
22.42N 72.54E

118 E11 **Nădlac Ger.** Nadlak, Hung.
Nagylak. Arad, W Romania
46.10N 20.47E

Nădlac see Nădlac

76 H6 **Nador** prev. Villa Nador.
NE Morocco 35.15N 2.56W

147 S9 **Nadqan, Qalamat var.** Nadgan.
well E Saudi Arabia 23.10N 50.08E

161 Q20 **Nadur** Gozo, N Malta 36.03N 14.18E

197 J13 **Nadurá** Vanua Levu, N Vanua
Levu, N Fiji 16.275 179.10E

118 I7 **Nadvirna Pol.** Nadwórna, Rus.
Nadvornaya. Ivano-Frankivs'ka
Oblast', W Ukraine 48.27N 24.30E

129 J8 **Nadvoitsy** Respublika Kareliya,
NW Russian Federation
63.52N 34.17E

Nadvornaya/Nadwórna see
Nadvirna

126 Gg9 **Nadym** Yamalo-Nenetskiy
Avtonomnyy Okrug, N Russian
Federation 65.25N 72.40E

158 J8 **Nāhan** Himāchal Pradesh,
NW India 30.33N 77.18E

42 J9 **Nahang, Rūd-e** see Nihing
6.36S 146.45E

110 H4 **Näfels** Glarus, NE Switzerland
47.06N 9.04E

146 C10 **Nahariyya** var. Nahariya,
Nahavand. N Israel 33.01N 35.04E

148 L6 **Nahāvand var.** Nehavend.
Hamadān, W Iran 34.13N 48.21E

103 F19 **Nahe** ↔ SW Germany

79 U14 **Nah-Iarmhidhe** see Westmeath

201 O13 **Nahoi, Cape** see Cumberland,
Cape

65 H16 **Nahuel Huapí, Lago**
◆ W Argentina

25 W7 **Nahunta** Georgia, SE USA
31.11N 81.58W

42 J6 **Naica** Chihuahua, N Mexico
27.53N 105.30W

9 U15 **Naicam** Saskatchewan, S Canada
52.26N 104.30W

170 E14 **Naiman Qi** Nei Mongol Zizhiqu,
N China 42.51N 120.41E

164 M4 **Naimin Bulak** spring NW China
44.57N 90.29E

11 P6 **Nain** Newfoundland, NE Canada
56.33N 61.41W

149 N8 **Na'īn** Esfahān, C Iran 32.52N 53.04E

---

158 K10 **Naini Tāl** Uttar Pradesh, N India
29.22N 79.25E

160 I10 **Nainpur** Madhya Pradesh,
C India 22.25N 80.10E

197 I14 **Nairai** island N Fiji

98 J8 **Nairn** N Scotland, UK 57.36N 3.51W

98 I8 **Nairn** cultural region NE Scotland,
UK

83 I19 **Nairobi ●** (Kenya) Nairobi Area,
S Kenya 1.16S 36.49E

83 I19 **Nairobi ✕** Nairobi Area, S Kenya
1.21S 37.01E

84 P13 **Nairoto** Cabo Delgado,
NE Mozambique 12.22S 39.05E

120 G3 **Naissaar** island N Estonia

Naissus see Niš

197 K13 **Naitaba var.** Naitauba; prev.
Naitamba. island Lau Group, E Fiji

Naitamba island/**Naitauba** see Naitaba

83 I19 **Naivasha** Rift Valley, SW Kenya
0.435 36.25E

83 H19 **Naivasha, Lake** ◎ SW Kenya

Najaf see An Najaf

149 N8 **Najafābād** var. Nejafabad.
Eşfahān, C Iran 32.37N 51.22E

147 N7 **Najd** var. Nejd. cultural region
C Saudi Arabia

107 O4 **Nájera** La Rioja, N Spain
42.25N 2.45W

107 P4 **Najerilla** ↔ N Spain

158 J9 **Najībābād** Uttar Pradesh, N India
29.37N 78.19E

169 Y11 **Najima** see Fukuoka

169 Y11 **Najin** NE North Korea
42.13N 130.15E

145 T9 **Najranal Ḩasan** C Iraq
32.24N 44.13E

147 O13 **Najrān** var. Abā as Su'ūd. Najrān,
S Saudi Arabia 17.31N 44.08E

147 P12 **Najrān off.** Minţaqat al Najrān. ◆
province S Saudi Arabia

170 Bb12 **Nakadōri-jima** island Gotō-rettō,
SW Japan

171 Kk15 **Naka-gawa** ↔ Honshū, S Japan
46.19N 142.04E

172 P14 **Nago** Okinawa, Okinawa,
SW Japan 26.36N 127.58E

160 K9 **Nagod** Madhya Pradesh, C India
24.36N 80.35E

161 J26 **Nagoda** Southern Province, S Sri
Lanka 6.13N 80.21E

103 G22 **Nagold** Baden-Württemberg,
SW Germany 48.33N 8.43E

Nagorno-Karabakhskaya
Avtonomnaya Oblast see
Nagornyy Karabakh

126 Ll13 **Nagornyy** Respublika Sakha
(Yakutiya), NE Russian Federation
55.53N 124.58E

143 V12 **Nagornyy Karabakh** var.
Nagorno-Karabakhskaya
Avtonomnaya Oblast, Arm.
Lerrnayin Gharabakh, Az. Dağlıq
Qarabağ. former autonomous region
SW Azerbaijan

129 R13 **Nagorsk** Kirovskaya Oblast',
NW Russian Federation
58.18N 50.49E

171 Hh15 **Nagoya** Aichi, Honshū, SW Japan
35.10N 136.53E

160 I12 **Nāgpur** Mahārāshtra, C India
21.09N 79.06E

162 K10 **Nagqu Chin.** Na-ch'ii; prev. Hei-
ho. Xizang Zizhiqu, W China
31.30N 91.57E

158 J8 **Nāg Tibba Range** ▲ N India

47 O8 **Nagua** NE Dominican Republic
19.18N 69.48W

118 H8 **Nagyatád** Somogy, SW Hungary
46.14N 17.19E

Nagybánya see Baia Mare

Nagybecskerek see Zrenjanin

Nagydisznód see Cisnădie

Nagyenyed see Aiud

113 N21 **Nagykálló** Szabolcs-Szatmár-
Bereg, E Hungary 47.49N 21.47E

113 G25 **Nagykanizsa** Ger. Grosskanizsa.
Zala, SW Hungary 46.27N 17.00E

82 J8 **Nagykőrös** Pest, C Hungary
47.24N 19.43E

113 K23 **Nagykikinda** see Kikinda

Nagyküküllő see Târnava Mare

113 J23 **Nagymihály** see Michalovce

113 J22 **Nagyőce** see Revúca

123 J16 **Nagysomkút** see Şomcuta Mare

197 J13 **Nagyszalonta** see Salonta

118 I7 **Nagyszeben** see Sibiu

129 T12 **Nagyszentmiklós** see Sânnicolau
Mare

113 J8 **Nagyszőllős** see Vynohradiv

Nagyszombat see Trnava

Nagytapolcsány see Topol'čany

Nagyvárad see Oradea

172 Oo15 **Naha** Okinawa, Okinawa,
SW Japan 26.10N 127.40E

158 J8 **Nāhan** Himāchal Pradesh,
NW India 30.33N 77.18E

144 F8 **Nahariyya** var. Nahariya.
N Israel 33.01N 35.04E

178 H15 **Nakhon Si Thammarat** var.
Nagara Sridharmaraj, Nakhon
Sithammaraj. Nakhon Si
Thammarat, SW Thailand
8.24N 99.58E

Nakhon Sithammaraj see
Nakhon Si Thammarat

145 V3 **Nakhrash** SE Iraq 31.13N 47.24E

8 I9 **Nakina** British Columbia,
W Canada 59.12N 132.48W

112 H9 **Nakło nad Notecią Ger.** Nakel.
Kujawsko-pomorskie, C Poland
53.07N 17.34E

41 P13 **Naknek** Alaska, USA
58.45N 157.01W

158 H10 **Nakodar** Punjab, NW India
31.06N 75.31E

84 L10 **Nakonde** Northern, NE Zambia
9.22S 32.45E

Nakorn Pathom see Nakhon
Pathom

97 H24 **Nakskov** Storstrøm, SE Denmark
54.50N 11.05E

169 Y15 **Naktong-gang** var. Nakdong, Jap. Rakutō-kō. ♣ C South Korea

83 H18 **Nakuru** Rift Valley, SW Kenya 0.16S 36.04E

83 H19 **Nakuru, Lake** ☺ Rift Valley, C Kenya

9 O17 **Nakusp** British Columbia, SW Canada 50.13N 117.48W

155 N15 **Näl** ♣ W Pakistan

168 M7 **Nalayh** Töv, C Mongolia 47.48N 107.17E

159 V12 **Nalbāri** Assam, NE India 26.36N 91.49E

65 G19 **Nalcayec, Isla** island Archipiélago de los Chonos, S Chile

131 N15 **Nal'chik** Kabardino-Balkarskaya Respublika, SW Russian Federation 43.29N 43.39E

161 I16 **Nalgonda** Andhra Pradesh, C India 17.04N 79.15E

159 S14 **Nalitabari** Dhaka, N Bangladesh 24.19N 87.52E

159 U14 **Nalitabari** Dhaka, N Bangladesh 26.90N 90.10E

161 I17 **Nallamala Hills** ▲ E India

142 G12 **Nallıhan** Ankara, NW Turkey 40.12N 31.22E

106 K2 **Nalón** ♣ NW Spain

178 Gg3 **Nalong** Kachin State, N Myanmar 24.42N 97.27E

77 N8 **Nālūt** NW Libya 31.52N 10.58E

176 Uu12 **Nama** Pulau Manawaka, E Indonesia 4.07S 131.22E

201 Q16 **Nama** island C Micronesia

85 O16 **Namacurra** Zambézia, NE Mozambique 17.31S 37.03E

196 F9 **Namai Bay** bay Babeldaob, N Palau

31 W2 **Namakan Lake** ☺ Canada/USA

149 O6 **Namak, Daryācheh-ye** marsh N Iran

149 T6 **Namak, Kavīr-e** salt pan NE Iran

178 H6 **Namaklwe** Shan State, E Myanmar 19.45N 99.01E

Namaksār, Kowl-e/Namakzār, Daryācheh-ye see Namakzar

154 I5 **Namakzar Pash.** Daryācheh-ye Namakzar, Kowl-e Namaksār. marsh Afghanistan/Iran

176 W13 **Namalau** Pulau Jursian, E Indonesia 5.50S 134.43E

83 I20 **Namanga** Rift Valley, S Kenya 2.33S 36.48E

153 S10 **Namangan** Namangan Wiloyati, E Uzbekistan 40.59N 71.33E

Namanganskaya Oblast' see Namangan Wiloyati

153 R10 **Namangan Wiloyati** Rus. Namanganskaya Oblast'. ♦ province E Uzbekistan

85 Q14 **Namapa** Nampula, NE Mozambique 13.43S 39.48E

85 C21 **Namaqualand** physical region S Namibia

83 G18 **Namatanai** C Uganda 1.01N 32.58E

195 P10 **Namatanai** New Ireland, NE PNG 3.42S 152.28E

85 I14 **Nambala** Central, C Zambia 15.06S 27.03E

83 J23 **Namanje** Lindi, SE Tanzania 8.37S 38.21E

176 Ww9 **Namber** Irian Jaya, E Indonesia 0.58S 134.51E

85 G16 **Nambiya** Ngamiland, N Botswana 18.09S 23.08E

191 V2 **Nambour** Queensland, E Australia 26.43S 152.54E

191 V6 **Nambucca Heads** New South Wales, SE Australia 30.37S 153.00E

165 N15 **Nam Co** ☺ W China

178 I5 **Năm Cum** Lai Châu, N Vietnam 22.37N 103.12E

Namdik see Namorik Atoll

178 J6 **Nam Đinh** Nam Ha, N Vietnam 20.25N 106.12E

175 Tt11 **Namea, Tanjung** headland Pulau Seram, E Indonesia

101 I20 **Namêche** Namur, SE Belgium 50.29N 5.02E

32 J4 **Namekagon Lake** ☺ Wisconsin, N USA

196 F10 **Namekakl Passage** passage Babeldaob, N Palau

Namen see Namur

85 P15 **Nametil** Nampula, NE Mozambique 15.46S 39.21E

169 X14 **Nam-gang** ♣ C North Korea

169 X19 **Nam-gang** ♣ S South Korea

169 Y17 **Namhae-do Jap.** Nankai-tō. island S South Korea

Namhoi see Foshan

85 C19 **Namib Desert** desert W Namibia

85 A15 **Namibe Port.** Moçâmedes, Mossâmedes. Namibe, SW Angola 15.10S 12.09E

85 A15 **Namibe** ♦ province SE Angola

85 C18 **Namibia off.** Republic of Namibia, var. South West Africa, Afr. Suidwes-Afrika, Ger. Deutsch-Südwestafrika; prev. German Southwest Africa, South-West Africa. ♦ republic S Africa

117 O17 **Namibia Plain** undersea feature S Atlantic Ocean

171 Ll14 **Namie** Fukushima, Honshū, C Japan 37.29N 140.58E

171 Mm8 **Namioka** Aomori, Honshū, C Japan 40.43N 140.34E

42 I5 **Namiquipa** Chihuahua, N Mexico 29.15N 107.25W

165 P15 **Namjagbarwa Feng** ▲ W China 29.38N 95.00E

175 Ss11 **Namlea** Pulau Buru, E Indonesia 3.12S 127.06E

164 L16 **Namling** Xizang Zizhiqu, W China 29.40N 88.58E

Namnetes see Nantes

178 I8 **Nam Ngum** ♣ C Laos

Namo see Namu Atoll

191 R5 **Namoi River** ♣ New South Wales, SE Australia

201 Q17 **Namoluk Atoll** atoll Mortlock Islands, C Micronesia

201 O15 **Namonuito Atoll** atoll Caroline Islands, C Micronesia

201 T9 **Namorik Atoll** var. Namdik. atoll Ralik Chain, S Marshall Islands

178 I6 **Nam Ou** ♣ N Laos

54 M14 **Nampa** Idaho, NW USA 43.32N 116.33W

78 M11 **Nampala** Ségou, W Mali 15.21N 5.32W

169 W14 **Namp'o** SW North Korea 38.45N 125.25E

85 P15 **Nampula** Nampula, NE Mozambique 15.09S 39.13E

85 P15 **Nampula off.** Província de Nampula. ♦ province NE Mozambique

169 W13 **Namsan-ni** NW North Korea 40.25N 125.01E

95 E15 **Namsos** Nord-Trøndelag, C Norway 64.28N 11.31E

95 F14 **Namskogan** Nord-Trøndelag, C Norway 64.57N 13.04E

186 M10 **Namsy** Respublika Sakha (Yakutiya), NE Russian Federation 62.42N 129.30E

178 H6 **Nam Teng** ♣ E Myanmar

178 I6 **Nam Tha** ♣ N Laos

178 Gg4 **Namtu** Shan State, E Myanmar 23.04N 97.25E

8 J15 **Namu** British Columbia, SW Canada 51.46N 127.49W

201 T9 **Namu Atoll** var. Namo. atoll Ralik Chain, C Marshall Islands, C Pacific Ocean

197 K15 **Namuka-i-lau** island Lau Group, E Fiji

85 O15 **Namuli, Mont** ▲ NE Mozambique 15.15S 37.33E

85 P14 **Namuno** Cabo Delgado, N Mozambique 13.36S 38.52E

101 I20 **Namur** Dut. Namen. Namur, SE Belgium 50.28N 4.52E

101 H21 **Namur** Dut. Namen. ♦ province S Belgium

85 D17 **Namutoni** Kunene, N Namibia 18.47S 16.48E

169 Y16 **Namwon** Jap. Nangen. S South Korea 35.24N 127.20E

178 Mm14 **Namyit Island** island S Spratly Islands

113 H14 **Namysłów** Ger. Namslau. Opolskie, S Poland 51.05N 17.41E

178 Hh7 **Nan** var. Muang Nan. Nan, NW Thailand 18.47N 100.46E

72 Nn7 **Nanae** Hokkaidō, NE Japan 41.55N 140.40E

81 I14 **Nana-Grébizi** ♦ prefecture N Central African Republic

8 L17 **Nanaimo** Vancouver Island, British Columbia, SW Canada 49.07N 123.58W

40 O9 **Nanakuli** Haw. Nānākuli. Oahu, Hawaii, USA, C Pacific Ocean 21.23N 158.09W

81 G15 **Nana-Mambéré** ♦ prefecture W Central African Republic

167 R13 **Nan'an** Fujian, SE China 24.57N 118.22E

191 U2 **Nanango** Queensland, E Australia 26.42S 151.58E

171 I12 **Nanao** Ishikawa, Honshū, SW Japan 37.02N 136.57E

167 Q14 **Nan'ao Dao** island S China

171 I11 **Nanatsu-shima** island SW Japan

58 F8 **Nanay, Río** ♣ NE Peru

166 J8 **Nanbu** Sichuan, C China 31.19N 106.02E

84 P13 **Nancha** Heilongjiang, NE China 47.09N 129.16E

167 P10 **Nanchang** var. Nan-ch'ang, Nanch'ang-hsien. Jiangxi, S China 28.38N 115.57E

Nanch'ang-hsien see Nanchang

167 P11 **Nancheng** Jiangxi, S China 27.37N 116.37E

Nan-ching see Nanjing

166 I10 **Nanchong** Sichuan, China 30.46N 106.03E

166 J10 **Nanchuan** Chongqing Shi, C China 29.06N 107.13E

105 T5 **Nancy** Meurthe-et-Moselle, NE France 48.69N 6.10E

193 A22 **Nanda** South Island, NZ

158 L9 **Nanda Devi** ▲ NW India 30.27N 80.00E

44 J11 **Nandaime** Granada, SW Nicaragua 11.46N 86.03W

166 K13 **Nandan** Guangxi Zhuangzu Zizhiqu, S China 25.03N 107.31E

161 H14 **Nanded** Mahārāshtra, C India 19.10N 77.21E

170 G15 **Nandeln** Hyōgo, Awaji-shima, SW Japan 34.19N 134.53E

191 S5 **Nandewar Range** ▲ New South Wales, SE Australia

Nandi see Nadi

86 E13 **Nanding He** ♣ China/Vietnam

Nándorhey see Oţelu Roşu

160 I10 **Nandurbār** Mahārāshtra, W India 21.22N 74.18E

Nanduri see Naduri

161 I17 **Nandyāl** Andhra Pradesh, E India 15.30N 78.28E

167 P11 **Nanfeng** Jiangxi, S China 27.15N 116.30E

Nang see Nang Xian

81 E15 **Nanga Eboko** Centre, C Cameroon 4.37N 12.21E

155 W4 **Nanga Parbat** ▲ India/Pakistan 35.15N 74.36E

174 L8 **Nangapinoh** Borneo, C Indonesia 0.21S 111.43E

155 R5 **Nangarhār** ♦ province E Afghanistan

174 M8 **Nangaserawai** var. Nangah Serawai. Borneo, C Indonesia 0.19S 112.25E

174 L9 **Nangatayap** Borneo, C Indonesia 1.30S 110.33E

Nangen see Namwŏn

105 P5 **Nangis** Seine-et-Marne, N France 48.36N 3.02E

169 X13 **Nangnim-sanmaek** ▲ North Korea

169 O4 **Nangong** Hebei, E China 37.24N 115.24E

165 Q14 **Nangqên** Qinghai, C China 32.05N 96.28E

178 I11 **Nang Rong** Buri Ram, E Thailand 14.37N 102.48E

165 O16 **Nang Xian** var. Nang. Xizang Zizhiqu, W China 29.04N 93.03E

166 F12 **Nanhua** Yunnan, SW China 25.15N 101.15E

166 J12 **Naniwa** see Ōsaka

161 G20 **Nanjangūd** Karnātaka, W India 12.07N 76.40E

166 Q8 **Nanjing** var. Nan-ching, Nanking; prev. Chianning, Chian-ning, Kiang-ning. Jiangsu, E China 32.03N 118.46E

167 O12 **Nankai-tō** see Namhae-do

167 N13 **Nanjiang** Jiangxi, S China 25.40N 114.40E

166 L15 **Nan Ling** ▲ S China

201 P13 **Nan Ling** ▲ S China

166 K15 **Nanliu Jiang** ♣ S China

206 M15 **Nan Madol** ruins Temwen Island, E Micronesia

166 H13 **Nanning** var. Nan-ning; prev. Yung-ning. Guangxi Zhuangzu Zizhiqu, S China 22.49N 108.19E

58 M11 **Nānpāra** Uttar Pradesh, N India 27.51N 81.30E

167 Q12 **Nanping** var. Nan-p'ing; prev. Yenping. Fujian, SE China 26.40N 118.07E

166 I7 **Nanping** Sichuan, C China 33.25N 104.05E

167 R12 **Nanri Dao** island SE China

172 Q13 **Nansei-shotō** Eng. Ryukyu Islands. island group SW Japan

Nansei Syotō Trench see Ryukyu Trench

207 T10 **Nansen Basin** undersea feature Arctic Ocean

207 T10 **Nansen Cordillera** var. Arctic-Mid Oceanic Ridge, Nansen Ridge. undersea feature Arctic Ocean

Nansen Ridge see Nansen Cordillera

133 T9 **Nan Shan** ▲ C China

179 Nn14 **Nanshan Island** island E Spratly Islands

Nansha Qundao see Spratly Islands

10 K3 **Nantais, Lac** ☺ Quebec, NE Canada

105 N5 **Nanterre** Hauts-de-Seine, N France 48.52N 2.13E

104 I8 **Nantes** Bret. Naoned; anc. Condivincum, Namnetes. Loire-Atlantique, NW France 47.12N 1.31W

12 G17 **Nanticoke** Ontario, S Canada 42.49N 80.04W

20 H13 **Nanticoke** Pennsylvania, NE USA 41.12N 76.00W

23 Y4 **Nanticoke River** ♣ Delaware/Maryland, NE USA

9 O17 **Nanton** Alberta, SW Canada 50.21N 113.46W

167 S8 **Nantong** Jiangsu, E China 32.00N 120.52E

167 S13 **Nant'ou** W Taiwan 23.54N 120.33E

105 S10 **Nantua** Ain, E France 46.10N 5.34E

21 Q13 **Nantucket** Nantucket Island, Massachusetts, NE USA 41.15N 70.05W

21 Q13 **Nantucket Island** island Massachusetts, NE USA

21 Q13 **Nantucket Sound** sound Massachusetts, NE USA

84 P13 **Nantulo** Cabo Delgado, N Mozambique 12.30S 39.03E

201 O12 **Nanuh** Pohnpei, E Micronesia

197 K13 **Nanuku Passage** channel NE Fiji

202 D6 **Nanumaga** var. Nanumanga. atoll NW Tuvalu

Nanumanga see Nanumaga

202 D5 **Nanumea Atoll** atoll NW Tuvalu

61 O19 **Nanuque** Minas Gerais, SE Brazil 17.49S 40.21N

175 Ss4 **Nanusa, Kepulauan** island group N Indonesia

169 U4 **Nanweng He** ♣ NE China

166 I10 **Nanxi** Sichuan, C China 28.54N 104.58E

167 N10 **Nanxian** var. Nan Xian. Hunan, S China 29.23N 112.18E

167 N7 **Nanyang** var. Nan-yang. Henan, C China 32.58N 112.29E

167 P6 **Nanyang Hu** ☺ E China

171 L13 **Nan'yō** Yamagata, Honshū, SW Japan 38.05N 140.07E

83 I18 **Nanyuki** Central, C Kenya 0.01N 37.04E

178 Jj4 **Nanzhang** Hubei, C China 31.47N 111.48E

107 T11 **Nao, Cabo de la** headland E Spain 38.43N 0.13E

10 M9 **Naococane, Lac** ☺ Quebec, E Canada

159 S14 **Naogaon** Rajshahi, NW Bangladesh 24.49N 88.58E

Naokot see Naukot

197 C12 **Naone** Maewo, C Vanuatu 15.03S 168.06E

Naoned see Nantes

191 S11 **Narooma** New South Wales, SE Australia 36.16S 150.08E

Narova see Narva

154 E14 **Nárowal** Punjab, E Pakistan 32.04N 74.54E

191 Q7 **Narowlya** Rus. Narovlya. Homyel'skaya Voblasts', SE Belarus 51.49N 29.30E

95 J17 **Närpes** Fin. Närpiö. Länsi-Suomi, W Finland 62.28N 21.19E

Närpiö see Närpes

191 S5 **Narrabri** New South Wales, SE Australia 30.21S 149.48E

79 V15 **Narrandera** New South Wales, SE Australia 34.46S 146.32E

191 Q4 **Narran Lake** ☺ New South Wales, SE Australia

46 H2 **Narran River** ♣ New South Wales/Queensland, SE Australia

188 J13 **Narrogin** Western Australia 32.52S 117.16E

191 Q7 **Narromine** New South Wales, SE Australia 32.16S 148.15E

23 R6 **Narrows** Virginia, NE USA 37.19N 80.48W

205 M15 **Narsarasuaq** ✕ Kitaa, S Greenland 61.07N 45.03W

160 I10 **Narsimhapur** Madhya Pradesh, C India 22.58N 79.15E

159 U15 **Narsingdi** var. Narsinghdi. Dhaka, C Bangladesh 23.55N 90.40E

159 H9 **Narsinghgarh** Madhya Pradesh, C India 23.45N 77.04E

159 Q11 **Nart** Nei Mongol Zizhiqu, N China 42.54N 115.35E

179 P11 **Nartès, Gjol i/Nartès, Laguna e** see Nartès, Liqeni i

96 M11 **Nartès, Liqeni i** var. Nartès, Gjol i. ☺ SW Albania

131 F17 **Nartháki** ▲ C Greece 39.12N 22.24E

24 J9 **Napoleonville** Louisiana, S USA 29.55N 91.01W

109 K17 **Napoli** Eng. Naples, Ger. Neapel; anc. Neapolis. Campania, S Italy 40.52N 14.15E

109 J18 **Napoli, Golfo di** gulf S Italy

59 F7 **Napo, Río** ♣ Ecuador/Peru

203 W9 **Napuka** island Îles Tuamotu, C French Polynesia

148 J3 **Naqadeh** Āžarbāyjān-e Bākhtarī, NW Iran 36.57N 45.20E

120 J3 **Naqnah** E Iraq 34.13N 45.33E

145 U6 **Naqnah** E Iraq 34.13N 45.33E

171 H15 **Nara** Nara, Honshū, SW Japan 34.40N 135.49E

78 L11 **Nara** Koulikoro, W Mali 15.04N 7.19W

171 Gg16 **Nara off.** Nara-ken. ♦ prefecture Honshū, SW Japan

155 R14 **Nāra Canal** irrigation canal S Pakistan

191 K11 **Naradhan** New South Wales, SE Australia 33.37S 146.19E

158 I9 **Naradhivas** see Narathiwat

54 W8 **Narayanganj** see Narathiwat

59 Q9 **Naranjos** Santa Cruz, E Bolivia

43 Q2 **Naranjos** Veracruz-Llave, E Mexico 21.20N 97.42W

165 Q6 **Naran Sebstein Bulag** spring NW China 42.40N 96.58E

149 X12 **Narāq** Eşfahān, C Iran

170 Bb12 **Narao** Nagasaki, Nakadōri-jima, SW Japan 32.40N 129.03E

161 J16 **Narasaraopet** Andhra Pradesh, E India 16.16N 80.06E

164 J5 **Narat** Xinjiang Uygur Zizhiqu, W China 43.19N 84.01E

178 Hh17 **Narathiwat** var. Naradhivas. Narathiwat, SW Thailand 6.25N 101.48E

39 V10 **Nara Visa** New Mexico, SW USA 35.35N 103.06W

Nārāyāni see Gandak

Narbada see Narmada

105 P16 **Narbonne** anc. Narbo Martius. Aude, S France 43.11N 3.00E

Narborough Island see Fernandina, Isla

106 J2 **Narcea** ♣ NW Spain

158 J9 **Nardaranagar** Uttar Pradesh, N India 30.10N 78.21E

66 G11 **Nares Plain** var. Nares Abyssal Plain. undersea feature NW Atlantic Ocean

207 P10 **Nares Strait** Dan. Nares Stræde. strait Canada/Greenland

Nares Stræde see Nares Strait

9 P10 **Narew** ♣ E Poland

161 N7 **Nargund** Karnātaka, S India 15.43N 75.23E

27 X5 **Nash** Texas, SW USA 33.26N 94.04W

160 E13 **Nāshik** prev. Nāsik. Mahārāshtra, W India 20.04N 73.48E

58 E7 **Nashino, Río** ♣ Ecuador/Peru

31 W2 **Nashua** Iowa, C USA 42.57N 92.32W

35 W7 **Nashua** Montana, NW USA 48.06N 106.16W

21 O10 **Nashua** New Hampshire, NE USA 42.45N 71.26W

29 S13 **Nashville** Arkansas, C USA 33.57N 93.51W

33 W14 **Nashville** Georgia, SE USA 31.12N 83.15W

32 L16 **Nashville** Illinois, N USA 38.20N 89.22W

35 O14 **Nashville** Indiana, N USA 39.13N 86.15W

23 V9 **Nashville** North Carolina, SE USA 35.58N 77.58W

22 J8 **Nashville** state capital Tennessee, S USA 36.10N 86.48W

22 J9 **Nashville** ✕ Tennessee, S USA 36.06N 86.44W

66 H10 **Nashville Seamount** undersea feature NW Atlantic Ocean

114 H9 **Našice** Osijek-Baranja, E Croatia 45.29N 18.05E

110 M11 **Nasielsk** Mazowieckie, C Poland 52.33N 20.46E

95 K18 **Näsijärvi** ☺ SW Finland

81 Q9 **Nasir** Upper Nile, SE Sudan 8.37N 33.06E

155 Q12 **Nasīrābād** Baluchistān, SW Pakistan 28.29N 68.24E

154 K15 **Nasīrābād** Baluchistān, SW Pakistan 36.16S 150.08E

Nasirabad see Mymensingh

197 I14 **Nasiri** see Ahvāz

38 F9 **Naso** Sicilia, Italy, C Mediterranean Sea 38.07N 14.46E

43 N6 **Nass** ♣ British Columbia, SW Canada

81 I20 **Nasarawa** Nassarawa, C Nigeria 8.33N 7.42E

46 H2 **Nassau** ● (Bahamas) New Providence, N Bahamas 25.03N 77.20W

46 H2 **Nassau** ✕ New Providence, C Bahamas 25.03N 77.30W

197 J14 **Nassau** island N Cook Islands

196 M9 **Nassau Sound** Florida, SE USA

110 L7 **Nassereith** Tirol, W Austria 47.19N 10.51E

97 L19 **Nässjö** Jönköping, S Sweden 57.39N 14.40E

101 K22 **Nassogne** Luxembourg, SE Belgium 50.09N 5.19E

159 M19 **Nastapoka Islands** island group Nunavut, C Canada

95 M19 **Nastola** Etelä-Suomi, S Finland 60.57N 25.55E

171 Ll14 **Nasu-dake** ▲ Honshū, S Japan 37.07N 139.57E

106 K10 **Nava del Rey** Castilla-León, N Spain 41.20N 5.05W

171 P14 **Nasushiobara** Tochigi, Honshū, S Japan 36.55N 140.02E

106 L6 **Nava del Rey** Castilla-León, N Spain 41.20N 5.05W

114 G9 **Našice**

159 P14 **Nasirabad**

92 J13 **Nazca Plate** tectonic feature

200 Oo11 **Nazca Ridge** undersea feature E Pacific Ocean

172 R12 **Naze** var. Nase. Kagoshima, Amami-ōshima, SW Japan 28.21N 129.30E

144 G9 **Nazerat** var. Natsrat, Ar. En Nazira, Eng. Nazareth. Northern, N Israel 32.42N 35.18E

143 Ff13 **Nazıyyevsk** Omskaya Oblast', C Russian Federation 55.35N 71.13E

84 J11 **Nchanga** Copperbelt, C Zambia 12.36S 27.52E

84 J11 **Ncheu** see Ntcheu

155 P6 **Nāvar, Dasht-e** Pash. Dasht-i-Nawar. desert C Afghanistan 3.30N 75.06W

127 Q5 **Navarin, Mys** headland NE Russian Federation 62.18N 179.06E

65 I25 **Navarino, Isla** island S Chile

107 Q4 **Navarra** Eng./Fr. Navarre. ♦ autonomous community N Spain

107 P4 **Navarrete** La Rioja, N Spain 42.25N 2.34W

63 C20 **Navarro** Buenos Aires, E Argentina 35.01S 59.18W

107 O12 **Navas de San Juan** Andalucía, S Spain 38.10N 3.19W

27 V10 **Navasota** Texas, SW USA 30.23N 96.05W

27 U9 **Navasota River** ♣ Texas, SW USA

46 I9 **Navassa Island** ◇ US unincorporated territory C West Indies

121 L19 **Navasyolki** Rus. Novosëlki. Homyel'skaya Voblasts', SE Belarus 52.24N 28.27E

121 H17 **Navayel'nya** Pol. Nowojelnia, Rus. Novoyel'nya. Hrodzyenskaya Voblasts', W Belarus 53.26N 25.36E

176 Yy11 **Naver** Irian Jaya, E Indonesia 3.27S 139.45E

120 H5 **Navesti** ♣ C Estonia

106 J2 **Navia** Asturias, N Spain 43.33N 6.43W

106 J2 **Navia** ♣ NW Spain

61 J21 **Naviraí** Mato Grosso do Sul, SW Brazil 23.05S 54.13W

197 G14 **Naviti** island Yasawa Group, NW Fiji

130 I6 **Navlya** Bryanskaya Oblast', W Russian Federation 52.47N 34.28E

197 J13 **Navoalevu** Vanua Levu, N Fiji 16.22S 179.28E

55 R12 **Navobod** Rus. Navabad, Novabad. C Tajikistan 39.00N 70.06E

55 P13 **Navobod** Rus. Navabad. W Tajikistan 38.37N 68.42E

Navoi see Nawoiy

Navoiyskaya Oblast' see Nawoiy Wiloyati

42 G7 **Navojoa** Sonora, NW Mexico 27.04N 109.28W

Navolat see Navolato

42 H9 **Navolato** var. Navolat. Navolat, C Mexico 24.46N 167.42W

197 C12 **Navonda** Ambae, C Vanuatu 15.21S 167.58E

Navpaktos see Náfpaktos

Návplion see Náfplio

79 P4 **Navrongo** N Ghana 10.54N 1.03W

160 D12 **Navsāri** var. Nausari. Gujarāt, W India 20.55N 72.55E

197 J15 **Navua** Viti Levu, W Fiji 18.13S 178.10E

148 H4 **Nawá** Dar'ā, S Syria 32.52N 36.03E

159 S14 **Nawabshah** see Nawābshāh

159 S14 **Nawabganj** Rajshahi, NW Bangladesh 24.36N 88.17E

159 S14 **Nawābganj** Uttar Pradesh, N India 26.52N 82.09E

155 Q15 **Nawābshāh** var. Nawabshah. Sind, S Pakistan 26.15N 68.25E

159 P14 **Nawada** Bihār, N India 24.54N 85.33E

158 H11 **Nawalgarh** Rājasthān, N India 27.51N 75.16E

Nawāl, Sabkhat an see Noual, Sebkhet en

Nawar, Dasht-i- see Nāvar, Dasht-e

178 Gg4 **Nawnghkio** var. Nawngkio. Shan State, C Myanmar 22.21N 96.48E

Nawngkio see Nawnghkio

152 M11 **Nawoiy** Rus. Navoi, Wiloyati, C Uzbekistan 40.05N 65.22E

152 K8 **Nawoiy Wiloyati** Rus. Navoiyskaya Oblast'. ♦ province N Uzbekistan

143 U13 **Naxçıvan** Rus. Nakhichevan'. SW Azerbaijan 39.13N 45.24E

166 I10 **Naxi** Sichuan, China 28.48N 105.25E

117 K21 **Náxos** var. Naxos. Náxos, Kykládes, Greece, Aegean Sea 37.06N 25.22E

117 K21 **Náxos** island Kykládes, Greece, Aegean Sea

42 K9 **Nayarit** ♦ state C Mexico

197 K14 **Nayau** island Lau Group, E Fiji

149 S8 **Näy Band** Khorāsān, E Iran 32.26N 57.30E

172 Pp4 **Nayoro** Hokkaidō, NE Japan 44.21N 142.27E

106 F9 **Nazaré** var. Nazare. Leiria, C Portugal 39.36N 9.04W

26 M4 **Nazareth** Texas, SW USA 34.32N 102.06W

181 O8 **Nazareth Bank** undersea feature W Indian Ocean

226 Hh14 **Nazarovo** Krasnoyarskiy Kray, S Russian Federation 56.00N 89.33E

42 K9 **Nazas** Durango, C Mexico 25.16N 104.04W

59 F16 **Nazca** Ica, S Peru 14.52S 75.01W

1 L15 **Nazca Plate** tectonic feature

Ndaghamcha, Sebkra de see Te-n-Dghâmcha, Sebkhet

83 G21 **Ndala** Tabora, C Tanzania
4.45S 33.15E

84 B11 **N'Dalatando** *Port.* Salazar, Vila Salazar. Cuanza Norte, NW Angola
9.18S 14.48E

79 S14 **Ndali** C Benin *9.52N 2.44E*

83 E18 **Ndde** SW Uganda *0.11S 30.04E*

80 J13 **Ndélé** Bamingui-Bangoran, N Central African Republic
8.24N 20.40E

81 E19 **Ndendé** Ngounié, S Gabon
2.21S 11.19E

81 E20 **Ndindi** Nyanga, S Gabon
3.46S 11.06E

80 G11 **Ndjamena** *var.* N'Djamena; *prev.* Fort-Lamy. ● (Chad) Chari-Baguirmi, W Chad *12.08N 15.01E*

80 G11 **Ndjamena** ✈ Chari-Baguirmi, W Chad *12.09N 15.00E*

81 D18 **Ndjolé** Moyen-Ogooué, W Gabon *0.07S 10.45E*

84 J13 **Ndola** Copperbelt, C Zambia *12.58S 28.35E*

Ndrhamcha, Sebkha de see Te-n-Dghâmcha, Sebkhet

81 L15 **Ndu** Orientale, N Dem. Rep. Congo (Zaire) *4.46N 22.54E*

83 H21 **Nduguti** Singida, C Tanzania *4.19S 34.40E*

195 X16 **Nduindui** Guadalcanal, C Solomon Islands *9.46S 159.54E*

Nduke see Kolombangara

117 F16 **Néa Anchíalos** *var.* Nea Anhialos, Néa Ankhíalos. Thessalía, C Greece *39.18N 22.49E*

Nea Anhialos/Néa Ankhíalos see Néa Anchíalos

117 H18 **Néa Artáki** Évvoia, C Greece *38.31N 23.39E*

99 F15 **Neagh, Lough** ⊚ E Northern Ireland, UK

34 F7 **Neah Bay** Washington, NW USA *48.21N 124.39W*

117 J22 **Nea Kaméni** *island* Kykládes, Greece, Aegean Sea

189 O8 **Neale, Lake** ⊚ Northern Territory, C Australia

190 G2 **Neales River** *seasonal river* South Australia

117 G14 **Néa Moudanía** *var.* Nea Moudhaniá. Kentrikí Makedonía, N Greece *40.15N 23.19E*

Néa Moudhanía see Néa Moudanía

118 K10 **Neamt** ◆ *county* NE Romania

Neapel see Napoli

117 D14 **Neápoli** *prev.* Neápolis. Dytikí Makedonía, N Greece *40.18N 21.23E*

117 K25 **Neápoli** Kríti, Greece, E Mediterranean Sea *35.15N 25.37E*

117 G22 **Neápoli** Pelopónnisos, S Greece *36.29N 23.05E*

Neápolis see Napoli, Italy

Neápolis see Nablus, West Bank

Neápolis see Neápoli, Greece

40 D16 **Near Islands** *island group* Aleutian Islands, Alaska, USA

99 J21 **Neath** S Wales, UK *51.39N 3.48W*

116 H13 **Néa Zíchni** *var.* Néa Zíchni; *prev.* Néa Zíkhna. Kentrikí Makedonía, NE Greece *41.02N 23.51E*

Néa Zíkhna/Néa Zíkhni see Néa Zíchni

44 C5 **Nebaj** Quiché, W Guatemala *15.25N 91.05W*

79 P13 **Nebbou** S Burkina *11.22N 1.49W*

152 B11 **Nebitdag** Balkanskiy Velayat, W Turkmenistan *39.33N 54.19E*

56 M13 **Neblina, Pico da** ▲ NW Brazil *0.49N 66.31W*

128 I13 **Nebolchi** Novgorodskaya Oblast', W Russian Federation *59.08N 33.19E*

38 L4 **Nebo, Mount** ▲ Utah, W USA *39.47N 111.46W*

30 L14 **Nebraska** *off.* State of Nebraska; also known as Blackwater State, Cornhusker State, Tree Planters State. ◆ *state* C USA

31 S16 **Nebraska City** Nebraska, C USA *40.38N 95.52W*

109 K23 **Nebrodi, Monti** *var.* Monti Caronie. ▲ Sicilia, Italy, C Mediterranean Sea

8 L14 **Nechako** ⊷ British Columbia, SW Canada

31 Q2 **Neche** North Dakota, N USA *48.57N 97.33W*

27 V8 **Neches** Texas, SW USA *31.51N 95.28W*

27 W8 **Neches River** ⊷ Texas, SW USA

103 H20 **Neckar** ⊷ SW Germany

103 H20 **Neckarsulm** Baden-Württemberg, SW Germany *49.12N 9.13E*

199 K5 **Necker Island** *island* C British Virgin Islands

183 U3 **Necker Ridge** *undersea feature* N Pacific Ocean

63 D23 **Necochea** Buenos Aires, E Argentina *38.33S 58.42W*

106 H2 **Neda** Galicia, NW Spain *43.28N 8.09W*

117 E20 **Nédas** ⊷ S Greece

27 Y11 **Nederland** Texas, SW USA *29.58N 93.59W*

Nederland see Netherlands

100 K12 **Neder Rijn** *Eng.* Lower Rhine. ⊷ C Netherlands

101 L16 **Nederweert** Limburg, SE Netherlands *51.16N 5.45E*

97 G16 **Nedre Tokke** ⊚ S Norway

119 S3 **Nedryhaylov** *Rus.* Nedrigaylov. Sums'ka Oblast', NE Ukraine *50.51N 33.52E*

100 O11 **Neede** Gelderland, E Netherlands *52.07N 6.36E*

35 U3 **Needle Mountain** ▲ Wyoming, C USA *44.03N 109.33W*

37 V4 **Needles** California, W USA *34.50N 114.37W*

99 M24 **Needles, The** *rocks* Isle of Wight, S England, UK

112 O7 **Neembucú** *off.* Departamento de Neembucú. ◆ *department* SW Paraguay

32 M7 **Neenah** Wisconsin, N USA *44.09N 88.26W*

9 W16 **Neepawa** Manitoba, S Canada *50.13N 99.28W*

101 K16 **Neerpelt** Limburg, NE Belgium *51.13N 5.27E*

76 M6 **Nefta** ✈ W Tunisia *34.03N 8.05E*

---

130 L15 **Neftegorsk** Krasnodarskiy Kray, SW Russian Federation *44.21N 39.40E*

131 U3 **Neftekamsk** Respublika Bashkortostan, W Russian Federation *56.06N 54.12E*

131 O14 **Neftekumsk** Stavropol'skiy Kray, SW Russian Federation *44.45N 45.00E*

125 G11 **Nefteyugansk** Khanty-Mansiyskiy Avtonomnyy Okrug, C Russian Federation *61.07N 72.18E*

84 C10 **Neftezavodsk** see Seydi

84 C10 **Negage** *var.* N'Gage. Uíge, NW Angola *7.46S 15.27E*

175 N16 **Negara** Bali, Indonesia *8.21S 114.34E*

175 N10 **Negara** Borneo, C Indonesia *2.40S 115.04E*

Negara Brunei Darussalam see Brunei

35 N4 **Negaunee** Michigan, N USA *46.30N 87.36W*

83 J15 **Negēlē** *var.* Negelli, *It.* Neghelli. Oromo, C Ethiopia *5.13N 39.43E*

Negelli see Negēlē

Negeri Pahang Darul Makmur see Pahang

Negeri Selangor Darul Ehsan see Selangor

174 H5 **Negeri Sembilan** *var.* Negri Sembilan. ◆ *state* Peninsular Malaysia

94 P3 **Negerpynten** *headland* S Svalbard *77.15N 22.40E*

Negev see HaNegev

Neghelli see Negēlē

118 I12 **Negoiu** *var.* Negoiul. ▲ S Romania *45.34N 24.34E*

Negoiul see Negoiu

84 P13 **Negomane** *var.* Negomano. Cabo Delgado, N Mozambique *11.25S 38.32E*

Negomano see Negomane

161 J25 **Negombo** Western Province, SW Sri Lanka *7.13N 79.51E*

Negoreleye see Nyeharelaye

114 P12 **Negotin** Serbia, E Yugoslavia *44.13N 22.31E*

115 P19 **Negotino** C FYR Macedonia *41.29N 22.06E*

58 A10 **Negra, Punta** *headland* NW Peru *6.03S 81.08W*

106 G3 **Negreira** Galicia, NW Spain *42.54N 8.46W*

118 L10 **Negrești** Vaslui, E Romania *46.49N 27.28E*

118 H8 **Negrești-Oaș** *Hung.* Avasfelsőfalu; *prev.* Negrești. Satu Mare, NE Romania *47.56N 23.21E*

46 H12 **Negril** W Jamaica *18.21N 78.21W*

Negri Sembilan see Negeri Sembilan

12 D7 **Nemegosenda** ⊷ Ontario, S Canada

12 D8 **Nemegosenda Lake** ⊚ Ontario, S Canada

65 K15 **Negro, Río** ⊷ E Argentina

64 N7 **Negro, Río** ⊷ N Argentina

59 N7 **Negro, Río** ⊷ E Bolivia

64 O5 **Negro, Río** ⊷ C Paraguay

50 F6 **Negro, Río** ⊷ N South America

63 E18 **Negro, Río** ⊷ Brazil/Uruguay

Negro, Río see Sico Tinto, Río, Honduras

Negro, Río see Chixoy, Río, Guatemala/Mexico

179 Q14 **Negros** *island* C Philippines

118 M15 **Negru Vodă** Constanța, SE Romania *43.47N 28.10E*

11 P13 **Neguac** New Brunswick, SE Canada *47.16N 65.04W*

12 B7 **Negwazu, Lake** ⊚ Ontario, S Canada

Négyfalu see Săcele

34 F10 **Nehalem** Oregon, NW USA *45.42N 123.55W*

34 F10 **Nehalem River** ⊷ Oregon, NW USA

Nehávand see Nahávand

149 V9 **Nehbandān** Khorāsān, E Iran *31.33N 60.01E*

169 V6 **Nehe** Heilongjiang, NE China

200 Si12 **Neiafu** 'Uta Vava'u, N Tonga *18.36S 173.58W*

47 N9 **Neiba** *var.* Neyba. SW Dominican Republic *18.27N 71.28W*

Néid, Carn Uí see Mizen Head

94 M9 **Neiden** Finnmark, N Norway *69.40N 29.22E*

Néifinn see Nephin

105 S10 **Neige, Crêt de la** ▲ E France *46.18N 5.58E*

181 O16 **Neiges, Piton des** ▲ C Réunion *21.04S 55.28E*

13 R9 **Neiges, Rivière des** ⊷ Quebec, SE Canada

166 I10 **Neijiang** Sichuan, C China *29.31N 105.03E*

32 K6 **Neillsville** Wisconsin, N USA *44.34N 90.36W*

Nei Monggol Zizhiqu/Nei Mongol see Nei Monggol Zizhiqu

169 Q10 **Nei Mongol Gaoyuan** *plateau* NE China

169 O12 **Nei Mongol Zizhiqu** *var.* Nei Mongol, *Eng.* Inner Mongolia, Inner Mongolian Autonomous Region; *prev.* Nei Monggol Zizhiqu. ◆ *autonomous region* N China

167 O4 **Neiqiu** Hebei, E China *37.22N 114.34E*

Neiriz see Neyrīz

103 Q16 **Neisse** *Cz.* Lužická Nisa, *Ger.* Lausitzer Neisse, *Pol.* Nisa, Nysa Łużycka. ⊷ C Europe

56 E11 **Neiva** Huila, S Colombia *2.58N 75.15W*

166 M7 **Neixiang** Henan, C China *33.07N 111.49E*

9 V9 **Nejanilini Lake** ⊚ Manitoba, C Canada

Nejd see Najd

82 J13 **Nek'emtē** *var.* Lakemti, Nakamti. Oromo, C Ethiopia *9.06N 36.31E*

129 M9 **Nekhayevskiy** Volgogradskaya Oblast', SW Russian Federation *50.25N 41.44E*

35 K7 **Nekoosa** Wisconsin, N USA *44.19N 89.54W*

---

97 M24 **Neksø** Bornholm, E Denmark *55.04N 15.05E*

117 C16 **Nekyomanteío** *ancient monument* Ípeiros, W Greece *39.13N 20.31E*

106 H7 **Nelas** Viseu, N Portugal *40.31N 7.52W*

128 H16 **Nelidovo** Tverskaya Oblast', W Russian Federation *56.13N 32.45E*

31 P13 **Neligh** Nebraska, C USA *42.07N 98.01W*

127 N12 **Nel'kan** Khabarovskiy Kray, E Russian Federation *57.44N 136.09E*

94 M10 **Nellim** *var.* Nellimö, *Lapp.* Njellim. Lappi, N Finland *68.49N 28.18E*

Nellimö see Nellim

161 J18 **Nellore** Andhra Pradesh, E India *14.29N 80.00E*

127 O16 **Nel'ma** Khabarovskiy Kray, SE Russian Federation *47.43N 139.08E*

85 F15 **Nelson** Santa Fe, C Argentina *31.16S 60.45W*

9 O17 **Nelson** British Columbia, SW Canada *49.29N 117.13W*

193 I14 **Nelson** South Island, NZ *41.16S 173.16E*

99 L17 **Nelson** NW England, UK *53.51N 2.13W*

31 N7 **Nelson** Nebraska, C USA *40.12N 98.04W*

193 J14 **Nelson** ◆ *unitary authority* South Island, NZ

9 X12 **Nelson** ⊷ Manitoba, C Canada

191 U8 **Nelson Bay** New South Wales, SE Australia *32.45S 152.09E*

190 K13 **Nelson, Cape** *headland* Victoria, SE Australia *38.25S 141.33E*

194 M15 **Nelson, Cape** *headland* S PNG *8.57S 149.19E*

65 G23 **Nelson, Estrecho** *strait* SE Pacific Ocean

9 W12 **Nelson House** Manitoba, C Canada *55.49N 98.51W*

32 J4 **Nelson Lake** ⊚ Wisconsin, N USA

35 T14 **Nelsonville** Ohio, N USA *39.27N 82.13W*

29 S2 **Nelsoon River** ⊷ Iowa/Missouri, C USA

85 K21 **Nelspruit** Mpumalanga, NE South Africa *25.28S 30.58E*

78 L10 **Néma** Hodh ech Chargui, SE Mauritania *16.31N 7.12W*

120 D13 **Neman** *Ger.* Ragnit. Kaliningradskaya Oblast', W Russian Federation *55.01N 22.00E*

86 I9 **Neman** *Bel.* Nyoman, *Ger.* Memel, *Lith.* Nemunas, *Pol.* Niemen, *Rus.* Neman. ⊷ NE Europe

117 F19 **Neméa** Pelopónnisos, S Greece *37.49N 22.40E*

Nemausus see Nîmes

12 D7 **Nemegosenda** ⊷ Ontario, S Canada

Neman see Arras

Nemirov see Nemyriv

172 R7 **Nemuro** Hokkaidō, NE Japan *43.19N 145.34E*

172 R7 **Nemuro-hantō** *peninsula* Hokkaidō, NE Japan

172 R6 **Nemuro-kaikyō** *strait* Japan/Russian Federation

118 H5 **Nemyriv** *var.* Nemirov. L'vivs'ka Oblast', NW Ukraine *50.07N 23.27E*

117 N7 **Nemyriv** *Rus.* Nemirov. Vinnyts'ka Oblast', C Ukraine *48.57N 28.51E*

99 D19 **Nenagh** *Ir.* an tAonach. C Ireland *52.52N 8.12W*

41 R9 **Nenana** Alaska, USA *64.33N 149.05W*

41 R9 **Nenana River** ⊷ Alaska, USA

195 W8 **Nendö** *var.* Swallow Island. *island* Santa Cruz Islands, E Solomon Islands

29 Q7 **Nene** ⊷ E England, UK

129 R4 **Nenetskiy Avtonomnyy Okrug** ◆ *autonomous district* NW Russian Federation

203 W11 **Nengonengo** *atoll* Îles Tuamotu, C French Polynesia

169 U6 **Nen Jiang** *var.* Nonni. NE China

169 V6 **Nenjiang** Heilongjiang, NE China *49.10N 125.18E*

201 P16 **Neoch** *atoll* Caroline Islands, C Micronesia

117 D18 **Neochóri** Dytikí Ellás, C Greece *38.23N 21.14E*

29 Q7 **Neodesha** Kansas, C USA

31 S14 **Neola** Iowa, C USA *41.27N 95.40W*

117 M19 **Néon Karlovási** *var.* Néon Karlovásion. Sámos, Dodekánisos, Greece, Aegean Sea *37.48N 26.42E*

Néon Karlovásion see Néon Karlovási

117 E16 **Néon Monastíri** Thessalía, C Greece *39.22N 21.55E*

29 R8 **Neosho** Missouri, C USA *36.52N 94.22W*

29 Q7 **Neosho River** ⊷ Kansas/Oklahoma, C USA

117 N17 **Nepa** ⊷ C Russian Federation

159 N10 **Nepal** ◆ *monarchy* S Asia

158 M11 **Nepalganj** Mid Western, SW Nepal *28.04N 81.37E*

12 L13 **Nepean** Ontario, SE Canada *29.31N 75.33W*

38 L4 **Nephi** Utah, W USA *39.43N 111.49W*

99 B16 **Nephin** *Ir.* Néifinn. ▲ N Ireland *54.00N 9.21W*

69 T9 **Nepoko** ⊷ NE Dem. Rep. Congo (Zaire)

20 K10 **Neptune** New Jersey, NE USA *40.13N 74.03W*

190 G10 **Neptune Islands** *island group* South Australia

111 W4 **Nerác** Lot-et-Garonne, SW France *44.07N 0.21E*

113 I17 **Neratovice** *Ger.* Neratowitz. Středočeský Kraj, C Czech Republic *50.16N 14.31E*

Neratowitz see Neratovice

126 L15 **Nercha** ⊷ S Russian Federation

126 L15 **Nerchinsk** Chitinskaya Oblast', S Russian Federation *52.01N 116.25E*

126 L16 **Nerchinskiy Zavod** Chitinskaya Oblast', S Russian Federation *51.13N 119.25E*

128 M15 **Nerekhta** Kostromskaya Oblast', NW Russian Federation *57.27N 40.33E*

128 H10 **Nereta** Aizkraukle, S Latvia *56.12N 25.18E*

108 K13 **Nereto** Abruzzo, C Italy *42.49N 13.50E*

115 H15 **Neretva** ⊷ Bosnia and Herzegovina/Croatia

117 C17 **Nerikós** *ruins* Lefkáda, Iónioi Nísoi, Greece, C Mediterranean Sea *38.48N 20.43E*

120 B12 **Neringa** *Ger.* Nidden; *prev.* Nida. Neringa, SW Lithuania *55.19N 21.00E*

120 I13 **Neris** *Bel.* Viliya, *Pol.* Wilia; *prev. Pol.* Wilija. ⊷ Belarus/Lithuania

Neris see Viliya

107 N15 **Nerja** Andalucía, S Spain *36.45N 3.34W*

128 L16 **Nerl'** ⊷ W Russian Federation

176 Vv13 **Nerong, Selat** *strait* Kepulauan Kai, E Indonesia

107 P12 **Nerpio** Castilla-La Mancha, C Spain *38.08N 2.18W*

106 J13 **Nerva** Andalucía, S Spain *37.39N 6.31W*

95 I14 **Neryungri** Respublika Sakha (Yakutiya), NE Russian Federation *56.37N 124.19E*

100 L4 **Nes** Friesland, N Netherlands *53.28N 5.46E*

96 G13 **Nesbyen** Buskerud, S Norway

94 L2 **Neskaupstadhur** Austurland, E Iceland *65.08N 13.45W*

94 F13 **Nesna** Nordland, C Norway *66.12N 13.01E*

28 K5 **Ness City** Kansas, C USA *38.27N 99.54W*

110 H7 **Nesslau** Sankt Gallen, NE Switzerland *47.13N 9.12E*

98 I9 **Ness, Loch** ⊚ N Scotland, UK

116 H12 **Néstos** *Bul.* Mesta, *Turk.* Kara Su. ⊷ Bulgaria/Greece *see also* Mesta

97 C14 **Nesttun** Hordaland, S Norway *60.19N 5.16E*

144 F9 **Nesvizh** see Nyasvizh

144 F9 **Netanya** *var.* Natanya, Nathanya. Central, C Israel *32.19N 34.51E*

100 I9 **Netherlands** *off.* Kingdom of the Netherlands, *var.* Holland, *Dut.* Koninkrijk der Nederlanden, Nederland. ◆ *monarchy* NW Europe

47 S9 **Netherlands Antilles** *prev.* Dutch West Indies. ◇ *Dutch autonomous region* S Caribbean Sea

Netherlands East Indies see Indonesia

Netherlands Guiana see Suriname

Netherlands New Guinea see Irian Jaya

118 L4 **Netishyn** Khmel'nyts'ka Oblast', W Ukraine *50.20N 26.38E*

144 E11 **Netivot** Southern, S Israel

109 O21 **Neto** ⊷ S Italy

16 N2 **Nettilling Lake** ⊚ Baffin Island, Nunavut, N Canada

31 V3 **Nett Lake** ⊚ Minnesota, N USA

109 I16 **Nettuno** Lazio, C Italy *41.26N 12.40E*

Netum see Noto

43 U16 **Netzahualcóyotl, Presa** ⊚ SE Mexico

Netze see Noteć

Neu Amerika see Puławy

114 L10 **Neubetsche** see Novi Bečej

Neubidschow see Nový Bydžov

102 N9 **Neubrandenburg** Mecklenburg-Vorpommern, NE Germany *53.33N 13.16E*

110 C8 **Neuburg an der Donau** Bayern, S Germany *48.43N 11.10E*

110 C8 **Neuchâtel** *Ger.* Neuenburg. Neuchâtel, W Switzerland *46.58N 6.55E*

110 C8 **Neuchâtel** *Ger.* Neuenburg. ◆ *canton* W Switzerland

110 C8 **Neuchâtel, Lac de** *Ger.* Neuenburger See. ⊚ W Switzerland

102 L10 **Neue Elde** *canal* N Germany

102 G9 **Neuenburg** see Neuchâtel

83 E17 **Neuenburg an der Elbe** see Nymburk

Neuenburger See see Neuchâtel, Lac de

110 F7 **Neuenhof** Aargau, N Switzerland *47.27N 8.17E*

102 H11 **Neuenland** ✈ (Bremen) Bremen, NW Germany *53.03N 8.46E*

Neuenstadt see La Neuveville

103 C18 **Neuerburg** Rheinland-Pfalz, W Germany *50.00N 6.13E*

101 K24 **Neufchâteau** Luxembourg, SE Belgium *49.49N 5.25E*

104 M3 **Neufchâteau** Vosges, NE France *48.22N 5.41E*

127 Oo16 **Neufchâtel-en-Bray** Seine-Maritime, N France *49.44N 1.26E*

111 S3 **Neufelden** Oberösterreich, N Austria *48.21N 14.05E*

Neugradisk see Nova Gradiška

110 G6 **Neuhaus** see Jindřichův Hradec

110 G6 **Neuhausen** *var.* Neuhausen am Rheinfall. Schaffhausen, N Switzerland *47.41N 8.37E*

Neuhausen am Rheinfall see Neuhausen

103 I17 **Neuhof** Hessen, C Germany *50.26N 9.34E*

90 G10 **Neukuhren** see Pionerskiy

108 I17 **Neulengbach** Niederösterreich, NE Austria *48.12N 15.89E*

130 M14 **Nevinnomyssk** Stavropol'skiy Kray, SW Russian Federation *44.42N 41.59E*

---

142 J14 **Nevşehir** *var.* Nevsehir. Nevşehir, C Turkey *38.37N 34.43E*

142 J14 **Nevşehir** *var.* Nevsehir. ◆ *province* C Turkey

Nevshehr see Nevşehir

125 Ee11 **Nev'yansk** Sverdlovskaya Oblast', C Russian Federation *57.26N 60.15E*

83 J25 **Newala** Mtwara, SE Tanzania *10.58S 39.18E*

35 P16 **New Albany** Indiana, N USA *38.16N 85.49W*

24 M2 **New Albany** Mississippi, S USA *34.29N 89.00W*

31 Y11 **New Albin** Iowa, C USA *43.30N 91.17W*

57 U8 **New Amsterdam** E Guyana *6.17N 57.30W*

191 Q4 **New Angledool** New South Wales, SE Australia *29.06S 147.54E*

23 Y2 **Newark** Delaware, NE USA *39.40N 75.45W*

20 K14 **Newark** New Jersey, NE USA *40.42N 74.12W*

18 G10 **Newark** New York, NE USA *43.01N 77.04W*

35 T13 **Newark** Ohio, N USA *40.03N 82.24W*

37 W5 **Newark Lake** ⊚ Nevada, W USA

18 G10 **Newark-on-Trent** *var.* Newark. C England, UK *53.04N 0.49W*

24 M7 **New Augusta** Mississippi, S USA *31.12N 89.03W*

21 P12 **New Bedford** Massachusetts, NE USA *41.37N 70.55W*

34 G11 **Newberg** Oregon, NW USA *45.18N 122.58W*

23 X10 **New Bern** North Carolina, SE USA *35.07N 77.03W*

23 F8 **Newbern** Tennessee, S USA *36.06N 89.15W*

35 P4 **Newberry** Michigan, N USA *46.21N 85.30W*

23 Q12 **Newberry** South Carolina, SE USA *34.16N 81.37W*

21 P15 **New Bloomfield** Pennsylvania, NE USA *40.24N 77.08W*

27 X5 **New Boston** Texas, SW USA *33.27N 94.25W*

27 S11 **New Braunfels** Texas, SW USA *29.43N 98.09W*

32 Q13 **New Bremen** Ohio, N USA *40.26N 84.22E*

99 F18 **Newbridge** *Ir.* An Droichead Nua. C Ireland *53.10N 6.48W*

20 B14 **New Brighton** Pennsylvania, NE USA *40.44N 80.18W*

20 M12 **New Britain** Connecticut, NE USA *41.37N 72.45W*

195 N13 **New Britain** *island* E PNG

199 Hh9 **New Britain Trench** *undersea feature* W Pacific Ocean

20 J15 **New Brunswick** New Jersey, NE USA *40.29N 74.27W*

13 V8 **New Brunswick** *Fr.* Nouveau-Brunswick. ◆ *province* SE Canada

20 K13 **Newburgh** New York, NE USA *41.30N 74.00W*

21 P10 **Newburyport** Massachusetts, NE USA *42.49N 70.53W*

79 T14 **New Bussa** Niger, W Nigeria *9.50N 4.32E*

197 J4 **New Caledonia** *var.* Kanaky, *Fr.* Nouvelle-Calédonie. ◇ *French overseas territory* SW Pacific Ocean

197 H5 **New Caledonia** *island* SW Pacific Ocean

183 O10 **New Caledonia Basin** *undersea feature* W Pacific Ocean

191 T8 **Newcastle** New South Wales, SE Australia *32.55S 151.46E*

11 O14 **Newcastle** New Brunswick, SE Canada *47.01N 65.36W*

12 I15 **Newcastle** Ontario, SE Canada *43.55N 78.35W*

99 C20 **Newcastle** *Ir.* An Caisleán Nua. SW Ireland *52.25N 9.04W*

85 K23 **Newcastle** KwaZulu/Natal, E South Africa *27.45S 29.59E*

99 J16 **Newcastle** *Ir.* An Caisleán Nua. SE Northern Ireland, UK *54.12N 5.54W*

35 P13 **New Castle** Indiana, N USA *39.55N 85.21W*

21 L5 **New Castle** Kentucky, S USA *38.22N 85.09W*

29 N11 **New Castle** Oklahoma, C USA *35.15N 97.36W*

18 B14 **New Castle** Pennsylvania, NE USA *40.59N 80.19W*

37 N6 **New Castle** Utah, W USA *37.40N 113.31W*

21 S6 **New Castle** Virginia, NE USA *37.29N 80.06W*

21 T8 **Newcastle** Wyoming, C USA *43.52N 104.13W*

37 V14 **Nevada** Iowa, C USA *42.01N 93.27W*

29 R6 **Nevada** Missouri, C USA *37.50N 94.21W*

37 R5 **Nevada** *off.* State of Nevada; also known as Battle Born State, Sagebrush State, Silver State. ◆ *state* W USA

37 P6 **Nevada City** California, W USA *39.15N 121.02W*

128 G16 **Nevel'** Pskovskaya Oblast', W Russian Federation *56.01N 29.54E*

127 Oo16 **Nevel'sk** Ostrov Sakhalin, Sakhalinskaya Oblast', SE Russian Federation *46.41N 141.54E*

111 S3 **Neufelden** Oberösterreich, N Austria *48.43N 14.01E*

130 M14 **Neva** ⊷ NW Russian Federation

128 G16 **Newcomerstown** Ohio, N USA *40.16N 81.36W*

105 P9 **Nevers** *anc.* Noviodunum. Nièvre, C France *47.00N 3.09E*

20 I12 **Nevertire** New South Wales, SE Australia *31.52S 147.42E*

191 Q6 **Nevertire** New South Wales, SE Australia *31.52S 147.42E*

Nevesinje see Nevesinje

128 J17 **Nevesinje** Republika Srpska, S Bosnia and Herzegovina

9 O17 **New Denver** British Columbia, SW Canada *49.58N 117.21W*

30 J7 **Newell** South Dakota, N USA *44.42N 103.25W*

85 K9 **Newell, Lake** ⊚ Alberta, SW Canada

30 L6 **New Leipzig** North Dakota, N USA *46.21N 101.54W*

12 H5 **New Liskeard** Ontario, S Canada *47.31N 79.40W*

22 L5 **Newllano** Louisiana, C USA *31.06N 93.16W*

20 N13 **New London** Connecticut, NE USA *41.21N 72.04W*

31 Y15 **New London** Iowa, C USA *40.55N 91.24W*

29 T4 **New London** Missouri, C USA *39.35N 91.24W*

32 L9 **New London** Wisconsin, N USA *44.24N 88.45W*

31 Q4 **New Madrid** Missouri, C USA *36.35N 89.31W*

188 J8 **Newman** Western Australia *23.18S 119.45E*

204 M13 **Newman Island** Antarctica

12 H15 **Newmarket** Ontario, S Canada *44.03N 79.26W*

99 P20 **Newmarket** E England, UK *52.17N 0.28E*

21 P10 **Newmarket** New Hampshire, NE USA *43.04N 70.53W*

3 U4 **New Market** Virginia, NE USA *38.39N 78.40W*

23 R2 **New Martinsville** West Virginia, NE USA *39.37N 80.48W*

35 U14 **New Matamoras** Ohio, N USA *39.32N 81.04W*

36 I6 **New Meadows** Idaho, NW USA *44.57N 116.16W*

28 R12 **New Mexico** *off.* State of New Mexico; also known as Land of Enchantment, Sunshine State. ◆ *state* SW USA

155 V6 **New Mirpur** Sind, SE Pakistan *33.09N 73.42E*

157 T17 **New Moore Island** island E India

25 S4 **Newnan** Georgia, SE USA *33.22N 84.48W*

191 P17 **New Norfolk** Tasmania, SE Australia *42.46S 147.01E*

22 I9 **New Orleans** Louisiana, S USA *29.57N 90.04W*

22 I9 **New Orleans** ✈ Louisiana, S USA *29.59N 90.17W*

20 K9 **New Paltz** New York, NE USA *41.44N 74.04W*

35 U12 **New Philadelphia** Ohio, N USA *40.29N 81.27W*

192 K10 **New Plymouth** Taranaki, North Island, NZ 39.04S 174.06E
99 M24 **Newport** S England, UK 50.42N 1.18W
99 K22 **Newport** SE Wales, UK 51.35N 3.00W
29 W10 **Newport** Arkansas, C USA 35.36N 91.16W
35 N13 **Newport** Indiana, N USA 39.52N 87.24W
31 W9 **Newport** Minnesota, N USA 44.52N 93.00W
34 F12 **Newport** Oregon, NW USA 44.38N 124.03W
21 O13 **Newport** Rhode Island, NE USA 41.29N 71.17W
23 O9 **Newport** Tennessee, S USA 35.58N 83.11W
21 N6 **Newport** Vermont, NE USA 44.55N 72.13W
34 M7 **Newport** Washington, NW USA 48.08N 117.05W
23 X7 **Newport News** Virginia, NE USA 36.58N 76.25W
99 N20 **Newport Pagnell** SE England, UK 52.04N 0.43W
25 U12 **New Port Richey** Florida, SE USA 28.14N 82.42W
31 V9 **New Prague** Minnesota, N USA 44.32N 93.34W
46 H3 **New Providence** island N Bahamas
99 H24 **Newquay** SW England, UK 50.27N 5.03W
99 I20 **New Quay** SW Wales, UK 52.13N 4.22W
31 V10 **New Richland** Minnesota, N USA 43.53N 93.29W
13 X7 **New-Richmond** Quebec, SE Canada 48.10N 65.54W
35 R15 **New Richmond** Ohio, N USA 38.57N 84.16W
32 I5 **New Richmond** Wisconsin, N USA 45.09N 92.31W
54 G1 **New River** ⊷ N Belize
57 T12 **New River** ⊷ SE Guyana
23 R6 **New River** ⊷ West Virginia, NE USA
44 G1 **New River Lagoon** ⊙ N Belize
24 J8 **New Roads** Louisiana, S USA 30.42N 91.26W
20 L14 **New Rochelle** New York, NE USA 40.55N 73.44W
31 O4 **New Rockford** North Dakota, N USA 47.40N 99.08W
99 P23 **New Romney** SE England, UK 50.58N 0.57E
99 F20 **New Ross** Ir. An tIúr. SE Ireland 52.24N 6.55W
99 F16 **Newry** Ir. An tIúr. SE Northern Ireland, UK 54.10N 6.19W
30 M5 **New Salem** North Dakota, N USA 46.51N 101.24W
New Sarum see Salisbury
31 W14 **New Sharon** Iowa, C USA 41.28N 92.39W
New Siberian Islands see Novosibirskiye Ostrova
25 X11 **New Smyrna Beach** Florida, SE USA 29.01N 80.55W
191 O7 **New South Wales** ◆ state SE Australia
41 O13 **New Stuyahok** Alaska, USA 59.27N 95.18W
23 N8 **New Tazewell** Tennessee, S USA 36.26N 83.36W
40 M12 **Newtok** Alaska, USA 60.56N 164.37W
25 S7 **Newton** Georgia, SE USA 31.18N 84.20W
31 W14 **Newton** Iowa, C USA 41.42N 93.03W
29 N6 **Newton** Kansas, C USA 38.03N 97.20W
21 O11 **Newton** Massachusetts, NE USA 42.19N 71.10W
24 M5 **Newton** Mississippi, S USA 32.19N 89.09W
20 J14 **Newton** New Jersey, NE USA 41.03N 74.45W
23 R9 **Newton** North Carolina, SE USA 35.40N 81.13W
27 Y9 **Newton** Texas, SW USA 30.51N 93.45W
99 J24 **Newton Abbot** SW England, UK 50.33N 3.34W
98 K13 **Newton St Boswells** SE Scotland, UK 55.34N 2.40W
99 I14 **Newton Stewart** SW Scotland, UK 54.58N 4.30W
94 O2 **Newtontoppen** ▲ C Svalbard 78.57N 17.34E
30 K3 **New Town** North Dakota, N USA 47.58N 102.30W
99 I20 **Newtown** E Wales, UK 52.31N 3.19W
99 G15 **Newtownabbey** Ir. Baile na Mainistreach. E Northern Ireland, UK 54.40N 5.57W
99 G15 **Newtownards** Ir. Baile Nua na hArda. SE Northern Ireland, UK 54.36N 5.40W
31 U10 **New Ulm** Minnesota, N USA 44.20N 94.28W
30 K10 **New Underwood** South Dakota, N USA 44.05N 102.46W
27 V10 **New Waverly** Texas, SW USA 30.32N 95.28W
20 K14 **New York** New York, NE USA 40.44N 73.57W
20 G10 **New York** ◆ state NE USA
X13 **New York Mountains** ▲ California, W USA
192 K12 **New Zealand** abbrev. NZ. ◆ commonwealth republic SW Pacific Ocean
129 O15 **Neya** Kostromskaya Oblast', NW Russian Federation 58.19N 43.51E
Neyba see Neiba
149 Q12 **Neyriz** var. Neiriz, Niriz. Fārs, S Iran 29.13N 54.18E
149 T4 **Neyshābūr** var. Nishapur. Khorāsān, NE Iran 36.14N 58.46E
161 J21 **Neyveli** Tamil Nādu, SE India 11.36N 79.25E
Nezhin see Nizhyn
35 N10 **Nezperce** Idaho, NW USA 46.14N 116.15W
22 H8 **Nezpique, Bayou** ⊷ Louisiana, S USA
196 W14 **Ngabordamlu, Tanjung** headland Pulau Trangan, E Indonesia 6.58S 134.13E

79 Y13 **Ngadda** ⊷ NE Nigeria
N'Gage see Negage
193 G16 **Ngahere** West Coast, South Island, NZ 42.22S 171.29E
79 Z12 **Ngala** Borno, NE Nigeria 12.19N 14.11E
85 G17 **Ngamiland** ◆ district N Botswana
164 K16 **Ngamring** Xizang Zizhiqu, W China 29.16N 87.10E
83 K19 **Ngangerabeli Plain** plain SE Kenya
164 I14 **Ngangla Ringco** ⊙ W China
164 G13 **Nganglong Kangri** ▲ W China 32.55N 81.00E
164 K15 **Ngangzê Co** ⊙ W China
81 F14 **Ngaoundéré** var. N'Gaoundéré. Adamaoua, N Cameroon 7.19N 13.34E
83 E20 **Ngara** Kagera, NW Tanzania 2.30S 30.40E
196 F8 **Ngardmau Bay** bay Babeldaob, N Palau
196 F7 **Ngaregur** island Palau Islands, N Palau
192 L7 **Ngaruawahia** Waikato, North Island, NZ 37.41S 175.09E
192 N11 **Ngaruroro** ⊷ North Island, NZ
202 I16 **Ngatangiia** Rarotonga, S Cook Islands 21.13S 159.43W
192 M6 **Ngatea** Waikato, North Island, NZ 37.16S 175.29E
177 F8 **Ngathainggyaung** Irrawaddy, SW Myanmar 17.22N 95.04E
Ngatik see Ngetik Atoll
Ngau see Gau
174 Ll15 **Ngawi** Jawa, S Indonesia 7.22S 111.22E
196 C7 **Ngcheangel** var. Kayangel Islands. island Palau Islands, N Palau
196 E10 **Ngchemiangel** Babeldaob, N Palau
196 C8 **Ngeaur** var. Angaur. island Palau Islands, S Palau
196 E10 **Ngerkeai** Babeldaob, N Palau
196 F9 **Ngermechau** Babeldaob, N Palau 7.34N 134.39E
196 C8 **Ngeruktabel** prev. Urukthapel. island Palau Islands, S Palau
196 F8 **Ngetbong** Babeldaob, N Palau 7.37N 134.34E
201 T17 **Ngetik Atoll** var. Ngatik; prev. Los Jardines. atoll Caroline Islands, E Micronesia
196 E10 **Ngetkip** Babeldaob, N Palau
Nggamea see Qamea
195 V15 **Nggatokae** island New Georgia Islands, NW Solomon Islands
85 C16 **N'Giva** var. Ondjiva, Port. Vila Pereira de Eça. Cunene, S Angola 17.01S 15.41E
81 G20 **Ngo** Plateaux, SE Congo 2.28S 15.43E
178 Jj7 **Ngoc Lac** Thanh Hoa, N Vietnam 20.06N 105.21E
81 G17 **Ngoko** ⊷ Cameroon/Congo
176 W14 **Ngoni, Tanjung** headland Maluku, Kepulauan Aru, SE Indonesia 6.10S 134.04E
83 H19 **Ngorengore** Rift Valley, SW Kenya 1.01S 35.26E
165 Q11 **Ngoring Hu** ⊙ C China
Ngorolaka see Banifing
83 H20 **Ngorongoro Crater** crater N Tanzania 3.10S 35.34E
81 D19 **Ngounié** off. Province de la Ngounié, var. La Ngounié. ◆ province S Gabon
81 D19 **Ngounié** ⊷ Congo/Gabon
80 H10 **Ngoura** var. N'Goura. Chari-Baguirmi, W Chad 12.52N 16.27E
80 G10 **Ngouri** var. N'Gouri; prev. Fort-Millot. Lac, W Chad 13.40N 15.24E
79 Y10 **Ngourti** Diffa, E Niger 15.22N 13.13E
79 Y11 **Nguigmi** var. N'Guigmi. Diffa, SE Niger 14.16N 13.07E
Nguimbo see Lumbala N'Guimbo
197 C14 **Ngulu Atoll** atoll Caroline Islands, W Micronesia
N'Gunza see Sumbe
175 N16 **Ngurah Rai** × (Bali) Bali, S Indonesia
79 W12 **Nguru** Yobe, NE Nigeria 12.55N 10.30E
Ngwaketze see Southern
85 I16 **Ngweze** ⊷ S Zambia
85 M17 **Nhamatanda** Sofala, C Mozambique 19.16S 34.10E
60 G12 **Nhamundá, Rio** var. Jamundá, Yamundá. ⊷ N Brazil
62 J7 **Nhandeara** São Paulo, S Brazil 20.40S 50.03W
84 D12 **N'Harea** var. N'Harea, Nhareia. Bié, W Angola 11.28S 16.57E
Nhareia see Nharêa
178 Kk13 **Nha Trang** Khanh Hoa, S Vietnam 12.15N 109.10E
190 L11 **Nhill** Victoria, SE Australia 36.21S 141.38E
85 L22 **Nhlangano** prev. Goedgegun. SW Swaziland 27.01S 31.11E
189 S1 **Nhulunbuy** Northern Territory, N Australia 12.15S 136.46E
79 N10 **Niafounké** Tombouctou, W Mali 15.54N 3.58W
35 N5 **Niagara** Wisconsin, N USA 45.45N 87.57W
12 H16 **Niagara** ⊷ Ontario, S Canada
12 G15 **Niagara Escarpment** hill range Ontario, S Canada
12 H16 **Niagara Falls** Ontario, S Canada 43.04N 79.06W
20 D9 **Niagara Falls** New York, NE USA 43.06N 79.04W
16 Pp17 **Niagara Falls** waterfall Canada/USA 43.04N 79.04W
K12 **Niagassola** var. Nyagassola. Haute-Guinée, NE Guinea 12.22N 9.07W
79 R12 **Niamey ●** (Niger) Niamey, SW Niger 13.28N 2.06E
79 R12 **Niamey** × Niamey, SW Niger 13.28N 2.14E
79 R14 **Niamtougou** N Togo 9.49N 1.07E
79 O16 **Niangara** Orientale, NE Dem. Rep. Congo (Zaire) 3.45N 27.54E
79 O10 **Niangay, Lac** ⊙ E Mali
79 N14 **Niangoloko** SW Burkina 10.15N 4.53W
29 U5 **Niangua River** ⊷ Missouri, C USA

81 O17 **Nia-Nia** Orientale, NE Dem. Rep. Congo (Zaire) 1.26N 27.38E
21 O17 **Niantic** Connecticut, NE USA 41.19N 72.11W
169 U7 **Nianzishan** Heilongjiang, NE China 47.31N 122.52E
173 N4 **Nias, Pulau** island W Indonesia
84 O13 **Niassa** off. Província da Niassa. ◆ province N Mozambique
203 U10 **Niau** island Îles Tuamotu, C French Polynesia
97 G20 **Nibe** Nordjylland, N Denmark 56.58N 9.39E
201 Q8 **Nibok** N Nauru 0.31S 166.55E
118 C10 **Nīca** Liepāja, W Latvia 56.21N 21.03E
44 J9 **Nicaragua** off. Republic of Nicaragua. ◆ republic Central America
44 K11 **Nicaragua, Lago de** var. Cocibolca, Gran Lago, Eng. Lake Nicaragua. ⊙ S Nicaragua
Nicaragua, Lake see Nicaragua, Lago de
66 D11 **Nicaraguan Rise** undersea feature NW Caribbean Sea
Nicaria see Ikaría
105 V15 **Nicastro** Calabria, SW Italy 38.58N 16.19E
Nice It. Nizza; anc. Nicaea. Alpes-Maritimes, SE France 43.43N 7.13E
Nice see Côte d'Azur
Nicephorium see Ar Raqqah
10 M9 **Nichicun, Lac** ⊙ Quebec, E Canada
170 C17 **Nichinan** var. Nitinan. Miyazaki, Kyūshū, SW Japan 31.36N 131.22E
42 L10 **Nicholas Channel** channel N Cuba
Nicholas II Land see Severnaya Zemlya
155 U2 **Nicholas Range** Pash. Selseleh-ye Kūh-e Vākhān, Taj. Qatorkühi Vakhon. ▲ Afghanistan/Tajikistan
22 M6 **Nicholasville** Kentucky, S USA 37.52N 84.34W
46 G2 **Nichols Town** Andros Island, NW Bahamas 25.07N 78.01W
23 U12 **Nichols** South Carolina, SE USA 34.13N 79.09W
57 U9 **Nickerie** ◆ district NW Suriname
57 V9 **Nickerie River** ⊷ NW Suriname
157 P22 **Nicobar Islands** island group India, E Indian Ocean
118 L9 **Nicolae Bălcescu** Botoșani, NE Romania 47.33N 26.52E
13 P11 **Nicolet** Quebec, SE Canada 46.13N 72.37W
13 Q12 **Nicolet** ⊷ Quebec, SE Canada
35 Q4 **Nicolet, Lake** ⊙ Michigan, N USA
31 U10 **Nicollet** Minnesota, N USA 44.16N 94.11W
63 F19 **Nico Pérez** Florida, S Uruguay 33.35S 55.10W
Nicopolis see Nikopol, Bulgaria
Nicopolis see Nikopol', Greece
124 R12 **Nicosia** Gk. Lefkosía, Turk. Lefkoşa. ● (Cyprus) C Cyprus 35.10N 33.22E
109 K24 **Nicosia** Sicilia, Italy, C Mediterranean Sea 37.45N 14.24E
109 N22 **Nicotera** Calabria, SW Italy 38.33N 15.55E
44 K13 **Nicoya** Guanacaste, NW Costa Rica 10.06S 85.26W
44 L14 **Nicoya, Golfo de** gulf W Costa Rica
44 L14 **Nicoya, Península de** peninsula NW Costa Rica
Nictheroy see Niterói
113 L15 **Nida** ⊷ S Poland
Nida see Neringa
Nidaros see Trondheim
110 D8 **Nidau** Bern, W Switzerland 47.07N 7.15E
103 H17 **Nidda** ⊷ W Germany
97 F17 **Nidelva** ⊷ S Norway
126 J10 **Nidym** Evenkiyskiy Avtonomnyy Okrug, N Russian Federation 64.08N 99.52E
112 L9 **Nidzica** Ger. Niedenburg. Warmińsko-Mazurskie, NE Poland, 53.22N 20.27E
102 H6 **Niebüll** Schleswig-Holstein, N Germany 54.47N 8.51E
85 S10 **Niederanven** Luxembourg, C Luxembourg 49.39N 6.15E
105 V4 **Niederbronn-les-Bains** Bas-Rhin, NE France 48.57N 7.37E
Niederdonau see Niederösterreich
111 U5 **Niederlausitz** Eng. Lower Lusatia. physical region E Germany
103 P14 **Niederösterreich** off. Niederösterreich, Eng. Lower Austria, Ger. Niederdonau; prev. Lower Danube. ◆ state NE Austria
102 G12 **Niedersachsen** Eng. Lower Saxony, Fr. Basse-Saxe. ◆ state NW Germany
79 D17 **Niefang** var. Sevilla de Niefang. NW Equatorial Guinea 1.52N 10.12E
85 G23 **Niekerkshoop** Northern Cape, W South Africa 29.21S 22.49E
85 M14 **Niel** Antwerpen, N Belgium 51.07N 4.19E
Niélé see Niellé
85 M14 **Niellé** var. Nielé. N Ivory Coast 10.12S 5.37W
81 O22 **Niemba** Katanga, E Dem. Rep. Congo (Zaire) 5.58S 28.24E
113 G15 **Niemcza** Ger. Nimptsch. Dolnośląskie, SW Poland 50.45N 16.52E
Niemen see Neman
94 J13 **Niemisel** Norrbotten, N Sweden 66.00N 22.00E
113 H15 **Niemodlin** Ger. Falkenberg. Opolskie, S Poland 50.37N 17.45E
78 M13 **Niéna** Sikasso, SW Mali 11.24N 6.20W
102 H12 **Nienburg** Niedersachsen, N Germany 52.37N 9.12E
102 N13 **Niepłotz** ⊷ NE Germany
113 L16 **Niepołomice** Małopolskie, S Poland 50.02N 20.12E
103 D14 **Niers** ⊷ Germany/Netherlands
103 Q15 **Niesky** Lus. Niska. Sachsen, E Germany 51.16N 14.49E

Nieśwież see Nyasvizh
Nieuport see Nieuwpoort
100 O8 **Nieuw-Amsterdam** Drenthe, NE Netherlands 52.43N 6.52E
57 W9 **Nieuw Amsterdam** Commewijne, NE Suriname 5.52N 55.04W
101 M14 **Nieuw-Bergen** Limburg, SE Netherlands 51.36N 6.04E
100 O7 **Nieuw-Buinen** Drenthe, NE Netherlands 52.57N 6.55E
100 J12 **Nieuwegein** Utrecht, C Netherlands 52.03N 5.06E
100 P6 **Nieuwe Pekela** Groningen, NE Netherlands 53.04N 6.58E
100 P5 **Nieuweschans** Groningen, NE Netherlands 53.10N 7.10E
Nieuw Guinea see New Guinea
100 I11 **Nieuwkoop** Zuid-Holland, C Netherlands 52.09N 4.46E
100 M9 **Nieuwleusen** Overijssel, E Netherlands 52.34N 6.16E
100 J11 **Nieuw-Loosdrecht** Utrecht, C Netherlands 52.12N 5.07E
57 U9 **Nieuw Nickerie** Nickerie, NW Suriname 05.52N 57.00W
100 P5 **Nieuwolda** Groningen, NE Netherlands 53.15N 6.58E
181 B17 **Nieuwpoort** var. Nieuport. West-Vlaanderen, W Belgium 51.07N 2.45E
101 G14 **Nieuw-Vossemeer** Noord-Brabant, S Netherlands 51.34N 4.12E
100 P7 **Nieuw-Weerdinge** Drenthe, NE Netherlands 52.51N 7.00E
42 L10 **Nieves** Zacatecas, C Mexico 24.00N 102.57W
66 O11 **Nieves, Pico de las** ▲ Gran Canaria, Islas Canarias, Spain, NE Atlantic Ocean 27.58N 15.34W
105 P8 **Nièvre** ◆ department C France
Niewenstar see Neustadt an der Weinstrasse
142 J15 **Niğde** Niğde, C Turkey 37.58N 34.42E
142 J15 **Niğde** ◆ province C Turkey
Nigel see Nile
193 C24 **Nightcaps** Southland, South Island, NZ 45.58S 168.03E
12 F7 **Night Hawk Lake** ⊙ Ontario, S Canada
117 M19 **Nightingale Island** island S Tristan da Cunha, S Atlantic Ocean
40 M12 **Nightmute** Alaska, USA 60.28N 164.43W
116 G13 **Nigrita** Kentrikí Makedonía, NE Greece 40.54N 23.30E
154 J15 **Nihing** Per. Rūd-e Nahang. ⊷ SW Afghanistan
203 V10 **Nihiru** atoll Îles Tuamotu, C French Polynesia
171 L13 **Nihommatsu** var. Nihommatsu, Nihonmatu. Fukushima, Honshū, C Japan 37.35N 140.22E
64 I12 **Nihuil, Embalse del** ⊙ W Argentina
171 K12 **Niigata** Niigata, Honshū, C Japan 37.55N 139.01E
171 K13 **Niigata** off. Niigata-ken. ◆ prefecture Honshū, C Japan
170 F15 **Niihama** Ehime, Shikoku, SW Japan 33.57N 133.15E
38 A8 **Nii'hau** island Hawaii, USA, C Pacific Ocean
170 E13 **Nii-jima** island E Japan
170 Fj13 **Niimi** Okayama, Honshū, SW Japan 35.00N 133.27E
171 Kk13 **Niitsu** var. Niitu. Niigata, Honshū, C Japan 37.48N 139.06E
Niitu see Niitsu
107 P15 **Níjar** Andalucía, S Spain 36.57N 2.13W
100 K11 **Nijkerk** Gelderland, C Netherlands 52.13N 5.30E
101 H16 **Nijlen** Antwerpen, N Belgium 51.10N 4.40E
100 L13 **Nijmegen** Ger. Nimwegen; anc. Noviomagus. Gelderland, SE Netherlands 51.49N 5.52E
100 N10 **Nijverdal** Overijssel, E Netherlands 52.22N 6.28E
202 G12 **Nikao** Rarotonga, S Cook Islands
Nikaria see Ikaría
128 I2 **Nikel'** Murmanskaya Oblast', NW Russian Federation 69.24N 30.12E
175 Rr17 **Nikiniki** Timor, S Indonesia 10.00S 124.30E
133 Q15 **Nikitin Seamount** undersea feature E Indian Ocean 5.48S 84.48E
79 S14 **Nikki** E Benin 9.55S 3.12E
171 Kk15 **Nikkō** var. Nikko. Tochigi, Honshū, S Japan 36.45N 139.37E
41 P10 **Nikolai** Alaska, USA 63.00N 154.22W
Nikolaiken see Mikołajki
Nikolainkaupunki see Länsi-Suomi
151 U15 **Nikolayevka** Almaty, SE Kazakhstan 43.99N 77.10E
151 O6 **Nikolayevka** Severnyy Kazakhstan, N Kazakhstan
131 P14 **Nikolayevsk** Volgogradskaya Oblast', SW Russian Federation 50.03N 45.30E
Nikolayevsk see Pugachev
179 P11 **Nikolayevsk-na-Amure** Khabarovskiy Kray, SE Russian Federation 53.04N 140.39E
131 O6 **Nikol'sk** Penzenskaya Oblast', W Russian Federation 53.46N 46.03E
131 R11 **Nikol'sk** Vologodskaya Oblast', NW Russian Federation 59.35N 45.31E
Nikol'sk see Ussuriysk

40 K17 **Nikolski** Umnak Island, Alaska, USA 52.56N 168.52W
Nikol'skiy see Satpayev
131 V7 **Nikol'skoye** Orenburgskaya Oblast', W Russian Federation 52.01N 55.48E
Nikol'sk-Ussuriyskiy see Ussuriysk
116 J7 **Nikopol** anc. Nicopolis. Pleven, N Bulgaria 43.43N 24.55E
119 S9 **Nikopol'** Dnipropetrovs'ka Oblast', SE Ukraine 47.34N 34.23E
117 C17 **Nikopol'** anc. Nicopolis. site of ancient city Ípeiros, W Greece 39.01N 20.43E
142 M12 **Niksar** Tokat, N Turkey 40.35N 36.54E
149 V14 **Nikshahr** Sīstān va Balūchestān, SE Iran 26.15N 60.10E
115 I15 **Nikšić** Montenegro, SW Yugoslavia 42.46N 18.56E
203 R4 **Nikumaroro** prev. Gardner Island, Kemins Island. atoll Phoenix Islands, C Kiribati
203 P3 **Nikunau** var. Nukunau; prev. Byron Island. atoll Tungaru, W Kiribati
Nīl, Nahr an see Nile
37 X16 **Niland** California, W USA 33.14N 115.31W
82 G8 **Nile** former province NW Uganda
77 W7 **Nile** Delta delta N Egypt
Nile see Neyrīz
69 T3 **Nile** Ar. Nahr an Nil. ⊷ N Africa
69 T3 **Nile Fan** undersea feature E Mediterranean Sea
35 O11 **Niles** Michigan, N USA 41.49N 86.15W
35 V11 **Niles** Ohio, N USA 41.10N 80.46W
161 F20 **Nileswaram** Kerala, SW India 12.18N 75.07E
12 K10 **Nilgiri, Lac** ⊙ Quebec, SE Canada
164 I5 **Nilka** Xinjiang Uygur Zizhiqu, NW China 43.46N 82.33E
161 G21 **Nilambūr** Kerala, SW India 11.16N 76.15E
160 F9 **Nimach** Madhya Pradesh, C India 24.30N 74.51E
158 G14 **Nimbāhera** Rājasthān, N India 24.37N 74.45E
78 L15 **Nimba, Monts** var. Nimba Mountains. ▲ W Africa
Nimba Mountains see Nimba, Monts
Nimburg see Nymburk
105 Q15 **Nîmes** anc. Nemausus, Nismes. Gard, S France 43.49N 4.19E
191 R11 **Nimmitabel** New South Wales, SE Australia 36.34S 149.18E
205 R11 **Nimrod Glacier** glacier Antarctica
154 K8 **Nimrūz** var. Nimroze; prev. Nīmrūz. ◆ province SW Afghanistan
83 F16 **Nimule** Eastern Equatoria, S Sudan 3.33N 32.06E
161 C23 **Nine Degree Channel** channel India/Maldives
20 G9 **Ninemile Point** headland New York, NE USA 43.31N 76.22W
181 S8 **Nineteast Ridge** undersea feature E Indian Ocean
191 P13 **Ninety Mile Beach** beach Victoria, SE Australia
192 I2 **Ninety Mile Beach** beach North Island, NZ
23 P12 **Ninety Six** South Carolina, SE USA 34.10N 82.01W
8 H7 **Nisling** ⊷ Yukon Territory, W Canada
169 Y9 **Ning'an** Heilongjiang, NE China 44.20N 129.28E
167 S9 **Ningbo** var. Ning-po, Yin-hsien; prev. Ninghsien. Zhejiang, SE China 29.54N 121.33E
118 M10 **Nisporeni** Rus. Nisporeny. W Moldova 47.04N 28.10E
Nisporeny see Nisporeni
167 U12 **Ningde** Fujian, SE China 26.48N 119.33E
167 P12 **Ningdu** Jiangxi, S China 26.28N 115.58E
194 E12 **Ningerum** Western, SW PNG 5.43S 141.09E
167 R9 **Ningguo** Anhui, E China 30.33N 118.58E
167 S9 **Ninghai** Zhejiang, SE China 29.19N 121.22E
Ning-hsia see Ningxia
Ninghsien see Ningbo
166 J15 **Ningming** Guangxi Zhuangzu Zizhiqu, S China 22.07N 106.43E
166 H11 **Ningnan** Sichuan, C China 26.59N 102.49E
Ning-po see Ningbo
166 J5 **Ningxia** off. Ningxia Huizu Zizhiqu, var. Ning-hsia, Ningsia, Eng. Ningsia Hui, Ningsia Hui Autonomous Region. ◆ autonomous region N China
166 J5 **Ningxian** Gansu, N China 35.30N 108.04E
205 T10 **Ninnis Glacier** glacier Antarctica
172 N10 **Ninohe** Iwate, Honshū, C Japan 40.17N 141.18E
101 F18 **Ninove** Oost-Vlaanderen, C Belgium 50.49N 4.01E
179 P11 **Ninoy Aquino** × (Manila) Luzon, N Philippines 14.26N 121.00E
Nio see Íos
31 P12 **Niobrara** Nebraska, C USA 42.43N 97.59W
30 M12 **Niobrara River** ⊷ Nebraska/Wyoming, C USA
81 I20 **Nioki** Bandundu, W Dem. Rep. Congo (Zaire) 2.44S 17.42E
78 M11 **Niono** Ségou, C Mali 13.56N 5.57W
78 K11 **Nioro** var. Nioro du Sahel. Kayes, W Mali 15.13N 9.38W

78 G11 **Nioro du Rip** SW Senegal 13.44N 15.48W
Nioro du Sahel see Nioro
104 K10 **Niort** Deux-Sèvres, W France 46.21N 0.24W
180 H14 **Nioumachoua** Mohéli, S Comoros 12.22S 43.43E
194 G12 **Nipa** Southern Highlands, W PNG 6.12S 143.29E
9 U14 **Nipawin** Saskatchewan, C Canada 53.21N 103.55W
10 D12 **Nipigon** Ontario, S Canada 49.01N 88.15W
10 D11 **Nipigon, Lake** ⊙ Ontario, S Canada
12 G11 **Nipissing, Lake** ⊙ Ontario, S Canada
37 P13 **Nipomo** California, W USA 35.02N 120.28W
Nippon see Japan
144 R6 **Niqniqiyah, Sabkha¹ an** ⊙ C Syria
64 I9 **Niquivil** San Juan, W Argentina 30.25S 68.42W
176 Yy10 **Nirabotong** Irian Jaya, E Indonesia 2.35S 140.08E
171 J16 **Nirasaki** Yamanashi, Honshū, S Japan 35.43N 138.24E
Niriz see Neyrīz
161 I14 **Nirmal** Andhra Pradesh, C India 19.04N 78.21E
159 Q13 **Nirmāli** Bihār, NE India 26.19N 86.35E
115 O14 **Niš** Eng. Nish, Ger. Nisch; anc. Naissus. Serbia, SE Yugoslavia 43.19N 21.52E
106 H9 **Nisa** Portalegre, C Portugal 39.31N 7.39W
115 O14 **Niš** ⊷ SE Yugoslavia
147 P4 **Nişāb** Al Ḩudūd ash Shamālīyah, N Saudi Arabia 29.11N 44.43E
147 Q15 **Nişāb** var. Anşāb. SW Yemen 14.24N 46.47E
115 P14 **Nišava** Bul. Nishava. ⊷ Bulgaria/Yugoslavia see also Nishava
Nisch/Nish see Niš
172 Nn5 **Niseko** Hokkaidō, NE Japan 42.50N 140.43E
116 G9 **Nishava** var. Nišava. ⊷ Bulgaria/Yugoslavia see also Nišava
120 L11 **Nishcha** Rus. Nishcha. ⊷ N Belarus
172 Qq7 **Nishibetsu-gawa** ⊷ Hokkaidō, NE Japan
170 E13 **Nishinfö** ...
Nishapur see Neyshābūr
116 G9 **Nishava** var. Nišava. ⊷ Bulgaria/Yugoslavia see also Nišava
170 Bb17 **Nishinoomote** Kagoshima, Tanega-shima, SW Japan 30.42N 130.59E
172 Ss16 **Nishino-shima** Eng. Rosario. island Ogasawara-shotō, SE Japan
171 Hh16 **Nishio** var. Nisio. Aichi, Honshū, SW Japan 34.52N 137.01E
170 C13 **Nishi-Sonogi-hantō** peninsula Kyūshū, SW Japan
171 Gg14 **Nishiwaki** var. Nisiwaki. Hyōgo, Honshū, SW Japan 35.02N 134.57E
147 U14 **Nishtūn** SE Yemen 15.47N 52.08E
Nisiros see Nísyros
Nisiwaki see Nishiwaki
115 O14 **Niška Banja** Serbia, SE Yugoslavia 43.18N 22.01E
13 O15 **Niskibi** ⊷ Ontario, C Canada
113 O15 **Nisko** Podkarpackie, SE Poland 50.31N 22.09E
8 H7 **Nisling** ⊷ Yukon Territory, W Canada
101 H22 **Nismes** Namur, S Belgium 50.04N 4.31E
Nismes see Nîmes
118 M10 **Nisporeni** Rus. Nisporeny. W Moldova 47.04N 28.10E
97 K20 **Nissan** ⊷ S Sweden
195 R11 **Nissan Island** island Green Islands, NE PNG
97 F16 **Nisser** ⊙ S Norway
119 M17 **Nissum Bredning** inlet NW Denmark
31 U6 **Nistru** see Dniester
46.31N 94.17W
Nisyros see Nísyros
117 M22 **Nísyros** var. Nisiros. island Dodekánisos, Greece, Aegean Sea
118 H8 **Nītaure** Cēsis, C Latvia 57.05N 25.12E
61 P10 **Niterói** prev. Nictheroy. Rio de Janeiro, SE Brazil 22.54S 43.06W
98 E16 **Nith** ⊷ Ontario, S Canada
98 J13 **Nith** ⊷ S Scotland, UK
Nitianan see Nichinan
113 I21 **Nitra** Ger. Neutra, Hung. Nyitra. Nitriansky Kraj, SW Slovakia 48.19N 18.04E
113 I20 **Nitra** Ger. Neutra, Hung. Nyitra. ⊷ W Slovakia
113 I21 **Nitriansky Kraj** ◆ region SW Slovakia
23 Q5 **Nitro** West Virginia, SE USA 38.24N 81.50W
125 F11 **Nitsa** ⊷ C Russian Federation
97 H14 **Nittedal** Akershus, S Norway 60.08N 10.45E
172 O6 **Niuatobutabu** see Niuatoputapu
200 S11 **Niuatoputapu** var. Niuatobutabu; prev. Keppel Island. island N Tonga
200 Q15 **Niu'Aunofa** headland Tongatapu, S Tonga 21.03S 175.19W
202 B16 **Niue** ◇ self-governing territory in free association with NZ S Pacific Ocean
202 B16 **Niulakita** var. Nurakita. atoll S Tuvalu
202 E6 **Niutao** island NW Tuvalu
95 L15 **Nivala** Oulu, C Finland 63.56N 25.00E
104 I15 **Nive** ⊷ SW France
79 R8 **Nivelles** Wallon Brabant, C Belgium 50.36N 4.04E
97 C17 **Niverville, Lac** ⊙ Quebec, SE Canada

29 T7 **Nixa** Missouri, C USA 37.02N 93.17W
37 R5 **Nixon** Nevada, W USA 39.48N 119.24W
27 S12 **Nixon** Texas, SW USA 29.16N 97.45W
Niya see Minfeng
152 K12 **Niyazov** Lebapskiy Velayat, NE Turkmenistan 39.13N 63.16E
161 H14 **Nizāmābād** Andhra Pradesh, C India 18.40N 78.04E
161 H15 **Nizām Sāgar** ⊙ C India
129 N16 **Nizhegorodskaya Oblast'** ◆ province W Russian Federation
126 K14 **Nizhneangarsk** Respublika Buryatiya, S Russian Federation 55.47N 109.39E
Nizhnegorskiy see Nyzhn'ohirs'kyy
131 S4 **Nizhnekamsk** Respublika Tatarstan, W Russian Federation 55.36N 51.45E
131 U3 **Nizhnekamskoye Vodokhranilishche** ⊟ W Russian Federation
127 O5 **Nizhnekolymsk** Respublika Sakha (Yakutiya), NE Russian Federation 68.32N 161.00E
127 N16 **Nizhne Leninskoye** Yevreyskaya Avtonomnaya Oblast', SE Russian Federation 47.50N 132.00E
126 H14 **Nizhneudinsk** Irkutskaya Oblast', S Russian Federation 54.48N 98.51E
126 Gg11 **Nizhnevartovsk** Khanty-Mansiyskiy Avtonomnyy Okrug, C Russian Federation 60.57N 76.40E
126 Ll6 **Nizhneyansk** Respublika Sakha (Yakutiya), NE Russian Federation 71.25N 135.59E
131 Q11 **Nizhniy Baskunchak** Astrakhanskaya Oblast', SW Russian Federation 48.15N 46.49E
126 M11 **Nizhniy Bestyakh** Respublika Sakha (Yakutiya), NE Russian Federation 61.55N 130.07E
131 O6 **Nizhniy Lomov** Penzenskaya Oblast', W Russian Federation 53.32N 43.39E
131 P3 **Nizhniy Novgorod** prev. Gor'kiy. Nizhegorodskaya Oblast', W Russian Federation 56.17N 43.59E
129 T8 **Nizhniy Odes** Respublika Komi, NW Russian Federation 63.42N 54.58E
Nizhniy Pyandzh see Panji Poyon
125 Ee11 **Nizhniy Tagil** Sverdlovskaya Oblast', C Russian Federation 57.57N 59.51E
129 T9 **Nizhnyaya-Omra** Respublika Komi, NW Russian Federation 62.46N 55.54E
129 P5 **Nizhnyaya Pesha** Nenetskiy Avtonomnyy Okrug, NW Russian Federation 66.54N 47.37E
125 F11 **Nizhnyaya Tavda** Tyumenskaya Oblast', C Russian Federation 57.41N 65.58E
126 Jj12 **Nizhnyaya Tunguska** Eng. Lower Tunguska. ⊷ N Russian Federation
119 Q3 **Nizhyn** Rus. Nezhin. Chernihivs'ka Oblast', NE Ukraine 51.03N 31.54E
142 M17 **Nizip** Gaziantep, S Turkey 37.01N 37.46E
147 X8 **Nizwá** var. Nazwāh. NE Oman 22.50N 57.27E
Nizza see Nice
108 C9 **Nizza Monferrato** Piemonte, NE Italy 44.47N 8.22E
Njávdám see Näätämöjoki
Njellim see Nellim
83 H24 **Njombe** Iringa, S Tanzania 9.19S 34.46E
98 G9 **Njombe** ⊷ C Tanzania
94 J10 **Njunis** ▲ N Norway 68.47N 19.24E
95 H17 **Njurunda** Västernorrland, C Sweden 62.15N 17.24E
96 H11 **Njutånger** Gävleborg, C Sweden 61.37N 17.04E
81 E16 **Nkambe** North-West, NW Cameroon 6.34N 10.43E
81 F21 **Nkata Bay** var. Nkhata Bay. Northern, N Malawi 11.36S 34.17E
85 J17 **Nkayi** prev. Jacob. La Bouenza, S Congo 4.10S 13.17E
84 D16 **Nkayi** Matabeleland North, W Zimbabwe 19.02S 28.55E
79 E17 **Nkhata Bay** var. Nkata Bay. Northern, N Malawi 11.36S 34.17E
83 E22 **Nkonde** Kigoma, N Tanzania 6.16S 30.17E
81 D15 **Nkongsamba** var. N'Kongsamba. Littoral, W Cameroon 4.58N 9.52E
85 E16 **Nkurenkuru** Okavango, N Namibia 17.39S 18.37E
78 Q15 **Nkwanta** E Ghana 8.18N 0.27E
178 H1 **Nmai Hka** var. Me Hka. ⊷ N Myanmar
Noardwâlde see Noordwolde
41 N7 **Noatak** Alaska, USA 67.34N 162.58W
41 N7 **Noatak River** ⊷ Alaska, USA
Nobeji see Noheji
61 N18 **Nobres** Mato Grosso, W Brazil 14.43S 56.15W
109 N21 **Nocera Terinese** Calabria, S Italy 39.03N 16.10E
43 Q16 **Nochixtlán** var. Asunción Nochixtlán. Oaxaca, SE Mexico 17.28N 97.18W
27 S5 **Nocona** Texas, SW USA 33.47N 97.43W
29 Q2 **Nodales, Bahía de los** bay S Argentina
29 Q2 **Nodaway River** ⊷ Iowa/Missouri, C USA
29 R8 **Noel** Missouri, C USA 36.33N 94.29W
97 C17 **Nærbø** Rogaland, S Norway 58.40N 5.39E
97 J24 **Næstved** Storstrøm, SE Denmark 55.12N 11.47E

◆ COUNTRY　◇ DEPENDENT TERRITORY　◈ ADMINISTRATIVE REGION　▲ MOUNTAIN　⊵ VOLCANO　⊙ LAKE
● COUNTRY CAPITAL　○ DEPENDENT TERRITORY CAPITAL　× INTERNATIONAL AIRPORT　▲ MOUNTAIN RANGE　⊷ RIVER　⊟ RESERVOIR

42 H3 **Nogales** Chihuahua, NW Mexico 18.49N 97.12W
42 F3 **Nogales** Sonora, NW Mexico 31.16N 110.52W
38 M17 **Nogales** Arizona, SW USA 31.20N 110.55W
Nogal Valley see Dooxo Nugaaleed
104 K15 **Nogaro** Gers, S France 43.46N 0.01W
112 J7 **Nogat** ☴ N Poland
170 D12 **Nogata** Fukuoka, Kyūshū, SW Japan 33.42N 130.43E
131 P15 **Nogayskaya Step'** steppe SW Russian Federation
104 M6 **Nogent-le-Rotrou** Eure-et-Loir, C France 48.19N 0.49E
105 O4 **Nogent-sur-Oise** Oise, N France 49.16N 2.28E
105 P6 **Nogent-sur-Seine** Aube, N France 48.30N 3.31E
126 I10 **Noginsk** Evenkiyskiy Avtonomnyy Okrug, N Russian Federation 64.28N 91.09E
130 L3 **Noginsk** Moskovskaya Oblast', W Russian Federation 55.51N 38.23E
127 O14 **Nogliki** Ostrov Sakhalin, Sakhalinskaya Oblast', SE Russian Federation 51.44N 143.14E
171 I14 **Nōgōhaku-san** ▲ Honshū, SW Japan 35.46N 136.36E
168 D5 **Nogoonnuur** Bayan-Ölgiy, NW Mongolia 49.31N 89.48E
63 C18 **Nogoyá** Entre Ríos, E Argentina 32.25S 59.49W
113 K21 **Nógrád** ♦ county N Hungary
Nógrád Megye. ♦ county N Hungary
107 U5 **Noguera Pallaresa** ☴ NE Spain
107 U4 **Noguera Ribagorçana** ☴ NE Spain
172 N9 **Noheji** var. Nobeji. Aomori, Honshū, C Japan 40.51N 141.07E
103 E19 **Nohfelden** Saarland, SW Germany 49.35N 7.08E
40 A8 **Nohili Point** headland Kauai, Hawaii, USA, C Pacific Ocean 22.03N 159.48W
106 G3 **Noia** Galicia, NW Spain 42.48N 8.52W
105 N16 **Noire, Montagne** ▲ S France
13 P12 **Noire, Rivière** ☴ Quebec, SE Canada
12 J10 **Noire, Rivière** ☴ Quebec, SE Canada
Noire, Rivière see Black River
104 G6 **Noires, Montagnes** ▲ NW France
104 H8 **Noirmoutier-en-l'Île** Vendée, NW France 47.00N 2.15W
104 H8 **Noirmoutier, Île de** island NW France
171 J17 **Nojima-zaki** headland Honshū, S Japan 34.54N 139.54E
195 W8 **Noka** Nendö, E Solomon Islands 10.42S 165.57E
85 G17 **Nokaneng** Ngamiland, NW Botswana 19.40S 22.12E
95 L18 **Nokia** Länsi-Suomi, W Finland 61.28N 23.30E
154 K11 **Nok Kundi** Baluchistān, SW Pakistan 28.49N 62.39E
32 L14 **Nokomis** Illinois, N USA 39.18N 89.17W
32 K5 **Nokomis, Lake** ☒ Wisconsin, N USA
80 G9 **Nokou** Kanem, W Chad 14.36N 14.45E
197 B12 **Nokuku** Espiritu Santo, N Vanuatu 14.56S 166.34E
97 J18 **Nol** Västra Götaland, S Sweden 57.55N 12.03E
81 H16 **Nola** Sangha-Mbaéré, SW Central African Republic 3.28N 16.05E
27 P7 **Nolan** Texas, SW USA 32.15N 100.15W
129 R15 **Nolinsk** Kirovskaya Oblast', NW Russian Federation 57.34N 49.54E
97 B19 **Nólsoy** Dan. Nolsø Island Faeroe Islands 61.59N 6.39W
194 F12 **Nomad** Western, SW Papau New Guinea 6.11S 142.13E
170 B15 **Noma-zaki** headland Kyūshū, SW Japan 31.24N 130.07E
42 K10 **Nombre de Dios** Durango, C Mexico 23.51N 104.13W
44 I5 **Nombre de Dios, Cordillera** ▲ N Honduras
40 M9 **Nome** Alaska, USA 64.30N 165.24W
31 Q6 **Nome** North Dakota, N USA 46.39N 97.49W
40 M9 **Nome, Cape** headland Alaska, USA 64.25N 165.00W
Nōmi-jima see Nishi-Nōmi-jima
12 M11 **Nominingue, Lac** ☒ Quebec, SE Canada
Nomoi Islands see Mortlock Islands
170 Bb13 **Nomo-zaki** headland Kyūshū, SW Japan 32.34N 129.45E
200 S13 **Nomuka** island Nomuka Group, C Tonga
200 S14 **Nomuka Group** island group W Tonga
201 Q15 **Nomwin Atoll** atoll Hall Islands, C Micronesia
15 I8 **Nonacho Lake** ☒ Northwest Territories, NW Canada
Nondaburi see Nonthaburi
41 P12 **Nondalton** Alaska, USA 59.58N 154.51W
169 V10 **Nong'an** Jilin, NE China
178 I10 **Nong Bua Khok** Nakhon Ratchasima, C Thailand 15.23N 101.51E
178 I9 **Nong Bua Lamphu** Udon Thani, E Thailand 17.11N 102.27E
178 J7 **Nông Hèt** Xiangkhoang, N Laos 19.27N 104.02E
Nongkaya see Nong Khai
178 I8 **Nong Khai** var. Mi Chai, Nongkaya. Nong Khai, E Thailand 17.52N 102.43E
178 Gg15 **Nong Met** Surat Thani, SW Thailand 9.27N 99.09E
85 L22 **Nongoma** KwaZulu/Natal, E South Africa 27.54S 31.40E
178 Hh10 **Nong Phai** Phetchabun, C Thailand 15.58N 101.02E
159 U13 **Nongstoin** Meghálaya, NE India 25.34N 91.19E
85 C19 **Nonidas** Erongo, N Namibia 22.36S 14.40E
Nonni see Nen Jiang

42 I7 **Nonoava** Chihuahua, N Mexico 27.24N 106.18W
203 O3 **Nonouti** prev. Sydenham Island. atoll Tungaru, W Kiribati
178 Hh11 **Nonthaburi** var. Nondaburi, Nontha Buri. Nonthaburi, C Thailand 13.55N 100.33E
104 L11 **Nontron** Dordogne, SW France 45.34N 0.41E
189 P1 **Noonamah** Northern Territory, N Australia 12.46S 131.08E
30 K2 **Noonan** North Dakota, N USA 48.51N 102.57W
101 E14 **Noord-Beveland** var. North Beveland. Island SW Netherlands
101 J14 **Noord-Brabant** Eng. North Brabant. ♦ province S Netherlands
100 H7 **Noorder Haaks** spit NW Netherlands
100 H9 **Noord-Holland** Eng. North Holland. ♦ province NW Netherlands
Noordhollands Kanaal see Noordhollandskanaal
100 H8 **Noordhollands Kanaal** var. Noordhollands Kanaal. canal NW Netherlands
Noord-Kaap see Northern Cape
100 L8 **Noordoostpolder** island N Netherlands
47 P16 **Noordpunt** headland Curaçao, C Netherlands Antilles 12.21N 69.08W
100 I8 **Noord-Scharwoude** Noord-Holland, NW Netherlands 52.42N 4.48E
Noordwes see North-West
100 G11 **Noordwijk aan Zee** Zuid-Holland, W Netherlands 52.15N 4.25E
100 H11 **Noordwijkerhout** Zuid-Holland, W Netherlands 52.16N 4.30E
100 M7 **Noordwolde** Fris. Noardwâlde. Friesland, N Netherlands 52.54N 6.10E
Noordzee see North Sea
100 H10 **Noordzee-Kanaal** canal NW Netherlands
95 K18 **Noormarkku** Swe. Norrmark. Länsi-Suomi, W Finland 61.34N 21.54E
41 N8 **Noorvik** Alaska, USA 66.50N 161.01W
8 J17 **Nootka Sound** inlet British Columbia, W Canada
84 A9 **Nóqui** Zaire, NW Angola 5.53S 13.26E
97 L15 **Nora** Örebro, C Sweden 59.31N 15.01E
153 Q13 **Norak** Rus. Nurek. W Tajikistan 38.23N 69.13E
16 P14 **Noranda** Quebec, SE Canada 48.16N 79.03W
31 W12 **Nora Springs** Iowa, C USA 43.08N 93.00W
97 M14 **Norberg** Västmanland, C Sweden 60.04N 15.34E
12 K13 **Norcan Lake** ☒ Ontario, SE Canada
207 R12 **Nord** Avannaarsua, N Greenland 81.38N 12.51W
80 F13 **Nord** Eng. North. ♦ province N Cameroon
105 P2 **Nord** ♦ department N France
94 P1 **Nordaustlandet** island NE Svalbard
97 G24 **Nordborg** Ger. Nordburg. Sønderjylland, SW Denmark 55.04N 9.40E
Nordburg see Nordborg
97 F23 **Nordby** Ribe, W Denmark 55.27N 8.25E
9 P15 **Nordegg** Alberta, SW Canada 52.27N 116.06W
102 E9 **Norden** Niedersachsen, NW Germany 53.36N 7.12E
102 G10 **Nordenham** Niedersachsen, NW Germany 53.28N 8.27E
126 I4 **Nordenshel'da, Arkhipelag** island group N Russian Federation
94 O3 **Nordenskiold Land** physical region W Svalbard
102 E9 **Norderney** island NW Germany
102 J9 **Norderstedt** Schleswig-Holstein, N Germany 53.42N 9.58E
96 C11 **Nordfjord** physical region S Norway
96 D11 **Nordfjord** fjord S Norway
96 D11 **Nordfjordeid** Sogn og Fjordane, S Norway 61.54N 6.01E
94 G11 **Nordfold** Nordland, C Norway 67.48N 15.16E
Nordfriesische Inseln see North Frisian Islands
102 H7 **Nordfriesland** cultural region N Germany
103 K15 **Nordhausen** Thüringen, C Germany 51.31N 10.48E
96 C13 **Nordhordland** physical region S Norway
102 E12 **Nordhorn** Niedersachsen, NW Germany 52.27N 7.04E
94 J1 **Nordhurfjördhur** Vestfirdhir, NW Iceland 66.01N 21.33W
94 J2 **Nordhurland Eystra** ♦ region N Iceland
94 J2 **Nordhurland Vestra** ♦ region N Iceland
180 H16 **Nord, Île du** island Inner Islands, NE Seychelles
97 F20 **Nordjylland** off. Nordjyllands Amt. ♦ county N Denmark
94 K7 **Nordkapp** Eng. North Cape. headland N Norway 71.10N 25.42E
94 K7 **Nordkapp** headland N Svalbard 80.31N 19.58E
81 N19 **Nordkivu** off. Région du Nord Kivu. ♦ region E Dem. Rep. Congo (Zaire)
Norske Havet see Norwegian Sea
94 G12 **Nordland** ♦ county C Norway
103 J20 **Nördlingen** Bayern, S Germany 48.49N 10.28E
95 K15 **Nordmaling** Västerbotten, N Sweden 63.34N 19.30E
97 K15 **Nordmark** Värmland, C Sweden 59.52N 14.04E
Nord, Mer du see North Sea
96 F8 **Nordmøre** physical region S Norway

102 I8 **Nord-Ostee-Kanal** canal N Germany
1 **Nordostrundingen** headland NE Greenland 81.30N 10.00W
81 D14 **Nord-Ouest** Eng. North-West. ♦ province NW Cameroon
Nord-Ouest, Territoires du see Northwest Territories
105 N2 **Nord-Pas-de-Calais** ♦ region N France
103 F19 **Nordpfälzer Bergland** ▲ W Germany
Nord, Pointe see Fatua, Pointe
197 H5 **Nord, Province** ♦ province C New Caledonia
103 D14 **Nordrhein-Westfalen** Eng. North Rhine-Westphalia, Fr. Rhénanie du Nord-Westphalie. ♦ state W Germany
Nordsee/Nordsjøen/ Nordsøen see North Sea
102 H7 **Nordstrand** island N Germany
95 E15 **Nord-Trøndelag** ♦ county C Norway
99 E18 **Nore** Ir. An Fheoir. ☴ S Ireland
31 Q14 **Norfolk** Nebraska, C USA 42.01N 97.25W
23 X7 **Norfolk** Virginia, NE USA 36.51N 76.17W
99 P19 **Norfolk** cultural region E England, UK
199 B11 **Norfolk Island** ♢ Australian external territory SW Pacific Ocean
183 P9 **Norfolk Ridge** undersea feature W Pacific Ocean
29 U8 **Norfolk Lake** ☒ Arkansas/Missouri, C USA
100 N6 **Norg** Drenthe, NE Netherlands 53.04N 6.28E
Norge see Norway
97 D14 **Norheimsund** Hordaland, S Norway 60.22N 6.09E
27 S16 **Norias** Texas, SW USA 26.47N 97.45W
171 J14 **Norikura-dake** ▲ Honshū, S Japan 36.06N 137.33E
126 I8 **Noril'sk** Taymyrskiy (Dolgano-Nenetskiy) Avtonomnyy Okrug, N Russian Federation 69.21N 88.01E
12 I13 **Norland** Ontario, SE Canada 44.46N 78.48W
23 V8 **Norlina** North Carolina, SE USA 36.26N 78.11W
32 L13 **Normal** Illinois, N USA 40.30N 88.59W
29 N11 **Norman** Oklahoma, C USA 35.13N 97.27W
Norman see Tulita
195 O16 **Normanby Island** island SE PNG
60 G9 **Normandia** Roraima, N Brazil 3.57N 59.39W
104 L5 **Normandie** Eng. Normandy. cultural region N France
104 J5 **Normandie, Collines de** hill range NW France
Normandy see Normandie
27 V9 **Normangee** Texas, SW USA 31.01N 96.06W
23 Q10 **Norman, Lake** ☒ North Carolina, SE USA
46 K13 **Norman Manley** ✈ (Kingston) E Jamaica 17.55N 76.46W
189 U5 **Norman River** ☴ Queensland, NE Australia
189 U4 **Normanton** Queensland, NE Australia 17.48S 141.01E
15 Gg5 **Norman Wells** Northwest Territories, NW Canada 65.18N 126.42W
10 H12 **Normétal** Quebec, SE Canada 48.98N 79.22W
9 V15 **Norquay** Saskatchewan, S Canada 51.51N 102.04W
96 I1 **Norra Dellen** ☒ C Sweden
95 G15 **Norråker** Jämtland, C Sweden 64.25N 15.40E
95 N12 **Norrala** Gävleborg, C Sweden 61.22N 17.04E
94 G13 **Norra Storfjället** ▲ N Sweden 65.57N 15.15E
94 H13 **Norrbotten** ♦ county N Sweden
97 G23 **Nørre Aaby** var. Nørre Åby C Denmark 55.28N 9.52E
97 G23 **Nørre Åby** var. Nørre Aaby. Fyn, C Denmark 55.28N 9.52E
97 I24 **Nørre Alslev** Storstrøm, SE Denmark 54.54N 11.52E
97 E23 **Nørre Nebel** Ribe, W Denmark 55.44N 8.16E
97 G20 **Nørresundby** Nordjylland, N Denmark 57.05N 9.55E
23 N8 **Norris Lake** ☒ Tennessee, S USA
20 I15 **Norristown** Pennsylvania, NE USA 40.07N 75.20W
97 N17 **Norrköping** Östergötland, S Sweden 58.34N 16.10E
96 N13 **Norrsundet** Gävleborg, C Sweden 60.55N 17.09E
97 P15 **Norrtälje** Stockholm, C Sweden 59.45N 18.42E
188 L12 **Norseman** Western Australia 32.16S 121.45E
95 N16 **Norsjö** Västerbotten, N Sweden 64.55N 19.30E
126 Mm15 **Norsk** Amurskaya Oblast', SE Russian Federation 52.20N 129.57E
197 O13 **Norsup** Malekula, C Vanuatu 16.05S 167.24E
203 V15 **Norte, Cabo** headland Easter Island, Chile, E Pacific Ocean 27.03S 109.24W
56 F7 **Norte de Santander** off. Departamento de Norte de Santander. ♦ province N Colombia
63 E21 **Norte, Punta** headland E Argentina 36.17S 56.46W
23 R13 **North** South Carolina, SE USA 33.37N 81.06W
1 **North America** continent
1 N12 **North American Basin** undersea feature W Sargasso Sea
1 C5 **North American Plate** tectonic feature
20 M11 **North Amherst** Massachusetts, NE USA 42.24N 72.31W
99 N20 **Northampton** C England, UK 52.13N 0.54W
99 M20 **Northamptonshire** cultural region C England, UK
157 P18 **North Andaman** island Andaman Islands, India, NE Indian Ocean
23 Q13 **North Augusta** South Carolina, SE USA 33.30N 81.58W
181 W8 **North Australian Basin** Fr. Bassin Nord de I' Australie. undersea feature E Indian Ocean
35 R11 **North Baltimore** Ohio, N USA 41.10N 83.40W
9 T15 **North Battleford** Saskatchewan, S Canada 52.46N 108.19W
12 H11 **North Bay** Ontario, S Canada 46.19N 79.28W
10 H6 **North Belcher Islands** island group Belcher Islands, Nunavut, C Canada
31 R15 **North Bend** Nebraska, C USA 41.27N 96.46W
34 E14 **North Bend** Oregon, NW USA 43.24N 124.13W
98 K12 **North Berwick** SE Scotland, UK 56.03N 2.44W
North Beveland see Noord-Beveland
21 O9 **Northfield** New Hampshire, NE USA 43.26N 71.34W
29 N11 **Northfield** Minnesota, N USA 44.27N 93.10W
191 P5 **North Bourke** New South Wales, SE Australia 30.03S 145.56E
North Brabant see Noord-Brabant
190 F2 **North Branch Neales** seasonal river South Australia
46 M6 **North Caicos** island NW Turks and Caicos Islands
28 L10 **North Canadian River** ☴ Oklahoma, C USA
35 U12 **North Canton** Ohio, N USA 40.52N 81.24W
11 R13 **North, Cape** headland Cape Breton Island, Nova Scotia, SE Canada 47.06N 60.24W
192 I1 **North, Cape** headland North Island, NZ 34.23S 173.02E
195 N9 **North Cape** headland New Ireland, NE PNG 2.33S 150.48E
20 J17 **North Cape May** New Jersey, NE USA 38.59N 74.57W
10 C9 **North Caribou Lake** ☒ Ontario, C Canada
23 U10 **North Carolina** off. State of North Carolina; also known as Old North State, Tar Heel State, Turpentine State. ♦ state SE USA
35 Q14 **North College Hill** Ohio, N USA 39.13N 84.33W
27 O8 **North Concho River** ☴ Texas, SW USA
21 O8 **North Conway** New Hampshire, NE USA 44.03N 71.06W
29 V14 **North Crossett** Arkansas, C USA 33.10N 91.56W
30 L4 **North Dakota** off. State of North Dakota; also known as Flickertail State, Peace Garden State, Sioux State. ♦ state N USA
North Devon Island see Devon Island
99 L22 **North Downs** hill range SE England, UK
20 C11 **North East** Pennsylvania, NE USA 42.13N 79.49W
85 I18 **North East** ♦ district NE Botswana
117 G15 **North East Bay** bay Ascension Island, C Atlantic Ocean
40 L10 **Northeast Cape** headland Saint Lawrence Island, Alaska, USA 63.16N 168.50W
178 Mm13 **Northeast Cay** SE PNG
83 J17 **North Eastern** ♦ province Kenya
North East Frontier Agency/North East Frontier Agency of Assam see Arunáchal Pradesh
117 E25 **North East Island** island E Falkland Islands
201 V11 **Northeast Island** island Chuuk, C Micronesia
46 L12 **North East Point** headland E Jamaica 18.09N 76.19W
46 L6 **North East Point** headland Great Inagua, S Bahamas 21.18N 73.01W
46 K5 **North East Point** headland Acklins Island, SE Bahamas 22.43N 73.50W
203 Z2 **Northeast Point** headland Kiritimati, E Kiribati 10.22S 105.45E
46 H2 **Northeast Providence Channel** channel N Bahamas
103 J14 **Northeim** Niedersachsen, C Germany 51.42N 10.00E
31 X14 **North English** Iowa, C USA 41.28N 81.47W
144 G8 **Northern** ♦ district N Israel
83 K14 **Northern** ♦ region N Malawi
194 L15 **Northern** ♦ province S PNG
85 J20 **Northern** ♦ province NE South Africa; prev. Northern Transvaal. ♦ province NE South Africa
82 B13 **Northern Bahr el Ghazal** ♦ state SW Sudan
82 D9 **Northern Border Region** see Al Ḥudūd ash Shamālīyah

85 F24 **Northern Cape** off. Northern Cape Province, Afr. Noord-Kaap. ♦ province W South Africa
202 K14 **Northern Cook Islands** island group N Cook Islands
82 B8 **Northern Darfur** ♦ state NW Sudan
Northern Dvina see Severnaya Dvina
99 F14 **Northern Ireland** var. The Six Counties. political division UK
82 D9 **Northern Kordofan** ♦ state C Sudan
197 K14 **Northern Lau Group** island group Lau Group, NE Fiji
196 K3 **Northern Mariana Islands** ♢ US commonwealth territory W Pacific Ocean
161 J23 **Northern Province** ♦ province N Sri Lanka
Northern Rhodesia see Zambia
Northern Sporades see Vóreioi Sporádes
190 D1 **Northern Territory** ♦ territory N Australia
Northern Transvaal see Northern
Northern Ural Hills see Severnyye Uvaly
86 I9 **North European Plain** plain N Europe
29 V2 **North Fabius River** ☴ Missouri, C USA
117 D24 **North Falkland Sound** sound N Falkland Islands
31 V9 **North Field** Minnesota, N USA 44.27N 93.10W
183 Q8 **North Fiji Basin** undersea feature N Coral Sea
99 Q22 **North Foreland** headland SE England, UK 51.22N 1.26E
37 P6 **North Fork American River** ☴ California, W USA
41 R7 **North Fork Chandalar River** ☴ Alaska, USA
30 K7 **North Fork Grand River** ☴ North Dakota/South Dakota, N USA
23 O6 **North Fork Kentucky River** ☴ Kentucky, S USA
41 Q7 **North Fork Koyukuk River** ☴ Alaska, USA
41 Q10 **North Fork Kuskokwim River** ☴ Alaska, USA
28 K11 **North Fork Red River** ☴ Oklahoma/Texas, SW USA
28 K3 **North Fork Solomon River** ☴ Kansas, C USA
25 W14 **North Fort Myers** Florida, SE USA 26.40N 81.52W
35 P5 **North Fox Island** island Michigan, N USA
102 G6 **North Frisian Islands** var. North Frisian Islands Ger. Nordfriesische Inseln. island group N Germany
207 N9 **North Geomagnetic Pole** pole Arctic Ocean 78.30N 69.00W
20 M13 **North Haven** Connecticut, NE USA 41.25N 72.51W
192 J5 **North Head** headland North Island, NZ 36.23S 174.01E
20 L6 **North Hero** Vermont, NE USA 44.49N 73.14W
10 H9 **North Highlands** California, W USA 38.40N 121.25W
35 N10 **North Chicago** Illinois, N USA 42.19N 87.50W
205 Y10 **North Highlands Glacier** glacier Antarctica
157 K21 **North Huvadhu Atoll** var. Gaafu Alifu Atoll. atoll S Maldives
117 A24 **North Island** island N Falkland Islands
192 I9 **North Island** island N NZ
23 I14 **North Island** island South Carolina, SE USA
35 O11 **North Judson** Indiana, N USA 41.12N 86.44W
North Kazakhstan see Severnyy Kazakhstan
35 V10 **North Kingsville** Ohio, N USA 41.54N 80.41W
169 Y13 **North Korea** off. Democratic People's Republic of Korea, Kor. Chosŏn-minjujuŭi-inmin-kanghwaguk. ♦ republic E Asia
159 X11 **North Lakhimpur** Assam, NE India 27.14N 94.00E
192 J3 **Northland** ♦ region North Island, NZ
199 J12 **Northland Plateau** undersea feature S Pacific Ocean
37 X11 **North Las Vegas** Nevada, W USA 36.12N 115.07W
35 O11 **North Liberty** Indiana, N USA 41.36N 86.22W
31 X14 **North Liberty** Iowa, C USA 41.45N 91.36W
29 U12 **North Little Rock** Arkansas, C USA 34.46N 92.15W
30 M13 **North Loup River** ☴ Nebraska, C USA
157 K18 **North Maalhosmadulu Atoll** var. North Maalhosmadulu Atoll, Raa Atoll. atoll N Maldives
35 U10 **North Madison** Ohio, N USA 41.48N 81.03W
35 P12 **North Manchester** Indiana, N USA 41.00N 85.45W
46 K5 **North Manitou Island** island Michigan, N USA
28 L10 **North Mankato** Minnesota, N USA 44.11N 94.03W
25 Z15 **North Miami** Florida, SE USA 25.54N 80.11W
157 K18 **North Miladummadulu Atoll** atoll N Maldives
98 I7 **North Minch** see Minch, The
25 W15 **North Naples** Florida, SE USA 26.13N 81.47W
183 P8 **North New Hebrides Trench** undersea feature W Pacific Ocean
27 O5 **North New River Canal** ☴ Florida, SE USA
157 I18 **North Nilandhe Atoll** var. Faafu Atoll. atoll C Maldives
35 S10 **North Ogden** Utah, W USA 41.18N 111.57W
37 S10 **North Palisade** ▲ California, W USA 37.06N 118.31W

201 U11 **North Pass** passage Chuuk Islands, C Micronesia
30 M15 **North Platte** Nebraska, C USA 41.07N 100.46W
35 X17 **North Platte River** ☴ C USA
117 G14 **North Point** headland Ascension Island, C Atlantic Ocean
180 I16 **North Point** headland Mahé, NE Seychelles 4.22S 55.28E
35 S13 **North Point** headland Michigan, N USA 45.01N 83.16W
23 P7 **North Point** headland Michigan, N USA 36.55N 82.37W
41 N9 **North Pole** Alaska, USA 64.42N 147.09W
207 R9 **North Pole** pole Arctic Ocean 90.00N 0.00W
25 O4 **Northport** Alabama, S USA 33.13N 87.34W
29 Q2 **Northport** Kansas, C USA 39.25N 95.19W
25 W14 **North Port** Florida, SE USA 27.03S 82.15W
34 L6 **Northport** Washington, NW USA 48.54N 117.48W
20 L13 **North Powder** Oregon, NW USA 45.00N 117.56W
31 V14 **North Raccoon River** ☴ Iowa, C USA
35 S11 **North Ridgeville** Ohio, N USA 41.14N 82.37W
21 P7 **North Riding** cultural region E England, UK
98 K4 **North Rona** island NW Scotland, UK
98 K4 **North Ronaldsay** island NE Scotland, UK
38 L2 **North Salt Lake** Utah, W USA 40.51N 111.54W
9 P15 **North Saskatchewan** ☴ Alberta/Saskatchewan, S Canada
37 X5 **North Schell Peak** ▲ Nevada, W USA 39.25N 114.34W
North Scotia Ridge see South Georgia Ridge
12 F16 **North Sea** Dan. Nordsøen, Dut. Noordzee, Fr. Mer du Nord, Ger. Nordsee, Nor. Nordsjøen; prev. German Ocean, Lat. Mare Germanicum. sea NW Europe
37 T6 **North Shoshone Peak** ▲ Nevada, W USA 39.08N 117.28W
North Siberian Lowland/North Siberian Plain see Severo-Sibirskaya Nizmennost'
31 R13 **North Sioux City** South Dakota, N USA 42.31N 96.28W
98 K4 **North Sound, The** sound N Scotland, UK
191 T4 **North Star** New South Wales, SE Australia 28.55S 150.25E
192 K9 **North Star State** see Minnesota
191 V3 **North Stradbroke Island** island Queensland, E Australia
North Sulawesi see Sulawesi Utara
North Sumatra see Sumatera Utara
12 D17 **North Sydenham** ☴ Ontario, S Canada
20 H9 **North Syracuse** New York, NE USA 43.07N 76.07W
192 K9 **North Taranaki Bight** gulf North Island, NZ
10 H9 **North Twin Island** island Nunavut, C Canada
98 E8 **North Uist** island NW Scotland, UK
99 L14 **Northumberland** cultural region N England, UK
189 Y7 **Northumberland Isles** island group Queensland, NE Australia
11 Q14 **Northumberland Strait** strait SE Canada
34 G14 **North Umpqua River** ☴ Oregon, NW USA
47 Q13 **North Union** Saint Vincent, Saint Vincent and the Grenadines 13.16N 61.07W
8 L17 **North Vancouver** British Columbia, SW Canada 49.21N 123.04W
20 K9 **Northville** New York, NE USA 43.13N 74.08W
99 Q19 **North Walsham** E England, UK 52.49N 1.22E
41 T10 **Northway** Alaska, USA 62.57N 141.56W
85 G21 **North-West** off. North-West Province, Afr. Noordwes. ♦ province South Africa
North-West see Nord-Ouest
16 I6 **Northwest Atlantic Mid-Ocean Canyon** undersea feature N Atlantic Ocean
188 G8 **North West Cape** headland Western Australia 21.48S 114.10E
40 J9 **Northwest Cape** headland Saint Lawrence Island, Alaska, USA 63.46N 171.45W
84 H13 **North Western** ♦ province W Zambia
161 J24 **North Western Province** ♦ province W Sri Lanka
155 U4 **North-West Frontier Province** ♦ province NW Pakistan
98 H8 **North West Highlands** ▲ N Scotland, UK
199 Hh4 **Northwest Pacific Basin** undersea feature N Pacific Ocean
203 Y2 **Northwest Point** headland Kiritimati, E Kiribati 10.22S 105.34E
27 O5 **Northwest Providence Channel** channel N Bahamas
11 Q8 **North West River** Newfoundland, E Canada 53.30N 60.10W
15 I5 **Northwest Territories** Fr. Territories du Nord-Ouest. ♦ territory NW Canada
99 K18 **Northwich** C England, UK 53.16N 2.31W
27 S5 **North Wichita River** ☴ Texas, SW USA
20 I17 **North Wildwood** New Jersey, NE USA 39.00N 74.48W
10 G14 **North Windham** Maine, NE USA 43.50N 70.26W
21 P8 **North Wilkesboro** North Carolina, SE USA 36.09N 81.09W
10 Q6 **Northwind Plain** undersea feature Arctic Ocean
31 V11 **Northwood** Iowa, C USA 43.26N 93.13W

31 Q4 **Northwood** North Dakota, N USA 47.43N 97.34W
99 M15 **North York Moors** moorland N England, UK
27 V9 **North Zulch** Texas, SW USA 30.54N 96.06W
28 K2 **Norton** Kansas, C USA 39.49N 99.53W
35 S13 **Norton** Ohio, N USA 40.25N 83.04W
23 P7 **Norton** Virginia, NE USA 36.55N 82.37W
41 N9 **Norton Bay** bay Alaska, USA
35 O9 **Norton Shores** Michigan, N USA 43.10N 86.15W
40 M10 **Norton Sound** inlet Alaska, USA
29 Q2 **Nortonville** Kansas, C USA 39.25N 95.19W
104 I8 **Nort-sur-Erdre** Loire-Atlantique, NW France 47.27N 1.30W
205 N2 **Norvegia, Cape** headland Antarctica 71.15S 12.15W
20 M13 **Norwalk** Connecticut, NE USA 41.08N 73.28W
31 V14 **Norwalk** Iowa, C USA 41.28N 93.40W
35 S11 **Norwalk** Ohio, N USA 41.14N 82.37W
21 P7 **Norway** Maine, NE USA 44.13N 70.30W
35 N5 **Norway** Michigan, N USA 45.47N 87.54W
95 E17 **Norway** off. Kingdom of Norway, Nor. Norge. ♦ monarchy N Europe
9 X13 **Norway House** Manitoba, C Canada 53.58N 97.49W
207 R16 **Norwegian Basin** undersea feature W Norwegian Sea
86 D6 **Norwegian Sea** Nor. Norske Havet. sea NE Atlantic Ocean
207 S17 **Norwegian Trench** undersea feature NE North Sea
12 F16 **Norwich** Ontario, S Canada 42.57N 80.37W
99 Q19 **Norwich** E England, UK 52.37N 1.18E
21 N13 **Norwich** Connecticut, NE USA 41.30N 72.02W
20 I11 **Norwich** New York, NE USA 42.31N 75.31W
31 U9 **Norwood** Minnesota, C USA 44.46N 93.55W
35 Q15 **Norwood** Ohio, N USA 39.07N 84.27W
12 H11 **Nosbonsing, Lake** ☒ Ontario, S Canada
Nösen see Bistriţa
172 P1 **Noshappu-misaki** headland Hokkaidō, NE Japan 45.26N 141.38E
171 M9 **Noshiro** var. Noshiro; prev. Noshirominato. Akita, Honshū, C Japan 40.10N 140.01E
Noshirominato/Nosiro see Noshiro
119 Q3 **Nosivka** Rus. Nosovka. Chernihivs'ka Oblast', NE Ukraine 50.55N 31.37E
69 T14 **Nosop** var. Nossob, Nossop. ☴ Botswana/Namibia
129 S4 **Nosovaya** Nenetskiy Avtonomnyy Okrug, NW Russian Federation 68.12N 54.33E
Nosovka see Nosivka
149 V11 **Noşratābād** Sīstān va Balūchestān, E Iran 29.53N 59.57E
97 J18 **Nossebro** Västra Götaland, S Sweden 58.12N 12.42E
98 K6 **Noss Head** headland N Scotland, UK 58.29N 3.03W
85 E20 **Nossob** ☴ E Namibia
Nossob/Nossop see Nosop
180 J2 **Nosy Be** ▲ Antsiranana, N Madagascar 23.36S 47.36E
180 J6 **Nosy Varika** Fianarantsoa, SE Madagascar 20.36S 48.31E
12 L10 **Notawassi** ☴ Quebec, SE Canada
12 L11 **Notawassi, Lac** ☒ Quebec, SE Canada
38 J5 **Notch Peak** ▲ Utah, W USA 39.08N 113.24W
110 G10 **Noteć** Ger. Netze. ☴ NW Poland
Nóties Sporádes see Dodekánisos
117 J22 **Nótion Aigaíon** Eng. Aegean South. ♦ region E Greece
117 H18 **Nótios Evvoïkós Kólpos** gulf E Greece
117 B16 **Nótio Stenó Kérkyras** strait W Greece
109 L25 **Noto** var. Netum. Sicilia, Italy, C Mediterranean Sea 36.52N 15.04E
171 J12 **Noto** Ishikawa, Honshū, SW Japan 37.18N 137.11E
97 J13 **Notodden** Telemark, S Norway 59.33N 9.15E
109 L25 **Noto, Golfo di** gulf Sicily, Italy, C Mediterranean Sea
171 J12 **Noto-hantō** peninsula Honshū, SW Japan
172 Qg5 **Notoro-ko** ☒ Hokkaidō, NE Japan
11 T11 **Notre Dame Bay** bay Newfoundland, E Canada
13 P6 **Notre-Dame-de-Lorette** Quebec, SE Canada 49.05N 72.24W
12 L11 **Notre-Dame-de-Pontmain** Quebec, SE Canada 46.13N 75.37W
13 T8 **Notre-Dame-du-Lac** Quebec, SE Canada 47.36N 68.48W
12 L12 **Notre-Dame-du-Nord** Quebec, SE Canada 48.48N 71.27W
79 R14 **Notsé** S Togo 6.53N 1.09E
172 R7 **Notsuke-suidō** strait Japan/Russian Federation
172 R7 **Notsuke-zaki** headland Hokkaidō, NE Japan 43.33N 145.18E
12 G14 **Nottawasaga** ☴ Ontario, S Canada
12 G14 **Nottawasaga Bay** lake bay Ontario, S Canada
10 G13 **Nottaway** ☴ Quebec, SE Canada
25 U3 **Nottely Lake** ☒ Georgia, SE USA
97 H16 **Nottersøy** island S Norway
99 M19 **Nottingham** C England, UK 52.58N 1.10W
16 N5 **Nottingham Island** island Nunavut, NE Canada

◆ COUNTRY ◇ DEPENDENT TERRITORY ◉ ADMINISTRATIVE REGION ▲ MOUNTAIN ▲ VOLCANO ☒ LAKE
◆ COUNTRY CAPITAL ○ DEPENDENT TERRITORY CAPITAL ✕ INTERNATIONAL AIRPORT ▲ MOUNTAIN RANGE ☴ RIVER ☒ RESERVOIR

99 N18 **Nottinghamshire** cultural region
C England, UK

23 V7 **Nottoway** Virginia, NE USA
37.07N 78.03 W

23 V7 **Nottoway River** ᴧ Virginia,
NE USA

78 G7 **Nouâdhibou** prev. Port-Étienne.
Dakhlet Nouâdhibou,
W Mauritania 20.54N 17.01W

78 G7 **Nouâdhibou ✕** Dakhlet
Nouâdhibou, W Mauritania
20.59N 17.02W

78 F7 **Nouâdhibou, Dakhlet** prev.
Baie du Lévrier. bay W Mauritania

78 F7 **Nouâdhibou, Râs** prev. Cap
Blanc. headland NW Mauritania
20.48N 17.03W

78 G9 **Nouakchott ●** (Mauritania)
Nouakchott District,
SW Mauritania 18.09N 15.58W

78 G9 **Nouakchott ✕** Trarza,
SW Mauritania 18.18N 15.54W

123 K13 **Noual, Sebkhet en** var. Sabkhat
an Nawāl. salt flat C Tunisia

78 G8 **Nouâmghâr** var. Nouamrhar.
Dakhlet Nouâdhibou,
W Mauritania 19.22N 16.31W

Nouamrhar see Nouâmghâr

Noua Sulita see Novoselytsya

197 I7 **Nouméa ○** (New Caledonia)
Province Sud, S New Caledonia
22.13S 166.29E

81 E15 **Noun** ᴧ C Cameroon

79 N12 **Nouna** W Burkina 12.43N 3.54W

85 H24 **Noupoort** Northern Cape,
C South Africa 31.10S 24.57E

Nouveau-Brunswick see New
Brunswick

Nouveau-Comptoir see
Wemindji

13 T4 **Nouvel, Lacs ⊚** Quebec,
SE Canada

13 W7 **Nouvelle** Quebec, SE Canada
48.07N 66.16W

13 W7 **Nouvelle** Quebec, SE Canada

Nouvelle-Calédonie see New
Caledonia

Nouvelle Écosse see Nova Scotia

105 R3 **Nouzonville** Ardennes, N France

153 Q11 **Nov Rus.** Nau. NW Tajikistan
40.10N 69.16E

61 I21 **Nova Alvorada** Mato Grosso do
Sul, SW Brazil 21.25S 54.19W

Novabad see Navobod

113 D19 **Nová Bystřice Ger.** Neubistritz.
Budějovický Kraj, S Czech
Republic 48.59N 15.05E

118 H13 **Novaci** Gorj, SW Romania
45.08N 23.39E

Nova Civitas see Neustadt an der
Weinstrasse

Novaesium see Neuss

62 H10 **Nova Esperança** Paraná, S Brazil
23.09S 52.13W

108 H11 **Novafeltria** Marche, C Italy

62 Q9 **Nova Friburgo** Rio de Janeiro,
SE Brazil 22.16S 42.34W

84 D12 **Nova Gaia** var. Cambundi-
Catembo. Malanje, NE Angola
10.03S 17.31E

111 S12 **Nova Gorica** W Slovenia
45.57N 13.40E

114 G10 **Nova Gradiška Ger.** Neugradisk,
Hung. Újgradiška. Brod-Posavina,
NE Croatia 45.15N 17.23E

62 K7 **Nova Granada** São Paulo,
S Brazil 20.33S 49.19W

62 O10 **Nova Iguaçu** Rio de Janeiro,
SE Brazil 22.31S 44.04W

119 S10 **Nova Kakhovka Rus.** Novaya
Kakhovka. Khersons'ka Oblast',
SE Ukraine 46.45N 33.19E

Nová Karvinná see Karviná

Nova Lamego see Gabú

Nova Lisboa see Huambo

114 C11 **Novalja** Lika-Senj, W Croatia
44.33N 14.53E

121 M14 **Novalukoml' Rus.** Novolukoml'.
Vitsyebskaya Voblasts', N Belarus
54.40N 29.09E

Nova Mambone see Mambone

85 P16 **Nova Naburi** Zambézia,
NE Mozambique 16.47S 38.55E

119 Q9 **Nova Odesa** var. Novaya Odessa.
Mykolayivs'ka Oblast', S Ukraine
47.18N 31.45E

62 H10 **Nova Olímpia** Paraná, S Brazil
23.28S 53.12W

63 I15 **Nova Prata** Rio Grande do Sul,
S Brazil 28.45S 51.37W

12 H12 **Novar** Ontario, S Canada
45.26N 79.14W

108 C7 **Novara** anc. Novaria. Piemonte,
NW Italy 45.27N 8.36E

Novaria see Novara

119 P7 **Novarkanels'k** Kirovohrads'ka
Oblast', C Ukraine 48.39N 30.48E

11 P15 **Nova Scotia Fr.** Nouvelle Écosse.
◆ province SE Canada

1 M9 **Nova Scotia** physical region
SE Canada

36 M8 **Novato** California, W USA
38.06N 122.35W

199 J8 **Nova Trough** undersea feature
W Pacific Ocean

118 L7 **Nova Ushtsya** Khmel'nyts'ka
Oblast', W Ukraine 48.50N 27.16E

85 M17 **Nova Vanduzi** Manica,
C Mozambique 18.54S 33.18E

119 U5 **Nova Vodolaha Rus.** Novaya
Vodolaga. Kharkivs'ka Oblast',
E Ukraine 49.43N 35.48E

Nova Chara Chitinskaya
Oblast', C Russian Federation
56.45N 117.58E

125 J14 **Novaya Igirma** Irkutskaya
Oblast', C Russian Federation
57.08N 103.52E

Novaya Kakhovka see Nova
Kakhovka

150 E10 **Novaya Kazanka** Zapadnyy
Kazakhstan, W Kazakhstan
48.57N 49.94E

128 I12 **Novaya Ladoga** Leningradskaya
Oblast', NW Russian Federation
60.03N 32.15E

125 Ee10 **Novaya Lyalya** Sverdlovskaya
Oblast', C Russian Federation
59.01N 60.32E

131 R5 **Novaya Malykla** Ul'yanovskaya
Oblast', W Russian Federation
54.13N 49.55E

Novaya Odessa see Nova Odesa

126 M4 **Novaya Sibir', Ostrov** island
Novosibirskiye Ostrova,
NE Russian Federation

Novaya Vodolaga see Nova
Vodolaha

121 P17 **Novaya Yel'nya** Rus. Novaya
Yel'ya. Mahilyowskaya Voblasts',
E Belarus 53.16N 31.13E

125 G4 **Novaya Zemlya** island group
N Russian Federation

Novaya Zemlya Trough see East
Novaya Zemlya Trough

116 K10 **Nova Zagora** Sliven, C Bulgaria
42.29N 26.00E

107 S12 **Novelda** País Valenciano, E Spain
38.24N 0.45W

113 H19 **Nové Mesto nad Váhom Ger.**
Waagneustadtll, Hung. Vágújhely.
Trenčiansky Kraj, W Slovakia
48.45N 17.50E

113 F17 **Nové Město na Moravě Ger.**
Neustadtl in Mähren. Jihlavský
Kraj, C Czech Republic
49.34N 16.04E

Novesium see Neuss

113 I21 **Nové Zámky Ger.** Neuhäusel,
Hung. Érsekújvár. Nitriansky Kraj,
SW Slovakia 48.00N 18.10E

125 C6 **Novgorod** Novgorodskaya
Oblast', W Russian Federation
58.31N 31.15E

125 C6 **Novgorod-Severskiy** see
Novhorod-Sivers'kyy

125 C6 **Novgorodskaya Oblast' ◆**
province W Russian Federation

119 R8 **Novhorodka** Kirovohrads'ka
Oblast', C Ukraine 48.21N 32.38E

119 R2 **Novhorod-Sivers'kyy Rus.**
Novgorod-Severskiy. Chernihivs'ka
Oblast', NE Ukraine 52.00N 33.15E

35 R10 **Novi** Michigan, N USA
42.28N 83.28W

Novi see Novi Vinodolski

114 L9 **Novi Bečej** prev. Új-Becse,
Vološinovo, Ger. Neubetsche,
Hung. Törökbecse. Serbia,
N Yugoslavia 45.36N 20.09E

Novi Grad see Bosanski Novi

116 G9 **Novi Iskŭr** Sofiya-Grad,
W Bulgaria 42.46N 23.19E

108 C9 **Novi Ligure** Piemonte, NW Italy
44.46N 8.46E

101 L22 **Noville** Luxembourg, SE Belgium
50.04N 5.46E

204 I10 **Noville Peninsula** peninsula
Thurston Island, Antarctica

Noviodunum see Soissons,
Aisne, France

Noviodunum see Nevers, Nièvre,
France

Noviodunum see Nyon, Vaud,
Switzerland

Noviomagus see Lisieux, France

Noviomagus see Nijmegen,
Netherlands

116 M8 **Novi Pazar** Shumen, NE Bulgaria
43.21N 27.13E

115 M15 **Novi Pazar Turk.** Yenipazar.
Serbia, S Yugoslavia 43.09N 20.31E

114 K10 **Novi Sad Ger.** Neusatz, Hung.
Újvidék. Serbia, N Yugoslavia
45.16N 19.49E

119 T6 **Novi Sanzhary** Poltavs'ka
Oblast', C Ukraine 49.21N 34.18E

114 H12 **Novi Travnik** prev. Pučarevo.
Federacija Bosna I Hercegovina,
C Bosnia and Herzegovina
44.12N 17.39E

114 B10 **Novi Vinodolski** var. Novi.
Primorje-Gorski Kotar,
NW Croatia 45.08N 14.46E

60 H14 **Novo Airão** Amazonas, N Brazil
2.06S 61.19W

131 N14 **Novoaleksandrovsk**
Stavropol'skiy Kray, SW Russian
Federation 44.41N 43.01E

Novoalekseyevka see Zhobda

126 H14 **Novoaltaysk** Altayskiy Kray,
S Russian Federation 53.22N 83.58E

131 N9 **Novoanninskiy** Volgogradskaya
Oblast', SW Russian Federation
50.29N 42.41E

60 F13 **Novo Aripuanã** Amazonas,
NW Brazil 5.04S 60.19W

119 Y6 **Novoaydar** Luhans'ka Oblast',
E Ukraine 49.00N 39.00E

119 X9 **Novoazovs'k Rus.** Novoazovsk.
Donets'ka Oblast', E Ukraine
47.07N 38.06E

126 Mm16 **Novobureyskiy** Amurskaya
Oblast', SE Russian Federation
49.42N 129.46E

131 U7 **Novocheboksarsk**
Chuvashskaya Respublika,
W Russian Federation 56.07N 47.32E

131 R5 **Novocheremshansk**
Ul'yanovskaya Oblast', W Russian
Federation 54.23N 50.14E

130 L12 **Novocherkassk** Rostovskaya
Oblast', SW Russian Federation
47.23N 40.00E

131 R6 **Novodevich'ye** Samarskaya
Oblast', W Russian Federation
53.33N 48.51E

125 M8 **Novodvinsk** Arkhangel'skaya
Oblast', NW Russian Federation
64.22N 40.48E

Novograd-Volynskiy see
Novhorod-Volyns'kyy

Novogrudok see Navahrudak

63 I15 **Novo Hamburgo** Rio Grande do
Sul, S Brazil 29.42S 51.07W

61 H16 **Novo Horizonte** Mato Grosso,
W Brazil 11.19S 57.11W

62 K8 **Novo Horizonte** São Paulo,
S Brazil 21.27S 49.14W

118 M4 **Novhrad-Volyns'kyy Rus.**
Novograd-Volynskiy.
Zhytomyrs'ka Oblast', N Ukraine
50.33N 27.31E

150 L14 **Novokazalinsk** see Ayteke Bi

130 M8 **Novokhopersk** Voronezhskaya
Oblast', W Russian Federation
51.09N 41.34E

131 R6 **Novokuybyshevsk** Samarskaya
Oblast', W Russian Federation
53.06N 49.56E

126 H14 **Novokuznetsk** prev. Stalinsk.
Kemerovskaya Oblast', S Russian
Federation 53.45N 87.12E

205 R1 **Novolazarevskaya** Russian
research station‡, Antarctica
70.42S 11.31E

Novolukoml' see Novalukoml'

111 V12 **Novo mesto Ger.** Rudolfswert;
prev. Ger. Neustadtl. SE Slovenia
45.48N 15.09E

130 K14 **Novomikhaylovskiy**
Krasnodarskiy Kray, SW Russian
Federation 44.18N 38.49E

114 L8 **Novo Miloševo** Serbia,
N Yugoslavia 45.43N 20.20E

Novomirgorod see
Novomyrhorod

130 L5 **Novomoskovsk** Tul'skaya
Oblast', W Russian Federation
54.04N 38.22E

119 U7 **Novomoskovs'k Rus.**
Novomoskovsk. Dnipropetrovs'ka
Oblast', E Ukraine 48.37N 35.13E

119 P7 **Novomykolayivka** Zaporiz'ka
Oblast', SE Ukraine 47.58N 35.54E

119 Q7 **Novomyrhorod Rus.**
Novomirgorod. Kirovohrads'ka
Oblast', S Ukraine 48.46N 31.39E

126 I12 **Novonazimovo** Krasnoyarskiy
Kray, C Russian Federation
59.30N 90.45E

131 N8 **Novonikolayevskiy**
Volgogradskaya Oblast',
SW Russian Federation
50.55N 42.24E

131 P10 **Novonikol'skoye**
Volgogradskaya Oblast',
SW Russian Federation
49.23N 45.06E

131 X7 **Novoorsk** Orenburgskaya
Oblast', W Russian Federation
51.22N 58.59E

126 M13 **Novopokrovskaya**
Krasnodarskiy Kray, SW Russian
Federation 45.58N 40.43E

119 Y5 **Novopolotsk** Luhans'ka Oblast',
E Ukraine 49.33N 39.07E

131 R8 **Novorepnoye** Saratovskaya
Oblast', W Russian Federation
51.04N 48.34E

130 K14 **Novorossiysk** Krasnodarskiy
Kray, SW Russian Federation
44.49N 37.37E

Novorossiyskiy see
Novorossiyskoye

150 J10 **Novorossiyskoye** prev.
Novorossiyskiy. Aktyubinsk,
NW Kazakhstan 51.13N 57.57E

126 Jj6 **Novorybnaya** Taymyrskaya
(Dolgano-Nenetskiy) Avtonomnyy
Okrug, N Russian Federation
72.48N 105.49E

128 F15 **Novorzhev** Pskovskaya Oblast',
W Russian Federation 57.01N 29.19E

119 S12 **Novoselitsa** see Novoselytsya

Novoselivs'ke Respublika Krym,
S Ukraine 45.26N 33.37E

116 G6 **Novo Selo** Vidin, NW Bulgaria
44.08N 22.48E

115 M14 **Novo Selo** Serbia, S Yugoslavia
43.39N 20.54E

118 K8 **Novoselytsya Rom.** Nouă Suliţă,
Rus. Novoselitsa. Chernivets'ka
Oblast', W Ukraine 48.13N 26.18E

131 U7 **Novosergiyevka** Orenburgskaya
Oblast', W Russian Federation
52.04N 53.40E

130 L11 **Novoshakhtinsk** Rostovskaya
Oblast', SW Russian Federation
47.48N 39.51E

126 Gg14 **Novosibirsk** Novosibirskaya
Oblast', C Russian Federation
55.04N 83.04E

125 G13 **Novosibirskaya Oblast' ◆**
province C Russian Federation

126 M4 **Novosibirskiye Ostrova Eng.**
New Siberian Islands. island group
N Russian Federation

130 K6 **Novosil'** Orlovskaya Oblast',
W Russian Federation 53.00N 37.59E

128 G18 **Novosokol'niki** Pskovskaya
Oblast', W Russian Federation
56.21N 30.07E

131 N9 **Novospasskoye** Ul'yanovskaya
Oblast', W Russian Federation
53.08N 47.48E

131 X8 **Novotroitsk** Orenburgskaya
Oblast', W Russian Federation
51.09N 58.18E

Novotroitskoye see Brlik,
Kazakhstan

Novotroitskoye see
Novotroyits'ke, Ukraine

119 T11 **Novotroyits'ke Rus.**
Novotroitskoye. Khersons'ka
Oblast', S Ukraine 46.21N 34.21E

Novoukrainka see
Novoukrayinka

119 Q8 **Novoukrayinka Rus.**
Novoukrainka. Kirovohrads'ka
Oblast', C Ukraine 48.19N 31.33E

131 Q5 **Novoul'yanovsk** Ul'yanovskaya
Oblast', W Russian Federation
54.10N 48.19E

131 W8 **Novouralets** Orenburgskaya
Oblast', W Russian Federation
51.19N 56.52E

81 E16 **Nsimaleen ✕** Centre, C Cameroon

118 I4 **Novovolyns'k Rus.** Novovolynsk.
Volyns'ka Oblast', NW Ukraine
50.46N 24.09E

119 S9 **Novovorontsovka** Khersons'ka
Oblast', S Ukraine 47.28N 33.55E

153 Y7 **Novovoznesenovka** Issyk-
Kul'skaya Oblast', E Kyrgyzstan
42.36N 78.44E

129 R14 **Novovyatsk** Kirovskaya Oblast',
NW Russian Federation
58.30N 49.42E

119 O6 **Novoyel'nya** see Navayel'nya

121 I19 **Novozhytov** Vinnyts'ka
Oblast', W Ukraine 49.16N 29.31E

130 H6 **Novozybkov** Bryanskaya Oblast',
W Russian Federation 52.36N 31.58E

114 F9 **Novska** Sisak-Moslavina,
NE Croatia 45.20N 16.58E

Nový Bohumín see Bohumín

113 D15 **Nový Bor Ger.** Haida; prev. Bor.
České Lípy, Hajda. Liberecký Kraj,
N Czech Republic 50.46N 14.32E

113 L18 **Nový Bydžov Ger.** Neubidschow.
Hradecký Kraj, N Czech Republic
50.15N 15.27E

121 G18 **Nový Dvor Rus.** Novyy Dvor.
Hrodzyenskaya Voblasts', W
Belarus 52.49N 24.22E

113 I17 **Nový Jičín Ger.** Neutitschein.
Ostravský Kraj, E Czech Republic
49.36N 18.00E

9 V9 **Nueltin Lake** ⊚
Manitoba/Nunavut, C Canada

120 K12 **Novy Pahost Rus.** Novyy Pogost.
Vitsyebskaya Voblasts',
NW Belarus 55.30N 27.28E

119 R9 **Novyy Buh Rus.** Novyy Bug.
Mykolayivs'ka Oblast', S Ukraine
47.39N 32.31E

119 Q4 **Novyy Bykiv** Chernihivs'ka
Oblast', N Ukraine 50.36N 31.39E

Novyy Dvor see Nový Dvor

Novyye Aneny see Anenii Noi

131 P7 **Novyy Burasy** Saratovskaya
Oblast', W Russian Federation
52.10N 46.00E

130 K8 **Novyy Oskol** Belgorodskaya
Oblast', W Russian Federation
50.43N 37.53E

Novyy Pogost see Novy Pahost

131 R2 **Novyy Tor"yal** Respublika Mariy
El, W Russian Federation
56.59N 48.53E

126 K14 **Novyy Uoyan** Respublika
Buryatiya, S Russian Federation
56.06N 111.27E

126 Gg9 **Novyy Urengoy** Yamalo-
Nenetskiy Avtonomnyy Okrug,
N Russian Federation 66.06N 76.25E

127 N15 **Novyy Urgal** Khabarovskiy Kray,
E Russian Federation 51.02N 132.45E

Novyy Uzen' see Zhanaozen

125 G12 **Novyy Vasyugan** Tomskaya
Oblast', C Russian Federation
58.28N 76.19E

113 N16 **Nowa Dęba** Podkarpackie,
SE Poland 50.31N 21.53E

113 G15 **Nowa Ruda Ger.** Neurode.
Dolnośląskie, SW Poland
50.34N 16.30E

112 F12 **Nowa Sól** var. Nowasól, Ger.
Neusalz an der Oder. Lubuskie,
W Poland 51.47N 15.42E

29 Q8 **Nowata** Oklahoma, C USA
36.42N 95.38W

148 M6 **Nowbarān** Markazī, W Iran
34.01N 49.51E

112 J8 **Nowe Kujawsko-pomorskie,**
C Poland 53.39N 18.43E

112 K9 **Nowe Miasto Lubawskie Ger.**
Neumark. Warmińsko-Mazurskie,
NE Poland 53.24N 19.36E

112 L13 **Nowe Miasto nad Pilicą**
Mazowieckie, C Poland
51.37N 20.34E

112 D8 **Nowe Warpno Ger.** Neuwarp.
Zachodniopomorskie, NW Poland
53.52N 14.12E

56 N9 **Nowgong** see Nagaon

112 E8 **Nowogard** var. Nowógard, Ger.
Naugard. Zachodniopomorskie,
NW Poland 53.41N 15.09E

112 N9 **Nowogród** Podlaskie, NE Poland
53.14N 21.52E

Nowogródek see Navahrudak

113 E14 **Nowogrodziec Ger.** Naumburg
am Queis. Dolnośląskie,
SW Poland 51.12N 15.24E

Nowojelnia see Navayel'nya

Nowo-Minsk see Mińsk
Mazowiecki

35 V13 **Nowood River** ᴧ Wyoming,
C USA

Nowo-Święciany see
Švenčionėliai

191 S10 **Nowra-Bomaderry** New South
Wales, SE Australia 34.51S 150.41E

155 T5 **Nowshera** var. Naushahra,
Naushara. North-West Frontier
Province, NE Pakistan 34.00N 72.00E

112 J7 **Nowy Dwór Gdański Ger.**
Tiegenhof. Pomorskie, N Poland
54.12N 19.03E

112 L11 **Nowy Dwór Mazowiecki**
Mazowieckie, C Poland
52.25N 20.43E

113 M17 **Nowy Sącz Ger.** Neu Sandec.
Małopolskie, S Poland 49.36N 20.41E

113 L18 **Nowy Targ Ger.** Neumark.
Małopolskie, S Poland 49.28N 20.68E

112 F11 **Nowy Tomyśl** var. Nowy Tomysl.
Wielkopolskie, C Poland
52.18N 16.07E

154 M7 **Now Zād** var. Nauzad. Helmand,
S Afghanistan 32.22N 64.31E

25 N4 **Noxubee River** ᴧ
Alabama/Mississippi, S USA

126 Gg10 **Noyabr'sk** Yamalo-Nenetskiy
Avtonomnyy Okrug, N Russian
Federation 63.08N 75.19E

104 L8 **Noyant** Maine-et-Loire,
NW France 47.28N 0.08W

41 X14 **Noyes Island** island Alexander
Archipelago, Alaska, USA

105 O3 **Noyon** Oise, N France 49.35N 3.00E

104 I7 **Nozay** Loire-Atlantique,
NW France 47.34N 1.36W

84 L12 **Nsando** Northern, NE Zambia
10.25S 31.14E

85 N16 **Nsanje** Southern, S Malawi
16.57S 35.10E

79 Q17 **Nsawam** SE Ghana 5.46N 0.19W

81 E16 **Nsimalen** see Nsimaleen

84 K12 **Nseluka** Northern, NE Zambia
10.35S 29.58E

84 H13 **Nsombo** Northern, N Zambia
12.22S 24.57E

85 N14 **Ntcheu** var. Ncheu. Central,
S Malawi 14.49S 34.37E

81 D17 **Ntem** prev. Campo, Kampo.
ᴧ Cameroon/Equatorial Guinea

85 I14 **Ntemwa** North Western,
NW Zambia 14.08S 26.13E

Ntlenyana, Mount see Thabana
Ntlenyana

81 J19 **Ntomba, Lac** var. Lac Tumba.
⊚ NW Dem. Rep. Congo (Zaire)

81 I20 **Ntungamo** SW Uganda
0.54S 30.16E

81 E18 **Ntusi** SW Uganda 0.03N 31.11E

84 H18 **Ntwetwe Pan** salt lake
NE Botswana

83 D19 **Nuasjärvi ⊚** C Finland

70 I9 **Nuba Mountains** C Sudan

118 G10 **Nubian Desert** desert NE Sudan

79 X14 **Numan** Adamawa, E Nigeria
9.26N 11.58E

171 K14 **Numata** Gunma, Honshū, S Japan
36.39N 139.00E

172 Oo4 **Numata** Hokkaidō, NE Japan
43.48N 141.55E

27 R14 **Nueces River** ᴧ Texas, SW USA

101 K15 **Nuenen** Noord-Brabant,
S Netherlands 51.29N 5.36E

64 G6 **Nuestra Señora, Bahía** bay
N Chile

63 D14 **Nuestra Señora Rosario de
Caa Catí** Corrientes,
NE Argentina 27.48S 57.42W

56 J9 **Nueva Antioquia** Vichada,
E Colombia 6.04N 69.30W

54 N4 **Nueva Caceres** see Naga

43 O7 **Nueva Ciudad Guerrera**
Tamaulipas, C Mexico 26.32N 99.13W

57 N4 **Nueva Esparta** off. Estado Nueva
Esparta. ◆ state NE Venezuela

46 C5 **Nueva Gerona** Isla de la
Juventud, S Cuba 21.49N 82.49W

44 M11 **Nueva Guadalupe** San Miguel,
E El Salvador 13.30N 88.21W

44 F6 **Nueva Guinea** Región
Autónoma Atlántico Sur,
SE Nicaragua 11.40N 84.22W

63 D19 **Nueva Helvecia** Colonia,
SW Uruguay 34.16S 57.52W

58 C13 **Nueva, Isla** island S Chile

42 M14 **Nueva Italia** Michoacán de
Ocampo, SW Mexico 19.01N 102.06W

58 D6 **Nueva Loja** var. Lago Agrio.
Sucumbíos, NE Ecuador
0.05N 76.41W

44 F6 **Nueva Ocotepeque** prev.
Ocotepeque. Ocotepeque,
W Honduras 14.25N 89.11W

63 D19 **Nueva Palmira** Colonia,
SW Uruguay 33.52S 58.25W

43 N6 **Nueva Rosita** Coahuila de
Zaragoza, NE Mexico
27.58N 101.10W

44 E7 **Nueva San Salvador** prev. Santa
Tecla. La Libertad, SW El Salvador
13.42N 89.18W

44 J7 **Nueva Segovia ◆** department
NW Nicaragua

Nueva Tabarca see Plana, Isla

Nueva Villa de Padilla see
Nuevo Padilla

63 B21 **Nueve de Julio** Buenos Aires,
E Argentina 35.29S 60.52W

46 H6 **Nuevitas** Camagüey, E Cuba
21.34N 77.18W

63 D18 **Nuevo Berlín** Río Negro,
W Uruguay 32.58S 58.03W

42 I4 **Nuevo Casas Grandes**
Chihuahua, N Mexico
30.23N 107.53W

45 T14 **Nuevo Chagres** Colón,
C Panama 9.13N 80.03W

43 W15 **Nuevo Coahuila** Campeche,
E Mexico 17.52N 90.46W

65 K17 **Nuevo, Golfo** gulf S Argentina

43 O7 **Nuevo Laredo** Tamaulipas,
NE Mexico 27.29N 99.31W

43 N8 **Nuevo León ◆** state NE Mexico

43 P10 **Nuevo Padilla** var. Nueva Villa
de Padilla. Tamaulipas, C Mexico
24.01N 98.48W

58 E6 **Nuevo Rocafuerte** Napo,
E Ecuador 0.55S 75.25W

168 G6 **Nuga** Dzavhan, W Mongolia
48.17N 95.07E

82 O13 **Nugaal** off. Gobolka Nugaal. ◆
region N Somalia

193 E24 **Nugget Point** headland South
Island, NZ 46.26S 169.49E

195 R9 **Nuguria Islands** island group
E PNG

192 P10 **Nuhaka** Hawke's Bay, North
Island, NZ 39.03S 177.43E

144 M10 **Nuhaydayn, Wādī an** dry
watercourse W Iraq

54 N4 **Nui Jiang** see Salween

190 G7 **Nui Atoll** atoll W Tuvalu
34.38S 135.36E

Nukha see Şäki

127 O9 **Nukh Yablonevyy, Gora** ▲
E Russian Federation
60.26N 151.45E

195 T13 **Nukiki** Choiseul Island,
NW Solomon Islands 6.45S 94.30E

194 F10 **Nuku** Sandaun, NW PNG
3.40S 142.29E

200 R15 **Nuku** island Tongatapu Group,
S Tonga

200 Qq15 **Nuku'alofa** Tongatapu, S Tonga
21.09S 175.13W

200 Qq15 **Nuku'alofa ✕** (Tonga) Tongatapu,
S Tonga 21.07S 175.13W

202 G12 **Nukufetau Atoll** atoll C Tuvalu

202 F7 **Nukufetau Atoll** atoll Wallis and
Futuna

202 G12 **Nukuhifala** island E Wallis and
Futuna

203 W7 **Nuku Hiva** island Îles Marquises,
NE French Polynesia

191 L19 **Nuku Hiva Island** island Îles
Marquises, N French Polynesia

202 F9 **Nukulaelae Atoll** var.
Nukulailai. atoll E Tuvalu

Nukulailai see Nukulaelae Atoll

202 G11 **Nukuloa** island N Wallis and
Futuna

195 W10 **Nukumanu Islands** prev.
Tasman Group. island group
NE PNG

Nukunau see Nikunau

85 F17 **Nxaunxau** Ngamiland,
NW Botswana 18.57S 21.18E

41 N12 **Nyac** Alaska, USA 61.00N 159.56W

125 FJ10 **Nyagan'** Khanty-Mansiyskiy
Avtonomnyy Okrug, N Russian
Federation 62.10N 65.32E

Nyainqêntanglha Feng see
Nyainqêntanglha Shan

158 M11 **Nyainqêntanglha Shan**
▲ W China

80 B11 **Nyala** Southern Darfur, W Sudan
12.01N 24.49E

85 H25 **Nyamandhlovu** Matabeleland
North, W Zimbabwe 19.50S 28.15E

81 I19 **Nyamtumbo** Ruvuma,
S Tanzania 10.35S 36.07E

85 M17 **Nyanda** see Masvingo

125 R14 **Nyandoma** Arkhangel'skaya
Oblast', NW Russian Federation
61.39N 40.09E

83 C15 **Numatinna** ᴧ W Sudan

171 J17 **Numazu** Shizuoka, Honshū,
S Japan 35.05N 138.52E

97 F14 **Numedalen** valley S Norway

97 G14 **Numedalslågen** ᴧ var. Laagen.
ᴧ S Norway

95 L19 **Nummela** Etelä-Suomi, S Finland
60.21N 24.19E

191 O11 **Numurkah** Victoria, SE Australia
36.04S 145.28E

206 L16 **Nunap Isua** var.
Uummannarsuaq, Dan. Kap
Farvel, Eng. Cape Farewell.
headland S Greenland 59.57N 44.27W

5 K5 **Nunavut ◆** Territory N Canada

56 H9 **Nunchia** Casanare, C Colombia
5.37N 72.13W

9 M20 **Nuneaton** C England, UK
52.31N 1.28W

159 W14 **Nungba** Manipur, NE India

40 L12 **Nunivak Island** island Alaska,
USA

158 I5 **Nun Kun** ▲ NW India
34.01N 76.04E

100 L10 **Nunspeet** Gelderland,
E Netherlands 52.21N 5.45E

109 C18 **Nuoro** Sardegna, Italy,
C Mediterranean Sea 40.19N 9.19E

77 R12 **Nuqayy, Jabal** hill range S Libya

56 C9 **Nuquí** Chocó, W Colombia
5.43N 77.16W

149 O4 **Nūr Māzandarān, N Iran
36.34N 52.01E

151 Q9 **Nura** N Kazakhstan

149 N11 **Nūrābād** Fārs, C Iran 30.07N 51.30E

Nurata see Nurota

Nurata, Khrebet see Nurota
Tizmasi

142 L17 **Nur Dağları ▲** S Turkey

Nurek see Norak

Nuremberg see Nürnberg

142 M15 **Nurhak** Kahramanmaraş,
S Turkey 37.57N 37.21E

190 J9 **Nuriootpa** South Australia
34.28S 139.00E

131 S5 **Nurlat** Respublika Tatarstan,
W Russian Federation 54.26N 50.48E

95 N15 **Nurmes** Itä-Suomi, E Finland
63.31N 29.08E

103 K20 **Nürnberg** Eng. Nuremberg.
Bayern, S Germany 49.27N 11.04E

103 K20 **Nürnberg ✕** Bayern, SE Germany
49.29N 11.04E

152 M10 **Nurota Rus.** Nurata. Nawoiy
Wiloyati, C Uzbekistan
40.35N 65.43E

153 N10 **Nurota Tizmasi Rus.** Khrebet
Nuratau. ▲ C Uzbekistan

155 T8 **Nürpur** Punjab, E Pakistan
31.54N 71.55E

191 P6 **Nurri, Mount** hill New South
Wales, SE Australia 31.42S 146.03E

27 T13 **Nursery** Texas, SW USA

176 Vv12 **Nusa Tenggara Eng.** Lesser
Sunda Islands. island group East
Timor/ Indonesia

175 O15 **Nusa Tenggara Barat off.**
Propinsi Nusa Tenggara Barat, Eng.
West Nusa Tenggara. ◆ province
S Indonesia

175 Q17 **Nusa Tenggara Timur off.**
Propinsi Nusa Tenggara Timur,
Eng. East Nusa Tenggara. ◆
province S Indonesia

176 E11 **Nu Shan** ▲ SW China

155 N11 **Nushki** Baluchistān, SW Pakistan
29.33N 66.01E

114 J7 **Nuštar** Vukovar-Srijem, E Croatia
45.20N 18.48E

101 T10 **Nuth** Limburg, SE Netherlands
50.55N 5.52E

102 N13 **Nuthe** ᴧ NE Germany

106 I5 **Nuuk var.** Núk, Dan. Godthaab,
Godthåb. ○ (Greenland) Kitaa,
SW Greenland 64.15N 51.34W

94 L13 **Nuupas** Lappi, NW Finland
66.29N 26.19E

203 O7 **Nuupere, Pointe** headland
Moorea, W French Polynesia
17.34S 149.46W

203 O7 **Nuuroa, Pointe** headland Tahiti,
W French Polynesia

168 M8 **Nüürst Töv, C Mongolia
47.44N 108.22E

161 K25 **Nuwara Eliya** var. Nuwara.
Central Province, Sri Lanka
6.58N 80.46E

190 E7 **Nuyts Archipelago** island group
South Australia

83 F17 **Nxaunxau** Ngamiland,
NW Botswana 18.57S 21.18E

85 M16 **Nyanga** off. Province de la
Nyanga, var. La Nyanga. ◆ province
SW Gabon

126 Kk11 **Nyurba** Respublika Sakha
(Yakutiya), NE Russian Federation
63.17N 118.14E

81 D20 **Nyanga** off. Province de la
Nyanga, var. La Nyanga. ◆ province
SW Gabon

81 E20 **Nyanga** ᴧ Congo/Gabon

83 F20 **Nyantakara** Kagera,
NW Tanzania 3.04S 31.22E

83 G19 **Nyanza ◆** province W Kenya

83 E21 **Nyanza-Lac** S Burundi
4.16S 29.38E

70 J14 **Nyasa, Lake** var. Lake Malawi;
prev. Lago Nyassa. ◊ E Africa

**Nyasaland/Nyasaland
Protectorate** see Malawi

Nyassa, Lago see Nyasa, Lake

121 J17 **Nyasvizh Pol.** Nieśwież, Rus.
Nesvizh. Minskaya Voblasts',
C Belarus 53.13N 26.41E

177 G8 **Nyaunglebin** Pegu,
SW Myanmar 17.58N 96.43E

177 N5 **Nyaung-u** Magwe, C Myanmar

97 H24 **Nyborg** Fyn, C Denmark
55.19N 10.49E

95 N11 **Nybro** Kalmar, S Sweden
56.45N 15.54E

121 J16 **Nyeharelaye Rus.** Negoreloye.
Minskaya Voblasts', C Belarus
53.36N 27.05E

205 W3 **Nye Mountains** ▲ Antarctica

83 J19 **Nyeri** Central, C Kenya 0.25S 36.55E

120 M11 **Nyeshcharda, Vozyera** ⊚
⊙ N Belarus

94 O2 **Ny-Friesland** physical region
N Svalbard

97 L14 **Nyhammar** Dalarna, C Sweden
60.19N 14.56E

166 F7 **Nyíkög Qu** ᴧ C China

164 L14 **Nyima** Xizang Zizhiqu, W China
31.53N 87.50E

85 L14 **Nyimba** Eastern, E Zambia
14.33S 30.49E

165 P16 **Nyingchi** Xizang Zizhiqu,
W China 29.34N 94.22E

113 O21 **Nyírbátor** Szabolcs-Szatmár-
Bereg, E Hungary 47.49N 22.06E

113 N21 **Nyíregyháza** Szabolcs-Szatmár-
Bereg, NE Hungary 47.57N 21.43E

Nyíro see Ewaso Ng'iro

Nyitra see Nitra

Nyitrabánya see Handlová

95 K16 **Nykarleby Fin.** Uusikaarlepyy.
Länsi-Suomi, W Finland
63.22N 22.30E

97 J25 **Nykøbing** Storstrøm,
SE Denmark 54.46N 11.52E

97 I22 **Nykøbing** Vestsjælland,
C Denmark 55.55N 11.40E

97 F21 **Nykøbing** Viborg, NW Denmark
56.48N 8.52E

97 N17 **Nyköping** Södermanland,
S Sweden 58.45N 17.03E

97 L15 **Nykroppa** Värmland, C Sweden
59.37N 14.18E

85 J20 **Nylstroom** Northern, NE South
Africa 24.39S 28.23E

191 P7 **Nymagee** New South Wales,
SE Australia 32.06S 146.19E

191 V5 **Nymboida** New South Wales,
SE Australia 29.57S 152.45E

191 U5 **Nymboida River** ᴧ New South
Wales, SE Australia

113 D16 **Nymburk var.** Neuenburg an der
Elbe, Ger. Nimburg. Středočeský
Kraj, C Czech Republic 50.12N 15.00E

97 O16 **Nynäshamn** Stockholm,
C Sweden 58.54N 17.55E

191 Q6 **Nyngan** New South Wales,
SE Australia 31.36S 147.07E

Nyoman see Neman

110 A10 **Nyon** Ger. Neuss, anc.
Noviodunum. Vaud,
SW Switzerland 46.24N 6.15E

105 S14 **Nyons** Drôme, E France
44.22N 5.06E

81 D16 **Nyong** ᴧ SW Cameroon

81 E14 **Nyos, Lac** Eng. Lake Nyos.
⊚ NW Cameroon

129 U11 **Nyrob var.** Nyrov. Permskaya
Oblast', NW Russian Federation
60.41N 56.42E

Nyrov see Nyrob

113 H15 **Nysa Ger.** Neisse. Opolskie,
S Poland 50.28N 17.20E

34 M13 **Nyssa** Oregon, NW USA
43.52N 116.59W

97 I25 **Nysted** Storstrøm, SE Denmark
54.40N 11.41E

129 U14 **Nytva** Permskaya Oblast',
NW Russian Federation
57.56N 55.22E

129 U12 **Nyukhcha** Arkhangel'skaya
Oblast', NW Russian Federation
63.32N 46.29E

119 U12 **Nyzhn'ohirs'kyy Rus.**
Nizhnegorskiy. Respublika Krym,
S Ukraine 45.29N 34.42E

83 G21 **Nzega** Tabora, C Tanzania
4.13S 33.10E

78 K15 **Nzérékoré** Guinée-Forestière,
SE Guinea 7.45N 8.49W

84 A10 **N'Zeto** Port. Ambrizete. Zaire,
NW Angola 7.13S 12.52E

81 M24 **Nzilo, Lac** Fr. Lac
Delcommune. ◊ SE Dem. Rep.
Congo (Zaire)

---

31 O11 **Oacoma** South Dakota, N USA
43.49N 99.25W

◆ COUNTRY ◊ DEPENDENT TERRITORY ◆ ADMINISTRATIVE REGION ▲ MOUNTAIN ◼ VOLCANO ⊚ LAKE
● COUNTRY CAPITAL ○ DEPENDENT TERRITORY CAPITAL ✕ INTERNATIONAL AIRPORT ▲ MOUNTAIN RANGE ᴧ RIVER ⊡ RESERVOIR

31 N9 **Oahe Dam** *dam* South Dakota, N USA 44.27N 100.24W
30 M9 **Oahe, Lake** ⊠ North Dakota/South Dakota, N USA
40 C9 **Oahu** *Haw.* O'ahu. *island* Hawaii, USA, C Pacific Ocean
172 Qq6 **O-Akan-dake** ▲ Hokkaidō, NE Japan 43.26N 144.06E
190 K8 **Oakbank** South Australia 33.07S 140.36E
21 P13 **Oak Bluffs** Martha's Vineyard, New York, NE USA 41.25N 70.32W
38 K4 **Oak City** Utah, W USA 39.22N 112.19W
39 R3 **Oak Creek** Colorado, C USA 40.16N 106.57W
37 P8 **Oakdale** California, W USA 37.46N 120.51W
24 H8 **Oakdale** Louisiana, S USA 30.49N 92.39W
31 P7 **Oakes** North Dakota, N USA 46.08N 98.05W
24 J4 **Oak Grove** Louisiana, S USA 32.51N 91.25W
99 N19 **Oakham** C England, UK 52.40N 0.45W
34 H7 **Oak Harbor** Washington, NW USA 48.17N 122.38W
23 R5 **Oak Hill** West Virginia, NE USA 37.59N 81.09W
37 N8 **Oakland** California, W USA 37.48N 122.16W
31 T15 **Oakland** Iowa, C USA 41.18N 95.22W
21 Q7 **Oakland** Maine, NE USA 44.32N 69.43W
23 T3 **Oakland** Maryland, NE USA 39.24N 79.24W
31 R14 **Oakland** Nebraska, C USA 41.50N 96.28W
35 N11 **Oak Lawn** Illinois, N USA 41.43N 87.45W
35 P16 **Oakley** Idaho, NW USA 42.13N 113.54W
28 I4 **Oakley** Kansas, C USA 39.06N 100.51W
35 N10 **Oak Park** Illinois, N USA 41.53N 87.46W
9 X16 **Oak Point** Manitoba, S Canada 50.23N 97.00W
34 G13 **Oakridge** Oregon, NW USA 43.45N 122.27W
22 M9 **Oak Ridge** Tennessee, S USA 36.01N 84.12W
192 K10 **Oakura** Taranaki, North Island, NZ 39.07S 173.58E
24 L7 **Oak Vale** Mississippi, S USA 31.26N 89.57W
12 G16 **Oakville** Ontario, S Canada 43.27N 79.40W
27 V8 **Oakwood** Texas, SW USA 31.34N 95.51W
193 F22 **Oamaru** Otago, South Island, NZ 45.10S 170.51E
98 F13 **Oa, Mull of** *headland* W Scotland, UK 55.35N 6.20W
175 Q7 **Oan** Sulawesi, N Indonesia 1.16N 121.25E
193 J17 **Oaro** Canterbury, South Island, NZ 42.29S 173.30E
37 X2 **Oasis** Nevada, W USA 41.01N 114.29W
205 S15 **Oates Land** *physical region* Antarctica
191 P17 **Oatlands** Tasmania, SE Australia 42.21S 147.23E
38 I11 **Oatman** Arizona, SW USA 35.03N 114.19W
43 R16 **Oaxaca** *var.* Oaxaca de Juárez; *prev.* Antequera. Oaxaca, SE Mexico 17.04N 96.40W
43 Q16 **Oaxaca** ◆ *state* SE Mexico
125 G8 **Ob'** ♣ C Russian Federation
12 G9 **Obabika Lake** ⊠ Ontario, S Canada
Obagan see Ubagan
120 M12 **Obal'** *Rus.* Obol'. Vitsyebskaya Voblasts', N Belarus 55.22N 29.16E
81 E16 **Obala** Centre, SW Cameroon 4.09N 11.31E
12 C6 **Oba Lake** ⊠ Ontario, S Canada
115 H14 **Obama** Fukui, Honshū, SW Japan 35.29N 135.42E
98 H11 **Oban** W Scotland, UK 56.25N 5.28W
Oban see Halfmoon Bay
171 Ll2 **Obanazawa** Yamagata, Honshū, C Japan 38.40N 140.21E
Obando see Puerto Inírida
106 I4 **O Barco** *var.* El Barco, El Barco de Valdeorras, O Barco de Valdeorras. Galicia, NW Spain 42.24N 7.00W
O Barco de Valdeorras see O Barco
Obbia see Hobyo
95 J16 **Obbola** Västerbotten, N Sweden 63.42N 20.18E
Obbrovazzo see Obrovac
Obchuga see Abchuha
Obdorsk see Salekhard
Óbecse see Bečej
120 I11 **Obeliai** Rokiškis, NE Lithuania 55.57N 25.47E
62 F13 **Oberá** Misiones, NE Argentina 27.28S 55.07W
110 E8 **Oberburg** Bern, W Switzerland 47.00N 7.37E
111 Q9 **Oberdrauburg** Salzburg, S Austria 46.45N 12.59E
Oberglogau see Głogówek
111 W4 **Ober Grafendorf** Niederösterreich, NE Austria 48.09N 15.33E
103 E15 **Oberhausen** Nordrhein-Westfalen, W Germany 51.28N 6.52E
Oberhollabrunn see Tulln
Oberlaibach see Vrhnika
103 Q15 **Oberlausitz** *physical region* E Germany
28 J2 **Oberlin** Kansas, C USA 39.48N 100.31W
24 H8 **Oberlin** Louisiana, S USA 30.37N 92.45W
35 T11 **Oberlin** Ohio, NE USA 41.17N 82.13W
105 U5 **Obernai** Bas-Rhin, NE France 48.28N 7.30E
111 R4 **Obernberg-am-Inn** Oberösterreich, N Austria 48.19N 13.02E
Oberndorf see Oberndorf am Neckar
103 G23 **Oberndorf am Neckar** *var.* Oberndorf. Baden-Württemberg, SW Germany 48.18N 8.32E

111 Q5 **Oberndorf bei Salzburg** Salzburg, NW Austria 47.57N 12.57E
Oberneustadtl see Kysucké Nové Mesto
191 S8 **Oberon** New South Wales, SE Australia 33.42S 149.50E
111 Q4 **Oberösterreich** *off.* Land Oberösterreich, *Eng.* Upper Austria. ◆ *state* NW Austria
Oberpahlen see Põltsamaa
191 M19 **Oberpfälzer Wald** ▲ SE Germany
111 Y6 **Oberpullendorf** Burgenland, E Austria 47.32N 16.30E
Oberradkersburg see Gornja Radgona
103 G18 **Oberursel** Hessen, W Germany 50.12N 8.34E
111 Q8 **Obervellach** Salzburg, S Austria 46.56N 13.10E
111 X7 **Oberwart** Burgenland, SE Austria 47.18N 16.12E
Oberwischau see Vișeu de Sus
111 T7 **Oberwölz** *var.* Oberwölz-Stadt. Steiermark, SE Austria 47.12N 14.20E
Oberwölz-Stadt see Oberwölz
35 S13 **Obetz** Ohio, N USA 39.52N 82.57W
Ob', Gulf of see Obskaya Guba
56 G8 **Obia** Santander, C Colombia 6.16N 73.18W
60 H12 **Óbidos** Pará, NE Brazil 1.52S 55.30W
106 F10 **Óbidos** Leiria, C Portugal 39.21N 9.09W
Obidovichi see Abidavichy
153 Q13 **Obigarm** W Tajikistan 38.42N 69.34E
172 P7 **Obihiro** Hokkaidō, NE Japan 42.55N 143.09E
153 P13 **Obikiik** W Tajikistan 38.07N 68.36E
115 N16 **Obilić** Serbia, S Yugoslavia 42.50N 20.57E
131 O12 **Obil'noye** Respublika Kalmykiya, SW Russian Federation 47.31N 44.24E
22 F8 **Obion** Tennessee, S USA 36.15N 89.11W
22 F8 **Obion River** ♣ Tennessee, S USA
175 T9 **Obi, Pulau** *island* Maluku, E Indonesia
175 Oo4 **Obira** Hokkaidō, NE Japan 44.01N 141.39E
175 T9 **Obi, Selat** *strait* Maluku, E Indonesia
131 N11 **Oblivskaya** Rostovskaya Oblast', SW Russian Federation 48.34N 42.31E
127 N16 **Obluch'ye** Yevreyskaya Avtonomnaya Oblast', SE Russian Federation 48.59N 131.18E
130 K4 **Obninsk** Kaluzhskaya Oblast', W Russian Federation 55.06N 36.40E
116 J8 **Obnova** Pleven, N Bulgaria 43.26N 25.04E
81 N15 **Obo** Haut-Mbomou, E Central African Republic 5.20N 26.28E
82 M11 **Obock** E Djibouti 11.57N 43.09E
Obol' see Obal'
Obolyanka see Abalyanka
176 Vv11 **Obome** Irian Jaya, E Indonesia 3.42S 133.21E
112 G11 **Oborniki** Wielkopolskie, C Poland 52.38N 16.48E
81 G19 **Obouya** Cuvette, C Congo 0.55S 15.40E
130 J8 **Oboyan'** Kurskaya Oblast', W Russian Federation 51.12N 36.15E
128 M9 **Obozerskiy** Arkhangel'skaya Oblast', NW Russian Federation 63.26N 40.20E
114 L11 **Obrenovac** Serbia, N Yugoslavia 44.39N 20.12E
114 D12 **Obrovac** *It.* Obbrovazzo. Zadar, SW Croatia 44.12N 15.40E
Obrovo see Abrova
37 Q3 **Observation Peak** ▲ California, W USA 40.48N 120.07W
126 H7 **Obskaya Guba** *Eng.* Gulf of Ob'. *gulf* N Russian Federation
181 N13 **Ob' Tablemount** *undersea feature* S Indian Ocean 50.16S 51.59E
181 T10 **Ob' Trench** *undersea feature* E Indian Ocean
79 P16 **Obuasi** S Ghana 6.15N 1.36W
119 P5 **Obukhiv** *Rus.* Obukhov. Kyyivs'ka Oblast', N Ukraine 50.05N 30.37E
Obukhov see Obukhiv
129 U14 **Obva** ♣ NW Russian Federation
119 V10 **Obytichna Kosa** *spit* SE Ukraine
119 V10 **Obytichna Zatoka** *gulf* SE Ukraine
107 O3 **Oca** ♣ N Spain
25 W10 **Ocala** Florida, SE USA 29.11N 82.08W
42 M7 **Ocampo** Coahuila de Zaragoza, NE Mexico 27.18N 102.24W
56 G7 **Ocaña** Norte de Santander, N Colombia 8.16N 73.21W
107 N9 **Ocaña** Castilla-La Mancha, C Spain 39.57N 3.30W
106 H4 **O Carballiño** *Cast.* Carballino. Galicia, NW Spain 42.26N 8.04W
39 T9 **Ocate** New Mexico, SW USA 36.09N 105.03W
56 D14 **Occidental, Cordillera** ▲ W Colombia
56 D14 **Occidental, Cordillera** ▲ W S America
23 Q6 **Oceana** West Virginia, NE USA 37.41N 81.37W
23 Z4 **Ocean City** Maryland, NE USA 38.20N 75.05W
20 J17 **Ocean City** New Jersey, NE USA 39.15N 74.33W
8 K15 **Ocean Falls** British Columbia, SW Canada 52.24N 127.42W
Ocean Island see Kure Atoll
Ocean Island see Banaba
66 I9 **Oceanographer Fracture Zone** *tectonic feature* NW Atlantic Ocean
37 U17 **Oceanside** California, W USA 33.12N 117.22W
24 M9 **Ocean Springs** Mississippi, S USA 30.24N 88.49W
Ocean State see Rhode Island
72 Q9 **O C Fisher Lake** ⊠ Texas, SW USA

119 Q10 **Ochakiv** *Rus.* Ochakov. Mykolayivs'ka Oblast', S Ukraine 46.36N 31.33E
Ochakov see Ochakiv
143 Q9 **Och'amch'ire** *Rus.* Ochamchira. W Georgia 42.45N 41.30E
Ochamchira see Och'amch'ire
129 T15 **Ochansk** Permskaya Oblast', NE Russian Federation 57.54N 54.40E
117 I19 **Óchi** ▲ Évvoia, C Greece 38.03N 24.27E
172 R8 **Ochiishi-misaki** *headland* Hokkaidō, NE Japan 43.10N 145.29E
25 S9 **Ochlockonee River** ♣ Florida/Georgia, SE USA
46 K12 **Ocho Rios** E Jamaica 18.24N 77.06W
Ochrida see Ohrid
103 J19 **Ochrida, Lake** see Ohrid, Lake
25 U7 **Ocilla** Georgia, SE USA 31.35N 83.15W
96 N13 **Ockelbo** Gävleborg, C Sweden 60.51N 16.46E
Ocker see Oker
97 I19 **Öckerö** Västra Götaland, S Sweden 57.43N 11.32E
25 U6 **Ocmulgee River** ♣ Georgia, SE USA
118 H11 **Ocna Mureș** *Hung.* Marosújvár; *prev.* Ocna Mureșului; *prev. Hung.* Marosújvárakna. Alba, C Romania 46.25N 23.52E
Ocna Mureșului see Ocna Mureș
118 H11 **Ocna Sibiului** *Ger.* Salzburg, *Hung.* Vizakna. Sibiu, C Romania 45.52N 23.59E
118 L7 **Ocnița** *Rus.* Oknitsa. N Moldova 48.25N 27.30E
25 U4 **Oconee, Lake** ⊠ Georgia, SE USA
25 U5 **Oconee River** ♣ Georgia, SE USA
32 M9 **Oconomowoc** Wisconsin, N USA 43.06N 88.29W
32 M6 **Oconto** Wisconsin, N USA 44.55N 87.52W
32 M6 **Oconto Falls** Wisconsin, N USA 44.52N 88.06W
32 M6 **Oconto River** ♣ Wisconsin, N USA
43 V16 **Ocosingo** Chiapas, SE Mexico 16.51N 92.06W
44 J8 **Ocotal** Nueva Segovia, NW Nicaragua 13.38N 86.27W
44 F6 **Ocotepeque** ◆ *department* W Honduras
Ocotepeque see Nueva Ocotepeque
42 L13 **Ocotlán** Jalisco, SW Mexico 20.18N 102.45W
43 R16 **Ocotlán** *var.* Ocotlán de Morelos. Oaxaca, SE Mexico 16.49N 96.49W
Ocotlán de Morelos see Ocotlán
43 V16 **Ocozocuautla** Chiapas, SE Mexico 16.43N 93.19W
23 Y10 **Ocracoke Island** North Carolina, SE USA
104 I3 **Octeville** Manche, N France 49.37N 1.39W
October Revolution Island see Oktyabr'skoy Revolyutsii, Ostrov
45 R17 **Ocú** Herrera, S Panama 7.55N 80.47W
85 Q14 **Ocua** Cabo Delgado, NE Mozambique 13.37S 39.44E
56 M5 **Ocumare del Tuy** *var.* Ocumare. Miranda, N Venezuela 10.07N 66.46W
79 P17 **Oda** SE Ghana 5.54N 1.01W
170 F12 **Ōda** *var.* Oda. Shimane, Honshū, SW Japan 35.09N 132.31E
94 K3 **Óðáðahraun** *lava flow* C Iceland
176 Y14 **Odammun** ♣ Irian Jaya, E Indonesia
171 Mm9 **Ōdate** Akita, Honshū, C Japan 40.18N 140.34E
171 J16 **Odawara** Kanagawa, Honshū, S Japan 35.13N 139.07E
97 D14 **Odda** Hordaland, S Norway 60.03N 6.34E
97 G22 **Odder** Århus, C Denmark 55.58N 10.10E
Oddur see Xuddur
31 T13 **Odebolt** Iowa, C USA 42.19N 95.15W
106 H14 **Odeleite** Faro, S Portugal 37.01N 7.28W
27 Q4 **Odell** Texas, SW USA 34.19N 99.24W
27 T14 **Odem** Texas, SW USA 27.57N 97.34W
106 F13 **Odemira** Beja, S Portugal 37.34N 8.37W
142 C14 **Ödemiş** İzmir, SW Turkey 38.10N 27.58E
85 G7 **Odendaalsrus** Free State, C South Africa 27.52S 26.42E
100 I9 **Odense** Fyn, C Denmark 55.24N 10.22E
103 H19 **Odenwald** ▲ W Germany
86 H10 **Oder** *Cz./Pol.* Odra. ♣ C Europe
Oderberg see Bohumín
102 P11 **Oderbruch** *wetland* Germany/Poland
102 P13 **Oder-Havel-Kanal** *canal* NE Germany
Oderhellen see Odorheiu Secuiesc
102 P13 **Oder-Spree-Kanal** *canal* NE Germany
108 I7 **Oderzo** Veneto, NE Italy 45.48N 12.33E
124 Pp4 **Odesa** *Rus.* Odessa. Odes'ka Oblast', SW Ukraine 46.28N 30.43E
97 L18 **Odeshög** Östergötland, S Sweden 58.13N 14.40E
119 Q10 **Odes'ka Oblast'** *var.* Odessa, *Rus.* Odesskaya Oblast'. ◆ *province* SW Ukraine

Odesskaya Oblast' see Odes'ka Oblast'
85 Ff13 **Odesskoye** Omskaya Oblast', C Russian Federation 54.15N 72.45E
Odessus see Varna
104 F6 **Odet** ♣ NW France
106 I14 **Odiel** ♣ SW Spain
78 L14 **Odienné** NW Ivory Coast 9.32N 7.34W
179 Pp12 **Odiongan** Tablas Island, C Philippines 12.23N 122.01E
118 L12 **Odobești** Vrancea, E Romania 45.46N 27.06E
112 H13 **Odolanów** *Ger.* Adelnau. Wielkopolskie, C Poland 51.35N 17.42E
61 N14 **Oeiras** Piauí, E Brazil 07.00S 42.07W
106 F11 **Oeiras** Lisboa, C Portugal 38.40N 9.18W
103 G14 **Oelde** Nordrhein-Westfalen, W Germany 51.49N 8.09E
30 J11 **Oelrichs** South Dakota, N USA 43.08N 103.13W
103 M17 **Oelsnitz** Sachsen, E Germany 50.22N 12.12E
Oels/Oels in Schlesien see Oleśnica
31 X12 **Oelwein** Iowa, C USA 42.40N 91.54W
Oeniadae see Oiniádes
203 N17 **Oeno Island** *atoll* Pitcairn Islands, C Pacific Ocean
Oesel see Saaremaa
103 L7 **Oetz** *var.* Ötz. Tirol, W Austria 47.13N 10.56E
143 P11 **Of** Trabzon, NE Turkey 40.57N 40.16E
32 K15 **O'Fallon** Illinois, N USA 38.35N 89.54W
29 W4 **O'Fallon** Missouri, C USA 38.54N 90.31W
109 N16 **Ofanto** ♣ S Italy
99 D18 **Offaly** *Ir.* Ua Uíbh Fhailí; *prev.* King's County. *cultural region* C Ireland
103 H18 **Offenbach** *var.* Offenbach am Main. Hessen, W Germany 50.06N 8.46E
Offenbach am Main see Offenbach
103 F22 **Offenburg** Baden-Württemberg, SW Germany 48.28N 7.57E
190 C2 **Officer Creek** *seasonal river* South Australia
Oficina María Elena see María Elena
Oficina Pedro de Valdívia see Pedro de Valdívia
117 K22 **Ofidoússa** *island* Kykládes, Greece, Aegean Sea
Ofiral see Sharm el Sheikh
97 H10 **Ofotfjorden** *fjord* N Norway
198 D8 **Ofu** *island* Manua Islands, E American Samoa
171 Mm12 **Ōfunato** Iwate, Honshū, C Japan 39.04N 141.41E
171 M10 **Oga** Akita, Honshū, C Japan 39.54N 139.48E
Ogaadeen see Ogadēn
171 M11 **Ogachi** Akita, Honshū, C Japan 39.03N 140.26E
171 M11 **Ogachi-tōge** *pass* Honshū, C Japan 39.00N 140.20E
83 N14 **Ogadēn** *Som.* Ogaadeen. *plateau* Ethiopia/Somalia
171 M10 **Oga-hantō** *peninsula* Honshū, C Japan
171 Hh14 **Ogaki** Gifu, Honshū, SW Japan 35.21N 136.35E
30 L15 **Ogallala** Nebraska, C USA 41.09N 101.43W
174 I12 **Ogan, Air** ♣ Sumatera, W Indonesia
172 T16 **Ogasawara-shotō** *Eng.* Bonin Islands. *island group* SE Japan
12 I9 **Ogascanane, Lac** ⊠ Quebec, SE Canada
172 N9 **Ogawara-ko** ⊠ Honshū, C Japan
79 T15 **Ogbomosho** Oyo, W Nigeria 8.10N 4.16E
31 Q13 **Ogden** Iowa, C USA 42.03N 94.01W
38 L2 **Ogden** Utah, W USA 41.11N 111.58W
20 I6 **Ogdensburg** New York, NE USA 44.42N 75.25W
197 L16 **Ogea Driki** *island* Lau Group, E Fiji
197 L16 **Ogea Levu** *island* Lau Group, E Fiji
25 W5 **Ogeechee River** ♣ Georgia, SE USA
Oger see Ogre
152 F6 **Oghiyon Shūrkhogi** *wetland* NW Uzbekistan
171 K12 **Ogi** Niigata, Sado, C Japan 37.49N 138.16E
8 H5 **Ogilvie** Yukon Territory, NW Canada 63.34N 139.43W
8 H4 **Ogilvie** ♣ Yukon Territory, NW Canada
8 G4 **Ogilvie Mountains** ▲ Yukon Territory, NW Canada
37 R14 **Ojai** California, W USA 34.25N 119.15W
96 K13 **Öje** Dalarna, C Sweden 60.50N 14.42E
95 J14 **Öjebyn** Norrbotten, N Sweden 65.20N 21.24E
42 M9 **Ojinaga** Chihuahua, N Mexico 29.30N 104.25W
171 K13 **Ojiya** *var.* Oziya. Niigata, Honshū, C Japan 37.18N 138.47E
42 M11 **Ojo Caliente** *var.* Ojocaliente. Zacatecas, C Mexico 22.39N 102.17W
42 J6 **Ojo de Liebre, Laguna** *var.* Laguna Scammon, Scammon
57 I7 **Ojos del Salado, Cerro** ▲ W Argentina 27.04S 68.34W
Oksu see Oqsu

107 R7 **Ojos Negros** Aragón, NE Spain 40.43N 1.30W
10 C10 **Ogoki** ♣ Ontario, S Canada
10 D11 **Ogoki Lake** ⊠ Ontario, C Canada
168 K10 **Ögöömör** Ömnögovï, S Mongolia 43.47N 104.31E
81 E19 **Ogooué** ♣ Congo/Gabon
81 E18 **Ogooué-Ivindo** *off.* Province de l'Ogooué-Ivindo, *var.* L'Ogooué-Ivindo. ◆ *province* NE Gabon
81 E19 **Ogooué-Lolo** *off.* Province de l'Ogooué-Lolo, *var.* L'Ogooué-Lolo. ◆ *province* C Gabon
81 C19 **Ogooué-Maritime** *off.* Province de l'Ogooué-Maritime, *var.* L'Ogooué-Maritime. ◆ *province* W Gabon
170 Cc13 **Ogōri** Fukuoka, Kyūshū, SW Japan 33.25N 130.30E
170 Dd13 **Ōgōri** Yamaguchi, Honshū, SW Japan 34.05N 131.20E
116 H7 **Ogosta** ♣ NW Bulgaria
114 Q9 **Ogražden** *Bul.* ▲ Bulgaria/FYR Macedonia *see also* Ograzhden
116 G12 **Ograzhden** *Mac.* Ogražden. ▲ Bulgaria/FYR Macedonia *see also* Ogražden
120 G9 **Ogre** *Ger.* Oger. Ogre, C Latvia 56.49N 24.36E
120 H9 **Ogre** ♣ C Latvia
114 C10 **Ogulin** Karlovac, NW Croatia 45.15N 15.13E
79 S16 **Ogun** ◆ *state* SW Nigeria
152 A12 **Ogurdzhaly, Ostrov** *Turkm.* Ogurjaly Adasy. *island* W Turkmenistan
Ogurjaly Adasy see Ogurdzhaly, Ostrov
79 U16 **Ogwashi-Uku** Delta, S Nigeria 6.08N 6.38E
193 B23 **Ohai** Southland, South Island, NZ 45.56S 167.59E
153 Q10 **Ohangaron** *Rus.* Akhangaran. Toshkent Wiloyati, E Uzbekistan 40.56N 69.37E
153 Q10 **Ohangaron** *Rus.* Akhangaran. ♣ E Uzbekistan
85 E17 **Ohangwena** ◆ *district* N Namibia
171 K17 **Ōhara** Chiba, Honshū, S Japan 35.14N 140.19E
32 M10 **O'Hare** ✕ (Chicago) Illinois, N USA 41.59N 87.56W
172 Nn8 **Ōhata** Aomori, Honshū, C Japan 41.23N 141.09E
192 L13 **Ohau** Manawatu-Wanganui, North Island, NZ 40.40S 175.15E
193 E20 **Ohau, Lake** ⊠ South Island, NZ
101 J20 **Ohey** Namur, SE Belgium 50.26N 5.07E
102 L12 **Ohre** *Ger.* Eger. ♣ Czech Republic/Germany
102 K13 **Ohre** ♣ NW Germany
114 L7 **Ohrid** *Turk.* Ochrida, Ohri. SW FYR Macedonia 41.07N 20.48E
115 M15 **Ohrid, Lake** *var.* Lake Ochrida, *Alb.* Liqeni i Ohrit, *Mac.* Ohridsko Ezero. ♦ Albania/FYR Macedonia
Ohridsko Ezero/Ohrit, Liqeni i see Ohrid, Lake
192 L9 **Ohura** Manawatu-Wanganui, North Island, NZ 38.51S 174.58E
60 J9 **Oiapoque** Amapá, E Brazil 3.54N 51.46W
60 J10 **Oiapoque, Rio** *var.* Fleuve l'Oyapok, Oyapok. ♣ Brazil/French Guiana *see also* Oyapok, Fleuve l'
13 O9 **Oies, Île aux** *island* Quebec, SE Canada
94 L13 **Oijärvi** Oulu, C Finland 65.37N 26.04E
95 J14 **Oikarainen** Lappi, N Finland 66.30N 25.46E
20 C13 **Oil City** Pennsylvania, NE USA 41.25N 79.42W
37 R13 **Oildale** California, W USA 35.25N 119.01W
Oilean Ciarraí see Castleisland
Oil Islands see Chagos Archipelago
117 L18 **Oinoússes** *island* E Greece
Oitf, Inis see Inisheer
101 N10 **Oirschot** Noord-Brabant, S Netherlands 51.30N 5.18E
105 P3 **Oise** ◆ *department* N France
105 N2 **Oise** ♣ N France
101 J14 **Oisterwijk** Noord-Brabant, S Netherlands 51.35N 5.12E
170 D14 **Ōita** Ōita, Kyūshū, SW Japan 33.15N 131.34E
170 D14 **Ōita** *off.* Ōita-ken. ◆ *prefecture* Kyūshū, SW Japan
117 E17 **Oítí** ▲ C Greece 38.48N 22.12E
172 O6 **Oiwake** Hokkaidō, NE Japan 42.57N 141.49E
172 Q4 **Oja** ♣ N Spain
82 H6 **Oko, Wadi** ♣ NE Sudan
170 J14 **Okaya** Nagano, Honshū, S Japan 36.04N 138.02E
96 K13 **Öje** Dalarna, C Sweden 60.50N 14.42E
95 J14 **Öjebyn** Norrbotten, N Sweden 65.20N 21.24E
9 Q16 **Okotoks** Alberta, SW Canada 50.43N 113.57W

150 M8 **Oktyabr'skiy** Kostanay, N Kazakhstan 40.43N 1.30W
194 E11 **Ok Tedi** Western, W PNG
171 G7 **Oktwin** Pegu, C Myanmar 18.46N 96.21E
131 R6 **Oktyabr'skiy** Samarskaya Oblast', W Russian Federation 53.11N 48.40E
85 L9 **Okahandja** Otjozondjupa, C Namibia 21.52S 101.40W
Oktyabr'sk see Kandyagash
129 N12 **Oktyabr'skiy** Arkhangel'skaya Oblast', NW Russian Federation 61.03N 43.16E
85 D18 **Okaihau** Northland, North Island, NZ 35.15S 173.47E
127 T5 **Oktyabr'skiy** Kamchatskaya Oblast', E Russian Federation 52.35N 94.18E
85 B13 **Okakarara** Otjozondjupa, N Namibia 20.34S 17.24E
131 T5 **Oktyabr'skiy** Respublika Bashkortostan, W Russian Federation 54.28N 53.29E
11 P5 **Okak Islands** *island group* Newfoundland, E Canada
131 O11 **Oktyabr'skiy** Volgogradskaya Oblast', SW Russian Federation 48.00N 43.35E
9 N17 **Okanagan** ♣ British Columbia, SW Canada
Oktyabr'skiy see Aktsyabrski
9 N17 **Okanagan Lake** ⊠ British Columbia, SW Canada
131 O17 **Oktyabr'skoye** Orenburgskaya Oblast', W Russian Federation
34 K6 **Okanogan River** ♣ Washington, NW USA
131 V7 **Oktyabr'skoye** Orenburgskaya Oblast', W Russian Federation
114 I13 **Okapa** Eastern Highlands, C PNG 6.22S 145.29E
126 J6 **Oktyabr'skoy Revolyutsii, Ostrov** *Eng.* October Revolution Island. *island* Severnaya Zemlya, N Russian Federation
85 C17 **Okakuejo** Kunene, N Namibia 19.09S 15.57E
170 C15 **Okuchi** *var.* Ōkuti. Kagoshima, Kyūshū, SW Japan 32.03N 130.36E
155 V9 **Okāra** Punjab, E Pakistan 30.49N 73.31E
171 Mm5 **Okushiri-tō** *var.* Okusiri Tō. *island* NE Japan
152 L9 **Okaputa** Otjozondjupa, N Namibia 20.09S 16.55E
28 M10 **Okarche** Oklahoma, C USA 35.43N 97.58W
Okusiri Tō see Okushiri-tō
79 S15 **Okuta** Kwara, W Nigeria 9.18N 3.09E
85 B14 **Okaukuejo** Kunene, N Namibia 19.09S 15.57E
Okuti see Okuchi
85 F19 **Okwa** *var.* Chapman's. ♣ Botswana/Namibia
170 Ff14 **Okayama** Okayama, Honshū, SW Japan 34.40N 133.54E
127 O10 **Ola** Magadanskaya Oblast', E Russian Federation 59.36N 151.18E
170 Ff14 **Okayama** *off.* Okayama-ken. ◆ *prefecture* Honshū, SW Japan
29 T11 **Ola** Arkansas, C USA 35.01N 93.13W
171 I16 **Okazaki** Aichi, Honshū, SW Japan 34.58N 137.10E
Ola see Ala
112 M12 **Okęcie** ✕ (Warszawa) Mazowieckie, C Poland 52.08N 20.57E
37 T11 **Olacha Peak** ▲ California, W USA 36.15N 118.07W
25 Y13 **Okeechobee** Florida, SE USA 27.14N 80.49W
94 J1 **Ólafsfjördhur** Nordhurland Eystra, N Iceland 66.04N 18.36W
25 Y14 **Okeechobee, Lake** ⊠ Florida, SE USA
94 H3 **Ólafsvík** Vesturland, W Iceland 64.52N 23.45W
28 M9 **Okeene** Oklahoma, C USA 36.07N 98.19W
Oláhbrettye see Bretea-Română
25 V8 **Okefenokee Swamp** *wetland* Georgia, SE USA
Oláhszentgyörgy see Sângeorz-Băi
99 J24 **Okehampton** SW England, UK 50.44N 4.00W
37 T11 **Olancha** California, W USA 36.16N 118.00W
29 P10 **Okemah** Oklahoma, C USA 35.25N 96.18W
44 J5 **Olanchito** Yoro, C Honduras 15.27N 86.37W
79 U16 **Okene** Kogi, S Nigeria 7.31N 6.15E
44 J6 **Olancho** ◆ *department* E Honduras
102 K13 **Oker** *var.* Ocker. ♣ NW Germany
97 O20 **Öland** *island* S Sweden
103 J14 **Oker** ♣ C Germany
97 O19 **Ölands norra udde** *headland* S Sweden 57.21N 17.06E
127 O13 **Okha** Ostrov Sakhalin, SE Russian Federation 53.33N 142.55E
97 N22 **Ölands södra udde** *headland* S Sweden 56.12N 16.26E
129 U15 **Okhansk** *var.* Ochansk. Permskaya Oblast', NW Russian Federation 57.44N 55.20E
190 K7 **Olary** South Australia 32.18S 140.16E
127 Nn10 **Okhotka** ♣ E Russian Federation
29 R4 **Olathe** Kansas, C USA 38.52N 94.49W
127 Nn11 **Okhotsk** Khabarovskiy Kray, E Russian Federation 59.21N 143.14E
63 C22 **Olavarría** Buenos Aires, E Argentina 36.57S 60.19W
199 I2 **Okhotsk, Sea of** *sea* NW Pacific Ocean
94 O2 **Ólav V Land** *physical region* E Svalbard
119 T4 **Okhtyrka** *Rus.* Akhtyrka. Sums'ka Oblast', NE Ukraine 50.19N 34.54E
113 H14 **Oława** *Ger.* Ohlau. Dolnośląskie, SW Poland 50.57N 17.18E
199 Gg6 **Oki-Daitō Ridge** *undersea feature* W Pacific Ocean
109 D17 **Olbia** *prev.* Terranova Pausania. Sardegna, Italy, C Mediterranean Sea 40.55N 9.30E
85 E23 **Okiep** Northern Cape, W South Africa 29.39S 17.53E
46 G5 **Old Bahama Channel** *channel* Bahamas/Cuba
170 Ff11 **Oki-kaikyō** *strait* SW Japan
Old Bay State/Old Colony State see Massachusetts
172 P15 **Okinawa** Okinawa, SW Japan 26.19N 127.46E
8 H2 **Old Crow** Yukon Territory, NW Canada 67.34N 139.55W
172 Oo14 **Okinawa** *var.* Okinawa-ken. ◆ *prefecture* Okinawa, SW Japan
Old Dominion see Virginia
172 Q14 **Okinawa** *island* SW Japan
Olderkeep see Oldeberkoop
170 Dd15 **Okino-erabu-jima** *island* Nansei-shotō, SW Japan
100 M7 **Oldeberkoop** *Fris.* Olderkeep. Friesland, N Netherlands 52.55N 6.07E
170 Ff11 **Oki-shotō** *var.* Oki-guntō. *island group* SW Japan
100 L10 **Oldebroek** Gelderland, E Netherlands 52.27N 5.53E
79 T16 **Okitipupa** Ondo, SW Nigeria 6.33N 4.43E
100 L8 **Oldemarkt** Overijssel, N Netherlands 52.49N 5.58E
79 T8 **Okkan** Pegu, SW Myanmar 17.31N 95.51E
96 E11 **Olden** Sogn og Fjordane, C Norway 61.52N 6.44E
150 N10 **Oklahoma** *off.* State of Oklahoma; also known as The Sooner State. ◆ *state* C USA
102 G10 **Oldenburg** Niedersachsen, NW Germany 53.09N 8.13E
28 N11 **Oklahoma City** *state capital* Oklahoma, C USA 35.28N 97.31W
102 K8 **Oldenburg** Schleswig-Holstein, N Germany 54.17N 10.55E
Old Line State see Maryland
Old North State see North Carolina
27 Q7 **Oklaunion** Texas, SW USA 34.07N 99.07W
100 P10 **Oldenzaal** Overijssel, E Netherlands 52.19N 6.52E
83 I17 **Ol Doinyo Lengeyo** ▲ C Kenya
9 Q15 **Olds** Alberta, SW Canada 51.49N 114.06W
29 P10 **Okmulgee** Oklahoma, C USA 35.37N 95.57W
20 J8 **Old Forge** New York, NE USA 43.42N 74.59W
Oknitsa see Ocnița
Old Goa see Goa
21 O6 **Old Speck Mountain** ▲ Maine, NE USA 44.34N 70.55W
24 M3 **Okolona** Mississippi, S USA 34.00N 88.45W
9 L17 **Oldham** N England, UK 53.36N 2.00W
21 S6 **Old Town** Maine, NE USA 44.54N 68.39W
172 Q4 **Okoppe** Hokkaidō, NE Japan 44.28N 143.08E
41 Q14 **Old Harbor** Kodiak Island, Alaska, USA 57.12N 153.18W
9 T17 **Old Wives Lake** ⊠ Saskatchewan, S Canada
168 J7 **Öldziyt** Arhangay, C Mongolia 48.30N 101.25E
46 J13 **Old Harbour** W Jamaica 17.55N 77.06W
169 N10 **Öldziyt** Dornogovi, SE Mongolia 44.21N 109.10E
99 C22 **Old Head of Kinsale** *Ir.* An Seancheann. *headland* SW Ireland 51.36N 8.33W
94 H6 **Oleai** *var.* San Jose. Saipan, S Northern Mariana Islands
22 J8 **Old Hickory Lake** ⊠ Tennessee, S USA
20 M12 **Olean** New York, NE USA 42.04N 78.24W
112 O7 **Olecko** *Ger.* Treuburg. Warmińsko-Mazurskie, NE Poland 54.01N 22.28E
127 O6 **Oleggio** Piemonte, NE Italy 45.36N 8.37E
126 L13 **Olëkma** ♣ C Russian Federation
126 L12 **Olëkminsk** Respublika Sakha (Yakutiya), NE Russian Federation 60.25N 120.25E

◆ COUNTRY ◇ DEPENDENT TERRITORY ▲ ADMINISTRATIVE REGION ▲ MOUNTAIN ▲ VOLCANO ⊠ LAKE
● COUNTRY CAPITAL ○ DEPENDENT TERRITORY CAPITAL ✕ INTERNATIONAL AIRPORT ▲ MOUNTAIN RANGE ♣ RIVER ⊠ RESERVOIR

119 W7 **Oleksandrivka** Donets'ka Oblast', E Ukraine 48.42N 36.56E
119 R7 **Oleksandrivka** Rus. Aleksandrivka. Kirovohrads'ka Oblast', C Ukraine 48.58N 32.13E
119 Q9 **Oleksandrivka** Mykolayivs'ka Oblast', S Ukraine 47.42N 31.17E
119 S7 **Oleksandriya** Rus. Aleksandriya. Kirovohrads'ka Oblast', C Ukraine 48.42N 33.07E
95 B20 **Ølen** Hordaland, S Norway 59.36N 5.48E
128 J4 **Olenegorsk** Murmanskaya Oblast', NW Russian Federation
126 K8 **Oleněk** Respublika Sakha (Yakutiya), NE Russian Federation 68.28N 112.18E
126 J9 **Oleněk** ≈ NE Russian Federation
126 K6 **Oleněkskiy Zaliv** bay N Russian Federation
126 K6 **Olenitsa** Murmanskaya Oblast', NW Russian Federation 66.27N 35.21E
104 I11 **Oléron, Île d'** island W France
113 H14 **Oleśnica** Ger. Oels, Oels in Schlesien. Dolnośląskie, SW Poland 51.13N 17.19E
113 I15 **Olesno** Ger. Rosenberg. Opolskie, S Poland 50.53N 18.23E
118 M3 **Olevs'k** Rus. Olevsk. Zhytomyrs'ka Oblast', N Ukraine 51.12N 27.38E
127 N18 **Ol'ga** Primorskiy Kray, SE Russian Federation 43.41N 135.06E
Olga, Mount see Kata Tjuta
94 P2 **Olgastretet** strait E Svalbard
168 D5 **Ölgiy** Bayan-Ölgiy, W Mongolia 48.57N 89.59E
97 T12 **Ølgod** Ribe, W Denmark 55.43N 8.37E
106 H14 **Olhão** Faro, S Portugal 37.01N 7.49W
95 L14 **Olhava** Oulu, C Finland 65.28N 25.25E
114 B12 **Olib** It. Ulbo. island W Croatia
85 B16 **Olifa** Kunene, NW Namibia 17.25S 14.27E
85 E20 **Olifants** var. Elephant River. ≈ E Namibia
85 E25 **Olifants** var. Elefantes. ≈ SW South Africa
85 G22 **Olifantshoek** Northern Cape, N South Africa 27.57S 22.46E
196 L15 **Olimarao Atoll** atoll Caroline Islands, C Micronesia
Ólimbos see Ólympos
Olimpo see Fuerte Olimpo
61 Q15 **Olinda** Pernambuco, E Brazil 08.00S 34.51W
Olinthos see Ólynthos
85 I20 **Oliphants Drift** Kgatleng, SE Botswana 24.13S 26.52E
Olisipo see Lisboa
Olita see Alytus
107 Q4 **Olite** Navarra, N Spain 42.28N 1.40W
64 K10 **Oliva** Córdoba, C Argentina 32.03S 63.34W
107 T11 **Oliva** País Valenciano, E Spain 38.55N 0.09W
106 I12 **Oliva de la Frontera** Extremadura, W Spain 38.16N 6.54W
Olivares see Olivares de Júcar
64 H9 **Olivares, Cerro de** ▲ N Chile 30.25S 69.52W
107 P9 **Olivares de Júcar** var. Olivares. Castilla-La Mancha, C Spain 39.45N 2.21W
24 L1 **Olive Branch** Mississippi, S USA 34.58N 89.49W
23 O5 **Olive Hill** Kentucky, S USA 38.18N 83.10W
37 Q6 **Olivehurst** California, W USA 39.05N 121.33W
106 F4 **Oliveira de Azeméis** Aveiro, N Portugal 40.49N 8.28W
106 I11 **Olivenza** Extremadura, W Spain 38.40N 7.06W
9 N17 **Oliver** British Columbia, SW Canada 49.10N 119.37W
105 N7 **Olivet** Loiret, C France 47.52N 1.53E
31 Q12 **Olivet** South Dakota, N USA 43.13N 97.40W
31 T9 **Olivia** Minnesota, N USA 44.46N 94.59W
193 C20 **Olivine Range** ▲ South Island, NZ
110 H10 **Olivone** Ticino, S Switzerland 46.32N 8.57E
Ölkeyek see Ul'kayak
131 O9 **Ol'khovka** Volgogradskaya Oblast', SW Russian Federation 49.54N 44.36E
113 K16 **Olkusz** Małopolskie, S Poland 50.16N 19.31E
24 I6 **Olla** Louisiana, S USA 31.54N 92.14W
64 I4 **Ollagüe, Volcán** var. Oyahue, Volcán Oyahue. ▲ N Chile 21.25S 68.10W
201 U13 **Ollan** island Chuuk, C Micronesia
196 F7 **Ollei** Babeldaob, N Palau 7.43N 134.37E
Ollius see Oglio
110 C10 **Ollon** Vaud, W Switzerland 46.19N 7.00E
153 Q10 **Olmaliq** Rus. Almalyk. Toshkent Wiloyati, E Uzbekistan 40.51N 69.39E
106 M6 **Olmedo** Castilla-León, N Spain 41.16N 4.40W
58 B10 **Olmos** Lambayeque, W Peru 6.00S 79.43W
Olmütz see Olomouc
32 M15 **Olney** Illinois, N USA 38.43N 88.05W
27 R5 **Olney** Texas, SW USA 33.22N 98.45W
97 L22 **Olofström** Blekinge, S Sweden 56.16N 14.31E
195 Y15 **Olomburi** Malaita, N Solomon Islands 9.00S 161.09E
113 H17 **Olomouc** Ger. Olmütz, Pol. Ołomuniec. Olomoucký Kraj, E Czech Republic 49.36N 17.13E
113 H18 **Olomoucký Kraj** ◆ region E Czech Republic
124 C6 **Olonets** Respublika Kareliya, NW Russian Federation 60.58N 33.01E
171 P10 **Olongapo** off. Olongapo City. Luzon, N Philippines 14.52N 120.16E

104 J16 **Oloron-Ste-Marie** Pyrénées-Atlantiques, SW France 43.12N 0.34W
198 D4§ **Olosega** island Manua Islands, E American Samoa
107 W4 **Olot** Cataluña, NE Spain 42.10N 2.30E
152 K12 **Olot** Rus. Alat. Bukhoro Wiloyati, S Uzbekistan 39.22N 63.42E
114 I12 **Olovo** Federacija Bosna I Hercegovina, E Bosnia and Herzegovina 44.08N 18.35E
126 K16 **Olovyannaya** Chitinskaya Oblast', S Russian Federation 50.58N 115.24E
127 O06 **Oloy** ≈ NE Russian Federation
103 F16 **Olpe** Nordrhein-Westfalen, W Germany 51.01N 7.51E
111 N8 **Olperer** ▲ SW Austria 47.03N 11.36E
Olshanka see Vil'shanka
Ol'shany see Al'shany
Olsnitz see Murska Sobota
100 M10 **Olst** Overijssel, E Netherlands 52.19N 6.06E
112 L8 **Olsztyn** Ger. Allenstein. Warmińsko-Mazurskie, NE Poland, 53.46N 20.28E
112 L8 **Olsztynek** Ger. Hohenstein in Ostpreussen. Warmińsko-Mazurskie, NE Poland, 53.34N 20.16E
118 I14 **Olt** ◆ county SW Romania
118 I14 **Olt** var. Oltul, Ger. Alt. ≈ S Romania
110 E7 **Olten** Solothurn, NW Switzerland 47.20N 7.51E
118 K14 **Olteniţa** prev. Eng. Oltenitsa, anc. Constantiola. Călăraşi, SE Romania 44.04N 26.40E
Oltenitsa see Olteniţa
118 H14 **Oltet** ≈ S Romania
26 M4 **Olton** Texas, SW USA 34.10N 102.07W
143 R12 **Oltu** Erzurum, NE Turkey 40.34N 41.58E
Oltul see Olt
152 G7 **Oltynkŭl** Qoraqalpoghiston Respublikasi, NW Uzbekistan 43.04N 58.51E
167 O15 **Oluan Pi** Eng. Cape Olwanpi. headland S Taiwan 21.57N 120.48E
Ólubló see Stará Ľubovňa
143 R11 **Olur** Erzurum, NE Turkey 40.49N 42.07E
106 L15 **Olvera** Andalucía, S Spain 36.56N 5.15W
Ol'viopol' see Pervomays'k
Olwanpi, Cape see Oluan Pi
17 G2 **Olympia** state capital Washington, NW USA 47.02N 122.54W
117 D20 **Olympía** Dytikí Ellás, S Greece 37.39N 21.36E
190 H5 **Olympic Dam** South Australia 30.25S 136.56E
34 F7 **Olympic Mountains** ▲ Washington, NW USA
124 R12 **Ólympos** var. Troodos, Eng. Mount Olympus. ▲ C Cyprus 34.55N 32.49E
117 F15 **Ólympos** var. Ólimbos, Eng. Mount Olympus. ▲ N Greece 40.04N 22.24E
117 L17 **Ólympos** ▲ Lésvos, E Greece 39.03N 26.20E
17 G1 **Olympus, Mount** ▲ Washington, NW USA 47.48N 123.42W
Olympus, Mount see Ólympos
117 C20 **Ólynthos** var. Olinthos; anc. Olynthus. site of ancient city Kentrikí Makedonía, N Greece 40.16N 23.21E
Olynthus see Ólynthos
119 O3 **Olyshivka** Chernihivs'ka Oblast', N Ukraine 51.13N 31.19E
127 Q7 **Olyutorskiy, Mys** headland E Russian Federation 59.56N 170.22E
127 Pp7 **Olyutorskiy Zaliv** bay E Russian Federation
194 F11 **Om** ≈ W PNG
133 S6 **Oma** ≈ N Russian Federation
164 I13 **Oma** Xizang Zizhiqu, W China 32.30N 83.13E
172 N8 **Ōma** Aomori, Honshū, C Japan 41.31N 140.54E
129 P6 **Oma** ≈ NW Russian Federation
171 J14 **Omachi** var. Ōmati. Nagano, S Japan 36.33N 137.49E
171 I17 **Omae-zaki** headland Honshū, S Japan 34.18N 138.12E
171 M11 **Omagari** Akita, Honshū, C Japan 39.27N 140.28E
99 E15 **Omagh Ir.** An Ómaigh. W Northern Ireland, UK 54.36N 7.18W
31 S15 **Omaha** Nebraska, C USA 41.14N 95.57W
147 W10 **Oman** off. Sultanate of Oman, Ar. Salṭanat 'Umān; prev. Muscat and Oman. ◆ monarchy SW Asia
133 O10 **Oman Basin** var. Bassin d'Oman. undersea feature N Indian Ocean
133 N10 **Oman, Gulf of** Ar. Khalīj 'Umān. gulf N Arabian Sea
192 J3 **Omapere** Northland, North Island, NZ 35.32S 173.24E
193 E20 **Omarama** Canterbury, South Island, NZ 44.29S 169.57E
114 F11 **Omarska** Republika Srpska, NW Bosnia and Herzegovina 44.53N 16.52E
85 C18 **Omaruru** Erongo, NW Namibia 21.28S 15.56E
85 C19 **Omaruru** ≈ W Namibia
85 E17 **Omatako** ≈ N Namibia
Ōmati see Ōmachi
85 E18 **Omawewozonyanda** Omaheke, E Namibia 21.30S 19.34E
172 N7 **Oma-zaki** headland Honshū, C Japan 41.32N 140.53E
175 Rr16 **Ombai, Selat** strait Nusa Tenggara, S Indonesia
85 C16 **Ombalantu** Omusati, N Namibia 17.35S 14.58E
81 H15 **Ombella-Mpoko** ◆ prefecture S Central African Republic
85 B17 **Ombombo** Kunene, NW Namibia 18.43S 13.55E
79 E16 **Omboué** Ogooué-Maritime, W Gabon 1.37S 9.19E

108 G13 **Ombrone** ≈ C Italy
82 F9 **Omdurman** var. Umm Durmān. Khartoum, C Sudan 15.37N 32.28E
171 Jj16 **Ōme** Tōkyō, Honshū, S Japan 35.48N 139.10E
108 C6 **Omegna** Piemonte, NE Italy 45.48N 8.25E
191 P12 **Omeo** Victoria, SE Australia 37.05S 147.36E
144 F11 **'Omer** Southern, C Israel 31.16N 34.51E
43 P16 **Ometepec** Guerrero, S Mexico 16.39N 98.22W
44 K11 **Ometepe, Isla de** island S Nicaragua
Om Hager see Om Hajer
82 I10 **Om Hajer** var. Om Hager. SW Eritrea 14.19N 36.46E
171 H14 **Omi-Hachiman** var. Ōmihaciman. Shiga, Honshū, SW Japan 35.10N 134.40E
115 F14 **Omiš** It. Almissa. Split-Dalmacija, S Croatia 43.27N 16.41E
114 B10 **Omišalj** Primorje-Gorski Kotar, NW Croatia 45.10N 14.33E
170 D12 **Ōmi-shima** island SW Japan
85 D19 **Omitara** Khomas, C Namibia 22.18S 18.01E
43 O16 **Omitlán, Río** ≈ S Mexico
41 X14 **Ommaney, Cape** headland Baranof Island, Alaska, USA 56.10N 134.40W
100 N9 **Ommen** Overijssel, E Netherlands 52.31N 6.25E
168 K11 **Ömnögovi** ◆ province S Mongolia
203 X7 **Omoa** Fatu Hiva, NE French Polynesia 10.30S 138.40E
Omo Botego see Omo Wenz
Ómoldova see Moldova Veche
127 O6 **Omolon** Chukotskiy Avtonomnyy Okrug, NE Russian Federation 65.11N 96.33E
127 O7 **Omolon** ≈ NE Russian Federation
126 L8 **Omoloy** ≈ NE Russian Federation
171 M10 **Omono-gawa** ≈ Honshū, C Japan
83 I14 **Omo Wenz** var. Omo Botego. ≈ Ethiopia/Kenya
125 Ff13 **Omsk** Omskaya Oblast', C Russian Federation 55.00N 73.22E
125 Ff12 **Omskaya Oblast'** ◆ province C Russian Federation
127 O8 **Omsukchan** Magadanskaya Oblast', E Russian Federation 62.25N 155.22E
172 N2 **Ōmu** Hokkaidō, NE Japan 44.34N 142.55E
112 M9 **Omulew** ≈ NE Poland
118 J12 **Omul, Vârful** prev. Vîrful Omu. ▲ C Romania 45.24N 25.26E
85 D16 **Omundaungilo** Ohangwena, N Namibia 17.28S 16.39E
170 C13 **Ōmura** Nagasaki, Kyūshū, SW Japan 32.55N 129.54E
85 D17 **Omusati** ◆ district N Namibia
170 Cc13 **Ōmuta** Fukuoka, Kyūshū, SW Japan 33.02N 130.26E
129 S14 **Omutninsk** Kirovskaya Oblast', NW Russian Federation 58.37N 52.08E
31 W11 **Onaga** Kansas, C USA 39.29N 96.06W
32 L3 **Onalaska** Wisconsin, N USA 43.52N 91.14W
25 X5 **Onancock** Virginia, NE USA 37.42N 75.45W
32 M12 **Onaping Lake** ◎ Ontario, S Canada
32 K3 **Onaqui** Illinois, N USA 40.39N 88.00W
13 R6 **Onatchiway, Lac** ◎ Québec, SE Canada
36 L4 **Onawa** Iowa, C USA 42.01N 96.06W
172 Pp7 **Ombetsu** var. Ombetsu. ≈ Hokkaidō, NE Japan 43.22N 143.54E
190 J7 **Oncócua** Cunene, SW Angola 16.37S 13.23E
107 S9 **Onda** País Valenciano, E Spain 39.58N 0.15W
113 N18 **Ondava** ≈ NE Slovakia
79 U16 **Ondjiva** see N'Giva
79 N16 **Ondo** Ondo, SW Nigeria 7.07N 4.50E
79 N16 **Ondo** ◆ state SW Nigeria
169 N8 **Öndörhaan** Hentiy, E Mongolia 47.20N 110.42E
85 D18 **Ondundazondgonda** Otjozondjupa, N Namibia 20.28S 18.00E
101 B16 **One and Half Degree Channel** channel S Maldives
197 L15 **Oneata** island Lau Group, E Fiji
128 L9 **Onega** Arkhangel'skaya Oblast', NW Russian Federation 63.54N 37.58E
125 Dd6 **Onega** ≈ NW Russian Federation
Onega Bay see Onezhskaya Guba
Onega, Lake see Onezhskoye Ozero
20 I10 **Oneida** New York, NE USA 43.05N 75.39W
22 I10 **Oneida** Tennessee, S USA 36.30N 84.30W
20 I10 **Oneida Lake** ◎ New York, NE USA
31 P13 **O'Neill** Nebraska, C USA 42.28N 98.37W
127 Pp13 **Onekotan, Ostrov** island Kuril'skiye Ostrova, SE Russian Federation
20 J11 **Oneonta** Alabama, S USA 33.57N 86.28W
20 J11 **Oneonta** New York, NE USA 42.27N 75.04W
202 I16 **Oneroa** island S Cook Islands
118 K11 **Oneşti** Hung. Onyest; prev. Gheorghe Gheorghiu-Dej. Bacău, E Romania 46.13N 26.46E
200 Qq15 **Onevai** island Tongatapu Group, S Tonga
110 A11 **Onex** Genève, SW Switzerland 46.11N 6.04E
128 K7 **Onezhskaya Guba** Eng. Onega Bay. bay NW Russian Federation 64.35N 36.55E
125 Ee6 **Onezhskoye Ozero** Eng. Lake Onega. ◎ NW Russian Federation
85 B17 **Ongandjera** Omusati, N Namibia 17.49S 15.06E
192 N12 **Ongaonga** Hawke's Bay, North Island, NZ 39.57S 176.21E

168 K9 **Ongi** Dundgovĭ, C Mongolia 45.27N 103.58E
168 K8 **Ongi** Övörhangay, C Mongolia 46.30N 102.18E
169 W14 **Ongjin** SW North Korea 37.55N 125.21E
161 J17 **Ongole** Andhra Pradesh, E India 15.33N 80.03E
168 K8 **Ongon** Övörhangay, C Mongolia 46.58N 103.45E
101 I21 **Onhaye** Namur, S Belgium 50.15N 4.51E
177 G8 **Ohn** Pegu, SW Myanmar 17.02N 96.28E
143 S9 **Oni** N Georgia 42.36N 43.13E
31 N9 **Onida** South Dakota, N USA 44.42N 100.03W
171 H14 **Oni-Hachiman** var. Ōmihaciman. Shiga, Honshū, SW Japan 33.10N 132.37E
170 E15 **Onigajō-yama** ▲ Shikoku, SW Japan
180 H7 **Onilahy** ≈ S Madagascar
79 U16 **Onitsha** Anambra, S Nigeria 6.09N 6.48E
171 Gg14 **Ōno** Hyōgo, Honshū, SW Japan 34.51N 134.56E
197 I15 **Ono** island SW Fiji
171 I14 **Ōno** Fukui, Honshū, SW Japan 35.59N 136.29E
170 D12 **Ōnoda** Yamaguchi, Honshū, SW Japan 33.59N 131.10E
197 I17 **Ono-i-lau** island SE Fiji
170 Cc13 **Ōnojō** var. Ōnozyō. Fukuoka, Kyūshū, SW Japan 33.30N 130.30E
126 K16 **Onokhoy** Respublika Buryatiya, S Russian Federation 51.51N 108.17E
170 F14 **Onomichi** var. Onomiti. Hiroshima, Honshū, SW Japan 34.25N 133.13E
Onomiti see Onomichi
169 O7 **Onon Gol** ≈ N Mongolia
57 N6 **Onoto** Anzoátegui, NE Venezuela 9.36N 65.10W
203 O3 **Onotoa** prev. Clerk Island. atoll Tungaru, W Kiribati
79 V17 **Onne** Rivers, S Nigeria 4.39N 7.37E
128 F16 **Onoch** Pskovskaya Oblast', W Russian Federation 56.42N 28.39E
112 L13 **Onpoczno** Łódzkie, C Poland 51.24N 20.18E
113 H15 **Onpole** Ger. Oppeln. Opolskie, S Poland 50.40N 17.55E
113 H15 **Onpolskie** ◆ province S Poland
150 G13 **Opornyy** Mangistau, SW Kazakhstan 46.09N 54.32E
Oporto see Porto
Oposhnya see Opishnya
85 B17 **Opowo** Kunene, N Namibia 18.06S 13.52E
152 H6 **Oqqal'a** var. Akkala, Rus. Karakala. Qoraqalpoghiston Respublikasi, NW Uzbekistan 43.43N 59.25E
153 V13 **Oqsu** Rus. Oksu. ≈ SE Tajikistan
153 P14 **Oqtogh, Qatorkŭhi** Rus. ▲ SW Tajikistan
152 M11 **Oqtosh** Rus. Aktash. Samarqand Wiloyati, C Uzbekistan 39.23N 65.45E
153 N11 **Oqtow Tizmasi** Rus. Khrebet Aktau. ▲ C Uzbekistan 32 J12 **Oquawka** Illinois, N USA 40.55N 90.55W
150 J10 **Or' Kaz.** Or. ≈ Kazakhstan/Russian Federation
38 M15 **Oracle** Arizona, SW USA 32.36N 110.46W
129 Q13 **Oparino** Kirovskaya Oblast', NW Russian Federation 59.52N 48.14E
12 H8 **Opasatica, Lac** ◎ Québec, SE Canada
114 B9 **Opatija** It. Abbazia. Primorje-Gorski Kotar, NW Croatia 45.18N 14.15E
113 N15 **Opatów** Świętokrzyskie, C Poland 50.45N 21.27E
113 I17 **Opava** Ger. Troppau. Ostravský Kraj, E Czech Republic 49.55N 17.53E
113 H16 **Opava** Ger. Oppa. ≈ NE Czech Republic
Opazova see Stara Pazova
194 L14 **Ope** ≈ S PNG
Opeepeeswag Lake ◎ Ontario, S Canada
25 R5 **Opelika** Alabama, S USA 32.39N 85.22W
24 I8 **Opelousas** Louisiana, S USA 30.31N 92.04W
195 O11 **Open Bay** bay New Britain, E PNG
12 H2 **Opeongo Lake** ◎ Ontario, SE Canada
101 K17 **Opglabbeek** Limburg, NE Belgium 51.04N 5.39E
35 W6 **Opheim** Montana, NW USA 48.50N 106.24W
41 P9 **Ophir** Alaska, USA 63.08N 94.31W
81 N18 **Opienge** Orientale, E Dem. Rep. Congo (Zaire) 0.15N 27.25E
193 G22 **Opihi** ≈ South Island, NZ
10 J9 **Opinaca** ≈ Québec, C Canada
10 J10 **Opinaca, Réservoir** ◎ Québec, C Canada
119 T5 **Opishnya** Rus. Oposhnya. Poltavs'ka Oblast', NE Ukraine 49.56N 34.36E
100 I8 **Opmeer** Noord-Holland, NW Netherlands 52.43N 4.55E
79 V17 **Opobo** Akwa Ibom, S Nigeria 4.30N 7.37E
128 F16 **Opochka** Pskovskaya Oblast', W Russian Federation 56.42N 28.39E

129 Q13 ...

12 G15 **Orangeville** Ontario, S Canada 43.55N 80.06W
38 M5 **Orangeville** Utah, W USA 39.14N 111.03W
44 G1 **Orange Walk** Orange Walk, N Belize 18.06N 88.33W
44 F2 **Orange Walk** ◆ district NW Belize
102 N17 **Oranienburg** Brandenburg, NE Germany 52.46N 13.15E
100 O7 **Oranjekanaal** canal NE Netherlands
85 D23 **Oranjemund** var. Orangemund; prev. Orange Mouth. Karas, SW Namibia 28.33S 16.27E
47 N7 **Oranjestad** ● (Aruba) W Aruba 12.31N 70.00W
Oranje Vrystaat see Free State
176 W9 **Oransbari** Irian Jaya, E Indonesia 1.18S 134.16E
Orany see Varėna
84 H18 **Orapa** Central, C Botswana 21.18S 25.22E
144 F9 **'Or 'Aqiva** var. Or Akiva. Haifa, W Israel 32.40N 34.58E
114 I10 **Orašje** Federacija Bosna I Hercegovina, N Bosnia and Herzegovina 45.01N 18.42E
118 G12 **Orăştie** Ger. Broos, Hung. Szászváros. Hunedoara, W Romania 45.49N 23.10E
113 K18 **Oraşul Stalin** see Braşov
113 K18 **Orava** Hung. Árva, Pol. Orawa. ◎ N Slovakia
95 K16 **Oravais Fin.** Oravainen. Länsi-Suomi, W Finland 63.18N 22.25E
118 F13 **Oraviţa** Ger. Orawitza, Hung. Oravicabánya. Caraş-Severin, SW Romania 45.01N 21.43E
Orawa see Orava
193 B24 **Orawia** Southland, South Island, NZ 46.03S 167.49E
Orawiţza see Oraviţa
105 P16 **Orb** ≈ S France
109 B19 **Orba** ≈ NW Italy
164 H12 **Orba Co** ◎ W China
110 B9 **Orbe** Vaud, W Switzerland 46.42N 6.28E
109 G14 **Orbetello** Toscana, C Italy 42.27N 11.14E
191 P14 **Orbost** Victoria, SE Australia 37.44S 148.28E
97 O14 **Örbyhus** Uppsala, C Sweden 60.19N 17.40E
204 I1 **Orcadas** Argentinian research station South Orkney Islands, Antarctica 60.37S 44.48W
106 L9 **Orcera** Andalucía, S Spain 38.20N 2.36W
35 P9 **Orchard Homes** Montana, NW USA 46.52N 114.01W
39 P5 **Orchard Mesa** Colorado, C USA 39.02N 108.33W
20 D10 **Orchard Park** New York, NE USA 42.46N 78.44W
197 V10 **Orchid Island** see Lan Yü
117 G18 **Orchómenos** var. Orhomenos, Orkhómenos; prev. Skripón, anc. Orchomenus. Stereá Elláa, C Greece 38.29N 22.56E
Orchomenus see Orchómenos
85 B17 **Orco** ≈ NW Italy
105 R8 **Or, Côte d'** physical region C France
31 O14 **Ord** Nebraska, C USA 41.36N 98.55W
188 L2 **Ord** ≈ Western Australia
38 M15 **Ord Mountain** ▲ California, W USA 34.41N 116.46W
85 D23 **Ordos Desert** see Mu Us Shamo
143 N11 **Ordu** anc. Cotyora. Ordu, N Turkey 41.00N 37.52E
143 V14 **Ordubad** SW Azerbaijan 38.55N 46.00E
39 U6 **Ordway** Colorado, C USA 38.13N 103.45W
39 U6 **Ordzhonikidze** Dnipropetrovs'ka Oblast', E Ukraine 47.39N 34.09E
158 K13 **Ordzhonikidze** see Vladikavkaz, Russian Federation
Ordzhonikidze see Yenakiyeve, Ukraine
Ordzhonikidzeabad see Kofarnihon
76 J1 **Orealla** E Guyana 5.13N 57.17W
115 G15 **Orebić** It. Sabbioncello. Dubrovnik-Neretva, S Croatia 42.58N 17.10E
97 M16 **Örebro** Örebro, C Sweden 59.18N 15.12E
97 L16 **Örebro** ◆ county C Sweden
27 W6 **Ore City** Texas, SW USA 32.48N 94.43W
30 L10 **Oregon** Illinois, N USA 42.02N 89.19W
23 U4 **Oregon** Ohio, N USA 41.38N 83.29W
34 H13 **Oregon** off. State of Oregon; also known as Beaver State, Sunset State, Valentine State, Webfoot State. ◆ state NW USA
34 G11 **Oregon City** Oregon, NW USA 45.21N 122.36W
97 P14 **Öregrund** Uppsala, C Sweden 60.19N 18.30E
125 G11 **Orekhov** see Orikhiv
125 L13 **Orekhovo-Zuyevo** Moskovskaya Oblast', W Russian Federation 55.46N 39.01E
26 I6 **Orel** ≈ W China
130 J6 **Orël** Orlovskaya Oblast', W Russian Federation 52.57N 36.06E

106 L11 **Orellana, Embalse de** ◎ W Spain
38 L3 **Orem** Utah, W USA 40.18N 111.41W
Ore Mountains see Erzgebirge/Krušné Hory
131 V7 **Orenburg** prev. Chkalov. Orenburgskaya Oblast', W Russian Federation 51.54N 55.11E
131 V7 **Orenburgskaya Oblast'** ◆ province W Russian Federation 51.54N 55.11E
131 T7 **Orenburgskaya Oblast'** ◆ province W Russian Federation
Orense see Ourense
196 C2 **Oreor** var. Koror. ● (Palau) Oreor, N Palau 7.21N 134.28E
196 C2 **Oreor** var. Koror. island N Palau
193 B24 **Orepuki** Southland, South Island, NZ 46.17S 167.45E
116 L12 **Orestiáda** prev. Orestiás. Anatolikí Makedonía kai Thráki, NE Greece 41.30N 26.31E
Orestiás see Orestiáda
Øresund/Öresund see Sound, The
193 C23 **Oreti** ≈ South Island, NZ
192 L5 **Orewa** Auckland, North Island, NZ 36.36S 174.42E
176 Y14 **Oreyabo** Irian Jaya, E Indonesia 6.57S 139.05E
117 A25 **Orford, Capo** headland West Falkland, Falkland Islands 52.00S 61.04W
46 B7 **Órganos, Sierra de los** ▲ W Cuba
39 R15 **Organ Peak** ▲ New Mexico, SW USA 32.17N 106.34W
107 N9 **Orgaz** Castilla-La Mancha, C Spain 39.39N 3.52W
168 I6 **Orgil** Hövsgöl, C Mongolia 49.31N 99.19E
107 O15 **Orgiva** var. Órgiva. Andalucía, S Spain 36.54N 3.25W
168 I9 **Örgön** Bayanhongor, C Mongolia 44.43N 100.21E
119 N9 **Orhei** var. Orheiu, Rus. Orgeyev. N Moldova 47.25N 28.48E
Orheiu see Orhei
107 R3 **Orhy, Pic d'/Orhy, Pic d'Orhy.** ▲ France/Spain see also Orhy 42.55N 1.01W
Orhomenos see Orchómenos
168 L6 **Orhon Gol** ≈ N Mongolia
104 J16 **Orhy, Pic d'/Orhy, Pic d'Orhy.** ▲ France/Spain see also Orhi 43.00N 1.00W
36 L2 **Orick** California, W USA 41.16N 124.03W
34 L6 **Orient** Washington, NW USA 48.51N 118.14W
50 D6 **Oriental, Cordillera** ▲ Bolivia/Peru
50 D6 **Oriental, Cordillera** ▲ C Colombia
50 C6 **Oriental, Cordillera** ▲ C Peru
65 M15 **Oriente** Buenos Aires, E Argentina 38.45S 60.37W
107 V12 **Orihuela** País Valenciano, E Spain 38.05N 0.56W
119 V9 **Orikhiv** Rus. Orekhov. Zaporiz'ka Oblast', SE Ukraine 47.32N 35.48E
115 K22 **Orikum** var. Orikumi, Vlorë, SW Albania 40.20N 19.28E
119 V6 **Oril'** Rus. Orel. ≈ E Ukraine
12 H13 **Orillia** Ontario, S Canada 44.36N 79.25W
95 M19 **Orimattila** Etelä-Suomi, S Finland 60.51N 25.46E
35 Y15 **Orin** Wyoming, C USA 43.05N 105.10W
49 R2 **Orinoco, Río** ≈ Colombia/Venezuela
49 R2 **Orinoco** Western, SW PNG 8.53S 143.13E
32 K11 **Orion** Illinois, N USA 41.21N 90.22W
31 Q5 **Oriska** North Dakota, N USA 46.54N 97.46W
159 P17 **Orissa** ◆ state NE India
72 E3 **Orissaar** see Orissaare
142 M11 **Orissaare** Ger. Orissaar. Saaremaa, W Estonia 58.33N 23.05E
D19 **Oristano** Sardegna, Italy, C Mediterranean Sea 39.54N 8.34E
D19 **Oristano, Golfo di** gulf Sardegna, Italy, C Mediterranean Sea
56 D13 **Orito** Putumayo, SW Colombia 0.41N 76.48W
95 L18 **Orivesi** Häme, SW Finland 61.39N 24.21E
95 N17 **Orivesi** ◎ Länsi-Suomi, SE Finland
60 H12 **Oriximiná** Pará, NE Brazil 1.45S 55.49W
43 Q14 **Orizaba** Veracruz-Llave, E Mexico 18.55N 97.57W
Orizaba, Volcán Pico de var. Citlaltépetl. ▲ S Mexico 19.00N 97.15W
19 **Ørje** Østfold, S Norway
115 G15 **Orjen** ▲ Bosnia and Herzegovina/Yugoslavia
Orjiva see Orgiva
98 G8 **Orkanger** Sør-Trøndelag, S Norway 63.18N 9.51E
97 K22 **Örkelljunga** Skåne, S Sweden 56.17N 13.17E
95 I15 **Orkdalen** valley S Norway
Orkhaniye see Botevgrad
98 H9 **Orkla** ≈ S Norway
Orkney see Orkney Islands
117 J22 **Orkney Deep** undersea feature Scotia Sea/Weddell Sea
98 J4 **Orkney Islands** var. Orkney, Orkneys. island group N Scotland, UK
Orkneys see Orkney Islands
26 K8 **Orla** Texas, SW USA 31.48N 103.55W
37 N5 **Orland** California, W USA 39.43N 122.11W
25 X11 **Orlando** Florida, SE USA 28.32N 81.22W
25 X12 **Orlando** Florida, SE USA 28.24N 81.16W
109 K23 **Orlando, Capo d'** headland Sicilia, Italy, C Mediterranean Sea 38.10N 14.44E

● COUNTRY   ◇ DEPENDENT TERRITORY   ✕ ADMINISTRATIVE REGION   ▲ MOUNTAIN   ☒ VOLCANO   ◎ LAKE
● COUNTRY CAPITAL   ◎ DEPENDENT TERRITORY CAPITAL   ✕ INTERNATIONAL AIRPORT   ▲ MOUNTAIN RANGE   ≈ RIVER   ◎ RESERVOIR

105 N6 **Orlau** see Orlová
Orléanais cultural region C France
36 L2 **Orleans** California, W USA 41.16N 123.36W
21 Q12 **Orleans** Massachusetts, NE USA 41.48N 69.57W
105 N7 **Orléans** anc. Aurelianum. Loiret, C France 47.54N 1.52E
13 R10 **Orléans, Île d'** island Quebec, SE Canada
Orléansville see Chlef
113 F16 **Orlice** Ger. Adler. ≈ NE Czech Republic
126 Ii5 **Orlik** Respublika Buryatiya, S Russian Federation 52.32N 99.36E
129 Q14 **Orlov** prev. Khalturin. Kirovskaya Oblast', NW Russian Federation 58.34N 48.57E
113 I17 **Orlová** Ger. Orlau, Pol. Orłowa. Ostravský Kraj, E Czech Republic 49.52N 18.25E
Orlov, Mys see Orlovskiy, Mys
130 I6 **Orlovskaya Oblast'** ◆ province W Russian Federation
128 M5 **Orlovskiy, Mys** var. Mys Orlov. headland NW Russian Federation 67.14N 41.17E
Orłowa see Orlová
105 O5 **Orly** ✈ (Paris) Essonne, N France 48.43N 2.24E
121 G16 **Orlya** Rus. Orlya. Hrodzyenskaya Voblasts', W Belarus 53.30N 24.58E
116 M7 **Orlyak** prev. Makenzen, Trubchular, Rom. Trupcilar. Dobrich, NE Bulgaria 43.39N 27.21E
154 L16 **Ormāra** Baluchistān, SW Pakistan 25.14N 64.36E
179 Qq13 **Ormoc** off. Ormoc City, var. MacArthur. Leyte, C Philippines 11.02N 124.35E
25 X10 **Ormond Beach** Florida, SE USA 29.16N 81.04W
111 X10 **Ormož** Ger. Friedau. NE Slovenia 46.24N 16.09E
12 J13 **Ormsby** Ontario, SE Canada 44.52N 77.45W
99 K17 **Ormskirk** NW England, UK 53.34N 2.54W
Ormsö see Vormsi
13 N13 **Ormstown** Quebec, SE Canada 45.08N 73.57W
Ormuz, Strait of see Hormuz, Strait of
105 T8 **Ornans** Doubs, E France 47.06N 6.06E
104 K5 **Orne** ◆ department N France
104 K5 **Orne** ≈ N France
94 G12 **Ørnes** Nordland, C Norway 66.51N 13.43E
112 L7 **Orneta** Warmińsko-Mazurskie, NE Poland 54.07N 20.10E
97 P16 **Ornö** Stockholm, C Sweden 59.03N 18.28E
39 Q3 **Orno Peak** ▲ Colorado, C USA 40.06N 107.06W
95 I16 **Örnsköldsvik** Västernorrland, C Sweden 63.16N 18.45E
169 X13 **Oro** E North Korea 39.59N 127.27E
47 T6 **Orocovis** C Puerto Rico 18.13N 66.22W
56 H10 **Orocué** Casanare, E Colombia 4.46N 71.22W
79 N13 **Orodara** SW Burkina 11.00N 4.54W
107 S4 **Oroel, Peña de a** ▲ N Spain 42.30N 0.31W
35 N10 **Orofino** Idaho, NW USA 46.28N 116.15W
168 J9 **Orog Nuur** ⊚ S Mongolia
37 U14 **Oro Grande** California, W USA 34.36N 117.19W
39 S15 **Orogrande** New Mexico, SW USA 32.24N 106.04W
203 Q7 **Orohena, Mont** ▲ Tahiti, W French Polynesia 17.37S 149.27W
Orolaunum see Arlon
Orol Dengizi see Aral Sea
201 S15 **Oroluk Atoll** atoll Caroline Islands, C Micronesia
82 J13 **Oromo** ◆ region C Ethiopia
11 O15 **Oromocto** New Brunswick, SE Canada 45.49N 66.28W
203 S4 **Orona** prev. Hull Island. atoll Phoenix Islands, C Kiribati
203 V17 **Orongo** ancient monument Easter Island, Chile, E Pacific Ocean 27.10S 109.27W
144 I3 **Orontes** var. Ononte, Ar. Nahr el Aassi, Nahr al 'Āṣī. ≈ SW Asia
106 L9 **Oropesa** Castilla-La Mancha, C Spain 39.55N 5.10W
107 T8 **Oropesa** País Valenciano, E Spain 40.06N 0.07E
Oropeza see Cochabamba
169 U5 **Oroqen Zizhiqi** Nei Mongol Zizhiqu, N China 50.34N 123.40E
179 Qq15 **Oroquieta** var. Oroquieta City. Mindanao, S Philippines 8.27N 123.46E
42 I8 **Oro, Río del** ≈ C Mexico
61 O14 **Orós, Açude** ⊞ E Brazil
109 D18 **Orosei, Golfo di** gulf Tyrrhenian Sea, C Mediterranean Sea
113 M24 **Orosháza** Békés, SE Hungary 46.33N 20.40E
Orosirá Rodhópis see Rhodope Mountains
113 I22 **Oroszlány** Komárom-Esztergom, W Hungary 47.31N 18.19E
196 B16 **Orote Peninsula** peninsula W Guam
127 O9 **Orotukan** Magadanskaya Oblast', E Russian Federation 62.18N 150.46E
37 O5 **Oroville** California, W USA 39.29N 121.35W
34 K6 **Oroville** Washington, NW USA 48.56N 119.25W
37 O5 **Oroville, Lake** ⊞ California, W USA
1 G15 **Orozco Fracture Zone** tectonic feature E Pacific Ocean
66 I7 **Orphan Knoll** undersea feature NW Atlantic Ocean 51.00N 47.00W
31 V3 **Orr** Minnesota, N USA 48.03N 92.48W
97 M21 **Orrefors** Kalmar, S Sweden 56.48N 15.45E
190 I7 **Orroroo** South Australia 32.46S 138.38E
35 T12 **Orrville** Ohio, N USA 40.50N 81.45W
96 L12 **Orsa** Dalarna, C Sweden 61.07N 14.40E
121 O14 **Orsha** Rus. Orsha. Vitsyebskaya Voblasts', NE Belarus 54.30N 30.25E

131 Q2 **Orshanka** Respublika Mariy El, W Russian Federation 56.54N 47.54E
110 C11 **Orsières** Valais, SW Switzerland 46.00N 7.09E
125 Dd13 **Orsk** Orenburgskaya Oblast', W Russian Federation 51.13N 58.34E
118 F13 **Orşova** Ger. Orschowa, Hung. Orsova. Mehedinţi, SW Romania 44.42N 22.22E
96 D10 **Ørsta** Møre og Romsdal, S Norway 62.12N 6.07E
97 O15 **Örsundsbro** Uppsala, C Sweden 59.45N 17.19E
142 D16 **Ortaca** Muğla, SW Turkey 36.49N 28.43E
109 M16 **Orta Nova** Puglia, SE Italy 41.19N 15.43E
142 I17 **Orta Toroslar** ▲ S Turkey
56 E11 **Ortega** Tolima, W Colombia 3.57N 75.10W
106 H1 **Ortegal, Cabo** headland NW Spain 43.46N 7.54W
104 J15 **Orthez** Pyrénées-Atlantiques, SW France 43.29N 0.46W
59 K14 **Orthon, Río** ≈ N Bolivia
62 J10 **Ortigueira** Paraná, S Brazil 24.10S 50.55W
106 H1 **Ortigueira** Galicia, NW Spain 43.40N 7.50W
108 H5 **Ortisei** Ger. Sankt-Ulrich. Trentino-Alto Adige, N Italy 46.35N 11.42E
42 F6 **Ortíz** Sonora, NW Mexico 28.18N 110.40W
56 L5 **Ortíz** Guárico, N Venezuela 9.37N 67.17W
Ortler see Ortles
108 F5 **Ortles** Ger. Ortler. ▲ N Italy 46.29N 10.33E
109 K14 **Ortona** Abruzzo, C Italy 42.21N 14.24E
31 R8 **Ortonville** Minnesota, N USA 45.18N 96.26N
153 W8 **Orto-Tokoy** Issyk-Kul'skaya Oblast', NE Kyrgyzstan 42.20N 76.03E
95 I15 **Örträsk** Västerbotten, N Sweden 64.10N 19.00E
102 J12 **Örtze** ≈ NW Germany
Oruba see Aruba
148 I3 **Orūmīyeh** var. Rizaiyeh, Urmia, Urmiyeh; prev. Reẓa'īyeh. Āžarbāyjān-e Bākhtarī, NW Iran 37.33N 45.06E
148 J3 **Orūmīyeh, Daryācheh-ye** var. Matianus, Sha Hi, Urumi Yeh, Eng. Lake Urmia; prev. Daryācheh-ye Reẓa'īyeh. ⊚ NW Iran
59 K19 **Oruro** Oruro, W Bolivia 17.57S 67.05W
59 J19 **Oruro** ◆ department W Bolivia
97 I18 **Orust** island S Sweden
Oruzgán/Orūzgān see Ürüzgān
108 H13 **Orvieto** anc. Velsuna. Umbria, C Italy 42.43N 12.07E
204 K7 **Orville Coast** physical region Antarctica
116 H7 **Oryakhovo** Vratsa, NW Bulgaria 43.43N 23.58E
Oryokko see Yalu
119 R5 **Orzhytsya** Poltav's'ka Oblast', C Ukraine 49.48N 32.40E
112 M9 **Orzyc** Ger. Orschütz. ≈ NE Poland
112 N8 **Orzysz** Ger. Arys. Warmińsko-Mazurskie, NE Poland 53.49N 21.54E
96 I10 **Os** Hedmark, S Norway 62.29N 11.14E
97 C14 **Os** Hordaland, S Norway 60.10N 5.30E
129 U15 **Osa** Permskaya Oblast', NW Russian Federation 57.16N 55.22E
31 W11 **Osage** Iowa, C USA 43.16N 92.48W
29 U5 **Osage** Iowa, C USA 38.09N 92.37W
29 P5 **Osage Beach** Missouri, C USA 38.37N 95.49W
29 U7 **Osage City** Kansas, C USA 38.37N 95.49W
29 U7 **Osage Fork River** ≈ Missouri, C USA
29 U5 **Osage River** ≈ Missouri, C USA
171 Gg15 **Ōsaka** hist. Naniwa. Ōsaka, Honshū, SW Japan 34.38N 135.27E
171 Gg15 **Ōsaka** off. Osaka-fu, var. Ōsaka Hu. ◆ urban prefecture Honshū, SW Japan
Ōsaka-fu/Osaka Hu see Ōsaka
153 R10 **Osakarovka** Karaganda, C Kazakhstan 50.37N 72.49E
170 G15 **Ōsaka-wan** bay SW Japan
31 T7 **Osakis** Minnesota, N USA 45.51N 95.08W
45 N16 **Osa, Península de** peninsula S Costa Rica
62 M10 **Osasco** São Paulo, S Brazil 23.31S 46.46W
29 R5 **Osawatomie** Kansas, C USA 38.30N 94.57W
28 L3 **Osborne** Kansas, C USA 39.25N 98.41W
181 S8 **Osborn Plateau** undersea feature E Indian Ocean
97 L21 **Osby** Skåne, S Sweden 56.24N 14.00E
Osca see Huesca
94 N2 **Oscar II Land** physical region W Svalbard
27 Y10 **Osceola** Arkansas, C USA 35.40N 89.58W
31 V15 **Osceola** Iowa, C USA 41.01N 93.45W
29 S8 **Osceola** Missouri, C USA 38.03N 93.42W
31 Q15 **Osceola** Nebraska, C USA 41.09N 97.28W
103 N15 **Oschatz** Sachsen, E Germany 51.17N 13.10E
102 K13 **Oschersleben** Sachsen-Anhalt, C Germany 52.02N 11.14E
35 R7 **Oscoda** Michigan, N USA 44.25N 83.19W
Ösel see Saaremaa
96 E9 **Osen** Sør-Trøndelag, S Norway 64.17N 10.29E
97 O14 **Österbybruk** Uppsala, C Sweden 60.13N 17.55E
97 M19 **Österbymo** Östergötland, S Sweden 57.49N 15.15E
96 K12 **Österdälälven** ≈ C Sweden
95 F16 **Österdalen** valley S Norway
97 L18 **Östergötland** ◆ county S Sweden
102 H10 **Osterholz-Scharmbeck** Niedersachsen, NW Germany 53.13N 8.46E
Östermark see Teuva
Östermyra see Seinäjoki

126 Ii12 **Osharovo** Evenkiyskiy Avtonomnyy Okrug, N Russian Federation 60.16N 98.20E
12 H15 **Oshawa** Ontario, SE Canada 43.54N 78.50W
171 Mm13 **Oshika-hantō** peninsula Honshū, C Japan
85 C16 **Oshikango** Ohangwena, N Namibia 17.28S 15.54E
170 Gg12 **Ō-shima** island SW Japan
171 M7 **Ō-shima** island NE Japan
171 Jj9 **Ō-shima** island S Japan
170 G17 **Ō-shima** island SW Japan
172 N6 **Oshima-hantō** ▲ Hokkaidō, NE Japan
85 D17 **Oshivelo** Otjikoto, N Namibia 18.37S 17.10E
30 K14 **Oshkosh** Nebraska, C USA 41.25N 102.21W
32 M7 **Oshkosh** Wisconsin, N USA 44.01N 88.31W
Oshmyany see Ashmyany
Osh Oblasty see Oshskaya Oblast'
79 T16 **Oshogbo** Osun, W Nigeria 7.42N 4.31E
153 S11 **Oshskaya Oblast'** Kir. Osh Oblasty. ◆ province SW Kyrgyzstan
81 J20 **Oshwe** Bandundu, C Dem. Rep. Congo (Zaire) 3.24S 19.31E
114 I9 **Osijek** prev. Osiek, Osjek, Ger. Esseg, Hung. Eszék. Osijek-Baranja, E Croatia 45.33N 18.40E
114 I9 **Osijek-Baranja** off. Osječko-Baranjska Županija. ◆ province E Croatia
130 L8 **Osinki** Kemerovskaya Oblast', S Russian Federation 53.30N 87.25E
126 J14 **Osinovka** Irkutskaya Oblast', C Russian Federation 56.19N 101.55E
112 N9 **Osinovka** Irkutskaya Oblast', C Russian Federation 56.19N 101.55E
112 N9 **Ostrołęka** Ger. Wiesenhof, Rus. Ostrolenka. Mazowieckie, C Poland 53.06N 21.33E
113 A16 **Osinki** Schlackenwerth. Karlovarský Kraj, W Czech Republic 50.18N 12.53E
114 N11 **Osipaonica** Serbia, NE Yugoslavia 44.34N 21.00E
Osipenko see Berdyans'k
Osipovichi see Asipovichy
114 I9 **Osječko-Baranjska Županija** see Osijek-Baranja
Osjek see Osijek
31 W15 **Oskaloosa** Iowa, C USA 41.17N 92.38W
29 Q4 **Oskaloosa** Kansas, C USA 39.13N 95.18W
97 N20 **Oskarshamn** Kalmar, S Sweden 57.16N 16.25E
97 J21 **Oskarström** Halland, S Sweden 56.48N 13.00E
12 M8 **Oskélaneo** Quebec, SE Canada 48.06N 75.12W
Öskemen see Ust'-Kamenogorsk
Oskil see Oskol
119 W5 **Oskol** Ukr. Oskil. ≈ Russian Federation/Ukraine
95 D20 **Oslo** prev. Christiania, Kristiania. ● (Norway) Oslo, S Norway 59.54N 10.43E
95 D21 **Oslo** ◆ county S Norway
95 D21 **Oslofjorden** fjord S Norway
161 G15 **Osmānābād** Mahārāshtra, C India 18.09N 76.06E
142 J11 **Osmancık** Çorum, N Turkey 40.58N 34.49E
142 L16 **Osmaniye** Osmaniye, S Turkey 37.04N 36.15E
97 O16 **Osmo** Stockholm, C Sweden 58.58N 17.55E
120 E3 **Osmussaar** island W Estonia
102 G13 **Osnabrück** Niedersachsen, NW Germany 52.08N 7.42E
112 D11 **Osno Lubuskie** Ger. Drossen. Lubuskie, W Poland 52.28N 14.51E
115 P19 **Osogovske Planine/Osogovski Planina, Mac.** Osogovski Planini. ▲ Bulgaria/FYR Macedonia
Osogovske Planine/Osogovski Planina/Osogovski Planini see Osogov Mountains
172 N8 **Osore-yama** ▲ Honshū, C Japan 41.18N 141.06E
Oşorhei see Târgu Mureş
63 I16 **Osório** Rio Grande do Sul, S Brazil 29.52S 50.16W
65 G16 **Osorno** Los Lagos, C Chile 40.38S 73.44W
106 M4 **Osorno** Castilla-León, N Spain 42.24N 4.22W
9 N17 **Osoyoos** British Columbia, SW Canada 49.01N 119.31W
56 J6 **Ospino** Portuguesa, N Venezuela 9.16N 69.25W
100 K13 **Oss** Noord-Brabant, S Netherlands 51.46N 5.31E
85 H11 **Ossa** S Portugal 38.43N 7.33W
117 F15 **Óssa** ▲ C Greece
25 X6 **Ossabaw Island** island Georgia, SE USA
25 X6 **Ossabaw Sound** sound Georgia, SE USA
191 O16 **Ossa, Mount** ▲ Tasmania, SE Australia 41.55S 146.03E
115 D15 **Ossa, Serra d'** ▲ SE Portugal
79 U16 **Osse** ≈ S Nigeria
32 J6 **Osseo** Wisconsin, N USA 44.33N 91.13W
111 S9 **Ossiacher See** ⊚ S Austria
21 N9 **Ossining** New York, NE USA 41.10N 73.50W
96 I12 **Ossjøen** ⊚ S Norway
127 P9 **Ossora** Koryakskiy Avtonomnyy Okrug, E Russian Federation 59.16N 163.01E
128 I15 **Ostashkov** Tverskaya Oblast', W Russian Federation 57.08N 33.10E
102 H9 **Oste** ≈ NW Germany
Ostee see Baltic Sea
Ostend/Ostende see Oostende
119 P3 **Oster** Chernihivs'ka Oblast', N Ukraine 50.51N 30.52E
Osterburg see Saint-Ulrich

118 F12 **Oţelu Roşu** Ger. Ferdinandsberg, Hung. Nándorhegy. Caras-Severin, SW Romania 45.30N 22.21E
193 E21 **Otematata** Canterbury, South Island, NZ 44.37S 170.12E
120 I6 **Otepää** Ger. Odenpäh. Valgamaa, SE Estonia 58.04N 26.31E
34 K9 **Othello** Washington, NW USA 46.49N 119.10W
117 A15 **Othonoí** island Iónioi Nísoi, Greece, C Mediterranean Sea
117 F17 **Óthrys** var. Othris. ▲ C Greece
79 R11 **Oti** N Togo
42 K10 **Otinapa** Durango, C Mexico 24.01N 104.58W
193 G17 **Otira** West Coast, South Island, NZ 42.51S 171.32E
3 V3 **Otis** Colorado, C USA 40.09N 102.57W
85 C16 **Otish, Monts** ▲ Quebec, E Canada
85 C17 **Otjikondo** Kunene, N Namibia 19.48S 15.28E
85 C17 **Otjikoto** var. Oshikoto. ◆ district N Namibia
85 E18 **Otjinene** Omaheke, NE Namibia 21.10S 18.43E
85 C18 **Otjiwarongo** Otjozondjupa, N Namibia 20.28S 16.36E
85 D18 **Otjosundu** var. Otjosundu. Otjozondjupa, N Namibia 21.19S 17.51E
85 D18 **Otjozondjupa** ◆ district C Namibia
114 C11 **Otočac** Lika-Senj, W Croatia 44.52N 15.13E
172 Pp6 **Otofuke-gawa** ≈ Hokkaidō, NE Japan
168 M14 **Otog Qi** Nei Mongol Zizhiqu, N China 39.05N 107.58E
172 Pp3 **Otoineppu** Hokkaidō, NE Japan 44.43N 142.13E
114 J10 **Otok** Vukovar-Srijem, E Croatia 45.10N 18.52E
113 A16 **Otovice** Schlackenwerth. Karlovarský Kraj, W Czech Republic 50.18N 12.53E
192 L8 **Otorohanga** Waikato, North Island, NZ 38.10S 175.13E
10 D9 **Otoskwin** ≈ Ontario, C Canada
170 F15 **Otoyo** Kōchi, Shikoku, SW Japan 33.45N 133.42E
97 E16 **Otra** ≈ S Norway
109 R19 **Otranto** Puglia, SE Italy 40.08N 18.28E
109 Q18 **Otranto, Canale d'** see Otranto, Strait of
109 Q18 **Otranto, Strait of** It. Canale d'Otranto. strait Albania/Italy
35 P10 **Otsego** Michigan, N USA 42.27N 85.42W
35 Q6 **Otsego Lake** ⊚ (Canada) Ontario, SE Canada 45.24N 75.40W
21 L12 **Otselic River** ≈ New York, NE USA
3 L11 **Ottawa** Illinois, N USA 41.21N 88.50W
29 Q5 **Ottawa** Kansas, C USA 38.37N 95.16W
35 R12 **Ottawa** Ohio, N USA 41.01N 84.03W
13 L14 **Ottawa** var. Uplands. ✈ Ontario, SE Canada 45.19N 75.40W
12 M12 **Ottawa** Fr. Outaouais. ≈ Ontario/Quebec, SE Canada
10 I4 **Ottawa Islands** island group Nunavut, C Canada
96 G11 **Otta** Oppland, S Norway 61.46N 9.33E
201 U13 **Otta** island Chuuk, C Micronesia
96 F11 **Otta** ≈ S Norway
201 U13 **Otta Pass** passage Chuuk Islands, C Micronesia
97 J22 **Ottarp** Skåne, S Sweden 55.55N 12.55E
13 L14 **Ottawa** ● (Canada) Ontario, SE Canada 45.24N 75.40W
33 L11 **Ottawa, Illinois, N USA** 41.21N 88.50W
76 F7 **Ottignies** Wallon Brabant, C Belgium 50.40N 4.34E
103 L23 **Ottobrunn** Bayern, SE Germany 48.02N 11.40E
194 I12 **Otto, Mount** ▲ C PNG 5.54S 145.24E
31 X15 **Ottumwa** Iowa, C USA 41.00N 92.24W
79 R16 **Ouidah** Eng. Whydah, Wida. S Benin 6.22N 2.07E
76 H7 **Oujda** Ar. Oudjda, Ujda. NE Morocco 34.45N 1.53W
78 I7 **Oujeft** Adrar, C Mauritania 20.05N 13.00W
95 L15 **Oulainen** Oulu, C Finland 64.14N 24.50E
95 L14 **Oulujärvi** Swe. Uleträsk. ⊚ C Finland
95 L14 **Oulujoki** Swe. Uleälv. ≈ C Finland
95 L14 **Oulunsalo** Oulu, C Finland 64.55N 25.26E
108 A8 **Oulx** Piemonte, NE Italy 45.05N 6.41E
78 K7 **Ouâdane** var. Ouadane, Adrar, C Mauritania 20.57N 11.34W
80 K13 **Ouadda** Haute-Kotto, N Central African Republic 8.02N 22.22E
80 K13 **Ouaddaï** off. Préfecture du Ouaddaï, var. Ouadaï, Wadaï. ◆ prefecture SE Chad

---

79 P13 **Ouagadougou** ✈ C Burkina 12.21N 1.27W
79 O12 **Ouahigouya** NW Burkina 13.31N 2.19W
Ouahran see Oran
81 J14 **Ouaka** ◆ prefecture C Central African Republic
81 J15 **Ouaka** ≈ S Central African Republic
78 M9 **Oualâta** var. Oualata. Hodh ech Chargui, SE Mauritania 17.18N 7.00W
79 R11 **Ouallam** var. Ouallam. Tillabéri, W Niger 14.13N 2.07E
180 H4 **Ouanani** Mohéli, S Comoros 12.19S 94.37E
57 Z10 **Ouanary** E French Guiana 4.10N 51.40W
80 L13 **Ouanda Djallé** Vakaga, NE Central African Republic 8.53N 22.47E
81 N14 **Ouango** Haut-Mbomou, SE Central African Republic 5.57N 25.57E
81 L15 **Ouango** Mbomou, S Central African Republic 4.19N 22.30E
79 N14 **Ouangolodougou** var. Wangolodougou. N Ivory Coast 9.58N 5.09W
180 I13 **Ouani** Anjouan, SE Comoros
81 M15 **Ouara** ≈ E Central African Republic
78 K7 **Ouarâne** desert C Mauritania
13 O11 **Ouareau** ≈ Quebec, SE Canada
76 K7 **Ouargla** var. Wargla. NE Algeria 32.00N 5.16E
76 F8 **Ouarzazate** S Morocco 30.54N 6.55W
79 Q11 **Ouatagouna** Gao, E Mali 15.06N 0.41E
76 G6 **Ouazzane** var. Ouezzane, Ar. Wazan, Wazzan. N Morocco 34.52N 5.34W
98 E8 **Outer Hebrides** var. Western Isles. island group NW Scotland, UK
32 K3 **Outer Island** island Apostle Islands, Wisconsin, N USA
37 S16 **Outer Santa Barbara Passage** passage California, SW USA
106 G3 **Outes** Galicia, NW Spain 42.50N 8.54W
85 C18 **Outjo** Kunene, N Namibia 20.06S 16.06E
9 T16 **Outlook** Saskatchewan, S Canada 51.30N 107.03W
95 N16 **Outokumpu** Itä-Suomi, E Finland 62.43N 29.01E
98 M2 **Out Skerries** island group NE Scotland, UK
203 J5 **Ouvéa** island Îles Loyauté, NE New Caledonia
105 U14 **Ouvèze** ≈ SE France
190 L9 **Ouyen** Victoria, SE Australia 35.06S 142.18E
41 Q14 **Ouzinkie** Kodiak Island, Alaska, USA 57.54N 152.27W
143 O13 **Ovacık** Tunceli, E Turkey 39.22N 39.13E
108 C9 **Ovada** Piemonte, NE Italy 44.41N 8.39E
197 I14 **Ovalau** island C Fiji
64 G9 **Ovalle** Coquimbo, N Chile 30.33S 71.16W
85 C17 **Ovamboland** physical region N Namibia
56 L10 **Ovana, Cerro** ▲ S Venezuela 4.41N 66.54W
106 G7 **Ovar** Aveiro, N Portugal 40.52N 8.37W
116 L10 **Ovcharitsa, Yazovir** ⊞ SE Bulgaria
56 E6 **Ovejas** Sucre, NW Colombia 9.30N 75.15W
103 E16 **Overath** Nordrhein-Westfalen, W Germany 50.55N 7.16E
100 F13 **Overflakkee** island SW Netherlands
101 H19 **Overijse** Vlaams Brabant, C Belgium 50.46N 4.31E
100 N10 **Overijssel** ◆ province E Netherlands
100 M9 **Overijssels Kanaal** canal E Netherlands
95 K13 **Överkalix** Norrbotten, N Sweden 66.19N 22.49E
99 R4 **Overland Park** Kansas, C USA 38.57N 94.40W
100 L14 **Overloon** Noord-Brabant, SE Netherlands 51.35N 5.54E
101 K16 **Overpelt** Limburg, NE Belgium 51.13N 5.24E
37 Y10 **Overton** Nevada, W USA 36.32N 114.25W
27 W7 **Overton** Texas, SW USA 32.16N 94.58W
94 K13 **Övertorneå** Norrbotten, N Sweden 66.22N 23.38E
27 W18 **Overum** Kalmar, S Sweden 57.59N 16.18E
94 G13 **Överuman** ⊚ N Sweden
168 H6 **Övögdiy** Dzavhan, C Mongolia 48.10N 97.49E
119 P11 **Ovidiopol'** Odes'ka Oblast', SW Ukraine 46.15N 30.27E
118 M14 **Ovidiu** Constanţa, SE Romania 44.16N 28.34E
47 N10 **Oviedo** SW Dominican Republic 17.46N 71.22W
106 K2 **Oviedo** anc. Asturias. Asturias, NW Spain 43.21N 5.49W
106 K2 **Oviedo** ◆ Asturias, N Spain 43.21N 5.49W
Ovilava see Wels
120 D7 **Oviši** Ventspils, NW Latvia 57.34N 21.43E
169 P10 **Övörhangay** ◆ province C Mongolia
95 C14 **Øvre Årdal** Sogn og Fjordane, S Norway 61.17N 7.44E
95 M14 **Øvre Fryken** ⊚ C Sweden
94 J11 **Övre Soppero** Norrbotten, N Sweden 68.05N 21.40E
119 N3 **Ovruch** Zhytomyrs'ka Oblast', N Ukraine 51.19N 28.50E
168 J9 **Övt Övörhangay, C Mongolia** 46.46N 102.18E
193 E24 **Owaka** Otago, South Island, NZ 46.27S 169.42E
171 Gg12 **Owase** Mie, Honshū, SW Japan 34.04N 136.10E
29 P9 **Owasso** Oklahoma, C USA 36.16N 95.51W
31 V10 **Owatonna** Minnesota, N USA 44.04N 93.13W

◆ COUNTRY ● COUNTRY CAPITAL ◇ DEPENDENT TERRITORY ○ DEPENDENT TERRITORY CAPITAL ▲ ADMINISTRATIVE REGION ✈ INTERNATIONAL AIRPORT ▲ MOUNTAIN ▲ MOUNTAIN RANGE ≈ RIVER ▧ VOLCANO ⊚ LAKE ⊞ RESERVOIR

181 O4 **Owen Fracture Zone** *tectonic feature* W Arabian Sea
193 H15 **Owen, Mount** ▲ South Island, NZ 41.32S 172.33E
193 H15 **Owen River** Tasman, South Island, NZ 41.40S 172.28E
46 D8 **Owen Roberts** ✈ Grand Cayman, Cayman Islands 19.15N 81.22W
22 I6 **Owensboro** Kentucky, S USA 37.46N 87.06W
37 T11 **Owens Lake** *salt flat* California, W USA
12 F14 **Owen Sound** Ontario, S Canada 44.34N 80.55W
12 F13 **Owen Sound** ◎ Ontario, S Canada
37 T10 **Owens River** ♒ California, W USA
194 K15 **Owen Stanley Range** ▲ S PNG
29 V5 **Owensville** Missouri, C USA 38.21N 91.30W
22 M4 **Owenton** Kentucky, S USA 38.33N 84.51W
79 U17 **Owerri** Imo, S Nigeria 5.19N 7.07E
192 M10 **Owhango** Manawatu-Wanganui, North Island, NZ 39.01S 175.22E
23 N5 **Owingsville** Kentucky, S USA 38.10N 83.42W
152 K10 **Owminzatovo-Toshi** *Rus.* Gory Auminzatau. ▲ N Uzbekistan
79 T16 **Owo** Ondo, SW Nigeria 7.10N 5.31E
35 R9 **Owosso** Michigan, N USA 43.00N 84.10W
37 V1 **Owyhee** Nevada, W USA 41.57N 116.07W
34 L14 **Owyhee, Lake** ◎ Oregon, NW USA
34 L15 **Owyhee River** ♒ Idaho/Oregon, NW USA
94 K1 **Oxarfjördhur** *var.* Axarfjördhur. *fjord* N Iceland
96 K12 **Oxberg** Dalarna, C Sweden 61.07N 14.10E
9 V17 **Oxbow** Saskatchewan, S Canada 49.16S 102.12W
97 O17 **Oxelösund** Södermanland, S Sweden 58.40N 17.10E
193 H18 **Oxford** Canterbury, South Island, NZ 43.18S 172.10E
99 M21 **Oxford** *Lat.* Oxonia. S England, UK 51.46N 1.15W
25 Q3 **Oxford** Alabama, S USA 33.36N 85.50W
24 L2 **Oxford** Mississippi, S USA 34.23N 89.30W
31 N16 **Oxford** Nebraska, C USA 40.15N 99.37W
20 I11 **Oxford** New York, NE USA 42.21N 75.39W
23 U8 **Oxford** North Carolina, SE USA 36.18N 78.35W
35 Q14 **Oxford** Ohio, N USA 39.30N 84.45W
20 H16 **Oxford** Pennsylvania, NE USA 39.46N 75.57W
9 X12 **Oxford House** Manitoba, C Canada 54.55N 95.13W
31 Y13 **Oxford Junction** Iowa, C USA 41.58N 90.57W
9 X12 **Oxford Lake** ◎ Manitoba, C Canada
99 M21 **Oxfordshire** *cultural region* S England, UK
Oxia *see* Oxyá
43 X12 **Oxkutzcab** Yucatán, SE Mexico 20.18N 89.25W
37 R15 **Oxnard** California, W USA 34.12N 119.10W
Oxonia *see* Oxford
12 I12 **Oxtongue** ♒ Ontario, SE Canada
Oxus *see* Amu Darya
117 E15 **Oxyá** *var.* Oxia. ▲ C Greece 38.46N 21.56E
171 I13 **Oyabe** Toyama, Honshū, SW Japan 36.41N 136.53E
Oyahue/Oyahue, Volcán *see* Ollagüe, Volcán
171 K16 **Oyama** Tochigi, Honshū, SW Japan 36.19N 139.64E
49 U5 **Oyapock** ♒ E French Guiana
57 Z10 **Oyapok, Baie de L'** *bay* Brazil/French Guiana
57 Z11 **Oyapok, Fleuve l'** *var.* Oyapock, Rio Oiapoque. ♒ Brazil/French Guiana *see also* Oiapoque, Rio
81 E17 **Oyem** Woleu-Ntem, N Gabon 1.34N 11.31E
9 R16 **Oyen** Alberta, SW Canada 51.19N 110.28W
97 I15 **Øyeren** ◎ S Norway
168 G6 **Oygon** Dzavhan, N Mongolia 46.57N 96.33E
98 I7 **Oykel** ♒ N Scotland, UK
127 N9 **Oymyakon** Respublika Sakha (Yakutiya), NE Russian Federation 63.28N 142.22E
81 H19 **Oyo** Cuvette, C Congo 1.05S 15.55E
79 S15 **Oyo** Oyo, W Nigeria 7.51N 3.57E
79 S15 **Oyo** ♦ *state* SW Nigeria
58 D13 **Oyón** Lima, C Peru 10.39S 76.46W
51 S10 **Oyonnax** Ain, E France
152 L10 **Oyoqghitma** *Rus.* Ayakagytma. Bukhoro Wiloyati, C Uzbekistan 40.37N 64.26E
152 M9 **Oyoqquduq** *Rus.* Ayakkuduk. Nawoiy Wiloyati, N Uzbekistan 41.16N 65.12E
34 F9 **Oysterville** Washington, NW USA 46.33N 124.03W
97 D14 **Øystese** Hordaland, S Norway 60.22N 6.13E
153 U10 **Oy-Tal** Oshskaya Oblast', SW Kyrgyzstan 40.23N 74.04E
153 T10 **Oy-Tal** ♒ SW Kyrgyzstan
151 S16 **Oytal** Zhambyl, S Kazakhstan 42.50N 73.21E
Oyyl *see* Uil
179 Qq15 **Ozamiz** Mindanao, S Philippines 8.09N 123.51E
25 R7 **Ozark** Alabama, S USA 31.27N 85.38W
29 T8 **Ozark** Arkansas, C USA 35.29N 93.49W
29 T8 **Ozark** Missouri, C USA 37.01N 93.12W
29 T6 **Ozarks, Lake of the** ◎ Missouri, C USA

199 Jj11 **Ozbourn Seamount** *undersea feature* W Pacific Ocean 26.00S 174.49W
113 L20 **Ózd** Borsod-Abaúj-Zemplén, NE Hungary 48.14N 20.18E
114 D11 **Ozeblin** ▲ C Croatia 44.37N 15.52E
127 Pp2 **Ozernovskiy** Kamchatskaya Oblast', E Russian Federation 51.28N 94.32E
150 M7 **Ózernoe** *var.* Ozérnyy. N Kazakhstan 53.27N 63.10E
Ozérnyy *see* Ózernoe
117 D18 **Ozerós, Límni** ◎ W Greece
121 D14 **Ozersk** *prev.* Darkehnen, *Ger.* Angerapp. Kaliningradskaya Oblast', W Russian Federation 54.23N 21.59E
130 L4 **Ozery** Moskovskaya Oblast', W Russian Federation 54.51N 38.37E
Özgön *see* Uzgen
109 C17 **Ozieri** Sardegna, Italy, C Mediterranean Sea 40.34N 9.01E
113 I15 **Ozimek** *Ger.* Malapane. Opolskie, S Poland 50.41N 18.16E
131 R8 **Ozinki** Saratovskaya Oblast', W Russian Federation 51.16N 49.45E
Oziya *see* Ojiya
27 O10 **Ozona** Texas, SW USA 30.42N 101.12W
112 J12 **Ozorków** *Rus.* Ozorkov. Łódz, C Poland 51.58N 19.16E
170 E14 **Ōzu** Ehime, Shikoku, SW Japan 33.31N 132.31E
143 R10 **Ozurget'i** *prev.* Makharadze. W Georgia 41.57N 42.01E

—— P ——

101 J17 **Paal** Limburg, NE Belgium 51.03N 5.08E
197 C13 **Paama** *island* C Vanuatu
206 M14 **Paamiut** *var.* Pâmiut, *Dan.* Frederikshåb. Kitaa, S Greenland 61.49N 49.52W
178 Gg9 **Pa-an** Karen State, S Myanmar 16.51N 97.37E
103 L22 **Paar** ♒ SE Germany
85 L26 **Paarl** Western Cape, SW South Africa 33.45S 18.58E
95 L15 **Paavola** Oulu, C Finland 64.34N 25.15E
31 Q5 **Pabbay** *island* NW Scotland, UK
175 P12 **Pabbiring, Kepulauan** *island group* C Indonesia
159 T15 **Pabna** Rajshahi, W Bangladesh 23.98N 89.15E
111 U4 **Pabneukirchen** Oberösterreich, N Austria 48.19N 14.48E
120 H13 **Pabradé** *Pol.* Podbrodzie. Švenčionys, SE Lithuania 54.58N 25.43E
58 L13 **Pacahuaras, Río** ♒ N Bolivia
58 B11 **Pacaraima, Sierra/Pacaraim, Serra** *see* Pakaraima Mountains
58 B11 **Pacasmayo** La Libertad, W Peru 7.27S 79.34W
44 D6 **Pacaya, Volcán de** ▲ S Guatemala 14.19N 90.36W
117 K23 **Pachiá** *island* Kykládes, Greece, Aegean Sea
109 L26 **Pachino** Sicilia, Italy, C Mediterranean Sea 36.43N 15.06E
58 F12 **Pachitea, Río** ♒ C Peru
160 I11 **Pachmarhi** Madhya Pradesh, C India 22.36N 78.18E
124 N4 **Páchna** *var.* Pakhna. SW Cyprus 34.47N 32.48E
117 H25 **Páchnes** ▲ Kríti, Greece, E Mediterranean Sea 35.19N 24.00E
56 F9 **Pacho** Cundinamarca, C Colombia 5.07N 74.11W
160 F12 **Páchora** Mahārāshtra, C India 20.52N 75.28E
43 P13 **Pachuca** *var.* Pachuca de Soto. Hidalgo, C Mexico 20.05N 98.46W
Pachuca de Soto *see* Pachuca
29 W5 **Pacific** Missouri, C USA 38.28N 90.44W
199 Jj15 **Pacific-Antarctic Ridge** *undersea feature* S Pacific Ocean
34 F8 **Pacific Beach** Washington, NW USA 47.09N 124.12W
37 N10 **Pacific Grove** California, W USA 36.35N 121.54W
31 S15 **Pacific Junction** Iowa, C USA 41.01N 95.48W
198-199 **Pacific Ocean** *ocean*
134-139 **Pacific Plate** *tectonic feature*
115 J15 **Pačir** ▲ SW Yugoslavia 43.19N 19.07E
190 L5 **Packsaddle** New South Wales, SE Australia 30.42S 141.55E
34 H9 **Packwood** Washington, NW USA 46.37N 121.38W
Padalung *see* Phatthalung
173 G9 **Padang** Sumatera, W Indonesia 01.00S 100.21E
174 Hh5 **Padang Endau** Pahang, Peninsular Malaysia 2.38N 103.37E
Padangpandjang *see* Padangpanjang
173 G8 **Padangpanjang** *prev.* Padangpandjang. Sumatera, W Indonesia 0.30S 100.25E
173 F7 **Padangsidempuan** *see* Padangsidimpoean
Padangsidimpoean *prev.* Padangsidempuan. Sumatera, W Indonesia 1.23N 99.15E
128 I9 **Padany** Respublika Kareliya, NW Russian Federation 63.18N 33.20E
95 M18 **Padasjoki** Etelä-Suomi, S Finland 61.20N 25.20E
59 M22 **Padauca** Tarija, S Bolivia 21.52S 64.46W
103 H14 **Paderborn** Nordrhein-Westfalen, NW Germany 51.43N 8.45E
Padeşul/Padeş, Vírful *see* Padeş, Vírful
118 F12 **Padeş, Vírful** *var.* Padeşul; *prev.* Vírful Padeş. ▲ W Romania 45.20N 22.50E
114 L10 **Padinska Skela** Serbia, N Yugoslavia 44.58N 20.25E
Padma *see* Brahmaputra
159 S14 **Padma** *var.* Ganges. Bangladesh/India *see also* Ganges

108 H8 **Padova** *Eng.* Padua; *anc.* Patavium. Veneto, NE Italy 45.24N 11.52E
84 A10 **Padrão, Ponta do** *headland* NW Angola 6.06S 12.18E
27 T16 **Padre Island** *island* Texas, SW USA
106 G3 **Padrón** Galicia, NW Spain 42.45N 8.40W
120 K13 **Padsvilye** *Rus.* Podsvil'ye. Vitsyebskaya Voblasts', N Belarus 55.10N 27.58E
190 K11 **Padthaway** South Australia 36.39S 140.30E
22 G7 **Paducah** Kentucky, S USA 37.09N 88.32W
27 P4 **Paducah** Texas, SW USA 33.59N 100.19W
107 N15 **Padul** Andalucía, S Spain 37.01N 3.37W
203 P8 **Paea** Tahiti, W French Polynesia 17.40S 149.34W
193 J14 **Paekakariki** Wellington, North Island, NZ 41.40S 174.58E
169 X11 **Paektu-san** *var.* Baitou Shan. ▲ China/North Korea 42.00N 128.03W
169 V15 **Paengnyŏng-do** *island* NW South Korea
192 M7 **Paeroa** Waikato, North Island, NZ 37.22S 175.39E
56 D12 **Páez** Cauca, SW Colombia 2.37N 76.00W
123 Mm4 **Páfos** *var.* Paphos. W Cyprus 34.46N 32.25E
123 Mm4 **Páfos** ✈ SW Cyprus 34.46N 32.25E
85 L19 **Pafúri** Gaza, SW Mozambique 22.51S 31.27E
114 C12 **Pag** *It.* Pago. Lika-Senj, W Croatia 44.26N 15.01E
114 B11 **Pag** *It.* Pago. *island* Zadar, SW Croatia
179 Qq16 **Pagadian** Mindanao, S Philippines 7.47N 123.22E
173 G11 **Pagai Selatan, Pulau** *island* Kepulauan Mentawai, W Indonesia
173 Ff10 **Pagai Utara, Pulau** *island* Kepulauan Mentawai, W Indonesia
196 K4 **Pagan** *island* C Northern Mariana Islands
117 G16 **Pagasitikós Kólpos** *gulf* E Greece
38 L8 **Page** Arizona, SW USA 36.54N 111.28W
31 Q5 **Page** North Dakota, N USA 47.09N 97.33W
120 D13 **Pagégiai** *Ger.* Pogegen. Šilutė, SW Lithuania 55.08N 21.54E
23 S11 **Pageland** South Carolina, SE USA 34.46N 80.23W
83 G16 **Pager** ♒ NE Uganda
155 Q5 **Paghman** Kābul, E Afghanistan 34.33N 68.55E
196 C16 **Pago Bay** *bay* E Guam, W Pacific Ocean
117 M20 **Pagóndas** *var.* Pagóndhas. Sámos, Dodekánisos, Greece, Aegean Sea 37.40N 26.49E
Pagóndhas *see* Pagóndas
198 C8 **Pago Pago** ◎ (American Samoa) Tutuila, W American Samoa 14.16S 170.43W
39 R8 **Pagosa Springs** Colorado, C USA 37.13N 107.01W
40 H12 **Pahala** *var.* Pāhala. Hawaii, USA, C Pacific Ocean 19.12N 155.28W
174 H4 **Pahang** *off.* Negeri Pahang Darul Makmur. ♦ *state* Peninsular Malaysia
Pahang *see* Pahang, Sungai
174 Hh5 **Pahang, Sungai** *var.* Pahang, Sungei Pahang. ♒ Peninsular Malaysia
155 S8 **Pahārpur** North-West Frontier Province, NW Pakistan 32.06N 71.00E
193 B24 **Pahia Point** *headland* South Island, NZ 46.19S 167.42E
192 M13 **Pahiatua** Manawatu-Wanganui, North Island, NZ 40.30S 175.48E
40 H12 **Pahoa** *Haw.* Pāhoa. Hawaii, USA, C Pacific Ocean 19.28N 154.55W
25 Y14 **Pahokee** Florida, SE USA 26.49N 80.40W
37 X9 **Pahranagat Range** ▲ Nevada, W USA
37 W11 **Pahrump** Nevada, W USA 36.11N 115.58W
178 H7 **Pai** Mae Hong Son, NW Thailand 19.19N 98.25E
40 F10 **Paia** *Haw.* Pā'ia. Maui, Hawaii, USA, C Pacific Ocean 20.54N 94.22W
Pai-ch'eng *see* Baicheng
120 H4 **Paide** *Ger.* Weissenstein. Järvamaa, N Estonia 58.54N 25.36E
99 J23 **Paignton** SW England, UK 50.25N 3.34W
192 K3 **Paihia** Northland, North Island, NZ 35.18S 174.06E
95 L19 **Päijänne** ◎ S Finland
116 F13 **Päijät** ♒ N Greece
59 M21 **Paila, Río** ♒ C Bolivia
178 H12 **Pailin** Bătdâmbâng, W Cambodia 12.51N 102.34E
56 F6 **Pailitas** Cesar, N Colombia 8.58N 73.37W
40 F9 **Pailolo Channel** *channel* Hawaii, USA, C Pacific Ocean
95 K19 **Paimio** *Swe.* Pemar. Länsi-Suomi, W Finland 60.27N 22.42E
171 O17 **Paimi-saki** *var.* Paimi. *headland* triomote-jima, SW Japan 24.18N 123.40E
104 G5 **Paimpol** Côtes d'Armor, NW France 48.46N 3.03W
161 H22 **Paimganj** Tamil Nādu, S India 10.30N 77.24E
173 G8 **Painan** Sumatera, W Indonesia 1.22S 100.33E
65 G23 **Paine, Cerro** ▲ S Chile 51.01S 72.57W
35 U11 **Painesville** Ohio, N USA 41.43N 81.15W
38 M10 **Painted Desert** *desert* Arizona, SW USA
27 O10 **Painted Rock** Texas, SW USA 31.30N 99.51W
23 O6 **Paintsville** Kentucky, S USA 37.49N 82.48W
Paisance *see* Piacenza

98 I12 **Paisley** W Scotland, UK 55.49N 4.25W
34 I15 **Paisley** Oregon, NW USA 42.40N 120.31W
107 R10 **País Valenciano** *var.* Valencia, *Cat.* València; *anc.* Valentia. ♦ *autonomous community* NE Spain
107 O15 **País Vasco** *Basq.* Euskadi, *Eng.* The Basque Country, *Sp.* Provincias Vascongadas. ♦ *autonomous community* N Spain
58 A9 **Paita** Piura, NW Peru 5.07S 81.07W
197 J7 **Paita** Province Sud, S New Caledonia 22.05S 166.18E
175 O1 **Paitan, Teluk** *bay* Sabah, East Malaysia
94 H7 **Paiva, Rio** ♒ N Portugal
94 K12 **Pajala** Norrbotten, N Sweden 67.13N 23.19E
106 K3 **Pajares, Puerto de** *pass* NW Spain
56 G9 **Pajárito** Boyacá, C Colombia 5.18N 72.43W
56 G4 **Pajaro** La Guajira, S Colombia 11.41N 72.37W
57 Q10 **Pakaraima Mountains** *var.* Serra Pacaraim, Sierra Pacaraima. ▲ South America
178 Hh11 **Pak Chong** Nakhon Ratchasima, C Thailand 14.38N 101.22E
127 Pp7 **Pakhachi** Koryakskiy Avtonomnyy Okrug, E Russian Federation 60.36N 168.59E
153 O11 **Pakhtakor** Jizzakh Wiloyati, C Uzbekistan 40.21N 67.54E
201 U16 **Pakin Atoll** *atoll* Caroline Islands, E Micronesia
155 Q12 **Pakistan** *off.* Islamic Republic of Pakistan; *var.* Islami Jamhuriya e Pakistan. ♦ *republic* S Asia
Pakistan, Islami Jamhuriya e *see* Pakistan
178 I8 **Pak Lay** *var.* Muang Pak Lay. Xaignabouli, C Laos 18.06N 101.21E
177 Ff5 **Pakokku** Magwe, C Myanmar 21.19N 95.04E
112 I10 **Pakość** *Ger.* Pakosch. Kujawski-pomorskie, C Poland 52.47N 18.03E
Pakosch *see* Pakość
155 V10 **Pākpattan** Punjab, E Pakistan 30.19N 73.27E
178 H16 **Pak Phanang** *var.* Ban Pak Phanang. Nakhon Si Thammarat, SW Thailand 8.19N 100.10E
114 G9 **Pakrac** *Hung.* Pakrácz. Požega-Slavonija, NE Croatia 45.26N 17.09E
120 F11 **Pakruojis** Pakruojis, N Lithuania 55.59N 23.50E
113 J24 **Paks** Tolna, S Hungary 46.37N 18.51E
155 R6 **Paktīā** ♦ *province* SE Afghanistan
155 Q7 **Paktīkā** ♦ *province* SE Afghanistan
175 Pp9 **Pakuli** Sulawesi, C Indonesia 1.14S 119.55E
83 F17 **Pakwach** NW Uganda 2.28N 31.28E
178 Ii8 **Pakxan** *var.* Muang Pakxan, Pak Sane. Bolikhamxai, C Laos 18.27N 103.38E
178 Jj10 **Pakxé** *var.* Paksé. Champasak, S Laos 15.09N 105.49E
80 G12 **Pala** Mayo-Kébbi, SW Chad 9.22N 14.54E
63 A17 **Palacios** Santa Fe, C Argentina 30.43S 61.37W
27 V13 **Palacios** Texas, SW USA 28.42N 96.13W
107 X9 **Palafrugell** Cataluña, NE Spain 41.55N 3.10E
109 L24 **Palagonia** Sicilia, Italy, C Mediterranean Sea 37.19N 14.45E
115 E17 **Palagruža** *It.* Pelagosa. *island* SW Croatia
117 G20 **Palaiá Epídavros** Peloponnísos, S Greece 37.38N 23.09E
124 Nn3 **Palaichóri** *var.* Palekhori. C Cyprus 34.55N 33.06E
117 H25 **Palaióchora** Kríti, Greece, E Mediterranean Sea 35.14N 23.37E
117 A15 **Palaiolastrítsa** *religious building* Kérkyra, Iónioi Nísoi, Greece, C Mediterranean Sea
117 J19 **Palaiópoli** Ándros, Kykládes, Greece, Aegean Sea 37.49N 24.49E
105 N5 **Palaiseau** Essonne, N France 48.40N 2.13E
Palakkad *see* Pālghāt
160 N11 **Pāla Laharha** Orissa, E India 21.27N 85.14E
85 G19 **Palamakoloi** Ghanzi, C Botswana 23.10S 22.22E
117 E16 **Palamás** Thessalía, C Greece 39.28N 22.04E
107 X5 **Palamós** Cataluña, NE Spain 41.51N 3.06E
84 Q12 **Palamuse** *Ger.* Sankt-Bartholomäi. Jõgevamaa, E Estonia 58.40N 26.34E
191 P9 **Palana** Tasmania, SE Australia 39.48S 147.54E
127 P9 **Palana** Koryakskiy Avtonomnyy Okrug, E Russian Federation 59.04N 159.58E
120 C11 **Palanga** *Ger.* Polangen. Palanga, NW Lithuania 55.56N 21.03E
149 V10 **Palangān, Kūh-e** ▲ E Iran
175 Mm10 **Palangkaraya** *prev.* Palangkaraja. Borneo, C Indonesia 2.16S 113.55E
62 O13 **Palanās** Paraná, S Brazil 26.29S 52.00W
161 H22 **Palani** Tamil Nādu, S India 10.30N 77.24E
160 D9 **Pālanpur** Gujarāt, W India 24.12N 72.28E
85 G19 **Palapye** Central, SE Botswana 22.37S 27.06E
161 H3 **Palas de Rei** Galicia, NW Spain 42.52N 7.51W
25 Y12 **Palm Bay** Florida, SE USA 28.01N 80.35W
127 O10 **Palatka** Magadanskaya Oblast', E Russian Federation 60.09N 150.33E
196 J6 **Palau** *var.* Belau. ♦ *republic* W Pacific Ocean

133 Y14 **Palau Islands** *var.* Palau. *island group* N Palau
198 Aa8 **Palauli Bay** *bay* Savai'i, Samoa, C Pacific Ocean
179 Oo15 **Palaw** Tenasserim, S Myanmar 12.57N 98.39E
179 Oo15 **Palawan** *island* W Philippines
198 F7 **Palawan Trough** *undersea feature* S South China Sea
179 P10 **Palayan City** Luzon, N Philippines 15.34N 121.34E
161 H23 **Pālayankottai** Tamil Nādu, SE India 8.44N 77.45E
109 L25 **Palazzola Acreide** *anc.* Acrae. Sicilia, Italy, C Mediterranean Sea 37.04N 14.54E
120 G3 **Paldiski** *prev.* Baltiski, *Eng.* Baltic Port, *Ger.* Baltischport. Harjumaa, NW Estonia 59.20N 24.04E
114 I13 **Pale** Republika Srpska, E Bosnia and Herzegovina 43.49N 18.35E
Palekhori *see* Palaichóri
175 Qq7 **Paleleh, Pegunungan** ▲ Sulawesi, N Indonesia
174 I11 **Palembang** Sumatera, W Indonesia 2.58S 104.45E
65 G18 **Palena** Los Lagos, S Chile 43.40S 71.49W
65 G18 **Palena, Río** ♒ S Chile
106 M5 **Palencia** *anc.* Palantia, Pallantia. Castilla-León, NW Spain 41.01N 4.31W
106 M3 **Palencia** ♦ *province* Castilla-León, N Spain
37 X15 **Palen Dry Lake** ◎ California, W USA
29 V2 **Palenque** Chiapas, SE Mexico 17.37N 92.03W
43 V15 **Palenque** *var.* Ruinas de Palenque. *ruins* Chiapas, SE Mexico 17.31N 91.58W
47 O9 **Palenque, Punta** *headland* S Dominican Republic 18.13N 70.08W
Palenque, Ruinas de *see* Palenque
Palerme *see* Palermo
109 I23 **Palermo** *Fr.* Palerme; *anc.* Panhormus, Panormus. Sicilia, Italy, C Mediterranean Sea 38.07N 13.22E
37 N9 **Palo Alto** California, W USA 37.26N 122.08W
27 V8 **Palestine** Texas, SW USA 31.44N 95.38W
27 V7 **Palestine, Lake** ◎ Texas, SW USA
109 I15 **Palestrina** Lazio, C Italy 41.49N 12.53E
177 F5 **Paletwa** Chin State, W Myanmar 21.30N 92.51E
161 G21 **Pālghāt** *var.* Palakkad; *prev.* Pulicat. Kerala, SW India 10.46N 76.42E
109 I15 **Palombara Sabina** Lazio, C Italy 42.04N 12.45E
107 S13 **Palos, Cabo de** *headland* SE Spain 37.38N 0.42W
201 O12 **Palikir** ◎ (Micronesia) Pohnpei, E Micronesia 6.58N 158.13E
106 I14 **Palos de la Frontera** Andalucía, S Spain 37.13N 6.52W
62 G11 **Palotina** Paraná, S Brazil 24.16S 53.49W
34 M9 **Palouse** Washington, NW USA 46.54N 117.04W
34 L9 **Palouse River** ♒ Washington, NW USA
37 Y16 **Palo Verde** California, W USA 33.25N 114.43W
59 E16 **Palpa** Ica, W Peru 14.33S 75.09W
97 M16 **Pålsboda** Örebro, C Sweden 59.04N 15.21E
95 M15 **Paltamo** Oulu, C Finland 64.25N 27.49E
175 Pp9 **Palu** *prev.* Paloe. Sulawesi, C Indonesia 0.54S 119.52E
143 P14 **Palu** Elâzığ, E Turkey 38.43N 39.55E
175 Q16 **Palu, Pulau** *island* S Indonesia
175 P8 **Palu, Teluk** *bay* Sulawesi, C Indonesia
158 I11 **Palwal** Haryāna, N India 28.15N 77.18E
79 Q13 **Pama** NE Burkina 11.13N 0.46E
180 J14 **Pamandzi** ✈ (Mamoudzou) Petite-Terre, E Mayotte
149 R11 **Pā Mazār** Kermān, C Iran
85 N19 **Pambarra** Inhambane, SE Mozambique 24.04S 34.44E
51 M16 **Pamiers** Ariège, S France 43.07N 1.36E
153 T14 **Pamir** *var.* Daryā-ye Pāmīr, *Taj.* Dar''yoi Pomir. ♒ Afghanistan/Tajikistan *see also* Pāmīr, Daryā-ye
Pamir/Pāmīr, Daryā-ye *see* Pamirs
153 U15 **Pāmīr, Daryā-ye** *var.* Pamir, *Taj.* Dar''yoi Pomir. ♒ Afghanistan/Tajikistan *see also* Pamir
Pāmīr-e Khord *see* Little Pamir
133 Q8 **Pamirs** *Pash.* Daryā-ye Pāmīr, *Rus.* Pamir. ▲ C Asia
Pāmiut *see* Paamiut
23 X10 **Pamlico River** ♒ North Carolina, SE USA
23 Y10 **Pamlico Sound** *sound* North Carolina, SE USA
26 M4 **Pampa** Texas, SW USA 35.31N 100.58W
59 N18 **Pampa Aullagas, Lago** ◎ C Bolivia
62 C11 **Pampa Húmeda** *grassland* E Argentina
58 A10 **Pampas** Huancavelica, C Peru 12.22S 74.52W
63 B21 **Pampas** *plain* C Argentina
56 E7 **Pamplona** Norte de Santander, N Colombia 7.24N 72.37W
107 Q3 **Pamplona** *Basq.* Iruña; *prev.* Pampeluna, *anc.* Pompaelo. Navarra, N Spain 42.49N 1.39W
116 I11 **Pamporovo** *prev.* Vasil Kolarov. Smolyan, S Bulgaria 41.39N 24.45E
142 D15 **Pamukkale** Denizli, W Turkey 37.51N 29.13E
23 W5 **Pamunkey River** ♒ Virginia, NE USA
158 K5 **Pamzal** Jammu and Kashmir, NW India 34.16N 78.49E
32 L14 **Pana** Illinois, N USA 39.23N 89.04W
43 Y11 **Panabá** Yucatán, SE Mexico 21.18N 88.15W
117 E19 **Panachaïkó** ▲ S Greece
52 F11 **Panache Lake** ◎ Ontario, S Canada
116 I10 **Panagyurishte** Pazardzhik, C Bulgaria 42.30N 24.10E
174 T14 **Panaitan, Pulau** *island* S Indonesia
174 Ii4 **Panaitan, Selat** *strait* Jawa, SW Indonesia
117 D18 **Panaitolikó** ▲ C Greece
161 E17 **Panaji** *var.* Pangim, Panjim, New Goa. Goa, W India 15.31N 73.52E
45 T14 **Panama** *off.* Republic of Panama. ♦ *republic* Central America
45 T15 **Panamá** *var.* Ciudad de Panamá, *Eng.* Panama City. ● (Panama) Panamá, C Panama 8.57N 79.33W
45 U14 **Panamá** *off.* Provincia de Panamá. ♦ *province* E Panama
45 U15 **Panamá, Bahía de** *bay* N Gulf of Panama
200 Oo8 **Panama Basin** *undersea feature* E Pacific Ocean
45 T14 **Panama Canal** *canal* E Panama
25 R9 **Panama City** Florida, SE USA 30.09N 85.39W
45 T15 **Panama City** ✈ Panamá, C Panama 9.02N 79.24W
25 Q9 **Panama City Beach** Florida, SE USA 30.10N 85.48W
45 T17 **Panamá, Golfo de** *var.* Gulf of Panama. *gulf* S Panama
Panama, Gulf of *see* Panamá, Golfo de
Panama, Isthmus of *see* Panamá, Istmo de
45 T15 **Panamá, Istmo de** *Eng.* Isthmus of Panama; *prev.* Isthmus of Darien. *isthmus* E Panama
37 U11 **Panamint Range** ▲ California, W USA
109 L22 **Panarea, Isola** *island* Isole Eolie, S Italy
109 O8 **Panaro** ♒ N Italy
179 Q14 **Panay Gulf** *gulf* C Philippines
179 P13 **Panay Island** *island* C Philippines
37 W7 **Pancake Range** ▲ Nevada, W USA
114 M11 **Pančevo** *Ger.* Pantschowa, *Hung.* Pancsova. Serbia, N Yugoslavia 44.52N 20.39E
115 M15 **Pančićev Vrh** ▲ SW Yugoslavia 43.16N 20.49E
118 L12 **Panciu** Vrancea, E Romania 45.54N 27.07E
118 F10 **Pâncota** *Hung.* Pankota; *prev.* Pincota. Arad, W Romania 46.19N 21.45E
85 N20 **Panda** Inhambane, SE Mozambique 24.04S 34.44E
176 X9 **Pandaidori, Kepulauan** *island group* E Indonesia
27 N11 **Pandale** Texas, SW USA 30.09N 101.34W
174 H7 **Pandang, Pulau** *island* W Indonesia
174 Kk9 **Pandang Tikar, Pulau** *island* W Indonesia
63 F20 **Pan de Azúcar** Maldonado, S Uruguay 34.45S 55.13W
120 H11 **Pandélys** Rokiškis, NE Lithuania 56.04N 25.18E
161 F15 **Pandharpur** Mahārāshtra, W India 17.42N 75.24E
190 J1 **Pandie Pandie** South Australia 26.06S 139.26E
175 Pp9 **Pandiri** Sulawesi, C Indonesia 1.32S 120.47E
63 F20 **Pando** Canelones, S Uruguay 34.43S 55.58W
59 I14 **Pando** ♦ *department* N Bolivia
199 Ii10 **Pandora Bank** *undersea feature* W Pacific Ocean
97 G20 **Pandrup** Nordjylland, N Denmark 57.13N 9.42E
174 H2 **Pandu** Assam, NE India 26.08N 91.37E
81 J15 **Pangala** Equateur, NW Dem. Rep. Congo (Zaire) 5.03N 19.14E
61 L15 **Panelas** Mato Grosso, W Brazil 9.06S 60.41W
120 G12 **Panevėžys** Panevėžys, C Lithuania 55.44N 24.21E
Panfilov *see* Zharkent
131 N9 **Panfilovo** Volgogradskaya Oblast', SW Russian Federation 50.25N 42.55E
200 Ss13 **Pangai** Lifuka, C Tonga 19.49S 174.22W
116 H3 **Pangala** Le Pool, S Congo 3.26S 14.36E
83 I21 **Pangani** Tanga, E Tanzania 5.27S 39.00E
83 I21 **Pangani** ♒ NE Tanzania
195 U13 **Pangboche** Choiseul Island, NW Solomon Islands 7.00S 95.05E
81 N20 **Pangi** Maniema, E Dem. Rep. Congo (Zaire) 3.12S 26.39E
172 F5 **Pangia** Southern Highlands, W PNG 6.18S 144.12E
173 F4 **Pangkalanbrandan** Sumatera, W Indonesia 4.00N 98.15E
Pangkalanbun *see* Pangkalanbuun
174 Ll10 **Pangkalanbuun** *prev.* Pangkalanbun. Borneo, C Indonesia 2.43S 111.37E
173 F5 **Pangkalpinang** Pulau Bangka, W Indonesia 2.04S 106.09E
9 U17 **Pangman** S Canada 49.37N 104.33W

◆ COUNTRY ◇ DEPENDENT TERRITORY ◈ ADMINISTRATIVE REGION ▲ MOUNTAIN ☒ VOLCANO ◎ LAKE
● COUNTRY CAPITAL ◇ DEPENDENT TERRITORY CAPITAL ✈ INTERNATIONAL AIRPORT ▲ MOUNTAIN RANGE ♒ RIVER ☒ RESERVOIR

307

Pang-Nga see Phang-Nga
16 Nn2 Pangnirtung Baffin Island, Nunavut, NE Canada 66.04N 65.45W
158 K6 Pangong Tso var. Bangong Co. ◇ China/India see also Bangong Co
38 K7 Panguitch Utah, W USA 37.49N 112.26W
195 S12 Panguna Bougainville Island, NE PNG 6.22S 155.19E
179 Pp17 Pangutaran Group island group Sulu Archipelago, SW Philippines
27 N2 Panhandle Texas, SW USA 35.18N 101.23W
Panhormus see Palermo
176 X12 Paniai, Danau ◉ Irian Jaya, E Indonesia
81 L21 Pania-Mutombo Kasai Oriental, C Dem. Rep. Congo (Zaire) 5.09S 23.49E
Panicherevo see Dolno Panicherevo
197 H5 Panié, Mont ▲ C New Caledonia 20.33S 164.41E
158 I10 Pānīpat Haryāna, N India 29.18N 77.00E
153 Q14 Panj Rus. Pyandzh; prev. Kirovabad. SW Tajikistan 37.39N 69.55E
153 P15 Panj Rus. Pyandzh. ✦ Afghanistan/Tajikistan
155 O5 Panjāb Bāmiān, C Afghanistan 34.21N 67.00E
153 O12 Panjakent Rus. Pendzhikent. W Tajikistan 39.28N 67.33E
154 L14 Panjgur Baluchistān, SW Pakistan 26.58N 64.05E
Panjim see Panaji
169 U12 Panjin Liaoning, NE China 41.11N 122.05E
153 P14 Panji Poyon Rus. Nizhniy Pyandzh. SW Tajikistan 37.14N 68.32E
155 Q4 Panjshir ✦ E Afghanistan
Pankota see Pâncota
79 W14 Pankshin Plateau, C Nigeria 9.21N 9.27E
169 Y10 Pan Ling ▲ N China
Panlong Jiang see Lô, Sông
160 J9 Panna Madhya Pradesh, C India 24.43N 80.10E
101 M16 Panningen Limburg, SE Netherlands 51.19N 5.58E
155 R13 Pāno Āqil Sind, SE Pakistan 27.56N 69.16E
124 Nn3 Páno Léfkara S Cyprus 34.52N 33.18E
124 N3 Páno Panagía var. Pano Panayia. W Cyprus 34.55N 32.38E
Pano Panayia see Páno Panagía
Panopolis see Akhmim
31 U14 Panora Iowa, C USA 41.41N 94.21W
62 I8 Panorama São Paulo, S Brazil 21.22S 51.51W
117 I24 Pánormos Kríti, Greece, E Mediterranean Sea 35.24N 24.42E
Panormus see Palermo
169 W11 Panshi Jilin, NE China 42.50N 126.06E
61 H19 Pantanal var. Pantanalmato-Grossense. swamp SW Brazil
Pantanalmato-Grossense see Pantanal
63 H16 Pantano Grande Rio Grande do Sul, S Brazil 30.12S 52.24W
175 R16 Pantar, Pulau island Kepulauan Alor, S Indonesia
23 X9 Pantego North Carolina, SE USA 35.34N 76.39E
109 G25 Pantelleria anc. Cossyra, Cosyra. Sicilia, Italy, C Mediterranean Sea 36.47N 12.00E
109 G25 Pantelleria, Isola di island Sicilia, Italy, C Mediterranean Sea
Pante Macassar/Pante Makassar see Pante Macassar
175 Rr17 Pante Makasar var. Pante Macassar, Pante Makassar. W East Timor 9.10S 124.27E
158 K10 Pantnagar Uttar Pradesh, N India 29.00N 79.28E
117 A15 Pantokrátoras ▲ Kérkyra, Iónioi Nísoi, Greece, C Mediterranean Sea 39.45N 19.51E
Pantschowa see Pančevo
179 Rr16 Pantukan Mindanao, S Philippines 7.10N 125.55E
43 P11 Pánuco Veracruz-Llave, E Mexico 22.01N 98.10W
43 P11 Pánuco, Río ✦ C Mexico
166 I12 Panxian Guizhou, S China 25.45N 104.28E
173 G7 Panyabungan Sumatera, N Indonesia 0.55N 99.30E
79 W14 Panyam Plateau, C Nigeria 9.28N 9.13E
163 N13 Panzhihua prev. Dukou, Tu-k'ou. Sichuan, C China 26.35N 101.41E
81 I22 Panzi Bandundu, SW Dem. Rep. Congo (Zaire) 7.10S 17.55E
44 E5 Panzós Alta Verapaz, E Guatemala 15.21N 89.40W
Pao-chi/Paoki see Baoji
Pao-king see Shaoyang
109 N20 Paola Calabria, SW Italy 39.21N 16.03E
123 Jj17 Paola E Malta 35.52N 14.30E
29 R5 Paola Kansas, C USA 38.34N 94.52W
35 O15 Paoli Indiana, N USA 38.31N 86.26W
197 D14 Paonangisu Éfaté, C Vanuatu 17.33S 168.23E
175 Tt11 Paoni var. Pauni. Pulau Seram, E Indonesia 2.48S 129.03E
39 Q5 Paonia Colorado, C USA 38.52N 107.35W
203 O7 Paopao Moorea, W French Polynesia 17.28S 149.48W
Pao-shan see Baoshan
Pao-ting see Baoding
Pao-t'ou/Paotow see Baotou
81 H14 Paoua Ouham-Pendé, W Central African Republic 7.22N 16.25E
Pap see Pop
113 H23 Pápa Veszprém, W Hungary 47.19N 17.27E
44 J12 Papagayo, Golfo de gulf NW Costa Rica
40 H11 Papaikou var. Pāpa'ikou. Hawaii, USA, C Pacific Ocean 19.45N 155.06W
43 R15 Papaloapan, Río ✦ SE Mexico
192 L6 Papakura Auckland, North Island, NZ 37.03S 174.57E

43 Q13 Papantla var. Papantla de Olarte. Veracruz-Llave, E Mexico 20.27N 97.21W
Papantla de Olarte see Papantla
203 P8 Papara Tahiti, W French Polynesia 17.45S 149.33W
192 K4 Paparoa Northland, North Island, NZ 36.06S 174.12E
193 G16 Paparoa Range ▲ South Island, NZ
117 K20 Pápas, Akrotírio headland Ikaría, Dodekánisos, Greece, Aegean Sea 37.31N 25.58E
98 L2 Papa Stour island NE Scotland, UK
192 L6 Papatoetoe Auckland, North Island, NZ 36.58S 174.52E
193 E25 Papatowai Otago, South Island, NZ 46.33S 169.33E
98 K4 Papa Westray island NE Scotland, UK
203 T10 Papeete ◉ (French Polynesia) Tahiti, W French Polynesia 17.31S 149.34W
102 F11 Papenburg Niedersachsen, NW Germany 53.04N 7.24E
100 H13 Papendrecht Zuid-Holland, SW Netherlands 51.49N 4.42E
203 Q7 Papenoo Tahiti, W French Polynesia 17.28S 149.25W
203 Q7 Papenoo Rivière ✦ Tahiti, W French Polynesia
203 N7 Papetoai Moorea, W French Polynesia 17.28S 149.52W
94 L3 Papey island E Iceland
42 H5 Papigochic, Río ✦ NW Mexico
120 E10 Papilė Akmenė, NW Lithuania 56.08N 22.51E
31 S15 Papillion Nebraska, C USA 41.09N 96.02W
13 T5 Papinachois ✦ Quebec, SE Canada
194 I15 Papua, Gulf of gulf S PNG
194 H13 Papua New Guinea off. Independent State of Papua New Guinea; prev. Territory of Papua and New Guinea, abbrev. PNG. ◆ commonwealth republic NW Melanesia
199 H10 Papua Plateau undersea feature N Coral Sea
114 G9 Papuk ▲ NE Croatia
177 G8 Papun Karen State, S Myanmar 18.04N 97.25E
44 L14 Paquera Puntarenas, W Costa Rica 9.52N 84.55W
57 V9 Pará ◆ district N Suriname
60 I13 Pará off. Estado do Pará. ◆ state NE Brazil
Pará see Belém
126 M12 Parabel' Tomskaya Oblast', C Russian Federation 58.54N 80.45E
188 I8 Paraburdoo Western Australia 23.07S 117.40E
62 L8 Pardo, Rio ✦ S Brazil
113 E16 Pardubice Ger. Pardubitz. Pardubický Kraj, C Czech Republic 50.01N 15.46E
113 E17 Pardubický Kraj ◆ region C Czech Republic
Pardubitz see Pardubice
121 F16 Parechcha Pol. Porzecze, Rus. Porech'ye. Hrodzyenskaya Voblasts', W Belarus 53.51N 24.07E
61 F17 Parecis, Chapada dos var. Serra dos Parecis. ▲ W Brazil
Parecis, Serra dos see Parecis, Chapada dos
106 M4 Paredes de Nava Castilla-León, N Spain 42.09N 4.42W
114 N13 Paraćin Serbia, C Yugoslavia 43.51N 21.25E
12 K8 Paradis Quebec, SE Canada 48.13N 76.36W
192 I1 Parengarenga Harbour inlet North Island, NZ
13 N8 Parent Quebec, SE Canada 47.55N 74.36W
104 J14 Parentis-en-Born Landes, SW France 44.22N 1.04W
Parenzo see Poreč
193 G20 Pareora Canterbury, South Island, NZ 44.28S 171.12E
175 P12 Parepare Sulawesi, C Indonesia 4.00S 119.40E
117 B16 Párga Ípeiros, W Greece 39.18N 20.19E
95 K20 Pargas Swe. Parainen. Länsi-Suomi, W Finland 60.18N 22.19E
66 O5 Pargo, Ponta do headland Madeira, Portugal, NE Atlantic Ocean 32.48N 17.16W
57 N6 Pariaguán Anzoátegui, NE Venezuela 8.51N 64.43W
47 X17 Paria, Gulf of var. Golfo de Paria. gulf Trinidad and Tobago/Venezuela
59 I15 Pariamanu, Río ✦ E Peru
38 L8 Paria River ✦ Utah, W USA
42 M14 Parícutin, Volcán ▮ C Mexico 19.25N 102.20W
45 P16 Parida, Isla island SW Panama
58 T8 Parika NE Guyana 6.51N 58.25W
95 O18 Parikkala Etelä-Suomi, S Finland 61.33N 29.33E
60 G13 Parima, Serra var. Sierra Parima. ▲ Brazil/Venezuela see also Parima, Sierra
57 N11 Parima, Sierra var. Serra Parima. ▲ Brazil/Venezuela see also Parima, Serra
59 F17 Parinacochas, Laguna ◉ SW Peru
64 A9 Pariñas, Punta headland NW Peru 4.45S 81.22W
60 H12 Parintins Amazonas, N Brazil 2.37S 56.45W
105 O5 Paris anc. Lutetia, Lutetia Parisiorum, Parisii. ● (France) Paris, N France 48.52N 2.19E
33 Y2 Paris Kiritimati, E Kiribati 1.55N 95.90W
29 S11 Paris Arkansas, C USA 35.17N 93.43W
35 S16 Paris Idaho, NW USA 42.14N 111.24W
35 N14 Paris Illinois, N USA 39.36N 87.42W
20 M5 Paris Kentucky, S USA 38.12N 84.15W
21 V3 Paris Missouri, C USA 39.28N 92.00W
22 H8 Paris Tennessee, S USA 36.18N 88.19W
25 V5 Paris Texas, SW USA 33.40N 95.33W
Parisii see Paris
21 R15 Parita Herrera, S Panama 7.59N 80.31W
45 S16 Parita, Bahía de bay S Panama

57 W8 Paramaribo ● (Suriname) Paramaribo, N Suriname 5.52N 55.13W
57 W9 Paramaribo ✦ district N Suriname
57 W9 Paramaribo ✈ Paramaribo, N Suriname 5.52N 55.13W
58 C13 Paramonga Lima, W Peru 10.40S 77.51W
127 Pp13 Paramushir, Ostrov island SE Russian Federation
117 C16 Paramythiá var. Paramithiá. Ípeiros, W Greece 39.28N 20.31E
64 M10 Paraná Entre Ríos, E Argentina 31.48S 60.29W
62 H11 Paraná off. Estado do Paraná. ◆ state S Brazil
49 U11 Paraná var. Alto Paraná. ✦ C South America
62 K12 Paranaíba, Rio ✦ E Brazil
63 C19 Paraná Ibicuy, Río ✦ E Argentina
61 H15 Paranaíta Mato Grosso, W Brazil 9.35S 57.01W
62 H9 Paranapanema, Rio ✦ S Brazil
62 K11 Paranapiacaba, Serra do ▲ S Brazil
62 H9 Paranavaí Paraná, S Brazil 23.03S 52.25W
149 N5 Parandak Markazī, W Iran 35.19N 50.40E
116 I12 Paranéstio Anatolikí Makedonía kai Thráki, NE Greece 41.16N 24.31E
203 W11 Paraoa ◇ atoll Îles Tuamotu, C French Polynesia
192 L13 Paraparaumu Wellington, North Island, NZ 40.55S 175.01E
59 N20 Parapeti, Río ✦ SE Bolivia
56 L10 Paraque, Cerro ▲ W Venezuela 6.00S 67.00W
160 I11 Parasiya Madhya Pradesh, C India 22.11N 78.47E
117 M23 Paraspóri, Akrotírio headland Kárpathos, SE Greece 35.54N 27.15E
62 O10 Parati Rio de Janeiro, SE Brazil 23.15S 44.42W
61 K14 Parauapebas Pará, N Brazil 6.03S 49.48W
105 Q10 Paray-le-Monial Saône-et-Loire, C France 46.27N 4.07E
160 G13 Parbatsar see Parvatsar
160 G13 Parbhani Mahārāshtra, C India 19.16N 76.51E
102 L10 Parchim Mecklenburg-Vorpommern, N Germany 53.25N 11.51E
112 P13 Parczew Lubelskie, E Poland 51.39N 22.59E
62 L8 Pardo, Rio ✦ S Brazil

60 I11 Paru, Rio ✦ N Brazil
161 M14 Pārvatipuram Andhra Pradesh, E India 17.01N 81.47E
158 G12 Parvatsar prev. Parbatsar. Rājasthān, N India 26.52N 74.49E
155 Q5 Parwān Per. Parván. ◆ province E Afghanistan
164 I15 Paryang Xizang Zizhiqu, W China 30.06N 83.53E
121 M18 Parychy Rus. Parichi. Homyel'skaya Voblasts', SE Belarus 52.48N 29.25E
85 J21 Parys Free State, C South Africa 26.51S 27.28E
37 T15 Pasadena California, W USA 34.09N 118.08W
27 W11 Pasadena Texas, SW USA 29.41N 95.12W
58 B8 Pasaje El Oro, SW Ecuador 3.19S 79.49W
179 R17 Pasar Volcano ▮ Mindanao, S Philippines 4.30N 125.52E
143 T9 P'asanauri N Georgia 42.21N 44.40E
173 G10 Pasapuat Pulau Pagai Utara, W Indonesia 2.36S 99.58E
178 Gg7 Pasawng Kayah State, C Myanmar 18.50N 97.16E
31 T5 Pas Rapids Minnesota, N USA 46.55N 93.03W
31 Q11 Parkston South Dakota, N USA 43.24N 97.58W
8 L17 Parksville Vancouver Island, British Columbia, SW Canada 49.13N 124.13W
59 S3 Parkview Mountain ▲ Colorado, C USA 40.19N 106.08W
107 N8 Parla Madrid, C Spain 40.13N 3.48W
31 S8 Parle, Lac qui ◉ Minnesota, N USA
117 F20 Parlía Tyroú Pelopónnisos, S Greece 37.17N 22.53E
161 G14 Parli Vaijnāth Mahārāshtra, C India 18.52N 76.36E
108 F9 Parma Emilia-Romagna, N Italy 44.49N 10.19E
33 T11 Parma Ohio, N USA 41.24N 81.43W
60 N13 Parnaíba var. Parnahyba. Piauí, E Brazil 2.58S 41.46W
117 J14 Parnassós ▲ C Greece
60 N13 Parnaíba, Rio ✦ NE Brazil
61 F18 Parnassós ✦ C Greece
193 J17 Pāsighāt Arunāchal Pradesh, NE India 28.08N 95.13E
159 X10 Parnassus Canterbury, South Island, NZ 42.41S 173.18E
190 H10 Parndana South Australia 35.48S 137.13E
117 H19 Párintha ▲ C Greece
117 F21 Párnon ▲ S Greece
120 G5 Pärnu Ger. Pernau, Latv. Pērnava; prev. Rus. Pernov. Pärnumaa, SW Estonia 58.23N 24.31E
120 G6 Pärnu var. Pärnu Jõgi, Ger. Pernau. ✦ SW Estonia
120 G5 Pärnu-Jaagupi Ger. Sankt-Jakobi. Pärnumaa, SW Estonia 58.36N 24.30E
Pärnu Jõgi see Pärnu
120 G5 Pärnu Laht Ger. Pernauer Bucht. bay SW Estonia
120 F5 Pärnumaa off. Pärnu Maakond. ◆ province SW Estonia
159 T11 Paro ✈ Bhutan 27.22N 89.31E
159 T11 Paro ✈ (Thimphu) W Bhutan 27.22N 89.31E
193 G17 Paro West Coast, South Island, NZ 42.31S 171.10E
169 X14 P'aro-ho var. Hwach'ŏn-chŏsuji. ◉ N South Korea
191 N6 Paroo seasonal river New South Wales/Queensland, SE Australia
117 J21 Páros Páros, Kykládes, Greece, Aegean Sea 37.04N 25.09E
117 J21 Páros island Kykládes, Greece, Aegean Sea
46 J7 Parowan Utah, W USA 37.50N 112.49W
9 U14 Parpaillon ▲ SE France
11 Y7 Parpan Graubünden, S Switzerland 46.46N 9.32E
64 G13 Parral Maule, C Chile 36.07S 71.47W
Parral see Hidalgo del Parral
191 T9 Parramatta New South Wales, SE Australia 33.49S 150.58E
42 M8 Parras var. Parras de la Fuente. Coahuila de Zaragoza, NE Mexico 25.26N 102.07W
Parras de la Fuente see Parras
44 M14 Parrita Puntarenas, S Costa Rica 9.33N 84.20W
12 G13 Parry Island Ontario, S Canada
207 O9 Parry Islands island group Nunavut, NW Canada
12 L12 Parry Sound Ontario, S Canada 45.21N 80.03W
109 L26 Passero, Capo headland Sicilia, Italy, C Mediterranean Sea 36.40N 15.08E
112 F7 Parsęta Ger. Persante. ✦ NW Poland
30 L3 Parshall North Dakota, N USA 47.57N 102.07W
27 Q3 Parsons Kansas, C USA 37.20N 95.15W
22 H9 Parsons Tennessee, S USA 35.39N 88.07W
23 H15 Parsons West Virginia, NE USA 39.06N 79.40W
Parsonstown see Birr
102 P11 Parsteiner See ◉ NE Germany
109 I24 Partanna Sicilia, Italy, C Mediterranean Sea 37.43N 12.54E
110 J8 Partenen Graubünden, E Switzerland 46.58N 10.01E
104 K9 Parthenay Deux-Sèvres, W France 46.39N 0.13W
97 J19 Partille Västra Götaland, S Sweden 57.43N 12.12E
109 I23 Partinico Sicilia, Italy, C Mediterranean Sea 38.03N 13.07E
113 I20 Partizánske prev. Šimonovany; Hung. Simony. Trenčiansky Kraj, W Slovakia 48.39N 18.22E
60 H12 Paru de Oeste, Rio ✦ N Brazil
190 K9 Paruna South Australia 34.45S 140.43E

39 O8 Pastora Peak ▲ Arizona, SW USA 36.48N 109.10W
107 O8 Pastrana Castilla-La Mancha, C Spain 40.24N 2.55W
174 M15 Pasuruan prev. Pasoeroean. Jawa, C Indonesia 7.37S 112.43E
120 F11 Pasvalys Pasvalys, N Lithuania 56.03N 24.42E
113 K21 Pásztó Nógrád, N Hungary 47.57N 19.41E
201 U12 Pata var. Patta. atoll Chuuk Islands, C Micronesia
38 M16 Patagonia Arizona, SW USA 31.32N 110.45W
65 H20 Patagonia physical region Argentina/Chile
160 D9 Patan Gujarāt, W India 23.51N 72.10E
160 J10 Pātan Madhya Pradesh, C India 23.54N 79.40E
175 T18 Patani Pulau Halmahera, E Indonesia 0.19N 128.46E
118 K13 Pătârlagele prev. Pătîrlagele. Buzău, SE Romania 45.19N 26.21E
61 P15 Paulo Afonso Bahia, E Brazil 9.21S 38.13W
190 I5 Patawarta Hill ▲ South Australia 30.57S 138.42E
190 L10 Patchewollock Victoria, SE Australia 35.24S 142.11E
192 K11 Patea Taranaki, North Island, NZ 39.48S 174.35E
192 K11 Patea ✦ North Island, NZ
79 U15 Pategi Kwara, C Nigeria 8.39N 5.46E
83 K20 Pate Island var. Patta Island. island SE Kenya
107 S10 Paterna País Valenciano, E Spain 39.30N 0.24W
203 N11 Pascua, Isla de var. Rapa Nui, Eng. Easter Island. island E Pacific Ocean
65 G21 Pascua, Río de la ✦ S Chile
105 N1 Pas-de-Calais ◆ department N France
102 P10 Pasewalk Mecklenburg-Vorpommern, NE Germany 53.30N 13.58E
9 T10 Pasfield Lake ◉ Saskatchewan, C Canada
Pasha-liman see Bashi Channel
120 J11 Pasinler Erzurum, NE Turkey 39.59N 41.40E
100 N6 Paterswolde Drenthe, NE Netherlands 53.07N 6.32E
158 H7 Pathānkot Himāchal Pradesh, N India 32.16N 75.43E
Pathein see Bassein
35 W15 Pathfinder Reservoir ◉ Wyoming, C USA
178 Hh1 Pathum Thani var. Patumdhani, Prathum Thani. Pathum Thani, C Thailand 14.03N 100.28E
174 L14 Pati Jawa, C Indonesia 6.45S 111.00E
56 C12 Patía var. El Bordo. Cauca, SW Colombia 2.06N 77.02W
158 I9 Patiāla var. Puttiala. Punjab, NW India 30.21N 76.27E
56 B12 Patía, Río ✦ SW Colombia
196 D15 Pati Point headland NE Guam 13.36N 144.39E
58 C13 Pativilca Lima, W Peru 10.40S 77.52W
145 S13 Patnos Ağrı, E Turkey 39.13N 42.52E
45 H12 Pato Branco Paraná, S Brazil 26.15S 52.40W
58 D8 Patoka ◉ Indiana, N USA
94 I3 Patonniva Lapp. Buoddobohki. Lappi, N Finland 69.44N 27.01E
155 W7 Pasrūr Punjab, E Pakistan 32.12N 74.42E
2 M1 Passage Island island Michigan, N USA
15 I1 Passage Point headland Banks Island, Northwest Territories, NW Canada 73.31N 115.12W
15 O4 Pastos Bons Maranhão, E Brazil
64 G13 Patos de Minas var. Patos. Minas Gerais, NE Brazil 18.34S 46.31W
63 I17 Patos, Lagoa dos lagoon S Brazil
64 J9 Patquía La Rioja, C Argentina 30.01S 66.54W
117 E19 Pátra Eng. Patras; prev. Pátrai. Dytikí Ellás, S Greece 38.13N 21.45E
Pátrai/Patras see Pátra
117 I25 Paximádia island SE Greece
Pax Julia see Beja
117 B16 Paxoí island Iónioi Nísoi, Greece, C Mediterranean Sea
41 S10 Paxson Alaska, USA 62.58N 145.27W
35 M13 Paxton Illinois, N USA 40.27N 88.06W
34 J10 Patterson Louisiana, S USA 29.41N 91.18W
193 C25 Paterson Inlet inlet Stewart Island, NZ
116 J8 Pavlikeni Veliko Tŭrnovo, N Bulgaria 43.15N 25.20E
151 T8 Pavlodar Pavlodar, NE Kazakhstan 52.21N 76.58E
151 S9 Pavlodar var. Pavlodarskaya Oblast', Kaz. Pavlodar Oblysy. ◆ province NE Kazakhstan Pavlodar Oblysy/Pavlodarskaya Oblast' see Pavlodar
174 L14 Pati Jawa, C Indonesia 6.45S 111.00E
119 U7 Pavlohrad Rus. Pavlograd. Dnipropetrovs'ka Oblast', E Ukraine 48.32N 35.50E
Pavlof Harbour see Pauloff Harbor
151 R9 Pavlovka Akmola, C Kazakhstan 51.22N 72.35E
131 V4 Pavlovka Respublika Bashkortostan, W Russian Federation 55.28N 56.36E
131 Q7 Pavlovka Ul'yanovskaya Oblast', W Russian Federation 52.49N 47.08E
131 N3 Pavlovo Nizhegorodskaya Oblast', W Russian Federation 55.59N 43.03E
130 L9 Pavlovsk Voronezhskaya Oblast', W Russian Federation 50.26N 40.08E
130 L13 Pavlovskaya Krasnodarskiy Kray, SW Russian Federation 46.06N 39.52E
119 S7 Pavlysh Kirovohrads'ka Oblast', C Ukraine 48.54N 33.28E
108 F10 Pavullo nel Frignano Emilia-Romagna, C Italy 44.19N 10.52E
29 P8 Pawhuska Oklahoma, C USA 36.40N 96.20W
23 U13 Pawleys Island South Carolina, SE USA 33.27N 79.07W
178 G7 Pawn ✦ C Myanmar
35 N9 Pawnee Illinois, S USA 39.35N 89.34W
29 O9 Pawnee Oklahoma, C USA 36.18N 96.47W
31 S17 Pawnee City Nebraska, C USA 40.06N 96.09W
28 M5 Pawnee River ✦ Kansas, C USA
31 S10 Paw Paw Michigan, N USA 42.12N 85.53W
31 S10 Paw Paw Lake Michigan, N USA 42.12N 86.16W
19 N12 Pawtucket Rhode Island, NE USA 41.52N 71.22W
Pax Augusta see Badajoz
117 I25 Paximádia island SE Greece
Pax Julia see Beja
117 B16 Paxoí island Iónioi Nísoi, Greece, C Mediterranean Sea
41 S10 Paxson Alaska, USA 62.58N 145.27W
35 M13 Paxton Illinois, N USA 40.27N 88.06W
119 V2 Pay-Khoy, Khrebet ▲ NW Russian Federation
Payne see Kangirsuk
10 K4 Payne, Lac ◉ Quebec, NE Canada
31 T8 Paynesville Minnesota, N USA 45.23N 94.42W
174 M4 Payong, Tanjung headland East Malaysia 3.46N 113.27E
Payo Obispo see Chetumal
63 D18 Paysandú Paysandú, W Uruguay 32.21S 58.04W
63 D17 Paysandú ◆ department W Uruguay
104 I7 Pays de la Loire ◆ region NW France
38 L12 Payson Arizona, SW USA 34.13N 111.19W

38 *L4* **Payson** Utah, W USA
40.02N 111.43W

129 *W4* **Payyer, Gora** ▲ NW Russian
Federation 66.49N 64.33E
Payzawat *see* Jiashi

143 *Q11* **Pazar** Rize, NE Turkey
41.10N 40.52E

142 *F10* **Pazarbaşı Burnu** *headland*
NW Turkey 41.12N 30.18E

142 *M16* **Pazarcık** Kahramanmaraş,
S Turkey 37.31N 37.19E

116 *I10* **Pazardzhik** *prev.* Tatar
Pazardzhik. Pazardzhik, C Bulgaria
42.11N 24.21E

66 *H11* **Pazardzhik** ◆ *province* C Bulgaria
42.11N 24.21E

56 *N9* **Paz de Ariporo** Casanare,
E Colombia 5.51N 71.52W

114 *A10* **Pazin** Ger. Mitterburg, *It.* Pisino.
Istra, NW Croatia 45.14N 13.56E

44 *D7* **Paz, Río** ❧
El Salvador/Guatemala

115 *O18* **Pčinja** ❧ S FYR Macedonia

200 *Oq15* **Pea** Tongatapu, S Tonga
21.10S 175.14W

29 *O6* **Peabody** Kansas, C USA
38.10N 97.06W

9 *O12* **Peace** ❧ Alberta/British
Columbia, W Canada
Peace Garden State *see* North
Dakota

9 *Q10* **Peace Point** Alberta, C Canada
59.11N 112.12W

9 *O12* **Peace River** Alberta, W Canada
56.15N 117.18W

25 *W13* **Peace River** ❧ Florida, SE USA

9 *N17* **Peachland** British Columbia,
SW Canada 49.44N 119.48W

38 *J10* **Peach Springs** Arizona, SW USA
35.33N 113.27W
Peach State *see* Georgia

25 *S4* **Peachtree City** Georgia, SE USA
33.24N 84.36W

201 *X11* **Peacock Point** *point* SE Wake
Island 19.16N 166.39E

99 *M18* **Peak District** *physical region*
C England, UK

191 *Q7* **Peak Hill** New South Wales,
SE Australia 32.39S 148.12E

117 *S15* **Peak, The** ▲ C Ascension Island

107 *O13* **Peal de Becerro** Andalucía,
S Spain 37.55N 3.07W

201 *X11* **Peale Island** *island* N Wake Island

39 *O6* **Peale, Mount** ▲ Utah, USA
38.26N 109.13W

41 *O4* **Peard Bay** *bay* Alaska, USA

25 *Q7* **Pea River** ❧ Alabama/Florida,
S USA

27 *W11* **Pearland** Texas, SW USA
29.33N 95.17W

40 *D9* **Pearl City** Oahu, Hawaii, USA,
C Pacific Ocean 21.24N 95.58W

40 *D9* **Pearl Harbor** *inlet* Oahu, Hawaii,
USA, C Pacific Ocean
Pearl Islands *see* Perlas,
Archipiélago de las
Pearl Lagoon *see* Perlas,
Laguna de

24 *M5* **Pearl River**
❧ Louisiana/Mississippi, S USA

27 *Q13* **Pearsall** Texas, SW USA
28.53N 99.05W

25 *U7* **Pearson** Georgia, SE USA
31.18N 82.51W

27 *P4* **Pease River** ❧ Texas, SW USA
34.55N 85.31W

10 *I7* **Peawanuk** Ontario, C Canada
54.55N 85.31W

85 *P16* **Pebane** Zambézia,
NE Mozambique 17.13S 38.10E

117 *C23* **Pebble Island** *island* N Falkland
Islands

117 *C23* **Pebble Island Settlement**
Pebble Island, N Falkland Islands
51.19S 59.40W

115 *L16* **Peć** *Alb.* Pejë, *Turk.* İpek. Serbia,
S Yugoslavia 42.40N 20.19E

27 *R8* **Pecan Bayou** ❧ Texas, SW USA

24 *H10* **Pecan Island** Louisiana, USA
29.39N 92.36W

62 *L12* **Peças, Ilha das** *island* S Brazil

32 *L10* **Pecatonica River**
❧ Illinois/Wisconsin, N USA

110 *G10* **Peccia** Ticino, S Switzerland
46.24N 8.39E
Pechenegi *see* Pechenihy
Pechenezhskoye
Vodokhranilishche
see Pechenizьke Vodokhvyshche

128 *I2* **Pechenga** *Fin.* Petsamo.
Murmanskaya Oblast',
NW Russian Federation
69.34N 31.14E

119 *V5* **Pechenihy** *Rus.* Pechenegi.
Kharkivs'ka Oblast', E Ukraine
49.49N 36.57E

119 *V5* **Pechenizьke**
Vodoskhovyshche *Rus.*
Pechenezhskoye
Vodokhranilishche. ◻ E Ukraine

129 *U7* **Pechora** Respublika Komi,
NW Russian Federation
65.08N 57.09E

129 *R6* **Pechora** ❧ NW Russian
Federation
Pechora Bay *see* Pechorskaya
Guba
Pechora Sea *see* Pechorskoye
More

129 *S3* **Pechorskaya Guba** *Eng.* Pechora
Bay. *bay* NW Russian Federation

125 *FJ6* **Pechorskoye More** *Eng.* Pechora
Sea. *sea* NW Russian Federation

118 *E11* **Pecica** *Ger.* Petschka, *Hung.*
Ópécska. Arad, W Romania
46.09N 21.06E

26 *K8* **Pecos** Texas, SW USA
31.25N 103.30W

27 *N11* **Pecos River** ❧ New
Mexico/Texas, SW USA

113 *I25* **Pécs** *Ger.* Fünfkirchen; *anc.*
Sopianae. Baranya, SW Hungary
46.04N 18.11E

45 *T17* **Pedasí** Los Santos, S Panama
7.30N 80.02W
Pedde *see* Pedja

45 *O17* **Pedder, Lake** ◻ Tasmania,
SE Australia

45 *M10* **Pedernales** SW Dominican
Republic

57 *Q5* **Pedernales** Delta Amacuro,
NE Venezuela 9.58N 62.14W

27 *R10* **Pedernales River** ❧ Texas,
SW USA

64 *H6* **Pedernales, Salar de** *salt lake*
N Chile
Pedhoulás *see* Pedoulás

57 *X11* **Pédima** *var.* Malavate. SW French
Guiana 3.15N 54.07W

190 *F1* **Pedirka** South Australia
26.41S 135.11E

175 *T7* **Pediwang** Pulau Halmahera,
E Indonesia 1.29N 127.57E

120 *I5* **Pedja** *var.* Pedja Jõgi, *Ger.* Pedde.
❧ E Estonia
Pedja Jõgi *see* Pedja

124 *N3* **Pedoulás** *var.* Pedhoulas.
W Cyprus 34.58N 32.51E

61 *N18* **Pedra Azul** Minas Gerais,
NE Brazil 16.01S 41.16W

106 *I3* **Pedrafita, Porto de** *var.* Puerto
de Piedrafita. *pass* NW Spain
42.43N 7.01W

78 *E9* **Pedra Lume** Sal, NE Cape Verde
16.46N 22.54W

45 *P16* **Pedregal** Chiriquí, W Panama
8.21N 82.26W

56 *J4* **Pedregal** Falcón, N Venezuela
11.01N 70.06W

42 *L9* **Pedriceña** Durango, C Mexico
25.08N 103.46W

62 *L11* **Pedro Barros** São Paulo, S Brazil
24.12S 47.22W

41 *Q13* **Pedro Bay** Alaska, USA
59.47N 154.06W

64 *H4* **Pedro de Valdivia** *var.* Oficina
Pedro de Valdivia. Antofagasta,
N Chile 22.33S 69.37W

64 *P4* **Pedro Juan Caballero**
Amambay, E Paraguay 22.33S 55.40W

65 *L15* **Pedro Luro** Buenos Aires,
E Argentina 39.26S 62.40W

107 *O10* **Pedro Muñoz** Castilla-La
Mancha, C Spain 39.25N 2.55W

161 *J22* **Pedro, Point** *headland* NW Sri
Lanka 9.54N 80.08E

190 *K9* **Pedrina** South Australia
34.56S 140.56E

98 *J13* **Peebles** SE Scotland, UK
55.39N 3.14W

35 *S15* **Peebles** Ohio, N USA
38.57N 83.23W

98 *J12* **Peebles** *cultural region* SE Scotland,
UK

20 *K13* **Peekskill** New York, NE USA
41.17N 73.54W

98 *F14* **Peel** W Isle of Man 54.13N 4.04W

14 *FJ4* **Peel** ❧ Northwest
Territories/Yukon Territory,
NW Canada

15 *I1* **Peel Point** *headland* Victoria
Island, Northwest Territories,
NW Canada 73.22N 114.33W

15 *K1* **Peel Sound** *passage* Nunavut,
N Canada

102 *N9* **Peene** ❧ NE Germany

101 *K17* **Peer** Limburg, NE Belgium
51.08N 5.28E

12 *H4* **Pefferlaw** Ontario, S Canada
44.18N 79.11W

193 *I18* **Pegasus Bay** *bay* South Island,
NZ

123 *Mm3* **Pégeia** *var.* Peyia. SW Cyprus
34.52N 32.24E

111 *V7* **Peggau** Steiermark, SE Austria
47.11N 15.20E

103 *L19* **Pegnitz** Bayern, SE Germany
49.45N 11.33E

103 *L19* **Pegnitz** ❧ SE Germany

107 *T11* **Pego** País Valenciano, E Spain
38.51N 0.07W

177 *G8* **Pegu** *var.* Bago. Pegu,
SW Myanmar 17.18N 96.31E

177 *G7* **Pegu** ◆ *division* S Myanmar

176 *W7* **Pegun, Pulau** *island* Kepulauan
Mapia, E Indonesia

201 *N13* **Pehleng** Pohnpei, E Micronesia

116 *M12* **Pehlivanköy** Kırklareli,
NW Turkey 41.21N 26.56E

79 *R14* **Péhonko** C Benin 10.14N 1.57E

63 *B21* **Pehuajó** Buenos Aires,
E Argentina 35.48S 61.52W

102 *J13* **Pei-ching** *see* Beijing/Beijing Shi
Peine Niedersachsen, C Germany
52.19N 10.14E
Pei-p'ing *see* Beijing/Beijing Shi
Peipsi Järv/Peipus-See *see*
Peipus, Lake

120 *J5* **Peipus, Lake** *Est.* Peipsi Järv, *Ger.*
Peipus-See, *Rus.* Chudskoye Ozero.
◻ Estonia/Russian Federation

117 *H19* **Peiraiás** *prev.* Piraiévs, *Eng.*
Piraeus. Attikí, C Greece
37.56N 23.39E
Peisern *see* Pyzdry

62 *J8* **Peixe, Rio do** ❧ S Brazil

61 *I16* **Peixoto de Azevedo** Mato
Grosso, W Brazil 10.18S 55.03W

174 *J8* **Pejantan, Pulau** *island*
W Indonesia
Pejé *see* Peć

114 *N12* **Pek** ❧ E Yugoslavia

178 *I7* **Pèk** *var.* Xieng Khouang; *prev.*
Xiangkhoang, Xiangkhoang,
N Laos 19.19N 103.23E

174 *Kk14* **Pekalongan** Jawa, C Indonesia
6.54S 109.37E

174 *Gg7* **Pekanbaru** *var.* Pakanbaru.
Sumatera, W Indonesia
0.31N 101.27E

32 *L12* **Pekin** Illinois, N USA
40.34N 89.38W
Peking *see* Beijing/Beijing Shi
**Pelabohan Kelang/Pelabuhan
Kelang** *see* Pelabuhan Klang

173 *G5* **Pelabuhan Klang** *var.* Kuala
Pelabohan Kelang, Pelabohan
Kelang, Pelabuhan Kelang, Port
Klang, Port Swettenham. Selangor,
Peninsular Malaysia 2.57N 101.24E

174 *Kk14* **Pelabuhan Ratu, Teluk** *bay*
Jawa, SW Indonesia

123 *L12* **Pelagie, Isole** *island group*
SW Italy
Pelagosa *see* Palagruža

24 *L5* **Pelahatchie** Mississippi, S USA
32.19N 89.48W

175 *N11* **Pelaihari** *var.* Pleihari. Borneo,
C Indonesia 3.48S 114.45E

105 *U14* **Pelat, Mont** ▲ SE France
44.16N 6.46E

118 *F12* **Peleaga, Vârful** *prev.* Vîrful
Peleaga. ▲ W Romania 45.23N 22.52E
Peleaga, Vîrful *see* Peleaga,
Vârful

126 *K12* **Peleduy** Respublika Sakha
(Yakutiya), NE Russian Federation
59.39N 112.36E

12 *C18* **Pelee Island** *island* Ontario,
S Canada

45 *Q11* **Pelée, Montagne**
▲ N Martinique 14.47N 61.10W

12 *D18* **Pelee, Point** *headland* Ontario,
S Canada 41.56N 82.30W

175 *R9* **Pelei** Pulau Peleng, N Indonesia
1.26S 123.27E

175 *R9* **Peleliu** *see* Beliliou

175 *Qq9* **Peleng, Pulau** *island* Kepulauan
Banggai, N Indonesia

175 *Qq9* **Peleng, Selat** *strait* Sulawesi,
C Indonesia

25 *T7* **Pelham** Georgia, SE USA
31.07N 84.09W

113 *E18* **Pelhřimov** *Ger.* Pilgram.
Jihlavský Kraj, C Czech Republic
49.25N 15.13E

41 *W13* **Pelican** Chicagof Island, Alaska,
USA 57.52N 136.05W

203 *Z3* **Pelican Lagoon** ◻ Kiritimati,
E Kiribati

31 *U6* **Pelican Lake** ◻ Minnesota,
N USA

31 *V3* **Pelican Lake** ◻ Minnesota,
N USA

32 *L5* **Pelican Lake** ◻ Wisconsin,
N USA

46 *G1* **Pelican Point** Grand Bahama
Island, N Bahamas 26.39N 78.09W

85 *B19* **Pelican Point** *headland*
W Namibia 22.55S 14.25E

31 *S6* **Pelican Rapids** Minnesota,
N USA 46.34N 96.04W
Pelican State *see* Louisiana

9 *U13* **Pelican Narrows** Saskatchewan,
C Canada 55.11N 102.51W

117 *L18* **Pelinnaío** ▲ Chíos, E Greece
38.31N 26.01E

117 *L16* **Pelinnaion** *anc.* Pelinnaio
Pelinnaíon *anc.* Pelinnaeum; *anc.*
Thessalía, C Greece 39.33N 21.45E

115 *N20* **Pelister** ▲ SW FYR Macedonia
41.00N 21.12E

115 *G15* **Pelješac** *peninsula* S Croatia

94 *M12* **Pelkosenniemi** Lappi,
NE Finland 67.06N 27.30E

31 *S3* **Pella** Iowa, C USA 41.24N 92.55W

116 *F13* **Pélla** *site of ancient city* Kentrikí
Makedonía, N Greece 40.46N 22.35E

25 *Q3* **Pell City** Alabama, S USA
33.35N 86.17W

63 *A22* **Pellegrini** Buenos Aires,
E Argentina 36.16S 63.07W

94 *K12* **Pello** Lappi, NW Finland
66.47N 24.00E

102 *G7* **Pellworm** *island* N Germany

8 *H6* **Pelly** ❧ Yukon Territory,
NW Canada

15 *L3* **Pelly Bay** Nunavut, N Canada
68.37N 89.45W

8 *I8* **Pelly Mountains** ▲ Yukon
Territory, W Canada
Pélmonostor *see* Beli Manastir

59 *P13* **Pelona Mountain** ▲ New
Mexico, SW USA 33.40N 108.06W

42 *M13* **Pelotas** Rio Grande do Sul,
S Brazil 31.45S 52.19W

63 *K14* **Pelotas, Rio** ❧ S Brazil

125 *F10* **Pelym** ❧ C Russian Federation

21 *R4* **Pemadumcook Lake** ◻ Maine,
NE USA

174 *Kk14* **Pemalang** Jawa, C Indonesia
6.52S 109.07E

174 *K7* **Pemangkat** *var.* Pamangkat.
Borneo, C Indonesia 1.11N 109.00E
Pemar *see* Paimio

173 *Ff5* **Pematangsiantar** Sumatera,
W Indonesia 2.58N 99.01E

85 *Q14* **Pemba** *prev.* Port Amelia, Porto
Amelia. Cabo Delgado,
NE Mozambique 13.00S 40.30E

83 *J22* **Pemba** ◆ *region* E Tanzania

83 *J21* **Pemba** *island* E Tanzania

85 *Q14* **Pemba, Baía de** *inlet*
NE Mozambique

83 *J21* **Pemba Channel** *channel*
E Tanzania

188 *I14* **Pemberton** Western Australia
34.27S 116.09E

9 *M16* **Pemberton** British Columbia,
SW Canada 50.19N 122.49W

31 *Q2* **Pembina** North Dakota, N USA
48.58N 97.14W

31 *Q2* **Pembina** ❧ Canada/USA

9 *P15* **Pembina** ❧ Alberta, SW Canada

176 *Xx15* **Pembre** Irian Jaya, E Indonesia
7.49S 138.01E

12 *K12* **Pembroke** Ontario, SE Canada
45.49N 77.07W

99 *H21* **Pembroke** SW Wales, UK
51.40N 4.55W

25 *W6* **Pembroke** Georgia, SE USA
32.09N 81.35W

23 *U11* **Pembroke** North Carolina,
SE USA 34.40N 79.12W

23 *R7* **Pembroke** Virginia, SE USA
37.19N 80.38W

99 *H21* **Pembrokeshire** *cultural region*
SW Wales, UK

45 *S15* **Peña Blanca, Cerro**
▲ C Panama 8.36N 80.39W

106 *K8* **Peña de Francia, Sierra de la**
▲ W Spain

106 *G6* **Peñafiel** *var.* Peñafiel. Porto,
N Portugal 41.12N 8.16W

106 *M6* **Peñafiel** Castilla-León, N Spain
41.36N 4.07W

107 *S8* **Peñagolosa** ▲ E Spain
40.10N 0.15W

107 *N7* **Peñalara, Pico de** ▲ C Spain
40.51N 3.57W

175 *Nn5* **Penambo, Banjaran** *var.*
Banjaran Tama Abu, Penambo
Range. ▲ Indonesia/Malaysia
Penambo Range *see* Penambo,
Banjaran

43 *O10* **Peña Nevada, Cerro**
▲ C Mexico 23.46N 99.52W

62 *J8* **Penápolis** São Paulo, S Brazil
21.23S 50.02W
Penang *see* Pinang, Pulau,
Peninsular Malaysia
Penang *see* George Town

106 *C7* **Peñaranda de Bracamonte**
Castilla-León, N Spain 40.54N 5.13W

107 *S8* **Peñarroya** ▲ E Spain 40.24N 0.42W

106 *L12* **Peñarroya-Pueblonuevo**
Andalucía, S Spain 38.21N 5.16W

99 *K22* **Penarth** S Wales, UK 51.27N 3.10W

106 *K1* **Peñas, Cabo de** *headland* N Spain
43.39N 5.52W

65 *F20* **Penas, Golfo de** *gulf* S Chile
Pen-ch'i *see* Benxi

81 *H14* **Pendé** *var.* Logone Oriental.
❧ Central African Republic/Chad

78 *I14* **Pendembu** E Sierra Leone
9.06N 12.12W

31 *R13* **Pender** Nebraska, C USA
42.06N 96.42W

34 *K4* **Pendleton** Oregon, NW USA
45.40N 118.47W

34 *M7* **Pend Oreille, Lake** ◻ Idaho,
NW USA

34 *M7* **Pend Oreille River**
❧ Idaho/Washington, NW USA
Pendzhikent *see* Panjakent
Peneius *see* Pineiós

106 *G8* **Penela** Coimbra, N Portugal
40.01N 8.22W

12 *G13* **Penetanguishene** Ontario,
S Canada 44.45N 79.55W

157 *H15* **Penganga** ❧ C India

167 *T12* **P'enghua Yu** *island* N Taiwan

81 *M21* **Penge** Kasai Oriental, C Dem.
Rep. Congo (Zaire) 5.29S 24.38E
**Penghu Archipelago/P'enghu
Ch'üntao/Penghu Islands** *see*
P'enghu/Liehtao

167 *R14* **P'enghu Liehtao** *var.* P'enghu
Ch'üntao, Penghu Islands, *Eng.*
Penghu Archipelago, Pescadores,
Jap. Hoko-guntō, Hoko-shotō.
island group W Taiwan

167 *R14* **Penghu Shuidao/P'enghu
Shuitao** *see* Pescadores Channel

167 *R14* **Penglai** *var.* Dengzhou.
Shandong, E China 37.48N 120.43E
Peng-pu *see* Bengbu
Penhsihu *see* Benxi
Penibético, Sistema *see* Béticos,
Sistemas

106 *F10* **Peniche** Leiria, W Portugal
39.21N 9.22W

98 *K11* **Penicuik** SE Scotland, UK
55.49N 3.13W

175 *Nn16* **Penida, Nusa** *island* S Indonesia

107 *T8* **Peñíscola** País Valenciano,
E Spain 40.22N 0.24E

42 *M13* **Pénjamo** Guanajuato, C Mexico
20.20N 101.35W
Penki *see* Benxi

104 *F7* **Penmarch, Pointe de** *headland*
NW France 47.46N 4.34W

109 *L15* **Penna, Punta della** *headland*
C Italy 42.10N 14.43E

118 *L7* **Penne** Abruzzo, C Italy
42.28N 13.57E

161 *J18* **Penner** *var.* Penner. ❧ C India

190 *J10* **Penneshaw** South Australia
35.45S 137.57E

20 *C14* **Penn Hills** Pennsylvania,
NE USA 40.28N 79.52W

20 *I16* **Penns Grove** New Jersey,
NE USA 39.42N 75.28W

20 *I16* **Pennsville** New Jersey, NE USA
39.37N 75.29W

20 *E14* **Pennsylvania** *off.*
Commonwealth of Pennsylvania;
also known as The Keystone State.
◆ *state* NE USA

20 *G10* **Penn Yan** New York, NE USA
42.39N 77.02W

128 *H16* **Peno** Tverskaya Oblast',
W Russian Federation 56.55N 32.44E

21 *R7* **Penobscot Bay** *bay* Maine,
NE USA

21 *S5* **Penobscot River** ❧ Maine,
NE USA

190 *K12* **Penola** South Australia
37.24S 140.50E

190 *F7* **Penong** South Australia
31.57S 133.01E

45 *R14* **Penonomé** Coclé, C Panama
8.29N 80.21W

202 *L13* **Penrhyn** *atoll* N Cook Islands

199 *Kk10* **Penrhyn Basin** *undersea feature*
C Pacific Ocean

191 *S9* **Penrith** New South Wales,
SE Australia 33.45S 150.48E

99 *K15* **Penrith** NW England, UK
54.40N 4.33W

25 *O9* **Pensacola** Florida, SE USA
30.25N 87.13W

25 *O9* **Pensacola Bay** *bay* Florida,
SE USA

205 *N7* **Pensacola Mountains**
▲ Antarctica

190 *M12* **Penshurst** Victoria, SE Australia
37.54S 142.19E

197 *C12* **Pentecost** *Fr.* Pentecôte. *island*
C Vanuatu

13 *O7* **Pentecôte, Lac** ◻ Quebec,
SE Canada
Pentecôte *see* Pentecost

15 *Gg16* **Penticton** British Columbia,
SW Canada 49.28N 119.37W

96 *J6* **Pentland Firth** *strait* N Scotland,
UK

98 *J12* **Pentland Hills** *hill range*
S Scotland, UK

155 *H17* **Penukonda** Andhra Pradesh,
E India 14.04N 77.38E

155 *H18* **Penu** Pulau Taliabu, E Indonesia
1.43S 125.09E

173 *G7* **Penwegon** Pegu, C Myanmar
18.13N 96.34E

26 *M8* **Penwell** Texas, SW USA
31.45N 102.32W

99 *J21* **Pen y Fan** ▲ SE Wales, UK
51.52N 3.25W

99 *L16* **Pen-y-ghent** ▲ N England, UK
54.11N 2.15W

175 *T12* **Penyu, Kepulauan** *island group*
E Indonesia

131 *N6* **Penza** Penzenskaya Oblast',
W Russian Federation 53.11N 45.00E

99 *G25* **Penzance** SW England, UK
50.07N 5.33W

131 *N6* **Penzenskaya Oblast'** ◆ *province*
W Russian Federation

127 *P7* **Penzhina** ❧ E Russian
Federation

127 *N17* **Penzhinskaya Guba** *bay*
E Russian Federation

38 *K13* **Peoria** Arizona, SW USA
33.34N 112.14W

32 *L12* **Peoria** Illinois, N USA
40.41N 89.35W

32 *L12* **Peoria Heights** Illinois, N USA
40.45N 89.34W

35 *N11* **Peotone** Illinois, N USA
41.19N 87.47W

20 *J11* **Pepacton Reservoir** ◻ New
York, NE USA

78 *I15* **Pepel** W Sierra Leone 8.39N 13.04W

32 *J6* **Pepin, Lake** ◻
Minnesota/Wisconsin, N USA

101 *L20* **Pepinster** Liège, E Belgium
50.34N 5.49E

115 *L20* **Peqin** *var.* Peqini. Elbasan,
C Albania 41.03N 19.46E
Peqini *see* Peqin

174 *Gg4* **Perak** ◆ *state* Peninsular Malaysia

174 *Gg4* **Perak, Sungai** ❧ Peninsular
Malaysia

107 *N7* **Perales del Alfambra** Aragón,
NE Spain 40.38N 1.00W

117 *C15* **Pérama** *var.* Perama. Ípeiros,
W Greece 39.42N 20.51E

94 *M13* **Parä-Posio** Lappi, NE Finland
66.10N 27.56E

13 *Z6* **Percé** Quebec, SE Canada
48.31N 64.15W

13 *Z6* **Percé, Rocher** *island* Quebec,
SE Canada

104 *L5* **Perche, Collines de** ▲ N France

203 *W16* **Percival Lakes** *lakes* Western
Australia

107 *T3* **Perdido, Monte** ▲ NE Spain
42.41N 0.04E

25 *O8* **Perdido River**
❧ Alabama/Florida, S USA

104 *F7* **Penmarch, Pointe de** *headland*
NW France 47.46N 4.34W
Perece Vela Basin *see* West
Mariana Basin

119 *L6* **Perechyn** Zakarpats'ka Oblast',
W Ukraine 48.45N 22.28E

56 *C10* **Pereira** Risaralda, W Colombia
4.52N 75.48W

58 *E12* **Pereira Barreto** São Paulo,
S Brazil 20.37S 51.07W

61 *G15* **Pereirinha** Pará, N Brazil
8.18S 57.30W

131 *N10* **Perelazovskiy** Volgogradskaya
Oblast', SW Russian Federation
49.10N 42.30E

131 *S7* **Perelyub** Saratovskaya Oblast',
W Russian Federation 51.52N 50.19E
Peremyshl *see* Podkarpackie

35 *P7* **Pere Marquette River**
❧ Michigan, N USA
Pennine Chain *see* Pennines

99 *L15* **Pennines** *see* Pennine Chain.
▲ N England, UK
Pennines, Alpes *see* Pennine
Alps

23 *O8* **Peremyshl'** *var.* Podkarpackie
Oblast', SW Russian Federation
Peremyshlyany L'viv's'ka Oblast',
W Ukraine 49.42N 24.33E

118 *L9* **Pereshchepyne** Dnipropetrovs'ka
Oblast', E Ukraine
48.38N 34.54E
Pereshchepino *see*
Pereshchepyne
Pereshchepyne *Rus.*
Pereshchepino. Dnipropetrovs'ka
Oblast', E Ukraine 48.38N 34.54E

128 *L16* **Pereslavl'-Zalesskiy**
Yaroslavskaya Oblast', W Russian
Federation 56.42N 38.45E
Pereval's'k *see* Podkarpackie
Pereval's'k Luhans'ka Oblast',
E Ukraine 48.28N 38.54E

119 *Y7* **Pereval's'k** Luhans'ka Oblast',
E Ukraine 48.28N 38.54E

131 *U7* **Perevolotskiy** Orenburgskaya
Oblast', W Russian Federation
51.52N 54.05E

127 *N16* **Pereyaslavka** Khabarovskiy Kray,
SE Russian Federation
47.49N 134.56E
Pereyaslav-Khmel'nitskiy *see*
Pereyaslav-Khmel'nyts'kyy

119 *Q5* **Pereyaslav-Khmel'nyts'kyy**
Rus. Pereyaslav-Khmel'nitskiy.
Kyyivs'ka Oblast', N Ukraine
50.04N 31.28E

117 *T5* **Perg** Oberösterreich, N Austria
48.15N 14.39E

63 *B19* **Pergamino** Buenos Aires,
E Argentina 33.53S 60.37W

108 *G6* **Pergine Valsugana** *Ger.* Persen.
Trentino-Alto Adige, N Italy
46.04N 11.13E

31 *S6* **Perham** Minnesota, N USA
46.35N 95.34W

95 *L16* **Perho** Länsi-Suomi, W Finland
63.15N 24.25E

118 *E11* **Periam** *Ger.* Perjamosch, *Hung.*
Perjámos. Timiş, W Romania
46.01N 20.54E

13 *O5* **Péribonca** ❧ Quebec,
SE Canada

10 *L11* **Péribonca, Lac** ◻ Quebec,
SE Canada

13 *O5* **Péribonca, Petite Rivière**
❧ Quebec, SE Canada

42 *G7* **Perico** Jujuy, N Argentina
24.25N 65.07W

42 *G7* **Pericos** Sinaloa, C Mexico
25.04N 107.40W

13 *O6* **Périgueux** *anc.* Vesuna.
Dordogne, SW France 45.12N 0.41E

56 *C6* **Perijá, Serranía de**
▲ Colombia/Venezuela

117 *H17* **Peristéra** *island* Vóreioi Sporádes,
Greece, Aegean Sea

58 *C12* **Perito Moreno** Santa Cruz,
S Argentina 46.35S 71.00W

161 *G22* **Periyār** *var.* Periyar. ❧ SW India

161 *H22* **Periyār Lake** ◻ S India

63 *D15* **Perugorría** Corrientes,
NE Argentina 29.21S 58.34W

62 *M11* **Peruíbe** São Paulo, S Brazil
24.18S 47.01W

161 *B21* **Perumalpar** *reef* India, N Indian
Ocean
Perusia *see* Perugia

101 *D20* **Péruwelz** Hainaut, SW Belgium
50.30N 3.34E

143 *R17* **Pervari** Siirt, SE Turkey
37.58N 42.30E

131 *O4* **Pervomaysk** Nizhegorodskaya
Oblast', W Russian Federation
54.52N 43.49E

119 *X7* **Pervomays'k** Luhans'ka Oblast',
E Ukraine 48.37N 38.36E

119 *P8* **Pervomays'k** *prev.* Ol'viopol'.
Mykolayivs'ka Oblast', S Ukraine
48.01N 30.51E

119 *S12* **Pervomays'ke** Respublika Krym,
S Ukraine 45.43N 33.49E

131 *V7* **Pervomayskiy** Orenburgskaya
Oblast', W Russian Federation
54.54N 56.58E

130 *M6* **Pervomayskiy** Tambovskaya
Oblast', W Russian Federation
53.15N 40.20E

119 *V6* **Pervomays'kyy** Kharkivs'ka
Oblast', E Ukraine 49.24N 36.12E

125 *E10* **Pervoural'sk** Sverdlovskaya
Oblast', C Russian Federation
56.59N 59.58E

127 *Pp12* **Pervyy Kuril'skiy Proliv** *strait*
E Russian Federation

101 *I19* **Perwez** Wallon Brabant,
C Belgium 50.39N 4.49E

108 *I11* **Pesaro** *anc.* Pisaurum. Marche,
C Italy 43.55N 12.52E

37 *N9* **Pescadero** California, W USA
37.15N 122.23W
Pescadores *see* P'enghu Liehtao

167 *S14* **Pescadores Channel** *var.*
Penghu Shuidao, P'enghu Shuitao.
channel W Taiwan

109 *K14* **Pescara** *anc.* Aternum, Ostia
Aterni. Abruzzo, C Italy
42.28N 14.13E

109 *K15* **Pescara** ❧ C Italy

118 *I11* **Pescia** Toscana, C Italy
43.54N 10.40E

110 *C8* **Peseux** Neuchâtel, W Switzerland
46.58N 6.52E

129 *P6* **Pesha** ❧ NW Russian Federation

155 *T5* **Peshāwar** North-West Frontier
Province, N Pakistan 34.00N 71.33E

155 *T6* **Peshāwar** ✈ North-West Frontier
Province, N Pakistan 34.01N 71.40E

115 *M19* **Peshkopi** *var.* Peshkopia,
Peshkopija. Dibër, NE Albania
41.40N 20.25E
Peshkopia/Peshkopija *see*
Peshkopi

116 *I11* **Peshtera** Pazardzhik,
C Bulgaria 42.01N 24.18E

35 *N6* **Peshtigo** Wisconsin, N USA
45.04N 87.43W

35 *N6* **Peshtigo River** ❧ Wisconsin,
N USA
Peski *see* Pyeski

129 *S13* **Peskovka** Kirovskaya Oblast',
NW Russian Federation
58.54N 52.17E

105 *S8* **Pesmes** Haute-Saône, E France
47.17N 5.33E

106 *H6* **Peso da Régua** *var.* Pêso da
Regua. Vila Real, N Portugal
41.10N 7.46W

42 *F5* **Pesqueira** Sonora, NW Mexico
29.22N 110.52W

104 *J13* **Pessac** Gironde, SW France
44.46N 0.42W

113 *J23* **Pest** *off.* Pest Megye. ◆ *county*
C Hungary

128 *J14* **Pestovo** Novgorodskaya Oblast',
W Russian Federation 58.37N 35.48E

42 *M15* **Petacalco, Bahía** *bay* W Mexico
Petach-Tikva/Petah Tiqwa *see*
Petah Tiqwa

144 *F10* **Petah Tiqwa** *var.* Petach-Tikva,
Petah Tiqva, Petakh Tikva. Tel
Aviv, C Israel 32.04N 34.52E
Petah Tikva *see* Petah Tiqwa

95 *L13* **Petäjävesi** Länsi-Suomi, W
Finland 62.16N 25.10E

24 *M7* **Petal** Mississippi, S USA
31.21N 89.15W

117 *I19* **Petalioí** *island* C Greece

117 *H19* **Petalión, Kólpos** *gulf* E Greece

117 *J18* **Pétalo** ▲ Ándros, Kykládes,
Greece, Aegean Sea 37.51N 24.50E

36 *M8* **Petaluma** California, W USA
38.15N 122.37W

101 *L25* **Pétange** Luxembourg,
SW Luxembourg 49.33N 5.52E

56 *M5* **Petare** Miranda, N Venezuela
10.29N 66.47W

43 *R17* **Petatlán** Guerrero, S Mexico
17.31N 101.16W

12 *J12* **Petawawa** Ontario, SE Canada
45.53N 77.16W

12 *J11* **Petawawa** ❧ Ontario, SE Canada

172 *D2* **Petchaburi** *see* Phetchaburi

188 *D2* **Petén** *off.* Departamento del
Petén. ◆ *department* N Guatemala

12 *L13* **Peten** Ontario, S Canada
44.54N 76.15W

44 *D2* **Petén Itzá, Lago** *var.* Lago de
Flores. ◻ N Guatemala

32 *K7* **Petenwell Lake** ◻ Wisconsin,
N USA

12 *J14* **Peterborough** Ontario,
SE Canada 44.19N 78.19W

99 *O20* **Peterborough** *prev.*
Medeshamstede. E England, UK
52.34N 0.15W

21 *N10* **Peterborough** New Hampshire,
NE USA 42.51N 71.54W

98 *I5* **Peterhead** NE Scotland, UK
57.30N 1.46W

199 *Mm16* **Peter I Island** ◇ *Norwegian
dependency* Antarctica

204 *H9* **Peter I Øy** *var.* Peter I Island.
island Antarctica
Peter I Øy *see* Peter I Island

99 *M14* **Peterlee** N England, UK
54.45N 1.18W
Peterlingen *see* Payerne

207 *P14* **Petermann Bjerg**
▲ C Greenland 73.16N 27.59W

◆ COUNTRY
● COUNTRY CAPITAL
◇ DEPENDENT TERRITORY
○ DEPENDENT TERRITORY CAPITAL
◆ ADMINISTRATIVE REGION
✕ INTERNATIONAL AIRPORT
▲ MOUNTAIN
▲ MOUNTAIN RANGE
▲ VOLCANO
❧ RIVER
◻ LAKE
◻ RESERVOIR

309

9 S12 **Peter Pond Lake**
◎ Saskatchewan, C Canada
41 X13 **Petersburg** Mytkof Island,
Alaska, USA 56.43N 132.51W
32 K13 **Petersburg** Illinois, N USA
40.01N 89.52W
35 N16 **Petersburg** Indiana, N USA
38.30N 87.16W
31 Q3 **Petersburg** North Dakota,
N USA 47.59N 97.59W
27 N5 **Petersburg** Texas, SW USA
33.52N 101.36W
23 V7 **Petersburg** Virginia, NE USA
37.13N 77.24W
23 T4 **Petersburg** West Virginia,
NE USA 38.59N 79.07W
102 H12 **Petershagen** Nordrhein-
Westfalen, NW Germany
52.22N 8.58E
57 S9 **Peters Mine** var. Peter's Mine.
N Guyana 6.13N 59.18W
109 O21a **Petilia Policastro** Calabria,
SW Italy 39.07N 16.48E
46 M9 **Pétionville** S Haiti 18.29N 72.16W
47 X6 **Petit-Bourg** Basse Terre,
C Guadeloupe 16.11N 61.34W
13 Y5 **Petit-Cap** Quebec, SE Canada
49.00N 64.26W
47 Y6 **Petit Cul-de-Sac Marin** bay
C Guadeloupe
10 K7 **Petite Rivière de la Baleine**
⋄ Quebec, NE Canada
46 M9 **Petite-Rivière-de-l'Artibonite**
C Haiti 19.06N 72.28W
181 X16 **Petite Rivière Noire, Piton de
la** ▲ C Mauritius
15 O2 **Petite-Rivière-St-François**
Quebec, SE Canada 47.18N 70.34W
46 L9 **Petit-Goâve** S Haiti 18.23N 72.51W
Petitjean see Sidi-Kacem
11 N10 **Petit Lac Manicouagan**
⋄ Quebec, E Canada
21 T7 **Petit Manan Point** headland
Maine, NE USA 44.23N 67.54W
Petit Mécatina, Rivière du see
Little Mecatina
9 N10 **Petitot** ⋄ Alberta/British
Columbia, W Canada
47 S12 **Petit Piton** ▲ SW Saint Lucia
13.49N 61.03W
Petit-Popo see Aného
Petit St-Bernard, Col du see
Little Saint Bernard Pass
11 O8 **Petitsikapau Lake**
⋄ Newfoundland, E Canada
94 L11 **Petkula** Lappi, N Finland
67.40N 26.43E
43 X12 **Peto** Yucatán, SE Mexico
20.09N 88.55W
64 G10 **Petorca** Valparaíso, C Chile
32.13S 70.49W
35 Q5 **Petoskey** Michigan, N USA
45.51N 88.03W
144 G14 **Petra** archaeological site Ma'ān,
W Jordan 30.19N 35.25E
Petra see Wādī Mūsā
117 F14 **Pétras, Sténa** pass N Greece
40.12N 22.15E
127 Nn18 **Petra Velikogo, Zaliv** bay
SE Russian Federation
Petrel see Petrer
12 K15 **Petre, Point** headland Ontario,
SE Canada 43.49N 77.09W
107 S12 **Petrer** var. Petrel. País Valenciano,
E Spain 38.28N 0.46W
129 U11 **Petretsovo** Permskaya Oblast',
NW Russian Federation
61.22N 57.21E
116 G12 **Petrich** Blagoevgrad, SW Bulgaria
41.24N 23.12E
197 H3 **Petrie, Récif** reef N New
Caledonia
39 N11 **Petrified Forest** prehistoric site
Arizona, SW USA 35.10N 109.49W
Petrikau see Piotrków Trybunalski
Petrikov see Pyetrykaw
118 H12 **Petrila** Hung. Petrilla.
Hunedoara, W Romania
45.27N 23.25E
Petrilla see Petrila
114 E9 **Petrinja** Sisak-Moslavina,
C Croatia 45.27N 16.14E
Petroaleksandrovsk see Türtkül
Petröcz see Bački Petrovac
128 G12 **Petrodvorets** Fin. Pietarhovi.
Leningradskaya Oblast',
NW Russian Federation
59.52N 29.52E
Petrograd see Sankt-Peterburg
Petrokov see Piotrków Trybunalski
56 G6 **Petrólea** Norte de Santander,
NE Colombia 8.30N 72.34W
12 D16 **Petrolia** Ontario, S Canada
42.54N 82.07W
21 S4 **Petrolia** Texas, SW USA
34.00N 98.13W
61 O15 **Petrolina** Pernambuco, E Brazil
9.22S 40.30W
47 T6 **Petrona, Punta** headland
C Puerto Rico 17.57N 66.23W
Petropavl see Petropavlovsk
119 V7 **Petropavlivka** Dnipropetrovs'ka
Oblast', E Ukraine 48.28N 36.28E
151 P6 **Petropavlovsk** Kaz. Petropavl.
Severnyy Kazakhstan,
N Kazakhstan 54.46N 69.06E
127 Pp11 **Petropavlovsk-Kamchatskiy**
Kamchatskaya Oblast', E Russian
Federation 53.03N 158.43E
62 P9 **Petrópolis** Rio de Janeiro,
SE Brazil 22.30S 43.28W
118 H12 **Petroşani** var. Petroşeni, Ger.
Petroschen, Hung. Petrozsény.
Hunedoara, W Romania
45.25N 23.22E
Petroschen/Petroşeni see
Petroşani
Petroskoi see Petrozavodsk
Petrovac/Petrováca see Bački
Petrovac
115 J17 **Petrovac na Moru** Montenegro,
SW Yugoslavia 42.11N 19.00E
119 S8 **Petrove** Kirovohrads'ka Oblast',
C Ukraine 48.22N 33.12E
115 O18 **Petrovec** C FYR Macedonia
41.57N 21.37E
Petrovgrad see Zrenjanin
131 P7 **Petrovsk** Saratovskaya Oblast',
W Russian Federation 52.20N 45.23E
128 J9 **Petrovskiy Yam** Respublika
Kareliya, NW Russian Federation
63.19N 35.14E

126 K16 **Petrovsk-Zabaykal'skiy**
Chitinskaya Oblast', S Russian
Federation 51.15N 108.36E
131 P9 **Petrov Val** Volgogradskaya
Oblast', SW Russian Federation
50.10N 45.16E
128 J11 **Petrozavodsk** Fin. Petroskoi.
Respublika Kareliya, NW Russian
Federation 61.46N 34.19E
Petrozsény see Petroşani
85 D20 **Petrusdal** Hardap, C Namibia
23.42S 17.23E
119 T7 **Petrykivka** Dnipropetrovs'ka
Oblast', E Ukraine 48.44N 34.42E
Petsamo see Pechenga
Petschka see Pecica
Pettau see Ptuj
111 S5 **Pettenbach** Oberösterreich,
C Austria 47.58N 14.03E
27 S3 **Pettus** Texas, SW USA
28.34N 97.49W
125 F13 **Petukhovo** Kurganskaya Oblast',
C Russian Federation 55.04N 67.49E
Petuna see Songyuan
111 R4 **Peuerbach** Oberösterreich,
N Austria 48.19N 13.45E
64 G12 **Peumo** Libertador, C Chile
34.22S 71.58W
173 Ee3 **Peusangan, Krueng**
⋄ Sumatera, NW Indonesia
127 O4 **Pevek** Chukotskiy Avtonomnyy
Okrug, NE Russian Federation
69.40N 170.19E
29 X5 **Pevely** Missouri, C USA
38.16N 90.24W
Peyia see Pégeia
104 J13 **Peyrehorade** Landes, SW France
43.33N 1.04W
105 P16 **Pézenas** Hérault, S France
43.28N 3.25E
113 H20 **Pezinok** Ger. Bösing, Hung.
Bazin. Bratislavský Kraj,
W Slovakia 48.16N 17.15E
103 L22 **Pfaffenhofen an der Ilm**
Bayern, SE Germany 48.31N 11.30E
110 G7 **Pfäffikon** Schwyz, C Switzerland
47.11N 8.46E
103 F20 **Pfälzer Wald** hill range
W Germany
103 N22 **Pfarrkirchen** Bayern,
SE Germany 48.25N 12.56E
103 G21 **Pforzheim** Baden-Württemberg,
SW Germany 48.52N 8.42E
103 H21 **Pfullendorf** Baden-
Württemberg, S Germany
47.55N 9.16E
110 K8 **Pfunds** Tirol, W Austria
46.56N 10.30E
103 G19 **Pfungstadt** Hessen, W Germany
49.48N 8.36E
85 L20 **Phalaborwa** Northern, NE South
Africa 23.50S 31.08E
158 E11 **Phalodi** Rājasthān, NW India
27.06N 72.22E
158 E12 **Phalsund** Rājasthān, NW India
26.20N 71.55E
161 E15 **Phaltan** Mahārāshtra, W India
18.01N 74.31E
178 Hh7 **Phan** var. Muang Phan. Chiang
Rai, NW Thailand 19.34N 99.43E
178 H14 **Phangan, Ko** island SW Thailand
178 Gg15 **Phang-Nga** var. Pang-Nga,
Phangnga. Phangnga,
SW Thailand 8.28N 98.31E
Phan Rang/Phanrang see Phan
Rang-Thap Cham
178 Kk13 **Phan Rang-Thap Cham** var.
Phanrang, Phan Rang, Phan Rang
Thap Cham. Ninh Thuận,
S Vietnam 11.34N 109.00E
178 Kk14 **Phan Ri** Bình Thuận, S Vietnam
11.11N 108.31E
178 Kk14 **Phan Thiết** Bình Thuận,
S Vietnam 10.55N 108.06E
Pharnacia see Giresun
27 S17 **Pharr** Texas, SW USA
26.11N 98.10W
178 H10 **Pharus** see Hvar
178 Hh7 **Phatthalung** var. Padalung,
Patalung. Phatthalung,
SW Thailand 7.33N 100.04E
178 Hh7 **Phayao** var. Muang Phayao.
Phayao, NW Thailand 19.10N 99.55E
9 U10 **Phelps Lake** ⋄ Saskatchewan,
C Canada
23 X9 **Phelps Lake** ⋄ North Carolina,
SE USA
25 R5 **Phenix City** Alabama, S USA
32.28N 85.00W
178 Jj8 **Pheo** Quang Bình, C Vietnam
17.42N 105.58E
Phet Buri see Phetchaburi
178 H12 **Phetchabun** var. Bejraburi,
Petchaburi, Phet Buri. Phetchaburi,
W Thailand 13.04N 99.58E
178 Hh9 **Phichit** var. Bichitra, Muang
Phichit, Pichit. Phichit, C Thailand
16.28N 100.21E
24 M5 **Philadelphia** Mississippi, S USA
32.45N 89.06W
20 I7 **Philadelphia** New York, NE USA
44.10N 75.40W
20 I16 **Philadelphia** Pennsylvania,
NE USA 40.00N 75.10W
20 I16 **Philadelphia** × Pennsylvania,
NE USA 39.51N 75.13W
Philadelphia see 'Ammān
30 L10 **Philip** South Dakota, N USA
44.02N 101.39W
101 L20 **Philippeville** Namur, S Belgium
50.12N 4.33E
Philippeville see Skikda
23 S3 **Philippi** West Virginia, NE USA
39.09N 80.02W
205 Y9 **Philippi Glacier** glacier
Antarctica
198 G7 **Philippine Basin** undersea
feature W Pacific Ocean
133 X12 **Philippine Plate** tectonic feature
179 Q13 **Philippines** off. Republic of
the Philippines. ◆ republic
SE Asia
133 X13 **Philippines** island group W Pacific
Ocean
179 S12 **Philippine Sea** sea W Pacific
Ocean
198 G7 **Philippine Trench** undersea
feature W Philippine Sea
85 H23 **Philippolis** Free State, C South
Africa 30.16S 25.16E
Philippopolis see Plovdiv,
Bulgaria

47 V9 **Philippopolis** see Shahbā', Syria
Philipsburg Sint Maarten,
N Netherlands Antilles
17.58N 63.02W
35 P10 **Philipsburg** Montana, NW USA
46.19N 113.17W
41 R6 **Philip Smith Mountains**
▲ Alaska, USA
158 H8 **Phillaur** Punjab, N India
31.01N 75.49E
191 N13 **Phillip Island** island Victoria,
SE Australia
27 N2 **Phillips** Texas, SW USA
35.39N 101.21W
32 K5 **Phillips** Wisconsin, N USA
45.42N 90.22W
28 K3 **Phillipsburg** Kansas, C USA
39.45N 99.19W
20 I14 **Phillipsburg** New Jersey, NE USA
40.41N 75.09W
23 S7 **Philpott Lake** ⋄ Virginia,
NE USA
Phintias see Licata
178 Hh9 **Phitsanulok** var. Bisnulok,
Muang Phitsanulok, Pitsanulok.
Phitsanulok, C Thailand
16.49N 100.15E
Phlórina see Flórina
178 J13 **Phnom Penh** see Phnum Penh
178 J13 **Phnum Penh** var. Phnom Penh.
● (Cambodia) Phnum Penh,
S Cambodia 11.34N 104.55E
178 J12 **Phnum Tbêng Meanchey**
Preăh Vihéar, N Cambodia
13.45N 104.58E
38 K13 **Phoenix** state capital Arizona,
SW USA 33.27N 112.04W
42 F9 **Phoenix** × Arizona, SW USA
33.25N 112.00W
203 R3 **Phoenix Island** see Rawaki
Phoenix Islands island group
C Kiribati
20 I15 **Phoenixville** Pennsylvania,
NE USA 40.07N 75.31W
85 K22 **Phofung** var. Mont-aux-Sources.
▲ N Lesotho 28.47S 28.52E
178 I10 **Phon** Khon Kaen, E Thailand
15.47N 102.36E
178 I15 **Phôngsali** var. Phong Saly.
Phôngsali, N Laos 21.40N 102.04E
Phong Saly see Phôngsali
12 G11 **Phonë** Ontario, S Canada
12 H15 **Phônhông** C Laos 18.29N 102.26E
178 J14 **Phô Rang** Lao Cai, N Vietnam
22.12N 104.27E
178 Gg10 **Phra Chedi Sam Ong**
Kanchanaburi, W Thailand
15.18N 98.26E
178 Hh8 **Phrae** var. Muang Phrae, Prae.
Phrae, NW Thailand 18.07N 100.09E
178 I5 **Phra Nakhon Si Ayutthaya** see
Ayutthaya
177 G15 **Phra Thong, Ko** island
SW Thailand
Phu Cương see Thu Dâu Một
177 Gg16 **Phuket** var. Bhuket, Puket, Mal.
Ujung Salang; prev. Junkseylon,
Salang. Phuket, SW Thailand
7.52N 98.22E
177 G16 **Phuket** × Phuket, SW Thailand
8.03N 98.16E
177 G16 **Phuket, Ko** island SW Thailand
160 N12 **Phulabāni** prev. Phulbani. Orissa,
E India 20.30N 84.18E
178 Kk10 **Phu Lôc** Th.a Thiên-Huê,
C Vietnam 16.13N 107.53E
Phulbani see Phulabāni
12 K15 **Picton** Ontario, SE Canada
178 Ii3 **Phumĭ Bânăm** Prey Vêng,
S Cambodia 11.14N 105.18E
178 Ii3 **Phumĭ Chôâm** Kâmpóng Spœ,
SW Cambodia 11.42N 103.58E
178 Ii3 **Phumĭ Kalêng** Stœng Trêng,
NE Cambodia 13.57N 106.17E
178 Ii2 **Phumĭ Kâmpóng Trábêk** prev.
Phum Kompong Trabek. Kâmpóng
Thum, C Cambodia 13.06N 105.16E
178 Ii2 **Phumĭ Koŭk Kduŏch**
Bătdâmbâng, NW Cambodia
13.16N 103.08E
178 Ii3 **Phumĭ Labăng** Rôtânôkiri,
NE Cambodia 13.51N 107.01E
178 Ii1 **Phumĭ Moŭng** Siêmréab,
NW Cambodia 13.45N 103.35E
178 Ii3 **Phumĭ Prâmaŏy** Poŭthisăt,
W Cambodia 12.13N 103.05E
178 I14 **Phumĭ Sâmĭt** Kaôh Kŏng,
SW Cambodia 10.54N 103.09E
178 Ii1 **Phumĭ Sâmraông** prev. Phum
Samrong. Siêmréab,
S Cambodia 14.09N 103.30E
Phum Kompong Trabek see
Phumĭ Kâmpóng Trábêk
Phum Samrong see Phumĭ
Sâmraông
178 Jj11 **Phu My** Bình Định, C Vietnam
14.09N 109.05E
178 J15 **Phung Hiệp** Cần Thơ, S Vietnam
9.49N 105.48E
159 T12 **Phuntsholing** SW Bhutan
26.52N 89.25E
178 J15 **Phước Long** Minh Hai, S Vietnam
9.27N 105.25E
178 Ii14 **Phu Quốc, Đao** var. Phu Quoc
Island. island S Vietnam
Phu Quoc Island see Phu Quôc,
Đao
178 J6 **Phu Thọ** Vinh Phu, N Vietnam
21.22N 105.13E
Phu Vinh see Tra Vinh
201 T13 **Piaanu Pass** passage Chuuk
Islands, C Micronesia
108 E8 **Piacenza** Fr. Paisance; anc.
Placentia. Emilia-Romagna, N Italy
45.01N 9.42E
109 K14 **Pianella** Abruzzo, C Italy
42.24N 14.05E
109 M15 **Pianosa, Isola** island Archipelago
Toscano, C Italy
176 V12 **Piar** Irian Jaya, E Indonesia
2.49S 132.46E
113 H20 **Piešťany** Ger. Pistyan, Hung.
Pöstyén. Trnavský, W Slovakia
48.36N 17.48E
111 X5 **Piesting** ⋄ E Austria

118 I15 **Piatra** Teleorman, S Romania
43.49N 25.10E
118 L10 **Piatra-Neamţ** Hung.
Karácsonkő. Neamţ, NE Romania
46.54N 26.23E
108 P1 **Piauhy** see Piauí
61 N15 **Piauí** off. Estado do Piauí; prev.
Piauhy. ◆ state E Brazil
61 N7 **Piave** ⋄ NE Italy
109 K24 **Piazza Armerina** var. Chiazza.
Sicilia, Italy, C Mediterranean Sea
37.22N 14.22E
83 G14 **Pibor** Texas, SW USA
Pibor Amh. Pibor Wenz.
⋄ Ethiopia/Sudan
83 G14 **Pibor Post** Jonglei, SE Sudan
6.49N 33.06E
Pibor Wenz see Pibor
83 K11 **Picacho Butte** ▲ Arizona,
C Poland 50.30N 20.31E
42 D4 **Picachos, Cerro** ▲ NW Mexico
29.15N 114.04W
105 O4 **Picardie** Eng. Picardy. ◇ region
N France
Picardy see Picardie
24 L8 **Picayune** Mississippi, S USA
30.31N 89.40W
85 G14 **Piccolo San Bernardo, Colle
di** see Little Saint Bernard Pass
8 A23 **Pichanal** Salta, N Argentina
23.22S 64.11W
153 P12 **Picher** Oklahoma, C USA
36.59N 94.49W
64 K5 **Pichidegua** Libertador, C Chile
34.25S 72.00W
58 B6 **Pichincha** ▲ SE Finland
58 C6 **Pichincha** ◆ province N Ecuador
64 K3 **Pichit** see Phichit
43 U15 **Pichucalco** Chiapas, SE Mexico
17.32N 93.07W
24 L5 **Pickens** Mississippi, S USA
32.52N 89.58W
23 O11 **Pickens** South Carolina, SE USA
34.52N 82.42W
12 G11 **Pickerel** ⋄ Ontario, S Canada
12 H15 **Pickering** Ontario, S Canada
43.50N 79.03W
99 N16 **Pickering** N England, UK
54.14N 0.46W
35 S13 **Pickerington** Ohio, N USA
39.52N 82.45W
10 C10 **Pickle Lake** Ontario, S Canada
51.30N 90.10W
31 P12 **Pickstown** South Dakota, N USA
43.02N 98.31W
25 V6 **Pickton** Texas, SW USA
33.01N 95.19W
25 N1 **Pickwick Lake** ⋄ S USA
66 N2 **Pico** var. Ilha do Pico. island
Azores, Portugal, NE
Atlantic Ocean
65 J19 **Pico de Salamanca** Chubut,
SE Argentina 45.26S 67.26W
61 N6 **Pico, Ilha do** see Pico
61 I10 **Picos** Piauí, E Brazil 7.04S 41.24W
65 I20 **Pico Truncado** Santa Cruz,
S Argentina 46.49S 68.01W
191 S9 **Picton** New South Wales,
SE Australia 34.12S 150.36E
193 K14 **Picton** Marlborough, South
Island, NZ 41.18S 174.00E
65 H15 **Pícún Leufú, Arroyo**
⋄ SW Argentina
161 K25 **Pidalion** see Gkréko, Akrotíri
158 L10 **Pidolos** Matese Campania,
S Italy 41.20N 14.30E
29 X7 **Piedmont** Missouri, C USA
37.09N 90.42W
23 P11 **Piedmont** South Carolina,
SE USA 34.42N 82.27W
23 U13 **Piedmont** Ohio, N USA
Piedmont see Piemonte
106 M11 **Piedrabuena** Castilla-La
Mancha, C Spain 39.00N 4.10W
Piedrafita, Puerto de see
Pedrafita, Porto de
106 L8 **Piedrahíta** Castilla-León,
C Spain 40.27N 5.19W
43 N6 **Piedras Negras** var. Ciudad
Porfirio Díaz. Coahuila de
Zaragoza, NE Mexico
28.40N 100.31W
63 E23 **Piedras, Punta** headland
E Argentina 35.27S 57.04W
59 I14 **Piedras, Río de las** ⋄ E Peru
113 J16 **Piekary Śląskie** Śląskie, S Poland
50.23N 19.01E
95 M17 **Pieksämäki** Isä-Suomi, E Finland
62.18N 27.10E
111 Y3 **Pielach** ⋄ NE Austria
95 M16 **Pielavesi** Itä-Suomi, C Finland
63.13N 26.45E
95 N16 **Pielinen** var. Pielisjärvi.
⋄ E Finland
Pielisjärvi see Pielinen
108 A8 **Piemonte** Eng. Piedmont. ◇
region NW Italy
113 I14 **Pieniny** S Poland
113 E14 **Pieńsk** Ger. Penzig.
SW Poland 51.14N 15.03E
31 Q13 **Pierce** Nebraska, C USA
42.11N 97.31W
31 R14 **Pierce** Saskatchewan,
C Canada
31 N10 **Pierre** state capital South Dakota,
N USA 44.22N 100.17W
104 K13 **Pierrefitte-Nestalas** Hautes-
Pyrénées, S France 42.58N 0.04W
105 R14 **Pierrelatte** Drôme, E France
44.22N 4.40E
13 P11 **Pierreville** Quebec, SE Canada
46.05N 72.48W
39 N15 **Pierz** Minnesota, N USA
60 H13 **Pimenta** Pará, N Brazil
5.25N 56.17W
58 B11 **Piment** Lambayeque, W Peru
6.51S 79.52W
31 N7 **Piéria** ▲ N Greece
31 N10 **Pierre** var. Pina. SW Belarus
41.28N 0.31N
104 J13 **Pina Rus.** Pina. ⋄ SW Belarus
42 E2 **Pinacate, Sierra del**
▲ NW Mexico 31.45N 113.33W
65 D20 **Pináculo, Cerro** ▲ S Argentina
50.46S 72.07W
203 X11 **Pinaki** atoll Îles Tuamotu,
E French Polynesia
39 N15 **Pinaleno Mountains**
▲ Arizona, SW USA
179 Pp12 **Pinamalayan** Mindoro,
N Philippines 13.00N 121.30E
174 Kk8 **Pinang** Borneo, C Indonesia
0.36N 109.10W
173 G3 **Pinang** var. Penang. ◆ state
Peninsular Malaysia

Pinang see Pinang, Pulau,
Peninsular Malaysia
Pinang see George Town
173 G3 **Pinang, Pulau** var. Penang,
Pinang; prev. Prince of Wales
Island. island Peninsular Malaysia
109 K24 **Pietraperzia** Sicilia, Italy,
SW Italy 38.35N 29.22E
109 N22 **Pietra Spada, Passo della** pass
SW Italy 38.31N 16.20E
85 K24 **Piet Retief** Mpumalanga, E South
Africa 27.00S 30.49E
118 I9 **Pietrosul, Mount** ▲ Luzon,
N Philippines 15.07N 120.21E
Pietrosul, Vârful prev. Virful
Pietrosu. ▲ N Romania
47.36N 24.39E
118 J10 **Pietrosul, Vârful** prev. Virful
Pietrosu. ▲ N Romania
47.06N 25.09E
Pietrosu, Virful see Pietrosul,
Vârful
108 I6 **Pieve di Cadore** Veneto, NE Italy
46.27N 12.22E
12 C18 **Pigeon Bay** lake bay Ontario,
S Canada
29 X8 **Piggott** Arkansas, C USA
36.22N 90.11W
85 L21 **Piggs Peak** NW Swaziland
25.58S 31.16E
8 A23 **Pigüé** Buenos Aires, E Argentina
37.37S 62.24W
43 O12 **Piguicas** ▲ C Mexico
21.08N 99.37W
200 Qq15 **Piha Passage** passage S Tonga
95 N18 **Pihkva Järv** see Pskov, Lake
95 U8 **Pihlava** Länsi-Suomi, W Finland
61.33N 21.36E
95 N15 **Pihtipudas** Länsi-Suomi, W
Finland 63.20N 25.37E
42 L14 **Pihuamo** Jalisco, SW Mexico
19.16N 103.21W
201 U11 **Piis Moen** var. Pis. atoll Chuuk
Islands, C Micronesia
43 U17 **Pijijiapán** Chiapas, SE Mexico
15.39N 93.13W
100 G12 **Pijnacker** Zuid-Holland,
W Netherlands 52.01N 4.25E
44 H5 **Pijol, Pico** ▲ NW Honduras
15.07N 87.39W
128 I13 **Pikaar** see Bikar Atoll
128 H13 **Pikalevo** Leningradskaya Oblast',
NW Russian Federation
59.33N 34.04E
196 M15 **Pikelot** island Caroline Islands,
C Micronesia
32 M5 **Pike River** ⋄ Wisconsin, N USA
39 T5 **Pikes Peak** ▲ Colorado, C USA
38.51N 105.06W
23 P6 **Pikeville** Kentucky, S USA
37.28N 82.31W
22 L9 **Pikeville** Tennessee, S USA
35.36N 85.11W
Pikinni see Bikini Atoll
81 H18 **Pikounda** La Sangha, C Congo
0.30N 16.43E
112 G9 **Piła** Ger. Schneidemühl.
Wielkopolskie, C Poland
53.09N 16.43E
7 P1 **Pico Fracture Zone** tectonic
feature NW Atlantic Ocean
61 U10 **Pilão Arcado** Bahia, E Brazil
9.49S 42.33W
63 D20 **Pilar** Buenos Aires, E Argentina
34.28S 58.55W
64 N7 **Pilar** var. Villa del Pilar.
Ñeembucú, S Paraguay
26.55S 58.19W
81 N6 **Pilcomayo, Río** ⋄ C South
America
153 R12 **Piles** see Pýlos
Pilgram see Pelhřimov
179 Q11 **Pili** Luzon, N Philippines
13.31N 123.15E
158 L10 **Pilibhīt** Uttar Pradesh, N India
28.37N 79.48E
112 J12 **Pilica** ⋄ C Poland
117 G16 **Pílio** ▲ C Greece
113 J22 **Pilisvörösvár** Pest, N Hungary
47.36N 18.54E
117 G15 **Pillar Bay** bay Ascension Island,
C Atlantic Ocean
191 P17 **Pillar, Cape** headland Tasmania,
SE Australia 43.13S 147.58E
Pillau see Baltiysk
191 R5 **Pilliga** New South Wales,
SE Australia 30.22S 148.53E
H8 **Pilón** Granma, E Cuba
19.54N 77.20W
24 N7 **Pineville** Kentucky, S USA
36.45N 83.42W
155 T5 **Pineville** Louisiana, S USA
31.19N 92.25E
29 R8 **Pineville** Missouri, C USA
36.35N 94.22W
23 R10 **Pineville** North Carolina, USA
35.04N 80.53W
23 Q6 **Pineville** West Virginia, USA
37.34N 81.32W
35 V8 **Piney Buttes** physical region
Montana, USA
169 W9 **Ping, Mae Nam** ⋄ W Thailand
166 H14 **Pingban** var. Pingbian Miaozu
Zizhixian. Yunnan, SW China
22.51N 103.28E
Pingbian Miaozu
Zizhixian see Pingban
163 S9 **Pingdingshan** Henan, C China
33.52N 113.19E
167 R4 **Pingdu** Shandong, E China
36.48N 119.56E
201 W16 **Pingelap Atoll** atoll Caroline
Islands, E Micronesia
167 Q13 **Pingnan** Fujian, SE China
24.30N 117.19E
167 O13 **Pingquan** Hebei, E China
41.01N 118.34E
31 P5 **Pingree** North Dakota, N USA
47.07N 98.54W
Pingsiang see Pingxiang
179 P12 **Pinamalayan** Mindoro,
N Philippines
166 I8 **Pingwu** Sichuan, C China
32.33N 104.32E

166 J15 **Pingxiang** Guangxi Zhuangzu
Zizhiqu, S China 22.03N 106.43E
167 O11 **Pingxiang** var. P'ing-hsiang;
prev. Pingsiang. Jiangxi, S China
27.42N 113.49E
167 R4 **Pingyang** Zhejiang, SE China
27.46N 120.37E
167 P5 **Pingyi** Shandong, E China
35.30N 117.37E
167 P5 **Pingyin** Shandong, E China
36.18N 116.24E
62 H13 **Pinhalzinho** Santa Catarina,
S Brazil 26.53S 52.57W
62 L10 **Pinhão** Paraná, S Brazil
25.46S 51.32W
63 H17 **Pinheiro Machado** Rio Grande
do Sul, S Brazil 31.34S 53.22W
106 F7 **Pinhel** Guarda, N Portugal
40.46N 7.03W
195 R10 **Pinipel Island** island Green
Islands, NE PNG
173 E8 **Pini, Pulau** island Kepulauan
Batu, W Indonesia
111 X7 **Pinka** ⋄ SE Austria
111 X7 **Pinkafeld** Burgenland, SE Austria
47.18N 16.09E
Pinkiang see Harbin
8 M12 **Pink Mountain** British
Columbia, W Canada
57.01N 122.26W
177 S13 **Pinlebu** Sagaing, N Myanmar
24.02N 95.21E
40 I12 **Pinnacle Island** island Alaska,
USA
188 I12 **Pinnacles, The** tourist site
Western Australia
190 K10 **Pinnaroo** South Australia
35.17S 140.54E
103 L12 **Pinneberg** Schleswig-Holstein,
N Germany 53.40N 9.49E
117 I15 **Pínnes, Akrotírio** headland
N Greece 40.06N 24.19E
37 R14 **Pinos, Isla de** see Juventud, Isla
de
37 R14 **Pinos, Mount** ▲ California,
W USA 34.48N 119.07W
107 R12 **Pinoso** País Valenciano, E Spain
38.25N 1.01W
107 N14 **Pinos-Puente** Andalucía, S Spain
37.16N 3.46W
43 U16 **Pinotepa Nacional** var. Santiago
Pinotepa Nacional. Oaxaca,
SE Mexico 16.19N 98.02W
116 N13 **Pins, Île des** var. Kunyé. island
E New Caledonia
121 L20 **Pinsk Pol.** Pińsk. Brestskaya
Voblasts', SW Belarus 52.07N 26.07E
12 D18 **Pins, Pointe aux** headland
Ontario, S Canada 42.14N 81.53W
59 B16 **Pinta, Isla** var. Abingdon. island
Galapagos Islands, Ecuador,
E Pacific Ocean
129 N8 **Pinyug** Kirovskaya Oblast',
NW Russian Federation
60.12N 47.45E
37 R17 **Pioche** Nevada, W USA
37.54N 114.27W
108 G11 **Piombino** Toscana, C Italy
42.54N 10.30E
1 **Pioneer Fracture Zone** tectonic
feature NE Pacific Ocean
126 J12 **Pioner, Ostrov** island Severnaya
Zemlya, N Russian Federation
120 A13 **Pionerskiy** Ger. Neukuhren.
Kaliningradskaya Oblast',
W Russian Federation 54.57N 20.16E
112 M13 **Pionki** Mazowieckie, C Poland
51.28N 21.27E
192 L9 **Piopio** Waikato, North Island, NZ
43.01N 102.33W
112 K13 **Piotrków Trybunalski** Ger.
Petrikau, Rus. Petrokov. Łodzkie,
C Poland 51.25N 19.42E
158 F12 **Pipār Road** Rājasthān, N India
26.25N 73.28E
117 I16 **Píperi** island Vóreioi Sporádes,
Greece, Aegean Sea
31 S10 **Pipestone** Minnesota, N USA
44.00N 96.19W
12 C11 **Pipestone** ⋄ Ontario, S Canada
8 E21 **Pipinas** Buenos Aires,
E Argentina 35.31S 57.19W
155 T7 **Piplán** prev. Liaqatabad. Punjab,
E Pakistan 32.19N 71.21E
13 S7 **Pipmuacan, Réservoir**
⋄ Quebec, SE Canada
Piqan see Shanshan
35 R12 **Piqua** Ohio, N USA 40.08N 84.14W
107 P5 **Piqueras, Puerto de** pass
N Spain 42.04N 2.35W
62 J13 **Piquiri, Rio** ⋄ S Brazil
62 I9 **Piracicaba** São Paulo, S Brazil
22.45S 47.33W
58 I11 **Piraeus/Piraiévs** see Peiraías
62 I7 **Piraju** São Paulo, S Brazil
23.12S 49.24W
62 L8 **Pirajuí** São Paulo, S Brazil
21.58S 49.27W
65 G12 **Pirámide, Cerro** ▲ S Chile
49.06S 73.32W
111 R13 **Piran It.** Pirano. SW Slovenia
45.31N 13.36E
64 N6 **Pirané** Formosa, N Argentina
25.42S 59.06W
61 J13 **Piranhas** Goiás, S Brazil
16.24S 51.51W
Pirano see Piran
61 M19 **Pirapòzinho** Minas Gerais, NE Brazil
17.19S 44.54W
62 J13 **Pirapòzinho** São Paulo, S Brazil
22.17S 51.31W
6 I3 **Pirarajá** Lavalleja, S Uruguay
34.43S 54.45W
76 I6 **Pirassununga** São Paulo, S Brazil
21.58S 47.23W
47 V6 **Pirata, Monte** ▲ E Puerto Rico
18.06N 65.33W
62 I13 **Piratuba** Santa Catarina, S Brazil
27.25S 51.46W
62 I15 **Pirdop** prev. Srednogorie. Sofiya,
116 I9 **Pirdop** prev. Srednogorie. Sofiya,
W Bulgaria 42.43N 24.09E
203 P7 **Pirea** Tahiti, W French Polynesia
61 K18 **Pirenópolis** Goiás, S Brazil
15.48S 49.00W
145 Q14 **Pīr Panjāl Range** ▲ NE India
148 I4 **Pīrāpòzinho** Āzarbāyjān-e
Bākhtari, NW Iran 36.46N 45.10E
159 S13 **Pirganj** Rajshahi,
NW Bangladesh 25.51N 88.25E

◆ COUNTRY ◇ DEPENDENT TERRITORY ▲ ADMINISTRATIVE REGION ▲ MOUNTAIN ⊠ VOLCANO ◎ LAKE
● COUNTRY CAPITAL ○ DEPENDENT TERRITORY CAPITAL × INTERNATIONAL AIRPORT ▲ MOUNTAIN RANGE ⋄ RIVER ⊡ RESERVOIR

Pirgi see Pyrgí
Pírgos see Pýrgos
63 F20 Piriápolis Maldonado, S Uruguay 34.51S 55.15W
116 G11 Pirin ▲▲ SW Bulgaria
Pirineos see Pyrenees
60 N13 Piripiri Piauí, E Brazil 4.15S 41.46W
120 H4 Pirita var. Pirita Jõgi. ≈ NW Estonia
Pirita Jõgi see Pirita
56 J6 Píritu Portuguesa, N Venezuela 9.21N 69.16W
95 L18 Pirkkala Länsi-Suomi, W Finland 61.27N 23.47E
103 F20 Pirmasens Rheinland-Pfalz, SW Germany 49.12N 7.36E
103 P16 Pirna Sachsen, E Germany 50.57N 13.56E
Piroe see Piru
115 Q15 Pirot Serbia, SE Yugoslavia 43.12N 22.34E
158 H6 Pīr Panjāl Range ▲▲ NE India
45 W16 Pirre, Cerro ▲ SE Panama 7.54N 77.42W
143 Y11 Pirsaat Rus. Pirsagat. ≈ E Azerbaijan
Pirsagat see Pirsaat
149 V11 Pīr Shūrān, Selseleh-ye ▲▲ SE Iran
94 M12 Pirttikoski Lappi, N Finland 66.20N 27.08E
Pirttikylä see Pörtom
175 T11 Piru prev. Piroe. Pulau Seram, E Indonesia 3.01S 128.10E
Piryatin see Pyryatyn
Pis see Piis Moen
108 F11 Pisa var. Pisae. Toscana, C Italy 43.43N 10.22E
Pisae see Pisa
176 Uu10 Pisang, Kepulauan island group E Indonesia
176 Xx11 Pisapa Irian Jaya, E Indonesia 3.25S 137.04E
201 V12 Pisar atoll Chuuk Islands, C Micronesia
Pisaurum see Pesaro
12 M10 Piscatosine, Lac ◎ Quebec, SE Canada
111 W7 Pischeldorf Steiermark, SE Austria 47.11N 15.48E
Pischk see Simeria
109 L19 Pisciotta Campania, S Italy 40.07N 15.13E
59 E16 Pisco Ica, SW Peru 13.46S 76.12W
118 G9 Pişcolt Hung. Piskolt. Satu Mare, NW Romania 47.34N 22.18E
59 E16 Pisco, Río ≈ E Peru
113 C18 Písek Budějovický Kraj, S Czech Republic 49.18N 14.07E
35 R14 Pisgah Ohio, N USA 39.19N 84.22W
164 F9 Pishan var. Guma. Xinjiang Uygur Zizhiqu, NW China 37.36N 78.45E
119 N8 Pishchanka Vinnyts'ka Oblast', C Ukraine 48.12N 28.52E
115 K21 Pishë Fier, SW Albania 40.40N 19.22E
149 X14 Pishīn Sīstān va Balūchestān, SE Iran 26.05N 61.46E
155 O9 Pishīn North-West Frontier Province, NW Pakistan 30.39N 66.52E
155 N11 Pishin Lora var. Psein Lora, Pash. Pseyn Bowr. ≈ SW Pakistan
Pishma see Pizhma
Pishpek see Bishkek
175 Q13 Pising Pulau Kabaena, C Indonesia 5.07S 121.50E
Pisino see Pazin
Piski see Simeria
Piskolt see Pişcolt
153 Q9 Piskom Rus. Pskem. ≈ E Uzbekistan
Piskom Tizmasi see Pskemskiy Khrebet
37 P13 Pismo Beach California, W USA 35.08N 120.38W
79 P12 Pissila C Burkina 13.07N 0.51W
64 H8 Pissis, Monte ▲ N Argentina 27.45S 68.43W
43 X12 Piste Yucatán, E Mexico 20.40N 88.34W
109 O18 Pisticci Basilicata, S Italy 40.22N 16.33E
108 F11 Pistoia anc. Pistoria, Pistoriæ. Toscana, C Italy 43.57N 10.52E
34 E15 Pistol River Oregon, NW USA 42.13N 124.23W
Pistoria/Pistoriæ see Pistoia
13 U5 Pistuacanis ≈ Quebec, SE Canada
Pistyan see Piešt'any
106 M5 Pisuerga ≈ N Spain
112 N8 Pisz Ger. Johannisburg. Warmińsko-Mazurskie, NE Poland 53.37N 21.49E
78 I13 Pita Moyenne-Guinée, NW Guinea 11.04N 12.15W
56 D12 Pitalito Huila, S Colombia 1.51N 76.01W
62 I11 Pitanga Paraná, S Brazil 24.45S 51.43W
190 M9 Pitarpunga Lake salt lake New South Wales, SE Australia
199 M11 Pitcairn Island island S Pitcairn Islands
199 M11 Pitcairn Islands ◇ UK dependent territory C Pacific Ocean
95 J14 Piteå Norrbotten, N Sweden 65.19N 21.30E
94 I13 Piteälven ≈ N Sweden
118 I13 Piteşti Argeş, S Romania 44.53N 24.49E
Pithagorio see Pythagóreio
188 I12 Pithara Western Australia 30.31S 116.38E
105 N6 Pithiviers Loiret, C France 48.10N 2.15E
158 L9 Pithorāgarh Uttar Pradesh, N India 29.34N 80.12E
196 B16 Piti W Guam 13.28N 144.42E
108 G13 Pitigliano Toscana, C Italy 42.38N 11.40E
42 F3 Pitiquito Sonora, NW Mexico 30.39N 112.00W
Pitkäranta see Pitkyaranta
40 M11 Pitkas Point Alaska, USA 62.01N 163.17W
128 H11 Pitkyaranta Fin. Pitkäranta. Respublika Kareliya, NW Russian Federation 61.34N 31.27E
98 J10 Pitlochry C Scotland, UK 56.46N 3.48W
20 I16 Pitman New Jersey, NE USA 39.43N 75.06W
114 G8 Pitomača Virovitica-Podravina, NE Croatia 45.57N 17.14E
37 O2 Pit River ≈ California, W USA
65 G15 Pitrufquén Araucanía, S Chile 38.58S 72.40W
Pitsanulok see Phitsanulok
Pitschen see Byczyna
Pitsunda see Bichvint'a
111 X6 Pitten ≈ E Austria
8 J14 Pitt Island island British Columbia, W Canada
Pitt Island see Makin
24 M3 Pittsboro Mississippi, S USA 33.55N 89.20W
23 T9 Pittsboro North Carolina, SE USA 35.46N 79.21W
29 R7 Pittsburg Kansas, C USA 37.24N 94.42W
27 W6 Pittsburg Texas, SW USA 33.00N 94.58W
20 B14 Pittsburgh Pennsylvania, NE USA 40.26N 80.00W
32 J14 Pittsfield Illinois, N USA 39.36N 90.48W
21 R6 Pittsfield Maine, NE USA 44.46N 69.22W
20 L11 Pittsfield Massachusetts, NE USA 42.27N 73.15W
191 U3 Pittsworth Queensland, E Australia 27.43S 151.36E
64 I8 Pituil La Rioja, NW Argentina 28.33S 67.24W
58 A10 Piura Piura, NW Peru 5.11S 80.41W
58 A9 Piura off. Departamento de Piura. ◐ department NW Peru
37 S13 Piute Peak ▲ California, W USA 35.27N 118.24W
115 J15 Piva ≈ SW Yugoslavia
119 V5 Pivdenne Kharkivs'ka Oblast', E Ukraine 49.52N 36.04E
119 P8 Pivdennyy Buh Rus. Yuzhnyy Bug. ≈ S Ukraine
56 F5 Pivijay Magdalena, N Colombia 10.28N 74.37W
111 T13 Pivka prev. Šent Peter, Ger. Sankt Peter, It. San Pietro del Carso. SW Slovenia 45.41N 14.12E
119 U13 Pivnichno-Kryms'kyy Kanal canal S Ukraine
115 J15 Pivsko Jezero ◎ SW Yugoslavia
113 M18 Piwniczna Małopolskie, S Poland 49.26N 20.43E
37 R12 Pixley California, W USA 35.58N 119.18W
129 Q15 Pizhma var. Pishma. ≈ NW Russian Federation
11 U13 Placentia Newfoundland, SE Canada 47.12N 53.58W
Placentia see Piacenza
11 U13 Placentia Bay inlet Newfoundland, SE Canada
179 Qq12 Placer Masbate, N Philippines 11.54N 123.54E
37 P7 Placerville California, W USA 38.42N 120.48W
46 F5 Placetas Villa Clara, C Cuba 22.18N 79.40W
115 Q18 Plačkovica ▲▲ E FYR Macedonia
38 L2 Plain City Utah, W USA 41.18N 112.05W
24 G4 Plain Dealing Louisiana, S USA 32.54N 93.42W
35 O14 Plainfield Indiana, N USA 39.42N 86.18W
20 K14 Plainfield New Jersey, NE USA 40.37N 74.25W
35 O8 Plains Montana, NW USA 47.27N 114.52W
26 L6 Plains Texas, SW USA 33.11N 102.49W
31 X10 Plainview Minnesota, N USA 44.10N 92.10W
31 Q13 Plainview Nebraska, C USA 42.21N 97.47W
27 N4 Plainview Texas, SW USA 34.12N 101.43W
28 K4 Plainville Kansas, C USA 39.13N 99.18W
117 L25 Pláka, Akrotírio headland Kríti, Greece, E Mediterranean Sea 35.10N 26.19E
117 J15 Pláka, Akrotírio headland Límnos, E Greece 40.02N 25.25E
115 N19 Plakenska Planina ▲▲ SW FYR Macedonia
46 K5 Plana Cays islets SE Bahamas
107 S12 Plana, Isla var. Nueva Tabarca. island E Spain
61 L18 Planaltina Goiás, S Brazil 15.34S 47.27W
85 O14 Planalto Moçambicano plateau N Mozambique
114 N10 Plandište Serbia, NE Yugoslavia 45.13N 21.07E
102 N13 Plane ≈ NE Germany
56 E6 Planeta Rica Córdoba, NW Colombia 8.24N 75.39W
31 P11 Plankinton South Dakota, N USA 43.43N 98.28W
32 M11 Plano Illinois, N USA 41.39N 88.32W
27 U6 Plano Texas, SW USA 33.01N 96.42W
25 W12 Plant City Florida, SE USA 28.01N 82.06W
24 J9 Plaquemine Louisiana, S USA 30.17N 91.13W
106 K9 Plasencia Extremadura, W Spain 40.01N 6.04W
112 P7 Plaska Podlaskie, NE Poland 53.55N 23.18E
114 C10 Plaški Karlovac, C Croatia 45.04N 15.21E
115 N19 Plasnica SW FYR Macedonia 41.28N 21.07E
125 E12 Plast Chelyabinskaya Oblast', C Russian Federation 54.24N 60.51E
11 N14 Plaster Rock New Brunswick, SE Canada 46.55N 67.24W
109 J24 Platani anc. Halycus. ≈ Sicilia, Italy, C Mediterranean Sea
117 G17 Plataniá Thessalía, C Greece 39.09N 23.15E
117 G24 Plátanos Kríti, Greece, E Mediterranean Sea 35.27N 23.34E
117 H18 Plata, Río de la var. River Plate. estuary Argentina/Uruguay
79 V15 Plateau ◐ state C Nigeria
81 G19 Plateaux var. Région des Plateaux. ◐ province C Congo
94 P1 Platen, Kapp headland NE Svalbard 80.30N 22.46E
Plate, River see Plata, Río de la
101 G22 Plate Taille, Lac de la var. L'Eau d'Heure. ◎ SE Belgium
Plathe see Płoty
41 N13 Platinum Alaska, USA 59.00N 161.49W
56 F5 Plato Magdalena, N Colombia 9.47N 74.46W
31 O11 Platte South Dakota, N USA 43.20N 98.51W
29 R3 Platte City Missouri, C USA 39.22N 94.46W
Plattensee see Balaton
29 R3 Platte River ≈ Iowa/Missouri, USA
31 Q15 Platte River ≈ Nebraska, C USA
39 T3 Platteville Colorado, C USA 40.13N 104.49W
32 K9 Platteville Wisconsin, N USA 42.44N 90.27W
103 N21 Plattling Bayern, SE Germany 48.45N 12.52E
29 R3 Plattsburg Missouri, C USA 39.34N 94.27W
20 L6 Plattsburgh New York, NE USA 44.42N 73.28W
31 S15 Plattsmouth Nebraska, C USA 41.00N 95.52W
103 M17 Plauen var. Plauen im Vogtland. Sachsen, E Germany 50.30N 12.08E
Plauen im Vogtland see Plauen
102 M10 Plauer See ◎ NE Germany
115 L16 Plav Montenegro, SW Yugoslavia 42.36N 19.57E
120 I10 Pļaviņas Ger. Stockmannshof. Aizkraukle, S Latvia 56.37N 25.40E
130 K5 Plavsk Tul'skaya Oblast', W Russian Federation 53.42N 37.21E
43 Z12 Playa del Carmen Quintana Roo, E Mexico 20.37N 87.04W
42 J12 Playa Los Corchos Nayarit, SW Mexico 22.31N 105.28W
39 P16 Playas Lake ◎ New Mexico, SW USA
43 S15 Playa Vicenté Veracruz-Llave, SE Mexico 17.42N 95.01W
178 K11 Plây Cu var. Pleiku. Gia Lai, C Vietnam 13.57N 108.01E
30 L3 Plaza North Dakota, N USA 48.00N 102.00W
65 I15 Plaza Huincul Neuquén, C Argentina 38.54S 69.10W
38 L3 Pleasant Grove Utah, W USA 40.21N 111.44W
31 V14 Pleasant Hill Iowa, C USA 41.34N 93.31W
29 R4 Pleasant Hill Missouri, C USA 38.47N 94.16W
Pleasant Island see Nauru
38 K13 Pleasant, Lake ⊡ Arizona, SW USA
21 P8 Pleasant Mountain ▲ Maine, NE USA 44.01N 70.47W
29 R5 Pleasanton Kansas, C USA 38.09N 94.43W
27 R12 Pleasanton Texas, SW USA 28.58N 98.28W
193 G20 Pleasant Point Canterbury, South Island, NZ 44.16S 171.09E
21 R5 Pleasant River ≈ Maine, NE USA
20 J17 Pleasantville New Jersey, NE USA 39.22N 74.31W
105 N12 Pléaux Cantal, C France 45.08N 2.10E
113 B19 Plechý Ger. Plöckenstein. ▲ Austria/Czech Republic 48.45N 13.50E
Pleebo see Plibo
Pleihari see Pelaihari
Pleiku see Plây Cu
103 M16 Pleisse ≈ E Germany
Plencia see Plentzia
192 O7 Plenty, Bay of bay North Island, NZ
35 Y6 Plentywood Montana, NW USA 48.46N 104.33W
107 O2 Plentzia var. Plencia. País Vasco, N Spain 43.25N 2.56W
104 H5 Plérin Côtes d'Armor, NW France 48.33N 2.46W
128 M10 Plesetsk Arkhangel'skaya Oblast', NW Russian Federation 62.40N 40.14E
Pleshchenitsy see Plyeshchanitsy
Pleskau see Pskov
Pleskauer See see Pskov, Lake
Pleskava see Pskov
114 E8 Pleso International ✕ (Zagreb) Zagreb, NW Croatia 45.45N 16.00E
Pless see Pszczyna
13 Q11 Plessisville Quebec, SE Canada 46.14N 71.45W
112 H12 Pleszew Wielkopolskie, C Poland 51.54N 17.46E
10 L10 Plétipi, Lac ◎ Quebec, SE Canada
103 F15 Plettenberg Nordrhein-Westfalen, W Germany 51.13N 7.52E
116 I8 Pleven prev. Plevna. Pleven, N Bulgaria 43.25N 24.36E
116 I8 Pleven ◐ province N Bulgaria
Plevlja/Plevlje see Pljevlja
Plevna see Pleven
Plezzo see Bovec
Pliberk see Bleiburg
78 L17 Plibo var. Pleebo. SE Liberia 4.37N 7.40W
124 Oo13 Pliny Trench undersea feature C Mediterranean Sea
120 K13 Plisa Rus. Plissa. Vitsyebskaya Voblasts', N Belarus 55.12N 27.58E
Plissa see Plisa
114 D11 Plitvica Selo Lika-Senj, W Croatia 44.53N 15.36E
114 D11 Plješevica ▲▲ C Croatia
115 K14 Pljevlja prev. Plevlja, Plevlje. Montenegro, N Yugoslavia 43.21N 19.21E
Ploça see Ploçë
Plocce see Ploče
115 G15 Ploče It. Plocce; prev. Kardeljevo. Dubrovnik-Neretva, SE Croatia 43.02N 17.25E
115 K22 Ploçë var. Ploça. Vlorë, SW Albania 40.24N 19.41E
112 K11 Płock Ger. Plozk. Mazowieckie, C Poland 52.31N 19.40E
111 O10 Plöcken Pass Ger. Plöckenpass, It. Passo di Monte Croce Carnico. pass SW Austria 46.36N 12.55E
Plöckenstein see Plechý
101 B19 Ploegsteert Hainaut, W Belgium 50.45N 2.52E
104 H6 Ploërmel Morbihan, NW France 47.57N 2.24W
Ploești see Ploiești
118 K13 Ploiești prev. Ploești. Prahova, SE Romania 44.56N 26.03E
117 L17 Plomári prev. Plomárion. Lésvos, E Greece 38.58N 26.24E
Plomárion see Plomári
105 O12 Plomb du Cantal ▲ C France 45.03N 2.48E
191 V6 Plomer, Point headland New South Wales, SE Australia 31.19S 153.00E
102 J8 Plön Schleswig-Holstein, N Germany 54.10N 10.25E
112 L11 Płońsk Mazowieckie, C Poland 52.37N 20.22E
121 J20 Plotnitsa Rus. Plotnitsa. Brestskaya Voblasts', SW Belarus 52.03N 26.39E
112 E8 Płoty Ger. Plathe. Zachodniopomorskie, NW Poland 53.48N 15.16E
104 G7 Plouay Morbihan, NW France 47.54N 3.14W
113 D15 Ploučnice Ger. Polzen. ≈ NE Czech Republic
116 I10 Plovdiv prev. Eumolpias, anc. Evmolpia, Philippopolis, Lat. Trimontium. Plovdiv, C Bulgaria 42.08N 24.47
116 J11 Plovdiv ◐ province C Bulgaria
32 L6 Plover Wisconsin, N USA 44.30N 89.33W
Plozk see Płock
29 U11 Plumerville Arkansas, C USA 35.09N 92.38W
21 P10 Plum Island island Massachusetts, NE USA
34 M9 Plummer Idaho, NW USA 47.19N 116.54W
85 J18 Plumtree Matabeleland South, SW Zimbabwe 20.27S 27.49E
120 D11 Plungė Plungé, W Lithuania 55.55N 21.53E
115 J15 Plužine Montenegro, SW Yugoslavia 43.08N 18.49E
121 K14 Plyeshchanitsy Rus. Pleshchenitsy. Minskaya Voblasts', N Belarus 54.25N 27.49E
47 V10 Plymouth ○ (Montserrat) SW Montserrat 16.39N 62.11W
99 I24 Plymouth SW England, UK 50.22N 4.10W
35 O11 Plymouth Indiana, N USA 41.19N 86.19W
21 P12 Plymouth Massachusetts, NE USA 41.57N 70.40W
21 N8 Plymouth New Hampshire, NE USA 43.43N 71.39W
23 X9 Plymouth North Carolina, SE USA 35.52N 76.45W
32 M8 Plymouth Wisconsin, N USA 43.48N 87.58W
99 J20 Plynlimon ▲ C Wales, UK 52.27N 3.48W
128 G14 Plyussa Pskovskaya Oblast', W Russian Federation 58.27N 29.21E
113 B17 Plzeň Ger. Pilsen, Pol. Pilzno. Plzeňský Kraj, W Czech Republic 49.44N 13.22E
113 B17 Plzeňský Kraj ◐ region W Czech Republic
112 F11 Pniewy Ger. Pinne. Wielkolpolskie, C Poland 52.31N 16.14E
108 D8 Po ≈ N Italy
79 P13 Pô S Burkina 11.10N 1.10W
44 M13 Poás, Volcán ⛰ NW Costa Rica 10.12N 84.12W
79 S16 Pobè S Benin 7.00N 2.41E
127 N8 Pobeda, Gora ▲ NE Russian Federation 65.28N 145.44E
Pobeda Peak see Pobedy, Pik/Tomur Feng
153 Z7 Pobedy, Pik var. Pobeda Peak, Chin. Tomur Feng. ▲ China/Kyrgyzstan see also Tomur Feng 42.02N 80.02E
112 H11 Pobiedziska Ger. Pudewitz. Wielkolpolskie, C Poland 52.30N 17.19E
Po, Bocche del see Po, Foci del
29 W9 Pocahontas Arkansas, C USA 36.15N 90.58W
31 U12 Pocahontas Iowa, C USA 42.44N 94.40W
35 Q15 Pocatello Idaho, NW USA 42.52N 112.27W
178 J13 Pochentong ✕ (Phnum Penh) Phnum Penh, S Cambodia 11.24N 104.52E
130 I6 Pochep Bryanskaya Oblast', W Russian Federation 52.56N 33.20E
130 H4 Pochinok Smolenskaya Oblast', W Russian Federation 54.20N 32.29E
43 R17 Pochutla var. San Pedro Pochutla. Oaxaca, SE Mexico 15.44N 96.27W
64 J6 Pocitos, Salar var. Salar Quiróm. salt lake NW Argentina
103 O22 Pocking Bayern, SE Germany 48.22N 13.17E
195 R17 Pocklington Reef reef SE PNG
199 Hh9 Pocklington Trough undersea feature W Pacific Ocean
29 R11 Pocola Oklahoma, C USA 35.13N 94.28W
23 Y5 Pocomoke City Maryland, NE USA 38.04N 75.34W
61 L21 Poços de Caldas Minas Gerais, NE Brazil 21.48S 46.33W
128 H14 Podberez'ye Novgorodskaya Oblast', NW Russian Federation 58.42N 31.22E
Podbrodzie see Pabradė
129 U8 Podcher'ye Respublika Komi, NW Russian Federation 63.55N 57.34E
113 E16 Poděbrady Ger. Podiebrad. Středočeský Kraj, C Czech Republic 50.09N 15.06E
176 Yy10 Podena, Kepulauan island group E Indonesia
130 L9 Podgorenskiy Voronezhskaya Oblast', W Russian Federation 50.22N 39.43E
115 J17 Podgorica prev. Titograd. Montenegro, SW Yugoslavia 42.25N 19.16E
115 K17 Podgorica ✕ Montenegro, SW Yugoslavia 42.22N 19.16E
111 T13 Podgrad SW Slovenia 45.31N 14.09E
Podiebrad see Poděbrady
118 M5 Podil's'ka Vysochina plateau W Ukraine
Podium Anicensis see le Puy
126 Ii11 Podkamennaya Tunguska Eng. Stony Tunguska. ≈ C Russian Federation
113 N17 Podkarpackie ◐ province SE Poland
Pod Klošter see Arnoldstein
64 O9 Podlaskie ◐ province NE Poland
131 Q8 Podlesnoye Saratovskaya Oblast', W Russian Federation 51.51N 47.03E
130 K4 Podol'sk Moskovskaya Oblast', W Russian Federation 55.24N 37.30E
78 H10 Podor N Senegal 16.40N 14.57W
129 P12 Podosinovets Kirovskaya Oblast', NW Russian Federation 60.15N 47.06E
128 I12 Podporozh'ye Leningradskaya Oblast', NW Russian Federation 60.52N 34.00E
Podravska Slatina see Slatina, Croatia
114 J13 Podromanlja Republika Srpska, SE Bosnia & Herzegovina 43.55N 18.46E
Podsvil'ye see Padsvillye
118 L9 Podu Iloaiei prev. Podul Iloaiei. Iaşi, NE Romania 47.13N 27.16E
115 N15 Podujevo Serbia, S Yugoslavia 42.56N 21.13E
Podul Iloaiei see Podu Iloaiei
Podunajská Rovina see Little Alföld
128 M12 Podyuga Arkhangel'skaya Oblast', NW Russian Federation 61.04N 40.46E
58 A9 Poechos, Embalse ◎ NW Peru
57 W10 Poeketi Sipaliwini, E Suriname
102 L8 Poel island N Germany
85 M20 Poelela, Lagoa ◎ S Mozambique
Poerwodadi see Purwodadi
Poetovio see Ptuj
85 E23 Pofadder Northern Cape, W South Africa 29.03S 19.25E
108 I9 Po, Foci del var. Bocche del Po. ≈ NE Italy
118 E12 Pogăniş ≈ W Romania
Pogegen see Pagėgiai
108 G12 Poggibonsi Toscana, C Italy 43.28N 11.09E
109 I14 Poggio Mirteto Lazio, C Italy 42.17N 12.42E
111 V4 Pöggstall Niederösterreich, N Austria 48.19N 15.10E
118 L13 Pogoanele Buzău, SE Romania 44.55N 27.00E
Pogónion see Delvináki
115 M21 Pogradec var. Pogradeci. Korçë, SE Albania 40.54N 20.40E
Pogradeci see Pogradec
127 N18 Pogranichnyy Primorskiy Kray, SE Russian Federation 44.18N 131.33E
40 M16 Pogromni Volcano ⛰ Unimak Island, Alaska, USA 54.34N 164.41W
169 Z15 P'ohang Jap. Hokō. E South Korea 36.01N 129.20E
13 T9 Pohénégamook, Lac ◎ Quebec, SE Canada
95 L20 Pohja Swe. Pojo. Etelä-Suomi, SW Finland 60.07N 23.30E
Pohjanlahti see Bothnia, Gulf of
201 U16 Pohnpei ◆ state E Micronesia
201 O12 Pohnpei ✕ Pohnpei, E Micronesia
201 O12 Pohnpei prev. Ponape Ascension Island. island E Micronesia
113 F19 Pohořelice Ger. Pohrlitz. Brněnský Kraj, SE Czech Republic 48.58N 16.30E
111 V10 Pohorje Ger. Bacher. ▲▲ N Slovenia
119 N6 Pohrebyshche Vinnyts'ka Oblast', C Ukraine 49.31N 29.16E
Pohrlitz see Pohořelice
175 Qq9 Poh, Teluk bay Sulawesi, C Indonesia
167 P9 Po Hu ⊡ E China
118 G15 Poiana Mare Dolj, S Romania 43.55N 23.01E
Poictiers see Poitiers
131 N6 Poim Penzenskaya Oblast', W Russian Federation 53.03N 43.11E
197 I6 Poindimié Province Nord, C New Caledonia 20.55S 165.18E
Poindo see Lhünzhub
119 Y13 Poinsett, Cape headland Antarctica 65.35S 113.00E
31 R9 Poinsett, Lake ◎ South Dakota, N USA
24 I10 Point Au Fer Island island Louisiana, S USA
41 X14 Point Baker Prince of Wales Island, Alaska, USA 56.19N 133.31W
27 U13 Point Comfort Texas, SW USA 28.40N 96.33W
Point de Galle see Galle
46 K10 Pointe à Gravois headland SW Haiti 18.00N 73.53W
24 L10 Pointe a la Hache Louisiana, S USA 29.34N 89.48W
47 Y6 Pointe-à-Pitre Grande Terre, C Guadeloupe 16.15N 61.31W
13 U7 Pointe-au-Père Quebec, SE Canada 48.31N 68.27W
13 V5 Pointe-aux-Anglais Quebec, SE Canada 49.40N 67.09W
47 T10 Pointe Du Cap headland N Saint Lucia 14.06N 60.56W
81 E21 Pointe-Noire Le Kouilou, S Congo 4.46S 11.52E
47 X6 Pointe Noire Basse Terre, W Guadeloupe 16.13N 61.47W
81 E21 Pointe-Noire ✕ Le Kouilou, S Congo 4.45S 11.55E
47 U15 Point Fortin Trinidad, Trinidad and Tobago 10.09N 61.41W
40 M6 Point Hope Alaska, USA 68.21N 166.48W
41 N5 Point Lay Alaska, USA 69.42N 162.57W
20 B16 Point Marion Pennsylvania, NE USA 39.44N 79.53W
20 K16 Point Pleasant New Jersey, NE USA 40.04N 74.00W
23 P4 Point Pleasant West Virginia, NE USA 38.53N 82.07W
47 R14 Point Salines ✕ (St.George's) SW Grenada 12.00N 61.47W
104 L9 Poitiers prev. Poictiers, anc. Limonum. Vienne, W France 46.34N 0.19E
104 K9 Poitou cultural region W France
104 K10 Poitou-Charentes ◐ region W France
105 N3 Poix-de-Picardie Somme, N France 49.47N 1.58E
Pojo see Pohja
39 S10 Pojoaque New Mexico, SW USA 35.52N 106.01W
158 E11 Pokaran Rājasthān, NW India 26.55N 71.55E
191 R4 Pokataroo New South Wales, SE Australia 29.37S 148.43E
121 P18 Pokats' Rus. Pokot'. ≈ SE Belarus
31 V5 Pokegama Lake ◎ Minnesota, N USA
192 L6 Pokeno Waikato, North Island, NZ 37.15S 175.01E
159 O11 Pokhara Western, C Nepal 28.13N 84.00E
131 T6 Pokhvistnevo Samarskaya Oblast', W Russian Federation 53.38N 52.07E
57 W10 Pokigron Sipaliwini, C Suriname 4.25N 55.24W
94 L10 Pokka Lapp. Bohkká. Lappi, N Finland 68.10N 25.45E
81 N16 Poko Orientale, NE Dem. Rep. Congo (Zaire) 3.07N 26.51E
Pokot' see Pokats'
Po-ko-to Shan see Bogda Shan
153 S7 Pokrovka Talasskaya Oblast', NW Kyrgyzstan 42.45N 71.33E
Pokrovka see Kyzyl-Suu
126 M11 Pokrovsk Respublika Sakha (Yakutiya), NE Russian Federation 61.40N 129.25E
119 V8 Pokrovs'ke Rus. Pokrovskoye. Dnipropetrovs'ka Oblast', E Ukraine 47.58N 36.15E
Pokrovskoye see Pokrovs'ke
Pola see Pula
39 N10 Polacca Arizona, SW USA 35.49N 110.21W
106 L2 Pola de Laviana Asturias, N Spain 43.15N 5.33W
106 K2 Pola de Lena Asturias, N Spain 43.10N 5.49W
106 L2 Pola de Siero Asturias, N Spain 43.24N 5.39W
203 Y3 Poland Kiritimati, E Kiribati 1.52N 95.33W
112 H12 Poland off. Republic of Poland, var. Polish Republic, Pol. Polska, Rzeczpospolita Polska; prev. Pol. Polska Rzeczpospolita Ludowa, Polish People's Republic. ◆ republic C Europe
Polangen see Palanga
112 G7 Polanów Ger. Pollnow. Zachodniopomorskie, NW Poland 54.07N 16.38E
142 H13 Polatlı Ankara, C Turkey 39.34N 32.07E
120 L12 Polatsk Rus. Polotsk. Vitsyebskaya Voblasts', N Belarus 55.28N 28.46E
112 F8 Połczyn-Zdrój Ger. Bad Polzin. Zachodniopomorskie, NW Poland 53.43N 16.02E
152 I16 Polekhatum prev. Pul'-I-Khatum. Akhalskiy Velayat, S Turkmenistan 36.01N 61.08E
155 Q3 Pol-e Khomrī var. Pul-i-Khumri. Baghlān, NE Afghanistan 35.55N 68.45E
207 S10 Pole Plain undersea feature Arctic Ocean
149 P5 Pol-e Safīd var. Pol-e-Sefid, Pul-i-Sefid. Māzandarān, N Iran 36.10N 53.03E
Pol-e-Sefid see Pol-e Safīd
120 B13 Polessk Ger. Labiau. Kaliningradskaya Oblast', W Russian Federation 54.52N 21.06E
Polesskoye see Polis'ke
175 P11 Polewali Sulawesi, C Indonesia 3.25S 119.22E
116 G11 Polezhan ▲ SW Bulgaria 41.42N 23.28E
80 F13 Poli Nord, N Cameroon 8.42N 13.09E
Poli see Pólis
109 M19 Policastro, Golfo di gulf S Italy
112 D8 Police Ger. Politz. Zachodniopomorskie, NW Poland 53.34N 14.34E
180 I17 Police, Pointe headland Mahé, NE Seychelles 4.48S 55.31E
117 L17 Políchnitos var. Polihnitos, Polikhnitos. Lésvos, E Greece 39.04N 26.10E
Poligiros see Polýgyros
109 P17 Polignano a Mare Puglia, SE Italy 40.58N 17.13E
105 S9 Poligny Jura, E France 46.51N 5.42E
Polihnitos see Políchnitos
Polikastro/Políkastron see Polýkastro
Polikhnitos see Políchnitos
116 K8 Polikrayshte Veliko Tŭrnovo, N Bulgaria 43.12N 25.38E
179 Pp10 Polillo Islands island group N Philippines
111 Q9 Polinik ▲ SW Austria 46.54N 13.10E
123 Mm3 Polis var. Poli. W Cyprus 35.02N 32.27E
Polish People's Republic see Poland
Polish Republic see Poland
119 O3 Polis'ke Rus. Polesskoye. Kyyivs'ka Oblast', N Ukraine 51.15N 29.27E
109 N22 Polistena Calabria, SW Italy 38.25N 16.04E
Politz see Police
Políyiros see Polýgyros
31 V14 Polk City Iowa, C USA 41.46N 93.42W
112 F13 Polkowice Ger. Heerwegen. Dolnośląskie, SW Poland 51.31N 16.04E
161 G22 Pollāchi Tamil Nādu, SE India 10.38N 77.00E
111 W7 Pöllau Steiermark, SE Austria 47.18N 15.46E
201 T13 Polle atoll Chuuk Islands, C Micronesia
Pollnow see Polanów
31 N7 Pollock South Dakota, N USA 45.53N 100.15W
94 L8 Polmak Finnmark, N Norway 70.01N 28.04E
32 L10 Polo Illinois, N USA 41.59N 89.34W
200 Qq15 Poloa island Tongatapu Group, N Tonga
44 E5 Polochic, Río ≈ C Guatemala
Pologi see Polohy
119 V9 Polohy Rus. Pologi. Zaporiz'ka Oblast', SE Ukraine 47.29N 36.18E
12 M10 Polonais, Lac des ◎ Quebec, SE Canada
63 G20 Polonio, Cabo headland E Uruguay 34.22S 53.46W
161 K24 Polonnaruwa North Central Province, C Sri Lanka 7.55N 81.01E
118 L5 Polonne Rus. Polonnoye. Khmel'nyts'ka Oblast', NW Ukraine 50.10N 27.30E
Polonnoye see Polonne
Polotsk see Polatsk
111 T7 Pöls var. Pölsbach. ≈ E Austria
Pölsbach see Pöls
Polska/Polska, Rzeczpospolita/Polska Rzeczpospolita Ludowa see Poland
116 L10 Polski Gradets Stara Zagora, C Bulgaria 42.12N 26.06E
116 K8 Polsko Kosovo Ruse, N Bulgaria 43.26N 25.40E
35 P8 Polson Montana, NW USA 47.41N 114.09W
119 T6 Poltava Poltavs'ka Oblast', NE Ukraine 49.33N 34.32E
Poltava see Poltavs'ka Oblast'
119 R5 Poltavs'ka Oblast' var. Poltava, Rus. Poltavskaya Oblast'. ◐ province NE Ukraine
Poltavskaya Oblast' see Poltavs'ka Oblast'
Poltoratsk see Ashgabat
120 I5 Põltsamaa Ger. Oberpahlen. Jõgevamaa, E Estonia 58.40N 25.58E
120 I4 Põltsamaa var. Põltsamaa Jõgi. ≈ C Estonia
Põltsamaa Jõgi see Põltsamaa
125 F10 Polunochnoye Sverdlovskaya Oblast', C Russian Federation 60.56N 60.15E
125 G8 Poluy ≈ N Russian Federation
120 J6 Põlva Ger. Põlwe. Põlvamaa, SE Estonia 58.03N 27.05E
95 N16 Polvijärvi Itä-Suomi, E Finland 62.52N 29.19E
Pölwe see Põlva
117 I22 Polýaigos island Kykládes, Greece, Aegean Sea
117 I22 Polyaígou Folégandrou, Stenó strait Kykládes, Greece, Aegean Sea
128 J3 Polyarnyy Murmanskaya Oblast', NW Russian Federation 69.10N 33.21E
129 W5 Polyarnyy Ural ▲▲ NW Russian Federation
117 G14 Polýgyros var. Poligiros, Políyiros. Kentrikí Makedonía, N Greece 40.21N 23.27E
116 F13 Polýkastro var. Polikastro; prev. Políkastron. Kentrikí Makedonía, N Greece 41.01N 22.33E
199 Kk9 Polynesia island group C Pacific Ocean
117 J15 Polýochni site of ancient city Límnos, E Greece 39.51N 25.21E
43 Y13 Polyuc Quintana Roo, E Mexico
111 V10 Polzela C Slovenia 46.18N 15.04E
Polzen see Ploučnice
176 Ww9 Pom Irian Jaya, E Indonesia 1.34S 135.39E
58 D12 Pomabamba Ancash, C Peru 8.51S 77.13W
193 D23 Pomahaka ≈ South Island, NZ
108 F12 Pomarance Toscana, C Italy 43.19N 10.53E
106 G9 Pombal Leiria, C Portugal 39.55N 8.37W
78 D9 Pombas Santo Antão, NW Cape Verde 17.09N 25.02W
85 N19 Pomene Inhambane, SE Mozambique 22.57S 35.34E
112 G8 Pomerania cultural region Germany/Poland
112 D7 Pomeranian Bay Ger. Pommersche Bucht, Pol. Zatoka Pomorska. bay Germany/Poland
35 T15 Pomeroy Ohio, N USA 39.01N 82.01W
34 L10 Pomeroy Washington, NW USA 46.28N 117.36W
119 Q8 Pomichna Kirovohrads'ka Oblast', C Ukraine 48.07N 31.25E
195 O12 Pomio New Britain, E PNG 5.28S 151.29E
153 Pomir, Dar"yoi see Pamir/Pāmīr, Daryā-ye
29 T6 Pomme de Terre Lake ⊡ Missouri, C USA
31 S8 Pomme de Terre River ≈ Minnesota, N USA
Pommersche Bucht see Pomeranian Bay
37 T15 Pomona California, W USA 34.03N 117.45W
116 N9 Pomorie Burgas, E Bulgaria 42.31N 27.39E
Pomorska, Zatoka see Pomeranian Bay
112 H8 Pomorskie ◐ province N Poland
129 Q4 Pomorskiy Proliv strait NW Russian Federation
129 T10 Pomozdino Respublika Komi, NW Russian Federation 62.11N 54.13E
Pompaelo see Pamplona
175 Q9 Pompangeo, Pegunungan ▲▲ Sulawesi, C Indonesia
25 Z15 Pompano Beach Florida, SE USA 26.14N 80.06W
109 K18 Pompei Campania, S Italy 40.45N 14.27E
35 V10 Pompeys Pillar Montana, NW USA 45.58N 107.55W
Ponape Ascension Island see Pohnpei
31 R13 Ponca Nebraska, C USA 42.34N 96.42W
29 O8 Ponca City Oklahoma, C USA 36.42N 97.05W
47 T6 Ponce C Puerto Rico 18.01N 66.36W
25 X10 Ponce de Leon Inlet inlet Florida, SE USA
24 K8 Ponchatoula Louisiana, S USA 30.26N 90.26W
28 M8 Pond Creek Oklahoma, C USA 36.40N 97.48W
161 J20 Pondicherry var. Puduchcheri, Fr. Pondichéry. Pondicherry, SE India 11.58N 79.49E
157 J20 Pondicherry var. Puduchcheri, Fr. Pondichéry. ◐ union territory India
Pondichéry see Pondicherry
207 N11 Pond Inlet Baffin Island, Nunavut, NE Canada 72.37N 77.56W
197 I6 Ponérihouen Province Nord, C New Caledonia 21.04S 165.24E

106 J4 **Ponferrada** Castilla-León, NW Spain 42.33N 6.34W

192 N13 **Pongaroa** Manawatu-Wanganui, North Island, NZ 40.36S 176.08E

178 I12 **Pong Nam Ron** Chantaburi, S Thailand 12.55N 102.15E

83 C14 **Pongo** ✍ S Sudan

158 I7 **Pong Reservoir** ☒ N India

113 N14 **Poniatowa** Lubelskie, E Poland 51.10N 22.04E

178 J13 **Pönley** Kâmpóng Chhnăng, C Cambodia 12.26N 104.25E

161 I20 **Ponnaiyār** ✍ SE India

9 Q15 **Ponoka** Alberta, SW Canada 52.42N 113.33W

131 U6 **Ponomarevka** Orenburgskaya Oblast', W Russian Federation 53.16N 54.10E

174 L15 **Ponorogo** Jawa, C Indonesia 7.51S 111.30E

128 M5 **Ponoy** Murmanskaya Oblast', NW Russian Federation 67.00N 41.06E

125 E5 **Ponoy** ✍ NW Russian Federation

104 K11 **Pons** Charente-Maritime, W France 45.31N 0.31W
Pons see Ponts
Pons Aelii see Newcastle upon Tyne
Pons Vetus see Pontevedra

101 G20 **Pont-à-Celles** Hainaut, S Belgium 50.31N 4.21E

104 K16 **Pontacq** Pyrénées-Atlantiques, SW France 43.10N 0.07W

66 P3 **Ponta Delgada** São Miguel, Azores, Portugal, NE Atlantic Ocean 37.28N 25.40W

66 P3 **Ponta Delgada** ✈ São Miguel, Azores, Portugal, NE Atlantic Ocean

66 N2 **Ponta do Pico** ▲ Pico, Azores, Portugal, NE Atlantic Ocean 38.28N 28.25W

62 J11 **Ponta Grossa** Paraná, S Brazil 25.07S 50.09W

105 S5 **Pont-à-Mousson** Meurthe-et-Moselle, NE France 48.55N 6.03E

105 T9 **Pontarlier** Doubs, E France 46.54N 6.19E

108 G11 **Pontassieve** Toscana, C Italy 43.46N 11.28E

104 L4 **Pont-Audemer** Eure, N France 49.22N 0.31E

24 K9 **Pontchartrain, Lake** ☒ Louisiana, S USA

104 I8 **Pontchâteau** Loire-Atlantique, NW France 47.26N 2.04W

105 R10 **Pont-de-Vaux** Ain, E France 46.25N 4.57E

106 G4 **Ponteareas** Galicia, NW Spain 42.10N 8.29W

108 J6 **Pontebba** Friuli-Venezia Giulia, NE Italy 46.32N 13.18E

106 G4 **Ponte Caldelas** Galicia, NW Spain 42.22N 8.30W

109 J16 **Pontecorvo** Lazio, C Italy 41.27N 13.40E

106 G5 **Ponte da Barca** Viana do Castelo, N Portugal 41.48N 8.25W

106 G5 **Ponte de Lima** Viana do Castelo, N Portugal 41.46N 8.34W

108 F11 **Pontedera** Toscana, C Italy 43.40N 10.37E

106 H10 **Ponte de Sor** Portalegre, C Portugal 39.15N 8.01W

106 H2 **Pontedeume** Galicia, NW Spain 43.22N 8.09W

108 F6 **Ponte di Legno** Lombardia, N Italy 46.16N 10.31E

9 T17 **Ponteix** Saskatchewan, S Canada 49.43N 107.22W

61 N20 **Ponte Nova** Minas Gerais, NE Brazil 20.25S 42.54W

61 G18 **Pontes e Lacerda** Mato Grosso, W Brazil 15.13S 59.21W

106 G4 **Pontevedra** anc. Pons Vetus. Galicia, NW Spain 42.25N 8.39W

106 G3 **Pontevedra** ◆ province Galicia, NW Spain

106 G4 **Pontevedra, Ría de** estuary NW Spain

32 M12 **Pontiac** Illinois, N USA 40.51N 88.37W

35 R9 **Pontiac** Michigan, N USA 42.38N 83.17W

174 Kk8 **Pontianak** Borneo, C Indonesia 0.04S 109.16E

109 I16 **Pontino, Agro** plain C Italy
Pontisarae see Pontoise

104 H6 **Pontivy** Morbihan, NW France 48.04N 2.58E

104 F6 **Pont-l'Abbé** Finistère, NW France 47.52N 4.13W

105 N4 **Pontoise** anc. Briva Isarae, Cergy-Pontoise, Pontisarae. Val-d'Oise, N France 49.03N 2.04E

9 W13 **Ponton** Manitoba, C Canada 54.36N 99.02W

104 J5 **Pontorson** Manche, N France 48.33N 1.31W

24 M2 **Pontotoc** Mississippi, S USA 34.15N 89.00W

27 R9 **Pontotoc** Texas, SW USA 30.52N 98.57W

108 E10 **Pontremoli** Toscana, C Italy 44.24N 9.55E

110 J10 **Pontresina** Graubünden, S Switzerland 46.29N 9.52E

107 U5 **Ponts** var. Pons. Cataluña, NE Spain 41.55N 1.12E

105 R14 **Pont-St-Esprit** Gard, S France 44.15N 4.37E

99 K21 **Pontypool** Wel. Pontypŵl. SE Wales, UK 51.43N 3.01W

99 J22 **Pontypridd** S Wales, UK 51.37N 3.22E
Pontypŵl see Pontypool

45 R17 **Ponuga** Veraguas, SE Panama 7.50N 80.58W

192 L6 **Ponui Island** island N NZ

121 X14 **Ponya** Rus. Ponya. ✍ N Belarus

109 I17 **Ponziane, Isole** island C Italy

190 F7 **Poochera** South Australia 32.45S 134.51E

99 I24 **Poole** England, UK 50.43N 1.58W

27 S6 **Poolville** Texas, SW USA 33.00N 97.55W
Poona see Pune

190 M8 **Pooncarie** New South Wales, SE Australia 33.26S 142.37E

15 K9 **Poopelloe Lake** seasonal lake New South Wales, SE Australia

57 K19 **Poopó** Oruro, C Bolivia 18.22S 66.58W

59 K19 **Poopó, Lago** var. Lago Pampa Aullagas. ☒ W Bolivia

192 L3 **Poor Knights Islands** island N NZ

41 P10 **Poorman** Alaska, USA 64.05N 155.34W

190 E3 **Pootnoura** South Australia 28.31S 134.09E

153 R10 **Pop** Rus. Pap. Namangan Wiloyati, E Uzbekistan 40.49N 71.06E

119 X7 **Popasna** Rus. Popasnaya. Luhans'ka Oblast', E Ukraine 48.37N 38.24E
Popasnaya see Popasna

56 D12 **Popayán** Cauca, SW Colombia 2.27N 76.31W

101 B18 **Poperinge** West-Vlaanderen, W Belgium 50.52N 2.43E

126 K7 **Popigay** Taymyrskiy (Dolgano-Nenetskiy) Avtonomnyy Okrug, N Russian Federation 71.54N 110.45E

126 J7 **Popigay** ✍ N Russian Federation

119 O5 **Popil'nya** Zhytomyrs'ka Oblast', N Ukraine 49.57N 29.24E

190 K8 **Popiltah Lake** seasonal lake New South Wales, SE Australia

35 X7 **Poplar** Montana, NW USA 48.06N 105.12W

9 Y14 **Poplar** ✍ Manitoba, C Canada

29 X8 **Poplar Bluff** Missouri, C USA 36.45N 90.23W

35 X6 **Poplar River** ✍ Montana, NW USA

43 P14 **Popocatépetl** ▲ S Mexico 18.59N 98.37W

174 Ll16 **Popoh** Jawa, S Indonesia 8.13S 111.50E

81 B21 **Popokabaka** Bandundu, SW Dem. Rep. Congo (Zaire) 5.43S 16.35E

109 J15 **Popoli** Abruzzo, C Italy 42.09N 13.51E

195 X16 **Popomanaseu, Mount** ▲ Guadalcanal, C Solomon Islands 9.40S 96.01E

194 L15 **Popondetta** Northern, S PNG 8.45S 148.15E

114 P9 **Popovača** Sisak-Moslavina, NE Croatia 45.35N 16.37E

116 J10 **Popovitsa** Türgovishte, C Bulgaria 42.08N 25.04E

116 L8 **Popovo** Türgovishte, N Bulgaria 43.19N 26.13E
Popovo see Iskra
Popper see Poprad

113 L19 **Popple River** ✍ Wisconsin, N USA

113 L19 **Poprad** Ger. Deutschendorf, Hung. Poprád. Prešovský Kraj, E Slovakia 49.03N 20.16E

113 L18 **Poprad** Ger. Popper, Hung. Poprád. ✍ Poland/Slovakia

113 L19 **Poprad-Tatry** ✈ (Poprad) Prešovský Kraj, E Slovakia 49.04N 20.21E

23 X7 **Poquoson** Virginia, NE USA 37.07N 76.21W

149 V8 **Porāli** ✍ SW Pakistan

192 N12 **Porangahau** Hawke's Bay, North Island, NZ 40.19S 176.36E

61 K17 **Porangatu** Goiás, C Brazil 13.28S 49.13W

121 G18 **Porazava** Pol. Porozow, Rus. Porozovo. Hrodzyenskaya Voblasts', W Belarus 52.57N 24.24E

160 A11 **Porbandar** Gujarāt, W India 21.40N 69.40E

8 I13 **Porcher Island** island British Columbia, SW Canada

106 M13 **Porcuna** Andalucía, S Spain 37.52N 4.12W

12 F7 **Porcupine** Ontario, S Canada 48.31N 81.07W

66 M6 **Porcupine Bank** undersea feature N Atlantic Ocean

9 V15 **Porcupine Hills** ▲ Manitoba/Saskatchewan, S Canada

32 L3 **Porcupine Mountains** hill range Michigan, N USA

66 M7 **Porcupine Plain** undersea feature E Atlantic Ocean

14 F4 **Porcupine River** ✍ Canada/USA

108 I7 **Pordenone** anc. Portenau. Friuli-Venezia Giulia, NE Italy 45.58N 12.39E

116 F9 **Pore** Casanare, E Colombia 5.42N 71.58W

114 A9 **Poreč** It. Parenzo. Istra, NW Croatia 45.16N 13.36E

62 J9 **Porecatu** Paraná, S Brazil 22.46S 51.22W
Porech'ye see Parechcha

131 P4 **Poretskoye** Chuvashskaya Respublika, W Russian Federation 55.12N 46.20E

79 Q13 **Porga** N Benin 11.04N 0.58E

194 G12 **Porgera** Enga, W PNG 5.27S 143.09E

95 K18 **Pori** Swe. Björneborg. Länsi-Suomi, W Finland 61.28N 21.49E

193 I14 **Porirua** Wellington, North Island, NZ 41.08S 174.50E

94 J12 **Porjus** Norrbotten, N Sweden 66.55N 19.51E

128 G14 **Porkhov** Pskovskaya Oblast', W Russian Federation 57.46N 29.26E

57 O4 **Porlamar** Nueva Esparta, NE Venezuela 10.57N 63.53W

104 I8 **Pornic** Loire-Atlantique, NW France 47.07N 2.07W

79 J13 **Poroma** Southern Highlands, W PNG 6.15S 143.34E

127 Oo15 **Poronaysk** Ostrov Sakhalin, Sakhalinskaya Oblast', SE Russian Federation 49.15N 143.00E

115 G20 **Póros** Póros, S Greece 37.30N 23.29E

115 C19 **Póros** Kefallinía, Iónioi Nísoi, Greece, C Mediterranean Sea 38.09N 20.45E

115 H20 **Póros** island S Greece

83 G2 **Poroto Mountains** ▲ SW Tanzania

114 B10 **Porozina** Primorje-Gorski Kotar, NW Croatia 45.07N 14.17E
Porozow/Porozovo see Porazava

205 X15 **Porpoise Bay** bay Antarctica

117 U25 **Porpoise Point** headland NE Ascension Island 7.54S 14.22W

117 C25 **Porpoise Point** headland East Falkland, Falkland Islands 51.19S 59.18W

110 C6 **Porrentruy** Jura, NW Switzerland 47.25N 7.06E

108 F10 **Porretta Terme** Emilia-Romagna, C Italy 44.10N 11.01E

106 G4 **Porriño** Galicia, NW Spain 42.10N 8.37W

94 L7 **Porsangen** fjord N Norway

94 K8 **Porsangerhalvøya** peninsula N Norway

97 G16 **Porsgrunn** Telemark, S Norway 59.07N 9.37E

142 E13 **Porsuk Çayı** ✍ C Turkey
Porsy see Boldumsaz

56 N18 **Portachuelo** Santa Cruz, C Bolivia 17.20S 63.24W

97 F15 **Portadown** Ir. Port An Dúnáin. S Northern Ireland, UK 54.25N 6.27W

97 G15 **Portadown Ir.** Port An Dúnáin.
Portaferry see Port An Dúnáin

25 P10 **Portage** Michigan, N USA 42.12N 85.34W

20 D15 **Portage** Pennsylvania, NE USA 40.23N 78.40W

32 K8 **Portage** Wisconsin, N USA 43.33N 89.28E

32 M3 **Portage Lake** ☒ Michigan, N USA

9 X16 **Portage la Prairie** Manitoba, S Canada 49.58N 98.19W

35 R11 **Portage River** ✍ Ohio, N USA

29 Y8 **Portageville** Missouri, C USA 36.25N 89.42W

30 L2 **Portal** North Dakota, N USA 48.57N 102.33W

8 L17 **Port Alberni** Vancouver Island, British Columbia, SW Canada 49.10N 124.49W

12 E15 **Port Albert** Ontario, S Canada 43.51N 81.42W

106 H10 **Portalegre** anc. Ammaia, Amoea. Portalegre, E Portugal 39.16N 7.25W

106 H10 **Portalegre** ◆ district C Portugal

39 V12 **Portales** New Mexico, SW USA 34.11N 103.19W

41 X14 **Port Alexander** Baranof Island, Alaska, USA 56.15N 134.38W

85 G25 **Port Alfred** Eastern Cape, S South Africa 33.36S 26.55E

8 J16 **Port Alice** Vancouver Island, British Columbia, SW Canada 50.22N 127.24W

24 J8 **Port Allen** Louisiana, S USA 30.27N 91.12W
Port Amelia see Pemba
Port An Dúnáin see Portadown

34 G7 **Port Angeles** Washington, NW USA 48.07N 123.25W

46 L12 **Port Antonio** NE Jamaica 18.10N 76.27W

117 D16 **Pórta Panagiá** religious building Thessalía, C Greece 39.28N 21.37E

27 T14 **Port Aransas** Texas, SW USA 27.49N 97.03W

99 C18 **Portarlington Ir.** Cúil an tSúdaire. C Ireland 53.10N 7.10W

191 P17 **Port Arthur** Tasmania, SE Australia 43.09S 147.51E

27 Y11 **Port Arthur** Texas, SW USA 29.55N 93.55W

98 G12 **Port Askaig** W Scotland, UK 55.51N 6.06W

190 I7 **Port Augusta** South Australia 32.29S 137.45E

46 J9 **Port-au-Prince** ● (Haiti) C Haiti 18.33N 72.19W

46 J9 **Port-au-Prince** ✈ E Haiti 18.38N 72.13N

24 I8 **Port Barre** Louisiana, S USA 30.33N 91.57W

157 Q19 **Port Blair** Andaman and Nicobar Islands, SE India 11.40N 92.43E

107 X4 **Port Bolívar** Texas, SW USA 29.21N 94.45W

107 X4 **Portbou** Cataluña, NE Spain 42.26N 3.10E

107 X4 **Port Bouet** ✈ (Abidjan) SE Ivory Coast 5.17N 3.55W

190 I8 **Port Broughton** South Australia 33.39S 137.55E

10 M12 **Port Burwell** Ontario, S Canada 42.37N 80.47W

10 L12 **Port Burwell** Quebec, NE Canada 60.25N 64.49W

190 M13 **Port Campbell** Victoria, SE Australia 38.37S 143.00E

13 V4 **Port-Cartier** Quebec, SE Canada 50.00N 66.55W

193 F23 **Port Chalmers** Otago, South Island, NZ 45.46S 170.37E

25 W14 **Port Charlotte** Florida, SE USA 27.00N 82.07W

40 L9 **Port Clarence** Alaska, USA 65.15N 166.51W

31 S11 **Port Clinton** Ohio, N USA 41.30N 82.56W

12 H17 **Port Colborne** Ontario, S Canada 42.51N 79.16W

13 Y7 **Port-Daniel** Quebec, SE Canada 48.10N 64.58W
Port Darwin see Darwin

191 O17 **Port Davey** headland Tasmania, SE Australia 43.19S 145.54E

41 X12 **Port Douglas** Queensland, NE Australia 16.32S 145.27E

8 J13 **Port Edward** British Columbia, SW Canada 54.10N 130.12W

85 L22 **Port Edward** KwaZulu-Natal, SE South Africa 31.03S 30.13E

61 J14 **Portel** Pará, NE Brazil 1.58S 50.45W

106 H12 **Portel** Évora, S Portugal 38.18N 7.42W

12 E14 **Port Elgin** Ontario, S Canada 44.26N 81.24W

85 I25 **Port Elizabeth** Eastern Cape, S South Africa 33.58S 25.36E

47 Y14 **Port Elizabeth** Bequia, Saint Vincent and the Grenadines 13.01N 61.15W

85 I25 **Port Elizabeth** Eastern Cape, S South Africa 33.58S 25.36E

98 F13 **Port Ellen** W Scotland, UK 55.37N 6.12W

98 H16 **Port Erin** SW Isle of Man 54.05N 4.47W

9 G18 **Porters Pass** pass South Island, NZ 43.18S 171.45E

85 E25 **Porterville** Western Cape, SW South Africa 33.03S 19.00E

37 R12 **Porterville** California, W USA 36.03N 119.03W
Port-Étienne see Nouâdhibou

190 L13 **Port Fairy** Victoria, SE Australia 38.24S 142.13E

192 M4 **Port Fitzroy** Great Barrier Island, Auckland, NE NZ 36.10S 175.21E
Port Florence see Kisumu
Port-Francqui see Ilebo

81 C18 **Port-Gentil** Ogooué-Maritime, W Gabon 0.40S 8.49E

190 I7 **Port Germein** South Australia 33.02S 138.01E

24 J6 **Port Gibson** Mississippi, S USA 31.57N 90.58W

41 Q13 **Port Graham** Alaska, USA 59.21N 151.49W

79 U17 **Port Harcourt** Rivers, S Nigeria 4.43N 7.02E

8 J16 **Port Hardy** Vancouver Island, British Columbia, SW Canada 50.40N 127.30W

11 R14 **Port Hawkesbury** Cape Breton Island, Nova Scotia, SE Canada 45.36N 61.22W

188 I6 **Port Hedland** Western Australia 20.22S 118.40E

41 O15 **Port Heiden** Alaska, USA 56.54N 158.40W

8 G10 **Pôrto Camargo** Paraná, S Brazil 23.25S 53.47W

27 U13 **Port O'Connor** Texas, SW USA 28.26N 96.26W

99 I19 **Porthmadog** var. Portmadoc. NW Wales, UK 52.55N 4.07W

12 I15 **Port Hope** Ontario, S Canada 43.56N 78.16W

11 S9 **Port Hope Simpson** Newfoundland, E Canada 52.30N 56.18W

62 G10 **Pôrto Camargo** Paraná, S Brazil 23.25S 53.47W

117 C24 **Port Howard Settlement** West Falkland, Falkland Islands 51.38S 59.54W

35 T9 **Port Huron** Michigan, N USA 42.58N 82.25W

109 K17 **Portici** Campania, S Italy 40.48N 14.19E

143 Y13 **Port-Ilıç Rus.** Port Il'ich. SE Azerbaijan 38.54N 48.59E
Port Il'ich see Port-Ilıç

106 G14 **Portimão** var. Vila Nova de Portimão. Faro, S Portugal 37.07N 8.31W

27 T17 **Port Isabel** Texas, SW USA 26.04N 97.13W

20 I13 **Port Jervis** New York, NE USA 41.22N 74.39W

57 S7 **Port Kaituma** NW Guyana 7.42N 59.52W

130 K12 **Port Katon** Rostovskaya Oblast', SW Russian Federation 46.52N 38.46E

191 S9 **Port Kembla** New South Wales, SE Australia 34.29S 150.53E

190 F8 **Port Kenny** South Australia 33.09S 134.38E
Port Láirge see Waterford

191 S8 **Portland** New South Wales, SE Australia 33.24S 150.00E

190 L13 **Portland** Victoria, SE Australia 38.21S 141.37E

192 K4 **Portland** Northland, North Island, NZ 35.48S 174.19E

35 Q13 **Portland** Indiana, N USA 40.25N 84.58W

21 P8 **Portland** Maine, NE USA 43.40N 70.16W

31 Q9 **Portland** Michigan, N USA 42.51N 84.52W

30 J3 **Portland** North Dakota, N USA 47.29N 97.22W

34 G11 **Portland** Oregon, NW USA 45.31N 122.40W

22 J5 **Portland** Tennessee, S USA 36.34N 86.31W

27 T14 **Portland** Texas, SW USA 27.52N 97.19W

34 G11 **Portland** ✈ Oregon, NW USA 45.36N 122.34W

190 L13 **Portland Bay** bay Victoria, SE Australia

46 K13 **Portland Bight** bay S Jamaica

99 J24 **Portland Bill** var. Bill of Portland. headland S England, UK 50.31N 2.28W
Portland, Bill of see Portland Bill

191 P15 **Portland, Cape** headland Tasmania, SE Australia 40.45S 147.58E

8 J12 **Portland Inlet** inlet British Columbia, W Canada 54.48N 130.14W

192 M12 **Portland Island** island E NZ

117 F15 **Portland Point** headland SW Ascension Island 7.59N 14.25W

46 J13 **Portland Point** headland S Jamaica 17.42N 77.10W

105 F16 **Port-la-Nouvelle** Aude, S France 43.01N 3.04E
Portlaoighise see Port Laoise

99 E18 **Port Laoise** var. Portlaoise, Ir. Portlaoighise; prev. Maryborough. C Ireland 53.01N 7.16W

27 U13 **Port Lavaca** Texas, SW USA 28.36N 96.39W

190 G9 **Port Lincoln** South Australia 34.43S 135.49E

78 H15 **Port Loko** W Sierra Leone 8.49N 12.49W

117 E24 **Port Louis** East Falkland, Falkland Islands 51.31S 58.07W

173 Y15 **Port-Louis** Grande Terre, N Guadeloupe 16.26N 61.31W

181 X16 **Port Louis** ● (Mauritius) NW Mauritius 20.10S 57.30E
Port Louis see Scarborough
Port-Lyautey see Kénitra

190 J7 **Port MacDonnell** South Australia 38.04S 140.40E

191 U7 **Port Macquarie** New South Wales, SE Australia 31.25S 152.55E
Port-Madoc see Porthmadog
Port Mahon see Mahón

8 K16 **Port McNeill** Vancouver Island, British Columbia, SW Canada 50.34N 127.06W

99 H16 **Port Maria** C Jamaica 18.21N 76.53W

85 J25 **Port Elizabeth** Eastern Cape, S South Africa 33.58S 25.36E

98 F13 **Port Ellen** W Scotland, UK 55.37N 6.12W

194 I16 **Port Moresby** ● (PNG) Central/National Capital District, SW PNG 9.28S 147.11E
Port Natal see Durban

190 L13 **Port Neill** South Australia 34.06S 136.19E

13 R6 **Portneuf** ✍ Quebec, SE Canada

13 R6 **Portneuf, Lac** ☒ Quebec, SE Canada

84 D23 **Port Nolloth** Northern Cape, W South Africa 29.18S 16.58E

20 J12 **Port Norris** New Jersey, NE USA 39.13N 75.00W
Port-Nouveau-Québec see Kangiqsualujjuaq

106 G6 **Porto Eng.** Oporto; anc. Portus Cale. Porto, NW Portugal 41.09N 8.37W

106 G6 **Porto var.** Pôrto. ◆ district N Portugal

106 G6 **Porto** ✈ Porto, W Portugal 41.15N 8.45W

62 G6 **Porto Alegre** var. Pôrto Alegre. state capital Rio Grande do Sul, S Brazil 30.03S 51.10W

82 A11 **Porto Alexandre** see Tombua

82 A12 **Porto Amboim** Cuanza Sul, NW Angola 10.43S 13.49E
Porto Amélia see Pemba
Porto Bello see Portobelo

45 T14 **Portobelo** var. Porto Bello, Puerto Bello. Colón, N Panama 9.32N 79.40W

63 G14 **Pôrto de Mós** see Porto de Moz

60 J12 **Porto de Moz** var. Pôrto de Mós. Pará, NE Brazil 1.45S 52.15W

61 H16 **Porto do Moniz** Madeira, Portugal, NE Atlantic Ocean 32.50N 17.10W
Porto Edda see Sarandë

109 I24 **Porto Empedocle** Sicilia, Italy, C Mediterranean Sea 37.18N 13.31E

61 H20 **Porto Esperança** Mato Grosso do Sul, SW Brazil 19.36S 57.24W

108 E13 **Portoferraio** Toscana, C Italy 42.48N 10.18E

98 G6 **Port of Ness** NW Scotland, UK 58.29N 6.15W

47 U7 **Port-of-Spain** ● (Trinidad and Tobago) Trinidad, Trinidad and Tobago 10.39N 61.30W
Port of Spain see Piarco

108 I7 **Porto, Golfe de** gulf Corse, France, C Mediterranean Sea
Porto Grande see Mindelo

108 I7 **Portogruaro** Veneto, NE Italy 45.46N 12.49E

37 P7 **Portola** California, W USA 39.48N 120.28W

37 B12 **Port-Olry** Espíritu Santo, C Vanuatu 15.03S 167.04E

95 M19 **Pörtom Fin.** Pirttikylä. Länsi-Suomi, W Finland 62.42N 21.40E
Port Omna see Portumna

61 K16 **Porto Murtinho** Mato Grosso do Sul, SW Brazil 21.42S 57.52W

61 K16 **Porto Nacional** Tocantins, C Brazil 10.40S 48.19W

79 S16 **Porto-Novo** ● (Benin) S Benin 6.28N 2.37E

25 X10 **Port Orange** Florida, SE USA 29.06N 80.59W

34 E15 **Port Orchard** Washington, NW USA 47.32N 122.38W
Pôrto Re see Kraljevica

34 E15 **Port Orford** Oregon, NW USA 42.45N 124.30W
Pôrto Rico see Puerto Rico

108 L13 **Porto San Giorgio** Marche, C Italy 43.10N 13.47E

108 L13 **Porto San Stefano** Toscana, C Italy 42.26N 11.09E

66 P5 **Porto Santo** var. Vila Baleira. Porto Santo, Madeira, Portugal, NE Atlantic Ocean 33.04N 16.19W

66 Q5 **Porto Santo** ✈ Porto Santo, Madeira, Portugal, NE Atlantic Ocean 33.04N 16.19W

66 P5 **Porto Santo** var. Ilha do Porto Santo. island Madeira, Portugal, NE Atlantic Ocean

62 I16 **Porto São José** Paraná, S Brazil 22.43S 53.10W

61 O19 **Porto Seguro** Bahia, E Brazil 16.25S 39.07W

109 B17 **Porto Torres** Sardegna, Italy, C Mediterranean Sea 40.49N 8.22E

105 Y16 **Porto-Vecchio** Corse, France, C Mediterranean Sea 41.35N 9.17E

59 E15 **Porto Velho** var. Velho. state capital Rondônia, W Brazil 8.45S 63.54W

56 A6 **Portoviejo** var. Puertoviejo. Manabí, W Ecuador 1.02S 80.31W

193 B26 **Port Pegasus** bay Stewart Island, NZ

12 H15 **Port Perry** Ontario, SE Canada 44.08N 78.57W

191 N12 **Port Phillip Bay** harbor Victoria, SE Australia

190 I8 **Port Pirie** South Australia 33.10S 138.01E

98 G9 **Portree** N Scotland, UK 57.24N 6.11W
Port Rex see East London

46 K13 **Port Royal** E Jamaica 17.56N 76.49W

23 R15 **Port Royal** South Carolina, SE USA 32.22N 80.41W

23 R15 **Port Royal Sound** inlet South Carolina, SE USA
Port Rush see Portrush

97 E14 **Portrush Ir.** Port Rois. N Northern Ireland, UK 55.12N 6.40W

75 W7 **Port Said Ar.** Bûr Sa'îd. N Egypt 31.16N 32.18E

25 R9 **Port Saint Joe** Florida, SE USA 29.49N 85.18W

25 Y11 **Port Saint John** Florida, SE USA 28.28N 80.46W

85 I24 **Port St.Johns** Eastern Cape, S South Africa 31.38S 29.32E

23 W5 **Port St-Louis-du-Rhône** Bouches-du-Rhône, SE France 43.23N 4.48E

27 R12 **Poteet** Texas, SW USA 29.02N 98.34W

117 E24 **Port Salvador** inlet East Falkland, Falkland Islands

117 D24 **Port San Carlos** East Falkland, Falkland Islands 51.30S 58.58W

85 K24 **Port Shepstone** KwaZulu-Natal, E South Africa 30.45S 30.24E

99 N24 **Portsmouth** S England, UK 50.48N 1.04W

31 S15 **Portsmouth** Ohio, N USA 38.43N 83.00W

23 X7 **Portsmouth** Virginia, NE USA 36.50N 76.18W

12 E17 **Port Stanley** Ontario, S Canada 42.39N 81.12W

117 B25 **Port Stephens** inlet West Falkland, Falkland Islands

117 B25 **Port Stephens Settlement** West Falkland, Falkland Islands 52.09S 60.58W

82 I7 **Port Sudan** Red Sea, NE Sudan 19.37N 37.13E

24 L10 **Port Sulphur** Louisiana, S USA 29.28N 89.41W

99 J22 **Port Talbot** S Wales, UK 51.36N 3.46W

94 L11 **Portuphahan Tekojärvi** ☒ N Finland

34 G7 **Port Townsend** Washington, NW USA 48.07N 122.45W

106 H9 **Portugal** off. Republic of Portugal. ◆ republic SW Europe

107 O2 **Portugalete** País Vasco, N Spain 43.19N 3.01W

56 J6 **Portuguesa** off. Estado Portuguesa. ◆ state N Venezuela
Portuguese East Africa see Mozambique
Portuguese Guinea see Guinea-Bissau
Portuguese Timor see East Timor
Portuguese West Africa see Angola

99 D18 **Portumna Ir.** Port Omna. W Ireland 53.06N 8.13W
Portus Cale see Porto
Portus Magnus see Almería
Portus Magnis see Mahón

105 P17 **Port-Vendres var.** Port Vendres. Pyrénées-Orientales, S France 42.31N 3.06E

190 H9 **Port Victoria** South Australia 34.34S 137.31E

197 C14 **Port-Vila var.** Vila. ● (Vanuatu) Éfaté, C Vanuatu 17.45S 168.21E

190 I9 **Port Wakefield** South Australia 34.13S 138.10E

55 N8 **Port Washington** Wisconsin, N USA 43.22N 87.54W

59 J14 **Porvenir** Pando, NW Bolivia 11.15S 68.43W

65 L22 **Porvenir** Magallanes, S Chile 53.18S 70.22W

97 N16 **Porvoo Swe.** Borgå. Etelä-Suomi, S Finland 60.25N 25.40E

178 J3 **Pôuthisăt prev.** Pursat. Pôuthisăt, W Cambodia 12.31N 103.55E

178 I13 **Pôuthisăt, Stœng prev.** Pursat. ✍ W Cambodia

104 J9 **Pouzauges** Vendée, NW France 46.47N 0.54W

104 J9 **Po, Valle del** see Po Valley

108 F8 **Po Valley** It. Valle del Po. valley N Italy

113 I19 **Považská Bystrica Ger.** Waagbistritz, Hung. Vágbeszterce. Trenčiansky Kraj, W Slovakia 49.07N 18.26E

110 J10 **Poschiavino** ✍ Italy/Switzerland

110 J10 **Poschiavo Ger.** Puschlav. Graubünden, S Switzerland 46.19N 10.02E

114 D12 **Posedarje** Zadar, SW Croatia 44.12N 15.27E

113 H18 **Posen** see Poznań

59 M19 **Poshekhon'ye** Yaroslavskaya Oblast', W Russian Federation 58.31N 39.07E

94 M13 **Posio** Lappi, NE Finland 66.06N 28.16E

61 O19 **Posjo Sulawesi, C Indonesia** 1.22S 120.45E

175 Po9 **Poso Danau** ☒ Sulawesi, C Indonesia

143 R9 **Posof** Ardahan, NE Turkey 41.31N 42.44E

175 Po9 **Poso, Sungai** ✍ Sulawesi, C Indonesia

27 R6 **Possum Kingdom Lake** ☒ Texas, SW USA

10 I7 **Poste-de-la-Baleine** Quebec, NE Canada 55.12N 77.54W

101 N14 **Posterholt** Limburg, SE Netherlands 51.07N 6.01E

85 G22 **Postmasburg** Northern Cape, N South Africa 28.19S 23.04E

110 E7 **Pôsto Diuarum** see Campo de Diauarum

111 T12 **Postojna Ger.** Adelsberg, It. Postumia. SW Slovenia 45.48N 14.12E
Postumia see Postojna

23 V6 **Postville** Iowa, C USA 43.04N 91.34W

35 V13 **Poston** Arizona, SW USA 33.59N 114.24W

115 C16 **Posušje** Federacija Bosna I Hercegovina, SW Bosnia & Herzegovina 43.29N 17.22E

23 V8 **Poteau** Oklahoma, C USA 35.03N 94.37W

117 G14 **Poteídaia** site of ancient city Kentrikí Makedonía, N Greece 40.12N 23.21E
Potentia see Potenza

109 M18 **Potenza** anc. Potentia. Basilicata, S Italy 40.40N 15.49E

193 A24 **Poteriteri, Lake** ☒ South Island, NZ

106 M2 **Potes** Cantabria, N Spain 43.10N 4.40W

85 J20 **Potgietersrus** Northern Transvaal, NE South Africa 24.09S 28.58E

85 J20 **Poth** Texas, SW USA 29.04N 98.04W

34 J7 **Potholes Reservoir** ☒ Washington, NW USA

143 Q9 **P'ot'i** W Georgia 42.10N 41.42E

79 X13 **Potiskum** Yobe, NE Nigeria 11.38N 11.07E
Potkozarje see Ivanjska

34 M9 **Potlatch** Idaho, NW USA 46.55N 116.51W

35 N9 **Pot Mountain** ▲ Idaho, NW USA 46.44N 115.24W

115 H14 **Potoci** Federacija Bosna I Hercegovina, SE Bosnia & Herzegovina 43.24N 17.52E

23 V3 **Potomac River** ✍ NE USA

29 W6 **Potosi** Missouri, C USA 37.56N 90.47W

59 L20 **Potosi** Potosí, S Bolivia 19.34S 65.51W

59 L19 **Potosí** ◆ department SW Bolivia

44 H9 **Potosí** Chinandega, NW Nicaragua 12.58N 87.30W

59 K21 **Potosí** ◆ department SW Bolivia

64 H7 **Potrerillos** Atacama, N Chile 26.25S 70.09W

44 H5 **Potrerillos** Cortés, NW Honduras 15.12N 87.57W

102 N12 **Potsdam** Brandenburg, NE Germany 52.24N 13.04E

19 K8 **Potsdam** New York, NE USA 44.40N 74.58W

111 X5 **Pottendorf** Niederösterreich, E Austria 47.53N 16.22E

111 X5 **Pottenstein** Niederösterreich, E Austria 47.55N 16.06E

197 G4 **Pott, Île** island Îles Belep, W New Caledonia

20 I15 **Pottstown** Pennsylvania, NE USA 40.15N 75.39W

20 H14 **Pottsville** Pennsylvania, NE USA 40.40N 76.10W

161 L25 **Pottuvil** Eastern Province, SE Sri Lanka 6.52N 81.49E

155 U6 **Potwar Plateau** plateau NE Pakistan

104 J7 **Pouancé** Maine-et-Loire, W France 47.46N 1.11W

197 N16 **Pouébo** Province Nord, C New Caledonia 20.40S 164.02E

197 N16 **Pouembout** Province Nord, W New Caledonia 21.09S 164.52E

13 R6 **Poulin de Courval, Lac** ☒ Quebec, SE Canada

20 I9 **Poultney** Vermont, NE USA 43.31N 73.12W

197 N16 **Poum** Province Nord, W New Caledonia 20.15S 164.03E

61 L21 **Pouso Alegre** Minas Gerais, NE Brazil 22.13S 45.55W

198 Bb8 **Poutasi** Upolu, SE Samoa 14.05S 171.43W

79 J9 **Poutrincourt, Lac** ☒ Quebec, SE Canada

12 H11 **Powassan** Ontario, S Canada 46.04N 79.21W

37 R12 **Poway** California, W USA 32.57N 117.02W

33 S14 **Powder River** Wyoming, C USA 43.01N 106.57W

35 Y10 **Powder River** ✍ Montana/Wyoming, NW USA

34 L12 **Powder River** ✍ Oregon, NW USA

33 V14 **Powder River Pass** pass Wyoming, C USA 44.08N 107.03W

33 N15 **Powell** Wyoming, C USA 44.45N 108.45W

117 I22 **Powell Basin** undersea feature NW Weddell Sea

38 M8 **Powell, Lake** ☒ Utah, W USA

39 M4 **Powell, Mount** ▲ Colorado, C USA 39.25S 106.20W

8 L17 **Powell River** British Columbia, SW Canada 49.54N 124.34W

32 K5 **Powers** Michigan, N USA 45.40N 87.29W

30 J2 **Powers Lake** North Dakota, N USA 48.33N 102.37W

23 V6 **Powhatan** Virginia, NE USA 37.32N 77.55W

31 X12 **Powhatan Point** Ohio, N USA 39.49N 80.49W

99 I20 **Powys** cultural region E Wales, UK

197 I6 **Poya** Province Nord, C New Caledonia 21.19S 165.07E

167 P10 **Poyang Hu** ☒ S China

126 M16 **Poyarkovo** Amurskaya Oblast', SE Russian Federation 49.37N 128.40E

32 L7 **Poygan, Lake** ☒ Wisconsin, N USA

114 N11 **Požarevac Ger.** Passarowitz. Serbia, NE Yugoslavia 44.37N 21.11E

43 Q13 **Poza Rica** var. Poza Rica de Hidalgo. Veracruz-Llave, E Mexico 20.33N 97.27W
Poza Rica de Hidalgo see Poza Rica
114 L13 **Požega** Prev. Slavonska Požega; Ger. Poschega, Hung. Pozsega. Požega-Slavonija, NE Croatia 43.19N 17.42E
114 H9 **Požega-Slavonija** off. Požeško-Slavonska Županija. ⬖ province NE Croatia
129 U13 **Pozhva** Komi-Permyatskiy Avtonomnyy Okrug, NW Russian Federation 59.07N 56.04E
112 G11 **Poznań** Ger. Posen, Posnania. Wielkolpolskie, C Poland 52.24N 16.56E
107 O13 **Pozo Alcón** Andalucía, S Spain 37.43N 2.55W
64 H3 **Pozo Almonte** Tarapacá, N Chile 20.13S 69.48W
106 L12 **Pozoblanco** Andalucía, S Spain 38.22N 4.47W
107 Q11 **Pozo Cañada** Castilla-La Mancha, C Spain 38.49N 1.45W
64 N5 **Pozo Colorado** Presidente Hayes, C Paraguay 23.25S 58.51W
65 J20 **Pozos, Punta** headland S Argentina 47.55S 65.46W
Pozsega see Požega
Pozsony see Bratislava
57 N5 **Pozuelos** Anzoátegui, NE Venezuela 10.10N 64.39W
109 L26 **Pozzallo** Sicilia, Italy, C Mediterranean Sea 36.43N 14.51E
109 K17 **Pozzuoli** anc. Puteoli. Campania, S Italy 40.49N 14.07E
79 P17 **Pra** ≈ S Ghana
Prabumulih see Perabumulih
113 C19 **Prachatice** Ger. Prachatitz. Budějovický Kraj, S Czech Republic 49.01N 14.00E
Prachatitz see Prachatice
178 Hh11 **Prachin Buri** var. Prachinburi. Prachin Buri, C Thailand 14.05N 101.19E
Prachuab Girikhand see Prachuap Khiri Khan
178 H13 **Prachuap Khiri Khan** var. Prachuab Girikhand. Prachuap Khiri Khan, SW Thailand 11.50N 99.45E
113 H16 **Praděd** Ger. Altvater. ▲ NE Czech Republic 50.06N 17.14E
56 D11 **Pradera** Valle del Cauca, SW Colombia 3.24N 76.19W
105 O17 **Prades** Pyrénées-Orientales, S France 42.36N 2.22E
61 O19 **Prado** Bahia, SE Brazil 17.13S 39.15W
56 E11 **Prado** Tolima, C Colombia 3.38N 74.57W
Prado del Ganso see Goose Green
Prae see Phrae
Prag/Praga/Prague see Praha
29 O10 **Prague** Oklahoma, C USA 35.29N 96.40W
113 D16 **Praha** Eng. Prague, Ger. Prag, Pol. Praga. ● (Czech Republic) Středočeský Kraj, NW Czech Republic 50.06N 14.25E
118 J13 **Prahova** ⬖ county SE Romania
118 J13 **Prahova** ≈ S Romania
78 E10 **Praia** ● (Cape Verde) Santiago, S Cape Verde 14.55N 23.31W
85 M21 **Praia do Bilene** Gaza, S Mozambique 25.18S 33.10E
85 M20 **Praia do Xai-Xai** Gaza, S Mozambique 25.04S 33.43E
118 J10 **Praid** Hung. Parajd. Harghita, C Romania 46.33N 25.06E
28 J3 **Prairie Dog Creek** ≈ Kansas/Nebraska, C USA
32 J9 **Prairie du Chien** Wisconsin, N USA 43.01N 91.07W
29 S9 **Prairie Grove** Arkansas, C USA 35.58N 94.19W
35 P10 **Prairie River** ≈ Michigan, N USA
Prairie State see Illinois
27 V11 **Prairie View** Texas, SW USA 30.05N 95.59W
178 Ii11 **Prakhon Chai** Buri Ram, E Thailand 14.36N 103.04E
111 R4 **Pram** ≈ N Austria
111 S4 **Prambachkirchen** Oberösterreich, N Austria 48.18N 13.50E
120 H2 **Prangli** island N Estonia
160 J13 **Prānhita** ≈ C India
180 I15 **Praslin** island Inner Islands, NE Seychelles
117 O23 **Prasonísi, Akrotírio** headland Ródos, Dodekánisos, Greece, Aegean Sea 35.53N 27.46E
113 I14 **Praszka** Opolskie, S Poland 51.05N 18.29E
121 M18 **Pratasy** Rus. Protasy. Homyel'skaya Voblasts', SE Belarus 52.48N 29.04E
178 I10 **Prathai** Nakhon Ratchasima, E Thailand 15.31N 102.42E
Prathet Thai see Thailand
Prathum Thani see Pathum Thani
65 F21 **Prat, Isla** island S Chile
108 G11 **Prato** Toscana, C Italy 43.52N 11.04E
105 O17 **Prats-de-Mollo-la-Preste** Pyrénées-Orientales, S France 42.25N 2.28E
28 L6 **Pratt** Kansas, C USA 37.38N 98.44W
110 E6 **Pratteln** Basel-Land, NW Switzerland 47.31N 7.42E
199 L2 **Pratt Seamount** undersea feature N Pacific Ocean 56.09N 142.30W
25 P5 **Prattville** Alabama, S USA 32.27N 86.27W
Praust see Pruszcz Gdański
116 M7 **Pravda** prev. Dogrular. Silistra, NE Bulgaria 43.53N 26.58E
121 B14 **Pravdinsk** Ger. Friedland. Kaliningradskaya Oblast', W Russian Federation 54.26N 21.01E
106 K2 **Pravia** Asturias, N Spain 43.29N 6.06W
120 L12 **Prazaroki** Rus. Prozoroki. Vitsyebskaya Voblasts', N Belarus 55.16N 28.11E
Prázsmár see Prejmer
178 J11 **Preăh Vihéar** Preăh Vihéar, N Cambodia 13.57N 104.48E
118 J12 **Predeal** Hung. Predeál. Brașov, C Romania 45.30N 25.31E
111 S8 **Predlitz** Steiermark, SE Austria 47.04N 13.54E
9 V15 **Preeceville** Saskatchewan, S Canada 51.58N 102.40W
Preenkuln see Priekule
104 K6 **Pré-en-Pail** Mayenne, NW France 48.27N 0.15W
111 T4 **Pregarten** Oberösterreich, N Austria 48.21N 14.31E
56 H7 **Pregonero** Táchira, NW Venezuela 8.01N 71.45W
120 J10 **Preiļi** Ger. Preli. Preiļi, SE Latvia 56.17N 26.52E
118 J12 **Prejmer** Ger. Tartlau, Hung. Prázsmár. Brașov, S Romania 45.42N 25.49E
115 J16 **Prekornica** ⛰ SW Yugoslavia
Preli see Preiļi
Prëmet see Përmet
102 M12 **Premnitz** Brandenburg, NE Germany 52.33N 12.22E
27 S15 **Premont** Texas, SW USA 27.21N 98.07W
115 H14 **Prenj** ⛰ S Bosnia and Herzegovina
Prenjas/Prenjasi see Përrenjas
24 L7 **Prentiss** Mississippi, S USA 31.36N 89.52W
Preny see Prienai
102 O10 **Prenzlau** Brandenburg, NE Germany 53.19N 13.52E
126 Jj12 **Preobrazhenka** Irkutskaya Oblast', C Russian Federation 60.01N 108.00E
177 Ee9 **Preparis Island** island SW Myanmar
Prerau see Přerov
113 H18 **Přerov** Ger. Prerau. Olomoucký Kraj, E Czech Republic 49.27N 17.27E
Preschau see Prešov
12 M14 **Prescott** Ontario, SE Canada 44.43N 75.33W
38 K12 **Prescott** Arizona, SW USA 34.33N 112.26W
29 T13 **Prescott** Arkansas, C USA 33.48N 93.22W
34 L10 **Prescott** Washington, NW USA 46.17N 118.21W
32 H6 **Prescott** Wisconsin, N USA 44.46N 92.45W
193 A24 **Preservation Inlet** inlet South Island, NZ
114 O7 **Preševo** Serbia, SE Yugoslavia 42.20N 21.38E
31 N10 **Presho** South Dakota, N USA 43.54N 100.03W
60 M13 **Presidente Dutra** Maranhão, E Brazil 5.16S 44.30W
62 I8 **Presidente Epitácio** São Paulo, S Brazil 21.45S 52.07W
64 N5 **Presidente Hayes** off. Departamento de Presidente Hayes. ⬖ department C Paraguay
62 I9 **Presidente Prudente** São Paulo, S Brazil 22.09S 51.24W
Presidente Stroessner see Ciudad del Este
Presidente Vargas see Itabira
62 I8 **Presidente Venceslau** São Paulo, S Brazil 21.52S 51.51W
199 L11 **President Thiers Seamount** undersea feature C Pacific Ocean 24.39S 145.50W
26 J11 **Presidio** Texas, SW USA 29.33N 104.22W
Preslav see Veliki Preslav
113 M19 **Prešov** var. Preschau, Ger. Eperies, Hung. Eperjes. Prešovský Kraj, E Slovakia 49.00N 21.13E
113 M19 **Prešovský Kraj** ⬖ region E Slovakia
115 N20 **Prespa, Lake** Alb. Liqen i Prespës, Gk. Límni Megáli Préspa, Limni Prespa, Mac. Prespansko Ezero, Serb. Prespansko Jezero. ◎ SE Europe
Prespa, Limni/Prespansko Ezero/Prespansko Jezero/Prespës, Liqen i see Prespa, Lake
21 S2 **Presque Isle** Maine, NE USA 46.40N 68.01W
20 B11 **Presque Isle** headland Pennsylvania, NE USA 42.09N 80.06W
Pressburg see Bratislava
79 P17 **Prestea** SW Ghana 5.22N 2.07W
113 B17 **Přeštice** Ger. Pschestitz. Plzeňský Kraj, W Czech Republic 49.36N 13.19E
99 K17 **Preston** NW England, UK 53.46N 2.42W
25 S6 **Preston** Georgia, SE USA 32.08N 84.35W
35 R16 **Preston** Idaho, NW USA 42.06N 111.52W
31 Z13 **Preston** Iowa, C USA 42.03N 90.24W
31 X11 **Preston** Minnesota, N USA 43.41N 92.06W
23 O6 **Prestonsburg** Kentucky, S USA 37.40N 82.46W
98 I13 **Prestwick** W Scotland, UK 55.30N 4.39W
85 J21 **Pretoria** var. Epitoli, Tshwane. ● (South Africa-administrative capital) Gauteng, NE South Africa 25.40S 28.11E
Pretoria-Witwatersrand-Vereeniging see Gauteng
Pretusha see Pretushë
115 M21 **Pretushë** var. Pretusha. Korçë, SE Albania 40.50N 20.45E
Preussisch Eylau see Bagrationovsk
Preussisch-Stargard see Starogard Gdański
Preußisch Holland see Pasłęk
117 C17 **Préveza** Ípeiros, W Greece 38.58N 20.43E
39 V3 **Prewitt Reservoir** ⊠ Colorado, C USA
178 J13 **Prey Vêng** Prey Vêng, S Cambodia 11.30N 105.19E
150 M12 **Priaral'skiye Karakumy, Peski** desert SW Kazakhstan
126 L16 **Priargunsk** Chitinskaya Oblast', S Russian Federation 50.25N 119.12E
40 K14 **Pribilof Islands** island group Alaska, USA
115 K14 **Priboj** Serbia, W Yugoslavia 43.34N 19.33E
113 C17 **Příbram** Ger. Pibrans. Středočeský Kraj, W Czech Republic 49.40N 14.01E
38 M4 **Price** Utah, W USA 39.34N 110.48W
39 N5 **Price River** ≈ Utah, W USA
25 N8 **Prichard** Alabama, S USA 30.44N 88.04W
27 R8 **Priddy** Texas, SW USA 31.39N 98.30W
107 P8 **Priego** Castilla-La Mancha, C Spain 40.25N 2.19W
106 M14 **Priego de Córdoba** Andalucía, S Spain 37.27N 4.12W
120 C10 **Priekule** Ger. Preenkuln. Liepāja, SW Latvia 56.26N 21.36E
120 C12 **Priekulė** Ger. Prökuls. Gargždai, W Lithuania 55.36N 21.16E
121 F14 **Prienai** Pol. Preny. Prienai, S Lithuania 54.37N 23.56E
85 G23 **Prieska** Northern Cape, C South Africa 29.40S 22.45E
34 M7 **Priest Lake** ⊠ Idaho, NW USA
34 M7 **Priest River** Idaho, NW USA 48.10N 117.02W
106 M3 **Prieta, Peña** ▲ N Spain 43.01N 4.42W
42 J10 **Prieto, Cerro** ▲ C Mexico 24.10N 105.21W
113 J19 **Prievidza** var. Priewitz, Ger. Priwitz, Hung. Privigye. Trenčiansky Kraj, C Slovakia 48.48N 18.37E
Priewitz see Prievidza
114 F10 **Prijedor** Republika Srpska, NW Bosnia & Herzegovina 45.00N 16.43E
115 K14 **Prijepolje** Serbia, W Yugoslavia 43.24N 19.39E
Prikaspiyskaya Nizmennost' see Caspian Depression
115 O19 **Prilep** Turk. Perlepe. S FYR Macedonia 41.21N 21.33E
110 B9 **Prilly** Vaud, SW Switzerland 46.32N 6.36E
Priluki see Pryluky
64 L10 **Primero, Río** ≈ C Argentina
31 S12 **Primghar** Iowa, C USA 43.05N 95.37W
114 B9 **Primorje-Gorski Kotar** off. Primorsko-Goranska Županija. ⬖ province NW Croatia
120 A13 **Primorsk** Ger. Fischhausen. Kaliningradskaya Oblast', W Russian Federation 54.45N 20.00E
128 G12 **Primorsk** Fin. Koivisto. Leningradskaya Oblast', NW Russian Federation 60.20N 28.39E
Primorsk/Primorskoye see Prymors'k
127 Nn17 **Primorskiy Kray** prev. Eng. Maritime Territory. ⬖ territory SE Russian Federation
116 N10 **Primorsko** prev. Keupriya. Burgas, E Bulgaria 42.15N 27.45E
130 K13 **Primorsko-Akhtarsk** Krasnodarskiy Kray, SW Russian Federation 46.03N 38.44E
119 U13 **Primors'kyy** Respublika Krym, S Ukraine 45.09N 35.33E
115 D14 **Primošten** Šibenik-Knin, S Croatia 43.34N 15.57E
9 R13 **Primrose Lake** ◎ Saskatchewan, C Canada
9 T14 **Prince Albert** Saskatchewan, S Canada 53.08N 105.43W
85 G25 **Prince Albert** Western Cape, SW South Africa 33.13S 22.03E
15 I1 **Prince Albert Peninsula** peninsula Victoria Island, Northwest Territories, NW Canada
15 I3 **Prince Albert Sound** inlet Northwest Territories, N Canada
15 Mm2 **Prince Charles Island** island Nunavut, NE Canada
205 W6 **Prince Charles Mountains** ⛰ Antarctica
Prince-Édouard, Île-du see Prince Edward Island
180 M13 **Prince Edward Fracture Zone** tectonic feature SW Indian Ocean
11 P14 **Prince Edward Island** Fr. Île-du Prince-Édouard. ⬖ province SE Canada
11 Q14 **Prince Edward Island** Fr. Île-du Prince-Édouard. island SE Canada
181 M12 **Prince Edward Islands** island group S South Africa
23 X4 **Prince Frederick** Maryland, NE USA 38.32N 76.33W
8 M14 **Prince George** British Columbia, SW Canada 53.55N 122.49W
23 W6 **Prince George** Virginia, NE USA 37.13N 77.13W
207 O8 **Prince Gustaf Adolf Sea** sea Nunavut, N Canada
207 Q3 **Prince of Wales, Cape** headland Alaska, USA 65.39N 168.12W
189 V1 **Prince of Wales Island** island Queensland, E Australia
15 Jj1 **Prince of Wales Island** island Queen Elizabeth Islands, Nunavut, NW Canada
41 Y14 **Prince of Wales Island** island Alexander Archipelago, Alaska, USA
Prince of Wales Island see Pinang, Pulau
15 I1 **Prince of Wales Strait** strait Northwest Territories, N Canada
207 O8 **Prince Patrick Island** island Parry Islands, Northwest Territories, NW Canada
15 Kk1 **Prince Regent Inlet** channel Nunavut, N Canada
8 J13 **Prince Rupert** British Columbia, SW Canada 54.18N 130.16W
Prince's Island see Príncipe
23 Y5 **Princess Anne** Maryland, NE USA 38.12N 75.48W
205 R1 **Princess Astrid Kyst** physical region Antarctica
189 W2 **Princess Charlotte Bay** bay Queensland, NE Australia
205 W7 **Princess Elizabeth Land** physical region Antarctica
8 J14 **Princess Royal Island** island British Columbia, SW Canada
47 U15 **Princes Town** Trinidad, Trinidad and Tobago 10.16N 61.22W
9 N17 **Princeton** British Columbia, SW Canada 49.25N 120.34W
32 L11 **Princeton** Illinois, N USA 41.22N 89.27W
35 N16 **Princeton** Indiana, N USA 38.21N 87.33W
31 Z14 **Princeton** Iowa, C USA 41.40N 90.21W
22 H7 **Princeton** Kentucky, S USA 37.06N 87.52W
31 V8 **Princeton** Minnesota, N USA 45.34N 93.34W
29 S1 **Princeton** Missouri, C USA 40.24N 93.34W
20 J15 **Princeton** New Jersey, NE USA 40.21N 74.39W
23 R6 **Princeton** West Virginia, NE USA 37.22N 81.06W
41 S12 **Prince William Sound** inlet Alaska, USA
69 P9 **Príncipe** var. Príncipe Island, Eng. Prince's Island. island N Sao Tome and Principe
Príncipe Island see Príncipe
34 I13 **Prineville** Oregon, NW USA 44.18N 120.50W
30 J11 **Pringle** South Dakota, N USA 43.34N 103.34W
27 N1 **Pringle** Texas, SW USA 35.55N 101.28W
101 H14 **Prinsenbeek** Noord-Brabant, S Netherlands 51.36N 4.42E
100 L6 **Prinses Margriet Kanaal** canal N Netherlands
205 T2 **Prinsesse Ragnhild Kyst** physical region Antarctica
205 U2 **Prins Harald Kyst** physical region Antarctica
94 N2 **Prins Karls Forland** island W Svalbard
45 N8 **Prinzapolka** Región Autónoma Atlántico Norte, NE Nicaragua 13.19N 83.34W
44 L8 **Prinzapolka, Río** ≈ NE Nicaragua
125 Ff9 **Priob'ye** Khanty-Mansiyskiy Avtonomnyy Okrug, N Russian Federation 62.25N 65.36E
106 H1 **Prior, Cabo** headland NW Spain 43.33N 8.21W
31 V9 **Prior Lake** Minnesota, N USA 44.42N 93.25W
128 H11 **Priozersk** Fin. Käkisalmi. Leningradskaya Oblast', NW Russian Federation 61.02N 30.07E
121 J20 **Pripet** Bel. Prypyats', Ukr. Pryp"yat'. ≈ Belarus/Ukraine
121 J20 **Pripet Marshes** wetland Belarus/Ukraine
Prishtinë see Priština
130 J8 **Pristen'** Kurskaya Oblast', W Russian Federation 51.15N 36.47E
115 N16 **Priština** Alb. Prishtinë. Serbia, S Yugoslavia 42.39N 21.09E
102 M10 **Pritzwalk** Brandenburg, NE Germany 53.10N 12.11E
105 R13 **Privas** Ardèche, E France 44.45N 4.34E
109 I16 **Priverno** Lazio, C Italy 41.28N 13.10E
Privigye see Prievidza
114 C12 **Privlaka** Zadar, SW Croatia 44.15N 15.07E
128 M15 **Privolzhsk** Ivanovskaya Oblast', NW Russian Federation 57.24N 41.16E
131 P7 **Privolzhskaya Vozvyshennost'** var. Volga Uplands. ⛰ W Russian Federation
131 P8 **Privolzhskoye** Saratovskaya Oblast', W Russian Federation 51.08N 45.57E
Priwitz see Prievidza
131 N13 **Priyutnoye** Respublika Kalmykiya, SW Russian Federation 46.08N 43.33E
115 M17 **Prizren** Alb. Prizreni. Serbia, S Yugoslavia 42.13N 20.46E
Prizreni see Prizren
109 I24 **Prizzi** Sicilia, Italy, C Mediterranean Sea 37.43N 13.25E
115 P18 **Probištip** NE FYR Macedonia 42.00N 22.06E
174 M15 **Probolinggo** Jawa, C Indonesia 7.45S 113.12E
Probstberg see Wyszków
113 F14 **Prochowice** Ger. Parchwitz. Dolnośląskie, SW Poland 51.15N 16.22E
31 W5 **Proctor** Minnesota, N USA 46.46N 92.13W
27 R8 **Proctor** Texas, SW USA 31.57N 98.25W
27 R8 **Proctor Lake** ⊠ Texas, SW USA
161 I18 **Proddatūr** Andhra Pradesh, E India 14.46N 78.39E
106 H9 **Proença-a-Nova** Castelo Branco, C Portugal 39.45N 7.55W
97 I24 **Præstø** Storstrøm, SE Denmark 55.07N 12.03E
101 I21 **Profondeville** Namur, SE Belgium 50.22N 4.52E
43 W11 **Progreso** Yucatán, SE Mexico 21.14N 89.40W
126 Mm16 **Progress** Amurskaya Oblast', SE Russian Federation 49.40N 129.30E
131 O15 **Prokhladnyy** Kabardino-Balkarskaya Respublika, SW Russian Federation 43.48N 44.02E
Prokletije see North Albanian Alps
126 H14 **Prokop'yevsk** Kemerovskaya Oblast', S Russian Federation 53.56N 86.48E
Prökuls see Priekulė
115 O15 **Prokuplje** Serbia, SE Yugoslavia 43.15N 21.35E
128 H14 **Proletariy** Novgorodskaya Oblast', W Russian Federation 58.24N 31.40E
130 M12 **Proletarsk** Rostovskaya Oblast', SW Russian Federation 46.42N 41.48E
130 J8 **Proletarskiy** Belgorodskaya Oblast', W Russian Federation 50.48N 35.46E
177 Ff7 **Prome** var. Pyè. Pegu, C Myanmar 18.49N 95.13E
62 J8 **Promissão** São Paulo, S Brazil 21.33S 49.51W
62 J8 **Promissão, Represa de** ⊠ S Brazil
129 V4 **Promyshlennyy** Respublika Komi, NW Russian Federation 67.36N 63.59E
121 O16 **Pronya** Rus. Pronya. ≈ E Belarus
8 M11 **Prophet River** British Columbia, W Canada 58.07N 122.39W
32 K11 **Prophetstown** Illinois, N USA 41.40N 89.56W
61 P16 **Propriá** Sergipe, E Brazil 10.15S 36.51W
105 X16 **Propriano** Corse, France, C Mediterranean Sea 41.41N 8.54E
Prościejów see Prostějov
Proskurov see Khmel 'nyts'kyy
116 H12 **Prosotsáni** Anatolikí Makedonía kai Thráki, NE Greece 41.10N 23.58E
179 Rr15 **Prosperidad** Mindanao, S Philippines 8.36N 125.54E
34 J10 **Prosser** Washington, NW USA 46.12N 119.46W
Prossnitz see Prostějov
113 G18 **Prostějov** Ger. Prossnitz, Pol. Prościejów. Olomoucký Kraj, E Czech Republic 49.28N 17.07E
205 V8 **Prosyana** Dnipropetrovs'ka Oblast', E Ukraine 48.07N 36.22E
113 L16 **Proszowice** Małopolskie, S Poland 50.12N 20.15E
Protasy see Pratasy
180 J11 **Protea Seamount** undersea feature SW Indian Ocean 36.49S 18.04E
117 D21 **Próti** island S Greece
116 N8 **Provadiya** Varna, E Bulgaria 43.10N 27.28E
105 S15 **Provence** prev. Marseille-Marignane. ✕ (Marseille) Bouches-du-Rhône, SE France 43.25N 5.15E
105 T14 **Provence** cultural region SE France
105 T14 **Provence-Alpes-Côte d'Azur** ⬖ region SE France
22 H6 **Providence** Kentucky, S USA 37.23N 87.47W
21 N12 **Providence** state capital Rhode Island, NE USA 41.50N 71.26W
38 L1 **Providence** Utah, W USA 41.42N 111.49W
Providence see Fort Providence
69 X10 **Providence Atoll** var. Providence. atoll S Seychelles
12 D12 **Providence Bay** Manitoulin Island, Ontario, S Canada 45.39N 82.16W
25 R6 **Providence Canyon** valley Alabama/Georgia, S USA
24 I5 **Providence, Lake** ◎ Louisiana, S USA
37 X13 **Providence Mountains** ⛰ California, W USA
46 L6 **Providenciales** island W Turks and Caicos Islands
127 Q4 **Provideniya** Chukotskiy Avtonomnyy Okrug, NE Russian Federation 64.22N 173.14W
21 Q12 **Provincetown** Massachusetts, NE USA 42.01N 70.10W
105 P5 **Provins** Seine-et-Marne, N France 48.34N 3.18E
38 L3 **Provo** Utah, W USA 40.13N 111.39W
9 R15 **Provost** Alberta, SW Canada 52.24N 110.16W
114 G13 **Prozor** Federacija Bosna I Hercegovina, SW Bosnia & Herzegovina 43.46N 17.38E
Prozoroki see Prazaroki
62 I11 **Prudentópolis** Paraná, S Brazil 25.12S 50.58W
41 R5 **Prudhoe Bay** Alaska, USA 70.16N 148.18W
41 R4 **Prudhoe Bay** bay Alaska, USA
113 H16 **Prudnik** Ger. Neustadt, Neustadt in Oberschlesien. Opolskie, S Poland 50.19N 17.34E
121 J16 **Prudy** Rus. Prudy. Minskaya Voblasts', C Belarus 53.48N 26.32E
103 D18 **Prüm** Rheinland-Pfalz, W Germany 50.15N 6.27E
103 D18 **Prüm** ≈ W Germany
Prusa see Bursa
112 J7 **Pruszcz Gdański** Ger. Praust. Pomorskie, N Poland 54.16N 18.36E
112 M12 **Pruszków** Ger. Kaltdorf. Mazowieckie, C Poland 52.09N 20.49E
118 K8 **Prut** Ger. Pruth. ≈ E Europe
Pruth see Prut
110 L8 **Prutz** Tirol, W Austria 47.07N 10.42E
Prużana see Pruzhany
121 G19 **Pruzhany** Pol. Prużana. Brestskaya Voblasts', SW Belarus 52.33N 24.28E
128 I11 **Pryazha** Respublika Kareliya, NW Russian Federation 61.42N 33.39E
119 U10 **Pryazovs'ke** Zaporiz'ka Oblast', SE Ukraine 46.43N 35.39E
Prychornomors'ka Nyzovyna see Black Sea Lowland
Prydniprovs'ka Nyzovyna/Prydnyaprowskaya Nizina see Dnieper Lowland
205 Y7 **Prydz Bay** bay Antarctica
119 R4 **Pryluky** Rus. Priluki. Chernihivs'ka Oblast', NE Ukraine 50.34N 32.23E
119 V10 **Prymors'k** Rus. Primorsk; prev. Primorskoye. Zaporiz'ka Oblast', SE Ukraine 46.43N 36.19E
29 Q9 **Pryor** Oklahoma, C USA 36.18N 95.18W
35 U11 **Pryor Creek** ≈ Montana, NW USA
Pryp"yat'/Prypyats' see Pripet
112 M10 **Przasnysz** Mazowieckie, C Poland 53.01N 20.53E
113 K14 **Przedbórz** Łodzkie, S Poland 51.04N 19.51E
113 P17 **Przemyśl** Rus. Peremyshl. Podkarpackie, SE Poland 49.46N 22.46E
113 O16 **Przeworsk** Podkarpackie, SE Poland 50.04N 22.30E
Przheval'sk see Karakol
112 L13 **Przysucha** Mazowieckie, SE Poland 51.22N 20.36E
117 H18 **Psachná** var. Psahna, Psakhná. Évvoia, C Greece 38.34N 23.40E
Psahna/Psakhná see Psachná
117 K18 **Psará** island E Greece
117 I16 **Psathoúra** island Vóreioi Sporádes, Greece, Aegean Sea
Pschestitz see Přeštice
Psein Lora see Pishin Lora
119 S5 **Psël** ≈ Russian Federation/Ukraine
117 M21 **Psérimos** island Dodekánisos, Greece, Aegean Sea
Pseyn Bowr see Pishin Lora
Pskem see Piskom
153 R8 **Pskemskiy Khrebet** Uzb. Piskom Tizmasi. ⛰ Kyrgyzstan/Uzbekistan
128 F14 **Pskov** Ger. Pleskau, Latv. Pleskava. Pskovskaya Oblast', W Russian Federation 58.31N 31.15E
120 K6 **Pskov, Lake** Est. Pihkva Järv, Ger. Pleskauer See, Rus. Pskovskoye Ozero. ◎ Estonia/Russian Federation
128 F15 **Pskovskaya Oblast'** ⬖ province W Russian Federation
Pskovskoye Ozero see Pskov, Lake
114 G9 **Psunj** ⛰ NE Croatia
113 J17 **Pszczyna** Ger. Pless. Śląskie, S Poland 49.58N 18.56E
Ptačnik/Ptacsnik see Vtáčnik
117 D17 **Ptéri** ▲ C Greece 39.08N 21.32E
Ptich' see Ptsich
117 E14 **Ptolemaïda** prev. Ptolemaïs. Dytikí Makedonía, N Greece 40.31N 21.40E
Ptolemaïs see Ptolemaïda, Greece
Ptolemaïs see 'Akko, Israel
123 Gg10 **Ptolemy Seamounts** undersea feature C Mediterranean Sea
121 M19 **Ptsich** Rus. Ptich'. Homyel'skaya Voblasts', SE Belarus 52.10N 28.49E
121 M18 **Ptsich** Rus. Ptich'. ≈ SE Belarus
111 X10 **Ptuj** Ger. Pettau; anc. Poetovio. NE Slovenia 46.26N 15.53E
194 E9 **Pua** ≈ NW PNG
63 A23 **Puán** Buenos Aires, E Argentina 37.34S 62.45W
198 B7 **Pu'apu'a** Savai'i, C Samoa 13.31S 172.09W
198 A7 **Puava, Cape** headland Savai'i, NW Samoa
58 F12 **Pucallpa** Ucayali, C Peru 8.21S 74.33W
59 J17 **Pucarani** La Paz, NW Bolivia 16.18S 68.28W
Pučarevo see Novi Travnik
163 U12 **Pucheng** Fujian, SE China 27.54N 118.34E
166 L6 **Pucheng** Shaanxi, C China 34.55N 109.28E
129 N16 **Puchezh** Ivanovskaya Oblast', W Russian Federation 56.58N 41.08E
113 I19 **Púchov** Hung. Puhó. Trenčiansky Kraj, W Slovakia 49.06N 18.19E
118 J13 **Pucioasa** Dâmbovița, S Romania 45.04N 25.22E
112 I6 **Puck** Pomorskie, N Poland 54.43N 18.24E
32 L8 **Puckaway Lake** ◎ Wisconsin, N USA
65 G15 **Pucón** Araucanía, S Chile 39.18S 71.52W
95 M14 **Pudasjärvi** Oulu, C Finland 65.19N 27.01E
154 L8 **Pūdeh Tal, Shelleh-ye** ≈ SW Afghanistan
131 S1 **Pudem** Udmurtskaya Respublika, NW Russian Federation 58.18N 52.08E
Pudewitz see Pobiedziska
128 K11 **Pudozh** Respublika Kareliya, NW Russian Federation 61.48N 36.30E
99 M17 **Pudsey** N England, UK 53.48N 1.40W
Puduchcheri see Pondicherry
157 H21 **Pudukkottai** Tamil Nādu, SE India 10.22N 78.46E
176 Z10 **Pue** Irian Jaya, E Indonesia 2.42S 140.36E
43 P14 **Puebla** var. Puebla de Zaragoza. Puebla, S Mexico 19.02N 98.12W
43 P15 **Puebla** ⬖ state S Mexico
106 L11 **Puebla de Alcocer** Extremadura, W Spain 38.58N 5.13W
Puebla de Don Fabrique see Puebla de Don Fadrique
107 P13 **Puebla de Don Fadrique** var. Puebla de Don Fabrique. Andalucía, S Spain 37.58N 2.25W
106 J11 **Puebla de la Calzada** Extremadura, W Spain 38.54N 6.37W
106 J5 **Puebla de Sanabria** Castilla-León, N Spain 42.04N 6.37W
106 I4 **Puebla de Trives** see A Pobla de Trives
Puebla de Zaragoza see Puebla
39 T6 **Pueblo** Colorado, C USA 38.15N 104.36W
39 N10 **Pueblo Colorado Wash** valley Arizona, SW USA
63 C16 **Pueblo Libertador** Corrientes, NE Argentina 30.13S 59.22W
42 J10 **Pueblo Nuevo** Durango, C Mexico 23.24N 105.21W
44 J8 **Pueblo Nuevo** Estelí, NW Nicaragua 13.24N 86.26W
56 J3 **Pueblo Nuevo** Falcón, N Venezuela 11.58N 69.57W
44 B6 **Pueblo Nuevo Tiquisate** var. Tiquisate. Escuintla, SW Guatemala 14.16N 91.21W
43 Q11 **Pueblo Viejo, Laguna de** lagoon E Mexico
65 J14 **Puelches** La Pampa, C Argentina 38.08S 65.56W
106 L14 **Puente-Genil** Andalucía, S Spain 37.23N 4.45W
107 Q3 **Puente la Reina** Navarra, N Spain 42.40N 1.49W
106 L12 **Puente Nuevo, Embalse de** ⊠ S Spain
59 D14 **Puente Piedra** Lima, W Peru 11.49S 77.01W
166 F14 **Pu'er** Yunnan, SW China 23.09N 100.57E
47 V6 **Puerca, Punta** headland E Puerto Rico 18.13N 65.36W
39 R12 **Puerco, Rio** ≈ New Mexico, SW USA
59 J17 **Puerto Acosta** La Paz, W Bolivia 15.33S 69.15W
65 G19 **Puerto Aisén** Aisén, S Chile 45.24S 72.42W
43 R17 **Puerto Ángel** Oaxaca, SE Mexico 15.39N 96.29W
Puerto Argentino see Stanley
43 T17 **Puerto Arista** Chiapas, SE Mexico 15.55N 93.47W
45 O16 **Puerto Armuelles** Chiriquí, SW Panama 8.16N 82.51W
Puerto Arrecife see Arrecife
56 D14 **Puerto Asís** Putumayo, SW Colombia 0.27N 76.27W
56 L9 **Puerto Ayacucho** Amazonas, SW Venezuela 5.44N 67.36W
59 C18 **Puerto Ayora** Galapagos Islands, Ecuador, E Pacific Ocean 0.45S 90.19W
59 C18 **Puerto Baquerizo Moreno** var. Baquerizo Moreno. Galapagos Islands, Ecuador, E Pacific Ocean 0.54S 89.37W
44 G4 **Puerto Barrios** Izabal, E Guatemala 15.42N 88.34W
Puerto Bello see Portobelo
56 F8 **Puerto Berrío** Antioquia, C Colombia 6.25N 74.27W
56 F9 **Puerto Boyaca** Boyacá, C Colombia 5.58N 74.36W
56 K4 **Puerto Cabello** Carabobo, N Venezuela 10.27N 68.02W
45 N7 **Puerto Cabezas** var. Bilwi. Región Autónoma Atlántico Norte, NE Nicaragua 14.04N 83.22W
56 L9 **Puerto Carreño** Vichada, E Colombia 6.08N 67.30W
56 E4 **Puerto Colombia** Atlántico, N Colombia 10.58N 74.57W
44 H4 **Puerto Cortés** Cortés, NW Honduras 15.49N 87.55W
56 J4 **Puerto Cumarebo** Falcón, N Venezuela 11.28N 69.21W
Puerto de Cabras see Puerto del Rosario
57 Q5 **Puerto de Hierro** Sucre, NE Venezuela 10.40N 62.03W
66 O11 **Puerto de la Cruz** Tenerife, Islas Canarias, Spain, NE Atlantic Ocean 28.24N 16.33W
66 Q11 **Puerto del Rosario** var. Puerto de Cabras. Fuerteventura, Islas Canarias, Spain, NE Atlantic Ocean 28.28N 13.52W
65 J20 **Puerto Deseado** Santa Cruz, SE Argentina 47.46S 65.52W
42 F8 **Puerto Escondido** Baja California Sur, W Mexico 25.49N 111.20W
43 R17 **Puerto Escondido** Oaxaca, SE Mexico 15.48N 96.57W
62 G12 **Puerto Esperanza** Misiones, NE Argentina 26.01S 54.39W
58 D6 **Puerto Francisco de Orellana** var. Coca. Napo, N Ecuador 0.27S 76.57W
56 H10 **Puerto Gaitán** Meta, C Colombia 4.19N 72.07W
Puerto Gallegos see Río Gallegos
62 G12 **Puerto Iguazú** Misiones, NE Argentina 25.39S 54.34W
58 F12 **Puerto Inca** Huánuco, N Peru 9.21S 74.55W
56 L11 **Puerto Inírida** var. Obando. Guainía, E Colombia 3.48N 67.54W
44 K13 **Puerto Jesús** Guanacaste, NW Costa Rica 10.08N 85.26W
43 Z11 **Puerto Juárez** Quintana Roo, SE Mexico 21.06N 86.46W
57 N5 **Puerto La Cruz** Anzoátegui, NE Venezuela 10.13N 64.40W
56 E14 **Puerto Leguízamo** Putumayo, S Colombia 0.07S 74.51W
45 N5 **Puerto Lempira** Gracias a Dios, E Honduras 15.13N 83.48W
Puerto Libertad see La Libertad
56 I11 **Puerto Limón** Meta, E Colombia 4.00N 71.09W
56 D13 **Puerto Limón** Putumayo, SW Colombia 1.01N 76.30W
Puerto Limón see Limón
107 N11 **Puertollano** Castilla-La Mancha, C Spain 38.40N 4.07W
65 K17 **Puerto Lobos** Chubut, SE Argentina 42.00S 64.58W
56 I3 **Puerto López** La Guajira, N Colombia 11.54N 71.21W
107 Q14 **Puerto Lumbreras** Murcia, SE Spain 37.34N 1.49W
43 V17 **Puerto Madero** Chiapas, SE Mexico 14.43N 92.25W
65 K17 **Puerto Madryn** Chubut, S Argentina 42.45S 65.01W
Puerto Magdalena see Bahía Magdalena
59 J15 **Puerto Maldonado** Madre de Dios, E Peru 12.37S 69.10W
Puerto Masachapa see Masachapa
Puerto México see Coatzacoalcos
65 G17 **Puerto Montt** Los Lagos, C Chile 41.28S 72.57W
43 Z12 **Puerto Morelos** Quintana Roo, SE Mexico 20.47N 86.54W
56 L10 **Puerto Nariño** Vichada, E Colombia 4.57N 67.51W
65 H23 **Puerto Natales** Magallanes, S Chile 51.42S 72.28W
45 X15 **Puerto Obaldía** San Blas, NE Panama 8.37N 77.25W
46 H6 **Puerto Padre** Las Tunas, E Cuba 21.13N 76.34W
56 L9 **Puerto Páez** Apure, C Venezuela 6.10N 67.30W
42 E3 **Puerto Peñasco** Sonora, NW Mexico 31.21N 113.32W
57 N5 **Puerto Píritu** Anzoátegui, NE Venezuela 10.02N 65.02W
47 N8 **Puerto Plata** var. San Felipe de Puerto Plata. N Dominican Republic 19.46N 70.42W
47 N8 **Puerto Plata** ✕ N Dominican Republic 19.43N 70.43W
Puerto Presidente Stroessner see Ciudad del Este
179 Oo14 **Puerto Princesa** off. Puerto Princesa City. Palawan, W Philippines 9.48N 118.43E
Puerto Princesa City see Puerto Princesa
Puerto Príncipe see Camagüey
Puerto Quellón see Quellón
62 F13 **Puerto Rico** Misiones, NE Argentina 26.48S 54.58W
59 K14 **Puerto Rico** Pando, N Bolivia 11.09S 67.28W
56 E12 **Puerto Rico** Caquetá, S Colombia 1.53N 75.08W
47 U5 **Puerto Rico** off. Commonwealth of Puerto Rico; prev. Porto Rico. ◇ US commonwealth territory C West Indies
66 F11 **Puerto Rico** island C West Indies
66 G11 **Puerto Rico Trench** undersea feature NE Caribbean Sea
56 I8 **Puerto Rondón** Arauca, E Colombia 6.16N 71.05W
Puerto San José see San José
65 J21 **Puerto San Julián** var. San Julián. Santa Cruz, SE Argentina 49.14S 67.40W
65 I22 **Puerto Santa Cruz** var. Santa Cruz. Santa Cruz, SE Argentina 50.05S 68.31W

◆ Country ● Country Capital ◇ Dependent Territory ○ Dependent Territory Capital ⬖ Administrative Region ✕ International Airport ▲ Mountain ⛰ Mountain Range Volcano ≈ River ◎ Lake ⊠ Reservoir

59 Q20 **Puerto Sauce** *see* Juan L.Lacaze
59 Q20 **Puerto Suárez** Santa Cruz, E Bolivia 18.58S 57.47W
56 D13 **Puerto Umbría** Putumayo, SW Colombia 0.52N 76.31W
42 J13 **Puerto Vallarta** Jalisco, SW Mexico 20.36N 105.15W
65 G16 **Puerto Varas** Los Lagos, C Chile 41.24S 72.55W
44 M13 **Puerto Viejo** Heredia, NE Costa Rica 10.27N 84.00W
Puertoviejo *see* Portoviejo
59 B18 **Puerto Villamil** *var.* Villamil. Galapagos Islands, Ecuador, E Pacific Ocean 0.57S 91.00W
56 F8 **Puerto Wilches** Santander, N Colombia 7.19N 73.55W
65 H20 **Pueyrredón, Lago** *var.* Lago Cochrane. ◎ S Argentina
131 R7 **Pugachëv** Saratovskaya Oblast', W Russian Federation 52.06N 48.50E
131 T3 **Pugachëv** Udmurtskaya Respublika, NW Russian Federation 56.38N 53.03E
34 H8 **Puget Sound** *sound* Washington, NW USA
109 O17 **Puglia** *var.* Le Puglie, *Eng.* Apulia. ◆ *region* SE Italy
109 N17 **Puglia, Canosa di** *anc.* Canusium. Puglia, SE Italy 41.13N 16.04E
120 I6 **Puhja** *Ger.* Kawelecht. Tartumaa, SE Estonia 58.19N 26.19E
Puhó *see* Púchov
107 V4 **Puigcerdà** Cataluña, NE Spain 42.25N 1.53E
105 N17 **Puigmal** *see* Puigmal d'Err
105 N17 **Puigmal d'Err** *var.* Puigmal. ▲ S France 42.24N 2.07E
78 I16 **Pujehun** S Sierra Leone 7.22N 11.43W
Puka *see* Pukë
193 E20 **Pukaki, Lake** ◎ South Island, NZ
40 F10 **Pukalani** Maui, Hawaii, USA, C Pacific Ocean 20.50N 94.20W
202 J13 **Pukapuka** *atoll* N Cook Islands
203 X9 **Pukapuka** *atoll* Îles Tuamotu, E French Polynesia
Pukari Neem *see* Purekkari Neem
203 X11 **Pukarua** *var.* Pukaruha. *atoll* Îles Tuamotu, E French Polynesia
Pukaruha *see* Pukarua
12 A7 **Pukaskwa** ᴧ Ontario, S Canada
9 V12 **Pukatawagan** Manitoba, C Canada 55.46N 101.13W
203 X16 **Pukatíkei, Maunga** ▲ Easter Island, Chile, E Pacific Ocean
190 C1 **Pukatja** *var.* Ernabella. South Australia 26.18S 132.13E
169 Y12 **Pukch'ŏng** E North Korea 40.13N 128.19E
115 L18 **Pukë** *var.* Puka. Shkodër, N Albania 42.03N 19.53E
192 L6 **Pukekohe** Auckland, North Island, NZ 37.12S 174.54E
192 L7 **Pukemiro** Waikato, North Island, NZ 37.37S 175.02E
202 D12 **Puke, Mont** ▲ Île Futuna, W Wallis and Futuna
Puket *see* Phuket
193 C20 **Puketeraki Range** ▲ South Island, NZ
192 N13 **Puketoi Range** ▲ North Island, NZ
193 F21 **Pukeuri Junction** Otago, South Island, NZ 45.01S 171.01E
121 L16 **Pukhavichy** *Rus.* Pukhovichi. Minskaya Voblasts', C Belarus 53.30N 28.15E
Pukhovichi *see* Pukhavichy
128 M10 **Puksoozero** Arkhangel'skaya Oblast', NW Russian Federation 62.37N 40.29E
114 A10 **Pula** *It.* Pola; *prev.* Pulj. Istra, NW Croatia 44.53N 13.51E
169 U14 **Pulandian** *var.* Xinjin. Liaoning, NE China 39.25N 121.58E
169 T14 **Pulandian Wan** *bay* NE China
179 Rr15 **Pulangi** ᴧ Mindanao, S Philippines
201 O15 **Pulap Atoll** *atoll* Caroline Islands, C Micronesia
20 H9 **Pulaski** New York, NE USA 43.34N 76.06W
22 I10 **Pulaski** Tennessee, S USA 35.11N 87.00W
23 R7 **Pulaski** Virginia, NE USA 37.03N 80.46W
176 Yy13 **Pulau, Sungai** ᴧ Irian Jaya, E Indonesia
112 N13 **Puławy** *Ger.* Neu Amerika. Lubelskie, E Poland 51.25N 21.56E
23 E16 **Pulheim** Nordrhein-Westfalen, W Germany 51.00N 6.48E
Pulicat *see* Pālghāt
161 J19 **Pulicat Lake** *lagoon* SE India
194 M12 **Pulie** ᴧ New Britain, C PNG
Pul'-I-Khatum *see* Polekhatum
Pul-i-Khumri *see* Pol-e Khomrī
Pul-i-Sefid *see* Pol-e Safīd
Pulj *see* Pula
111 W2 **Pulkau** ᴧ NE Austria
95 L15 **Pulkkila** Oulu, C Finland 64.14N 25.52E
125 Cc6 **Pul'kovo** ✈ (Sankt-Peterburg) Leningradskaya Oblast', NW Russian Federation 60.06N 30.23E
34 M9 **Pullman** Washington, NW USA 46.43N 117.10W
110 H10 **Pully** Vaud, SW Switzerland
42 F7 **Púlpita, Punta** *headland* W Mexico 26.30N 111.28W
112 M10 **Pułtusk** Mazowieckie, C Poland 52.41N 21.04E
164 H10 **Pulu** Xinjiang Uygur Zizhiqu, W China 36.36N 81.33E
143 P13 **Pülümür** Tunceli, E Turkey 39.30N 39.54E
201 N10 **Pulusuk** *island* Caroline Islands, C Micronesia
201 N16 **Puluwat Atoll** *atoll* Caroline Islands, C Micronesia
27 N11 **Pumpville** Texas, SW USA 30.55N 101.43W
203 P7 **Punaauia** *var.* Hakapehi. Tahiti, W French Polynesia 17.37S 149.37W
58 S8 **Punã, Isla** *island* SW Ecuador
193 G16 **Punakaiki** West Coast, South Island, NZ 42.07S 171.21E
159 T11 **Púnakha** E Bhutan 27.37N 89.49E

59 L18 **Punata** Cochabamba, C Bolivia 17.33S 65.52W
161 E14 **Pune** *prev.* Poona. Mahārāshtra, W India 18.31N 73.52E
85 M17 **Pungoè, Rio** *var.* Púnguè, Pungwe. ᴧ C Mozambique
23 X10 **Pungo River** ᴧ North Carolina, SE USA
Púnguè/Pungwe *see* Pungoè, Rio
81 N19 **Punia** Maniema, E Dem. Rep. Congo 1.28S 26.25E
64 H8 **Punilla, Sierra de la** ▲ W Argentina
167 P14 **Puning** Guangdong, S China 23.18N 116.12E
64 G10 **Punitaqui** Coquimbo, C Chile 30.49S 71.13W
158 N18 **Punjab** ◆ *state* NW India
155 T9 **Punjab** *prev.* West Punjab, Western Punjab. ◆ *province* E Pakistan
133 Q9 **Punjab Plains** *plain* N India
95 O17 **Punkaharju** *var.* Punkasalmi. Isä-Suomi, E Finland 61.45N 29.21E
59 L18 **Puno** Puno, SE Peru 15.52S 70.03W
59 H17 **Puno** *off.* Departamento de Puno. ◆ *department* S Peru
63 B24 **Punta Alta** Buenos Aires, E Argentina 38.53S 62.00W
65 H24 **Punta Arenas** *prev.* Magallanes. Magallanes, S Chile 53.10S 70.55W
47 T6 **Punta Chame** Panamá, C Panama 8.39N 79.42W
59 G17 **Punta Colorada** Arequipa, SW Peru 16.17S 73.59W
42 F9 **Punta Coyote** Baja California Sur, W Mexico
64 G8 **Punta de Díaz** Atacama, N Chile 28.03S 70.36W
63 D20 **Punta del Este** Maldonado, S Uruguay 34.58S 54.58W
65 K17 **Punta Delgada** Chubut, SE Argentina 42.46S 63.40W
57 S5 **Punta de Mata** Monagas, NE Venezuela 9.43N 63.39W
57 O4 **Punta de Piedras** Nueva Esparta, NE Venezuela 10.54N 64.06W
44 F4 **Punta Gorda** Toledo, SE Belize 16.07N 88.47W
45 N11 **Punta Gorda** Región Autónoma Atlántico Sur, SE Nicaragua 11.31N 83.46W
25 Y14 **Punta Gorda** Florida, SE USA 26.55N 82.03W
44 M11 **Punta Gorda, Río** ᴧ SE Nicaragua
64 H6 **Punta Negra, Salar de** *salt lake* N Chile
42 D5 **Punta Prieta** Baja California, NW Mexico 28.55N 114.10W
44 L13 **Puntarenas** Puntarenas, W Costa Rica 9.57N 84.49W
44 L13 **Puntarenas** *off.* Provincia de Puntarenas. ◆ *province* W Costa Rica
56 J4 **Punto Fijo** Falcón, N Venezuela 11.42N 70.13W
107 S4 **Puntón de Guara** ▲ N Spain 42.18N 0.13W
20 C10 **Punxsutawney** Pennsylvania, NE USA 40.55N 78.57W
95 M14 **Puolanka** Oulu, C Finland 64.51N 27.42E
59 L18 **Pupuya, Nevado** ▲ W Bolivia 15.04S 69.01W
167 Q10 **Puqi** Hubei, C China 29.45N 113.51E
59 F16 **Puquio** Ayacucho, S Peru 14.43S 74.06W
29 N11 **Pur** ᴧ N Russian Federation
29 N11 **Purcell** Oklahoma, C USA 35.00N 97.21W
107 P14 **Purcell Mountains** ▲ British Columbia, SW Canada
107 P14 **Purchena** Andalucía, S Spain 37.21N 2.21W
29 S8 **Purdy** Missouri, C USA 36.49N 93.55W
120 I2 **Purekkari Neem** *prev.* Pukari Neem. *headland* N Estonia 59.33N 24.49E
39 U7 **Purgatoire River** ᴧ Colorado, C USA
111 V3 **Purgstall** *see* Purgstall an der Erlauf
111 V3 **Purgstall an der Erlauf** *var.* Purgstall. Niederösterreich, NE Austria 48.01N 15.08E
160 O13 **Puri** *var.* Jagannath. Orissa, E India 19.52N 85.49E
131 X4 **Purkersdorf** Niederösterreich, NE Austria 48.13N 16.12E
100 I9 **Purmerend** Noord-Holland, C Netherlands 52.30N 4.55E
157 G16 **Pūrna** ᴧ C India
159 R13 **Purnea** *see* Pūrnia
159 R13 **Pūrnia** *prev.* Purnea. Bihār, NE India 25.46N 87.28E
128 M6 **Purnululu** ◎ NW Australia
128 K10 **Puror** ᴧ NW Russian Federation
128 I6 **Pursat** *see* Poŭthísát, Poŭthísát, W Cambodia
128 I6 **Pursat** *see* Poŭthísát, Stoeng, W Cambodia
156 L13 **Purulia** *see* Puruliya
156 L13 **Puruliya** *prev.* Purulia. West Bengal, NE India 23.19N 86.24E
49 G7 **Purus, Río** ᴧ Brazil/Peru
194 G15 **Pūrvik** *Dan.* Paasila *island* SW PNG
95 N17 **Puruvesi** ◎ SE Finland
24 L7 **Puryis** Mississippi, S USA 31.08N 89.24W
116 J11 **Pūrvomay** *prev.* Borisovgrad. Plovdiv, C Bulgaria 42.06N 25.14E
174 J14 **Purwakarta** *prev.* Poerwakarta. Jawa, C Indonesia 6.30S 107.25E
174 L14 **Purwodadi** *prev.* Poerwodadi. Jawa, C Indonesia 7.04S 110.52E
174 K15 **Purwokerto** *prev.* Poerwokerto. Jawa, C Indonesia 7.25S 109.13E
174 Kk15 **Purworejo** *prev.* Poerworedjo. Jawa, C Indonesia 7.45S 110.04E
22 H8 **Puryear** Tennessee, S USA 36.25N 88.21W
160 H13 **Pusad** Mahārāshtra, C India 19.56N 77.40E
169 X16 **Pusan** *off.* Pusan-gwangyŏk, *var.* Busan, *Jap.* Fusan. SE South Korea 35.11N 129.04E

173 Ee4 **Pusatgajo, Pegunungan** ▲ Sumatera, NW Indonesia
Puschlav *see* Poschiavo
131 Q8 **Pushkin** *prev.* Tsarskoye Selo
Pushkin *see* Tsarskoye Selo
131 Q8 **Pushkino** Saratovskaya Oblast', W Russian Federation 51.09N 47.00E
113 M22 **Püspökladány** Hajdú-Bihar, E Hungary 47.19N 21.06E
120 J3 **Püssi** *Ger.* Isenhof. Ida-Virumaa, NE Estonia 59.21N 27.05E
118 I5 **Pustomyty** L'vivs'ka Oblast', W Ukraine 49.43N 23.55E
128 F16 **Pustoshka** Pskovskaya Oblast', W Russian Federation 56.21N 29.16E
Pusztakalán *see* Cǎlan
178 H1 **Putao** *prev.* Fort Hertz. Kachin State, N Myanmar 27.22N 97.24E
192 M8 **Putaruru** Waikato, North Island, NZ 38.02S 175.49E
167 R12 **Putian** Fujian, SE China 25.28N 119.01E
109 O17 **Putignano** Puglia, SE Italy 40.51N 17.07E
Putivl' *see* Putyvl'
43 S10 **Putla** *var.* Putla de Guerrero. Oaxaca, SE Mexico 16.54N 97.55W
Putla de Guerrero *see* Putla
21 N12 **Putnam** Connecticut, NE USA 41.56N 71.52W
27 Q7 **Putnam** Texas, SW USA 32.22N 99.11W
20 M10 **Putney** Vermont, NE USA 42.59N 72.30W
113 L20 **Putnok** Borsod-Abaúj-Zemplén, NE Hungary 48.18N 20.25E
Putorana, Gory/Putorana Mountains *see* Putorana, Plato
126 I18 **Putorana, Plato** *var.* Gory Putorana, *Eng.* Putorana Mountains. ▲ N Russian Federation
64 H2 **Putre** Tarapacá, N Chile 18.11S 69.30W
161 I24 **Puttalam** North Western Province, W Sri Lanka 8.01N 79.54E
161 I24 **Puttalam Lagoon** *lagoon* W Sri Lanka
101 H17 **Putte** Antwerpen, C Belgium 51.04N 4.39E
100 K11 **Putten** Gelderland, C Netherlands 52.15N 5.36E
102 K7 **Puttgarden** Schleswig-Holstein, N Germany 54.30N 11.12E
Puttiala *see* Patiāla
103 D20 **Püttlingen** Saarland, SW Germany 49.16N 6.52E
56 D14 **Putumayo** *off.* Intendencia del Putumayo. ◆ *province* S Colombia
50 E7 **Putumayo, Río** *var.* Río Içá. ᴧ NW South America *see also* Içá, Rio
174 K8 **Putus, Tanjung** *headland* Borneo, N Indonesia 0.27S 109.04E
118 J8 **Putyla** Chernivets'ka Oblast', W Ukraine 47.59N 25.04E
119 S3 **Putyvl'** *Rus.* Putivl'. Sums'ka Oblast', NE Ukraine 51.21N 33.52E
95 M18 **Puula** ◎ SE Finland
95 N18 **Puumala** Isä-Suomi, E Finland 61.31N 28.12E
120 I5 **Puurmani** *Ger.* Talkhof. Jõgevamaa, E Estonia 58.36N 26.17E
101 G17 **Puurs** Antwerpen, C Belgium 51.04N 4.16E
Pu'uUla'ula *see* Red Hill
40 A8 **Puuwai** Niihau, Hawaii, USA, C Pacific Ocean 21.54N 96.11W
10 J7 **Puvirnituq** *prev.* Povungnituk. Quebec, NE Canada 60.10N 77.19W
34 H8 **Puyallup** Washington, NW USA 47.11N 122.17W
167 O5 **Puyang** Henan, C China 35.42N 115.03E
167 R9 **Puyang Jiang** *var.* Tsien Tang. ᴧ SE China
105 O11 **Puy-de-Dôme** ◆ *department* C France
105 N15 **Puylaurens** Tarn, S France 43.33N 2.01E
104 M13 **Puy-l'Évêque** Lot, S France 44.31N 1.10E
105 N17 **Puymorens, Col de** *pass* S France 42.33N 1.50E
58 C7 **Puyo** Pastaza, C Ecuador 1.30S 77.58W
193 A24 **Puysegur Point** *headland* South Island, NZ 46.09S 166.38E
154 J8 **Pūzak, Hāmūn-e Pash.** Hāmūn-i-Puzak. ◎ SW Afghanistan
147 U13 **Puzak, Hāmūn-i-** *see* Pūzak, Hāmūn-e
83 J23 **Pwani** *Eng.* Coast. ◆ *region* E Tanzania
81 O23 **Pweto** Katanga, SE Dem. Rep. Congo (Zaire) 8.29S 28.57E
99 I19 **Pwllheli** NW Wales, UK 52.53N 4.22W
201 O14 **Pwok** Pohnpei, E Micronesia
126 Gg10 **Pyakupur** ᴧ N Russian Federation
128 M6 **Pyalitsa** Murmanskaya Oblast', NW Russian Federation 66.16N 39.55E
128 K10 **Pyal'ma** Respublika Kareliya, NW Russian Federation 62.24N 35.56E
128 I6 **Pyandzh** *see* Panj
177 O9 **Pyaozero, Ozero** ◎ NW Russian Federation
177 F9 **Pyapon** Irrawaddy, SW Myanmar 16.17N 95.40E
121 J15 **Pyarshai** *Rus.* Pershay. Minskaya Voblasts', C Belarus 54.02N 26.44E
126 I6 **Pyasina** ᴧ N Russian Federation
116 I10 **Pyasŭchnik, Yazovir** ◎ C Bulgaria
125 B13 **Pyatigorsk** Stavropol'skiy Kray, SW Russian Federation 44.01N 43.06E
119 S7 **Pyatykhatki** P"yatykhatky. Dnipropetrovs'ka Oblast', E Ukraine 48.23N 33.43E
177 G6 **Pyawbwe** Mandalay, C Myanmar 20.39N 96.04E
131 T3 **Pychas** Udmurtskaya Respublika, NW Russian Federation 56.30N 52.33E
177 F6 **Pyechin** Chin State, W Myanmar 20.01N 93.36E
121 G17 **Pyeski** *Rus.* Peski. Hrodzyenskaya Voblasts', W Belarus 53.12N 24.37E

121 L19 **Pyetrykaw** *Rus.* Petrikov. Homyel'skaya Voblasts', SE Belarus 52.07N 28.30E
95 M16 **Pyhäjärvi** ◎ C Finland
95 O17 **Pyhäjärvi** ◎ SW Finland
95 L15 **Pyhäjoki** Oulu, W Finland
95 L15 **Pyhäjoki** ᴧ W Finland
95 M15 **Pyhäntä** Oulu, C Finland 64.07N 26.19E
95 M16 **Pyhäsalmi** Oulu, C Finland 63.38N 26.00E
95 N16 **Pyhäselkä** ◎ SE Finland
95 M19 **Pyhtää** *Swe.* Pyttis. Etelä-Suomi, S Finland 60.29N 26.40E
177 G6 **Pyinmana** Mandalay, C Myanmar 19.45N 96.12E
117 N24 **Pýles** *var.* Píles. Kárpathos, SE Greece 35.31N 27.08E
117 D21 **Pýlos** *var.* Pílos. Pelopónnisos, S Greece 36.55N 21.42E
20 B12 **Pymatuning Reservoir** ◎ Ohio/Pennsylvania, NE USA
169 X15 **P'yŏngt'aek** SW South Korea 37.00N 127.04E
169 V14 **P'yŏngyang** *var.* P'yŏngyang-si, *Eng.* Pyongyang. ● (North Korea) SW North Korea 39.04N 125.46E
37 Q4 **P'yŏngyang-si** *see* P'yŏngyang
39 P15 **Pyramid Mountains** ▲ New Mexico, SW USA
39 R5 **Pyramid Peak** ▲ Colorado, C USA 39.04N 106.57W
117 D17 **Pyramída** *var.* Piramida. ▲ C Greece 39.08N 21.18E
150 J23 **Pyrenaei Montes** *see* Pyrenees
104 J16 **Pyrénées** *Fr.* Pyrénées, *Sp.* Pirineos; *anc.* Pyrenaei Montes. ▲ SW Europe
104 J16 **Pyrénées-Atlantiques** ◆ *department* SW France
105 N17 **Pyrénées-Orientales** ◆ *department* S France
117 L19 **Pýrgi** *var.* Pirgi. Chíos, E Greece 38.13N 26.01E
117 D19 **Pýrgos** *var.* Pírgos. Dytikí Ellás, S Greece 37.40N 21.27E
119 R4 **Pyryatyn** *Rus.* Piryatin. Poltavs'ka Oblast', NE Ukraine 50.13N 32.31E
112 D9 **Pyrzyce** *Ger.* Pyritz. Zachodniopomorskie, NW Poland 53.09N 14.52E
128 F15 **Pytalovo** *Latv.* Abrene; *prev.* Jaunlatgale. Pskovskaya Oblast', W Russian Federation 57.06N 27.55E
117 M20 **Pythagóreio** *var.* Pithagorio. Sámos, Dodekánisos, Greece, Aegean Sea 37.42N 26.57E
12 L11 **Pythonga, Lac** ◎ Quebec, SE Canada
177 G7 **Pyttegga** ▲ S Norway 62.13N 7.40E
Pyttis *see* Pyhtää
177 G7 **Pyu** Pegu, C Myanmar 18.28N 96.25E
177 G8 **Pyuntaza** Pegu, SW Myanmar 17.51N 96.43E
159 N11 **Pyuthan** Mid Western, W Nepal 28.09N 82.50E
112 H12 **Pyzdry** *Ger.* Peisern. Wielkopolskie, C Poland 52.10N 17.42E

Q

144 H13 **Qā' al Jafr** ◎ S Jordan
207 O11 **Qaanaaq** *var.* Qânâq, *Dan.* Thule. Avannaarsua, N Greenland 77.34N 69.44W
144 G7 **Qabb Eliâs** E Lebanon 33.46N 35.49E
Qâbil *see* Al Qâbil
Qabirri *see* Iori
Qabis *see* Gabès
155 Q10 **Qābis, Khalīj** *see* Gabès, Golfe de
104 M13 **Qacentina** *see* Constantine
145 S14 **Qabr Hūd** C Yemen 16.02N 49.36E
154 L4 **Qaddāhīyah** S Iraq 31.43N 44.28E
149 O4 **Qā'emshahr** *prev.* 'Aliābad, Shāhi. Māzandarān, N Iran 36.31N 52.49E
149 U7 **Qā'en** *var.* Qayen, Qāyen. Khorāsān, E Iran 33.43N 59.07E
147 U13 **Qafa** *spring/well* SW Oman 17.46N 52.55E
169 V9 **Qagan Nur** ◎ NE China
169 Q11 **Qagan Nur** ◎ NE China
164 H13 **Qagan Us** *see* Dulan
164 H13 **Qagcaka** Xizang Zizhiqu, W China 32.31N 81.52E
165 Q10 **Qaidam He** ᴧ C China
162 L8 **Qaidam Pendi** *basin* C China
Qain *see* Qā'en
145 U3 **Qala Áhangarán** *see* Chaghcharān
145 U3 **Qalāat Sikkar** *see* Al Qal'at
Qala Diza *see* Qal'at Dīzah
Qala Nau *see* Qal'eh-ye Now
147 W2 **Qalansīyah** Suquţrā, S Yemen 12.40N 53.30E
145 N9 **Qala Panja** *see* Qal'eh-ye Panjeh
145 N9 **Qala Shāhar** *see* Qal'eh Shahr
Qalāt *see* Kalāt
145 W3 **Qal'at al Aḥmad** E Iraq 32.24N 46.46E
147 N19 **Qal'at Bīshah** 'Asīr, SW Saudi Arabia 19.59N 42.38E
144 H4 **Qal'at Burzay** Ḩamāh, W Syria 35.37N 36.16E
145 N5 **Qal'at Dīzah** *see* Qalā Diza
145 V10 **Qal'at Ḩusayn** E Iraq 32.09N 46.46E
145 V11 **Qal'at Majnūnah** S Iraq 31.39N 45.44E
145 X3 **Qal'at Şāliḥ** *var.* Qal'ah Şālih. E Iraq 31.30N 47.24E
145 W7 **Qal'at Sukkar** SE Iraq 31.52N 46.04E
Qalba Zhotasy *see* Kalbinskiy Khrebet
145 U5 **Qal'eh Bīābān** Fārs, S Iran 143 T9 ...
155 N4 **Qal'eh Shahr** Pash. Qala Shāhar. Sar-e Pol, N Afghanistan 35.34N 65.38E

154 L4 **Qal'eh-ye Now** *var.* Qala Nau. Bādghīs, NW Afghanistan 40.03N 48.56E
155 T2 **Qal'eh-ye Panjeh** *var.* Qala Panja. Badakhshān, NE Afghanistan 36.56N 72.15E
147 U14 **Qamar Bay** *see* Qamar, Ghubbat al
147 V13 **Qamar, Jabal al** ▲ SW Oman
165 R14 **Qamba** Xizang Zizhiqu, W China 31.09N 97.09E
197 K13 **Qamea** *prev.* Nggamea. *island* N Fiji
77 R7 **Qaminis** NE Libya 31.48N 20.04E
143 T9 **Qamishly** *see* Al Qāmishlī
Qānāq *see* Qaanaaq
82 Q11 **Qandala** Bari, NE Somalia 11.30N 50.00E
144 L14 **Qantari** Ar Raqqah, N Syria 36.24N 39.16E
143 V13 **Qapıçiğ Dağı** *Rus.* Gora Kapydzhik. ▲ SW Azerbaijan 39.18N 46.00E
164 H5 **Qapqal** *var.* Qapqal Xibe Zizhixian. Xinjiang Uygur Zizhiqu, NW China 43.46N 81.09E
Qapqal Xibe Zizhixian *see* Qapqal
39 P15 **Qapshagay Böyeni** *see* Kapchagayskoye Vodokhranilishche
145 T4 **Qara Anjīr** N Iraq 35.30N 44.37E
Qarabagh *see* Qarah Bāgh
Qarabóget *see* Karaboget
Qarabulaq *see* Karabulak
Qarabutaq *see* Karabutak
Qaraghandy/Qaraghandy Oblysy *see* Karaganda
145 U4 **Qara Gol** NE Iraq 36.21N 45.38E
Qārah *see* Qāra
154 J4 **Qarah Bāgh** *var.* Qarabāgh. Herāt, NW Afghanistan 35.06N 61.33E
144 G7 **Qaraoun, Lac de** Buḩayrat al Qir'awn. ◎ S Lebanon
Qaraoy *see* Karaoy
Qaraqoyym *see* Karakoyyn, Ozero
Qarasū *see* Karasu
Qaratal *see* Karatal
Qarataū *see* Karatau, Khrebet
Qarataū *see* Karatau, Zhambyl, Kazakhstan
Qaraton *see* Karaton
82 P13 **Qardho** *var.* Kardh. *It.* Gardo. Bari, N Somalia 9.34N 49.30E
148 M4 **Qareh Chāy** ᴧ N Iran
148 K2 **Qareh Sū** ᴧ NW Iran
148 L3 **Qariateine** *see* Al Qaryatayn
Qarkilik *see* Ruoqiang
159 N11 **Qarluq** Mid Western, W Nepal 28.09N 82.50E
153 O13 **Qarokŭl** *Rus.* Karakul'. Sukhondaryo Wiloyati, S Uzbekistan 38.17N 67.39E
153 O13 **Qarokŭl** *Rus.* Ozero Karakul'. ◎ E Tajikistan
153 T12 **Qarqan** *see* Qiemo
164 K9 **Qarqan He** ᴧ NW China
Qarqannah, Juzur *see* Kerkenah, Îles de
155 O1 **Qarqaraly** *see* Karkaralinsk
152 M12 **Qarqin** Jowzjān, N Afghanistan 37.25N 66.03E
Qars *see* Kars
Qarsaqbay *see* Karsakpay
152 M12 **Qarshi** *Rus.* Karshi; *prev.* Bek-Budi. Qashqadaryo Wiloyati, S Uzbekistan 38.54N 65.48E
153 L12 **Qarshi Chŭli** *Rus.* Karshinskaya Step. *grassland* S Uzbekistan
152 M13 **Qarshi Kanali** *Rus.* Karshinskiy Kanal. *canal* S Uzbekistan
Qaryatayn *see* Al Qaryatayn
Qashqadaryo Wiloyati *Rus.* Kashkadar'inskaya Oblast'. ◆ *province* S Uzbekistan
207 N13 **Qasigiannguit** *var.* Qasigiánguit, *Dan.* Christianshâb. Kitaa, C Greenland 68.42N 50.49W
148 M8 **Qāşim, Minţaqat** *see* Al Qaşim
145 R8 **Qaşr 'Amīj** C Iraq 33.30N 41.52E
145 R9 **Qaşr Darwīshah** C Iraq 32.36N 43.27E
148 J6 **Qaşr-e Shīrīn** Kermānshāh, W Iran 34.31N 45.35E
77 O7 **Qaşr Farāfra** W Egypt 27.00N 27.58E
148 M8 **Qaşşim** *see* Al Qaşim
Qa'tabah *see* Kataba
144 H7 **Qaţanā** *var.* Katana. Dimashq, S Syria 33.27N 36.04E
149 N15 **Qatar** *off.* State of Qatar, *Ar.* Dawlat Qatar. ◆ *monarchy* SW Asia
153 R13 **Qatrūyeh** Fārs, S Iran 143 T9 ...
147 S12 **Qattara Depression/Qaţţārah, Munkhafaḑ al** *see* Qaţţāra, Munkhafaḑ al
147 N13 **Qaţţāra, Monkhafad var.** Qattara Depression. *desert* NW Egypt
144 I5 **Qaţţīnah, Buḩayrat** *see* Ḩimş, Buḩayrat
Qaydār *see* Qeydār
Qāyen *see* Qā'en
153 T12 **Qayroqqum** *Rus.* Kayrakkum. Sughd, NW Tajikistan
153 Q10 **Qayroqqum, Obanbori** *Rus.* Kayrakkumskoye Vodokhranilishche. ◎ NW Tajikistan
153 O13 **Qazangöldağ** ▲ SW Tajikistan 39.52N 46.13E
148 L3 **Qazaniyeh** *var.* Dhū Shaykh. E Iraq 33.39N 45.33E
Qazaqstan/Qazaqstan Respublikasy *see* Kazakhstan

143 Y12 **Qazimämmäd** *Rus.* Kazi Magomed. SE Azerbaijan 40.03N 48.56E
148 M4 **Qazris** *see* Cáceres
197 M5 **Qazvīn** *prev.* Kazvin. Qazvīn, N Iran 36.16N 50.00E
77 X10 **Qazvīn** ◆ *province* N Iran
115 L23 **Qeleulevu Lagoon** *lagoon* NE Fiji
Qena *see* Qinā; *anc.* Caene, Caenepolis. E Egypt 26.12N 32.49E
207 N13 **Qeparo** Vlorë, S Albania 40.04N 19.49E
Qeqertarssuaq *var.*
Qeqertarsuaq, Dan. Godhavn. Kitaa, Greenland 69.27N 52.54W
206 M13 **Qeqertarsuaq** *island* W Greenland
207 N13 **Qeqertarsuaq** *island* Disko Bugt. *inlet* W Greenland
206 M13 **Qeqertarsuup Tunua** *Dan.* Disko Bugt. *inlet* W Greenland
153 V14 **Qeshm** Hormozgān, S Iran 26.58N 56.16E
149 R14 **Qeshm** *var.* Jazīreh-ye Qeshm, Qeshm Island. *island* S Iran
149 R14 **Qeshm Island/Qeshm, Jazīreh-ye** *see* Qeshm
148 L4 **Qeydār** *var.* Qaydār. Zanjān, NW Iran 36.50N 47.40E
148 K5 **Qezel Owzan** *var.* Ki Zil Uzen, Qi Zil Uzun. ᴧ NW Iran
148 L5 **Qezel Owzan, Rūd-e** *see* Qezel Owzan
Qian *see* Guizhou
169 V9 **Qian Gorlo** *see* Qian Gorlos
169 V9 **Qian Gorlos** *var.* Qian Gorlo, Qian Gorlos Mongolzu Zizhixian, Qianguozhen. Jilin, NE China 45.06N 124.48E
Qian Gorlos Mongolzu Zizhixian/Qianguozhen *see* Qian Gorlos
167 N9 **Qianjiang** Hubei, C China 30.26N 112.55E
166 K10 **Qianjiang** Sichuan, C China 29.30N 108.45E
166 L14 **Qian Jiang** ᴧ S China
166 G9 **Qianning** *var.* Gartar. Sichuan, C China 30.33N 101.22E
169 U13 **Qian Shan** ▲ NE China
166 H10 **Qianwei** Sichuan, C China 29.15N 103.52E
166 J11 **Qianxi** Guizhou, S China 27.00N 106.01E
165 Q7 **Qianxian** Shaanxi, C China 34.37N 96.43E
164 H10 **Qiemo** *var.* Qarqan. Xinjiang Uygur Zizhiqu, NW China 38.09N 85.30E
166 J10 **Qijiang** Chongqing Shi, C China 29.06N 106.35E
165 N5 **Qijiaojing** Xinjiang Uygur Zizhiqu, NW China 43.28N 91.34E
155 P9 **Qila Saifullāh** Baluchistan, SW Pakistan 30.45N 68.08E
165 S9 **Qilian** Qinghai, C China 38.09N 100.08E
139 Nn10 **Qilian Shan** *var.* Kilien Mountains. ▲ N China
207 O13 **Qimusseriarsuaq** *Dan.* Melville Bugt, *Eng.* Melville Bay. *bay* NW Greenland
Qinā *see* Qena
165 Q7 **Qin'an** Gansu, C China 34.49N 105.50E
169 W7 **Qing** *see* Qinghai
164 K9 **Qingan** Heilongjiang, NE China 46.53N 127.29E
167 R5 **Qingdao** *var.* Ching-Tao, Ch'ing-tao, Tsingtao, Tsintao, *Ger.* Tsingtau. Shandong, E China 36.30N 120.55E
169 V8 **Qinggang** Heilongjiang, NE China 46.40N 126.04E
165 S10 **Qinghai** *var.* Chinghai, Koko Nor, Qing, Qinghai Sheng, Tsinghai. ◆ *province* C China
165 S10 **Qinghai Hu** *var.* Ch'ing Hai, Tsing Hai, Mong. Koko Nor. ◎ C China
Qinghai Sheng *see* Qinghai
164 M3 **Qinghe** *var.* Qinggil. Xinjiang Uygur Zizhiqu, NW China 46.42N 90.19E
164 I6 **Qingjian** Shaanxi, C China 37.10N 110.09E
169 W7 **Qing Jiang** ᴧ C China
166 I12 **Qinglong** var. Liancheng. Guizhou, S China 25.49N 105.10E
166 H8 **Qinglong** Hebei, E China 40.24N 118.57E
166 M12 **Qingyang** *see* Jinjiang
169 U7 **Qingyuan** Liaoning, NE China 42.04N 124.56E
167 Q4 **Qingzhou** *prev.* Yidu. Shandong, E China 36.46N 118.23E
181 X16 **Qin He** ᴧ C China
166 H10 **Qinhuangdao** Hebei, E China 39.57N 119.31E
175 P10 **Qin Xian** *see* Qinxian
13 C20 **Qinxian** *var.* Qin Xian. Shanxi, C China 36.46N 112.42E
169 N3 **Qinzhou** Guangxi Zhuangzu Zizhiqu, S China 22.09N 108.36E
31 X13 **Qionghai** *prev.* Jiaji. Hainan, S China 19.12N 110.26E
161 J1 **Qionglai** Sichuan, C China 30.26N 103.27E
142 H8 **Qionglai Shan** ▲ C China
149 Q1 **Qiongzhou Haixia** *var.* Hainan Strait. *strait* S China
169 U7 **Qiqihar** *var.* Ch'i-ch'i-ha-erh, Tsitsihar; *prev.* Lungkiang. Heilongjiang, NE China 47.23N 124.00E
149 Q12 **Qir** Fārs, S Iran 28.27N 53.04E
164 H10 **Qira** Xinjiang Uygur Zizhiqu, NW China 37.04N 80.45E
155 P15 **Qiryat Aḩmad** *var.* Kiryat Aḩmad. Sind, SE Pakistan 26.19N 68.06E

144 F11 **Qiryat Gat** *var.* Kiryat Gat. Southern, C Israel 31.37N 34.46E
144 G8 **Qiryat Shemona** Northern, N Israel 33.13N 35.34E
147 U14 **Qishn** SE Yemen 15.28N 51.43E
144 G9 **Qishon, Naḩal** ᴧ N Israel
162 K5 **Qita Ghazzah** *see* Gaza Strip
162 K5 **Qitai** Xinjiang Uygur Zizhiqu, NW China 44.00N 89.33E
169 Y8 **Qitaihe** Heilongjiang, NE China 45.45N 130.53E
147 W12 **Qitbīt, Wādī** *dry watercourse* S Oman
167 Q5 **Qixian** *var.* Qi Xian, Zhaoge. Henan, C China 35.34N 114.10E
Qīzān *see* Jīzān
Qizil Orda *see* Kyzylorda
153 V14 **Qizil Qum/Qizilqum** *see* Kyzyl Kum
153 V14 **Qizilrabot** *Rus.* Kyzylrabot. SE Tajikistan 37.28N 74.43E
152 J10 **Qizilrawbe** *Rus.* Kyzylrabat. Bukhoro Wiloyati, C Uzbekistan 40.35N 62.09E
145 S4 **Qi Zil Uzun** *see* Qezel Owzan
152 G7 **Qizil Yār** N Iraq 35.26N 44.12E
Qoraqalpoghiston Respublikasi, W Uzbekistan 42.59N 59.21E
152 H6 **Qoghaly** *see* Kugaly
149 N16 **Qogir Feng** *see* K2
149 N16 **Qom** *var.* Kum, Qum, Qom, N Iran 34.43N 50.53E
149 N16 **Qom** ◆ *province* N Iran
149 Q10 **Qomisheh** *see* Shahrezā
Qomolangma Feng *see* Everest, Mount
148 M7 **Qomsheh** *see* Shahrezā
Qomul *see* Hami
Qondūz *see* Kunduz
Qongyrat *see* Konyrat
Qoqek *see* Tacheng
169 V9 **Qorabowur Kirlari** *see* Karabowur', Uval
Qoradaryo *see* Karadar'ya
152 G6 **Qorajar** *Rus.* Karadzhar. Qoraqalpoghiston Respublikasi, NW Uzbekistan 43.34N 58.35E
152 K12 **Qorako'l** *Rus.* Karakul'. Bukhoro Wiloyati, C Uzbekistan 39.27N 63.45E
152 E5 **Qoraŭzak** *Rus.* Karaŭzyak. Qoraqalpoghiston Respublikasi, NW Uzbekistan 43.07N 60.03E
Qorghanbuy *see* Kurgal'dzhino
Qornet es Saouda ▲ NE Lebanon 36.06N 34.06E
152 L12 **Qorowulbozor** *Rus.* Karaulbazar. Bukhoro Wiloyati, C Uzbekistan 39.28N 64.49E
148 M5 **Qorveh** *var.* Qerveh, Qurveh. Kordestān, W Iran 35.07N 47.46E
Qosshaghyl *see* Koschagyl
Qostanay/Qostanay Oblysy *see* Kostanay
149 P12 **Qoubaïyât** Fārs, S Iran 28.52N 53.40E
149 R13 **Qoubaïyât var.** Al Qubayyât. N Lebanon 37.00N 34.30E
144 H6 **Qoussantina** *see* Constantine
152 H6 **Qowowuyag** *see* Cho Oyu
152 H6 **Qozoqdaryo** *Rus.* Kazakdar'ya. Qoraqalpoghiston Respublikasi, NW Uzbekistan 43.36N 59.47E
21 N11 **Quabbin Reservoir** ◎ Massachusetts, NE USA
102 F12 **Quakenbrück** Niedersachsen, NW Germany 52.41N 7.57E
20 I15 **Quakertown** Pennsylvania, NE USA 40.26N 75.17W
190 M10 **Quambatook** Victoria, SE Australia 35.52S 143.28E
27 Q4 **Quanah** Texas, SW USA 34.19N 99.45W
178 Kk11 **Quang Ngai** *var.* Quangngai, Quang Nghia. Quang Ngai, C Vietnam 15.09N 108.49E
178 X9 **Quang Nghia** *see* Quang Ngai
178 X9 **Quang Tri** Quang Tri, C Vietnam 16.42N 107.15E
Quan Long *see* Ca Mau
152 H6 **Quanshuigou** China/India 35.40N 79.28E
158 H4 **Quanshan** *see* Dedu
167 R13 **Quanshuihe** Guizhou, S China 25.49N 105.05E
166 M12 **Quanzhou** Guangxi Zhuangzu Zizhiqu, S China 25.59N 111.01E
9 V16 **Qu'Appelle** ᴧ Saskatchewan, S Canada
10 M3 **Quaqtaq** *prev.* Koartac. Quebec, NE Canada 60.49N 69.30W
63 E16 **Quaraí** Rio Grande do Sul, S Brazil 30.24S 56.24N
61 H24 **Quaraí, Rio** ᴧ Brazil/Uruguay *see also* Cuareim, Río
175 P10 **Quarles, Pegunungan** ▲ Sulawesi, C Indonesia
109 C20 **Quartano** ◎ Sardegna
109 C20 **Quartu Sant' Elena** Sardegna, Italy, C Mediterranean Sea 39.15N 9.12E
31 X13 **Quasqueton** Iowa, C USA 42.23N 91.45W
181 X16 **Quatre Bornes** W Mauritius 20.15S 57.28E
180 I17 **Quatre Bornes** Mahé, NE Seychelles
143 X10 **Quba** *Rus.* Kuba. N Azerbaijan 41.22N 48.30E
149 S3 **Qubba** *see* Ba'qūbah
149 U7 **Qūchān** *var.* Kuchan. Khorāsān, NE Iran 37.12N 58.28E
191 N10 **Queanbeyan** New South Wales, SE Australia 35.24S 149.14E
13 K6 **Quebec** Quebec, SE Canada 46.49N 71.15W
13 K6 **Quebec** ◆ *province* SE Canada

63 D17 **Quebracho** Paysandú, W Uruguay 31.58S 57.52W

◆ COUNTRY   ◇ DEPENDENT TERRITORY   ◆ ADMINISTRATIVE REGION   ▲ MOUNTAIN   ᴧ VOLCANO   ◎ LAKE
● COUNTRY CAPITAL   ○ DEPENDENT TERRITORY CAPITAL   ✕ INTERNATIONAL AIRPORT   ▲ MOUNTAIN RANGE   ᴧ RIVER   ◙ RESERVOIR

103 K14 **Quedlinburg** Sachsen-Anhalt, C Germany 51.48N 11.09E
144 H10 **Queen Alia** ✈ ('Ammān) 'Ammān, C Jordan
8 L16 **Queen Bess, Mount** ▲ British Columbia, SW Canada 51.15N 124.29W
8 I14 **Queen Charlotte** British Columbia, SW Canada 53.18N 132.04W
117 B24 **Queen Charlotte Bay** *bay* West Falkland, Falkland Islands
8 H14 **Queen Charlotte Islands** *Fr.* Îles de la Reine-Charlotte. *island group* British Columbia, SW Canada
8 I15 **Queen Charlotte Sound** *sea area* British Columbia, W Canada
8 J16 **Queen Charlotte Strait** *strait* British Columbia, W Canada
29 U1 **Queen City** Missouri, C USA 40.24N 92.34W
27 X5 **Queen City** Texas, SW USA 33.09N 94.09W
207 O9 **Queen Elizabeth Islands** *Fr.* Îles de la Reine-Élisabeth. *island group* Nunavut, N Canada
205 Y10 **Queen Mary Coast** *physical region* Antarctica
117 N24 **Queen Mary's Peak** ▲ C Tristan da Cunha
206 M8 **Queen Maud Gulf** *gulf* Arctic Ocean
205 P11 **Queen Maud Mountains** ▲▲ Antarctica
Queen's County *see* Laois
189 U7 **Queensland** ◈ *state* N Australia
199 Hh10 **Queensland Plateau** *undersea feature* N Coral Sea
191 O16 **Queenstown** Tasmania, SE Australia 42.06S 145.33E
193 C22 **Queenstown** Otago, South Island, NZ 45.03S 168.41E
85 I24 **Queenstown** Eastern Cape, S South Africa 31.52S 26.50E
Queenstown *see* Cobh
34 F8 **Queets** Washington, NW USA 47.31N 124.19W
63 D18 **Queguay Grande, Río** ≈ W Uruguay
61 O16 **Queimadas** Bahia, E Brazil 10.58S 39.37W
84 D11 **Quela** Malanje, NW Angola 9.18S 17.07E
85 O16 **Quelimane** *var.* Kilimane, Kilmain, Quilimane. Zambézia, NE Mozambique 17.52S 36.51E
65 G18 **Quellón** *var.* Puerto Quellón. Los Lagos, S Chile 43.05S 73.38W
Quelpart *see* Cheju-do
39 P12 **Quemado** New Mexico, SW USA 34.18N 108.29W
27 O12 **Quemado** Texas, SW USA 28.58N 100.36W
46 K7 **Quemado, Punta de** *headland* E Cuba 20.13N 74.07W
Quemoy *see* Chinmen Tao
64 K13 **Quemú Quemú** La Pampa, E Argentina 36.03S 63.30W
161 E17 **Quepem** Goa, W India 15.13N 74.03E
44 M14 **Quepos** Puntarenas, S Costa Rica 9.28N 84.10W
Que Que *see* Kwekwe
63 D23 **Quequén** Buenos Aires, E Argentina 38.30S 58.43W
63 D23 **Quequén Grande, Río** ≈ E Argentina
63 C23 **Quequén Salado, Río** ≈ E Argentina
Quera *see* Chur
43 N13 **Querétaro** Querétaro de Arteaga, C Mexico 20.36N 100.24W
42 F4 **Querobabi** Sonora, NW Mexico 30.03N 111.02W
44 M13 **Quesada** *var.* Ciudad Quesada, San Carlos. Alajuela, N Costa Rica 10.17N 84.24W
107 O13 **Quesada** Andalucía, S Spain 37.52N 3.05W
167 O7 **Queshan** Henan, C China 32.48N 114.03E
8 M15 **Quesnel** British Columbia, SW Canada 52.58N 122.30W
39 S9 **Questa** New Mexico, SW USA 36.41N 105.37W
104 H7 **Questembert** Morbihan, NW France 47.39N 2.24W
59 K22 **Quetena, Río** ≈ SW Bolivia
155 O10 **Quetta** Baluchistān, SW Pakistan 30.15N 67.00E
Quetzalcoalco *see* Coatzacoalcos
Quetzaltenango *see* Quezaltenango
58 B6 **Quevedo** Los Ríos, C Ecuador 1.01S 79.27W
44 B6 **Quezaltenango** *var.* Quetzaltenango. Quezaltenango, W Guatemala 14.48N 91.27W
44 A2 **Quezaltenango** *off.* Departamento de Quezaltenango, *var.* Quetzaltenango. ◈ *department* SW Guatemala
44 E6 **Quezaltepeque** Chiquimula, SE Guatemala 14.38N 89.25W
179 O15 **Quezon** Palawan, W Philippines 9.13N 118.01E
179 P10 **Quezon City** Luzon, N Philippines 14.39N 121.01E
167 P5 **Qufu** Shandong, E China 35.37N 117.05E
84 B12 **Quibala** Cuanza Sul, NW Angola 10.44S 14.58E
84 B11 **Quibaxe** *var.* Quibaxi. Cuanza Norte, NW Angola 8.30S 14.36E
Quibaxi *see* Quibaxe
56 D9 **Quibdó** Chocó, W Colombia 5.40N 76.37W
104 G7 **Quiberon** Morbihan, NW France 47.30N 3.07W
104 G7 **Quiberon, Baie de** *bay* NW France
56 J5 **Quíbor** Lara, N Venezuela 9.55N 69.34W
44 C4 **Quiché** *off.* Departamento del Quiché. ◈ *department* W Guatemala
101 E21 **Quiévrain** Hainaut, S Belgium 50.25N 3.40E
42 I9 **Quila** Sinaloa, C Mexico 24.24N 107.11W
85 B14 **Quilengues** Huíla, SW Angola 14.03S 14.03E
Quilimane *see* Quelimane
59 G15 **Quillabamba** Cusco, C Peru 12.48S 72.42W
59 L18 **Quillacollo** Cochabamba, C Bolivia 17.23S 66.15W
64 H4 **Quillagua** Antofagasta, N Chile 21.33S 69.32W
105 N17 **Quillan** Aude, S France 42.52N 2.10E
9 U15 **Quill Lakes** ◎ Saskatchewan, S Canada
64 G11 **Quillota** Valparaíso, C Chile 32.54S 71.16W
161 G23 **Quilon** *var.* Kolam, Kollam. Kerala, SW India 8.57N 76.36E
189 V9 **Quilpie** Queensland, C Australia 26.39S 144.15E
155 O4 **Quil-Qala** Bāmīān, N Afghanistan 35.13N 67.02E
64 L7 **Quimilí** Santiago del Estero, C Argentina 27.38S 62.25W
59 O19 **Quimome** Santa Cruz, E Bolivia 17.45S 61.15W
104 F6 **Quimper** *anc.* Quimper Corentin. Finistère, NW France 48.00N 4.05W
Quimper Corentin *see* Quimper
104 G7 **Quimperlé** Finistère, NW France 47.52N 3.33W
34 F8 **Quinault** Washington, NW USA 47.27N 123.53W
34 F8 **Quinault River** ≈ Washington, NW USA
37 P5 **Quincy** California, W USA 39.55N 120.57W
25 S8 **Quincy** Florida, SE USA 30.35N 84.34W
32 I13 **Quincy** Illinois, N USA 39.56N 91.24W
21 O11 **Quincy** Massachusetts, NE USA 42.15N 71.00W
34 J9 **Quincy** Washington, NW USA 47.13N 119.51W
56 E10 **Quindío** *off.* Departamento del Quindío. ◈ *province* C Colombia
56 E10 **Quindío, Nevado del** ▲ C Colombia 4.42N 75.25W
64 J10 **Quines** San Luis, C Argentina 32.15S 65.46W
41 N13 **Quinhagak** Alaska, USA 59.45N 161.55W
78 G13 **Quinhámel** W Guinea-Bissau 11.52N 15.52W
Qui Nhon/Quinhon *see* Quy Nhơn
Quinindé *see* Rosa Zárate
27 U6 **Quinlan** Texas, SW USA 32.54N 96.08W
63 H17 **Quinta** Rio Grande do Sul, S Brazil 32.05S 52.18W
107 O10 **Quintanar de la Orden** Castilla-La Mancha, C Spain 39.36N 3.03W
43 X13 **Quintana Roo** ◈ *state* SE Mexico
107 S6 **Quinto** Aragón, NE Spain 41.25N 0.31W
110 G10 **Quinto** Ticino, S Switzerland 46.32N 8.44E
29 Q11 **Quinton** Oklahoma, C USA 35.07N 95.22W
64 K12 **Quinto, Río** ≈ C Argentina
84 A10 **Quinzau** Zaire, NW Angola 6.50S 12.48E
12 H8 **Quinze, Lac des** ◎ Quebec, SE Canada
85 B15 **Quipungo** Huíla, C Angola 14.49S 14.29E
64 G13 **Quirihue** Bío Bío, C Chile 36.15S 72.34W
84 D12 **Quirima** Malanje, NW Angola 10.51S 18.06E
191 T6 **Quirindi** New South Wales, SE Australia 31.29S 150.40E
57 P5 **Quiriquire** Monagas, NE Venezuela 9.58N 63.13W
12 D10 **Quirke Lake** ◎ Ontario, S Canada
63 B21 **Quiroga** Buenos Aires, E Argentina 35.18S 61.22W
106 I4 **Quiroga** Galicia, NW Spain 42.28N 7.15W
Quiróm, Salar *see* Pocitos, Salar
58 B9 **Quiroz, Río** ≈ NW Peru
84 Q13 **Quissanga** Cabo Delgado, NE Mozambique 12.21S 40.31E
85 M20 **Quissico** Inhambane, S Mozambique 24.42S 34.43E
27 O4 **Quitaque** Texas, SW USA 34.22N 101.03W
84 Q13 **Quiterajo** Cabo Delgado, NE Mozambique 11.37S 40.22E
25 T6 **Quitman** Georgia, SE USA 30.46N 83.33W
24 M6 **Quitman** Mississippi, S USA 32.02N 88.43W
27 V6 **Quitman** Texas, SW USA 32.48N 95.27W
58 C6 **Quito** ● (Ecuador) Pichincha, N Ecuador 0.13S 78.30W
Quito *see* Mariscal Sucre
60 P13 **Quixadá** Ceará, E Brazil 4.57S 39.04W
85 Q15 **Quixaxe** Nampula, NE Mozambique 15.15S 40.07E
166 J9 **Qu Jiang** ≈ C China
167 R10 **Qu Jiang** ≈ SE China
167 N13 **Qujiang** *prev.* Maba. Guangdong, S China 24.47N 113.34E
166 H12 **Qujing** Yunnan, SW China 25.39N 103.52E
Qulan *see* Kulan
152 L10 **Quljuqtov-Toghi** *Rus.* Gory Kul'dzhuktau. ▲▲ C Uzbekistan
Qulsary *see* Kul'sary
Qulyndy Zhazyghy *see* Kulunda Steppe
Qum *see* Qom
Qumälisch *see* Lubartów
165 P11 **Qumar He** ≈ C China
165 Q12 **Qumarlêb** Qinghai, C China 34.06N 95.54E
Qumisheh *see* Shahreẕā
153 O14 **Qumqurghon** *Rus.* Kumkurgan. Surkhondaryo Wiloyati, S Uzbekistan 37.54N 67.31E
Qunaytirah/Qunayṭirah, Muḥāfaẓat al/Qunaytra *see* Al Qunayṭirah
152 G7 **Qŭnghirot** *Rus.* Kungrad. Qoraqalpoghiston Respublikasi, NW Uzbekistan 43.01N 58.49E
201 V12 **Quoi** *island* Chuuk, C Micronesia
15 Kk6 **Quoich** ≈ Nunavut, NE Canada
85 E26 **Quoin Point** *headland* SW South Africa 34.48S 19.39E
190 I7 **Quorn** South Australia 32.22S 138.03E
153 R10 **Qŭqon** *var.* Khokand, *Rus.* Kokand. Farghona Wiloyati, E Uzbekistan 40.34N 70.55E
Qurein *see* Al Kuwayt
153 P14 **Qŭrghonteppa** *Rus.* Kurgan-Tyube. SW Tajikistan 37.51N 68.42E
Qurlurtuuq *see* Kugluktuk
Qurveh *see* Qorveh
Qusair *see* Quseir
143 X10 **Qusar** *Rus.* Kusary. NE Azerbaijan 41.26N 48.27E
Quşayr *see* Al Quşayr
77 Y10 **Quseir** *var.* Al Quşayr, Qusair. E Egypt 26.05N 34.16E
148 I2 **Qūshchī** Āẕarbāyjān-e Bākhtarī, N Iran 37.58N 45.04E
153 N11 **Qŭshrabot** *Rus.* Kushrabat. Samarqand Wiloyati, C Uzbekistan 40.15N 66.40E
Qusmuryn *see* Kushmurun, Kostanay, Kazakhstan
Qusmuryn *see* Kushmurun, Ozero, Kazakhstan
Quṭayfah/Qutayfe/Quteife *see* Al Quṭayfah
Quthing *see* Moyeni
Quwair *see* Guwēr
153 S10 **Quwasoy** *Rus.* Kuvasay. Farghona Wiloyati, E Uzbekistan 40.17N 71.53E
Qu Xian *see* Quzhou
165 N16 **Qüxü** Xizang Zizhiqu, W China 29.25N 90.48E
178 Kk13 **Quy Chanh** Ninh Thuân, S Vietnam 11.28N 108.53E
178 Kk12 **Quy Nhơn** *var.* Quinhon, Qui Nhon. Binh Đinh, C Vietnam 13.46N 109.10E
153 O11 **Qŭytosh** *Rus.* Koytash. Jizzakh Wiloyati, C Uzbekistan 40.13N 67.19E
167 R10 **Quzhou** *var.* Qu Xian. Zhejiang, SE China 28.55N 118.54E
Qyteti Stalin *see* Kuçovë
Qyzylorda/Qyzylorda Oblysy *see* Kyzylorda
Qyzyltū *see* Kishkenekol'
Qyzylzhar *see* Kyzylzhar

R

111 R4 **Raab** Oberösterreich, N Austria 48.19N 13.40E
111 X8 **Raab** *Hung.* Rába. ≈ Austria/Hungary *see also* Rába
Raab *see* Győr
111 V2 **Raabs an der Thaya** Niederösterreich, E Austria 48.51N 15.28E
95 L14 **Raahe** *Swe.* Brahestad. Oulu, W Finland 64.42N 24.30E
100 M10 **Raalte** Overijssel, E Netherlands 52.22N 6.16E
101 I14 **Raamsdonksveer** Noord-Brabant, S Netherlands 51.42N 4.54E
94 L12 **Raanujärvi** Lappi, NW Finland 66.39N 24.40E
98 G9 **Raasay** *island* NW Scotland, UK
120 H3 **Raasiku** *Ger.* Rasik. Harjumaa, NW Estonia 59.22N 25.12E
114 B11 **Rab** *It.* Arbe. Primorje-Gorski Kotar, NW Croatia 44.46N 14.46E
114 B11 **Rab** *It.* Arbe. *island* NW Croatia
175 P16 **Raba** Sumbawa, S Indonesia 8.30S 118.46E
113 G22 **Rába** *Ger.* Raab. ≈ Austria/Hungary *see also* Raab
114 A10 **Rabac** Istra, NW Croatia 45.03N 14.09E
106 I2 **Rábade** Galicia, NW Spain 42.07N 7.37W
82 F10 **Rabak** White Nile, C Sudan 13.12N 32.43E
194 M16 **Rabaraba** Milne Bay, SE PNG 10.02S 149.53E
104 K16 **Rabastens-de-Bigorre** Hautes-Pyrénées, S France 43.22N 0.10E
123 Jj17 **Rabat** W Malta 35.51N 14.25E
76 F6 **Rabat** *var.* al Dar al Baida. ● (Morocco) NW Morocco 34.01N 6.51W
Rabat *see* Victoria
195 P10 **Rabaul** New Britain, E PNG 4.13S 152.10E
Rabbah Ammon/Rabbath Ammon *see* 'Ammān
30 K8 **Rabbit Creek** ≈ South Dakota, N USA
12 H10 **Rabbit Lake** ◎ Ontario, S Canada
197 K13 **Rabi** *prev.* Rambi. *island* N Fiji
146 K9 **Rābigh** Makkah, W Saudi Arabia 22.51N 39.00E
44 D5 **Rabinal** Baja Verapaz, C Guatemala 15.05N 90.23W
173 Ee6 **Rabi, Pulau** *island* NW Indonesia, East Indies
113 L17 **Rabka** Małopolskie, S Poland 49.37N 20.00E
161 F16 **Rabkavi** Karnātaka, W India 16.40N 75.03E
Râbniṭa *see* Rîbniṭa
111 Y6 **Rabnitz** ≈ E Austria
128 J7 **Rabocheostrovsk** Respublika Kareliya, NW Russian Federation 64.58N 34.46E
25 U1 **Rabun Bald** ▲ Georgia, SE USA 34.58N 83.18W
77 S11 **Rabyānah** SE Libya 24.07N 21.58E
77 S11 **Rabyānah, Ramlat** *var.* Rebiana Sand Sea, Şaḥrā' Rabyānah. *desert* SE Libya
Rabyānah, Şaḥrā' *see* Rabyānah, Ramlat
118 L11 **Răcăciuni** Bacău, E Romania 46.20N 27.00E
109 J24 **Racalmuto** Sicilia, Italy, C Mediterranean Sea 37.25N 13.43E
118 J14 **Răcari** Dâmboviṭa, SE Romania 44.37N 25.43E
Răcari *see* Durankulak
118 F13 **Răcășdia** *Hung.* Rakasd. Caraș-Severin, SW Romania 44.58N 21.36E
108 B9 **Racconigi** Piemonte, NE Italy 44.45N 7.41E
35 T15 **Raccoon Creek** ≈ Ohio, N USA
11 V13 **Race, Cape** *headland* Newfoundland, E Canada 46.40N 53.05W
24 K10 **Raceland** Louisiana, S USA 29.43N 90.36W
21 Q12 **Race Point** *headland* Massachusetts, NE USA 42.03N 70.14W
178 J15 **Rach Gia** Kiên Giang, S Vietnam 10.01N 105.04E
178 J14 **Rach Gia, Vinh** *bay* S Vietnam
78 J8 **Rachid** Tagant, C Mauritania 18.48N 11.40W
112 L10 **Raciąż** Mazowieckie, C Poland 52.46N 20.04E
113 I16 **Racibórz** *Ger.* Ratibor. Śląskie, S Poland 50.06N 18.13E
35 N9 **Racine** Wisconsin, N USA 42.42N 87.49W
12 D7 **Racine Lake** ◎ Ontario, S Canada
113 J23 **Ráckeve** Pest, C Hungary 47.07N 18.57E
Rácz-Becse *see* Bečej
147 O15 **Radā'** *var.* Ridā'. W Yemen 14.24N 44.49E
115 O15 **Radan** ▲ SE Yugoslavia 42.59N 21.31E
65 J19 **Rada Tilly** Chubut, SE Argentina 45.54S 67.33W
118 K8 **Rădăuṭi** *Ger.* Radautz, *Hung.* Rádóc. Suceava, N Romania 47.49N 25.58E
118 L8 **Rădăuṭi-Prut** Botoșani, NE Romania 48.14N 26.47E
Radautz *see* Rădăuṭi
Radbusa *see* Radbuza
113 A17 **Radbuza** *Ger.* Radbusa. ≈ SE Czech Republic
22 K6 **Radcliff** Kentucky, S USA 37.50N 85.57W
145 O2 **Radd, Wādī ar** *dry watercourse* N Syria
97 H16 **Råde** Østfold, S Norway 59.21N 10.52E
111 V11 **Radeče** *Ger.* Ratschach. C Slovenia 46.01N 15.10E
Radein *see* Radenci
118 J4 **Radekhiv** *Pol.* Radziechów, *Rus.* Radekhov. L'vivs'ka Oblast', W Ukraine 50.17N 24.39E
Radekhov *see* Radekhiv
111 X9 **Radenci** *Ger.* Radein; *prev.* Radinci. NE Slovenia 46.36N 16.02E
111 S9 **Radenthein** Kärnten, S Austria 46.48N 13.42E
23 R7 **Radford** Virginia, NE USA 37.07N 80.34W
160 C9 **Rādhanpur** Gujarāt, W India 23.52N 71.49E
Radinci *see* Radenci
131 Q6 **Radishchevo** Ul'yanovskaya Oblast', W Russian Federation 52.49N 47.54E
10 I9 **Radisson** Quebec, E Canada 53.47N 77.35W
9 P16 **Radium Hot Springs** British Columbia, SW Canada 50.39N 116.09W
118 F11 **Radna** *Hung.* Máriaradna. Arad, W Romania 46.04N 21.40E
116 K10 **Radnevo** Stara Zagora, C Bulgaria 42.18N 25.57E
99 J20 **Radnor** *cultural region* E Wales, UK
Radnót *see* Iernut
Rádóc *see* Rădăuṭi
103 H24 **Radolfzell am Bodensee** Baden-Württemberg, S Germany 47.43N 8.58E
112 M13 **Radom** Mazowieckie, C Poland 51.23N 21.07E
118 I14 **Radomirești** Olt, S Romania 44.06N 25.00E
113 K14 **Radomsko** *Rus.* Novoradomsk. Łodzkie, C Poland 51.04N 19.25E
119 N4 **Radomyshl'** Zhytomyrs'ka Oblast', N Ukraine 50.30N 29.16E
115 P19 **Radoviš** *prev.* Radovište. E FYR Macedonia 41.39N 22.26E
Radovište *see* Radoviš
96 B13 **Radøy** *island* S Norway
111 R7 **Radstadt** Salzburg, NW Austria 47.24N 13.31E
190 E8 **Radstock, Cape** *headland* South Australia 33.11S 134.18E
121 G15 **Radun'** *Rus.* Radun'. Hrodzyenskaya Voblasts', W Belarus 54.03N 25.00E
126 Gg11 **Raduzhnyy** Khanty-Mansiyskiy Avtonomnyy Okrug, C Russian Federation 62.03N 77.28E
120 F11 **Radviliškis** Radviliškis, N Lithuania 55.48N 23.32E
9 U17 **Radville** Saskatchewan, S Canada 49.28N 104.19W
146 K7 **Raḍwá, Jabal** ▲ W Saudi Arabia 24.31N 38.21E
113 P16 **Radymno** Podkarpackie, SE Poland 49.57N 22.49E
118 J5 **Radyvyliv** Rivnens'ka Oblast', NW Ukraine 50.07N 25.12E
Radziechów *see* Radekhiv
112 I11 **Radziejów** Kujawskio-pomorskie, C Poland 52.36N 18.33E
112 O12 **Radzyń Podlaski** Lubelskie, E Poland 51.48N 22.36E
15 Hh4 **Rae** ≈ Nunavut, NW Canada
158 M13 **Rāe Bareli** Uttar Pradesh, N India 26.13N 81.13E
Rae-Edzo *see* Edzo
23 T11 **Raeford** North Carolina, SE USA 34.58N 79.13W
101 M19 **Raeren** Liège, E Belgium 50.42N 6.06E
15 Kk3 **Rae Strait** *strait* Nunavut, N Canada
192 L11 **Raetihi** Manawatu-Wanganui, North Island, NZ 39.28S 175.16E
203 U13 **Raevavae** *var.* Raivavae. *island* Îles Australes, SW French Polynesia
Rafa *see* Rafah
64 M10 **Rafaela** Santa Fe, E Argentina 31.16S 61.25W
144 E11 **Rafah** *var.* Rafa, Rafaḥ, *Heb.* Rafiah, Raphiah. SW Gaza Strip 31.17N 34.18E
81 L15 **Rafaï** Mbomou, SE Central African Republic 5.01N 23.51E
147 O4 **Rafḥah** Al Ḥudūd ash Shamālīyah, N Saudi Arabia 29.40N 43.28E
Rafiah *see* Rafah
149 R10 **Rafsanjān** Kermān, C Iran 30.25N 56.00E
82 B13 **Raga** Western Bahr el Ghazal, SW Sudan 8.28N 25.40E
21 S8 **Ragged Island** *island* Maine, NE USA
46 I5 **Ragged Island Range** *island group* S Bahamas
192 L7 **Raglan** Waikato, North Island, NZ 37.49S 174.52E
24 G8 **Ragley** Louisiana, S USA 30.31N 93.13W
Ragnit *see* Neman
109 K25 **Ragusa** Sicilia, Italy, C Mediterranean Sea 36.55N 14.42E
Ragusa *see* Dubrovnik
Ragusavecchia *see* Cavtat
175 Qq12 **Raha** Pulau Muna, C Indonesia 4.49S 122.43E
121 N17 **Rahachow** *Rus.* Rogachëv. Homyel'skaya Voblasts', SE Belarus 53.03N 30.04E
69 U6 **Rahad** *var.* Nahr ar Rahad. ≈ W Sudan
Rahad, Nahr ar *see* Rahad
Rahaeng *see* Tak
144 F11 **Rahat** Southern, C Israel 31.20N 34.43E
146 L8 **Rahaṭ, Ḥarrat** *lavaflow* W Saudi Arabia
155 S12 **Rahīmyār Khān** Punjab, SE Pakistan 28.27N 70.21E
97 I14 **Råholt** Akershus, S Norway 60.16N 11.10E
Rahovec *see* Orahovac
203 S10 **Raiatea** *island* Îles Sous le Vent, W French Polynesia
161 H16 **Rāichūr** Karnātaka, C India 16.15N 77.19E
Raidestos *see* Tekirdağ
159 S13 **Rāiganj** West Bengal, NE India 25.37N 88.10E
160 M11 **Raigarh** Madhya Pradesh, C India 21.55N 83.24E
175 Q18 **Raijua, Selat** *strait* Nusa Tenggara, S Indonesia
191 O16 **Railton** Tasmania, SE Australia 41.24S 146.28E
38 L8 **Rainbow Bridge** *natural arch* Utah, W USA
25 Q3 **Rainbow City** Alabama, S USA 33.57N 86.02W
9 N11 **Rainbow Lake** Alberta, W Canada 58.30N 119.24W
23 R5 **Rainelle** West Virginia, NE USA 37.57N 80.46W
34 G10 **Rainier** Oregon, NW USA 46.05N 122.55W
34 H9 **Rainier, Mount** ⛰ Washington, NW USA 46.51N 121.45W
25 Q2 **Rainsville** Alabama, S USA 34.29N 85.51W
10 B11 **Rainy Lake** ◎ Canada/USA
10 A11 **Rainy River** Ontario, C Canada 48.43N 94.33W
Raippaluoto *see* Replot
160 K12 **Raipur** Madhya Pradesh, C India 21.16N 81.42E
160 H10 **Raisen** Madhya Pradesh, C India 23.21N 77.49E
13 N13 **Raisin** ≈ Ontario, SE Canada
35 R11 **Raisin, River** ≈ Michigan, N USA
Raivavae *see* Raevavae
155 W9 **Rāiwind** Punjab, E Pakistan 31.13N 74.10E
176 U9 **Raja Ampat, Kepulauan** *island group* E Indonesia
161 L16 **Rājahmundry** Andhra Pradesh, E India 17.05N 81.42E
161 I18 **Rājampet** Andhra Pradesh, E India 14.09N 79.10E
Rajang *see* Rajang, Batang
174 Mm6 **Rajang, Batang** *var.* Rajang. ≈ East Malaysia
155 S11 **Rājanpur** Punjab, E Pakistan 29.07N 70.19E
161 H23 **Rājapālaiyam** Tamil Nādu, SE India 9.25N 77.36E
158 E12 **Rājasthān** ◈ *state* NW India
159 T15 **Rajbari** Dhaka, C Bangladesh 23.46N 89.39E
159 R12 **Rajbiraj** Eastern, E Nepal 26.34N 86.52E
160 G9 **Rājgarh** Madhya Pradesh, C India 24.01N 76.42E
158 H10 **Rājgarh** Rājasthān, NW India 28.37N 75.25E
159 P14 **Rājgīr** Bihār, N India 25.01N 85.25E
112 O8 **Rajgród** Podlaskie, NE Poland 53.43N 22.40E
160 L12 **Rājim** Madhya Pradesh, C India 20.57N 81.58E
114 C11 **Rajinac, Mali** ▲ W Croatia 44.47N 15.04E
160 B10 **Rājkot** Gujarāt, W India 22.18N 70.46E
159 R14 **Rājmahal** Bihār, NE India 25.03N 87.49E
159 Q14 **Rājmahāl Hills** *hill range* N India
160 K12 **Rāj Nāndgaon** Madhya Pradesh, C India 21.06N 81.01E
158 I8 **Rājpura** Punjab, NW India 30.30N 76.36E
159 S14 **Rajshahi** *prev.* Rampur Boalia. Rajshahi, W Bangladesh 24.24N 88.40E
159 S13 **Rajshahi** ◈ *division* NW Bangladesh
202 K13 **Rakahanga** *atoll* N Cook Islands
193 H19 **Rakaia** Canterbury, South Island, NZ 43.45S 172.02E
193 G19 **Rakaia** ≈ South Island, NZ
158 H3 **Rakaposhi** ▲ N India 36.06N 74.31E
Rakasd *see* Răcășdia
174 Ii14 **Rakata, Pulau** *var.* Pulau Krakatau. *island* S Indonesia
164 K16 **Raka Zangbo** ≈ W China
147 U10 **Rakbah, Qalamat ar** *well* SE Saudi Arabia 20.37N 52.45E
Rakhine State *see* Arakan State
118 I8 **Rakhiv** Zakarpats'ka Oblast', W Ukraine 48.05N 24.15E
147 V13 **Rakhyūt** SW Oman 16.41N 53.09E
197 I14 **Rakiraki** Viti Levu, W Fiji 17.21S 178.11E
Rakka *see* Ar Raqqah
120 I4 **Rakke** Lääne-Virumaa, NE Estonia 58.58N 26.14E
97 I16 **Rakkestad** Østfold, S Norway 59.25N 11.19E
112 F12 **Rakoniewice** *Ger.* Rakwitz. Wielkolpolskie, C Poland 52.09N 16.10E
Rakonitz *see* Rakovník
85 H18 **Rakops** Central, C Botswana 21.01S 24.23E
113 C16 **Rakovník** *Ger.* Rakonitz. Středočeský Kraj, W Czech Republic 50.07N 13.43E
116 J10 **Rakovski** Plovdiv, C Bulgaria 42.16N 24.58E
Rakutō-kō *see* Naktong-gang
120 I3 **Rakvere** *Ger.* Wesenberg. Lääne-Virumaa, N Estonia 59.21N 26.19E
Rakwitz *see* Rakoniewice
24 L6 **Raleigh** Mississippi, S USA 32.01N 89.30W
23 U9 **Raleigh** *state capital* North Carolina, SE USA 35.46N 78.38W
23 Y11 **Raleigh Bay** *bay* North Carolina, SE USA
23 U9 **Raleigh-Durham** ✈ North Carolina, SE USA 35.54N 78.45W
201 S6 **Ralik Chain** *island group* Ralik Chain, W Marshall Islands
27 N5 **Ralls** Texas, SW USA 33.40N 101.23W
20 G13 **Ralston** Pennsylvania, NE USA 41.29N 76.57W
147 O16 **Ramādah** W Yemen 13.35N 43.50E
Ramadi *see* Ar Ramādī
107 N2 **Ramales de la Victoria** Cantabria, N Spain 43.15N 3.28W
144 F10 **Ramallah** C West Bank 31.53N 34.49E
63 C19 **Ramallo** Buenos Aires, E Argentina 33.30S 60.01W
161 H20 **Rāmanagaram** Karnātaka, E India 12.45N 77.16E
161 I23 **Rāmanāthapuram** Tamil Nādu, SE India 9.22N 78.52E
160 N12 **Rāmapur** Orissa, E India 21.48N 84.00E
161 I14 **Rāmāreddi** *var.* Kāmāreddi, Kamareddy. Andhra Pradesh, C India 18.19N 78.23E
144 F10 **Ramat Gan** Tel Aviv, W Israel 32.04N 34.48E
105 T6 **Rambervillers** Vosges, NE France 48.15N 6.50E
Rambi *see* Rabi
105 N5 **Rambouillet** Yvelines, N France 48.39N 1.49E
194 L8 **Rambutyo Island** *island* N PNG
159 Q12 **Ramechhap** Central, C Nepal 27.19N 86.04E
191 R12 **Rame Head** *headland* Victoria, SE Australia 37.48S 149.30E
130 L4 **Ramenskoye** Moskovskaya Oblast', W Russian Federation 55.31N 38.24E
128 J15 **Rameshki** Tverskaya Oblast', W Russian Federation 57.21N 36.05E
159 P14 **Rāmgarh** Bihār, N India 23.37N 85.31E
158 D11 **Rāmgarh** Rājasthān, NW India 27.26N 70.35E
148 M9 **Rāmhormoz** *var.* Ram Hormuz, Ramuz. Khūzestān, SW Iran 31.17N 49.37E
Ram Hormuz *see* Rāmhormoz
Ram, Jebel *see* Ramm, Jabal
144 F10 **Ramla** *var.* Ramle, Ramleh, *Ar.* Er Ramle. Central, C Israel 31.55N 34.52E
Ramle/Ramleh *see* Ramla
144 F14 **Ramm, Jabal** *var.* Jebel Ram. ▲ SW Jordan 29.34N 35.24E
158 K10 **Rāmnagar** Uttar Pradesh, N India 29.22N 79.07E
97 N15 **Ramnäs** Västmanland, C Sweden 59.46N 16.16E
Râmnicul-Sărat *see* Râmnicu Sărat
118 L12 **Râmnicu Sărat** *prev.* Râmnicul-Sărat, Rîmnicu-Sărat. Buzău, E Romania 45.24N 27.06E
118 I13 **Râmnicu Vâlcea** *prev.* Rîmnicu Vîlcea. Vâlcea, C Romania 45.04N 24.32E
Ramokgwebana *see* Ramokgwebane
85 J18 **Ramokgwebane** *var.* Ramokgwebana. Central, NE Botswana 20.36S 27.39E
130 L7 **Ramon'** Voronezhskaya Oblast', W Russian Federation 51.51N 39.18E
37 V17 **Ramona** California, W USA 33.02N 116.52W
58 A10 **Ramón, Laguna** ◎ NW Peru
12 G7 **Ramore** Ontario, S Canada 48.26N 80.19W
42 M11 **Ramos** San Luis Potosí, C Mexico 22.48N 101.55W
43 N8 **Ramos Arizpe** Coahuila de Zaragoza, NE Mexico 25.34N 100.58W
42 J9 **Ramos, Río de** ≈ C Mexico
41 R8 **Rampart** Alaska, USA 65.30N 150.10W
14 G5 **Ramparts** ≈ Northwest Territories, NW Canada
158 K10 **Rāmpur** Uttar Pradesh, N India 28.48N 79.03E
160 F9 **Rāmpura** Madhya Pradesh, C India 24.34N 75.25E
Rampur Boalia *see* Rajshahi
177 F6 **Ramree Island** *island* W Myanmar
147 W6 **Rams** *var.* Ar Rams. Ra's al Khaymah, NE UAE 25.52N 56.01E
149 N4 **Rāmsar** *prev.* Sakhtsar. Māzandarān, N Iran 36.55N 50.39E
95 H16 **Ramsele** Västernorrland, N Sweden 63.33N 16.35E
23 T9 **Ramseur** North Carolina, SE USA 35.43N 79.39W
99 I16 **Ramsey** NE Isle of Man 54.19N 4.24W
99 I16 **Ramsey Bay** *bay* NE Isle of Man
12 E9 **Ramsey Lake** ◎ Ontario, S Canada
99 Q22 **Ramsgate** SE England, UK 51.19N 1.25E
96 M10 **Ramsjö** Gävleborg, C Sweden 62.10N 15.40E
160 I12 **Rāmtek** Mahārāshtra, C India 21.28N 79.28E
Ramtha *see* Ar Ramthā
194 H11 **Ramu** ≈ N PNG
Ramuz *see* Rāmhormoz
120 G12 **Ramygala** Panevėžys, C Lithuania 55.30N 24.18E
94 G12 **Rana** ≈ C Norway
158 H14 **Rāna Pratāp Sāgar** ◎ N India
175 O2 **Ranau** Sabah, East Malaysia 5.55N 116.43E
174 I12 **Ranau, Danau** ◎ Sumatera, W Indonesia
64 H12 **Rancagua** Libertador, C Chile 34.10S 70.45W
101 G22 **Rance** Hainaut, S Belgium 50.09N 4.16E
104 H6 **Rance** ≈ NW France
62 J9 **Rancharia** São Paulo, S Brazil 22.13S 50.53W
159 P15 **Rānchi** Bihār, N India 23.22N 85.19E
63 D21 **Ranchos** Buenos Aires, E Argentina 35.31S 58.22W
39 S9 **Ranchos De Taos** New Mexico, SW USA 36.21N 105.36W
65 G16 **Ranco, Lago** ◎ C Chile
97 C16 **Randaberg** Rogaland, S Norway 59.00N 5.38E
31 U7 **Randall** Minnesota, N USA 46.05N 94.30W
109 L23 **Randazzo** Sicilia, Italy, C Mediterranean Sea 37.52N 14.57E
97 G21 **Randers** Århus, C Denmark 56.28N 10.03E
94 I12 **Randijaure** ◎ N Sweden
23 T9 **Randleman** North Carolina, SE USA 35.49N 79.48W
21 O11 **Randolph** Massachusetts, NE USA 42.09N 71.02W
31 Q13 **Randolph** Nebraska, C USA 42.25N 97.05W
38 M1 **Randolph** Utah, W USA 41.40N 111.10W
102 P9 **Randow** ≈ NE Germany
97 H14 **Randsfjorden** ◎ S Norway
94 K13 **Råneå** Norrbotten, N Sweden 65.52N 22.17E
95 F15 **Ranemsletta** Nord-Trøndelag, C Norway 64.36N 11.55E
78 H10 **Ranérou** C Senegal 15.17N 14.00W
193 E22 **Ranfurly** Otago, South Island, NZ 45.07S 170.06E
178 Hh17 **Rangae** Narathiwat, SW Thailand 6.19N 101.45E
159 V16 **Rangamati** Chittagong, SE Bangladesh 22.40N 92.10E
192 I2 **Rangauru Bay** *bay* North Island, NZ
21 P6 **Rangeley** Maine, NE USA 44.58N 70.37W
39 O4 **Rangely** Colorado, C USA 40.05N 108.48W
27 R7 **Ranger** Texas, SW USA 32.28N 98.40W
12 C9 **Ranger Lake** Ontario, S Canada 46.51N 83.34W
12 C9 **Ranger Lake** ◎ Ontario, S Canada
159 V12 **Rangia** Assam, NE India 26.27N 91.34E
193 I18 **Rangiora** Canterbury, South Island, NZ 43.19S 172.33E
203 T9 **Rangiroa** *atoll* Îles Tuamotu, W French Polynesia
192 N9 **Rangitaiki** ≈ North Island, NZ
193 F19 **Rangitata** ≈ South Island, NZ
192 M12 **Rangitikei** ≈ North Island, NZ
192 L6 **Rangitoto Island** *island* N NZ
Rangkasbitoeng *see* Rangkasbitung
174 Ii14 **Rangkasbitung** *prev.* Rangkasbitoeng. Jawa, SW Indonesia 6.21S 106.12E
178 Hh9 **Rang, Khao** ▲ C Thailand 16.13N 99.03E
153 V13 **Rangkŭl** *Rus.* Rangkul'. SE Tajikistan 38.30N 74.24E
Rangkul' *see* Rangkŭl
Rangoon *see* Yangon
159 T13 **Rangpur** Rajshahi, N Bangladesh 25.46N 89.20E
174 Hh7 **Rangsang, Pulau** *island* W Indonesia
161 F18 **Rānibennur** Karnātaka, W India 14.36N 75.39E
159 R15 **Rāniganj** West Bengal, NE India 23.34N 87.13E
155 Q13 **Rānīpur** Sind, SE Pakistan 27.16N 68.34E
Rāniyah *see* Ṛānya
27 N9 **Rankin** Texas, SW USA 31.12N 101.56W
15 L7 **Rankin Inlet** Nunavut, C Canada 62.52N 92.13W
191 P8 **Rankins Springs** New South Wales, SE Australia 33.51S 146.16E
Rankovićevo *see* Kraljevo
110 I7 **Rankweil** Vorarlberg, W Austria 47.16N 9.40E
Rann *see* Brežice
131 T8 **Ranneye** Orenburgskaya Oblast', W Russian Federation 51.28N 52.29E
98 I10 **Rannoch, Loch** ◎ C Scotland, UK
203 U17 **Rano Kau** *var.* Rano Kao. *crater* Easter Island, Chile, E Pacific Ocean 27.10S 109.25W
178 Gg14 **Ranong** Ranong, SW Thailand 9.58N 98.40E
195 T14 **Ranongga** *var.* Ghanongga. *island* NW Solomon Islands
203 W16 **Rano Raraku** *ancient monument* Easter Island, Chile, E Pacific Ocean 27.07S 109.18W
176 W9 **Ransiki** Irian Jaya, E Indonesia 1.27S 134.12E
94 K12 **Rantajärvi** Norrbotten, N Sweden 66.45N 23.39E
95 N17 **Rantasalmi** Isä-Suomi, SE Finland 62.02N 28.22E
175 N11 **Rantau** Borneo, C Indonesia 2.55S 115.09E
174 Hh7 **Rantau, Pulau** *var.* Pulau Tebingtinggi. *island* W Indonesia
175 Pp11 **Rantepao** Sulawesi, C Indonesia 2.58S 119.58E
32 M13 **Rantoul** Illinois, N USA 40.19N 88.08W
95 L15 **Rantsila** Oulu, C Finland 64.31N 25.40E
94 L13 **Ranua** Lappi, NW Finland 65.55N 26.34E
145 T3 **Ṛānya** *var.* Rāniyah. NE Iraq 36.15N 44.52E
163 X3 **Raohe** Heilongjiang, NE China 46.49N 134.00E
76 H9 **Raoui, Erg er** *desert* W Algeria
199 L11 **Rapa** *island* Îles Australes, S French Polynesia
203 V14 **Rapa Iti** *island* Îles Australes, SW French Polynesia
108 D10 **Rapallo** Liguria, NW Italy 44.21N 9.13E
Rapa Nui *see* Pascua, Isla de
Raphiah *see* Rafah
23 V5 **Rapidan River** ≈ Virginia, NE USA
30 J10 **Rapid City** South Dakota, N USA 44.04N 103.13W
13 P8 **Rapide-Blanc** Quebec, SE Canada 47.48N 72.57W
12 I8 **Rapide-Deux** Quebec, SE Canada 47.56N 78.33W
120 K6 **Räpina** *Ger.* Rappin. Põlvamaa, SE Estonia 58.06N 27.27E
120 G4 **Rapla** *Ger.* Rappel. Raplamaa, NW Estonia 59.00N 24.46E
120 G4 **Raplamaa** *off.* Rapla Maakond. ◈ *province* NW Estonia
23 X6 **Rappahonnock River** ≈ Virginia, NE USA

◆ Country
● Country Capital
◇ Dependent Territory
○ Dependent Territory Capital
◈ Administrative Region
✈ International Airport
▲ Mountain
▲▲ Mountain Range
⛰ Volcano
≈ River
◎ Lake
⊞ Reservoir

Rappel see Rapla

110 G7 Rapperswil Sankt Gallen, NE Switzerland 47.13N 8.49E

Rappin see Räpina

159 N12 Räpti N India

59 K16 Rapulo, Río ↔ E Bolivia

Raqqah/Raqqah, Muḥāfaẓat al see Ar Raqqah

20 J8 Raquette Lake ◎ New York, NE USA

20 J6 Raquette River ↔ New York, NE USA

203 V10 Raraka atoll Îles Tuamotu, C French Polynesia

203 V10 Raroia atoll Îles Tuamotu, C French Polynesia

202 H15 Rarotonga ✈ Rarotonga, S Cook Islands, C Pacific Ocean 21.15S 159.45W

202 H16 Rarotonga island S Cook Islands, C Pacific Ocean

153 P12 Rarz W Tajikistan 39.23N 68.43E

145 N2 Ras al 'Ayn var. Ras al 'Ain. Al Ḥasakah, N Syria 36.52N 40.04E

144 H3 Ra's al Basīṭ Al Lādhiqīyah, W Syria 35.57N 35.55E

147 R5 Ras al-Hafgī see Ra's al Khafjī

147 R5 Ra's al Khafjī var. Ra's al-Hafgī. Ash Sharqīyah, NE Saudi Arabia 28.22N 48.29E

Ras al-Khaimah/Ras al Khaimah see Ra's al Khaymah

149 R15 Ra's al Khaymah var. Ras al Khaimah. Ra's al Khaymah, NE UAE 25.44N 55.54E

149 R15 Ra's al Khaymah var. Ras al-Khaimah. ✈ Ra's al Khaymah, NE UAE 25.37N 55.51E

144 G13 Ra's an Naqb Ma'ān, S Jordan 30.00N 35.29E

63 B26 Rasa, Punta headland E Argentina 40.50S 62.15W

176 W10 Rasawi Irian Jaya, E Indonesia 2.04S 134.02E

Râşcani see Rîşcani

82 J10 Ras Dashen Terara ▲ N Ethiopia 13.12N 38.09E

157 K19 Rasdu Atoll atoll C Maldives

120 E12 Raseiniai Raseiniai, C Lithuania 55.23N 23.06E

77 X8 Rās Ghārib E Egypt 28.16N 33.01E

168 D6 Rashaant Bayan-Ölgiy, W Mongolia 47.48N 90.45E

168 L10 Rashaant Dundgovĭ, C Mongolia 44.54N 106.32E

168 J6 Rashaant Hövsgöl, N Mongolia 49.08N 101.27E

145 Y11 Rashid E Iraq 31.15N 47.31E

77 V7 Rashid Eng. Rosetta. N Egypt 31.24N 30.25E

148 M3 Rasht var. Resht. Gīlān, NW Iran 37.18N 49.37E

145 S2 Rashwān N Iraq 36.28N 43.54E

115 M15 Raška Serbia, C Yugoslavia 43.18N 20.37E

121 P15 Rasna Rus. Ryasna. Mahilyowskaya Voblasts', E Belarus 54.01N 31.12E

118 J12 Râşnov prev. Rîşno, Rozsnyó, Hung. Barcarozsnyó. Braşov, C Romania 45.34N 25.27E

120 L11 Rasony Rus. Rossony. Vitsyebskaya Voblasts', N Belarus 55.55N 28.51E

Ra's Shamrah see Ugarit

131 N7 Rasskazovo Tambovskaya Oblast', W Russian Federation 52.42N 41.45E

121 O16 Rasta ↔ E Belarus

Rastadt see Rastatt

Rastäne see Ar Rastān

147 S6 Ra's Tannūrah Eng. Ras Tanura. Ash Sharqīyah, NE Saudi Arabia 26.44N 50.04E

Ras Tanura see Ra's Tannūrah

103 G21 Rastatt var. Rastadt. Baden-Württemberg, SW Germany 48.52N 8.12E

Rastenburg see Kętrzyn

155 V7 Rasūlnagar Punjab, E Pakistan 32.19N 73.51E

201 U6 Ratak Chain island group Ratak Chain, E Marshall Islands

121 K15 Ratamka Rus. Ratomka. Minskaya Voblasts', C Belarus 53.57N 27.23E

95 G17 Råtan Jämtland, C Sweden 62.28N 14.34E

158 G11 Ratangarh Rājasthān, NW India 28.01N 74.39E

Rat Buri see Ratchaburi

178 H17 Ratchaburi var. Rat Buri. Ratchaburi, W Thailand 13.30N 99.49E

31 W15 Rathbun Lake ◎ Iowa, C USA

Ráth Caola see Rathkeale

177 F5 Rathedaung Arakan State, W Myanmar 20.30N 92.48E

102 M12 Rathenow Brandenburg, NE Germany 52.34N 12.20E

99 C19 Rathkeale Ir. Ráth Caola. SW Ireland 52.31N 8.55W

98 F13 Rathlin Island Ir. Reachlainn. island N Northern Ireland, UK

99 C20 Ráthluirc Ir. An Ráth. SW Ireland 52.22N 8.44W

Ratibor see Racibórz

Ratisbon/Ratisbona/ Ratisbonne see Regensburg

Rätische Alpen see Rhaetian Alps

40 E17 Rat Island island Aleutian Islands, Alaska, USA

40 E17 Rat Islands island group Aleutian Islands, Alaska, USA

160 F10 Ratlam prev. Rutlam. Madhya Pradesh, C India 23.23N 75.03E

161 D15 Ratnāgiri Mahārāshtra, W India 17.00N 73.20E

161 K26 Ratnapura Sabaragamuwa Province, S Sri Lanka 6.40N 80.25E

118 J2 Ratne Rus. Ratno. Volyns'ka Oblast', NW Ukraine 51.40N 24.33E

Ratno see Ratne

Ratomka see Ratamka

39 U8 Raton New Mexico, SW USA 36.54N 104.27W

145 S12 Ratqah, Wādī ar dry watercourse W Iraq

Ratschach see Radeče

178 H17 Rattaphum Songkhla, SW Thailand 7.07N 100.16E

28 L6 Rattlesnake Creek ↔ Kansas, C USA

96 L13 Rättvik Dalarna, C Sweden 60.53N 15.12E

102 K9 Ratzeburg Mecklenburg-Vorpommern, N Germany 53.41N 10.48E

102 K9 Ratzeburger See ◎ N Germany

8 J10 Ratz, Mount ▲ British Columbia, SW Canada 57.22N 132.17W

63 D22 Rauch Buenos Aires, E Argentina 36.47S 59.06W

43 U16 Raudales Chiapas, SE Mexico

Raudhatain see Ar Rawḍatayn

Raudnitz an der Elbe see Roudnice nad Labem

94 K1 Raufarhöfn Nordhurland Eystra, NE Iceland 66.26N 15.57W

96 H13 Raufoss Oppland, S Norway 60.43N 10.36E

Raukawa see Cook Strait

192 Q8 Raukumara ▲ North Island, NZ 37.46S 178.07E

199 J12 Raukumara Plain undersea feature N Coral Sea

192 P8 Raukumara Range ▲ North Island, NZ

160 N11 Raulakela var. Raurkela; prev. Rourkela. Orissa, E. India 22.13N 84.52E

97 F15 Rauland Telemark, S Norway 59.41N 7.57E

95 J19 Rauma Swe. Raumo. Länsi-Suomi, W Finland 61.09N 21.30E

96 F10 Rauma ↔ S Norway

Raumo see Rauma

120 H8 Rauna Cēsis, C Latvia 57.19N 25.34E

174 Mm16 Raung, Gunung ▲ Jawa, S Indonesia 8.00S 114.07E

97 J22 Rauna Skåne, S Sweden 56.01N 12.48E

172 R6 Rausu Hokkaidō, NE Japan 44.00N 145.06E

172 R6 Rausu-dake ▲ Hokkaidō, NE Japan 44.06N 145.04E

95 M17 Rautalampi Itä-Suomi, C Finland 62.37N 26.48E

95 N16 Rautavaara Itä-Suomi, C Finland 63.28N 28.15E

118 M9 Răuţel C Moldova

95 O18 Rautjärvi Etelä-Suomi, SE Finland 61.23N 29.23E

Rautu see Sosnovo

203 V11 Ravahere atoll Îles Tuamotu, C French Polynesia

109 J25 Ravanusa Sicilia, Italy, C Mediterranean Sea 37.16N 13.57E

149 S9 Rāvar Kermān, C Iran 31.15N 56.51E

153 Q11 Ravat Oshskaya Oblast', SW Kyrgyzstan 39.54N 70.06E

20 K11 Ravena New York, NE USA 42.28N 73.49W

108 H10 Ravenna Emilia-Romagna, N Italy 44.28N 12.15E

31 O15 Ravenna Nebraska, C USA 41.01N 98.54W

35 U1 Ravenna Ohio, N USA 41.09N 81.14W

103 I24 Ravensburg Baden-Württemberg, S Germany 47.46N 9.37E

189 W4 Ravenshoe Queensland, NE Australia 17.29S 145.28E

188 K13 Ravensthorpe Western Australia 33.37S 120.03E

23 Q4 Ravenswood West Virginia, NE USA 38.57N 81.45W

155 U9 Rāvi ↔ India/Pakistan

114 C9 Ravna Gora Primorje-Gorski Kotar, NW Croatia 45.20N 14.54E

111 U10 Ravne na Koroškem Ger. Gutenstein. N Slovenia 46.33N 14.57E

145 P6 Rāwah N Iraq 34.32N 41.54E

203 T4 Rawaki prev. Phoenix Island. atoll Phoenix Islands, C Kiribati

155 U6 Rāwalpindi Punjab, NE Pakistan 33.38N 73.06E

112 L13 Rawa Mazowiecka Łódzkie, C Poland 51.46N 20.15E

145 T2 Rawāndiz var. Rawandoz, Rawāndūz. N Iraq 36.37N 44.31E

Rawandoz/Rawāndūz see Rawāndiz

176 Vv9 Rawarra ↔ Irian Jaya, E Indonesia

176 V9 Rawas Irian Jaya, E Indonesia 1.07S 132.12E

145 O4 Rawdah ◊ E Syria

112 G13 Rawicz Ger. Rawitsch. Wielkopolskie, C Poland 51.37N 16.51E

Rawitsch see Rawicz

188 M11 Rawlinna Western Australia 30.05S 125.35E

35 W16 Rawlins Wyoming, C USA 41.47N 107.14W

65 J21 Rawson Chubut, SE Argentina 43.22S 65.01W

165 R16 Rawu Xizang Zizhiqu, W China 29.30N 96.42E

159 N12 Raxaul Bihār, N India 26.58N 84.51E

30 K3 Ray North Dakota, N USA 48.19N 103.11W

174 M9 Raya, Bukit ▲ Borneo, N Indonesia 0.40S 112.40E

161 I18 Rāyachoti Andhra Pradesh, E India 14.03N 78.43E

161 H17 Rāyadrug var. Rāyagarha. Orissa, E India 19.11N 83.22E

161 M14 Rāyagarha prev. Rāyadrug. Orissa, E India 19.11N 83.22E

144 H7 Rayak var. Rayaq, Fr. Rayak. E Lebanon 33.51N 36.03E

Rayaq see Rayak

145 X12 Rāyat E Iraq 36.39N 44.56E

174 J10 Raya, Tanjung headland Pulau Bangka, W Indonesia 1.49S 106.04E

11 R13 Ray, Cape headland Newfoundland, E Canada 47.59N 59.15W

126 Mm16 Raychikhinsk Amurskaya Oblast', SE Russian Federation 49.47N 129.19E

131 U9 Rayevskiy Respublika Bashkortostan, W Russian Federation 54.04N 54.58E

9 Q17 Raymond Alberta, SW Canada 49.30N 112.40W

34 K6 Raymond Mississippi, S USA 32.15N 90.25W

34 F9 Raymond Washington, NW USA 46.41N 123.43W

191 T8 Raymond Terrace New South Wales, SE Australia 32.46S 151.45E

27 T17 Raymondville Texas, SW USA 26.27N 97.45W

9 U16 Raymore Saskatchewan, S Canada 51.24N 104.34W

190 I8 Ray Mountains ▲ Alaska, USA 33.34S 138.13E

24 H9 Rayne Louisiana, S USA

43 O12 Rayón San Luis Potosí, C Mexico 21.54N 99.33W

42 G4 Rayón Sonora, NW Mexico 29.42N 110.33W

178 Hh12 Rayong Rayong, S Thailand 12.01N 101.16E

27 T5 Ray Roberts, Lake ◎ Texas, SW USA

20 E15 Raystown Lake ◎ Pennsylvania, NE USA

147 V13 Raysūt SW Oman 16.58N 54.01E

29 R4 Raytown Missouri, C USA 39.00N 94.27W

24 I5 Rayville Louisiana, S USA 32.29N 91.45W

148 L5 Razan Hamadān, W Iran 35.22N 48.58E

145 X9 Razāzah, Buḥayrat ar var. Baḥr al Milḥ. ◎ C Iraq

116 L8 Razboyna ▲ E Bulgaria 42.54N 26.31E

Razdan see Hrazdan

Razdolnoye see Rozdol'ne

Razelm, Lacul see Razim, Lacul

116 L8 Razga E Iraq 36.25N 45.06E

116 L8 Razgrad Razgrad, NE Bulgaria 43.33N 26.31E

116 L8 Razgrad ◊ province N Bulgaria

119 N13 Razim, Lacul prev. Lacul Razelm. lagoon NW Black Sea

116 G11 Razlog Blagoevgrad, SW Bulgaria 41.52N 23.28E

120 K10 Razna ◎ SE Latvia

104 E6 Raz, Pointe du headland NW France 48.04N 4.52W

Reachlainn see Rathlin Island

Reachrainn see Lambay Island

99 N22 Reading S England, UK 51.28N 0.58W

20 H15 Reading Pennsylvania, NE USA 40.19N 75.55W

50 C7 Real, Cordillera ▲ C Ecuador

64 K12 Realicó La Pampa, C Argentina 35.01S 64.13W

27 S13 Realitos Texas, SW USA 27.26N 98.31W

110 G9 Realp Uri, C Switzerland 46.36N 8.32E

178 Hh12 Reăng Kesei Bătdâmbâng, W Cambodia 12.57N 103.15E

203 T11 Reao atoll Îles Tuamotu, E French Polynesia

Reate see Rieti

188 L11 Rebecca, Lake ◎ Western Australia

Rebiana Sand Sea see Rabyānah, Ramlat

128 H8 Reboly Respublika Kareliya, NW Russian Federation 63.51N 30.49E

172 P3 Rebun Rebun-tō, NE Japan 45.19N 141.02E

172 P3 Rebun-suidō strait E Sea of Japan

172 P1 Rebun-tō island NE Japan

108 J12 Recanati Marche, C Italy 43.23N 13.34E

Rechitsa see Rechytsa

111 Y7 Rechnitz Burgenland, SE Austria 47.18N 16.26E

121 J20 Rechytsa Rus. Rechitsa. Brestskaya Voblasts', SW Belarus 51.51N 26.49E

121 O19 Rechytsa Rus. Rechitsa. Homyel'skaya Voblasts', SE Belarus 52.22N 30.25E

61 Q15 Recife prev. Pernambuco. state capital Pernambuco, E Brazil 8.06S 34.52W

85 I26 Recife, Cape Afr. Kaap Recife. headland S South Africa 34.03S 25.37E

85 I26 Recife, Kaap see Recife, Cape

180 I16 Récifs, Îles aux island Inner Islands, NE Seychelles

103 E14 Recklinghausen Nordrhein-Westfalen, W Germany 51.37N 7.12E

102 M8 Recknitz ↔ NE Germany

101 K23 Recogne Luxembourg, SE Belgium 49.56N 5.20E

63 C15 Reconquista Santa Fe, C Argentina 29.10S 59.41W

205 O6 Recovery Glacier glacier Antarctica

61 G15 Recreio Mato Grosso, W Brazil 8.13S 58.15W

29 X9 Rector Arkansas, S USA 36.15N 90.17W

112 F9 Recz Ger. Reetz Neumark. Zachodniopomorskie, NW Poland 53.16N 15.32E

99 D17 Ree, Lough Ir. Loch Rí. ◎ C Ireland

101 L24 Redange var. Redange-sur-Attert. Diekirch, W Luxembourg 49.46N 5.52E

Redange-sur-Attert see Redange

20 C13 Redbank Creek ↔ Pennsylvania, NE USA

11 S9 Red Bay Quebec, E Canada 51.40N 56.27W

25 N2 Red Bay Alabama, S USA 34.26N 88.08W

37 N4 Red Bluff California, W USA 40.09N 122.14W

26 J8 Red Bluff Reservoir ◎ New Mexico/Texas, SW USA

32 K16 Red Bud Illinois, S USA 38.12N 89.59W

31 P9 Red Cedar River ↔ Wisconsin, N USA

9 T17 Redcliff Alberta, SW Canada 50.06N 110.48W

85 K17 Redcliff Midlands, C Zimbabwe 19.01S 29.43E

190 L9 Red Cliffs Victoria, SE Australia 34.21S 142.12E

31 P17 Red Cloud Nebraska, C USA 40.05N 98.31W

9 P15 Red Deer ↔ Alberta, SW Canada 52.15N 113.48W

9 Q16 Red Deer ↔ Alberta, SW Canada

9 T15 Red Deer ↔ Saskatchewan, C Canada

41 O11 Red Devil Alaska, USA 61.45N 95.18W

9 R16 Redding California, W USA 40.33N 122.26W

99 L20 Redditch W England, UK 52.19N 1.55W

31 P9 Redfield South Dakota, N USA 44.51N 98.31W

26 J12 Redford Texas, SW USA 29.31N 104.19W

47 V13 Redhead Trinidad, Trinidad and Tobago 10.48N 60.56W

190 I8 Red Hill South Australia 33.34S 138.13E

40 F10 Red Hill Haw. Pu'u Ula'ula. ▲ Maui, Hawaii, USA, C Pacific Ocean 20.42N 94.16W

28 K7 Red Hills hill range Kansas, C USA

11 T7 Red Indian Lake ◎ Newfoundland, E Canada

128 J16 Redkino Tverskaya Oblast', W Russian Federation 56.41N 36.07E

10 A10 Red Lake Ontario, C Canada 51.00N 93.55W

38 I10 Red Lake salt flat Arizona, SW USA

31 S4 Red Lake Falls Minnesota, N USA 47.52N 96.16W

31 R4 Red Lake River ↔ Minnesota, N USA

37 U15 Redlands California, W USA 34.03N 117.10W

20 G16 Red Lion Pennsylvania, NE USA 39.53N 76.36W

35 U11 Red Lodge Montana, NW USA 45.11N 109.15W

34 H13 Redmond Oregon, NW USA 44.16N 121.10W

38 L5 Redmond Utah, W USA 39.00N 111.51W

34 H9 Redmond Washington, NW USA 47.40N 122.07W

31 T15 Red Oak Iowa, C USA 41.00N 95.10W

20 K12 Red Oaks Mill New York, NE USA 41.39N 73.52W

104 I7 Redon Ille-et-Vilaine, NW France 47.39N 2.04W

47 W10 Redonda island SW Antigua and Barbuda

106 G4 Redondela Galicia, NW Spain 42.16N 8.36W

106 H11 Redondo Évora, S Portugal 38.39N 7.33W

41 Q12 Redoubt Volcano ▲ Alaska, USA 60.29N 152.44W

9 Y16 Red River ↔ Canada/USA

133 U12 Red River var. Yuan, Chin. Yuan Jiang, Vtn. Sông Hông Hà. ↔ China/Vietnam

27 W4 Red River ↔ S USA

24 H7 Red River ↔ Louisiana, S USA

32 M6 Red River ↔ Wisconsin, N USA

39 N11 Ré, Île de island W France

Red Rock, Lake see Red Rock Reservoir

31 W14 Red Rock Reservoir var. Lake Red Rock. ◎ Iowa, C USA

194 J15 Redscar Bay bay S PNG

77 Y9 Red Sea anc. Sinus Arabicus. sea Africa/Asia

23 T1 Red Springs North Carolina, SE USA 34.49N 79.10W

15 Gg6 Redstone ↔ Northwest Territories, NW Canada

9 V17 Redvers Saskatchewan, S Canada 49.31N 101.33W

79 P13 Red Volta var. Nazinon, Fr. Volta Rouge. ↔ Burkina/Ghana

9 Q15 Redwater Alberta, SW Canada 53.57N 113.06W

30 M16 Red Willow Creek ↔ Nebraska, C USA

31 W9 Red Wing Minnesota, N USA 44.33N 92.31W

37 N9 Redwood City California, W USA 37.29N 122.13W

31 T9 Redwood Falls Minnesota, N USA 44.33N 95.07W

198 Ff7 Reed Bank undersea feature S South China Sea

32 M8 Reed City Michigan, W USA 43.52N 85.30W

30 K6 Reeder North Dakota, N USA 46.03N 102.55W

37 R11 Reedley California, W USA 36.35N 119.27W

34 G12 Reedsport Oregon, NW USA 43.42N 124.06W

34 E13 Reedsburg Wisconsin, S USA 43.33N 90.03W

195 X8 Reef Islands island group Santa Cruz Islands, E Solomon Islands

193 H16 Reefton West Coast, South Island, NZ 42.07S 171.52E

22 P8 Reelfoot Lake ◎ Tennessee, C USA

97 N17 Rekarne Västmanland, C Sweden 59.25N 16.04

Reengus see Ringas

76 I7 Reese River ↔ Nevada, W USA

100 M8 Reest ↔ E Netherlands

143 N13 Refahiye Erzincan, C Turkey 39.54N 38.45E

25 U16 Reform Alabama, S USA 33.22N 88.01W

190 I7 Reftele Jönköping, S Sweden 57.10N 13.34E

27 T6 Refugio Texas, SW USA 28.18N 97.16W

112 E6 Rega ↔ NW Poland

Regar see Tursunzoda

103 O21 Regen ↔ SE Germany

103 M20 Regen ↔ SE Germany

103 M21 Regensburg Eng. Ratisbon, Fr. Ratisbonne; hist. Ratisbona; anc. Castra Regina, Reginum. Bayern, SE Germany 49.01N 12.06E

103 M21 Regenstauf Bayern, SE Germany 49.06N 12.07E

77 O5 Reggane C Algeria 26.45N 0.10E

100 N9 Regge ↔ E Netherlands

Reggio see Reggio di Calabria

Reggio nell' Emilia see Reggio di Calabria

Reggio di Calabria see Reggio nell' Emilia

109 M23 Reggio di Calabria var. Reggio Calabria, Gk. Rhegion; anc. Regium, Rhegium. Calabria, SW Italy 38.06N 15.39E

108 F9 Reggio nell' Emilia var. Reggio Emilia, abbrev. Reggio; anc. Regium Lepidum. Emilia-Romagna, N Italy 44.42N 10.37E

118 I10 Reghin Ger. Sächsisch-Reen, Hung. Szászrégen; prev. Reghinul Săsesc, Reghinul Sásesc. Mureş, C Romania 46.46N 24.40E

Reghinul Săsesc/Reghinul Sásesc see Reghin

9 U16 Regina Saskatchewan, S Canada 50.25N 104.39W

9 U16 Regina ✈ Saskatchewan, S Canada 50.25N 104.43W

57 Z10 Régina E French Guiana 4.19N 52.07W

9 U16 Regina Beach Saskatchewan, S Canada 50.44N 105.03W

Reginum see Regensburg

Registan see Rīgestān

43 J20 Registro São Paulo, S Brazil 24.30S 47.49W

Regium see Reggio di Calabria

Regium Lepidum see Reggio nell' Emilia

103 K19 Regnitz var. Rednitz. ↔ SE Germany

42 G12 Regocijo Durango, W Mexico 23.35N 104.34W

106 H12 Reguengos de Monsaraz Évora, S Portugal 38.25N 7.31W

103 M18 Rehau Bayern, SE Germany 50.15N 12.03E

85 D19 Rehoboth Hardap, C Namibia 23.18S 17.03E

Rehoboth/Rehovoth see Reḥovot

82 F11 Rehoboth Beach Delaware, NE USA 38.42N 75.03W

144 F10 Reḥovot var. Rehoboth, Rekhovot, Rehovoth. Central, C Israel 31.54N 34.49E

83 J20 Rei spring/well S Kenya 3.24S 39.18E

111 C18 Reichenau var. Rychnov nad Kněžnou, Czech Republic

Reichenau see Bogatynia, Poland

103 M17 Reichenbach var. Reichenbach im Vogtland. Sachsen, E Germany 50.36N 12.18E

Reichenbach see Dzierżoniów

Reichenbach im Vogtland see Reichenbach

Reichenberg see Liberec

189 O11 Reid Western Australia 30.48S 128.24E

23 V6 Reidsville Georgia, SE USA 32.05N 82.07W

23 T8 Reidsville North Carolina, SE USA 36.21N 79.39W

Reifnitz see Ribnica

99 O22 Reigate SE England, UK 51.13N 0.13W

104 I10 Ré, Île de island W France

39 N15 Reiley Peak ▲ Arizona, SW USA 32.24N 110.09W

105 Q4 Reims Eng. Rheims; anc. Durocortorum, Remi. Marne, N France 49.16N 4.01E

65 G23 Reina Adelaida, Archipiélago island group S Chile

47 O12 Reina Beatrix ✈ (Oranjestad) ◆ C Aruba 12.30N 69.57W

110 E6 Reinach Aargau, N Switzerland 47.15N 8.12E

110 E6 Reinach Basel-Land, NW Switzerland 47.30N 7.36E

66 O1 Reina Sofía ✈ (Tenerife) Tenerife, Islas Canarias, Spain, NE Atlantic Ocean

31 W13 Reinbeck Iowa, C USA 42.19N 92.36W

102 J10 Reinbek Schleswig-Holstein, N Germany 53.31N 10.15E

9 U12 Reindeer ↔ Saskatchewan, C Canada

9 U11 Reindeer Lake ◎ Manitoba/Saskatchewan, C Canada

95 J16 Replot Fin. Raippaluoto. island W Finland

Reppen see Rzepin

Reps see Rupea

29 T7 Republic Missouri, C USA 37.07N 93.28W

34 K7 Republic Washington, NW USA 48.39N 118.44W

29 N3 Republican River ↔ Kansas/Nebraska, C USA

15 L4 Repulse Bay Northwest Territories, N Canada 66.34N 86.19W

58 P9 Requena Loreto, NE Peru 5.05 73.47W

107 R10 Requena País Valenciano, E Spain 39.28N 1.07W

105 R13 Réquista Aveyron, S France 44.00N 2.31E

142 M12 Reşadiye Tokat, N Turkey 40.24N 37.19E

115 X8 Resen Turk. Resne. SW FYR Macedonia 41.07N 21.00E

Resen see Resen

57 Z10 Rémire NE French Guiana 4.52N 52.16W

131 N13 Remontnoye Rostovskaya Oblast', SW Russian Federation 46.35N 43.38E

176 V13 Remu Maluku, E Indonesia 0.36S 130.45E

9 W17 Reston Manitoba, S Canada 49.33N 101.03W

9 H11 Restoule Lake ◎ Ontario, S Canada

63 F13 Retamosa Meta, C Colombia

27.27S 58.55W

118 F22 Reşiţa Ger. Reschitza, Hung. Resicabánya. Caraş-Severin, W Romania 45.13N 21.58E

8 H8 Rémigny, Lac ◎ Quebec, SE Canada

103 M21 Regenstauf Bayern, SE Germany 49.06N 12.07E

57 Z10 Rémire NE French Guiana 4.52N 52.16W

100 N9 Regge ↔ E Netherlands

44 A1 Retalhuleu off. Departamento de Retalhuleu. ◆ department SW Guatemala

99 N18 Retford C England, UK 53.18N 0.52W

105 Q3 Rethel Ardennes, N France 49.31N 4.22E

Rethimno/Réthimnon see Réthymno

117 I25 Réthymno var. Rethimno; prev. Réthimnon. Kríti, Greece, E Mediterranean Sea 35.21N 24.28E

101 J21 Retie Antwerpen, N Belgium 51.18N 5.05E

113 J21 Rétság Nógrád, N Hungary 47.57N 19.07E

111 W2 Retz Niederösterreich, NE Austria 48.46N 15.58E

181 N15 Réunion off. La Réunion. ◇ French overseas department W Indian Ocean

132 L17 Réunion island W Indian Ocean

107 U6 Reus Cataluña, E Spain 41.10N 1.06E

101 J15 Reusel Noord-Brabant, S Netherlands 51.21N 5.10E

110 F7 Reuss ↔ NW Switzerland

Reutel see Ciuhru

103 H22 Reutlingen Baden-Württemberg, S Germany 48.30N 9.13E

110 L7 Reutte Tirol, W Austria 47.30N 10.45E

101 M16 Reuver Limburg, SE Netherlands 51.16N 6.04E

30 K7 Reva South Dakota, N USA 45.30N 103.03W

128 K4 Revda Murmanskaya Oblast', NW Russian Federation 67.57N 34.29E

125 Ee11 Revda Sverdlovskaya Oblast', C Russian Federation 56.48N 59.42E

105 N16 Revel Haute-Garonne, S France 43.28N 1.57E

Reval/Revel' see Tallinn

9 O16 Revelstoke British Columbia, SW Canada 51.01N 118.12W

45 N13 Reventazón, Río ↔ E Costa Rica

108 G9 Revere Lombardia, N Italy 45.03N 11.07E

41 Y14 Revillagigedo Island island Alexander Archipelago, Alaska, USA

105 K13 Revin Ardennes, N France 49.57N 4.39E

94 L3 Reydharfjördhur Austurland, E Iceland 65.02N 14.12W

59 K16 Reyes Beni, NW Bolivia 14.17S 67.18W

37 N8 Reyes, Point headland California, W USA 37.58N 123.01W

56 B12 Reyes, Punta headland SW Colombia 2.43S 78.07W

142 L17 Reyhanlı Hatay, S Turkey 36.16N 36.33E

45 P15 Rey, Isla del island Archipiélago de las Perlas, SE Panama

94 J6 Reykhólar Vestfirðhir, W Iceland 65.28N 22.12W

94 J4 Reykjahlídh Nordhurland Eystra, NE Iceland 65.37N 16.54W

207 O16 Reykjanes Basin var. Irminger Basin. undersea feature N Atlantic Ocean

207 O15 Reykjanes Ridge undersea feature N Atlantic Ocean

94 I5 Reykjavík var. Reikjavik. ● (Iceland) Höfudhborgarsvaedhi, W Iceland 64.07N 21.54W

20 D13 Reynoldsville Pennsylvania, NE USA 41.04N 78.51W

43 P8 Reynosa Tamaulipas, C Mexico 26.03N 98.19W

Reza'iyeh see Orūmīyeh

Reza'īyeh, Daryācheh-ye see Orūmīyeh, Daryācheh-ye

104 I8 Rezé Loire-Atlantique, NW France 47.10N 1.36W

120 K10 Rēzekne Ger. Rositten; prev. Rus. Rezhitsa. Rēzekne, SE Latvia 56.31N 27.22E

Rezhitsa see Rēzekne

119 N9 Rezina NE Moldova 47.44N 28.58E

116 N11 Rezovo Turk. Rezve. Burgas, E Bulgaria 42.00N 28.00E

116 N11 Rezovo Reka Turk. Rezve Deresi. ↔ Bulgaria/Turkey see also Rezve Deresi

Rezve see Rezovo

116 N11 Rezve Deresi Bul. Rezovska Reka. ↔ Bulgaria/Turkey see also Rezovska Reka

Rhadames see Ghādāmis

Rhaedestus see Tekirdağ

110 J10 Rhaetian Alps Fr. Alpes Rhétiques, Ger. Rätische Alpen, It. Alpi Retiche. ▲ C Europe

Rhätikon ↔ C Europe

103 G14 Rheda-Wiedenbrück Nordrhein-Westfalen, W Germany 51.51N 8.19E

100 M12 Rheden Gelderland, E Netherlands 52.01N 6.03E

Rhegion/Rhegium see Reggio di Calabria

Rheims see Reims

101 H15 Rhein see Rhine

103 F15 Rheinbach Nordrhein-Westfalen, W Germany 50.37N 6.57E

102 F13 Rheine var. Rhine in Westfalen. Nordrhein-Westfalen, NW Germany 52.16N 7.27E

Rheine in Westfalen see Rheine
Rheinfeld see Rheinfelden
103 F24 **Rheinfelden** Baden-Württemberg, S Germany 47.34N 7.46E
110 E6 **Rheinfelden** var. Rheinfeld. Aargau, N Switzerland 47.33N 7.46E
103 E17 **Rheinisches Schiefergebirge** var. Rhine State Uplands, Eng. Rhenish Slate Mountains. ▲ W Germany
103 D18 **Rheinland-Pfalz** Eng. Rhineland-Palatinate, Fr. Rhénanie-Palatinat. ◆ state W Germany
103 G18 **Rhein/Main** ✕ (Frankfurt am Main) Hessen, W Germany 50.03N 8.33E
Rhénanie du Nord-Westphalie see Nordrhein-Westfalen
Rhénanie-Palatinat see Rheinland-Pfalz
100 K12 **Rhenen** Utrecht, C Netherlands 51.57N 5.34E
Rhenish Slate Mountains see Rheinisches Schiefergebirge
Rhétiques, Alpes see Rhaetian Alps
102 N10 **Rhin** ♒ NE Germany
Rhin see Rhine
86 F19 **Rhine** Dut. Rijn, Fr. Rhin, Ger. Rhein. ♒ W Europe
32 L5 **Rhinelander** Wisconsin, N USA 45.39N 89.22W
Rhineland-Palatinate see Rheinland-Pfalz
Rhine State Uplands see Rheinisches Schiefergebirge
102 N11 **Rhinkanal** canal NE Germany
83 F17 **Rhino Camp** NW Uganda 2.58N 31.24E
76 D7 **Rhir, Cap** headland W Morocco 30.40N 9.54W
108 D7 **Rho** Lombardia, N Italy 45.31N 9.01E
21 N12 **Rhode Island** off. State of Rhode Island and Providence Plantations; also known as Little Rhody, Ocean State. ◆ state NE USA
21 O13 **Rhode Island** island Rhode Island, NE USA
21 O13 **Rhode Island Sound** sound Maine/Rhode Island, NE USA
Rhodes see Ródos
Rhode-Saint-Genèse see Sint-Genesius-Rode
86 L14 **Rhodes Basin** undersea feature E Mediterranean Sea
Rhodesia see Zimbabwe
116 I12 **Rhodope Mountains** var. Rodhópi Óri, Bul. Rhodope Planina, Rodopi, Gk. Orosirá Rodhópis, Turk. Dospad Dagh. ▲ Bulgaria/Greece
Rhodope Planina see Rhodope Mountains
Rhodos see Ródos
103 I18 **Rhön** ▲ C Germany
105 Q10 **Rhône** ◆ department E France
150 C12 **Rhône** ♒ France/Switzerland
105 Q11 **Rhône-Alpes** ◆ region E France
123 J6 **Rhône Fan** undersea feature W Mediterranean Sea
100 G13 **Rhoon** Zuid-Holland, SW Netherlands 51.52N 4.25E
98 O9 **Rhum** var. Rum. island N Scotland, UK
Rhuthun see Ruthin
99 I18 **Rhyl** NE Wales, UK 53.19N 3.28W
61 K18 **Rialma** Goiás, S Brazil 15.22S 49.35W
106 L3 **Riaño** Castilla-León, N Spain 42.59N 5.00W
107 O9 **Riansáres** ♒ C Spain
158 H6 **Riasi** Jammu and Kashmir, NW India 33.03N 74.51E
174 Gg7 **Riau** off. Propinsi Riau. ◆ province W Indonesia
174 I10 **Riau Archipelago** see Riau, Kepulauan
174 I10 **Riau, Kepulauan** var. Riau Archipelago, Dut. Riouw-Archipel. island group W Indonesia
107 O6 **Riaza** Castilla-León, N Spain 41.17N 3.29W
106 L3 **Riaza** ♒ N Spain
79 K17 **Riba** spring/well NE Kenya 1.56N 40.38E
106 H4 **Ribadavia** Galicia, NW Spain 42.16N 8.07W
106 J2 **Ribadeo** Galicia, NW Spain 43.31N 7.04W
106 L2 **Ribadesella** Asturias, N Spain 43.27N 5.04W
106 G10 **Ribatejo** former province C Portugal
149 Q8 **Rībār-e Rīzāb** Yazd, C Iran
85 P15 **Ribáuè** Nampula, N Mozambique 14.56S 38.19E
99 K17 **Ribble** ♒ NW England, UK
97 F23 **Ribe** Ribe, W Denmark 55.19N 8.46E
97 F23 **Ribe** off. Ribe Amt, var. Ripen. ◆ county W Denmark
106 G3 **Ribeira** Galicia, NW Spain 42.33N 9.01W
66 O5 **Ribeira Brava** Madeira, Portugal, NE Atlantic Ocean 32.39N 17.04W
60 P3 **Ribeirão Preto** São Paulo, S Brazil 21.09S 47.48W
62 L14 **Ribeira, Rio** ♒ S Brazil
109 I24 **Ribera** Sicilia, Italy, C Mediterranean Sea 37.31N 13.16E
59 L14 **Riberalta** Beni, N Bolivia 11.00S 66.06W
107 W4 **Ribes de Freser** Cataluña, NE Spain 42.18N 2.10E
32 L5 **Rib Mountain** ▲ Wisconsin, N USA 44.54N 89.41W
111 U12 **Ribnica** Ger. Reifnitz. S Slovenia 45.46N 14.40E
119 N9 **Ribniţa** var. Râbniţa, Rus. Rybnitsa. NE Moldova 47.46N 29.01E
102 M8 **Ribnitz-Damgarten** Mecklenburg-Vorpommern, NE Germany 54.14N 12.25E

113 D16 **Říčany** Ger. Ritschan. Středočeský Kraj, W Czech Republic 49.58N 14.39E
31 U7 **Rice** Minnesota, N USA 45.42N 94.10W
32 L5 **Rice Lake** Wisconsin, N USA 45.31N 91.43W
12 I15 **Rice Lake** ⊜ Ontario, SE Canada
25 V3 **Rice Lake** ⊜ Ontario, S Canada
25 V3 **Richard B.Russell Lake** ⊠ Georgia, SE USA
27 U6 **Richardson** Texas, SW USA 32.57N 96.43W
9 R11 **Richardson** ♒ Alberta, C Canada
8 I3 **Richardson Mountains** ▲ Yukon Territory, NW Canada
193 C21 **Richardson Mountains** ▲ South Island, NZ
44 F3 **Richardson Peak** ▲ SE Belize 16.34N 88.46W
78 G10 **Richard Toll** N Senegal 16.27N 15.44W
30 L5 **Richardton** North Dakota, N USA 46.52N 102.19W
12 F13 **Rich, Cape** headland Ontario, S Canada 44.42N 80.37W
104 L8 **Richelieu** Indre-et-Loire, C France 47.01N 0.18E
35 P15 **Richfield** Idaho, NW USA 43.03N 114.11W
38 K5 **Richfield** Utah, W USA 38.46N 112.06W
20 J10 **Richfield Springs** New York, NE USA 42.52N 74.57W
20 M6 **Richford** Vermont, NE USA 44.59N 72.37W
29 R6 **Rich Hill** Missouri, C USA 38.06N 94.22W
11 P14 **Richibucto** New Brunswick, SE Canada 46.42N 64.54W
110 E8 **Richisau** Glarus, NE Switzerland 47.00N 8.54E
25 S6 **Richland** Georgia, SE USA 32.05N 84.40W
29 U6 **Richland** Missouri, C USA 37.51N 92.24W
27 U8 **Richland** Texas, SW USA 31.55N 96.25W
34 K10 **Richland** Washington, NW USA 46.17N 119.16W
32 K8 **Richland Center** Wisconsin, N USA 43.18N 90.22W
23 W11 **Richlands** North Carolina, SE USA 34.52N 77.33W
23 Q7 **Richlands** Virginia, NE USA 37.05N 81.47W
27 R9 **Richland Springs** Texas, SW USA 31.16N 98.56W
191 S8 **Richmond** New South Wales, SE Australia 33.36S 150.43E
8 L17 **Richmond** British Columbia, SW Canada 49.07N 123.09W
12 L13 **Richmond** Ontario, SE Canada 45.12N 75.49W
13 Q12 **Richmond** Quebec, SE Canada 45.39N 72.07W
193 I14 **Richmond** Tasman, South Island, NZ 41.24S 173.04E
37 N8 **Richmond** California, W USA 37.57N 122.22W
35 Q4 **Richmond** Indiana, N USA 39.48N 84.52W
22 M6 **Richmond** Kentucky, S USA 37.45N 84.17W
29 S4 **Richmond** Missouri, C USA 39.16N 93.58W
27 V11 **Richmond** Texas, SW USA 29.34N 95.45W
38 L1 **Richmond** Utah, W USA 41.55N 111.51W
23 W6 **Richmond** state capital Virginia, NE USA 37.33N 77.27W
12 H15 **Richmond Hill** Ontario, S Canada 43.51N 79.24W
193 J15 **Richmond Range** ▲ South Island, NZ
29 S12 **Rich Mountain** ▲ Arkansas, C USA 34.37N 94.17W
33 S13 **Richwood** Ohio, N USA 40.25N 83.18W
23 R5 **Richwood** West Virginia, NE USA 38.13N 80.31W
106 K5 **Ricobayo, Embalse de** ⊠ NW Spain
Ricomagus see Riom
Rida' see Radā'
100 H13 **Ridderkerk** Zuid-Holland, SW Netherlands 51.52N 4.34E
35 N16 **Riddle** Idaho, NW USA 42.07N 116.09W
34 F14 **Riddle** Oregon, NW USA 42.57N 123.21W
12 L13 **Rideau** ♒ Ontario, SE Canada
37 T12 **Ridgecrest** California, W USA 35.37N 117.40W
18 D13 **Ridgefield** Connecticut, NE USA 41.16N 73.30W
24 K5 **Ridgeland** Mississippi, S USA 32.25N 90.07W
23 R15 **Ridgeland** South Carolina, SE USA 32.28N 80.58W
22 F8 **Ridgely** Tennessee, S USA 36.15N 89.29W
12 D17 **Ridgetown** Ontario, S Canada 42.27N 81.52W
23 R12 **Ridgeway** Ridgeway South Carolina, SE USA 34.17N 80.56W
20 D13 **Ridgway** var. Ridgeway. Pennsylvania, NE USA
9 W16 **Riding Mountain** ▲ Manitoba, S Canada
Ried see Ried im Innkreis
111 R4 **Ried im Innkreis** var. Ried. Oberösterreich, NW Austria 48.13N 13.28E
111 X8 **Riegersburg** Steiermark, SE Austria 47.03N 15.52E
110 E6 **Riehen** Basel-Stadt, NW Switzerland 47.34N 7.39E
94 J9 **Riehppegáisá** var. Rieppe. N Norway 69.38N 21.31E
101 K18 **Riemst** Limburg, NE Belgium 50.49N 5.35E
Rieppe see Riehppegáisá
103 O15 **Riesa** Sachsen, E Germany 51.18N 13.18E
63 H24 **Riesco, Isla** island S Chile
109 K25 **Riesi** Sicilia, Italy, C Mediterranean Sea 37.16N 14.04E

85 F25 **Riet** ♒ SW South Africa
85 I23 **Riet** ♒ SW South Africa
120 D11 **Rietavas** Plungė, W Lithuania 55.43N 21.56E
85 F19 **Rietfontein** Omaheke, E Namibia 21.54S 20.57E
109 I14 **Rieti** anc. Reate. Lazio, C Italy 42.22N 12.49E
86 D14 **Rif** var. Er Rif, Er Riff, Riff. ▲ N Morocco
39 Q4 **Rifle** Colorado, C USA 39.30N 107.46W
35 V7 **Rifle River** ♒ Michigan, N USA
83 H18 **Rift Valley** ◆ province Kenya
Rift Valley see Great Rift Valley
120 F9 **Rīga** Eng. Riga. ● (Latvia) Rīga, C Latvia 56.57N 24.07E
149 U22 **Rīgān** Kermān, SE Iran 28.39N 59.01E
Rigas Jūras Līcis see Riga, Gulf of
13 N12 **Rigaud** ♒ Ontario/Quebec, SE Canada
35 R14 **Rigby** Idaho, NW USA 43.40N 111.54W
154 M10 **Rīgestān** var. Registan. desert region S Afghanistan
34 M11 **Riggins** Idaho, NW USA 45.24N 116.18W
11 R8 **Rigolet** Newfoundland, NE Canada 51.10N 58.25W
80 G9 **Rig-Rig** Kanem, W Chad 14.19N 14.19E
120 F4 **Riguldi** Läänemaa, W Estonia 59.07N 23.34E
95 L19 **Riihimäki** Etelä-Suomi, S Finland 60.45N 24.45E
205 O2 **Riiser-Larsen Ice Shelf** ice shelf Antarctica
205 U2 **Riiser-Larsen Peninsula** peninsula Antarctica
117 P22 **Riiser-Larsen Sea** sea Antarctica
42 D2 **Riíto** Sonora, NW Mexico 32.06N 114.57W
114 B9 **Rijeka** Ger. Sankt Veit am Flaum, It. Fiume, Slvn. Reka; anc. Tarsatica. Primorje-Gorski Kotar, NW Croatia 45.20N 14.25E
101 I14 **Rijen** Noord-Brabant, S Netherlands 51.34N 4.55E
101 H15 **Rijkevorsel** Antwerpen, N Belgium 51.23N 4.43E
Rijn see Rhine
100 G11 **Rijnsburg** Zuid-Holland, W Netherlands 52.12N 4.27E
100 N10 **Rijssen** Overijssel, E Netherlands 52.19N 6.30E
100 G12 **Rijswijk** Eng. Ryswick. Zuid-Holland, W Netherlands 52.04N 4.22E
94 I9 **Riksgränsen** Norrbotten, N Sweden 68.24N 18.15E
172 Q6 **Rikubetsu** Hokkaidō, NE Japan 43.30N 143.43E
171 Mm12 **Rikuzen-Takata** Iwate, Honshū, C Japan 39.01N 141.37E
29 O4 **Riley** Kansas, C USA 39.18N 96.49W
101 I17 **Rillaar** Vlaams Brabant, C Belgium 50.58N 4.58E
116 L11 **Rí, Loch** see Ree, Lough
79 T12 **Rima** ♒ N Nigeria
114 I9 **Rilska Reka** ♒ W Bulgaria
147 N7 **Rimah, Wādī ar** var. Wādī ar Rummah. dry watercourse C Saudi Arabia
Rimaszombat see Rimavská Sobota
203 R12 **Rimatara** island Îles Australes, SW French Polynesia
113 L20 **Rimavská Sobota** Ger. Gross-Steffelsdorf, Hung. Rimaszombat. Banskobystrický Kraj, C Slovakia 48.24N 20.01E
9 R15 **Rimbey** Alberta, SW Canada 52.39N 114.10W
97 P17 **Rimbo** Stockholm, C Sweden 59.43N 18.21E
97 H21 **Rimforsa** Östergötland, S Sweden 58.06N 15.40E
108 I8 **Rimini** anc. Ariminum. Emilia-Romagna, N Italy 44.03N 12.33E
Rimnicu-Sărat see Râmnicu Sărat
Rimnicu Vîlcea see Râmnicu Vâlcea
155 Y3 **Rimo Muztāgh** ▲ India/Pakistan
13 U7 **Rimouski** Quebec, SE Canada 48.25N 68.31W
164 M16 **Rinbung** Xizang Zizhiqu, W China 29.15N 89.40E
168 I5 **Rinchinlhümbe** Hövsgöl, N Mongolia 51.06N 99.40E
64 I5 **Rincón, Cerro** ▲ N Chile 24.01S 67.19W
106 M15 **Rincón de la Victoria** Andalucía, S Spain 36.43N 4.18W
Rincón del Bonete, Lago Artificial de see Río Negro, Embalse del
107 Q4 **Rincón de Soto** La Rioja, N Spain 42.15N 1.49W
96 G8 **Rindal** Møre og Romsdal, S Norway 63.02N 9.09E
117 J20 **Rineia** island Kykládes, Greece, Aegean Sea
158 H11 **Ringas** prev. Reengus, Ringus. Rājasthān, N India 27.18N 75.27E
97 H24 **Ringe** Fyn, C Denmark 55.13N 10.30E
96 H11 **Ringebu** Oppland, S Norway 61.31N 10.09E
25 R1 **Ringgold** Georgia, SE USA 34.55N 85.06W
24 H2 **Ringgold** Louisiana, S USA 32.19N 93.16W
27 S5 **Ringgold** Texas, SW USA 33.47N 97.56W
61 J19 **Ringkøbing** Ringkøbing, W Denmark 56.04N 8.22E
61 J19 **Ringkøbing** off. Ringkøbing Amt. ◆ county W Denmark

97 E22 **Ringkøbing Fjord** fjord W Denmark
35 S10 **Ringling** Montana, NW USA 46.15N 110.48W
29 N13 **Ringling** Oklahoma, C USA 34.12N 97.35W
96 H13 **Ringsaker** Hedmark, S Norway 60.54N 10.45E
97 I23 **Ringsted** Vestsjælland, E Denmark 55.28N 11.48E
94 I9 **Ringus** see Ringas
20 K13 **Ringvassøya** island N Norway 41.06N 74.15W
102 H13 **Ringwood** New Jersey, NE USA 31.11N 52.10N 9.04E
Rinn Dúain see Hook Head
81 Niedersachsen, NW Germany 52.10N 9.04E
117 E18 **Rinteln** Niedersachsen, NW Germany 52.10N 9.04E
58 C7 **Río Dytikí Ellás,** S Greece 38.18N 21.48E
22 M7 **Riobamba** Chimborazo, C Ecuador 1.38S 78.40W
61 C16 **Rio Bonito** Rio de Janeiro, SE Brazil 22.42S 42.38W
63 H18 **Rio Branco** state capital Acre, W Brazil 9.58S 67.49W
Rio Branco Cerro Largo, NE Uruguay 32.34S 53.21W
Río Branco, Território de see Roraima
43 P8 **Río Bravo** Tamaulipas, C Mexico 25.57N 98.03W
65 G16 **Río Bueno** Los Lagos, C Chile 40.19S 72.55W
57 P5 **Río Caribe** Sucre, NE Venezuela 10.40N 63.07W
56 M5 **Río Chico** Miranda, N Venezuela 10.18N 66.00W
65 H18 **Río Cisnes** Aisén, S Chile 44.29S 71.15W
62 L9 **Rio Claro** São Paulo, S Brazil 22.23S 47.31W
47 V14 **Rio Claro** Trinidad, Trinidad and Tobago 10.18N 61.10W
56 J5 **Río Claro** Lara, N Venezuela 9.54N 69.22W
65 K15 **Río Colorado** Río Negro, E Argentina 39.04S 64.04W
64 K11 **Río Cuarto** Córdoba, C Argentina 33.06S 64.20W
61 N9 **Rio de Janeiro** var. Rio. state capital Rio de Janeiro, SE Brazil 22.52S 43.16W
62 P9 **Rio de Janeiro** off. Estado do Rio de Janeiro. ◆ state SE Brazil
45 R17 **Río de Jesús** Veraguas, S Panama 7.57N 81.09W
36 K3 **Rio Dell** California, W USA 40.30N 124.07W
62 K13 **Rio do Sul** Santa Catarina, S Brazil 27.15S 49.37W
65 J24 **Río Gallegos** var. Gallegos, Puerto Gallegos. Santa Cruz, S Argentina 51.39S 69.21W
63 I18 **Rio Grande** ♒ Texas, S USA
65 J24 **Río Grande** Tierra del Fuego, S Argentina 53.45S 67.46W
42 L10 **Río Grande** Zacatecas, C Mexico 23.48N 103.03W
44 J9 **Río Grande** León, NW Nicaragua 12.57N 86.31W
47 V9 **Río Grande** E Puerto Rico 18.22N 65.49W
27 V12 **Río Grande City** Texas, SW USA 26.22N 98.49W
61 J17 **Rio Grande do Norte** off. Estado do Rio Grande do Norte. ◆ state E Brazil
63 G15 **Rio Grande do Sul** off. Estado do Rio Grande do Sul. ◆ state S Brazil
117 M17 **Rio Grande Fracture Zone** tectonic feature C Atlantic Ocean
117 L18 **Rio Grande Gap** undersea feature S Atlantic Ocean
Rio Grande Plateau see Rio Grande Rise
117 J18 **Rio Grande Rise** var. Rio Grande Plateau. undersea feature SW Atlantic Ocean
56 G7 **Ríohacha** La Guajira, N Colombia 11.22N 72.46W
45 N8 **Río Hato** Coclé, C Panama 8.22N 80.09W
27 T7 **Rio Hondo** Texas, SW USA 26.14N 97.34W
58 D10 **Rioja** San Martín, N Peru 5.59S 77.05W
43 Y11 **Río Lagartos** Yucatán, SE Mexico 21.34N 88.07W
105 P11 **Riom** anc. Ricomagus. Puy-de-Dôme, C France 45.54N 3.06E
106 F10 **Rio Maior** Santarém, C Portugal 39.19N 8.55W
105 O12 **Riom-ès-Montagnes** Cantal, C France 45.15N 2.39E
62 J12 **Rio Negro** Paraná, S Brazil 26.06S 49.46W
62 I15 **Río Negro** off. Provincia de Río Negro. ◆ province C Argentina
63 D18 **Río Negro** ◆ department W Uruguay
49 V12 **Río Negro, Embalse del** var. Lago Artificial de Rincón del Bonete. ⊠ C Uruguay
58 D13 **Rinconada** La Rioja, N Spain 42.15N 1.49W
107 Q4 **Rincón de Soto** La Rioja, N Spain
155 Y5 **Ríopar** Castilla-La Mancha, C Spain 38.30N 2.26W
63 H16 **Rio Pardo** Rio Grande do Sul, S Brazil 29.41S 52.25W
37 R11 **Río Rancho Estates** New Mexico, SW USA 35.14N 106.40W
44 L13 **Río San Juan** ◆ department S Nicaragua
56 D8 **Ríosucio** Caldas, W Colombia 5.25N 75.43W
56 C7 **Ríosucio** Chocó, NW Colombia 7.24N 77.09W
42 K10 **Río Tercero** Córdoba, C Argentina 32.12S 64.03W
57 S15 **Río Tocuyo** Lara, NW Venezuela 10.12N 69.58W
61 J19 **Río Verde** Goiás, C Brazil 17.49S 50.55W

43 O12 **Río Verde** var. Rioverde. San Luis Potosí, C Mexico 21.58N 100.00W
37 O8 **Río Vista** California, W USA 38.09N 121.42W
114 M11 **Ripanj** Serbia, N Yugoslavia 44.37N 20.30E
108 J13 **Ripatransone** Marche, C Italy 43.00N 13.45E
24 H4 **Ripley** Mississippi, S USA 34.43N 88.57W
35 R13 **Ripley** Ohio, N USA 38.45N 83.51W
22 F9 **Ripley** Tennessee, S USA 35.45N 89.31W
23 Q4 **Ripley** West Virginia, NE USA 38.49N 81.42W
99 M16 **Ripon** N England, UK 54.07N 1.31W
32 M7 **Ripon** Wisconsin, N USA 43.52N 88.48W
109 L24 **Riposto** Sicilia, Italy, C Mediterranean Sea 37.43N 15.13E
101 L14 **Rips** Noord-Brabant, SE Netherlands 51.31N 5.49E
56 D9 **Risaralda** off. Departamento de Risaralda. ◆ province C Colombia
118 L8 **Rîşcani** var. Râşcani, Rus. Ryshkany. NW Moldova 47.55N 27.31E
158 J9 **Rishikesh** Uttar Pradesh, N India 30.06N 78.16E
172 R2 **Rishiri-suidō** strait E Sea of Japan
172 Oo2 **Rishiri-tō** var. Risiri Tō. island NE Japan
172 P2 **Rishiri-yama** ▲ Rishiri-tō, NE Japan 45.11N 141.11E
27 R7 **Rising Star** Texas, SW USA 32.06N 98.57W
35 Q15 **Rising Sun** Indiana, N USA 38.58N 84.52W
Risiri Tō see Rishiri-tō
104 I4 **Risle** ♒ N France
29 Q9 **Rison** Arkansas, C USA 33.57N 92.11W
97 C17 **Risør** Aust-Agder, S Norway 58.43N 9.13E
94 I7 **Risøyhamn** Nordland, C Norway 69.00N 15.37E
103 I23 **Riss** ♒ S Germany
120 G4 **Risti** Ger. Kreuz. Läänemaa, W Estonia 59.03N 24.11E
13 V8 **Ristigouche** ♒ Quebec, SE Canada
95 N18 **Ristiina** Isä-Suomi, E Finland 61.31N 27.15E
95 N14 **Ristijärvi** Oulu, C Finland 64.30N 28.15E
196 C14 **Ritidian Point** headland N Guam 13.39N 144.51E
Ritschan see Říčany
37 R9 **Ritter, Mount** ▲ California, W USA 37.40N 119.10W
35 T12 **Rittman** Ohio, N USA 40.58N 81.46W
34 L9 **Ritzville** Washington, NW USA 47.07N 118.22W
23 W8 **Riva** see Riva del Garda
25 J23 **Rivadavia** Buenos Aires, E Argentina 35.28S 62.58W
108 F7 **Riva del Garda** var. Riva. Trentino-Alto Adige, N Italy 45.54N 10.50E
108 B8 **Rivarolo Canavese** Piemonte, W Italy 45.19N 7.42E
44 K11 **Rivas** Rivas, SW Nicaragua 11.25N 85.49W
44 I11 **Rivas** ◆ department SW Nicaragua
105 R11 **Rive-de-Gier** Loire, E France 45.31N 4.36E
63 A22 **Rivera** Buenos Aires, E Argentina 37.13S 63.13W
63 F16 **Rivera** Rivera, NE Uruguay 30.54S 55.31W
63 F15 **Rivera** ◆ department NE Uruguay
37 P9 **Riverbank** California, W USA 37.43N 120.56W
78 K17 **River Cess** SW Liberia 5.28N 9.31W
30 M4 **Riverdale** North Dakota, N USA 47.29N 101.22W
32 I6 **River Falls** Wisconsin, N USA 44.52N 92.38W
9 T16 **Riverhurst** Saskatchewan, S Canada 50.52N 106.49W
191 O10 **Riverina** physical region New South Wales, SE Australia
9 W16 **River Nile** ◆ state NE Sudan
65 F19 **Rivero, Isla** island Archipiélago de los Chonos, S Chile
9 W16 **Rivers** Manitoba, S Canada 50.01N 100.13W
79 V17 **Rivers** ◆ state S Nigeria
193 D23 **Riverside** Canterbury, South Island, NZ 45.54S 168.44E
85 F26 **Riverside** Cape Town, SW South Africa 34.04S 21.15E
37 U15 **Riverside** California, W USA 33.57N 117.24W
27 W9 **Riverside** Texas, SW USA 30.51N 95.23W
39 U3 **Riverside Reservoir** ⊠ Colorado, C USA
8 K15 **Rivers Inlet** British Columbia, SW Canada 51.43N 127.19W
8 K15 **Rivers Inlet** inlet British Columbia, SW Canada
9 X15 **Riverton** Manitoba, S Canada 51.00N 97.00W
193 C24 **Riverton** Southland, South Island, NZ 46.22S 168.02E
32 K9 **Riverton** Illinois, N USA 39.50N 89.31W
38 L3 **Riverton** Utah, W USA 40.32N 111.57W
35 V16 **Riverton** Wyoming, C USA 43.01N 108.22W
12 G10 **River Valley** Ontario, S Canada 46.36N 80.09W
11 P14 **Riverview** New Brunswick, SE Canada 46.03N 64.46W
105 O17 **Rivesaltes** Pyrénées-Orientales, S France 42.46N 2.48E
21 X8 **Riverhead** New York, NE USA 40.55N 72.40W
190 M9 **Riverina** ◆ Southern Province
107 P11 **Riverton** Mountain ▲ Nunavut, Kalar, Alaska, USA 60.01N 166.15W
190 J9 **Riverview** Queensland, E Australia 23.31S 150.31E
183 S8 **Robbie Ridge** undersea feature W Pacific Ocean
23 N10 **Robbins** North Carolina, SE USA 35.25N 79.35W
191 N15 **Robbins Island** island Tasmania, SE Australia
23 N10 **Robbinsville** North Carolina, SE USA 35.19N 83.48W
191 O10 **Robe** South Australia 37.11S 139.48E
23 W9 **Robersonville** North Carolina, SE USA 35.49N 77.15W
85 F26 **Robertson** Western Cape, SW South Africa 34.04S 19.52E
37 O15 **Robertson** California, W USA 33.57N 117.24W
95 J15 **Robertsfors** Västerbotten, N Sweden 64.12N 20.49E
29 V6 **Robert Lee** Texas, SW USA 31.54N 100.29W
37 S9 **Roberts Creek Mountain** ▲ Nevada, W USA 39.54N 116.16W
29 R8 **Robert S.Kerr Reservoir** ⊠ Oklahoma, C USA
85 F26 **Robertson** Western Cape, SW South Africa 33.48S 19.52E
204 H4 **Robertson Island** island Antarctica
78 K17 **Robertsport** W Liberia 6.45N 11.15W
191 J16 **Robertstown** South Australia 34.05S 139.04E
13 P7 **Roberval** Quebec, SE Canada 48.31N 72.13W
8 K15 **Robeson Channel** strait Canada/Greenland
190 J8 **Robinson** physical region New South Wales, SE Australia
31 N15 **Robinson** Illinois, N USA 38.59N 87.44W
200 Oo12 **Róbinson Crusoe, Isla** island Islas Juan Fernández, Chile, E Pacific Ocean
188 B9 **Robinson Range** ▲ Western Australia
194 L16 **Robinson River** Central, S PNG 10.05 148.51E
190 M9 **Robinvale** Victoria, SE Australia 34.37S 142.45E
37 S11 **Robinson** Arizona, W USA 35.06N 114.36W
190 M9 **Robinvale** Victoria, SE Australia
107 O11 **Robledo** Castilla-La Mancha, C Spain 38.45N 2.27W
23 Z14 **Robersonville** Florida, SE USA 26.46N 80.03W
13 Q10 **Rivière-à-Pierre** Quebec, SE Canada 46.59N 72.12W
13 T9 **Rivière-Bleue** Quebec, SE Canada 47.25N 69.01W

13 T8 **Rivière-du-Loup** Quebec, SE Canada 47.49N 69.31W
181 Y15 **Rivière du Rempart** NE Mauritius 20.06S 57.40E
47 R12 **Rivière-Pilote** S Martinique 14.29N 60.54W
181 O17 **Rivière St-Étienne, Point de la** headland SE Réunion
11 S10 **Rivière-St-Paul** Quebec, SE Canada 51.26N 57.52W
27 P6 **Rivière Sèche** see Bel Air
118 K4 **Rivne** Pol. Równe, Rus. Rovno. Rivnens'ka Oblast', NW Ukraine 50.37N 26.15E
22 F9 **Rivne** see Rivnens'ka Oblast'
118 K3 **Rivnens'ka Oblast'** var. Rivne, Rus. Rovenskaya Oblast'. ◆ province NW Ukraine
108 B8 **Rivoli** Piemonte, NW Italy 45.04N 7.31E
165 Q14 **Riwoqê** Xizang Zizhiqu, W China 31.14N 96.30E
101 H19 **Rixensart** Wallon Brabant, C Belgium 50.43N 4.31E
143 P11 **Rize** Rize, NE Turkey 41.02N 40.33E
143 P11 **Rize** prev. Çoruh. ◆ province NE Turkey
167 R5 **Rizhao** Shandong, E China 35.23N 119.31E
Rizhskiy Zaliv see Riga, Gulf of
Rizokarpaso/Rizokárpason see Dipkarpaz
109 O21 **Rizzuto, Capo** headland S Italy 38.54N 17.05E
97 F15 **Rjukan** Telemark, S Norway 59.52N 8.37E
97 D16 **Rjuven** ▲ S Norway
78 H9 **Rkîz** Trarza, W Mauritania 16.49N 15.19W
97 I14 **Roa** Oppland, S Norway 60.17N 10.37E
107 N5 **Roa** Castilla-León, N Spain 41.42N 3.55W
11 P13 **Rocher Percé** island Rocher Percé, Quebec, C Canada
13 V3 **Rochers Ouest, Rivière aux** ♒ Quebec, SE Canada
99 U13 **Rob Roy** island NW Solomon Islands
9 S17 **Robsart** Saskatchewan, S Canada 49.22N 109.15W
9 N15 **Robson, Mount** ▲ British Columbia, SW Canada 53.09N 119.16W
27 T14 **Robstown** Texas, SW USA 27.47N 97.40W
27 P6 **Roby** Texas, SW USA 32.44N 100.22W
106 E11 **Roca, Cabo da** headland C Portugal 38.47N 9.32W
43 S14 **Rocadas** see Xangongo
49 X6 **Roca Partida, Punta** headland C Mexico 18.43N 95.11W
109 L18 **Rocas, Atol das** island E Brazil
5 S13 **Roca d'Aspide.** Campania, S Italy 40.25N 15.12E
109 K15 **Roccaraso** Abruzzo, C Italy 41.51N 14.05E
108 H10 **Rocca San Casciano** Emilia-Romagna, C Italy 44.06N 11.51E
108 G13 **Roccastrada** Toscana, C Italy 43.00N 11.09E
63 G20 **Rocha** Rocha, E Uruguay 34.30S 54.22W
63 G20 **Rocha** ◆ department E Uruguay
99 L17 **Rochdale** NW England, UK 53.37N 2.09W
104 L11 **Rochechouart** Haute-Vienne, C France 45.49N 0.49E
101 J22 **Rochefort** Namur, S Belgium 50.10N 5.13E
104 J11 **Rochefort** var. Rochefort sur Mer. Charente-Maritime, W France 45.57N 0.58W
Rochefort sur Mer see Rochefort
129 N10 **Rochegda** Arkhangel'skaya Oblast', NW Russian Federation 62.37N 43.21E
32 L10 **Rochelle** Illinois, N USA 41.54N 89.03W
27 Q9 **Rochelle** Texas, SW USA 31.13N 99.10W
98 F9 **Roag, Loch** inlet NW Scotland, UK
35 O12 **Rochester** Indiana, N USA 41.03N 86.13W
31 W10 **Rochester** Minnesota, N USA 44.01N 92.28W
21 P5 **Rochester** New Hampshire, NE USA 43.18N 70.58W
21 T5 **Rochester** New York, NE USA 43.09N 77.37W
35 P5 **Rochester Hills** Michigan, N USA 42.39N 83.04W
66 M6 **Rockall** island UK, N Atlantic Ocean
66 L6 **Rockall Bank** undersea feature N Atlantic Ocean
86 B8 **Rockall Rise** undersea feature N Atlantic Ocean
86 C9 **Rockall Trough** undersea feature N Atlantic Ocean
37 O5 **Rock Creek** ♒ Nevada, W USA
25 T10 **Rockdale** Texas, SW USA 30.39N 96.58W
205 N12 **Rockefeller Plateau** plateau Antarctica
32 K11 **Rock Falls** Illinois, N USA 41.46N 89.41W
25 Q5 **Rockford** Alabama, S USA 32.53N 86.11W
32 L10 **Rockford** Illinois, N USA 42.16N 89.05W
13 S6 **Rock Forest** Quebec, SE Canada 45.21N 71.53W
9 T9 **Rockglen** Saskatchewan, S Canada 49.10N 105.57W
189 Y8 **Rockhampton** Queensland, E Australia 23.31S 150.31E
23 R11 **Rock Hill** South Carolina, SE USA 34.55N 80.59W
188 I13 **Rockingham** Western Australia 32.16S 115.21E
23 R10 **Rockingham** North Carolina, SE USA 34.56N 79.46W
191 N15 **Rock Island** Tasmania, SE Australia
32 J10 **Rock Island** Illinois, N USA 41.30N 90.34W
27 U12 **Rock Island** Texas, SW USA 29.31N 96.33W
12 C10 **Rock Lake** Ontario, S Canada 46.25N 83.49W
30 K8 **Rock Lake** North Dakota, N USA 48.45N 99.12W
12 M12 **Rockland** Ontario, SE Canada 45.33N 75.16W
21 R7 **Rockland** Maine, NE USA 44.08N 69.06W
190 M9 **Rocklands Reservoir** ⊠ Victoria, SE Australia
37 O7 **Rocklin** California, W USA 38.48N 121.13W
25 R3 **Rockmart** Georgia, SE USA 34.00N 85.02W
31 Q4 **Rock Port** Missouri, C USA 40.26N 95.33W
37 W7 **Rockport** Washington, NW USA 48.28N 121.36W
33 S11 **Rock Rapids** Iowa, C USA 43.27N 96.10W
32 K11 **Rock River** ♒ Illinois/Wisconsin, N USA
35 O16 **Rock Sound** Eleuthera Island, C Bahamas 24.51N 76.09W
35 U17 **Rock Springs** Wyoming, C USA 41.35N 109.12W
27 P11 **Rocksprings** Texas, SW USA 30.01N 100.14W
57 T9 **Rockstone** C Guyana 5.58N 58.33W
33 S12 **Rock Valley** Iowa, C USA 43.12N 96.17W
23 W3 **Rockville** Maryland, NE USA 39.04N 77.04W

◆ COUNTRY ◇ DEPENDENT TERRITORY ◉ ADMINISTRATIVE REGION ▲ MOUNTAIN ⊻ VOLCANO ⊜ LAKE
● COUNTRY CAPITAL ○ DEPENDENT TERRITORY CAPITAL ✕ INTERNATIONAL AIRPORT ▲ MOUNTAIN RANGE ♒ RIVER ⊠ RESERVOIR

317

27 U6 **Rockwall** Texas, SW USA *32.55N 96.27W*
31 U13 **Rockwell City** Iowa, C USA *42.24N 94.37W*
35 S10 **Rockwood** Michigan, N USA *42.04N 83.15W*
22 M9 **Rockwood** Tennessee, S USA *35.52N 84.41W*
27 Q8 **Rockwood** Texas, SW USA *31.29N 99.23W*
39 U6 **Rocky Ford** Colorado, C USA *38.03N 103.45W*
12 D9 **Rocky Island Lake** ◎ Ontario, S Canada
23 V9 **Rocky Mount** North Carolina, SE USA *35.56N 77.47W*
23 S7 **Rocky Mount** Virginia, NE USA *37.00N 79.53W*
35 Q8 **Rocky Mountain** ▲ Montana, NW USA *47.45N 112.46W*
9 P15 **Rocky Mountain House** Alberta, SW Canada *52.24N 114.52W*
39 T3 **Rocky Mountain National Park** *national park* Colorado, C USA
2 E12 **Rocky Mountains** ***var.*** Rockies, ***Fr.*** Montagnes Rocheuses. ▲▲ Canada/USA
44 H1 **Rocky Point** *headland* NE Belize *18.21N 88.04W*
85 A17 **Rocky Point** *headland* NW Namibia *19.01S 12.27E*
97 F14 **Rødberg** Buskerud, S Norway *60.16N 9.00E*
97 I25 **Rødby** Storstrøm, SE Denmark *54.42N 11.24E*
97 I25 **Rødbyhavn** Storstrøm, SE Denmark *54.39N 11.24E*
11 T10 **Roddickton** Newfoundland, SE Canada *50.51N 56.03W*
97 F23 **Rødding** Sønderjylland, SW Denmark *55.22N 9.04E*
97 M22 **Rödeby** Blekinge, S Sweden *56.16N 15.34E*
100 N6 **Roden** Drenthe, NE Netherlands *53.07N 6.25E*
64 H9 **Rodeo** San Juan, W Argentina *30.12S 69.06W*
105 O14 **Rodez** ***anc.*** Segodunum. Aveyron, S France *44.21N 2.34E*
Rodholívos *see* Rodolívos
Rodhópi Óri *see* Rhodope Mountains
Ródhos/Rodi *see* Ródos
109 N15 **Rodi Gargancio** Puglia, SE Italy *41.54N 15.51E*
103 N20 **Roding** Bayern, SE Germany *49.12N 12.30E*
115 J19 **Rodinit, Kepi i** *headland* W Albania *41.35N 19.27E*
118 I9 **Rodnei, Munții** ▲▲ N Romania
192 L4 **Rodney, Cape** *headland* North Island, NZ *36.16S 174.48E*
40 L9 **Rodney, Cape** *headland* Alaska, USA *64.39N 166.24W*
128 M16 **Rodniki** Ivanovskaya Oblast', W Russian Federation *57.04N 41.45E*
121 Q16 **Rodnya** ***Rus.*** Rodnya. Mahilyowskaya Voblasts', E Belarus *53.30N 32.12E*
Rodó *see* José Enrique Rodó
116 H13 **Rodolívos** ***var.*** Rodholívos. Kentrikí Makedonía, NE Greece *40.55N 23.59E*
Rodopi *see* Rhodope Mountains
117 O22 **Ródos** ***var.*** Ródhos, ***Eng.*** Rhodes, ***It.*** Rodi. Ródos, Dodekánisos, Greece, Aegean Sea *36.25N 28.13E*
117 O22 **Ródos** ***var.*** Ródhos, ***Eng.*** Rhodes, ***It.*** Rodi; ***anc.*** Rhodos. *island* Dodekánisos, Greece, Aegean Sea
Rodosto *see* Tekirdağ
61 A14 **Rodrigues** Amazonas, W Brazil *6.50S 73.45W*
181 P8 **Rodrigues** ***var.*** Rodriquez. *island* E Mauritius
Rodriquez *see* Rodrigues
Rodunma *see* Roanne
188 I7 **Roebourne** Western Australia *20.49S 117.04E*
85 J20 **Roedtan** Northern, NE South Africa *24.37S 29.04E*
100 H11 **Roelofarendsveen** Zuid-Holland, W Netherlands *52.12N 4.37E*
Roepat *see* Rupat, Pulau
Roer *see* Rur
101 M16 **Roermond** Limburg, SE Netherlands *51.12N 6.00E*
101 C18 **Roeselare** ***Fr.*** Roulers; ***prev.*** Rousselaere. West-Vlaanderen, W Belgium *50.57N 3.07E*
15 Ll5 **Roes Welcome Sound** *strait* Nunavut, N Canada
Roeteng *see* Ruteng
Rofreit *see* Rovereto
Rogachëv *see* Rahachow
59 L15 **Rogagua, Laguna** ◎ NW Bolivia
97 C16 **Rogaland** ◈ *county* S Norway
27 Y9 **Roganville** Texas, SW USA *30.49N 93.54W*
111 W11 **Rogaška Slatina** ***Ger.*** Rohitsch-Sauerbrunn; ***prev.*** Rogatec-Slatina. E Slovenia *46.13N 15.38E*
Rogatec-Slatina *see* Rogaška Slatina
114 J13 **Rogatica** Republica Srpska, SE Bosnia & Herzegovina *43.50N 18.55E*
Rogatin *see* Rohatyn
95 F17 **Rogen** ◎ C Sweden
29 S9 **Rogers** Arkansas, C USA *36.19N 94.07W*
31 P5 **Rogers** North Dakota, N USA *47.03N 98.12W*
27 T9 **Rogers** Texas, SW USA *30.53N 97.10W*
35 R5 **Rogers City** Michigan, N USA *45.25N 83.49W*
Roger Simpson Island *see* Abemama
37 T14 **Rogers Lake** *salt flat* California, W USA
23 Q8 **Rogers, Mount** ▲ Virginia, NE USA *36.39N 81.32W*
35 O16 **Rogerson** Idaho, NW USA *42.11N 114.36W*
9 O16 **Rogers Pass** *pass* British Columbia, SW Canada *51.18N 117.36W*
23 O8 **Rogersville** Tennessee, S USA *36.26N 83.01W*
101 L16 **Roggel** Limburg, SE Netherlands *51.16N 5.55E*
Roggeveen *see* Roggewein, Cabo
200 Nn12 **Roggeveen Basin** *undersea feature* E Pacific Ocean
203 X16 **Roggewein, Cabo** ***var.*** Roggeveen. *headland* Easter Island, Chile, E Pacific Ocean *27.07S 109.15W*
105 Y13 **Rogliano** Corse, France, C Mediterranean Sea *42.58N 9.25E*
109 N21 **Rogliano** Calabria, SW Italy *39.09N 16.18E*
94 G12 **Rognan** Nordland, C Norway *67.04N 15.21E*
102 K10 **Rögnitz** ≈ N Germany
Rogozhina/Rogozhinë *see* Rrogozhinë
112 G10 **Rogoźno** Wielkopolskie, C Poland *52.46N 16.57E*
34 E15 **Rogue River** ≈ Oregon, NW USA
118 I6 **Rohatyn** ***Rus.*** Rogatin. Ivano-Frankivs'ka Oblast', W Ukraine *49.25N 24.35E*
201 O14 **Rohi** Pohnpei, E Micronesia
Rohitsch-Sauerbrunn *see* Rogaška Slatina
155 Q13 **Rohri** Sind, SE Pakistan *27.40N 68.52E*
158 I10 **Rohtak** Haryāna, N India *28.55N 76.32E*
Roi Ed *see* Roi Et
178 Ii10 **Roi Et** ***var.*** Muang Roi Et, Roi Ed. Roi Et, E Thailand *16.04N 103.37E*
203 U9 **Roi Georges, Îles du** *island group* Îles Tuamotu, C French Polynesia
159 Y10 **Roing** Arunāchal Pradesh, NE India *28.06N 95.46E*
120 E7 **Roja** Talsi, NW Latvia *57.31N 22.44E*
63 B20 **Rojas** Buenos Aires, E Argentina *34.13S 60.41W*
155 R12 **Rojhān** Punjab, E Pakistan *28.44N 70.01E*
43 Q12 **Rojo, Cabo** *headland* C Mexico *21.33N 97.19W*
47 Q10 **Rojo, Cabo** *headland* W Puerto Rico *17.57N 67.10W*
173 G7 **Rokan Kanan, Sungai** ≈ Sumatera, W Indonesia
173 G7 **Rokan Kiri, Sungai** ≈ Sumatera, W Indonesia
Rokha *see* Rokhah
155 R4 **Rokhah** ***var.*** Rokha. Kāpīsā, E Afghanistan *35.16N 69.28E*
120 I11 **Rokiškis** Rokiškis, NE Lithuania *55.58N 25.34E*
172 Nn9 **Rokkasho** Aomori, Honshū, C Japan *40.59N 141.22E*
113 B17 **Rokycany** ***Ger.*** Rokytzan. Plzeňský Kraj, NW Czech Republic *49.45N 13.36E*
119 P6 **Rokytne** Kyyivs'ka Oblast', N Ukraine *49.40N 30.29E*
118 L3 **Rokytne** Rivnens'ka Oblast', NW Ukraine *51.19N 27.09E*
Rokytzan *see* Rokycany
164 L11 **Rola Co** ◎ W China
31 V13 **Roland** Iowa, C USA *42.10N 93.30W*
97 D15 **Røldal** Hordaland, S Norway *59.52N 6.49E*
100 O7 **Rolde** Drenthe, NE Netherlands *52.58N 6.39E*
31 O2 **Rolette** North Dakota, N USA *48.39N 99.50W*
29 V6 **Rolla** Missouri, C USA *37.57N 91.46W*
31 O2 **Rolla** North Dakota, N USA *48.51N 99.37W*
110 A10 **Rolle** Vaud, W Switzerland *46.27N 6.19E*
189 X8 **Rolleston** Queensland, E Australia *24.30S 148.36E*
193 H19 **Rolleston** Canterbury, South Island, NZ *43.34S 172.24E*
193 G18 **Rolleston Range** ▲▲ South Island, NZ
12 H8 **Rollet** Quebec, SE Canada *47.56N 79.14W*
24 J4 **Rolling Fork** Mississippi, S USA *32.54N 90.52W*
22 L6 **Rolling Fork** ≈ Kentucky, S USA
12 J11 **Rolphton** Ontario, SE Canada *46.09N 77.43W*
Röm *see* Rømø
189 X10 **Roma** Queensland, E Australia *26.36S 148.53E*
109 I15 **Roma** ***Eng.*** Rome. ● (Italy) Lazio, C Italy *41.52N 12.30E*
97 P19 **Roma** Gotland, SE Sweden *57.31N 18.28E*
23 T14 **Romain, Cape** *headland* South Carolina, SE USA *33.00N 79.21W*
11 P11 **Romaine** ≈ Newfoundland/Quebec, E Canada
27 R17 **Roma Los Saenz** Texas, SW USA *26.24N 99.01W*
116 H8 **Roman** Vratsa, NW Bulgaria *43.09N 23.56E*
118 L10 **Roman** ***Hung.*** Románvásár. Neamț, NE Romania *46.46N 26.55E*
66 M13 **Romanche Fracture Zone** *tectonic feature* E Atlantic Ocean
63 C15 **Romang** Santa Fe, C Argentina *29.30S 59.46W*
175 T15 **Romang, Pulau** ***var.*** Pulau Roma. *island* Kepulauan Damar, E Indonesia
175 Ss15 **Romang, Selat** *strait* Nusa Tenggara, S Indonesia
118 J11 **Romania** ***Bul.*** Rumŭniya, ***Ger.*** Rumänien, ***Hung.*** Románia, ***Rom.*** România, ***SCr.*** Rumunjska, ***Ukr.*** Rumuniya; ***prev.*** Republica Socialistă România, Roumania, Rumania, Socialist Republic of Romania, ***Rom.*** Romînia. ◆ *republic* SE Europe
119 T14 **Roman-Kash** ▲ S Ukraine *44.37N 34.13E*
25 W16 **Romano, Cape** *headland* Florida, SE USA *25.51N 81.40W*
46 G5 **Romano, Cayo** *island* C Cuba
126 Kk15 **Romanovka** Respublika Buryatiya, S Russian Federation *53.10N 112.34E*
131 N8 **Romanovka** Saratovskaya Oblast', W Russian Federation *51.45N 42.45E*
110 I6 **Romanshorn** Thurgau, NE Switzerland *47.33N 9.21E*
105 R12 **Romans-sur-Isère** Drôme, E France *45.03N 5.03E*
201 U12 **Romanum** *island* Chuuk, C Micronesia
Románvásár *see* Roman
41 S5 **Romanzof Mountains** ▲▲ Alaska, USA
Roma, Pulau *see* Romang, Pulau
105 S4 **Rombas** Moselle, NE France *49.15N 6.04E*
176 Xx10 **Rombebai, Danau** ◎ Irian Jaya, E Indonesia
25 R2 **Rome** Georgia, SE USA *34.01N 85.01W*
20 I9 **Rome** New York, NE USA *43.13N 75.28W*
Rome *see* Roma
35 S9 **Romeo** Michigan, N USA *42.48N 83.00W*
Römerstadt *see* Rýmařov
Rometan *see* Romitan
105 P5 **Romilly-sur-Seine** Aube, N France *48.31N 3.43E*
Romînia *see* Romania
152 L11 **Romitan** ***Rus.*** Rometan. Bukhoro Wiloyati, C Uzbekistan *39.56N 64.21E*
23 U3 **Romney** West Virginia, NE USA *39.20N 78.45W*
119 S4 **Romny** Sums'ka Oblast', NE Ukraine *50.45N 33.30E*
97 E24 **Rømø** ***Ger. Rom.*** *island* SW Denmark
119 S5 **Romodan** Poltavs'ka Oblast', NE Ukraine *50.00N 33.20E*
131 P5 **Romodanovo** Respublika Mordoviya, W Russian Federation *54.25N 45.24E*
Romorantin *see* Romorantin-Lanthenay
105 N8 **Romorantin-Lanthenay** ***var.*** Romorantin. Loir-et-Cher, C France *47.22N 1.43E*
174 Hh5 **Rompin, Sungai** ≈ Peninsular Malaysia
96 F9 **Romsdal** *physical region* S Norway
96 F10 **Romsdalen** *valley* S Norway
96 E9 **Romsdalsfjorden** *fjord* S Norway
35 P8 **Ronan** Montana, NW USA *47.31N 114.06W*
61 M14 **Roncador** Maranhão, E Brazil *5.48S 45.08W*
195 W12 **Roncador Reef** *reef* N Solomon Islands
61 J17 **Roncador, Serra do** ▲▲ C Brazil
23 S6 **Ronceverte** West Virginia, NE USA *37.45N 80.27W*
109 H14 **Ronciglione** Lazio, C Italy *42.16N 12.15E*
106 L15 **Ronda** Andalucía, S Spain *36.45N 5.10W*
96 G11 **Rondane** ▲▲ S Norway
106 L15 **Ronda, Serranía de** ▲▲ S Spain
97 H22 **Rønde** Århus, C Denmark *56.18N 10.28E*
Rōņdik *see* Rongrik Atoll
61 E16 **Rondônia** ***off.*** Estado de Rondônia; ***prev.*** Território de Rondônia. ◈ *state* W Brazil
61 I18 **Rondonópolis** Mato Grosso, W Brazil *16.28S 54.37W*
96 G11 **Rondslottet** ▲ S Norway *61.54N 9.48E*
97 P20 **Ronehamn** Gotland, SE Sweden *57.10N 18.30E*
166 L13 **Rong'an** ***var.*** Chang'an, Rongan. Guangxi Zhuangzu Zizhiqu, S China *25.13N 109.19E*
201 R4 **Rongelap Atoll** ***var.*** Rōņļap. *atoll* Ralik Chain, NW Marshall Islands
Rongerik *see* Rongrik Atoll
166 L13 **Rong Jiang** ≈ S China
166 K12 **Rongjiang** ***prev.*** Guzhou. Guizhou, S China *25.59N 108.27E*
Rong, Kas *see* Rŭng, Kaôh
178 Hh8 **Rong Kwang** Phrae, NW Thailand *18.19N 100.18E*
201 T4 **Rongrik Atoll** ***var.*** Rōņdik, Rongerik. *atoll* Ralik Chain, N Marshall Islands
201 X2 **Rongrong** *island* SE Marshall Islands
166 L13 **Rongshui** ***var.*** Rongshui Miaozu Zizhixian. Guangxi Zhuangzu Zizhiqu, S China *25.08N 109.15E*
Rongshui Miaozu Zizhixian *see* Rongshui
120 I6 **Rõngu** ***Ger.*** Ringen. Tartumaa, SE Estonia *58.10N 26.17E*
166 L15 **Rongxian** ***var.*** Rong Xian. Guangxi Zhuangzu Zizhiqu, S China *22.52N 110.33E*
Roniu *see* Ronui, Mont
201 N13 **Ronkiti** Pohnpei, E Micronesia *6.48N 158.10E*
Rōņļap *see* Rongelap Atoll
97 L24 **Rønne** Bornholm, E Denmark *55.07N 14.43E*
97 M22 **Ronneby** Blekinge, S Sweden *56.12N 15.18E*
204 J7 **Ronne Entrance** *inlet* Antarctica
204 L6 **Ronne Ice Shelf** *ice shelf* Antarctica
101 E19 **Ronse** ***Fr.*** Renaix. Oost-Vlaanderen, SW Belgium *50.45N 3.36E*
203 R8 **Ronui, Mont** ***var.*** Roniu. Tahiti, W French Polynesia *17.49S 149.12W*
32 K14 **Roodhouse** Illinois, N USA *39.28N 90.22W*
85 C19 **Rooibank** Erongo, W Namibia *23.04S 14.34E*
Rooke Island *see* Umboi Island
117 N24 **Rookery Point** *headland* NE Tristan da Cunha *37.03S 12.15W*
176 W10 **Roon, Pulau** *island* E Indonesia
181 V7 **Roo Rise** *undersea feature* E Indian Ocean
158 J9 **Roorkee** Uttar Pradesh, N India *29.51N 77.54E*
101 H15 **Roosendaal** Noord-Brabant, S Netherlands *51.31N 4.28E*
27 P10 **Roosevelt** Texas, SW USA *30.28N 100.06W*
39 N3 **Roosevelt** Utah, W USA *40.18N 109.59W*
49 T8 **Roosevelt** ≈ W Brazil
205 O13 **Roosevelt Island** *island* Antarctica
8 L10 **Roosevelt, Mount** ▲ British Columbia, W Canada *58.28N 125.22W*
9 P17 **Roosville** British Columbia, SW Canada *48.59N 115.03W*
31 X10 **Root River** ≈ Minnesota, N USA
113 N16 **Ropczyce** Podkarpackie, SE Poland *50.03N 21.36E*
189 Q3 **Roper Bar** Northern Territory, N Australia *14.45S 134.30E*
26 M5 **Ropesville** Texas, SW USA *33.24N 102.09W*
104 K14 **Roquefort** Landes, SW France *44.01N 0.18W*
63 C21 **Roque Pérez** Buenos Aires, E Argentina *35.25S 59.24W*
60 E10 **Roraima** ***off.*** Estado de Roraima; ***prev.*** Território de Rio Branco, Território de Roraima. ◈ *state* N Brazil
60 F9 **Roraima, Mount** ▲ N South America *5.10N 60.36W*
176 X10 **Rori** Irian Jaya, E Indonesia *1.44S 136.49E*
Ro Ro Reef *see* Malolo Barrier Reef
96 I9 **Røros** Sør-Trøndelag, S Norway *62.37N 11.25E*
110 I7 **Rorschach** Sankt Gallen, NE Switzerland *47.28N 9.30E*
95 E14 **Rørvik** Nord-Trøndelag, C Norway *64.52N 11.13E*
121 G17 **Ros'** ***Rus.*** Ross'. Hrodzyenskaya Voblasts', W Belarus *53.20N 24.25E*
121 G17 **Ros'** ***Rus.*** Ross'. ≈ W Belarus
119 O6 **Ros'** ≈ N Ukraine
46 K7 **Rosa, Lake** ◎ Great Inagua, S Bahamas
34 M9 **Rosalia** Washington, NW USA *47.14N 117.22W*
203 W15 **Rosalia, Punta** *headland* Easter Island, Chile, E Pacific Ocean *27.04S 109.19W*
47 P12 **Rosalie** E Dominica *15.22N 61.15W*
37 T14 **Rosamond** California, W USA *34.51N 118.09W*
37 S14 **Rosamond Lake** *salt flat* California, W USA
63 B18 **Rosario** Santa Fe, C Argentina *32.56S 60.38W*
42 J11 **Rosario** Sinaloa, C Mexico *23.00N 105.51W*
42 G6 **Rosario** Sonora, NW Mexico *27.53N 109.18W*
64 O6 **Rosario** San Pedro, C Paraguay *24.26S 57.06W*
63 E20 **Rosario** Colonia, SW Uruguay *34.19S 57.18W*
56 H5 **Rosario** Zulia, NW Venezuela *10.18N 72.19W*
Rosario *see* Rosarito
42 B4 **Rosario, Bahía del** *bay* NW Mexico
64 K6 **Rosario de la Frontera** Salta, N Argentina *25.50S 65.00W*
63 C18 **Rosario del Tala** Entre Ríos, E Argentina *32.19S 59.10W*
63 F16 **Rosário do Sul** Rio Grande do Sul, S Brazil *30.15S 54.55W*
61 H18 **Rosário Oeste** Mato Grosso, W Brazil *14.49S 56.25W*
42 E7 **Rosarito** Baja California, NW Mexico *26.27N 111.37W*
42 B1 **Rosarito** ***var.*** Rosario. Baja California, NW Mexico *32.25N 117.03W*
42 E7 **Rosarito** Baja California Sur, W Mexico *26.28N 111.40W*
106 L9 **Rosarito, Embalse del** ▣ W Spain
109 N22 **Rosarno** Calabria, SW Italy *38.28N 15.58E*
58 B5 **Rosa Zárate** ***var.*** Quinindé. Esmeraldas, SW Ecuador *0.18N 79.28W*
Roscianum *see* Rossano
31 O8 **Roscoe** South Dakota, N USA *45.24N 99.19W*
27 P7 **Roscoe** Texas, SW USA *32.27N 100.32W*
104 F5 **Roscoff** Finistère, NW France *48.43N 4.00W*
Ros Comáin *see* Roscommon
99 C17 **Roscommon** ***Ir.*** Ros Comáin. C Ireland *53.37N 8.10W*
35 Q7 **Roscommon** Michigan, N USA *44.30N 84.34W*
99 C17 **Roscommon** ***Ir.*** Ros Comáin. *cultural region* C Ireland
Ros. Cré *see* Roscrea
99 D19 **Roscrea** ***Ir.*** Ros. Cré. C Ireland *52.57N 7.46W*
47 X12 **Roseau** ***prev.*** Charlotte Town. ● (Dominica) SW Dominica *15.16N 61.22W*
31 S2 **Roseau** Minnesota, N USA *48.51N 95.45W*
181 Y16 **Rose Belle** SE Mauritius *20.24S 57.36E*
191 O16 **Rosebery** Tasmania, SE Australia *41.51S 145.33E*
23 U11 **Roseboro** North Carolina, SE USA *34.58N 78.31W*
27 T9 **Rosebud** Texas, SW USA *31.04N 96.58W*
35 W10 **Rosebud Creek** ≈ Montana, NW USA
34 F14 **Roseburg** Oregon, NW USA *43.13N 123.20W*
24 J3 **Rosedale** Mississippi, S USA *33.51N 91.01W*
101 H21 **Rosée** Namur, S Belgium *50.15N 4.43E*
57 U8 **Rose Hall** E Guyana *6.14N 57.30W*
181 X16 **Rose Hill** W Mauritius *20.13S 57.28E*
82 H12 **Roseires, Reservoir** ***var.*** Lake Rusayris. ▣ E Sudan
Rosenau *see* Rožnov pod Radhoštěm, Czech Republic
Rosenau *see* Rožňava, Slovakia
27 V11 **Rosenberg** Texas, SW USA *29.33N 95.48W*
Rosenberg *see* Olesno, Poland
Rosenberg *see* Ružomberok, Slovakia
102 I10 **Rosengarten** Niedersachsen, N Germany *53.24N 9.53E*
103 M24 **Rosenheim** Bayern, S Germany *47.51N 12.07E*
Rosenhof *see* Zilupe
107 X4 **Roses** Cataluña, NE Spain *42.15N 3.10E*
107 X4 **Roses, Golf de** *gulf* NE Spain
109 K14 **Roseto degli Abruzzi** Abruzzo, C Italy *42.39N 14.01E*
9 S16 **Rosetown** Saskatchewan, S Canada *51.34N 107.58W*
Rosetta *see* Rashîd
37 O7 **Roseville** California, W USA *38.44N 121.16W*
32 J12 **Roseville** Illinois, N USA *40.42N 90.40W*
31 V8 **Roseville** Minnesota, N USA *45.00N 93.09W*
31 R7 **Rosholt** South Dakota, N USA *45.51N 96.42W*
108 F12 **Rosignano Marittimo** Toscana, C Italy *43.24N 10.28E*
118 I14 **Roșiori de Vede** Teleorman, S Romania *44.06N 25.00E*
116 K8 **Rositsa** ≈ N Bulgaria
Rositten *see* Rēzekne
97 J23 **Roskilde** Roskilde, E Denmark *55.39N 12.07E*
97 I23 **Roskilde** ***off.*** Roskilde Amt. ◈ *county* E Denmark
Ros Láir *see* Rosslare
130 H5 **Roslavl'** Smolenskaya Oblast', W Russian Federation *53.59N 32.57E*
34 I8 **Roslyn** Washington, NW USA *47.13N 120.52W*
101 K14 **Rosmalen** Noord-Brabant, S Netherlands *51.43N 5.21E*
Ros Mhic Thriúin *see* New Ross
115 P19 **Rosoman** C FYR Macedonia *41.31N 21.55E*
104 F6 **Rosporden** Finistère, NW France *47.58N 3.54W*
193 F17 **Ross** West Coast, South Island, NZ *42.54S 170.51E*
8 J7 **Ross** ≈ Yukon Territory, W Canada
Ross' *see* Ros'
98 H8 **Ross and Cromarty** *cultural region* N Scotland, UK
109 O20 **Rossano** ***anc.*** Roscianum. Calabria, SW Italy *39.34N 16.37E*
24 L5 **Ross Barnett Reservoir** ▣ Mississippi, S USA
9 W16 **Rossburn** Manitoba, S Canada *50.42N 100.49W*
12 H13 **Rosseau** Ontario, S Canada *45.15N 79.38W*
12 H13 **Rosseau, Lake** ◎ Ontario, S Canada
195 R17 **Rossel Island** ***prev.*** Yela Island. *island* SE PNG
205 P12 **Ross Ice Shelf** *ice shelf* Antarctica
11 P16 **Rossignol, Lake** ◎ Nova Scotia, SE Canada
85 C19 **Rössing** Erongo, W Namibia *22.27S 14.52E*
205 Q14 **Ross Island** *island* Antarctica
Rossitten *see* Rybachiy
Rossiyskaya Federatsiya *see* Russian Federation
9 N17 **Rossland** British Columbia, SW Canada *49.03N 117.49W*
99 F20 **Rosslare** ***Ir.*** Ros Láir. SE Ireland *52.15N 6.22W*
99 F20 **Rosslare Harbour** Wexford, SE Ireland *52.15N 6.19W*
103 M14 **Rosslau** Sachsen-Anhalt, E Germany *51.52N 12.15E*
78 G10 **Rosso** Trarza, SW Mauritania *16.36N 15.49W*
105 X14 **Rosso, Cap** *headland* Corse, France, C Mediterranean Sea *42.25N 8.22E*
95 H16 **Rosson** Jämtland, C Sweden *63.54N 16.21E*
99 K21 **Ross-on-Wye** W England, UK *51.55N 2.34W*
Rossony *see* Rasony
130 L9 **Rossosh'** Voronezhskaya Oblast', W Russian Federation *50.09N 39.34E*
189 Q7 **Ross River** Northern Territory, N Australia *23.36S 134.30E*
8 J7 **Ross River** Yukon Territory, W Canada *61.57N 132.26W*
205 O15 **Ross Sea** *sea* Antarctica
94 G13 **Røssvatnet** ◎ C Norway
25 R1 **Rossville** Georgia, SE USA *34.59N 85.22W*
Rostak *see* Ar Rustāq
149 P14 **Rostāq** Hormozgān, S Iran *26.48N 53.50E*
119 N5 **Rostavytsya** ≈ N Ukraine
9 T15 **Rosthern** Saskatchewan, S Canada *52.40N 106.19W*
102 M8 **Rostock** Mecklenburg-Vorpommern, NE Germany *54.04N 12.07E*
128 L16 **Rostov** Yaroslavskaya Oblast', W Russian Federation *57.11N 39.19E*
Rostov *see* Rostov-na-Donu
130 L12 **Rostov-na-Donu** ***var.*** Rostov, ***Eng.*** Rostov-on-Don. Rostovskaya Oblast', SW Russian Federation *47.16N 39.45E*
Rostov-on-Don *see* Rostov-na-Donu
130 L10 **Rostovskaya Oblast'** ◈ *province* SW Russian Federation
95 J14 **Rosvik** Norrbotten, N Sweden *65.26N 21.48E*
25 S3 **Roswell** Georgia, SE USA *34.01N 84.21W*
39 U14 **Roswell** New Mexico, SW USA *33.23N 104.31W*
96 K12 **Rot** Dalarna, C Sweden *61.16N 14.04E*
103 I23 **Rot** ≈ S Germany
106 J15 **Rota** Andalucía, S Spain *36.39N 6.20W*
196 K9 **Rota** *island* S Northern Mariana Islands
27 P6 **Rotan** Texas, SW USA *32.51N 100.28W*
Rotcher Island *see* Tamana
102 I11 **Rotenburg** Niedersachsen, NW Germany *53.06N 9.25E*
Rotenburg *see* Rotenburg an der Fulda
103 I16 **Rotenburg an der Fulda** ***var.*** Rotenburg. Thüringen, C Germany *51.00N 9.43E*
103 L18 **Roter Main** ≈ E Germany
103 K20 **Roth** Bayern, SE Germany *49.15N 11.06E*
103 G16 **Rothaargebirge** ▲▲ W Germany
Rothenburg *see* Rothenburg ob der Tauber
103 J20 **Rothenburg ob der Tauber** ***var.*** Rothenburg. Bayern, S Germany *49.23N 10.10E*
204 H6 **Rothera** *UK research station* Antarctica *67.28S 68.31W*
193 I17 **Rotherham** Canterbury, South Island, NZ *42.42S 172.56E*
99 M17 **Rotherham** N England, UK *53.25N 1.19W*
98 H12 **Rothesay** W Scotland, UK *55.49N 5.03W*
110 E7 **Rothrist** Aargau, N Switzerland *47.18N 7.54E*
204 H6 **Rothschild Island** *island* Antarctica
175 Qq18 **Roti, Pulau** *island* S Indonesia
175 R18 **Roti, Selat** *strait* Nusa Tenggara, S Indonesia
191 O8 **Roto** New South Wales, SE Australia *33.04S 145.27E*
192 N8 **Rotoiti, Lake** ◎ North Island, NZ
Rotomagus *see* Rouen
109 N19 **Rotondella** Basilicata, S Italy *40.12N 16.30E*
105 X15 **Rotondo, Monte** ▲ Corse, France, C Mediterranean Sea *42.15N 9.03E*
193 I15 **Rotoroa, Lake** ◎ South Island, NZ
192 N8 **Rotorua** Bay of Plenty, North Island, NZ *38.09S 176.13E*
192 N8 **Rotorua, Lake** ◎ North Island, NZ
103 N22 **Rott** ≈ SE Germany
110 F10 **Rotten** ≈ S Switzerland
111 T6 **Rottenmann** Steiermark, E Austria *47.31N 14.18E*
100 H12 **Rotterdam** Zuid-Holland, SW Netherlands *51.55N 4.30E*
20 K10 **Rotterdam** New York, NE USA *42.46N 73.57W*
97 M21 **Rottnen** ◎ S Sweden
100 N4 **Rottumeroog** *island* Waddeneilanden, NE Netherlands
100 N4 **Rottumerplaat** *island* Waddeneilanden, NE Netherlands
103 G23 **Rottweil** Baden-Württemberg, S Germany *48.10N 8.37E*
203 O7 **Rotui, Mont** Moorea, W French Polynesia *17.30S 149.49W*
105 P1 **Roubaix** Nord, N France *50.42N 3.10E*
113 C15 **Roudnice nad Labem** ***Ger.*** Raudnitz an der Elbe. Ústecký Kraj, NW Czech Republic *50.25N 14.13E*
104 M4 **Rouen** ***anc.*** Rotomagus. Seine-Maritime, N France *49.25N 1.04E*
176 Y11 **Rouffaer Reserves** *reserve* Irian Jaya, E Indonesia
13 N10 **Rouge, Rivière** ≈ Quebec, SE Canada
22 J6 **Rough River** ≈ Kentucky, S USA
22 J6 **Rough River Lake** ▣ Kentucky, S USA
Rouhaïbé *see* Ar Ruḩaybah
104 K11 **Rouillac** Charente, W France *45.46N 0.04W*
Roulers *see* Roeselare
Roumania *see* Romania
181 Y15 **Round Island** ***var.*** Île Ronde. *island* NE Mauritius
12 J12 **Round Lake** ◎ Ontario, SE Canada
37 U7 **Round Mountain** Nevada, W USA *38.42N 117.04W*
27 R10 **Round Mountain** Texas, SW USA *30.25N 98.20W*
191 U5 **Round Mountain** ▲ New South Wales, SE Australia *30.22S 152.13E*
27 S10 **Round Rock** Texas, SW USA *30.30N 97.40W*
35 U10 **Roundup** Montana, NW USA *46.27N 108.32W*
57 Y10 **Roura** NE French Guiana *4.45N 52.18W*
Rourkela *see* Rāulakela
98 J4 **Rousay** *island* N Scotland, UK
Rousselaere *see* Roeselare
105 O17 **Roussillon** *cultural region* S France
13 V7 **Routhierville** Quebec, SE Canada *48.09N 67.07W*
101 K25 **Rouvroy** Luxembourg, SE Belgium *49.33N 5.28E*
12 I7 **Rouyn-Noranda** Quebec, SE Canada *48.16N 79.01W*
Rouyuanchengzi *see* Huachi
94 L12 **Rovaniemi** Lappi, N Finland *66.28N 25.40E*
108 E7 **Rovato** Lombardia, N Italy *45.34N 10.03E*
129 N11 **Rovdino** Arkhangel'skaya Oblast', NW Russian Federation *61.36N 42.28E*
Roven'ki *see* Roven'ky
119 Y8 **Roven'ky** ***var.*** Roven'ki. Luhans'ka Oblast', E Ukraine *48.04N 39.19E*
Rovenskaya Oblast' *see* Rivnens'ka Oblast'
Rovenskaya Sloboda *see* Rovyenskaya Slabada
108 G7 **Rovereto** ***Ger.*** Rofreit. Trentino-Alto Adige, N Italy *45.52N 11.03E*
178 J12 **Rôviĕng Tbong** Preăh Vihéar, N Cambodia *13.18N 105.06E*
Rovigno *see* Rovinj
108 H8 **Rovigo** Veneto, NE Italy *45.04N 11.48E*
114 A10 **Rovinj** ***It.*** Rovigno. Istra, NW Croatia *45.06N 13.39E*
56 E10 **Rovira** Tolima, C Colombia *4.15N 75.15W*
Rovno *see* Rivne
131 P9 **Rovnoye** Saratovskaya Oblast', W Russian Federation *50.43N 46.03E*
84 Q12 **Rovuma, Rio** ***var.*** Ruvuma. ≈ Mozambique/Tanzania ***see also*** Ruvuma
121 O19 **Rovyenskaya Slabada** ***Rus.*** Rovenskaya Sloboda. Homyel'skaya Voblasts', SE Belarus *52.12N 30.19E*
191 R5 **Rowena** New South Wales, SE Australia *29.51S 148.55E*
23 T11 **Rowland** North Carolina, SE USA *34.32N 79.17W*
15 M1 **Rowley** ≈ Baffin Island, Nunavut, NE Canada
15 M2 **Rowley Island** *island* Nunavut, NE Canada
181 W8 **Rowley Shoals** *reef* NW Australia
179 Pp12 **Roxas** Mindoro, N Philippines *12.36N 121.29E*
179 Q13 **Roxas City** Panay Island, C Philippines *11.33N 122.43E*
23 U8 **Roxboro** North Carolina, SE USA *36.23N 78.58W*
193 D23 **Roxburgh** Otago, South Island, NZ *45.32S 169.18E*
98 K13 **Roxburgh** *cultural region* SE Scotland, UK
190 H5 **Roxby Downs** South Australia *30.29S 136.56E*
97 M17 **Roxen** ◎ S Sweden
27 V5 **Roxton** Texas, SW USA *33.33N 95.43W*
13 P12 **Roxton-Sud** Quebec, SE Canada *45.30N 72.35W*
35 U8 **Roy** Montana, NW USA *47.19N 108.55W*
39 U10 **Roy** New Mexico, SW USA *35.56N 104.12W*
99 E17 **Royal Canal** ***Ir.*** An Chanáil Ríoga. *canal* C Ireland
32 L1 **Royale, Isle** *island* Michigan, N USA
39 S6 **Royal Gorge** *valley* Colorado, C USA
99 M20 **Royal Leamington Spa** ***var.*** Leamington, Leamington Spa. C England, UK *52.18N 1.31W*
99 O23 **Royal Tunbridge Wells** ***var.*** Tunbridge Wells. SE England, UK *51.07N 0.16E*
26 L9 **Royalty** Texas, SW USA *31.21N 102.51W*
104 J11 **Royan** Charente-Maritime, W France *45.37N 1.01W*
117 B24 **Roy Cove Settlement** West Falkland, Falkland Islands *51.31S 60.22W*
105 O3 **Roye** Somme, N France *49.42N 2.46E*
97 H15 **Røyken** Buskerud, S Norway *59.47N 10.21E*
95 F14 **Røyrvik** Nord-Trøndelag, C Norway *64.53N 13.30E*
27 U6 **Royse City** Texas, SW USA *32.58N 96.19W*
99 O21 **Royston** E England, UK *52.05N 0.01W*
25 U2 **Royston** Georgia, SE USA *34.17N 83.06W*
116 L10 **Roza** ***prev.*** Gyulovo. Yambol, E Bulgaria *42.29N 26.30E*
115 L16 **Rožaje** Montenegro, SW Yugoslavia *42.51N 20.11E*
112 M10 **Różan** Mazowieckie, C Poland *52.36N 21.27E*
119 O10 **Rozdil'na** Odes'ka Oblast', SW Ukraine *46.51N 30.03E*
119 S12 **Rozdol'ne** ***Rus.*** Razdolnoye. Respublika Krym, S Ukraine *45.45N 33.27E*
151 Q9 **Rozhdestvenka** Akmola, C Kazakhstan *50.51N 71.25E*
118 I6 **Rozhnyativ** Ivano-Frankivs'ka Oblast', W Ukraine *48.58N 24.07E*
118 J3 **Rozhyshche** Volyns'ka Oblast', NW Ukraine *50.54N 25.16E*
Roznau am Radhost *see* Rožnov pod Radhoštěm
113 L19 **Rožňava** ***Ger.*** Rosenau, ***Hung.*** Rozsnyó. Košický Kraj, E Slovakia *48.40N 20.31E*
118 K10 **Roznov** Neamț, NE Romania *46.46N 26.33E*
113 I18 **Rožnov pod Radhoštěm** ***Ger.*** Rosenau, Roznau am Radhost. Zlínský Kraj, E Czech Republic *49.28N 18.09E*
Rózsahegy *see* Ružomberok
Rozsnyó *see* Râșnov, Romania
Rozsnyó *see* Rožňava, Slovakia
115 K18 **Rranxë** Shkodër, NW Albania *41.58N 19.27E*
115 L18 **Rrëshen** ***var.*** Rresheni, Rrshen. Lezhë, C Albania *41.46N 19.54E*
Rresheni *see* Rrëshen
Rrogozhina *see* Rrogozhinë
115 K20 **Rrogozhinë** ***var.*** Rogozhina, Rogozhinë, Rrogozhina. Tiranë, W Albania *41.04N 19.40E*
Rrshen *see* Rrëshen
114 O13 **Rtanj** ▲ E Yugoslavia *43.45N 21.54E*
131 O7 **Rtishchevo** Saratovskaya Oblast', W Russian Federation *52.16N 43.46E*
192 N12 **Ruahine Range** ***var.*** Ruarine. ▲▲ North Island, NZ
193 L14 **Ruamahanga** ≈ North Island, NZ
Ruanda *see* Rwanda
192 M10 **Ruapehu, Mount** ▲ North Island, NZ *39.15S 175.33E*
193 C25 **Ruapuke Island** *island* SW NZ
Ruarine *see* Ruahine Range
192 O9 **Ruatahuna** Bay of Plenty, North Island, NZ *38.38S 176.56E*
192 Q8 **Ruatoria** Gisborne, North Island, NZ *37.54S 178.18E*
192 K4 **Ruawai** Northland, North Island, NZ *36.08S 174.03E*
13 N8 **Ruban** ≈ Quebec, SE Canada
83 I22 **Rubeho Mountains** ▲▲ C Tanzania
172 Q5 **Rubeshibe** Hokkaidō, NE Japan *43.49N 143.37E*
Rubezhnoye *see* Rubizhne
115 L18 **Rubik** Lezhë, C Albania *41.46N 19.48E*
56 H7 **Rubio** Táchira, W Venezuela *7.42N 72.22W*
119 X6 **Rubizhne** ***Rus.*** Rubezhnoye. Luhans'ka Oblast', E Ukraine *49.01N 38.22E*
83 F20 **Rubondo Island** *island* N Tanzania
126 Gg15 **Rubtsovsk** Altayskiy Kray, S Russian Federation *51.34N 81.10E*

◆ Country
● Country Capital
◇ Dependent Territory
○ Dependent Territory Capital
◈ Administrative Region
✕ International Airport
▲ Mountain
▲▲ Mountain Range
Volcano
≈ River
◎ Lake
▣ Reservoir

41 P9 **Ruby** Alaska, USA 64.44N 155.29W
37 W3 **Ruby Dome** ▲ Nevada, W USA 40.35N 115.25W
37 W4 **Ruby Lake** ◎ Nevada, W USA
37 W4 **Ruby Mountains** ▲ Nevada, W USA
35 Q12 **Ruby Range** ▲ Montana, NW USA
120 C10 **Rucava** Liepāja, SW Latvia 56.09N 21.10E
Rūdān see Dehbārez
Rudelstadt see Ciechanowiec
Rudensk see Rudzyensk
121 G14 **Rūdiškės** Trakai, S Lithuania 54.31N 24.49E
97 H24 **Rudkøbing** Fyn, C Denmark 54.57N 10.43E
127 N17 **Rudnaya Pristan'** Primorskiy Kray, SE Russian Federation 44.19N 135.42E
151 N14 **Rudnichnyy** Kaz. Rüdnichnyy. Almaty, SE Kazakhstan 44.39N 78.57E
129 S13 **Rudnichnyy** Kirovskaya Oblast', NW Russian Federation 59.37N 52.28E
116 N9 **Rudnik** Varna, E Bulgaria 42.57N 27.46E
Rudnya see Rudnyy
130 H4 **Rudnya** Smolenskaya Oblast', W Russian Federation 54.55N 31.10E
131 O8 **Rudnya** Volgogradskaya Oblast', SW Russian Federation 50.54N 44.27E
150 M7 **Rudnyy** var. Rudny. Kostanay, N Kazakhstan 53.00N 63.05E
126 Hh1 **Rudol'fa, Ostrov** island Zemlya Frantsa-Iosifa, NW Russian Federation
83 H16 **Rudolf, Lake** var. Lake Turkana. ◎ N Kenya
Rudolfswert see Novo mesto
103 L17 **Rudolstadt** Thüringen, C Germany 50.43N 11.19E
33 Q4 **Rudyard** Michigan, N USA 46.15N 84.36W
35 S7 **Rudyard** Montana, NW USA 48.33N 110.37W
121 K16 **Rudzyensk** Rus. Rudensk. Minskaya Voblasts', C Belarus 53.36N 27.52E
106 L6 **Rueda** Castilla-León, N Spain 41.24N 4.58W
116 F14 **Ruen** ▲ Bulgaria/FYR Macedonia 42.10N 22.31E
82 G10 **Rufa'a** Gezira, C Sudan 14.49N 33.21E
104 L10 **Ruffec** Charente, W France 46.01N 0.10E
23 R14 **Ruffin** South Carolina, SE USA 33.00N 80.48W
83 J23 **Rufiji** ≈ E Tanzania
63 A20 **Rufino** Santa Fe, C Argentina 34.15S 62.40W
78 H7 **Rufisque** W Senegal 14.44N 17.18W
85 K14 **Rufunsa** Lusaka, C Zambia 15.03S 29.36E
113 H17 **Rugāji** Balvi, E Latvia 57.01N 27.07E
167 R7 **Rugao** Jiangsu, E China 32.25N 120.39E
99 M20 **Rugby** C England, UK 52.22N 1.18W
31 N3 **Rugby** North Dakota, N USA 48.24N 100.00W
102 N7 **Rügen** headland NE Germany 54.25N 13.21E
167 N7 **Ru He** ≈ C China
83 E19 **Ruhengeri** NW Rwanda 1.39S 29.16E
102 M10 **Ruhner Berg** hill N Germany 53.17N 12.00E
120 F7 **Ruhnu** var. Ruhnu Saar, Swe. Runö. island SW Estonia
Ruhnu Saar see Ruhnu
103 G15 **Ruhr** ≈ W Germany
93 W6 **Ruhr Valley** industrial region W Germany
167 S11 **Rui'an** var. Rui an. Zhejiang, SE China 27.48N 120.36E
167 P10 **Ruichang** Jiangxi, S China 29.46N 115.37E
26 J11 **Ruidosa** Texas, SW USA 30.00N 104.40W
39 S14 **Ruidoso** New Mexico, SW USA 33.19N 105.40W
167 P12 **Ruijin** Jiangxi, S China 25.52N 116.01E
166 D13 **Ruili** Yunnan, SW China 24.04N 97.49E
100 N8 **Ruinen** Drenthe, NE Netherlands 52.46N 6.19E
101 D17 **Ruiselede** West-Vlaanderen, W Belgium 51.03N 3.21E
66 P5 **Ruivo de Santana, Pico** ▲ Madeira, Portugal, NE Atlantic Ocean 32.46N 16.57W
42 J12 **Ruiz** Nayarit, SW Mexico 21.00N 105.09W
56 E10 **Ruiz, Nevado del** ≋ W Colombia 4.52S 75.22W
124 J9 **Rujaylah, Ḥarrat ar** salt lake N Jordan
Rujen see Rūjiena
120 H7 **Rūjiena** Est. Ruhja, Ger. Rujen. Valmiera, N Latvia 57.54N 25.22E
81 J18 **Ruki** ≈ W Dem. Rep. Congo (Zaire)
83 E22 **Rukwa** ◇ region SW Tanzania
83 F23 **Rukwa, Lake** ◎ SE Tanzania
27 P6 **Rule** Texas, SW USA 33.10N 99.53W
24 K3 **Ruleville** Mississippi, S USA 33.43N 90.33W
Rum see Rhum
114 K10 **Ruma** Serbia, N Yugoslavia 45.02N 19.51E
Rumadīya see Ar Ramādī
147 Q7 **Rumāḥ** Ar Riyāḍ, C Saudi Arabia 25.35N 47.09E
150 M7 **Rumaitha** see Ar Rumaythah
Rumania/Rumänien see Romania
Rumänisch-Sankt-Georgen see Sângeorz-Băi
145 X14 **Rumaylah** SE Iraq 30.16N 47.22E
145 P2 **Rumaylah, Wādī** dry watercourse NE Syria
119 V10 **Rumbai** Irian Jaya, E Indonesia 2.44S 132.04E

83 E14 **Rumbek** El Buhayrat, S Sudan 6.49N 29.42E
176 W10 **Rumberpon, Pulau** island E Indonesia
Rumburg see Rumburk
113 D14 **Rumburk** Ger. Rumburg. Ústecký Kraj, NW Czech Republic 50.59N 14.31E
46 J4 **Rum Cay** island C Bahamas
101 M26 **Rumelange** Luxembourg, S Luxembourg 49.28N 6.01E
101 D20 **Rumes** Hainaut, SW Belgium 50.33N 3.19E
21 P7 **Rumford** Maine, NE USA 44.31N 70.31W
112 F16 **Rumia** Pomorskie, N Poland 54.57N 18.25E
115 J17 **Rumija** ▲ SW Yugoslavia
105 T11 **Rumilly** Haute-Savoie, E France 45.52N 5.57E
145 O6 **Rūmiyah** N Iraq 34.28N 41.17E
Rummah, Wādī ar see Rimah, Wādī ar
Rummelsburg in Pommern see Miastko
172 Oo4 **Rumoi** Hokkaidō, NE Japan 43.55N 141.37E
84 M12 **Rumphi** var. Rumpi. Northern, N Malawi 11.00S 33.51E
Rumpi see Rumphi
31 V7 **Rum River** ≈ Minnesota, N USA
196 F16 **Rumung** island Caroline Islands, W Micronesia
193 G16 **Runanga** West Coast, South Island, NZ 42.24S 171.15E
192 P7 **Runaway, Cape** headland North Island, NZ 37.33S 177.59E
99 K18 **Runcorn** C England, UK 53.19N 2.43W
120 K10 **Rundāni** Ludza, E Latvia 56.19N 27.51E
85 L18 **Runde** var. Lundi. ≈ SE Zimbabwe
95 O16 **Rundu** var. Runtu. Okavango, NE Namibia 17.55S 19.45E
95 I16 **Rundvik** Västerbotten, N Sweden 63.31N 19.22E
83 G20 **Runere** Mwanza, N Tanzania 3.06S 33.18E
27 S3 **Runge** Texas, SW USA 28.52N 97.42W
178 I14 **Runga, Kaôh** prev. Kas Rong. island SW Cambodia
81 O16 **Rungu** Orientale, N Dem. Rep. Congo (Zaire) 3.09N 27.58E
83 F23 **Rungwa** Rukwa, W Tanzania 7.18S 31.40E
83 G22 **Rungwa** Singida, C Tanzania 6.54S 33.33E
46 M13 **Runn** ◎ C Sweden
26 M4 **Running Water Draw** valley New Mexico/Texas, SW USA
Runö see Ruhnu
Runtu see Rundu
201 U12 **Ruo** island Caroline Islands, C Micronesia
164 G12 **Ruoqiang** var. Jo-ch'iang. Uigh. Charkhlik, Charkhliq, Qarkilik. Xinjiang Uygur Zizhiqu, NW China 38.58N 88.07E
165 S7 **Ruo Shui** ≈ N China
94 L8 **Ruostefjelbma** var. Rustefjelbma Finnmark, N Norway 70.25N 28.10E
95 L18 **Ruovesi** Länsi-Suomi, W Finland 61.58N 24.04E
114 B9 **Rupa** Primorje-Gorski Kotar, NW Croatia 45.29N 14.15E
190 M11 **Rupanyup** Victoria, SE Australia 36.38S 142.37E
174 H6 **Rupat, Pulau** prev. Roepat. island W Indonesia
174 G6 **Rupat, Selat** strait Sumatera, W Indonesia
118 J11 **Rupea** Ger. Reps, Hung. Kőhalom; prev. Cohalm. Braşov, C Romania 46.01N 25.13E
101 G17 **Rupel** ≈ N Belgium
35 P15 **Rupella** see La Rochelle
35 P15 **Rupert** Idaho, NW USA 42.37N 113.40W
23 Q10 **Rutherfordton** North Carolina, SE USA 35.22N 81.57W
21 T5 **Rupert** West Virginia, NE USA 37.57N 80.40W
Rupert House see Fort Rupert
10 J10 **Rupert, Rivière de** ≈ Quebec, C Canada
204 M13 **Ruppert Coast** physical region Antarctica
102 N11 **Ruppiner Kanal** canal NE Germany
191 N11 **Rupununi River** ≈ S Guyana
103 D16 **Rur** Dut. Roer. ≈ Germany/Netherlands
60 H13 **Rurópolis Presidente Medici** Pará, N Brazil 4.05S 55.26W
203 S12 **Rurutu** island Îles Australes, SW French Polynesia
Rusaddir see Melilla
85 L17 **Rusape** Manicaland, E Zimbabwe 18.31S 32.07E
Rusayris, Lake see Roseires, Reservoir
Ruschuk/Rusçuk see Ruse
116 K7 **Ruse** var. Ruschuk, Rustchuk, Turk. Ruschuk. Ruse, N Bulgaria 43.49N 25.58E
116 L7 **Ruse** ◇ province N Bulgaria
111 W10 **Ruše** NE Slovenia 46.31N 15.30E
116 K7 **Rusenski Lom** ≈ N Bulgaria
99 G17 **Rush** Ir. An Ros. E Ireland 53.31N 6.06W
167 S4 **Rushan** var. Xiacun. Shandong, E China 36.57N 121.33E
Rushan see Rūshon
Rushankiy Khrebet see Rushon, Qatorkŭhi
31 V7 **Rush City** Minnesota, N USA 45.41N 92.56W
31 X10 **Rush Creek** ≈ Colorado, C USA
31 X10 **Rushford** Minnesota, N USA 43.48N 91.45W
160 N13 **Rushikulya** ≈ E India
32 M7 **Rush Lake** ◎ Ontario, S Canada
30 J10 **Rush Lake** ◎ Wisconsin, N USA
30 M12 **Rushmore, Mount** ▲ South Dakota, N USA 43.52N 103.27W
153 S13 **Rūshon** Rus. Rushan. S Tajikistan 37.59N 71.32E
153 S14 **Rushon, Qatorkŭhi** Rus. Rushankiy Khrebet. ▲ SE Tajikistan

28 M12 **Rush Springs** Oklahoma, C USA 34.46N 97.57W
47 V15 **Rushville** Trinidad, Trinidad and Tobago 10.07N 61.03W
32 J13 **Rushville** Illinois, N USA 40.07N 90.33W
30 K12 **Rushville** Nebraska, C USA 42.41N 102.28W
191 O11 **Rushworth** Victoria, SE Australia 36.36S 145.03E
27 W8 **Rusk** Texas, SW USA 31.48N 95.09W
95 I14 **Rusksele** Västerbotten, N Sweden 64.49N 18.55E
120 C12 **Rusnė** Šilutė, W Lithuania 55.18N 21.19E
116 M10 **Rusokastrenska Reka** ≈ E Bulgaria
Russ see Melilla
111 X3 **Russbach** ≈ NE Austria
9 V16 **Russell** Manitoba, S Canada 50.46N 101.16W
192 K2 **Russell** Northland, North Island, NZ 35.17S 174.07E
28 L4 **Russell** Kansas, C USA 38.54N 98.51W
23 O4 **Russell** Kentucky, S USA 38.31N 82.41W
195 W15 **Russell Islands** island group C Solomon Islands
22 L7 **Russell Springs** Kentucky, S USA 37.02N 85.03W
25 O2 **Russellville** Alabama, S USA 34.30N 87.43W
9 T11 **Russellville** Arkansas, S USA 35.16N 93.07W
22 J7 **Russellville** Kentucky, S USA 36.51N 86.53W
103 G18 **Rüsselsheim** Hessen, W Germany 50.00N 8.25E
94 K8 **Russenes** Finnmark, N Norway 70.29N 24.58E
Russia see Russian Federation
Russian America see Alaska
127 N17 **Russian Federation** off. Russian Federation, var. Russia, Latv. Krievija, Rus. Rossiyskaya Federatsiya. ◆ republic Asia/Europe
41 N11 **Russian Mission** Alaska, USA 61.48N 161.23W
36 M7 **Russian River** ≈ California, W USA
204 I8 **Russkaya** Russian research station Antarctica 74.45S 135.24W
126 H3 **Russkaya Gavan'** Novaya Zemlya, Arkhangel'skaya Oblast', N Russian Federation 76.13N 62.48E
126 J4 **Russkiy, Ostrov** island N Russian Federation
111 Y5 **Rust** Burgenland, E Austria 47.48N 16.42E
143 U10 **Rust'avi** SE Georgia 41.36N 45.00E
23 T7 **Rustburg** Virginia, NE USA 37.16N 79.04W
Rustchuk see Ruse
85 I21 **Rustefjelbma** see Ruostefjelbma
85 I21 **Rustenburg** North-West, N South Africa 25.40S 27.15W
24 H5 **Ruston** Louisiana, S USA 32.31N 92.38W
83 E21 **Rutana** SE Burundi 4.01S 30.01E
64 I4 **Rutana, Volcán** ▲ N Chile 22.43S 67.52W
Rutanzige, Lake see Edward, Lake
106 M14 **Rute** Andalucía, S Spain 37.19N 4.22W
175 Pp16 **Ruteng** prev. Roeteng. Flores, C Indonesia 8.34S 120.28E
204 L8 **Rutford Ice Stream** ice feature Antarctica
37 X6 **Ruth** Nevada, W USA 39.15N 115.00W
103 G15 **Rüthen** Nordrhein-Westfalen, W Germany 51.30N 8.28E
12 D17 **Rutherford** Ontario, S Canada 42.39N 82.06W
99 J18 **Ruthin** Wel. Rhuthun. NE Wales, UK 37.57N 80.40W
110 G7 **Rüti** Zürich, N Switzerland 47.16N 8.51E
Rutlam see Ratlām
20 M9 **Rutland** Vermont, NE USA 43.32N 72.58W
99 N19 **Rutland** cultural region C England, UK
23 N8 **Rutledge** Tennessee, S USA 36.16N 83.31W
164 G12 **Rutog** var. Rutok. Xizang Zizhiqu, W China 33.27N 79.43E
Rutok see Rutog
81 P19 **Rutshuru** Nord Kivu, E Dem. Rep. Congo (Zaire) 1.13S 29.27E
100 L8 **Rutten** Flevoland, N Netherlands 52.49N 5.44E
131 Q17 **Rutul** Respublika Dagestan, SW Russian Federation 41.35N 47.30E
95 U13 **Ruukki** Oulu, C Finland 64.40N 25.35E
100 N11 **Ruurlo** Gelderland, E Netherlands 52.04N 6.27E
149 S15 **Ru'ūs al Jibāl** headland Oman/UAE
145 X7 **Ru'ūs aṭ Ṭiwāl, Jabal** ▲ W Syria
83 I25 **Ruvuma** var. Rio Rovuma. ◇ region SE Tanzania
83 I25 **Ruvuma** var. Rio Rovuma. ≈ Mozambique/Tanzania see also Rovuma, Rio
144 L9 **Ruwais** see Ar Ruways
147 Z10 **Ruwayshid, Wadi ar** dry watercourse NE Jordan
81 P18 **Ruwenzori** ≈ Uganda/Dem. Rep. Congo (Zaire)
147 Y8 **Ruwī** NE Oman 23.33N 58.31E
116 F9 **Ruy** ▲ Bulgaria/Yugoslavia
83 E20 **Ruya, Lake** see Ruyi, Rio
131 P5 **Ruyigi** E Burundi 3.28S 30.19E
131 P5 **Ruzayevka Respublika** Mordoviya, W Russian Federation 54.04N 44.54E
121 G18 **Ruzhany** Rus. Ruzhany. Brestskaya Voblasts', SW Belarus 52.52N 24.53E

116 I10 **Rŭzhevo Konare** var. Rŭzhevo Konare. Plovdiv, C Bulgaria 42.16N 24.58E
116 G7 **Ruzhin** see Ruzhyn
119 N5 **Ruzhintsi** Vidin, NW Bulgaria 43.38N 22.50E
Ruzhyn Rus. Ruzhin. Zhytomyrs'ka Oblast', N Ukraine 49.42N 29.03E
113 K19 **Ružomberok** Ger. Rosenberg, Hung. Rózsahegy. Žilinský Kraj, N Slovakia 49.03N 19.18E
113 C16 **Ruzyně** × (Praha) Praha, C Czech Republic 50.06N 14.16E
83 D19 **Rwanda** off. Rwandese Republic; prev. Ruanda. ◆ republic C Africa
Rwandese Republic see Rwanda
97 G22 **Ry** Århus, C Denmark 56.06N 9.46E
Ryasna see Rasna
130 I7 **Ryazan'** Ryazanskaya Oblast', W Russian Federation 54.37N 39.37E
130 I5 **Ryazanskaya Oblast'** ◇ province W Russian Federation
130 M6 **Ryazhsk** Ryazanskaya Oblast', W Russian Federation 53.42N 40.09E
120 B13 **Rybachiy** Ger. Rossitten. Kaliningradskaya Oblast', W Russian Federation 55.09N 20.49E
128 J2 **Rybachiy, Poluostrov** peninsula NW Russian Federation
128 L15 **Rybach'ye** see Balykchy
Rybinsk prev. Andropov. Yaroslavskaya Oblast', W Russian Federation 58.02N 38.52E
128 K14 **Rybinskoye Vodokhranilishche** Eng. Rybinsk Reservoir, Rybinsk Sea. ◎ W Russian Federation
Rybinsk Reservoir/Rybinsk Sea see Rybinskoye Vodokhranilishche
113 I16 **Rybnik** Śląskie, S Poland 50.05N 18.30E
113 F16 **Rybnitsa** see Rîbniţa
Rychnov nad Kněžnou Ger. Reichenau. Hradecký Kraj, N Czech Republic 50.10N 16.17E
112 I12 **Rychwał** Wielkopolskie, C Poland 52.04N 18.09E
9 O13 **Rycroft** Alberta, W Canada 55.45N 118.42W
97 L21 **Ryd** Kronoberg, S Sweden 56.27N 14.44E
97 L20 **Rydaholm** Jönköping, S Sweden 56.57N 14.19E
204 I8 **Rydberg Peninsula** peninsula Antarctica
99 T13 **Rye** SE England, UK 50.57N 0.42E
35 T10 **Ryegate** Montana, NW USA 46.21N 109.12W
37 S3 **Rye Patch Reservoir** ◎ Nevada, W USA
97 D15 **Ryfylke** physical region S Norway
97 H16 **Rygge** Østfold, S Norway
112 N13 **Ryki** Lubelskie, E Poland 51.37N 21.57E
130 I7 **Rykovo** see Yenakiyeve
171 Kk16 **Ryl'sk** Kurskaya Oblast', W Russian Federation 51.34N 34.41E
191 S8 **Rylstone** New South Wales, SE Australia 32.48S 149.58E
113 H17 **Rýmařov** Ger. Römerstadt. Ostravský Kraj, E Czech Republic 49.57N 17.13E
150 E11 **Ryn-Peski** desert W Kazakhstan
171 K12 **Ryōtsu** var. Ryōtu. Niigata, Sado, C Japan 38.02N 138.23E
Ryōtu see Ryōtsu
112 K10 **Rypin** Kujawsko-pomorskie, C Poland 53.03N 19.25E
Ryshkany see Rîşcani
Ryssel see Lille
97 M24 **Rytterknægten** hill E Denmark 50.10N 14.53E
171 Kk16 **Ryūgasaki** Ibaraki, Honshū, S Japan 35.54N 140.11E
198 G5 **Ryukyu Trench** var. Nansei Syotō Trench. undersea feature S East China Sea
112 D11 **Rzepin** Ger. Reppen. Lubuskie, W Poland 52.20N 14.50E
113 N16 **Rzeszów** Podkarpackie, SE Poland 50.03N 22.00E
128 I16 **Rzhev** Tverskaya Oblast', W Russian Federation 56.16N 34.21E
119 P5 **Rzhishchev** Rus. Rzhishchev. Kyyivs'ka Oblast', N Ukraine 49.58N 31.01E

—————— S ——————

144 E11 **Sa'ad** Southern, W Israel 31.27N 34.31E
111 P7 **Saalach** ≈ W Austria
103 L14 **Saale** ≈ C Germany
103 L17 **Saalfeld** var. Saalfeld an der Saale. Thüringen, C Germany 50.39N 11.22E
Saalfeld see Zalewo
Saalfeld an der Saale see Saalfeld
110 C8 **Saane** ≈ W Switzerland
100 N11 **Saar** Fr. Sarre. ≈ France/Germany see also Sarre
103 E20 **Saarbrücken** Fr. Sarrebruck. Saarland, SW Germany 49.13N 7.01E
Saarburg see Sarrebourg
120 D6 **Sääre** var. Sjar. Saaremaa, W Estonia 57.55N 22.03E
120 E6 **Saaremaa** Ger. Oesel, Ösel; prev. Saaremaa. island W Estonia
94 L12 **Saarenkylä** Lappi, N Finland 66.31N 25.51E
94 L17 **Saarijärvi** Länsi-Suomi, W Finland 62.42N 25.16E
95 M10 **Saariselkä** Lapp. Suoločielgi. Lappi, N Finland 68.25N 27.25E
94 M10 **Saariselkä** hill range NE Finland
103 D20 **Saarland** Fr. Sarre. ◇ state SW Germany
103 D20 **Saarlautern** see Saarlouis

171 Tj16 **Sagamihara** Kanagawa, Honshū, S Japan 35.32N 139.23E
171 Jj16 **Sagami-nada** inlet SW Japan
171 Jj17 **Sagami-wan** bay SW Japan
Sagan see Żagań
31 Y3 **Saganaga Lake** ◎ Minnesota, N USA
161 F18 **Sāgar** Karnātaka, W India 14.09N 75.02E
160 I9 **Sāgar** prev. Saugor. Madhya Pradesh, C India 23.52N 78.46E
13 S8 **Sagard** Quebec, SE Canada 48.01N 70.03W
179 Qq13 **Sagay** Negros, C Philippines 10.54N 123.26E
149 V11 **Sāghand** Yazd, C Iran 32.33N 55.12E
21 N14 **Sag Harbor** Long Island, New York, NE USA 40.59N 72.15W
Saghez see Saqqez
33 R8 **Saginaw** Michigan, N USA 43.25N 83.57W
33 R8 **Saginaw Bay** lake bay Michigan, N USA
153 X7 **Sagiz** Atyrau, W Kazakhstan 48.12N 54.55E
66 H6 **Saglek Bank** undersea feature W Labrador Sea
11 P5 **Saglek Bay** bay SW Labrador Sea
105 X15 **Saglouc/Sagluk** see Salluit
107 P13 **Sagonne, Golfe de** gulf Corse, France, C Mediterranean Sea
106 F14 **Sagra** ▲ S Spain 37.59N 2.33W
106 F14 **Sagres** Faro, S Portugal 37.01N 8.55W
39 S7 **Saguache** Colorado, C USA 38.05N 106.05W
46 J7 **Sagua de Tánamo** Holguín, E Cuba 20.34N 75.14W
46 E5 **Sagua la Grande** Villa Clara, C Cuba 22.48N 80.06W
13 R7 **Saguenay** ≈ Quebec, SE Canada
76 C9 **Saguia al Hamra** var. As Saqia al Hamra. ≈ N Western Sahara
Sagunt/Saguntum see Sagunto
107 S9 **Sagunto** var. Sagunt, Ar. Murviedro; anc. Saguntum. País Valenciano, E Spain 39.40N 0.16W
144 H10 **Şaḥāb** 'Ammān, N Jordan 31.52N 36.00E
56 E6 **Sahagún** Córdoba, NW Colombia 8.57N 75.26W
106 L4 **Sahagún** Castilla-León, N Spain 42.22N 5.01W
147 X8 **Saham** N Oman 24.06N 56.52E
77 U9 **Sahara** desert Libya/Algeria
77 U9 **Sahara el Gharbīya** var. Aş Şaḥrā' al Gharbīyah, Eng. Western Desert. desert C Egypt
77 X9 **Sahara el Sharqīya** var. Aş Şaḥrā' ash Sharqīyah, Eng. Arabian Desert, Eastern Desert. desert E Egypt
158 J9 **Saharan Atlas** see Atlas Saharien
158 I9 **Sahāranpur** Uttar Pradesh, N India 29.58N 77.33E
66 L10 **Saharan Seamounts** var. Saharan Seamounts. undersea feature E Atlantic Ocean 25.00N 20.00W
159 S13 **Saharsa** Bihār, NE India 25.54N 86.36E
159 R14 **Sāhibganj** Bihār, NE India 25.14N 87.38E
145 Q7 **Sahīlīyah** C Iraq 33.43N 42.42E
144 J4 **Sāḥiliyah, Jibāl as** ▲ NW Syria
116 M13 **Şahin** Tekirdağ, NW Turkey 41.01N 26.51E
155 U8 **Sāhīwāl** Punjab, E Pakistan 31.57N 72.22E
155 U9 **Sāhīwāl** prev. Montgomery. Punjab, E Pakistan 30.40N 73.04E
147 W11 **Şaḥmah, Ramlat as** desert C Oman
145 T13 **Şaḥrā' al Ḥijārah** desert S Iraq
42 H5 **Sahuaripa** Sonora, NW Mexico 29.02N 109.14W
38 M16 **Sahuarita** Arizona, SW USA 31.24N 110.55W
42 L13 **Sahuayo** var. Sahuayo de José Mariá Morelos; prev. Sahuayo de Díaz, Sahuayo de Porfirio Díaz, Michoacán de Ocampo, SW Mexico 20.04N 102.44W
Sahuayo de Díaz/Sahuayo de José Mariá Morelos/Sahuayo de Porfirio Díaz see Sahuayo
181 W8 **Sahul Shelf** undersea feature N Timor Sea
178 Hh17 **Sai Buri** Pattani, SW Thailand 6.42N 101.37E
76 I6 **Saïda** NW Algeria 34.49N 0.10E
144 G7 **Saïda** var. Şaydā, Sayida; anc. Sidon. W Lebanon 33.20N 35.24E
Sa'īdābād see Sirjān
82 B13 **Sa'id Bundas** Western Bahr el Ghazal, SW Sudan 8.24N 24.53E
194 J12 **Saidor** Madang, N PNG 5.37S 146.28E
159 S13 **Saidpur** var. Syedpur. Rajshahi, NW Bangladesh 25.48N 89.00E
110 C7 **Saignelégier** Jura, NW Switzerland 47.18N 7.03E
170 G11 **Saigō** Shimane, Dōgo, SW Japan 36.12N 133.18E
Saigon see Hồ Chi Minh
168 I12 **Saihan Toroi** Nei Mongol Zizhiqu, N China 42.46N 100.41W
168 I12 **Saihan Tal** see Sonid Youqi
Sai Hun see Syr Darya
94 M11 **Saija** Lappi, NE Finland 67.07N 28.46E
170 Ee14 **Saijō** Ehime, Shikoku, SW Japan 33.55N 133.10E
170 Dd15 **Saiki** Ōita, Kyūshū, SW Japan 32.57N 131.52E
176 Uu9 **Saileen** Irian Jaya, E Indonesia 1.14S 130.56E
95 N18 **Saimaa** ◎ SE Finland
95 N18 **Saimaa Canal** Fin. Saimaan Kanava, Rus. Saymenskiy Kanal. canal Finland/Russian Federation
Saimaan Kanava see Saimaa Canal
43 L10 **Sain Alto** Zacatecas, C Mexico 23.35N 103.14W
98 L12 **St Abb's Head** headland SE Scotland, UK 55.54N 2.07W

319

9 Y16 **St.Adolphe** Manitoba, S Canada 49.39N 96.55W

105 O15 **St-Affrique** Aveyron, S France 43.57N 2.52E

13 Q10 **St-Agapit** Quebec, SE Canada 46.33N 71.25W

99 O21 **St Albans** anc. Verulamium. E England, UK 51.46N 0.21W

20 L6 **St Albans** Vermont, NE USA 44.49N 73.07W

23 Q5 **Saint Albans** West Virginia, NE USA 38.21N 81.47W

St Alban's Head var. St Aldhelm's Head

9 Q14 **St.Albert** Alberta, SW Canada 53.37N 113.37W

99 M24 **St.Aldhelm's Head** var. St.Alban's Head. headland S England, UK 50.34N 2.04W

13 S8 **St-Alexandre** Quebec, SE Canada 47.39N 69.36W

13 O11 **St-Alexis-des-Monts** Quebec, SE Canada 46.30N 73.08W

105 P2 **Saint-Amand-les-Eaux** Nord, N France 50.27N 3.25E

105 O9 **St-Amand-Montrond** var. St-Amand-Mont-Rond. Cher, C France 46.43N 2.28E

13 Q7 **St-Ambroise** Quebec, SE Canada 48.35N 71.19W

181 P16 **St-André** NE Réunion

12 M12 **St-André-Avellin** Quebec, SE Canada 45.45N 75.04W

104 K12 **St-André-de-Cubzac** Gironde, SW France 45.01N 0.26W

98 K11 **St Andrews** E Scotland, UK 56.20N 2.48W

25 Q9 **Saint Andrews Bay** bay Florida, SE USA

25 W7 **Saint Andrew Sound** sound Georgia, SE USA

Saint Anna Trough see Svyataya Anna Trough

46 J11 **St.Ann's Bay** C Jamaica 18.25N 77.12W

11 T10 **St.Anthony** Newfoundland, SE Canada 51.21N 55.34W

35 R13 **Saint Anthony** Idaho, NW USA 43.56N 111.38W

190 M11 **Saint Arnaud** Victoria, SE Australia 36.39S 143.15E

193 I15 **St.Arnaud Range** ▲ South Island, NZ

13 T8 **St-Arsène** Quebec, SE Canada 47.55N 69.21W

11 R10 **St-Augustin** Quebec, E Canada 51.13N 58.39W

25 X9 **Saint Augustine** Florida, SE USA 29.54N 81.19W

99 H24 **St Austell** SW England, UK 50.21N 4.46W

105 T4 **St-Avold** Moselle, NE France 49.06N 6.43E

105 N17 **St-Barthélemy** ◆ S France

104 L17 **St-Béat** Haute-Garonne, S France 42.55N 0.39E

99 I15 **St Bees Head** headland NW England, UK 54.30N 3.39W

181 P16 **St-Benoit** E Réunion

105 T13 **St-Bonnet** Hautes-Alpes, SE France 44.41N 6.04E

St.Botolph's Town see Boston

99 G21 **St Brides Bay** inlet SW Wales, UK

104 H5 **St-Brieuc** Côtes d'Armor, NW France 48.31N 2.45W

104 H5 **St-Brieuc, Baie de** bay NW France

104 L7 **St-Calais** Sarthe, NW France 47.55N 0.48E

13 Q10 **St-Casimir** Quebec, SE Canada 46.40N 72.05W

12 H16 **St.Catharines** Ontario, S Canada 43.10N 79.15W

47 S14 **St.Catherine, Mount** ▲ N Grenada 12.10N 61.41W

66 C11 **St Catherine Point** headland E Bermuda

25 X6 **Saint Catherines Island** island Georgia, SE USA

99 M24 **St Catherine's Point** headland S England, UK 50.34N 1.17W

105 N13 **St-Céré** Lot, S France 44.52N 1.53E

110 A10 **St.Cergue** Vaud, SW Switzerland 46.25N 6.10E

105 R11 **St-Chamond** Loire, E France 45.28N 4.31E

35 S16 **Saint Charles** Idaho, NW USA 42.05N 111.23W

49 X4 **Saint Charles** Missouri, C USA 38.48N 90.28W

105 P13 **St-Chély-d'Apcher** Lozère, S France 44.51N 3.16E

Saint Christopher-Nevis see Saint Kitts and Nevis

33 S9 **St Clair** Michigan, N USA 42.49N 82.29W

12 D17 **St.Clair** ≈ Canada/USA

191 O17 **St.Clair, Lake** ◎ Tasmania, SE Australia

12 C17 **St.Clair, Lake** var. Lac à l'eau Claire. ◎ Canada/USA

33 S10 **Saint Clair Shores** Michigan, N USA 42.30N 82.53W

105 S10 **St-Claude** anc. Condate. Jura, E France 46.22N 5.52E

47 X6 **St-Claude** Basse Terre, SW Guadeloupe 16.01N 61.41W

31 S7 **Saint Cloud** Florida, SE USA 28.15N 81.15W

31 U8 **Saint Cloud** Minnesota, N USA 45.33N 94.09W

47 T9 **Saint Croix** island S Virgin Islands (US)

32 J4 **Saint Croix Flowage** ◎ Wisconsin, N USA

21 T5 **Saint Croix River** ≈ Canada/USA

31 W7 **Saint Croix River** ≈ Minnesota/Wisconsin, N USA

47 S14 **St.David's** SE Grenada 12.01N 61.40W

99 H21 **St David's** SW Wales, UK 51.53N 5.16W

99 G21 **St David's Head** headland SW Wales, UK 51.54N 5.19W

66 C12 **St David's Island** island E Bermuda

181 O16 **St-Denis** ◎ (Réunion) NW Réunion 20.55S 14.33E

105 U4 **St-Dié** Vosges, NE France 48.16N 6.57E

105 R5 **St-Dizier** anc. Desiderii Fanum. Haute-Marne, N France 48.39N 5.00E

13 N11 **St-Donat** Quebec, SE Canada 46.16N 74.12W

13 N11 **Ste-Adèle** Quebec, SE Canada 45.58N 74.10W

180 I16 **Sainte Anne** Inner Islands, NE Seychelles

9 Y16 **Ste.Anne** Manitoba, S Canada 49.40N 96.40W

47 R12 **Ste-Anne** Grande Terre, E Guadeloupe 16.13N 61.22W

47 Y6 **Ste-Anne** SE Martinique 14.25N 60.53W

12 Q10 **Ste-Anne** Quebec, SE Canada 50.36N 5.20E

13 W6 **Ste-Anne-des-Monts** Quebec, SE Canada 49.07N 66.28W

12 M10 **Ste-Anne-du-Lac** Quebec, SE Canada 46.51N 75.20W

13 U4 **Ste-Anne, Lac** ◎ Quebec, SE Canada

13 S10 **Ste-Apolline** Quebec, SE Canada 46.47N 70.15W

13 U7 **Ste-Blandine** Quebec, SE Canada 48.22N 68.27W

13 R10 **Ste-Claire** Quebec, SE Canada 46.36N 70.40W

13 Q10 **Ste-Croix** Quebec, SE Canada 46.36N 71.42W

110 B8 **Ste.Croix** Vaud, SW Switzerland 46.49N 6.31E

105 P14 **Ste-Énimie** Lozère, S France 44.21N 3.26E

49 Y6 **Sainte Genevieve** Missouri, C USA 37.57N 90.01W

105 S12 **St-Égrève** Isère, E France 45.15N 5.40E

13 S7 **Ste-Maguerite Nord-Est** ≈ SE Canada

13 R7 **Ste-Marguerite** ≈ SE Canada

13 V4 **Ste-Marguerite, Pointe** headland Quebec, SE Canada 50.01N 66.43W

13 V3 **Ste-Marguesite** ≈ Quebec, SE Canada

13 R10 **Ste-Marie** Quebec, SE Canada 46.28N 71.00W

47 Q11 **Ste-Marie** NE Martinique 14.48N 61.01W

181 P16 **Ste-Marie** NE Réunion

105 U6 **Ste-Marie-aux-Mines** Haut-Rhin, NE France 48.16N 7.12E

10 J14 **Ste-Marie, Lac** ◎ S Canada

180 K4 **Sainte Marie, Nosy** island E Madagascar

104 L8 **Ste-Maure-de-Touraine** Indre-et-Loire, C France 47.06N 0.38E

105 R4 **Ste-Menehould** Marne, NE France 49.06N 4.54E

Ste-Perpétue see Ste-Perpétue-de-l'Islet

13 S9 **Ste-Perpétue-de-l'Islet** var. Ste-Perpétue. Quebec, SE Canada 47.02N 69.54W

47 X11 **Ste-Rose** Basse Terre, N Guadeloupe 16.20N 61.41W

181 P16 **Ste-Rose** E Réunion

9 W15 **Ste.Rose du Lac** Manitoba, S Canada 51.04N 99.31W

104 J11 **Saintes** anc. Mediolanum. Charente-Maritime, W France 45.45N 0.37W

47 X7 **Saintes, Canal des** channel SW Guadeloupe

Saintes, Îles des see les Saintes

181 P16 **Ste-Suzanne** N Réunion

13 P10 **Ste-Thècle** Quebec, SE Canada 46.48N 72.31W

105 Q12 **St-Étienne** Loire, E France 45.25N 4.22E

104 M4 **St-Étienne-du-Rouvray** Seine-Maritime, N France 49.22N 1.07E

Saint Eustatius see Sint Eustatius

12 M11 **St-Véronique** Quebec, SE Canada 46.30N 74.58W

13 T7 **St-Fabien** Quebec, SE Canada 48.19N 68.51W

13 P7 **St-Félicien** Quebec, SE Canada 48.39N 72.28W

13 O11 **St-Félix-de-Valois** Quebec, SE Canada 46.10N 73.25W

105 Y14 **St-Florent** Corse, France, C Mediterranean Sea 42.41N 9.19E

105 Y14 **St-Florent, Golfe de** gulf Corse, France, C Mediterranean Sea

105 P6 **St-Florentin** Yonne, C France 48.00N 3.46E

105 N9 **St-Florent-sur-Cher** Cher, C France 47.00N 2.13E

105 P12 **St-Flour** Cantal, C France 45.01N 3.04E

28 H2 **Saint Francis** Kansas, C USA 39.45N 101.31W

84 X10 **St Francis, Cape** headland S South Africa 34.11S 24.45E

49 X7 **Saint Francis River** ≈ Arkansas/Missouri, C USA

24 J8 **Saint Francisville** Louisiana, S USA 30.46N 91.22W

13 Q12 **St-François** ◎ Quebec, SE Canada

47 Y6 **St-François** Grande Terre, E Guadeloupe 16.15N 61.16W

13 R11 **St-François, Lac** ◎ Quebec, SE Canada

49 X7 **Saint Francois Mountains** ▲ Missouri, C USA

13 T10 **St-Gaudens** Haute-Garonne, S France 43.07N 0.43E

13 Q8 **St-Gédéon** Quebec, SE Canada 45.51N 70.36W

189 X10 **Saint George** Queensland, E Australia 28.04S 148.39E

66 B12 **St George** N Bermuda 32.24N 64.42W

40 K15 **Saint George** Saint George Island, Alaska, USA 56.34N 169.30W

13 S14 **Saint George** South Carolina, SE USA 33.10N 80.34W

38 J8 **Saint George** Utah, W USA 37.06N 113.35W

11 R12 **St.George, Cape** headland Newfoundland, E Canada 48.26N 59.17W

195 P11 **St.George, Cape** headland New Ireland, NE PNG 4.49S 152.52E

40 J15 **Saint George Island** Pribilof Islands, Alaska, USA

25 S10 **Saint George Island** island Florida, SE USA

101 J19 **Saint-Georges** Liège, E Belgium 50.36N 5.20E

13 R11 **St-Georges** Quebec, SE Canada 46.07N 70.40W

57 Z11 **St-Georges** E French Guiana 3.55N 51.49W

47 R14 **St.George's** ◎ (Grenada) SW Grenada 12.03N 61.45W

11 R12 **St.George's Bay** inlet Newfoundland, E Canada 48.26N 59.17W

99 G21 **Saint George's Channel** channel Ireland/Wales, W Europe

195 P10 **St.George's Channel** channel NE PNG

66 B11 **St George's Island** island E Bermuda

101 I21 **Saint-Gérard** Namur, S Belgium 50.20N 4.47E

St-Germain see St-Germain-en-Laye

13 P12 **St-Germain-de-Grantham** Quebec, SE Canada 45.49N 72.32W

105 N5 **St-Germain-en-Laye** var. St-Germain. Yvelines, N France 48.52N 2.04E

104 H8 **St-Gildas, Pointe du** headland NW France 47.09N 2.25W

105 R15 **St-Gilles** Gard, S France 43.41N 4.24E

104 I9 **St-Gilles-Croix-de-Vie** Vendée, W France 46.41N 1.55W

181 O16 **St-Gilles-les-Bains** W Réunion 21.01S 55.13E

104 M16 **St-Girons** Ariège, S France 42.58N 1.07E

Saint Gotthard see Szentgotthárd

110 G9 **Gotthard Tunnel** tunnel Ticino, S Switzerland

99 H22 **St Govan's Head** headland SW Wales, UK 51.35N 4.55W

36 M7 **Saint Helena** California, W USA 38.29N 122.30W

67 F24 **Saint Helena** ◇ UK dependent territory C Atlantic Ocean

69 O12 **Saint Helena** island C Atlantic Ocean

85 E25 **St.Helena Bay** bay SW South Africa

67 M16 **Saint Helena Fracture Zone** tectonic feature C Atlantic Ocean

36 M7 **Saint Helena, Mount** ▲ California, W USA 38.40N 122.39W

23 S15 **Saint Helena Sound** inlet South Carolina, SE USA

32 J7 **Saint Helen, Lake** ◎ Michigan, N USA

191 Q16 **Saint Helens** Tasmania, SE Australia 41.21S 148.15E

96 W18 **St Helens** NW England, UK 53.28N 2.43W

34 G10 **Saint Helens** Oregon, NW USA 45.55N 122.51W

34 H10 **Saint Helens, Mount** ▲ Washington, NW USA 46.24N 121.49W

34 J6 **St Helier** ◎ (Jersey) S Jersey, Channel Islands 49.12N 2.07W

13 S9 **St-Hilarion** Quebec, SE Canada 47.34N 70.24W

101 K22 **Saint-Hubert** Luxembourg, SE Belgium 50.02N 5.23E

13 T8 **St-Hubert** Quebec, SE Canada 47.46N 69.15W

13 P12 **St-Hyacinthe** Quebec, SE Canada 45.37N 72.57W

St.Iago de la Vega see Spanish Town

33 Q4 **Saint Ignace** Michigan, N USA 45.52N 84.43W

13 O10 **St-Ignace-du-Lac** Quebec, SE Canada 46.43N 73.49W

12 D12 **St.Ignace Island** island Ontario, S Canada

98 K8 **St.Imier** Bern, N Switzerland 47.09N 6.55E

99 G25 **St Ives** SW England, UK 50.12N 5.28W

31 U10 **Saint James** Minnesota, N USA 43.58N 94.40W

10 I15 **St.James, Cape** headland Graham Island, British Columbia, SW Canada 51.57N 131.04W

13 Q10 **St-Jean** var. St-Jean-sur-Richelieu. Quebec, SE Canada 45.15N 73.16W

57 X9 **St-Jean** NW French Guiana 5.21N 54.09W

13 R8 **St-Jean, Lac** ◎ Quebec, SE Canada

104 K11 **Saint-Jean-d'Angély** Charente-Maritime, W France 45.57N 0.31W

105 N7 **St-Jean-de-Braye** Loiret, C France 47.54N 1.56E

104 I16 **St-Jean-de-Luz** Pyrénées-Atlantiques, SW France 43.24N 1.40W

105 T12 **St-Jean-de-Maurienne** Savoie, E France 45.16N 6.21E

104 I5 **St-Jean-de-Monts** Vendée, NW France 46.45N 2.00W

105 Q14 **St-Jean-du-Gard** Gard, S France 44.06N 3.49E

13 Q7 **St-Jean, Lac** ◎ Quebec, SE Canada

104 I16 **St-Jean-Pied-de-Port** Pyrénées-Atlantiques, SW France 43.10N 1.13W

13 S9 **St-Jean-Port-Joli** Quebec, SE Canada 47.13N 70.16W

13 N12 **St-Jérôme** Quebec, SE Canada 45.51N 70.36W

45 T5 **Saint Jo** Texas, SW USA 33.42N 97.33W

11 O15 **St.John** New Brunswick, SE Canada 45.16N 66.03W

28 K16 **Saint John** Kansas, C USA 37.59N 98.44W

78 K16 **Saint John** ≈ C Liberia

47 T9 **Saint John** island C Virgin Islands (US)

24 I6 **Saint John, Lake** ◎ Louisiana, S USA

21 Q2 **Saint John** Fr. Saint-John. ≈ Canada/USA

47 W10 **St John's** ● (Antigua and Barbuda) Antigua, Antigua and Barbuda 17.06N 61.50W

11 V12 **St.John's** Newfoundland, E Canada 47.34N 52.40W

39 O12 **Saint Johns** Arizona, SW USA 34.28N 109.22W

33 Q9 **Saint Johns** Michigan, N USA 42.58N 84.31W

11 V12 **St.John's** ★ Newfoundland, E Canada 47.52N 52.45W

47 N12 **St.Joseph** Dominica 15.24N 61.25W

181 P17 **St.Joseph** S Réunion

24 J6 **Saint Joseph** Louisiana, S USA 31.56N 91.14W

33 O10 **Saint Joseph** Michigan, N USA 42.04N 86.30W

49 R3 **Saint Joseph** Missouri, C USA 39.45N 94.49W

22 J10 **Saint Joseph** Tennessee, S USA 35.02N 87.29W

24 R9 **Saint Joseph Bay** bay Florida, SE USA

13 S13 **St-Joseph-de-Beauce** Quebec, SE Canada 46.20N 70.52W

10 C10 **St.Joseph, Lake** ◎ Ontario, C Canada

33 N10 **Saint Joseph River** ≈ N USA

12 C11 **Saint Joseph's Island** island Ontario, S Canada

13 N11 **St-Jovite** Quebec, SE Canada 46.07N 74.35W

123 Jj16 **St Julian's** N Malta 35.55N 14.29E

St-Julien see St-Julien-en-Genevois

57 S10 **St-Julien-en-Genevois** var. St-Julien. Haute-Savoie, E France 46.07N 6.06E

104 M11 **St-Junien** Haute-Vienne, C France 45.52N 0.54E

105 Q11 **St-Just-St-Rambert** Loire, E France 45.30N 4.12E

98 D8 **St Kilda** NW Scotland, UK

47 U10 **Saint Kitts** island Saint Kitts and Nevis

47 U10 **Saint Kitts and Nevis** off. Federation of Saint Christopher and Nevis, var. Saint Christopher-Nevis. ◆ commonwealth republic E West Indies

105 X6 **St.Laurent** Manitoba, S Canada 50.20N 97.55W

St-Laurent see St-Laurent-du-Maroni

13 S9 **St-Pacôme** Quebec, SE Canada 47.22N 69.56W

57 X9 **St-Laurent-du-Maroni** var. St-Laurent. NW French Guiana 5.28N 54.03W

St-Laurent, Fleuve see St.Lawrence

104 J12 **Saint-Médard** Gironde, SW France 45.11N 0.56W

11 N12 **St.Lawrence** Fr. Fleuve St-Laurent. ≈ Canada/USA

11 Q12 **St.Lawrence, Gulf of** gulf NW Atlantic Ocean

40 K10 **Saint Lawrence Island** island Alaska, USA

12 M14 **Saint Lawrence River** ≈ Canada/USA

101 L25 **Saint-Léger** Luxembourg, SE Belgium 49.36N 5.39E

11 N14 **St.Léonard** New Brunswick, SE Canada 47.16N 67.55W

13 P11 **St-Léonard** Quebec, SE Canada 46.06N 72.18W

181 O17 **St-Leu** W Réunion 21.09S 55.16E

104 J4 **St-Lô** anc. Briovera, Laudus. Manche, N France 49.07N 1.08W

T15 **St.Louis** Saskatchewan, S Canada 52.50N 105.43W

105 V7 **St-Louis** Haut-Rhin, NE France 47.34N 7.34E

181 O17 **St-Louis** S Réunion

78 G10 **Saint Louis** NW Senegal 15.58N 16.30W

49 X4 **Saint Louis** Missouri, C USA 38.38N 90.15W

31 W5 **Saint Louis River** ≈ Minnesota, N USA

105 T7 **St-Loup-sur-Semouse** Haute-Saône, E France 47.53N 6.15E

13 O12 **St-Luc** Quebec, SE Canada 45.19N 73.18W

85 L22 **St.Lucia** KwaZulu/Natal, E South Africa 28.22S 32.25E

47 X13 **Saint Lucia** ◆ commonwealth republic SE West Indies

85 L22 **St.Lucia, Cape** headland E South Africa 28.29S 32.42E

47 Y13 **Saint Lucia Channel** channel Martinique/Saint Lucia

25 Y14 **Saint Lucie Canal** canal Florida, SE USA

25 Z13 **Saint Lucie Inlet** inlet Florida, SE USA

98 L2 **St Magnus Bay** bay N Scotland, UK

104 K10 **St-Maixent-l'École** Deux-Sèvres, W France 46.24N 0.13W

104 I5 **St-Malo** Manitoba, S Canada 49.16N 96.03W

104 I5 **St-Malo** Ille-et-Vilaine, NW France 48.39N 2.00W

104 H4 **St-Malo, Golfe de** gulf NW France

98 K5 **St-Marc** N Haiti 19.05N 72.42W

46 L9 **St-Marc, Canal de** channel W Haiti

105 P10 **St-Marcel, Mont** ▲ S French Guiana 2.32N 53.00W

13 S11 **St-Marcellin-le-Mollard** Isère, E France 45.12N 5.18E

98 K5 **St Margaret's Hope** NE Scotland, UK 58.49N 2.57W

34 M9 **Saint Maries** Idaho, NW USA 47.19N 116.37W

25 Y9 **Saint Marks** Florida, SE USA 33.42N 97.33W

110 D11 **St.Martin** Valais, SW Switzerland 46.09N 7.27E

Saint Martin see Sint Maarten

33 O5 **Saint Martin Island** island Michigan, N USA

24 I9 **Saint Martinville** Louisiana, S USA 30.09N 91.51W

193 E20 **St.Mary, Mount** ▲ South Island, NZ 44.16S 169.42E

194 K14 **St.Mary, Mount** ▲ S PNG 8.06S 147.00E

190 I6 **Saint Mary Peak** ▲ South Australia 31.25S 138.39E

191 Q16 **Saint Marys** Tasmania, SE Australia 41.34S 148.13E

12 E16 **St.Marys** Ontario, S Canada 43.15N 81.08W

40 M11 **Saint Marys** Alaska, USA 62.03N 163.10W

25 W8 **Saint Marys** Georgia, SE USA 30.44N 81.30W

28 L3 **Saint Marys** Kansas, C USA 39.09N 96.00W

33 Q4 **Saint Marys** Ohio, N USA 40.31N 84.22W

23 R3 **Saint Marys** West Virginia, NE USA 39.23N 81.11W

25 W8 **Saint Marys River** ≈ Florida/Georgia, SE USA

33 Q4 **Saint Marys River** ≈ N USA

104 D6 **St-Mathieu, Pointe** headland NW France 48.17N 4.56W

40 J12 **Saint Matthew Island** island Alaska, USA

23 R13 **Saint Matthews** South Carolina, SE USA 33.40N 80.46W

St.Matthew's Island see Zadetkyi Kyun

194 M8 **St.Matthias Group** island group NE PNG

110 C11 **St.Maurice** Valais, SW Switzerland 46.09N 7.28E

13 P9 **St-Maurice** ≈ Quebec, SE Canada

104 J13 **St-Médard-en-Jalles** Gironde, SW France 44.54N 0.43W

41 N10 **Saint Michael** Alaska, USA 63.28N 162.02W

13 N10 **St-Michel** see Mikkeli

13 N10 **St-Michel-des-Saints** Quebec, SE Canada 46.39N 73.54W

105 S5 **St-Mihiel** Meuse, NE France 48.57N 5.33E

110 J10 **St.Moritz** Ger. Sankt Moritz, Rmsch. San Murezzan. Graubünden, SE Switzerland 46.30N 9.50E

104 M8 **St-Nazaire** Loire-Atlantique, NW France 47.16N 2.12W

Saint Nicholas see São Nicolau

Saint-Nicolas see Sint-Niklaas

105 N1 **St-Omer** Pas-de-Calais, N France 50.45N 2.15E

104 J11 **Saintonge** cultural region W France

13 S9 **St-Pacôme** Quebec, SE Canada 47.22N 69.56W

13 S10 **St-Pamphile** Quebec, SE Canada 46.57N 69.46W

13 R10 **St-Pascal** Quebec, SE Canada 47.25N 69.51W

12 J1 **St-Patrice, Lac** ◎ Quebec, SE Canada

9 R14 **St.Paul** Alberta, SW Canada 54.00N 111.18W

181 O16 **St-Paul** NW Réunion

40 K14 **St Paul** Saint Paul Island, Alaska, USA 57.08N 170.13W

31 V8 **Saint Paul** state capital Minnesota, N USA 45.00N 93.10W

31 P15 **Saint Paul** Nebraska, C USA 41.12N 98.26W

23 P7 **Saint Paul** Virginia, NE USA 36.53N 82.18W

79 Q17 **Saint Paul, Cape** headland S Ghana 5.44N 0.55E

105 O17 **St-Paul-de-Fenouillet** Pyrénées-Orientales, S France 42.49N 2.28E

67 N14 **Saint Paul Fracture Zone** tectonic feature E Atlantic Ocean

40 J14 **Saint Paul Island** island Pribilof Islands, Alaska, USA

104 J15 **St-Paul-lès-Dax** Landes, SW France 43.45N 1.01W

23 U11 **Saint Pauls** North Carolina, SE USA 34.45N 78.56W

203 R16 **St.Paul's Point** headland Pitcairn Island, Pitcairn Islands

31 U10 **Saint Peter** Minnesota, N USA 44.18N 93.59W

99 L26 **St Peter Port** ◎ (Guernsey) C Guernsey, Channel Islands 49.28N 2.33W

25 V13 **Saint Petersburg** Florida, SE USA 27.46N 82.37W

Saint Petersburg see Sankt-Peterburg

25 V13 **Saint Petersburg Beach** Florida, SE USA 27.43N 82.43W

181 P17 **St-Philippe** SE Réunion

47 Q11 **St-Pierre** NW Martinique 14.44N 61.10W

181 O17 **St-Pierre** SW Réunion

11 S13 **St-Pierre and Miquelon** Fr. Îles St-Pierre et Miquelon. ◇ French territorial collectivity NE North America

13 S9 **St-Pierre, Lac** ◎ Quebec, SE Canada

104 F5 **St-Pol-de-Léon** Finistère, NW France 48.42N 4.00W

105 O2 **St-Pol-sur-Ternoise** Pas-de-Calais, N France 50.22N 2.21E

105 O16 **St-Pons-de-Thomières** var. St.Pons. Hérault, S France 43.28N 2.48E

105 P10 **St-Pourçain-sur-Sioule** Allier, C France 46.19N 3.16E

13 S11 **St-Prosper** Quebec, SE Canada 46.14N 70.28W

105 P3 **St-Quentin** Aisne, N France 49.51N 3.17E

13 R10 **St-Raphaël** Quebec, SE Canada 46.47N 70.46W

105 U15 **St-Raphaël** Var, SE France 43.25N 6.46E

13 Q10 **St-Raymond** Quebec, SE Canada 46.53N 71.49W

35 Q9 **Saint Regis** Montana, NW USA 47.18N 115.06W

20 J7 **Saint Regis River** ≈ New York, NE USA

105 R15 **St-Rémy-de-Provence** Bouches-du-Rhône, SE France 43.48N 4.49E

13 V6 **St-René-de-Matane** Quebec, SE Canada 48.42N 67.22W

104 M9 **St-Savin** Vienne, W France 46.34N 0.53E

13 S8 **St-Siméon** Quebec, SE Canada 47.49N 69.55W

35 X7 **Saint Simons Island** island Georgia, SE USA

203 Y2 **Saint Stanislas Bay** bay Kiritimati, E Kiribati

11 O15 **St.Stephen** New Brunswick, SE Canada 45.12N 67.18W

41 X12 **Saint Terese** Alaska, USA 58.28N 134.46W

12 E17 **St.Thomas** Ontario, S Canada 42.46N 81.12W

31 Q2 **Saint Thomas** North Dakota, N USA 48.37N 97.28W

47 T9 **Saint Thomas** island W Virgin Islands (US)

Saint Thomas see São Tomé, Sao Tome and Principe

Saint Thomas see Charlotte Amalie, Virgin Islands (US)

13 P10 **St-Tite** Quebec, SE Canada 46.42N 72.32W

105 U16 **St-Tropez** Var, SE France 43.16N 6.39E

St Ubes see Setúbal

104 L3 **St-Valéry-en-Caux** Seine-Maritime, N France 49.53N 0.42E

105 Q9 **St-Vallier** Saône-et-Loire, C France 46.39N 4.19E

108 B7 **St-Vincent** Valle d'Aosta, NW Italy 45.47N 7.42E

47 Q14 **Saint Vincent** island S Saint Vincent and the Grenadines

47 W14 **Saint Vincent and the Grenadines** ◆ commonwealth republic SE West Indies

47 N15 **St Vincent, Cape** see São Vicente, Cabo de

104 I15 **St-Vincent-de-Tyrosse** Landes, SW France 43.39N 1.16W

190 I9 **Saint Vincent, Gulf** gulf South Australia

25 R10 **Saint Vincent Island** island Florida, SE USA

47 T12 **Saint Vincent Passage** passage Saint Lucia/Saint Vincent and the Grenadines

191 N18 **Saint Vincent, Point** headland Tasmania, SE Australia 43.19S 145.50E

Saint-Vith see Sankt-Vith

9 S14 **St.Walburg** Saskatchewan, S Canada 53.37N 109.12W

St Wolfgangsee see Wolfgangsee

104 M11 **St-Yrieix-la-Perche** Haute-Vienne, C France 45.31N 1.12E

Saint Yves see Setúbal

13 Y5 **St-Yvon** Quebec, SE Canada 49.09N 64.51W

196 H5 **Saipan** island ● (Northern Mariana Islands) S Northern Mariana Islands

196 H6 **Saipan Channel** channel S Northern Mariana Islands

196 H6 **Saipan International Airport** ★ Saipan, S Northern Mariana Islands

13 G9 **Sais** ★ (Fès) C Morocco

96 G6 **Sais** 33.58N 4.48W

Saishū see Cheju-do

Saishū see Cheju-do

13 Q10 **Saison** ≈ SW France

13 P17 **Sai, Sungai** ≈ Borneo, N Indonesia

171 Jj16 **Saitama** off. Saitama-ken. ◆ prefecture Honshū, S Japan

170 C16 **Saito** Miyazaki, Kyūshū, SW Japan 32.07N 131.23E

144 J15 **Saiyid Abid** Sind, SE Pakistan 25.50N 68.32E

146 M3 **Sakākāh** Al Jawf, NW Saudi Arabia 29.55N 40.10E

13 M20 **Sakajószentpéter** Borsod-Abaúj-Zemplén, NE Hungary 48.13N 20.43E

85 F24 **Sak** ≈ SW South Africa 30.09N 20.27E

183 J8 **Saka** Coast, E Kenya 4.39S 39.27E

178 I11 **Sa Kaeo** Prachin Buri, C Thailand 13.47N 102.03E

171 Gg15 **Sakai** Osaka, Honshū, SW Japan 34.34N 135.28E

170 E12 **Sakai** Fukui, Honshū, SW Japan 36.09N 136.13E

170 Ff12 **Sakaiminato** Tottori, Honshū, SW Japan 35.34N 133.12E

170 L4 **Sakakawea, Lake** ◎ North Dakota, N USA

10 J9 **Sakami** ≈ Quebec, C Canada

81 O26 **Sakania** Katanga, SE Dem. Rep. Congo (Zaire) 12.43S 28.34E

152 I14 **Sakar-Chaga** Turkm. Sakarchäge. Maryysky Velayat, C Turkmenistan 37.40N 61.33E

Sakarchäge see Sakar-Chaga

Sak'art'velo see Georgia

142 F11 **Sakarya** province NW Turkey

142 F12 **Sakarya Nehri** ≈ NW Turkey

150 K13 **Saksaul'skiy** var. Saksaul'skoye. Kaz. Sekseûîl. Kyzylorda, S Kazakhstan 47.07N 61.06E

Saksaul'skoye see Saksaul'skiy

170 E12 **Sakata** Yamagata, Honshū, C Japan 38.54N 139.51E

126 L9 **Sakha (Yakutiya), Respublika** var. Respublika Yakutiya, Yakutiya, Eng. Yakutia. ◆ autonomous republic NE Russian Federation

Sakhalin see Sakhalin, Ostrov

160 O14 **Sakhalin, Ostrov** var. Sakhalin. island SE Russian Federation

127 P14 **Sakhalinskaya Oblast'** ◆ province SE Russian Federation

127 Nn13 **Sakhalinskiy Zaliv** gulf E Russian Federation

Sakhnovshchina see Sakhnovshchyna

119 O6 **Sakhnovshchyna** Rus. Sakhnovshchina. Kharkivs'ka Oblast', E Ukraine 49.08N 35.51E

Sakhon Nakhon see Sakon Nakhon

Sakhtsar see Rämsar

Saki see Saky

13 W10 **Şäki** Rus. Sheki; prev. Nukha. NW Azerbaijan 41.09N 47.10E

120 E13 **Šakiai** Ger. Schaken. Šakiai, S Lithuania 54.57N 23.04E

179 Z16 **Sakirmethe** Irian Jaya, E Indonesia 8.36S 140.55E

172 Oo16 **Sakishima-shotō** var. Sakisima Syotô. island group SW Japan

Sakiz see Saqqez

Sakiz-Adasi see Chíos

161 F19 **Sakleshpur** Karnātaka, S India 12.58N 75.45E

178 I9 **Sakon Nakhon** var. Muang Sakon Nakhon, Sakhon Nakhon. Sakon Nakhon, E Thailand 17.10N 104.07E

155 P15 **Sakrand** Sind, SE Pakistan 26.10N 68.13E

85 F24 **Sak River** Afr. Sakrivier. Northern Cape, W South Africa 30.49S 20.24E

Sakrivier see Sak River

150 K13 **Saksaul'skoye** prev. Saksaul'skiy, Kaz. Sekseûîl. Kyzylorda, S Kazakhstan 47.07N 61.06E

97 I25 **Sakskøbing** Storstrøm, SE Denmark 54.48N 11.39E

171 Ij15 **Saku** Nagano, Honshū, S Japan 36.15N 138.28E

171 K16 **Sakura** Chiba, Honshū, S Japan 35.42N 140.10E

119 S13 **Saky** Rus. Saki. Respublika Krym, S Ukraine 45.08N 33.36E

78 E9 **Sal** island Ilhas de Barlavento, NE Cape Verde

131 N12 **Sal** ≈ SW Russian Federation

113 I21 **Sal'a** Hung. Sellye, Vágsellye. Nitriansky Kraj, SW Slovakia 48.08N 17.55E

97 N15 **Sala** Västmanland, C Sweden 59.55N 16.37E

44 C2 **Salabangka, Kepulauan** island group N Indonesia

63 D14 **Salada** Corrientes, NE Argentina 28.15S 58.40W

63 C21 **Saladillo** Buenos Aires, E Argentina 35.40S 59.49W

28 T9 **Saladillo, Río** ≈ C Argentina

45 T5 **Salado** Texas, SW USA 30.57N 97.32W

J16 **Salado, Arroyo** ≈ SE Argentina

29 Q12 **Salado, Río** ≈ New Mexico, SW USA

62 D21 **Salado, Río** ≈ E Argentina

63 C21 **Salado, Río** ≈ C Argentina

43 N7 **Salado, Río** ≈ NE Mexico

149 N6 **Salafchegân** var. Sarafjagän. Qom, N Iran 34.28N 50.28E

79 Q15 **Salaga** C Ghana 8.31N 0.37W

198 A2 **Sala'ilua** Savai'i, W Samoa 13.39S 172.33W

85 H20 **Salajwe** Kweneng, SE Botswana 23.40S 24.46E

80 H9 **Salal** Kanem, W Chad 14.48N 17.12E

82 I6 **Salala** Red Sea, NE Sudan 21.16N 36.16E

147 V13 **Salālah** SW Oman 17.01N 54.03E

44 D5 **Salamá** Baja Verapaz, C Guatemala 15.06N 90.18W

44 J6 **Salamá** Olancho, C Honduras 14.48N 86.34W

64 G10 **Salamanca** Coquimbo, C Chile 31.49S 70.58W

43 N13 **Salamanca** Guanajuato, C Mexico 20.33N 101.06W

106 K7 **Salamanca** anc. Helmantica, Salmantica. Castilla-León, NW Spain 40.58N 5.40W

20 D11 **Salamanca** New York, NE USA 42.09N 78.43W

106 J7 **Salamanca** ◆ province Castilla-León, W Spain

65 J19 **Salamanca, Pampa de** plain S Argentina

80 J12 **Salamat** off. Préfecture du Salamat. ◆ prefecture SE Chad

80 J12 **Salamat, Bahr** ≈ S Chad

56 F5 **Salamina** Magdalena, N Colombia 10.28N 74.46W

117 G19 **Salamína** var. Salamís. Salamína, C Greece 37.58N 23.28E

117 G19 **Salamína** island C Greece

Salamís see Salamína

144 I5 **Salamiyah** var. As Salamiyah. Ḥamāh, W Syria 35.01N 37.01E

33 P12 **Salamonie Lake** ◎ Indiana, N USA

33 P12 **Salamonie River** ≈ Indiana, N USA

Salang see Phuket

198 B8 **Salani** Upolu, SE Samoa 14.00S 171.33W

120 C11 **Salantai** Kretinga, NW Lithuania 56.05N 21.36E

106 K2 **Salas** Asturias, N Spain 43.25N 6.15W

107 O5 **Salas de los Infantes** Castilla-León, N Spain 42.01N 3.17W

104 M16 **Salat** ≈ S France

201 V13 **Salat** island Chuuk, C Micronesia

174 L15 **Salatiga** Jawa, C Indonesia 7.15S 110.34E

201 V13 **Salat Pass** passage W Pacific Ocean

131 V6 **Salavat** Respublika Bashkortostan, W Russian Federation 53.20N 55.54E

58 C12 **Salaverry** La Libertad, W Peru 8.15S 78.57W

176 Uu9 **Salawati, Pulau** *island* E Indonesia

200 Nn11 **Sala y Gomez** *island* Chile, E Pacific Ocean

Sala y Gomez Fracture Zone *see* Sala y Gomez Ridge

200 O11 **Sala y Gomez Ridge** *var.* Sala y Gomez Fracture Zone. *tectonic feature* SE Pacific Ocean

63 A22 **Salazar** Buenos Aires, E Argentina 36.19S 62.10W

56 G7 **Salazar** Norte de Santander, N Colombia 7.46N 72.46W

Salazar *see* N'Dalatando

181 P16 **Salazie** Réunion 21.01S 55.33E

105 N8 **Salbris** Loir-et-Cher, C France 47.25N 2.02E

59 G15 **Salcantay, Nevado** ▲ C Peru 13.21S 72.31W

47 O8 **Salcedo** N Dominican Republic 19.21N 70.23W

41 S9 **Salcha River** ✍ Alaska, USA

121 H15 **Šalčininkai** Šalčininkai, SE Lithuania 54.19N 25.26E

Saldae *see* Béjaïa

56 E11 **Saldaña** Tolima, C Colombia 3.57N 75.01W

106 M4 **Saldaña** Castilla-León, N Spain 42.31N 4.43W

85 E25 **Saldanha** Western Cape, SW South Africa 33.00S 17.56E

Salduba *see* Zaragoza

63 B23 **Saldungaray** Buenos Aires, E Argentina 38.15S 61.45W

120 D9 **Saldus** *Ger.* Frauenburg. Saldus, W Latvia 56.40N 22.29E

191 P13 **Sale** Victoria, SE Australia 38.06S 147.06E

76 F6 **Salé** NW Morocco 34.07N 6.40W

76 F6 **Salé** ✈ (Rabat) W Morocco 33.69N 6.30W

Salehābād *see* Andimeshk

174 Ii10 **Saleh, Air** ✍ Sumatera, W Indonesia

175 Oo16 **Saleh, Teluk** *bay* Nusa Tenggara, S Indonesia

125 G8 **Salekhard** *prev.* Obdorsk. Yamalo-Nenetskiy Avtonomnyy Okrug, N Russian Federation 66.33N 66.35E

198 B7 **Sālelologa** Savai'i, C Samoa 13.42S 172.10W

161 H21 **Salem** Tamil Nādu, SE India 11.37N 78.07E

49 V9 **Salem** Arkansas, C USA 36.21N 91.49W

32 L15 **Salem** Illinois, N USA 38.37N 88.57W

33 P15 **Salem** Indiana, N USA 38.37N 86.06W

21 P11 **Salem** Massachusetts, NE USA 42.30N 70.51W

49 V6 **Salem** New Jersey, NE USA 37.39N 91.32W

20 I16 **Salem** New Jersey, NE USA 39.33N 75.26W

33 U12 **Salem** Ohio, N USA 40.52N 80.51W

34 G12 **Salem** *state capital* Oregon, NW USA 44.57N 123.01W

31 Q11 **Salem** South Dakota, N USA 43.43N 97.23W

38 L4 **Salem** Utah, W USA 40.03N 111.40W

23 S7 **Salem** Virginia, NE USA 37.16N 80.00W

23 R3 **Salem** West Virginia, NE USA 39.15N 80.32W

109 H23 **Salemi** Sicilia, Italy, C Mediterranean Sea 37.48N 12.48E

Salemy *see* As Sālimī

96 K12 **Sälen** Dalarna, C Sweden 61.11N 13.14E

109 Q18 **Salentina, Campi** Puglia, SE Italy 40.22N 18.01E

109 Q18 **Salentina, Penisola** *peninsula* SE Italy

109 L18 **Salerno** *anc.* Salernum. Campania, S Italy 40.40N 14.43E

109 L18 **Salerno, Golfo di** *Eng.* Gulf of Salerno. *gulf* S Italy

Salerno, Gulf of *see* Salerno, Golfo di

Salernum *see* Salerno

99 K17 **Salford** NW England, UK 53.30N 2.16W

Salgir *see* Salhyr

113 K21 **Salgótarján** Nógrád, N Hungary 48.06N 19.46E

61 O15 **Salgueiro** Pernambuco, E Brazil 8.04S 39.04W

96 C13 **Salhus** Hordaland, S Norway 60.30N 5.15E

119 T12 **Salhyr** *Rus.* Salgir. ✍ S Ukraine

175 S4 **Salibabu, Pulau** *island* N Indonesia

39 S6 **Salida** Colorado, C USA 38.29N 105.57W

104 J15 **Salies-de-Béarn** Pyrénées-Atlantiques, SW France 43.28N 0.55W

224 C14 **Salihli** Manisa, W Turkey 38.28N 28.07E

121 K18 **Salihorsk** *Rus.* Soligorsk. Minskaya Voblasts', S Belarus 52.48N 27.31E

121 K18 **Salihorskáye Vodaskhovishcha** ☑ C Belarus

85 N14 **Salima** Central, C Malawi 13.44S 34.21E

177 Ff5 **Salin** Magwe, W Myanmar 20.30N 94.40E

49 N4 **Salina** Kansas, C USA 38.50N 97.36W

38 L5 **Salina** Utah, W USA 38.57N 111.54W

45 S17 **Salina Cruz** Oaxaca, SE Mexico 16.10N 95.12W

109 L22 **Salina, Isola** Isole Eolie, S Italy

46 J5 **Salina Point** *headland* Acklins Island, SE Bahamas 22.10N 74.16W

42 A7 **Salinas** Guayas, W Ecuador 2.15S 80.54W

42 M11 **Salinas** *var.* Salinas de Hidalgo. San Luis Potosí, C Mexico 22.36N 101.41W

47 T6 **Salinas** C Puerto Rico 17.58N 66.18W

37 O10 **Salinas** California, W USA 36.40N 121.40W

Salinas, Cabo de *see* Salines, Cap de ses

Salinas de Hidalgo *see* Salinas

84 A13 **Salinas, Ponta das** *headland* W Angola 12.50S 12.57E

47 O10 **Salinas, Punta** *headland* S Dominican Republic 18.11N 70.32W

Salinas, Río *see* Chixoy, Río

37 O11 **Salinas River** ✍ California, W USA

24 H6 **Saline Lake** ☑ Louisiana, S USA

27 R17 **Salineno** Texas, SW USA 26.29N 99.06W

49 V14 **Saline River** ✍ Arkansas, C USA

32 M11 **Saline River** ✍ Illinois, N USA

107 X10 **Salines, Cap de ses** *var.* Cabo de Salinas. *headland* Mallorca, Spain, W Mediterranean Sea 39.15N 3.03E

Salisbury *see* Mazsalaca

47 O12 **Salisbury** *var.* Baroui. W Dominica 15.25N 61.27W

99 M23 **Salisbury** *var.* New Sarum. S England, UK 51.04N 1.48W

23 Y4 **Salisbury** Maryland, NE USA 38.21N 75.36W

49 T3 **Salisbury** Missouri, C USA 39.25N 92.48W

23 S9 **Salisbury** North Carolina, SE USA 35.40N 80.28W

Salisbury *see* Harare

16 N5 **Salisbury Island** *island* Nunavut, NE Canada

99 L23 **Salisbury, Lake** *see* Bisina, Lake

99 L23 **Salisbury Plain** *plain* S England, UK

23 R14 **Salkehatchie River** ✍ South Carolina, SE USA

144 I9 **Salkhad** As Suwaydā', SW Syria 32.28N 36.42E

94 M12 **Salla** Lappi, NE Finland 66.49N 28.40E

105 U11 **Sallanches** Haute-Savoie, E France 45.55N 6.37E

107 V5 **Sallent** Cataluña, NE Spain 41.48N 1.52E

63 A22 **Salliqueló** Buenos Aires, E Argentina 36.42S 62.52W

49 R10 **Sallisaw** Oklahoma, C USA 35.29N 94.47W

82 I7 **Sallom** Red Sea, NE Sudan 19.30N 37.11E

10 J2 **Salluit** *prev.* Saglouc, Sagluk. Quebec, NE Canada 62.10N 75.40W

77 T7 **Salūm** *var.* As Sallūm. NW Egypt 31.31N 25.09E

158 F14 **Salūm** Rājasthān, N India 24.16N 74.04E

77 T7 **Salūm, Gulf of Ar.** Khalīj as Salūm. *gulf* Egypt/Libya

175 Q7 **Salumpaga** Sulawesi, N Indonesia 1.18N 120.58E

161 M14 **Salūr** Andhra Pradesh, E India 18.31N 83.14E

57 Y9 **Salut, Îles du** *island group* N French Guiana

108 A9 **Saluzzo** *Fr.* Saluces; *anc.* Saluciae. Piemonte, NW Italy 44.39N 7.28E

65 F23 **Salvación, Bahía** *bay* S Chile

61 P17 **Salvador** *prev.* São Salvador. Bahia, E Brazil 12.58S 38.28W

67 E24 **Salvador** East Falkland, Falkland Islands 51.28S 58.22W

24 K10 **Salvador, Lake** ☑ Louisiana, S USA

56 D13 **Salvaleón de Higüey** *see* Higüey

106 F10 **Salvaterra de Magos** Santarém, C Portugal 39.01N 8.46W

43 N13 **Salvatierra** Guanajuato, C Mexico 20.13N 100.52W

107 P3 **Salvatierra Basq.** Agurain. País Vasco, N Spain 42.52N 2.22W

178 H5 **Salwa/Salwah** *see* As Salwá

177 I13 **Salween Bur.** Thanlwin, *Chin.* Nu Chiang, Nu Jiang. ✍ SE Asia

143 Y12 **Salyan** Rus. Sal'yany. SE Azerbaijan 39.36N 48.57E

159 N11 **Salyan** *var.* Sallyana. Mid Western, W Nepal 28.22N 82.10E

Sal'yany *see* Salyan

23 O6 **Salyersville** Kentucky, S USA 37.44N 83.01W

111 V6 **Salza** ✍ E Austria

111 Q7 **Salzach** ✍ Austria/Germany

111 Q6 **Salzburg** *anc.* Juvavum. Salzburg, N Austria 47.48N 13.03E

111 O8 **Salzburg off.** Land Salzburg. ◆ *state* C Austria

Salzburg *see* Ocna Sibiului

111 Q7 **Salzburger Kalkalpen Eng.** Salzburg Alps. ▲ C Austria

102 J13 **Salzgitter** *prev.* Wenstedt-Salzgitter. Niedersachsen, C Germany 52.06N 10.24E

103 G14 **Salzkotten** Nordrhein-Westfalen, W Germany 51.40N 8.36E

102 K11 **Salzwedel** Sachsen-Anhalt, N Germany 52.51N 11.10E

158 D11 **Sām** Rājasthān, NW India 26.49N 70.30E

56 G9 **Šamac** *see* Bosanski Šamac

94 N3 **Samá** Boyacá, C Colombia 78.12N 12.11E

144 I3 **Samachique** Chihuahua, N Mexico 27.16N 107.28W

95 F14 **Salsbruket** Nord-Trøndelag, C Norway 64.49N 11.48E

130 M13 **Sal'sk** Rostovskaya Oblast', SW Russian Federation 46.30N 41.30E

108 E9 **Salsomaggiore Terme** Emilia-Romagna, N Italy 44.49N 9.58E

109 K25 **Salso** ✍ Sicilia, Italy, C Mediterranean Sea

109 J25 **Salso** ✍ Sicilia, Italy, C Mediterranean Sea

64 J3 **Salta** Salta, NW Argentina 24.47S 65.23W

99 Q17 **Salta off.** Provincia de Salta. ◆ *province* N Argentina

98 I24 **Saltash** SW England, UK 50.24N 4.13W

26 J8 **Salt Basin** *basin* Texas, SW USA

9 V16 **Saltcoats** Saskatchewan, S Canada 51.06N 102.12W

46 K4 **Salt Cay** *island* SE Bahamas

142 K17 **Salt Draw** ✍ Texas, SW USA

99 F21 **Saltee Islands** *island group* SE Ireland

155 P3 **Saltfjorden** *inlet* C Norway

26 I8 **Salt Flat** Texas, SW USA

49 N8 **Salt Fork Arkansas River** ✍ Oklahoma, C USA 36.35N 105.05W

33 T13 **Salt Fork Lake** ☑ Ohio, N USA

28 J11 **Salt Fork Red River** ✍ Oklahoma/Texas, C USA

97 J23 **Saltholm** *island* E Denmark

43 N8 **Saltillo** Coahuila de Zaragoza, NE Mexico 25.30N 101.00W

190 L5 **Salt Lake** *salt lake* New South Wales, SE Australia

39 V15 **Salt Lake** ☑ New Mexico, SW USA

38 K2 **Salt Lake City** *state capital* Utah, W USA 40.44N 111.54W

63 C20 **Salto** Buenos Aires, E Argentina 34.18S 60.17W

63 D17 **Salto** Salto, N Uruguay 31.25S 57.58W

63 D17 **Salto** ◆ *department* N Uruguay

64 Q6 **Salto del Guairá** Canindeyú, E Paraguay 24.06S 54.22W

63 D17 **Salto Grande, Embalse de** *var.* Lago de Salto Grande. ☑ Argentina/Uruguay

Salto Grande, Lago de *see* Salto Grande, Embalse de

37 W16 **Salton Sea** ☑ California, W USA

62 I12 **Salto Santiago, Represa de** ☑ S Brazil

155 U7 **Salt Range** ▲ E Pakistan

38 M13 **Salt River** ✍ Arizona, SW USA

22 L5 **Salt River** ✍ Kentucky, S USA

49 V3 **Salt River** ✍ Missouri, C USA

97 F17 **Saltrød** Aust-Agder, S Norway 58.28N 8.49E

97 P16 **Saltsjöbaden** Stockholm, C Sweden 59.15N 18.19E

94 G12 **Saltstraumen** Nordland, C Norway 67.16N 14.42E

23 Q7 **Saltville** Virginia, NE USA 36.52N 81.48W

23 Q12 **Saluda** South Carolina, SE USA 34.00N 81.47W

23 X6 **Saluda** Virginia, NE USA 37.36N 76.36W

23 Q12 **Saluda River** ✍ South Carolina, SE USA

175 R10 **Salue Timpuas, Selat** *var.* Selat Banggai. *strait* N Banda Sea

77 T7 **Salūm** *var.* As Sallūm. NW Egypt 31.31N 25.09E

179 R12 **Samar** *island* C Philippines

131 S6 **Samara** *prev.* Kuybyshev. Samarskaya Oblast', W Russian Federation 53.11N 50.15E

131 S6 **Samara** ✈ Samarskaya Oblast', W Russian Federation 53.11N 50.27E

131 T7 **Samara** ✍ W Russian Federation

119 V7 **Samara** ✍ E Ukraine

195 N17 **Samarai** Milne Bay, SE PNG 10.37S 150.39E

144 G9 **Samarang** *see* Semarang

144 G9 **Samarian Hills** *hill range* N Israel

56 L9 **Samariapo** Amazonas, C Venezuela 5.13N 67.47W

175 O8 **Samarinda** Borneo, C Indonesia 0.30S 117.09E

Samarkand *see* Samarqand

Samarkandskaya Oblast' *see* Samarqand Wiloyati

Samarkandski/Samarkandskoye *see* Temirtau

153 N11 **Samarobriva** *see* Amiens

152 M11 **Samarqand Rus.** Samarkand. Samarqand Wiloyati, C Uzbekistan 39.39N 66.55E

152 M11 **Samarqand Wiloyati Rus.** Samarkandskaya Oblast'. ◆ *province* C Uzbekistan

145 S6 **Sāmarrā'** C Iraq 34.13N 43.52E

131 R7 **Samarskaya Oblast' prev.** Kuybyshevskaya Oblast'. ◆ *province* W Russian Federation

159 Q13 **Samastipur** Bihār, N India 25.52N 85.46E

78 L14 **Samatiguila** NW Ivory Coast 9.51N 7.36W

121 Q17 **Samatsevichy Rus.** Samotevichi. Mahilyowskaya Voblasts', E Belarus 53.12N 31.49E

143 Y11 **Samaxı Rus.** Shemakha. C Azerbaijan 40.31N 48.54E

Samawa *see* As Samāwah

158 H6 **Samba** Jammu and Kashmir, NW India 32.31N 75.07E

81 K18 **Samba** Equateur, NW Dem. Rep. Congo 2.21N 21.01E

81 N21 **Samba** Maniema, E Dem. Rep. Congo (Zaire) 4.40S 26.22E

175 Oo6 **Sambaliung, Pegunungan** ▲ Borneo, N Indonesia

160 M11 **Sambalpur** Orissa, E India 21.28N 83.04E

69 X12 **Sambao** ✍ W Madagascar

174 Kk7 **Sambas, Sungai** ✍ Borneo, N Indonesia

180 K2 **Sambava** Antsirañana, NE Madagascar 14.16S 50.10E

176 Ww9 **Samberi** Irian Jaya, E Indonesia 1.07S 135.54E

158 J10 **Sambhal** Uttar Pradesh, N India 28.34N 78.34E

158 H12 **Sāmbhar Salt Lake** ☑ N India

109 N21 **Sambiase** Calabria, SW Italy 38.58N 16.16E

118 H5 **Sambir prev.** Sambor. *L'viv'ska Oblast'*, NW Ukraine 49.29N 23.09E

84 C13 **Samba** Huambo, C Angola 13.07S 16.04E

Sambor *see* Sambir

63 E21 **Samborombón, Bahía** *bay* NE Argentina

63 C20 **Sambre** ◆ Belgium/France

101 M20 **Sambre** ✍ Belgium/France

45 V16 **Sambú, Río** ✍ SE Panama

169 Z14 **Samch'ŏk Jap.** Sanchoku. NE South Korea 37.21N 129.12E

163 J21 **Samch'ŏnp'o** *see* Sach'ŏn

110 J10 **Same** Kilimanjaro, NE Tanzania 4.02S 37.46E

83 K22 **Samfya** Luapula, N Zambia 11.25S 29.30E

147 W13 **Samhān, Jabal** ▲ SW Oman

117 C18 **Sámi** Kefallinía, Iónioi Nísoi, Greece, C Mediterranean Sea 38.15N 20.39E

196 H6 **Sami** Saipan, S Northern Mariana Islands

58 F10 **Samíria, Río** ✍ N Peru

143 V11 **Şämkir Rus.** Shamkhor. NW Azerbaijan 40.51N 46.03E

126 J7 **Sam, Nam Vtn.** Sông Chu. ✍ Laos/Vietnam

Samnān *see* Semnān

Sam Neua *see* Xam Nua

77 P10 **Samnū** C Libya 27.19N 15.01E

198 Bb7 **Samoa off.** Independent State of Samoa, *var.* Sāmoa; *prev.* Western Samoa ◆ *monarchy* W Polynesia

198 C8 **Samoa** *island group* American /Samoa

183 T9 **Samoa Basin** *undersea feature* W Pacific Ocean

Sāmoa-i-Sisifo *see* Sāmoa

114 D8 **Sambor** Zagreb, N Croatia 45.48S 15.38E

116 H10 **Samokov** *var.* Samakov. Sofiya, W Bulgaria 42.19N 23.34E

113 H21 **Šamorín Ger.** Sommerein, *Hung.* Somorja. Trnavský Kraj, W Slovakia 48.01N 17.18E

117 M19 **Sámos prev.** Limín Vathéos. Sámos, Dodekánisos, Greece, Aegean Sea 37.46N 26.58E

117 M20 **Sámos** *island* Dodekánisos, Greece, Aegean Sea

116 K13 **Samosch** *see* Someş

173 Ff5 **Samosir, Pulau** *island*

Samotevichi *see* Samatsevichy

117 K14 **Samothráki** Samothráki, NE Greece 40.28N 25.31E

117 J14 **Samothráki anc.** Samothrace. *island* NE Greece

117 A15 **Samothráki** *var.* Samothrace; *anc.* Samothrace, Samothraki. *island* Iónioi Nísoi, Greece, C Mediterranean Sea

174 M10 **Sampit** Borneo, C Indonesia 2.30S 112.30E

174 M10 **Sampit, Sungai** ✍ Borneo, N Indonesia

27 X8 **San Augustine** Texas, SW USA 31.31N 94.06W

147 T13 **Sanāw** *var.* Sanaw. NE Yemen 18.00N 51.00E

126 J7 **Sam, Nam Vtn.** Sông Chu. ✍ Laos/Vietnam

195 P11 **Sampun** New Britain, E PNG 5.19S 152.06E

81 N24 **Sampwe** Katanga, S Dem. Rep. Congo (Zaire) 9.17S 27.22E

27 X8 **Sam Rayburn Reservoir** ☑ Texas, SW USA

175 T13 **Samsang** Xizang Zizhiqu, W China

97 H22 **Samsø** *island* E Denmark

97 H23 **Samsø Bælt** *channel* E Denmark

142 J7 **Sâm Sơn** Thanh Hoa, N Vietnam 19.43N 105.52E

142 L11 **Samsun** ✈ W Turkey

142 L11 **Samsun anc.** Amisus. Samsun, N Turkey 41.16N 36.22E

142 L11 **Samsun** ◆ *province* N Turkey

143 R9 **Samtredia** W Georgia 42.09N 42.20E

61 E15 **Samuel, Represa de** ☑ W Brazil

178 H15 **Samui, Ko** *island* SW Thailand

155 U9 **Samundari** *see* Samundri

155 U9 **Samundri var.** Samundari. Punjab, E Pakistan 31.04N 72.58E

143 X10 **Samur** ✍ Azerbaijan/Russian Federation

143 Y11 **Samur-Abşeron Kanalı Rus.** Samur-Apsheronskiy Kanal. *canal* E Azerbaijan

Samur-Apsheronskiy Kanal *see* Samur-Abşeron Kanalı

178 Hh1 **Samut Prakan** *var.* Muang Samut Prakan, Paknam. Samut Prakan, C Thailand 13.39N 100.13E

178 Hh1 **Samut Sakhon** *var.* Maha Chai, Samut Sakorn, Tha Chin. Samut Sakhon, C Thailand 13.31N 100.15E

178 Hh1 **Samut Sakorn** *see* Samut Sakhon

178 Hh1 **Samut Songhram prev.** Meklong. Samut Songkhram, SW Thailand 13.25N 100.01E

79 N12 **San Ségou, C Mali** 13.18N 4.51W

113 O15 **San** ✍ SE Poland

147 O15 **San'ā' Eng.** Sana. ● (Yemen) W Yemen 15.24N 44.13E

114 F11 **Sana** ✍ NW Bosnia and Herzegovina

82 O12 **Sanaag off.** Gobolka Sanaag. ◆ *region* N Somalia

116 J8 **Sanadinovo** Pleven, N Bulgaria 43.33N 25.00E

205 P1 **Sanae** South African research station Antarctica 70.19S 1.31W

145 V10 **Sanāf, Hawr as** ☑ S Iraq

81 C15 **Sanaga** ✍ C Cameroon

56 D12 **San Agustín** Huila, SW Colombia 1.52N 76.13W

179 S16 **San Agustin, Cape** *headland* Mindanao, S Philippines 6.17N 126.12E

39 Q13 **San Agustín, Plains of** *plain* New Mexico, SW USA

40 M16 **Sanak Islands** *island group* Aleutian Islands, Alaska, USA

200 P11 **San Ambrosio, Isla Eng.** San Ambrosio Island. *island* W Chile

200 P11 **San Ambrosio Island** *see* San Ambrosio, Isla

175 S10 **San Antonio** Pulau Sanana, E Indonesia 2.04S 125.58E

175 S10 **Sanana, Pulau** *island* Maluku, E Indonesia

148 K5 **Sanandaj prev.** Sinneh. Kordestān, W Iran 35.47.01E

37 P8 **San Andreas** California, W USA 38.10N 120.40W

2 C13 **San Andreas Fault** *fault* W USA

56 G8 **San Andrés** Santander, C Colombia 72.52N 72.52W

63 C20 **San Andrés de Giles** Buenos Aires, E Argentina 34.25S 59.27W

39 R14 **San Andres Mountains** ▲ New Mexico, USA

169 Z14 **Samch'ŏk Jap.** Sanchoku. NE South Korea 37.21N 129.12E

43 S15 **San Andrés Tuxtla var.** Tuxtla. Veracruz-Llave, E Mexico 18.27N 95.18W

27 P8 **San Angelo** Texas, SW USA 31.27N 100.26W

109 A20 **Sant'Antioco, Isola di** *island* W Italy

44 F4 **San Antonio** Toledo, S Belize 16.13N 89.02W

64 G11 **San Antonio** Valparaíso, C Chile 33.35S 71.34W

196 H6 **San Antonio** Saipan, S Northern Mariana Islands

39 R13 **San Antonio** New Mexico, SW USA 33.53N 106.52W

27 R12 **San Antonio** Texas, SW USA 29.25N 98.29W

56 M11 **San Antonio** Amazonas, S Venezuela 3.31N 66.46W

56 I7 **San Antonio** Barinas, C Venezuela 7.24N 71.28W

57 O5 **San Antonio** Monagas, NE Venezuela 10.03S 63.45W

57 S12 **San Antonio** ✈ Texas, SW USA 29.31N 98.11W

San Antonio *see* San Antonio del Táchira

107 V11 **San Antonio Abad** Eivissa, Spain, W Mediterranean Sea 38.58N 1.18E

27 U13 **San Antonio Bay** *inlet* Texas, SW USA

42 F4 **San Antonio, Cabo** *headland* E Argentina 36.45S 56.40W

44 A5 **San Antonio, Cabo de** *headland* W Cuba 21.51N 84.58W

107 T11 **San Antonio, Cabo de** *headland* E Spain 38.50N 0.09E

175 O9 **Sancoins** Cher, C France 46.49N 3.00E

195 Z17 **San Cristóbal** *var.* Makira. *island* SE Solomon Islands

56 H7 **San Cristóbal** Táchira, W Venezuela 7.46N 72.15W

190 K9 **Sandalwood** South Australia 34.51S 140.13E

Sandalwood Island *see* Sumba, Pulau

96 D11 **Sandane** Sogn og Fjordane, S Norway 61.46N 6.13E

116 G12 **Sandanski prev.** Sveti Vrach. Blagoevgrad, SW Bulgaria 41.36N 23.18E

78 J11 **Sandaré** Kayes, W Mali 14.36N 10.22W

97 J19 **Sandared** Västra Götaland, S Sweden 57.43N 12.46E

96 N12 **Sandarne** Gävleborg, C Sweden 61.15N 17.15E

194 E10 **Sandaun prev.** West Sepik. ◆ *province* NW PNG

18 K4 **Sanday** *island* NE Scotland, UK

179 N14 **Sand Cay** *island* W Spratly Islands

33 S5 **Sand Creek** ✍ Indiana, N USA

97 H15 **Sande** Vestfold, S Norway 59.34N 10.13E

97 H16 **Sandefjord** Vestfold, S Norway 59.07N 10.13E

179 O15 **Sandégué** E Ivory Coast 7.58N 3.33W

79 P14 **Sandema** N Ghana 10.42N 1.17W

39 O11 **Sanders** Arizona, SW USA 35.13N 109.21W

26 M11 **Sanderson** Texas, SW USA 30.10N 102.23W

25 U4 **Sandersville** Georgia, SE USA 32.58N 82.48W

94 H4 **Sandgerdhi** Sudhurland, SW Iceland 64.01N 22.42W

30 K14 **Sand Hills** ▲ Nebraska, C USA

25 S14 **Sandia** Texas, SW USA 27.59N 97.52W

25 T17 **San Diego** California, SW USA 32.43N 117.09W

25 S14 **San Diego** Texas, SW USA 27.45N 98.14W

142 F14 **Sandıklı** Afyon, W Turkey 38.28N 30.16E

158 L12 **Sandila** Uttar Pradesh, N India 27.05N 80.37E

174 Gg12 **Sanding, Selat** *strait* W Indonesia

32 J3 **Sand Island** *island* Apostle Islands, Wisconsin, N USA

97 C16 **Sandnes** Rogaland, S Norway 58.51N 5.45E

94 F13 **Sandnessjøen** Nordland, C Norway 66.00N 12.37E

81 L24 **Sandoa** Katanga, S Dem. Rep. Congo (Zaire) 9.39S 22.58E

113 N15 **Sandomierz Rus.** Sandomir. Świętokrzyskie, C Poland 50.42N 21.44E

Sandomir *see* Sandomierz

56 C13 **Sandoná** Nariño, SW Colombia 1.13N 77.29W

108 I7 **San Dona di Piave** Veneto, NE Italy 45.37N 12.34E

128 K14 **Sandovo** Tverskaya Oblast', W Russian Federation 58.26N 36.30E

177 Ff7 **Sandoway** Arakan State, W Myanmar 18.28N 94.19E

99 M24 **Sandown** S England, UK 50.39N 1.11W

97 B19 **Sandoy Dan.** Sandø Island Faeroe Islands 61.52N 6.51W

41 N16 **Sand Point** Popof Island, Alaska, USA 55.20N 160.40W

67 N24 **Sand Point** *headland* E Tristan da Cunha

33 R7 **Sand Point** *headland* Michigan, N USA 43.54N 83.24W

29 M7 **Sandpoint** Idaho, NW USA 48.16N 116.33W

95 H14 **Sandsele** Västerbotten, N Sweden 65.16N 17.40E

10 I14 **Sandspit** Moresby Island, British Columbia, SW Canada

49 P9 **Sand Springs** Oklahoma, C USA 36.08N 96.06W

31 W7 **Sandstone** Minnesota, N USA 46.07N 92.51W

38 K15 **Sand Tank Mountains** ▲ Arizona, SW USA

33 S8 **Sandusky** Michigan, N USA 43.24N 82.47W

33 S11 **Sandusky** Ohio, N USA 41.27N 82.42W

33 S12 **Sandusky River** ✍ Ohio, N USA

85 D22 **Sandverhaar** Karas, S Namibia 26.49S 17.25E

97 L24 **Sandvig** Bornholm, E Denmark 55.51N 14.45E

97 H15 **Sandvika** Akershus, S Norway 59.54N 10.28E

96 M11 **Sandviken** Gävleborg, C Sweden 60.37N 16.49E

32 M11 **Sandwich** Illinois, N USA 41.39N 88.37W

Sandwich Island *see* Éfaté

Sandwich Islands *see* Hawaiian Islands

159 V16 **Sandwip Island** *island* SE Bangladesh

9 U12 **Sandy Bay** Saskatchewan, C Canada 55.31N 102.14W

191 N16 **Sandy Cape** *headland* Tasmania, SE Australia 41.27S 144.43E

178 Mm14 **Sandy Cay** *island* NW Spratly Islands

38 L3 **Sandy City** Utah, W USA 40.36N 111.53W

33 O5 **Sandy Creek** ✍ Ohio, N USA

23 O5 **Sandy Hook** Kentucky, S USA 38.09N 83.05W

20 K15 **Sandy Hook** *headland* New Jersey, NE USA 40.27N 73.59W

152 J15 **Sandykachi Turkm.** Sandykgachy. Maryyskiy Velayat, S Turkmenistan 36.34N 62.28E

152 J15 **Sandykgachy** *see* Sandykachi

Sandykly, Peski *desert* E Turkmenistan

9 Q13 **Sandy Lake** Alberta, C Canada 55.50N 113.30W

10 B8 **Sandy Lake** Ontario, C Canada 53.00N 93.25W

10 B8 **Sandy Lake** ☑ Ontario, C Canada

25 S3 **Sandy Springs** Georgia, SE USA 33.57N 84.23W

26 H8 **San Elizario** Texas, SW USA *31.35N 106.16W*
101 L25 **Sanem** Luxembourg, SW Luxembourg *49.33N 5.55E*
44 K5 **San Esteban** Olancho, C Honduras *15.18N 85.45W*
107 O6 **San Esteban de Gormaz** Castilla-León, N Spain *41.34N 3.13W*
42 E5 **San Esteban, Isla** *island* NW Mexico
San Eugenio/San Eugenio del Cuareim *see* Artigas
64 H11 **San Felipe** ***var.*** San Felipe de Aconcagua. Valparaíso, C Chile *32.45S 70.42W*
42 D3 **San Felipe** Baja California, NW Mexico *31.02N 114.55W*
42 N12 **San Felipe** Guanajuato, C Mexico *21.27N 101.12W*
56 K5 **San Felipe** Yaracuy, NW Venezuela *10.25N 68.40W*
46 B5 **San Felipe, Cayos de** *island group* W Cuba
San Felipe de Aconcagua *see* San Felipe
San Felipe de Puerto Plata *see* Puerto Plata
39 R11 **San Felipe Pueblo** New Mexico, SW USA *35.25N 106.27W*
San Feliú de Guixols *see* Sant Feliu de Guíxols
200 Oo11 **San Félix, Isla** ***Eng.*** San Felix Island. *island* W Chile
San Felix Island *see* San Félix, Isla
56 L11 **San Fernanado de Atabapo** Amazonas, S Venezuela *4.00N 67.42W*
42 C4 **San Fernando** ***var.*** Misión San Fernando. Baja California, NW Mexico *29.58N 115.14W*
43 P9 **San Fernando** Tamaulipas, C Mexico *24.51N 98.09W*
179 P9 **San Fernando** Luzon, N Philippines *16.45N 120.21E*
179 P10 **San Fernando** Luzon, N Philippines *15.01N 120.41E*
106 J16 **San Fernando** ***prev.*** Isla de León. Andalucía, S Spain *36.28N 6.12W*
47 U14 **San Fernando** Trinidad, Trinidad and Tobago *10.16N 61.27W*
37 S15 **San Fernando** California, W USA *34.16N 118.26W*
56 L7 **San Fernando** ***var.*** San Fernando de Apure. Apure, C Venezuela *7.54N 67.28W*
San Fernando de Apure *see* San Fernando
64 L8 **San Fernando del Valle de Catamarca** ***var.*** Catamarca. Catamarca, NW Argentina *28.28S 65.46W*
San Fernando de Monte Cristi *see* Monte Cristi
43 P9 **San Fernando, Río** ≈ C Mexico
25 X11 **Sanford** Florida, SE USA *28.48N 81.16W*
21 P9 **Sanford** Maine, NE USA *43.26N 70.46W*
23 T10 **Sanford** North Carolina, SE USA *35.28N 79.10W*
27 N2 **Sanford** Texas, SW USA *35.42N 101.31W*
41 T10 **Sanford, Mount** ▲ Alaska, USA *62.21N 144.12W*
44 G8 **San Francisco** ***var.*** Gotera, San Francisco Gotera. Morazán, E El Salvador *13.40N 88.06W*
45 R16 **San Francisco** Veraguas, C Panama *8.14N 80.58W*
179 Pp11 **San Francisco** ***var.*** Aurora. Luzon, N Philippines *13.22N 122.31E*
37 L8 **San Francisco** California, W USA *37.46N 122.25W*
56 H5 **San Francisco** Zulia, NW Venezuela *10.36N 71.39W*
36 M8 **San Francisco** ✈ California, W USA *37.37N 122.23W*
37 N9 **San Francisco Bay** *bay* California, W USA
63 C24 **San Francisco de Bellocq** Buenos Aires, E Argentina *38.42S 60.01W*
42 I6 **San Francisco de Borja** Chihuahua, N Mexico *27.57N 106.42W*
44 J6 **San Francisco de la Paz** Olancho, C Honduras *14.55N 86.13W*
42 J7 **San Francisco del Oro** Chihuahua, N Mexico *26.52N 105.49W*
42 M12 **San Francisco del Rincón** Jalisco, SW Mexico *20.57N 101.54W*
47 O8 **San Francisco de Macorís** C Dominican Republic *19.15N 70.15W*
San Francisco de Satipo *see* Satipo
San Francisco Gotera *see* San Francisco
San Francisco Telixtlahuaca *see* Telixtlahuaca
109 K23 **San Fratello** Sicilia, Italy, C Mediterranean Sea *38.00N 14.35E*
San Fructuoso *see* Tacuarembó
84 C12 **Sanga** Cuanza Sul, NW Angola *11.10S 15.27E*
58 C5 **San Gabriel** Carchi, N Ecuador *0.37N 77.49W*
165 S15 **Sa'ngain** Xizang Zizhiqu, W China *30.46N 98.45E*
160 E13 **Sangamner** Mahārāshtra, W India *19.37N 74.18E*
158 H12 **Sāngāner** Rājasthān, N India *26.48N 75.48E*
Sangan, Koh-i- *see* Sangān, Kūh-e
155 N6 **Sangān, Kūh-e** ***Pash.*** Koh-i-Sangan. ▲▲ C Afghanistan
126 Ll10 **Sangar** Respublika Sakha (Yakutiya), NE Russian Federation *63.48N 127.37E*
175 O8 **Sangasanga** Borneo, C Indonesia *0.36S 117.12E*
105 N1 **Sangatte** Pas-de-Calais, N France *50.56N 1.41E*
109 B19 **San Gavino Monreale** Sardegna, Italy, C Mediterranean Sea *39.33N 8.47E*
59 D16 **Sangayan, Isla** *island* W Peru
32 L14 **Sangchris Lake** ⊠ Illinois, N USA
175 P15 **Sangeang, Pulau** *island* S Indonesia
118 I10 **Sângeorgiu de Pădure** ***prev.*** Erdăt-Sângeorz, Sîngeorgiu de Pădure, ***Hung.*** Erdőszentgyörgy. Mureș, C Romania *46.27N 24.49E*
118 I9 **Sângeorz-Băi** ***var.*** Singeroz Băi, ***Ger.*** Rumänisch-Sankt-Georgen, ***Hung.*** Oláhszentgyőrgy; ***prev.*** Sîngeorz-Băi. Bistrița-Năsăud, N Romania *47.24N 24.40E*
37 R10 **Sanger** California, W USA *36.42N 119.33W*
27 T5 **Sanger** Texas, SW USA *33.21N 97.10W*
Sângerei *see* Sîngerei
103 L15 **Sangerhausen** Sachsen-Anhalt, C Germany *51.28N 11.18E*
47 S6 **San Germán** W Puerto Rico *18.05N 67.02W*
San Germano *see* Cassino
167 N2 **Sanggan He** ≈ E China
175 Oo16 **Sanggar, Teluk** *bay* Nusa Tenggara, S Indonesia
174 L8 **Sanggau** Borneo, C Indonesia *0.07N 110.34E*
81 H16 **Sangha** ≈ Central African Republic/Congo
81 G16 **Sangha-Mbaéré** ◈ *prefecture* SW Central African Republic
155 Q15 **Sānghar** Sind, SE Pakistan *26.10N 68.58E*
117 F22 **Sangiás** ▲ S Greece *36.39N 22.24E*
Sangihe, Kepulauan *see* Sangir, Kepulauan
175 S4 **Sangihe, Pulau** ***var.*** Sangir. *island* N Indonesia
56 G8 **San Gil** Santander, C Colombia *6.34N 73.07W*
108 F12 **San Gimignano** Toscana, C Italy *43.30N 11.00E*
154 M8 **Sangīn** ***var.*** Sangin. Helmand, S Afghanistan *32.03N 64.49E*
109 O21 **San Giovanni in Fiore** Calabria, SW Italy *39.16N 16.42E*
109 M16 **San Giovanni Rotondo** Puglia, SE Italy *41.43N 15.43E*
108 G12 **San Giovanni Valdarno** Toscana, C Italy *43.34N 11.31E*
Sangir *see* Sangihe, Pulau
175 Rr6 **Sangir, Kepulauan** ***var.*** Kepulauan Sangihe. *island group* N Indonesia
168 K9 **Sangiyn Dalay** Dundgovĭ, C Mongolia *45.59N 104.58E*
168 H9 **Sangiyn Dalay** Govĭ-Altay, C Mongolia *45.12N 97.51E*
168 K11 **Sangiyn Dalay** Ömnögovĭ, S Mongolia *42.50N 105.04E*
168 K8 **Sangiyn Dalay** Övörhangay, C Mongolia *46.35N 103.18E*
169 Y15 **Sangju** ***Jap.*** Shōshū. C South Korea *36.26N 128.09E*
178 Ii11 **Sangkha** Surin, E Thailand *14.36N 103.43E*
175 Oo7 **Sangkulirang** Borneo, N Indonesia *1.00N 117.56E*
175 Oo7 **Sangkulirang, Teluk** *bay* Borneo, N Indonesia
161 E16 **Sāngli** Mahārāshtra, W India *16.55N 74.37E*
81 E16 **Sangmélima** Sud, S Cameroon *2.57N 11.55E*
37 V15 **San Gorgonio Mountain** ▲ California, W USA *34.06N 116.50W*
39 T8 **Sangre de Cristo Mountains** ▲▲ Colorado/New Mexico, C USA
63 A20 **San Gregorio** Santa Fe, C Argentina *34.18S 62.01W*
63 F18 **San Gregorio de Polanco** Tacuarembó, C Uruguay *32.37S 55.49W*
47 V14 **Sangre Grande** Trinidad, Trinidad and Tobago *10.35N 61.07W*
165 N16 **Sangri** Xizang Zizhiqu, W China *29.17N 92.01E*
158 H9 **Sangrūr** Punjab, NW India *30.16N 75.52E*
46 I11 **Sangster** ***off.*** Sir Donald Sangster International Airport, ***var.*** Montego Bay. ✈ (Montego Bay) W Jamaica *18.30N 77.54W*
61 G17 **Sangue, Rio do** ≈ W Brazil
107 R4 **Sangüesa** Navarra, N Spain *42.34N 1.16W*
63 C16 **San Gustavo** Entre Ríos, E Argentina *30.40S 59.22W*
Sangyuan *see* Wuqiao
42 C6 **San Hipólito, Punta** *headland* W Mexico *26.57N 114.00W*
25 W15 **Sanibel** Sanibel Island, Florida, SE USA *26.27N 82.01W*
25 V15 **Sanibel Island** *island* Florida, SE USA
62 F13 **San Ignacio** Misiones, NE Argentina *27.13S 55.29W*
44 F2 **San Ignacio** ***prev.*** Cayo, El Cayo. Cayo, W Belize *17.09N 89.02W*
59 L16 **San Ignacio** Beni, N Bolivia *14.54S 65.34W*
59 O18 **San Ignacio** Santa Cruz, E Bolivia *16.27S 60.57W*
44 M14 **San Ignacio** ***var.*** San Ignacio de Acosta. San José, W Costa Rica *9.46N 84.10W*
42 E6 **San Ignacio** Baja California Sur, W Mexico *27.18N 112.51W*
42 J10 **San Ignacio** Sinaloa, W Mexico *23.55N 106.25W*
58 B9 **San Ignacio** Cajamarca, N Peru *5.03S 79.03W*
San Ignacio de Acosta *see* San Ignacio
42 D7 **San Ignacio, Laguna** *lagoon* W Mexico
10 I6 **Sanikiluaq** Belcher Islands, Nunavut, C Canada *55.16N 77.44W*
179 Pp9 **San Ildefonso Peninsula** *peninsula* Luzon, N Philippines
Saniquillie *see* Sanniquellie
63 D20 **San Isidro** Buenos Aires, E Argentina *34.28S 58.31W*
45 N14 **San Isidro** ***var.*** San Isidro de El General. San José, SE Costa Rica *9.21N 83.42W*
San Isidro de El General *see* San Isidro
56 E5 **San Jacinto** Bolívar, N Colombia *9.52N 75.10W*
37 U16 **San Jacinto** California, W USA *33.47N 116.58W*
37 V15 **San Jacinto Peak** ▲ California, W USA *33.48N 116.40W*
63 F14 **San Javier** Misiones, NE Argentina *27.49S 55.06W*
63 C16 **San Javier** Santa Fe, C Argentina *30.34S 59.58W*
107 S13 **San Javier** Murcia, SE Spain *37.49N 0.49W*
63 D18 **San Javier** Río Negro, W Uruguay *32.40S 58.07W*
63 C16 **San Javier, Río** ≈ C Argentina
Sanjiang *see* Jinping
166 L12 **Sanjiang** ***var.*** Guyi, Sanjiang Dongzu Zizhixian. Guangxi Zhuangzu Zizhiqu, S China *25.49N 109.31E*
Sanjiang Dongzu Zizhixian *see* Sanjiang
171 Kk13 **Sanjō** ***var.*** Sanzyô. Niigata, Honshū, C Japan *37.39N 139.00E*
59 M15 **San Joaquín** Beni, N Bolivia *13.03S 64.47W*
57 O6 **San Joaquín** Anzoátegui, NE Venezuela *9.21N 64.30W*
37 O9 **San Joaquin River** ≈ California, W USA
37 P10 **San Joaquin Valley** *valley* California, W USA
63 A18 **San Jorge** Santa Fe, C Argentina *31.49S 61.49W*
195 W15 **San Jorge** *island* N Solomon Islands
42 D3 **San Jorge, Bahía de** *bay* NW Mexico
San Jorge, Isla de *see* Weddell Island
65 J19 **San Jorge, Golfo** ***var.*** Gulf of San Jorge. *gulf* S Argentina
San Jorge, Gulf of *see* San Jorge, Golfo
196 K8 **San Jose** Tinian, S Northern Mariana Islands *15.00S 145.38E*
179 Pp12 **San Jose** Mindoro, N Philippines *12.20N 121.07E*
37 N9 **San Jose** California, W USA *37.18N 121.53W*
63 F14 **San José** Misiones, NE Argentina *27.46S 55.46W*
59 P19 **San José** ***var.*** San José de Chiquitos. Santa Cruz, E Bolivia *14.13S 68.04W*
44 M14 **San José** ● (Costa Rica) San José, C Costa Rica *9.55N 84.05W*
44 C7 **San José** ***var.*** Puerto San José. Escuintla, S Guatemala *13.55N 90.48W*
42 G6 **San José** Sonora, NW Mexico *27.31N 110.09W*
107 U11 **San José** Eivissa, Spain, W Mediterranean Sea *38.55N 1.18E*
56 H5 **San José** Zulia, NW Venezuela *9.58N 72.22W*
44 M14 **San José** ***off.*** Provincia de San José. ◈ *province* W Costa Rica
63 E19 **San José** ◈ *department* S Uruguay
44 M13 **San José** ✈ Alajuela, C Costa Rica *10.03N 84.12W*
San José *see* San José del Guaviare, Colombia
San José *see* San José de Mayo, S Uruguay
179 P9 **San Jose City** Luzon, N Philippines *15.49N 120.57E*
179 Pp13 **San Jose de Buenavista** Panay Island, C Philippines *10.44N 122.00E*
San José de Cúcuta *see* Cúcuta
63 D16 **San José de Feliciano** Entre Ríos, E Argentina *30.21S 58.47W*
57 O6 **San José de Guanipa** ***var.*** El Tigrito. Anzoátegui, NE Venezuela *8.54N 64.10W*
64 I9 **San José de Jáchal** San Juan, W Argentina *30.15S 68.46W*
42 G10 **San José del Cabo** Baja California Sur, W Mexico *23.01N 109.40W*
56 G12 **San José del Guaviare** ***var.*** San José. Guaviare, S Colombia *2.34N 72.37W*
63 E20 **San José de Mayo** ***var.*** San José. San José, S Uruguay *34.19S 56.42W*
56 I10 **San José de Ocuné** Vichada, E Colombia *4.10N 70.21W*
43 O9 **San José de Raíces** Nuevo León, NE Mexico *24.32N 100.15W*
65 K17 **San José, Golfo** *gulf* E Argentina
42 F9 **San José, Isla** *island* W Mexico
45 U16 **San José, Isla** *island* SE Panama
27 U14 **San Jose Island** *island* Texas, SW USA
64 I10 **San Juan** San Juan, W Argentina *31.36S 68.26W*
47 N9 **San Juan** ***var.*** San Juan de la Maguana. C Dominican Republic *18.46N 71.13W*
59 E17 **San Juan** Ica, S Peru *15.22S 75.08W*
47 U5 **San Juan** ○ (Puerto Rico) NE Puerto Rico *18.28N 66.06W*
64 H10 **San Juan** ***off.*** Provincia de San Juan. ◈ *province* W Argentina
47 U5 **San Juan** ***var.*** Luis Muñoz Marin. ✈ NE Puerto Rico *18.27N 66.05W*
San Juan *see* San Juan de los Morros
64 O7 **San Juan Bautista** Misiones, S Paraguay *26.39S 57.08W*
37 O10 **San Juan Bautista** California, W USA *36.50N 121.34W*
San Juan Bautista *see* Villahermosa
San Juan Bautista Cuicatlán *see* Cuicatlán
San Juan Bautista Tuxtepec *see* Tuxtepec
81 C17 **San Juan, Cabo** *headland* S Equatorial Guinea *1.09N 9.25E*
107 S12 **San Juan de Alicante** País Valenciano, E Spain *38.25N 0.27W*
56 H7 **San Juan de Colón** Táchira, NW Venezuela *8.01N 72.16W*
42 L9 **San Juan de Guadalupe** Durango, C Mexico *25.12N 100.50W*
San Juan de la Maguana *see* San Juan
56 G4 **San Juan del Cesar** La Guajira, N Colombia *10.45N 73.00W*
42 L15 **San Juan de Lima, Punta** *headland* SW Mexico *18.34N 103.40W*
44 I8 **San Juan de Limay** Estelí, NW Nicaragua *13.10N 86.36W*
45 N12 **San Juan del Norte** ***var.*** Greytown. Río San Juan, SE Nicaragua *10.54N 83.42W*
56 K4 **San Juan de los Cayos** Falcón, N Venezuela *11.06N 68.25W*
42 M12 **San Juan de los Lagos** Jalisco, C Mexico *21.15N 102.15W*
56 L5 **San Juan de los Morros** ***var.*** San Juan. Guárico, N Venezuela *9.52N 67.22W*
42 K9 **San Juan del Río** Durango, C Mexico *25.12N 100.50W*
43 O13 **San Juan del Río** Querétaro de Arteaga, C Mexico *20.21N 100.01W*
44 J11 **San Juan del Sur** Rivas, SW Nicaragua *11.14N 85.52W*
56 M9 **San Juan de Manapiare** Amazonas, S Venezuela *5.15N 66.04W*
42 E7 **San Juanico** Baja California Sur, W Mexico
42 D7 **San Juanico, Punta** *headland* W Mexico *26.01N 112.17W*
34 G6 **San Juan Islands** *island group* Washington, NW USA
42 I6 **San Juanito** Chihuahua, N Mexico
42 I12 **San Juanito, Isla** *island* C Mexico
39 R8 **San Juan Mountains** ▲▲ Colorado, C USA
56 E5 **San Juan Nepomuceno** Bolívar, NW Colombia *9.57N 75.06W*
46 E5 **San Juan, Pico** ▲ C Cuba *21.58N 80.10W*
203 W15 **San Juan, Punta** *headland* Easter Island, Chile, E Pacific Ocean *27.03S 109.22W*
44 M12 **San Juan, Río** ≈ Costa Rica/Nicaragua
43 S15 **San Juan, Río** ≈ SE Mexico
39 O8 **San Juan River** ≈ Colorado/Utah, W USA
San Julián *see* Puerto San Julián
63 B17 **San Justo** Santa Fe, C Argentina *30.46S 60.31W*
111 W5 **Sankt Aegyd-am-Neuwalde** Niederösterreich, E Austria *47.51N 15.34E*
111 U9 **Sankt Andrä** ***Slvn.*** Šent Andraž. Kärnten, S Austria *46.46N 14.49E*
Sankt Andrä *see* Szentendre
Sankt Anna *see* Sântana
110 K8 **Sankt Anton-am-Arlberg** Vorarlberg, W Austria *47.08N 10.11E*
103 E16 **Sankt Augustin** Nordrhein-Westfalen, W Germany *50.46N 7.10E*
Sankt-Bartholomäi *see* Palamuse
103 F24 **Sankt Blasien** Baden-Württemberg, SW Germany *47.43N 8.09E*
111 R3 **Sankt Florian am Inn** Oberösterreich, N Austria *48.24N 13.27E*
110 I7 **Sankt Gallen** ***var.*** St.Gallen, ***Eng.*** Saint Gall, ***Fr.*** St-Gall. Sankt Gallen, NE Switzerland *47.25N 9.22E*
110 H8 **Sankt Gallen** ***var.*** St.Gallen, Eng, Saint Gall, ***Fr.*** St-Gall. ◈ *canton* NE Switzerland
110 J8 **Sankt Gallenkirch** Vorarlberg, W Austria *47.00N 10.59E*
111 Q5 **Sankt Georgen** Salzburg, N Austria *47.59N 12.57E*
Sankt Georgen *see* Đurđevac, Croatia
Sankt-Georgen *see* Sfântu Gheorghe, Romania
111 R6 **Sankt Gilgen** Salzburg, NW Austria *47.46N 13.21E*
Sankt Gotthard *see* Szentgotthárd
103 E20 **Sankt Ingbert** Saarland, SW Germany *49.16N 7.07E*
Sankt-Jakobi *see* Viru-Jaagupi, Lääne-Virumaa, Estonia
Sankt-Jakobi *see* Pärnu-Jaagupi, Pärnumaa, Estonia
Sankt Johann *see* Sankt Johann in Tirol
111 T7 **Sankt Johann am Tauern** Steiermark, E Austria *47.20N 14.27E*
111 Q7 **Sankt Johann im Pongau** Salzburg, NW Austria *47.22N 13.13E*
111 P6 **Sankt Johann in Tirol** ***var.*** Sankt Johann. Tirol, W Austria *47.31N 12.25E*
Sankt-Johannis *see* Järva-Jaani
110 L8 **Sankt Leonhard** Tirol, W Austria *47.05N 10.53E*
Sankt Margarethen *see* Sankt Margarethen im Burgenland
111 Y5 **Sankt Margarethen im Burgenland** ***var.*** Sankt Margarethen. Burgenland, E Austria *47.49N 16.37E*
Sankt Martin *see* Martin
111 X8 **Sankt Martin an der Raab** Burgenland, SE Austria *46.59N 16.12E*
111 U7 **Sankt Michael in Obersteiermark** Steiermark, SE Austria *47.21N 14.59E*
Sankt Michel *see* Mikkeli
Sankt Moritz *see* St.Moritz
110 E11 **Sankt Niklaus** Valais, S Switzerland *46.09N 7.48E*
111 S7 **Sankt Nikolai** ***var.*** Sankt Nikolai im Sölktal. Steiermark, SE Austria *47.18N 14.04E*
Sankt Nikolai im Sölktal *see* Sankt Nikolai
111 U9 **Sankt Paul** ***var.*** Sankt Paul im Lavanttal. Kärnten, S Austria *46.42N 14.53E*
Sankt Paul im Lavanttal *see* Sankt Paul
Sankt Peter *see* Pivka
111 W9 **Sankt Peter am Ottersbach** Steiermark, SE Austria *46.49N 15.48E*
128 J13 **Sankt-Peterburg** ***prev.*** Leningrad, Petrograd, ***Eng.*** Saint Petersburg, ***Fin.*** Pietari. Leningradskaya Oblast', NW Russian Federation *59.55N 30.25E*
102 H8 **Sankt Peter-Ording** Schleswig-Holstein, N Germany *54.18N 8.37E*
111 V4 **Sankt Pölten** Niederösterreich, N Austria *48.14N 15.37E*
111 W7 **Sankt Ruprecht** ***var.*** Sankt Ruprecht an der Raab. Steiermark, SE Austria *47.10N 15.41E*
Sankt Ruprecht an der Raab *see* Sankt Ruprecht
Sankt-Ulrich *see* Ortisei
111 T4 **Sankt Valentin** Niederösterreich, C Austria *48.09N 14.30E*
Sankt Veit am Flaum *see* Rijeka
111 T9 **Sankt Veit an der Glan** ***Slvn.*** Šent Vid. Kärnten, S Austria *46.46N 14.22E*
101 M21 **Sankt-Vith** ***var.*** Saint-Vith. Liège, E Belgium *50.16N 6.07E*
103 E20 **Sankt Wendel** Saarland, SW Germany *49.28N 7.10E*
111 R6 **Sankt Wolfgang** Salzburg, NW Austria *47.43N 13.30E*
81 K21 **Sankuru** ≈ C Dem. Rep. Congo (Zaire)
42 D8 **San Lázaro, Cabo** *headland* W Mexico *24.46N 112.15W*
143 O16 **Şanlıurfa** ***prev.*** Sanli Urfa, Urfa, ***anc.*** Edessa. Şanlıurfa, S Turkey *37.07N 38.45E*
143 O16 **Şanlıurfa** ***prev.*** Urfa. ◈ *province* SE Turkey
143 O16 **Şanlıurfa Yaylası** *plateau* SE Turkey
63 B18 **San Lorenzo** Santa Fe, C Argentina *32.37S 60.48W*
59 M21 **San Lorenzo** Tarija, S Bolivia *21.27S 64.47W*
58 C5 **San Lorenzo** Esmeraldas, N Ecuador *1.15N 78.51W*
44 H8 **San Lorenzo** Valle, S Honduras *13.25N 87.27W*
107 N8 **San Lorenzo de El Escorial** ***var.*** El Escorial. Madrid, C Spain *40.36N 4.07W*
42 E5 **San Lorenzo, Isla** *island* NW Mexico
59 C14 **San Lorenzo, Isla** *island* W Peru
65 G20 **San Lorenzo, Monte** ▲ S Argentina *47.40S 72.12W*
42 I9 **San Lorenzo, Río** ≈ C Mexico
106 J15 **Sanlúcar de Barrameda** Andalucía, S Spain *36.46N 6.21W*
106 J14 **Sanlúcar la Mayor** Andalucía, S Spain *37.24N 6.13W*
42 F11 **San Lucas** Baja California Sur, NW Mexico *22.49N 109.52W*
42 E6 **San Lucas** ***var.*** Cabo San Lucas. Baja California Sur, W Mexico *27.13N 112.15W*
42 G11 **San Lucas, Cabo** ***var.*** San Lucas Cape. *headland* W Mexico *22.52N 109.55W*
San Lucas Cape *see* San Lucas, Cabo
64 J11 **San Luis** San Luis, C Argentina *33.18S 66.18W*
44 E4 **San Luis** Petén, NE Guatemala *16.16N 89.27W*
42 D2 **San Luis** ***var.*** San Luis Río Colorado. Sonora, NW Mexico *32.25N 114.48W*
44 M7 **San Luis** Región Autónoma Atlántico Norte, NE Nicaragua *13.58N 84.10W*
38 H15 **San Luis** Arizona, SW USA *32.27N 114.45W*
39 T8 **San Luis** Colorado, C USA *37.09N 105.24W*
56 J4 **San Luis** Falcón, N Venezuela *11.08N 69.36W*
64 J11 **San Luis** ***off.*** Provincia de San Luis. ◈ *province* C Argentina
43 N12 **San Luis de la Paz** Guanajuato, C Mexico *21.15N 100.33W*
42 K8 **San Luis del Cordero** Durango, C Mexico *25.25N 104.09W*
42 D4 **San Luis, Isla** *island* NW Mexico
44 E6 **San Luis Jilotepeque** Jalapa, SE Guatemala *14.36N 89.40W*
59 M16 **San Luis, Laguna de** ◉ NW Bolivia
37 P13 **San Luis Obispo** California, W USA *35.16N 120.39W*
39 R7 **San Luis Peak** ▲ Colorado, C USA *37.59N 106.55W*
43 N11 **San Luis Potosí** San Luis Potosí, C Mexico *22.09N 100.57W*
43 N11 **San Luis Potosí** ◈ *state* C Mexico
37 O10 **San Luis Reservoir** ⊠ California, W USA
San Luis Río Colorado *see* San Luis
39 S8 **San Luis Valley** *basin* Colorado, C USA
109 C19 **Sanluri** Sardegna, Italy, C Mediterranean Sea *39.34N 8.54E*
63 D23 **San Manuel** Buenos Aires, E Argentina *37.46S 58.49W*
38 M15 **San Manuel** Arizona, SW USA *32.36N 110.37W*
108 F11 **San Marcello Pistoiese** Toscana, C Italy *44.03N 10.46E*
109 N20 **San Marco Argentano** Calabria, SW Italy *39.31N 16.07E*
56 E6 **San Marcos** Sucre, N Colombia *8.37N 75.12W*
44 M14 **San Marcos** San José, C Costa Rica *9.39N 84.00W*
44 B5 **San Marcos** San Marcos, W Guatemala *14.57N 91.46W*
44 F6 **San Marcos** Ocotepeque, SW Honduras *14.23N 88.57W*
43 O16 **San Marcos** Guerrero, S Mexico *16.47N 99.29W*
27 S11 **San Marcos** Texas, SW USA *29.52N 97.56W*
44 A5 **San Marcos** ***off.*** Departamento de San Marcos. ◈ *department* W Guatemala
San Marcos de Arica *see* Arica
42 E6 **San Marcos, Isla** *island* W Mexico
108 H11 **San Marino** ● (San Marino) C San Marino *43.53N 12.27E*
108 I11 **San Marino** ***off.*** Republic of San Marino. ◆ *republic* S Europe
64 I11 **San Martín** Mendoza, C Argentina *33.04S 68.28W*
56 F11 **San Martín** Meta, C Colombia *3.43N 73.42W*
58 D11 **San Martín** ***off.*** Departamento de San Martín. ◈ *department* C Peru
204 I5 **San Martín** *Argentinian research station* Antarctica *68.18S 67.03W*
65 H16 **San Martín de los Andes** Neuquén, W Argentina *40.10S 71.22W*
106 M8 **San Martín de Valdeiglesias** Madrid, C Spain *40.21N 4.24W*
65 G21 **San Martín, Lago** ***var.*** Lago O'Higgins. ◉ S Argentina
108 H6 **San Martino di Castrozza** Trentino-Alto Adige, N Italy *46.16N 11.50E*
59 N16 **San Martín, Río** ≈ N Bolivia
San Martín Texmelucan *see* Texmelucan
37 N9 **San Mateo** California, W USA *37.33N 122.19W*
57 O6 **San Mateo** Anzoátegui, NE Venezuela *9.34N 64.30W*
44 B4 **San Mateo Ixtatán** Huehuetenango, W Guatemala *15.48N 91.30W*
59 Q18 **San Matías** Santa Cruz, E Bolivia *16.19S 58.23W*
65 K16 **San Matías, Golfo** ***var.*** Gulf of San Matías. *gulf* E Argentina
San Matías, Gulf of *see* San Matías
13 O8 **Sanmaur** Quebec, SE Canada *47.52N 73.47W*
167 T10 **Sanmen Wan** *bay* E China
166 M6 **Sanmenxia** ***var.*** Shan Xian. Henan, C China *34.46N 111.16E*
Sânmiclăuș Mare *see* Sânnicolau Mare
63 D14 **San Miguel** Corrientes, NE Argentina *28.02S 57.38W*
59 L16 **San Miguel** Beni, N Bolivia *16.43S 61.06W*
44 G8 **San Miguel** San Miguel, SE El Salvador *13.27N 88.10W*
42 L6 **San Miguel** Coahuila de Zaragoza, N Mexico *29.10N 101.28W*
42 J9 **San Miguel** ***var.*** San Miguel de Cruces. Durango, C Mexico *24.25N 105.55W*
45 U16 **San Miguel** Panamá, SE Panama *8.26N 78.57W*
37 P12 **San Miguel** California, W USA *35.45N 120.42W*
44 B9 **San Miguel** ◈ *department* E El Salvador
43 N13 **San Miguel de Allende** Guanajuato, C Mexico *20.54N 100.46W*
San Miguel de Cruces *see* San Miguel
San Miguel de Ibarra *see* Ibarra
63 D21 **San Miguel del Monte** Buenos Aires, E Argentina *35.25S 58.49W*
64 J7 **San Miguel de Tucumán** ***var.*** Tucumán. Tucumán, N Argentina *26.46S 65.15W*
45 V16 **San Miguel, Golfo de** *gulf* S Panama
37 P15 **San Miguel Island** *island* California, W USA
44 L11 **San Miguelito** Río San Juan, S Nicaragua *11.22N 84.52W*
45 T15 **San Miguelito** Panamá, C Panama *8.58N 79.31W*
59 N18 **San Miguel, Río** ≈ E Bolivia
58 D6 **San Miguel, Río** ≈ Colombia/Ecuador
42 I7 **San Miguel, Río** ≈ N Mexico
44 G8 **San Miguel, Volcán de** ⛰ SE El Salvador *13.27N 88.18W*
167 Q12 **Sanming** Fujian, SE China *26.10N 117.37E*
108 F11 **San Miniato** Toscana, C Italy *43.40N 10.53E*
San Murezzan *see* St.Moritz
Sannār *see* Sennar
109 M15 **Sannicandro Garganico** Puglia, SE Italy *41.49N 15.31E*
42 H6 **San Nicolás** Sonora, NW Mexico *28.31N 109.24W*
63 C19 **San Nicolás de los Arroyos** Buenos Aires, E Argentina *33.17S 60.12W*
37 R16 **San Nicolas Island** *island* Channel Islands, California, W USA
Sânnicolaul-Mare *see* Sânnicolau Mare
118 E11 **Sânnicolau Mare** ***var.*** Sânnicolaul-Mare, ***Hung.*** Nagyszentmiklós; ***prev.*** Sânmiclăuș Mare, Sînnicolau Mare. Timiș, W Romania *46.05N 20.37E*
126 Ll5 **Sannikova, Proliv** *strait* NE Russian Federation
78 K16 **Sanniquellie** ***var.*** Saniquillie. NE Liberia *7.24N 8.45W*
172 N9 **Sannohe** Aomori, Honshū, C Japan *40.23N 141.16E*
Sanntaler Alpen *see* Kamnikšo-Savinjske Alpe
171 K15 **Sano** Tochigi, Honshū, S Japan *36.19N 139.26E*
113 O17 **Sanok** Podkarpackie, SE Poland *49.31N 22.14E*
56 E5 **San Onofre** Sucre, NW Colombia *9.45N 75.33W*
59 K21 **San Pablo** Potosí, S Bolivia *21.43S 66.37W*
179 P11 **San Pablo** ***off.*** San Pablo City. Luzon, N Philippines *14.04N 121.16E*
San Pablo Balleza *see* Balleza
37 N8 **San Pablo Bay** *bay* California, W USA
42 C6 **San Pablo, Punta** *headland* W Mexico *27.12N 114.30W*
45 R16 **San Pablo, Río** ≈ C Panama
179 Q11 **San Pascual** Burias Island, C Philippines *13.06N 122.59E*
123 Jj16 **San Pawl il-Bahar** ***Eng.*** Saint Paul's Bay. E Malta *35.57N 14.24E*
63 C19 **San Pedro** Buenos Aires, E Argentina *33.37S 59.42W*
64 K5 **San Pedro** Jujuy, N Argentina *24.13S 64.51W*
62 G13 **San Pedro** Misiones, NE Argentina *26.37S 54.12W*
44 H1 **San Pedro** Corozal, NE Belize *17.58N 87.55W*
42 L8 **San Pedro** ***var.*** San Pedro de las Colonias. Coahuila de Zaragoza, NE Mexico *25.47N 102.57W*
64 O5 **San Pedro** San Pedro, SE Paraguay *24.04S 57.03W*
64 O6 **San Pedro** ***off.*** Departamento de San Pedro. ◈ *department* C Paraguay
79 N16 **San Pedro** ✈ (Yamoussoukro) C Ivory Coast *6.49N 5.14W*
46 G6 **San Pedro** ≈ C Cuba
San Pedro *see* San Pedro del Pinatar
78 M17 **San-Pédro** S Ivory Coast *4.45N 6.37W*
44 D5 **San Pedro Carchá** Alta Verapaz, C Guatemala *15.30N 90.12W*
37 S16 **San Pedro Channel** *channel* California, W USA
64 I5 **San Pedro de Atacama** Antofagasta, N Chile *22.52S 68.10W*
San Pedro de Durazno *see* Durazno
42 G5 **San Pedro de la Cueva** Sonora, NW Mexico *29.16N 109.46W*
San Pedro de las Colonias *see* San Pedro
58 B11 **San Pedro de Lloc** La Libertad, NW Peru *7.27S 79.34W*
107 S13 **San Pedro del Pinatar** ***var.*** San Pedro. Murcia, SE Spain *37.49N 0.46W*
47 P9 **San Pedro de Macorís** SE Dominican Republic *18.28N 69.19W*
42 C3 **San Pedro Mártir, Sierra** ▲▲ NW Mexico
San Pedro Pochutla *see* Pochutla
44 D2 **San Pedro, Río** ≈ Guatemala/Mexico
42 K10 **San Pedro, Río** ≈ C Mexico
106 J10 **San Pedro, Sierra de** ▲▲ W Spain
44 G5 **San Pedro Sula** Cortés, NW Honduras *15.25N 88.01W*
San Pedro Tapanatepec *see* Tapanatepec
64 I4 **San Pedro, Volcán** ▲ N Chile *21.46S 68.13W*
108 E7 **San Pellegrino Terme** Lombardia, N Italy *45.53N 9.42E*
27 T16 **San Perlita** Texas, SW USA *26.30N 97.38W*
San Pietro *see* Supetar
San Pietro del Carso *see* Pivka
109 A20 **San Pietro, Isola di** *island* W Italy
34 K7 **Sanpoil River** ≈ Washington, NW USA
171 L12 **Sanpoku** ***var.*** Sampoku. Niigata, Honshū, C Japan *38.32N 139.33E*
42 C3 **San Quintín** Baja California, NW Mexico *30.21N 115.58W*
42 B3 **San Quintín, Bahía de** *bay* NW Mexico
42 B3 **San Quintín, Cabo** *headland* NW Mexico *30.22N 116.01W*
64 I12 **San Rafael** Mendoza, W Argentina *34.43S 68.15W*
43 N9 **San Rafael** Nuevo León, NE Mexico *25.01N 100.33W*
36 M8 **San Rafael** California, W USA *37.58N 122.31W*
39 Q11 **San Rafael** New Mexico, SW USA *35.03N 107.52W*
56 H4 **San Rafael** ***var.*** El Moján. Zulia, NW Venezuela *10.58N 71.45W*
44 J8 **San Rafael del Norte** Jinotega, NW Nicaragua *13.10N 86.10W*
44 J10 **San Rafael del Sur** Managua, SW Nicaragua *11.51N 86.24W*
38 M5 **San Rafael Knob** ▲ Utah, W USA *38.46N 110.45W*
37 Q14 **San Rafael Mountains** ▲▲ California, W USA
44 M13 **San Ramón** Alajuela, C Costa Rica *10.04N 84.27W*
59 E14 **San Ramón** Junín, C Peru *11.08S 75.19W*
63 F19 **San Ramón** Canelones, S Uruguay *34.18S 55.55W*
64 K5 **San Ramón de la Nueva Orán** Salta, N Argentina *23.07S 64.19W*
59 O16 **San Ramón, Río** ≈ E Bolivia
108 B11 **San Remo** Liguria, NW Italy *43.48N 7.46E*
56 J3 **San Román, Cabo** *headland* NW Venezuela *12.10N 70.01W*
63 C15 **San Roque** Corrientes, NE Argentina *28.34S 58.45W*
196 I4 **San Roque** Saipan, S Northern Mariana Islands *15.15S 85.46E*
106 K16 **San Roque** Andalucía, S Spain *36.13N 5.22W*
27 R9 **San Saba** Texas, SW USA *31.12N 98.43W*
27 Q9 **San Saba River** ≈ Texas, SW USA
63 D17 **San Salvador** Entre Ríos, E Argentina *31.37S 58.30W*
44 F7 **San Salvador** ● (El Salvador) San Salvador, SW El Salvador *13.42N 89.12W*
44 A10 **San Salvador** ◈ *department* C El Salvador
44 F8 **San Salvador** ✈ La Paz, S El Salvador *13.27N 89.04W*
46 K4 **San Salvador** ***prev.*** Watlings Island. *island* E Bahamas
64 J5 **San Salvador de Jujuy** ***var.*** Jujuy. Jujuy, N Argentina *24.10S 65.19W*
44 F7 **San Salvador, Volcán de** ⛰ C El Salvador *13.58N 89.14W*
79 Q14 **Sansanné-Mango** ***var.*** Mango. N Togo *10.21N 0.28E*
47 S5 **San Sebastián** W Puerto Rico *18.21N 67.00W*
65 J24 **San Sebastián, Bahía** *bay* S Argentina
Sansenhō *see* Sach'ŏn
108 H12 **Sansepolcro** Toscana, C Italy *43.34N 12.12E*
109 M16 **San Severo** Puglia, SE Italy *41.40N 15.22E*
114 F11 **Sanski Most** Federacija Bosna I Hercegovina, NW Bosnia & Herzegovina *44.43N 16.40E*
176 Ww9 **Sansundi** Irian Jaya, E Indonesia *0.42S 135.48E*
106 K11 **Santa Amalia** Extremadura, W Spain *39.00N 6.01W*
62 F13 **Santa Ana** Misiones, NE Argentina *27.22S 55.34W*
59 L16 **Santa Ana** Beni, N Bolivia *13.43S 65.37W*
44 E7 **Santa Ana** Santa Ana, NW El Salvador *13.58N 89.34W*
42 F4 **Santa Ana** Sonora, NW Mexico *30.33N 111.07W*
37 T16 **Santa Ana** California, W USA *33.45N 117.52W*

◆ Country
● Country Capital
◇ Dependent Territory
○ Dependent Territory Capital
◈ Administrative Region
✈ International Airport
▲ Mountain
▲▲ Mountain Range
⛰ Volcano
≈ River
◉ Lake
⊠ Reservoir

57 N6 **Santa Ana** Nueva Esparta, NE Venezuela 9.15N 64.39W

44 A9 **Santa Ana** ✦ *department* NW El Salvador

Santa Ana de Coro *see* Coro

44 E7 **Santa Ana, Volcán de** *var.* La Matepec. ▲ W El Salvador 13.49N 89.36W

42 J7 **Santa Barbara** Chihuahua, N Mexico 26.46N 105.46W

37 Q14 **Santa Bárbara** California, W USA 34.24N 119.40W

44 G6 **Santa Bárbara** Santa Bárbara, W Honduras 14.57N 88.15W

56 L11 **Santa Bárbara** Amazonas, S Venezuela 3.55N 67.06W

56 I7 **Santa Bárbara** Barinas, W Venezuela 7.48N 71.10W

44 F5 **Santa Bárbara** ✦ *department* NW Honduras

Santa Bárbara *see* Iscuandé

37 Q15 **Santa Barbara Channel** *channel* California, W USA

Santa Bárbara de Samaná *see* Samaná

37 R16 **Santa Barbara Island** *island* Channel Islands, California, W USA

56 E5 **Santa Catalina** Bolívar, N Colombia 10.34N 75.22W

45 R15 **Santa Catalina** Bocas del Toro, W Panama 8.46N 81.18W

37 T17 **Santa Catalina, Gulf of** *gulf* California, W USA

42 F8 **Santa Catalina, Isla** *island* W Mexico

37 S16 **Santa Catalina Island** *island* Channel Islands, California, W USA

43 N8 **Santa Catarina** Nuevo León, NE Mexico 25.39N 100.30W

62 H13 **Santa Catarina** *off.* Estado de Santa Catarina. ✦ *state* S Brazil

Santa Catarina de Tepehuanes *see* Tepehuanes

62 L13 **Santa Catarina, Ilha de** *island* S Brazil

47 Q16 **Santa Catherina** Curaçao, C Netherlands Antilles 12.07N 68.46W

46 E5 **Santa Clara** Villa Clara, C Cuba 22.25N 78.00W

37 N9 **Santa Clara** California, W USA 37.20N 121.57W

38 J8 **Santa Clara** Utah, W USA 37.07N 113.39W

Santa Clara *see* Santa Clara de Olimar

63 F18 **Santa Clara de Olimar** *var.* Santa Clara. Cerro Largo, NE Uruguay 32.54S 54.55W

63 A17 **Santa Clara de Saguier** Santa Fe, C Argentina 31.21S 61.49W

Santa Coloma *see* Santa Coloma de Gramanet

107 X5 **Santa Coloma de Farners** *var.* Santa Coloma de Farnés. Cataluña, NE Spain 41.52N 2.39E

Santa Coloma de Farnés *see* Santa Coloma de Farners

107 W6 **Santa Coloma de Gramanet** *var.* Santa Coloma. Cataluña, NE Spain 41.28N 2.13E

106 G2 **Santa Comba** Galicia, NW Spain 43.01N 8.49W

Santa Comba *see* Uaco Cungo

106 H8 **Santa Comba Dão** Viseu, N Portugal 40.22N 8.07W

84 C10 **Santa Cruz** Uíge, NW Angola 6.56S 16.25E

59 N19 **Santa Cruz** *var.* Santa Cruz de la Sierra. Santa Cruz, C Bolivia 17.49S 63.10W

64 G12 **Santa Cruz** Libertador, C Chile 34.39S 71.16W

44 K13 **Santa Cruz** Guanacaste, W Costa Rica 10.15N 85.34W

46 I12 **Santa Cruz** W Jamaica 18.03N 77.41W

66 P6 **Santa Cruz** Madeira, Portugal, NE Atlantic Ocean 32.43N 16.46W

37 N10 **Santa Cruz** California, W USA 36.58N 122.01W

65 H20 **Santa Cruz** *off.* Provincia de Santa Cruz. ✦ *province* S Argentina

59 O18 **Santa Cruz** ✒ *department* E Bolivia

Santa Cruz *see* Viru-Viru

Santa Cruz *see* Puerto Santa Cruz

Santa Cruz Barillas *see* Barillas

61 O18 **Santa Cruz Cabrália** Bahia, E Brazil 16.16S 39.03W

Santa Cruz de El Seibo *see* El Seibo

66 N11 **Santa Cruz de la Palma** La Palma, Islas Canarias, Spain, NE Atlantic Ocean 28.40N 17.46W

Santa Cruz de la Sierra *see* Santa Cruz

107 O9 **Santa Cruz de la Zarza** Castilla-La Mancha, C Spain 39.59N 3.10W

44 C5 **Santa Cruz del Quiché** Quiché, W Guatemala 15.01N 91.08W

107 N8 **Santa Cruz del Retamar** Castilla-La Mancha, C Spain 40.07N 4.13W

Santa Cruz del Seibo *see* El Seibo

56 G7 **Santa Cruz del Sur** Camagüey, C Cuba 20.44N 78.00W

37 O11 **Santa Cruz de Mudela** Castilla-La Mancha, C Spain 38.37N 3.27W

66 Q11 **Santa Cruz de Tenerife** Tenerife, Islas Canarias, Spain, NE Atlantic Ocean 28.28N 16.15W

66 P11 **Santa Cruz de Tenerife** ✦ *province* Islas Canarias, Spain, NE Atlantic Ocean

62 K9 **Santa Cruz do Rio Pardo** São Paulo, S Brazil 22.52S 49.37W

63 H15 **Santa Cruz do Sul** Rio Grande do Sul, S Brazil 29.42S 52.25W

59 C17 **Santa Cruz, Isla** *var.* Indefatigable Island, Isla Chávez. *island* Galapagos Islands, Ecuador, E Pacific Ocean

37 Q15 **Santa Cruz Island** *island* California, W USA

195 X8 **Santa Cruz Islands** *island group* E Solomon Islands

65 I22 **Santa Cruz, Río** ✒ S Argentina

38 L15 **Santa Cruz River** ✒ Arizona, SW USA

63 C17 **Santa Elena** Entre Ríos, E Argentina 30.58S 59.46W

44 F2 **Santa Elena** Cayo, W Belize 17.08N 89.04W

37 R16 **Santa Elena** Texas, SW USA 26.43N 98.30W

58 A7 **Santa Elena, Bahía de** *bay* W Ecuador

57 R10 **Santa Elena de Uairén** Bolívar, E Venezuela 4.40N 61.03W

44 K12 **Santa Elena, Península** *peninsula* NW Costa Rica

58 A7 **Santa Elena, Punta** *headland* W Ecuador 2.11S 81.00W

106 L11 **Santa Eufemia** Andalucía, S Spain 38.36N 4.54W

109 N21 **Santa Eufemia, Golfo di** *gulf* S Italy

109 N21 **Santa Eufemia Lamezia Terme** Calabria, SE Italy 38.54N 16.13E

120 S4 **Santa Eulalia de Gállego** Aragón, NE Spain 42.16N 0.46W

107 V11 **Santa Eulalia del Río** Eivissa, Spain, W Mediterranean Sea 39.00N 1.33E

63 B17 **Santa Fe** Santa Fe, C Argentina 31.36S 60.46W

107 N14 **Santa Fe** Andalucía, S Spain 37.10N 3.43W

39 S10 **Santa Fe** state capital New Mexico, SW USA 35.41N 105.56W

63 B15 **Santa Fe** *off.* Provincia de Santa Fe. ✒ *province* C Argentina

Santa Fé de Bogotá *see* Bogotá

46 C6 **Santa Fé** var. La Fe. Isla de la Juventud, W Cuba 21.39N 82.45W

45 R16 **Santa Fé** Veraguas, C Panama 8.28N 81.03W

Santa Fé do Sul São Paulo, S Brazil 20.13S 50.56W

62 J7 **Santa Fé do Sul** São Paulo, S Brazil 20.13S 50.56W

59 B18 **Santa Fe, Isla** *var.* Barrington Island. *island* Galapagos Islands, Ecuador, E Pacific Ocean

25 V9 **Santa Fe River** ✒ Florida, SE USA

61 M15 **Santa Filomena** Piauí, E Brazil 9.06S 45.52W

42 G10 **Santa Genoveva** ▲ W Mexico 23.07N 109.56W

159 S14 **Santahar** Rajshahi, NW Bangladesh 24.45N 89.03E

62 G11 **Santa Helena** Paraná, S Brazil 24.53S 54.19W

56 J5 **Santa Inés** Lara, N Venezuela 10.37N 69.18W

65 Z24 **Santa Inés, Isla** *island* S Chile

64 J13 **Santa Isabel** La Pampa, C Argentina 36.11S 66.59W

45 U14 **Santa Isabel** Colón, N Panama 9.31N 79.12W

195 W14 **Santa Isabel** *var.* Bughotu. *island* N Solomon Islands

Santa Isabel *see* Malabo

80 D11 **Santa Isabel do Rio Negro** Amazonas, NW Brazil 0.40S 64.55W

63 C15 **Santa Lucía** Corrientes, NE Argentina 28.58S 59.05W

59 I17 **Santa Lucía** Puno, S Peru 15.45S 70.34W

63 F20 **Santa Lucía** *var.* Santa Lucia. Canelones, S Uruguay 34.25S 56.25W

44 B6 **Santa Lucía Cotzumalguapa** Escuintla, SW Guatemala 14.20N 91.00W

109 L23 **Santa Lucia del Mela** Sicilia, Italy, C Mediterranean Sea 38.07N 15.16E

37 O11 **Santa Lucia Range** ▲ California, W USA

42 D9 **Santa Margarita, Isla** *island* W Mexico

63 G15 **Santa Maria** Rio Grande do Sul, S Brazil 29.40S 53.48W

37 P13 **Santa Maria** California, W USA 34.56N 120.25W

66 Q4 **Santa Maria** × Santa Maria, Azores, Portugal, NE Atlantic Ocean

66 P3 **Santa Maria** *island* Azores, Portugal, NE Atlantic Ocean

Santa Maria *see* Gaua

64 J7 **Santa María** Catamarca, N Argentina 26.38S 66.01W

42 G9 **Santa María, Bahía** *bay* W Mexico

85 L21 **Santa Maria, Cabo de** *headland* S Mozambique 26.05S 32.58E

106 G15 **Santa Maria, Cabo de** *headland* S Portugal 36.57N 7.55W

46 J4 **Santa Maria, Cape** *headland* Long Island, C Bahamas 23.40N 75.20W

109 J17 **Santa Maria Capua Vetere** Campania, S Italy 41.04N 14.15E

61 M17 **Santa Maria da Vitória** Bahia, E Brazil 13.25S 44.09W

57 N9 **Santa María de Erebato** Bolívar, SE Venezuela 5.09N 64.49W

106 G7 **Santa Maria da Feira** Aveiro, N Portugal 40.55N 8.31W

57 N6 **Santa María de Ipire** Guárico, C Venezuela 8.51N 65.21W

Santa María del Buen Aire *see* Buenos Aires

42 J8 **Santa María del Oro** Durango, C Mexico 25.57N 105.22W

43 N12 **Santa María del Río** San Luis Potosí, C Mexico 21.48N 100.42W

Santa María di Castellabate *see* Castellabate

109 Q20 **Santa Maria di Leuca, Capo** *headland* SE Italy 39.48N 18.21E

110 K10 **Santa-Maria-im-Münstertal** Graubünden, SE Switzerland 46.36N 10.25E

59 A16 **Santa María, Isla** *var.* Isla Floreana, Charles Island. *island* Galapagos Islands, Ecuador, E Pacific Ocean

42 J3 **Santa María, Laguna de** ☉ N Mexico

63 F15 **Santa Maria, Río** ✒ S Brazil

45 R16 **Santa María, Río** ✒ C Panama

38 J12 **Santa Maria River** ✒ Arizona, SW USA

109 G15 **Santa Marinella** Lazio, C Italy 42.01N 11.51E

56 F4 **Santa Marta** Magdalena, N Colombia 11.13N 74.13W

106 J11 **Santa Marta** Extremadura, W Spain 38.37N 6.39W

Santa Maura *see* Lefkáda

37 S15 **Santa Monica** California, W USA 34.01N 118.29W

118 F10 **Sântana** *Ger.* Sankt Anna, *Hung.* Újszentanna; *prev.* Sintana. Arad, W Romania 46.19N 21.30E

63 F16 **Santana, Coxilha de** *hill range* S Brazil

83 H16 **Santana da Boa Vista** Rio Grande do Sul, S Brazil 30.52S 53.03W

63 F16 **Santana do Livramento** *prev.* Livramento. Rio Grande do Sul, S Brazil 30.52S 55.30W

107 N2 **Santander** Cantabria, N Spain 43.28N 3.48W

56 F8 **Santander** *off.* Departamento de Santander. ✒ *province* C Colombia

Santander Jiménez *see* Jiménez

Sant'Andrea *see* Svetac

109 B20 **Sant'Antioco** Sardegna, Italy, C Mediterranean Sea 39.03S 8.28E

106 J13 **Santa Olalla del Cala** Andalucía, S Spain 37.54N 6.13W

37 R15 **Santa Paula** California, W USA 34.21N 119.03W

38 L4 **Santaquin** Utah, W USA 39.58N 111.46W

60 I12 **Santarém** Pará, N Brazil 2.25S 54.40W

106 G10 **Santarém** *anc.* Scalabis. Santarém, W Portugal 39.13N 8.40W

106 G10 **Santarém** ✦ *district* C Portugal

46 F4 **Santaren Channel** *channel* W Bahamas

56 K10 **Santa Rita** Vichada, E Colombia 4.51N 68.27W

196 B16 **Santa Rita** SW Guam

44 H5 **Santa Rita** Cortés, NW Honduras 15.10N 87.54W

42 E9 **Santa Rita** Baja California Sur, W Mexico 27.28N 100.33W

56 H5 **Santa Rita** Zulia, NW Venezuela 10.33N 71.31W

61 I19 **Santa Rita de Araguaia** Goiás, S Brazil 17.17S 53.13W

63 D14 **Santa Rita de Cassia** *see* Cássia

63 G14 **Santa Rosa** Corrientes, NE Argentina 28.18S 58.04W

64 K13 **Santa Rosa** La Pampa, C Argentina 36.37S 64.15W

63 G14 **Santa Rosa** Rio Grande do Sul, S Brazil 27.49S 54.28W

60 E10 **Santa Rosa** Roraima, N Brazil 3.41N 62.29W

59 I16 **Santa Rosa** Puno, S Peru 14.38S 70.48W

37 N7 **Santa Rosa** California, W USA 38.26N 122.42W

39 U11 **Santa Rosa** New Mexico, SW USA 34.54N 104.43W

57 O6 **Santa Rosa** Anzoátegui, NE Venezuela 9.36N 64.16W

44 A3 **Santa Rosa** ✦ *Departamento de* SE Guatemala

Santa Rosa *see* Santa Rosa de Copán

65 J15 **Santa Rosa, Bajo de** *basin* E Argentina

44 F6 **Santa Rosa de Copán** *var.* Santa Rosa. Copán, W Honduras 14.46N 88.48W

56 E8 **Santa Rosa de Osos** Antioquia, C Colombia 6.40N 75.27W

37 Q15 **Santa Rosa Island** *island* California, W USA

25 O9 **Santa Rosa Island** *island* Florida, SE USA

56 K6 **Santa Rosalía** Portuguesa, NW Venezuela 9.01N 69.02W

196 C15 **Santa Rosa, Mount** ▲ NE Guam

37 V16 **Santa Rosa Mountains** ▲ California, W USA

37 T2 **Santa Rosa Range** ▲ Nevada, W USA

54 M8 **Santa Sylvina** Chaco, N Argentina 27.49S 61.07W

44 B19 **Santa Teresa** Santa Fe, C Argentina 33.30S 60.45W

61 O20 **Santa Teresa** Espírito Santo, SE Brazil 19.51S 40.49W

109 M23 **Santa Teresa di Riva** Sicilia, Italy, C Mediterranean Sea 38.00N 15.25E

63 E21 **Santa Teresita** Buenos Aires, E Argentina 36.34S 56.43W

63 H19 **Santa Vitória do Palmar** Rio Grande do Sul, S Brazil 33.31S 53.25W

37 Q14 **Santa Ynez River** ✒ California, W USA

Sant Carles de la Ràpida *see* Sant Carles de la Ràpita

107 U7 **Sant Carles de la Ràpita** *var.* Sant Carles de la Ràpida. Cataluña, NE Spain 40.37N 0.36E

107 W5 **Sant Celoni** Cataluña, NE Spain 41.39N 2.05E

37 U17 **Santee** California, W USA 32.50N 116.58W

23 T13 **Santee** ✒ South Carolina, SE USA

42 K15 **San Telmo, Punta** *headland* SW Mexico 18.19N 103.30W

109 O17 **Santeramo in Colle** Puglia, SE Italy 40.46N 16.45E

107 X5 **Sant Feliu de Guíxols** *var.* San Feliú de Guixols. Cataluña, NE Spain 41.46N 3.01E

107 W6 **Sant Feliú de Llobregat** Cataluña, NE Spain 41.21N 2.00E

108 C7 **Santhià** Piemonte, NE Italy 45.21N 8.11E

63 F15 **Santiago** Rio Grande do Sul, S Brazil 29.10S 54.52W

64 H11 **Santiago** *var.* Gran Santiago. ● (Chile) Santiago, C Chile 33.30S 70.40W

47 N9 **Santiago** *var.* Santiago de los Caballeros. N Dominican Republic 19.27N 70.42W

42 G10 **Santiago** Baja California Sur, W Mexico 23.32N 109.45W

43 O8 **Santiago** Nuevo León, NE Mexico 25.22N 100.09W

45 R16 **Santiago** Veraguas, S Panama 8.06N 80.58W

59 E16 **Santiago** Ica, SW Peru 14.13S 75.43W

106 G3 **Santiago** *var.* Santiago de Compostela, *Eng.* Compostela; *anc.* Campus Stellae. Galicia, NW Spain 42.52N 8.33W

44 C2 **Santiago** Baja California, NW Mexico 31.18N 116.12W

196 M6 **Santiago** Saipan, N Northern Mariana Islands

64 H11 **Santiago** *off.* Región Metropolitana de Santiago, *var.* Metropolitan. ✦ *region* C Chile

64 H11 **Santiago** × Santiago, C Chile 33.27S 70.40W

106 G3 **Santiago** × Galicia, NW Spain

78 D10 **Santiago** ✒ El Salvador

59 E15 **Santiago de Cañete** *var.* Cañete. Lima, W Peru 13.04S 76.25W

56 E12 **San Vicente de la Barquera** Cantabria, N Spain 43.22N 4.24W

Santiago *see* Santiago de Cuba, Cuba

Santiago *see* Grande de Santiago, Río, Mexico

42 B6 **Santiago Atitlán** Sololá, SW Guatemala 14.36N 91.13W

44 G16 **Santiago, Cerro** ▲ W Panama 8.27N 81.42W

Santiago de Compostela *see* Santiago

46 I8 **Santiago de Cuba** *var.* Santiago. Santiago de Cuba, E Cuba 20.01N 75.50W

Santiago de Guayaquil *see* Guayaquil

64 K8 **Santiago del Estero** Santiago del Estero, C Argentina 27.51S 64.15W

63 A15 **Santiago del Estero** *off.* Provincia de Santiago del Estero. ✒ *province* N Argentina

44 I8 **Santiago de los Caballeros** Sinaloa, W Mexico 25.33N 107.22W

Santiago de los Caballeros *see* Santiago, Dominican Republic

Santiago de los Caballeros *see* Guatemala, Guatemala

44 F8 **Santiago de María** Usulután, SE El Salvador 13.48N 88.28W

106 F12 **Santiago do Cacém** Setúbal, S Portugal 38.01N 8.42W

42 J12 **Santiago Ixcuintla** Nayarit, C Mexico 21.49N 105.07W

Santiago Jamiltepec *see* Jamiltepec

26 L11 **Santiago Mountains** ▲ Texas, SW USA

42 J9 **Santiago Papasquiaro** Durango, C Mexico 25.03N 105.25W

Santiago Pinotepa Nacional *see* Pinotepa Nacional

58 C8 **Santiago, Río** ✒ N Peru

42 M10 **San Tiburcio** Zacatecas, C Mexico 24.07N 101.28W

107 N2 **Santillana** Cantabria, N Spain 43.24N 4.06W

56 I5 **San Timoteo** Zulia, NW Venezuela 9.49N 71.04W

Santi Quaranta *see* Sarandë

Santíssima Trinidad *see* Chilung

107 O12 **Santisteban del Puerto** Andalucía, S Spain 38.15N 3.10W

107 U7 **Sant Jordi, Golf de** *gulf* NE Spain

107 T8 **Sant Mateu** País Valenciano, E Spain 40.28N 0.10E

27 S7 **Santo** Texas, SW USA 32.35N 98.06W

Santo *see* Espíritu Santo

56 M10 **Santo Amaro, Ilha de** *island* SE Brazil 23.41S 46.29W

63 G14 **Santo Ângelo** Rio Grande do Sul, S Brazil 28.16S 54.15W

78 C9 **Santo Antão** *island* Ilhas de Barlavento, N Cape Verde

62 J10 **Santo Antônio da Platina** Paraná, S Brazil 23.20S 50.05W

60 C13 **Santo Antônio do Içá** Amazonas, N Brazil 3.04S 67.55W

45 E5 **Santo Corazón, Río** ✒ E Bolivia

47 O9 **Santo Domingo** *prev.* Ciudad Trujillo. ● (Dominican Republic) SE Dominican Republic 18.30N 69.57W

42 E8 **Santo Domingo** Baja California Sur, W Mexico 25.31N 111.54W

42 M10 **Santo Domingo** San Luis Potosí, C Mexico 23.18N 101.42W

44 L10 **Santo Domingo** Chontales, S Nicaragua 12.15N 85.06W

107 P4 **Santo Domingo de la Calzada** La Rioja, N Spain 42.25N 2.57W

58 B6 **Santo Domingo de los Colorados** Pichincha, NW Ecuador 0.16S 79.11W

Santo Domingo Tehuantepec *see* Tehuantepec

57 O6 **San Tomé** Anzoátegui, NE Venezuela 8.54N 64.14W

San Tomé de Guayana *see* Ciudad Guayana

107 R13 **Santomera** Murcia, SE Spain 38.03N 1.05W

107 O2 **Santoña** Cantabria, N Spain 43.27N 3.28W

Santorin/Santorini *see* Thíra

62 M10 **Santos** São Paulo, S Brazil 23.55S 46.22W

62 N10 **Santos Plateau** *undersea feature* SW Atlantic Ocean

106 G6 **Santo Tirso** Porto, N Portugal 41.20N 8.25W

42 B2 **Santo Tomás** Baja California, NW Mexico 31.31N 116.25W

44 M12 **Santo Tomás** Chontales, S Nicaragua 12.04N 85.01W

59 I16 **Santo Tomás, Punta** *headland* NW Mexico 31.30N 116.40W

58 C8 **Santo Tomás, Río** ✒ C Peru

59 B18 **Santo Tomás, Volcán** ▲ Galapagos Islands, Ecuador, E Pacific Ocean 0.46S 91.01W

63 F14 **Santo Tomé** Corrientes, NE Argentina 28.31S 56.03W

Santo Tomé de Guayana *see* Ciudad Guayana

100 H10 **Santpoort** Noord-Holland, W Netherlands 19.27N 70.42W

42 G10 **Santiago** Baja California Sur, W Mexico 23.32N 109.45W

Santurce *see* Santurtzi

107 O2 **Santurtzi** *var.* Santurce, Santurzi. País Vasco, N Spain 43.19N 3.03W

Santurzi *see* Santurtzi

65 G20 **San Valentín, Cerro** ▲ S Chile 46.36S 73.17W

44 F8 **San Vicente** San Vicente, C El Salvador 13.37N 88.44W

44 C2 **San Vicente** Baja California, NW Mexico 31.18N 116.12W

44 B9 **San Vicente** ✦ *department* E El Salvador

106 J10 **San Vicente de Alcántara** Extremadura, W Spain 39.21N 7.07W

107 N2 **San Vicente de Barakaldo** *var.* Baracaldo. País Vasco, N Spain 43.16N 2.58W

59 E15 **San Vicente de Cañete** *var.* Cañete. Lima, W Peru 13.04S 76.25W

56 E12 **San Vicente del Caguán** Caquetá, S Colombia 2.07N 74.46W

44 F8 **San Vicente, Volcán de** ▲ C El Salvador 13.34N 88.50W

45 O15 **San Vito** Puntarenas, SE Costa Rica 8.49N 82.58W

108 I7 **San Vito al Tagliamento** Friuli-Venezia Giulia, NE Italy 45.54N 12.55E

109 H23 **San Vito, Capo** *headland* Sicilia, C Mediterranean Sea 38.11N 12.41E

109 P18 **San Vito dei Normanni** Puglia, SE Italy 40.40N 17.42E

166 L17 **Sanya** var. Ya Xian. Hainan, S China 18.17N 109.32E

85 I17 **Sanyati** N Zimbabwe

27 Q6 **San Ygnacio** Texas, SW USA 27.04N 99.25W

126 Ll12 **Sanyakhtakh** Respublika Sakha (Yakutiya), NE Russian Federation 60.34N 124.09E

84 C10 **Sanza Pombo** Uíge, NW Angola 07.20S 16.00E

Sanzyō *see* Sanjō

106 G14 **São Bartolomeu de Messines** Faro, S Portugal 37.12N 8.16W

62 M10 **São Bernardo do Campo** São Paulo, S Brazil 23.41S 46.29W

63 F15 **São Borja** Rio Grande do Sul, S Brazil 28.34S 56.01W

106 H14 **São Brás de Alportel** Faro, S Portugal 37.09N 7.55W

62 M10 **São Caetano do Sul** São Paulo, S Brazil 23.37S 46.34W

62 L9 **São Carlos** São Paulo, S Brazil 22.01S 47.52W

61 P16 **São Cristóvão** Sergipe, E Brazil 10.58S 37.10W

61 F15 **São Francisco de Assis** Rio Grande do Sul, S Brazil 29.31S 55.07W

60 K13 **São Félix** Pará, NE Brazil 6.43S 51.55W

São Félix *see* São Félix do Araguaia

61 J16 **São Félix do Araguaia** *var.* São Félix. Mato Grosso, W Brazil 11.36S 50.40W

61 J14 **São Félix do Xingu** Pará, NE Brazil 6.37S 51.58W

62 Q9 **São Fidélis** Rio de Janeiro, SE Brazil 21.37S 41.45W

78 D10 **São Filipe** Fogo, S Cape Verde 14.52N 24.28W

62 K12 **São Francisco do Sul** Santa Catarina, S Brazil 26.16S 48.39W

61 P16 **São Francisco, Ilha de** *island* S Brazil

61 P16 **São Francisco, Rio** ✒ E Brazil

63 G16 **São Gabriel** Rio Grande do Sul, S Brazil 30.17S 54.17W

62 P10 **São Gonçalo** Rio de Janeiro, SE Brazil 22.48S 43.02W

83 H23 **São Hill** Iringa, S Tanzania 8.19S 35.10E

62 R9 **São João da Barra** Rio de Janeiro, SE Brazil 21.39S 41.04W

106 G7 **São João da Madeira** Aveiro, N Portugal 40.52N 8.28W

60 M12 **São João de Cortês** Maranhão, E Brazil 2.30S 44.27W

61 N15 **São João del Rei** Minas Gerais, NE Brazil 21.07S 44.15W

61 N14 **São João do Piauí** Piauí, E Brazil 8.21S 42.13W

61 M14 **São João dos Patos** Maranhão, E Brazil 6.28S 43.43W

60 C11 **São Joaquim** Amazonas, NW Brazil 0.08S 67.10W

63 J14 **São Joaquim** Santa Catarina, S Brazil 28.20S 49.55W

62 L7 **São Joaquim da Barra** São Paulo, S Brazil 20.36S 47.50W

66 N2 **São Jorge** Azores, Portugal, NE Atlantic Ocean

63 K14 **São José** Santa Catarina, S Brazil 27.34S 48.39W

62 M8 **São José do Rio Pardo** São Paulo, S Brazil 21.35S 46.52W

62 K8 **São José do Rio Preto** São Paulo, S Brazil 20.49S 49.19W

62 N10 **São Jose dos Campos** São Paulo, S Brazil 23.10S 45.53W

63 I17 **São Lourenço do Sul** Rio Grande do Sul, S Brazil 31.25S 51.58W

61 P15 **São Luís** Roraima, N Brazil 1.11N 60.15W

61 M12 **São Luís** state capital Maranhão, NE Brazil 2.34S 44.16W

61 F14 **São Luís, Ilha de** *island* NE Brazil

63 F14 **São Luiz Gonzaga** Rio Grande do Sul, S Brazil 28.22S 55.00W

106 I10 **São Mamede** ▲ C Portugal 39.18N 7.19W

São Mandol *see* São Manuel, Rio

59 U8 **São Manuel** ✒ S Brazil

61 G13 **São Manuel, Rio** var. São Mandol, Teles Pirés. ✒ C Brazil

117 H17 **Sarakíniko, Akrotírio** *headland* Évvoia, C Greece 38.46N 23.43E

117 I18 **Sarakinó** *island* Vóreioi Sporádes, Greece, Aegean Sea

131 W2 **Saraktash** Orenburgskaya Oblast', W Russian Federation 51.45N 56.23E

32 L15 **Sara, Lake** ☉ Illinois, N USA

25 N8 **Saraland** Alabama, S USA 30.49N 88.04W

57 V9 **Saramacca** ✦ *district* N Suriname

57 V10 **Saramacca Rivier** ✒ C Suriname

177 G2 **Saramati** ▲ N Myanmar 25.46N 95.01E

151 R12 **Saran'** *Kaz.* Saran. Karaganda, C Kazakhstan 49.46N 73.01E

20 K7 **Saranac Lake** New York, NE USA 44.18N 74.06W

20 K7 **Saranac River** ✒ New York, NE USA

115 L23 **Sarandë** *var.* Saranda, *It.* Porto Edda; *prev.* Santi Quaranta. Vlorë, S Albania 39.53N 19.59E

63 H14 **Sarandi** Rio Grande do Sul, S Brazil 27.57S 52.58W

63 F19 **Sarandí del Yí** Durazno, C Uruguay 33.18S 55.37W

63 F19 **Sarandí Grande** Florida, S Uruguay 33.43S 56.19W

179 R17 **Sarangani Islands** *island group* S Philippines

131 P5 **Saransk** Respublika Mordoviya, W Russian Federation 54.10N 45.09E

117 C14 **Sarantáporos** ✒ N Greece

116 H9 **Sarantsi** Sofiya, W Bulgaria 42.43N 23.46E

131 T3 **Sarapul** Udmurtskaya Respublika, NW Russian Federation 56.26N 53.52E

144 I3 **Saráqeb** *Fr.* Saráqeb. Idlib, N Syria 35.52N 36.48E

56 J5 **Sarare** Lara, N Venezuela 9.46N 69.10W

57 O10 **Sararína** Amazonas, S Venezuela 4.10N 64.31W

149 S10 **Sar Ashk** Kermán, C Iran

25 V13 **Sarasota** Florida, SE USA 27.20N 82.31W

119 O11 **Sarata** Odes'ka Oblast', SW Ukraine 46.01N 29.40E

118 I10 **Sărățel** *Hung.* Szeretfalva. Bistrița-Năsăud, N Romania 47.02N 24.24E

27 X10 **Saratoga** Texas, SW USA 30.19N 94.31W

20 K10 **Saratoga Springs** New York, NE USA 43.04N 73.47W

131 P8 **Saratov** Saratovskaya Oblast', W Russian Federation 51.33N 45.57E

131 P8 **Saratovskaya Oblast'** ✦ *province* W Russian Federation

131 Q7 **Saratovskoye Vodokhranilishche** ☒ W Russian Federation

194 K12 **Sarawaget Range** *var.* Saruwaged Range. ▲ C PNG

174 M5 **Sarawak** ✦ *state* East Malaysia

Sarawak *see* Kuching

145 U6 **Sarāy** *var.* Saráı. It. Iraq 34.06N 45.06E

142 D10 **Saray** Tekirdağ, NW Turkey 41.27N 27.54E

78 J2 **Saré Sé** Senegal 12.49N 11.45W

149 W14 **Sarbāz** Sīstán va Balūchestán, SE Iran 26.31N 61.15E

149 U8 **Sarbīsheh** Khorāsān, E Iran 32.34N 59.49E

113 J24 **Sárbogárd** Fejér, C Hungary 37.04N 94.07W

149 S7 **Sarcoxie** Missouri, C USA 37.04N 94.07W

158 I11 **Sārda** *Nep.* Kāli ✒ India/Nepal

158 G10 **Sardārshahr** Rājasthān, NW India 28.24N 74.32E

109 C18 **Sardegna** *Eng.* Sardinia. ✦ *region* Italy, C Mediterranean Sea

109 A18 **Sardegna** *Eng.* Sardinia. *island* Italy, C Mediterranean Sea

44 K13 **Sardinal** Guanacaste, NW Costa Rica 10.30N 85.38W

56 G7 **Sardinata** Norte de Santander, N Colombia 8.05N 72.50W

123 K8 **Sardinia-Corsica Trough** *undersea feature* Tyrrhenian Sea, C Mediterranean Sea

24 L2 **Sardis** Mississippi, S USA 34.25N 89.55W

24 L2 **Sardis Lake** ☒ Mississippi, S USA

49 P2 **Sardis Lake** ☒ Oklahoma, C USA

94 H17 **Sarek** ▲ N Sweden

155 N3 **Sar-e Pol** *var.* Sar-i-Pul. Sar-e Pol, N Afghanistan 36.16N 65.55E

155 O3 **Sar-e Pol** ✦ *province* N Afghanistan

148 I6 **Sar-e Pol-e Zahāb** *var.* Sar-e Pol, Sar-i-Pul. Kermānshāh, W Iran 34.28N 45.52E

55 T13 **Sarez, Kŭli** *Rus.* Sarezskoye Ozero. ☉ SE Tajikistan

Sarezskoye Sea *see* Sarez, Kŭli

66 G10 **Sargasso Sea** ✒ W Atlantic Ocean

155 U8 **Sargodha** Punjab, NE Pakistan 32.06N 72.47E

79 I14 **Sarh** *prev.* Fort-Archambault. Moyen-Chari, S Chad 9.07N 18.22E

149 P4 **Sārī** *var.* Sari, Sārı. Māzandarān, N Iran 36.36N 53.04E

117 M21 **Saría** *island* SE Greece

42 F3 **Saric** Sonora, NW Mexico 31.07N 111.22W

196 K6 **Sarigan** *island* C Northern Mariana Islands

142 D14 **Sarıgöl** Manisa, SW Turkey

145 T6 **Sārīhah** E Iraq 34.34N 44.38E

143 R12 **Sarıkamış** Kars, NE Turkey 40.18N 42.34E

174 L6 **Sarikei** Sarawak, East Malaysia 2.07N 111.30E

153 U12 **Sarikol Range** *Rus.* Sarykol'skiy Khrebet. ▲ China/Tajikistan

189 Y7 **Sarina** Queensland, NE Australia 21.34S 149.12E

◆ COUNTRY ◇ DEPENDENT TERRITORY ◆ ADMINISTRATIVE REGION ▲ MOUNTAIN ⊠ VOLCANO ☉ LAKE
● COUNTRY CAPITAL ○ DEPENDENT TERRITORY CAPITAL × INTERNATIONAL AIRPORT ▲ MOUNTAIN RANGE ✒ RIVER ☒ RESERVOIR

323

Sarine see La Sarine

107 S5 Sariñena Aragón, NE Spain
41.46N 0.10W

153 O13 Sariosiyo Rus. Sariasiya.
Surkhondaryo Wiloyati,
S Uzbekistan 38.25N 67.51E

Sar-i-Pul see Sar-e Pol,
Afghanistan

Sar-i Pul see Sar-e Pol-e Žaháb,
Iran

Sariqamish Küli see
Sarykamyshskoye Ozero

155 V1 Sari Qül Rus. Ozero Zurkul', Taj.
Zürköl. ⊚ Afghanistan/Tajikistan
see also Zürkül

77 Q12 Sarir Tibisti var. Serir Tibesti.
desert S Libya

27 S15 Sarita Texas, SW USA
27.12N 97.48W

169 W14 Sariwŏn SW North Korea
38.30N 125.52E

116 H12 Sarıyer İstanbul, NW Turkey
41.10N 29.03E

99 L26 Sark Fr. Sercq. island Channel
Islands

113 N24 Sarkad Rom. Sârcad. Békés,
SE Hungary 46.42N 21.21E

151 W14 Sarkand Almaty, SW Kazakhstan
45.25N 79.53E

Sarkani see Krasnogorskoye

158 D11 Sarkāri Tala Rājasthān, NW India
27.59N 70.52E

142 G15 Şarkikaraağaç var. Şarki
Karaağaç. Isparta, SW Turkey
38.04N 31.22E

142 L13 Şarkışla Sivas, C Turkey
39.21N 36.27E

142 C11 Şarköy Tekirdağ, NW Turkey
40.37N 27.07E

Sárköz see Livada

Sarlat see Sarlat-la-Canéda

104 M13 Sarlat-la-Canéda var. Sarlat.
Dordogne, SW France 44.54N 1.12E

111 S3 Sarleinsbach Oberösterreich,
N Austria 48.33N 13.55E

176 Y10 Sarmi Irian Jaya, E Indonesia
1.51S 138.45E

65 I19 Sarmiento Chubut, S Argentina
45.37S 69.06W

65 M25 Sarmiento, Monte ▲ S Chile
54.28S 70.49W

96 J11 Särna Dalarna, C Sweden
61.40N 13.10E

110 F8 Sarnen Obwalden, C Switzerland
46.54N 8.15E

110 F9 Sarner See ⊚ C Switzerland

12 D16 Sarnia Ontario, S Canada
42.57N 82.22W

118 L3 Sarny Rivnens'ka Oblast',
NW Ukraine 51.20N 26.34E

175 Q10 Saroako Sulawesi, C Indonesia
2.31S 121.18E

120 L13 Sarochyna Rus. Sorochino.
Vitsyebskaya Voblasts', N Belarus
55.12N 28.45E

174 Hh10 Sarolangun Sumatera,
W Indonesia 2.17S 102.39E

172 Q5 Saroma Hokkaidō, NE Japan
44.01N 143.43E

172 Q5 Saroma-ko ⊚ Hokkaidō,
NE Japan

Saronic Gulf see Saronikós
Kólpos

117 H20 Saronikós Kólpos Eng. Saronic
Gulf. gulf S Greece

108 D7 Saronno Lombardia, N Italy
45.37N 9.01E

142 B11 Saros Körfezi gulf NW Turkey

113 N20 Sárospatak Borsod-Abaúj-
Zemplén, NE Hungary 48.18N 21.30E

131 P12 Sarpa Respublika Kalmykiya,
SW Russian Federation
47.00N 45.42E

131 P12 Sarpa, Ozero ⊚ SW Russian
Federation

115 M18 Šar Planina ▲ FYR
Macedonia/Yugoslavia

97 I16 Sarpsborg Østfold, S Norway
59.16N 11.07E

145 U5 Sarqalā N Iraq

105 U4 Sarralbe Moselle, NE France
49.02N 7.01E

Sarre see Saar, France/Germany

Sarre see Saarland, Germany

105 U5 Sarrebourg Ger. Saarburg.
Moselle, NE France 48.43N 7.03E

Sarrebruck see Saarbrücken

105 U4 Sarreguemines prev.
Saargemünd. Moselle, NE France
49.06N 7.04E

106 I3 Sarria Galicia, NW Spain
42.46N 7.25W

107 S8 Sarrión Aragón, NE Spain
40.09N 0.49W

44 F4 Sarstoon Sp. Río Sarstún.
⊷ Belize/Guatemala
Sarstún, Río see Sarstoon

126 M9 Sartang ⊷ NE Russian
Federation

105 X16 Sartène Corse, France,
C Mediterranean Sea 41.37N 8.58E

104 K7 Sarthe ◆ department NW France

104 K7 Sarthe ⊷ N France

117 H15 Sárti Kentrikí Makedonía,
N Greece 40.04N 23.59E

172 Pp2 Sarufutsu Hokkaidō, NE Japan
45.20N 142.03E

172 Oo7 Saru-gawa ⊷ Hokkaidō,
NE Japan

Saruhan see Manisa

158 G9 Sarüpsar Rājasthān, NW India
29.25N 73.49E

143 U13 Sārur prev. Il'ichevsk.
SW Azerbaijan 39.30N 44.59E

Saruwaged Range see Sarawaget
Range

Sarvani see Marneuli

113 G23 Sárvár Vas, W Hungary
47.14N 16.57E

149 P11 Sarvestān Fārs, S Iran
29.16N 53.13E

176 X9 Sarwon Irian Jaya, E Indonesia
0.58S 136.08E

151 P17 Saryagach Kaz. Saryaghash.
Yuzhnyy Kazakhstan, S Kazakhstan
41.28N 69.10E

Saryaghash see Saryagach

Saryarqa see Kazakhskiy
Melkosopochnik

153 W8 Sary-Bulak Narynskaya Oblast',
C Kyrgyzstan 41.56N 75.44E

153 U10 Sary-Bulak Oshskaya Oblast',
SW Kyrgyzstan 40.49N 73.44E

119 S14 Sarych, Mys headland S Ukraine
44.23N 33.44E

153 Z7 Sary-Dzhaz var. Aksu He.
⊷ China/Kyrgyzstan see also Aksu
He

151 T14 Sarysesik-Atyrau, Peski desert
E Kazakhstan

150 G13 Sarykamys Kaz. Saryqamys.
Mangistau, SW Kazakhstan
45.58N 53.30E

152 F8 Sarykamyshskoye Ozero Uzb.
Sariqamish Küli. salt lake
Kazakhstan/Uzbekistan
Sarykol'skiy Khrebet see Sarikol
Range

150 M10 Sarykopa, Ozero
⊚ C Kazakhstan

151 V15 Saryozek Kaz. Saryözek. Almaty,
SE Kazakhstan 44.22N 77.57E
Saryqamys see Sarykamys

151 S13 Saryshagan Kaz. Saryshahan.
Zhezkazgan, SE Kazakhstan
46.03N 73.36E
Saryshahan see Saryshagan

151 O13 Sarysu ⊷ S Kazakhstan

153 T11 Sary-Tash Oshskaya Oblast',
SW Kyrgyzstan 39.43N 73.13E

152 J15 Saryyazynskoye
Vodokhranilishche
⊚ S Turkmenistan

108 E10 Sarzana Liguria, NW Italy
44.07N 9.59E

196 B17 Sasalaguan, Mount ▲ S Guam

159 O14 Sasarām Bihār, N India
24.58N 84.01E

195 W14 Sasari, Mount ▲ Santa Isabel,
N Solomon Islands 8.09S 159.32E

170 C12 Sasebo Nagasaki, Kyūshū,
SW Japan 33.10N 129.42E

12 I9 Saseginaga, Lac ⊚ Quebec,
SE Canada
Saseno see Sazan

9 R13 Saskatchewan ❖ province
SW Canada

9 U14 Saskatchewan
⊷ Manitoba/Saskatchewan,
C Canada

9 T15 Saskatoon Saskatchewan,
S Canada 52.10N 106.40W

9 T15 Saskatoon ✕ Saskatchewan,
S Canada 52.15N 107.05W

126 K7 Saskylakh Respublika Sakha
(Yakutiya), NE Russian Federation
71.56N 114.07E

44 L7 Saslaya, Cerro ▲ N Nicaragua
13.52N 85.06W

40 G17 Sasmik, Cape headland Tanaga
Island, Alaska, USA 51.36N 105.55W

121 N19 Sasnovy Bor Rus. Sosnovyy Bor.
Homyel'skaya Voblasts', SE Belarus
52.31N 29.37E

131 N15 Sasovo Ryazanskaya Oblast',
W Russian Federation 54.19N 41.54E

27 S12 Saspamco Texas, SW USA
29.13N 98.18W

111 W9 Sass var. Sassbach. ⊷ SE Austria

78 M17 Sassandra S Ivory Coast
4.58N 6.07W

78 M17 Sassandra var. Ibo, Sassandra
Fleuve. ⊷ S Ivory Coast
Sassandra Fleuve see Sassandra

109 B17 Sassari Sardegna, Italy,
C Mediterranean Sea 40.43N 8.33E
Sassbach see Sass

100 H11 Sassenheim Zuid-Holland,
W Netherlands 52.13N 4.31E
Sassmacken see Valdemārpils

102 O7 Sassnitz Mecklenburg-
Vorpommern, NE Germany
54.32N 13.39E

101 E16 Sas van Gent Zeeland,
SW Netherlands 51.13N 3.48E

151 W12 Sasykkol', Ozero
⊚ E Kazakhstan

119 O12 Sasyk Kunduk, Ozero
⊚ SW Ukraine

78 J12 Satadougou Kayes, SW Mali
12.40N 11.25W

107 V11 Sa Talaiassa ▲ Eivissa, Spain,
W Mediterranean Sea 38.55N 1.17E

170 B17 Sata-misaki headland Kyūshū,
SW Japan 31.00N 130.39E

28 I7 Satanta Kansas, C USA
37.23N 102.00W

161 E15 Sātāra Mahārāshtra, W India
17.40N 73.58E

198 Aa7 Sātaua Savai'i, NW Samoa
13.25S 172.40W

196 M16 Satawal island Caroline Islands,
C Micronesia

201 R17 Satawan Atoll atoll Mortlock
Islands, C Micronesia

25 Y12 Satellite Beach Florida, SE USA
28.10N 80.35W

97 M14 Säter Dalarna, C Sweden
60.21N 15.45E
Sathmar see Satu Mare

25 V7 Satilla River ⊷ Georgia, SE USA

59 F14 Satipo var. San Francisco de
Satipo. Junín, C Peru 11.13S 74.40W

125 E11 Satka Chelyabinskaya Oblast',
C Russian Federation 55.08N 58.54E

159 S17 Satkhira Khulna, SW Bangladesh
22.43N 89.06E

160 K9 Satna prev. Sutna. Madhya
Pradesh, C India 24.33N 80.49E

105 R11 Satolas ✕ (Lyon) Rhône, E France
45.44N 5.01E

113 N20 Sátoraljaújhely Borsod-Abaúj-
Zemplén, NE Hungary 48.24N 21.39E

151 O12 Satpayev prev. Nikol'skiy.
Zhezkazgan, C Kazakhstan
47.59N 67.27E

160 G11 Sātpura Range ▲ C India

170 Bb16 Satsuma-hantō peninsula Kyūshū,
SW Japan

178 Hh12 Sattahip var. Ban Sattahip, Ban
Sattahipp. Chon Buri, S Thailand
12.41N 100.55E

94 L11 Sattanen Lappi, NE Finland
67.35N 26.35E

118 H9 Satulung Hung. Kővárhosszúfalu.
Maramureş, N Romania
47.34N 23.25E
Satul see Satun

160 C11 Satura Germania
Satul-Vechi see Staro Selo

118 F11 Sǎvârşin Hung. Soborsin; prev.
Sǎvîrşin. Arad, W Romania
46.00N 22.12E

142 L13 Savaştepe Balıkesir, W Turkey
39.19N 27.37E

85 N18 Save Inhambane, E Mozambique
21.07S 34.35E

104 L16 Save ⊷ S France

85 L17 Save var. Sabi.
⊷ Mozambique/Zimbabwe see
also Sabi
Save see Sava

79 R15 Save SE Benin 8.04N 2.28E

148 M6 Sāveh Markazī, W Iran
35.03N 50.21E

118 L8 Săveni Botoşani, NE Romania
47.57N 26.49E

105 N16 Saverdun Ariège, S France
43.15N 1.34E

105 U5 Saverne var. Zabern; anc. Tres
Tabernae. Bas-Rhin, NE France
48.45N 7.22E

121 O21 Savichy Rus. Savichi.
Homyel'skaya Voblasts', SE Belarus
51.37N 30.19E

108 B9 Savigliano Piemonte, NW Italy
44.39N 7.39E

121 Q16 Savichy Rus. Savichi.
Mahilyowskaya Voblasts', E Belarus
53.28N 31.46E
Savinski see Savinskiy

125 Dd6 Savinskiy var. Savinski.
Arkhangel'skaya Oblast',
NW Russian Federation
62.54N 40.07E

108 N11 Savio ⊷ C Italy
Sǎvîrşin see Sǎvârşin

207 O11 Savissivik var. Savigsvik.
Avannaarsua, S Greenland
76.09N 65.24W

95 N18 Savitaipale Etelä-Suomi,
S Finland 61.12N 27.43E

115 J15 Šavnik Montenegro,
SW Yugoslavia 42.57N 19.04E

195 W15 Savo island C Solomon Islands

110 I9 Savognin Graubünden,
S Switzerland 46.34N 9.35E

105 T12 Savoie ◆ department E France

108 C10 Savona Liguria, NW Italy
44.18N 8.28E

95 M17 Savonlinna Swe. Nyslott. Itä-
Suomi, SE Finland 61.51N 28.55E

95 N17 Savonranta Itä-Suomi, SE Finland
62.10N 29.10E

40 L9 Savoonga Saint Lawrence Island,
Alaska, USA 63.40N 170.29W

32 M13 Savory Illinois, N USA
40.03N 88.15W

119 O8 Savran' Odes'ka Oblast',
SW Ukraine 48.07N 30.00E

143 R11 Savşat Artvin, NE Turkey
41.15N 42.20E

97 L19 Sävsjö Jönköping, S Sweden
57.25N 14.40E

94 M11 Savukoski Lappi, NE Finland
67.17N 28.14E

197 J13 Savusavu Vanua Levu, N Fiji
16.47S 179.21E

175 Q17 Savu Sea Ind. Laut Sawu. sea
S Indonesia

85 H17 Savute Chobe, N Botswana
18.33S 24.06E

145 N7 Şawāb ‘Uqlat well W Iraq
33.57N 40.04E

144 M7 Sawāb, Wādī as dry watercourse
W Iraq 37.16N 0.04W

158 H13 Sawāi Mādhopur Rājasthān,
N India 26.00N 76.22E

175 Tt10 Sawai, Teluk bay Pulau Seram,
E Indonesia

79 N16 Sawakin ⊘ NE England, UK
54.16N 0.24W
Sawakin see Suakin

178 H17 Sawang Daen Din Sakon
Nakhon, E Thailand 17.28N 103.27E

178 H8 Sawankhalok var. Swankalok.
Sukhothai, NW Thailand
17.19N 99.49E

171 Kk17 Sawara Chiba, Honshū, S Japan
35.52N 140.29E

171 Jj12 Sawasaki-bana headland Sado,
C Japan 37.48N 138.11E

39 R7 Sawatch Range ▲ Colorado,
C USA

147 N12 Sawdā', Jabal ▲ SW Saudi Arabia
18.15N 42.26E

77 P9 Sawdā', Jabal as ▲ C Libya
Sawdirī see Sodiri

176 W9 Saweba, Tanjung headland Irian
Jaya, E Indonesia 0.41S 133.59E

99 F14 Sawel Mountain ▲ C Northern
Ireland, UK 54.49N 7.04W
Sawhāj see Sohâg

79 O15 Sawla N Ghana 9.14N 2.26W

153 P11 Sawot Rus. Savat. Sirdaryo
Wiloyati, E Uzbekistan 40.03N 68.35E

147 N12 Şawqirah var. Suqrah. S Oman
18.16N 56.34E

147 N12 Şawqirah, Dawḥat var. Ghubbat
Sawqirah, Sukra Bay, Suqrah Bay.
bay S Oman
Şawqirah, Ghubbat see
Sawqirah, Dawḥat

191 V5 Sawtell New South Wales,
SE Australia 30.22S 153.04E

191 V5 Sawtooth Mountains ▲
Shebshi Mountains

144 K7 Şawt, Wādī aş dry watercourse
S Syria

175 O13 Sawu, Kepulauan var. Kepulauan
Savu. island group S Indonesia

175 Qq18 Sawu, Laut see Savu Sea

102 I10 Sawu, Pulau var. Pulau Savu.
island Kepulauan Sawu, S Indonesia

11 N8 Sax País Valenciano, E Spain
38.33N 0.49W
Saxe see Sachsen

100 C11 Saxon Valais, SW Switzerland
46.07N 7.09E

37 S5 Say Niamey, SW Niger 13.02N 2.22E

13 V7 Sayabec Quebec, SE Canada
48.33N 67.42W
Sayaboury see Xaignabouli

118 F11 Sayak Kaz. Sayaq. Zhezkazgan,
E Kazakhstan 46.54N 77.17E

59 J14 Sayán Lima, W Peru 11.06S 77.09W

126 Hh15 Sayanogorsk Respublika
Khakasiya, S Russian Federation
53.07N 91.08E

126 J15 Sayansk Irkutskaya Oblast',
S Russian Federation 54.06N 102.10E

133 T6 Sayanskiy Khrebet ▲ S Russian
Federation
Sayaq see Sayak

152 K13 Sayat Lebapskiy Velayat,
E Turkmenistan 38.44N 63.51E

44 D3 Sayaxché Petén, N Guatemala
16.31N 90.10W
Şayda/Sayida see Saïda

147 T15 Şaydā E Yemen 15.18N 51.15E

31 U14 Saylorville Lake ⊚ Iowa, C USA

169 N10 Saynshand Dornogovĭ,
SE Mongolia 44.51N 110.07E

168 J11 Saynshand Ömnögovĭ,
S Mongolia 43.30N 102.08E

168 F7 Sayn-Ust Govi-Altay, W Mongolia
47.23N 94.19E

144 J7 Sayqal, Baḩr ⊚ S Syria

164 H4 Sayram He ⊷ NW China

28 K11 Sayre Oklahoma, C USA
35.17N 99.38W

20 J9 Sayre Pennsylvania, NE USA
41.57N 76.30W

20 K15 Sayreville New Jersey, NE USA
40.27N 74.19W

153 N13 Sayrob Rus. Sayrab.
Surkhondaryo Wiloyati,
S Uzbekistan 38.03N 66.94E

42 L13 Sayula Jalisco, SW Mexico
19.52N 103.36V

147 R14 Say 'ūn var. Saywūn. C Yemen
15.52N 48.31E

150 GX4 Say-Utës Kaz. Say-Ötesh.
Mangistau, SW Kazakhstan
44.20N 53.32E

10 K16 Sayward Vancouver Island, British
Columbia, SW Canada
50.20N 126.01W
Saywūn see Say 'ūn
Sayyāl see As Sayyāl

145 U8 Sayyid 'Abid var. Saiyid Abid.
E Iraq 37.54N 45.07E

115 J22 Sazan var. Ishulli i Sazanit, It.
Saseno. island SW Albania

113 E17 Sázava var. Sazau, Ger. Sazawa.
⊷ C Czech Republic

128 J14 Sazonovo Vologodskaya Oblast',
NW Russian Federation
59.04N 35.10E

104 G6 Scaër Finistère, NW France
48.00N 3.40W

99 J15 Scafell Pike ▲ NW England, UK
54.26N 3.10W
Scalabis see Santarém

98 M2 Scalloway N Scotland, UK
60.10N 1.17W

40 M11 Scammon Bay Alaska, USA
61.50N 165.34W
Scammon Lagoon/Scammon,
Laguna see Ojo de Liebre, Laguna

86 F7 Scandinavia geophysical region
NW Europe
Scania see Skåne

98 K5 Scapa Flow sea basin N Scotland,
UK

109 K26 Scaramia, Capo headland Sicilia,
Italy, C Mediterranean Sea
36.46N 14.29E

12 H15 Scarborough Ontario, SE Canada
43.46N 79.14W

99 N16 Scarborough N England, UK
54.16N 0.24W

47 Z16 Scarborough prev. Port Louis.
Tobago, Trinidad and Tobago
11.10N 60.45W

193 I17 Scargill Canterbury, South Island,
NZ 42.57S 172.57E

98 E7 Scarpanto see Kárpathos
Scarpanto Strait see Karpathou,
Stenó

109 G25 Scauri Sicilia, Italy,
C Mediterranean Sea 36.45N 12.06E
Scealg, Bá na see Ballinskelligs
Bay
Scebeli see Shebeli

102 K9 Schaalsee ⊚ N Germany

101 G18 Schaerbeek Brussels, C Belgium
50.51N 4.21E

110 G6 Schaffhausen Fr. Schaffhouse.
Schaffhausen, N Switzerland
47.42N 8.37E

110 G6 Schaffhausen Fr. Schaffhouse. ◆
canton N Switzerland
Schaffhouse see Schaffhausen

100 I8 Schagen Noord-Holland,
NW Netherlands 52.46N 4.46E
Schaken see Šakiai

102 G9 Schalkhaar Overijssel,
NE Netherlands 52.16N 6.10E

111 R3 Schärding Oberösterreich,
N Austria 48.27N 13.26E

102 G9 Scharhörn island NW Germany

102 F10 Scharteberg ▲ W Germany
50.33N 6.42W

110 E8 Schattdorf ⊘ C Switzerland
46.52N 8.38E

103 I22 Schatzberg ▲ SW Germany
48.48N 9.48E

100 K11 Scherpenzeel Gelderland,
C Netherlands 52.07N 5.30E

100 G11 Scheveningen Zuid-Holland,
W Netherlands 52.07N 4.18E

100 G12 Schiedam Zuid-Holland,
SW Netherlands 51.55N 4.25E

101 M24 Schieren Diekirch,
NE Luxembourg 49.49N 6.06E

100 M4 Schiermonnikoog Fris.
Skiermûntseach. Friesland,
N Netherlands 53.28N 6.09E

100 M4 Schiermonnikoog Fris.
Skiermûntseach. island
Waddeneilanden, N Netherlands

101 K14 Schijndel Noord-Brabant,
S Netherlands 51.37N 5.27E

101 H16 Schilde Antwerpen, N Belgium
51.13N 4.34E
Schillen see Zhilino

105 V5 Schiltigheim Bas-Rhin,
NE France 48.37N 7.46E

100 H10 Schiphol ✕ (Amsterdam) Noord-
Holland, C Netherlands 52.18N 4.48E

103 I24 Schippenbeil see Sępopol

Schiria see Şiria

113 D22 Schíza island SW Greece

183 U3 Schjetman Reef reef Antarctica

111 R7 Schlackenwerth see Ostrov

Schladming Steiermark,
SE Austria 47.23N 13.37E
Schlan see Slaný

102 I7 Schlei inlet N Germany

102 H8 Schleiden Nordrhein-Westfalen,
W Germany 50.31N 6.30E

31 T13 Schleinitz Range ▲ New Ireland,
E PNG

102 H8 Schleswig-Holstein ◆ state
N Germany 54.31N 9.34E

103 N14 Schleswig Schleswig-Holstein,
N Germany 54.31N 9.34E

Schlettstadt see Sélestat

110 F7 Schlieren Zürich, N Switzerland
47.22N 8.27E

115 J22 Schlochau see Człuchów

Schloppe see Człopa

103 J18 Schlüchtern Hessen, C Germany
50.19N 9.27E

103 J17 Schmalkalden Thüringen,
C Germany 50.42N 10.26E

111 N7 Schmida ⊷ NE Austria

67 P19 Schmidt-Ott Seamount var.
Schmitt-Ott Seamount, Schmitt-
Ott Tablemount. undersea feature
SW Indian Ocean 39.37S 13.00E

111 Y4 Schmiegel see Śmigiel

Schmitt-Ott
Seamount/Schmitt-Ott
Tablemount see Schmidt-Ott
Seamount

13 V3 Schneeberg ▲ NE Austria

103 M18 Schneeberg ▲ Veliki Snežnik

Schneeberg see Veliki Snežnik

102 L9 Schnee-Eifel see Schneifel

Schneekoppe see Sněžka

Schneidemühl see Piła

102 L9 Schneifel var. Schnee-Eifel. plateau
W Germany

13 U3 Schnelle Körös/Schnelle
Kreisch see Crişul Repede

109 K26 Schnelle Körös/Schnelle
Kreisch see Crişul Repede

102 I11 Schneverdingen var.
Schneverdingen (Wümme).
Niedersachsen, NW Germany
53.07N 9.48E

Schneverdingen (Wümme) see
Schneverdingen

20 K10 Schodack New York, NE USA
42.40N 74.19W

20 K11 Schoharie Creek ⊷ New York,
NE USA

117 J21 Schoinoússa island Kykládes,
Greece, Aegean Sea

102 L13 Schönebeck Sachsen-Anhalt,
C Germany 52.01N 11.45E

102 O12 Schöneck see Skarszewy

103 K24 Schönefeld ✕ (Berlin) Berlin,
NE Germany 52.23N 13.29E

102 K13 Schöningen Niedersachsen,
C Germany 52.07N 10.58E

Schönlanke see Trzcianka

Schönsee see Kowalewo
Pomorskie

33 P10 Schoolcraft Michigan, N USA
42.07N 85.39W

100 O8 Schoonebeek Drenthe,
NE Netherlands 52.39N 6.57E

100 I12 Schoonhoven Zuid-Holland,
C Netherlands 51.57N 4.51E

100 H8 Schoorl Noord-Holland,
NW Netherlands 52.42N 4.40E

100 M10 Schalkhaar see Šakiai

103 F24 Schopfheim Baden-
Württemberg, S Germany
47.49N 10.54E

103 I21 Schorndorf Baden-Württemberg,
SW Germany 48.49N 9.31E

102 F10 Schortens Niedersachsen,
NW Germany 53.31N 7.57E

101 H16 Schoten var. Schooten.
Antwerpen, N Belgium 51.15N 4.30E

191 T7 Schouten Island Tasmania,
SE Australia

194 W9 Schouten Islands island group
NW PNG

100 F13 Schouwen island SW Netherlands

47 O12 Schœlcher W Martinique
14.37N 61.06W

111 Y4 Schrems Niederösterreich,
N Austria 48.47N 15.01E

103 L22 Schrobenhausen Bayern,
SE Germany 48.33N 11.16E

20 K10 Schroon Lake ⊚ New York,
NE USA

110 J8 Schruns Vorarlberg, W Austria
47.04N 9.54E

Schubin see Szubin

27 U11 Schulenburg Texas, SW USA
29.40N 96.54W

Schuls see Scuol

110 E8 Schüpfheim Luzern,
C Switzerland 47.02N 7.23E

37 S6 Schurz Nevada, W USA
38.55N 118.48W

103 I24 Schussen ⊷ S Germany

Schüttenhofen see Sušice

31 K5 Schuyler Nebraska, C USA
41.25N 97.04W

20 L10 Schuylerville New York, NE USA
43.05N 73.34W

103 K20 Schwabach Bayern, SE Germany
49.19N 11.01E

103 I23 Schwabenalb see Schwäbische Alb

103 I23 Schwäbische Alb var.
Schwabenalb. ▲ S Germany

103 J22 Schwäbisch Gmünd var.
Gmünd. Baden-Württemberg,
SW Germany 48.48N 9.48E

103 I21 Schwäbisch Hall var. Hall.
Baden-Württemberg, SW Germany
49.07N 9.43E

103 H16 Schwalm ⊷ C Germany

111 V9 Schwanberg Steiermark,
SE Austria 46.46N 15.12E

103 H18 Schwanden Glarus, E Switzerland
47.02N 9.04E

103 M20 Schwandorf Bayern, SE Germany
49.19N 12.07E

111 S5 Schwanenstadt Oberösterreich,
NW Austria 48.03N 13.45E

174 M9 Schwaner, Pegunungan
▲ Borneo, N Indonesia

111 W5 Schwarza ⊷ E Austria

103 D17 Schwarzach ⊷ S Austria

103 M20 Schwarzach Cz. Černice.
⊷ Czech Republic/Germany

195 P9 Schleinitz Range ▲ New Ireland,
E PNG

Schwarzach see Schwarzach im
Pongau, Austria

Schwarzach see Svratka,
Czech Republic

111 Q7 Schwarzach im Pongau var.
Schwarzach. Salzburg, NW Austria
47.19N 13.09E

Schwarzawa see Svratka

103 N14 Schwarze Elster ⊷ E Germany

Schwarze Körös see Crişul Negru

110 D9 Schwarzenburg Bern,
W Switzerland 46.51N 7.28E

85 D21 Schwarzrand ▲ S Namibia

103 G23 Schwarzwald Eng. Black Forest.
▲ SW Germany

Schwarzwasser see Wda

41 P7 Schwatka Mountains ▲ Alaska,
USA

111 N7 Schwaz Tirol, W Austria
47.21N 11.43E

111 Y4 Schwechat Niederösterreich,
NE Austria 48.09N 16.28E

111 Y4 Schwechat ✕ (Wien) Wien,
E Austria 48.04N 16.31E

102 P11 Schwedt Brandenburg,
NE Germany 53.04N 14.16E

103 D19 Schweich Rheinland-Pfalz,
SW Germany 49.49N 6.44E

13 V3 Schweidnitz see Świdnica

103 J18 Schweinfurt Bayern, SE Germany
50.03N 10.13E

Schweiz see Switzerland

102 L9 Schwerin Mecklenburg-
Vorpommern, N Germany
53.37N 11.25E

102 L9 Schweriner See ⊚ N Germany

103 F15 Schwerte Nordrhein-Westfalen,
W Germany 51.27N 7.34E

Schwiebus see Świebodzin

102 P13 Schwielochsee ⊚ NE Germany

Schwihau see Švihov

110 G8 Schwyz var. Schwiz, Schwyz,
C Switzerland 47.01N 8.39E

110 G8 Schwyz var. Schwiz. ◆ canton
C Switzerland

12 J11 Schyan ⊷ Quebec, SE Canada

Schyl see Jiu

109 I24 Sciacca Sicilia, Italy,
C Mediterranean Sea 37.30N 13.05E

109 L26 Sciasciamana see Shashemenē

109 L26 Scicli Sicilia, Italy,
C Mediterranean Sea 36.48N 14.43E

99 F25 Scilly, Isles of island group
SW England, UK

113 H17 Ścinawa Ger. Steinau an der Elbe.
Dolnośląskie, SW Poland
51.22N 16.27E

38 L5 Scio see Chíos

Scioto River ⊷ Ohio, N USA

35 X6 Scipio Utah, W USA
39.15N 112.06W

191 T7 Scobey Montana, NW USA
48.47N 105.25W

193 I17 Scone New South Wales,
SE Australia 32.02S 150.51E

98 H11 Scotland national region UK

23 W8 Scotland Neck North Carolina,
SE USA 36.07N 77.25W

205 K24 Scott Base NZ research station
Antarctica 77.52S 167.18E

47 Q15 Scott, Cape headland Vancouver
Island, British Columbia,
SW Canada 50.43N 128.24W

25 V14 Scott City Kansas, US USA
38.28N 100.54W

49 V7 Scott City Missouri, C USA
37.13N 89.31W

49 Y14 Scott Coast physical region
Antarctica

36 K3 Scotia California, W USA
40.04N 124.07W

49 V14 Scotia Plate tectonic feature

49 V15 Scotia Ridge undersea feature
S Atlantic Ocean

204 F10 Scotia Sea sea SW Atlantic Ocean

31 Q12 Scotland South Dakota, N USA
43.09N 97.43W

27 R5 Scotland Texas, SW USA
33.37N 98.27W

205 W8 Scott Island island
Antarctica

110 J8 Scottdale Pennsylvania, NE USA
40.05N 79.35W

205 Y11 **Scott Glacier** *glacier* Antarctica
205 Q17 **Scott Island** *island* Antarctica
28 L11 **Scott, Mount** ▲ Oklahoma, USA 34.52N 98.34W
34 G15 **Scott, Mount** ▲ Oregon, NW USA 42.53N 122.06W
36 M1 **Scott River** ≈ California, W USA
30 I13 **Scottsbluff** Nebraska, C USA 41.52N 103.40W
25 Q2 **Scottsboro** Alabama, S USA 34.40N 86.01W
33 P15 **Scottsburg** Indiana, N USA 38.42N 85.46W
191 P16 **Scottsdale** Tasmania, SE Australia 41.13S 147.30E
38 L13 **Scottsdale** Arizona, SW USA 33.30N 111.54W
47 O12 **Scotts Head Village** *var.* Cachacrou. S Dominica 15.12N 61.22W
199 Jj17 **Scott Shoal** *undersea feature* S Pacific Ocean
22 K7 **Scottsville** Kentucky, S USA 36.45N 86.11W
31 U14 **Scranton** Iowa, C USA 42.01N 94.33W
20 I13 **Scranton** Pennsylvania, NE USA 41.25N 75.40W
194 G10 **Screw** ≈ NW PNG
31 R14 **Scribner** Nebraska, C USA 41.40N 96.40W
Scrobesbyrig' *see* Shrewsbury
12 I14 **Scugog** ≈ Ontario, SE Canada
12 I14 **Scugog, Lake** ◎ Ontario, SE Canada
99 N17 **Scunthorpe** E England, UK 53.34N 0.39W
110 K9 **Scuol** *Ger.* Schuls. Graubünden, E Switzerland 46.51N 10.21E
Scupi *see* Skopje
Scutari *see* Shkodër
115 K17 **Scutari, Lake** *Alb.* Liqeni i Shkodrës, *SCr.* Skadarsko Jezero. ◎ Albania/Yugoslavia
Scyros *see* Skýros
Scythopolis *see* Bet She'an
27 U13 **Seadrift** Texas, SW USA 28.25N 96.42W
23 Y4 **Seaford** *var.* Seaford City. Delaware, NE USA 38.38N 75.36W
Seaford City *see* Seaford
12 E15 **Seaforth** Ontario, S Canada 43.33N 81.25W
26 M6 **Seagraves** Texas, SW USA 32.56N 102.33W
9 X9 **Seal** ≈ Manitoba, C Canada
190 M10 **Sea Lake** Victoria, SE Australia 35.34S 142.51E
85 G26 **Seal, Cape** *headland* S South Africa 34.06S 23.18E
67 D26 **Sea Lion Islands** *island group* SE Falkland Islands
21 S8 **Seal Island** *island* Maine, NE USA
27 V11 **Sealy** Texas, SW USA 29.46N 96.09W
37 X12 **Searchlight** Nevada, W USA 35.27N 114.54W
49 V11 **Searcy** Arkansas, C USA 35.15N 91.44W
21 R7 **Searsport** Maine, NE USA 44.28N 68.54W
37 N10 **Seaside** California, W USA 36.36N 121.51W
34 F10 **Seaside** Oregon, NW USA 45.57N 123.55W
20 K16 **Seaside Heights** New Jersey, NE USA 39.56N 74.03W
34 H8 **Seattle** Washington, NW USA 47.34N 122.19W
34 H9 **Seattle-Tacoma** ✕ Washington, NW USA 47.04N 122.27W
193 J16 **Seaward Kaikoura Range** ▲ South Island, NZ
44 J9 **Sébaco** Matagalpa, W Nicaragua 12.50N 86.04W
21 P8 **Sebago Lake** ◎ Maine, NE USA
176 V11 **Sebakor, Teluk** *bay* Irian Jaya, E Indonesia
Sebangan, Sungai *see* Sebangau Besar, Sungai
174 M11 **Sebangan, Teluk** *bay* Borneo, C Indonesia
174 Mm11 **Sebanganu, Teluk** *bay* Borneo, C Indonesia
174 Mm11 **Sebangau Besar, Sungai** *var.* Sungai Sebangan. ≈ Borneo, N Indonesia
174 I8 **Sebanglea, Pulau** *island* W Indonesia
Sebaste/Sebastia *see* Sivas
25 Y12 **Sebastian** Florida, SE USA 27.55N 80.31W
42 C5 **Sebastián Vizcaíno, Bahía** *bay* NW Mexico
21 R6 **Sebasticook Lake** ◎ Maine, NE USA
36 M7 **Sebastopol** California, W USA 38.22N 122.50W
Sebastopol *see* Sevastopol'
175 Oo4 **Sebatik, Pulau** *island* N Indonesia
21 R5 **Sebec Lake** ◎ Maine, NE USA
78 K12 **Sébékoro** Kayes, W Mali 13.00N 9.03W
Sebenico *see* Šibenik
42 G6 **Seberi, Cerro** ▲ NW Mexico 27.49N 110.18W
118 H11 **Sebeş** *Ger.* Mühlbach, *Hung.* Szászsebes; *prev.* Sebeşu Săsesc. Alba, W Romania 45.57N 23.34E
Sebes-Körös *see* Crişul Repede
Sebeşu Săsesc *see* Sebeş
33 R8 **Sebewaing** Michigan, N USA 43.43N 83.27W
128 F16 **Sebezh** Pskovskaya Oblast', W Russian Federation 56.19N 28.31E
143 N12 **Şebinkarahisar** Giresun, N Turkey 40.19N 38.25E
118 F11 **Sebiş** *Hung.* Borossebes. Arad, W Romania 46.21N 22.09E
Sebkra Azz el Matti *see* Azzel Matti, Sebkha
21 Q4 **Seboomook Lake** ◎ Maine, NE USA
76 G6 **Sebou** *var.* Sebu. ≈ N Morocco
22 I6 **Sebree** Kentucky, S USA 37.34N 87.30W
25 X13 **Sebring** Florida, SE USA 27.30N 81.26W
Sebta *see* Ceuta
Sebu *see* Sebou
175 Nn11 **Sebuku, Pulau** *island* N Indonesia
175 Oo4 **Sebuku, Teluk** *bay* Borneo, N Indonesia
176 Vv10 **Sebyar** ≈ Irian Jaya, E Indonesia
108 F10 **Secchia** ≈ N Italy
10 L17 **Sechelt** British Columbia, SW Canada 49.25N 123.37W
58 C12 **Sechín, Río** ≈ W Peru
58 A10 **Sechura, Bahía de** *bay* NW Peru
193 A22 **Secretary Island** *island* SW NZ
161 I15 **Secunderābād** *var.* Sikandarabad. Andhra Pradesh, C India 17.30N 78.33E
59 L17 **Sécure, Río** ≈ C Bolivia
120 D10 **Seda** Mažeikiai, NW Lithuania 56.10N 22.04E
49 T5 **Sedalia** Missouri, C USA 38.42N 93.13W
105 R3 **Sedan** Ardennes, N France 49.42N 4.55E
49 P7 **Sedan** Kansas, C USA 37.07N 96.11W
107 N3 **Sedano** Castilla-León, N Spain 42.43N 3.43W
106 H10 **Seda, Ribeira de** *stream* C Portugal
193 K15 **Seddon** Marlborough, South Island, NZ 41.41S 174.04E
193 H15 **Seddonville** West Coast, South Island, NZ 41.34S 171.59E
149 U7 **Sedeh** Khorāsān, E Iran 33.18N 59.12E
144 E11 **Sederot** Southern, S Israel 31.31N 34.34E
67 B23 **Sedge Island** *island* NW Falkland Islands
78 G12 **Sédhiou** SW Senegal 12.39N 15.33W
9 U16 **Sedley** Saskatchewan, S Canada 50.06N 103.51W
Sedlez *see* Siedlce
125 G12 **Sedel'nikovo** Omskaya Oblast', C Russian Federation 56.54N 75.24E
119 Q2 **Sedniv** Chernihivs'ka Oblast', N Ukraine 51.39N 31.34E
38 L11 **Sedona** Arizona, SW USA 34.52N 111.45W
Sedunum *see* Sion
120 F12 **Šeduva** Radviliškis, N Lithuania 55.45N 23.46E
147 Y8 **Seeb** *var.* Muscat Sīb Airport. ✕ (Masqaṭ) NE Oman 23.36N 58.27E
Seeb *see* As Sīb
110 M7 **Seefeld-in-Tirol** Tirol, W Austria 47.19N 11.16E
85 E22 **Seeheim Noord** Karas, S Namibia 26.49S 17.50E
Seeland *see* Sjælland
205 N9 **Seelig, Mount** ▲ Antarctica 81.45S 102.15W
Seeonee *see* Seoni
168 E6 **Seer** Hovd, W Mongolia 48.18N 92.37E
104 L5 **Sées** Orne, N France 48.36N 0.11E
103 J14 **Seesen** Niedersachsen, C Germany 51.54N 10.10E
Seesker Höhe *see* Szeskie Wzgórza
102 J10 **Seevetal** Niedersachsen, N Germany 53.24N 10.01E
111 V6 **Seewiesen** Steiermark, E Austria 47.37N 15.16E
142 J13 **Şefaatli** *var.* Kızılkoca. Yozgat, C Turkey 39.31N 34.45E
155 N3 **Sefid, Darya-ye** *Pash.* Āb-i-Safed. ≈ N Afghanistan
154 K5 **Sefīd Kūh, Selseleh-ye** *Eng.* Paropamisus Range. ▲ W Afghanistan
76 G6 **Sefrou** N Morocco 33.51N 4.49W
193 E19 **Sefton, Mount** ▲ South Island, NZ 43.43S 169.58E
176 U10 **Segaf, Kepulauan** *island group* E Indonesia
175 Oo3 **Segama, Sungai** ≈ East Malaysia
174 Hh6 **Segamat** Johor, Peninsular Malaysia 2.30N 102.48E
79 S13 **Ségbana** NE Benin 10.55N 3.42E
Segestica *see* Sisak
Segesvár *see* Sighişoara
176 Uu9 **Seget** Irian Jaya, E Indonesia 1.21S 131.04E
Segewold *see* Sigulda
128 J9 **Segezha** Respublika Kareliya, NW Russian Federation 63.39N 34.24E
Seghedin *see* Szeged
Segna *see* Senj
109 I16 **Segni** Lazio, C Italy 41.41N 13.02E
Segodunum *see* Rodez
107 S9 **Segorbe** País Valenciano, E Spain 39.51N 0.30W
78 M12 **Ségou** *var.* Segu. Ségou, C Mali 13.25N 6.12W
78 M12 **Ségou** ◆ *region* SW Mali
56 E8 **Segovia** Antioquia, N Colombia 7.06N 74.42W
107 N7 **Segovia** Castilla-León, C Spain 40.57N 4.07W
106 M6 **Segovia** ◆ *province* Castilla-León, N Spain
Segoviao Wangkí *see* Coco, Río
128 J9 **Segozero, Ozero** ◎ NW Russian Federation
107 U5 **Segre** ≈ NE Spain
104 J7 **Segré** Maine-et-Loire, NW France 47.40N 0.51W
Segu *see* Ségou
40 I17 **Seguam Island** *island* Aleutian Islands, Alaska, USA
40 I17 **Seguam Pass** *strait* Aleutian Islands, Alaska, USA
79 Y7 **Séguédine** Agadez, NE Niger 20.12N 13.03E
78 M15 **Séguéla** W Ivory Coast 8.01N 6.38W
27 S11 **Seguin** Texas, SW USA 29.34N 97.58W
40 E17 **Segula Island** *island* Aleutian Islands, Alaska, USA
64 K10 **Segundo, Río** ≈ C Argentina
107 Q12 **Segura** ≈ S Spain
107 P13 **Segura, Sierra de** ▲ S Spain
85 G18 **Sehithwa** Ngamiland, N Botswana 20.28S 22.43E
160 H10 **Sehore** Madhya Pradesh, C India 23.12N 77.07E
195 O16 **Sehulea** Normanby Island, S PNG 9.55S 151.10E
155 P15 **Sehwān** Sind, SE Pakistan 26.27N 67.46E
111 V8 **Seiersberg** Steiermark, SE Austria 47.01N 15.22E
28 L9 **Seiling** Oklahoma, C USA 36.09N 98.55W
105 S9 **Seille** ≈ E France
101 J20 **Seilles** Namur, SE Belgium 50.31N 5.12E
95 K17 **Seinäjoki** *Swe.* Östermyra. Länsi-Suomi, W Finland 62.45N 22.54E
10 B12 **Seine** ≈ Ontario, S Canada
104 M4 **Seine** ≈ N France
104 K4 **Seine, Baie de la** *bay* N France
Seine, Banc de la *see* Seine Seamount
105 O5 **Seine-et-Marne** ◆ *department* N France
104 L3 **Seine-Maritime** ◆ *department* N France
86 B14 **Seine Plain** *undersea feature* E Atlantic Ocean
86 B15 **Seine Seamount** *var.* Banc de la Seine. *undersea feature* E Atlantic Ocean 33.45N 14.25W
104 E6 **Sein, Île de** *island* NW France
176 Y12 **Seinma** Irian Jaya, E Indonesia 4.10S 138.54E
Seisbierrum *see* Sexbierum
111 U5 **Seitenstetten Markt** Niederösterreich, C Austria 48.03N 14.41E
Seiyu *see* Chŏnju
97 H22 **Sejerø** *island* E Denmark
112 P7 **Sejny** Podlaskie, NE Poland 54.09N 23.21E
174 Ii13 **Sekampung, Way** ≈ Sumatera, SW Indonesia
83 G20 **Seke** Shinyanga, N Tanzania 3.16S 33.31E
171 I15 **Seki** Gifu, Honshū, SW Japan 35.25N 136.51E
167 U12 **Sekibi-sho** *island* China/Japan/Taiwan
172 Pp5 **Sekihoku-tōge** *pass* Hokkaidō, NE Japan 43.40N 143.10E
Sekondi *see* Sekondi-Takoradi
79 P17 **Sekondi-Takoradi** *var.* Sekondi. S Ghana 4.55N 1.45W
82 J11 **Sek'ot'a** Amhara, N Ethiopia 12.41N 39.05E
Seksëúil *see* Saksaul'skiy
34 I9 **Selah** Washington, NW USA 46.39N 120.31W
174 Gg5 **Selangor** *var.* Negeri Selangor Darul Ehsan. ◐ *state* Peninsular Malaysia
Selânik *see* Thessaloníki
174 Hh7 **Selapanjang** Pulau Rantau, W Indonesia 1.00N 102.44E
178 Ii10 **Selaphum** Roi Et, E Thailand 16.00N 103.54E
176 Uu16 **Selaru, Pulau** *island* Kepulauan Tanimbar, E Indonesia
176 Vv11 **Selassi** Irian Jaya, E Indonesia 3.16S 132.50E
173 G3 **Selatan, Selat** *strait* Peninsular Malaysia
41 N8 **Selawik** Alaska, USA 66.36N 160.00W
41 N8 **Selawik Lake** ◎ Alaska, USA
175 Pp13 **Selayar, Selat** *strait* Sulawesi, C Indonesia
97 C14 **Selbjørnsfjorden** *fjord* S Norway
96 H8 **Selbusjøen** ◎ S Norway
99 M17 **Selby** N England, UK 53.49N 1.06W
31 N8 **Selby** South Dakota, N USA 45.30N 100.01W
23 Z4 **Selbyville** Delaware, NE USA 38.28N 75.12W
142 B15 **Selçuk** *var.* Akıncılar. İzmir, SW Turkey 37.55N 27.21E
41 Q13 **Seldovia** Alaska, USA 59.26N 151.42W
109 M18 **Sele** *anc.* Silarius. ≈ S Italy
44 B5 **Selegua, Río** ≈ W Guatemala
133 X7 **Selemdzha** ≈ SE Russian Federation
133 U7 **Selenga** *Mong.* Selenge Mörön. ≈ Mongolia/Russian Federation
168 K6 **Selenge** Bulgan, N Mongolia 49.34N 104.18E
168 J6 **Selenge** Hövsgöl, N Mongolia 49.25N 101.30E
81 I19 **Selenge** Bandundu, W Dem. Rep. Congo (Zaire) 1.58S 18.10E
168 L6 **Selenge** ◐ *province* N Mongolia
Selenge Mörön *see* Selenga
126 Jj16 **Selenginsk** Respublika Buryatiya, S Russian Federation 52.00N 106.40E
Selenica *see* Selenicë
115 K22 **Selenicë** *var.* Selenica. Vlorë, SW Albania 40.32N 19.38E
126 M7 **Selennyakh** ≈ NE Russian Federation
102 J8 **Selenter See** ◎ N Germany
Sele Sound *see* Soela Väin
105 U6 **Sélestat** *Ger.* Schlettstadt. Bas-Rhin, NE France 48.16N 7.28E
Selety *see* Sileti
Seleucia *see* Silifke
94 I4 **Selfoss** Sudhurland, SW Iceland 63.56N 20.59W
30 M7 **Selfridge** North Dakota, N USA 46.01N 100.52W
78 I15 **Seli** ≈ N Sierra Leone
78 I11 **Sélibabi** *var.* Sélibaby. Guidimaka, S Mauritania 15.13N 12.10W
Sélibaby *see* Sélibabi
Selidovka/Selidovo *see* Selydove
128 I15 **Seliger, Ozero** ◎ W Russian Federation
38 J11 **Seligman** Arizona, SW USA 35.20N 112.56W
49 S8 **Seligman** Missouri, C USA 36.31N 93.56W
82 E6 **Selima Oasis** *oasis* N Sudan 21.22N 29.19E
78 L13 **Sélingué, Lac de** ⊟ S Mali
Selinoûs *see* Kréstena
20 G14 **Selinsgrove** Pennsylvania, NE USA 40.47N 76.51W
Selishche *see* Syelishcha
128 I16 **Selizharovo** Tverskaya Oblast', W Russian Federation 56.50N 33.24E
96 C10 **Selje** Sogn og Fjordane, S Norway 62.02N 5.22E
9 X16 **Selkirk** Manitoba, S Canada 50.10N 96.52W
98 K13 **Selkirk** SE Scotland, UK 55.35N 2.48W
98 K13 **Selkirk** *cultural region* SE Scotland, UK
9 O16 **Selkirk Mountains** ▲ British Columbia, SW Canada
200 Oo12 **Selkirk Rise** *undersea feature* SE Pacific Ocean
117 F21 **Sellasía** Pelopónnisos, S Greece 37.14N 22.24E
46 M9 **Selle, Pic de la** *var.* La Selle. ▲ SE Haiti 18.18N 71.55W
104 M8 **Selles-sur-Cher** Loir-et-Cher, C France 47.16N 1.31E
38 K16 **Sells** Arizona, SW USA 31.54N 111.52W
Sellye *see* Sal'a
25 P5 **Selma** Alabama, S USA 32.24N 87.01W
37 Q11 **Selma** California, W USA 36.33N 119.37W
22 G10 **Selmer** Tennessee, S USA 35.10N 88.35W
181 N17 **Sel, Pointe au** *headland* W Réunion
Selseleh-ye Kūh-e Vākhān *see* Nicholas Range
131 S2 **Selty** Udmurtskaya Respublika, NW Russian Federation 57.19N 52.09E
Selukwe *see* Shurugwi
64 L9 **Selva** Santiago del Estero, N Argentina 29.46S 62.01W
9 T9 **Selwyn Lake** ◎ Northwest Territories/Saskatchewan, C Canada
8 K6 **Selwyn Mountains** ▲ Yukon Territory, NW Canada
189 T6 **Selwyn Range** ▲ Queensland, C Australia
119 W8 **Selydove** *var.* Selidovka, *Rus.* Selidovo. Donets'ka Oblast', SE Ukraine 48.06N 37.16E
Selzaete *see* Zelzate
Seman *see* Semanit, Lumi i
174 Ii13 **Semangka, Teluk** *bay* Sumatera, SW Indonesia
174 I13 **Semangka, Way** ≈ Sumatera, SW Indonesia
115 D22 **Semanit, Lumi i** *var.* Seman. ≈ W Albania
174 Kk14 **Semarang** *var.* Samarang. Jawa, C Indonesia 6.58S 110.28E
174 Kk6 **Sematan** Sarawak, East Malaysia 1.49N 109.43E
175 Qq17 **Semau, Pulau** *island* S Indonesia
175 Nn8 **Semayang, Danau** ◎ Borneo, N Indonesia
175 O4 **Sembakung, Sungai** ≈ Borneo, N Indonesia
81 G17 **Sembé** La Sangha, NW Congo 1.37N 14.34E
Semberong *see* Semberong, Sungai
174 Hh6 **Semberong, Sungai** *var.* Semberong. ≈ Peninsular Malaysia
174 M10 **Sembulu, Danau** ◎ Borneo, N Indonesia
Semendria *see* Smederevo
119 R1 **Semenivka** Chernihivs'ka Oblast', N Ukraine 52.10N 32.37E
119 S6 **Semenivka** *Rus.* Semenovka. Poltavs'ka Oblast', NE Ukraine 49.36N 33.11E
131 O3 **Semenov** Nizhegorodskaya Oblast', W Russian Federation 56.47N 44.27E
Semenovka *see* Semenivka
174 M16 **Semeru, Gunung** *var.* Mahameru. ▲ Jawa, S Indonesia 8.01S 112.53E
Semey *see* Semipalatinsk
Semezhevo *see* Syemyezhava
130 L7 **Semiluki** Voronezhskaya Oblast', W Russian Federation 51.46N 39.00E
35 W16 **Seminoe Reservoir** ⊟ Wyoming, C USA
49 O11 **Seminole** Oklahoma, C USA 35.13N 96.40W
26 M6 **Seminole** Texas, SW USA 32.43N 102.39W
25 S8 **Seminole, Lake** ⊟ Florida/Georgia, SE USA
150 M8 **Semiozernoye** Kostanay, N Kazakhstan 52.22N 64.06E
151 V9 **Semipalatinsk** *Kaz.* Semey. Vostochnyy Kazakhstan, E Kazakhstan 50.25N 80.16E
149 O9 **Semīrom** *var.* Samirum. Eşfahān, C Iran 31.19N 51.49E
40 F17 **Semisopochnoi Island** *island* Aleutian Islands, Alaska, USA
174 Ll7 **Semitau** Borneo, C Indonesia 0.30N 111.58E
83 E18 **Semliki** ≈ Uganda/Dem. Rep. Congo (Zaire)
149 P5 **Semnān** *var.* Samnān. Semnān, N Iran 35.37N 53.21E
149 Q5 **Semnān** *off.* Ostān-e Semnān. ◐ *province* N Iran
101 K24 **Semois** ≈ SE Belgium
110 E8 **Sempacher See** ◎ C Switzerland
Sena *see* Vila de Sena
32 L12 **Senachwine Lake** ◎ Illinois, N USA
61 O14 **Senador Pompeu** Ceará, E Brazil 5.30S 39.25W
Sena Gallica *see* Senigallia
61 C15 **Sena Madureira** Acre, W Brazil 9.04S 68.40W
161 L25 **Senanayake Samudra** ◎ E Sri Lanka
85 G15 **Senanga** Western, SW Zambia 16.09S 23.16E
49 Y9 **Senath** Missouri, C USA 36.07N 90.09W
24 L2 **Senatobia** Mississippi, S USA 34.37N 89.58W
170 C15 **Sendai** Kagoshima, Kyūshū, SW Japan 31.48N 130.16E
171 M13 **Sendai** Miyagi, Honshū, C Japan 38.16N 140.52E
170 Bb15 **Sendai-gawa** ≈ Kyūshū, SW Japan
171 M14 **Sendai-wan** *bay* E Japan
103 J23 **Senden** Bayern, S Germany 48.18N 10.04E
160 F11 **Sendhwa** Madhya Pradesh, C India 21.38N 75.04E
113 H21 **Senec** *Ger.* Wartberg, *Hung.* Szenc; *prev.* Szempcz. Bratislavský Kraj, W Slovakia 48.14N 17.24E
49 P3 **Seneca** Kansas, C USA 39.47N 96.04W
49 R8 **Seneca** Missouri, C USA 36.50N 94.36W
34 K13 **Seneca** Oregon, NW USA 44.06N 118.57W
23 O11 **Seneca** South Carolina, SE USA 34.41N 82.57W
20 G11 **Seneca Lake** ◎ New York, NE USA
33 U13 **Senecaville Lake** ⊟ Ohio, N USA
78 G11 **Senegal** *off.* Republic of Senegal, *Fr.* Sénégal. ◆ *republic* W Africa
78 H9 **Senegal** *Fr.* Sénégal. ≈ W Africa
33 O4 **Seney Marsh** *wetland* Michigan, N USA
103 P14 **Senftenberg** Brandenburg, E Germany 51.31N 14.01E
84 L11 **Senga Hill** Northern, NE Zambia 9.26S 31.12E
164 G13 **Sênggê Zangbo** ≈ W China
176 Z11 **Senggi** Irian Jaya, E Indonesia 3.26S 140.46E
131 R5 **Sengiley** Ul'yanovskaya Oblast', W Russian Federation 53.54N 48.51E
65 I19 **Senguerr, Río** ≈ S Argentina
85 J16 **Sengwa** ≈ C Zimbabwe
Senia *see* Senj
113 H19 **Senica** *Ger.* Senitz, *Hung.* Szenice. Trnavský Kraj, W Slovakia 48.40N 17.22E
Seniça *see* Sjenica
108 J11 **Senigallia** *anc.* Sena Gallica. Marche, C Italy 43.43N 13.13E
142 F15 **Senirkent** Isparta, SW Turkey 38.07N 30.34E
Senitz *see* Senica
114 C10 **Senj** *Ger.* Zengg, *It.* Segna; *anc.* Senia. Lika-Senj, NW Croatia 44.58N 14.55E
169 O10 **Senj** Dornogovĭ, SE Mongolia 44.34N 110.58E
94 H9 **Senja** *prev.* Senjen. *island* N Norway
Senjen *see* Senja
167 U12 **Senkaku-shotō** *island group* SW Japan
143 R12 **Şenkaya** Erzurum, NE Turkey 40.33N 42.16E
85 I16 **Senkobo** Southern, S Zambia 17.34S 25.57E
105 O4 **Senlis** Oise, N France 49.13N 2.33E
178 K13 **Senmonorom** Môndól Kiri, E Cambodia 12.27N 107.12E
82 G10 **Sennar** *var.* Sannār. Sinnar, C Sudan 13.31N 33.37E
Senno *see* Syanno
Senones *see* Sens
111 W11 **Senovo** E Slovenia 46.01N 15.24E
105 P6 **Sens** *anc.* Agendicum, Senones. Yonne, C France 48.12N 3.16E
Sensburg *see* Mrągowo
178 J12 **Sên, Stœng** ≈ C Cambodia
44 F7 **Sensuntepeque** Cabañas, NE El Salvador 13.52N 88.37W
114 L8 **Senta** *Hung.* Zenta. Serbia, N Yugoslavia 45.57N 20.04E
Šent Andraž *see* Sankt Andrä
176 Z10 **Sentani, Danau** ◎ Irian Jaya, E Indonesia
30 J5 **Sentinel Butte** ▲ North Dakota, N USA 46.52N 103.50W
8 M13 **Sentinel Peak** ▲ British Columbia, W Canada 54.51N 122.02W
61 N16 **Sento Sé** Bahia, E Brazil 9.51S 41.56W
Šent Peter *see* Pivka
Šent Vid *see* Sankt Veit an der Glan
194 E10 **Senu** ≈ NW PNG
Seo de Urgel *see* La See d'Urgel
160 I7 **Seondha** Madhya Pradesh, C India 26.09N 78.46E
160 J11 **Seoni** *prev.* Seeonee. Madhya Pradesh, C India 22.07N 79.33E
Seoul *see* Sŏul
192 I13 **Separation Point** *headland* South Island, NZ 40.46S 172.58E
175 O7 **Sepasu** Borneo, N Indonesia 0.44N 117.38E
194 F10 **Sepik** ≈ Indonesia/PNG
Sepone *see* Muang Xépôn
112 M7 **Sępopol** *Ger.* Schippenbeil. Warmińsko-Mazurskie, NE Poland, 54.16N 21.09E
118 F10 **Şepreuş** *Hung.* Seprős. Arad, W Romania 46.34N 21.44E
Seprős *see* Şepreuş
Şepşi-Sângeorz/Sepsiszentgyörgy *see* Sfântu Gheorghe
13 W4 **Sept-Îles** Quebec, SE Canada 50.11N 66.19W
107 N6 **Sepúlveda** Castilla-León, N Spain 41.18N 3.45W
174 Ii12 **Seputih, Way** ≈ Sumatera, SW Indonesia
106 K8 **Sequeros** Castilla-León, N Spain 40.31N 6.04W
106 L5 **Sequillo** ≈ NW Spain
34 G7 **Sequim** Washington, NW USA 48.04N 123.06W
37 S11 **Sequoia National Park** *national park* California, W USA
143 Q14 **Şerafettin Dağları** ▲ E Turkey
131 N10 **Serafimovich** Volgogradskaya Oblast', SW Russian Federation 49.34N 42.43E
175 R6 **Serai** Sulawesi, N Indonesia 1.45N 124.58E
101 K19 **Seraing** Liège, E Belgium 50.37N 5.31E
Séraitang *see* Baima
Serajgonj *see* Shirajganj Ghat
152 I15 **Serakhs** *var.* Saragt. Akhalskiy Velayat, S Turkmenistan 36.33N 61.10E
176 X10 **Serami** Irian Jaya, E Indonesia 2.11S 136.46E
Seram, Laut *see* Ceram Sea
Serampore/Serampur *see* Shrīrāmpur
175 Tt11 **Seram, Pulau** *var.* Serang, *Eng.* Ceram. *island* Maluku, E Indonesia
174 J14 **Serang** Jawa, C Indonesia 6.07S 106.09E
Serang *see* Seram, Pulau
174 Kk6 **Serasan, Pulau** *island* Kepulauan Natuna, W Indonesia
174 Kk6 **Serasan, Selat** *strait* Indonesia/Malaysia
114 M12 **Serbia** *Ger.* Serbien, *Serb.* Srbija. ◐ *republic* Yugoslavia
Serbien *see* Serbia
Sercq *see* Sark
Serdica *see* Sofiya
131 O7 **Serdobsk** Penzenskaya Oblast', W Russian Federation 52.30N 44.16E
151 X9 **Serebryansk** Vostochnyy Kazakhstan, E Kazakhstan 49.43N 83.16E
126 Ll13 **Serebryanyy Bor** Respublika Sakha (Yakutiya), NE Russian Federation 56.40N 124.46E
113 H20 **Sered'** *Hung.* Szered. Trnavský Kraj, SW Slovakia 48.17N 17.44E
119 S1 **Seredyna-Buda** Sums'ka Oblast', NE Ukraine 52.09N 34.00E
120 E13 **Seredžius** Jurbarkas, C Lithuania 55.04N 23.24E
142 I14 **Şereflikoçhisar** Ankara, C Turkey 38.55N 33.31E
108 D7 **Seregno** Lombardia, N Italy 45.39N 9.12E
105 P7 **Serein** ≈ C France
174 H5 **Seremban** Negeri Sembilan, Peninsular Malaysia 2.42N 101.54E
83 H20 **Serengeti Plain** *plain* N Tanzania
84 K13 **Serenje** Central, E Zambia 13.12S 30.15E
Seres *see* Sérres
118 J5 **Seret** ≈ W Ukraine
Seret/Sereth *see* Siret
117 I21 **Serfopoúla** *island* Kykládes, Greece, Aegean Sea
131 P4 **Sergach** Nizhegorodskaya Oblast', W Russian Federation 55.31N 45.29E
31 S13 **Sergeant Bluff** Iowa, C USA 42.24N 96.19W
169 P7 **Sergelen** Dornod, NE Mongolia 48.31N 114.01E
169 O9 **Sergelen** Sühbaatar, E Mongolia 46.12N 111.48E
173 F4 **Sergeulangit, Pegunungan** ▲ Sumatera, NW Indonesia
126 I4 **Sergeya Kirova, Ostrova** *island* N Russian Federation
Sergeyevichi *see* Syarhyeyevichy
151 O7 **Sergeyevka** Severnyy Kazakhstan, N Kazakhstan 53.51N 67.17E
Sergiopol *see* Ayagoz
61 P16 **Sergipe** *off.* Estado de Sergipe. ◐ *state* E Brazil
130 L3 **Sergiyev Posad** Moskovskaya Oblast', W Russian Federation 56.21N 38.10E
128 K5 **Sergozero, Ozero** ◎ NW Russian Federation
174 L7 **Serian** Sarawak, East Malaysia 1.10N 110.34E
174 J13 **Seribu, Kepulauan** *island group* S Indonesia
117 I21 **Sérifos** *anc.* Seriphos. *island* Kykládes, Greece, Aegean Sea
117 I21 **Sérifou, Stenó** *strait* SE Greece
142 F16 **Serik** Antalya, SW Turkey 36.55N 31.06E
108 E7 **Serio** ≈ N Italy
Seriphos *see* Sérifos
Serir Tibesti *see* Sarīr Tībistī
131 S5 **Sernovodsk** Samarskaya Oblast', W Russian Federation 53.56N 51.16E
131 R2 **Sernur** Respublika Mariy El, W Russian Federation 56.55N 49.09E
112 M11 **Serock** Mazowieckie, C Poland 52.30N 21.03E
63 B18 **Serodino** Santa Fe, C Argentina 32.33S 60.52W
Seroei *see* Serui
107 P14 **Serón** Andalucía, S Spain 37.20N 2.28W
101 E14 **Serooskerke** Zeeland, SW Netherlands 51.42N 3.52E
107 T6 **Seròs** Cataluña, NE Spain 41.27N 0.24E
125 F10 **Serov** Sverdlovskaya Oblast', C Russian Federation 59.42N 60.31E
85 I19 **Serowe** Central, SE Botswana 22.25S 26.43E
106 H13 **Serpa** Beja, S Portugal 37.55N 7.36W
Serpa Pinto *see* Menongue
190 A4 **Serpentine Lakes** *salt lake* South Australia
47 T15 **Serpent's Mouth, The** *Sp.* Boca de la Serpiente. *strait* Trinidad and Tobago/Venezuela
Serpiente, Boca de la *see* Serpent's Mouth, The
130 K4 **Serpukhov** Moskovskaya Oblast', W Russian Federation 54.54N 37.25E
62 K13 **Serra do Mar** ▲ S Brazil
Sérrai *see* Sérres
109 N22 **Serra San Bruno** Calabria, SW Italy 38.32N 16.18E
105 S14 **Serres** Hautes-Alpes, SE France 44.26N 5.42E
116 H13 **Sérres** *var.* Seres; *prev.* Sérrai. Kentrikí Makedonía, NE Greece 41.04N 23.34E
64 J9 **Serrezuela** Córdoba, C Argentina 30.37S 65.25W
61 O16 **Serrinha** Bahia, E Brazil 11.37S 38.55W
61 M19 **Serro** *var.* Sêrro. Minas Gerais, NE Brazil 18.37S 43.22W
Sêrro *see* Serro
Sert *see* Siirt
Sertá *see* Sertã
106 H9 **Sertã** *var.* Sertá. Castelo Branco, C Portugal 39.48N 8.04W
62 L8 **Sertãozinho** São Paulo, S Brazil 21.04S 47.55W
166 F7 **Sêrtar** Sichuan, C China 32.18N 100.18E
176 X10 **Serui** *prev.* Seroei. Irian Jaya, E Indonesia 1.52S 136.15E
85 J19 **Serule** Central, E Botswana 21.58S 27.18E
174 Ll10 **Seruyan, Sungai** *var.* Sungai Pembuang. ≈ Borneo, N Indonesia
117 E14 **Sérvia** Dytikí Makedonía, N Greece 40.12N 22.01E
166 E7 **Sêrxü** Sichuan, C China 32.54N 98.06E
126 Mm15 **Seryshevo** Amurskaya Oblast', SE Russian Federation 51.03N 128.16E
Sesana *see* Sežana
175 Nn5 **Sesayap, Sungai** ≈ Borneo, N Indonesia
Sesdlets *see* Siedlce
81 N17 **Sese** Orientale, N Dem. Rep. Congo (Zaire) 2.13N 25.52E
83 F18 **Sese Islands** *island group* S Uganda
175 T9 **Sesepe** Pulau Obi, E Indonesia 1.26S 127.55E
85 H16 **Sesheke** *var.* Sesheko. Western, SE Zambia 17.27S 24.19E
Sesheko *see* Sesheke
108 C8 **Sesia** *anc.* Sessites. ≈ NW Italy
106 F11 **Sesimbra** Setúbal, S Portugal 38.25N 9.06W
117 N22 **Seskliō** *island* Dodekánisos, Greece, Aegean Sea
32 L16 **Sesser** Illinois, N USA 38.05N 89.03W
Sessites *see* Sesia
108 G11 **Sesto Fiorentino** Toscana, C Italy 43.49N 11.12E
108 E7 **Sesto San Giovanni** Lombardia, N Italy 45.31N 9.13E
108 A8 **Sestriere** Piemonte, NE Italy 45.00N 6.54E
108 D10 **Sestri Levante** Liguria, NW Italy 44.16N 9.22E
109 C20 **Sestu** Sardegna, Italy, C Mediterranean Sea 39.15N 9.06E
114 E8 **Sesvete** Zagreb, N Croatia 45.50N 16.03E
120 G12 **Šeta** Kėdainiai, C Lithuania 55.17N 24.16E
Setabis *see* Xàtiva
172 N5 **Setana** Hokkaidō, NE Japan 42.27N 139.52E
105 Q16 **Sète** *prev.* Cette. Hérault, S France 43.24N 3.42E
60 J11 **Sete Ilhas** Amapá, NE Brazil 1.06N 52.06W
61 L20 **Sete Lagoas** Minas Gerais, NE Brazil 19.28S 44.15W
62 G10 **Sete Quedas, Ilha das** *island* S Brazil
94 I10 **Setermoen** Troms, N Norway 68.51N 18.19E
97 E17 **Setesdal** *valley* S Norway
45 W16 **Setetule, Cerro** ▲ SE Panama 7.51N 77.37W
23 Q5 **Seth** West Virginia, NE USA 38.06N 81.40W
Setia *see* Sezze
76 K5 **Sétif** *var.* Stif. N Algeria 36.10N 5.24E
171 I15 **Seto** Aichi, Honshū, SW Japan 35.13N 137.03E
170 F14 **Seto-naikai** *Eng.* Inland Sea. *sea* S Japan
172 Qq13 **Setouchi** *var.* Setoushi. Kagoshima, Amami-Ō-shima, SW Japan 44.19N 142.58E
76 F6 **Settat** W Morocco 33.03N 7.37W
81 D20 **Setté Cama** Ogooué-Maritime, SW Gabon 2.31S 9.46E
9 W13 **Setting Lake** ◎ Manitoba, C Canada
99 L16 **Settle** N England, UK 54.04N 2.17W
201 Y12 **Settlement** E Wake Island 19.16N 166.37E
106 F11 **Setúbal** *Eng.* Saint Ubes, Saint Yves. Setúbal, W Portugal 38.31N 8.54W
106 F11 **Setúbal** ◐ *district* S Portugal
106 F12 **Setúbal, Baía de** *bay* W Portugal
Setul *see* Satun
10 B10 **Seul, Lac** ◎ Ontario, S Canada
105 R8 **Seurre** Côte d'Or, C France 47.00N 5.09E
143 U11 **Sevan** C Armenia 40.31N 44.55E
143 V12 **Sevana Lich** *Eng.* Lake Sevan, *Rus.* Ozero Sevan. ◎ E Armenia
Sevan, Lake/Sevan, Ozero *see* Sevana Lich
79 N11 **Sévaré** Mopti, C Mali 14.30N 4.08W
119 S14 **Sevastopol'** *Eng.* Sebastopol. Respublika Krym, S Ukraine 44.36N 33.33E
27 R14 **Seven Sisters** Texas, SW USA 27.57N 98.34W
8 K13 **Seven Sisters Peaks** ▲ British Columbia, SW Canada 54.57N 128.10W
101 M15 **Sevenum** Limburg, SE Netherlands 51.25N 6.01E
105 P14 **Séverac-le-Château** Aveyron, S France 44.18N 3.03E
12 H13 **Severn** ≈ Ontario, S Canada
99 L21 **Severn** *Wel.* Hafren. ≈ England/Wales, UK
129 O11 **Severnaya Dvina** *var.* Northern Dvina. ≈ NW Russian Federation
131 N16 **Severnaya Osetiya-Alaniya, Respublika** *Eng.* North Ossetia; *prev.* Respublika Severnaya Osetiya, Severo-Osetinskaya SSR. ◐ *autonomous republic* SW Russian Federation
Severnaya Osetiya, Respublika *see* Severnaya Osetiya-Alaniya Respublika
125 F9 **Severnaya Sos'va** ≈ N Russian Federation
127 N17 **Severnaya Zemlya** *var.* Nicholas II Land. *island group* N Russian Federation
131 T5 **Severnoye** Orenburgskaya Oblast', W Russian Federation 54.03N 52.31E
37 S3 **Severn Troughs Range** ▲ Nevada, W USA
129 W3 **Severnyy** Respublika Komi, NW Russian Federation 67.37N 64.12E
150 I13 **Severnyy Chink Ustyurta** ▲ W Kazakhstan
129 Q13 **Severnyy Uval** *var.* Northern Ural Hills. *hill range* NW Russian Federation
151 O6 **Severnyy Kazakhstan** *off.* Severo-Kazakhstanskaya Oblast', *var.* North Kazakhstan, *Kaz.* Soltüstik Qazaqstan Oblysy. ◐ *province* N Kazakhstan
129 V9 **Severnyy Ural** ▲ NW Russian Federation
Severo-Alichurskiy Khrebet *see* Alichuri Shimolí, Qatorkŭhi
126 K14 **Severobaykal'sk** Respublika Buryatiya, S Russian Federation 55.39N 109.17E
Severodonetsk *see* Syeverodonets'k

◆ Country ● Country Capital ◇ Dependent Territory ○ Dependent Territory Capital ◐ Administrative Region ✕ International Airport ▲ Mountain ▲ Mountain Range Volcano ≈ River ◎ Lake ⊟ Reservoir

128 M8 **Severodvinsk** *prev.* Molotov, Sudostroy. Arkhangel'skaya Oblast', NW Russian Federation 64.31N 39.50E
Severo-Kazakhstanskaya Oblast' *see* Severnyy Kazakhstan
127 Pp13 **Severo-Kuril'sk** Sakhalinskaya Oblast', SE Russian Federation 50.38N 155.57E
128 J3 **Severomorsk** Murmanskaya Oblast', NW Russian Federation 69.00N 33.15E
Severo-Osetinskaya SSR *see* Severnaya Osetiya-Alaniya, Respublika
126 J6 **Severo-Sibirskaya Nizmennost'** *var.* North Siberian Plain, *Eng.* North Siberian Lowland. *lowlands* N Russian Federation
125 Ee10 **Severoural'sk** Sverdlovskaya Oblast', C Russian Federation 60.09N 59.58E
126 J12 **Severo-Yeniseyskiy** Krasnoyarskiy Kray, C Russian Federation 60.29N 93.13E
130 M11 **Severskiy Donets** *Ukr.* Sivers'kyy Donets'. River Russian Federation/Ukraine *see also* Sivers'kyy Donets'
94 M9 **Sevettijärvi** Lappi, N Finland 69.31N 28.40E
38 M5 **Sevier Bridge Reservoir** Reservoir Utah, W USA
38 J4 **Sevier Desert** *plain* Utah, W USA
38 J5 **Sevier Lake** Lake Utah, W USA
23 N9 **Sevierville** Tennessee, S USA 35.52N 83.33W
106 J14 **Sevilla** *Eng.* Seville; *anc.* Hispalis. Andalucía, SW Spain 37.24N 5.58W
106 J13 **Sevilla** Administrative region *province* Andalucía, SW Spain
Sevilla de Niefang *see* Niefang
45 O16 **Sevilla, Isla** *island* SW Panama
Seville *see* Sevilla
116 J9 **Sevlievo** Gabrovo, N Bulgaria 43.01N 25.07E
Sevluš/Sevlyush *see* Vynohradiv
111 V11 **Sevnica** *Ger.* Lichtenwald. E Slovenia 46.00N 15.20E
130 J7 **Sevsk** Bryanskaya Oblast', W Russian Federation 52.03N 34.31E
78 J15 **Sewa** River E Sierra Leone
41 R12 **Seward** Alaska, USA 60.06N 149.26W
31 R15 **Seward** Nebraska, C USA 40.52N 97.06W
8 G8 **Seward Glacier** *glacier* Yukon Territory, W Canada
207 Q3 **Seward Peninsula** *peninsula* Alaska, USA
Seward's Folly *see* Alaska
64 H12 **Sewell** Libertador, C Chile 34.03S 70.16W
100 K5 **Sexbierum** *Fris.* Seisbierrum. Friesland, N Netherlands 53.13N 5.28E
9 O13 **Sexsmith** Alberta, W Canada 55.18N 118.45W
43 W13 **Seybaplaya** Campeche, SE Mexico 19.39N 90.36W
181 N6 **Seychelles** *off.* Republic of Seychelles. Country *republic* W Indian Ocean
69 Z9 **Seychelles** *island group* NE Seychelles
181 N6 **Seychelles Bank** *var.* Le Banc des Seychelles. *undersea feature* W Indian Ocean
Seychelles, Le Banc des *see* Seychelles Bank
180 H17 **Seychellois, Morne** Mountain Mahé, NE Seychelles
94 L2 **Seydhisfjördhur** Austurland, E Iceland 65.15N 14.00W
152 J12 **Seydi** *prev.* Neftezavodsk. Lebapskiy Velayat, E Turkmenistan 39.30N 62.52E
142 G16 **Seydişehir** Konya, SW Turkey 37.25N 31.51E
142 J13 **Seyfe Gölü** Lake C Turkey
Seyhan *see* Adana
142 K16 **Seyhan Barajı** Reservoir S Turkey
142 K17 **Seyhan Nehri** River S Turkey
142 F13 **Seyitgazi** Eskişehir, W Turkey 39.27N 30.42E
130 J7 **Seym** River W Russian Federation
119 S3 **Seym** River N Ukraine
127 O9 **Seymchan** Magadanskaya Oblast', E Russian Federation 62.54N 152.27E
116 N12 **Seymen** Tekirdağ, NW Turkey 41.06N 27.56E
191 O11 **Seymour** Victoria, SE Australia 37.01S 145.10E
85 I25 **Seymour** Eastern Cape, S South Africa 32.33S 26.48E
31 W16 **Seymour** Iowa, C USA 40.40N 93.07W
49 U7 **Seymour** Missouri, C USA 37.09N 92.46W
27 Q5 **Seymour** Texas, SW USA 33.35N 99.15W
116 M12 **Şeytan Deresi** River NW Turkey
111 S12 **Sežana** *It.* Sesana. SW Slovenia 45.42N 13.52E
105 P5 **Sézanne** Marne, N France 48.43N 3.41E
109 I16 **Sezze** *anc.* Setia. Lazio, C Italy 41.28N 13.04E
117 H25 **Sfákia** Kríti, Greece, E Mediterranean Sea 35.12N 24.05E
117 D21 **Sfaktiria** *island* S Greece
118 J11 **Sfântu Gheorghe** *Ger.* Sankt-Georgen, *Hung.* Sepsiszentgyörgy; *prev.* Şepşi-Sângeorz, Sfintu Gheorghe. Covasna, C Romania 45.52N 25.49E
119 N13 **Sfântu Gheorghe, Braţul** *var.* Gheorghe Braţul. River E Romania
77 N6 **Sfax** *Ar.* Şafāqis. E Tunisia 34.45N 10.45E
77 N6 **Sfax** International airport E Tunisia 34.43N 10.37E
Sfintu Gheorghe *see* Sfântu Gheorghe
100 H13 **'s-Gravendeel** Zuid-Holland, SW Netherlands 51.48N 4.36E
100 F11 **'s-Gravenhage** *var.* Den Haag, *Eng.* The Hague, *Fr.* La Haye. Country capital (Netherlands-seat of government) Zuid-Holland, W Netherlands 52.07N 4.16E
100 G12 **'s-Gravenzande** Zuid-Holland, W Netherlands 52.00N 4.10E
Shaan/Shaanxi Sheng *see* Shaanxi
165 X11 **Shaanxi** *var.* Shaan, Shaanxi Sheng, Shan-hsi, Shenshi, Shensi. Administrative region *province* C China
Shaartuz *see* Shahrtuz
Shabani *see* Zvishavane
83 N17 **Shabeellaha Dhexe** *off.* Gobolka Shabeellaha Dhexe. Administrative region *region* E Somalia
83 L17 **Shabeellaha Hoose** *off.* Gobolka Shabeellaha Hoose. Administrative region *region* S Somalia
Shabeelle, Webi *see* Shebeli
116 O7 **Shabla** Dobrich, NE Bulgaria 43.33N 28.31E
116 O7 **Shabla, Nos** *headland* NE Bulgaria 43.30N 28.36E
11 N9 **Shabogama Lake** Lake Newfoundland, E Canada
81 N20 **Shabunda** Sud Kivu, E Dem. Rep. Congo (Zaire) 2.42S 27.19E
147 Q15 **Shabwah** C Yemen 15.09N 46.46E
164 F8 **Shache** *var.* Yarkant. Xinjiang Uygur Zizhiqu, NW China 38.27N 77.16E
Shacheng *see* Huailai
205 R12 **Shackleton Coast** *physical region* Antarctica
205 Z10 **Shackleton Ice Shelf** *ice shelf* Antarctica
Shaddādī *see* Ash Shadādah
30 K7 **Shadehill Reservoir** Reservoir South Dakota, N USA
125 Ee12 **Shadrinsk** Kurganskaya Oblast', C Russian Federation 56.08N 63.18E
33 O12 **Shafer, Lake** Reservoir Indiana, N USA
37 R13 **Shafter** California, W USA 35.27N 119.15W
26 J11 **Shafter** Texas, SW USA 29.49N 104.18W
99 L23 **Shaftesbury** S England, UK 51.01N 2.12W
193 F22 **Shag** River South Island, NZ
151 V9 **Shagan** River E Kazakhstan
41 O11 **Shageluk** Alaska, USA 62.40N 159.33W
126 I16 **Shagonar** Respublika Tyva, S Russian Federation 51.31N 93.06E
193 F22 **Shag Point** *headland* South Island, NZ 45.28S 170.50E
150 J12 **Shagyray, Plato** *plain* SW Kazakhstan
Shāhābād *see* Eslāmābād
174 H5 **Shah Alam** Selangor, Peninsular Malaysia 3.01N 101.31E
119 O12 **Shahany, Ozero** Lake SW Ukraine
144 H9 **Shahbā'** *anc.* Philippopolis. As Suwaydā', S Syria 32.49N 36.37E
Shahbān *see* Ad Dayr
155 P17 **Shāhbandar** Sind, SE Pakistan 23.59N 67.54E
155 P13 **Shāhdād Kot** Sind, SW Pakistan 27.49N 67.49E
149 T10 **Shahdād, Namakzār-e** *salt pan* E Iran
155 Q15 **Shāhdādpur** Sind, SE Pakistan 25.55N 68.40E
160 K10 **Shahdol** Madhya Pradesh, C India 23.19N 81.25E
167 N7 **Sha He** River C China
Shahepu *see* Linze
159 N13 **Shāhganj** Uttar Pradesh, N India 26.03N 82.40E
158 C11 **Shāhgarh** Rājasthān, NW India 27.07N 69.55E
Sha Hi *see* Orūmīyeh, Daryācheh-ye, Iran
Shāhī *see* Qā'emshahr, Māzandarān, Iran
145 Q6 **Shāhimah** *var.* Shahma. C Iraq 34.21N 42.19E
Shahjahanabad *see* Delhi
158 L11 **Shāhjahānpur** Uttar Pradesh, N India 27.52N 79.55E
Shahma *see* Shāhimah
155 U7 **Shāhpur** Punjab, E Pakistan 32.15N 72.31E
Shāhpur *see* Shāhpur Chākar
158 G13 **Shāhpura** Rājasthān, N India 25.37N 75.01E
155 Q15 **Shāhpur Chākar** *var.* Shāhpur. Sind, SE Pakistan 26.09N 68.40E
154 M5 **Shahrak** Ghowr, C Afghanistan 34.09N 64.16E
149 Q11 **Shahr-e Bābak** Kermān, C Iran 30.07N 55.04E
149 N8 **Shahr-e Kord** *var.* Shahr Kord. Chahār Maḩall va Bakhtīārī, C Iran 32.19N 50.52E
149 O9 **Shahreẕā** *var.* Qomisheh, Qumisheh, Shahriza; *prev.* Qomsheh. Eşfahān, C Iran 32.01N 51.51E
153 S10 **Shahrikhon** *Rus.* Shakhrikhan. Andijon Wiloyati, E Uzbekistan 40.42N 72.03E
153 N12 **Shahrisabz** *Rus.* Shakhrisabz. Qashqadaryo Wiloyati, S Uzbekistan 39.01N 66.45E
153 P11 **Shahriston** *Rus.* Shakhristan. NW Tajikistan 39.45N 68.47E
Shahriza *see* Shahreẕā
Shahr-i-Zabul *see* Zābol
Shahr Kord *see* Shahr-e Kord
153 P14 **Shahrtuz** *Rus.* Shaartuz. SW Tajikistan 37.13N 68.05E
149 Q4 **Shāhrūd** *prev.* Emāmrūd, Emāmshahr. Semnān, N Iran 36.30N 54.59E
Shahsavār/Shahsawar *see* Tonekābon
Shaidara *see* Step' Nardara
Shaikh Ābid *see* Shaykh 'Ābid
Shaikh Fāris *see* Shaykh Fāris
Shaikh Najm *see* Shaykh Najm
144 K5 **Shā'īr, Jabal** Mountain C Syria 34.51N 37.49E
160 G10 **Shājāpur** Madhya Pradesh, C India 23.27N 76.21E
82 J8 **Shakal, Ras** *headland* NE Sudan 18.04N 38.34E
Shakhdarinskiy Khrebet *see* Shokhdara, Qatorkŭhi
Shakhrikhan *see* Shahrikhon
Shakhrisabz *see* Shahrisabz
Shakhristan *see* Shahriston
119 X8 **Shakhtars'k** *Rus.* Shakhtërsk. Donets'ka Oblast', SE Ukraine 48.02N 38.18E
127 O15 **Shakhtersk** Ostrov Sakhalin, Sakhalinskaya Oblast', SE Russian Federation 49.10N 142.09E
Shakhtërsk *see* Shakhtars'k
151 R10 **Shakhtinsk** Karaganda, C Kazakhstan 49.40N 72.37E
130 L11 **Shakhty** Rostovskaya Oblast', SW Russian Federation 47.45N 40.14E
131 P2 **Shakhun'ya** Nizhegorodskaya Oblast', W Russian Federation 57.42N 46.36E
79 S15 **Shaki** Oyo, W Nigeria 8.37N 3.25E
83 J15 **Shakiso** Oromo, C Ethiopia 5.33N 38.48E
119 X8 **Shakmars'k** Donets'ka Oblast', E Ukraine 48.04N 38.22E
31 V9 **Shakopee** Minnesota, N USA 44.48N 93.31W
172 Nn5 **Shakotan-hantō** *peninsula* Hokkaidō, NE Japan
172 O4 **Shakotan-misaki** *headland* Hokkaidō, NE Japan 43.22N 140.28E
41 N9 **Shaktoolik** Alaska, USA 64.19N 161.05W
83 J14 **Shala Hāyk'** Lake C Ethiopia
128 M10 **Shalakusha** Arkhangel'skaya Oblast', NW Russian Federation 62.16N 40.16E
151 U8 **Shalday** Pavlodar, NE Kazakhstan 51.57N 78.51E
131 P16 **Shali** Chechenskaya Respublika, SW Russian Federation 43.03N 45.55E
147 W12 **Shalīm** *var.* Shelim. S Oman 18.07N 55.39E
Shaliuhe *see* Gangca
150 F9 **Shalkar, Ozero** *prev.* Chelkar, Ozero. Lake W Kazakhstan
23 V12 **Shallotte** North Carolina, SE USA 33.58N 78.21W
27 N5 **Shallowater** Texas, SW USA 33.41N 102.00W
128 K11 **Shal'skiy** Respublika Kareliya, NW Russian Federation 61.45N 36.02E
166 F9 **Shaluli Shan** Mountain range C China
83 F22 **Shama** River C Tanzania
9 Z11 **Shamattawa** Manitoba, C Canada 55.52N 92.04W
10 F8 **Shamattawa** River Ontario, C Canada
Shām, Bādiyat ash *see* Syrian Desert
Shamiya *see* Ash Shāmīyah
147 X8 **Shām, Jabal ash** *var.* Jebel Sham. Mountain NW Oman 23.21N 57.08E
Sham, Jebel *see* Shām, Jabal ash
Shamkhor *see* Şämkir
20 G14 **Shamokin** Pennsylvania, NE USA 40.47N 76.33W
27 P2 **Shamrock** Texas, SW USA 35.12N 100.15W
Sha'nabi, Jabal ash *see* Chambi, Jebel
145 Y12 **Shanāwah** E Iraq 30.57N 47.25E
165 T8 **Shandan** Gansu, N China 38.43N 101.12E
Shandi *see* Shendi
167 Q5 **Shandong** *var.* Lu, Shandong Sheng, Shantung. Administrative region *province* E China
167 R4 **Shandong Bandao** *var.* Shantung Peninsula. *peninsula* E China
Shandong Peninsula *see* Shandong Bandao
Shandong Sheng *see* Shandong
145 U8 **Shandrūkh** E Iraq 33.20N 45.19E
85 J17 **Shangani** River W Zimbabwe
167 O15 **Shangchuan Dao** *island* S China
Shangchuankou *see* Minhe
169 P12 **Shangdu** Nei Mongol Zizhiqu, N China 41.32N 113.33E
167 O11 **Shanggao** Jiangxi, S China 28.16N 114.52E
167 S8 **Shanghai** *var.* Shang-hai. Shanghai Shi, E China 31.13N 121.28E
Shanghai *see* Shanghai Shi
167 S8 **Shanghai Shi** *var.* Hu, Shanghai. Administrative region *municipality* E China
167 P13 **Shanghang** Fujian, SE China 25.02N 116.21E
166 K14 **Shanglin** Guangxi Zhuangzu Zizhiqu, S China 23.25N 108.31E
85 G15 **Shangombo** Western, W Zambia 16.21S 22.12E
167 O6 **Shangqiu** *var.* Zhuji. Henan, C China 34.29N 115.39E
167 Q10 **Shangrao** Jiangxi, S China 28.27N 117.57E
167 S9 **Shangyu** *var.* Baiguan. Zhejiang, SE China 30.03N 120.52E
169 X9 **Shangzhi** Heilongjiang, NE China 45.11N 127.58E
166 L7 **Shangzhou** *var.* Shang Xian. Shaanxi, C China 33.51N 109.55E
169 W9 **Shanhetun** Heilongjiang, NE China 44.42N 127.12E
Shan-hsi *see* Shaanxi, China
Shan-hsi *see* Shanxi, China
165 O6 **Shankou** Xinjiang Uygur Zizhiqu, W China 42.01N 94.07E
192 M13 **Shannon** Manawatu-Wanganui, North Island, NZ 40.33S 175.25E
99 B19 **Shannon** International airport W Ireland 52.42N 8.57W
99 C17 **Shannon** *Ir.* An tSionainn. River W Ireland
178 H6 **Shan Plateau** *plateau* E Myanmar
164 M6 **Shanshan** *var.* Piqan. Xinjiang Uygur Zizhiqu, NW China 42.53N 90.18E
Shansi *see* Shanxi
178 Gg5 **Shan State** Administrative region *state* E Myanmar
Shantar Islands *see* Shantarskiye Ostrova
127 N13 **Shantarskiye Ostrova** *Eng.* Shantar Islands. *island group* E Russian Federation
167 Q14 **Shantou** *var.* Shan-t'ou, Swatow. Guangdong, S China 23.22N 116.39E
Shantung *see* Shandong
Shantung Peninsula *see* Shandong Bandao
169 O14 **Shanxi** *var.* Jin, Shan-hsi, Shansi, Shanxi Sheng. Administrative region *province* C China
Shan Xian *see* Sanmenxia
167 P6 **Shanxian** *var.* Shan Xian. Shandong, E China 34.51N 116.05E
Shanxi Sheng *see* Shanxi
166 L7 **Shanyang** Shaanxi, C China 33.35N 109.48E
167 O13 **Shaoguan** *var.* Shao-kuan, *Cant.* Kukong; *prev.* Ch'u-chiang. Guangdong, S China 24.56N 113.37E
Shao-kuan *see* Shaoguan
167 Q11 **Shaowu** Fujian, SE China 27.24N 117.26E
167 S9 **Shaoxing** Zhejiang, SE China 30.01N 120.34E
166 M12 **Shaoyang** *prev.* Tangdukou. Hunan, S China 26.54N 111.14E
166 M11 **Shaoyang** *var.* Baoqing, Shao-yang; *prev.* Pao-king. Hunan, S China 27.18N 111.33E
98 K5 **Shapinsay** *island* NE Scotland, UK
129 S4 **Shapkina** River NW Russian Federation
Shapūr *see* Salmās
164 M4 **Shaqiuhe** Xinjiang Uygur Zizhiqu, W China 45.00N 88.52E
145 T2 **Shaqlāwa** *var.* Shaqlāwah. E Iraq 36.24N 44.21E
Shaqlāwah *see* Shaqlāwa
144 I8 **Shaqqā** As Suwaydā', S Syria 32.55N 36.42E
147 P7 **Shaqrā'** Ar Riyāḑ, C Saudi Arabia 25.10N 45.08E
Shaqrā *see* Shuqrah
151 W10 **Shar** *var.* Charsk. Vostochnyy Kazakhstan, E Kazakhstan 49.33N 81.03E
155 O6 **Sharan** Urūzgān, SE Afghanistan 33.28N 66.19E
Sharaqpur *see* Sharqpur
Sharbaqty *see* Shcherbakty
147 X12 **Sharbatāt** S Oman 17.57N 56.14E
147 X12 **Sharbithāt, Ras** *var.* Ra's Sharbatāt. *headland* S Oman 17.55N 56.30E
12 K14 **Sharbot Lake** Ontario, SE Canada 44.45N 76.46W
151 P17 **Shardara** *var.* Chardara. Yuzhnyy Kazakhstan, S Kazakhstan 41.17N 68.03E
Shardara Dalasy *see* Step' Nardara
168 F8 **Sharga** Govĭ-Altay, W Mongolia 46.16N 95.32E
168 H6 **Sharga** Hövsgöl, N Mongolia 49.33N 98.36E
188 M7 **Sharhorod** Vinnyts'ka Oblast', C Ukraine 48.46N 28.05E
168 K10 **Sharhulsan** Ömnögovĭ, S Mongolia 44.43N 104.06E
172 Qq6 **Shari** Hokkaidō, NE Japan 43.54N 144.42E
Shari *see* Chari
145 T6 **Shārī, Buḩayrat** Lake C Iraq
Sharjah *see* Ash Shāriqah
120 K12 **Sharkawshchyna** *var.* Sharkowshchyna, *Pol.* Szarkowszczyzna, *Rus.* Sharkovshchina. Vitsyebskaya Voblasts', NW Belarus 55.21N 27.27E
188 G9 **Shark Bay** *bay* Western Australia
147 Y9 **Sharkh** E Oman 21.19N 59.04E
Sharkovshchina/Sharkowshchyna *see* Sharkawshchyna
131 U6 **Sharlyk** Orenburgskaya Oblast', W Russian Federation 52.52N 54.45E
Sharm ash Shaykh *see* Sharm el Sheikh
77 Y9 **Sharm el Sheikh** *var.* Ofiral, Sharm ash Shaykh. E Egypt 27.51N 34.16E
20 B13 **Sharon** Pennsylvania, NE USA 41.12N 80.28W
28 H4 **Sharon Springs** Kansas, C USA 38.54N 101.45W
33 Q14 **Sharonville** Ohio, N USA 39.16N 84.24W
Sharourah *see* Sharūrah
31 O10 **Sharpe, Lake** Reservoir South Dakota, N USA
Sharqī, Al Jabal ash/Sharqi, Jebel esh *see* Anti-Lebanon
Sharqīyah, Al Minṭaqah ash *see* Ash Sharqīyah
144 I6 **Sharqīyat an Nabk, Jabal** Mountain range W Syria
155 W8 **Sharqpur** *var.* Sharaqpur. Punjab, E Pakistan 31.27N 74.10E
147 Q13 **Sharūrah** *var.* Sharourah. Najrān, S Saudi Arabia 17.29N 47.04E
129 O14 **Shar'ya** Kostromskaya Oblast', NW Russian Federation 58.22N 45.30E
151 V15 **Sharyn** *var.* Charyn. River SE Kazakhstan
Sharyn *see* Charyn
85 J18 **Shashe** Central, NE Botswana 21.25S 27.28E
85 J18 **Shashe** *var.* Shashi. River Botswana/Zimbabwe
83 J14 **Shashemenē** *var.* Shashemenne, Shashhamana, *It.* Sciasciamana. Oromo, C Ethiopia 7.16N 38.38E
Shashemenne/Shashhamana *see* Shashemenē
Shashi *see* Shashe
Shashi/Sha-shih/Shasi *see* Jingzhou
37 N3 **Shasta Lake** Reservoir California, W USA
37 N2 **Shasta, Mount** Mountain California, W USA 41.24N 122.11W
131 O4 **Shatki** Nizhegorodskaya Oblast', W Russian Federation 55.09N 44.04E
152 J13 **Shatlyk** Maryyskiy Velayat, C Turkmenistan 37.55N 61.00E
Shatra *see* Ash Shaṭrah
121 K17 **Shatsk** *Rus.* Shatsk. Minskaya Voblasts', C Belarus 53.25N 27.44E
131 N5 **Shatsk** Ryazanskaya Oblast', W Russian Federation 54.02N 41.38E
28 J9 **Shattuck** Oklahoma, C USA 36.16N 99.52W
151 P16 **Shaul'der** Yuzhnyy Kazakhstan, S Kazakhstan 42.49N 68.22E
9 S17 **Shaunavon** Saskatchewan, S Canada 49.37N 108.22W
Shavat *see* Showot
164 K4 **Shawan** Xinjiang Uygur Zizhiqu, NW China 44.19N 85.34E
12 G12 **Shawanaga** Ontario, S Canada 45.29N 80.16W
32 M6 **Shawano** Wisconsin, N USA 44.46N 88.36W
32 M6 **Shawano Lake** Lake Wisconsin, N USA
13 P10 **Shawinigan** *prev.* Shawinigan Falls. Quebec, SE Canada 46.35N 72.45W
Shawinigan Falls *see* Shawinigan
13 P10 **Shawinigan-Sud** Quebec, SE Canada 46.30N 72.43W
144 J5 **Shawmarīyah, Jabal ash** Mountain range C Syria
49 O11 **Shawnee** Oklahoma, C USA 35.19N 96.55W
12 K12 **Shawville** Quebec, SE Canada 45.37N 76.31W
Shaykh *see* Ash Shakk
145 W9 **Shaykh 'Ābid** *var.* Shaikh Ābid. E Iraq 32.40N 46.09E
145 Y10 **Shaykh Fāris** *var.* Shaikh Fāris. E Iraq 32.06N 47.39E
145 T7 **Shaykh Ḩātim** E Iraq 33.28N 44.15E
Shaykh, Jabal ash *see* Hermon, Mount
145 X10 **Shaykh Najm** *var.* Shaikh Najm. E Iraq 32.04N 46.54E
145 W9 **Shaykh Sa'd** E Iraq 32.35N 46.16E
153 T14 **Shazud** SE Tajikistan 37.45N 72.22E
121 N18 **Shchadryn** *Rus.* Shchedrin. Homyel'skaya Voblasts', SE Belarus 52.55N 29.32E
121 H18 **Shchara** River SW Belarus
Shchedrin *see* Shchadryn
Shcheglovsk *see* Kemerovo
130 K5 **Shchëkino** Tul'skaya Oblast', W Russian Federation 53.57N 37.33E
129 S7 **Shchel'yayur** Respublika Komi, NW Russian Federation 65.19N 53.27E
151 U8 **Shcherbakty** *Kaz.* Sharbaqty. Pavlodar, E Kazakhstan 52.28N 78.00E
130 K7 **Shchigry** Kurskaya Oblast', W Russian Federation 51.53N 36.49E
Shchitkovichi *see* Shchytkavichy
119 Q2 **Shchors** Chernihivs'ka Oblast', N Ukraine 51.49N 31.58E
119 T8 **Shchors'k** Dnipropetrovs'ka Oblast', E Ukraine 48.20N 34.10E
Shchuchin *see* Shchuchyn
151 Q7 **Shchuchinsk** *prev.* Shchuchye. Severnyy Kazakhstan, N Kazakhstan 52.56N 70.09E
Shchuchye *see* Shchuchinsk
121 G16 **Shchuchyn** *Pol.* Szczuczyn Nowogródzki, *Rus.* Shchuchin. Hrodzyenskaya Voblasts', W Belarus 53.38N 24.48E
121 K17 **Shchytkavichy** *Rus.* Shchitkovichi. Minskaya Voblasts', C Belarus 53.13N 27.58E
126 H15 **Shebalino** Respublika Altay, S Russian Federation 51.16N 85.41E
130 J9 **Shebekino** Belgorodskaya Oblast', W Russian Federation 50.25N 36.54E
Shebelē Wenz, Wabē *see* Shebeli
83 L14 **Shebeli** *Amh.* Wabē Shebelē Wenz, *It.* Scebeli, *Som.* Webi Shabeelle. River Ethiopia/Somalia
115 M20 **Shebenikut, Maja e** Mountain E Albania 41.13N 20.27E
155 N2 **Sheberghān** *var.* Shibarghān, Shiberghan, Shiberghān. Jowzjān, N Afghanistan 36.40N 65.45E
150 F14 **Shebir** Mangistau, SW Kazakhstan 44.52N 52.01E
33 N8 **Sheboygan** Wisconsin, N USA 43.46N 87.43W
79 X15 **Shebshi Mountains** *var.* Schebschi Mountains. Mountain range E Nigeria
Shechem *see* Nablus
Shedadi *see* Ash Shadādah
11 P14 **Shediac** New Brunswick, SE Canada 46.13N 64.34W
130 L15 **Shedok** Krasnodarskiy Kray, SW Russian Federation 44.12N 40.49E
82 N12 **Sheekh** Woqooyi Galbeed, N Somalia 10.01N 45.21E
40 M11 **Sheenjek River** River Alaska, USA
98 D13 **Sheep Haven** *Ir.* Cuan na gCaorach. *inlet* N Ireland
37 X10 **Sheep Range** Mountain range Nevada, W USA
100 M13 **'s-Heerenberg** Gelderland, E Netherlands 51.52N 6.15E
99 P22 **Sheerness** SE England, UK 51.27N 0.45E
11 Q15 **Sheet Harbour** Nova Scotia, SE Canada 44.55N 62.31W
193 H18 **Sheffield** Canterbury, South Island, NZ 43.22S 172.01E
99 M18 **Sheffield** N England, UK 53.22N 1.30W
25 O2 **Sheffield** Alabama, S USA 34.46N 87.42W
31 V12 **Sheffield** Iowa, C USA 42.53N 93.13W
27 N10 **Sheffield** Texas, SW USA 30.42N 101.49W
65 H22 **Shehuen, Río** River S Argentina
Shekhem *see* Nablus
155 V8 **Shekhūpura** Punjab, NE Pakistan 31.43N 73.58E
Sheki *see* Şäki
128 L14 **Sheksna** Vologodskaya Oblast', NW Russian Federation 59.11N 38.32E
127 O4 **Shelagskiy, Mys** *headland* NE Russian Federation 70.04N 170.39E
49 V3 **Shelbina** Missouri, C USA 39.41N 92.01W
11 P16 **Shelburne** Nova Scotia, SE Canada 43.46N 65.19W
12 G14 **Shelburne** Ontario, S Canada 44.04N 80.12W
35 R7 **Shelby** Montana, NW USA 48.30N 111.52W
23 Q10 **Shelby** North Carolina, SE USA 35.17N 81.32W
33 S12 **Shelby** Ohio, N USA 40.52N 82.39W
32 L14 **Shelbyville** Illinois, N USA 39.24N 88.47W
33 P14 **Shelbyville** Indiana, N USA 39.31N 85.46W
22 L5 **Shelbyville** Kentucky, S USA 38.12N 85.13W
49 V2 **Shelbyville** Missouri, C USA 39.48N 92.02W
22 J10 **Shelbyville** Tennessee, S USA 35.28N 86.27W
27 X8 **Shelbyville** Texas, SW USA 31.42N 94.03W
32 L14 **Shelbyville, Lake** Reservoir Illinois, N USA
31 S12 **Sheldon** Iowa, C USA 43.10N 95.51W
40 M11 **Sheldons Point** Alaska, USA 62.31N 165.03W
126 J16 **Shelekhov** Irkutskaya Oblast', C Russian Federation 52.04N 104.03E
Shelekhov Gulf *see* Shelikhova, Zaliv
127 Oo9 **Shelikhova, Zaliv** *Eng.* Shelekhov Gulf. *gulf* E Russian Federation
41 P14 **Shelikof Strait** *strait* Alaska, USA
Shelim *see* Shalīm
9 T14 **Shellbrook** Saskatchewan, S Canada 53.13N 106.24W
30 L3 **Shell Creek** River North Dakota, N USA
Shellif *see* Chelif, Oued
24 I10 **Shell Keys** *island group* Louisiana, S USA
32 J4 **Shell Lake** Wisconsin, N USA 45.43N 91.55W
31 W12 **Shell Rock** Iowa, C USA 42.42N 92.34W
193 C26 **Shelter Point** *headland* Stewart Island, NZ 47.04S 168.13E
20 L13 **Shelton** Connecticut, NE USA 41.19N 73.06W
34 G8 **Shelton** Washington, NW USA 47.13N 123.06W
Shemakha *see* Şamaxı
151 W9 **Shemonaikha** Vostochnyy Kazakhstan, E Kazakhstan 50.39N 81.51E
131 Q4 **Shemursha** Chuvashskaya Respublika, W Russian Federation 54.57N 47.27E
40 D16 **Shemya Island** *island* Aleutian Islands, Alaska, USA
31 T16 **Shenandoah** Iowa, C USA 40.46N 95.23W
23 U4 **Shenandoah** Virginia, NE USA 38.26N 78.34W
23 U4 **Shenandoah Mountains** *ridge* West Virginia, NE USA
23 V3 **Shenandoah River** River West Virginia, NE USA
79 W15 **Shendam** Plateau, C Nigeria 8.52N 9.30E
82 G8 **Shendi** *var.* Shandī. River Nile, NE Sudan 16.40N 33.22E
78 I15 **Shenge** SW Sierra Leone 7.54N 12.54W
151 U15 **Shengel'dy** Almaty, SE Kazakhstan 44.04N 77.31E
115 K18 **Shëngjin** *var.* Shëngjini. Lezhë, NW Albania 41.49N 19.34E
Shëngjini *see* Shëngjin
Shengking *see* Liaoning
Sheng Xian/Shengxian *see* Shengzhou
167 S9 **Shengzhou** *var.* Shengxian, Sheng Xian. Zhejiang, SE China 29.36N 120.47E
Shenking *see* Liaoning
129 N11 **Shenkursk** Arkhangel'skaya Oblast', NW Russian Federation 62.10N 42.58E
166 L3 **Shenmu** Shaanxi, C China 38.49N 110.27E
115 L19 **Shën Noj i Madh** Mountain C Albania 41.23N 20.07E
Shenshi/Shensi *see* Shaanxi
169 V12 **Shenyang** *Chin.* Shen-yang, *Eng.* Moukden, Mukden; *prev.* Fengtien. Liaoning, NE China 41.49N 123.25E
167 O15 **Shenzhen** Guangdong, S China 22.39N 114.02E
160 G8 **Sheopur** Madhya Pradesh, C India 25.40N 76.42E
118 L5 **Shepetivka** *Rus.* Shepetovka. Khmel'nyts'ka Oblast', NW Ukraine 50.12N 27.01E
Shepetovka *see* Shepetivka
27 W10 **Shepherd** Texas, SW USA 30.30N 95.00W
197 D14 **Shepherd Islands** *island group* C Vanuatu
22 K5 **Shepherdsville** Kentucky, S USA 37.59N 85.43W
191 O11 **Shepparton** Victoria, SE Australia 36.25S 145.25E
99 P22 **Sheppey, Isle of** *island* SE England, UK
Sherabad *see* Sherobod
99 L23 **Sherborne** S England, UK 50.57N 2.30W
78 H16 **Sherbro Island** *island* SW Sierra Leone
13 Q12 **Sherbrooke** Quebec, SE Canada 45.23N 71.54W
31 T11 **Sherburn** Minnesota, N USA 43.39N 94.43W
80 H6 **Sherda** Borkou-Ennedi-Tibesti, N Chad 20.04N 16.48E
82 G7 **Shereik** River Nile, N Sudan 18.43N 33.37E
130 K3 **Sheremet'yevo** International airport (Moskva) Moskovskaya Oblast', W Russian Federation 56.05N 37.10E
159 P14 **Shergāti** Bihār, N India 24.35N 84.51E
49 U12 **Sheridan** Arkansas, C USA 34.18N 92.24W
35 W12 **Sheridan** Wyoming, C USA 44.47N 106.59W
190 G8 **Sheringa** South Australia 33.51S 135.13E
126 L16 **Sherlovaya Gora** Chitinskaya Oblast', S Russian Federation 50.26N 116.09E
27 U5 **Sherman** Texas, SW USA 33.39N 96.34W
204 J10 **Sherman Island** *island* Antarctica
21 S4 **Sherman Mills** Maine, NE USA 45.51N 68.23W
31 O15 **Sherman Reservoir** Reservoir Nebraska, C USA
153 N14 **Sherobod** *Rus.* Sherabad. Surkhondaryo Wiloyati, S Uzbekistan 37.43N 66.59E
153 O13 **Sherobod** *Rus.* Sherabad. River S Uzbekistan
159 T14 **Sherpur** Dhaka, N Bangladesh 25.00N 90.01E
39 T4 **Sherrelwood** Colorado, C USA 39.49N 105.00W
101 J14 **'s-Hertogenbosch** *Fr.* Bois-le-Duc, *Ger.* Herzogenbusch. Noord-Brabant, S Netherlands 51.40N 5.19E
30 M2 **Sherwood** North Dakota, N USA 48.55N 101.36W
9 Q14 **Sherwood Park** Alberta, SW Canada 53.34N 113.04W
58 F13 **Sheshea, Río** River E Peru
149 T5 **Sheshtamad** Khorāsān, NE Iran 36.03N 57.45E
31 S10 **Shetek, Lake** Lake Minnesota, N USA
98 M2 **Shetland Islands** *island group* NE Scotland, UK
150 F14 **Shetpe** Mangistau, SW Kazakhstan 44.06N 52.03E
160 C11 **Shetrunji** River W India
Shevchenko *see* Aktau
119 W5 **Shevchenkove** Kharkivs'ka Oblast', E Ukraine 49.40N 37.13E
83 H14 **Shewa Gīmīra** Southern, S Ethiopia 7.12N 35.49E
167 Q9 **Shexian** *var.* Huicheng, She Xian. Anhui, E China 29.52N 118.27E
167 R6 **Sheyang** *prev.* Hede. Jiangsu, E China 33.49N 120.13E
31 O4 **Sheyenne** North Dakota, N USA 47.49N 99.08W
31 P4 **Sheyenne River** River North Dakota, N USA
98 G7 **Shiant Islands** *island group* NW Scotland, UK
127 Pp14 **Shiashkotan, Ostrov** *island* Kuril'skiye Ostrova, SE Russian Federation
33 R9 **Shiawassee River** River Michigan, N USA
147 R14 **Shibām** C Yemen 15.49N 48.24E
Shibarghān *see* Sheberghān
171 Kk12 **Shibata** *var.* Sibata. Niigata, Honshū, C Japan 37.57N 139.19E
172 Qq7 **Shibecha** Hokkaidō, NE Japan 43.19N 144.34E
Shiberghan/Shiberghān *see* Sheberghān
172 Qq6 **Shibetsu** Hokkaidō, N Japan 43.40N 145.10E
172 Pp4 **Shibetsu** *var.* Sibetu. Hokkaidō, NE Japan 44.12N 142.23E
172 P5 **Shibetsu** *var.* Sibetu. Hokkaidō, NE Japan 43.57N 142.13E
Shibh Jazīrat Sīnā' *see* Sinai
Shibīn al Kawm *see* Shibīn el Kôm
77 W8 **Shibîn el Kôm** *var.* Shibīn al Kawm. N Egypt 30.33N 30.59E
149 O13 **Shīb, Kūh-e** Mountain S Iran
10 D8 **Shibogama Lake** Lake Ontario, C Canada
Shibotsu-jima *see* Zelënyy, Ostrov
171 K14 **Shibukawa** *var.* Sibukawa. Gunma, Honshū, S Japan 36.31N 138.58E
170 Bb16 **Shibushi** Kagoshima, Kyūshū, SW Japan 31.27N 131.05E
170 C17 **Shibushi-wan** *bay* SW Japan
172 N9 **Shichinohe** Aomori, Honshū, C Japan 40.40N 141.07E
201 U13 **Shichiyo Islands** *island group* Chuuk, C Micronesia
Shickshock Mountains *see* Chic-Chocs, Monts
151 S9 **Shiderti** River N Kazakhstan
151 S8 **Shiderty** Pavlodar, NE Kazakhstan 51.43N 74.34E
98 G10 **Shiel, Loch** Lake N Scotland, UK
171 H15 **Shiga** *off.* Shiga-ken, *var.* Siga. Administrative region *prefecture* Honshū, SW Japan
Shigatse *see* Xigazê
147 U13 **Shiḩan** *oasis* NE Yemen 17.46N 52.25E
Shih-chia-chuang/Shihmen *see* Shijiazhuang
164 K4 **Shihezi** Xinjiang Uygur Zizhiqu, NW China 44.20N 85.59E
Shiichi *see* Shyichy
115 K19 **Shijak** *var.* Shijaku. Durrës, W Albania 41.21N 19.34E
Shijaku *see* Shijak
167 O4 **Shijiazhuang** *var.* Shih-chia-chuang; *prev.* Shihmen. Hebei, E China 38.04N 114.28E
172 Nn7 **Shikabe** Hokkaidō, NE Japan 42.03N 140.45E
155 Q13 **Shikārpur** Sind, S Pakistan 27.59N 68.39E
201 V12 **Shiki Islands** *island group* Chuuk, C Micronesia
170 Ee15 **Shikoku** *var.* Sikoku. *island* SW Japan
199 Gg6 **Shikoku Basin** *var.* Sikoku Basin. *undersea feature* N Philippine Sea
170 Ee15 **Shikoku-sanchi** Mountain range Shikoku, SW Japan
172 S7 **Shikotan, Ostrov** *Jap.* Shikotan-tō. *island* NE Russian Federation
Shikotan-tō *see* Shikotan, Ostrov
172 O6 **Shikotsu-ko** *var.* Sikotu Ko. Lake Hokkaidō, NE Japan
83 N15 **Shilabo** Somali, E Ethiopia 6.05N 44.48E
131 X7 **Shil'da** Orenburgskaya Oblast', W Russian Federation 51.46N 59.48E
145 V3 **Shīler, Āw-e** River E Iraq
159 S12 **Shiliguri** *prev.* Siliguri. West Bengal, NE India 26.45N 88.24E
126 Kk16 **Shilka** Chitinskaya Oblast', S Russian Federation 51.52N 115.49E
133 V7 **Shilka** River S Russian Federation
20 H15 **Shillington** Pennsylvania, NE USA 40.18N 75.57W
159 V13 **Shillong** Meghālaya, NE India 25.36N 91.54E
130 M5 **Shilovo** Ryazanskaya Oblast', W Russian Federation 54.18N 40.53E
170 C14 **Shimabara** *var.* Simabara. Nagasaki, Kyūshū, SW Japan 32.48N 130.19E
170 C14 **Shimabara-wan** *bay* SW Japan
171 Ii17 **Shimada** *var.* Simada. Shizuoka, Honshū, S Japan 34.49N 138.10E
170 Ee12 **Shimane** *off.* Shimane-ken, *var.* Simane. Administrative region *prefecture* Honshū, SW Japan

Country ◆ Country Capital ● Dependent Territory ◇ Dependent Territory Capital ○ Administrative Region International Airport Mountain ▲ Mountain Range Volcano River Lake Reservoir

170 F11 **Shimane-hantō** *peninsula* Honshū, SW Japan

126 M15 **Shimanovsk** Amurskaya Oblast', SE Russian Federation 52.00N 127.36E

Shimbir Berris *see* Shimbiris

82 O12 **Shimbiris** *var.* Shimbir Berris. ▲ N Somalia 10.43N 47.10E

172 P6 **Shimizu** Hokkaidō, NE Japan 42.58N 142.54E

171 Ii16 **Shimizu** *var.* Simizu. Shizuoka, Honshū, S Japan 35.01N 138.28E

158 I8 **Shimla** *prev.* Simla. Himāchal Pradesh, N India 31.07N 77.09E

Shimminato *see* Shinminato

171 J18 **Shimoda** *var.* Simoda. Shizuoka, Honshū, S Japan 34.49N 138.09E

171 Kk16 **Shimodate** *var.* Simodate. Ibaraki, Honshū, S Japan 36.18N 139.57E

161 F18 **Shimoga** Karnātaka, W India 13.55N 75.31E

170 Bb14 **Shimo-jima** *island* SW Japan

170 B15 **Shimo-Koshiki-jima** *island* SW Japan

83 J21 **Shimoni** Coast, S Kenya 4.40S 39.22E

170 D12 **Shimonoseki** *var.* Simonoseki; *hist.* Akamagaseki, Bakan. Yamaguchi, Honshū, SW Japan 33.57N 130.54E

171 Kk16 **Shimotsuma** *var.* Simotuma. Ibaraki, Honshū, S Japan 36.10N 139.58E

128 G14 **Shimsk** Novgorodskaya Oblast', NW Russian Federation 58.12N 30.43E

171 Jj14 **Shinano-gawa** *var.* Sinano Gawa. ⊘ Honshū, C Japan

147 W7 **Shinās** N Oman 24.45N 56.24E

154 J6 **Shindand** Farāh, W Afghanistan 33.19N 62.09E

Shinei *see* Hsinying

27 T12 **Shiner** Texas, SW USA 29.25N 97.10W

178 Gg1 **Shingbwiyang** Kachin State, N Myanmar 26.40N 96.14E

151 W11 **Shingozha** Vostochnyy Kazakhstan, E Kazakhstan 47.46N 80.38E

171 Gg17 **Shingū** *var.* Singū. Wakayama, Honshū, SW Japan 33.40N 135.57E

12 F8 **Shining Tree** Ontario, S Canada 47.36N 81.12W

170 F12 **Shinji-ko** *var.* Sinzi-ko. ⊗ Honshū, SW Japan

171 Ll12 **Shinjō** *var.* Sinzyō. Yamagata, Honshū, C Japan 38.46N 140.16E

98 I7 **Shin, Loch** ⊗ N Scotland, UK

171 Ii13 **Shinminato** *var.* Shimminato, Sinminato. Toyama, Honshū, SW Japan 36.46N 137.04E

171 Dd12 **Shinnanyō** *var.* Shin-Nan'yō, Sinn'anyō. Yamaguchi, Honshū, SW Japan 34.04N 131.43E

23 S3 **Shinnston** West Virginia, NE USA 39.22N 80.19W

144 I6 **Shinshār** *Fr.* Chinnchâr. Ḥimṣ, W Syria 34.36N 36.45E

171 Ii16 **Shinshiro** *var.* Sinsiro. Aichi, Honshū, SW Japan 34.52N 137.29E

Shinshū *see* Chinju

172 P6 **Shintoku** Hokkaidō, NE Japan 43.03N 142.50E

83 G20 **Shinyanga** Shinyanga, NW Tanzania 3.40S 33.25E

83 G20 **Shinyanga** ◆ *region* N Tanzania

171 M13 **Shiogama** *var.* Siogama. Miyagi, Honshū, C Japan 38.19N 140.59E

171 J15 **Shiojiri** *var.* Sioziri. Nagano, Honshū, S Japan 36.07N 137.57E

170 G17 **Shiono-misaki** *headland* Honshū, SW Japan 33.25N 135.45E

171 Ll15 **Shioya-zaki** *headland* Honshū, C Japan 37.00N 140.57E

116 J9 **Shipchenski Prokhod** *pass* C Bulgaria 42.46N 25.21E

166 G14 **Shiping** Yunnan, SW China 23.45N 102.23E

11 P13 **Shippagan** *var.* Shippegan. New Brunswick, SE Canada 47.45N 64.43W

Shippegan *see* Shippagan

20 F15 **Shippensburg** Pennsylvania, NE USA 40.03N 77.31W

39 O9 **Ship Rock** ▲ New Mexico, SW USA 36.41N 108.50W

39 P9 **Shiprock** New Mexico, SW USA 36.47N 108.41W

13 R6 **Shipshaw** ⊘ Quebec, SE Canada

127 Pp11 **Shipunskiy, Mys** *headland* E Russian Federation 53.04N 159.57E

166 K7 **Shiquan** Shaanxi, C China 33.06N 108.10E

126 Hh14 **Shira** Respublika Khakasiya, C Russian Federation 54.35N 89.58E

170 G16 **Shirahama** Wakayama, Honshū, SW Japan 33.40N 135.21E

159 T14 **Shirajganj Ghat** *var.* Serajgonj, Serajganj. Rajshahi, C Bangladesh 24.25N 89.40E

171 Mm7 **Shirakami-misaki** *headland* NE Japan 41.26N 140.10E

171 L14 **Shirakawa** *var.* Sirakawa. Fukushima, Honshū, C Japan 37.07N 140.11E

171 Ii13 **Shirakawa** Gifu, Honshū, SW Japan 36.17N 136.52E

171 K14 **Shirane-san** ▲ Honshū, S Japan 36.44N 139.21E

171 J16 **Shirane-san** ▲ Honshū, S Japan 35.39N 138.13E

172 Pp7 **Shiranuka** Hokkaidō, NE Japan 42.57N 144.01E

172 O6 **Shiraoi** Hokkaidō, NE Japan 42.34N 141.24E

205 N12 **Shirase Coast** *physical region* Antarctica

172 Pp5 **Shirataki** Hokkaidō, NE Japan 43.55N 143.14E

149 O11 **Shīrāz** *var.* Shīrāz. Fārs, S Iran 29.37N 52.34E

85 N15 **Shire** *var.* Chire. ⊘ Malawi/Mozambique

168 G7 **Shiree** Dzavhan, W Mongolia 47.30N 96.48E

169 O9 **Shireet** Sühbaatar, SE Mongolia 45.33N 112.19E

172 R6 **Shiretoko-hantō** *headland* Hokkaidō, NE Japan 44.06N 145.07E

172 R5 **Shiretoko-misaki** *headland* Hokkaidō, NE Japan 44.20N 145.19E

131 N5 **Shiringushi** Respublika Mordoviya, W Russian Federation 53.50N 42.49E

154 M3 **Shīrīn Tagāb** Fāryāb, N Afghanistan 36.49N 65.01E

155 N2 **Shīrīn Tagāb** ⊘ N Afghanistan

172 Nn8 **Shiriya-zaki** *headland* Honshū, C Japan 41.24N 141.27E

150 I12 **Shirkala, Gryada** *plain* W Kazakhstan

171 Ll13 **Shiroishi** *var.* Siroisi. Miyagi, Honshū, C Japan 38.01N 140.37E

Shirokoye *see* Shyroke

171 K12 **Shirone** *var.* Sirone. Niigata, Honshū, C Japan 49.00N 31.27E

171 I14 **Shirotori** Gifu, Honshū, SW Japan 35.53N 136.52E

171 J13 **Shirouma-dake** ▲ Honshū, S Japan 36.46N 137.46E

207 T1 **Shirshov Ridge** *undersea feature* W Bering Sea

Shirshūtūr *see* Shirshyutyur, Peski

152 K12 **Shirshyutyur, Peski** *Turkm.* Shirshūtūr. *desert* E Turkmenistan

149 T3 **Shīrvān** *var.* Shirwān. Khorāsān, NE Iran 37.25N 57.55E

Shirwa, Lake *see* Chilwa, Lake

Shirwān *see* Shīrvān

165 N5 **Shisanjianfang** Xinjiang Uygur Zizhiqu, W China 43.10N 91.15E

40 M16 **Shishaldin Volcano** ▲ Unimak Island, Alaska, USA 54.45N 163.58W

Shishchitsy *see* Shyshchytsy

40 M8 **Shishmaref** Alaska, USA 66.15N 166.04W

Shisur *see* Ash Shiṣar

171 Ii16 **Shitara** Aichi, Honshū, SW Japan 35.06N 137.33E

158 D12 **Shiv** Rājasthān, NW India 26.10N 71.13E

157 E15 **Shivājī Sāgar** *prev.* Konya Reservoir ☒ W India

160 H8 **Shivpuri** Madhya Pradesh, C India 25.26N 77.41E

38 J9 **Shivwits Plateau** *plain* Arizona, SW USA

Shiwalik Range *see* Siwalik Range

166 M8 **Shiyan** Hubei, C China 32.39N 110.48E

166 H13 **Shizong** Yunnan, SW China 24.49N 103.59E

171 Mm13 **Shizugawa** Miyagi, Honshū, NE Japan 38.40N 141.26E

165 W8 **Shizuishan** *var.* Dawukou. Ningxia, N China 39.04N 106.22E

172 Oo7 **Shizunai** Hokkaidō, NE Japan 42.20N 142.24E

171 Ii16 **Shizuoka** *var.* Sizuoka. Shizuoka, Honshū, S Japan 34.58N 138.20E

171 Ii16 **Shizuoka** *off.* Shizuoka-ken, *var.* Sizuoka. ◆ *prefecture* Honshū, S Japan

Shklov *see* Shklow

121 N15 **Shklow** *Rus.* Shklov. Mahilyowskaya Voblasts', E Belarus 54.13N 30.16E

115 K18 **Shkodër** *var.* Shkodra, *It.* Scutari, *SCr.* Skadar. Shkodër, NW Albania 42.03N 19.31E

115 K17 **Shkodër** ◆ *district* NW Albania

Shkodra *see* Shkodër

Shkodrës, Liqeni i *see* Scutari, Lake

Shkumbin/Shkumbin *see* Shkumbini, Lumi i

115 L20 **Shkumbinit, Lumi i** *var.* Shkumbin, Shkumbin. ⊘ C Albania

Shligigh, Cuan *see* Sligo Bay

126 Ii2 **Shmidta, Ostrov** *island* Severnaya Zemlya, N Russian Federation

191 S10 **Shoalhaven River** ⊘ New South Wales, SE Australia

9 W16 **Shoal Lake** Manitoba, S Canada 50.28N 100.36W

33 O15 **Shoals** Indiana, N USA 38.40N 86.46W

170 F13 **Shōbara** *var.* Syōbara. Hiroshima, Honshū, SW Japan 34.50N 132.58E

170 Ff14 **Shōdo-shima** *island* SW Japan

171 Ii13 **Shō-gawa** ⊘ Honshū, SW Japan

126 J3 **Shokal'skogo, Proliv** *strait* N Russian Federation

172 Oo4 **Shokanbetsu-dake** ▲ Hokkaidō, NE Japan 43.43N 141.33E

153 T14 **Shokhdara, Qatorkūhi** *Rus.* Shakhdarinskiy Khrebet. ▲ SE Tajikistan

151 N9 **Sholaksay** Kostanay, N Kazakhstan 51.43N 64.45E

Sholāpur *see* Solāpur

Sholdaneshty *see* Şoldăneşti

151 P17 **Sholakorgan** *var.* Chulakkurgan. Yuzhnyy Kazakhstan, S Kazakhstan 43.48N 69.12E

Shoqpar *see* Chokpar

161 G21 **Shoranūr** Kerala, SW India 10.53N 76.06E

161 G20 **Shorāpur** Karnātaka, C India 16.34N 76.48E

32 M11 **Shorewood** Illinois, N USA 41.31N 88.12W

Shorkazakhly, Solonchak *see* Kazakhlyshor, Solonchak

151 Q9 **Shortandy** Akmola, C Kazakhstan

Shortepa/Shor Tepe *see* Shūr Tappeh

195 S12 **Shortland Island** *var.* Alu. *island* Shortland Islands, NW Solomon Islands

195 T13 **Shortland Islands** *island group* NW Solomon Islands

172 P3 **Shosanbetsu** *var.* Shosanbetsu. Hokkaidō, NE Japan 44.31N 141.47E

35 O15 **Shoshone** Idaho, NW USA 42.56N 114.24W

37 T6 **Shoshone Mountains** ▲ Nevada, W USA

35 U12 **Shoshone River** ⊘ Wyoming, C USA

85 I19 **Shoshong** Central, SE Botswana 23.03S 26.33E

35 V14 **Shoshoni** Wyoming, C USA 43.13N 108.06W

Shōshū *see* Sangju

119 S2 **Shostka** Sums'ka Oblast', NE Ukraine 51.51N 33.30E

193 Q2 **Shotover** ⊘ South Island, NZ

39 N12 **Show Low** Arizona, SW USA 34.15N 110.01W

Show Me State *see* Missouri

152 H9 **Showot** *Rus.* Shavat. Khorazm Wiloyati, W Uzbekistan 41.41N 60.13E

129 O4 **Shoyna** Nenetskiy Avtonomnyy Okrug, NW Russian Federation 67.50N 44.09E

128 M11 **Shozhma** Arkhangel'skaya Oblast', NW Russian Federation 61.57N 40.10E

119 Q7 **Shpola** Cherkas'ka Oblast', C Ukraine 49.00N 31.27E

Shqipëria/Shqipërisë, Republika e *see* Albania

24 G5 **Shreveport** Louisiana, S USA 32.31N 93.45W

99 K19 **Shrewsbury** *hist.* Scrobesbyrig'. W England, UK 52.43N 2.45W

158 D11 **Shri Mohangarh** *prev.* Sri Mohangorh. Rājasthān, NW India 27.16N 71.18E

159 S16 **Shrīrāmpur** *prev.* Serampore, Serampur. West Bengal, NE India 22.43N 88.19E

99 K19 **Shropshire** *cultural region* W England, UK

151 S16 **Shu** *Kaz.* Shū. Zhambyl, SE Kazakhstan 43.34N 73.40E

Shū *see* Chu

166 G13 **Shuangbai** Yunnan, SW China 24.45N 101.38E

169 W9 **Shuangcheng** Heilongjiang, NE China 45.20N 126.21E

166 E14 **Shuangjiang** Yunnan, SW China 23.28N 99.43E

169 U10 **Shuangliao** *var.* Zhengjiatun. Jilin, NE China 43.31N 123.32E

169 Y7 **Shuangyashan** *var.* Shuang-ya-shan. Heilongjiang, NE China 46.37N 131.10E

147 W12 **Shu'aymiah** *var.* Shu'aymiyah. S Oman 17.55N 55.39E

150 I10 **Shubarkuduk** *Kaz.* Shubarqudyq. Aktyubinsk, W Kazakhstan 49.10N 56.28E

Shubarqudyq *see* Shubarkuduk

151 N12 **Shubar-Tengiz, Ozero** ⊗ C Kazakhstan

41 S5 **Shublik Mountains** ▲ Alaska, USA

Shubrā al Khaymah *see* Shubrā el Kheima

124 Qq16 **Shubrā el Kheima** *var.* Shubrā al Khaymah. N Egypt 30.06N 31.15E

164 E8 **Shufu** Xinjiang Uygur Zizhiqu, NW China 39.18N 75.43E

153 S14 **Shughnon, Qatorkūhi** *Rus.* Shugnanskiy Khrebet. ▲ SE Tajikistan

Shugnan, Qatorkūhi/Shugnanskiy Khrebet *see* Shughnon, Qatorkūhi

117 Q6 **Shu He** ⊘ E China

Shuiding *see* Huocheng

Shuiji *see* Laixi

Shū-Ile Taūlary *see* Chu-Iliyskiye Gory

155 T10 **Shujāābād** Punjab, E Pakistan 29.52N 71.22E

169 W9 **Shulan** Jilin, NE China 44.28N 126.57E

164 E8 **Shule** Xinjiang Uygur Zizhiqu, NW China 39.19N 76.06E

Shuleh *see* Shule He

165 Q8 **Shule He** *var.* Shuleh, Sulo. ⊘ C China

32 K9 **Shullsburg** Wisconsin, N USA 42.37N 90.12W

41 N16 **Shumagin Islands** *island group* Alaska, USA

152 G7 **Shumanay** Qoraqalpoghiston Respublikasi, W Uzbekistan 42.42N 58.56E

116 M8 **Shumen** Shumen, NE Bulgaria 43.17N 26.57E

116 M8 **Shumen** ◆ *province* NE Bulgaria

131 P4 **Shumerlya** Chuvashskaya Respublika, W Russian Federation 55.31N 46.24E

125 Ee12 **Shumikha** Kurganskaya Oblast', C Russian Federation 55.12N 63.09E

120 M12 **Shumilina** *Rus.* Shumilino. Vitsyebskaya Voblasts', NE Belarus 55.19N 29.33E

Shumilino *see* Shumilina

127 Pp12 **Shumshu, Ostrov** *island* SE Russian Federation

118 K5 **Shums'k** Ternopil's'ka Oblast', W Ukraine 50.06N 26.04E

41 O7 **Shungnak** Alaska, USA 66.53N 157.08W

Shunsen *see* Ch'unch'ŏn

166 M3 **Shuo Xian** Shanxi, NE China 39.19N 112.25E

Shuozhou *see* Shuo Xian

167 N3 **Shuozhou** *var.* Shuoxian; *prev.* Shuo Xian. Shanxi, C China 39.19N 112.25E

147 P16 **Shuqrah** *var.* Shaqrā. SW Yemen 13.25N 45.43E

Shurab *see* Shūrob

153 O14 **Shūrchi** *Rus.* Shurchi. Surkhondaryo Wiloyati, S Uzbekistan 37.58N 67.40E

153 R11 **Shūrob** *Rus.* Shurab. NW Tajikistan 40.02N 70.31E

149 T10 **Shūr, Rūd-e** ⊘ E Iran

155 O2 **Shūr Tappeh** *var.* Shortepa, Shor Tepe. Balkh, N Afghanistan 37.22N 66.49E

85 K17 **Shurugwi** *prev.* Selukwe. Midlands, C Zimbabwe 19.40S 30.00E

148 L8 **Shūsh** *anc.* Susa, *Bibl.* Shushan. Khūzestān, SW Iran 32.11N 48.13E

148 L9 **Shūshtar** *var.* Shushtar. Khūzestān, SW Iran 32.03N 48.51E

Shushter/Shustar *see* Shūshtar

148 K8 **Shūsh, Hawr ash** *var.* Hawr as Suwayqiyah. ⊗ E Iraq

145 V9 **Shuwayjah, Hawr ash** *var.* Hawr as Suwayqiyah. ⊗ E Iraq

128 M16 **Shuya** Ivanovskaya Oblast', W Russian Federation 56.51N 41.24E

41 Q14 **Shuyak Island** *island* Alaska, USA

177 G4 **Shwebo** Sagaing, C Myanmar 22.34N 95.42E

177 Ff7 **Shwedaung** Pegu, W Myanmar 18.43N 95.11E

177 G8 **Shwegyin** Pegu, SW Myanmar 17.55N 96.58E

178 Gg4 **Shweli** *Chin.* Longchuan Jiang. ⊘ Myanmar/China

177 G6 **Shwemyo** Mandalay, C Myanmar 22.05N 93.36E

Shyghys Qazaqstan Oblysy *see* Vostochnyy Kazakhstan

Shyghys Qongyrat *see* Shygys Konyrat

151 T12 **Shygys Konyrat** *var.* Vostochno-Kounradskiy, *Kaz.* Shyghys Qongyrat. Karaganda, C Kazakhstan 47.01N 75.05E

121 M19 **Shyichy** *Rus.* Shiichi. Homyel'skaya Voblasts', SE Belarus 52.15N 29.13E

151 Q17 **Shymkent** *prev.* Chimkent. Yuzhnyy Kazakhstan, S Kazakhstan 42.19N 69.36E

Shynggyrlaū *see* Chingirlau

119 S9 **Shyroke** *Rus.* Shirokoye. Dnipropetrovs'ka Oblast', E Ukraine 47.37N 33.15E

119 O9 **Shyryayeve** Odes'ka Oblast', SW Ukraine 47.21N 30.11E

119 S5 **Shyshaky** Poltavs'ka Oblast', C Ukraine 49.54N 34.00E

121 K17 **Shyshchytsy** *Rus.* Shishchitsy. Minskaya Voblasts', C Belarus 53.12N 27.33E

155 Y3 **Siachen Muztāgh** ▲ NE Pakistan

Siadehan *see* Tākestān

79 N23 **Siaderno** Calabria, SW Italy 38.17N 16.19E

148 I1 **Sīāh Range** ▲ W Pakistan

155 W7 **Siāh Chashmeh** Āzarbāyjān-e Bākhtarī, N Iran 39.01N 44.22E

155 W7 **Siālkot** Punjab, NE Pakistan 32.31N 74.33E

194 K12 **Sialum** Morobe, C PNG 6.05S 147.33E

Siam *see* Thailand

Siam, Gulf of *see* Thailand, Gulf of

Sian *see* Xi'an

Siang *see* Brahmaputra

Siangtan *see* Xiangtan

174 J5 **Siantan, Pulau** *island* Kepulauan Anambas, W Indonesia

56 H11 **Siare, Río** ⊘ C Colombia

179 Rr13 **Siargao Island** *island* S Philippines

194 K12 **Siassi** Umboi Island, C PNG 5.34S 147.50E

117 D14 **Siátista** Dytikí Makedonía, N Greece 40.16N 21.34E

177 Ff4 **Siatlai** Chin State, W Myanmar 22.05N 93.36E

179 Q15 **Siaton** Negros, C Philippines 9.03S 123.03E

120 F11 **Šiauliai** *Ger.* Schaulen. Šiauliai, N Lithuania 55.57N 23.21E

175 S5 **Siau, Pulau** *island* N Indonesia

85 J15 **Siavonga** Southern, SE Zambia 16.28S 28.45E

Siazan' *see* Siyäzän

109 N20 **Sibari** Calabria, S Italy 39.45N 16.26E

Sibata *see* Shibata

131 X6 **Sibay** Respublika Bashkortostan, W Russian Federation 52.39N 58.39E

95 M19 **Sibbo** *Fin.* Sipoo. Etelä-Suomi, S Finland 60.23N 25.15E

114 D13 **Šibenik** *It.* Sebenico. Šibenik-Knin, S Croatia 43.43N 15.54E

Šibenik *see* Šibenik-Knin

114 E13 **Šibenik-Knin** *off.* Šibenska Županija. *var.* Šibenik. ◆ *province* S Croatia

Šibenska Županija *see* Šibenik-Knin

Siberia *see* Sibir'

112 O12 **Sibir'** *var.* Siberia. *physical region* NE Russian Federation

81 F20 **Sibiti** S Congo 3.40S 13.19E

83 G21 **Sibiti** ⊘ E Tanzania

118 I12 **Sibiu** *Ger.* Hermannstadt, *Hung.* Nagyszeben. Sibiu, C Romania 45.48N 24.08E

118 I11 **Sibiu** ◆ *county* C Romania

31 S11 **Sibley** Iowa, C USA 43.24N 95.45W

173 Ff6 **Sibolga** Sumatera, W Indonesia 1.42N 98.48E

Sibolga, Teluk *see* Tapanuli

112 J13 **Sibolga, Teluk** *var.* Teluk Tapanuli. *bay* Sumatera, W Indonesia

112 K10 **Sibu** Sarawak, East Malaysia 2.18N 111.49E

Sibukawa *see* Shibukawa

44 G2 **Sibun** ⊘ E Belize

81 I15 **Sibut** *prev.* Fort-Sibut. Kémo, S Central African Republic 5.44N 19.07E

179 P17 **Sibutu** *island* SW Philippines

179 P17 **Sibutu Passage** *passage* SW Philippines

179 Q12 **Sibuyan Island** *island* C Philippines

179 Q12 **Sibuyan Sea** *sea* C Philippines

131 J26 **Sibylla Island** *island* N Marshall Islands

166 H9 **Sichuan** *var.* Chuan, Sichuan Sheng, Ssu-ch'uan, Szechuan, Szechwan. ◆ *province* C China

166 I9 **Sichuan Pendi** *basin* C China

105 S16 **Sicie, Cap** *headland* SE France 43.03N 5.50E

109 J24 **Sicilia** *Eng.* Sicily; *anc.* Trinacria. ◆ *region* Italy, C Mediterranean Sea

109 M24 **Sicilia** *Eng.* Sicily; *anc.* Trinacria. *island* Italy, C Mediterranean Sea

Sicilian Channel *see* Sicily, Strait of

Sicily *see* Sicilia

109 H24 **Sicily, Strait of** *var.* Sicilian Channel. *strait* C Mediterranean Sea

44 K5 **Sico Tinto, Río** *var.* Río Negro. ⊘ N Honduras

59 H16 **Sicuani** Cusco, S Peru 14.18S 71.16W

114 D12 **Šid** Serbia, NW Yugoslavia 45.07N 19.13E

117 A15 **Sidári** Kérkyra, Iónioi Nísoi, Greece, C Mediterranean Sea 39.47N 19.43E

174 K8 **Sidas** Borneo, C Indonesia 0.24N 109.46E

100 O5 **Siddeburen** Groningen, NE Netherlands 53.15N 6.52E

160 D9 **Siddhapur** *prev.* Siddhpur, Sidhpur. Gujarāt, W India 23.57N 72.28E

161 I15 **Siddipet** Andhra Pradesh, C India 18.10N 78.54E

Sidhirókastron *see* Sidirókastro

Sidhpur *see* Siddhapur

Sidi al Hāni', Sabkhat *see* Sidi el Hani, Sebkhet

160 L9 **Sidhi** Madhya Pradesh, C India 24.23N 81.23E

79 N14 **Sidéradougou** SW Burkina 10.39N 4.16W

Siders *see* Sierre

77 U7 **Sidi Barráni** NW Egypt 31.33N 25.54E

76 I6 **Sidi Bel Abbès** *var.* Sidi bel Abbès, Sidi-Bel-Abbès. NW Algeria 35.12N 0.42W

76 E7 **Sidi-Bennour** W Morocco 32.39N 8.28W

76 M6 **Sidi Bouzid** *var.* Gammouda, Sidi Bu Zayd. C Tunisia 35.05N 9.20E

123 K12 **Sidi el Hani, Sebkhet de** *var.* Sabkhat Sīdī al Hāni'. *salt flat* NE Tunisia

76 D8 **Sidi-Ifni** SW Morocco 29.33N 10.04W

76 G6 **Sidi-Kacem** *prev.* Petitjean. N Morocco 34.21N 5.46W

116 G12 **Sidirókastro** *prev.* Sidhirókastron. Kentrikí Makedonía, NE Greece 41.13N 23.24E

204 L12 **Sidley, Mount** ▲ Antarctica 76.39S 124.48W

31 S16 **Sidney** Iowa, C USA 40.45N 95.39W

35 Y7 **Sidney** Montana, NW USA 47.42N 104.10W

30 J15 **Sidney** Nebraska, C USA 41.09N 102.57W

20 I11 **Sidney** New York, NE USA 42.18N 75.21W

33 R13 **Sidney** Ohio, N USA 40.16N 84.09W

25 T2 **Sidney Lanier, Lake** ☒ Georgia, SE USA

Sidon *see* Saïda

129 X4 **Sidorovsk** Yamalo-Nenetskiy Avtonomnyy Okrug, N Russian Federation 66.34N 82.12E

Sidra *see* Surt

Sidra/Sidra, Gulf of *see* Surt, Khalīj, N Libya

78 K13 **Sidra** *var.* Surt, N Libya

Siebenbürgen *see* Transylvania

Sieben Dörfer *see* Săcele

112 O12 **Siedlce** *Ger.* Sedlez, *Rus.* Sedlets. Mazowieckie, C Poland 52.10N 22.18E

103 E16 **Sieg** ⊘ W Germany

103 F16 **Siegen** Nordrhein-Westfalen, W Germany 50.52N 8.01E

111 X4 **Sieghartskirchen** Niederösterreich, E Austria 48.13N 16.01E

112 O11 **Siemiatycze** Podlaskie, NE Poland 52.27N 22.51E

178 Fj11 **Siĕmpang** Stœ̆ng Trĕng, NE Cambodia 14.07N 106.24E

178 Ii12 **Siĕmréab** *prev.* Siemreap. Siĕmréab, NW Cambodia 13.21N 103.49E

Siemreap *see* Siĕmréab

108 G7 **Siena** *Fr.* Sienne; *anc.* Saena Julia. Toscana, C Italy 43.19N 11.19E

Sienne *see* Siena

94 K12 **Sieppijärvi** Lappi, NW Finland 67.09N 23.58E

112 J13 **Sieradz** Sieradz, C Poland 51.35N 18.43E

112 K10 **Sierpc** Mazowieckie, C Poland 52.51N 19.43E

26 J9 **Sierra Blanca** Texas, SW USA 31.10N 105.21W

39 S14 **Sierra Blanca Peak** ▲ New Mexico, SW USA 33.22N 105.48W

37 P5 **Sierra City** California, W USA 39.34N 120.35W

65 G21 **Sierra Colorada** Río Negro, S Argentina 40.37S 67.48W

62 H5 **Sierra del Nevado** ▲ W Argentina

65 G15 **Sierra Grande** Río Negro, E Argentina 41.40S 65.22W

78 G15 **Sierra Leone** *off.* Republic of Sierra Leone. ◆ *republic* W Africa

66 L13 **Sierra Leone Basin** *undersea feature* E Atlantic Ocean

66 K8 **Sierra Leone Fracture Zone** *tectonic feature* E Atlantic Ocean

Sierra Leone Ridge *see* Sierra Leone Rise

66 L13 **Sierra Leone Rise** *var.* Sierra Leone Ridge, Sierra Leone Schwelle. *undersea feature* E Atlantic Ocean

Sierra Leone Schwelle *see* Sierra Leone Rise

45 U17 **Sierra Madre** ▲ Mexico/Guatemala

109 J24 **Sierra Madre** ▲ Luzon, N Philippines

39 R2 **Sierra Madre** ▲ Colorado/Wyoming, C USA

2 H15 **Sierra Madre del Sur** ▲ S Mexico

2 G13 **Sierra Madre Occidental** *var.* Western Sierra Madre. ▲ C Mexico

2 H13 **Sierra Madre Oriental** *var.* Eastern Sierra Madre. ▲ C Mexico

46 H8 **Sierra Maestra** ▲ E Cuba

42 L7 **Sierra Mojada** Coahuila de Zaragoza, C Mexico 27.13N 103.42W

107 O14 **Sierra Nevada** ▲ S Spain

37 P6 **Sierra Nevada** ▲ W USA

56 F4 **Sierra Nevada de Santa Marta** ▲ NE Colombia

44 K5 **Sierra Río Tinto** ▲ NE Honduras

26 K3 **Sierra Vieja** ▲ Texas, SW USA

39 N16 **Sierra Vista** Arizona, SW USA 31.33N 110.18W

110 D10 **Sierre** *Ger.* Siders. Valais, SW Switzerland 46.18N 7.33E

38 L16 **Sierrita Mountains** ▲ Arizona, SW USA

Siete Moai *see* Ahu Akivi

78 K5 **Sīfié** W Ivory Coast 7.58N 6.55W

117 I21 **Sífnos** *var.* Siphnos. *island* Kykládes, Greece, Aegean Sea

117 I21 **Sífnou, Stenó** *strait* SE Greece

79 N23 **Sigatoka** *prev.* Singatoka. Viti Levu, W Fiji 18.10S 105.30E

105 P16 **Sigean** Aude, S France 43.01N 2.58E

Sighet *see* Sighetu Marmaţiei

118 I8 **Sighetu Marmaţiei** *var.* Sighet, Sighetul Marmaţiei, *Hung.* Máramarossziget. Maramureş, N Romania 47.55N 23.52E

Sighetul Marmaţiei *var.* Sighet, Sighetu Marmaţiei, *Hung.*

118 I11 **Sighişoara** *Ger.* Schässburg, *Hung.* Segesvár. Mureş, C Romania 46.12N 24.48E

173 E3 **Sigli** Sumatera, W Indonesia 5.21N 95.55E

94 J1 **Siglufjördhur** Nordhurland Vestra, N Iceland 66.09N 18.55W

103 H23 **Sigmaringen** Baden-Württemberg, S Germany 48.04N 9.12E

103 N20 **Signalberg** ▲ SE Germany 49.30N 12.34E

38 I13 **Signal Peak** ▲ Arizona, SW USA 33.20N 114.03W

Signan *see* Xi'an

204 H1 **Signy** UK research station South Orkney Islands, Antarctica 60.27S 45.35W

31 X15 **Sigourney** Iowa, C USA 41.19N 92.12W

117 K17 **Sígri, Akrotírio** *headland* Lésvos, E Greece 39.12N 25.49E

9 P15 **Sigsbee Deep** *see* Mexico Basin

49 N2 **Sigsbee Escarpment** *undersea feature* N Gulf of Mexico

58 C8 **Sígsig** Azuay, S Ecuador 3.04S 78.49W

97 O15 **Sigtuna** Stockholm, C Sweden 59.36N 17.43E

44 P4 **Siguatepeque** Comayagua, W Honduras 14.38N 87.52W

34 I14 **Sigüenza** Castilla-La Mancha, C Spain 41.04N 2.37W

107 P7 **Sigües** Aragón, NE Spain 42.39N 1.01W

78 W3 **Siguiri** Haute-Guinée, NE Guinea 11.25N 9.07W

120 G8 **Sigulda** *Ger.* Segewold. Riga, C Latvia 57.08N 24.51E

79 Q7 **Sihanoukville** *var.* Kâmpóng Saôm

Sihlsee ⊗ NW Switzerland

95 M16 **Siikainen** Länsi-Suomi, W Finland 61.52N 21.49E

95 M16 **Siilinjärvi** Itä-Suomi, C Finland 63.04N 27.40E

143 R15 **Siirt** *var.* Sert; *anc.* Tigranocerta. Siirt, SE Turkey 37.55N 41.55E

143 R15 **Siirt** ◆ *province* SE Turkey

195 Z15 **Sikaiana** *var.* Stewart Islands. *island group* W Solomon Islands

Sikandarabad *see* Secunderābād

158 J11 **Sikandra Rao** Uttar Pradesh, N India 27.42N 78.22E

8 M11 **Sikanni Chief** British Columbia, W Canada 57.16N 122.44W

8 M11 **Sikanni Chief** ⊘ British Columbia, W Canada

158 H11 **Sīkar** Rājasthān, N India 27.39N 75.09E

78 M13 **Sikasso** Sikasso, S Mali 11.21N 5.42W

78 M13 **Sikasso** ◆ *region* SW Mali

178 Gg3 **Sikaw** Kachin State, C Myanmar 23.50N 97.04E

29 Y7 **Sikeston** Missouri, C USA 36.53N 89.35W

127 O16 **Sikhote-Alin', Khrebet** ▲ SE Russian Federation

165 J22 **Síkinos** *island* Kykládes, Greece, Aegean Sea

159 S11 **Sikkim** *Tib.* Denjong. ◆ *state* N India

111 I26 **Siklós** Baranya, SW Hungary 45.51N 18.18E

Sikondo *see* Shikoku

83 G14 **Sikongo** Western, W Zambia 15.03S 22.07E

Sikou/Sikourío *see* Sykourío

126 L7 **Siktyakh** Respublika Sakha (Yakutiya), NE Russian Federation 69.45N 124.42E

120 D12 **Šilalė** Šiauliai, W Lithuania 55.29N 22.10E

108 G5 **Silandro** *Ger.* Schlanders. Trentino-Alto Adige, N Italy 46.39N 10.55E

43 N12 **Silao** Guanajuato, C Mexico 20.55N 101.24W

Silarius *see* Sele

179 Q13 **Silay** *off.* Silay City. Negros, C Philippines 10.49N 122.58E

159 W14 **Silchar** Assam, NE India 24.49N 92.48E

110 G9 **Silenen** Uri, C Switzerland 46.49N 8.39E

23 T9 **Siler City** North Carolina, SE USA 35.43N 79.27W

35 U1 **Silesia** Montana, NW USA

112 F13 **Silesia** *physical region* SW Poland

76 K12 **Silet** S Algeria 22.43N 4.51E

151 R8 **Sileti** *var.* Selety. ⊘ N Kazakhstan

Siletitengiz *see* Siletiteniz, Ozero

151 R7 **Siletiteniz, Ozero** *Kaz.* Siletitengiz. ⊗ N Kazakhstan

180 H16 **Silhouette** *island* Inner Islands, SE Seychelles

142 I17 **Silifke** *anc.* Seleucia. İçel, S Turkey 36.22N 33.57E

162 J10 **Siliguri** *see* Shiliguri

198 Aa7 **Silisili** ▲ Savai'i, C Samoa 13.37S 172.26W

116 M6 **Silistra** *var.* Silistria; *anc.* Durostorum. Silistra, NE Bulgaria 44.07N 27.16E

116 M7 **Silistra** ◆ *province* NE Bulgaria

Silistria *see* Silistra

142 D10 **Silivri** İstanbul, NW Turkey 41.04N 28.15E

96 L13 **Siljan** ⊗ C Sweden

97 G22 **Silkeborg** Århus, C Denmark 56.10N 9.34E

110 M8 **Sill** ⊘ W Austria

107 S10 **Silla** País Valenciano, E Spain 39.22N 0.25W

64 I13 **Sillajguay, Cordillera** ▲ N Chile 19.45S 68.39W

120 K3 **Sillamäe** *Ger.* Sillamäggi. Ida-Virumaa, NE Estonia 59.23N 27.45E

Sillamäggi *see* Sillamäe

Sillein *see* Žilina

111 P9 **Sillian** Tirol, W Austria

114 B10 **Šilo** Primorje-Gorski Kotar, NW Croatia 45.09N 14.39E

49 R9 **Siloam Springs** Arkansas, C USA 36.11N 94.32W

27 X10 **Silsbee** Texas, SE USA 30.21N 94.10W

149 W15 **Silūp, Rūd-e** ⊘ SE Iran

120 C12 **Silute** *Ger.* Heydekrug. Šilutė, W Lithuania 55.21N 21.29E

143 Q15 **Silvan** Diyarbakır, SE Turkey 38.08N 41.00E

110 J10 **Silvaplana** Graubünden, S Switzerland 46.27N 9.45E

Silva Porto *see* Kuito

60 I9 **Silva, Recife do** *reef* E Brazil

160 D12 **Silvassa** Dādra and Nagar Haveli, W India 20.13S 73.03E

31 X4 **Silver Bay** Minnesota, N USA 47.17N 91.15W

39 P15 **Silver City** New Mexico, SW USA 32.46N 108.16W

20 D10 **Silver Creek** New York, NE USA 42.32N 79.10W

39 N12 **Silver Creek** ⊘ Arizona, SW USA

44 P4 **Silver Lake** Kansas, C USA 39.06N 95.51W

34 I14 **Silver Lake** Oregon, NW USA 43.07N 121.04W

37 T9 **Silver Peak Range** ▲ Nevada, W USA

39 W3 **Silver Spring** Maryland, NE USA 38.59N 77.01W

Silver State *see* Nevada

Silver State *see* Colorado

39 Q7 **Silverton** Colorado, SW USA 37.48N 107.39W

20 K16 **Silverton** New Jersey, NE USA 40.00N 74.09W

34 I14 **Silverton** Oregon, NW USA 45.00N 122.46W

27 N4 **Silverton** Texas, SW USA 34.28N 101.19W

106 G14 **Silves** Faro, S Portugal 37.10N 8.25W

56 D12 **Silvia** Cauca, SW Colombia 2.36N 76.24W

110 J9 **Silvrettagruppe** ▲ Austria/Switzerland

Sily-Vajdej *see* Vulcan

110 I7 **Silz** Tirol, W Austria 47.17N 11.00E

180 J13 **Sima** Anjouan, SE Comoros 12.10S 44.18E

Simabara *see* Shimabara

Simada *see* Shimada

Simane *see* Shimane

121 L20 **Simanichy** *Rus.* Simonichi. Homyel'skaya Voblasts', SE Belarus 51.53N 28.04E

166 F14 **Simao** Yunnan, SW China 22.30N 101.06E

159 P12 **Simara** Central, C Nepal 27.14N 85.00E

12 M7 **Simard, Lac** ⊗ Quebec, SE Canada

143 V14 **Simav** Kütahya, W Turkey 39.04N 28.58E

142 Q13 **Simav** ⊘ NW Turkey

81 L18 **Simba** Orientale, N Dem. Rep. Congo (Zaire) 0.46N 22.54E

194 H11 **Simbai** Madang, N PNG 5.12S 144.33E

195 O9 **Simberi Island** *island* Tabar Islands, N PNG

82 F17 **Simbirsk** *prev.* Ul'yanovsk

116 K11 **Simeonovgrad** *prev.* Maritsa. Khaskovo, S Bulgaria 42.03N 25.36E

118 G11 **Simeria** *Ger.* Pischk, *Hung.* Piski. Hunedoara, W Romania 45.51N 23.00E

109 L24 **Simeto** ✦ Sicilia, Italy,
C Mediterranean Sea

173 E5 **Simeulue, Pulau** island
NW Indonesia

119 T13 **Simferopol'** Respublika Krym,
S Ukraine 44.55N 33.05E

119 T13 **Simferopol'** ✕ Respublika Krym,
S Ukraine 44.55N 34.04E

Simi see **Sými**

158 M9 **Simikot** Far Western, NW Nepal
30.02N 81.49E

56 F7 **Simití** Bolívar, N Colombia
7.57N 73.57W

116 G11 **Simitla** Blagoevgrad, SW Bulgaria
41.57N 23.06E

37 S15 **Simi Valley** California, W USA
34.17N 118.52W

Simizu see **Shimizu**

Simla see **Shimla**

**Şimlãul Silvaniei/Şimleul
Silvaniei** see **Şimleu Silvaniei**

118 G9 **Şimleu Silvaniei** Hung.
Szilágysomlyó; prev. Şimlãul
Silvaniei, Şimleul Silvaniei. Sãlaj,
NW Romania 47.12N 22.48E

Simmer see **Simmerbach**

103 E19 **Simmerbach** ✦
✦ W Germany

103 F18 **Simmern** Rheinland-Pfalz,
W Germany 50.00N 7.30E

24 I7 **Simmesport** Louisiana, S USA
30.58N 91.48W

121 F14 **Simnas** Alytus, S Lithuania
54.23N 23.40E

94 L13 **Simo** Lappi, NW Finland
65.40N 25.04E

Simoda see **Shimoda**

Simodate see **Shimodate**

94 M13 **Simojärvi** ⊙ N Finland

94 L13 **Simojoki** ✦ NW Finland

43 U15 **Simojovel** var. Simojovel de
Allende. Chiapas, SE Mexico
17.12N 92.42W

Simojovel de Allende see
Simojovel

58 B7 **Simón Bolívar** var. Guayaquil.
✕ (Guayaquil) Guayas, W Ecuador
2.16S 79.54W

56 L5 **Simón Bolívar** ✕ (Caracas)
Distrito Federal, N Venezuela

Simonichi see **Simanichy**

12 M10 **Simon, Lac** ⊙ Quebec, SE Canada

Simonoseki see **Shimonoseki**

Šimonovany see **Partizánske**

Simonstad see **Simon's Town**

85 E26 **Simon's Town** var. Simonstad.
Western Cape, SW South Africa
34.12S 18.25E

Simony see **Partizánske**

Simotuma see **Shimotsuma**

173 F6 **Simpangkaman, Sungai**
✦ Sumatera, NW Indonesia

173 F5 **Simpangkiri, Sungai**
✦ Sumatera, NW Indonesia

Simpeln see **Simplon**

101 M18 **Simpelveld** Limburg,
SE Netherlands 50.49N 5.58E

110 E11 **Simplon** var. Simpeln. Valais,
SW Switzerland 46.13N 8.01E

110 E11 **Simplon Pass** pass S Switzerland
46.18N 8.01E

108 C6 **Simplon Tunnel** tunnel
Italy/Switzerland

Simpson see **Fort Simpson**

190 G1 **Simpson Desert** desert Northern
Territory/South Australia

8 J9 **Simpson Peak** ▲ British
Columbia, W Canada
59.43N 131.29W

15 L3 **Simpson Peninsula** peninsula
Nunavut, NE Canada

23 P1 **Simpsonville** South Carolina,
SE USA 34.44N 82.15W

97 L23 **Simrishamn** Skåne, S Sweden
55.34N 14.20E

127 Pp15 **Simushir, Ostrov** island
Kuril'skiye Ostrova, SE Russian
Federation

Sinä'/Sinai Peninsula see **Sinai**

173 Ee6 **Sinabang** Sumatera, W Indonesia
2.27N 96.24E

83 N15 **Sina Dhaqa** Galguduud,
C Somalia 5.21N 46.21E

77 X8 **Sinai** var. Sinai Peninsula, Ar.
Shibh Jazīrat Sīnã', Sīnā'. physical
region NE Egypt

118 J12 **Sinaia** Prahova, SE Romania
45.19N 25.33E

196 B16 **Sinajana** C Guam 13.28N 144.45E

42 H8 **Sinaloa** ✦ state C Mexico

56 H4 **Sinamaica** Zulia, NW Venezuela
11.06N 71.52W

169 X14 **Sinan-ni** SE North Korea
38.37N 127.43E

Sinano Gawa see **Shinano-gawa**

Sinãwan see **Sinãwin**

77 N8 **Sinãwin** var. Sinãwan. NW Libya
31.00N 10.37E

85 J16 **Sinazongwe** Southern, S Zambia
17.11S 27.20E

177 Ff6 **Sinbaungwe** Magwe, W Myanmar
19.43N 95.10E

177 Ff5 **Sinbyugyun** Magwe, W Myanmar
20.37N 94.40E

56 E6 **Since** Sucre, NW Colombia
9.15N 75.12W

56 E6 **Sincelejo** Sucre, NW Colombia
9.16N 75.22W

177 F5 **Sinchaingbyin** var. Zullapara.
Arakan State, W Myanmar
20.51N 92.32E

25 U4 **Sinclair, Lake** ⊠ Georgia,
SE USA

8 M14 **Sinclair Mills** British Columbia,
SW Canada 54.03N 121.37W

178 Mm15 **Sin Cowe East Island** island
S Spratly Islands

178 Mm15 **Sin Cowe Island** island
SW Spratly Islands

155 Q14 **Sind** var. Sindh. ✦ province
SE Pakistan

160 I8 **Sind** ✦ N India

97 H19 **Sindal** Nordjylland, N Denmark
57.28N 10.13E

179 Q15 **Sindangan** Mindanao,
S Philippines 8.09N 122.59E

81 D19 **Sindara** Ngounié, W Gabon
1.07S 10.40E

158 E13 **Sindari** prev. Sindri. Rãjasthãn,
N India 25.34N 71.57E

175 Q16 **Sindeh, Teluk** bay Nusa Tenggara,
C Indonesia

116 N8 **Sindel** Varna, E Bulgaria
43.07N 27.35E

103 H22 **Sindelfingen** Baden-
Württemberg, SW Germany
48.43N 9.01E

161 G16 **Sindgi** Karnãtaka, C India
17.01N 76.22E

Sindh see **Sind**

120 G8 **Sindi** Ger. Zintenhof. Pärnumaa,
SW Estonia 58.25N 24.40E

142 C13 **Sındırgı** Balıkesir, W Turkey
39.15N 28.10E

79 N14 **Sindou** SW Burkina 10.34N 5.04W

Sindri see **Sindari**

155 T9 **Sind Sãgar Doãb** desert E Pakistan

130 M11 **Sinegorskiy** Rostovskaya Oblast',
SW Russian Federation
48.01N 40.52E

127 O9 **Sinegor'ye** Magadanskaya Oblast',
E Russian Federation 62.04N 150.33E

116 O12 **Sinekli** İstanbul, NW Turkey
41.13N 28.11E

106 F12 **Sines** Setúbal, S Portugal
37.58N 8.52W

106 F12 **Sines, Cabo de** headland
S Portugal 37.57N 8.55W

94 L12 **Sinettä** Lappi, NW Finland
66.39N 25.25E

195 P11 **Sinewit, Mount** ▲ New Britain,
C PNG 4.42S 151.58E

82 G11 **Singa** var. Sinja, Sinjah. Sinnar,
E Sudan 13.07N 33.54E

80 J12 **Singa** Moyen-Chari, S Chad
9.52N 19.31E

Singan see **Xi'an**

174 I7 **Singapore** ● (Singapore)
S Singapore 1.17N 103.48E

174 I7 **Singapore** off. Republic of
Singapore. ◆ republic SE Asia

174 I7 **Singapore Strait** var. Strait of
Singapore, Mal. Selat Singapura.
strait Indonesia/Singapore

**Singapore, Strait of/Singapura,
Selat** see Singapore Strait

175 N16 **Singaraja** Bali, C Indonesia
8.06S 115.04E

Singatoka see **Sigatoka**

178 H10 **Sing Buri** var. Singhaburi. Sing
Buri, C Thailand 14.55N 100.21E

103 H24 **Singen** Baden-Württemberg,
S Germany 47.46N 8.49E

Singeorgiu de Pãdure see
Sângeorgiu de Pãdure

Singeorz-Bãi/Singerzo Bãi see
Sângeorz-Bãi

118 M9 **Sîngerei** var. Sîngerei; prev.
Lazovsk. N Moldova 47.38N 28.08E

83 N1 **Singhaburi** see Sing Buri

83 H21 **Singida** Singida, C Tanzania
4.45S 34.48E

83 G22 **Singida** ◆ region C Tanzania

Singidunum see **Beograd**

177 G2 **Singkaling Hkamti** Sagaing,
N Myanmar 26.00N 95.43E

175 Pp12 **Singkang** Sulawesi, C Indonesia
4.09S 119.58E

174 Gg8 **Singkarak, Danau** ⊚ Sumatera,
W Indonesia

174 K7 **Singkawang** Borneo, C Indonesia
0.57N 108.57E

173 F6 **Singkep, Pulau** island Kepulauan
Lingga, W Indonesia

191 T7 **Singleton** New South Wales,
SE Australia 32.36S 151.10E

Singora see **Songkhla**

Singü see **Shingü**

109 D17 **Siniscola** Sardegna, Italy,
C Mediterranean Sea 40.34N 9.42E

115 F14 **Sinj** Split-Dalmacija, SE Croatia
43.41N 16.37E

Sinja/Sinjah see **Singa**

Sinjajevina see **Sinjavina**

145 R13 **Sinjãr** NW Iraq 36.19N 41.51E

145 P2 **Sinjãr, Jabal** ▲ N Iraq

115 K15 **Sinjavina** var. Sinjajevina.
▲ SW Yugoslavia

82 I7 **Sinkat** Red Sea, NE Sudan
18.52N 36.51E

**Sinkiang/Sinkiang Uighur
Autonomous Region** see
Xinjiang Uygur Zizhiqu

Sinmal a see **Tãrnãveni**

169 N13 **Sinmi-do** island NW North Korea

101 I18 **Sinn** ✦ C Germany

57 Y9 **Sinnamarie** see Sinnamary

57 Y9 **Sinnamary** var. Sinnamarie.
N French Guiana 5.23N 52.57W

82 G11 **Sinnar** ◆ state E Sudan

Sinneh see **Sanandaj**

20 E13 **Sinnemahoning Creek**
✦ Pennsylvania, NE USA

Sinnicolau Mare see Sânnicolau
Mare

Sino/Sinoe see **Greenville**

161 I12 **Sinoe, Lacul** prev. Lacul Sinoie.
lagoon SE Romania

119 N14 **Sinoie, Lacul** see Lacul Sinoe.

47 V9 **Sint Eustatius** Eng. Saint
Eustatius. island N Netherlands
Antilles

101 G19 **Sint-Genesius-Rode** Fr. Rhode-
Saint-Genèse. Vlaams Brabant,
C Belgium 50.45N 4.21E

101 F16 **Sint-Gillis-Waas** Oost-
Vlaanderen, N Belgium 51.14N 4.08E

101 H17 **Sint-Katelijne-Waver**
Antwerpen, C Belgium 51.05N 4.31E

101 E18 **Sint-Lievens-Houtem** Oost-
Vlaanderen, NW Belgium
50.55N 3.52E

47 V9 **Sint Maarten** Eng. Saint Martin.
island N Netherlands Antilles

101 F14 **Sint Maartensdijk** Zeeland,
SW Netherlands 51.33N 4.05E

101 I19 **Sint-Martens-Voeren** Fr.
Fouron-Saint-Martin. Limburg,
NE Belgium 50.46N 5.49E

101 J14 **Sint-Michielsgestel** Noord-
Brabant, S Netherlands 51.37N 5.21E

47 O6 **Sint-Miclaas** see Gheorgheni

176 Ww17 **Siriwo** ✦ Irian Jaya, E Indonesia

101 F16 **Sint Nicolaas** S Aruba
12.25N 69.52W

101 F16 **Sint-Niklaas** Fr. Saint-Nicolas.
Oost-Vlaanderen, N Belgium
51.10N 4.09E

101 K14 **Sint-Oedenrode** Noord-Brabant,
S Netherlands 51.34N 5.28E

27 T14 **Sinton** Texas, SW USA
28.02N 97.30W

101 G14 **Sint Philipsland** Zeeland,
SW Netherlands 51.37N 4.11E

101 G16 **Sint-Pieters-Leeuw** Vlaams
Brabant, C Belgium 50.46N 4.16E

106 F12 **Sintra** prev. Cintra. Lisboa,
W Portugal 38.48N 9.22W

101 K14 **Sint-Truiden** Fr. Saint-Trond.
Limburg, NE Belgium 50.48N 5.13E

101 H14 **Sint Willebrord** Noord-Brabant,
S Netherlands 51.33N 4.34E

169 V11 **Sinŭiju** W North Korea
40.08N 124.33E

82 P13 **Sini'il Nugaal, NE Somalia
8.33N 49.05E

Sinus Aelaniticus see Aqaba,
Gulf of

Sinus Gallicus see Lion, Golfe du

Sinyang see **Xinyang**

121 I18 **Sinyawka** var. Sinyawka.
Minskaya Voblasts', SW Belarus
52.57N 26.29E

126 Ll11 **Sinyaya** ✦ NE Russian
Federation

Sinying see Hsinying

Sinyukha see **Synyukha**

Sinzi-ko see Shinji-ko

Sinzyõ see **Shinjõ**

113 L14 **Sió** ✦ W Hungary

179 Q16 **Siocon** Mindanao, S Philippines
7.37N 122.09E

113 L14 **Siófok** Somogy, C Hungary
46.54N 18.03E

Siogama see Shiogama

85 G15 **Sioma** Western, SW Zambia
16.41S 23.34E

110 D11 **Sion** Ger. Sitten; anc. Sedunum.
Valais, SW Switzerland 46.15N 7.23E

105 O11 **Sioule** ✦ C France

31 S12 **Sioux Center** Iowa, C USA
43.04N 96.10W

31 R13 **Sioux City** Iowa, C USA
42.30N 96.24W

31 R13 **Sioux Falls** South Dakota, N USA
43.33N 96.45W

10 B11 **Sioux Lookout** Ontario,
S Canada 50.07N 91.54W

31 T12 **Sioux Rapids** Iowa, C USA
42.53N 95.09W

Sioux State see North Dakota

Sioziri see Shiojiri

179 Q14 **Sipalay** Negros, C Philippines
9.46N 122.25E

57 V11 **Sipaliwini** ◆ district S Suriname

47 U15 **Siparia** Trinidad, Trinidad and
Tobago 10.07N 61.33W

Siphnos see **Sífnos**

169 V11 **Siping** var. Ssu-p'ing, Szeping;
prev. Ssu-p'ing-chieh. Jilin,
NE China 43.09N 124.22E

9 X12 **Sipiwesk** Manitoba, C Canada
55.28N 97.16W

9 W13 **Sipiwesk Lake** ⊚ Manitoba,
C Canada

205 O11 **Siple Coast** physical region
Antarctica

204 K12 **Siple Island** island Antarctica

204 K13 **Siple, Mount** ▲ Siple Island,
Antarctica 73.25S 126.24W

Sipoo see **Sibbo**

114 G12 **Sipovo** Republika Srpska,
W Bosnia and Herzegovina
44.16N 17.05E

25 O4 **Sipsey River** ✦ Alabama, S USA

173 Hh9 **Sipura, Pulau** island W Indonesia

2 G16 **Siqueiros Fracture Zone** tectonic
feature E Pacific Ocean

44 J10 **Siquia, Río** ✦ SE Nicaragua

179 Q14 **Siquijor Island** island
C Philippines

45 N13 **Siquirres** Limón, E Costa Rica
10.06N 83.33W

56 J5 **Siquisique** Lara, N Venezuela
10.36N 69.38W

161 G19 **Sira** Karnãtaka, W India
13.46N 76.54E

97 C16 **Sira** ✦ S Norway

178 Hh12 **Siracha** var. Ban Si Racha, Si
Racha. Chon Buri, S Thailand
13.10N 100.57E

109 L25 **Siracusa** Eng. Syracuse. Sicilia,
Italy, C Mediterranean Sea
37.04N 15.16E

159 N14 **Sirãjganj** see Shirajganj Ghat

9 N14 **Sir Alexander, Mount** ▲ British
Columbia, W Canada
54.00N 120.33W

143 O12 **Şiran** Gümüşhane, NE Turkey
40.12N 39.07E

79 O12 **Sirba** ✦ E Burkina

149 O17 **Şir Bani Yãs** island W UAE

97 D17 **Sirdalsvatnet** ⊚ S Norway

Sir Darya/Sirdaryo see Syr Darya

153 O11 **Sirdaryo Wiloyati** Rus.
Syrdar'inskaya Oblast'. ◆ province
E Uzbekistan

**Sir Donald Sangster
International Airport** see
Sangster

194 H13 **Sirebi** ✦ S PNG

189 S3 **Sir Edward Pellew Group** island
group Northern Territory,
NE Australia

118 K8 **Siret** Ger. Sereth, Hung. Szeret.
Suceava, N Romania 47.55N 26.04E

118 K8 **Siret** var. Siretul, Ger. Sereth, Rus.
Seret, Ukr. Siret.
✦ Romania/Ukraine

Siretul see **Siret**

146 K3 **Sirḩãn, Wãdì as** dry watercourse
Jordan/Saudi Arabia

158 I8 **Sirhind** Punjab, N India
30.39N 76.28E

118 F11 **Şiria** Ger. Schiria. Arad,
W Romania 46.16N 21.37E

149 S14 **Sirik** Hormozgãn, SE Iran
26.31N 57.06E

178 Hh8 **Sirikit Reservoir** ⊠ N Thailand

60 N2 **Sirituba, Ilha** island NE Brazil

176 Ww17 **Siriwo** ✦ Irian Jaya, E Indonesia

149 R11 **Sirjãn** prev. Sa'ìdãbãd. Kermãn,
S Iran 29.28N 55.39E

190 H9 **Sir Joseph Banks Group** island
group South Australia

94 K11 **Sirkka** Lappi, N Finland
67.49N 24.48E

143 R16 **Sirna** see **Sýrna**

143 S16 **Sirnach** ▲ ◆ province SE Turkey
37.33N 42.27E

161 J14 **Sironcha** Mahãrãshtra, C India
18.51N 80.03E

Sirone see **Shirone**

120 M12 **Sirotina** Rus. Sirotino.
Vitsyebskaya Voblasts', N Belarus
55.22N 29.34E

158 H9 **Sirsa** Haryãna, NW India
29.31N 75.04E

181 V17 **Sir Seewoosagur Ramgoolam**
✕ (Port Louis) SE Mauritius

161 E18 **Sirsi** Karnãtaka, W India
14.46N 74.49E

190 A2 **Sir Thomas, Mount** ▲ South
Australia 27.09S 129.49E

105 T16 **Six-Fours-les-Plages** Var,
SE France 43.04N 5.49E

167 Q7 **Sixian** var. Si Xian. Anhui,
E China 33.28N 117.52E

24 J9 **Six Mile Lake** ⊚ Louisiana,
S USA

145 V3 **Siyãh Güz** E Iraq 35.49N 45.45E

161 L25 **Siyãmbalanduwa** Uva Province,
SE Sri Lanka 6.54N 81.31E

143 Y10 **Siyãzãn** Rus. Siazan'.
NE Azerbaijan 41.04N 49.04E

Sizebolu see Sozopol

Sizuoka see Shizuoka

Sjar see **Sãre**

115 L15 **Sjenica** Turk. Seniça. Serbia,
SW Yugoslavia 43.16N 20.01E

96 G13 **Sjoa** ✦ S Norway

97 K23 **Sjöbo** Skåne, S Sweden
55.37N 13.45E

96 F11 **Sjælland** Eng. Zealand, Ger.
Seeland. island E Denmark

96 E9 **Sjøholt** Møre og Romsdal,
S Norway 62.28N 6.49E

97 F22 **Sjørn** Ringkøbing, W Denmark
55.55N 8.30E

97 F22 **Sjern Å** var. Skjern Aa.
✦ W Denmark

97 I22 **Skjern Aa** see Skjern Å

94 J8 **Skjerstad** Nordland, C Norway
67.14N 15.06E

119 R11 **Skjervøy** Troms, N Norway
70.01N 20.57E

94 I9 **Skjold** Troms, N Norway
69.03N 19.18E

113 I17 **Skoczów Śląskie, S Poland
49.47N 18.46E

97 I24 **Skælskør** Vestsjælland,
E Denmark 55.16N 11.18E

111 T11 **Škofja Loka** Ger. Bischoflack.
NW Slovenia 46.12N 14.16E

96 N12 **Skog** Gävleborg, C Sweden
61.10N 16.49E

97 K16 **Skoghall** Värmland, C Sweden
59.19N 13.30E

33 N10 **Skokie** Illinois, N USA
42.01N 87.43W

118 H6 **Skole** L'viv'ska Oblast', W Ukraine
49.04N 23.29E

117 H17 **Skóllis** ▲ S Greece 37.58N 21.33E

178 J13 **Skon** Kâmpóng Cham,
C Cambodia 12.56N 104.36E

117 H17 **Skópelos** Skópelos, Vóreioi
Sporádes, Greece, Aegean Sea
39.07N 23.44E

117 H17 **Skópelos** island Vóreioi Sporádes,
Greece, Aegean Sea

113 P14 **Skopin** Ryazanskaya Oblast',
W Russian Federation 53.46N 39.37E

115 N18 **Skopje** var. Usküb, Turk. Üsküb;
prev. Skoplje, anc. Scupi. ● (FYR
Macedonia) N FYR Macedonia
42.01N 21.27E

115 O18 **Skopje** ✕ N FYR Macedonia
41.58N 21.35E

Skoplje see Skopje

Skórcz Ger. Skurz. Pomorskie,
N Poland 53.46N 18.43E

Skorodnoye see Skarodnaye

117 H17 **Skántzoúra** island Vóreioi
Sporádes, Greece, Aegean Sea

95 M18 **Skorped** Västernorrland,
C Sweden 63.22N 17.56E

112 G11 **Skórzewo** Zachodniopomorskie,
NW Poland 54.22N 16.43E

113 G19 **Sławno** see Slavonia

96 I16 **Skreia** Oppland, S Norway
60.37N 11.00E

130 J11 **Skripón** see Orchómenos

113 O16 **Skrudaliena** Daugvapils,
SE Latvia 55.50N 26.42E

120 D12 **Skrunda** Kuldīga, W Latvia
56.39N 22.01E

97 C16 **Skudeneshavn** Rogaland,
S Norway 59.09N 5.16E

85 L20 **Skukuza** Mpumalanga, NE South
Africa 24.54S 31.33E

99 B22 **Skull** Ir. An Scoil. SW Ireland
51.32N 9.33W

24 L3 **Skuna River** ✦ Mississippi,
S USA

31 X15 **Skunk River** ✦ Iowa, C USA

28 C10 **Skuodas** Ger. Schoden, Pol.
Szkudy. Skuodas, NW Lithuania
56.16N 21.30E

97 K23 **Skurup** Skåne, S Sweden
55.28N 13.30E

116 H8 **Skǔf** ✦ NW Bulgaria

96 O13 **Skutskär** Uppsala, C Sweden

97 B19 **Skúvoy** Dan. Skuø island Faeroe
Islands 61.46N 6.49W

Skvira see Skvyra

119 O5 **Skvyra** Rus. Skvira. Kyyivs'ka
Oblast', N Ukraine 49.44N 29.40E

41 Q11 **Skwentna** Alaska, USA
61.56N 151.03W

112 E11 **Skwierzyna** Ger. Schwerin.
Lubuskie, W Poland 52.36N 15.27E

38 K13 **Sky Harbour** ✕ (Phoenix)
Arizona, SW USA 33.26N 112.00W

98 G9 **Skye, Isle of** island NW Scotland,
UK

34 I8 **Skykomish** Washington,
NW USA 47.40N 121.20W

65 F19 **Skylge** see Terschelling

65 H24 **Skyring, Peninsula** peninsula
S Chile

65 H24 **Skyring, Seno** inlet S Chile

117 H17 **Skyropoúla** var. Skiropoula. island
Vóreioi Sporádes, Greece, Aegean
Sea

117 I17 **Skýros** var. Skíros. Skýros, Vóreioi
Sporádes, Greece, Aegean Sea
38.55N 24.34E

117 I17 **Skýros** var. Skíros; anc. Scyros.
island Vóreioi Sporádes, Greece,
Aegean Sea

120 J12 **Slabodka** Rus. Slobodka.
Vitsyebskaya Voblasts', NW Belarus
55.42N 27.10E

97 I23 **Slagelse** Vestsjælland, E Denmark
55.25N 11.21E

95 I14 **Slagnäs** Norrbotten, N Sweden
65.35N 18.12E

174 Kk15 **Slamet, Gunung** ▲ Jawa,
S Indonesia 7.12S 109.13E

41 S10 **Slana** Alaska, USA 62.46N 144.00W

99 F20 **Slaney** Ir. An tSláine.
✦ SE Ireland

118 J13 **Slănic** Prahova, SE Romania
45.13N 25.58E

118 K11 **Slănic Moldova** Bacãu, E Romania
46.12N 26.23E

115 H16 **Slano** Dubrovnik-Neretva,
SE Croatia 42.47N 17.54E

128 F13 **Slantsy** Leningradskaya Oblast',
NW Russian Federation
59.06N 28.00E

113 C16 **Slaný** Ger. Schlan. Střední Čechy,
NW Czech Republic 50.13N 14.04E

118 K8 **Ślaskie** ◆ province S Poland

10 C10 **Slate Falls** Ontario, S Canada
51.11N 91.32W

49 T4 **Slater** Missouri, C USA
39.13N 93.04W

114 F9 **Slater** Lake Alberta, SW Canada
55.16N 114.46W

40 P13 **Slave Lake** Alberta, SW Canada

70 G11 **Slave Coast** coastal region W Africa

8 J9 **Slave** ✦ Alberta/Northwest
Territories, C Canada

125 O14 **Slavgorod** Altayskiy Kray,
S Russian Federation 52.55N 78.46E

Slavgorod see Slawharad

Slavonia see Slavonija

114 G9 **Slavonija** Eng. Slavonia, Ger.
Slawonien, Hung. Szlavónia,
Szlavonország. cultural region
NE Croatia

Slavonska Požega see Požega

114 H10 **Slavonski Brod** Ger. Brod, Hung.
Bród; prev. Brod, Brod na Savi.
Brod-Posavina, NE Croatia
45.09N 18.00E

118 L4 **Slavuta** Khmel'nyts'ka Oblast',
W Ukraine 50.18N 26.52E

119 P2 **Slavutych** Chernihiv's'ka Oblast',
N Ukraine 51.31N 30.47E

127 R14 **Slavyanka** Primorskiy Kray,
SE Russian Federation
42.46N 131.19E

116 J6 **Slavyanovo** Pleven, N Bulgaria
43.28N 24.52E

Slavyansk see Slov"yans'k

130 N12 **Slavyansk-na-Kubani**
Krasnodarskiy Kray, SW Russian
Federation 45.16N 38.09E

121 N20 **Slavyechna** Rus. Slovechna.
✦ Belarus/Ukraine

121 O16 **Slawharad** Rus. Slavgorod.
Mahilyowskaya Voblasts', E Belarus
53.27N 31.00E

112 G12 **Sławno** Zachodniopomorskie,
NW Poland 54.22N 16.43E

Slawonien see Slavonija

31 U6 **Slayton** Minnesota, N USA
43.59N 95.45W

99 L18 **Sleaford** E England, UK
52.58N 0.27W

99 A20 **Slea Head** Ir. Ceann Sléibhe.
headland SW Ireland 52.05N 10.25W

98 F7 **Sleat, Sound of** strait
NW Scotland, UK

6 L6 **Sledyuki** see Slyedzyuki

10 I5 **Sleeper Islands** island group
Nunavut, C Canada

31 T7 **Sleepy Eye** Minnesota, N USA
44.18N 94.43W

41 O11 **Sleetmute** Alaska, USA
61.42N 157.10W

99 E18 **Slíebhe, Ceann** see Slea Head

Slěmãní see As Sulaymãnīyah

205 O1 **Slessor Glacier** glacier Antarctica

24 L9 **Slidell** Louisiana, S USA 30.16N 89.46W

20 K12 **Slide Mountain** ▲ New York, NE USA 42.00N 74.23W

100 I13 **Sliedrecht** Zuid-Holland, C Netherlands 51.49N 4.45E

123 Jj16 **Sliema** N Malta 35.54N 14.31E

99 G16 **Slieve Donard** ▲ SE Northern Ireland, UK 54.10N 5.57W

Sligeach see Sligo

99 D16 **Sligo** Ir. Sligeach. NW Ireland 54.16N 8.28W

99 C16 **Sligo** Ir. Sligeach. cultural region NW Ireland

99 C15 **Sligo Bay** Ir. Cuan Shligigh. inlet NW Ireland

20 B13 **Slippery Rock** Pennsylvania, NE USA 41.02N 80.02W

97 P19 **Slite** Gotland, SE Sweden 57.37N 18.46E

116 L9 **Sliven** var. Slivno. Sliven, C Bulgaria 42.41N 26.15E

116 L10 **Sliven** ◆ province C Bulgaria

116 G9 **Slivnitsa** Sofiya, W Bulgaria 42.51N 23.01E

Slivno see Sliven

116 L7 **Slivo Pole** Ruse, N Bulgaria 43.57N 26.15E

31 S13 **Sloan** Iowa, C USA 42.13N 96.13W

37 X12 **Sloan** Nevada, W USA 35.56N 115.13W

Slobodka see Slabodka

129 R14 **Slobodskoy** Kirovskaya Oblast', NW Russian Federation 58.43N 50.12E

Slobozeya see Slobozia

119 O10 **Slobozia** Rus. Slobodzeya. E Moldova 46.45N 29.42E

118 L14 **Slobozia** Ialomiţa, SE Romania 44.34N 27.22E

100 O5 **Slochteren** Groningen, NE Netherlands 53.13N 6.48E

121 H17 **Slonim** Pol. Słonim, Rus. Slonim. Hrodzyenskaya Voblasts', W Belarus 53.04N 25.21E

100 K7 **Sloter Meer** ⊚ N Netherlands

Slot, The see New Georgia Sound

99 N22 **Slough** S England, UK 51.31N 0.36W

113 J20 **Slovakia** off. Slovenská Republika, Ger. Slowakei, Hung. Szlovákia, Slvk. Slovensko. ◆ republic C Europe

Slovak Ore Mountains see Slovenské rudohorie

Slovechna see Slavvechna

111 S12 **Slovenia** off. Republic of Slovenia, Ger. Slowenien, Slvn. Slovenija. ◆ republic SE Europe

Slovenia see Slovenia

111 V10 **Slovenj Gradec** Ger. Windischgraz. N Slovenia 46.29N 15.05E

111 W10 **Slovenska Bistrica** Ger. Windischfeistritz. NE Slovenia 46.21N 15.27E

Slovenská Republika see Slovakia

111 W10 **Slovenske Konjice** E Slovenia 46.21N 15.28E

113 K20 **Slovenské rudohorie** Eng. Slovak Ore Mountains, Ger. Slowakisches Erzgebirge, Ungarisches Erzgebirge. ▲ C Slovakia

Slovensko see Slovakia

119 Y7 **Slov'yanoserbs'k** Luhans'ka Oblast', E Ukraine 48.41N 39.00E

119 W6 **Slov"yans'k** Rus. Slavyansk. Donets'ka Oblast', E Ukraine 48.51N 37.38E

Slowakei see Slovakia

Slowakisches Erzgebirge see Slovenské rudohorie

Slowenien see Slovenia

112 D11 **Słubice** Ger. Frankfurt. Lubuskie, W Poland 52.19N 14.34E

121 K19 **Sluch** Rus. Sluch'. ◆ C Belarus

118 L4 **Sluch** ◆ NW Ukraine

101 D16 **Sluis** Zeeland, SW Netherlands 51.18N 3.22E

114 D10 **Slunj** Hung. Szluin. Karlovac, C Croatia 45.06N 15.35E

112 I11 **Słupca** Wielkopolskie, C Poland 52.16N 17.54E

112 G6 **Słupia** Ger. Stolpe. ◆ NW Poland

112 G6 **Słupsk** Ger. Stolp. Pomorskie, N Poland 54.27N 17.01E

121 K18 **Slutsk** Rus. Sluck. Minskaya Voblasts', S Belarus 53.01N 27.31E

121 O16 **Slyedzyuki** Rus. Sledyuki. Mahilyowskaya Voblasts', E Belarus 53.34N 30.19E

99 A17 **Slyne Head** Ir. Ceann Léime. headland W Ireland 53.25N 10.11W

126 J16 **Slyudyanka** Irkutskaya Oblast', S Russian Federation 51.36N 103.28E

49 U14 **Smackover** Arkansas, C USA 33.21N 92.43W

97 L20 **Småland** cultural region S Sweden

97 K20 **Smålandsstenar** Jönköping, S Sweden 57.10N 13.24E

11 O8 **Smallwood Reservoir** ⊠ Newfoundland, S Canada

121 N14 **Smalyany** Rus. Smolyany. Vitsyebskaya Voblasts', NE Belarus 54.36N 30.04E

121 L15 **Smalyavichy** Rus. Smolevichi. Minskaya Voblasts', C Belarus 54.01N 28.09E

76 C9 **Smara** var. Es Semara. N Western Sahara 26.65N 11.43W

121 J14 **Smarhon'** Pol. Smorgonie, Rus. Smorgon'. Hrodzyenskaya Voblasts', W Belarus 54.28N 26.24E

114 M11 **Smederevo** Ger. Semendria. Serbia, N Yugoslavia 44.40N 20.55E

114 M12 **Smederevska Palanka** Serbia, C Yugoslavia 44.23N 20.55E

118 L13 **Smeeni** Buzău, SE Romania 45.00N 26.52E

99 D16 **Smela** see Smila

113 J22 **Śmigiel** Ger. Schmiegel. Wielkopolskie, C Poland 52.02N 16.33E

119 Q6 **Smila** Rus. Smela. Cherkas'ka Oblast', C Ukraine 49.15N 31.54E

100 N7 **Smilde** Drenthe, NE Netherlands 52.57N 6.28E

9 S16 **Smiley** Saskatchewan, S Canada 51.40N 109.24W

27 T12 **Smiley** Texas, SW USA 29.16N 97.38W

Smilten see Smiltene

120 I8 **Smiltene** Ger. Smilten. Valka, N Latvia 57.25N 25.53E

127 O14 **Smirnykh** Ostrov Sakhalin, Sakhalinskaya Oblast', SE Russian Federation 49.47N 142.48E

9 Q13 **Smith** Alberta, W Canada 55.06N 113.57W

41 P4 **Smith Bay** bay Alaska, USA

10 I3 **Smith, Cape** headland Quebec, NE Canada 60.50N 78.06W

28 L3 **Smith Center** Kansas, C USA 39.46N 98.46W

8 K13 **Smithers** British Columbia, SW Canada 54.45N 127.10W

23 V10 **Smithfield** North Carolina, SE USA 35.30N 78.20W

38 L1 **Smithfield** Utah, W USA 41.50N 111.49W

23 X7 **Smithfield** Virginia, NE USA 36.41N 76.38W

10 I3 **Smith Island** island Nunavut, C Canada

Smith Island see Sumisu-jima

22 H7 **Smithland** Kentucky, S USA 37.06N 88.24W

23 T7 **Smith Mountain Lake** var. Leesville Lake. ⊠ Virginia, NE USA

36 L1 **Smith River** California, W USA 41.54N 124.09W

33 R9 **Smith River** ✍ Montana, NW USA

12 L13 **Smiths Falls** Ontario, SE Canada 44.54N 76.01W

35 N13 **Smith Ferry** Idaho, NW USA 44.19N 116.04W

22 K7 **Smiths Grove** Kentucky, S USA 37.01N 86.14W

191 N15 **Smithton** Tasmania, SE Australia 40.54S 145.06E

20 L14 **Smithtown** Long Island, New York, NE USA 40.52N 73.13W

22 K9 **Smithville** Tennessee, S USA 35.57N 85.48W

27 T11 **Smithville** Texas, SW USA 30.04N 97.32W

Šmohor see Hermagor

37 Q4 **Smoke Creek Desert** desert Nevada, W USA

9 O14 **Smoky** ✍ Alberta, W Canada

190 E7 **Smoky Bay** South Australia 32.22S 133.57E

191 V6 **Smoky Cape** headland New South Wales, SE Australia 30.54S 153.06E

28 L4 **Smoky Hill River** ✍ Kansas, C USA

28 L4 **Smoky Hills** hill range Kansas, C USA

9 Q14 **Smoky Lake** Alberta, SW Canada 54.07N 112.25W

96 E8 **Smøla** island W Norway

130 M4 **Smolensk** Smolenskaya Oblast', W Russian Federation 54.48N 32.07E

130 H4 **Smolenskaya Oblast'** ◆ province W Russian Federation

Smolensk-Moscow Upland see Smolensko-Moskovskaya Vozvyshennost'

130 J3 **Smolensko-Moskovskaya Vozvyshennost'** var. Smolensk-Moscow Upland. ▲ W Russian Federation

Smolevichi see Smalyavichy

117 C15 **Smólikas** ▲ W Greece 40.06N 20.54E

116 I12 **Smolyan** prev. Pashmakli. Smolyan, S Bulgaria 41.33N 24.46E

116 I12 **Smolyan** ◆ province S Bulgaria

Smolyany see Smalyany

35 S15 **Smoot** Wyoming, C USA 42.37N 110.55W

10 G12 **Smooth Rock Falls** Ontario, S Canada 49.16N 81.37W

Smorgon'/Smorgonie see Smarhon'

97 J23 **Smygehamn** Skåne, S Sweden 55.19N 13.25E

204 I7 **Smyley Island** island Antarctica

23 Y3 **Smyrna** Delaware, NE USA 39.18N 75.36W

25 S3 **Smyrna** Georgia, SE USA 33.52N 84.30W

22 J9 **Smyrna** Tennessee, S USA 36.00N 86.30W

Smyrna see İzmir

176 W10 **Snabai** Irian Jaya, E Indonesia 1.45S 134.14E

99 I16 **Snaefell** ▲ C Isle of Man 54.15N 4.29W

94 H3 **Snaefellsjökull** ▲ W Iceland 64.51N 23.51W

8 J4 **Snake** ✍ Yukon Territory, NW Canada

31 O8 **Snake Creek** ✍ South Dakota, N USA

191 P13 **Snake Island** island Victoria, SE Australia

37 Y6 **Snake Range** ▲ Nevada, W USA

34 X14 **Snake River** ✍ NW USA

31 V6 **Snake River** ✍ Minnesota, N USA

30 L12 **Snake River** ✍ Nebraska, C USA

35 Q14 **Snake River Plain** plain Idaho, NW USA

15 I7 **Snare** ✍ Northwest Territories, NW Canada

95 F15 **Snåsa** Nord-Trøndelag, C Norway 64.15N 12.23E

23 O8 **Sneedville** Tennessee, S USA 36.31N 83.13W

100 K6 **Sneek** Friesland, N Netherlands 53.01N 5.40E

55 G8 **Snežnik** ▲ SW Slovenia 45.35N 14.29E

72 G8 **Snezhnogorsk** Taymyrskiy (Dolgano-Nenetskiy) Avtonomnyy Okrug, N Russian Federation 68.06N 87.37E

Snezhnoye see Snizhne

113 G15 **Sněžka** Ger. Schneekoppe. ▲ N Czech Republic 50.42N 15.55E

112 N8 **Śniardwy, Jezioro** Ger. Spirdingsee. ⊚ NE Poland

Śnieczkus see Visaginas

119 R10 **Snihurivka** Mykolayivs'ka Oblast', S Ukraine 47.05N 32.48E

118 I5 **Snilov** (L'viv) L'vivs'ka Oblast', W Ukraine 49.49N 23.59E

113 O19 **Snina** Hung. Szinna. Prešovský Kraj, E Slovakia 49.00N 22.10E

119 Y8 **Snizhne** Rus. Snezhnoye. Donets'ka Oblast', SE Ukraine 48.01N 38.46E

94 J3 **Snæfellur** ▲ E Iceland 64.38N 19.18W

96 G10 **Snøhetta** ▲ S Norway 62.22N 9.08E

94 G12 **Snøtinden** ▲ C Norway 66.39N 13.50E

99 I18 **Snowdon** ▲ NW Wales, UK 53.04N 4.04W

99 I18 **Snowdonia** ▲ NW Wales, UK

15 I8 **Snowdrift** ✍ Northwest Territories, NW Canada

Snowdrift see Łutselk'e

39 N2 **Snowflake** Arizona, SW USA 34.30N 110.04W

23 Y5 **Snow Hill** Maryland, NE USA 38.10N 75.23W

23 W10 **Snow Hill** North Carolina, SE USA 35.25N 77.40W

204 H3 **Snowhill Island** island Antarctica

9 V13 **Snow Lake** Manitoba, C Canada 54.55N 100.01W

39 R5 **Snowmass Mountain** ▲ Colorado, C USA 39.07N 107.04W

20 M10 **Snow, Mount** ▲ Vermont, NE USA

36 M5 **Snow Mountain** ▲ California, W USA 39.44N 123.01W

Snow Mountains see Maoke, Pegunungan

25 S7 **Snowshoe Peak** ▲ Montana, NW USA 48.15N 115.44W

190 I8 **Snowtown** South Australia 33.49S 138.13E

38 K1 **Snowville** Utah, W USA 41.59N 112.42W

37 X3 **Snow Water Lake** ⊚ Nevada, W USA

191 Q11 **Snowy Mountains** ▲ New South Wales/Victoria, SE Australia

191 Q12 **Snowy River** ✍ New South Wales/Victoria, SE Australia

46 K5 **Snug Corner** Acklins Island, SE Bahamas 22.31N 73.51W

178 J13 **Snuol** Krâchéh, E Cambodia 12.04N 106.25E

118 J7 **Snyatyn** Rus. Snyatyn. Ivano-Frankivs'ka Oblast', W Ukraine 48.28N 25.33E

28 L12 **Snyder** Oklahoma, C USA 34.37N 98.56W

27 O6 **Snyder** Texas, SW USA 32.43N 100.54W

180 H3 **Soalala** Mahajanga, W Madagascar 16.04S 45.21E

180 J4 **Soanierana-Ivongo** Toamasina, E Madagascar 16.52S 49.34E

175 Ss7 **Soasiu** var. Tidore. Pulau Tidore, E Indonesia 0.40N 127.25E

56 G8 **Soatá** Boyacá, C Colombia 6.14N 72.42W

180 I5 **Soavinandriana** Antananarivo, C Madagascar 19.09S 46.43E

176 Yy12 **Soba** Irian Jaya, E Indonesia 4.18S 139.11E

79 V13 **Soba** Kaduna, C Nigeria 10.58N 8.06E

169 Y16 **Sobaek-sanmaek** ▲ S South Korea

82 F13 **Sobat** ✍ E Sudan

176 Z12 **Sobger, Sungai** ✍ Irian Jaya, E Indonesia

176 W10 **Sobiei** Irian Jaya, E Indonesia 2.31S 134.30E

130 M3 **Sobinka** Vladimirskaya Oblast', W Russian Federation 56.00N 39.55E

131 S7 **Sobolevo** Orenburgskaya Oblast', W Russian Federation 51.57N 51.42E

170 D15 **Sobo-san** ▲ Kyūshū, SW Japan 32.50N 131.16E

113 G14 **Sobótka** Dolnośląskie, SW Poland 50.53N 16.48E

61 O15 **Sobradinho** Bahia, E Brazil 9.33S 40.56W

179 R13 **Sobradinho, Barragem de** see Sobradinho, Represa de

61 O16 **Sobradinho, Represa de** var. Barragem de Sobradinho. ⊠ E Brazil

60 O13 **Sobral** Ceará, E Brazil 3.45S 40.19W

107 T4 **Sobrarbe** physical region NE Spain

111 N10 **Soča** It. Isonzo. ✍ Italy/Slovenia

112 L11 **Sochaczew** Mazowieckie, C Poland 52.15N 20.15E

130 L15 **Sochi** Krasnodarskiy Kray, SW Russian Federation 43.34N 39.46E

116 G13 **Sochós** var. Sohos, Sokhós. Kentriki Makedonía, N Greece 40.49N 23.22E

203 R11 **Société, Archipel de la** var. Archipel de Tahiti, Îles de la Société, Eng. Society Islands. island group W French Polynesia

Société, Îles de la/Society Islands see Société, Archipel de la

23 T11 **Society Hill** South Carolina, SE USA 34.28N 79.54W

183 W9 **Society Ridge** undersea feature C Pacific Ocean

64 I5 **Socompa, Volcán** ▲ N Chile 24.18S 68.03W

57 E5 **Soconusco, Sierra de** see Sierra Madre

56 G8 **Socorro** Santander, C Colombia 6.25N 73.13W

37 R12 **Socorro** New Mexico, SW USA 33.58N 106.55W

42 C3 **Socotra** island Suqutrā

59 FJ15 **Sóc Trăng** var. Khanh Hung. Soc Trăng, S Vietnam 9.36N 105.58E

107 P10 **Socuéllamos** Castilla-La Mancha, C Spain 39.16N 2.48W

37 W13 **Soda Lake** salt flat California, W USA

94 L11 **Sodankylä** Lappi, N Finland 67.25N 26.34E

35 R15 **Soda Springs** Idaho, NW USA 42.39N 111.36W

22 L10 **Soddy Daisy** Tennessee, S USA 35.14N 85.11W

Soddo/Soddu see Sodo

97 N14 **Söderfors** Uppsala, C Sweden 60.22N 17.19E

96 N12 **Söderhamn** Gävleborg, C Sweden 61.19N 17.10E

97 N17 **Söderköping** Östergötland, S Sweden 58.28N 16.19E

97 N17 **Södermanland** ◆ county C Sweden

97 O16 **Södertälje** Stockholm, C Sweden 59.10N 17.39E

82 D10 **Sodiri** var. Sawdirī, Sodari. Northern Kordofan, C Sudan 14.22N 29.06E

83 I14 **Sodo** var. Soddo, Soddu. Southern, S Ethiopia 6.49N 37.43E

96 N11 **Södra Dellen** ⊚ C Sweden

95 M19 **Södra Vi** Kalmar, S Sweden 57.45N 15.45E

20 G9 **Sodus Point** headland New York, NE USA 43.16N 76.59W

175 Rr17 **Soe** prev. Soë. Timor, C Indonesia 9.51S 124.28E

174 J14 **Soekarno-Hatta** ✕ (Jakarta) Jawa, S Indonesia 6.19N 117.10E

120 E5 **Soela Väin** prev. Eng. Sele Sound, Ger. Dagden-Sund, Soëla-Sund. strait W Estonia

Soemba see Sumba, Pulau

Soembawa see Sumbawa

Soemenep see Sumenep

Soengaipenoeh see Sungaipenuh

Soerabaja see Surabaya

93 G14 **Soest** Nordrhein-Westfalen, W Germany 51.34N 8.07E

100 J11 **Soest** Utrecht, C Netherlands 52.10N 5.19E

102 F11 **Soeste** ✍ NW Germany

100 J11 **Soesterberg** Utrecht, C Netherlands 52.07N 5.16E

57 E16 **Sofádes** var. Sofádhes. Thessalía, C Greece 39.19N 22.06E

Sofádhes see Sofádes

85 N18 **Sofala** Sofala, C Mozambique 20.04S 34.43E

85 N17 **Sofala** ◆ province C Mozambique

85 N18 **Sofala, Baia de** bay E Mozambique

180 J3 **Sofia** seasonal river NW Madagascar

Sofia see Sofiya

117 G19 **Sofikó** Pelopónnisos, S Greece 37.46N 23.04E

Sofi-Kurgan see Sopu-Korgon

116 G10 **Sofiya** var. Sophia, Eng. Sofia; Lat. Serdica. ● (Bulgaria) Sofiya-Grad, W Bulgaria 42.42N 23.20E

112 I10 **Sofiya** × Sofiya-Grad, W Bulgaria 42.42N 23.26E

116 H9 **Sofiya** ◆ province W Bulgaria

57 S4 **Sofía** ◆ municipality W Bulgaria

119 S8 **Sofiyevka** see Sofiivka

Sofiyivka var. Sofiyevka. Dnipropetrovs'ka Oblast', E Ukraine 48.03N 33.52E

127 Nn14 **Sofiysk** Khabarovskiy Kray, SE Russian Federation 51.31N 139.46E

127 N14 **Sofiysk** Khabarovskiy Kray, SE Russian Federation 52.20N 133.37E

128 I6 **Sofporog** Respublika Kareliya, NW Russian Federation 65.48N 31.30E

172 Ss15 **Sōfu-gan** island Izu-shotō, SE Japan

56 G9 **Sog** see Sog Xian

56 G9 **Sogamoso** Boyacá, C Colombia 5.43N 72.55W

142 I11 **Soğanlı Çayı** ✍ N Turkey

96 E12 **Søgne** physical region S Norway

129 U13 **Sogn og Fjordane**, N Norway 61.12N 7.06E

Sogndalsfjøra see Sogndal

95 C14 **Sogndal** Vest-Agder, S Norway 58.05N 7.48E

95 C14 **Sognefjorden** fjord NE North Sea

96 C12 **Sogn Og Fjordane** ◆ county S Norway

179 R13 **Sogod** Leyte, C Philippines 10.25N 125.00E

168 I11 **Sogo Nur** ⊚ N China

169 T12 **Sogruma** Qinghai, W China 32.31N 100.52E

169 S17 **Sŏgwip'o** S South Korea 33.11N 126.33E

162 K10 **Sog Xian** var. Sog. Xizang Zizhiqu, W China 31.52N 93.40E

77 X10 **Sohâg** var. Sawhâj, Suliag, C Egypt 26.27N 31.43E

66 H9 **Sohar** see Şuḩār

102 H7 **Sohm Plain** undersea feature NW Atlantic Ocean

101 E13 **Sohos** see Sochós

Sohrau see Żory

101 K15 **Soignies** Hainaut, SW Belgium 50.34N 4.04E

127 Nn15 **Soila** Xizang Zizhiqu, W China 30.40N 136.42E

125 S4 **Soissons** anc. Augusta Suessionum, Noviodunum. Aisne, N France 49.22N 3.19E

127 N11 **Sōja** Okayama, Honshū, SW Japan 34.41N 133.45E

158 I10 **Sojat** Rājasthān, N India 25.55N 73.43E

169 V18 **Sŏjosŏn-man** inlet W North Korea

118 I4 **Sokal'** Rus. Sokal. L'vivs'ka Oblast', NW Ukraine 50.28N 24.16E

169 V14 **Sokch'o** N South Korea 38.07N 128.33E

142 J15 **Söke** Aydın, SW Turkey 37.45N 27.24E

81 M24 **Sokele** Katanga, SE Dem. Rep. Congo (Zaire) 9.54S 24.38E

153 R11 **Sokh** Taj. Sokh. ✍ Kyrgyzstan/Uzbekistan

143 Q8 **Sokh** see Sükh

Sokhós see Sochós

143 Q8 **Sokhumi** Rus. Sukhumi. NW Georgia 43.01N 41.01E

115 O14 **Sokobanja** Serbia, E Yugoslavia 43.39N 21.51E

79 X6 **Sokodé** C Togo 8.58N 1.10E

127 O10 **Sokol** Magadanskaya Oblast', E Russian Federation 59.51N 150.56E

128 M13 **Sokol** Vologodskaya Oblast', NW Russian Federation 59.26N 40.09E

112 P9 **Sokółka** Podlaskie, NE Poland 53.24N 23.30E

78 M1 **Sokolo** Ségou, W Mali 14.43N 6.02W

113 A16 **Sokolov** Ger. Falkenau an der Eger; prev. Falknov nad Ohří. Karlovarský Kraj, W Czech Republic 50.10N 12.38E

113 O16 **Sokołów Małopolski** Podkarpackie, SE Poland 50.12N 22.07E

112 O11 **Sokołów Podlaski** Mazowieckie, E Poland 52.25N 22.14E

78 U11 **Sokone** W Senegal 13.52N 16.22W

79 T12 **Sokoto** Sokoto, NW Nigeria 13.05N 5.15E

79 S12 **Sokoto** ◆ state NW Nigeria

79 S12 **Sokoto** ✍ NW Nigeria

147 **Sokotra** see Suquṭrā

153 U7 **Sokotra** Chuvaskaya Oblast', N Kyrgyzstan 52.53N 74.19E

118 L7 **Sokyryany** Chernivets'ka Oblast', W Ukraine 48.28N 27.25E

97 C16 **Sola** Rogaland, S Norway 58.52N 5.37E

197 C10 **Sola** Vanua Lava, N Vanuatu 13.51S 167.34E

197 L22 **Solaísvesborg** Blekinge, S Sweden 56.04N 14.34E

176 Y15 **Solaka** Irian Jaya, E Indonesia 7.52S 138.45E

158 I8 **Solan** Himāchal Pradesh, N India 30.54N 77.06E

193 A25 **Solander Island** island SW NZ

161 F15 **Solāpur** var. Sholāpur. Mahārāshtra, W India 17.42N 75.54E

95 H16 **Solberg** Västernorrland, C Sweden 63.48N 17.40E

113 H18 **Solca** Ger. Solka. Suceava, N Romania 47.40N 25.49E

107 O16 **Sol, Costa del** coastal region S Spain

108 F5 **Solda** Trentino-Alto Adige, N Italy 46.33N 10.35E

119 N9 **Şoldăneşti** Rus. Sholdaneshty. N Moldova 47.49N 28.45E

110 L8 **Sölden** Tirol, W Austria 46.58N 11.01E

49 P3 **Soldier Creek** ✍ Kansas, C USA

41 R12 **Soldotna** Alaska, USA 60.29N 151.03W

112 I10 **Solec Kujawski** Kujawsko-pomorskie, C Poland 53.04N 18.09E

63 B16 **Soledad** Santa Fe, C Argentina 30.37S 60.52W

57 E4 **Soledad** Atlántico, N Colombia 10.54N 74.48W

37 O11 **Soledad** California, W USA 36.25N 121.19W

57 O7 **Soledad** Anzoátegui, NE Venezuela 8.10N 63.31W

Soledad see East Falkland

Soledad, Isla see East Falkland

105 Y15 **Soledade** Rio Grande do Sul, S Brazil 28.49S 52.30W

35 P7 **Somers** Montana, NW USA 48.04N 114.16W

49 N4 **Solomon** Kansas, C USA 38.55N 97.22W

195 U16 **Solomon Islands** prev. British Solomon Islands Protectorate. ◆ commonwealth republic W Pacific Ocean

195 T12 **Solomon Islands** island group PNG/Solomon Islands

28 M3 **Solomon River** ✍ Kansas, C USA

199 Hh9 **Solomon Sea** sea W Pacific Ocean

33 U11 **Solon** Ohio, N USA 41.23N 81.26W

119 T8 **Solone** Dnipropetrovs'ka Oblast', E Ukraine 48.11N 34.21E

175 R16 **Solor, Kepulauan** island group S Indonesia

130 M4 **Solotcha** Ryazanskaya Oblast', W Russian Federation 54.43N 39.50E

110 D7 **Solothurn** Fr. Soleure. NW Switzerland 47.12N 7.28E

110 D7 **Solothurn** Fr. Soleure. ◆ canton NW Switzerland

128 J7 **Solovetskiye Ostrova** island group NW Russian Federation

107 V5 **Solsona** Cataluña, NE Spain 42.00N 1.31E

115 E14 **Šolta** It. Solta. island S Croatia

148 L4 **Solţānābād** see Kāshmar

148 L4 **Solţānīyeh** Zanjān, NW Iran 36.24N 48.49E

102 I11 **Soltau** Niedersachsen, NW Germany 52.58N 9.49E

128 G14 **Sol'tsy** Novgorodskaya Oblast', W Russian Federation 58.09N 30.22E

Soltüstik Qazaqstan Oblysy see Severnyy Kazakhstan

Solun see Thessaloníki

115 O19 **Solunska Glava** ▲ C FYR Macedonia 41.43N 21.24E

99 J15 **Solway Firth** inlet England/Scotland, UK

84 I13 **Solwezi** North Western, NW Zambia 12.10S 26.22E

171 LJ14 **Sōma** Fukushima, Honshū, C Japan 37.49N 140.52E

142 C13 **Soma** Manisa, W Turkey 39.11N 27.34E

83 M4 **Somali** ◆ region E Ethiopia

83 O15 **Somalia** off. Somali Democratic Republic, Som. Jamuuriyada Demuqraadiga Soomaaliyeed, Soomaaliya; prev. Italian Somaliland, Somaliland Protectorate. ◆ republic E Africa

181 N4 **Somali Basin** undersea feature W Indian Ocean

69 J8 **Somali Plain** undersea feature W Indian Ocean

114 J8 **Sombor** Hung. Zombor. Serbia, NW Yugoslavia 45.46N 19.07E

113 H20 **Sombreffe** Namur, S Belgium 50.32N 4.39E

42 L10 **Sombrerete** Zacatecas, C Mexico 23.36N 103.46W

47 P9 **Sombrero** island N Anguilla

157 Q21 **Sombrero Channel** channel Nicobar Islands, India

118 H9 **Şomcuta Mare** Hung. Nagysomkút; prev. Somcuţa Mare. Maramureş, N Romania 47.28N 23.30E

126 H7 **Somero** Länsi-Suomi, W Finland 60.37N 23.30E

20 L15 **Someren** Noord-Brabant, SE Netherlands 51.22N 5.42E

95 L19 **Somero** Länsi-Suomi, W Finland 60.37N 23.30E

35 P5 **Somers** Montana, NW USA 48.04N 114.16W

46 A12 **Somerset** var. Somerset Village. W Bermuda 32.18N 64.52W

33 Q5 **Somerset** Colorado, C USA 38.55N 107.27W

22 M7 **Somerset** Kentucky, S USA 37.05N 84.36W

21 O12 **Somerset** Massachusetts, NE USA 41.46N 71.07W

K23 **Somerset** cultural region SW England, UK

46 A12 **Somerset East** Somerset-Oos

21 P9 **Somersworth** New Hampshire, NE USA 43.15N 70.52W

38 H15 **Somerton** Arizona, SW USA 32.36N 114.42W

20 J11 **Somerville** New Jersey, NE USA 40.34N 74.36W

22 H9 **Somerville** Tennessee, S USA 35.14N 89.21W

27 U10 **Somerville** Texas, SW USA 30.21N 96.31W

27 T10 **Somerville Lake** ⊠ Texas, SW USA

Someş/Somesch/Someşul see Szamos

197 J13 **Somosomo** Taveuni, N Fiji 16.46S 179.57W

44 I9 **Somotillo** Chinandega, NW Nicaragua 13.01N 86.54W

44 I8 **Somoto** Madriz, NW Nicaragua 13.28N 86.36W

112 I11 **Sompolno** Wielkopolskie, C Poland 52.24N 18.30E

107 S3 **Somport** var. Puerto de Somport, Fr. Col de Somport. pass France/Spain see also Somport, Col du var. Puerto de Somport, Sp. Somport; anc. Summus Portus. pass France/Spain see also Somport 42.47N 0.33W

104 J17 **Somport, Col du** var. Puerto de Somport, Sp. Somport; anc. Summus Portus. pass France/Spain see also Somport

101 J15 **Son** Noord-Brabant, S Netherlands 51.32N 5.34E

97 H15 **Son** Akershus, S Norway 59.31N 10.42E

160 L9 **Son** ✍ C India

45 R16 **Soná** Veraguas, W Panama 08.00N 81.20W

160 M12 **Sonapur** prev. Sonepur. Orissa, E India 20.49N 83.58E

176 Vv10 **Sonar** Irian Jaya, E Indonesia 2.31S 133.01E

97 G24 **Sønderborg** Ger. Sonderburg. Sønderjylland, SW Denmark 54.55N 9.48E

Sonderburg see Sønderborg

97 F24 **Sønderjylland** off. Sønderjyllands Amt. ◆ county SW Denmark

103 K15 **Sondershausen** Thüringen, C Germany 51.22N 10.52E

Søndre Strømfjord see Kangerlussuaq

108 E6 **Sondrio** Lombardia, N Italy 46.10N 9.52E

Sone see Son

Sonepur see Sonapur

59 K22 **Sonequera** ▲ S Bolivia 22.06S 67.10W

84 I13 **Solwezi** North Western, NW Zambia 12.10S 26.22E

59 K22 **Sonequera** ▲ S Bolivia

142 C13 **Soma** Manisa, W Turkey

83 H25 **Songea** Ruvuma, S Tanzania 10.42S 35.39E

176 Z11 **Songgato, Sungai** ✍ Irian Jaya, E Indonesia

169 X10 **Songhua Hu** ⊚ NE China

169 Y7 **Songhua Jiang** var. Sungari. ✍ NE China

167 S8 **Songjiang** Shanghai Shi, E China 31.01N 121.13E

167 Y3 **Sŏngjin** see Kimch'aek

178 H17 **Songkhla** var. Songkla, Mal. Singora. Songkhla, SW Thailand 7.12N 100.34E

Songkla see Songkhla

169 S9 **Song Ling** ▲ NE China

169 W14 **Songnim** SW North Korea 38.43N 125.40E

84 B10 **Songo** Uíge, NW Angola 7.30S 14.55E

85 M15 **Songo** Tete, NW Mozambique 15.35S 32.43E

81 F21 **Songolo** Bas-Congo, SW Dem. Rep. Congo (Zaire) 5.40S 14.04E

168 M7 **Songpan** prev. Sungpu. Sichuan, C China 32.49N 103.39E

167 S4 **Songsri** Fujian, SE China 27.33N 118.68E

166 M6 **Songxian** var. Song Xian. Henan, C China 34.11N 112.04E

167 R10 **Songyin** Zhejiang, SE China 28.29N 119.27E

169 V9 **Songyuan** var. Fu-yü, Petuna; prev. Fuyu. Jilin, NE China 45.10N 124.52E

169 P11 **Sonid Youqi** var. Saihon Tal. Nei Mongol Zizhiqu, N China 42.45N 112.36E

169 P11 **Sonid Zuoqi** Nei Mongol Zizhiqu, N China 43.49N 113.36E

158 I10 **Sonipat** Haryāna, N India 29.00N 77.01E

95 M13 **Sonkajärvi** Itä-Suomi, C Finland 63.40N 27.30E

178 J6 **Son La** Son La, N Vietnam 21.19N 103.55E

155 O16 **Sonmiāni** Baluchistān, S Pakistan 25.24N 66.37E

155 O16 **Sonmiāni Bay** bay S Pakistan

103 K18 **Sonneberg** Thüringen, C Germany 50.22N 11.10E

103 N22 **Sonntagshorn** ▲ Austria/Germany 47.40N 12.42E

42 E3 **Sonoita, Rio** var. Río Sonoyta. ✍ Mexico/USA

37 N7 **Sonoma** California, W USA 38.16N 122.28W

37 T3 **Sonoma Peak** ▲ Nevada, W USA 40.50N 117.34W

37 P8 **Sonora** California, W USA 37.58N 120.22W

39 H15 **Sonora** Arizona, SW USA 32.36N 114.42W

27 O10 **Sonora** Texas, SW USA 30.31N 100.40W

42 E5 **Sonora** ◆ state NW Mexico

37 X17 **Sonora**

Sonoran Desert var. Desierto de Altar. desert Mexico/USA see also Altar, Desierto de

42 G5 **Sonora, Río** ✍ NW Mexico

42 E2 **Sonoyta** var. Sonoita. Sonora, NW Mexico 31.49N 112.50W

148 K6 **Sonqor** var. Sunqur. Kermānshāh, W Iran 34.45N 47.39E

107 N9 **Sonseca** var. Sonseca con Casalgordo. Castilla-La Mancha, C Spain 39.40N 3.58W

Sonseca con Casalgordo see Sonseca

56 E9 **Sonsón** N Colombia 5.41N 75.15W

44 A7 **Sonsonate** Sonsonate, W El Salvador 13.43N 89.43W

44 A9 **Sonsonate** ◆ department SW El Salvador

196 A10 **Sonsorol Islands** island group S Palau

114 J9 **Sonta** Hung. Szond; prev. Szonta. Serbia, NW Yugoslavia 45.34N 19.06E

◆ COUNTRY ◇ DEPENDENT TERRITORY ◆ ADMINISTRATIVE REGION ▲ MOUNTAIN ⊠ VOLCANO ⊚ LAKE
● COUNTRY CAPITAL ○ DEPENDENT TERRITORY CAPITAL ✕ INTERNATIONAL AIRPORT ▲ MOUNTAIN RANGE ✍ RIVER ⊠ RESERVOIR

329

Sơn Tây var. Sontay. Ha Tây,
N Vietnam 21.06N 105.31E

Sonthofen Bayern, S Germany
47.31N 10.16E

Soochow see Suzhou

Soomaaliya/Soomaaliyeed,
Jamuuriyada Demuqraadiga
see Somalia

Soome Laht see Finland, Gulf of

Sooner State see Oklahoma

Soperton Georgia, SE USA
32.22N 82.35W

Sop Hao Houaphan, N Laos
20.33N 104.25E

Sophia see Sofiya

Sopi Pulau Morotai, E Indonesia
2.36N 128.32E

Sopianae see Pécs

Sopinusa Irian Jaya, E Indonesia
3.31S 132.55E

Sopi, Tanjung headland Pulau
Morotai, N Indonesia 2.39N 128.34E

Sopo var. W Sudan

Sopockinie/Sopotskin see
Sapotskino

Sopot Plovdiv, C Bulgaria
42.40N 24.45E

Sopot Ger. Zoppot. Pomorskie,
N Poland 54.25N 18.33E

Sop Prap var. Ban Sop Prap.
Lampang, NW Thailand
17.55N 99.19E

Sopron Ger. Ödenburg. Györ-
Moson-Sopron, NW Hungary
47.40N 16.34E

Sopu-Korgon var. Sofi-Kurgan.
Oshskaya Oblast', SW Kyrgyzstan
40.03N 73.30E

Sopur Jammu and Kashmir,
NW India 34.19N 74.28E

Sora Lazio, C Italy 41.43N 13.37E

Sorada Orissa, E India
19.46N 84.28E

Söraker Västernorrland,
C Sweden 62.31N 17.31E

Sorata La Paz, W Bolivia
15.49S 68.39W

Sorau/Sorau in der
Niederlausitz see Zary

Sorbas Andalucía, S Spain
37.06N 2.06W

Sord/Sórd Choluim Chille see
Swords

Sorel Quebec, SE Canada
46.02N 73.06W

Sorell Tasmania, SE Australia
42.49S 147.34E

Sorell, Lake ⊚ Tasmania,
SE Australia

Soresina Lombardia, N Italy
45.16N 9.51E

Sørfjorden fjord S Norway

Sörfjärden Gävleborg, C Sweden
61.45N 17.00E

Sorgues Vaucluse, SE France
44.00N 4.52E

Sorgun Yozgat, C Turkey
39.49N 35.10E

Soria Castilla-León, N Spain
41.46N 2.26W

Soria ◆ province Castilla-León,
N Spain

Soriano Soriano, SW Uruguay
33.25S 58.21W

Soriano ◆ department
SW Uruguay

Sørkapp headland SW Svalbard
76.34N 16.33E

Sorkh, Küh-e ▲ NE Iran

Soro var Ghazal, Bahr el

Sorø Vestsjælland, E Denmark
55.25N 11.34E

Soroca Rus. Soroki. N Moldova
48.10N 28.18E

Sorocaba São Paulo, S Brazil
23.28S 47.27W

Sorochino see Sorochyna

Sorochinsk Orenburgskaya
Oblast', W Russian Federation
52.26N 53.10E

Soroki see Soroca

Sorol atoll Caroline Islands,
W Micronesia

Sorong Irian Jaya, E Indonesia
0.49S 131.16E

Soroti C Uganda 1.42N 33.37E

Sørøy see Sørøya

Sørøya var. Sørøy. island
N Norway

Sorraia, Rio ⋧ C Portugal

Sørreisa Troms, N Norway
69.08N 18.09E

Sorrento anc. Surrentum.
Campania, S Italy 40.37N 14.22E

Sorsele Västerbotten, N Sweden
65.31N 17.34E

Sorsogon Luzon, N Philippines
12.57N 124.04E

Sort Cataluña, NE Spain
42.25N 1.07E

Sortavala Respublika Kareliya,
NW Russian Federation
61.43N 30.40E

Sortino Sicilia, Italy,
C Mediterranean Sea 37.10N 15.01E

Sør-Trøndelag ◆ county S Norway

Sørumsand Akershus, S Norway
59.58N 11.13E

Sõrve Säär headland SW Estonia
57.54N 22.02E

Sösdala Skåne, S Sweden
56.02N 13.42E

Sos del Rey Católico Aragón,
NE Spain 42.30N 1.13W

Sösel Västerbotten, N Sweden
65.31N 17.34E

Sosna ⋧ W Russian Federation

Sosneado, Cerro ▲ W Argentina
34.44S 69.52W

Sosnogorsk Respublika Komi,
NW Russian Federation
63.33N 53.55E

Sosnovets Respublika Kareliya,
NW Russian Federation
64.25N 34.23E

Sosnovets see Sosnowiec

Soochow see Suzhou

Sosnovka Chuvashskaya
Respublika, W Russian Federation
56.18N 47.14E

Sosnovka Kirovskaya Oblast',
NW Russian Federation
56.15N 51.20E

Sosnovka Murmanskaya Oblast',
NW Russian Federation
66.28N 40.31E

Sosnovka Tambovskaya Oblast',
W Russian Federation 53.14N 41.19E

Sosnovo Fin. Rautu.
Leningradskaya Oblast',
NW Russian Federation
60.30N 30.13E

Sosnovo-Ozerskoye Respublika
Buryatiya, S Russian Federation
52.34N 111.36E

Sosnovyy Bor see Sasnovy Bor

Sosnowiec Ger. Sosnowitz, Rus.
Sosnovets. Śląskie, S Poland
50.16N 19.07E

Sosnovets see Sosnowiec

Sosnytsya Chernihivs'ka Oblast',
N Ukraine 51.31N 32.30E

Sos'va Sverdlovskaya Oblast',
C Russian Federation 59.13N 61.58E

Sotará, Volcán ℞ S Colombia
2.04N 76.40W

Sotavento, Ilhas de var. Leeward
Islands. island group S Cape Verde

Sotkamo Oulu, C Finland
64.05N 28.30E

Soto la Marina Tamaulipas,
C Mexico 23.46N 98.12W

Soto la Marina, Río
⋧ C Mexico

Sotuta Yucatán, SE Mexico
20.34N 89.00W

Souanké La Sangha, NW Congo
2.03N 14.01E

Soubré S Ivory Coast 5.49N 6.34W

Soúda var. Soúdha, Eng. Suda.
Kríti, Greece, E Mediterranean Sea
35.28N 24.04E

Soúdha see Soúda

Soueida var. As Suwaydā'

Souflí prev. Souflion. Anatolikí
Makedonía kai Thráki, NE Greece
41.12N 26.17E

Souflion see Souflí

Soufrière W Saint Lucia
13.51N 61.03W

Soufrière ℞ Basse Terre,
S Guadeloupe 16.03N 61.39W

Souillac Lot, S France 44.53N 1.29E

Souillac S Mauritius 20.31S 57.31E

Souk Ahras NE Algeria
36.14N 8.00E

Souk-el-Arba-Rharb var. Souk
el Arba du Rharb, Souk-el-Arba-
du-Rharb, Souk-el-Arba-el-Rhab.
NW Morocco 34.38N 6.00W

Soukhné see As Sukhnah

Sŏul off. Sŏul-t'ŭkpyŏlsi, Eng.
Seoul, Jap. Keijŏ; prev. Kyŏngsŏng.
● (South Korea) NW South Korea
37.30N 126.57E

Soulac-sur-Mer Gironde,
SW France 45.31N 1.06W

Soumagne Liège, E Belgium
50.36N 5.48E

Sound Beach Long Island, New
York, NE USA 40.56N 72.58W

Sound, The Dan. Øresund, Swe.
Öresund. strait Denmark/Sweden

Soúnio, Akrotírio headland
C Greece 37.39N 24.01E

Soûr var. Şür; anc. Tyre.
SW Lebanon 33.18N 35.30E

Sources, Mont-aux- see Phofung

Soure Coimbra, N Portugal
40.04N 8.37W

Souris Manitoba, S Canada
49.37N 100.16W

Souris Prince Edward Island,
SE Canada 46.22N 62.16W

Souris River var. Mouse River.
⋧ Canada/USA

Sour Lake Texas, SW USA
30.08N 94.24W

Sourpi Thessalía, C Greece
39.07N 22.54E

Sousel Portalegre, C Portugal
38.57N 7.40W

Sousse var. Süsah. NE Tunisia
35.45N 10.37E

South see Sud

South Africa off. Republic of
South Africa, Afr. Suid-Afrika.
◆ republic S Africa

South America continent

South American Plate tectonic
feature

Southampton hist. Hamwih, Lat.
Clausentum. S England, UK
50.54N 1.22W

Southampton Long Island, New
York, NE USA 40.52N 72.22W

Southampton Island island
Nunavut, NE Canada

South Andaman island
Andaman Islands, India, NE Indian Ocean

South Aulatsivik Island island
Newfoundland, E Canada

South Australia ◆ state
S Australia

South Australian Abyssal Plain
see Australian-Antarctic Basin

South Australian Basin
undersea feature SW Indian Ocean

South Australian Plain var.
undersea feature SE Indian Ocean

South Bend Indiana, N USA
41.40N 86.15W

South Bend Texas, SW USA
32.58N 98.39W

South Bend Washington,
NW USA 46.38N 123.48W

South Beveland see Zuid-
Beveland

South Borneo see Kalimantan
Selatan

South Boston Virginia, NE USA
36.42N 78.54W

South Branch Neales seasonal
river South Australia

South Branch Potomac River
⋧ West Virginia, NE USA

Southbridge Canterbury, South
Island, NZ 43.49S 172.17E

Southbridge Massachusetts,
NE USA 42.04N 72.01W

South Bruny Island island
Tasmania, SE Australia

South Burlington Vermont,
NE USA 44.27N 73.08W

South Caicos island S Turks and
Caicos Islands

South Cape see Ka Lae

South Carolina off. State of South
Carolina; also known as The
Palmetto State. ◆ state SE USA

South Carpathians see Carpaţii
Meridionali

South Celebes see Sulawesi
Selatan

South Charleston West Virginia,
NE USA 38.22N 81.42W

South China Sea undersea
feature SE South China Sea

South China Sea Chin. Nan Hai,
Ind. Laut China Selatan, Vtn. Biển
Đông. sea SE Asia

South Dakota off. State of South
Dakota; also known as The Coyote
State, Sunshine State. ◆ state N USA

South Daytona Florida, SE USA
29.09N 81.01W

South Domingo Pueblo New
Mexico, SW USA 35.28N 106.24W

South Downs hill range
SE England, UK

South East ◆ district SE Botswana

South East Bay bay Ascension
Island, C Atlantic Ocean

South East Cape headland
Tasmania, SE Australia
43.36S 146.52E

South-East Celebes see Sulawesi
Tenggara

Southeast Indian Ridge undersea
feature Indian Ocean/Pacific Ocean

Southeast Island see Tagula
Island

Southeast Pacific Basin var.
Belling Hausen Mulde. undersea
feature SE Pacific Ocean

South East Point headland
SE Ascension Island

South East Point headland
Victoria, S Australia 39.10S 146.21E

South East Point headland
Kiritimati, NE Kiribati
1.42N 157.10W

South East Point headland
Mayaguana, SE Bahamas
22.15N 72.44W

South-East Sulawesi see Sulawesi
Tenggara

Southend Saskatchewan,
S Canada 56.19N 103.13W

Southend-on-Sea E England, UK
51.33N 0.43E

Southern ◆ region S Botswana

Southern ◆ district S Israel

Southern ◆ region S Malawi

Southern ◆ province S Zambia

Southern Alps ▲ South Island,
NZ

Southern Cook Islands island
group S Cook Islands

Southern Cross Western
Australia 31.17S 119.15E

Southern Darfur ◆ state
W Sudan

Southern Highlands ◆ province
W PNG

Southern Indian Lake
⊚ Manitoba, C Canada

Southern Kordofan ◆ state
C Sudan

Southern Lau Group island group
Lau Group, SE Fiji

Southern Ocean ocean

Southern Pines North Carolina,
SE USA 35.10N 79.23W

Southern Province ◆ province
S Sri Lanka

Southern Uplands ▲ S Scotland,
UK

Southern Urals see Yuzhnyy Ural

Southery E England, UK

South Esk River ⋧ Tasmania,
SE Australia

South Fabius River ⋧ Missouri,
C USA

Southfield Michigan, N USA
42.28N 83.12W

South Fiji Basin undersea feature
S Pacific Ocean

South Foreland headland
SE England, UK 51.08N 1.22E

South Fork American River
⋧ California, W USA

South Fork Grand River
⋧ North/South Dakota, N USA

South Fork Kern River
⋧ California, W USA

South Fork Koyukuk River
⋧ Alaska, USA

South Fork Kuskokwim River
⋧ Alaska, USA

South Fork Republican River
⋧ C USA

South Fork Solomon River
⋧ Kansas, C USA

South Fox Island island
Michigan, N USA

South Fulton Tennessee, S USA
36.28N 88.53W

South Geomagnetic Pole pole
Antarctica 78.30S 111.00E

South Georgia island S
Georgia and the South Sandwich
Islands, SW Atlantic Ocean

South Georgia and the South
Sandwich Islands ◇ UK
dependent territory SW
Atlantic Ocean

South Georgia Ridge var. North
Scotia Ridge. undersea
feature SW Atlantic Ocean

South Goulburn Island island
Northern Territory, N Australia

South Hatia Island island
SE Bangladesh

South Haven Michigan, N USA
42.24N 86.16W

South Hill Virginia, NE USA
36.43N 78.07W

South Holland see Zuid-Holland

South Holston Lake
⊞ Tennessee/Virginia, S USA

South Honshu Ridge undersea
feature S Pacific Ocean

South Hutchinson Kansas,
C USA 38.01N 97.56W

South Huvadhu Atoll var. Gaafu
Dhaalu Atoll. atoll S Maldives

South Indian Basin undersea
feature Indian Ocean/Pacific Ocean

South Indian Lake Manitoba,
C Canada 56.48N 98.55W

South Island island NW Kenya

South Island island S NZ

South Jason island Jason Islands,
NW Falkland Islands

South Kalimantan see
Kalimantan Selatan

South Kazakhstan see Yuzhnyy
Kazakhstan

South Korea off. Republic of
Korea, Kor. Taehan Min'guk.
◆ republic E Asia

South Lake Tahoe California,
W USA 38.56N 119.57W

South Loup River ⋧ Nebraska,
C USA

South Maalhosmadulu Atoll
var. Baa Atoll. atoll N Maldives

South Maitland ⋧ Ontario,
S Canada

South Makassar Basin undersea
feature S Java Sea

South Manitou Island island
Michigan, N USA

South Miladummadulu Atoll
atoll N Maldives

South Mills North Carolina,
SE USA 36.28N 76.18W

South Nahanni ⋧ Northwest
Territories, NW Canada

South Naknek Alaska, USA
58.39N 157.01W

South Nation ⋧ Ontario,
SE Canada

South Negril Point headland
W Jamaica 18.14N 78.21W

South Nilandhe Atoll var.
Dhaalu Atoll. atoll C Maldives

South Ogden Utah, W USA
41.09N 111.58W

Southold Long Island, New York,
NE USA 41.03N 72.24W

South Orkney Islands island
group Antarctica

South Ossetia former autonomous
region SW Georgia

South Pacific Basin see
Southwest Pacific Basin

South Paris Maine, NE USA
44.14N 70.29W

South Pass Wyoming, C USA
42.20N 108.55W

South Pass passage Chuuk Islands,
C Micronesia

South Pittsburg Tennessee,
S USA 35.00N 85.42W

South Platte River
⋧ Colorado/Nebraska, C USA

South Point Ohio, N USA
38.25N 82.35W

South Point headland S Ascension
Island

South Point headland Michigan,
N USA 44.51N 83.17W

South Point see Ka Lae

South Pole pole Antarctica
90.00S 0.00E

Southport Tasmania, SE Australia
43.26S 146.57E

Southport NW England, UK
53.39N 3.01W

Southport North Carolina,
SE USA 33.55N 78.01W

South Portland Maine, NE USA
43.38N 70.14W

South River Ontario, S Canada
45.48N 79.21W

South River ⋧ North Carolina,
SE USA

South Ronaldsay island
NE Scotland, UK

South Salt Lake Utah, W USA
40.42N 111.52W

South Sandwich Islands island
group E South Georgia and South
Sandwich Islands

South Sandwich Trench
undersea feature SW Atlantic Ocean

South Saskatchewan
⋧ Alberta/Saskatchewan,
S Canada

South Scotia Ridge undersea
feature S Scotia Sea

South Seal ⋧ Manitoba,
C Canada

South Shetland Islands island
group Antarctica

South Shetland Trough
undersea feature Atlantic
Ocean/Pacific Ocean

South Shields NE England, UK
55.00N 1.25W

South Sioux City Nebraska,
C USA 42.99N 96.24W

South Solomon Trench
undersea feature W Pacific Ocean

South Stradbroke Island island
Queensland, E Australia
33.55N 85.27W

South Sulawesi see Sulawesi
Selatan

South Sumatra see Sumatera

South Taranaki Bight bight
SE New Zealand

South Tasmania Plateau see
Tasman Plateau

South Tucson Arizona, SW USA
32.11N 110.56W

South Twin Island island
Nunavut, C Canada

South Uist island NW Scotland,
UK

South-West see Sud-Ouest

South-West Africa/South West
Africa see Namibia

South West Bay bay Ascension
Island, C Atlantic Ocean

South West Cape headland
Tasmania, SE Australia
43.34S 146.01E

South West Cape headland
Stewart Island, NZ 47.15S 167.28E

Southwest Cape headland Saint
Lawrence Island, Alaska, USA
63.19N 171.27W

Southwest Cay island NW Spratly
Islands

Southwest Indian Ocean Ridge
see Southwest Indian Ridge

Southwest Indian Ridge var.
Southwest Indian Ocean Ridge.
undersea feature SW Indian Ocean

Southwest Pacific Basin var.
South Pacific Basin. undersea
feature SE Pacific Ocean

Southwest Point headland Great
Abaco, N Bahamas 25.50N 77.12W

South West Point headland
Kiritimati, NE Kiribati 1.52N 157.34E

South West Point headland
SW Saint Helena 16.00S 5.48W

South Wichita River ⋧ Texas,
SW USA

Southwold E England, UK
52.15N 1.36E

South Yarmouth Massachusetts,
NE USA 41.38N 70.09W

Sovata Harg. Szováta. Mureş,
C Romania 46.36N 25.04E

Soverato Calabria, SW Italy
38.40N 16.31E

Sovetabad see Ghafurov

Sovetsk Ger. Tilsit.
Kaliningradskaya Oblast',
W Russian Federation 55.04N 21.52E

Sovetsk Kirovskaya Oblast',
NW Russian Federation
57.37N 49.02E

Sovetskaya Rostovskaya Oblast',
SW Russian Federation
49.00N 42.09E

Sovetskaya Gavan' Khabarovskiy
Kray, SE Russian Federation
48.54N 140.19E

Sovetskiy Khanty-Mansiyskiy
Avtonomnyy Okrug, C Russian
Federation 61.20N 63.34E

Sovetskoye see Ketchenery

Sovet"yab prev. Sovet"yap.
Akhalskiy Velayat, S Turkmenistan
36.29N 61.13E

Sovet"yap see Sovet"yab

Sovyets'kyy Respublika Krym,
S Ukraine 45.20N 34.54E

Sowa Pan salt lake N Botswana

Sowa Pan see Sua. Central,
NE Botswana 20.35S 26.18E

Soweto Gauteng, NE South Africa
26.08S 27.53E

Sōya-kaikyō see La Perouse Strait

Sōya-misaki headland Hokkaidō,
NE Japan 45.31N 141.55E

Soyana ⋧ NW Russian
Federation

Soyang, Mys var. Mys Suz. headland
NW Turkmenistan 41.47N 52.27E

Soyo Zaire, NW Angola
6.07S 12.19E

Soyra ▲ C Eritrea 14.46N 39.29E

Sozaq see Suzak

Sozh Rus. Sozh. ⋧ NE Europe

Sozopol prev. Sizebolu anc.
Apollonia. Burgas, E Bulgaria
42.25N 27.41E

Spa Liège, E Belgium 50.29N 5.52E

Spaatz Island Antarctica

Space Launching Centre space
station Kzylorda, S Kazakhstan
45.50N 63.20E

Spain off. Kingdom of Spain, Sp.
España; anc. Hispania, Iberia, Lat.
Hispana. ◆ monarchy SW Europe

Spalato see Split

Spalding E England, UK
52.48N 0.06W

Spanish Ontario, S Canada
46.12N 82.21W

Spanish Fork Utah, W USA
40.09N 111.40W

Spanish Point headland
C Bermuda 32.18N 64.49W

Spanish River ⋧ Ontario,
S Canada

Spanish Town hist. St.Iago de la
Vega. C Jamaica 18.00N 76.57W

Spánta, Akrotírio headland Kríti,
Greece, E Mediterranean Sea
35.40N 23.44E

Sparks Nevada, W USA
39.32N 119.45W

Sparnacum see Épernay

Sparreholm Södermanland,
C Sweden 59.04N 16.51E

Sparta Georgia, SE USA
33.16N 82.58W

Sparta Illinois, N USA
38.07N 89.42W

Sparta Michigan, N USA
43.09N 85.42W

Sparta North Carolina, SE USA
36.34N 81.21W

Sparta Tennessee, S USA
35.55N 85.27W

Sparta Wisconsin, N USA
43.57N 90.49W

Sparta see Spárti

Spartanburg South Carolina,
SE USA 34.56N 81.57W

Spárti Eng. Sparta. Pelopónnisos,
S Greece 37.04N 22.25E

Spartel, Cap headland N Morocco
35.49N 5.55W

Spartivento, Capo headland
Sardegna, Italy, C Mediterranean
Sea 38.52N 8.50E

Sparwood British Columbia,
SW Canada 49.43N 114.49W

Spas-Demensk Kaluzhskaya
Oblast', W Russian Federation
54.22N 34.16E

Spas-Klepiki Ryazanskaya
Oblast', W Russian Federation
55.08N 40.15E

Spasovo see Kulen Vakuf

Spassk-Dal'niy Primorskiy Kray,
SE Russian Federation
44.34N 132.52E

Spassk-Ryazanskiy Ryazanskaya
Oblast', W Russian Federation
54.25N 40.21E

Spáta Attikí, C Greece 37.58N 23.55E

Spáta, Akrotírio headland Kríti,
Greece, E Mediterranean Sea
35.42N 23.43E

Spatrjan see Paternion

Spearfish South Dakota, N USA
44.29N 103.51W

Spearman Texas, SW USA
36.12N 101.11W

Speedwell Island island
S Falkland Islands

Speedwell Island Settlement
S Falkland Islands 52.13S 59.40W

Speery Island island
S Saint Helena

Speightstown NW Barbados
13.13N 59.37W

Spello Umbria, C Italy
43.00N 12.41E

Spenard Alaska, USA
61.09N 150.03W

Spence Bay see Taloyoak

Spencer Indiana, N USA
39.18N 86.46W

Spencer Iowa, C USA
43.09N 95.07W

Spencer Nebraska, C USA
42.52N 98.42W

Spencer North Carolina, SE USA
35.41N 80.26W

Spencer Tennessee, S USA
35.46N 85.27W

Spencer West Virginia, NE USA
38.48N 81.21W

Spencer Wisconsin, C USA
44.46N 90.17W

Spencer, Cape headland South
Australia 35.17S 136.52E

Spencer Gulf gulf South Australia

Spencerport New York, NE USA
43.11N 77.48W

Spencerville Ohio, N USA
40.42N 84.21W

Spercheiáda var. Sperhiada,
Sperkhiás. Stereá Ellás, C Greece
38.54N 22.07E

Spercheiós ⋧ C Greece

Sperhiada see Spercheiáda

Sperillen ⊚ S Norway

Sperkhiás see Spercheiáda

Spessart hill range C Germany

Spétsai see Spétses

Spétses prev. Spétsai. Spétses,
S Greece 37.16N 23.09E

Spétses island S Greece

Spey ⋧ NE Scotland, UK

Speyer Eng. Spires; anc. Civitas
Nemetum, Spira. Rheinland-Pfalz,
SW Germany 49.19N 8.25E

Speyerbach ⋧ W Germany

Spezzano Albanese Calabria,
SW Italy 39.40N 16.17E

Spice Islands see Maluku

Spiekeroog island NW Germany

Spielfeld Steiermark, SE Austria
46.43S 15.36E

Spiess Seamount undersea
feature S Atlantic Ocean 53.00S 2.00W

Spiez Bern, W Switzerland
46.42N 7.40E

Spijkenisse Zuid-Holland,
SW Netherlands 51.52N 4.19E

Spike Mountain ▲ Alaska, USA
67.42N 141.39W

Spili Kríti, Greece,
E Mediterranean Sea 35.12N 24.33E

Spillern ▲ W Switzerland

Spilsby E England, UK
53.09N 0.06E

Spin Būldak Kandahār,
S Afghanistan 31.01N 66.22E

Spira see Speyer

Spirding See see Śniardwy, Jezioro

Spires see Speyer

Spirit Lake Iowa, C USA
43.25N 95.06W

Spirit River Alberta, W Canada
55.46N 118.51W

Spiritwood Saskatchewan,
S Canada 53.18N 107.33W

Spiro Oklahoma, C USA
35.14N 94.37W

Spišská Nová Ves Ger. Neudorf,
Zipser Neudorf, Hung. Igló.
Košický Kraj, E Slovakia

Spitak NW Armenia 40.51N 44.17E

Spitsbergen island NW Svalbard

Spittal see Spittal an der Drau

Spittal an der Drau var. Spittal.
Kärnten, S Austria 46.48N 13.30E

Spitz Niederösterreich, NE Austria
48.24N 15.22E

Spjelkavik Møre og Romsdal,
S Norway 62.28N 6.22E

Splendora Texas, SW USA
30.13N 95.09W

Split It. Spalato. Split-Dalmacija,
S Croatia 43.31N 16.27E

Split ✕ Split-Dalmacija, S Croatia
43.33N 16.19E

Split-Dalmacija off. Splitsko-
Dalmatinska Županija. ◆ province
S Croatia

Split Lake ⊚ Manitoba, C Canada

Splitsko-Dalmatinska Županija
see Split-Dalmacija

Splügen Graubünden,
S Switzerland 46.59N 9.18E

Spodnji Dravograd see
Dravograd

Spofford Texas, SW USA
29.10N 100.24W

Špogi Daugavpils, SE Latvia
56.03N 26.47E

Spokane Washington, NW USA
47.39N 117.25W

Spokane River ⋧ Washington,
NW USA

Spoleto Umbria, C Italy
42.43N 12.43E

Spooner Wisconsin, N USA
45.51N 91.49W

Spoon River ⋧ Illinois, N USA

Spotsylvania Virginia, NE USA
38.13N 77.31W

Sprague Washington, NW USA
47.19N 117.55W

Spratly Island island
SW Spratly Islands

Spratly Islands Chin. Nansha
Qundao. ◇ disputed territory SE Asia

Spray Oregon, NW USA
44.50N 119.38W

Spreča ⋧ N Bosnia and
Herzegovina

Spree ⋧ E Germany

Spreewald wetland NE Germany

Spremberg Brandenburg,
E Germany 51.34N 14.22E

Spring Texas, SW USA
30.03N 95.24W

Spring Arbor Michigan, N USA
42.12N 84.33W

Springbok Northern Cape,
W South Africa 29.38S 17.56E

Spring City Pennsylvania,
NE USA 40.10N 75.33W

Spring City Tennessee, S USA
35.41N 84.51W

Spring City Utah, W USA
39.28N 111.30W

Spring Creek Nevada, W USA
40.45N 115.40W

Springdale Arkansas, C USA
36.11N 94.07W

Springdale Ohio, N USA
39.17N 84.29W

Springe Niedersachsen,
N Germany 52.13N 9.33E

Springer New Mexico, SW USA
36.21N 104.35W

Springerville Arizona, SW USA
34.07N 109.16W

Springfield Colorado, C USA
37.24N 102.36W

Springfield Georgia, SE USA
32.21N 81.20W

Springfield state capital Illinois,
N USA 39.48N 89.38W

Springfield Kentucky, S USA
37.41N 85.13W

Springfield Massachusetts,
NE USA 42.06N 72.32W

Springfield Minnesota,
N USA 44.15N 94.58W

Springfield Missouri, C USA
37.13N 93.18W

Springfield Ohio, N USA
39.55N 83.48W

Springfield Oregon, NW USA
44.03N 123.01W

Springfield South Dakota, N USA
42.51N 97.54W

Springfield Tennessee, S USA
36.30N 86.53W

Springfield Vermont, NE USA
43.18N 72.27W

Springfield, Lake ⊞ Illinois,
N USA

Spring Garden NE Guyana
6.58N 58.34W

Spring Green Wisconsin, N USA
43.10N 90.02W

Spring Grove Minnesota, N USA
43.33N 91.38W

Springhill Louisiana, S USA
33.01N 93.27W

Spring Hill Florida, SE USA
28.28N 82.36W

Spring Hill Kansas, C USA
38.44N 94.49W

Springhill Nova Scotia,
SE Canada 45.40N 64.04W

Spring Lake North Carolina,
SE USA 35.10N 78.58W

Springlake Texas, SW USA
34.13N 102.18W

Spring Mountains ▲ Nevada,
W USA

Spring Point West Falkland,
Falkland Islands 51.49S 60.27W

Spring River ⋧
Arkansas/Missouri, C USA

Spring River ⋧
Missouri/Oklahoma, C USA

Springs Gauteng, NE South Africa
26.15S 28.26E

Springs Junction West Coast,
South Island, NZ 42.20S 172.10E

Springsure Queensland,
E Australia 24.08S 148.06E

Spring Valley Minnesota, N USA
43.41N 92.23W

Spring Valley New York, NE USA
41.10N 73.58W

Springview Nebraska, C USA
42.48N 99.45W

330

◆ COUNTRY ◇ DEPENDENT TERRITORY ◆ ADMINISTRATIVE REGION ▲ MOUNTAIN ℞ VOLCANO ⊚ LAKE
● COUNTRY CAPITAL ○ DEPENDENT TERRITORY CAPITAL ✕ INTERNATIONAL AIRPORT ▲ MOUNTAIN RANGE ⋧ RIVER ⊞ RESERVOIR

Springville New York, NE USA 42.27N 78.52W — 20 D11
Springville Utah, W USA 40.10N 111.36W — 38 L3
Sprottau see Szprotawa
Sproule, Pointe headland Quebec, SE Canada 49.47N 67.02W — 13 V4
Spruce Grove Alberta, SW Canada 53.36N 113.55W — 9 Q14
Spruce Knob ▲ West Virginia, NE USA 38.40N 79.37W — 23 T4
Spruce Mountain ▲ Nevada, W USA 40.33N 114.46W — 37 X3
Spruce Pine North Carolina, SE USA 35.55N 82.03W — 23 P9
Spui ♦ SW Netherlands — 100 G13
Spulico, Capo headland S Italy 39.57N 16.38E — 101 O19
Spur Texas, SW USA 33.28N 100.51W — 27 O5
Spurn Head headland E England, UK 53.34N 0.06E — 99 O17
Spy Namur, S Belgium 50.29N 4.43E — 101 H20
Spydeberg Østfold, S Norway 59.36N 11.04E — 97 J15
Spy Glass Point headland South Island, NZ 59.36N 11.04E — 193 J17
Squamish British Columbia, SW Canada 49.40N 123.10W — 8 L17
Squam Lake ◎ New Hampshire, NE USA — 21 O8
Squa Pan Mountain ▲ Maine, NE USA 46.36N 68.09W — 21 S2
Squaw Harbor Unga Island, Alaska, USA 55.12N 160.41W — 41 N16
Squaw Island island Ontario, S Canada — 12 I11
Squillace, Golfo di gulf S Italy — 109 O22
Squinzano Puglia, SE Italy 40.25N 18.03E — 109 Q18
Sragen Jawa, C Indonesia 7.24S 111.01E — 174 L15
Sráid na Cathrach see Milltown Malbay
Srālau Stœng Trêng, N Cambodia 14.03N 105.46E — 178 Jj11
Srath an Urláir see Stranorlar
Srbac Republika Srpska, N Bosnia & Herzegovina 45.06N 17.33E — 114 G10
Srbinje see Foča
Srbobran see Donji Vakuf
Srě Âmběl Kaôh Kong, SW Cambodia 11.07N 103.46E — 114 K9
Srebrenica Republika Srpska, E Bosnia & Herzegovina 44.04N 19.18E — 114 K13
Srebrenik Federacija Bosna I Hercegovina, E Bosnia & Herzegovina 44.42N 18.30E — 114 I11
Sredets prev. Grudovo. Burgas, E Bulgaria 42.21N 27.13E — 116 M10
Sredets prev. Syulemeshlii. Stara Zagora, C Bulgaria 42.16N 25.40E — 116 K10
Sredetska Reka ♒ SE Bulgaria — 116 M10
Sredinnyy Khrebet ▲ E Russian Federation — 127 P9
Sredishte Rom. Beibunar; prev. Knyazhevo. Dobrich, NE Bulgaria 43.51N 27.30E — 116 N7
Sredna Gora ▲ C Bulgaria — 116 I10
Srednekolymsk Respublika Sakha (Yakutiya), NE Russian Federation 67.28N 153.52E — 127 N7
Srednerusskaya Vozvyshennost' Eng. Central Russian Upland. ▲ W Russian Federation — 130 K7
Srednesibirskoye Ploskogor'ye var. Central Siberian Uplands, Eng. Central Siberian Plateau. ▲ N Russian Federation — 128 I9
Sredniy Ural ▲ NW Russian Federation — 129 V13
Srê Khtŭm Môndól Kiri, E Cambodia 12.10N 106.52E — 178 Jj13
Śrem Wielkopolskie, C Poland 52.07N 17.00E — 113 G17
Sremska Mitrovica prev. Mitrovica, Ger. Mitrowitz. Serbia, NW Yugoslavia 44.58N 19.37E — 114 K10
Srêng, Stœng ♒ NW Cambodia — 178 Ii13
Srê Noy Siěmréab, NW Cambodia 13.47N 104.03E — 178 Ii11
Srepok, Sông see Srêpôk, Tônle
Srêpôk, Tônle var. Sông Srepok. ♒ Cambodia/Vietnam — 178 K12
Sretensk Chitinskaya Oblast', S Russian Federation 52.14N 117.33E — 126 L15
Sri Aman Sarawak, East Malaysia 1.13N 111.25E — 174 L17
Sribne Chernihivs'ka Oblast', N Ukraine 50.40N 32.55E — 119 R4
Sri Jayawardanapura var. Sri Jayawardenepura; prev. Kotte. ● (Sri Lanka) Western Province, W Sri Lanka 6.54N 79.58E — 161 I25
Srikākulam Andhra Pradesh, E India 18.18N 83.54E — 161 M14
Sri Lanka off. Democratic Socialist Republic of Sri Lanka; prev. Ceylon. ♦ republic S Asia — 161 I25
Sri Lanka island S Asia — 138 Mm15
Srimangal Chittagong, E Bangladesh 24.19N 91.40E — 159 V14
Sri Mohangorh see Shri Mohangarh
Srinagar Jammu and Kashmir, N India 34.06N 74.50E — 158 H5
Srinagarind Reservoir ◎ W Thailand — 178 H10
Sringeri Karnātaka, W India 13.25N 75.13E — 161 F19
Sri Pada Eng. Adam's Peak. ▲ S Sri Lanka 6.49N 80.25E — 161 K25
Sri Saket see Si Sa Ket
Środa Śląska Ger. Neumarkt. Dolnośląskie, SW Poland 51.10N 16.30E — 113 G14
Środa Wielkopolska Wielkopolskie, C Poland 52.13N 17.16E — 113 H12
Srpska Kostajnica see Bosanska Kostajnica
Srpska, Republika ♦ republic Bosnia & Herzegovina — 115 G14

Srpski Brod see Bosanski Brod
Ssu-ch'uan see Sichuan
Ssu-p'ing/Ssu-p'ing-chieh see Siping
Stablo see Stavelot
Stabroek Antwerpen, N Belgium 51.21N 4.22E — 101 G15
Stackeln see Strenči
Stack Skerry island N Scotland, UK — 98 I5
Stad peninsula S Norway — 96 C10
Stade Niedersachsen, NW Germany 53.36N 9.28E — 111 R3
Stadl-Paura Oberösterreich, NW Austria 48.14N 13.51E — 111 R3
Stadolichy Rus. Stadolichi. Homyel'skaya Voblasts', SE Belarus 51.43N 28.30E — 121 L20
Stadskanaal Groningen, NE Netherlands 53.00N 6.55E — 100 P7
Stadtallendorf Hessen, C Germany 50.49N 9.01E — 103 H16
Stadtbergen Bayern, S Germany 48.21N 10.50E — 103 K23
Stäfa Zürich, NE Switzerland 47.14N 8.42E — 110 G7
Staffanstorp Skåne, S Sweden 55.37N 13.13E — 97 K23
Staffelstein Bayern, C Germany 50.05N 11.00E — 103 K18
Stafford C England, UK 52.48N 2.07W — 99 L19
Stafford Kansas, C USA 37.57N 98.36W — 28 L6
Stafford Virginia, NE USA 38.24N 77.22W — 23 W4
Staffordshire cultural region C England, UK — 99 L19
Stafford Springs Connecticut, NE USA 41.57N 72.18W — 21 N12
Stágira Kentrikí Makedonía, N Greece 40.31N 23.46E — 117 H14
Staicele Limbaži, N Latvia 57.51N 24.44E — 120 G7
Staierdorf-Anina see Anina
Stainz Steiermark, SE Austria 46.55N 15.18E — 111 V8
Stájerlakanina see Anina
Stakhanov Luhans'ka Oblast', E Ukraine 48.30N 38.42E — 119 Y7
Stalden Valais, SW Switzerland 46.12N 7.55E — 110 E11
Stalin see Varna
Stalinabad see Dushanbe
Stalingrad see Volgograd
Staliniri see Ts'khinvali
Stalino see Donets'k
Stalinobod see Dushanbe
Stalinov Štít see Gerlachovský štít
Stalinsk see Novokuznetsk
Stalinskaya Oblast' see Donets'ka Oblast'
Stalinski Zaliv see Varnenski Zaliv
Stalin, Yazovir see Iskŭr, Yazovir
Stalowa Wola Podkarpackie, SE Poland 50.34N 22.01E — 113 N15
Stamboliyski Popovitsa Plovdiv, C Bulgaria 42.08N 24.36E — 116 I11
Stamboliyski, Yazovir ◎ N Bulgaria — 116 J8
Stamford E England, UK 52.39N 0.32W — 99 N19
Stamford Connecticut, NE USA 41.03N 73.31W — 20 L14
Stamford Texas, SW USA 32.55N 99.49W — 27 P6
Stamford, Lake ◎ Texas, SW USA — 27 Q6
Stampa Graubünden, SE Switzerland 46.21N 9.35E — 110 I10
Stampalia see Astypálaia
Stamps Arkansas, C USA 33.22N 93.30W — 49 T14
Stamsund Nordland, C Norway 68.00N 13.49E — 94 G11
Stanberry Missouri, C USA 40.12N 94.33W — 49 R2
Stancomb-Wills Glacier glacier Antarctica — 205 O3
Standerton Mpumalanga, E South Africa 26.54S 29.15E — 85 K21
Standish Michigan, N USA 43.58N 83.58W — 33 R7
Stanford Kentucky, S USA 37.30N 84.39W — 21 N6
Stanford Montana, NW USA 47.08N 110.15W — 33 S9
Stånga Gotland, SE Sweden 57.16N 18.30E — 97 P19
Stange Hedmark, S Norway 60.43N 11.12E — 96 I13
Stanger KwaZulu/Natal, E South Africa 29.18S 31.17E — 85 L23
Stanimaka see Asenovgrad
Stanislaus River ♒ California, W USA — 37 P8
Stanislav see Ivano-Frankivs'k
Stanislavskaya Oblast' see Ivano-Frankivs'ka Oblast'
Stanisławów see Ivano-Frankivs'k
Stanke Dimitrov see Dupnitsa
Stanley Tasmania, SE Australia 40.48S 145.18E — 191 O15
Stanley var. Port Stanley, Puerto Argentino ○ (Falkland Islands) East Falkland, Falkland Islands 51.45S 57.55W — 67 E24
Stanley Idaho, NW USA 44.12N 114.58W — 35 O13
Stanley North Dakota, N USA 48.19N 102.23W — 30 L1
Stanley var. Sichuan 38.34N 78.30W — 23 U4
Stanley Wisconsin, N USA 44.58N 90.54W — 32 J6
Stanley Pool var. Pool Malebo. ◎ Congo/Dem. Rep. Congo (Zaire) — 81 G21
Stanley Reservoir ◎ S India — 161 H20
Stann Creek ♦ district SE Belize — 44 G3
Stann Creek see Kisangani — 44 G3
Stanovoy Khrebet ▲ SE Russian Federation — 127 N17
Stans Unterwalden, C Switzerland 46.57N 8.22E — 110 F8
Stansted ✕ (London) Essex, E England, UK 51.53N 0.16E — 99 Q21

Stanthorpe Queensland, E Australia 28.35S 151.52E — 191 U4
Stanton Kentucky, S USA 37.51N 83.51W — 23 N6
Stanton Michigan, N USA 43.17N 85.01W — 33 Q8
Stanton Nebraska, C USA 41.57N 97.13W — 31 Q14
Stanton North Dakota, N USA 47.19N 101.22W — 30 L5
Stanton Texas, SW USA 32.07N 101.47W — 27 N7
Stanwood Washington, NW USA 48.14N 122.22W — 34 H7
Stanychno-Luhans'ke Luhans'ka Oblast', E Ukraine 48.39N 39.30E — 119 Y7
Stanzach Tirol, W Austria 47.24N 10.36E — 110 K7
Staphorst Overijssel, E Netherlands 52.37N 6.12E — 100 M9
Staples Ontario, S Canada 42.09N 82.34W — 12 D18
Staples Minnesota, N USA 46.21N 94.47W — 31 T6
Stapleton Nebraska, C USA 41.28N 100.30W — 30 M14
Star Texas, SW USA 31.27N 98.16W — 27 S8
Starachowice Świętokrzyskie, C Poland 51.04N 21.02E — 113 M14
Stará Ľubovňa Ger. Altlublau, Hung. Ólubló. Prešovský Kraj, E Slovakia 49.18N 20.40E — 113 M18
Stara Pazova Ger. Altpasua, Hung. Ópazova. Serbia, N Yugoslavia 44.59N 20.10E — 114 L10
Stara Planina see Balkan Mountains — 23 T5
Stara Reka ♒ C Bulgaria — 116 I9
Stara Synyava Khmel'nyts'ka Oblast', W Ukraine 49.39N 27.38E — 118 M5
Stara Vyzhivka Volyns'ka Oblast', NW Ukraine 51.27N 24.25E — 118 I2
Staraya Byelitsa Rus. Staraya Belitsa. Vitsyebskaya Voblasts', NE Belarus 54.42N 29.37E — 121 M14
Staraya Mayna Ul'yanovskaya Oblast', W Russian Federation 54.36N 48.57E — 131 R5
Staraya Rudnya Rus. Staraya Rudnya. Homyel'skaya Voblasts', SE Belarus 52.50N 30.15E — 121 O18
Staraya Russa Novgorodskaya Oblast', W Russian Federation 57.59N 31.18E — 128 H14
Stara Zagora Lat. Augusta Trajana. Stara Zagora, C Bulgaria 42.26N 25.39E — 116 K10
Stara Zagora ♦ province C Bulgaria — 116 K10
Starbuck Minnesota, N USA 45.36N 95.31W — 31 S8
Starbuck Island prev. Volunteer Island. island E Kiribati — 203 W4
Star City Arkansas, C USA 33.56N 91.50W — 49 V13
Staretina ▲ W Bosnia and Herzegovina — 114 F13
Stargard in Pommern see Stargard Szczeciński
Stargard Szczeciński Ger. Stargard in Pommern. Zachodniopomorskie, NW Poland 53.19N 15.01E — 112 E9
Star Harbour harbor San Cristobal, SE Solomon Islands — 195 Z17
Stari Bečej see Bečej
Stari Grad It. Cittavecchia. Split-Dalmacija, S Croatia 43.11N 16.36E — 115 F15
Staring, Teluk var. Teluk Wawosungu. bay Sulawesi, C Indonesia — 175 Qq12
Staritsa Tverskaya Oblast', W Russian Federation 56.28N 34.51E — 128 J16
Starke Florida, SE USA 29.56N 82.07W — 25 V9
Starkville Mississippi, S USA 33.27N 88.49W — 24 M4
Star Mountains Ind. Pegunungan ▲ Indonesia/PNG — 194 E11
Starnberg Bayern, SE Germany 48.00N 11.19E — 103 L23
Starnberger See ◎ SE Germany — 103 L24
Starobesheve Donets'ka Oblast', E Ukraine 47.45N 38.01E — 119 X8
Starobil's'k Rus. Starobel'sk. Luhans'ka Oblast', E Ukraine 49.16N 38.55E — 119 Y6
Starobin see Starobyn
Starobyn Rus. Starobin. Minskaya Voblasts', S Belarus 52.43N 27.28E — 121 K18
Starodub Bryanskaya Oblast', W Russian Federation 52.30N 32.56E — 130 H6
Starogard Gdański Ger. Preussisch-Stargard. Pomorskie, N Poland 53.57N 18.29E — 112 I8
Staroikan Yuzhnyy Kazakhstan, S Kazakhstan 43.09N 68.34E — 151 P16
Starokonstantinov see Starokostyantyniv
Starokostyantyniv Rus. Starokonstantinov. Khmel'nyts'ka Oblast', NW Ukraine 49.43N 27.12E — 118 L5
Starominskaya Krasnodarskiy Kray, SW Russian Federation 46.33N 39.05E — 130 K12
Staro Selo Rom. Satul-Vechi; prev. Startsevo. Silistra, NE Bulgaria 43.58N 26.32E — 116 L7
Staroshcherbinovskaya Krasnodarskiy Kray, SW Russian Federation 46.36N 38.42E — 130 K12
Starosubkhangulovo Respublika Bashkortostan, W Russian Federation 53.05N 57.22E — 131 V6
Star Peak ▲ Nevada, USA 40.31N 118.09W — 57 S4
Star-Smil see Staro Selo
Start Point headland SW England, UK 50.13N 3.38W — 99 J15
Startsyo see Kirawsk
Starum see Stavoren
Staryya Darohi Rus. Staryye Dorogi. Minskaya Voblasts', S Belarus 53.01N 28.12E — 121 L18
Staryye Dorogi see Staryya Darohi — 101 L24

Staryye Zyatsy Udmurtskaya Respublika, NW Russian Federation 57.22N 52.42E — 131 T2
Staryy Krym Respublika Krym, S Ukraine 45.03N 35.06E — 119 U13
Staryy Oskol Belgorodskaya Oblast', W Russian Federation 51.21N 37.52E — 130 K8
Staryy Sambir L'vivs'ka Oblast', W Ukraine 49.27N 23.00E — 118 H6
Stassfurt var. Staßfurt. Sachsen-Anhalt, C Germany 51.51N 11.34E — 103 L14
Staszów Świętokrzyskie, C Poland 50.34N 21.08E — 113 M15
State Center Iowa, C USA 42.01N 93.09W — 31 W13
State College Pennsylvania, NE USA 40.48N 77.52W — 20 E14
Staten Island island New York, NE USA — 20 K15
Staten Island see Estados, Isla de los
Statenville Georgia, SE USA 30.42N 83.00W — 25 U8
Statesboro Georgia, SE USA 32.28N 81.46W — 25 W5
States, The see United States of America
Statesville North Carolina, SE USA 35.46N 80.53W — 23 R9
Stathelle Telemark, S Norway 59.01N 9.40E — 97 L19
Staunton Illinois, N USA 39.00N 89.47W — 32 K15
Staunton Virginia, NE USA 38.09N 79.04W — 23 T5
Stavanger Rogaland, S Norway 58.58N 5.43E — 97 C16
Stavelot Dut. Stablo. Liège, E Belgium 50.24N 5.55E — 101 L21
Stavern Vestfold, S Norway 58.58N 10.01E — 97 G16
Stavers Island see Vostok Island
Stavoren Fris. Starum. Friesland, N Netherlands 52.52N 5.22E — 100 J7
Stavropol prev. Voroshilovsk. Stavropol'skiy Kray, SW Russian Federation 45.02N 41.57E — 130 M14
Stavropol' see Tol'yatti
Stavropol'skaya Vozvyshennost' ▲ SW Russian Federation — 130 M14
Stavropol'skiy Kray ♦ territory SW Russian Federation — 130 M14
Stavrós Kentrikí Makedonía, N Greece 40.39N 23.42E — 117 H14
Stavrós, Akrotírio headland Kríti, Greece, E Mediterranean Sea 35.25N 24.57E — 117 J24
Stavrós, Akrotírio headland Náxos, Kykládes, Greece, Aegean Sea 37.12N 25.32E — 117 K21
Stavroúpoli prev. Stavroúpolis. Anatolikí Makedonía kai Thráki, NE Greece 41.12N 24.42E — 116 I12
Stavroúpolis see Stavroúpoli
Stavyshche Kyyivs'ka Oblast', N Ukraine 49.23N 30.10E — 119 O6
Stawell Victoria, SE Australia 37.03S 142.47E — 190 M11
Stawiski Podlaskie, NE Poland 53.22N 22.08E — 112 N9
Stayner Ontario, S Canada 44.25N 80.05W — 12 G14
Steamboat Springs Colorado, C USA 40.28N 106.51W — 39 R3
Stearns Kentucky, S USA 36.39N 84.27W — 22 M8
Stebbins Alaska, USA 63.30N 162.15W — 41 N10
Steeg Tirol, W Austria 47.15N 10.18E — 110 K7
Steele Missouri, C USA 36.04N 89.49W — 49 Y9
Steele North Dakota, N USA 46.51N 99.55W — 31 N5
Steele Island land Antarctica — 204 J5
Steeleville Illinois, N USA 38.00N 89.39W — 32 K16
Steelville Missouri, C USA 37.58N 91.21W — 49 W6
Steenbergen Noord-Brabant, S Netherlands 51.34N 4.13E — 101 G14
Steenkool see Bintuni
Steen River Alberta, W Canada 59.37N 117.16W — 9 O10
Steenwijk Overijssel, N Netherlands 52.46N 6.07E — 100 M8
Steeple Jason island Jason Islands, NW Falkland Islands — 67 A23
Steep Point headland Western Australia 26.09S 113.10E — 182 J8
Ştefăneşti Botoşani, NE Romania 47.43N 27.15E — 118 L9
Stefanie, Lake see Ch'ew Bahir — 5 J1
Stefansson Island island Nunavut, N Canada — 203 J16
Ştefan Vodă Rus. Suvorovo. SE Moldova 46.33N 29.40E — 119 O10
Steffen, Cerro ▲ S Chile 44.27S 71.42W — 65 H8
Steffisburg Bern, C Switzerland 46.46N 7.37E — 110 D9
Steg Tirol, W Austria 47.15N 10.18E — 110 K7
Steier see Steyr
Steierdorf/Steierdorf-Anina see Anina
Steiermark off. Land Steiermark, Eng. Styria. ♦ state C Austria — 111 T7
Steigerwald hill range C Germany — 103 J19
Stein Limburg, SE Netherlands 50.58N 5.45E — 101 L17
Stein see Stein an der Donau, Austria
Stein see Kamnik, Slovenia
Steinach Tirol, W Austria 47.07N 11.30E — 110 M8
Steinamanger see Szombathely
Stein an der Donau var. Stein. Niederösterreich, NE Austria 48.24N 15.35E — 111 W3
Steinau an der Elbe see Ścinawa
Steinbach Manitoba, S Canada 49.31N 96.40W — 5 J6
Steiner Alpen see Kamniško-Savinjske Alpe

Steinhuder Meer ◎ NW Germany — 102 H12
Steinkjer Nord-Trøndelag, C Norway 64.01N 11.28E — 95 E15
Stejarul see Karapelit
Stekene Oost-Vlaanderen, NW Belgium 51.13N 4.04E — 101 F16
Stellenbosch Western Cape, SW South Africa 33.55N 18.49E — 85 E26
Stellendam Zuid-Holland, SW Netherlands 51.48N 4.01E — 100 F13
Steller, Mount ▲ Alaska, USA 60.36N 142.49W — 41 T12
Steyerlak-Anina see Anina — 111 T5
Steyr var. Steier. Oberösterreich, N Austria 48.02N 14.26E — 111 T5
Steyr ♒ NW Austria — 111 T5
Stickney South Dakota, N USA 43.24N 98.23W — 31 P11
Stiens Friesland, N Netherlands 53.15N 5.45E — 100 L5
Stif see Sétif
Stigler Oklahoma, C USA 35.15N 95.07W — 49 Q11
Stigliano Basilicata, S Italy 40.24N 16.13E — 109 N18
Stigtomta Södermanland, C Sweden 58.48N 16.46E — 97 N17
Stikine ♒ British Columbia, W Canada — 8 I11
Stensjön Jönköping, S Sweden 57.36N 14.42E — 97 L19
Stilida/Stilís see Stylida
Stillwater Minnesota, N USA 45.03N 92.48W — 31 W8
Stillwater Oklahoma, C USA 36.07N 97.02W — 49 O9
Stillwater Range ▲ Nevada, W USA — 37 S5
Stillwater Reservoir ◎ New York, NE USA — 20 I8
Stilo, Punta headland S Italy 38.27N 16.36E — 109 O22
Stilwell Oklahoma, C USA 35.48N 94.37W — 49 R10
Štimlje Serbia, S Yugoslavia 42.27N 21.03E — 115 N17
Stinnett Texas, SW USA 35.49N 101.26W — 27 N1
Stirling Virginia, NE USA 39.03N 78.10W — 23 V3
Stirling cultural region C Scotland, UK — 11 S7
Stirling Range ▲ Western Australia — 188 J14
Stjørdal Nord-Trøndelag, C Norway 63.27N 10.57E — 97 E16
Stochod see Stokhid — 103 H24
Step' Nardara Kaz. Shardara Dalasy; prev. Shaidara. grassland S Kazakhstan — 151 P17
Stepanakert see Xankändi
Stepenitz ♒ N Germany — 102 K9
Stephan South Dakota, N USA 44.12N 99.25W — 31 O10
Stephen Minnesota, N USA 48.27N 96.54W — 31 R3
Stephens Arkansas, C USA 33.25N 93.04W — 49 T14
Stephens, Cape headland D'Urville Island, Marlborough, SW NZ 40.42S 173.56E — 192 J13
Stephens City Virginia, NE USA 39.03N 78.10W — 23 V3
Stephens Creek New South Wales, SE Australia 31.51S 141.30E — 190 L6
Stephens Island island C NZ — 192 K13
Stephenson Michigan, USA 45.27N 87.36W — 33 N5
Stephenville Newfoundland, SE Canada 48.33N 58.33W — 11 S7
Stephenville Texas, SW USA 32.12N 98.13W — 27 S7
Stepnogorsk Akmola, C Kazakhstan 52.04N 72.18E — 151 R8
Stepnoye Stavropol'skiy Kray, SW Russian Federation 44.18N 44.34E — 131 O15
Stepnyak Severnyy Kazakhstan, N Kazakhstan 52.52N 70.49E — 151 Q8
Steps Point headland W American Samoa 14.22S 170.46W — 198 C9
Stereá Ellás Eng. Greece Central. ♦ region C Greece — 117 F17
Sterkspruit Eastern Cape, SE South Africa 30.28S 27.24E — 85 J24
Sterlibashevo Respublika Bashkortostan, W Russian Federation 53.19N 55.12E — 131 U6
Sterling Alaska, USA 60.32N 150.51W — 41 R12
Sterling Colorado, C USA 40.37N 103.12W — 39 V3
Sterling Illinois, N USA 41.47N 89.42W — 32 K11
Sterling Kansas, C USA 38.12N 98.12W — 28 M5
Sterling City Texas, SW USA 31.50N 100.58W — 111 X3
Sterling Heights Michigan, N USA 42.34N 83.01W — 26 M10
Sterling Park Virginia, NE USA 39.00N 77.24W — 23 W3
Sterling Reservoir ◎ Colorado, C USA — 39 V2
Sterlington Louisiana, C USA 32.42N 92.05W — 24 I5
Sterlitamak Respublika Bashkortostan, W Russian Federation 53.39N 56.00E — 131 U6
Šternberk Ger. Sternberg. Olomoucký Kraj, E Czech Republic 49.45N 17.19E — 113 H17
Stêroh Suquţrā, S Yemen 12.21N 53.50E — 147 V17
Steszew Wielkopolskie, C Poland 52.16N 16.41E — 112 G11
Steubenville Ohio, N USA 40.21N 80.37W — 99 V13
Stevenage E England, UK 51.55N 0.13W — 99 O21
Stevenson Alabama, S USA 34.52N 85.50W — 25 Q1
Stevenson Washington, NW USA 45.43N 121.54W — 34 H1
Stevenson Creek seasonal river South Australia — 190 E1
Stevenson Entrance strait Alaska, USA — 41 Q13
Stevens Point Wisconsin, N USA 44.31N 89.33W — 32 L6
Stevens Village Alaska, USA 66.01N 149.02W — 41 R8
Stevensville Montana, NW USA 46.30N 114.05W — 35 P10
Stevns Klint headland E Denmark 55.15N 12.25E — 99 E25
Stewart British Columbia, W Canada 55.58N 129.52W — 8 J12
Stewart ♒ Yukon Territory, NW Canada — 8 J6
Stewart Crossing Yukon Territory, NW Canada 63.22N 136.31W — 8 I6
Stewart, Isla island S Chile — 65 H8
Stewart Island island S NZ — 193 B25

Stewart, Mount ▲ Queensland, E Australia 20.11S 145.29E — 189 W6
Stewart River Yukon Territory, NW Canada 63.17N 139.24W — 8 H6
Stewartsville Missouri, C USA 39.45N 94.30W — 49 R3
Stewart Valley Saskatchewan, S Canada 50.34N 107.47W — 9 S16
Stewartville Minnesota, N USA 43.51N 92.29W — 31 W10
Steyerlak-Anina see Anina — 31 W10
Steyr var. Steier. Oberösterreich, N Austria 48.02N 14.26E — 111 T5
Steyr ♒ NW Austria — 111 T5
Stickney South Dakota, N USA 43.24N 98.23W — 31 P11
Stiens Friesland, N Netherlands 53.15N 5.45E — 100 L5
Stif see Sétif
Stigler Oklahoma, C USA 35.15N 95.07W — 49 Q11
Stigliano Basilicata, S Italy 40.24N 16.13E — 109 N18
Stigtomta Södermanland, C Sweden 58.48N 16.46E — 97 N17
Stigtomta — 97 N17
Stilida/Stilís see Stylida
Stillwater Minnesota, N USA 45.03N 92.48W — 31 W8
Stillwater Oklahoma, C USA 36.07N 97.02W — 49 O9
Stillwater Range ▲ Nevada, W USA — 37 S5
Stillwater Reservoir ◎ New York, NE USA — 20 I8
Stilo, Punta headland S Italy 38.27N 16.36E — 109 O22
Stilwell Oklahoma, C USA 35.48N 94.37W — 49 R10
Štimlje Serbia, S Yugoslavia 42.27N 21.03E — 115 N17
Stinnett Texas, SW USA 35.49N 101.26W — 27 N1
Stirling Virginia, NE USA 39.03N 78.10W — 23 V3
Stirling Creek New South Wales, SE Australia 31.51S 141.30E — 190 L6
Stirling Island island C NZ — 192 K13
Stockach Baden-Württemberg, S Germany 47.51N 9.01E — 103 H24
Stockdale Texas, SW USA 29.14N 97.57W — 27 S12
Stockerau Niederösterreich, NE Austria 48.24N 16.13E — 111 X3
Stockholm ● (Sweden) Stockholm, C Sweden 59.16N 18.03E — 95 H20
Stockholm ♦ county C Sweden — 97 O15
Stockmannshof see Pļaviņas
Stockport NW England, UK 53.25N 2.10W — 99 L18
Stocks Seamount undersea feature C Atlantic Ocean — K15
Stockton California, W USA 37.55N 121.19W — 37 P15
Stockton Kansas, C USA 39.25N 99.17W — 28 L3
Stockton Missouri, C USA 37.42N 93.48W — 49 S6
Stockton Island island Apostle Islands, Wisconsin, N USA — 32 K3
Stockton Lake ◎ Missouri, C USA — 49 S7
Stockton-on-Tees var. Stockton on Tees. N England, UK 54.34N 1.19W — 9 M15
Stockton Plateau plain Texas, SW USA — 26 M10
Stockville Nebraska, C USA 40.30N 100.21W — 31 N12
Stöde Västernorrland, C Sweden 62.27N 16.34E — 95 M16
Stœng Trêng prev. Stung Treng. Stœng Trêng, NE Cambodia 13.31N 105.58E — 178 Jj12
Stogovo Karaorman ▲ W FYR Macedonia — 115 M19
Stoke-on-Trent var. Stoke. C England, UK 53.00N 2.10W — 99 L19
Stokes Point headland Tasmania, SE Australia — 190 M15
Stokhid Pol. Stochód, Rus. Stokhod. ♒ NW Ukraine — 118 J2
Stokhod see Stokhid
Stokkseyri Suðurland, SW Iceland 63.49N 21.00W — 95 G10
Stokmarknes Nordland, C Norway 68.33N 14.54E — 94 H11
Stol see Veliki Krš
Stolac Federacija Bosna I Hercegovina, S Bosnia and Herzegovina 43.04N 17.58E — 115 H15
Stolberg var. Stolberg im Rheinland. Nordrhein-Westfalen, W Germany 50.46N 6.13E — 103 D16
Stolberg im Rheinland see Stolberg
Stolbovoy, Ostrov island NE Russian Federation — 126 L5
Stolbtsy see Stowbtsy
Stolin Rus. Stolin. Brestskaya Voblasts', SW Belarus 51.52N 26.51E — 121 J20
Stolp see Słupsk
Stolpe see Słupia
Stolpmünde see Ustka
Stómio Thessalía, C Greece 39.51N 22.45E — 117 F15
Stonehaven NE Scotland, UK 56.58N 2.13W — 98 L10
Stonehenge ancient monument Wiltshire, S England, UK 51.12N 1.54W — 99 M23
Stonecliffe Ontario, SE Canada 46.12N 77.58W — 123 G10

Stone Mountain ▲ Georgia, SE USA 33.48N 84.10W — 25 T3
Stonewall Manitoba, S Canada 50.07N 97.19W — 9 X16
Stonewood West Virginia, NE USA 39.15N 80.18W — 23 S3
Stoney Point Ontario, S Canada 42.18N 82.32W — 12 D17
Stonglandseidet Troms, N Norway 69.03N 17.06E — 94 H10
Stonybeach Bay bay Tristan da Cunha, SE Atlantic Ocean — 67 N25
Stony Creek ♒ California, W USA — 37 N5
Stonyhill Point headland S Tristan da Cunha — 67 N25
Stony Lake ◎ Ontario, SE Canada — 12 I14
Stony Plain Alberta, SW Canada 53.31N 114.04W — 9 Q14
Stony Point North Carolina, SE USA 35.51N 81.04W — 23 R9
Stony Point headland New York, NE USA 43.50N 76.18W — 20 G8
Stony Rapids Saskatchewan, C Canada 59.13N 105.48W — 9 T10
Stony River Alaska, USA 61.48N 156.37W — 41 P11
Stony Tunguska see Podkamennaya Tunguska
Stooping ♒ Ontario, C Canada — 10 G10
Stör ♒ N Germany — 102 I9
Storå Örebro, S Sweden 59.43N 15.10E — 97 M15
Stora Gla ◎ C Sweden — 97 J16
Stora Le Nor. Store Le. ◎ Norway/Sweden — 97 I16
Stora Lulevatten ◎ N Sweden — 94 I12
Storavan ◎ N Sweden — 94 H13
Storby Åland, SW Finland 60.12N 19.33E — 95 L20
Stordalen Møre og Romsdal, S Norway 62.22N 7.00E — 96 E10
Storebælt var. Store Bælt, Eng. Great Belt, Storebelt. channel Baltic Sea/Kattegat — H23
Storebro Kalmar, S Sweden 57.36N 15.30E — 97 M19
Store Heddinge Storstrøm, SE Denmark 55.19N 12.24E — 97 J24
Store Le see Stora Le
Støren Sør-Trøndelag, S Norway 63.01N 10.16E — 95 E16
Store Sotra island S Norway — 97 B14
Storfjorden fjord S Norway — 94 I4
Storfors Värmland, C Sweden 59.33N 14.16E — 97 L15
Storforshei Nordland, C Norway 66.25N 14.25E — 94 G13
Storhammer see Hamar
Störkanal canal N Germany — 102 L10
Storlien Jämtland, C Sweden 63.18N 12.10E — 95 F16
Storm Bay inlet Tasmania, SE Australia — 191 P17
Storm Lake Iowa, C USA 42.38N 95.12W — 31 T12
Storm Lake ◎ Iowa, C USA — 31 S13
Stornoway NW Scotland, UK 53.25N 2.10W — 98 G7
Storøya island N Svalbard — 94 J1
Storozhevsk Respublika Komi, NW Russian Federation 61.56N 52.18E — 129 S10
Storozhinets see Storozhynets'
Storozhynets' Ger. Storozynetz, Rom. Storojinet, Rus. Storozhinets. Chernivets'ka Oblast', W Ukraine 48.09N 25.40E — 118 K8
Storozynetz see Storozhynets'
Storrs Connecticut, NE USA 41.48N 72.15W — 21 N12
Storsjøen ◎ S Norway — 96 H11
Storsjön ◎ C Sweden — 96 N13
Storsjön ◎ C Sweden — 95 I15
Storslett Troms, N Norway 69.45N 21.03E — 94 J9
Storsteinnes Troms, N Norway 69.13N 19.14E — 96 J8
Storstrøm off. Storstrøms Amt. ♦ county SE Denmark — 97 J24
Storsylen ▲ S Norway 63.07N 12.10E — 95 J15
Storsund Norrbotten, SW Sweden 65.36N 20.40E — 95 F16
Stortoppen ▲ N Sweden 67.28N 17.56E — 97 H11
Storuman Västerbotten, N Sweden 65.04N 17.10E — 95 H14
Storuman ◎ N Sweden — 94 J9
Storvik Gävleborg, C Sweden 60.37N 16.30E — 96 N13
Storvreta Uppsala, C Sweden 59.58N 17.42E — 97 O14
Story City Iowa, C USA 42.10N 93.36W — 31 V13
Stoughton Saskatchewan, S Canada 49.40N 103.01W — 9 L23
Stoughton Massachusetts, NE USA 42.07N 71.06W — 21 O11
Stoughton Wisconsin, N USA 42.56N 89.12W — 32 L9
Stour ♒ E England, UK — 99 L23
Stour ♒ S England, UK — 99 P21
Stour ♒ S England, UK — 99 N23
Stover Missouri, C USA 38.26N 92.59W — 49 T6
Støvring Nordjylland, N Denmark 56.55N 9.52E — 97 I13
Stowbtsy Pol. Stołbce, Rus. Stowbtsy. Minskaya Voblasts', C Belarus 53.27N 26.44E — 121 J17
Stowell Texas, SW USA 29.47N 94.21W — 27 X11
Stowmarket E England, UK 52.04N 0.54E — 99 P20
Stozher Dobrich, NE Bulgaria 43.27N 27.49E — 116 N8
Strabane Ir. An Srath Bán. W Northern Ireland, UK 68.09N 17.12E — 99 E14
Strabo Trench undersea feature C Mediterranean Sea — 123 G10
Strafford Missouri, C USA 37.16N 93.07W — 49 T7
Strahan Tasmania, SE Australia 42.10S 145.18E — 191 N17

◆ COUNTRY   ● COUNTRY CAPITAL   ○ DEPENDENT TERRITORY   ○ DEPENDENT TERRITORY CAPITAL   ◇ ADMINISTRATIVE REGION   ✕ INTERNATIONAL AIRPORT   ▲ MOUNTAIN   ▲ MOUNTAIN RANGE   ☆ VOLCANO   ♒ RIVER   ◎ LAKE   ▨ RESERVOIR

331

113 C18 **Strakonice** Ger. Strakonitz. Budějovický Kraj, S Czech Republic 49.13N 13.55E

Strakonitz see Strakonice

102 N8 **Stralsund** Mecklenburg-Vorpommern, NE Germany 54.18N 13.06E

101 L16 **Stramproy** Limburg, SE Netherlands 51.12N 5.43E

85 E26 **Strand** Western Cape, SW South Africa 34.06S 18.49E

96 E10 **Stranda** Møre og Romsdal, S Norway 62.18N 6.55E

99 G15 **Strangford Lough** Ir. Loch Cuan. inlet E Northern Ireland, UK

97 N16 **Strängnäs** Södermanland, C Sweden 59.22N 17.01E

99 E14 **Stranorlar** Ir. Srath an Urláir. NW Ireland 54.48N 7.46W

99 H14 **Stranraer** S Scotland, UK 54.54N 5.01W

9 U16 **Strasbourg** Saskatchewan, S Canada 51.04N 104.58W

105 V5 **Strasbourg** Ger. Strassburg; anc. Argentoratum. Bas-Rhin, NE France 48.34N 7.45E

39 U4 **Strasburg** Colorado, C USA 39.42N 104.13W

31 N7 **Strasburg** North Dakota, N USA 46.07N 100.10W

33 U12 **Strasburg** Ohio, N USA 40.35N 81.31W

23 U3 **Strasburg** Virginia, NE USA 38.59N 78.21W

119 N10 **Strășeni** var. Strasheny. C Moldova 47.07N 28.37E

Strasheny see Strășeni

111 T8 **Strassburg** Kärnten, S Austria 46.54N 14.21E

Strassburg see Strasbourg, France **Strassburg** see Aiud, Romania

101 M25 **Strassen** Luxembourg, S Luxembourg 49.37N 6.04E

111 R5 **Strasswalchen** Salzburg, C Austria 47.59N 13.19E

12 F16 **Stratford** Ontario, S Canada 43.22N 81.00W

192 K10 **Stratford** Taranaki, North Island, NZ 39.20S 174.15E

37 Q11 **Stratford** California, W USA 36.10N 119.47W

31 V13 **Stratford** Iowa, C USA 42.16N 93.55W

49 O12 **Stratford** Oklahoma, C USA 34.48N 96.57W

27 N1 **Stratford** Texas, SW USA 36.20N 102.04W

32 K6 **Stratford** Wisconsin, N USA 44.53N 90.13W

Stratford see Stratford-upon-Avon

99 M20 **Stratford-upon-Avon** var. Stratford. C England, UK 52.12N 1.40W

191 O17 **Strathgordon** Tasmania, SE Australia 42.49S 146.04E

9 Q16 **Strathmore** Alberta, SW Canada 51.04N 113.19W

37 R11 **Strathmore** California, W USA 36.07N 119.04W

12 E16 **Strathroy** Ontario, S Canada 42.57N 81.40W

98 I6 **Strathy Point** headland N Scotland, UK 58.36N 4.04W

39 W4 **Stratton** Colorado, C USA 39.16N 102.34W

21 P6 **Stratton** Maine, NE USA 45.08N 70.25W

20 M10 **Stratton Mountain** ▲ Vermont, NE USA 43.05N 72.55W

103 N21 **Straubing** Bayern, SE Germany 48.52N 12.34E

102 O12 **Strausberg** Brandenburg, E Germany 52.34N 13.52E

34 K13 **Strawberry Mountain** ▲ Oregon, NW USA 44.18N 118.43W

31 X12 **Strawberry Point** Iowa, C USA 42.40N 91.31W

38 M3 **Strawberry Reservoir** ⊠ Utah, W USA

38 M4 **Strawberry River** ⋙ Utah, W USA

27 R7 **Strawn** Texas, SW USA 32.33N 98.30W

115 P17 **Straža** ▲ Bulgaria/FYR Macedonia 42.16N 22.13E

113 I19 **Strážov** Hung. Sztrazsó. ▲ NW Slovakia 48.59N 18.29E

190 F7 **Streaky Bay** South Australia 32.49S 134.13E

190 E7 **Streaky Bay** bay South Australia

32 L12 **Streator** Illinois, N USA 41.07N 88.50W

Streckenbach see Świdnik

113 C17 **Středočeský Kraj** ♦ region C Czech Republic

Strednogorie see Pirdop

31 O6 **Streeter** North Dakota, N USA 46.37N 99.23W

27 U8 **Streetman** Texas, SW USA 31.52N 96.19W

118 G13 **Strehaia** Mehedinţi, SW Romania 44.37N 23.10E

Strehlen see Strzelin

113 L16 **Strelcha** Pazardzhik, C Bulgaria 42.28N 24.21E

126 I13 **Strelka** Krasnoyarskiy Kray, C Russian Federation 58.04N 92.54E

128 L6 **Strel'na** ⋙ NW Russian Federation

120 H7 **Strenči** Ger. Stackeln. Valka, N Latvia 57.38N 25.42E

110 K8 **Strengen** Tirol, W Austria 47.07N 10.25E

108 C6 **Stresa** Piemonte, NE Italy 45.52N 8.32E

Streshin see Streshyn

121 N18 **Streshyn** Rus. Streshin. Homyel'skaya Voblasts', SE Belarus 52.42N 30.08E

97 B18 **Streymoy** Dan. Strømø Island Faeroe Islands 62.11N 7.18W

126 Gg11 **Strezhevoy** Tomskaya Oblast', C Russian Federation 60.39N 77.37E

97 G23 **Strib** Fyn, C Denmark 55.33N 9.46E

113 A17 **Stříbro** Ger. Mies. Plzeňský Kraj, W Czech Republic 49.44N 12.55E

194 E13 **Strickland** ⋙ SW PNG

---

65 H21 **Strobel, Lago** ☺ S Argentina

63 B25 **Stroeder** Buenos Aires, E Argentina 40.10S 62.34W

117 C20 **Strofádes** island Iónioi Nísoi, Greece, C Mediterranean Sea

Strofiliá see Strofyliá

117 G17 **Strofyliá** var. Strofiliá. Évvoia, C Greece 38.49N 23.25E

102 O10 **Strom** ⋙ NE Germany

109 L22 **Stromboli** ⋙ Isola Stromboli, SW Italy 38.48N 15.13E

109 L22 **Stromboli, Isola** island Isole Eolie, S Italy

98 H9 **Stromeferry** N Scotland, UK 57.20N 5.34W

98 J5 **Stromness** N Scotland, UK 58.57N 3.18W

96 N11 **Strömsbruk** Gävleborg, C Sweden 61.52N 17.19E

31 Q15 **Stromsburg** Nebraska, C USA 41.06N 97.36W

97 K21 **Strömsnäsbruk** Kronoberg, S Sweden 56.34N 13.45E

97 I17 **Strömstad** Västra Götaland, S Sweden 58.55N 11.10E

95 G16 **Strömsund** Jämtland, C Sweden 63.51N 15.34E

95 G15 **Ströms Vattudal** valley N Sweden

49 V14 **Strong** Arkansas, C USA 33.06N 92.21W

109 O21 **Strongoli** Calabria, SW Italy 39.17N 17.03E

33 T11 **Strongsville** Ohio, N USA 41.18N 81.50W

117 Q23 **Strongylí** var. Strongili. island SE Greece

98 K5 **Stronsay** island NE Scotland, UK

99 L21 **Stroud** C England, UK 51.45N 2.15W

49 O10 **Stroud** Oklahoma, C USA 35.45N 96.37W

20 I14 **Stroudsburg** Pennsylvania, NE USA 40.59N 75.12W

97 F21 **Struer** Ringkøbing, W Denmark 56.28N 8.37E

115 M20 **Struga** SW FYR Macedonia 41.11N 20.41E

Strugi-Kranyse see Strugi-Krasnyye

128 M32 **Strugi-Krasnyye** var. Strugi-Kranyse. Pskovskaya Oblast', W Russian Federation 58.19N 29.09E

116 G11 **Struma** Gk. Strymónas. ⋙ Bulgaria/Greece see also Strymónas

99 G22 **Strumble Head** headland SW Wales, UK 52.01N 5.05W

115 Q19 **Strumešnitsa** | Mac. Strumica. ⋙ Bulgaria/FYR Macedonia

115 Q19 **Strumica** E FYR Macedonia 41.27N 22.39E

116 G11 **Strumyani** Blagoevgrad, SW Bulgaria 41.41N 23.13E

33 V12 **Struthers** Ohio, N USA 41.03N 80.36W

116 I10 **Stryama** ⋙ C Bulgaria

116 G13 **Strymónas** Bul. Struma. ⋙ Bulgaria/Greece see also Struma

117 H14 **Strymonikós Kólpos** gulf N Greece

118 I6 **Stryy** L'vivs'ka Oblast', NW Ukraine 49.16N 23.51E

118 H6 **Stryy** ⋙ W Ukraine

113 F14 **Strzegom** Ger. Striegau. Wałbrzych, SW Poland 50.58N 16.19E

112 E10 **Strzelce Krajeńskie** Ger. Friedeberg Neumark. Lubuskie, W Poland 52.52N 15.30E

113 I15 **Strzelce Opolskie** Ger. Gross Strehlitz. Opolskie, S Poland 50.31N 18.19E

190 K3 **Strzelecki Creek** seasonal river South Australia

190 J3 **Strzelecki Desert** desert South Australia

113 G15 **Strzelin** Ger. Strehlen. Dolnośląskie, SW Poland 50.46N 17.03E

112 I11 **Strzelno** Kujawsko-pomorskie, C Poland 52.38N 18.11E

113 N17 **Strzyżów** Podkarpackie, SE Poland 49.52N 21.46E

25 Y13 **Stuart** Florida, SE USA 27.12N 80.15W

31 U13 **Stuart** Iowa, C USA 41.30N 94.19W

31 O13 **Stuart** Nebraska, C USA 42.36N 99.08W

23 S8 **Stuart** Virginia, NE USA 36.38N 80.16W

8 L13 **Stuart** ♦ British Columbia, W Canada

41 N10 **Stuart Island** island Alaska, USA

8 L13 **Stuart Lake** ⊠ British Columbia, SW Canada

193 B22 **Stuart Mountains** ▲ South Island, NZ

190 F3 **Stuart Range** hill range South Australia

Stubaital see Neustift im Stubaital

97 J24 **Stubbekøbing** Storstrøm, SE Denmark 54.52N 12.04E

47 P14 **Stubbs** Saint Vincent, Saint Vincent and the Grenadines 13.08N 61.09W

111 V6 **Stübming** ⋙ E Austria

116 J11 **Studen Kladenets, Yazovir** ⊠ S Bulgaria

193 G21 **Studholme** Canterbury, South Island, NZ 44.44S 171.07E

Stuhlweissenburg see Székesfehérvár

Stuhm see Sztum

10 C7 **Stull Lake** ⊠ Ontario, C Canada

130 L4 **Stung Treng** see Stœng Trêng

10 U4 **Stupino** Moskovskaya Oblast', W Russian Federation 54.54N 38.06E

10 U4 **Sturgeon** Missouri, C USA 39.13N 92.16W

12 G10 **Sturgeon** ⋙ Ontario, S Canada

33 N6 **Sturgeon Bay** Wisconsin, N USA 44.51N 87.21W

12 G11 **Sturgeon Falls** Ontario, S Canada 46.22N 79.57W

10 C12 **Sturgeon Lake** ⊠ Ontario, S Canada

---

32 M3 **Sturgeon River** ⋙ Michigan, N USA

22 H6 **Sturgis** Kentucky, S USA 37.33N 87.58W

33 P11 **Sturgis** Michigan, N USA 41.48N 85.25W

30 J9 **Sturgis** South Dakota, N USA 44.24N 103.30W

114 D10 **Šturlić** Federacija Bosna I Hercegovina, NW Bosnia and Herzegovina 45.03N 15.47E

113 J22 **Štúrovo** Hung. Párkány; prev. Parkan. Nitriansky Kraj, SW Slovakia 47.49N 18.44E

190 L4 **Sturt, Mount** hill New South Wales, SE Australia 29.30S 141.41E

189 P4 **Sturt Plain** plain Northern Territory, N Australia

189 P9 **Sturt Stony Desert** desert South Australia

85 J25 **Stutterheim** Eastern Cape, S South Africa 32.34S 27.25E

103 H21 **Stuttgart** Baden-Württemberg, SW Germany 48.47N 9.12E

49 W12 **Stuttgart** Arkansas, C USA 34.30N 91.33W

94 H2 **Stykkishólmur** Vesturland, W Iceland 65.03N 22.43W

117 F17 **Stylída** var. Stilida, Stilís. Stereá Ellás, C Greece 38.55N 22.37E

118 K2 **Styr** Rus. Styr'. ⋙ Belarus/Ukraine

117 I19 **Stýra** var. Stira. Évvoia, C Greece 38.10N 24.13E

Styria see Steiermark

Su see Jiangsu

Sue see S Sudan

175 S17 **Suai** W East Timor 9.19S 125.16E

56 G9 **Suaita** Santander, C Colombia 6.07N 73.30W

42 I7 **Suakin** var. Sawakin. Red Sea, NE Sudan 19.06N 37.17E

167 T13 **Suao** Jap. Suō. N Taiwan 24.33N 121.48E

Suao see Suau

42 G6 **Suaqui Grande** Sonora, NW Mexico 28.22N 109.52W

63 A14 **Suardi** Santa Fe, C Argentina 30.31S 61.58W

56 D11 **Suárez** Cauca, SW Colombia 2.55N 76.40W

195 N17 **Suau** var. Suao. Suao Island, SE PNG 10.44S 150.18E

120 G12 **Subačius** Kupiškis, NE Lithuania 55.45N 24.55E

174 Ji14 **Subang** prev. Soebang. Jawa, C Indonesia 6.31S 107.45E

174 Gg5 **Subang** ✈ (Kuala Lumpur) Pahang, Peninsular Malaysia

133 S10 **Subansiri** ⋙ NE India

120 E11 **Subate** Daugavpils, SE Latvia

145 N5 **Subaykhān** Dayr az Zawr, E Syria 34.47N 40.38E

165 P8 **Subei** var. Dangchengwan, Subei Mongolzu Zizhixian. Gansu, C China 39.33N 94.50E

Subei Mongolzu Zizhixian see Subei

174 K5 **Subi Besar, Pulau** island Kepulauan Natuna, W Indonesia

28 I7 **Sublette** Kansas, C USA 37.26N 100.48W

114 K8 **Subotica** Ger. Maria-Theresiopel, Hung. Szabadka. Serbia, N Yugoslavia 46.06N 19.40E

118 K9 **Suceava** Ger. Suczawa, Hung. Szucsava. Suceava, NE Romania 47.40N 26.15E

118 J9 **Suceava** ♦ county NE Romania

118 K9 **Suceava** Ger. Suczawa. ⋙ N Romania

114 D12 **Sučević** Zadar, SW Croatia 44.13N 16.04E

113 K17 **Sucha Beskidzka** Małopolskie, S Poland 49.39N 19.11E

113 M14 **Suchedniów** Świętokrzyskie, C Poland 51.01N 20.49E

44 A2 **Suchitepéquez** off. Departamento de Suchitepéquez. ♦ department SW Guatemala

Su-chou see Suzhou

Suchow see Suzhou, Jiangsu, China

Suchow see Xuzhou, Jiangsu, China

99 D17 **Suck** ⋙ C Ireland

194 M16 **Suckling, Mount** ▲ S PNG 9.36S 149.00E

59 L19 **Sucre** hist. Chuquisaca, La Plata. • (Bolivia-legal capital) Chuquisaca, S Bolivia 18.52S 65.24W

54 E6 **Sucre** Santander, N Colombia 8.50N 74.22W

56 C11 **Sucre** Manabí, W Ecuador 1.21S 80.27W

56 E6 **Sucre** off. Departamento de Sucre. ♦ department N Colombia

57 O5 **Sucre** off. Estado Sucre. ♦ state NE Venezuela

58 D6 **Sucumbíos** ♦ province NE Ecuador

115 G15 **Sućuraj** Split-Dalmacija, S Croatia

60 K10 **Sucuriju** Amapá, NE Brazil 1.30N 50.00W

61 E16 **Sud** Eng. South. ♦ province S Cameroon

128 K13 **Suda** ⋙ NW Russian Federation

119 U13 **Sudak** Respublika Krym, S Ukraine 44.51N 34.55E

26 M4 **Sudan** Texas, SW USA 34.04N 102.31W

82 C10 **Sudan** off. Republic of Sudan, Ar. Jumhuriyat as-Sudan; prev. Anglo-Egyptian Sudan.

Sudan see Mali

Sudanese Republic see Mali

Sudan, Jumhuriyat as- see Sudan

12 G13 **Sudbury** Ontario, S Canada 46.29N 80.59W

99 P21 **Sudbury** E England, UK 52.04N 0.43E

44 G8 **Sud, Canal de** see Gonâve, Canal de la

---

82 E13 **Sudd** swamp region N Sudan

102 K10 **Sude** ⋙ N Germany

113 E15 **Sudest Island** see Tagula Island

113 E15 **Sudeten** var. Sudetes, Sudetic Mountains, Cz./Pol. Sudety. ▲ Czech Republic/Poland

Sudetes see Sudeten

Sudetic Mountains/Sudety see Sudeten

94 G1 **Suðureyri** Vestfirðir, NW Iceland 66.08N 23.31W

82 A3 **Suður** ♦ region S Iceland

44 H6 **Sudurho** see S Sulaymāniyah

97 B19 **Suðuroy** Dan. Suderø Island Faeroe Islands 61.60N 6.29W

176 Xs12 **Sudirman, Pegunungan** ▲ Irian Jaya, E Indonesia

128 M15 **Sudislavl'** Kostromskaya Oblast', NW Russian Federation 57.55N 41.45E

Südkarpaten see Carpaţii Meridionali

81 N20 **Sud Kivu** off. Région Sud Kivu. ♦ region E Dem. Rep. Congo (Zaire)

Südliche Morava see Južna Morava

102 E12 **Süd-Nord-Kanal** canal NW Germany

130 M3 **Sudogda** Vladimirskaya Oblast', W Russian Federation 55.58N 40.57E

Sudostroy see Severodvinsk

81 C15 **Sud-Ouest** Eng. South-West. ♦ province W Cameroon

181 X17 **Sud Ouest, Pointe** headland SW Mauritius 20.27S 57.18E

197 J7 **Sud, Province** ♦ province S New Caledonia

130 J8 **Sudzha** Kurskaya Oblast', W Russian Federation 51.12N 35.19E

83 D15 **Sue** ⋙ S Sudan

107 S10 **Sueca** Pais Valenciano, E Spain 39.13N 0.19W

116 I10 **Süedinenie** Plovdiv, C Bulgaria 42.14N 24.36E

Suero see Alzira

77 X8 **Suez** Ar. As Suways, El Suweis. NE Egypt 29.58N 32.33E

77 W7 **Suez Canal** Ar. Qanāt as Suways. canal NE Egypt

77 X8 **Suez, Gulf of** Ar. Khalīj as Suways. gulf NE Egypt

9 R17 **Suffield** Alberta, SW Canada 50.15N 111.05W

23 X3 **Suffolk** Virginia, NE USA 36.43N 76.34W

99 P20 **Suffolk** cultural region E England, UK

148 J2 **Şūfiān** Āzarbāyjān-e Khāvari, N Iran 38.15N 45.58E

33 N12 **Sugar Creek** ⋙ Illinois, N USA

32 L13 **Sugar Creek** ⋙ Illinois, N USA

33 R3 **Sugar Island** island Michigan, N USA

27 V11 **Sugar Land** Texas, SW USA 29.37N 95.37W

21 P6 **Sugarloaf Mountain** ▲ Maine, NE USA 45.01N 70.18W

67 G24 **Sugar Loaf Point** headland N Saint Helena 15.54S 5.43W

127 O8 **Sugoy** ⋙ E Russian Federation

164 F7 **Sugun** Xinjiang Uygur Zizhiqu, W China 39.46N 76.45E

153 U11 **Sugut, Gora** ▲ SW Kyrgyzstan 39.52N 73.36E

175 O2 **Sugut, Sungai** ⋙ East Malaysia

126 08 **Suhai Hu** ☺ C China

168 K14 **Suhait** Nei Mongol Zizhiqu, N China 39.29N 105.11E

147 X7 **Şuḥār** var. Sohar. N Oman 24.23N 56.43E

168 L6 **Sühbaatar** Selenge, N Mongolia 50.11N 106.14E

169 P9 **Sühbaatar** ♦ province E Mongolia

103 K17 **Suhl** Thüringen, C Germany 50.37N 10.43E

110 F7 **Suhr** Aargau, N Switzerland 47.22N 8.04E

110 F7 **Suhr** ⋙ N Switzerland

167 O12 **Suichuan** Jiangxi, S China 26.18N 114.31E

168 I13 **Suid-Afrika** see South Africa

166 L4 **Suide** Shaanxi, C China 37.30N 110.10E

169 W8 **Suifenhe** Heilongjiang, NE China 44.22N 131.12E

169 W8 **Suihua** Heilongjiang, NE China 46.40N 127.00E

167 Q6 **Suining** Jiangsu, E China 33.54N 117.58E

166 I9 **Suining** Sichuan, C China 30.31N 105.33E

105 Q4 **Suippes** Marne, N France 49.08N 4.31E

99 E20 **Suir** Ir. An tSiúir. ⋙ S Ireland

171 Gg15 **Suita** Osaka, Honshū, SW Japan 34.39N 135.27E

166 L16 **Suixi** Guangdong, S China 21.22N 110.13E

Sui Xian see Suizhou

169 T13 **Suizhong** Liaoning, NE China 40.19N 120.20E

167 N8 **Suizhou** prev. Sui Xian. Hubei, C China 31.46N 113.20E

155 P17 **Sujāwal** Sind, SE Pakistan

174 Ji14 **Sukabumi** prev. Soekaboemi. Jawa, C Indonesia 6.55S 106.55E

174 Kk9 **Sukadana, Teluk** bay Borneo, W Indonesia

171 LJi9 **Sukagawa** Fukushima, Honshū, N Japan 37.16N 140.19E

151 O15 **Sukarnapura** see Jayapura

Sukarno, Puntjak see Jaya, Puncak

153 R11 **Sükh Rus.** Sokh. Farghona Wiloyati, E Uzbekistan 39.56N 71.10E

116 J12 **Sŭkh** see Sokh

130 J5 **Sukhinichi** Kaluzhskaya Oblast', W Russian Federation 54.06N 35.22E

133 Q4 **Sukhona** var. Tot'ma. ⋙ NW Russian Federation

178 H9 **Sukhothai** var. Sukotai. Sukhothai, W Thailand 17.01N 99.51E

115 E16 **Sukhumi** see Sokhumi

Sukkertoppen see Maniitsoq

155 Q13 **Sukkur** Sind, SE Pakistan 27.44N 68.46E

Sukotai see Sukhothai

129 V15 **Sukra Bay** see Şawqirah, Dawḥat

170 E16 **Suksun** Permskaya Oblast', NW Russian Federation 57.10N 57.27E

96 B12 **Sukumo** Kōchi, Shikoku, SW Japan 32.55N 132.42E

119 R5 **Sula** island S Norway

44 H6 **Sula** ⋙ N Ukraine

155 S10 **Sulaco, Río** ⋙ NW Honduras

131 Q16 **Sulaimaniya** see As Sulaymāniyah

131 Q16 **Sulaimān Range** ▲ C Pakistan

175 Rr10 **Sulak** Respublika Dagestan, SW Russian Federation 43.19N 47.28E

142 I12 **Sulak** ⋙ SW Russian Federation

175 R17 **Sula, Kepulauan** island group C Indonesia

98 F5 **Sula Sgeir** island NW Scotland, UK

175 Q9 **Sulawesi** Eng. Celebes. island C Indonesia

175 P11 **Sulawesi Selatan** off. Propinsi Sulawesi Selatan, Eng. South Celebes, South Sulawesi. ♦ province C Indonesia

175 Q9 **Sulawesi Tengah** off. Propinsi Sulawesi Tengah, Eng. Central Celebes, Central Sulawesi. ♦ province N Indonesia

175 Q11 **Sulawesi Tenggara** off. Propinsi Sulawesi Tenggara, Eng. South-East Celebes, South-East Sulawesi. ♦ province C Indonesia

175 Qq7 **Sulawesi Utara** off. Propinsi Sulawesi Utara, Eng. North Celebes, North Sulawesi. ♦ province N Indonesia

145 T5 **Sulaymān Beg** N Iraq 34.48N 44.54E

97 D15 **Suldalsvatnet** ☺ S Norway

112 E12 **Sulechów** Ger. Züllichau. Lubuskie, W Poland 52.04N 15.37E

112 E11 **Sulęcin** Lubuskie, W Poland 52.29N 15.06E

79 U14 **Suleja** Niger, C Nigeria 9.15N 7.10E

113 K14 **Sulejów** Łodzkie, S Poland 51.21N 19.57E

98 I5 **Sule Skerry** island N Scotland, UK

98 J6 **Suliag** see Sohâg

78 A9 **Sulima** S Sierra Leone 6.58N 11.34W

119 O13 **Sulina** Tulcea, SE Romania 45.07N 29.40E

119 N13 **Sulina, Braţul** ⋙ SE Romania

102 H12 **Sulingen** Niedersachsen, NW Germany 52.40N 8.48E

94 H12 **Sulitjelma** Nordland, C Norway 67.10N 16.16E

58 A9 **Sulitjelma** Piura, NW Peru 4.54S 80.42W

25 N3 **Sulligent** Alabama, S USA 33.54N 88.07W

32 M14 **Sullivan** Illinois, N USA 39.36N 88.36W

33 N15 **Sullivan** Indiana, N USA 39.04N 87.24W

49 W5 **Sullivan** Missouri, C USA 38.12N 91.09W

98 M1 **Sullom Voe** NE Scotland, UK 60.24N 1.09W

105 O7 **Sully-sur-Loire** Loiret, C France 47.46N 2.21E

109 K15 **Sulmona** anc. Sulmo. Abruzzo, C Italy 42.03N 13.55E

44 F7 **Sulmul** El Salvador/Honduras

116 M11 **Süloğlu** Edirne, NW Turkey 41.46N 26.55E

24 G9 **Sulphur** Louisiana, S USA 30.14N 93.22W

49 O12 **Sulphur** Oklahoma, C USA 34.30N 96.58W

30 K9 **Sulphur Creek** ⋙ South Dakota, N USA

27 V6 **Sulphur Draw** ⋙ Texas, SW USA

27 V6 **Sulphur River** ⋙ Arkansas/Texas, SW USA

23 U6 **Sulphur Springs** Texas, SW USA 33.09N 95.36W

24 M6 **Sulphur Springs Draw** ⋙ Texas, SW USA

128 J8 **Sultan** Ontario, S Canada

145 Q15 **Sultan Dağları** ▲ C Turkey

116 N13 **Sultanköy** Tekirdağ, NW Turkey 41.01N 27.58E

179 R16 **Sultan Kudarat** var. Nuling. Mindanao, S Philippines 7.20N 124.16E

158 M13 **Sultānpur** Uttar Pradesh, N India 26.15N 82.04E

179 Pp17 **Sulu Archipelago** island group SW Philippines

198 F7 **Sulu Basin** undersea feature SE South China Sea

179 W14 **Sülüktü** see Sulyukta

169 W14 **Sulu Sea** Ind. Laut Sulu. sea

175 Pp1 **Sulutobe** Kaz. Sülútóbe. Kzylorda, S Kazakhstan

21 N9 **Sulyukta** Kir. Sülükti. Oshskaya Oblast', SW Kyrgyzstan 39.57N 69.30E

22 M8 **Sulz** see Sulz am Neckar

35 R6 **Sulz am Neckar** var. Sulz. Baden-Württemberg, SW Germany 48.21N 8.37E

103 G22 **Sulzbach-Rosenberg** Bayern, SE Germany 49.30N 11.46E

23 X8 **Sulzberger Bay** bay Antarctica

205 N13 **Sumaco** see Sonqor

20 G4 **Sumartin** Split-Dalmacija, S Croatia 43.17N 16.52E

34 G4 **Sumas** Washington, NW USA 49.00N 122.15W

169 W13 **Sumatera Barat** off. Propinsi Sumatera Barat, Eng. West Sumatra. ♦ province W Indonesia

---

173 G9 **Sumatera Barat** off. Propinsi Sumatera Barat, Eng. West Sumatra. ♦ province W Indonesia

174 Hh1 **Sumatera Selatan** off. Propinsi Sumatera Selatan, Eng. South Sumatra. ♦ province W Indonesia

173 Fj6 **Sumatera Utara** off. Propinsi Sumatera Utara, Eng. North Sumatra. ♦ province W Indonesia

Sumatra see Sumatera

112 H23 **Sumava** Eng. Bohemian Forest. ▲ SW Czech Republic

Sumayl see Summêl

145 U7 **Summa' al Muḥammad** E Iraq 33.34N 45.06E

174 Ji4 **Sumba, Selat** strait Jawa/Sumatera, SW Indonesia

175 P17 **Sumba, Pulau** Eng. Sandalwood Island; prev. Soemba. island Nusa Tenggara, C Indonesia

152 D12 **Sumbar** ⋙ W Turkmenistan

175 P16 **Sumba, Selat** strait Nusa Tenggara, S Indonesia

175 O16 **Sumbawa** prev. Soembawa. island Nusa Tenggara, C Indonesia 8.30S 117.25E

175 O16 **Sumbawabesar** Sumbawa, W Tanzania 7.57S 31.36E

83 F23 **Sumbawanga** Rukwa, W Tanzania 7.57S 31.36E

84 B12 **Sumbe** prev. N'Gunza, Port. Novo Redondo. Cuanza Sul, W Angola 11.12S 13.49E

98 M3 **Sumburgh Head** headland NE Scotland, UK 59.51N 1.16W

113 H23 **Sümeg** Veszprém, W Hungary 47.00N 17.13E

82 C12 **Summêl** Southern Darfur, S Sudan 9.49N 27.39E

174 Mm14 **Summenep** prev. Soemenep. Pulau Madura, C Indonesia 7.01S 113.51E

175 Q8 **Summer Island** island Michigan, N USA

34 H15 **Summer Lake** ☺ Oregon, NW USA

9 N17 **Summerland** British Columbia, SW Canada 49.34N 119.45W

11 P14 **Summerside** Prince Edward Island, SE Canada 46.24N 63.46W

23 R5 **Summersville** West Virginia, NE USA 38.16N 80.51W

23 R5 **Summersville Lake** ⊠ West Virginia, NE USA

25 S13 **Summerton** South Carolina, SE USA 33.36N 80.21W

25 W13 **Summerville** Georgia, SE USA 34.28N 85.21W

23 S14 **Summerville** South Carolina, SE USA 33.01N 80.10W

41 R10 **Summit** Alaska, USA 63.21N 148.50W

37 V4 **Summit Mountain** ▲ Nevada, W USA 39.23N 116.25W

39 R8 **Summit Peak** ▲ Colorado, C USA 37.21N 106.42W

Summus Portus see Somport, Col du

31 X12 **Sumner** Iowa, C USA 42.51N 92.05W

24 K3 **Sumner** Mississippi, S USA 33.58N 90.22W

193 H17 **Sumner, Lake** ☺ South Island, NZ

39 U12 **Sumner, Lake** ⊠ New Mexico, SW USA

171 Kk13 **Sumon-dake** ▲ Honshū, C Japan

170 G15 **Sumoto** Hyōgo, Awaji-shima, SW Japan 34.18N 134.52E

113 G17 **Šumperk** Ger. Mährisch-Schönberg. Olomoucký Kraj, E Czech Republic 49.59N 16.58E

44 F7 **Sumpul** El Salvador/Honduras

143 Z11 **Sumqayıt** Rus. Sumgait. E Azerbaijan 40.33N 49.41E

143 Y11 **Sumqayıtçay** Rus. Sumgait. ⋙ E Azerbaijan

153 R9 **Sumsar** Dzhalal-Abadskaya Oblast', W Kyrgyzstan 41.12N 71.16E

119 S3 **Sums'ka Oblast'** var. Sumy, Rus. Sumskaya Oblast'. ♦ province NE Ukraine

128 J8 **Sumskiy Posad** Respublika Kareliya, NW Russian Federation 64.12N 35.22E

23 X10 **Sumter** South Carolina, SE USA 33.55N 80.20W

119 T3 **Sumy** Sums'ka Oblast', NE Ukraine 50.54N 34.48E

Sumy see Sums'ka Oblast'

165 Q15 **Sumzom** Xizang Zizhiqu, W China 29.45N 96.13E

129 R15 **Suna** Kirovskaya Oblast', NW Russian Federation 57.53N 50.04E

120 Oo5 **Sunagawa** Hokkaidō, NE Japan 43.30N 141.55E

178 M13 **Sunāmganj** Chittagong, NE Bangladesh 25.03N 91.26E

179 Pp17 **Sunan** var. Hongwan, Sunan Yugurzu Zizhixian. Gansu, N China 38.39N 99.51E

169 W14 **Sunan** ✈ (P'yŏngyang) SW North Korea 39.12N 125.41E

Sunan Yugurzu Zizhixian see Sunan

169 W14 **Sunch'ŏn** SW North Korea 39.28N 125.58E

175 Y16 **Sunch'ŏn** Jap. Junten. S South Korea 34.56N 127.28E

---

38 K13 **Sun City** Arizona, SW USA

21 O3 **Suncook** New Hampshire, NE USA 43.07N 71.25W

Sunda Islands see Greater Sunda Islands

35 Z12 **Sundance** Wyoming, C USA 44.24N 104.22W

159 T17 **Sundarbans** wetland Bangladesh/India

160 M11 **Sundargarh** Orissa, E India 22.07N 84.01E

174 Ji4 **Sunda, Selat** strait Jawa/Sumatera, SW Indonesia

133 U15 **Sunda Shelf** undersea feature S South China Sea

Sunda Trench see Java Trench

133 U17 **Sunda Trench** undersea feature E Indian Ocean

97 O16 **Sundbyberg** Stockholm, C Sweden 59.22N 17.58E

99 M14 **Sunderland** var. Wearmouth. NE England, UK 54.55N 1.22W

103 F15 **Sundern** Nordrhein-Westfalen, W Germany 51.19N 8.00E

142 F12 **Sündiken Dağları** ▲ C Turkey

26 M5 **Sundown** Texas, SW USA 33.27N 102.29W

9 P16 **Sundre** Alberta, SW Canada 51.49N 114.46W

12 G11 **Sundridge** Ontario, S Canada 45.46N 79.25W

95 H17 **Sundsvall** Västernorrland, C Sweden 62.22N 17.19E

174 Ji13 **Sungaibuntu** Sumatera, SW Indonesia 4.04S 105.37E

174 Gg9 **Sungaidareh** Sumatera, W Indonesia 0.58S 101.30E

178 Hh17 **Sungai Ko-Lok** var. Sungai Ko-Lok. Narathiwat, SW Thailand 6.01N 101.58E

174 Gg10 **Sungaipenuh** prev. Soengaipenoeh. Sumatera, W Indonesia 2.00S 101.28E

174 Kk8 **Sungaipinyuh** Borneo, C Indonesia 0.16N 109.06E

Sungari see Songhua Jiang

Sungaria see Dzungaria

Sungei Pahang see Pahang, Sungai

178 Hh8 **Sung Men** Phrae, NW Thailand 17.59N 100.07E

85 M15 **Sungo** Tete, NW Mozambique 16.31S 33.68E

116 M9 **Sungurlare** Burgas, E Bulgaria 42.47N 26.46E

142 J12 **Sungurlu** Çorum, N Turkey 40.10N 34.22E

114 F9 **Sunja** Sisak-Moslavina, C Croatia 45.22N 16.33E

159 Q12 **Sun Koshi** ⋙ E Nepal

159 S12 **Sundalen** valley S Norway

96 P9 **Sunndalsøra** Møre og Romsdal, S Norway

97 K15 **Sunne** Värmland, C Sweden 59.52N 14.30E

97 O15 **Sunnersta** Uppsala, C Sweden 59.46N 17.40E

96 C11 **Sunnfjord** physical region S Norway

97 C15 **Sunnhordland** physical region S Norway

96 D10 **Sunnmøre** physical region S Norway

39 N4 **Sunnyside** Utah, W USA 39.33N 110.23W

34 L10 **Sunnyside** Washington, NW USA 46.01N 119.58W

37 N8 **Sunnyvale** California, W USA 37.22N 122.02W

32 L8 **Sun Prairie** Wisconsin, N USA 43.12N 89.12W

27 N1 **Sunray** Texas, SW USA 36.01N 101.49W

24 J8 **Sunset** Louisiana, S USA 30.24N 92.04W

27 S5 **Sunset** Texas, SW USA 33.24N 97.43W

Sunset State see Oregon

189 Z10 **Sunshine Coast** cultural region Queensland, E Australia

Sunshine State see Florida, USA

Sunshine State see New Mexico, USA

Sunshine State see South Dakota, USA

126 Kk11 **Suntar** Respublika Sakha (Yakutiya), NE Russian Federation 62.10N 117.35E

41 R10 **Suntrana** Alaska, USA 63.54N 148.58W

154 J15 **Suntsar** Baluchistān, SW Pakistan 25.30N 62.03E

79 O16 **Sunyani** W Ghana 7.22N 2.18W

95 M17 **Suolahti** Länsi-Suomi, W Finland 62.32N 25.51E

Suoločielgi see Saariselkä

Suomenlahti see Finland, Gulf of

Suomen Tasavalta/Suomi see Finland

95 N14 **Suomussalmi** Oulu, E Finland 64.54N 29.05E

170 H13 **Suō-nada** SW Japan

95 M17 **Suonenjoki** Itä-Suomi, C Finland 62.36N 27.06E

191 N12 **Sunbury** Victoria, SE Australia 37.36S 144.42E

178 Ji13 **Suŏng** Kâmpóng Cham, C Cambodia 11.53N 105.41E

128 I10 **Suoyarvi** Respublika Kareliya, NW Russian Federation 62.01N 32.24E

79 S14 **Supe** Lima, W Peru 10.49S 77.42W

3 V7 **Supérieur, Lac** see Superior, Lake

38 M14 **Superior** Arizona, SW USA 33.17N 111.06W

35 O9 **Superior** Montana, NW USA 47.11N 114.53W
31 P17 **Superior** Nebraska, C USA 40.01N 98.04W
32 I3 **Superior** Wisconsin, N USA 46.41N 92.03W
43 S17 **Superior, Laguna** lagoon S Mexico
33 N2 **Superior, Lake** Fr. Lac Supérieur. ☺ Canada/USA
38 L13 **Superstition Mountains** ▲ Arizona, SW USA
115 F14 **Supetar** It. San Pietro. Split-Dalmacija, S Croatia 43.22N 16.34E
178 H11 **Suphan Buri** var. Supanburi. Suphan Buri, W Thailand 14.28N 100.10E
176 W9 **Supiori, Pulau** island E Indonesia
196 K2 **Supply Reef** reef N Northern Mariana Islands
205 O7 **Support Force Glacier** glacier Antarctica
143 R10 **Sup'sa** var. Supsa. ◈ W Georgia
45 W12 **Sūq 'Abs** see 'Abs
142 H4 **Sūq ash Shuyūkh** SE Iraq 30.52N 46.28E
167 Q6 **Suqian** Jiangsu, E China 33.57N 118.18E
Suqrah see Şawqirah
Suqrah Bay see Şawqirah, Dawḩat
147 V16 **Suqutra** var. Sokotra, Eng. Socotra. island SE Yemen
147 Z8 **Şūr** NE Oman 22.32N 59.33E
131 P5 **Şūr** Penzenskaya Oblast', W Russian Federation 53.23N 45.03E
131 P4 **Sura** ◈ W Russian Federation
155 N12 **Sūrāb** Baluchistān, SW Pakistan 28.28N 66.15E
Surabaia see Surabaya
174 M15 **Surabaya** prev. Soerabaja, Surabaja. Jawa, C Indonesia 7.13S 112.45E
97 N15 **Surahammar** Västmanland, C Sweden 59.43N 16.13E
174 L15 **Surakarta** Eng. Solo; prev. Soerakarta. Jawa, S Indonesia 7.31S 110.49E
Surakhany see Suraxanı
179 R17 **Surallah** Mindanao, S Philippines 6.16N 124.46E
143 S10 **Surami** C Georgia 41.59N 43.36E
149 X13 **Sūrān** Sīstān va Balūchestān, SE Iran 27.18N 61.58E
113 I21 **Šurany** Hung. Nagysurány. Nitriansky Kraj, SW Slovakia 48.05N 18.10E
160 D12 **Sūrat** Gujarāt, W India 21.10N 72.54E
Suratdhani see Surat Thani
158 G9 **Sūratgarh** Rājasthān, NW India 29.19N 73.58E
178 Gg15 **Surat Thani** var. Suratdhani. Surat Thani, SW Thailand 9.09N 99.19E
121 Q16 **Suraw** Rus. Surov. ◈ E Belarus
143 Z11 **Suraxanı** Rus. Surakhany. E Azerbaijan 40.25N 49.59E
147 Y11 **Surayr** E Oman 19.55N 57.46E
144 K2 **Suraysāt** Ḩalab, N Syria 36.42N 38.01E
120 O12 **Surazh** Rus. Surazh. Vitsyebskaya Voblasts', NE Belarus 55.24N 30.46E
130 H6 **Surazh** Bryanskaya Oblast', W Russian Federation 53.04N 32.29E
203 V17 **Sur, Cabo** headland Easter Island, Chile, E Pacific Ocean 27.10S 109.25W
114 L11 **Surčin** Serbia, N Yugoslavia 44.48N 20.19E
118 H9 **Surduc** Hung. Szurduk. Sălaj, NW Romania 47.13N 23.19E
115 P16 **Surdulica** Serbia, SE Yugoslavia 42.43N 22.10E
101 L24 **Sûre** var. Sauer. ◈ W Europe see also Sauer
160 C10 **Surendranagar** Gujarāt, W India 22.43N 71.43E
20 K16 **Surf City** New Jersey, NE USA 39.21N 74.24W
191 V3 **Surfers Paradise** Queensland, E Australia 27.54S 153.18E
23 U13 **Surfside Beach** South Carolina, SE USA 33.36N 78.58W
104 J10 **Surgères** Charente-Maritime, W France 46.07N 0.44W
125 G11 **Surgut** Khanty-Mansiyskiy Avtonomnyy Okrug, C Russian Federation 61.13N 73.28E
126 Hh10 **Surgutikha** Krasnoyarskiy Kray, N Russian Federation 64.44N 87.13E
100 M6 **Surhuisterveen** Friesland, N Netherlands 53.10N 6.10E
107 V5 **Súria** Cataluña, NE Spain 41.49N 1.45E
179 P10 **Sūrīān** Fārs, S Iran
159 J15 **Sūriāpet** Andhra Pradesh, C India 17.10N 79.42E
179 R14 **Surigao** Mindanao, S Philippines 9.43N 125.31E
178 Ii11 **Surin** Surin, E Thailand 14.52N 103.28E
57 U11 **Suriname** off. Republic of Suriname, var. Surinam; prev. Dutch Guiana, Netherlands Guiana. ◆ republic N South America
Sūriya/Sūriyah, Al-Jumhūrīyah al-'Arabīyah as - see Syria
Surkhab, Darya-i- see Kahmard, Daryā-ye
Surkhandar'inskaya Oblast' see Surkhondaryo Wiloyati
Surkhandar'ya see Surkhondaryo
Surkhet see Birendranagar
153 R12 **Surkhob** ◈ C Tajikistan
153 P13 **Surkhondaryo** Rus. Surkhandar'ya. ◈ S Tajikistan/Uzbekistan
153 N13 **Surkhondaryo Wiloyati** Rus. Surkhandar'inskaya Oblast'. ◆ province E Uzbekistan
81 P11 **Sürmene** Trabzon, NE Turkey 40.55N 40.03E
Surov see Suraw

131 N11 **Surovikino** Volgogradskaya Oblast', SW Russian Federation 48.39N 42.46E
126 Jj14 **Surovo** Irkutskaya Oblast', S Russian Federation 55.45N 105.31E
37 N11 **Sur, Point** headland California, W USA 36.18N 121.54W
197 F3 **Surprise, Ile** island N New Caledonia
63 E22 **Sur, Punta** headland E Argentina 50.58S 69.10W
Surrentum see Sorrento
30 M3 **Surrey** North Dakota, N USA 48.13N 101.05W
99 O22 **Surrey** cultural region SE England, UK
23 X7 **Surry** Virginia, NE USA 37.08N 76.49W
110 F8 **Sursee** Luzern, W Switzerland 47.10N 8.07E
131 P6 **Sursk** Penzenskaya Oblast', W Russian Federation 53.06N 45.46E
131 P5 **Surskoye** Ul'yanovskaya Oblast', W Russian Federation 54.28N 46.47E
77 P8 **Surt** var. Sidra, Sirte. N Libya 31.13N 16.34E
97 I19 **Surte** Västra Götaland, S Sweden 57.49N 12.01E
77 Q8 **Surt, Khalij** Eng. Gulf of Sidra, Gulf of Sirti, Sidra. gulf N Libya
94 I5 **Surte** island S Iceland
143 N17 **Suruç** Şanlıurfa, S Turkey 36.58N 38.24E
171 Ii17 **Suruga-wan** bay SE Japan
174 Hh10 **Surulangun** Sumatera, W Indonesia 2.36S 102.43E
Süs see Susch
108 A8 **Susa** Piemonte, NE Italy 45.09N 7.01E
170 E12 **Susa** Yamaguchi, Honshū, SW Japan 34.35N 131.34E
Susa see Shūsh
115 E16 **Susac** It. Cazza. island SW Croatia
170 Ee15 **Susaki** Kōchi, Shikoku, SW Japan 33.22N 133.13E
130 G17 **Susami** Wakayama, Honshū, SW Japan 33.33N 135.32E
148 K9 **Süsangerd** var. Susangird. Khūzestān, SW Iran 31.40N 48.06E
Susangird see Süsangerd
37 P4 **Susanville** California, W USA 40.25N 120.39W
110 J9 **Susch** var. Süs. Graubünden, SE Switzerland 46.45N 10.04E
143 N12 **Suşehri** Sivas, N Turkey 40.10N 38.06E
113 B18 **Sušice** Ger. Schüttenhofen. Plzeňský Kraj, W Czech Republic 49.13N 13.31E
41 R11 **Susitna** Alaska, USA 61.32N 150.30W
41 R11 **Susitna River** ◈ Alaska, USA
131 Q3 **Suslonger** Respublika Mariy El, W Russian Federation 56.18N 48.16E
107 N14 **Suspiro del Moro, Puerto del** pass S Spain 37.04N 3.39W
20 H16 **Susquehanna River** ◈ New York/Pennsylvania, NE USA
11 O15 **Sussex** New Brunswick, SE Canada 45.43N 65.31W
20 J13 **Sussex** New Jersey, NE USA 41.12N 74.34W
23 W7 **Sussex** Virginia, NE USA 36.54N 77.16W
99 O23 **Sussex** cultural region S England, UK
191 S10 **Sussex Inlet** New South Wales, SE Australia 35.10S 150.35E
101 L17 **Susteren** Limburg, SE Netherlands 51.04N 5.49E
8 K12 **Sustut Peak** ▲ British Columbia, W Canada 56.25N 126.34W
127 Nn9 **Susuman** Magadanskaya Oblast', E Russian Federation 62.46N 148.07E
196 H6 **Susupe** Saipan, S Northern Mariana Islands
142 D12 **Susurluk** Balıkesir, NW Turkey 39.55N 28.10E
116 M13 **Susuzmüsellim** Tekirdağ, NW Turkey 41.04N 27.03E
142 F15 **Sütçüler** Isparta, SW Turkey 37.31N 31.00E
118 L13 **Suţeşti** Brăila, SE Romania 45.13N 27.26E
85 F25 **Sutherland** Western Cape, SW South Africa 32.22S 20.42E
30 L15 **Sutherland** Nebraska, C USA 41.09N 101.07W
98 I7 **Sutherland** cultural region N Scotland, UK
193 B21 **Sutherland Falls** waterfall South Island, NZ 44.49S 167.32E
34 F14 **Sutherlin** Oregon, NW USA 43.23N 123.18W
155 V10 **Sutlej** ◈ India/Pakistan
Sutna see Satna
35 P7 **Sutter Creek** California, W USA 38.23N 120.49W
41 R11 **Sutton** Alaska, USA 61.42N 148.53W
31 Q16 **Sutton** Nebraska, C USA 40.36N 97.52W
23 R4 **Sutton** West Virginia, NE USA 38.39N 80.42W
10 F8 **Sutton** ◈ Ontario, S Canada
99 M19 **Sutton Coldfield** C England, UK 52.34N 1.48W
23 R4 **Sutton Lake** ◈ West Virginia, NE USA
13 P13 **Sutton, Monts** hill range Quebec, SE Canada
10 F8 **Sutton Ridges** ▲ Ontario, C Canada
172 Nn5 **Suttsu** Hokkaidō, NE Japan 42.46N 140.12E
41 P15 **Sutwik Island** island Alaska, USA 56.31N 157.11W
168 K7 **Süüj** Bulgan, C Mongolia 47.49N 104.06E
120 H5 **Suure-Jaani** Ger. Gross-Sankt-Johannis. Viljandimaa, S Estonia 58.34N 25.26E
120 J7 **Suur Munamägi** var. Munamägi, Ger. Eier-Berg. ▲ SE Estonia
120 F5 **Suur Väin** Ger. Grosser Sund. strait W Estonia
153 U8 **Suusamyr** Chuyskaya Oblast', C Kyrgyzstan 42.07N 73.51E

197 II3 **Suva ●** (Fiji) Viti Levu, W Fiji 18.08S 178.30E
197 II3 **Suva ✈** Viti Levu, C Fiji 18.01S 178.30E
115 N18 **Suva Gora** ▲ W FYR Macedonia
120 H11 **Suvainiškis** Rokiškis, NE Lithuania 56.09N 25.15E
Suvalkai/Suvalki see Suwałki
115 P15 **Suva Planina** ▲ SE Yugoslavia
115 M17 **Suva Reka** Serbia, S Yugoslavia 42.23N 20.50E
130 K5 **Suvorov** Tul'skaya Oblast', W Russian Federation 54.08N 36.33E
119 N12 **Suvorove** Odes'ka Oblast', SW Ukraine 45.35N 28.58E
171 J15 **Suwa** Nagano, Honshū, S Japan 36.01N 138.07E
121 A14 **Suwałki** see As Suwayq
Suwaira see Aş Şuwayrah
112 O7 **Suwałki** Lith. Suvalkai, Rus. Suvalki. Podlaskie, NE Poland 54.06N 22.55E
178 Ii10 **Suwannaphum** Roi Et, E Thailand 15.36N 103.46E
25 V8 **Suwannee River** ◈ Florida/Georgia, SE USA
Suwār see Aş Şuwār
202 K14 **Suwarrow** atoll N Cook Islands
149 R16 **Suwaydān** var. Sweiham. Abū Ẕaby, E UAE 24.30N 55.18E
121 M17 **Suwayqiyah, Hawr as** see Shuwayjah, Hawr ash
Suways, Khalij as see Suez, Gulf of
Suways, Qanāt as see Suez Canal
Suweida see Aş Suwaydā'
Suweon see Suwŏn
169 X15 **Suwŏn** var. Suweon, Jap. Suigen. NW South Korea 37.17N 127.03E
149 W14 **Sūzā** Hormozgān, S Iran 26.49N 56.04E
151 P15 **Suzak** Kaz. Sozaq. Yuzhnyy Kazakhstan, S Kazakhstan 44.09N 68.28E
130 M3 **Suzdal'** Vladimirskaya Oblast', W Russian Federation 56.27N 40.29E
167 P7 **Suzhou** var. Su Xian. Anhui, E China 33.39N 116.56E
167 R8 **Suzhou** var. Soochow, Su-chou, Suchow; prev. Wuhsien. Jiangsu, E China 31.22N 120.34E
171 J12 **Suzu** Ishikawa, Honshū, SW Japan 37.24N 137.12E
171 Hh16 **Suzuka** Mie, Honshū, SW Japan 34.51N 136.31E
171 Jj14 **Suzuka** var. Suzaka. Nagano, Honshū, S Japan 36.39N 138.16E
171 J12 **Suzu-misaki** headland Honshū, SW Japan 37.31N 137.19E
113 F17 **Svabinov** Ger. Zwittau. Pardubický Kraj, C Czech Republic 49.44N 16.27E
119 S6 **Svatovods'ka** Rus. Svetlovodsk. Kirovohrads'ka Oblast', C Ukraine 49.04N 33.15E
Svizzera see Switzerland
126 Mm15 **Svobodnyy** Amurskaya Oblast', SE Russian Federation 51.24N 128.05E
120 N13 **Svorove** see Svyatoye
94 G11 **Svolvær** Nordland, C Norway 68.15N 14.29E
113 F18 **Svratka** Ger. Schwarzach, Schwarzawa. ◈ SE Czech Republic 49.26N 22.07E
207 U10 **Svyataya Anna Trough** var. Saint Anna Trough. undersea feature N Kara Sea
92 J6 **Svyatoy Nos, Mys** headland N Russian Federation 68.07N 39.49E
126 M5 **Svyatoy Nos, Mys** headland NE Russian Federation 72.49N 140.45E
121 N18 **Svyetlahorsk** Rus. Svetlogorsk. Homyel'skaya Voblasts', SE Belarus 52.39N 29.43E
99 P19 **Swaffham** E England, UK 52.38N 0.40E
25 V5 **Swainsboro** Georgia, SE USA 32.36N 82.19W
85 C19 **Swakop** ◈ W Namibia
85 C19 **Swakopmund** Erongo, W Namibia 22.40S 14.34E
99 M15 **Swale** ◈ N England, UK
101 N15 **Swallow Island** see Nendö
101 M16 **Swalmen** Limburg, SE Netherlands 51.13N 6.01E
10 G8 **Swan** ◈ Ontario, C Canada
99 L24 **Swanage** S England, UK 50.37N 1.59W
190 M10 **Swan Hill** Victoria, SE Australia 35.21S 143.34E
9 P13 **Swan Hills** Alberta, SW Canada 54.40N 116.19W
67 D24 **Swan Island** island E Falkland Islands
Swankalok see Sawankhalok
31 U10 **Swan Lake** ◈ Minnesota, C USA
9 Y10 **Swanquarter** North Carolina, SE USA 35.23N 76.17W
190 J9 **Swan Reach** South Australia 34.39S 139.35E
9 R11 **Swan River** Manitoba, S Canada 52.06N 101.16W
191 P13 **Swan Islands, SE Australia** 42.09S 148.03E
7 S1 **Swans Island** island Maine, NE USA
30 L17 **Swanson Lake** ◈ Nebraska, C USA
33 R11 **Swanton** Ohio, N USA 41.35N 83.53W
112 G11 **Swarzędz** Poznań, C Poland 52.24N 17.05E
65 L22 **Swaziland** off. Kingdom of Swaziland. ◆ monarchy S Africa

95 G18 **Sweden** off. Kingdom of Sweden, Swe. Sverige. ◆ monarchy N Europe
Swedru see Agona Swedru
34 G12 **Sweet Home** Oregon, NW USA 44.24N 122.44W
27 T12 **Sweet Home** Texas, SW USA 29.21N 97.04W
49 T4 **Sweet Springs** Missouri, C USA 38.57N 93.24W
22 M10 **Sweetwater** Tennessee, S USA 35.36N 84.27W
27 P7 **Sweetwater** Texas, SW USA 32.28N 100.24W
35 V15 **Sweetwater River** ◈ Wyoming, C USA
85 F26 **Swellendam** Western Cape, SW South Africa 34.01S 20.25E
113 G15 **Świdnica** Ger. Schweidnitz. Wałbrzych, SW Poland 50.51N 16.28E
113 O14 **Świdnik** Ger. Streckenbach. Lubelskie, E Poland 51.13N 22.39E
112 F8 **Świdwin** Ger. Schivelbein. Zachodniopomorskie, NW Poland 53.46N 15.43E
113 F15 **Świebodzice** Ger. Freiburg in Schlesien, Swiebodzice. Wałbrzych, SW Poland 50.54N 16.22E
112 E11 **Świebodzin** Ger. Schwiebus. Lubuskie, W Poland 52.15N 15.30E
112 I9 **Świecie** Ger. Schwertberg. Kujawsko-pomorskie, N Poland 53.24N 18.24E
113 L15 **Świętokrzyskie** ◆ province C Poland
9 T16 **Swift Current** Saskatchewan, S Canada 50.16N 107.49W
100 K9 **Swifterbant** Flevoland, C Netherlands 52.36N 5.33E
117 L22 **Swilly, Lough** Ir. Loch Súilí. inlet N Ireland
99 M22 **Swindon** S England, UK 51.34N 1.46W
112 D8 **Swinemünde** see Świnoujście
112 D8 **Świnoujście** Ger. Swinemünde. Zachodniopomorskie, NW Poland 53.54N 14.12E
121 J9 **Swintsowy Rudnik** see Svintsovyy Rudnik
110 E9 **Switzerland** off. Swiss Confederation, Fr. La Suisse, Ger. Schweiz, It. Svizzera; anc. Helvetia. ◆ federal republic C Europe
99 F17 **Swords** It. Sord, Sórd Choluim Chille. E Ireland 53.28N 6.13W
20 H13 **Swoyersville** Pennsylvania, NE USA 41.18N 75.48W
128 I10 **Syamozero, Ozero** ◈ NW Russian Federation
128 M13 **Syamzha** Vologodskaya Oblast', NW Russian Federation 60.02N 41.09E
112 G11 **Szamotuły** Poznań, C Poland 52.35N 16.35E
120 N13 **Syanno** Rus. Senno. Vitsyebskaya Voblasts', NE Belarus 54.48N 29.44E
121 K16 **Syarhyeyevichy** Rus. Sergeyevichi. Minskaya Voblasts', C Belarus 53.33N 27.55E
128 I12 **Syas'stroy** Leningradskaya Oblast', NW Russian Federation 60.05N 32.37E
32 M10 **Sycamore** Illinois, N USA 41.59N 88.41W
130 J3 **Sychëvka** Smolenskaya Oblast', W Russian Federation 55.52N 34.19E
113 H14 **Syców** Ger. Gross Wartenberg. Dolnośląskie, SW Poland 51.18N 17.42E
12 E17 **Sydenham** ◈ Ontario, S Canada
Sydenham Island see Nonouti
191 T9 **Sydney** state capital New South Wales, SE Australia 33.55S 151.10E
11 R14 **Sydney** Cape Breton Island, Nova Scotia, SE Canada 46.10N 60.10W
202 I9 **Sydney Island** see Manra
11 R14 **Sydney Mines** Cape Breton Island, Nova Scotia, SE Canada 46.14N 60.19W
121 K18 **Syelishcha** Rus. Selishche. Minskaya Voblasts', C Belarus 53.01N 27.25E
121 J18 **Syemyezhava** Rus. Semezhevo. Minskaya Voblasts', C Belarus 52.57N 27.01E
176 X6 **Syene** see Aswān
Syeverodonets'k Rus. Severodonetsk. Luhans'ka Oblast', E Ukraine 48.58N 38.28E
96 D10 **Sykkylven** More og Romsdal, S Norway 62.22N 6.34E
117 F15 **Sykoúri** var. Sikoúri; prev. Sikoúrion. Thessalía, C Greece 39.46N 22.34E
91 R11 **Syktyvkar** prev. Ust'-Sysol'sk. Respublika Komi, NW Russian Federation 61.42N 50.45E
25 Q4 **Sylacauga** Alabama, S USA 33.10N 86.15W
96 J9 **Sylene** Swe. Sylen Norway/Sweden 63.00N 12.14E
159 V14 **Sylhet** Chittagong, NE Bangladesh 24.52N 91.51E
102 G6 **Sylt** island NW Germany
23 O10 **Sylva** North Carolina, SE USA 35.22N 83.13W
23 S4 **Sylva** ◈ NW Russian Federation
25 W5 **Sylvania** Georgia, SE USA 32.45N 81.38W
33 R11 **Sylvania** Ohio, N USA 41.43N 83.42W
9 Q15 **Sylvan Lake** Alberta, SW Canada 52.18N 114.02W
33 T13 **Sylvan Pass** pass Wyoming, C USA 44.29N 100.03W

25 T7 **Sylvester** Georgia, SE USA 31.31N 83.50W
27 P6 **Sylvester** Texas, SW USA 32.42N 100.15W
8 L11 **Sylvia, Mount** ▲ British Columbia, W Canada 58.58N 111.58W
126 Hh12 **Sym** ◈ C Russian Federation
117 N22 **Sými** var. Simi. island Dodekánisos, Greece, Aegean Sea
119 U8 **Synel'nykove** Dnipropetrovs'ka Oblast', E Ukraine 48.18N 35.31E
129 U6 **Synya** Respublika Komi, NW Russian Federation 65.21N 58.01E
119 P7 **Synyukha** Rus. Sinyukha. ◈ S Ukraine
205 V2 **Syowa** Japanese research station Antarctica
28 H6 **Syracuse** Kansas, C USA 38.00N 101.43W
31 S16 **Syracuse** Nebraska, C USA 40.39N 96.11W
20 H10 **Syracuse** New York, NE USA 43.03N 76.09W
Syracuse see Siracusa
Syrdar'inskaya Oblast' see Sirdaryo Wiloyati
Syrdariya see Syr Darya
151 S7 **Syr Darya** var. Sai Hun, Sir Darya, Syrdariya, Kaz. Syrdariya, Rus. Syrdar'ya, Uzb. Sirdaryo; anc. Jaxartes. ◈ C Asia
150 L14 **Syr Darya** ◈ C Asia
153 P10 **Syrdar'ya** Sirdaryo Wiloyati, E Uzbekistan 40.46N 68.34E
144 J6 **Syria** off. Syrian Arab Republic, var. Siria, Syrie, Ar. Al-Jumhūrīyah al-'Arabīyah as-Sūrīyah, Sūrīya. ◆ republic SW Asia
144 L9 **Syrian Desert** Ar. Al Ḩamad, Bādiyat ash Shām. desert SW Asia
Syrie see Syria
117 L22 **Sýrna** var. Sirna. island Kykládes, Greece, Aegean Sea
117 I20 **Sýros** var. Síros. island Kykládes, Greece, Aegean Sea
95 M18 **Sysmä** Etelä-Suomi, S Finland 61.28N 25.37E
129 R12 **Sysola** ◈ NW Russian Federation
Syulemeshlii see Sredets
131 S2 **Syumsi** Udmurtskaya Respublika, NW Russian Federation 57.07N 51.35E
116 K10 **Syuyab** see Aq Beshim
119 U12 **Syvash, Zaliv** see Syvash, Zaliv Syvash. inlet S Ukraine
21 Q6 **Syzran'** Samarskaya Oblast', W Russian Federation 53.10N 48.22E
113 N21 **Szabadka** see Subotica
113 N21 **Szabolcs-Szatmár-Bereg** off. Szabolcs-Szatmár-Bereg Megye. ◈ county E Hungary
112 G10 **Szamocin** Ger. Samotschin. Wielkopolskie, C Poland 53.02N 17.04E
118 H8 **Szamos** var. Someş, Someşul, Ger. Samosch, Somesch. ◈ Hungary/Romania
Szamosújvár see Gherla
112 G11 **Szamotuły** Poznań, C Poland 52.35N 16.35E
113 M24 **Szarvas** Békés, SE Hungary 46.52N 20.32E
Szászmagyarós see Măieruş
Szászrégen see Reghin
Szászsebes see Sebeş
Szászváros see Orăştie
Szatmárrémeti see Satu Mare
Száva see Sava
113 P15 **Szczebrzeszyn** Lubelskie, E Poland 50.43N 23.00E
112 D9 **Szczecin** Eng./Ger. Stettin. Zachodniopomorskie, NW Poland 53.25N 14.31E
112 G8 **Szczecinek** Ger. Neustettin. Zachodniopomorskie, NW Poland 53.25N 16.42E
112 I7 **Szczeciński, Zalew** var. Stettiner Haff, Ger. Oderhaff. bay Germany/Poland
113 K15 **Szczuczyn Podlaskie, NE Poland** 50.38N 19.46E
112 N8 **Szczuczyn Nowogródzki** see Shchuchyn
113 M8 **Szczytno** Ger. Ortelsburg. Warmińsko-Mazurskie, NE Poland 53.33N 21.00E
113 K21 **Szechuan/Szechwan** see Sichuan
113 K17 **Szécsény** Nógrád, N Hungary 48.04N 19.31E
113 N23 **Szeged** Ger. Szegedin, Rom. Seghedin. Csongrád, SE Hungary 46.16N 20.06E
113 F21 **Szegedin** see Szeged
113 N23 **Szeghalom** Békés, SE Hungary 5.04S 32.49E
113 Q13 **Székelyhid** see Săcueni
113 J23 **Székesfehérvár** Ger. Stuhlweissenberg; anc. Alba Regia. Fejér, W Hungary 47.13N 18.24E
Szeklerburg see Miercurea-Ciuc
Szekler Neumarkt see Târgu Secuiesc
113 J25 **Szekszárd** Tolna, S Hungary 33.10N 86.15W
119 J22 **Szempcz/Szenc** see Senec
96 J9 **Szenice** see Senica
Szentágota see Agnita
179 P8 **Szentendre** Pest, N Hungary 47.40N 19.04E
146 J4 **Szentes** Csongrád, SE Hungary 46.41N 20.16E
146 J5 **Szentgotthárd** Eng. Saint Gotthard, Slov. Monošter. Vas, W Hungary 46.57N 16.18E
197 B12 **Szentgyörgy** see Đurđevac
Szenttamás see Srbobran
113 N3 **Széphely** see Jebel
113 J25 **Szeping** see Siping
Szered see Sered'
85 L22 **Szerencs** Borsod-Abaúj-Zemplén, NE Hungary 48.10N 21.10E

Szeret see Siret
Szeretfalva see Sărăţel
112 N7 **Szeskie Wzgórza** Ger. Seesker Höhe. hill NE Poland 54.15N 22.19E
113 H25 **Szigetvár** Baranya, SW Hungary 46.03N 17.47E
Szilágysomlyó see Şimleu Silvaniei
Szinna see Snina
Sziszek see Sisak
Szitás-Keresztúr see Cristuru Secuiesc
113 E15 **Szklarska Poręba** Ger. Schreiberhau. Dolnośląskie, SW Poland 50.50N 15.30E
Szkudy see Skuodas
Szlatina see Slatina, Croatia
Szlavonia/Szlavonország see Slavonija
Szlovákia see Slovakia
Szluin see Slunj
113 L23 **Szolnok** Jász-Nagykun-Szolnok, C Hungary 47.10N 20.12E
Szolyva see Svalyava
113 G23 **Szombathely** Ger. Steinamanger; anc. Sabaria, Savaria. Vas, W Hungary 47.13N 16.37E
Szond/Szonta see Sonta
Szováta see Sovata
112 F13 **Szprotawa** Ger. Sprottau. Lubuskie, W Poland 51.33N 15.31E
Sztálinváros see Dunaújváros
Sztrazsó see Strážov
112 J8 **Sztum** Ger. Stuhm. Pomorskie, N Poland 53.54N 19.01E
112 H10 **Szubin** Ger. Schubin. Kujawsko-pomorskie, W Poland 53.04N 17.49E
Szucsava see Suceava
Szurduk see Surduc
113 M14 **Szydłowiec** Ger. Schlelau. Mazowieckie, C Poland 51.15N 20.51E

━━━ **T** ━━━

Taalintehdas see Dalsbruk
179 P11 **Taal, Lake** ☺ Luzon, NW Philippines
Taastrup see Tåstrup
113 I24 **Tab** Somogy, W Hungary 46.40N 18.01E
179 Q11 **Tabaco** Luzon, N Philippines 13.22N 123.42E
194 M7 **Tabalo** Mussau Island, NE PNG 1.22S 149.37E
106 K5 **Tabanera** Castilla-León, N Spain 41.49N 5.57W
195 P9 **Tabar Island** island Tabar Islands, N PNG
195 P9 **Tabar Islands** island group NE PNG
Tabariya, Bahrat see Tiberias, Lake
149 S7 **Ţabas** var. Golshan. Khorāsān, C Iran
45 P15 **Tabasará, Serranía de** ▲ W Panama
43 U15 **Tabasco** ◆ state SE Mexico
Tabasco see Grijalva, Río
131 Q2 **Tabashino** Respublika Mariy El, W Russian Federation 57.00N 47.47E
60 B13 **Tabatinga** Amazonas, N Brazil 4.13S 69.43W
76 G9 **Tabelbala** N Algeria 29.22N 3.01W
9 Q17 **Taber** Alberta, SW Canada 49.48N 112.09W
176 W14 **Taberfane** Pulau Trangan, E Indonesia 6.14S 134.08E
97 L19 **Täberg** Jönköping, S Sweden 57.42N 14.04E
Tabiauea see Tabiteuea
194 H12 **Tabibuga** Western Highlands, C PNG 5.32S 144.37E
203 O3 **Tabiteuea** prev. Drummond Island. atoll Tungaru, W Kiribati
179 Q12 **Tablas Island** C Philippines
179 Pp12 **Tablas Strait** strait C Philippines
194 M16 **Table Bay** bay SE PNG
192 Q10 **Table Cape** headland North Island, NZ 39.07S 178.00E
11 S13 **Table Mountain** ▲ Newfoundland, E Canada
181 P17 **Table, Pointe de la** headland SE Réunion 21.19S 55.49E
49 S8 **Table Rock Lake** ☺ Arkansas/Missouri, C USA
38 K14 **Table Top** ▲ Arizona, SW USA 32.45N 112.07W
194 J13 **Tabletop, Mount** ▲ C PNG 6.51S 146.00E
126 Mm5 **Tabor** Respublika Sakha (Yakutiya), NE Russian Federation 71.14N 150.23E
31 T6 **Tabor** Iowa, C USA 40.54N 95.40W
113 D18 **Tábor** Ger. Budějovický kraj, S Czech Republic 49.25N 14.40E
83 F21 **Tabora** Tabora, W Tanzania 5.04S 32.49E
83 F21 **Tabora** ◆ region C Tanzania
23 U12 **Tabor City** North Carolina, SE USA 34.10N 78.53W
153 Q10 **Taboshar** NW Tajikistan 40.37N 69.33E
78 L18 **Tabou** var. Tabu. S Ivory Coast 4.28N 7.19W
148 J2 **Tabrīz** var. Tebriz; anc. Tauris. Āzarbāyjān-e Khāvarī, NW Iran 38.05N 46.18E
Tabu see Tabou
203 W1 **Tabuaeran** prev. Fanning Island. atoll Line Islands, E Kiribati
194 E11 **Tabubil** Western, SW PNG 5.13S 141.13E
179 P8 **Tabuk** Luzon, N Philippines 17.26N 121.25E
146 J4 **Tabūk** Tabūk, NW Saudi Arabia 28.25N 36.33E
146 J5 **Tabūk** off. Minţaqat Tabūk. ◆ province NW Saudi Arabia
197 B12 **Tabwemasana, Mount** ▲ Espiritu Santo, W Vanuatu 15.22S 166.44E
97 O15 **Täby** Stockholm, C Sweden 59.28N 18.04E
43 N14 **Tacámbaro** Michoacán de Ocampo, C Mexico 19.12N 101.27W

44 A5 **Tacaná, Volcán**
🌋 Guatemala/Mexico *15.07N 92.06W*

45 X16 **Tacarcuna, Cerro ▲** SE Panama
8.08N 77.15W

Tachau *see* Tachov

164 J3 **Tacheng** *var.* Qoqek. Xinjiang Uygur Zizhiqu, NW China
46.45N 82.55E

56 H7 **Táchira** *off.* Estado Táchira.
◆ *state* W Venezuela

167 T13 **Tachoshui** N Taiwan *24.26N 121.43E*

113 A17 **Tachov** *Ger.* Tachau. Plzeňský Kraj, W Czech Republic *49.48N 12.37E*

179 R13 **Tacloban** *off.* Tacloban City. Leyte, C Philippines *11.15N 124.59E*

59 I19 **Tacna** Tacna, SE Peru *18.00S 70.15W*

59 H18 **Tacna** off. Departamento de Tacna.
◆ *department* S Peru

34 H8 **Tacoma** Washington, NW USA
47.15N 122.26W

20 L11 **Taconic Range ▲** NE USA

64 L6 **Taco Pozo** Formosa, N Argentina
25.35S 63.15W

59 M20 **Tacsara, Cordillera de ▲**
S Bolivia

63 F17 **Tacuarembó** *prev.* San Fructuoso. Tacuarembó, C Uruguay
31.42S 56.00W

63 E18 **Tacuarembó** ◆ *department*
C Uruguay

63 F17 **Tacuarembó, Río ♦** C Uruguay

85 I14 **Taculi** North Western,
NW Zambia *14.17S 26.51E*

179 R16 **Tacurong** Mindanao,
S Philippines *6.42N 124.40E*

171 Kk13 **Tadamu-gawa ♦** Honshū,
C Japan

79 N8 **Tadek ♦** NW Niger

76 J9 **Tademaït, Plateau du** *plateau*
C Algeria

197 K16 **Tadine** Province des Îles Loyauté, E New Caledonia *21.33S 167.54E*

82 L11 **Tadjoura** E Djibouti *11.47N 42.51E*

82 M11 **Tadjoura, Golfe de** *Eng.* Gulf of Tajura. *inlet* E Djibouti

Tadmor/Tadmur *see* Tudmur

9 W10 **Tadoule Lake** ⊚ Manitoba,
C Canada

13 S8 **Tadoussac** Quebec, SE Canada
48.07N 69.55W

161 H14 **Tādpatri** Andhra Pradesh, E India
14.55N 77.58E

Tadzhikabad *see* Tojikobod

Tadzhikistan *see* Tajikistan

169 Y14 **T'aebaek-sanmaek ▲** E South Korea

169 V15 **Taechŏng-do** *island* NW South Korea

169 X13 **Taedong-gang ♦** C North Korea

169 Y16 **Taegu** *off.* Taegu-gwangyŏksi, *var.* Daegu, *Jap.* Taikyū. SE South Korea *35.55N 128.32E*

Taehan-haehyŏp *see* Korea Strait

Taehan Min'guk *see* South Korea

169 V15 **Taejŏn** *off.* Taejŏn-gwangyŏksi, *Jap.* Taiden. C South Korea

200 T11 **Tafahi** *island* N Tonga

107 Q4 **Tafalla** Navarra, N Spain
42.31N 1.40W

77 M12 **Tafassâsset, Oued ♦** SE Algeria

79 W7 **Tafassâsset, Ténéré du** *desert*
N Niger

57 U11 **Tafelberg ▲** S Suriname
3.55N 56.09W

99 J21 **Taff ♦** SE Wales, UK

Tafila/Ṭafilah, Muḥāfaẓat *see*
Aṭ Ṭafīlah

79 N18 **Tafiré** N Ivory Coast *9.04N 5.10W*

148 M6 **Tafresh** Markazī, W Iran
34.40N 50.00E

149 O9 **Taft** Yazd, C Iran *31.48N 54.10E*

37 R13 **Taft** California, W USA
35.08N 119.27W

27 T14 **Taft** Texas, SW USA *27.58N 97.24W*

149 W12 **Taftān, Kūh-e ▲** SE Iran
28.38N 61.06E

37 R13 **Taft Heights** California, W USA
35.06N 119.29W

201 Y14 **Tafunsak** Kosrae, E Micronesia
5.21N 162.58E

198 Aa8 **Tāga** Savai'i, SW Samoa
13.46S 172.31W

155 O6 **Tagāb** Kāpīsā, E Afghanistan
33.52N 66.22E

41 O8 **Tagagawik River ♦** Alaska, USA

171 M13 **Tagajō** *var.* Tagazyō. Miyagi, Honshū, C Japan *38.21N 141.02E*

130 K12 **Taganrog** Rostovskaya Oblast', SW Russian Federation
47.10N 38.54E

130 K12 **Taganrog, Gulf of** *Rus.* Taganrogskiy Zaliv, *Ukr.* Tahanroz'ka Zatoka. *gulf* Russian Federation/Ukraine

Taganrogskiy Zaliv *see* Taganrog, Gulf of

78 J8 **Tagant ♦** *region* C Mauritania

154 M14 **Tagas** Baluchistān, SW Pakistan
27.09N 64.36E

170 D13 **Tagawa** Fukuoka, Kyūshū, SW Japan *33.37N 130.46E*

179 R11 **Tagaytay** Luzon, N Philippines
14.04N 120.55E

Tagazyō *see* Tagajō

179 Q14 **Tagbilaran** *var.* Tagbilaran City. Bohol, C Philippines *9.41N 123.54E*

108 B10 **Taggia** Liguria, NW Italy
43.52N 7.48E

79 V9 **Taghouaji, Massif de ▲** C Niger
17.13N 8.37E

109 J15 **Tagliacozzo** Lazio, C Italy
42.03N 13.15E

108 J7 **Tagliamento ♦** NE Italy

179 R14 **Tagolo Point** *headland* Mindanao, S Philippines
8.30N 124.45E

61 N3 **Tagow Bāy** *var.* Bai. Sar-e Pol, N Afghanistan *35.41N 66.01E*

Tagtabazar *see* Takhtabazar

61 L14 **Taguatinga** Tocantins, C Brazil
12.16S 46.25W

195 V17 **Tagula Island** *prev.* Southeast Island, Sudest Island. *island* SE PNG

179 R15 **Tagum** Mindanao, S Philippines
7.22N 125.51E

56 C7 **Tagún, Cerro** *elevation* Colombia/Panama *7.57N 77.13W*

---

107 P7 **Tagus** *Port.* Rio Tejo, *Sp.* Río Tajo.
♦ Portugal/Spain

66 M9 **Tagus Plain** *undersea feature*
E Atlantic Ocean

203 S10 **Tahaa** *island* Îles Sous le Vent,
W French Polynesia

203 U10 **Tahanea** *atoll* Îles Tuamotu,
C French Polynesia

Tahanroz'ka Zatoka *see*
Taganrog, Gulf of

171 I16 **Tahara** Aichi, Honshū, SW Japan
34.40N 137.15E

76 K12 **Tahat ▲** SE Algeria *23.15N 5.34E*

73 I4 **Ta He ♦** NE China

169 U4 **Tahe** Heilongjiang, NE China
52.21N 124.42E

168 Q9 **Tahilt** Govĭ-Altay, W Mongolia
45.20N 96.42E

203 T10 **Tahiti** *island* Îles du Vent,
W French Polynesia

Tahiti, Archipel de *see* Société,
Archipel de la

120 E4 **Tahkuna nina** *headland*
W Estonia *59.06N 22.35E*

154 K12 **Tāhlāb ♦** W Pakistan

154 K12 **Tāhlāb, Dasht-i** *desert*
SW Pakistan

49 N10 **Tahlequah** Oklahoma, C USA
35.55N 94.58W

37 Q6 **Tahoe City** California, W USA
39.09N 120.09W

37 P6 **Tahoe, Lake** ⊚ California/Nevada,
W USA

15 N6 **Tahoena** *see* Tahuna

34 F8 **Taholah** Washington, NW USA
47.19N 124.17W

79 U11 **Tahoua** Tahoua, W Niger
14.52N 5.18E

79 T11 **Tahoua** ◆ *department* W Niger

33 J9 **Tahquamenon Falls** *waterfall*
Michigan, N USA *46.34N 85.14W*

33 P4 **Tahquamenon River**
♦ Michigan, N USA

8 K17 **Tahsis** Vancouver Island, British Columbia, SW Canada
49.42N 126.31W

75 T7 **Tahta** *see* Tahkta

142 L15 **Tahtalı Dağları ▲** C Turkey

59 I14 **Tahuamanu, Río ♦** Bolivia/Peru

58 F13 **Tahuania, Río ♦** E Peru

203 X7 **Tahuata** *island* Îles Marquises,
NE French Polynesia

175 S6 **Tahulandang, Pulau** *island*
N Indonesia

175 S5 **Tahuna** *prev.* Tahoena. Pulau Sangihe, N Indonesia *3.33N 125.33E*

176 Y10 **Tahun, Danau** *see* Tahun, Danau

78 L17 **Taï** SW Ivory Coast *5.52N 7.28W*

167 P5 **Tai'an** Shandong, E China
36.13N 117.12E

203 R8 **Taiarapu, Presqu'île de** *peninsula* Tahiti,
W French Polynesia

Taïbad *see* Täybad

166 K7 **Taibai Shan ▲** C China
33.57N 107.31E

107 Q12 **Taibilla, Sierra de ▲** S Spain

169 Q12 **Taibus Qi** *var.* Baochang. Nei Mongol Zizhiqu, N China
41.55N 115.22E

167 S13 **Taichū** *see* T'aichung

167 S13 **T'aichung** *Jap.* Taichū; *prev.* Taiwan. C Taiwan *24.09N 120.40E*

Taiden *see* Taejŏn

193 E23 **Taieri ♦** South Island, NZ

117 E21 **Taïgetos ▲** S Greece

167 N4 **Taihang Shan ▲** C China

192 M11 **Taihape** Manawatu-Wanganui, North Island, NZ *39.41S 175.46E*

167 O7 **Taihe** Anhui, E China
33.14N 115.35E

167 O12 **Taihe** Jiangxi, S China
26.50N 114.49E

167 Q6 **Taihu** Anhui, E China
30.26N 116.13E

167 K8 **Tai Hu** ⊚ E China

167 P9 **Taihu** Anhui, E China
30.26N 116.13E

188 B17 **Taikang** Henan, C China
34.06N 114.53E

172 Y7 **Taiki** Hokkaidō, NE Japan
42.29N 143.15E

177 M12 **Taikkyi** Yangon, SW Myanmar
17.16N 95.55E

Taikyū *see* Taegu

169 U8 **Tailai** Heilongjiang, NE China
46.28N 123.24E

173 Ff10 **Taileleo** Pulau Siberut,
W Indonesia *1.45S 99.06E*

190 J10 **Tailem Bend** South Australia
35.20S 139.33E

98 I3 **Tain** N Scotland, UK *57.49N 4.04W*

167 S14 **T'ainan** *Jap.* Tainan; *prev.* Dainan. S Taiwan *23.00N 120.11E*

117 E22 **Taínaro, Akrotírio** *headland*
S Greece *36.41N 22.28E*

167 Q11 **Taining** Fujian, SE China
26.55N 117.13E

203 N7 **Taiohae** *prev.* Madisonville. Nuku Hiva, NE French Polynesia
8.55S 140.04W

167 S13 **T'aipei** *Jap.* Taihoku; *prev.* Daihoku. ◆ (Taiwan) N Taiwan
25.01N 121.28E

173 Gg3 **Taiping** Perak, Peninsular Malaysia *4.54N 100.42E*

168 K3 **Taiping** *see* Chongzuo

172 X5 **Taisei** Hokkaidō, NE Japan
42.13N 139.52E

170 F12 **Taisha** Shimane, Honshū,
SW Japan *35.23N 132.40E*

111 R4 **Taiskirchen** Oberösterreich,
NW Austria *48.15N 13.33E*

65 F20 **Taitao, Península de** *peninsula*
S Chile

167 S14 **Taitō** *Jap.* Taitō. S Taiwan
22.49N 121.04E

172 Y6 **Taitung** Hokkaidō, NE Japan
44.10N 143.09E

94 I3 **Taivassalo** Länsi-Suomi, W Finland *60.35N 21.36E*

167 S14 **Taiwan** *off.* Republic of China, *var.* Formosa, Formo'sa. ◆ *republic* E Asia

139 Q11 **Taiwan** *var.* Formosa. *island* E Asia

Taiwan *see* T'aichung

T'aiwan Haihsia/Taiwan Haixia *see* Taiwan Strait

Taiwan Shan *see* Chungyang Shanmo

167 R13 **Taiwan Strait** *var.* Formosa Strait, *Chin.* T'aiwan Haihsia, Taiwan Haixia. *strait* China/Taiwan

167 N4 **Taiyuan** *var.* T'ai-yuan, T'ai-yüan, Yangku. Shanxi, C China
37.48N 112.33E

167 R7 **Taizhou** Jiangsu, E China
32.36N 119.52E

167 S10 **Taizhou** *prev.* Haimen, Jiaojiang. Zhejiang, SE China *28.36N 121.19E*

167 K16 **Taizhou** *see* Linhai

147 I16 **Ta'izz** SW Yemen *13.36N 44.04E*

147 O16 **Ta'izz** × SW Yemen *13.41N 44.13E*

77 P12 **Tajarhī** SW Libya *24.21N 14.28E*

153 P13 **Tajikistan** *off.* Republic of Tajikistan, *Rus.* Tadzhikistan, *Taj.* Jumhurii Tojikiston; *prev.* Tadzhik S.S.R. ◆ *republic* C Asia

Tajik S.S.R. *see* Tajikistan

171 Kk14 **Tajima** Fukushima, Honshū,
C Japan *37.10N 139.46E*

Tajo, Río *see* Tagus

44 B5 **Tajumulco, Volcán ▲**
▲ W Guatemala *15.04N 91.50W*

107 P7 **Tajuña ♦** C Spain

Tajura, Gulf of *see* Tadjoura,
Golfe de

178 N19 **Tak** *var.* Rahaeng. Tak, W Thailand
16.51N 99.07E

201 U4 **Taka Atoll** *var.* Tōke. *atoll* Ratak Chain, N Marshall Islands

171 L16 **Takahagi** Ibaraki, Honshū,
S Japan *36.43N 140.40E*

170 Ff13 **Takahashi** *var.* Takahasi. Okayama, Honshū, SW Japan
34.48N 133.37E

170 F13 **Takahashi-gawa ♦** Honshū,
SW Japan

Takahasi *see* Takahashi

201 P12 **Takaieu Island** *island*
E Micronesia

192 I13 **Takaka** Tasman, South Island, NZ
40.52S 172.49E

175 P13 **Takalar** Sulawesi, C Indonesia
5.28S 119.24E

170 Ff14 **Takamatsu** *var.* Takamatu. Kagawa, Shikoku, SW Japan
34.18N 133.58E

Takamatu *see* Takamatsu

170 Cc14 **Takamori** Kumamoto, Kyūshū, SW Japan *32.50N 131.08E*

170 Cc16 **Takanabe** Miyazaki, Kyūshū, SW Japan *32.13N 131.31E*

175 O16 **Takan, Gunung ▲** Pulau Sumba, S Indonesia *8.52S 117.32E*

171 M4 **Takanosu** Akita, C Japan
40.13N 140.23E

171 Ii13 **Takaoka** Toyama, Honshū,
SW Japan *36.43N 137.01E*

192 N12 **Takapau** Hawke's Bay, North Island, NZ *40.01S 176.21E*

203 U9 **Takapoto** *atoll* Îles Tuamotu,
C French Polynesia

192 L5 **Takapuna** Auckland, North Island, NZ *36.48S 174.45E*

171 Gg14 **Takarazuka** Hyōgo, Honshū, SW Japan *34.48N 135.18E*

203 U9 **Takaroa** *atoll* Îles Tuamotu,
C French Polynesia

171 Jj15 **Takasaki** Gunma, Honshū,
S Japan *36.20N 139.00E*

170 Gg15 **Takatsuki** *var.* Takatuki. Ōsaka, Honshū, SW Japan *34.50N 135.36E*

Takatuki *see* Takatsuki

171 Ii4 **Takayama** Gifu, Honshū,
SW Japan *36.09N 137.17E*

170 Ff14 **Takefu** *var.* Takehu. Fukui, Honshū, SW Japan *35.58N 136.11E*

170 C13 **Takeo** Saga, Kyūshū, SW Japan
33.12N 130.01E

Takeo *see* Takêv

170 C13 **Take-shima** *island* Nansei-shotō, SW Japan

178 M5 **Taketan** *var.* Takistan; *prev.* Siadehan. Qazvin, N Iran
36.02N 49.36E

170 D14 **Taketa** Ōita, Kyūshū, SW Japan
32.56N 131.21E

178 Hh10 **Tak Fah** Nakhon Sawan,
C Thailand

145 V13 **Takhādid** *well* S Iraq *29.56N 44.33E*

155 R3 **Takhār ♦** *province* NE Afghanistan

152 H8 **Takhiatash** *see* Takhiatosh

152 H8 **Takhiatosh** *Rus.* Takhiatash. Qoraqalpogiston Respublikasi, W Uzbekistan *42.27N 59.26E*

7 K19 **Ta Khmau** Kândal, S Cambodia
11.30N 104.59E

152 H9 **Takhta** Dashta. Tahta. Dashkhovuzskiy Velayat, N Turkmenistan *41.40N 59.51E*

152 I15 **Takhtabazar** *var.* Tagtabazar. Maryyskiy Velayat, S Turkmenistan *35.57N 62.49E*

151 O8 **Takhtabrod** Severnyy Kazakhstan, N Kazakhstan
52.35N 67.37E

148 L5 **Takhtakŭpir** *Rus.* Takhtakupyr. Qoraqalpogiston Respublikasi, NW Uzbekistan *43.04N 60.23E*

148 M8 **Takht-e Shāh, Kūh-e ▲** C Iran

79 V12 **Takiéta** Zinder, S Niger
13.43N 8.33E

15 I5 **Takijuq Lake** ⊚ Nunavut,
NW Canada

172 X4 **Takikawa** Hokkaidō, NE Japan
43.34N 141.54E

172 Y4 **Takinoue** Hokkaidō, NE Japan
44.10N 143.09E

193 B23 **Takitimu Mountains ▲** South Island, NZ

172 N10 **Takizawa** Iwate, Honshū,
C Japan *39.45N 141.06E*

8 L13 **Takla Lake** ⊚ British Columbia, SW Canada

Takla Makan Desert *see*
Taklimakan Shamo

164 H9 **Taklimakan Shamo** *Eng.* Takla Makan Desert. *desert* NW China

178 Jj12 **Takôk** Môndól Kiri, E Cambodia
12.37N 106.30E

175 P9 **Takolekaju, Pegunungan ▲** Sulawesi, N Indonesia

41 P10 **Takotna** Alaska, USA
62.59N 156.03W

Takow *see* Kaohsiung

126 Kk14 **Taksimo** Respublika Buryatiya, S Russian Federation *56.18N 114.53E*

170 Cc13 **Taku** Saga, Kyūshū, SW Japan
33.17N 130.07E

8 I10 **Taku ♦** British Columbia,
W Canada

105 G15 **Takua Pa** *var.* Ban Takua Pa. Phangnga, SW Thailand
8.47N 98.16E

79 N13 **Takum** Taraba, E Nigeria
7.16N 10.00E

203 V10 **Takume** *atoll* Îles Tuamotu,
C French Polynesia

202 L16 **Takutea** *island* S Cook Islands
19.49S 158.18W

195 N11 **Takutea** *see* Tayu

121 L18 **Tal'** *Rus.* Tal'. Minskaya Voblasts', S Belarus *52.52N 27.59E*

42 L13 **Tala** Jalisco, C Mexico
20.39N 103.45W

63 F19 **Tala** Canelones, S Uruguay
34.24S 55.45W

144 H5 **Talabriga** *see* Aveiro, Portugal

Talabriga *see* Talavera de la Reina, Spain

121 N24 **Talachyn** *Rus.* Tolochin. Vitsyebskaya Voblasts', NE Belarus *54.25N 29.42E*

155 U7 **Talagang** Punjab, E Pakistan
32.55N 72.23E

161 I23 **Talaimannar** Northern Province, NW Sri Lanka *9.07N 79.45E*

119 R3 **Talalayivka** Chernihivs'ka Oblast', N Ukraine *50.51N 33.09E*

45 O15 **Talamanca, Cordillera de ▲**
▲ S Costa Rica

58 A9 **Talara** Piura, NW Peru
4.31S 81.17W

106 L11 **Talarrubias** Extremadura,
W Spain *39.03N 5.13W*

153 S8 **Talas** Talasskaya Oblast',
NW Kyrgyzstan *42.29N 72.21E*

153 S8 **Talas** ♦ NW Kyrgyzstan

195 N11 **Talasea** New Britain, E PNG
5.19S 150.02E

153 S8 **Talas Oblasty** *see* Talasskaya Oblast'

153 S8 **Talasskaya Oblast'** *Kir.* Talas Oblasty. ♦ *province* NW Kyrgyzstan

153 S8 **Talasskiy Alatau, Khrebet ▲**
▲ Kazakhstan/Kyrgyzstan

79 U12 **Talata Mafara** Zamfara,
NW Nigeria *12.33N 6.05E*

175 Ss4 **Talaud, Kepulauan** *island group* N Indonesia

106 M9 **Talavera de la Reina** *anc.* Caesarobriga, Talabriga. Castilla-La Mancha, C Spain *39.58N 4.49W*

106 J11 **Talavera la Real** Extremadura, W Spain *38.52N 6.46W*

194 L12 **Talawe, Mount ▲** New Britain, C PNG *5.30S 148.24E*

25 S5 **Talbotton** Georgia, SE USA
32.40N 84.32W

191 R7 **Talbragar River ♦** New South Wales, SE Australia

64 G4 **Talca** Maule, C Chile *35.28S 71.42W*

64 F13 **Talcahuano** Bío Bío, C Chile
36.43S 73.07W

160 N12 **Tālcher** Orissa, E India
20.57N 85.13E

27 W5 **Talco** Texas, SW USA *33.21N 95.06W*

151 V14 **Taldykorgan** *Kaz.* Taldyqorghan; *prev.* Taldy-Kurgan. Almaty, SE Kazakhstan *45.00N 78.23E*

Taldy-Kurgan/Taldyqorghan
see Taldykorgan

153 T7 **Taldy-Suu** Issyk-Kul'skaya Oblast', E Kyrgyzstan *42.49N 78.33E*

153 U10 **Taldy-Suu** Oshskaya Oblast', SW Kyrgyzstan *40.33N 73.52E*

Tal-e Khosravi *see* Yāsūj

200 Ss14 **Taleki Tonga** *island* Otu Tolu Group, C Tonga

200 Ss13 **Taleki Vavu'u** *island* Otu Tolu Group, C Tonga

104 J13 **Talence** Gironde, SW France
44.49N 0.35W

151 U16 **Talgar** *Kaz.* Talghar. Almaty, SE Kazakhstan *43.25N 77.07E*

Talghar *see* Talgar

175 Rr10 **Taliabu, Pulau** *island* Kepulauan Sula, C Indonesia

117 L22 **Taliarós, Akrotírio** *headland* Astypálaia, Kykládes, Greece, Aegean Sea *36.31N 26.18E*

Ta-lien *see* Dalian

49 U12 **Talihina** Oklahoma, C USA
34.45N 95.03W

143 T12 **Talimardzhan** *see* Tollimarjon

148 R7 **Talin** *see* T'alin

143 T12 **T'alin** *Rus.* Talin; *prev.* Verin T'alin. W Armenia *40.23N 43.51E*

83 K17 **Tali Post** Bahr el Gabel, S Sudan
5.55N 30.43E

148 M5 **Taliq, Rūd-e** *see* Tālūqān

148 L5 **Talış Dağları** *see* Talish Mountains

Talish Dağları, Per. Kūhhā-ye Ţavālesh, *Rus.* Talyshskiye Gory.
▲ Azerbaijan/Iran

181 X16 **Talkaikajuju Sredslovvskaya Oblast'** C Russian Federation *56.58N 63.34E*

107 S16 **Taliboang Sumbawa, C Indonesia**
8.45S 116.55E

121 L17 **Tal'ka** Minskaya Voblasts', C Belarus *53.22N 28.22E*

41 P10 **Talkeetna** Alaska, USA
62.19N 150.06W

41 P9 **Talkeetna Mountains ▲** Alaska, USA

59 I17 **Talkhof** *see* Puurmani

94 M8 **Tálknafjördhur** Vestfirdhir, NW Iceland *65.38N 23.50W*

145 O12 **Tall 'Abtah** N Iraq *35.52N 42.40E*

144 M2 **Tall Abyaḍ** *var.* Tell Abiad. Ar Raqqah, N Syria *36.42N 38.56E*

---

25 Q4 **Talladega** Alabama, S USA
33.26N 86.06W

25 S8 **Tallahassee** *prev.* Muskogean. *state capital* Florida, SE USA
30.26N 84.16W

25 S8 **Tallahatchie River ♦**
♦ Mississippi, S USA

Tall al Abyaḍ *see* At Tall al Abyaḍ

145 W12 **Tall al Laḥm** S Iraq *30.46N 46.22E*

191 P11 **Tallangatta** Victoria, SE Australia
36.15S 147.13E

105 T13 **Tallard** Hautes-Alpes, SE France
44.30N 6.04E

145 Q3 **Tall ash Sha'ir** N Iraq

25 R4 **Tallassee** Alabama, S USA
32.32N 85.53W

145 R4 **Tall 'Azbah** NW Iraq *35.47N 43.13E*

144 I5 **Tall Bīsah** Ḥimş, W Syria
34.49N 36.43E

145 R3 **Tall Ḥassūnah** N Iraq *36.05N 43.10E*

145 Q2 **Tall Ḥuqnah** *var.* Tell Huqnah.
N Iraq *36.33N 42.34E*

23 H4 **Tallin** *see* Tallinn

120 G3 **Tallinn** *Ger.* Reval, *Rus.* Tallin; *prev.* Revel. ◆ (Estonia) Harjumaa, NW Estonia *59.25N 24.42E*

120 H3 **Tallinn** × Harjumaa, NW Estonia
59.23N 24.52E

144 H5 **Tall Kalakh** *var.* Tell Kalakh.
Ḥimş, C Syria *34.40N 36.18E*

145 R2 **Tall Kayf** NW Iraq *36.30N 43.07E*

145 P2 **Tall Kūchak** *see* Tall Kūshik

145 P2 **Tall Kūshik** *var.* Tall Kūchak.
Al Ḥasakah, E Syria *36.48N 42.01E*

33 R9 **Tallmadge** Ohio, N USA
41.06N 81.26W

22 J9 **Tallulah** Louisiana, S USA
32.22N 91.12W

145 Q2 **Tall Zāhir** N Iraq *36.51N 42.29E*

126 H14 **Tal'menka** Altayskiy Kray,
S Russian Federation *53.55N 83.26E*

126 I8 **Talnakh** Taymyrskiy (Dolgano-Nenetskiy) Avtonomnyy Okrug, N Russian Federation *69.26N 88.26E*

119 P7 **Tal'noye** *Rus.* Tal'noye. Cherkas'ka Oblast', C Ukraine *48.54N 30.39E*

Tal'noye *see* Tal'ne

82 L12 **Talodi** Southern Kordofan,
C Sudan *10.40N 30.25E*

127 O10 **Talon** Magadanskaya Oblast',
E Russian Federation *59.47N 148.46E*

12 H11 **Talon, Lake** ⊚ Ontario, S Canada

155 R2 **Tāloqān** *var.* Taliq-an. Takhār,
NE Afghanistan *36.41N 69.33E*

130 M8 **Talovaya** Voronezhskaya Oblast', W Russian Federation *51.07N 40.46E*

175 Qq10 **Talowa, Teluk** *bay* Sulawesi,
C Indonesia

15 Kk3 **Taloyoak** *prev.* Spence Bay. Nunavut, N Canada *69.30N 93.25W*

27 Q8 **Talpa** Texas, SW USA
31.46N 99.42W

42 K13 **Talpa de Allende** Jalisco,
C Mexico *20.22N 104.51W*

25 S5 **Talquin, Lake** ⊚ Florida, SE USA

168 H9 **Talsen** *see* Talsi

120 E8 **Talsi** *Ger.* Talsen. Talsi, NW Latvia
57.14N 22.34E

64 G4 **Taltal** Antofagasta, N Chile
25.22S 70.27W

15 I5 **Taltson** ♦ Northwest Territories,
NW Canada

174 M8 **Taluk** Sumatera, W Indonesia
0.30S 101.36E

94 J8 **Talvik** Finnmark, N Norway
70.02N 22.58E

190 M7 **Talyawalka Creek ♦** New South Wales, SE Australia

148 L5 **Talyshskiye Gory** *see* Talish Mountains

120 I4 **Talsalu** *Ger.* Talsse. Lääne-Virumaa, NE Estonia *59.10N 26.07E*

111 S8 **Tamsweg** Salzburg, SW Austria
47.08N 13.48E

177 Ff3 **Tamu** Sagaing, N Myanmar
24.11N 94.21E

196 C15 **Tamuning** NW Guam
13.29N 144.47E

191 T6 **Tamworth** New South Wales, SE Australia *31.07S 150.54E*

99 M19 **Tamworth** C England, UK
52.39N 1.40W

94 L3 **Tana** Finnmark, N Norway
70.10N 28.06E

83 N18 **Tana** SE Kenya

170 G17 **Tana** Wakayama, Honshū,
SW Japan *33.43N 135.22E*

41 T10 **Tanacross** Alaska, USA
63.30N 143.21W

94 L3 **Tanafjorden** *fjord* N Norway

40 L7 **Tanaga Island** *island* Aleutian Islands, Alaska, USA

40 L7 **Tanaga Volcano ▲** Tanaga Island, Alaska, USA *51.53N 178.08W*

109 M18 **Tanagro ♦** S Italy

82 H11 **T'ana Hāyk'** *Eng.* Lake Tana.
⊚ NW Ethiopia

173 F8 **Tanahbela, Pulau** *island* Kepulauan Batu, W Indonesia *0.23S 98.23E*

175 P8 **Tanahjampea, Pulau** *island* C Indonesia

173 Ff8 **Tanahmasa, Pulau** *island* Kepulauan Batu, W Indonesia
0.12S 98.27E

173 I11 **Tanahputih** Sumatera,
W Indonesia *1.55S 100.51E*

158 L10 **Tanakpur** Uttar Pradesh, N India
29.04N 80.06E

175 O6 **Tanala** ♦ SE China

189 P5 **Tanambao** NE China

Taniantaweng Shan *see*
Tanggula Shan

180 H10 **Tanami Desert** *desert* Northern Territory, N Australia

41 Q9 **Tanana** Alaska, USA
65.12N 152.00W

Tananarive *see* Antananarivo

41 Q9 **Tanana River ♦** Alaska, USA

97 C16 **Tananger** Rogaland, S Norway
58.55N 5.34E

196 H5 **Tanapag** Saipan, S Northern Mariana Islands *16.45S 34.14E*

196 H5 **Tanapag, Puetton** *bay* Saipan,
S Northern Mariana Islands
1.09S 120.30E

108 C9 **Tanaro ♦** N Italy

42 M14 **Tancítaro, Cerro ▲** C Mexico
19.16N 102.25W

159 N12 **Tānda** Uttar Pradesh, N India

79 O15 **Tanda** E Ivory Coast *7.48N 3.10W*

179 Rr14 **Tandag** Mindanao, S Philippines
9.00N 126.13E

118 L14 **Ţăndărei** Ialomiţa, SE Romania
44.39N 27.40E

65 N14 **Tandil** Buenos Aires, E Argentina
37.18S 59.10W

155 Q16 **Tando Allāhyār** Sind, SE Pakistan
25.30N 68.43E

155 Q17 **Tando Bāgo** Sind, SE Pakistan
24.48N 68.58E

155 P16 **Tando Muhammad Khān** Sind, SE Pakistan *25.07N 68.40E*

190 I7 **Tandou Lake** *seasonal lake* New South Wales, SE Australia

96 L11 **Tandsjöborg** Gävleborg,
C Sweden *61.40N 14.42E*

161 K15 **Tāndūr** Andhra Pradesh, C India
17.16N 77.37E

170 Bb17 **Tanega-shima** *island* Nansei-shotō, SW Japan

172 N10 **Tanegachi** Iwate, Honshū, C Japan *40.23N 141.42E*

78 H8 **Tanezrouft** *desert* Algeria/Mali

83 J21 **Tanga** Tanga, E Tanzania
5.07S 39.04E

83 J22 **Tanga** ♦ *region* E Tanzania

159 T14 **Tangail** Dhaka, C Bangladesh

195 Q9 **Tanga Islands** *island group*
NE PNG

161 K28 **Tangalla** Southern Province, S Sri Lanka *6.01N 80.46E*

Tanganyika and Zanzibar *see*
Tanzania

70 I13 **Tanganyika, Lake** ⊚ E Africa

58 E7 **Tangarana, Río ♦** N Peru

195 W16 **Tangaroa, Mauna ▲** Easter Island, Chile, E Pacific Ocean

203 V16 **Tangaroa, Maunga ▲** Easter Island, Chile, E Pacific Ocean

76 G5 **Tanger** *var.* Tangier, Tanger, *Fr./Ger.* Tangerk, *Sp.* Tánger; *anc.* Tingis. NW Morocco *35.49N 5.48W*

174 I2 **Tangerang** Jawa, C Indonesia
6.13S 106.36E

Tangerk *see* Tanger

102 M12 **Tangermünde** Sachsen-Anhalt, C Germany *52.35N 11.57E*

162 K10 **Tanggula Shan** *var.* Dangla, Tangla Range. ▲ W China

165 N13 **Tanggula Shan ▲** W China
33.18N 91.10E

Tanggulashan *see* Tuotuoheyan

162 K10 **Tanggula Shankou** *pass* W China

79 N7 **Tanghe** Henan, C China
32.40N 112.49E

155 T5 **Tangi** North-West Frontier Province, NW Pakistan
34.18N 71.42E

23 Y5 **Tangier** *see* Tanger

23 Y5 **Tangier Island** *island* Virginia, NE USA

Tangiers *see* Tanger

24 K8 **Tangipahoa River ♦** Louisiana, S USA

171 H13 **Tango-hantō** *peninsula* Honshū, SW Japan

162 I10 **Tangra Yumco** *var.* Tangro Tso.
⊚ W China

163 I7 **Tangro Tso** *see* Tangra Yumco

179 Qq15 **Tangub** *var.* Tangub City. Mindanao, S Philippines
8.07N 123.42E

79 N8 **Tangwang He ♦** NE China

169 X7 **Tangyuan** Heilongjiang,
NE China *46.45N 129.52E*

94 M11 **Tanhua** Lappi, N Finland
67.31N 27.30E

176 Uu16 **Tanimbar, Kepulauan** *island group* Maluku, E Indonesia

Tanintharyi *see* Tenasserim

133 T15 **Tanjay** E Iraq

179 N10 **Tanjore** *see* Thanjāvūr

175 N10 **Tanjung** *prev.* Tandjoeng. Borneo, C Indonesia *2.07S 115.22E*

175 O6 **Tanjungbatu** Borneo,
N Indonesia *2.19N 118.03E*

174 J11 **Tanjungkarang** *see* Bandarlampung

173 G14 **Tanjungpandan** *prev.* Tandjoengpandan. Pulau Belitung, W Indonesia *2.43S 107.36E*

174 J11 **Tanjungpinang** *prev.* Tandjoengpinang. Pulau Bintan, W Indonesia *0.55N 104.27E*

175 O6 **Tanjungredeb** *var.* Tanjungredep; *prev.* Tandjoengredeb. Borneo, C Indonesia *2.09N 117.28E*

◆ COUNTRY   ◇ DEPENDENT TERRITORY   ◈ ADMINISTRATIVE REGION   ▲ MOUNTAIN   ▲ VOLCANO   ⊚ LAKE
● COUNTRY CAPITAL   ○ DEPENDENT TERRITORY CAPITAL   ✕ INTERNATIONAL AIRPORT   ▲ MOUNTAIN RANGE   ♦ RIVER   ▨ RESERVOIR

Tanjungredep *see* Tanjungredeb
155 *S8* **Tānk** North-West Frontier Province, NW Pakistan 32.13N 70.28E
197 *H26* **Tanna** *island* S Vanuatu
95 *F17* **Tännäs** Jämtland, C Sweden 62.27N 12.40E
Tannenhof *see* Krynica
110 *K7* **Tannheim** Tirol, W Austria 47.30N 10.32E
Tannu-Tuva *see* Tyva, Respublika
175 *R10* **Tano** Pulau Taliabu, E Indonesia 1.51S 124.55E
79 *O17* **Tano** ✔ S Ghana
158 *D10* **Tanot** Rājasthān, NW India 27.49N 70.21E
79 *V11* **Tanout** Zinder, C Niger 14.58N 8.54E
43 *P12* **Tanquián** San Luis Potosí, C Mexico 21.38N 98.39W
79 *R13* **Tansarga** E Burkina 11.51N 1.51E
178 *Jj14* **Tan Son Nhat** ✈ (Hồ Chi Minh) Tây Ninh, S Vietnam 10.52N 106.38E
77 *V8* **Tanta** *var.* Tantā, Tantā. N Egypt 30.42N 31.00E
76 *D9* **Tan-Tan** SW Morocco 28.30N 11.10W
43 *P12* **Tantoyuca** Veracruz-Llave, E Mexico 21.18N 98.12W
158 *J12* **Tāntpur** Uttar Pradesh, N India 26.51N 77.28E
Tan-tung *see* Dandong
40 *M12* **Tanunak** Alaska, USA 60.35N 165.15W
177 *Ff5* **Ta-nyaung** Magwe, W Myanmar 20.49N 94.40E
178 *J5* **Tân Yên** Tuyên Quang, N Vietnam 22.08N 104.58E
83 *F22* **Tanzania** *off.* United Republic of Tanzania, *Swa.* Jamhuri ya Muungano wa Tanzania; *prev.* German East Africa, Tanganyika and Zanzibar. ◆ *republic* E Africa
Tanzania, Jamhuri ya Muungano wa *see* Tanzania
169 *U9* **Tao'an** *var.* Taoan, Taonan. Jilin, NE China 45.19N 122.46E
169 *T8* **Tao'er He** ✔ NE China
165 *U11* **Tao He** ✔ C China
T'aon-an *see* Baicheng
Taongi *see* Bokaak Atoll
109 *M23* **Taormina** *anc.* Tauromenium. Sicilia, Italy, C Mediterranean Sea 37.54N 15.18E
39 *S9* **Taos** New Mexico, SW USA 36.24N 105.34W
Taoudenit *see* Taoudenni
79 *O6* **Taoudenni** *var.* Taoudenit. Tombouctou, N Mali 22.46N 3.54W
76 *G6* **Taounate** N Morocco 34.34N 4.35W
167 *S13* **T'aoyüan** *Jap.* Tōen. N Taiwan 25.00N 121.15E
120 *I3* **Tapa** *Ger.* Taps. Lääne-Virumaa, NE Estonia 59.15N 26.00E
43 *V17* **Tapachula** Chiapas, SE Mexico 14.53N 92.18W
Tapaiu *see* Gvardeysk
61 *H14* **Tapajós, Rio** *var.* Tapajóz. ✔ NW Brazil
Tapajóz *see* Tapajós, Rio
63 *C21* **Tapalqué** *var.* Tapalquén. Buenos Aires, E Argentina 36.21S 60.01W
Tapalquén *see* Tapalqué
Tapanahoni Rivier *see* Tapanahony Rivier
57 *W11* **Tapanahony Rivier** *var.* Tapanahoni. ✔ E Suriname
43 *T16* **Tapanatepec** *var.* San Pedro Tapanatepec. Oaxaca, SE Mexico 16.23N 94.09W
193 *D23* **Tapanui** Otago, South Island, NZ 45.58S 169.16E
61 *E14* **Tapanuli, Teluk** *see* Sibolga, Teluk
61 *E14* **Tapauá** Amazonas, N Brazil 5.42S 64.15W
49 *R7* **Tapauá, Rio** ✔ W Brazil
193 *I14* **Tapawera** Tasman, South Island, NZ 41.24S 172.50E
63 *I16* **Tapes** Rio Grande do Sul, S Brazil 30.40S 51.25W
78 *H11* **Tāpi** *prev.* Tāpti. ✔ W India
106 *J2* **Tapia de Casariego** Asturias, N Spain 43.34N 6.55W
58 *F10* **Tapiche, Río** ✔ N Peru
178 *Gg15* **Tapi, Mae Nam** *var.* Luang. ✔ SW Thailand
194 *K14* **Tapini** Central, S PNG 8.20S 146.57E
Tapirapecó, Serra *see* Tapirapecó, Sierra
57 *N13* **Tapirapecó, Sierra** *Port.* Serra Tapirapecó. ▲ Brazil/Venezuela
79 *R13* **Tapoa** ✔ Benin/Niger
196 *H5* **Tapochau, Mount** ▲ Saipan, S Northern Mariana Islands
113 *H24* **Tapolca** Veszprém, W Hungary 46.54N 17.28E
23 *X5* **Tappahannock** Virginia, NE USA 37.55N 76.51W
31 *U13* **Tappan Lake** ☐ Ohio, N USA
171 *Mm7* **Tappi-zaki** *headland* Honshū, C Japan 41.15N 140.19E
Taps *see* Tapa
Tāpti *see* Tāpi
193 *J16* **Tapuaenuku** ▲ South Island, NZ 42.00S 173.39E
179 *Pp17* **Tapul Group** *island group* Sulu Archipelago, SW Philippines
60 *E11* **Tapurucuará** *var.* Tapuruquara. Amazonas, NW Brazil 0.175 65.00W
Tapuruquara *see* Tapurucuará
198 *C9* **Taputapu, Cape** *headland* Tutuila, W American Samoa 14.19S 170.51W
147 *W13* **Tāqah** S Oman 17.04N 54.24E
145 *T3* **Taqtaq** N Iraq 35.54N 44.36E
63 *J15* **Taquara** Rio Grande do Sul, S Brazil 29.40S 50.46W
61 *H19* **Taquari, Rio** ✔ C Brazil
62 *L8* **Taquaritinga** São Paulo, S Brazil 21.22S 48.29W
125 *G12* **Tara** Omskaya Oblast', C Russian Federation 56.54N 74.17E
85 *O16* **Tara** Southern, S Zambia 16.54S 26.45E
115 *I15* **Tara** ✔ SW Yugoslavia
114 *I13* **Tara** ✔ W Yugoslavia
79 *X15* **Taraba** ◆ *state* E Nigeria
79 *X15* **Taraba** ✔ E Nigeria

---

77 *O7* **Ţarābulus** *var.* Ţarābulus al Gharb, *Eng.* Tripoli. ● (Libya) NW Libya 32.54N 13.10E
77 *O7* **Ţarābulus** ✈ NW Libya 32.97N 13.07E
Ţarābulus/Ţarābulus ash Shām *see* Tripoli
Ţarābulus al Gharb *see* Ţarābulus
107 *O7* **Taracena** Castilla-La Mancha, C Spain 40.39N 3.07W
119 *N12* **Taraclia** *Rus.* Tarakilya. S Moldova 45.55N 28.40E
145 *V10* **Tarād al Kahf** SE Iraq 31.58N 45.58E
191 *R10* **Tarago** New South Wales, SE Australia 35.04S 149.40E
79 *V11* **Taraju** Jawa, S Indonesia 7.27S 107.58E
176 *Vv11* **Tarak** Irian Jaya, E Indonesia 3.21S 132.43E
175 *O5* **Tarakan** Borneo, C Indonesia 3.19N 117.37E
175 *O5* **Tarakan, Pulau** *island* N Indonesia
Tarakilya *see* Taraclia
172 *Pp16* **Tarama-jima** *island* Sakishima-shotō, SW Japan
192 *K10* **Taranaki** *off.* Taranaki Region. ◆ *region* North Island, NZ
192 *K10* **Taranaki, Mount** *var.* Egmont. ▲ North Island, NZ 39.16S 174.04E
107 *O9* **Tarancón** Castilla-La Mancha, C Spain 40.01N 3.01W
98 *E7* **Taransay** *island* NW Scotland, UK
109 *P18* **Taranto** *var.* Tarentum. Puglia, SE Italy 40.30N 17.10E
109 *O19* **Taranto, Golfo di** *Eng.* Gulf of Taranto. *gulf* S Italy
Taranto, Gulf of *see* Taranto, Golfo di
64 *G3* **Tarapacá** *off.* Región de Tarapacá. ◆ *region* N Chile
195 *Y16* **Tarapaina** Maramasike Island, N Solomon Islands 9.28S 161.24E
58 *D10* **Tarapoto** San Martín, N Peru 6.31S 76.24W
144 *M6* **Ţaraq an Na'jah** *hill range* E Syria
144 *M6* **Ţaraq Sīdawī** *hill range* E Syria
105 *Q11* **Tarare** Rhône, E France 45.54N 4.26E
Tararite de Litera *see* Tamarite de Litera
192 *M13* **Tararua Range** ▲ North Island, NZ
192 *M13* **Tarascon** Bouches-du-Rhône, SE France 43.48N 4.39E
104 *M17* **Tarascon-sur-Ariège** Ariège, S France 42.51N 1.36E
119 *P6* **Tarashcha** Kyyivs'ka Oblast', N Ukraine 49.34N 30.31E
59 *L18* **Tarata** Cochabamba, C Bolivia 17.34S 66.04W
59 *I18* **Tarata** Tacna, S Peru 17.30S 70.00W
202 *H2* **Taratai** atoll Tungaru, W Kiribati
61 *B15* **Tarauacá** Acre, W Brazil 8.06S 70.45W
61 *B15* **Tarauacá, Río** ✔ NW Brazil
205 *Q8* **Taravao** Tahiti, W French Polynesia 17.43S 149.19W
203 *R8* **Taravao, Baie de** *bay* Tahiti, W French Polynesia
203 *Q8* **Taravao, Isthme de** *isthmus* Tahiti, W French Polynesia
105 *X16* **Taravo** ✔ Corse, France, C Mediterranean Sea
202 *J3* **Tarawa** ▲ Tarawa, W Kiribati 0.52S 169.31E
202 *H2* **Tarawa** *atoll* Tungaru, W Kiribati
192 *N10* **Tarawera** Hawke's Bay, North Island, NZ 39.03S 176.34E
192 *N8* **Tarawera, Lake** ☐ North Island, NZ
192 *N8* **Tarawera, Mount** ▲ North Island, NZ 38.13S 176.29E
107 *S8* **Tarayuela** ▲ N Spain 40.28N 0.22W
151 *R16* **Taraz** *prev.* Aulie Ata, Auliye-Ata, Dzhambul, Zhambyl. Zhambyl, S Kazakhstan 42.55N 71.27E
107 *Q5* **Tarazona** Aragón, NE Spain 41.54N 1.43W
107 *Q10* **Tarazona de la Mancha** Castilla-La Mancha, C Spain 39.16N 1.55W
76 *E8* **Tarbagatay, Khrebet** ▲ China/Kazakhstan
98 *J8* **Tarbat Ness** *headland* N Scotland, UK 57.51N 3.48W
98 *H12* **Tarbela Reservoir** ☐ N Pakistan
98 *F7* **Tarbert** W Scotland, UK 55.52N 5.25W
98 *F7* **Tarbert** Western Isles, NW Scotland, UK 57.53N 6.48W
104 *K16* **Tarbes** *anc.* Bigorra. Hautes-Pyrénées, S France 43.13N 0.04E
23 *W9* **Tarboro** North Carolina, SE USA 35.54N 77.32W
107 *T7* **Tarca** *see* Torysa
107 *T7* **Tarcento** Friuli-Venezia Giulia, NE Italy 46.13N 13.13E
190 *F5* **Tarcoola** South Australia 30.44S 134.33E
107 *S5* **Tardienta** Aragón, NE Spain 41.58N 0.31W
104 *L11* **Tardoire** ✔ W France
191 *N12* **Taree** New South Wales, SE Australia 31.55S 152.28E
94 *K12* **Tärendö** Norrbotten, N Sweden 67.10N 22.40E
Tarentum *see* Taranto
76 *C9* **Tarfaya** SW Morocco 27.56N 12.55W
118 *J13* **Târgovişte** *prev.* Tîrgovişte. Dâmboviţa, S Romania 44.54N 25.28E
118 *M12* **Târgu Bujor** *prev.* Tîrgu Bujor. Galaţi, E Romania 45.52N 27.55E
118 *H13* **Târgu Cărbuneşti** *prev.* Tîrgu Cărbuneşti. Gorj, SW Romania 44.57N 23.31E
118 *L9* **Târgu Frumos** *prev.* Tîrgu Frumos. Iaşi, NE Romania 47.12N 27.00E
118 *H13* **Târgu Jiu** *prev.* Tîrgu Jiu. Gorj, W Romania 45.03N 23.16E
118 *J15* **Târgu Lăpuş** *prev.* Tîrgu Lăpuş. Maramureş, N Romania 47.28N 23.54E
120 *I5* **Târgu-Neamţ** *see* Târgu-Neamţ

---

144 *H5* **Ţarţūs** *Fr.* Tartouss; *anc.* Tortosa. Ţarţūs, W Syria 34.55N 35.52E
144 *H5* **Ţarţūs** *off.* Muḩāfaz̧at Ţarţūs, *var.* Tartous, Tartus. ◆ *governorate* W Syria
125 *G13* **Tarumizu** Kagoshima, Kyūshū, SW Japan 31.30N 130.40E
130 *K4* **Tarusa** Kaluzhskaya Oblast', W Russian Federation 54.45N 37.10E
173 *G9* **Tarusan** Sumatera, W Indonesia 1.13S 100.22E
119 *N11* **Tarutyne** Odes'ka Oblast', SW Ukraine 46.11N 29.09E
168 *I7* **Tarvagatyn Nuruu** ▲ N Mongolia
107 *T7* **Tarvisio** Friuli-Venezia Giulia, NE Italy 46.31N 13.33E
Tarvisium *see* Treviso
59 *O16* **Tarvo, Río** ✔ E Bolivia
12 *G8* **Tarzwell** Ontario, S Canada 48.00N 79.58W
43 *K5* **Tasajera, Sierra de la** ▲ N Mexico
149 *P17* **Ţarīf** Abū Z̧aby, C UAE 24.01N 53.46E
106 *K16* **Tarifa** Andalucía, S Spain 36.01N 5.36W
86 *C14* **Tarifa, Punta de** *headland* SW Spain 36.01N 5.39W
59 *M21* **Tarija** Tarija, S Bolivia 21.33S 64.42W
59 *M21* **Tarija** ◆ *department* S Bolivia
147 *R14* **Tarīm** S Yemen 16.00N 48.50E
83 *G19* **Tarime** Mara, N Tanzania 1.19S 34.24E
Tarim Basin *see* Tarim Pendi
165 *H8* **Tarim He** ✔ NW China
165 *H8* **Tarim Pendi** *Eng.* Tarim Basin. *basin* NW China
155 *N7* **Tarīn Kowt** *var.* Terinkot. Urūzgān, C Afghanistan 32.37N 65.52E
155 *Pp10* **Taripa** Sulawesi, C Indonesia 1.51S 120.46E
176 *Z11* **TarIturu, Sungai** *prev.* Idenburg-rivier. ✔ Irian Jaya, E Indonesia
119 *Q12* **Tarkhankut, Mys** *headland* S Ukraine 45.20N 32.32E
54 *Q1* **Tarkio** Missouri, C USA 40.25N 95.24W
125 *U11* **Tarko-Sale** Yamalo-Nenetskiy Avtonomnyy Okrug, N Russian Federation 64.55N 77.34E
79 *P17* **Tarkwa** S Ghana 5.16N 1.58W
179 *P10* **Tarlac** Luzon, N Philippines 15.29N 120.34E
97 *F22* **Tarm** Ringkøbing, W Denmark 55.55N 8.31E
59 *E14* **Tarma** Junín, C Peru 11.25S 75.43W
105 *N15* **Tarn** ◆ *department* S France
104 *M15* **Tarn** ✔ S France
113 *L22* **Tarna** ✔ C Hungary
94 *G13* **Tärnaby** Västerbotten, N Sweden 65.43N 15.19E
155 *P8* **Tarnak Rūd** ✔ SE Afghanistan
118 *J11* **Târnava Mare** *Ger.* Grosse Kokel, *Hung.* Nagy-Küküllő; *prev.* Tîrnava Mare. ✔ S Romania
118 *I11* **Târnava Mică** *Ger.* Kleine Kokel, *Hung.* Kis-Küküllő; *prev.* Tîrnava Mică. ✔ C Romania
118 *I11* **Târnăveni** *Ger.* Marteskirch, Martinskirch, *Hung.* Dicsöszentmárton; *prev.* Sinmartin, Tîrnăveni. Mureş, C Romania 46.19N 24.16E
113 *P18* **Tarnica** ▲ SE Poland 49.05N 22.43E
113 *N15* **Tarnobrzeg** Podkarpackie, SE Poland 50.34N 21.40E
129 *N12* **Tarnogskiy Gorodok** Vologodskaya Oblast', NW Russian Federation 60.28N 43.45E
113 *M16* **Tarnów** Małopolskie, SE Poland 50.01N 20.58E
Tarnowice/Tarnowitz *see* Tarnowskie Góry
113 *J16* **Tarnowskie Góry** *var.* Tarnowice, Tarnowskie Gory, *Ger.* Tarnowitz. Śląskie, S Poland 50.27N 18.52E
97 *N14* **Tärnsjö** Västmanland, C Sweden 60.10N 16.57E
79 *O8* **Taro** ✔ NW Italy
195 *Q10* **Taron** New Ireland, NE PNG 4.22S 153.04E
76 *E8* **Taroudannt** *var.* Taroudant. SW Morocco 30.31N 8.50W
Taroudant *see* Taroudannt
25 *V12* **Tarpon, Lake** ☐ Florida, SE USA
25 *V12* **Tarpon Springs** Florida, SE USA 28.09N 82.45W
104 *G14* **Tarquinia** *anc.* Tarquinii; *hist.* Corneto. Lazio, C Italy 42.22N 11.45E
Tarquinii *see* Tarquinia
Tarraco *see* Tarragona
78 *D10* **Tarrafal** Santiago, S Cape Verde 15.16N 23.45W
107 *V6* **Tarragona** *anc.* Tarraco. Cataluña, E Spain 41.07N 1.15E
107 *T7* **Tarragona** ◆ *province* Cataluña, NE Spain
191 *O17* **Tarraleah** Tasmania, SE Australia 42.11S 146.29E
61 *M15* **Tasso Fragoso** Maranhão, E Brazil 8.25S 45.53W
97 *J23* **Tástrup** *var.* Taastrup. København, E Denmark 55.39N 12.19E
151 *O9* **Tasty-Taldy** Akmola, C Kazakhstan 50.45N 66.35E
61 *O10* **Tatajuba** Sulawesi, C Indonesia

---

121 *M17* **Tatarka** *Rus.* Tatarka. Mahilyowskaya Voblasts', E Belarus 53.15N 28.49E
142 *D15* **Tatar Pazardzhik** *see* Pazardzhik
125 *G13* **Tatarsk** Novosibirskaya Oblast', C Russian Federation 55.08N 75.58E
127 *O15* **Tatarskiy Proliv** *Eng.* Tatar Strait. *strait* SE Russian Federation
131 *R4* **Tatarstan, Respublika** *prev.* Tatarskaya ASSR. ◆ *autonomous republic* W Russian Federation
Tatar Strait *see* Tatarskiy Proliv
195 *O9* **Tatau Island** *island* Tabar Islands, N PNG
175 *P8* **Tate** Sulawesi, N Indonesia 0.12S 119.44E
171 *Jj17* **Tateyama** Chiba, Honshū, S Japan 35.00N 139.51E
171 *J14* **Tate-yama** ▲ Honshū, S Japan 36.27N 137.32E
147 *N11* **Tathlith** 'Asīr, S Saudi Arabia 19.37N 43.31E
147 *O11* **Tathlīth, Wādī** *dry watercourse* S Saudi Arabia
191 *R11* **Tathra** New South Wales, SE Australia 36.44S 149.58E
131 *P8* **Tatishchevo** Saratovskaya Oblast', W Russian Federation 51.43N 45.35E
41 *S12* **Tatitlek** Alaska, USA 60.49N 146.29W
8 *L15* **Tatla Lake** British Columbia, SW Canada 51.54N 124.39W
124 *O2* **Tatlisu** *Gk.* Akanthoú. N Cyprus 35.22N 33.44E
9 *Z10* **Tatnam, Cape** *headland* Manitoba, C Canada 57.16N 91.03W
113 *K18* **Tatra Mountains** *Ger.* Tatra, *Hung.* Tátra, *Pol./Svk.* Tatry. ▲ Poland/Slovakia
Tatra/Tátra *see* Tatra Mountains
Tatry *see* Tatra Mountains
171 *I16* **Tatsuno** var. Tatuno. Hyōgo, Honshū, SW Japan 34.51N 134.33E
151 *S16* **Tatti** *var.* Tatty. Zhambyl, S Kazakhstan 43.10N 73.22E
Tatty *see* Tatti
62 *L10* **Tatui** São Paulo, S Brazil 23.21S 47.49W
39 *V14* **Tatum** New Mexico, SW USA 33.15N 103.19W
27 *X7* **Tatum** Texas, SW USA 32.19N 94.31V
97 *H14* **Tåsinge** *island* C Denmark
10 *M5* **Tasiujaq** Quebec, E Canada 58.43N 69.58W
127 *N9* **Taskan** Magadanskaya Oblast', E Russian Federation 63.10N 150.03E
151 *Q14* **Taskesken** Vostochnyy Kazakhstan, E Kazakhstan
142 *J10* **Taşköprü** Kastamonu, N Turkey 41.30N 34.12E
155 *N9* **Taskuduk, Peski** *see* Goshquduq Qum
143 *S13* **Taşlıçay** Ağrı, E Turkey 39.37N 43.22E
193 *H14* **Tasman** *off.* Tasman District. ◆ *unitary authority* South Island, NZ
193 *W10* **Tasman** ✔ C French Polynesia
199 *I14* **Tasman Basin** *var.* East Australian Basin. *undersea feature* S Tasman Sea
193 *I14* **Tasman Bay** *inlet* South Island, NZ
199 *Hh14* **Tasman Fracture Zone** *tectonic feature* S Indian Ocean
193 *E19* **Tasman Glacier** *glacier* South Island, NZ
Tasman Group *see* Nukumanu Islands
191 *N15* **Tasmania** *prev.* Van Diemen's Land. ◆ *state* SE Australia
191 *Q16* **Tasmania** *island* SE Australia
193 *H14* **Tasman Mountains** ▲ South Island, NZ
191 *P17* **Tasman Peninsula** *peninsula* Tasmania, SE Australia
199 *Hh13* **Tasman Plain** *undersea feature* W Tasman Sea
199 *Hh14* **Tasman Plateau** *var.* South Tasmania Plateau. *undersea feature* SW Tasman Sea
199 *I14* **Tasman Sea** *sea* SW Pacific Ocean
118 *G9* **Tăşnad** *Ger.* Trestenberg, Trestendorf, *Hung.* Tasnád. Satu Mare, NW Romania 47.30N 22.33E
142 *I11* **Taşova** Amasya, N Turkey 40.45N 36.19E
79 *T10* **Tassara** Tahoua, W Niger 16.40N 5.34E
181 *D10* **Tassialouc, Lac** ☐ Quebec, C Canada
76 *L11* **Tassili-n-Ajjer** *plateau* E Algeria
76 *K14* **Tassili ta-n-Ahaggar** *var.* Tassili du Hoggar. *plateau* S Algeria
Tassili du Hoggar *see* Tassili ta-n-Ahaggar
120 *D12* **Tauragė** *Ger.* Tauroggen. Tauragė, SW Lithuania 55.15N 22.17E
120 *D12* **Tauragė** ◆ *province* W Lithuania
92 *N7* **Tauranga** Bay of Plenty, North Island, NZ 37.41S 176.09E
10 *O10* **Taureau, Réservoir** ☐ Quebec, SE Canada
109 *N22* **Taurianova** Calabria, SW Italy 38.22N 16.01E
Tauris *see* Tabriz
192 *I2* **Tauroa Point** *headland* North Island, NZ 35.09S 173.02E
Tauroggen *see* Tauragė
Tauromenium *see* Taormina
Taurus Mountains *see* Toros Dağları
Taus *see* Domažlice
Taúshyq *see* Tauchik
57 *R5* **Tauste** Aragón, NE Spain 41.55N 1.15W
203 *V16* **Tautira** Tahiti, W French Polynesia 17.45S 149.10W

---

151 *P7* **Tayynsha** *prev.* Krasnoarmeysk. Severnyy Kazakhstan, N Kazakhstan 53.52N 69.51E
126 *H9* **Taz** ✔ N Russian Federation
76 *G6* **Taza** NE Morocco 34.13N 4.06W
145 *T4* **Tāza Khurmātū** E Iraq 35.18N 44.21E
171 *M10* **Tazawa-ko** ☐ Honshū, C Japan
23 *N8* **Tazewell** Tennessee, S USA 36.27N 83.34V
23 *Q7* **Tazewell** Virginia, NE USA 37.06N 81.31W
77 *S11* **Tāzirbū** SE Libya 25.43N 21.16E
41 *S11* **Tazlina Lake** ☐ Alaska, USA
126 *H7* **Tazovskaya Guba** *Eng.* Bay of Taz. *bay* N Russian Federation
126 *H8* **Tazovskiy** Yamalo-Nenetskiy Avtonomnyy Okrug, N Russian Federation 67.33N 78.21E
143 *U10* **T'bilisi** *Eng.* Tiflis. ● (Georgia) SE Georgia 41.43N 44.49E
81 *E14* **Tchabal Mbabo** ▲ NW Cameroon 7.12N 12.16E
Tchad *see* Chad
Tchad, Lac *see* Chad, Lake
79 *S15* **Tchaourou** E Benin 8.55N 2.39E
81 *E20* **Tchibanga** Nyanga, S Gabon 2.49S 11.00E
Tchien *see* Zwedru
77 *Z6* **Tchigaï, Plateau du** ▲ NE Niger
79 *V9* **Tchighozérine** Agadez, C Niger 17.15N 7.48E
79 *T10* **Tchin-Tabaradene** Tahoua, W Niger 15.57N 5.49E
80 *G13* **Tcholliré** Nord, NE Cameroon 8.48N 14.00E
Tchongking *see* Chongqing
24 *K4* **Tchula** Mississippi, S USA 33.10N 90.13V
112 *I7* **Tczew** *Ger.* Dirschau. Pomorskie, N Poland 54.05N 18.46E
118 *I10* **Teaca** *Ger.* Teckendorf, *Hung.* Teke; *prev. Ger.* Teckendorf. Bistriţa-Năsăud, N Romania 46.55N 24.30E
25 *O04* **Teacapán** Sinaloa, C Mexico 22.33N 105.44W
202 *A10* **Teafuafou** *island* Funafuti Atoll, C Tuvalu
27 *U8* **Teague** Texas, SW USA 31.37N 96.16W
203 *R9* **Teahupoo** Tahiti, W French Polynesia 17.51S 149.15W
202 *H15* **Te Aiti Point** *headland* Rarotonga, S Cook Islands 21.10S 59.46W
67 *D24* **Teal Inlet** East Falkland, Falkland Islands 51.34S 58.25W
193 *B22* **Te Anau** Southland, South Island, NZ 45.24S 167.44E
193 *B22* **Te Anau, Lake** ☐ South Island, NZ
43 *U15* **Teapa** Tabasco, SE Mexico 17.36N 92.57W
192 *Q7* **Te Araroa** Gisborne, North Island, NZ 37.37S 178.21E
192 *M7* **Te Aroha** Waikato, North Island, NZ 37.33S 175.41E
Teate *see* Chieti
202 *A9* **Te Ava Fuagea** *channel* Funafuti Atoll, SE Tuvalu
202 *B8* **Te Ava I Te Lape** *channel* Funafuti Atoll, SE Tuvalu
202 *B9* **Te Ava Pua Pua** *channel* Funafuti Atoll, SE Tuvalu
192 *M8* **Te Awamutu** Waikato, North Island, NZ 37.59S 175.19E
176 *Xx9* **Teba** Irian Jaya, E Indonesia 1.27S 137.54E
106 *L15* **Teba** Andalucía, S Spain 36.59N 4.54W
130 *M5* **Teberda** Karachayevo-Cherkesskaya Respublika, SW Russian Federation 43.28N 41.45E
76 *M6* **Tébessa** NE Algeria 35.21N 8.06E
33 *B22* **Tebicuary, Río** ✔ S Paraguay
174 *H11* **Tebingtinggi** Sumatera, W Indonesia 3.33S 103.00E
173 *Ff5* **Tebingtinggi** Sumatera, N Indonesia 3.19N 99.07E
173 *Ff5* **Tebingtinggi, Pulau** *see* Rantau, Pulau
Tebriz *see* Tabriz
143 *U9* **Tebulos Mt'a** *Rus.* Gora Tebulosmta. ▲ Georgia/Russian Federation 42.33N 45.21E
Tebulosmta, Gora *see* Tebulos Mt'a
43 *Q14* **Tecamachalco** Puebla, S Mexico 18.52N 97.43W
42 *B1* **Tecate** Baja California, NW Mexico 32.33N 116.37W
142 *M13* **Tecer Dağları** ▲ C Turkey
105 *P16* **Techimann** W Ghana 7.35N 1.56W
119 *N15* **Techirghiol** Constanţa, SE Romania 44.03N 28.37E
76 *A12* **Techla** *var.* Western Sahara 21.39N 14.57W
65 *H14* **Tecka, Sierra de** ▲ SW Argentina
Teckendorf *see* Teaca
43 *N16* **Tecolotlán** Jalisco, SW Mexico 20.14N 104.01W
42 *K13* **Tecomán** Colima, SW Mexico 18.52N 103.54W
42 *G13* **Tecopa** California, W USA 35.51N 116.14W
42 *G5* **Tecoripa** Sonora, NW Mexico 28.36N 109.57W
42 *J11* **Tecpan** *var.* Tecpan de Galeana. Guerrero, S Mexico 17.11N 100.39W
Tecpan de Galeana *see* Tecpan
42 *J11* **Tecuala** Nayarit, C Mexico 22.24N 105.30W
118 *L12* **Tecuci** Galaţi, E Romania 45.49N 27.27E
31 *R10* **Tecumseh** Michigan, N USA 42.00N 83.57W
31 *S16* **Tecumseh** Nebraska, C USA 40.22N 96.12W
29 *O11* **Tecumseh** Oklahoma, C USA 35.15N 96.56W
194 *E12* **Tedi** ✔ W PNG
152 *J14* **Tedzhen** *Per.* Harīrūd, *Turkm.* Tejen. ✔ Afghanistan/Iran *see also* Harīrūd

---

◆ **COUNTRY** ○ **DEPENDENT TERRITORY** ◇ **ADMINISTRATIVE REGION** ▲ **MOUNTAIN** ☈ **VOLCANO** ☐ **LAKE**
● **COUNTRY CAPITAL** ○ **DEPENDENT TERRITORY CAPITAL** ✈ **INTERNATIONAL AIRPORT** ▲ **MOUNTAIN RANGE** ✔ **RIVER** ☐ **RESERVOIR**

152 H15 **Tedzhenstroy** *Turkm.* Tejenstroy. Akhalskiy Velayat, S Turkmenistan 36.57N 60.49E
168 I7 **Teel** Arhangay, C Mongolia 48.01N 100.30E
99 L15 **Tees** ♒ N England, UK
12 E15 **Teeswater** Ontario, S Canada 44.00N 81.17W
202 A10 **Tefala** *atoll* Funafuti Atoll, C Tuvalu
60 D13 **Tefé** Amazonas, N Brazil 3.24S 64.45W
76 K11 **Tefedest** ▲ S Algeria
142 E16 **Tefenni** Burdur, SW Turkey 37.19N 29.45E
60 D13 **Tefé, Rio** ♒ NW Brazil
174 Kk14 **Tegal** Jawa, C Indonesia 6.52S 109.07E
102 O12 **Tegel** ✕ (Berlin) Berlin, NE Germany 52.33N 13.16E
101 M15 **Tegelen** Limburg, SE Netherlands 51.19N 6.09E
103 L24 **Tegernsee** ⊚ SE Germany
109 M18 **Teggiano** Campania, S Italy 40.25N 15.28E
79 U14 **Tegina** Niger, C Nigeria 10.06N 6.10E
197 B10 **Tegua** *island* Torres Islands, N Vanuatu
44 I7 **Tegucigalpa** ● (Honduras) Francisco Morazán, SW Honduras 14.04N 87.10W
44 H7 **Tegucigalpa** ✕ Central District, C Honduras 14.03N 87.20W
   Tegucigalpa *see* Central District, Honduras
   Tegucigalpa *see* Francisco Morazán, Honduras
79 U9 **Teguidda-n-Tessoumt** Agadez, C Niger 17.27N 6.40E
66 Q11 **Teguise** Lanzarote, Islas Canarias, Spain, NE Atlantic Ocean 29.04N 13.37W
126 Hh13 **Tegul'det** Tomskaya Oblast', C Russian Federation 57.16N 87.58E
79 S13 **Tehachapi** California, W USA 35.07N 118.27W
79 S13 **Tehachapi Mountains** ▲ California, W USA
   Tehama *see* Tīhāmah
79 O14 **Téhini** NE Ivory Coast 9.36N 3.40W
149 N5 **Tehrān** *var.* Teheran. ● (Iran) Tehrān, N Iran 35.43N 51.26E
149 N6 **Tehrān** *off.* Ostān-e Tehrān, *var.* Tehran. ♦ *province* N Iran
158 X9 **Tehri** Uttar Pradesh, N India 30.12N 78.28E
   Tehri *see* Tikamgarh
43 Q15 **Tehuacán** Puebla, S Mexico 18.28N 97.24W
43 S17 **Tehuantepec** *var.* Santo Domingo Tehuantepec. Oaxaca, SE Mexico 16.18N 95.13W
43 S17 **Tehuantepec, Golfo de** *var.* Gulf of Tehuantepec. *gulf* S Mexico
   Tehuantepec, Golfo de *see* Tehuantepec, Golfo de
   Tehuantepec, Isthmus of *see* Tehuantepec, Istmo de
43 T16 **Tehuantepec, Istmo de** *var.* Isthmus of Tehuantepec. *isthmus* SE Mexico
2 I6 **Tehuantepec Ridge** *undersea feature* E Pacific Ocean
43 S16 **Tehuantepec, Río** ♒ SE Mexico
203 W10 **Tehuata** *atoll* Îles Tuamotu, C French Polynesia
66 O11 **Teide, Pico de** ▲ Gran Canaria, Islas Canarias, Spain, NE Atlantic Ocean 28.16N 16.39W
99 I21 **Teifi** ♒ SW Wales, UK
82 B9 **Teiga Plateau** *plateau* W Sudan
99 J24 **Teignmouth** SW England, UK 50.34N 3.29W
   Teisen *see* Chech'ŏn
111 L23 **Teiuş** *Ger.* Dreikirchen, *Hung.* Tövis. Alba, C Romania 46.12N 23.40E
175 N16 **Tejakula** Bali, C Indonesia 8.09S 115.19E
   Tejen *see* Harīrūd/Tedzhen
   Tejenstroy *see* Tedzhenstroy
37 S14 **Tejon Pass** *pass* California, W USA 34.46N 118.49W
   Tejo, Rio *see* Tagus
43 O14 **Tejupilco** *var.* Tejupilco de Hidalgo. México, S Mexico 18.55N 100.10W
   Tejupilco de Hidalgo *see* Tejupilco
192 P7 **Te Kaha** Bay of Plenty, North Island, NZ 37.45S 105.42E
31 N14 **Tekamah** Nebraska, C USA 41.46N 96.13W
192 Q3 **Te Kao** Northland, North Island, NZ 34.39S 172.57E
193 F20 **Tekapo** South Island, NZ
193 F19 **Tekapo, Lake** ⊚ South Island, NZ
192 P9 **Te Karaka** Gisborne, North Island, NZ 38.30S 105.52E
192 L7 **Te Kauwhata** Waikato, North Island, NZ 37.22S 175.07E
43 X12 **Tekax** *var.* Tekax de Álvaro Obregón. Yucatán, SE Mexico 20.07N 89.10W
   Tekax de Álvaro Obregón *see* Tekax
   Tekax/Tekendorf *see* Teaca
142 A14 **Teke Burnu** *headland* W Turkey
116 M12 **Teke Deresi** ♒ NW Turkey
152 D10 **Tekeli, Gory** *hill range* NW Turkmenistan
151 V14 **Tekeli** Almaty, SE Kazakhstan 44.49N 78.46E
151 R7 **Teke, Ozero** ⊚ N Kazakhstan
134 I5 **Tekes** Xinjiang Uygur Zizhiqu, NW China 43.15N 81.43E
151 W16 **Tekes** Almaty, SE Kazakhstan 42.40N 80.01E
   Tekes *see* Tekes He
164 H5 **Tekes He** *Rus.* Tekes. ♒ China/Kazakhstan
82 I10 **Tekezē** *var.* Takkaze. ♒ Eritrea/Ethiopia
142 C10 **Tekirdağ** *It.* Rodosto; *anc.* Bisanthe, Raidestos, Rhaedestus. Tekirdağ, NW Turkey 40.58N 27.31E

142 C10 **Tekirdağ** ♦ *province* NW Turkey
161 N14 **Tekkali** Andhra Pradesh, E India 18.37N 84.15E
117 K15 **Tekke Burnu** *Turk.* Ilyasbaba Burnu. *headland* NW Turkey 40.03N 26.12E
143 Q13 **Tekman** Erzurum, NE Turkey 39.39N 41.31E
34 M9 **Tekoa** Washington, NW USA 47.13N 117.05W
202 H16 **Te Kou** ▲ Rarotonga, S Cook Islands 21.13S 159.46W
175 R9 **Teku** Sulawesi, N Indonesia 0.46S 123.25E
192 L9 **Te Kuiti** Waikato, North Island, NZ 38.21S 175.09E
44 H4 **Tela** Atlántida, NW Honduras 15.43N 87.27W
144 F12 **Telalim** Southern, N Israel 30.58N 34.47E
   Telanaipura *see* Jambi
143 O10 **T'elavi** E Georgia 41.55N 45.29E
144 F10 **Tel Aviv** ✕ *district* W Israel
144 F10 **Tel Aviv-Yafo** *see* Tel Aviv-Yafo
   Tel Aviv-Yafo *var.* Tel Aviv-Yafo. Tel Aviv, C Israel 32.04N 34.46E
144 F10 **Tel Aviv-Yafo** ✕ Tel Aviv, C Israel 32.04N 34.45E
113 E18 **Telč** *Ger.* Teltsch. Jihlavský Kraj, C Czech Republic 49.10N 15.28E
194 E11 **Telefomin** Sandaun, NW PNG 5.05S 141.40E
8 I10 **Telegraph Creek** British Columbia, W Canada 57.55N 131.10W
202 B10 **Telele** *island* Funafuti Atoll, C Tuvalu
175 P12 **Telen, Sungai** ♒ Borneo, C Indonesia
118 I15 **Teleorman** ♦ *county* S Romania
118 I14 **Teleorman** ♒ S Romania
27 V5 **Telephone** Texas, SW USA 33.48N 96.00W
37 U14 **Telescope Peak** ▲ California, W USA 36.09N 117.03W
   Teles Pirés *see* São Manuel, Rio
99 L19 **Telford** C England, UK 52.42N 2.28W
110 L7 **Telfs** Tirol, W Austria 47.19N 11.04E
44 I9 **Telica** León, NW Nicaragua 12.29N 86.51W
78 I13 **Telica, Río** ♒ C Honduras
85 O14 **Telire, Río** ♒ Costa Rica/Panama
116 I8 **Telish** *prev.* Azizie. Pleven, N Bulgaria 43.20N 24.16E
43 N16 **Telixtlahuaca** *var.* San Francisco Telixtlahuaca. Oaxaca, SE Mexico 17.18N 96.54W
8 K13 **Telkwa** British Columbia, SW Canada 54.39N 126.51W
27 P4 **Tell** Texas, SW USA 34.18N 100.20W
   Tell Abiad *see* Tall Abyaḍ
   Tell Abiad/Tell Abyaḍ *see* At Tall al Abyaḍ
33 O16 **Tell City** Indiana, N USA 37.56N 86.45W
40 M9 **Teller** Alaska, USA 65.15N 166.21W
161 F20 **Tellicherry** *var.* Thalassery. Kerala, SW India 11.48N 75.30E
22 M10 **Tellico Plains** Tennessee, S USA 35.19N 84.18W
   Tell Kalakh *see* Tall Kalakh
   Tell Mardīkh *see* Ebla
56 C11 **Tello** Huila, C Colombia 3.06N 75.07W
   Tell Shedadi *see* Ash Shadādah
39 Q7 **Telluride** Colorado, C USA 37.56N 107.48W
   Tel'man/Tel'mansk *see* Gubadag
119 X9 **Tel'manove** Donets'ka Oblast', E Ukraine 47.24N 38.03E
168 H6 **Telmen Nuur** ⊚ NW Mongolia
   Teloekbetoeng *see* Bandarlampung
43 O10 **Teloloápan** Guerrero, S Mexico 18.21N 99.54W
   Telo Martius *see* Toulon
117 N5 **Telpoziz, Gora** ▲ NW Russian Federation 62.52N 59.15E
   Telschen *see* Telšiai
65 J13 **Telsen** Chubut, S Argentina 42.27S 66.59W
120 D11 **Telšiai** *Ger.* Telschen. Telšiai, NW Lithuania 55.59N 22.21E
   Teltsch *see* Telč
   Telukbetung *see* Bandarlampung
173 F7 **Telukdalam** Pulau Nias, W Indonesia 0.34N 97.47E
12 H9 **Temagami** Ontario, S Canada 47.03N 79.47W
12 H9 **Temagami, Lake** ⊚ Ontario, S Canada
202 H16 **Te Manga** ▲ Rarotonga, S Cook Islands 21.13S 159.45W
   Temanggoeng *see* Temanggung
174 Kk15 **Temanggung** *prev.* Temanggoeng. Jawa, C Indonesia
203 W12 **Tematangi** *atoll* Îles Tuamotu, S French Polynesia
43 X11 **Temax** Yucatán, SE Mexico 21.10N 88.55W
176 X12 **Tembagapura** Irian Jaya, E Indonesia 4.10S 137.18E
133 U3 **Tembenchi** ♒ N Russian Federation
57 P6 **Tembladora** Monagas, NE Venezuela 9.01N 62.38W
107 N9 **Tembleque** Castilla-La Mancha, C Spain 39.40N 3.30W
78 M14 **Tembo Aluma** N Ivory Coast 10.25N 6.25W
109 J11 **Temecula** California, W USA 33.29N 117.09W
174 Gg3 **Temengor, Tasik** ⊚ Peninsular Malaysia

114 L9 **Temerin** Serbia, N Yugoslavia 45.25N 19.54E
   Temes/Temesch *see* Tamiš
   Temeschburg/Temeschwar *see* Timişoara
   Temes-Kubin *see* Kovin
   Temesvár/Temeswar *see* Timişoara
176 V9 **Teminabuan** *prev.* Teminaboean. Irian Jaya, E Indonesia 1.30S 131.58E
151 P17 **Temirlanovka** Yuzhnyy Kazakhstan, S Kazakhstan 42.36N 69.15E
151 R10 **Temirtau** *prev.* Samarkandski, Samarkandskoye. Karaganda, C Kazakhstan 50.04N 72.55E
12 H10 **Témiscaming** Quebec, SE Canada 46.40N 79.04W
   Témiscamingue, Lac *see* Timiskaming, Lake
13 T8 **Témiscouata, Lac** ⊚ Quebec, SE Canada
131 N5 **Temnikov** Respublika Mordoviya, W Russian Federation 54.39N 43.09E
203 T13 **Temoe** *island* Îles Gambier, E French Polynesia
191 Q9 **Temora** New South Wales, SE Australia 34.28S 147.33E
42 H7 **Temósachic** Chihuahua, W Mexico 27.16N 108.15W
42 I5 **Temósachic** Chihuahua, N Mexico 28.55N 107.42W
195 W8 **Temotu** *off.* Temotu Province. ♦ *province* E Solomon Islands
38 L14 **Tempe** Arizona, SW USA 33.24N 111.54W
175 P12 **Tempe, Danau** ⊚ Sulawesi, C Indonesia
108 B11 **Tempio Pausania** Sardegna, Italy, C Mediterranean Sea 40.55N 9.07E
44 K12 **Tempisque, Río** ♒ NW Costa Rica
27 T9 **Temple** Texas, SW USA 31.06N 97.22W
102 O12 **Templehof** ✕ (Berlin) Berlin, NE Germany 52.28N 13.24E
99 D19 **Templemore** *Ir.* An Teampall Mór. C Ireland 52.48N 7.49W
102 O11 **Templin** Brandenburg, NE Germany 53.07N 13.31E
43 P12 **Tempoal** *var.* Tempoal de Sánchez. Veracruz-Llave, E Mexico 21.27N 98.21W
   Tempoal de Sánchez *see* Tempoal
43 P13 **Tempoal, Río** ♒ C Mexico
85 E14 **Tempué** Moxico, C Angola 13.36S 18.56E
130 J14 **Temryuk** Krasnodarskiy Kray, SW Russian Federation 45.15N 37.26E
101 G17 **Temse** Oost-Vlaanderen, N Belgium 51.07N 4.13E
65 O14 **Temuco** Araucanía, C Chile 38.45S 72.37W
193 G20 **Temuka** Canterbury, South Island, NZ 44.13S 171.16E
201 P13 **Temwen Island** *island* E Micronesia
58 C6 **Tena** Napo, C Ecuador 1.00S 77.48W
54 W13 **Tenabo** Campeche, E Mexico 20.01N 90.12W
   Tenaghau *see* Aola
27 X7 **Tenaha** Texas, SW USA 31.56N 94.14W
41 X13 **Tenake** Chicagof Island, Alaska, USA 57.46N 135.13W
161 K16 **Tenali** Andhra Pradesh, E India 16.13N 80.36E
   Tenan *see* Ch'ŏnan
43 O10 **Tenancingo** *var.* Tenencingo de Degollado. México, S Mexico 18.57N 99.39W
203 X12 **Tenararo** *island* Groupe Actéon, SE French Polynesia
178 Gg12 **Tenasserim** Tenasserim, S Myanmar 12.06N 98.55E
178 H11 **Tenasserim** *var.* Tanintharyi. ♦ *division* S Myanmar
100 O5 **Ten Boer** Groningen, NE Netherlands 53.16N 6.42E
99 I21 **Tenby** SW Wales, UK 51.40N 4.43W
82 K11 **Tendaho** Afar, NE Ethiopia 11.39N 40.59E
105 V14 **Tende** Alpes Maritimes, SE France 44.04N 7.34E
   Ten Degree Channel *strait* Andaman and Nicobar Islands, India, E Indian Ocean
82 I7 **Tendelti** White Nile, E Sudan 13.01N 31.55E
78 I8 **Te-n-Dghâmcha, Sebkhet** *var.* Sebkha de Ndrhamcha, Sebkra de Ndaghamcha. *salt lake* W Mauritania
171 Ll12 **Tendō** Yamagata, Honshū, C Japan 38.22N 140.22E
74 H6 **Tendrara** NE Morocco 33.06N 1.58W
119 Q11 **Tendriv's'ka Kosa** *spit* S Ukraine
119 Q11 **Tendriv's'ka Zatoka** *gulf* S Ukraine
79 N12 **Ténenkou** Mopti, C Mali 14.28N 4.55W
79 W9 **Ténéré** *physical region* C Niger
79 W9 **Ténéré, Erg du** *desert* C Niger
66 O11 **Tenerife** *island* Islas Canarias, Spain, NE Atlantic Ocean
78 J5 **Ténès** NW Algeria 36.30N 1.18E
175 Oo15 **Tengah, Kepulauan** *island group* C Indonesia
175 Q14 **Tenggarong** Borneo, C Indonesia 0.23S 117.00E
168 F13 **Tenggar Shamo** *desert* N China
174 I4 **Tenggul, Pulau** *island* Peninsular Malaysia
   Tengiz Köl *see* Tengiz, Ozero
151 P9 **Tengiz, Ozero** *Kaz.* Tengiz Köl. *salt lake* C Kazakhstan
78 M14 **Tengréla** *var.* Tingréla. N Ivory Coast 10.25N 6.25W
166 M14 **Tengxian** *var.* Teng Xian. Guangxi Zhuangzu Zizhiqu, S China 23.24N 110.49E

204 H2 **Teniente Rodolfo Marsh** *Chilean research station* South Shetland Islands, Antarctica 61.57S 58.23W
34 G9 **Tenino** Washington, NW USA 46.51N 122.51W
114 I9 **Tenja** Osijek-Baranja, E Croatia 45.30N 18.45E
196 B16 **Tenjo, Mount** ▲ W Guam
161 H23 **Tenkasi** Tamil Nādu, SE India 8.58N 77.22E
81 N24 **Tenke** Katanga, SE Dem. Rep. Congo (Zaire) 10.34S 26.12E
   Tenke *see* Tinca
126 M7 **Tenkeli** Respublika Sakha (Yakutiya), NE Russian Federation 70.09N 140.39E
49 R10 **Tenkiller Ferry Lake** ⊚ Oklahoma, C USA
79 Q13 **Tenkodogo** S Burkina 11.43N 0.19W
189 Q5 **Tennant Creek** Northern Territory, C Australia 19.40S 134.16E
22 G9 **Tennessee** *off.* State of Tennessee; also known as The Volunteer State. ♦ *state* SE USA
39 R5 **Tennessee Pass** *pass* Colorado, C USA 39.21N 106.18W
22 H10 **Tennessee River** ♒ S USA
25 N2 **Tennessee Tombigbee Waterway** *canal* Alabama/Mississippi, S USA
101 K22 **Tenneville** Luxembourg, SE Belgium 50.05N 5.31E
94 M11 **Teno** *var.* Tenojoki, *Lapp.* Dealnu, *Nor.* Tana. ♒ Finland/Norway *see also* Tana
94 L9 **Teno** *var.* Tenojoki, *Lapp.* Dealnu, *Nor.* Tana. ♒ Finland/Norway *see also* Tana
   Tenojoki *see* Tana/Teno
175 Nn3 **Tenom** Sabah, East Malaysia 5.07N 115.57E
43 V15 **Tenosique** *var.* Tenosique de Pino Suárez. Tabasco, SE Mexico 17.30N 91.24W
   Tenosique de Pino Suárez *see* Tenosique
171 H15 **Tenri** Nara, Honshū, SW Japan 34.36N 135.51E
171 I16 **Tenryū** Shizuoka, Honshū, SW Japan 34.54N 137.48E
172 ii15 **Tenryū-gawa** ♒ Honshū, C Japan
25 I6 **Tensas River** ♒ Louisiana, S USA
25 O4 **Tensaw River** ♒ Alabama, S USA
175 Pp10 **Tentena** *var.* Tenteno. Sulawesi, C Indonesia 1.46S 120.40E
   Tenten *var.* Tentena
191 U4 **Tenterfield** New South Wales, SE Australia 29.04S 152.02E
25 X16 **Ten Thousand Islands** *island group* Florida, SE USA
62 H9 **Teodoro Sampaio** São Paulo, S Brazil 22.30S 52.13W
61 N9 **Teófilo Otoni** *var.* Theophilo Ottoni. Minas Gerais, NE Brazil 17.52S 41.31W
118 K6 **Teofipol'** Khmel'nyts'ka Oblast', W Ukraine 50.00N 26.22E
203 Q8 **Teohatu** Tahiti, W French Polynesia
43 P14 **Teotihuacán** *ruins* México, S Mexico 19.49N 98.48W
   Teotilán del Camino *var.* Teotitlán. Oaxaca, S Mexico 18.05N 97.04W
   Teotitlán del Camino *see* Teotilán
127 O17 **Tepa** Île Uvea, E Wallis and Futuna 13.19S 176.09W
203 P8 **Tepace, Récif** *reef* Tahiti, W French Polynesia
42 L14 **Tepalcatepec** Michoacán de Ocampo, SW Mexico 19.10N 102.49W
202 A16 **Tepa Point** *headland* SW Niue 19.07S 169.55E
43 N9 **Tepatitlán** *var.* Tepatitlán de Morelos. Jalisco, SW Mexico 20.54N 102.45W
   Tepatitlán de Morelos *see* Tepatitlán
42 J9 **Tepehuanes** *var.* Santa Catarina de Tepehuanes. Durango, C Mexico 25.18N 105.43W
   Tepelena *see* Tepelenë
115 L22 **Tepelenë** *var.* Tepelena. *It.* Tepeleni. Gjirokastër, S Albania 40.18N 20.00E
   Tepeleni *see* Tepelenë
43 N12 **Tepic** Nayarit, C Mexico 21.29N 104.54W
113 C16 **Teplice** *Ger.* Teplitz; *prev.* Teplice-Šanov, Teplitz-Schönau. Ústecký Kraj, NW Czech Republic 50.37N 13.48E
   Teplice-Sanov/Teplitz/Teplitz-Schönau *see* Teplice
119 O7 **Teplyk** Vinnyts'ka Oblast', C Ukraine 48.40N 29.46E
203 W9 **Tepoto** *island* Îles du Désappointement, C French Polynesia
94 I11 **Tepsa** Lappi, N Finland
202 B8 **Tepuka** *atoll* Funafuti Atoll, C Tuvalu
192 N7 **Te Puke** Bay of Plenty, North Island, NZ 37.48S 176.20E
42 L9 **Tequila** Jalisco, SW Mexico 20.52N 103.48W
43 O13 **Tequisquiapan** Querétaro de Arteaga, C Mexico 20.34N 99.52W
35 Q4 **Terace** Idaho, NW USA 43.49N 112.25W
96 M7 **Tera** ♒ N Spain
79 V9 **Téra** Tillabéri, W Niger 14.01N 0.48E
35 X9 **Terry** Montana, NW USA 46.46N 105.16W
28 D11 **Terral** Oklahoma, C USA 33.55N 97.54W
110 I7 **Teufen** Sankt Gallen, NE Switzerland 47.24N 9.24E
42 L12 **Teul** *var.* Teul de Gonzáles Ortega. Zacatecas, C Mexico 21.30N 103.28W
   Terranova di Sicilia *see* Gela
   Terranova Pausania *see* Olbia
109 B21 **Teulada** Sardegna, Italy, C Mediterranean Sea 38.58N 8.46E
   Terrassa *Cast.* Tarrasa. Cataluña, E Spain 41.34N 2.01E
   Teul de Gonzáles Ortega *see* Teul
5 X16 **Teulon** Manitoba, S Canada 50.20N 97.14W
44 I7 **Teupasenti** El Paraíso, S Honduras 14.14N 86.43W
172 Oo3 **Teuri-tō** *island* NE Japan
102 G13 **Teutoburger Wald** *Eng.* Teutoburg Forest. *hill range* NW Germany
   Teutoburg Forest *see* Teutoburger Wald
74 G5 **Tevua, Swe.** Östermark. Länsi-Suomi, W Finland 62.28N 21.45E
109 W5 **Terre Eng.** Tiber. ♒ C Italy
144 G9 **Teveraya** *var.* Tiberias, Tverya. Northern, N Israel 32.47N 35.32E
98 K11 **Teviot** ♒ SE Scotland, UK
116 I9 **Tevli** *see* Tewli
125 Ff12 **Tevriz** Omskaya Oblast', C Russian Federation 57.30N 72.13E
178 I7 **Thar Pärkar** *prov?* SE Pakistan
193 R24 **Te Waewae Bay** *bay* South Island, NZ

106 H11 **Tera, Ribeira de** ♒ S Portugal
193 K14 **Terawhiti, Cape** *headland* North Island, NZ 41.17S 174.36E
100 N12 **Terborg** Gelderland, E Netherlands 51.55N 6.22E
143 P13 **Tercan** Erzincan, NE Turkey 39.46N 40.22E
66 O2 **Terceira** ✕ Terceira, Azores, Portugal, NE Atlantic Ocean 38.43N 27.13W
66 O2 **Terceira** *var.* Ilha Terceira. *island* Azores, Portugal, NE Atlantic Ocean
   Terceira, Ilha *see* Terceira
118 K6 **Terebovlya** Ternopil's'ka Oblast', W Ukraine 49.18N 25.43E
131 O15 **Terek** ♒ SW Russian Federation
153 R9 **Terek-Say** Dzhalal-Abadskaya Oblast', W Kyrgyzstan 41.28N 71.06E
174 Hh3 **Terengganu** *var.* Trengganu. ♦ *state* Peninsular Malaysia
131 X7 **Terensay** Orenburgskaya Oblast', W Russian Federation 51.35N 59.28E
60 N13 **Teresina** *var.* Therezina. *state capital* Piauí, NE Brazil 5.09S 42.46W
62 P9 **Teresópolis** Rio de Janeiro, SE Brazil 22.25S 42.59W
112 P12 **Terespol** Lubelskie, E Poland 52.05N 23.37E
203 V16 **Tereveka, Maunga** ▲ Easter Island, Chile, E Pacific Ocean 27.04S 109.22W
45 O14 **Teribe, Río** ♒ NW Panama
128 K3 **Teriberka** Murmanskaya Oblast', NW Russian Federation 69.10N 35.18E
   Terijoki *see* Zelenogorsk
   Terinkot *see* Tarīn Kowt
   Terisaqqan *see* Tersakkan
26 K2 **Terlingua** Texas, SW USA 29.18N 103.36W
26 K1 **Terlingua Creek** ♒ Texas, SW USA
64 K7 **Termas de Río Hondo** Santiago del Estero, N Argentina 27.28S 64.52W
142 M11 **Terme** Samsun, N Turkey 41.11N 36.58E
   Termez *see* Termiz
109 J23 **Termini Imerese** *anc.* Thermae Himerenses. Sicilia, Italy, C Mediterranean Sea 38.01N 13.55E
43 V14 **Términos, Laguna de** *lagoon* SE Mexico
79 X10 **Termit-Kaoboul** Zinder, C Niger 15.34N 11.31E
153 O14 **Termez** *Rus.* Termez. Surkhondaryo Wiloyati, S Uzbekistan 37.17N 67.12E
109 L15 **Termoli** Molise, C Italy 42.00N 14.58E
100 M8 **Ternunde** *see* Dendermonde
100 P5 **Termunten** Groningen, NE Netherlands 53.18N 7.02E
175 S7 **Ternate** Pulau Ternate, E Indonesia 0.50N 127.20E
175 Si7 **Ternate, Pulau** *island* E Indonesia
111 T5 **Ternberg** Oberösterreich, N Austria 47.57N 14.22E
101 E15 **Terneuzen** *var.* Neuzen. Zeeland, SW Netherlands 51.19N 3.49E
127 O17 **Terney** Primorskiy Kray, SE Russian Federation 45.03N 136.43E
109 I14 **Terni** *anc.* Interamna Nahars. Umbria, C Italy 42.34N 12.37E
111 X6 **Ternitz** Niederösterreich, E Austria 47.43N 16.01E
118 K6 **Ternivka** Dnipropetrovs'ka Oblast', E Ukraine 48.30N 36.05E
118 K6 **Ternopil' Pol.** Tarnopol, *Rus.* Ternopol'. Ternopil's'ka Oblast', W Ukraine 49.32N 25.37E
118 I6 **Ternopil's'ka Oblast'** *var.* Ternopil', *Rus.* Ternopol'skaya Oblast'. ♦ *province* NW Ukraine
   Ternopol'/Ternopol'skaya Oblast' *see* Ternopil'/Ternopil's'ka Oblast'
127 Oo15 **Terpeniya, Mys** *headland* Ostrov Sakhalin, SE Russian Federation 48.37N 144.40E
127 Oo15 **Terpeniya, Zaliv** *inlet* Ostrov Sakhalin, SE Russian Federation
24 V13 **Térraba, Río** *var.* Grande de Térraba, Río
8 J13 **Terrace** British Columbia, W Canada 54.34N 128.31W
3 J12 **Terrace Bay** Ontario, S Canada 48.46N 87.06W
109 I16 **Terracina** Lazio, C Italy 41.17N 13.13E
95 N7 **Terråk** Troms, N Norway 65.03N 12.22E
28 D11 **Terral** Oklahoma, C USA 33.55N 97.54W
110 I7 **Teufen** Sankt Gallen, NE Switzerland 47.24N 9.24E
109 H16 **Terralba** Sardegna, Italy, C Mediterranean Sea 39.47N 8.35E

80 H10 **Tersef** Chari-Baguirmi, C Chad 12.55N 16.49E
153 X8 **Terskey Ala-Too, Khrebet** ▲ Kazakhstan/Kyrgyzstan
   Terter *see* Tärtär
107 R8 **Teruel** *anc.* Turba. Aragón, E Spain 40.21N 1.06W
107 R8 **Teruel** ♦ *province* Aragón, E Spain
116 M7 **Tervel** *prev.* Kurtbunar, *Rom.* Curtbunar. Dobrich, NE Bulgaria 43.45N 27.25E
95 M16 **Tervo** Itä-Suomi, C Finland 62.57N 26.48E
94 L13 **Tervola** Lappi, NW Finland 66.04N 24.49E
   Tervueren *see* Tervuren
101 H18 **Tervuren** *var.* Tervueren. Vlaams Brabant, C Belgium 50.48N 4.28E
114 H11 **Tešanj** Federacija Bosna I Hercegovina, N Bosnia and Herzegovina 44.37N 18.00E
85 M19 **Tesenane** Inhambane, S Mozambique 22.48S 34.02E
82 I9 **Teseney** *var.* Tesseneï. W Eritrea 15.05N 36.42E
   Teshekpuk Lake ⊚ Alaska, USA
168 K6 **Teshig** Bulgan, N Mongolia
172 Q6 **Teshikaga** Hokkaidō, NE Japan 43.29N 144.27E
172 P2 **Teshio** Hokkaidō, NE Japan 44.49N 141.46E
172 P2 **Teshio-gawa** *var.* Tesio Gawa. ♒ Hokkaidō, NE Japan
172 P3 **Teshio-sanchi** ▲ Hokkaidō, NE Japan
   Tešín *see* Cieszyn
168 F5 **Tesiyn Gol** *var.* Tes-Khem. ♒ Mongolia/Russian Federation *see also* Tes-Khem
133 T7 **Tes-Khem** *var.* Tesiyn Gol. ♒ Mongolia/Russian Federation *see also* Tesiyn Gol
114 H11 **Teslić** Republika Srpska, N Bosnia and Herzegovina 44.35N 17.50E
8 I9 **Teslin** Yukon Territory, W Canada 60.12N 132.44W
8 I9 **Teslin** ♒ British Columbia/Yukon Territory, W Canada
79 Q8 **Tessalit** Kidal, NE Mali 20.11N 0.58E
79 S12 **Tessaoua** Maradi, S Niger 13.43N 7.59E
101 K17 **Tessenderlo** Limburg, NE Belgium 51.04N 5.04E
   Tesseneï *see* Teseney
12 L7 **Tessier, Lac** ⊚ Quebec, SE Canada
   Tessin *see* Ticino
99 M23 **Test** ♒ S England, UK
29 U3 **Testama** *see* Tõstamaa
57 P4 **Testigos, Islas los** *island group* N Venezuela
39 S10 **Tesuque** New Mexico, SW USA 35.45N 105.55W
105 O17 **Têt** *var.* Tet. ♒ S France
56 G5 **Tetas, Cerro de las** ▲ NW Venezuela 9.58N 73.00W
85 M15 **Tete** NE Mozambique 16.14S 33.34E
85 M15 **Tete** *off.* Província de Tete. ♦ *province* NW Mozambique
9 N15 **Tête Jaune Cache** British Columbia, SW Canada 52.52N 119.22W
192 O8 **Te Teko** Bay of Plenty, North Island, NZ 38.03S 176.48E
129 U15 **Tetepare** *island* New Georgia Islands, NW Solomon Islands
118 M5 **Teteriv** *Rus.* Teterev. ♒ N Ukraine
102 M9 **Teterow** Mecklenburg-Vorpommern, NE Germany 53.46N 12.34E
116 I9 **Teteven** Lovech, N Bulgaria 42.58N 24.16E
203 T10 **Tetiaroa** *atoll* Îles du Vent, W French Polynesia
107 P14 **Tetica de Bacares** ▲ S Spain 37.15N 2.31W
   Tetiyev *see* Tetiyiv
119 O6 **Tetiyiv** *Rus.* Tetiyev. Kyyivs'ka Oblast', N Ukraine 49.21N 29.37E
41 T10 **Tetlin** Alaska, USA 63.08N 142.31W
35 R8 **Teton River** ♒ Montana, NW USA
85 I20 **Tétouan** see Tetouan
74 G5 **Tétouan** *var.* Tetouan, Tetuán. N Morocco 35.33N 5.22W
114 V13 **Tetovo** *Alb.* Tetova, Tetovë, *Turk.* Kalkandelen. NW FYR Macedonia 42.01N 20.58E
192 M6 **Tetrázio** ▲ S Greece
   Tetschen *see* Děčín
   Tetuán *see* Tétouan
203 P8 **Tetufera, Mont** ▲ Tahiti, W French Polynesia 17.40S 149.25W
116 J13 **Tetyushi** Respublika Tatarstan, W Russian Federation 54.55N 48.46E
147 S9 **Thamūd** N Yemen 17.17N 49.57E
110 I7 **Teufen** Sankt Gallen, NE Switzerland 47.24N 9.24E
44 L12 **Teul** *var.* Teul de Gonzáles Ortega. Zacatecas, C Mexico 21.30N 103.28W
85 I20 **Thamaga** Kweneng, SE Botswana 24.40S 25.31E

99 L21 **Tewkesbury** C England, UK 51.58N 2.09W
121 F19 **Tewli** *Rus.* Tevli. Brestskaya Voblasts', SW Belarus 52.20N 24.13E
165 U12 **Têwo** *var.* Dêngkagoin. Gansu, C China 34.05N 103.15E
27 U3 **Texana, Lake** ⊚ Texas, SW USA
49 S14 **Texarkana** Arkansas, C USA 33.25N 94.03W
27 X5 **Texarkana** Texas, SW USA 33.25N 94.03W
27 N9 **Texas** *off.* State of Texas; also known as The Lone Star State. ♦ *state* S USA
27 W12 **Texas City** Texas, SW USA 29.22N 94.54W
43 P9 **Texcoco** México, C Mexico 19.31N 98.52W
100 I6 **Texel** *island* Waddeneilanden, NW Netherlands
28 H8 **Texhoma** Oklahoma, C USA 36.30N 101.46W
27 N1 **Texhoma** Texas, SW USA 36.30N 101.46W
39 W12 **Texico** New Mexico, SW USA 34.23S 103.03W
43 O13 **Texmelucan** *var.* San Martín Texmelucan. Puebla, S Mexico 19.13N 98.25W
49 O13 **Texoma, Lake** ⊚ Oklahoma/Texas, C USA
27 N9 **Texon** Texas, SW USA 31.13N 101.42W
126 I12 **Teya** Krasnoyarskiy Kray, C Russian Federation 60.27N 92.46E
85 J23 **Teyateyaneng** NW Lesotho
128 M16 **Teykovo** Ivanovskaya Oblast', W Russian Federation 56.49N 40.31E
128 M16 **Teza** ♒ W Russian Federation
43 Q13 **Teziutlán** Puebla, S Mexico 19.49N 97.22E
159 W12 **Tezpur** Assam, NE India 26.39N 92.47E
15 L8 **Tha-Anne** ♒ Nunavut, NE Canada
85 K23 **Thabana Ntlenyana** *var.* Thabantshonyana, Mount Ntlenyana. ▲ E Lesotho 29.26S 29.16E
   Thabantshonyana *see* Thabana Ntlenyana
85 J22 **Thaba Putsoa** ▲ C Lesotho 29.48S 27.46E
178 J6 **Tha Bo** Nong Khai, E Thailand 17.52N 102.34E
105 T14 **Thabor, Pic du** ▲ E France 45.07N 6.34E
   Tha Chin *see* Samut Sakhon
177 G7 **Thagaya** Pegu, C Myanmar 19.19N 96.16E
178 J10 **Thai, Ao** *see* Thailand, Gulf of
178 I6 **Thai Binh** Thai Binh, N Vietnam 20.27N 106.19E
178 J6 **Thai Hoa** Nghê An, N Vietnam 19.21N 105.26E
178 Hh10 **Thailand** *off.* Kingdom of Thailand, *Th.* Prathet Thai; *prev.* Siam. ♦ *monarchy* SE Asia
178 J10 **Thailand, Gulf of** *var.* Gulf of Siam, *Th.* Ao Thai, *Vtn.* Vinh Thai Lan. *gulf* SE Asia
   Thai Lan, Vinh *see* Thailand, Gulf of
178 J6 **Thai Nguyên** Bac Thai, N Vietnam 21.36N 105.49E
178 J9 **Thakhèk** *prev.* Muang Khammouan. Khammouan, C Laos 17.24N 104.50E
159 S13 **Thakurgaon** Rajshahi, NW Bangladesh 26.04N 88.34E
155 S6 **Thal** North-West Frontier Province, NW Pakistan 33.24N 70.31E
105 U16 **Thalang** Phuket, SW Thailand 08.00N 98.21E
   Thalassery *see* Tellicherry
178 H10 **Thalat Khae** Nakhon Ratchasima, C Thailand 15.15N 102.24E
110 I7 **Thalgau** Salzburg, NW Austria 47.49N 13.19E
108 G8 **Thalwil** Zürich, NW Switzerland 47.18N 8.34E
85 I20 **Thamaga** Kweneng, SE Botswana 24.40S 25.31E
   Thamarid *see* Thamarīt
147 V13 **Thamarīt** *var.* Thamarid, Thumrayt. SW Oman 17.39N 54.01E
147 P16 **Thamar, Jabal** ▲ SW Yemen 13.46N 45.32E
192 M6 **Thames** Waikato, North Island, NZ 37.10S 175.33E
12 D17 **Thames** ♒ Ontario, S Canada
99 O22 **Thames** ♒ S England, UK
192 M6 **Thames, Firth of** *gulf* North Island, NZ
12 D17 **Thamesville** Ontario, S Canada 42.33S 81.58W
147 R13 **Thamūd** N Yemen 17.17N 49.57E
178 Gg9 **Thanbyuzayat** Mon State, S Myanmar 15.58N 97.43E
158 I9 **Thānesar** Haryāna, NW India 29.58N 76.51E
178 J7 **Thanh Hoa** Thanh Hoa, N Vietnam 19.49N 105.48E
   Thanintari Taungdan *see* Bilauktaung Range
161 J21 **Thanjāvūr** *prev.* Tanjore. Tamil Nādu, SE India 10.46N 79.09E
   Thanlwin *see* Salween
105 U14 **Thann** Haut-Rhin, NE France 47.51N 7.04E
178 H16 **Tha Nong Phrom** Phatthalung, SW Thailand 7.24N 100.04E
178 H13 **Thap Sakae** *var.* Thap Sakau. Prachuap Khiri Khan, SW Thailand 11.30N 99.34E
   Thap Sakau *see* Thap Sakae
100 L10 **'t Harde** Gelderland, E Netherlands 52.25N 5.52E
144 G9 **Tharad** Gujarāt, W India 24.25N 71.40E
158 D11 **Thar Desert** *var.* Great Indian Desert, Indian Desert. *desert* India/Pakistan
189 V10 **Thargomindah** Queensland, C Australia 28.00S 143.47E
156 D11 **Thar Pärkar** *prov?* SE Pakistan
145 S7 **Tharthār al Furāt, Qanāt ath** *canal* C Iraq

◆ COUNTRY   ◇ DEPENDENT TERRITORY   ✦ ADMINISTRATIVE REGION   ▲ MOUNTAIN   ☒ VOLCANO   ⊚ LAKE
● COUNTRY CAPITAL   ○ DEPENDENT TERRITORY CAPITAL   ✕ INTERNATIONAL AIRPORT   ▲ MOUNTAIN RANGE   ♒ RIVER   ⊡ RESERVOIR

145 R7 **Tharthār, Buḥayrat ath** ☒ C Iraq
145 R5 **Tharthār, Wādī ath** *dry watercourse* N Iraq
178 Gg14 **Tha Sae** Chumphon, SW Thailand
178 H15 **Tha Sala** Nakhon Si Thammarat, SW Thailand 8.43N 99.54E
116 I13 **Thásos** Thásos, E Greece 40.46N 24.43E
117 I14 **Thásos** *island* E Greece
39 N14 **Thatcher** Arizona, SW USA 32.47N 109.46W
178 Ij5 **That Khê** *var.* Tráng Dinh. Lang Son, N Vietnam 22.15N 106.26E
178 Gg9 **Thaton** Mon State, S Myanmar 16.55N 97.19E
178 J9 **That Phanom** Nakhon Phanom, E Thailand 16.52N 104.41E
178 Ii10 **Tha Tum** Surin, E Thailand 15.18N 103.39E
105 P16 **Thau, Bassin de** *var.* Étang de Thau, *anc.* S France
Thau, Étang de *see* Thau, Bassin de
177 G3 **Thaungdut** Sagaing, N Myanmar 24.25N 94.45E
178 Gg8 **Thaungyin** *Th.* Mae Nam Moei. ↝ Myanmar/Thailand
178 J9 **Tha Uthen** Nakhon Phanom, E Thailand 17.31N 104.34E
111 W2 **Thaya** *var.* Dyje. ↝ Austria/Czech Republic *see also* Dyje
49 V8 **Thayer** Missouri, C USA 36.31N 91.34W
177 Ff7 **Thayetmyo** Magwe, C Myanmar 19.19N 95.10E
35 S15 **Thayne** Wyoming, C USA 42.54N 111.01W
177 G6 **Thazi** Mandalay, C Myanmar 20.49N 96.04E
Thebes *see* Thíva
46 L5 **The Carlton** *var.* Abraham Bay. Mayaguana, SE Bahamas 22.21N 72.56W
47 O14 **The Crane** *var.* Crane. S Barbados 13.06N 59.26W
34 I11 **The Dalles** Oregon, NW USA 45.36N 121.10W
30 M14 **Thedford** Nebraska, C USA 41.58N 100.34W
The Hague *see* 's-Gravenhage
114 **Theiss** *see* Tisa/Tisza
115 Jj6 **Thelon** ↝ Northwest Territories/Nunavut, N Canada
9 V15 **Theodore** Saskatchewan, S Canada 51.25N 103.01W
25 N8 **Theodore** Alabama, S USA 30.33N 88.10W
38 L13 **Theodore Roosevelt Lake** ☒ Arizona, SW USA
Theodosia *see* Feodosiya
Theophilo Ottoni *see* Teófilo Otoni
15 K13 **The Pas** Manitoba, C Canada 53.49N 101.09W
33 T14 **The Plains** Ohio, N USA 39.22N 82.07W
Thera *see* Thíra
180 H17 **Therese, Île** *island* Inner Islands, NE Seychelles
Therezina *see* Teresina
117 L20 **Thérma** Ikaría, Dodekánisos, Greece, Aegean Sea 37.37N 26.18E
Thermae Himerenses *see* Termini Imerese
Thermae Pannonicae *see* Baden
Thermaic Gulf/Thermaicus Sinus *see* Thermaïkós Kólpos
123 Gg10 **Thermaïkós Kólpos** *Eng.* Thermaic Gulf; *anc.* Thermaicus Sinus. *gulf* N Greece
Thermía *see* Kýthnos
117 L17 **Thérmis** Lésvos, E Greece 39.08N 26.32E
117 E18 **Thérmo** Dytikí Ellás, C Greece 38.32N 21.42E
35 V14 **Thermopolis** Wyoming, C USA 43.39N 108.12W
191 P10 **The Rock** New South Wales, SE Australia 35.18S 147.07E
205 O5 **Theron Mountains** ▲ Antarctica
117 E18 **Thespiés** Stereá Ellás, C Greece 38.18N 23.08E
117 E16 **Thessalía** *Eng.* Thessaly. ◊ *region* C Greece
2 C10 **Thessalon** Ontario, S Canada 46.15N 83.32W
117 G14 **Thessaloníki** *Eng.* Salonica, Salonika, *SCr.* Solun, *Turk.* Selânik. Kentrikí Makedonía, N Greece 40.37N 22.58E
117 G14 **Thessaloníki ✈** Kentrikí Makedonía, N Greece 40.30N 22.58E
Thessaly *see* Thessalía
86 B12 **Theta Gap** *undersea feature* E Atlantic Ocean
99 P20 **Thetford** E England, UK 52.25N 0.45E
13 R11 **Thetford-Mines** Quebec, SE Canada 46.07N 71.16W
115 K17 **Theth** *var.* Thethi. Shkodër, N Albania 42.25N 19.45E
Thethi *see* Theth
101 L20 **Theux** Liège, E Belgium 50.33N 5.48E
47 V9 **The Valley** ○ (Anguilla) E Anguilla 18.12N 63.00W
49 N10 **The Village** Oklahoma, C USA 35.33N 97.33W
57 W10 **The Woodlands** Texas, SW USA 30.09N 95.27E
Thiamis *see* Thýamis
Thian Shan *see* Tien Shan
24 J9 **Thibodaux** Louisiana, S USA 29.48N 90.49W
31 S3 **Thief Lake** ☒ Minnesota, N USA
31 S3 **Thief River** ↝ Minnesota, C USA
31 S3 **Thief River Falls** Minnesota, N USA 48.07N 96.10W
Thiele *see* La Thielle

157 K18 **Thikombia** *see* Cikobia
Tiladhunmathi Atoll *var.* Tiladunmati Atoll. *atoll* N Maldives
159 T11 **Thimphu** *var.* Thimbu; *prev.* Tashi Chho Dzong. ● (Bhutan) W Bhutan 27.28N 89.37E
94 H2 **Thingeyri** Vestfirdhir, NW Iceland 65.52N 23.28W
94 I3 **Thingvellir** Sudhurland, SW Iceland 64.15N 21.06W
197 J6 **Thio** Province Sud, C New Caledonia 21.37S 166.13E
105 T4 **Thionville** Ger. Diedenhofen. Moselle, NE France 49.22N 6.10E
117 K22 **Thíra** Thíra, Kykládes, Greece, Aegean Sea 36.25N 25.26E
117 K22 **Thíra** *prev.* Santorini, Santoríni, *anc.* Thera. *island* Kykládes, Greece, Aegean Sea
117 J22 **Thirasía** *island* Kykládes, Greece, Aegean Sea
99 M16 **Thirsk** N England, UK 54.06N 1.16W
12 F12 **Thirty Thousand Islands** *island* group Ontario, S Canada
Thiruvananthapuram *see* Trivandrum
97 F20 **Thisted** Viborg, NW Denmark 56.58N 8.42E
94 L1 **Thistil Fjord** *see* Thistilfjördhur
Thistilfjördhur *var.* Thistil Fjord. *fjord* NE Iceland
190 G9 **Thistle Island** *island* South Australia
Thithia *see* Cicia
179 N14 **Thitu Island** *island* NW Spratly Islands
Thiukhaoluang Phrahang *see* Luang Prabang Range
117 G18 **Thíva** *Eng.* Thebes; *prev.* Thívai. Stereá Ellás, C Greece 38.19N 23.19E
Thívai *see* Thíva
104 M12 **Thiviers** Dordogne, SW France 45.24N 0.54E
94 J4 **Thjórsá** ↝ C Iceland
15 L9 **Thlewiaza** ↝ Nunavut, NE Canada
15 J9 **Thoa** ↝ Northwest Territories, NW Canada
101 G14 **Tholen** Zeeland, SW Netherlands 51.31N 4.13E
101 F14 **Tholen** *island* SW Netherlands
28 L10 **Thomas** Oklahoma, C USA 35.44N 98.45W
23 T3 **Thomas** West Virginia, NE USA 39.09N 79.28E
49 U3 **Thomas Hill Reservoir** ☒ Missouri, C USA
25 S5 **Thomaston** Georgia, SE USA 32.53N 84.19W
21 R7 **Thomaston** Maine, NE USA 44.06N 69.10W
27 T12 **Thomaston** Texas, SW USA 28.56N 97.07W
25 O6 **Thomasville** Alabama, S USA 31.54N 87.42W
25 S8 **Thomasville** Georgia, SE USA 30.49N 83.57W
23 S9 **Thomasville** North Carolina, SE USA 35.52N 80.04W
37 N5 **Thomes Creek** ↝ California, W USA
9 W12 **Thompson** Manitoba, C Canada 55.45N 97.54W
31 R4 **Thompson** North Dakota, N USA 47.45N 97.07W
2 F8 **Thompson** ↝ Alberta/British Columbia, SW Canada
35 O8 **Thompson Falls** Montana, NW USA 47.36N 115.20W
31 Q10 **Thompson, Lake** ☒ South Dakota, N USA
36 M3 **Thompson Peak** ▲ California, W USA 41.00N 123.01W
49 S2 **Thompson River** ↝ Missouri, C USA
193 A22 **Thompson Sound** *sound* South Island, NZ
15 Hh1 **Thomsen** ↝ Banks Island, Northwest Territories, NW Canada
25 V4 **Thomson** Georgia, SE USA 33.28N 82.30W
105 T10 **Thonon-les-Bains** Haute-Savoie, E France 46.22N 6.30E
105 O15 **Thoré** *var.* Thore. ↝ S France
39 P11 **Thoreau** New Mexico, SW USA 35.24N 108.13W
Thorenburg *see* Turda
94 J3 **Thórisvatn** ☒ C Iceland
94 P4 **Thor, Kapp** *headland* S Svalbard 76.25N 25.01E
126 J14 **Thorlákshöfn** Sudhurland, SW Iceland 63.51N 21.24W
Thorn *see* Toruń
29 T10 **Thorndale** Texas, SW USA 30.36N 97.12W
12 H10 **Thorne** Ontario, S Canada 46.38N 79.04W
99 J14 **Thornhill** S Scotland, UK 55.13N 3.46W
27 U8 **Thornton** Texas, SW USA 31.24N 96.34W
Thornton Island *see* Millennium Island
12 H16 **Thorold** Ontario, S Canada 43.07N 79.15W
34 H1 **Thorp** Washington, NW USA 47.03N 120.40W
205 S3 **Thorshavnheiane** *physical region* Antarctica
Thórshöfn *see* Tórshavn
94 L1 **Thórshöfn** Nordhurland Eystra, NE Iceland 66.09N 15.18W
Thospitis *see* Van Gölü
178 J14 **Thôt Nôt** Cân Thơ, S Vietnam 10.16N 105.51E
165 U9 **Thouars** Deux-Sèvres, W France 46.58N 0.13W
159 X14 **Thoubal** Manipur, NE India 24.40N 94.00E
104 K9 **Thouet** ↝ W France
20 H7 **Thoune** *see* Thun
13 R11 **Thousand Islands** *island* Canada/USA
37 S15 **Thousand Oaks** California, W USA 34.10N 118.50W
116 L12 **Thrace** *cultural region* SE Europe
116 J13 **Thracian Sea** *Gk.* Thrakikó Pélagos; *anc.* Thracium Mare. *sea* Greece/Turkey

35 R11 **Thracian Mare/Thrakikó Pélagos** *see* Thracian Sea
Thrá Lí, Bá *see* Tralee Bay
9 Q16 **Three Forks** Montana, NW USA 45.53N 111.34W
191 N15 **Three Hills** Alberta, SW Canada 51.43N 113.15W
192 H1 **Three Hummock Island** *island* Tasmania, SE Australia
Three Kings Islands *island group* N NZ
183 P10 **Three Kings Rise** *undersea feature* W Pacific Ocean
79 O18 **Three Points, Cape** *headland* S Ghana 4.43N 2.03W
27 S13 **Three Rivers** Michigan, N USA 41.56N 85.37W
85 G24 **Three Rivers** Texas, SW USA 28.27N 98.10W
34 H13 **Three Sisters** Western Cape, South Africa 31.51S 23.04E
195 Z16 **Three Sisters** ▲ Oregon, NW USA 44.08N 121.46W
27 Q6 **Three Sisters Islands** *island group* SE Solomon Islands
Thrissur *see* Trichūr
188 M10 **Throssell, Lake** *salt lake* Western Australia
117 K25 **Thrýptis** ▲ Kríti, Greece, E Mediterranean Sea 35.06N 25.51E
178 Ij14 **Thu Dâu Môt** *var.* Phu Cương. Sông Bé, S Vietnam 10.58N 106.40E
178 Ij6 **Thu Do ✈** (Ha Nôi) Ha Nôi, N Vietnam 21.13N 105.46E
101 G21 **Thuin** Hainaut, S Belgium 50.21N 4.18E
155 Q12 **Thul** Sind, SE Pakistan 28.13N 68.49E
Thule *see* Qaanaaq
85 J18 **Thuli** *var.* Tuli. ↝ S Zimbabwe
110 D9 **Thumrayt** *var.* Thamarit
110 D9 **Thun** *Fr.* Thoune. Bern, W Switzerland 46.46N 7.37E
10 C12 **Thunder Bay** Ontario, S Canada 48.27N 89.12W
32 M1 **Thunder Bay** *lake bay* S Canada
33 R6 **Thunder Bay** *lake bay* Michigan, N USA
33 R6 **Thunder Bay River** ↝ Michigan, N USA
49 N11 **Thunderbird, Lake** ☒ Oklahoma, C USA
30 L8 **Thunder Butte Creek** ↝ South Dakota, N USA
110 D9 **Thuner See** ☒ C Switzerland
178 H16 **Thung Song** *var.* Cha Mai. Nakhon Si Thammarat, SW Thailand 8.10N 99.40E
110 H7 **Thur** ↝ N Switzerland
110 G6 **Thurgau** *Fr.* Thurgovie. ◊ *canton* NE Switzerland
Thurgovie *see* Thurgau
Thuringe *see* Thüringen
110 J7 **Thüringen** Vorarlberg, W Austria 47.12N 9.48E
103 J17 **Thüringen** *Eng.* Thuringia, *Fr.* Thuringe. ◊ *state* C Germany
103 J17 **Thüringer Wald** *Eng.* Thuringian Forest. ▲ C Germany
Thuringia *see* Thüringen
Thuringian Forest *see* Thüringer Wald
99 D19 **Thurles** *Ir.* Durlas. S Ireland 52.40N 7.49W
21 W2 **Thurmont** Maryland, NE USA 39.36N 77.22W
Thurø *see* Thurø By
97 H24 **Thurø By** *var.* Thurø. Fyn, C Denmark 55.03N 10.43E
98 J6 **Thurso** Quebec, SE Canada 45.36N 75.13W
204 I10 **Thurso** N Scotland, UK 58.34N 3.31W
110 I9 **Thurston Island** *island* Antarctica
110 I9 **Thusis** Graubünden, S Switzerland 46.40N 9.27E
117 C15 **Thýamis** *var.* Thiamis. ↝ W Greece
97 E21 **Thyborøn** *var.* Tyborøn. Ringkøbing, W Denmark 56.40N 8.12E
117 L20 **Thýmaina** *island* Dodekánisos, Greece, Aegean Sea
85 N15 **Thyolo** *var.* Cholo. Southern, S Malawi 16.03S 35.11E
191 U6 **Ti, Ba** *var.* Tashi, SE Australia 31.14S 151.51E
186 J14 **Tiandong** *var.* Pingma. Guangxi Zhuangzu Zizhiqu, S China 23.37N 107.06E
167 O3 **Tianjin** *var.* Tientsin. Tianjin Shi, E China 39.12N 117.06E
167 P3 **Tianjin Shi** *var.* Jin, Tianjin, T'ien-ching, Tientsin. ◊ *municipality* E China
155 I10 **Tianjun** *var.* Xinyuan. Qinghai, C China 37.16N 99.03E
166 J13 **Tianlin** *prev.* Leli. Guangxi Zhuangzu Zizhiqu, S China 24.27N 106.03E
Tian Shan *see* Tien Shan
165 W11 **Tianshui** Gansu, C China 34.33N 105.51E
156 I7 **Tianshuihai** Xinjiang Uygur Zizhiqu, W China 35.16N 79.30E
167 S10 **Tiantai** Zhejiang, SE China 29.11N 121.01E
166 J14 **Tianyang** Guangxi Zhuangzu Zizhiqu, S China 23.45N 106.54E
165 U9 **Tianzhu** *var.* Tianzhu Zangzu Zizhixian. Gansu, C China 37.01N 103.04E
Tianzhu Zangzu Zizhixian *see* Tianzhu
203 Q7 **Tiarei** Tahiti, W French Polynesia 17.31S 149.19W
76 J6 **Tiaret** *var.* Tihert. NW Algeria 35.23N 1.18E
191 N17 **Tiassalé** S Ivory Coast 5.54N 4.49W
197 Bb8 **Ti'avea** Upolu, SE Samoa 13.58S 171.30W
62 J11 **Tibagi** *var.* Tibají. Paraná, S Brazil 24.28S 50.49W

62 J10 **Tibagi, Rio** *var.* Rio Tibají. ↝ S Brazil
Tibají *see* Tibagi
Tibají, Rio *see* Tibagi, Rio
145 Q9 **Tibal, Wādī** *dry watercourse* S Iraq
56 G9 **Tibaná** Boyacá, C Colombia 5.19N 73.25W
81 F14 **Tibati** Adamaoua, N Cameroon 6.28N 12.37E
78 K15 **Tibé, Pic de** ▲ SE Guinea 8.39N 8.58W
Tiber *see* Tivoli, Italy
Tiber *see* Tevere, Italy
Tiberias *see* Teverya
144 G8 **Tiberias, Lake** *var.* Chinnereth, Sea of Bahr Tabariya, Sea of Galilee, *Ar.* Bahrat Tabariya, *Heb.* Yam Kinneret. ☒ N Israel
69 Q5 **Tibesti** *var.* Tibesti Massíf, *Ar.* Tibistī. ▲ N Africa
Tibesti Massif *see* Tibesti
Tibetan Autonomous Region *see* Xizang Zizhiqu
Tibet, Plateau of *see* Qingzang Gaoyuan
Tibistī *see* Tibesti
2 K7 **Tiblemont, Lac** ☒ Quebec, SE Canada
145 X9 **Ţīb, Nahr aţ** ↝ S Iraq
190 L14 **Tibooburra** New South Wales, SE Australia 29.24S 142.01E
97 N18 **Tibro** Västra Götaland, S Sweden 58.25N 14.10E
42 E5 **Tiburón, Isla** *var.* Isla del Tiburón. *island* NW Mexico
25 W14 **Tice** Florida, SE USA 26.40N 81.49W
78 K9 **Tichît** *var.* Tichitt. Tagant, C Mauritania 18.25N 9.31W
Tichitt *see* Tichît
110 G11 **Ticha, Yazovir** ☒ NE Bulgaria
108 D8 **Ticino** *Fr./Ger.* Tessin. ◊ *canton* S Switzerland
110 H11 **Ticino** *Fr./Ger.* Tessin. ↝ SW Switzerland
43 X12 **Ticul** Yucatán, SE Mexico 20.21N 89.29W
97 K18 **Tidaholm** Västra Götaland, S Sweden 58.12N 13.55E
128 I8 **Tiksha** Respublika Kareliya, NW Russian Federation 64.07N 32.31E
Tidjikdja *see* Tidjikja
78 J8 **Tidjikja** *var.* Tidjikdja; *prev.* Fort-Cappolani. Tagant, C Mauritania 18.30N 11.24W
Tidore *see* Soasiu
175 Ss7 **Tidore, Pulau** *island* E Indonesia
79 N16 **Tidra, Île** *see* Et Tidra
79 N16 **Tiébissou** *var.* Tiebissou. N Ivory Coast 7.10N 5.10W
169 V11 **Tiefa** Liaoning, NE China 42.25N 123.39E
110 I9 **Tiefencastel** Graubünden, S Switzerland 46.40N 9.33E
101 H15 **Tiegenhof** *see* Nowy Dwór Gdański
100 K13 **Tiel** Gelderland, C Netherlands 51.54N 5.04E
169 W7 **Tieli** Heilongjiang, NE China 46.57N 128.01E
169 V11 **Tieling** *var.* T'ieh-ling. Liaoning, NE China 42.19N 123.52E
31 Q14 **Tilden** Nebraska, C USA 42.03N 97.49W
12 L8 **Tilden** Texas, SW USA 28.26N 98.32W
78 M12 **Tilia** *var.* Tilemsi. ↝ C Mali
78 I6 **Tilimsen** *see* Tlemcen
64 J17 **Tilio Martius** *see* Toulon
79 R11 **Tillabéri** *var.* Tillabéry. Tillabéri, W Niger 14.12N 1.25E
79 R11 **Tillabéri** ◊ *department* SW Niger
34 F11 **Tillamook** Oregon, NW USA 45.27N 123.50W
34 F11 **Tillamook Bay** *inlet* Oregon, NW USA
155 J17 **Tillanchāng Dwip** *island* Nicobar Islands, India, NE Indian Ocean
97 N15 **Tillberga** Västmanland, C Sweden 59.52N 16.39E
Tillberg *see* Dyleń
25 S10 **Tillery, Lake** ☒ North Carolina, SE USA
79 T10 **Tillia** Tahoua, W Niger 16.13N 4.51E
25 N8 **Tillmans Corner** Alabama, S USA 30.35N 88.10W
2 F17 **Tillsonburg** Ontario, S Canada 42.51N 80.41W
91 N22 **Tilpa** New South Wales, SE Australia 30.56S 144.24E
191 N5 **Tilpa** New South Wales, SE Australia
Tilsit *see* Sovetsk
31 N13 **Tilton** Illinois, N USA 40.06N 87.39W
130 K7 **Tim** Kurskaya Oblast', W Russian Federation 51.39N 37.11E
56 G9 **Timaná** Huila, S Colombia 1.56N 75.57W
Timan Ridge *see* Timanskiy Kryazh
128 I4 **Timanskiy Kryazh** *Eng.* Timan Ridge. *ridge* NW Russian Federation 65.45N 58.36E
193 F22 **Timaru** Canterbury, South Island, NZ 44.23S 171.15E
96 F9 **Tingvoll** Møre og Romsdal, S Norway 62.55N 8.13E
Timashevo Samarskaya Oblast', W Russian Federation 53.22N 51.13E
130 K13 **Timashevsk** Krasnodarskiy Kray, SW Russian Federation 45.37N 38.57E
Timbaki/Timbákion *see* Tympáki
24 U7 **Timbalier Bay** *bay* Louisiana, S USA

161 J19 **Tirupati** Andhra Pradesh, E India 13.39N 79.25E
161 I20 **Tiruppattūr** Tamil Nādu, SE India 12.28N 78.31E
161 H21 **Tiruppūr** Tamil Nādu, SW India 11.04N 77.19E
161 I20 **Tiruvannāmalai** Tamil Nādu, SE India 12.13N 79.07E
114 L10 **Tisa** Ger. Theiss, Hung. Tisza, Rus. Tissa, Ukr. Tysa. ≈ SE Europe see also Tisza
Tischnowitz see Tišnov
9 U14 **Tisdale** Saskatchewan, S Canada 52.51N 104.01W
49 O13 **Tishomingo** Oklahoma, C USA 34.14N 96.40W
97 M17 **Tisnaren** ◎ S Sweden
113 F18 **Tišnov** Ger. Tischnowitz. Brněnský Kraj, SE Czech Republic 49.21N 16.24E
Tissa see Tisa/Tisza
76 J6 **Tissemsilt** N Algeria 35.37N 1.48E
159 S12 **Tista** ≈ NE India
114 L8 **Tisza** Ger. Theiss, Rom./Slvn./SCr. Tisa, Rus. Tissa, Ukr. Tysa. ≈ SE Europe see also Tisa
113 L23 **Tiszaföldvár** Jász-Nagykun-Szolnok, E Hungary 47.00N 20.16E
113 M22 **Tiszafüred** Jász-Nagykun-Szolnok, E Hungary 47.34N 20.45E
113 L23 **Tiszakécske** Bács-Kiskun, C Hungary 46.55N 20.04E
113 M21 **Tiszaújváros** prev. Leninváros. Borsod-Abaúj-Zemplén, NE Hungary 47.55N 21.03E
113 L23 **Tiszavasvári** Szabolcs-Szatmár-Bereg, NE Hungary 47.57N 21.24E
Titibu see Chichibu
59 I17 **Titicaca, Lake** ◎ Bolivia/Peru
202 H17 **Titikaveka** Rarotonga, S Cook Islands 21.16S 159.45W
160 M13 **Titilāgarh** Orissa, E India 20.18N 83.09E
174 Gg4 **Titiwangsa, Banjaran** ▲ Peninsular Malaysia
Titograd see Podgorica
Titose see Chitose
Titova Mitrovica see Kosovska Mitrovica
Titovo Užice see Užice
115 M18 **Titov Vrv** ▲ NW FYR Macedonia 41.58N 20.49E
96 F7 **Titran** Sør-Trøndelag, S Norway 63.40N 8.20E
33 Q8 **Tittabawassee River** ≈ Michigan, N USA
118 J13 **Titu** Dâmbovița, S Romania 44.40N 25.31E
81 M16 **Titule** Orientale, N Dem. Rep. Congo (Zaire) 3.19N 25.23E
25 X11 **Titusville** Florida, SE USA 28.34N 80.48W
20 C12 **Titusville** Pennsylvania, NE USA 41.36N 79.39W
78 G11 **Tivaouane** W Senegal 14.59N 16.50W
115 I17 **Tivat** Montenegro, SW Yugoslavia 42.25N 18.43E
12 E14 **Tiverton** Ontario, S Canada 44.15N 81.31W
99 J23 **Tiverton** SW England, UK 50.54N 3.30W
21 O12 **Tiverton** Rhode Island, NE USA 41.38N 71.10W
109 I15 **Tivoli** anc. Tíber. Lazio, C Italy 41.58N 12.45E
27 U13 **Tivoli** Texas, SW USA 28.26N 96.54W
176 W11 **Tiwarra** Irian Jaya, E Indonesia 2.54S 133.52E
147 Z8 **Tiwi** NE Oman 22.43N 59.20E
Tiworo see Tioro, Selat
176 Ww12 **Tiyo, Pegunungan** ▲ Irian Jaya, E Indonesia
43 Y11 **Tizimín** Yucatán, SE Mexico 21.10N 88.09W
76 K5 **Tizi Ouzou** var. Tizi-Ouzou. N Algeria 36.44N 4.06E
76 D8 **Tiznit** SW Morocco 29.43N 9.39W
115 I14 **Tjentište** Republika Srpska, SE Bosnia and Herzegovina 43.25N 18.42E
Tjepoe/Tjepu see Cepu
100 L7 **Tjeukemeer** ◎ N Netherlands
Tjiamis see Ciamis
Tjiandjoer see Cianjur
Tjilatjap see Cilacap
Tjiledoeg see Ciledug
97 F23 **Tjæreborg** Ribe, W Denmark 55.28N 8.34E
97 I18 **Tjörn** island S Sweden
94 O3 **Tjuvfjorden** fjord S Svalbard
Tkvarcheli see Tqvarch'eli
42 L8 **Tlahualilo** Durango, N Mexico 26.06N 103.25W
43 P14 **Tlalnepantla** México, C Mexico 19.37N 99.09W
43 Q13 **Tlapacoyán** Veracruz-Llave, E Mexico 19.57N 97.18W
43 P16 **Tlapa de Comonfort** Guerrero, S Mexico 17.33N 98.33W
42 L13 **Tlaquepaque** Jalisco, C Mexico 20.36N 103.19W
43 P14 **Tlascala** see Tlaxcala
43 P14 **Tlaxcala** var. Tlaxcala, Tlaxcala de Xicohténcatl. Tlaxcala, C Mexico 19.17N 98.15W
43 P14 **Tlaxcala** ◆ state S Mexico
Tlaxcala de Xicohténcatl see Tlaxcala
43 Q16 **Tlaxco** var. Tlaxco de Morelos. Tlaxcala, S Mexico 19.37N 98.07W
Tlaxco de Morelos see Tlaxco
43 Q16 **Tlaxiaco** var. Santa María Asunción Tlaxiaco. Oaxaca, S Mexico 17.18N 97.42W
76 I6 **Tlemcen** var. Tilimsen, Tlemsen. NW Algeria 34.52N 1.21W
Tlemsen see Tlemcen
144 L4 **Tlété Ouâte Rharbi, Jebel** ▲ N Syria
118 J7 **Tlumach** Ivano-Frankivs'ka Oblast', W Ukraine 48.53S 25.00E
131 P17 **Tlyarata** Respublika Dagestan, SW Russian Federation 42.10N 46.10E
118 K10 **Toaca, Vârful** prev. Vîrful Toaca. ▲ NE Romania 46.58N 25.55E
Toaca, Vîrful see Toaca, Vârful

197 C13 **Toak** Ambrym, C Vanuatu 16.21S 168.16E
180 J4 **Toamasina** var. Tamatave. Toamasina, E Madagascar 18.10S 49.22E
180 J4 **Toamasina** ◆ province E Madagascar
180 J4 **Toamasina** ✕ Toamasina, E Madagascar 18.10S 49.22E
23 X6 **Toano** Virginia, NE USA 37.22N 76.46W
203 U10 **Toau** atoll Îles Tuamotu, C French Polynesia
47 T6 **Toa Vaca, Embalse** ⊟ C Puerto Rico
64 K13 **Toay** La Pampa, C Argentina 36.43S 64.22W
165 R14 **Toba** Xizang Zizhiqu, W China 31.16N 97.37E
171 Hh6 **Toba** Mie, Honshū, SW Japan 34.55N 136.46E
173 Ff5 **Toba, Danau** ◎ Sumatera, W Indonesia
155 Q9 **Toba Kākar Range** ▲ NW Pakistan
175 T10 **Tobalai, Selat** strait Maluku, E Indonesia
175 Q9 **Tobamawu** Sulawesi, N Indonesia 1.16S 121.42E
107 U2 **Tobarra** Castilla-La Mancha, C Spain 38.36N 1.40W
155 U9 **Toba Tek Singh** Punjab, E Pakistan 30.54N 72.30E
175 T6 **Tobelo** Pulau Halmahera, E Indonesia 1.45N 127.58E
12 E12 **Tobermory** Ontario, S Canada 45.13N 81.39W
98 G10 **Tobermory** W Scotland, UK 56.37N 6.12W
172 Oo5 **Tōbetsu** Hokkaidō, NE Japan 43.12N 141.28E
188 M6 **Tobin Lake** ◎ Western Australia
9 U14 **Tobin Lake** ◎ Saskatchewan, C Canada
37 T4 **Tobin, Mount** ▲ Nevada, W USA 40.25N 117.28W
171 L10 **Tobi-shima** island C Japan
174 J11 **Toboali** Pulau Bangka, W Indonesia 2.57S 106.25E
150 M8 **Tobol** Kaz. Tobyl. Kostanay, N Kazakhstan 52.42N 62.36E
150 L8 **Tobol** Kaz. Tobyl. ≈ Kazakhstan/Russian Federation
125 F11 **Tobol'sk** Tyumenskaya Oblast', C Russian Federation 58.15N 68.12E
Tobruch/Tobruk see Ṭubruq
129 R3 **Tobseda** Nenetskiy Avtonomnyy Okrug, NW Russian Federation 68.37N 52.24E
Tobyl see Tobol
129 Q6 **Tobysh** ≈ NW Russian Federation
56 F10 **Tocaima** Cundinamarca, C Colombia 4.30N 74.37W
61 K16 **Tocantins** ◆ state C Brazil
61 K15 **Tocantins, Rio** ≈ N Brazil
25 T2 **Toccoa** Georgia, SE USA 34.34N 83.19W
171 K15 **Tochigi** var. Totigi. Tochigi, Honshū, S Japan 36.24N 139.42E
171 Kk15 **Tochigi** off. Tochigi-ken, var. Totigi. ◆ prefecture Honshū, S Japan
171 K13 **Tochio** var. Totio. Niigata, Honshū, C Japan 37.27N 139.00E
97 M18 **Töcksfors** Värmland, C Sweden 59.30N 11.49E
44 J5 **Tocoa** Colón, N Honduras 15.36N 86.01W
64 H4 **Tocopilla** Antofagasta, N Chile 22.06S 70.08W
64 I4 **Tocorpuri, Cerro de** ▲ Bolivia/Chile 22.25S 67.52W
191 O10 **Tocumwal** New South Wales, SE Australia 35.53S 145.35E
54 K4 **Tocuyo de La Costa** Falcón, NW Venezuela 11.02N 68.27W
158 K13 **Toda Rāisingh** Rājasthān, N India 26.01N 75.34E
108 H3 **Todi** Umbria, C Italy 42.46N 12.25E
110 G9 **Tödi** ▲ NE Switzerland 46.52N 8.53E
176 Uu9 **Todio** Irian Jaya, E Indonesia 0.46S 130.50E
172 N12 **Todoga-saki** headland Honshū, C Japan 39.33N 142.02E
61 P17 **Todos os Santos, Baía de** bay E Brazil
42 F10 **Todos Santos** Baja California Sur, W Mexico 23.26N 110.14W
35 B2 **Todos Santos, Bahía de** bay NW Mexico
Toeban see Tuban
158 E6 **Toekang Besi Eilanden** see Tukangbesi, Kepulauan
Töen see T'aoyüan
193 D25 **Toetoes Bay** bay South Island, NZ
9 Q14 **Tofield** Alberta, SW Canada 53.22N 112.39W
8 K17 **Tofino** Vancouver Island, British Columbia, SW Canada 49.04N 125.51W
201 X10 **Tofol** Kosrae, E Micronesia
97 J20 **Tofta** Halland, S Sweden 57.10N 12.19E
97 F23 **Tofte** Buskerud, S Norway 59.31N 10.33E
171 J15 **Toftlund** Sønderjylland, SW Denmark 55.12N 9.04E
200 U13 **Tofua** island Ha'apai Group, C Tonga
197 B10 **Toga** island Torres Islands, N Vanuatu
82 L12 **Togdheer** off. Gobolka Togdheer. ◆ region NW Somalia
Toghyzaq see Toguzak
171 Ii12 **Togi** Ishikawa, Honshū, SW Japan 37.06N 136.43E
168 D6 **Togiak** Alaska, USA 59.03N 160.31W
175 Qq8 **Togian, Kepulauan** island group C Indonesia
79 Q15 **Togo** off. Togolese Republic; prev. French Togoland. ◆ republic W Africa
168 F8 **Tögrög** Dzavhan-Gol, SW Mongolia 45.51N 95.04E

168 E7 **Tögrög** Hovd, W Mongolia 47.24N 92.06E
Togton-heyan see Tuotuoheyan
150 L7 **Toguzak** Kaz. Toghyzaq. ≈ Kazakhstan/Russian Federation
39 P10 **Tohatchi** New Mexico, SW USA 35.51N 108.45W
203 O7 **Tohiea, Mont** ◬ Moorea, W French Polynesia 17.33S 149.48W
95 O17 **Tohmajärvi** Itä-Suomi, E Finland 62.12N 30.19E
143 N14 **Tohma Çayı** ≈ C Turkey
95 L16 **Toholampi** Länsi-Suomi, W Finland 63.46N 24.15E
168 M10 **Tōhöm** Dornogovĭ, SE Mongolia 44.25N 108.18E
25 X12 **Tohopekaliga, Lake** ◎ Florida, SE USA
171 I17 **Toi** Shizuoka, Honshū, S Japan 34.55N 138.45E
202 H15 **Toi** N Niue 18.57S 169.51W
95 L19 **Toijala** Länsi-Suomi, W Finland 61.09N 23.51E
175 Qq9 **Toima** Sulawesi, N Indonesia 0.48S 122.21E
170 C17 **Toi-misaki** headland Kyūshū, SW Japan 31.21N 131.18E
175 Rr17 **Toineke** Timor, S Indonesia 10.06S 124.22E
37 U6 **Toiyabe Range** ▲ Nevada, W USA
Tojarví, Inis see Inishturk
Tojikiston, Jumhurii see Tajikistan
153 R12 **Tojikobod** Rus. Tadzhikabad. C Tajikistan 39.08N 70.45E
170 F13 **Tōjō** Hiroshima, Honshū, SW Japan 34.54N 133.15E
41 T10 **Tok** Alaska, USA 63.20N 142.59W
172 P5 **Tokachi-dake** ▲ Hokkaidō, NE Japan 43.24N 142.41E
172 P7 **Tokachi-gawa** var. Tokati Gawa. ≈ Hokkaidō, NE Japan
171 Hh6 **Tōkai** Aichi, Honshū, SW Japan 35.01N 136.51E
113 N21 **Tokaj** Borsod-Abaúj-Zemplén, NE Hungary 48.07N 21.25E
171 Jj13 **Tōkamachi** Niigata, Honshū, C Japan 37.08N 138.46E
193 D25 **Tokanui** Southland, South Island, NZ 46.33S 169.01E
82 I11 **Tokar** var. Ṭawkar. Red Sea, NE Sudan 18.27N 37.40E
142 L12 **Tokat** Tokat, N Turkey 40.19N 36.34E
142 L12 **Tokat** ◆ province N Turkey
169 X15 **Tŏkchŏk-gundo** island group NW South Korea
Tōke see Taka Atoll
202 J9 **Tokelau** ◇ NZ overseas territory W Polynesia
Tokerebes see Trebišov
Tokhtamyshbek see Tükhtamish
26 M6 **Tokio** Texas, SW USA 33.09N 102.31W
Tokio see Tōkyō
201 W14 **Toki Point** point NW Wake Island 19.19N 166.36E
Tokkuztara see Gongliu
153 V7 **Tokmak** Kir. Tokmok. Chuyskaya Oblast', N Kyrgyzstan 42.49N 75.18E
119 V9 **Tokmak** var. Velykyy Tokmak. Zaporiz'ka Oblast', SE Ukraine 47.13S 35.42E
Tokmok see Tokmak
192 Q8 **Tokomaru Bay** Gisborne, North Island, NZ 38.10S 178.18E
171 Hh16 **Tokoname** Aichi, Honshū, SW Japan 34.54N 136.49E
172 Q5 **Tokoro** Hokkaidō, NE Japan 44.06N 144.03E
192 M8 **Tokoroa** Waikato, North Island, NZ 38.14S 175.52E
172 Q6 **Tokoro-gawa** ≈ Hokkaidō, NE Japan
78 K14 **Tokounou** Haute-Guinée, C Guinea 9.43N 9.46W
40 M7 **Toksook Bay** Alaska, USA 60.33N 165.01W
Toksu see Xinhe
Toksum see Toksun
164 L6 **Toksun** var. Toksum. Xinjiang Uygur Zizhiqu, NW China 42.46N 88.45E
153 W7 **Toktogul** Talasskaya Oblast', NW Kyrgyzstan 41.51N 72.56E
153 T9 **Toktogul'skoye Vodokhranilishche** ⊟ W Kyrgyzstan
Tokmush see Tükhtamish
200 Ss12 **Toku** island Vava'u Group, N Tonga
172 Qq14 **Tokunoshima** Kagoshima, Tokuno-shima, SW Japan
172 Q14 **Tokuno-shima** island Nansei-shotō, SW Japan
170 Ff15 **Tokushima** var. Tokusima. Tokushima, Shikoku, SW Japan 34.04N 134.28E
170 F15 **Tokushima** off. Tokushima-ken, var. Tokusima. ◆ prefecture Shikoku, SW Japan
Tokusima see Tokushima
171 Jj16 **Tokuyama** Yamaguchi, Honshū, SW Japan 34.04N 131.48E
171 K16 **Tōkyō** var. Tokio. ● (Japan) Tōkyō, Honshū, S Japan 35.40N 139.45E
171 J15 **Tōkyō** off. Tōkyō-to. ◆ capital district Honshū, S Japan
151 O7 **Tokyrau** ≈ C Kazakhstan
155 O3 **Tokzār** Pash. Tukzār. Sar-e Pol, N Afghanistan 35.47N 66.28E
201 U12 **Tol** atoll Chuuk Islands, C Micronesia
87 O10 **Tôlañaro** prev. Faradofay, Fort-Dauphin. Toliara, SE Madagascar 25.01S 46.59E
79 N9 **Tolbo** Bayan-Ölgiy, W Mongolia 48.25N 90.13W
Tolbukhin see Dobrich
62 G1 **Toledo** Paraná, S Brazil 24.45S 53.41W
56 G8 **Toledo** Norte de Santander, N Colombia 7.16N 72.28W
107 N9 **Toledo** City. Cebu, C Philippines 10.23N 123.39E

107 N9 **Toledo** anc. Toletum. Castilla-La Mancha, C Spain 39.52N 4.01W
32 M14 **Toledo** Illinois, N USA 39.16N 88.15W
31 W13 **Toledo** Iowa, C USA 42.00N 92.34W
33 R11 **Toledo** Ohio, N USA 41.39N 83.33W
34 F12 **Toledo** Oregon, NW USA 44.37N 123.56W
34 Q9 **Toledo** Washington, NW USA 46.27N 122.49W
44 F4 **Toledo** ◆ district S Belize
106 M9 **Toledo** ◆ province Castilla-La Mancha, C Spain
27 Y7 **Toledo Bend Reservoir** ⊟ Louisiana/Texas, SW USA
106 M10 **Toledo, Montes de** ▲ C Spain
108 J12 **Tolentino** Marche, C Italy 43.08N 13.17E
Toletum see Toledo
96 H10 **Tolga** Hedmark, S Norway 62.25N 11.00E
164 J3 **Toli** Xinjiang Uygur Zizhiqu, NW China 45.55N 83.33E
180 N7 **Toliara** var. Tuléar; prev. Tuléar. Toliara, SW Madagascar 23.19S 43.40E
180 N7 **Toliara** ◆ province SW Madagascar
Toliary see Toliara
56 D11 **Tolima** off. Departamento del Tolima. ◆ province C Colombia
175 Pp7 **Tolitoli** Sulawesi, C Indonesia 1.04N 120.49E
97 L23 **Tollarp** Skåne, S Sweden 55.55N 14.00E
102 N9 **Tollense** ≈ NE Germany
102 N10 **Tollensesee** ◎ NE Germany
38 K13 **Tolleson** Arizona, SW USA 33.25N 112.15W
152 M13 **Tollimarjon** Rus. Talimardzhan. Qashqadaryo Wiloyati, S Uzbekistan 38.22N 65.31E
108 J6 **Tolmezzo** Friuli-Venezia Giulia, NE Italy 46.27N 13.01E
111 S11 **Tolmin** Ger. Tolmein, It. Tolmino. W Slovenia 46.12N 13.39E
Tolmino see Tolmin
113 J25 **Tolna** Ger. Tolnau. Tolna, S Hungary 46.25N 18.46E
113 I24 **Tolna** off. Tolna Megye. ◆ county SW Hungary
81 I20 **Tolo** Bandundu, W Dem. Rep. Congo (Zaire) 2.57S 18.35E
202 D12 **Toloke** Île Futuna, W Wallis and Futuna
32 M13 **Tolono** Illinois, N USA 39.59N 88.16W
107 Q3 **Tolosa** País Vasco, N Spain 43.09N 2.04W
Tolosa see Toulouse
175 Qq10 **Tolo, Teluk** bay Sulawesi, C Indonesia
41 R9 **Tolovana River** ≈ Alaska, USA
127 Oo10 **Tolstoy, Mys** headland E Russian Federation 59.12N 155.04E
65 G15 **Toltén** Araucanía, C Chile 39.13S 73.10W
56 E6 **Tolú** Sucre, NW Colombia 9.31N 75.34W
43 O14 **Toluca** var. Macouria. N French Guiana 05.00N 52.28W
201 N13 **Toluca** var. Toluca de Lerdo. México, S Mexico 19.19N 99.40W
Toluca de Lerdo see Toluca
43 O14 **Toluca, Nevado de** ▲ C Mexico 19.05N 99.45W
175 Qq10 **Tolvana** ≈ C Indonesia

152 L9 **Tomditow-Toghi** ▲ N Uzbekistan
64 G13 **Tomé** Bío Bío, C Chile 36.39S 72.53W
60 L12 **Tomé-Açu** Pará, NE Brazil 2.25S 48.09W
97 L23 **Tomelilla** Skåne, S Sweden 55.33N 14.00E
107 O10 **Tomelloso** Castilla-La Mancha, C Spain 39.09N 3.01W
12 H10 **Tomiko, Lake** ◎ Ontario, S Canada
79 N12 **Tominian** Ségou, C Mali 13.18N 4.39W
175 Pp8 **Tomini, Gulf of** var. Teluk Tomini; prev. Teluk Gorontalo. bay Sulawesi, C Indonesia
108 J12 **Tomini, Teluk** see Tomini, Gulf of
171 M14 **Tomioka** Fukushima, Honshū, C Japan 37.19N 140.57E
172 Jj15 **Tomioka** Gunma, Honshū, S Japan 36.15N 138.51E
126 L12 **Tommot** Respublika Sakha (Yakutiya), NE Russian Federation 58.56N 126.24E
175 Rr7 **Tomohon** Sulawesi, N Indonesia 1.19N 124.49E
115 L21 **Tomorrit, Mali i** ▲ S Albania 40.43N 20.12E
9 S17 **Tompkins** Saskatchewan, S Canada 50.03N 108.49W
22 K8 **Tompkinsville** Kentucky, S USA 36.42N 85.41W
175 Pp7 **Tompo** Sulawesi, N Indonesia 0.56N 120.16E
188 I8 **Tom Price** Western Australia 22.48S 117.48E
126 H13 **Tomsk** Tomskaya Oblast', C Russian Federation 56.30N 85.04E
126 Gg12 **Tomskaya Oblast'** ◆ province C Russian Federation
20 K16 **Toms River** New Jersey, NE USA 39.56N 74.09W
28 L12 **Tom Steed Reservoir** var. Tom Steed Lake. ⊟ Oklahoma, C USA
176 Vv10 **Tomu** Irian Jaya, E Indonesia 2.07S 133.01E
194 F13 **Tomu** ≈ W PNG
194 H6 **Tomur Feng** var. Pik Pobedy, Pobeda Peak. ▲ China/Kyrgyzstan see also Pobedy, Pik 42.02N 80.07E
201 N13 **Tomworahlang** Pohnpei, E Micronesia
43 O17 **Tonalá** Chiapas, SE Mexico 16.03N 93.43W
108 F6 **Tonale, Passo del** pass N Italy 46.16N 10.37E
171 I13 **Tonami** Toyama, Honshū, SW Japan 36.39N 136.57E
65 G15 **Tonantins** Amazonas, W Brazil 2.58S 67.30W
34 K6 **Tonasket** Washington, NW USA 48.41N 119.27W
57 Y9 **Tonate** var. Macouria. N French Guiana 05.00N 52.28W
43 O14 **Tonawanda** New York, NE USA 43.00N 78.51W
175 Rr7 **Tondano, Danau** ◎ Sulawesi, N Indonesia 1.19N 124.55E
106 H7 **Tondela** Viseu, N Portugal 40.31N 8.04W
97 F24 **Tønder** Ger. Tondern. Sønderjylland, SW Denmark 54.57N 8.52E
Tondern see Tønder
171 K16 **Tone-gawa** ≈ Honshū, S Japan
149 N4 **Tonekābon** var. Shahsavar, Tonkābon; prev. Shahsavār. Māzandarān, N Iran 36.49N 51.51E
Tonezh see Tonyezh
200 S15 **Tonga** off. Kingdom of Tonga, var. Friendly Islands. ◆ monarchy SW Pacific Ocean
183 R9 **Tonga** island group SW Pacific Ocean
85 K23 **Tongaat** KwaZulu/Natal, E South Africa 29.31S 31.09E
167 Q13 **Tong'an** var. Tong an. Fujian, SE China 24.43N 118.07E
49 Q4 **Tonganoxie** Kansas, C USA 39.06N 95.05W
44 Y13 **Tongass National Forest** reserve Alaska, USA
200 S14 **Tongatapu** island Tongatapu Group, S Tonga
200 R15 **Tongatapu Group** island group S Tonga
183 S9 **Tonga Trench** undersea feature S Pacific Ocean
166 N8 **Tongbai Shan** ▲ C China
167 P8 **Tongcheng** Anhui, SE China 31.16N 117.00E
166 L6 **Tongchuan** Shaanxi, C China 35.10N 109.03E
166 L12 **Tongdao** var. Tongdao Dongzu Zizhixian; prev. Shuangjiang. Hunan, S China 26.06N 109.46E
165 T11 **Tongde** Qinghai, C China

169 U10 **Tongliao** Nei Mongol Zizhiqu, N China 43.37N 122.15E
167 Q9 **Tongling** Anhui, E China 30.54N 117.51E
167 R9 **Tonglu** Zhejiang, SE China 29.49N 119.37E
197 D14 **Tongoa** island Shepherd Islands, S Vanuatu
64 G9 **Tongoy** Coquimbo, C Chile 30.20S 71.28W
166 L11 **Tongren** Guizhou, S China 27.43N 109.10E
165 T11 **Tongren** Qinghai, C China 35.51N 101.58E
Tongres see Tongeren
159 O11 **Tongsa** var. Tongsa Dzong. C Bhutan 27.33N 90.30E
Tongsa Dzong see Tongsa
165 P12 **Tongshan** see Xuzhou
166 I7 **Tongshi** Hainan, S China 18.37N 109.34E
98 I6 **Tongue** N Scotland, UK
35 X10 **Tongue River** ≈ Montana, NW USA
35 W11 **Tongue River Resevoir** ⊟ Montana, NW USA
165 V11 **Tongwei** Gansu, C China
165 W9 **Tongxin** Ningxia, N China 37.00N 105.41E
169 U9 **Tongyu** var. Tonggou. Jilin, NE China 44.49N 123.08E
166 J11 **Tongzi** Guizhou, S China 28.07N 106.49E
42 G5 **Tónichi** Sonora, NW Mexico 28.34N 109.33W
83 D14 **Tonj** Warab, SW Sudan 7.18N 28.40E
158 H13 **Tonk** Rājasthān, N India 26.10N 75.49E
178 Ii12 **Tônlé Sap** Eng. Great Lake. ◎ W Cambodia
104 L14 **Tonneins** Lot-et-Garonne, SW France 44.23N 0.21E
105 Q7 **Tonnerre** Yonne, C France 47.50N 4.00E
37 S8 **Tonopah** Nevada, W USA 38.04N 117.13W
170 Ff14 **Tonoshō** Okayama, Shōdo-shima, SW Japan 34.29N 134.10E
45 S7 **Tonosí** Los Santos, S Panama 7.23N 80.25W
97 F16 **Tønsberg** Vestfold, S Norway 59.16N 10.25E
41 T1 **Tonsina** Alaska, USA 61.39N 145.10W
97 D17 **Tonstad** Vest-Agder, S Norway 58.40N 6.42E
200 S14 **Tonumea** island Nomuka Group, W Tonga
143 O1 **Tonya** Trabzon, NE Turkey 40.52N 39.16E
121 K29 **Tonyezh** Rus. Tonezh. Homyel'skaya Voblasts', SE Belarus 51.49N 27.48E
38 L3 **Tooele** Utah, W USA 40.31N 112.18W
126 L15 **Toora-Khem** Respublika Tyva, S Russian Federation 52.25N 96.01E
191 O5 **Tooraweenah** New South Wales, SE Australia 30.29S 145.25E
85 K25 **Toorberg** ▲ S South Africa 32.02S 24.02E
120 G5 **Tootsi** Pärnumaa, SW Estonia 58.35N 24.46E
191 U3 **Toowoomba** Queensland, E Australia 27.34S 151.54E
49 Q4 **Topeka** state capital Kansas, C USA 39.06N 95.05W
113 M18 **Topla** Hung. Toplya. ≈ NE Slovakia
126 H14 **Topki** Kemerovskaya Oblast', S Russian Federation 55.12N 85.40E
118 J10 **Topliţa** Ger. Töplitz, Hung. Maroshévíz; prev. Topliţa Română, Hung. Oláh-Toplicza, Toplicza. Harghita, C Romania 46.57N 25.22E
Topliţa Română/Töplitz see Topliţa
Toplya see Topla
42 I8 **Topolobampo** Sinaloa, C Mexico 25.37N 109.02W
118 I13 **Topoloveni** Argeş, S Romania 44.49N 25.01E
116 L11 **Topolovgrad** prev. Kavakli. Khaskovo, S Bulgaria 42.06N 26.20E
Topolya see Bačka Topola
128 I6 **Topozero, Ozero** ◎ NW Russian Federation
34 J4 **Toppenish** Washington, NW USA 46.22N 120.18W
189 P4 **Top Springs Roadhouse** Northern Territory, N Australia 16.37S 131.49E
201 U11 **Tora** island Chuuk, C Micronesia
Toraigh see Tory Island
201 U11 **Tora Island Pass** passage Chuuk Islands, C Micronesia
149 U6 **Torbat-e Ḥeydarīyeh** var. Turbat-i-Haidari. Khorāsān, NE Iran 35.18N 59.12E
149 V7 **Torbat-e Jām** var. Turbat-i-Jam. Khorāsān, NE Iran 35.16N 60.36E
41 Q11 **Torbert, Mount** ▲ Alaska, USA 61.30N 152.15W
33 P6 **Torch Lake** ◎ Michigan, N USA
169 W11 **Torch River** ≈ Heilongjiang, NE China 46.00N 128.45E
169 X8 **Torch River** Jilin, NE China 41.43N 125.56E
97 K13 **Töre** Norrbotten, N Sweden 65.55N 22.40E
97 L17 **Töreboda** Västra Götaland, S Sweden 58.40N 14.07E
97 J21 **Torekov** Skåne, S Sweden 56.25N 12.34E
106 M9 **Torelló** Cataluña, NE Spain
149 O3 **Torell Land** physical region SW Svalbard
178 K7 **Torez** Donets'ka Oblast', SE Ukraine 48.02N 38.45E

103 N14 **Torgau** Sachsen, E Germany 51.34N 13.01E
Torgay Ústirti see Turgayskaya Stolovaya Strana
Torghay see Turgay
97 N22 **Torhamn** Blekinge, S Sweden 56.04N 15.49E
101 C17 **Torhout** West-Vlaanderen, W Belgium 51.04N 3.06E
108 B8 **Torino** Eng. Turin. Piemonte, NW Italy 45.03N 7.39E
172 Q13 **Tori-shima** island Izu-shotō, SE Japan
83 F16 **Torit** Eastern Equatoria, S Sudan 4.27N 32.31E
195 O11 **Toriu** New Britain, E PNG 4.39S 151.42E
154 M4 **Torkestān, Selseleh-ye Band-e** var. Bandi-i Turkistan. ▲ NW Afghanistan
106 L7 **Tormes** ≈ W Spain
Tornacum see Tournai
Torneå see Tornio
94 K12 **Torneå** | var. Torniojoki, Fin. Torniojoki. ≈ Finland/Sweden
11 O4 **Torneträsk** ◎ N Sweden
Torngat Mountains see Torngat Mountains
26 H8 **Tornillo** Texas, SW USA 31.26N 106.06W
94 K13 **Tornio** Swe. Torneå. Lappi, NW Finland 65.50N 24.17E
Torniojoki/Torniojoki see Tornealven
81 R23 **Tornquist** Buenos Aires, E Argentina 38.05S 62.13W
57 G6 **Toro** Castilla-León, N Spain 41.31N 5.24W
64 H9 **Toro, Cerro del** ▲ N Chile 29.10S 69.43W
79 R2 **Torodi** Tillabéri, SW Niger 13.05N 1.46E
195 S12 **Torokina** Bougainville Island, NE PNG 6.12S 155.04E
113 L23 **Törökszentmiklós** Jász-Nagykun-Szolnok, E Hungary 47.10N 20.25E
44 G7 **Torola, Río** ≈ El Salvador/Honduras
Toronaíos, Kólpos see Kassándras, Kólpos
12 H15 **Toronto** Ontario, S Canada 43.42N 79.25W
33 V12 **Toronto** Ohio, N USA 40.27N 80.36W
Toronto see Lester B.Pearson
49 P6 **Toronto Lake** ⊟ Kansas, C USA
37 V16 **Toro Peak** ▲ California, SW USA 33.31N 116.25W
128 H16 **Toropets** Tverskaya Oblast', W Russian Federation 56.29N 31.37E
83 G8 **Tororo** E Uganda 0.42N 34.12E
142 H16 **Toros Dağları** Eng. Taurus Mountains. ▲ S Turkey
191 N13 **Torquay** Victoria, SE Australia 38.21S 144.18E
99 J24 **Torquay** SW England, UK 50.28N 3.30W
106 M5 **Torquemada** Castilla-León, N Spain 42.02N 4.17W
37 S16 **Torrance** California, W USA 33.49N 118.19W
106 H8 **Torre, Alto da** ▲ C Portugal 40.21N 7.31W
109 L17 **Torre Annunziata** Campania, S Italy 40.45N 14.27E
107 S10 **Torreblanca** País Valenciano, E Spain 40.13N 0.12E
106 L15 **Torrecilla** ▲ S Spain 36.38N 4.54W
107 P6 **Torrecilla en Cameros** La Rioja, N Spain 42.18N 2.33W
106 M9 **Torredelcampo** Andalucía, S Spain 37.45N 3.52W
109 L17 **Torre del Greco** Campania, S Italy 40.46N 14.22E
106 I6 **Torre de Moncorvo** var. Moncorvo, Tôrre de Moncorvo. Bragança, N Portugal 41.10N 7.03W
106 J9 **Torrejoncillo** Extremadura, W Spain 39.54N 6.28W
107 O8 **Torrejón de Ardoz** Madrid, C Spain 40.27N 3.28W
107 N7 **Torrelaguna** Madrid, C Spain 40.49N 3.33W
106 M2 **Torrelavega** Cantabria, N Spain 43.21N 4.03W
109 L16 **Torremaggiore** Puglia, SE Italy 41.42N 15.17E
106 M15 **Torremolinos** Andalucía, S Spain 36.37N 4.30W
190 I6 **Torrens, Lake** salt lake South Australia
Torrent/Torrent de l'Horta see Torrente
107 S10 **Torrente** var. Torrent, Torrent de l'Horta. País Valenciano, E Spain 39.26N 0.28W
42 L8 **Torreón** Coahuila de Zaragoza, NE Mexico 25.47N 103.21W
107 R13 **Torre-Pacheco** Murcia, SE Spain 37.43N 0.57W
108 A8 **Torre Pellice** Piemonte, NE Italy 44.49N 7.12E
107 O13 **Torreperogil** Andalucía, S Spain 38.01N 3.16W
63 H15 **Torres** Rio Grande do Sul, S Brazil 29.19S 49.44W
Torrès, Îles see Torres Islands
197 B10 **Torres Islands** Fr. Îles Torrès. island group N Vanuatu
106 G9 **Torres Novas** Santarém, C Portugal 39.28N 8.31W
189 V1 **Torres Strait** strait Australia/PNG
106 F10 **Torres Vedras** Lisboa, C Portugal 39.04N 9.15W
107 S13 **Torrevieja** País Valenciano, E Spain 37.59N 0.41W
194 H11 **Torricelli Mountains** ▲ NW PNG
98 G8 **Torridon, Loch** inlet NW Scotland, UK
108 D9 **Torriglia** Liguria, NW Italy 44.31N 9.08E
106 M9 **Torrijos** Castilla-La Mancha, C Spain 39.58N 4.18W
20 L12 **Torrington** Connecticut, NE USA 41.48N 73.07W

◆ COUNTRY ◇ DEPENDENT TERRITORY ◆ ADMINISTRATIVE REGION ▲ MOUNTAIN ☒ VOLCANO ◎ LAKE
● COUNTRY CAPITAL ○ DEPENDENT TERRITORY CAPITAL ✕ INTERNATIONAL AIRPORT ▲ MOUNTAIN RANGE ≈ RIVER ⊟ RESERVOIR

35 Z15 **Torrington** Wyoming, C USA 42.04N 104.10W
95 F16 **Torröjen** var. Torrön. ◎ C Sweden
Torrön see Torröjen
107 N15 **Torrox** Andalucía, S Spain 36.45N 3.58W
96 N13 **Torsåker** Gävleborg, C Sweden 60.31N 16.30E
97 N21 **Torsås** Kalmar, S Sweden 56.24N 16.00E
97 J14 **Torsby** Värmland, C Sweden 60.07N 13.00E
97 N16 **Torshälla** Södermanland, C Sweden 59.25N 16.28E
97 B19 **Tórshavn Dan.** Thorshavn *Dependent territory capital* Faeroe Islands 62.02N 6.47W
Torshiz see Kāshmar
47 T9 **Tortola** *island* C British Virgin Islands
108 D9 **Tortona** *anc.* Dertona. Piemonte, NW Italy 44.54N 8.52E
109 L23 **Tórtoli** Sardegna, Italy, C Mediterranean Sea 38.01N 14.49E
107 U7 **Tortosa** *anc.* Dertosa. Cataluña, E Spain 40.49N 0.31E
Tortosa see Ţarţūs
107 U7 **Tortosa, Cap** *headland* E Spain 40.43N 0.52E
46 L8 **Tortue, Île de la** *var.* Tortuga Island. *island* N Haiti
57 Y10 **Tortue, Montagne** ▲ C French Guiana
Tortuga, Isla see La Tortuga, Isla
Tortuga Island see Tortue, Île de la
56 C11 **Tortugas, Golfo** *gulf* W Colombia
47 T5 **Tortuguero, Laguna** *lagoon* N Puerto Rico
143 Q12 **Tortum** Erzurum, NE Turkey 40.15N 41.30E
Torugart, Pereval see Turugart Shankou
143 O12 **Torul** Gümüşhane, NE Turkey 40.34N 39.18E
112 J10 **Toruń** *Ger.* Thorn. Toruń, Kujawsko-pomorskie, C Poland 53.01N 18.36E
97 K20 **Torup** Halland, S Sweden 56.57N 13.04E
120 I6 **Tõrva** *Ger.* Tõrwa. Valgamaa, S Estonia 58.00N 25.54E
Tõrwa see Tõrva
98 D13 **Tory Island** *Ir.* Toraigh. *island* NW Ireland
113 N19 **Torysa** *Hung.* Tarca. ♒ NE Slovakia
Tȍrzburg see Bran
128 J16 **Torzhok** Tverskaya Oblast', W Russian Federation 57.04N 34.55E
170 Ee15 **Tosa** Kōchi, Shikoku, SW Japan 33.28N 133.25E
170 E16 **Tosa-Shimizu** var. Tosasimizu. Kōchi, Shikoku, SW Japan 32.46N 132.55E
Tosasimizu see Tosa-Shimizu
170 Ee16 **Tosa-wan** *bay* SW Japan
85 H21 **Tosca** North-West, N South Africa 25.51S 23.56E
108 F12 **Toscana** *Eng.* Tuscany. ♦ *region* C Italy
109 E14 **Toscano, Archipelago** *Eng.* Tuscan Archipelago. *island group* C Italy
108 G10 **Tosco-Emiliano, Appennino** *Eng.* Tuscan-Emilian Mountains. ▲ C Italy
Tōsei see Tungshih
171 J18 **To-shima** *island* Izu-shotō, SE Japan
153 Q9 **Toshkent** *Eng./Rus.* Tashkent. ● (Uzbekistan) Toshkent Wiloyati, E Uzbekistan 41.19N 69.17E
153 Q9 **Toshkent** ✈ Toshkent Wiloyati, E Uzbekistan 41.13N 69.15E
153 P9 **Toshkent Wiloyati** *Rus.* Tashkentskaya Oblast'. ♦ *province* E Uzbekistan
128 H13 **Tosno** Leningradskaya Oblast', NW Russian Federation 59.34N 30.48E
165 Q10 **Toson Hu** ◎ C China
168 H6 **Tosontsengel** Dzavhan, NW Mongolia 48.42N 98.14E
Tosquduq Qumlari see Goshqudug Qum
107 U4 **Tossal de l'Orri** Llorri. ▲ NE Spain 42.24N 1.15E
63 A15 **Tostado** Santa Fe, C Argentina 29.14S 61.43W
120 F6 **Tõstamaa** *Ger.* Testama. Pärnumaa, SW Estonia 58.19N 23.58E
102 I10 **Tostedt** Niedersachsen, NW Germany 53.16N 9.42E
142 J11 **Tosya** Kastamonu, N Turkey 41.01N 34.01E
97 F15 **Totak** ◎ S Norway
107 R13 **Totana** Murcia, SE Spain 37.45N 1.30W
96 H13 **Toten** *physical region* S Norway
85 G18 **Toteng** Ngamiland, C Botswana 20.19S 22.57E
104 M3 **Tôtes** Seine-Maritime, N France 49.40N 1.02E
Totigi see Tochigi
Totio see Tochio
Totis see Tata
201 U13 **Totiw** *island* Chuuk, C Micronesia
129 N13 **Tot'ma** var. Totma. Vologodskaya Oblast', NW Russian Federation 59.58N 42.42E
Tot'ma see Sukhona
57 V9 **Totness** Coronie, N Suriname 5.51N 56.19W
44 C5 **Totonicapán** Totonicapán, W Guatemala 14.54N 91.18W
44 A2 **Totonicapán** *off.* Departamento de Totonicapán. ♦ *department* W Guatemala
63 B18 **Totoras** Santa Fe, C Argentina 32.34S 61.10W
197 K15 **Totoya** *island* S Fiji
191 Q7 **Tottenham** New South Wales, SE Australia 32.16S 147.23E
171 G23 **Tottori** Tottori, Honshū, SW Japan
170 FJ13 **Tottori** *off.* Tottori-ken. ♦ *prefecture* Honshū, SW Japan
78 I6 **Touâjîl** Tiris Zemmour, N Mauritania 22.03N 12.39W

78 L15 **Touba** W Ivory Coast 8.16N 7.40W
78 G11 **Touba** W Senegal 14.55N 15.53W
76 E7 **Toubkal, Jbel** ▲ W Morocco 31.00N 7.50W
34 K10 **Touchet** Washington, NW USA 46.03N 118.40W
105 P7 **Toucy** Yonne, C France 47.45N 3.18E
79 O12 **Tougan** W Burkina 13.06N 3.03W
78 L7 **Touggourt** NE Algeria 33.07N 6.04E
79 Q12 **Tougouri** N Burkina 13.22N 0.25W
78 J13 **Tougué** Moyenne-Guinée, NW Guinea 11.28N 11.48W
78 K12 **Toukoto** Kayes, W Mali 13.24N 9.52W
105 S5 **Toul** Meurthe-et-Moselle, NE France 48.40N 5.54E
78 L16 **Toulépleu** var. Touloblli. W Ivory Coast 6.37N 8.27W
167 S14 **Touliu** *Jap.* Taiwan 23.44N 120.27E
13 U3 **Toulnustouc** ♒ Quebec, SE Canada
Touloblli see Toulépleu
105 T16 **Toulon** *anc.* Telo Martius, Tilio Martius. Var, SE France 43.07N 5.55E
32 K12 **Toulon** Illinois, N USA 41.04N 89.54W
104 M15 **Toulouse** *anc.* Tolosa. Haute-Garonne, S France 43.36N 1.24E
104 M15 **Toulouse** ✈ Haute-Garonne, S France 43.38N 1.19E
79 N16 **Toumodi** C Ivory Coast 6.34N 5.01W
76 G9 **Tounassine, Hamada** *hill range* W Algeria
177 G7 **Toungoo** Pegu, C Myanmar 18.57N 96.25E
104 L8 **Touraine** *cultural region* C France
105 P1 **Tourane** see Đà Nẵng
106 F2 **Touriñán, Cabo** *headland* NW Spain 43.02N 9.20W
78 J6 **Tourine** Tiris Zemmour, N Mauritania 22.22N 11.49W
104 J3 **Tourlaville** Manche, N France 49.39N 1.34W
101 D19 **Tournai** var. Tournay, *Dut.* Doornik; *anc.* Tornacum. Hainaut, SW Belgium 50.36N 3.24E
104 L16 **Tournay** Hautes-Pyrénées, S France 43.10N 0.16E
Tournay see Tournai
105 R12 **Tournon** Ardèche, E France 45.04N 4.49E
105 R9 **Tournus** Saône-et-Loire, C France 46.33N 4.53E
61 Q14 **Touros** Rio Grande do Norte, E Brazil 5.10S 35.28W
104 L8 **Tours** *anc.* Caesarodunum, Turoni. Indre-et-Loire, C France 47.23N 0.42E
191 Q17 **Tourville, Cape** *headland* Tasmania, SE Australia 42.09S 148.20E
168 L8 **Töv** ♦ *province* C Mongolia
56 H7 **Tovar** Mérida, N W Venezuela 8.21N 71.45W
130 L5 **Tovarkovskiy** Tul'skaya Oblast', W Russian Federation 53.41N 38.18E
Tovil'-Dora see Tavildara
Tővis see Teiuş
143 V11 **Tovuz** *Rus.* Tauz. W Azerbaijan 40.58N 45.41E
172 N9 **Towada** Aomori, Honshū, C Japan 40.36N 141.11E
172 N9 **Towada-ko** var. Towada Ko. ◎ Honshū, C Japan
192 K3 **Towai** Northland, North Island, NZ 35.29S 174.06E
20 I7 **Towanda** Pennsylvania, NE USA 41.45N 76.25W
31 W4 **Tower** Minnesota, N USA 47.48N 92.16W
175 Pp8 **Towera** Sulawesi, N Indonesia 0.29S 120.01E
188 M13 **Tower Island** see Genovesa, Isla
37 U11 **Tower Peak** ▲ Western Australia 33.23S 123.27E
31 N3 **Towne Pass** California, W USA
35 R10 **Towner** North Dakota, N USA 48.20N 100.27W
189 X6 **Townsend** Montana, NW USA 46.19N 111.31W
175 Q10 **Townsville** Queensland, NE Australia 19.15S 146.49E
154 K4 **Towoeti Meer** see Towuti, Danau
23 X3 **Towraghoudi** Herāt, NW Afghanistan 35.12N 62.19E
175 Q11 **Towson** Maryland, NE USA 39.22N 76.33W
172 No6 **Towuti, Danau** *Dut.* Towoeti Meer. ◎ Sulawesi, C Indonesia
171 Ii13 **Toxkan He** see Ak-say
26 K9 **Toyah** Texas, SW USA 31.18N 103.47W
172 No6 **Tōya-ko** ◎ Hokkaidō, NE Japan
171 Ii13 **Toyama** Toyama, Honshū, SW Japan 36.41N 137.12E
171 Ii13 **Toyama** *off.* Toyama-ken. ♦ *prefecture* Honshū, SW Japan
171 Ii13 **Toyama-wan** *bay* SW Japan
170 F16 **Tōyo** Ehime, Shikoku, SW Japan 33.28N 133.13E
170 Ee14 **Tōyo** Kōchi, Shikoku, SW Japan 33.28N 134.15E
171 Hh16 **Toyohashi** var. Toyohasi. Aichi, Honshū, SW Japan 34.45N 137.22E
Toyohara see Yuzhno-Sakhalinsk
Toyohasi see Toyohashi
171 Hh16 **Toyokawa** Aichi, Honshū, SW Japan 34.48N 137.23E
171 Gg13 **Toyooka** Hyōgo, Honshū, SW Japan 35.33N 134.48E
172 Kk12 **Toyosaka** Niigata, Honshū, SW Japan 35.04N 137.09E
171 Hh13 **Toyoshina** Nagano, Honshū, SW Japan 36.18N 137.54E
171 Pp2 **Toyotomi** Hokkaidō, NE Japan 45.07N 141.45E
171 Dd12 **Toyoura** Yamaguchi, Honshū, SW Japan 34.09N 130.55E
Toytepa see Tuytepa
76 M6 **Tozeur** var. Tawzar. W Tunisia 34.00N 8.09E
41 Q8 **Tozi, Mount** ▲ Alaska, USA 65.45N 151.01W

143 Q9 **Tqvarch'eli** *Rus.* Tkvarcheli. NW Georgia 42.51N 41.42E
Tráblous see Tripoli
143 O11 **Trabzon** *Eng.* Trebizond; *anc.* Trapezus. Trabzon, NE Turkey 41.00N 39.43E
143 O11 **Trabzon** *Eng.* Trebizond. ♦ *province* NE Turkey
11 P13 **Tracadie** New Brunswick, SE Canada 47.31N 64.57W
13 O1 **Tracy** Quebec, SE Canada 45.59N 73.07W
37 O8 **Tracy** California, W USA 37.43N 121.27W
31 S10 **Tracy** Minnesota, N USA 44.14N 95.37W
22 K9 **Tracy City** Tennessee, S USA 35.15N 85.44W
108 D7 **Tradate** Lombardia, N Italy 45.43N 8.57E
86 F6 **Traena Bank** *undersea feature* E Norwegian Sea
31 W13 **Traer** Iowa, C USA 42.11N 92.28W
106 J16 **Trafalgar, Cabo de** *headland* SW Spain 36.10N 6.03W
Traiectum ad Mosam/Traiectum Tungorum see Maastricht
9 O17 **Tráigh Mhór** see Tramore
Trail British Columbia, SW Canada 49.04N 117.46W
60 B11 **Traíra, Serra do** ▲ NW Brazil
111 V5 **Traisen** Niederösterreich, NE Austria 48.03N 15.37E
111 W4 **Traisen** ♒ NE Austria
111 X4 **Traiskirchen** Niederösterreich, NE Austria 48.01N 16.18E
Trajani Portus see Civitavecchia
Trajectum ad Rhenum see Utrecht
121 H14 **Trakai** *Ger.* Traken, *Pol.* Troki. Trakai, SE Lithuania 54.39N 24.58E
Traken *Ir.* Trá Lí. SW Ireland 52.16N 9.42W
99 A20 **Tralee** *Ir.* Trá Lí. SW Ireland 52.16N 9.42W
99 A20 **Tralee Bay** *Ir.* Bá Thrá Lí. *bay* SW Ireland
Trá Lí see Tralee
Trälleborg see Trelleborg
Tralles see Aydın
63 J16 **Tramandaí** Rio Grande do Sul, S Brazil 30.01S 50.11W
110 C7 **Tramelan** Bern, W Switzerland 47.13N 7.07E
99 G18 **Tramore** *Ir.* Tráigh Mhór, Trá Mhór. S Ireland 52.10N 7.10W
97 I18 **Tranås** Jönköping, S Sweden 58.03N 15.00E
64 J7 **Trancas** Tucumán, N Argentina 26.10S 65.19W
106 I7 **Trancoso** Guarda, N Portugal 40.46N 7.21W
97 H22 **Tranebjerg** Århus, C Denmark 55.51N 10.36E
97 K19 **Tranemo** Västra Götaland, S Sweden 57.30N 13.19E
178 Gg16 **Trang** Trang, S Thailand 7.33N 99.36E
176 W14 **Trangan, Pulau** *island* Kepulauan Aru, E Indonesia
Tràng Dinh see Thất Khê
191 Q7 **Trangie** New South Wales, SE Australia 32.01S 147.58E
96 K12 **Trängslet** Dalarna, C Sweden 61.22N 13.43E
108 I7 **Trani** Puglia, SE Italy 41.16N 16.24E
63 F17 **Tranqueras** Rivera, NE Uruguay 31.13S 55.45W
65 G15 **Tranqui, Isla** *island* S Chile
41 V6 **Trans-Alaska pipeline** *oil pipeline* Alaska, USA
205 Q10 **Transantarctic Mountains** ▲ Antarctica
Transcarpathian Oblast see Zakarpats'ka Oblast'
Transylvania see Transylvania
Transilvaniei, Alpi see Carpaţii Meridionali
Transjordan see Jordan
180 L11 **Transkei Basin** *undersea feature* SW Indian Ocean
127 N17 **Trans-Siberian Railway** *Railroad* Russian Federation
Transsylvanische Alpen/Transylvanian Alps see Carpaţii Meridionali
96 K12 **Transtrand** Dalarna, C Sweden 61.06N 13.19E
118 G10 **Transylvania** *Eng.* Ardeal, Transilvania, *Ger.* Siebenbürgen, *Hung.* Erdély. *cultural region* NW Romania
178 Ji15 **Tra Ôn** Vinh Long, S Vietnam 9.58N 105.58E
109 H23 **Trapani** *anc.* Drepanum. Sicilia, Italy, C Mediterranean Sea 38.02N 12.31E
178 Ji12 **Trâpeăng Vêng** Kâmpóng Thum, C Cambodia 12.37N 104.58E
Trapezus see Trabzon
116 F9 **Trapoklovo** Sliven, C Bulgaria 42.40N 26.36E
191 P13 **Traralgon** Victoria, SE Australia 38.15S 146.35E
78 H9 **Trarza** ♦ *region* SW Mauritania
108 H12 **Trasimeno, Lago** *Eng.* Lake of Perugia, *anc.* Trasimenus. ◎ C Italy
Trasimeno, Lago see Trasimeno, Lago
97 J20 **Tråslövsläge** Halland, S Sweden 57.02N 12.18E
106 I6 **Trás-os-Montes** see Cucumbi
106 I6 **Trás-os-Montes e Alto Douro** *former province* N Portugal
178 H10 **Trat** *var.* Bang Phra. Trat, S Thailand 12.16N 102.30E
98 C8 **Trá Tholl, Inis** see Inishtrahull
Traí see Trogir
111 T4 **Traun** Oberösterreich, N Austria 48.14N 14.13E
111 S5 **Traun** ♒ N Austria
103 N23 **Traunreut** Bayern, SE Germany 47.58N 12.36E
111 S5 **Traun, Lake** see Traunsee
Traytepa see Tuytepa
76 M6 **Trautenau** see Trutnov

23 P11 **Travelers Rest** South Carolina, SE USA 34.58N 82.26W
190 L8 **Travellers Lake** *seasonal lake* New South Wales, SE Australia
33 P6 **Traverse City** Michigan, N USA 44.45N 85.37W
31 R7 **Traverse, Lake** ◎ Minnesota/South Dakota, N USA
193 I16 **Travers, Mount** ▲ South Island, NZ 42.01S 172.46E
9 P7 **Travers Reservoir** ⊟ Alberta, SW Canada
178 Ji15 **Tra Vinh** *var.* Phu Vinh. Tra Vinh, S Vietnam 9.57N 106.19E
27 S10 **Travis, Lake** ⊟ Texas, SW USA
114 H12 **Travnik** Federacija Bosna I Hercegovina, C Bosnia and Herzegovina 44.14N 17.40E
111 V11 **Trbovlje** *Ger.* Trifail. C Slovenia 46.09N 15.03E
25 V13 **Treasure Island** Florida, SE USA 27.46N 82.46W
Treasure State see Montana
195 S14 **Treasury Islands** *island group* NW Solomon Islands
108 D9 **Trebbia** *anc.* Trebia. ♒ NW Italy
102 N8 **Trebel** ♒ NE Germany
105 O16 **Trèbes** Aude, S France 43.12N 2.25E
Trebia see Trebbia
113 F18 **Třebíč** *Ger.* Trebitsch. Jihlavský Kraj, S Czech Republic 49.13N 15.52E
115 I16 **Trebinje** Republika Srpska, S Bosnia and Herzegovina 42.42N 18.19E
115 H16 **Trebišnica** ♒ S Bosnia and Herzegovina
113 N20 **Trebišov** *Hung.* Tőketerebes. Košický Kraj, E Slovakia 48.36N 21.44E
Trebitsch see Třebíč
Trebizond see Trabzon
Trebnitz see Trzebnica
111 V12 **Trebnje** SE Slovenia 45.54N 15.01E
113 D19 **Třeboň** *Ger.* Wittingau. Budějovický Kraj, S Czech Republic 49.00N 14.46E
106 J15 **Trebujena** Andalucía, S Spain 36.52N 6.10W
102 I7 **Treene** ♒ N Germany
111 S9 **Tree Planters State** see Nebraska
111 T6 **Treffen** Kärnten, S Austria 46.40N 13.51E
104 G5 **Tréguier** Côtes d'Armor, NW France 48.50N 3.12W
63 G18 **Treinta y Tres, y Tres, E** Uruguay 33.12S 54.19W
63 F18 **Treinta y Tres** ♦ *department* E Uruguay
116 F9 **Treklyanska Reka** ♒ W Bulgaria
175 R10 **Treko, Kepulauan** *island group* N Indonesia
104 K8 **Trélazé** Maine-et-Loire, NW France 47.27N 0.28W
65 K17 **Trelew** Chubut, SE Argentina 43.13S 65.15W
97 K23 **Trelleborg** var. Trälleborg. Skåne, S Sweden 55.22N 13.10E
115 P15 **Trem** ▲ SE Yugoslavia 43.10N 22.12E
13 N11 **Tremblant, Mont** ▲ Quebec, SE Canada 46.13N 74.34W
101 H17 **Tremelo** Vlaams Brabant, C Belgium 50.59N 4.34E
109 M15 **Tremiti, Isole** *island group* SE Italy
32 K12 **Tremont** Illinois, N USA 40.30N 89.31W
38 L1 **Tremonton** Utah, W USA 41.42N 112.09W
107 U4 **Tremp** Cataluña, NE Spain 42.09N 0.53E
32 J7 **Trempealeau** Wisconsin, N USA 44.01N 91.25W
13 P8 **Trenche** ♒ Quebec, SE Canada
13 O7 **Trenche, Lac** ◎ Quebec, SE Canada
117 D18 **Trichonída, Límni** ◎ C Greece
161 G22 **Trichūr** var. Thrissur. Kerala, SW India 10.31N 76.13E
191 O8 **Trida** New South Wales, SE Australia 33.02S 145.03E
27 S1 **Trident Peak** ▲ Nevada, W USA 41.52N 118.22W
63 A21 **Tridentum/Trient** see Trento
111 T6 **Trieben** Steiermark, SE Austria 47.30N 14.30E
99 N18 **Trent** ♒ C England, UK
108 F5 **Trentino-Alto Adige** *prev.* Venezia Tridentina. ♦ *region* N Italy
108 G6 **Trento** *Eng.* Trent, *Ger.* Trient; *anc.* Tridentum. Trentino-Alto Adige, N Italy 46.04N 11.07E
12 J15 **Trenton** Ontario, SE Canada 44.06N 77.36W
25 V10 **Trenton** Florida, SE USA 29.36N 82.49W
25 R1 **Trenton** Georgia, SE USA 34.52N 85.27W
33 S10 **Trenton** Michigan, N USA 42.08N 83.10W
29 W3 **Trenton** Missouri, C USA 40.04N 93.37W
30 M17 **Trenton** Nebraska, C USA 40.10N 101.00W
20 J11 **Trenton** *state capital* New Jersey, NE USA 40.13N 74.44W
23 W10 **Trenton** North Carolina, SE USA 35.03N 77.20W
22 G9 **Trenton** Tennessee, S USA 35.58N 88.56W
38 L1 **Trenton** Utah, W USA 41.53N 111.57W
Trentschin see Trenčín
Treptow an der Rega see Trzebiatów
63 C23 **Tres Arroyos** Buenos Aires, E Argentina 38.21S 60.16W
115 I15 **Três Cachoeiras** Rio Grande do Sul, S Brazil 29.21S 49.48W
45 O14 **Tres Cruces, Cerro** ▲ SE Mexico 15.28N 92.27W
103 N23 **Tres Cruces, Cordillera** ▲ W Bolivia
115 I14 **Treska** ♒ NW FYR Macedonia
115 I14 **Treskavica** ▲ SE Bosnia and Herzegovina

61 J20 **Três Lagoas** Mato Grosso do Sul, SW Brazil 20.46S 51.43W
42 H12 **Tres Marías, Islas** *island group* C Mexico
61 M19 **Tres Marias, Represa** ⊟ SE Brazil
61 F20 **Tres Montes, Península** *headland* S Chile 46.49S 75.29W
107 O3 **Trespaderne** Castilla-León, N Spain 42.46N 3.24W
62 G13 **Três Passos** Rio Grande do Sul, S Brazil 27.28S 53.55W
63 A23 **Tres Picos, Cerro** ▲ E Argentina 38.10S 61.54W
65 G17 **Tres Picos, Cerro** ▲ SW Argentina 42.22S 71.51W
62 I12 **Três Pinheiros** Paraná, S Brazil 25.25S 51.57W
61 J23 **Três Pontas** Minas Gerais, SE Brazil 21.35S 45.18W
62 P9 **Tres Puntas, Cabo** see Manabique, Punta
62 P9 **Três Rios** Rio de Janeiro, SE Brazil 22.06S 43.15W
Tres Tabernae see Saverne
Trestenberg/Trestendorf see Tăşnad
43 R15 **Tres Valles** Veracruz-Llave, SE Mexico 18.14N 96.03W
96 I17 **Tretten** Oppland, S Norway 61.19N 10.19E
Treuburg see Olecko
103 K21 **Treuchtlingen** Bayern, S Germany 48.57N 10.55E
102 N13 **Treuenbrietzen** Brandenburg, E Germany 52.06N 12.52E
97 F16 **Treungen** Telemark, S Norway 59.00N 8.34E
65 I17 **Trevelín** Chubut, SW Argentina 43.02S 71.27W
108 I13 **Treves/Trèves** see Trier
108 E7 **Trevi** Umbria, C Italy 42.52N 12.46E
111 V12 **Treviglio** Lombardia, N Italy 45.31N 9.34E
106 J4 **Trevinca, Peña** ▲ NW Spain 42.10N 6.49W
107 P3 **Treviño** Castilla-León, N Spain 42.45N 2.41W
108 I7 **Treviso** *anc.* Tarvisium. Veneto, NE Italy 45.40N 12.15E
99 G24 **Trevose Head** *headland* SW England, UK 50.33N 5.03W
191 P17 **Triabunna** Tasmania, SE Australia 42.33S 147.55E
23 W4 **Triangle** Virginia, NE USA 38.30N 77.17W
85 L18 **Triangle** Masvingo, SE Zimbabwe 20.58S 31.28E
117 L23 **Tría Nísia** *island* Kykládes, Greece, Aegean Sea
99 G24 **Triberg** see Triberg im Schwarzwald
175 R10 **Triberg im Schwarzwald** var. Triberg. Baden-Württemberg, SW Germany 48.07N 8.13E
159 V15 **Tribhuvan** ✈ (Kathmandu) Central, C Nepal
56 C9 **Tribugá, Golfo de** *gulf* W Colombia
189 V8 **Tribulation, Cape** *headland* Queensland, NE Australia 16.14S 145.48E
110 M8 **Tribulaun** ▲ SW Austria 46.59N 11.18E
9 U17 **Tribune** Saskatchewan, S Canada 49.16N 103.50W
28 H5 **Tribune** Kansas, C USA 38.28N 101.45W
109 N18 **Tricarico** Basilicata, S Italy 40.37N 16.09E
109 Q20 **Tricase** Puglia, SE Italy 39.56N 18.21E
13 O6 **Trichonída, Límni** see Trichonída, Límni
Trichinopoly see Tiruchchirāppalli
117 D18 **Trichonída, Límni** ◎ C Greece
161 G22 **Trichūr** var. Thrissur. Kerala, SW India 10.31N 76.13E
191 O8 **Trida** New South Wales, SE Australia 33.02S 145.03E
27 S1 **Trident Peak** ▲ Nevada, W USA 41.52N 118.22W
Tridentum/Trient see Trento
111 T6 **Trieben** Steiermark, SE Austria 47.30N 14.30E
103 D19 **Trier** *Eng.* Treves, *Fr.* Trèves; *anc.* Augusta Treverorum. Rheinland-Pfalz, SW Germany 49.45N 6.39E
108 K7 **Trieste** *Slvn.* Trst. Friuli-Venezia Giulia, NE Italy 45.39N 13.45E
108 J8 **Trieste, Golfo di/Triest, Golf von** see Trieste, Gulf of
76 H5 **Trieste, Gulf of** *Cro.* Tršćanski Zaljev, *Ger.* Golf von Triest, *It.* Golfo di Trieste, *Slvn.* Tržaški Zaliv. *gulf* S Europe
111 W4 **Triesting** ♒ W Austria
Trifail see Trbovlje
111 S10 **Trifeşti** Iaşi, NE Romania 47.08N 27.32E
111 S10 **Triglav** *It.* Tricorno. ▲ NW Slovenia 46.22N 13.40E
106 I14 **Trigueros** Andalucía, S Spain 37.24N 6.49W
117 E16 **Tríkala** *prev.* Trikkala. Thessalía, C Greece 39.33N 21.46E
117 E17 **Trikeríotis** ♒ C Greece
Trikkala see Tríkala
Trikomo/Tríkomon see Iskele
99 F17 **Trim** *Ir.* Baile Átha Troim. E Ireland 53.34N 6.50W
110 E7 **Trimbach** Solothurn, NW Switzerland 47.21N 7.49E
111 Q5 **Trimmelkam** Oberösterreich, N Austria 48.02N 12.58E
31 U11 **Trimont** Minnesota, C USA 43.45N 94.42W
Trimontium see Plovdiv
Trinacria see Sicilia
161 K24 **Trincomalee** var. Trinkomali, Eastern Province, NE Sri Lanka 8.34N 81.13E
45 X18 **Tres Cruces, Cerro** ▲ SE Mexico 15.28N 92.27W
49 V13 **Trindade Spur** *undersea feature* W Atlantic Ocean
4 U12 **Trina** California, W USA 35.46N 117.21W

59 M16 **Trinidad** Beni, N Bolivia 14.52S 64.54W
56 H9 **Trinidad** Casanare, E Colombia 5.25N 71.39W
46 E6 **Trinidad** Sancti Spíritus, C Cuba 21.48N 80.00W
39 U8 **Trinidad** Colorado, C USA 37.10N 104.31W
63 E19 **Trinidad** Flores, S Uruguay 33.34S 56.54W
47 Y17 **Trinidad** *island* C Trinidad and Tobago
47 Y16 **Trinidad and Tobago** *off.* Republic of Trinidad and Tobago. ♦ *republic* SE West Indies
65 E23 **Trinidad, Golfo** *gulf* S Chile
109 N16 **Trinitapoli** Puglia, SE Italy 41.22N 16.06E
57 X10 **Trinité, Montagnes de la** ▲ C French Guiana
11 U12 **Trinity Bay** *inlet* Newfoundland, E Canada
41 P15 **Trinity Islands** *island group* Alaska, USA
37 N2 **Trinity Mountains** ▲ California, W USA
37 S4 **Trinity Peak** ▲ Nevada, W USA 40.13N 118.43W
37 S5 **Trinity Range** ▲ Nevada, W USA
37 N2 **Trinity River** ♒ California, W USA
27 V8 **Trinity River** ♒ Texas, SW USA
181 Y15 **Triolet** NW Mauritius 20.04S 57.31E
109 O20 **Trinitò, Capo** *headland* SE Italy 39.37N 16.46E
118 L11 **Trotuş** ♒ E Romania
46 M8 **Trou-du-Nord** N Haiti 19.34N 71.57W
27 W7 **Troup** Texas, SW USA 32.08N 95.07W
15 H8 **Trout** ♒ Northwest Territories, NW Canada
25 N8 **Trout Creek** Montana, NW USA 47.51N 115.40W
34 H10 **Trout Lake** Washington, NW USA 45.59N 121.33W
10 I9 **Trout Lake** ◎ Ontario, S Canada
35 T12 **Trout Peak** ▲ Wyoming, C USA 44.36N 109.32W
104 L4 **Trouville** Calvados, N France 49.21N 0.07E
99 L22 **Trowbridge** S England, UK 51.19N 2.13W
25 Q6 **Troy** Alabama, S USA 31.48N 85.58W
49 Q3 **Troy** Kansas, C USA 39.46N 95.05W
49 W4 **Troy** Missouri, C USA 38.58N 90.58W
20 L10 **Troy** New York, NE USA 42.73N 73.37W
23 S10 **Troy** North Carolina, SE USA 35.23N 79.58W
33 R13 **Troy** Ohio, N USA 40.02N 84.12W
27 T9 **Troy** Texas, SW USA 31.12N 97.18W
116 I9 **Troyan** Lovech, N Bulgaria 42.53N 24.43E
116 I9 **Troyanski Prokhod** *pass* N Bulgaria 42.48N 24.38E
151 N6 **Troyebratskiy** Severnyy Kazakhstan, N Kazakhstan 54.21N 66.07E
105 Q5 **Troyes** *anc.* Augustobona Tricassium. Aube, N France 48.18N 4.04E
119 X5 **Troyits'ke** Luhans'ka Oblast', E Ukraine 49.55N 38.18E
37 W7 **Troy Peak** ▲ Nevada, W USA 38.18N 115.27W
115 G15 **Trpanj** Dubrovnik-Neretva, S Croatia 43.00N 17.18E
Trščanski Zaljev see Trieste, Gulf of
Trst see Trieste
115 N14 **Trstenik** Serbia, C Yugoslavia 43.38N 21.01E
130 I6 **Trubchevsk** Bryanskaya Oblast', W Russian Federation 52.33N 33.45E
Trubchular see Orlyak
39 S10 **Truchas Peak** ▲ New Mexico, SW USA 35.58N 105.39W
149 P16 **Trucial Coast** *physical region* C UAE
Trucial States see United Arab Emirates
37 Q6 **Truckee** California, W USA 39.18N 120.10W
37 R5 **Truckee River** ♒ Nevada, W USA
131 Q13 **Trudfront** Astrakhanskaya Oblast', SW Russian Federation 45.56N 47.42E
12 I9 **Truite, Lac à la** ◎ Quebec, SE Canada
44 K4 **Trujillo** Colón, NE Honduras 15.55N 86.00W
58 C12 **Trujillo** La Libertad, NW Peru 8.04S 79.02W
106 K10 **Trujillo** Extremadura, W Spain 39.28N 5.53W
56 I6 **Trujillo** N W Venezuela 9.19N 70.37W
56 I6 **Trujillo** *off.* Estado Trujillo. ♦ *state* W Venezuela
Truk see Chuuk
Truk Islands see Chuuk Islands
31 N10 **Truman** Minnesota, C USA 43.49N 94.26W
49 X10 **Trumann** Arkansas, C USA 35.40N 90.30W
38 J9 **Trumbull, Mount** ▲ Arizona, SW USA 36.22N 113.09W
116 F9 **Trŭn** Pernik, W Bulgaria 42.51N 22.37E
191 Q8 **Trundle** New South Wales, SE Australia 32.55S 147.43E
133 U13 **Trung Phân** *physical region* S Vietnam
11 Q15 **Truro** Nova Scotia, SE Canada 45.16N 63.12W
99 H25 **Truro** SW England, UK 50.16N 5.03W
25 W3 **Truscott** Texas, SW USA 33.43N 99.48W
118 K9 **Truşeşti** Botoşani, NE Romania 47.45N 27.01E

◆ COUNTRY ◇ DEPENDENT TERRITORY ▲ ADMINISTRATIVE REGION ▲ MOUNTAIN ▲ VOLCANO ◎ LAKE
● COUNTRY CAPITAL ◉ DEPENDENT TERRITORY CAPITAL ✕ INTERNATIONAL AIRPORT ▲ MOUNTAIN RANGE ♒ RIVER ⊟ RESERVOIR

339

118 H6 **Truskavets'** L'viv'ska Oblast', W Ukraine 49.15N 23.30E
97 H22 **Trustrup** Århus, C Denmark 56.20N 10.46E
8 M11 **Trutch** British Columbia, W Canada 57.42N 123.00W
39 Q14 **Truth Or Consequences** New Mexico, SW USA 33.07N 107.15W
113 F15 **Trutnov** Ger. Trautenau. Hradecký Kraj, NE Czech Republic 50.34N 15.52E
105 P13 **Truyère** ✍ C France
116 K9 **Tryavna** Lovech, N Bulgaria 42.52N 25.30E
30 M14 **Tryon** Nebraska, C USA 41.31N 100.56W
96 I11 **Trysilelva** ✍ S Norway
114 D10 **Tržac** Federacija Bosna I Hercegovina, NW Bosnia and Herzegovina 44.58N 15.48E
Tržaski Zaliv see Trieste, Gulf of
112 G10 **Trzcianka** Ger. Schönlanke. Piła, Wielkopolskie, C Poland 53.01N 16.24E
112 E7 **Trzebiatów** Ger. Treptow an der Rega. Zachodniopomorskie, NW Poland 54.04N 15.14E
113 G14 **Trzebnica** Ger. Trebnitz. Dolnośląskie, SW Poland 51.18N 17.03E
111 T10 **Tržič** Ger. Neumarktl. NW Slovenia 46.22N 14.17E
Trzynietz see Třinec
Tsabong see Tshabong
168 G7 **Tsagaanchuluut** Dzavhan, C Mongolia 47.06N 96.40E
169 P7 **Tsagaanders** Dornod, NE Mongolia 48.03N 114.16E
169 S8 **Tsagaannuur** Dornod, E Mongolia 47.30N 118.45E
168 G8 **Tsagaan-Olom** Govĭ-Altay, C Mongolia 46.42N 96.30E
168 J8 **Tsagaan-Ovoo** Övörhangay, C Mongolia 45.57N 101.25E
168 D5 **Tsagaantüngi** Bayan-Ölgiy, W Mongolia 49.06N 90.26E
131 P12 **Tsagan Aman** Respublika Kalmykiya, SW Russian Federation 47.37N 46.43E
25 V11 **Tsala Apopka Lake** ☒ Florida, SE USA
Tsamkong see Zhanjiang
Tsangpo see Brahmaputra
168 L9 **Tsant** Dundgovĭ, C Mongolia 46.16N 106.55E
85 G17 **Tsau** Ngamiland, NW Botswana 20.08S 22.29E
180 I4 **Tsaratanana** Mahajanga, C Madagascar 16.46S 47.40E
116 N10 **Tsarevo** prev. Michurin. Burgas, E Bulgaria 42.10N 27.51E
Tsarigrad see İstanbul
Tsaritsyn see Volgograd
128 G13 **Tsarskoye Selo** prev. Pushkin. Leningradskaya Oblast', NW Russian Federation 59.42N 30.24E
119 T7 **Tsarychanka** Dnipropetrovs'ka Oblast', E Ukraine 48.56N 34.29E
85 H21 **Tsatsu** Southern, S Botswana 25.21S 24.45E
83 J20 **Tsavo** Coast, S Kenya 2.58S 38.28E
85 E21 **Tsawisis** Karas, S Namibia 26.18S 18.07E
Tschakathurn see Čakovec
Tschaslau see Čáslav
Tschenstochau see Częstochowa
Tschernembl see Črnomelj
30 K6 **Tschida, Lake** ☒ North Dakota, N USA
Tschorna see Mustvee
85 I17 **Tsebanana** Central, NE Botswana 19.50S 26.29E
Tsefat see Zefat
168 G8 **Tseel** Govĭ-Altay, SW Mongolia 45.45N 95.54E
130 M13 **Tselina** Rostovskaya Oblast', SW Russian Federation 46.31N 41.01E
Tselinograd see Astana
Tselinogradskaya Oblast' see Akmola
168 J6 **Tsengel** Hövsgöl, N Mongolia 49.29N 101.09E
168 E7 **Tsenher** Hovd, W Mongolia 47.07N 92.04E
152 E12 **Tsentral'nyye Nizmennyye Garagumy** Turkm. Mençezi Garagum. desert C Turkmenistan
85 E21 **Tses** Karas, S Namibia 25.54S 18.09E
85 G14 **Tseshebi** ✍ C Namibia
168 E7 **Tsetsegnuur** Hovd, W Mongolia 46.36N 93.16E
168 J7 **Tsetserleg** Arhangay, C Mongolia 47.28N 101.19E
79 R16 **Tsévié** S Togo 6.25N 1.13E
85 G21 **Tshabong** var. Tsabong. Kgalagadi, SW Botswana 26.01S 22.24E
85 G20 **Tshane** Kgalagadi, SW Botswana 24.02S 21.54E
Tshangalele, Lac see Lufira, Lac de Retenue de la
85 H17 **Tshauxaba** Central, C Botswana 19.56S 25.09E
81 F21 **Tshela** Bas-Congo, W Dem. Rep. Congo 4.55S 13.01E
81 K22 **Tshibala** Kasai Occidental, S Dem. Rep. Congo 6.53S 22.01E
81 J22 **Tshibala** Kasai Occidental, SW Dem. Rep. Congo 6.23S 20.47E
81 L22 **Tshilenge** Kasai Oriental, S Dem. Rep. Congo 6.16S 23.48E
81 L24 **Tshimbalanga** Katanga, S Dem. Rep. Congo 9.42S 23.04E
81 L22 **Tshimbulu** Kasai Oriental, S Dem. Rep. Congo 6.27S 22.54E
Tshiumbe see Chiumbe
81 M21 **Tshofa** Kasai Oriental, S Dem. Rep. Congo 5.13S 25.13E
81 K18 **Tshuapa** ✍ C Dem. Rep. Congo (Zaire)
Tshwane see Pretoria
116 G7 **Tsibritsa** ✍ NW Bulgaria
116 I12 **Tsigansko Gradishte** ▲ Bulgaria/Greece 41.24N 24.41E

14 G3 **Tsiigehtchic** prev. Arctic Red River. Northwest Territories, NW Canada 67.24N 133.40W
129 Q7 **Tsil'ma** ✍ NW Russian Federation
121 J17 **Tsimkavichy** Rus. Timkovichi. Minskaya Voblasts', C Belarus 53.04N 26.58E
130 M11 **Tsimlyansk** Rostovskaya Oblast', SW Russian Federation 47.39N 42.05E
131 N11 **Tsimlyanskoye Vodokhranilishche** var. Tsimlyansk Vodoskhovshche, Eng. Tsimlyansk Reservoir. ☒ SW Russian Federation
131 **Tsimlyansk Reservoir** see Tsimlyanskoye Vodokhranilishche
131 **Tsimlyansk Vodoskhovshche** see Tsimlyanskoye Vodokhranilishche
Tsinan see Jinan
Tsing Hai see Qinghai Hu, China
Tsinghai see Qinghai, China
Tsingtao/Tsingtau see Qingdao
Tsingyuan see Baoding
Tsinkiang see Quanzhou
Tsintao see Qingdao
85 D17 **Tsintsabis** Otjikoto, N Namibia 18.44S 17.57E
180 H8 **Tsiombe** var. Tsihombe. Toliara, S Madagascar
126 Kk14 **Tsipa** ✍ S Russian Federation
180 H5 **Tsiribihina** ✍ W Madagascar
180 I5 **Tsiroanomandidy** Antananarivo, C Madagascar 18.43S 46.01E
201 U13 **Tsis** island Chuuk, C Micronesia
Tsitsihar see Qiqihar
131 Q3 **Tsivil'sk** Chuvashskaya Respublika, W Russian Federation 55.51N 47.30E
143 T9 **Ts'khinvali** prev. Staliniri. C Georgia 42.12N 43.58E
121 J19 **Tsna** ✍ SW Belarus
128 L15 **Tsna** var. Zna. ✍ W Russian Federation
168 K11 **Tsoohor** Ömnögovĭ, S Mongolia 43.15N 104.04E
171 H16 **Tsu** var. Tu. Mie, Honshū, SW Japan 34.40N 136.30E
171 K13 **Tsubame** var. Tubame. Niigata, Honshū, C Japan 37.39N 138.55E
171 Ii3 **Tsubata** Ishikawa, Honshū, SW Japan 36.43N 136.42E
172 Q6 **Tsubetsu** Hokkaidō, NE Japan 43.43N 144.01E
171 Kk16 **Tsuchiura** var. Tutiura. Ibaraki, Honshū, S Japan 36.03N 140.09E
172 N7 **Tsugaru-kaikyō** strait N Japan
171 Kk13 **Tsugawa** Niigata, Honshū, C Japan 37.40N 139.26E
172 Oo5 **Tsukigata** Hokkaidō, NE Japan 43.18N 141.37E
170 Dd15 **Tsukumi** var. Tukumi. Ōita, Kyūshū, SW Japan 33.02N 131.51E
168 E5 **Tsul-Ulaan** Bayan-Ölgiy, W Mongolia 48.51N 91.13E
85 D17 **Tsumeb** Otjikoto, N Namibia 19.13S 17.42E
85 F17 **Tsumkwe** Otjozondjupa, NE Namibia 19.35S 20.26E
170 Cc16 **Tsuno** Miyazaki, Kyūshū, SW Japan 32.43N 131.32E
171 H14 **Tsuruga** var. Turuga. Fukui, Honshū, SW Japan 35.38N 136.01E
170 F15 **Tsurugi-san** ▲ Shikoku, SW Japan 33.51N 134.04E
170 Dd15 **Tsurumi-zaki** headland Kyūshū, SW Japan 32.55N 132.03E
171 L11 **Tsuruoka** var. Turuoka. Yamagata, Honshū, C Japan 38.43N 139.48E
171 Hh15 **Tsushima** var. Tusima. Aichi, Honshū, SW Japan 35.10N 136.45E
170 C10 **Tsushima** var. Tsushima-tō, Tusima. island group SW Japan
Tsushima-tō see Tsushima
170 E12 **Tsuwano** Shimane, Honshū, SW Japan 34.28N 131.43E
170 Ff13 **Tsuyama** var. Tuyama. Okayama, Honshū, SW Japan 35.03N 133.57E
85 G19 **Tswaane** Ghanzi, SW Botswana
121 N16 **Tsyakhtsin** Rus. Tekhtin. Mahilyowskaya Voblasts', E Belarus
121 P19 **Tsyerakhowka** Rus. Terekhovka. Homyel'skaya Voblasts', SE Belarus 52.13N 31.24E
121 J17 **Tsyeshawlya** Rus. Cheshevlya, Tseshevlya. Brestskaya Voblasts', SW Belarus 53.13N 25.49E
119 R10 **Tsyurupyns'k** Rus. Tsyurupyns'k. Khersons'ka Oblast', S Ukraine 46.34N 32.42E
Tu see Tsu
194 H13 **Tua** ✍ C PNG
Tuaim see Tuam
192 L6 **Tuakau** Waikato, North Island, NZ 37.16S 174.56E
99 C17 **Tuam** Ir. Tuaim. W Ireland 53.31N 8.49W
193 K14 **Tuamarina** Marlborough, South Island, NZ 41.27S 174.00E
175 R13 **Tuamba** see Tuambe
Tuamotu, Archipel des see Tuamotu, Îles
199 N10 **Tuamotu Fracture Zone** tectonic feature E Pacific Ocean
203 W9 **Tuamotu, Îles** var. Archipel des Tuamotu, Dangerous Archipelago, Tuamotu Islands. island group N French Polynesia
Tuamotu Islands see Tuamotu, Îles
183 X10 **Tuamotu Ridge** undersea feature C Pacific Ocean
178 Ii5 **Tuan Giao** Lai Châu, N Vietnam 21.34N 103.24E
179 P8 **Tuao** Luzon, N Philippines 17.42N 121.25E
202 B15 **Tuapa** Niue 18.56S 169.58W
45 N7 **Tuapi** Región Autónoma Atlántico Norte, NE Nicaragua 14.10N 83.18W
130 K15 **Tuapse** Krasnodarskiy Kray, SW Russian Federation 44.07N 39.07E
175 Nn2 **Tuaran** Sabah, East Malaysia 6.12N 116.12E
106 I2 **Tua, Rio** ✍ N Portugal

198 B7 **Tuasivi** Savai'i, C Samoa 13.37S 172.07W
193 B24 **Tuatapere** Southland, South Island, NZ 46.09S 167.43E
38 M9 **Tuba City** Arizona, SW USA 36.08N 111.14W
144 H11 **Ţūbah, Qaşr aţ** castle Ma'ān, C Jordan 31.32N 36.39E
Tubame see Tsubame
174 U14 **Tuban** prev. Toeban. Jawa, C Indonesia 6.55S 112.01E
147 O16 **Tuban, Wādī** dry watercourse SW Yemen
63 K14 **Tubarão** Santa Catarina, S Brazil 28.29S 49.00W
100 O10 **Tubbergen** Overijssel, E Netherlands 52.25N 6.46E
Tubeke see Tubize
103 H22 **Tübingen** var. Tuebingen. Baden-Württemberg, SW Germany 48.31N 9.04E
131 W6 **Tubinsk** Respublika Bashkortostan, W Russian Federation 52.48N 58.18E
101 G19 **Tubize** Dut. Tubeke. Wallon Brabant, C Belgium 50.43N 4.14E
78 J16 **Tubmanburg** NW Liberia 6.50N 10.53W
179 Qq15 **Tubod** Mindanao, S Philippines 7.58N 123.46E
77 T7 **Ţubruq** Eng. Tobruk, It. Tobruch. NE Libya 32.04N 23.58E
203 T13 **Tubuai** island Îles Australes, SW French Polynesia
Tubuai, Îles/Tubuai Islands see Australes, Îles
42 F3 **Tubutama** Sonora, NW Mexico 30.51N 111.31W
56 K4 **Tucacas** Falcón, N Venezuela 10.46N 68.19W
61 P16 **Tucano** Bahia, E Brazil 10.52S 38.48W
59 P19 **Tucavaca, Río** ✍ E Bolivia
112 H8 **Tuchola** Kujawsko-pomorskie, C Poland 53.36N 17.49E
113 M17 **Tuchów** Małopolskie, SE Poland 49.53N 21.04E
25 S3 **Tucker** Georgia, SE USA 33.53N 84.10W
49 W10 **Tuckerman** Arkansas, C USA 35.43N 91.12W
28 B12 **Tucker's Town** E Bermuda 32.19N 64.42W
38 M15 **Tucson** Arizona, SW USA 32.13N 111.00W
64 J7 **Tucumán** off. Provincia de Tucumán. ◆ province N Argentina
Tucumán see San Miguel de Tucumán
39 V11 **Tucumcari** New Mexico, SW USA 35.10N 103.43W
60 H13 **Tucunaré** Pará, N Brazil 5.15S 55.49W
57 Q6 **Tucupita** Delta Amacuro, NE Venezuela 9.01N 62.04W
60 K13 **Tucuruí, Represa de** ☒ NE Brazil
112 F9 **Tuczno** Zachodniopomorskie, NW Poland 53.12N 16.08E
Tuddo see Tudu
107 Q5 **Tudela** Basq. Tutera; anc. Tutela. Navarra, N Spain 42.04N 1.37W
106 M6 **Tudela de Duero** Castilla-León, N Spain 41.35N 4.34W
144 K6 **Tudmur** var. Tadmur, Tamar. Gk. Palmyra; Bibl. Tadmor. Ḥimṣ, C Syria 34.36N 38.15E
120 J4 **Tudu** Lääne-Virumaa, NE Estonia 59.12N 26.52E
126 H16 **Tuekta** Respublika Altay, S Russian Federation 50.51N 85.52E
106 I5 **Tuela, Rio** ✍ N Portugal
159 X12 **Tuensang** Nāgāland, NE India 26.16N 94.45E
142 L15 **Tufanbeyli** Adana, C Turkey 38.16N 36.13E
194 M15 **Tufi** Northern, S PNG 9.04S 149.15E
199 L3 **Tufts Plain** undersea feature N Pacific Ocean
69 V14 **Tugela** ✍ E South Africa
23 P6 **Tug Fork** ✍ S USA
41 P15 **Tugidak Island** island Trinity Islands, Alaska, USA
179 P8 **Tuguegarao** Luzon, N Philippines 17.36N 121.47E
127 N13 **Tugur** Khabarovskiy Kray, SE Russian Federation 53.43N 137.00E
167 O13 **Tui** var. Grand Balé. ✍ W Burkina
59 J16 **Tuichi** ✍ W Bolivia
201 Q11 **Tuineje** Fuerteventura, Islas Canarias, Spain, NE Atlantic Ocean 28.18N 14.03W
45 X16 **Tuira, Río** ✍ SE Panama
Tuisarkan see Tūysarkān
191 Q10 **Tujiabu** see Yongxiu
131 W5 **Tukan** Respublika Bashkortostan, W Russian Federation 53.58N 57.29E
175 R13 **Tukangbesi, Kepulauan** Dut. Toekang Besi Eilanden. island group C Indonesia
153 V13 **Tükhtamish** Rus. Toktomush, prev. Tokhtamyshbek. SE Tajikistan 37.51N 74.41E
192 I12 **Tükrah** NE Libya 32.28N 20.36E
14 G2 **Tuktoyaktuk** Northwest Territories, NW Canada
173 Fj6 **Tuktuk** Pulau Samosir, W Indonesia 2.32N 98.42E
120 E9 **Tukums** Ger. Tuckum. Tukums, W Latvia 56.58N 23.12E
83 G24 **Tukuyu** prev. Neu-Langenburg. Mbeya, S Tanzania 9.13S 33.39E
161 K21 **Tükzār** var. Tokzār ▲
Tula de Allende see Tula

165 N10 **Tulage** see Tula
195 X15 **Tulaghi** var. Tulagi. Florida Islands, C Solomon Islands 9.04S 160.09E
Tulagi see Tulaghi
43 P13 **Tulancingo** Hidalgo, C Mexico 20.34N 98.24W
37 R11 **Tulare** California, W USA 36.12N 119.21W
37 P9 **Tulare** South Dakota, N USA 44.43N 98.29W
37 Q12 **Tulare Lake Bed** salt flat California, W USA
39 S14 **Tularosa** New Mexico, SW USA 33.04N 106.01W
39 P13 **Tularosa Mountains** ▲ New Mexico, SW USA
39 S15 **Tularosa Valley** basin New Mexico, SW USA
8 E25 **Tulbagh** Western Cape, SW South Africa 33.16S 19.09E
58 C5 **Tulcán** Carchi, N Ecuador 0.44N 77.43W
119 N13 **Tulcea** Tulcea, E Romania 45.11N 28.48E
119 N13 **Tulcea** ◆ county SE Romania
119 N7 **Tul'chyn** Rus. Tul'chin. Vinnyts'ka Oblast', C Ukraine 48.40N 28.48E
35 O1 **Tulélake** California, W USA 41.57N 121.30W
125 J10 **Tulghes** Hung. Gyergyótölgyes. Harghita, C Romania 46.57N 25.46E
N120 **Tul'govichi** see Tul'havichy
N120 **Tul'havichy** Rus. Tul'govichi. Homyel'skaya Voblasts', SE Belarus 51.45N 29.41E
27 N4 **Tulia** Texas, SW USA 34.32N 101.45W
25 Gg6 **Tulita** prev. Fort Norman, Norman. Northwest Territories, NW Canada 64.55N 125.25W
22 J10 **Tullahoma** Tennessee, S USA 35.21N 86.12W
191 N12 **Tullamarine** ✈ (Melbourne) Victoria, SE Australia 37.40S 144.46E
191 Q7 **Tullamore** New South Wales, SE Australia 32.39S 147.35E
99 E18 **Tullamore** Ir. Tulach Mhór. C Ireland 53.16N 7.30W
105 O12 **Tulle** anc. Tutela. Corrèze, C France 45.16N 1.46E
111 X3 **Tulln** var. Oberhollabrunn. Niederösterreich, NE Austria 48.19N 16.01E
111 W4 **Tulln** ✍ NE Austria
99 D20 **Tullos** Louisiana, S USA 31.48N 92.19W
99 F19 **Tullow** Ir. An Tullach. SE Ireland 52.48N 6.43W
189 W5 **Tully** Queensland, NE Australia 18.03S 145.55E
128 J3 **Tuloma** ✍ NW Russian Federation
116 K10 **Tulovo** Stara Zagora, C Bulgaria 42.34N 25.34E
49 P9 **Tulsa** Oklahoma, C USA 36.09N 95.59W
159 N11 **Tulsipur** Mid Western, W Nepal 28.01N 82.22E
37 P9 **Tuolumne River** see Tuolumne River
Tuong Buong see Tương Đương
178 J7 **Tul'skiy** Respublika Adygeya, SW Russian Federation 44.26N 40.12E
194 K8 **Tulu** Manus Island, N PNG 1.58S 146.50E
56 D10 **Tuluá** Valle del Cauca, W Colombia 4.01N 76.16W
118 M12 **Tulucești** Galaţi, E Romania 45.58N 28.03E
41 N12 **Tuluksak** Alaska, USA 61.06N 160.57W
43 Z12 **Tulum, Ruinas de** ruins Quintana Roo, SE Mexico 20.13N 87.24W
126 Ii15 **Tulun** Irkutskaya Oblast', S Russian Federation 54.28N 100.18E
174 Ll15 **Tulungagung** prev. Toeloengagoeng. Jawa, C Indonesia 8.03S 111.54E
195 S11 **Tulun Islands** var. Kilinailau Islands; prev. Carteret Islands. island group NE PNG
130 M4 **Tuma** Ryazanskaya Oblast', W Russian Federation 55.09N 40.27E
56 B12 **Tumaco** Nariño, SW Colombia 1.51N 78.46W
56 B12 **Tumaco, Bahía de** bay SW Colombia
Tuman-gang see Tumen
54 H11 **Tumba, Río** ✍ N Nicaragua
97 O16 **Tumba** Stockholm, C Sweden 59.12N 17.49E
Tumba, Lac see Ntomba, Lac
174 M9 **Tumbangsenamang** Borneo, C Indonesia 1.16S 112.12E
191 Q10 **Tumbarumba** New South Wales, SE Australia 35.47S 148.03E
58 A8 **Tumbes** Tumbes, NW Peru 3.33S 80.27W
58 A9 **Tumbes** ◆ department NW Peru
21 P5 **Tumbledown Mountain** ▲ Maine, NE USA 45.27N 70.28W
9 N13 **Tumbler Ridge** British Columbia, W Canada 55.06N 120.51W
178 I12 **Tumbôt, Phnum** ▲ W Cambodia
190 G9 **Tumby Bay** South Australia 34.23S 136.05E
169 N10 **Tumen** Jilin, NE China 42.58N 129.52E
169 Y11 **Tumen** Chin. Tumen Jiang, Kor. Tuman-gang, Rus. Tumyn'tszyan. ✍ E Asia
57 Q8 **Tumeremo** Bolívar, E Venezuela 7.19N 61.28W
161 G14 **Tumkūr** Karnātaka, W India 13.19N 77.06E
115 V5 **Tummel** ✍ C Scotland, UK
196 B15 **Tumon Bay** bay W Guam
43 O11 **Tula** Tamaulipas, C Mexico 22.59N 99.43W
60 P14 **Tumuc-Humac Mountains** var. Serra Tumucumaque. ▲ N South America

191 Q10 **Tumut** New South Wales, SE Australia 35.19S 148.12E
Tumyn'tszyan see Tumen
Tün see Ferdows
47 U14 **Tunapuna** Trinidad, Trinidad and Tobago 10.38N 61.23W
62 K11 **Tunas** Paraná, S Brazil 24.57S 49.05W
Tunbridge Wells see Royal Tunbridge Wells
116 L11 **Tunca Nehri** Bul. Tundzha. ✍ Bulgaria/Turkey see also Tundzha
143 O14 **Tunceli** var. Kalan. Tunceli, E Turkey 39.07N 39.34E
143 O14 **Tunceli** ◆ province C Turkey
158 J22 **Tündla** Uttar Pradesh, N India 27.13N 78.13E
83 I25 **Tunduru** Ruvuma, S Tanzania 11.07S 37.21E
116 L10 **Tundzha** Turk. Tunca Nehri. ✍ Bulgaria/Turkey see also Tunca Nehri
161 H17 **Tungabhadra** ✍ S India
161 F17 **Tungabhadra Reservoir** ☒ S India
203 P2 **Tungaru** prev. Gilbert Islands. island group W Kiribati
179 Q16 **Tungawan** Mindanao, S Philippines 7.33N 122.22E
174 Hh9 **Tungkal** ✍ Sumatera, W Indonesia
167 Q16 **Tungsha Tao** Chin. Dongsha Qundao, Eng. Pratas Island. island S Taiwan
167 S13 **Tungshih** Jap. Tōsei. N Taiwan 24.13N 120.54E
14 G7 **Tungsten** Northwest Territories, NW Canada 62.00N 128.09W
167 Q16 **Tung-t'ing Hu** see Dongting Hu
58 A13 **Tungurahua** ◆ province C Ecuador
97 F14 **Tunhovdfjorden** ☒ S Norway
24 K2 **Tunica** Mississippi, S USA 34.40N 90.22W
77 N5 **Tunis** var. Tūnis. ● (Tunisia) NE Tunisia 36.52N 10.10E
77 N5 **Tunis, Golfe de** Ar. Khalij Tūnis. gulf NE Tunisia
77 N6 **Tunisia** off. Republic of Tunisia, Ar. Al Jumhūrīyah at Tūnisīyah, Fr. République Tunisienne. ◆ republic N Africa
Tūnisīyah, Al Jumhūrīyah at see Tunisia
Tūnis, Golfe de see Tunis, Golfe de
56 G9 **Tunja** Boyacá, C Colombia 5.33N 73.22W
95 F14 **Tunnsjøen** ☒ C Norway
41 S9 **Tununak** Alaska, USA 60.21N 162.40W
153 U8 **Tunuk** Chuyskaya Oblast', N Kyrgyzstan 42.11N 73.55E
11 Q6 **Tunungayualok Island** island Newfoundland, E Canada
64 H11 **Tunuyán** Mendoza, W Argentina 33.28S 69.01W
64 I11 **Tunuyán, Río** ✍ W Argentina
207 P14 **Tunu** ◆ province E Greenland
Tunxi see Huangshan
37 P9 **Tuolumne River** ✍ California, W USA
Tuong Buong see Tương Đương
178 J7 **Tương Đương** var. Hòa Bình. Nghệ An, N Vietnam 19.14N 104.30E
166 I13 **Tuoniang Jiang** ✍ S China
165 N12 **Tuotuo He** ✍ C China
165 O12 **Tuotuoheyan** var. Tanggulashan, Togton-heyan. Qinghai, C China 34.13N 92.25E
Tüp see Tyup
62 J9 **Tupã** São Paulo, S Brazil 21.57S 50.28W
203 S10 **Tupai** var. Motu Iti. atoll Îles Sous le Vent, W French Polynesia
63 G15 **Tupanciretã** Rio Grande do Sul, S Brazil 29.06S 53.48W
28 M2 **Tupelo** Mississippi, S USA 34.15N 88.42W
126 L14 **Tupik** Chitinskaya Oblast', S Russian Federation 54.21N 119.56E
61 K18 **Tupiraçaba** Goiás, S Brazil 14.33S 48.40W
59 L21 **Tupiza** Potosí, S Bolivia 21.27S 65.45W
9 N13 **Tupper** British Columbia, W Canada 55.30N 119.59W
20 J8 **Tupper Lake** ☒ New York, NE USA
64 H11 **Tupungato, Volcán** ▲ W Argentina 33.27S 69.42W
169 T9 **Tuquan** Nei Mongol Zizhiqu, N China 45.21N 121.36E
56 C13 **Túquerres** Nariño, SW Colombia 1.06N 77.37W
159 U13 **Tura** Meghālaya, NE India 25.33N 90.14E
126 J10 **Tura** Evenkiyskiy Avtonomnyy Okrug, N Russian Federation 64.19N 100.16E
146 M10 **Turabah** Makkah, W Saudi Arabia
57 O8 **Turagua, Cerro** ▲ C Venezuela 6.59N 64.44W
192 L12 **Turakina** Manawatu-Wanganui, North Island, NZ 40.03S 175.13E
193 I19 **Turakirae Head** headland North Island, NZ 41.26S 174.54E
192 G9 **Turama** ✍ S PNG
194 G15 **Turan** Respublika Tyva, S Russian Federation 52.11N 93.40E
192 M10 **Turangi** Waikato, North Island, NZ 39.01S 175.46E
152 F11 **Turan Lowland** var. Turan Plain, Kaz. Turan Oypaty, Rus. Turanskaya Nizmennost', Turk. Turan Pesligi, Uzb. Turon Pasttekisligi. plain C Asia
Turan Oypaty/Turan Pesligi/Turan Plain/Turanskaya Nizmennost' see Turan Lowland

146 L2 **Ţurayf** Al Ḩudūd ash Shamālīyah, NW Saudi Arabia 31.43N 38.39E
191 Q10 **Turball** New South Wales, SE Australia 35.19S 148.12E
56 E5 **Turbaco** Bolívar, N Colombia 10.19N 75.25W
154 K15 **Turbat** Baluchistān, SW Pakistan 26.02N 62.56E
Turbat-i-Ḩeydariyeh see Torbat-e Ḩeydarīyeh
Turbat-i-Jam see Torbat-e Jām
56 D7 **Turbo** Antioquia, NW Colombia 8.06N 76.43W
Turčiansky Svätý Martin see Martin
118 H10 **Turda** Ger. Thorenburg, Hung. Torda. Cluj, NW Romania 46.34N 23.49E
112 I12 **Turek** Wielkopolskie, C Poland 52.01N 18.30E
95 L19 **Turenki** Etelä-Suomi, S Finland 60.55N 24.37E
152 J10 **Turfan** see Turpan
151 R8 **Turgay** Kaz. Torghay. Akmola, N Kazakhstan 51.43N 72.46E
151 N10 **Turgay** Kaz. Torgay. ✍ C Kazakhstan
150 M8 **Turgayskaya Stolovaya Strana** Kaz. Torgay Üstirti. plateau Kazakhstan/Russian Federation
116 L8 **Turgel** see Türi
66 L8 **Türgovishte** prev. Eski Dzhumaya. Türgovishte, N Bulgaria 43.15N 26.33E
142 C14 **Türgovishte** ◆ province N Bulgaria
142 J12 **Turgutlu** Manisa, W Turkey 38.30N 27.43E
120 H4 **Türi** Ger. Turgel. Järvamaa, N Estonia 58.49N 25.25E
107 S9 **Turia** ✍ E Spain
60 M12 **Turiaçu** Maranhão, E Brazil 1.40S 45.22W
Turin see Torino
118 I3 **Turiys'k** Volyns'ka Oblast', NW Ukraine 51.05N 24.31E
126 K15 **Turka** Respublika Buryatiya, S Russian Federation 53.02N 108.19E
118 H6 **Turka** L'viv'ska Oblast', W Ukraine 49.07N 23.01E
94 K12 **Turkana, Lake** see Rudolf, Lake
151 Q12 **Turkestan** Kaz. Türkistan. Yuzhnyy Kazakhstan, S Kazakhstan 43.18N 68.18E
153 Q12 **Turkestan Range** Rus. Turkestanskiy Khrebet. ▲ C Asia
Turkestanskiy Khrebet see Turkestan Range
123 M23 **Túrkeve** Jász-Nagykun-Szolnok, E Hungary 47.07N 20.48E
27 O4 **Turkey** Texas, SW USA 34.23N 100.54W
142 H14 **Turkey** off. Republic of Turkey, Turk. Türkiye Cumhuriyeti. ◆ republic SW Asia
29 M9 **Turkey Creek** Oklahoma, C USA
31 T9 **Turkey Mountains** ▲ New Mexico, SW USA
31 X1 **Turkey River** ✍ Iowa, C USA
131 N7 **Türki** Saratovskaya Oblast', W Russian Federation 52.04N 43.16E
124 Nn2 **Turkish Republic of Northern Cyprus** ◇ disputed territory Cyprus
Turkistan see Turkestan
Turkistan, Bandi-i see Torkestān
Türkiye Cumhuriyeti see Turkey
Türkmen Aylagy see Turkmenbashi Aylagy
152 A10 **Turkmenbashi** prev. Krasnovodsk. Balkanskiy Velayat, W Turkmenistan 40.00N 53.04E
Türkmengala see Turkmen-kala
152 G13 **Türkmengala** off. Turkmen-kala; prev. Turkmenskaya Soviet Socialist Republic. ◆ republic C Asia
152 A11 **Turkmenistan** off. Turkm. Türkmenistan.
176 Z14 **Türkmenskiy Zaliv** Turkm. Türkmen Aylagy. lake gulf W Turkmenistan
142 L16 **Türkoğlu** Kahramanmaraş, S Turkey 37.24N 36.49E
46 M7 **Turks and Caicos Islands** ◇ UK dependent territory N West Indies
46 G10 **Turks and Caicos Islands** island group N West Indies
47 N6 **Turks Islands** island group SE Turks and Caicos Islands
95 K19 **Turku** Swe. Åbo. Länsi-Suomi, W Finland 60.27N 22.16E
49 P9 **Turley** Oklahoma, C USA 36.14N 95.58W
49 P9 **Turlock** California, W USA 37.29N 120.52W
105 O2 **Turmantas** Zarasai, NE Lithuania 55.41N 26.27E
57 L5 **Turmero** Aragua, N Venezuela 10.14N 67.48E
192 N13 **Turnagain, Cape** headland North Island, NZ 40.30S 176.36E
44 H2 **Turneffe Islands** island group E Belize
20 L11 **Turners Falls** Massachusetts, NE USA 42.36N 72.31W
9 P16 **Turner Valley** Alberta, SW Canada 50.43N 114.19W
101 I16 **Turnhout** Antwerpen, N Belgium 51.19N 4.57E
113 E15 **Turnov** Ger. Turnau. Liberecký Kraj, N Czech Republic 50.36N 15.10E

118 I15 **Türnovo** see Veliko Türnovo
118 I15 **Turnu Măgurele** var. Turnu-Măgurele. Teleorman, S Romania 43.43N 24.52E
Turnu Severin see Drobeta-Turnu Severin
Turóczszentmárton see Martin
Turoni see Tours
152 **Turan Pasttekisligi** see Turan Lowland
Turov see Turaw
Turpakkala see Turpoqqal'a
164 M6 **Turpan** var. Turfan. Xinjiang Uygur Zizhiqu, NW China 42.54N 89.06E
Turpan Depression see Turpan Pendi
164 M6 **Turpan Pendi** Eng. Turpan Depression. depression NW China
164 M5 **Turpan Zhan** Xinjiang Uygur Zizhiqu, W China 43.10N 89.06E
Turpentine State see North Carolina
153 P13 **Turpoqqal'a** Rus. Turpakkala. Khorazm Wiloyati, W Uzbekistan 40.52N 62.00E
46 K8 **Turquino, Pico** ▲ E Cuba 19.54N 76.55W
49 Y10 **Turrell** Arkansas, C USA 35.22N 90.13W
45 N14 **Turrialba** Cartago, E Costa Rica 9.52N 83.40W
98 K8 **Turriff** NE Scotland, UK 57.32N 2.28W
145 V7 **Turşaq** E Iraq 33.27N 45.47E
Turshiz see Kāshmar
Tursunzade see Tursunzoda
153 P13 **Tursunzoda** Rus. Tursunzade; prev. Regar. W Tajikistan 38.30N 68.18E
168 J4 **Turt** Hövsgöl, N Mongolia 51.30N 100.40E
152 I9 **Türtkül** Rus. Turtkul'; prev. Petroaleksandrovsk. Qoraqalpog'iston Respublikasi, W Uzbekistan 41.34N 61.00E
31 O9 **Turtle Creek** ✍ South Dakota, N USA
32 K4 **Turtle Flambeau Flowage** ☒ Wisconsin, N USA
9 S14 **Turtleford** Saskatchewan, S Canada 53.21N 108.51W
30 M4 **Turtle Lake** North Dakota, N USA 47.31N 100.53W
94 K12 **Turtola** Lappi, NW Finland 66.39N 23.55E
126 Ij10 **Turu** ✍ N Russian Federation
Turuga see Tsuruga
153 V10 **Turugart Pass** Rus. China/Kyrgyzstan 40.33N 74.04E
164 E7 **Turugart Shankou** var. Pereval Torugart. pass China/Kyrgyzstan 40.33N 75.21E
126 Hh9 **Turukhan** ✍ N Russian Federation
126 J9 **Turukhansk** Krasnoyarskiy Kray, N Russian Federation 65.50N 87.48E
145 N3 **Ţurumbah** well NE Syria 36.09N 40.24E
Turuoka see Tsuruoka
150 H14 **Turush** Mangistau, SW Kazakhstan 45.24N 56.02E
62 K7 **Turvo, Rio** ✍ S Brazil
118 J2 **Tur''ya** Pol. Turja, Rus. Tur'ya. ✍ NW Ukraine
31 X1 **Tuscaloosa** Alabama, S USA
23 O4 **Tuscaloosa, Lake** ☒ Alabama, S USA
Tuscan Archipelago see Toscano, Archipelago
Tuscan-Emilian Mountains see Tosco-Emiliano, Appennino
Tuscany see Toscana
37 V2 **Tuscarora** Nevada, W USA 41.16N 116.13W
22 F15 **Tuscarora Mountain** ridge Pennsylvania, NE USA
32 M14 **Tuscola** Illinois, N USA 39.46N 88.19W
27 P7 **Tuscola** Texas, SW USA 32.12N 99.48W
23 S13 **Tuscumbia** Alabama, S USA 34.43N 87.42W
94 O4 **Tusenøyane** island group S Svalbard
150 K13 **Tushybas, Zaliv** prev. Zaliv Paskevicha. lake gulf SW Kazakhstan
Tusima see Tsushima
25 Q5 **Tuskegee** Alabama, S USA 32.25N 85.41W
96 E8 **Tustna** island S Norway
41 H12 **Tustumena Lake** ☒ Alaska, USA
112 K13 **Tuszyn** Łódzkie, C Poland 51.36N 19.51E
143 S13 **Tutak** Ağrı, E Turkey 39.34N 42.48E
193 C20 **Tutamoe Range** ▲ North Island, NZ
Tutasev see Tutayev
128 L15 **Tutayev** var. Tutasev. Yaroslavskaya Oblast', W Russian Federation 57.51N 39.29E
161 H23 **Tuticorin** Tamil Nādu, SE India 8.48N 78.10E
115 L15 **Tutin** Serbia, S Yugoslavia 43.00N 20.20E
192 O10 **Tutira** Hawke's Bay, North Island, NZ 39.14S 176.53E
126 Ii10 **Tutonchany** Evenkiyskiy Avtonomnyy Okrug, N Russian Federation 64.13N 93.49E
116 L6 **Tutrakan** Silistra, NE Bulgaria 44.03N 26.38E
31 O11 **Tuttle** North Dakota, N USA 47.07N 99.58W
29 S12 **Tuttle Creek Lake** ☒ Kansas, C USA
103 H23 **Tuttlingen** Baden-Württemberg, SW Germany 47.58N 8.49E
175 S16 **Tutualã** E West Timor 8.23S 127.12E

◆ COUNTRY   ◇ DEPENDENT TERRITORY   ◆ ADMINISTRATIVE REGION   ▲ MOUNTAIN   ⛰ VOLCANO   ☒ LAKE
● COUNTRY CAPITAL   ○ DEPENDENT TERRITORY CAPITAL   ✈ INTERNATIONAL AIRPORT   ▲ MOUNTAIN RANGE   ✍ RIVER   ☒ RESERVOIR

198 Cc9 **Tutuila** island W American Samoa
85 I18 **Tutume** Central, E Botswana 20.27S 26.58E
41 N7 **Tututalak Mountain** ▲ Alaska, USA 67.51N 161.27W
24 K3 **Tutwiler** Mississippi, S USA 34.00N 90.25W
168 L8 **Tuul Gol** ✍ N Mongolia
95 I16 **Tuupovaara** Itä-Suomi, E Finland 62.30N 30.40E
Tuva see Tyva, Respublika
202 E7 **Tuvalu** prev. Ellice Islands. ◆ commonwealth republic SW Pacific Ocean
197 L17 **Tuvana-i-Colo** prev. Tuvana-i-Tholo. island Lau Group, SE Fiji
197 L18 **Tuvana-i-Ra** island Lau Group, SE Fiji
Tuvana-i-Tholo see Tuvana-i-Colo
Tuvinskaya ASSR see Tyva, Respublika
197 L14 **Tuvuca** prev. Tuvutha. island Lau Group, E Fiji
Tuvutha see Tuvuca
147 P9 **Ṭuwayq, Jabal** ▲ C Saudi Arabia
144 H13 **Ṭuwayyil ash Shiḥāq** desert S Jordan
9 U16 **Tuxford** Saskatchewan, S Canada 50.33N 105.32W
178 K13 **Tu Xoay** Đặc Lắc, S Vietnam 12.18N 107.33E
42 L14 **Tuxpan** Jalisco, C Mexico 19.33N 103.21W
42 J12 **Tuxpan** Nayarit, C Mexico 21.57N 105.12W
43 Q12 **Tuxpan** var. Tuxpan de Rodríguez Cano. Veracruz-Llave, E Mexico 20.58N 97.22W
Tuxpan de Rodríguez Cano see Tuxpan
43 R15 **Tuxtepec** var. San Juan Bautista Tuxtepec. Oaxaca, S Mexico 18.01N 96.05W
43 U16 **Tuxtla** var. Tuxtla Gutiérrez. Chiapas, SE Mexico 16.43N 93.03W
Tuxtla see San Andrés Tuxtla
Tuxtla Gutiérrez see Tuxtla
Tuyama see Tsuyama
178 Ji5 **Tuyên Quang** Tuyên Quang, N Vietnam 21.48N 105.10E
178 K14 **Tuy Hoa** Bình Thuận, S Vietnam 11.03N 108.12E
178 Kk12 **Tuy Hoa** Phú Yên, S Vietnam 13.01N 109.15E
131 U5 **Tuymazy** Respublika Bashkortostan, W Russian Federation 54.36N 53.40E
148 L6 **Tūysarkān** var. Tuiserkan, Tuyserkân. Hamadān, W Iran 34.31N 48.30E
Tuyserkān see Tūysarkān
153 Q10 **Tuytepa** Rus. Toytepa. Toshkent Wiloyati, E Uzbekistan 41.04N 69.22E
151 W16 **Tuyuk** Kaz. Tuyyq. Almaty, SE Kazakhstan 43.07N 79.24E
Tuyyq see Tuyuk
142 I14 **Tüz Gölü** ◎ C Turkey
129 Q15 **Tuzha** Kirovskaya Oblast', NW Russian Federation 57.37N 48.02E
115 K17 **Tuzi** Montenegro, SW Yugoslavia 42.22N 19.21E
145 T5 **Tūz Khurmātū** N Iraq 34.55N 44.37E
114 I11 **Tuzla** Federacija Bosna I Hercegovina, NE Bosnia and Herzegovina 44.33N 18.40E
119 N15 **Tuzla** Constanța, SE Romania 43.58N 28.38E
143 T12 **Tuzluca** Iğdır, NE Turkey 40.01N 43.39E
97 J20 **Tvååker** Halland, S Sweden 57.04N 12.25E
97 F17 **Tvedestrand** Aust-Agder, S Norway 58.37N 8.55E
128 J16 **Tver'** prev. Kalinin. Tverskaya Oblast', W Russian Federation 56.52N 35.52E
Tverya see Teverya
130 I15 **Tverskaya Oblast'** ◆ province W Russian Federation
128 I15 **Tvertsa** ✍ W Russian Federation
112 H13 **Twardogóra** Ger. Festenberg. Dolnośląskie, SW Poland 51.21N 17.27E
12 J14 **Tweed** Ontario, SE Canada 44.28N 77.19W
98 K13 **Tweed** ✍ England/Scotland, UK
100 O7 **Tweede-Exloërmond** Drenthe, NE Netherlands 52.55N 6.55E
191 V3 **Tweed Heads** New South Wales, SE Australia 28.10S 153.32E
100 M11 **Twello** Gelderland, E Netherlands 52.13N 6.07E
37 W15 **Twentynine Palms** California, W USA 34.08N 116.03W
27 P9 **Twin Buttes Reservoir** ◙ Texas, SW USA
35 O15 **Twin Falls** Idaho, NW USA 42.33N 114.27W
41 N13 **Twin Hills** Alaska, USA 59.06N 160.21W
9 O16 **Twin Lakes** Alberta, W Canada 57.46N 117.30W
35 O12 **Twin Peaks** ▲ Idaho, NW USA 44.37N 114.24W
193 I14 **Twins, The** ▲ South Island, NZ 41.14S 172.38E
31 S5 **Twin Valley** Minnesota, N USA 47.15N 96.15W
102 G11 **Twistringen** Niedersachsen, NW Germany 52.48N 8.39E
193 E20 **Twizel** Canterbury, South Island, NZ 44.15S 170.06E
31 X5 **Two Harbors** Minnesota, N USA 47.01N 91.40W
9 R14 **Two Hills** Alberta, SW Canada 53.40N 111.43W
33 N7 **Two Rivers** Wisconsin, N USA 44.07N 87.33W
118 H8 **Tyachiv** Zakarpats'ka Oblast', W Ukraine 48.02N 23.35E
119 R6 **Tyas'myn** ✍ N Ukraine
25 X6 **Tybee Island** Georgia, SE USA 32.00N 80.51W
Tyborøn see Thyborøn

113 J16 **Tychy** Ger. Tichau. Śląskie, S Poland 50.12N 19.01E
113 O16 **Tyczyn** Podkarpackie, SE Poland 49.58N 22.03E
96 I18 **Tydal** Sør-Trøndelag, S Norway 63.01N 11.36E
117 H24 **Tyflos** ✍ Kríti, Greece, E Mediterranean Sea
23 S3 **Tygart Lake** ◙ West Virginia, NE USA
126 M15 **Tygda** Amurskaya Oblast', SE Russian Federation 53.07N 126.12E
23 U7 **Tyger River** ✍ South Carolina, SE USA
34 I11 **Tygh Valley** Oregon, NW USA 45.15N 121.12W
96 F12 **Tyin** ◎ S Norway
31 S10 **Tyler** Minnesota, N USA 44.16N 96.07W
27 W7 **Tyler** Texas, SW USA 32.21N 95.18W
27 W7 **Tyler, Lake** ◙ Texas, SW USA
24 K7 **Tylertown** Mississippi, S USA 31.90N 90.08W
Tylos see Bahrain
126 Gg12 **Tym** ✍ C Russian Federation
117 C15 **Týmfi** var. Timfi. ▲ W Greece 39.58N 20.51E
117 E17 **Tymfristós** var. Timfristos. ▲ C Greece 28.57N 21.49E
127 O14 **Tymovskoye** Ostrov Sakhalin, Sakhalinskaya Oblast', SE Russian Federation 50.36N 142.45E
117 J25 **Tympáki** var. Timbaki; prev. Timbákion. Kríti, Greece, E Mediterranean Sea 35.04N 24.46E
126 Ll14 **Tynda** Amurskaya Oblast', SE Russian Federation 55.09N 124.43E
31 Q12 **Tyndall** South Dakota, N USA 42.57N 97.52W
99 L14 **Tyne** ✍ N England, UK
99 M14 **Tynemouth** NE England, UK 55.01N 1.24W
99 L14 **Tyneside** cultural region NE England, UK
96 H10 **Tynset** Hedmark, S Norway 61.45N 10.48E
41 Q12 **Tyonek** Alaska, USA 61.04N 151.08W
Tyôsi see Chōshi
Tyras see Dniester, Moldova/Ukraine
Tyras see Bilhorod-Dnistrovs'kyy, Ukraine
Tyre see Soûr
97 C13 **Tyrifjorden** ◎ S Norway
97 K22 **Tyringe** Skåne, S Sweden 56.09N 13.34E
127 N15 **Tyrma** Khabarovskiy Kray, SE Russian Federation 50.00N 132.04E
Tyrnau see Trnava
117 F15 **Týrnavos** var. Tírnavos. Thessalía, C Greece 39.45N 22.18E
131 N14 **Tyrnyauz** Kabardino-Balkarskaya Respublika, SW Russian Federation 43.19N 42.55E
20 O4 **Tyrone** Pennsylvania, NE USA 40.41N 78.12W
99 E15 **Tyrone** cultural region W Northern Ireland, UK
190 M10 **Tyrrell, Lake** salt lake Victoria, SE Australia
86 H4 **Tyrrhenian Basin** undersea feature Tyrrhenian Sea, C Mediterranean Sea
123 L9 **Tyrrhenian Sea** It. Mare Tirreno. sea N Mediterranean Sea
Tysa see Tisa/Tisza
118 J7 **Tysmenytsya** Ivano-Frankivs'ka Oblast', W Ukraine 48.54N 24.50E
97 C14 **Tysnesøya** island S Norway
97 C14 **Tysse** Hordaland, S Norway 60.23N 5.46E
97 D14 **Tyssedal** Hordaland, S Norway 60.07N 6.36E
97 J17 **Tystberga** Södermanland, C Sweden 58.51N 17.15E
120 E12 **Tytuvénai** Kelmé, C Lithuania 55.36N 23.14E
150 D14 **Tyub-Karagan, Mys** headland SW Kazakhstan 44.40N 50.19E
153 V8 **Tyugel'-Say** Narynskaya Oblast', C Kyrgyzstan 41.57N 74.40E
125 Ff13 **Tyukalinsk** Omskaya Oblast', C Russian Federation 55.56N 72.02E
131 V7 **Tyul'gan** Orenburgskaya Oblast', W Russian Federation 52.27N 56.08E
125 F11 **Tyumen'** Tyumenskaya Oblast', C Russian Federation 57.10N 65.28E
125 Ff10 **Tyumenskaya Oblast'** ◆ province C Russian Federation
126 Kk10 **Tyung** ✍ NE Russian Federation
153 Y7 **Tyup** Kir. Tüp. Issyk-Kul'skaya Oblast', NE Kyrgyzstan 42.43N 78.18E
126 Ii16 **Tyva, Respublika** prev. Tannu-Tuva, Tuva, Tuvinskaya ASSR. ◆ autonomous republic C Russian Federation
119 N7 **Tyvriv** Vinnyts'ka Oblast', C Ukraine 49.01N 28.28E
99 J23 **Tywi** ✍ S Wales, UK
99 I19 **Tywyn** W Wales, UK 52.34N 4.06W
85 K20 **Tzaneen** Northern, NE South Africa 23.49S 30.09E
Tzekung see Zigong
43 U17 **Tzucacab** Yucatán, SE Mexico 20.04N 89.03W

—— U ——

84 B12 **Uaco Cungo** var. Waku Kungo, Port. Santa Comba. Cuanza Sul, C Angola 11.21S 15.04E
UAE see United Arab Emirates
203 X7 **Ua Huka** island Îles Marquises, NE French Polynesia
60 E10 **Uaiacás** Roraima, N Brazil 3.28N 63.13W
Uamba see Wamba
203 W7 **Ua Pu** island Îles Marquises, NE French Polynesia
83 O7 **Uar Garas** spring/well SW Somalia 1.19N 41.22E

60 G12 **Uatumã, Rio** ✍ C Brazil
Ua Uíbh Fhailí see Offaly
60 C11 **Uaupés, Rio** var. Río Vaupés. ✍ Brazil/Colombia see also Vaupés, Río
151 X9 **Uba** ✍ E Kazakhstan
151 N6 **Ubagan** Kaz. Obagan.
195 N12 **Ubai** New Britain, E PNG 5.38S 150.45E
81 J15 **Ubangi** Fr. Oubangui. ✍ C Africa
Ubangi-Shari see Central African Republic
118 M3 **Ubarts'** Ukr. Ubort'. ✍ Belarus/Ukraine see also Ubort'
56 F9 **Ubaté** Cundinamarca, C Colombia 5.19N 73.49W
62 N10 **Ubatuba** São Paulo, S Brazil 23.24S 45.06W
155 N6 **Ubauro** Sind, SE Pakistan 28.07N 69.43E
179 Qq14 **Ubay** Bohol, C Philippines 10.02N 124.29E
105 U14 **Ubaye** ✍ SE France
Ubayid, Wadi al see Ubayyiḍ, Wādī al
145 N8 **Ubaylah** N Iraq 33.06N 40.13E
145 O10 **Ubayyiḍ, Wādī al** var. Wadi al Ubayid. dry watercourse SW Iraq
100 L13 **Ubbergen** Gelderland, E Netherlands 51.49N 5.54E
170 Dd13 **Ube** Yamaguchi, Honshū, SW Japan 33.56N 131.14E
107 O13 **Úbeda** Andalucía, S Spain 38.01N 3.22W
111 V7 **Ubelbach** var. Markt-Ubelbach. Steiermark, SE Austria 47.13N 15.15E
61 L20 **Uberaba** Minas Gerais, SE Brazil 19.46S 47.57W
61 K19 **Uberaba, Laguna** ◎ E Bolivia
61 K19 **Uberlândia** Minas Gerais, SE Brazil 18.16S 48.16W
103 H24 **Überlingen** Baden-Württemberg, S Germany 47.46N 9.10E
79 O16 **Ubiaja** Edo, S Nigeria 6.39N 6.23E
106 K3 **Ubiña, Peña** ▲ NW Spain 43.01N 5.58W
59 J17 **Ubinas, Volcán** ℞ S Peru 16.16S 70.49W
Ubol Rajadhani/Ubol Ratchathani see Ubon Ratchathani
178 Ii9 **Ubolratna Reservoir** ◙ C Thailand
178 J10 **Ubon Ratchathani** var. Muang Ubon, Ubol Rajadhani, Ubol Ratchathani, Udon Ratchathani. E Thailand 15.15N 104.49E
121 L20 **Ubort'** Bel. Ubarts'. ✍ Belarus/Ukraine see also Ubarts'
106 K13 **Ubrique** Andalucía, S Spain 36.42N 5.27W
81 M18 **Ubundu** Orientale, C Dem. Rep. Congo (Zaire) 0.24S 25.30E
143 N13 **Ucar** Rus. Udzhary. C Azerbaijan 40.31N 47.40E
58 C11 **Ucayali** off. Departamento de Ucayali. ◆ department E Peru
58 F10 **Ucayali, Río** ✍ C Peru
Uccle see Ukkel
152 J13 **Uch-Adzhi** Turkm. Üchajy. Maryyskiy Velayat, C Turkmenistan 38.06N 62.44E
131 X4 **Uchaly** Respublika Bashkortostan, W Russian Federation 54.19N 59.33E
151 W13 **Ucharal** Kaz. Üsharal. Almaty, E Kazakhstan 46.07N 80.55E
170 Ci7 **Uchinoura** Kagoshima, Kyūshū, SW Japan 31.16N 131.04E
172 Nn6 **Uchiura-wan** bay NW Pacific Ocean
Uchkuduk see Uchquduq
152 K8 **Uchquduq** Rus. Uchkuduk. Navoiy Wiloyati, N Uzbekistan 42.12N 63.27E
153 S9 **Uchqurghon** Rus. Uchkurghan. Namangan Wiloyati, E Uzbekistan 41.06N 72.04E
152 G6 **Uchsay** see Uchsoy
152 D10 **Uchtagan, Peski** Turkm. Uchtagan Gumy. desert NW Turkmenistan
126 Mm12 **Uchur** ✍ E Russian Federation
102 O10 **Uckermark** cultural region E Germany
8 K17 **Ucluelet** Vancouver Island, British Columbia, SW Canada 48.58N 125.28W
126 Ii14 **Uda** ✍ S Russian Federation
126 Mm13 **Uda** ✍ E Russian Federation
126 K9 **Udachnyy** Respublika Sakha (Yakutiya), NE Russian Federation 66.27N 112.18E
161 G21 **Udagamandalam** var. prev. Ootacamund. Tamil Nādu, SW India 11.28N 76.42E
158 F14 **Udaipur** prev. Oodeypore. Rājasthān, N India 24.41N 73.40E
149 N16 **'Udayd, Khawr al** var. Khor al Udeid. inlet Qatar/Saudi Arabia
114 D11 **Udbina** Lika-Senj, W Croatia 44.33N 15.64E
97 I18 **Uddevalla** Västra Götaland, S Sweden 58.19N 11.55E
94 H13 **Uddjaur** var. Uddjaur.
Udeid, Khor al see 'Udayd, Khawr al
101 K14 **Uden** Noord-Brabant, SE Netherlands 51.40N 5.37E
100 L13 **Udenhout** Noord-Brabant, S Netherlands 51.37N 5.09E
161 G21 **Udgīr** Mahārāshtra, C India 18.23N 77.06E
Udhagamandalam see Udagamandalam

158 H6 **Udhampur** Jammu and Kashmir, NW India 32.55N 75.07E
145 X14 **'Udhaybah, 'Uqlat al** well S Iraq 29.46N 46.50E
108 I7 **Udine** anc. Utina. Friuli-Venezia Giulia, NE Italy 46.04N 13.10E
183 T14 **Udintsev Fracture Zone** tectonic feature S Pacific Ocean
Udipi see Udupi
Udjak see Ilok
'Ujmān see 'Ajmān
Ujohradze see Moldova Nouă
Udmurtia see Udmurtskaya Respublika
131 S2 **Udmurtskaya Respublika** Eng. Udmurtia. ◆ autonomous republic NW Russian Federation
128 J15 **Udomlya** Tverskaya Oblast', W Russian Federation 57.53N 34.59E
160 L11 **Udon Thani** var. Ban Mak Khaeng, Udorndhani. Udon Thani, N Thailand 17.25N 102.45E
Udorndhani see Udon Thani
201 U12 **Udot** atoll Chuuk Islands, C Micronesia
127 N13 **Udskaya Guba** bay E Russian Federation
161 E19 **Udupi** var. Udipi. Karnātaka, SW India 13.18N 74.46E
36 L6 **Udzhary** see Ucar
34 K12 **Ueda** var. Uyeda. Nagano, Honshū, S Japan 36.25N 138.14E
81 L16 **Uele** var. Welle. ✍ NE Dem. Rep. Congo (Zaire)
Uele (upper course) see Uolo, Río, Equatorial Guinea/Gabon
Uele (upper course) see Kibali, Dem. Rep. Congo (Zaire)
127 N13 **Uelen** Chukotskiy Avtonomnyy Okrug, NE Russian Federation 66.01N 169.52W
102 J11 **Uelzen** N Germany 52.58N 10.34E
171 H15 **Ueno** Mie, Honshū, SW Japan 34.45N 136.09E
131 V4 **Ufa** Respublika Bashkortostan, W Russian Federation 54.46N 56.02E
131 V4 **Ufa** ✍ W Russian Federation
152 A10 **Ufra** Balkanskiy Velayat, NW Turkmenistan 40.00N 53.05E
120 D8 **Ugāle** Ventspils, NW Latvia 57.16N 21.58E
83 F17 **Uganda** off. Republic of Uganda. ◆ republic E Africa
144 G4 **Ugarit** Ar. Ra's Shamrah. site of ancient city Al Lādhiqīyah, NW Syria 35.34N 35.45E
41 O14 **Ugashik** Alaska, USA 57.30N 157.24W
109 O10 **Ugento** Puglia, SE Italy 39.53N 18.09E
107 O15 **Ugíjar** Andalucía, S Spain 36.58N 3.03W
105 T11 **Ugine** Savoie, E France 45.45N 6.25E
127 O15 **Uglegorsk** Ostrov Sakhalin, Sakhalinskaya Oblast', SE Russian Federation 49.05N 142.06E
128 L15 **Uglich** Yaroslavskaya Oblast', W Russian Federation 57.33N 38.23E
128 I14 **Uglovka** var. Okulovka. Novgorodskaya Oblast', W Russian Federation 58.24N 33.15E
127 Pp5 **Ugol'nyye Kopi** Chukotskiy Avtonomnyy Okrug, NE Russian Federation 64.43N 105.46E
130 I4 **Ugra** ✍ W Russian Federation
153 V9 **Ugyut** Narynskaya Oblast', C Kyrgyzstan 41.22N 74.49E
127 Nn13 **Ul'banskiy Zaliv** strait E Russian Federation
113 H19 **Uherské Hradiště** Ger. Ungarisch-Hradisch. Zlínský Kraj, E Czech Republic 49.04N 17.26E
113 H19 **Uherský Brod** Ger. Ungarisch-Brod. Zlínský Kraj, E Czech Republic 49.01N 17.37E
Uhorshchyna see Hungary
33 T13 **Uhrichsville** Ohio, N USA 40.23N 81.21W
171 H15 **Uji** var. Uzi. Kyōto, Honshū, SW Japan 34.52N 135.47E

170 Aa16 **Uji-guntō** island Nansei-shotō, SW Japan
83 J21 **Ujiji** Kigoma, W Tanzania 4.55S 29.39E
160 Q10 **Ujjain** prev. Ujain. Madhya Pradesh, C India 23.10N 75.49E
175 P13 **Ujungpandang** var. Macassar, Makassar; prev. Makasar. Sulawesi, C Indonesia 5.09S 119.28E
Ujung Salang see Phuket
160 L11 **Ukái Reservoir** ◙ W India
83 G19 **Ukara Island** island N Tanzania
83 F19 **Ukerewe Island** island N Tanzania
159 X13 **Ukhrul** Manipur, NE India 25.07N 94.24E
129 S9 **Ukhta** Respublika Komi, NW Russian Federation 63.30N 53.47E
36 L6 **Ukiah** California, W USA 39.07N 123.14W
34 K12 **Ukiah** Oregon, NW USA 45.06N 118.57W
101 G18 **Ukkel** Fr. Uccle. Brussels, C Belgium 50.47N 4.18E
120 G3 **Ukmergė** Pol. Wiłkomierz.
118 L6 **Ukraine** off. Ukraine, Rus. Ukraina, Ukr. Ukrayina; prev. Ukrainian Soviet Socialist Republic, Ukrainskaya S.S.R. ◆ republic SE Europe
Ukrainskaya S.S.R/Ukrayina see Ukraine
84 B13 **Uku** Cuanza Sul, NW Angola 11.25S 14.18E
170 Bb12 **Uku-jima** island Gotō-rettō, SW Japan
85 F20 **Ukwi** Kgalagadi, SW Botswana 23.41S 20.26E
120 M13 **Ula** Rus. Ulla. Vitsyebskaya Voblasts', N Belarus 55.13N 29.15E
142 C16 **Ula** Muğla, SW Turkey 37.07N 28.25E
120 M13 **Ula** Rus. Ulla. ✍ N Belarus
168 L7 **Ulaanbaatar** Eng. Ulan Bator. ● (Mongolia) Töv, C Mongolia 47.54N 106.57E
169 N8 **Ulaan-Ereg** Hentiy, E Mongolia 46.50N 109.39E
168 S5 **Ulaangom** Uvs, NW Mongolia 49.56N 92.06E
168 L7 **Ulaantolgoy** Hovd, W Mongolia 46.39N 92.50E
168 I8 **Ulaan-Uul** Bayankhongor, C Mongolia 46.03N 100.52E
169 O10 **Ulaan-Uul** Dornogovi, SE Mongolia 44.21N 111.06E
165 R10 **Ulan** Qinghai, C China 36.59N 98.21E
Ulan Bator see Ulaanbaatar
168 L13 **Ulan Buh Shamo** desert N China
127 O15 **Uleglegorsk** Ostrov Sakhalin, Sakhalinskaya Oblast', SE Russian Federation 49.05N 142.06E
169 T8 **Ulanhot** Nei Mongol Zizhiqu, N China 46.02N 122.00E
131 Q14 **Ulan Khol** Respublika Kalmykiya, SW Russian Federation 45.27N 46.48E
168 M13 **Ulan-Ude** prev. Verkhneudinsk. Respublika Buryatiya, S Russian Federation 51.55N 107.40E
165 N12 **Ulan Ul Hu** ◎ C China
195 Z16 **Ulawa Island** island SE Solomon Islands
144 J7 **'Ulayyāniyah, Bi'r al** well S Syria 34.01N 38.06E
127 Nn13 **Ul'banskiy Zaliv** strait E Russian Federation
119 O7 **Ulbo** see Olib
115 J18 **Ulcinj** Montenegro, SW Yugoslavia 41.56N 19.14E
169 O7 **Uldz** Hentiy, NE Mongolia 48.47N 112.01E
169 O7 **Uleåborg** see Oulu
Uleälv see Oulujoki
176 Ww12 **Uleåträsk** see Oulujärvi
33 T13 **Ulëz** var. Uleza. Dibër, C Albania 41.42N 19.52E
Uleza see Ulëz
97 F22 **Ulfborg** Ringkøbing, W Denmark 56.16N 8.21E
100 N13 **Ulft** Gelderland, E Netherlands 51.52N 6.22E
200 Ss13 **Ulimang** Babeldaob, N Palau 7.46N 96.53E
69 T10 **Ulindi** ✍ W Dem. Rep. Congo (Zaire)
196 H14 **Ulithi Atoll** atoll Caroline Islands, W Micronesia
114 N10 **Uljma** Serbia, NE Yugoslavia 45.04N 21.08E
150 L11 **Ul'kayak** Kaz. Ölkeyek. ✍ C Kazakhstan
151 Q7 **Ul'ken-Karoy, Ozero** ◎ N Kazakhstan
108 J7 **Umbria** ◆ region C Italy
Ülkenözen see Bol'shoy Uzen'
Ülkenqobda see Bol'shaya Khobda
106 G3 **Ulla** ✍ NW Spain
Ulla see Ula
191 S10 **Ulladulla** New South Wales, SE Australia 35.22S 150.28E
159 T14 **Ulapara** Rajshahi, W Bangladesh 24.19N 89.34E
98 H7 **Ullapool** N Scotland, UK 57.54N 5.10W
85 K23 **Ulmazi** KwaZulu/Natal, E South Africa 29.58S 30.50E
107 T7 **Ulldecona** Cataluña, NE Spain 50.07N 12.45E
94 I9 **Ullsfjorden** fjord N Norway
147 U12 **Ullung-do** Jap. Utsuryō. island NE South Korea 37.29N 130.53E
103 J22 **Ulm** Baden-Württemberg, S Germany 48.24N 9.58E
35 P7 **Ulm** Montana, NW USA 47.27N 111.32W
191 V3 **Ulmarra** New South Wales, SE Australia 29.37S 153.06E

118 K13 **Ulmeni** Buzău, C Romania 45.08N 26.43E
118 K14 **Ulmeni** Călăraşi, S Romania 44.08N 26.43E
44 L7 **Ulmukhuás** Región Autónoma Atlántico Norte, NE Nicaragua 14.21N 84.34W
196 C8 **Ulong** var. Aulong. island Palau Islands, N Palau
85 N14 **Ulongué** var. Ulongwé. Tete, NW Mozambique 14.42S 34.21E
Ulongwé see Ulongué
97 K19 **Ulricehamn** Västra Götaland, S Sweden 57.57N 13.25E
100 N5 **Ulrum** Groningen, NE Netherlands 53.24N 6.20E
169 Z16 **Ulsan** Jap. Urusan. SE South Korea 35.33N 129.19E
96 D13 **Ulsteinvik** Møre og Romsdal, S Norway 62.19N 5.52E
99 D15 **Ulster** ◆ province Northern Ireland, UK/Ireland
175 S3 **Ulu** Pulau Siau, N Indonesia 2.46N 125.22E
126 Ll12 **Ulu** Respublika Sakha (Yakutiya), NE Russian Federation 60.18N 127.27E
Ulua, Río see Úlua, Río
44 H5 **Úlua, Río** ✍ NW Honduras
142 D12 **Uludağ** ▲ NW Turkey 40.08N 29.13E
142 E12 **Uludağ** ▲ NW Turkey
Ulugh Muztag see Muztag Feng
164 D7 **Uliugqat** Xinjiang Uygur Zizhiqu, W China 39.45N 74.10E
142 J16 **Ulukışla** Niğde, S Turkey 37.33N 34.28E
201 O15 **Ulul** island Caroline Islands, C Micronesia
85 L22 **Ulundi** KwaZulu/Natal, E South Africa 28.18S 31.26E
164 M3 **Ulungur He** ✍ NW China
164 K2 **Ulungur Hu** ◎ NW China
189 P8 **Uluru** var. Ayers Rock. rocky outcrop Northern Territory, C Australia 25.20S 130.59E
99 K19 **Ulverston** NW England, UK 54.12N 3.07W
191 O16 **Ulverstone** Tasmania, SE Australia 41.13S 146.09E
96 D13 **Ulvik** Hordaland, S Norway 60.34N 6.53E
95 J18 **Ulvila** Länsi-Suomi, W Finland 61.25N 21.55E
119 O8 **Ulyanivka** Rus. Ul'yanovka. Kirovohrads'ka Oblast', C Ukraine 48.18N 30.15E
131 Q5 **Ul'yanovsk** prev. Simbirsk. Ul'yanovskaya Oblast', W Russian Federation 54.16N 48.21E
131 Q5 **Ul'yanovskaya Oblast'** ◆ province W Russian Federation
151 S10 **Ul'yanovskiy** Karaganda, C Kazakhstan 50.04N 73.45E
Ul'yanovskiy Kanal see Ul'yanow Kanali
152 M13 **Ul'yanow Kanali** Rus. Ul'yanovskiy Kanal. canal Turkmenistan/Uzbekistan
Ulyshylanshyq see Uly-Zhylanshyk
28 M3 **Ulysses** Kansas, C USA 37.34N 101.21W
151 O12 **Ulytau, Gory** ▲ C Kazakhstan
126 K14 **Ulyunkhan** Respublika Buryatiya, S Russian Federation 54.48N 111.01E
151 N11 **Uly-Zhylanshyk** Kaz. Ulyshylanshyq. ✍ C Kazakhstan
114 A9 **Umag** It. Umago. Istra, NW Croatia 45.25N 13.32E
Umago see Umag
201 V13 **Uman** island Chuuk Islands, C Micronesia
119 O7 **Uman'** Rus. Uman. Cherkas'ka Oblast', C Ukraine 48.45N 30.10E
43 W12 **Umán** Yucatán, SE Mexico 20.51N 89.45W
41 N16 **Umanak/Umanaq** see Uummannaq
'Umān, Khalīj see Oman, Gulf of
'Umān, Salṭanat see Oman
176 Ww12 **Umari** Irian Jaya, E Indonesia 4.18S 135.22E
160 K10 **Umaria** Madhya Pradesh, C India 23.31N 80.48E
155 R16 **Umar Kot** Sind, SE Pakistan 25.22N 69.46E
196 B17 **Umatac** SW Guam 13.17N 144.40E
Umatac Bay bay SW Guam
145 S6 **Umayqah** C Iraq 34.32N 43.45E
128 J5 **Umba** Murmanskaya Oblast', NW Russian Federation 66.39N 34.24E
82 A12 **Umbelasha** ✍ W Sudan
108 H12 **Umbertide** Umbria, C Italy 43.16N 12.21E
63 H15 **Umberto** var. Humberto. Santa Fe, C Argentina 30.52S 61.19W
194 K11 **Umboi Island** var. Rooke Island. island C PNG
128 J7 **Umbozero, Ozero** ◎ NW Russian Federation
108 I7 **Umbria** ◆ region C Italy
Umbrian-Machigian Mountains see Umbro-Marchigiano, Appennino
108 H11 **Umbro-Marchigiano, Appennino** Eng. Umbrian-Machigian Mountains. ▲ C Italy
95 H14 **Umeälven** ✍ N Sweden
95 J16 **Umeå** Västerbotten, N Sweden 63.50N 20.15E

144 J3 **Umm 'Āmūd** Ḥalab, N Syria 35.57N 37.39E
147 Y10 **Umm ar Ruşaş** var. Umm Ruşayş. W Oman 20.26N 58.48E
147 X9 **Umma Samin** salt flat C Oman
147 V9 **Umm az Zumūl** oasis E Saudi Arabia 22.39N 54.45E
82 A9 **Umm Buru** Western Darfur, W Sudan 15.01N 23.36E
82 A12 **Umm Dafag** Southern Darfur, W Sudan 10.28N 23.19E
144 F9 **Umm el Fahm** Haifa, N Israel 32.30N 35.06E
82 F9 **Umm Inderab** Northern Kordofan, C Sudan 15.18N 31.56E
82 C10 **Umm Keddada** Northern Darfur, W Sudan 13.36N 26.42E
146 J7 **Umm Lajj** Tabūk, W Saudi Arabia 25.01N 37.19E
145 Y13 **Umm Qasr** SE Iraq 30.01N 47.55E
82 F11 **Umm Ruwaba** var. Umm Ruwābah, Um Ruwāba. Northern Kordofan, C Sudan 12.54N 31.13E
149 N16 **Umm Sa'id** var. Musay'id. E Qatar 24.57N 51.31E
144 K10 **Umm Ṭuways, Wādī** dry watercourse W Jordan
40 J17 **Umnak Island** island Aleutian Islands, Alaska, USA
34 F13 **Umpqua River** ✍ Oregon, NW USA
84 D13 **Umpulo** Bié, C Angola 12.43S 17.42E
160 I12 **Umred** Mahārāshtra, C India 20.54N 79.19E
145 Y10 **Um Sawān, Hawr** ◎ S Iraq
Um Ruwāba see Umm Ruwaba
Umtali see Mutare
85 J24 **Umtata** Eastern Cape, SE South Africa 31.35S 28.47E
79 V17 **Umuahia** Abia, SW Nigeria 5.30N 7.33E
62 H10 **Umuarama** Paraná, S Brazil 23.45S 53.19W
Um Sawān see Mvuma
85 K18 **Umzingwani** ✍ S Zimbabwe
114 D11 **Una** ✍ Bosnia and Herzegovina/Croatia
114 E12 **Una** ✍ W Bosnia and Herzegovina
25 T6 **Unadilla** Georgia, SE USA 32.15N 83.44W
20 I10 **Unadilla River** ✍ New York, NE USA
61 L18 **Unaí** Minas Gerais, SE Brazil 16.24S 46.49W
41 N10 **Unalakleet** Alaska, USA 63.52N 160.47W
40 K17 **Unalaska Island** island Aleutian Islands, Alaska, USA
193 I16 **Una, Mount** ▲ South Island, NZ 42.12S 172.34E
84 L13 **Unango** Niassa, N Mozambique 12.45S 35.28E
Unao see Unnāo
94 L12 **Unari** Lappi, N Finland 67.07N 25.37E
147 O6 **'Unayzah** var. Anaiza. Al Qaşīm, C Saudi Arabia 26.03N 44.00E
144 L10 **'Unayzah, Jabal** ▲ Jordan/Saudi Arabia 32.09N 39.10E
Unci see Almería
59 X9 **Uncia** Potosí, C Bolivia 18.30S 66.29W
39 Q7 **Uncompahgre Peak** ▲ Colorado, C USA 38.04N 107.27W
39 P6 **Uncompahgre Plateau** plain Colorado, C USA
30 M4 **Unden** ◎ S Sweden
30 M4 **Underwood** North Dakota, N USA 47.25N 101.09W
176 Uu11 **Undur** Pulau Seram, E Indonesia
130 H6 **Unecha** Bryanskaya Oblast', W Russian Federation 52.51N 32.38E
41 N16 **Unga Island** island Alaska, USA 55.14N 160.34W
191 P8 **Ungarie** New South Wales, SE Australia 33.39S 146.54E
Ungarisch-Brod see Uherský Brod
Ungarisch-Hradisch see Uherské Hradiště
Ungarisches Erzgebirge see Slovenské rudohorie
Ungary see Hungary
10 M4 **Ungava Bay** bay Quebec, E Canada
10 I2 **Ungava, Péninsule d'** peninsula Quebec, SE Canada
Ungeny see Ungheni
118 M9 **Ungheni** Rus. Ungeny. W Moldova 47.13N 27.48E
Unguja see Zanzibar
Üngüz Angyrsyndaky Garagum see Zaungukskiye Garagumy
152 H11 **Unguz, Solonchakovyye Vpadiny** salt marsh C Turkmenistan
Ungvár see Uzhhorod
62 I12 **União da Vitória** Paraná, S Brazil 26.13S 51.04W
113 G17 **Uničov** Ger. Mährisch-Neustadt. Olomoucký Kraj, E Czech Republic 49.46N 17.05E
112 I11 **Uniejów** Łódzkie, C Poland 51.58N 18.46E
40 L16 **Unije** island Aleutian Islands, Alaska, USA
40 L16 **Unimak Island** island Aleutian Islands, Alaska, USA
40 L16 **Unimak Pass** strait Aleutian Islands, Alaska, USA
29 U5 **Union** Missouri, C USA 38.27N 91.01W
34 K12 **Union** Oregon, NW USA 45.12N 117.51W
23 R6 **Union** South Carolina, SE USA 34.40N 81.35W
23 Q3 **Union** West Virginia, USA 37.33N 80.33W
33 Q13 **Unión, Bahía** bay E Argentina
98 [ ] **Union City** Indiana, N USA 40.12N 84.50W

◆ COUNTRY   ● COUNTRY CAPITAL   ◇ DEPENDENT TERRITORY   ○ DEPENDENT TERRITORY CAPITAL   ◆ ADMINISTRATIVE REGION   ✈ INTERNATIONAL AIRPORT   ▲ MOUNTAIN   ▲ MOUNTAIN RANGE   ℞ VOLCANO   ✍ RIVER   ◎ LAKE   ◙ RESERVOIR

33 Q10 **Union City** Michigan, N USA 42.03N 85.06W
20 C12 **Union City** Pennsylvania, NE USA 41.54N 79.51W
22 G8 **Union City** Tennessee, S USA 36.25N 89.01W
34 G14 **Union Creek** Oregon, NW USA 42.54N 122.26W
85 G25 **Uniondale** Western Cape, SW South Africa 33.40S 23.07E
42 K13 **Unión de Tula** Jalisco, SW Mexico 19.58N 104.20W
32 M9 **Union Grove** Wisconsin, N USA 42.39N 88.03W
47 Y15 **Union Island** *island* S Saint Vincent and the Grenadines
48 K5 **Union Reefs** *reef* SW Mexico
2 D7 **Union Seamount** *undersea feature* NE Pacific Ocean 49.34N 132.45W
25 Q6 **Union Springs** Alabama, S USA 32.08N 85.43W
22 H6 **Uniontown** Kentucky, S USA 37.46N 87.55W
20 C16 **Uniontown** Pennsylvania, NE USA 39.54N 79.43W
49 T1 **Unionville** Missouri, C USA 40.28N 93.00W
147 V8 **United Arab Emirates** *Ar.* Al Imārāt al ʿArabīyah al Muttaḥidah, *abbrev.* UAE; *prev.* Trucial States. ◆ *federation* SW Asia
United Arab Republic *see* Egypt
99 H14 **United Kingdom** *off.* UK of Great Britain and Northern Ireland, *abbrev.* UK. ◆ *monarchy* NW Europe
United Mexican States *see* Mexico
United Provinces *see* Uttar Pradesh
18 L9 **United States of America** *off.* United States of America, *var.* America, The States, *abbrev.* U.S., USA. ◆ *federal republic*
128 J10 **Unitsa** Respublika Kareliya, NW Russian Federation 62.31N 34.31E
9 S15 **Unity** Saskatchewan, S Canada 52.27N 109.10W
Unity State *see* Wahda
107 Q8 **Universales, Montes** ⛰ C Spain
49 X4 **University City** Missouri, C USA 38.40N 90.19W
197 B13 **Unmet** Malekula, C Vanuatu 16.09S 167.16E
103 F15 **Unna** Nordrhein-Westfalen, W Germany 51.31N 7.40E
158 L12 **Unnāo** *prev.* Unao. Uttar Pradesh, N India 26.31N 80.30E
197 D15 **Unpongkor** Erromango, S Vauautu 18.48S 169.01E
Unruhstadt *see* Kargowa
98 M1 **Unst** *island* NE Scotland, UK
103 K16 **Unstrut** ≈ C Germany
Unterdrauburg *see* Dravograd
Unterlimbach *see* Lendava
103 L23 **Unterschleissheim** Bayern, SE Germany 48.16N 11.34E
103 H24 **Untersee** ◎ Germany/Switzerland
102 O10 **Unterueckersee** ◎ NE Germany
110 F9 **Unterwalden** ◈ *canton* C Switzerland
57 N12 **Unturán, Sierra de** ⛰ Brazil/Venezuela
165 N11 **Unuli Horog** Qinghai, W China 35.10N 91.49E
142 M11 **Ünye** Ordu, W Turkey 41.07N 37.14E
Unza *see* Unzha
129 O14 **Unzha** *var.* Unza. ≈ NW Russian Federation
81 E17 **Uolo, Río** *var.* Eyo (lower course), Mbini, Uele (upper course), Woleu; *prev.* Benito. ≈ Equatorial Guinea/Gabon
57 Q10 **Uonán** Bolívar, SE Venezuela 4.33N 62.10W
167 T12 **Uotsuri-shima** *island* China/Japan/Taiwan
171 J13 **Uozu** Toyama, Honshū, SW Japan 36.48N 137.23E
44 L12 **Upala** Alajuela, NW Costa Rica 10.52N 85.00W
57 P7 **Upata** Bolívar, E Venezuela 8.01N 62.25W
81 M23 **Upemba, Lac** ◎ SE Dem. Rep. Congo (Zaire)
207 O12 **Upernavik** *var.* Upernivik. Kitaa, C Greenland 73.06N 55.42W
Upernivik *see* Upernavik
85 F22 **Upington** Northern Cape, W South Africa 28.24S 21.13E
Uplands *see* Ottawa
198 Bb8 **Upolu** *island* SE Samoa
40 G11 **Upolu Point** *headland* Hawaii, USA, C Pacific Ocean 20.15N 155.51W
Upper Austria *see* Oberösterreich
Upper Bann *see* Bann
12 M13 **Upper Canada Village** *tourist site* Ontario, SE Canada 44.57N 75.04W
20 I16 **Upper Darby** Pennsylvania, NE USA 39.57N 75.15W
30 L2 **Upper Des Lacs Lake** ▣ North Dakota, N USA
193 L14 **Upper Hutt** Wellington, North Island, NZ 41.08S 174.58E
31 X11 **Upper Iowa River** ≈ Iowa, C USA
34 H15 **Upper Klamath Lake** ◎ Oregon, NW USA
36 M6 **Upper Lake** California, W USA 39.07N 122.53W
37 Q1 **Upper Lake** ◎ California, W USA
8 K9 **Upper Liard** Yukon Territory, W Canada 60.01N 128.59W
99 E16 **Upper Lough Erne** ◎ SW Northern Ireland, UK
82 F12 **Upper Nile** ◈ *state* E Sudan
31 T3 **Upper Red Lake** ◎ Minnesota, N USA
33 S12 **Upper Sandusky** Ohio, N USA 40.49N 83.16W
Upper Volta *see* Burkina
97 O15 **Upplandsväsby** *var.* Upplands Väsby. Stockholm, C Sweden 59.28N 17.49E
97 O15 **Uppsala** Uppsala, C Sweden 59.52N 17.37E
97 O14 **Uppsala** ◈ *county* C Sweden
40 J12 **Upright Cape** *headland* Saint Matthew Island, Alaska, USA 60.19N 172.15W
22 K6 **Upton** Kentucky, S USA 37.25N 85.53W
35 Y13 **Upton** Wyoming, C USA 44.06N 104.37W
147 N7 **ʿUqlat aş Şuqūr** Al Qaşīm, W Saudi Arabia 25.51N 42.12E
Uqturpan *see* Wushi
56 C7 **Urabá, Golfo de** *gulf* NW Colombia
Uracas *see* Farallon de Pajaros
Uradar'ya *see* Ŭradaryo
153 N13 **Ŭradaryo** *Rus.* Uradar'ya. ≈ S Uzbekistan
168 M13 **Urad Qianqi** *var.* Xishanzui. Nei Mongol Zizhiqu, N China 40.43N 108.41E
171 Jj17 **Uraga-suidō** *strait* S Japan
172 Pp7 **Urahoro** Hokkaidō, NE Japan 42.47N 143.41E
172 Oo8 **Urakawa** Hokkaidō, NE Japan 42.11N 142.42E
131 X6 **Ural** *Kaz.* Zayyq. ≈ Kazakhstan/Russian Federation
191 T6 **Uralla** New South Wales, SE Australia 30.39S 151.30E
Ural Mountains *see* Ural'skiye Gory
150 F8 **Ural'sk** *Kaz.* Oral. Zapadnyy Kazakhstan, NW Kazakhstan 51.12N 51.17E
Ural'skaya Oblast' *see* Zapadnyy Kazakhstan
131 W5 **Ural'skiye Gory** *var.* Ural'skiy Khrebet, *Eng.* Ural Mountains. ⛰ Kazakhstan/Russian Federation
Ural'skiy Khrebet *see* Ural'skiye Gory
144 I3 **Urām aş Şughrá** Ḩalab, N Syria 36.10N 36.55E
191 P10 **Urana** New South Wales, SE Australia 35.22S 146.16E
9 S10 **Uranium City** Saskatchewan, C Canada 59.30N 108.46W
60 F10 **Uraricoera** Roraima, N Brazil 3.26N 60.54W
49 S5 **Uraricoera, Rio** ≈ N Brazil
Ura-Tyube *see* Ŭroteppa
171 K16 **Urawa** Saitama, Honshū, S Japan 35.51N 139.37E
125 F10 **Uray** Khanty-Mansiyskiy Avtonomnyy Okrug, C Russian Federation 60.07N 64.38E
147 R7 **ʿUray'irah** Ash Sharqīyah, E Saudi Arabia 25.59N 48.51E
32 M13 **Urbana** Illinois, N USA 40.06N 88.12W
33 R13 **Urbana** Ohio, N USA 40.04N 83.46W
31 V14 **Urbandale** Iowa, C USA 41.37N 93.42W
108 I11 **Urbania** Marche, C Italy 43.40N 12.33E
176 Uu8 **Urbinasopon** Irian Jaya, E Indonesia 0.19S 131.12E
108 I11 **Urbino** Marche, C Italy 43.45N 12.38E
59 H16 **Urcos** Cusco, S Peru 13.45S 71.37W
150 D10 **Urda** Zapadnyy Kazakhstan, W Kazakhstan 48.52N 47.31E
107 N10 **Urda** Castilla-La Mancha, C Spain 39.25N 3.43W
168 E7 **Urdgol** Hovd, W Mongolia 47.39N 92.46E
Urdunn *see* Jordan
151 X12 **Urdzhar** *Kaz.* Ürzhar. Vostochnyy Kazakhstan, E Kazakhstan 47.06N 81.37E
99 L16 **Ure** ≈ N England, UK
121 K18 **Urechcha** *Rus.* Urech'ye. Minskaya Voblasts', S Belarus 52.57N 27.54E
Urech'ye *see* Urechcha
131 P2 **Uren'** Nizhegorodskaya Oblast', W Russian Federation 57.29N 45.47E
126 H9 **Urengoy** Yamalo-Nenetskiy Avtonomnyy Okrug, N Russian Federation 65.52N 78.42E
192 K10 **Urenui** Taranaki, North Island, NZ 38.59S 174.25E
197 B10 **Ureparapara** *island* Banks Islands, N Vanuatu
42 G5 **Ures** Sonora, NW Mexico 29.25N 110.24W
Urfa *see* Şanlıurfa
152 H9 **Urganch** *Rus.* Urgench; *prev.* Novo-Urgench. Khorazm Wiloyati, W Uzbekistan 41.39N 60.32E
Urgench *see* Urganch
142 J14 **Ürgüp** Nevşehir, C Turkey 38.39N 34.55E
153 O12 **Urgut** Samarqand Wiloyati, C Uzbekistan 39.25N 67.15E
164 K3 **Urho** Xinjiang Uygur Zizhiqu, W China 46.04N 84.51E
158 G5 **Uri** Jammu and Kashmir, NW India 34.04N 74.03E
110 G9 **Uri** ◈ *canton* C Switzerland
56 F11 **Uribe** Meta, C Colombia 3.01N 74.33W
56 H4 **Uribia** La Guajira, N Colombia 11.45N 72.19W
118 G12 **Uricani** *Hung.* Hobicaurikány. Hunedoara, SW Romania 45.18N 23.03E
59 M21 **Uriondo** Tarija, S Bolivia 21.40S 64.37W
42 I7 **Urique** Chihuahua, N Mexico 27.16N 107.51W
42 I7 **Urique, Río** ≈ N Mexico
58 E9 **Urituyacu, Río** ≈ N Peru
151 N7 **Uritskiy** Kostanay, N Kazakhstan 53.21N 65.27E
100 K8 **Urk** Flevoland, N Netherlands 52.40N 5.34E
142 B14 **Urla** İzmir, W Turkey 38.19N 26.46E
118 K13 **Urlați** Prahova, SE Romania 44.58N 26.15E
131 V4 **Urman** Respublika Bashkortostan, W Russian Federation 54.53N 56.52E
153 P12 **Urmetan** W Tajikistan 39.27N 68.13E
Urmia *see* Orūmīyeh
Urmia, Lake *see* Orūmīyeh, Daryācheh-ye
Urmiyeh *see* Orūmīyeh
115 N17 **Uroševac** *Alb.* Ferizaj. Serbia, S Yugoslavia 42.23N 21.09E
153 P11 **Ŭroteppa** *Rus.* Ura-Tyube. NW Tajikistan 39.54N 68.57E
56 D8 **Urrao** Antioquia, W Colombia 6.16N 76.10W
Ursat'yevskaya *see* Khowos
168 I11 **Urt** Ömnögovĭ, S Mongolia 43.16N 101.00E
131 X7 **Urtazym** Orenburgskaya Oblast', W Russian Federation 52.12N 58.48E
61 K18 **Uruaçu** Goiás, C Brazil 14.37S 49.06W
42 M14 **Uruapan** *var.* Uruapan del Progreso. Michoacán de Ocampo, SW Mexico 19.25N 102.04W
Uruapan del Progreso *see* Uruapan
59 G15 **Urubamba, Cordillera** ⛰ C Peru
59 G14 **Urubamba, Río** ≈ C Peru
60 G12 **Urucará** Amazonas, N Brazil 2.30S 57.45W
63 E16 **Uruguaiana** Rio Grande do Sul, S Brazil 29.45S 57.04W
Uruguai, Rio *see* Uruguay
63 E18 **Uruguay** *off.* Oriental Republic of Uruguay; *prev.* La Banda Oriental. ◆ *republic* E South America
63 E15 **Uruguay** *var.* Rio Uruguai, Río Uruguay. ≈ E South America
Uruguay, Río *see* Uruguay
Urumchi *see* Ürümqi
164 L5 **Ürümqi** *var.* Tihwa, Urumchi, Urumqi, Urumtsi, Wu-lu-k'o-mu-shi, Wu-lu-mu-ch'i; *prev.* Ti-hua. *autonomous region capital* Xinjiang Uygur Zizhiqu, NW China 43.52N 87.31E
Urumtsi *see* Ürümqi
Urundi *see* Burundi
191 V6 **Urunga** New South Wales, SE Australia 30.33S 152.58E
196 C15 **Uruno Point** *headland* NW Guam 13.37N 144.49E
127 P15 **Urup, Ostrov** *island* Kuril'skiye Ostrova, SE Russian Federation
147 P11 **ʿUruq al Mawārid** *desert* S Saudi Arabia
Urusan *see* Ulsan
131 T5 **Urussu** Respublika Tatarstan, W Russian Federation 54.34N 53.23E
192 K10 **Uruti** Taranaki, North Island, NZ 38.57S 174.32E
59 K19 **Uru Uru, Lago** ◎ W Bolivia
57 P9 **Uruyén** Bolívar, SE Venezuela 5.40N 62.25W
155 O7 **Ūrūzgān** *var.* Oruzgān, Orūzgān. Urūzgān, C Afghanistan 32.58N 66.39E
155 N6 **Urūzgān** *Per.* Orūzgān. ◈ *province* C Afghanistan
172 P6 **Uryū-gawa** ≈ Hokkaidō, NE Japan
172 P4 **Uryū-ko** ◎ Hokkaidō, NE Japan
131 N8 **Uryupinsk** Volgogradskaya Oblast', SW Russian Federation 50.51N 41.59E
Ürzhar *see* Urdzhar
129 R16 **Urzhum** Kirovskaya Oblast', NW Russian Federation 57.09N 49.56E
118 K13 **Urziceni** Ialomița, SE Romania 44.43N 26.39E
U.S./USA *see* United States of America
170 D13 **Usa** Ōita, Kyūshū, SW Japan 33.32N 131.20E
121 L16 **Usa** *Rus.* Usa. ≈ C Belarus
129 T6 **Usa** ≈ NW Russian Federation
142 E14 **Uşak** *prev.* Ushak. Uşak, W Turkey 38.42N 29.25E
142 D14 **Uşak** *var.* Ushak. ◈ *province* W Turkey
85 C19 **Usakos** Erongo, W Namibia 22.01S 15.31E
83 J21 **Usambara Mountains** ⛰ NE Tanzania
83 G23 **Usangu Flats** *wetland* SW Tanzania
67 D24 **Usborne, Mount** ▲ East Falkland, Falkland Islands 51.34S 58.57W
102 O8 **Usedom** *island* NE Germany
101 M24 **Useldange** Diekirch, C Luxembourg 49.46N 5.58E
120 L12 **Ushacha** *Rus.* Ushacha. Vitsyebskaya Voblasts', N Belarus 55.31N 28.30E
Ushachi *see* Ushachy
120 L13 **Ushachy** *Rus.* Ushachi. Vitsyebskaya Voblasts', N Belarus 55.09N 28.37E
Ushak *see* Uşak
126 I2 **Ushakova, Ostrov** *island* Severnaya Zemlya, N Russian Federation
Ushant *see* Ouessant, Île d'
Üsharal *see* Ucharal
170 Bb14 **Ushibuka** *var.* Usibuka. Kumamoto, Shimo-jima, SW Japan 32.13N 130.01E
Ushi Point *see* Sabaneta, Puntan
151 V14 **Ushtobe** *Kaz.* Üshtöbe. Almaty, SE Kazakhstan 45.15N 77.58E
65 I25 **Ushuaia** Tierra del Fuego, S Argentina 54.48S 68.19W
41 R10 **Usibelli** Alaska, USA 63.54N 148.41W
Usibuka *see* Ushibuka
194 I12 **Usino** Madang, N PNG 5.40S 145.31E
129 U6 **Usinsk** Respublika Komi, NW Russian Federation 66.00N 57.37E
99 K22 **Usk** *Wel.* Wysg. ≈ SE Wales, UK
Uskočke Planine/Uskokengebirge *see* Gorjanci/Žumberačko Gorje
Uskoplje *see* Gornji Vakuf
Üsküb *see* Skopje
116 M11 **Üsküpdere** Kırklareli, NW Turkey 41.41N 27.21E
130 L7 **Usman'** Lipetskaya Oblast', W Russian Federation 52.04N 39.41E
120 D8 **Usmas Ezers** ◎ NW Latvia
129 U13 **Usol'ye** Permskaya Oblast', NW Russian Federation 59.27N 56.33E
126 J15 **Usol'ye-Sibirskoye** Irkutskaya Oblast', C Russian Federation 52.48N 103.40E
43 T16 **Uspanapa, Río** ≈ SE Mexico
151 R11 **Uspenskiy** Zhezkazgan, C Kazakhstan 48.45N 72.46E
105 O11 **Ussel** Corrèze, C France 45.33N 2.18E
169 Z6 **Ussuri** *var.* Usuri, Wusuri, *Chin.* Wusuli Jiang. ≈ China/Russian Federation
127 Nn18 **Ussuriysk** *prev.* Nikol'sk, Nikol'sk-Ussuriyskiy, Voroshilov. Primorskiy Kray, SE Russian Federation 43.48N 131.58E
142 J10 **Usta Burnu** *headland* N Turkey 41.58N 34.30E
155 P13 **Usta Muhammad** Baluchistān, SW Pakistan 28.07N 68.00E
126 K15 **Ust'-Barguzin** Respublika Buryatiya, S Russian Federation 53.28N 109.00E
127 P12 **Ust'-Bol'sheretsk** Kamchatskaya Oblast', E Russian Federation 52.48N 156.12E
131 N9 **Ust'-Buzulukskaya** Volgogradskaya Oblast', SW Russian Federation 50.12N 42.06E
113 C16 **Ústecký Kraj** ◈ *region* NW Czech Republic
110 G7 **Uster** Zürich, NE Switzerland 47.20N 8.40E
109 I22 **Ustica, Isola d'** *island* S Italy
126 J13 **Ust'-Ilimsk** Irkutskaya Oblast', C Russian Federation 57.57N 102.30E
113 C15 **Ústí nad Labem** *Ger.* Aussig. Ústecký Kraj, NW Czech Republic 50.40N 14.04E
113 F17 **Ústí nad Orlicí** *Ger.* Wildenschwert. Pardubický Kraj, E Czech Republic 49.58N 16.24E
Ustinov *see* Izhevsk
115 J14 **Ustiprača** Republika Srpska, SE Bosnia and Herzegovina 43.42N 19.03E
125 Ff12 **Ust'-Ishim** Omskaya Oblast', C Russian Federation 57.42N 70.58E
112 G6 **Ustka** *Ger.* Stolpmünde. Pomorskie, N Poland 54.34N 16.50E
127 Pp10 **Ust'-Kamchatsk** Kamchatskaya Oblast', E Russian Federation 56.13N 162.28E
151 X9 **Ust'-Kamenogorsk** *Kaz.* Öskemen. Vostochnyy Kazakhstan, E Kazakhstan 49.58N 82.36E
127 Oo10 **Ust'-Khayryuzovo** Koryakskiy Avtonomnyy Okrug, E Russian Federation 57.07N 156.37E
126 H16 **Ust'-Koksa** Respublika Altay, S Russian Federation 50.15N 85.45E
129 S11 **Ust'-Kulom** Respublika Komi, NW Russian Federation 61.42N 53.42E
126 Jj14 **Ust'-Kut** Irkutskaya Oblast', C Russian Federation 56.49N 105.31E
126 M7 **Ust'-Kuyga** Respublika Sakha (Yakutiya), NE Russian Federation 69.59N 135.27E
130 L14 **Ust'-Labinsk** Krasnodarskiy Kray, SW Russian Federation 44.40N 40.46E
126 Mm11 **Ust'-Maya** Respublika Sakha (Yakutiya), NE Russian Federation 60.27N 134.28E
127 N9 **Ust'-Nera** Respublika Sakha (Yakutiya), NE Russian Federation 64.28N 143.01E
126 Ll13 **Ust'-Nyukzha** Amurskaya Oblast', S Russian Federation 56.30N 121.32E
126 Kk6 **Ust'-Olenëk** Respublika Sakha (Yakutiya), NE Russian Federation 73.03N 119.34E
127 O10 **Ust'-Omchug** Magadanskaya Oblast', E Russian Federation 61.07N 149.17E
126 Jj15 **Ust'-Ordynskiy** Ust'-Ordynskiy Buryatskiy Avtonomnyy Okrug, S Russian Federation 52.49N 104.42E
126 J15 **Ust'-Ordynskiy Buryatskiy Avtonomnyy Okrug** ◈ *autonomous district* S Russian Federation
129 N8 **Ust'-Pinega** Arkhangel'skaya Oblast', NW Russian Federation 64.09N 41.55E
126 Hh8 **Ust'-Port** Taymyrskiy (Dolgano-Nenetskiy) Avtonomnyy Okrug, N Russian Federation 69.42N 84.25E
116 L11 **Ustrem** *prev.* Vakav. Yambol, E Bulgaria 42.01N 26.28E
113 O18 **Ustrzyki Dolne** Podkarpackie, SE Poland 49.26N 22.36E
Ust'-Sysol'sk *see* Syktyvkar
129 R7 **Ust'-Tsil'ma** Respublika Komi, NW Russian Federation 65.25N 52.09E
Ust Urt *see* Ustyurt Plateau
129 O11 **Ust'ya** ≈ NW Russian Federation
119 R8 **Ustynivka** Kirovohrads'ka Oblast', C Ukraine 47.58N 32.32E
150 H15 **Ustyurt Plateau** *var.* Ust Urt, *Uzb.* Ustyurt Platosi. *plateau* Kazakhstan/Uzbekistan
Ustyurt Platosi *see* Ustyurt Plateau
128 K14 **Ustyuzhna** Vologodskaya Oblast', NW Russian Federation 58.50N 36.25E
164 J4 **Usu** Xinjiang Uygur Zizhiqu, NW China 44.27N 84.37E
175 Q10 **Usu** Sulawesi, C Indonesia 2.34S 120.58E
170 Dd14 **Usuki** Ōita, Kyūshū, SW Japan 33.07N 131.46E
44 G8 **Usulután** Usulután, SE El Salvador 13.19N 88.26W
44 B9 **Usulután** ◈ *department* SE El Salvador
43 W16 **Usumacinta, Río** ≈ Guatemala/Mexico
Usumbura *see* Bujumbura
Usuri *see* Ussuri
176 X12 **Uta** Irian Jaya, E Indonesia 4.28S 136.03E
38 K5 **Utah** *off.* State of Utah; also known as Beehive State, Mormon State. ◈ *state* W USA
38 L3 **Utah Lake** ◎ Utah, W USA
Utaidhani *see* Uthai Thani
95 M14 **Utajärvi** Oulu, C Finland 64.45N 26.25E
Utamboni *see* Mitemele, Río
Utaradit *see* Uttaradit
173 G3 **Utara, Selat** *strait* Peninsular Malaysia
172 P5 **Utashinai** *var.* Utasinai. Hokkaidō, NE Japan 43.32N 142.03E
Utasinai *see* Utashinai
176 X12 **Uta, Sungai** ≈ Irian Jaya, E Indonesia
200 Ss12 **ʿUta Vavaʿu** *island* Vavaʿu Group, N Tonga
39 V9 **Ute Creek** ≈ New Mexico, SW USA
120 H12 **Utena** Utena, E Lithuania 55.30N 25.34E
39 V10 **Ute Reservoir** ▣ New Mexico, SW USA
178 H10 **Uthai Thani** *var.* Muang Uthai Thani, Udayadhani, Utaidhani. Uthai Thani, W Thailand 15.22N 100.03E
155 O15 **Uthal** Baluchistān, SW Pakistan 25.53N 66.37E
20 I10 **Utica** New York, NE USA 43.06N 75.15W
107 R10 **Utiel** País Valenciano, E Spain 39.33N 1.13W
9 O13 **Utikuma Lake** ◎ Alberta, W Canada
44 I4 **Utila, Isla de** *island* Islas de la Bahía, N Honduras
Utina *see* Udine
61 O17 **Utinga** Bahia, E Brazil 12.05S 41.07W
Utirik *see* Utrik Atoll
97 M22 **Utlängan** *island* S Sweden
119 U11 **Utlyuts'kyy Lyman** *bay* S Ukraine
170 Cc14 **Uto** Kumamoto, Kyūshū, SW Japan 32.40N 130.37E
97 P16 **Utö** Stockholm, C Sweden 58.55N 18.19E
27 Q12 **Utopia** Texas, SW USA 29.30N 99.31W
100 J11 **Utrecht** *Lat.* Trajectum ad Rhenum. Utrecht, C Netherlands 52.06N 5.07E
85 K22 **Utrecht** KwaZulu/Natal, E South Africa 27.34S 30.19E
100 I11 **Utrecht** ◈ *province* C Netherlands
106 K14 **Utrera** Andalucía, S Spain 37.10N 5.46W
201 V4 **Utrik Atoll** *var.* Utirik, Utrōk, Utrōnk. *atoll* Ratak Chain, N Marshall Islands
Utrōk/Utrōnk *see* Utrik Atoll
97 B16 **Utsira** *island* SW Norway
94 L8 **Utsjoki** *var.* Ohcejohka. Lappi, N Finland 69.54N 27.01E
171 Kk15 **Utsunomiya** *var.* Utunomiya. Tochigi, Honshū, S Japan 36.36N 139.52E
131 P13 **Utta** Respublika Kalmykiya, SW Russian Federation 46.22N 46.03E
178 Hh8 **Uttaradit** *var.* Utaradit. Uttaradit, N Thailand 17.37N 100.04E
158 J8 **Uttarkāshi** Uttar Pradesh, N India 30.45N 78.19E
158 K11 **Uttar Pradesh** *prev.* United Provinces, United Provinces of Agra and Oudh. ◈ *state* N India
47 T5 **Utuado** C Puerto Rico 18.16N 66.43W
164 K3 **Utubulak** Xinjiang Uygur Zizhiqu, W China 46.49N 86.15E
41 N5 **Utukok River** ≈ Alaska, USA
Utunomiya *see* Utsunomiya
195 X9 **Utupua** *island* Santa Cruz Islands, E Soloman Islands
150 G9 **Utva** ≈ NW Kazakhstan
201 Y15 **Utwe** Kosrae, E Micronesia
201 X15 **Utwe Harbor** *harbor* Kosrae, E Micronesia
168 J7 **Uubulan** Arhangay, C Mongolia 48.37N 101.58E
120 G6 **Uulu** Pärnumaa, SW Estonia 58.15N 24.31E
207 N13 **Uummannaq** *var.* Umanak, Umanaq. Kitaa, C Greenland 70.37N 52.25W
Uummannarsuaq *see* Nunap Isua
168 E4 **Üüreg Nuur** ◎ NW Mongolia
Uusikaarlepyy *see* Nykarleby
95 J19 **Uusikaupunki** *Swe.* Nystad. Länsi-Suomi, W Finland 60.48N 21.25E
131 S2 **Uva** Udmurtskaya Respublika, NW Russian Federation 56.41N 52.15E
115 L14 **Uvac** ≈ W Yugoslavia
27 Q12 **Uvalde** Texas, SW USA 29.13N 99.49W
161 K25 **Uva Province** ◈ *province* SE Sri Lanka
121 O18 **Uvaravichy** *Rus.* Uvarovichi. Homyel'skaya Voblasts', SE Belarus 52.36N 30.43E
56 J11 **Uvá, Río** ≈ E Colombia
Uvarovichi *see* Uvaravichy
131 N7 **Uvarovo** Tambovskaya Oblast', W Russian Federation 51.58N 42.13E
125 Ff11 **Uvat** Tyumenskaya Oblast', C Russian Federation 59.11N 68.37E
202 G12 **Uvea, Île** *island* N Wallis and Futuna
83 E21 **Uvinza** Kigoma, W Tanzania 5.04S 30.24E
81 O20 **Uvira** Sud Kivu, E Dem. Rep. Congo (Zaire) 3.24S 29.04E
168 E5 **Uvs** ◈ *province* NW Mongolia
168 F5 **Uvs Nuur** *var.* Ozero Ubsu-Nur. ◎ Mongolia/Russian Federation
170 E15 **Uwa** Ehime, Shikoku, SW Japan 33.22N 132.29E
170 E15 **Uwajima** *var.* Uwazima. Ehime, Shikoku, SW Japan 33.13N 132.32E
82 B5 **ʿUwaynāt, Jabal al** *var.* Jebel Uweinat. ▲ Libya/Sudan 21.51N 25.01E
Uwazima *see* Uwajima
Uweinat, Jebel *see* ʿUwaynāt, Jabal al
176 Z14 **Uwimmerah, Sungai** ≈ Irian Jaya, E Indonesia
12 H14 **Uxbridge** Ontario, S Canada 44.07N 79.07W
Uxellodunum *see* Issoudun
168 M15 **Uxin Qi** Nei Mongol Zizhiqu, N China 38.29N 108.48E
43 X12 **Uxmal, Ruinas** *ruins* Yucatán, SE Mexico 20.20N 89.46W
133 Q5 **Uy** ≈ Kazakhstan/Russian Federation
150 K15 **Uyaly** Kzylorda, S Kazakhstan 44.22N 61.16E
126 Mm7 **Uyandina** ≈ NE Russian Federation
126 I14 **Uyar** Krasnoyarskiy Kray, S Russian Federation 55.48N 94.12E
168 L10 **Üydzen** Ömnögovĭ, S Mongolia 44.08N 106.48E
Uyeda *see* Ueda
126 Hh3 **Uyedineniya, Ostrov** *island* N Russian Federation
79 V17 **Uyo** Akwa Ibom, S Nigeria 5.00N 7.57E
168 D8 **Üyönch** Hovd, W Mongolia 46.04N 92.05E
151 Q15 **Uyuk** Zhambyl, S Kazakhstan 43.46N 70.55E
147 V13 **ʿUyūn** SW Oman 17.12N 53.46E
59 K20 **Uyuni** Potosí, W Bolivia 20.26S 66.48W
59 J20 **Uyuni, Salar de** *wetland* SW Bolivia
152 I9 **Uzbekistan** *off.* Republic of Uzbekistan. ◆ *republic* C Asia
164 D8 **Uzbel Shankou** *Rus.* Pereval Kyzyl-Dzhiik. *pass* China/Tajikistan 38.33N 73.46E
121 J17 **Uzda** *Rus.* Uzda. Minskaya Voblasts', C Belarus 53.29N 27.10E
105 N12 **Uzerche** Corrèze, C France 45.24N 1.35E
105 R14 **Uzès** Gard, S France 44.00N 4.25E
153 T10 **Uzgen** *Kir.* Özgön. Oshskaya Oblast', SW Kyrgyzstan 40.42N 73.17E
119 O3 **Uzh** ≈ N Ukraine
Uzhgorod *see* Uzhhorod
118 G7 **Uzhhorod** *Rus.* Uzhgorod; *prev.* Ungvár. Zakarpats'ka Oblast', W Ukraine 48.36N 22.19E
126 Hh14 **Uzhur** Krasnoyarskiy Kray, S Russian Federation 55.18N 89.36E
Uzi *see* Uji
114 K13 **Užice** *prev.* Titovo Užice. Serbia, W Yugoslavia 43.52N 19.51E
Uzin *see* Uzyn
130 L5 **Uzlovaya** Tul'skaya Oblast', W Russian Federation 54.01N 38.15E
110 H7 **Uznach** Sankt Gallen, NE Switzerland 47.12N 9.00E
151 U16 **Uzunagach** Almaty, SE Kazakhstan 43.07N 76.19E
142 B10 **Uzunköprü** Edirne, NW Turkey 41.15N 26.42E
120 D11 **Užventis** Kelmė, C Lithuania 55.49N 22.38E
119 P5 **Uzyn** *Rus.* Uzin. Kyyivs'ka Oblast', N Ukraine 49.52N 30.28E

V

Vääksy *see* Asikkala
85 H23 **Vaal** ≈ C South Africa
95 M14 **Vaala** Oulu, C Finland 64.34N 26.49E
95 N19 **Vaalimaa** Etelä-Suomi, SE Finland 60.34N 27.49E
101 M19 **Vaals** Limburg, SE Netherlands 50.46N 6.01E
95 J16 **Vaasa** *Swe.* Vasa; *prev.* Nikolainkaupunki. Vaasa, W Finland 63.07N 21.39E
100 L10 **Vaassen** Gelderland, E Netherlands 52.18N 5.58E
120 G11 **Vabalninkas** Biržai, NE Lithuania 55.59N 24.45E
Vabkent *see* Wobkent
113 J22 **Vác** *Ger.* Waitzen. Pest, N Hungary 47.46N 19.07E
63 I14 **Vacaria** Rio Grande do Sul, S Brazil 28.30S 50.57W
37 N7 **Vacaville** California, W USA 38.21N 121.59W
105 R15 **Vaccarès, Étang de** ◎ SE France
46 L10 **Vache, Île à** *island* SW Haiti
181 Y16 **Vacoas** W Mauritius 20.18S 57.28E
34 G10 **Vader** Washington, NW USA 46.23N 122.58W
96 D12 **Vadheim** Sogn og Fjordane, S Norway 61.12N 5.48E
124 O3 **Vadili** *Gk.* Vatilí. C Cyprus 35.09N 33.39E
160 D11 **Vadodara** *prev.* Baroda. Gujarāt, W India 22.19N 73.13E
94 M8 **Vadsø** *Fin.* Vesisaari. Finnmark, N Norway 70.07N 29.47E
97 L17 **Vadstena** Östergötland, S Sweden 58.25N 14.55E
110 I8 **Vaduz** ● (Liechtenstein) W Liechtenstein 47.07N 9.31E
Vág *see* Váh
129 N12 **Vaga** ≈ NW Russian Federation
96 G11 **Vågåmo** Oppland, S Norway 61.52N 9.06E
114 D12 **Vaganski Vrh** ▲ W Croatia 44.24N 15.32E
97 A19 **Vágar** *Dan.* Vågø *Island* Faeroe Islands 62.03N 7.19W
Vágbeszterce *see* Považská Bystrica
97 L19 **Vaggeryd** Jönköping, S Sweden 57.30N 14.10E
195 U14 **Vaghena** *var.* Wagina. *island* NW Solomon Islands
97 O16 **Vagnhärad** Södermanland, C Sweden 58.57N 17.31E
106 G7 **Vagos** Aveiro, N Portugal 40.33N 8.42W
Vágsellye *see* Sal'a
94 H10 **Vågsfjorden** *fjord* N Norway
96 C10 **Vågsøy** *island* S Norway
Vágújhely *see* Nové Mesto nad Váhom
113 I21 **Váh** *Ger.* Waag, *Hung.* Vág. ≈ W Slovakia
95 K16 **Vähäkyrö** Länsi-Suomi, W Finland 63.04N 22.04E
203 X11 **Vahitahi** *atoll* Îles Tuamotu, E French Polynesia
Vaidei *see* Vulcan
24 L4 **Vaiden** Mississippi, S USA 33.19N 89.42W
161 I23 **Vaigai** ≈ SE India
203 V16 **Vaihu** Easter Island, Chile, E Pacific Ocean 27.10S 109.22W
120 I6 **Väike Emajõgi** ≈ S Estonia
120 I4 **Väike-Maarja** *Ger.* Klein-Marien. Lääne-Virumaa, NE Estonia 59.07N 26.13E
Väike-Salatsi *see* Mazsalaca
39 R4 **Vail** Colorado, C USA 39.36N 106.20W
200 Qq15 **Vaina** Tongatapu, S Tonga 21.12S 175.10W
120 E5 **Väinameri** *prev.* Muhu Väin, *Ger.* Moon-Sund. *sea* E Baltic Sea
95 N18 **Vainikkala** Etelä-Suomi, SE Finland 60.54N 28.18E
120 D10 **Vaiņode** Liepāja, SW Latvia 56.25N 21.52E
161 H23 **Vaippār** ≈ SE India
203 W11 **Vairaatea** *atoll* Îles Tuamotu, C French Polynesia
203 R8 **Vairao** Tahiti, W French Polynesia 17.48S 149.16W
105 R14 **Vaison-la-Romaine** Vaucluse, SE France 44.15N 5.04E
202 G11 **Vaitupu** Île Uvea, E Wallis and Futuna 13.13S 176.09W
202 F7 **Vaitupu** *atoll* C Tuvalu
Vajdahunyad *see* Hunedoara
Vajdej *see* Vulcan
80 K12 **Vakaga** ◈ *prefecture* NE Central African Republic
116 H10 **Vakarel** Sofiya, W Bulgaria 42.34N 23.43E
Vakav *see* Ustrem
143 O11 **Vakfıkebir** Trabzon, NE Turkey 41.03N 39.19E
126 H11 **Vakh** ≈ C Russian Federation
Vakhon, Qatorkŭhi *see* Nicholas Range
153 P14 **Vakhsh** SW Tajikistan 37.46N 68.48E
153 Q12 **Vakhsh** ≈ SW Tajikistan
131 P1 **Vakhtan** Nizhegorodskaya Oblast', W Russian Federation 58.00N 46.43E
96 C13 **Vaksdal** Hordaland, S Norway 60.28N 5.45E
129 O8 **Vaksha** ≈ NW Russian Federation
195 O15 **Vakuta Island** *island* Kiriwina Islands, SE PNG
Valachia *see* Wallachia
110 D11 **Valais** *Ger.* Wallis. ◈ *canton* SW Switzerland
115 M21 **Valamarës, Mali i** ▲ SE Albania 40.48N 20.31E
131 S2 **Valamaz** Udmurtskaya Respublika, NW Russian Federation 57.36N 52.07E
115 Q19 **Valandovo** SE FYR Macedonia 41.20N 22.33E
113 I18 **Valašské Meziříčí** *Ger.* Wallachisch-Meseritsch, *Pol.* Wałeckie Międzyrzecze. Zlínský Kraj, E Czech Republic 49.28N 17.57E
117 I17 **Valáxa** *island* Vóreioi Sporádes, Greece, Aegean Sea
97 K16 **Vålberg** Värmland, C Sweden 59.24N 13.12E
118 H12 **Vâlcea** *prev.* Vîlcea. ◈ *county* SW Romania
65 J16 **Valcheta** Río Negro, E Argentina 40.42S 66.07W
13 P12 **Valcourt** Quebec, SE Canada 45.28N 72.18W
Valdai Hills *see* Valdayskaya Vozvyshennost'
106 M3 **Valdavia** ≈ N Spain
128 I15 **Valday** Novgorodskaya Oblast', W Russian Federation 57.56N 33.19E
128 I15 **Valdayskaya Vozvyshennost'** *var.* Valdai Hills. *hill range* W Russian Federation
106 L9 **Valdecañas, Embalse de** ▣ W Spain
120 E8 **Valdemārpils** *Ger.* Sassmacken. Talsi, NW Latvia 57.23N 22.36E
97 N18 **Valdemarsvik** Östergötland, S Sweden 58.13N 16.34E
107 N8 **Valdemoro** Madrid, C Spain 40.12N 3.40W
107 O11 **Valdepeñas** Castilla-La Mancha, C Spain 38.46N 3.24W
106 L5 **Valderaduey** ≈ NE Spain
106 L5 **Valderas** Castilla-León, N Spain 42.04N 5.27W
107 T7 **Valderrobres** *var.* Vall-de-roures. Aragón, NE Spain 40.52N 0.07E
65 K17 **Valdés, Península** *peninsula* SE Argentina
41 S11 **Valdez** Alaska, USA 61.07N 146.21W
58 C5 **Valdéz** *var.* Limones. Esmeraldas, NW Ecuador 1.17N 78.56W
Valdia *see* Weldiya
105 U11 **Val d'Isère** Savoie, E France 45.23N 7.03E
65 G15 **Valdivia** Los Lagos, C Chile 39.49S 73.12W
Valdivia Bank *see* Valdivia Seamount
67 P17 **Valdivia Seamount** *var.* Valdivia Bank. *undersea feature* E Atlantic Ocean 26.15S 6.25E
105 N4 **Val-d'Oise** ◈ *department* N France
12 J8 **Val-d'Or** Quebec, SE Canada 48.05N 77.42W
25 U8 **Valdosta** Georgia, SE USA 30.49N 83.16W
96 G13 **Valdres** *physical region* S Norway
34 L13 **Vale** Oregon, NW USA 43.58N 117.14W
118 F9 **Valea lui Mihai** *Hung.* Érmihályfalva. Bihor, NW Romania 47.31N 22.08E
9 N15 **Valemount** British Columbia, SW Canada 52.46N 119.17W
61 O17 **Valença** Bahia, E Brazil 13.22S 39.06W
106 F4 **Valença do Minho** Viana do Castelo, N Portugal 42.01N 8.37W
61 N14 **Valença do Piauí** Piauí, E Brazil 6.25S 41.46W
105 N8 **Valençay** Indre, C France 47.10N 1.31E
105 R13 **Valence** *anc.* Valentia, Valentia Julia, Ventia. Drôme, E France 44.55N 4.54E
107 S10 **Valencia** País Valenciano, E Spain 39.28N 0.24W
56 K5 **Valencia** Carabobo, N Venezuela 10.11N 68.02W
107 R10 **Valencia** *Cat.* València. ◈ *province* País Valenciano, E Spain

◆ Country
● Country Capital
◇ Dependent Territory
○ Dependent Territory Capital
◈ Administrative Region
✖ International Airport
▲ Mountain
⛰ Mountain Range
♨ Volcano
≈ River
◎ Lake
▣ Reservoir

107 S10 **Valencia ✈** Valencia, E Spain
València/Valencia see País Valenciano
106 I10 **Valencia de Alcántara** Extremadura, W Spain 39.25N 7.13W
106 I4 **Valencia de Don Juan** Castilla-León, N Spain 42.16N 5.31W
107 U9 **Valencia, Golfo de** var. Gulf of Valencia, gulf E Spain
Valencia, Gulf of see Valencia, Golfo de
99 A21 **Valencia Island** Ir. Dairbhre. island SW Ireland
105 P2 **Valenciennes** Nord, N France 50.21N 3.31E
118 K13 **Vălenii de Munte** Prahova, SE Romania 45.10N 26.01E
Valentia see Valence, France
Valentia see País Valenciano
Valentia Julia see Valence
105 T8 **Valentigney** Doubs, E France 47.27N 6.49E
30 M12 **Valentine** Nebraska, C USA 42.52N 100.31W
26 J10 **Valentine** Texas, SW USA 30.35N 104.30W
Valentine State see Oregon
108 C8 **Valenza** Piemonte, NW Italy 45.01N 8.37E
96 I13 **Våler** Hedmark, S Norway 60.39N 11.52E
56 I6 **Valera** Trujillo, NW Venezuela 9.21N 70.37W
199 K13 **Valerie Guyot** undersea feature S Pacific Ocean 33.00S 164.00W
Valetta see Valletta
120 I7 **Valga** Ger. Walk, Latv. Valka. Valgamaa, S Estonia 57.48N 26.04E
120 I7 **Valgamaa** off. Valga Maakond. ◆ province S Estonia
45 Q15 **Valiente, Península** peninsula NW Panama
105 X16 **Valinco, Golfe de** gulf Corse, France, C Mediterranean Sea
114 L12 **Valjevo** Serbia, W Yugoslavia 44.16N 19.54E
Valjok see Väljohka
120 I7 **Valka** Ger. Walk. Valka, N Latvia 57.48N 26.01E
Valka see Valga
95 L18 **Valkeakoski** Länsi-Suomi, W Finland 61.16N 24.04E
95 M19 **Valkeala** Etelä-Suomi, S Finland 60.55N 26.49E
101 L18 **Valkenburg** Limburg, SE Netherlands 50.52N 5.46E
101 K15 **Valkenswaard** Noord-Brabant, S Netherlands 51.21N 5.28E
121 G15 **Valkininkai** Varėna, S Lithuania 54.22N 24.51E
119 U5 **Valky** Kharkivs'ka Oblast', E Ukraine 49.51N 35.40E
43 Y12 **Valladolid** Yucatán, SE Mexico 20.39N 88.13W
106 M5 **Valladolid** Castilla-León, NW Spain 41.39N 4.45W
106 L5 **Valladolid** ◆ province Castilla-León, N Spain
105 U15 **Vallauris** Alpes-Maritimes, SE France 43.34N 7.03E
Vall-de-roures see Valderrobres
107 S9 **Vall d'Uxó** País Valenciano, E Spain 39.49N 0.15W
97 E16 **Valle** Aust-Agder, S Norway 59.13N 7.33E
107 N2 **Valle** Cantabria, N Spain 43.14N 4.16W
44 H8 **Valle** ◆ department S Honduras
107 N8 **Vallecas** Madrid, C Spain 40.22N 3.37W
39 Q8 **Vallecito Reservoir** ☐ Colorado, C USA
108 A7 **Valle d'Aosta** ◆ region NW Italy
43 O14 **Valle de Bravo** México, S Mexico 19.19N 100.08W
57 N5 **Valle de Guanape** Anzoátegui, N Venezuela 9.49N 65.34W
56 M6 **Valle de La Pascua** Guárico, N Venezuela 9.15N 66.00W
56 B11 **Valle del Cauca** off. Departamento del Valle del Cauca. ◆ province W Colombia
43 N13 **Valle de Santiago** Guanajuato, C Mexico 20.21N 101.13W
42 J7 **Valle de Zaragoza** Chihuahua, N Mexico 27.25N 105.50W
56 G5 **Valledupar** Cesar, N Colombia 10.31N 73.16W
78 G9 **Vallée de Ferlo** ❧ NW Senegal
59 M19 **Vallegrande** Santa Cruz, C Bolivia 18.30S 64.06W
43 P8 **Valle Hermoso** Tamaulipas, C Mexico 25.39N 97.49W
37 N8 **Vallejo** California, W USA 38.07N 122.16W
64 G8 **Vallenar** Atacama, N Chile 28.35S 70.44W
97 O15 **Vallentuna** Stockholm, C Sweden 59.31N 18.04E
123 L12 **Valletta** prev. Valetta. ● (Malta) E Malta 35.54N 14.30E
49 N6 **Valley Center** Kansas, C USA 37.49N 97.22W
31 Q5 **Valley City** North Dakota, N USA 46.57N 97.58W
34 I15 **Valley Falls** Oregon, NW USA 42.28N 120.16W
Valleyfield see Salaberry-de-Valleyfield
23 S4 **Valley Head** West Virginia, NE USA 38.33N 80.01W
27 T8 **Valley Mills** Texas, SW USA 31.39N 97.27W
77 W10 **Valley of the Kings** ancient monument E Egypt 23.41N 32.30E
31 T9 **Valley Springs** South Dakota, N USA 43.34N 96.28W
22 K5 **Valley Station** Kentucky, S USA 38.06N 85.52W
9 O14 **Valleyview** Alberta, W Canada 55.01N 117.16W
27 T5 **Valley View** Texas, SW USA 33.27N 97.08W
63 C21 **Vallimanca, Arroyo** ❧ E Argentina
95 L9 **Väljohka** var. Valjok Finnmark, N Norway 69.39N 25.52E
109 M19 **Vallo della Lucania** Campania, S Italy 40.13N 15.15E

110 B9 **Vallorbe** Vaud, W Switzerland 46.43N 6.21E
107 V6 **Valls** Cataluña, NE Spain 41.18N 1.15E
96 N11 **Vallsta** Gävleborg, C Sweden 61.30N 16.25E
96 N12 **Vallvik** Gävleborg, C Sweden 61.10N 17.15E
9 T17 **Val Marie** Saskatchewan, S Canada 49.15N 107.43W
120 H7 **Valmiera** Est. Volmari, Ger. Wolmar. Valmiera, N Latvia 57.33N 25.26E
107 N3 **Valnera** ▲ N Spain 43.08N 3.39W
104 J3 **Valognes** Manche, N France 49.31N 1.28W
Valona see Vlorë
Valona Bay see Vlorës, Gjiri i
106 G6 **Valongo** var. Valongo de Gaia. Porto, N Portugal 41.10N 8.30W
Valongo de Gaia see Valongo
106 M5 **Valoria la Buena** Castilla-León, N Spain 41.48N 4.33W
121 J15 **Valozhyn** Pol. Wołożyn, Rus. Volozhin. Minskaya Voblasts', C Belarus 54.07N 26.31E
106 I5 **Valpaços** Vila Real, N Portugal 41.36N 7.16W
25 P8 **Valparaiso** Florida, SE USA 30.30N 86.28W
33 N11 **Valparaiso** Indiana, N USA 41.28N 87.04W
64 G11 **Valparaíso** Valparaíso, C Chile 33.04S 71.38W
42 L11 **Valparaíso** Zacatecas, C Mexico 22.49N 103.28W
64 G11 **Valparaíso** off. Región de Valparaíso. ◆ region C Chile
Valpo see Valpovo
114 I9 **Valpovo** Hung. Valpo. Osijek-Baranja, E Croatia 45.40N 18.25E
105 R14 **Valréas** Vaucluse, SE France 44.22N 5.00E
Vals see Vals-Platz
160 D12 **Valsād** prev. Bulsar. Gujarāt, W India 20.40N 72.55E
Valsbaai see False Bay
176 Uu10 **Vals Pisang, Kepulauan** island group E Indonesia
110 H9 **Vals-Platz** var. Vals. Graubünden, S Switzerland 46.39N 9.09E
176 Xx16 **Vals, Tanjung** headland Irian Jaya, SE Indonesia 8.25S 137.34E
95 N15 **Valtimo** Itä-Suomi, E Finland 63.39N 28.49E
117 D17 **Váltou** ▲ C Greece
131 O12 **Valuyevka** Rostovskaya Oblast', SW Russian Federation 46.48N 43.49E
130 K9 **Valuyki** Belgorodskaya Oblast', W Russian Federation 50.11N 38.07E
38 L2 **Val Verda** Utah, W USA 40.51N 111.53W
66 I13 **Valverde** Hierro, Islas Canarias, Spain, NE Atlantic Ocean 27.48N 17.55W
106 J13 **Valverde del Camino** Andalucía, S Spain 37.34N 6.45W
97 G23 **Vamdrup** Vejle, C Denmark 55.25N 9.16E
96 L13 **Vämhus** Dalarna, C Sweden 61.07N 14.30E
95 K18 **Vammala** Länsi-Suomi, W Finland 61.19N 22.55E
Vámosudvarhely see Odorheiu Secuiesc
143 S14 **Van** Van, E Turkey 38.30N 43.22E
27 V7 **Van** Texas, SW USA 32.31N 95.38W
143 T14 **Van** ◆ province E Turkey
143 T11 **Vanadzor** prev. Kirovakan. N Armenia 40.49N 44.28E
27 U5 **Van Alstyne** Texas, SW USA 33.25N 96.34W
35 W10 **Vananda** Montana, NW USA 46.22N 106.58W
118 I11 **Vânători** Hung. Héjjasfalva; prev. Vinători. Mureş, C Romania 46.13N 24.56E
203 W12 **Vanavana** atoll Îles Tuamotu, SE French Polynesia
126 J12 **Vanavara** Evenkiyskiy Avtonomnyy Okrug, C Russian Federation 60.19N 102.19E
13 Q8 **Van Bruyssel** Quebec, SE Canada 47.56N 72.08W
29 R10 **Van Buren** Arkansas, C USA 35.26N 94.21W
21 S1 **Van Buren** Maine, NE USA 47.07N 67.57W
49 W7 **Van Buren** Missouri, C USA 37.00N 91.00W
21 T5 **Vanceboro** Maine, NE USA 45.31N 67.25W
23 W10 **Vanceboro** North Carolina, SE USA 35.18N 77.06W
23 O4 **Vanceburg** Kentucky, S USA 38.32N 83.18W
129 T3 **Vanch** see Vanj
8 L17 **Vancouver** British Columbia, SW Canada 49.13N 123.06W
34 G10 **Vancouver** Washington, NW USA 45.38N 122.39W
8 K16 **Vancouver ✈** British Columbia, SW Canada
8 K16 **Vancouver Island** island British Columbia, SW Canada
Vanda see Vantaa
32 L15 **Vandalia** Illinois, N USA 38.57N 89.05W
33 R14 **Vandalia** Ohio, N USA 39.53N 84.11W
27 T2 **Vanderbilt** Texas, SW USA 28.45N 96.37W
33 Q7 **Vandercook Lake** Michigan, N USA 42.11N 84.23W
8 K14 **Vanderhoof** British Columbia, SW Canada 53.54N 124.00W
20 K8 **Vanderwhacker Mountain** ▲ New York, NE USA 43.54N 74.06W
189 P1 **Van Diemen Gulf** gulf Northern Territory, N Australia
189 N12 **Van Diemen's Land** see Tasmania
120 H5 **Vändra** Ger. Fennern; prev. Vana-Vändra. Pärnumaa, SW Estonia 58.40N 25.02E

Vandsburg see Więcbork
36 L4 **Van Duzen River** ❧ California, W USA
120 F13 **Vandžiogala** Kaunas, C Lithuania 55.07N 23.55E
43 N10 **Vanegas** San Luis Potosí, C Mexico 23.53N 100.55W
Vaner, Lake see Vänern
97 K17 **Vänern Eng.** Lake Vaner; prev. Lake Vener. ☐ S Sweden
97 J18 **Vänersborg** Västra Götaland, S Sweden 58.16N 12.22E
96 F12 **Vang** Oppland, S Norway 61.07N 8.34E
180 I7 **Vangaindrano** Fianarantsoa, SE Madagascar 23.21S 47.34E
143 S14 **Van Gölü, Lake** var. anc. Thospitis. salt lake E Turkey
195 V15 **Vangunu** island New Georgia Islands, NW Solomon Islands
26 J9 **Van Horn** Texas, SW USA 31.03N 104.51W
195 X9 **Vanikolo** var. Vanikoro. island Santa Cruz Islands, E Solomon Islands
Vanikoro see Vanikolo
194 E9 **Vanimo** Sandaun, NW PNG 2.43S 141.22E
127 O15 **Vanino** Khabarovskiy Kray, SE Russian Federation 49.10N 140.18E
153 S13 **Vāniyambādi** ❧ SW India
159 S12 **Vanj Rus.** Vanch. S Tajikistan 38.22N 71.27E
118 G14 **Vânju Mare** prev. Vînju Mare. Mehedinţi, SW Romania 44.25N 22.52E
127 P3 **Vankarem** Chukotskiy Avtonomnyy Okrug, NE Russian Federation 67.48N 176.11W
96 L13 **Vankleek Hill** Ontario, SE Canada 45.32N 74.39W
95 L14 **Van, Lake** see Van Gölü
95 I15 **Vännäs** Västerbotten, N Sweden 63.54N 19.43E
95 I15 **Vännäsby** Västerbotten, N Sweden 63.55N 19.53E
104 H7 **Vannes** anc. Dariorigum. Morbihan, NW France 47.40N 2.45W
94 I8 **Vannøya** island N Norway
105 T12 **Vanoise, Massif de la** ▲ E France
176 Xx10 **Van Rees, Pegunungan** ▲ Irian Jaya, E Indonesia
85 E24 **Vanrhynsdorp** Western Cape, South Africa 31.33S 18.42E
96 L13 **Vansbro** Dalarna, C Sweden 60.31N 14.15E
97 D18 **Vanse** Vest-Agder, S Norway 58.04N 6.40E
15 M14 **Vansittart Island** island Nunavut, NE Canada
94 M20 **Vantaa Swe.** Vanda. Etelä-Suomi, S Finland 60.18N 25.01E
95 L19 **Vantaa ✈** (Helsinki) Etelä-Suomi, S Finland 60.18N 25.01E
34 J4 **Vantage** Washington, NW USA 46.55N 119.55W
197 K14 **Vanua Balavu** prev. Vanua Mbalavu. island Lau Group, E Fiji
197 C10 **Vanua Lava** island Banks Islands, N Vanuatu
197 I9 **Vanua Levu** island N Fiji
197 I12 **Vanua Levu Barrier Reef** reef C Fiji
197 B10 **Vanuatu** off. Republic of Vanuatu; prev. New Hebrides. ◆ republic SW Pacific Ocean
183 P8 **Vanuatu** island group SW Pacific Ocean
197 K15 **Vanua Vatu** island Lau Group, E Fiji
33 Q12 **Van Wert** Ohio, N USA 40.52N 84.34W
197 K7 **Vao Province Sud, S New Caledonia 22.40S 167.29E
119 N7 **Vapnyarka** Vinnyts'ka Oblast', C Ukraine 48.32N 28.44E
105 T15 **Var** ◆ department SE France
105 U14 **Var** ❧ SE France
97 J18 **Vara** Västra Götaland, S Sweden 58.16N 12.57E
117 H25 **Varadinska Županija** see Varaždin
120 J10 **Varakļāni** Madona, C Latvia 56.36N 26.40E
108 B7 **Varallo** Piemonte, NE Italy 45.51N 8.16E
149 O5 **Varāmīn** var. Veramin. Tehrān, N Iran 35.19N 51.40E
159 N14 **Vārānasi** prev. Banaras, Benares, hist. Kasi. Uttar Pradesh, N India 25.20N 83.00E
129 T3 **Varandey** Nenetskiy Avtonomnyy Okrug, NW Russian Federation 68.48N 57.54E
118 M10 **Vărāncău** see Ştei
97 O15 **Varangerbotn** Finnmark, N Norway 70.09N 28.28E
94 N8 **Varangerfjorden** fjord N Norway 70.00N 30.00E
94 M8 **Varangerhalvøya** peninsula N Norway
Varannó see Vranov nad Topľou
120 J13 **Varapayeva Rus.** Voropayevo. Vitsyebskaya Voblasts', NW Belarus 55.09N 27.13E
94 H11 **Varāzdin Ger.** Warasdin, Hung. Varasd. Varaždin, N Croatia 46.18N 16.20E
114 E7 **Varaždin** off. Varadinska Županija. ◆ province N Croatia
108 C8 **Varazze** Liguria, NW Italy 44.21N 8.35E
97 J19 **Varberg** Halland, S Sweden 57.06N 12.15E
115 Q19 **Vardar Gk.** Axiós. ❧ FYR Macedonia/Greece see also Axiós
97 F23 **Varde** Ribe, W Denmark 55.38N 8.29E
143 V12 **Vardenis** E Armenia 40.11N 45.43E
94 N8 **Vardø Fin.** Vuoreija. Finnmark, N Norway 70.22N 31.04E
117 E18 **Vardoúsia** ▲ C Greece
Vareia see Logroño

102 G10 **Varel** Niedersachsen, NW Germany 53.24N 8.07E
121 G15 **Varėna Pol.** Orany. Varėna, S Lithuania 54.13N 24.35E
13 O12 **Varennes** Quebec, SE Canada 45.42N 73.25W
105 P10 **Varennes-sur-Allier** Allier, C France 46.17N 3.24E
114 I12 **Vareš** Federacija Bosna I Hercegovina, E Bosnia and Herzegovina 44.12N 18.19E
108 D7 **Varese** Lombardia, N Italy 45.49N 8.49E
94 K3 **Vatnajökull** glacier SE Iceland
97 P15 **Vättö** Stockholm, C Sweden 59.48N 18.55E
97 J18 **Vårgårda** Västra Götaland, S Sweden 58.00N 12.49E
125 F12 **Vargashi** Kurganskaya Oblast', C Russian Federation 55.22N 65.39E
97 J18 **Vårgön** Västra Götaland, S Sweden 58.21N 12.22E
97 C17 **Varhaug** Rogaland, S Norway 58.37N 5.39E
95 N17 **Varkaus** Itä-Suomi, C Finland 62.19N 27.49E
94 J2 **Varmahlíð** Norðhurland Vestra, N Iceland 65.32N 19.33W
97 J15 **Värmland** ◆ county C Sweden
97 K16 **Värmlandsnäs** peninsula S Sweden
116 N8 **Varna** prev. Stalin, anc. Odessus. Varna, E Bulgaria 43.13N 27.55E
116 N8 **Varna** ✈ Varna, E Bulgaria 43.16N 27.52E
116 N8 **Varna** ◆ province E Bulgaria
97 L20 **Värnamo** Jönköping, S Sweden 57.10N 14.03E
116 N8 **Varnenski Zaliv** prev. Stalinski Zaliv. bay E Bulgaria
116 N8 **Varnensko Ezero** estuary E Bulgaria
120 D11 **Varniai** Telšiai, W Lithuania 55.45N 22.22E
Varnoús see Baba
113 D14 **Varnsdorf Ger.** Warnsdorf. Ústecký Kraj, N Czech Republic 50.55N 14.34E
113 I23 **Várpalota** Veszprém, W Hungary 47.13N 18.07E
Varshava see Warszawa
120 K6 **Värska** Põlvamaa, SE Estonia 57.58N 27.37E
100 N12 **Varsseveld** Gelderland, E Netherlands 51.55N 6.28E
117 D19 **Vartholomió** prev. Vartholomión. Dytikí Ellás, S Greece 37.52N 21.12E
Vartholomión see Vartholomió
143 Q14 **Varto** Muş, E Turkey 39.10N 41.28E
97 K18 **Vartofta** Västra Götaland, S Sweden 58.06N 13.40E
95 O17 **Vārtsilä** Itä-Suomi, E Finland 62.10N 30.35E
Vartsilya see Vyartsilya
119 R4 **Varva** Chernihivs'ka Oblast', NE Ukraine 50.31N 32.43E
61 I14 **Várzea Grande** Mato Grosso, SW Brazil 15.39S 56.07W
108 D9 **Varzi** Lombardia, N Italy 44.51N 9.13E
Varzimanor Ayni see Ayní
128 K5 **Varzuga** ❧ NW Russian Federation
105 P8 **Varzy** Nièvre, C France 47.22N 3.22E
113 G23 **Vas Alt.** Vas Megye. ◆ county W Hungary
Vasa see Vaasa
202 A9 **Vasafua** island Funafuti Atoll, C Tuvalu
113 O17 **Vásárosnamény** Szabolcs-Szatmár-Bereg, E Hungary 48.07N 22.19E
106 H13 **Vasco, Ribeira de** ❧ S Portugal
118 G10 **Vaşcău Hung.** Vaskoh. Bihor, NE Romania 46.26N 22.30E
Vascongadas, Provincias see País Vasco
Vashess Bay see Vaskess Bay
Vāsht see Khāsh
117 G14 **Vasilikí** Kentriki Makedonía, NE Greece 40.28N 23.57E
117 C18 **Vasilikí** Lefkáda, Iónioi Nísoi, Greece, C Mediterranean Sea 38.36N 20.37E
117 K25 **Vasiliki** Kríti, Greece, E Mediterranean Sea 35.04N 25.49E
121 G16 **Vasilishki Pol.** Wasiliszki, Rus. Vasilishki. Hrodzyenskaya Voblasts', W Belarus 53.46N 24.51E
Vasil Kolarov see Pamporovo
Vasil'kov see Vasyl'kiv
121 N19 **Vasilyevichy Rus.** Vasilevichi. Homyel'skaya Voblasts', SE Belarus 52.15N 29.49E
203 Y3 **Vaskess Bay** var. Vashess Bay. bay Kiritimati, E Kiribati
Vaskoh see Vaşcău
118 M10 **Vaslui** Vaslui, E Romania 46.38N 27.44E
118 L11 **Vaslui** ◆ county NE Romania
33 R8 **Vassar** Michigan, N USA 43.22N 83.34W
97 J16 **Vassdalseggi** ▲ S Norway 59.47N 7.07E
62 P9 **Vassouras** Rio de Janeiro, SE Brazil 22.24S 43.38W
94 H11 **Vastenjaure** ◇ N Sweden
97 N15 **Västerås** Västmanland, C Sweden 59.37N 16.33E
95 G15 **Västerbotten** ◆ county N Sweden
97 O16 **Västerhaninge** Stockholm, C Sweden 59.07N 18.06E
97 N19 **Västervik** Kalmar, S Sweden 57.44N 16.40E
97 M15 **Västmanland** ◆ county C Sweden
109 L15 **Vasto** anc. Histonium. Abruzzo, C Italy 42.07N 14.40E
97 J19 **Västra Götaland** ◆ county S Sweden
97 J19 **Västra Silen** ◇ S Sweden
113 G23 **Vasvár** Vas. Eisenburg. Vas, W Hungary 47.04N 16.46E
116 M7 **Vetkilski** Shumen, NE Bulgaria 43.33N 27.19E

119 O5 **Vasyl'kiv Rus.** Vasil'kov. Kyyivs'ka Oblast', N Ukraine 50.10N 30.18E
126 Gg12 **Vasyugan** ❧ C Russian Federation
105 N8 **Vatan** Indre, C France 47.06N 1.49E
109 G15 **Vaté** see Efate
109 G15 **Vatican City off.** Vatican City State. ♦ papal state S Europe
109 M22 **Vaticano, Capo** headland S Italy 38.37N 15.49E
Vatili see Vadili
94 K3 **Vatnajökull** glacier SE Iceland
13 N12 **Vaudreuil** Quebec, SE Canada 45.24N 74.01W
39 T12 **Vaughn** New Mexico, SW USA 34.36N 105.12W
56 J14 **Vaupés** off. Comisaría del Vaupés. ◆ province SE Colombia
56 J13 **Vaupés, Río** var. Río Uaupés. ❧ Brazil/Colombia see also Uaupés, Rio
105 Q15 **Vauvert** Gard, S France 43.42N 4.16E
9 T17 **Vauxhall** Alberta, SW Canada 50.04N 112.09W
101 K23 **Vaux-sur-Sûre** Luxembourg, SE Belgium 49.55N 5.34E
180 J4 **Vatovavy** ❧ E Madagascar
200 Ss12 **Vava'u Group** island group N Tonga
78 M16 **Vavoua** W Ivory Coast 7.22N 6.28W
131 S2 **Vavozh** Udmurtskaya Respublika, NW Russian Federation 56.48N 51.53E
161 K23 **Vavuniya** Northern Province, N Sri Lanka 8.45N 80.30E
121 G17 **Vawkavysk Pol.** Wołkowysk, Rus. Volkovysk. Hrodzyenskaya Voblasts', W Belarus 53.10N 24.28E
121 F17 **Vawkavysk Wzvyshsha Rus.** Volkovyskiye Vysoty. hill range W Belarus
97 P15 **Vaxholm** Stockholm, C Sweden 59.43N 18.57E
97 L21 **Växjö** var. Vexiö. Kronoberg, S Sweden 56.52N 14.49E
129 T1 **Vaygach, Ostrov** island NW Russian Federation
143 V13 **Vayk'** var. Azizbekov. SE Armenia 39.41N 45.28E
129 P8 **Vazhgort** prev. Chasovo. Respublika Komi, NW Russian Federation 64.06N 46.58E
47 V10 **V.C.Bird ✈** (St John's) Antigua, Antigua and Barbuda 17.07N 61.49W
97 C16 **Veavågen** Rogaland, S Norway 59.18N 5.13E
31 Q7 **Veblen** South Dakota, N USA 45.50N 97.17W
100 N9 **Vecht Ger.** Vechte. ❧ Germany/Netherlands see also Vechte
102 G12 **Vechta** Niedersachsen, NW Germany 52.44N 8.16E
102 E12 **Vechte Dut.** Vecht. ❧ Germany/Netherlands see also Vecht
120 I8 **Vecpiebalga** Cēsis, C Latvia 57.03N 25.47E
120 G9 **Vecumnieki** Bauska, C Latvia 56.36N 24.30E
Vedavati see Hagari
97 J20 **Veddige** Halland, S Sweden 57.16N 12.19E
118 J15 **Vedea** ❧ S Romania
131 P16 **Vedeno** Chechenskaya Respublika, SW Russian Federation 42.57N 46.02E
197 H14 **Ve Drala Reef** reef N Fiji
100 I6 **Veendam** Groningen, NE Netherlands 53.04N 6.52E
100 K12 **Veenendaal** Utrecht, C Netherlands 52.03N 5.33E
101 E14 **Veere** Zeeland, SW Netherlands 51.33N 3.40E
26 M2 **Vega** Texas, SW USA 35.14N 102.25W
94 E13 **Vega** island C Norway
47 T5 **Vega Baja** C Puerto Rico 18.27N 66.23W
40 D17 **Vega Point** headland Kiska Island, Alaska, USA 51.49N 105.19E
97 J17 **Vegår** ◇ S Norway
101 K14 **Veghel** Noord-Brabant, S Netherlands 51.37N 5.33E
115 D14 **Véglia** see Krk
117 F16 **Vegoritís, Límni** ◇ N Greece
9 Q14 **Vegreville** Alberta, SW Canada 53.30N 112.01W
97 K21 **Veinge** Halland, S Sweden 56.33N 13.04E
63 B21 **Veintincinco de Mayo** var. 25 de Mayo. Buenos Aires, E Argentina 35.27S 60.11W
65 G14 **Veinticinco de Mayo** La Pampa, C Argentina 37.45S 67.40W
115 J19 **Vejsiejai** Lazdijai, S Lithuania 54.06N 23.41E
97 F23 **Vejen** Ribe, W Denmark 55.28N 9.09E
106 K16 **Vejer de la Frontera** Andalucía, S Spain 36.15N 5.58W
97 G23 **Vejle** Vejle, C Denmark 55.43N 9.33E
97 F23 **Vejle Amt.** ◆ county C Denmark
116 M7 **Vekilski** Shumen, NE Bulgaria 43.33N 27.19E

56 G3 **Vela, Cabo de la** headland NE Colombia 12.13N 72.13W
Vela Goa see Goa
115 F15 **Vela Luka** Dubrovnik-Neretva, S Croatia 42.58N 16.43E
63 G9 **Velázquez** Rocha, E Uruguay 34.04S 54.16W
109 E15 **Velbert** Nordrhein-Westfalen, W Germany 51.19N 7.03E
111 S9 **Velden** Kärnten, S Austria 46.37N 13.59E
101 K15 **Veldhoven** Noord-Brabant, S Netherlands 51.24N 5.24E
114 C11 **Veldes** see Bled
116 N11 **Veleka** ❧ SE Bulgaria
111 V10 **Velenje Ger.** Wöllan. N Slovenia 46.21N 15.07E
202 E12 **Vele, Pointe** headland Île Futuna, S Wallis and Futuna
115 O16 **Veles Turk.** Köprülü. C FYR Macedonia 41.43N 21.49E
115 M20 **Velesta** S FYR Macedonia 41.16N 20.37E
Velestíno see Velestíno
117 F16 **Velestíno** var. Velestíno. Thessalía, C Greece 39.22N 22.43E
Velestíno see Velestíno
Velevshchina see Vyelyewshchyna
56 F9 **Vélez** Santander, C Colombia 6.01N 73.37W
107 Q13 **Vélez Blanco** Andalucía, S Spain 37.43N 2.07W
106 M17 **Vélez de la Gomera, Peñon de** island group S Spain
107 N15 **Vélez-Málaga** Andalucía, S Spain 36.46N 4.06W
107 Q13 **Vélez Rubio** Andalucía, S Spain 37.39N 2.04W
Velha Goa see Goa
Velho see Porto Velho
114 E8 **Velika Gorica** Zagreb, N Croatia 45.44N 16.04E
114 C9 **Velika Kikinda** see Kikinda
114 D10 **Velika Kladuša** Federacija Bosna I Hercegovina, NW Bosnia and Herzegovina 45.10N 15.48E
114 N11 **Velika Morava** var. Glavn'a Morava, Morava, Ger. Grosse Morava. ❧ C Yugoslavia
114 N12 **Velika Plana** Serbia, C Yugoslavia 44.20N 21.01E
111 U10 **Velika Raduha** ▲ N Slovenia 46.24N 14.46E
127 Pp5 **Velikaya** ❧ NE Russian Federation
128 F15 **Velikaya** ❧ W Russian Federation
Velikaya Berestovitsa see Vyalikaya Byerastavitsa
Velikaya Lepetikha see Velyka Lepetykha
Veliki Bečkerek see Zrenjanin
114 F12 **Veliki Krš** var. Stol. ▲ E Yugoslavia 44.10N 22.09E
116 L8 **Veliki Preslav** prev. Preslav. Shumen, NE Bulgaria 43.09N 26.46E
114 B9 **Veliki Risnjak** ▲ NW Croatia 45.30N 14.31E
111 T13 **Veliki Snežnik Ger.** Schneeberg, It. Monte Nevoso. ▲ SW Slovenia 45.36N 14.25E
114 J13 **Veliki Stolac** ▲ E Bosnia and Herzegovina 43.55N 19.15E
Veliki Bor see Vyaliki Bor
128 G16 **Velikiye Luki** Pskovskaya Oblast', W Russian Federation 56.19N 30.27E
129 P12 **Velikiy Ustyug** Vologodskaya Oblast', NW Russian Federation 60.46N 46.18E
114 N11 **Veliko Gradište** Serbia, NE Yugoslavia 44.46N 21.28E
161 I18 **Velikonda Range** ▲ SE India
116 K9 **Veliko Tŭrnovo** prev. Tirnovo, Trnovo, Turnovo. Veliko Tŭrnovo, N Bulgaria 43.04N 25.40E
66 K8 **Veliko Tŭrnovo** ◆ province N Bulgaria
Velikovec see Völkermarkt
129 R5 **Velikovisochnoye** Nenetskiy Avtonomnyy Okrug, NW Russian Federation 67.13N 52.00E
78 H11 **Vélingara** S Senegal 15.00N 14.39W
78 H11 **Vélingara** S Senegal 13.12N 14.04W
116 H11 **Velingrad** Pazardzhik, C Bulgaria 42.01N 24.00E
130 I13 **Velizh** Smolenskaya Oblast', W Russian Federation 55.30N 31.06E
113 F16 **Velká Deštná** var. Deitná, Grosskoppe, Ger. Deschnaer Koppe. ▲ NE Czech Republic 50.18N 16.25E
113 F18 **Velké Meziříčí Ger.** Grossmeseritsch. Jihlavský Kraj, C Czech Republic 49.22N 16.01E
Vel'ký Krtíš Banskobystrický Kraj, C Slovakia 48.13N 19.21E
195 T14 **Vella Lavella** var. Mbilua. island New Georgia Islands, NW Solomon Islands
109 I15 **Velletri** Lazio, C Italy 41.43N 12.43E
97 K23 **Vellinge** Skåne, S Sweden 55.29N 13.00E
161 I21 **Vellore** Tamil Nādu, SE India 12.55N 79.09E
Veloca see Viana do Castelo
117 G21 **Velopoúla** island S Greece
100 M12 **Veluwe** island S Greece
100 L12 **Veluwemeer** lake channel C Netherlands
30 M3 **Velva** North Dakota, N USA 48.03N 100.55W
115 O18 **Velvendós** var. Velvendos. Dytikí Makedonía, N Greece 40.15N 22.04E
119 S5 **Velyka Bahachka** Poltavs'ka Oblast', C Ukraine 49.46N 33.44E
119 S9 **Velyka Lepetykha Rus.** Velikaya Lepetikha. Khersons'ka Oblast', S Ukraine 47.10N 33.55E
119 O10 **Velyka Mykhaylivka** Odes'ka Oblast', SW Ukraine 47.07N 29.49E
119 W8 **Velyka Novosilka** Donets'ka Oblast', SE Ukraine 47.49N 36.49E

119 S9 **Velyka Oleksandrivka** Khersons'ka Oblast', S Ukraine 47.17N 33.16E
119 T4 **Velyka Pysarivka** Sums'ka Oblast', NE Ukraine 50.25N 35.28E
118 G6 **Velykyy Bereznyy** Zakarpats'ka Oblast', W Ukraine 48.54N 22.27E
119 W4 **Velykyy Burluk** Kharkivs'ka Oblast', E Ukraine 50.04N 37.25E
181 P7 **Velykyy Tokmak** see Tokmak
Vema Fracture Zone tectonic feature W Indian Ocean
67 P18 **Vema Seamount** undersea feature SW Indian Ocean 31.37S 8.19E
95 P17 **Vemdalen** Jämtland, C Sweden 62.26N 13.50E
97 N19 **Vena Kalmar, S Sweden 57.31N 16.00E
43 N11 **Venado** San Luis Potosí, C Mexico 22.54N 101.06W
64 L11 **Venado Tuerto** Entre Ríos, E Argentina 33.45S 61.57W
63 A19 **Venado Tuerto** Santa Fe, C Argentina 33.46S 61.57W
109 K13 **Venafro** Molise, C Italy 41.28N 14.03E
57 Q9 **Venamo, Cerro** ▲ E Venezuela 5.56N 61.25W
108 B8 **Venaria** Piemonte, NW Italy 45.07N 7.38E
105 U15 **Vence** Alpes-Maritimes, SE France 43.45N 7.07E
106 H5 **Venda Nova** Vila Real, N Portugal 41.40N 7.58W
106 G11 **Vendas Novas** Évora, S Portugal 38.40N 8.27W
104 J9 **Vendée** ◆ department NW France
105 Q6 **Vendeuvre-sur-Barse** Aube, NE France 48.08N 4.17E
104 M7 **Vendôme** Loir-et-Cher, C France 47.48N 1.04E
Venedig see Venezia
108 I8 **Veneta, Laguna** lagoon NE Italy
Venetia see Venezia
41 T5 **Venetie** Alaska, USA 67.00N 146.25W
108 I8 **Veneto** var. Venezia Euganea. ◆ region NE Italy
116 M7 **Venets** Shumen, NE Bulgaria 43.33N 26.56E
130 L5 **Venev** Tul'skaya Oblast', W Russian Federation 54.18N 38.16E
108 I8 **Venezia Eng.** Venice, Fr. Venise, Ger. Venedig; anc. Venetia. NE Italy 45.25N 12.19E
Venezia Euganea see Veneto
Venezia, Golfo di see Venice, Gulf of
Venezia Tridentina see Trentino-Alto Adige
56 L4 **Venezuela off.** Republic of Venezuela; prev. Estados Unidos de Venezuela, United States of Venezuela. ♦ republic N South America
Venezuela, Cordillera de see Costa, Cordillera de la
56 I4 **Venezuela, Golfo de Eng.** Gulf of Maracaibo, Gulf of Venezuela. gulf NW Venezuela
Venezuela, Gulf of see Venezuela, Golfo de
66 F11 **Venezuelan Basin** undersea feature E Caribbean Sea
161 D16 **Vengurla** Mahārāshtra, W India 15.55N 73.39E
41 O15 **Veniaminof, Mount** ▲ Alaska, USA 56.12N 159.24W
25 V14 **Venice** Florida, SE USA 27.06N 82.27W
24 L10 **Venice** Louisiana, S USA 29.15N 89.20W
Venice see Venezia
108 J8 **Venice, Gulf of It.** Golfo di Venezia, Slvn. Beneški Zaliv. gulf N Adriatic Sea
Venise see Venezia
96 K13 **Venjan** Dalarna, C Sweden 60.58N 13.55E
96 J13 **Venjansjön** ◇ C Sweden
161 J18 **Venkatagiri** Andhra Pradesh, E India 14.00N 79.39E
101 M15 **Venlo** prev. Venloo. Limburg, SE Netherlands 51.22N 6.10E
Venloo see Venlo
97 E18 **Vennesla** Vest-Agder, S Norway 58.15N 7.58E
109 M17 **Venosa** anc. Venusia. Basilicata, S Italy 40.57N 15.49E
Venoste, Alpi see Ötztaler Alpen
101 M14 **Venray** var. Venraij. Limburg, SE Netherlands 51.31N 5.58E
120 C8 **Venta Ger.** Windau. ❧ Latvia/Lithuania
Venta Belgarum see Winchester
42 G9 **Ventana, Punta Arena de la** var. Punta de la Ventana. headland W Mexico 24.03N 109.49W
Ventana, Punta de la see Ventana, Punta Arena de la
Ventana, Sierra de la hill range E Argentina
Ventia see Valence
203 S11 **Vent, Îles du** var. Windward Islands. island group Archipel de la Société, W French Polynesia
203 R10 **Vent, Îles Sous le** var. Leeward Islands. island group Archipel de la Société, W French Polynesia
108 B11 **Ventimiglia** Liguria, NW Italy 43.46N 7.37E
99 M24 **Ventnor** England, UK 50.36N 1.10W
20 J17 **Ventnor City** New Jersey, NE USA 39.19N 74.27W
105 S14 **Ventoux, Mont** ▲ SE France 44.10N 5.16E
120 C8 **Ventspils Ger.** Windau. Ventspils, NW Latvia 57.22N 21.34E
56 M10 **Venturi, Río** ❧ S Venezuela
37 R15 **Ventura, CA** USA 34.15N 119.14W
190 F8 **Venus Bay** South Australia 33.15S 134.42E
Venusia see Venosa

● COUNTRY   ◇ DEPENDENT TERRITORY   ♦ ADMINISTRATIVE REGION   ▲ MOUNTAIN   ☐ VOLCANO   ◇ LAKE
○ COUNTRY CAPITAL   ○ DEPENDENT TERRITORY CAPITAL   ✈ INTERNATIONAL AIRPORT   ▲ MOUNTAIN RANGE   ❧ RIVER   ☐ RESERVOIR

203 P7 **Vénus, Pointe** var. Pointe Tataaihoa. headland Tahiti, W French Polynesia 17.28S 149.28W
43 V16 **Venustiano Carranza** Chiapas, SE Mexico 16.24N 92.04W
43 N7 **Venustiano Carranza, Presa** ⊠ NE Mexico
63 B15 **Vera** Santa Fe, C Argentina 29.28S 60.10W
107 Q14 **Vera** Andalucía, S Spain 37.15N 1.51W
65 K18 **Vera, Bahía** bay E Argentina
43 R14 **Veracruz** var. Veracruz Llave. Veracruz-Llave, E Mexico 19.09N 96.09W
43 Q13 **Veracruz-Llave** var. Veracruz. ◈ state E Mexico
45 Q16 **Veraguas** off. Provincia de Veraguas. ◈ province W Panama
Veramin see Varāmīn
160 B12 **Verāval** Gujarāt, W India 20.54N 70.22E
108 C6 **Verbania** Piemonte, NW Italy 45.55N 8.34E
109 N20 **Verbicaro** Calabria, SW Italy 39.44N 15.51E
110 D11 **Verbier** Valais, SW Switzerland 46.06N 7.14E
Vercellae see Vercelli
108 C8 **Vercelli** anc. Vercellae. Piemonte, NW Italy 45.19N 8.25E
105 S13 **Vercors** physical region E France
95 E16 **Verdalsøra** Nord-Trøndelag, C Norway 63.46N 11.27E
Verde, Cabo see Cape Verde
46 J5 **Verde, Cape** headland Long Island, C Bahamas 22.51N 75.50W
106 M2 **Verde, Costa** coastal region N Spain
Verde Grande, Río/Verde Grande y de Belem, Río see Verde, Río
102 H11 **Verden** Niedersachsen, NW Germany 52.55N 9.13E
61 J19 **Verde, Rio** ≈ SE Brazil
59 P16 **Verde, Río** ≈ Bolivia/Brazil
42 M12 **Verde, Río** var. Río Verde Grande, Río Verde Grande y de Belem. ≈ C Mexico
43 Q16 **Verde, Río** ≈ SE Mexico
38 L13 **Verde River** ≈ Arizona, SW USA
Verdhikoúsa/Verdhikoússa see Verdikoússa
49 Q8 **Verdigris River** ≈ Kansas/Oklahoma, C USA
117 E15 **Verdikoússa** var. Verdhikoúsa, Verdhikoússa. Thessalía, C Greece 39.46N 21.58E
105 S15 **Verdon** ≈ SE France
13 O12 **Verdun** Quebec, SE Canada 45.27N 73.36W
105 S4 **Verdun** var. Verdun-sur-Meuse; anc. Verodunum. Meuse, NE France 49.09N 5.25E
Verdun-sur-Meuse see Verdun
85 J21 **Vereeniging** Gauteng, NE South Africa 26.40S 27.55E
Veremeyki see Vyeramyeyki
129 T14 **Vereshchagino** Permskaya Oblast', NW Russian Federation 58.06N 54.38E
78 G14 **Verga, Cap** headland W Guinea 10.12N 14.27W
63 G18 **Vergara** Treinta y Tres, E Uruguay 32.58S 53.54W
110 G11 **Vergeletto** Ticino, S Switzerland 46.13N 8.34E
20 L8 **Vergennes** Vermont, NE USA 44.09N 73.13W
Veria see Véroia
106 I5 **Verín** Galicia, NW Spain 41.55N 7.25W
Verin T'alin see T'alin
120 K6 **Veriora** Põlvamaa, SE Estonia 57.57N 27.23E
119 T7 **Verkhivtseve** Dnipropetrovs'ka Oblast', E Ukraine 48.27N 34.15E
131 W3 **Verkhiye Kigi** Respublika Bashkortostan, W Russian Federation 55.25N 58.40E
Verkhnedvinsk see Vyerkhnyadzvinsk
126 Hh11 **Verkhneimbatsk** Krasnoyarskiy Kray, N Russian Federation 63.06N 88.03E
128 I3 **Verkhnetulomskiy** Murmanskaya Oblast', NW Russian Federation 68.37N 31.46E
128 I3 **Verkhnetulomskoye Vodokhranilishche** ⊠ NW Russian Federation
Verkhneudinsk see Ulan-Ude
126 L11 **Verkhnevilyuysk** Respublika Sakha (Yakutiya), NE Russian Federation 63.44N 119.59E
131 W5 **Verkhniy Avzyan** Respublika Bashkortostan, W Russian Federation 53.31N 57.26E
131 Q11 **Verkhniy Baskunchak** Astrakhanskaya Oblast', SW Russian Federation 48.14N 46.43E
119 T9 **Verkhniy Rohachyk** Khersons'ka Oblast', S Ukraine 47.16N 34.16E
126 Ll12 **Verkhnyaya Amga** Respublika Sakha (Yakutiya), NE Russian Federation 59.34N 127.07E
129 V6 **Verkhnyaya Inta** Respublika Komi, NW Russian Federation 65.55N 60.07E
126 Ii6 **Verkhnyaya Taymyra** ≈ N Russian Federation
129 O10 **Verkhnyaya Toyma** Arkhangel'skaya Oblast', NW Russian Federation 62.12N 44.57E
130 K6 **Verkhov'ye** Orlovskaya Oblast', W Russian Federation 52.49N 37.20E
118 I8 **Verkhovyna** Ivano-Frankivs'ka Oblast', W Ukraine 48.09N 24.48E
126 M8 **Verkhoyansk** Respublika Sakha (Yakutiya), NE Russian Federation 67.27N 133.27E
126 L8 **Verkhoyanskiy Khrebet** ▲▲ NE Russian Federation
119 T7 **Verkhn'odniprovs'k** Dnipropetrovs'ka Oblast', E Ukraine 48.40N 34.17E
103 G14 **Verl** Nordrhein-Westfalen, NW Germany 51.52N 8.30E
94 N1 **Verlegenhuken** headland N Svalbard 80.03N 16.15E
84 A9 **Vermelha, Ponta** headland NW Angola 5.40S 12.09E
105 P7 **Vermenton** Yonne, C France 47.40N 3.43E
9 R14 **Vermilion** Alberta, SW Canada 53.21N 110.52W
33 T11 **Vermilion** Ohio, N USA 41.25N 82.21W
24 I10 **Vermilion Bay** bay Louisiana, S USA
31 V4 **Vermilion Lake** ◎ Minnesota, N USA
12 F9 **Vermilion River** ≈ Ontario, S Canada
32 L12 **Vermilion River** ≈ Illinois, N USA
31 R12 **Vermillion** South Dakota, N USA 42.46N 96.55W
31 R12 **Vermillion River** ≈ South Dakota, N USA
13 O9 **Vermillon, Rivière** ≈ Quebec, SE Canada
117 E14 **Vérmio** ▲▲ N Greece
20 L8 **Vermont** off. State of Vermont; also known as The Green Mountain State. ◈ state NE USA
115 K16 **Vermosh** var. Vermoshi. Shkodër, N Albania 42.37N 19.42E
Vermoshi see Vermosh
39 O3 **Vernal** Utah, W USA 40.27N 109.31W
12 G11 **Verner** Ontario, S Canada 46.24N 80.04W
104 M5 **Verneuil-sur-Avre** Eure, N France 48.44N 0.55E
116 D13 **Vérno** ▲▲ N Greece
9 N17 **Vernon** British Columbia, SW Canada 50.16N 119.19W
104 M4 **Vernon** Eure, N France 49.04N 1.28E
25 N3 **Vernon** Alabama, S USA 33.45N 88.06W
33 P15 **Vernon** Indiana, N USA 38.58N 85.39W
27 Q4 **Vernon** Texas, SW USA 34.10N 99.16W
34 G10 **Vernonia** Oregon, NW USA 45.51N 123.11W
12 G12 **Vernon, Lake** ◎ Ontario, S Canada
24 G7 **Vernon Lake** ⊠ Louisiana, S USA
25 Y13 **Vero Beach** Florida, SE USA 27.38N 80.24W
Verőcze see Virovitica
Verodunum see Verdun
117 E14 **Véroia** var. Veria, Vérroia, Turk. Karaferiye. Kentrikí Makedonía, N Greece 40.31N 22.14E
108 E8 **Verolanuova** Lombardia, N Italy 45.20N 10.06E
12 K14 **Verona** Ontario, SE Canada 44.30N 76.42W
108 G8 **Verona** Veneto, NE Italy 45.26N 11.00E
31 P6 **Verona** North Dakota, N USA 46.19N 98.03W
32 L9 **Verona** Wisconsin, N USA 42.59N 89.33W
63 E20 **Verónica** Buenos Aires, E Argentina 35.25S 57.16W
24 J9 **Verret, Lake** ◎ Louisiana, S USA
Vérroia see Véroia
195 P10 **Verron Range** ▲▲ New Ireland, NE PNG
105 N5 **Versailles** Yvelines, N France 48.48N 2.07E
33 P15 **Versailles** Indiana, N USA 39.04N 85.16W
22 M5 **Versailles** Kentucky, S USA 38.03N 84.43W
49 U5 **Versailles** Missouri, C USA 38.25N 92.50W
33 Q13 **Versailles** Ohio, N USA 40.13N 84.28W
Versecz see Vršac
110 A10 **Versoix** Genève, SW Switzerland 46.16N 6.10E
13 Z6 **Verte, Pointe** headland Quebec, SE Canada 48.36N 64.10W
113 I22 **Vértes** ▲▲ NW Hungary
46 G6 **Vertientes** Camagüey, C Cuba 21.15N 78.09W
116 G13 **Vertískos** ▲▲ N Greece
104 I8 **Vertou** Loire-Atlantique, NW France 47.10N 1.28W
Verulamium see St Albans
101 L19 **Verviers** Liège, E Belgium 50.36N 5.52E
105 Y14 **Vescovato** Corse, France, C Mediterranean Sea 42.30N 9.27E
101 L20 **Vesdre** ≈ E Belgium
119 U10 **Vesele** Rus. Veseloye. Zaporiz'ka Oblast', S Ukraine 47.00N 34.52E
113 D18 **Veselí nad Lužnicí** var. Weseli an der Lainsitz, Ger. Frohenbruck. Budějovický Kraj, S Czech Republic 49.11N 14.40E
116 M9 **Veselinovo** Shumen, E Bulgaria 43.01N 27.02E
130 L12 **Veselovskoye Vodokhranilishche** ⊠ SW Russian Federation
Veseloye see Vesele
119 Q9 **Veselynove** Mykolayivs'ka Oblast', S Ukraine 47.21N 31.15E
Veseya see Vyasyeya
130 M10 **Veshenskaya** Rostovskaya Oblast', SW Russian Federation 49.37N 41.43E
131 Q5 **Veshkayma** Ul'yanovskaya Oblast', W Russian Federation 54.04N 47.06E
Vesisaari see Vadsø
Vesontio see Besançon
105 T7 **Vesoul** anc. Vesulium, Vesulum. Haute-Saône, E France 47.37N 6.09E
97 J20 **Vessigebro** Halland, S Sweden 56.58N 12.40E
97 D17 **Vest-Agder** ◈ county S Norway
25 P4 **Vestavia Hills** Alabama, S USA 33.27N 86.47W
86 F6 **Vesterålen** island NW Norway
94 G10 **Vesterålen** island group N Norway
89 V3 **Vestervig** Viborg, NW Denmark 56.46N 8.19E
94 H2 **Vestfirdhir** ◈ region NW Iceland
94 G11 **Vestfjorden** fjord C Norway
97 G16 **Vestfold** ◈ county S Norway
97 B18 **Vestmanna** Dan. Vestmanhavn Faeroe Islands 62.09N 7.11W
94 I4 **Vestmannaeyjar** Sudhurland, S Iceland 63.26N 20.13W
96 E9 **Vestnes** Møre og Romsdal, S Norway 62.39N 7.00E
97 I23 **Vestsjælland** off. Vestsjællands Amt. ◈ county E Denmark
94 H3 **Vesturland** ◈ region W Iceland
94 G11 **Vestvågøya** island C Norway
Vesulium/Vesulum see Vesoul
Vesuna see Périgueux
109 K17 **Vesuvio** Eng. Vesuvius. Ж S Italy 40.48N 14.29E
Vesuvius see Vesuvio
128 K14 **Ves'yegonsk** Tverskaya Oblast', W Russian Federation 58.40N 37.13E
113 I23 **Veszprém** Ger. Veszprim. Veszprém, W Hungary 47.06N 17.54E
113 H23 **Veszprém** off. Veszprém Megye. ◈ county W Hungary
Veszprim see Veszprém
Vetka see Vyetka
97 M19 **Vetlanda** Jönköping, S Sweden 57.25N 15.04E
131 P1 **Vetluga** Nizhegorodskaya Oblast', W Russian Federation 57.51N 45.45E
129 P14 **Vetluga** ≈ NW Russian Federation
129 O14 **Vetluzhskiy** Kostromskaya Oblast', NW Russian Federation 58.21N 45.25E
131 P2 **Vetluzhskiy** Nizhegorodskaya Oblast', W Russian Federation 57.10N 45.07E
109 H14 **Vetralla** Lazio, C Italy 42.18N 12.03E
116 M9 **Vetren** prev. Zhitarovo. Burgas, E Bulgaria 42.38N 27.22E
116 M8 **Vetrino** Varna, E Bulgaria 43.19N 27.26E
Vetrino see Vyetryna
126 Ii6 **Vetrovaya, Gora** ▲ N Russian Federation 73.54N 95.00E
Vetter, Lake see Vättern
108 J13 **Vettore, Monte** ▲ C Italy 42.49N 13.15E
101 A17 **Veurne** var. Furnes. West-Vlaanderen, W Belgium 51.04N 2.40E
33 Q15 **Vevay** Indiana, N USA 38.45N 85.07W
110 C10 **Vevey** Ger. Vivis; anc. Vibiscum. Vaud, SW Switzerland 46.28N 6.51E
Vexiö see Växjö
105 S13 **Veynes** Hautes-Alpes, SE France 44.33N 5.51E
105 N11 **Vézère** ≈ W France
116 I9 **Vezhen** ▲ C Bulgaria 42.45N 24.22E
142 K11 **Vezirköprü** Samsun, N Turkey 41.09N 35.27E
59 J18 **Viacha** La Paz, W Bolivia 16.40S 68.16W
49 R10 **Vian** Oklahoma, C USA 35.30N 94.56W
Viana de Castelo see Viana do Castelo
106 H12 **Viana do Alentejo** Évora, S Portugal 38.20N 8.00W
106 I4 **Viana do Bolo** Galicia, NW Spain 42.10N 7.06W
106 G5 **Viana do Castelo** var. Viana de Castelo; anc. Velobriga. Viana do Castelo, NW Portugal 41.40N 8.49W
106 G5 **Viana do Castelo** var. Viana de Castelo. ◈ district N Portugal
100 J12 **Vianen** Zuid-Holland, C Netherlands 52.00N 5.06E
178 Ii8 **Viangchan** Eng./Fr. Vientiane. ● (Laos) C Laos 17.57N 102.38E
178 I6 **Viangphoukha** var. Vieng Pou Kha. Louang Namtha, N Laos 20.41N 101.03E
106 K13 **Viar** ≈ SW Spain
108 E11 **Viareggio** Toscana, C Italy 43.52N 10.15E
105 O14 **Viaur** ≈ S France
Vibiscum see Vevey
97 G21 **Viborg** Viborg, NW Denmark 56.28N 9.25E
31 R12 **Viborg** South Dakota, N USA 43.10N 97.04W
97 F21 **Viborg** off. Viborg Amt. ◈ county NW Denmark
109 N22 **Vibo Valentia** prev. Monteleone di Calabria; anc. Hipponium. Calabria, SW Italy 38.40N 16.06E
107 W5 **Vic** var. Vich; anc. Ausa, Vicus Ausonensis. Cataluña, NE Spain 41.55N 2.16E
104 K16 **Vic-en-Bigorre** Hautes-Pyrénées, S France 43.22N 0.03E
42 K10 **Vicente Guerrero** Durango, C Mexico 23.30N 104.24W
43 P10 **Vicente Guerrero, Presa** var. Presa de las Adjuntas. ⊠ NE Mexico
Vicentia see Vicenza
108 G8 **Vicenza** anc. Vicentia. Veneto, NE Italy 45.31N 11.31E
Vich see Vic
56 J10 **Vichada** off. Comisaría del Vichada. ◈ province E Colombia
56 K10 **Vichada, Río** ≈ E Colombia
63 G17 **Vichadero** Rivera, NE Uruguay 31.45S 54.40W
Vichegda see Vychegda
128 M16 **Vichuga** Ivanovskaya Oblast', W Russian Federation 57.13N 41.51E
105 P10 **Vichy** Allier, C France 46.08N 3.26E
28 K9 **Vici** Oklahoma, C USA 36.09N 99.18W
33 P10 **Vicksburg** Michigan, N USA 42.07N 85.31W
24 J5 **Vicksburg** Mississippi, S USA 32.21N 90.52W
105 O12 **Vic-sur-Cère** Cantal, C France 45.00N 2.36E
31 X14 **Victor** Iowa, C USA 41.45N 92.18W
61 I21 **Víctor** Mato Grosso do Sul, SW Brazil 21.39S 53.21W
190 I10 **Victor Harbor** South Australia 35.33S 138.37E
63 C18 **Victoria** Entre Ríos, E Argentina 32.36S 60.12W
8 L17 **Victoria** Vancouver Island, British Columbia, SW Canada 48.25N 123.22W
47 R14 **Victoria** NW Grenada 12.11N 61.42W
44 H6 **Victoria** Yoro, NW Honduras 15.01N 87.28W
123 J16 **Victoria** var. Rabat. Gozo, NW Malta 36.02N 14.14E
118 I12 **Victoria** Ger. Viktoriastadt. Braşov, C Romania 45.43N 24.40E
180 H17 **Victoria** ● (Seychelles) Mahé, SW Seychelles 4.37S 28.28E
27 U13 **Victoria** Texas, SW USA 28.47N 96.58W
191 N12 **Victoria** ◈ state SE Australia
182 K7 **Victoria** ≈ Western Australia
Victoria see Labuan, East Malaysia
Victoria see Masvingo, Zimbabwe
Victoria Bank see Vitória Seamount
9 Y15 **Victoria Beach** Manitoba, S Canada 50.40N 96.30W
Victoria de Durango see Durango
Victoria de las Tunas see Las Tunas
85 I16 **Victoria Falls** Matabeleland North, W Zimbabwe 17.55S 25.48E
85 I16 **Victoria Falls** ✕ Matabeleland North, W Zimbabwe 18.03S 25.48E
85 I16 **Victoria Falls** waterfall Zambia/Zimbabwe 18.03S 25.50E
Victoria Falls see Iguaçu, Salto do
65 F19 **Victoria, Isla** island Archipiélago de los Chonos, S Chile
15 J2 **Victoria Island** island Northwest Territories/Nunavut, NW Canada
190 L8 **Victoria, Lake** ◎ New South Wales, SE Australia
70 I12 **Victoria, Lake** var. Victoria Nyanza. ◎ E Africa
205 S13 **Victoria Land** physical region Antarctica
177 F5 **Victoria, Mount** ▲ W Myanmar 21.13N 93.53E
197 I14 **Victoria, Mount** ▲ Viti Levu, W Fiji 17.37S 178.00E
194 K15 **Victoria, Mount** ▲ S PNG 8.51S 147.36E
83 F17 **Victoria Nile** var. Somerset Nile. ≈ C Uganda
Victoria Nyanza see Victoria, Lake
44 G3 **Victoria Peak** ▲ SE Belize 16.50N 88.38W
193 H16 **Victoria Range** ▲▲ South Island, NZ
189 O3 **Victoria River** ≈ Northern Territory, N Australia
189 P3 **Victoria River Roadhouse** Northern Territory, N Australia 15.37S 131.07E
13 Q11 **Victoriaville** Quebec, SE Canada 46.03N 71.55W
Victoria-Wes see Victoria West
85 G24 **Victoria West** Afr. Victoria-Wes. Northern Cape, W South Africa 31.22S 23.06E
64 J13 **Victorica** La Pampa, C Argentina 36.14S 65.21W
205 T3 **Victor, Mount** ▲ Antarctica 72.49S 33.01E
37 U14 **Victorville** California, W USA 34.32N 117.17W
64 G9 **Vicuña** Coquimbo, N Chile 30.00S 70.44W
64 K11 **Vicuña Mackenna** Córdoba, C Argentina 33.52S 64.25W
Vicus Ausonensis see Vic
Vicus Elbii see Viterbo
35 X7 **Vida** Montana, NW USA 47.52N 105.30W
25 V6 **Vidalia** Georgia, SE USA 32.13N 82.24W
24 J7 **Vidalia** Louisiana, S USA 31.34N 91.25W
97 F22 **Videbæk** Ringkøbing, C Denmark 56.07N 8.37E
62 I13 **Videira** Santa Catarina, S Brazil 27.00S 51.08W
118 J14 **Videle** Teleorman, S Romania 44.15N 25.27E
Videm-Krško see Krško
Vídeň see Wien
106 H12 **Vidigueira** Beja, S Portugal 38.12N 7.48W
116 J9 **Vidima** ≈ N Bulgaria
116 G7 **Vidin** anc. Bononia. Vidin, NW Bulgaria 44.00N 22.50E
116 F8 **Vidin** ◈ province NW Bulgaria
160 H10 **Vidisha** Madhya Pradesh, C India 23.30N 77.49E
27 Y10 **Vidor** Texas, SW USA 30.07N 94.01W
97 L20 **Vidöstern** ◎ S Sweden
94 J13 **Vidsel** Norrbotten, N Sweden 65.49N 20.31E
120 H9 **Vidzemes Augstiene** ▲▲ C Latvia
120 J12 **Vidzy** Rus. Vidzy. Vitsyebskaya Voblasts', NW Belarus 55.22N 26.37E
65 L16 **Viedma** Río Negro, E Argentina 40.50S 62.57W
65 H22 **Viedma, Lago** ◎ S Argentina
47 O11 **Vieille Case** var. Itassi. N Dominica 15.36N 61.24W
106 M2 **Vieja, Peña** ▲ N Spain 43.09N 4.47W
42 E4 **Viejo, Cerro** ▲ NW Mexico 30.16N 112.18W
58 B9 **Viejo, Cerro** ▲ N Peru 4.54S 79.24W
120 E10 **Viekšniai** Akmenė, NW Lithuania 56.14N 22.33E
107 U3 **Vielha** var. Viella. Cataluña, NE Spain 42.40N 0.46E
Viella see Vielha
101 L21 **Vielsalm** Luxembourg, E Belgium 50.16N 5.55E
Vieng Pou Kha see Viangphoukha
25 T6 **Vienna** Georgia, SE USA 32.05N 83.48W
32 L17 **Vienna** Illinois, N USA 37.22N 88.51W
49 V5 **Vienna** Missouri, C USA 38.11N 91.57W
23 Q3 **Vienna** West Virginia, NE USA 39.19N 81.33W
Vienna see Wien, Austria
Vienna see Vienne, France
105 R11 **Vienne** anc. Vienna. Isère, E France 45.31N 4.52E
104 L10 **Vienne** ◈ department W France
104 L9 **Vienne** ≈ W France
Vientiane see Viangchan
Vientos, Paso de los see Windward Passage
47 V6 **Vieques** var. Isabel Segunda. E Puerto Rico 18.08N 65.27W
47 V6 **Vieques, Isla de** island E Puerto Rico
47 V6 **Vieques, Pasaje de** passage E Puerto Rico
47 V5 **Vieques, Sonda de** sound E Puerto Rico
Vierdörfer see Săcele
95 M15 **Vieremä** Itä-Suomi, C Finland 63.42N 27.02E
101 M14 **Vierlingsbeek** Noord-Brabant, SE Netherlands 51.36N 6.01E
103 G20 **Viernheim** Hessen, W Germany 49.31N 8.34E
103 D15 **Viersen** Nordrhein-Westfalen, W Germany 51.15N 6.24E
110 G8 **Vierwaldstätter See** Eng. Lake of Lucerne. ◎ C Switzerland
105 N8 **Vierzon** Cher, C France 47.13N 2.04E
42 L8 **Viesca** Coahuila de Zaragoza, NE Mexico 25.25N 102.45W
120 H10 **Viesīte** Ger. Eckengraf. Jēkabpils, S Latvia 56.21N 25.30E
109 N15 **Vieste** Puglia, SE Italy 41.52N 16.10E
178 Jj9 **Vietnam** off. Socialist Republic of Vietnam, Vtn. Công Hoa Xa Hôi Chu Nghia Viêt Nam. ◆ republic SE Asia
178 J5 **Viêt Quang** Ha Giang, N Vietnam 22.24N 104.48E
Vietri see Viêt Tri
178 Jj6 **Viêt Tri** var. Vietri. Vinh Phu, N Vietnam 21.19N 105.25E
32 L4 **Vieux Desert, Lac** ◎ Michigan/Wisconsin, N USA
47 Y13 **Vieux Fort** S Saint Lucia 13.43N 60.57W
47 X6 **Vieux-Habitants** Basse Terre, SW Guadeloupe 16.03N 61.45W
121 G14 **Vievis** Kaišiadorys, S Lithuania 54.46N 24.51E
179 P8 **Vigan** Luzon, N Philippines 17.34N 102.21E
108 D8 **Vigevano** Lombardia, N Italy 45.19N 8.51E
109 N18 **Viggiano** Basilicata, S Italy 40.21N 15.54E
60 L12 **Vigia** Pará, NE Brazil 0.49S 48.07W
43 Y12 **Vigía Chico** Quintana Roo, SE Mexico 19.49N 87.31W
47 T11 **Vigie** ✕ (Castries) NE Saint Lucia 14.01N 60.59W
104 K17 **Vignemale** var. Pic de Vignemale. ▲ France/Spain 42.48N 0.06W
Vignemale, Pic de see Vignemale
108 G10 **Vignola** Emilia-Romagna, C Italy 44.28N 11.00E
106 G4 **Vigo** Galicia, NW Spain 42.15N 8.43W
106 G4 **Vigo, Ría de** estuary NW Spain
96 D9 **Vigra** island S Norway
97 C17 **Vigrestad** Rogaland, S Norway 58.34N 5.42E
95 L15 **Vihanti** Oulu, C Finland 64.28N 25.00E
155 U10 **Vihāri** Punjab, E Pakistan 30.03N 72.31E
104 K8 **Vihiers** Maine-et-Loire, NW France 47.09N 0.37W
113 O19 **Vihorlat** ▲ E Slovakia 48.54N 22.09E
95 L19 **Vihti** Etelä-Suomi, S Finland 60.25N 24.16E
Viipuri see Vyborg
95 M16 **Viitasaari** Länsi-Suomi, W Finland 63.05N 25.52E
120 K3 **Viivikonna** Ida-Virumaa, NE Estonia 59.19N 27.40E
161 K16 **Vijayawāda** prev. Bezwada. Andhra Pradesh, SE India 16.34N 80.40E
Vijosa/Vijosë see Aóos, Albania/Greece
Vijosa/Vijosë see Vjosës, Lumi i, Albania/Greece
Vik see Vikøyri
94 J4 **Vík** Sudhurland, S Iceland 63.25N 18.58W
96 L13 **Vika** Dalarna, C Sweden 60.55N 14.30E
94 L12 **Vikajärvi** Lappi, N Finland 66.37N 26.10E
96 L13 **Vikarbyn** Dalarna, C Sweden 60.57N 15.00E
97 J22 **Viken** Skåne, S Sweden 56.09N 12.36E
97 L17 **Viken** ◎ C Sweden
97 G15 **Vikersund** Buskerud, S Norway 59.58N 9.58E
126 J14 **Vikhorevka** Irkutskaya Oblast', S Russian Federation 56.09N 99.09E
116 G11 **Vikhren** ▲ SW Bulgaria 41.45N 23.24E
9 R15 **Viking** Alberta, SW Canada 53.07N 111.49W
86 E7 **Viking Bank** undersea feature N North Sea
97 M14 **Vikmanshyttan** Dalarna, C Sweden 60.19N 15.55E
96 D12 **Vikøyri** var. Vik. Sogn og Fjordane, S Norway 61.04N 6.34E
95 H17 **Viksjö** Västernorrland, C Sweden 62.45N 17.30E
Viktoriastadt see Victoria
Vila see Port-Vila
Vila Arriaga see Bibala
Vila Artur de Paiva see Cubango
Vila Bela da Santissima Trindade see Mato Grosso
60 B12 **Vila Bittencourt** Amazonas, NW Brazil 1.25S 69.24W
Vila da Ponte see Cubango
66 O2 **Vila da Praia da Vitória** Terceira, Azores, Portugal, NE Atlantic Ocean 38.43N 27.04W
Vila de Aljustrel see Cangamba
Vila de Almoster see Chiange
Vila de João Belo see Xai-Xai
Vila de Macia see Macia
Vila de Manhiça see Manhiça
Vila de Manica see Manica
Vila de Mocímboa da Praia see Mocímboa da Praia
85 N16 **Vila de Sena** var. Sena. Sofala, C Mozambique 17.25S 34.59E
106 F14 **Vila do Bispo** Faro, S Portugal 37.04N 8.52W
106 G6 **Vila do Conde** Porto, NW Portugal 41.21N 8.45W
Vila do Maio see Maio
66 P3 **Vila do Porto** Santa Maria, Azores, Portugal, NE Atlantic Ocean 36.57N 25.10W
85 K15 **Vila do Zumbo** prev. Vila do Zumbu, Zumbo. Tete, NW Mozambique 15.36S 30.30E
Vila do Zumbu see Vila do Zumbo
106 I6 **Vila Flor** var. Vila Flôr. Bragança, N Portugal 41.18N 7.09W
107 V6 **Vilafranca del Penedès** var. Villafranca del Panadés. Cataluña, NE Spain 41.21N 1.42E
106 F10 **Vila Franca de Xira** var. Vilafranca de Xira. Lisboa, C Portugal 38.57N 8.58W
Vila Gago Coutinho see Lumbala N'Guimbo
106 G3 **Vilagarcía de Arousa** var. Villagarcía de Arosa. Galicia, NW Spain 42.34N 8.45W
Vila General Machado see Camacupa
Vila Henrique de Carvalho see Saurimo
104 I7 **Vilaine** ≈ NW France
Vila João de Almeida see Chibia
120 K8 **Viļaka** Ger. Marienhausen. Balvi, NE Latvia 57.11N 27.42E
106 I2 **Vilalba** Galicia, NW Spain 43.16N 7.40W
Vila Marechal Carmona see Uíge
Vila Mariano Machado see Ganda
180 G3 **Vilanandro, Tanjona** headland W Madagascar 16.10S 44.27E
Vilanculos see Vilankulo
120 J10 **Viļāni** Rēzekne, E Latvia 56.33N 26.55E
85 N19 **Vilankulo** var. Vilanculos. Inhambane, E Mozambique 22.01S 35.19E
Vila Norton de Matos see Balombo
106 G6 **Vila Nova de Famalicão** var. Vila Nova de Famalicao. Braga, N Portugal 41.24N 8.31W
106 I6 **Vila Nova de Foz Côa** var. Vila Nova de Fozcôa. Guarda, N Portugal 41.04N 7.09W
106 F6 **Vila Nova de Gaia** Porto, NW Portugal 41.07N 8.37W
Vila Nova de Portimão see Portimão
107 V6 **Vilanova i La Geltrú** Cataluña, NE Spain 41.15N 1.42E
Vila Pereira de Eça see N'Giva
106 H6 **Vila Pouca de Aguiar** Vila Real, N Portugal 41.30N 7.37W
106 H6 **Vila Real** var. Vila Rial. Vila Real, N Portugal 41.16N 7.45W
106 H6 **Vila Real** ◈ district N Portugal
107 T9 **Vila-real de los Infantes** var. Villarreal. País Valenciano, E Spain 39.55N 0.07W
106 H14 **Vila Real de Santo António** Faro, S Portugal 37.12N 7.25W
106 J7 **Vilar Formoso** Guarda, N Portugal 40.37N 6.49W
Vila Rial see Vila Real
61 J15 **Vila Rica** Mato Grosso, W Brazil 9.52S 50.44W
Vila Robert Williams see Caála
Vila Salazar see N'Dalatando
Vila Serpa Pinto see Menongue
Vila Teixeira da Silva see Bailundo
Vila Teixeira de Sousa see Luau
106 H9 **Vila Velha de Ródão** Castelo Branco, C Portugal 39.39N 7.40W
106 G5 **Vila Verde** Braga, N Portugal 41.39N 8.27W
106 H11 **Vila Viçosa** Évora, S Portugal 38.46N 7.25W
59 G15 **Vilcabamba, Cordillera de** ▲▲ C Peru
Vîlcea see Vâlcea
126 Hh1 **Vil'cheka, Zemlya** Eng. Wilczek Land. island Zemlya Frantsa-Iosifa, NW Russian Federation
97 F22 **Vildbjerg** Ringkøbing, C Denmark 56.12N 8.46E
Vileyka see Vilyeyka
95 H15 **Vilhelmina** Västerbotten, N Sweden 64.37N 16.40E
61 F17 **Vilhena** Rondônia, W Brazil 12.40S 60.07W
117 G19 **Vília** Attikí, C Greece 38.09N 23.21E
Viliya see Viliya
121 I14 **Viliya** Lith. Neris, Rus. Viliya. ≈ W Belarus
Viliya see Neris
120 H5 **Viljandi** Ger. Fellin. Viljandimaa, S Estonia 58.22N 25.34E
120 H5 **Viljandimaa** off. Viljandi Maakond. ◈ province SW Estonia
121 E14 **Vilkaviškis** Pol. Wyłkowyszki. Vilkaviškis, SW Lithuania 54.39N 23.03E
120 F13 **Vilkija** Kaunas, C Lithuania 55.02N 23.36E
207 V9 **Vil'kitskogo, Proliv** strait N Russian Federation
Vilkovo see Vylkove
59 L21 **Villa Abecia** Chuquisaca, S Bolivia 21.01S 65.12W
43 N5 **Villa Acuña** var. Ciudad Acuña. Coahuila de Zaragoza, NE Mexico 29.17N 100.57W
42 J4 **Villa Ahumada** Chihuahua, N Mexico 30.37N 106.30W
47 O9 **Villa Altagracia** C Dominican Republic 18.37N 70.11W
58 L13 **Villa Bella** Beni, N Bolivia 10.21S 65.25W
106 J3 **Villablino** Castilla-León, N Spain 42.55N 6.21W
56 K6 **Villa Bruzual** Portuguesa, N Venezuela 9.19N 69.06W
107 O9 **Villacañas** Castilla-La Mancha, C Spain 39.37N 3.19W
107 O12 **Villacarrillo** Andalucía, S Spain 38.07N 3.04W
106 M7 **Villacastín** Castilla-León, N Spain 40.46N 4.25W
Villa Cecilia see Ciudad Madero
111 S9 **Villach** Slvn. Beljak. Kärnten, S Austria 46.36N 13.49E
109 B20 **Villacidro** Sardegna, Italy, C Mediterranean Sea 39.27N 8.43E
Villa Concepción see Concepción
106 L4 **Villada** Castilla-León, N Spain 42.15N 4.58W
42 M10 **Villa de Cos** Zacatecas, C Mexico 23.20N 102.20W
56 L5 **Villa de Cura** var. Cura. Aragua, N Venezuela 10.00N 67.30W
Villa del Nevoso see Ilirska Bistrica
Villa del Pilar see Pilar
106 M13 **Villa del Río** Andalucía, S Spain 37.58N 4.16W
Villa de Méndez see Méndez
44 H6 **Villa de San Antonio** Comayagua, W Honduras 14.24N 87.37W
107 N4 **Villadiego** Castilla-León, N Spain 42.31N 4.01W
107 T8 **Villafames** País Valenciano, E Spain 40.07N 0.03W
43 U16 **Villa Flores** Chiapas, SE Mexico 16.12N 93.16W
106 J3 **Villafranca del Bierzo** Castilla-León, N Spain 42.36N 6.49W
107 S8 **Villafranca del Cid** País Valenciano, E Spain 40.25N 0.15W
106 J11 **Villafranca de los Barros** Extremadura, W Spain 38.34N 6.19W
107 N10 **Villafranca de los Caballeros** Castilla-La Mancha, C Spain 39.25N 3.21W
Villafranca del Panadés see Vilafranca del Penedès
108 F8 **Villafranca di Verona** Veneto, NE Italy 45.22N 10.51E
109 J23 **Villafrati** Sicilia, Italy, C Mediterranean Sea 37.53N 13.30E
Villagarcía de Arosa see Vilagarcía de Arousa
43 O9 **Villagrán** Tamaulipas, C Mexico 24.28N 99.30W
63 C17 **Villaguay** Entre Ríos, E Argentina 31.55S 59.01W
64 O6 **Villa Hayes** Presidente Hayes, S Paraguay 25.04S 57.25W
43 U15 **Villahermosa** prev. San Juan Bautista. Tabasco, SE Mexico 17.56N 92.50W
107 O11 **Villahermosa** Castilla-La Mancha, C Spain 38.46N 2.52W
66 O11 **Villahermoso** Gomera, Islas Canarias, Spain, NE Atlantic Ocean 38.46N 2.52W
Villa Hidalgo see Hidalgo
107 T12 **Villajoyosa** var. La Vila Jojosa. País Valenciano, E Spain 38.31N 0.13W
Villa Juárez see Juárez
Villalba see Collado Villalba
43 N8 **Villaldama** Nuevo León, NE Mexico 26.29N 100.27W
106 L5 **Villalón de Campos** Castilla-León, N Spain 42.04N 5.03W
63 A25 **Villalonga** Buenos Aires, E Argentina 39.55S 62.34W
106 L5 **Villalpando** Castilla-León, N Spain 41.51N 5.25W
42 K9 **Villa Madero** var. Francisco I.Madero. Durango, C Mexico 24.27N 104.11W
43 O9 **Villa Mainero** Tamaulipas, C Mexico 24.32N 99.39W
Villamaña see Villamañán
106 L4 **Villamañán** var. Villamaña. Castilla-León, N Spain 42.19N 5.34W
64 L10 **Villa María** Córdoba, C Argentina 32.22S 63.15W
63 C17 **Villa María Grande** Entre Ríos, E Argentina 31.39S 59.54W
59 K21 **Villa Martín** Potosí, SW Bolivia 20.48S 67.36W
106 K15 **Villamartin** Andalucía, S Spain 36.50N 5.39W
64 J8 **Villa Mazán** La Rioja, NW Argentina 28.43S 66.25W
Villa Mercedes see Mercedes
Villamil see Puerto Villamil
Villa Nador see Nador
56 G5 **Villanueva** La Guajira, N Colombia 10.37N 72.58W
44 H5 **Villanueva** Cortés, NW Honduras 15.17N 87.58W
42 L11 **Villanueva** Zacatecas, C Mexico 22.24N 102.52W
44 I9 **Villa Nueva** Chinandega, NW Nicaragua 12.58N 86.46W
39 T11 **Villanueva** New Mexico, SW USA 35.18N 105.20W
106 M12 **Villanueva de Córdoba** Andalucía, S Spain 38.19N 4.37W
107 O12 **Villanueva del Arzobispo** Andalucía, S Spain 38.10N 3.00W
106 K11 **Villanueva de la Serena** Extremadura, W Spain 38.58N 5.48W
106 L5 **Villanueva del Campo** Castilla-León, N Spain 41.58N 5.25W
107 O11 **Villanueva de los Infantes** Castilla-La Mancha, C Spain 38.45N 3.01W
63 C14 **Villa Ocampo** Santa Fe, C Argentina 28.28S 59.22W
42 J8 **Villa Ocampo** Durango, C Mexico 26.26N 105.28W
42 J7 **Villa Orestes Pereyra** Durango, C Mexico 26.30N 105.38W
107 N3 **Villarcayo** Castilla-León, N Spain 42.55N 3.34W
106 L5 **Villardefrades** Castilla-León, N Spain 41.43N 5.15W
107 S9 **Villar del Arzobispo** País Valenciano, E Spain 39.43N 0.49W
107 Q6 **Villaroya de la Sierra** Aragón, NE Spain 41.28N 1.46W
Villarreal see Vila-real de los Infantes
64 P6 **Villarrica** Guairá, SE Paraguay 25.45S 56.28W
65 G15 **Villarrica, Volcán** Ж S Chile 39.28S 71.57W
107 P10 **Villarrobledo** Castilla-La Mancha, C Spain 39.16N 2.36W
107 N10 **Villarrubia de los Ojos** Castilla-La Mancha, C Spain 39.13N 3.36W

◆ Country
● Country Capital
◇ Dependent Territory
○ Dependent Territory Capital
◈ Administrative Region
✕ International Airport
▲ Mountain
▲▲ Mountain Range
Ж Volcano
≈ River
◎ Lake
⊠ Reservoir

20 J17 **Villas** New Jersey, NE USA 39.01N 74.54W

107 O3 **Villasana de Mena** Castilla-León, N Spain 43.04N 3.16W

109 M23 **Villa San Giovanni** Calabria, S Italy 38.12N 15.39E

63 D18 **Villa San José** Entre Ríos, E Argentina 32.12S 58.15W

Villa Sanjurjo see Al-Hoceima

107 P6 **Villasayas** Castilla-León, N Spain 41.19N 2.36W

109 C20 **Villasimius** Sardegna, Italy, C Mediterranean Sea 39.10N 9.30E

43 N6 **Villa Unión** Coahuila de Zaragoza, NE Mexico 28.18N 100.43W

42 K10 **Villa Unión** Durango, C Mexico 23.58N 104.01W

42 J10 **Villa Unión** Sinaloa, C Mexico 23.13N 106.10W

64 K12 **Villa Valeria** Córdoba, C Argentina 34.21S 64.55W

107 N8 **Villaverde** Madrid, C Spain 40.21N 3.43W

56 F10 **Villavicencio** Meta, C Colombia 4.09N 73.37W

106 L2 **Villaviciosa** Asturias, N Spain 43.28N 5.25W

106 L12 **Villaviciosa de Córdoba** Andalucía, S Spain 38.04N 5.00W

59 L2 **Villazón** Potosí, S Bolivia 22.04S 65.34W

12 J8 **Villebon, Lac** ⊚ Quebec, SE Canada

Ville de Kinshasa see Kinshasa

104 J5 **Villedieu-les-Poêles** Manche, N France 48.51N 1.12W

Villefranche see Villefranche-sur-Saône

105 N16 **Villefranche-de-Lauragais** Haute-Garonne, S France 43.24N 1.42E

105 N14 **Villefranche-de-Rouergue** Aveyron, S France 44.21N 2.01E

105 R10 **Villefranche-sur-Saône** var. Villefranche. Rhône, E France 46.00N 4.40E

12 H9 **Ville-Marie** Quebec, SE Canada 47.21N 79.25W

104 M15 **Villemur-sur-Tarn** Haute-Garonne, S France 43.50N 1.32E

107 S11 **Villena** País Valenciano, E Spain 38.39N 0.52W

Villeneuve-d'Agen see Villeneuve-sur-Lot

104 L13 **Villeneuve-sur-Lot** var. Villeneuve-d'Agen; hist. Gajac. Lot-et-Garonne, SW France 44.24N 0.43E

105 P6 **Villeneuve-sur-Yonne** Yonne, C France 48.04N 3.21E

24 H8 **Ville Platte** Louisiana, S USA 30.41N 92.16W

105 R11 **Villeurbanne** Rhône, E France 45.46N 4.54E

103 G23 **Villingen-Schwenningen** Baden-Württemberg, S Germany 48.04N 8.27E

31 T15 **Villisca** Iowa, C USA 40.55N 94.58W

Villmanstrand see Lappeenranta

Vilna see Vilnius

121 H14 **Vilnius** Pol. Wilno, Ger. Wilna; prev. Rus. Vilna. ● (Lithuania) Vilnius, SE Lithuania 54.41N 25.19E

121 H14 **Vilnius** × Vilnius, SE Lithuania 54.33N 25.17E

119 S7 **Vil'nohirs'k** Dnipropetrovs'ka Oblast', E Ukraine 48.31N 34.01E

119 U8 **Vil'nyans'k** Zaporiz'ka Oblast', SE Ukraine 47.56N 35.22E

95 L17 **Vilppula** Länsi-Suomi, W Finland 62.01N 24.30E

103 M20 **Vils** ♠ SE Germany

120 C5 **Vilsandi Saar** island W Estonia

119 P8 **Vil'shanka** Rus. Olshanka. Kirovohrads'ka Oblast', C Ukraine 48.12N 30.54E

103 O22 **Vilshofen** Bayern, SE Germany 48.36N 13.10E

161 J20 **Viluppuram** Tamil Nādu, SE India 12.54N 79.40E

115 I16 **Vilusi** Montenegro, SW Yugoslavia 42.44N 18.34E

101 G18 **Vilvoorde** Fr. Vilvorde. Vlaams Brabant, C Belgium 50.55N 4.25E

Vilvorde see Vilvoorde

121 J14 **Vilyeyka** Pol. Wilejka, Rus. Vileyka. Minskaya Voblasts', NW Belarus 54.30N 26.54E

126 Kk11 **Vilyuy** ♠ NE Russian Federation

126 L10 **Vilyuysk** Respublika Sakha (Yakutiya), NE Russian Federation 63.42N 121.20E

126 K11 **Vilyuyskoye Vodokhranilishche** ⊡ NE Russian Federation

106 G2 **Vimianzo** Galicia, NW Spain 43.06N 9.03W

97 M19 **Vimmerby** Kalmar, S Sweden 57.40N 15.49E

104 L5 **Vimoutiers** Orne, C France 48.56N 0.10E

95 L16 **Vimpeli** Länsi-Suomi, W Finland 63.10N 23.49E

81 G14 **Vina** ♠ Cameroon/Chad

64 G11 **Viña del Mar** Valparaíso, C Chile 33.01S 71.34W

21 R8 **Vinalhaven** island Maine, NE USA

107 T8 **Vinaròs** País Valenciano, E Spain 40.28N 0.28E

33 N15 **Vincennes** Indiana, N USA 38.42N 87.30W

205 Y12 **Vincennes Bay** bay Antarctica

27 O7 **Vincent** Texas, SW USA 32.30N 101.10W

97 H24 **Vindeby** Fyn, C Denmark 54.55N 11.09E

95 I15 **Vindeln** Västerbotten, N Sweden 64.10N 19.45E

97 I22 **Vinderup** Ringkøbing, C Denmark 56.28N 8.48E

Vindhya Mountains see Vindhya Range

159 N14 **Vindhya Range** var. Vindhya Mountains. ▲ N India

Vindobona see Wien

22 K6 **Vine Grove** Kentucky, S USA 37.48N 85.58W

---

20 J17 **Vineland** New Jersey, NE USA 39.29N 75.01W

118 E11 **Vinga** Arad, W Romania 46.00N 21.14E

97 M16 **Vingåker** Södermanland, C Sweden 59.01N 15.59E

178 Jj8 **Vinh** Nghệ An, N Vietnam 18.42N 105.40E

178 K9 **Vinh** Linh Quang Tri, C Vietnam 17.02N 107.03E

178 Jj14 **Vinh Loi** see Bac Liêu

115 Q18 **Vinh Long** var. Vinhlong. Vinh Long, S Vietnam 10.15N 105.58E

Vinhlong see Vinh Long

113 V13 **Vinica** NE FYR Macedonia 41.53N 22.30E

116 G8 **Vinica** SE Slovenia 45.28N 15.12E

49 Q8 **Vinita** Oklahoma, C USA 36.38N 95.09W

100 I11 **Vinkeveen** Utrecht, C Netherlands 52.13N 4.55E

118 L6 **Vin'kivtsi** Khmel'nyts'ka Oblast', W Ukraine 49.02N 27.13E

114 I10 **Vinkovci** Ger. Winkowitz, Hung. Vinkovce. Vukovar-Srijem, E Croatia 45.18N 18.45E

Vinkovce see Vinkovci

Vinnitsa see Vinnytsya

118 M7 **Vinnitskaya Oblast'** var. Vinnytsya, Rus. Vinnitskaya Oblast'. ♦ province C Ukraine

119 N6 **Vinnytsya** Rus. Vinnitsa. Vinnyts'ka Oblast', C Ukraine 49.14N 28.30E

119 N6 **Vinnyts'ka Oblast'** × Vinnyts'ka Oblast', N Ukraine 49.13N 28.40E

Vinogradov see Vynohradiv

118 K12 **Vintilă Vodă** Buzău, SE Romania 45.28N 26.44E

31 X13 **Vinton** Iowa, C USA 42.10N 92.01W

24 F9 **Vinton** Louisiana, S USA 30.13N 93.37W

161 J17 **Vinukonda** Andhra Pradesh, E India 16.03N 79.41E

85 E23 **Vioolsdrif** Northern Cape, W South Africa 28.50S 17.38E

84 M13 **Viphya Mountains** ▲ C Malawi

179 Qq11 **Virac** Catanduanes Island, N Philippines 13.39N 124.17E

128 K8 **Virandozero** Respublika Kareliya, NW Russian Federation

143 P16 **Viranşehir** Şanlıurfa, SE Turkey 37.13N 39.31E

160 D13 **Virār** Mahārāshtra, W India 19.30N 72.48E

37 U17 **Virden** Manitoba, S Canada 49.49N 100.57W

32 K14 **Virden** Illinois, N USA 39.30N 89.46W

104 J5 **Virdois** see Virrat

104 J4 **Vire** Calvados, N France 48.49N 0.52W

104 J4 **Vire** ♠ N France

85 A15 **Virei** Namibe, SW Angola 15.43S 12.54E

Vîrful Moldoveanu see Vârful Moldoveanu

116 I8 **Virful** ♠ NW Bulgaria

37 R5 **Virginia Peak** ▲ Nevada, W USA 39.46N 119.26W

47 U9 **Virgin Gorda** island C British Virgin Islands

109 H14 **Virginia** Free State, C South Africa 28.04S 26.51E

114 I10 **Virginia** Illinois, N USA 39.57N 90.12W

31 W4 **Virginia** Minnesota, N USA 47.31N 92.32W

23 T6 **Virginia** off. Commonwealth of Virginia; also known as Mother of Presidents, Mother of States, Old Dominion. ♦ state NE USA

23 Y7 **Virginia Beach** Virginia, NE USA 36.51N 75.58W

35 R14 **Virginia City** Montana, NW USA 45.17N 111.54W

37 Q6 **Virginia City** Nevada, W USA 39.19N 119.39W

12 H8 **Virginiatown** Ontario, S Canada 48.09N 79.35W

Virgin Islands see British Virgin Islands

47 T9 **Virgin Islands (US)** var. Virgin Islands of the United States; prev. Danish West Indies. ♦ US unincorporated territory E West Indies

47 T9 **Virgin Passage** passage Puerto Rico/Virgin Islands (US)

37 Y10 **Virgin River** ♠ Nevada/Utah, W USA

Virihaur see Virihaure

94 H12 **Virihaure** var. Virihaur. ⊚ N Sweden

178 Jj11 **Virôchey** Rôtânôkiri, NE Cambodia 13.58N 106.49E

95 N19 **Virolahti** Etelä-Suomi, S Finland 60.33N 27.37E

32 J8 **Viroqua** Wisconsin, N USA 43.33N 90.53W

114 G8 **Virovitica** Ger. Virovititz, Hung. Veröcze; prev. Ger. Werowitz. Virovitica-Podravina, NE Croatia 45.49N 17.25E

114 G8 **Virovitica-Podravina** off. Virovitičko-Podravinska Županija. ♦ province NE Croatia

Virovititz see Virovitica

115 J17 **Virpazar** Montenegro, SW Yugoslavia 42.15N 19.06E

95 L17 **Virrat** Swe. Virdois. Länsi-Suomi, W Finland 62.13N 23.49E

97 N15 **Virserum** Kalmar, S Sweden 57.17N 15.18E

Virtsu see Vänatori

114 G8 **Viru** La Libertad, C Peru 8.27S 78.44W

161 H23 **Virudhunagar** see Virudunagar
Virudunagar var. Virudhunagar. Tamil Nādu, SE India 9.34N 77.57E

120 I3 **Viru-Jaagupi** Ger. Sankt-Jakobi. Lääne-Virumaa, NE Estonia 59.13N 26.28E

59 N14 **Viru-Viru** var. Santa Cruz. ✈ (Santa Cruz) Santa Cruz, C Bolivia 17.49S 63.12W

115 E15 **Vis** It. Lissa; anc. Issa. island S Croatia

Vis see Fish

120 I12 **Visaginas** prev. Sneičkus. Ignalina, E Lithuania 55.36N 26.22E

161 M15 **Visakhapatnam** Andhra Pradesh, SE India 17.45N 83.19E

37 R11 **Visalia** California, W USA 36.19N 119.19W

Vişău see Vişeu

179 Qq12 **Visayan Sea** sea C Philippines

97 P19 **Visby** Ger. Wisby. Gotland, SE Sweden 57.37N 18.19E

207 N9 **Viscount Melville Sound** prev. Melville Sound. sound Northwest Territories/Nunavut, N Canada

101 L19 **Visé** Liège, E Belgium 50.43N 5.42E

114 K13 **Višegrad** Republika Srpska, E Bosnia and Herzegovina 43.46N 19.18E

60 L12 **Viseu** Pará, NE Brazil 1.10S 46.09W

106 H7 **Viseu** prev. Vizeu. Viseu, N Portugal 40.40N 7.55W

106 H7 **Viseu** var. Vizeu. ♦ district N Portugal

118 I8 **Vişeu** Hung. Visó; prev. Vişău. ♠ NW Romania

118 I8 **Vişeu de Sus** var. Vişeul de Sus, Ger. Oberwischau, Hung. Felsővisó. Maramureş, N Romania 47.43N 23.24E

Vişeul de Sus see Vişeu de Sus

129 R10 **Vishera** ♠ NW Russian Federation

97 J19 **Viskafors** Västra Götaland, S Sweden 57.37N 12.49E

97 J20 **Visland** Kronoberg, S Sweden 56.46N 14.30E

97 L21 **Vislanda** Kronoberg, S Sweden 56.46N 14.30E

Vislinskiy Zaliv see Vistula Lagoon

Visó see Vişeu

114 H13 **Visoko** Federacija Bosna I Hercegovina, C Bosnia and Herzegovina 43.58N 18.12E

114 A59 **Viso, Monte** ▲ NW Italy 44.42N 7.04E

110 E10 **Visp** Valais, SW Switzerland 46.18N 7.52E

110 E10 **Vissefjärda** Kalmar, S Sweden 56.31N 15.34E

97 M21 **Vissefjärda** Kalmar, S Sweden 56.31N 15.34E

102 I11 **Visselhövede** Niedersachsen, NW Germany 52.58N 9.36E

97 G23 **Vissenbjerg** Fyn, C Denmark 55.22N 10.07E

37 U17 **Vista** California, W USA 33.12N 117.14W

60 C11 **Vista Alegre** Amazonas, NW Brazil 1.23N 68.13W

116 J13 **Vistonída, Límni** ⊚ NE Greece

121 A14 **Vistula** see Wisła

121 A14 **Vistula Lagoon** Ger. Frisches Haff, Pol. Zalew Wiślany, Rus. Vislinskiy Zaliv. lagoon Poland/Russian Federation

116 I8 **Vit** ♠ NW Bulgaria

Vitebsk see Vitsyebsk

Vitebskaya Oblast' see Vitsyebskaya Voblasts'

109 H14 **Viterbo** anc. Vicus Elbii. Lazio, C Italy 42.25N 12.07E

114 H10 **Vitez** Federacija Bosna I Hercegovina, C Bosnia and Herzegovina 44.08N 17.47E

178 J15 **Vi Thanh** Cần Thơ, S Vietnam 9.45N 105.28E

Viti see Fiji

194 K12 **Vitiaz Strait** strait NE PNG

106 J7 **Vitigudino** Castilla-León, N Spain 41.00N 6.26W

197 H15 **Viti Levu** island W Fiji

126 Kk14 **Vitim** ♠ C Russian Federation

126 Kk13 **Vitimskiy** Irkutskaya Oblast', C Russian Federation 58.12N 113.10E

111 V2 **Vitis** Niederösterreich, N Austria 48.45N 15.09E

61 O20 **Vitória** Espírito Santo, SE Brazil 20.19S 40.21W

61 N18 **Vitória Bank** see Vitória Seamount

Vitória da Conquista Bahia, E Brazil 14.52S 40.52W

107 P3 **Vitoria-Gasteiz** var. Vitoria, Eng. Vittoria. País Vasco, N Spain 42.51N 2.40W

67 J16 **Vitória Seamount** var. Victoria Bank, Vitoria Bank. undersea feature C Atlantic Ocean 18.48S 37.24W

114 H13 **Vitorog** ▲ SW Bosnia and Herzegovina 44.06N 17.03E

104 J6 **Vitré** Ille-et-Vilaine, NW France 48.07N 1.12W

105 R5 **Vitry-le-François** Marne, N France 48.43N 4.36E

120 N13 **Vitsi** see Vérnon

94 H4 **Vittangi** Norrbotten, N Sweden 67.40N 21.39E

114 D13 **Vodice** Šibenik-Knin, S Croatia 43.46N 15.46E

116 K7 **Vodňany** Rus, N Bulgaria 43.33N 25.49E

114 C10 **Vodnjan** It. Dignano d'Istria. Istra, NW Croatia 44.57N 13.51E

129 Q9 **Vodnyy** Respublika Komi, NW Russian Federation 63.31N 53.21E

79 Q17 **Vodskov** Nordjylland, N Denmark 57.04N 10.03E

94 H4 **Vogar** Suðurland, SW Iceland 63.58N 22.20W

---

199 Jj7 **Vityaz Seamount** undersea feature C Pacific Ocean 13.30N 173.15W

183 Q7 **Vityaz Trench** undersea feature W Pacific Ocean

110 G8 **Vitznau** Luzern, W Switzerland 47.01N 8.28E

106 I1 **Viveiro** Galicia, NW Spain 43.39N 7.34W

107 S9 **Viver** País Valenciano, E Spain 39.55N 0.36W

105 Q13 **Viverais, Monts du** ♠ C France

126 H10 **Viviers** ♠ W Russian Federation

24 F4 **Vivian** Louisiana, S USA 32.52N 93.59W

31 N10 **Vivian** South Dakota, N USA 43.55N 100.16W

105 R13 **Viviers** Ardèche, E France 44.31N 4.40E

85 K19 **Vivo** Northern, NE South Africa 22.58S 29.13E

104 L10 **Vivonne** Vienne, W France 46.25N 0.15E

197 G14 **Viwa** island Yasawa Group, NW Fiji

Vizakna see Ocna Sibiului

107 Q2 **Vizcaya** Basq. Bizkaia. ♦ province País Vasco, N Spain

55 Q6 **Vizcaya, Golfo de** see Biscay, Bay of

142 C10 **Vize** Kırklareli, NW Turkey 41.33N 27.49E

126 I2 **Vize, Ostrov** island Severnaya Zemlya, N Russian Federation 79.55N 76.57E

161 M15 **Vizianagaram** var. Vizianagram. Andhra Pradesh, E India 18.07N 83.25E

Vizianagram see Vizianagaram

129 R11 **Vizille** Isère, E France 45.05N 5.46E

129 R11 **Vizinga** Respublika Komi, NW Russian Federation 61.06N 50.09E

118 M13 **Viziru** Brăila, SE Romania 45.00N 27.43E

115 Q16 **Vjosës, Lumi i** var. Vijosa, Vijosë, Gk. Aóos. ♠ Albania/Greece see also Aóos

101 H18 **Vlaams Brabant** ♦ province C Belgium

100 G12 **Vlaanderen** see Flanders

118 F10 **Vlaardingen** Zuid-Holland, SW Netherlands 51.55N 4.21E

115 P16 **Vlădeasa, Vârful** prev. Vîrful Vlădeasa. ▲ NW Romania 46.45N 22.46E

Vlădeasa, Virful see Vlădeasa, Vârful

115 P16 **Vladičin Han** Serbia, SE Yugoslavia 42.44N 22.04E

131 O16 **Vladikavkaz** prev. Dzaudzhikau, Ordzhonikidze. Respublika Severnaya Osetiya, SW Russian Federation 43.05N 44.41E

130 M3 **Vladimir** Vladimirskaya Oblast', W Russian Federation 56.09N 40.21E

150 M7 **Vladimirovka** Kostanay, N Kazakhstan 53.28N 64.01E

Vladimirovka see Yuzhno-Sakhalinsk

130 L3 **Vladimirskaya Oblast'** ♦ province W Russian Federation

130 J3 **Vladimirskiy Tupik** Smolenskaya Oblast', W Russian Federation 55.45N 33.25E

Vladimir-Volynskiy see Volodymyr-Volyns'kyy

127 Nn18 **Vladivostok** Primorskiy Kray, SE Russian Federation 43.09N 131.52E

119 U3 **Vladyslavivka** Respublika Krym, S Ukraine 45.09N 35.25E

100 P6 **Vlagtwedde** Groningen, NE Netherlands 53.01N 7.07E

114 J12 **Vlajna** see Kukavica

114 J12 **Vlasenica** Republika Srpska, E Bosnia and Herzegovina 44.10N 18.57E

115 G12 **Vlašić** ▲ C Bosnia and Herzegovina 44.18N 17.40E

113 D17 **Vlašim** Ger. Wlaschim. Středočeský Kraj, C Czech Republic 49.42N 14.54E

115 P15 **Vlasotince** Serbia, SE Yugoslavia 42.58N 22.07E

126 L9 **Vlasovo** Respublika Sakha (Yakutiya), NE Russian Federation 70.41N 134.49E

100 I11 **Vleuten** Utrecht, C Netherlands 52.07N 5.01E

119 N8 **Vlieland** Fris. Flylân. island Waddeneilanden, N Netherlands

100 I5 **Vlieland** Fris. Flylân. island Waddeneilanden, N Netherlands

100 I5 **Vliestroom** strait NW Netherlands

101 J14 **Vlijmen** Noord-Brabant, S Netherlands 51.42N 5.13E

101 D15 **Vlissingen** Eng. Flushing, Fr. Flessingue. Zeeland, SW Netherlands 51.25N 3.34E

115 K22 **Vlodava** see Włodawa

115 K22 **Vlonë/Vlora** see Vlorë

115 K22 **Vlorë** prev. Vlonë, It. Valona, Vlora. Vlorë, SW Albania 40.27N 19.31E

115 K22 **Vlorë, Gjiri i** var. Valona Bay. bay SW Albania

113 C16 **Vltava** Ger. Moldau. ♠ W Czech Republic

130 K3 **Vnukovo** × (Moskva) Gorod Moskva, W Russian Federation 55.30N 36.52E

27 Q9 **Vnukovo** see Novi Bečej

111 R5 **Vöcklabruck** Oberösterreich, NW Austria 48.01N 13.37E

114 D13 **Vodice** Šibenik-Knin, S Croatia 43.46N 15.46E

---

195 N16 **Vogel, Cape** headland SE PNG 9.42S 150.04E

Vogelkop see Doberai, Jazirah

79 X5 **Vogel Peak** prev. Dim lang. ▲ E Nigeria 8.16N 11.14E

103 H17 **Vogelsberg** ▲ C Germany

108 D8 **Voghera** Lombardia, N Italy 44.58N 9.01E

114 I13 **Vogošća** Federacija Bosna I Hercegovina, SE Bosnia and Herzegovina 43.55N 18.20E

103 M17 **Vogtland** historical region E Germany

129 V12 **Vogul'skiy Kamen', Gora** ▲ NW Russian Federation 60.10N 58.41E

197 N6 **Voh** Province Nord, C New Caledonia 20.57S 164.41E

180 H8 **Vohémar** see Iharaña

Vohimena, Tanjona Fr. Cap Sainte Marie. headland S Madagascar 25.20S 45.06E

180 J6 **Vohipeno** Fianarantsoa, SE Madagascar 22.21S 47.51E

120 H5 **Võhma** Ger. Wöchma. Viljandimaa, S Estonia 58.37N 25.34E

83 J20 **Voi** Coast, S Kenya 3.22S 38.34E

78 K15 **Voinjama** N Liberia 8.23N 9.48W

105 S12 **Voiron** Isère, E France 45.22N 5.34E

111 V8 **Voitsberg** Steiermark, SE Austria 47.04N 15.09E

97 F24 **Vojens** Ger. Woyens. Sønderjylland, SW Denmark 55.15N 9.19E

114 K9 **Vojvodina** Ger. Wojwodina. Region N Yugoslavia

13 S6 **Volant** ⊚ Quebec, SE Canada

45 P15 **Volcán** var. Hato del Volcán. Chiriquí, W Panama 8.45N 82.38W

96 D10 **Volda** Møre og Romsdal, S Norway 62.07N 6.04E

100 J9 **Volendam** Noord-Holland, C Netherlands 52.30N 5.04E

128 L15 **Volga** Yaroslavskaya Oblast', W Russian Federation 57.56N 38.23E

31 Q10 **Volga** South Dakota, N USA 44.19N 96.55W

125 Cc11 **Volga** ♠ NW Russian Federation

Volga-Baltic Waterway see Volgo-Baltiyskiy Kanal

131 **Volga Hills/Volga Uplands** see Privolzhskaya Vozvyshennost'

128 L13 **Volgo-Baltiyskiy Kanal** Eng. Volga-Baltic Waterway. canal NW Russian Federation

130 M12 **Volgodonsk** Rostovskaya Oblast', SW Russian Federation 47.34N 42.03E

131 O10 **Volgograd** prev. Stalingrad, Tsaritsyn. Volgogradskaya Oblast', SW Russian Federation 48.42N 44.28E

131 N9 **Volgogradskaya Oblast'** ♦ province SW Russian Federation

131 P10 **Volgogradskoye Vodokhranilishche** ⊡ W Russian Federation

103 J19 **Volkach** Bayern, C Germany 49.51N 10.15E

111 U9 **Völkermarkt** Slvn. Velikovec. Kärnten, S Austria 46.39N 14.37E

128 K12 **Volkhov** Leningradskaya Oblast', NW Russian Federation 59.56N 32.19E

103 D20 **Völklingen** Saarland, SW Germany 49.15N 6.51E

128 K12 **Volkovysk** see Vawkavysk

Volkovyskiye Vysoty see Vawkavyskaye Wzvyshsha

85 K22 **Volksrust** Mpumalanga, E South Africa 27.18S 29.53E

100 L8 **Vollenhove** Overijssel, N Netherlands 52.40N 5.58E

L16 **Volma** Rus. Volma. ♠ C Belarus

121 L16 **Volmari** see Valmiera

114 W9 **Volnovakha** Donets'ka Oblast', E Ukraine 47.36N 37.31E

118 K6 **Volochys'k** Khmel'nyts'ka Oblast', W Ukraine 49.32N 26.14E

119 06 **Volodarka** Kyyivs'ka Oblast', N Ukraine 49.31N 29.55E

119 W9 **Volodars'ke** Donets'ka Oblast', E Ukraine 47.11N 37.19E

131 R13 **Volodarskiy** Astrakhanskaya Oblast', SW Russian Federation 46.23N 48.39E

Volodarskoye see Saumalkol'

128 L14 **Volodymerets'** Rivnens'ka Oblast', NW Ukraine 51.24N 25.52E

118 I3 **Volodymyr-Volyns'kyy** Pol. Włodzimierz, Rus. Vladimir-Volynskiy. Volyns'ka Oblast', NW Ukraine 50.51N 24.19E

128 L12 **Vologda** Vologodskaya Oblast', W Russian Federation 59.10N 39.55E

128 L12 **Vologodskaya Oblast'** ♦ province NW Russian Federation

130 K3 **Volokolamsk** Moskovskaya Oblast', W Russian Federation 56.03N 35.57E

130 K9 **Volokonovka** Belgorodskaya Oblast', W Russian Federation 50.30N 37.54E

117 I17 **Vólos** Thessalía, C Greece 39.21N 22.58E

128 M11 **Voloshka** Arkhangel'skaya Oblast', NW Russian Federation 61.19N 40.06E

Volosinovo see Novi Bečej

118 H7 **Volovets'** Zakarpats'ka Oblast', W Ukraine 48.42N 23.12E

116 K7 **Volovo** Rus, N Bulgaria 43.33N 25.49E

130 L7 **Vol'sk** Saratovskaya Oblast', W Russian Federation 52.04N 47.19E

79 Q17 **Volta** ♠ SE Ghana

79 P16 **Volta Blanche** see White Volta

79 Q17 **Volta, Lake** ⊚ SE Ghana

Volta Noire see Black Volta

62 O9 **Volta Redonda** Rio de Janeiro, SE Brazil 22.31S 44.04W

Volta Rouge see Red Volta

---

108 F12 **Volterra** anc. Volaterrae. Toscana, C Italy 43.25N 10.51E

109 K17 **Volturno** ♠ S Italy

115 I15 **Volujak** ▲ SW Yugoslavia

103 H17 **Volkelsberg** see ...

108 D8 **Voghera** Lombardia, N Italy

67 F24 **Volunteer Point** headland East Falkland, Falkland Islands 51.31S 57.43W

Volunteer State see Tennessee

116 H13 **Vólvi, Límni** ⊚ N Greece

118 I3 **Volyn'ka Oblast'** var. Volyn, Rus. Volynskaya Oblast'. ♦ province NW Ukraine

Volynskaya Oblast' see Volyn'ka Oblast'

197 H6 **Voh** Province Nord, C New Caledonia

131 Q3 **Volzhsk** Respublika Mariy El, W Russian Federation 55.53N 48.21E

131 O10 **Volzhskiy** Volgogradskaya Oblast', SW Russian Federation 48.48N 44.40E

180 I7 **Vondrozo** Fianarantsoa, SE Madagascar 22.49S 47.19E

116 K9 **Voneshta Voda** Veliko Tŭrnovo, N Bulgaria 42.55N 25.40E

41 P10 **Von Frank Mountain** ▲ Alaska, USA 63.41N 154.29W

117 C17 **Vónitsa** Dytikí Ellás, W Greece 38.55N 20.52E

120 J6 **Võnnu** Ger. Wendau. Tartumaa, SE Estonia 58.15N 27.08E

100 G12 **Voorburg** Zuid-Holland, W Netherlands 52.04N 4.22E

100 H11 **Voorschoten** Zuid-Holland, W Netherlands 52.07N 4.25E

100 M11 **Voorst** Gelderland, E Netherlands 52.10N 6.10E

100 K11 **Voorthuizen** Gelderland, C Netherlands 52.12N 5.36E

94 K2 **Vopnafjördhur** Austurland, E Iceland 65.45N 14.51W

94 L2 **Vopnafjördhur** bay E Iceland

Vora see Vorë

121 H15 **Voranava** Pol. Werenów, Rus. Voronovo. Hrodzyenskaya Voblasts', W Belarus 54.10N 25.21E

110 I8 **Vorarlberg** off. Land Vorarlberg. ♦ state W Austria

111 X7 **Vorau** Steiermark, E Austria 47.22N 15.55E

100 N11 **Vorden** Gelderland, E Netherlands 52.07N 6.18E

110 H9 **Vorderrhein** ♠ SE Switzerland

94 J2 **Vordhufell** ▲ N Iceland 65.42N 18.45W

97 J24 **Vordingborg** Storstrøm, SE Denmark 55.01N 11.55E

115 K19 **Vorë** var. Vora. Tiranë, W Albania 41.23N 19.37E

117 H17 **Vóreioi Sporádes** var. Vórioi Sporádhes, Eng. Northern Sporades. island group E Greece

117 K19 **Voríon Aigaíon** Eng. Aegean North. ♦ region SE Greece

117 G18 **Voreiós Evvoïkós Kólpos** gulf E Greece

207 S16 **Voring Plateau** undersea feature N Norwegian Sea

Vórioi Sporádhes see Vóreioi Sporádes

129 W9 **Vorkuta** Respublika Komi, NW Russian Federation 67.27N 64.00E

97 J11 **Vorma** ♠ S Norway

120 E4 **Vorma** var. Vormsi Saar, Ger. Worms, Swed. Ormsö. island W Estonia

Vormsi Saar see Vorma

126 Hh12 **Vorogovo** Krasnoyarskiy Kray, C Russian Federation 61.01N 89.25E

131 N7 **Vorona** ♠ W Russian Federation

130 L7 **Voronezh** Voronezhskaya Oblast', W Russian Federation 51.39N 39.13E

130 L7 **Voronezh** ♠ W Russian Federation

130 M3 **Voronezhskaya Oblast'** ♦ province W Russian Federation

Voronovitsya see Voronovytsya

130 J7 **Voronovo** see Voranava

119 N6 **Voronovytsya** Rus. Voronovitsya. Vinnyts'ka Oblast', C Ukraine 49.06N 28.49E

126 Hh7 **Vorontsovo** Taymyrskiy (Dolgano-Nenetskiy) Avtonomnyy Okrug, N Russian Federation 71.45N 83.31E

115 K22 **Vran** ▲ SW Bosnia and Herzegovina 43.35N 17.30E

118 K12 **Vrancea** ♦ county E Romania

153 T14 **Vrang** SE Tajikistan 37.03N 72.26E

127 Oo2 **Vrangelya, Ostrov** Eng. Wrangel Island. island NE Russian Federation

114 K13 **Vranica** ▲ C Bosnia and Herzegovina 43.57N 17.43E

115 O16 **Vranje** Serbia, SE Yugoslavia 42.33N 21.55E

113 N19 **Vranov nad Topl'ou** var. Vranov, Hung. Varannó. Prešovský Kraj, E Slovakia 48.54N 21.40E

116 H8 **Vratsa** Vratsa, NW Bulgaria 43.13N 23.33E

116 F10 **Vrattsa** prev. Mirovo. Kyustendil, W Bulgaria 42.16N 22.39E

114 G11 **Vrbanja** ♠ N Bosnia and Herzegovina

114 K9 **Vrbas** Serbia, N Yugoslavia 45.34N 19.39E

114 G13 **Vrbas** ♠ N Bosnia and Herzegovina

114 E8 **Vrbovec** Zagreb, N Croatia 45.53N 16.24E

114 C9 **Vrbovsko** Primorje-Gorski Kotar, NW Croatia 45.22N 15.06E

113 E15 **Vrchlabí** Ger. Hohenelbe. Hradecký Kraj, NE Czech Republic 50.37N 15.37E

85 J21 **Vrede** Free State, E South Africa 27.25S 29.10E

102 F13 **Vreden** Nordrhein-Westfalen, NW Germany 52.01N 6.50E

85 D25 **Vredenburg** Western Cape, SW South Africa 32.55S 18.00E

85 E25 **Vredendal** Western Cape, SW South Africa 31.40S 18.30E

101 I17 **Vorst** Antwerpen, N Belgium 51.06N 5.01E

130 K3 **Vorstershoop** North-West, N South Africa 25.46S 22.57E

120 H6 **Vörtsjärv** Ger. Wirz-See. ⊚ SE Estonia

120 J7 **Võru** Ger. Werro. Võrumaa, SE Estonia 57.51N 27.00E

120 I7 **Võrumaa** off. Võru Maakond. ♦ province SE Estonia

143 V13 **Vorotan** Az. Bärgusad. ♠ Armenia/Azerbaijan

119 S3 **Vorozhba** Sums'ka Oblast', NE Ukraine 51.09N 34.16E

119 T5 **Vorskla** ♠ Russian Federation/Ukraine

101 I17 **Vorst** Antwerpen, N Belgium 51.06N 5.01E

85 G21 **Vorstershoop** North-West, N South Africa 25.46S 22.57E

120 E5 **Vormsi Saar** Ger. Worms. ⊡ Estonia

120 I7 **Võru** Ger. Werro. Võrumaa, SE Estonia

130 L4 **Voskresensk** Moskovskaya Oblast', W Russian Federation 55.19N 38.42E

131 P2 **Voskresenskoye** Nizhegorodskaya Oblast', W Russian Federation 57.00N 45.33E

131 N9 **Voskresenskoye** Respublika Bashkortostan, W Russian Federation 53.07N 56.02E

96 D13 **Voss** Hordaland, S Norway 60.37N 6.25E

96 D13 **Vóslvi, Límni** ⊚ N Greece

101 I16 **Vosselaar** Antwerpen, N Belgium 51.19N 4.55E

96 D13 **Vosso** ♠ S Norway

Vostochno-Kazakhstanskaya Oblast' see Shygys Konyrat

151 T12 **Vostochno-Kounradskiy** Kaz. Shyghys Qongyrat. Zhezkazgan, C Kazakhstan 47.01N 75.05E

127 N4 **Vostochno-Sibirskoye More** Eng. East Siberian Sea. sea Arctic Ocean

151 X10 **Vostochnyy Kazakhstan** off. Vostochno-Kazakhstanskaya Oblast', var. East Kazakhstan, Kaz. Shyghys Qazaqstan Oblysy. ♦ province E Kazakhstan

Vostochnyy Sayan see Eastern Sayans

205 U10 **Vostok Island** see Vostok Island

205 U10 **Vostok** Russian research station Antarctica 77.18S 105.32E

203 X5 **Vostok Island** var. Vostok Island; prev. Stavers Island. island Line Islands, SE Kiribati

131 T2 **Votkinsk** Udmurtskaya Respublika, NW Russian Federation 57.04N 54.00E

129 U15 **Votkinskoye Vodokhranilishche** var. Votkinsk Reservoir. ⊡ NW Russian Federation

Votkinsk Reservoir see Votkinskoye Vodokhranilishche

62 J7 **Votuporanga** São Paulo, S Brazil 20.25S 49.52W

106 M7 **Vouga, Rio** ♠ N Portugal

117 L14 **Voúrinos** ▲ N Greece

117 G24 **Voúxa, Akrotírio** headland Kríti, Greece, E Mediterranean Sea

105 R4 **Vouziers** Ardennes, N France 49.24N 4.42E

119 V7 **Vovcha** Rus. Volchya. ♠ E Ukraine

119 V4 **Vovchans'k** Rus. Volchansk. Kharkivs'ka Oblast', E Ukraine 50.19N 36.54E

105 N6 **Voves** Eure-et-Loir, C France 48.18N 1.39E

81 M14 **Vovodo** ♠ S Central Africa Republic

96 M12 **Voxna** Gävleborg, C Sweden 61.20N 15.34E

96 L11 **Voxnan** ♠ C Sweden

116 F7 **Voynishka Reka** ♠ NW Bulgaria

129 T9 **Voyvozh** Respublika Komi, NW Russian Federation 62.54N 54.52E

128 M12 **Vozhega** Vologodskaya Oblast', NW Russian Federation

128 L12 **Vozhe, Ozero** ⊚ NW Russian Federation

119 Q9 **Voznesens'k** Rus. Voznesensk. Mykolayivs'ka Oblast', S Ukraine 47.33N 31.22E

128 J12 **Voznesen'ye** Leningradskaya Oblast', NW Russian Federation 61.00N 35.24E

150 J14 **Vozrozhdeniya, Ostrov** Uzb. Vozrojdeniye Oroli. island Kazakhstan/Uzbekistan

97 Q9 **Vrå** var. Vraa. Nordjylland, N Denmark 57.21N 9.57E

Vraa see Vrå

116 H9 **Vrachesh** Sofiya, NW Bulgaria 42.52N 23.45E

117 C19 **Vrachionás** ▲ Zákynthos, Iónioi Nísioi, Greece, C Mediterranean Sea 37.49N 20.43E

119 P8 **Vradiyivka** Mykolayivs'ka Oblast', S Ukraine 47.51N 30.37E

118 K12 **Vrancea** ♦ county E Romania

153 T14 **Vrang** SE Tajikistan

127 Oo2 **Vrangelya, Ostrov** Eng. Wrangel Island. island NE Russian Federation

114 H13 **Vranica** ▲ C Bosnia and Herzegovina 43.57N 17.43E

115 O16 **Vranje** Serbia, SE Yugoslavia 42.33N 21.55E

113 N19 **Vranov nad Topl'ou** var. Vranov, Hung. Varannó. Prešovský Kraj, E Slovakia 48.54N 21.40E

116 H8 **Vratsa** Vratsa, NW Bulgaria 43.13N 23.33E

116 F10 **Vrattsa** prev. Mirovo. Kyustendil, W Bulgaria 42.16N 22.39E

114 H9 **Vrbanja** ♠ N Bosnia and Herzegovina

114 K9 **Vrbas** Serbia, N Yugoslavia 45.34N 19.39E

114 G13 **Vrbas** ♠ N Bosnia and Herzegovina

114 E8 **Vrbovec** Zagreb, N Croatia 45.53N 16.24E

114 C9 **Vrbovsko** Primorje-Gorski Kotar, NW Croatia 45.22N 15.06E

113 E15 **Vrchlabí** Ger. Hohenelbe. Hradecký Kraj, NE Czech Republic 50.37N 15.37E

85 J21 **Vrede** Free State, E South Africa 27.25S 29.10E

102 F13 **Vreden** Nordrhein-Westfalen, NW Germany 52.01N 6.50E

85 D25 **Vredenburg** Western Cape, SW South Africa 32.55S 18.00E

115 G15 **Vrgorac** prev. Vrhgorac. Split-Dalmacija, SE Croatia 43.10N 17.24E

Vrhgorac see Vrgorac

◆ COUNTRY ○ COUNTRY CAPITAL
◇ DEPENDENT TERRITORY ○ DEPENDENT TERRITORY CAPITAL
◆ ADMINISTRATIVE REGION ✕ INTERNATIONAL AIRPORT
▲ MOUNTAIN ▲ MOUNTAIN RANGE
♦ VOLCANO ♠ RIVER
⊚ LAKE ⊡ RESERVOIR

111 T12 **Vrhnika** Ger. Oberlaibach. W Slovenia 45.57N 14.18E
161 I21 **Vriddhāchalam** Tamil Nādu, SE India 11.33N 79.18E
100 N6 **Vries** Drenthe, NE Netherlands 53.04N 6.34E
100 O10 **Vriezenveen** Overijssel, E Netherlands 52.25N 6.39E
97 L20 **Vrigstad** Jönköping, S Sweden 57.19N 14.30E
110 H9 **Vrin** Graubünden, S Switzerland 46.40N 9.06E
114 E13 **Vrlika** Split-Dalmacija, S Croatia 43.54N 16.24E
115 M14 **Vrnjačka Banja** Serbia, C Yugoslavia 43.36N 20.55E
Vrondádhes/Vrondádo see Vrontádos
117 L18 **Vrontádos** var. Vrondádo; prev. Vrondádhes. Chíos, E Greece 38.25N 26.07E
100 N9 **Vroomshoop** Overijssel, E Netherlands 52.28N 6.34E
114 N10 **Vršac** Ger. Werschetz, Hung. Versecz. Serbia, NE Yugoslavia 45.08N 21.17E
114 M10 **Vršački Kanal** canal N Yugoslavia
85 H21 **Vryburg** North-West, N South Africa 26.57S 24.43E
85 K22 **Vryheid** KwaZulu/Natal, E South Africa 27.45S 30.48E
113 I18 **Vsetín** Ger. Wsetin. Zlínský Kraj, E Czech Republic 49.21N 17.57E
113 J20 **Vtáčnik** Hung. Madaras, Ptacsnik; ▲ W Slovakia 48.38N 18.38E
Vuadil′ see Wodil
Vuanggava see Vuaqava
197 K15 **Vuaqava** prev. Vuanggava. island Lau Group, SE Fiji
116 I11 **Vŭcha** ♣ SW Bulgaria
115 N16 **Vučitrn** Serbia, S Yugoslavia 42.49N 21.00E
101 J14 **Vught** Noord-Brabant, S Netherlands 51.37N 5.19E
119 W8 **Vuhledar** Donets′ka Oblast′, E Ukraine 47.48N 37.11E
114 I9 **Vuka** ♣ E Croatia
115 K17 **Vukël** var. Vukli. Shkodër, N Albania 42.29N 19.39E
Vukli see Vukël
114 J9 **Vukovar** Hung. Vukovár. Vukovar-Srijem, E Croatia 45.18N 18.45E
114 I10 **Vukovar-Srijem** off. Vukovarsko-Srijemska Županija. ♦ province E Croatia
129 U8 **Vuktyl** Respublika Komi, NW Russian Federation 63.49N 57.07E
9 Q17 **Vulcan** Alberta, SW Canada 50.27N 113.12W
118 G12 **Vulcan** Ger. Wulkan, Hung. Zsilyvajdevulkán; prev. Crivadia Vulcanului, Vaidei, Hung. Sily-Vajdej, Vajdej. Hunedoara, W Romania 45.22N 23.16E
118 M12 **Vulcăneşti** Rus. Vulkaneshty. S Moldova 45.41N 28.25E
109 L22 **Vulcano, Isola** island Isole Eolie, S Italy
116 G7 **Vŭlchedrŭm** Montana, NW Bulgaria 43.42N 23.25E
116 N8 **Vŭlchidol** prev. Kurt-Dere. Varna, E Bulgaria 43.25N 27.33E
Vulkaneshty see Vulcăneşti
38 J13 **Vulture Mountains** ▲ Arizona, SW USA
178 K14 **Vung Tau** prev. Fr. Cape Saint Jacques, Cap Saint-Jacques. Ba Ria-Vung Tau, S Vietnam 10.21N 107.04E
197 I15 **Vunisea** Kadavu, SE Fiji 19.04S 178.09E
Vuohčču see Vuotso
95 N15 **Vuokatti** Oulu, C Finland 64.08N 28.16E
95 M15 **Vuolijoki** Oulu, C Finland 64.09N 27.00E
94 J13 **Vuollerim** Norrbotten, N Sweden 66.24N 20.36E
Vuoreija see Vardø
94 L10 **Vuotso** Lapp. Vuohčču. Lappi, N Finland 68.04N 27.05E
116 J11 **Vŭrbitsa** var. Filevo. Khaskovo, S Bulgaria 42.02N 25.25E
116 J12 **Vŭrbitsa** ♣ S Bulgaria
131 Q4 **Vurnary** Chuvashskaya Respublika, W Russian Federation 55.30N 46.59E
116 G8 **Vŭrshets** Montana, NW Bulgaria 43.14N 23.20E
121 F17 **Vyalikaya Byerastavitsa** Pol. Brzostowica Wielka, Rus. Bol′shaya Berëstovitsa; prev. Velikaya Berestovitsa. Hrodzyenskaya Voblasts′, SW Belarus 53.12N 24.03E
121 N20 **Vyaliki Bor** Rus. Velikiy Bor. Homyel′skaya Voblasts′, SE Belarus 52.01N 29.54E
121 J18 **Vyaliki Rozhan** Rus. Bol′shoy Rozhan. Minskaya Voblasts′, S Belarus 52.46N 27.07E
128 H10 **Vyartsilya** Fin. Värtsilä. Respublika Kareliya, NW Russian Federation 62.07N 30.43E
121 K17 **Vyasyeya** Rus. Veseya. Minskaya Voblasts′, C Belarus 53.04N 27.40E
129 R15 **Vyatka** ♣ NW Russian Federation
Vyatka see Kirov
129 S16 **Vyatskiye Polyany** Kirovskaya Oblast′, NW Russian Federation 56.15N 51.06E
127 Nn16 **Vyazemskiy** Khabarovskiy Kray, SE Russian Federation 47.28N 134.39E
130 I4 **Vyaz′ma** Smolenskaya Oblast′, W Russian Federation 55.09N 34.20E
131 N3 **Vyazniki** Vladimirskaya Oblast′, W Russian Federation 56.15N 42.06E
129 O8 **Vyazovka** Volgogradskaya Oblast′, SW Russian Federation 50.57N 43.57E
121 J14 **Vyazyn′** Rus. Vyazyn′. Minskaya Voblasts′, NW Belarus 54.25N 27.10E
128 G11 **Vyborg** Fin. Viipuri. Leningradskaya Oblast′, NW Russian Federation 60.44N 28.47E
129 P11 **Vychegda** ♣ NW Russian Federation

126 Jj16 **Vydrino** Respublika Buryatiya, S Russian Federation 51.22N 104.34E
121 L14 **Vyelyewshchyna** Rus. Velevshchina. Vitsyebskaya Voblasts′, N Belarus 54.44N 28.33E
121 P16 **Vyeramyeyki** Rus. Veremeyki. Mahilyowskaya Voblasts′, E Belarus 53.46N 31.18E
120 K11 **Vyerkhnyadzvinsk** Rus. Verkhnedvinsk. Vitsyebskaya Voblasts′, N Belarus 55.46N 27.55E
121 P18 **Vyetka** Rus. Vetka. Homyel′skaya Voblasts′, SE Belarus 52.34N 31.13E
120 L12 **Vyetryna** Rus. Vetrino. Vitsyebskaya Voblasts′, N Belarus 55.24N 28.28E
Vygonovskoye, Ozero see Vyhanawskaye, Vozyera
128 J9 **Vygozero, Ozero** ◉ NW Russian Federation
Vyhanashchanskaye Vozyera see Vyhanawskaye, Vozyera
121 I18 **Vyhanawskaye, Vozyera** var. Vyhanashchanskaye Vozyera, Rus. Ozero Vygonovskoye. ◉ SW Belarus
131 N4 **Vyksa** Nizhegorodskaya Oblast′, W Russian Federation 55.21N 42.10E
119 O12 **Vylkove** Rus. Vilkovo. Odes′ka Oblast′, SW Ukraine 45.24N 29.37E
118 H8 **Vym′** ♣ NW Russian Federation
100 K12 **Vynohradiv** Cz. Sevluš, Hung. Nagyszőllős, Rus. Vinogradov; prev. Sevlyush. Zakarpats′ka Oblast′, W Ukraine 48.09N 23.01E
128 G13 **Vyritsa** Leningradskaya Oblast′, NW Russian Federation 59.25N 30.20E
99 J19 **Vyrnwy** Wel. Afon Efyrnwy. ♣ E Wales, UK
151 X9 **Vyshe Ivanovskiy Belak, Gora** ▲ E Kazakhstan 50.16N 83.46E
119 P4 **Vyshhorod** Kyyivs′ka Oblast′, N Ukraine 50.36N 30.28E
128 I15 **Vyshniy Volochek** Tverskaya Oblast′, W Russian Federation 57.37N 34.33E
113 G18 **Vyškov** Ger. Wischau. Brněnský Kraj, SE Czech Republic 49.16N 16.58E
113 F17 **Vysoké Mýto** Ger. Hohenmauth. Pardubický Kraj, C Czech Republic 49.58N 16.08E
119 S9 **Vysokopillya** Khersons′ka Oblast′, S Ukraine 47.28N 33.30E
130 K3 **Vysokovsk** Moskovskaya Oblast′, W Russian Federation 56.12N 36.42E
128 K12 **Vytegra** Vologodskaya Oblast′, NW Russian Federation 60.59N 36.27E
118 J8 **Vyzhnytsya** Chernivets′ka Oblast′, W Ukraine 48.14N 25.10E

W

79 O14 **Wa** NW Ghana 10.07N 2.28W
Waadt see Vaud
Waag see Váh
Waagbistritz see Povážská Bystrica
Waagneustadtl see Nové Mesto nad Váhom
83 M16 **Waajid** Gedo, SW Somalia 3.37N 43.19E
100 L13 **Waal** ♣ S Netherlands
197 G4 **Waala** Province Nord, W New Caledonia 19.46S 163.41E
101 I14 **Waalwijk** Noord-Brabant, S Netherlands 51.42N 5.04E
101 E16 **Waarschoot** Oost-Vlaanderen, NW Belgium 51.09N 3.35E
194 G12 **Wabag** Enga, W PNG 5.28S 143.40E
13 N7 **Wabano** ♣ Quebec, SE Canada
9 P11 **Wabasca** ♣ Alberta, SW Canada
33 P12 **Wabash** Indiana, N USA 40.46N 85.48W
31 X9 **Wabasha** Minnesota, N USA 44.22N 92.01W
33 N13 **Wabash River** ♣ N USA
12 C7 **Wabatongushi Lake** ◉ Ontario, S Canada
83 L15 **Wabē Gestro Wenz** ♣ SE Ethiopia
12 B9 **Wabos** Ontario, S Canada 46.48N 84.06W
9 W13 **Wabowden** Manitoba, C Canada 54.57N 98.37W
112 J9 **Wąbrzeźno** Kujawsko-pomorskie, N Poland 53.18N 18.55E
194 G14 **Wabuda Island** island SW PNG
23 U12 **Waccamaw River** ♣ South Carolina, SE USA
25 U11 **Waccasassa Bay** bay Florida, SE USA
101 F16 **Wachtebeke** Oost-Vlaanderen, NW Belgium 51.10N 3.52E
27 T8 **Waco** Texas, SW USA 31.33N 97.09W
28 M3 **Waconda Lake** var. Great Elder Reservoir. ◉ Kansas, C USA
Wad see Ouaddaï
74 H6 **Wad al-Hajarah** see Guadalajara
171 Gg13 **Wadayama** Hyōgo, Honshū,
82 D10 **Wad Banda** Western Kordofan, C Sudan 13.07N 27.55E
77 Y9 **Waddān** NW Libya 29.10N 16.07E
Waddeneilanden Eng. West Frisian Islands. island group N Netherlands
100 J6 **Waddenzee** var. Wadden Zee. sea SE North Sea
8 L16 **Waddington, Mount** ▲ British Columbia, SW Canada 51.17N 125.16W
100 H12 **Waddinxveen** Zuid-Holland, C Netherlands 52.03N 4.37E
7 U5 **Wadena** Saskatchewan, S Canada 51.57N 103.48W
31 T6 **Wadena** Minnesota, N USA 46.27N 95.07W
110 G7 **Wädenswil** Zürich, N Switzerland 47.13N 8.39E
23 S11 **Wadesboro** North Carolina, SE USA 34.58N 80.04W
161 G16 **Wādī** Karnātaka, C India 17.00N 76.58E

144 G10 **Wādī as Sīr** var. Wadi es Sīr. ′Ammān, NW Jordan 31.57N 35.49E
Wadi es Sīr see Wādī as Sīr
82 F5 **Wadi Halfa** var. Wādī Ḥalfā′. Northern, N Sudan 21.46N 31.16E
144 G13 **Wādī Mūsā** var. Petra. Ma′ān, S Jordan 30.19N 35.28E
25 V4 **Wadley** Georgia, SE USA 32.52N 82.24W
82 G10 **Wad Medani** see Wad Madani
82 F10 **Wad Nimr** White Nile, C Sudan 14.31N 32.10E
172 Q14 **Wadomari** Kagoshima, Okinoerabu-jima, SW Japan 27.25N 128.40E
113 K17 **Wadowice** Małopolskie, S Poland 49.52N 19.30E
37 R5 **Wadsworth** Nevada, W USA 39.39N 119.16W
33 T12 **Wadsworth** Ohio, N USA 41.01N 81.43W
27 T11 **Waelder** Texas, SW USA 29.42N 97.16W
Waereghem see Waregem
169 U13 **Wafangdian** var. Fuxian, Fu Xian. Liaoning, NE China 39.36N 122.00E
175 S11 **Waflia** Pulau Buru, E Indonesia 3.09S 126.05E
Wagadugu see Ouagadougou
100 K12 **Wageningen** Gelderland, SE Netherlands 51.58N 5.40E
57 V9 **Wageningen** Nickerie, NW Suriname 5.43N 56.45W
115 L15 **Wager Bay** inlet Nunavut, N Canada
176 Y10 **Wageseri** Irian Jaya, E Indonesia 1.48S 138.19E
191 P10 **Wagga Wagga** New South Wales, SE Australia 35.10S 147.22E
188 J13 **Wagin** Western Australia 33.16S 117.25E
Wagina see Vaghena
110 H8 **Wägitaler See** ◉ SW Switzerland
31 P12 **Wagner** South Dakota, N USA 43.04N 98.17W
29 V10 **Wagoner** Oklahoma, C USA 35.57N 95.22W
39 U10 **Wagon Mound** New Mexico, SW USA 36.00N 104.42W
34 J14 **Wagontire** Oregon, NW USA 43.15N 119.51W
112 H10 **Wągrowiec** Wielkopolskie, NW Poland 52.47N 17.10E
155 U6 **Wāh** Punjab, NE Pakistan 33.49N 72.43E
176 U10 **Wahai** Pulau Seram, E Indonesia 2.48S 129.28E
175 O7 **Wahau, Sungai** ♣ Borneo, C Indonesia
Wahaybah, Ramlat Al see Wahībah, Ramlat Āl
82 D13 **Wahda** var. Unity State. ♦ state S Sudan
40 D9 **Wahiawa** Haw. Wahiawā. Oahu, Hawaii, USA, C Pacific Ocean 21.30N 158.01W
Wahībah, Ramlat Ahl see Wahībah, Ramlat Āl
147 Y9 **Wahībah, Ramlat Āl** var. Ramlat Ahl Wahībah, Ramlat Al Wahybah, Eng. Wahībah Sands. desert N Oman
Wahībah Sands see Wahībah, Ramlat Āl
103 E16 **Wahn** ✈ (Köln) Nordrhein-Westfalen, W Germany 50.51N 7.09E
31 R15 **Wahoo** Nebraska, C USA 41.12N 96.37W
31 R6 **Wahpeton** North Dakota, N USA 46.16N 96.36W
Wahran see Oran
38 J6 **Wah Wah Mountains** ▲ Utah, W USA
40 D9 **Waialua** Oahu, Hawaii, USA, C Pacific Ocean 21.34N 158.07W
40 D9 **Waianae** Haw. Wai′anae. Oahu, Hawaii, USA, C Pacific Ocean 21.26N 158.11W
192 Q8 **Waiapu** ♣ North Island, NZ
193 I17 **Waiau** Canterbury, South Island, NZ 42.39S 173.03E
193 I17 **Waiau** ♣ South Island, NZ
193 B23 **Waiau** ♣ South Island, NZ
103 H21 **Waiblingen** Baden-Württemberg, S Germany 48.49N 9.19E
192 L6 **Waihi** Waikato, North Island, NZ 37.24S 175.49E
193 C20 **Waihou** ♣ North Island, NZ
23 U9 **Wake Forest** North Carolina, SE USA 35.58N 78.30W
175 P17 **Waikabubak** prev. Waikaboebak. Pulau Sumba, C Indonesia 9.40S 119.25E
175 P13 **Waikaia** ♣ South Island, NZ
193 D23 **Waikaka** Southland, South Island, NZ 45.55S 168.59E
192 L13 **Waikanae** Wellington, North Island, NZ 40.52S 175.04E
192 M7 **Waikare, Lake** ◉ North Island, NZ
192 O9 **Waikaremoana, Lake** ◉ North Island, NZ
192 L7 **Waikato** off. Waikato Region. ♦ region North Island, NZ
192 M9 **Waikato** ♣ North Island, NZ
190 J9 **Waikerie** South Australia 34.12S 139.58E
193 F23 **Waikouaiti** Otago, South Island, NZ 45.36S 170.39E
192 L6 **Waikato** Waikato, North Island, NZ 37.24S 175.49E
201 Y11 **Waikiki** Oahu, Hawaii, USA, C Pacific Ocean 21.16N 157.49W
40 G12 **Wailuku** Maui, Hawaii, USA, C Pacific Ocean 20.53N 156.30W

193 H18 **Waimakariri** ♣ South Island, NZ
40 D9 **Waimanalo Beach** Oahu, Hawaii, USA, C Pacific Ocean 21.20N 157.42W
193 G15 **Waimangaroa** West Coast, South Island, NZ 41.41S 171.49E
193 G21 **Waimate** Canterbury, South Island, NZ 44.45S 171.03E
40 G11 **Waimea** var. Kamuela. Hawaii, USA, C Pacific Ocean 20.01N 155.39W
40 D9 **Waimea** var. Maunawai. Oahu, Hawaii, USA, C Pacific Ocean 21.38N 158.03W
40 B8 **Waimea** Kauai, Hawaii, USA, C Pacific Ocean 21.57N 159.39W
101 M20 **Waimes** Liège, E Belgium 50.25N 6.10E
100 J11 **Wainganga** var. Wain River. ♣ C India
Waingapo see Waingapu
175 Pp17 **Waingapu** prev. Waingapoe. Pulau Sumba, C Indonesia 9.40S 120.16E
57 S7 **Waini** N Guyana
57 S7 **Waini Point** headland NW Guyana 8.24N 59.48W
9 R15 **Wainwright** Alberta, SW Canada 52.49N 110.51W
41 O5 **Wainwright** Alaska, USA 70.38N 160.02W
192 M11 **Waiouru** Manawatu-Wanganui, North Island, NZ 39.27S 175.40E
176 X11 **Waipa** Irian Jaya, E Indonesia 3.47S 136.16E
192 P9 **Waipaoa** ♣ North Island, NZ
193 D25 **Waipapa Point** headland South Island, NZ 46.39S 168.51E
193 I18 **Waipara** Canterbury, South Island, NZ 43.03S 172.44E
192 N12 **Waipawa** Hawke's Bay, North Island, NZ 39.57S 176.35E
192 K4 **Waipu** Northland, North Island, NZ 35.59S 174.25E
192 N12 **Waipukurau** Hawke's Bay, North Island, NZ 40.01S 176.33E
176 Vv13 **Wair** Pulau Kai Besar, E Indonesia 5.16S 133.09E
192 N9 **Wairakei** var. Wairakei. Waikato, North Island, NZ 38.37S 176.05E
193 M14 **Wairarapa, Lake** ◉ North Island, NZ
193 J15 **Wairau** ♣ South Island, NZ
192 P10 **Wairoa** Hawke's Bay, North Island, NZ 39.03S 105.25E
192 P10 **Wairoa** ♣ North Island, NZ
192 J4 **Wairoa** ♣ North Island, NZ
192 N9 **Waitahanui** Waikato, North Island, NZ 38.48S 176.04E
192 M6 **Waitakaruru** Waikato, North Island, NZ 37.14S 175.22E
193 F21 **Waitaki** ♣ South Island, NZ
192 K10 **Waitara** Taranaki, North Island, NZ 39.00S 174.14E
192 M7 **Waitoa** Waikato, North Island, NZ 37.36S 175.37E
192 L8 **Waitomo Caves** Waikato, North Island, NZ 38.17S 175.06E
192 L11 **Waitotara** Taranaki, North Island, NZ 39.49S 174.43E
192 L11 **Waitotara** ♣ North Island, NZ
34 L10 **Waitsburg** Washington, NW USA 46.16N 118.09W
Waitzen see Vác
40 D9 **Waiuku** Auckland, North Island, NZ 37.15S 174.44E
171 J12 **Wajima** var. Wazima. Ishikawa, Honshū, SW Japan 37.21N 136.53E
83 K14 **Wajir** North Eastern, NE Kenya 1.43N 40.04E
83 I14 **Waka** Southern, SW Ethiopia 7.12N 37.19E
81 I17 **Waka** Equateur, NW Dem. Rep. Congo (Zaire) 1.04N 20.11E
12 D9 **Wakami Lake** ◉ Ontario, S Canada
170 G13 **Wakasa** Tottori, Honshū, SW Japan 35.18N 134.25E
171 H13 **Wakasa-wan** bay C Japan
193 C22 **Wakatipu, Lake** ◉ South Island, NZ
9 T15 **Wakaw** Saskatchewan, S Canada 52.40N 105.45W
170 J12 **Wakayama** Wakayama, Honshū, SW Japan 34.12N 135.09E
170 G16 **Wakayama** off. Wakayama-ken. ♦ prefecture Honshū, SW Japan
28 K4 **Wa Keeney** Kansas, C USA 39.01N 99.52W
193 I14 **Wakefield** Tasman, South Island, NZ 41.24S 173.03E
99 M17 **Wakefield** N England, UK 53.42N 1.28W
49 O4 **Wakefield** Kansas, C USA 39.12N 97.00W
32 L4 **Wakefield** Michigan, N USA 46.27N 89.55W
34 L10 **Walla Walla** Washington, NW USA 46.03N 118.20W
47 V9 **Wallblake** ✈ (The Valley) C Anguilla 18.12N 63.02W
201 Y11 **Wake Island** ◇ US unincorporated territory NW Pacific Ocean
201 Y12 **Wake Island** atoll NW Pacific Ocean
201 X12 **Wake Lagoon** lagoon Wake Island, NW Pacific Ocean
177 F9 **Wakema** Irrawaddy, SW Myanmar 16.36N 95.10E
Wakhan see Khandūd
170 Fj15 **Waki** Tokushima, Shikoku, SW Japan 34.04N 134.10E
172 N8 **Wakinosawa** Aomori, Honshū, C Japan 41.08N 140.47E
172 Pp1 **Wakkanai** Hokkaidō, NE Japan 45.26N 141.43E
85 K22 **Wakkerstroom** Mpumalanga, E South Africa 27.21S 30.10E
191 N10 **Wakool** New South Wales, SE Australia 35.30S 144.22E
Wakra see Al Wakrah
Waku Kungo see Uaco Cungo

195 S12 **Wakunai** Bougainville Island, NE PNG 5.52S 155.13E
Walachei/Walachia see Wallachia
175 Pp12 **Walanae, Sungai** ♣ Sulawesi, C Indonesia
161 K26 **Walawe Ganga** ♣ S Sri Lanka
113 F15 **Wałbrzych** Ger. Waldenburg, Waldenburg in Schlesien. Dolnośląskie, SW Poland 50.44N 16.19E
191 T6 **Walcha** New South Wales, SE Australia 31.01S 151.38E
101 D14 **Walcheren** island SW Netherlands
31 Z14 **Walcott** Iowa, C USA 41.34N 90.46W
35 W16 **Walcott** Wyoming, C USA 41.46N 106.46W
101 M20 **Walcourt** Namur, S Belgium 50.15N 4.26E
112 G9 **Wałcz** Ger. Deutsch Krone. Zachodniopomorskie, NW Poland 53.16N 16.28E
110 H7 **Wald** Zürich, N Switzerland 47.16N 8.54E
111 U3 **Waldaist** ♣ N Austria
188 I9 **Waldburg Range** ▲ Western Australia
39 R3 **Walden** Colorado, C USA 40.43N 106.16W
80 K13 **Walden** New York, NE USA 41.35N 74.09W
Waldenburg/Waldenburg in Schlesien see Wałbrzych
9 T15 **Waldheim** Saskatchewan, S Canada 52.40N 106.35W
Waldia see Weldiya
103 M23 **Waldkraiburg** Bayern, SE Germany 48.10N 12.23E
49 T14 **Waldo** Arkansas, C USA 33.21N 93.18W
25 S9 **Waldo** Florida, SE USA 29.47N 82.07W
21 R7 **Waldoboro** Maine, NE USA 44.06N 69.22W
34 F12 **Waldport** Oregon, NW USA 44.25N 124.04W
49 S11 **Waldron** Arkansas, C USA 34.54N 94.05W
205 Y13 **Waldron, Cape** headland Antarctica 66.08S 116.00E
103 F24 **Waldshut-Tiengen** Baden-Württemberg, S Germany 47.37N 8.13E
175 Qq9 **Walea, Selat** strait Sulawesi, C Indonesia
Wałeckie Międzyrzecze see Valašské Meziříčí
110 H8 **Walensee** ◉ NW Switzerland
110 I8 **Wales** Alaska, USA 65.36N 168.02W
99 J20 **Wales** Wel. Cymru. national region UK
115 L13 **Wales Island** island Nunavut, NE Canada
79 P14 **Walewale** N Ghana 10.21N 0.48W
101 M24 **Walferdange** Luxembourg, C Luxembourg 49.39N 6.07E
191 Q5 **Walgett** New South Wales, SE Australia 30.02S 148.13E
204 K10 **Walgreen Coast** physical region Antarctica
31 Q2 **Walhalla** North Dakota, N USA 48.55N 97.55W
23 O11 **Walhalla** South Carolina, SE USA 34.45N 83.03W
81 O19 **Walikale** Nord Kivu, E Dem. Rep. Congo (Zaire) 1.25S 28.03E
194 G9 **Walis Island** island NW PNG
Walk see Valga, Estonia
Walk see Valka, Latvia
31 U5 **Walker** Minnesota, N USA 47.06N 94.35W
13 V4 **Walker, Lac** ◉ Quebec, SE Canada
37 S7 **Walker Lake** ◉ Nevada, W USA
37 R6 **Walker River** ♣ Nevada, W USA
30 K10 **Wall** South Dakota, N USA 43.58N 102.12W
181 U9 **Wallaby Plateau** undersea feature E Indian Ocean
35 N8 **Wallace** Idaho, NW USA 47.28N 115.55W
23 V11 **Wallace** North Carolina, SE USA 34.42N 77.59W
9 T15 **Wallace** Saskatchewan, S Canada 52.40N 105.45W
12 D17 **Wallaceburg** Ontario, S Canada 42.36N 82.22W
24 F5 **Wallace Lake** ◉ Louisiana, S USA
9 P13 **Wallace Mountain** ▲ Alberta, W Canada 54.50N 115.57W
Wallachia var. Walachia, Ger. Walachei, Rom. Valachia. cultural region S Romania
Wallachisch-Meseritsch see Valašské Meziříčí
191 U4 **Wallangarra** New South Wales, SE Australia 28.56S 151.57E
190 I8 **Wallaroo** South Australia 33.56S 137.38E
34 L10 **Walla Walla** Washington, NW USA 46.03N 118.20W
47 V9 **Wallblake** ✈ (The Valley) C Anguilla 18.12N 63.02W
103 H19 **Walldorf** Baden-Württemberg, SW Germany 49.34N 9.22E
102 F12 **Wallenhorst** Niedersachsen, NW Germany 52.21N 8.01E
Wallenthal see Haţeg
111 S4 **Wallern** Oberösterreich, N Austria 48.13N 13.58E
Wallern see Wallern im Burgenland
111 Z5 **Wallern im Burgenland** var. Wallern. Burgenland, E Austria 47.43N 16.56E
20 M9 **Wallingford** Vermont, NE USA 43.27N 72.56W
27 V11 **Wallis** Texas, SW USA 29.37N 96.04W
Wallis see Valais
199 Jj10 **Wallis and Futuna** Fr. Territoire de Wallis et Futuna. ♦ French overseas territory C Pacific Ocean
201 Y11 **Wallis Island** var. Île Uvea. island N Wallis and Futuna
202 H11 **Wallis, Îles** island group N Wallis and Futuna

101 H19 **Wallon Brabant** ♦ province C Belgium
33 S5 **Walloon Lake** ◉ Michigan, N USA
34 K10 **Wallula** Washington, NW USA 46.03N 118.54W
34 K10 **Wallula, Lake** ◉ Washington, NW USA
23 S8 **Walnut Cove** North Carolina, SE USA 36.18N 80.08W
37 N8 **Walnut Creek** California, W USA 37.52N 122.04W
26 L5 **Walnut Creek** ♣ Kansas, C USA
49 W9 **Walnut Ridge** Arkansas, C USA 36.06N 90.56W
27 S7 **Walnut Springs** Texas, SW USA 32.05N 97.42W
190 L10 **Walpeup** Victoria, SE Australia 35.09S 142.01E
197 I2 **Walpole, Île** island SE New Caledonia
41 N13 **Walrus Islands** island group Alaska, USA
99 L19 **Walsall** C England, UK 52.34N 1.58W
39 T7 **Walsenburg** Colorado, C USA 37.37N 104.46W
8 S17 **Walsh** Alberta, SW Canada 49.58N 110.03W
39 W7 **Walsh** Colorado, C USA 37.20N 102.17W
102 I11 **Walsrode** Niedersachsen, NW Germany 52.52N 9.36E
23 R14 **Walterboro** South Carolina, SE USA 32.54N 80.40W
Walter F. George Lake see Walter F. George Reservoir
25 R6 **Walter F. George Reservoir** var. Walter F. George Lake. ◉ Alabama/Georgia, SE USA
28 M2 **Walters** Oklahoma, C USA 34.21N 98.18W
103 J16 **Waltershausen** Thüringen, C Germany 50.53N 10.33E
31 N10 **Walters Shoal** var. Walters Shoals. reef S Madagascar
Walters Shoals see Walters Shoal
24 M3 **Walthall** Mississippi, S USA 33.36N 89.16W
12 L9 **Walton** Kentucky, S USA 38.52N 84.36W
20 J11 **Walton** New York, NE USA 42.10N 75.07W
81 O20 **Walungu** Sud Kivu, E Dem. Rep. Congo (Zaire) 2.40S 28.37E
Walvisbaai see Walvis Bay
85 C19 **Walvis Bay** Afr. Walvisbaai. Erongo, NW Namibia 22.59S 14.33E
85 B19 **Walvis Bay** bay NW Namibia
Walvish Ridge see Walvis Ridge
191 V4 **Walvis Ridge** var. Walvish Ridge. undersea feature E Atlantic Ocean
103 H15 **Wamal** Irian Jaya, E Indonesia 8.00S 139.06E
79 V15 **Wamba** Nassarawa, C Nigeria 8.57N 8.35E
81 O17 **Wamba** Orientale, NE Dem. Rep. Congo (Zaire) 2.10N 27.58E
81 H22 **Wamba** var. Uamba. ♣ Angola/Dem. Rep. Congo (Zaire)
49 P4 **Wamego** Kansas, C USA 39.12N 96.18W
37 P13 **Wampú, Río** ♣ E Honduras
176 Xx16 **Wan** Irian Jaya, E Indonesia 8.15S 138.00E
Wan see Anhui
191 N4 **Wanaaring** New South Wales, SE Australia 29.42S 144.07E
193 D21 **Wanaka** Otago, South Island, NZ 44.42S 169.09E
193 D20 **Wanaka, Lake** ◉ South Island, NZ
12 F9 **Wanapitei** ♣ Ontario, S Canada
12 F10 **Wanapitei Lake** ◉ Ontario, S Canada
20 K14 **Wanaque** New Jersey, NE USA 41.02N 74.17W
176 Y9 **Wanau** Irian Jaya, E Indonesia 1.20S 132.40E
193 F22 **Wanbrow, Cape** headland South Island, NZ 45.07S 170.59E
Wanchuan see Zhangjiakou
176 X11 **Wandai** var. Komeyo. Irian Jaya, E Indonesia 3.35S 136.15E
169 Z8 **Wanda Shan** ▲ NE China
207 R11 **Wandel Sea** sea Arctic Ocean
176 D13 **Wanding** var. Wandingzhen. Yunnan, SW China 24.12N 98.05E
Wandingzhen see Wanding
176 Z14 **Wandoti** Irian Jaya, E Indonesia 6.08S 140.47E
101 H20 **Wanfercée-Baulet** Hainaut, S Belgium 50.27N 4.37E
192 F9 **Wangerooge** see Wangeroog
Wangeroog see Wangerooge
178 J10 **Warin Chamrap** Ubon Ratchathani, E Thailand 15.10N 104.51E
102 F9 **Wangerooge** island NW Germany
176 Ww13 **Warilau, Pulau** island Kepulauan Aru, E Indonesia 5.19S 134.33E
192 L12 **Wanganui** Manawatu-Wanganui, North Island, NZ 39.56S 175.02E
192 L11 **Wanganui** ♣ North Island, NZ
191 P11 **Wangaratta** Victoria, SE Australia 36.22S 146.16E
166 J8 **Wangcang** var. Fengjiaba. Sichuan, C China 32.15N 106.16E
167 S9 **Wangpan Yang** sea E China
169 Y10 **Wangqing** Jilin, NE China 43.19N 129.42E
Wangda see Zogang
103 I24 **Wangen im Allgäu** Baden-Württemberg, S Germany 47.40N 9.49E
102 F9 **Wangerooge** island NW Germany
166 J13 **Wangmo** var. Fuxing. Guizhou, S China 25.10N 106.01E

Wankie see Hwange
Wanki, Río see Coco, Río
83 N17 **Wanlaweyn** var. Wanle Weyn, It. Uanle Uen. Shabeellaha Hoose, SW Somalia 2.36N 44.47E
Wanle Weyn see Wanlaweyn
188 I12 **Wanneroo** Western Australia 31.37S 115.43E
166 L17 **Wanning** Hainan, S China 18.55N 110.27E
178 I8 **Wanon Niwat** Sakon Nakhon, E Thailand 17.39N 103.45E
161 H16 **Wanparti** Andhra Pradesh, C India 16.19N 78.06E
Wansen see Wiązów
166 L11 **Wanshan** Guizhou, S China 27.45N 109.12E
101 M4 **Wanssum** Limburg, SE Netherlands 51.31N 6.04E
192 M13 **Wanstead** Hawke's Bay, North Island, NZ 40.09S 176.31E
169 K9 **Wanxian** Chongqing Shi, C China 30.48N 108.21E
196 F16 **Wanyaan** Yap, Micronesia
166 K8 **Wanyuan** Sichuan, C China 32.04N 108.07E
167 O11 **Wanzai** Jiangxi, S China 28.06N 114.27E
101 J20 **Wanze** Liège, E Belgium 50.32N 5.16E
33 N12 **Wapakoneta** Ohio, N USA 40.34N 84.11W
10 D7 **Wapaseese** ♣ Ontario, C Canada
34 L10 **Wapato** Washington, NW USA 46.27N 120.25W
31 Y15 **Wapello** Iowa, C USA 41.10N 91.13W
194 H12 **Wapenamanda** Enga, W PNG 5.36S 143.51E
9 N13 **Wapiti** ♣ Alberta/British Columbia, SW Canada
49 X7 **Wappapello Lake** ◉ Missouri, C USA
80 J16 **Wappingers Falls** New York, NE USA 41.36N 73.54W
31 X13 **Wapsipinicon River** ♣ Iowa, C USA
194 G14 **Wapumba Island** island SW PNG
12 L9 **Wapus** ♣ Quebec, SE Canada
166 H7 **Waqên** Sichuan, C China
23 Q7 **War** West Virginia, NE USA 37.18N 81.39W
82 D13 **Warab** C S Sudan 8.13N 28.52E
82 D13 **Warab** ♦ state SW Sudan
161 J15 **Warangal** Andhra Pradesh, C India 18.00N 79.35E
Warasdin see Varaždin
191 O16 **Waratah** Tasmania, SE Australia 41.28S 145.34E
191 O14 **Waratah Bay** bay Victoria, SE Australia
103 R15 **Warburg** Nordrhein-Westfalen, W Germany 51.30N 9.10E
190 I1 **Warburton Creek** seasonal river South Australia
188 M9 **Warburton** Western Australia 26.11S 126.18E
101 M20 **Warche** ♣ E Belgium
155 P5 **Wardag** var. Wardak, Per. Vardak. ♦ province E Afghanistan
Wardak see Wardag
34 K9 **Warden** Washington, NW USA 46.58N 119.02W
160 I12 **Wardha** Mahārāshtra, W India 20.40N 78.40E
194 L14 **Ward Hunt, Cape** headland PNG 8.03S 148.15E
195 N16 **Ward Hunt Strait** strait PNG
123 J16 **Wardija, Ras il-** var. Wardija Point. headland Gozo, NW Malta 36.03N 14.11E
Wardija Point see Wardija, Ras il-
154 N4 **Wardīyah** N Iraq 36.18N 41.45E
193 E19 **Ward, Mount** ▲ South Island, NZ 43.49S 169.54E
8 L11 **Ware** British Columbia, W Canada 57.25N 125.40W
101 D18 **Waregem** var. Waereghem. West-Vlaanderen, W Belgium 50.52N 3.25E
101 J19 **Waremme** Liège, E Belgium 50.40N 5.15E
102 N10 **Waren** Mecklenburg-Vorpommern, NE Germany 53.31N 12.42E
176 X10 **Waren** Irian Jaya, E Indonesia 2.13S 136.21E
103 F14 **Warendorf** Nordrhein-Westfalen, W Germany 51.57N 8.00E
23 P12 **Ware Shoals** South Carolina, SE USA 34.24N 82.15W
100 N4 **Warffum** Groningen, NE Netherlands 53.24N 6.34E
83 O15 **Wargalo** Mudug, E Somalia 6.06N 47.40E
152 M12 **Warganza** Rus. Varganzi. Qashqadaryo Wiloyati, S Uzbekistan 38.10N 66.00E
Wargla see Ouargla
191 T4 **Warialda** New South Wales, SE Australia 29.34S 150.35E
178 J10 **Warin Chamrap** Ubon Ratchathani, E Thailand 15.10N 104.51E
176 Ww13 **Warilau, Pulau** island Kepulauan Aru, E Indonesia 5.19S 134.33E

100 H8 **Warmenhuizen** Noord-Holland, NW Netherlands 52.43N 4.45E

112 L8 **Warmińsko-Mazurkie** ◇ province NW Netherlands

37 R12 **Wasco** California, W USA 35.34N 119.20W

33 V10 **Waseca** Minnesota, N USA 44.04N 93.30W

12 H13 **Washago** Ontario, S Canada 44.46N 78.48W

21 S2 **Washburn** Maine, NE USA 46.46N 68.08W

30 M5 **Washburn** North Dakota, N USA 47.15N 101.02W

32 K3 **Washburn** Wisconsin, N USA 46.40N 90.52W

33 S14 **Washburn Hill** hill Ohio, N USA 39.10N 83.25W

160 H13 **Wāshīm** Mahārāshtra, C India 20.06N 77.08E

99 M14 **Washington** NE England, UK 54.54N 1.31W

25 U3 **Washington** Georgia, SE USA 33.44N 82.44W

32 L12 **Washington** Illinois, N USA 40.42N 89.24W

33 N15 **Washington** Indiana, N USA 38.40N 87.10W

31 X15 **Washington** Iowa, C USA 41.18N 91.41W

49 O3 **Washington** Kansas, C USA 39.46N 97.03W

49 W5 **Washington** Missouri, C USA 38.31N 91.01W

25 X9 **Washington** North Carolina, SE USA 35.33N 77.03W

20 B15 **Washington** Pennsylvania, NE USA 40.10N 80.16W

27 V10 **Washington** Texas, SW USA 30.18N 96.08W

38 J8 **Washington** Utah, W USA 37.07N 113.30W

23 V4 **Washington** Virginia, NE USA 38.40N 78.10W

34 I9 **Washington** off. State of Washington; also known as Chinook State, Evergreen State. ◇ state NW USA

Washington see Washington Court House

33 S14 **Washington Court House** var. Washington. Ohio, NE USA 39.31N 83.25W

23 W4 **Washington DC** ● (USA) District of Columbia, NE USA 38.54N 77.02W

33 O5 **Washington Island** island Wisconsin, N USA

Washington Island see Teraina

21 O7 **Washington, Mount** ▲ New Hampshire, NE USA 44.16N 71.18W

28 M11 **Washita River** ↔ Oklahoma/Texas, C USA

99 O18 **Wash, The** inlet E England, UK

34 L9 **Washtucna** Washington, NW USA 46.44N 118.19W

112 P9 **Wasilków** Podlaskie, NE Poland 53.12N 23.15E

41 R11 **Wasilla** Alaska, USA 61.34N 149.26W

57 U9 **Wasjabo** Sipaliwini, NW Suriname 5.09N 57.09W

9 X11 **Waskaiowaka Lake** ◎ Manitoba, C Canada

9 T14 **Waskesiu Lake** Saskatchewan, C Canada 53.55N 106.04W

27 X7 **Waskom** Texas, SW USA 32.28N 94.03W

112 G13 **Wąsosz** Dolnośląskie, SW Poland 51.36N 16.30E

44 M6 **Waspam** var. Waspán. Región Autónoma Atlántico Norte, NE Nicaragua 14.40N 84.04W

Waspán see Waspam

172 P4 **Wassamu** Hokkaidō, NE Japan 44.01N 142.25E

110 G9 **Wassen** Uri, C Switzerland 46.42N 8.34E

100 G11 **Wassenaar** Zuid-Holland, W Netherlands 52.07N 4.24E

101 N24 **Wasserbillig** Grevenmacher, E Luxembourg 49.43N 6.30E

Wasserburg see Wasserburg am Inn

103 M23 **Wasserburg am Inn** var. Wasserburg. Bayern, SE Germany 48.02N 12.12E

103 I17 **Wasserkuppe** ▲ C Germany 50.30N 9.55E

105 R5 **Wassy** Haute-Marne, N France 48.32N 4.54E

175 Pp12 **Watampone** var. Bone. Sulawesi, C Indonesia 4.31S 120.15E

175 S11 **Watawa** Pulau Buru, E Indonesia 3.36S 127.13E

Watenstedt-Salzgitter see Salzgitter

20 M13 **Waterbury** Connecticut, NE USA 41.33N 73.01W

23 R11 **Wateree Lake** ◎ South Carolina, SE USA

23 R12 **Wateree River** ↔ South Carolina, SE USA

99 E20 **Waterford** Ir. Port Láirge. S Ireland 52.15N 7.07W

33 S9 **Waterford** Michigan, N USA 42.42N 83.24W

99 E20 **Waterford** Ir. Port Láirge. cultural region S Ireland

99 E21 **Waterford Harbour** Ir. Cuan Phort Láirge. inlet S Ireland

100 G13 **Wateringen** Zuid-Holland, W Netherlands 52.01N 4.16E

101 G19 **Waterloo** Wallon Brabant, C Belgium 50.43N 4.24E

12 D15 **Waterloo** Ontario, S Canada 43.28N 80.31W

13 P16 **Waterloo** Québec, SE Canada 45.20N 72.28W

32 K9 **Waterloo** Illinois, N USA 38.20N 90.09W

31 X13 **Waterloo** Iowa, C USA 42.31N 92.16W

20 H8 **Waterloo** New York, NE USA 42.54N 76.51W

32 L4 **Watersmeet** Michigan, N USA 46.16N 89.10W

25 V9 **Watertown** Florida, SE USA 30.11N 82.36W

20 I8 **Watertown** New York, NE USA 43.57N 75.55W

31 R9 **Watertown** South Dakota, N USA 44.54N 97.06W

32 M8 **Watertown** Wisconsin, N USA 43.12N 88.44W

24 L3 **Water Valley** Mississippi, S USA 34.09N 89.37W

49 O3 **Waterville** Kansas, C USA 39.41N 96.45W

21 S4 **Waterville** Maine, NE USA 44.34N 69.40W

31 V10 **Waterville** Minnesota, N USA 44.13N 93.34W

20 I10 **Waterville** New York, NE USA 42.55N 75.18W

12 E16 **Watford** Ontario, S Canada 42.57N 81.51W

99 N21 **Watford** SE England, UK 51.39N 0.24W

30 K4 **Watford City** North Dakota, N USA 47.48N 103.16W

147 X12 **Waṭīf** S Oman 18.34N 56.31E

20 G11 **Watkins Glen** New York, NE USA 42.26N 76.52W

Watlings Island see San Salvador

176 V13 **Watnil** Pulau Kai Kecil, E Indonesia 5.45S 132.39E

28 M10 **Watonga** Oklahoma, C USA 35.50N 98.24W

39 T10 **Watrous** New Mexico, SW USA 35.48N 104.58W

9 T16 **Watrous** Saskatchewan, S Canada 51.40N 105.28W

81 P16 **Watsa** Orientale, NE Dem. Rep. Congo (Zaire) 3.00N 29.31E

33 N12 **Watseka** Illinois, N USA 40.46N 87.44W

81 J19 **Watsikengo** Equateur, C Dem. Rep. Congo (Zaire) 0.49S 20.34E

190 C5 **Watson** South Australia 30.32S 131.28E

9 U15 **Watson** Saskatchewan, S Canada 52.13N 104.30W

205 O14 **Watson Escarpment** ▲ Antarctica

8 K9 **Watson Lake** Yukon Territory, W Canada 60.04N 128.46W

37 N10 **Watsonville** California, W USA 36.53N 121.43W

178 I8 **Wattay** ✈ (Viangchan) Viangchan, C Laos 18.03N 102.36E

111 N7 **Wattens** Tirol, W Austria 47.18N 11.37E

22 M9 **Watts Bar Lake** ◎ Tennessee, S USA

110 H7 **Wattwil** Sankt Gallen, NE Switzerland 47.19N 9.04E

176 Uu12 **Watubela, Kepulauan** island group E Indonesia

103 N24 **Watzmann** ▲ SE Germany 47.32N 12.56E

194 J13 **Wau** Morobe, C PNG 7.18S 146.38E

83 D14 **Wau** var. Wāw. Western Bahr el Ghazal, S Sudan 7.43N 28.01E

31 Q8 **Waubay** South Dakota, N USA 45.19N 97.18W

31 Q8 **Waubay Lake** ◎ South Dakota, N USA

191 U7 **Wauchope** New South Wales, SE Australia 31.30S 152.46E

25 W13 **Wauchula** Florida, SE USA 27.33N 81.48W

32 M10 **Wauconda** Illinois, N USA 42.15N 88.08W

190 J7 **Waukaringa** South Australia 32.19S 139.27E

31 N10 **Waukegan** Illinois, N USA 42.21N 87.50W

32 M9 **Waukesha** Wisconsin, N USA 43.01N 88.13W

31 X11 **Waukon** Iowa, C USA 43.16N 91.28W

32 L8 **Waunakee** Wisconsin, N USA 43.13N 89.28E

32 L7 **Waupaca** Wisconsin, N USA 44.21N 89.04W

32 M8 **Waupun** Wisconsin, N USA 43.40N 88.43W

28 M13 **Waurika** Oklahoma, C USA 34.10N 98.00W

28 M12 **Waurika Lake** ◎ Oklahoma, C USA

32 L6 **Wausau** Wisconsin, N USA 44.58N 89.40W

33 R11 **Wauseon** Ohio, N USA 41.33N 84.08W

32 L7 **Wautoma** Wisconsin, N USA 44.04N 89.16W

32 M9 **Wauwatosa** Wisconsin, N USA 43.03N 88.03W

24 L9 **Waveland** Mississippi, S USA 30.17N 89.22W

99 Q20 **Waveney** ↔ E England, UK

192 L11 **Waverley** Taranaki, North Island, NZ 39.46S 174.37E

31 W12 **Waverly** Iowa, C USA 42.43N 92.28W

49 T4 **Waverly** Missouri, C USA 39.12N 93.31W

49 Q4 **Waverly** Nebraska, C USA 40.56N 96.27W

20 H11 **Waverly** New York, NE USA 42.00N 76.33W

23 U6 **Waverly** Tennessee, S USA 36.04N 87.47W

23 V7 **Waverly** Virginia, NE USA 37.02N 77.06W

101 H19 **Wavre** Wallon Brabant, C Belgium 50.43N 4.37E

177 G8 **Waw** Pegu, SW Myanmar 17.25N 96.40E

12 D12 **Wawa** Ontario, S Canada 47.59N 84.43W

79 T14 **Wāwā** Niger, W Nigeria 9.52N 4.33E

77 U11 **Wāwa, Río** var. Rio Huahua. ↔ NE Nicaragua

194 G13 **Wawoi** ↔ SW PNG

176 Uu12 **Wawosungu, Teluk** see Staring, Teluk

27 T7 **Waxahachie** Texas, SW USA 32.23N 96.51W

164 I9 **Waxxari** Xinjiang Uygur Zizhiqu, NW China 38.43N 87.11E

197 N12 **Waya** island Yasawa Group, NW Fiji

32 K5 **Waycross** Georgia, SE USA 31.12N 82.21W

188 K10 **Way, Lake** ◎ Western Australia

33 P9 **Wayland** Michigan, N USA 42.40N 85.39W

31 R13 **Wayne** Nebraska, C USA 42.13N 97.01W

20 K14 **Wayne** New Jersey, NE USA 40.57N 74.16W

23 P5 **Wayne** West Virginia, NE USA 38.13N 82.26W

25 V4 **Waynesboro** Georgia, SE USA 33.04N 82.01W

24 M7 **Waynesboro** Mississippi, S USA 31.40N 88.39W

22 H10 **Waynesboro** Tennessee, S USA 35.19N 87.45W

23 U5 **Waynesboro** Virginia, NE USA 38.04N 78.53W

20 B16 **Waynesburg** Pennsylvania, NE USA 39.51N 80.10W

49 U6 **Waynesville** Missouri, C USA 37.49N 92.12W

23 O10 **Waynesville** North Carolina, SE USA 35.29N 82.59W

28 L8 **Waynoka** Oklahoma, C USA 36.36N 98.51W

Wazan see Ouazzane

Wazima see Wajima

155 V7 **Wazīrābād** Punjab, NE Pakistan 32.28N 74.04E

Wazzan see Ouazzane

112 I8 **Wda** var. Czarna Woda, Ger. Schwarzwasser. ↔ N Poland

197 K6 **Wé** Province des Îles Loyauté, E New Caledonia 20.55S 167.15E

99 O23 **Weald, The** lowlands SE England, UK

194 E15 **Weam** Western, SW PNG 8.33S 141.10E

99 L15 **Wear** ↔ N England, UK

28 L10 **Wearmouth** see Sunderland

27 S6 **Weatherford** Oklahoma, C USA 35.31N 98.42W

36 M3 **Weatherford** Texas, SW USA 32.45N 97.48W

49 R7 **Weaverville** California, W USA 40.42N 122.57W

199 G9 **Webb City** Missouri, C USA 37.07N 94.28W

Weber Basin undersea feature S Ceram Sea

20 F9 **Webfoot State** see Oregon

22 M9 **Webster** New York, NE USA 43.13N 77.25W

31 Q8 **Webster** South Dakota, N USA 45.19N 97.31W

31 X13 **Webster City** Iowa, C USA 42.28N 93.49W

49 X5 **Webster Groves** Missouri, C USA 38.32N 90.20W

23 S4 **Webster Springs** var. Addison. West Virginia, NE USA 38.27N 80.24W

175 T8 **Weda, Teluk** bay Pulau Halmahera, E Indonesia

67 B25 **Weddell Island** var. Isla San Jose. Island W Falkland Islands

67 K22 **Weddell Plain** undersea feature SW Atlantic Ocean

67 K23 **Weddell Sea** sea SW Atlantic Ocean

67 B25 **Weddell Settlement** Weddell Island, W Falkland Islands 52.52S 60.54W

190 M11 **Wedderburn** Victoria, SE Australia 36.26S 143.37E

102 I9 **Wedel** Schleswig-Holstein, N Germany 53.34N 9.42E

94 N3 **Wedel Jarlsberg Land** physical region SW Svalbard

102 I12 **Wedemark** Niedersachsen, NW Germany 52.33N 9.43E

8 M17 **Wedge Mountain** ▲ British Columbia, SW Canada 50.10N 122.43W

25 R4 **Wedowee** Alabama, S USA 33.16N 85.28W

176 Vv13 **Weduar** Pulau Kai Besar, E Indonesia 5.55S 132.51E

176 Vv14 **Weduar, Tanjung** headland Pulau Kai Besar, SE Indonesia 5.58S 132.49E

37 N2 **Weed** California, W USA 41.26N 122.24W

23 Q12 **Weedon Centre** Québec, SE Canada 45.40N 71.28W

20 E13 **Weedville** Pennsylvania, NE USA 41.15N 78.28W

102 F10 **Weener** Niedersachsen, NW Germany 53.09N 7.19E

31 S16 **Weeping Water** Nebraska, C USA 40.52N 96.08W

101 L16 **Weert** Limburg, SE Netherlands 51.15S 5.43E

100 I10 **Weesp** Noord-Holland, C Netherlands 52.18N 5.03E

191 S5 **Wee Waa** New South Wales, SE Australia 30.16S 149.27E

112 N7 **Wegorzewo** Ger. Angerburg. Warmińsko-Mazurskie, NE Poland 54.12N 21.49E

112 E9 **Wegorzyno** Ger. Wangerin. Zachodniopomorskie, NW Poland 53.34N 15.35E

112 N11 **Wegrów** Ger. Bingerau. Mazowieckie, E Poland 52.24N 22.01E

100 N5 **Wehe-Den Hoorn** Groningen, NE Netherlands 53.20N 6.29E

100 M12 **Wehl** Gelderland, E Netherlands 51.58N 6.13E

173 S7 **Weh, Pulau** island NW Indonesia

167 U7 **Wei** see Weifang

167 P3 **Weichang** prev. Zhuizishan. Hebei, E China 41.55N 117.45E

77 U9 **Weichsel** see Wisła

33 M16 **Weida** Thüringen, C Germany 50.46N 12.05E

20 M19 **Weiden** see Weiden in der Oberpfalz

103 M19 **Weiden in der Oberpfalz** var. Weiden. Bayern, SE Germany 49.40N 12.10E

167 Q4 **Weifang** var. Wei, Wei-fang; prev. Weihsien. Shandong, E China 36.43N 119.10E

167 S4 **Weihai** Shandong, E China 37.30N 122.04E

166 K6 **Wei He** ↔ C China

Weihsien see Weifang

23 W7 **Weilburg** Hessen, W Germany 50.31N 8.18E

103 K24 **Weilheim** Bayern, SE Germany 47.50N 11.09E

191 P4 **Weilmoringle** New South Wales, SE Australia 29.13S 146.51E

103 L16 **Weimar** Thüringen, C Germany 50.58N 11.19E

27 U11 **Weimar** Texas, SW USA 29.42N 96.46W

159 L6 **Weinan** Shaanxi, C China 34.30N 109.30E

110 H6 **Weinfelden** Thurgau, NE Switzerland 47.33N 9.09E

103 I24 **Weingarten** Baden-Württemberg, S Germany 47.49N 9.37E

103 G20 **Weinheim** Baden-Württemberg, SW Germany 49.33N 8.40E

166 H11 **Weining** var. Weining Yizu Huizu Miaozu Zizhixian. Guizhou, S China 26.51N 104.16E

Weining Yizu Huizu Miaozu Zizhixian see Weining

189 V2 **Weipa** Queensland, NE Australia 12.43S 142.01E

9 Y11 **Weir River** Manitoba, C Canada 56.44N 94.06W

23 R1 **Weirton** West Virginia, NE USA 40.25N 80.35W

34 M13 **Weiser** Idaho, NW USA 44.15N 116.58W

166 F12 **Weishan** Yunnan, SW China 25.22N 100.19E

167 P6 **Weishan Hu** ◎ E China

103 M15 **Weisse Elster** Eng. White Elster. ↔ Czech Republic/Germany

Weisse Körös/Weisse Kreisch see Crişul Alb

110 L7 **Weissenbach am Lech** Tirol, W Austria 47.27N 10.39E

103 K21 **Weissenburg** Bayern, SE Germany 49.02N 10.58E

Weissenburg see Wissembourg, France

Weissenburg see Alba Iulia, Romania

103 M15 **Weissenfels** var. Weißenfels. Sachsen-Anhalt, C Germany 51.12N 11.58E

111 R9 **Weissensee** ◎ S Austria

110 E11 **Weissenstein** see Paide

Weisshorn ▲ SW Switzerland 46.06N 7.43E

Weisskirchen see Bela Crkva

25 R3 **Weiss Lake** ◎ Alabama, S USA

103 Q14 **Weisswasser** Lus. Běla Woda. Sachsen, E Germany 51.30N 14.37E

101 M22 **Weiswampach** Diekirch, N Luxembourg 50.07N 6.04E

111 U2 **Weitra** Niederösterreich, N Austria 48.41N 14.54E

167 O4 **Weixian** var. Wei Xian. Hebei, E China 36.58N 115.15E

165 V11 **Weiyuan** Gansu, C China 35.07N 104.12E

166 F14 **Weiyuan Jiang** ↔ SW China

111 W7 **Weiz** Steiermark, SE Austria 47.13N 15.37E

166 K16 **Weizhou Dao** island S China 6.34N 38.28E

112 I6 **Wejherowo** Pomorskie, NW Poland 54.36N 18.12E

49 Q8 **Welch** Oklahoma, C USA 36.52N 95.06W

26 M6 **Welch** Texas, SW USA 32.52N 102.06W

23 Q6 **Welch** West Virginia, NE USA 37.25N 81.37W

47 O14 **Welchman Hall** C Barbados 13.10N 59.34W

82 J11 **Weldiya** var. Waldia, It. Valdia. Amhara, N Ethiopia 11.45N 39.39E

23 W8 **Weldon** North Carolina, SE USA 36.25N 77.36W

27 V9 **Weldon** Texas, SW USA 31.00N 95.33W

101 M19 **Welkenraedt** Liège, E Belgium 50.40N 5.58E

199 L2 **Welker Seamount** undersea feature N Pacific Ocean

85 I22 **Welkom** Free State, C South Africa 27.58S 26.43E

12 H16 **Welland** Ontario, S Canada 43.58N 79.13W

12 G16 **Welland** ↔ Ontario, S Canada

99 O19 **Welland** ↔ C England, UK

12 H17 **Welland Canal** canal Ontario, S Canada

161 K25 **Wellawaya** Uva Province, SE Sri Lanka 6.43N 81.07E

189 T4 **Wellesley Islands** island group Queensland, N Australia

101 J22 **Wellin** Luxembourg, SE Belgium 50.05N 5.05E

99 N20 **Wellingborough** C England, UK 52.19N 0.42W

191 R7 **Wellington** New South Wales, SE Australia 32.34S 148.55E

12 J15 **Wellington** Ontario, SE Canada 43.57N 77.24W

193 L14 **Wellington** ● (NZ) Wellington, North Island, NZ 41.16S 174.46E

3 N14 **Wellington** Somalí, E Ethiopia 6.59N 45.20E

85 E26 **Wellington** Western Cape, SW South Africa 33.39S 19.00E

39 T2 **Wellington** Colorado, C USA 40.42N 105.00W

49 N7 **Wellington** Kansas, C USA 37.16N 97.22W

37 R7 **Wellington** Nevada, W USA 38.45N 119.22W

33 P7 **Wellington** Texas, SW USA 34.51N 100.12W

38 M4 **Wellington** Utah, W USA 39.31N 110.45W

193 M14 **Wellington** off. Wellington Region. ◇ region North Island, NZ

65 F22 **Wellington, Isla** var. Wellington. island S Chile

191 P12 **Wellington, Lake** ◎ Victoria, SE Australia

31 X14 **Wellman** Iowa, C USA 41.27N 91.50W

26 K6 **Wellman** Texas, SW USA 33.03N 102.26W

99 K22 **Wells** SW England, UK 51.13N 2.39W

31 V11 **Wells** Minnesota, N USA 43.45N 93.43W

37 X2 **Wells** Nevada, W USA 41.06N 114.57W

27 W8 **Wells** Texas, SW USA 31.28N 94.54W

20 F12 **Wellsboro** Pennsylvania, NE USA 41.43N 77.39W

23 R1 **Wellsburg** West Virginia, NE USA 40.16N 80.36W

192 K4 **Wellsford** Auckland, North Island, NZ 36.17S 174.30E

188 I9 **Wells, Lake** ◎ Western Australia

189 N4 **Wells, Mount** ▲ Western Australia 17.20S 128.35E

99 P18 **Wells-next-the-Sea** E England, UK 52.58N 0.49E

33 T15 **Wellston** Ohio, N USA 39.07N 82.31W

49 O10 **Wellston** Oklahoma, C USA 35.41N 97.03W

23 V12 **Wellsville** New York, NE USA 42.06N 77.55W

38 L1 **Wellsville** Utah, W USA 41.38N 111.55W

38 I14 **Wellton** Arizona, SW USA 32.40N 114.09W

111 S4 **Wels** anc. Ovilava. Oberösterreich, N Austria 48.10N 14.01E

101 K15 **Welschap** ✈ (Eindhoven) Noord-Brabant, S Netherlands 51.27N 5.22E

102 P10 **Welse** ↔ NE Germany

24 H9 **Welsh** Louisiana, S USA 30.12N 92.49W

99 K19 **Welshpool** Wel. Y Trallwng. E Wales, UK 52.38N 3.06W

99 O21 **Welwyn Garden City** SE England, UK 51.48N 0.13W

81 K18 **Wema** Equateur, NW Dem. Rep. Congo (Zaire) 0.25S 21.33E

83 G21 **Wembere** ↔ C Tanzania

9 N13 **Wembley** Alberta, C Canada 55.07N 119.12W

10 I9 **Wemindji** prev. Nouveau-Comptoir, Paint Hills. Québec, C Canada 53.00N 78.42W

101 G18 **Wemmel** Vlaams Brabant, C Belgium 50.54N 4.18E

34 J8 **Wenatchee** Washington, NW USA 47.49N 120.48W

166 M17 **Wenchang** Hainan, S China 19.34N 110.46E

167 R11 **Wencheng** prev. Daxue. Zhejiang, SE China 27.48N 120.03E

79 P16 **Wenchi** W Ghana 7.45N 2.01W

Wen-chou/Wenchow see Wenzhou

166 H8 **Wenchuan** prev. Wei Xian. Hebei, E China 31.29N 103.39E

Wendau see Võnnu

167 S4 **Wendeng** Shandong, E China 37.14N 122.06E

81 J14 **Wendo** Southern, S Ethiopia 6.34N 38.28E

38 J2 **Wendover** Utah, W USA 40.41N 114.00W

12 D9 **Wenebegon** ↔ Ontario, S Canada

12 D8 **Wenebegon Lake** ◎ Ontario, S Canada

110 E9 **Wengen** Bern, W Switzerland 46.36S 7.55E

167 O13 **Wengyuan** prev. Longxian. Guangdong, S China 24.22N 114.06E

201 P15 **Weno** prev. Moen. atoll Chuuk, C Micronesia

201 V12 **Weno** prev. Moen. island Chuuk Islands, C Micronesia

164 N13 **Wenquan** Qinghai, C China 33.16N 91.43E

164 G8 **Wenquan** var. Arixang. Xinjiang Uygur Zizhiqu, NW China 45.01N 81.02E

166 H14 **Wenshan** Yunnan, SW China 23.22N 104.21E

164 F6 **Wensu** Xinjiang Uygur Zizhiqu, W China 41.15N 80.18E

190 D14 **Wentworth** New South Wales, SE Australia 34.04S 141.53E

165 V12 **Wenxian** var. Wen Xian. Gansu, C China 32.57N 104.42E

167 S10 **Wenzhou** var. Wen-chou, Wenchow. Zhejiang, SE China 28.02N 120.36E

36 L4 **Weott** California, W USA 40.19N 123.57W

101 L20 **Wépion** Namur, SE Belgium 50.24N 4.53E

102 I11 **Werbellinsee** ◎ NE Germany

101 L21 **Werbomont** Liège, E Belgium 50.22S 5.43E

85 G20 **Werda** Kgalagadi, S Botswana 25.13S 23.16E

Werder see Virtsu

83 N14 **Werdēr** Somalí, E Ethiopia 6.59N 45.20E

Werenów see Voranava

176 U13 **Weri** Irian Jaya, E Indonesia 3.10S 132.39E

103 I19 **Werkendam** Noord-Brabant, S Netherlands 51.48N 4.54E

103 J18 **Wernberg-Köblitz** Bayern, SE Germany 49.31N 12.10E

103 J18 **Werneck** Bayern, C Germany 50.00N 10.06E

103 L15 **Wernigerode** Sachsen-Anhalt, C Germany 51.51N 10.48E

55 C12 **Werowitz** see Virovitica

103 I17 **Werra** ↔ C Germany

190 K12 **Werribee** Victoria, SE Australia 37.55S 144.39E

191 T6 **Werris Creek** New South Wales, SE Australia 31.22S 150.40E

25 O14 **Werrvcq** see Võru

167 T9 **Wertach** ↔ S Germany

3 T3 **Wertheim** Baden-Württemberg, SW Germany 49.45N 9.31E

101 H18 **Wervershoof** Noord-Holland, NW Netherlands 52.43S 5.09E

101 J18 **Wervicq** see Wervik

101 C18 **Wervik** var. Wervicq, Werwick. W Belgium 50.46N 3.03E

Werwick see Wervik

103 D14 **Wesel** Nordrhein-Westfalen, W Germany 51.40N 6.37E

Weseli an der Lainsitz see Veselí nad Lužnicí

Wesenberg see Rakvere

102 H12 **Weser** ↔ NW Germany

Wes-Kaap see Western Cape

27 S17 **Weslaco** Texas, SW USA 26.09N 97.59W

12 J13 **Weslemkoon Lake** ◎ Ontario, SE Canada

189 R1 **Wessel Islands** island group Northern Territory, N Australia

31 P9 **Wessington** South Dakota, N USA 44.27N 98.43W

31 P10 **Wessington Springs** South Dakota, N USA 44.02N 98.33E

27 T8 **West** Texas, SW USA 31.48N 97.05W

West see Ouest

32 M9 **West Allis** Wisconsin, N USA 43.01N 88.00W

190 E8 **Westall, Point** headland South Australia 32.54S 134.04E

West Antarctica see Lesser Antarctica

12 G11 **West Arm** Ontario, S Canada 46.16N 80.25W

West Azerbaijan see Āz̄arbāyjān-e Gharbī

9 F10 **Westbank** British Columbia, SW Canada 49.51N 119.37W

12 E11 **West Bay** Manitoulin Island, Ontario, S Canada 45.48N 82.09W

99 U20 **West Bay** bay Louisiana, S USA

32 M8 **West Bend** Wisconsin, N USA 43.25S 88.13W

159 R16 **West Bengal** ◇ state NE India

West Borneo see Kalimantan Barat

31 Y14 **West Branch** Iowa, C USA 41.40N 91.21W

33 R7 **West Branch** Michigan, N USA 44.16N 84.13W

20 F13 **West Branch Susquehanna River** ↔ Pennsylvania, NE USA

99 L20 **West Bromwich** C England, UK 52.28N 1.59W

21 P8 **Westbrook** Maine, NE USA 43.42N 70.21E

31 T10 **Westbrook** Minnesota, N USA 44.02N 95.26W

31 Y15 **West Burlington** Iowa, C USA 40.49N 91.09W

98 I2 **West Burra** island NE Scotland, UK

31 W10 **West Concord** Minnesota, N USA 44.09N 92.54W

31 V14 **West Des Moines** Iowa, C USA 41.33N 93.42W

39 Q6 **West Elk Peak** ▲ Colorado, C USA 38.43N 107.12W

46 F1 **West End** Grand Bahama Island, N Bahamas 26.30 78.55W

46 F1 **West End Point** headland Grand Bahama Island, N Bahamas 26.40N 78.58W

100 O7 **Westerbork** Drenthe, NE Netherlands 52.49N 6.36E

100 N3 **Westereems** strait Germany/Netherlands

100 O9 **Westerhaar-Vriezenveensewijk** Overijssel, NE Netherlands 52.28N 6.38E

102 G6 **Westerland** Schleswig-Holstein, N Germany 54.54N 8.19E

101 I17 **Westerlo** Antwerpen, N Belgium 51.05N 4.55E

21 N13 **Westerly** Rhode Island, NE USA 41.22N 71.45W

83 G13 **Western** ◇ province W Kenya

159 N11 **Western** ◇ zone C Nepal

194 F14 **Western** ◇ province SW PNG

195 T14 **Western** off. Western Province. ◇ province NW Solomon Islands

85 G15 **Western** ◇ province SW Zambia

188 K8 **Western Australia** ◇ state W Australia

82 A13 **Western Bahr el Ghazal** ◇ state SW Sudan

Western Bug see Bug

85 F25 **Western Cape** var. Western Cape Province, Afr. Wes-Kaap. ◇ province SW South Africa

82 A11 **Western Darfur** ◇ state W Sudan

Western Desert see Sahara el Gharbiya

120 G9 **Western Dvina** Bel. Dzvina, Ger. Düna, Latv. Daugava, Rus. Zapadnaya Dvina. ↔ W Europe

83 D15 **Western Equatoria** ◇ state SW Sudan

161 E16 **Western Ghats** ▲ SW India

194 G12 **Western Highlands** ◇ province C PNG

Western Isles see Outer Hebrides

82 C12 **Western Kordofan** ◇ state C Sudan

27 J3 **Westernport** Maryland, NE USA 39.29N 79.03W

161 J26 **Western Province** ◇ province SW Sri Lanka

76 B10 **Western Sahara** ◇ disputed territory N Africa

Western Samoa see Samoa

Western Sayans see Zapadnyy Sayan

Western Scheldt see Westerschelde

Western Sierra Madre see Madre Occidental, Sierra

101 E15 **Westerschelde** Eng. Western Scheldt; prev. Honte. inlet S North Sea

33 S13 **Westerville** Ohio, N USA 40.07N 82.55W

◆ COUNTRY
● COUNTRY CAPITAL
◇ DEPENDENT TERRITORY
○ DEPENDENT TERRITORY CAPITAL
◆ ADMINISTRATIVE REGION
✕ INTERNATIONAL AIRPORT
▲ MOUNTAIN
▲ MOUNTAIN RANGE
☈ VOLCANO
↔ RIVER
◎ LAKE
◙ RESERVOIR

103 *F17* **Westerwald** ◆ W Germany

67 *C25* **West Falkland** *var.* Gran Malvina,
Isla Gran Malvina. *island*
W Falkland Islands

31 *R5* **West Fargo** North Dakota, N USA
46.49N 96.51W

196 *M15* **West Fayu Atoll** *atoll* Caroline
Islands, C Micronesia

20 *C11* **Westfield** New York, NE USA
42.18N 79.34W

32 *L7* **Westfield** Wisconsin, N USA
43.56N 89.31W

West Flanders *see* West-
Vlaanderen

49 *S10* **West Fork** Arkansas, C USA
35.55N 94.11W

31 *P16* **West Fork Big Blue River**
Nebraska, C USA

31 *U12* **West Fork Des Moines River**
Iowa/Minnesota, C USA

27 *S5* **West Fork Trinity River**
Texas, SW USA

32 *L16* **West Frankfort** Illinois, N USA
37.54N 88.55W

100 *I8* **West-Friesland** *physical region*
NW Netherlands

West Frisian Islands *see*
Waddeneilanden

21 *T5* **West Grand Lake** ⊗ Maine,
NE USA

20 *M12* **West Hartford** Connecticut,
NE USA 41.44N 72.45W

20 *M13* **West Haven** Connecticut,
NE USA 41.16N 72.57W

49 *X12* **West Helena** Arkansas, C USA
34.33N 90.38W

30 *M2* **Westhope** North Dakota, N USA
48.54N 101.01W

205 *Y8* **West Ice Shelf** *ice shelf* Antarctica

49 *R2* **West Indies** *island group* SE North
America

West Irian *see* Irian Jaya

West Java *see* Jawa Barat

38 *L3* **West Jordan** Utah, W USA
40.37N 111.55W

West Kalimantan *see* Kalimantan
Barat

101 *D14* **Westkapelle** Zeeland,
SW Netherlands 51.32N 3.26E

33 *O13* **West Lafayette** Indiana, N USA
40.24N 86.54W

33 *T13* **West Lafayette** Ohio, N USA
40.16N 81.45W

West Lake *see* Kagera

31 *Y14* **West Liberty** Iowa, C USA
41.34N 91.15W

23 *O5* **West Liberty** Kentucky, S USA
38.04N 83.22W

Westliche Morava *see* Zapadna
Morava

15 *I13* **Westlock** Alberta, SW Canada
54.12N 113.49W

12 *E17* **West Lorne** Ontario, S Canada
42.36N 81.34W

98 *J12* **West Lothian** *cultural region*
S Scotland, UK

101 *H16* **Westmalle** Antwerpen, N Belgium
51.18N 4.40E

199 *H6* **West Mariana Basin** *var.* Perece
Vela Basin. *undersea feature*
W Pacific Ocean

99 *E17* **Westmeath** *Ir.* An Iarmhí, Na h-
Iarmhidhe. *cultural region* C Ireland

49 *Y11* **West Memphis** Arkansas, C USA
35.09N 90.11W

23 *W2* **Westminster** Maryland, NE USA
39.34N 77.00W

23 *O11* **Westminster** South Carolina,
SE USA 34.39N 83.06W

24 *I5* **West Monroe** Louisiana, S USA
32.31N 92.09W

20 *D15* **Westmont** Pennsylvania, NE USA
40.16N 78.55W

49 *O3* **Westmoreland** Kansas, C USA
39.23N 96.30W

37 *W17* **Westmorland** California, W USA
33.02N 115.37W

194 *L11* **West New Britain** ◆ *province*
E PNG

West New Guinea *see* Irian Jaya

85 *K18* **West Nicholson** Matabeleland
South, S Zimbabwe 21.06S 29.23E

31 *T14* **West Nishnabotna River**
⚌ Iowa, C USA

183 *P11* **West Norfolk Ridge** *undersea
feature* W Pacific Ocean

27 *P12* **West Nueces River** ⚌ Texas,
SW USA

West Nusa Tenggara *see* Nusa
Tenggara Barat

31 *T11* **West Okoboji Lake** ⊗ Iowa,
C USA

35 *R16* **Weston** Idaho, NW USA
42.01N 119.29W

23 *R4* **Weston** West Virginia, NE USA
39.02N 80.28W

99 *J22* **Weston-super-Mare**
SW England, UK 51.21N 2.58W

25 *Z14* **West Palm Beach** Florida,
SE USA 26.43N 80.03W

West Papua *see* Irian Jaya

196 *E9* **West Passage** *passage* Babeldaob,
N Palau

25 *O9* **West Pensacola** Florida, SE USA
30.25N 87.16W

49 *V8* **West Plains** Missouri, C USA
36.43N 91.51W

37 *P7* **West Point** California, W USA
38.21N 120.33W

25 *R5* **West Point** Georgia, SE USA
32.52N 85.10W

24 *M3* **West Point** Mississippi, S USA
33.36N 88.40W

31 *R14* **West Point** Nebraska, C USA
41.50N 96.42W

23 *X6* **West Point** Virginia, NE USA
37.31N 76.48W

190 *G10* **West Point** *headland* South
Australia 35.01S 135.58E

67 *B24* **Westpoint Island Settlement**
Westpoint Island, NW Falkland
Islands 51.21S 60.40W

25 *R4* **West Point Lake**
⊠ Alabama/Georgia, SE USA

99 *B16* **Westport** *Ir.* Cathair na Mart.
W Ireland 53.48N 9.31W

193 *G15* **Westport** West Coast, South
Island, NZ 41.46S 171.37E

34 *F10* **Westport** Oregon, NW USA
46.07N 123.22W

34 *F9* **Westport** Washington, NW USA
46.53N 124.06W

33 *S15* **West Portsmouth** Ohio, N USA
38.45N 83.01W

West Punjab *see* Punjab

9 *V14* **Westray** Manitoba, C Canada
53.30N 101.19W

98 *J4* **Westray** *island* NE Scotland, UK

12 *F9* **Westree** Ontario, S Canada
47.25N 81.32W

39 *T4* **West Ridge** *cultural region*
N England, UK

32 *J7* **West River** *see* Xi Jiang

32 *I7* **West Salem** Wisconsin, N USA
43.54N 91.04W

25 *O2* **Wheeler Lake** ⊠ Alabama,
S USA

37 *Y6* **Wheeler Peak** ▲ Nevada, W USA
39.00N 114.17W

39 *T7* **Wheeler Peak** ▲ New Mexico,
SW USA 36.34N 105.25W

33 *S15* **Wheelersburg** Ohio, N USA
38.43N 82.51W

23 *R7* **Wheeling** West Virginia, NE USA
40.03N 80.43W

99 *L16* **Whernside** ▲ N England, UK
54.13N 2.27W

190 *F9* **Whidbey, Point** *headland* South
Australia 34.36S 135.08E

188 *I7* **Whim Creek** Western Australia
20.51S 117.54E

8 *L17* **Whistler** British Columbia,
SW Canada 50.07N 122.57W

33 *W8* **Whitakers** North Carolina,
SE USA 36.06N 77.43W

12 *H15* **Whitby** Ontario, S Canada
43.53N 78.54W

99 *N15* **Whitby** N England, UK
54.28N 0.37W

8 *G6* **White** ⚌ Yukon Territory,
W Canada

11 *T11* **White Bay** *bay* Newfoundland,
E Canada

22 *I8* **White Bluff** Tennessee, S USA
36.06N 87.13W

30 *J6* **White Butte** ▲ North Dakota,
N USA 46.23N 103.18W

21 *R5* **White Cap Mountain** ▲ Maine,
NE USA 45.33N 69.15W

24 *J9* **White Castle** Louisiana, S USA
30.10N 91.09W

190 *M5* **White Cliffs** New South Wales,
SE Australia 30.52S 143.04E

33 *P8* **White Cloud** Michigan, N USA
43.34N 85.46W

9 *P14* **Whitecourt** Alberta, SW Canada
54.10N 115.37W

27 *O2* **White Deer** Texas, SW USA
35.26N 101.10W

White Elster *see* Weisse Elster

26 *M5* **Whiteface** Texas, SW USA
33.36N 102.36W

20 *K7* **Whiteface Mountain** ▲ New
York, NE USA 44.22N 73.54W

31 *W3* **Whiteface Reservoir**
⊠ Minnesota, N USA

35 *O7* **Whitefish** Montana, NW USA
48.24N 114.20W

33 *N9* **Whitefish Bay** Wisconsin, N USA
43.39N 87.54W

33 *Q3* **Whitefish Bay** *lake bay*
Canada/USA

12 *E11* **Whitefish Falls** Ontario,
S Canada 46.06N 81.42W

12 *B7* **Whitefish Lake** ⊗ Ontario,
S Canada

31 *U6* **Whitefish Lake** ⊗ Minnesota,
C USA

33 *Q3* **Whitefish Point** *headland*
Michigan, N USA 46.46N 84.57W

33 *O4* **Whitefish River** ⚌ Michigan,
N USA

27 *O4* **Whiteflat** Texas, SW USA
34.06N 100.55W

49 *V12* **White Hall** Arkansas, C USA
34.18N 92.05W

32 *K14* **White Hall** Illinois, N USA
39.26N 90.24W

33 *O8* **Whitehall** Michigan, N USA
43.24N 86.21W

20 *L9* **Whitehall** New York, NE USA
43.33N 73.24W

33 *S13* **Whitehall** Ohio, N USA
39.58N 82.53W

32 *J7* **Whitehall** Wisconsin, N USA
44.22N 91.19W

39 *J15* **Whitehaven** NW England, UK
54.33N 3.34W

8 *I8* **Whitehorse** *territory capital* Yukon
Territory, W Canada 60.40N 135.07W

192 *O7* **White Island** *island* NE NZ

12 *K13* **White Lake** ⊗ Ontario,
SE Canada

24 *H10* **White Lake** ⊗ Louisiana, S USA

195 *N12* **Whiteman Range** ▲ New
Britain, E PNG

24 *L8* **Whitesand** ⚌ Texas, SW USA

26 *M2* **White Sands** Arkansas, SW USA

38 *K13* **Wickenburg** Arizona, SW USA
33.57N 112.43W

24 *L8* **Wickett** Texas, SW USA
31.34N 103.00W

188 *I7* **Wickham** Western Australia
20.40S 117.11E

191 *N7* **Wickham, Cape** *headland*
Tasmania, SE Australia
39.36S 143.55E

23 *G7* **Wickliffe** Kentucky, S USA
37.14N 89.16W

99 *G19* **Wicklow** *Ir.* Cill Mhantáin.
E Ireland 52.58N 6.03W

99 *F19* **Wicklow** *Ir.* Cill Mhantáin. *cultural
region* E Ireland

99 *G19* **Wicklow Head** *Ir.* Ceann Chill
Mhantáin. *headland* E Ireland
52.57N 6.00W

99 *F18* **Wicklow Mountains** *Ir.* Sléibhte
Chill Mhantáin. ▲ E Ireland

12 *H10* **Wicksteed Lake** ⊗ Ontario,
S Canada

Wida *see* Ouidah

67 *G15* **Wideawake Airfield**
✈ (Georgetown) SW Ascension
Island

Wilia/Wilja *see* Neris

20 *I3* **Wilkes Barre** Pennsylvania,
NE USA 41.15N 75.49W

33 *R9* **Wilkesboro** North Carolina,
SE USA

205 *W15* **Wilkes Coast** *physical region*
Antarctica

201 *N12* **Wilkes Island** *island* N Wake
Island

205 *X12* **Wilkes Land** *physical region*
Antarctica

113 *L17* **Wieliczka** Małopolskie, S Poland
50.00N 20.02E

112 *G12* **Wielkopolskie** ◆ *province*
C Poland

113 *J14* **Wieluń** Sieradz, C Poland
51.13N 18.33E

111 *X4* **Wien** *Eng.* Vienna, *Hung.* Bécs,
Slvk. Vídeň, *Slvn.* Dunaj; *anc.*
Vindobona. ● (Austria) Wien,
NE Austria 48.13N 16.22E

111 *X4* **Wien** ◆ *state* NE Austria

White Sea *see* Beloye More

**White Sea-Baltic Canal/White
Sea Canal** *see* Belomorsko-
Baltiyskiy Kanal

65 *I25* **Whiteside, Canal** *channel* S Chile

35 *S10* **White Sulphur Springs**
Montana, NW USA 46.33N 110.54W

23 *R6* **White Sulphur Springs** West
Virginia, NE USA 37.48N 80.18W

22 *J2* **Whitesville** Kentucky, S USA
37.40N 86.48W

34 *I10* **White Swan** Washington,
NW USA 46.22N 120.46W

23 *U12* **Whiteville** North Carolina,
SE USA 34.20N 78.42W

22 *J2* **Whiteville** Tennessee, S USA
35.19N 89.09W

79 *Q13* **White Volta** *var.* Nakambé, *Fr.*
Volta Blanche. ⚌ Burkina/Ghana

32 *M9* **Whitewater** Wisconsin, N USA
42.51N 88.43W

39 *P14* **Whitewater Baldy** ▲ New
Mexico, SW USA 33.18N 108.38W

25 *X17* **Whitewater Bay** *bay* Florida,
SE USA

33 *Q14* **Whitewater River**
⚌ Indiana/Ohio, N USA

9 *V16* **Whitewood** South Dakota,
N USA 44.27N 103.38W

99 *L16* **Whithorn** S Scotland, UK
54.43N 4.26W

192 *M6* **Whitianga** Waikato, North Island,
NZ 36.49S 175.42E

21 *N11* **Whitinsville** Massachusetts,
NE USA 42.06N 71.40W

22 *M8* **Whitley City** Kentucky, S USA
36.40N 84.28W

23 *Q11* **Whitmire** South Carolina,
SE USA 34.30N 81.36W

33 *O2* **Whitmore Lake** Michigan,
N USA 42.26N 83.44W

205 *N9* **Whitmore Mountains**
▲ Antarctica

12 *I13* **Whitney** Ontario, SE Canada
45.29N 78.11W

27 *T8* **Whitney** Texas, SW USA
31.56N 97.20W

28 *S8* **Whitney, Lake** ⊠ Texas, SW USA

37 *S11* **Whitney, Mount** ▲ California,
W USA 37.45N 119.55W

189 *Y6* **Whitsunday Group** *island group*
Queensland, E Australia

27 *S6* **Whitt** Texas, SW USA
32.55N 98.01W

31 *U12* **Whittemore** Iowa, C USA
43.03N 94.25W

41 *R12* **Whittier** Alaska, USA
60.46N 148.40W

37 *T15* **Whittier** California, W USA
33.58N 118.01W

85 *I25* **Whittlesea** Eastern Cape, S South
Africa 32.08S 26.51E

22 *K10* **Whitwell** Tennessee, S USA
35.12N 85.31W

190 *H7* **Whyalla** South Australia
33.04S 137.34E

Whydah *see* Ouidah

12 *F13* **Wiarton** Ontario, S Canada
44.44N 81.09W

175 *Q11* **Wiau** Sulawesi, C Indonesia
3.08S 121.22E

113 *H15* **Wiązów** Ger. Wansen.
Dolnośląskie, SW Poland
50.49N 17.13E

35 *Y8* **Wibaux** Montana, NW USA
46.57N 104.11W

49 *N6* **Wichita** Kansas, C USA
37.41N 97.20W

27 *R5* **Wichita Falls** Texas, SW USA
33.54N 98.29W

28 *L11* **Wichita Mountains**
▲ Oklahoma, C USA

27 *R5* **Wichita River** ⚌ Texas, SW USA

98 *K6* **Wick** N Scotland, UK 58.25N 3.06W

113 *N13* **Whiteriver** Arizona, SW USA
33.50N 109.57W

27 *U1* **White River Lake** ⊠ Texas,
SW USA

34 *H11* **White Salmon** Washington,
NW USA 45.43N 121.29W

20 *I10* **Whitesboro** Texas, SW USA
43.07N 75.17W

27 *T5* **Whitesboro** Texas, SW USA
33.39N 96.54W

23 *O7* **Whitesburg** Kentucky, S USA
37.16N 82.55W

192 *K6* **Whatipu** Auckland, North Island,
NZ 37.17S 174.44E

35 *V1* **Wheatland** Wyoming, C USA
42.03N 104.57W

12 *D18* **Wheatley** Ontario, S Canada
42.06N 82.27W

32 *M10* **Wheaton** Illinois, N USA
41.52N 88.06W

31 *R7* **Wheaton** Minnesota, N USA
45.48N 96.30W

34 *F11* **Wharton** Texas, SW USA
29.19N 96.08W

181 *U8* **Wharton Basin** *var.* West
Australian Basin. *undersea feature*
E Indian Ocean

193 *E18* **Whataroa** West Coast, South
Island, NZ 43.16S 170.19E

25 *Hh7* **Wha Ti** *prev.* Lac La Martre.
Northwest Territories, W Canada
63.10N 117.12W

204 *I6* **Wilkins Ice Shelf** *ice shelf*
Antarctica

190 *D4* **Wilkinsons Lakes** *salt lake* South
Australia

Wilkomierz *see* Ukmergė

190 *K11* **Willalooka** South Australia
36.24S 140.20E

34 *G12* **Willamette River** ⚌ Oregon,
NW USA

191 *O8* **Willandra Billabong Creek**
seasonal river New South Wales,
SE Australia

34 *F9* **Willapa Bay** *inlet* Washington,
NW USA

49 *T7* **Willard** Missouri, C USA
37.18N 93.25W

39 *S12* **Willard** New Mexico, SW USA
34.36N 106.01W

33 *S12* **Willard** Ohio, N USA
41.03N 82.43W

38 *L1* **Willard** Utah, W USA

101 *G17* **Willebroek** Antwerpen,
C Belgium 51.04N 4.22E

47 *P16* **Willemstad** ○ (Netherlands
Antilles) Curaçao, Netherlands
Antilles 12.06N 68.54W

101 *G14* **Willemstad** Noord-Brabant,
S Netherlands 51.40N 4.27E

9 *S17* **William** ⚌ Saskatchewan,
C Canada

25 *O6* **William "Bill" Dannelly
Reservoir** ⊠ Alabama, S USA

190 *G3* **William Creek** South Australia
28.55S 136.23E

189 *T15* **William, Mount** ▲ South
Australia

33 *K11* **Williams** Arizona, SW USA
35.15N 112.11W

31 *X14* **Williamsburg** Iowa, C USA
41.39N 92.00W

22 *M8* **Williamsburg** Kentucky, S USA
36.43N 84.06W

33 *R15* **Williamsburg** Ohio, N USA
39.04N 84.02W

23 *X6* **Williamsburg** Virginia, NE USA
37.16N 76.41W

8 *L17* **Williams Lake** British Columbia,
SW Canada 52.07N 122.09W

23 *P6* **Williamson** West Virginia,
NE USA 37.40N 82.16W

33 *O13* **Williamsport** Indiana, N USA
40.18N 87.18W

20 *G13* **Williamsport** Pennsylvania,
NE USA 41.13N 76.59W

23 *W9* **Williamston** North Carolina,
SE USA 35.51N 77.03W

23 *Q11* **Williamston** South Carolina,
SE USA 34.37N 82.28W

22 *M4* **Williamstown** Kentucky, S USA
38.38N 84.33W

20 *L10* **Williamstown** Massachusetts,
NE USA 42.41N 73.11W

20 *J16* **Willingboro** New Jersey, NE USA
40.01N 74.52W

9 *Q14* **Willingdon** Alberta, SW Canada
53.49N 112.08W

27 *W10* **Willis** Texas, SW USA
30.25N 95.28W

110 *F8* **Willisau** Luzern, W Switzerland
47.07N 8.00E

85 *F24* **Williston** Northern Cape,
W South Africa 31.19S 20.52E

25 *V10* **Williston** Florida, SE USA
29.23N 82.27W

30 *J3* **Williston** North Dakota, N USA
48.07N 103.37W

23 *Q13* **Williston** South Carolina, SE USA
33.24N 81.25W

8 *L12* **Williston Lake** ⊠ British
Columbia, W Canada

36 *L5* **Willits** California, W USA
39.24N 123.22W

31 *T8* **Willmar** Minnesota, N USA
45.07N 95.02W

8 *K11* **Will, Mount** ▲ British Columbia,
W Canada 57.31N 128.48W

33 *T11* **Willoughby** Ohio, N USA
41.38N 81.24W

9 *U17* **Willow Bunch** Saskatchewan,
S Canada 49.30N 105.40W

34 *J11* **Willow Creek** ⚌ Oregon,
NW USA

41 *R11* **Willow Lake** Alaska, USA
61.44N 150.02W

15 *H7* **Willowlake** ⚌ Northwest
Territories, NW Canada

85 *H25* **Willowmore** Eastern Cape,
S South Africa 33.18S 23.30E

32 *L5* **Willow Reservoir** ⊠ Wisconsin,
N USA

36 *M5* **Willows** California, W USA
39.28N 122.12W

49 *V7* **Willow Springs** Missouri, C USA
36.59N 91.58W

190 *I7* **Wilmington** South Australia
32.42S 138.08E

23 *Y2* **Wilmington** Delaware, NE USA
39.45N 75.33W

33 *R14* **Wilmington** Ohio, N USA
39.27N 83.49W

22 *M6* **Wilmore** Kentucky, S USA
37.51N 84.39W

31 *R8* **Wilmot** South Dakota, N USA
45.24N 96.51W

39 *Q7* **Wilna/Wilno** *see* Vilnius

23 *R9* **Wilson** North Carolina, S USA
35.42N 77.54W

27 *N5* **Wilson** Texas, SW USA
33.21N 101.44W

190 *A7* **Wilson Bluff** *headland* South
Australia/Western Australia
31.41S 129.01E

37 *Y7* **Wilson Creek Range** ▲ Nevada,
W USA

25 *O1* **Wilson Lake** ⊠ Alabama, S USA

28 *M4* **Wilson Lake** ⊠ Kansas, C USA

39 *P7* **Wilson, Mount** ▲ Colorado,
C USA 37.50N 107.59W

191 *P13* **Wilsons Promontory** *peninsula*
Victoria, SE Australia

31 *V14* **Wilton** Iowa, C USA 41.35N 91.01W

21 *P7* **Wilton** Maine, NE USA

30 *M5* **Wilton** North Dakota, N USA

99 *L22* **Wiltshire** *cultural region* S England,
UK

101 *M23* **Wiltz** Diekirch, NW Luxembourg
49.58N 5.55E

188 *K9* **Wiluna** Western Australia
26.34S 120.14E

101 *M23* **Wilwerwiltz** Diekirch,
NE Luxembourg 49.59N 6.00E

31 *P5* **Wimbledon** North Dakota,
N USA 47.08N 98.25W

44 *K7* **Wina** *var.* Güina. Jinotega,
N Nicaragua 13.58N 85.14W

33 *O12* **Winamac** Indiana, N USA
41.03N 86.37W

83 *G19* **Winam Gulf** *var.* Kavirondo Gulf.
gulf SW Kenya

85 *I22* **Winburg** Free State, C South
Africa 28.31S 27.01E

21 *N10* **Winchendon** Massachusetts,
NE USA 42.41N 72.01W

12 *M13* **Winchester** Ontario, SE Canada
45.07N 75.19W

99 *M23* **Winchester** *hist.* Wintanceaster,
Lat. Venta Belgarum. S England,
UK 51.04N 1.19W

34 *M10* **Winchester** Idaho, NW USA
46.13N 116.35W

32 *J14* **Winchester** Illinois, N USA
39.37N 90.28W

33 *Q13* **Winchester** Indiana, N USA
40.09N 84.58W

22 *M5* **Winchester** Kentucky, S USA
37.59N 84.10W

21 *N10* **Winchester** New Hampshire,
NE USA 42.46N 72.21W

22 *K10* **Winchester** Tennessee, S USA
35.11N 86.06W

23 *V3* **Winchester** Virginia, NE USA
39.11N 78.10W

101 *L22* **Wincrange** Diekirch,
NW Luxembourg 50.03N 5.55E

8 *I5* **Wind** ⚌ Yukon Territory,
NW Canada

191 *S8* **Windamere, Lake** ⊠ New South
Wales, SE Australia

Windau *see* Ventspils, Latvia

Windau *see* Venta,
Latvia/Lithuania

20 *D15* **Windber** Pennsylvania, NE USA
40.12N 78.47W

25 *T3* **Winder** Georgia, SE USA
33.59N 83.43W

99 *K15* **Windermere** NW England, UK
54.24N 2.54W

12 *C7* **Windermere Lake** ⊗ Ontario,
S Canada

33 *U11* **Windham** Ohio, N USA
41.14N 81.03W

85 *D19* **Windhoek** Ger. Windhuk.
● (Namibia) Khomas, C Namibia
22.34S 17.06E

85 *D20* **Windhoek** ✈ Khomas, C Namibia
22.31S 17.04E

Windhuk *see* Windhoek

13 *O8* **Windigo** Quebec, SE Canada
47.45N 73.19W

13 *O8* **Windigo** ⚌ Quebec, SE Canada

111 *T6* **Windischfeistritz** *see* Slovenska
Bistrica

Windischgarsten
Oberösterreich, W Austria
47.42N 14.21E

Windischgraz *see* Slovenj Gradec

39 *T16* **Wind Mountain** ▲ New Mexico,
SW USA 32.01N 105.35W

31 *T10* **Windom** Minnesota, N USA
43.52N 95.07W

39 *Q7* **Windom Peak** ▲ Colorado,
C USA 37.37N 107.35W

189 *U9* **Windorah** Queensland,
C Australia 25.25S 142.40E

39 *O10* **Window Rock** Arizona, SW USA
35.40N 109.03W

33 *N9* **Wind Point** *headland* Wisconsin,
N USA 42.46N 87.46W

1 *P15* **Wind River** ⚌ Wyoming, C USA

13 *S9* **Windsor** Nova Scotia,
SE Canada 44.58N 64.13W

12 *C17* **Windsor** Ontario, S Canada
42.18N 83.00W

13 *Q12* **Windsor** Quebec, SE Canada
45.34N 72.00W

99 *N22* **Windsor** S England, UK
51.29N 0.39W

39 *T3* **Windsor** Colorado, C USA
40.28N 104.54W

20 *M12* **Windsor** Connecticut, NE USA
41.51N 72.38W

49 *T5* **Windsor** Missouri, C USA
38.31N 93.31W

23 *X9* **Windsor** North Carolina, SE USA
36.00N 76.57W

20 *M12* **Windsor Locks** Connecticut,
NE USA 41.55N 72.37W

27 *R5* **Windthorst** Texas, SW USA
33.34N 98.26W

47 *Z14* **Windward Islands** *island group*
E West Indies

Windward Islands *see* Vent, Îles
du, Archipel de la Société, French
Polynesia

Windward Islands *see*
Barlavento, Ilhas de, Cape Verde

46 *K8* **Windward Passage** *Sp.* Paso de
los Vientos. *channel* Cuba/Haiti

25 *O3* **Winfield** Alabama, S USA
33.55N 87.49W

31 *Y15* **Winfield** Iowa, C USA
41.07N 91.26W

49 *O7* **Winfield** Kansas, C USA
37.14N 97.00W

23 *Q4* **Winfield** West Virginia, NE USA
38.30N 81.54W

31 N5 **Wing** North Dakota, N USA *47.06N 100.16W*
191 U7 **Wingham** New South Wales, SE Australia *31.52S 152.24E*
10 G16 **Wingham** Ontario, S Canada *43.54N 81.19W*
35 T8 **Winifred** Montana, NW USA *47.33N 109.26W*
10 E8 **Winisk** ≈ Ontario, S Canada
10 E9 **Winisk Lake** ◎ Ontario, C Canada
26 L8 **Wink** Texas, SW USA *31.45N 103.09W*
38 M14 **Winkelman** Arizona, SW USA *32.59N 110.46W*
9 X17 **Winkler** Manitoba, S Canada *49.12N 97.55W*
111 Q9 **Winklern** Tirol, W Austria *46.54N 12.54E*
Winkowitz *see* Vinkovci
34 G9 **Winlock** Washington, NW USA *46.29N 122.56W*
79 P17 **Winneba** SE Ghana *5.22N 0.37W*
31 U11 **Winnebago** Minnesota, N USA *43.46N 94.10W*
31 R13 **Winnebago** Nebraska, C USA *42.14N 96.28W*
32 M7 **Winnebago, Lake** ◎ Wisconsin, N USA
32 M7 **Winneconne** Wisconsin, N USA *44.07N 88.44W*
37 T3 **Winnemucca** Nevada, W USA *40.58N 117.43W*
37 R4 **Winnemucca Lake** ◎ Nevada, W USA
103 H21 **Winnenden** Baden-Württemberg, SW Germany *48.52N 9.22E*
31 N11 **Winner** South Dakota, N USA *43.22N 99.51W*
35 U9 **Winnett** Montana, NW USA *47.00N 108.18W*
12 I9 **Winneway** Quebec, SE Canada *47.35N 78.33W*
24 H6 **Winnfield** Louisiana, S USA *31.55N 92.38W*
31 U4 **Winnibigoshish, Lake** ◎ Minnesota, N USA
27 X11 **Winnie** Texas, SW USA *29.49N 94.22W*
9 Y16 **Winnipeg** Manitoba, S Canada *49.52N 97.10W*
9 X16 **Winnipeg** ✈ Manitoba, S Canada *49.56N 97.16W*
2 J8 **Winnipeg** ≈ Manitoba, S Canada
9 X16 **Winnipeg Beach** Manitoba, S Canada *50.25N 96.59W*
9 W14 **Winnipeg, Lake** ◎ Manitoba, C Canada
9 W15 **Winnipegosis** Manitoba, S Canada *51.36N 99.59W*
9 W15 **Winnipegosis, Lake** ◎ Manitoba, C Canada
21 O8 **Winnipesaukee, Lake** ◎ New Hampshire, NE USA
24 I6 **Winnsboro** Louisiana, S USA *32.09N 91.43W*
23 R12 **Winnsboro** South Carolina, SE USA *34.22N 81.05W*
27 W6 **Winnsboro** Texas, SW USA *33.01N 95.16W*
31 X10 **Winona** Minnesota, N USA *44.03N 91.37W*
24 L4 **Winona** Mississippi, S USA *33.30N 89.42W*
49 W7 **Winona** Missouri, C USA *37.00N 91.19W*
27 W7 **Winona** Texas, SW USA *32.29N 95.10W*
20 M7 **Winooski River** ≈ Vermont, NE USA
100 P6 **Winschoten** Groningen, NE Netherlands *53.09N 7.03E*
102 J10 **Winsen** Niedersachsen, N Germany *53.22N 10.13E*
38 M11 **Winslow** Arizona, SW USA *35.01N 110.42W*
21 Q7 **Winslow** Maine, NE USA *44.33N 69.35W*
20 M12 **Winsted** Connecticut, NE USA *41.55N 73.03W*
34 F14 **Winston** Oregon, NW USA *43.07N 123.24W*
23 S9 **Winston Salem** North Carolina, SE USA *36.06N 80.14W*
100 N5 **Winsum** Groningen, NE Netherlands *53.19N 5.37E*
Wintanceaster *see* Winchester
25 W11 **Winter Garden** Florida, SE USA *28.34N 81.35W*
8 J16 **Winter Harbour** Vancouver Island, British Columbia, SW Canada *50.28N 128.03W*
25 W12 **Winter Haven** Florida, SE USA *28.01N 81.43W*
25 X11 **Winter Park** Florida, SE USA *28.36N 81.20W*
27 P8 **Winters** Texas, SW USA *31.57N 99.57W*
31 U15 **Winterset** Iowa, C USA *41.19N 94.00W*
100 O12 **Winterswijk** Gelderland, E Netherlands *51.58N 6.43E*
110 G6 **Winterthur** Zürich, NE Switzerland *47.30N 8.43E*
31 U9 **Winthrop** Minnesota, N USA *44.32N 94.22W*
34 J7 **Winthrop** Washington, NW USA *48.28N 120.13W*
189 V7 **Winton** Queensland, E Australia *22.22S 143.04E*
193 C24 **Winton** Southland, South Island, NZ *46.08S 168.19E*
23 X8 **Winton** North Carolina, SE USA *36.22N 76.56W*
103 K15 **Wipper** ≈ C Germany
103 K14 **Wipper** ≈ C Germany
Wipper *see* Wieprza
176 Vv9 **Wiriagar, Sungai** ≈ Irian Jaya, E Indonesia
190 G6 **Wirraminna** South Australia *31.10S 136.13E*
190 F4 **Wirrida** South Australia *29.34S 134.33E*
190 F7 **Wirrulla** South Australia *32.27S 134.33E*
Wirsitz *see* Wyrzysk
Wirz-See *see* Võrtsjärv
99 O19 **Wisbech** E England, UK *52.39N 0.08E*
Wisby *see* Visby
21 Q8 **Wiscasset** Maine, NE USA *44.01N 69.40W*
Wischau *see* Vyškov
32 J5 **Wisconsin** ***off.*** State of Wisconsin; also known as The Badger State. ◆ *state* N USA
32 L8 **Wisconsin Dells** Wisconsin, N USA *43.37N 89.43W*
32 L8 **Wisconsin, Lake** ▣ Wisconsin, N USA
32 L7 **Wisconsin Rapids** Wisconsin, N USA *44.24N 89.49W*
32 L7 **Wisconsin River** ≈ Wisconsin, N USA
35 P11 **Wisdom** Montana, NW USA *45.36N 113.27W*
23 P7 **Wise** Virginia, NE USA *36.58N 82.34W*
41 Q7 **Wiseman** Alaska, USA *67.24N 150.06W*
98 J12 **Wishaw** W Scotland, UK *55.46N 3.55W*
31 O6 **Wishek** North Dakota, N USA *46.12N 99.33W*
34 I11 **Wishram** Washington, NW USA *45.40N 120.53W*
113 J17 **Wisła** Śląskie, S Poland *49.39N 18.49E*
112 K11 **Wisła** ***Eng.*** Vistula, ***Ger.*** Weichsel. ≈ C Poland
Wiślany, Zalew *see* Vistula Lagoon
113 M16 **Wisłoka** ≈ SE Poland
102 L9 **Wismar** Mecklenburg-Vorpommern, N Germany *53.54N 11.28E*
31 R14 **Wisner** Nebraska, C USA *41.59N 96.54W*
105 V4 **Wissembourg** ***var.*** Weissenburg. Bas-Rhin, NE France *49.03N 7.57E*
32 J6 **Wissota, Lake** ▣ Wisconsin, N USA
99 O18 **Witham** ≈ E England, UK
99 O17 **Withernsea** E England, UK *53.45N 0.00W*
39 Q13 **Withington, Mount** ▲ New Mexico, SW USA *33.52N 107.29W*
25 U8 **Withlacoochee River** ≈ Florida/Georgia, SE USA
112 H11 **Witkowo** Wielkopolskie, C Poland *52.27N 17.49E*
99 M21 **Witney** S England, UK *51.47N 1.30W*
103 E15 **Witten** Nordrhein-Westfalen, W Germany *51.25N 7.19E*
103 N14 **Wittenberg** Sachsen-Anhalt, E Germany *51.52N 12.39E*
32 L6 **Wittenberg** Wisconsin, N USA *44.53N 89.20W*
102 L11 **Wittenberge** Brandenburg, N Germany *52.58N 11.45E*
105 U7 **Wittenheim** Haut-Rhin, NE France *47.49N 7.19E*
188 I7 **Wittenoom** Western Australia *22.17S 118.22E*
Wittingau *see* Třeboň
102 K12 **Wittingen** Niedersachsen, C Germany *52.42N 10.43E*
103 E18 **Wittlich** Rheinland-Pfalz, SW Germany *49.59N 6.54E*
102 F9 **Wittmund** Niedersachsen, NW Germany *53.34N 7.46E*
102 M10 **Wittstock** Brandenburg, NE Germany *53.09N 12.28E*
194 M11 **Witu Islands** *island group* E PNG
112 O7 **Wiżajny** Podlaskie, NE Poland *54.21N 22.51E*
57 W10 **W.J. van Blommesteinmeer** ▣ E Suriname
112 L11 **Wkra** ***Ger.*** Soldau. ≈ C Poland
112 I6 **Władysławowo** Pomorskie, N Poland *54.48N 18.25E*
Wlaschim *see* Vlašim
113 E14 **Wleń** ***Ger.*** Lähn. Dolnośląskie, SW Poland *51.00N 15.39E*
112 J11 **Włocławek** ***Ger./Rus.*** Vlotslavsk. Kujawskio-pomorskie, C Poland *52.39N 19.02E*
112 P13 **Włodawa** ***Rus.*** Vlodava. Lubelskie, SE Poland *51.33N 23.31E*
Włodzimierz *see* Volodymyr-Volyns'kyy
113 K15 **Włoszczowa** Świętokrzyskie, C Poland *50.51N 19.58E*
85 C19 **Wlotzkasbaken** Erongo, W Namibia *22.25S 14.30E*
152 L11 **Wobkent** ***Rus.*** Vabkent. Bukhoro Wiloyati, C Uzbekistan *40.01N 64.25E*
13 R12 **Woburn** Quebec, SE Canada *45.22N 70.52W*
21 O11 **Woburn** Massachusetts, NE USA *42.28N 71.09W*
Wocheiner Feistritz *see* Bohinjska Bistrica
Wöchma *see* Võhma
153 S11 **Wodil** ***var.*** Vuadil'. Farghona Wiloyati, E Uzbekistan *40.10N 71.43E*
189 V14 **Wodonga** Victoria, SE Australia *36.10S 146.55E*
113 I17 **Wodzisław Śląski** ***Ger.*** Loslau. Śląskie, S Poland *49.59N 18.27E*
100 I11 **Woerden** Zuid-Holland, C Netherlands *52.06N 4.54E*
100 I8 **Wognum** Noord-Holland, NW Netherlands *52.40N 5.01E*
Wohlau *see* Wołów
110 F7 **Wohlen** Aargau, NW Switzerland *47.21N 8.16E*
205 R2 **Wohlthat Mountains** ▲▲ Antarctica
176 W9 **Woinui, Selat** *strait* Irian Jaya, E Indonesia
194 K15 **Woitape** Central, S PNG *8.35S 147.15E*
Wōjjā *see* Wotje Atoll
Wojwodina *see* Vojvodina
176 W13 **Wokam, Pulau** *island* Kepulauan Aru, E Indonesia
99 N22 **Woking** SE England, UK *51.19N 0.34W*
Woldenberg Neumark *see* Dobiegniew
196 K15 **Woleai Atoll** *atoll* Caroline Islands, W Micronesia
Woleu *see* Uolo, Río
81 E17 **Woleu-Ntem** ***off.*** Province du Woleu-Ntem, ***var.*** Le Woleu-Ntem. ◆ *province* W Gabon
34 F15 **Wolf Creek** Oregon, NW USA *42.40N 123.22W*
28 K9 **Wolf Creek** ≈ Oklahoma/Texas, SW USA
39 R7 **Wolf Creek Pass** *pass* Colorado, C USA *37.28N 106.48W*
21 O9 **Wolfeboro** New Hampshire, NE USA *43.34N 71.10W*
27 U5 **Wolfe City** Texas, SW USA *33.22N 96.04W*
12 L15 **Wolfe Island** *island* Ontario, SE Canada
103 M14 **Wolfen** Sachsen-Anhalt, E Germany *51.40N 12.16E*
102 J13 **Wolfenbüttel** Niedersachsen, C Germany *52.10N 10.31E*
111 T4 **Wolfern** Oberösterreich, N Austria *48.06N 14.16E*
111 Q6 **Wolfgangsee** ***var.*** Abersee, St Wolfgangsee. ◎ N Austria
41 P9 **Wolf Mountain** ▲ Alaska, USA *65.20N 154.08W*
35 X7 **Wolf Point** Montana, NW USA *48.04N 105.40W*
24 L8 **Wolf River** ≈ Mississippi, S USA
32 M7 **Wolf River** ≈ Wisconsin, N USA
111 U9 **Wolfsberg** Kärnten, SE Austria *46.49N 14.49E*
102 K12 **Wolfsburg** Niedersachsen, N Germany *52.25N 10.46E*
59 B17 **Wolf, Volcán** ⛰ Galapagos Islands, Ecuador, E Pacific Ocean *0.01N 91.22W*
102 O8 **Wolgast** Mecklenburg-Vorpommern, NE Germany *54.03N 13.47E*
110 F8 **Wolhusen** Luzern, W Switzerland *47.04N 8.06E*
112 D8 **Wolin** ***Ger.*** Wollin. Zachodniopomorskie, NW Poland *53.52N 14.34E*
111 Y3 **Wolkersdorf** Niederösterreich, NE Austria *48.24N 16.31E*
Wołkowysk *see* Vawkavysk
Wöllan *see* Velenje
15 I2 **Wollaston, Cape** *headland* Victoria Island, Northwest Territories, NW Canada *71.00N 118.21W*
65 J25 **Wollaston, Isla** *island* S Chile
9 U11 **Wollaston Lake** Saskatchewan, C Canada *58.04N 103.37W*
9 T10 **Wollaston Lake** ◎ Saskatchewan, C Canada
15 I3 **Wollaston Peninsula** *peninsula* Victoria Island, Northwest Territories/Nunavut, NW Canada
Wollin *see* Wolin
191 S9 **Wollongong** New South Wales, SE Australia *34.25S 150.52E*
Wolmar *see* Valmiera
102 L13 **Wolmirstedt** Sachsen-Anhalt, C Germany *52.15N 11.37E*
112 M11 **Wołomin** Mazowieckie, C Poland *52.21N 21.15E*
112 G3 **Wołów** ***Ger.*** Wohlau. Dolnośląskie, SW Poland *51.21N 16.39E*
Wołożyn *see* Valozhyn
12 G11 **Wolseley Bay** Ontario, S Canada *46.05N 80.16W*
31 P10 **Wolsey** South Dakota, N USA *44.22N 98.28W*
112 F12 **Wolsztyn** Wielkopolskie, W Poland *52.06N 16.06E*
100 M7 **Wolvega** ***Fris.*** Wolvegea. Friesland, N Netherlands *52.53N 6.00E*
Wolvegea *see* Wolvega
99 K19 **Wolverhampton** C England, UK *52.36N 2.07W*
Wolverine State *see* Michigan
101 G18 **Wolvertem** Vlaams Brabant, C Belgium *50.55N 4.19E*
101 H16 **Wommelgem** Antwerpen, N Belgium *51.12N 4.31E*
176 W11 **Wondiwoi, Pegunungan** ▲▲ Irian Jaya, E Indonesia
194 J13 **Wonenara** ***var.*** Wonerara. Eastern Highlands, C PNG *6.46S 145.54E*
Wonerara *see* Wonenara
Wongalara Lake *see* Wongalarroo Lake
191 N6 **Wongalarroo Lake** ***var.*** Wongalara Lake. *seasonal lake* New South Wales, SE Australia
169 Y15 **Wŏnju** ***Jap.*** Genshū. N South Korea *37.21N 127.57E*
8 M12 **Wonowon** British Columbia, W Canada *56.46N 121.54W*
169 X13 **Wŏnsan** SE North Korea *39.11N 127.21E*
191 O13 **Wonthaggi** Victoria, SE Australia *38.37S 145.39E*
25 N2 **Woodall Mountain** ▲ Mississippi, S USA *34.47N 88.14W*
25 W7 **Woodbine** Georgia, SE USA *30.58N 81.43W*
31 S14 **Woodbine** Iowa, C USA *41.44N 95.42W*
20 J17 **Woodbine** New Jersey, NE USA *39.12N 74.47W*
23 W4 **Woodbridge** Virginia, NE USA *38.39N 77.14W*
191 V4 **Woodburn** New South Wales, SE Australia *29.07S 153.23E*
34 G11 **Woodburn** Oregon, NW USA *45.08N 122.51W*
22 K9 **Woodbury** Tennessee, S USA *35.49N 86.04W*
191 V5 **Wooded Bluff** *headland* New South Wales, SE Australia *29.24S 153.22E*
191 V3 **Woodenbong** New South Wales, SE Australia *28.24S 152.39E*
37 R11 **Woodlake** California, W USA *36.24N 119.06W*
37 N7 **Woodland** California, W USA *38.39N 121.46W*
21 T5 **Woodland** Maine, NE USA *45.10N 67.25W*
34 G10 **Woodland** Washington, NW USA *45.54N 122.44W*
39 T5 **Woodland Park** Colorado, C USA *38.59N 105.03W*
195 P15 **Woodlark Island** ***var.*** Murua Island. *island* SE PNG
Woodle Island *see* Kuria
9 T17 **Wood Mountain** ▲▲ Saskatchewan, S Canada
32 K15 **Wood River** Illinois, N USA *38.51N 90.06W*
31 P16 **Wood River** Nebraska, C USA *40.48N 98.33W*
41 R9 **Wood River** ≈ Alaska, USA
41 O13 **Wood River Lakes** *lakes* Alaska, USA
190 C1 **Woodroffe, Mount** ▲ South Australia *26.19S 131.42E*
23 P11 **Woodruff** South Carolina, SE USA *34.44N 82.02W*
32 K4 **Woodruff** Wisconsin, N USA *45.55N 89.41W*
27 T14 **Woodsboro** Texas, SW USA *28.14N 97.19W*
33 U13 **Woodsfield** Ohio, N USA *39.45N 81.07W*
189 P4 **Woods, Lake** ◎ Northern Territory, N Australia
9 Z16 **Woods, Lake of the** ***Fr.*** Lac des Bois. ◎ Canada/USA
27 Q6 **Woodson** Texas, SW USA *33.00N 99.01W*
11 N14 **Woodstock** New Brunswick, SE Canada *46.10N 67.37W*
12 F16 **Woodstock** Ontario, S Canada *43.07N 80.46W*
32 M10 **Woodstock** Illinois, N USA *42.18N 88.27W*
20 M9 **Woodstock** Vermont, NE USA *43.37N 72.33W*
23 U4 **Woodstock** Virginia, NE USA *38.51N 78.28W*
21 N8 **Woodsville** New Hampshire, NE USA *44.07N 72.01W*
192 M12 **Woodville** Manawatu-Wanganui, North Island, NZ *40.21S 175.58E*
24 J7 **Woodville** Mississippi, S USA *31.06N 91.18W*
27 X9 **Woodville** Texas, SW USA *30.46N 94.25W*
28 K9 **Woodward** Oklahoma, C USA *36.25N 99.23W*
31 O5 **Woodworth** North Dakota, N USA *47.06N 99.19W*
176 Y12 **Woogi** Irian Jaya, E Indonesia *3.59S 138.45E*
176 Ww9 **Wool** Irian Jaya, E Indonesia *1.38S 135.34E*
191 V5 **Woolgoolga** New South Wales, E Australia *30.04S 153.09E*
190 H6 **Woomera** South Australia *31.12S 136.52E*
21 O12 **Woonsocket** Rhode Island, NE USA *41.58N 71.27W*
31 P10 **Woonsocket** South Dakota, N USA *44.03N 98.16W*
33 T12 **Wooster** Ohio, N USA *40.48N 81.56W*
82 L12 **Woqooyi Galbeed** ***off.*** Gobolka Woqooyi Galbeed. ◆ *region* NW Somalia
110 E8 **Worb** Bern, C Switzerland *46.54N 7.36E*
85 F26 **Worcester** Western Cape, SW South Africa *33.40S 19.22E*
99 L20 **Worcester** ***hist.*** Wigorna Ceaster. W England, UK *52.10N 2.13W*
21 N11 **Worcester** Massachusetts, NE USA *42.17N 71.48W*
99 L20 **Worcestershire** *cultural region* C England, UK
34 H16 **Worden** Oregon, NW USA *42.04N 121.50W*
111 O6 **Wörgl** Tirol, W Austria *47.28N 12.04E*
176 Ww14 **Workai, Pulau** *island* Kepulauan Aru, E Indonesia
99 J15 **Workington** NW England, UK *54.39N 3.33W*
100 K7 **Workum** Friesland, N Netherlands *52.58N 5.25E*
35 V13 **Worland** Wyoming, C USA *44.01N 107.57W*
Wormatia *see* Worms
101 N25 **Wormeldange** Grevenmacher, E Luxembourg *49.37N 6.25E*
100 I9 **Wormer** Noord-Holland, C Netherlands *52.30N 4.49E*
103 G19 **Worms** ***anc.*** Augusta Vangionum, Borbetomagus, Wormatia. Rheinland-Pfalz, SW Germany *49.37N 8.22E*
Worms *see* Vormsi
103 K21 **Wörnitz** ≈ S Germany
103 G21 **Wörth** Rheinland-Pfalz, SW Germany *49.04N 8.16E*
27 U8 **Wortham** Texas, SW USA *31.47N 96.27W*
111 S9 **Worther See** ◎ S Austria
99 O23 **Worthing** SE England, UK *50.48N 0.22W*
31 S11 **Worthington** Minnesota, N USA *43.37N 95.36W*
33 S13 **Worthington** Ohio, N USA *40.05N 83.01W*
37 W8 **Worthington Peak** ▲ Nevada, W USA *37.57N 115.32W*
176 Y12 **Wosi** Irian Jaya, E Indonesia *3.55S 138.54E*
176 W11 **Wosimi** Irian Jaya, E Indonesia *2.44S 134.34E*
201 R5 **Wotho Atoll** ***var.*** Wōtto. *atoll* Ralik Chain, W Marshall Islands
201 V5 **Wotje Atoll** ***var.*** Wōjjā. *atoll* Ratak Chain, E Marshall Islands
Wotoe *see* Wotu
Wottawa *see* Otava
Wōtto *see* Wotho Atoll
175 Pp10 **Wotu** ***prev.*** Wotoe. Sulawesi, C Indonesia *2.34S 120.46E*
100 K11 **Woudenberg** Utrecht, C Netherlands *52.04N 5.25E*
100 I13 **Woudrichem** Noord-Brabant, S Netherlands *51.49N 5.00E*
45 N8 **Wounta** ***var.*** Huaunta. Región Autónoma Atlántico Norte, NE Nicaragua *13.33N 83.31W*
175 R12 **Wowoni, Pulau** *island* C Indonesia
175 Qq12 **Wowoni, Selat** *strait* Sulawesi, C Indonesia
83 J17 **Woyamdero Plain** *plain* E Kenya
Woyens *see* Vojens
Wozrojdeniye Oroli *see* Vozrozhdeniya, Ostrov
Wrangel Island *see* Vrangelya, Ostrov
41 Y13 **Wrangell** Wrangell Island, Alaska, USA *56.28N 132.22W*
40 C15 **Wrangell, Cape** *headland* Attu Island, Alaska, USA *52.55N 172.28E*
41 S11 **Wrangell, Mount** ▲ Alaska, USA *62.00N 144.01W*
41 T11 **Wrangell Mountains** ▲▲ Alaska, USA
207 S7 **Wrangel Plain** *undersea feature* Arctic Ocean
98 H6 **Wrath, Cape** *headland* N Scotland, UK *58.37N 5.01W*
39 W3 **Wray** Colorado, C USA *40.01N 102.12W*
46 K13 **Wreck Point** *headland* C Jamaica *17.50N 76.55W*
85 C23 **Wreck Point** *headland* W South Africa *28.52S 16.17E*
25 V4 **Wrens** Georgia, SE USA *33.12N 82.23W*
99 K18 **Wrexham** NE Wales, UK *53.03N 3.00W*
49 R13 **Wright City** Oklahoma, C USA *34.03N 95.00W*
204 J12 **Wright Island** *island* Antarctica
11 N9 **Wright, Mont** ▲ Quebec, E Canada *52.36N 67.40W*
27 X5 **Wright Patman Lake** ▣ Texas, SW USA
38 M16 **Wrightson, Mount** ▲ Arizona, SW USA *31.42N 110.51W*
25 U5 **Wrightsville** Georgia, SE USA *32.43N 82.43W*
23 W12 **Wrightsville Beach** North Carolina, SE USA *34.12N 77.48W*
37 T15 **Wrightwood** California, W USA *34.21N 117.37W*
15 Gg7 **Wrigley** Northwest Territories, W Canada *63.16N 123.39W*
113 G14 **Wrocław** ***Eng./Ger.*** Breslau. Dolnośląskie, SW Poland *51.06N 17.01E*
112 F10 **Wronki** ***Ger.*** Fronicken. Wielkopolskie, NW Poland *52.42N 16.21E*
112 H11 **Września** Wielkolpolskie, C Poland *52.19N 17.33E*
112 F12 **Wschowa** Lubuskie, W Poland *51.48N 16.18E*
Wsetin *see* Vsetín
188 I12 **Wubin** Western Australia *30.05S 116.43E*
169 W9 **Wuchang** Heilongjiang, NE China *44.55N 127.13E*
Wuchang *see* Wuhan
Wu-chou/Wuchow *see* Wuzhou
166 M16 **Wuchuan** ***var.*** Meilu. Guangdong, S China *21.30N 110.40E*
166 K10 **Wuchuan** ***prev.*** Duru. Guizhou, S China *28.40N 108.04E*
169 O13 **Wuchuan** Nei Mongol Zizhiqu, N China *41.04N 111.28E*
169 V6 **Wudalianchi** Heilongjiang, NE China *48.40N 126.06E*
165 O11 **Wudaoliang** Qinghai, C China *35.16N 93.03E*
147 Q13 **Wuday'ah** *spring/well* S Saudi Arabia *17.03N 47.06E*
79 V13 **Wudil** Kano, N Nigeria *11.46N 8.49E*
166 G12 **Wuding** Yunnan, SW China *25.31N 102.24E*
166 L4 **Wuding He** ≈ C China
190 G8 **Wudinna** South Australia *33.06S 135.30E*
163 P10 **Wudu** Gansu, C China *33.22N 105.01E*
166 L9 **Wufeng** Hubei, C China *30.09N 110.31E*
167 O11 **Wugong Shan** ▲▲ S China
163 P7 **Wuhai** Nei Mongol Zizhiqu, N China *39.40N 106.48E*
167 O9 **Wuhan** ***var.*** Han-kou, Han-k'ou, Hanyang, Wuchang, Wu-han; ***prev.*** Hankow. Hubei, C China *30.34N 114.19E*
167 Q7 **Wuhe** Anhui, E China *33.10N 117.50E*
Wuhsi/Wu-hsi *see* Wuxi
Wuhsien *see* Suzhou
167 Q8 **Wuhu** ***var.*** Wu-na-mu. Anhui, E China *31.22N 118.25E*
Wūjae *see* Ujae Atoll
166 K11 **Wu Jiang** ≈ C China
Wujlān *see* Ujelang Atoll
79 W15 **Wukari** Taraba, E Nigeria *7.51N 9.49E*
166 H11 **Wulian Feng** ▲▲ SW China
166 F13 **Wuliang Shan** ▲▲ SW China
176 U15 **Wuliaru, Pulau** *island* Kepulauan Tanimbar, E Indonesia
166 K11 **Wuling Shan** ▲▲ S China
111 Y5 **Wulka** ≈ E Austria
Wulkan *see* Vulcan
111 T3 **Wullowitz** Oberösterreich, N Austria *48.37N 14.27E*
Wu-lu-k'o-mu-shi/Wu-lu-mu-ch'i *see* Ürümqi
81 D14 **Wum** Nord-Ouest, NE Cameroon *6.24N 10.04E*
166 H12 **Wumeng Shan** ▲▲ SW China
166 K14 **Wuming** Guangxi Zhuangzu Zizhiqu, S China *22.55N 108.16E*
102 I10 **Wümme** ≈ NW Germany
Wu-na-mu *see* Wuhu
176 Y11 **Wunen** Irian Jaya, E Indonesia *3.40S 138.31E*
10 D9 **Wunnummin Lake** ◎ Ontario, C Canada
82 D13 **Wun Rog** Warab, S Sudan *09.00N 28.20E*
103 M18 **Wunsiedel** Bayern, E Germany *50.02N 12.00E*
102 I12 **Wunstorf** Niedersachsen, NW Germany *52.25N 9.25E*
177 G3 **Wuntho** Sagaing, N Myanmar *23.52N 95.43E*
103 F15 **Wupper** ≈ W Germany
103 E15 **Wuppertal** ***prev.*** Barmen-Elberfeld. Nordrhein-Westfalen, W Germany *51.16N 7.12E*
166 K5 **Wuqi** Shaanxi, C China *36.55N 108.13E*
167 P4 **Wuqiao** ***var.*** Sangyuan. Hebei, E China *37.40N 116.21E*
103 L23 **Würm** ≈ SE Germany
79 T12 **Wurno** Sokoto, NW Nigeria *13.15N 5.24E*
103 I19 **Würzburg** Bayern, SW Germany *49.48N 9.55E*
103 N15 **Wurzen** Sachsen, E Germany *51.21N 12.48E*
166 L9 **Wu Shan** ▲▲ C China
164 G7 **Wushi** ***var.*** Uqturpan. Xinjiang Uygur Zizhiqu, NW China *41.07N 79.09E*
Wusih *see* Wuxi
67 N18 **Wüst Seamount** *undersea feature* S Atlantic Ocean *32.00S 0.06E*
Wusuli Jiang/Wusuri *see* Ussuri
167 N3 **Wutai Shan** ▲ C China *39.00N 114.00E*
166 H10 **Wutongqiao** Sichuan, C China *29.24N 103.54E*
165 P6 **Wutongwozi Quan** *spring* NW China *42.30N 95.21E*
194 E9 **Wutung** Sandaun, NW PNG *2.39S 141.01E*
101 H15 **Wuustwezel** Antwerpen, N Belgium *51.24N 4.34E*
194 G8 **Wuvulu Island** *island* NW PNG
165 U9 **Wuwei** ***var.*** Liangzhou. Gansu, C China *38.02N 102.30E*
167 R8 **Wuxi** ***var.*** Wuhsi, Wu-hsi, Wusih. Jiangsu, E China *31.34N 120.19E*
Wuxing *see* Huzhou
166 L14 **Wuxuan** Guangxi Zhuangzu Zizhiqu, S China *23.40N 109.41E*
166 K11 **Wuyang He** ≈ S China
169 X6 **Wuyiling** Heilongjiang, NE China *48.36N 129.24E*
163 T12 **Wuyi Shan** ▲▲ SE China
167 Q11 **Wuyishan** ***prev.*** Chong'an. Fujian, SE China *27.48N 118.03E*
168 M13 **Wuyuan** Nei Mongol Zizhiqu, N China *41.04N 108.15E*
166 L17 **Wuzhi Shan** ▲ S China *18.52N 109.36E*
165 W8 **Wuzhong** Ningxia, N China *37.58N 106.09E*
166 M14 **Wuzhou** ***var.*** Wu-chou, Wuchow. Guangxi Zhuangzu Zizhiqu, S China *23.30N 111.19E*
20 H12 **Wyalusing** Pennsylvania, NE USA *41.40N 76.13W*
190 M10 **Wycheproof** Victoria, SE Australia *36.06S 143.13E*
99 K21 **Wye** ***Wel.*** Gwy. ≈ England/Wales, UK
Wyłkowyszki *see* Vilkaviškis
99 P19 **Wymondham** E England, UK *52.29N 1.10E*
31 R17 **Wymore** Nebraska, C USA *40.07N 96.39W*
190 E5 **Wynbring** South Australia *30.34S 133.27E*
189 N3 **Wyndham** Western Australia *15.28S 128.07E*
31 R6 **Wyndmere** North Dakota, N USA *46.16N 97.07W*
49 X11 **Wynne** Arkansas, C USA *35.13N 90.47W*
49 N12 **Wynnewood** Oklahoma, C USA *34.39N 97.09W*
191 O15 **Wynyard** Tasmania, SE Australia *40.57S 145.33E*
9 U15 **Wynyard** Saskatchewan, S Canada *51.46N 104.10W*
35 V11 **Wyola** Montana, NW USA *45.07N 107.23W*
190 A4 **Wyola Lake** *salt lake* South Australia
33 P9 **Wyoming** Michigan, N USA *42.54N 85.42W*
35 V14 **Wyoming** ***off.*** State of Wyoming; also known as The Equality State. ◆ *state* C USA
35 S15 **Wyoming Range** ▲▲ Wyoming, C USA
191 T8 **Wyong** New South Wales, SE Australia *33.18S 151.27E*
112 G9 **Wyrzysk** ***Ger.*** Wirsitz. Wielkopolskie, C Poland *53.09N 17.15E*
Wysg *see* Usk
112 O10 **Wysokie Mazowieckie** Łomża, E Poland *52.54N 22.34E*
112 M11 **Wyszków** ***Ger.*** Probstberg. Mazowieckie, C Poland *52.36N 21.27E*
112 L11 **Wyszogród** Mazowieckie, C Poland *52.24N 20.14E*
23 R7 **Wytheville** Virginia, NE USA *36.57N 81.05W*

X

82 Q12 **Xaafuun** ***It.*** Hafun. Bari, NE Somalia *10.25N 51.17E*
82 Q12 **Xaafuun, Raas** ***var.*** Ras Hafun. *headland* NE Somalia *10.36N 51.09E*
Xábia *see* Jávea
44 C4 **Xaclbal, Río** ***var.*** Xalbal. ≈ Guatemala/Mexico
143 Y10 **Xaçmaz** ***Rus.*** Khachmas. N Azerbaijan *41.26N 48.46E*
82 O12 **Xadeed** ***var.*** Haded. *physical region* N Somalia
165 O14 **Xagquka** Xizang Zizhiqu, W China *31.46N 92.46E*
178 I6 **Xai** ***var.*** Muang Xay, Muong Sai. Oudômxai, N Laos *20.41N 102.00E*
164 F10 **Xaidulla** Xinjiang Uygur Zizhiqu, W China *36.27N 77.46E*
178 I7 **Xaignabouli** ***prev.*** Muang Xaignabouri, ***Fr.*** Sayaboury. Xaignabouli, N Laos *19.16N 101.43E*
178 J7 **Xai Lai Leng, Phou** ▲ Laos/Vietnam *19.13N 104.09E*
164 L15 **Xainza** Xizang Zizhiqu, W China *30.54N 88.36E*
164 L16 **Xaitongmoin** Xizang Zizhiqu, W China *29.27N 88.13E*
85 M20 **Xai-Xai** ***prev.*** João Belo, Vila de João ***Bel.*** Gaza, S Mozambique *25.00S 33.37E*
Xalbal *see* Xaclbal, Río
82 P13 **Xalin** Nugaal, N Somalia *9.16N 49.00E*
178 J6 **Xam Nua** ***var.*** Sam Neua. Houaphan, N Laos *20.24N 104.03E*
84 D11 **Xá-Muteba** ***Port.*** Cinco de Outubro. Lunda Norte, NE Angola *9.31S 17.46E*
85 C16 **Xangongo** ***Port.*** Rocadas. Cunene, SW Angola *16.41S 14.58E*
143 W12 **Xankändi** ***Rus.*** Khankendi; ***prev.*** Stepanakert. SW Azerbaijan *39.50N 46.44E*
143 V11 **Xanlar** ***Rus.*** Khanlar. NW Azerbaijan *40.37N 46.18E*
116 J13 **Xánthi** Anatolikí Makedonía kai Thráki, NE Greece *41.09N 24.54E*
62 H13 **Xanxerê** Santa Catarina, S Brazil *26.52S 52.25W*
83 O15 **Xarardheere** Mudug, E Somalia *4.45N 47.54E*
133 W8 **Xar Moron** ≈ NE China
Xarra *see* Xarrë
115 L23 **Xarrë** ***var.*** Xarra. Vlorë, S Albania *39.45N 20.01E*
84 D12 **Xassengue** Lunda Sul, NW Angola *10.28S 18.32E*
107 S11 **Xátiva** ***var.*** Jativa; ***anc.*** Setabis. País Valenciano, E Spain *39.00N 0.32W*
Xauen *see* Chefchaouen
62 K10 **Xavantes, Represa de** ***var.*** Represa de Chavantes. ▣ S Brazil
164 I7 **Xayar** Xinjiang Uygur Zizhiqu, W China *41.16N 82.52E*
Xäzär Dänizi *see* Caspian Sea
178 Jj9 **Xé Bangfai** ≈ C Laos
178 Jj10 **Xé Banghiang** ***var.*** Bang Hieng. ≈ S Laos
Xêgar *see* Tingri
33 R14 **Xenia** Ohio, N USA *39.40N 83.55W*
Xeres *see* Jeréz de la Frontera
117 E15 **Xeriás** ≈ C Greece
117 G17 **Xeró** ▲ Évvoia, C Greece *38.52N 23.18E*
85 H18 **Xhumo** Central, C Botswana *21.13S 24.37E*
167 N15 **Xiachuan Dao** *island* S China
Xiacun *see* Rushan
Xiaguan *see* Dali
165 U11 **Xiahe** ***var.*** Labrang. Gansu, C China *35.12N 102.28E*
167 Q13 **Xiamen** ***var.*** Hsia-men; ***prev.*** Amoy. Fujian, SE China *24.28N 118.04E*
166 L6 **Xi'an** ***var.*** Changan, Sian, Signan, Siking, Singan, Xian. Shaanxi, C China *34.16N 108.54E*
166 L10 **Xianfeng** Hubei, C China *29.39N 109.07E*
Xiang *see* Hunan
167 N7 **Xiangcheng** Henan, C China *33.52N 113.29E*
166 F10 **Xiangcheng** ***prev.*** Qagchêng. Sichuan, C China *28.52N 99.45E*
166 M8 **Xiangfan** ***var.*** Xiangyang. Hubei, C China *32.07N 112.00E*
167 N10 **Xiang Jiang** ≈ S China
Xiangkhoang *see* Pèk
178 Ii7 **Xiangkhoang, Plateau de** ***var.*** Plain of Jars. *plateau* N Laos
167 N11 **Xiangtan** ***var.*** Hsiang-t'an, Siangtan. Hunan, S China *27.52N 112.54E*
167 N11 **Xiangxiang** Hunan, S China *27.50N 112.31E*
Xiangyang *see* Xiangfan
167 S10 **Xianju** Zhejiang, SE China *28.53N 120.41E*
166 F8 **Xianshui He** ≈ C China
167 N9 **Xiantao** ***var.*** Mianyang. Hubei, C China *30.19N 113.31E*
167 R10 **Xianxia Ling** ▲▲ SE China
166 K6 **Xianyang** Shaanxi, C China *34.23N 118.40E*
164 L5 **Xiaocaohu** Xinjiang Uygur Zizhiqu, W China *45.43N 90.07E*
169 W6 **Xiao Hinggan Ling** ***Eng.*** Lesser Khingan Range. ▲▲ NE China
166 M6 **Xiao Shan** ▲▲ C China
166 M12 **Xiao Shui** ≈ S China
167 P6 **Xiaoxian** ***var.*** Xiao Xian. Anhui, E China *34.12N 116.55E*
166 G11 **Xichang** Sichuan, C China *27.52N 102.16E*
43 P11 **Xicoténcatl** Tamaulipas, C Mexico *22.59N 98.54W*
Xieng Khouang *see* Pèk
Xieng Ngeun *see* Muong Xiang Ngeun
165 X10 **Xifeng** Gansu, C China *35.46N 107.35E*
166 J11 **Xifeng** Guizhou, S China *27.15N 106.44E*
Xigang *see* Helan
164 L16 **Xigazê** ***var.*** Jih-k'a-tse, Shigatse, Xigaze. Xizang Zizhiqu, W China *29.18N 88.49E*
166 I8 **Xi He** ≈ C China
165 W11 **Xihe** Gansu, C China *34.00N 105.24E*
Xihuachi *see* Heshui
165 Q7 **Xijan Quan** *spring* NW China *40.52N 96.31E*
165 W10 **Xiji** Ningxia, N China *36.02N 105.33E*
166 M14 **Xi Jiang** ***var.*** Hsi Chiang, ***Eng.*** West River. ≈ S China
166 K15 **Xijin Shuiku** ▣ S China
Xilaganí *see* Xylaganí
166 I13 **Xilin** ***prev.*** Bada. Guangxi Zhuangzu Zizhiqu, S China *24.30N 105.00E*
169 Q10 **Xilinhot** ***var.*** Silinhot. Nei Mongol Zizhiqu, N China *43.58N 116.06E*
Xilokastro *see* Xylókastro
Xin *see* Xinjiang Uygur Zizhiqu
167 R10 **Xin'anjiang Shuiku** ▣ SE China
Xin'anzhen *see* Xinyi
169 Q7 **Xin Barag Youqi** ***var.*** Altan Emel. Nei Mongol Zizhiqu, N China *48.37N 116.40E*
169 R7 **Xin Barag Zuoqi** ***var.*** Amgalang. Nei Mongol Zizhiqu, N China *48.12N 118.15E*
169 W12 **Xinbin** Liaoning, NE China *41.39N 125.04E*
167 O7 **Xincai** Henan, C China *32.46N 114.54E*
165 V8 **Xincheng** ***var.*** Yinchuanzhan. Ningxia, N China *38.27N 106.04E*
167 O13 **Xinfeng** Jiangxi, S China *25.30N 114.52E*
167 O14 **Xinfengjiang Shuiku** ▣ S China
169 T13 **Xingcheng** Liaoning, NE China *40.38N 120.47E*
84 E11 **Xinge** Lunda Norte, NE Angola *9.44S 19.10E*
167 P12 **Xingguo** Jiangxi, S China *26.25N 115.22E*
165 S11 **Xinghai** Qinghai, C China *35.12N 102.28E*

◆ Country
● Country Capital
◇ Dependent Territory
○ Dependent Territory Capital
◆ Administrative Region
✈ International Airport
▲ Mountain
▲▲ Mountain Range
⛰ Volcano
≈ River
◎ Lake
▣ Reservoir

167 R7 **Xinghua** Jiangsu, E China
32.54N 119.48E
Xingkai Hu see Khanka, Lake

167 P13 **Xingning** Guangdong, S China
24.13N 115.38E

166 I13 **Xingren** Guizhou, S China
25.25N 105.07E

167 O4 **Xingtai** Hebei, E China
37.07N 114.28E

61 J14 **Xingu, Rio** ~ C Brazil

165 P6 **Xingxingxia** Xinjiang Uygur
Zizhiqu, NW China 41.48N 95.01E

166 I13 **Xingyi** Guizhou, S China
25.04N 104.51E

164 I6 **Xinhe** var. Toksu. Xinjiang Uygur
Zizhiqu, NW China 41.34N 82.30E
Xin Hot see Abag Qi

165 T10 **Xining** var. Hsining, Hsi-ning,
Sining. province capital Qinghai,
C China 36.37N 101.46E

167 O4 **Xinji** prev. Shulu. Hebei, E China
37.55N 115.14E

167 P10 **Xinjian** Jiangxi, S China
28.42N 115.43E
Xinjiang see Xinjiang Uygur
Zizhiqu

168 D8 **Xinjiang Uygur Zizhiqu** var.
Sinkiang, Sinkiang Uighur
Autonomous Region, Xin,
Xinjiang. ◆ autonomous region
NW China

166 H9 **Xinjin** Sichuan, C China
30.24N 103.48E
Xinjin see Pulandian

169 U12 **Xinmin** Liaoning, NE China
41.58N 122.51E

166 M12 **Xinning** Hunan, S China
26.34N 110.57E
Xinpu see Lianyungang

167 P5 **Xinwen** prev. Suncun. Shandong,
E China 35.49N 117.36E
Xin Xian see Xinzhou

167 N6 **Xinxiang** Henan, C China
35.13N 113.48E

167 O8 **Xinyang** var. Hsin-yang, Sinyang.
Henan, C China 32.09N 114.04E

167 Q6 **Xinyi** var. Xin'anzhen. Jiangsu,
E China 34.25N 118.19E

167 Q6 **Xinyi He** ~ E China

167 O11 **Xinyu** Jiangxi, S China
27.51N 115.00E

164 I5 **Xinyuan** var. Künes. Xinjiang
Uygur Zizhiqu, NW China
43.25N 83.12E
Xinyuan see Tianjun

168 M14 **Xinzhao Shan** ▲ N China
39.37N 107.51E

167 N3 **Xinzhou** var. Xin Xian. Shanxi,
C China 38.24N 112.43E

106 H4 **Xinzo de Limia** Galicia,
NW Spain 42.04N 7.45W
Xions see Książ Wielkopolski

167 O7 **Xiping** Henan, C China
33.22N 114.00E

165 T11 **Xiqing Shan** ▲ C China

61 N16 **Xique-Xique** Bahia, E Brazil
10.46S 42.43W

117 E14 **Xirovoúni** ▲ N Greece
40.31N 21.58E
Xishanzui see Urad Qianqi

166 J11 **Xishui** Guizhou, S China
28.24N 106.09E

167 O9 **Xishui** Hubei, C China
30.29N 115.13E

169 R10 **Xi Ujimqin Qi** var. Bayan Ul Hot.
Nei Mongol Zizhiqu, N China
44.31N 117.36E

166 K11 **Xiushan** Sichuan, C China
28.23N 108.52E

167 O10 **Xiu Shui** ~ S China

164 J16 **Xixabangma Feng** ▲ W China
28.25N 85.47E

166 M7 **Xixia** Henan, C China
33.19N 111.25E
Xixón see Gijona
Xixona see Jijona
Xizang see Xizang Zizhiqu
Xizang Gaoyuan see Qingzang
Gaoyuan

166 E9 **Xizang Zizhiqu** var. Thibet,
Tibetan Autonomous Region,
Xizang, Eng. Tibet. ◆ autonomous
region W China

169 U14 **Xizhong Dao** island N China
Xolotlán see Managua, Lago de

165 N9 **Xorkol** Xinjiang Uygur Zizhiqu,
NW China 38.45N 91.07E

43 X14 **Xpujil** Quintana Roo, E Mexico
18.30N 89.24W
Xuanchou see Xuanzhou

178 Jj9 **Xuân Ðuc** Quang Bình,
C Vietnam 17.19N 106.38E

166 L9 **Xuan'en** Hubei, C China
30.03N 109.26E

166 K8 **Xuanhan** Sichuan, C China
31.25N 107.41E

167 O2 **Xuanhua** Hebei, C China
40.37N 115.04E

167 P4 **Xuanhui He** ~ E China

167 Q8 **Xuanzhou** var. Xuancheng.
Anhui, E China 30.59N 118.43E

167 N7 **Xuchang** Henan, C China
34.03N 113.48E

143 X10 **Xudat** Rus. Khudat.
NE Azerbaijan 41.37N 48.39E

83 M16 **Xuddur** var. Hudur, It. Oddur.
Bakool, SW Somalia 4.06N 43.47E

82 O13 **Xudun** Nugaal, N Somalia
9.12N 47.41E

166 L11 **Xuefeng Shan** ▲ S China

44 F2 **Xunantunich** ruins Cayo,
W Belize 17.06N 89.10W

169 W6 **Xun He** ~ NE China

166 L7 **Xun He** ~ C China

164 L14 **Xun Jiang** ~ S China

169 W5 **Xunke** Heilongjiang, NE China
49.36N 128.25E

166 L14 **Xunwu** Jiangxi, SE China
24.58N 115.37E

167 O3 **Xushui** Hebei, E China
39.01N 115.37E

---

116 K13 **Xylaganí** var. Xilaganí. Anatolikí
Makedonía kai Thráki, NE Greece
40.58N 25.27E

117 F19 **Xylókastro** var. Xilokastro.
Pelopónnisos, S Greece
38.04N 22.36E

Y

166 H9 **Ya'an** var. Yaan. Sichuan, C China
30.00N 102.57E

190 L10 **Yaapeet** Victoria, SE Australia
35.48S 142.03E

81 D15 **Yabassi** Littoral, W Cameroon
4.30N 9.58E

83 J15 **Yabélo** Oromo, C Ethiopia
4.53N 38.00E

172 Pp5 **Yabetsu-gawa** var. Yūbetsu-gawa.
~ Hokkaidō, NE Japan

116 H9 **Yablanitsa** Lovech Oblast,
W Bulgaria 43.02N 24.04E

45 N7 **Yablis** Región Autónoma Atlántico
Norte, NE Nicaragua 14.02N 83.44W

126 Kk16 **Yablonovyy Khrebet**
▲ S Russian Federation

168 J14 **Yabrai Shan** ▲ NE China

47 U6 **Yabucoa** E Puerto Rico
18.03N 65.52W

197 K14 **Yacata** Island Lau Group, E Fiji

166 J11 **Yachi He** ~ S China

34 H10 **Yacolt** Washington, NW USA
45.49N 122.22W

56 M10 **Yacuaray** Amazonas, S Venezuela
4.12N 66.30W

59 M22 **Yacuíba** Tarija, S Bolivia
22.03S 63.40W

59 K16 **Yacuma, Río** ~ C Bolivia

161 H16 **Yādgīr** Karnātaka, C India
16.46N 77.09E

23 R8 **Yadkin River** ~ North Carolina,
SE USA

23 R9 **Yadkinville** North Carolina,
SE USA 36.07N 80.39W

131 P13 **Yadrin** Chuvashskaya Respublika,
W Russian Federation 55.55N 46.10E

197 I13 **Yadua** prev. Yandua. Island NW Fiji

172 Oo17 **Yaeyama-shotō** var. Yaegama-
shotō. island group SW Japan

77 O8 **Yafran** NW Libya 32.04N 12.31E

197 L15 **Yagasa Cluster** island group Lau
Group, E Fiji

172 Oo3 **Yagashiri-tō** island NE Japan

67 H21 **Yaghan Basin** undersea feature
SE Pacific Ocean

127 Nn9 **Yagodnoye** Magadanskaya
Oblast', E Russian Federation
62.37N 149.18E
Yagotin see Yahotyn

80 L17 **Yagoua** Extrême-Nord,
NE Cameroon 10.22N 15.13E

165 Qq11 **Yagradagzê Shan** ▲ C China
35.06N 95.41E
Yaguachi see Yaguachi Nuevo

58 B7 **Yaguachi Nuevo** var. Yaguachi.
Guayas, W Ecuador 2.06S 79.43W
Yaguarón, Río see Jaguarão, Rio

171 I16 **Yahagi-gawa** ~ Honshū,
SW Japan

119 Q1 **Yahorlyts'kyy Lyman** bay
S Ukraine

119 Q5 **Yahotyn** Rus. Yagotin. Kyyivs'ka
Oblast', N Ukraine 50.15N 31.48E

42 L12 **Yahualica** Jalisco, SW Mexico
21.12N 102.52W

81 L17 **Yahuma** Orientale, N Dem. Rep.
Congo (Zaire) 1.12N 23.00E

142 K15 **Yahyalı** Kayseri, C Turkey
38.07N 35.22E

178 Gg15 **Yai, Khao** ▲ SW Thailand
8.45N 99.32E

171 Kk15 **Yaita** Tochigi, Honshū, S Japan
36.47N 139.54E

171 Ii17 **Yaizu** Shizuoka, Honshū, S Japan
34.52N 138.19E

166 G9 **Yajiang** Sichuan, C China
30.05N 100.57E

121 O14 **Yakawlyevichi** Rus. Yakovlevichi.
Vitsyebskaya Voblasts', NE Belarus
54.21N 30.29E

169 S6 **Yakeshi** Nei Mongol Zizhiqu,
N China 49.16N 120.42E

34 I9 **Yakima** Washington, NW USA
46.36N 120.30W

34 J10 **Yakima River** ~ Washington,
NW USA

116 G7 **Yakimovo** Montana, NW Bulgaria
43.39N 23.21E
Yakkabag see Yakkabogh

153 N12 **Yakkabogh** Rus. Yakkabag.
Qashqadaryo Wiloyati,
S Uzbekistan 38.57N 66.35E

154 L12 **Yakmach** Baluchistān,
SW Pakistan 28.48N 63.48E

79 O12 **Yako** W Burkina 12.58N 2.15W

41 W13 **Yakobi Island** island Alexander
Archipelago, Alaska, USA

81 K16 **Yakoma** Equateur, N Dem. Rep.
Congo (Zaire) 4.04N 22.22E

116 H11 **Yakoruda** Blagoevgrad,
SW Bulgaria 42.01N 23.40E
Yakovlevichi see Yakawlyevichi

131 T2 **Yakshur-Bod'ya** Udmurtskaya
Respublika, NW Russian
Federation 57.10N 53.10E

172 N6 **Yakumo** Hokkaidō, NE Japan
42.18N 140.15E

170 B17 **Yaku-shima** island Nansei-shotō,
SW Japan

41 V12 **Yakutat** Alaska, USA
59.33N 139.43W

41 U12 **Yakutat Bay** inlet Alaska, USA

127 N12 **Yakutia/Yakutiya/Yakutiya,
Respublika** see Sakha (Yakutiya),
Respublika

126 M11 **Yakutsk** Respublika Sakha
(Yakutiya), NE Russian Federation
62.10N 129.49E

178 Hh17 **Yala** Yala, SW Thailand
6.31N 101.19E

190 D6 **Yalata** South Australia
31.30S 131.53E

33 S9 **Yale** Michigan, N USA
43.07N 82.45W

188 I11 **Yalgoo** Western Australia
28.23S 116.43E

116 M19 **Yalıköy** İstanbul, NW Turkey
41.29N 28.19E

81 L14 **Yalinga** Haute-Kotto, C Central
African Republic 6.47N 23.09E

121 M17 **Yalizava** Rus. Yelizovo.
Mahilyowskaya Voblasts', E Belarus
53.24N 29.01E

46 L13 **Yallahs Hill** ▲ E Jamaica
17.53N 76.31W

24 L3 **Yalobusha River** ~ Mississippi,
S USA

81 N15 **Yaloké** Ombella-Mpoko,
W Central African Republic
5.15N 17.12E

166 E7 **Yalong Jiang** ~ C China

142 M11 **Yalova** Yalova, NW Turkey
40.40N 29.16E

142 M11 **Yalova** ◆ province NW Turkey
Yaloveny see Ialoveni

125 F12 **Yalpug, Ozero** see Yalpuh, Ozero

119 N12 **Yalpuh, Ozero** Rus. Ozero Yalpug.
⊗ SW Ukraine

119 T14 **Yalta** Respublika Krym, S Ukraine
44.30N 34.09E

169 W12 **Yalu Chin.** Yalu Jiang, Jap.
Oryokko, Kor. Amnok-kang.
~ China/North Korea
Yalu Jiang see Yalu

125 F12 **Yalutorovsk** Tyumenskaya
Oblast', C Russian Federation
56.36N 66.09E

142 F14 **Yalvaç** Isparta, SW Turkey
38.16N 31.09E

172 N12 **Yamada** Iwate, Honshū, N Japan
39.27N 141.56E

170 Cc14 **Yamaga** Kumamoto, Kyūshū,
SW Japan 33.01N 130.42E

171 L12 **Yamagata** Yamagata, Honshū,
C Japan 38.15N 140.19E

171 Ll12 **Yamagata** off. Yamagata-ken. ◆
prefecture Honshū, C Japan

170 Bb16 **Yamaga** Kagoshima, Kyūshū,
SW Japan 31.12N 130.37E

170 Dd12 **Yamaguchi** var. Yamaguti.
Yamaguchi, Honshū, SW Japan
34.10N 131.26E

170 Dd12 **Yamaguchi** off. Yamaguchi-ken,
var. Yamaguti. ◆ prefecture Honshū,
SW Japan
Yamaguti see Yamaguchi

129 X5 **Yamalo-Nenetskiy
Avtonomnyy Okrug** ◆
autonomous district N Russian
Federation

126 Gg6 **Yamal, Poluostrov** peninsula
N Russian Federation

171 J16 **Yamanashi** off. Yamanashi-ken,
var. Yamanasi. ◆ prefecture Honshū,
S Japan
Yamanasi see Yamanashi
Yamaniyah, Al Jumhūrīyah al see
Yemen

131 W5 **Yamantau** ▲ W Russian
Federation 53.11N 57.30E

126 K16 **Yamarovka** Chitinskaya Oblast',
S Russian Federation 50.36N 110.25E
Yamasaki see Yamazaki

13 P2 **Yamassa** ▲ Quebec, SE Canada

171 Jj17 **Yamato** Kanagawa, Honshū,
S Japan 35.30N 139.25E

199 Gg4 **Yamato Ridge** undersea feature
E Sea of Japan

170 G14 **Yamazaki** var. Yamasaki. Hyōgo,
Honshū, SW Japan 35.00N 134.31E

191 V5 **Yamba** New South Wales,
SE Australia 29.28S 153.22E

83 D16 **Yambio** var. Yambiyo. Western
Equatoria, S Sudan 4.34N 28.21E
Yambiyo see Yambio

116 L10 **Yambol** Turk. Yanboli. Yambol,
E Bulgaria 42.28N 26.30E

116 M11 **Yambol** ◆ province E Bulgaria

81 M17 **Yambuya** Orientale, N Dem. Rep.
Congo (Zaire) 1.22N 24.21E

176 Uu15 **Yamdena, Pulau** prev. Jamdena.
island Kepulauan Tanimbar,
E Indonesia

170 Cc13 **Yame** Fukuoka, Kyūshū, SW Japan
33.12N 130.31E

177 G6 **Yamethin** Mandalay, C Myanmar
20.25N 96.08E

194 G11 **Yaminbot** East Sepik, NW PNG
4.30S 143.44E

171 L15 **Yamizo-san** ▲ Honshū, C Japan
36.56N 140.14E

189 U9 **Yamma Yamma, Lake**
⊗ Queensland, C Australia

78 M16 **Yamoussoukro** ● (Ivory Coast)
C Ivory Coast 6.51N 5.21W

39 P3 **Yampa River** ~ Colorado,
C USA

119 S2 **Yampil'** Sums'ka Oblast',
NE Ukraine 51.57N 33.49E

118 M8 **Yampil'** Vinnyts'ka Oblast',
C Ukraine 48.15N 28.18E

127 Oo10 **Yamsk** Magadanskaya Oblast',
E Russian Federation 59.33N 154.04E

158 J8 **Yamuna** prev. Jumna. ~ N India

158 I9 **Yamunānagar** Haryāna, N India
30.07N 77.16E

151 U8 **Yamyshevo** Pavlodar,
NE Kazakhstan 51.49N 77.28E

165 N16 **Yamzho Yumco** ⊗ W China

126 L6 **Yana** ~ NE Russian Federation

195 P15 **Yanaba Island** island SE PNG

170 C13 **Yanagawa** Fukuoka, Kyūshū,
SW Japan 33.08N 130.23E

170 E13 **Yanai** Yamaguchi, Honshū,
SW Japan 33.56N 132.08E

151 L16 **Yanam** var. Yanaon. Pondicherry,
E India 16.45N 82.16E

126 L5 **Yan'an** var. Yan. Shaanxi,
C China 36.34N 109.26E
Yanaon see Yanam

131 U3 **Yanaul** Respublika Bashkortostan,
W Russian Federation 56.15N 54.57E

220 O12 **Yanavichy** Rus. Yanovichi.
Vitsyebskaya Voblasts', NE Belarus
55.16N 30.42E

146 K8 **Yanbu' al Baḥr** Al Madīnah,
W Saudi Arabia 24.06N 38.03E
Yanboli see Yambol

23 T8 **Yanceyville** North Carolina,
SE USA 36.24N 79.20W

167 R7 **Yancheng** Jiangsu, E China
33.27N 120.10E

165 W8 **Yanchi** Ningxia, N China
37.49N 107.24E

166 L5 **Yanchuan** Shaanxi, C China
36.54N 110.04E

191 O10 **Yanco Creek** seasonal river New
South Wales, SE Australia

191 O6 **Yanda Creek** seasonal river New
South Wales, SE Australia

190 K4 **Yandama Creek** seasonal river
New South Wales/South Australia

167 S11 **Yandang Shan** ▲ SE China

197 G3 **Yandé, Île** island Îles Belep,
W New Caledonia
Yandua see Yadua

165 O6 **Yandun** Xinjiang Uygur Zizhiqu,
W China 42.24N 94.07E

78 L13 **Yanfolila** Sikasso, SW Mali
11.08N 8.12W

81 M18 **Yangambi** Orientale, N Dem. Rep.
Congo (Zaire) 0.46N 24.24E

164 M15 **Yangbajain** Xizang Zizhiqu,
W China 30.04N 90.34E
Yangchow see Yangzhou

166 M15 **Yangchun** Guangdong, S China
22.16N 111.49E

167 N2 **Yanggao** Shanxi, C China
40.24N 113.51E
Yanggeta see Yaqeta
Yangiabad see Yangiobod
Yangibazar see Dzhany-Bazar,
Kyrgyzstan

161 F16 **Yangishlak** see Yangiqishloq
153 M13 **Yangi-Nishon** Rus. Yang-Nishan.
Qashqadaryo Wiloyati,
S Uzbekistan 38.37N 65.39E

153 Q9 **Yangiobod** Rus. Yangiabad.
Toshkent Wiloyati, E Uzbekistan
41.10N 70.10E

153 O10 **Yangiqishloq** Rus. Yangikishlak.
Jizzakh Wiloyati, C Uzbekistan
40.27N 67.06E

153 P9 **Yangiyer** Sirdaryo Wiloyati,
E Uzbekistan 40.19N 68.48E

153 P9 **Yangiyŭl** Rus. Yangiyul'. Toshkent
Wiloyati, E Uzbekistan 41.12N 69.05E

166 M15 **Yangjiang** Guangdong, S China
21.52N 111.55E
Yangku see Taiyuan
Yang-Nishan see Yangi-Nishon

177 G9 **Yangon** Eng. Rangoon.
● (Myanmar) Yangon, S Myanmar
16.49N 96.10E

177 G8 **Yangon** Eng. Rangoon. ◆ division
SW Myanmar

166 M17 **Yangpu Gang** harbor Hainan,
S China

167 N4 **Yangquan** Shanxi, C China
37.52N 113.28E

167 N15 **Yangshan** Guangdong, S China
24.32N 112.36E

178 Kk13 **Yang Sin, Chu** ▲ S Vietnam
12.23N 108.25E
Yangtze see Chang Jiang, C China
Yangtze Kiang see Chang Jiang

167 N8 **Yangzhou** var. Yangchow. Jiangsu,
E China 32.22N 119.24E

165 L5 **Yan He** ~ C China

169 Y10 **Yanji** Jilin, NE China 42.53N 129.31E
Yanji see Longjing

31 Q12 **Yankton** South Dakota, N USA
42.52N 97.24W

126 M6 **Yannina** see Ioánnina
**Yano-Indigirskaya
Nizmennost'** plain NE Russian
Federation
Yanovichi see Yanavichy

161 K24 **Yan Oya** ~ N Sri Lanka

134 K6 **Yanqi** var. Yanqi Huizu Zizhixian.
Xinjiang Uygur Zizhiqu,
NW China 42.04N 86.32E
Yanqi Huizu Zizhixian see Yanqi

167 P2 **Yan Shan** ▲ E China

167 Q10 **Yanshan** Jiangxi, S China
28.17N 117.47E

166 H14 **Yanshan** prev. Hekou. Yunnan,
SW China 23.36N 104.20E

169 X8 **Yanshou** Heilongjiang, NE China
45.27N 128.19E

126 L6 **Yanskiy Zaliv** bay N Russian
Federation

191 O4 **Yantabulla** New South Wales,
SE Australia 29.22S 145.00E

167 R4 **Yantai** var. Yan-t'ai; prev. Chefoo,
Chih-fu. Shandong, E China
37.30N 121.22E

120 A13 **Yantarnyy** Ger. Palmnicken.
Kaliningradskaya Oblast',
W Russian Federation 54.53N 19.59E

167 J10 **Yantra** Gabrovo, N Bulgaria
42.58N 25.19E

116 K9 **Yantra** ~ N Bulgaria

166 G11 **Yanyuan** Sichuan, C China
27.30N 101.22E

167 P5 **Yanzhou** Shandong, E China
35.34N 116.52E

81 E16 **Yaoundé** var. Yaunde.
● (Cameroon) Centre, S Cameroon
3.51N 11.31E

196 I14 **Yap** ◆ state W Micronesia

198 F16 **Yap** island Caroline Islands,
W Micronesia

59 M18 **Yapacani, Río** ~ C Bolivia

176 Ww12 **Yapa Kopra** Irian Jaya,
E Indonesia 4.18S 135.05E
Yapan see Yapen, Selat

176 U16 **Yapanskoye More** see Japan,
Sea of

79 F17 **Yapei** N Ghana 9.13N 1.09W

10 M10 **Yapeitso, Mont** ▲ Quebec,
E Canada 52.18N 70.24W

176 X10 **Yapen, Pulau** prev. Japen. island
E Indonesia

176 X9 **Yapen, Selat** var. Yapan. strait
Irian Jaya, E Indonesia

63 E15 **Yapeyú** Corrientes, NE Argentina
29.28S 56.49W

142 I11 **Yapraklı** Çankırı, N Turkey
40.45N 33.46E

182 M3 **Yap Trench** var. Yap Trough.
undersea feature SE Philippine Sea
Yap Trough see Yap Trench

176 C14 **Yapurá, Río** see Japurá, Rio,
Brazil/Colombia
Yapurá see Japurá, Rio,
Brazil/Colombia

42 G6 **Yaqui** Sonora, NW Mexico
27.21N 109.59W

34 G4 **Yaquina Bay** bay Oregon,
NW USA

42 G6 **Yaqui, Río** ~ NW Mexico

176 Y14 **Yar** channel Irian Jaya, E Indonesia

56 K5 **Yaracuy** off. Estado Yaracuy. ◆
state NW Venezuela

152 E13 **Yaradzhi** Turkm. Yarajy.
Akhalskiy Velayat, C Turkmenistan
38.12N 57.40E
Yarajy see Yaradzhi

129 Q15 **Yaransk** Kirovskaya Oblast',
NW Russian Federation
57.18N 47.52E

99 Q9 **Yare** ~ E England, UK

129 S9 **Yarega** Respublika Komi,
NW Russian Federation
63.04N 53.28E

118 I7 **Yaremcha** Ivano-Frankivs'ka
Oblast', W Ukraine 48.27N 24.34E

171 I4 **Yarí, Río** ~ SW Colombia

56 K5 **Yaritagua** Yaracuy, N Venezuela
10.04N 69.07W

164 E9 **Yarkand** see Yarkant He
Yarkant see Shache

155 U3 **Yarkant He** var. Yarkand.
~ NW China
Yarkhün ~ NW Pakistan
Yarlung Zangbo Jiang see
Brahmaputra

118 L6 **Yarlovtsi** Khmel'nyts'ka
Oblast', W Ukraine 49.13N 26.53E

169 T11 **Yar Moron** ~ N China

11 O16 **Yarmouth** Nova Scotia,
SE Canada 43.53N 66.08W
Yarmouth see Great Yarmouth
Yaroslav see Jarosław

128 L15 **Yaroslavl'** Yaroslavskaya Oblast',
W Russian Federation 57.38N 39.52E

128 Kk12 **Yaroslavskaya Oblast'** ◆ province
W Russian Federation

190 J4 **Yarram** Victoria, SE Australia
38.36S 146.40E

191 O1 **Yarrawonga** Victoria, SE Australia
36.04S 145.58E

190 L4 **Yarriarraburra Swamp** wetland
New South Wales, SE Australia

126 Gg8 **Yar-Sale** Yamalo-Nenetskiy
Avtonomnyy Okrug, N Russian
Federation 66.51N 70.42E

126 J12 **Yartsevo** Krasnoyarskiy Kray,
C Russian Federation 60.15N 90.09E

130 J4 **Yartsevo** Smolenskaya Oblast',
W Russian Federation 55.03N 32.46E

56 E8 **Yarumal** Antioquia,
NW Colombia 6.58N 75.25W

197 H13 **Yasawa** Yasawa Group,
NW Fiji

197 G13 **Yasawa Group** island group
NW Fiji

79 U17 **Yashi** Katsina, N Nigeria
12.21N 7.56E

79 S14 **Yashikera** Kwara, W Nigeria
9.40N 3.19E

153 T14 **Yashilkŭl** Rus. Ozero Yashil'kul'.
⊗ SE Tajikistan
Yashil'kul', Ozero see Yashilkŭl

171 L11 **Yashima** Akita, Honshū, C Japan
39.10N 140.10E

170 Dd14 **Ya-shima** island SW Japan

170 L14 **Yashiro-jima** island SW Japan

131 P13 **Yashkul'** Respublika Kalmykiya,
SW Russian Federation
46.09N 45.22E

152 F13 **Yashlyk** Akhalskiy Velayat,
C Turkmenistan 37.46N 58.51E
Yasinovataya see Yasynuvata

116 N10 **Yasna Polyana** Burgas,
SE Bulgaria 42.18N 27.35E

126 M14 **Yasnyy** Amurskaya Oblast',
SE Russian Federation
53.03N 127.52E

178 J10 **Yasothon** Yasothon, E Thailand
15.46N 104.12E

191 R10 **Yass** New South Wales,
SE Australia 34.52S 148.55E

170 J4 **Yasugi** Shimane, Honshū,
SW Japan 35.25N 133.14E

149 N10 **Yāsūj** var. Yasuj; prev. Tal-e
Khosravī. Kohkīlūyeh va Būyer
Aḥmadī, C Iran 30.40N 51.34E

142 M11 **Yasun Burnu** headland N Turkey
41.07N 37.40E

115 G13 **Yasynuvata** Rus. Yasinovataya.
Donets'ka Oblast', SE Ukraine
48.04N 37.56E

142 C15 **Yatağan** Muğla, SW Turkey
37.22N 28.07E

171 M9 **Yatate-tōge** pass Honshū, S Japan
40.25N 140.36E

197 J2 **Yaté** Province Sud, S New
Caledonia 22.10S 166.56E

29 P6 **Yates Center** Kansas, C USA
37.52N 95.43W

193 B21 **Yates Point** headland South Island,
NZ 44.30S 167.49E

15 Kk7 **Yathkyed Lake** ⊗ Nunavut,
NE Canada

81 M18 **Yatolema** Orientale, N Dem. Rep.
Congo (Zaire) 0.25N 24.34E

171 J15 **Yatsuga-take** ▲ Honshū, S Japan
35.58N 138.22E

170 C14 **Yatsushiro** var. Yatusiro.
Kumamoto, Kyūshū, SW Japan
32.30N 130.34E

170 C15 **Yatsushiro-kai** bay SW Japan

144 H11 **Yatta** var. Yuta. S West Bank
31.29N 35.10E

83 J20 **Yatta Plateau** plateau SE Kenya
Yatusiro see Yatsushiro

57 F17 **Yauca, Río** ~ SW Peru

47 N16 **Yauco** W Puerto Rico 18.02N 66.51W

176 Xx9 **Yauke** Irian Jaya, E Indonesia
1.34S 137.56E

81 L17 **Yaunde** see Yaoundé

81 D15 **Yavan** see Yovon

176 Y14 **Yavari Mirim, Río** ~ NE Peru
Yavari see Javari, Rio

56 K5 **Yaracuy** off. Estado Yaracuy. ◆
state NW Venezuela

160 I13 **Yavatmāl** Mahārāshtra, C India
20.22N 78.10E

56 M9 **Yavi, Cerro** ▲ C Venezuela
5.43N 65.51W

45 W16 **Yaviza** Darién, SE Panama
8.09N 77.40W

144 F10 **Yavne** Central, W Israel
31.52N 34.45E

118 H5 **Yavoriv Pol.** Jaworów, Rus.
Yavorov. L'vivs'ka Oblast',
NW Ukraine 49.57N 23.21E

118 I7 **Yavorov** see Yavoriv

170 E15 **Yawatahama** Ehime, Shikoku,
SW Japan 33.27N 132.24E
Ya Xian see Sanya

142 L17 **Yayladağı** Hatay, S Turkey
35.55N 36.03E

129 V13 **Yayva** Permskaya Oblast',
NW Russian Federation
59.19N 57.15E

129 V12 **Yayva** ~ NW Russian Federation

149 Q8 **Yazd** var. Yezd. Yazd, C Iran
31.55N 54.22E

149 Q8 **Yazd** off. Ostān-e Yazd, var. Yezd.
◆ province C Iran
Yazdān see Yazdān

149 R7 **Yazgulom, Qatorkŭhi** see
Yazgulemskiy Khrebet

153 S13 **Yazgulemskiy Khrebet**
▲ S Tajikistan

24 K5 **Yazoo City** Mississippi, S USA
32.51N 90.24W

24 K5 **Yazoo River** ~ Mississippi,
S USA

147 O15 **Yemen** off. Republic of
Yemen, Ar. Al Jumhūrīyah
al Yamanīyah, Al Yaman. ◆ republic
SW Asia

118 M4 **Yemil'chyne** Zhytomyrs'ka
Oblast', N Ukraine 50.51N 27.49E

128 M10 **Yemtsa** Arkhangel'skaya Oblast',
NW Russian Federation
63.04N 40.18E

128 M10 **Yemtsa** ~ NW Russian
Federation

129 R10 **Yemva** prev. Zheleznodorozhnyy.
Respublika Komi, NW Russian
Federation 62.38N 50.58E

79 U17 **Yenagoa** Bayelsa, S Nigeria
4.58N 6.16E

191 X7 **Yenakiyeve** Rus. Yenakiyevo; prev.
Ordzhonikidze, Rykovo. Donets'ka
Oblast', E Ukraine 48.13N 38.13E
Yenakiyevo see Yenakiyeve

177 Ff6 **Yenangyaung** Magwe,
W Myanmar 20.28N 94.54E

177 Jj5 **Yên Bai** Yên Bai, N Vietnam
21.43N 104.54E

191 P9 **Yenda** New South Wales,
SE Australia 34.16S 146.15E

176 W10 **Yende** Irian Jaya, E Indonesia
2.19S 134.34E

79 U14 **Yendi** NE Ghana 9.23N 0.02W

164 E8 **Yengisar** Xinjiang Uygur Zizhiqu,
NW China 38.50N 76.10E

124 O3 **Yeniboğaziçi** var. Ayios Seryios,
Gk. Ágios Sérgios. E Cyprus
35.10N 33.53E

124 Oo2 **Yenierenköy** var. Yialousa, Gk.
Agialoúsa. NE Cyprus 35.32N 34.12E
Yenipazar see Novi Pazar

142 E12 **Yenişehir** Bursa, NW Turkey
40.16N 29.37E

126 Hh8 **Yenisey** ~ Mongolia/Russian
Federation

130 L4 **Yeniseysk** Krasnoyarskiy Kray,
C Russian Federation 58.23N 92.06E

207 W10 **Yeniseyskiy Zaliv** var. Yenisei
Bay. bay N Russian Federation

126 Hh8 **Yeniseyskiy Zaliv** see Yeniseyskiy
Zaliv

15 Hh8 **Yellowknife** territory capital
Northwest Territories, W Canada
62.30N 114.28W

15 I7 **Yellowknife** ~ Northwest
Territories, NW Canada

194 F10 **Yellow River** ~ NW PNG

25 P8 **Yellow River**
~ Alabama/Florida, S USA

32 I4 **Yellow River** ~ Wisconsin,
N USA

32 J6 **Yellow River** ~ Wisconsin,
N USA

32 K7 **Yellow River** ~ Wisconsin,
N USA
Yellow River see Huang He

163 V8 **Yellow Sea** Chin. Huang Hai, Kor.
Hwang-Hae. sea E Asia

35 T13 **Yellowstone Lake** ⊗ Wyoming,
C USA

35 V9 **Yellowstone National Park**
national park Wyoming, NW USA

35 X8 **Yellowstone River**
~ Montana/Wyoming, NW USA

98 L12 **Yell Sound** strait N Scotland, UK

49 U9 **Yellville** Arkansas, C USA
36.13N 92.40W

126 Hh11 **Yeloguy** ~ C Russian Federation

152 J14 **Yëloten** prev. Iolotan', Turkm.
Yolöten. Maryyskiy Velayat,
S Turkmenistan 37.18N 62.21E

121 M20 **Yel'sk** Rus. Yel'sk. Homyel'skaya
Voblasts', SE Belarus 51.49N 29.09E

79 T13 **Yelwa** Kebbi, W Nigeria
10.52N 4.46E

125 Ee12 **Yemanzhelinsk** Chelyabinskaya
Oblast', C Russian Federation
54.43N 61.08E

23 R15 **Yemassee** South Carolina, SE USA
32.41N 80.51W

119 X7 **Yenotayevka** Astrakhanskaya
Oblast', SW Russian Federation
47.16N 47.01E

125 Ee11 **Yekaterinburg** prev. Sverdlovsk.
Sverdlovskaya Oblast', C Russian
Federation 56.52N 60.34E
Yekaterinodar see Krasnodar
Yekaterinoslav see
Dnipropetrovs'k

126 Mm15 **Yekaterinoslavka** Amurskaya
Oblast', SE Russian Federation
50.23N 129.03E

131 O7 **Yekaterinovka** Saratovskaya
Oblast', W Russian Federation
52.01N 44.11E

78 K16 **Yekepa** NE Liberia 7.34N 8.31W

131 T3 **Yelabuga** Respublika Tatarstan,
W Russian Federation 55.46N 52.07E

130 M5 **Yelan'** Volgogradskaya Oblast',
SW Russian Federation 50.54N 43.40E
Yelanets' see Kuryk

131 Q9 **Yelanets'** Rus. Yelanets.
Mykolayivs'ka Oblast', S Ukraine
47.40N 31.51E

125 G13 **Yelanka** Novosibirskaya Oblast',
C Russian Federation 55.38N 75.23E

143 T12 **Yerevan** Eng. Erivan. ● (Armenia)
C Armenia 40.12N 44.31E

143 U12 **Yerevan** × C Armenia
40.07N 44.34E

151 R9 **Yereymentau** var. Jermentau,
Yermentau, Kaz. Ereymentaū.
Akmola, C Kazakhstan 51.37N
73.10E

131 O12 **Yergeni** hill range SW Russian
Federation
Yeriho see Jericho

37 R6 **Yerington** Nevada, W USA
38.58N 119.10W

142 J13 **Yerköy** Yozgat, C Turkey
39.39N 34.28E

116 L13 **Yerlisu** Edirne, NW Turkey
Yermak see Aksu

151 R9 **Yermentau** Kaz. Ereymentaū,
Jermentau. Akmola, C Kazakhstan
51.37N 73.10E

151 R9 **Yermentau, Gory**
~ C Kazakhstan

129 R5 **Yermitsa** Respublika Komi,
NW Russian Federation
66.57N 52.15E

37 V14 **Yermo** California, W USA
34.54N 116.49W

101 F15 **Yerseke** Zeeland, SW Netherlands 51.30N 4.03E
131 Q8 **Yershov** Saratovskaya Oblast', W Russian Federation 51.18N 48.16E
129 P9 **Yërtom** Respublika Komi, NW Russian Federation 63.27N 47.52E
58 D13 **Yerupaja, Nevado** ▲ C Peru 10.23S 76.58W
Yerushalayim see Jerusalem
107 R4 **Yesa, Embalse de** ▧ NE Spain
151 V15 **Yesik** Kaz. Esik; prev. Issyk. Almaty, SE Kazakhstan 43.23N 77.31E
151 O8 **Yesil'** Kaz. Esil. Akmola, C Kazakhstan 51.58N 66.22E
142 K15 **Yeşilhisar** Kayseri, C Turkey 38.22N 35.07E
142 L11 **Yeşilırmak** anc. Iris. ≈ N Turkey
39 U12 **Yeso** New Mexico, SW USA 34.25N 104.36W
Yeso see Hokkaidō
131 N15 **Yessentuki** Stavropol'skiy Kray, SW Russian Federation 44.06N 42.51E
126 J9 **Yessey** Evenkiyskiy Avtonomnyy Okrug, N Russian Federation 68.18N 101.49E
107 P12 **Yeste** Castilla-La Mancha, C Spain 38.21N 2.18W
Yesuj see Yāsūj
191 T4 **Yetman** New South Wales, SE Australia 28.56S 150.47E
78 L4 **Yetti** physical region N Mauritania
177 G4 **Ye-u** Sagaing, C Myanmar 22.49N 95.25E
104 H9 **Yeu, Île d'** island NW France
Yevlakh see Yevlax
143 W11 **Yevlax** Rus. Yevlakh. C Azerbaijan 40.36N 47.09E
119 S13 **Yevpatoriya** Respublika Krym, S Ukraine 45.12N 33.22E
125 B17 **Yevreyskaya Avtonomnaya Oblast'** Eng. Jewish Autonomous Oblast. ◈ autonomous province SE Russian Federation
130 K12 **Yeya** ≈ SW Russian Federation
164 I10 **Yeyik** Xinjiang Uygur Zizhiqu, W China 36.43N 83.13E
130 K12 **Yeysk** Krasnodarskiy Kray, SW Russian Federation 46.41N 38.15E
Yezd see Yazd
Yezerishche see Yezyaryshcha
Yezo see Hokkaidō
120 N11 **Yezyaryshcha** Rus. Yezerishche. Vitsyebskaya Voblasts', NE Belarus 55.49N 29.58E
Yialí see Gyalí
Yialousa see Yenierenköy
169 V7 **Yi'an** Heilongjiang, NE China 47.52N 125.13E
Yiannitsá see Giannitsá
166 I10 **Yibin** Sichuan, C China 28.47N 104.36E
164 K13 **Yibug Caka** ◎ W China
166 M9 **Yichang** Hubei, C China 30.37N 111.02E
166 L5 **Yichuan** Shaanxi, C China 36.05N 110.02E
163 W3 **Yichun** Heilongjiang, NE China 47.40N 129.10E
169 X6 **Yichun** var. I-ch'un. Heilongjiang, NE China 47.39N 128.54E
167 O11 **Yichun** Jiangxi, S China 27.45N 114.22E
Yidu see Qingzhou
196 C15 **Yigo** NE Guam 13.33N 144.52E
167 Q5 **Yi He** ≈ E China
169 X8 **Yilan** Heilongjiang, NE China 46.18N 129.36E
142 C9 **Yıldız Dağları** ▲▲ NW Turkey
142 L13 **Yıldızeli** Sivas, N Turkey 39.52N 36.37E
169 U4 **Yilehuli Shan** ▲▲ NE China
169 S7 **Yimin He** ≈ NE China
165 W8 **Yinchuan** var. Yinch'uan, Yin-ch'uan, Yinchwan. Ningxia, N China 38.30N 106.19E
Yinchuanzhan see Xincheng
Yinchwan see Yinchuan
Yindu He see Indus
167 N14 **Yingde** Guangdong, S China 24.08N 113.21E
167 O7 **Ying He** ≈ C China
169 U13 **Yingkou** var. Ying-k'ou, Yingkow; prev. Newchwang, Niuchwang. Liaoning, NE China 40.38N 122.17E
Yingkow see Yingkou
167 P9 **Yingshan** Hubei, C China 30.45N 115.41E
Yingshan see Guangshui
167 Q10 **Yingtan** Jiangxi, S China 28.17N 117.03E
Yin-hsien see Ningbo
164 H5 **Yining** var. I-ning, Uigh. Gulja, Kuldja. Xinjiang Uygur Zizhiqu, NW China 43.53N 81.18E
166 K11 **Yinjiang** Guizhou, S China 28.22N 108.07E
177 Ff4 **Yinmabin** Sagaing, C Myanmar 22.04N 94.57E
169 N13 **Yin Shan** ▲▲ N China
Yin-tu Ho see Indus
165 P15 **Yi'ong Zangbo** ≈ W China
Yioúra see Gyáros
83 J14 **Yirga 'Alem** It. Irgalem. Southern, S Ethiopia 6.43N 38.24E
63 E19 **Yí, Río** ≈ C Uruguay
83 E14 **Yirol** El Buhayrat, S Sudan 6.34N 30.33E
Yirshi see Yirxie
169 S8 **Yirxie** prev. Yirshi. Nei Mongol Zizhiqu, N China 47.16N 119.51E
167 Q5 **Yishui** Shandong, E China 35.49N 118.39E
Yisrael/Yisra'el see Israel
Yíthion see Gýtheio
Yitiaoshan see Jingtai
169 W10 **Yitong** Jilin, NE China 43.22N 125.19E
165 P5 **Yiwu** var. Aratürük. Xinjiang Uygur Zizhiqu, NW China 43.16N 94.38E
169 U12 **Yiwulü Shan** ▲▲ N China
169 T12 **Yi Xian** Liaoning, NE China 41.29N 121.21E
167 N10 **Yiyang** Hunan, S China 28.39N 112.19E
167 Q10 **Yiyang** Jiangxi, S China 28.23N 117.24E
167 N13 **Yizhang** Hunan, S China 25.24N 112.55E
95 K19 **Yläne** Länsi-Suomi, W Finland 60.50N 22.25E
95 L14 **Yli-Ii** Oulu, C Finland 65.21N 25.55E
95 L14 **Ylikiiminki** Oulu, C Finland 65.00N 26.10E
94 N13 **Yli-Kitka** ◎ NE Finland
95 K17 **Ylistaro** Länsi-Suomi, W Finland 62.58N 22.30E
94 K13 **Ylitornio** Lappi, NW Finland 66.16N 23.39E
95 L15 **Ylivieska** Oulu, W Finland 64.04N 24.30E
95 L18 **Ylöjärvi** Länsi-Suomi, W Finland 61.31N 23.37E
97 N17 **Yngaren** ◎ C Sweden
27 T12 **Yoakum** Texas, SW USA 29.17N 97.09W
79 X13 **Yobe** ◈ state NE Nigeria
172 Nn4 **Yobetsu-dake** 🌋 Hokkaidō, NE Japan 43.15N 140.27E
176 Xx10 **Yobi** Irian Jaya, E Indonesia 1.42S 138.09E
82 L11 **Yoboki** C Djibouti 11.30N 42.04E
170 C12 **Yobuko** Saga, Kyūshū, SW Japan 33.31N 129.50E
24 M4 **Yockanookany River** ≈ Mississippi, S USA
24 L2 **Yocona River** ≈ Mississippi, S USA
176 Yy15 **Yodom** Irian Jaya, E Indonesia 7.12S 139.24E
174 Kk15 **Yogyakarta** prev. Djokjakarta, Jogjakarta, Jokyakarta. Jawa, C Indonesia 7.48S 110.24E
174 Kk16 **Yogyakarta** off. Daerah Istimewa Yogyakarta, var. Djokjakarta, Jogjakarta, Jokyakarta. ◈ autonomous district S Indonesia
172 O5 **Yoichi** Hokkaidō, NE Japan 43.11N 140.45E
44 G6 **Yojoa, Lago de** ◎ NW Honduras
81 G16 **Yokadouma** Est, SE Cameroon 3.25N 15.06E
171 H15 **Yōkaichi** var. Yôkaiti. Shiga, Honshū, SW Japan 35.07N 136.10E
Yôkaiti see Yōkaichi
171 H15 **Yokkaichi** var. Yokkaiti. Mie, Honshū, SW Japan 34.58N 136.36E
Yokkaiti see Yokkaichi
81 E15 **Yoko** Centre, C Cameroon 5.28N 12.19E
172 Qq12 **Yokoate-jima** island Nansei-shotō, SW Japan
172 N9 **Yokohama** Aomori, Honshū, C Japan 41.04N 141.14E
171 Jj16 **Yokohama** Kanagawa, Honshū, S Japan 35.26N 139.37E
171 Jj17 **Yokosuka** Kanagawa, Honshū, S Japan 35.15N 139.39E
170 F12 **Yokota** Shimane, Honshū, SW Japan 35.10N 133.03E
171 M11 **Yokote** Akita, Honshū, C Japan 39.19N 140.33E
172 Nn7 **Yokotsu-dake** ▲ Hokkaidō, NE Japan 41.54N 140.48E
79 Y14 **Yola** Adamawa, E Nigeria 9.07N 12.24E
81 L19 **Yolombo** Equateur, C Dem. Rep. Congo (Zaire) 1.36S 23.13E
Yolöten see Yëloten
176 W10 **Yomber** Irian Jaya, E Indonesia 2.04S 134.22E
172 T16 **Yome-jima** island Ogasawara-shotō, SE Japan
78 K16 **Yomou** Guinée-Forestière, SE Guinea 7.30N 9.13W
176 Y15 **Yomuka** Irian Jaya, E Indonesia 7.25S 138.36E
196 C16 **Yona** E Guam 13.24N 144.46E
170 Ff12 **Yonago** Tottori, Honshū, SW Japan 35.30N 134.15E
172 O17 **Yonaguni** Okinawa, SW Japan 24.29N 123.00E
172 Nn16 **Yonaguni-jima** island Nansei-shotō, SW Japan
172 Pp14 **Yonaha-dake** ▲ Okinawa, SW Japan 26.43N 128.13E
169 X14 **Yonan** SW North Korea 37.50N 126.15E
171 L13 **Yonezawa** Yamagata, Honshū, C Japan 37.54N 140.06E
167 Q12 **Yong'an** var. Yongan. Fujian, SE China 25.58N 117.25E
165 T9 **Yongchang** Gansu, N China 38.15N 101.55E
167 P7 **Yongcheng** Henan, C China 33.55N 116.21E
169 Z15 **Yŏngch'ŏn** Jap. Eisen. SE South Korea 35.56N 128.55E
166 J10 **Yongchuan** Chongqing Shi, C China 29.27N 105.56E
165 U10 **Yongdeng** Gansu, C China 35.58N 103.27E
133 W9 **Yongding He** ≈ E China
167 P11 **Yongfeng** Jiangxi, S China 27.19N 115.22E
164 L5 **Yongfengqu** Xinjiang Uygur Zizhiqu, W China 43.28N 87.09E
166 L13 **Yongfu** Guangxi Zhuangzu Ziziqu, S China 24.57 109.59E
169 X13 **Yŏnghŭng** E North Korea 39.30N 127.13E
165 U10 **Yongjing** Gansu, C China 36.00N 103.30E
169 Y15 **Yŏngju** Jap. Eishū. C South Korea 36.48N 128.37E
Yongning see Xuyong
166 E12 **Yongping** Yunnan, SW China 25.30N 99.28E
166 G12 **Yongren** Yunnan, SW China 26.11N 101.49E
166 L10 **Yongshun** var. Lingxi. Hunan, S China 29.01N 109.48E
167 P10 **Yongxiu** var. Tujiabu. Jiangxi, S China 29.08N 115.47E
166 M12 **Yongzhou** Hunan, S China 26.12N 111.36E
20 K14 **Yonkers** New York, NE USA 40.56N 73.51W
105 Q7 **Yonne** ◈ department C France
105 P6 **Yonne** ≈ C France
56 H9 **Yopal** var. El Yopal. Casanare, C Colombia 5.19N 72.19W
164 E8 **Yopurga** var. Yukuriawat. Xinjiang Uygur Zizhiqu, NW China 39.13N 76.44E
188 J12 **York** Western Australia 31.55S 116.52E
99 M16 **York** anc. Eboracum, Eburacum. N England, UK 53.58N 1.04W
25 N5 **York** Alabama, S USA 32.29N 88.18W
31 Q15 **York** Nebraska, C USA 40.52N 97.35W
20 G16 **York** Pennsylvania, NE USA 39.55N 76.42W
23 R11 **York** South Carolina, SE USA 34.59N 81.14W
12 J13 **York** ≈ Ontario, SE Canada
13 X6 **York** ≈ Quebec, SE Canada
189 V1 **York, Cape** headland Queensland, NE Australia 10.40S 142.36E
190 I9 **Yorke Peninsula** peninsula South Australia
190 I9 **Yorketown** South Australia 35.01S 137.38E
21 P9 **York Harbor** Maine, NE USA 43.10N 70.37W
23 X6 **York River** ≈ Virginia, NE USA
99 M16 **Yorkshire** cultural region N England, UK
99 L16 **Yorkshire Dales** physical region N England, UK
9 V16 **Yorkton** Saskatchewan, S Canada 51.12N 102.28W
27 T12 **Yorktown** Texas, SW USA 28.58N 97.30W
23 X6 **Yorktown** Virginia, NE USA 37.13N 76.29W
32 M11 **Yorkville** Illinois, N USA 41.38N 88.27W
44 I5 **Yoro** Yoro, C Honduras 15.06N 87.09W
44 H5 **Yoro** ◈ department N Honduras
172 Pp14 **Yoron-jima** island Nansei-shotō, SW Japan
79 N13 **Yorosso** Sikasso, S Mali 12.18N 4.44W
37 R8 **Yosemite National Park** national park California, W USA
170 Ff14 **Yoshii-gawa** ≈ Honshū, SW Japan
170 Ff15 **Yoshino-gawa** var. Yosino Gawa. ≈ Shikoku, SW Japan
131 Q3 **Yoshkar-Ola** Respublika Mariy El, W Russian Federation 56.37N 47.53E
Yosino Gawa see Yoshino-gawa
176 Y15 **Yos Sudarso, Pulau** var. Pulau Dolak, Pulau Kolepom; prev. Jos Sudarso. island E Indonesia
176 Z10 **Yos Sudarso, Teluk** bay Irian Jaya, E Indonesia
169 Y17 **Yŏsu** Jap. Reisui. S South Korea 34.45N 127.40E
172 Nn5 **Yotei-zan** ▲ Hokkaidō, NE Japan 42.50N 140.46E
99 D21 **Youghal** Ir. Eochaill. S Ireland 51.57N 7.49W
99 D21 **Youghal Bay** Ir. Cuan Eochaille. inlet S Ireland
20 C15 **Youghiogheny River** ≈ Pennsylvania, NE USA
166 K14 **You Jiang** ≈ S China
191 Q9 **Young** New South Wales, SE Australia 34.19S 148.19E
9 T15 **Young** Saskatchewan, S Canada 51.44N 105.44W
63 E18 **Young** Río Negro, W Uruguay 32.43S 57.36W
190 G5 **Younghusband, Lake** salt lake South Australia
190 J10 **Younghusband Peninsula** peninsula South Australia
192 Q10 **Young Nicks Head** headland North Island, NZ 39.38S 105.03E
193 D20 **Young Range** ▲▲ South Island, NZ
203 Q15 **Young's Rock** island Pitcairn Island, Pitcairn Islands
9 R16 **Youngstown** Alberta, SW Canada 51.31N 111.12W
33 V12 **Youngstown** Ohio, N USA 41.06N 80.39W
165 N9 **Youshashan** Qinghai, C China 38.12N 90.58E
Youth, Isle of see Juventud, Isla de la
79 N11 **Youvarou** Mopti, C Mali 15.19N 4.15W
166 K10 **Youyang** Sichuan, C China 28.48N 108.48E
169 Y7 **Youyi** Heilongjiang, NE China 46.51N 131.54E
153 P13 **Yovon** Rus. Yavan. SW Tajikistan 38.19N 69.02E
142 J13 **Yozgat** Yozgat, C Turkey 39.49N 34.48E
142 K13 **Yozgat** ◈ province C Turkey
64 O6 **Ypacaraí** var. Ypacaray. Central, S Paraguay 25.22S 57.16W
Ypacaray see Ypacaraí
64 P5 **Ypané, Río** ≈ C Paraguay
Ypres see Ieper
116 I13 **Ypsário** var. Ipsario. ▲ Thásos, E Greece 40.43N 24.39E
33 R10 **Ypsilanti** Michigan, N USA 42.12N 83.36W
36 M1 **Yreka** California, W USA 41.43N 122.38W
Yrendagüé see General Eugenio A. Garay
195 N8 **Ysabel Channel** channel N PNG
12 K8 **Yser, Lac** ◎ Quebec, SE Canada
153 Y8 **Yshtyk** Issyk-Kul'skaya Oblast', E Kyrgyzstan 41.34N 78.21E
Yssel see IJssel
105 Q12 **Yssingeaux** Haute-Loire, C France 45.09N 4.07E
97 K23 **Ystad** Skåne, S Sweden 55.25N 13.51E
Ysyk-Köl see Balykchy, Kyrgyzstan
Ysyk-Köl see Issyk-Kul', Ozero, Kyrgyzstan
Ysyk-Köl Oblasty see Issyk-Kul'skaya Oblast'
98 L8 **Ythan** ≈ NE Scotland, UK
Y Trallwng see Welshpool
96 C13 **Ytre Arna** Hordaland, S Norway 60.28N 5.25E
96 B12 **Ytre Sula** island S Norway
95 G17 **Ytterhogdal** Jämtland, C Sweden 62.10N 14.55E
126 M10 **Ytyk-Kyuyel'** Respublika Sakha (Yakutiya), NE Russian Federation 62.22N 133.37E
Yu see Henan
Yuan Jiang see Red River
167 S13 **Yüanlin** Jap. Inrin. C Taiwan 23.57N 120.33E
167 N3 **Yuanping** Shanxi, C China 38.26N 112.42E
167 O11 **Yuan Shui** ≈ S China
170 G16 **Yuasa** Wakayama, Honshū, SW Japan 34.00N 135.08E
194 H10 **Yuat** ≈ N PNG
37 O6 **Yuba City** California, W USA 39.07N 121.40W
172 Oo6 **Yūbari** Hokkaidō, NE Japan 43.09N 141.00E
172 P6 **Yūbari-sanchi** ▲▲ Hokkaidō, NE Japan
37 O6 **Yuba River** ≈ California, W USA
82 H13 **Yubdo** Oromo, C Ethiopia 9.05N 35.28E
172 Q5 **Yūbetsu** Hokkaidō, NE Japan 44.12N 143.34E
Yūbetsu-gawa see Yabetsu-gawa
43 X12 **Yucatán** ◈ state SE Mexico
49 O3 **Yucatan Basin** var. Yucatan Deep. undersea feature N Caribbean Sea
Yucatán, Canal de see Yucatan Channel
43 Y10 **Yucatan Channel** Sp. Canal de Yucatán. channel Cuba/Mexico
Yucatan Deep see Yucatan Basin
Yucatan Peninsula see Yucatán, Península de
43 X13 **Yucatán, Península de** Eng. Yucatan Peninsula. peninsula Guatemala/Mexico
38 I11 **Yucca** Arizona, SW USA 34.49N 114.06W
37 V15 **Yucca Valley** California, W USA 34.06N 116.30W
167 P4 **Yucheng** Shandong, E China 37.01N 116.37E
167 N4 **Yuci** Shanxi, C China 37.34N 112.45E
133 X5 **Yudoma** ≈ E Russian Federation
167 P12 **Yudu** Jiangxi, C China 26.02N 115.24E
Yue see Guangdong
166 M12 **Yuecheng Ling** ▲▲ S China
189 P7 **Yuendumu** Northern Territory, N Australia 22.19S 131.51E
166 H10 **Yuexi** Sichuan, C China 28.50N 102.36E
167 N10 **Yueyang** Hunan, S China 29.24N 113.08E
129 U14 **Yug** Permskaya Oblast', NW Russian Federation 57.49N 56.08E
129 P13 **Yug** ≈ NW Russian Federation
127 N11 **Yugorenok** Respublika Sakha (Yakutiya), NE Russian Federation 59.46N 137.36E
125 F10 **Yugorsk** Khanty-Mansiyskiy Avtonomnyy Okrug, C Russian Federation 61.17N 63.25E
125 G6 **Yugorskiy Poluostrov** peninsula NW Russian Federation
114 M13 **Yugoslavia** off. Federal Republic of Yugoslavia, SCr. Jugoslavija, Savezna Republika Jugoslavija. ◆ federal republic SE Europe
152 K14 **Yugo-Vostochnyye Garagumy** prev. Yugo-Vostochnyye Karakumy. desert E Turkmenistan
Yugo-Vostochnyye Karakumy see Yugo-Vostochnyye Garagumy
167 S10 **Yuhuan Dao** island SE China
166 L14 **Yu Jiang** ≈ S China
127 Nn7 **Yukagirskoye Ploskogor'ye** plateau NE Russian Federation
120 L11 **Yukhavichy** Rus. Yukhovichi. Vitsyebskaya Voblasts', N Belarus 56.02N 28.39E
130 J4 **Yukhnov** Kaluzhskaya Oblast', W Russian Federation 54.43N 35.15E
Yukhovichi see Yukhavichy
81 J20 **Yuki** var. Yuki Kengunda. Bandundu, W Dem. Rep. Congo (Zaire) 3.52S 19.32E
Yuki Kengunda see Yuki
28 M10 **Yukon** Oklahoma, C USA 35.30N 97.45W
2 F4 **Yukon** ≈ Canada/USA
Yukon see Yukon Territory
41 S7 **Yukon Flats** salt flat Alaska, USA
14 F5 **Yukon Territory** var. Yukon, Fr. Territoire du Yukon. ◈ territory NW Canada
143 T16 **Yüksekova** Hakkâri, SE Turkey 37.34N 44.16E
126 Jj11 **Yukta** Evenkiyskiy Avtonomnyy Okrug, C Russian Federation 63.16N 106.04E
170 Dd13 **Yukuhashi** var. Yukuhasi. Fukuoka, Kyūshū, SW Japan 33.41N 131.00E
Yukuhasi see Yukuhashi
Yukuriawat see Yopurga
129 O9 **Yula** ≈ NW Russian Federation
189 P8 **Yulara** Northern Territory, N Australia 25.15S 130.57E
131 W6 **Yuldybayevo** Respublika Bashkortostan, W Russian Federation 52.22N 57.55E
25 W8 **Yulee** Florida, SE USA 30.37N 81.36W
164 K7 **Yuli** var. Lopnur. Xinjiang Uygur Zizhiqu, NW China 41.24N 86.12E
167 T14 **Yüli** C Taiwan 23.23N 121.18E
166 L15 **Yulin** Guangxi Zhuangzu Zizhiqu, S China 22.37N 110.07E
166 L4 **Yulin** Shaanxi, C China 38.22N 109.47E
167 T14 **Yüli Shan** ▲ E Taiwan 23.22N 121.13E
166 F11 **Yulongxue Shan** ▲ SW China 27.09N 100.10E
38 H14 **Yuma** Arizona, SW USA 32.40N 114.38W
39 W3 **Yuma** Colorado, C USA 40.07N 102.43W
56 K5 **Yumare** Yaracuy, N Venezuela 10.37N 68.40W
65 G14 **Yumbel** Bío Bío, C Chile 37.07S 72.33W
81 N19 **Yumbi** Maniema, E Dem. Rep. Congo (Zaire) 1.13S 26.13E
165 R8 **Yumen** var. Laojunmiao, Yümen. Gansu, N China 39.49N 97.46E
165 Q7 **Yumenzhen** Gansu, N China 40.15N 97.03E
164 J3 **Yumin** Xinjiang Uygur Zizhiqu, NW China 46.14N 82.52E
Yun see Yunnan
142 G14 **Yunak** Konya, W Turkey 38.49N 31.42E
47 O8 **Yuna, Río** ≈ E Dominican Republic
40 I17 **Yunaska Island** island Aleutian Islands, Alaska, USA
166 M6 **Yuncheng** Shanxi, C China 35.07N 110.45E
59 L18 **Yungas** physical region E Bolivia
Yungki see Jilin
Yung-ning see Nanning
166 I12 **Yun Gui Gaoyuan** plateau SW China
Yunjinghong see Jinghong
166 M15 **Yunkai Dashan** ▲▲ S China
Yunki see Jilin
166 E11 **Yun Ling** ▲▲ SW China
167 N9 **Yunmeng** Hubei, C China 30.59N 113.44E
163 N14 **Yunnan** var. Yun, Yunnan Sheng, Yünnan, Yun-nan. ◈ province SW China
Yunnan see Kunming
Yunnan Sheng see Yunnan
170 Cc15 **Yunomae** Kumamoto, Kyūshū, SW Japan 32.16N 131.00E
167 N8 **Yun Shui** ≈ C China
190 J7 **Yunta** South Australia 32.37S 139.33E
167 Q14 **Yunxiao** Fujian, SE China 23.56N 117.16E
166 K9 **Yunyang** Sichuan, C China 31.03N 109.43E
200 Nn10 **Yupanqui Basin** undersea feature E Pacific Ocean
Yuratishki see Yuratsishki
121 I15 **Yuratsishki** Pol. Juraciszki, Rus. Yuratishki. Hrodzyenskaya Voblasts', W Belarus 54.01N 25.55E
Yurev see Tartu
126 H14 **Yurga** Kemerovskaya Oblast', S Russian Federation 55.42N 84.59E
58 E10 **Yurimaguas** Loreto, N Peru 5.54S 76.07W
131 P3 **Yurino** Respublika Mariy El, W Russian Federation 56.19N 46.15E
43 N13 **Yuriria** Guanajuato, C Mexico 20.12N 101.09W
129 T13 **Yurla** Komi-Permyatskiy Avtonomnyy Okrug, NW Russian Federation 59.18N 54.19E
Yuruá, Río see Juruá, Rio
116 M13 **Yürük** Tekirdağ, NW Turkey 40.58N 27.09E
164 G10 **Yurungkax He** ≈ W China
129 Q14 **Yur'ya** var. Jarja. Kirovskaya Oblast', NW Russian Federation 59.01N 49.22E
Yur'yev see Tartu
129 N16 **Yur'yevets** Ivanovskaya Oblast', W Russian Federation 57.19N 43.01E
130 M3 **Yur'yev-Pol'skiy** Vladimirskaya Oblast', W Russian Federation 56.28N 39.39E
119 V7 **Yur"yivka** Dnipropetrovs'ka Oblast', E Ukraine 48.45N 36.01E
126 K6 **Yuryung-Khaya** Respublika Sakha (Yakutiya), NE Russian Federation 72.45N 113.32E
44 I7 **Yuscarán** El Paraíso, S Honduras 13.58N 86.48W
167 P12 **Yu Shan** ▲▲ S China
128 I7 **Yushkozero** Respublika Kareliya, NW Russian Federation 64.46N 32.13E
165 R13 **Yushu** Qinghai, C China 33.03N 97.00E
131 P12 **Yusta** Respublika Kalmykiya, SW Russian Federation 47.06N 46.16E
128 I10 **Yustozero** Respublika Kareliya, NW Russian Federation 62.44N 33.31E
143 Q11 **Yusufeli** Artvin, NE Turkey 40.49N 41.31E
170 E15 **Yusuhara** Kōchi, Shikoku, SW Japan 33.22N 132.52E
129 T14 **Yus'va** Permskaya Oblast', NW Russian Federation 58.48N 54.59E
Yuta see Yatta
167 P2 **Yutian** Hebei, E China 39.52N 117.43E
164 H10 **Yutian** var. Keriya. Xinjiang Uygur Zizhiqu, NW China 36.49N 81.31E
64 K5 **Yuto** Jujuy, NW Argentina 23.35S 64.28W
64 P7 **Yuty** Caazapá, S Paraguay 26.28S 56.11W
166 G13 **Yuxi** Yunnan, SW China 24.22N 102.28E
167 O2 **Yuxian** prev. Yu Xian. Hebei, E China 39.50N 114.33E
171 M11 **Yuzawa** Akita, Honshū, C Japan 39.11N 140.29E
129 N16 **Yuzha** Ivanovskaya Oblast', W Russian Federation 56.34N 42.00E
Yuzhno-Alichurskiy Khrebet see Alichuri Janubí, Qatorkŭhi
Yuzhno-Kazakhstanskaya Oblast' see Yuzhnyy Kazakhstan
127 Oo15 **Yuzhno-Sakhalinsk** Jap. Toyohara; prev. Vladimirovka. Ostrov Sakhalin, Sakhalinskaya Oblast', SE Russian Federation 46.58N 142.45E
131 P14 **Yuzhno-Sukhokumsk** Respublika Dagestan, SW Russian Federation 44.43N 45.32E
125 Ee12 **Yuzhnoural'sk** Chelyabinskaya Oblast', C Russian Federation 54.28N 61.13E
126 I13 **Yuzhno-Yeniseyskiy** Krasnoyarskiy Kray, C Russian Federation 58.40N 94.49E
151 Z10 **Yuzhnyy Altay, Khrebet** ▲▲ E Kazakhstan
Yuzhnyy Bug see Pivdennyy Buh
151 O15 **Yuzhnyy Kazakhstan** off. Yuzhno-Kazakhstanskaya Oblast', Eng. South Kazakhstan, Kaz. Ongtüstik Qazaqstan Oblysy; prev. Chimkentskaya Oblast'. ◈ province S Kazakhstan
127 Oo10 **Yuzhnyy, Mys** headland E Russian Federation 57.44N 156.49E
131 W6 **Yuzhnyy Ural** var. Southern Urals. ▲▲ W Russian Federation
165 V10 **Yuzhong** Gansu, C China 35.52N 104.09E
Yuzhou see Chongqing
105 N5 **Yvelines** ◈ department N France
110 B9 **Yverdon** var. Yverdon-les-Bains, Ger. Iferten; anc. Eborodunum. Vaud, W Switzerland 46.46N 6.37E
Yverdon-les-Bains see Yverdon
104 M3 **Yvetot** Seine-Maritime, N France 49.37N 0.48E
Ylanly see Il'yaly

Z

153 T12 **Zaalayskiy Khrebet** Taj. Qatorkŭhi Pasi Oloy. ▲▲ Kyrgyzstan/Tajikistan
Zaamin see Zomin
Zaandam see Zaanstad
100 I10 **Zaanstad** prev. Zaandam. Noord-Holland, C Netherlands 52.27N 4.49E
Zabadani see Az Zabdānī
121 L18 **Zabalatstsye** Rus. Zabolot'ye. Homyel'skaya Voblasts', SE Belarus 52.38N 28.35E
114 L9 **Žabalj** Ger. Josefsdorf, Hung. Zsablya; prev. Józseffalva;. Serbia, N Yugoslavia 45.22N 20.01E
Zāb aş Şaghīr, Nahraz see Little Zab
126 L16 **Zabaykal'sk** Chitinskaya Oblast', S Russian Federation 49.37N 117.19E
Zāb-e Kūchek, Rūdkhāneh-ye see Little Zab
Zabeln see Sabile
Zabéré see Zabré
Zabern see Saverne
147 N16 **Zabīd** W Yemen 14.00N 43.00E
147 O16 **Zabīd, Wādī** dry watercourse SW Yemen
Żabinka see Zhabinka
Ząbkowice see Ząbkowice Śląskie
113 G15 **Ząbkowice Śląskie** var. Ząbkowice, Ger. Frankenstein, Frankenstein in Schlesien. Dolnośląskie, SW Poland 50.34N 16.48E
112 P10 **Zabłudów** Podlaskie, NE Poland 53.00N 23.21E
114 D8 **Zabok** Krapina-Zagorje, N Croatia 46.00N 15.48E
149 W9 **Zābol** var. Shahr-i-Zabul, Zabul; prev. Nasratabad. Sīstān va Balūchestān, E Iran 31.00N 61.32E
Zābol see Zābul
149 W13 **Zābolī** Sīstān va Balūchestān, SE Iran 27.09N 61.31E
Zabolot'ye see Zabalatstsye
79 Q13 **Zabré** var. Zabéré. S Burkina 11.13N 0.34W
113 G17 **Zábřeh** Ger. Hohenstadt. Olomoucký Kraj, E Czech Republic 49.52N 16.52E
113 J16 **Zabrze** Ger. Hindenburg, Hindenburg in Oberschlesien. Śląskie, S Poland 50.19N 18.52E
155 O7 **Zābul** Per. Zābol. ◈ province SE Afghanistan
Zabul see Zābol
44 E6 **Zacapa** Zacapa, E Guatemala 14.59N 89.32W
44 A3 **Zacapa** off. Departamento de Zacapa. ◈ department E Guatemala
42 M14 **Zacapú** Michoacán de Ocampo, SW Mexico 19.49N 101.52W
43 V14 **Zacatal** Campeche, SE Mexico 18.37N 91.52W
42 M11 **Zacatecas** Zacatecas, C Mexico 22.45N 102.33W
42 L10 **Zacatecas** ◈ state C Mexico
44 F8 **Zacatecoluca** La Paz, S El Salvador 13.28N 88.51W
43 P15 **Zacatepec** Morelos, S Mexico 18.40N 99.11W
43 Q13 **Zacatlán** Puebla, S Mexico 19.54N 97.59W
150 F8 **Zachagansk** Zapadnyy Kazakhstan, NW Kazakhstan 51.04N 51.13E
117 D20 **Zacháro** var. Zaharo, Zakháro. Dytikí Ellás, S Greece 37.28N 21.40E
24 J8 **Zachary** Louisiana, S USA 30.39N 91.09W
119 U6 **Zachepylivka** Kharkivs'ka Oblast', E Ukraine 49.13N 35.15E
Zachist'ye see Zachystsye
112 E9 **Zachodniopomorskie** ◈ province NW Poland
121 L14 **Zachystsye** Rus. Zachist'ye. Minskaya Voblasts', NW Belarus 54.24N 28.45E
42 L13 **Zacoalco** var. Zacoalco de Torres. Jalisco, SW Mexico 20.12N 103.31W
Zacoalco de Torres see Zacoalco
43 P13 **Zacualtipán** Hidalgo, C Mexico 20.39N 98.42W
114 C12 **Zadar** It. Zara; anc. Iader. Zadar, W Croatia 44.06N 15.14E
114 C12 **Zadar** off. Zadarsko-Kninska Županija prev. Zadar-Knin. ◈ province SW Croatia
Zadar-Knin see Zadar
177 G14 **Zadetkyi Kyun** var. St. Matthew's Island. island Mergui Archipelago, S Myanmar
69 Q9 **Zadié** var. Djadié. ≈ NE Gabon
165 Q13 **Zadoi** Qinghai, C China 32.56N 95.21E
130 L7 **Zadonsk** Lipetskaya Oblast', W Russian Federation 52.25N 38.55E
77 X8 **Za'farâna** E Egypt 29.06N 32.34E
155 W7 **Zafarwāl** Punjab, E Pakistan 32.19N 74.52E
124 P1 **Zafer Burnu** var. Cape Andreas, Cape Apostolas Andreas, Gk. Akrotíri Apostólou Andréa. headland NE Cyprus 35.42N 34.34E
109 J23 **Zafferano, Capo** headland Sicilia, Italy, C Mediterranean Sea 38.06N 13.31E
116 M7 **Zafirovo** Silistra, NE Bulgaria 44.00N 26.51E
117 L23 **Zaforá** island Kykládes, Greece, Aegean Sea
106 J12 **Zafra** Extremadura, W Spain 38.25N 6.27W
112 E13 **Żagań** var. Zagań, Żegań, Ger. Sagan. Lubuskie, W Poland 51.37N 15.18E
120 F10 **Žagarė** Pol. Žagory. Joniškis, N Lithuania 56.22N 23.16E
77 W7 **Zagazig** var. Az Zaqāzīq. N Egypt 30.35N 31.31E
76 M5 **Zaghouan** var. Zaghwān. NE Tunisia 36.26N 10.05E
Zaghwān see Zaghouan
117 G16 **Zagorá** Thessalía, C Greece 39.27N 23.06E
Zagorod'ye see Zaharoddzye
Žagory see Žagarė
Zágráb see Zagreb
114 E8 **Zagreb** Ger. Agram, Hung. Zágráb. ● (Croatia) Zagreb, N Croatia 45.48N 15.57E
114 E8 **Zagreb** prev. Grad Zagreb. ◈ province NC Croatia
148 L7 **Zāgros, Kūhhā-ye** Eng. Zagros Mountains. ▲▲ W Iran
Zagros Mountains see Zāgros, Kūhhā-ye
114 O12 **Žagubica** Serbia, E Yugoslavia 44.13N 21.47E
113 L22 **Zagyva** ≈ N Hungary
Zaharo see Zacháro
121 G19 **Zaharoddzye** Rus. Zagorod'ye. physical region SW Belarus
149 W11 **Zāhedān** var. Zahidan; prev. Duzdab. Sīstān va Balūchestān, SE Iran 29.31N 60.51E
Zahidan see Zāhedān
Zaḩlah see Zahlé
144 H7 **Zahlé** var. Zaḩlah. C Lebanon 33.51N 35.54E
Zähmet see Zakhmet
113 O20 **Záhony** Szabolcs-Szatmár-Bereg, NE Hungary 48.26N 22.10E
147 N13 **Z̧ahrān** 'Asīr, S Saudi Arabia 17.47N 43.27E
145 R12 **Z̧ahrat al Baţn** hill range S Iraq
123 I12 **Zahrez Chergui** var. Zahrez Chergúi. marsh N Algeria
131 S4 **Zainsk** Respublika Tatarstan, W Russian Federation 55.12N 52.01E
84 A10 **Zaire** prev. Congo. ◈ province NW Angola
Zaire see Congo (Democratic Republic of)
Zaire see Congo (river)
114 P13 **Zaječar** Serbia, E Yugoslavia 43.54N 22.16E
85 L18 **Zaka** Masvingo, E Zimbabwe 20.19S 31.27E
126 J16 **Zakamensk** Respublika Buryatiya, S Russian Federation 50.18N 102.57E
118 G7 **Zakarpats'ka Oblast'** Eng. Transcarpathian Oblast, Rus. Zakarpatskaya Oblast'. ◈ province W Ukraine
Zakarpatskaya Oblast' see Zakarpats'ka Oblast'
Zakataly see Zaqatala
Zakháro see Zacháro
Zakhidnyy Buh/Zakhodni Buh see Bug
152 J14 **Zakhmet** Turkm. Zähmet. Maryyskiy Velayat, C Turkmenistan 37.48N 62.33E
145 Q1 **Zākhō** var. Zākhū. N Iraq 37.09N 42.40E
Zākhū see Zākhō
Zákinthos see Zákynthos
113 L18 **Zakopane** Małopolskie, S Poland 49.17N 19.57E
80 J12 **Zakouma** Salamat, S Chad 10.47N 19.51E
117 L25 **Zákros** Kríti, Greece, E Mediterranean Sea 35.06N 26.12E
117 C19 **Zákynthos** var. Zákinthos. Zákynthos, W Greece 37.46N 20.54E
117 C20 **Zákynthos** var. Zákinthos, It. Zante. island Iónioi Nísoi, Greece, C Mediterranean Sea
117 C19 **Zakýnthou, Porthmós** strait SW Greece
113 G24 **Zala** off. Zala Megye. ◈ county W Hungary
113 G24 **Zala** ≈ W Hungary
144 M4 **Zalābiyah** Dayr az Zawr, C Syria 35.39N 39.51E
113 G24 **Zalaegerszeg** Zala, W Hungary 46.51N 16.49E
106 K11 **Zalamea de la Serena** Extremadura, W Spain 38.38N 5.37W
106 J13 **Zalamea la Real** Andalucía, S Spain 37.40N 6.40W
169 U7 **Zalantun** var. Butha Qi. Nei Mongol Zizhiqu, N China 47.57N 122.43E
126 J15 **Zalari** Irkutskaya Oblast', S Russian Federation 53.31N 102.10E
113 G23 **Zalaszentgrót** Zala, SW Hungary 46.57N 17.04E
Zalatna see Zlatna
118 G9 **Zalău** Ger. Waltenberg, Hung. Zilah; prev. Ger. Zillenmarkt. Sălaj, NW Romania 47.10N 23.03E
111 V10 **Žalec** Ger. Sachsenfeld. C Slovenia 46.15N 15.08E
119 S9 **Zalenodol's'k** Dnipropetrovs'ka Oblast', E Ukraine 47.31N 33.56E
112 K8 **Zalewo** Ger. Saalfeld. Warmińsko-Mazurskie, NE Poland 53.54N 19.39E
147 N9 **Z̧alim** Makkah, W Saudi Arabia 22.46N 42.12E
82 A11 **Zalingei** var. Zalinje. Western Darfur, W Sudan 12.51N 23.28E
Zalinje see Zalingei
118 K7 **Zalishchyky** Ternopil's'ka Oblast', W Ukraine 48.40N 25.43E
Zallah see Zillah
100 J13 **Zaltbommel** Gelderland, C Netherlands 51.49N 5.15E
128 H15 **Zaluch'ye** Novgorodskaya Oblast', NW Russian Federation 57.40N 31.45E
Zamak see Zamakh
147 Q14 **Zamakh** var. Zamak. N Yemen 16.25N 47.35E
142 K15 **Zamantı Irmağı** ≈ C Turkey

◆ Country
● Country Capital
◇ Dependent Territory
○ Dependent Territory Capital
◈ Administrative Region
✈ International Airport
▲ Mountain
▲▲ Mountain Range
🌋 Volcano
≈ River
◎ Lake
▧ Reservoir

Zambesi/Zambeze *see* Zambezi
85 G14 **Zambezi** North Western, W Zambia *13.33S 23.07E*
85 K15 **Zambezi** *var.* Zambesi, *Port.* Zambeze. ≈ S Africa
85 O15 **Zambézia** *off.* Província da Zambézia. ◐ *province* C Mozambique
85 I14 **Zambia** *off.* Republic of Zambia; *prev.* Northern Rhodesia. ◆ *republic* S Africa
179 Q16 **Zamboanga** *off.* Zamboanga City. Mindanao, S Philippines *6.56N 122.03E*
56 E5 **Zambrano** Bolívar, N Colombia *9.45N 74.49W*
112 N10 **Zambrów** Łomża, E Poland *52.59N 22.14E*
85 L14 **Zambué** Tete, NW Mozambique *15.03S 30.49E*
79 T13 **Zamfara** ≈ NW Nigeria
58 C9 **Zamora** Zamora Chinchipe, S Ecuador *4.05S 78.58W*
106 K6 **Zamora** Castilla-León, NW Spain *41.30N 5.45W*
106 K5 **Zamora** ◐ *province* Castilla-León, NW Spain
Zamora *see* Barinas
58 A13 **Zamora Chinchipe** ◐ *province* S Ecuador
42 M13 **Zamora de Hidalgo** Michoacán de Ocampo, SW Mexico *20.00N 102.18W*
113 P15 **Zamość** *Rus.* Zamoste. Lubelskie, E Poland *50.43N 23.16E*
Zamoste *see* Zamość
166 G7 **Zamtang** *prev.* Gamda. Sichuan, C China *32.19N 100.55E*
77 O8 **Zamzam, Wādī** *dry watercourse* NW Libya
81 F20 **Zanaga** La Lékoumou, S Congo *2.49S 13.52E*
43 T16 **Zanatepec** Oaxaca, SE Mexico *16.28N 94.24W*
107 P9 **Záncara** ≈ C Spain
Zancle *see* Messina
164 G14 **Zanda** Xizang Zizhiqu, W China *31.28N 79.49E*
100 H10 **Zandvoort** Noord-Holland, W Netherlands *52.22N 4.31E*
41 P8 **Zane Hills** *hill range* Alaska, USA
33 T13 **Zanesville** Ohio, N USA *39.55N 82.01W*
Zanga *see* Hrazdan
148 L4 **Zanjān** *var.* Zenjan, Zinjan. Zanjān, NW Iran *36.40N 48.30E*
148 L4 **Zanjān** *off.* Ostān-e Zanjān, *var.* Zenjan, Zinjan. ◐ *province* NW Iran
Zante *see* Zákynthos
83 J22 **Zanzibar** Zanzibar, E Tanzania *6.10S 39.12E*
83 J22 **Zanzibar** ◐ *region* E Tanzania
83 J22 **Zanzibar** *Swa.* Unguja. *island* E Tanzania
83 J22 **Zanzibar Channel** *channel* E Tanzania
171 Ll13 **Zaō-san** ▲ Honshū, C Japan *38.06N 140.27E*
167 N8 **Zaoyang** Hubei, C China *32.11N 112.42E*
126 I14 **Zaozernyy** Krasnoyarskiy Kray, S Russian Federation *55.53N 94.37E*
167 Q6 **Zaozhuang** Shandong, E China *34.52N 117.37E*
30 L4 **Zap** North Dakota, N USA *47.18N 101.55W*
114 L13 **Zapadna Morava** *Ger.* Westliche Morava. ≈ C Yugoslavia
128 H16 **Zapadnaya Dvina** Tverskaya Oblast', W Russian Federation *56.16N 32.03E*
Zapadnaya Dvina *see* Western Dvina
126 H10 **Zapadno-Sibirskaya Ravnina** *Eng.* West Siberian Plain. *plain* C Russian Federation
Zapadnyy Bug *see* Bug
150 E9 **Zapadnyy Kazakhstan** *off.* Zapadno-Kazakhstanskaya Oblast', *Eng.* West Kazakhstan, *Kaz.* Batys Qazaqstan Oblysy; *prev.* Ural'skaya Oblast'. ◐ *province* NW Kazakhstan
126 Hh15 **Zapadnyy Sayan** *Eng.* Western Sayans. ▲▲ S Russian Federation
65 H15 **Zapala** Neuquén, W Argentina *38.54S 70.06W*
64 I4 **Zapaleri, Cerro** *var.* Cerro Sapaleri. ▲ N Chile *22.51S 67.10W*
27 Q16 **Zapata** Texas, SW USA *26.54N 99.16W*
46 D5 **Zapata, Península de** *peninsula* W Cuba
63 G19 **Zapicán** Lavalleja, S Uruguay *33.31S 54.55W*
67 J19 **Zapiola Ridge** *undersea feature* SW Atlantic Ocean
67 L19 **Zapiola Seamount** *undersea feature* S Atlantic Ocean *38.15S 26.15W*
128 I2 **Zapolyarnyy** Murmanskaya Oblast', NW Russian Federation *69.24N 30.53E*
119 U8 **Zaporizhzhya** *Rus.* Zaporozh'ye; *prev.* Aleksandrovsk. Zaporiz'ka Oblast', SE Ukraine *47.46N 35.12E*
Zaporizhzhya *see* Zaporiz'ka Oblast'
119 U9 **Zaporiz'ka Oblast'** *var.* Zaporizhzhya, *Rus.* Zaporozhskaya Oblast'. ◐ *province* SE Ukraine
Zaporozhskaya Oblast' *see* Zaporiz'ka Oblast'
Zaporozh'ye *see* Zaporizhzhya
42 L14 **Zapotiltic** Jalisco, SW Mexico *19.35N 103.25W*
164 G13 **Zapug** Xizang Zizhiqu, W China
143 V10 **Zaqatala** *Rus.* Zakataly. NW Azerbaijan *41.38N 46.37E*
165 P13 **Zaqên** Qinghai, W China *33.22N 94.31E*
165 Q13 **Za Qu** ≈ C China
142 M13 **Zara** Sivas, C Turkey *39.55N 37.43E*
Zara *see* Zadar
Zarafshan *see* Zarafshon
153 P12 **Zarafshon** *Rus.* Zeravshan. W Tajikistan *39.12N 68.36E*
152 L9 **Zarafshon** *Rus.* Zarafshan. Nawoiy Wiloyati, N Uzbekistan *41.33N 64.09E*
Zarafshon *see* Zeravshan
153 O12 **Zarafshon, Qatorkŭhi** *Rus.* Zeravshanskiy Khrebet, *Uzb.* Zarafshon Tizmasi. ▲▲ Tajikistan/Uzbekistan
Zarafshon Tizmasi *see* Zarafshon, Qatorkŭhi
56 E7 **Zaragoza** Antioquia, N Colombia *7.30N 74.52W*
42 I5 **Zaragoza** Chihuahua, N Mexico *29.36N 107.41W*
43 N6 **Zaragoza** Coahuila de Zaragoza, NE Mexico *28.30N 100.52W*
43 O10 **Zaragoza** Nuevo León, NE Mexico *23.59N 99.49W*
107 R5 **Zaragoza** *Eng.* Saragossa; *anc.* Caesaraugusta, Salduba. Aragón, NE Spain *41.39N 0.54W*
107 R6 **Zaragoza** ◐ *province* Aragón, NE Spain
107 R5 **Zaragoza** ✕ Aragón, NE Spain *41.37N 0.52W*
149 S10 **Zarand** Kermān, C Iran *30.49N 56.34E*
154 J9 **Zaranj** Nīmrūz, SW Afghanistan *30.59N 61.54E*
120 I11 **Zarasai** Zarasai, E Lithuania *55.44N 26.17E*
64 N12 **Zárate** *prev.* General José F.Uriburu. Buenos Aires, E Argentina *34.06S 59.03W*
107 Q2 **Zarautz** *var.* Zarauz. País Vasco, N Spain *43.16N 2.10W*
Zarauz *see* Zarautz
Zaravecchia *see* Biograd na Moru
Zarāyīn *see* Zarēn
130 L4 **Zaraysk** Moskovskaya Oblast', W Russian Federation *54.48N 38.54E*
57 N6 **Zaraza** Guárico, N Venezuela *9.21N 65.19W*
Zarbdar *see* Zarbdor
153 P11 **Zarbdor** *Rus.* Zarbdar. Jizzakh Wiloyati, C Uzbekistan *40.04N 68.10E*
148 M8 **Zard Kūh** ▲ SW Iran *32.19N 50.03E*
128 I5 **Zarechensk** Murmanskaya Oblast', NW Russian Federation *66.39N 31.27E*
155 Q7 **Zareh Sharan** Paktīkā, E Afghanistan *33.07N 68.46E*
41 Y14 **Zarembo Island** *island* Alexander Archipelago, Alaska, USA
145 V4 **Zarēn** *var.* Zarāyīn. E Iraq *35.16N 45.43E*
155 Q7 **Zarghūn Shahr** *var.* Katawaz. Paktīkā, SE Afghanistan *32.40N 68.19E*
79 V13 **Zaria** Kaduna, C Nigeria *11.06N 7.42E*
118 K2 **Zarichne** Rivnens'ka Oblast', NW Ukraine *51.49N 26.09E*
126 H14 **Zarinsk** Altayskiy Kray, S Russian Federation *53.34N 85.22E*
118 J12 **Zărneşti** *Hung.* Zernest. Braşov, C Romania *45.34N 25.18E*
117 J25 **Zarós** Kríti, Greece, E Mediterranean Sea *35.07N 24.54E*
102 O9 **Zarow** ≈ NE Germany
Zarqa/Zarqā', Muḩāfaẓat az *see* Az Zarqā'
113 G20 **Záruby** ▲ W Slovakia *48.30N 17.24E*
58 B8 **Zaruma** El Oro, SW Ecuador *3.41S 79.32W*
112 E13 **Żary** *Ger.* Sorau, Sorau in der Niederlausitz. Lubuskie, W Poland *51.43N 15.09E*
56 D10 **Zarzal** Valle del Cauca, W Colombia *4.22N 76.03W*
44 I7 **Zarzalar, Cerro** ▲ S Honduras *14.15N 86.49W*
158 I5 **Zāskār** ≈ NE India
158 I5 **Zāskār Range** ▲▲ NE India
121 K15 **Zaslawye** Minskaya Voblasts', C Belarus *54.01N 27.16E*
118 K7 **Zastavna** Chernivets'ka Oblast', W Ukraine *48.30N 25.51E*
113 B16 **Žatec** *Ger.* Saaz. Ústecký Kraj, NW Czech Republic *50.19N 13.32E*
Zaumgarten *see* Chrzanów
152 G10 **Zaungukskiye Garagumy** *Turkm.* Üngüz Angyrsyndaky Garagum. *desert* N Turkmenistan
27 X9 **Zavalla** Texas, SW USA *31.09N 94.25W*
101 H18 **Zaventem** Vlaams Brabant, C Belgium *50.52N 4.28E*
101 H18 **Zaventem** ✕ (Brussel/Bruxelles) Vlaams Brabant, C Belgium *50.55N 4.28E*
Zavertse *see* Zawiercie
116 L7 **Zavet** Razgrad, NE Bulgaria *43.46N 26.40E*
131 O12 **Zavetnoye** Rostovskaya Oblast', SW Russian Federation *47.10N 43.54E*
162 M3 **Zavhan Gol** ≈ W Mongolia
114 H12 **Zavidovići** Federacija Bosna I Hercegovina, N Bosnia and Herzegovina *44.26N 18.07E*
126 Mm16 **Zavitinsk** Amurskaya Oblast', SE Russian Federation *50.23N 128.57E*
125 F12 **Zavodoukovsk** Tyumenskaya Oblast', C Russian Federation *56.27N 66.37E*
Zawia *see* Az Zāwiyah
113 K15 **Zawiercie** *Rus.* Zavertse. Śląskie, S Poland *50.28N 19.24E*
77 P11 **Zawīlah** *var.* Zuwaylah, *It.* Zueila. C Libya *26.10N 15.07E*
144 I4 **Zāwīyah, Jabal az** ▲▲ NW Syria
111 Y3 **Zaya** ≈ NE Austria
177 G8 **Zayatkyi** Pegu, C Myanmar *17.48N 96.27E*
151 Y11 **Zaysan** Vostochnyy Kazakhstan, E Kazakhstan *47.28N 84.48E*
Zaysan Köl *see* Zaysan, Ozero
151 Y11 **Zaysan, Ozero** *Kaz.* Zaysan Köl. ◎ E Kazakhstan
165 R16 **Zayü** *var.* Gyigang. Xizang Zizhiqu, W China *28.36N 97.25E*
Zayyq *see* Ural
46 F6 **Zaza** ≈ C Cuba
118 K5 **Zbarazh** Ternopil's'ka Oblast', W Ukraine *49.40N 25.47E*
118 J5 **Zboriv** Ternopil's'ka Oblast', W Ukraine *49.40N 25.07E*
113 F18 **Zbraslav** Brněnský Kraj, SE Czech Republic *49.13N 16.19E*
118 K6 **Zbruch** ≈ W Ukraine
Žďár *see* Žďár nad Sázavou
113 F17 **Žďár nad Sázavou** *Ger.* Saar in Mähren; *prev.* Žďár. Jihlavský Kraj, C Czech Republic *49.34N 15.55E*
118 K4 **Zdolbuniv** *Pol.* Zdolbunów, *Rus.* Zdolbunov. Rivnens'ka Oblast', NW Ukraine *50.33N 26.15E*
Zdolbunov/Zdolbunów *see* Zdolbuniv
112 J13 **Zduńska Wola** Sieradz, C Poland *51.37N 18.57E*
119 O4 **Zdvizh** ≈ N Ukraine
Zdzięciół *see* Dzyatlava
113 I16 **Zdzieszowice** *Ger.* Odertal. Opolskie, S Poland *50.24N 18.06E*
Zealand *see* Sjælland
196 K6 **Zealandia Bank** *undersea feature* C Pacific Ocean
65 H20 **Zeballos, Monte** ▲ S Argentina *47.04S 71.32W*
85 K20 **Zebediela** Northern, NE South Africa *24.16S 29.21E*
115 L18 **Zebë, Mal** *var.* Mali i Zebës. ▲ NE Albania *41.57N 20.16E*
Zebës, Mali i *see* Zebë, Mal
23 V9 **Zebulon** North Carolina, SE USA *35.49N 78.19W*
114 K8 **Žednik** *Hung.* Bácsjózseffalva. Serbia, N Yugoslavia *45.58N 19.40E*
101 C15 **Zeebrugge** West-Vlaanderen, NW Belgium *51.19N 3.13E*
191 N16 **Zeehan** Tasmania, SE Australia *41.54S 145.19E*
101 L14 **Zeeland** Noord-Brabant, SE Netherlands *51.42N 5.40E*
31 N7 **Zeeland** North Dakota, N USA *45.57N 99.49W*
101 E14 **Zeeland** ◐ *province* SW Netherlands
85 I21 **Zeerust** North-West, N South Africa *25.33S 26.04E*
100 K10 **Zeewolde** Flevoland, C Netherlands *52.19N 5.31E*
144 G8 **Ẕefat** *var.* Safed, Tsefat, *Ar.* Safad. Northern, N Israel *32.57N 35.27E*
Żegań *see* Żagań
Zehden *see* Cedynia
102 O11 **Zehdenick** Brandenburg, NE Germany *52.58N 13.19E*
Zē-i Bādīnān *see* Great Zab
Zeiden *see* Codlea
152 M14 **Zeidskoye Vodokhranilishche** ⊡ E Turkmenistan
Zē-i Kōya *see* Little Zab
189 P7 **Zeil, Mount** ▲ Northern Territory, C Australia *23.31S 132.41E*
100 J12 **Zeist** Utrecht, C Netherlands *52.04N 5.15E*
103 M16 **Zeitz** Sachsen-Anhalt, E Germany *51.03N 12.07E*
165 T11 **Zêkog** Qinghai, C China *35.03N 101.30E*
Zelaya Norte *see* Atlántico Norte, Región Autónoma
Zelaya Sur *see* Atlántico Sur, Región Autónoma
101 F17 **Zele** Oost-Vlaanderen, NW Belgium *51.04N 4.01E*
112 N12 **Żelechów** Lubelskie, E Poland *51.49N 21.57E*
115 H14 **Zelena Glava** ▲ SE Bosnia and Herzegovina *43.32N 17.55E*
115 I14 **Zelengora** ▲▲ S Bosnia and Herzegovina
128 I5 **Zelenoborskiy** Murmanskaya Oblast', NW Russian Federation *66.52N 32.25E*
131 R3 **Zelenodol'sk** Respublika Tatarstan, W Russian Federation *55.52N 48.49E*
128 G12 **Zelenogorsk** *Fin.* Terijoki. Leningradskaya Oblast', NW Russian Federation *60.08N 30.06E*
130 K3 **Zelenograd** Moskovskaya Oblast', W Russian Federation *56.02N 37.08E*
120 B13 **Zelenogradsk** *Ger.* Cranz, Kranz. Kaliningradskaya Oblast', W Russian Federation *54.57N 20.30E*
131 O15 **Zelenokumsk** Stavropol'skiy Kray, SW Russian Federation *44.22N 43.48E*
172 Rr7 **Zelënyy, Ostrov** *var.* Shibotsu-jima. *island* NE Russian Federation
Železna Kapela *see* Eisenkappel
Železna Vrata *see* Demir Kapija
114 L11 **Železniki** Serbia, N Yugoslavia *44.45N 20.23E*
100 N12 **Zelhem** Gelderland, E Netherlands *52.00N 6.21E*
115 N18 **Želino** NW FYR Macedonia *42.00N 21.06E*
115 M14 **Željin** ▲▲ C Yugoslavia
103 K17 **Zella-Mehlis** Thüringen, C Germany *50.40N 10.40E*
111 P7 **Zell am See** *var.* Zell-am-See. Salzburg, S Austria *47.19N 12.48E*
111 N7 **Zell am Ziller** Tirol, W Austria *47.13N 11.52E*
Zelle *see* Celle
111 W2 **Zellerndorf** Niederösterreich, NE Austria *48.40N 15.57E*
111 U7 **Zeltweg** Steiermark, S Austria *47.10N 14.43E*
121 G17 **Zel'va** *Pol.* Zelwa. Hrodzyenskaya Voblasts', W Belarus *53.09N 24.49E*
120 H13 **Želva** Ukmergė, C Lithuania *55.13N 25.07E*
Zelwa *see* Zel'va
101 E16 **Zelzate** *var.* Selzaete. Oost-Vlaanderen, NW Belgium *51.12N 3.49E*
120 E11 **Žemaičių Aukštumas** *physical region* W Lithuania
120 C12 **Žemaičių Naumiestis** Šilutė, SW Lithuania *55.22N 21.39E*
Zembin *see* Zyembin
131 N6 **Zemetchino** Penzenskaya Oblast', W Russian Federation *53.31N 42.35E*
81 M15 **Zémio** Haut-Mbomou, E Central African Republic *5.04N 25.07E*
43 R16 **Zempoaltepec, Cerro** ▲ SE Mexico *17.04N 95.54W*
101 G17 **Zemst** Vlaams Brabant, C Belgium *50.59N 4.28E*
114 L11 **Zemun** Serbia, N Yugoslavia *44.51N 20.24E*
Zendajan *see* Zendeh Jan
154 J5 **Zendeh Jan** *var.* Zendajan, Zindajān. Herāt, NW Afghanistan *34.55N 61.53E*
Zengg *see* Senj
114 H12 **Zenica** Federacija Bosna I Hercegovina, C Bosnia and Herzegovina *44.12N 17.52E*
Zenjan *see* Zanjān
Zen'kov *see* Zin'kiv
Zenshū *see* Chŏnju
Zenta *see* Senta
170 Ff14 **Zentsūji** *var.* Zentûzi. Kagawa, Shikoku, SW Japan *34.13N 133.45E*
Zentûzi *see* Zentsūji
84 B11 **Zenza do Itombe** Cuanza Norte, NW Angola *9.22S 14.10E*
114 H12 **Žepče** Federacija Bosna I Hercegovina, N Bosnia and Herzegovina *44.26N 18.00E*
25 W12 **Zephyrhills** Florida, SE USA *28.13N 82.10W*
199 J10 **Zephyr Reef** *reef* Antarctica
164 F9 **Zepu** *var.* Poskam. Xinjiang Uygur Zizhiqu, NW China *38.10N 77.18E*
153 Q12 **Zeravshan** *Taj./Uzb.* Zarafshon. ≈ Tajikistan/Uzbekistan
Zeravshan *see* Zarafshon
Zeravshanskiy Khrebet *see* Zarafshon, Qatorkŭhi
103 M14 **Zerbst** Sachsen-Anhalt, E Germany *51.57N 12.05E*
151 P8 **Zerenda** Severnyy Kazakhstan, N Kazakhstan *52.55N 69.09E*
112 H12 **Żerków** Wielkopolskie, C Poland *52.03N 17.33E*
110 E11 **Zermatt** Valais, SW Switzerland *46.00N 7.44E*
Zernest *see* Zărneşti
110 J9 **Zernez** Graubünden, SE Switzerland *46.42N 10.06E*
130 L12 **Zernograd** Rostovskaya Oblast', SW Russian Federation *46.52N 40.13E*
Zestafoni *see* Zestap'oni
143 S9 **Zestap'oni** *Rus.* Zestafoni. C Georgia *42.09N 43.00E*
100 H12 **Zestienhoven** ✕ (Rotterdam) Zuid-Holland, SW Netherlands *51.57N 4.30E*
115 J16 **Zeta** ≈ SW Yugoslavia
15 J2 **Zeta Lake** ◎ Victoria Island, Nunavut, N Canada
100 L12 **Zetten** Gelderland, SE Netherlands *51.55N 5.43E*
103 M17 **Zeulenroda** Thüringen, C Germany *50.40N 11.58E*
102 H10 **Zeven** Niedersachsen, NW Germany *53.17N 9.16E*
100 M12 **Zevenaar** Gelderland, SE Netherlands *51.55N 6.04E*
101 H14 **Zevenbergen** Noord-Brabant, S Netherlands *51.39N 4.36E*
126 M14 **Zeya** Amurskaya Oblast', SE Russian Federation *53.51N 126.53E*
133 X6 **Zeya** ≈ SE Russian Federation
126 **Zeya Reservoir** *see* Zeyskoye Vodokhranilishche
149 T11 **Zeynalābād** Kermān, C Iran *29.55N 57.28E*
126 M14 **Zeyskoye Vodokhranilishche** *Eng.* Zeya Reservoir. ⊡ SE Russian Federation
106 H8 **Zêzere, Rio** ≈ C Portugal
Zgerzh *see* Zgierz
144 H6 **Zgharta** N Lebanon *34.24N 35.54E*
112 K12 **Zgierz** *Ger.* Neuhof, *Rus.* Zgerzh. Łódź, C Poland *51.55N 19.19E*
113 E14 **Zgorzelec** *Ger.* Görlitz. Dolnośląskie, SW Poland *51.10N 15.00E*
Zhabdün *see* Zhongba
121 F19 **Zhabinka** *Pol.* Żabinka, *Rus.* Zhabinka. Brestskaya Voblasts', SW Belarus *52.12N 24.01E*
Zhaggo *see* Luhuo
165 R15 **Zhag'yab** Xizang Zizhiqu, W China *30.42N 97.33E*
150 L9 **Zhailma** *Kaz.* Zhayylma. Kostanay, N Kazakhstan *51.34N 61.39E*
151 V16 **Zhalanash** Almaty, SE Kazakhstan *43.07N 78.40E*
Zhalashash *see* Dzhalagash
151 S7 **Zhalauly, Ozero** ◎ NE Kazakhstan
150 E9 **Zhalpaktal** *prev.* Furmanovo. Zapadnyy Kazakhstan, W Kazakhstan *49.40N 49.27E*
121 G16 **Zhaludok** *Rus.* Zheludok. Hrodzyenskaya Voblasts', W Belarus *53.36N 24.58E*
Zhaman-Akkol', Ozero *see* Akkol', Ozero
151 R16 **Zhambyl** *see* Taraz
151 Q14 **Zhambyl** *off.* Zhambylskaya Oblast', *Kaz.* Zhambyl Oblysy; *prev.* Dzhambulskaya Oblast'. ◐ *province* S Kazakhstan
Zhambyl Oblysy/Zhambylskaya Oblast' *see* Zhambyl
Zhamo *see* Bomi
151 S12 **Zhamshy** ≈ C Kazakhstan
150 M15 **Zhanadar'ya** Kzylorda, S Kazakhstan *44.41N 64.39E*
151 O15 **Zhanakorgan** *Kaz.* Zhangaqorghan. Kzylorda, S Kazakhstan *43.57N 67.14E*
165 N16 **Zhanang** Xizang Zizhiqu, W China *29.15N 91.19E*
151 T12 **Zhanaortalyk** Zhezkazgan, C Kazakhstan *47.31N 75.42E*
150 F15 **Zhanaozen** *Kaz.* Zhangaözen, *prev.* Novyy Uzen'. Mangistau, W Kazakhstan *43.21N 52.50E*
151 Q16 **Zhanatas** Zhambyl, S Kazakhstan *43.33N 69.40E*
Zhangaözen *see* Zhanaozen
Zhangaqazaly *see* Ayteke Bi
Zhangaqorghan *see* Zhanakorgan
167 O2 **Zhangbei** Hebei, E China *41.13N 114.43E*
Zhangdian *see* Zibo
169 X9 **Zhangguangcai Ling** ▲▲ NE China
151 W10 **Zhangiztobe** Vostochnyy Kazakhstan, E Kazakhstan *49.16N 81.16E*
165 W11 **Zhangjiachuan** Gansu, N China *34.55N 106.25E*
166 L10 **Zhangjiajie** *var.* Dayong. Hunan, S China *29.10N 110.22E*
167 O2 **Zhangjiakou** *var.* Changkiakow, Zhang-chia-k'ou, *Eng.* Kalgan; *prev.* Wanchuan. Hebei, E China *40.48N 114.51E*
167 Q13 **Zhangping** Fujian, SE China *25.21N 117.29E*
167 Q13 **Zhangpu** Fujian, SE China *24.07N 117.36E*
169 U11 **Zhangwu** Liaoning, NE China *42.21N 122.32E*
165 S8 **Zhangye** Gansu, N China *38.59N 100.27E*
167 Q13 **Zhangzhou** Fujian, SE China *24.31N 117.40E*
169 W6 **Zhan He** ≈ NE China
Zhänibek *see* Dzhanybek
166 L16 **Zhanjiang** *var.* Chanchiang, Chan-chiang, *Cant.* Tsamkong, *Fr.* Fort-Bayard. Guangdong, S China *21.10N 110.19E*
Zhansügirov *see* Dzhansugurov
169 V8 **Zhaodong** Heilongjiang, NE China *46.03N 125.58E*
Zhaoge *see* Qixian
166 H11 **Zhaojue** Sichuan, C China *28.03N 102.50E*
167 N14 **Zhaoqing** Guangdong, S China *23.07N 112.26E*
164 H5 **Zhaosu** *var.* Mongolküre. Xinjiang Uygur Zizhiqu, NW China *43.09N 81.07E*
166 H11 **Zhaotong** Yunnan, SW China *27.17N 103.42E*
169 V9 **Zhaoyuan** Heilongjiang, NE China *45.30N 125.04E*
169 V9 **Zhaozhou** Heilongjiang, NE China *45.40N 125.16E*
151 X13 **Zharbulak** Vostochnyy Kazakhstan, E Kazakhstan *46.05N 82.06E*
164 J15 **Zhari Namco** ◎ W China
150 I12 **Zharkamys** *Kaz.* Zharqamys. Aktyubinsk, W Kazakhstan *47.58N 56.33E*
151 W15 **Zharkent** *prev.* Panfilov. Almaty, SE Kazakhstan *44.10N 80.01E*
128 H17 **Zharkovskiy** Tverskaya Oblast', W Russian Federation *55.51N 32.19E*
151 W11 **Zharma** Vostochnyy Kazakhstan, E Kazakhstan *48.48N 80.55E*
150 F14 **Zharmysh** Mangistau, SW Kazakhstan *44.12N 52.27E*
Zharqamys *see* Zharkamys
120 L13 **Zhary** *Rus.* Zhary. Vitsyebskaya Voblasts', N Belarus *55.04N 28.40E*
Zhaslyk *see* Jasliq
126 Ll11 **Zhatay** Respublika Sakha (Yakutiya), NE Russian Federation *62.07N 129.42E*
164 J14 **Zhaxi Co** ◎ W China
Zhayylma *see* Zhailma
Zhdanov *see* Beyläqan, Azerbaijan
Zhdanov *see* Mariupol', Ukraine
Zhe *see* Zhejiang
167 R10 **Zhejiang** *var.* Che-chiang, Chekiang, Zhe, Zhejiang Sheng. ◐ *province* SE China
Zhejiang Sheng *see* Zhejiang
151 S7 **Zhelezinka** Pavlodar, N Kazakhstan *53.35N 75.16E*
121 C14 **Zheleznodorozhnyy** *Ger.* Gerdauen. Kaliningradskaya Oblast', W Russian Federation *54.21N 21.17E*
126 J13 **Zheleznodorozhnyy** Irkutskaya Oblast', C Russian Federation *57.54N 102.40E*
Zheleznodorozhnyy *see* Yemva
130 J7 **Zheleznogorsk** Kurskaya Oblast', W Russian Federation *52.22N 35.21E*
126 Jj14 **Zheleznogorsk-Ilimskiy** Irkutskaya Oblast', S Russian Federation *56.38N 103.53E*
131 N15 **Zheleznovodsk** Stavropol'skiy Kray, SW Russian Federation *44.12N 43.01E*
Zhëltyye Vody *see* Zhovti Vody
Zheludok *see* Zhaludok
Zhem *see* Emba
166 K7 **Zhenba** Shaanxi, C China *32.42N 107.55E*
166 I13 **Zhenfeng** Guizhou, S China *25.27N 105.38E*
Zhengjiatun *see* Shuangliao
165 X10 **Zhengning** Gansu, N China *35.28N 108.21E*
169 Q12 **Zhengxiangbai Qi** Nei Mongol Zizhiqu, N China *42.10N 114.57E*
167 N6 **Zhengzhou** *var.* Ch'eng-chou, Chengchow; *prev.* Chenghsien. Henan, C China *34.45N 113.37E*
167 R8 **Zhenjiang** *var.* Chenkiang. Jiangsu, E China *32.12N 119.30E*
169 U9 **Zhenlai** Jilin, NE China *45.52N 123.11E*
166 I11 **Zhenxiong** Yunnan, SW China *27.31N 104.52E*
166 K11 **Zhenyuan** *prev.* Wuyang. Guizhou, S China *27.07N 108.33E*
167 R11 **Zherong** Fujian, SE China *27.16N 119.54E*
Zhetiqara *see* Dzhetygara
150 F15 **Zhetybay** Mangistau, SW Kazakhstan *43.30N 52.09E*
151 P17 **Zhetysay** *var.* Dzhetysay. Yuzhnyy Kazakhstan *40.45N 68.18E*
166 M11 **Zhexi Shuiku** ⊡ C China
151 O12 **Zhezdy** Zhezkazgan, C Kazakhstan *48.06N 67.01E*
151 O12 **Zhezkazgan** *Kaz.* Zhezqazghan; *prev.* Dzhezkazgan. Zhezkazgan, C Kazakhstan *47.48N 67.43E*
Zhezqazghan *see* Zhezkazgan
166 M9 **Zhicheng** Hubei, C China *30.21N 111.27E*
Zhidachov *see* Zhydachiv
165 Q12 **Zhidoi** Qinghai, C China *33.55N 95.39E*
126 Jj15 **Zhigalovo** Irkutskaya Oblast', S Russian Federation *54.47N 105.00E*
126 L9 **Zhigansk** Respublika Sakha (Yakutiya), NE Russian Federation *66.45N 123.20E*
131 R6 **Zhigulevsk** Samarskaya Oblast', W Russian Federation *53.24N 49.30E*
120 D13 **Zhilino** *Ger.* Schillen. Kaliningradskaya Oblast', W Russian Federation *54.55N 21.54E*
131 O8 **Zhirnovsk** Volgogradskaya Oblast', SW Russian Federation *51.01N 44.49E*
Zhitarovo *see* Vetren
Zhitkovichi *see* Zhytkavichy
131 P10 **Zhitkur** Volgogradskaya Oblast', SW Russian Federation *49.00N 46.16E*
Zhitomir *see* Zhytomyr
Zhitomirskaya Oblast' *see* Zhytomyrs'ka Oblast'
130 J5 **Zhizdra** Kaluzhskaya Oblast', W Russian Federation *53.38N 34.39E*
121 N18 **Zhlobin** Homyel'skaya Voblasts', SE Belarus *52.52N 30.01E*
Zhmerinka *see* Zhmerynka
118 M7 **Zhmerynka** *Rus.* Zhmerinka. Vinnyts'ka Oblast', C Ukraine *49.01N 28.01E*
155 R9 **Zhob** *var.* Fort Sandeman. Baluchistān, SW Pakistan *31.22N 69.25E*
155 R8 **Zhob** ≈ C Pakistan
150 I10 **Zhobda** *prev.* Novoalekseyevka. Aktyubinsk, W Kazakhstan *50.10N 55.39E*
Zhodino *see* Zhodzina
121 L15 **Zhodzina** *Rus.* Zhodino. Minskaya Voblasts', C Belarus *54.05N 28.19E*
126 Mm3 **Zhokhova, Ostrov** *island* Novosibirskiye Ostrova, NE Russian Federation
Zholkev/Zholkva *see* Zhovkva
Zholsaly *see* Dzhusaly
Zhondor *see* Jondor
164 I15 **Zhongba** *var.* Zhabdün. Xizang Zizhiqu, W China *29.37N 84.11E*
166 F11 **Zhongdian** Yunnan, SW China *27.48N 99.40E*
Zhonghua Renmin Gongheguo *see* China
165 V9 **Zhongning** Ningxia, N China *37.25N 105.40E*
167 N15 **Zhongshan** Guangdong, S China *22.30N 113.19E*
205 X7 **Zhongshan** *Chinese research station* Antarctica *69.23S 76.34E*
166 M6 **Zhongtiao Shan** ▲▲ C China
165 V9 **Zhongwei** Ningxia, N China *37.31N 105.10E*
166 K9 **Zhongxian** *var.* Zhong Xian. Chongqing Shi, C China *30.16N 108.03E*
167 N9 **Zhongxiang** Hubei, C China *31.12N 112.35E*
167 O7 **Zhoukou** *var.* Zhoukouzhen. Henan, C China *33.37N 114.34E*
Zhoukouzhen *see* Zhoukou
167 S9 **Zhoushan** Zhejiang, S China
167 S9 **Zhoushan Qundao** *Eng.* Zhoushan Islands. *island group* SE China
118 I5 **Zhovkva** *Pol.* Żółkiew, *Rus.* Zholkev, Zholkva; *prev.* Nesterov. L'vivs'ka Oblast', NW Ukraine *50.03N 24.00E*
119 S7 **Zhovti Vody** *Rus.* Zhëltyye Vody. Dnipropetrovs'ka Oblast', E Ukraine *48.24N 33.30E*
119 Q10 **Zhovtneve** *Rus.* Zhovtnevoye. Mykolayivs'ka Oblast', S Ukraine *46.50N 32.00E*
Zhovtnevoye *see* Zhovtneve
116 K9 **Zhrebchevo, Yazovir** ⊡ C Bulgaria
169 V13 **Zhuanghe** Liaoning, NE China *39.43N 122.57E*
165 W11 **Zhuanglang** Gansu, C China *35.06N 106.21E*
151 P15 **Zhuantobe** *Kaz.* Zhūantöbe. Yuzhnyy Kazakhstan, S Kazakhstan *44.45N 68.50E*
167 Q5 **Zhucheng** Shandong, E China *35.58N 119.24E*
165 V12 **Zhugqu** Gansu, C China *33.51N 104.14E*
167 N15 **Zhuhai** Guangdong, S China *22.16N 113.30E*
Zhuizishan *see* Weichang
Zhuji *see* Shangqiu
130 I5 **Zhukovka** Bryanskaya Oblast', W Russian Federation *53.33N 33.48E*
167 O3 **Zhuozhou** *prev.* Zhuo Xian. Hebei, E China *39.22N 115.40E*
168 L14 **Zhuozi Shan** ▲ N China *39.28N 106.58E*
Zhuravichi *see* Zhuravichy
121 O17 **Zhuravichy** *Rus.* Zhuravichi. Homyel'skaya Voblasts', SE Belarus *53.15N 30.29E*
151 Q8 **Zhuravlevka** Akmola, N Kazakhstan *51.56N 69.56E*
119 Q4 **Zhurivka** Kyyivs'ka Oblast', N Ukraine *50.28N 31.48E*
150 J11 **Zhuryn** Aktyubinsk, W Kazakhstan *49.13N 57.36E*
151 T15 **Zhusandala, Step'** *grassland* SE Kazakhstan
166 L8 **Zhushan** Hubei, C China *32.11N 110.05E*
167 N11 **Zhuzhou** Hunan, S China *27.52N 112.52E*
118 I6 **Zhydachiv** *Pol.* Żydaczów, *Rus.* Zhidachov. L'vivs'ka Oblast', NW Ukraine *49.22N 24.09E*
Zhympīty *see* Dzhambeyty
121 K19 **Zhytkavichy** *Rus.* Zhitkovichi. Homyel'skaya Voblasts', SE Belarus *52.13N 27.52E*
119 N4 **Zhytomyr** *Rus.* Zhitomir. Zhytomyrs'ka Oblast', NW Ukraine *50.17N 28.39E*
Zhytomyr *see* Zhytomyrs'ka Oblast'
118 M4 **Zhytomyrs'ka Oblast'** *var.* Zhytomyr, *Rus.* Zhitomirskaya Oblast'. ◐ *province* N Ukraine
159 U15 **Zia** ✕ (Dhaka) Dhaka, C Bangladesh
113 J20 **Žiar nad Hronom** *var.* Sväty Kríž nad Hronom, *Ger.* Heiligenkreuz, *Hung.* Garamszentkereszt. Banskobystrický Kraj, C Slovakia *48.36N 18.52E*
167 Q4 **Zibo** *var.* Zhangdian. Shandong, E China *36.51N 118.01E*
166 L4 **Zichang** *prev.* Wayaobu. Shaanxi, C China *37.08N 109.40E*
Zichenau *see* Ciechanów
113 G15 **Ziębice** *Ger.* Münsterberg in Schlesien. Dolnośląskie, SW Poland *50.37N 17.01E*
Ziebingen *see* Cybinka
Ziegenhais *see* Głuchołazy
112 E12 **Zielona Góra** *Ger.* Grünberg, Grünberg in Schlesien, Grüneberg. Lubuskie, W Poland *51.55N 15.30E*
101 F14 **Zierikzee** Zeeland, SW Netherlands *51.39N 3.55E*
166 I10 **Zigong** *var.* Tzekung. Sichuan, C China *29.19N 104.48E*
78 G12 **Ziguinchor** SW Senegal *12.33N 16.19W*
43 N16 **Zihuatanejo** Guerrero, S Mexico *17.39N 101.33W*
Zilah *see* Zalău
131 W7 **Zilair** Respublika Bashkortostan, W Russian Federation *52.12N 57.15E*
142 L12 **Zile** Tokat, N Turkey *40.18N 35.52E*
113 J18 **Žilina** *Ger.* Sillein, *Hung.* Zsolna. Žilinský Kraj, N Slovakia *49.13N 18.43E*
113 J19 **Žilinský Kraj** ◐ *region* N Slovakia
77 Q9 **Zillah** *var.* Zallah. C Libya *28.30N 17.33E*
Zillenmarkt *see* Zalău
111 N7 **Ziller** ≈ W Austria
Zillertal Alps *see* Zillertaler Alpen
111 N8 **Zillertaler Alpen** *Eng.* Zillertal Alps, *It.* Alpi Aurine. ▲▲ Austria/Italy
120 K10 **Zilupe** *Ger.* Rosenhof. Ludza, E Latvia *56.10N 28.06E*
126 J15 **Zima** Irkutskaya Oblast', S Russian Federation *53.57N 101.57E*
43 O13 **Zimapán** Hidalgo, C Mexico *20.42N 99.23W*
85 I16 **Zimba** Southern, S Zambia *17.16S 26.10E*
85 J17 **Zimbabwe** *off.* Republic of Zimbabwe; *prev.* Rhodesia. ◆ *republic* S Africa
118 H10 **Zimbor** *Hung.* Magyarzsombor. Sălaj, NW Romania *47.00N 23.16E*
Zimmerbude *see* Svetlyy
118 J15 **Zimnicea** Teleorman, S Romania *43.39N 25.21E*
116 L9 **Zimnitsa** Yambol, E Bulgaria *42.34N 26.37E*
131 N12 **Zimovniki** Rostovskaya Oblast', SW Russian Federation *47.07N 42.28E*
Zindajān *see* Zendeh Jan
79 V12 **Zinder** Zinder, S Niger *13.46N 9.01E*
79 W11 **Zinder** ◐ *department* S Niger
79 P12 **Ziniaré** C Burkina *12.35N 1.21W*
Zinjan *see* Zanjān
147 P16 **Zinjibār** SW Yemen *13.07N 45.22E*
119 T4 **Zin'kiv** *var.* Zen'kov. Poltavs'ka Oblast', NE Ukraine *50.11N 34.21E*
Zinov'yevsk *see* Kirovohrad
Zintenhof *see* Sindi
33 N10 **Zion** Illinois, N USA *42.27N 87.49W*
56 F10 **Zipaquirá** Cundinamarca, C Colombia *5.03N 74.01W*
Zipser Neudorf *see* Spišská Nová Ves
113 H23 **Zirc** Veszprém, W Hungary *47.16N 17.52E*
115 D14 **Žirje** *It.* Zuri. *island* S Croatia
Zirknitz *see* Cerknica
110 M7 **Zirl** Tirol, W Austria *47.16N 11.16E*
103 K20 **Zirndorf** Bayern, SE Germany *49.27N 10.57E*
166 M11 **Zi Shui** ≈ C China
111 Y3 **Zistersdorf** Niederösterreich, NE Austria *48.31N 16.45E*
43 O14 **Zitácuaro** Michoacán de Ocampo, SW Mexico *19.28N 100.21W*
103 Q16 **Zittau** Sachsen, E Germany *50.53N 14.48E*
114 I12 **Živinice** Federacija Bosna I Hercegovina, E Bosnia and Herzegovina *44.26N 18.39E*
Ziwa Magharibi *see* Kagera
83 J14 **Ziway Hāyk'** ◎ C Ethiopia
167 N12 **Zixing** Hunan, S China *26.01N 113.25E*
131 W7 **Ziyanchurino** Orenburgskaya Oblast', W Russian Federation *51.36N 56.58E*

◆ Country
● Country Capital
◇ Dependent Territory
○ Dependent Territory Capital
◐ Administrative Region
✕ International Airport
▲ Mountain
▲▲ Mountain Range
Volcano
≈ River
◎ Lake
⊡ Reservoir

166 K8 **Ziyang** Shaanxi, C China *32.33N 108.27E*
113 I20 **Zlaté Moravce** *Hung.* Aranyosmarót. Nitriansky Kraj, SW Slovakia *48.22N 18.22E*
114 K13 **Zlatibor** ▲ W Yugoslavia
116 L9 **Zlati Voyvoda** Sliven, E Bulgaria *42.36N 26.13E*
118 G11 **Zlatna** *Ger.* Kleinschlatten, *Hung.* Zalatna; *prev. Ger.* Goldmarkt. Alba, C Romania *46.07N 23.10E*
116 I8 **Zlatna Panega** Lovech, N Bulgaria *43.07N 24.09E*
116 N8 **Zlatni Pyasŭtsi** Dobrich, NE Bulgaria *43.19N 28.03E*
125 E11 **Zlatoust** Chelyabinskaya Oblast', C Russian Federation *55.12N 59.33E*
113 M19 **Zlatý Stôl** *Ger.* Goldener Tisch, *Hung.* Aranyosasztal. ▲ C Slovakia *48.45N 20.39E*
115 P18 **Zletovo** NE FYR Macedonia *42.00N 22.14E*
113 H18 **Zlín** *prev.* Gottwaldov. Zlínský Kraj, SE Czech Republic *49.13N 17.40E*
113 H19 **Zlínský Kraj** ◐ *region* E Czech Republic
77 O7 **Zlīṭan** W Libya *32.28N 14.34E*
112 F9 **Złocieniec** *Ger.* Falkenburg in Pommern. Zachodniopomorskie, NW Poland *53.31N 16.01E*
112 J13 **Złoczew** Sieradz, S Poland *51.24N 18.36E*
Złoczów *see* Zolochiv
113 F14 **Złotoryja** *Ger.* Goldberg. Dolnośląskie, W Poland *51.07N 15.57E*
112 G9 **Złotów** Wielkopolskie, NW Poland *53.22N 17.01E*
126 Gg15 **Zmeinogorsk** Altayskiy Kray, S Russian Federation *51.07N 82.16E*
112 G13 **Żmigród** *Ger.* Trachenberg. Dolnośląskie, SW Poland *51.31N 16.55E*
130 J6 **Zmiyevka** Orlovskaya Oblast', W Russian Federation *52.39N 36.20E*
119 V5 **Zmiyiv** Kharkivs'ka Oblast', E Ukraine *49.40N 36.22E*
Zna *see* Tsna
Znaim *see* Znojmo
130 M7 **Znamenka** Tambovskaya Oblast', W Russian Federation *52.24N 42.28E*
Znamenka *see* Znam"yanka
121 C14 **Znamensk** *Ger.* Wehlau. Kaliningradskaya Oblast', W Russian Federation *54.37N 21.13E*
125 Ff12 **Znamenskoye** Omskaya Oblast', C Russian Federation *57.09N 73.40E*
119 R7 **Znam"yanka** *Rus.* Znamenka. Kirovohrads'ka Oblast', C Ukraine *48.41N 32.39E*
112 H10 **Żnin** Kujawsko-pomorskie, C Poland *52.50N 17.40E*
113 F19 **Znojmo** *Ger.* Znaim. Brněnský Kraj, S Czech Republic *48.52N 16.04E*
81 N16 **Zobia** Orientale, N Dem. Rep. Congo (Zaire) *2.57N 25.55E*
85 N15 **Zóbuè** Tete, NW Mozambique *15.28S 34.25E*
100 G12 **Zoetermeer** Zuid-Holland, W Netherlands *52.04N 4.30E*
110 E7 **Zofingen** Aargau, N Switzerland *47.16N 7.56E*
165 R15 **Zogang** *var.* Wangda. Xizang Zizhiqu, W China *29.41N 97.54E*
108 E7 **Zogno** Lombardia, N Italy *45.49N 9.42E*
148 M10 **Zohreh, Rūd-e** ↝ SW Iran
166 H7 **Zoigê** Sichuan, C China *33.44N 102.57E*
Żółkiew *see* Zhovkva
110 D8 **Zollikofen** Bern, W Switzerland *47.01N 7.28E*
Zolochev *see* Zolochiv
119 U4 **Zolochiv** *Rus.* Zolochev. Kharkivs'ka Oblast', E Ukraine *50.16N 35.58E*
118 J5 **Zolochiv** *Pol.* Złoczów, *Rus.* Zolochev. L'vivs'ka Oblast', W Ukraine *49.48N 24.51E*
119 X7 **Zolote** *Rus.* Zolotoye. Luhans'ka Oblast', E Ukraine *48.42N 38.33E*
119 Q6 **Zolotonosha** Cherkas'ka Oblast', C Ukraine *49.39N 32.04E*
Zolotoye *see* Zolote
Zólyom *see* Zvolen
85 N15 **Zomba** Southern, S Malawi *15.22S 35.22E*
Zombor *see* Sombor
101 D17 **Zomergem** Oost-Vlaanderen, NW Belgium *51.07N 3.31E*
153 P11 **Zomin** *Rus.* Zaamin. Jizzakh Wiloyati, C Uzbekistan *39.56N 68.16E*
81 I15 **Zongo** Equateur, N Dem. Rep. Congo (Zaire) *4.17N 18.42E*
142 G10 **Zonguldak** Zonguldak, NW Turkey *41.25N 31.46E*
142 H10 **Zonguldak** ◐ *province* NW Turkey
101 K17 **Zonhoven** Limburg, NE Belgium *50.58N 5.22E*
148 J2 **Zonūz** Āzarbāyjān-e Khāvarī, NW Iran *38.31N 45.54E*
105 Y16 **Zonza** Corse, France, C Mediterranean Sea *41.49N 9.13E*
Zoppot *see* Sopot
Zorgho *see* Zorgo
79 Q13 **Zorgo** *var.* Zorgho. C Burkina *12.15N 0.37W*
106 K10 **Zorita** Extremadura, W Spain *39.18N 5.42W*
153 U14 **Zorkŭl** *Rus.* Ozero Zorkul'. ⊚ SE Tajikistan
Zorkul', Ozero *see* Zorkŭl
58 A8 **Zorritos** Tumbes, N Peru *3.40S 80.36W*
113 J16 **Żory** *var.* Zory, *Ger.* Sohrau. Śląskie, S Poland *50.03N 18.41E*
78 K15 **Zorzor** N Liberia *7.46N 9.28W*
101 E18 **Zottegem** Oost-Vlaanderen, NW Belgium *50.52N 3.49E*
79 R15 **Zou** ↝ S Benin
80 H6 **Zouar** Borkou-Ennedi-Tibesti, N Chad *20.24N 16.28E*
78 J6 **Zouérat** *var.* Zouérate, Zouîrât. Tiris Zemmour, N Mauritania *22.44N 12.29W*
Zouérate *see* Zouérat
Zoug *see* Zug
Zouîrât *see* Zouérat
78 M16 **Zoukougbeu** C Ivory Coast *9.47N 6.50W*
100 M5 **Zoutkamp** Groningen, NE Netherlands *53.22N 6.17E*
101 J18 **Zoutleeuw** *Fr.* Leau. Vlaams Brabant, C Belgium *50.49N 5.06E*
114 L9 **Zrenjanin** *prev.* Petrovgrad, Veliki Bečkerek, *Ger.* Grossbetschkerek, *Hung.* Nagybecskerek. Serbia, N Yugoslavia *45.22N 20.24E*
114 E10 **Zrinska Gora** ▲ C Croatia
Zsablya *see* Žabalj
103 N16 **Zschopau** ↝ E Germany
Zsebely *see* Jebel
Zsibó *see* Jibou
Zsil/Zsily *see* Jiu
Zsilyvajdevulkán *see* Vulcan
Zsolna *see* Žilina
Zsombolya *see* Jimbolia
Zsupanya *see* Županja
57 N7 **Zuata** Anzoátegui, NE Venezuela *8.21N 65.12W*
107 N14 **Zubia** Andalucía, S Spain *37.10N 3.36W*
67 P16 **Zubov Seamount** *undersea feature* E Atlantic Ocean *20.45S 8.45E*
128 I16 **Zubtsov** Tverskaya Oblast', W Russian Federation *56.10N 34.34E*
110 M8 **Zuckerhütl** ▲ SW Austria *46.57N 11.07E*
Zueila *see* Zawīlah
78 M16 **Zuénoula** C Ivory Coast *7.25N 6.03W*
107 S5 **Zuera** Aragón, NE Spain *41.52N 0.46W*
147 V13 **Ẓufār** *Eng.* Dhofar. *physical region* SW Oman
110 G8 **Zug** *Fr.* Zoug. Zug, C Switzerland *47.10N 8.30E*
110 G8 **Zug** *Fr.* Zoug. ◐ *canton* C Switzerland
143 R9 **Zugdidi** W Georgia *42.30N 41.52E*
110 G8 **Zuger See** ⊚ NW Switzerland
103 K25 **Zugspitze** ▲ S Germany *47.25N 10.58E*
101 E15 **Zuid-Beveland** *var.* South Beveland. *island* SW Netherlands
100 K10 **Zuidelijk-Flevoland** *polder* C Netherlands
Zuider Zee *see* IJsselmeer
100 G12 **Zuid-Holland** *Eng.* South Holland. ◐ *province* W Netherlands
100 N5 **Zuidhorn** Groningen, NE Netherlands *53.15N 6.25E*
100 O6 **Zuidlaardermeer** ⊚ NE Netherlands
100 O6 **Zuidlaren** Drenthe, NE Netherlands *53.06N 6.40E*
101 K14 **Zuid-Willemsvaart Kanaal** *canal* S Netherlands
100 N8 **Zuidwolde** Drenthe, NE Netherlands *52.40N 6.25E*
Zuitaizi *see* Kangxian
107 O14 **Zújar** Andalucía, S Spain *37.33N 2.52W*
106 L11 **Zújar** ↝ W Spain
106 L11 **Zújar, Embalse del** ⊞ W Spain
82 J9 **Zula** E Eritrea *15.19N 39.40E*
56 G6 **Zulia** *off.* Estado Zulia. ◐ *state* NW Venezuela
Zullapara *see* Sinchaingbyin
Züllichau *see* Sulechów
107 P3 **Zumárraga** País Vasco, N Spain *43.04N 2.19W*
114 D8 **Žumberačko Gorje** *var.* Gorjanči, Uskocke Planine, Žumberak, *Ger.* Uskokengebirge; *prev.* Sichelbuger Gebirge. ▲ Croatia/Slovenia *see also* Gorjanci
Žumberak *see* Gorjanci/Žumberačko Gorje
204 K7 **Zumberge Coast** *coastal feature* Antarctica
Zumbo *see* Vila do Zumbo
31 W10 **Zumbro Falls** Minnesota, N USA *44.15N 92.25W*
31 W10 **Zumbro River** ↝ Minnesota, N USA
31 W10 **Zumbrota** Minnesota, N USA *44.18N 92.37W*
101 H15 **Zundert** Noord-Brabant, S Netherlands *51.28N 4.40E*
Zungaria *see* Dzungaria
79 U14 **Zungeru** Niger, C Nigeria *9.49N 6.10E*
39 O11 **Zuni** New Mexico, SW USA *35.03N 108.52W*
39 P11 **Zuni Mountains** ▲ New Mexico, SW USA
166 J11 **Zunyi** Guizhou, S China *27.40N 106.55E*
166 J15 **Zuo Jiang** ↝ China/Vietnam
110 J9 **Zuoz** Graubünden, SE Switzerland *46.37N 9.58E*
114 I10 **Županja** *Hung.* Zsupanya. Vukovar-Srijem, E Croatia *45.03N 18.42E*
115 M17 **Žur** Serbia, S Yugoslavia *42.10N 20.37E*
131 T2 **Zura** Udmurtskaya Respublika, NW Russian Federation *57.36N 53.19E*
145 V8 **Zurbāṭiyah** E Iraq *33.13N 46.07E*
Zuri *see* Žirje
110 F7 **Zürich** *Eng./Fr.* Zurich, *It.* Zurigo. Zürich, N Switzerland *47.22N 8.33E*
110 G6 **Zürich** *Eng./Fr.* Zurich. ◐ *canton* N Switzerland
Zurich, Lake *see* Zürichsee
110 G7 **Zürichsee** *Eng.* Lake Zurich. ⊚ NE Switzerland
Zurigo *see* Zürich
155 V1 **Zŭrkŭl** *Pash.* Sarī Qūl, *Rus.* Ozero Zurkul'. ⊚ Afghanistan/Tajikistan *see also* Sarī Qūl
Zurkul', Ozero *see* Sarī Qūl/Zŭrkŭl
112 K10 **Żuromin** Mazowieckie, N Poland *53.00N 19.54E*
110 J8 **Zürs** Vorarlberg, W Austria *47.11N 10.11E*
79 T13 **Zuru** Kebbi, W Nigeria *11.28N 5.13E*
110 F6 **Zurzach** Aargau, N Switzerland *47.33N 8.21E*
103 J22 **Zusam** ↝ S Germany
100 M11 **Zutphen** Gelderland, E Netherlands *52.09N 6.12E*
77 N7 **Zuwārah** NW Libya *32.55N 12.06E*
Zuwaylah *see* Zawīlah
129 R14 **Zuyevka** Kirovskaya Oblast', NW Russian Federation *58.24N 51.08E*
167 N10 **Zuzhou** Hunan, S China *27.52N 113.00E*
Zvenigorodka *see* Zvenyhorodka
119 P6 **Zvenyhorodka** *Rus.* Zvenigorodka. Cherkas'ka Oblast', C Ukraine *49.05N 30.58E*
126 Jj14 **Zvezdnyy** Irkutskaya Oblast', C Russian Federation *56.43N 106.22E*
85 K18 **Zvishavane** *prev.* Shabani. Matabeleland South, S Zimbabwe *20.19S 30.01E*
113 J20 **Zvolen** *Ger.* Altsohl, *Hung.* Zólyom. Banskobystrický Kraj, C Slovakia *48.35N 19.06E*
114 J12 **Zvornik** E Bosnia and Herzegovina *44.24N 19.07E*
100 M5 **Zwaagwesteinde** *Fris.* De Westerein. Friesland, N Netherlands *53.16N 6.07E*
100 H10 **Zwanenburg** Noord-Holland, C Netherlands *52.22N 4.43E*
100 L8 **Zwarte Meer** ⊚ N Netherlands
100 M9 **Zwarte Water** ↝ N Netherlands
100 M8 **Zwartsluis** Overijssel, E Netherlands *52.39N 6.04E*
78 L17 **Zwedru** *var.* Tchien. E Liberia *6.04N 8.07W*
100 O8 **Zweeloo** Drenthe, NE Netherlands *52.48N 6.45E*
103 E20 **Zweibrücken** *Fr.* Deux-Ponts; *Lat.* Bipontium. Rheinland-Pfalz, SW Germany *49.15N 7.22E*
110 D9 **Zweisimmen** Fribourg, W Switzerland *46.33N 7.22E*
103 M15 **Zwenkau** Sachsen, E Germany *51.11N 12.19E*
111 V3 **Zwettl** Wien, NE Austria *48.28N 14.16E*
111 T3 **Zwettl an der Rodl** Oberösterreich, N Austria *48.28N 14.16E*
101 D18 **Zwevegem** West-Vlaanderen, W Belgium *50.48N 3.19E*
103 M17 **Zwickau** Sachsen, E Germany *50.43N 12.31E*
103 O21 **Zwiesel** Bayern, SE Germany *49.02N 13.14E*
100 H13 **Zwijndrecht** Zuid-Holland, SW Netherlands *51.49N 4.39E*
103 N16 **Zwikauer Mulde** ↝ E Germany
Zwischenwässern *see* Medvode
Zwittau *see* Svitavy
112 N13 **Zwoleń** Mazowieckie, SE Poland *51.21N 21.37E*
100 M9 **Zwolle** Overijssel, E Netherlands *52.31N 6.06E*
24 G6 **Zwolle** Louisiana, S USA *31.37N 93.38W*
112 K12 **Żychlin** Łodzkie, C Poland *52.13N 19.43E*
Żydaczów *see* Zhydachiv
121 L14 **Zyembin** *Rus.* Zembin. Minskaya Voblasts', C Belarus *54.25N 28.14E*
Zyôetu *see* Jōetsu
112 L12 **Żyrardów** Mazowieckie, C Poland *52.03N 20.27E*
127 Nn8 **Zyryanka** Respublika Sakha (Yakutiya), NE Russian Federation *65.45N 150.43E*
151 Y9 **Zyryanovsk** Vostochnyy Kazakhstan, E Kazakhstan *49.45N 84.16E*
113 J17 **Żywiec** *Ger.* Bäckermühle Schulzenmühle. Śląskie, S Poland *49.43N 19.10E*

Picture Credits

DORLING KINDERSLEY *would like to express their thanks to the following individuals, companies and institutions for their help in preparing this Atlas.*

Earth Resources Mapping Ltd., *Egham, Surrey*
Brian Groombridge, World Conservation Monitoring Centre, *Cambridge*
The British Library, *London*
British Library of Political and Economic Science, *London*
The British Museum, *London*
The City Business Library, *London*
King's College, *London*
National Meteorological Library and Archive, *Bracknell, Berkshire*
The Printed Word, *London*
The Royal Geographical Society, *London*
University of London Library
Paul Beardmore
Philip Boyes
Hayley Crockford
Alistair Dougal
Nick Drake
Reg Grant
Louise Keane
Zoe Livesley
Laura Porter
Andy Summers
Jeff Eidenshink
Chris Hornby
Rachelle Smith
Ray Pinchard
Robert Meisner
Fiona Strawbridge
Wim Jenkins

T = top, B = bottom, A=above, L = left, R = right, C = center

Adams Picture Library: 150CLA; **Ardea London Ltd**: K Ghana 268C; M Iljima 238TC; R Waller 264TR; **Aspect Picture Library**: P Carmichael 235CRB, 288TR; Tompkinson 340TRB; **Axiom**: C Bradley 264CA, 285CA; Jim Holmes xviiiCRA, xxviiiBCA, xxxiCRB, 268TCR, 302TL, 302BC; J Morris 129TL, 129CRB; J Spaull 232BL; **Bath Spa University College/GeoTechnologies**: xxBL, xxviiiBCR, 166BR, 166TL; **Bridgeman Art Library, London/New York**: Collection of the Earl of Pembroke, Wilton House xxivBCA; **British Antarctic Survey**: 352BL; **British National Space Centre**: xiiiCRB, 119TL, 315BR, 348; **The J. Allan Cash Photolibrary**: 10BC, 108CL, 123CLB, 124CL, 126CLB, 129BR, 132BC, 151BL, 189BR, 244BCL, 249TL, 276CR, 322BR, 325TR; **Bruce Coleman Ltd**: 150BC, 168CL, 172TC; S Alden 344BR; Atlantide xxxTCR, 244BR; E Bjurstrom 249BR; S Bond 164CRB; Thomas Buchholz xixCL, 160TR, 226TCL; J Burton xxviiiC; J Cancalosi 325TRB; B J Coates xxixBLA, 344BC; A Compost xxviiCBR; B Coleman 113TL; B & C Colhoun 4TR, 68CB; Dr S Coyne 85TL; Gerald Cubitt xxTCB, 303BCL, 322TR, 332TR; P Davey xxxiCLB, 215BL; N Devore 347CBL; S J Doylee xxviCRR; Halle Flygare xxiCR; Jeff Foott Productions xxvii CRB, 11CRA; MPL Fogden 27CB; Michael Freeman 155BRA; Paul van Gaalen 150TR; G Gualco 248C; B Henderson 346CR; Dr C Henneghein 123C; HPH Photography, HVan den Berg 123CR; C Hughs BCL; Johnny Johnson 73CR, 355TR; Janos Jurka 155CA; SC Kaufman 52C; Stephen J Krasemann 61TR; Harad Lange 10TRB, 122CA; C Lockwood 60BC; LC Marigo xxviBCA, xxxiCLA. 91CRA, 105BR; M McCoy 337TR; D Meredith 5CR; John Murray xixC, 323BR; Orion Press 302TR; Orion Services & Trading Co. Inc. 301TR; C Ott 28BL; Dr Eckart Pott 24C, 76CL, 151C, 157TL, 350CLB; F Prenzel 343C, 346CB; Marie Read 80BR, 81CRB; H Reinhard xxxiCR, xxxiTR, 350BR; L Lee Rue III 269BCL; John Shaw xxiiiTL; K N Swenson 350BC; Peter Terry 201CR; N Tomalin 96BCL; Peter Ward 136TC; S Widstrand 101TR; Konad Wothe 155C, 313TCL; Jonathan T Wright 227BR; **Colorific**: Black Star/L Mulvehil 280CL; Black Star/R Rogers 101BR; Black Star/J Rupp 289BCR; Camera Tres/C Meyer 105BRA; Robert Caputo/Matrix 136CL; J Hill 205CLB; M Koene 97TR; M Yamashita 280BL, 309CA; **Comstock**: 188CRB; D Cousens: 261CRA; **Sue Cunningham Photographic**: 93CR; **James Davis Travel Photography**: 15CA, 33BC, 91TLB, 100BCR, 101CLA, 109BL, 157BC, 176TR, 214CB, 284BC, 323CRA, 341BR; Sogn & Fjordane 160TC; **Deutsche Forschungsanstalt für Luft-und Raumfahrt**: 66TL, 174TR, 175TL, 175BL, 191BL, 352CL, 352CR, 352CRB; **George Dunnit**: 222CA; **Environmental Picture Library**: Chris Westwood 226C; **European Space Agency**: xiiTRB, xiiiCRB, 118TL, 315BR; **Eye Ubiquitous**: L Fordyce 14CLA; L Johnstone 8CRA, 52BLA, 56CB; S Miller xxvCA; Mike Southern 127BLA; **Ffotograff**: C Aithie 235CL; N Tapsell 284CL; **Geoscience Features**: xxBRA, xxBCRA, 176CL, 188BC, 219BR; Solar Film 116TC; **Robert Harding Picture Library**: xxiTC, xxviiiCR, xxxivC, 4TLB, 5CA, 19CR, 19CRB, 69BC, 72CRA, 92BL, 161BR, 169CR, 200CR, 218BL, 236CLA, 252CB, 253TL, 261TR, 280TR, 303CA, 307BR; Paul G Adam 15TCB; David Atchison-Jones 124BLA; Julia Bayne 126BCL; Bildagentur Schuster 140CR; C Bowman 92BR, 95CA, 112CL, 124CRL; C Campbell xxvBCA; G Corrigan 285CRB, 289CRB; R Cundy 123BR; Delu 137BC; A Durand 193BR; Financial Times 252BR; R Frerck 93BL; Tony Gervis 5BCL, 9CR; Ian Griffiths xxxivCL, 133TL; T Hall 307CRA; D Harney 252CA; Gavin Hellier xixCR, 233BL; F Jackson 241BCR; Franz Joseph Land 217TR; P Koch 245TR; Y Marcoux 26BR; Siroua Massif xixBCA; Andrew Mills 152CLB; Louise Murray 200TR; G Renner 128CB, 350C; C Rennie 90CL, 204BR; Rolf Richardson 208CL; P Van Riel 90BR; Ellen Rooney 222TR; Sassoon xxviiiCL, 264CLB; P Scholey 320TR; M Short 241TL; E Simanor xxxiCR; V Southwell 245CR; James Strachan 80TR, 193BL, 234BCR; C Tokeley 238CLA; A C Waltham 289C; Tony Waltham xxiBLA, xxviCLLL, 244CRB; Westlight 69CR; N Wheeler 245BL; A Woolfitt 161BRA; **Paul Harris**: 218TR, 304TC; **Hutchison Library**: 8BL, 238BCL; P Collomb 241CR; C Dodwell 237TR; N Durrell McKenna xxxBCRA; Sarah Errington 124BCL; P Hellyer 252BC; J Horner xxxTC; R Ian Lloyd 232CRA; J Nowell 233CLB, 253TC; A Zvoznikov xxviCL; **Image Bank**: 151BR; J Banagan 340BCA; A Becker xxviiiBCA; M Isy-Schwart 341CL, 346C; K Forest 293TR; P Hendrie 304BC; M Khansa 216BR; Lomeo xxxviiTCR; T Madison 307CR; C Molyneux xxviCRRR; K Mori 346TC; Carlos Navajas xxxiTR; Ocean Images Inc. 344CLB; Joseph van Os xxiTCR; Steve Proehl 8CL; Terje Rakke xxiiiTC 116CL; M Reitz 354CA; M Romanelli 307BL; G A Rossi 269BCR, 320BLA; Bernard Roussel 189TL; Steve Satusheck xxiBCRA; Stock Photos/J M Spielman xxviiiTRL; **Images Colour Library**: xxviCLL, 5BR, 33BR, 69TL, 84TL, 112TC, 155BR, 176CLB, 177CR, 268CL, 301CL, 302TRB, 324CA; **Impact Photos**: C Bluntzer 280BR; Cosmos/G Buthaud 117BC; Sarah Franklin 226BL; A le Garsmeur 235CRA; A Indge xxxiTC; J & G Andrews 336BL; Colin Jones xxxvCB, 124BL; V Nemirousky 241BR; Jeremy Nicholl 132TCR; C Penn 343BR; Geray Sweeney xxiBRA, 354CRB, 354TR; **JVZ Picture Library**: 233TC; **Frank Lane Picture Agency**: xxvTCR, xxviiBLA, 157TR; A Christiansen 104CRA; J Holmes xviiiBL; S McCutcheon 5C; Silvestris 313TCR; D Smith xxviBCLA; W Wisniewsli 218TL, 351BR; **Magnum**: Abbas 145CR, 240CA; S Franklin 232CRB; David Hurn 6BCL; P Jones-Griffiths 341BL; Hiroji Kubota xxxBCLA, 280CLB; Fred Maver xxBLA; Steve McCurry 127CL, 239BCR; G Rodger 128TR; Chris Steele Perkins 126BL; **Mallin Space Science Systems/NASA/JPL**: xvBC; **Mountain Camera/ John Cleare**: 273TR; C Monteath 273CR; **NASA Dryden Flight Center**: xBL; **NASA Goddard Spaceflight Center**: xxBCR, xxiiiBL, 74CL, 79CR, 206TL, 291TR, 294BL, 311CL; **NASA Jet Propulsion Laboratory**: xxviBCR, xxxiBL, xxxiBCR, xxxvBL, 187CL, 199TL,310CL; **National Aeronautical and Space Administration**: xiCLA, xxiCL; **National Oceanic and Atmospheric Administration**: xCAL, xxivBL; **National Reconnaissance Office**: xiiCLB, xiiBL, xxxivBL; **NOAA National Geophysical Data Center**: xiiiBRI, xxxviBL; **Natural Environment Research Council**: xiiiCRB, 119TL, 315BR; **Natural Science Photos**: M Andera 192C; **Nature Photographers**: EA Janes 196CL; **Network Photographers Ltd.**: Christian Sappa/Rapho 209BL; **N.H.P.A.**:N J Dennis xxviiCL; D Heuchlin xxxviiCLA; Stephen Krasemann 19BL, 45BR, 72TC; K Schafer 91CB; R Tidman 288CLB; D Tomlinson 257CR; M Wendler 90TR; **Nottingham Trent University**: Tony Waltham xviiiCLA, xixBRA; **Novosti**: 256BLA; **Oxford Scientific Films**: D Allan xxviTR; M Brown 248BL; M Colbeck 261CA; Hjalmar R Bardarson xxxiiBCA; D Bown xxviiCBLL; W Faidley 5TL; L Gould xxviiTRB; D Guravich xxviiTR; P Hammerschmidy/Okapia 151CLA; M Hill 101TL, 351TR; C Menteath 238TR; J Netherton 4CRB; S Osolinski 144CA; Richard Packwood 126CA; M Pitts 323TC; Norbetr Rosing xxviiCBL, 11TR, 355BR; D Simonson 101C; Survival Anglia/C Catton 241TR; R Toms xxviiBRA; Konrad Wothe xxiCLA, xxvBLA; **Panos Pictures**: B Aris 239C; P Barker xxviiiBRA; T Bolstao 273BR; N Cooper 144CB, 273TC; J-L Dugast 307CB, 308BC; Jeremy Hartley 127CA, 154CL; J Holmes 265BC; James Morris 132CLB; M Rose 260TR; D Sansoni 277CL; C Stowers 293TL; **Edward Parker**: 91CLB, 91TL; **Pictor International**: xviiiBRA, xixCRB, xxiiiTCL, xxivCL, 5CLA, 29BR, 36CRB, 36TR, 41BCA, 41CL, 48CB, 49BC, 56CA, 61TRB, 64BR, 64CR, 64BC, 72CL, 72CB, 81CL, 113BR, 117TC, 144CL, 145CLB, 169BR, 185CLA, 307TCR, 308BR, 309CR, 324CLB, 333TL; **Pictures Colour Library**: xxvBCLA, xxviBRA, xxxiBCLA, 8BR, 19TR, 24TC, 27TR, 33TL, 36BL, 44C, 44CLA, 49TR, 60TRB, 68BC, 77CA, 81CRA, 122BL, 154TCB, 160BL, 169BL, 184CA, 185CR, 185CB, 185BR, 205BL, 300BC, 301BR, 344CL; **Planet Earth Pictures**: 346BL; D Barrett 264CB, 332CA; R Coomber 27BL; G Douwma 312BR; E Edmonds 313BR; HC Heap 216TR; J Lythgoe 354BL; A Mounter 235BCR, 312CR; Mike Potts 8CA; P Scoones xxivTR; J Walencik 192TR; J Waters 95BCL; **Planetary Visions Limited**: xCRA, xiCRA, xiiTR, xiiiTR, xiiiCL, xiiiCLB, xiiiCLB, xiiiBL, xxiiBR, xxviBL, 22TL, 356TL, 356BL, 357BR; **Popperfoto**: Reuters/J Drake xxxviCLA; **Rex Features**: 300CR; Antelope xxxviCLB; M Friedel xxvCR; I McIlgorm xxxivCBR; J Shelley xxivCR; Sipa Press xxxivCR; Sipa Press/Alix xxxivCBL; Sipa Press/Chamussy 320BLA; **Russia & Republics Photolibrary**: Mark Wadlow 208CR, 209CL, 222CL, 222BC, 223TL, 223BR, 226TCR; **Rutherford Appleton Laboratory**: xiiCRB, 119TL, 315BR; **Science Photo Library**: CNES, 1990 Distribution Spot Image 235BL; Earth Satellite Corporation xxiiiTRB, xxxvCR, 91BCL; F Gohier xvCR; John Heseltine xxTCB; Keith Kent xixCLB; Peter Menzell xixBLA; NASA. xivBCA; David Parker xviiiBCA; University of Cambridge Collection Air Pix 151CLB; RJ Wainscoat, Peter Arnold, Inc. xvBCA; D Weintraub xvBLA; **South American Pictures**: 101BL, 112TR; R Francis 94BL; Guyana Space Centre 92TR; T Morrison 91BL, 91CRB, 92CR, 94TR, 96TR, 108BL, 109C; **Southampton Oceanography Centre**: xxiiBLA, xiiCRB, 119TL, 315BR; **Sovofoto/Eastfoto**: xxxviCBR; **Sovzond/SSC Satellitbild**: 171BL; **Spectrum Colour Library**: 92BC, 288BC; J King 257BR; **Frank Spooner Pictures/Gamma**: 48CRB; E Baitel xxxviBCLA; Bernstein xxxvCL; Contrast 196CR; Diard/Photo News 197CL; Liaison/C Hires xxxviTCB; Liaison/Nickelsberg xxxviTR; Liaison /Vogel 238BL; D Marleen 197TL; Novosti 204CA; P Piel xxxivCA; N Quidu 233CL; H Stucke 340CA, 342CLB; Torrengo /Figaro 136BR; Art Zamur 197BL; **Still Pictures**: Chris Caldicot 133TC; A Crump 347CL; M & C Denis-Huot xxviBLA, 136CR, 141BL; M Edwards xvCRL, 95BL, 116CR, 123BLA, 277BR; J Frebet 95CLB; H Giradet 95TC; M Gunther 215BC; E Parker 94CL; R Seitre 235CA, 236TL, 236BL; **Tony Stone Images**: xxxTR, 6CA, 9BL, 9CL, 15CRB, 53TR, 73BR, 104C, 165BC, 173BR, 184TR, 189CRB, 189CL, 239BR, 300CLB, 301C, 324CB, 325BR, 342BC, 344TR; Glen Allison 32TR, 57CRB; Doug Armand 18TCB; D Austen 324TR, 336CR, 337CL; J Beatty 128CL; O Benn xxxBRA; K Biggs xxvTL; R Bradbury 84BR; R A Butcher xxxTE; J Callahan xxxiCRA; P Chesley 333BCL, 342C; Willard Clay 56BL, 57CRA; Joe Cornish 164BL, 185TL; Cosmo Condina 77CB; Tony Craddock xxviiiTR; Phil Degginger 68CLB; Demetrio 7BR; N DeVore xxviiiBCA; A Diesendruck 108BR; Shaun Egan 151CRA, 164BR; R Elliot xxviBCRA; S Elmore 33C; F Frerck 214TR; John Garrett 127CR; Sylvain Grandadam 18BR; R Grosskopf 52BL; D Hanson 180BC; C Harvey 123TL; G Hellier 192BL, 302CR; S Huber 177CRB; D Hughs xxvBR; Amulf Husmo 155TR; Gary Irvine 57BC; J Jangoux 104CL; G Johnson 236CLB; Donald Johnston xxiTR; Alan Kehr 197C; Raphael Koskas xxTR; J Lamb 164CRA; J Lawrence 129CRA; Lester Lefkowitz 9CA; M Lewis 85CLA; S Mayman 97BR; Murray & Associates 85CR; G Norways 180CA; N Parfitt xxxiCLL, 122TCR, 141TL; R Passmore 215TR; Nigel Press xxBCA; Ed Pritchard 152CA, 154CLR; Tom Raymond 37BL; L Resnick 128BR; M Rogers 140BR; A Sacks 52TCB; Charlotte Saule 154CR; S Schulhof xxviiiTC; Pete Seaward 64CL; Mark Segal 60BL; V Shenai 272CL; Ron Sherman 48CL; H Sitton 240CR; R Smith xxixCLB, 100C; H Strand 91BR, 113TR; Stephen Studd 188CLA; Penny Tweedie 321CR; L Ulrich 29BL; M Vines 29TC; AB Wadham 108CR; J Warden 113CLB; Randy Wells 41CRA, 345BL; Gary Yeowell 64BL; **Telegraph Colour Library**: 109TCR, 281TL; J Sims 48TR; **Topham Picturepoint**: xxxvCBL, 235BCL, 237CR, 292BR, 304TR, 306BC; **Travel Ink**: Andrew Cowin 152TR; **Trifid Corporation**: xxxivBCL; **Trip**: 248BR, 256CA, 277CRA; B Ashe 285TR; D Cole 340BCL, 340CR; D Davis 153BL; I Deineko xxxvTR; Jerry Dennis 40BL; Dinodia 276CL; Eye Ubiquitous/L Fordyce 4CLB; A Gasson 265CR; W Jacobs 81TL, 96BL, 321BC, 322CLA, 333BCR, 343BL; P Kingsbury 196C; E Knight 109BR; K Knight 321BR; V Kolpakov 261BL; T Noorits 151TL, 209BR, 260CL; Richard Powers 77TR; N Ray 306CA; C Rennie 204CLB; V Sidoropolev 257TR; E Smith 329TL, 329BC; **UCL/NOAA**: Seymour Laxon/Dave McAdoo: xixBR; **United States Geological Survey**: 158TL; **University College London**: Dept of Geomatic Engineering: xiiiCRB, 66TL; Mullard Space Science Laboratory: 119TL, 315BR; **University of Ulster**: 167BL; **USGS EROS Data Center**: xBC, xBR, xiCR, xiCRB, xiiCR, xiiCRB, xiiBR, xiiiTL, xxivBC, xxvBC, xxviiBL, xxixBL, xxxiBC, xxxviBC, 35BR, 170CL, 170CR; **Woodfin Camp & Associates**: 156BLR; **World Pictures**: xixCA, xxiCRA, 26CRB, 40CL, 41BC, 44BL, 65BL, 76TR, 93TR, 125BR, 140TCR, 144TR, 145BL, 150BCR, 164TC, 168BL, 172CR, 173CR, 177BC, 181TC, 215CB, 216BL, 281BL, 289BCL, 292CLB, 312CLB, 312BC, 323BL, 328CB, 329C, 332CL, 333CR; **Zefa Picture Library**: xxBLRA, xxiBCLA, xxiiCL, 5CL, 15BC, 18TC, 27CA, 37TL, 40CRB, 45BL, 60TCR, 68BCR, 105BCL, 117TCL, 123CLA, 137TL, 141BR, 151CRB, 156C, 168C, 169TL, 172BL, 185TR, 208CRB, 214BL, 218CB, 222CLA, 300CA, 329TR; Anatol 197BR; Barone 200BL; Brandenburg 7C; A J Brown 84TR; H J Clauss 97CLB; Damm 125BC; Evert 156BL; W Felger 5BL; J Fields 347CRA; R Frerck 6BL; Konrad Heibig 201BR; G Heil 100BR; Heilman 52BC; Hunter 10C; Kitchen 10TR, 24BL, 24CL, 26TR; Dr Hans Kramarz 9BLA, 219CRA; Mechlio 277BL; Jose Fuste Raga 44TR; Rossenbach 181BR, 214CA; T Stewart 15TR, 33CR; Streichan 153TL; Sunak 96BR, 292TR; D H Teuffen 161TL; Bo Zaunders 76BC.

Additional Photography: Geoff Dann; Rob Reichenfeld; H Taylor; Jerry Young.

◆ COUNTRY ● COUNTRY CAPITAL ◇ DEPENDENT TERRITORY ○ DEPENDENT TERRITORY CAPITAL ◐ ADMINISTRATIVE REGION ✈ INTERNATIONAL AIRPORT ▲ MOUNTAIN ▲ MOUNTAIN RANGE ♜ VOLCANO ↝ RIVER ⊚ LAKE ⊞ RESERVOIR

Abyssal plain A broad plain found in the depths of the ocean, more than 10,000 ft (3,000 m) below sea level.

Air mass A huge, homogeneous mass of air, within which horizontal patterns of temperature and humidity are consistent. Air masses are separated by fronts.

Alluvial fan Large fan-shaped deposit of fine sediments deposited by a river as it emerges from a narrow, mountain valley onto a broad, open plain.

Alluvium Material deposited by rivers. Nowadays usually only applied to finer particles of silt and clay.

Anticline A geological fold that forms an arch shape, curving upward in the rock strata.

Aquifer A body of rock that can absorb water. .

Arête A thin, jagged mountain ridge that divides two adjacent cirques, found in regions where glaciation has occurred.

Artesian well A naturally occurring source of underground water, stored in an aquifer.

Atoll A ring-shaped island or coral reef often enclosing a lagoon of sea water.

Badlands A landscape that has been heavily eroded and dissected by rain-water, and which has little or no vegetation.

Back slope The gentler windward slope of a sand dune or gentler slope of a cuesta.

Bajos An alluvial fan deposited by a river at the base of mountains and hills that encircle desert areas.

Bar, coastal An offshore strip of sand or shingle, either above or below the water. Usually parallel to the shore but sometimes crescent-shaped or at an oblique angle.

Barchan A crescent-shaped sand dune, formed where wind direction is very consistent. The horns of the crescent point downwind and where there is enough sand the barchan is mobile.

Base level The level below which flowing water cannot erode the land.

Basement rock A mass of ancient rock often of PreCambrian age, covered by a layer of more recent sedimentary rocks. Commonly associated with shield areas.

Bedrock Solid, consolidated and relatively unweathered rock, found on the surface of the land or just below a layer of soil or weathered rock.

Bluff The steep bank of a meander, formed by the erosive action of a river.

Breccia A type of rock composed of sharp fragments, cemented by a fine-grained material such as clay.

Butte An isolated, flat-topped hill with steep or vertical sides, buttes are the eroded remnants of a former land surface.

Calcite Hexagonal crystals of calcium carbonate.

Caldera A huge volcanic vent, often containing a number of smaller vents, and sometimes a crater lake.

Carbonation Process whereby rocks are broken down by carbonic acid. Carbon dioxide in the air dissolves in rainwater, forming carbonic acid.

Castle kopje Hill or rock outcrop, especially in southern Africa, where steep sides, and a summit composed of blocks, give a castle-like appearance.

Cataracts A series of stepped waterfalls created as a river flows over a band of hard, resistant rock.

Chernozem A fertile soil, also known as "black earth" consisting of a layer of dark topsoil, rich in decaying vegetation, overlying a lighter chalky layer.

Confluence The point at which two rivers meet.

Continental drift The theory that the continents of today are fragments of one or more prehistoric supercontinents that have moved across the Earth's surface, creating ocean basins.

Continental shelf An area of the continental crust, below sea level, which slopes gently.

Continental slope A steep slope running from the edge of the continental shelf to the ocean floor.

Core The center of the Earth, consisting of a dense mass of iron and nickel.

Coulées A US / Canadian term for a ravine formed by river erosion.

Craton A large block of the Earth's crust which has remained stable for a long period of geological time. It is made up of ancient shield rocks.

Cretaceous A period of geological time beginning about 145 million years ago and lasting until c. 65 million years ago.

Crevasse A deep crack in a glacier.

Crust The hard, thin outer shell of the Earth. It floats on the mantle, which is softer and more dense.

Crystalline rock Rocks formed when molten magma crystallizes (igneous rocks) or when heat or pressure cause re-crystallization (metamorphic rocks).

Cuesta A hill which rises into a steep slope on one side but has a gentler gradient on its other slope.

Delta Low-lying, fan-shaped area at a river mouth, formed by the deposition of successive layers of sediment.

Denudation The combined effect of weathering, erosion, and mass movement, which, over long periods, exposes underlying rocks.

Deposition The laying down of material that has accumulated: after being eroded and then transported by wind, ice, or water; as organic remains, such as coal and coral; as the result of evaporation and chemical precipitation.

Depression 1 In climatic terms it is a large low pressure system; 2 a complex fold, producing a large valley, which incorporates both a syncline and an anticline.

Detritus Piles of rock deposited by an erosive agent such as a river or glacier.

Distributary A minor branch of a river, which does not rejoin the main stream, common at deltas.

Divide A US term describing the area of high ground separating two drainage basins.

Donga A steep-sided gully, resulting from erosion by a river or by floods.

Drainage basin The area drained by a single river system, its boundary is marked by a watershed or divide.

Drumlin A long, streamlined hillock composed of material deposited by a glacier. They often occur in groups known as swarms.

Earthflow The rapid movement of soil and other loose surface material down a slope, when saturated by water.

Ephemeral A nonpermanent feature, often used in connection with seasonal rivers or lakes in dry areas.

Epicenter The point on the Earth's surface directly above the underground origin or focus of an earthquake.

Erg An extensive area of sand dunes, particularly in the Sahara Desert.

Erosion The processes which wear away the surface of the land. Glaciers, wind, rivers, waves, and currents all carry debris that causes erosion.

Escarpment A steep slope at the margin of a level, upland surface. In a landscape created by folding, escarpments (or scarps) frequently lie behind a more gentle backward slope.

Esker A narrow, winding ridge of sand and gravel deposited by streams of water flowing beneath or at the edge of a glacier.

Erratic A rock transported by a glacier and deposited some distance from its place of origin.

Eustacy A world-wide fall or rise in ocean levels.

Exfoliation A kind of weathering whereby scalelike flakes of rock are peeled or broken off by the development of salt crystals in water within the rocks.

Extrusive rock Igneous rock formed when molten material (magma) pours forth at the Earth's surface and cools rapidly. It usually has a glassy texture.

Fault A fracture or crack in rock, where strains (tectonic movement) have caused blocks to move, vertically or laterally, relative to each other.

Ferrel cell A component in the global pattern of air circulation, which rises in the colder latitudes (60° N and S) and descends in warmer latitudes (30° N and S).

Fissure A deep crack in a rock or a glacier.

Fjord A deep, narrow inlet, created when the sea inundates the U-shaped valley created by a glacier.

Flash flood A sudden, short-lived rise in the water level of a river or stream, or surge of water down a dry river channel, or wadi, caused by heavy rainfall.

Floodplain The broad, flat part of a river valley, adjacent to the river itself, formed by sediment deposited during flooding.

Fold A bend in the rock strata of the Earth's crust, resulting from compression.

Frost shattering A form of weathering where water freezes in cracks, causing expansion. As temperatures fluctuate and the ice melts and refreezes, it eventually causes the rocks to shatter.

Geosyncline A concave fold (syncline) or large depression in the Earth's crust, extending hundreds of miles.

Geothermal energy Heat derived from hot rocks within the Earth's crust and resulting in hot springs, steam, or hot rocks at the surface.

Geyser A jet of steam and hot water that intermittently erupts from vents in the ground in areas that are, or were, volcanic.

Glaciation The growth of glaciers and ice sheets, and their impact on the landscape.

Glacier A body of ice moving down-slope under the influence of gravity and consisting of compacted and frozen snow.

Glacio-eustacy A worldwide change in the level of the oceans, caused when the formation of ice sheets takes up water or when their melting returns water to the ocean.

Glaciofluvial To do with glacial meltwater, the landforms it creates and its processes; erosion, transportation, and deposition.

Glacis A gentle slope or pediment.

Gondwanaland The supercontinent thought to have existed over 200 million years ago in the southern hemisphere.

Graben A block of rock let down between two parallel faults. Where the graben occurs within a valley, the structure is known as a rift valley.

Grease ice Slicks of ice that form in Antarctic seas, when ice crystals are bonded together by wind and wave action.

Groundwater Water that has seeped into the pores, cavities, and cracks of rocks or into soil and water held in an aquifer.

Gully A deep, narrow channel eroded in the landscape by ephemeral streams.

Guyot A small, flat-topped submarine mountain, formed as a result of subsidence which occurs during sea-floor spreading.

Hadley cell A large-scale component in the global pattern of air circulation. Warm air rises over the Equator and blows at high altitude toward the poles, sinking in subtropical regions (30° N and 30° S) and creating high pressure. The air then flows at the surface toward the Equator in the form of trade winds.

Hamada An Arabic word for a plateau of bare rock in a desert.

Hanging valley A tributary valley that ends suddenly, high above the bed of the main valley.

Headwards The action of a river eroding back upstream, as opposed to the normal process of downstream erosion. Headwards erosion is often associated with gullying.

Hoodos Pinnacles of rock that have been worn away by weathering in semiarid regions.

Horst A block of the Earth's crust that has been left upstanding by the sinking of adjoining blocks along fault lines.

Hot spot A region of the Earth's crust where high thermal activity occurs, often leading to volcanic eruptions.

Hydrolysis The chemical breakdown of rocks in reaction with water, forming new compounds.

Ice Age A period in the Earth's history when surface temperatures in the temperate latitudes were much lower and ice sheets expanded considerably. There have been ice ages from Pre-Cambrian times onward.

Ice cap A permanent dome of ice in highland areas.

Ice floe A large, flat mass of ice floating free on the ocean surface. It is usually formed after the breakup of winter ice by heavy storms.

Ice sheet A continuous, very thick layer of ice and snow. The term is usually used of ice masses which are continental in extent.

Ice shelf A floating mass of ice attached to the edge of a coast. The seaward edge is usually a sheer cliff up to 100 ft (30 m) high.

Ice wedge Massive blocks of ice up to 6.5 ft (2 m) wide at the top and extending 32 ft (10 m) deep.

Iceberg A large mass of ice in a lake or a sea, which has broken off from a floating ice sheet (an ice shelf) or from a glacier.

Igneous rock Rock formed when molten material, magma, from the hot, lower layers of the Earth's crust, cools, solidifies, and crystallizes, either within the Earth's crust (intrusive) or on the surface (extrusive).

Inselberg An isolated, steep-sided hill, rising from a low plain in semiarid and savannah landscapes.

Interglacial A period of global climate, between two ice ages, when temperatures rise and ice sheets and glaciers retreat.

Intraplate volcano A volcano that lies in the center of one of the Earth's tectonic plates, rather than, as is more common, at its edge.

Intrusion (intrusive igneous rock) Rock formed when molten material, magma, penetrates existing rocks below the Earth's surface before cooling and solidifying.

Isostacy The state of equilibrium that the Earth's crust maintains as its lighter and heavier parts float on the denser underlying mantle.

Isthmus A narrow strip of land connecting two larger landmasses or islands.

Joint A crack in a rock, formed where blocks of rock have not shifted relative to each other, as is the case with a fault. Joints are created by folding; by shrinkage in igneous rock as it cools or sedimentary rock as it dries out; and by the release of pressure in a rock mass when overlying materials are removed by erosion.

Kame A mound of stratified sand and gravel with steep sides, deposited in a crevasse by meltwater running over a glacier. When the ice retreats, this forms an undulating terrain of hummocks.

Karst A barren limestone landscape created by carbonic acid in streams and rainwater, in areas where limestone is close to the surface.

Kettle hole A round hollow formed in a glacial deposit by a detached block of glacial ice, which later melted. They can fill with water to form kettle-lakes.

Lagoon A shallow stretch of coastal salt-water behind a partial barrier such as a sandbank or coral reef. Also used to describe the water encircled by an atoll.

Laterite A hard red deposit left by chemical weathering in tropical conditions, and consisting mainly of oxides of iron and aluminum.

Latitude The angular distance from the Equator, to a given point on the Earth's surface. Imaginary lines of latitude running parallel to the Equator encircle the Earth, and are measured in degrees north or south of the Equator. The Equator is 0°, the poles 90° South and North respectively. Also called parallels.

Laurasia In the theory of continental drift, the northern part of the great supercontinent of Pangaea. Laurasia is said to consist of N America, Greenland and all of Eurasia north of the Indian subcontinent.

Lava The molten rock, magma, which erupts onto the Earth's surface through a volcano, or through a fault or crack in the Earth's crust.

Leaching The process whereby water dissolves minerals and moves them down through layers of soil or rock.

Levée A raised bank alongside the channel of a river. Levées are either human-made or formed in times of flood when the river overflows its channel, slows and deposits much of its sediment load.

Lithosphere The rigid, upper layer of the Earth, comprising the crust and the upper part of the mantle..

Loess Fertile, fine-grained, yellow deposits of unstratified silts and sands.

Longitude A division of the Earth which pinpoints how far east or west a given place is from the Prime Meridian (0°) which runs through the Royal Observatory at Greenwich, England (UK). Imaginary lines of longitude are drawn around the world from pole to pole. The world is divided into 360 degrees.

Longshore drift The movement of sand and silt along the coast, carried by waves hitting the beach at an angle.

Magma Undground, molten rock, which is very hot and highly charged with gas. It is generated at great pressure, at depths 10 miles (16 km) or more below the Earth's surface.

Mantle The layer of the Earth between the crust and the core. it is about 1,800 miles (2,900 km) thick.

Massif A single very large mountain or an area of mountains with uniform characteristics and clearly-defined boundaries.

Meltwater Water resulting from the melting of a glacier or ice sheet.

Mesa A broad, flat-topped hill, characteristic of arid regions.

Metamorphic rocks Rocks that have been altered from their original form, in terms of texture, composition, and structure by intense heat, pressure, or by the introduction of new chemical substances – or a combination of more than one of these.

Milankovitch hypothesis A theory suggesting that there are a series of cycles that slightly alter the Earth's position when rotating about the Sun.

Mistral A strong, dry, cold northerly or north-westerly wind, which blows from the Massif Central of France to the Mediterranean Sea.

Mohoroviâiá discontinuity (Moho) The structural divide at the margin between the Earth's crust and the mantle. On average it is 20 miles (35 km) below the continents and 6 miles (10 km) below the oceans.

Monsoon A wind that changes direction biannually. The change is caused by the reversal of pressure over landmasses and the adjacent oceans. Because the inflowing moist winds bring rain, the term monsoon is also used to refer to the rains themselves.

Moraine Debris, transported and deposited by a glacier or ice sheet in unstratified, mixed, piles of rock, boulders, pebbles, and clay.

Mountain-building The formation of fold mountains by tectonic activity. Also known as orogeny, mountain-building often occurs on the margin where two tectonic plates collide.

Nappe A mass of rocks that has been overfolded by repeated thrust faulting.

Oasis A fertile area in the midst of a desert, usually watered by an underground aquifer.

Oceanic ridge A mid-ocean ridge formed, according to the theory of plate tectonics, when plates drift apart and hot magma pours through to form new oceanic crust.

Onion-skin weathering The weathering away or exfoliation of a rock or outcrop by the peeling off of surface layers.

Outwash plain Glaciofluvial material (typically clay, sand, and gravel) carried beyond an ice sheet by meltwater streams, forming a broad, flat deposit.

Oxbow lake A crescent-shaped lake formed on a river floodplain when a river erodes the outside bend of a meander, making the neck of the meander narrower until the river cuts across the neck. The meander is cut off and is dammed off with sediment, creating an oxbow lake.

Oxidation A form of chemical weathering where oxygen dissolved in water reacts with minerals in rocks – particularly iron to form oxides.

Pack ice Ice masses more than 10 ft (3 m) thick that form on the sea surface and are not attached to a landmass.

Pancake ice Thin discs of ice, up to 8 ft (2.4 m) wide which form when slicks of grease ice are tossed together by winds and stormy seas.

Pangaea In the theory of continental drift, Pangaea is the original great land mass which, about 190 million years ago, began to split into Gondwanaland in the south and Laurasia in the north, separated by the Tethys Sea.

Pediment A gently-sloping ramp of bedrock below a steeper slope, often found at mountain edges in desert areas, but also in other climatic zones. Pediments may include depositional elements such as alluvial fans.

Periglacial Regions on the edges of ice sheets or glaciers or, more commonly, cold regions experiencing intense frost action, permafrost or both.

Permafrost Permanently frozen ground, typical of Arctic regions.

Permeable rocks Rocks through which water can seep, because they are either porous or cracked.

Phreatic eruption A volcanic eruption which occurs when lava combines with groundwater, superheating the water and causing a sudden emission of steam at the surface.

Pingo A dome of earth with a core of ice, found in tundra regions. Pingos are formed either when groundwater freezes and expands, pushing up the land surface, or when trapped, freezing water in a lake expands and pushes up lake sediments to form the pingo dome.

Placer A belt of mineral-bearing rock strata lying at or close to the Earth's surface, from which minerals can be easily extracted.

Plate, plate tectonics The study of tectonic plates, that helps to explain continental drift, mountain formation and volcanic activity. The movement of tectonic plates may be explained by the currents of rock rising and falling from within the Earth's mantle, as it heats up and then cools. The boundaries of the plates are known as plate margins and most mountains, earthquakes, and volcanoes occur at these margins. Constructive margins are moving apart; destructive margins are crunching together and conservative margins are sliding past one another.

Pleistocene A period of geological time spanning from about 5.2 million years ago to 1.6 million years ago.

Plutonic rock Igneous rocks found deep below the surface. They are coarse-grained because they cooled and solidified slowly.

Polje A long, broad depression found in karst (limestone) regions.

Polygonal patterning Typical ground patterning, found in areas where the soil is subject to severe frost action, often in periglacial regions.

Porosity A measure of how much water can be held within a rock or a soil.

PreCambrian The earliest period of geological time dating from over 570 million years ago.

Precipitation The fall of moisture from the atmosphere onto the surface of the Earth, whether as dew, hail, rain, sleet, or snow.

Pyramidal peak A steep, isolated mountain summit, formed when the back walls of three or more cirques are cut back and move toward each other. The cliffs around such a horned peak, or horn, are divided by sharp arêtes.

Pyroclasts Fragments of rock ejected during volcanic eruptions.

Quaternary The current period of geological time, which started about 1.6 million years ago.

Reg A large area of stony desert, where tightly-packed gravel lies on top of clayey sand. A reg is formed where the wind blows away the finer sand.

Resistance The capacity of a rock to resist denudation, by processes such as weathering and erosion.

Ria A flooded V-shaped river valley or estuary, flooded by a rise in sea level (eustacy) or sinking land. It is shorter than a fjord and gets deeper as it meets the sea.

Rift valley A long, narrow depression in the Earth's crust, formed by the sinking of rocks between two faults.

Roche moutonée A rock found in a glaciated valley. The side facing the flow of the glacier has been smoothed and rounded, while the other side has been left more rugged because the glacier, as it flows over it, has plucked out frozen fragments and carried them away.

Runoff Water draining from a land surface by flowing across it.

Sabkha The floor of an isolated depression that occurs in an arid environment – usually covered by salt deposits and devoid of vegetation.

Salt plug A rounded hill produced by the upward doming of rock strata caused by the movement of salt or other evaporite deposits under intense pressure.

Sastrugi Ice ridges formed by wind action. They lie parallel to the direction of the wind.

Scree Piles of rock fragments beneath a cliff or rock face, caused by mechanical weathering, especially frost shattering, where the expansion and contraction of freezing and thawing water within the rock, gradually breaks it up.

Sea-floor spreading The process whereby tectonic plates move apart, allowing hot magma to erupt and solidify.

Seamount An isolated, submarine mountain or hill, probably of volcanic origin.

Sediment Grains of rock transported and deposited by rivers, sea, ice, or wind.

Sedimentary rocks Rocks formed from the debris of preexisting rocks or of organic material. They are found in many environments on the ocean floor, on beaches, rivers, and deserts.

Seif A sand dune which lies parallel to the direction of the prevailing wind. Seifs form steep-sided ridges, sometimes extending for miles.

Selva A region of wet forest found in the Amazon Basin.

Shale (marine shale) A compacted sedimentary rock, with fine-grained particles. Marine shale is formed on the seabed. Fuel such as oil may be extracted from it.

Sheetwash Water that runs downhill in thin sheets without forming channels. It can cause sheet erosion.

Sheet erosion The washing away of soil by a thin film or sheet of water, known as sheetwash.

Shield A vast stable block of the Earth's crust, which has experienced little or no mountain-building.

Sinkhole A circular depression in a limestone region. They are formed by the collapse of an underground cave system or the chemical weathering of the limestone.

Slip face The steep leeward side of a sand dune or slope. Opposite side to a back slope.

Soil creep The very gradual downslope movement of rock debris and soil, under the influence of gravity. This is a type of mass movement.

Solifluction A kind of soil creep, where water in the surface layer has saturated the soil and rock debris which slips slowly downhill. It often happens where frozen top-layer deposits thaw, leaving frozen layers below them.

Spit A thin linear deposit of sand or shingle extending from the sea shore.

Stack A tall, isolated pillar of rock near a coastline, created as wave action erodes away the adjacent rock.

Strike-slip fault Occurs where plates move sideways past each other and blocks of rocks move horizontally in relation to each other, not up or down as in normal faults.

Subduction zone A region where two tectonic plates collide, forcing one beneath the other.

Submarine fan Deposits of silt and alluvium, carried by large rivers forming great fan-shaped deposits on the ocean floor.

Supercontinent A large continent that breaks up to form smaller continents or that forms when smaller continents merge.

Syncline A basin-shaped downfold in rock strata, created when the strata are compressed, for example where tectonic plates collide.

Tableland A highland area with a flat or gently undulating surface.

Tectonic plates Plates, or tectonic plates, are the rigid slabs which form the Earth's outer shell, the lithosphere. Eight big plates and several smaller ones have been identified.

Thermokarst Subsidence created by the thawing of ground ice in periglacial areas, creating depressions.

Till Unstratified glacial deposits or drift left by a glacier or ice sheet. Includes mixtures of clay, sand, gravel, and boulders.

Topography The typical shape and features of a given area such as land height and terrain.

Tombolo A large sand spit which attaches part of the mainland to an island.

Transform fault In plate tectonics, a fault of continental scale, occurring where two plates slide past each other, staying close together for example, the San Andreas Fault. The jerky, uneven movement creates earthquakes but does not destroy or add to the Earth's crust

Trench (oceanic trench) A long, deep trough in the ocean floor, formed, according to the theory of plate tectonics, when two plates collide and one dives under the other, creating a subduction zone.

Tropic of Cancer A line of latitude or imaginary circle round the Earth, lying at 23° 28' N.

Tropic of Capricorn A line of latitude or imaginary circle round the Earth, lying at 23° 28' S.

U-shaped valley A river valley that has been deepened and widened by a glacier. They are characteristically flat-bottomed and steep-sided and generally much deeper than river valleys.

V-shaped valley A typical valley eroded by a river in its upper course.

Wadi The dry bed left by a torrent of water. Also classified as a ephemeral stream, found in arid and semiarid regions, which are subject to sudden and often severe flash flooding.

Watershed The dividing line between one drainage basin an area where all streams flow into a single river system – and another. In the US, watershed also means the whole drainage basin of a single river system its catchment area.

Waterspout A rotating column of water in the form of cloud, mist, and spray which form on open water. Often has the appearance of a small tornado.

Weathering The decay and breakup of rocks at or near the Earth's surface, caused by water, wind, heat, or ice, organic material, or the atmosphere. Physical weathering includes the effects of frost and temperature changes. Biological weathering includes the effects of palnt roots, burrowing animals and the acids produced by animals, especially as they decay after death. Carbonation and hydrolysis are among many kinds of chemical weathering.

NORTH AMERICA

 **CANADA**
PAGES 8–16

 **UNITED STATES OF AMERICA**
PAGES 17–41

 **MEXICO**
PAGES 42–43

 **BELIZE**
PAGES 44–45

 **COSTA RICA**
PAGES 44–45

 **EL SALVADOR**
PAGES 44–45

 **GUATEMALA**
PAGES 44–45

 **HONDURAS**
PAGES 44–45

SOUTH AMERI

 **GRENADA**
PAGES 46–47

 **HAITI**
PAGES 46–47

 **JAMAICA**
PAGES 46–47

 **ST KITTS & NEVIS**
PAGES 46–47

 **ST LUCIA**
PAGES 46–47

 **ST VINCENT & THE GRENADINES**
PAGES 46–47

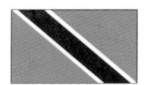 **TRINIDAD & TOBAGO**
PAGES 46–47

 **COLOMBIA**
PAGES 56–57

AFRICA

 **URUGUAY**
PAGES 62–63

 **CHILE**
PAGES 64–65

 **PARAGUAY**
PAGES 64–65

 **ALGERIA**
PAGES 76–77

 **EGYPT**
PAGES 76–77

 **LIBYA**
PAGES 76–77

 **MOROCCO**
PAGES 76–77

 **TUNISIA**
PAGES 76–77

 **LIBERIA**
PAGES 78–79

 **MALI**
PAGES 78–79

 **MAURITANIA**
PAGES 78–79

 **NIGER**
PAGES 78–79

 **NIGERIA**
PAGES 78–79

 **SENEGAL**
PAGES 78–79

 **SIERRA LEONE**
PAGES 78–79

 **TOGO**
PAGES 78–79

 **BURUNDI**
PAGES 82–83

 **DJIBOUTI**
PAGES 82–83

 **ERITREA**
PAGES 82–83

 **ETHIOPIA**
PAGES 82–83

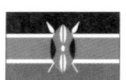 **KENYA**
PAGES 82–83

 **RWANDA**
PAGES 82–83

 **SOMALIA**
PAGES 82–83

 **SUDAN**
PAGES 82–83

EUROPE

 **SOUTH AFRICA**
PAGES 84–85

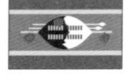 **SWAZILAND**
PAGES 84–85

 **ZAMBIA**
PAGES 84–85

 **ZIMBABWE**
PAGES 84–85

 **DENMARK**
PAGES 94–97

 **FINLAND**
PAGES 94–95

 **ICELAND**
PAGES 94–95

 **NORWAY**
PAGES 94–97

 **MONACO**
PAGES 104–105

 **ANDORRA**
PAGES 106–107

 **PORTUGAL**
PAGES 106–107

 **SPAIN**
PAGES 106–107

 **ITALY**
PAGES 108–109

 **SAN MARINO**
PAGES 108–109

 **VATICAN CITY**
PAGES 108–109

 **AUSTRIA**
PAGES 110–111

 **BOSNIA & HERZEGOVINA**
PAGES 114–115

 **CROATIA**
PAGES 114–115

 **MACEDONIA**
PAGES 114–115

 **YUGOSLAVIA**
PAGES 114–115

 **BULGARIA**
PAGES 116–117

 **GREECE**
PAGES 116–117

 **MOLDOVA**
PAGES 118–119

 **ROMANIA**
PAGES 118–119

ASIA

 **ARMENIA**
PAGES 142–143

 **AZERBAIJAN**
PAGES 142–143

 **GEORGIA**
PAGES 142–143

 **TURKEY**
PAGES 142–143/116–117

 **IRAQ**
PAGES 144–145

 **ISRAEL**
PAGES 144–145

 **JORDAN**
PAGES 144–145

 **LEBANON**
PAGES 144–145

 **IRAN**
PAGES 148–149

 **KAZAKHSTAN**
PAGES 150–151

KYRGYZSTAN
PAGES 152–153

 **TAJIKISTAN**
PAGES 152–153

 **TURKMENISTAN**
PAGES 152–153

UZBEKISTAN
PAGES 152–153

AFGHANISTAN
PAGES 154–155

PAKISTAN
PAGES 154–157

 **SOUTH KOREA**
PAGES 162–163/168–169

TAIWAN
PAGES 166–167

JAPAN
PAGES 170–172

BRUNEI
PAGES 173–176

INDONESIA
PAGES 173–176

MALAYSIA
PAGES 173–176

SINGAPORE
PAGES 173–176

 **MYANMAR**
PAGES 177–179

AUSTRALASIA & OCEANIA

MAURITIUS
PAGES 180–181

SEYCHELLES
PAGES 180–181

AUSTRALIA
PAGES 188–191

NEW ZEALAND
PAGES 192–193

PAPUA NEW GUINEA
PAGES 194–195

SOLOMON ISLANDS
PAGES 194–195

MARSHALL ISLANDS
PAGES 196/201

MICRONESIA
PAGES 196/201